BAILEY & SCOTT'S

DIAGNOSTIC
MICROBIOLOGY

ELSEVIER

evolve

∴ *To access your Student Resources, visit:*

http://evolve.elsevier.com/Forbes/

Evolve® Student Resources for *Forbes: Baily & Scott's Diagnostic Microbiology, Twelfth Edition,* offer the following features:

Student Resources

- **Weblinks**
  Links to places of interest on the web specifically for diagnostic microbiology.

- **Content Updates**
  Find out the latest information on relevant issues in the field.

BAILEY & SCOTT'S

DIAGNOSTIC MICROBIOLOGY

TWELFTH EDITION

Betty A. Forbes, PhD, D(ABMM), F(AAM)
Professor, Pathology and Medicine
Department of Pathology
Medical College of Virginia Campus
Virginia Commonwealth University Medical Center
Richmond, Virginia

Daniel F. Sahm, PhD, D(ABMM), F(AAM)
President
Eurofins: Anti-Infective Services
Herndon, Virginia

Alice S. Weissfeld, PhD, D(ABMM), F(AAM)
President, Microbiology Specialists Incorporated
Adjunct Assistant Professor
Department of Molecular Virology and Microbiology
Baylor College of Medicine
Houston, Texas

Photography by
Ernest A. Trevino, MT (ASCP)
Director of Operations
Microbiology Specialists Incorporated
Houston, Texas

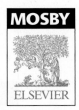

MOSBY

ELSEVIER

11830 Westline Industrial Drive
St. Louis, Missouri 63146

BAILEY & SCOTT'S DIAGNOSTIC MICROBIOLOGY, 12th Edition   ISBN 13 978-0-323-03065-6
ISBN 10 0-323-03065-3

BAILEY & SCOTT'S DIAGNOSTIC MICROBIOLOGY,   ISBN 13 978-0-8089-2364-0
INTERNATIONAL EDITION, 12th Edition   ISBN 10 0-8089-2364-1

Copyright © 2007, 2002, 1998, 1994, 1990, 1986, 1982, 1978, 1974, 1970, 1966, 1962 by Mosby, Inc., an affiliate of Elsevier Inc.

Notice

ISBN 13: 978-0-323-03065-6
ISBN 10: 0-323-03065-3

Publishing Director: Andrew Allen
Executive Editor: Loren Wilson
Developmental Editor: Ellen Wurm
Publishing Services Manager: Patricia Tannian
Senior Project Manager: Anne Altepeter
Senior Designer: Julia Dummitt

Printed in China
Last digit is the print number: 9 8 7 6 5 4 3

Pamela A. Fuselier, MT (ASCP)
Technical Specialist
Microbiology Specialists Incorporated
Houston, Texas

Lynne Shore Garcia, MS, MT, F(AAM)
Director, LSG & Associates
Santa Monica, California

Gary W. Procop, MD
Co-Director, Clinical Microbiology
Jackson Health System
Professor of Pathology
Miller School of Medicine
University of Miami
Miami, Florida

Glenn D. Roberts, MD
Consultant in Clinical Microbiology
Department of Clinical Microbiology
Mayo Clinic and Mayo Foundation
Professor of Microbiology and of Laboratory Medicine
Mayo Clinic College of Medicine
Rochester, Minnesota

Richard B. Thomson, Jr, PhD, ABMM, F(AAM)
Director, Clinical Microbiology and Virology
Department of Pathology and Laboratory Medicine
Evanston Northwestern Healthcare
Evanston, Illinois
Professor of Pathology
Northwestern University Feinberg
    School of Medicine
Chicago, Illinois

Mary K. York, PhD, ABMM
MKY Microbiology Consulting
Walnut Creek, California

To Elizabeth and Jonathan,
who are incredible and continue to inspire me and enrich my life;
to my sis and her family, who are always there for me;
to my family of friends, who keep me going; and finally, to my new friends
Olivia, Will, and all the others at The Crossroads for saving "my table"
each morning and supplying me with laughs and outstanding coffee
while working on this edition.
BAF

To my wife
Jan
and our sons
Nathan, Elias, Aaron, Zachary, and Matthew,
You are the Center of everything.
DFS

To my father, whose memory continues to inspire me;
to my mother, whose love has sustained me through life's rough times;
and to Paula Vance, my business partner, and everyone at Microbiology
Specialists Incorporated, who have been my rock
as we reinvented ourselves as clinical microbiologists.
ASW

This now twelfth edition of *Bailey and Scott's Diagnostic Microbiology* attests to the continued need for a comprehensive and functional publication to guide students and practitioners through the ever-changing field of clinical microbiology. Through seven decades, the many revisions of the original, concise laboratory manual, *Methods for Diagnostic Bacteriology,* by Isabelle G. Schaub and M. Kathleen Foley, have been mandatory additions to personal and laboratory bookshelves. The classic Bailey and Scott series grew from the efforts of Elvyn G. Scott and W. Robert Bailey, who joined Schaub and Foley for their fifth edition. Since then, a succession of authors has undertaken the increasingly daunting task of updating and expanding this work while adding the areas of parasitology, mycology, virology, and molecular diagnosis—disciplines now considered integral to the clinical microbiology laboratory.

The challenges that confront clinical microbiologists in this first part of the twenty-first century could not have been foreseen by the authors of the original text. They were not charged with the need to balance fiscal restraint with increasing demands for conducting expensive, albeit more rapid, automated and molecular diagnostic tests. Nor was vigilance for acts of bioterrorism (inflicted even in the earliest centuries of civilization) of everyday concern to them. In the absence of antimicrobial therapy, there was no need for these pioneers to follow strict criteria for performing antimicrobial susceptibility tests in the face of a growing number of antimicrobial agents active against an increasing number of potential pathogens, many of which formerly were considered normal flora. The original authors were unaware of the opportunistic infections of patients with AIDS and other immune-compromising conditions and the emerging and reemerging infectious diseases that have arisen, often as the consequence of ongoing political and social upheavals. The specter of antimicrobial agent–resistant microorganisms, with their expanding repertoire of resistance factors, had not yet appeared. To our benefit, however, the present advances in molecular techniques have permitted a better understanding of antimicrobial agent resistance factors and the relationships among certain groups of microorganisms, so that we can better control them and define their role in specific disease syndromes. These molecular advances have also given rise to highly specific methods for microbial identification and in some cases, methods for their direct detection in clinical specimens, even when they cannot be cultivated in vitro.

To address such issues as they affect traditional and modern practices in the field, Betty Ann Forbes, Daniel F. Sahm, and Alice Weissfeld again bring their expertise to this latest edition of *Bailey and Scott's Diagnostic Microbiology.* Each author has made contributions to update previous information and provide new concepts in his or her respective areas of expertise. Whether a beginning or experienced practitioner, the reader will appreciate the clearly written, practical, and informative approach that has been taken in each chapter. In three parts, the authors describe basic issues in medical microbiology; general principles for safety, specimen management, and diagnostic approaches, including for antimicrobial susceptibility testing; and bacterial identification. Of note are the extensive revisions of the chapter that addresses nucleic acid–based analytic methods for microbial identification and characterization and the addition of case studies for each chapter. The fourth through sixth parts are written by other experts in the fields of parasitology, mycology, and virology. Again, these parts have been revised and updated extensively to include, for example, coverage of SARS and avian influenza. The seventh part assists diagnosis by organizing the material in an organ system approach. In a new eighth part on clinical laboratory management, a chapter is included to guide the laboratory response to bioterrorism. New and updated tables, diagrams, and color photographs throughout the text enhance and facilitate comprehension of the subject matter.

As before, the authors have approached the subject from the point of view of a bench technologist confronted with a culture plate of microbial growth. A few direct observations (e.g., gram-stain reaction and morphology, growth on certain media) and standard rapid tests (e.g., oxidase, catalase) are used to place the organisms into major groupings from which identification then proceeds. The chapter sections guide workup of the culture. The chapters that deal with antimicrobial susceptibility testing are especially valuable for technologists in small clinical laboratories who often are requested to perform susceptibility tests on organisms for which no standardized method is available. Emphasis on clinical relevance and cost effectiveness are especially pertinent in the current health care climate.

The success of the previous editions written by these authors signifies that their approach is valued by clinical microbiologists at all levels of expertise. With its updated material and attention to cutting-edge concepts, this revision merits high regard and will no doubt quickly replace well-worn copies of the previous edition.

Josephine A. Morello, PhD
Professor Emerita of Pathology
Former Director of Hospitals Laboratories and
Clinical Microbiology Laboratories
University of Chicago Medical Center
Chicago, Illinois

This, the twelfth edition of *Bailey and Scott's Diagnostic Microbiology,* is the third consecutive edition that we have co-authored. Although we have learned much during our preparation of these three editions, the dynamics of infectious disease trends along with the technical developments available for diagnosing, treating, and controlling these diseases has continued to present major challenges. In meeting these challenges, our primary goal has always been to provide an updated and reliable reference text for practicing clinical microbiologists and technologists while also presenting this information in a format that supports the educational efforts of all those responsible for preparing others for a career in diagnostic microbiology. Admittedly this is not an easy task. We realize that in our efforts to achieve both purposes we have had to make some difficult decisions, the results of which may from time to time dissatisfy either the practitioners or the educators. Nonetheless, by carefully reviewing the compliments and the criticisms of our eleventh edition readers, we believe that in the twelfth edition we were able to achieve strong compromise for both a reference and a teaching text.

To align our goals with the reader's expectations and needs, we have kept the favorite features and made adjustments in response to important critical input from users of the text. Key features such as the glossary and the case studies have been updated. Importantly, more new case studies have been added. Also, the succinct presentation of each organism group's key laboratory, clinical, epidemiologic, and therapeutic features in tables and figures has been kept and updated, and new tables have been added. Regarding content, the major changes reflect the changes that the discipline of diagnostic microbiology continues to experience. The chapter that deals with molecular methods for identifying and characterizing microbial agents has been expanded and updated. Also, although the grouping of organisms into sections according to key features (e.g., Gram reaction, catalase or oxidase reaction, growth on MacConkey) has remained, changes regarding the genera and species discussed in these sections have been made. These changes, along with changes in organism nomenclature, were made to accurately reflect the changes that have occurred, and continue to occur, in bacterial taxonomy. Also, throughout the text the content has been enhanced with new photographs, line drawings, and color schemes for tables. Finally, although some classic methods for bacterial identification and characterization developed over the years (e.g., catalase, oxidase, Gram stain) still play a critical role in today's laboratory, others (e.g., IMViC profile) have given way to commercial identification systems. We realize that in a textbook such as this that balances the needs for practicing and teaching diagnostic microbiology, our selection of identification methods that received the most detailed attention may not always meet the needs of both groups. However, we have tried to be consistent in selecting those methods that reflect the most current and common practices of today's clinical microbiology laboratories.

Finally, in terms of organization, the twelfth edition is similar in many aspects to the eleventh edition, but some changes have been made. Part I contains three chapters that address the basic field of medical microbiology in terms of taxonomy, genetics, physiology, and host interactions. Part II comprises nine chapters that address the basic principles of clinical microbiology, including laboratory safety, microscopy, serology, molecular testing, and antimicrobial susceptibility testing. All of bacteriology is captured in Part III, which begins with three brief overview chapters on identification schemes and the organization and use of the subsequent organism-based chapters. Part III is then further organized into subsections based on key identification features, with supporting chapters for specific organism genera. Parts IV, V, and VI address, in total, the areas of parasitology, mycology, and virology, respectively. The 10 chapters in Part VII explore diagnostic microbiology from the human organ system perspective and emphasize clinical manifestations and diagnostic protocols. The twelfth edition ends with Part VIII, which uses four chapters to address important aspects of clinical laboratory management, including a new chapter regarding bioterrorism.

We offer various instructor ancillaries specifically geared for the twelfth edition. The **Evolve Resources** include an instructor's manual, a 900-question test bank, a laboratory manual, and an electronic image collection, all with instructor-only access. Student resources include weblinks and content updates.

A **Study Guide** is available and includes chapter objectives, summary of key points, review questions, and case studies.

We sincerely hope that clinical microbiology practitioners and educators alike find Bailey and Scott's Twelfth Edition to be a worthy and useful tool to support their professional activities.

ACKNOWLEDGMENTS

We acknowledge the help of our colleagues at Elsevier who guided us through this project: Anne Altepeter, senior project manager, and Ellen Wurm, developmental editor. Of course, none of this would have been possible without the extraordinary efforts of Liz Williams, who interpreted so much poor handwriting

while typing the manuscripts, and, of course, Loren Wilson, executive editor, and Jennifer Presley, editorial assistant, who watched out for all of the behind-the-scenes issues on our behalf.

Finally, we acknowledge and thank all the clinical microbiologists, scientists, clinicians, and educators who have over many years been our colleagues and who, through their contribution to the field and their communications with us, have made the writing of this and the previous editions possible.

Betty A. Forbes
Daniel F. Sahm
Alice S. Weissfeld

CONTENTS

PART VIII
Clinical Laboratory Management

P A R T I

Basic Medical Microbiology

CHAPTER 1

Microbial Taxonomy

Taxonomy is the area of biologic science comprising three distinct, but highly interrelated, disciplines that include classification, nomenclature, and identification. Applied to all living entities, taxonomy provides a consistent means to classify, name, and identify organisms. This consistency allows biologists worldwide to use a common label for every organism they study within their particular disciplines. The common language that taxonomy provides minimizes confusion about names and allows attention to center on more important scientific issues and phenomena. The importance of this contribution is not only realized in phylogeny (the evolutionary history of organisms) but also in virtually every other discipline of biology, including microbiology.

In diagnostic microbiology, classification, nomenclature, and identification of microorganisms play a central role in providing accurate and timely diagnosis of infectious diseases. Because of taxonomy's important contribution to diagnostic microbiology, some detailed discussion of the three areas that make up taxonomy follows.

CLASSIFICATION

Classification is the organization of microorganisms that share similar morphologic, physiologic, and genetic traits into specific groups, or taxa. The classification system is hierarchic and consists of the following taxa designations:

- Species
- Genus (composed of similar species)
- Family (composed of similar genera)
- Order (composed of similar families)
- Class (composed of similar orders)
- Division (composed of similar classes)
- Kingdom (composed of similar divisions)

SPECIES

Species is the most basic taxonomic group and can be defined as a collection of bacterial strains that share many common physiologic and genetic features and as a group differ notably from other bacterial species. Occasionally, taxonomic subgroups within a species, called **subspecies**, are recognized. Furthermore, designations such as biotype, serotype, or genotype may be given to groups below the subspecies levels that share specific, but relatively minor, characteristics. Although these subgroups may have some taxonomic importance and occasional practical utility, their usefulness in diagnostic microbiology is usually limited.

GENUS

Genus (pl. genera) is the next higher taxon and comprises different species that have several important features in common but differ sufficiently to still maintain their status as individual species. All bacterial species belong to a genus, and relegation of a species to a particular genus is based on various genetic and phenotypic characteristics shared among the species. However, microorganisms do not possess the multitude of physical features exhibited by higher organisms such as plants and animals. For instance, they rarely leave any fossil record and they exhibit a tremendous capacity to intermix genetic material among supposedly unrelated species and genera. For these reasons the ability to confidently establish microorganism relatedness so they may be classified in higher taxa beyond the genus level is difficult. Therefore, although grouping similar genera into common families and similar families into common orders and so on is used for classification of plants and animals, these higher taxa designations (i.e., division, class, order, family) are not usually useful for classifying bacteria.

NOMENCLATURE

Nomenclature, the naming of microorganisms according to established rules and guidelines, provides the accepted labels by which organisms are universally recognized. Because genus and species are the groups of most concern to microbiologists, the discussion of rules governing microbial nomenclature is limited to these two taxon designations. In this binomial (two-name) system of nomenclature, every organism is assigned a genus and species name of Latin or Greek derivation. In other words, every organism has a scientific "label" consisting of two parts: the genus designation, which is always capitalized, and the species designation, which is never capitalized. Both components are always used simultaneously and are printed in italics, or underlined in script. For example, the streptococci include *Streptococcus pneumoniae, Streptococcus pyogenes, Streptococcus agalactiae, Streptococcus bovis,* among others. Alternatively, the name may be abbreviated by using the upper case form of the first letter of the genus designation followed by a period (.) and the full species name, which is never abbreviated, such as *S. pneumoniae, S. pyogenes, S. agalactiae,* and *S. bovis.* Frequently an informal designation (e.g., staphylococci, streptococci, enterococci) may be used to label a particular group of organisms, but such designations are not capitalized or italicized.

As more information is gained regarding organism classification and identification, a particular species may be moved to a different genus or assigned a new genus name. The rules and criteria for these changes are beyond the scope of this chapter, but such changes are documented in the *International Journal for Systematic Bacteriology.* In the diagnostic laboratory, changes in nomenclature are phased in gradually so that physicians and laboratorians have ample opportunity to recognize that a familiar pathogen has been given a new name. This is usually accomplished by using the new genus designation while continuing to provide the previous designation in parentheses, for example, *Stenotrophomonas (Xanthomonas) maltophilia* or *Burkholderia (Pseudomonas) cepacia.*

IDENTIFICATION

Microbial **identification** is the process by which a microorganism's key features are delineated. Once those features are established, the profile is compared with those of other previously characterized microorganisms so that the organism in question can be classified within the most appropriate taxa (classification) and can be assigned an appropriate genus and species name (nomenclature); both are essential aspects of the role taxonomy plays in diagnostic microbiology and infectious diseases (Box 1-1).

IDENTIFICATION METHODS

A wide variety of methods and criteria are used to establish a microorganism's identity. These methods usually can be separated into either of two general categories: genotypic characteristics and phenotypic characteristics. **Genotypic characteristics** relate to an organism's genetic makeup, including the nature of the organism's genes and constituent nucleic acids (see Chapter 2 for more information about microbial genetics). **Phenotypic characteristics** are based on features beyond the genetic level and include readily observable characteristics and those characteristics that may require extensive analytic procedures to be detected. Examples of characteristics used as criteria for bacterial identification and classification are provided in Table 1-1. Modern microbial taxonomy usually uses a combination of several methods so that the microorganism is characterized as completely as possible and the most appropriate classification and naming of the organism is accomplished.

Although the criteria and examples given in Table 1-1 are in the context of microbial identification for classification purposes, the principles and practices of classification exactly parallel the approaches used in diagnostic microbiology for the identification and characterization of microorganisms encountered in the clinical setting. Fortunately, because of the previous efforts and accomplishments of microbial taxonomists, microbiologists do not have to use several burdensome classification and identification schemes to identify infectious agents. Instead, microbiologists extract key organism features on which to base their identification methods so that organism identity can be made in a timely and clinically useful manner (see Chapter 14). In a sense, diagnostic microbiology uses streamlined

Table 1-1   Identification Criteria and Characteristics for Microbial Classification

Phenotypic Criteria Examples	Principles
Macroscopic morphology	Characteristics of microbial growth patterns on artificial media as observed when inspected with the unaided eye. Examples of such characteristics include the size, texture, and pigmentation of bacterial colonies
Microscopic morphology	Size, shape, intracellular inclusions, cellular appendages, and arrangement of cells when observed with the aid of microscopic magnification
Staining characteristics	Ability of organism to reproducibly stain a particular color with the application of specific dyes and reagents. Staining is usually used in conjunction with microscopic morphology as part of the process of bacterial identification. For example, the Gram stain for bacteria is a critical criterion for identification
Environmental requirements	Ability of organism to grow at various temperatures, in the presence of oxygen and other gases, at various pH levels, or in the presence of other ions and salts such as NaCl
Nutritional requirements	Ability of organism to utilize various carbon and nitrogen sources as nutritional substrates when grown under specific environmental conditions
Resistance profiles	Exhibition of a characteristic inherent resistance to specific antibiotics, heavy metals, or toxins by certain microorganisms
Antigenic properties	Establishment of profiles of microorganisms by various serologic and immunologic methods that are useful for determining the relatedness among various microbial groups
Subcellular properties	Establishment of the molecular constituents of the cell that are typical of a particular taxon, or organism group, by various analytic methods. Some examples include cell wall components, components of the cell membrane, and enzymatic content of the microbial cell
Genotypic Criteria Examples	**Principles**
DNA base composition ratio	DNA comprises four bases (guanine, cytosine, adenine, and thymine). The extent to which the DNA from two organisms is made up of cytosine and guanine (i.e., G + C content) relative to their total base content can be used as indicator of relatedness, or lack thereof. For example, an organism with G + C content of 50% will not be closely related to an organism whose G + C content is 25%
Nucleic acid (DNA and RNA) base sequence analysis	The order of bases along a strand of DNA or RNA is known as the *base sequence*, and the extent to which sequences are similar (homologous) between two microorganisms can be determined directly or indirectly by various molecular methods. The degree of similarity in the sequences may be a measure of the degree of organism relatedness

versions of the phenotypic and genotypic approaches employed by taxonomists to identify microorganisms. This should not be taken to mean that the identification of all clinically relevant organisms is easy and straightforward, and this is also not meant to imply that microbiologists can only identify or recognize organisms that have already been characterized and named by taxonomists. Indeed, the clinical microbiology laboratory is well recognized as the place in which previously unknown or uncharacterized infectious agents are initially encountered and, as such, has an ever-increasing responsibility to be the sentinel for emerging etiologies of infectious diseases.

ADDITIONAL READING

Balows A, Turper HG, Dworkin M, et al, editors: *The prokaryotes: a handbook on the biology of bacteria: ecophysiology, isolation, identification, applications,* vols 1-4, New York, 1992, Springer Verlag.

Brock TD et al, editors: *Biology of microorganisms,* Englewood Cliffs, NJ, 1994, Prentice Hall.

Krieg NR, Holt JG, editors: *Bergey's manual of systematic bacteriology,* vol 1, Baltimore, 1984, Williams & Wilkins.

Garrity GM, editor: *Bergey's manual of systematic bacteriology,* ed 2, New York, 2005, Springer.

Microbial genetics, metabolism, and structure are the keys to microbial viability and survival. These processes involve numerous pathways that are widely varied, often complicated, and frequently interactive. Essentially, survival requires energy to fuel the synthesis of materials necessary for growth, propagation, and carrying out all other metabolic processes (Figure 2-1). Although the goal of survival is the same for all organisms, the strategies microorganisms use to accomplish this vary substantially.

Knowledge regarding genetic, metabolic, and structural characteristics of microorganisms provides the basis for understanding almost every aspect of diagnostic microbiology, including:

- The mechanism by which microorganisms cause disease
- Developing and implementing optimum techniques for microbial detection, cultivation, identification, and characterization
- Understanding antimicrobial action and resistance
- Developing and implementing tests for antimicrobial resistance detection
- Designing strategies for disease therapy and control

Microorganisms vary greatly in many genetic and physiologic aspects, and a detailed consideration of all these differences is beyond the scope of this textbook. Therefore, bacterial systems are used as a model to discuss microbial genetics, metabolism, and structure. Information regarding characteristics of fungi, viruses, and parasites are found in the chapters discussing these specific organism groups.

BACTERIAL GENETICS

Genetics, the process of heredity and variation, is the starting point from which all other cellular pathways, functions, and structures originate. The ability of a microorganism to maintain viability, adapt, multiply, and cause disease is founded in genetics. The three major aspects of microbial genetics that require discussion include:

- The structure and organization of genetic material
- Replication and expression of genetic information
- The mechanisms by which genetic information is changed and exchanged among bacteria

NUCLEIC ACID STRUCTURE AND ORGANIZATION

For all living entities, hereditary information resides or is encoded in **nucleic acids.** The two major classes include **deoxyribonucleic acid (DNA),** which is the most common macromolecule that encodes genetic information, and **ribonucleic acid (RNA).** In some forms, RNA encodes genetic information for various viruses; in other forms, RNA plays an essential role in several of the genetic processes to be discussed.

Nucleotide Structure and Sequence

DNA consists of deoxyribose sugars connected by phosphodiester bonds (Figure 2-2, A). The **bases** that are covalently linked to each deoxyribose sugar are the key to the genetic code within the DNA molecule. The four bases include two purines, adenine (A) and guanine (G), and the two pyrimidines, cytosine (C) and thymine (T) (Figure 2-3). In RNA, uracil replaces thymine. Taken together, the sugar, the phosphate, and a base form a single unit referred to as a **nucleotide.** DNA and RNA are nucleotide polymers (i.e., chains or strands), and the order of bases along a DNA or RNA strand is known as the **base sequence.** This sequence provides the information that specifies the proteins that will be synthesized by microbial cells (i.e., the sequence is the **genetic code**).

DNA Molecular Structure

The intact DNA molecule usually is composed of two nucleotide polymers. Each strand has a 5' and a 3' hydroxyl terminus (see Figure 2-2, B). The two strands run antiparallel, with the 5' terminus of one strand opposed to the 3' terminal of the other. The strands are also complementary as the adenine base of one strand always binds, via two hydrogen bonds, to the thymine base of the other strand, or vice versa. Likewise, the guanine base of one strand always binds by three hydrogen

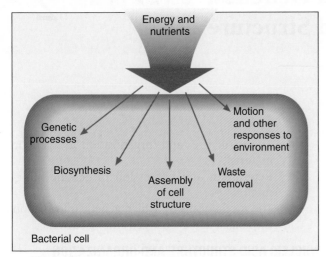

Figure 2-1 General overview of bacterial life processes.

bonds to the cytosine base of the other strand, or vice versa. The molecular restrictions of these base pairings, along with the conformation of the sugar-phosphate backbones oriented in antiparallel fashion, result in DNA having the unique structural conformation often referred to as a "twisted ladder" (see Figure 2-2, *B*). Additionally, the dedicated base pairs provide the format essential for consistent replication and expression of the genetic code. In contrast to DNA, RNA rarely exists as a double-stranded molecule and, whereas DNA carries the genetic code, the three major types of RNA **(messenger RNA [mRNA], transfer RNA [tRNA], and ribosomal RNA [rRNA])** play other key roles in gene expression.

Genes and the Genetic Code

A DNA sequence that encodes for a specific product (RNA or protein) is defined as a **gene.** Thousands of genes within an organism encode messages or blueprints for production, by gene expression, of specific protein and RNA products that play essential metabolic roles in the cell. All genes taken together within an organism comprise that organism's **genome.** The size of a gene and an entire genome is usually expressed in the number of base pairs (bp) present (e.g., kilobases [10^3 bases], megabases [10^6 bases]).

Certain genes are widely distributed among various organisms while others are limited to particular species. Also, the base pair sequence for individual genes may be highly conserved (i.e., show limited sequence differences among different organisms) or be widely variable. As discussed in Chapter 8, these similarities and differences in gene content and sequences are the basis for the development of molecular tests used to detect, identify, and characterize clinically relevant microorganisms.

Chromosomes

The genome is organized into discrete elements known as **chromosomes.** The set of genes within a given chromosome is arranged in a linear fashion, but the number of genes per chromosome is variable. Similarly, although the number of chromosomes per cell is consistent for a given species, this number varies considerably among species. For example, human cells contain 23 pairs (i.e., diploid) of chromosomes whereas bacteria contain a single, unpaired (i.e., haploid) chromosome.

The bacterial chromosome contains all genes essential for viability and exists as a double-stranded, closed circular, naked (i.e., not enclosed within a membrane) macromolecule. The molecule is extensively folded and twisted (i.e., supercoiled) so that it may be accommodated within the confines of the bacterial cell. The fact that the linearized, unsupercoiled chromosome of the bacterium *Escherichia coli* is about 1300 μm in length but fits within a 1 μm × 3 μm cell attests to the extreme compactness that the bacterial chromosome must achieve. For genes within the compacted chromosome to be expressed and replicated, unwinding or relaxation of the molecule is essential.

In contrast to the bacterial chromosome, the chromosomes of parasites and fungi number greater than one per cell, are linear, and are housed within a membrane structure known as the **nucleus.** This difference is a major criterion for classifying bacteria as prokaryotic organisms, while classifying fungi and parasites as eukaryotes. The genome of viruses may be referred to as a chromosome, but the DNA (or RNA) is contained within a protein coat rather than within a cell.

Nonchromosomal Elements of the Genome

Although the bacterial chromosome represents the majority of the genome, not all genes within a given cell are confined to the chromosome. Many genes are also located on plasmids and transposable elements. Both of these are able to replicate and encode information for the production of various cellular products. Although considered part of the bacterial genome, they are not as stable as the chromosome and may be lost during cellular replication, often without severe detrimental effects on the cell.

Plasmids exist as "miniature" chromosomes in being double-stranded, closed, circular structures with size ranges from 1 to 2 kilobases up to 1 megabase or more. The number of plasmids per bacterial cell varies extensively, and each plasmid is composed of several genes. Some genes encode products that mediate plasmid replication and transfer between bacterial cells, whereas others encode products that provide a survival edge such as determinants of antimicrobial

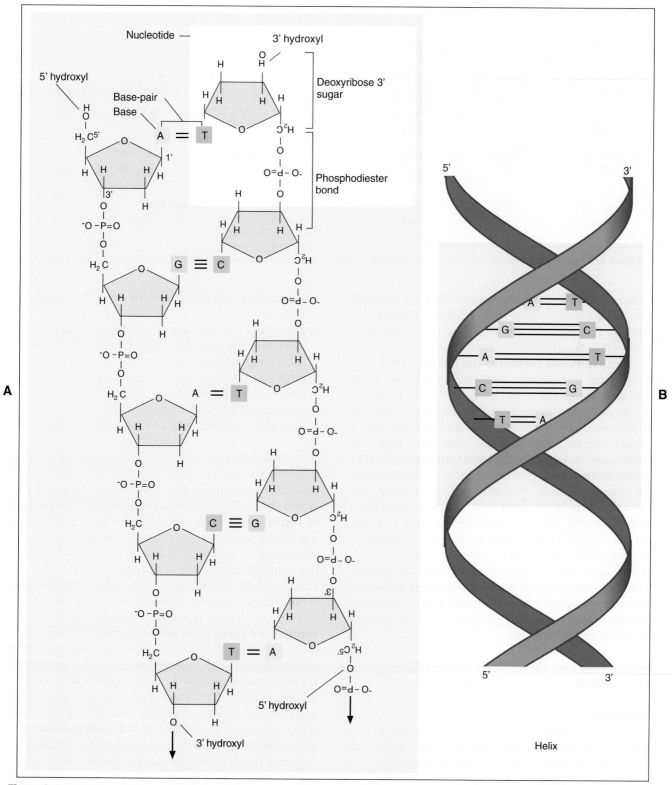

Figure 2-2   **A,** Molecular structure of DNA showing nucleotide structure, phosphodiester bond connecting nucleotides, and complementary pairing of bases (A, adenine; T, thymine; G, guanine; C, cytosine) between antiparallel nucleic acid strands. **B,** 5' and 3' antiparallel polarity and helical ("twisted ladder") configuration of DNA.

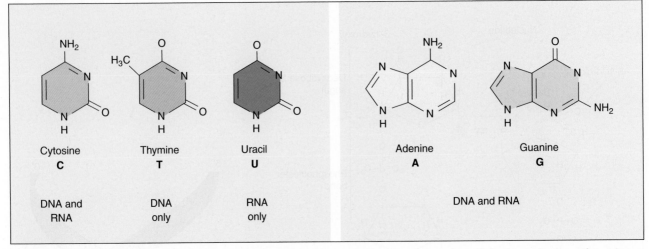

Figure 2-3   Molecular structure of nucleic acid bases. Pyrimidines: cytosine, thymine, and uracil. Purines: adenine and guanine.

resistance. Unlike most chromosomal genes, plasmid genes do not usually encode for products essential for viability. Plasmids, in whole or in part, may also become incorporated in the chromosome.

Transposable elements are pieces of DNA that move from one genetic element to another, from plasmid to chromosome or vice versa. Unlike plasmids, they do not exist as separate entities within the bacterial cell because they must either be incorporated into a plasmid or the chromosome. The two types of transposable elements are the **insertion sequence (IS)** and the **transposon.** Insertion sequences contain genes that simply encode for information required for movement among plasmids and chromosomes. Transposons contain genes for movement as well as genes that encode for other features such as drug resistance. Plasmids and transposable elements coexist with chromosomes in the cells of many bacterial species. These extrachromosomal elements play a key role in the exchange of genetic material throughout the bacterial microbiosphere, including genetic exchange among clinically relevant bacteria.

REPLICATION AND EXPRESSION OF GENETIC INFORMATION

Replication

Bacteria multiply by cell division that results in the production of two daughter cells from one parent cell. As part of this process, the genome must be replicated so that each daughter cell receives the same complement of functional DNA. Replication is a complex process that is mediated by various enzymes, such as DNA polymerase and cofactors; replication must occur quickly and accurately. For descriptive purposes, replication may be considered in four stages that are depicted together in Figure 2-4:

1. Unwinding or relaxation of the chromosome's supercoiled DNA
2. Unzipping, or disconnecting, the complementary strands of the parental DNA so that each may serve as a template (i.e., pattern) for synthesis of new DNA strands
3. Synthesis of the new DNA strands
4. Termination of replication with release of two identical chromosomes, one for each daughter cell

Relaxation of supercoiled chromosomal DNA is required so that enzymes and cofactors involved in replication can access the DNA molecule at the site where the replication process will originate (i.e., **origin of replication**). On exposure of the replication site (a specific sequence of approximately 300 base pairs that is recognized by several initiation proteins), unzipping of the complementary strands of parental DNA begins. Each parental strand serves as a template for the synthesis of a new complementary daughter strand. The site of active replication is referred to as the **replication fork;** there are two during the replication process. Each replication fork moves through the parent DNA molecule in opposite directions so that replication is a bidirectional process. Activity at each replication fork involves different cofactors and enzymes, with DNA polymerases playing a central role. Using each parental strand as a template, **DNA polymerases** add nucleotide bases to each growing daughter strand in a sequence that is complementary to the base sequence of the template (parent) strand. The complementary bases of each strand are then crosslinked. The new nucleotides can only be added to the 3' hydroxyl end of the growing strand so that synthesis for each daughter strand only occurs in a 5' to 3' direction.

Termination of replication occurs when the two replication forks meet, resulting in two complete

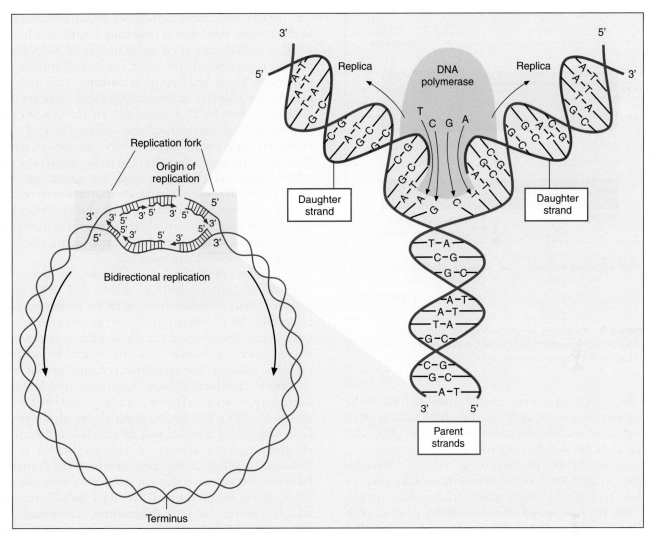

Figure 2-4   Bacterial DNA replication depicting bidirectional movement of two replication forks from origin of replication. Each parent strand serves as a template for production of a complementary daughter strand and, eventually, two identical chromosomes.

chromosomes, each containing two complementary strands, one of parental origin and one newly synthesized daughter strand. Although the time required for replication can vary among bacteria, the process generally takes approximately 40 minutes in rapidly growing bacteria such as *E. coli*. However, the replication time for a particular bacterial strain can vary depending on environmental conditions such as the availability of nutrients or the presence of toxic substances (e.g., antimicrobial agents).

Expression of Genetic Information

Gene expression is the processing of information encoded in genetic elements (i.e., chromosomes, plasmids, and transposons) that results in the production of biochemical products. The overall process is composed of two complex steps, transcription and translation, and requires various components, including a DNA template representing a single gene or cluster of genes, various

enzymes and cofactors, and RNA molecules of specific structure and function.

Transcription. Gene expression begins with transcription, which converts the DNA base sequence of the gene (i.e., the genetic code) into an mRNA (messenger RNA) molecule that is complementary to the gene's DNA sequence (Figure 2-5). Usually only one of the two DNA strands (the sense strand) encodes for a functional gene product, and this same strand is the template for mRNA synthesis.

RNA polymerase is the enzyme central to the transcription process. The enzyme is composed of four protein subunits (α [two copies], β, β') and a sigma (σ) factor. Sigma factor is loosely affiliated with the enzyme structure and identifies the appropriate site on the DNA template where transcription of mRNA is initiated. This initiation site is also known as the **promoter sequence.** The remainder of the enzyme

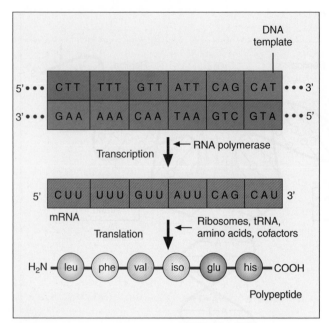

Figure 2-5   Overview of gene expression components; transcription for production of mRNA and translation for production of polypeptide (protein).

($\alpha_2\beta\beta'$) functions to open double-stranded DNA at the promoter sequence and use the DNA strand as a template to sequentially add ribonucleotides (ATP, GTP, UTP, and CTP) to form the growing mRNA strand.

Transcription proceeds in a 5′ to 3′ direction. However, in mRNA the thymine triphosphate (TTP) of DNA is replaced with uracil triphosphate (UTP). Synthesis of the single-stranded mRNA product ends when specific nucleotide base sequences on the DNA template are encountered. In some instances, termination of transcription may be facilitated by a rho cofactor, which can disrupt the mRNA-RNA polymerase-template DNA complex.

In bacteria, the mRNA molecules that result from the transcription process are **polycistronic,** that is, they encode for several gene products. Frequently, polycistronic mRNA may encode several genes whose products (proteins) are involved in a single or closely related cellular function. When a cluster of genes is under the control of a single promoter sequence, the gene group is referred to as an **operon.**

The transcription process not only produces mRNA but also tRNA and rRNA. All three types of RNA have key roles in protein synthesis.

Translation.   The next phase in gene expression, translation, involves protein **synthesis.** By this process the genetic code within mRNA molecules is translated into specific amino acid sequences that are responsible for protein structure, and hence, function (see Figure 2-5).

Before discussing translation, an understanding of the genetic code that is originally transcribed from DNA to mRNA and then translated from mRNA to protein is warranted. The code consists of triplets of nucleotide bases, referred to as **codons;** each codon encodes for a specific amino acid. Because there are 64 different codons for 20 amino acids, an amino acid can be encoded by more than one codon (Table 2-1). However, each codon specifies only one amino acid. Therefore, through translation the codon sequences in mRNA direct which amino acids are added and in what order. Translation ensures that proteins with proper structure and function are produced. Errors in the process can result in aberrant proteins that are nonfunctional, underscoring the need for translation to be well controlled and accurate.

To accomplish the task of translation, intricate interactions between mRNA, tRNA, and rRNA are required. Sixty different types of tRNA molecules are responsible for transferring different amino acids from intracellular reservoirs to the site of protein synthesis. These molecules, whose structure resembles an inverted t, contain one sequence recognition site for binding to specific three-base sequences (codons) on the mRNA molecule (Figure 2-6). A second site binds specific amino acids, the building blocks of proteins. Each amino acid is joined to a specific tRNA molecule via the enzymatic activity of aminoacyl-tRNA synthetases. Therefore, tRNA molecules have the primary function of using the codons of the mRNA molecule as the template for precisely delivering a specific amino acid for polymerization. **Ribosomes,** composed of rRNA and proteins, also are central to translation and provide the site at which translation occurs.

Translation, diagrammatically shown in Figure 2-6, involves three steps: initiation, elongation, and termination. Following termination, bacterial proteins often undergo posttranslational modifications as a final step in protein synthesis.

Initiation begins with the association of ribosomal subunits, mRNA, formylmethionine tRNA (f-met; carrying the initial amino acid of the protein to be synthesized), and various initiation factors (see Figure 2-6, *A*). Assembly of the complex begins at a specific 3- to 9-base sequence on the mRNA, referred to as the **ribosomal binding site,** or **RBS.** After the initial complex is formed, addition of individual amino acids begins.

Elongation involves tRNAs mediating the sequential addition of amino acids in a specific sequence that is dictated by the codon sequence of the mRNA molecule (see Figure 2-6, *B* to *C*, and Table 2-1). As the mRNA molecule threads through the ribosome in a 5′ to 3′ direction, peptide bonds are formed between adjacent amino acids still bound by their respective

Table 2-1   The Genetic Code as Expressed by Triplet Base Sequences of mRNA*

Codon	Amino Acid	Codon	Amino Acid	Codon	Amino Acid	Codon	Amino Acid
UUU	Phenylalanine	CUU	Leucine	GUU	Valine	AUU	Isoleucine
UUC	Phenylalanine	CUC	Leucine	GUC	Valine	AUC	Isoleucine
UUG	Leucine	CUG	Leucine	GUG	Valine	AUG (start)†	Methionine
UUA	Leucine	CUA	Leucine	GUA	Valine	AUA	Isoleucine
UCU	Serine	CCU	Proline	GCU	Alanine	ACU	Threonine
UCC	Serine	CCC	Proline	GCC	Alanine	ACC	Threonine
UCG	Serine	CCG	Proline	GCG	Alanine	ACG	Threonine
UCA	Serine	CCA	Proline	GCA	Alanine	ACA	Threonine
UGU	Cysteine	CGU	Arginine	GGU	Glycine	AGU	Serine
UGC	Cysteine	CGC	Arginine	GGC	Glycine	AGC	Serine
UGG	Tryptophan	CGG	Arginine	GGG	Glycine	AGG	Arginine
UGA	None (stop signal)	CGA	Arginine	GGA	Glycine	AGA	Arginine
UAU	Tyrosine	CAU	Histidine	GAU	Aspartic	AAU	Asparagine
UAC	Tyrosine	CAC	Histidine	GAC	Aspartic	AAC	Asparagine
UAG	None (stop signal)	CAG	Glutamine	GAG	Glutamic	AAG	Lysine
UAA	None (stop signal)	CAA	Glutamine	GAA	Glutamic	AAA	Lysine

Modified from Brock TD et al, editor: *Biology of microorganisms*, Englewood Cliffs, NJ, 1994, Prentice Hall.

*The codons in DNA are complementary to those given here. Thus, U here is complementary to the A in DNA, C is complementary to G, G to C, and A to T. The nucleotide on the left is at the 5′-end of the triplet.

†AUG codes for *N*-formylmethionine at the beginning of mRNAs of bacteria.

tRNA molecules in the P (peptide) and A (acceptor) sites of the ribosome. During the process, the forming peptide is moved to the P site and the most 5′ tRNA is released from the E (exit) site. This movement vacates the A site, which contains the codon specific for the next amino acid, so that the incoming tRNA-amino acid can join the complex (see Figure 2-6, *C*).

Because multiple proteins encoded on an mRNA strand can be translated at the same time, multiple ribosomes may be simultaneously associated with one mRNA molecule. Such an arrangement is referred to as a **polysome;** its appearance resembles a string of pearls.

Termination, the final step in translation, occurs when the ribosomal A site encounters a stop or nonsense codon that does not specify an amino acid (i.e., a "stop signal"; Table 2-1). At this point, the protein synthesis complex disassociates and the ribosomes are available for another round of translation. Following termination, most proteins must undergo some extent of modification such as folding or enzymatic trimming so that protein function, transportation, or incorporation into various cellular structures can be accomplished. This process is referred to as **posttranslational modification.**

Regulation and Control of Gene Expression

The vital role that gene expression and protein synthesis play in the survival of cells dictates that bacteria judiciously control these processes. The cell must regulate gene expression and control the activities of gene products so that a physiologic balance is maintained. Regulation and control are also key and highly complex mechanisms by which single-cell organisms are able to respond and adapt to environmental challenges, regardless of whether the challenges occur naturally or result from medical progress (e.g., antibiotics).

Regulation occurs at one of three levels of the gene expression and protein synthesis pathway: transcriptional, translational, and posttranslational. The most common is transcriptional level regulation. Because direct interactions with genes and their ability to be transcribed to mRNA are involved, transcriptional level regulation is also referred to as **genetic level control.** Genes that encode enzymes involved in biosynthesis **(anabolic enzymes)** and genes that encode enzymes for biodegradation **(catabolic enzymes)** will be used as examples of genetic level control.

In general, genes that encode anabolic enzymes for the synthesis of particular products are **repressed**

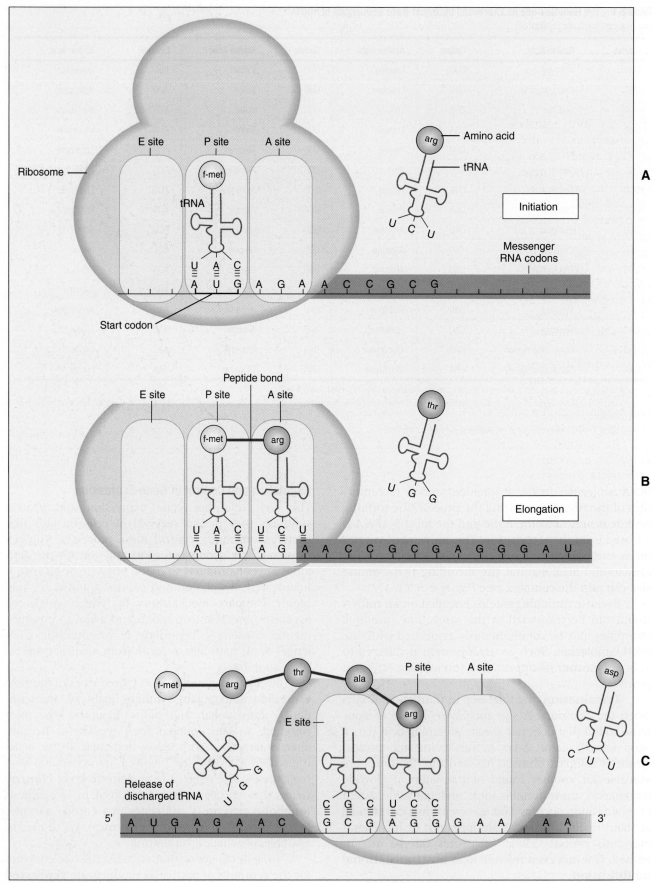

For legend see opposite page

Figure 2-6 Overview of translation in which mRNA serves as the template for the assembly of amino acids into polypeptides. The three steps include initiation **(A)**, elongation **(B** and **C)**, and termination (not shown).

(i.e., are not transcribed and therefore are not expressed) in the presence of those products. This strategy avoids waste and overproduction of products that are already present in sufficient supply. In this system the product acts as a co-repressor that forms a complex with a repressor molecule. In the absence of co-repressor product (i.e., gene product) transcription occurs (Figure 2-7, *A*). When present in sufficient quantity, the product forms a complex with the repressor. The complex then binds to a specific base region of the gene sequence known as the **operator region** (see Figure 2-7, *B*). This binding blocks RNA polymerase progression from the promoter sequence and inhibits transcription. As the supply of product (co-repressor) dwindles, an insufficient amount remains to form a complex with the repressor, the operator region is no longer bound, and transcription of the genes for the anabolic enzymes will commence and continue until a sufficient supply of end product is again available.

In contrast to repression, genes that encode catabolic enzymes are usually **induced,** that is, transcription only occurs when the substrate to be degraded by enzymatic action is present. Production of degradative enzymes in the absence of substrates would be a waste of cellular energy and resources. When the substrate is absent in an inducible system, a repressor binds to the operator sequence of the DNA and blocks transcription of the genes for the degradative enzymes (see Figure 2-7, *C*). In the presence of an inducer, which often is the target substrate for degradation, a complex is formed between inducer and repressor that results in the release of the repressor from the operator site, thus allowing transcription of the genes encoding for the appropriate catabolic enzymes (see Figure 2-7, *D*).

Certain genes are not regulated, that is, they are not under the control of inducers or repressors. These genes are referred to as **constitutive.** Because they usually encode for products that are essential for viability under almost all growth and environmental conditions, these genes are continuously expressed. Also, not all regulation occurs at the genetic level (i.e., the level of transcription). For example, the production of some enzymes may be controlled at the protein synthesis (i.e., translational) level. The activities of other enzymes that have already been synthesized may be regulated at a posttranslational level, that is, certain catabolic or anabolic metabolites may directly interact

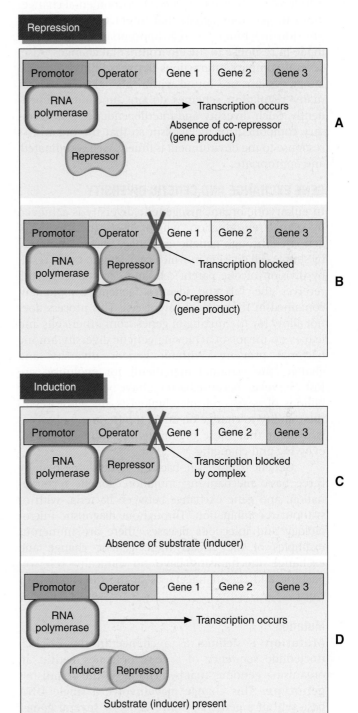

Figure 2-7 Transcriptional (i.e., genetic level) control of gene expression. Gene repression is depicted in **A** and **B**; induction is shown in **C** and **D**.

with enzymes to either increase or decrease their enzymatic activity.

Among different bacteria and even among different genes within the same bacterium, the mechanisms by which inducers and co-repressors are involved in gene regulation vary widely. Furthermore, bacterial cells have mechanisms to detect environmental changes and can generate signals that interact with gene expression machinery so that appropriate products are made in response to the environmental changes. Also, several complex interactions between different regulatory systems are found within a single cell so that many regulation schemes do not function independently. Such diversity and interdependence is a necessary component of metabolism so that the organism's response to the environment is timely, well coordinated, and appropriate.

GENE EXCHANGE AND GENETIC DIVERSITY

In eukaryotic organisms, genetic diversity is achieved by sexual reproduction, which allows the mixing of genomes through genetic exchange. Bacteria multiply by simple cell division in which two daughter cells result by division of one parent cell, and each daughter cell receives the full and identical genetic complement contained in the original parent cell. This process does not allow for the mixing of genes from other cells and leaves no means of achieving genetic diversity among bacterial progeny. Without genetic diversity and change, the essential ingredients for evolution are lost. However, because bacteria have been on earth for billions of years and microbiologists are witnesses to their ability to change on an all too frequent a basis, it is evident that these organisms are fully capable of evolving and changing.

Genetic change in bacteria is accomplished by three basic mechanisms: mutation, genetic recombination, and gene exchange between bacteria, with or without recombination. Throughout diagnostic microbiology and infectious diseases, there are numerous examples of the impact these genetic change and exchange mechanisms have on clinically relevant bacteria and the management of the infections they cause.

Mutation

Mutation is defined as a change in the original nucleotide sequence of a gene or genes within an organism's genome, that is, a change in the organism's **genotype.** This change may involve a single DNA base within a gene, an entire gene, or several genes. Mutational changes in the sequence may arise spontaneously, perhaps by an error made during DNA replication. Alternatively, mutations may be induced by chemical or physical factors (i.e., **mutagens**) in the

environment, or by biologic factors such as the introduction of foreign DNA into the cell. Changes in the DNA base sequence can result in changes in the base sequence of mRNA codons during transcription. This, in turn, can affect the types and sequences of amino acids that will be incorporated in proteins during translation.

Various outcomes may result from a mutation and are dependent on the site and extent of the mutation. For example, a mutation may be so devastating that it is lethal to the organism; the mutation therefore "dies" along with the organism. In other instances the mutation may be silent so that no changes in the organism's observable properties (i.e., the organism's **phenotype**) are detected. Alternatively, the mutation may result in a noticeable change in the organism's phenotype and the change may provide the organism with a survival advantage. This outcome, in Darwinian terms, is the basis for prolonged survival and evolution. Nonlethal mutations are considered stable if they are passed on from one generation to another as an integral part of the cell's genotype (i.e., genetic makeup). Additionally, genes that have undergone stable mutations may also be transferred to other bacteria by one of the mechanisms of gene exchange. In other instances, the mutation may be lost through repair mechanisms in the cell that restore the original phenotype and genotype, or be lost spontaneously during subsequent cycles of DNA replication.

Genetic Recombination

Besides mutations, bacterial genotypes can be changed through **recombination.** In this process, some segment of DNA that originated from one bacterial cell (i.e., donor) enters a second bacterial cell (i.e., recipient) and is exchanged with a DNA segment of the recipient's genome. This is also referred to as **homologous recombination** because the pieces of DNA that are exchanged usually have extensive homology or similarities in their nucleotide sequences. Recombination involves a number of binding proteins, with the RecA protein playing a central role (Figure 2-8, *A*). After recombination, the recipient DNA consists of one original, unchanged strand and the second from the donor DNA fragment that has been recombined.

Recombination is a molecular event that occurs frequently in many varieties of bacteria, including most of the clinically relevant species, and may involve any portion of the organism's genome. However, the recombinational event may go unnoticed unless the exchange of DNA results in a distinct alteration in the phenotype. Nonetheless, recombination is a major means by which bacteria may achieve genetic diversity and change.

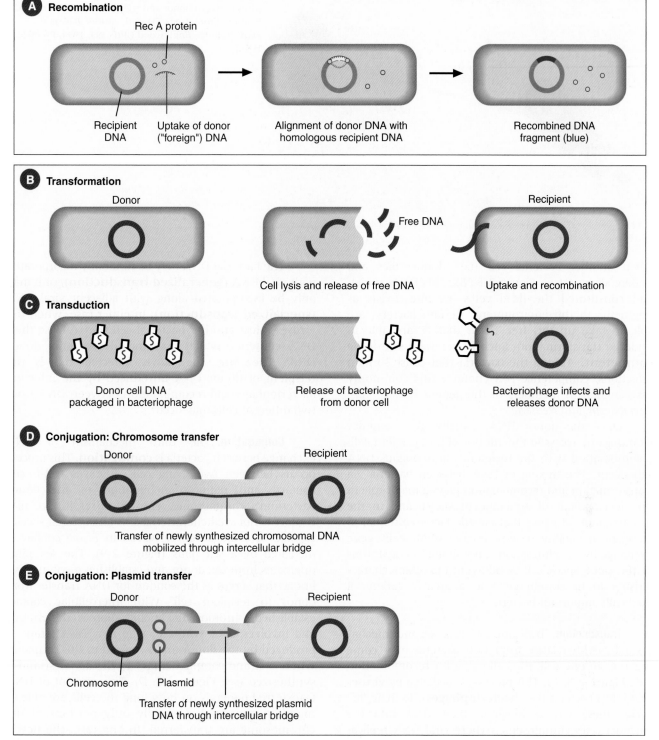

Figure 2-8   Genetic recombination (**A**). The mechanisms of gene exchange between bacteria: transformation (**B**), transduction (**C**), and conjugational transfer of chromosomal (**D**) and plasmid (**E**) DNA.

Gene Exchange

As just mentioned, an organism's opportunity for undergoing recombination depends on the acquisition of "foreign" DNA from a donor cell. The three mechanisms by which bacteria physically exchange DNA include transformation, transduction, and conjugation.

Transformation. Transformation involves recipient cell uptake of free DNA released into the environment

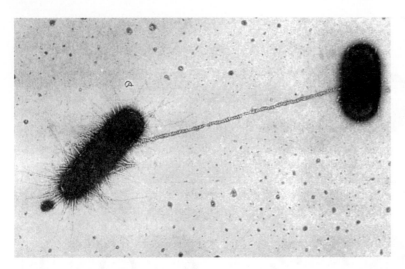

Figure 2-9 Photomicrograph of *Escherichia coli* sex pilus between donor and recipient cell. (Courtesy C. Brinton. From Brock TD et al, editors: *Biology of microorganisms*, Englewood Cliffs, NJ, 1994, Prentice Hall.)

when another bacterial cell (i.e., donor) dies and undergoes lysis (see Figure 2-8, *B*). This DNA, which had constituted the dead cell's genome, exists as fragments in the environment. Certain bacteria are able to take up this free DNA, that is, are able to undergo transformation. Such bacteria are said to be **competent.** Among the bacteria that cause human infections, competence is a characteristic commonly associated with members of the genera *Haemophilus, Streptococcus,* and *Neisseria.*

Once the donor DNA, usually as a singular strand, gains access to the interior of the recipient cell, recombination with the recipient's homologous DNA can occur. The mixing of DNA between bacteria via transformation and recombination plays a major role in the development of antibiotic resistance and in the dissemination of genes that encode factors essential to an organism's ability to cause disease. Additionally, gene exchange by transformation is not limited to organisms of the same species, thus allowing important characteristics to be disseminated to a greater variety of medically important bacteria.

Transduction. Transduction is a second mechanism by which DNA from two bacteria may come together in one cell, thus allowing for recombination (see Figure 2-8, *C*). This process is mediated by viruses that infect bacteria (i.e., **bacteriophages**). In their "life cycle" these viruses integrate their DNA into the bacterial cell's chromosome, where viral DNA replication and expression is directed. When the production of viral products is completed, viral DNA is excised (cut) from the bacterial chromosome and packaged within protein coats. The viruses are then released when the infected bacterial cell lyses. In transduction, the virus not only packages its own DNA but may also package a portion of the donor bacterium's DNA.

The bacterial DNA may be randomly incorporated with viral DNA **(generalized transduction),** or it may only be incorporated along with adjacent viral DNA **(specialized transduction).** In either case, when the viruses infect another bacterial cell they release their DNA contents, which may include bacterial donor DNA. Therefore, the newly infected cell is the recipient of donor DNA introduced by the infecting bacteriophage and recombination between DNA from two different cells may occur.

Conjugation. The third mechanism of DNA exchange between bacteria is **conjugation.** This process occurs between two living cells, involves cell-to-cell contact, and requires mobilization of the donor bacterium's chromosome. The nature of intercellular contact is not well characterized in all bacterial species capable of conjugation. However, in *E. coli* contact is mediated by a sex pilus (Figure 2-9). The sex pilus originates from the donor and establishes a conjugative bridge that serves as the conduit for DNA transfer from donor to recipient cell. With intercellular contact established, chromosomal mobilization is undertaken and involves DNA synthesis. One new DNA strand is produced by the donor and is passed to the recipient, where a strand complementary to the donor strand is synthesized (see Figure 2-8, *D*). The amount of DNA transferred depends on how long the cells are able to maintain contact, but usually only portions of the chromosome are transferred. In any case, the newly introduced DNA is then available to recombine with the recipient's chromosome.

In addition to chromosomal DNA, genes encoded in nonchromosomal genetic elements, such as plasmids and transposons, may be transferred by conjugation (see Figure 2-8, *E*). Not all plasmids are capable of conjugative transfer, but for those that are, the donor plasmid usually is replicated so that the donor retains

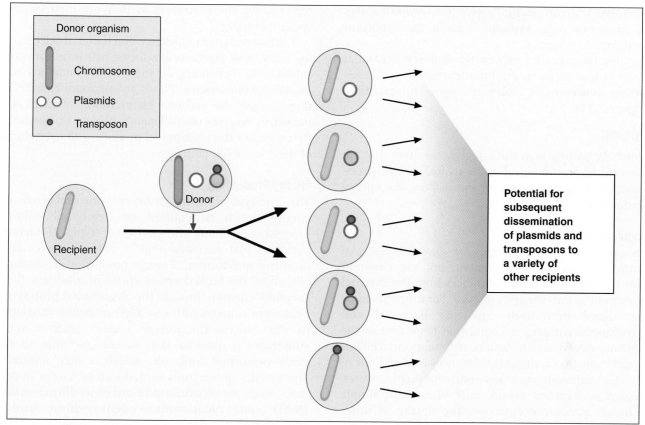

Figure 2-10   Pathways for bacterial dissemination of plasmids and transposons, together and independently.

a copy of the plasmid that is being transferred to the recipient. Plasmid DNA may also become incorporated into the host cell's chromosome.

In contrast to plasmids, transposons do not exist independently within the cell. Except when they are moving from one location to another, transposons must be incorporated into the chromosome and/or into plasmids. These elements are often referred to as "jumping genes," because of their ability to change location within, and even between, the genomes of bacterial cells. **Transposition** is the process by which these genetic elements excise from one genomic location and insert into another. Transposons carry genes whose products help mediate the transpositional process as well as genes that encode for some other characteristic such as antimicrobial resistance. In the cases of both plasmids and transposons, homologous recombination between the genes of these elements and the host bacterium's chromosomal DNA can occur but is not necessary. Therefore, these elements provide an alternative means for those organisms that cannot accommodate recombination to maintain foreign DNA in their genome.

Plasmids and transposons play a key role in genetic diversity and dissemination of genetic information among bacteria. Many characteristics that significantly alter the activities of clinically relevant bacteria are encoded and disseminated on these elements. Furthermore, as shown in Figure 2-10, the variety of strategies that bacteria can use to mix and match genetic elements provides them with a tremendous capacity to genetically adapt to environmental changes, including those imposed by human medical practices. A good example of this is the emergence and widespread dissemination of resistance to antimicrobial agents among clinically important bacteria. Bacteria have used their capacity for disseminating genetic information to establish resistance to most of the commonly used antibiotics (see Chapter 11 for more information regarding antimicrobial resistance mechanisms).

BACTERIAL METABOLISM

Fundamentally, bacterial metabolism involves all the cellular processes required for the organism's survival and replication. Familiarity with bacterial metabolism is essential for understanding bacterial interactions with human host cells, the mechanisms bacteria use to cause disease, and the basis of diagnostic microbiology, that is, the tests and strategies used for laboratory identification of infectious etiologies. Because metabolism is an extensive and complicated topic, this section focuses on processes most typical of medically important bacteria.

For the sake of clarity, metabolism is discussed in terms of four primary, but interdependent, processes: fueling, biosynthesis, polymerization, and assembly (Figure 2-11).

FUELING

Metabolic pathways in this category are those that are involved in acquisition of nutrients from the environment, production of precursor metabolites, and energy production.

Acquisition of Nutrients

Bacteria use various strategies for obtaining essential nutrients from the environment, but the common goal is to transport these substances from the external environment into the cell's interior. For nutrients to be internalized they must cross the **bacterial cell envelope,** a complex structure that helps protect the cell from environmental insults, maintains intracellular equilibrium, and transports substances into and out of the cell. Although some key nutrients such as water, oxygen, and carbon dioxide enter the cell by simple diffusion across the envelope, the uptake of other substances requires energy and selectivity by the cell's envelope.

Active transport is among the most common methods used for the uptake of nutrients such as certain sugars, most amino acids, organic acids, and many inorganic ions. The mechanism, which is driven by an energy-dependent pump, involves carrier molecules embedded in the membrane portion of the cell envelope. These carriers combine with the nutrients, transport them across the membrane, and release them within the cell. **Group translocation** is another transport mechanism that requires energy but differs from active transport in that the nutrient being transported undergoes chemical modification. Many sugars, purines, pyrimidines, and fatty acids are transported by this mechanism.

Production of Precursor Metabolites

Once inside the cell, many nutrients serve as the raw materials from which precursor metabolites for subsequent biosynthetic processes are produced. These metabolites, listed in Figure 2-11, are produced by three central pathways that include the Embden-Meyerhof-Parnas (EMP) pathway, the tricarboxylic acid (TCA) cycle, and the pentose phosphate shunt. These pathways and their relationship to one another are shown in Figure 2-12; not shown are the several alternative pathways (e.g., the Entner-Douderoff pathway) that play key roles in redirecting and

replenishing the precursors as they are used in subsequent processes.

The production efficiency of a bacterial cell resulting from these precursor-producing pathways can vary substantially, depending on the growth conditions and availability of nutrients. This is an important consideration because the accurate identification of medically important bacteria often depends heavily on methods that measure the presence of products and byproducts of these metabolic pathways.

Energy Production

The third type of fueling pathway is one that produces energy, which is required for nearly all cellular processes, including the two other types of fueling pathways just discussed, that is, nutrient uptake and precursor production. Energy production is accomplished by the breakdown of chemical substrates (i.e., chemical energy) through the degradative process of catabolism coupled with oxidation-reduction reactions. In this process the energy source molecule (i.e., **substrate**) is oxidized as it donates electrons to an electron-acceptor molecule, which is then reduced. The transfer of electrons is mediated by carrier molecules such as **nicotinamide-adenine-dinucleotide (NAD$^+$)** and **nicotinamide-adenine-dinucleotide-phosphate (NADP$^+$).** The energy released by the oxidation-reduction reaction is transferred to phosphate-containing compounds where high-energy phosphate bonds are formed. Adenosine triphosphate (ATP) is the most common of such molecules; the energy contained within this compound is eventually released by hydrolysis of ATP under controlled conditions. The release of this chemical energy and enzymatic activities specifically catalyze each biochemical reaction within the cell and drive nearly all cellular reactions.

The two general mechanisms for ATP production in bacterial cells are substrate-level phosphorylation and electron transport, also referred to as oxidative phosphorylation. In **substrate-level phosphorylation,** high-energy phosphate bonds produced by the central pathways are donated to adenosine diphosphate (ADP) to form ATP (see Figure 2-12). Additionally, pyruvate, a primary intermediate in the central pathways, serves as the initial substrate for several other pathways that also can generate ATP by substrate level phosphorylation. These other pathways constitute **fermentative metabolism,** which does not require oxygen and produces various end products, including alcohols, acids, carbon dioxide, and hydrogen. The specific fermentative pathways used, and hence end products produced, vary with different bacterial species. Detecting these products serves as an important basis

Precursor metabolites

- Glucose 6-phosphate
- Fructose 6-phosphate
- Pentose 5-phosphate
- Erythrose 4-phosphate
- 3-Phosphoglycerate
- Phosphoenolpyruvate
- Pyruvate
- Acetyl CoA
- α—Ketoglutarate
- Succinyl CoA
- Oxaloacetate

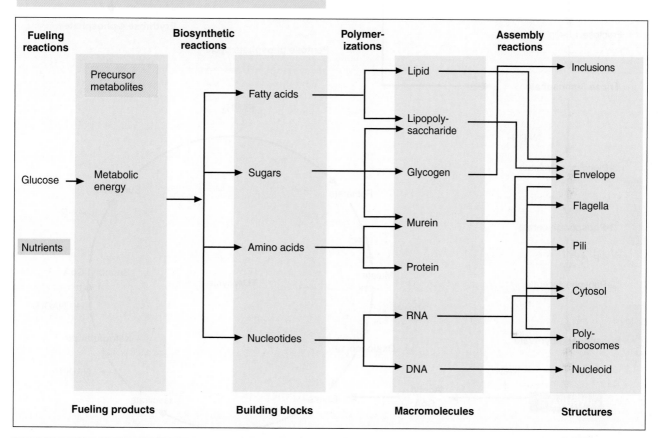

Figure 2-11   Overview of bacterial metabolism, which includes the processes of fueling, biosynthesis, polymerization, and assembly. (Modified from Niedhardt FC, Ingraham JL, Schaechter M, editors: *Physiology of the bacterial cell: a molecular approach,* Sunderland, Mass, 1990, Sinauer Associates.)

Nutrients

- Gases
  Carbon dioxide (CO_2)
  Oxygen (O_2)
  Ammonia (NH_3)
- Organic compounds, including amino acids
- Water (H_2O)
- Nitrate (NO^{3-})
- Phosphate (PO_4^{3-})
- Hydrogen sulfide (H_2S)
- Sulfate (SO_4^{2-})
- Potassium (K^+)
- Magnesium (Mg^{2+})
- Calcium (Ca^{2+})
- Sodium (Na^+)
- Iron (Fe^{3+})
Organic iron complexes

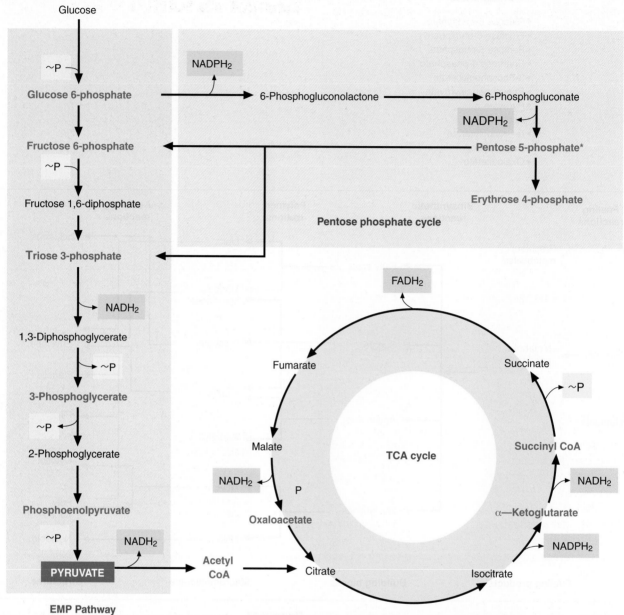

Figure 2-12   Overview diagram of the central metabolic pathways (Embden-Meyerhof-Parnas [EMP], the tricarboxylic acid [TCA] cycle, and the pentose phosphate shunt). Precursor metabolites (see also Figure 2-11) that are produced are highlighted in red; production of energy in the form of ATP (~P) by substrate-level phosphorylation is highlighted in yellow, and reduced carrier molecules for transport of electrons used in oxidative phosphorylation are highlighted in green.   (Modified from Niedhardt FC, Ingraham JL, Schaechter M, editors: *Physiology of the bacterial cell: a molecular approach,* Sunderland, Mass, 1990, Sinauer Associates.)

for laboratory identification of bacteria (see Chapter 7 for more information on the biochemical basis for bacterial identification).

Oxidative   phosphorylation   involves   an electron transport system that conducts a series of electron transfers from reduced carrier molecules such as NADH$_2$ and NADPH$_2$, produced in the central pathways (see Figure 2-12), to a terminal electron acceptor. The energy produced by the series of

oxidation-reduction reactions is used to generate ATP from ADP. When oxidative phosphorylation uses oxygen as the terminal electron acceptor, the process is known as **aerobic respiration. Anaerobic respiration** refers to processes that use final electron acceptors other than oxygen.

Knowledge regarding which mechanisms bacteria use to generate ATP is important for designing laboratory protocols for cultivating and identifying these organisms. For example, some bacteria solely depend on aerobic respiration and cannot grow in the absence of oxygen **(strictly aerobic bacteria).** Others can use either aerobic respiration or fermentation, depending on the availability of oxygen **(facultatively anaerobic bacteria).** For still others, oxygen is absolutely toxic **(strictly anaerobic bacteria).**

BIOSYNTHESIS

The fueling reactions essentially bring together all the raw materials needed to initiate and maintain all other cellular processes. The production of precursors and energy is mostly accomplished by catabolic processes that involve the degradation of substrate molecules. The three remaining primary pathways—biosynthesis, polymerization, and assembly—depend on anabolic metabolism, in which precursor compounds are put together for the creation of larger molecules (polymers) needed for assembly of cellular structures (see Figure 2-11).

Biosynthetic processes use the 12 precursor products in dozens of pathways to produce nearly 100 different building blocks such as amino acids, fatty acids, sugars, and nucleotides (see Figure 2-11). Many of these pathways are highly complex and interdependent, whereas other pathways are almost completely independent. In many cases, the enzymes that drive the individual pathways are encoded on a single mRNA molecule that has been cotranscribed from contiguous genes in the bacterial chromosome (i.e., an operon).

As discussed with the fueling pathways, bacterial genera and species vary extensively in their biosynthetic capabilities. Knowledge of these variations is necessary to ensure the design of optimal conditions for growing organisms under laboratory conditions. For example, some organisms may not be capable of synthesizing a particular amino acid that is a necessary building block for several essential proteins. Without the ability to synthesize the amino acid, the bacterium must obtain that building block from the environment. Similarly, if this organism is to be grown in the microbiology laboratory, this need must be fulfilled by supplying the amino acid in culture media.

POLYMERIZATION AND ASSEMBLY

Various anabolic reactions assemble (polymerize) the building blocks into macromolecules that include lipids, lipopolysaccharides, polysaccharides, proteins, and nucleic acids. This synthesis of macromolecules is driven by energy and enzymatic activity within the cell. Similarly, energy and enzymatic activities also drive the assembly of these various macromolecules into the component structures of the bacterial cell. That is, the cellular structures are the product of all the genetic and metabolic processes discussed.

STRUCTURE AND FUNCTION OF THE BACTERIAL CELL

Before discussing bacterial cellular structure, a general description of cells is useful. Based on key characteristics, all cells are classified into two basic types: prokaryotic and eukaryotic. Although these two cell types share many common features, there are many important differences in terms of structure, metabolism, and genetics.

EUKARYOTIC AND PROKARYOTIC CELLS

Among clinically relevant microorganisms, bacterial cells are prokaryotic, whereas cells of parasites and fungi are eukaryotic, as are those of plants and all higher animals. Viruses are acellular and depend on host cells for survival.

A notable characteristic of **eukaryotic cells,** such as those of parasites and fungi, is the presence of membrane-enclosed organelles that have specific cellular functions. Examples of these organelles and their respective functions include:

- Endoplasmic reticulum—process and transport proteins
- Golgi body—process substances for transport out of cell
- Mitochondria—generate energy (ATP)
- Lysosomes—provide environment for controlled enzymatic degradation of intracellular substances
- Nucleus—provide membrane enclosure for chromosomes

Additionally, eukaryotic cells have an infrastructure, or cytoskeleton, that provides support for the different organelles.

Prokaryotic cells, such as those of bacteria, do not contain organelles. All functions take place in the cytoplasm or cytoplasmic membrane of the cell. Prokaryotic and eukaryotic cell types also differ considerably at the macromolecular level regarding protein synthesis machinery, chromosomal organization, and gene expression. One notable structure present only in bacterial cells that is absent among eukaryotes is a cell wall composed of peptidoglycan. As is mentioned many times over in this text, this structure alone has immeasurable impact on the practice of diagnostic bacteriology and the management of bacterial diseases.

BACTERIAL MORPHOLOGY

Most clinically relevant bacterial species range in size from 0.25 to 1 μm in width and 1 to 3 μm in length, thus requiring microscopy for visualization (see Chapter 6 for more information on microscopy). Just as bacterial species and genera vary in their metabolic

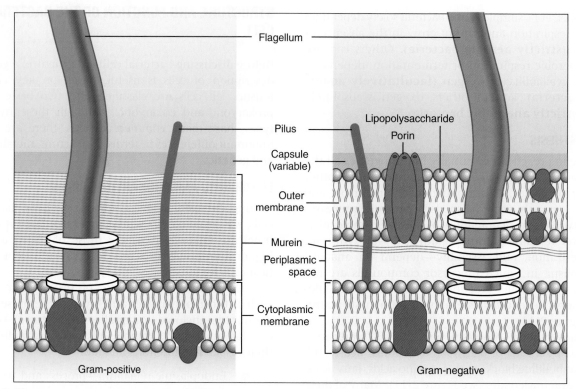

Figure 2-13   General structures of the gram-positive and gram-negative bacterial cell envelopes. The outer membrane and periplasmic space are only present in the gram-negative envelope. The murein layer is substantially more prominent in gram-positive envelopes. (Modified from Niedhardt FC, Ingraham JL, Schaechter M, editors: *Physiology of the bacterial cell: a molecular approach,* Sunderland, Mass, 1990, Sinauer Associates.)

processes, their cells also vary in size, morphology, cell-to-cell arrangements, and in the makeup of one of the most prominent cellular structures, the cell wall. In fact, cell wall differences provide the basis for the **Gram stain,** the most fundamental test used in bacterial identification schemes. This staining procedure separates almost all medically important bacteria into two general types: **gram-positive bacteria,** which stain a deep blue color, and **gram-negative bacteria,** which stain a pink to red color (see Figure 6-3). This simple but important color distinction is due to differences in the constituents of bacterial cell walls that influence the cell's ability to retain certain dyes, even after decolorization with alcohol.

Common bacterial cellular morphologies include **cocci** (round), **coccobacilli** (ovoid), and **bacillus** (rod-shaped), as well as fusiform (pointed-end), curved, or spiral shapes. Cellular arrangements are also noteworthy because cells may characteristically occur singly, in pairs, tetrads, clusters, or in chains (see Figure 6-4 for examples of bacterial staining and morphologies). The determination of Gram reaction coupled with cell size, morphology, and arrangement are essential aspects of bacterial identification.

BACTERIAL CELL COMPONENTS

Bacterial cell components can be divided into those that make up the cell envelope and its appendages and those associated with the cell's interior. Importantly, the cellular structures work together to function as a complex and integrated unit.

Cell Envelope

As shown in Figure 2-13, the outermost structure, the cell envelope, comprises:

- An outer membrane (in gram-negative bacteria only)
- A cell wall composed of the peptidoglycan macromolecule (also known as the *murein layer*)
- Periplasm (in gram-negative bacteria only)
- The cytoplasmic membrane, interior to which is the cytoplasm

Outer Membrane. Outer membranes are only found in gram-negative bacteria, and they function as the cell's initial barrier to the environment. These membranes serve as primary permeability barriers to hydrophilic and hydrophobic compounds, and they retain essential enzymes and other proteins located in the periplasmic space. The membrane is a bilayered structure composed of **lipopolysaccharide,** which gives the surface of gram-negative bacteria a net negative charge; it also plays a significant role in the ability of certain bacteria to cause disease.

Scattered throughout the lipopolysaccharide macromolecules are protein structures called **porins.** These water-filled structures control the passage of nutrients and other solutes, including antibiotics, through the outer membrane. The number and types of porins vary with bacterial species, and these differences can substantially influence the extent to which various substances pass through the outer membranes of different bacteria. In addition to porins, other proteins (murein lipoproteins) facilitate attachment of the outer membrane to the next deeper layer in the cell envelope, the cell wall.

Cell Wall (Murein Layer). The cell wall, also referred to as the **peptidoglycan,** or murein layer, is an essential structure found in nearly all clinically important bacteria. This structure gives the bacterial cell shape and strength to withstand changes in environmental osmotic pressures that would otherwise result in cell lysis. The murein layer also protects against mechanical disruption of the cell and offers some barrier to the passage of larger substances. Because this structure is essential for the survival of bacteria, its synthesis and structure have been the primary target for the development and design of several antimicrobial agents.

The structure of the cell wall is unique in nature and is composed of **disaccharide-pentapeptide** subunits. The disaccharides *N*-acetylglucosamine and *N*-acetylmuramic acid are the alternating sugar components (moieties), with the amino acid chain only linked to *N*-acetylmuramic acid molecules (Figure 2-14). Polymers of these subunits crosslink to one another via peptide bridges to form peptidoglycan sheets. Layers of these sheets are, in turn, crosslinked with one another to give a multilayered, crosslinked structure of considerable strength. Referred to as the *murein sacculus,* or *sack,* this peptidoglycan structure surrounds the entire cell.

A notable difference between the cell walls of gram-positive and gram-negative bacteria is that the peptidoglycan layer in gram-positive bacteria is substantially thicker (see Figure 2-13). Additionally, the cell wall of gram-positive bacteria contains **teichoic acids** (i.e., glycerol or ribitol phosphate polymers that are combined with various sugars, amino acids, and amino sugars). Some teichoic acids are linked to *N*-acetylmuramic acid, and others (e.g., lipoteichoic acids) are linked to the next underlying layer, the cellular membrane. Still other gram-positive

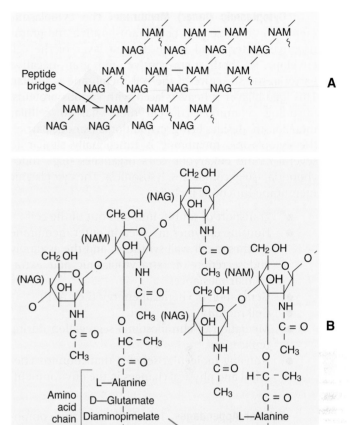

Figure 2-14   Peptidoglycan sheet **(A)** and subunit **(B)** structure. Multiple peptidoglycan layers compose the murein structure, and different layers are extensively crosslinked by peptide bridges (NAG, *N*-acetylglucosamine; NAM, *N*-acetylmuramic acid). Note that amino acid chains only derive from NAM. (Modified from Saylers AA, Whitt DD: *Bacterial pathogenesis: a molecular approach,* Washington, DC, 1994, American Society for Microbiology Press.)

bacteria (e.g., mycobacteria) fortify their murein layer with waxy substances, such as **mycolic acids,** that make their cells even more refractory to toxic substances, including acids.

Periplasmic Space. The periplasmic space is only found in gram-negative bacteria and is bounded by the internal surface of the outer membrane and the external surface of the cellular membrane. This area, which contains the murein layer, is not just an empty space but consists of gel-like substances that help to secure nutrients from the environment. This space also contains several enzymes that degrade macromolecules and detoxify environmental solutes, including antibiotics, that enter through the outer membrane.

Cytoplasmic (Inner) Membrane. The cytoplasmic membrane is present in both gram-positive and gram-negative bacteria and is the deepest layer of the cell envelope. In both gram-positive and gram-negative bacteria, the structure of the cell membrane is similar. This lipid bilayer is heavily laced with various proteins, including a number of enzymes vital to cellular metabolism. Besides being an additional osmotic barrier, the cytoplasmic membrane is functionally similar to several of the eukaryotic cell's organelles (e.g., mitochondria, Golgi complexes, lysosomes). The cytoplasmic membrane functions include:

- Transport of solutes into and out of the cell
- Housing enzymes involved in outer membrane synthesis, cell wall synthesis, and the assembly and secretion of extracytoplasmic and extracellular substances
- Generation of chemical energy (i.e., ATP)
- Cell motility
- Mediation of chromosomal segregation during replication
- Housing molecular sensors that monitor chemical and physical changes in the environment

Cellular Appendages. In addition to the components of the cell envelope proper, there are also cellular appendages (i.e., capsules, fimbriae, and flagella) that are associated with or proximal to this portion of the cell. The presence of these appendages, which can play a role in causing infections and in laboratory identification, varies among bacterial species and even among strains within the same species.

The **capsule** is immediately exterior to the murein layer of gram-positive bacteria and the outer membrane of gram-negative bacteria. Often referred to as the **slime layer,** the capsule is composed of high–molecular-weight polysaccharides whose production may depend on the environment and growth conditions surrounding the bacterial cell. The capsule does not function as an effective permeability barrier or add strength to the cell envelope but does protect bacteria from attack by cells of the human defense system. The capsule also facilitates and maintains bacterial colonization of biologic (e.g., teeth) and inanimate (e.g., prosthetic heart valves) surfaces through the formation of biofilms.

Fimbriae, or **pili,** are hairlike, proteinaceous structures that extend from the cell membrane into the external environment; some may be up to 2 μm in length. Two general types are known: common pili and sex pili. Common pili are adhesins that help bacteria attach to animal host cell surfaces, often as the first step in establishing infection. The sex pilus, well characterized in the gram-negative bacillus *E. coli,* serves as the conduit for the passage of DNA from donor to recipient during conjugation.

Flagella are complex structures, mostly composed of the protein flagellin, intricately embedded in the cell envelope. These structures are responsible for bacterial motility. Although not all bacteria are motile, motility plays an important role in survival and the ability of certain bacteria to cause disease. Depending on the bacterial species, flagella may be located at one end of the cell **(monotrichous flagella)** or at both ends of the cell **(lophotrichous flagella),** or the entire cell surface may be covered with flagella **(peritrichous flagella).**

Cell Interior

Those structures and substances that are bounded internally by the cytoplasmic membrane compose the cell interior and include the cytosol, polysomes, inclusions, the nucleoid, plasmids, and endospores.

The **cytosol,** where nearly all other functions not conducted by the cell membrane occur, contains thousands of enzymes and is the site of protein synthesis. The cytosol has a granular appearance caused by the presence of many polysomes (mRNA complexed with several ribosomes during translation and protein synthesis) and **inclusions** (i.e., storage reserve granules). The number and nature of the inclusions vary depending on the bacterial species and the nutritional state of the organism's environment. Two common types of granules include **glycogen,** a storage form of glucose, and **polyphosphate granules,** a storage form for inorganic phosphates that are microscopically visible in certain bacteria stained with specific dyes.

Unlike eukaryotic chromosomes, the bacterial chromosome is not enclosed within a membrane-bound nucleus. Instead the bacterial chromosome exists as a **nucleoid** in which the highly coiled DNA is intermixed with RNA, polyamines, and various proteins that lend structural support. At times, depending on the stage of cell division, there may be more than one chromosome per bacterial cell. **Plasmids** are the other genetic elements that exist independently in the cytosol, and their numbers may vary from none to several per bacterial cell.

The final bacterial structure to be considered is the **endospore.** Under adverse physical and chemical conditions, or when nutrients are scarce, some bacterial genera are able to form spores (i.e., **sporulate**). Sporulation involves substantial metabolic and structural changes in the bacterial cell. Essentially, the cell transforms from an actively metabolic and growing state to a dormant state, with a decrease in cytosol and a concomitant increase in the thickness and strength of the cell envelope. The spore state is maintained until favorable conditions for growth are again encountered. This survival tactic is demonstrated by a number of clinically relevant bacteria and frequently challenges our ability to thoroughly sterilize materials and foods for human use.

ADDITIONAL READING

Brock TD et al, editors: *Biology of microorganisms,* Englewood Cliffs, NJ, 1994, Prentice Hall.

Joklik WK et al, editors: *Zinsser microbiology,* Norwalk, Conn, 1992, Appleton & Lange.

Moat AG, Foster JW: *Microbial physiology,* New York, 1995, Wiley-Liss.

Neidhardt FC, Ingraham JL, Schaecter M, editors: *Physiology of the bacterial cell: a molecular approach,* Sunderland, Mass, 1990, Sinauer Associates.

Ryan KJ, editor: *Sherris medical microbiology: an introduction to infectious diseases,* Norwalk, Conn, 1994, Appleton & Lange.

Saylers AA, Whitt DD: *Bacterial pathogenesis: a molecular approach,* Washington, DC, 1994, American Society for Microbiology Press.

CHAPTER 3

Host-Microorganism Interactions

Interactions between humans and microorganisms are exceedingly complex and are far from being completely understood. What is known about the interactions between these two living entities plays an important role both in the practice of diagnostic microbiology and in the management of infectious diseases. Understanding these interactions is necessary for establishing methods to reliably isolate specific microorganisms from patient specimens and for developing effective treatment strategies. This chapter provides the framework for understanding the various aspects of host-microorganism interactions.

Humans and microorganisms inhabit the same planet, and their paths cross in many and varied ways so that interactions are inevitable. The most important point regarding host-microorganism interactions is that the relationships are always bidirectional in nature. The bias is to believe that humans dominate the relationship because humans use microorganisms in various settings, including the food and fermentation industry, as biologic insecticides for agriculture; to genetically engineer a multitude of products; and even for biodegrading our industrial wastes. However, as one realizes that microbial populations share the common goal of survival with humans, and that they have been successful at achieving that goal, then exactly which participant in the relationship is the user and which one is the used becomes much less clear. This is especially true when considering the microorganisms most closely associated with humans and human disease.

The complex relationships between human hosts and medically relevant microorganisms are best understood by considering the sequential steps in the development of microbial-host associations and subsequent development of infection and disease. The stages of interaction are shown in Figure 3-1 and include physical encounter between host and microorganism, colonization or survival of the microorganism on an internal (gastrointestinal, respiratory, or genitourinary tract) or external (skin) surface of the host, microbial entry, invasion, and dissemination to deeper tissues and organs of the human body, and resolution or outcome. Each stage is discussed from two perspectives: the human host's and the microorganism's.

THE ENCOUNTER BETWEEN HOST AND MICROORGANISM

THE HUMAN HOST'S PERSPECTIVE

Because microorganisms are found everywhere, human encounters are inevitable, but the means of encounter vary widely. Which microbial population a human is exposed to and the mechanism of exposure are often direct consequences of a person's activity. Certain activities carry different risks for an encounter and there is a wide spectrum of activities or situations over which a person may or may not have absolute control. For example, acquiring salmonellosis because one fails to thoroughly cook the holiday turkey is avoidable, whereas contracting tuberculosis as a consequence of living in conditions of extreme poverty and overcrowding may be less avoidable. The role that human activities play in the encounter process cannot be overstated because most of the crises associated with infectious disease could be avoided or greatly reduced if human behavior and living conditions could be changed.

Microbial Reservoirs and Transmission

Humans encounter microorganisms when they enter or are exposed to the same environment in which the microbial agents live or when the infectious agents are brought to the human host by indirect means. The environment, or place of origin, of the infecting agent is referred to as the **reservoir.** As shown in Figure 3-2, microbial reservoirs include humans, animals, water, food, air, and soil. The human host may acquire microbial agents by various means referred to as the **modes of transmission.** The mode of transmission is direct when the host directly contacts the microbial reservoir and is indirect when the host encounters the microorganism by an intervening agent of transmission.

The agents of transmission that bring the microorganism from the reservoir to the host may be a living entity, in which case they are called **vectors,** or they may be a nonliving entity referred to as a **vehicle** or **fomite.** Additionally, some microorganisms may have a single mode of transmission while others may spread by various methods. From a diagnostic microbiology perspective, knowledge about an infectious agent's mode of transmission is often important for

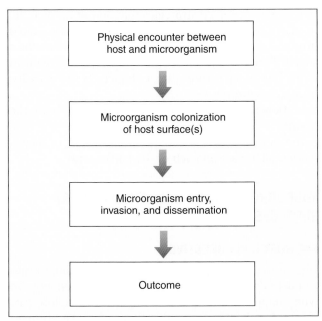

Figure 3-1 General stages of microbial-host interaction.

determining optimum specimens for organism isolation and for taking precautions that minimize the risk of laboratory-acquired infections (see Chapters 4 and 64 for more information regarding laboratory safety).

Humans as Microbial Reservoirs

Humans play a substantial role as microbial reservoirs. Indeed, the passage of a neonate from the sterile environment of the mother's womb through the birth canal, which is heavily colonized with various microbial agents, is a primary example of one human directly acquiring microorganisms from another human serving as the reservoir. This is the mechanism by which newborns first encounter microbial agents. Other examples in which humans serve as the microbial reservoir include acquisition of "strep" throat through touching; hepatitis by blood transfusions; gonorrhea, syphilis, and acquired immunodeficiency syndrome by sexual contact; tuberculosis by coughing; and the common cold through sneezing. Indirect transfer can occur when microorganisms from one individual contaminate a vehicle of transmission such as water (e.g., cholera) that is then ingested by another person. In the medical setting, indirect transmission of microorganisms from one human host to another via contaminated medical devices helps disseminate infections in hospitals. Hospital-acquired infections are referred to as **nosocomial infections.**

Animals as Microbial Reservoirs

From animal reservoirs infectious agents can be transmitted directly to humans such as through an animal bite (e.g., rabies) or indirectly such as through the bite of insect vectors that feed on both animals and humans (e.g., Lyme disease and Rocky Mountain spotted fever). Animals may also transmit infectious agents by water and food supplies. For example, beavers are often heavily colonized with parasites that cause infections of the human gastrointestinal tract. These parasites may be encountered and subsequently acquired when contaminated stream water is used by beaver and camper alike. Alternatively, animals used for human food carry numerous bacteria (e.g., *Salmonella* and *Campylobacter*) that, if not destroyed by appropriate cooking during preparation, can cause severe gastrointestinal illness.

Many other infectious diseases are encountered through direct or indirect animal contact, and informa-

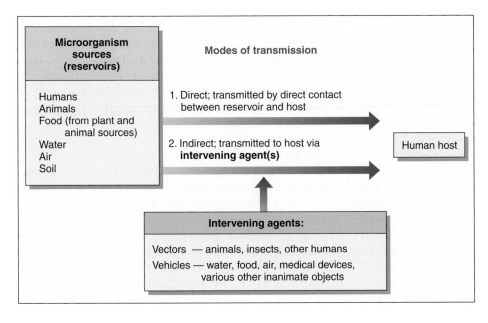

Figure 3-2 Summary of microbial reservoirs and modes of transmission to humans.

tion regarding a patient's exposure to animals is often a key component to diagnosing these infections. Some microorganisms primarily infect animal populations and on occasion accidentally encounter and infect humans. Such infectious diseases are classified as **zoonoses;** when a human infection results from such an encounter it is referred to as a **zoonotic infection.**

Insects as Vectors

The most common role of insects (arthropods) in the transmission of infectious diseases is as vectors rather than as reservoirs. A wide variety of arthropods transmit viral, parasitic, and bacterial diseases from animals to humans, whereas others transmit microorganisms between human hosts without an animal reservoir being involved. Malaria, a leading cause of death from infection, is a prime example of a disease that is maintained in the human population by the activities of an insect (i.e., mosquito) vector. Still other arthropods may themselves be agents of disease. These include organisms such as lice and scabies that are spread directly between humans and cause skin irritations but do not penetrate the body. Because they are able to survive on the skin of the host without usually gaining access to internal tissues, they are referred to as **ectoparasites.**

The Environment as a Microbial Reservoir

The soil and natural environmental debris are reservoirs for countless types of microorganisms. Therefore, it is not surprising that these also serve as reservoirs for microorganisms that can cause infections in humans. Many of the fungal agents (see Chapter 50) are acquired by inhalation of soil and dust particles containing the microorganisms (e.g., San Joaquin Valley fever). Other, nonfungal infections (e.g., tetanus) may result when microbial agents in the environment are introduced into the human body as a result of a penetrating wound.

THE MICROORGANISM'S PERSPECTIVE

Clearly, numerous activities can result in human encounters with many microorganisms. Because humans are engaged in all of life's complex activities, the tendency is to perceive the microorganism as having a passive role in the encounter process. However, this assumption is a gross oversimplification.

Microorganisms are also driven by survival, and the environments of the reservoirs they occupy must allow their metabolic and genetic needs to be fulfilled. Furthermore, most reservoirs are usually inhabited by hundreds, if not thousands, of different species. Yet human encounters with the reservoirs, either directly or indirectly, do not result in all species establishing an association with the human host. Although some species have evolved strategies that do not involve the human host to ensure survival, others have included humans to a greater or lesser extent as part of their survival tactics. Therefore, the latter type of organism often has mechanisms that enhance its chances for human encounter.

Depending on factors associated with both the human host and the microorganism involved, the encounter may have a beneficial, disastrous, or inconsequential impact on each of the participants.

MICROORGANISM COLONIZATION OF HOST SURFACES

THE HOST'S PERSPECTIVE

Once a microbe and the human host are brought into contact, the outcome of the encounter depends on what happens during each step of interaction (see Figure 3-1), beginning with colonization. The human host's role in microbial **colonization,** defined as the persistent survival of microorganisms on a surface of the human body, is dictated by the defenses that protect vital internal tissues and organs against microbial invasion. The first defenses are the external and internal body surfaces that are in relatively direct contact with the external environment and as such are the body areas with which microorganisms will initially associate. These surfaces include:

- Skin (including conjunctival epithelium covering the eye)
- Mucous membranes that line the mouth or oral cavity, the respiratory tract, the gastrointestinal tract, and the genitourinary tract

Because body surfaces are always present and they protect against all microorganisms, skin and mucous membranes are considered constant and nonspecific protection mechanisms. As is discussed in this text, other protective mechanisms are produced only in response to the presence of microbial agents (i.e., inducible defenses) and some are specifically directed to particular microorganisms (i.e., specific defenses).

Skin and Skin Structures

Skin serves as a physical and chemical barrier to microorganisms; its protective characteristics are summarized in Box 3-1 and Figure 3-3. The acellular, outermost layer of skin and tightly packed cellular layers underneath provide an impenetrable physical barrier to all microorganisms, unless damaged. Additionally, these layers continuously shed, thus dislodging bacteria that have attached to the outer layers. The skin is also a dry and cool environment that contrasts with the warm

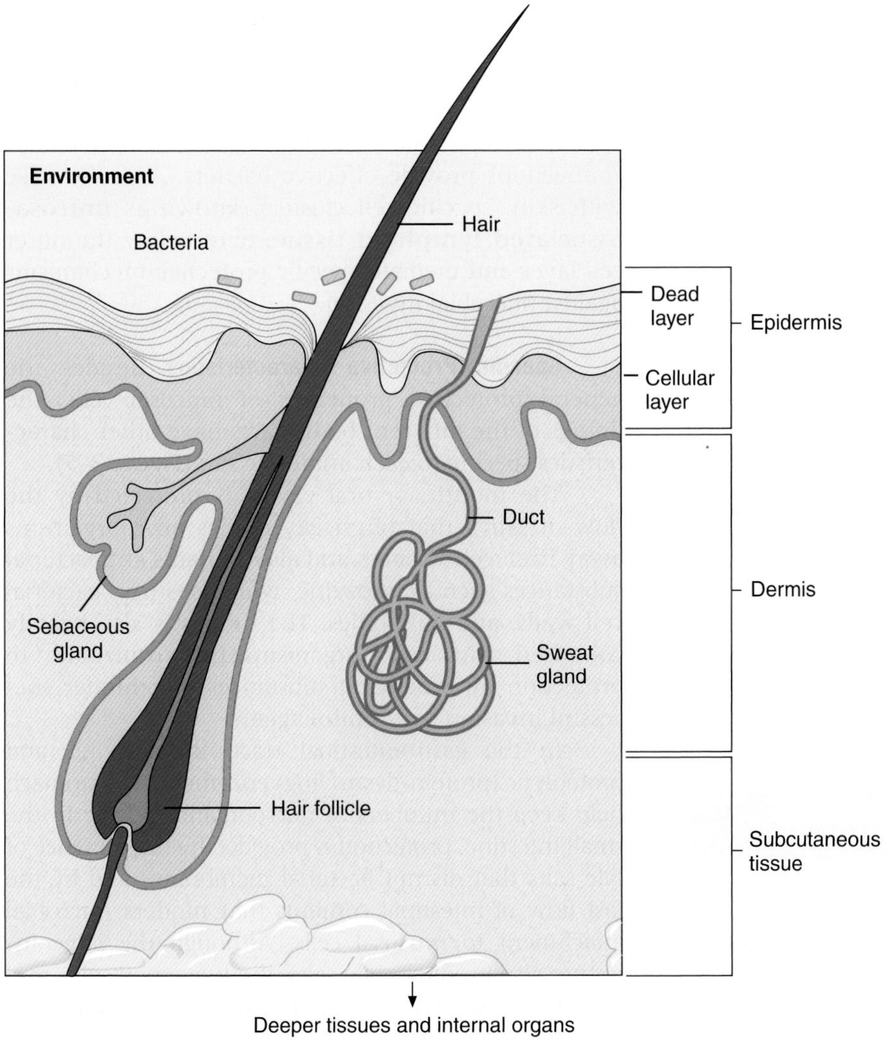

Figure 3-3   Skin and skin structures.

Deeper tissues and internal organs

BOX 3-1	Protective Characteristics of the Skin and Skin Structures

SKIN STRUCTURE	PROTECTIVE ACTIVITY
Outer (dermal) layers	Physical barrier to microbial penetration
	Sloughing of outer layers removes attached bacteria
	Provide dry, acidic, and cool conditions that limit bacterial growth
Hair follicles, sweat glands, sebaceous glands	Production of acids, alcohols, and toxic lipids that limit bacterial growth
Conjunctival epithelium covering the eyes	Flushing action of tears removes microorganisms
	Tears contain lysozyme that destroys bacterial cell wall
Skin-associated lymphoid tissue	Mediate specific and nonspecific protection mechanisms against microorganisms that penetrate outer tissue layers

and moist environment in which many microbial agents thrive.

The follicles and glands of the skin produce various natural antibacterial substances. However, many microorganisms can survive the conditions of the skin. These bacteria are known as *skin colonizers,* and they often produce substances that may be toxic and inhibit the growth of more harmful microbial agents. Beneath the outer layers of skin are various host cells that protect against organisms that breach the surface barriers. These cells, collectively known as **skin-associated lymphoid tissue,** mediate specific and nonspecific responses directed at controlling microbial invaders.

Mucous Membranes

Because cells that line the respiratory tract, gastrointestinal tract, and genitourinary tract are involved in numerous functions besides protection, they are not covered with a hardened acellular layer like that

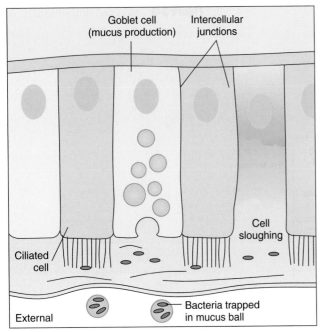

Figure 3-4 General features of mucous membranes highlighting protective features such as ciliated cells, mucus production, tight intercellular junctions, and cell sloughing.

BOX 3-2	Protective Characteristics of Mucous Membranes

MUCOUS MEMBRANE STRUCTURES	PROTECTIVE ACTIVITY
Mucosal cells	Rapid sloughing for bacterial removal Tight intercellular junctions prevent bacterial penetration
Goblet cells	Mucus production: Protective lubrication of cells Bacterial trapping Contains specific antibodies with specific activity against bacteria Provision of antibacterial substances to mucosal surface: Lysozyme: degrades bacterial cell wall Lactoferrin: competes for bacterial iron supply Lactoperoxidase: production of substances toxic to bacteria
Mucosa-associated lymphoid tissue	Mediates specific responses against bacteria that penetrate outer layer

found on the skin. However, the cells that compose these membranes still exhibit various protective characteristics (Box 3-2 and Figure 3-4).

General Protective Characteristics. A major protective component of mucous membranes is the mucus itself. This substance serves to trap bacteria before they can reach the outer surface of the cells, lubricates the

cells to prevent damage that may promote microbial invasion, and contains numerous specific (i.e., antibodies) and nonspecific antibacterial substances. In addition to mucous activity and flow mediated by cilia action, rapid cellular shedding and tight intercellular connections provide effective barriers. As is the case with skin, specific cell clusters, known as **mucosa-associated lymphoid tissue,** exist below the outer cell layer and mediate specific protective mechanisms against microbial invasion.

Specific Protective Characteristics. Besides the general protective properties of mucosal cells, the linings of the different body tracts have other characteristics specific to each anatomic site (Figure 3-5).

The mouth, or oral cavity, is protected by the flow of saliva that physically carries microorganisms away from cell surfaces and also contains antibacterial substances such as lysozyme, which destroys bacterial cell walls, and antibodies. The mouth is also heavily colonized with microorganisms that contribute to protection by producing substances that hinder successful invasion by harmful agents.

In the gastrointestinal tract, the low pH and proteolytic (protein-destroying) enzymes of the stomach help keep the numbers of microorganisms low. In the small intestine, protection is provided by the presence of bile salts that disrupt bacterial membranes and by the fast flow of intestinal contents that hinders microbial attachment to mucosal cells. Although the large intestine also produces bile salts, the movement of bowel contents is slower so that a higher concentration of microbial agents are able to attach to the mucosal cells and inhabit this portion of the gastrointestinal tract. As in the oral cavity, the high concentration of normal microbial inhabitants in the large bowel also contributes significantly to protection.

In the upper respiratory tract, nasal hairs keep out large airborne particles that may contain microorganisms. The cough-sneeze reflex also significantly contributes to the removal of potentially infective agents. The cells lining the trachea contain cilia (hairlike cellular projections) that move mucus and the microorganisms trapped within upward and away from delicate cells of the lungs (see Figure 3-4). These barriers are so effective that only inhalation of particles smaller than 2 to 3 μm have a chance of reaching the lungs.

In the female urogenital tract, the vaginal lining and the ectocervix are protected by heavy colonization with normal microbial inhabitants and a low pH. A thick mucous plug in the cervical opening is a substantial barrier that keeps microorganisms from ascending and invading the more delicate tissues of the uterus, fallopian tubes, and ovaries. The anterior

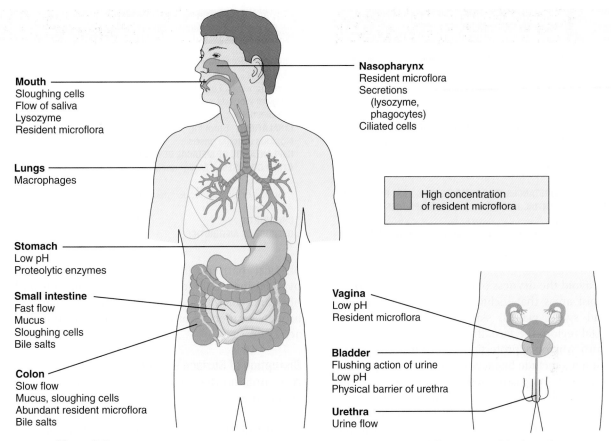

Mouth
Sloughing cells
Flow of saliva
Lysozyme
Resident microflora

Lungs
Macrophages

Stomach
Low pH
Proteolytic enzymes

Small intestine
Fast flow
Mucus
Sloughing cells
Bile salts

Colon
Slow flow
Mucus, sloughing cells
Abundant resident microflora
Bile salts

Nasopharynx
Resident microflora
Secretions
  (lysozyme,
  phagocytes)
Ciliated cells

High concentration
of resident microflora

Vagina
Low pH
Resident microflora

Bladder
Flushing action of urine
Low pH
Physical barrier of urethra

Urethra
Urine flow

Figure 3-5   Protective characteristics associated with the mucosal linings of different internal body surfaces.

urethra of males and females is naturally colonized with microorganisms, but stricture at the urethral opening provides a physical barrier that, combined with low urine pH and the flushing action of urination, protects against bacterial invasion of the bladder, ureters, and kidneys.

THE MICROORGANISM'S PERSPECTIVE

As previously discussed, microorganisms that inhabit many surfaces of the human body (see Figure 3-5) are referred to as **colonizers,** or **normal flora.** Some are **transient colonizers** because they merely survive, but do not multiply, on the surface and are frequently shed with the host cells. Others, called **resident flora,** not only survive but also thrive and multiply; their presence is more permanent.

The normal flora of body surfaces vary considerably with anatomic location. For example, environmental conditions, such as temperature and oxygen availability, vary considerably between the nasal cavity and small bowel. Only microorganisms with the metabolic capacity to survive under the circumstances that each anatomic location offers will be inhabitants of those particular body surfaces.

Knowledge of the normal flora of the human body is extremely important in diagnostic microbiology, especially for determining the clinical significance of microorganisms that are isolated from patient specimens. Normal flora organisms frequently are found in clinical specimens. This is a result of contamination of normally sterile specimens during collection or because the colonizing organism is actually involved in the infection. Microorganisms considered as normal colonizers of the human body and the anatomic locations they colonize are addressed in Part VII.

Microbial Colonization

Colonization may be the last step in the establishment of a long-lasting, mutually beneficial (i.e., **commensal**), or harmless, relationship between a colonizer and the human host. Alternatively, colonization may be the first step in the process of developing infection and disease. Whether colonization results in a harmless or harmful situation depends on the host and microorganism characteristics. In either case, successful initial colonization depends on the microorganism's ability to survive the conditions first encountered on the host surface (Box 3-3).

BOX 3-3   Microbial Activities Contributing to Colonization of Host Surfaces

SURVIVAL AGAINST ENVIRONMENTAL CONDITIONS: Localization in moist areas
Protection within ingested or inhaled debris
Expression of specific metabolic characteristics (e.g., salt tolerance)
ATTACHMENT AND ADHERENCE TO HOST CELL SURFACES: Pili
Adherence proteins
Biofilms
Various protein adhesins
MOTILITY
PRODUCTION OF SUBSTANCES THAT COMPETE WITH HOST FOR ACQUISITION OF ESSENTIAL NUTRIENTS (E.G., SIDEOPHORES FOR CAPTURE OF IRON)
ABILITY TO COEXIST WITH OTHER COLONIZING MICROORGANISMS

BOX 3-4   Factors Contributing to Disruption of Skin and Mucosal Surface

TRAUMA: Penetrating wounds
Abrasions
Burns (chemical and fire)
Surgical wounds
Needlesticks
INHALATION: Smoking
Noxious or toxic gases
IMPLANTATION OF MEDICAL DEVICES
OTHER DISEASES: Malignancies
Diabetes
Previous or simultaneous infections
Alcoholism
CHILDBIRTH
OVERUSE OF ANTIBIOTICS

To avoid the dryness of the skin, organisms often seek moist areas that include hair follicles, sebaceous (oil) and sweat glands, skin folds, underarms, the anogenital region, the face, the scalp, and areas around the mouth. Microbial protection against mucous surface environments include being embedded in food particles to survive oral and gastrointestinal conditions or being contained within airborne particles to aid in surviving the respiratory tract. Microorganisms also exhibit metabolic capabilities that help survival. For example, the ability of staphylococci to thrive in relatively high salt concentrations enhances their survival in and among the sweat glands of the skin.

Besides surviving the host's physical and chemical conditions, colonization also requires that microorganisms attach and adhere to host surfaces (Box 3-3). This can be particularly challenging in places such as the mouth and bowel, in which the surfaces are frequently washed by passing fluids. Pili, the rodlike projections of bacterial envelopes, various molecules (e.g., adherence proteins and adhesins), and biochemical complexes (e.g., biofilm) work together to enhance attachment of microorganisms to the host cell surface. For more information about pili, see Chapter 2.

In addition, microbial motility by flagella allows organisms to move around and actively seek optimum conditions. Finally, because no single microbial species is a lone colonizer, successful colonization also requires that a microorganism be able to coexist with other microbial genera and species.

MICROORGANISM ENTRY, INVASION, AND DISSEMINATION

THE HOST'S PERSPECTIVE

In most instances, to establish infection, microorganisms must penetrate or circumvent the host's physical barriers (i.e., skin or mucosal surfaces); overcoming these defensive barriers depends on both host and microbial factors. When these barriers are broken, numerous other host defensive strategies are activated.

Disruption of Surface Barriers

Any situation that disrupts the physical barrier of the skin and mucosa, alters the environmental conditions (e.g., loss of stomach acidity or dryness of skin), changes the functioning of surface cells, or alters the normal flora population can facilitate the penetration of microorganisms past the barriers and into deeper host tissues. Disruptive factors may vary from accidental or intentional (medical) trauma that result in surface destruction to the use of antibiotics that remove normal, protective, colonizing microorganisms (Box 3-4). Importantly, a number of these factors are related to medical activities and procedures.

Responses to Microbial Invasion of Deeper Tissues

Once surface barriers have been bypassed, the host responds to microbial presence in the underlying tissue in various ways. Some of these responses are nonspecific because they occur regardless of the type of invading organism, whereas other responses are more specific and involve the host's immune system. Both nonspecific and specific host responses are critical if the host is to survive. Without them, microorganisms would multiply and invade vital tissues and organs unchecked, resulting in severe damage to the host.

Nonspecific Responses. Some nonspecific responses are biochemical; others are cellular. Biochemical factors remove essential nutrients, such as iron, from tissues so that they cannot be used by invading microorganisms. Cellular responses are central to tissue and organ defenses, and the cells involved are known as *phagocytes.*

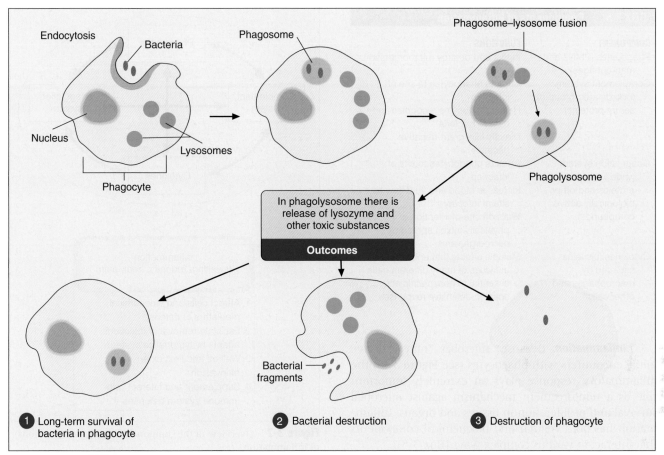

Figure 3-6   Overview of phagocyte activity and possible outcomes of phagocyte-bacterial interactions.

Phagocytes. **Phagocytes** are cells that ingest and destroy bacteria and other foreign particles. The two types of phagocytes are **polymorphonuclear leukocytes** (also known as **PMNs** or **neutrophils**) and **macrophages.** Phagocytes ingest bacteria by a process known as **endocytosis** and engulf them in a membrane-lined structure called a **phagosome** (Figure 3-6). The phagosome is then fused with a second structure, the lysosome. When the lysosome, which contains toxic chemicals and destructive enzymes, combines with the phagosome, the bacteria trapped within are neutralized and destroyed. This destructive process must be carried out inside membrane-lined structures; otherwise the noxious substances contained within would destroy the phagocyte itself. This is evident during the course of rampant infections when thousands of phagocytes exhibit "sloppy" ingestion of the microorganisms and toxic substances spill from the cells, thus damaging the surrounding host tissue.

Although both PMNs and macrophages are phagocytes, these cell types differ. PMNs develop in the bone marrow and spend their short lives, usually a day or less, circulating in blood and tissues. Widely dispersed in the body, PMNs usually are the first cells on the scene of bacterial invasion. Macrophages also develop in the bone marrow but first go through a cellular phase when they are called **monocytes.** During the monocyte stage, the cells circulate in the blood. When deposited in tissue or at a site of infection, monocytes mature into macrophages. In the absence of infection, macrophages usually reside in specific organs, such as the spleen, lymph nodes, liver, or lungs, where they may live for days to several weeks awaiting encounters with invading bacteria. In addition to the ingestion and destruction of bacteria, macrophages play an important role in mediating immune system defenses (see Specific Responses— The Immune System later in this chapter).

In addition to the inhibition of microbial proliferation by phagocytes and by biochemical substances such as lysozyme, microorganisms are "washed" from tissues by the flow of lymph fluid. The fluid carries infectious agents through the lymphatic system, where they are deposited in tissues and organs (e.g., lymph nodes and spleen) heavily populated with phagocytes. This process functions as an efficient filtration system.

BOX 3-5	**Components of Inflammation**
COMPONENT	**FUNCTIONS**
Phagocytes (PMNs and macrophages)	Ingest and destroy microorganisms
Complement system (coordinated group of serum proteins)	Attracts phagocytes to site of infection
	Helps phagocytes recognize and bind to bacteria
	Directly kills gram-negative bacteria
Coagulation system (wide variety of proteins and other biologically active compounds)	Attracts phagocytes to site of infection
	Increases blood and fluid flow to site of infection
	Walls off site of infection to physically inhibit spread of microorganisms
Cytokines (proteins secreted by macrophages and other cells)	Multiple effects that enhance the activities of many different cells essential to nonspecific and specific defensive responses

Inflammation. Because microbes may survive initial encounters with phagocytes (see Figure 3-6), the inflammatory response plays an extremely important role as a reinforcement mechanism against microbial survival and proliferation in tissues and organs. Inflammation has both cellular and biochemical components that interact in various complex ways (Box 3-5).

The **complement system** is composed of a coordinated group of proteins that are activated by the immune system or by the mere presence of invading microorganisms. On activation of this system, a cascade of biochemical events occurs that attracts and enhances the activities of more phagocytes. Because PMNs and macrophages are widely dispersed throughout the body, signals are needed to attract and concentrate these cells at the point of invasion. The complement system provides many of these signals.

Protective functions of the complement system also are enhanced by the **coagulation system,** which works to increase blood flow to the area of infection and can also effectively wall off the infection through the production of barrier substances.

Another key component of inflammation is a group of biochemicals known as **cytokines,** substances secreted by one type of cell that have substantial effects on the antiinfective activities of other cells.

The manifestations of inflammation are evident and familiar to most of us and include the following:

- Swelling—caused by increased flow of fluid and cells to the affected body site
- Redness—results from vasodilation of blood vessels at the infection site

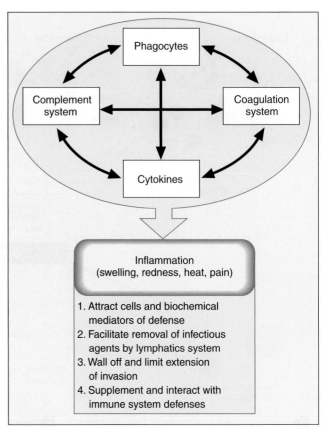

Figure 3-7 Overview of the components, signs, and functions of inflammation.

- Heat—results from increased temperature of affected tissue
- Pain—due to tissue damage and pressure from increased flow of fluid and cells

On a microscopic level, the presence of phagocytes at the infection site is an important observation in diagnostic microbiology. Microorganisms seen associated with these host cells are frequently identified as the cause of a particular infection. An overview of inflammation is shown in Figure 3-7.

SPECIFIC RESPONSES—THE IMMUNE SYSTEM

The immune system provides the human host with the ability to mount a specific protective response to the presence of a microorganism, a customized defense against the invading microorganism. In addition to this specificity, the immune system has a "memory" so that if a microorganism is encountered a second or third time, an immune-mediated defensive response is immediately available. It is important to remember that nonspecific (i.e., phagocytes, inflammation) and specific (i.e., the immune system) host defensive systems are interdependent in their efforts to limit the spread of infection.

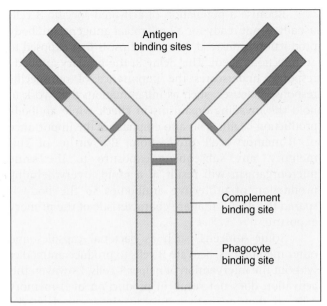

Figure 3-8   General structure of the IgG class antibody molecule.

Components of the Immune System

The central molecule of the immune response is the **antibody.** Antibodies, also referred to as **immunoglobulins,** are specific proteins produced by certain cells in response to the presence of foreign molecules known as **antigens.** In the case of infectious diseases, the antigens are components of the invading microorganism's structure that are usually composed of proteins or polysaccharides. Antibodies circulate in the serum portion of the host's blood and are present in secretions such as saliva. These molecules have two active areas: the **antigen binding site** and the **phagocyte binding site** (Figure 3-8).

There are five different classes of antibody: IgG, IgA, IgM, IgD, and IgE. Each class has distinctive molecular configurations. IgG, IgM, IgA, and IgE are most involved in combating infections. IgM is the first antibody produced when an invading microorganism is initially encountered; production of the most abundant antibody, IgG, follows. IgA is secreted in various body fluids and primarily protects those body surfaces lined with mucous membranes. Increased IgE is associated with various parasitic infections. As is discussed in Chapter 10, our ability to measure specific antibody production is a valuable tool for the laboratory diagnosis of infectious diseases.

Regarding the cellular components of the immune response, there are three major types of cells: B lymphocytes, T lymphocytes, and natural killer cells. The functions of these cells are summarized in Box 3-6. **B lymphocytes** originate from stem cells and develop into B cells in the bone marrow before being widely distributed to lymphoid tissues throughout the body. These cells primarily function as antibody producers. **T lymphocytes** also originate from bone marrow stem cells, but they mature in the thymus and either directly destroy infected cells or work with B cells to regulate antibody production. The development of **natural killer cells,** which destroy infected or malignant host cells, is uncertain. Each of the three cell types is strategically located within lymphoid tissue throughout the body to maximize the chances of encountering invading microorganisms that the lymphatic system drains from the infection site.

Two Arms of the Immune System. The immune system provides immunity that generally can be divided into two arms:

- **Antibody-mediated immunity** (or humoral immunity)
- **Cell-mediated immunity** (or cellular immunity)

Antibody-mediated immunity is centered on the activities of B cells and the production of antibodies. When B cells encounter microbial antigen, they become activated and a series of events are initiated. These events are mediated by the activities of helper T cells and the release of cytokines. Cytokines mediate an explosion in the number of B cells that recognize the antigen and the maturation of B cells into plasma cells that produce tremendous amounts of antibodies

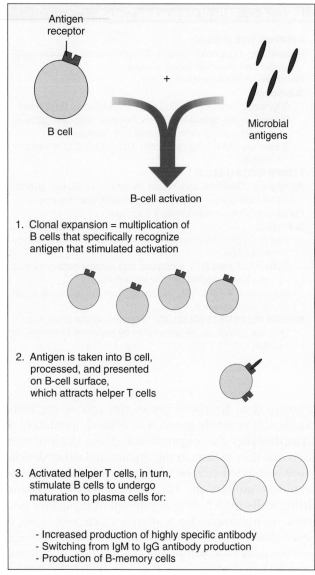

1. Clonal expansion = multiplication of B cells that specifically recognize antigen that stimulated activation

2. Antigen is taken into B cell, processed, and presented on B-cell surface, which attracts helper T cells

3. Activated helper T cells, in turn, stimulate B cells to undergo maturation to plasma cells for:

   - Increased production of highly specific antibody
   - Switching from IgM to IgG antibody production
   - Production of B-memory cells

Figure 3-9 Overview of B-cell activation that is central to antibody-mediated immunity.

specific for the antigen. The process also results in the production of B-memory cells (Figure 3-9).

Antibodies protect the host in a number of ways:

- Helping phagocytes ingest and kill microorganisms
- Neutralizing microbial toxins detrimental to host cells and tissues
- Promoting bacterial clumping (agglutination) that facilitates clearing from infection site
- Inhibiting bacterial motility
- Combining with microorganisms to activate the complement system and inflammatory response

Because a population of activated specific B cells usually is not ready for all microbial antigens, antibody production is delayed when the host is first exposed to an infectious agent. This delay in the primary antibody response underscores the importance of nonspecific response defenses, such as inflammation, that work to hold the invading organisms in check while antibody production begins. This also emphasizes the importance of B-memory cell production. By virtue of this memory, any subsequent exposure to the same microorganism will result in a rapid, overwhelming production of protective antibodies so the body is spared the delays that are characteristic of the primary exposure.

Some antigens, such as bacterial capsules and outer membranes, activate B cells to produce antibodies without the intervention of helper T cells. However, this activation does not result in production of B-memory cells so that on reexposure to the same bacterial antigens, there will be no rapid memory response on the part of the host.

The primary cells that mediate cell-mediated immunity are T lymphocytes that recognize and destroy human host cells infected with microorganisms. This function is extremely important for the destruction and elimination of infecting microorganisms (e.g., viruses, tuberculosis, some parasites, and fungi) that are able to survive within host cells where they are "hidden" from antibody action. Therefore, antibody-mediated immunity targets microorganisms outside of human cells while cell-mediated immunity targets microorganisms inside human cells. However, in many instances these two arms of the immune system overlap and work together.

Like B cells, T cells must be activated. Activation is accomplished by T-cell interactions with other cells that process microbial antigens and present them on their surface (e.g., macrophages and B cells). The responses of activated T cells are very different and depend on the subtype of T cell (Figure 3-10). Activated helper T cells work with B cells for antibody production (see Figure 3-9) and facilitate inflammation by releasing cytokines. Cytotoxic T cells directly interact with and destroy host cells that contain microorganisms. The T-cell subset, helper or cytotoxic cells, that is activated is controlled by an extremely complex series of molecular and genetic events known as the **major histocompatibility complex,** or **MHC,** which is a part of cells that present antigens to the T cells.

In summary, the host presents a spectrum of challenges to invading microorganisms, from the physical barriers of skin and mucous membranes to the interactive cellular and biochemical components of inflammation and the immune system. All these systems work together to minimize microbial invasion

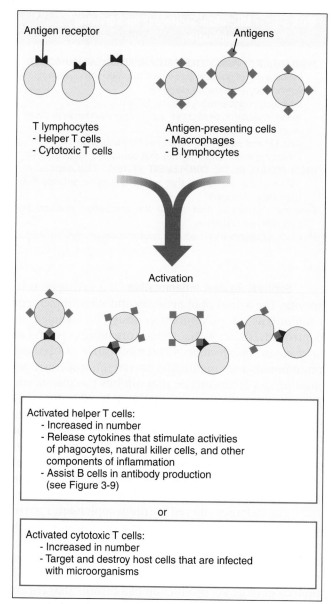

Figure 3-10   Overview of T-cell activation that is central to cell-mediated immunity.

multiplication of microorganisms that result in damage to the host. The extent and severity of the damage depend on many factors, including the microorganism's ability to cause disease, the body site of the infection, and the general health of the person infected. **Disease** results when the infection produces notable changes in human physiology that are often associated with damages to one or more of the body's organ systems.

Pathogens and Virulence

Microorganisms that cause infections and/or disease are called **pathogens,** and the characteristics that enable them to cause disease are referred to as **virulence factors.** Most virulence factors protect the organism against host attack or mediate damaging effects on host cells. The terms **pathogenicity** and **virulence** reflect the degree to which a microorganism is capable of causing disease. An organism of high pathogenicity is very likely to cause disease when encountered, whereas an organism of low pathogenicity is much less likely to cause infection. When disease does occur, highly virulent organisms often severely damage the human host. The degree of severity diminishes with diminishing virulence of the microorganism.

Because host factors play a role in the development of infectious diseases, the distinction between a pathogen and nonpathogen, or colonizer, is not always clear. For example, many organisms that colonize our skin usually do not cause disease (i.e., exhibit low pathogenicity) under normal circumstances. However, when damage to the skin occurs (Box 3-4), or when the skin is disrupted in some other way, these organisms can gain access to deeper tissues and establish an infection.

Organisms that only cause infection when one or more of the host's defense mechanisms are disrupted or malfunction are known as **opportunistic pathogens,** and the infections they cause are referred to as **opportunistic infections.** On the other hand, several pathogens known to cause serious infections can be part of a person's normal flora and never cause disease in that person. However, the same organism can cause life-threatening infection when transmitted to other persons. The reasons for these inconsistencies are not fully understood, but such widely different results undoubtedly involve complex interactions between microorganism and human. Recognizing and separating pathogens from nonpathogens presents one of the greatest challenges in interpreting diagnostic micro-biology laboratory results.

Microbial Virulence Factors

Virulence factors provide microorganisms with the capacity to avoid host defenses and damage host cells, tissues, and organs in a number of ways. Some

and avoid damage to vital tissues and organs that can result from the presence of infectious agents.

THE MICROORGANISM'S PERSPECTIVE

Given the complexities of the human host's defense systems, it is no wonder that microbial strategies designed to survive these systems are equally complex. Before considering the microorganism's perspective, a number of definitions and terms must be considered.

Colonization and Infection

Many of our body surfaces are colonized with a wide variety of microorganisms without apparent detriment. In contrast, an **infection** involves the growth and virulence factors are specific for certain pathogenic genera or species and substantial differences exist in the way bacteria, viruses, parasites, and fungi cause disease. Of importance, knowledge of a microorganism's capacity to cause specific types of infections plays a major role in developing the diagnostic microbiology procedures used for isolating and identifying microorganisms (see Part VII for more information regarding diagnosis by organ system).

Attachment. Whether humans encounter microorganisms via the air, through ingestion, or by direct contact, the first step of infection and disease development (a process referred to as **pathogenesis**) is microbial attachment to a surface (exceptions being instances in which the organisms are directly implanted by trauma or other means into deeper tissues).

Many of the microbial factors that facilitate attachment of pathogens are the same as those used by nonpathogen colonizers (Box 3-3). The difference between pathogens and colonizers is that pathogens do not always stop at colonization. Also, many pathogens are not part of the normal microbial flora so that their successful attachment also requires that they outcompete colonizers for a place on the body's surface. Often medical interventions such as the overuse of antimicrobial agents that can destroy much of the normal flora tilt the competition in favor of the invading organism.

Invasion. Once surface attachment has been secured, microbial invasion into subsurface tissues and organs (i.e., infection) is accomplished by traumatic factors such as those listed in Box 3-4, or by the direct action of an organism's virulence factors. Some microorganisms produce factors that force mucosal surface phagocytes (M cells) to ingest them and then release them unharmed into the tissue below the mucosal surface. Other organisms, such as staphylococci and streptococci, are not so subtle. These organisms produce an array of enzymes (e.g., hyaluronidases, nucleases, collagenases) that hydrolyze host proteins and nucleic acids that destroy host cells and tissues. This destruction allows the pathogen to "burrow" through minor openings in the outer surfaces and through deeper tissues. Once a pathogen has penetrated the body surface, strategies that allow microbial survival of the host's inflammatory and immune responses must be used. Alternatively, some pathogens cause disease from their site of attachment without further penetration. For example, in diseases such as diphtheria and whooping cough, the bacteria produce toxic substances that destroy surrounding tissues but the organisms themselves generally do not penetrate the mucosal surface they inhabit.

BOX 3-7	**Microbial Strategies for Surviving Inflammation**

AVOID KILLING BY PHAGOCYTES (POLYMORPHONUCLEAR LEUKOCYTES):
   Inhibit ability of phagocyte to ingest by producing capsule
Avoid phagocyte-mediated killing by:
   Inhibiting phagosome-lysosome fusion
   Being resistant to destructive agents (e.g., lysozyme) released by lysosomes
   Actively and rapidly multiplying within phagocyte
   Releasing toxins and enzymes that damage or kill phagocyte
AVOID EFFECTS OF THE COMPLEMENT SYSTEM: Use capsule to hide surface molecules that would otherwise activate the complement system
Produce substances that inhibit the processes involved in complement activation
Produce substances that destroy specific complement proteins

Survival Against Inflammation. If a pathogen is to survive, the action of phagocytes and the complement components of inflammation must be avoided or controlled (Box 3-7). Some organisms, such as *Streptococcus pneumoniae,* a common cause of bacterial pneumonia and meningitis, avoid phagocytosis by producing a large capsule that inhibits the phagocytic process. Other pathogens may not be able to avoid phagocytosis but are not effectively destroyed and are able to survive within phagocytes. This is the case for *Mycobacterium tuberculosis,* the bacterium that causes tuberculosis. Still other pathogens use toxins and enzymes to attack and destroy phagocytes before the phagocytes attack and destroy them.

The defenses offered by the complement system depend on a series of biochemical reactions triggered by specific microorganism molecular structures. Therefore, microbial avoidance of complement activation requires that the infecting agent either mask its activating molecules (e.g., via production of a capsule that covers bacterial surface antigens) or produce substances (e.g., enzymes) that disrupt critical biochemical components of the complement pathway.

Any single microorganism may produce various virulence factors and several may be expressed simultaneously. For example, while trying to avoid phagocytosis, an organism may also be excreting other enzymes and toxins that help destroy and penetrate tissue, and be producing other factors designed to interfere with the immune response. Microorganisms may also use host systems to their own advantage. For example, the lymphatic and blood circulation system used to carry pathogens away from the site of infection can also be used by surviving pathogens to become dispersed throughout the body.

Survival Against the Immune System. Microbial strategies to avoid the defenses of the immune system

BOX 3-8	Microbial Strategies for Surviving the Immune System

- Pathogen multiplies and invades so quickly that damage to host is complete before immune response can be fully activated, or organism's virulence is so great that the immune response is insufficient
- Pathogen invades and destroys cells involved in the immune response
- Pathogen survives, unrecognized, in host cells and avoids detection by immune system
- Pathogen covers its antigens with a capsule so that an immune response is not activated
- Pathogen changes antigens so that immune system is constantly fighting a primary encounter (i.e., the memory of the immune system is neutralized)
- Pathogen produces enzymes (proteases) that directly destroy or inactivate antibodies

BOX 3-9	Summary of Bacterial Toxins

ENDOTOXINS
- General toxin common to almost all gram-negative bacteria
- Composed of lipopolysaccharide portion of cell envelope
- Released when gram-negative bacterial cell is destroyed
- Effects on host include:
  Disruption of clotting, causing clots to form throughout body (i.e., disseminated intravascular coagulation [DIC])
  Fever
  Activation of complement and immune systems
  Circulatory changes that lead to hypotension, shock, and death

EXOTOXINS
- Most commonly associated with gram-positive bacteria
- Produced and released by living bacteria; do not require bacterial death for release
- Specific toxins target specific host cells; the type of toxin varies with the bacterial species
- Some kill host cells and help spread bacteria in tissues (e.g., enzymes that destroy key biochemical tissue components or specifically destroy host cell membranes)
- Some destroy or interfere with the specific intracellular activities (e.g., interruption of protein synthesis, interruption of internal cell signals, or interruption of neuromuscular system)

are outlined in Box 3-8. Again, a pathogen can use more that one strategy to avoid immune-mediated defenses, and microbial survival does not necessarily require devastation of the immune system. The pathogen may merely need to "buy" time to reach a safe area in the body or to be transferred to the next susceptible host. Also, microorganisms can avoid much of the immune response if they do not penetrate the surface layers. This strategy is the hallmark of diseases that are caused by microbial toxins.

Microbial Toxins. **Toxins** are biochemically active substances that are released by microorganisms and have a particular effect on host cells. Microorganisms use toxins to help them establish infections and multiply within the host. Alternatively, a pathogen may be restricted to a particular body site from which toxins are released to cause widespread problems throughout the body. Toxins also can cause human disease in the absence of the pathogens that produced them. This common mechanism of food poisoning that involves ingestion of preformed bacterial toxins is referred to as **intoxication,** a notable example of which is botulism.

Endotoxins and exotoxins are the two general types of bacterial toxins (Box 3-9). **Endotoxins** are released by gram-negative bacteria and can have devastating effects on the body's metabolism, the most serious being endotoxic shock, which often results in death. The effects of **exotoxins** produced by gram-positive bacteria tend to be more limited and specific than the effects of gram-negative endotoxins. The activities of exotoxins range from those enzymes produced by many staphylococci and streptococci that augment bacterial invasion by damaging host tissues and cells to those that have highly specific activities (e.g., diphtheria toxin that inhibits protein synthesis

or the cholera toxin that interferes with host cell signals). Examples of other highly active and specific toxins are those that cause botulism and tetanus by interfering with neuromuscular functions.

Genetics of Virulence: Pathogenicity Islands

The evolution and dissemination of many of the virulence factors discussed is now known to be largely facilitated by what are referred to as pathogenicity islands (PAIs). These are mobile genetic elements that contribute to the change and spread of virulence factors among bacterial populations of a variety of species. These genetic elements are thought to have evolved from lysogenic bacteriophages and plasmids and are spread by horizontal gene transfer (see Chapter 2 for information about bacterial genetics). PAIs are typically comprised of one or more virulence-associated genes and "mobility" genes (i.e., integrases and transposases) that mediate movement between various genetic elements (e.g., plasmids and chromosomes) and among different bacterial strains. In essence, PAIs facilitate the dissemination of virulence capabilities among bacteria in a manner similar to the mechanism diagrammed in Figure 2-10, which also facilitates dissemination of antimicrobial resistance genes (see Chapter 11).

The existence and function of PAIs must be kept in mind whenever considering the virulence factors discussed in this chapter and when considering pathogenesis in each bacterial organism discussed in Part III. It is important to appreciate how widely disseminated this genetic mechanism is among medically important bacteria. For example, PAIs have

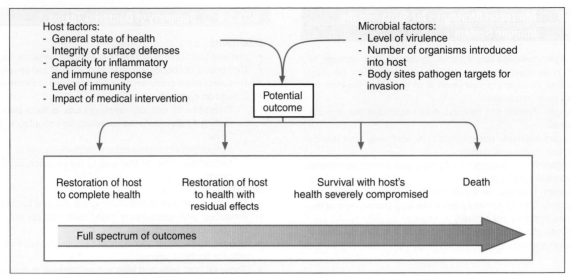

Figure 3-11 Possible outcomes of infections and infectious diseases.

been identified as playing a role in virulence for each of the following organisms (gram-positive and gram-negative alike):

Helicobacter pylori
Pseudomonas aeruginosa
Shigella spp.
Yersinia spp.
Vibrio cholerae
Salmonella spp.
E. coli (enteropathogenic, enterohemorrhagic, uropathogenic)
Neisseria spp.
Bacteroides fragilis
Listeria monocytogenes
Staphylococcus aureus
Streptococcus spp.
Enterococcus faecalis
Clostridium difficile

OUTCOME AND PREVENTION OF INFECTIOUS DISEASES

OUTCOME OF INFECTIOUS DISEASES

Given the complexities of host defenses and microbial virulence, it is not surprising that the factors determining outcome between these two living entities are also complicated. Basically, outcome depends on the state of the host's health, the virulence of the pathogen, and whether the host can clear the pathogen before infection and disease cause irreparable harm or death (Figure 3-11).

The time for a disease or infection to develop also depends on host and microbial factors. Infectious processes that develop quickly are referred to as **acute infections,** and those that develop and progress slowly, sometimes over a period of years, are known as **chronic infections.** Some pathogens, particularly certain viruses, can be clinically silent inside the body without any noticeable effect on the host before suddenly causing a severe and acute infection. During the silent phase, the infection is said to be **latent.** Again, depending on host and microbial factors, acute, chronic, or latent infections can result in any of the outcomes detailed in Figure 3-11.

Medical intervention can help the host fight the infection but usually is not instituted until after the host is aware that an infectious process is underway. The clues that an infection is occurring are known as the **signs** and **symptoms** of disease and result from host responses (e.g., inflammatory and immune responses) to the action of microbial virulence factors (Box 3-10). The signs and symptoms reflect the stages of infection. In turn, the stages of infection generally reflect the stages in host-microorganism interactions (Figure 3-12).

Whether medical procedures contribute to controlling or clearing an infection depends on key factors, including:

- The severity of the infection, which is determined by host and microbial interactions already discussed
- Accuracy in diagnosing the pathogens or pathogens causing the infection, which is the primary function of diagnostic microbiology
- Whether the patient receives appropriate treatment for the infection (which heavily depends on accurate diagnosis)

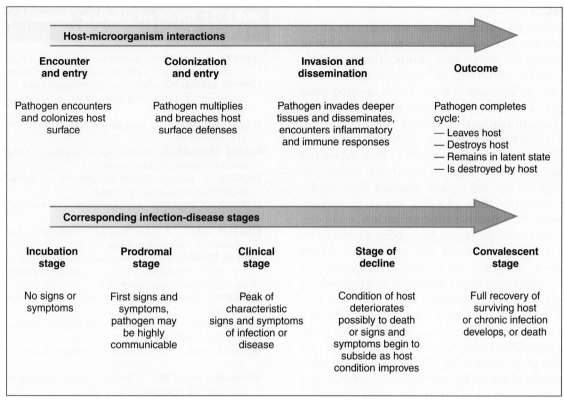

Figure 3-12 Host-microorganism interactions and stages of infection or disease.

BOX 3-10	Signs and Symptoms of Infection and Infectious Diseases

General or localized aches and pains
Headache
Fever
Swollen lymph nodes
Rashes
Redness and swelling
Cough and sneezes
Congestion of nasal and sinus passages
Sore throat
Nausea and vomiting
Diarrhea

BOX 3-11	Strategies for Prevention of Infectious Diseases

PREVENTING TRANSMISSION: Avoid direct contact with infected persons or take protective measures when direct contact is going to occur (e.g., wear gloves, wear condoms)
Block spread of airborne microorganisms by wearing masks or isolating persons with infections transmitted by air
Use sterile medical techniques
CONTROLLING MICROBIAL RESERVOIRS: Sanitation and disinfection
Sewage treatment
Food preservation
Water treatment
Control of pests and insect vector populations
MINIMIZING RISK BEFORE OR SHORTLY AFTER EXPOSURE: Immunization
Cleansing and use of antiseptics
Prophylactic use of antimicrobial agents

PREVENTION OF INFECTIOUS DISEASES

The treatment of an infection is often difficult and not always successful. Because much of the damage is already done before appropriate medical intervention is begun, the microorganisms gain too much of a "head start." Another strategy for combating infectious diseases is to stop infections before they even start (i.e., **disease prevention**). As discussed at the beginning of this chapter, the first step in any host-microorganism relationship is the encounter. Therefore, strategies to prevent disease involve interrupting encounters or minimizing the risk of infection when encounters do occur. As outlined in Box 3-11, interruption of encounters may be accomplished by preventing transmission of the infecting agents and by controlling or destroying reservoirs of human pathogens. Of interest, most of these measures do not really involve medical practices but rather social practices and policies.

Immunization

If encounters do occur, medical strategies exist for minimizing the risk for disease development. One of the most effective methods is **immunization.** This practice takes advantage of the specificity and memory of the immune system. There are two basic approaches to immunization: active and passive. With **active immunization,** modified antigens from pathogenic microorganisms are introduced into the body and cause an immune response. If or when the host encounters the pathogen in nature, the memory of the immune system will ensure minimal delay in the immune response, thus affording strong protection. Alternatively, with **passive immunization** antibodies against a particular pathogen have been produced in one host and are transferred to a second host where they provide temporary protection. The passage of maternal antibodies to the newborn is a key example of natural passive immunization. Active immunity is generally more long lasting because the immunized host's own immune response has been activated. However, for complex reasons, active immunity only has been successfully applied for relatively few infectious diseases, some of which include diphtheria, whooping cough (pertussis), tetanus, influenza, polio, smallpox, measles, hepatitis, and certain *S. pneumoniae* and *H. influenzae* infections.

Prophylactic antimicrobial therapy, the administration of antibiotics when the risk of developing an infection is high, is another common medical intervention for preventing infection.

Epidemiology

To prevent infectious diseases, information is required regarding the sources of pathogens, the method of transmission to and among humans, human risk factors for encountering the pathogen and developing infection, and factors that contribute to good and bad outcomes resulting from the encounter. **Epidemiology** is the science that characterizes these aspects of infectious diseases and monitors the effect diseases have on public health. By fully characterizing the circumstances associated with the acquisition and dissemination of infectious diseases, there is a better chance for preventing and eliminating these diseases. Additionally, many epidemiologic strategies that have been developed for public health also apply in hospitals for the control of hospital-based (i.e., nosocomial) infections (for more information on infection control in hospitals, see Chapter 64).

The field of epidemiology is broad and complex. Diagnostic microbiology laboratory personnel and hospital epidemiologists often work closely to investigate problems. Therefore, familiarity with certain epidemiologic terms and concepts is important (Box 3-12).

BOX 3-12   Definitions of Selected Epidemiologic Terms

CARRIER: A person who carries the etiologic agent but shows no apparent signs or symptoms of infection or disease

COMMON SOURCE: The etiologic agent responsible for an epidemic or outbreak originates from a single source or reservoir

DISEASE INCIDENCE: The number of diseased or infected persons in a population

DISEASE PREVALENCE: Percentage of diseased persons in a given population at a particular time

ENDEMIC: A disease constantly present at some rate of occurrence in a particular location

EPIDEMIC: A larger than normal number of diseased or infected individuals in a particular location

ETIOLOGIC AGENT: A microorganism responsible for causing infection or infectious disease

MODE OF TRANSMISSION: Means by which etiologic agents are brought in contact with the human host (e.g., infected blood, contaminated water, insect bite)

MORBIDITY: The state of disease and its associated effects on the host

MORBIDITY RATE: The incidence of a particular disease state

MORTALITY: Death resulting from disease

MORTALITY RATE: The incidence in which a disease results in death

NOSOCOMIAL INFECTION: Infection in which etiologic agent was acquired in a hospital

OUTBREAK: A larger than normal number of diseased or infected individuals that occurs over a relatively short period

PANDEMIC: An epidemic that spans the world

RESERVOIR: Origin of the etiologic agent or location from which they disseminate (e.g., water, food, insects, animals, other humans)

STRAIN TYPING: Laboratory-based characterization of etiologic agents designed to establish their relatedness to one another during a particular outbreak or epidemic

SURVEILLANCE: Any type of epidemiologic investigation that involves data collection for characterizing circumstances surrounding the incidence or prevalence of a particular disease or infection

VECTOR: A living entity (animal, insect, or plant) that transmits the etiologic agent

VEHICLE: A nonliving entity that is contaminated with the etiologic agent and as such is the mode of transmission for that agent

---

Because the central focus of epidemiology is on tracking and characterizing infections and infectious diseases, this field heavily depends on diagnostic microbiology. Epidemiologic investigations cannot proceed without first knowing the etiologic agents involved. Therefore, the procedures and protocols used in diagnostic microbiology to detect, isolate, and characterize human pathogens are not only essential for patient care but also play a central role in epidemiologic studies focused on disease prevention and the general improvement of public health. In fact, microbiologists who work in clinical laboratories are often the first to recognize patterns that suggest potential outbreaks or epidemics.

ADDITIONAL READING

Atlas RM: *Principles of microbiology,* St Louis, 1995, Mosby.

Brock TD et al, editors: *Biology of microorganisms,* Englewood Cliffs, NJ, 1994, Prentice Hall.

Dobrindt U et al: Genomic islands in pathogenic and environmental microorganisms, *Nat Rev Microbiol* 2:414, 2002.

Murray PR et al, editors: *Medical microbiology,* ed 5, St Louis, 2005, Mosby.

Ryan KJ, editor: *Sherris medical microbiology: an introduction to infectious diseases,* Norwalk, Conn, 1994, Appleton & Lange.

Salyers AA, Whitt DD: *Bacterial pathogenesis: a molecular approach,* Washington, DC, 1994, ASM Press.

Schaechter M, Medoff G, Eisenstein BI, editors: *Mechanisms of microbial disease,* ed 2, Baltimore, 1993, Williams & Wilkins.

Schmidt H, Hensel M: Pathogenicity islands in bacterial pathogenesis, *Clin Microbiol Rev* 17:14, 2004.

II PART

General Principles in Clinical Microbiology

Laboratory Safety CHAPTER 4

Microbiology laboratory safety practices were first published in 1913 in a textbook by Eyre.[2] It included admonitions such as the necessity to (1) wear gloves, (2) wash hands after working with infectious materials, (3) disinfect all instruments immediately after use, (4) use water to moisten specimen labels rather than the tongue, (5) disinfect all contaminated waste before discarding, and (6) report to appropriate personnel all accidents or exposures to infectious agents.

These guidelines are still incorporated into safety programs in the twenty-first century laboratory. In addition, safety programs have been expanded to include not only the proper handling of biologic hazards encountered in processing patient specimens or handling infectious microorganisms but also fire safety; electrical safety; the safe handling, storage, and disposal of chemicals and radioactive substances; and techniques for the safe lifting or moving of heavy objects. In areas of the country prone to natural disasters (e.g., earthquakes, hurricanes, snowstorms), safety programs also involve disaster preparedness plans that outline steps to take in an emergency. Although all microbiologists are responsible for their own health and safety, their institution and immediate supervisors are required to provide safety training to help familiarize microbiologists with known hazards and to avoid accidental exposure. Laboratory safety is considered such an integral part of overall laboratory services that federal law in the United States mandates preemployment safety training followed by at least quarterly safety in-services. Microbiologists should find very little reason to be afraid while performing duties if the safety regulations are internalized and followed without deviation. Investigation of the causes of accidents usually shows that accidents happen when individuals become sloppy in performing their duties or when they do not believe that they will be affected by departures from safety standards.

STERILIZATION AND DISINFECTION

Sterilization is a process whereby all forms of microbial life, including bacterial spores, are killed. Sterilization may be accomplished by physical or chemical means. **Disinfection** is a process whereby pathogenic organisms, but not necessarily all microorganisms or spores, are destroyed. As with sterilization, disinfection may be accomplished by physical or chemical methods.

Methods of Sterilization

The **physical** methods of sterilization include the following:

- Incineration
- Moist heat
- Dry heat
- Filtration
- Ionizing (gamma) radiation

Incineration is the most common method of treating infectious waste. Hazardous material is literally burned to ashes at temperatures of 870° to 980° C. Toxic air emissions and the presence of heavy metals in ash have limited the use of incineration in most large U.S. cities, however.

Moist heat (steam under pressure) is used to sterilize biohazardous trash and heat-stable objects; an autoclave is used for this purpose. An autoclave is essentially a large pressure cooker. Moist heat in the form of saturated steam under 1 atmosphere (15 psi [pounds per square inch]) of pressure causes the irreversible denaturation of enzymes and structural proteins. The most common type of steam sterilizer in the microbiology laboratory is the gravity displacement type shown in Figure 4-1. Steam enters at the top of the sterilizing chamber and, because steam is lighter than air, it displaces the air in the chamber and forces it out the bottom through the drain vent. The two common sterilization temperatures are 121° C (250° F) and 132° C (270° F). Items such as media, liquids, and instruments are usually autoclaved for 15 minutes at

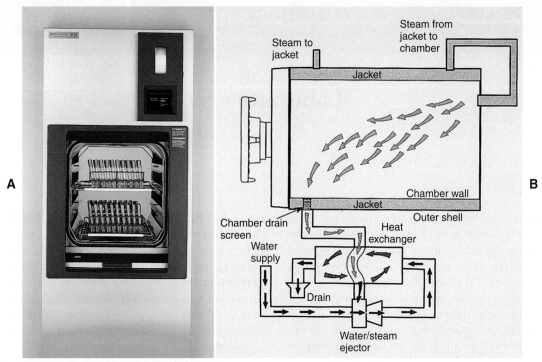

Figure 4-1 Gravity displacement type autoclave. **A,** Typical Eagle Century Series sterilizer for laboratory applications. **B,** Typical Eagle 3000 sterilizer piping diagram. The arrows show the entry of steam into the chamber and the displacement of air. (Courtesy AMSCO International Inc., a wholly owned subsidiary of STERIS Corp., Mentor, Ohio.)

121° C. Infectious medical waste, on the other hand, is often sterilized at 132° C for 30 to 60 minutes to allow penetration of the steam throughout the waste and the displacement of air trapped inside the autoclave bag. Moist heat is the fastest and simplest physical method of sterilization.

Dry heat requires longer exposure times (1.5 to 3 hours) and higher temperatures than moist heat (160° to 180° C). Dry-heat ovens are used to sterilize items such as glassware, oil, petrolatum, or powders. **Filtration** is the method of choice for antibiotic solutions, toxic chemicals, radioisotopes, vaccines, and carbohydrates, which are all heat-sensitive. Filtration of liquids is accomplished by pulling the solution through a cellulose acetate or cellulose nitrate membrane with a vacuum. Filtration of air is accomplished using high-efficiency particulate air (HEPA) filters designed to remove organisms larger than 0.3 μm from isolation rooms, operating rooms, and biological safety cabinets (BSCs). Ionizing radiation used in microwaves and radiograph machines are short wavelength and high-energy gamma rays. **Ionizing radiation** is used for sterilizing disposables such as plastic syringes, catheters, or gloves before use. The most common **chemical** sterilant is ethylene oxide (EtO), which is used in gaseous form for sterilizing heat-sensitive objects. Formaldehyde vapor and vapor-phase hydrogen per-

oxide (an oxidizing agent) have been used to sterilize HEPA filters in BSCs. Glutaraldehyde, which is sporicidal (kills spores) in 3 to 10 hours, is used for medical equipment such as bronchoscopes, because it does not corrode lenses, metal, or rubber. Peracetic acid, effective in the presence of organic material, has also been used for the surface sterilization of surgical instruments. The use of glutaraldehyde or peracetic acid is called **cold sterilization.**

METHODS OF DISINFECTION

Physical Methods of Disinfection
The three **physical** methods of disinfection are:

- Boiling at 100° C for 15 minutes, which kills vegetative bacteria
- Pasteurizing at 63° C for 30 minutes or 72° C for 15 seconds, which kills food pathogens
- Using nonionizing radiation such as ultraviolet (UV) light

UV rays are long wavelength and low energy. They do not penetrate well and organisms must have direct surface exposure, such as the working surface of a BSC, for this form of disinfection to work.

Chemical Methods of Disinfection

Chemical disinfectants comprise many classes, including the following:

- Alcohols
- Aldehydes
- Halogens
- Heavy metals
- Quaternary ammonium compounds
- Phenolics

When chemicals are used to destroy all life they are called chemical sterilants, or **biocides;** however, these same chemicals used for shorter periods are disinfectants. Disinfectants used on living tissue (skin) are called **antiseptics.**

A number of factors influence the activity of disinfectants such as the following:

- Types of organisms present
- Temperature and pH of process
- Number of organisms present (microbial load)
- Concentration of disinfectant
- Amount of organics (blood, mucus, pus) present
- Nature of surface to be disinfected (e.g., potential for corrosion; porous vs. nonporous surface)
- Length of contact time
- Type of water available (hard vs. soft)

Resistance to disinfectants varies with the type of microorganisms present. Bacterial spores such as *Bacillus* spp., are the most resistant, followed by mycobacteria (acid-fast bacilli); nonlipid viruses, for example, poliovirus; fungi; vegetative (nonsporulating) bacteria, for example, gram-negative rods; and, finally, lipid viruses, for example, herpes simplex virus, which are the most susceptible to the action of disinfectants. The Environmental Protection Agency (EPA) registers chemical disinfectants used in the United States and requires manufacturers to specify the activity level of each compound at the working dilution. Therefore, microbiologists who must recommend appropriate disinfectants should check the manufacturer's cut sheets (product information) for the classes of microorganisms that will be killed. Generally, the time necessary for killing microorganisms increases in direct proportion with the number of organisms (microbial load). This is particularly true of instruments contaminated with organic material such as blood, pus, or mucus. The organic material should be mechanically removed before chemical sterilization to decrease the microbial load. This is analogous to removing dried food from utensils before placing them in a dishwasher and is important for cold sterilization of instruments such as bronchoscopes. The type of water and its concentration in a solution are also important.

Hard water may reduce the rate of killing of microorganisms, and surprisingly, 70% ethyl alcohol is more effective as a disinfectant than 95% ethyl alcohol.

Ethyl or isopropyl alcohol is nonsporicidal (does not kill spores) and evaporates quickly. Therefore, either is best used on the skin as an antiseptic or on thermometers and injection vial rubber septa as a disinfectant. Because of their irritating fumes, the aldehydes (formaldehyde and glutaraldehyde) are generally not used as surface disinfectants. The halogens, especially chlorine and iodine, are frequently used as disinfectants. Chlorine is most often used in the form of sodium hypochlorite (NaOCl), the compound known as household bleach. The Centers for Disease Control and Prevention (CDC) recommends that tabletops be cleaned following blood spills with a 1:10 dilution of bleach. Iodine is prepared either as tincture with alcohol or as an iodophor coupled to a neutral polymer, for example, povidone-iodine. Both iodine compounds are widely employed antiseptics. In fact, 70% ethyl alcohol, followed by an iodophor, is the most common compound used for skin disinfection before drawing blood cultures or surgery. Because mercury is toxic to the environment, heavy metals containing mercury are no longer recommended, but an eyedrop solution containing 1% silver nitrate is still instilled in the eyes of newborns to prevent infections with *Neisseria gonorrhoeae.* Quaternary ammonium compounds are used to disinfect bench-tops or other surfaces in the laboratory. However, organic materials, such as blood, may inactivate heavy metals or quaternary ammonium compounds, thus limiting their utility. Finally, phenolics, such as the common laboratory disinfectant amphyl, are derivatives of carbolic acid (phenol). The addition of detergent results in a product that cleans and disinfects at the same time, and at concentrations between 2% and 5% these products are widely used for cleaning bench-tops.

The most important point to remember when working with biocides or disinfectants is to prepare a working solution of the compound *exactly* according to the manufacturer's package insert. Many people think that if the manufacturer says to dilute 1:200, they will be getting a stronger product of they dilute it 1:10. However, the ratio of water to active ingredient may be critical, and if sufficient water is not added, the free chemical for surface disinfection may not be released.

CHEMICAL SAFETY

In 1987, the United States Occupational Safety and Health Administration (OSHA) published the Hazard Communication Standard, which provides for certain institutional educational practices to ensure all labora-

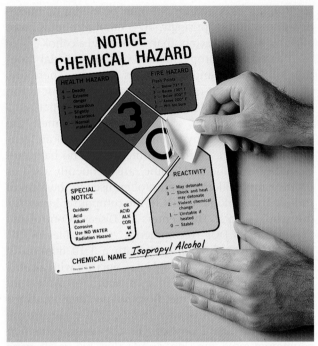

Figure 4-2 National Fire Protection Association diamond stating chemical hazard. This information can be customized as shown here for isopropyl alcohol by applying the appropriate self-adhesive polyester numbers to the corresponding color-coded hazard area. (Courtesy Lab Safety Supply Inc., Janesville, Wis.)

tory personnel have a thorough working knowledge of the hazards of chemicals with which they work. This standard has also been called the "employee right-to-know." It mandates that all hazardous chemicals in the workplace be identified and clearly marked with a National Fire Protection Association (NFPA) label stating the health risks, such as carcinogens (cause of cancer), mutagens (cause of mutations in DNA or RNA), or teratogens (cause of birth defects), and the hazard class, for example, corrosive (harmful to mucous membranes, skin, eyes, or tissues), poison, flammable, or oxidizing. Examples of these labels are shown in Figure 4-2.

Each laboratory should have a **chemical hygiene plan** that includes guidelines on proper labeling of chemical containers, manufacturer's material safety data sheets (MSDSs), and the written chemical safety training and retraining programs. Hazardous chemicals must be inventoried annually. In addition, laboratories are required to maintain a file of every chemical that they use and a corresponding MSDS. The manufacturer provides these MSDS sheets for every hazardous chemical; some manufacturers also provide letters for nonhazardous chemicals, such as saline, so these can be included with the other MSDSs. The MSDSs include information on the nature of the chemical, the precautions to take if the chemical is spilled, and disposal recommendations. The sections in the typical MSDS include:

- Substance name
- Name, address, and telephone number of manufacturer
- Hazardous ingredients
- Physical and chemical properties
- Fire and explosion data
- Toxicity
- Health effects and first aid
- Stability and reactivity
- Shipping data
- Spill, leak, and disposal procedures
- Personal protective equipment
- Handling and storage

Employees should become familiar with each MSDS so that they know where to look in the event of an emergency.

Fume hoods (Figure 4-3) are provided in the laboratory to prevent inhalation of toxic fumes. Fume hoods protect against chemical odor by exhausting air to the outside but are not HEPA-filtered to trap pathogenic microorganisms. It is important to remember that a BSC is *not* a fume hood. Work with toxic or noxious chemicals should always be done in a fume hood or when wearing a fume mask. Spills should be cleaned up using a fume mask, gloves, impervious (impenetrable to moisture) apron, and goggles. Acid and alkaline, flammable, and radioactive spill kits are available to assist in rendering any chemical spills harmless.

FIRE SAFETY

Fire safety is an important component of the laboratory safety program. Each laboratory is required to post fire evacuation plans that are essentially blueprints for finding the nearest exit in case of fire (Figure 4-4). Fire drills conducted quarterly or annually depending on local laws ensure that all personnel know what to do in case of fire. Exit ways should always remain clear of obstructions, and employees should be trained to use fire extinguishers. The local fire department is often an excellent resource for training in the types and use of fire extinguishers.

Type A fire extinguishers are used for trash, wood, and paper; type B are used for chemical fires; type C are used for electrical fires. Combination type ABC extinguishers are found in most laboratories, so personnel need not worry about which extinguisher to reach for in case of a fire. However, type C extinguishers, which contain carbon dioxide (CO_2) or another dry chemical to smother fire, are also used because this type of extinguisher will not damage equipment.

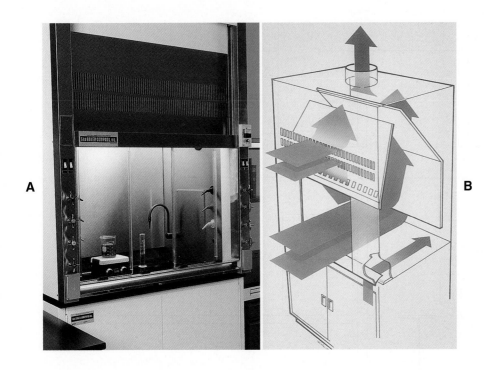

Figure 4-3 Fume hood. **A,** Model ChemGARD. **B,** Schematics. Arrows indicate air flow through cabinet to outside vent. (Courtesy The Baker Co., Sanford, Me.)

The important actions in case of fire and the order in which to perform tasks are remembered as the standard acronym RACE:

1. **R**escue any injured individuals.
2. **A**ctivate the fire alarm.
3. **C**ontain (smother) the fire, if feasible (close fire doors).
4. **E**xtinguish the fire, if possible.

ELECTRICAL SAFETY

Electrical cords should be checked regularly for fraying and replaced when necessary. All plugs should be the three-prong, grounded type. All sockets should be checked for electrical grounding and leakage at least annually. No extension cords should be used in the laboratory.

HANDLING OF COMPRESSED GASES

Compressed gas cylinders (CO_2, anaerobic gas mixture) contain pressurized gases and must be properly handled and secured. In cases where leaking cylinders have fallen, tanks have become missiles, resulting in loss of life and destruction of property. Therefore, gas tanks should be properly chained (Figure 4-5, *A*) and stored in well-ventilated areas. The metal cap, which is removed when the regulator is installed, should always be in place when a gas cylinder is not in use.

Cylinders should be transported chained to special dollies (Figure 4-5, *B*).

BIOSAFETY

Individuals are exposed in various ways to laboratory-acquired infections in microbiology laboratories. These involve the following:

- Rubbing the eyes or nose with contaminated hands
- Inhaling aerosols produced during centrifugation, vortexing, or spills of liquid cultures
- Accidentally ingesting microorganisms by putting pens or fingers in the mouth
- Suffering percutaneous inoculation, that is, being punctured by a needlestick

Risks from a microbiology laboratory may extend to adjacent laboratories and to families of those who work in the microbiology laboratory. For example, Blaser and Feldman[1] noted that 5 of 31 individuals who contracted typhoid fever from proficiency testing specimens did not work in a microbiology laboratory. Two patients were family members of a microbiologist who had worked with the *Salmonella typhi*, two were students whose afternoon class was in the laboratory where the organism had been cultured that morning, and one worked in an adjacent chemistry laboratory.

In the clinical microbiology laboratory, shigellosis, salmonellosis, tuberculosis, brucellosis, and hepatitis are

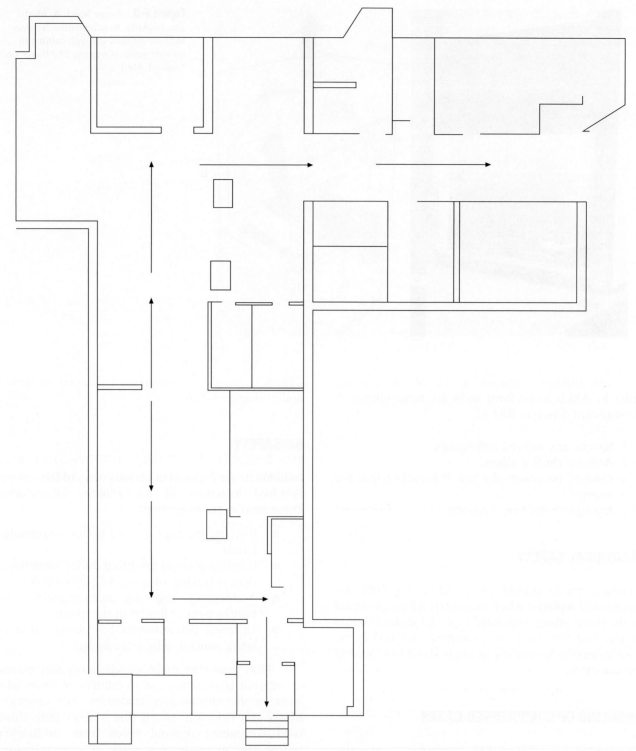

Figure 4-4   Fire evacuation plan. Arrows indicate quickest fire exits.

the five most frequently acquired laboratory infections. Viral agents transmitted through blood and body fluids cause most of the infections in nonmicrobiology laboratory workers and in health care workers in general. These include hepatitis B virus (HBV), hepatitis C virus (HCV), hepatitis D virus (HDV), and the human immunodeficiency virus (HIV). Of interest, males and younger employees (17 to 24 years old) are involved in more laboratory-acquired infections than females and older employees (45 to 64 years old).

Figure 4-5   **A,** Gas cylinders chained to the wall. **B,** Gas cylinder chained to dolly during transportation. (Courtesy Lab Safety Supply Inc., Janesville, Wis.)

EXPOSURE CONTROL PLAN

It is the legal responsibility of the laboratory director and supervisor to ensure that an **Exposure Control Plan** has been implemented and that the mandated safety guidelines are followed. The plan identifies tasks that are hazardous to employees and promotes employee safety through use of the following:

- Employee education and orientation
- Appropriate disposal of hazardous waste
- Standard (formerly Universal) Precautions
- Engineering controls and safe work practices, as well as appropriate waste disposal and use of BSCs
- Personal protective equipment (PPE) such as laboratory coats, shoe covers, gowns, gloves, and eye protection (goggles, face shields)
- Postexposure plan involving the investigation of all accidents and a plan to prevent recurrences

EMPLOYEE EDUCATION AND ORIENTATION

Each institution should have a safety manual that is reviewed by all employees and a safety officer who is knowledgeable about the risks associated with laboratory-acquired infections. The safety officer should provide orientation of new employees, as well as quarterly continuing education updates for all personnel. Initial training and all retraining should be documented in writing. Hand washing should be emphasized for all laboratory personnel. The mechanical action of rubbing the hands together and soaping under the fingernails is the most important part of the process; in the laboratory, products containing antibacterial agents do not prove to be any more effective than ordinary soap, unlike the situation in areas of the hospital such as the operating room.

All employees should also be offered, at no charge, the HBV vaccine and annual skin tests for tuberculosis. For those employees whose skin tests are already positive or those who have previously been vaccinated with BCG (Bacillus Calmette-Guerin), the employer should offer chest radiographs on employment although annual chest x-ray studies thereafter are no longer recommended by the CDC.

DISPOSAL OF HAZARDOUS WASTE

All materials contaminated with potentially infectious agents must be decontaminated before disposal. These

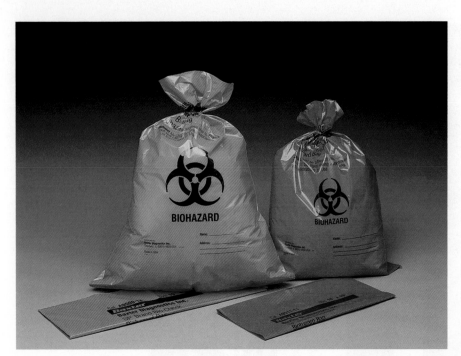

Figure 4-6 Autoclave bags. (Courtesy Allegiance Healthcare Corp., McGaw Park, Ill.)

include unused portions of patient specimens, patient cultures, stock cultures of microorganisms, and disposable sharp instruments, such as scalpels and syringes with needles. Infectious waste may be decontaminated by use of an autoclave, incinerator, or any one of several alternative waste-treatment methods. Some state or local municipalities permit blood, serum, urine, feces, and other body fluids to be carefully poured into a sanitary sewer. Infectious waste from microbiology laboratories is usually autoclaved on-site or sent for incineration, however.

In 1986, the EPA published a guide to hazardous waste reduction to limit the amount of hazardous waste generated and released into the environment. These regulations call for the following:

- Substituting less hazardous chemicals when possible, for example, the substitution of ethyl acetate for ether in ova and parasite concentrations and Hemo-de in place of xylene for trichrome stains
- Developing procedures that use less of a hazardous chemical, for example, the substitution of infrared technology for radioisotopes in blood culture instruments
- Segregating infectious waste from uncontaminated (paper) trash
- Substituting miniaturized systems for identification and antimicrobial susceptibility testing of potential pathogens to reduce the volume of chemical reagents and infectious waste

Recently, several alternative waste-treatment machines were developed to reduce the amount of waste buried in landfills. These systems combine mechanical shredding or compacting of the waste with either chemical (sodium hypochlorite, chlorine dioxide, peracetic acid), thermal (moist heat, dry heat), or ionizing radiation (microwaves, radio waves) decontamination. Most state regulations for these units require at least a sixfold reduction in vegetative bacteria, fungi, mycobacteria, and lipid-containing viruses and at least a fourfold reduction in bacterial spores.

Infectious waste (agar plates, tubes, reagent bottles) should be placed into two leak-proof, plastic bags for sturdiness (Figure 4-6); this is known as **double-bagging** in common laboratory jargon. Pipettes, swabs, and other glass objects should be placed into rigid cardboard containers (Figure 4-7) before disposal. Broken glass is placed in thick boxes lined with plastic biohazard bags (Figure 4-8); when full, the box is incinerated or autoclaved. Sharp objects, including scalpels and needles, are placed in Sharps containers (Figure 4-9), which are autoclaved or incinerated when full.

STANDARD PRECAUTIONS

In 1987, the CDC published guidelines known as *Universal Precautions,* to reduce the risk of HBV transmission in clinical laboratories and blood banks. In 1996, these safety recommendations became known as **Standard Precautions.** These precautions require

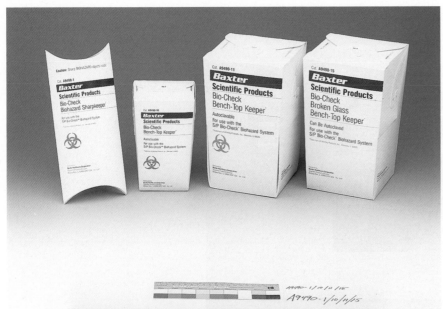

Figure 4-7 **A,** Various bench-top pipette discard containers. **B,** Bench-top serologic pipette discard container. (Courtesy Allegiance Healthcare Corp., McGaw Park, Ill.)

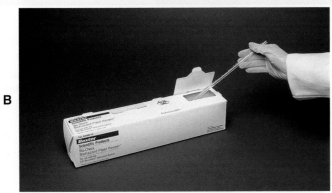

A

B

Figure 4-8 Cartons for broken glass. (Courtesy Lab Safety Supply Inc., Janesville, Wis.)

that blood and body fluids from every patient be treated as potentially infectious. The essentials of Standard Precautions and safe laboratory work practices are as follows:

- Do not eat, drink, smoke, or apply cosmetics (including lip balm).
- Do not insert or remove contact lenses.
- Do not bite nails or chew on pens.
- Do not mouth-pipette.
- Limit access to the laboratory to trained personnel only.
- Assume all patients are infectious for HIV or other blood-borne pathogens.
- Use appropriate barrier precautions to prevent skin and mucous membrane exposure, including wearing gloves at all times and masks, goggles, gowns, or aprons if there is a risk of splashes or droplet formation.
- Thoroughly wash hands and other skin surfaces after gloves are removed and immediately after any contamination.

Figure 4-9 Sharps containers. (Courtesy Lab Safety Supply Inc., Janesville, Wis.)

● Take special care to avoid injuries with sharp objects such as needles and scalpels.

The CDC's Standard Precautions should be followed for handling blood and body fluids, including all secretions and excretions (e.g., serum, semen, all sterile body fluids, saliva from dental procedures, and vaginal secretions) submitted to the microbiology laboratory. Standard Precautions do not apply to feces, nasal secretions, saliva (except in dental procedures), sputum, sweat, tears, urine, or vomitus unless they are grossly bloody. The consistent practice of Standard Precautions by health care workers handling all patient material will lessen the risks associated with such specimens. Mouth pipetting is strictly prohibited. Mechanical devices must be used for drawing all liquids into pipettes. Eating, drinking, smoking, and applying cosmetics are strictly forbidden in work areas. Food and drink must be stored in refrigerators in areas separate from the work area. All personnel should wash their hands with soap and water after removing gloves, after handling infectious material, and before leaving the laboratory area. In addition, it is good practice to store sera collected periodically from all health care workers so that, in the event of an accident, a seroconversion (acquisition of antibodies to an infectious agent) can be documented (see Chapter 10).

All health care workers should follow Standard Precautions whether working inside or outside the laboratory. When collecting specimens outside the laboratory, individuals should follow these guidelines:

● Wear gloves and a laboratory coat.
● Deal carefully with needles and lancets.
● Discard sharps in an appropriate puncture-resistant container.
● Never recap needles by hand; if necessary, special devices should be available for resheathing needles.

ENGINEERING CONTROLS

LABORATORY ENVIRONMENT

The biohazard symbol should be prominently displayed on laboratory doors and any equipment (refrigerators, incubators, centrifuges) that contain infectious material. The air-handling system of a microbiology laboratory should move air from lower to higher risk areas, never the reverse. Ideally, the microbiology laboratory should be under negative pressure, and air should not be recirculated after passing through microbiology (see Chapter 62 for a more detailed discussion of negative pressure in microbiology laboratories). The selected use of BSCs for procedures that generate infectious aerosols is critical to laboratory safety. Many infectious diseases, such as plague, tularemia, brucellosis, tuberculosis, and legionellosis, may be contracted by inhaling infectious particles, often present in a droplet of liquid. Because blood is a primary specimen that may contain infectious virus particles, subculturing blood cultures by puncturing the septum with a needle should be performed behind a barrier to protect the worker from droplets. Several other common procedures used to process

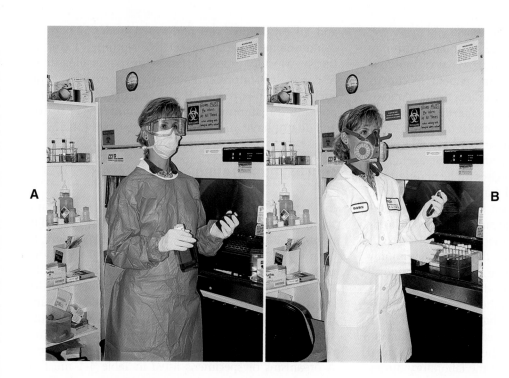

Figure 4-13 Personal protective equipment. **A,** Microbiologist wearing a laboratory gown, gloves, goggles, and face mask. **B,** Microbiologist wearing a laboratory coat, gloves, and respirator with HEPA filters (pink cartridges) for cleaning up spills of *Mycobacterium tuberculosis.*

and gloves that cover all potentially exposed skin on the arms are most beneficial. If the laboratory protective clothing becomes contaminated with body fluids or potential pathogens, it should be sterilized in an auto-clave immediately and cleaned before reusing. The institution or a uniform agency should clean laboratory coats; it is no longer permissible for microbiologists to launder their own coats. Alternatively, disposable gowns may be used. Obviously, laboratory workers who plan to enter an area of the hospital where patients at special risk of acquiring infection are present (e.g., intensive care units, the nursery, operating rooms, or areas in which immunosuppressive therapy is being admini-stered) should take every precaution to cover their street clothes with clean or sterile protective clothing appro-priate to the area being visited. Special impervious protective clothing is advisable for certain activities, such as working with radioactive substances or caustic chemicals. Solid-front gowns are indicated for those working with specimens being cultured for myco-bacteria. Unless large-volume spills of potentially infec-tious material are anticipated, impervious laboratory gowns are not necessary in most microbiology laboratories.

POSTEXPOSURE CONTROL

All laboratory accidents and potential exposures *must* be reported to the supervisor and safety officer, who will immediately arrange to send the individual to employee health or an outside occupational health

physician. Immediate medical care is of foremost importance; investigation of the accident should take place only after the employee has received appropriate care. If the accident is a needlestick injury, for example, the patient should be identified and the risk of the laboratorian acquiring a blood-borne infection should be assessed. The investigation helps the physician determine the need for prophylaxis, that is, HBIG (hepatitis B virus immunoglobulin) or an HBV booster immunization. The physician also is able to discuss the potential for disease transmission to family members following, for example, exposure to a patient with *Neisseria meningitidis.* Follow-up treatment should also be assessed including, for example, drawing additional sera at intervals of 6 weeks, 3 months, and 6 months for HIV testing. Finally, the safety committee or, at mini-mum, the laboratory director and safety officer should review the accident to determine whether it could have been prevented and to delineate measures to be taken to prevent future accidents. The investigation of the accident and corrective action should be documented in writing in an incident report.

CLASSIFICATION OF BIOLOGIC AGENTS BASED ON HAZARD

A CDC booklet titled *Classification of Etiological Agents on the Basis of Hazard* served as a reference for assessing the relative risks of working with various biologic agents until an updated CDC/NIH document was published titled *Biosafety in Microbiological and Biomedical Laboratories.* The fourth edition of this manual is currently available on the CDC website (http://www.cdc.gov/od/ohs/biosfty/biosfty.htm). In general, patient specimens pose a greater risk to laboratory workers than do microorganisms in culture, because the nature of etiologic agents in patient specimens is initially unknown. Biosafety Level 1 agents include those that have no known potential for infecting healthy people. These agents are used in laboratory teaching exercises for beginning-level students of microbiology. Level 1 agents include *Bacillus subtilis* and *Mycobacterium gordonae.* Precautions for working with Level 1 agents include standard good laboratory technique as described previously.

Biosafety Level 2 agents are those most commonly being sought in clinical specimens, and they include all the common agents of infectious disease, as well as HIV and several more unusual pathogens. For handling clinical specimens suspected of harboring any of these pathogens, Biosafety Level 2 precautions are sufficient. This level of safety includes the principles outlined previously plus limiting access to the laboratory during working procedures, training laboratory personnel in handling pathogenic agents, direction by competent supervisors, and performing aerosol-generating procedures in a BSC. Employers must offer hepatitis B vaccine to all employees determined to be at risk of exposure. *Bacillus anthracis* and *Yersinia pestis*, two organisms mentioned as possible bioterrorism agents, are actually listed as BSL-2 organisms.

Biosafety Level 3 procedures have been recommended for the handling of material suspected of harboring viruses unlikely to be encountered in a routine clinical laboratory, and for cultures growing *Mycobacterium tuberculosis,* the mold stages of systemic fungi, and for some other organisms when grown in quantities greater than that found in patient specimens. These precautions, in addition to those undertaken for Level 2, consist of laboratory design and engineering controls that contain potentially dangerous material by careful control of air movement and the requirement that personnel wear protective clothing and gloves, for instance. Persons working with Biosafety Level 3 agents should have baseline sera stored for comparison with acute sera that can be drawn in the event of unexplained illness. *Francisella tularensis* and *Brucella* spp. isolated from naturally occurring infections or infections following a bioterrorist event, are both BSL-3 pathogens. BSL-3 organisms are transmitted primarily by aerosols.

Biosafety Level 4 agents, which include only certain viruses of the arbovirus, arenavirus, or filovirus groups, none of which are commonly found in the United States, require the use of maximum containment facilities. Personnel and all materials must be decontaminated before leaving the facility,

and all procedures are performed under maximum containment (special protective clothing, Class III BSCs). Most of these facilities are public health or research laboratories. Potential bioterrorist agents such as smallpox require BSL-4 facilities. BSL-4 agents pose life-threatening risks, are transmitted via aerosols and do not have an available vaccine or therapy.

MAILING BIOHAZARDOUS MATERIALS

In March 2005 the requirements for packaging and shipping biologic material were significantly revised in response to the international community's desire to ensure safe and trouble-free shipment of infectious material while attempting to be more reasonable. Before this date, clinical specimens submitted for infectious disease diagnosis as well as isolates of any microorganism were considered an "infectious substance" and packaged and labeled under UN6.2 dangerous goods regulations. Now, only those category A organisms or specimens listed in Table 4-1 must be shipped as dangerous goods. The UN created a new category, UN3373, so that non-category A specimens or cultures can be packed and shipped as diagnostic or clinical specimens. In fact, the only category A infectious substance that most clinical laboratories would ship is a *Mycobacterium tuberculosis* culture. If the laboratory director is unsure whether a patient has symptoms of a category A agent it is prudent to ship the specimen as an infectious substance rather than a diagnostic specimen. Figure 4-14, *A*, shows triple packaging for both diagnostic/clinical or infectious substances in a pouch, and Figure 14-4, *B*, shows triple packaging for diagnostic, clinical, or infectious substances in a rigid bottle.

Packaging must meet requirements of IATA (The International Air Transport Association) and IACO (International Civil Aviation Organization). Packaging instructions are available in the annual IATA regulations under 602 (dangerous goods) or 650 (patient or clinical specimens). All air and ground shippers such as the United States Postal Service (USPS), the United States Department of Transportation (DOT) and Federal Express (Fed Ex) have adopted IATA standards.

Training in the proper packing and shipping of infectious material is a key feature of the regulations. Every institution that ships infectious materials, whether a hospital or POL (physician office laboratory), is required to have appropriately trained individuals; training may be obtained through carriers, package manufacturers, and special safety training organizations. The **shipper** is the individual (institution) who is ultimately responsible for safe and appropriate packaging. Any fines or penalties will be the shipper's responsibility.

Table 4-1   Examples of Infectious Substances Included in Category A*

UN Number and Proper Shipping Name	Microorganism
UN 2814—Infectious Substance Affecting Humans	*Bacillus anthracis* (cultures only)
	Brucella abortus (cultures only)
	Brucella melitensis (cultures only)
	Brucella suis (cultures only)
	Burkholderia mallei - Glanders (cultures only)
	Burkholderia pseudomallei – (cultures only)
	Chlamydia psittaci – avian strains (cultures only)
	Clostridium botulinum (cultures only)
	Coccidioides immitis (cultures only)
	Coxiella burnetii (cultures only)
	Crimean-Congo hemorrhagic fever virus
	Dengue virus (cultures only)
	Eastern equine encephalitis virus (cultures only)
	Escherichia coli, verotoxigenic (cultures only)
	Ebola virus
	Flexal virus
	Francisella tularensis (cultures only)
	Guanarito virus
	Hantaan virus
	Hantaviruses causing hemorrhagic fever with renal syndrome
	Hendra virus
	Hepatitis B virus (cultures only)
	Herpes B virus (cultures only)
	Human immunodeficiency virus (cultures only)
	Highly pathogenic avian influenza virus (cultures only)
	Japanese encephalitis virus (cultures only)
	Junin virus
	Kyasanur Forest disease virus
	Lassa virus
	Machupo virus
	Marburg virus
	Monkeypox virus
	Mycobacterium tuberculosis (cultures only)
	Nipah virus
	Omsk hemorrhagic fever virus
	Poliovirus (cultures only)
	Rabies virus (cultures only)
	Rickettsia prowazekii (cultures only)
	Rickettsia rickettsii (cultures only)
	Rift Valley fever virus (cultures only)
	Russian spring-summer encephalitis virus (cultures only)
	Sabia virus
	Shigella dysenteriae type 1 (cultures only)
	Tick-borne encephalitis virus (cultures only)
	Variola virus
	Venezuelan equine encephalitis virus (cultures only)
	West Nile virus (cultures only
	Yellow fever virus (cultures only)
	Yersinia pestis (cultures only)
UN 2900—Infectious Substance Affecting Animals	African swine fever virus (cultures only)
	Avian paramyxovirus type 1—Velogenic Newcastle disease virus (cultures only)
	Classical swine fever virus (cultures only)
	Foot and mouth disease virus (cultures only)
	Lumpy skin disease virus (cultures only)
	Mycoplasma mycoides—Contagious bovine pleuropneumonia (cultures only)
	Peste des petits ruminants virus (cultures only)
	Rinderpest virus (cultures only)
	Sheep-pox virus (cultures only)
	Goatpox virus (cultures only)
	Swine vesicular disease virus (cultures only)
	Vesicular stomatitis virus (cultures only)

Courtesy Saf T Pak, Edmonton, Alberta, Canada.

*This table was updated in March 2005. It is not exhaustive. Infectious substances, including new or emerging pathogens, that do not appear in the table but that meet the same criteria must be assigned to Category A. In addition, if there is doubt as to whether a substance meets the criteria it must be included in Category A.

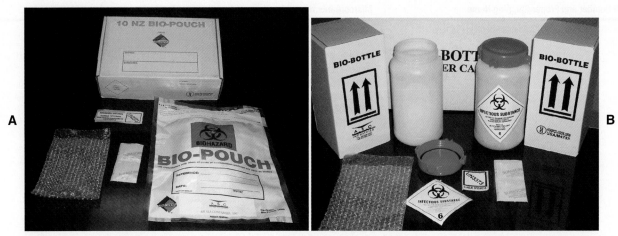

Figure 4-14   **A,** The Bio-Pouch (lower right-hand corner) is made of laminated, low-density polyethylene, which is virtually unbreakable. The label for shipping a diagnostic specimen is shown (UN 3373). **B,** The Bio-Bottle is made of high-density polyethylene and is used as the secondary container. This packaging is used for both infectious substances (with the Class 6 label shown) or for clinical and diagnostic specimens with the UN 3373 label. (Courtesy Air Sea Containers, Inc, Miami, Florida.)

Infectious specimens or isolates should be wrapped with absorbent material and placed within a plastic biohazard bag, called a primary receptacle. The primary receptacle is then inserted into a secondary container, most often a watertight hard plastic mailer. After the secondary container is capped, it is placed in an outer tertiary container constructed of fiberboard. This type of packaging is shown in Figure 4-14, *B*. A UN Class 6 label on the outer box confirms that the packaging meets all the required standards. A Shippers Declaration for Dangerous Goods Form must accompany the airbill or ground form. Diagnostic or clinical specimens are packaged similarly but a UN3373 label is attached to a different outer box and it is not necessary to fill out a Shippers Declaration.

Shippers should note that some carriers have additional requirements for coolant materials such as ice, dry ice, or liquid nitrogen. Because the shipper is liable for appropriate packaging, it is best to check with individual carriers in special circumstances and update your instructions yearly when the new IATA *Dangerous Goods Regulations* are published.

Shipping and packaging regulations from the Code of Federal Regulations can be found at http://hazmat.dot.gov and http://www.cdc.gov/od/ohs/biosfty/shipregs.htm. IATA regulations can be found at http://www.iata.org. The international importation or exportation of biological agents requires a permit from CDC. Information on obtaining a permit may be found at http://cdc.gov.

REFERENCES

1. Blaser MJ, Feldman RA: Acquisition of typhoid fever from proficiency testing specimens, *N Engl J Med* 303:1481, 1980.

2. Eyre JWH: *Bacteriologic technique*, Philadelphia, 1913, WB Saunders.

ADDITIONAL READING

Centers for Disease Control: Update: Universal precautions for prevention of transmission of human immunodeficiency virus, hepatitis B virus, and other blood-borne pathogens in health care settings, *MMWR Morb Mortal Wkly Rep* 37:377, 1988.

Centers for Disease Control: Recommendations for prevention of HIV transmission in health-care settings, *MMWR Morb Mortal Wkly Rep* 36:3S, 1987.

Denys GA, Gary LD, Snyder JW: Cumitech 40; Packing and shipping of diagnostic specimens and infectious substances, Sewell, DL, coordinating editor, 2003, Washington, DC, ASM Press.

Fleming DO, Hunt DL: *Biological safety: principles and practices,* ed 3 Washington, DC, 2000, ASM Press.

Hospital Infection Control Practices Advisory Committee: Guideline for isolation precautions in hospitals, *Infect Control Hosp Epidemiol* 17:53, 1996.

International Air Transport Association: Dangerous Goods Regulations, ed 46, Montreal, 2005, International Air Transport Association.

Jamison R, Noble MA, Proctor EM, et al: Cumitech 29, Laboratory safety in clinical microbiology, Smith JA, coordinating editor, 1996, Washington, DC, American Society for Microbiology.

National Committee for Clinical Laboratory Standards: Clinical laboratory safety; Approved guideline GP17-A2, Wayne, Pa, 2004, National Committee for Clinical Laboratory Standards.

National Committee for Clinical Laboratory Standards: Protection of laboratory workers from instrument biohazards and infectious disease transmitted by blood, body fluids and tissue; Approved guideline M29-A, Wayne,

Pa, 1997, National Committee for Clinical Laboratory Standards.

National Committee for Clinical Laboratory Standards: Protection of laboratory workers from occupationally acquired infections; Approved standard, M29-A2, Wayne, PA, ;2001, National Committee for Clinical Laboratory Standards.

Occupational Safety and Health Administration: Occupational exposure to blood-borne pathogens: final rule, *Federal Register* 56: 64175, 1991.

Occupational Safety and Health Administration: Occupational exposure to blood-borne pathogens: correction July 1, 1992, 29 CFR Part 1910. *Federal Register* 57:127: 29206, 1991.

Occupational Safety and Health Administration: Draft guidelines for preventing the transmission of tuberculosis in health care facilities, *Federal Register* 58:52810, 1993.

Sewell DL: Laboratory-associated infections and biosafety, *Clin Microbiol Rev* 8:389, 1995.

United States Department of Health and Human Services: Biosafety in microbiological and biomedical laboratories, ed 4, Washington, DC, 1999, US Government Printing Office.

United States Environmental Protection Agency: EPA guide for infectious waste management, Washington, DC, 1986, Publication EPA/530-SW-86-014, US Environmental Protection Agency.

CHAPTER 5

Specimen Management

In the late 1800s, the first clinical microbiology laboratories were organized to diagnose infectious diseases such as tuberculosis, typhoid fever, malaria, intestinal parasites, syphilis, gonorrhea, and diphtheria. Between 1860 and 1900, microbiologists such as Pasteur, Koch, and Gram developed the techniques for staining and the use of solid media for isolation of microorganisms that are still used in clinical laboratories today. Microbiologists continue to look for the same organisms that these laboratorians did, as well as a whole range of others that have been uncovered in the twentieth century, for example, *Legionella,* viral infections, nontuberculosis acid-fast bacteria, and fungal infections. Microbiologists work in public health laboratories, hospital laboratories, reference or independent laboratories, and physician office laboratories (POLs). Depending on the level of service of each facility, the type of testing differs, but in general a microbiologist will perform one or more of the following functions:

- Cultivation (growth), identification, and antimicrobial susceptibility testing of microorganisms
- Direct detection of infecting organisms by microscopy
- Direct detection of specific products of infecting organisms using chemical, immunologic, or molecular techniques
- Detection of antibodies produced by the patient in response to an infecting organism (serodiagnosis)

This chapter presents an overview of issues involved in infectious disease diagnostic testing. Many of these issues are covered in detail in separate chapters, which are cited.

GENERAL CONCEPTS FOR SPECIMEN COLLECTION AND HANDLING

Specimen collection and transportation are critical considerations, because any results the laboratory generates is limited by the quality of the specimen and its condition on arrival in the laboratory. Specimens should be obtained to preclude or minimize the possibility of introducing extraneous (contaminating) microorganisms that are not involved in the infectious process. This is a particular problem, for example, in specimens collected from mucous membranes that are already colonized with microorganisms that are part of an individual's endogenous or "normal" flora; these organisms are usually contaminants but may also be opportunistic pathogens. For example, the throats of hospitalized patients on ventilators may frequently be colonized with *Klebsiella pneumoniae;* although *K. pneumoniae* is not usually involved in cases of community-acquired pneumonia, it can cause a hospital-acquired respiratory infection in this subset of patients. Use of special techniques that bypass areas containing normal flora when this is feasible (e.g., covered brush bronchoscopy in critically ill patients with pneumonia) prevents many problems associated with false-positive results. Likewise, careful skin preparation before procedures such as blood cultures and spinal taps decreases the chance that organisms normally present on the skin will contaminate the specimen.

APPROPRIATE COLLECTION TECHNIQUES

Specimens should be collected during the acute (early) phase of an illness (or within 2 to 3 days for viral infections), and before antibiotics are administered, if possible. Swabs generally are poor specimens if tissue or needle aspirates can be obtained. It is the microbiologist's responsibility to provide clinicians with a collection manual or instruction cards listing optimal specimen collection techniques and transport information. Information for the nursing staff and clinicians should include:

- Safety considerations
- Selection of appropriate anatomic site
- Collection instructions including type of swab or transport medium
- Transportation instructions including time and temperature
- Labeling instructions including minimum patient demographic information
- Special instructions such as patient preparation

Instructions should be written, so specimens collected by the patient (e.g., urine, sputum, or stool) are handled properly. Most urine or stool collection kits contain instructions in several languages, but nothing substitutes for a concise set of verbal instructions. Similarly, when distributing kits for sputum collection, the microbiologist should be able to explain to the patient the difference between spitting in a cup (saliva) and producing good lower respiratory secretions from a deep cough (sputum). General collection information is shown in Table 5-1. An in-depth discussion of each type of specimen is found in Part VII.

SPECIMEN TRANSPORT

Ideally, specimens should be transported to the laboratory within 2 hours of collection. All specimen containers should be leak-proof, and the specimens should be transported within sealable, leak-proof, plastic bags with a separate section for paperwork; resealable bags or bags with a permanent seal are common for this purpose. Bags should be marked with a biohazard label (Figure 5-1). Many microorganisms are susceptible to environmental conditions such as the presence of oxygen (anaerobic bacteria), changes in temperature (*Neisseria meningitidis*), or changes in pH (*Shigella*). Thus, use of special preservatives or holding media for transportation of specimens delayed for more than 2 hours is important to ensure organism viability (survival).

SPECIMEN PRESERVATION

Preservatives, such as boric acid for urine or polyvinyl alcohol (PVA) and buffered formalin for stool for ova and parasite (O&P) examination, are designed to maintain the appropriate colony counts (for urines) or the integrity of trophozoites and cysts (for O&Ps), respectively. Other transport, or **holding**, media maintain the viability of microorganisms present in a specimen without supporting the growth of any of the organisms. This maintains the organisms in a state of suspended animation so that no organism overgrows another or dies out. Stuart's medium and Amie's medium are two common holding media. Sometimes charcoal is added to these media to absorb fatty acids present in the specimen that could kill fastidious (fragile) organisms such as *Neisseria gonorrhoeae* or *Bordetella pertussis*.

Anticoagulants are used to prevent clotting of specimens such as blood, bone marrow, and synovial fluid because microorganisms will otherwise be bound up in the clot. The type and concentration of anticoagulant is very important because many organisms are inhibited by some of these chemicals. Sodium polyanethol sulfonate (SPS) at a concentration of 0.025% (w/v) is usually used, because *Neisseria* spp. and some anaerobic bacteria are particularly sensitive to higher concentrations. Because the ratio of specimen to SPS is so important, it is necessary to have both large (adult-size) and small (pediatric-size) tubes

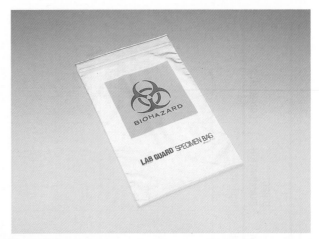

Figure 5-1   Specimen bag with biohazard label, separate pouch for paperwork, and self-seal. (Courtesy Allegiance Healthcare Corp., McGaw Park, Ill.)

available, so organisms in small amounts of bone marrow or synovial fluid are not overwhelmed by the concentration of SPS. Heparin is also a commonly used anticoagulant, especially for viral cultures, although it may inhibit growth of gram-positive bacteria and yeast. Citrate, EDTA (ethylenediaminetetraacetic acid), or other anticoagulants should not be used for microbiology, because their efficacy has not been demonstrated for a majority of organisms. It is the microbiologist's job to make sure the appropriate anticoagulant is used for each procedure. One generally should not specify a color ("yellow-top") tube for collection without specifying the anticoagulant (SPS), because at least one popular brand of collection tube (Vacutainer, Becton, Dickinson and Company) has a yellow-top tube with either SPS or ACD (trisodium citrate/citric acid/dextrose); ACD is not appropriate for use in microbiology.

Specimen Storage

If specimens cannot be processed as soon as they are received, they must be stored (see Table 5-1). Several storage methods are used (i.e., refrigerator temperature [4° C], ambient [room] temperature [22° C], body temperature [37° C], and freezer temperature [either –20° or –70° C]), depending on the type of transport media (if applicable) and the etiologic (infectious) agents sought. Specimens suspected of containing anaerobic bacteria, for example, should never be stored in the refrigerator, while cerebral spinal fluid (CSF) should always be kept at 37° C. Urine, stool, viral specimens, sputa, swabs, and foreign devices such as catheters should be stored at 4° C. Serum for serologic studies may be frozen for up to 1 week at –20° C, and tissues or specimens for long-term storage should be frozen at –70° C.

Table 5-1 Collection, Transport, Storage, and Processing of Specimens Commonly Submitted to a Microbiology Laboratory*

SPECIMEN	CONTAINER	PATIENT PREPARATION	SPECIAL INSTRUCTIONS	TRANSPORTATION TO LABORATORY	STORAGE BEFORE PROCESSING	PRIMARY PLATING MEDIA	DIRECT EXAMINATION	COMMENTS
Aqueous/vitreous fluid	Sterile, screw cap tube			Immediately/RT	Set up immediately on receipt	BA, CA, Mac, Sab, 7H10, thio	Gram/AO	
Abscess (also lesion, wound, pustule, ulcer)								
Superficial	Aerobic swab moistened with Stuart's or Amie's medium	Wipe area with sterile saline or 70% alcohol	Swab along leading edge of wound	Within 24 hrs/RT	24 hrs/RT	BA, CA, Mac	Gram	Add CNA if smear suggests mixed gram-positive and gram negative flora
Deep	Anaerobic transporter	Wipe area with sterile saline or 70% alcohol	Aspirate material from wall or excise tissue	Within 24 hrs/RT	24 hrs/RT	BA, CA, Mac, Ana	Gram	Wash any granules and "emulsify" in saline
Blood or Bone Marrow	Blood culture media set (aerobic and anaerobic bottle) or Vacutainer tube with SPS	Disinfect venipuncture site with 70% alcohol and disinfectant such as betadine	Draw blood at time of febrile episode; draw 2 sets from right and left arms; do not draw more than 3 sets in a 24-hr period; draw ≥20 ml/set (adults) or 1-20 ml/set (pediatric) depending on patient's weight	Within 2 hrs/RT	Must be incubated at 37° C on receipt in laboratory	Blood culture bottles	AO or Giemsa	Other considerations: brucellosis, tularemia, cellwall-deficient bacteria, leptospirosis, or AFB
Body Fluids								
Amniotic, abdominal, ascites (peritoneal), bile, joint (synovial), pericardial, pleural	Sterile, screw-cap tube or anaerobic transporter	Disinfect skin before aspirating specimen	Needle aspiration	Immediately/RT	Plate as soon as received	Use an aerobic and anaerobic blood culture bottle set for body fluids	Gram	May need to concentrate by centrifugation or filtration – stain and culture sediment
Bone	Sterile, screw-cap container	Disinfect skin before surgical procedure	Take sample from affected area for biopsy	Immediately/RT	Plate as soon as received	BA, CA, Mac, thio	Gram	May need to homogenize

Table 5-1   Collection, Transport, Storage, and Processing of Specimens Commonly Submitted to a Microbiology Laboratory*—cont'd

SPECIMEN	CONTAINER	PATIENT PREPARATION	SPECIAL INSTRUCTIONS	TRANSPORTATION TO LABORATORY	STORAGE BEFORE PROCESSING	PRIMARY PLATING MEDIA	DIRECT EXAMINATION	COMMENTS
Cerebrospinal Fluid								
	Sterile, screw-cap tube	Disinfect skin before aspirating specimen	Consider rapid testing (e.g., Gram stain; cryptococcal antigen)	Immediately/RT	6 hrs/37° C except for viruses, which can be held at 4° C for up to 3 days	BA, CA	Gram – best sensitivity by cytocentrifugation (may also want to do AO if cytocentrifuge not available)	Add thio for CSF collected from shunt
Ear								
Inner	Sterile, screw-cap tube or anaerobic transporter	Clean ear canal with mild soap solution before myringotomy (puncture of the ear drum)	Aspirate material behind drum with syringe if ear drum intact; use swab to collect material from ruptured ear drum	Immediately/RT	6 hrs/RT	BA, CA, Mac (add thio if prior antimicrobial therapy)	Gram	Add anaerobic culture plates for tympanocentesis specimens
Outer	Aerobic swab moistened with Stuart's or Amie's medium	Wipe away crust with sterile saline	Firmly rotate swab in outer canal	Within 24 hrs/RT	24 hrs/RT	BA, CA, Mac	Gram	
Eye								
Conjunctiva	Aerobic swab moistened with Stuart's or Amie's medium		Sample both eyes; use swab premoistened with sterile saline	Within 24 hrs/RT	24 hrs/RT	BA, CA, Mac	Gram, AO, histologic stains (e.g., Giemsa)	Other considerations: *Chlamydia trachomatis*, viruses, and fungi
Corneal scrapings	Bedside inoculation of BA, CA, Sab, 7H10, thio	Clinician should instill local anesthetic before collection		Immediately/RT	Must be incubated at 28° C (Sab) or 37° C (everything else) on receipt in laboratory	BA, CA, Sab, 7H10, Ana, thio	Gram/AO	Other considerations: *Acanthamoeba* spp., herpes simplex virus and other viruses, *Chlamydia trachomatis*; and fungi
Foreign Bodies								
IUD	Sterile, screw-cap container	Disinfect skin before removal		Immediately/RT	Plate as soon as received	Thio		

cont'd

Table 5-1 Collection, Transport, Storage, and Processing of Specimens Commonly Submitted to a Microbiology Laboratory*—cont'd

SPECIMEN	CONTAINER	PATIENT PREPARATION	SPECIAL INSTRUCTIONS	TRANSPORTATION TO LABORATORY	STORAGE BEFORE PROCESSING	PRIMARY PLATING MEDIA	DIRECT EXAMINATION	COMMENTS
IV catheters, pins, prosthetic valves	Sterile, screw-cap container	Disinfect skin before removal	Do not culture Foley catheters; IV catheters are cultured quantitatively by rolling the segment back and forth across agar with sterile forceps 4 times; ≥15 colonies are associated with clinical significance	Immediately/RT	Plate as soon as received	BA, thio		
GI Tract								
Gastric aspirate	Sterile, screw-cap tube	Collect in early AM before patient eats or gets out of bed	Most gastric aspirates are on infants or for AFB	Immediately/RT	Must be neutralized with sodium bicarbonate within 1 hr of collection	BA, CA, Mac, CNA	Gram/AO	Other considerations: AFB
Gastric biopsy	Sterile, screw-cap tube		Rapid urease test or culture for *Helicobacter pylori*	Immediately/4° C	Must be set up immediately	Skirrow's, BA		Other considerations: urea breath test
Rectal swab	Swab placed in enteric transport medium		Insert swab ≈ 2.5 cm past anal sphincter; feces should be visible on swab	Within 24 hrs/4° C	72 hrs/4° C	BA, Mac, XLD or HE, Campy	Methylene blue for fecal leukocytes	Other considerations: *Vibrio, Yersinia enterocolitica, Escherichia coli* O157:H7
Stool culture	Clean, leak-proof container; transfer feces to enteric transport medium if transport will exceed 1 hr		Routine culture should include *Salmonella, Shigella,* and *Campylobacter,* specify *Vibrio, Aeromonas, Plesiomonas, Yersinia, Escherichia coli* O157:H7, if needed	Within 24 hrs/4° C	72 hrs/4° C	BA, Mac, XLD or HE, Campy	Methylene blue for fecal leukocytes	Do not perform routine stool cultures for patients whose length of stay in the hospital exceeds 3 days and whose admitting diagnosis was not diarrhea; these patients should be tested for *Clostridium difficile*

Table 5-1   Collection, Transport, Storage, and Processing of Specimens Commonly Submitted to a Microbiology Laboratory*—cont'd

SPECIMEN	CONTAINER	PATIENT PREPARATION	SPECIAL INSTRUCTIONS	TRANSPORTATION TO LABORATORY	STORAGE BEFORE PROCESSING	PRIMARY PLATING MEDIA	DIRECT EXAMINATION	COMMENTS
O&P	O&P transporters (e.g., 10% formalin and PVA)	Collect 3 specimens every other day at a minimum for outpatients; hospitalized patients (inpatients) should have 3 specimens collected every day; specimens from inpatients hospitalized more than 3 days should be discouraged	Wait 7-10 days if patient has received antiparasitic compounds, barium, iron, Kaopectate, metronidazole, Milk of Magnesia, Pepto-Bismol, or tetracycline	Within 24 hrs/RT	Indefinitely/RT			
Genital Tract								
Female								
Bartholin cyst	Anaerobic transporter	Disinfect skin before collection	Aspirate fluid; consider chlamydia and GC culture	Within 24 hrs/RT	24 hrs/RT	BA, CA, Mac, TM, Ana	Gram	
Cervix	Swab moistened with Stuart's or Amie's medium	Remove mucus before collection of specimen	Do *not* use lubricant on speculum; use viral/chlamydial transport medium, if necessary; swab deeply into endocervical canal	Within 24 hrs/RT	24 hrs/RT	BA, CA, Mac, TM	Gram	
Cul-de-sac	Anaerobic transporter		Submit aspirate	Within 24 hrs/RT	24 hrs/RT	BA, CA, Mac, TM, Ana	Gram	
Endometrium	Anaerobic transporter		Surgical biopsy or transcervical aspirate via sheathed catheter	Within 24 hrs/RT	24 hrs/RT	BA, CA, Mac, TM, Ana	Gram	

cont'd

Table 5-1   Collection, Transport, Storage, and Processing of Specimens Commonly Submitted to a Microbiology Laboratory*—cont'd

SPECIMEN	CONTAINER	PATIENT PREPARATION	SPECIAL INSTRUCTIONS	TRANSPORTATION TO LABORATORY	STORAGE BEFORE PROCESSING	PRIMARY PLATING MEDIA	DIRECT EXAMINATION	COMMENTS
Urethra	Swab moistened with Stuart's or Amie's medium	Remove exudate from urethral opening	Collect discharge by massaging urethra against pubic symphysis or insert flexible swab 2-4 cm into urethra and rotate swab for 2 seconds; collect at least 1 hr after patient has urinated	Within 24 hrs/RT	24 hrs/RT	BA, CA, TM	Gram	Other considerations: *Chlamydia*, *Mycoplasma*
Vagina	Swab moistened with Stuart's or Amie's medium or JEMBEC transport system	Remove exudate	Swab secretions and mucous membrane of vagina	Within 24 hrs/RT	24 hrs/RT	Culture is not recommended for the diagnosis of bacterial vaginosis; inoculate selective medium for Group B *Streptococcus* (LIM broth) if indicated on pregnant women	Gram	Examine Gram stain for bacterial vaginosis, especially white blood cells, gram-positive rods indicative of *Lactobacillus*, and curved, gram-negative rods indicative of *Mobiluncus* spp.
Male								
Prostate	Swab moistened with Stuart's or Amie's medium *or sterile, screw-cap tube*	Clean glans with soap and water	Collect secretions on swab or in tube	Within 24 hrs/RT for swab; immediately if in tube/RT	Swab: 24 hrs/RT; tube: plate secretions immediately	BA, CA, Mac, TM, CNA	Gram	
Urethra	Swab moistened with Stuart's or Amie's medium or JEMBEC transport system		Insert flexible swab 2-4 cm into urethra and rotate for 2 seconds or collect discharge on JEMBEC transport system	Within 24 hrs/RT for swab; within 2 hrs for JEMBEC system	24 hrs/RT for swab; put JEMBEC at 37° C immediately on receipt in laboratory	BA, CA, TM	Gram	Other considerations: *Chlamydia*, *Mycoplasma*

Table 5-1 Collection, Transport, Storage, and Processing of Specimens Commonly Submitted to a Microbiology Laboratory*—cont'd

SPECIMEN	CONTAINER	PATIENT PREPARATION	SPECIAL INSTRUCTIONS	TRANSPORTATION TO LABORATORY	STORAGE BEFORE PROCESSING	PRIMARY PLATING MEDIA	DIRECT EXAMINATION	COMMENTS
Hair, Nails, or Skin Scrapings (for fungal culture)								
	Clean, screw-top tube	Nails or skin: wipe with 70% alcohol	Hair: collect hairs with intact shaft Nails: send clippings of affected area Skin: scrape skin at leading edge of lesion	Within 24 hrs/RT	Indefinitely/RT	Sab, IMAcg, Sabcg	CW	
Respiratory Tract								
Lower								
BAL, BB, BW	Sterile, screw-top container		Anaerobic culture appropriate only if sheathed (protected) catheter used	Within 24 hrs/RT	24 hrs/4° C	BA, CA, Mac	Gram and other special stains as requested, e.g., *Legionella* DFA, acid-fast stain	Other considerations: quantitative culture for BAL, AFB, *Legionella*, *Nocardia*, *Mycoplasma*, *Pneumocystis*, *cytomegalovirus*
Sputum, tracheal aspirate (suction)	Sterile, screw-top container	Sputum: have patient brush teeth and then rinse or gargle with water before collection	Sputum: have patient collect from deep cough; specimen should be examined for suitability for culture by Gram stain; induced sputa on pediatric or uncooperative patients may be watery because of saline nebulization	Within 24 hrs/RT	24 hrs/4° C	BA, CA, Mac	Gram and other special stains as requested, e.g., *Legionella* DFA, acid-fast stain	Other considerations: AFB, *Nocardia*

cont'd

Table 5-1 Collection, Transport, Storage, and Processing of Specimens Commonly Submitted to a Microbiology Laboratory*—cont'd

SPECIMEN	CONTAINER	PATIENT PREPARATION	SPECIAL INSTRUCTIONS	TRANSPORTATION TO LABORATORY	STORAGE BEFORE PROCESSING	PRIMARY PLATING MEDIA	DIRECT EXAMINATION	COMMENTS
Upper								
Nasopharynx	Swab moistened with Stuart's or Amie's medium		Insert flexible swab through nose into posterior nasopharynx and rotate for 5 seconds; specimen of choice for *Bordetella pertussis*	Within 24 hrs/RT	24 hrs/RT	BA, CA		Other considerations: add special media for *Corynebacterium diphtheriae*, pertussis, *Chlamydia*, and *Mycoplasma*
Pharynx (throat)	Swab moistened with Stuart's or Amie's medium		Swab posterior pharynx and tonsils; routine culture for group A *Streptococcus* (*S. pyogenes*) only	Within 24 hrs/RT	24 hrs/RT	BA		Other considerations: add special media for *C. diphtheriae*, *Neisseria gonorrhoeae*, and epiglottis (*Haemophilus influenzae*)
Tissue								
	Anaerobic transporter or sterile, screw-cap tube	Disinfect skin	Do not allow specimen to dry out; moisten with sterile, distilled water if not bloody	Within 24 hrs/RT	24 hrs/RT	BA, CA, Mac, Ana (thio, if indicated)	Gram	May need to homogenize
Urine								
Clean-voided midstream (CVS)	Sterile, screw-cap container	Females: clean area with soap and water, then rinse with water; hold labia apart and begin voiding in commode; after several mL have passed, collect		Within 24 hrs/4° C	24 hrs/4° C	BA, Mac	Gram or check for pyuria	Plate quantitatively at 1:1000; consider plating quantitatively at 1:100 if patient is female of childbearing age with white blood cells and possible acute urethral syndrome

Table 5-1 Collection, Transport, Storage, and Processing of Specimens Commonly Submitted to a Microbiology Laboratory*—cont'd

SPECIMEN	CONTAINER	PATIENT PREPARATION	SPECIAL INSTRUCTIONS	TRANSPORTATION TO LABORATORY	STORAGE BEFORE PROCESSING	PRIMARY PLATING MEDIA	DIRECT EXAMINATION	COMMENTS
		midstream Males: clean glans with soap and water, then rinse with water; retract foreskin; after several mL have passed, collect midstream						
Straight catheter (in and out)	Sterile, screw-cap container	Clean urethral area (soap and water) and rinse (water)	Insert catheter into bladder; allow first 15 mL to pass; then collect remainder	Within 2 hrs/4° C	24 hrs/4° C	BA, Mac	Gram or check for pyuria	Plate quantitatively at 1:100 and 1:1000
Indwelling catheter (Foley)	Sterile, screw-cap container	Disinfect catheter collection port	Aspirate 5-10 mL of urine with needle and syringe	Within 2 hrs/4° C	24 hrs/4° C	BA, Mac	Gram or check for pyuria	Plate quantitatively at 1:1000
Suprapubic aspirate	Sterile, screw-cap container or anaerobic transporter	Disinfect skin	Needle aspiration above the symphysis pubis through the abdominal wall into the full bladder	Immediately/RT	Plate as soon as received	BA, Mac, Ana	Gram or check for pyuria	Plate quantitatively at 1:100 and 1:1000

*Specimens for viruses, chlamydia, and mycoplasma are usually submitted in appropriate transport media at 4° C to stabilize respective microorganisms.
≈, approximately; *7H10*, Middlebrook 7H10 agar; *AFB*, acid-fast bacilli; *AM*, morning; *Ana*, anaerobic agars as appropriate (see Chapter 43); *AO*, acridine orange stain; *BA*, blood agar; *BAL*, bronchial alveolar lavage; *BB*, bronchial brush; *BW*, bronchial wash; *CA*, chocolate agar; *Campy*, selective *Campylobacter* agar; *CNA*, Columbia agar with colistin and nalidixic acid; *CW*, calcofluor white stain; *DFA*, direct fluorescent antibody stain; *GC*, *Neisseria gonorrhoeae*; transport using JEMBEC system with modified Thayer-Martin; *GI*, gastrointestinal; *Gram*, Gram stain; *HBT*, human blood-bilayer Tween agar; *HE*, Hektoen enteric agar; *hrs*, hours; *IMAcg*, inhibitory mold agar with chloramphenicol and gentamicin; *IUD*, intrauterine device; *Mac*, MacConkey agar; *mL*, milliliters; *O&P*, ova and parasite examination; *PVA*, polyvinyl alcohol; *RT*, room temperature; *Sab*, Sabouraud dextrose agar; *Sabcg*, Sabouraud dextrose agar with cycloheximide and gentamicin; *SPS*, sodium polyanethol sulfate; *thio*, thioglycollate broth; *TM*, Thayer-Martin agar; *XLD*, xylose lysine deoxycholate agar.

SPECIMEN LABELING

Specimens should be labeled at the very least with the patient's name, identifying number (hospital number) or birth date, and source. Enough information must be provided on the specimen label so that the specimen can be matched up with the requisition when it is received in the laboratory.

SPECIMEN REQUISITION

The specimen (or test) requisition is an order form that is sent to the laboratory along with a specimen. Often the requisition is a hard (paper) copy of the physician's orders and the patient demographic information (e.g., name and hospital number). Sometimes, however, if a hospital information system offers computerized order entry, the requisition is transported to the laboratory electronically. The requisition should contain as much information as possible regarding the patient history, diagnosis, and immunization record. This information helps the microbiologist to work up the specimen and determine which organisms are significant in the culture. A complete requisition should include the following:

- The patient's name
- Hospital number
- Age or date of birth
- Sex
- Collection date and time
- Ordering physician (may be UPIN number)
- Exact nature and source of the specimen
- Diagnosis (may be ICD-9-CM code)
- Immunization history
- Current antimicrobial therapy

REJECTION OF UNACCEPTABLE SPECIMENS

Criteria for specimen rejection should be set up and distributed to all clinical practitioners. In general, specimens are unacceptable if any of the following conditions apply:

- The information on the label does not match the information on the requisition (patient name or source of specimen is different).
- The specimen has been transported at the improper temperature.
- The specimen has not been transported in the proper medium (e.g., specimens for anaerobic bacteria submitted in aerobic transports).
- The quantity of specimen is insufficient for testing (the specimen is considered QNS [quantity not sufficient]).
- The specimen is leaking.

- The specimen transport time exceeds 2 hours postcollection and the specimen is not preserved.
- The specimen was received in a fixative (formalin) which, in essence, kills any microorganism present.
- The specimen has been received for anaerobic culture from a site known to have anaerobes as part of the normal flora (vagina, mouth).
- The specimen is dried up.
- Processing the specimen would produce information of questionable medical value (e.g., Foley catheter tip).

It is an important rule to always talk to the requesting physician or another member of the health care team before discarding unacceptable specimens. In some cases, such as mislabeling of a specimen or requisition, the person who collected the specimen and filled out the paperwork can come to the laboratory and correct the problem; identification of a mislabeled specimen or requisition should not be done over the telephone. Frequently, it may be necessary to do the best possible job on a less than optimal specimen, if the specimen would be impossible to collect again because the patient is taking antibiotics, the tissue was collected at surgery, or the patient would have to undergo a second invasive procedure (bone marrow or spinal tap). A notation regarding improper collection should be added to the final report in this instance, because only the primary caregiver is able to determine the validity of the results.

SPECIMEN PROCESSING

Depending on the site of testing (hospital, independent lab, POL) and how the specimens are transported to the laboratory (in-house courier or driver), microbiology samples may arrive in the laboratory in large numbers or as single tests. Although batch processing may be possible in large independent laboratories, most often hospital testing is performed as specimens arrive. When multiple specimens arrive at the same time, priority should be given to those that are most critical, such as cerebrospinal fluid (CSF), tissue, blood, and sterile fluids. Urine, throat, sputa, stool, or wound drainage specimens can be saved for later. Acid-fast, viral, and fungal specimens are usually batched for processing at one time. When a specimen is received with multiple requests but the amount of specimen is insufficient to do all of them, the microbiologist should call the clinician to prioritize the testing. On arrival in the laboratory, the time and date received should be recorded.

GROSS EXAMINATION OF SPECIMEN

All processing should begin with a gross examination of the specimen. Areas with blood or mucus should be located and sampled for culture and direct examination. Stool should be examined for evidence of barium (i.e., chalky white color), which would preclude O&P examination. Notations should be made on the handwritten or electronic workcard regarding the status of the specimen (e.g., bloody, cloudy, clotted) so that if more than one person works on the sample, all microbiologists working it up will know the results of the gross examination.

DIRECT MICROSCOPIC EXAMINATION

All appropriate specimens should have a direct microscopic examination. The direct examination serves several purposes. first, the quality of the specimen can be assessed; for example, sputa can be rejected that represent saliva and not lower respiratory tract secretions by quantitation of white blood cells or squamous epithelial cells present in the specimen. Second, the microbiologist and clinician can be given an early indication of what may be wrong with the patient (e.g., 4+ gram-positive cocci in clusters in an exudate). Third, the workup of the specimen can be guided by comparing what grows in culture to what was seen on smear. A situation in which three different morphotypes (cellular types) are seen on direct Gram stain but only two grow out in culture, for example, alerts the microbiologist to the fact that the third organism may be an anaerobic bacterium.

Direct examinations are usually not performed on throat, nasopharyngeal, or stool specimens but are indicated from most other sources.

The most common stain in bacteriology is the Gram stain, which helps to visualize rods, cocci, white blood cells, red blood cells, or squamous epithelial cells present in the sample. The most common direct fungal stains are KOH (potassium hydroxide), PAS (periodic-acid Schiff), and calcofluor white. The most common direct acid-fast stains are AR (auramine rhodamine), ZN (Ziehl-Neelsen), and Kinyoun. Chapter 6 describes the use of microscopy in clinical diagnosis in more detail.

SELECTION OF CULTURE MEDIA

Primary culture media are divided into several categories. The first are **nutritive** media, such as blood or chocolate agars. Nutritive media support the growth of a wide range of microorganisms and are considered nonselective because, theoretically, the growth of most organisms is supported. Nutritive media can also be **differential,** in that microorganisms can be distinguished on the basis of certain growth characteristics evident on the medium. Blood agar is considered both a nutritive and differential medium because it differentiates organisms based on whether they are alpha (α)-, beta (β)-, or gamma (γ)- hemolytic (Figure 5-2). **Selective** media support the growth of one group of organisms, but not another, by adding antimicrobials, dyes, or alcohol to a particular medium. MacConkey agar, for example, contains the dye crystal violet, which inhibits gram-positive organisms. Columbia agar with colistin and nalidixic acid (CNA) is a selective medium for gram-positive organisms because the antimicrobials colistin and nalidixic acid inhibit gram-negative organisms. Selective media can also be differential media if, in addition to their inhibitory activity, they differentiate between groups of organisms. MacConkey agar, for example, differentiates between lactose-fermenting and nonfermenting gram-negative rods by the color of the colonial growth (pink or clear, respectively); this is shown in Figure 5-3. In some cases (sterile body fluids, tissues or deep abscesses in a patient on antimicrobial therapy), **backup broth** (also called supplemental or enrichment broth) medium is inoculated, along with primary solid (agar) media, so small numbers of organisms present may be detected; this allows detection of anaerobes in aerobic cultures and organisms that may be damaged by either previous or concurrent antimicrobial therapy. Thioglycollate (thio) broth, brain-heart infusion broth (BHIB), and tryptic soy broth (TSB) are common backup broths.

Selection of media to inoculate for any given specimen is usually based on the organisms most likely to be involved in the disease process. For example, in determining what to set up for a CSF specimen, one considers the most likely pathogens that cause meningitis (*Streptococcus pneumoniae, Haemophilus influenzae, Neisseria meningitidis, Escherichia coli,* group B *Streptococcus*) and selects media that will support the growth of these organisms (blood and chocolate agar at a minimum). Likewise, if a specimen is collected from a source likely to be contaminated with normal flora, for example, an anal fistula (an opening of the surface of the skin near the anus that may communicate with the rectum), one might want to add a selective medium, such as CNA, to suppress gram-negative organisms and allow gram-positive organisms and yeast to be recovered.

Routine primary plating media and direct examinations for specimens commonly submitted to the microbiology laboratory are shown in Table 5-1. Chapter 7 on bacterial cultivation reemphasizes the strategies described here for selection and use of bacterial media.

SPECIMEN PREPARATION

Many specimens require some form of initial treatment before inoculation onto primary plating media.

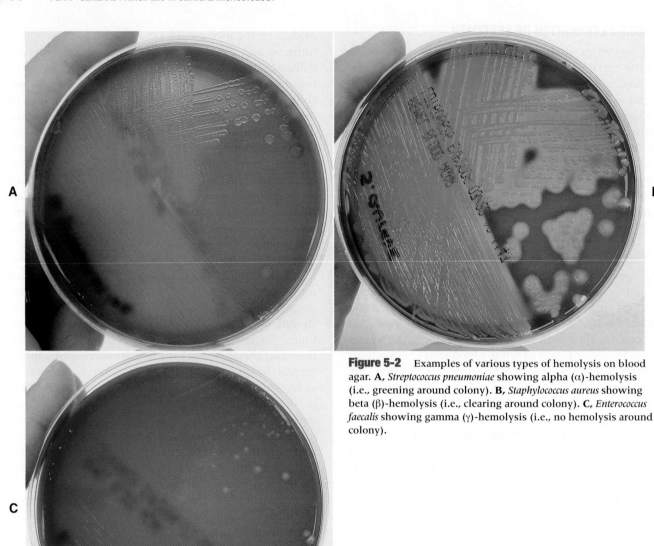

Figure 5-2 Examples of various types of hemolysis on blood agar. **A,** *Streptococcus pneumoniae* showing alpha (α)-hemolysis (i.e., greening around colony). **B,** *Staphylococcus aureus* showing beta (β)-hemolysis (i.e., clearing around colony). **C,** *Enterococcus faecalis* showing gamma (γ)-hemolysis (i.e., no hemolysis around colony).

Such procedures include **homogenization** (grinding) of tissue; **concentration** by centrifugation or filtration of large volumes of sterile fluids, such as ascites (peritoneal) or pleural (lung) fluids; or **decontamination** of specimens, such as those for legionellae or mycobacteria. Swab specimens are often vortexed (mixed) in 0.5 to 1 mL of saline or broth for 10 to 20 seconds to dislodge material from the fibers.

INOCULATION OF SOLID MEDIA

Specimens can be inoculated (plated) onto solid media either quantitatively by a dilution procedure or by means of a quantitative loop, or semiquantitatively

using an ordinary inoculating loop. Urine cultures and tissues from burn victims are plated quantitatively; everything else is usually plated semiquantitatively. Plates inoculated for quantitation are usually streaked with a 1:100 or 1:1000 loop, as shown in Figure 5-4, *A.* The original streak line is cross-struck with an ordinary inoculating loop to produce isolated, countable colonies. Plates inoculated for semiquantitation are usually struck out in four quadrants, as shown in Figure 5-4, *B.* The inoculum is applied by swabbing a dime-sized area or placing a drop of liquid specimen on the plate. The original inoculum is then cross-struck with an ordinary reuseable nichrome inoculating loop (quadrant one) or a sterile plastic single-use loop. The

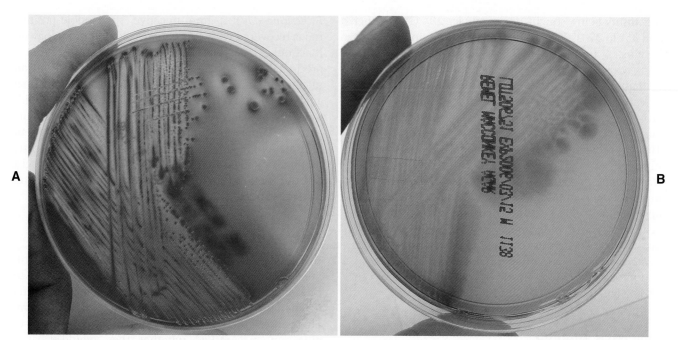

Figure 5-3 MacConkey agar. **A,** *Escherichia coli,* a lactose fermenter. **B,** *Pseudomonas aeruginosa,* a non–lactose fermenter.

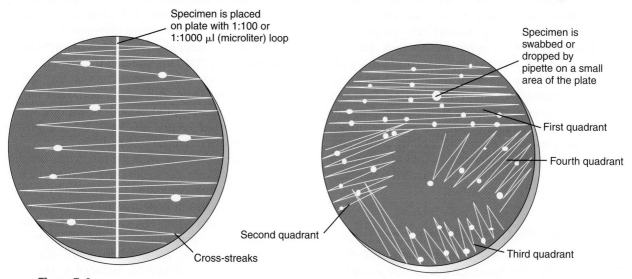

Figure 5-4 Methods of inoculating solid media. **A,** Streaking for quantitation. **B,** Streaking semiquantitatively.

loop is then flipped over or flamed and quadrant two is struck by pulling the loop through quadrant one a few times and streaking the rest of the area. The nichrome loop is then flamed again and quadrant three is streaked by going into quadrant two a few times and streaking the rest of the area. finally, quadrant four is streaked by pulling the loop over the rest of the agar without further flaming. This process is called **streaking for isolation,** because the microorganisms present in the specimen are successively diluted out as each quadrant is streaked until finally each morphotype is present as a single colony. Numbers of organisms present can subsequently be graded as 4+ (many, heavy growth) if growth is out

to the fourth quadrant, 3+ (moderate growth) if growth is out to the third quadrant, 2+ (few or light growth) if growth is in the second quadrant, and 1+ (rare) if growth is in the first quadrant. This tells the clinician the relative numbers of different organisms present in the specimen; such semiquantitative information is usually sufficient for the physician to be able to treat the patient.

INCUBATION CONDITIONS

Inoculated media are incubated under various temperatures and environmental conditions, depending on the organisms being sought, for example, 28° C for fungi and 35° to 37° C for most bacteria, viruses, and acid-fast bacillus. A number of different environmental conditions exist. **Aerobes** grow in ambient air, which contains 21% oxygen (O_2) and a small amount (0.03%) of carbon dioxide (CO_2). **Anaerobes** usually cannot grow in the presence of O_2 and the atmosphere in anaerobe jars, bags, or chambers is composed of 5% to 10% hydrogen (H_2), 5% to 10% CO_2, 80% to 90% nitrogen (N_2), and 0% O_2. **Capnophiles,** such as *Haemophilus influenzae* and *Neisseria gonorrhoeae,* require increased concentrations of CO_2 (5% to 10%) and approximately 15% O_2. This atmosphere can be achieved by a candle jar (3% CO_2) or a CO_2 incubator, jar, or bag. **Microaerophiles** *(Campylobacter jejuni, Helicobacter pylori)* grow under reduced O_2 (5% to 10%) and increased CO_2 (8% to 10%). This environment can also be obtained in specially designed jars or bags.

SPECIMEN WORKUP

One of the most important functions that a microbiologist performs is to decide what is clinically relevant regarding specimen workup. Considerable judgment is required to decide what organisms to look for and report. It is essential to recognize what constitutes indigenous (normal) flora and what constitutes a potential pathogen. Indiscriminate reporting of normal flora can contribute to unnecessary use of antibiotics and potential emergence of resistant organisms. Because organisms that are clinically relevant to identify and report vary by source, the microbiologist should know which ones cause disease at various sites. Part VII contains a detailed discussion of these issues.

EXTENT OF IDENTIFICATION REQUIRED

As health care continues to change, one of the most problematic issues for microbiologists is the extent of culture workup. Microbiologists still rely heavily on definitive identification, although shortcuts, including the use of limited identification procedures in some cases, are becoming commonplace in most clinical laboratories. Careful application of knowledge of the significance of various organisms in specific situations and thoughtful use of limited approaches will keep microbiology testing cost effective and the laboratory's workload manageable, while providing for optimum patient care.

Complete identification of a blood culture isolate, such as *Clostridium septicum* as opposed to a genus identification of *Clostridium* spp., will alert the clinician to the possibility of malignancy or other disease of the colon. At the same time, a presumptive identification of *Escherichia coli* if a gram-negative, spot indole-positive rod is recovered with appropriate colony morphology on MacConkey agar (flat, lactose-fermenting colony that is precipitating bile salts) is probably permissible from an uncomplicated urinary tract infection. In the final analysis, culture results should always be compared with the suspected diagnosis. The clinician should be encouraged to supply the microbiologist with all pertinent information (e.g., recent travel history, pet exposure, pertinent radiograph findings) so that the microbiologist can use the information to interpret culture results and plan appropriate strategies for workup.

COMMUNICATION OF LABORATORY FINDINGS

To fulfill their professional obligation to the patient, microbiologists must communicate their findings to those health care professionals responsible for treating the patient. This task is not as easy as it may seem. This is nicely illustrated in a study in which a group of physicians was asked whether they would treat a patient with a sore throat given two separate laboratory reports, that is, one that stated, "many group A *Streptococcus*" and one that stated, "few group A *Streptococcus*".[1] Although group A *Streptococcus (Streptococcus pyogenes)* is considered significant in any numbers in a symptomatic individual, the physicians said that they would treat the patient with many organisms but not the one with few organisms. Thus, although a pathogen (group A *Streptococcus*) was isolated in both cases, one word on the report (either "many" or "few") made a difference in how the patient would be handled.

In communicating with the physician, the microbiologist can avoid confusion and misunderstanding by not using jargon or abbreviations and by providing reports with clear-cut conclusions. The microbiologist should not assume that the clinician is fully familiar with laboratory procedures or the latest microbial taxonomic schemes. Thus, when appropriate, interpretive statements should be included in the written report along with the specific results. One such example would be the addition of a statement, such as "suggests contamination at collection," when more than three organisms are isolated from a clean-voided midstream urine specimen.

Laboratory newsletters should be used to provide physicians with material such as details of new procedures, nomenclature changes, and changes in usual antimicrobial susceptibility patterns of frequently isolated organisms. This last information, discussed in more detail in Chapter 12 is very useful to clinicians when selecting empiric therapy. Empiric therapy is based on the physician determining the most likely organism causing a patient's clinical symptoms and then selecting an antimicrobial that, in the past, has worked against that organism in a particular hospital or geographic area. Empiric therapy is used to start patients on treatment before the results of the patient's culture are known and may be critical to the patient's well-being in cases of life-threatening illnesses.

Positive findings should be telephoned to the clinician, and all verbal reports should be followed by written confirmation of results. Results should be legibly handwritten or generated electronically in the laboratory information system (LIS).

CRITICAL (PANIC) VALUES

Certain critical results *must* be communicated to the clinician immediately. Each clinical microbiology laboratory, in consultation with its medical staff, should prepare a list of these so-called panic values. Common panic values include:

- Positive blood cultures
- Positive spinal fluid Gram stain or culture
- *Streptococcus pyogenes* (group A *Streptococcus*) in a surgical wound
- Gram stain suggestive of gas gangrene (large box-car shaped gram-positive rods)
- Blood smear positive for malaria
- Positive cryptococcal antigen test
- Positive acid-fast stain
- Detection of a significant pathogen (e.g., *Legionella, Brucella,* vancomycin-resistant *Staphylococcus aureus*).

EXPEDITING RESULTS REPORTING— COMPUTERIZATION

Before widespread computerization of clinical microbiology laboratories, result reporting was accomplished by handwriting reports and having couriers deliver hard copies that were pasted into the patient's chart. Today, microbiology computer software is available that simplifies and speeds up this task.

CPUs (central processing units), disks, tape drives, controllers, printers, video terminals, communication ports, modems, and other types of hardware support running the software. The hardware and software together make up the complete LIS. Many LIS systems are, in turn, hooked up with a hospital information system (HIS). Between the HIS and LIS, most functions involved in ordering and reporting laboratory tests can be handled electronically. Order entry, patient identification, and specimen identification can be handled using the same type of bar coding that is commonly used in supermarkets. The LIS also takes care of results reporting and supervisory verification

of results, stores quality control data, allows easy test inquiries, and assists in test management reporting by storing, for example, the number of positive, negative, and unsatisfactory specimens. Most large systems also are capable of **interfacing** (communicating) with microbiology instruments to automatically download (transfer) and store data regarding positive cultures or antimicrobial susceptibility results. Results of individual organism antibiograms (patterns) can then be retrieved monthly so hospital-wide susceptibility patterns can be studied for emergence of resistant organisms or other epidemiologic information. Many vendors of laboratory information systems are now writing software for microbiology to adapt to personal computers (PCs) so that large CPUs may no longer be needed. This brings down the cost of microbiology systems so that even smaller laboratories are able to afford them. Today, small systems can be interfaced with printers or electronic facsimile machines (faxes) for quick and easy reporting and information retrieval. This further improves the quality of patient care.

REFERENCE

Lee A and McLean S: The laboratory report: a problem in communication between clinician and microbiologist? *Med J Aust* 2:858, 1977.

ADDITIONAL READING

Aller RD, Elevitch FR, editors: Laboratory and hospital information systems. In *Clinics in laboratory medicine,* Philadelphia, 1991, WB Saunders.

Isenberg HD, Schoenknecht FD, von Graevenitz A: *Cumitech 9: collection and processing of bacteriological specimens,* Rubin SJ, coordinating editor, Washington, DC, 1979, American Society for Microbiology.

Miller JM: Handbook of specimen collection and handling in microbiology, *CDC laboratory manual,* Atlanta, 1985, U.S. Department of Health and Human Services.

Morris AJ, Wilson SJ, Marx CE, et al: Clinical impact of bacteria and fungi recovered only from broth cultures, *J Clin Microbiol* 33:161–165.

National Committee for Clinical Laboratory Standards: Abbreviated identification of bacteria and yeast; Approved guideline M35-A, Wayne, Pa, 2002, National Committee for Clinical Laboratory Standards.

Pezzlo M: Specimen collection: the role of the clinical microbiologist, *Clinical microbiology updates* 3:1, Somerville, NJ, 1991, Hoechst-Roussel Pharmaceuticals Inc.

Thomson RB Jr., Miller JM: Specimen collection, transport, and processing: bacteriology. In Murray PR, Baron EJ, Jorgensen JH, et al, editors: *Manual of clinical microbiology,* ed 8, Washington, DC, 2003, ASM Press.

CHAPTER **6** — **Role of Microscopy**

T he basic flow of procedures involved in the laboratory diagnosis of infectious diseases is as follows:

1. Direct examination of patient specimens for the presence of etiologic agents
2. Growth and cultivation of the agents from these same specimens
3. Analysis of the cultivated organisms to establish their identification and other pertinent characteristics such as susceptibility to antimicrobial agents

For certain infectious diseases, this process may also include measuring the patient's immune response to the infectious agent.

Microscopy is the most common method used both for the detection of microorganisms directly in clinical specimens and for the characterization of organisms grown in culture (Box 6-1). **Microscopy** is defined as the use of a microscope to magnify (i.e., visually enlarge) objects too small to be visualized with the naked eye so that their characteristics are readily observable. Because most infectious agents cannot be detected with the unaided eye, microscopy plays a pivotal role in the laboratory. Microscopes and microscopic methods vary, but only those of primary use in diagnostic microbiology are discussed.

The method used to process patient specimens is dictated by the type and body source of specimen (see Part VII). Regardless of the method used, some portion of the specimen usually is reserved for microscopic examination. Specific stains or dyes applied to the specimens, combined with particular methods of microscopy, can detect etiologic agents in a rapid, relatively inexpensive, and productive way. Microscopy also plays a key role in the characterization of organisms that have been cultivated in the laboratory (for more information regarding cultivation of bacteria see Chapter 7).

The types of microorganisms to be detected, identified, and characterized determine the most appropriate types of microscopy to use. Table 6-1 outlines the four types of microscopy used in diagnostic microbiology and their relative utility for each of the four major microorganism groups. **Bright-field microscopy** (also known as **light microscopy**) and **fluorescence microscopy** have the widest use and application. Which microorganisms can be detected or identified by each microscopic method also depends on the methods used to highlight the microorganisms and their key characteristics. This enhancement is usually achieved using various dyes or stains.

BRIGHT-FIELD (LIGHT) MICROSCOPY

PRINCIPLES OF LIGHT MICROSCOPY

For light microscopy, visible light is passed through the specimen and then through a series of lenses that reflect the light in a manner that results in magnification of the organisms present in the specimen (Figure 6-1). The total magnification achieved is the product of the lenses used.

Magnification

In most light microscopes, the **objective lens,** which is closest to the specimen, magnifies objects 100× (times) and the **ocular lens,** which is nearest the eye, magnifies 10×. Using these two lenses in combination, organisms in the specimen are magnified 1000× their actual size when viewed through the ocular lens. Objective lenses of lower magnification are available so that those of 10×, 20×, and 40× magnification power can provide total magnifications of 100×, 200×, and 400×, respectively. Magnification of 1000× allows for the visualization of fungi, most parasites, and most bacteria but is not sufficient for observing viruses, which require magnification of 100,000× or more (see Electron Microscopy in this chapter).

Resolution

To optimize visualization, other factors besides magnification must be considered. **Resolution,** defined as the extent to which detail in the magnified object is maintained, is also essential. Without it everything would be magnified as an indistinguishable blur.

BOX 6-1 Applications of Microscopy in Diagnostic Microbiology

- Rapid preliminary organism identification by direct visualization in patient specimens
- Rapid final identification of certain organisms by direct visualization in patient specimens
- Detection of different organisms present in the same specimen
- Detection of organisms not easily cultivated in the laboratory
- Evaluation of patient specimens for the presence of cells indicative of inflammation (i.e., phagocytes) or contamination (i.e., squamous epithelial cells)
- Determination of an organism's clinical significance. Bacterial contaminants usually are not present in patient specimens at sufficiently high numbers ($\times 10^5$ cells/mL) to be seen by light microscopy
- Provide preculture information about which organisms might be expected to grow so that appropriate cultivation techniques are used
- Determine which tests and methods should be used for identification and characterization of cultivated organisms
- Provide a method for investigating unusual or unexpected laboratory test results

Table 6-1 Microscopy for Diagnostic Microbiology

Organism Group	Bright-Field Microscopy	Fluorescence Microscopy	Dark-Field Microscopy	Electron Microscopy
Bacteria	+	+	±	−
Fungi	+	+	−	−
Parasites	+	+	−	±
Viruses	−	+	−	±

+, Commonly used; ±, limited use; −, rarely used.

Therefore, **resolving power,** which is the closest distance between two objects that when magnified still allows the two objects to be distinguished from each other, is extremely important. The resolving power of most light microscopes allows bacterial cells to be distinguished from one another but usually does not allow bacterial structures, internal or external, to be detected.

To achieve the level of resolution desired with 1000× magnification, oil immersion must be used in conjunction with light microscopy. Immersion oil is used to fill the space between the objective lens and the

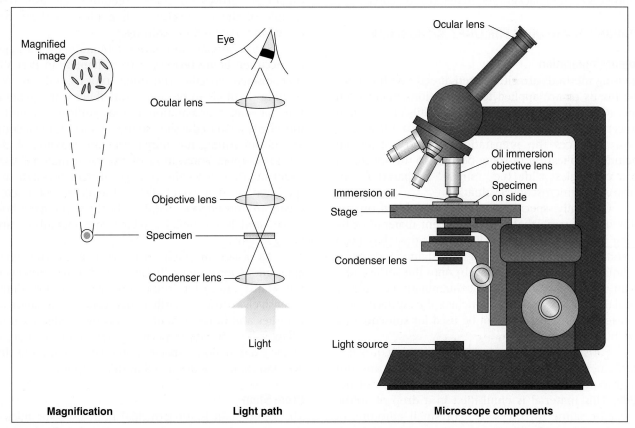

Figure 6-1 Principles of bright-field (light) microscopy. (Modified from Atlas RM: *Principles of microbiology,* St Louis, 1995, Mosby.)

glass slide onto which the specimen has been affixed. The oil enhances resolution by preventing light rays from dispersing and changing wavelength after passing through the specimen. A specific objective lens, the **oil immersion lens,** is designed for use with oil; this lens provides 100× magnification on most light microscopes.

Lower magnifications (i.e., 100× or 400×) may be used to locate specimen samples in certain areas on a microscope slide, or to observe microorganisms such as some fungi and parasites. The 1000× magnification provided by the combination of ocular and oil immersion lenses usually is required for optimal detection and characterization of bacteria.

Contrast

The third key component to light microscopy is **contrast,** which is needed to make objects stand out from the background. Because microorganisms are essentially transparent, owing to their microscopic dimensions and high water content, they cannot be easily detected among the background materials and debris in patient specimens. Lack of contrast is also a problem for the microscopic examination of microorganisms grown in culture. Contrast is most commonly achieved by staining techniques that highlight organisms and allow them to be differentiated from one another and from background material and debris.

STAINING TECHNIQUES FOR LIGHT MICROSCOPY

Smear Preparation

Staining methods are either used directly with patient specimens or are applied to preparations made from microorganisms grown in culture. Details of specimen processing are presented throughout Part VII and in most instances the preparation of every specimen includes application of some portion of the specimen to a clean glass slide (i.e., "smear" preparation) for subsequent microscopic evaluation.

Generally, specimen samples are placed on the slide using a swab that contains patient material or by using a pipette into which liquid specimen has been aspirated (Figure 6-2). Material to be stained is dropped (if liquid) or rolled (if on a swab) onto the surface of a clean, dry, glass slide. To avoid contamination of culture media, once a swab has touched the surface of a nonsterile slide, it should not be used for subsequently inoculating media.

For staining microorganisms grown in culture, a sterile needle may be used to transfer a small amount of growth from a solid medium to the surface of the slide. This material is emulsified in a drop of sterile water or saline on the slide. For small amounts of growth that might become lost in even a drop of saline, a sterile wooden applicator stick can be used to

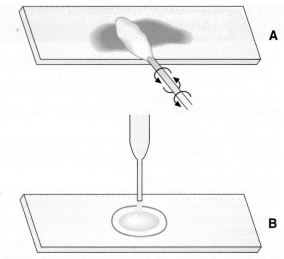

Figure 6-2   Smear preparations by swab roll (**A**) and pipette deposition (**B**) of patient specimen on glass microscope slide.

touch the growth; this material is then rubbed directly onto the slide, where it can be easily seen. The material placed on the slide to be stained is allowed to dry and is affixed to the slide by placing it on a slide warmer (60° C) for at least 10 minutes or by flooding it with 95% methanol for 1 minute. To examine organisms grown in liquid medium, an aspirated sample of the broth culture is applied to the slide, dried, and fixed before staining.

Smear preparation varies depending on the type of specimen being processed (see chapters in Part VII that discuss specific specimen types) and on the staining methods to be used. Nonetheless, the general rule for smear preparation is that sufficient material must be applied to the slide so that chances for detecting and distinguishing microorganisms are maximized. At the same time, application of excessive material that could interfere with the passage of light through the specimen or that could distort the details of microorganisms must be avoided. Finally, the staining method to be used is dictated by which microorganisms are being sought.

As listed in Table 6-1, light microscopy has applications for bacteria, fungi, and parasites. However, the stains used for these microbial groups differ extensively. Those primarily designed for examination of parasites and fungi by light microscopy are discussed in Chapters 49 and 50, respectively. The stains for microscopic examination of bacteria, the Gram stain and the acid-fast stains, are discussed in this chapter.

Gram Stain

The Gram stain is the principal stain used for microscopic examination of bacteria. Nearly all clinically important bacteria can be detected using this method, the only exceptions being those organisms that exist almost exclusively within host cells (e.g., chlamydia), those that lack a cell wall (e.g., mycoplasma and ureaplasma), and those of insufficient dimension to be resolved by light microscopy (e.g., spirochetes). First devised by Hans Christian Gram during the late nineteenth century, the Gram stain can be used to divide most bacterial species into two large groups: those that take up the basic dye, crystal violet (i.e., gram-positive bacteria), and those that allow the crystal violet dye to wash out easily with the decolorizer alcohol or acetone (i.e., gram-negative bacteria).

Procedure. Although modifications of the classic Gram stain that involve changes in reagents and timing exist, the principles and results are the same for all modifications. The classic Gram stain procedure entails fixing clinical material to the surface of the microscope slide either by heating or by using methanol. Methanol fixation preserves the morphology of host cells, as well as bacteria, and is especially useful for examining bloody specimen material. Slides are overlaid with 95% methanol for 1 minute; the methanol is allowed to run off, and the slides are air dried before staining. After fixation, the first step in the Gram stain is the application of the primary stain **crystal violet.** A mordant, **Gram's iodine,** is applied after the crystal violet to chemically bond the alkaline dye to the bacterial cell wall. The decolorization step distinguishes gram-positive from gram-negative cells. After **decolorization,** organisms that stain gram-positive retain the crystal violet and those that are gram-negative are cleared of crystal violet. Addition of the counterstain safranin will stain the clear gram-negative bacteria pink or red (Figure 6-3).

Principle. The difference in composition between gram-positive cell walls, which contain thick peptidoglycan with numerous teichoic acid cross-linkages, and gram-negative cell walls, which consist of a thinner layer of peptidoglycan, accounts for the Gram staining differences between these two major groups of bacteria. Presumably, the extensive teichoic acid cross-links contribute to the ability of gram-positive organisms to resist alcohol decolorization. Although the counterstain may be taken up by the gram-positive organisms, their purple appearance will not be altered.

Gram-positive organisms that have lost cell wall integrity because of antibiotic treatment, old age, or action of autolytic enzymes may allow the crystal violet to wash out with the decolorizing step and may appear gram-variable, with some cells staining pink and others staining purple. However, for identification purposes, these organisms are considered to be truly gram-positive. On the other hand, gram-negative bacteria rarely, if ever, retain crystal violet (e.g., appear

purple) if the staining procedure has been properly performed. Host cells, such as red and white blood cells (phagocytes), allow the crystal violet stain to wash out with decolorization and should appear pink on smears that have been correctly prepared and stained.

Gram Stain Examination. Once stained, the smear is examined using the oil immersion (1000× magnification) lens. When clinical material is Gram-stained (e.g., the **direct smear**), the slide is evaluated for the presence of bacterial cells as well as the Gram reactions, morphologies (e.g., cocci or bacilli), and arrangements (e.g., chains, pairs, clusters) of the cells seen (Figure 6-4). This information often provides a preliminary diagnosis regarding the infectious agents and frequently is used to direct initial therapies for the patient.

The direct smears should also be examined for the presence of inflammatory cells (e.g., phagocytes) that are key indicators of an infectious process. Noting the presence of other host cells, such as squamous epithelial cells in respiratory specimens, is also helpful because the presence of these cells may indicate contamination with organisms and cells from the mouth (for more information regarding interpretation of respiratory smears see Chapter 53). Observing background tissue debris and proteinaceous material, which generally stain gram-negative, also provides helpful information. For example, the presence of such material indicates that specimen material was adequately affixed to the slide. Therefore, the absence of bacteria or inflammatory cells on such a smear is "real" and not likely the result of loss of specimen during staining (Figure 6-5). Other ways that Gram stain evaluations of direct smears are used are discussed throughout the chapters of Part VII that deal with infections of specific body sites.

Several examples of Gram stains of direct smears are provided in Figure 6-6. Basically, whatever is observed is also recorded and is used to produce a laboratory report for the physician. The report typically includes:

- The presence of host cells and debris
- The Gram reactions, morphologies (e.g., cocci, bacilli, coccobacilli), and arrangement of bacterial cells present. **NOTE**: Reporting the absence of bacteria and host cells can be equally as important.
- Optionally, the relative amounts of bacterial cells (e.g., rare, few, moderate, many) may be provided. However, it is important to remember that to visualize bacterial cells by light microscopy, a minimum concentration of 10^5 cells per 1 mL of specimen is required. This is a large

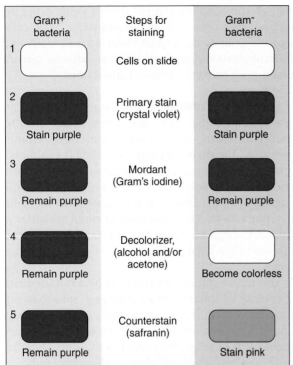

Gram+ bacteria	Steps for staining	Gram− bacteria
1	Cells on slide	1
2 **Stain purple**	Primary stain (crystal violet)	2 **Stain purple**
3 **Remain purple**	Mordant (Gram's iodine)	3 **Remain purple**
4 **Remain purple**	Decolorizer, (alcohol and/or acetone)	4 **Become colorless**
5 **Remain purple**	Counterstain (safranin)	5 **Stain pink**

1 Fix material on slide with methanol or heat. If slide is heat fixed, allow it to cool to the touch before applying stain.

2 Flood slide with crystal violet (purple) and allow it to remain on the surface without drying for 10 to 30 seconds. Rinse the slide with tap water, shaking off all excess.

3 Flood the slide with iodine to increase affinity of crystal violet and allow it to remain on the surface without drying for twice as long as the crystal violet was in contact with the slide surface (20 seconds of iodine for 10 seconds of crystal violet, for example). Rinse with tap water, shaking off all excess.

4 Flood the slide with decolorizer for 10 seconds and rinse off immediately with tap water. Repeat this procedure until the blue dye on longer runs off the slide with the decolorizer. Thicker smears require more prolonged decolorizing. Rinse with tap water and shake off excess.

5 Flood the slide with counterstain and allow it to remain on the surface without drying for 30 seconds. Rinse with tap water and gently blot the slide dry with paper towels or bibulous paper or air dry. For delicate smears, such as certain body fluids, air drying is the best method.

6 Examine microscopically under an oil immersion lens at 1000x for phagocytes, bacteria, and another cellular material.

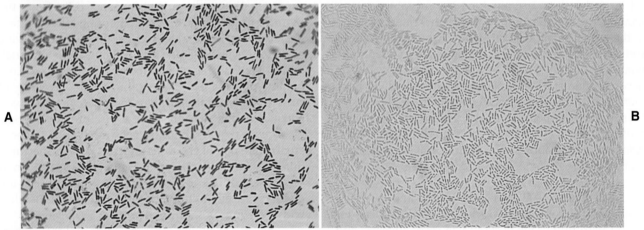

A   B

Figure 6-3   Gram stain procedures and principles. **A,** Gram-positive bacteria observed under oil immersion appear purple. **B,** Gram-negative bacteria observed under oil immersion appear pink. (Modified from Atlas RM: *Principles of microbiology,* St Louis, 1995, Mosby.)

number of bacteria for any normally sterile body site and to describe the quantity as rare or few based on microscopic observation may be understating their significance in a clinical specimen. On the other hand, noting the relative amounts seen on direct smear may be useful laboratory information to correlate smear results with the amount of growth observed subsequently from cultures.

Although Gram stain evaluation of direct smears is routinely used as an aid in the diagnosis of bacterial infections, unexpected but significant findings of other infectious etiologies may be detected and cannot be ignored. For example, fungal cells and elements

generally stain gram-positive, but they may take up the crystal violet poorly and appear gram-variable (e.g., both pink and purple) or gram-negative. Because infectious agents besides bacteria may be detected by Gram stain, any unusual cells or structures observed on the smear should be evaluated further before being dismissed as unimportant (Figure 6-7).

Gram Stain of Bacteria Grown in Culture. The Gram stain also plays a key role in the identification of bacteria grown in culture. Similar to direct smears, smears prepared from bacterial growth are evaluated for the bacterial cells' Gram reactions, morphologies, and arrangements (see Figure 6-4). If growth from

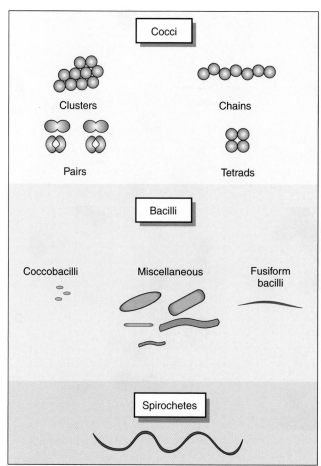

Figure 6-4   Examples of common bacterial cellular morphologies, Gram staining reactions, and cellular arrangements.

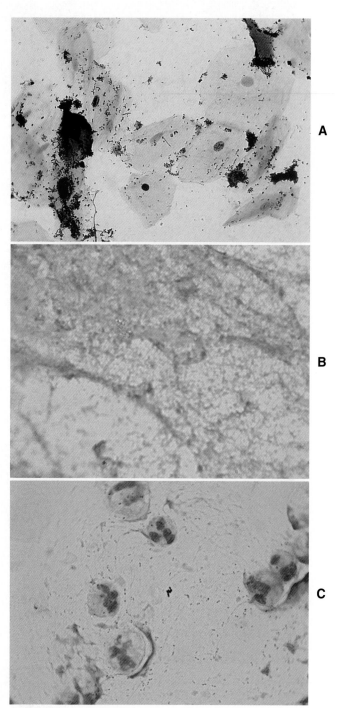

Figure 6-5   Gram stains of direct smears showing squamous cells and bacteria (**A**), proteinaceous debris alone (**B**), and proteinaceous debris with polymorphonuclear leukocytes and bacteria (**C**).

more than one specimen is to be stained on the same slide, a wax pencil may be used to create divisions. Drawing a "map" of such a slide so that different Gram stain results can be recorded in an organized fashion is helpful (Figure 6-8). The smear results will be used to determine subsequent testing for identifying and characterizing the organisms isolated from the patient specimen.

Acid-Fast Stains

The acid-fast stain is the other commonly used stain for light-microscopic examination of bacteria.

Principle. The acid-fast stain is specifically designed for a subset of bacteria whose cell walls contain long-chain fatty (mycolic) acids. Mycolic acids render the cells resistant to decolorization, even with acid alcohol decolorizers. Thus, these bacteria are referred to as being **acid-fast.** Although these organisms may stain slightly or poorly as gram-positive, the acid-fast stain takes full advantage of the waxy content of the cell walls to

maximize detection. Mycobacteria are the most commonly encountered acid-fast bacteria, typified by *Mycobacterium tuberculosis*, the etiologic agent of tuberculosis. Bacteria lacking cell walls fortified with mycolic acids cannot resist decolorization with acid alcohol and

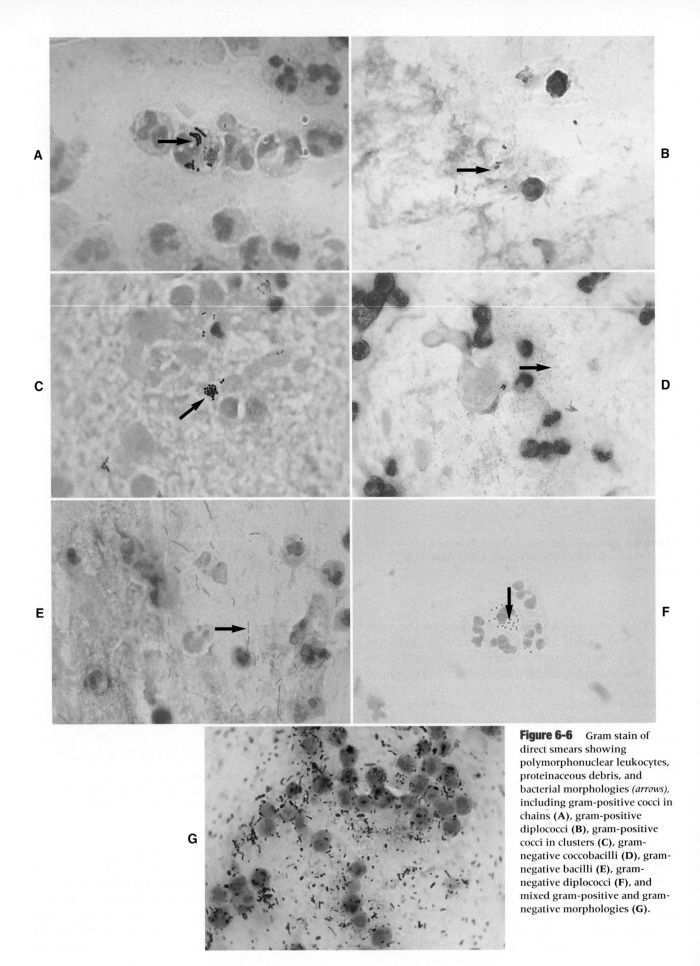

Figure 6-6 Gram stain of direct smears showing polymorphonuclear leukocytes, proteinaceous debris, and bacterial morphologies *(arrows),* including gram-positive cocci in chains (**A**), gram-positive diplococci (**B**), gram-positive cocci in clusters (**C**), gram-negative coccobacilli (**D**), gram-negative bacilli (**E**), gram-negative diplococci (**F**), and mixed gram-positive and gram-negative morphologies (**G**).

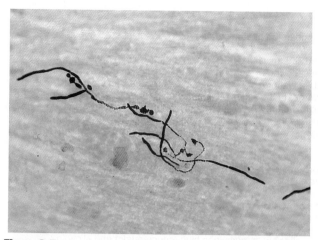

Figure 6-7 Gram stains of direct smears can reveal infectious etiologies other than bacteria, such as the yeast *Candida tropicalis*.

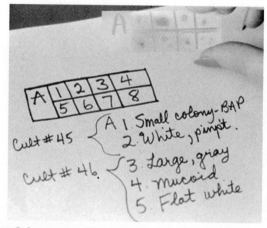

Figure 6-8 Example of a slide map for staining several bacterial colony samples on a single slide.

are categorized as being **non–acid-fast,** a trait typical of most other clinically relevant bacteria. However, some degree of acid-fastness is a characteristic of a few nonmycobacterial bacteria, such as *Nocardia* spp., and coccidian parasites, such as *Cryptosporidium* spp.

Procedure. The classic acid-fast staining method, **Ziehl-Neelsen,** is outlined in Figure 6-9. The procedure requires heat to allow the **primary stain (carbolfuchsin)** to enter the wax-containing cell wall. A modification of this procedure, the **Kinyoun** acid-fast method, is more commonly used today. Because of a higher concentration of phenol in the primary stain solution, heat is not required for the intracellular penetration of carbolfuchsin. This modification, which is presented in more detail in Chapter 45, is referred to as the "cold" method. Another

modification of the acid-fast stain that is used for identifying certain nonmycobacterial species is described and discussed in Part III, Section 14. When the acid-fast–stained smear is read with 1000× magnification, acid-fast–positive organisms stain red. Depending on the type of counterstain used (e.g., **methylene blue** or **malachite green**), other microorganisms, host cells, and debris stain a blue to blue-green color (Figures 6-9 and 6-10).

As with the Gram stain, the acid-fast stain is used to detect acid-fast bacteria (e.g., mycobacteria) directly in clinical specimens and provide preliminary identification information for suspicious bacteria grown in culture. Because mycobacterial infections are much less common than infections caused by other non–acid-fast bacteria, the acid-fast stain is only performed on specimens from patients highly suspected of having a mycobacterial infection. That is, Gram staining is a routine part of most bacteriology procedures while acid-fast staining is reserved for specific situations. Similarly, the acid-fast stain is only applied to bacteria grown in culture when it is suspected based on other characteristics that mycobacteria are present (for more information regarding identification of mycobacteria see Chapter 45).

PHASE CONTRAST MICROSCOPY

Instead of using a stain to achieve the contrast necessary for observing microorganisms, altering microscopic techniques to enhance contrast offers another approach. Phase contrast microscopy is one such contrast-enhancing technique. By this method, beams of light pass through the specimen and are partially deflected by the different densities or thicknesses (i.e., refractive indices) of the microbial cells or cell structures in the specimen. The greater the refractive index of an object, the more the beam of light is slowed, which results in decreased light intensity. These differences in light intensity translate into differences that provide contrast. Therefore, phase microscopy translates differences in phases within the specimen into differences in light intensities that result in contrast among objects within the specimen being observed.

Microscopy that depends on staining microorganisms only allows dead organisms to be observed. Because staining is not part of phase contrast microscopy, this method offers the advantage of allowing observation of viable microorganisms. The method is not commonly used in most aspects of diagnostic microbiology, but it is used to identify medically important fungi grown in culture (for more information regarding the use of phase contrast microscopy for fungal identification see Chapter 50).

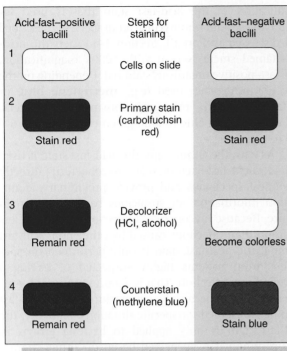

Acid-fast–positive bacilli	Steps for staining	Acid-fast–negative bacilli
1 Stain red	Cells on slide	1 Stain red
2 Stain red	Primary stain (carbolfuchsin red)	2 Stain red
3 Remain red	Decolorizer (HCl, alcohol)	3 Become colorless
4 Remain red	Counterstain (methylene blue)	4 Stain blue

1 Fix smears on heated surface (60° C for at least 10 minutes).

2 Flood smears with carbolfuchsin (primary stain) and heat to almost boiling by performing the procedure on an electrically heated platform or by passing the flame of a Bunsen burner underneath the slides on a metal rack. The stain on the slides should steam. Allow slides to sit for 5 minutes after heating; do not allow them to dry out. Wash the slides in distilled water (note: tap water may contain acid-fast bacilli). Drain off excess liquid.

3 Flood slides with 3% HCl in 95% ethanol (decolorizer) for approximately 1 minute. Check to see that no more red color runs off the surface when the slide is tipped. Add a bit more decolorizer for very thick slides or those that continue to "bleed" red dye. Wash thoroughly with water and remove the excess.

4 Flood slides with methylene blue (counterstain) and allow to remain on surface of slides for 1 minute. Wash with distilled water and stand slides upright on paper towels to air dry. Do not blot dry.

5 Examine microscopically, screening at 400× magnification and confirm all suspicious (i.e., red) organisms at 1000× magnification using an oil-immersion lens.

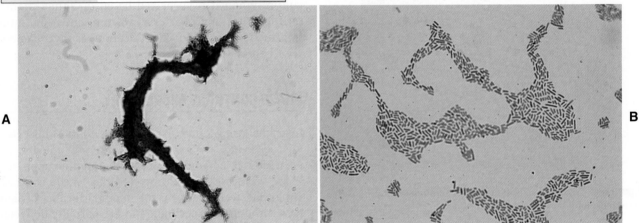

Figure 6-9 The Ziehl-Neelsen acid-fast stain procedures and principles. **A,** Acid-fast positive bacilli. **B,** Acid-fast negative bacilli. (Modified from Atlas RM: *Principles of microbiology,* St Louis, 1995, Mosby.)

FLUORESCENT MICROSCOPY

PRINCIPLE OF FLUORESCENT MICROSCOPY

Certain dyes, called **fluors** or **fluorochromes,** can be raised to a higher energy level after absorbing ultraviolet (excitation) light. When the dye molecules return to their normal, lower energy state, they release excess energy in the form of visible (fluorescent) light. This process is called **fluorescence,** and microscopic methods have been developed to exploit the enhanced contrast and detection that this phenomenon provides.

Figure 6-11 depicts diagrammatically the principle of fluorescent microscopy in which the excitation light is emitted from above (epifluorescence). An excitation filter passes light of the desired wavelength to excite the fluorochrome that has been used to stain the specimen.

A barrier filter in the objective lens prevents the excitation wavelengths from damaging the eyes of the observer. When observed through the ocular lens, fluorescing objects appear brightly lit against a dark background.

The color of the fluorescent light depends on the dye and light filters used. For example, use of the fluorescent dyes acridine orange, auramine, and fluorescein isothiocyanate (FITC) requires blue excitation light, exciter filters that select for light in the 450- to 490-λ wavelength range, and a barrier filter for 515 λ. Calcofluor white, on the other hand, requires violet excitation light, an exciter filter that selects for light in the 355- to 425-λ wavelength range, and a barrier filter for 460 λ. Which dye is used often depends on which organism is being sought and the fluorescent method used. The intensity of the contrast obtained

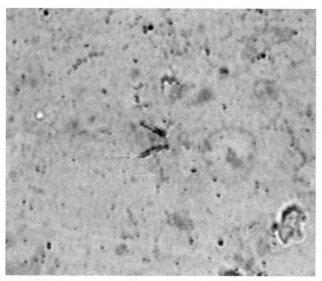

Figure 6-10   Acid-fast stain of direct smear to show acid-fast bacilli staining deep red *(arrow A)* and non–acid-fast bacilli and host cells staining blue with the counterstain methylene blue *(arrow B).*

with fluorescent microscopy is an advantage it has over the use of chromogenic dyes (e.g., crystal violet and safranin of the Gram stain) and light microscopy.

STAINING TECHNIQUES FOR FLUORESCENT MICROSCOPY

Based on the composition of the fluorescent stain reagents, fluorescent staining techniques may be divided into two general categories: **fluorochroming,** in which a fluorescent dye is used alone, and **immunofluorescence,** in which fluorescent dyes have been linked (conjugated) to specific antibodies. The principal differences between these two methods are diagrammed in Figure 6-12.

Fluorochroming

In fluorochroming a direct chemical interaction occurs between the fluorescent dye and a component of the bacterial cell; this interaction is the same as occurs with the stains used in light microscopy. The difference is that use of a fluorescent dye enhances contrast and amplifies the observer's ability to detect stained cells tenfold greater than would be observed by light microscopy. For example, a minimum concentration of at least 10^5 organisms per milliliter of specimen is required for visualization by light microscopy, whereas by fluorescent microscopy that number decreases to 10^4 per milliliter. The most

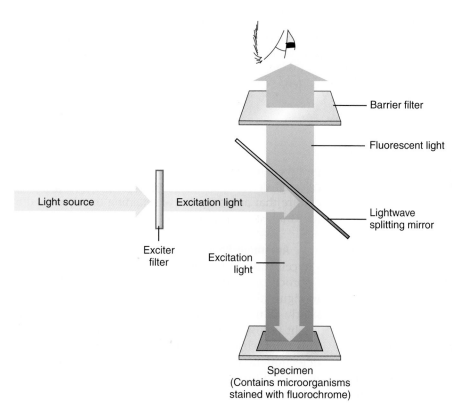

Figure 6-11   Principle of fluorescent microscopy. Microorganisms in a specimen are stained with a fluorescent dye. On exposure to excitation light, organisms are visually detected by the emission of fluorescent light by the dye with which they have been stained (i.e., fluorochroming) or "tagged" (i.e., immunofluorescence).

Barrier filter

Fluorescent light

Light source

Excitation light

Lightwave splitting mirror

Exciter filter

Excitation light

Specimen
(Contains microorganisms stained with fluorochrome)

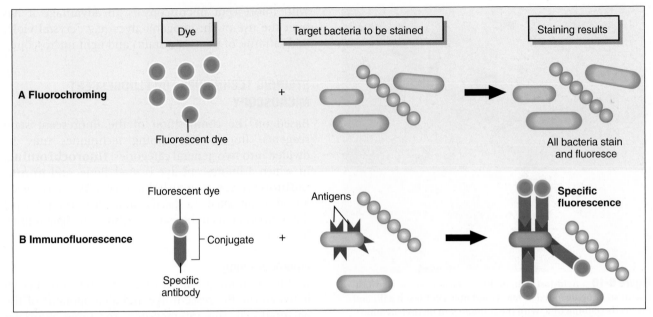

Figure 6-12   Principles of fluorochroming and immunofluorescence. Fluorochroming **(A)** involves nonspecific staining of any bacterial cell with a fluorescent dye. Immunofluorescence **(B)** uses antibodies labeled with fluorescent dye (i.e., a conjugate) to specifically stain a particular bacterial species.

Procedure 6-1

ACRIDINE ORANGE STAIN

PRINCIPLE
Acridine orange, a vital stain, will intercalate with nucleic acid, changing the dye's optical characteristics so that it will fluoresce bright orange under ultraviolet light. All nucleic acid–containing cells will fluoresce orange (see Figure 6-13).

METHOD
1. Fix slide, either in methanol or with heat, as described previously for Gram stain.
2. Flood slide with acridine orange stain (available from various commercial suppliers). Allow stain to remain on surface of slide for 2 minutes without drying.

3. Rinse with tap water and allow moisture to drain from slide and air dry.
4. Examine the slide using fluorescent microscopy.

EXPECTED RESULTS
Bacteria will fluoresce bright orange against a green-fluorescing or dark background. The nuclei of host cells may also fluoresce.

---

common fluorochroming methods used in diagnostic microbiology include acridine orange stain, auramine-rhodamine stain, and calcofluor white stain.

Acridine Orange.   The fluorochrome acridine orange binds to nucleic acid. This staining method (Procedure 6-1) can be used to confirm the presence of bacteria in blood cultures when Gram stain results are difficult to interpret or when the presence of bacteria is highly suspected but none are detected using light microscopy. Because acridine orange stains all nucleic acids, it is nonspecific. Therefore, all microorganisms and host cells will stain and give a bright orange fluorescence. Although this stain can be used to enhance detection, it does not discriminate between gram-negative and gram-

positive bacteria. The stain is also used for detection of cell wall–deficient bacteria (e.g., mycoplasmas) grown in culture that are incapable of retaining the dyes used in the Gram stain (Figure 6-13).

Auramine-Rhodamine.   The waxy mycolic acids in the cell walls of mycobacteria have an affinity for the fluorochromes **auramine** and **rhodamine**. As shown in Figure 6-14, these dyes will nonspecifically bind to nearly all mycobacteria. The mycobacterial cells appear bright yellow or orange against a greenish background. This fluorochroming method can be used to enhance detection of mycobacteria directly in patient specimens and for initial characterization of cells grown in culture.

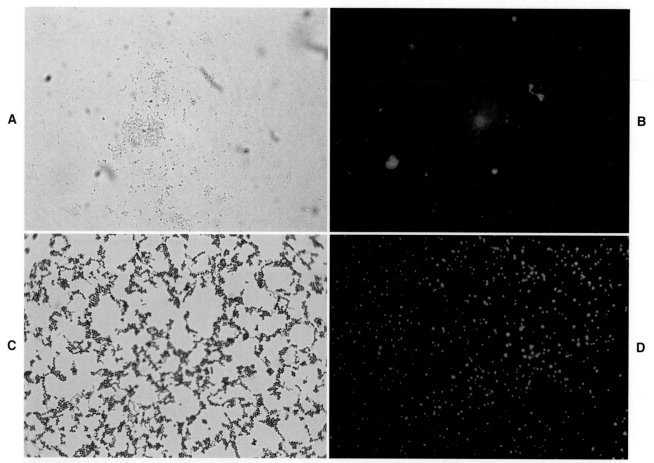

Figure 6-13   Comparison of acridine orange fluorochroming and Gram stain. Gram stain of mycoplasma demonstrates inability to distinguish these cell wall–deficient organisms from amorphous gram-negative debris (**A**). Staining the same specimen with acridine orange confirms the presence of nucleic acid–containing organisms (**B**). Gram stain distinguishes between gram-positive and gram-negative bacteria (**C**), but all bacteria stain the same with the nonspecific acridine orange dye (**D**).

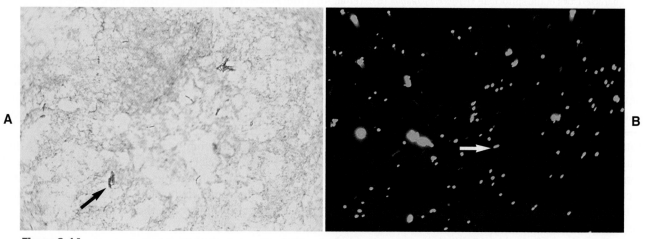

Figure 6-14   Comparison of the Ziehl-Neelsen–stained (**A**) and auramine-rhodamine–stained (**B**) *Mycobacterium* spp. *(arrows).*

Calcofluor White.   The cell walls of fungi will bind the stain calcofluor white, which greatly enhances fungal visibility in tissue and other specimens. This fluorochrome is commonly used to directly detect fungi in clinical material and to observe subtle characteristics of fungi grown in culture (for more information regarding the use of calcofluor white for the laboratory diagnosis of fungal infections, see Chapter 50).

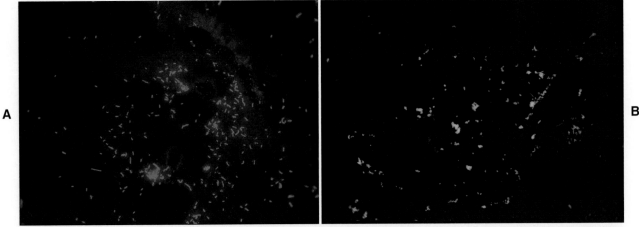

Figure 6-15　Immunofluoresence stains of *Legionella* spp. **(A)** and *Bordetella pertussis* **(B)** used for identification.

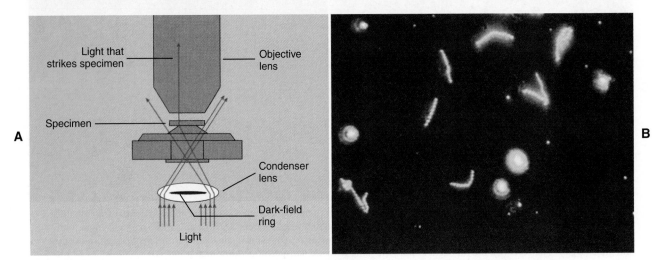

Figure 6-16　Dark-field microscopy. Principle **(A)** and dark-field photomicrograph showing the tightly coiled characteristics of the spirochete *Treponema pallidum* **(B)**.　(From Atlas RM: *Principles of microbiology,* St Louis, 1995, Mosby.)

Immunofluorescence

As discussed in Chapter 3, antibodies are molecules that have high specificity for interacting with microbial antigens. That is, antibodies specific for an antigen characteristic of a particular microbial species will only combine with that antigen. Therefore, if antibodies are conjugated (chemically linked) to a fluorescent dye, the resulting **dye-antibody conjugate** can be used to detect, or "tag," specific microbial agents (see Figure 6-12). When "tagged," the microorganisms become readily detectable by fluorescent microscopy. Thus, immunofluorescence combines the amplified contrast provided by fluorescence with the specificity of antibody-antigen binding.

This method is used to directly examine patient specimens for bacteria that are difficult or slow to grow (e.g., *Legionella* spp., *Bordetella pertussis,* and *Chlamydia trachomatis*) or to identify organisms already grown in culture. FITC, which emits an intense, apple green fluorescence, is the fluorochrome most commonly used for conjugation to antibodies (Figure 6-15). Immunofluorescence is also used in virology (Chapter 51) and to some extent in parasitology (Chapter 49).

Two additional types of microscopy, dark-field microscopy and electron microscopy, are not commonly used to diagnose infectious diseases. However, because of their importance in the detection and characterization of certain microorganisms, they are discussed here.

DARK-FIELD MICROSCOPY

Dark-field microscopy is similar to phase contrast microscopy in that it involves the alteration of microscopic technique rather than the use of dyes or stains to achieve contrast. By the dark-field method,

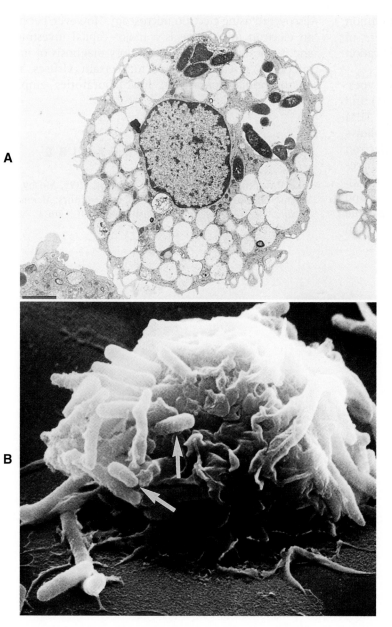

Figure 6-17   **A,** Transmission electron micrograph showing *Escherichia coli* cells internalized by a human mast cell *(arrows).* **B,** Scanning electron micrograph of *E. coli* interacting with the surface of human mast cell *(arrows).*   (**A** and **B** Courtesy SN Abraham, Washington University School of Medicine, St Louis.)

the condenser does not allow light to pass directly through the specimen but directs the light to hit the specimen at an oblique angle (Figure 6-16, *A*). Only light that hits objects, such as microorganisms in the specimen, will be deflected upward into the objective lens for visualization. All other light that passes through the specimen will miss the objective, thus making the background a dark field.

This method has greatest utility for detecting certain bacteria directly in patient specimens that, because of their thin dimensions, cannot be seen by light microscopy and, because of their physiology, are difficult to grow in culture. Dark-field microscopy is used to detect spirochetes, the most notorious of which is the bacterium *Treponema pallidum,* the causative agent

of syphilis (for more information regarding spirochetes, see Chapter 48). As shown in Figure 6-16, *B,* spirochetes viewed using dark-field microscopy will appear extremely bright against a black field. The use of dark-field microscopy in diagnostic microbiology has substantially decreased with the advent of reliable serologic techniques for the diagnosis of syphilis.

ELECTRON MICROSCOPY

The electron microscope uses electrons instead of light to visualize small objects and, instead of lenses, the electrons are focused by electromagnetic fields and form an image on a fluorescent screen, like a television screen. Because of the substantially increased resolution this technology allows, magnifications in excess of 100,000× compared with the 1000× magnification provided by light microscopy are achieved.

Electron microscopes are of two general types: the **transmission electron microscope (TEM)** and the **scanning electron microscope (SEM).** TEM passes the electron beam through objects and allows visualization of internal structures. SEM uses electron beams to scan the surface of objects and provides three-dimensional views of surface structures (Figure 6-17). These microscopes are powerful research tools, and many new morphologic features of bacteria, bacterial components, fungi, viruses, and parasites have been discovered using electron microscopy. However, because an electron microscope is a major capital investment and is not needed for the laboratory diagnosis of most infectious diseases (except for certain viruses and microsporidian parasites), few laboratories employ this method.

ADDITIONAL READING

Atlas RM: *Principles of microbiology,* St Louis, 1995, Mosby.

Murray PR, Baron EJ, Jorgensen JH, et al, editors: *Manual of clinical microbiology,* ed 8, Washington, DC, 2003, ASM Press.

Traditional Cultivation and Identification

CHAPTER **7**

Direct laboratory methods such as microscopy provide preliminary information about the bacteria involved in an infection, but bacterial growth is usually required for definitive identification and characterization. This chapter presents the various principles and methods required for bacterial cultivation and identification.

PRINCIPLES OF BACTERIAL CULTIVATION

This section focuses on the principles and practices of bacterial cultivation, which has three main purposes:

- To grow and isolate all bacteria present in a clinical specimen
- To determine which of the bacteria that grow are most likely causing infection and which are likely contaminants or colonizers
- To obtain sufficient growth of clinically relevant bacteria to allow identification and characterization

Cultivation is the process of growing microorganisms in culture by taking bacteria from the infection site (i.e., the **in vivo** environment) by some means of specimen collection and growing them in the artificial environment of the laboratory (i.e., the **in vitro** environment). Once grown in culture, most bacterial populations are easily observed without microscopy and are present in sufficient quantities to allow laboratory identification procedures to be performed.

The successful transition from the in vivo to the in vitro environment requires that the nutritional and environmental growth requirements of bacterial pathogens be met. The in vivo to in vitro transition is not necessarily easy for bacteria. In vivo they are utilizing various complex metabolic and physiologic pathways developed for survival on or within the human host. Then, relatively suddenly, they are exposed to the artificial in vitro environment of the laboratory. The bacteria must adjust to survive and multiply in vitro. Of importance, their survival depends on the availability of essential nutrients and appropriate environmental conditions.

Although growth conditions can be met for most known bacterial pathogens, the needs of certain clinically relevant bacteria are not sufficiently understood to allow for the ready development of in vitro growth conditions. Examples include *Treponema pallidum* (the causative agent of syphilis) and *Mycobacterium leprae* (the causative agent of leprosy).

NUTRITIONAL REQUIREMENTS

As discussed in Chapter 2, bacteria have numerous nutritional needs that include different gases, water, various ions, nitrogen, sources for carbon, and energy. The latter two requirements are most commonly provided by carbohydrates (e.g., sugars and their derivatives) and proteins.

General Concepts of Culture Media

In the laboratory, nutrients are incorporated into culture media on or in which bacteria are grown. If a culture medium meets a bacterial cell's growth requirements, then that cell will multiply to sufficient numbers to allow visualization by the unaided eye. Of course, bacterial growth after inoculation also requires that the medium be placed in optimal environmental conditions.

Because different pathogenic bacteria have different nutritional needs, various types of culture media have been developed for use in diagnostic microbiology. For certain bacteria the needs are relatively complex, and exceptional media components must be used for growth. Bacteria with such requirements are said to be **fastidious.** Alternatively, the nutritional needs of most clinically important bacteria are relatively basic and straightforward. These bacteria are considered **nonfastidious.**

Phases of Growth Media

Growth media are used in either of two phases: liquid **(broth)** or solid **(agar).** In some instances (e.g., certain blood culture methods), a biphasic medium that contains both a liquid and a solid phase may be used.

In broth media, nutrients are dissolved in water, and bacterial growth is indicated by a change in the broth's appearance from clear to turbid (i.e., cloudy). The turbidity, or cloudiness, of the broth is due to light deflected by bacteria present in the culture (Figure 7-1). The more bacteria growth, the greater the turbidity. At least 10^6 bacteria per milliliter of broth are needed for turbidity to be detected with the unaided eye.

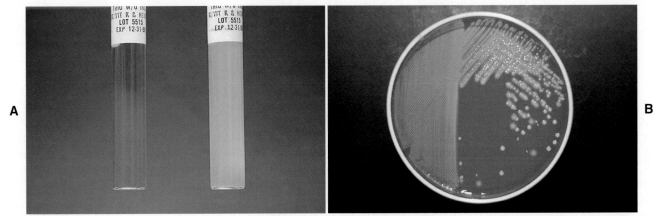

Figure 7-1 **A,** Clear broth indicating no bacterial growth *(left)* and turbid broth indicating bacterial growth *(right).* **B,** Individual bacterial colonies growing on the agar surface following incubation.

Solid media are made by adding a solidifying agent to the nutrients and water. **Agarose,** the most common solidifying agent, has the unique property of melting at high temperatures (≥95° C) but resolidifying only after its temperature falls below 50° C. Addition of agar allows a solid medium to be prepared by heating to an extremely high temperature, which is required for sterilization, and cooling to 55° to 60° C for distribution into petri dishes. On further cooling, the agarose-containing medium forms a stable solid gel referred to as **agar.** The petri dish containing the agar is referred to as the **agar plate.** Different agar media usually are identified according to the major nutritive components of the medium (e.g., sheep blood agar, bile esculin agar, xylose-lysine-desoxycholate agar).

With appropriate incubation conditions, each bacterial cell inoculated onto the agar medium surface will proliferate to sufficiently large numbers to be observable with the unaided eye (see Figure 7-1). The resulting bacterial population is considered to be derived from a single bacterial cell and is known as a **colony.** In other words, all bacterial cells within a single colony are the same genus and species, having identical genetic and phenotypic characteristics (i.e., are a single **clone**). Bacterial cultures derived from a single colony or clone are considered **"pure."** Pure cultures are required for subsequent procedures used to identify and characterize bacteria. The ability to select pure (individual) colonies is one of the first and most important steps required for bacterial identification and characterization.

Media Classifications and Functions

Media are categorized according to their function and use. In diagnostic bacteriology there are four general categories of media: **enrichment, supportive, selective,** and **differential.**

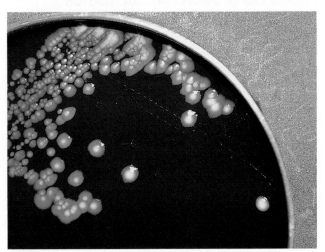

Figure 7-2 Growth of *Legionella pneumophila* on the enrichment medium buffered charcoal-yeast extract (BCYE) agar, used specifically to grow this bacterial genus.

Enrichment media contain specific nutrients required for the growth of particular bacterial pathogens that may be present alone or with other bacterial species in a patient specimen. This media type is used to enhance the growth of a particular bacterial pathogen from a mixture of organisms by using nutrient specificity. One example of such a medium is buffered charcoal-yeast extract agar, which provides L-cysteine and other nutrients required for the growth of *Legionella pneumophila,* the causative agent of legionnaires' disease (Figure 7-2).

Supportive media contain nutrients that support growth of most nonfastidious organisms without giving any particular organism a growth advantage. **Selective media** contain one or more agents that are inhibitory to all organisms except those being sought. In other words, these media select for the growth of

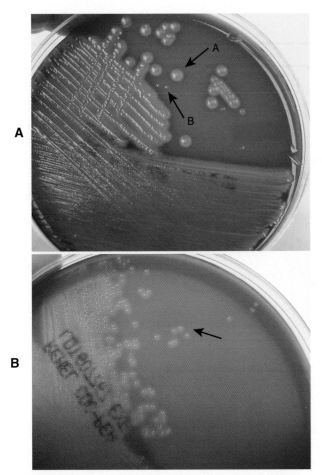

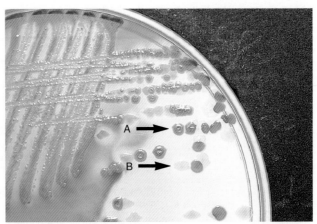

Figure 7-4 Differential capabilities of MacConkey agar as gram-negative bacilli capable of fermenting lactose appear deep purple *(arrow A)*, whereas those not able to ferment lactose appear light pink or relatively colorless *(arrow B)*.

Figure 7-3 **A,** Heavy mixed growth of the gram-negative bacillus *Escherichia coli (arrow A)* and the gram-positive coccus *Enterococcus* spp. *(arrow B)* on the nonselective medium sheep blood agar (SBA). **B,** The selective medium phenylethyl-alcohol agar (PEA) only allows the enterococci to grow *(arrow)*.

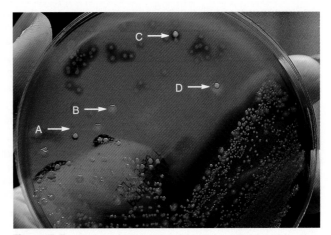

Figure 7-5 Different colony morphologies exhibited on sheep blood agar by various bacteria, including alpha-hemolytic streptococci *(arrow A)*, gram-negative bacilli *(arrow B)*, beta-hemolytic streptococci *(arrow C)*, and *Staphylococcus aureus (arrow D)*.

certain bacteria to the disadvantage of others. Inhibitory agents used for this purpose include dyes, bile salts, alcohols, acids, and antibiotics. An example of a selective medium is phenylethyl alcohol agar, which inhibits the growth of aerobic and facultatively anaerobic gram-negative rods and allows gram-positive cocci to grow (Figure 7-3).

Differential media employ some factor (or factors) that allows colonies of one bacterial species or type to exhibit certain metabolic or culture characteristics that can be used to distinguish them from other bacteria growing on the same agar plate. One commonly used differential medium is MacConkey agar, which differentiates between gram-negative bacteria that can and cannot ferment the sugar lactose (Figure 7-4).

Of importance, many media used in diagnostic bacteriology provide more than one function. For example, MacConkey agar is both differential and selective because it will not allow most gram-positive bacteria to grow. Another example is sheep blood

agar. This is the most commonly used supportive medium for diagnostic bacteriology because it allows many organisms to grow. However, in many ways this agar is also differential because the appearance of colonies produced by certain bacterial species is readily distinguishable (Figure 7-5).

Summary of Artificial Media for Routine Bacteriology

Various broth and agar media that have enrichment, selective, and/or differential capabilities and are used frequently for routine bacteriology are listed alphabetically in Table 7-1. Anaerobic bacteriology (Section 13), mycobacteriology (Section 14), and mycology (Chapter 50) use similar media strategies; details regarding these media are provided in the appropriate chapters.

Table 7-1 Plating Media for Routine Bacteriology

Medium	Components/Comments	Primary Purpose
Bile esculin agar (BEA)	Nutrient agar base with ferric citrate. Hydrolysis of esculin by group D streptococci imparts a brown color to medium; sodium desoxycholate inhibits many bacteria	Differential isolation and presumptive identification of group D streptococci and enterococci
Bile esculin azide agar with vancomycin	Contains azide to inhibit gram-negative bacteria, vancomycin to select for resistant gram-positive bacteria, and bile esculin to differentiate enterococci from other vancomycin-resistant bacteria that may grow	Selective and differential for cultivation of vancomycin-resistant enterococci from clinical and surveillance specimens
Blood agar	Trypticase soy agar, *Brucella* agar, or beef heart infusion with 5% sheep blood	Cultivation of fastidious microorganisms, determination of hemolytic reactions
Bordet-Gengou agar	Potato-glycerol–based medium enriched with 15%-20% defibrinated blood. Contaminants inhibited by methicillin (final concentration of 2.5 μm/mL)	Isolation of *Bordetella pertussis*
Buffered charcoal-yeast extract agar (BCYE)	Yeast extract, agar, charcoal, and salts supplemented with l-cysteine HCl, ferric pyrophosphate, ACES buffer, and α-ketoglutarate	Enrichment for *Legionella* spp.
Buffered charcoal-yeast extract (BCYE) agar with antibiotics	BCYE supplemented with polymyxin B, vancomycin, and ansamycin, to inhibit gram-negative bacteria, gram-positive bacteria, and yeast, respectively	Enrichment and selection for *Legionella* spp.
Campy-blood agar	Contains vancomycin (10 mg/L), trimethoprim (5 mg/L), polymixin B (2500 U/L), amphotericin B (2 mg/L), and cephalothin (15 mg/L) in a *Brucella* agar base with sheep blood	Selective for *Campylobacter* spp.
Campylobacter thioglycollate broth	Thioglycollate broth supplemented with increased agar concentration and antibiotics	Selective holding medium for recovery of *Campylobacter* spp.
Cefoperazone, vancomycin, amphotericin (CVA) medium	Blood-supplemented enrichment medium containing cefoperazone, vancomycin, and amphotericin to inhibit growth of most gram-negative bacteria, gram-positive bacteria, and yeast, respectively	Selective medium for isolation of *Campylobacter* spp.
Cefsulodin- irgasan- novobiocin (CIN) agar	Peptone base with yeast extract, mannitol, and bile salts. Supplemented with cefsulodin, irgasan, and novobiocin; neutral red and crystal violet indicators	Selective for *Yersinia* spp.; may be useful for isolation of *Aeromonas* spp.
Chocolate agar	Peptone base, enriched with solution of 2% hemoglobin or IsoVitaleX (BBL)	Cultivation of *Haemophilus* spp. and pathogenic *Neisseria* spp.
Columbia colistin-nalidixic acid (CNA) agar	Columbia agar base with 10 mg colistin per liter, 15 mg nalidixic acid per liter, and 5% sheep blood	Selective isolation of gram-positive cocci
Cystine-tellurite blood agar	Infusion agar base with 5% sheep blood. Reduction of potassium tellurite by *Corynebacterium diphtheriae* produces black colonies	Isolation of *C. diphtheriae*
Eosin methylene blue (EMB) agar (Levine)	Peptone base with lactose and sucrose. Eosin and methylene blue as indicators	Isolation and differentiation of lactose-fermenting and non–lactose-fermenting enteric bacilli
Gram-negative broth (GN)	Peptone base broth with glucose and mannitol. Sodium citrate and sodium desoxycholate act as inhibitory agents	Selective (enrichment) liquid medium for enteric pathogens
Hektoen enteric (HE) agar	Peptone base agar with bile salts, lactose, sucrose, salicin, and ferric ammonium citrate. Indicators include bromthymol blue and acid fuchsin	Differential, selective medium for the isolation and differentiation of *Salmonella* and *Shigella* spp. from other gram-negative enteric bacilli
MacConkey agar	Peptone base with lactose. Gram-positive organisms inhibited by crystal violet and bile salts. Neutral red as indicator	Isolation and differentiation of lactose fermenting and non–lactose-fermenting enteric bacilli
MacConkey sorbitol agar	A modification of MacConkey agar in which lactose has been replaced with d-sorbitol as the primary carbohydrate	For the selection and differentiation of *E. coli* O157:H7 in stool specimens

Table 7-1 Plating Media for Routine Bacteriology—cont'd

Medium	Components/Comments	Primary Purpose
Mannitol salt agar	Peptone base, mannitol, and phenol red indicator. Salt concentration of 7.5% inhibits most bacteria	Selective isolation of staphylococci
New York City (NYC) agar	Peptone agar base with cornstarch, supplemented with yeast dialysate, 3% hemoglobin, and horse plasma. Antibiotic supplement includes vancomycin (2 μg/mL), colistin (5.5 μg/mL), amphotericin B (1.2 μg/mL), and trimethoprim (3 μg/mL)	Selective for *Neisseria gonorrhoeae*
Phenylethyl alcohol (PEA) agar	Nutrient agar base. Phenylethanol inhibits growth of gram-negative organisms	Selective isolation of gram-positive cocci and anaerobic gram-negative bacilli
Regan Lowe	Charcoal agar supplemented with horse blood, cephalexin, and amphotericin B	Enrichment and selective medium for isolation of *Bordetella pertussis*
Salmonella-Shigella (SS) agar	Peptone base with lactose, ferric citrate, and sodium citrate. Neutral red as indicator; inhibition of coliforms by brilliant green and bile salts	Selective for *Salmonella* and *Shigella* spp.
Schaedler agar	Peptone and soy protein base agar with yeast extract, dextrose, and buffers. Addition of hemin, l-cystine, and 5% blood enriches for anaerobes	Nonselective medium for the recovery of anaerobes and aerobes
Selenite broth	Peptone base broth. Sodium selenite toxic for most *Enterobacteriaceae*	Enrichment of isolation of *Salmonella* spp.
Skirrow agar	Peptone and soy protein base agar with lysed horse blood. Vancomycin inhibits gram-positive organisms; polymyxin B and trimethoprim inhibit most gram-negative organisms	Selective for *Campylobacter* spp.
Streptococcal selective agar (SSA)	Contains crystal violet, colistin, and trimethoprim-sulfamethoxazole in 5% sheep blood agar base	Selective for *Streptococcus pyogenes* and *Streptococcus agalactiae*
Tetrathionate broth	Peptone base broth. Bile salts and sodium thiosulfate inhibit gram-positive organisms and *Enterobacteriaceae*	Selective for *Salmonella* and *Shigella* spp.
Thayer-Martin agar	Blood agar base enriched with hemoglobin and supplement B; contaminating organisms inhibited by colistin, nystatin, vancomycin, and trimethoprim	Selective for *N. gonorrhoeae* and *N. meningitidis*
Thioglycollate broth	Pancreatic digest of casein, soy broth, and glucose enrich growth of most microorganisms.	Supports growth of anaerobes, aerobes, microaerophilic, and fastidious microorganisms
Thiosulfate citrate-bile salts (TCBS) agar	Peptone base agar with yeast extract, bile salts, citrate, sucrose, ferric citrate, and sodium thiosulfate. Bromthymol blue acts as indicator	Selective and differential for vibrios
Todd-Hewitt broth supplemented with antibiotics	Todd-Hewitt, an enrichment broth for streptococci, is supplemented with nalidixic acid and gentamicin or colistin for greater selectivity Thioglycollate and agar reduce redox potential	Selection and enrichment for *Streptococcus agalactiae* in female genital specimens
Trypticase soy broth (TSB)	All-purpose enrichment broth that can support the growth of many fastidious and nonfastidious bacteria	Enrichment broth used for subculturing various bacteria from primary agar plates
Xylose lysine desoxycholate (XLD) agar	Yeast extract agar with lysine, xylose, lactose, sucrose, and ferric ammonium citrate. Sodium desoxycholate inhibits gram-positive organisms; phenol red as indicator	Isolation and differentiation of *Salmonella* and *Shigella* spp. from other gram-negative enteric bacilli

Of the dozens of available media, only those most commonly used for routine diagnostic bacteriology are summarized in this discussion. Part VII discusses which media should be used to culture bacteria from various clinical specimens. Similarly, other chapters throughout Part III discuss media used to identify and characterize specific organisms.

Brain-Heart Infusion. Brain-heart infusion (BHI) is a nutritionally rich medium used to grow various microorganisms, either as a broth or as an agar, with or without added blood. Key ingredients include infusion from several animal tissue sources, added peptone (protein), phosphate buffer, and a small concentration of dextrose. The carbohydrate provides a readily accessible source of energy for many bacteria. BHI broth is often used as a major component of the media developed for culturing a patient's blood for bacteria (see Chapter 52), establishing bacterial identification, and for certain tests to determine bacterial susceptibility to antimicrobial agents (see Chapter 12).

Chocolate Agar. Chocolate agar is essentially the same as blood agar except that during preparation the red blood cells are lysed when added to molten agar base. This lysis releases intracellular nutrients such as hemoglobin, hemin ("X" factor), and the coenzyme nicotinamide adenine dinucleotide (NAD or "V" factor) into the agar for utilization by fastidious bacteria. Red blood cell lysis gives the medium a chocolate-brown color from which this agar gets its name. The most common bacterial pathogens that require this enriched medium for growth include *Neisseria gonorrhoeae,* the causative agent of gonorrhea, and *Haemophilus* spp., which cause infections usually involving the respiratory tract and middle ear. Neither of these species are able to grow on sheep blood agar.

Columbia CNA with Blood. Columbia agar base is a nutritionally rich formula containing three peptone sources and 5% defibrinated sheep blood. This supportive medium can also be used to help differentiate bacterial colonies based on the hemolytic reactions they produce. *CNA* refers to the antibiotics colistin (C) and nalidixic acid (NA) that are added to the medium to suppress the growth of most gram-negative organisms while allowing gram-positive bacteria to grow, thus conferring a selective property to this medium.

Gram-Negative (GN) Broth. A selective broth, gram-negative (GN) broth is used for the cultivation of gastrointestinal pathogens (i.e., *Salmonella* spp. and *Shigella* spp.) from stool specimens and rectal swabs. The broth contains several active ingredients, including sodium citrate and sodium desoxycholate (a bile salt) that inhibit gram-positive organisms and the early multiplication of gram-negative, nonenteric pathogens. To optimize its selective nature, GN broth should be subcultured 6 to 8 hours after initial inoculation and incubation. After this time, the nonenteric pathogens begin to overgrow the pathogens.

Hektoen Enteric (HE) Agar. Hektoen enteric (HE) agar contains bile salts and dyes (bromthymol blue and acid fuchsin) to selectively slow the growth of most nonpathogenic gram-negative bacilli found in the gastrointestinal tract and allow *Salmonella* spp. and *Shigella* spp. to grow. The medium is also differential because many nonenteric pathogens that do grow will appear as orange to salmon-colored colonies. This colony appearance results from the organism's ability to ferment the lactose in the medium, resulting in the production of acid, which lowers the medium's pH and causes a change in the pH indicator bromthymol blue. *Salmonella* spp. and *Shigella* spp. do not ferment lactose, so no color change occurs and their colonies maintain the original blue-green color of the medium. As an additional differential characteristic, the medium contains ferric ammonium citrate, an indicator for the detection of H_2S, so that H_2S-producing organisms, such as *Salmonella* spp., can be visualized as colonies exhibiting a black precipitate (Figure 7-6).

MacConkey Agar. MacConkey agar is the most frequently used primary selective and differential agar. This medium contains crystal violet dye to inhibit the growth of gram-positive bacteria and fungi, and allows many types of gram-negative bacilli to grow. The pH indicator, neutral red, provides this medium with a differential capacity. Bacterial fermentation of lactose results in acid production, which decreases medium pH and causes the neutral red indicator to give bacterial colonies a pink to red color. Non–lactose-fermenters, such as *Shigella* spp., remain colorless and translucent (see Figure 7-4).

Phenylethyl Alcohol (PEA) Agar. Phenylethyl alcohol (PEA) agar is essentially sheep blood agar that is supplemented with phenylethyl alcohol to inhibit the growth of gram-negative bacteria. Five percent sheep blood in PEA provides nutrients for common gram-positive cocci such as enterococci, streptococci, and staphylococci (see Figure 7-3).

Sheep Blood Agar. Most bacteriology specimens are inoculated to sheep blood agar plates because this medium supports all but the most fastidious clinically significant bacteria. Additionally, the colony morphologies that commonly encountered bacteria exhibit on this medium are familiar to most clinical microbiologists.

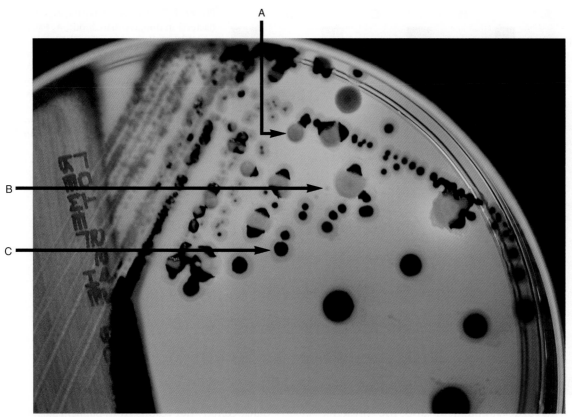

Figure 7-6   Differential capabilities of HE agar for lactose-fermenting, gram-negative bacilli (e.g., *Escherichia coli, arrow A*), non–lactose-fermenters (e.g., *Shigella* spp., *arrow B*), and H₂S producers (e.g., *Salmonella* spp., *arrow C*).

The medium consists of a base containing a protein source (e.g., tryptones), soybean protein digest (containing a slight amount of natural carbohydrate), sodium chloride, agar, and 5% sheep blood.

Certain bacteria produce extracellular enzymes that lyse red blood cells in the agar **(hemolysis).** This activity can result in complete clearing of the red blood cells around the bacterial colony **(beta hemolysis)** or in only partial lysis of the cells to produce a greenish discoloration around the colony **(alpha hemolysis).** Other bacteria have no effect on the red blood cells, and no halo is produced around the colony **(gamma** or **nonhemolysis).** Microbiologists often use colony morphology and the degree or absence of hemolysis as criteria for determining what additional steps will be necessary for identification of a bacterial isolate. To read the hemolytic reaction on a blood agar plate accurately, the technologist must hold the plate up to the light and observe the plate with the light coming from behind (i.e., transmitted light).

Thayer-Martin Agar. Thayer-Martin agar is an enrichment and selective medium for the isolation of *Neisseria gonorrhoeae,* the causative agent of gonorrhea, and *Neisseria meningitidis,* a life-threatening cause of

meningitis. The enrichment portion of the medium is the basal components and the chocolatized blood, while the addition of antibiotics provides a selective capacity. The antibiotics include colistin to inhibit other gram-negative bacteria, vancomycin to inhibit gram-positive bacteria, and nystatin to inhibit yeast. The antimicrobial trimethoprim is also added to inhibit *Proteus* spp., which tend to swarm over the agar surface and mask the detection of individual colonies of the two pathogenic *Neisseria* spp. A further modification, **Martin-Lewis agar,** substitutes ansamycin for nystatin and has a higher concentration of vancomycin.

Thioglycollate Broth. Thioglycollate broth is the enrichment broth most frequently used in diagnostic bacteriology. The broth contains many nutrient factors, including casein, yeast and beef extracts, and vitamins, to enhance the growth of most medically important bacteria. Other nutrient supplements, an oxidation-reduction indicator (resazurin), dextrose, vitamin K1, and hemin have been used to modify the basic thioglycollate formula. In addition, this medium contains 0.075% agar to prevent convection currents from carrying atmospheric oxygen throughout the

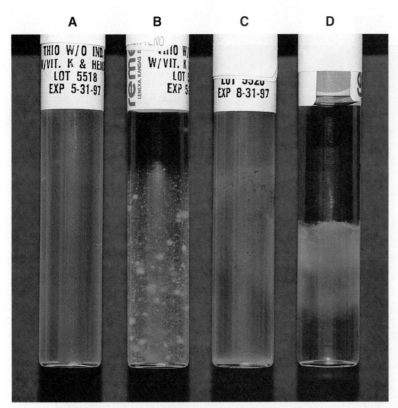

Figure 7-7   Growth characteristics of various bacteria in thioglycollate broth. **A,** Facultatively anaerobic gram-negative bacilli (i.e., those that grow in the presence or absence of oxygen) grow throughout broth. **B,** Gram-positive cocci grow as "puff balls." **C,** Strictly aerobic organisms (i.e., those that require oxygen for growth), such as *Pseudomonas aeruginosa,* grow toward the top of the broth. **D,** Strictly anaerobic organisms (i.e., those that do not grow in the presence of oxygen) grow in the bottom of the broth.

broth. This agar supplement and the presence of thioglycolic acid, which acts as a reducing agent to create an anaerobic environment deeper in the tube, allows anaerobic bacteria to grow.

Gram-negative, facultatively anaerobic bacilli (i.e., those that can grow in the presence or absence of oxygen) generally produce diffuse, even growth throughout the broth, whereas gram-positive cocci frequently grow as discrete "puffballs." Strict aerobic bacteria (i.e., require oxygen for growth), such as *Pseudomonas* spp., tend to grow toward the surface of the broth, whereas strict anaerobic bacteria (i.e., those that cannot grow in the presence of oxygen) grow at the bottom of the broth (Figure 7-7).

Xylose-Lysine-Desoxycholate (XLD) Agar.

As with HE agar, xylose-lysine-desoxycholate (XLD) agar is selective and differential for *Shigella* spp. and *Salmonella* spp. The salt, sodium desoxycholate, inhibits many gram-negative bacilli that are not enteric pathogens and inhibits gram-positive organisms. A phenol red indicator in the medium detects increased acidity from carbohydrate (i.e., lactose, xylose, and sucrose) fermentation. Enteric pathogens, such as *Shigella* spp., do not ferment these carbohydrates, so their colonies remain colorless (i.e., the same approximate pink to red color of the uninoculated medium). Colonies of *Salmonella* spp. are also colorless on XLD, because of the decarboxylation of lysine, which results in a pH increase that causes the pH indicator to turn red. These colonies often exhibit a

black center that results from *Salmonella* spp. producing H$_2$S. Several of the nonpathogens ferment one or more of the sugars and produce yellow colonies (Figure 7-8).

Preparation of Artificial Media

Nearly all media are commercially available as ready-to-use agar plates or tubes of broth. If media are not purchased, laboratory personnel can prepare agars and broths using dehydrated powders that are reconstituted in water (distilled or deionized) according to manufacturer's recommendations. Generally, media are reconstituted by dissolving a specified amount of media powder, which usually contains all necessary components, in water. Boiling is often required to dissolve the powder, but specific manufacturer's instructions printed in media package inserts should be followed exactly. Most media require sterilization so that only bacteria from patient specimens will grow and not those that are contaminants from water or the powdered media. Broth media are distributed to individual tubes before sterilization. Agar media are usually sterilized in large flasks or bottles capped with either plastic screw caps or plugs before being placed in an autoclave.

Media Sterilization.

The timing of **autoclave sterilization** should start from the moment the temperature reaches 121° C and usually requires a minimum of 15 minutes. Once the sterilization cycle is completed, molten agar is allowed to cool to approximately 50° C before being distributed to individual petri

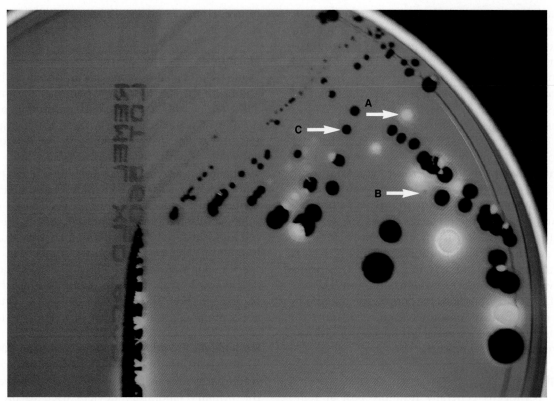

Figure 7-8   Differential capabilities of xylose-lysine-desoxycholate (XLD) agar for lactose-fermenting, gram-negative bacilli (e.g., *Escherichia. coli, arrow A*), non–lactose-fermenters (e.g., *Shigella* spp., *arrow B*), and H_2S producers (e.g., *Salmonella* spp., *arrow C*).

plates (usually 25 mL of molten agar per plate). If other ingredients are to be added (e.g., supplements such as sheep blood or specific vitamins, nutrients, or antibiotics), they should be incorporated when the molten agar has cooled, just before distribution to plates.

Delicate media components that cannot withstand steam sterilization by autoclaving (e.g., serum, certain carbohydrate solutions, certain antibiotics, and other heat-labile substances) can be sterilized by **membrane filtration.** Passage of solutions through membrane filters with pores ranging in size from 0.2 to 0.45 μm in diameter will not remove viruses but does effectively remove most bacterial and fungal contaminants. Finally, all media, whether purchased or prepared, must be subjected to stringent quality control before being used in the diagnostic setting (for more information regarding quality control see Chapter 63).

Cell Cultures. Although most bacteria grow readily on artificial media, certain pathogens require factors provided only by living cells. These bacteria are **obligate intracellular parasites** that require viable host cells for propagation. Although all viruses are obligate intracellular parasites, chlamydiae, rickettsiae, and rickettsiae-like organisms are bacterial pathogens that require living cells for cultivation.

The cultures for growth of these bacteria comprise layers of living cells growing on the surface of a solid matrix such as the inside of a glass tube or the bottom of a plastic flask. The presence of bacterial pathogens within the cultured cells is detected by specific changes in the cells' morphology. Alternatively, specific stains, composed of antibody conjugates, may be used to detect bacterial antigens within the cells. Cell cultures may also detect certain bacterial toxins (e.g., *Clostridium difficile* cytotoxin). Cell cultures are most commonly used in diagnostic virology. Cell culture maintenance and inoculation is addressed in Chapter 51.

ENVIRONMENTAL REQUIREMENTS

Optimizing the environmental conditions to support the most robust growth of clinically relevant bacteria is as important as meeting the organism's nutritional needs for in vitro cultivation. The four most critical environmental factors to consider include oxygen and carbon dioxide (CO_2) availability, temperature, pH, and moisture content of medium and atmosphere.

Oxygen and Carbon Dioxide Availability

Most clinically relevant bacteria are either aerobic, facultatively anaerobic, or strictly anaerobic. **Aerobic**

bacteria use oxygen as a terminal electron acceptor and grow well in room air. Most clinically significant aerobic organisms are actually **facultatively anaerobic,** being able to grow in the presence (i.e., aerobically) or absence (i.e., anaerobically) of oxygen. However, some bacteria, such as *Pseudomonas* spp., members of the Neisseriaceae family, *Brucella* spp., *Bordetella* spp., and *Francisella* spp., are strictly aerobic and cannot grow in the absence of oxygen. Other aerobic bacteria require only low levels of oxygen and are referred to as being **microaerophilic,** or microaerobic. **Anaerobic bacteria** are unable to use oxygen as an electron acceptor, but some aerotolerant strains will still grow slowly and poorly in the presence of oxygen. Oxygen is inhibitory or lethal for strictly anaerobic bacteria.

In addition to oxygen, the availability of CO_2 is important for growth of certain bacteria. Organisms that grow best with higher CO_2 concentrations (i.e., 5% to 10% CO_2) than is provided in room air are referred to as being **capnophilic.** For some bacteria, a 5% to 10% CO_2 concentration is essential for successful cultivation from patient specimens.

Temperature

Bacterial pathogens generally multiply best at temperatures similar to those of internal human host tissues and organs (i.e., 37° C). Therefore, cultivation of most medically important bacteria is done using incubators with temperatures maintained in the 35° to 37° C range. For others, an incubation temperature of 30° C (i.e., the approximate temperature of the body's surface) may be preferable, but such bacteria are encountered relatively infrequently so that use of this incubation temperature occurs only when dictated by special circumstances.

Recovery of certain organisms can be enhanced by incubation at other temperatures. For example, the gastrointestinal pathogen *Campylobacter jejuni* grows at 42° C, whereas many other pathogens and nonpathogens cannot. Therefore, incubation at this temperature can be used as an enrichment procedure. Other bacteria, such as *Listeria monocytogenes* and *Yersinia enterocolitica,* can grow at 0° C, but grow best at temperatures between 20° and 40° C. Cold enrichment has been used to enhance the recovery of these organisms in the laboratory.

pH

The pH scale is a measure of the hydrogen ion concentration of an organism's environment, with a pH value of 7 being neutral. Values less than 7 indicate the environment is **acidic;** values greater than 7 indicate **alkaline** conditions. Most clinically relevant bacteria prefer a near neutral pH range, from 6.5 to 7.5. Commercially prepared media are buffered in this range so that checking their pH is rarely necessary.

Moisture

Water is provided as a major constituent of both agar and broth media. However, when media are incubated at the temperatures used for bacterial cultivation, a large portion of water content can be lost by evaporation. Loss of water from media can be deleterious to bacterial growth in two ways: (1) less water is available for essential bacterial metabolic pathways and (2) with a loss of water there is a relative increase in the solute concentration of the media. An increased solute concentration can osmotically shock the bacterial cell and cause lysis. In addition, increased atmospheric humidity enhances the growth of certain bacterial species. For these reasons, measures, such as sealing agar plates to trap moisture or using humidified incubators, are taken to ensure appropriate moisture levels are maintained throughout the incubation period.

Methods for Providing Optimum Incubation Conditions

Although heating blocks and temperature-controlled water baths may be used occasionally, **incubators** are the primary laboratory devices used to provide the environmental conditions required for cultivating microorganisms. The conditions of incubators can be altered to accommodate the type of organisms to be grown. This section focuses on the incubation of routine bacteriology cultures. Conditions for growing anaerobic bacteria (Section 13), mycobacteria (Section 14), fungi (Chapter 50), and viruses (Chapter 51) are covered in other areas of the text.

Once inoculated with patient specimens, most media are placed in incubators with temperatures maintained between 35° and 37° C and humidified atmospheres that contain 3% to 5% CO_2. Incubators containing room air may be used for some media, but the lack of increased CO_2 may hinder the growth of certain bacteria.

Various atmosphere-generating systems are commercially available and are used instead of CO_2-generating incubators. For example, a self-contained culture medium and a compact CO_2-generating system can be used for culturing fastidious organisms such as *Neisseria gonorrhoeae.* A tablet of sodium bicarbonate is dissolved by the moisture created within an airtight plastic bag and releases sufficient CO_2 to support growth of the pathogen. As an alternative to commercial systems, a **candle jar** can also generate a CO_2 concentration of approximately 3% and has historically been used as a common method for cultivating certain fastidious bacteria. The burning candle, which is placed in a container of inoculated agar plates that is subsequently sealed, uses up just enough oxygen before it goes out (from lack of oxygen) to lower the oxygen tension and produce CO_2 and water by combustion. Other atmosphere-generating systems are available to create conditions optimal for cultivating specific bacterial pathogens (e.g., *Campylobacter* spp. and anaerobic bacteria).

Finally, the duration of incubation required for obtaining good bacterial growth depends on the organisms being cultured. Most bacteria encountered in routine bacteriology will grow within 24 to 48 hours, if not sooner. Certain anaerobic bacteria may require longer incubation, and mycobacteria frequently take weeks before detectable growth occurs.

BACTERIAL CULTIVATION

The process of bacterial cultivation involves the use of optimal artificial media and incubation conditions to isolate and identify the bacterial etiologies of an infection as rapidly and as accurately as possible.

ISOLATION OF BACTERIA FROM SPECIMENS

As discussed in detail throughout Part VII, the cultivation of bacteria from infections at various body sites is accomplished by inoculating processed specimens directly onto artificial media. The media are summarized in Table 7-1 and incubation conditions are selected for their ability to support the growth of the bacteria most likely to be involved in the infectious process.

To enhance the growth, isolation, and selection of etiologic agents, specimen inocula are usually spread over the surface of plates in a standard pattern so that individual bacterial colonies are obtained and semiquantitative analysis can be performed. A commonly used streaking technique is illustrated in Figure 7-9. Using this method, the relative numbers of organisms in the original specimen can be estimated based on the growth of colonies past the original area of inoculation. To enhance isolation of bacterial colonies, the loop should be flamed for sterilization between streaking each subsequent quadrant.

Streaking plates inoculated with a measured amount of specimen, such as when a calibrated loop is used to quantify colony-forming units (CFUs) in urine cultures, is accomplished by spreading the inoculum evenly over the entire agar surface (Figure 7-10). This facilitates counting colonies by ensuring that individual bacterial cells will be well dispersed over the agar surface.

Evaluation of Colony Morphologies

Initial evaluation of colony morphologies on the primary plating media is extremely important. Labora-

torians can provide physicians with early preliminary information regarding the patient's culture results. This information also is important for deciding which subsequent steps to take for definitive organism identification and characterization.

Type of Media Supporting Bacterial Growth. As previously discussed, different media are used to recover particular bacterial pathogens so that determining which media support growth is a clue to the type of organism isolated (e.g., growth on MacConkey agar indicates the organism is a gram-negative bacillus). The incubation conditions that support growth may also be a preliminary indicator of which bacteria have been isolated (e.g., aerobic vs. anaerobic bacteria).

Relative Quantities of Each Colony Type. The predominance of a bacterial isolate is often used as one of the criteria, along with direct smear results, organism virulence, and the body site from which the culture was obtained, for establishing the organism's clinical significance.

Colony Characteristics. Noting key features of a bacterial colony is important for any bacterial identification; success or failure of subsequent identification procedures often depends on the accuracy of these observations. Criteria frequently used to characterize bacterial growth include the following:

- Colony size (usually measured in millimeters or described in relative terms such as pinpoint, small, medium, large)
- Colony pigmentation
- Colony shape (includes form, elevation, and margin of the colony [Figure 7-11])
- Colony surface appearance (e.g., glistening, opaque, dull, transparent)
- Changes in agar media resulting from bacterial growth (e.g., hemolytic pattern on blood agar, changes in color of pH indicators, pitting of the agar surface; for examples see Figures 7-3 through 7-8)
- Odor (certain bacteria produce distinct odors that can be helpful in preliminary identification)

Many of these criteria are somewhat subjective, and the adjectives and descriptive terms used may vary among different laboratories. Regardless of the terminology used, nearly every laboratory's protocol for bacterial identification begins with some agreed-upon colony description of the commonly encountered pathogens.

Although careful determination of colony appearance is important, it is unwise to place total confidence on colony morphology for preliminary identification.

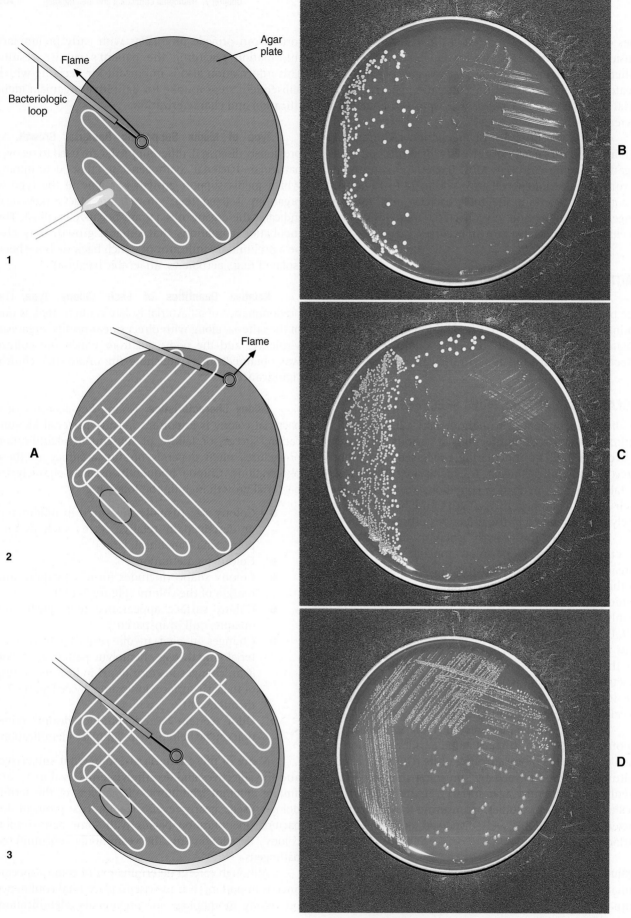

For legend see opposite page.

Figure 7-9 **A,** Dilution streak technique for isolation and semiquantitation of bacterial colonies. **B,** Actual plates show sparse, or 1+ bacterial growth that is limited to the first quadrant. **C,** Moderate, or 2+ bacterial growth that extends to the second quadrant. **D,** Heavy, or 3+ to 4+ bacterial growth that extends to the fourth quadrant.

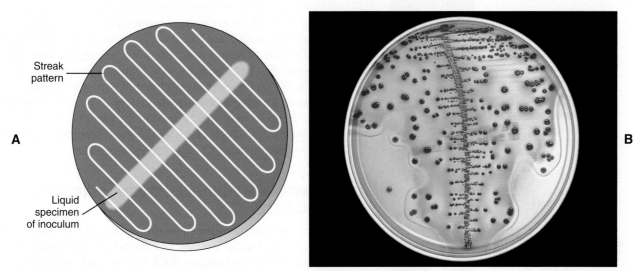

Streak pattern

A

Liquid specimen of inoculum

B

Figure 7-10 **A,** Streaking pattern using a calibrated loop for enumeration of bacterial colonies grown from a liquid specimen such as urine. **B,** Actual plate shows well-isolated and dispersed bacterial colonies for enumeration obtained with the calibrated loop streaking technique.

Bacteria of one species often exhibit colony characteristics that are nearly indistinguishable from those of many other species. Additionally, bacteria of the same species exhibit morphologic diversity. For example, certain colony characteristics may be typical of a given species, but different strains of that species may have different morphologies.

Gram Stain and Subcultures. Isolation of individual colonies during cultivation not only is important for examining morphologies and characteristics but also is necessary for timely performance of Gram stains and subcultures.

The Gram stain and microscopic evaluation of cultured bacteria are used with colony morphology to decide which identification steps are needed. To avoid confusion, organisms from a single colony are stained. In many instances, staining must be performed with each different colony morphology that is observed on the primary plate. In other cases, staining may not be necessary because growth on a particular selective agar provides dependable evidence of the organism's Gram stain morphology (e.g., gram-negative bacilli essentially are the only clinically relevant bacteria that grow well on MacConkey agar).

Following characterization of growth on primary plating media, all subsequent procedures for definitive identification require the use of pure cultures (i.e.,

cultures containing one strain of a single species). If sufficient inocula for testing can be obtained from the primary media, then a subculture is not necessary, except as a precaution to obtain more of the etiologic agent if needed and to ensure that a pure inoculum has been used for subsequent tests (i.e., a "purity" check). However, frequently the primary media do not yield sufficient amounts of bacteria in pure culture and a subculture step is required (Figure 7-12).

Using a sterile loop, a portion of an isolated colony is taken and transferred to the surface of a suitable enrichment medium that is then incubated under conditions optimal for the organism. When making transfers for subculture it is beneficial to flame the inoculating loop between streaks to each area on the agar surface. This avoids overinoculation of the subculture media and ensures individual colonies will be obtained. Once a pure culture is available in a sufficient amount, an inoculum for subsequent identification procedures can be prepared.

PRINCIPLES OF IDENTIFICATION

Microbiologists use various methods to identify organisms cultivated from patient specimens. Although many of the principles and issues about bacterial identification discussed in this chapter are generally applicable to most

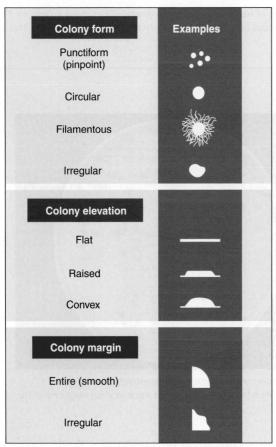

Colony form	Examples
Punctiform (pinpoint)	
Circular	
Filamentous	
Irregular	
Colony elevation	
Flat	
Raised	
Convex	
Colony margin	
Entire (smooth)	
Irregular	

Figure 7-11 Colony morphologic features and descriptive terms for commonly encountered bacterial colonies.

clinically relevant bacteria, specific information regarding particular organism groups is covered in the appropriate chapters in Part III.

The importance of accurate bacterial identification cannot be overstated because identity is central to diagnostic bacteriology issues, including:

- Determining the clinical significance of a particular pathogen (e.g., is the isolate a pathogen or a contaminant?)
- Guiding physician care of the patient
- Determining whether laboratory testing for detection of antimicrobial resistance is warranted
- Determining the type of antimicrobial therapy that is appropriate
- Determining whether the antimicrobial susceptibility profiles are unusual or aberrant for a particular bacterial species
- Determining whether the infecting organism is a risk for other patients in the hospital, the public, or laboratory workers (i.e., is the organism one that may pose problems for infection control, public health, or laboratory safety?)

The identification of a bacterial isolate requires analysis of information gathered from laboratory tests that provide characteristic profiles of bacteria. The tests and the order in which they are used for organism identification are often referred to as an **identification scheme.** Identification schemes can be classified into one of two categories: (1) those that are based on genotypic characteristics of bacteria and (2) those that are based on phenotypic characteristics. Certain schemes rely on both genotypic and phenotypic characteristics. Additionally, some tests, such as the Gram stain, are an integral part of many schemes used for identifying a wide variety of bacteria, whereas other tests may only be used in the identification scheme for a single species such as the fluorescent antibody test for identification of *Legionella pneumophila.*

ORGANISM IDENTIFICATION USING GENOTYPIC CRITERIA

Genotypic identification methods involve characterization of some portion of a bacterium's genome using molecular techniques for DNA or RNA analysis. This usually involves detecting the presence of a gene, or a part thereof, or an RNA product that is specific for a particular organism. In principle, the presence of a specific gene or a particular nucleic acid sequence is interpreted as a definitive identification of the organism. The genotypic approach is highly specific and often very sensitive. With the ever-expanding list of molecular techniques being developed, the genetic approach to organism identification will continue to grow and become more integrated into diagnostic microbiology laboratory protocols (for more information regarding molecular methods, see Chapter 8).

ORGANISM IDENTIFICATION USING PHENOTYPIC CRITERIA

Phenotypic criteria are based on observable physical or metabolic characteristics of bacteria, that is, identification is through analysis of gene products rather than through the genes themselves. The phenotypic approach is the classic approach to bacterial identification, and most identification strategies are still based on bacterial phenotype. Delineation of some characteristics may require subcellular analysis involving sophisticated instrumentation (e.g., high-performance liquid chromatography [HPLC] to analyze cell wall components. For more information about these techniques, see Chapter 8). Other characterizations are based on the antigenic makeup of the organisms and involve techniques based on antigen-antibody interactions (for more information regarding immunologic diagnosis of infectious diseases, see Chapter 10). However, most of the phenotypic

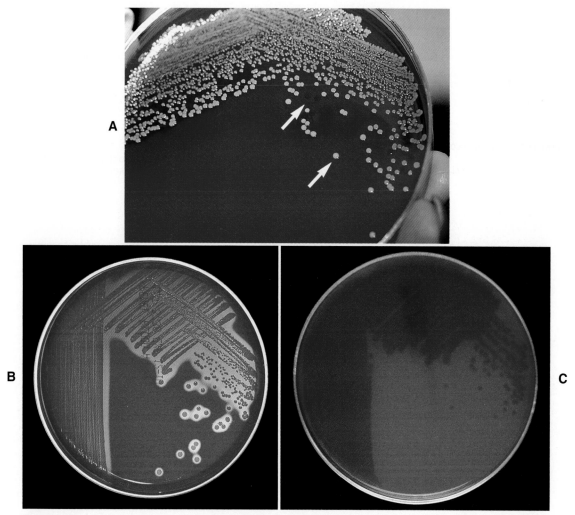

Figure 7-12   Mixed bacterial culture on sheep blood agar (**A**) requires subculture of individually distinct colonies *(arrows)* to obtain pure cultures of *Staphylococcus aureus* (**B**) and *Streptococcus pneumoniae* (**C**).

characterizations used in diagnostic bacteriology are based on tests that establish a bacterial isolate's morphology and metabolic capabilities. The most commonly used phenotypic criteria include:

- Microscopic morphology and staining characteristics
- Macroscopic (colony) morphology
- Environmental requirements for growth
- Resistance or susceptibility to antimicrobial agents
- Nutritional requirements and metabolic capabilities

Microscopic Morphology and Staining Characteristics

Microscopic evaluation of bacterial cellular morphology, as facilitated by the Gram stain or other enhancing methods discussed in Chapter 6, provides the most basic and important information on which final identification

strategies are based. For this reason, a Gram stain of bacterial growth from isolated colonies on various media is usually the first step in any identification scheme. Based on these findings, most clinically relevant bacteria can be divided into four distinct groups: gram-positive cocci, gram-negative cocci, gram-positive bacilli, and gram-negative bacilli (Figure 7-13). Some bacterial species are morphologically indistinct and are described as "gram-negative coccobacilli," "gram-variable bacilli," or **pleomorphic** (i.e., exhibiting various shapes). Still other morphologies include curved and/or rods and spirals.

Even without staining, examination of a wet preparation of bacterial colonies under oil immersion (1000× magnification) can provide clues as to possible identity. For example, a wet preparation prepared from a translucent, alpha-hemolytic colony on blood agar may reveal cocci in chains, a strong indication that the bacteria are probably streptococci. Also, the

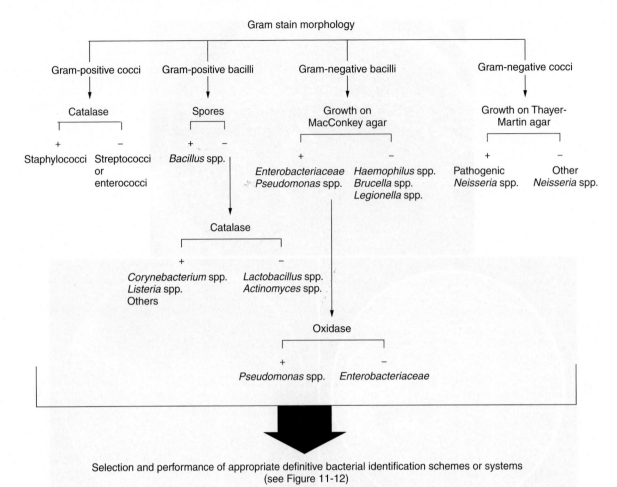

Figure 7-13   Flowchart example of a bacterial identification scheme.

presence of yeast, whose colonies can closely mimic bacterial colonies but whose cells are generally much larger, can be determined (Figure 7-14).

In most instances, identification schemes for final identification are based on the cellular morphologies and staining characteristics of bacteria. To illustrate, an abbreviated identification flowchart for commonly encountered bacteria is shown in Figure 7-13 (more detailed identification schemes are presented throughout Part III); this flowchart simply illustrates how information about microorganisms is integrated into subsequent identification schemes that are usually based on the organism's nutritional requirements and metabolic capabilities. In certain cases, staining characteristics alone are used to definitively identify a bacterial species. Examples of this are mostly restricted to the use of fluorescent-labeled specific antibodies and fluorescent microscopy to identify organisms such as *Legionella pneumophila* and *Bordetella pertussis*.

Macroscopic (Colony) Morphology

Evaluation of colony morphology includes considering colony size, shape, color (pigment), surface appearance,

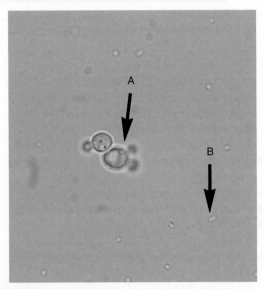

Figure 7-14   Microscopic examination of a wet preparation demonstrates the size difference between most yeast cells, such as those of *Candida albicans (arrow A)*, and bacteria, such as *Staphylococcus aureus (arrow B)*.

and any changes that colony growth produces in the surrounding agar medium (e.g., hemolysis of blood in blood agar plates).

Although these characteristics usually are not sufficient for establishing a final or definitive identification, the information gained provides preliminary information necessary for determining what identification procedures should follow. However, it is unwise to place too much confidence on colony morphology alone for preliminary identification of isolates. Microorganisms often grow as colonies whose appearance is not that different from many other species, especially if the colonies are relatively young (i.e., less than 14 hours old). Therefore, unless colony morphology is distinctive or unless growth occurs on a particular selective medium, other characteristics must be included in the identification scheme.

Environmental Requirements for Growth

Environmental conditions required for growth can be used to supplement other identification criteria. However, as with colony morphologies, this information alone is not sufficient for establishing a final identification. The ability to grow in particular incubation atmospheres most frequently provides insight about the organism's potential identity. For example, organisms growing only in the bottom of a tube containing thioglycollate broth are not likely to be strictly aerobic bacteria, thus eliminating these types of bacteria from the list of identification possibilities. Similarly, anaerobic bacteria can be discounted in the identification schemes for organisms that grow on blood agar plates incubated in an ambient (room) atmosphere. An organism's requirement, or preference, for increased carbon dioxide concentrations can provide hints for the identification of other bacteria such as *Streptococcus pneumoniae, Haemophilus influenzae,* and *Neisseria gonorrhoeae.*

In addition to atmosphere, the ability to survive or even thrive in temperatures that exceed or are well below the normal body temperature of 37° C may be helpful for organism identification. The growth of *Campylobacter jejuni* at 42° C and the ability of *Yersinia enterocolitica* to survive at 0° C are two examples.

Resistance or Susceptibility to Antimicrobial Agents

The ability of an organism to grow in the presence of certain antimicrobial agents or specific toxic substances is widely used to establish preliminary identification information. This is accomplished by using agar media supplemented with inhibitory substances or antibiotics (for examples, see Table 7-1) or by directly measuring an organism's resistance to antimicrobial agents that may be used to treat infections (for more information regarding antimicrobial susceptibility testing, see Chapter 12)

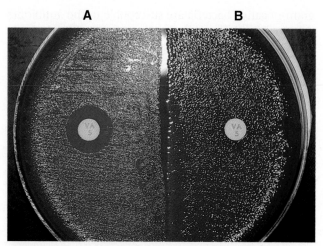

Figure 7-15   **A,** Zone of growth inhibition around the 5-mg vancomycin disk is indicative of a gram-positive bacterium. **B,** The gram-negative organism is not inhibited by this antibiotic, and growth extends to the edge of the disk.

As discussed earlier in this chapter, most clinical specimens are inoculated to several media, including some selective or differential agars. Therefore, the first clue to identification of an isolated colony is the nature of the media on which the organism is growing. For example, with rare exceptions, only gram-negative bacteria grow well on MacConkey agar. Alternatively, other agar plates, such as Columbia agar with CNA, support the growth of gram-positive organisms to the exclusion of most gram-negative bacilli. Certain agar media can be used to differentiate even more precisely than simply separating gram-negative and gram-positive bacteria. Whereas chocolate agar will support the growth of all *Neisseria* spp., the antibiotic-supplemented Thayer-Martin formulation will almost exclusively support the growth of the pathogenic species *N. meningitidis* and *N. gonorrhoeae.*

Directly testing a bacterial isolate's susceptibility to a particular antimicrobial agent may be a very useful part of an identification scheme. Many gram-positive bacteria (with a few exceptions, such as certain enterococci, lactobacilli, *Leuconostoc,* and *Pediococcus* spp.) are susceptible to vancomycin, an antimicrobial agent that acts on the bacterial cell wall. In contrast, most clinically important gram-negative bacteria are resistant to vancomycin. Therefore, when organisms with uncertain Gram stain results are encountered, susceptibility to vancomycin can be used to help establish the organism's Gram "status." Any zone of inhibition around a vancomycin-impregnated disk after overnight incubation is usually indicative of a gram-positive bacterium (Figure 7-15). With few exceptions (e.g., certain *Chryseobacterium, Moraxella,* or *Acinetobacter* spp. isolates may be vancomycin susceptible), truly gram-negative bacteria are resistant to vancomycin. Conversely, most gram-negative bacteria are susceptible to the antibiotics colistin or polymyxin, whereas gram-positive bacteria are frequently resistant to these agents.

Nutritional Requirements and Metabolic Capabilities

Determining the nutritional and metabolic capabilities of a bacterial isolate is the most common approach used for determining the genus and species of an organism. The methods available for making these determinations share many commonalties but also have some important differences. In general, all methods use a combination of tests to establish the enzymatic capabilities of a given bacterial isolate as well as the isolate's ability to grow or survive the presence of certain inhibitors (e.g., salts, surfactants, toxins, and antibiotics).

Establishing Enzymatic Capabilities.

As discussed in Chapter 2, enzymes are the driving force in bacterial metabolism. Because enzymes are genetically encoded, the enzymatic content of an organism is a direct reflection of the organism's genetic makeup, which, in turn, is specific for individual bacterial species.

Types of Enzyme-Based Tests.

In diagnostic bacteriology, enzyme-based tests are designed to measure the presence of one specific enzyme or a complete metabolic pathway that may contain several different enzymes. Although the specific tests most useful for the identification of particular bacteria are discussed in Part III, some examples of tests commonly used to characterize a wide spectrum of bacteria are reviewed here.

Single Enzyme Tests.

Several tests are commonly used to determine the presence of a single enzyme. These tests usually provide rapid results because they can be performed on organisms already grown in culture. Of importance, these tests are easy to perform and interpret and often play a key role in the identification scheme. Although most single enzyme tests do not yield sufficient information to provide species identification, they are used extensively to determine which subsequent identification steps should be followed. For example, the catalase test can provide pivotal information and is commonly used in schemes for gram-positive identifications. The oxidase test is of comparable importance in identification schemes for gram-negative bacteria (see Figure 7-13).

Catalase Test.

The enzyme **catalase** catalyzes the release of water and oxygen from hydrogen peroxide (H_2O_2 + catalase = $H_2O + O_2$); its presence is determined by direct analysis of a bacterial culture (see Procedure 13-8). The rapid production of bubbles (effervescence) when bacterial growth is mixed with a hydrogen peroxide solution is interpreted as a positive test (i.e., the presence of catalase). Failure to produce effervescence or weak effervescence is interpreted as negative. If the bacterial inoculum is inadvertently contaminated with red blood cells when the test inoculum is collected from a sheep blood agar plate, weak production of bubbles may occur, but this should not be interpreted as a positive test.

Because the catalase test is key to the identification scheme of many gram-positive organisms, interpretation must be done carefully. For example, staphylococci are catalase-positive, whereas streptococci and enterococci are negative; similarly, the catalase reaction differentiates *Listeria monocytogenes* and corynebacteria (catalase-positive) from other gram-positive, non–spore-forming bacilli (see Figure 7-13).

Oxidase Test.

Cytochrome oxidase participates in electron transport and in the nitrate metabolic pathways of certain bacteria. The test for the presence of oxidase can be performed by flooding bacterial colonies on the agar surface with the reagent 1% tetramethyl-*p*-phenylenediamine dihydrochloride. Alternatively, a sample of the bacterial colony can be rubbed onto filter paper impregnated with the reagent (see Procedure 13-33). If an iron-containing wire is used to transfer growth, a false-positive reaction may result; therefore, platinum wire or wooden sticks are recommended. Certain organisms may show slight positive reactions after the initial 10 seconds have passed; such results are not considered definitive.

The test is initially used for differentiating between groups of gram-negative bacteria. Among the commonly encountered gram-negative bacilli Enterobacteriaceae, *Stenotrophomonas maltophilia*, and *Acinetobacter* spp. are oxidase-negative, whereas many other bacilli, such as *Pseudomonas* spp. and *Aeromonas* spp., are positive (see Figure 7-13). The oxidase test is also a key reaction for the identification of *Neisseria* spp. (oxidase-positive).

Indole Test.

Bacteria that produce the enzyme **tryptophanase** are able to degrade the amino acid tryptophan into pyruvic acid, ammonia, and indole. Indole is detected by combining with an indicator, aldehyde (1% paradimethylaminocinnamaldehyde), that results in a blue color formation (see Procedure 13-20). This test is used in numerous identification schemes, especially to presumptively identify *Escherichia coli*, the gram-negative bacillus most commonly encountered in diagnostic bacteriology.

Urease Test.

Urease hydrolyzes the substrate urea into ammonia, water, and carbon dioxide. The presence of the enzyme is determined by inoculating an organism to broth or agar that contains urea as the primary carbon source and detecting the production of ammonia (see Procedure 13-41). Ammonia increases the pH of the medium so its presence is readily detected using a pH indicator. Change in medium pH is a common indicator of metabolic process and, because **pH indicators** change color with increases (alkalinity) or decreases (acidity) in the medium's pH, they are commonly used in many identification test schemes. The urease test helps identify certain species of Enterobacteriaceae, such as *Proteus* spp., and other important bacteria such as *Corynebacterium urealyticum* and *Helicobacter pylori*.

PYR Test.
The enzyme L-pyrroglutamyl-aminopeptidase hydrolyzes the substrate L-pyrrolidonyl-β-naphthylamide (PYR) to produce a β-naphthylamine. When the β-naphthylamine combines with a cinnamaldehyde reagent, a bright red color is produced (see Procedure 13-36). The PYR test is particularly helpful in identifying gram-positive cocci such as *Streptococcus pyogenes* and *Enterococcus* spp., which are positive, whereas other streptococci are negative.

Hippurate Hydrolysis.
Hippuricase is a constitutive enzyme that hydrolyzes the substrate hippurate to produce the amino acid glycine. Glycine is detected by oxidation with ninhydrin reagent that results in the production of a deep purple color (see Procedure 13-19). The hippurate test is most frequently used in the identification of *Streptococcus agalactiae*, *Campylobacter jejuni*, and *Listeria monocytogenes*.

Tests for Presence of Metabolic Pathways.
Several identification schemes are based on determining what metabolic pathways an organism uses and the substrates processed by these pathways. In contrast to single enzyme tests, these pathways may involve several interactive enzymes. The presence of an end product resulting from these interactions is measured in the testing system. Assays for metabolic pathways can be classified into three general categories: carbohydrate oxidation and fermentation, amino acid degradation, and single substrate utilizations.

Oxidation and Fermentation Tests.
As discussed in Chapter 2, bacteria use various metabolic pathways to produce biochemical building blocks and energy. For most clinically relevant bacteria, this involves utilization of carbohydrates (e.g., sugar or sugar derivatives) and protein substrates. Determining whether substrate utilization is an oxidative or fermentative process is important for the identification of several different bacteria.

Oxidative processes require oxygen; fermentative ones do not. The clinical laboratory determines how an organism utilizes a substrate by observing whether acid byproducts are produced in the presence or absence of oxygen. In most instances, the presence of acid byproducts is detected by a change in the pH indicator incorporated into the medium. The color changes that occur in the presence of acid depend on the type of pH indicator used.

Oxidation-fermentation determinations are usually accomplished using a special medium (**oxidative-fermentative [O-F] medium**) that contains low concentrations of peptone and a single carbohydrate substrate such as glucose. The organism to be identified is inoculated into two glucose **O-F tubes,** one of which is then overlaid with mineral oil as a barrier to oxygen. Common pH indicators used for O-F tests, and the color changes they undergo with acidic conditions, include **bromcresol purple,** which changes from purple to yellow; **Andrade's acid fuchsin** indicator, which changes from pale yellow to pink; **phenol red,** which changes from red to yellow; and **bromthymol blue,** which changes from green to yellow.

As shown in Figure 7-16, when acid production is detected in both tubes, the organism is identified as a **glucose fermenter** because fermentation can occur with or without oxygen. If acid is only detected in the open, aerobic tube, the organism is characterized as a **glucose-oxidizer.** As a third possibility, some bacteria do not use glucose as a substrate and no acid is detected in either tube (a **nonutilizer**). The glucose fermentative or oxidative capacity is generally used to separate organisms into major groups (e.g., Enterobacteriaceae are fermentative; *Pseudomonas* spp. are oxidative). However, the utilization pattern for several other carbohydrates (e.g., lactose, sucrose, xylose, maltose) is often needed to help identify an organism's genus and species.

Amino Acid Degradation.
Determining the ability of bacteria to produce enzymes that either deaminate, dihydrolyze, or decarboxylate certain amino acids is often used in identification schemes. The amino acid substrates most often tested include lysine, ornithine, arginine, and phenylalanine. (The indole test for tryptophan cleavage is presented earlier in this chapter.)

Decarboxylases cleave the carboxyl group from amino acids so that amino acids are converted into amines; lysine is converted to cadaverine, and ornithine is converted to putrescine. Because amines increase medium pH, they are readily detected by color changes in a pH indictor indicative of alkalinity. Decarboxylation is an anaerobic process that requires an acid environment for activation. The most common medium used for this test is **Moeller decarboxylase base,** whose components include glucose, the amino acid substrate

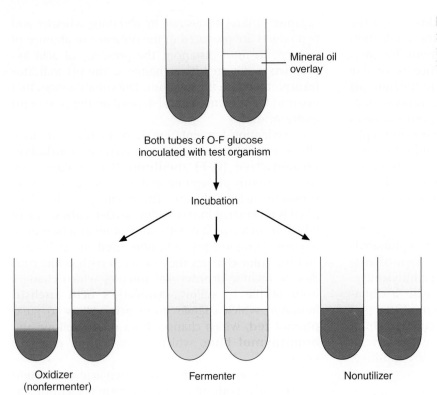

Both tubes of O-F glucose
inoculated with test organism

Mineral oil
overlay

Incubation

Oxidizer
(nonfermenter)

Fermenter

Nonutilizer

Figure 7-16   Principle of glucose oxidative-
fermentation (O-F) test. Fermentation patterns
shown in O-F tubes are examples of oxidative,
fermentative, and nonutilizing bacteria.

of interest (i.e., lysine, ornithine, or arginine), and a pH indicator.

Organisms are inoculated into the tube medium that is then overlaid with mineral oil to ensure anaerobic conditions. (See Chapter 13) Early during incubation, bacteria utilize the glucose and produce acid, resulting in a yellow coloration of the pH indicator. Organisms that can decarboxylate the amino acid then begin to attack that substrate and produce the amine product, which increases the pH and changes the indicator back from yellow to purple (if bromcresol purple is the pH indicator used; red if phenol red is the indicator). Therefore, after overnight incubation, a positive test is indicated by a purple color and a negative test (i.e., lack of decarboxylase activity) is indicated by a yellow color. With each amino acid tested, a control tube of the glucose-containing broth base without amino acid is inoculated. This standard's color is compared with that of the tube containing the amino acid following incubation.

Because it is a two-step process, the breakdown of arginine is more complicated than that of lysine or ornithine. Arginine is first **dehydrolyzed** to citrulline, which is subsequently converted to ornithine. Ornithine is then decarboxylated to putrescine, which results in the same pH indicator changes as just outlined for the other amino acids.

Unlike decarboxylation, **deamination** of the amino acid phenylalanine occurs in air. The presence

of the end product (phenylpyruvic acid) is detected by the addition of 10% ferric chloride, which results in the development of a green color. Agar slant medium is commercially available for this test.

Single Substrate Utilization. Whether an organism can grow in the presence of a single nutrient or carbon source provides useful identification information. Such tests entail inoculation of organisms to a medium that contains a single source of nutrition (e.g., citrate, malonate, or acetate) and, after incubation, observing the medium for growth. Growth is determined by observing the presence of bacterial colonies or by using a pH indicator to detect end products of metabolic activity.

Establishing Inhibitor Profiles. The ability of a bacterial isolate to grow in the presence of one or more inhibitory substances can provide valuable identification information. Examples regarding the use of inhibitory substances are presented earlier in this chapter.

In addition to the information gained from using inhibitory media and/or antimicrobial susceptibility testing, other more specific tests may be incorporated into bacterial identification schemes. Because most of these tests are used to identify a particular group of bacteria, their protocols and principles are discussed in the appropriate chapters in Part III. A few examples of such tests include:

1. Selection and inoculation of tests

> • Number and type of tests selected depend on type of organism to be identified, clinical significance of isolates, and availability of reliable methods
> • Identification systems must be inoculated with pure cultures

2. Incubation for substrate utilization

> • Duration depends on whether bacterial multiplication is or is not required for substrate utilization (i.e., growth-based test vs. a non–growth-based test)

3. Detection of metabolic activity (substrate utilization)

> • Colorimetry, fluorescence, or turbidity are used to detect products of substrate utilization
> • Detection is done visually or with the aid of various photometers

4. Analysis of metabolic profiles

> • Involves conversion of substrate utilization profile to a numeric code (see Figure 7-18)
> • Computer-assisted comparison of numeric code with extensive taxonomic data base provides most likely identification of the bacterial isolate
> • For certain organisms for which identification is based on a few tests, extensive testing and analysis are not routinely needed

Figure 7-17 Four basic components of bacterial identification schemes and systems.

- Growth in the presence of various NaCl concentrations (identification of enterococci and *Vibrio* spp.)
- Susceptibility to optochin and solubility in bile (identification of *Streptococcus pneumoniae*)
- Ability to hydrolyze esculin in the presence of bile (identification of enterococci)
- Ethanol survival (identification of *Bacillus* spp.)

PRINCIPLES OF PHENOTYPE-BASED IDENTIFICATION SCHEMES

As shown in Figure 7-13, growth characteristics, microscopic morphologies, and single test results are used to categorize most bacterial isolates into general groups. However, the definitive identification to species requires use of schemes designed to produce metabolic profiles of the organisms. Identification systems usually consist of four major components (Figure 7-17):

- Selection and inoculation of a set (i.e., **battery**) of specific metabolic substrates and growth inhibitors
- Incubation to allow substrate utilization to occur or to allow growth inhibitors to act
- Determination of metabolic activity that occurred during incubation
- Analysis of metabolic profiles and comparison with established profile databases for known bacterial species to establish definitive identification

SELECTION AND INOCULATION OF IDENTIFICATION TEST BATTERY

The number and types of tests that are selected for inclusion in a battery depends on various factors, including the type of bacteria to be identified, the clinical significance of the bacterial isolate, and the availability of reliable testing methods.

Type of Bacteria to Be Identified

Certain organisms have such unique features that relatively few tests are required to establish identity. For example, *Staphylococcus aureus* is essentially the only gram-positive coccus that appears microscopically in clusters, is catalase-positive, and produces coagulase. Therefore, identification of this common pathogen usually requires the use of only two tests coupled with colony and microscopic morphology. In contrast, identification of most clinically relevant gram-negative bacilli, such as those of the *Enterobacteriaceae* family, requires establishing metabolic profiles often involving 20 or more tests.

Clinical Significance of the Bacterial Isolate

Although a relatively large number of tests may be required to identify a particular bacterial species, the number of tests actually inoculated may depend on the clinical significance of an isolate. For instance, if a gram-negative bacillus is mixed with five other bacterial species in a urine culture, it is likely to be a contaminant. In this setting, multiple tests to establish species identity are not warranted and should not routinely be performed. However, if this same organism is isolated in pure culture from cerebrospinal fluid, the full battery of tests required for definitive identification should be performed.

Availability of Reliable Testing Methods

With the complicated medical procedures that patients are subjected to and the increasing population of immunocompromised patients, uncommon or unusual bacteria are isolated. Because of the unusual nature that some of these bacteria exhibit, reliable testing methods and identification criteria may not be established in most clinical laboratories. In these instances, only the genus of the organism may be identified (e.g., *Bacillus* spp.), or identification may not go beyond a description of the organism's microscopic morphology (e.g., gram-positive, pleomorphic bacilli, or gram-variable, branching organism). When such bacteria are encountered and are thought to be clinically significant, they should be sent to a reference laboratory whose personnel are experienced in identifying unusual organisms.

Although the number of tests included in an identification battery may vary and different identification systems may require various inoculation techniques, the one common feature of all systems is the requirement for inoculation with a pure culture. Inoculation with a mixture of bacteria produces mixed and often uninterpretable results. To expedite identification, cultivation strategies (described earlier in this chapter) should focus on obtaining pure cultures as soon as possible. Furthermore, controls should be run with most identification systems as a check for purity of the culture used to inoculate the system.

INCUBATION FOR SUBSTRATE UTILIZATION

The time required to obtain bacterial identification depends heavily on the length of incubation needed before the test result is available. In turn, the duration of incubation depends on whether the test is measuring metabolic activity that requires bacterial growth or whether the assay is measuring the presence of a particular enzyme or cellular product that can be detected without the need for bacterial growth.

Conventional Identification

Because the **generation time** (i.e., the time required for a bacterial population to double) for most clinically relevant bacteria is 20 to 30 minutes, growth-based tests usually require hours of incubation before the presence of an end product can be measured. Many conventional identification schemes require 18 to 24 hours of incubation, or longer, before the tests can be accurately interpreted. Although the conventional approach has been the standard for most bacterial identification schemes, the desire to produce results and identifications in a more timely fashion has resulted in the development of rapid identification strategies.

Rapid Identification

In the context of diagnostic bacteriology, the term *rapid* is relative. In some instances a rapid method is one that provides a result the same day that the test was inoculated. Alternatively, the definition may be more precise whereby "rapid" is only used to describe tests that provide results within 4 hours of inoculation.

Two general approaches have been developed to obtain more rapid identification results. One has been to vary the conventional testing approach by decreasing the test substrate medium volume and increasing the concentration of bacteria in the inoculum. Several conventional methods, such as carbohydrate fermentation profiles, use this strategy for more rapid results.

The second approach uses unique or unconventional substrates. Particular substrates are chosen, based on their ability to detect enzymatic activity at all times. That is, detection of the enzyme does not depend on multiplication of the organism (i.e., not a growth-based test) so that delays caused by depending on bacterial growth are minimized. The catalase, oxidase, and PYR tests discussed previously are examples of such tests, but many others are available as part of commercial testing batteries.

Still other rapid identification schemes are based on antigen-antibody reactions, such as latex agglutination tests, that are commonly used to quickly and easily identify certain beta-hemolytic streptococci and *S. aureus* (for more information regarding these test formats, see Chapter 10).

DETECTION OF METABOLIC ACTIVITY

The accuracy of an identification scheme heavily depends on the ability to reliably detect whether a bacterial isolate has utilized the substrates composing the identification battery. The sensitivity and strength of the detection signal can also contribute to how rapidly results are available. No matter how quickly an organism may metabolize a particular substrate, if the end products are slowly or weakly detected, the ultimate production of results will still be "slow."

Detection strategies for determining the end products of different metabolic pathways use one of the following: colorimetry, fluorescence, and turbidity.

Colorimetry

Several identification systems measure color change to detect the presence of metabolic end products. Most frequently the color change is produced by pH indicators in the media. Depending on the byproducts to be measured and the testing method, additional reagents may need to be added to the reaction before result interpretation. An alternative to the use of pH indicators is the oxidation-reduction potential indicator tetrazolium violet. Organisms are inoculated into wells that contain a single, utilizable carbon source. Metabolism of that substrate generates electrons that reduce the tetrazolium violet, resulting in production of a purple color (positive reaction) that can be spectrophotometrically detected. In a third approach, the substrates themselves may be chromogenic so that when they are "broken down" by the organism the altered substrate produces a color.

Some commercial systems use a miniaturized modification of conventional biochemical batteries, with the color change being detectable with the unaided eye. Alternatively, in certain automated systems, a photoelectric cell measures the change in the wavelength of light transmitted through miniaturized growth cuvettes or wells, thus eliminating the need for direct visual interpretation by laboratory personnel. Additionally, a complex combination of dyes and filters may be used to enhance and broaden the scope of substrates and color changes that can be used in such systems. These combinations hasten identification and increase the variety of organisms that can be reliably identified.

Fluorescence

There are two basic strategies for using fluorescence to measure metabolic activity. In one approach, substrate-fluorophore complexes are used. If a bacterial isolate processes the substrate, the fluorophore is released and assumes a fluorescent configuration. Alternatively, pH changes resulting from metabolic activity can be measured by changes in fluorescence of certain fluorophore markers. In these pH-driven, fluorometric reactions, pH changes result in either the fluorophore becoming fluorescent or, in other instances, fluorescence being quenched or lost. To detect fluorescence, ultraviolet light of appropriate wavelength is focused on the reaction mixture and a special kind of photometer, a fluorometer, measures fluorescence.

Turbidity

Turbidity measurements are not commonly used for bacterial identifications but do have widespread application for determining growth in the presence of specific growth inhibitors, including antimicrobial agents, and for detecting bacteria present in certain clinical specimens.

Turbidity is the ability of particles in suspension to refract and deflect light rays passing through the suspension such that the light is reflected back into the eyes of the observer. The **optical density** (OD), a measurement of turbidity, is determined in a spectrophotometer. This instrument compares the amount of light that passes through the suspension (the **percent transmittance**) with the amount of light that passes through a control suspension without particles. A photoelectric sensor, or photometer, converts the light that impinges on its surface to an electrical impulse, which can be quantified. A second type of turbidity measurement is obtained by nephelometry, or light scatter. In this case, the photometers are placed at angles to the suspension, and the scattered light, generated by a laser or incandescent bulb, is measured. The amount of light scattered depends on the number and size of the particles in suspension.

ANALYSIS OF METABOLIC PROFILES

The metabolic profile obtained with a particular bacterial isolate is essentially the phenotypic fingerprint, or signature, of that organism. Typically, the profile is recorded as a series of pluses (+) for positive reactions and minuses (–) for negative or nonreactions (Figure 7-18). Although this profile by itself provides little information, microbiologists can compare the profile with an extensive identification database to establish the identity of that specific isolate.

Identification Databases

Reference databases are available for clinical use. These databases are maintained by manufacturers of identification systems and are based on the continuously

	Test/ substrate	Test results (– or +)	Binary code conversion (0 or 1)	Octal value	Octal score	Octal triplet total	Octal profile
1	ONPG	+	1	× 1	1		
2	Arginine dihydrolase	–	0	× 2	0	5	
3	Lysine decarboxylase	+	1	× 4	4		
4	Ornithine decarboxylase	+	1	× 1	1		
5	Citrate utilization	–	0	× 2	0	1	
6	H$_2$S production	–	0	× 4	0		
7	Urea hydrolysis	–	0	× 1	0		
8	Tryptophane deaminase	–	0	× 2	0	4	
9	Indole production	+	1	× 4	4		
10	VP test	–	0	× 1	0		
11	Gelatin hydrolysis	–	0	× 2	0	4	5144572 (*E. coli*)
12	Glucose fermentation	+	1	× 4	4		
13	Mannitol fermentation	+	1	× 1	1		
14	Inositol fermentation	–	0	× 2	0	5	
15	Sorbitol fermentation	+	1	× 4	4		
16	Rhamnose fermentation	+	1	× 1	1		
17	Sucrose fermentation	+	1	× 2	2	7	
18	Melibiose fermentation	+	1	× 4	4		
19	Amygdalin fermentation	–	0	× 1	0		
20	Arabinose fermentation	+	1	× 2	2	2	
21	Oxidase production	–	0	× 4	0		

*As derived from API 20E (bioMérieux, Inc.) for identification of *Enterobacteriaceae*.

Figure 7-18   Example of converting a metabolic profile to an octal profile for bacterial identification.

updated taxonomic status of clinically relevant bacteria. Although microbiologists typically do not establish and maintain their own databases, an overview of the general approach provides background information.

The first step in developing a database is to accumulate many bacterial strains of the same species. Each strain is inoculated to an identical battery of metabolic tests to generate a positive-negative test profile. The cumulative results of each test are expressed as a percentage of each genus or species that possesses that characteristic. For example, suppose that 100 different known *E. coli* strains and 100 known *Shigella* spp. strains are tested in four biochemicals, yielding the results illustrated in Table 7-2. In reality, many more strains and tests would be performed. However, the principle—to generate a database for each species that contains the percentage probability for a positive result with each test in the battery—is the same.

Manufacturers develop databases for each of the identification systems they produce for diagnostic use (e.g., *Enterobacteriaceae*, gram-positive cocci, non-fermentative gram-negative bacilli). Because the data are based on organism "behavior" in a particular commercial system, the databases cannot and should

Table 7-2   Generation and Use of Genus-Identification Database Probability: Percent Positive Reactions for 100 Known Strains

	Biochemical Parameter			
Organism	Lactose	Sucrose	Indole	Ornithine
Escherichia	91	49	99	63
Shigella	1	1	38	20

not be applied to interpret profiles obtained by other testing methods.

Furthermore, most databases are established with the assumption that the isolate to be identified has been appropriately characterized using adjunctive tests. For example, if an *S. aureus* isolate is mistakenly tested using a system for identification of *Enterobacteriaceae*, the database will not identify the gram-positive coccus because the results obtained will only be compared with data available for enteric bacilli. This underscores the importance of accurately performing preliminary tests and observations, such as colony and Gram stain morphologies, before selecting a particular identification battery.

Use of the Database to Identify Unknown Isolates

Once a metabolic profile has been obtained with a bacterial isolate of unknown identity, the profile must be converted to a numeric code that will facilitate comparison of the unknown's phenotypic fingerprint with the appropriate database.

To exemplify this step in the identification process, a **binary code conversion system** that uses the numerals 0 and 1 to represent negative and positive metabolic reactions, respectively, is used as an example (although other strategies are now used). As shown in Figure 7-18, using binary code conversion, a 21-digit binomial number (e.g., 101100001001101111010, as read from top to bottom in the figure) is produced from the test result. This number is then used in an **octal code conversion** scheme to produce a mathematic number (**octal profile** [see Figure 7-18]). The octal profile number is used to generate a numerical profile distinctly related to a specific bacterial species. As shown in Figure 7-18, the octal profile for the unknown organism is 5144572. This profile would then be compared with database profiles to determine the most likely identity of the organism. In this example, the octal profile indicates the unknown organism is *E. coli*.

Confidence in Identification. Once metabolic profiles have been translated into numeric scores, the probability that a correct correlation with the database has been made must be established, that is, how confident can the laboratorian be in knowing that a correct identification has been made. This is accomplished by establishing the percentage probability, which is usually provided as part of most commercially available identification database schemes.

For example, unknown organism X is tested against the four biochemicals listed in Table 7-2 and yields results as follows: lactose (+), sucrose (+), indole (−), and ornithine (+). Based on the results of each test, the percentage of known strains in the database that produced positive results are used to calculate the percentage probability that strain X is a member of one of the two genera (*Escherichia* or *Shigella*) given in the example (Table 7-3). Therefore, if 91% of *Escherichia* spp. are lactose-positive (see Table 7-2), the probability that X is a species of *Escherichia* based on lactose alone is 0.91. If 38% of *Shigella* spp. are indole positive (see Table 7-2), then the probability that X is a species of *Shigella* based on indole alone is 0.62 (1.00 [all *Shigella*] −0.38 [percent positive *Shigella*] = 0.62 [percent of all *Shigella* that are indole negative]). The probabilities of the individual tests are then multiplied to achieve a calculated likelihood that X is one of these two genera. In this example, X is more likely to be a species of *Escherichia*, with a probability of 357:1 (1 divided by

Table 7-3 Generation and Use of Genus-Identification Database Probability: Probability That Unknown Strain X is Member of Known Genus Based on Results of Each Individual Parameter Tested

Organism	Biochemical Parameter			
	Lactose	Sucrose	Indole	Ornithine
X	+	+	−	+
Escherichia	0.91	0.49	0.01	0.63
Shigella	0.01	0.01	0.62	0.20

Probability that X is *Escherichia* = 0.91 × 0.49 × 0.01 × 0.63 = 0.002809.
Probability that X is *Shigella* = 0.01 × 0.01 × 0.62 × 0.20 = 0.000012.

0.0028; see Table 7-3). This is still a very unlikely probability for correct identification, but only four parameters were tested, and the indole result was atypical. As more parameters are added to the formula, the importance of just one test decreases and the overall pattern prevails.

With many organisms being tested for 20 or more reactions, computer-generated databases provide the probabilities. As more organisms are included in the database, the genus and species designations and probabilities become more precise. Also, with more profiles in a database, the unusual patterns can be more readily recognized and, in some cases, new or unusual species may be discovered.

The most common commercial suppliers of multi-component identification systems are driven by patent information technology and data management systems that automatically provide analysis and outcome of the metabolic process and identification.

COMMERCIAL IDENTIFICATION SYSTEMS

ADVANTAGES AND EXAMPLES OF COMMERCIAL SYSTEM DESIGNS

Commercially available identification systems have largely replaced compilations of conventional test media and substrates prepared in-house for bacterial identification. This replacement has mostly come about because the design of commercial systems has continuously evolved to maximize the speed and optimize the convenience with which all four identification components shown in Figure 7-17 can be achieved. Because laboratory workload has increased, conventional methodologies have had difficulty competing with the advantages of convenience and updated databases offered by commercial systems. Table 13-1 lists and describes the most common manual and automated bacterial identification systems available.

Some of the simplest multitest commercial systems consist of a conventional format that can be

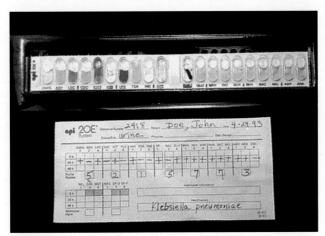

Figure 7-19   Biochemical test panel (API; bioMérieux, Inc., Hazelwood, Mo). The test results obtained with the substrates in each cupule are recorded, and an organism identification code is calculated by octal code conversion on the form provided. The octal profile obtained then is matched with an extensive database to establish organism identification.

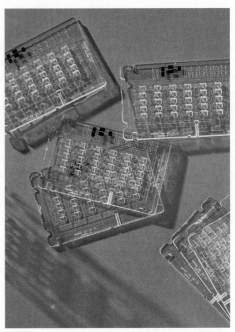

Figure 7-20   Plastic cards composed of multiple wells containing dried substrates that are reconstituted by inoculation with a bacterial suspension (bioMérieux, Inc., Hazelwood, Mo). Test results in the card wells are automatically read by the manufacturer's reading device.

inoculated once to yield more than one result. By combining reactants, for example, one substrate can be used to determine indole and nitrate results; indole and motility results; motility, indole, and ornithine decarboxylase; or other combinations. Alternatively, conventional tests have been assembled in smaller volumes and packaged so that they can be inoculated easily with one manipulation instead of several. When used in conjunction with a computer-generated database, species identifications are made relatively easily.

Another approach is to have substrates dried in plastic cupules that are arranged in series on strips into which a suspension of the test organism is placed (Figure 7-19). For some of these systems, use of a heavy inoculum or use of substrates whose utilization is not dependent on extended bacterial multiplication allows results to be available after 4 to 6 hours of incubation.

Still other identification battery formats have been designed to more fully automate several aspects of the identification process. One example of this is use of "cards" that are substantially smaller than most microtitre trays or cupule strips (Figure 7-20). Analogous to the microtitre tray format, these cards contain dried substrates in tiny wells that are resuspended on inoculation.

Commercial systems are often categorized as either being **automated** or **manual.** As shown in Table 13-1, various aspects of an identification system can be automated and these usually include, in whole or in part, the inoculation steps, the incubation and reading of tests, and the analysis of results. However, no strict criteria exist that state how many aspects

must be automated for a whole system to be classified as automated. Therefore, whether a system is considered automated can be controversial. Furthermore, regardless of the lack or level of automation, the selection of an identification system ultimately depends on system accuracy and reliability, whether the system meets the needs of the laboratory, and limitations imposed by laboratory financial resources.

Overview of Commercial Systems

Various multitest bacterial identification systems (as listed in Table 13-1) are commercially available for use in diagnostic microbiology laboratories, and the four basic identification components outlined in Figure 7-17 are common to them all. However, different systems vary in their approach to each component. The most common variations involve:

- Types and formats of tests included in the test battery
- Method of inoculation (manual or automated)
- Required length of incubation for substrate utilization. This usually depends on whether utilization requires bacterial growth.
- Method for detecting substrate utilization and whether detection is manual or automated
- Method of interpreting and analyzing results (manual or computer assisted), and if computer

assisted, the extent to which assistance is automated

The general features of some commercial identification systems are summarized in Table 13-1. More specific information is available from the manufacturers.

CHROMATOGRAPHY

Chromotagrphy refers to procedures used to separate and characterize substances based on their size, ionic charge, or solubility in particular solvents. This technique has many variations that can be applied for specific purposes.

PRINCIPLES

Chromatography involves two phases: the **mobile phase** and the **stationary phase.** The mobile phase contains and carries the sample to be analyzed through or across the stationary phase. The stationary phase maintains conditions necessary for separating various substances (i.e., the **analytes**) within the sample being studied. The mobile phase may be gas or liquid. The stationary phase may also be liquid or solid and is usually housed within a column of some design. Specific chromatographic methods are named according to the phase used. For example, **gas-liquid chromatography (GLC)** refers to any chromatographic method that uses gas as the mobile phase and liquid as the stationary phase.

Regardless of the phases used, chromatographic principles are generally the same. A sample is mixed by injection with the mobile phase, which carries the sample through a column containing the stationary phase. Depending on the nature of the solid phase, different analytes within the sample will be retained within the column based on size, ionic, charge, or analyte solubility in the mobile phase. As the mobile phase continues to move through the column, analytes with different characteristics will have different **retention times** and thus will elute (come out) from the column at different times. Eluted analytes pass through a detector that generates a signal. The signal is then translated into an electronic signal that is recorded and used to produce a chromatogram.

The **chromatogram** is the plot of elution peaks produced by each analyte in the sample. With the inclusion of reliable controls and standards, retention times are used to definitively identify specific analytes. The size of each peak can be used to determine the amount of analyte present in each sample.

APPLICATIONS

Many substances, including proteins, carbohydrates, organic acids, fatty acids, and mycolic acids, may be detected by chromatography. Therefore, chromatographic methods can provide analytic, phenotypic fingerprints (i.e., chromatograms) regarding the types and amounts of these substances within a bacterial cell. Comparison of chromatograms obtained with organisms of unknown identity with those of known identity is used to identify an organism from a clinical specimen. The chromatographic approach has greatest application when conventional phenotypic-based methods for identification are poor or nonexistent.

ADDITIONAL READING

Atlas RM, Parks LC, editors: *Handbook of microbiological media,* Boca Raton, Fla, 1993, CRC Press.

Clinical Laboratory Standards and Institute (National Committee for Clinical Laboratory Standards): *Abbreviated identification of bacteria and yeast;* Approved Guidelines M35-A, Wayne, PA, 2002, NCCLS.

Murray PR, Baron EJ, Jorgensen JH, et al, editors: *Manual of Clinical Microbiology,* ed 8, Washington, DC, 2003, ASM Press.

O'Hara CM, Weinstein MP, Miller JM: Manual and automated systems for detection and identification of microorganisms. In Murray PR, Baron EJ, Jorgensen JH, et al, editors: *Manual of clinical microbiology,* ed 8, Washington, DC, 2003, ASM Press.

York, MK, Traylor MM, Hardy J, et al: Biochemical tests for the identification of aerobic bacteria. In Isenberg HD, editor: *Clinical microbiology procedures handbook,* Washington, DC, 2004, ASM Press.

CHAPTER 8

Nucleic Acid–Based Analytic Methods for Microbial Identification and Characterization

The principles of bacterial cultivation and identification discussed in Chapter 7 focus on phenotypic methods. These methods analyze readily observable bacterial traits and "behavior." Although these strategies are the mainstay of diagnostic bacteriology, notable limitations are associated with the use of phenotypic methods. These limitations are as follows:

- Inability to grow certain fastidious pathogens
- Inability to maintain viability of certain pathogens in specimens during transport to laboratory
- Extensive delay in cultivation and identification of slowly growing pathogens
- Lack of reliable methods to identify certain organisms grown in vitro
- Use of considerable time and resources in establishing the presence and identity of pathogens in specimens

The explosion in molecular biology over the past 20 years has provided alternatives to phenotype-based strategies used in clinical microbiology. These alternatives have the potential to avert some of the aforementioned limitations. The detection and manipulation of nucleic acids (DNA and RNA) allows microbial genes to be examined directly (i.e., **genotypic methods**) rather than by analysis of their products such as enzymes (i.e., **phenotypic methods**). Additionally, nonnucleic acid–based analytic methods that detect phenotypic traits not detectable by conventional strategies (e.g., cell wall components) have been developed to enhance bacterial detection, identification, and characterization. For the laboratory diagnosis of infectious diseases to remain timely and effective, strategies that integrate conventional, nucleic acid–based, and analytic techniques must continue to evolve.

Several methods that analyze microbial DNA or RNA can detect, identify, and characterize infectious etiologies. Although technical aspects may differ, all molecular procedures involve the direct manipulation and analysis of genes, in whole or in part, rather than the analysis of gene products. Furthermore, because nucleic acids are common to all living entities, most methods are adaptable for the diagnosis of viral, fungal, parasitic, or bacterial infections. This chapter discusses the general principles and applications of molecular diagnostics.

OVERVIEW OF MOLECULAR METHODS

Because molecular diagnostic tests are based on the consistent and somewhat predictable nature of DNA and RNA, understanding these methods requires a basic understanding of nucleic acid composition and structure. Therefore, a review of the section titled, "Nucleic Acid Structure and Organization" in Chapter 2 may be helpful.

The molecular methods to be discussed are classified into one of three categories: hybridization, amplification, and sequencing and enzymatic digestion of nucleic acids.

NUCLEIC ACID HYBRIDIZATION METHODS

Hybridization methods are based on the ability of two nucleic acid strands that have complementary base sequences (i.e., are **homologous**) to specifically bond with each other and form a double-stranded molecule, or **duplex** or **hybrid.** This duplex formation is driven by the consistent manner in which the base adenine always bonds to thymine, while the bases guanine and cytosine always form a bonding pair (see Figure 2-2). Because hybridization requires nucleic acid sequence homology, a positive hybridization reaction between two nucleic acid strands, each from a different source, indicates genetic relatedness between the two organisms that donated each of the nucleic acid strands for the hybridization reaction.

Hybridization assays require that one nucleic acid strand (the **probe**) originate from an organism of known identity and the other strand (the **target**) originate from an unknown organism to be detected or identified (Figure 8-1). Positive hybridization identifies the unknown organism as being the same as the probe-source organism. With a negative hybridization test, the organism remains undetected or unidentified. The single-stranded nucleic acid components used in hybridization may be either RNA or DNA so that DNA-DNA, DNA-RNA, and even RNA-RNA duplexes may form,

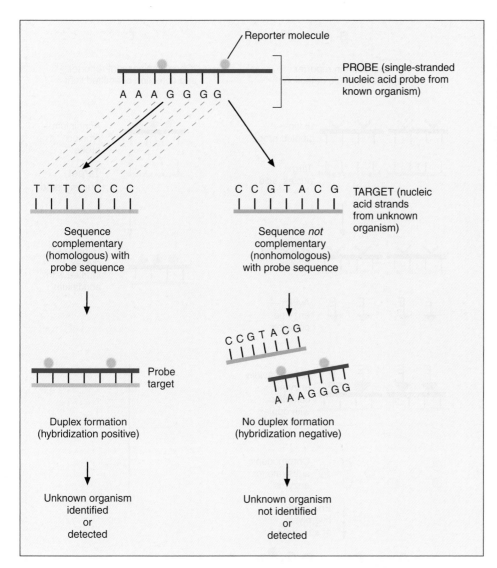

Figure 8-1   Principles of nucleic acid hybridization. Identification of unknown organism is established by positive hybridization (i.e., duplex formation) between a nucleic acid strand from the known organism (i.e., the probe), and a target nucleic acid strand from the organism to be identified. Failure to hybridize indicates lack of homology between probe and target nucleic acid.

depending on the specific design of the hybridization assay.

Hybridization Steps and Components

The basic steps in a hybridization assay include:

1. Production and labeling of single-stranded probe nucleic acid
2. Preparation of single-stranded target nucleic acid
3. Mixture and hybridization of target and probe nucleic acid
4. Detection of hybridization

Production and Labeling of Probe Nucleic Acid   In keeping with the requirement of complementation for hybridization, the probe design (i.e., probe length and its sequence of nucleic acid bases) depends on the sequence of the intended target nucleic acid. Therefore, selection and design of a probe depends on the

intended use. For example, if a probe is to be used for recognizing only gram-positive bacteria, the probe's nucleic acid sequence needs to be specifically complementary to a nucleic acid sequence common only to gram-positive bacteria and not to gram-negative bacteria. Even more specific probes can be designed to identify a particular bacterial genus or species, a virulence, or an antibiotic resistance gene that may only be present in certain strains within a given species.

In the past, probes were produced through a labor-intensive process that involved recombinant DNA and cloning techniques with the piece of nucleic acid of interest. More recently, probes are chemically synthesized using instrumentation, a service that is widely available commercially. This greatly facilitates probe development because the user need only supply the manufacturer with the nucleotide base sequence of the desired probe. The base sequence of potential

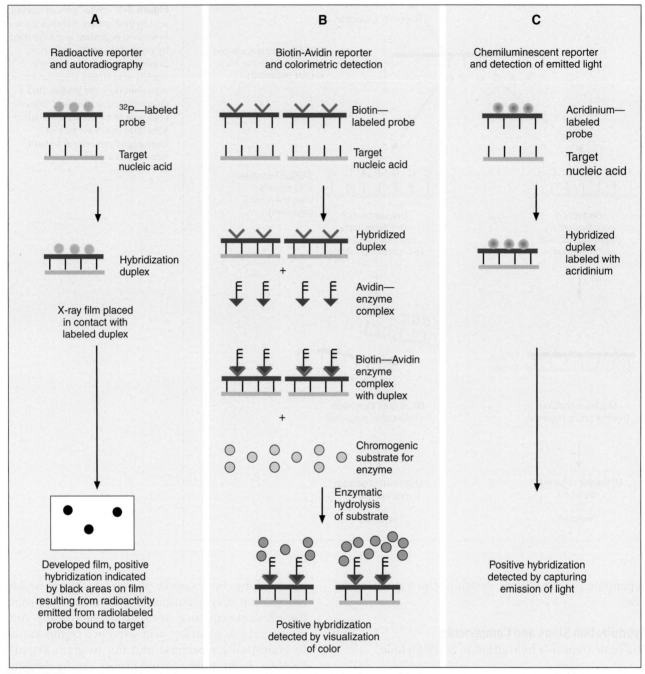

Figure 8-2 Reporter molecule labeling of nucleic acid probes and principles of hybridization detection. Use of probes labeled with a radioactive reporter, with hybridization detected by autoradiography (**A**); probes labeled with biotin-avidin reporter, with hybridization detected by a colorimetric assay (**B**); probes labeled with chemiluminescent reporter (i.e., acridinium), with hybridization detected by a luminometer to detect emitted light (**C**).

target genes or gene fragments for probe design also is easily accessed using computer on-line services for gene sequence information (e.g., GENBANK, National Center for Biological Information). In short, the design and production of nucleic acid probes is now relatively easy. Although probes may be hundreds to thousands of bases long, oligonucleotide (i.e., 20 to 50 bases long) probes are usually sufficient for detection of most clinically relevant targets.

All hybridization tests must have a means to detect hybridization. This is accomplished with the use of a **"reporter"** molecule that chemically forms a complex with the single-stranded probe DNA. Probes may be labeled with a variety of such molecules, but most commonly radioactive (e.g., ^{32}P, ^{125}I, or ^{35}S), biotin-avidin, digoxigenin, or chemiluminescent labels are used (Figure 8-2).

With the use of radioactively labeled probes, hybridization is detected by the emission of radioactivity from the probe-target mixture (see Figure 8-2, *A*). Although this is a highly sensitive method for detecting hybridization, the difficulties of working with radioactivity in a clinical microbiology laboratory have limited the use of this type of label in the diagnostic setting.

Biotinylation is a nonradioactive alternative for labeling probes and involves the chemical incorporation of biotin into probe DNA. Biotin-labeled probe-target nucleic acid duplexes are detected using avidin, a biotin-binding protein that has been conjugated with an enzyme such as horseradish peroxidase. When a chromogenic substrate is added, the peroxidase produces a colored product that can be detected visually or spectrophotometrically (see Figure 8-2, *B*)

Other nonradioactive labels are based on principles similar to biotinylation. For example, with digoxigenin-labeled probes, hybridization is detected using anti-digoxigenin antibodies that have been conjugated with an enzyme. Successful duplex formation means the enzyme is present so that with the addition of a chromogenic substrate color production is interpreted as positive hybridization. Alternatively, the antibody may be conjugated with fluorescent dyes that can be directly detected without the need for an enzymatic reaction to produce a colored or fluorescent end product.

Chemiluminescent reporter molecules can be directly chemically linked to the nucleic acid probe without using a conjugated antibody. These molecules (e.g., acridinium) emit light so hybridization between a chemiluminescent-labeled probe and target nucleic acid can be detected using a luminometer (see Figure 8-2, *C*). The chemiluminescent approach is used in one commercially available hybridization system (Gen-Probe, San Diego, Calif).

Preparation of Target Nucleic Acid. Because hybridization is driven by complementary binding of homologous nucleic acid sequences between probe and target, target nucleic acid must be single-stranded and its base sequence integrity maintained. Failure to meet these requirements will result in negative hybridization reactions that are due to factors other than the absence of microbial target nucleic acid (i.e., false-negative results).

Because the relatively rigorous procedures for releasing nucleic acid from the target microorganism can be deleterious to the molecule's structure, obtaining target nucleic acid and maintaining its appropriate conformation and sequence can be difficult. The steps in

target preparation vary depending on the organism source of the nucleic acid and the nature of the environment from which the target organism is being prepared (i.e., laboratory culture media; fresh clinical material, such as fluid, tissue, and stool; and fixed or preserved clinical material). Generally, target preparation steps involve enzymatic and/or chemical destruction of the microbial envelope to release target nucleic acid, stabilization of target nucleic acid to preserve structural integrity, and, if the target is DNA, denaturation to a single strand, which is necessary for binding to complementary probe nucleic acid.

Mixture and Hybridization of Target and Probe. Designs for mixing target and probe nucleic acids are discussed later, but some general concepts regarding the hybridization reaction require consideration.

The ability of the probe to bind the correct target depends on the extent of base sequence homology between the two nucleic acid strands and the environment in which probe and target are brought together. Environmental conditions set the **stringency** for a hybridization reaction, and the degree of stringency can determine the outcome of the reaction. Hybridization stringency is most affected by:

- Salt concentration in the hybridization buffer (stringency increases as salt concentration decreases)
- Temperature (stringency increases as temperature increases)
- Concentration of destabilizing agents (stringency increases with increasing concentrations of formamide or urea)

With greater stringency, a higher degree of base-pair complementarity is required between probe and target to obtain successful hybridization (i.e., less tolerance for deviations in base sequence). Under less stringent conditions, strands with less base-pair complementarity (i.e., strands having a higher number of mismatched base pairs within the sequence) may still hybridize. Therefore, as stringency increases, the specificity of hybridization increases and as stringency decreases, specificity decreases. For example, under high stringency a probe specific for a target sequence in *Streptococcus pneumoniae* may only bind to target prepared from this species (high specificity), but under low stringency the same probe may bind to targets from various streptococcal species (lower specificity). Therefore, to ensure accuracy in hybridization, reaction conditions must be carefully controlled.

Detection of Hybridization. The method of detecting hybridization depends on the reporter molecule used for labeling probe nucleic acid and on the hybridization format (see Figure 8-2). Detection of hybridization using radioactively labeled probes is done by exposing the reaction mixture to radiograph film (i.e., **autoradiography**). Hybridization with non-radioactively labeled probes is detected using colorimetry, fluorescence, or chemiluminescence, and detection can be somewhat automated using spectrophotometers, fluorometers, or luminometers, respectively. The more commonly used nonradioactive detection systems (e.g., digoxigenin, chemiluminescence) are able to detect approximately 10^4 target nucleic acid sequences per hybridization reaction.

Hybridization Formats

Hybridization reactions can be done using either a solution format or solid support format.

Solution Format. In the solution format, probe and target nucleotide strands are put together in a liquid reaction mixture that facilitates duplex formation, so hybridization occurs substantially faster than with the use of a solid support format. However, before detection of hybridization can be accomplished some method must be used to separate hybridized, labeled probe from nonhybridized, labeled probe (i.e., "background noise"). Separation methods include enzymatic (e.g., S1 nuclease) digestion of single-stranded probe and precipitation of hybridized duplexes, hydroxyapatite or charged magnetic microparticles that preferentially bind duplexes, or chemical destruction of the reporter molecule (e.g., acridinium dye) that is attached to unhybridized probe nucleic acid. After the duplexes have been "purified" from the reaction mixture and the background noise minimized, hybridization detection can proceed by the method appropriate for the type of reporter molecule used to label the probe (Figure 8-3).

Solid Support Format. Either probe or target nucleic acids can form a complex to a solid support and still be capable of forming duplexes with complementary strands. Various solid support materials and common solid formats exist, including filter hybridizations, southern hybridizations, sandwich hybridizations, and in situ hybridizations.

Filter hybridizations have several variations. By one approach, the target sample, which can be previously purified DNA, the microorganism containing the target DNA, or the clinical specimen that contains the microorganism of interest, is affixed to a membrane (e.g., nitrocellulose or nylon fiber filters). To be able to identify specimens, samples are usually oriented on the membrane using a template or grid. The membrane is then processed to release target DNA from the microorganism and denature it to a single strand. A solution containing labeled probe nucleic acid is used

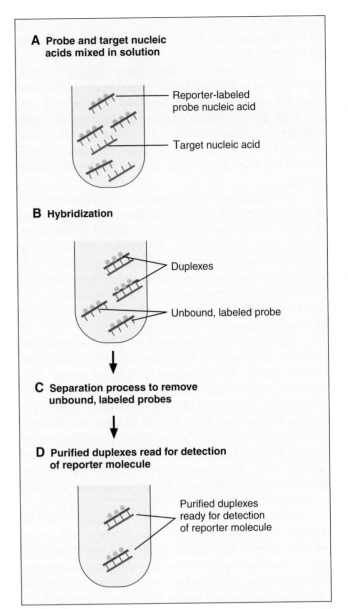

Figure 8-3 Principle of the solution hybridization format.

to "flood" the membrane and allow for hybridization to occur. After a series of incubations and washings to remove unbound probe, the membrane is processed for detection of duplexes (Figure 8-4, *A*). An advantage of this method is that a single membrane can hold several samples for exposure to the same probe.

Southern hybridization is another method that uses membranes as the solid support, but in this instance the nucleic acid target has been purified from the organisms and digested with specific enzymes to produce several fragments of various sizes (see Figure 8-4, *B;* also see the section titled, "Enzymatic Digestion and Electrophoresis of Nucleic Acids" later in this chapter). By **gel electrophoresis,** the nucleic acid fragments, which carry a net negative charge, are

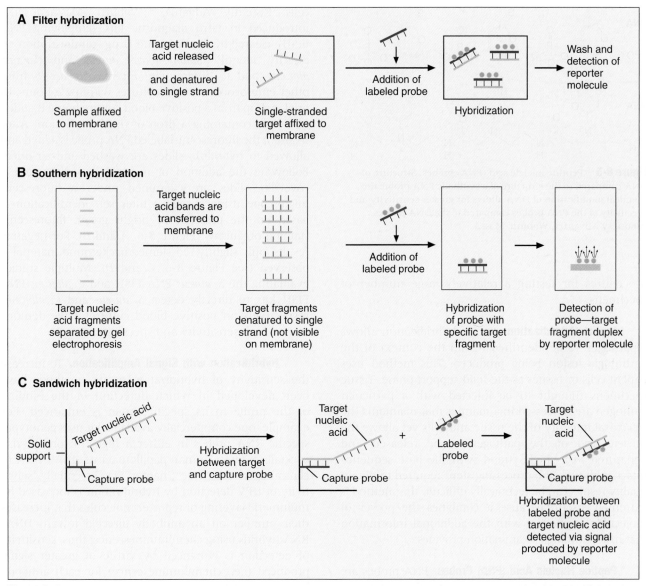

Figure 8-4 Principle of solid support hybridization formats. **A,** Filter hybridization. **B,** Southern hybridization. **C,** Sandwich hybridization.

subjected to an electrical field that forces them to migrate through an agarose matrix. Because fragments of different sizes migrate through the porous agarose at different rates, they can be separated over the time they are exposed to the electrical field. When electrophoresis is complete, the nucleic acid fragments are stained with the fluorescent dye **ethidium bromide** so that fragment "banding patterns" may be visualized on exposure of the gel to ultraviolet (UV) light. For Southern hybridization, the target nucleic acid bands are transferred to a membrane that is then exposed to probe nucleic acid. This method allows the determination of which specific target nucleic acid fragment is carrying the base sequence of interest, but the labor intensity of the

procedure precludes its common use in most diagnostic settings.

With **sandwich hybridizations** two probes are used. One probe is attached to the solid support, is not labeled, and via hybridization "captures" the target nucleic acid from the sample to be tested. The presence of this duplex is then detected using a labeled second probe that is specific for another portion of the target sequence (see Figure 8-4, *C*). Sandwiching the target between two probes decreases nonspecific reactions but requires a greater number of processing and washing steps. For such formats, plastic microtiter wells coated with probes have replaced filters as the solid support material, thereby facilitating these multiple-step

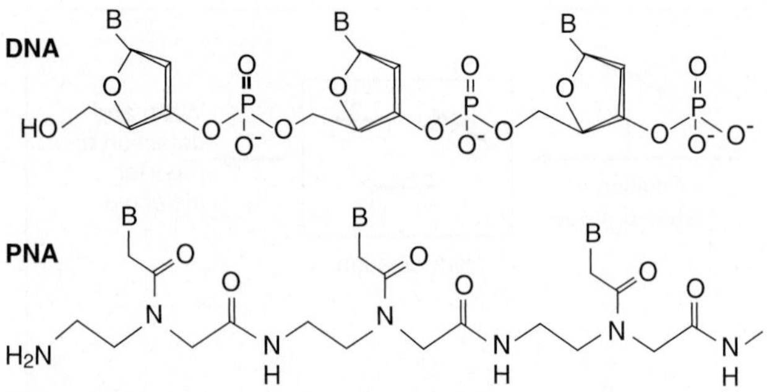

Figure 8-5 Peptide nucleic acid (PNA) probes. Structure of DNA compared to the structure of a synthetic PNA probe; the chemical modification of DNA allows for greater sensitivity and specificity of the PNA probes compared to the DNA probes. (Courtesy AdvanDx, Woburn, Mass.)

procedures for testing a relatively large number of specimens.

In Situ Hybridization. In situ hybridization allows a pathogen to be identified within the context of the pathologic lesion being produced. This method uses patient cells or tissues as the solid support phase. Tissue specimens thought to be infected with a particular pathogen are processed in a manner that maintains the structural integrity of the tissue and cells yet allows the nucleic acid of the pathogen to be released and denatured to a single strand with the base sequence intact. Although the processing steps required to obtain quality results can be technically difficult, this method is extremely useful because it combines the power of molecular diagnostics with the additional information that histopathologic examination provides.

Peptide Nucleic Acid (PNA) Probes. PNA probes are synthetic pieces of DNA that possess unique chemical characteristics in which the negatively charged sugar-phosphate backbone of DNA is replaced by a neutral polyamide backbone of repetitive units (Figure 8-5). Individual nucleotide bases can be attached to this neutral backbone, which then allows the PNA probe to hybridize to complementary nucleic acid targets. Because of the synthetic structure of the backbone, these probes have improved hybridization characteristics, providing faster and more specific results than traditional DNA probes. In addition, these probes are not degraded by ubiquitous enzymes such as nucleases and proteases, thereby providing longer shelf-life in diagnostic applications. PNA FISH is a novel fluorescence in situ hybridization (FISH) technique using PNA probes targeting species-specific rRNA sequences. Upon penetration of the microbial cell wall, the fluorescent-labeled PNA probes hybridize to multicopy rRNA sequences within the microorganisms, resulting in fluorescent

cells. Recently, AdvanDx (Woburn, Massachusetts) introduced in vitro diagnostic kits (employing PNA FISH), cleared by the Food and Drug Administration, to directly identify *Staphylococcus aureus* and *Candida albicans*, and to differentiate *Enterococcus faecalis* from other enterococci in blood cultures within 2½ hours. In brief, a drop from a positive blood culture bottle is added to a slide containing a drop of fixative solution. After fixation, the fluorescent-labeled PNA probe is added and allowed to hybridize; slides are washed and air-dried. Following the addition of mounting medium and a coverslip, slides are examined under a fluorescent microscope using a special filter set. Identification is based on the presence of bright green, fluorescent-staining organisms (Figure 8-6, *A* and *B*). For negative results, only slightly red-stained background material is observed (see Figure 8-6, *C* and *D*). Multiple studies evaluating the *S. aureus* PNA FISH and *C. albicans* PNA FISH kits to directly detect *S. aureus* and *C. albicans*, respectively, in positive blood cultures have demonstrated high sensitivity and specificity.

Hybridization with Signal Amplification. To increase the sensitivity of hybridization assays, methods have been developed in which detection of the binding of the probe to its specific target is enhanced. For example, one commercially available kit uses genotype-specific RNA probes in either a high-risk or low-risk cocktail to detect human papillomavirus (HPV) DNA in clinical specimens (see Chapter 51). Essentially, sensitivity of HPV detection by hybridization is increased by multimeric layering of reporter molecules that increases their number on an antibody directed toward DNA-RNA hybrids using chemiluminescence; thus, sensitivity of detection is enhanced by virtue of greater signal produced (i.e., chemiluminescence) for each antibody bound to target.

AMPLIFICATION METHODS—PCR-BASED

Although hybridization methods are highly specific for organism detection and identification, they are limited by their sensitivity, that is, without sufficient target nucleic acid in the reaction, false-negative results occur. Therefore, many hybridization methods require "amplifying" target nucleic acid by growing target organisms to greater numbers in culture. The requirement for cultivation detracts from the potential speed advantage that molecular methods can offer. Therefore, the development of molecular amplification techniques not dependent on organism multiplication has contributed greatly to circumvent the speed problem while enhancing sensitivity and maintaining specificity. For purposes of discussion, amplification methods are divided into two major categories: those

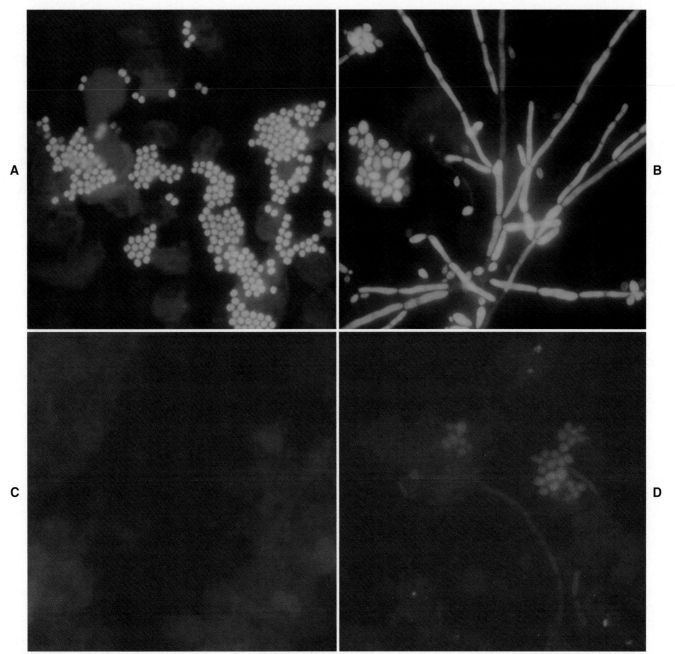

Figure 8-6   Using a fluorescent-tagged peptide nucleic acid (PNA) probe in conjunction with fluorescent in situ hybridization (FISH), *Staphylococcus aureus* (**A**) or *Candida albicans* (**B**) was directly identified in blood cultures within $2^{1}/_{2}$ hours. A drop from the positive blood culture bottle is added to a slide containing a drop of fixative solution, which keeps the cells intact. After fixation, the appropriate fluorescent-labeled PNA probe is added. The PNA probe penetrates the microbial cell wall and hybridizes to the rRNA. Slides are examined under a fluorescent microscope. If the specific target is present, bright green, fluorescent-staining organisms will be present. Blood cultures negative for either *S. aureus* (**C**) or *C. albicans* (**D**) by PNA FISH technology are shown. (Courtesy AdvanDx, Woburn, Mass.)

methods that use polymerase chain reaction (PCR) technology and those assays that are not PCR-based.

Overview of PCR and Derivations

The most widely used target nucleic acid amplification method is the **polymerase chain reaction (PCR).** This method combines the principles of complementary nucleic acid hybridization with those of nucleic acid replication that are applied repeatedly through numerous cycles. By this method, a single copy of a nucleic acid target, often undetectable by standard hybridization methods, is multiplied to 10^7 or more copies within a relatively short period. This provides ample target that can be readily detected by numerous methods.

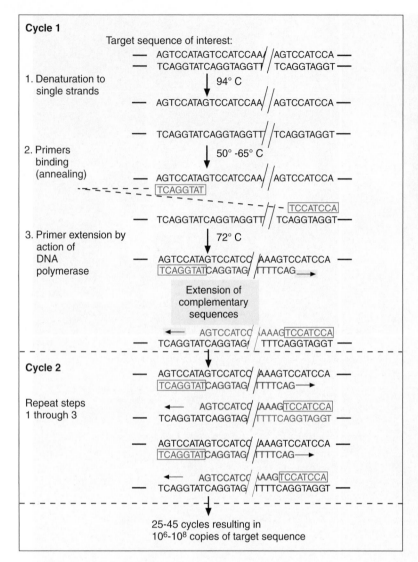

Figure 8-7 Overview of polymerase chain reaction. The target sequence is denatured to single strands, primers specific for each target strand sequence are added, and DNA polymerase catalyzes the addition of deoxynucleotides to extend and produce new strands complementary to each of the target sequence strands (*cycle 1*). The second cycle begins by both double-stranded products of cycle 1 being denatured and subsequently serving as targets for more primer annealing and extension by DNA polymerase. After 25 to 30 cycles, at least 10^7 copies of target DNA may be produced. (Modified from Ryan KJ, Champoux JJ, Drew WL, et al: *Medical microbiology: an introduction to infectious diseases,* Norwalk, Conn, 1994, Appleton & Lange.)

Conventional PCR involves 25 to 50 repetitive cycles, with each cycle comprising three sequential reactions: denaturation of target nucleic acid, primer annealing to single-stranded target nucleic acid, and extension of primer-target duplex.

Extraction and Denaturation of Target Nucleic Acid.

For PCR, nucleic acid is first **extracted** (released) from the organism or a clinical sample that potentially contains the target organism by heat, chemical, or enzymatic methods. Numerous manual methods are available to accomplish this task including a variety of commercially available kits that will extract either RNA or DNA, depending on the specific target of interest. In addition, other commercially available kits are designed to extract nucleic acids from specific types of clinical specimens such as blood or tissues. Most recently, automated instruments have been introduced for the extraction of nucleic acid

from various sources (e.g., bacteria, viruses, tissue, blood).

Once extracted, target nucleic acid is then added to the reaction mix that contains all the necessary components for PCR to occur (primers, covalent ions, buffer, and enzyme) and then placed into a thermal cycler to undergo amplification (Figure 8-7). For PCR to begin, target nucleic acid must first be in the single-stranded conformation for the second reaction, primer annealing, to occur. Denaturation to a single strand, which is not necessary for RNA targets, is accomplished by heating to 94° C (see Figure 8-7). Of note, for many PCR procedures, especially those involving commonly encountered bacterial pathogens, disruption of the organism to release DNA is done in one step by heating the sample to 94° C.

Primer Annealing.

Primers are short, single-stranded sequences of nucleic acid (i.e., oligonucleotides usually 20 to 30 nucleotides long) that are selected to specifically hybridize **(anneal)** to a particular nucleic acid target, essentially functioning like probes. As noted for hybridization tests, the abundance of available gene sequence data allows for the design of primers specific for a number of microbial pathogens and their virulence or antibiotic resistance genes. Thus, primer nucleotide sequence design depends on the intended target, such as genus-specific genes, species-specific genes, virulence genes, or antibiotic-resistance genes.

Primers are designed to be used in pairs that flank the target sequence of interest (see Figure 8-7). When the primer pair is mixed with the denatured target DNA, one primer anneals to a specific site at one end of the target sequence of one target strand while the other primer anneals to a specific site at the opposite end of the other, complementary target strand. Usually, primers are designed so that the distance between them on the target DNA is 50 to 1000 base pairs. The annealing process is conducted at 50° to 58° C, or higher. Once the duplexes are formed the last step in the cycle, which mimics the DNA replication process, begins.

Extension of Primer-Target Duplex. Annealing of primers to target sequences provides the necessary template format that allows DNA polymerase to add nucleotides to the 3′ **terminus** (end) of each primer and produce by extension a sequence complementary to the target template (see Figure 8-7). *Taq* polymerase is the enzyme commonly used for primer extension, which occurs at 72° C. This enzyme is used because of its ability to function efficiently at elevated temperatures and withstand the denaturing temperature of 94° C through several cycles. The ability to allow primer annealing and extension to occur at elevated temperatures without detriment to the polymerase increases the stringency of the reaction, thus decreasing the chance for amplification of nontarget nucleic acid (i.e., nonspecific amplification).

All three reaction components of PCR occur in the same tube that contains a mixture of target nucleic acid, primers, components to optimize polymerase activity (i.e., buffer, cation [$MgCl_2$], salt), and deoxynucleotides. To maintain continuous reaction cycles programmable **thermal cyclers** are used. These cyclers hold the reaction vessel and carry the PCR mixture through each reaction step at the precise temperature and for the optimal duration.

As shown in Figure 8-7, for each target sequence originally present in the PCR mixture, two double-stranded fragments containing the target sequence are produced after one cycle. At the beginning of the second cycle of PCR, denaturation then produces four templates to which the primers will anneal. Following extension at the end of the second cycle, there will be four double-

stranded fragments containing target nucleic acid. Therefore, with completion of each cycle there is a doubling of target nucleic acid and after 30 to 40 cycles 10^7 to 10^8 target copies will be present in the reaction mixture.

Although it is possible to detect one copy of a pathogen's gene in a sample or patient specimen by PCR technology, detection is dependent on the ability of the primers to locate and anneal to the single target copy and on optimum PCR conditions. Nonetheless, PCR has proved to be a powerful amplification tool to enhance the sensitivity of molecular diagnostic techniques.

Detection of PCR Products. The specific PCR amplification product containing the target nucleic acid of interest is referred to as the **amplicon.** Because PCR produces an amplicon in substantial quantities, any of the basic methods previously described for detecting hybridization can be adopted for detecting specific amplicons. Basically this involves using a labeled probe that is specific for the target sequence within the amplicon. Therefore, solution or solid-phase formats may be used with reporter molecules that generate radioactive, colorimetric, fluorometric, or chemiluminescent signals. Probe-based detection of amplicons serves two purposes: it allows visualization of the PCR product and it provides specificity by ensuring that the amplicon is the target sequence of interest and not the result of nonspecific amplification.

When the reliability of PCR for a particular amplicon has been well established, hybridization-based detection may not always be necessary. Confirming the presence of an amplicon of the expected size may be sufficient. This is commonly accomplished by subjecting a portion of the PCR mixture, after amplification, to gel electrophoresis. After electrophoresis, the gel is stained with ethidium bromide to visualize the amplicon and, using molecular-weight–size markers, the presence of amplicons of appropriate size (i.e., the size of target sequence amplified depends on the primers selected for PCR) is confirmed (Figure 8-8).

Derivations of the PCR Method. The powerful amplification capacity of PCR has prompted the development of several modifications that enhance the utility of this methodology, particularly in the diagnostic setting. Specific examples include multiplex PCR, nested PCR, quantitative PCR, RT-PCR, arbitrary primed PCR, and PCR for nucleotide sequencing.

Multiplex PCR is a method by which more than one primer pair is included in the PCR mixture. This approach offers a couple of notable advantages. First, strategies that include internal controls for PCR can be developed. For example, one primer pair can be directed at sequences present in all clinically relevant

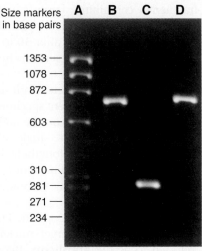

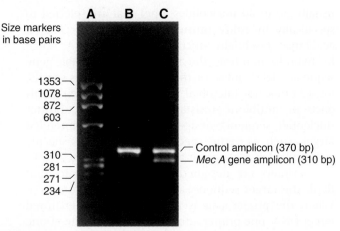

Figure 8-8 Use of ethidium bromide–stained agarose gels to determine the size of PCR amplicons for identification. Lane A shows molecular-size markers, with the marker sizes indicated in base pairs. Lanes B, C, and D contain PCR amplicons typical of the enterococcal vancomycin-resistance genes *vanA* (783 kb), *vanB* (297 kb), and *vanC1* (822 kb), respectively.

Figure 8-9 Ethidium bromide–stained gels containing amplicons produced by multiplex PCR. Lane A shows molecular-size markers, with the marker sizes indicated in base pairs. Lanes B and C show amplicons obtained with multiplex PCR consisting of control primers and primers specific for the staphylococcal methicillin-resistance gene *mecA*. The presence of only the control amplicon (370 bp) in Lane B indicates that PCR was successful, but the strain on which the reaction was performed did not contain *mecA*. Lane C shows both the control and the *mecA* (310 bp) amplicons, indicating that the reaction was successful and that the strain tested carries the *mecA* resistance gene.

bacteria (i.e., the control or universal primers) and the second primer pair can be directed at a sequence specific for the particular gene of interest (i.e., the test primers). The control amplicon should always be detectable after PCR, and its absence would indicate that PCR conditions were not met and the test would require repeating. When the control amplicon is detected, the absence of the test amplicon can be more confidently interpreted to indicate the absence of target nucleic acid in the specimen rather than a failure of the PCR system (Figure 8-9).

Another advantage of multiplex PCR is the ability to search for different targets using one reaction. Primer pairs directed at sequences specific for different organisms or genes can be put together so that the use of multiple reaction vessels can be minimized. For example, multiplexed PCR assays containing primers to detect viral agents that cause meningitis or encephalitis (e.g., herpes simplex virus, enterovirus, West Nile virus) have been used in a single reaction tube. A limitation of multiplex PCR is that mixing different primers can cause some interference in the amplification process so that optimizing conditions can be difficult, especially as the number of different primer pairs used increases.

Nested PCR involves the sequential use of two primer sets. The first set is used to amplify a target sequence. The amplicon obtained is then used as the target sequence for a second amplification using primers internal to those of the first amplicon. The advantage of this approach is extreme sensitivity and confirmed specificity without the need for using probes. Because production of the second amplicon requires the presence of the first amplicon, production of the second amplicon automatically verifies the accuracy of the first amplicon. The problem encountered with nested PCR is that the procedure requires open manipulations of amplified DNA that is readily, albeit inadvertently, aerosolized and capable of contaminating other reaction vials.

Arbitrary primed PCR uses short primers that are not specifically complementary to a particular sequence of a target DNA. Although they are not specifically directed, their short sequence (approximately 10 nucleotides) ensures that they will randomly anneal to multiple sites in a chromosomal sequence. On cycling, the multiple annealing sites result in the amplification of multiple fragments of different sizes. Theoretically, strains that have similar nucleotide sequences will have similar annealing sites and thus will produce amplified fragments (i.e., amplicons) of similar sizes. Therefore, by comparing fragment migration patterns following agarose gel electrophoresis, strains or isolates can be judged to be the same, similar, or unrelated.

Quantitative PCR is an approach that combines the power of PCR for the detection and identification of infectious agents with the ability to quantitate the actual number of targets originally in the clinical specimen. The ability to quantitate "infectious burden" has tremendous implications for studying and understanding the disease state (e.g., acquired immunodeficiency syndrome [AIDS]), the prognosis of certain infections, and the effectiveness of antimicrobial therapy.

The PCR methods discussed thus far have focused on amplification of a DNA target. **Reverse transcription PCR (RT-PCR)** amplifies an RNA target. Because many clinically important viruses have genomes composed of RNA rather than DNA (e.g., human immunodeficiency virus [HIV], hepatitis B virus), the ability to amplify RNA greatly facilitates laboratory-based diagnostic testing for these infectious agents. The unique step to this procedure is the use of the enzyme **reverse transcriptase** that directs synthesis of DNA from the viral RNA template, usually within 30 minutes. Once the DNA has been produced, relatively routine PCR technology is applied to obtain amplification.

Real-Time PCR

Most conventional PCR-based tests used in clinical laboratories were developed in-house and needed to be performed in dedicated spaces to control or reduce cross contamination that was always a threat for producing false-positive results. Conventional PCR assays also require multiple manipulations including initial amplification of target nucleic acid, detection of amplified product by gel electrophoresis, and then confirmation by an alternative method such as Southern blotting or chemiluminescence techniques. In general, a conventional PCR assay would require a minimum of at least 4 to 6 hours from the time that extracted nucleic acid is placed into a thermal cycler to begin amplification to subsequent product detection.

Recently, small, automated instruments that combine target nucleic acid amplification with qualitative or quantitative measurement of amplified product have become commercially available. These instruments are noteworthy for four reasons: First, these new instruments combine thermocycling or target DNA amplification with the ability to detect amplified target by fluorescently labeled probes as the hybrids are formed (i.e., detection of amplicon in real time). Second, because both amplification and product detection can be accomplished in one reaction vessel without ever opening, the major concern of cross contamination of samples with amplified product associated with conventional PCR assays is greatly lessened. Third, these instruments are not only able to measure amplified product **(amplicon)** as it is made, but because of this capability, they are also able to quantitate the amount of product and thereby determine the number of copies of target in the original specimen. And fourth, the amount of time to complete a real-time PCR assay is significantly decreased compared to conventional PCR-based assays because the time required for the post-PCR detection of

amplified product is eliminated by the use of fluorescent probes. Also, some systems are able to perform rapid thermal cycling based on instrument design, detecting product in as little as 20 to 30 minutes.

Several instruments (also referred to as *platforms*) are available for amplification in conjunction with real-time detection of PCR-amplified products; examples are shown in Figure 8-10. Each instrument has unique features that permit some flexibility such that a clinical laboratory can fulfill its specific needs in terms of specimen capacity, number of targets simultaneously detected, detection format, and time for analysis (Table 8-1). Nevertheless, all instruments have amplification (i.e., thermal cycling) capability as well as an excitation or light source, an emission detection source, and a computer interface to selectively monitor the formation of amplified product.

As with conventional PCR, nucleic acid must first be extracted from the clinical specimen before real-time amplification. In principle, real-time amplification is accomplished in the same manner as previously described for conventional PCR-based assays in which denaturation of double-stranded nucleic acid followed by primer annealing and extension (elongation) are performed in one cycle. However, it is the detection process that discriminates real-time PCR from conventional PCR assays. In real-time PCR assays, accumulation of amplicon is monitored as it is generated. Monitoring of amplified target is made possible by the labeling of primers, oligonucleotide probes (oligoprobes) or amplicons with molecules capable of fluorescing. These labels produce a change in fluorescent signal that is measured by the instrument following their direct interaction with or hybridization to the amplicon. This signal is related to the amount of amplified product present during each cycle and increases as the amount of specific amplicon increases.

Currently, a range of fluorescent chemistries are used for amplicon detection; the more commonly used chemistries can be divided into two categories: (1) those that involve the nonspecific binding of a fluorescent dye (e.g., SYBER Green I) to double-stranded DNA and (2) fluorescent probes that bind specifically to the target of interest. SYBER Green I chemistry is based on the binding of SYBER Green I to a site referred to as the **DNA minor groove** (where the strand backbones of DNA are closer together on one side of the helix than on the other) that is present in all double-stranded DNA molecules. Once bound, fluorescence of this dye increases more than 100-fold. Therefore, as the amount of double-stranded amplicon increases, the fluorescent signal or output increases proportionally and can be measured by the instrument during the elongation stage of amplification. However, a major disadvantage of this particular

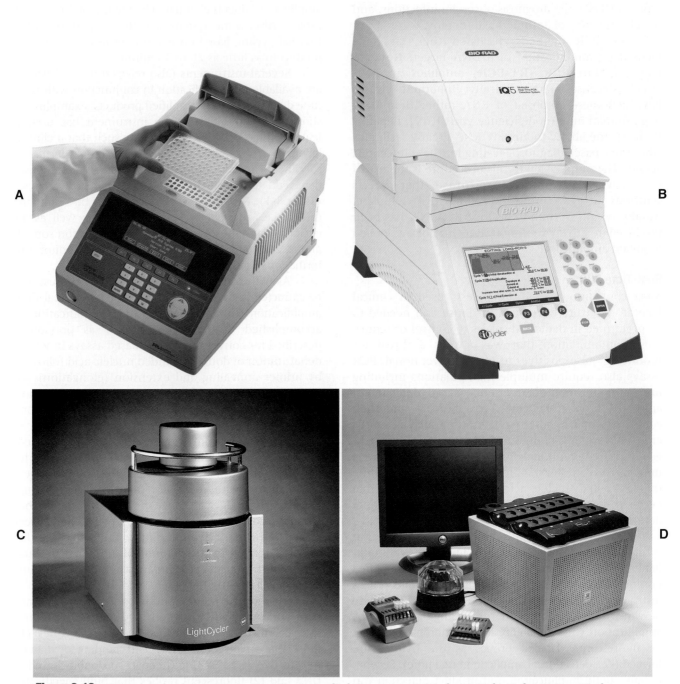

Figure 8-10   Examples of real-time PCR instruments. **A,** Applied Biosystems. **B,** iCycler. **C,** Light Cycler. **D,** SmartCycler. (**A** courtesy Applied Biosystems, Foster, Calif; **B** courtesy Bio-Rad Laboratories, Hercules, Calif; **C** courtesy Roche Applied Science, Indianapolis, Ind; **D** courtesy Cepheid, Sunnyvale, Calif.)

means of detection is that it can detect both specific and nonspecific amplified products.

Three different chemistries commonly employed to detect amplicon in real time involve additional fluorescence-labeled oligonucleotides or probes. Sufficient amounts of fluorescence are only released either after cleavage of the probe (hydrolysis probes) or during

hybridization of one (molecular beacon) or two oligonucleotides (hybridization probes) to the amplicon. These fluorescent chemistries for detection of amplified product are overviewed in Figure 8-11. Introduction of these additional probes increases the specificity of the PCR product. Also, there are real-time PCR instruments (e.g., Light Cycler, Roche Applied Science, Indianapolis,

Table 8-1   Examples of Commercially Available Real-Time Polymerase Chain Reaction (PCR) Instruments

Instrument*	Manufacturer	Detection Format	Turn-Around Time[†]	Ability to Detect Multiple Targets	Maximum No. of Reactions per Run
Applied Biosystems 7300 and 7500 Real-Time PCR Systems	Applied Biosystems Foster, Calif	Hydrolysis probe (TaqMan) or molecular beacons	40 min-approx 2 hr	Yes	96[‡]
iCycler IQ Real-Time PCR Detection System	Bio-Rad Laboratories Hercules, Calif	Hydrolysis probe (TaqMan), hybridization probes (FRET), or molecular beacons	30 min-1.5 hr	Yes	96
Light Cycler	Roche Applied Science Indianapolis, Ind	Hydrolysis probe (TaqMan), hybridization probes (FRET), or molecular beacons	20 min-1 hr	Yes	32
Smart Cycler	Cepheid Sunnyvale, Clif	Hydrolysis probe (TaqMan), hybridization probes (FRET), or molecular beacons	20 min-1 hr	Yes	16

*Different thermocycler models are available in most instances.
[†]Does not include specimen preparation.
[‡]Applied Biosystems 7900HT-Real-Time PCR System has 384 wells.

Ind) that have the ability to detect multiple targets (multiplex PCR) by using different probes labeled with specific fluorescent dyes that each have their own unique emission spectra.

Some real-time PCR instruments also have the ability to perform melting curve analysis. This type of analysis of amplified products confirms the identification (i.e., specificity) of the amplified products and/or identifies nonspecific products. Melting curve analysis can be performed with assays using hybridization probes and molecular beacons, but not hydrolysis probes because hydrolysis probes are destroyed during the amplification process. The underlying basis of melting curve analysis is the ability of the double-stranded DNA to become single-stranded upon heating (referred to as *melting* or *denaturation*). The **melting temperature,** or **Tm,** is the temperature at which the DNA becomes single-stranded ("melts") and is dependent on its base sequence (stretches of double-stranded DNA with more cytosines and guanines require more heat [energy] to break the three hydrogen bonds between these two bases in contrast to adenine and thymidine base pairing, which has only two hydrogen bonds). By definition, the **Tm** is the temperature at which 50% of the DNA is single-stranded. Because the Tm of the probe from its target is specific, being primarily based on probe-target base composition, amplification products can be confirmed as correct by its melting characteristics or Tm. Of significance, the Tm can also be used to distinguish base pair differences (e.g., genotypes, mutations, or polymorphisms) in target DNA, thus forming the basis for many genetic testing assays because base pair mismatches due to mutations

lower the Tm. With respect to real-time PCR assays, because fluorescence of single-labeled probes is reversible by breaking the hydrogen bonds between the probe and target (i.e., denaturation), the Tm can be determined by measuring fluorescence. In real-time PCR thermal cyclers, melting curve analysis is performed once amplification is finished. The temperature of the reaction vessel is lowered below the established annealing temperature of the hybridization probe or molecular beacon; this step allows the probe or beacon to anneal to its target as well as other similar DNA sequences in the reaction. As the temperature is slowly raised, the hybridization probes or molecular beacon that were hybridized to the target will separate (melt) and the fluorescent signal will decrease (Figure 8-12).

Finally, as with conventional PCR, real-time PCR assays also have the ability to quantitate the amount of target in a clinical sample. For quantitative analysis, amplification curves are evaluated. As previously discussed, amplification is monitored either through the fluorescence of double-stranded DNA–specific dyes like SYBR Green 1 or by sequence-specific probes; thus during amplification, a curve is generated. During real-time PCR, there are at least three distinct phases for these curves: (1) an initial lag phase in which no product is detected, (2) an exponential phase of amplified product detected, and a (3) plateau phase. The number of targets in the original specimen can be determined with precision when the number of cycles needed for the signal to achieve an arbitrary threshold (the portion of the curve where the signal begins to increase exponentially or logarithmically) is determined. This segment of the real-time PCR cycle is within the

A

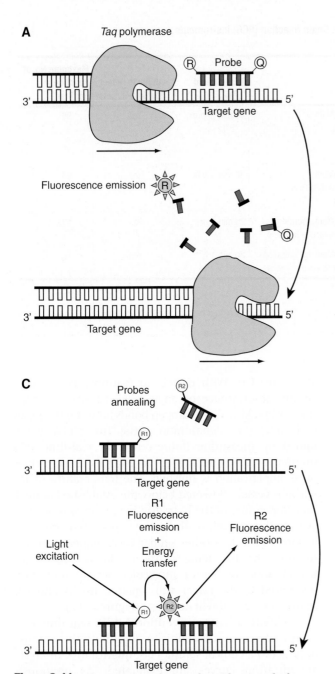

B

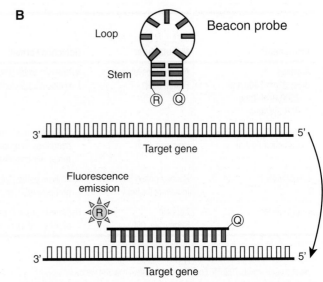

(b). **B,** Molecular beacon. Molecular beacons are hairpin-shaped molecules with an internally quenched fluorophore that fluoresces once the beacon probe binds to the amplified target and the quencher is no longer in close proximity to the fluorophore. These probes are designed such that the loop portion of the molecule is a sequence complementary to the target of interest *(a).* The "stem" portion of the beacon probe is formed by the annealing of complementary arm sequences on the respective ends of the probe sequence. In addition, a fluorescent moiety *(R)* and a quencher moiety *(Q)* at opposing ends of the probe are attached *(a).* The stem portion of the probe keeps the fluorescent and quencher moieties in close proximity to one another such that the fluorescence of the fluorophore remains quenched. When it encounters a target molecule with a complementary sequence, the molecular beacon undergoes a spontaneous conformational change that forces the stem apart, thereby causing the fluorophore and quencher to move away from each other and leading to restoration of fluorescence *(b).*
C, Fluorescent resonant energy transfer (FRET) or hybridization probes. Two different hybridization probes are used, one carrying a fluorescent reporter moiety at its 3' end (designated R1) and the other probe carrying a fluorescent dye at its 5' end (designated R2) *(a).* These two oligonucleotide probes are designed such that they hybridize to amplified DNA target in a head-to-tail arrangement in very close proximity to one another. The first dye (R1) is excited by a filtered light source and emits a fluorescent light at a slightly longer wavelength. Because the two dyes are in close proximity to each other, the energy emitted from R1 excites R2 attached to the second hybridization probe, which emits fluorescent light at an even longer wavelength *(b).* This energy transfer is referred to as *FRET.* By selecting an appropriate detection channel on the instrument, the intensity of light emitted from R2 is filtered and measured. (Modified with permission from Mocellin S, Rossi CR, Pilati P, et al: Quantitative real-time PCR: a powerful ally in cancer research, *Trends Mol Med* 9:189, 2003.)

Figure 8-11 Fluorogenic probes (probe with an attached **fluorophore,** a fluorescent molecule that can absorb light energy and then is elevated to an excited state that is released as fluorescence in the absence of a quencher) commonly used for detection of amplified product in real-time PCR assays.
A, Hydrolysis probe. In addition to the specific primers for amplification, an oligonucleotide probe with a reporter fluorescent dye *(R)* and a quencher dye *(Q)* at its 5' and 3' ends, respectively, is added to the reaction mix. During the extension phase, the **quencher** (molecule that can accept energy from a fluorophore and then dissipate the energy so that no fluorescence results) can only quench the reporter fluorescence when the two dyes are close to each other, which is only the case with an intact probe *(a).* Once amplification occurs and the fluorogenic probe binds to amplified product, bound probe is degraded by the 5'-3' exonuclease activity of *Taq* polymerase; thus quenching is no longer possible and fluorescence is emitted and then measured

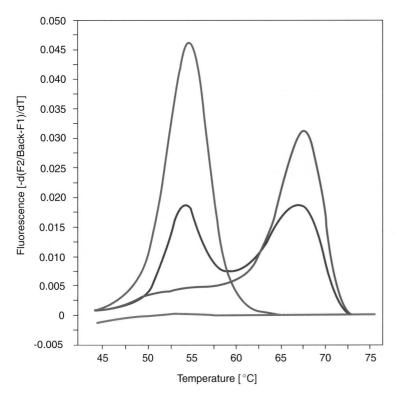

Figure 8-12 Melting curve analyses performed using the LightCycler HSV1/2 Detection Kit. DNA was extracted and subjected to real-time PCR using the LightCycler to detect the presence of HSV DNA. Following amplification, melting curve analysis was performed in which amplified product was cooled to below 55° C and then the temperature slowly raised. The Tm is the temperature at which half of the DNA is single-stranded and is specific for the sequence of the particular DNA product. The specific melting temperature is determined at 640 nm (channel F2 on the cycler) for the clinical samples and the positive and negative controls. For illustration purposes, melting curve analyses are "overlaid" relative to one another in this figure for three clinical samples and the HSV-1 and HSV-2 positive or "template" control. The clinical specimens containing HSV-1 DNA (red line) or HSV-2 (green line) result in a melting peak at 54° C (the Tm) or 67° C (the Tm), respectively. The LightCycler positive or template control containing HSV-1 and HSV-2 DNA, displayed as a purple line, shows two peaks at 54° C and 67° C, respectively. The clinical sample that is negative (brown line) for both HSV-1 and HSV-2 shows no peaks.

linear amplification portion of the reaction where conditions are optimal and fluorescence accumulates in proportion to the amplicon.

With most instrument analyses, the value used for quantitative measurement is the PCR cycle number in which the fluorescence reaches a threshold value of 10 times the standard deviation of baseline fluorescence emission; this cycle number is referred to as the **threshold cycle (CT)** or **crossing point** and is inversely proportional to the starting amount of target present in the clinical sample (see Mackay in Additional Reading). In other words, the C_T is the cycle number in which the fluorescent signal rises above background (the threshold value previously defined) and is dependent on the amount of target at the beginning of the reaction. Thus, to quantitate the target in a clinical specimen, a standard curve is generated in which known amounts of target are prepared and then subjected to real-time PCR along with the clinical sample containing an unknown amount of target. A standard curve is generated using the C_T values for each of the known amounts of target amplified. By taking the C_T value of the clinical specimen and using the standard curve, the amount of target in the original sample can then be determined (Figure 8-13). Quantitative nucleic acid methods are used to monitor response to therapy, detect the development of drug resistance, and predict disease progression.

The introduction of commercially available analyte-specific reagents (ASRs) followed soon after the introduction of real-time PCR. ASRs represent a new regulatory approach by the FDA in which reagents (e.g., antibodies, specific receptor proteins, ligands, oligonucleotides such as DNA or RNA probes or primers, and many reagents used in in-house PCR assays) in this broad category can be used in multiple diagnostic applications. ASR-labeled reagents carry the "For Research Use Only" label, and the manufacturer is prohibited from promoting any applications for these reagents or providing recipes for using the reagents. On a cautionary note, because they are not FDA-cleared, ASR assays cannot be reimbursed by Medicare carriers. Because rulings vary on a state-by-state basis, one is advised to check into Medicare reimbursement before development and introduction of an ASR assay. The high-complexity CLIA laboratory must take full responsibility for developing, validating, and offering the diagnostic assay that employs these reagents. This new regulation essentially allows for new diagnostic methods to become available more quickly, particularly those methods targeted toward smaller patient populations. Importantly, because good manufacturing practices are mandated, ASRs provide more standardized products for the performance of amplification assays. ASRs are available for a number of organisms such as beta-hemolytic group A and B streptococci, methicillin-resistant *S. aureus* (MRSA), *Bordetella pertussis,* vancomycin-resistant enterococci, hepatitis A virus, and Epstein-Barr virus. Finally, molecular kits such as the IDI-MRSA and IDI–StrepB (GeneOhm Sciences, San Diego, Calif) for

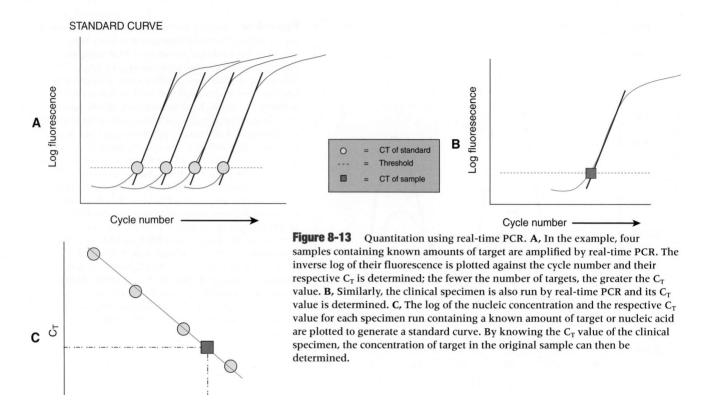

STANDARD CURVE

Figure 8-13 Quantitation using real-time PCR. **A,** In the example, four samples containing known amounts of target are amplified by real-time PCR. The inverse log of their fluorescence is plotted against the cycle number and their respective C_T is determined; the fewer the number of targets, the greater the C_T value. **B,** Similarly, the clinical specimen is also run by real-time PCR and its C_T value is determined. **C,** The log of the nucleic concentration and the respective C_T value for each specimen run containing a known amount of target or nucleic acid are plotted to generate a standard curve. By knowing the C_T value of the clinical specimen, the concentration of target in the original sample can then be determined.

the direct detection in clinical specimens of MRSA and beta-hemolytic group B streptococci, respectively, have recently received FDA clearance. The GeneOhm kits are presently approved for real-time amplification using the SmartCycler (Cepheid).

AMPLIFICATION METHODS: NON–PCR-BASED

Although PCR was developed first and there are numerous PCR-based assays, rapid, sensitive, and specific detection of infectious agents by nucleic acid amplification can be achieved by a number of methods other than PCR. These amplification formats can be divided into two broad categories: those that amplify the signal used to detect the target nucleic acid and those that directly amplify the target nucleic acid but are not PCR-based. Examples of signal amplification methods used in infectious disease diagnostics are listed in Table 8-2. One such system is the **ligase** chain reaction (LCR). LCR uses two pairs of probes that span the target sequence of interest. Once annealed to the target sequence, a space remains between the probes that is enzymatically closed using a ligase (i.e., a ligation reaction). On heating, the joined probes are released as a single strand that is complementary to the target nucleic acid; these newly synthesized strands are then used as the template for subsequent cycles of probe annealing and ligations.

Table 8-2 Examples of Commercially Available Signal Amplification Methods

Method	Manufacturer
Branched DNA (bDNA)	Bayer Diagnostics, Tarrytown, NY
Invader assays	Third Wave Molecular Diagnostics, Madison, Wis
Signal-mediated amplification of RNA (SMART)	Cytocell Technologies, Ltd, Cambridge, UK
Ligase chain reaction (LCR)	Abbott Laboratories, Abbott Park, Ill

Through the process, probe DNA is amplified to a level readily detectable using assays similar to those described for the biotin-avidin system. One commercial system (Abbott Laboratories, Abbott Park, Ill) uses an LCR approach to detect *Chlamydia trachomatis* and *Neisseria gonorrhoeae* in urine or other specimens.

Rather than amplifying the probe, the signal used to detect hybridization between the amplicon and probe can be amplified by increasing the number of reporter molecules per probe. This is accomplished by attaching multiple signal-generating molecules to the probe that recognizes the target DNA. Thus, several reporter molecules are associated with each probe-target duplex. The potential of this approach is exemplified by the development of branched probes. These probes are complex nucleic acid constructions involving capture probes, extender probes, and multi-copy DNA sequences that can accommodate thousands of reporter molecules per probe-target duplex. This method for detecting target nucleic acid is exquisitely sensitive.

A number of other non-PCR-based technologies have been successfully used to detect a variety of infectious agents (Table 8-3). As with PCR, these applications are able to amplify DNA, RNA, mRNA, and rRNA targets; to have multiplex capabilities; and to be qualitative or quantitative. To learn more about these alternative target amplification methods, refer to additional reading and articles authored by Ginocchio.

SEQUENCING AND ENZYMATIC DIGESTION OF NUCLEIC ACIDS

The nucleotide sequence of a microorganism's genome is the blueprint for that organism. Therefore, molecular methods that elucidate some part of a pathogen's genomic sequence provide a powerful tool for diagnostic microbiology. Other methods, either used independently or in conjunction with hybridization or amplification procedures, can provide nucleotide sequence information to detect, identify, and characterize clinically relevant microorganisms. These methods include nucleic acid sequencing and enzymatic digestion and electrophoresis of nucleic acids.

Nucleic Acid Sequencing

Nucleic acid sequencing involves methods that determine the exact nucleotide sequence of a gene or gene fragment obtained from an organism. Although explaining the technology involved is beyond the scope of this text, nucleic acid sequencing will powerfully affect clinical microbiology for some time to come. To illustrate, nucleotide sequences obtained from a microorganism can be compared with an ever-growing gene sequence database for:

- Detecting and classifying previously unknown human pathogens
- Identifying various known microbial pathogens and their subtypes
- Determining which specific nucleotide changes resulting from mutations are responsible for antibiotic resistance
- Establishing the relatedness between isolates of the same species

Before the development of rapid and automated methods, DNA sequencing was a laborious task only undertaken in the research setting. However, determining the sequence of nucleotides in a segment of nucleic acid from an infectious agent can be done rapidly using amplified target from the organism and an automated DNA sequencer. Because sequence information can now be rapidly produced, DNA sequencing has entered the arena of diagnostic microbiology. Identification of microorganisms using PCR in conjunction with automated sequencing is slowly making its way into clinical microbiology laboratories; presently, such molecular analyses are limited for the most part to research-oriented laboratories. It is becoming quite clear that combinations of phenotypic and genotypic characterization are most successful in identifying a variety of microorganisms for which identification is difficult such as the speciation of *Nocardia,* mycobacteria, and organisms that commercial automated instruments fail to identify or correctly identify. Recently, Perkin Elmer Applied Biosystems Division (now Applera, Foster City, Calif) has introduced MicroSeq kit-based reagents in conjunction with automated sequencing that allows analysis of a sequence of either the bacterial 16S rRNA gene or the D2 expansion segment region of the nuclear large-subunit rRNA gene of fungi. Of significance, the MicroSeq sequence libraries contain accurate and rigorously verified sequence data; an important component for successful sequencing in the identification of organisms is an accurate and complete sequence database. In addition, the ability to create customized libraries for specific sequences of interest is possible by the availability of flexible software.

High-Density DNA Probes

An alternative to gel-based sequencing has been the introduction of the high-density oligonucleotide probe array. This technology was developed recently by Affymetrix (Santa Clara, Calif). The method relies upon the hybridization of a fluorescent-labeled nucleic acid target to large sets of oligonucleotides synthesized at precise locations on a miniaturized glass substrate or "chip." The hybridization pattern of the probe to the various oligonucleotides is then used to gain primary structure information about the target (Figure 8-14). This technology has been applied to a broad range of nucleic acid sequence analysis problems including pathogen identification and classification, polymorphism detection, and drug resistance mutations for viruses (e.g., HIV) as well as bacteria.

Enzymatic Digestion and Electrophoresis of Nucleic Acids

Enzymatic digestion and electrophoresis of DNA fragments are less exacting methods of identifying and characterizing microorganisms than is nucleic acid sequencing. However, enzyme digestion-electrophoresis

Table 8-3 Examples of Non–Polymerase Chain Reaction–Based Nucleic Amplification Tests

Amplification Method	Manufacturer/Name	Method Overview	Examples of Commercially Available Assays	Additional Comments
Nucleic acid sequence-based amplification (NASBA)	bioMérieux Inc. NucliSens technology: nucleic acid release, extraction, NASBA amplification, product detection.	(1) Isothermal amplification achieved through coordination of 3 enzymes (avian mycloblastosis, RNAseH, T7 RNA polymerase) in conjunction with 2 oligonucleotide primers specific for the target sequence. (2) Amplification based on primer extension and RNA transcription.	NucliSens HIV-1 QT NucliSens CMV pp67 NucliSens EasyQ HIV-1 NucliSens EasyQ enterovirus	(1) Can be adapted to real-time format using molecular beacons (2) Can develop in-house assays (3) Automated extraction available (NucliSens extractor) (4) Easy Q System = incubator, analyzer, and computer
Transcription-mediated amplification (TMA)	Gen-Probe Inc.: Sample processing, amplification, target detection by hybridization protection or dual kinetic assays for *Chlamydia trachomatis* and *Neisseria gonorrhoeae*. Also, ASRs for hepatitis C virus (HCV), Bayer Inc., Tarrytown, NY; Gen-Probe/Chiron Corp.: TMA for screening donated blood products for HIV-1 and HCV.	(1) Autocatalytic, isothermal amplification utilizing reverse transcriptase and T7 RNA polymerase and 2 primers complementary to the target. (2) Exponential extension of RNA (up to 10 billion amplicons within 10 minutes).	Gen-Probe: *Mycobacterium tuberculosis* Direct Test; APTIMA Combo 2 for dual detection of *C. trachomatis* and *N. gonorrhoeae*; Bayer ASR reagents for HCV; Gen-Probe/Chiron: Procleix HIV-1/HCV	(1) Second-generation TMA assays of Gen-Probe better at removing interfering substances • Less labor-intensive • Uses target capture after sample lysis using an intermediate capture oligomer • TMA performed directly on captured target (2) Automated system for TMA-based assays: TIGRIS DTS system (Gen-Probe) • Instrument handles specimen processing through amplification and detection • 500 tests in 9 hours; 3.5 hours to first result (3) Real-time TMA-based assays using molecular beacons under development by Gen-Probe that will be compatible with a variety of thermal cyclers (e.g., iCycler, ABI Prism)
Standard displacement amplification (SDA)	BD ProbeTec ET System: SDA coupled with homogeneous real-time detection.	(1) Isothermal process in which single-stranded target is first generated. (2) Exponential amplification of target.	BDProbe Tec ET System for *C. trachomatis* and *N. gonorrhoeae*; Panel assays for *Mycoplasma pneumoniae*, *Chlamydophila pneumoniae*, and *L. pneumoniae*, *Chlamydiaceae*; Assay that detects *C. trachomatis*, *C. pneumophila*, and *C. psittaci*; BD ProbeTec *M. tuberculosis* Direct	(1) Reagents dried in separate disposable microwell strips (2) All assays have internal control to monitor for inhibition (3) Automated system for sample processing: BD Viper Sample Processor

procedures still provide valuable information for the diagnosis and control of infectious diseases.

Enzymatic digestion of DNA is accomplished using any of a number of enzymes known as **restriction endonucleases.** Each specific endonuclease rec-ognizes a specific nucleotide sequence (usually four to eight nucleotides in length) known as the enzyme's **recognition,** or **restriction, site.** Once the recognition site is located, the enzyme catalyzes the digestion of the nucleic acid strand at that site,

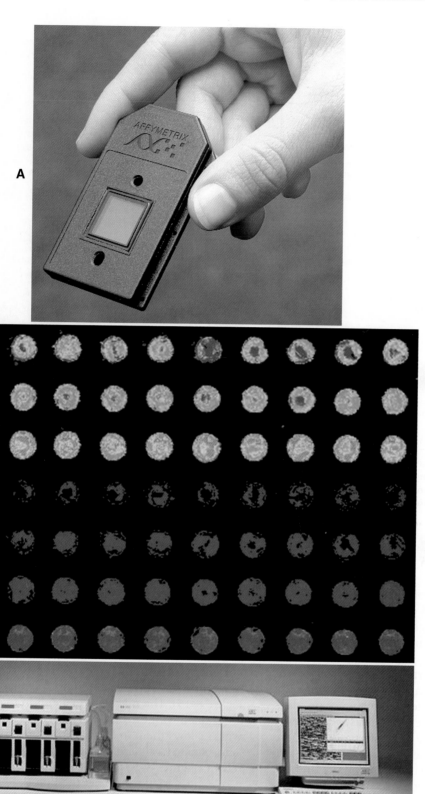

Figure 8-14   Overview of high-density DNA probes. High-density oligonucleotide arrays are created using light-directed chemical synthesis that combines photolithography and solid-phase chemical synthesis. Because of this sophisticated process, more than 500 to as many as a million different oligonucleotide probes may be formed on a chip; an array is shown in **A**. Nucleic acid is extracted from a sample and then hybridized within seconds to the probe array in a GeneChip Fluidics Station. The hybridized array (**B**) is scanned using a laser confocal fluorescent microscope that looks at each site (i.e., probe) on the chip and the intensity of hybridization is analyzed using imaging processing software. The Affymetrix system is shown in part **C**. From left to right is the GeneChip Fluidics Station, scanner, and computer system for data. (Courtesy Affymetrix, Santa Clara, Calif.)

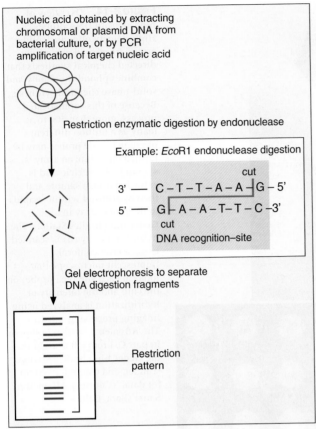

Figure 8-15   DNA enzymatic digestion and gel electrophoresis to separate DNA fragments resulting from the digestion. An example of a nucleic acid recognition site and enzymatic cut produced by *Eco*R1, a commonly used endonuclease, is shown in the inset.

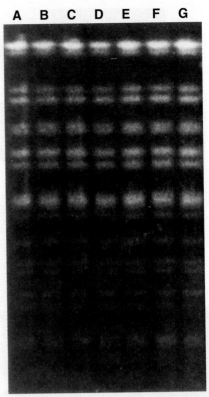

Figure 8-16   Restriction fragment length polymorphisms of vancomycin-resistant *Enterococcus faecalis* isolates in lanes A through G as determined by pulsed-field gel electrophoresis. All isolates appear to be the same strain.

causing a break, or **cut,** in the nucleic acid strand (Figure 8-15).

The number and size of fragments produced by enzymatic digestion depend on the length of nucleic acid being digested (the longer the strand, the greater the likelihood of more recognition sites and thus more fragments), the nucleotide sequence of the strand being digested, and the particular enzyme used for digestion. For example, enzymatic digestion of a bacterial plasmid whose nucleotide sequence provides several recognition sites for endonuclease A, but only rare sites for endonuclease B, will produce more fragments with endonuclease A. Additionally, the size of the fragments produced will depend on the number of nucleotides between each of endonuclease A's recognition sites present on the nucleic acid being digested.

The DNA used for digestion is obtained by various methods. A target sequence may be obtained by amplification via PCR, in which case the length of the DNA to be digested is relatively short (e.g., 50 to 1000 bases). Alternatively, specific procedures may be used to

cultivate the organism of interest to large numbers (e.g., 10^{10} bacterial cells) from which plasmid DNA, chromosomal DNA, or total cellular DNA may be isolated and purified for endonuclease digestion.

Following digestion, fragments are subjected to agarose gel electrophoresis, which allows them to be separated according to their size differences as previously described for Southern hybridization (see Figure 8-4, *B*). During electrophoresis all nucleic acid fragments of the same size comigrate as a single band. For many digestions, electrophoresis results in the separation of several different fragment sizes (see Figure 8-15). The nucleic acid bands in the agarose gel are stained with the fluorescent dye ethidium bromide, which allows them to be visualized on exposure to UV light. Stained gels are photographed for a permanent record (Figures 8-16 to 8-18).

One variation of this method, known as **ribotyping,** involves enzymatic digestion of chromosomal DNA followed by Southern hybridization using probes for genes that encode ribosomal RNA. Because all bacteria contain ribosomal genes, a hybridization pattern will be obtained with almost any isolate, but the pattern will vary depending on the arrangement of genes in a particular strain's genome.

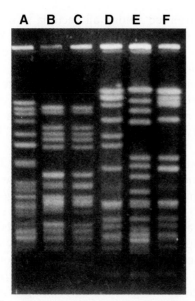

Figure 8-17   Although antimicrobial susceptibility profiles indicated that several methicillin-resistant *S. aureus* isolates were the same strain, restriction fragment length polymorphism analysis using pulsed-field gel electrophoresis (Lanes A through F) demonstrates that only isolates B and C were the same.

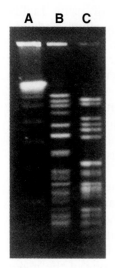

Figure 8-18   Restriction patterns generated by pulsed-field gel electrophoresis for two *Streptococcus pneumoniae* isolates, one that was susceptible to penicillin (Lane B) and one that was resistant (Lane C), from the same patient. Restriction fragment length polymorphism analysis indicates that the patient was infected with different strains. Molecular-size markers are shown in Lane A.

Regardless of the method, the process by which enzyme digestion patterns are analyzed is referred to as **restriction enzyme analysis (REA).** The patterns obtained after gel electrophoresis are referred to as **restriction patterns,** and differences between microorganism restriction patterns are known as **restriction fragment length polymorphisms (RFLPs).** Because RFLPs reflect differences or similarities in nucleotide sequences, REA methods can be used for organism identification and for establishing strain relatedness within the same species (see Figures 8-16 to 8-18).

APPLICATIONS OF NUCLEIC ACID–BASED METHODS

Categories for the application of molecular diagnostic microbiology methods are the same as those for conventional, phenotype-based methods:

- Direct detection of microorganisms in patient specimens
- Identification of microorganisms grown in culture
- Characterization of microorganisms beyond identification

DIRECT DETECTION OF MICROORGANISMS

Nucleic acid hybridization and target or probe amplification methods are the molecular techniques most commonly used for direct organism detection in clinical specimens.

Advantages and Disadvantages

When considering the advantages and disadvantages of molecular approaches to direct organism detection, comparison with the most commonly used conventional method (i.e., direct smears, culture, and microscopy) is helpful.

Specificity.   Both hybridization and amplification methods are driven by the specificity of a nucleotide sequence for a particular organism. Therefore, a positive assay not only indicates the presence of an organism but also provides the organism's identity, potentially precluding the need for follow-up culture. Although molecular methods may not be faster than microscopic smear examinations, the opportunity to avoid delays associated with culture can be a substantial advantage.

However, for many infectious agents, detection and identification are only part of the diagnostic requirement. Determination of certain characteristics, such as strain relatedness or resistance to antimicrobial agents, is often an important diagnostic or epidemiologic component that is not possible without the availability of culture. For this reason, most molecular direct detection methods target organisms for which antimicrobial susceptibility testing is not routinely needed (e.g., chlamydia) or for which reliable cultivation methods are not widely available (e.g., ehrlichia).

The high specificity of molecular techniques also presents a limitation in what can be detected with any one assay, that is, most molecular assays focus on detecting the presence of only one or two potential pathogens. Even if tests for those organisms are positive, the possibility of a mixed infection involving other organisms has not been ruled out. If the tests are negative, other procedures may be needed to determine whether additional pathogens are present. In contrast, smear examination and cultivation procedures can detect and identify a broader selection of possible infectious etiologies. Of importance, Gram-stained smear results are often needed to determine the clinical relevance of finding a particular organism upon culture or detection using molecular assays. However, given the rapid development of molecular technology, protocols that widen the spectrum of detectable organisms in any particular specimen are becoming available. ASRs for real-time PCR that can detect as many as six to seven organisms are commercially available. Finally, a concern always associated with any amplification-based assay is the possibility for cross contamination between samples and/or by amplified byproduct. Thus, it is of utmost importance for any laboratory performing these assays to employ measures to prevent false-positive results.

Sensitivity. Hybridization-based methods can have difficulty directly detecting organisms. The quantity of target nucleic acid may be insufficient, or the patient specimen may contain substances that interfere with or cross-react in the hybridization and signal-generating reactions. One approach developed by Gen-Probe (San Diego, Calif) to enhance sensitivity has been to use DNA probes targeted for bacterial ribosomal RNA, of which there are up to 10,000 copies per cell. Essentially, amplification is accomplished by the choice of a target that exists within the cell as multiple copies rather than as a single copy.

Amplification Techniques Enhance Sensitivity. As discussed with direct hybridization methods, patient specimens also may contain substances that interfere with or inhibit amplification reactions such as PCR. Nonetheless, the ability to amplify target or probe nucleic acid to readily detectable levels has provided a tremendous means for overcoming the lack of sensitivity characteristic of most direct hybridization methods.

Besides the potential for providing more reliable test results than direct hybridization (i.e., fewer false-negative results), amplification methods have other advantages that include:

- Ability to detect nonviable organisms that are not retrievable by cultivation-based methods
- Ability to detect and identify organisms that cannot be grown in culture or are extremely difficult to grow (e.g., hepatitis B virus and the agent of Whipple's disease)
- More rapid detection and identification of organisms that grow slowly (e.g., mycobacteria, certain fungi)
- Ability to detect previously unknown agents directly in clinical specimens by using **broad-range primers** (e.g., use of primers that will anneal to a region of target DNA that is conserved among all bacteria so an amplicon can be produced without previous knowledge of the bacterial species involved in the infection)
- Ability to quantitate infectious agent burden directly in patient specimens, an application that has particular importance for managing HIV and hepatitis B and hepatitis C infections.

Despite these significant advantages, limitations still exist, notably the ability to only find the organisms toward which the primers have been targeted. Additionally, no cultured organism is available if subsequent characterization beyond identification is necessary. As with hybridization, the first limitation may eventually be addressed using broad-range amplification methods that would be used to screen specimens for the presence of any organism (e.g., bacteria, fungi, parasite). Specimens that are positive by this test would then be processed further for a more specific diagnosis. The second limitation is more difficult to overcome and is one reason why culture methods will remain a major part of diagnostic microbiology for some time to come.

An interesting consequence of using highly sensitive amplification methods is the effect on clinical interpretation of results. For example, if a microbiologist detects organisms that are no longer viable, can he or she assume the organisms are or were involved in the infectious process being diagnosed? Also, amplification may detect microorganisms present in insignificant quantities as part of the patient's normal or transient flora, or as an established latent infection, that have nothing to do with the current disease state of the patient.

Finally, as previously mentioned, an underlying complication in the development and application of any direct detection method is that various substances in patient specimens can interfere with the reagents and conditions required for optimum hybridization or amplification. Specimen interference is one of the major issues that must be addressed in the design of any useful direct method for molecular diagnosis of infectious diseases.

Applications for Direct Molecular Detection of Microorganisms

Given their inherent advantages and disadvantages, molecular direct detection methods are most useful when:

- One or two pathogens cause the majority of infections (e.g., *Chlamydia trachomatis* and *Neisseria gonorrhoeae* as common agents of genitourinary tract infections)

- Further organism characterization, such as antimicrobial susceptibility testing, is not required for management of the infection (e.g., various viral agents)
- Either no reliable diagnostic methods exist or they are notably suboptimal (e.g., various bacterial, parasitic, viral, and fungal agents)
- Reliable diagnostic methods exist but are slow (e.g., *Mycobacterium tuberculosis*)
- Quantitation of infectious agent burden that influences patient management (e.g., AIDS) is desired

A large number and variety of commercially available molecular systems and products for the detection and identification of infectious organisms are now available. These include automated or semiautomated systems such as Roche Diagnostic System COBAS instrument which is capable of processing and assaying 22 specimens in $3\frac{1}{2}$ hours. Many of these systems and products are mentioned throughout this textbook. Additionally, many direct detection assays have been developed by diagnostic manufacturers and research laboratories associated with academic medical centers. Therefore, direct molecular diagnostic methods based on amplification will continue to expand and enhance our understanding and diagnosis of infectious diseases. However, as with any laboratory method, their ultimate utility and application will depend on their accuracy, potential impact on patient care, advantages over currently available methods, and resources required to establish and maintain their use in the diagnostic setting.

IDENTIFICATION OF MICROORGANISMS GROWN IN CULTURE

Once organisms are grown in culture, hybridization, amplification, or RFLP analysis may be used to establish identity. Because the target nucleic acid is already amplified via microbial cultivation, sensitivity is not usually a problem for molecular identification methods. Additionally, extensive nucleotide sequence data are available for most clinically relevant organisms so that probes and primers that are highly specific can be readily produced. With neither specificity nor sensitivity as problems in this setting, other criteria regarding the application of molecular identification methods must be considered.

The criteria that are often considered in comparing molecular and conventional methods for microbial identification include speed, accuracy, and cost. For slow-growing organisms, such as mycobacteria and fungi, growth-based identification schemes can take weeks to months to produce a result. Molecular-based methods can identify these microorganisms almost immediately after sufficient inoculum is available, clearly demonstrating a speed advantage over conventional methods. On the other hand, phenotypic-based methods used to identify frequently encountered bacteria, such as *S. aureus* and beta-hemolytic streptococci, can usually provide highly accurate results within minutes and are less costly and time-consuming than any currently available molecular method.

Although many of the phenotype-based identification schemes are highly accurate and reliable, in some situations phenotypic profiles may yield uncertain identifications and molecular methods are providing an alternative for establishing a definitive identification. This is especially the case when a common pathogen exhibits unusual phenotypic traits (e.g., optochin-resistant *S. pneumoniae*).

CHARACTERIZATION OF MICROORGANISMS BEYOND IDENTIFICATION

Situations exist in which characterizing a microbial pathogen beyond identification provides important information for patient management and public health. In such situations, knowledge regarding an organism's virulence, resistance to antimicrobial agents, or relatedness to other strains of the same species can be extremely important. Although various phenotypic methods have been able to provide some of this information, the development of molecular technologies has greatly expanded our ability to generate this information in the diagnostic setting. This is especially true with regard to antimicrobial resistance and strain relatedness.

Detection of Antimicrobial Resistance

Like all phenotypic traits, those that render microorganisms resistant to antimicrobial agents are encoded on specific genes (for more information regarding antimicrobial resistance mechanisms, see Chapter 11). Therefore, molecular methods for gene amplification or hybridization can be used to detect antimicrobial resistance. In many ways, phenotypic methods for resistance detection are reliable and are the primary methods for antimicrobial susceptibility testing (see Chapter 12). However, the complexity of emerging resistance mechanisms often challenges the ability of commonly used susceptibility testing methods to detect clinically important resistance to antimicrobial agents.

Methods such as PCR play a role in the detection of certain resistance profiles that may not always readily be detected by phenotypic methods. Two such examples include detection of the *van* genes, which mediate vancomycin resistance among enterococci (see Figure 8-8), and the *mec* gene, which encodes resistance among staphylococci to all currently available drugs of the beta-

Table 8-4 Examples of Methods to Determine Strain Relatedness

Method	Advantages/Limitations
Plasmid analysis	Simple to implement but cannot often discriminate because many bacterial species have few or no plasmids
Multilocus enzyme electrophoresis	Provides only an estimate of overall genetic relatedness and diversity (protein-based)
Multilocus sequence typing	Data are electronically portable and used as non–culture-based typing method; labor intensive and expensive
Pulsed-field gel electrophoresis	Highly discriminatory but it is difficult to resolve bands of similar size and interlaboratory reproducibility is limited
Randomly amplified polymorphic DNA	High discriminatory power but poor laboratory interlaboratory and intralaboratory reproducibility due to short random primer sequences and low PCR annealing temperatures
Repetitive sequence-based PCR; manual and automated	Manual system—useful for strain typing but low rates of interlaboratory reproducibility; suboptimal turn-around times (TATs). Automated system—increased reproducibility and decreased TATs
Ribotyping and PCR ribotyping	Difficult to distinguish among different subtypes

lactam class (see Figure 8-9). Undoubtedly, conventional and molecular methods will both continue to play key roles in the characterization of microbial resistance to antimicrobial agents.

Investigation of Strain Relatedness/Pulsed-Field Gel Electrophoresis

An important component to recognizing and controlling disease outbreaks inside or outside of a hospital is identification of the reservoir and mode of transmission of the infectious agents involved. This often requires establishing relatedness among the pathogens isolated during the outbreak. For example, if all the microbial isolates thought to be associated with a nosocomial infection outbreak are shown to be identical, or at least very closely related, then a common source or reservoir for those isolates should be sought. If the etiologic agents are not the same, other explanations for the outbreak must be investigated (see Chapter 64). Because each species of a microorganism comprises an almost limitless number of strains, identification of an organism to the species level is not sufficient for establishing relatedness. **Strain typing,** the process used to establish the relatedness among organisms belonging to the same species, is required.

Although phenotypic characteristics (e.g., bio-typing, serotyping, antimicrobial susceptibility profiles) historically have been used to type strains, these methods often are limited by their inability to consistently discriminate between different strains, their labor intensity, or their lack of reproducibility. In contrast, certain molecular methods do not have these limitations and have enhanced strain-typing capabilities. The molecular typing methods either directly compare nucleotide sequences between strains or produce results that indirectly reflect similarities in nucleotide sequences among "outbreak" organisms. Indirect methods usually involve enzymatic digestion and electrophoresis of microbial DNA to produce RFLPs for comparison and analysis.

Several molecular methods have been investigated for establishing strain relatedness. Examples of these methods are in Table 8-4. The method chosen primarily depends on the extent to which the following four criteria proposed by Maslow and colleagues are met:

- **Typeability:** the method's capacity to produce clearly interpretable results with most strains of the bacterial species to be tested
- **Reproducibility:** the method's capacity to repeatedly obtain the same typing profile result with the same bacterial strain
- **Discriminatory power:** the method's ability to produce results that clearly allow differentiation between unrelated strains of the same bacterial species
- **Practicality:** the method should be versatile, relatively rapid, inexpensive, technically simple, and provide readily interpretable results

The last criterion, *practicality*, is especially important for busy clinical microbiology laboratories that provide support for infection control and hospital epidemiology.

Among the molecular methods used for strain typing, pulsed-field gel electrophoresis (PFGE) meets most of Maslow's criteria for a good typing system and is frequently referred to as the microbial typing "gold standard" for the present. This method is applicable to most of the commonly encountered bacterial pathogens, particularly those frequently associated with nosocomial infections and outbreaks such as staphylococci (MRSA), enterococci (vancomycin-resistant enterococci), and gram-negative pathogens including *Escherichia coli, Klebsiella* spp., *Enterobacter* spp., and *Acinetobacter* spp. For

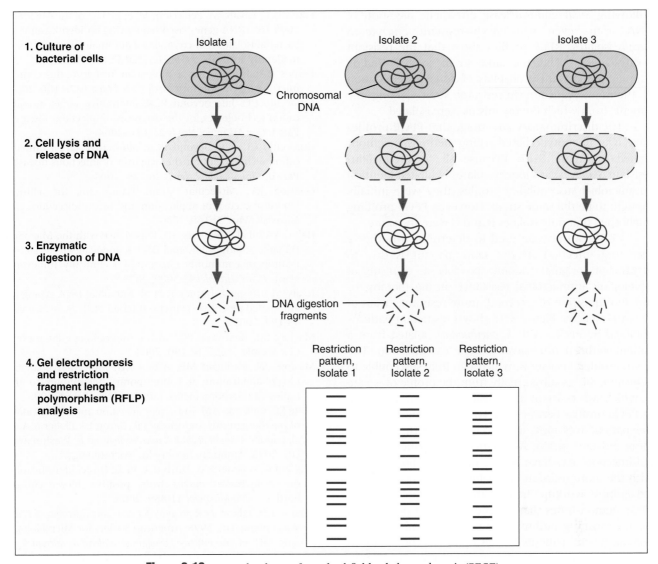

Figure 8-19   Procedural steps for pulsed-field gel electrophoresis (PFGE).

these reasons, PFGE has been widely accepted among microbiologists, infection control personnel, and infectious disease specialists as a primary laboratory tool for epidemiology.

The principle of PFGE is to use a specialized electrophoresis device to separate chromosomal fragments produced by enzymatic digestion of intact bacterial chromosomal DNA. Bacterial suspensions are first embedded in agarose plugs, where they are carefully lysed to release intact chromosomal DNA; the DNA is then digested using restriction endonuclease enzymes. Enzymes that have relatively few restriction sites on the genomic DNA are selected so that 10 to 20 DNA fragments ranging in size from 10 to 1000 kb are produced (Figure 8-19). Because of the large DNA fragment sizes produced, resolution of the banding patterns requires the use of a **pulsed electrical field** across

the agarose gel that subjects the DNA fragments to different voltages from varying angles at different time intervals.

Although comparison and interpretation of RFLP profiles produced by PFGE can be complex, the basic premise is that strains with the same or highly similar digestion profiles share substantial similarities in their nucleotide sequences and therefore are likely to be most closely related. For example, in Figure 8-18 isolates 1 and 2 have identical RFLP patterns, whereas isolate 3 has only 7 of its 15 bands in common with either isolates 1 or 2. Therefore, isolates 1 and 2 would be considered closely related, if not identical, whereas isolate 3 would not be considered related to the other two isolates.

One example of PFGE application for the investigation of an outbreak is shown in Figure 8-16.

Following *Sma*I endonuclease enzymatic digestion of DNA from seven vancomycin-resistant *Enterococcus faecalis* isolates, RFLP profiles show that the resistant isolates are probably the same strain. Such a finding strongly supports the probability of clonal dissemination of the same vancomycin-resistant strain among the patients from which the organisms were isolated.

The discriminatory advantage that PFGE profiles have over phenotype-based typing methods is demonstrated in Figure 8-17. Because all six methicillin-resistant *Staphylococcus aureus* isolates exhibited identical antimicrobial susceptibility profiles, they were initially thought to be the same strain. However, PFGE profiling established that only isolates B and C were the same.

PFGE can also be used to determine whether a recurring infection in the same patient is due to insufficient original therapy, possibly as a result of developing antimicrobial resistance during therapy, or due to acquisition of a second, more resistant, strain of the same species. Figure 8-18 shows restriction patterns obtained by PFGE, with *S. pneumoniae* isolated from a patient with an unresolved middle ear infection. The PFGE profile of isolate B, which was fully susceptible to penicillin, differs substantially from the profile of isolate C, which was resistant to penicillin. The clear difference in PFGE profiles between the two strains indicates that the patient was most likely reinfected with a second, more resistant, strain. Alternatively, the patient's original infection may have been a mixture of both strains, with the more resistant one being lost during the original culture workup. In any case, this application of PFGE demonstrates that the method is not only useful for investigating outbreaks or strain dissemination involving several patients but also gives us the ability to investigate questions regarding reinfections, treatment failures, and mixed infections involving more than one strain of the same species.

ADDITIONAL READING

Chapin K, Musgnug M: Evaluation of three rapid methods for the direct detection of *Staphylococcus aureus* from positive blood cultures, *J Clin Microbiol* 41:4324, 2003.

Cockerill FR: Application of rapid-cycle real-time polymerase chain reaction for diagnostic testing in the clinical microbiology laboratory, *Arch Pathol Med* 127:1112, 2003.

Fontana C, Favaro M, Pelliccioni M, et al: Use of the MicroSeq 5000 16S rRNA gene-based sequencing for identification of bacterial isolates that commercial automated systems failed to identify correctly, *J Clin Microbiol* 43:615, 2005.

Forbes BA: Introducing a molecular test into the clinical microbiology laboratory, *Arch Pathol Med* 127:1106, 2003.

Ginocchio CC: Life beyond PCR: alternative target amplification technologies for the diagnosis of infectious diseases, Part I, *Clin Microbiol Newls* 26:121, 2004.

Ginocchio CC: Life beyond PCR: alternative target amplification technologies for the diagnosis of infectious diseases, Part II, *Clin Microbiol Newls* 26:129, 2004.

Goering RV: Molecular strain typing for the clinical laboratory: current application and future direction, *Clin Microbiol Newsl* 22:169, 2000.

Hall L, Wohlfiel S, Roberts GD: Experience with the MicroSeq D2 large-subunit ribosomal DNA sequencing kit for identification of commonly encountered clinically important yeast, *J Clin Microbiol* 41:5009, 2003.

Healy M, Huong J, Bittner T, et al: Microbial DNA typing by automated repetitive-sequenced-based PCR, *J Clin Microbiol* 43:199, 2005.

Mackay IM: Real-time PCT in the microbiology laboratory, *Clin Microbiol Infect* 10:190, 2004.

Maslow JN, Mulligan ME, Arbeit RD: Molecular epidemiology: application of contemporary techniques to the typing of microorganisms, *Clin Infect Dis* 17:153, 1993.

Nolte FR, Caliendo AM: Molecular detection and identification of microorganisms. In Murray PR, Baron EJ, Pfaller MA, et al, editors: *Manual of clinical microbiology,* ed 8, Washington, DC, 2003, American Society for Microbiology.

Oliviera K, Brecher SM, Durbin A, et al: Direct identification of *Staphylococcus aureus* from positive blood culture bottles, *J Clin Microbiol* 41:889, 2003.

Persing DH, editor: *PCR protocols for emerging infectious diseases,* Washington, DC, 1996, American Society for Microbiology.

Persing DH et al, editors: *Diagnostic molecular microbiology: principles and applications,* Washington, DC, 1993, American Society for Microbiology.

Soll DR, Lockart SR, Pujol C, et al: Laboratory procedures for the epidemiologic analysis of microorganisms. In Murray PR, Baron EJ, Pfaller MA, et al, editors: *Manual of clinical microbiology,* ed 8, Washington, DC, 2003, American Society for Microbiology.

Tenover FC, Arbeit RD, Goering RV, et al: Interpreting chromosomal DNA restriction patterns produced by pulsed-field gel electrophoresis: criteria for bacterial strain typing, *J Clin Microbiol* 33:2233, 1995.

Immunochemical Methods Used for Organism Detection

Certain factors can hinder the diagnosis of an infectious disease by culture and biochemical techniques. These factors include the inability to cultivate an organism on artificial media, such as with *Treponema pallidum,* the agent of syphilis, or the fragility of an organism and its subsequent failure to survive transport to the laboratory, such as with respiratory syncytial virus and varicella-zoster virus. The fastidious nature of some organisms, such as leptospira or *Bartonella,* can result in long incubation periods before growth is evident, or administration of antimicrobial therapy before a specimen is obtained, such as with a patient who has received partial treatment, also can impede diagnosis. In these cases, detecting a specific product of the infectious agent in clinical specimens is very important because this product would not be present in the specimen in the absence of the agent. This chapter discusses the direct detection of microorganisms in patient specimens using immunochemical methods and the identification of microorganisms by these methods once they have been isolated on laboratory media. Chapter 10 discusses the diagnosis of infectious diseases by using some of these same immunochemical methods to detect antibodies produced in response to the presence of an infecting agent in patient serum.

PRODUCTION OF ANTIBODIES FOR USE IN LABORATORY TESTING

Immunochemical methods use antigens and antibodies as tools to detect microorganisms. **Antigens** are foreign substances, usually high–molecular-weight proteins or carbohydrates, that elicit the production of other proteins, called **antibodies,** in a human or animal host (see Chapter 3). Antibodies attach to the antigens and aid the host in removing the infectious agent (see Chapters 3 and 10). Antigens may be part of the physical structure of the pathogen, such as the bacterial cell wall, or they may be a chemical produced and released by the pathogen, such as an enzyme or toxin; each antigen is also called an **epitope.** Figure 9-1 shows the multiple epitopes of group A *Streptococcus (Streptococcus pyogenes).*

POLYCLONAL ANTIBODIES

Because one organism contains many different epitopes that the host will recognize as foreign and because the host usually responds by producing antibodies to each antigen, the host's serum will contain various **polyclonal antibodies.** Polyclonal antibodies used in immunodiagnosis are prepared by immunizing animals (usually rabbits, sheep, or goats) with an infectious agent and then isolating and purifying the resulting antibodies from the animal's serum. However, because individual animals produce different antibodies with different **idiotypes** (antigen-binding sites), the lack of uniformity in polyclonal antibody reagents requires that different lots be continually retested for specificity and avidity (strength of binding) in any given immunochemical test system.

MONOCLONAL ANTIBODIES

The ability to create an immortal cell line producing large quantities of a completely characterized and highly specific antibody, known as a **monoclonal antibody,** has revolutionized immunologic testing. Monoclonal antibodies are produced by the offspring (clones) of a single hybrid cell, the product of fusion of an antibody-producing plasma B cell and a malignant antibody-producing myeloma cell. One technique for the production of such a clone of cells, called **hybridoma cells,** is illustrated in Figure 9-2.

The process starts by immunizing a mouse with the antigen for which an antibody is to be produced. The animal responds by producing many antibodies to the epitope injected. The mouse's spleen, which contains antibody-producing plasma cells, is removed and emulsified so that antibody-producing cells can be separated and placed into individual wells of a microdilution tray. Because cells cannot remain viable in cell culture for very long, they must be fused together with cells that are able to survive and multiply in tissue culture, that is, the continuously propagating, or immortal cells, of multiple myeloma (a malignant tumor of antibody-producing plasma cells). The special myeloma tumor cells used for hybridoma production, however, possess a very important defect—they are deficient in

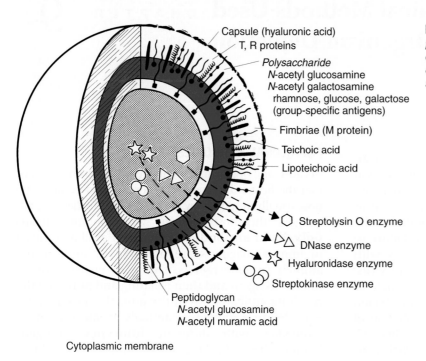

Capsule (hyaluronic acid)
T, R proteins
Polysaccharide
   N-acetyl glucosamine
   N-acetyl galactosamine
   rhamnose, glucose, galactose
   (group-specific antigens)
Fimbriae (M protein)
Teichoic acid
Lipoteichoic acid

Streptolysin O enzyme
DNase enzyme
Hyaluronidase enzyme
Streptokinase enzyme

Peptidoglycan
   N-acetyl glucosamine
   N-acetyl muramic acid

Cytoplasmic membrane

Figure 9-1   Group A *Streptococcus (Streptococcus pyogenes)* contains many antigenic structural components and produces various antigenic enzymes, each of which may elicit a specific antibody response from the infected host.

the enzyme hypoxanthine phosphoribosyltransferase. This defect leads to their inability to survive in a medium containing hypoxanthine, aminopterin, and thymidine (HAT medium). Antibody-producing spleen cells, however, possess the enzyme. Thus, fused hybridoma cells survive in the selective medium and can be recognized by their ability to grow indefinitely in the medium. Unfused antibody-producing lymphoid cells die after several multiplications in vitro because they are not immortal, and unfused myeloma cells die in the presence of the toxic enzyme substrates. The only surviving cells will be true hybrids. The growth medium supernatant from the microdilution tray wells in which the hybridoma cells are growing is then tested for the presence of the desired antibody. Many such cell lines are usually examined before a suitable antibody is found because it must be specific enough to bind only the type of antigen to be tested, but not so specific that it binds only the antigen from the particular strain with which the mouse was first immunized. When a good candidate antibody-producing cell is found, the hybridoma cells are either grown in cell culture in vitro or they are reinjected into the peritoneal cavities of many mice, where the cells multiply and produce large quantities of antibody in the ascitic (peritoneal) fluid that is formed. Ascitic fluid can be removed from mice many times over the animals' lifetimes, and antibody molecules will be identical to the original clone.

Polyclonal and monoclonal antibodies are both used in commercial systems to detect infectious agents.

PRINCIPLES OF IMMUNOCHEMICAL METHODS USED FOR ORGANISM DETECTION

Numerous immunologic methods are used for the rapid detection of bacteria, fungi, parasites, and viruses in patient specimens; the same reagents can often be used to identify these organisms grown in culture. The techniques fall under the following categories: precipitin tests, particle agglutination tests, immunofluorescence assays, and enzyme immunoassays.

PRECIPITIN TESTS

The classic method of detecting soluble antigen (i.e., antigen in solution) is Ouchterlony **double immunodiffusion.**

Double Immunodiffusion

In the double immunodiffusion method, small circular wells are cut out of agar or agarose, a gelatin-like matrix from seaweed through which molecules can readily diffuse, in Petri dishes. The patient specimen containing antigen is placed in one well, and antibody directed against the antigen being sought is placed in the adjacent well. Over a period of 18 to 24 hours, the antigen and antibody diffuse toward each other, producing a visible precipitin band at the point at which they meet, known as the **zone of equivalence.** This method is currently used to detect exoantigens produced by the systemic fungi to confirm their presence in culture (Figure 9-3). However, the technique is too slow

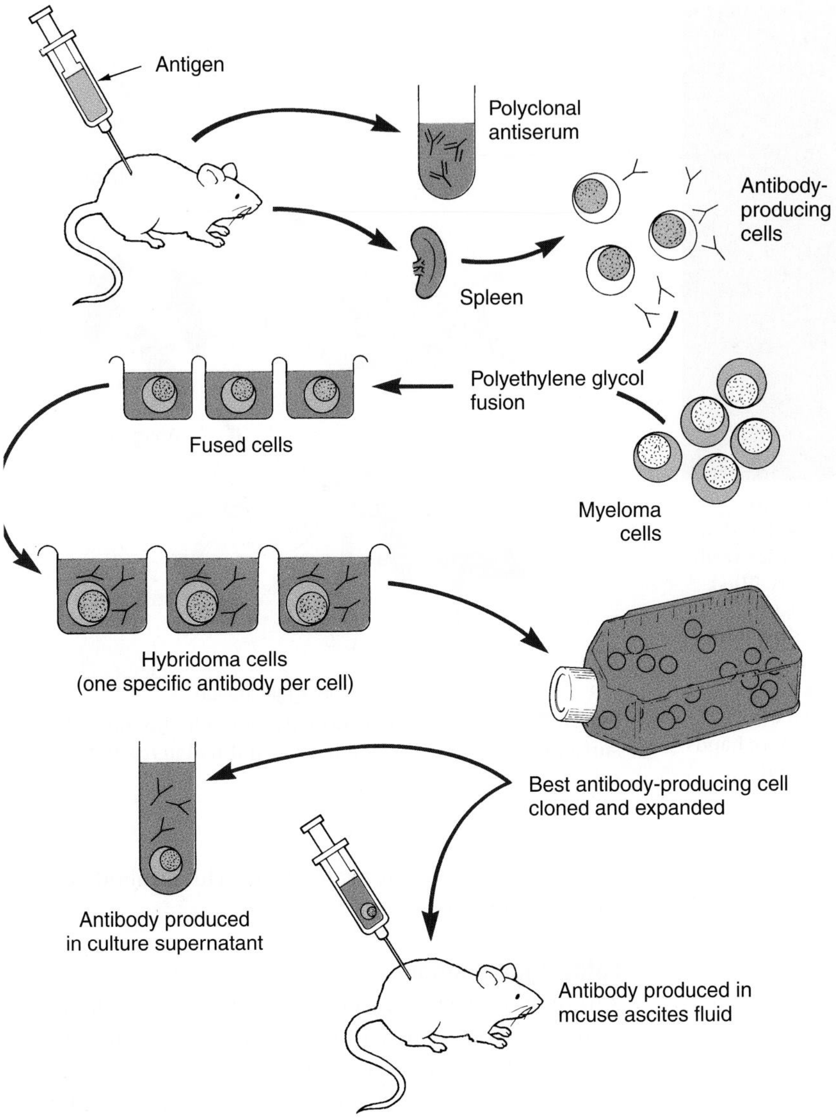

Figure 9-2   Production of a monoclonal antibody.

Antigen

Polyclonal antiserum

Antibody-producing cells

Spleen

Polyethylene glycol fusion

Myeloma cells

Fused cells

Hybridoma cells (one specific antibody per cell)

Best antibody-producing cell cloned and expanded

Antibody produced in culture supernatant

Antibody produced in mcuse ascites fluid

to be commonly used for antigen detection directly in patient specimens.

Counterimmunoelectrophoresis

Counterimmunoelectrophoresis (CIE) is a modification of the Ouchterlony method that speeds up migration of an antigen and antibody by applying an electrical current. With some exceptions (e.g., *Streptococcus pneumoniae* serotypes 7 and 14), most bacterial antigens are negatively charged in a slightly alkaline environment, whereas antibodies are neutral. This feature of bacterial antigens is exploited by CIE assays, in which solutions of antibody and sample fluid to be tested are placed in small wells cut into a slab of agarose on a glass surface (Figure 9-4). A paper or fiber wick is used to connect the two opposite sides of the agarose to troughs of slightly alkaline buffer, formulated for each antibody-antigen system. When an electrical current is applied

through the buffer, the negatively charged antigen molecules migrate toward the positive electrode and thus toward the wells filled with antibody. The neutrally charged antibodies are carried toward the negative electrode by the flow of the buffer. At some point between the wells, a zone of equivalence occurs, and the antigen-antibody complexes form a visible precipitin band. The entire procedure usually takes about 1 hour.

Any antigen for which antisera are available can be tested by CIE. The sensitivity appears to be less than that of particle agglutination, detecting approximately 0.01 to 0.05 mg/mL of antigen, which translates to about 10^3 organisms per milliliter of fluid. Bands are often difficult to see, and the agarose gel may require overnight washing in distilled water to remove nonspecific precipitin reactions. Testing positive and negative controls is especially critical, because sera may contain nonspecifically reacting agents that form nonstable

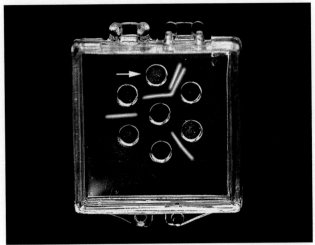

Figure 9-3 Exo-Antigen Identification System, Immuno-Mycologics, Inc., Norman, Okla. The center well is filled with a 50× concentrate of an unknown mold. The arrow identifies well 1; wells 2 to 6 are shown clockwise. Wells 1, 3, and 5 are filled with anti-*Histoplasma,* anti-*Blastomyces,* and anti-*Coccidioides* reference antisera, respectively. Wells 2, 4, and 6 are filled with *Histoplasma* antigen, *Blastomyces* antigen, and *Coccidioides* antigen, respectively. The unknown organism can be identified as *Histoplasma capsulatum* based on the formation of line(s) of identity (arc) linking the control band(s) with one or more bands formed between the unknown extract (center well) and the reference antiserum well (well 1).

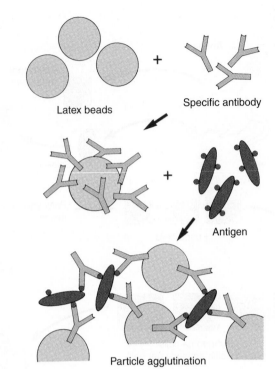

Figure 9-5 Alignment of antibody molecules bound to the surface of a latex particle and latex agglutination reaction.

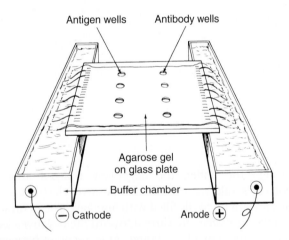

Figure 9-4 Apparatus for performing counterimmunoelectrophoresis.

complexes in the gel. Because of the initial capital outlay for the apparatus and the large quantities of antigen and antibody that must be used, CIE is more expensive than agglutination-based tests and is, therefore, no longer widely used for immunodiagnosis.

PARTICLE AGGLUTINATION

Numerous procedures have been developed to detect antigen via the agglutination (clumping) of an artificial carrier particle such as a latex bead with antibody bound to its surface.

Latex Agglutination

Antibody molecules can be bound in random alignment to the surface of latex (polystyrene) beads (Figure 9-5). Because the number of antibody molecules bound to each latex particle is large, the potential number of antigen-binding sites exposed is also large. Antigen present in a specimen being tested binds to the combining sites of the antibody exposed on the surfaces of the latex beads, forming cross-linked aggregates of latex beads and antigen. The size of the latex bead (0.8 µm or larger) enhances the ease with which the agglutination reaction is recognized. Levels of bacterial polysaccharides detected by latex agglutination have been shown to be as low as 0.1 ng/mL. Because the pH, osmolarity, and ionic concentration of the solution influence the amount of binding that occurs, conditions under which latex agglutination procedures are carried out must be carefully standardized. Additionally, some constituents of body fluids, such as rheumatoid factor, have been found to cause false-positive reactions in the latex agglutination systems available. To counteract this problem, it is recommended that all specimens be pretreated by boiling or with ethylenediaminetetraacetic acid (EDTA) before testing. Commercial test systems are usually performed on cardboard cards or glass slides; manu-

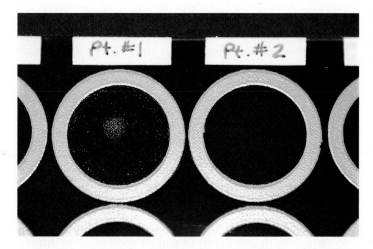

Figure 9-6 Cryptococcal Antigen Latex Agglutination System (CALAS), Meridian Diagnostics, Inc., Cincinnati, Ohio. Patient 1 shows positive agglutination; patient 2 is negative.

Figure 9-7 Streptex, Remel, Inc., Lenexa, Kan. Colony of beta-hemolytic *Streptococcus* agglutinates with group B *Streptococcus* (*Streptococcus agalactiae*) latex suspension.

facturer's recommendations should be followed precisely to ensure accurate results. Reactions are graded on a 1+ to 4+ scale, with 2+ usually the minimum amount of agglutination seen in a positive sample. Control latex (coated with antibody from the same animal species from which the specific antibody was made) is tested alongside the test latex. If the patient specimen or the culture isolate reacts with both the test and control latex, the test is considered nonspecific and therefore uninterpretable.

Latex tests are very popular in clinical laboratories to detect antigen to *Cryptococcus neoformans* in cerebrospinal fluid or serum (Figure 9-6) and to confirm the presence of beta-hemolytic *Streptococcus* from culture plates (Figure 9-7). Latex tests are also available to detect *Streptococcus agalactiae, Clostridium difficile* toxins A and B, and rotavirus.

Coagglutination

Similar to latex agglutination, coagglutination uses antibody bound to a particle to enhance visibility of the agglutination reaction between antigen and antibody. In this case the particles are killed and treated *S. aureus* organisms (Cowan I strain), which contain a large amount of an antibody-binding protein, protein A, in their cell walls. In contrast to latex particles, these staphylococci bind only the base of the heavy-chain portion of the antibody, leaving both antigen-binding ends free to form complexes with specific antigen (Figure 9-8). Several commercial suppliers have prepared coagglutination reagents for identification of streptococci, including Lancefield groups A, B, C, D, F, G, and N; *Streptococcus pneumoniae; Neisseria meningitidis; N. gonorrhoeae;* and *Haemophilus influenzae* types A to F grown in culture. The coagglutination reaction is highly

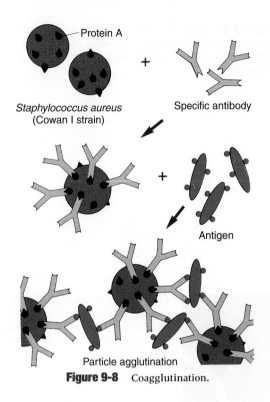

Figure 9-8   Coagglutination.

agglutinating particles in a reaction. By combining liposomes containing reactive molecules on their surfaces and latex particles that harbor antibody-binding sites on their surfaces, reagents are created that have the potential to transform a weak antigen-antibody particle agglutination reaction into a stronger, more easily visualized reaction (Figure 9-9). Liposomes have yet to reach their full potential as diagnostic reagents in clinical microbiology.

IMMUNOFLUORESCENT ASSAYS

Immunofluorescent assays are popular tests for detecting antigens in clinical laboratories. In these tests, antigenic determinants in patient specimens are immobilized and fixed onto glass slides with formalin, methanol, ethanol, or acetone. Monoclonal or polyclonal antibodies conjugated (attached) to fluorescent dyes are then applied to the specimen. After appropriate incubation, washing, and counterstaining (staining the background with a nonspecific fluorescent stain such as rhodamine or Evan's blue), the slide is viewed using a microscope equipped with a high-intensity light source (usually halogen) and filters to excite the fluorescent tag. Most kits used in clinical microbiology laboratories use fluorescein isothiocyanate (FITC) as the dye; FITC fluoresces a bright apple-green (Figure 9-10).

Fluorescent antibody tests are performed using either a direct (DFA) or indirect (IFA) technique (Figure 9-11). In the DFA, FITC is conjugated directly to the specific antibody. In the IFA, the antigen-specific antibody is unlabeled and a second antibody (usually raised against the animal species from which the antigen-specific antibody was harvested) is conjugated

specific but may not be as sensitive for detecting small quantities of antigen as latex agglutination. Thus, it is not usually used for direct antigen detection.

Liposome-Enhanced Latex Agglutination
Phospholipid molecules form small, closed vesicles under certain controlled conditions. These vesicles, consisting of a single lipid bilayer, are called **liposomes.** Molecules bound to the surface of liposomes act as

Figure 9-9   Diagram of liposome-latex agglutination reaction.

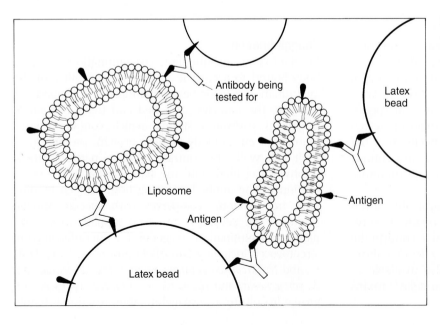

Figure 9-10 *Legionella* (Direct) Fluorescent Test System, Scimedx Corp., Denville, NJ. *Legionella pneumophila* serogroup 1 in sputum.

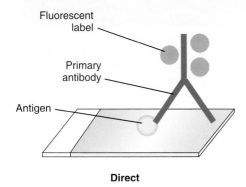

Direct

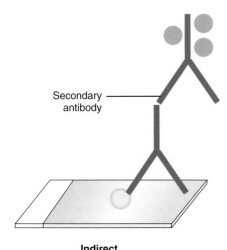

Indirect

Figure 9-11   Direct and indirect fluorescent antibody tests for antigen detection.

to the FITC. The IFA is a two-step, or sandwich, procedure. IFA is more sensitive than DFA, although DFA is faster because there is only one incubation step.

The major advantage of immunofluorescent microscopy assays is that they allow visual assessment of

the adequacy of a specimen. This is a major factor in their use in tests for chlamydial elementary bodies or respiratory syncytial virus (RSV) antigens. Microscopists can see if the specimen was taken from the columnar epithelial cells at the opening of the cervix in the case of the chlamydia DFA, or from the basal cells of the nasal epithelium in the case of RSV. Many individuals, however, consider it problematic that reading slides is completely subjective and that microbiologists must have extensive training to perform testing. Likewise, many individuals view the requirement for a fluorescent scope as an expensive luxury. Finally, fluorescence fades rapidly over time, which makes the archiving of slides difficult. Therefore, antibodies have been conjugated to other markers besides fluorescent dyes. These newer colorimetric labels use enzymes, such as horseradish peroxidase, alkaline phosphatase, and avidin-biotin, to detect the presence of antigen by converting a colorless substrate to a colored end product. Advantages of these tags are that they allow the preparation of permanent mounts because the reactions do not fade with storage and they can be detected using a simple light microscope.

Fluorescent antibody tests are commonly used to detect *Bordetella pertussis, Legionella pneumophila, Giardia, Cryptosporidium, Pneumocystis, Trichomonas,* herpes simplex virus, cytomegalovirus, varicella-zoster virus, RSV, adenovirus, influenza virus, and parainfluenza virus in clinical specimens.

ENZYME IMMUNOASSAYS

Enzyme immunoassay (EIA), or enzyme-linked immunosorbent assay (ELISA), systems were first developed during the 1960s. The basic test consists of antibodies bonded to enzymes; the enzymes remain able to catalyze a reaction yielding a visually discernible end product while attached to the antibodies. Furthermore, the antibody-binding sites remain free to react with their specific antigen. The use of enzymes as labels has several advantages. First, the enzyme itself is not changed during activity; it can catalyze the reaction of many substrate molecules, greatly amplifying the reaction and enhancing detection. Second, enzyme-conjugated antibodies are stable and can be stored for a relatively long time. Third, the formation of a colored end product allows direct observation of the reaction or automated spectrophotometric reading.

The use of monoclonal antibodies has helped increase the specificity of currently available ELISA systems. New ELISA systems are continually being developed for detection of etiologic agents or their products. In some instances, such as detection of RSV, HIV, and certain adenoviruses, ELISA systems may even be more sensitive than current culture methods.

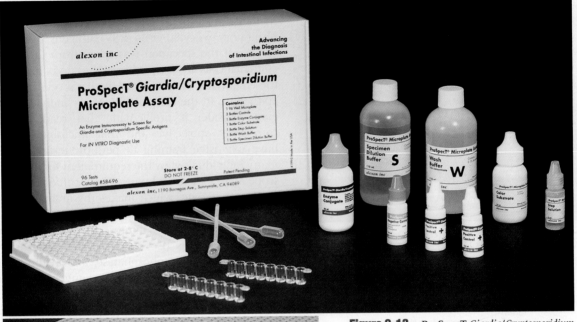

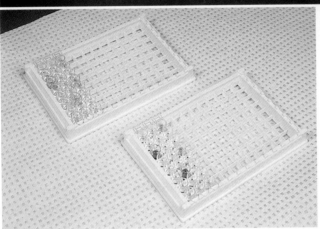

Figure 9-12 ProSpecT *Giardia/Cryptosporidium* Microplate Assay. **A,** Breakaway microwell cupules and kit components. **B,** Positive (yellow) and negative (blue) reactions. (Courtesy Remel, Inc., Lenexa, Kan.)

Solid-Phase Immunoassay

Most ELISA systems developed to detect infectious agents consist of antibody directed against the agent in question firmly fixed to a solid matrix, either the inside of the wells of a microdilution tray or the outside of a spherical plastic or metal bead or some other solid matrix (Figure 9-12). Such systems are called **solid-phase immunosorbent assays** (SPIA). If antigen is present in the fluid to be tested, stable antigen-antibody complexes form when the fluid is added to the matrix. Unbound antigen is thoroughly removed by washing, and a second antibody against the antigen being sought is then added to the system. This antibody has been complexed to an enzyme such as alkaline phosphatase or horseradish peroxidase. If the antigen is present on the solid matrix, it binds the second antibody, forming a sandwich with antigen in the middle. After washing has removed unbound, labeled antibody, the addition and hydrolysis of the enzyme substrate causes the color change and completes the reaction. The visually detectable end point appears wherever the enzyme is present (Figure 9-13). Because of the expanding nature of the reaction, even minute amounts of antigen (>1 ng/mL) can be detected. The system just described requires a specific enzyme-labeled antibody for each antigen tested. However, it is simpler to use an indirect assay in which a second, unlabeled antibody is used to bind to the antigen-antibody complex on the matrix. A third antibody, labeled with enzyme and directed against the non-variable Fc portion of the unlabeled second antibody, can then be used as the detection marker for many different antigen-antibody complexes (Figure 9-14). ELISA systems are important diagnostic tools for hepatitis B s (surface) and e antigens and HIV p24 protein, all indicators of early, active, acute infection.

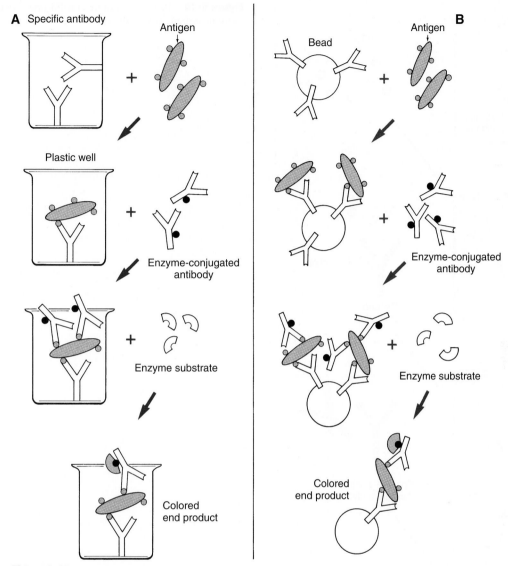

Figure 9-13   Principle of direct solid-phase enzyme immunosorbent assay (SPIA). **A,** Solid phase is microtiter well. **B,** Solid phase is bead.

Membrane-Bound SPIA

The flow-through and large surface area characteristic of nitrocellulose, nylon, or other membranes have been exploited to enhance the speed and sensitivity of ELISA reactions. The presence of absorbent material below the membrane serves to pull the liquid reactants through the membrane and helps to separate nonreacted components from the antigen-antibody complexes bound to the membrane; washing steps are also simplified. Membrane-bound SPIA systems are available for several viruses (Figure 9-15), group A beta-hemolytic streptococci antigen directly from throat swabs, and group B streptococcal antigen in vaginal secretions. In addition to their use in clinical laboratories, these assays are expected to become more prevalent for home testing systems.

OTHER IMMUNOASSAYS

Several other methods, including radioimmunoassay (RIA) and fluorescent immunoassay (FIA), are similar to ELISA except that radionucleotides (usually ^{125}I or ^{14}C) are substituted for enzymes in RIA and fluorochromes are substituted for enzymes in FIA. Although RIA was formerly the key method for antigen detection for numerous infectious agents, including hepatitis B virus, it has been largely replaced by ELISA testing, which does not require use of radioactive substances.

An optical immunoassay (OIA) has recently been introduced to detect group A streptococcal pharyngitis (Figure 9-16). It detects changes in the reflection of light occurring on the inert matrix after antigen and antibody have attached to it.

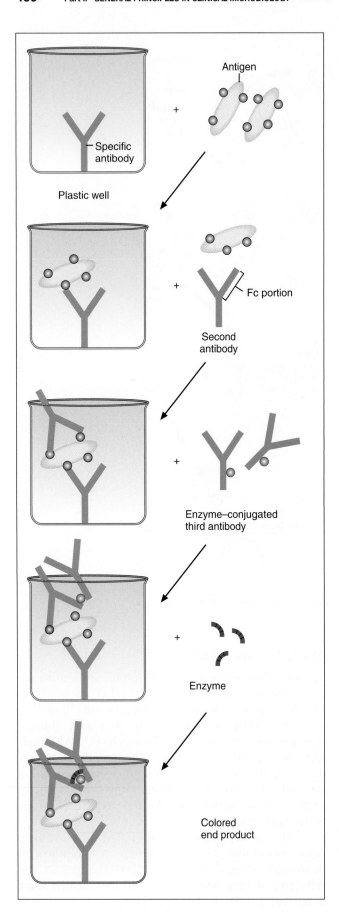

Figure 9-14 Principle of indirect solid-phase enzyme immunosorbent assay (SPIA).

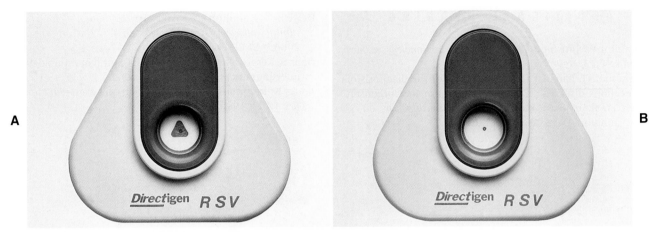

Figure 9-15 Directigen respiratory syncytial virus (RSV) membrane-bound cassette. **A,** Positive reaction. **B,** Negative reaction. (Courtesy Becton Dickinson Diagnostic Systems, Sparks, Md.)

Figure 9-16 Optical ImmunoAssay (OIA) for group A *Streptococcus.* **A,** Positive reaction. **B,** Negative reaction. **C,** Invalid reaction. (Courtesy BioStar, Inc., Boulder, Colo.)

ADDITIONAL READING

Benjamini E, Sunshine G, Leskowitz S: *Immunology: a short course*, ed 3, New York, 1996, Wiley-Liss.

Constatine NT, Lana DP: Immunoassays for the diagnosis of infections diseases. In Murray PR, Baron EJ, Jorgensen JH, et al, editors, *Manual of clinical microbiology*, ed 8, Washington, DC, 2003, ASM Press.

Detrick B, Hamilton RG, Folds JD: *Manual of molecular and clinical laboratory immunology*, ed 7, Washington, DC, 2006, ASM Press.

Gaur S, Kesarwala H, Gavai M, et al: Clinical immunology and infectious diseases, *Pediatr Clin North Am* 41:745, 1994.

James K: Immunoserology of infectious diseases, *Clin Microbiol Rev* 3:132, 1990.

Serologic Diagnosis of Infectious Diseases

An understanding of how immunochemical methods have been adapted as tools for serodiagnosis of infectious diseases requires a basic knowledge of the workings of the immune system. For further discussion of the host's response to foreign substances, see Chapter 3.

FEATURES OF THE IMMUNE RESPONSE

The host or patient has physical barriers, such as intact skin and ciliated epithelial cells, and chemical barriers, such as oils produced by the sebaceous glands and lysozyme found in tears and saliva, to prevent infections by foreign organisms. In addition, **natural immunity,** which is not specific, causes **chemotaxis,** the process by which phagocytes engulf organisms that enter the host and produce substances that attract white blood cells to the site of infection. **Acquired immunity** is the specific response of the host to an infecting organism.

The human specific immune responses are simplistically divided into the following two categories: cell-mediated and antibody-mediated.

Cell-mediated immune responses are carried out by special lymphocytes of the T (thymus-derived) class. T cells proliferate and differentiate into various effector T cells, including killer and helper cells. Killer cells, also known as **cytotoxic T lymphocytes,** specifically attack and kill microorganisms or host cells damaged or infected by these pathogens. Helper cells promote the maturation of B cells by producing activator cytokines that induce the B cells to produce antibodies and attach to and kill invading organisms. Although diagnosis of certain diseases may be aided by measuring the cell-mediated immune response to the pathogen, such tests entail skin tests performed by physicians or in vitro cell function assays performed by specially trained immunologists. These tests are usually not within the repertoire of clinical microbiology laboratories.

Antibody-mediated immune responses are those produced by specific proteins generated by lymphocytes of the B (bone marrow–derived) class. Because these proteins exhibit immunologic function and because they fold into a globular structure in the active state, they are also called **immunoglobulins.**

Antibodies are either secreted into the blood or lymphatic fluid (and sometimes other body fluids) by the B lymphocytes, or they remain attached to the surface of the lymphocyte or other cells. Because the cells involved in this category of immune response chiefly circulate in the blood, this type of immunity is also called **humoral immunity.** For purposes of determining whether an antibody has been produced against a particular infectious agent by a patient, the patient's serum (or occasionally the plasma) is tested for the presence of the antibody. The study of the diagnosis of disease by measuring antibody levels in serum is called **serology.**

CHARACTERISTICS OF ANTIBODIES

By a genetically determined mechanism, immunocompetent humans are able to produce antibodies specifically directed against almost all the antigens with which they might come into contact throughout their lifetimes and which the body recognizes as foreign. Antigens may be part of the physical structure of the pathogen, or they may be a chemical produced and released by the pathogen, such as an exotoxin. One pathogen may contain or produce many different antigens that the host will recognize as foreign, so that infection with one agent may cause a number of different antibodies to be produced. In addition, some antigenic determinants of a pathogen may not be available for recognition by the host until the pathogen has undergone a physical change. For example, until a pathogenic bacterium has been digested by a human polymorphonuclear leukocyte, certain antigens deep within the cell wall are not detected by the host immune system. Once the bacterium is broken down, these new antigens are revealed and antibodies can be produced against them. For this reason, a patient may produce different antibodies at different times during the course of a single disease. The immune response to an antigen also matures with continued exposure, and the antibodies produced against it become more specific and more **avid** (able to bind more tightly).

Antibodies function by (1) attaching to the surface of pathogens and making the pathogens more amenable to ingestion by phagocytic cells **(opsonizing antibodies);** (2) binding to and blocking surface receptors for host cells **(neutralizing antibodies);** or

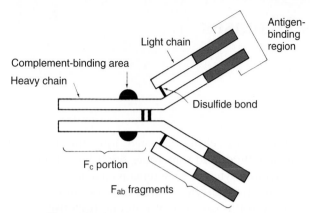

Figure 10-1   Structure of immunoglobulin G.

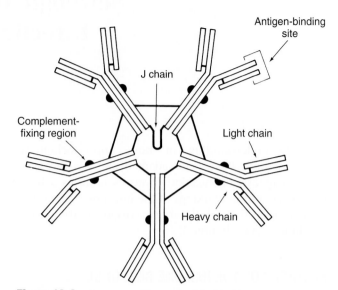

Figure 10-2   Structure of immunoglobulin M.

(3) attaching to the surface of pathogens and contributing to their destruction by the lytic action of complement **(complement-fixing antibodies).** Although routine diagnostic serologic methods are usually used to measure only two antibody classes, IgM and IgG, there are five different classes of antibodies, immunoglobulin G (IgG), immunoglobulin M (IgM), immunoglobulin A (IgA), immunoglobulin D (IgD), and immunoglobulin E (IgE). IgA is the predominant class of antibody in saliva, tears, and intestinal secretions. The role of IgD in infection is unknown, and IgE rises during infections by several parasites.

The basic structure of an antibody molecule comprises two mirror images, each composed of two identical protein chains (Figure 10-1). At the terminal ends are the antigen-binding sites, which specifically attach to the antigen against which the antibody was produced. Depending on the specificity of the antibody, antigens of some similarity, but not total identity, to the inducing antigen may also be bound; this is called a **cross reaction.** The complement-binding site is found in the center of the molecule in a structure that is similar for all antibodies of the same class. IgM is produced as a first response to many antigens, although the levels remain high only transiently. Thus, presence of IgM usually indicates recent or active exposure to antigen or infection. On the other hand, IgG antibody may persist long after infection has run its course. The IgM antibody type (Figure 10-2) consists of five identical proteins, with the basic antibody structures linked together at the bases, leaving 10 antigen-binding sites available. The second antibody class, IgG, consists of one basic antibody molecule with two binding sites (see Figure 10-1). The differences in the size and conformation between these two classes of immunoglobulins result in differences in activities and functions.

Features of the Humoral Immune Response Useful in Diagnostic Testing

Immunocompetent humans produce both IgM and IgG antibodies in response to most pathogens. In most cases, IgM is produced by a patient only after the first interaction with a given pathogen and is no longer detectable within a relatively short period afterward. For serologic diagnostic purposes, one important difference between IgG and IgM is that IgM cannot cross the placenta of pregnant women. Therefore, any IgM detected in the serum of a newborn baby must have been produced by the baby itself. The larger number of binding sites on IgM molecules can help to clear the offending pathogen more quickly, even though each individual antigen-binding site may not be the most efficient for attaching the antigen. Over time, the cells that were producing IgM switch to producing IgG.

IgG is often more specific for the antigen (more avid). The IgG has only two antigen-binding sites, but it can also bind complement. When IgG has bound to an antigen, the base of the molecule may be left projecting out in the environment. Structures on the IgG's base attract and bind the cell membranes of phagocytes, increasing the chances of engulfment and destruction of the pathogen by the host cells. A second encounter with the same pathogen usually induces only an IgG response. Because the B lymphocytes retain memory of this pathogen, they can respond more quickly and with larger numbers of antibodies than at the initial interaction. This enhanced response is called the **anamnestic response.** Because the B-cell memory is not perfect, occasional clones of memory cells will be stimulated by an antigen that is similar but not identical

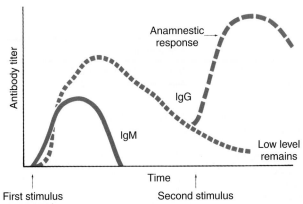

Figure 10-3 Relative humoral response to antigen stimulation over time.

to the original antigen; thus the anamnestic response may be polyclonal and nonspecific. For example, reinfection with cytomegalovirus may stimulate memory B cells to produce antibody against Epstein-Barr virus (another herpes family virus), which they encountered previously, in addition to antibody against cytomegalovirus. The relative humoral responses over time are diagrammed in Figure 10-3.

Interpretation of Serologic Tests

A central dogma of serology is the concept of rise in titer. The **titer** of antibody is the reciprocal of the highest dilution of the patient's serum in which the antibody is still detectable. Patients with large amounts of antibody have high titers, because antibody is still detectable at very high dilutions of serum. Serum for antibody levels should be drawn during the acute phase of the disease (when it is first discovered or suspected) and again during convalescence (usually at least 2 weeks later). These specimens are called **acute** and **convalescent sera.** For some infections, such as legionnaires' disease and hepatitis, titers may not rise until months after the acute infection, or they may never rise.

Patients with intact humoral immunity develop increasing amounts of antibody to a disease-causing pathogen over several weeks. If it is the patient's first encounter with the pathogen and the specimen has been obtained early enough, no or very low titers of antibody will be detected at the onset of disease. In the case of a second encounter, the patient's serum will usually contain measurable antibody during the initial phase of the disease and the level of the antibody will quickly increase, because of the anamnestic response. For most pathogens, an increase in the patient's titer of two doubling dilutions (e.g., from a positive result at 1:8 to a positive result at 1:32) is considered to be diagnostic of current infection. This is called a **fourfold rise in titer.**

Accurate results used for diagnosis of many infections are achieved only when acute and conva-

lescent sera are tested concurrently in the same test system, because variables inherent in the procedures and laboratory error can easily result in differences of one doubling (or twofold) dilution in the results obtained from even the same sample tested at the same time. Unfortunately, a certain proportion of infected patients may never show a rise in titer, necessitating the use of other diagnostic measures. Because the delay inherent in testing paired acute and convalescent sera results in diagnostic information that arrives too late to influence initial therapy, increasing numbers of early (IgM) serologic testing assays are being commercially evaluated. Moreover, it is sometimes more realistic to see a fourfold fall in titer between acute and convalescent sera when they are tested concurrently in the same system because sera may be collected late in the course of an infection in many cases when antibodies have already begun to decrease.

SERODIAGNOSIS OF INFECTIOUS DISEASES

With most diseases there exists a spectrum of responses in infected humans, such that a person may develop antibody from subclinical infection or after colonization by an agent without actually having symptoms of the disease. In these cases, the presence of antibody in a single serum specimen or a similar titer of antibody in paired sera may merely indicate past contact with the agent and cannot be used to accurately diagnose recent disease. Therefore, in the vast majority of serologic procedures for diagnosis of recent infection, testing both acute and convalescent sera is the method of choice. Except for detecting the presence of IgM, testing a single serum can be recommended only in certain cases, such as diagnosis of recent infection with *Mycoplasma pneumoniae* and viral influenza B, when high titers may indicate recent infection, or if the infecting or disease-causing agent is extremely rare, and people without disease or prior immunization would have no chance of developing an immune response, such as the case with rabies virus or the toxin of botulism.

The prevalence of antibody to an etiologic agent of disease in the population relates to the number of people who have come into contact with the agent, not the number who actually develop disease. For most diseases, only a small proportion of infected individuals actually develop symptoms; others develop antibodies that are protective without experiencing actual disease. In a number of circumstances serum is tested only to determine whether a patient is immune, that is, has antibody to a particular agent either in response to a past infection or to immunization. These tests can be performed with a single serum sample. Correlation of the results of such tests with the actual immune status of

Table 10-1   Tests Available for Serodiagnosis of Infectious Diseases

TEST	SERA NEEDED	INTERPRETATION	APPLICATION
IgM	Single, acute (collected at onset of illness)	Newborn, positive: in utero (congenital) infection Adult, positive: primary or current infection Negative: no infection or past infection	Newborn: STORCH* agents; other organisms Adults: any infectious agent
IgG	Acute and convalescent (collected 2-6 weeks after onset)	Positive: fourfold rise or fall in titer between acute and convalescent sera tested at the same time in the same test system Negative: no current infection or past infection, or patient is immunocompromised and cannot mount a humoral antibody response, or convalescent specimen collected before increase in IgG (Lyme disease, *Legionella*)	Any infectious agent
IgG	Single specimen collected between onset and convalescence	Adult, positive: adult evidence of infection at some unknown time except in certain cases in which a single high titer is diagnostic (rabies, *Legionella, Ehrlichia*) Newborn, positive: maternal antibodies that crossed the placenta Negative: patient has not been exposed to microorganism or patient has a congenital or acquired immune deficiency or specimen collected before increase in IgG (Lyme disease or *Legionella*)	Any infectious agent
Immune status evaluation	Single specimen collected at any time	Positive: previous exposure Negative: no exposure	Rubella testing for women of childbearing age, syphilis testing to obtain marriage license, cytomegalovirus testing transplant donor and recipient

*STORCH, Syphilis, toxoplasma, rubella virus, cytomegalovirus, herpes simplex virus.

individual patients must be performed to determine the level of detectable antibody present that corresponds to actual immunity to infection or reinfection in the host. For example, sensitive tests can detect the presence of very tiny amounts of antibody to the rubella virus. Certain people, however, may still be susceptible to infection with the rubella virus with such small amounts of circulating antibody, and a higher level of antibody may be required to ensure protection from disease.

Alternatively, depending on the etiologic agent, even low levels of antibody may protect a patient from pathologic effects of disease yet not prevent reinfection. For example, a person previously immunized with killed poliovirus vaccine who becomes infected with pathogenic poliovirus experiences multiplication of the virus in the gut and virus entry into the circulation, but damage to the central nervous system is blocked by humoral antibody in the circulation. As more sensitive testing methods are developed and these types of problems become more common, microbiologists will need to work closely with clinicians to develop guidelines for interpreting serologic test results as they relate to the immune status of individual patients. Moreover, patients may respond to an antigenic stimulus by producing antibodies that can cross-react with other antigens. These antibodies are nonspecific and thus may cause misinterpretation of serologic tests.

Table 10-1 lists the serologic tests available for immunodiagnosis of infectious diseases, the specimen required, interpretation of positive and negative test results, and examples of applications of each technique.

PRINCIPLES OF SEROLOGIC TEST METHODS

Antibodies can be detected in many ways. In some cases, antibodies to an agent may be detected in more than one way, but the different antibody detection tests may not be measuring the same antibody. For this reason, the presence of antibodies to a particular pathogen as detected by one method may not correlate with the presence of antibodies to the same agent detected by another test method. Moreover, different test methodologies have varying degrees of sensitivity in detecting antibodies even if they are present. However, because IgM is usually produced only during a patient's first exposure to an infectious agent, the detection of specific IgM can help the clinician a great deal in establishing a diagnosis. Most of the serologic test methods can be adapted for analysis of IgM.

SEPARATING IgM FROM IgG FOR SEROLOGIC TESTING

IgM testing is especially helpful for diseases that may have nonspecific clinical presentations, such as toxoplasmosis, or for those for which rapid therapeutic decisions may be required. For example, rubella infection in pregnant women can lead to serious consequences to the fetus such as cataracts, glaucoma, mental retardation, and deafness. Therefore, pregnant women who are exposed to rubella virus and develop a mild febrile illness can be tested for the presence of antirubella IgM. If positive, the option of elective termination of pregnancy can be offered. An additional reason to measure for IgM alone is for diagnosis of neonatal infections. Because IgG can readily cross the placenta, newborn babies carry titers of IgG nearly identical to those of their mothers during their first 2 to 3 months of life until they produce their own antibodies. Accurate serologic diagnosis of infection in neonates requires either demonstration of a rise in titer (which takes time to occur) or the detection of specific IgM directed against the putative agent. Because the IgM molecule does not cross the placental barrier, any IgM would have to be of fetal origin. Agents that are difficult to culture or those that adult females would be expected to have encountered during their lifetimes, such as *Treponema pallidum,* cytomegalovirus, herpesvirus, *Toxoplasma,* or rubella virus are those for which specific identification of fetal IgM is most often used. The names of some of these agents have been grouped together with the acronym **STORCH** (syphilis, *Toxoplasma,* rubella, cytomegalovirus, and herpes). These tests should be ordered separately, depending on the clinical illness of a newborn suspected of having one of these diseases. In many instances however, infected babies appear clinically well. Furthermore, in many cases serologic tests yield false-positive or false-negative results. Thus, multiple considerations, including patient history and clinical situation, must be employed in the serodiagnosis of neonatal infection, and culture in many cases is still the most reliable diagnostic method.

Several methods have been developed to measure only specific IgM in sera that may also contain specific IgG. In addition to using a labeled antibody specific for only IgM as the marker or the IgM capture sandwich assays, the immunoglobulins can be separated from each other by physical means. Centrifugation through a sucrose gradient, performed at very high speeds, has been used in the past to separate IgM, which has a greater molecular weight, from IgG.

Other available IgM separation systems operate based on the fact that certain proteins on the surface of staphylococci (protein A) and streptococci (protein G) bind the Fc portion of IgG. A simple centrifugation step then separates the particles and their bound immunoglobulins from the remaining mixture, which contains the bulk of the IgM. Other methods use antibodies to remove IgM from sera containing both IgG and IgM. An added bonus of IgM separation systems is that **rheumatoid factor,** IgM antibodies produced by some patients against their own IgG, often binds to the IgG molecules being removed from the serum; consequently, these IgM antibodies are removed along with the IgG. Rheumatoid factor can cause nonspecific and interfering results with many serologic tests, and its presence should be taken into account.

METHODS OF ANTIBODY DETECTION

Direct Whole Pathogen Agglutination Assays

The most basic tests for antibody detection are those that measure the antibody produced by a host to determinants on the surface of a bacterial agent in response to infection with that agent. Specific antibodies bind to surface antigens of the bacteria in a thick suspension and cause the bacteria to clump together in visible aggregates. Such antibodies are called **agglutinins,** and the test is called **bacterial agglutination.** Electrostatic and other forces influence the formation of aggregates in solutions so that certain conditions are usually necessary for good results. Because most bacterial surfaces exhibit a negative charge, they tend to repel each other. Performance of agglutination tests in sterile physiologic saline (0.85% sodium chloride in distilled water), which contains free positive ions, enhances the ability of antibody to cause aggregation of bacteria. Although bacterial agglutination tests can be performed on the surface of both glass slides and in test tubes, tube agglutination tests are often more sensitive, because a longer incubation period (allowing more antigen and antibody to interact) can be used. The small volume of liquid used for slide tests requires a rather rapid reading of the result, before the liquid can evaporate.

Examples of bacterial agglutination tests are the tests for antibodies to *Francisella tularensis* and *Brucella* spp., which are part of a panel called **febrile agglutinin tests.** Bacterial agglutination tests are often used to diagnose diseases in which the bacterial agent is difficult to cultivate in vitro. Diseases that can be diagnosed by this technique include tetanus, yersiniosis, leptospirosis, brucellosis, and tularemia. The reagents necessary to perform many of these tests are commercially available, singly or as complete systems. Because most laboratories are able to culture and identify the causative agent, agglutination tests for certain diseases, such as typhoid fever, are seldom used today. Furthermore, the typhoid febrile agglutinin test (called the **Widal test**) is often positive in patients with infections caused by other bacteria because of cross-reacting antibodies or a previous immunization against typhoid. Appropriate specimens from patients suspected of having typhoid fever should be cultured for the presence of salmonellae instead.

Whole cells of parasites, including *Plasmodium, Leishmania,* or *Toxoplasma gondii,* have also been used for direct detection of antibody by agglutination. In addition to using the actual infecting bacteria or parasites as the agglutinating particles for the detection of antibodies, certain bacteria may be agglutinated by antibodies produced against another infectious agent. Many patients infected with one of the rickettsiae produce antibodies that can nonspecifically agglutinate bacteria of the genus *Proteus.* The **Weil-Felix test** detects these cross-reacting antibodies. As newer, more specific serologic methods of diagnosing rickettsial disease become more widely available, however, the use of the *Proteus* agglutinating test is being discontinued.

Particle Agglutination Tests

Numerous serologic procedures have been developed to detect antibody via the agglutination of an artificial carrier particle with antigen bound to its surface. As noted in Chapter 9, similar systems employing artificial carriers coated with antibodies are commonly used for detection of microbial antigens. Either artificial carriers, such as latex particles or treated red blood cells, or biologic carriers, such as whole bacterial cells, can carry an antigen on their surface that will bind with antibody produced in response to that antigen when it was introduced to the host. The size of the carrier enhances the visibility of the agglutination reaction, and the artificial nature of the system allows the antigen bound to the surface to be extremely specific.

Results of particle agglutination tests are dependent on several factors, including the amount and avidity of antigen conjugated to the carrier, the time of incubation together with the patient's serum (or other source of antibody), and the microenvironment of the interaction (including pH and protein concentration). Commercial tests have been developed as systems, complete with their own diluents, controls, and containers. For accurate results, they should be used as units without modifications. If tests have been developed for use with cerebrospinal fluid, for example, they should not be used with serum unless the package insert or the technical representative has certified such usage.

Treated animal red blood cells have also been used as carriers of antigen for agglutination tests; these tests are called indirect **hemagglutination,** or passive hemagglutination tests, because it is not the antigens of the blood cells themselves, but rather the passively attached antigens, that are being bound by antibody. The most widely used of these tests are the microhemagglutination test for antibody to *T. pallidum* (MHA-TP, so called because it is performed in a microtiter plate), the hemagglutination treponemal test for syphilis (HATTS), the passive hemagglutination tests for antibody to extracellular antigens of streptococci, and the rubella indirect hemagglutination tests, all of which are available commercially. Certain reference laboratories, such as the Centers for Disease Control and Prevention (CDC), also perform indirect hemagglutination tests for antibodies to some clostridia, *Burkholderia pseudomallei, Bacillus anthracis, Corynebacterium diphtheriae, Leptospira,* and the agents of several viral and parasitic diseases.

Complete systems for the use of latex or other particle agglutination tests are available commercially for the accurate and sensitive detection of antibody to cytomegalovirus, rubella virus, varicella-zoster virus, the heterophile antibody of infectious mononucleosis, teichoic acid antibodies of staphylococci, antistreptococcal antibodies, mycoplasma antibodies, and others. Latex tests for antibodies to *Coccidioides, Sporothrix, Echinococcus,* and *Trichinella* are available, although they are not widely used because of the uncommon occurrence of the corresponding infection or its limited geographic distribution. Use of tests for *Candida* antibodies has not yet shown results reliable enough for accurate diagnosis of disease.

Flocculation Tests

In contrast to the aggregates formed when particulate antigens bind to specific antibody, the interaction of soluble antigen with antibody may result in the formation of a precipitate, a concentration of fine particles, usually visible only because the precipitated product is forced to remain in a defined space within a matrix. Two variations of the **precipitin test,** flocculation and counterimmunoelectrophoresis, are widely used for serologic studies.

In **flocculation tests** the precipitin end product forms macroscopically or microscopically visible clumps. The Venereal Disease Research Laboratory test, known as the **VDRL,** is the most widely used flocculation test. Patients infected with pathogenic treponemes, most commonly *T. pallidum,* the agent of syphilis, form an antibody-like protein called **reagin** that binds to the test antigen, cardiolipin-lecithin–coated cholesterol particles, causing the particles to flocculate. Because reagin is not a specific antibody directed against *T. pallidum* antigens, the test is not highly specific but it is a good screening test, detecting more than 99% of cases of secondary syphilis.

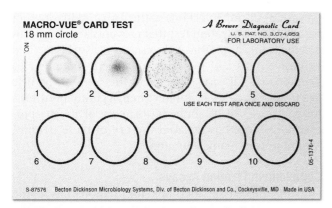

MACRO-VUE® CARD TEST
18 mm circle
FOR LABORATORY USE
USE EACH TEST AREA ONCE AND DISCARD

Figure 10-4 MACRO-VUE RPR card test. *R*, Reactive (positive) test; *N*, nonreactive (negative) test. (Courtesy Becton Dickinson Diagnostic Systems, Sparks, Md.)

The VDRL is the single most useful test available for testing cerebrospinal fluid in cases of suspected neurosyphilis, although it may be falsely positive in the absence of this disease. The performance of the VDRL test requires scrupulously clean glassware and exacting attention to detail, including numerous daily quality control checks. In addition, the reagents must be prepared fresh each time the test is performed, patients' sera must be inactivated by heating for 30 minutes at 56° C before testing, and the reaction must be read microscopically. For all these reasons, it has been supplanted in many laboratories by a qualitatively comparable test, the **rapid plasma reagin,** or **RPR, test.**

The RPR test is commercially available as a complete system containing positive and negative controls, the reaction card, and the prepared antigen suspension. The antigen, cardiolipin-lecithin–coated cholesterol with choline chloride, also contains charcoal particles to allow for macroscopically visible flocculation. Sera are tested without heating, and the reaction takes place on the surface of a specially treated cardboard card, which is then discarded (Figure 10-4). The RPR test is not recommended for testing cerebrospinal fluid. All procedures are standardized and clearly described in product inserts, and these procedures should be adhered to strictly. Overall, the RPR appears to be a more specific screening test for syphilis than the VDRL, and it is certainly easier to perform. Several modifications have been made, such as the use of dyes to enhance visualization of results.

Conditions and infections other than syphilis can cause a patient's serum to yield a positive result in the VDRL or RPR tests; these are called **biologic false-positive tests.** Autoimmune diseases, such as systemic lupus erythematosus and rheumatic fever; infectious mononucleosis; hepatitis; pregnancy; and old age have caused false-positive tests, so results of screening tests should be considered presumptive until confirmed with a specific treponemal test.

Counterimmunoelectrophoresis

Another variation of the classic precipitin test has been widely used to detect small amounts of antibody. This test takes advantage of the net electric charge of the antigens and antibodies being tested in a particular test buffer. Because the antigen and antibody being sought migrate toward each other in a semisolid matrix under the influence of an electrical current, the method is called **counterimmunoelectrophoresis** (CIE). The principle of this test is outlined in Chapter 9; the same methodology is used to identify specific antigen or antibody. When antigen and antibody meet in optimal proportions, a line of precipitation appears. Because all variables, such as buffer pH, type of gel or agarose matrix, amount of electrical current, amount and concentration of antigen and antibody, size of antigen and antibody inocula, and placement of these inocula, must be carefully controlled for maximum reactivity, CIE tests are difficult to develop and perform. Other methods for detection of antibody to infectious agents are more commonly used in most diagnostic laboratories.

Immunodiffusion Assays

Closely resembling the precipitin test is the Ouchterlony double immunodiffusion assay, which is widely used for detecting antibodies directed against fungal cell components. This test is described in Chapter 9. Whole-cell extracts or other antigens of the suspected fungus are placed in wells in an agarose plate, and the patient's serum and a control positive serum are placed in adjoining wells. If the patient has produced specific antibody against the fungus, precipitin lines will be visible within the agarose between the homologous (identical) antigen and antibody wells; their identity with similar lines from the control serum helps confirm the results. The type and thickness of the precipitin bands may have prognostic, as well as diagnostic, value. Antibodies against the pathogenic fungi *Histoplasma, Blastomyces, Coccidioides, Paracoccidioides,* and some opportunistic fungi are routinely detected by immunodiffusion. The test usually requires at least 48 hours, but additional time may be required to develop the bands.

Hemagglutination Inhibition Assays

Many human viruses can bind to surface structures on red blood cells from different species. For example, rubella virus particles can bind to human type O, goose, or chicken erythrocytes and cause agglutination of the red blood cells. Influenza and parainfluenza viruses agglutinate guinea pig, chicken, or human O ery-throcytes; many arboviruses agglutinate goose red blood cells; adenoviruses agglutinate rat or rhesus monkey cells; mumps virus binds red blood cells of monkeys; and herpesvirus and cytomegalovirus agglutinate sheep red blood cells. Serologic tests for the presence of antibodies to these viruses exploit the agglutinating properties of the virus particles. Patients' sera that have been treated with kaolin or heparin-magnesium chloride (to remove nonspecific inhibitors of red cell agglutina-tion and nonspecific agglutinins of the red cells) are added to a system that contains the virus suspected of causing disease. If antibodies to the virus are present, they will form complexes and block the binding sites on the viral surfaces. When the proper red cells are added to the solution, all of the virus particles will be bound by antibody, preventing the virus from aggluti-nating the red cells. Thus, the patient's serum is positive for hemagglutination-inhibiting antibodies. As for most serologic procedures, a fourfold increase in such titers is considered diagnostic. The hemagglutination inhibition tests for most agents are performed only at reference laboratories. Rubella antibodies, however, are often detected with this method in routine diagnostic labo-ratories. Several commercial rubella hemagglutination inhibition test systems are available.

Neutralization Assays

Antibody that inhibits the infectivity of a virus by blocking its host cell receptor site is called a **neu-tralizing antibody.** The serum to be tested is mixed with a suspension of infectious viral particles of the same type as those with which the patient is suspected of being infected. A control suspension of viruses is mixed with normal serum. The viral suspensions are then inoculated into a cell culture system that supports growth of the virus. The control cells will display evi-dence of viral infection. If the patient's serum contains antibody to the virus, that antibody will bind the viral particles and prevent them from invading the cells in culture. The antibody has neutralized the "infectivity" of the virus. These tests are technically demanding and time-consuming and are performed only in reference laboratories.

Antibodies to bacterial toxins and other extra-cellular products that display measurable activities can be tested in the same way. The ability of a patient's serum to neutralize the erythrocyte-lysing capability of streptolysin O, an extracellular enzyme produced by *S. pyogenes* during infection, has been used for many years as a test for previous streptococcal infection. After pharyngitis with streptolysin O–producing strains, most patients show a high titer of the antibody to streptolysin O, that is, antistreptolysin O (ASO) antibody. Strepto-cocci also produce the enzyme deoxyribonuclease B

(DNase B) during infections of the throat, skin, or other tissue. A neutralization test that prevents activity of this enzyme, the anti-DNase B test, has also been used extensively as an indicator of recent or previous streptococcal disease. However, the use of particle agglutination (latex or indirect hemagglutination) tests for the presence of antibody to many of the strepto-coccal enzymes has replaced the use of these neutral-ization tests in many laboratories.

Complement Fixation Assays

One of the classic methods for demonstrating the presence of antibody in a patient's serum has been the **complement fixation test** (CF). This test consists of two separate systems. The first (the test system) consists of the antigen suspected of causing the patient's disease and the patient's serum. The second (the indicator system) consists of a combination of sheep red blood cells, complement-fixing antibody (IgG) raised against the sheep red blood cells in another animal, and an exogenous source of complement (usually guinea pig serum). When these three components are mixed together in optimum concentrations, the antisheep erythrocyte antibody will bind to the surface of the red blood cells and the complement will then bind to the antigen-antibody complex, ultimately causing lysis (bursting) of the red blood cells. For this reason the antisheep red blood cell antibody is also called **hemol-ysin.** For the CF test these two systems are tested in sequence (Figure 10-5). The patient's serum is first added to the putative antigen; then the limiting amount of complement is added to the solution. If the patient's serum contains antibody to the antigen, the resulting antigen-antibody complexes will bind all of the complement added. In the next step the sheep red blood cells and the hemolysin (indicator system) are added. Only if the complement has not been bound by a complex formed with antibody from the patient's serum will the complement be available to bind to the sheep cell–hemolysin complexes and cause lysis. A positive result, meaning the patient does possess complement-fixing antibodies, is revealed by failure of the red blood cells to lyse in the final test system. Lysis of the indicator cells indicates lack of antibody and a negative CF test.

Although requiring many manipulations, at least 48 hours for both stages of the test to be completed, and often yielding nonspecific results, this test has been used over the years to detect many types of antibodies, particularly antiviral and antifungal. Many new systems provide for improved recovery of pathogens or their products and for more sensitive and less demanding procedures for detection of antibodies, including particle agglutination, indirect fluorescent antibody tests, and enzyme-linked immunosorbent assay (ELISA) proce-dures, and have gradually been introduced to replace

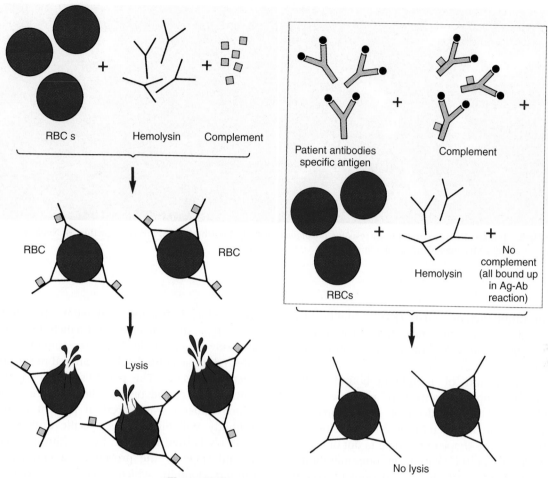

Figure 10-5   Complement fixation test.

the CF test. At this time, CF tests are performed chiefly for diagnosis of unusual infections and are done primarily in reference laboratories. This test is still probably the most common method for diagnosing infection caused by fungi, respiratory viruses, and arboviruses, as well as to diagnose Q fever. Laboratories without experience in performing these tests should not adopt complement fixation tests for routine diagnostic testing when other less demanding procedures are available.

Enzyme-Linked Immunosorbent Assays

ELISA technology for antibodies to infectious agents are sensitive and specific. As described more thoroughly in Chapter 9, the presence of specific antibody is detected by the ability of a second antibody conjugated to a colored or fluorescent marker to bind to the target antibody, which is bound to its homologous antigen. Various enzyme-substrate systems, including the use of avidin-biotin to bind marker substances, are also discussed in Chapter 9. The antigen to which the antibodies bind, if these antibodies are present in the

patient's sera, is either attached to the inside of the wells of a microtiter plate, adherent to a filter matrix, or bound to the surface of beads or plastic paddles. Advantages of ELISA tests are that they can be performed easily on many serum samples at the same time and that the colored or fluorescent end products are easily detected by instruments, removing the element of subjectivity inherent in so many serologic procedures that rely on a technologist's interpretation of a reaction. Disadvantages include the need for special equipment, the fairly long reaction times (often hours instead of minutes for particle agglutination tests), the relative end point of the test (which relies on measuring the amount of a visible end product that is not dependent on the original antigen-antibody reaction itself but on a second enzymatic reaction as compared with a directly quantitative result), and the requirement for batch processing to ensure that performance of the test is cost effective.

Commercial microdilution or solid-phase matrix systems are available to detect antibody specific for hepatitis virus antigens, herpes simplex viruses 1 and 2, respiratory syncytial virus (RSV), cytomegalovirus,

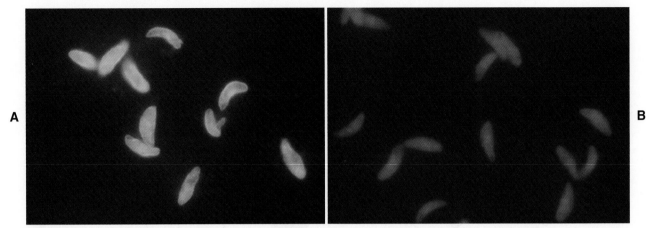

Figure 10-6   Indirect fluorescent antibody tests for *Toxoplasma gondii*, IgG antibodies. **A,** Positive reaction. **B,** Negative reaction. (Courtesy Meridian, Cincinnati, Ohio.)

human immunodeficiency virus (HIV), rubella virus (both IgG and IgM), mycoplasmas, chlamydiae, *Borrelia burgdorferi, Entamoeba histolytica, T. gondii,* and many other agents.

The introduction of membrane-bound ELISA components has improved sensitivity and ease of use dramatically. Slot-blot and dot-blot assays force the target antigen through a membrane filter, causing it to become affixed in the shape of the hole (a dot or a slot). Several antigens can be placed on one membrane. When test (patient) serum is layered onto the membrane, specific antibodies, if present, will bind to the corresponding dot or slot of antigen. Addition of a labeled second antibody and subsequent development of the label allows visual detection of the presence of antibodies based on the pattern of antigen sites. Cassette-based membrane-bound ELISA assays, designed for testing a single serum, can be performed rapidly (often within 10 minutes). Commercial kits to detect antibodies to *Helicobacter pylori, T. gondii,* and some other infectious agents are available. Accuracy of the results of tests using each of these formats is variable, however.

Antibody capture ELISAs are particularly valuable for detecting IgM in the presence of IgG. Anti-IgM antibodies are fixed to the solid phase, and thus only IgM antibodies, if present in the patient's serum, are bound. In a second step, specific antigen is added in a sandwich format and a second antigen-specific labeled antibody is finally added. Toxoplasmosis, rubella, and other infections are diagnosed using this technology, usually in research settings.

Indirect Fluorescent Antibody Tests and Other Immunomicroscopic Methods

A widely applied method to detect diverse antibodies is that of indirect fluorescent antibody determination

(IFA), which is described in detail in Chapter 9. For tests of this type, the antigen against which the patient makes antibody (e.g., whole *Toxoplasma* organisms or virus-infected tissue culture cells) is affixed to the surface of a microscope slide. The patient's serum to be tested is diluted and placed on the slide, covering the area in which antigen was placed. If present in the serum, antibody will bind to its specific antigen. Unbound antibody is then removed by washing the slide. In the second stage of the procedure, a conjugate of anti-human globulin, which may be directed specifically against IgG or IgM, and a dye that will fluoresce when exposed to ultraviolet light (e.g., fluorescein) is placed on the slide. This labeled marker for human antibody will bind to the antibody already bound to the antigen on the slide and will serve as a detector of binding of the antibody to the antigen when viewed under a fluorescence microscope (Figure 10-6). Commercially available test kits include the slides with the antigens, positive and negative control sera, diluent for the patients' sera, and the properly diluted conjugate. As with other commercial products, IFA systems should be used as units, without modifying the manufacturer's instructions. Currently, commercially available IFA tests include those for antibodies to *Legionella* species, *B. burgdorferi, T. gondii,* varicella-zoster virus, cytomegalovirus, Epstein-Barr virus capsid antigen, early antigen and nuclear antigen, herpes simplex viruses types 1 and 2, rubella virus, *M. pneumoniae, T. pallidum* (the **fluorescent treponemal antibody absorption test [FTA-ABS]**), and several rickettsiae. Most of these tests, if performed properly, give extremely specific and sensitive results. Proper interpretation of IFA tests requires experienced and technically competent technologists. These tests can be performed rapidly and are cost effective if only a few samples are tested.

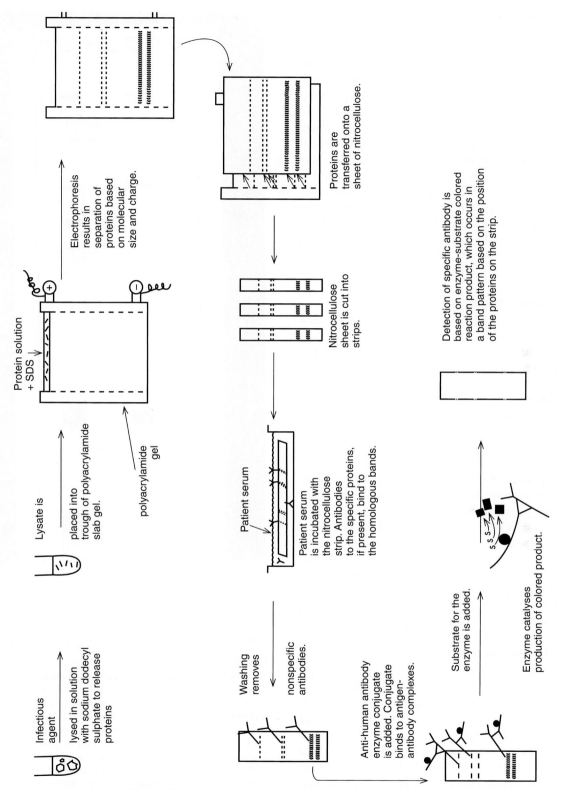

Figure 10-7   Diagram of Western blot immunoassay system.

Radioimmunoassays

Radioimmunoassay (RIA) is an automated method of detecting antibodies that is usually performed in a chemistry laboratory, rather than in a serology laboratory. RIA tests were originally used to detect antibody to hepatitis B viral antigens. Radioactively labeled antibody competes with the patient's unlabeled antibody for binding sites on a known amount of antigen. A reduction in radioactivity of the antigen–patient antibody complex compared with the radioactive counts measured in a control test with no antibody is used to quantitate the amount of patient antibody bound to the antigen. The development of new marker substances, such as ELISA systems, chemiluminescence, and fluorescence, resulted in the production of diagnostic tests as sensitive as RIA without the hazards and associated disposal problems of radioactive reagents.

Fluorescent Immunoassays

Because of the inconveniences associated with RIA (radioactive substances and expensive scintillation counters), fluorescent immunoassays (FIA) were developed. These tests, which use fluorescent dyes or molecules as markers instead of radioactive labels, are based on the same principle as RIA. The primary difference is that in RIA systems the competitive antibody is labeled with a radioisotope and in FIA the antigen is labeled with a compound that fluoresces under the appropriate light rays. Binding of patient antibody to a fluorescent-labeled antigen can reduce or quench the fluorescence, or binding can cause fluorescence by allowing conformational change in a fluorescent molecule. Measurement of fluorescence is thus a direct measurement of antigen-antibody binding, not dependent on a second marker system such as that in ELISA tests. Systems are commercially available to measure antibody developed against numerous infectious agents, as well as against self-antigens (autoimmune antibodies).

Western Blot Immunoassays

Requirements for the detection of very specific antibodies have driven the development of the Western blot immunoassay system (Figure 10-7). The method is based on the electrophoretic separation of major proteins of an infectious agent in a two-dimensional agarose gel matrix. A suspension of the organism against which antibodies are being sought is mechanically or chemically disrupted, and the solubilized antigen suspension is placed at one end of a polyacrylamide (polymer) gel. Under influence of an electrical current, the proteins migrate through the gel. Most bacteria or viruses contain several major proteins that can be recognized based on their position in the gel after electrophoresis. Smaller proteins travel faster and migrate

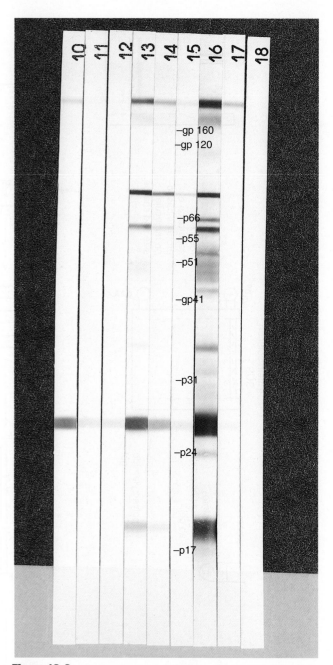

Figure 10-8   Human immunodeficiency virus type 1 (HIV-1) Western blot. Samples are characterized as positive, indeterminate, or negative based on the bands found to be present in significant intensity. A positive blot has any two or more of the following bands: p24, gp41, and/or gp120/160. An indeterminate blot contains some bands but not the definitive ones. A negative blot has no bands present. Lane 16 shows antibodies from a control serum binding to the virus-specific proteins (p) and glycoproteins (gp) transferred onto the nitrocellulose paper. (Courtesy Calypte Biomedical Corporation, Pleasanton, Calif.)

farther in the lanes of the gel than do the larger proteins. The protein bands are transferred from the gel to a nitrocellulose or other type of thin membrane, and the membrane is treated to immobilize the proteins. The membrane is then cut into many thin strips, each carrying the pattern of protein bands. When patient serum is layered over the strip, antibodies will bind to each of the protein components represented by a band on the strip. The pattern of antibodies present can be used to determine whether the patient is infected with the agent (Figure 10-8). Antibodies against microbes with numerous cross-reacting antibodies, such as *T. pallidum, B. burgdorferi,* herpes simplex virus types 1 and 2, and HIV, are identified more specifically using this technology than a method that tests for only one general antibody type. For example, CDC defines an ELISA or immunofluorescence assay as a first-line

Lyme disease antibody test but positive or equivocal results must be confirmed by a Western blot test.

ADDITIONAL READING

Benjamini E, Sunshine G, Leskowitz S: *Immunology: a short course,* ed 3, New York, 1996, Wiley-Liss.

Constantine NT, Lana DP: Immunoassays for the diagnosis of infectious diseases. In Murray PR, Baron EJ, Jorgensen JH, et al, editors: *Manual of clinical microbiology,* ed 8, Washington, DC, 2003, ASM Press.

Detrick B, Hamilton RG, Fold JD: *Manual of molecular and clinical laboratory immunology,* ed 7, Washington, DC, 2006, ASM Press.

Gaur S, Kesarwala H, Gavai M, et al: Clinical immunology and infectious diseases, *Pediatr Clin North Am* 41:745, 1994.

James K: Immunoserology of infectious diseases, *Clin Microbiol Rev* 3:132, 1990.

11 Principles of Antimicrobial Action and Resistance

Medical intervention in an infection primarily involves attempts to eradicate the infecting pathogen using substances that actively inhibit or kill it. Some of these substances are obtained and purified from other microbial organisms and are known as antibiotics. Others are chemically synthesized. Collectively, these natural and synthesized substances are referred to as *antimicrobial agents*. Depending on the type of organisms targeted, these substances are also known as antibacterial, antifungal, antiparasitic, or antiviral agents.

Because antimicrobial agents play a central role in the control and management of infectious diseases, some understanding of their mode of action and the mechanisms microorganisms deploy to circumvent antimicrobial activity is important, especially because diagnostic laboratories are expected to design and implement tests that measure a pathogen's response to antimicrobial activity (see Chapter 12). Much of what is discussed here regarding antimicrobial action and resistance is based on antibacterial agents, but the principles generally apply to almost all antiinfective agents. More information regarding antiparasitic, antifungal, and antiviral agents may be found in Parts IV, V, and VI, respectively.

ANTIMICROBIAL ACTION

PRINCIPLES

Several key steps must be completed for an antimicrobial agent to successfully inhibit or kill the infecting microorganism (Figure 11-1). First, the agent must be in an active form. This is ensured through the pharmacodynamic design of the drug, which takes into account the route through which the patient will receive the agent (e.g., orally, intramuscularly, intravenously). Second, the antibiotic must also be able to achieve sufficient levels or concentrations at the site of infection so that it has a chance to exert an antibacterial effect (i.e., be in anatomic approximation with the infecting bacteria). The ability to achieve adequate levels depends on the pharmacokinetic properties of the agent. Table 11-1 provides examples of various anatomic limitations that are characteristic of a few commonly used antibacterial agents. Some agents, such as ampicillin and ceftriaxone, achieve therapeutically effective levels in several body sites, whereas others, such as nitrofurantoin and norfloxacin, are limited to the urinary tract. Therefore, knowledge of the site of infection can substantially affect the selection of antimicrobial agent for therapeutic use.

The remaining steps in antimicrobial action relate to direct interactions between the antibacterial agent and the bacterial cell. When the antibiotic contacts the cell surface, adsorption results in the drug molecules maintaining contact with the cell surface. Next, because most targets for antibacterial agents are essentially intracellular, uptake of the antibiotic to some location within the bacterial cell is required. Once the antibiotic has achieved sufficient intracellular concentration, binding to a specific target occurs. This binding involves molecular interactions between the antimicrobial agent and one or more biochemical components that play an important role in the microorganism's cellular metabolism. Adequate binding of the target results in disruption of certain cellular processes leading to cessation of bacterial cell growth and, depending on the antimicrobial agent's mode of action, perhaps cell death. Antimicrobial agents that inhibit bacterial growth but generally do not kill the organism are known as *bacteriostatic* agents. Agents that usually kill target organisms are said to be *bactericidal* (Box 11-1).

The primary goal in the development and design of antimicrobial agents is to optimize a drug's ability to efficiently achieve all steps outlined in Figure 11-1 while minimizing toxic effects on human cells and physiology. Different antibacterial agents exhibit substantial specificity in terms of their bacterial cell targets, that is, their mode of action. For this reason, antimicrobial agents are frequently categorized according to their mode of action.

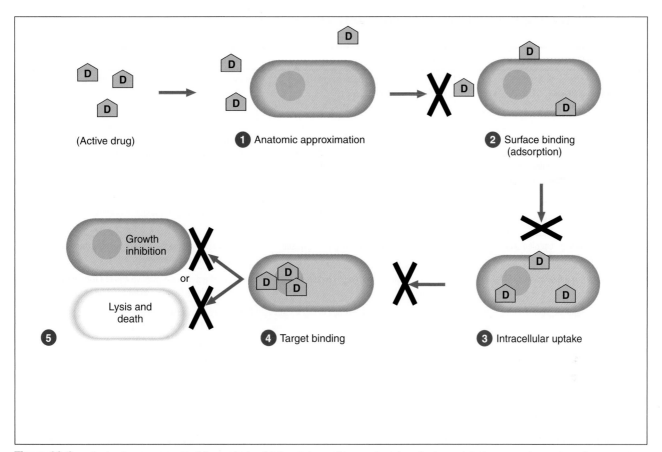

Figure 11-1 The basic steps required for antimicrobial activity and strategic points for bacterial circumvention or interference (marked by X) of antimicrobial action leading to resistance.

Table 11-1   Anatomic Distribution of Some Common Antibacterial Agents

	Serum-Blood*	Cerebrospinal Fluid	Urine
Ampicillin	+	+	+
Ceftriaxone	+	+	+
Vancomycin	+	±	+
Ciprofloxacin	+	±	+
Gentamicin	+	−	+
Clindamycin	+	−	−
Norfloxacin	−	−	+
Nitrofurantoin	−	−	+

*Serum-blood represents a general anatomic distribution.

+, Therapeutic levels generally achievable at that site; ±, therapeutic achievable levels moderate to poor; −, therapeutic levels generally not achievable at that site.

BOX 11-1   Bacteriostatic and Bactericidal Antibacterial Agents*

GENERALLY BACTERIOSTATIC
Chloramphenicol
Erythromycin and other macrolides
Clindamycin
Sulfonamides
Trimethoprim
Tetracyclines
Tigecycline
Linezolid
Quinupristin/dalfopristin
GENERALLY BACTERICIDAL
Aminoglycosides
Beta-lactams
Vancomycin
Daptomycin
Teicoplanin
Quinolones (e.g., ciprofloxacin, levofloxacin)
Rifampin
Metronidazole

*The bactericidal and bacteriostatic nature of an antimicrobial may vary depending on the concentration of the agent used and the bacterial species targeted.

MODE OF ACTION OF ANTIBACTERIAL AGENTS

Several potential antimicrobial targets exist within the bacterial cell, but those pathways or structures most frequently targeted include cell wall (peptidoglycan)

Table 11-2 Summary of Mechanisms of Action for Commonly Used Antibacterial Agents

Antimicrobial Class	Mechanism of Action	Spectrum of Activity
Aminoglycosides (e.g., gentamicin, tobramycin, amikacin, netilmicin, streptomycin, kanamycin)	Inhibit protein synthesis by binding to 30S ribosomal subunit	Gram-positive and gram-negative bacteria; not anaerobic bacteria
Beta-lactams (e.g., penicillin, ampicillin, mezlocillin, piperacillin, cefazolin, cefotetan, ceftriaxone, cefotaxime, ceftazidime, aztreonam, imipenem)	Inhibit cell wall synthesis by binding enzymes involved in peptidoglycan production (i.e., penicillin-binding proteins, or PBPs)	Both gram-positive and gram-negative bacteria, but spectrum may vary with the individual antibiotic
Chloramphenicol	Inhibits protein synthesis by binding 50S ribosomal subunit	Gram-positive and gram-negative bacteria
Fluoroquinolones (e.g., ciprofloxacin, ofloxacin, norfloxacin)	Inhibit DNA synthesis by binding DNA gyrases	Gram-positive and gram-negative bacteria, but spectrum may vary with individual antibiotic
Glycylglycines (e.g., tigecycline)	Inhibition of protein synthesis by binding to 30S ribosomal subunit	Wide spectrum of gram-positive and gram-negative species including those resistant to tetracycline
Ketolides (e.g., telithromycin)	Inhibition of protein synthesis by binding to 50S ribosomal subunit	Gram-positive cocci including certain macrolide-resistant strains and some fastidious gram-negatives (e.g., *H. influenzae* and *M. catarrhalis*)
Lipopeptides (e.g., daptomycin)	Binding and disruption of cell membrane	Gram-positive bacteria including those resistant to beta-lactams and glycopeptides
Nitrofurantoin	Exact mechanism uncertain; may have several bacterial enzyme targets and directly damage DNA	Gram-positive and gram-negative bacteria
Oxazolidinones (e.g., linezolid)	Bind to 50S ribosomal subunit to interfere with initiation of protein synthesis	Wide variety of gram-positive bacteria, including those resistant to other antimicrobial classes
Polymyxins (e.g., polymyxin B and colistin)	Disruption of cell membrane	Gram-negative bacteria, poor activity against most gram-positive bacteria
Rifampin	Inhibits RNA synthesis by binding DNA-dependent, RNA polymerase	Gram-positive and certain gram-negative (e.g., *N. meningitidis*) bacteria
Streptogramins (e.g., quinupristin/dalfopristin)	Inhibit protein synthesis by binding to two separate sites on the 50S ribosomal subunit	Primarily gram-positive bacteria
Sulfonamides	Interfere with folic acid pathway by binding the enzyme dihydropteroate synthase	Gram-positive and many gram-negative bacteria
Tetracycline	Inhibits protein synthesis by binding 30S ribosomal subunit	Gram-positive and gram-negative bacteria, and several intracellular bacterial pathogens (e.g., chlamydia)
Trimethoprim	Interferes with folic acid pathway by binding the enzyme dihydrofolate reductase	Gram-positive and many gram-negative bacteria

synthesis, the cell membrane, protein synthesis, and DNA and RNA synthesis. The different modes of antimicrobial action are summarized in Table 11-2.

Inhibitors of Cell Wall Synthesis

The bacterial cell wall, also known as the peptidoglycan, or murein layer, plays an essential role in the life of the bacterial cell. This fact, combined with the lack of a similar structure in human cells, has made the cell wall the focus of attention for the development of bactericidal agents that are relatively nontoxic for humans.

Beta-Lactam Antimicrobial Agents. Beta-lactam antibiotics are those that contain the four-membered, nitrogen-containing, beta-lactam ring at the core of their structure. The names and basic structures of commonly used beta-lactams are shown in Figure 11-2. This drug class comprises the largest group of antibacterial agents, and dozens of derivatives are available for clinical use. The popularity of these agents results from their bactericidal action and lack of toxicity to humans; also, their molecular structures can be manipulated to achieve greater activity for wider therapeutic applications.

Beta-lactam class	Examples	Base molecular structure
Penicillins	Penicillin Ampicillin Piperacillin Mezlocillin	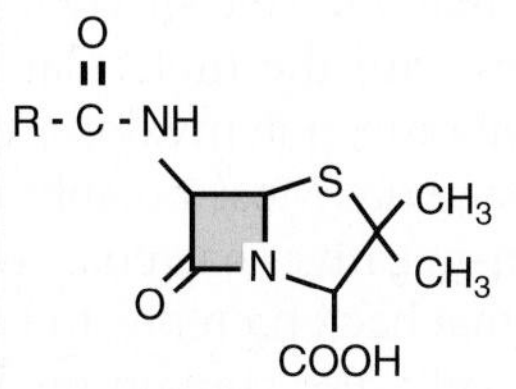
Cephalosporins	Cefazolin Cefuroxime Cefotetan Cefotaxime Ceftriaxone Ceftazidime Cefepime	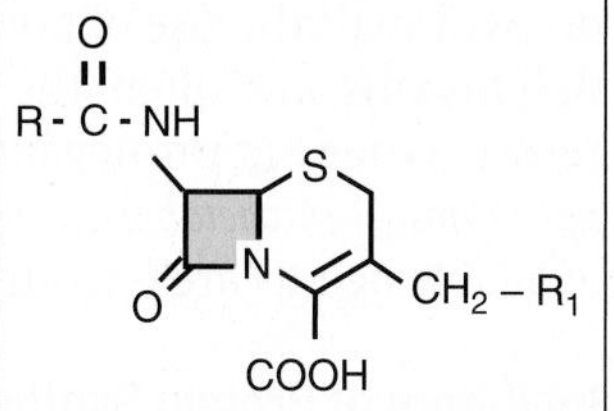
Monobactams	Aztreonam	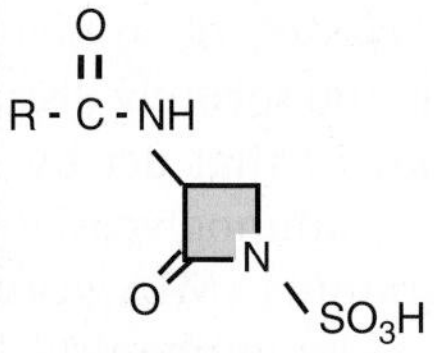
Carbapenems	Imipenem Meropenem	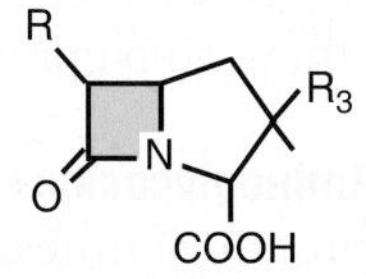

Figure 11-2 Basic structures and examples of commonly used beta-lactam antibiotics. The core beta-lactam ring is highlighted in yellow in each structure. (Modified from Salyers AA, Whitt DD, editors: *Bacterial pathogenesis: a molecular approach,* Washington, DC, 1994, ASM Press.)

The beta-lactam ring is key to the mode of action of these drugs that target and inhibit cell wall synthesis by binding the enzymes involved in synthesis. Most bacterial cells cannot survive once they have lost the capacity to produce and maintain their peptidoglycan layer. The enzymes essential for this function are anchored in the cell membrane and as a group are referred to as penicillin-binding proteins, or PBPs. Bacterial species may contain between four and six different types of PBPs. The PBPs involved in cell wall cross-linking (i.e., transpeptidases) are often the most critical for survival. When beta-lactams bind to these PBPs, cell wall synthesis is essentially halted. Death results from osmotic instability caused by faulty cell wall synthesis, or the binding of the beta-lactam to PBP may trigger a series of events that leads to autolysis and death of the cell.

Because nearly all clinically relevant bacteria have cell walls, beta-lactam agents act against a broad spectrum of gram-positive and gram-negative bacteria. However, because of differences among bacteria in their PBP content, natural structural characteristics (e.g., the outer membrane present in gram-negative but not gram-positive bacteria), and their common antimicrobial resistance mechanisms, the effectiveness of beta-lactams against different types of bacteria can vary widely. Therefore, any given beta-lactam drug will have a specific group or type of bacteria against which it is considered to have the greatest activity. The types of bacteria against which a particular antimicrobial agent does and does not have activity is referred to as that drug's *spectrum of activity*. Many factors contribute to an antibiotic's spectrum of activity. Knowledge of this spectrum of activity is key to many aspects of antimicrobial use and laboratory testing.

A common mechanism of bacterial resistance to beta-lactams is production of enzymes (i.e., beta-lactamases) that bind and hydrolyze these drugs. Therefore, beta-lactam combinations comprised of a beta-lactam with antimicrobial activity (e.g., ampicillin, amoxicillin, piperacillin) and a beta-lactam without activity that is capable of binding and inhibiting beta-lactamases (e.g., sulbactam, clavulanate, tazobactam) have been developed. The binding beta-lactam "ties up" the beta-lactamases produced by the bacteria and allows the other beta-lactam in the combination to exert its antimicrobial effect. Examples of these beta-lactam combinations include ampicillin/sulbactam, amoxicillin/clavulanate, and piperacillin/tazobactam. Such combinations are only effective against organisms that produce beta-lactamases that are bound by the inhibitor; they have little effect on resistance that is mediated by altered PBPs (see Mechanisms of Antibiotic Resistance section).

Glycopeptides. Glycopeptides are the other major class of antibiotics that inhibit bacterial cell wall synthesis; vancomycin is the most commonly used agent in this class. Vancomycin is a large molecule and functions differently from beta-lactam antibiotics (Figure 11-3). This agent does not bind to PBPs but does bind to precursors of cell wall synthesis. The binding interferes with the ability of the PBP enzymes, such as transpeptidases and transglycosylases, to incorporate the precursors into the growing cell wall. With the cessation of cell wall synthesis, cell growth stops and death often follows. Because vancomycin has a different mode of action, the resistance to beta-lactam agents by gram-positive bacteria does not generally hinder vancomycin activity. However, because of its relatively large size, vancomycin cannot penetrate the outer membrane of most gram-negative bacteria to

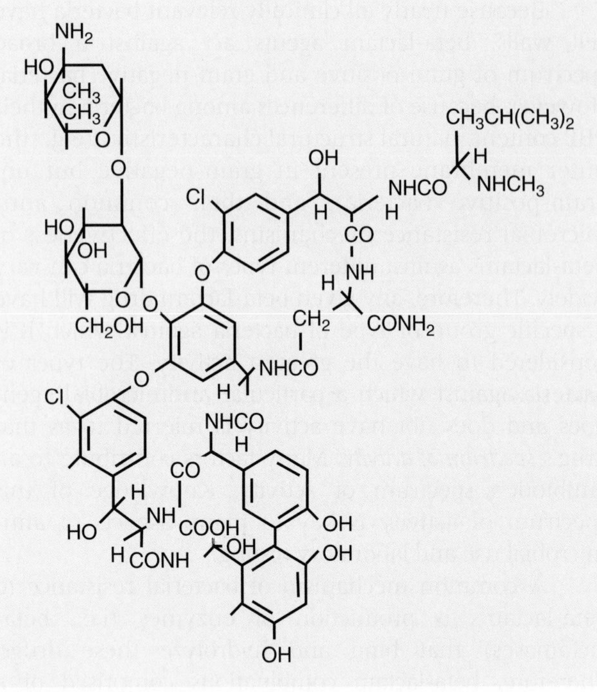

Figure 11-3 Structure of vancomycin, a non–beta-lactam antibiotic that inhibits cell wall synthesis. (Modified from Salyers AA, Whitt DD, editors: *Bacterial pathogenesis: a molecular approach,* Washington, DC, 1994, ASM Press.)

reach their cell wall precursor targets. Therefore, this agent is usually ineffective against gram-negative bacteria.

Several other cell wall–active antibiotics have been discovered and developed over the years, but toxicity to the human host has prevented their widespread clinical use. One example is bacitracin, which inhibits the recycling of certain metabolites required for maintaining peptidoglycan synthesis. Because of potential toxicity, bacitracin is usually only used as a topical antibacterial agent and internal consumption is generally avoided.

Inhibitors of Cell Membrane Function

Daptomycin, a lipopeptide agent that has recently been approved for clinical use is the most recent agent developed that exerts its antimicrobial effect by binding and disrupting the cell membrane of gram-positive bacteria. Although the exact mechanism of action is not fully understood, daptomycin has potent activity against gram-positive cocci, including those resistant to other agents such as beta-lactams and glycopeptides (e.g., methicillin-resistant *Staphylococcus aureus* [MRSA], vancomycin-resistant enterococci [VRE], and vancomycin-resistant *S. aureus* [VRSA]). Because of the molecule's size, daptomycin is unable to penetrate the outer membrane of gram-negative bacilli and thus is ineffective against that group of organisms.

Polymyxins (polymyxin B and colistin) are older agents that disrupt bacterial cell membranes. This disruption results in leakage of macromolecules and ions essential for cell survival. Because their effectiveness varies with the molecular makeup of the bacterial cell membrane, polymyxins are not equally effective against all bacteria. Most notably they are more active against gram-negative bacteria, while activity against gram-positive bacteria tends to be poor. Furthermore, human host cells also contain membranes so that toxicity risks do exist with the use of polymyxins. Although toxic, the polymyxins are often the antimicrobial agents of last resort when gram-negative bacilli (e.g. *Pseudomonas aeruginosa, Acinetobacter* spp.) resistant to all other available agents are encountered.

Inhibitors of Protein Synthesis

Several classes of antibiotics target bacterial protein synthesis and severely disrupt cellular metabolism. Antibiotic classes that act by inhibiting protein synthesis include aminoglycosides, macrolide-lincosamide-streptogramins (MLS group), ketolides (e.g., telithromycin) chloramphenicol, tetracyclines, glycylglycines (e.g., tigecycline), and oxazolidinones (e.g., linezolid). Although these antibiotics are generally categorized as protein synthesis inhibitors, the specific mechanisms by which they inhibit protein synthesis differ significantly.

Aminoglycosides. Aminoglycosides inhibit bacterial protein synthesis by binding to protein receptors on the organism's 30S ribosomal subunit. This process interrupts several steps, including initial formation of the protein synthesis complex, accurate reading of the mRNA code, and formation of the ribosomal-mRNA complex. The structure of a commonly used aminoglycoside, gentamicin, is given in Figure 11-4. Other available aminoglycosides include tobramycin, amikacin, netilmicin, streptomycin, and kanamycin. The spectrum of activity of aminoglycosides includes a wide variety of gram-negative and gram-positive bacteria. Aminoglycosides are often used in combination with cell wall–active antibiotics, such as beta-lactams or vancomycin, to achieve more rapid killing of certain bacteria. Anaerobic bacteria cannot take up these agents intracellularly so they are usually not inhibited by aminoglycosides.

Macrolide-Lincosamide-Streptogramin (MLS) Group. The most common antibiotics in the MLS group include the macrolides, such as erythromycin, azithromycin, and clarithromycin, and clindamycin (a lincosamide). Protein synthesis is inhibited by drug binding to receptors on the bacterial 50S ribosomal subunit and subsequent disruption of the growing peptide chain. Primarily because of uptake difficulties associated with gram-

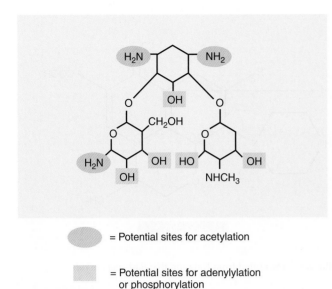

= Potential sites for acetylation

= Potential sites for adenylylation
or phosphorylation

Figure 11-4 Structure of the commonly used aminoglycoside gentamicin. Potential sites of modification by adenylylating, phosphorylating, and acetylating enzymes produced by bacteria are highlighted. (Modified from Salyers AA, Whitt DD, editors: *Bacterial pathogenesis: a molecular approach*, Washington, DC, 1994, ASM Press.)

negative outer membranes, the macrolides and clindamycin generally are not effective against most genera of gram-negative bacteria. However, they are effective against gram-positive bacteria. Newer agents include quinupristin/dalfopristin, which is a dual streptogramin that targets two sites on the 50S ribosomal subunit.

Ketolides. This group of compounds is made up of chemical derivatives related to erythromycin and other macrolides. As such, they too act by binding to the 50S ribosomal subunit and inhibiting protein synthesis. The key difference between the only currently available ketolide (i.e., telithromycin) and macrolides is that this ketolide maintains activity against most macrolide-resistant gram-positive organisms and does not induce a common macrolide resistance mechanism (i.e., the MLS$_B$ methylase) that alters the ribosomal target.

Oxazolidinones. Oxazolidinones, currently represented by linezolid, are a relatively new class of antibacterial agent available for clinical use. Linezolid is a synthetic agent that inhibits protein synthesis through a unique mechanism. Therefore, linezolid is not expected to be affected by resistance mechanisms that affect other drug classes.

Chloramphenicol. Chloramphenicol inhibits the addition of new amino acids to the growing peptide chain by binding to the 50S ribosomal subunit. This

antibiotic is highly active against a wide variety of gram-negative and gram-positive bacteria; however, its use has dwindled because of serious toxicity associated with it and the development of many other effective and safer agents, mostly of the beta-lactam class.

Tetracyclines. Tetracyclines inhibit protein synthesis by binding to the 30S ribosomal subunit so that incoming tRNA–amino acid complexes cannot bind to the ribosome, thus halting peptide chain elongation. Tetracyclines have a broad spectrum of activity that includes gram-negative bacteria, gram-positive bacteria, and several intracellular bacterial pathogens such as chlamydia, rickettsia, and rickettsia-like organisms. Similar to chloramphenicol, the development of several effective beta-lactams has caused a marked decrease in the use of these agents.

Glycylglycines. These agents are synthetic derivatives related to the tetracycline class and tigecycline is the first agent of this class to be approved for clinical use. Similar to the tetracyclines, tigecycline inhibits protein synthesis by binding to the 30S ribosomal subunit. However, tigecycline has the advantage of being refractory to the most common tetracycline resistance mechanisms expressed by gram-negative and gram-positive bacteria.

Inhibitors of DNA and RNA Synthesis

The primary antimicrobial agents that target DNA metabolism are the fluoroquinolones and metronidazole.

Fluoroquinolones. Fluoroquinolones, also often simply referred to as *quinolones,* are derivatives of nalidixic acid, an older antibacterial agent. The structures of two quinolones, ciprofloxacin and ofloxacin, are shown in Figure 11-5. These agents bind to and interfere with DNA gyrase enzymes involved in the regulation of bacterial DNA supercoiling, a process that is essential for DNA replication and transcription. The fluoroquinolones are potent bactericidal agents, and they have a broad spectrum of activity that includes gram-negative and gram-positive bacteria. However, the spectrum of activity can vary with each individual quinolone agent.

Metronidazole. The exact mechanism for metronidazole's antibacterial activity is uncertain, but it is believed to involve direct interactions between the activated drug and DNA that results in breakage of DNA strands. Activation of metronidazole requires reduction under conditions of low redox potential such as that found in anaerobic environments. Therefore, this agent is most potent against anaerobic bacteria, notably those that are gram-negative.

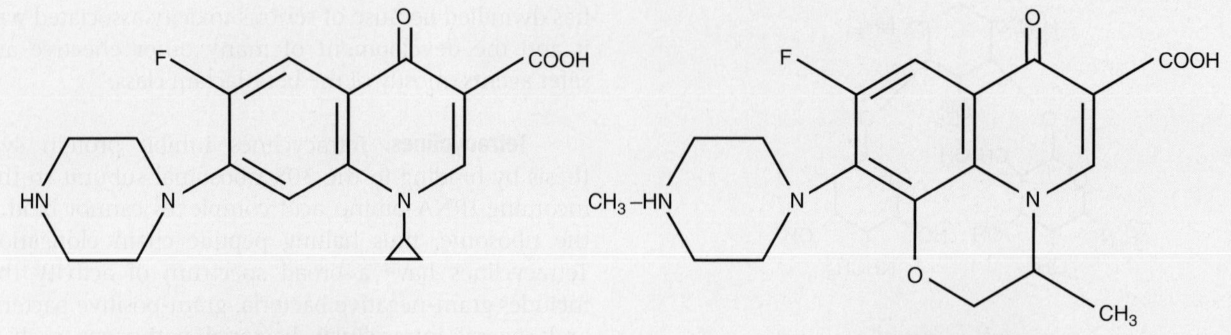

Figure 11-5 Structures of the fluoroquinolones ciprofloxacin and ofloxacin. (Modified from Katzung BG: *Basic and clinical pharmacology*, Norwalk, Conn, 1995, Appleton & Lange.)

Rifampin. Rifampin binds to the enzyme DNA-dependent RNA polymerase and inhibits synthesis of the RNA. Because rifampin does not penetrate the outer membrane of all gram-negative bacteria effectively, activity against these organisms is usually somewhat less than against gram-positive bacteria. Also, because spontaneous mutations that result in production of rifampin-insensitive RNA polymerases occur at relatively high frequencies, resistance can develop quickly. Therefore, rifampin is usually only used in combination with other antimicrobial agents.

Inhibitors of Other Metabolic Processes

Antimicrobial agents that target other bacterial processes other than those already discussed include sulfonamides, trimethoprim, and nitrofurantoin.

Sulfonamides. The folic acid pathway is used by bacteria to produce precursors important for DNA synthesis (Figure 11-6). Sulfonamides target and bind to one of the enzymes, dihydropteroate synthase, and disrupt the folic acid pathway. Several different sulfonamide derivatives are available for clinical use. These agents are active against a wide variety of gram-positive and gram-negative (except *P. aeruginosa*) bacteria.

Trimethoprim. Like sulfonamides, trimethoprim also targets the folic acid pathway. However, a different enzyme, dihydrofolate reductase, is inhibited (see Figure 11-6). Trimethoprim is active against several gram-positive and gram-negative species. Frequently, trimethoprim is combined with a sulfonamide (usually sulfamethoxazole) into a single formulation to produce an antibacterial agent that can simultaneously attack two targets on the same folic acid metabolic pathway. This drug combination can enhance activity against various bacteria and may help avoid the emergence of bacterial resistance to a single agent.

Nitrofurantoin. The mechanism of action of nitrofurantoin is not completely known. This agent may have several targets involved in bacterial protein and enzyme synthesis and the drug also may directly damage DNA. Nitrofurantoin is only used to treat urinary tract infections and has good activity against most of the gram-positive and gram-negative bacteria that cause infections at that site.

MECHANISMS OF ANTIBIOTIC RESISTANCE

PRINCIPLES

Successful bacterial resistance to antimicrobial action requires interruption or disturbance of one or more of the steps essential for effective antimicrobial action (see Figure 11-1). These disturbances or resistance mechanisms can come about in various ways, but the end result is partial or complete loss of antibiotic effectiveness. Different aspects concerning antimicrobial resistance mechanisms that are discussed include biologic vs. clinical antimicrobial resistance, environmentally mediated antimicrobial resistance, and microorganism-mediated antimicrobial resistance.

BIOLOGIC VS. CLINICAL RESISTANCE

Development of bacterial resistance to antimicrobial agents to which they were originally susceptible requires alterations in the cell's physiology or structure. Biologic

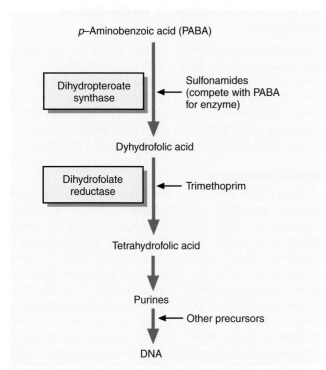

Figure 11-6 Bacterial folic acid pathway indicating the target enzymes for sulfonamide and trimethoprim activity. (Modified from Katzung BG: *Basic and clinical pharmacology*, Norwalk, Conn, 1995, Appleton & Lange.)

resistance refers to changes that result in the organism being less susceptible to a particular antimicrobial agent than has been previously observed. When antimicrobial susceptibility has been lost to such an extent that the drug is no longer effective for clinical use, the organism has achieved clinical resistance.

Of importance, biologic resistance and clinical resistance do not necessarily coincide. In fact, because most laboratory methods used to detect resistance are focused on detecting clinical resistance, microorganisms may undergo substantial changes in their levels of biologic resistance without notice. For example, for some time *Streptococcus pneumoniae*, a common cause of pneumonia and meningitis, could be inhibited by penicillin at concentrations of 0.03 μg/mL or less. The clinical laboratory focused on the ability to detect strains that required 2 μg/mL of penicillin or more for inhibition. This was the threshold for resistance that was believed to be required for interference with penicillin effectiveness. However, although no isolates were being detected that required more than 2 μg/mL of penicillin for inhibition, strains were developing biologic resistance that required penicillin concentrations 10 to 50 times higher than 0.03 μg/mL for inhibition.

From a clinical laboratory and public health perspective, it is important to realize that biologic development of antimicrobial resistance is an ongoing process.

Our inability to reliably detect all these processes with current laboratory procedures and criteria should not be misinterpreted as evidence that they are not occurring.

ENVIRONMENTALLY MEDIATED ANTIMICROBIAL RESISTANCE

Antimicrobial resistance is the result of nearly inseparable interactions among the drug, the microorganism, and the environment in which they are brought together. Characteristics of the antimicrobial agents, other than the mode and spectrum of activity, include important aspects of each drug's pharmacologic attributes. However these factors are beyond the scope of this text. Microorganism characteristics are discussed in subsequent sections of this chapter (see discussion of Microorganism-Mediated Antimicrobial Resistance). The impact that environment has on antimicrobial activity is considered here, and its importance cannot be overstated.

Environmentally mediated resistance is defined as resistance that directly results from physical or chemical characteristics of the environment that either directly alter the antimicrobial agent or alter the microorganism's normal physiologic response to the drug. Examples of environmental factors that mediate resistance include pH, anaerobic atmosphere, cation (e.g., Mg^{++} and Ca^{++}) concentrations, and thymidine content.

Several antibiotics are affected by the pH of the environment. For instance, the antibacterial activities of erythromycin and aminoglycosides diminish with decreasing pH, whereas the activity of tetracycline decreases with increasing pH.

Aminoglycoside-mediated shutdown of bacterial protein synthesis requires intracellular uptake across the cell membrane. Much of this aminoglycoside uptake is driven by oxidative processes, so that in the absence of oxygen, uptake, and hence activity, is substantially diminished.

Aminoglycoside activity is also affected by the concentration of cations such as Ca^{++} and Mg^{++} in the environment. This effect is most notable with *P. aeruginosa*. As shown in Figure 11-1, an important step in antimicrobial activity is the adsorption of the antibiotic to the bacterial cell surface. Aminoglycoside molecules have a net positive charge and, as is true for most gram-negative bacteria, the outer membrane of *P. aeruginosa* has a net negative charge. This electrostatic attraction facilitates attachment of the drug to the surface before internalization and subsequent inhibition of protein synthesis (Figure 11-7). However, calcium and magnesium cations compete with the aminoglycosides for negatively charged binding sites on the cell surface. If the positively charged calcium and magnesium ions outcompete aminoglycoside molecules for

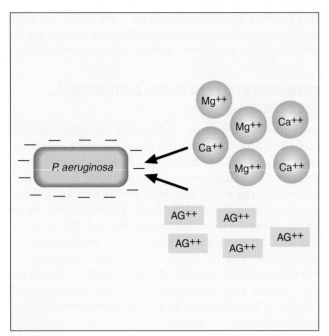

Figure 11-7 Cations (Mg^{++} and Ca^{++}) and aminoglycosides (AG^{++}) compete for the negatively charged binding sites on the outer membrane surface of *Pseudomonas aeruginosa*. Such competition is an example of the impact that environmental factors (e.g., cation concentrations) can have on the antibacterial activity of aminoglycosides.

these sites, less drug will be taken up and antimicrobial activity will be diminished. For this reason, aminoglycoside activity against *P. aeruginosa* tends to decrease as environmental cation concentrations increase.

The presence of certain metabolites or nutrients in the environment can also affect antimicrobial activity. For example, enterococci are able to use thymine and other exogenous folic acid metabolites to circumvent the activities of the sulfonamides and trimethoprim, which are folic acid pathway inhibitors (see Figure 11-6). In essence, if the environment supplies other metabolites that a microorganism can use, the activities of antibiotics that target pathways for producing those metabolites are greatly diminished, if not entirely lost. In the absence of the metabolites, full susceptibility to the antibiotics may be restored.

Information regarding environmentally mediated resistance is used to establish standardized testing methods that minimize the impact of environmental factors so that microorganism-mediated resistance (see the following discussion) is more accurately determined. Of importance, testing conditions are not established to recreate the in vivo physiology of infection but are set to optimize detection of resistance expressed by microorganisms. This is an extremely important point and a major reason why susceptibility testing results cannot be used to predict the clinical outcome of patients undergoing antimicrobial therapy.

MICROORGANISM-MEDIATED ANTIMICROBIAL RESISTANCE

Microorganism-mediated resistance refers to antimicrobial resistance that is due to genetically encoded traits of the microorganism and is the type of resistance that in vitro susceptibility testing methods are targeted to detect (see Chapter 12). Organism-based resistance can be divided into two subcategories: intrinsic or inherent resistance, and acquired resistance.

Intrinsic Resistance

Antimicrobial resistance resulting from the normal genetic, structural, or physiologic state of a microorganism is referred to as intrinsic resistance. Such resistance is considered to be a natural and consistently inherited characteristic that is associated with the vast majority of strains that constitute a particular bacterial group, genus, or species. Therefore, this is predictable resistance so that once the identity of the organism is known, so are certain aspects of its antimicrobial resistance profile. Box 11-2 lists common examples of intrinsic resistance and the underlying reason for the resistance. Intrinsic resistance profiles are useful for determining which antimicrobial agents should be included in the battery of drugs that will be tested against specific types of organisms. For example, referring to the information given in Box 11-2, aztreonam would not be included in antibiotic batteries tested against gram-positive cocci. Similarly, vancomycin would not be routinely tested against gram-negative bacilli. As discussed in Chapter 7, intrinsic resistance profiles are also useful markers to aid in the identification of certain bacteria or bacterial groups.

Acquired Resistance

Antibiotic resistance that results from altered cellular physiology and structure caused by changes in a microorganism's usual genetic makeup is known as *acquired resistance*. Unlike intrinsic resistance, acquired resistance may be a trait associated with only some strains of a particular organism group or species, but not others. Therefore, the presence of this type of resistance in any clinical isolate is unpredictable, and this unpredictability is the primary reason why laboratory methods to detect resistance are necessary.

Because acquired resistance mechanisms are all genetically encoded, the methods for acquisition basically are those that allow for gene change or exchange. Therefore, resistance may be acquired by:

- Successful genetic mutation
- Acquisition of genes from other organisms via gene transfer mechanisms
- A combination of mutational and gene transfer events

BOX 11-2   Examples of Intrinsic Resistance to Antibacterial Agents

Natural Resistance	Mechanism
Anaerobic bacteria vs. aminoglycosides	Lack of oxidative metabolism to drive uptake of aminoglycosides
Gram-positive bacteria vs. aztreonam (a beta-lactam)	Lack of penicillin-binding proteins (PBPs) that bind and are inhibited by this beta-lactam antibiotic
Gram-negative bacteria vs. vancomycin	Lack of uptake resulting from inability of vancomycin to penetrate outer membrane
Pseudomonas aeruginosa vs. sulfonamides, trimethoprim, tetracycline, or chloramphenicol	Lack of uptake resulting from inability of antibiotics to achieve effective intracellular concentrations
Klebsiella spp. vs. ampicillin (a beta-lactam) targets	Production of enzymes (beta-lactamases) that destroy ampicillin before the drug can reach the PBP
Aerobic bacteria vs. metronidazole	Inability to anaerobically reduce drug to its active form
Enterococci vs. aminoglycosides	Lack of sufficient oxidative metabolism to drive uptake of aminoglycosides
Enterococci vs. all cephalosporin antibiotics	Lack of PBPs that effectively bind and are inhibited by these beta-lactam
Lactobacilli and *Leuconostoc* vs. vancomycin	Lack of appropriate cell wall precursor target to allow vancomycin to bind and inhibit cell wall synthesis
Stenotrophomonas maltophilia vs. imipenem (a beta-lactam)	Production of enzymes (beta-lactamases) that destroy imipenem before the drug can reach the PBP targets

COMMON PATHWAYS FOR ANTIMICROBIAL RESISTANCE

Whether resistance is intrinsic or acquired, bacteria share similar pathways or strategies to effect resistance to antimicrobial agents. Of the pathways listed in Figure 11-8, those that involve enzymatic destruction or alteration of the antibiotic, decreased intracellular uptake or accumulation of drug, and altered antibiotic target are the most common. One or more of these pathways may be expressed by a single cell to successfully avert the action of one or more antibiotics.

Resistance to Beta-Lactam Antibiotics

Bacterial resistance to beta-lactams may be mediated by enzymatic destruction of the antibiotics, altered antibiotic targets, or decreased intracellular uptake of the drug (Table 11-3). All three pathways play an important role in clinically relevant antibacterial resistance, but bacterial destruction of beta-lactams by producing beta-lactamases is by far the most common method of resistance. As shown in Figure 11-9, beta-lactamases open the beta-lactam ring and the altered structure of the drug prohibits subsequent effective binding to PBPs so that cell wall synthesis is able to continue.

Staphylococci are the gram-positive bacteria that most commonly produce beta-lactamase; approximately 90% or more of clinical isolates are resistant to penicillin as a result of enzyme production. Rare isolates of enterococci also produce beta-lactamase. Gram-negative bacteria, including *Enterobacteriaceae, P. aeruginosa,* and *Acinetobacter* spp. produce dozens of different beta-lactamase types that mediate resistance to one or more of the beta-lactam antibiotics.

Although the basic mechanism shown in Figure 11-9 for beta-lactamase activity is the same for all types

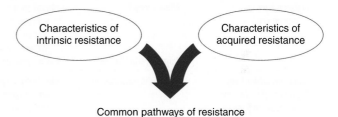

Common pathways of resistance

1. Enzymatic degradation or modification of the antimicrobial agent
2. Decreased uptake or accumulation of the antimicrobial agent
3. Altered antimicrobial target
4. Circumvention of the consequences of antimicrobial action
5. Uncoupling of antimicrobial agent-target interactions and subsequent effects on bacterial metabolism
6. Any combination of mechanisms 1 through 5

Figure 11-8   Overview of common pathways bacteria use to effect antimicrobial resistance.

of these enzymes, there are distinct differences. For example, beta-lactamases produced by gram-positive bacteria, such as staphylococci, are excreted into the surrounding environment where the hydrolysis of beta-lactams takes place before the drug can bind to PBPs in the cell membrane (Figure 11-10). In contrast, beta-lactamases produced by gram-negative bacteria remain intracellular in the periplasmic space where they are strategically positioned to hydrolyze beta-lactams as they transverse the outer membrane through water-filled, protein-lined porin channels (see Figure

Table 11-3 Summary of Resistance Mechanisms for Beta-Lactams, Vancomycin, Aminoglycosides, and Fluoroquinolones

Antimicrobial Class	Resistance Pathway	Specific Mechanism	Examples
Beta-lactams (e.g., penicillin, ampicillin, mezlocillin, piperacillin, cefazolin, cefotetan, ceftriaxone, cefotaxime, ceftazidime, aztreonam, imipenem)	Enzymatic destruction	Beta-lactamase enzymes destroy beta-lactam ring so antibiotic cannot bind to penicillin-binding protein (PBP) and interfere with cell wall synthesis (see Figure 11-9)	Staphylococcal resistance to penicillin. Resistance of *Enterobacteriaceae* and *Pseudomonas aeruginosa* to several penicillins, cephalosporins, and aztreonam
	Altered target	Mutational changes in original PBPs or acquisition of different PBPs that do not bind beta-lactams sufficiently to inhibit cell wall synthesis	Staphylococcal resistance to methicillin and other available beta-lactams. Penicillin and cephalosporin resistance in *Streptococcus pneumoniae* and viridans streptococci
	Decreased uptake	Porin channels (through which beta-lactams cross the outer membrane to reach PBP of gram-negative bacteria) change in number or character so that beta-lactam uptake is substantially diminished	*P. aeruginosa* resistance to imipenem
Glycopeptides (e.g., vancomycin)	Altered target	Alteration in the molecular structure of cell wall precursor components decreases binding of vancomycin so that cell wall synthesis is able to continue	Enterococcal and *Staphylococcus aureus* resistance to vancomycin
	Target overproduction	Excess peptidoglycan	Vancomycin-intermediate staphylococci
Aminoglycosides (e.g., gentamicin, tobramycin, amikacin, netilmicin, streptomycin, kanamycin)	Enzymatic modification	Modifying enzymes alter various sites on the aminoglycoside molecule so that the ability of drug to bind the ribosome and halt protein synthesis is greatly diminished or lost	Gram-positive and gram-negative resistance to aminoglycosides
	Decreased uptake	Porin channels (through which aminoglycosides cross the outer membrane to reach the ribosomes of gram-negative bacteria) change in number or character so that aminoglycoside uptake is substantially diminished	Aminoglycoside resistance in a variety of gram-negative bacteria
	Altered target	Mutational changes in ribosomal binding site diminishes ability of aminoglycoside to bind sufficiently and halt protein synthesis	Enterococcal resistance to streptomycin (may also be mediated by enzymatic modifications)
Quinolones (e.g., ciprofloxacin, ofloxacin, levofloxacin, norfloxacin, lomefloxacin)	Decreased uptake	Alterations in the outer membrane diminishes uptake of drug and/or activation of an "efflux" pump that removes quinolones before intracellular concentration sufficient for inhibiting DNA metabolism can be achieved	Gram-negative and staphylococcal (efflux mechanism only) resistance to various quinolones
	Altered target	Changes in the DNA gyrase subunits decrease ability of quinolones to bind this enzyme and interfere with DNA processes	Gram-negative and gram-positive resistance to various quinolones
Macrolides (e.g., erythromycin, azithromycin, clarithromycin)	Efflux	Pumps drug out of cell before target binding	Various streptococci and staphylococci
	Altered target	Enzymatic alteration of ribosomal target reduces drug binding	Various streptococci and staphylococci

11-10). Beta-lactamases also vary in their spectrum of substrates, that is, not all beta-lactams are susceptible to hydrolysis by every beta-lactamase. For example, staphylococcal beta-lactamase can readily hydrolyze penicillin and penicillin derivatives, such as ampicillin, mezlocillin, and piperacillin, but this enzyme cannot effectively hydrolyze many cephalosporins or imipenem.

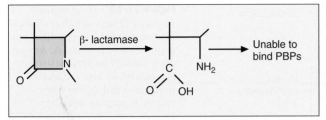

Figure 11-9 Mode of beta-lactamase enzyme activity. By cleaving the beta-lactam ring, the molecule can no longer bind to penicillin-binding proteins (PBPs) and is no longer able to inhibit cell wall synthesis. (Modified from Salyers AA, Whitt DD, editors: *Bacterial pathogenesis: a molecular approach,* Washington, DC, 1994, ASM Press.)

Various molecular alterations in the beta-lactam structure have been developed to protect the beta-lactam ring against enzymatic hydrolysis. This development has resulted in the production of more effective antibiotics in this class. For example, methicillin and closely related agents oxacillin and nafcillin are molecular derivatives of penicillin that by the nature of their structure are not susceptible to staphylococcal beta-lactamases. These agents are the mainstay of antistaphylococcal therapy. Similar strategies have been applied to develop penicillins and cephalosporins that are more resistant to the variety of beta-lactamases produced by gram-negative bacilli. Even with this strategy, it is important to note that among common gram-negative bacilli (e.g., *Enterobacteriaceae, P. aeruginosa,* and *Acinetobacter* spp.), the list of molecular types and numbers of beta-lactamases continues to emerge and diverge, thus challenging the effectiveness of currently available beta-lactam agents.

Another therapeutic strategy has been to combine two different beta-lactam drugs. One of the beta-lactams (the beta-lactamase inhibitor) avidly and irreversibly binds to the beta-lactamase and renders the enzyme incapable of hydrolysis, while the second beta-lactam, which is susceptible to beta-lactamase activity, exerts its antibacterial activity. Examples of beta-lactam/beta-lactamase inhibitor combinations include ampicillin/sulbactam, amoxicillin/clavulanic acid, and piperacillin/ tazobactam.

Altered targets also play a key role in clinically relevant beta-lactam resistance (see Table 11-3). By this pathway the organism changes, or acquires from another organism, genes that encode altered cell wall–synthesizing enzymes (i.e., PBPs). These "new" PBPs continue their function even in the presence of a beta-lactam antibiotic, usually because the beta-lactam lacks sufficient affinity for the altered PBP. This is the mechanism by which staphylococci are resistant to methicillin and all other beta-lactams (e.g., cephalosporins and imipenem). Therefore, strains that exhibit

this mechanism of resistance must be challenged with a non–beta-lactam agent, such as vancomycin, that acts on the cell wall. Changes in PBPs are also responsible for ampicillin resistance in *Enterococcus faecium* and in the widespread beta-lactam resistance observed in *S. pneumoniae* and viridans streptococci, organisms that to date have not been known to produce beta-lactamase.

Because gram-positive bacteria do not have outer membranes through which beta-lactams must pass before reaching their PBP targets, decreased uptake is not a pathway for beta-lactam resistance among these bacteria. However, diminished uptake can contribute significantly to beta-lactam resistance seen in gram-negative bacteria (see Figure 11-10). Changes in the number or characteristics of the outer membrane porins through which beta-lactams pass contribute to absolute resistance (e.g., *P. aeruginosa* resistance to imipenem). Additionally, porin changes combined with the presence of certain beta-lactamases in the periplasmic space may result in clinically relevant levels of resistance.

Resistance to Glycopeptides

To date, acquired, high-level resistance to vancomycin has been commonly encountered among enterococci, rarely among staphylococci, and not at all among streptococci. The mechanism involves the production of altered cell wall precursors that do not bind vancomycin with sufficient avidity to allow inhibition of peptidoglycan synthesizing enzymes. The altered targets are readily incorporated into the cell wall so that synthesis progresses as usual (see Table 11-3). A second mechanism of resistance to glycopeptides, described only among staphylococci to date, results in a lower level of resistance and is thought to be mediated by overproduction of the peptidoglycan layer, resulting in excessive binding of the glycopeptide molecule and diminished ability for the drug to exert its antibacterial effect.

Vancomycin is the only cell wall–inhibiting agent for use against gram-positive organisms that are resistant to all currently available beta-lactams (e.g., methicillin-resistant staphylococci and ampicillin-resistant enterococci). Therefore, the potential for vancomycin resistance to spread to other gram-positive genera poses a serious threat to public health. Resistance to vancomycin by enzymatic modification or destruction has not been described.

Resistance to Aminoglycosides

Analogous to beta-lactam resistance, aminoglycoside resistance is accomplished by enzymatic, altered target, or decreased uptake pathways (see Table 11-3). Gram-positive and gram-negative bacteria produce several

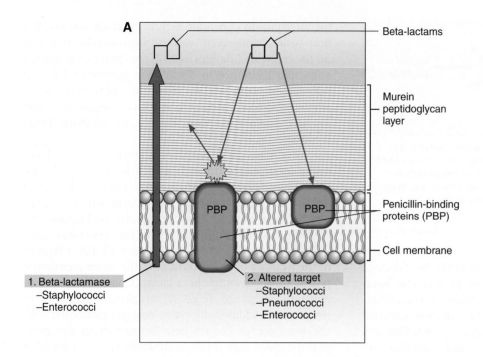

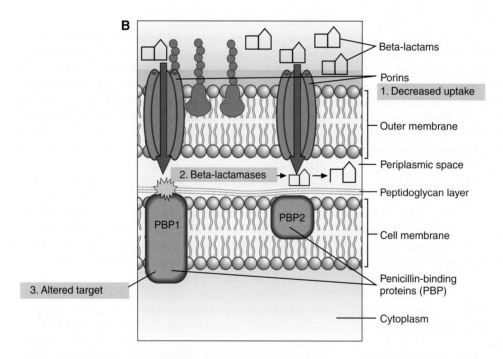

Figure 11-10 Diagrammatic summary of beta-lactam resistance mechanisms for gram-positive and gram-negative bacteria. **A,** Among gram-positive bacteria, resistance is mediated by beta-lactamase production and altered PBP targets. **B,** In gram-negative bacteria, resistance can also be mediated by decreased uptake through the outer membrane porins.

different aminoglycoside-modifying enzymes. Three general types of enzymes catalyze one of the following modifications of an aminoglycoside molecule (see Figure 11-4):

- Phosphorylation of hydroxyl groups
- Adenylylation of hydroxyl groups
- Acetylation of amine groups

Once an aminoglycoside has been modified, its affinity for binding to the 30S ribosomal subunit may be sufficiently diminished or totally lost so protein synthesis is able to continue unabated.

Aminoglycosides enter the gram-negative cell by passing through outer membrane porin channels. Therefore, porin alterations may also contribute to aminoglycoside resistance among these bacteria. Although some mutations that result in altered ribosomal targets have been described, the altered target pathway is thought to be a rare means for bacteria to achieve resistance to most commonly used aminoglycosides.

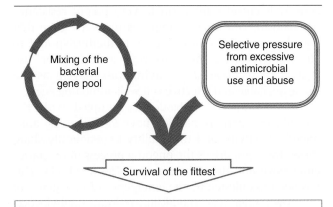

1. Emergence of "new" genes (e.g., methicillin-resistant staphylococci, vancomycin-resistant enterococci)
2. Spread of "old" genes to new hosts (e.g., penicillin-resistant *Neisseria gonorrhoeae*)
3. Mutations of "old" genes resulting in more potent resistance (e.g., beta-lactamase–mediated resistance to advanced cephalosporins in *Escherichia coli* and *Klebsiella* spp.)
4. Emergence of intrinsically resistant opportunistic bacteria (e.g., *Stenotrophomonas maltophilia*)

Figure 11-11 Factors contributing to the emergence and dissemination of antimicrobial resistance among bacteria.

Resistance to Quinolones

Enzymatic degradation or alteration of quinolones has not been fully described as a key pathway for resistance. Resistance is most frequently mediated by either decreased uptake or accumulation or by production of an altered target (see Table 11-3). Components of the gram-negative cellular envelope can limit quinolone access to the cell's interior site of DNA processing. Other bacteria, notably staphylococci, exhibit a mechanism by which the drug is "pumped" out of the cell after having entered, thus keeping the intracellular quinolone concentration sufficiently low to allow DNA processing to continue relatively unaffected. Therefore, this "efflux" process is a pathway of diminished accumulation of drug rather than of diminished uptake.

The primary quinolone resistance pathway involves mutational changes in the subunits of the DNA gyrases that are the target of quinolone activity. With a sufficient number or substantial type of changes, the gyrases no longer bind quinolones so DNA processing is able to continue.

Resistance to Other Antimicrobial Agents

Bacterial resistance mechanisms for other antimicrobial agents involve modifications or derivations of the recurring pathway strategies of enzymatic activity, altered target, or decreased uptake. These are summarized in Box 11-3.

EMERGENCE AND DISSEMINATION OF ANTIMICROBIAL RESISTANCE

The resistance pathways that have been discussed are not necessarily new mechanisms that have recently evolved among bacteria. By definition, antibiotics originate from microorganisms. Therefore, antibiotic resistance mechanisms always have been part of the evolution of bacteria as a means of survival among antibiotic-producing competitors. However, with the introduction of antibiotics into medical practice, clinically relevant bacteria have had to adopt resistance mechanisms as part of their survival strategy. With our use of antimicrobial agents, a survival of the fittest strategy has been used by bacteria to adapt to the pressures of antimicrobial attack (Figure 11-11).

All bacterial resistance strategies are encoded by one or more genes, and these resistance genes are readily shared between strains of the same species, between species of different genera, and even between more distantly related bacteria. When a resistance mechanism arises, either by mutation or gene transfer, in a particular bacterial strain or species, it is possible for this mechanism to then be passed on to other organisms using commonly described paths of genetic communication (see Figure 2-10). Therefore, resistance may spread to a wide variety of clinically important bacteria, and any single organism can acquire multiple genes and become resistant to the full spectrum of available antimicrobial agents. For example, there already exist strains of enterococci and *Pseudomonas aeruginosa* for which few effective therapeutic choices currently exist. Alternatively, multiple resistance may be mediated by a gene that encodes for a single very potent resistance mechanism. One such example is the *mecA* gene that encodes staphylococcal resistance to methicillin and to all other beta-lactams currently available for use against these organisms, leaving vancomycin as the single available and effective cell wall–inhibiting agent.

In summary, antibiotic use coupled with the formidable repertoire bacteria have for thwarting antimicrobial activity, and their ability to genetically share these strategies, drives the ongoing process of resistance emergence and dissemination (see Figure 11-11). This has been manifested by the emergence of new genes of unknown origin (e.g., methicillin-resistant staphylococci and vancomycin-resistant enterococci), the movement of old genes into new bacterial hosts (e.g., penicillin-resistant *N. gonorrhoeae*), mutations in familiar resistance genes that result in greater potency (e.g., beta-lactamase–mediated resistance to cephalosporins in *Escherichia coli*), and the emergence of new pathogens whose most evident virulence factor is intrinsic or natural resistance to many of the antimicrobial agents used in the hospital setting (e.g., *Stenotrophomonas maltophilia*).

The ongoing nature of resistance emergence and dissemination dictates that reliable laboratory procedures be used to detect resistance as an aid to managing the infected patient and as a means for monitoring changing resistance trends among clinically relevant bacteria.

ADDITIONAL READING

Bryan LE, editor: Microbial resistance to drugs. *Handb Exp Pharmacol* 91 Berlin, 1989, Springer-Verlag.

Courvalin P: Transfer of antibiotic resistance genes between gram-positive and gram-negative bacteria, *Antimicrob Agents Chemother* 38:1447, 1994.

Davies J: Inactivation of antibiotics and the dissemination of resistance genes, *Science* 264:375, 1994.

Lorian V, editor: *Antibiotics in laboratory medicine*, Philadelphia, 2005, Lippincott Williams &.Wilkins

Mandell GL, Bennett JE, Dolin R, editors: *Principles and practice of infectious diseases*, ed 6, New York, 2005, Elsevier Churchill Livingstone.

Neu HC: The crisis in antibiotic resistance, *Science* 257:1064, 1992.

Nikaido H: Prevention of drug access to bacterial targets: permeability barriers and active efflux, *Science* 264:382, 1994.

Rice LB, Sahm D, Bonomo RA: Mechanisms of resistance to antibacterial agents. In Murray PR, Baron EJ, Jorgensen JH, et al, editors: *Manual of clinical microbiology*, ed 8, Washington, DC, 2003, ASM Press.

Yao JD, Moellering RC: Antibacterial agents. In Murray PR, Baron EJ, Jorgensen JH, et al, editors: *Manual of clinical microbiology*, ed 8, Washington, DC, 2003, ASM Press.

Laboratory Methods and Strategies for Antimicrobial Susceptibility Testing

As discussed in Chapter 11, most clinically important bacteria are capable of acquiring and expressing resistance to antimicrobial agents commonly used to treat the infections they cause. Therefore, once an organism is isolated in the laboratory, characterization frequently includes tests to detect antimicrobial resistance. That is to say that in addition to the organism's identification, its antimicrobial susceptibility profile is often another key component of the clinical laboratory report that is produced for the physician. The procedures used to produce antimicrobial susceptibility profiles and detect resistance to agents that may be used therapeutically are referred to as *antimicrobial susceptibility testing methods*. The methods applied for profiling aerobic and facultatively anaerobic bacteria are the focus of this chapter; strategies for when and how these methods should be applied are also considered. Procedures for antimicrobial susceptibility testing of clinical isolates of anaerobic bacteria and mycobacteria are discussed in Chapters 42 and 45, respectively.

GOAL AND LIMITATIONS

The primary goal of antimicrobial susceptibility testing is to determine whether the bacterial etiology of concern is capable of expressing resistance to the antimicrobial agents that are potential choices for therapy. Because intrinsic resistance is usually well known for most organisms, testing to establish this resistance usually is not necessary and organism identification is sufficient. In essence, antimicrobial susceptibility tests are assays designed to determine the extent of an organism's acquired resistance for any clinically important organism whose antimicrobial susceptibility profile is unpredictable.

STANDARDIZATION

For laboratory tests to accurately determine organism-based resistances, the potential influence of environmental factors on antibiotic activity must be minimized (see Chapter 11). This is not to suggest that environmental resistance does not play a clinically relevant role, but the major focus of the in vitro tests is to measure an organism's expression of resistance. To control the

impact of environmental factors, the conditions for susceptibility testing are extensively standardized. Standardization serves three important purposes:

- To optimize bacterial growth conditions so that inhibition of growth can be attributed to the antimicrobial agent against which the organism is being tested and is not due to limitations of nutrient, temperature, or other environmental conditions that could hinder the organism's growth
- To optimize conditions for maintaining antimicrobial integrity and activity so that failure to inhibit bacterial growth can be attributed to organism-associated resistance mechanisms rather than to environmental drug inactivation
- To maintain reproducibility and consistency of results so that the same organism will produce the same resistance profile, regardless of which microbiology laboratory performs the test

Standard conditions for antimicrobial susceptibility testing methods have been established based on numerous laboratory investigations. Guidelines and recommendations for their use are published by the Clinical and Laboratory Standards Institute (CLSI), Subcommittee on Antimicrobial Susceptibility Testing. The CLSI documents that describe various methods for antimicrobial susceptibility testing are continuously updated and may be obtained by contacting CLSI, 940 W. Valley Road, Suite 1400, Wayne, Pa, 19087.

The components of antimicrobial susceptibility testing that are standardized and controlled include the following:

- Bacterial inoculum size
- Growth medium (most frequently a Mueller-Hinton base is used)
  - pH
  - Cation concentration
  - Blood and serum supplements
  - Thymidine content
- Incubation atmosphere
- Incubation temperature
- Incubation duration
- Antimicrobial concentrations tested

LIMITATIONS OF STANDARDIZATION

Although standardization of in vitro conditions is essential, the use of standard conditions imparts some limitations. Most notably, the laboratory test conditions in no way mimic the in vivo environment at the infection sites where the antimicrobial agent and bacteria will actually interact. Factors, such as bacterial inoculum size, pH, cation concentration, and oxygen tension (i.e., atmosphere), can differ substantially depending on the site of infection. Because of the lack of correlation between in vitro test conditions and the in vivo setting, antimicrobial susceptibility testing results cannot, and should not, be used as predictors of therapeutic outcome for the use of particular antimicrobial agents. Additionally, several other important factors that play key roles in patient outcome are not taken into account by susceptibility testing. Some of these factors include:

● Antibiotic diffusion in tissues and host cells
● Serum protein binding of antimicrobial agents
● Drug interactions and interference
● Status of patient defense and immune systems
● Multiple simultaneous illnesses
● Virulence and pathogenicity of infecting bacterium
● Site and severity of infection

Despite these limitations, it is known that antimicrobial resistance can substantially affect the morbidity and mortality of infected patients and the early and accurate recognition of resistant bacteria significantly aids in the selection of antimicrobial therapy and optimal management of patients. Thus, in vitro susceptibility testing provides valuable data that are used in conjunction with other diagnostic information to guide therapy of patients. Additionally, as discussed later in this chapter, in vitro susceptibility testing provides the data to track resistance trends among clinically relevant bacteria.

TESTING METHODS

PRINCIPLES

Three general methods are available to detect and evaluate antimicrobial susceptibility:

● Methods that directly measure the activity of one or more antimicrobial agents against a bacterial isolate
● Methods that directly detect the presence of a specific resistance mechanism in a bacterial isolate
● Special methods that measure complex antimicrobial-organism interactions

Which of these methods is used depends on factors that involve clinical need, accuracy, and convenience.

Given the complexities of antimicrobial resistance patterns, a laboratory may commonly use methods from one or more of these categories.

METHODS THAT DIRECTLY MEASURE ANTIMICROBIAL ACTIVITY

Methods that directly measure antimicrobial activity involve bringing the antimicrobial agents of interest and the infecting bacterium together in the same in vitro environment to determine the impact of the drug's presence on bacterial growth or viability. The level of impact on bacterial growth is measured and interpreted so that the organism's resistance or susceptibility to each agent tested can be reported to the clinician. Direct measures of antimicrobial activity are accomplished using:

● Conventional susceptibility testing methods such as broth dilution, agar dilution, and disk diffusion
● Commercial susceptibility testing systems
● Special screens and indicator tests

Conventional Testing Methods: General Considerations

Some general considerations apply to all three methods regarding inoculum preparation and selection of antimicrobial agents for testing.

Inoculum Preparation. Properly prepared inocula are key to any of the antimicrobial susceptibility testing methods. Inconsistencies in inoculum preparation often lead to inconsistencies and inaccuracies in susceptibility test results. The two important requirements for appropriate inoculum preparation include use of a pure culture and a standardized inoculum.

Interpretation of results obtained with a mixed inoculum is not reliable and failure to use a pure culture can substantially delay reporting of results. Pure inocula are obtained by selecting four to five colonies of the same morphology, inoculating them to a broth medium, and allowing the culture to achieve good active growth (i.e., mid-logarithmic phase), as indicated by observable turbidity in the broth. For most organisms this requires 3 to 5 hours of incubation. Alternatively, four to five colonies 16 to 24 hours of age may be selected from an agar plate and suspended in broth or 0.85% saline solution to achieve a turbid suspension.

Use of a standard inoculum size is as important as culture purity and is accomplished by comparison of the turbidity of the organism suspension with a turbidity standard. McFarland turbidity standards, prepared by mixing various volumes of 1% sulfuric acid and 1.175% barium chloride to obtain solutions with specific optical

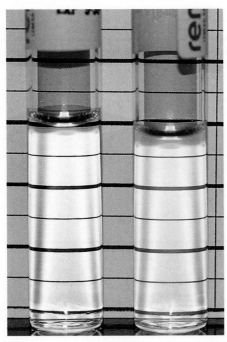

Figure 12-1 Bacterial suspension prepared to match the turbidity of the 0.5 McFarland turbidity standard. Matching this turbidity provides a bacterial inoculum concentration of 1 to 2 × 10^8 CFU/mL.

BOX 12-1 Criteria for Antimicrobial Battery Content and Use

ORGANISM IDENTIFICATION OR GROUP:
Antimicrobials to which the organism is intrinsically resistant are routinely excluded from the test battery (e.g., vancomycin vs. gram-negative bacilli). Similarly, certain antimicrobials were specifically developed for use against particular organisms but not others (e.g., ceftazidime for use against *Pseudomonas aeruginosa* but not against *Staphylococcus aureus*) and should only be included in the appropriate battery.

ACQUIRED RESISTANCE PATTERNS COMMON TO LOCAL MICROBIAL FLORA:
If resistance to a particular agent is common, the utility of the agent may be sufficiently limited so that routine testing is not warranted and only more potent antimicrobials are included in the test battery. Conversely, more potent agents may not need to be in the test battery if susceptibility to less potent agents is highly prevalent.

ANTIMICROBIAL SUSCEPTIBILITY TESTING METHOD USED:
Depending on the testing method, some agents do not reliably detect resistance and should not be included in the battery.

SITE OF INFECTION:
Antimicrobial agents, such as nitrofurantoin, that only achieve effective levels in the urinary tract should not be included in batteries tested against bacterial isolates from other body sites (i.e., must be able to achieve anatomic approximation). See Figure 11-1.

AVAILABILITY OF ANTIMICROBIAL AGENTS IN FORMULARY:
Antimicrobial agent test batteries are selected for their ability to detect bacterial resistance to agents that are used by the medical staff that the laboratory services.

densities, are most commonly used. The 0.5 McFarland standard, which is commercially available, provides an optical density comparable to the density of a bacterial suspension of 1.5 × 10^8 colony forming units (CFU)/mL. Pure cultures are grown or are directly prepared from agar plates to match the turbidity of the 0.5 McFarland standard (Figure 12-1). Matching turbidity using the unaided eye is facilitated by holding the bacterial suspension and McFarland tubes side by side and viewing them against a black-lined background. Alternatively, any one of various commercially available instruments that measure turbidity can be used to standardize inocula to match the 0.5 McFarland standard. If the bacterial suspension initially does not match the standard's turbidity, the suspension may be diluted or supplemented with more organisms as needed.

Selection of Antimicrobial Agents for Testing. The antimicrobial agents that are chosen for testing against a particular bacterial isolate are referred to as the **antimicrobial battery or panel.** A laboratory may use different testing batteries, but the content and application of each battery are based on various criteria. Although the criteria given in Box 12-1 affect decisions regarding battery content, the final decision is one that cannot and should not be made by the laboratory alone. Input from the medical staff (particularly infectious diseases specialists) and pharmacy is imperative.

The CLSI publishes up-to-date tables that list potential antimicrobial agents to include in batteries for testing against particular organisms or organism groups. Of particular interest is Table 1, titled, "Suggested Groupings of U.S. FDA-Approved Antimicrobial Agents That Should Be Considered for Routine Testing and Reporting on *Nonfastidious* Organisms by Clinical Microbiology Laboratories," and Table 1A, titled, "Suggested Groupings of U.S. FDA-Approved Antimicrobial Agents That Should Be Considered for Routine Testing and Reporting on *Fastidious* Organisms by Clinical Microbiology Laboratories." Because revisions can occur annually, up-to-date tables should be consulted when changes in testing batteries are being contemplated (see Additional Reading).

Further considerations about antibiotics that may be tested against particular organism groups are presented later in this chapter and in appropriate parts of the chapters in Part III. Most commonly, individual testing batteries are considered for each of the following organism groups:

- *Enterobacteriaceae*
- *Pseudomonas aeruginosa* and *Acinetobacter* spp.
- *Staphylococcus* spp.

Table 12-1 Summary of Broth Dilution Susceptibility Testing Conditions

Organism Groups	Test Medium	Inoculum Size (CFU/mL)	Incubation Conditions	Incubation Duration
Enterobacteriaceae	Mueller-Hinton	5×10^5	35° C; air	16-20 hr
Pseudomonas aeruginosa				
Enterococci				
Staphylococci to detect methicillin-resistant staphylococci	Mueller-Hinton plus 2% NaCl		30°-35° C; air	24 hr
Streptococcus pneumoniae and other streptococci	Mueller-Hinton plus 2%-5% lysed horse blood	5×10^5	35° C; air	20-24 hr
Haemophilus influenzae	*Haemophilus* test medium	5×10^5	35° C; air	20-24 hr
Neisseria meningitidis	Mueller-Hinton plus 2%-5% lysed horse blood	5×10^5	35° C; 5%-7% CO_2	24 hr

- *Enterococcus* spp.
- *Streptococcus* spp. (not including *S. pneumoniae*)
- *Streptococcus pneumoniae*
- *Haemophilus influenzae*
- *Neisseria gonorrhoeae*

Conventional Testing Methods: Broth Dilution

Broth dilution testing involves challenging the organism of interest with antimicrobial agents in a broth environment. Each antimicrobial agent is tested using a range of concentrations commonly expressed as μg (micrograms) of active drug/mL (milliliter) of broth (i.e., μg/mL). The concentration range tested for a particular drug depends on various criteria, including the concentration that is safely achievable in a patient's serum. Therefore, the concentration range tested will often vary from one drug to the next, depending on the pharmacologic properties of each. Additionally, the concentration range tested may be based on the level of drug that is needed to most reliably detect a particular underlying resistance mechanism. In this case, the test concentration for a drug may vary depending on the organism and its associated resistances that the test is attempting to detect. For example, to detect clinically significant resistance to cefotaxime in *S. pneumoniae,* the dilution scheme need only go as high as 2 μg/mL, but to detect cefotaxime resistance in *Escherichia coli,* the scheme must go up to 16 μg/mL or beyond.

Typically, the range of concentrations tested for each antibiotic are a series of doubling dilutions (e.g., 16, 8, 4, 2, 1, 0.5, 0.25 μg/mL); the lowest antimicrobial concentration that completely inhibits visible bacterial growth, as detected visually or via an automated or semiautomated method, is recorded as the **minimal inhibitory concentration** (**MIC**).

Procedures. The key features of broth dilution testing procedures are given in Table 12-1. Because

changes in procedural recommendations occur, the CLSI M07 series titled, "Methods for Dilution Antimicrobial Susceptibility Tests for Bacteria that Grow Aerobically" should be consulted annually.

Medium and Antimicrobial Agents. For any in vitro susceptibility testing method, it is necessary to alter certain conditions when testing particular types of organisms in order to optimize growth of some fastidious bacteria and facilitate expression of certain types of bacterial resistance. For example, Mueller-Hinton is the standard medium used for most broth dilution testing, and conditions of this medium (e.g., pH, cation concentration, thymidine content) are well controlled by commercial media manufacturers. However, media supplements or different media altogether are required to obtain good growth and reliable susceptibility profiles with relatively fastidious bacteria such as *S. pneumoniae* and *Haemophilus influenzae*. Also, although staphylococci are not fastidious organisms, media supplementation with NaCl is needed to enhance expression and detection of methicillin-resistant isolates (see Table 12-1).

Broth dilution testing is divided into two categories: microdilution and macrodilution. The principles of the tests are the same; the only difference is the volume of broth in which the test is performed. For microdilution testing, the total broth volume is 0.05 to 0.1 mL; for macrodilution testing, the broth volumes are usually 1 mL or greater. Because most susceptibility test batteries require testing several antibiotics at several different concentrations, the smaller volume used in microdilution allows this to be conveniently accomplished in a single microtitre tray format (Figure 12-2).

Use of test tubes as required by the macrodilution method becomes substantially cumbersome and labor intensive, especially because most laboratories must test

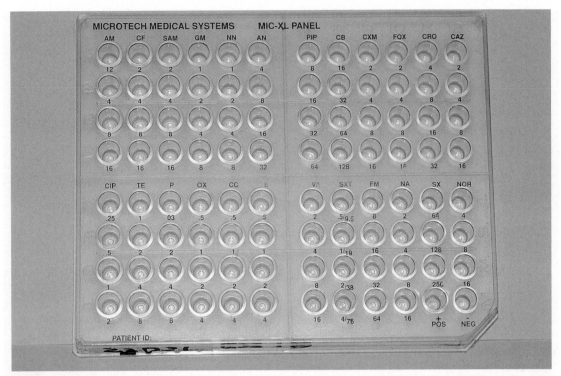

Figure 12-2   Microtitre tray for broth microdilution testing. Doubling dilutions of each antimicrobial agent in test broth occupies one vertical row of wells.

several bacterial isolates daily. For this reason, macrodilution is rarely used in most clinical laboratories and subsequent comments regarding broth dilution focus on the microdilution approach.

A key component to broth dilution testing is proper preparation and dilution of the antimicrobial agents that will be incorporated into the broth medium. Most laboratories that perform broth microdilution use commercially supplied microdilution panels in which the broth is already supplemented with appropriate antimicrobial concentrations. Therefore, antimicrobial preparation and dilution is not commonly done in most clinical laboratories, but details of this procedure are outlined in the CLSI M07-A6 document. In most instances, each antimicrobial agent is presented in the microtitre trays as a series of doubling twofold dilutions. To ensure against loss of antibiotic potency, the antibiotic microdilution panels are stored at −20° C or lower, if possible, and are thawed just before inoculation and use. Once thawed the panels should never be refrozen because doing so can result in substantial loss of antimicrobial potency. Alternatively, the antimicrobial agents may be lyophilized or freeze-dried with the medium or drug in each well and upon inoculation with the bacterial suspension the medium and drug are simultaneously reconstituted to the appropriate concentration.

Inoculation and Incubation. Standardized bacterial suspensions that match the turbidity of the 0.5 McFarland standard (i.e., 1.5×10^8 CFU/mL) usually serve as the starting point for dilutions that ultimately allow the final standard bacterial concentration of 5×10^5 CFU/mL in each microtitre well to be achieved. Of importance, the standard inoculum should be prepared from a fresh, overnight, pure culture of the test organism. Inoculation of the microdilution panel is readily accomplished using manual or automated multiprong inoculators that are calibrated to deliver the precise volume of inoculum to each well in the panel simultaneously (see Figure 12-2).

Inoculated trays are incubated under environmental conditions that optimize bacterial growth but do not interfere with antimicrobial activity (i.e., avoiding environmentally mediated results). For the most commonly tested bacteria (e.g., *Enterobacteriaceae, P. aeruginosa,* staphylococci, and enterococci) the environment is air at 35° C (see Table 12-1). Exceptions exist for the sake of testing more fastidious bacteria (e.g., *N. meningitidis* optimally requires 5% to 7% CO_2). Similarly, incubation durations for some organisms may need to be extended beyond the usual 16 to 20 hours (see Table 12-1). However, prolonged incubation times beyond recommended limits should be avoided because antimicrobial deterioration may

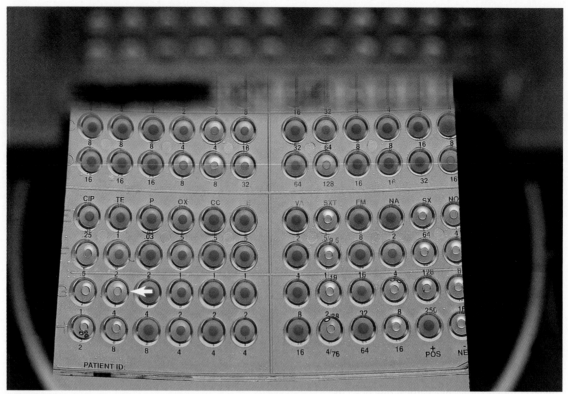

Figure 12-3 Bacterial growth profiles in a broth microdilution tray. The wells containing the lowest concentration of an antibiotic that completely inhibits visible growth *(arrow)* is recorded, in µg/mL, as the minimal inhibitory concentration (MIC).

result in false resistance interpretations. This is a primary factor that limits our ability to perform accurate testing with some slow-growing bacteria.

Reading and Interpretation of Results. Following incubation, the microdilution trays are examined for bacterial growth. Each tray should include a growth control that does not contain antimicrobial agent and a sterility control that was not inoculated. Once growth in the growth control and no growth in the sterility control wells have been confirmed, the growth profiles for each antimicrobial dilution can be established and the MIC determined. Detecting the presence of growth in microdilution wells is often augmented through the use of light boxes and reflecting mirrors. When a panel is placed in these devices, the presence of bacterial growth, which may be manifested as light to heavy turbidity or a button of growth on the well bottom, is more reliably visualized (Figure 12-3).

When the dilution series of each antibiotic is inspected, the microdilution well containing the lowest drug concentration that completely inhibited visible bacterial growth is recorded as the MIC. Once the MICs for the antimicrobials in the test battery for a particular organism have been recorded, they are usually translated into **interpretive categories of sus-**

ceptible, intermediate, or **resistant** (Box 12-2). The interpretive criteria for these categories are based on extensive studies that correlate MIC with serum achievable levels for each antimicrobial agent, particular resistance mechanisms, and successful therapeutic outcomes. The interpretive criteria for an array of antimicrobial agents are published in the CLSI M07 series document titled, "Methods for Dilution Antimicrobial Susceptibility Tests for Bacteria that Grow Aerobically (M100 supplements)." For example, using these standards, an isolate of *P. aeruginosa* with an imipenem MIC of less than or equal to 4 µg/mL would be classified as being susceptible, one with an MIC of 8 µg/mL would be considered intermediate, and one with an MIC of 16 µg/mL or greater would be classified as resistant to imipenem.

After the MICs are determined and their respective and appropriate interpretive categories assigned, the laboratory may report the MIC, the category, or both. Because the MIC alone will not provide most physicians with a meaningful interpretation of data, either the category result with or without the MIC is usually reported.

In some settings, the full range of antimicrobial dilutions is not used; only the concentrations that separate the categories of susceptible, intermediate,

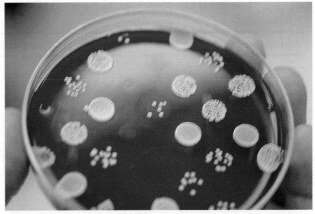

Figure 12-4 Growth pattern on an agar dilution plate. Each plate contains a single concentration of antibiotic, and growth is indicated by a spot on the agar surface. No spot is seen for isolates inhibited by the concentration of antibiotic incorporated into the agar of that particular plate.

and resistant are used. The specific concentrations that separate or define the different categories are known as **breakpoints,** and panels that only contain these antimicrobial concentrations are referred to as **breakpoint panels.** In this case, only category results are produced; precise MICs are not available because the full range of dilutions is not tested.

Advantages and Disadvantages. Broth dilution testing allows the option of providing both quantitative (i.e., MIC) and qualitative (i.e., category interpretation) results. Whether this is an advantage or not is controversial. On one hand, an MIC can be helpful in establishing the level of resistance of a particular bacterial strain and can substantially affect the decision to use certain antimicrobial agents. For example, the penicillin MIC for *S. pneumoniae* may determine whether penicillin or alternative agents will be used to treat a case of meningitis. On the other hand, for most antimicrobial susceptibility testing methods, a category report is sufficient so that once this determination is made the actual MIC data are superfluous. This is one reason why other methods (e.g., disk diffusion) that focus only on producing interpretive categories have been maintained in the clinical microbiology community.

Conventional Testing Methods: Agar Dilution

With agar dilution the antimicrobial concentrations and organisms to be tested are brought together on an agar-based medium rather than in broth. Each doubling dilution of an antimicrobial agent is incorporated into a single agar plate so that testing a series of six dilutions of one drug would require the use of six plates, plus one positive growth control plate without antibiotic. The standard conditions for agar dilution are given in Table 12-2 and, as shown in Figure 12-4, the surface of each plate is inoculated with 1×10^4 CFU. By this method, one or more bacterial isolates are tested per plate. After incubation, the plates are examined for growth and the MIC is the lowest concentration of an antimicrobial agent in agar that completely inhibits visible growth. The same MIC breakpoints and interpretive categories used for broth dilution are applied for interpretation of agar dilution methods. Similarly, test results may be reported as MICs only, as category only, or as both.

Preparing agar dilution plates (see CLSI M07-A6 series document titled, "Methods for Dilution Antimicrobial Susceptibility Tests for Bacteria That Grow Aerobically") is sufficiently labor intensive to preclude the use of this method in most clinical laboratories in which multiple antimicrobial agents must be tested, even though several isolates may be tested per plate. As with broth dilution, the standard medium is Mueller-Hinton, but supplements and substitutions are made as needed to facilitate growth of more fastidious organisms. (See Table 12-3 for agar media used for agar dilution and disk diffusion testing.) In fact, one advantage of this method is that it provides a means for determining MICs for *N. gonorrhoeae*, which does not grow sufficiently in broth to be tested by broth dilution methods.

Table 12-2 Summary of Agar Dilution Susceptibility Testing Conditions

Organism Groups	Test Medium	Inoculum Size (CFU/spot)	Incubation Conditions	Incubation Duration
Enterobacteriaceae	Mueller-Hinton	1×10^4	35° C; air	16-20 hr
Pseudomonas aeruginosa				
Enterococci				
Staphylococci to detect methicillin-resistant staphylococci	Mueller-Hinton plus 2% NaCl		30°-35° C; air	24 hr
Neisseria meningitidis	Mueller-Hinton plus 5% sheep blood	1×10^4	35° C; 5%-7% CO_2	24 hr
Streptococcus pneumoniae	Agar dilution not recommended method for testing this organism			
Other streptococci	Mueller-Hinton plus 5% sheep blood	1×10^4	35° C; air, CO_2 may be need for some isolates	20-24 hr
Neisseria gonorrhoeae	GC agar plus supplements	1×10^4	35° C; 5%-% CO_2	24 hr

Table 12-3 Summary of Disk Diffusion Susceptibility Testing Conditions

Organism Groups	Test Medium	Inoculum Size (CFU/mL)	Incubation Conditions	Incubation Duration
Enterobacteriaceae	Mueller-Hinton agar	Swab from 1.5×10^8	35° C; air	16-18 hr
Pseudomonas aeruginosa	Mueller-Hinton agar	Swab from 1.5×10^8 suspension	35° C; air	16-18 hr
Enterococci	Mueller-Hinton agar	Swab from 1.5×10^8 suspension	35° C; air	16-18 hr (24 hr for vancomycin)
Staphylococci to detect methicillin-resistant staphylococci	Mueller-Hinton agar	Swab from 1.5×10^8 suspension	30°-35° C; air	24 hr
Streptococcus pneumoniae and other streptococci	Mueller-Hinton agar plus 5% sheep blood	Swab from 1.5×10^8 suspension	35° C; 5%-7% CO_2	20-24 hr
Haemophilus influenzae	*Haemophilus* test medium	Swab from 1.5×10^8 suspension	35° C; 5%-7% CO_2	16-18 hr
Neisseria gonorrhoeae	GC agar plus supplements	Swab from 1.5×10^8 suspension	35° C; 5%-7% CO_2	20-24 hr

Conventional Testing Methods: Disk Diffusion

As more antimicrobial agents became available for treating bacterial infections, the limitations of the macrobroth dilution method became apparent. Before microdilution technology became widely available, a more practical and convenient method for testing multiple antimicrobial agents against bacterial strains was needed. Out of this need, the disk diffusion test was developed, spawned by the landmark study by Bauer, Kirby, Sherris, and Turck in 1966. These investigators standardized and correlated the use of antibiotic-impregnated filter paper disks (i.e., antibiotic disks) with MICs using many bacterial strains. Using the disk diffusion susceptibility test, antimicrobial resistance is detected by challenging bacterial isolates with antibiotic disks that are placed on the surface of an agar plate that has been seeded with a lawn of bacteria (Figure 12-5).

When disks containing a known concentration of antimicrobial agent are placed on the surface of a freshly inoculated plate, the agent immediately begins to diffuse and establish a concentration gradient around the paper disk. The highest concentration is closest to the disk. Upon incubation, the bacteria grow on the surface of the plate except where the antibiotic concentration in the gradient around each disk is sufficiently high to inhibit growth. Following incubation, the diameter of

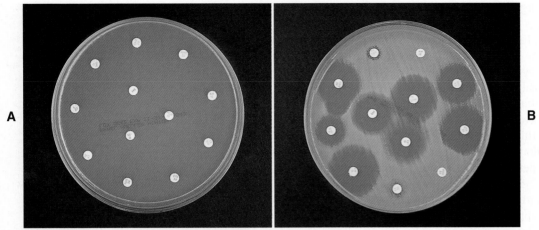

Figure 12-5   **A,** By the disk diffusion method, antibiotic disks are placed on the surface just after the agar surface was inoculated with the test organism. **B,** Zones of growth inhibition around various disks are apparent following 16 to 18 hours of incubation.

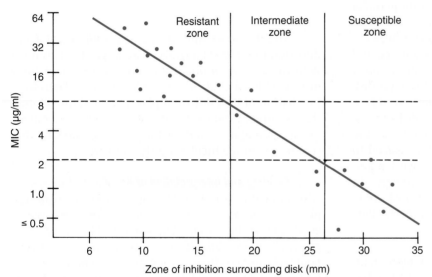

Figure 12-6   Example of a regression analysis plot to establish zone-size breakpoints for defining the susceptible, intermediate, and resistant categories for an antimicrobial agent. In this example, the maximum achievable serum concentration of the antibiotic is 8 μg/mL. Disk inhibition zones less than or equal to 18 mm in diameter indicate resistance, zones greater than or equal to 26 mm indicate susceptibility, and the intermediate category is indicated by zones ranging from 19 to 25 mm.

the zone of inhibition around each disk is measured in millimeters (see Figure 12-5).

To establish reference inhibitory zone–size breakpoints to define susceptible, intermediate, and resistant categories for each antimicrobial agent-bacterial species combination, hundreds of strains are tested. The inhibition zone sizes obtained are then correlated with MICs obtained by broth or agar dilution, and a regression analysis (i.e., line of best fit) is performed by plotting the zone size in millimeters against the MIC (Figure 12-6). As the MIC of the bacterial strains tested increase (i.e., more resistant bacterial strains), the corresponding inhibition zone sizes (i.e., diameters) decrease. Using Figure 12-6 to illustrate, horizontal lines are drawn from the MIC resistant breakpoint and the susceptible MIC breakpoint, 8 μg/mL

and 2 μg/mL, respectively. Where the horizontal lines intersect the regression line, vertical lines are drawn to delineate the corresponding inhibitory zone size breakpoints (in millimeters). Using this approach, zone size interpretive criteria have been established for most of the commonly tested antimicrobial agents and are published in the CLSI M02 series titled, "Performance Standards for Antimicrobial Disk Susceptibility Tests."

Procedures. The key features of disk diffusion testing procedures are summarized in Table 12-3, with more details and updates available through CLSI.

Medium and Antimicrobial Agents. Mueller-Hinton is the standard agar base medium for testing most bacterial organisms, with certain supplements and substitutions again required for testing more fastidious organisms. In addition to factors such as pH and cation content, the depth of the agar medium can also affect test accuracy and must be carefully controlled. Because antimicrobial agents diffuse in all directions from the surface of the agar plate, the thickness of the agar affects the antimicrobial drug concentration gradient. If the agar is too thick, zone sizes would be smaller; if too thin, the inhibition zones would be larger. For many laboratories that perform disk diffusion, commercial manufacturers are reliable sources for properly prepared and controlled Mueller-Hinton plates.

The appropriate concentration of drug for each disk is predetermined and set by the Food and Drug Administration. The disks are available from various commercial sources and should be kept at $-20°$ C in a desiccator until used. Thawed, unused disks may be stored at $4°$ to $8°$ C for up to a week. Inappropriate storage can lead to deterioration of the antimicrobial agents and result in misleading zone size results.

To ensure equal diffusion of the drug into the agar, the disks must be placed flat on the surface and be firmly applied to ensure adhesion. This is most easily accomplished by using any one of several disk dispensers that are available through commercial disk manufacturers. With these dispensers, all disks in the test battery are simultaneously delivered to the inoculated agar surface and are adequately spaced to minimize the chances for inhibition zone overlap and significant interactions between antimicrobials. In most instances, a maximum of 12 antibiotic disks may be applied to the surface of a single 150-mm Mueller-Hinton agar plate (see Figure 12-5).

Inoculation and Incubation.
Before disk placement, the plate surface is inoculated using a swab that has been submerged in a bacterial suspension standardized to match the turbidity of the 0.5 McFarland turbidity standard (i.e., 1.5×10^8 CFU/mL). The surface of the plate is swabbed in three directions to ensure an even and complete distribution of the inoculum over the entire plate. Within 15 minutes of inoculation, the antimicrobial agent disks are applied and the plates are inverted for incubation to avoid accumulation of moisture on the agar surface that can interfere with interpretation of test results.

For most organisms, incubation is at $35°$ C in air, but increased CO_2 is used when testing certain fastidious bacteria (see Table 12-3). Similarly, the incubation time may be increased beyond 16 hours to enhance detection of certain resistance patterns (e.g., methicillin resistance in staphylococci and vancomycin resistance in enterococci) and to ensure accurate results in general for certain fastidious organisms such as *N. gonorrhoeae*.

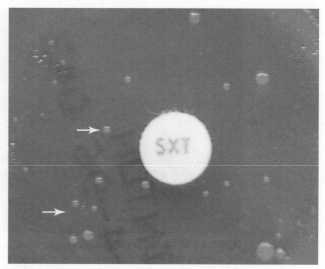

Figure 12-7 Disk diffusion plate that was inoculated with a mixed culture as evidenced by different colony morphologies *(arrows)* appearing throughout the lawn of growth.

The dynamics and timing of antimicrobial agent diffusion to establish a concentration gradient coupled with the growth of organisms over an 18- to 24-hour duration is critical for reliable results. Therefore, incubation of disk diffusion plates beyond the allotted time should be avoided and disk diffusion generally is not an acceptable method for testing organisms that require extended incubation times to grow.

Reading and Interpretation of Results.
Before results with individual antimicrobial agent disks are read, the plate is examined to confirm that a confluent lawn of good growth has been obtained (see Figure 12-5). If growth between inhibitory zones around each disk is poor and nonconfluent, then the test should not be interpreted but should be repeated. The lack of confluent growth may be due to insufficient inoculum. Alternatively, a particular isolate may have undergone mutation so that growth factors supplied by the standard susceptibility testing medium are no longer sufficient for supporting robust growth. In the latter case, medium supplemented with blood and/or incubation in CO_2 may enhance growth. However, caution in interpreting results is required when extraordinary measures are used to obtain good growth and the standard medium recommended for testing a particular type of organism is not used. Plates should also be examined for purity because mixed cultures are most evident as different colony morphologies scattered throughout the lawn of bacteria that is being tested (Figure 12-7). Mixed cultures require purification and repeat testing of the bacterial isolate of interest.

Using a dark background and reflected light (Figure 12-8), the plate is situated so that a ruler or

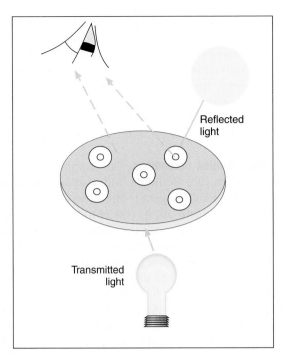

Figure 12-8   Examination of a disk diffusion plate by transmitted and reflected light.

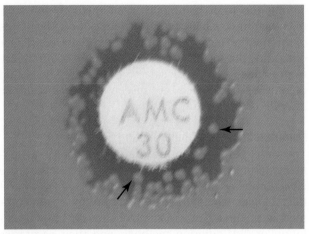

Figure 12-9   Individual bacterial colonies within a more obvious zone of inhibition *(arrows)*. This could indicate inoculation with a mixed culture. However, emergence of resistant mutants of the test isolate is a more likely reason for this growth pattern.

caliper may be used to measure inhibition zone diameters for each antimicrobial agent. Certain motile organisms, such as *Proteus* spp., may swarm over the surface of the plate and complicate clear interpretation of the zone boundaries. In these cases, the swarming haze is ignored and zones are measured at the point where growth is obviously inhibited. Similarly, hazes of bacterial growth may be observed when testing sulfonamides and trimethoprim as a result of the organism population going through several doubling generations before inhibition; the resulting haze of growth should be ignored for disk interpretation with these agents.

In instances not involving swarming organisms or the testing of sulfonamides and trimethoprim, hazes of growth that occur within more obvious inhibition zones should not be ignored. In many instances, this is the only way in which clinically relevant resistance patterns are manifested by certain bacterial isolates when tested using the disk diffusion method. Key examples in which this can occur include cephalosporin resistance among several species of *Enterobacteriaceae*, methicillin resistance in staphylococci, and vancomycin resistance in some enterococci. In fact, detection of the haze produced by some staphylococci and enterococci can best be accomplished using transmitted rather than reflected light. In these cases, the disk diffusion plates are held in front of the light source when methicillin and vancomycin inhibition zones are being read (see Figure 12-8). Still other significant resistances may be subtly

evident and appear as individual colonies within an obvious zone of inhibition (Figure 12-9). When such colonies are seen, purity of the test isolate must be confirmed. If purity is confirmed, the individual colonies are variants or resistant mutants of the same species and the test isolate should be considered resistant.

Once zone sizes are recorded, interpretive categories are assigned. Interpretive criteria for antimicrobial agent–organism combinations that can be tested by disk diffusion are provided in the CLSI-M2 series titled, "Performance Standards for Antimicrobial Disk Susceptibility Tests (M100 supplements)." Definitions for susceptible, intermediate, and resistant are the same as those used for dilution methods (see Box 12-2). For example, using the CLSI interpretive standards, an *E. coli* isolate that produces an ampicillin inhibition zone diameter of 13 mm or less is classified as resistant; if the zone is 14 to 16 mm, the isolate is considered intermediate to ampicillin; if the zone is 17 mm or greater, the organism is categorized as susceptible.

Unlike MICs, inhibition zone sizes are only used to produce a category interpretation and have no clinical utility in and of themselves. Therefore, when testing is performed by disk diffusion, only the category interpretation of susceptible, intermediate, or resistant is reported.

Advantages and Disadvantages.   One of the greatest advantages of the disk diffusion test is convenience and user friendliness. Up to 12 antimicrobial agents can be tested against one bacterial isolate with minimal use of extra materials and devices. Because results are generally accurate and most commonly encountered bacteria are reliably tested, the disk diffusion test is still among the most commonly used methods for antimicrobial susceptibility testing. The major disadvantages of this method include the lack of interpretive criteria for organisms not included in Table 12-3 and the inability to provide more precise data regarding the level of an organism's resistance or susceptibility as can be provided by MIC methods.

Commercial Susceptibility Testing Systems

The variety and widespread use of commercial susceptibility testing methods reflect the key role that resistance detection plays among the responsibilities of clinical microbiology laboratories. In many instances, the commercial methods are variations of the conventional dilution or disk diffusion methods, and their accuracies have been evaluated by comparison of results with those obtained by conventional methods. Additionally, many of the media and environmental conditions standardized for conventional methods are maintained with the use of commercial systems. The goal of detecting resistance is the same for all commercial methods, but the principles and practices that are applied to achieve that goal vary with respect to:

- Format in which bacteria and antimicrobial agents are brought together
- Extent of automation for inoculation, incubation, interpretation, and reporting
- Method used for detection of bacterial growth inhibition
- Speed with which results are produced
- Accuracy

Accuracy is an extremely important aspect of any susceptibility testing system and is addressed in more detail later in this chapter.

Broth Microdilution Methods. Several systems have been developed that provide microdilution panels already prepared and formatted according to the guidelines for conventional broth microdilution methods (e.g., BBL Sceptor, BD Microbiology Systems, Cockeysville, Md; Sensititre, Trek Diagnostics Systems, Inc., Westlake, Ohio; MicroScan touchSCAN-SR, Dade Behring, Inc., West Sacramento, Calif). These systems enable laboratories to perform broth microdilution without having to prepare their own panels.

The systems may differ to some extent regarding volume within the test wells, how inocula are prepared and added, the availability of different supplements for the testing of fastidious bacteria, the types of antimicrobial agents and the dilution schemes used, and the format of medium and antimicrobial agents (e.g., dry-lyophilized or frozen). Furthermore, the degree of automation for inoculation of the panels and the devices available for reading results vary among the different

products. In general, these commercial panels are designed to receive the standard inoculum and are incubated using conditions and durations recommended for conventional broth microdilution. They are growth-based systems that require overnight incubation, and CLSI interpretive criteria apply for interpretation of most results. Reading of these panels is frequently augmented by the availability of semiautomated reading devices.

Agar Dilution Derivations. One commercial system (Spiral Biotech Inc., Bethesda, Md) uses an instrument to apply antimicrobial agent to the surface of an already prepared agar plate in a concentric spiral fashion. Starting in the center of the plate, the instrument deposits the highest concentration of antibiotic and from that point drug application proceeds to the periphery of the plate. Diffusion of the drug in the agar establishes a concentration gradient from high (center of plate) to low (periphery of plate). Starting at the periphery of the plate, bacterial inocula are applied as a single streak perpendicular to the established gradient in a spoke-wheel fashion. Following incubation, the distance from where growth is noted at the edge of the plate to where growth is inhibited toward the center of the plate is measured (Figure 12-10). This value is used to calculate the MIC of the antimicrobial agent against each of the bacterial isolates streaked on the plate.

Diffusion in Agar Derivations. One test has been developed that combines the convenience of disk diffusion with the ability to generate MIC data. The Etest (AB BIODISK, Solna, Sweden) uses plastic strips; one side of the strip contains the antimicrobial agent concentration gradient and the other contains a numeric scale that indicates the drug concentration (Figure 12-11). Mueller-Hinton plates are inoculated as for disk diffusion and the strips are placed on the inoculum lawn. Several strips may be placed radially on the same plate so that multiple antimicrobials can be tested against a single isolate. Following overnight incubation, the plate is examined and the number present at the point at which the border of growth inhibition intersects the E-strip is taken as the MIC (see Figure 12-11). The same MIC interpretive criteria used for dilution methods as provided in CLSI guidelines are used with the E-test value to assign an interpretive category of susceptible, intermediate, or resistant. This method provides a means for producing MIC data in those situations in which the level of resistance can be clinically important (e.g., penicillin or cephalosporins against *S. pneumoniae*).

Another method (BIOMIC, Giles Scientific, Inc., New York) combines the use of conventional disk diffusion methodology with video digital analysis to

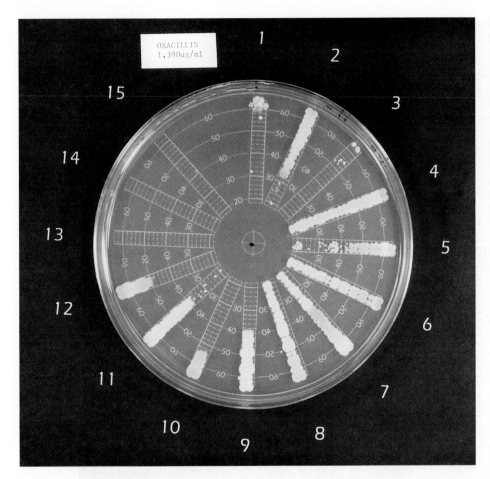

Figure 12-10 Growth patterns on a plate containing an antibiotic gradient (concentration decreases from center of the plate to the periphery) applied by the Spiral Gradient instrument. The distance from where growth is noted at the edge of the plate to where growth is inhibited toward the center of the plate is measured. This value is used in a formula to calculate the MIC of the antimicrobial agent against each of the bacterial isolates streaked on the plate. (Courtesy Spiral Biotech, Inc., Bethesda, Md.)

automate interpretation of inhibition zone sizes. Automated zone readings and interpretations are combined with computer software to produce MIC values and to allow for data manipulations and evaluations for detecting unusual resistance profiles and producing antibiogram reports.

Automated Antimicrobial Susceptibility Test Systems.

The automated antimicrobial susceptibility test systems available for use in the United States include the Vitek Legacy system and Vitek 2 Systems (bioMérieux, Inc., Hazelwood, Mo), the MicroScan WalkAway System (Dade International, Sacramento, Calif), and the Phoenix System (BD Microbiology Systems, Cockeysville, Md). These different systems vary with respect to the extent which inoculum preparation and inoculation are automated, the methods used to detect growth, and the algorithms used to interpret and assign MIC values and categorical (i.e., susceptible, intermediate, resistant) findings.

For example, the Vitek 2 AST inoculum is automatically introduced via a filling tube into a miniaturized plastic 64-well, closed card containing specified concentrations of antibiotics (Figure 12-12). Cards are incubated in a temperature-controlled compartment.

Optical readings are performed every 15 minutes to measure the amount of light transmitted through each well, including a growth control well. Algorithmic analysis of the growth kinetics in each well is performed by the system's software to derive the MIC data. The MIC results are validated with the Advanced Expert System (AES) software, a category interpretation is assigned, and the organism's antimicrobial resistance patterns are reported. Resistance detection is enhanced with the sophisticated AES software, which can recognize and report resistance patterns utilizing MICs. In summary, this system facilitates standardized susceptibility testing in a closed environment with validated results and recognition of an organism's antimicrobial resistance mechanism in 6 to 8 hours for most clinically relevant bacteria (Figure 12-13).

The WalkAway System uses the broth microdilution panel format that is manually inoculated with a multiprong device. Inoculated panels are placed in an incubator-reader unit, where they are incubated for the required length of time, and then the growth patterns are automatically read and interpreted. Depending on the microdilution tray used, bacterial growth may be detected spectrophotometrically or fluorometrically (Figure 12-14).

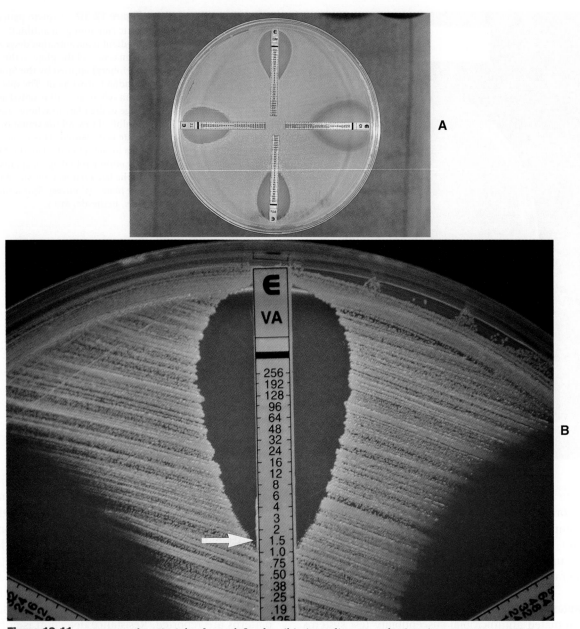

Figure 12-11 Etest uses the principle of a predefined antibiotic gradient on a plastic strip to generate an MIC value. It is processed like the disk diffusion. **A,** Individual antibiotic strips are placed on an inoculated agar surface. **B,** After incubation, the MIC is read where the growth/inhibition edge intersects the strip graduated with an MIC scale across 15 dilutions *(arrow).* Several antibiotic strips can be tested on a plate. (Courtesy AB BIODISK, Solna, Sweden.)

Spectrophotometrically analyzed panels require overnight incubation, and the growth patterns may be read manually as described for routine microdilution testing. Fluorometric analysis is based on the degradation of fluorogenic substrates by viable bacteria as the means for detecting bacterial inhibition by the antimicrobial agents. The fluorogenic approach can provide susceptibility results in 3.5 to 5.5 hours. Either full dilution schemes or breakpoint panels are available. In addition to speed and facilitating work flow, the auto-mated systems also provide increasingly powerful computer-based data management systems that can be used to evaluate the accuracy of results, manage larger databases, and interface with pharmacy areas to enhance the utility of antimicrobial susceptibility testing data.

The Phoenix System provides a convenient, albeit manual, gravity-based inoculation process. Growth is monitored in an automated fashion based on a redox indicator system and interpretation of results of augmented by a rules-based data management expert system.

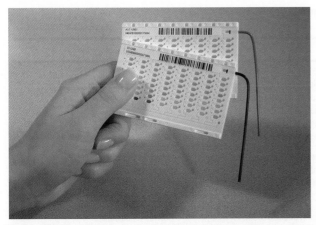

Figure 12-12 The Vitek 2 antimicrobial susceptibility test card contains 64 wells with multiple concentrations of up to 22 antibiotics. The antibiotic is rehydrated when the organism suspension is introduced into the card during the automated filling process. (Courtesy bioMérieux, Inc., Hazelwood, Mo.)

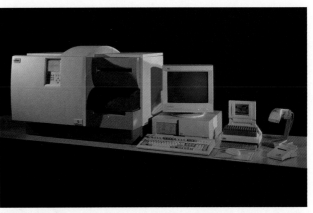

Figure 12-13 The components of the Vitek 2 System consist of the instrument housing; the sample processing and reader/incubator; the computer workstation, which provides data analysis, storage, and epidemiology reports; the Smart Carrier Station, which is the direct interface between the microbiologist on the bench and the instrument; and a bar code scanner to facilitate data entry. (Courtesy bioMérieux, Inc., Hazelwood, Mo.)

A   B

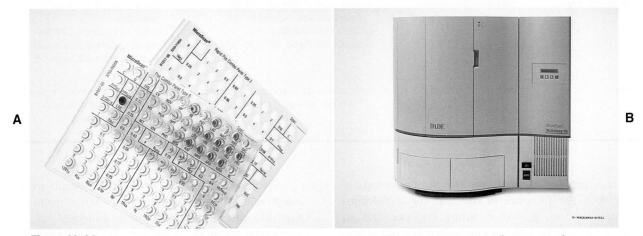

Figure 12-14 Microdilution tray format (**A**) used with the MicroScan WalkAway instrument (**B**) for automated incubation, reading, and interpretation of antimicrobial susceptibility tests. (Courtesy Dade International, Sacramento, Calif.)

Alternative Approaches for Enhancing Resistance Detection

Although the various conventional and commercial antimicrobial susceptibility testing methods provide accurate results in most instances, certain clinically relevant resistance mechanisms can be difficult to detect. In these instances supplemental tests and alternative approaches are needed to ensure reliable detection of resistance. Also, as new and clinically important resistance mechanisms emerge and are recognized, they will be a "lag time" during which conventional and commercial methods are being developed to ensure accurate detection of these new patterns. During such lag periods,

special tests may be used until more conventional or commercial methods become available. Key examples of such alternative approaches are discussed in this section.

Supplemental Testing Methods. Table 12-4 highlights some of the features of supplemental tests that may be used to enhance resistance detection. For certain strains of staphylococci, conventional and commercial systems may have difficulty detecting resistance to oxacillin and the related drugs methicillin and nafcillin. The oxacillin agar screen provides a backup test that may be used when other methods provide equivocal or

Table 12-4 **Supplemental Methods for Detection of Antimicrobial Resistance**

Test	Purpose	Conditions	Interpretation
Oxacillin agar screen	Detection of staphylococcal resistance to penicillinase-resistant penicillins (e.g., oxacillin, methicillin, or nafcillin)	Medium: Mueller-Hinton agar plus 6 μg oxacillin/mL plus 4% NaCL Inoculum: Swab or spot from 1.5×10^8 standard suspension Incubation: 30°-35° C 24 hr, up to 48 hr for non–*Staphylococcus aureus*	Growth = Resistance No growth = Susceptible
Vancomycin agar screen	Detection of enterococcal resistance to vancomycin	Medium: Brain-heart infusion agar plus 6 μg vancomycin/mL Inoculum: Spot of 10^5-10^6 CFU Incubation: 35° C, 24 hr	Growth = Resistance No growth = Susceptible
Aminoglycoside screens	Detection of acquired enterococcal high-level resistance to aminoglycosides that would compromise synergy with a cell wall–active agent such as ampicillin or vancomycin	Medium: Brain-heart infusion broth: 500 μg/mL gentamicin; 1000 μg/mL streptomycin Agar: 500 μg/mL gentamicin; 2000 μg/mL streptomycin Inoculum: Broth; 5×10^5 CFU/mL agar; 10^6 CFU/spot Incubation: 35° C, 24 hr; 48 hr for streptomycin, only if no growth at 24 hr	Growth = Resistance No growth = Susceptible
Oxacillin disk screen	Detection of *Streptococcus pneumoniae* resistance to penicillin	Medium: Mueller-Hinton agar plus 5% sheep blood plus 1 μg oxacillin disk Inoculum: as for disk diffusion Incubation: 5%-7% CO_2 35° C; 20-24 hr	Inhibition zone $\leq$20 mm; penicillin susceptible Inhibition zone <20 mm; penicillin resistant, intermediate, or susceptible. Further testing by MIC method is needed
"D" test	Differentiate clindamycin resistance among *S. aureus* resulting from efflux (*mcrA* or MLS$_B$)	Approximation clindamycin and erythromycin disk to look for blunting of clindamycin zone	Blunting of clindamycin zone to give "D" pattern, indicating inducible clindamycin resistance

uncertain profiles. Growth on the screen correlates highly with the presence of oxacillin (or methicillin) resistance, and no growth is strong evidence that an isolate is susceptible. This is an important determination; strains that are classified as resistant are considered resistant to all other currently available beta-lactam antibiotics so that therapy must include the use of vancomycin. The agar screen plates can be made in-house, and they are available commercially (e.g., REMEL, Lenexa, Kan; BBL, Cockeysville, Md). Additionally, other commercial tests designed to detect oxacillin resistance more rapidly (i.e., 4 hours) have been developed and may provide another approach to supplemental testing (e.g., Crystal MRSA ID System, BBL, Cockeysville, Md). In addition to the agar screen, use of the 30-μg cefoxitin disk has been developed to use by disk diffusion for assisting in the detection of oxacillin-resistant staphylococci (CLSI M100-15). By this method cefoxitin inhibitory zones of less than or equal to 24 mm indicates oxacillin resistance among staphylococci. The cefoxitin disk test is especially helpful in detecting oxacillin resistance among coagulase-negative staphylococci.

Similarly, detection of reduced staphylococcal susceptibility to vancomycin (i.e., MICs from 4 to 16 μg/mL) can be difficult by disk diffusion and some commercial methods. Although the therapeutic relevance of staphylococci with vancomycin MICs in this range is currently uncertain, this diminished susceptibility is outside the normal MIC range for fully susceptible strains; therefore, there is a need to have the capability of detecting this phenotype. The agar screen used for this purpose is outlined in Table 12-4 and is essentially the same as that outlined for enterococci, also in Table 12-4. Strains that grow on the screen should be tested by broth microdilution to obtain a definitive MIC value.

In the same fashion, detection of enterococcal resistance to vancomycin can be difficult by some conventional and commercial methods, and the agar screen can be helpful in confirming this resistance pattern (see Table 12-4). However, as a screen, not all enterococcal isolates that grow are resistant to vancomycin at clinically relevant levels. Therefore, strains that are detected using this method should be more fully characterized using a broth microdilution method to determine the isolate's MIC.

Aminoglycosides also play a key role in therapy for serious enterococcal infections, and acquired high-level resistance, which essentially destroys the therapeutic value of these drugs for combination therapy with ampicillin or vancomycin, is not readily detected by conventional methods. Therefore, screens (Table 12-4) that use high concentrations of aminoglycosides have been developed specifically for detecting this resistance and are available commercially (e.g., REMEL, Lenexa, Kan; or BBL, Cockeysville, Md).

With the emergence of penicillin resistance among *S. pneumoniae,* the penicillin disk diffusion test was not sufficiently sensitive to detect subtle but significant changes in susceptibility to this agent. To address this issue, the oxacillin disk screen described in Table 12-4 is useful but does have a notable limitation. Although organisms that give zones greater than or equal to 20 mm can be accurately characterized as penicillin susceptible, the penicillin susceptibility status of those with zones less than 20 mm remains uncertain and use of some method that produces an MIC value is required.

With regard to macrolide (e.g., erythromycin, azithromycin, clarithromycin) and lincosamide (e.g., clindamycin) resistance among staphylococci, interpretation of in vitro results can also be complicated by the different underlying mechanisms of resistance that have very different therapeutic implications. Isolates that produce a profile that demonstrates resistance to a macrolide (e.g., erythromycin) and susceptibility to clindamycin may do so as a result of two different resistance mechanisms. If this profile is the result of the efflux (*msrA* gene) mechanism, then the isolate can be considered susceptible to clindamycin. However, if this profile resulted from the inducible MLS$_B$ mechanism, which results in an altered ribosomal target, then clindamycin-resistant mutants may readily arise during therapy with this agent. Currently such strains should be reported as resistant to clindamycin. The "D" test that is used to discern between these two different resistance mechanisms is outlined in Table 12-4.

Undoubtedly, as complicated resistance mechanisms that require laboratory detection continue to emerge, screening and supplemental testing methods will continue to be developed. Some of these will be maintained as the primary method for detecting a particular resistance mechanism, while others may tend to fade away as adjustments in conventional and commercial procedures enhance resistance detection and preclude the need to use a supplemental test.

Predictor Antimicrobial Agents.   Another approach that may be used to ensure accuracy in resistance detection is the use of "predictor" antimicrobial agents in the test batteries. The basic premise of this approach is to use antimicrobial agents (i.e., **the predictor drugs**) that are the most sensitive indicators of certain resistance mechanisms. The profile obtained with such a battery of agents is then used to deduce the underlying resistance mechanism. A susceptibility report is subsequently produced based on the likely effect that the resistance mechanisms would have on the antimicrobials being considered for therapeutic management of the patient. Use of predictor drugs is not a new concept; this approach has already been used in various situations. For example:

- Staphylococcal resistance to oxacillin is used to determine and report resistance to all currently available beta-lactams, including penicillins, cephalosporins, and carbapenems (e.g., imipenem, meropenem, ertapenem)
- Enterococcal high-level gentamicin resistance predicts resistance to nearly all other currently available aminoglycosides, including amikacin, tobramycin, netilmicin, and kanamycin
- Enterococcal resistance to ampicillin predicts resistance to all penicillin derivatives

METHODS THAT DIRECTLY DETECT SPECIFIC RESISTANCE MECHANISMS

As an alternative to detecting resistance by measuring the effect of antimicrobial presence on bacterial growth, some strategies focus on assaying for the presence of a particular mechanism. When the presence or absence of the mechanism is established, the resistance profile of the organism can be generated without having to test several different antimicrobial agents. The utility of this approach, which can involve phenotypic and genotypic methods, depends on the presence of a particular resistance mechanism as being a sensitive and specific indicator of clinical resistance.

Phenotypic Methods

The most common phenotypic-based assays are those that test for the presence of beta-lactamase enzymes in the clinical bacterial isolate of interest. Less commonly used are tests to detect the chloramphenicol-modifying enzyme chloramphenicol acetyltransferase.

Beta-Lactamase Detection.   Beta-lactamases play a key role in bacterial resistance to beta-lactam agents, and detection of their presence can provide useful information (see Chapter 11). Various assays are available to detect beta-lactamases, but the most useful one for clinical laboratories is the chromogenic cephalosporinase test. Beta-lactamases exert their effect by opening the beta-lactam ring (see Figure 11-9). With the use of a chromogenic cephalosporin as the substrate, this process results in a colored product. One such assay

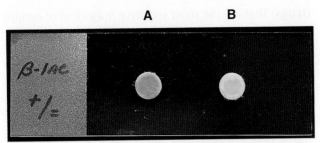

Figure 12-15 The chromogenic cephalosporin test allows direct detection of beta-lactamase production. When the beta-lactam ring of the cephalosporin substrate in the disk is hydrolyzed by the bacterial inoculum, a deep pink color is produced (**A**). Lack of color production indicates the absence of beta-lactamase (**B**).

is the cefinase disk (BD Microbiology Systems, cockeysville, Md) test shown in Figure 12-15.

Useful application of tests to directly detect beta-lactamase production is limited to those organisms for which the list of beta-lactams significantly affected by the enzyme are known. Furthermore, this list must include the beta-lactams commonly considered for therapeutic eradication of the organism. Examples of useful applications include detection of:

- *N. gonorrhoeae* resistance to penicillin
- *H. influenzae* resistance to ampicillin
- Staphylococcal resistance to penicillin

The actual utility of this approach even for the organisms listed is decreasing. As beta-lactamase–mediated resistance has become widespread among *N. gonorrhoeae, H. influenzae,* and staphylococci, other agents not affected by the beta-lactamases have become the antimicrobials of choice for therapy. Therefore, the need to know the beta-lactamase status of these bacterial species has become substantially less urgent. Whereas several *Enterobacteriaceae* and *P. aeruginosa* produce beta-lactamases, the effect of these enzymes on the various beta-lactams depends on which enzymes are produced. Therefore, even though such organisms would frequently produce a positive beta-lactamase assay, very little, if any, information regarding which antimicrobial agents are affected would be gained. Detection of beta-lactam resistance among these organisms is best accomplished using conventional and commercial systems for directly evaluating antimicrobial agent–organism interactions.

Chloramphenicol Acetyltransferase Detection. Chloramphenicol modification by chloramphenicol acetyltransferase (CAT) detection is only one mechanism by which bacteria may express resistance to this agent. This, coupled with the substantially diminished use of chloramphenicol in today's clinical settings, significantly limits the utility of this test. Commercial colorimetric

assays, such as that produced by REMEL (Lenexa, Kan), do provide a convenient method for establishing the presence of this enzyme. If positive, chloramphenicol resistance can be reported, but a negative test does not rule out resistance that may be mediated by other mechanisms such as decreased uptake.

Genotypic Methods

The genes that encode many of the clinically relevant acquired resistance mechanisms are known, as is all or part of their nucleotide sequences. This has allowed for the development of molecular methods involving nucleic acid hybridization and amplification for the study and detection of antimicrobial resistance (for more information regarding molecular methods for the characterization of bacteria, see Chapter 8). The ability to definitively determine the presence of a particular gene that encodes antimicrobial resistance has several advantages. However, as with any laboratory procedure, certain disadvantages and limitations also exist.

From a research and development perspective, molecular methods are extremely useful for more thoroughly characterizing the resistances of bacterial collections used to establish and evaluate conventional standards recommended by CLSI. Phenotype-based commercial (automated and nonautomated) susceptibility testing methods and systems can also be evaluated.

Molecular methods also may be directly applied in the clinical setting as an important backup resource to investigate and arbitrate equivocal results obtained by phenotypic methods. For example, the clinical importance of accurately detecting methicillin resistance among staphylococci coupled with the inconsistencies of phenotype-based methods is problematic. In doubtful situations, molecular detection of the *mec* gene that encodes methicillin resistance can be usefully applied to definitively establish an isolate's methicillin resistance. Similarly, doubt raised by equivocal phenotypic results obtained with potentially vancomycin-resistant enterococci can be definitively resolved by establishing the presence and classification of *van* genes that mediate this resistance.

Although molecular methods have been and will continue to be extremely important in antimicrobial resistance detection, numerous factors still complicate their use beyond supplementing phenotype-based susceptibility testing protocols. These factors include the following:

- Use of probes or oligonucleotides for specific resistance genes only allows those particular genes to be found. Resistance mediated by divergent genes or totally different mechanisms could be missed (i.e., the absence of one gene may not guarantee antimicrobial susceptibility).

- Phenotypic resistance to a level that is clinically significant for any one antimicrobial agent may be due to a culmination of processes that involve enzymatic modification of the antimicrobial, decreased uptake, altered affinity of the target for the drug, or some combination of these mechanisms (i.e., the presence of one gene does not guarantee resistance).

- The presence of a gene encoding resistance does not provide information regarding the status of the control genes necessary for expression of resistance. That is, although present, the genes may be silent or nonfunctional and the organism may be incapable of expressing the resistance encoded by the detected gene.

- From a clinical laboratory perspective, there is practical difficulty in adopting molecular methods specific for only a few resistance mechanisms when the vast majority of the susceptibility testing still will be accomplished using phenotypic-based methods.

Even though there are challenges to the widespread adoption of molecular methods for routine antimicrobial susceptibility testing, the significant contributions that this approach has made to resistance detection will continue to expand.

SPECIAL METHODS FOR COMPLEX ANTIMICROBIAL–ORGANISM INTERACTIONS

Certain in vitro tests have been developed to investigate aspects of antimicrobial activity not routinely addressed by commonly used susceptibility testing procedures. Specifically, these are tests designed to measure bactericidal activity (i.e., bacterial killing) or to measure the antibacterial effect of antimicrobial agents used in combination.

These tests are often labor intensive, fraught with the potential for technical problems, frequently difficult to interpret, and of uncertain clinical utility. For these reasons, their use should be substantially limited. If performed at all, they should be done with the availability of expert microbiology and infectious disease consultation.

Bactericidal Tests

Bactericidal tests are designed to determine the ability of antimicrobial agents to kill bacteria. The killing ability of most drugs is already known, and they are commonly classified as bacteriostatic or bactericidal agents. However, many variables, including the concentration of antimicrobial agent and the species of targeted organism, can influence this classification. For example, beta-lactams, such as penicillins, typically are

bactericidal against most gram-positive cocci but are usually only bacteriostatic against enterococci. If bactericidal tests are clinically appropriate, they should only be applied to evaluate antimicrobials usually considered to be bactericidal (e.g., beta-lactams and vancomycin) and not to those agents known to be bacteriostatic (e.g., macrolides).

Key clinical situations in which achieving bactericidal activity is of greatest clinical importance include severe and life-threatening infections, infections in the immunocompromised host, and infections in body sites where assistance from the patient's own defenses is minimal (e.g., endocarditis or osteomyelitis). Based on research trials in animal models and clinical trials in humans, the most effective therapy for these types of infections is often already known. However, occasionally the laboratory may be asked to substantiate that bactericidal activity is being achieved or is achievable. The methods available for this include minimal bactericidal concentration (MBC) testing, time-kill studies, and serumcidal testing. Regardless of which method is used, the need to interpret the results cautiously with the understanding of uncertain clinical correlation and the potential for substantial technical artifacts cannot be overemphasized.

Minimal Bactericidal Concentration. The MBC test involves continuation of the procedure for conventional broth dilution testing. After incubation and determination of the antimicrobial agent's MIC, an aliquot from each tube or well in the dilution series showing inhibition of visible bacterial growth is subcultured to an enriched agar medium (usually sheep blood agar). Following overnight incubation, the plates are examined and the CFUs counted. Knowing the volume of the aliquot sampled and the number of CFUs obtained, the number of viable cells per milliliter for each antimicrobial dilution well can be calculated. This number is compared with the known CFU/mL in the original inoculum. The antimicrobial concentration that resulted in a 99.9% reduction in CFU/mL compared with the organism concentration in the original inoculum is recorded as the MBC.

Although the clinical significance of MBC results is uncertain, applications of this information include considering whether treatment failure could be occurring because an organism's MBC exceeds the serum achievable level of the antimicrobial agent. Alternatively, if an antibiotic's MBC is greater than or equal to 32 times higher than the MIC, the organism may be tolerant to that drug. **Tolerance** is a phenomenon most commonly associated with bacterial resistance to beta-lactam antibiotics and reflects an organism's ability to be only inhibited by an agent that is usually bactericidal. Although the physiologic basis of tolerance has been studied in several bacterial species, the actual clinical relevance of this phenomenon has not been well established.

Time-Kill Studies. Another approach to examine bactericidal activity involves exposing a bacterial isolate to a concentration of antibiotic in a broth medium and measuring the rate of killing over a specified period. By this time-kill analysis, samples are taken from the antibiotic-broth solution immediately after the inoculum was added and at regular intervals afterward. Each time-sample is plated to agar plates; following incubation, CFU counts are performed as described for MBC testing. The number of viable bacteria from each sample is plotted over time so that the rate of killing can be determined. Generally, a 1000-fold decrease in the number of viable bacteria in the antibiotic-containing broth after a 24-hour period compared with the number of bacteria in the original inoculum is interpreted as bactericidal activity. Although time-kill analysis is frequently used in the research environment to study the in vitro activity of anti-microbial agents, the labor intensity and technical specifications of the procedure preclude its use in most clinical microbiology laboratories for the production of data used to manage a patient's infection.

Serum Bactericidal Test. The serum bactericidal test (SBT) is analogous to the MIC-MBC test except that the medium used is patient's serum that contains the therapeutic antimicrobial agents that the patient has been receiving. By using patient serum to detect bacteriostatic and bactericidal activity, the antibacterial impact of factors other than the antibiotics (e.g., antibodies and complement) also are observed.

For each test, two serum samples are required. One is collected just before the patient is to receive the next antimicrobial dose and is referred to as the **trough specimen.** The second sample is collected when the serum antimicrobial concentration is highest and is referred to as the **peak specimen.** When to collect the peak specimen varies with the pharmacokinetic pro-perties of the antimicrobial agents and the route through which they are being administered. Peak levels for intravenously, intramuscularly, and orally administered agents are generally obtained 30 to 60 minutes, 60 minutes, and 90 minutes after admin-istration, respectively. The trough and peak levels should be collected around the same dose and tested simultaneously.

Serial twofold dilutions of each specimen are prepared and inoculated with the bacterial isolate (final inoculum of 5×10^5 CFU/mL) that is causing the patient's infection. Dilutions are incubated overnight, and the highest dilution inhibiting visibly detectable growth is the **serumstatic titre** (e.g., 1:8, 1:16, 1:32).

Aliquots of known size are then taken from each dilution at or below the serumstatic titre (i.e., those dilutions that inhibited bacterial growth) and are plated on sheep blood agar plates. Following incubation, the CFUs per plate are counted and the serum dilution that results in a 99.9% reduction in the CFU/mL as compared with the original inoculum is recorded as the **serumcidal titre.** For example, if a bacterial isolate showed a serumstatic titre of 1:32, then the tubes containing dilutions of 1:2, 1:4, 1:8, 1:16, and 1:32 would be subcultured. If the 1:8 dilution was the highest dilution to yield a 99.9% decrease in CFUs, then the serumcidal titre would be recorded as 1:8.

The SBT was originally developed to help predict the clinical efficacy of antimicrobial therapy for staphylococcal endocarditis. Peak serumcidal titres of 1:32 to 1:64 or greater have been thought to correlate best with a positive clinical outcome. However, even though the test is performed in patient serum, there are still many differences not accounted for between the in vitro test environment and the in vivo site of infection. Therefore, although the test is used to evaluate whether effective bactericidal concentrations are being achieved, the predictive clinical value for staphylococcal endocarditis or any other infection caused by other bacteria is still uncertain.

Details regarding the performance of these bactericidal tests are provided in the CLSI document M26-A titled, "Methods for Determining Bactericidal Activity of Antimicrobial Agents."

Tests for Activity of Antimicrobial Combinations

Therapeutic management of bacterial infections often requires simultaneous use of more than one anti-microbial agent. Multiple therapies are used for rea-sons that include:

- Treating polymicrobial infections caused by organisms with different antimicrobial resis-tance profiles
- Achieving more rapid bactericidal activity than could be achieved with any single agent
- Achieving bactericidal activity against bacteria for which no single agent is lethal (e.g., enterococci)
- Minimizing the chance of resistant organisms emerging during therapy (e.g., *M. tuberculosis*)

Testing the effectiveness of antimicrobial combi-nations against a single bacterial isolate is referred to as **synergy testing.** When combinations are tested, three outcome categories are possible:

- **Synergy:** the activity of the antimicrobial combination is substantially greater than the activity of the single most active drug alone

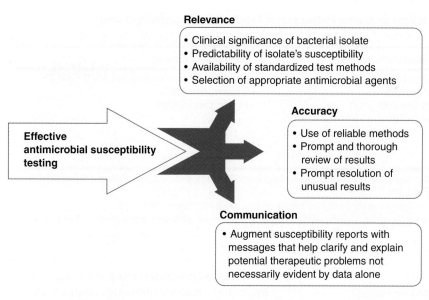

Figure 12-16   Goals of effective antimicrobial susceptibility testing strategies.

- **Indifference:** the activity of the combination is no better or worse than the single most active drug alone
- **Antagonism:** the activity of the combination is substantially less than the activity of the single most active drug alone (an interaction to be avoided)

The checkerboard assay and the time-kill assay are two basic methods for synergy testing. In the checkerboard method, MIC panels are set up that contain two antimicrobial agents serially diluted alone and in combination with each other. Following inoculation and incubation, the MICs obtained with the single agents and the various combinations are recorded. By calculating the MIC ratios obtained with single and combined agents, the drug combination in question is classified as synergistic, indifferent, or antagonistic.

Using the time-kill assay, the same procedure described for testing bactericidal activity is used except the killing curve obtained with a single agent is compared with the killing curve obtained with antimicrobial combinations. Synergy is indicated when the combination exhibits 100-fold or more greater killing than the most active single agent tested alone following 24 hours of incubation. Similar killing rates between the most active agent and the combination is interpreted as indifference. Antagonism is evidenced by the combination being less active than the most active single agent.

The decision to use more than one antimicrobial agent may be based on antimicrobial resistance profiles or identifications of particular bacterial pathogens reported by the clinical microbiology laboratory. However, the decision regarding which antimicrobial agents to combine should not rely on the results of complex synergy tests performed in the clinical laboratory. Most

clinically useful antimicrobial combinations have been investigated in a clinical research setting and are well described in the medical literature. These data should be used to guide the decision for combination therapy. The technical difficulties associated with performing and interpreting synergy tests, which at most would only be performed rarely in the clinical laboratory, precludes their reliable utility in the diagnostic setting.

LABORATORY STRATEGIES FOR ANTIMICROBIAL SUSCEPTIBILITY TESTING

The clinical microbiology laboratory is responsible for maximizing the positive impact that susceptibility testing information can have on the use of antimicrobial agents to therapeutically manage infectious diseases. However, meeting this responsibility is difficult because of demands for more efficient use of laboratory resources, the increasing complexities of important bacterial resistance profiles, and the continued expectations for high-quality results. To ensure quality in the midst of dwindling resources and expanding antimicrobial resistance, strategies for antimicrobial susceptibility testing must be carefully developed. These strategies should target relevance, accuracy, and communication as their goals (Figure 12-16).

RELEVANCE

Antimicrobial susceptibility testing should only be performed when there is sufficient potential for providing clinically useful and reliable information regarding those antimicrobial agents that are appropriate for the bacterial isolate in question. Therefore, for the sake of relevance, two questions must be addressed:

Table 12-5 Categorization of Bacteria According to Need for Routine Performance of Antimicrobial Susceptibility Testing*

Testing Commonly Required	Testing Occasionally Required†	Testing Rarely Required
Staphylococci *Streptococcus pneumoniae* Viridans streptococci‡ Enterococci *Enterobacteriaceae* *Pseudomonas aeruginosa* *Acinetobacter* spp.	*Haemophilus influenzae* *Neisseria gonorrhoeae* *Moraxella catarrhalis* Anaerobic bacteria	Beta-hemolytic streptococci (groups A, B, C, F, and G) *Neisseria meningitides* *Listeria monocytogenes*

*Based on the assumption that the organism is clinically significant. Table only includes those bacteria for which standardized testing procedures are available as outlined and recommended by the National Committee for Clinical Laboratory Standards.
†Testing only needed if an antimicrobial to which the organisms are frequently resistant is still considered for use (e.g., penicillin for *Neisseria gonorrhoeae*).
‡Viridans streptococci only require testing when implicated in endocarditis or isolated in pure culture from a normally sterile site with a strong suspicion of being clinically important.

- When should testing be performed?
- Which antimicrobial agents should be tested?

WHEN TO PERFORM A SUSCEPTIBILITY TEST

The first issue that must be resolved is whether antimicrobial susceptibility testing is appropriate for a particular isolate. Although the answer may not always be clear, the issue must always be addressed. The decision to perform susceptibility testing depends on the following criteria:

- Clinical significance of a bacterial isolate
- Predictability of a bacterial isolate's susceptibility to the antimicrobial agents most commonly used against them, often referred to as the **drugs of choice**
- Availability of reliable standardized methods for testing the isolate

Determining Clinical Significance

Performing tests and reporting antimicrobial susceptibility data on clinically insignificant bacterial isolates is a waste of resources and, more importantly, can mislead physicians, who depend on laboratory information to help establish the clinical significance of a bacterial isolate. Useful criteria for establishing the clinical importance of a bacterial isolate include:

- Detection and/or the abundance of the organism on direct Gram stain of a patient's specimen, preferably in the presence of white blood cells, and growth of an organism with the same morphology in culture
- Known ability of the bacterial species isolated to cause infection at the body site from which the specimen was obtained (see Part VII)
- Whether the organism is generally considered either an epithelial or mucosal colonizer or is generally considered a pathogen

- Body site from which the organism was isolated (normally sterile vs. normally colonized site)

Although these criteria are helpful and heavily depend on the capacity of the bacterial species isolated to cause disease, the final designation of clinical significance often still requires dialogue between the laboratorian and physician.

Reporting susceptibility results on organisms of doubtful clinical significance can be falsely interpreted by clinicians as an indicator of clinical significance. Therefore, using criteria such as those just listed should be a major feature of any laboratory's antimicrobial susceptibility testing strategy.

Predictability of Antimicrobial Susceptibility

If the organisms are clinically significant, what are the chances they could be resistant to the antimicrobial agents commonly used to eradicate them (i.e., the drugs of choice)? Unfortunately, the increasing dissemination of resistance among clinically relevant bacteria has diminished the list of bacteria whose antimicrobial susceptibility can be confidently predicted based on identification without the need to perform testing. Table 12-5 categorizes many of the commonly encountered bacteria according to the need to perform testing to detect resistance.

Acquired resistance to various antimicrobial agents dictates that susceptibility testing be performed on all clinically relevant isolates of several bacterial groups, genera, and species. For other organisms, such as *Haemophilus influenzae* and *Neisseria gonorrhoeae*, resistance to the original drugs of choice (ampicillin and penicillin) has become so widespread that more potent antibiotics (e.g., ceftriaxone) for which no resistance has been described to date have become the drugs of choice. Therefore, whereas testing used to be routinely indicated to detect ampicillin and penicillin resistance, testing for resistance to currently recommended antimicrobials for these organisms is not routinely necessary. The possible exception to this is the relatively recent emergence of fluoroquinolone resistance among *N. gonorrhoeae* that may warrant testing of clinical isolates.

One notable exception to the widespread emergence of resistance has been the absence of penicillin resistance among beta-hemolytic streptococci. Because susceptibility to penicillin is extremely predictable among these organisms, testing against penicillin provides little, if any, information that is not already provided by accurate organism identification. However, in instances in which the patient cannot tolerate penicillin, alternative agents, such as erythromycin, may be considered. Resistance among beta-hemolytic streptococci to this agent has been well documented, and susceptibility testing in this instance would be indicated.

The recommendations outlined in Table 12-5 are guidelines. In any clinical setting there will be exceptions that must be considered in consultation with physicians. Also, these guidelines are for providing data regarding the management of a single patient's infection. When susceptibility testing is performed as a means of gathering surveillance data for monitoring emerging resistance (see Accuracy and Antimicrobial Resistance Surveillance in this chapter) the guidelines may not necessarily apply.

Availability of Reliable Susceptibility Testing Methods

If a reliable, standardized method for testing a particular bacterial genus or species does not exist, then the ability to produce accurate and meaningful data is substantially compromised. Although standard methods exist for most of the commonly encountered bacteria (see Tables 12-1 to 12-3), clinically relevant isolates of bacterial species for which standard testing methods do not exist are encountered. In these instances, the dilemma stems from the conflict between the laboratory's urge to contribute in some way by providing data and the lack of confidence in producing interpretable and accurate information.

Many organisms not listed in Table 12-5 will grow on the media and under the conditions recommended for testing commonly encountered bacteria. However, the ability to grow and the ability to detect important antimicrobial resistances are not the same thing. For example, the gram-negative bacillus *Stenotrophomonas maltophilia* grows extremely well under most susceptibility testing conditions, but the results obtained with beta-lactam antibiotics can be widely variable and seriously misleading. This organism produces potent beta-lactamases that seriously compromise the effectiveness of most beta-lactams, yet certain isolates may appear susceptible by standard in vitro testing criteria.

Therefore, even though testing can provide an answer, substantial potential to obtain the wrong answer exists.

Given the uncertainty that surrounds testing of bacteria for which standardized methods are lacking, two approaches may be used. One is to not perform testing but rather to provide physicians with information based on clinical studies published in the medical literature about the antimicrobial agents generally accepted as the drugs of choice for the bacterial species in question. This approach is best handled when the laboratory medical director and infectious disease specialists are involved. The other option is to still provide the literature information but to also perform the test to the best of the laboratory's ability. In this case, results must be accompanied by a message that clearly indicates testing was performed by a nonstandardized method and results should be interpreted with caution. When such tests are undertaken, customized antimicrobial batteries that include the agents most commonly used to eradicate the bacterial species of interest need to be assembled and used. Recently CLSI has published the document M45 to provide guidelines for testing certain less frequently encountered bacteria.

SELECTION OF ANTIMICROBIAL AGENTS FOR TESTING

Selection of relevant antimicrobial agents is based on the criteria outlined in Box 12-1. These criteria should be carefully considered in consultation with the medical staff so that reports contain information that is most appropriate for the therapeutic management of the patient. Careful selection of antimicrobial agents avoids cluttering reports with superfluous information, minimizes the risk of confusing physicians, and substantially decreases the waste of time and resources in the clinical microbiology laboratory.

Antimicrobial agents that may be considered for inclusion in batteries to be tested against certain bacterial groups are provided in Table 12-6. The list is not exhaustive but is useful for illustrating some points regarding the development of relevant testing batteries. For example, with all the penicillins, cephalosporins, and other beta-lactam antibiotics available for testing, only penicillin and oxacillin need to be tested against staphylococci. The information gained with these two agents alone will reflect the general effectiveness of any other beta-lactam. In essence, these drugs are predictor agents as discussed earlier in this chapter. Similarly, ampicillin can be used alone as an indicator of enterococcal susceptibility to various penicillins and, because of intrinsic resistance, cephalosporins should never be tested against these organisms.

In contrast to the relatively few agents that may be included in testing batteries for gram-positive cocci,

Table 12-6   Selection of Antimicrobial Agents for Testing Against Common Bacterial Groups*

Antimicrobial Agents	Enterobacteriaceae	Pseudomonas aeruginosa	Staphylococci	Enterococci	Staphylococcus pneumoniae	Viridans Streptococci
Penicillins						
Penicillin	−	−	+	−	+	+
Oxacillin	−	−	+	−	−	−
Ampicillin	+	−	−	+	−	−
Piperacillin/tazobactam	+	+	−	−	−	−
Cephalosporins						
Cefazolin	+	−	−	−	−	−
Cefotetan	+	−	−	−	−	−
Ceftriaxone	+	−	−	−	+	+
Cefotaxime	+	−	−	−	+	+
Ceftazidime	+	+	−	−	−	−
Other beta-lactams						
Aztreonam	+	+	−	−	−	−
Imipenem	+	+	−	−	±	−
Glycopeptides						
Vancomycin	−	−	+	+	+	+
Aminoglycosides						
Gentamicin	+	+	±	+[†]	−	−
Tobramycin	+	+	−	−	−	−
Amikacin	+	+	−	−	−	−
Quinolones						
Ciprofloxacin	+	+	+	−	+	−
Levofloxacin	+	+	+	−	+	+
Other agents						
Erythromycin	−	−	+	−	+	+
Clindamycin	−	−	+	−	+	+
Trimethoprim-sulfamethoxazole	+	−	+	−	+	−±
Tigecycline	+	−	+	+	±	+
Daptomycin	−	−	+	+	±	±
Linezolid	−	−	+	+	±	+
Telithromycin	−	−	+	−	+	

+, May be selected for inclusion in testing batteries (not all agents with + need to be selected); ±, may be selected in certain situations; −, selection for testing is not necessary or not recommended.

*Not all available antimicrobial agents are included. Selection recommendation is based on nonurinary tract infections only.

[†]Gentamicin testing against enterococci requires use of high-concentration disks or a special screen (see Table 12-4).

several potential choices exist for use against gram-negative bacilli. This is mostly due to the commercial availability of several beta-lactams with similar activities against *Enterobacteriaceae* and the general inability of one beta-lactam to serve as a reliable predictor drug for other beta-lactams. For example, an organism resistant to cefazolin may or may not be resistant to cefotetan, and an organism resistant to cefotetan may or may not be resistant to ceftazidime. With the lack of potential for selecting a predictor drug in these instances, more agents must be tested. However, in some instances, overlap in activities does exist so that some duplication of effort can be avoided. For example, the spectrum of activity of ceftriaxone and cefotaxime is sufficiently similar to allow selection of only one for the testing battery.

Many scenarios exist in which the spectrum of activity and other criteria listed in Box 12-1 are considered for the sake of designing the most relevant and useful testing batteries. Again, it must be emphasized that consideration of these criteria must be done in close consultation with the medical staff.

ACCURACY

Susceptibility testing strategies focused on production of accurate results have two key components:

- Use of methods that produce accurate results
- The application of real-time review of results before reporting

USE OF ACCURATE METHODOLOGIES

The complexities of the various resistance mechanisms have left no one method (i.e., conventional, automated, or molecular) singularly sufficient for the detection of all clinically relevant resistance patterns. Therefore, the selection of testing methods and careful consideration of how different methods are most effectively used together is necessary to ensure accurate and reliable detection of resistance.

Microbiologists must be aware of the strengths and weaknesses of their primary susceptibility testing methods (e.g., disk diffusion, commercial broth micro-dilution, automation) for detecting relevant resistance patterns and know when adjunct or supplemental testing is necessary. This awareness is accomplished by reviewing studies published in peer-reviewed journals that focus on the performance of antimicrobial testing systems and by periodically challenging one's own system with organisms that have been thoroughly characterized with respect to their resistance profiles (e.g., proficiency testing programs). Furthermore, accurate and relevant testing not only means using various conventional methods, or even using a mixture of automated, conventional, and screening methods, but also encompasses the potential application of molecular techniques and predictor drug.

Testing of *S. pneumoniae* provides one example of the need to be aware of testing limitations and the importance of implementing supplemental tests. Not long ago, routine susceptibility testing of *S. pneumoniae* was considered unnecessary. However, with the emergence of beta-lactam resistance, testing has become imperative. As the need for testing emerged, the inability of conventional tests, such as penicillin disk diffusion, to detect resistance became apparent. Fortunately, a test that uses the penicillin derivative oxacillin was developed and widely used as a reliable screen for detecting nonsusceptibility to penicillin. However, this test is only a screen in that the level of penicillin resistance (i.e., the MIC) can vary greatly among nonsusceptible isolates and some strains that appear resistant by the screen may actually be susceptible. Because the level of resistance can affect therapeutic decisions, another method that allows for MIC determinations should be used for testing these organisms. Additionally, the emergence of cephalosporin (i.e., ceftriaxone or cefotaxime) resistance requires the use of tests that detect resistance to these agents.

Other important examples in which more than one method is required to obtain complete and accurate susceptibility testing data for certain organism groups or species include vancomycin-resistant enterococci, methicillin-resistant staphylococci, and extended-spectrum, beta-lactamase–producing *Enterobacteriaceae*. In addition, molecular methods also may be used in the clinical setting as an important backup resource to investigate and arbitrate equivocal results obtained by phenotypic methods. However, multiple testing protocols are not routinely necessary for every organism encountered. In most laboratories, one conventional or commercial method is likely to be the mainstay of testing, with adjunct testing used as a supplement when necessary.

REVIEW OF RESULTS

In addition to selecting one or more methods to accurately detect resistance, the strengths and weaknesses of these testing systems must be continuously monitored. This is primarily accomplished by carefully reviewing the susceptibility data that are produced daily. In the past, establishing and maintaining aggressive and effective monitoring programs often have been prohibitively labor intensive. Also, the effectiveness varied with the laboratorian's level of interest and knowledge. However, the speed and flexibility afforded by computerization of results review and reporting greatly facilitates the administration of such quality assurance programs, even in laboratories with modest resources. Effective computer programs may be a part of the general laboratory information system, or in some cases, such programs are available through the commercial susceptibility testing system being used. Because automated expert data review greatly facilitates the review process and enhances data accuracy, this feature should be seriously considered when selecting an antimicrobial susceptibility testing system.

Susceptibility profiles must be scrutinized manually or with the aid of computers according to what profiles are likely, somewhat likely, somewhat unlikely, and nearly impossible. This awareness not only pertains to profiles exhibited by organisms within a particular institution but also to those exhibited by clinically relevant bacteria in general. The unusual resistance profiles must be discovered and evaluated expeditiously to determine whether they are due to technical or clerical errors or are truly indicative of an emerging resistance problem. The urgency of making this determination is twofold. First, if the profile results from laboratory error, it must be corrected and the physician notified so that the patient is not subjected to ineffective or inappropriate antimicrobial therapy. Second, if the profile is valid and presents a threat to the patient and to others (e.g., the emergence of vancomycin-resistant staphylococci), the immediate notification of infection control and infectious disease personnel is warranted.

Components of Results-Review Strategies

Any laboratory-based strategy for monitoring the accuracy of results and the emergence of resistance must have two components:

Table 12-7   Examples of Susceptibility Testing Profiles Requiring Further Evaluation

Organism	Susceptibility Profile
Staphylococci	Vancomycin intermediate or resistant Clindamycin resistant; erythromycin susceptible Linezolid resistant Daptomycin-resistant
Viridans streptococci	Vancomycin intermediate or resistant
Streptococcus pneumoniae	Vancomycin intermediate or resistant
Beta-hemolytic streptococci	Penicillin intermediate or resistant
Enterobacteriaceae	Imipenem resistant
Enterobacter/Citrobacter/Serratia/Morganella/Providencia	Susceptible to ampicillin or cefazolin
Pseudomonas aeruginosa	Amikacin resistant; gentamicin or tobramycin susceptible
Stenotrophomonas maltophilia	Imipenem susceptible; trimethoprim/sulfamethoxazole resistant
Neisseria gonorrhoeae	Ceftriaxone resistant
Neisseria meningitidis	Penicillin resistant

Modified from Courvalin P: *Am Soc Microbiol News* 58:368, 1992.

- Data review—a mechanism for recognizing new or unusual susceptibility profiles
- Resolution—the application of protocols for determining whether an unusual profile is due to error (technical or clerical) or accurately reflects the emergence of a new resistance mechanism

Both components must be integrated into the review process to ensure efficient and timely use of resources.

Data Review. Recognition of unusual resistance profiles is primarily accomplished by carefully reviewing the susceptibility data produced daily. Examples of unusual susceptibility profiles for gram-positive and gram-negative bacteria are given in Table 12-7. The examples are a mixture of profiles that clearly demonstrate a likely error (i.e., clindamycin-resistant, erythromycin-susceptible staphylococci), profiles that have rarely been encountered but if observed require immediate attention (i.e., vancomycin resistance in staphylococci), or profiles that have been described but may not be common (i.e., imipenem resistance in *Enterobacteriaceae*).

The data review process for evaluation of profiles should not be the responsibility of a single person within the laboratory. Furthermore, the process requires checks and balances that do not impede the workflow or increase the time needed to get information to physicians. How this is established will vary depending on a particular laboratory's division of labor and workflow, but several key aspects to be considered include:

- The identification of the organism must be known. To evaluate the accuracy of a susceptibility profile, identification and susceptibility data must be simultaneously analyzed in a timely fashion. Without knowing the organism's identification, it is frequently difficult to determine whether the susceptibility profile is unusual.
- Susceptibility results should be analyzed and reported as early in the day as possible. The workflow should allow time for corrective action of errors found during data review so that corrected, or substantiated, results can be provided to physicians as soon as possible.
- Use of two or more tiers of data review. The first tier is at the bench level where technologists are simultaneously reading the results and evaluating an organism's susceptibility profile for appropriateness. When unusual profiles are found, the technologist should be able to initiate troubleshooting protocols (see Resolution). To avoid the release of erroneous and potentially dangerous information, results should not be reported at this point. Review at this level, which is greatly facilitated by automated expert review systems, maintains proficiency among technologists regarding the nuances of susceptibility profiles and important resistance patterns. The second tier is at the level of supervisory or laboratory director. The purpose of review at this level is to track and monitor the efficiency of the first tier, to take ultimate responsibility for the accuracy of results, to

provide constructive and educational feedback to the technologists performing the first-line review, and to provide guidance for resolution of the unusual profiles. Again, a computer-based review process that searches all reports for predefined unusual profiles (similar to those outlined in Table 12-7) can greatly enhance the efficiency and accuracy of the second level of review.

- The review process must be flexible and updated. Because bacterial capabilities for antimicrobial resistance profiles change, laboratory resistance detection systems can become outdated. Therefore, the list of unusual profiles requires periodic review and updating.

Resolution. The importance of having strategies for resolving unusual profiles cannot be overstated. However, developing detailed procedures for every contingency is not possible or practical. Most resolution strategies should focus on certain general approaches, with supervisory or laboratory director consultation always being among the options available to technologists. Although the steps taken to investigate and resolve an unusual profile often depend on the organism and antibiotics involved, most protocols for resolution should include one or more of the following approaches:

- Review data for possible clerical error.
- Determine whether susceptibility panel and identification system were inoculated with same isolate.
- Reexamine test panel or plate for reading error (e.g., misreading of actual zone of inhibition).
- Confirm purity of inoculum and proper inoculum preparation.
- For commercial systems, determine whether manufacturer's recommended procedures were followed.
- Confirm accuracy of organism identification.
- Confirm resistance by using a second method or screening test.

Often a quick review of the data recording and interpretation aspects, or purity of culture, will reveal the reason why an unusual profile was obtained. Other times more extensive testing, perhaps by more than one method, may be needed to establish the validity of an unusual or unexpected resistance profile.

ACCURACY AND ANTIMICROBIAL RESISTANCE SURVEILLANCE

Antimicrobial resistance surveillance involves tracking the susceptibility profiles produced by the bacteria that are encountered within a particular institution, within a specific geographic location (i.e., regionally, nationally, or internationally). For laboratories that serve a particular institution or group of institutions, periodic publication of an antibiogram report that contains susceptibility data is the extent of the surveillance offered. These reports, which may be further organized in various ways (e.g., according to hospital location, site of infection, outpatient vs. inpatient, duration of hospital stay), provide valuable information for monitoring emerging resistance trends among the local microbial flora. Such information is also helpful for establishing **empiric therapy** guidelines (i.e., therapy that is instituted before knowledge of the infecting organism's identification or its antimicrobial susceptibility profile), detecting areas of potential inappropriate or excessive antimicrobial use, and contributing data to larger, more extensive surveillance programs.

Data that have been validated through a results-review and resolution program not only enhance the reliability of laboratory reports for patient management but also strengthen the credibility of susceptibility data used for resistance surveillance and antibiogram profiling. Therefore, the need for each institution to scrutinize susceptibility profiles daily can be accomplished by establishing a results review and resolution format that ensures accuracy for patient management, detects emerging resistance patterns quickly, and ensures accuracy of the data that are included in summary antibiogram reports.

COMMUNICATION

Susceptibility testing profiles produced for each bacterial isolate are typically reported to the physician as a listing of the antimicrobial agents, with each agent accompanied by the category interpretation of susceptible, intermediate, or resistant. In most instances, this reporting approach is sufficient. However, as resistance profiles and their underlying mechanisms become more varied and complex, laboratory personnel must ensure that the significance of susceptibility data is clearly and accurately communicated to clinicians in a way that optimizes both patient care and antimicrobial use. In many situations, passively communicating only the susceptibility data to the physician without adding comments or appropriately amending the reports is no longer sufficient.

For example, methicillin-resistant staphylococci are to be considered cross-resistant to all beta-lactams, but in vitro results occasionally may indicate susceptibility to certain cephalosporins, beta-lactam/beta-lactamase inhibitor combinations, or imipenem. Simply reporting these findings without editing such profiles to reflect probable resistance to all beta-lactams would be seriously misleading. As another example, serious enterococcal infections often require combination therapy that includes both a cell wall–active agent (ampicillin or vancomycin) and an aminoglycoside (i.e., gentamicin). This important information would not be conveyed in a report that simply lists the agents and their interpretive category results. Such an approach can leave the false impression that a "susceptible" result for any single agent indicates that one drug used alone provides appropriate therapy. Therefore, an explanatory note that clearly states the recommended use of combination therapy should accompany the enterococcal susceptibility report.

To avoid misinterpretations that may result by providing only antimicrobial susceptibility data, strategies must consider those organism-antimicrobial combinations that may require reporting supplemental messages to the physician. Consultations with infectious disease specialists and other members of the medical staff is an important part of determining when such messages are needed and what the content should include. Finally, if a laboratory does not have the means to reliably relay these messages, either via the computer or by paper, then a policy of direct communication with the attending physician by telephone or in person should be established.

ADDITIONAL READING

Bauer AW, Kirby WM, Sherris JC, et al: Antibiotic susceptibility testing by a single disc method, *Am J Clin Pathol* 45:49, 1966.

Clinical and Laboratory Standards Institute: Methods for determining bactericidal activity of antimicrobial agents; tentative guideline M26-A, Villanova, Pa, CLSI.

Clinical and Laboratory Standards Institute: Methods for dilution antimicrobial susceptibility testing for bacteria that grow aerobically; M07-A7, Wayne, Pa, CLSI.

Clinical and Laboratory Standards Institute: Methods for antimicrobial dilution and disk susceptibility testing of infrequently isolated or fastidious bacteria; approved guideline; M45-A, Wayne, Pa, CLSI.

Clinical and Laboratory Standards Institute: Analysis and presentation of cumulative antimicrobial susceptibility test data; approved guideline; M39-A2, Wayne, Pa, CLSI.

Clinical and Laboratory Standards Institute: Performance standards for antimicrobial disk susceptibility testing; M02-A9, Wayne, Pa, CLSI.

Clinical and Laboratory Standards Institute: Performance standards for antimicrobial susceptibility testing; M100-S16, Wayne, Pa, CLSI.

Courvalin P: Interpretive reading of antimicrobial susceptibility tests, *Am Soc Microbiol News* 58:368, 1992.

Ferraro MJ, Jorgensen JM: Susceptibility testing instrumentation and computerized expert systems for data analysis and interpretation. In Murray PR, Baron EJ, Jorgensen JM, et al, editors: *Manual of clinical microbiology,* ed 8, Washington, DC, 1995, ASM Press.

Lorian V, editor: *Antibiotics in laboratory medicine,* Philadelphia, 2005, Lippincott Williams and Wilkins.

Rice LB, Sahm D, Bonomo RA: Mechanisms of resistance to antibacterial agents. In Murray PR, Baron EJ, Jorgensen JM, et al, editors: *Manual of clinical microbiology,* ed 8, Washington, DC, 1995, ASM Press.

Sahm DF, O'Brien TF: Detection and surveillance of antimicrobial resistance, *Trends Microbiol* 2:366, 1994.

Sahm DF: The role of clinical microbiology in the control and surveillance of antimicrobial resistance, *Am Soc Microbiol News* 62:25, 1996.

Steward CD, Raney PM, Morrell AK, et al: Testing for induction of clindamycin resistance in erythromycin-resistant isolates of *Staphylococcus aureus, J Clin Microbiol* 43:1716, 2005.

Yao JD, Moellering RC: Antimicrobial agents. In Murray PR, Baron EJ, Jorgensen JM, et al, editors: *Manual of clinical microbiology,* ed 8, Washington, DC, 1995, ASM Press.

PART III

Bacteriology

CHAPTER 13 **Overview of Bacterial Identification Methods and Strategies**

RATIONALE FOR A METHOD OF ORGANISM IDENTIFICATION

Deciding how to teach microbiology in a manner that is both comprehensive and yet understandable is difficult. Most microbiology text chapters are organized by genus name and provide no obvious approach as to how to work up a clinical isolate. Some texts have flowcharts containing algorithms for organism workup; many of these, however, are either too broad to be helpful (e.g., gram-positive vs. gram-negative bacilli) or too esoteric (e.g., cellular analysis of fatty acids) to be practical in routine clinical practice. Unfortunately, the student ends up simply memorizing seemingly unrelated bits of information about various organisms.

The chapters in Part III: Bacteriology are arranged to guide the student through the workup of a microorganism. To accomplish this, chapters have been grouped into subsections using results of basic microbiology procedures, such as the Gram stain, oxidase and catalase tests, and growth on common laboratory media, such as blood, chocolate, and MacConkey agars. Each chapter begins with a short description of the organisms covered in the chapter and an assessment of what the organisms have in common. Because microbiology is ultimately the identification of organisms based on common phenotypic traits shared with known members of the same genus or family, microbiologists "play the odds" every day by finding the best biochemical fit and assigning the most probable identification. For example, the gram-negative rod known as CDC group EF-4a may be considered with either the MacConkey-positive or -negative organisms because it grows on MacConkey agar 50% of the time. Therefore, although CDC group EF-4a has been arbitrarily assigned to the section on oxidase-positive, MacConkey-positive, gram-negative bacilli and coccobacilli in this text, it is also included in the discussion of oxidase-positive, MacConkey-negative, gram-negative bacilli and coccobacilli. This approach to the identification of microorganisms is similar to that used by clinical

microbiologists. Internalization of this method allows students entering the workforce to immediately be able to identify bacteria.

The following chapters help guide the determination of whether a clinical isolate is relevant and provide key biochemical characteristics necessary for organism identification, information on whether susceptibility testing is indicated, and information on the correct antimicrobial agents to use. Most of the procedures described in these chapters can be found at the end of this chapter. In this chapter, each procedure includes a photograph of positive and negative reactions so that the information discussed in the section on expected results can be easily visualized. Chapter 6 includes photographs of some commonly used bacteriologic stains. In addition, Table 13-1 lists several commonly used commercial identification systems for a variety of the microorganisms discussed in the following pages.

HOW TO USE PART III: BACTERIOLOGY

In most instances, the first information that a microbiologist uses in the identification process is the macroscopic description of the colony, or colony morphology. This includes the type of hemolysis (if any), pigment (if present), size, texture (opaque, translucent, or transparent), adherence to agar, pitting of agar, and many other characteristics. After careful observation of the colony, the Gram stain is used to separate the organism into one of a variety of broad categories based on Gram reaction and cellular morphology of gram-positive or gram-negative bacteria (e.g., gram-positive cocci, gram-negative rods). For gram-positive organisms, the catalase test should follow the Gram stain, and testing on gram-negative organisms should begin with the oxidase test. These simple tests plus growth on MacConkey agar, if the isolate is a gram-negative rod or coccobacillus, help the microbiologist assign the organism to one of the primary categories (organized here as subsections). Subsequent testing criteria that are outlined in each

Text continued on p. 222

Table 13-1   Examples of Commercial Identification Systems for Various Organisms

Organism Group	Type of System	Manufacturer	Incubation Time
Enterobacteriaceae	Manual:		
	API 20E	bioMérieux, Inc.*	24-48 hr
	API Rapid 20E	bioMérieux, Inc.	4 hr
	Crystal Enteric/Nonfermenter	Becton Dickinson Diagnostic Systems†	18 hr
	RapID ONE	Remel, Inc.‡	4 hr
	Automated:		
	GNI	bioMérieux, Inc.	4-13 hr
	GNI+	bioMérieux, Inc.	2-12 hr
	NEG ID Type 2	Dade MicroScan, Inc.§	15-42 hr
	Rapid NEG ID Type 3	Dade MicroScan, Inc.	2.5 hr
	Sensititre AP80	Trek Diagnostic Systems, Inc.¶	5-18 hr
Enterococcus spp. and	Manual:		
Streptococcus spp.	API 20 Strep	bioMérieux, Inc.	4-24 hr
	RapID STR	Remel, Inc.	4 hr
	Crystal Gram-Positive ID	Becton Dickinson Diagnostic Systems	18 hr
	Automated:		
	GPI	bioMérieux, Inc.	2-15 hr
	Pos ID2	Dade MicroScan, Inc.	18-48 hr
	Sensititre AP90	Trek Diagnostic Systems, Inc.	24 hr
Haemophilus spp.	Manual:		
	API NH	bioMérieux, Inc.	2 hr
	RapID NH	Remel, Inc.	4 hr
	NHI	bioMérieux, Inc.	4 hr
	Crystal *Neisseria/Haemophilus*	Becton Dickinson Diagnostic Systems	4 hr
	Automated:		
	HNID	Dade MicroScan, Inc.	4 hr
Neisseria spp. and *Moraxella catarrhalis*	Manual:		
	API NH	bioMérieux, Inc.	2 hr
	RapID NH	Remel, Inc.	4 hr
	NHI	bioMérieux, Inc.	4 hr
	Crystal *Neisseria/Haemophilus*	Becton Dickinson Diagnostic Systems	4 hr
	Automated:		
	HNID	Dade MicroScan, Inc.	4 hr
Nonenteric gram-negative rods	Manual:		
	API 20NE	bioMérieux, Inc.	24-48 hr
	Crystal Enteric/Nonfermenter	Becton Dickinson Diagnostic Systems	18-20 hr
	RapID NF Plus	Remel, Inc.	4 hr
	Automated:		
	GNI	bioMérieux, Inc.	2-18 hr
	NEG ID Type 2	Dade MicroScan, Inc.	15-42 hr
	Sensititre AP80	Trek Diagnostic Systems, Inc.	5-18 hr
Staphylococcus spp.	Manual:		
	API STAPH	bioMérieux, Inc.	24 hr
	Crystal Gram-Positive	Becton Dickinson Diagnostic Systems	18-24 hr
	Automated:		
	GPI	bioMérieux, Inc.	2-15 hr
	Pos ID2	Date MicroScan, Inc.	24-48 hr
Coryneform rods	Manual:		
	API Coryne	bioMérieux, Inc.	24 hr
	RapID CB Plus	Remel, Inc.	4 hr
	Crystal Gram-Positive	Becton Dickinson Diagnostic Systems	18-24 hr
	Automated:		
	GPI	bioMérieux, Inc.	2-15 hr

*St. Louis, Mo.: www.bioMerieux-Vitek.com
†Sparks, Md.; www.bectondickinson.com
‡Lenexa, Kan.; www.remelinc.com
§West Sacramento, Calif.; www.dadebehring.com
¶Westlake, Ohio; www.trekds.com

Procedure 13-1

ACETAMIDE UTILIZATION

PRINCIPLE

This test is used to determine the ability of an organism to use acetamide as the sole source of carbon. Bacteria that can grow on this medium deaminate acetamide to release ammonia. The production of ammonia results in a pH-driven color change of the medium from green to royal blue.

METHOD

1. Inoculate acetamide slant lightly with a needle using growth from an 18- to 24-hour culture. Do not inoculate from a broth culture, because the growth will be too heavy.
2. Incubate at 35° C for up to 7 days.

EXPECTED RESULTS

Positive: Deamination of the acetamide resulting in a blue color (Figure 13-1, *A*).

Negative: No color change (Figure 13-1, *B*).

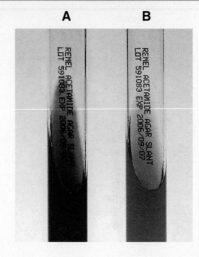

Figure 13-1
Acetamide utilization. **A**, Positive. **B**, Negative.

QUALITY CONTROL

Positive: *Pseudomonas aeruginosa*
Negative: *Stenotrophomonas maltophilia*

Procedure 13-2

ACETATE UTILIZATION

PRINCIPLE

This test is used to determine if an organism can use acetate as the sole source of carbon. If so, breakdown of the sodium acetate causes the pH of the medium to shift toward the alkaline range, turning the indicator from green to blue.

METHOD

1. With a straight inoculating needle, inoculate acetate slant lightly from an 18- to 24-hour culture. Do not inoculate from a broth culture, because the growth will be too heavy.
2. Incubate at 35° C for up to 7 days.

EXPECTED RESULTS

Positive: Medium becomes alkalinized (blue) because of the growth of the organism (Figure 13-2, *A*).

Negative: No growth or growth with no indicator change to blue (Figure 13-2, *B*).

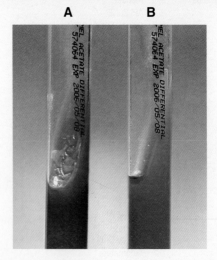

Figure 13-2
Acetate utilization. **A**, Positive. **B**, Negative.

QUALITY CONTROL

Positive: *Escherichia coli*
Negative: *Shigella flexneri*

Procedure 13-3

BACITRACIN TEST

PRINCIPLE
This test is used to determine the effect of a small amount of bacitracin (0.04 U) on an organism. *Streptococcus pyogenes* is inhibited by the small amount of bacitracin in the disk; other beta-hemolytic streptococci usually are not.

METHOD
1. Using an inoculating loop, streak two or three suspect colonies of a pure culture onto a blood agar plate.
2. Using heated forceps, place a bacitracin disk in the first quadrant (area of heaviest growth). Gently tap the disk to ensure adequate contact with the agar surface.
3. Incubate the plate for 18 to 24 hours at 35°C in ambient air.
4. Look for zone of inhibition around disk.

EXPECTED RESULTS
Positive: Any zone of inhibition around the disk (Figure 13-3, *A*).
Negative: No zone of inhibition (Figure 13-3, *B*).

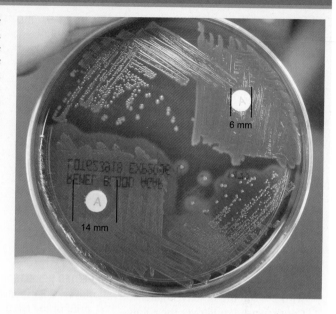

Figure 13-3 Any zone of inhibition is positive *(Streptococcus pyogenes);* growth up to the disk is negative *(Streptococcus algalatiae).*

QUALITY CONTROL
Positive: *Streptococcus pyogenes*
Negative: *Streptococcus agalactiae*

Procedure 13-4

BILE ESCULIN AGAR

PRINCIPLE
Gram-positive bacteria other than some streptococci and enterococci are inhibited by the bile in this medium. If streptococci and enterococci can grow in the presence of 40% bile and hydrolyze esculin, they subsequently turn the indicator, ferric ammonium citrate, a dark brown color. This dark brown color results from the combination of esculetin (the end product of esculin hydrolysis) with ferric ions to form a phenolic iron complex.

METHOD
1. Inoculate one to two colonies from an 18- to 24-hour culture onto the surface of the slant.
2. Incubate at 35° C in ambient air for 48 hours.

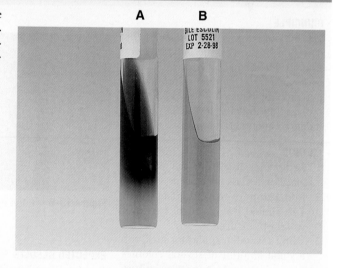

Figure 13-4 Bile esculin agar. **A,** Positive. **B,** Negative.

EXPECTED RESULTS
Positive: Blackening of the agar slant (Figure 13-4, *A*).
Negative: No blackening of medium (Figure 13-4, *B*).

QUALITY CONTROL
Positive: *Enterococcus faecalis*
Negative: *Streptococcus mitis*

Procedure 13-5

BILE SOLUBILITY TEST

PRINCIPLE

The bile solubility test differentiates *Streptococcus pneumoniae* (positive) from alpha-hemolytic streptococci (negative). Bile or a solution of a bile salt, such as sodium desoxycholate rapidly lyses pneumococcal colonies. Lysis depends on the presence of an intracellular autolytic enzyme. Bile salts lower the surface tension between the bacterial cell membrane and the medium, thus accelerating the organism's natural autolytic process.

METHOD

1. Place 1 to 2 drops of 10% sodium desoxycholate to the side of a young (13- to 24-hour), well-isolated colony growing on 5% sheep blood agar. **NOTE:** A tube test is performed with 2% sodium desoxycholate.
2. Gently wash liquid over the colony, without dislodging the colony from the agar.
3. Incubate the plate at 35° C in ambient air for 30 minutes.
4. Examine for lysis of colony.

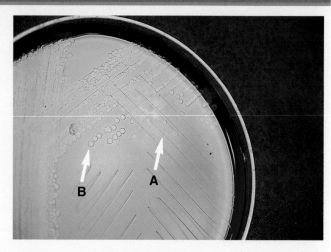

Figure 13-5 Bile solubility test. **A,** Colony lysed. **B,** Intact colony.

EXPECTED RESULTS

Positive: Colony disintegrates; an imprint of the lysed colony may remain within the zone (Figure 13-5, *A*).
Negative: Intact colonies (Figure 13-5, *B*).

QUALITY CONTROL

Positive: *Streptococcus pneumoniae*
Negative: *Enterococcus faecalis*

Procedure 13-6

BUTYRATE DISK

PRINCIPLE

The butyrate disk is a rapid test for the detection of the enzyme butyrate esterase in identifying *Moraxella (Branhamella) catarrhalis*. If bromo-chloro-indolyl butyrate impregnated in the disks is hydrolyzed by the enzyme, a blue-colored indigo compound is formed.

METHOD

1. Remove a disk from the vial and place on a glass microscope slide.
2. Add 1 drop of reagent grade water. This should leave a slight excess of water on the disk.
3. Using a wooden applicator stick, rub a small amount of several colonies from an 18-to 24-hour pure culture onto the disk.
4. Incubate at room temperature for up to 5 minutes.

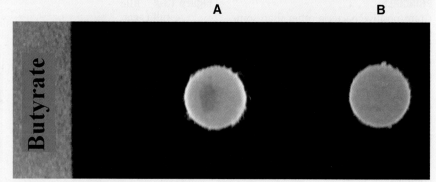

Figure 13-6   Butyrate disk. **A,** Positive. **B,** Negative.

EXPECTED RESULTS

Positive: Development of a blue color during 5-minute incubation period (Figurer 13-6, *A*).
Negative: No color change (Figure 13-6, *B*).

QUALITY CONTROL

Positive: *Moraxella catarrhalis*
Negative: *Neisseria gonorrhoeae*

Procedure 13-7

CAMP TEST

PRINCIPLE

Certain organisms (including group B streptococci) produce a diffusible extra-cellular protein (CAMP factor) that acts synergistically with the beta-lysin of *Staphylococcus aureus* to cause enhanced lysis of red blood cells.

METHOD

1. Streak a beta-lysin–producing strain of *S. aureus* down the center of a sheep blood agar plate.
2. Streak test organisms across the plate perpendicular to the *S. aureus* streak. (Multiple organisms can be tested on a single plate if they are 3 to 4 mm apart.)
3. Incubate overnight at 35° C in ambient air.

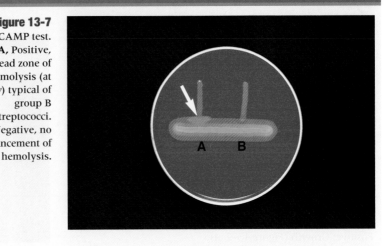

Figure 13-7 CAMP test. **A,** Positive, arrowhead zone of beta-hemolysis (at *arrow*) typical of group B streptococci. **B,** Negative, no enhancement of hemolysis.

EXPECTED RESULTS

Positive: Enhanced hemolysis is indicated by an arrowhead-shaped zone of beta-hemolysis at the juncture of the two organisms (Figure 13-7, *A*).

Negative: No enhancement of hemolysis (Figure 13-7, *B*).

QUALITY CONTROL

Positive: *Streptococcus agalactiae*
Negative: *Streptococcus pyogenes*

Procedure 13-8

CATALASE TEST

PRINCIPLE

The enzyme catalase mediates the breakdown of hydrogen peroxide (H_2O_2) into oxygen and water. The presence of the enzyme in a bacterial isolate is evident when a small inoculum is introduced into hydrogen peroxide (30% for the slide test), and the rapid elaboration of oxygen bubbles occurs. The lack of catalase is evident by a lack of or weak bubble production.

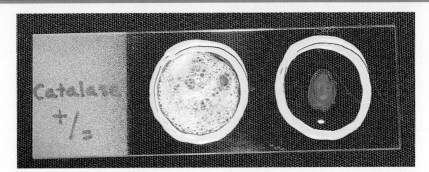

Figure 13-8 Catalase test. **A,** Positive. **B,** Negative.

METHOD

1. Use a loop or sterile wooden stick to transfer a small amount of colony growth to the surface of a clean, dry glass slide.
2. Place a drop of 30% hydrogen peroxide (H_2O_2) onto the medium.
3. Observe for the evolution of oxygen bubbles (Figure 13-8).

EXPECTED RESULTS

Positive: Copious bubbles produced (Figures 13-8, *A*).
Negative: No or few bubbles produced (Figures 13-8, *B*).
NOTE: Some organisms (enterococci) produce a peroxidase that slowly catalyzes the breakdown of (H_2O_2) and the test may appear weakly positive. This reaction is not a truly positive test.

QUALITY CONTROL

Positive: *Staphylococcus aureus*
Negative: *Streptococcus pyogenes*

Procedure 13-9

CETRIMIDE

PRINCIPLE
This test is used to determine the ability of an organism to grow in the presence of cetrimide, a toxic substance that inhibits the growth of many bacteria.

METHOD
1. Inoculate a cetrimide agar slant with 1 drop of an 18- to 24-hour culture in brain-heart infusion broth.
2. Incubate at 35° C for up to 7 days.
3. Examine the slant for bacterial growth.

EXPECTED RESULTS
Positive: Growth (Figure 13-9, *A*).
Negative: No growth (Figure 13-9, *B*).

QUALITY CONTROL
Positive: *Pseudomonas aeruginosa*
Negative: *Escherichia coli*

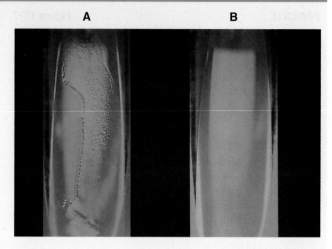

Figure 13-9
Cetrimide.
A, Positive.
B, Negative.

Procedure 13-10

CITRATE UTILIZATION

PRINCIPLE
This test is used to determine the ability of an organism to utilize sodium citrate as its only carbon source and inorganic ammonium salts as its only nitrogen source. Bacteria that can grow on this medium turn the bromthymol blue indicator from green to blue.

METHOD
1. Inoculate Simmons citrate agar lightly on the slant by touching the tip of a needle to a colony that is 18 to 24 hours old. **NOTE:** There is no need to stab into the butt of the tube. Do not inoculate from a broth culture, because the inoculum will be too heavy.
2. Incubate at 35° to 37° C for up to 7 days.

3. Observe for development of blue color, denoting alkalinization.

EXPECTED RESULTS
Positive: Growth on the medium, with or without a change in the color of the indicator. The color change of the indicator is due to acid or alkali production by the test organism as it grows on the medium. Growth usually results in the bromthymol blue indicator, turning from green to blue (Figure 13-10, *A*).
Negative: Absence of growth (Figure 13-10, *B*).

QUALITY CONTROL
Positive: *Klebsiella pneumoniae*
Negative: *Escherichia coli*

Figure 13-10   Citrate utilization.
A, Positive. **B,** Negative.

chapter are then used to more definitively identify the isolate.

The clinical diagnosis and the source of the specimen also can aid in determining which group of organisms to consider. For example, if a patient has endocarditis or the specimen source is blood, and a small, gram-negative rod is observed on Gram stain, the microbiologist should consider a group of gram-negative bacilli known as the *HACEK* bacteria for *Haemophilus aphrophilus*, *Actinobacillus actinomycetemcomitans*, *Cardiobacterium hominis*, *Eikenella corrodens*, and *Kingella* spp. Similarly, if the patient has suffered an animal bite, the

Procedure 13-11

COAGULASE TEST

PRINCIPLE

This test is used to differentiate *Staphylococcus aureus* (positive) from coagulase-negative staphylococci (negative). *S. aureus* produces two forms of coagulase: bound and free. Bound coagulase, or "clumping factor," is bound to the bacterial cell wall and reacts directly with fibrinogen. This results in an alteration of fibrinogen so that it precipitates on the staphylococcal cell, causing the cells to clump when a bacterial suspension is mixed with plasma. The presence of bound coagulase correlates well with free coagulase, an extracellular protein enzyme that causes the formation of a clot when *S. aureus* colonies are incubated with plasma. The clotting mechanism involves activation of a plasma coagulase-reacting factor (CRF), which is a modified or derived thrombin molecule, to form a coagulase-CRF complex. This complex in turn reacts with fibrinogen to produce the fibrin clot.

METHOD

A. Slide test
  1. Place a drop of coagulase plasma (preferably rabbit plasma with EDTA) on a clean, dry glass slide.
  2. Place a drop of distilled water or saline next to the drop of plasma as a control.
  3. With a loop, straight wire, or wooden stick, emulsify a portion of the isolated colony being tested in each drop, inoculating the water or saline first. Try to create a smooth suspension.
  4. Mix well with a wooden applicator stick.
  5. Rock the slide gently for 5 to 10 seconds.

EXPECTED RESULTS

Positive: Macroscopic clumping in 10 seconds or less in coagulated plasma drop and no clumping in saline or water drop (Figure 13-11, *A*, left side).

Negative: No clumping in either drop. **NOTE:** *All* negative slide tests *must*

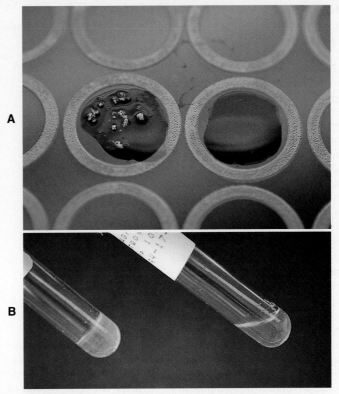

Figure 13-11   Coagulase test. **A,** Slide coagulase test for clumping factor. Left side is positive; right side is negative. **B,** Tube coagulase test for free coagulase. Tube on the left is positive, exhibiting clot. Tube on the right is negative.

be confirmed using the tube test (Figure 13-11, *B*, right side).

Equivocal: Clumping in both drops indicates that the organism autoagglutinates and is unsuitable for the slide coagulase test.

B. Tube test
  1. Emulsify several colonies in 0.5 mL of rabbit plasma (with EDTA) to give a milky suspension.
  2. Incubate tube at 35° C in ambient air for 4 hours.
  3. Check for clot formation. **NOTE:** Tests can be positive at 4 hours and then revert to negative after 24 hours.
  4. If negative at 4 hours, incubate at room temperature overnight and check again for clot formation.

EXPECTED RESULTS

Positive: Clot of any size (Figure 13-11, *A*, left side).

Negative: No clot (Figure 13-11, *B*, right side).

QUALITY CONTROL

Positive: *Staphylococcus aureus*

Negative: *Staphylococcus epidermidis*

Procedure 13-12

DECARBOXYLASE TESTS (MOELLER'S METHOD)

PRINCIPLE

This test measures the enzymatic ability of an organism to decarboxylate (or hydrolyze) an amino acid to form an amine. Decarboxylation, or hydrolysis, of the amino acid results in an alkaline pH change.

METHOD

A. Glucose nonfermenting organisms
  1. Prepare a very heavy suspension (≥McFarland No. 5 turbidity standard) in brain-heart infusion broth from young bacteria (18 to 24 hours old) growing on 5% sheep blood agar.
  2. Inoculate each of the three decarboxylase broths (arginine, lysine, and ornithine) and the control broth (no amino acid) with 4 drops of broth.
  3. Add a 4-mm layer of sterile mineral oil to each tube.
  4. Incubate the cultures at 35° C in ambient air for up to 7 days.
B. Glucose-fermenting organisms
  1. Inoculate tubes with 1 drop of an 18- to 24-hour brain-heart infusion broth culture.
  2. Add a 4-mm layer of sterile mineral oil to each tube.
  3. Incubate the cultures for 4 days at 35° C in ambient air.

Figure 13-12
Decarboxylase tests (Moeller's method).
A, Positive.
B, Negative.
C, Uninoculated tube.

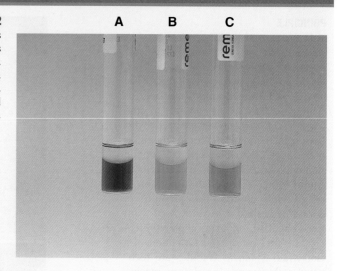

EXPECTED RESULTS

Positive: Alkaline (purple) color change compared with the control tube (Figure 13-12, *A*).

Negative: No color change or acid (yellow) color in test and control tube. Growth in the control tube. **NOTE:** The fermentation of dextrose in the medium causes the acid color change. It would not, however, mask the alkaline color change brought about by a positive decarboxylation reaction (Figure 13-12, *B*). Uninoculated tube is shown in Figure 13-12, *C*.

QUALITY CONTROL

Positive:

Lysine	*Klebsiella pneumoniae*
Ornithine	*Enterobacter cloacae*
Arginine	*Enterobacter cloacae*
Base	—

Negative:

Lysine	*Enterobacter cloacae*
Ornithine	*Klebsiella pneumoniae*
Arginine	*Klebsiella pneumoniae*
Base	*Klebsiella pneumoniae*

microbiologist should think of *Pasteurella multocida*, EF-4a, and EF-4b if the isolate is gram-negative and *Staphylococcus hyicus* and *Staphylococcus intermedius* if the organism is gram-positive.

The most important point to remember when attempting to group and identify any microorganism is that no piece of information should be overlooked. In this era of cost containment, it is more important than ever to take a stepwise approach to microbial identi-

fication, beginning with the simplest procedures and proceeding to more expensive definitive testing only if the clinical situation warrants further work. The days of going from a quick glance at a colony on a plate to a rapid identification system without a Gram stain, catalase, or oxidase test are over. Microbiologists are going back to basics. The real challenge is how to most efficiently and effectively bring all the information together in a cost-effective and clinically relevant manner.

Procedure 13-13

DNA HYDROLYSIS

PRINCIPLE
This test is used to determine the ability of an organism to hydrolyze DNA. The medium is pale green because of the DNA–methyl green complex. If the organism growing on the medium hydrolyses DNA, the green color fades and the colony is surrounded by a colorless zone.

METHOD
1. Inoculate the DNase agar with the organism to be tested and streak for isolation.
2. Incubate aerobically at 35° C for 13 to 24 hours.

EXPECTED RESULTS
Positive: When DNA is hydrolyzed, methyl green is released and combines with highly polymerized DNA at a pH of 7.5, turning the medium colorless around the test organism (Figure 13-13, *A* and *B*).

Negative: If there is no degradation of

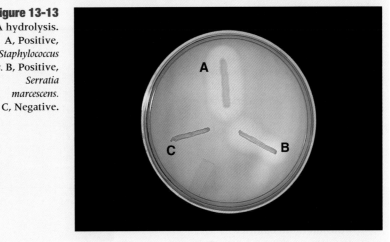

Figure 13-13
DNA hydrolysis.
A, Positive,
*Staphylococcus
aureus*. B, Positive,
*Serratia
marcescens*.
C, Negative.

DNA, the medium remains green (Figure 13-13, *C*).

QUALITY CONTROL
Positive: *Staphylococcus aureus*
Negative: *Staphylococcus epidermidis*

Procedure 13-14

ESCULIN HYDROLYSIS

PRINCIPLE
This test is used to determine whether an organism is able to hydrolyze the glycoside esculin.

METHOD
1. Inoculate the medium with 1 drop of a 24-hour broth culture.
2. Incubate at 35° C for up to 7 days.
3. Examine the slants for blackening and under the ultraviolet rays of a Wood's lamp for esculin hydrolysis.

EXPECTED RESULTS
Positive: Blackened medium (Figure 13-14, *A*), which would also show a loss of fluorescence under the Wood's lamp.

Negative: No blackening and no loss of fluorescence under Wood's lamp, or slight blackening with no loss of fluorescence under Wood's lamp. Uninoculated tube is shown in Figure 13-14, *B*.

Figure 13-14
Esculin hydrolysis.
A, Positive,
blackening of slant.
B, Uninoculated
tube.

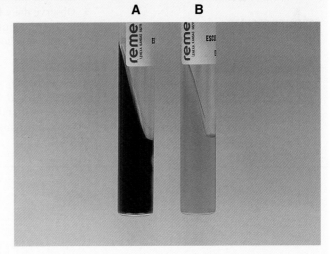

QUALITY CONTROL
Positive: *Klebsiella pneumoniae*
Negative: *Shigella flexneri*

Procedure 13-15

FERMENTATION MEDIA

PRINCIPLE

These media are used to determine the ability of an organism to ferment a specific carbohydrate that is incorporated in a basal medium, thereby producing acid with or without visible gas. A medium for *Enterobacteriaceae* and coryneforms and a medium for streptococci and enterococci are described.

METHOD

A. Peptone medium with Andrade's indicator (for enterics and coryneforms)
   1. Inoculate each tube with 1 drop of an 18- to 24-hour brain-heart infusion broth culture.
   2. Incubate at 35° C for up to 7 days in ambient air. **NOTE:** Tubes are only held 4 days for the organisms belonging to the *Enterobacteriaceae* family.
   3. Examine the tubes for acid (indicated by a pink color) and gas production.
   4. Tubes must show growth for the test to be valid. If after 24 hours of incubation there is no growth in the fermentation tubes or control, add 1 to 2 drops of sterile rabbit serum per 5 mL of fermentation broth to each tube.

EXPECTED RESULTS

Positive: Indicator change to pink with or without gas formation in Durham tube (Figure 13-15, *A*, left and middle).
Negative: Growth, but no change in color. Medium remains clear to straw-colored (Figure 13-15, *A*, right).
B. Heart infusion broth with bromcresol purple indicator (for streptococci and enterococci)

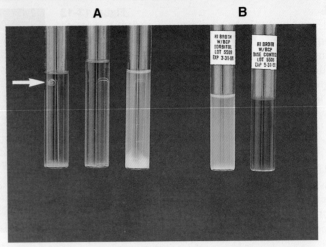

Figure 13-15   Fermentation media. **A,** Peptone medium with Andrade's indicator. The tube on the left ferments glucose with the production of gas (visible as a bubble [*arrow*] in the inverted [Durham] tube), the tube in the middle ferments glucose with no gas production, and the tube on the right does not ferment glucose. **B,** Heart infusion broth with bromcresol purple indicator. The tube on the left is positive; the tube on the right is negative.

   1. Inoculate each tube with 2 drops of an 18- to 24-hour brain-heart infusion broth culture.
   2. Incubate 4 days at 35° C in ambient air.
   3. Observe daily for a change of the bromcresol purple indicator from purple to yellow (acid).

EXPECTED RESULTS

Positive: Indicator change to yellow (Figure 13-15, *B*, left).
Negative: Growth, but no change in color. Medium remains purple (Figure 13-15, *B*, right).

QUALITY CONTROL

NOTE: Appropriate organisms depend on which carbohydrate has been added to the basal medium. An example is given for each type of medium.
A. Peptone medium with Andrade's indicator
Dextrose:
Positive, with gas: *Escherichia coli*
Positive, no gas: *Shigella flexneri*
Negative: *Pseudomonas aeruginosa*
B. Heart infusion broth with bromcresol purple indicator
Sorbitol:
Positive: *Streptococcus mutans*
Negative: *Streptococcus mitis*

Procedure 13-16

FLAGELLA STAIN (WET-MOUNT TECHNIQUE)

PRINCIPLE

A wet-mount technique[1] for staining bacterial flagella is simple and is useful when the number and arrangement of flagella are critical in identifying species of motile bacteria.

METHOD

1. Grow the organism to be stained at room temperature on blood agar for 16 to 24 hours.
2. Add a small drop of water to a microscope slide.
3. Dip a sterile inoculating loop into sterile water.
4. Touch the loopful of water to the colony margin briefly (this allows motile cells to swim into the droplet of water).
5. Touch the loopful of motile cells to the drop of water on the slide. **NOTE:** Agitating the loop in the droplet of water on the slide causes the flagella to shear off the cell.
6. Cover the faintly turbid drop of water on the slide with a cover slip. A proper wet mount has barely enough liquid to fill the space under a cover slip. Small air spaces around the edge are preferable.
7. Examine the slide immediately under 40× to 50× for motile cells. If motile cells are not seen, do not proceed with the stain.
8. If motile cells are seen, leave the slide at room temperature for 5 to 10 minutes. This allows time for the bacterial cells to adhere to either the glass slide or the cover slip.
9. Apply 2 drops of RYU flagella stain (Remel, Inc., Lenexa, Kan.) gently to the edge of the cover slip. The stain will flow by capillary action and mix with the cell suspension. Small air pockets around the edge of the wet mount are useful in aiding the capillary action.
10. After 5 to 10 minutes at room temperature, examine the cells for flagella.
11. Cells with flagella may be observed at 100× (oil) in the zone of opti-

mum stain concentration, about half way from the edge of the cover slip to the center of the mount.
12. Focusing the microscope on the cells attached to the coverslip rather than the cells attached to the slide facilitates visualization of the flagella. The precipitate from the stain is primarily on the slide rather than the cover slip.

EXPECTED RESULTS

Observe the slide and note the following:
1. Presence or absence of flagella
2. Number of flagella per cell
3. Location of flagella per cell

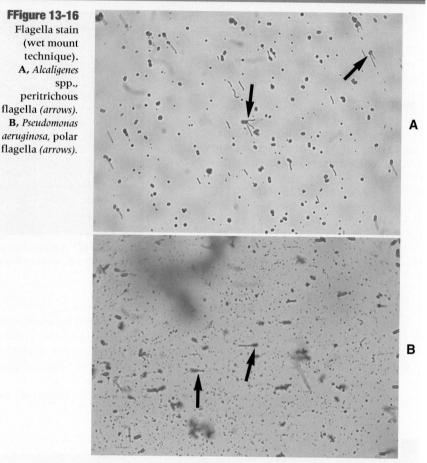

FFigure 13-16 Flagella stain (wet mount technique). **A,** *Alcaligenes* spp., peritrichous flagella *(arrows).* **B,** *Pseudomonas aeruginosa,* polar flagella *(arrows).*

   a. Peritrichous (Figure 13-16, *A*)

   b. Lophotrichous

   c. Polar   (Figure 13-16, *B*)

4. Amplitude of wavelength

   a. Short

   b. Long

5. Whether or not "tufted"

QUALITY CONTROL

Peritrichous: *Escherichia coli*
Polar: *Pseudomonas aeruginosa*
Negative: *Klebsiella pneumoniae*

Procedure 13-17

GELATIN HYDROLYSIS

PRINCIPLE

This test is used to determine the ability of an organism to produce proteolytic enzymes (gelatinases) that liquefy gelatin.

METHOD

1. Inoculate the gelatin deep with 4 to 5 drops of a 24-hour broth culture.
2. Incubate at 35° C in ambient air for up to 14 days. **NOTE:** Incubate the medium at 25° C if the organism grows better at 25° C than at 35° C.
3. Alternatively, inoculate the gelatin deep from a 24-hour-old colony by stabbing 4 to 5 times 1/2 inch into the medium.
4. Remove the gelatin tube daily from the incubator and place at 4° C to check for liquefaction. Do not invert or tip the tube, because sometimes the only discernible liquefaction will occur at the top of the deep where inoculation occurred.
5. Refrigerate an uninoculated control along with the inoculated tube. Liquefaction is determined only after the control has hardened (gelled).

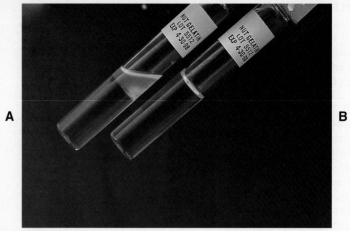

Figure 13-17 Gelatin hydrolysis. **A,** Positive; note liquefaction at top of tube. **B,** Uninoculated tube.

EXPECTED RESULTS

Positive: Partial or total liquefaction of the inoculated tube (the control tube must be completely solidified) at 4° C within 14 days (Figure 13-17, *A*).
Negative: Complete solidification of tube at 4° C (Figure 13-17, *B*).

QUALITY CONTROL

Positive: *Proteus vulgaris*
Negative: *Enterobacter aerogenes*

Procedure 13-18

GROWTH AT 42° C

PRINCIPLE

The test is used to determine the ability of an organism to grow at 42° C.

METHOD

1. Inoculate two tubes of trypticase soy agar (TSA) with a light inoculum by lightly touching a needle to the top of a single 13- to 24-hour-old colony and streaking the slant.
2. Immediately incubate one tube at 35° C and one at 42° C.
3. Record the presence of growth on each slant after 18 to 24 hours.

EXPECTED RESULTS

Positive: Good growth at both 35° and 42° C (Figure 13-18, *A*).
Negative: No growth at 42° C (Figure 13-18, *B*), but good growth at 35° C.

Figure 13-18
Growth at 42° C.
A, Positive, good growth.
B, Negative, no growth.

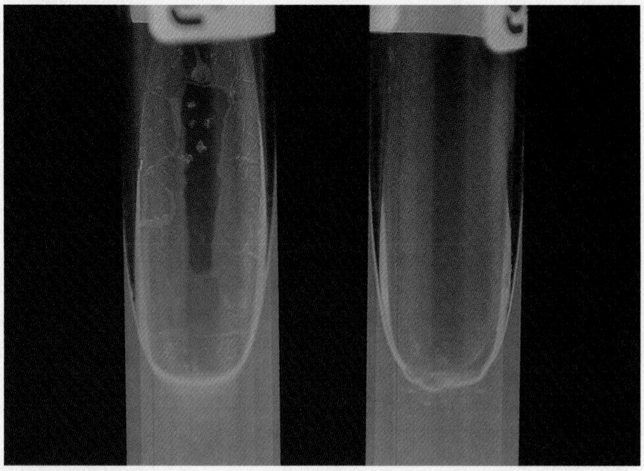

QUALITY CONTROL

Positive: *Pseudomonas aeruginosa*
Negative: *Pseudomonas fluorescens*

Procedure 13-19

HIPPURATE TEST

PRINCIPLE

The end products of hydrolysis of hippuric acid by hippuricase include glycine and benzoic acid. Glycine is deaminated by the oxidizing agent ninhydrin, which is reduced during the process. The end products of the ninhydrin oxidation react to form a purple-colored product. The test medium must contain only hippurate, because ninhydrin might react with any free amino acids present in growth media or other broths.

METHOD

1. Add 0.1 mL of sterile water to a 12 × 75 mm plastic test tube.
2. Make a heavy suspension of the organism to be tested.
3. Using heated forceps, place a rapid hippurate disk in the mixture.
4. Cap and incubate the tube for 2 hours at 35° C; use of a water bath is preferred.
5. Add 0.2 mL ninhydrin reagent and reincubate for an additional 15 to

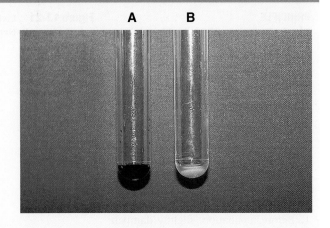

Figure 13-19
Hippurate test.
A, Positive.
B, Negative.

30 minutes. Observe the solution for development of a deep purple color.

EXPECTED RESULTS

Positive: Deep purple color (Figure 13-19, *A*).
Negative: Colorless or slightly yellow pink color (Figure 13-19, *B*).

QUALITY CONTROL

Positive: *Streptococcus agalactiae*
Negative: *Streptococcus pyogenes*

Procedure 13-20

INDOLE PRODUCTION

PRINCIPLE

The test is used to determine the ability of an organism to split tryptophan to form the compound indole. The spot indole test is described in Procedure 13-39.

METHOD

A. *Enterobacteriaceae*
   1. Inoculate tryptophane broth with 1 drop from a 24-hour brain-heart infusion broth culture.
   2. Incubate at 35° C in ambient air for 48 hours.
   3. Add 0.5 mL of Kovac's reagent to the broth culture.
B. Other gram-negative bacilli
   1. Inoculate tryptophane broth with 1 drop of a 24-hour broth culture.
   2. Incubate at 35° C in ambient air for 48 hours.
   3. Add 1 mL of xylene to the culture.
   4. Shake mixture vigorously to extract the indole and allow to stand until the xylene forms a layer on top of the aqueous phase.
   5. Add 0.5 mL of Ehrlich's reagent down the side of the tube.

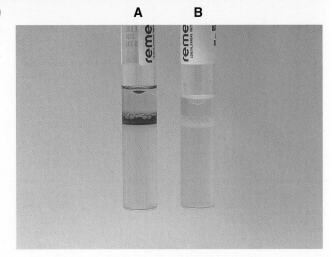

Figure 13-20
Indole production.
A, Positive.
B, Negative.

EXPECTED RESULTS

Positive: Pink- to wine-colored ring after addition of appropriate reagent (Figure 13-20, *A*).
Negative: No color change after the addition of the appropriate reagent (Figure 13-20, *B*).

QUALITY CONTROL

A. Kovac's method
   Positive: *Escherichia coli*
   Negative: *Klebsiella pneumoniae*
B. Ehrlich's method
   Positive: *Elizabethkingia meningoseptica*
   Negative: CDC group EO-2

Procedure 13-21

LAP TEST

PRINCIPLE
The LAP test is used to identify catalase-negative gram-positive cocci. The LAP disk is a rapid test for the detection of the enzyme leucine aminopeptidase. Leucine-β-naphthylamide impregnated disks serve as a substrate for the detection of leucine aminopeptidase. Following hydrolysis of the substrate by the enzyme, the resulting β-naphthylamine produces a red color upon the addition of cinnamaldehyde reagent.

METHOD
1. Before incubation, slightly dampen the LAP disk with reagent grade water. Do *not* supersaturate the disk.
2. Using a wooden applicator stick, rub a small amount of several colonies of an 18- to 24-hour pure culture onto a small area of the LAP disk.
3. Incubate at room temperature for 5 minutes.
4. After this incubation period, add 1 drop of cinnamaldehyde reagent.

Figure 13-21   LAP Test. **A**, Positive. **B**, Negative.

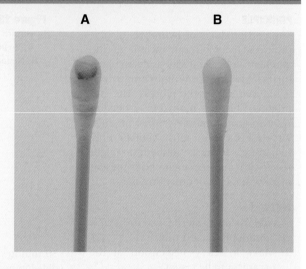

EXPECTED RESULTS
Positive: Development of a red color within 1 minute after adding cinnamaldehyde reagent (Figure 13-21, *A*).
Negative: No color change or development of a slight yellow color (Figure 13-21, *B*).

QUALITY CONTROL
Positive: *Enterococcus faecalis*
Negative: *Leuconostoc* sp.

Procedure 13-22

LITMUS MILK

PRINCIPLE

This test is used to determine an organism's ability to metabolize litmus milk. Fermentation of lactose is evidenced by the litmus turning pink as a result of acid production. If sufficient acid is produced, casein in the milk is coagulated, solidifying the milk. With some organisms, the curd shrinks and whey is formed at the surface. Some bacteria hydrolyze casein, causing the milk to become straw-colored and resemble turbid serum. Additionally, some organisms reduce litmus, in which case the medium becomes colorless in the bottom of the tube.

METHOD

1. Inoculate with 4 drops of a 24-hour broth culture.
2. Incubate at 35° to 37° C in ambient air.
3. Observe daily for 7 days for alkaline reaction (litmus turns blue), acid reaction (litmus turns pink), indicator reduction, acid clot, rennet clot, and peptonization. Multiple changes can occur over the observation period.
4. Record all changes.

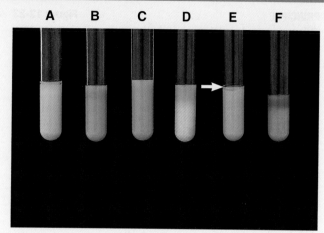

Figure 13-22
Litmus milk.
A, Acid reaction.
B, Alkaline reaction. **C,** No change.
D, Reduction of indicator. **E,** Clot. (Note separation of clear fluid from clot at arrow.)
F, Peptonization.

QUALITY CONTROL

Alkaline: *Alcaligenes faecalis*
Acid: *Enterococcus faecium*
Peptonization: *Burkholderia cepacia*

EXPECTED RESULTS

Appearance of Indicator (Litmus Dye)

COLOR	pH CHANGE TO . . .	RECORD
Pink, mauve (Figure 13-22, *A*)	Acid	Acid (A)
Blue (Figure 13-22, *B*)	Alkaline	Alkaline (K)
Purple (identical to uninoculated control) (Figure 13-22, *C*)	No change	No change
White (Figure 13-22, *D*)	Independent of pH change; result of reduction of indicator	Decolorized

Appearance of Milk

CONSISTENCY OF MILK	OCCURS WHEN pH IS . . .	RECORD
Coagulation or clot (Figure 13-22, *E*)	Acid or alkaline	Clot
Dissolution of clot with clear, grayish, watery fluid and a shrunken, insoluble pink clot (Figure 13-22, *F*)	Acid	Digestion
Dissolution of clot with clear, grayish, watery fluid and a shrunken, insoluble blue clot	Alkaline	Peptonization

Procedure 13-23

LYSIN IRON AGAR

PRINCIPLE

This test is used to determine whether a gram-negative rod decarboxylates or deaminates lysine and forms hydrogen sulfide (H_2S). Lysine iron agar (LIA) contains lysine, peptones, a small amount of glucose, ferric ammonium citrate, and sodium thiosulfate. When glucose is fermented, the butt of the medium becomes acidic (yellow). If the organism produces lysine decarboxylase, cadaverine is formed. Cadaverine neutralizes the organic acids formed by glucose fermentation, and the butt of the medium reverts to the alkaline state (purple). If the decarboxylase is not produced, the butt remains acidic (yellow). If oxidative deamination of lysine occurs, a compound is formed that, in the presence of ferric ammonium citrate and a coenzyme, flavin mononucleotide, forms a burgundy color on the slant. If deamination does not occur, the LIA slant remains purple.

METHOD

1. With a straight inoculating needle, inoculate LIA (Figure 13-23, *E*) by twice stabbing through the center of the medium to the bottom of the tube and then streaking the slant.

Figure 13-23
Lysine iron agar.
A, Alkaline slant/alkaline butt (K/K). **B,** Alkaline slant/alkaline butt, H_2S positive (K/K H_2S^+). **C,** Alkaline slant/acid butt (K/A). **D,** Red slant/acid butt (R/A). **E,** Uninoculated tube.

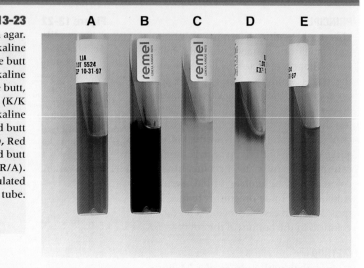

2. Cap the tube tightly and incubate at 35°C in ambient air for 18 to 24 hours.

EXPECTED RESULTS

1. Alkaline slant/alkaline butt (K/K) = lysine decarboxylation and no fermentation of glucose (Figure 13-23, *A*).
2. Alkaline slant/acid butt (K/A) = glucose fermentation (Figure 13-23, *C*).
3. **NOTE**: Patterns shown in 1 and 2 above can be accompanied by a black precipitate of ferrous sulfide (FeS), which indicates production of H_2S (Figure 13-23, *B*).
4. Red slant/acid butt (R/A) = lysine deamination and glucose fermentation (Figure 13-23, *D*).

QUALITY CONTROL

Alkaline slant and butt, H_2S positive: *Salmonella typhimurium*
Alkaline slant, acid butt: *Shigella flexneri*
Red slant, acid butt: *Proteus vulgaris*

Procedure 13-24

METHYL RED/VOGES-PROSKAUER (MRVP) TESTS

PRINCIPLE
This test is used to determine the ability of an organism to produce and maintain stable acid end products from glucose fermentation, to overcome the buffering capacity of the system, and to determine the ability of some organisms to produce neutral end products (e.g., acetyl-methylcarbinol or acetoin) from glucose fermentation.

METHOD
1. Inoculate MRVP broth with 1 drop from a 24-hour brain-heart infusion broth culture.
2. Incubate at 35° to 37° C for a minimum of 48 hours in ambient air. Tests should not be made with cultures incubated less than 48 hours, because the end products build up to detectable levels over time. If results are equivocal at 48 hours, repeat the tests with cultures incubated at 35° to 37° C for 4 to 5 days in ambient air; in such instances, duplicate tests should be incubated at 25° C.
3. Split broth into aliquots for MR test and VP test.

A. MR (methyl red) test
   1. Add 5 or 6 drops of methyl red reagent per 5 mL of broth.
   2. Read reaction immediately.

EXPECTED RESULTS
Positive: Bright red color indicative of mixed acid fermentation (Figure 13-24, *A*).
   Weakly positive: Red-orange color.
   Negative: Yellow color (Figure 13-24, *B*).
B. VP (Voges-Proskauer) test (Barritt's method) for gram-negative rods
   1. Add 0.6 mL (6 drops) of solution A (α-naphthol) and 0.2 mL (2 drops) of solution B (KOH) to 1 mL of MRVP broth.
   2. Shake well after addition of each reagent.
   3. Observe for 5 minutes.

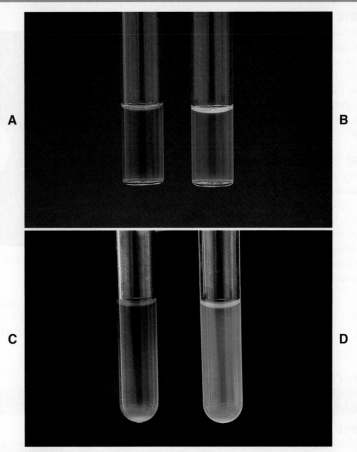

Figure 13-24   Methyl Red/Voges-Proskauer (MRVP) tests. A, Positive methyl red. B, Negative methyl red. C, Positive Voges-Proskauer. D, Negative Voges-Proskauer.

EXPECTED RESULTS
Positive: Red color indicative of acetoin production (Figure 13-24, *C*).
   Negative: Yellow color (Figure 13-24, *D*).
C. VP (Voges-Proskauer) test (Coblentz method) for streptococci
   1. Use 24-hour growth from blood agar plate to heavily inoculate 2 mL of MRVP broth.
   2. After 6 hours of incubation at 35° C in ambient air, add 1.2 mL (12 drops) of solution A (α-naphthol) and 0.4 mL (4 drops) solution B (40% KOH with creatine).

3. Shake the tube and incubate at room temperature for 30 minutes.

QUALITY CONTROL
A. Methyl red
   Positive: *Escherichia coli*
   Negative: *Enterobacter cloacae*
B. VP (Barritt's method)
   Positive: *Enterobacter cloacae*
   Negative: *Escherichia coli*
C. VP (Coblentz method)
   Positive: *Streptococcus mutans*
   Negative: *Streptococcus mitis*

Procedure 13-25

MICRODASE TEST

PRINCIPLE

The microdase test is a rapid method to differentiate *Staphylococcus* from *Micrococcus* by detection of the enzyme oxidase. In the presence of atmospheric oxygen, the oxidase enzyme reacts with the oxidase reagent and cytochrome C to form the colored compound, indophenol.

METHOD

1. Using a wooden applicator stick, rub a small amount of several colonies of an 18- to 24-hour pure culture grown on blood agar onto a small area of the microdase disk. **NOTE**: Do *not* rehydrate the disk before use.
2. Incubate at room temperature for 2 minutes.

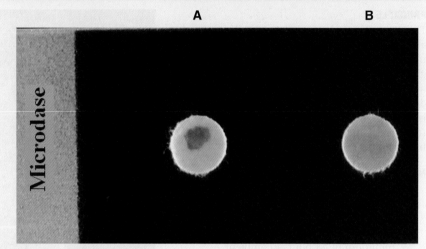

Figure 13-25   Microdase Test. **A**, Positive. **B**, Negative.

EXPECTED RESULTS

Positive: Development of blue to purple-blue color (Figure 13-25, *A*).
Negative: No color change (Figure 13-25, *B*).

QUALITY CONTROL

Positive: *Micrococcus luteus*
Negative: *Staphylococcus aureus*

Procedure 13-26

MOTILITY TESTING

PRINCIPLE

These tests are used to determine if an organism is motile. An organism must possess flagella to be motile.

METHOD

A. Hanging drop
1. Using a young (6- to 24-hour) actively growing 25° C broth culture, place 1 drop of the culture in the center of a 22 × 22-mm coverslip.
2. Place a small drop of immersion oil on each corner of the coverslip.
3. Invert the coverslip over the concavity of a depression slide.
4. Examine with the high dry (40×) objective.
B. Semisolid agar deep
1. Touch a straight needle to a colony of a young (18- to 24-hour) culture growing on agar medium.
2. Stab once to a depth of only ¹/₃ to ¹/₂ inch in the middle of the tube.
3. Incubate at 35° to 37° C and examine daily for up to 7 days.

Figure 13-26
Motility test.
A, Positive.
B, Negative.

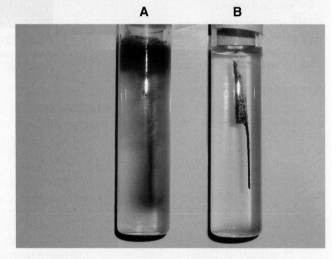

EXPECTED RESULTS

A. Hanging drop
Positive: In true motility, the organisms change position with respect to each other, often darting across the field.
Negative: In Brownian movement, the organisms may appear quite active, but they remain in the same relative position to other organisms or debris in the field.

B. Semisolid agar deep
Positive: Motile organisms will spread out into the medium from the site of inoculation (Figure 13-26, *A*).
Negative: Nonmotile organisms remain at the site of inoculation (Figure 13-26, *B*).

QUALITY CONTROL

Positive: *Escherichia coli*
Negative: *Klebsiella pneumoniae*

Procedure 13-27

MRS BROTH

PRINCIPLE

This test is used to determine whether an organism forms gas during glucose fermentation. Some *Lactobacillus* spp. and *Leuconostoc* sp. produce gas.

METHOD

1. Inoculate MRS broth with an 18- to 24-hour culture from agar or broth.
2. Incubate 24 to 48 hours at 35° C in ambient air.

EXPECTED RESULTS

Positive: Gas production indicated by a bubble in Durham tube (Figure 13-27, *A*).

Negative: No gas production (Figure 13-27, *B*).

QUALITY CONTROL

Positive: *Leuconostoc* spp.
Negative: *Pediococcus* spp.

Figure 13-27
MRS broth.
A, Positive, gas production by *Leuconostoc (arrow)*.
B, Negative, no gas production.

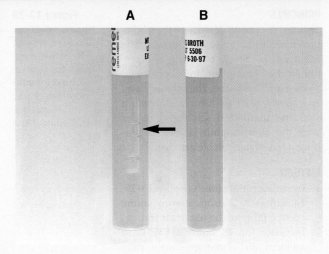

Procedure 13-28

MUG TEST

PRINCIPLE

Escherichia coli produces the enzyme beta-D-glucuronidase, which hydrolyzes beta-D-glucopyranosid-uronic derivatives to aglycons and D-glucuronic acid. The substrate 4-methylumbelliferyl-beta-D-glucuronide is impregnated in the disk and is hydrolyzed by the enzyme to yield the 4-methylumbelliferyl moiety, which fluoresces blue under long wavelength ultraviolet light.

METHOD

1. Wet the disk with one drop of water.
2. Using a wooden applicator stick, rub a portion of a colony from an 18- to 24-hour-old pure culture onto the disk.
3. Incubate at 35° C in a closed container for up to 2 hours.
4. Observe disk using a 366-nm ultraviolet light.

EXPECTED RESULTS

Positive: Electric blue fluorescence (Figure 13-28, *A*).

Negative: Lack of fluorescence (Figure 13-28, *B*).

QUALITY CONTROL

Positive: *Escherichia coli*
Negative: *Pseudomonas aeruginosa*

Figure 13-28
MUG Test.
A, Positive.
B, Negative.

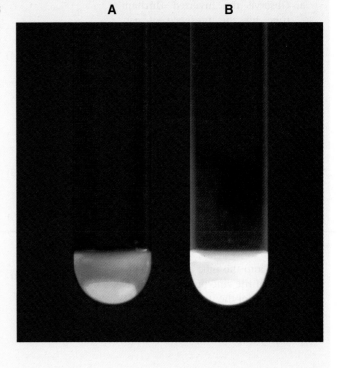

Procedure 13-29

NITRATE REDUCTION

PRINCIPLE

This test is used to determine the ability of an organism to reduce nitrate. The reduction of nitrate to nitrite is determined by adding sulfanilic acid and alpha-naphthylamine. The sulfanilic acid and nitrite react to form a diazonium salt. The diazonium salt then couples with the alpha-naphthylamine to produce a red, water-soluble azo dye.

METHOD

1. Inoculate nitrate broth (Figure 13-29, *D*) with 1 to 2 drops from a young broth culture of the test organism.
2. Incubate for 48 hours at 35° C in ambient air (some organisms may require longer incubation for adequate growth). Test these cultures 24 hours after obvious growth is detected or after a maximum of 7 days.
3. After a suitable incubation period, test the nitrate broth culture for the presence of gas, reduction of nitrate, and reduction of nitrite according to the following steps:
   a. Observe the inverted Durham tube for the presence of gas, indicated by bubbles inside the tube.
   b. Add 5 drops each of nitrate reagent solution A (sulfanilic acid) and B (alpha-naphthylamine). Observe for at least 3 minutes for a red color to develop.
   c. If no color develops, test further with zinc powder. Dip a wooden applicator stick into zinc powder and transfer only the amount that adheres to the stick to the nitrate broth culture to which solutions A and B have been added. Observe for at least 3 minutes for a red color to develop. Breaking the stick into the tube after the addition of the zinc provides a useful marker for the stage of testing.

Figure 13-29
Nitrate reduction.
A, Positive, no gas.
B, Positive, gas *(arrow).* **C,** Positive, no color after addition of zinc *(arrow).*
D, Uninoculated tube.

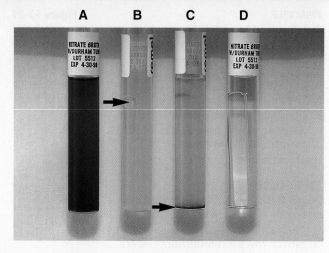

EXPECTED RESULTS

The nitrate reduction test is read for the presence or absence of three metabolic products: gas, nitrate (NO_3), and nitrite (NO_2). The expected results can be summarized as follows:

QUALITY CONTROL

NO_3^+, no gas: *Escherichia coli*
NO_3^+, gas: *Pseudomonas aeruginosa*
Negative: *Acinetobacter* spp.

Reaction	Gas	Color after addition of solutions A and B	Color after addition of zinc	Interpretation
$NO_3 \rightarrow NO_2$ (Figure 13-29, *A*)	None	Red	—	NO_3^+, no gas
$NO_3 \rightarrow NO_2$, partial nongaseous end products	None	Red	—	NO_3^+, no gas
$NO_3 \rightarrow NO_2$, gaseous end products (Figure 13-29, *B*)	Yes	Red	—	NO_3^+, gas$^+$
$NO_3 \rightarrow$ gaseous end product (Figure 13-29, *C*)	Yes	None	None	NO_3^+, NO_2^+, gas$^+$
$NO_3 \rightarrow$ nongaseous end products	None	None	None	NO_3^+, NO_2^+, no gas
$NO_3 \rightarrow$ no reaction	None	None	Red	Negative

Procedure 13-30

NITRITE REDUCTION

PRINCIPLE

This test is used to determine whether an organism can reduce nitrites to gaseous nitrogen or to other compounds containing nitrogen.

METHOD

1. Inoculate nitrite broth with 1 drop from a 24-hour broth culture.
2. Incubate for 48 hours at 35° C.
3. Examine 48-hour nitrite broth cultures for nitrogen gas in the inverted Durham tube and add 5 drops each of the nitrate reagents A and B to determine if nitrite is still present in the medium (reagents A and B are described under the nitrate reduction test).

EXPECTED RESULTS

Positive: No color change to red 2 minutes after the addition of the reagents and gas production observed in the Durham tube (Figure 13-30, *A*). **NOTE**: If broth does not become red and no gas production is observed, zinc dust is added to make sure the nitrite has not been oxidized to ni-

Figure 13-30
Nitrite reduction.
A, Positive, no color change after addition of zinc dust and gas in Durham tube *(arrow)*.
B, Negative.

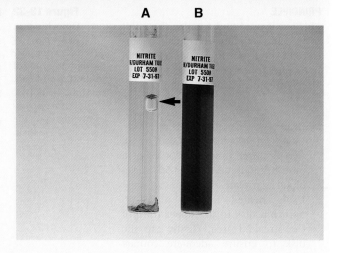

trate, thus invalidating the test. If oxidation has occurred, the mixture becomes red after the addition of zinc.
Negative: The broth becomes red after the addition of the reagents. No gas production is observed (Figure 13-30, *B*).

QUALITY CONTROL

Positive: *Alcaligenes faecalis*
Negative: *Alcaligenes piechaudii*

Procedure 13-31

ONPG (O-NITROPHENYL-β-D-GALACTOPYRANOSIDE) TEST

PRINCIPLE

This test is used to determine the ability of an organism to produce β-galactosidase, an enzyme that hydrolyzes the substrate ONPG to form a visible (yellow) product, orthonitrophenol.

METHOD

1. Aseptically suspend a loopful of organism in 0.85% saline.
2. Place an ONPG disk in the tube.
3. Incubate for 4 hours at 37° C in ambient air.
4. Examine tubes for a color change.

EXPECTED RESULTS

Positive: Yellow (presence of β-galactosidase) (Figure 13-31, *A*).
Negative: Colorless (absence of enzyme) (Figure 13-31, *B*).

QUALITY CONTROL

Positive: *Escherichia coli*
Negative: *Salmonella typhimurium*

Figure 13-31
OPNG test.
A, Positive.
B, Negative.

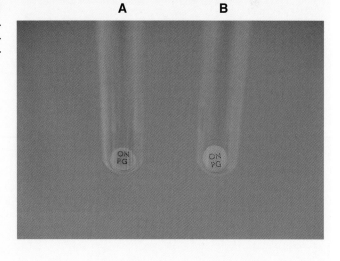

Procedure 13-32

OPTOCHIN TEST

PRINCIPLE

This test is used to determine the effect of optochin (ethylhydrocupreine hydrochloride) on an organism. Optochin lyses pneumococci (positive test), but alpha-streptococci are resistant (negative test).

METHOD

1. Using an inoculating loop, streak two or three suspect colonies of a pure culture onto half of a 5% sheep blood agar plate.
2. Using heated forceps, place an optochin disk in the upper third of the streaked area. Gently tap the disk to ensure adequate contact with the agar surface.
3. Incubate plate for 18 to 24 hours at 35° C in 5% CO_2. **NOTE:** Cultures do not grow as well in ambient air, and larger zones of inhibition occur.
4. Measure zone of inhibition in millimeters, including diameter of disk.

EXPECTED RESULTS

Positive: Zone of inhibition is 14 mm or greater in diameter, with 6-mm disk (Figure 13-32, *A*).

Figure 13-32
Optochin test. **A,** *Streptococcus pneumoniae* showing zone of inhibition >14 mm. **B,** Alpha-hemolytic *Streptococcus* growing up to the disk.

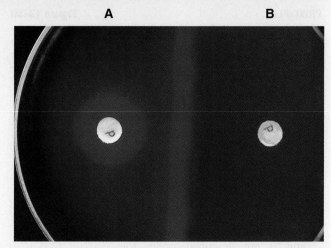

Negative: No zone of inhibition (Figure 13-32, *B*).
Equivocal: Any zone of inhibition less than 14 mm is questionable for pneumococci; the strain is identified as a pneumococcus only if it is bile-soluble.

QUALITY CONTROL

Positive: *Streptococcus pneumoniae*
Negative: *Streptococcus mitis*

Procedure 13-33

OXIDASE TEST (KOVAC'S METHOD)

PRINCIPLE
To determine the presence of bacterial cytochrome oxidase using the oxidation of the substrate tetramethyl-*p*-phenylenediamine dihydrochloride to indophenol, a dark purple-colored end product. A positive test (presence of oxidase) is indicated by the development of a dark purple color. No color development indicates a negative test and the absence of the enzyme.

METHOD
1. Moisten filter paper with the substrate (1% tetramethyl-*p*-phenylene-diamine dihydrochloride) or select a commercially available paper disk that has been impregnated with the substrate.
2. Use a platinum wire or wooden stick to remove a small portion of a bacterial colony (preferably not more than 24 hours old) from the agar surface and rub the sample on the filter paper or commercial disk. **NOTE:** Nickel-base alloy wires containing chromium and iron (nichrome) used to rub the colony paste onto the filter paper may cause false-positive results.
3. Observe inoculated area of paper or disk for a color change to deep blue or purple (Figure 13-33) within 10 seconds (timing is critical).

EXPECTED RESULTS
Positive: Development of a dark purple color within 10 seconds (Figure 13-33, *A*).
Negative: Absence of color (Figure 13-33, *B*).

QUALITY CONTROL
Positive: *Neisseria gonorrhoeae*
Negative: *Escherichia coli*

Figure 13-33   Oxidase test. **A,** Positive. **B,** Negative.

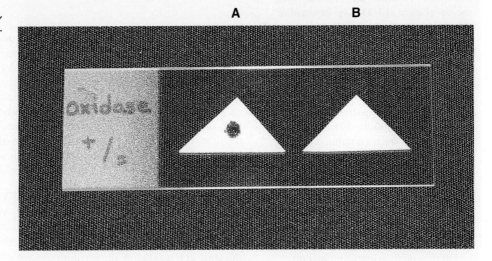

Procedure 13-34

OXIDATION/FERMENTATION (OF) MEDIUM (CDC METHOD)

PRINCIPLE

This test is used to determine whether an organism uses carbohydrate substrates to produce acid byproducts. Nonfermentative bacteria are routinely tested for their ability to produce acid from six carbohydrates (glucose, xylose, mannitol, lactose, sucrose, and maltose). In addition to the six tubes containing carbohydrates, a control tube containing the OF base without carbohydrate is also inoculated. Triple sugar iron agar (TSI) (see Procedure 13-40) is also used to determine whether an organism can ferment glucose. OF glucose is used to determine whether an organism ferments (Figure 13-34, *A*) or oxidizes (Figure 13-34, *B*) glucose. If no reaction occurs in either the TSI or OF glucose, the organism is considered a nonglucose utilizer (Figure 13-34, *C*).

METHOD

1. To determine whether acid is produced from carbohydrates, inoculate agar deeps, each containing a single carbohydrate, with bacterial growth from an 18- to 24-hour culture by stabbing a needle 4 to 5 times into the medium to a depth of 1 cm. **NOTE**: Two tubes of OF dextrose are usually inoculated; one is overlaid with either sterile melted petrolatum or sterile paraffin oil to detect fermentation.

2. Incubate the tubes at 35°C in ambient air for up to 7 days. **NOTE**: If screwcap tubes are used, loosen the caps during incubation to allow for air exchange. Otherwise, the control tube and tubes containing carbohydrates that are not oxidized might not become alkaline.

Figure 13-34

Oxidation/fermentation medium (CDC method).
A, Fermenter.
B, Oxidizer.
C, Nonutilizer.

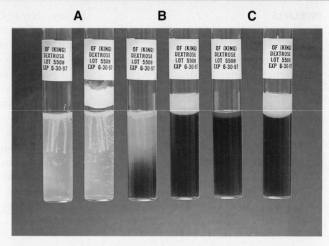

EXPECTED RESULTS

Positive: Acid production (A) is indicated by the color indicator changing to yellow in the carbohydrate-containing deep.

Weak-positive (A^w): Weak acid formation can be detected by comparing the tube containing the medium with carbohydrate with the inoculated tube containing medium with no carbohydrate. Most bacteria that can grow in the OF base produce an alkaline reaction in the control tube. If the color of the medium in a tube containing carbohydrate remains about the same as it was before the medium was inoculated and if the inoculated medium in the control tube becomes a deeper red (i.e., becomes alkaline), the culture being tested is considered weakly positive, assuming the amount of growth is about the same in both tubes.

Negative: Red or alkaline (K) color in the deep with carbohydrate equal to the color of the inoculated control tube.

No change (NC) or neutral (N): There is growth in the media, but neither the carbohydrate-containing media nor the control base turn alkaline (red).

NOTE: If the organism does not grow at all in the OF medium, mark the reaction as no growth (NG).

QUALITY CONTROL

Note: Appropriate organisms depend on which carbohydrate has been added to the basal medium. Glucose is used as an example.

Fermenter: *Escherichia coli*
Oxidizer: *Pseudomonas aeruginosa*
Nonutilizer: *Alcaligenes faecalis*

Procedure 13-35

PHENYLALANINE DEAMINASE

PRINCIPLE

This test is used to determine the ability of an organism to oxidatively deaminate phenylalanine to phenyl-pyruvic acid. The phenylpyruvic acid is detected by adding a few drops of 10% ferric chloride; a green-colored complex is formed between these two compounds.

METHOD

1. Inoculate phenylalanine slant with 1 drop of a 24-hour brain-heart infusion broth.
2. Incubate 18 to 24 hours (or until good growth is apparent) at 35° C in ambient air with cap loose.
3. After incubation, add 4 to 5 drops of 10% aqueous ferric chloride to the slant.

EXPECTED RESULTS

Positive: Green color develops on slant after ferric chloride is added (Figure 13-35, *A*).

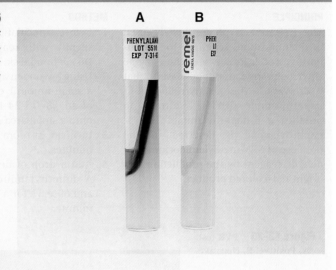

Figure 13-35
Phenylalanine deaminase.
A, Positive.
B, Negative.

Negative: Slant remains original color after the addition of ferric chloride (Figure 13-35, *B*).

QUALITY CONTROL

Positive: *Proteus vulgaris*
Negative: *Escherichia coli*

Procedure 13-36

PYR TEST

PRINCIPLE
The PYR test is predominately used in identification schemes for gram-positive cocci. Presence of the enzyme L-pyrroglutamylaminopeptidase that hydrolyzes the L-pyrrolidonyl-β-naphthylamide (PYR) substrate to produce a β-naphthylamine. The β-naphthylamine is detected in the presence of *N, N*-methylaminocinnamaldehyde reagent by the production of a bright red–colored product.

METHOD
1. Before inoculation, moisten disk slightly with reagent grade water. Do *not* flood disk.
2. Using a wooden applicator stick, rub a small amount of several colonies of an 18- to 24-hour pure culture onto a small area of the PYR disk.
3. Incubate at room temperature for 2 minutes.
4. Add a drop of detector reagent, *N, N*-dimethylaminocinnamaldehyde and observe for a red color within 1 minute.

EXPECTED RESULTS
Positive: Bright red color within 5 minutes (Figure 13-36, *A*).
Negative: No color change or an orange color (Figure 13-36, *B*).

QUALITY CONTROL
Positive: *Enterococcus faecalis*
Negative: *Streptococcus mitis*

Figure 13-36 PYR test.
A, Positive. **B**, Negative.

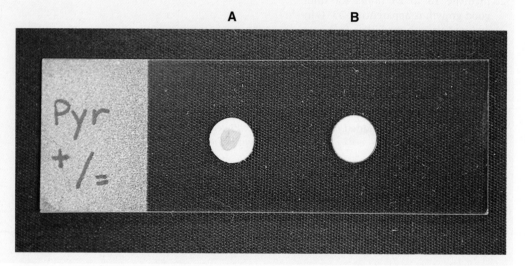

Procedure 13-37

PYRUVATE BROTH

PRINCIPLE

This test is used to determine the ability of an organism to utilize pyruvate. This aids in the differentiation between *Enterococcus faecalis* (positive) and *Enterococcus faecium* (negative).

METHOD

1. Lightly inoculate the pyruvate broth with an 18- to 24-hour culture of the organism from 5% sheep blood agar.
2. Incubate at 35° C in ambient air for 24 to 48 hours.

EXPECTED RESULTS

Positive: Indicator changes from green to yellow (Figure 13-37, *A*).

Negative: No color change; yellow-green indicates a weak reaction and should be regarded as negative (Figure 13-37, *B*).

Figure 13-37
Pyruvate broth.
A, Positive.
B, Negative.

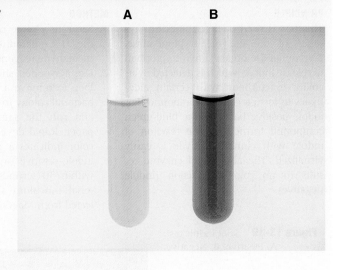

QUALITY CONTROL

Positive: *Enterococcus faecalis*
Negative: *Enterococcus faecium*

Procedure 13-38

SALT TOLERANCE TEST

PRINCIPLE

This test is used to determine the ability of an organism to grow in high concentrations of salt. It is used to differentiate enterococci (positive) from nonenterococci (negative). A heart infusion broth containing 6.5% NaCl is used as the test medium. This broth also contains a small amount of glucose and bromcresol purple as the indicator for acid production.

METHOD

1. Inoculate one to two colonies from an 18- to 24-hour culture into 6.5% NaCl broth.
2. Incubate tube at 35° C in ambient air for 48 hours.
3. Check daily for growth.

EXPECTED RESULTS

Positive: Visible turbidity in broth with or without color change from purple to yellow (Figure 13-38, *A*).

Figure 13-38
Salt tolerance test.
A, Positive.
B, Negative.

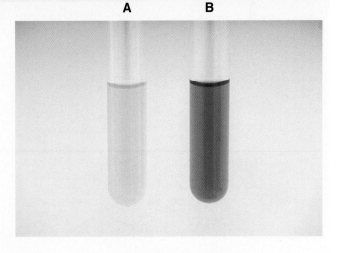

Negative: No turbidity and no color change (Figure 13-38, *B*).

QUALITY CONTROL

Positive: *Enterococcus faecalis*
Negative: *Streptococcus mitis*

Procedure 13-39

SPOT INDOLE TEST

PRINCIPLE

This test is used to determine the presence of the enzyme tryptophanase. Tryptophanase breaks down tryptophan to release indole, which is detected by its ability to combine with certain aldehydes to form a colored compound. For indole-positive bacteria, the blue-green compound formed by the reaction of indole with cinnamaldehyde is easily visualized. The absence of enzyme results in no color production (indole negative).

METHOD

1. Saturate a piece of filter paper with the 1% paradimethylaminocinnamaldehyde reagent.
2. Use a wooden stick or bacteriologic loop to remove a small portion of a bacterial colony from the agar surface and rub the sample on the filter paper. Rapid development of a blue color indicates a positive test. Most indole-positive organisms turn blue within 30 seconds. **NOTE:** The bacterial inoculum should not be selected from MacConkey agar because the color of lactose-fermenting colonies on this medium can interfere with test interpretation.

EXPECTED RESULTS

Positive: Development of a blue color within 20 seconds (Figure 13-39, *A*).
Negative: No color development or slightly pink color (Figure 13-39, *B*).

QUALITY CONTROL

Positive: *Escherichia coli*
Negative: *Enterobacter cloacae*

Figure 13-39   Spot indole test.
A, Positive. **B**, Negative.

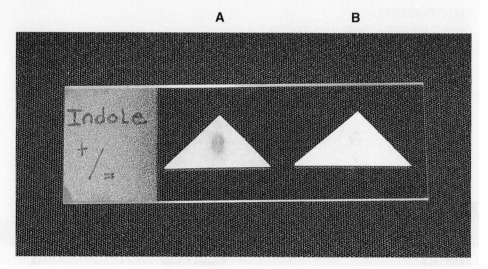

Procedure 13-40

TRIPLE SUGAR IRON AGAR (TSI)

PRINCIPLE

Triple sugar iron agar (TSI) is used to determine whether a gram-negative rod utilizes glucose and lactose or sucrose fermentatively and forms hydrogen sulfide (H_2S). TSI contains 10 parts lactose : 10 parts sucrose : 1 part glucose and peptone. Phenol red and ferrous sulfate serve as indicators of acidification and H_2S formation, respectively. When glucose is utilized by a fermentative organism, the entire medium becomes acidic (yellow) in 8 to 12 hours. The butt remains acidic after the recommended 18- to 24-hour incubation period because of the presence of organic acids resulting from the fermentation of glucose under anaerobic conditions in the butt of the tube. The slant, however, reverts to the alkaline (red) state because of oxidation of the fermentation products under aerobic conditions on the slant. This change is a result of the formation of CO_2 (carbon dioxide) and H_2O and the oxidation of peptones in the medium to alkaline amines. When, in addition to glucose, lactose and/or sucrose are fermented, the large amount of fermentation products formed on the slant will more than neutralize the alkaline amines and render the slant acidic (yellow), provided the reaction is read in 18 to 24 hours. Reactions in TSI should not be read beyond 24 hours of incubation, because aerobic oxidation of the fermentation products from lactose and/or sucrose does proceed and the slant will eventually revert to the alkaline state. The formation of CO_2 and H_2 (hydrogen gas) is indicated by the presence of bubbles or cracks in the agar or by separation of the agar

Figure 13-40
Triple sugar iron agar. **A,** Acid slant/acid butt with gas, no H_2S (A/A). **B,** Alkaline slant/acid butt, no gas, H_2S-positive (K/A H_2S^+). **C,** Alkaline slant/no change butt, no gas, no H_2S (K/NC). **D,** Uninoculated tube.

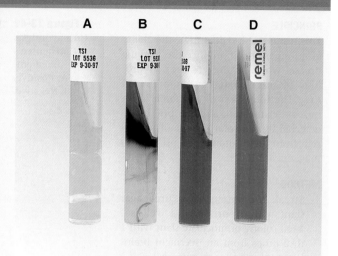

from the sides or bottom of the tube. The production of H_2S requires an acidic environment and is manifested by blackening of the butt of the medium in the tube.

METHOD

1. With a straight inoculation needle, touch the top of a well-isolated colony.
2. Inoculate TSI (Figure 13-40, *D*) by first stabbing through the center of the medium to the bottom of the tube and then streaking the surface of the agar slant.
3. Leave the cap on loosely and incubate the tube at 35° C in ambient air for 18 to 24 hours.

EXPECTED RESULTS

Alkaline slant/no change in the butt (K/NC) = glucose, lactose, and sucrose nonutilizer; this may also be

recorded as K/K (alkaline slant/alkaline butt) (Figure 13-40, *C*).

Alkaline slant/acid butt (K/A) = glucose fermentation only.

Acid slant/acid butt (A/A) = glucose, sucrose, and/or lactose fermenter (Figure 13-40, A)

NOTE: A black precipitate in the butt indicates production of ferrous sulfide and H_2S gas (H_2S^+) (Figure 13-40, *B*). Bubbles or cracks in the tube indicate the production of CO_2 or H_2. Drawing a circle around the A for the acid butt, that is, A/Ⓐ, usually indicates this means the organism ferments glucose and sucrose, glucose and lactose, or glucose, sucrose, and lactose, with the production of gas.

QUALITY CONTROL

A/Ⓐ: *Escherichia coli*
K/A H_2S^+: *Salmonella typhi*
K/NC: *Pseudomonas aeruginosa*

Procedure 13-41

UREA HYDROLYSIS (CHRISTENSEN'S METHOD)

PRINCIPLE

This test is used to determine the ability of an organism to produce the enzyme urease, which hydrolyzes urea. Hydrolysis of urea produces ammonia and CO_2. The formation of ammonia alkalinizes the medium, and the pH shift is detected by the color change of phenol red from light orange at pH 6.8 to magenta at pH 8.1.

METHOD

1. Streak the surface of a urea agar slant with a portion of a well-isolated colony or inoculate slant with 1 to 2 drops from an overnight brain-heart infusion broth culture.
2. Leave the cap on loosely and incubate tube at 35° C in ambient air for 48 hours to 7 days.

EXPECTED RESULTS

Positive: Change in color of slant from light orange to magenta (Figure 13-41, *A*).

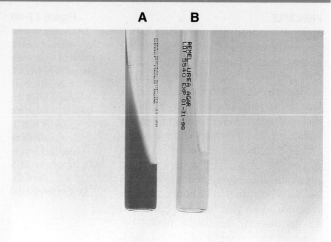

Figure 13-41
Urea hydrolysis (Christensen's method).
A, Positive.
B, Negative.

Negative: No color change (agar slant and butt remain light orange) (Figure 13-41, *B*).

QUALITY CONTROL

Positive: *Proteus vulgaris*
Negative: *Escherichia coli*

Procedure 13-42

X AND V FACTOR TEST

PRINCIPLE

Members of the genus *Haemophilus* require accessory growth factors in vitro. Some *Haemophilus* spp. require X factor (hemin) alone, V factor (NAD, nicotinamide-adenine denucleotide) alone, or a combination of both.

METHOD

1. Make a very light suspension (MacFarland 0.5) of the organism in sterile saline. **NOTE:** It is important not to carry over any X-factor contained in the medium that the organism is taken from. Therefore, a loop, *not* a swab, should be used to make the suspension.
2. Dip a sterile swab into the organism suspension. Roll the swab over the

entire surface of a trypticase soy-agar plate.
3. Place the X, V, and XV factor disks on the agar surface. If using separate disks, place them at least 4 to 5 cm apart.
4. Incubate overnight at 35° C in ambient air.

EXPECTED RESULTS

Positive: Growth around the XV disk only shows a requirement for both factors (Figure 13-42, *A*). Growth around the V disk, no growth around the X disk, and light growth around the XV disk shows a V factor requirement (Figure 13-42, *B*).
Negative: Growth over entire surface of the agar indicates no requirement

for either X or V factor (Figure 13-42, *C*).

QUALITY CONTROL

Positive: *Haemophilus influenzae* will show a halo of growth around the XV disk; the rest of the agar surface will show no growth. *Haemophilus parainfluenzae* will show a halo of growth around the XV and V disks.
Negative: *Haemophilus aphrophilus* will grow over the entire surface of the plate. Neither X, nor V, nor XV factors are necessary for growth.

Procedure 13-42 (cont'd)

X and V Factor Test

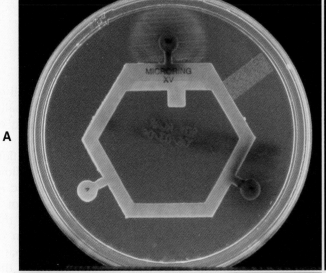

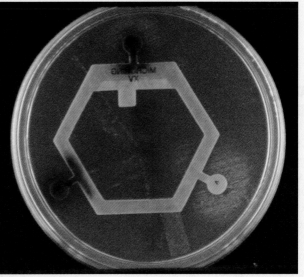

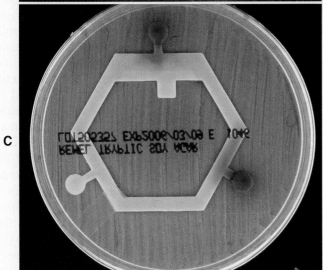

Figure 13-42   X and V factor test. **A,** Positive, growth around XV disk only. **B,** Positive, growth around V disk. **C,** Negative, growth over entire plate.

REFERENCE

1. Heimbrook ME, Wang WL, Campbell G: Staining bacterial flagella easily, *J Clin Microbiol* 26:2612, 1989.

ADDITIONAL READING

National Committee for Clinical Laboratory Standards: Abbreviated identification of bacteria and yeast; approved guideline M35-A, Wayne, Pa, 2002, NCCLS.

York MK, Traylor MM, Hardy J, et al: Biochemical tests for the identification of aerobic bacteria. In Isenberg HD, editor: *Clinical microbiology procedures handbook*, ed 2, Washington, DC, 2004, ASM Press.

CHAPTER 14

General Considerations and Applications of Information Provided In Bacterial Sections of Part III

RATIONALE FOR APPROACHING ORGANISM IDENTIFICATION

It is difficult to determine how most effectively to present and teach diagnostic microbiology in a way that is sufficiently comprehensive and yet not excessively cluttered with rare and seldom-needed facts about bacterial species that are uncommonly encountered. There are approximately 530 different bacterial species or taxa reported by clinical microbiology laboratories across the United States (Figure 14-1). Yet 95% of the bacterial identifications reported are distributed across only 27 of these taxa. This is an indication of how infrequently the other 500 or more taxa are identified and reported. Therefore, while the chapters in Part III: Bacteriology are intended to be comprehensive in terms of the variety of bacterial species presented, it is helpful to keep in perspective which taxa are most likely to be encountered in the clinical environment. The relative frequencies with which the common bacterial species and organism groups are reported in clinical laboratories are presented in Figure 14-2.

Historically, most microbiology text chapters are organized by genus name, but fail to provide information and processes that are needed to understand what is involved in getting from the clinical specimen to the correct genus. Many texts (including this one, see Chapter 15) provide flow charts containing algorithms or identification schemes for organism workup. Although these are helpful, one must be aware of the limitations of flow charts. In some cases they may be too general to be helpful, that is, they lack sufficient detail to be useful for discriminating among key microbial groups and species. In other cases, they may be too esoteric to be of practical use in routine clinical practice, such as identification schemes based on cellular analysis of fatty acid analysis. In addition, there are many other criteria that must be incorporated into the identification process that are too complex to be included in most flow charts. Thus flow charts are only one of many tools that must be used in the field of diagnostic microbiology. Also, as discussed later in this chapter, organism taxonomy and profiles continuously change so that flow charts that are too detailed are at risk of quickly becoming outdated. Furthermore, as is evident throughout the chapters in Part III, diagnostic microbiology is full of exceptions to rules, and flow charts are not usually constructed in a manner that readily captures many of the important exceptions.

To meet the challenges of bacterial identification processes beyond what can be portrayed in flow charts, the chapters in Part III have been arranged to guide the student through the entire workup of a microorganism, beginning with initial culture of the specimen that originally contained the organism of interest. In most instances, the first information that a microbiologist uses in the identification process is the macroscopic description of the colony, or colony morphology. This includes the type of hemolysis (if any), pigment (if present), size, texture (opaque, translucent, or transparent), adherence to agar, pitting of agar, and many other characteristics (see Chapter 7). After careful observation of the colony, the Gram stain is used to separate the organism into one of a variety of broad categories based on Gram reaction and cellular morphology of gram-positive or gram-negative bacteria (e.g., gram-positive cocci, gram-negative rods—Chapter 6). For gram-positive organisms, the catalase test should follow the Gram stain, and testing on gram-negative organisms should begin with the oxidase test. These simple tests plus growth on MacConkey agar, if the isolate is a gram-negative rod or coccobacillus, help the microbiologist assign the organism to one of the primary categories (organized here as subsections). Application of the various identification methods and systems outlined in Chapter 13 generate the data and criteria that are discussed in each chapter for the definitive identification of clinically important bacteria.

Because diagnostic microbiology is centered around the identification of organisms based on common phenotypic traits shared with known members of the same genus or family, microbiologists "play the odds" every day by finding the best biochemical "fit" and assigning the most probable identification. For example, the gram-negative rod known as CDC group EF-4a may be considered with either the MacConkey-positive or -negative organisms because it grows on MacConkey agar 50% of the time. Therefore, although CDC group EF-4a has been arbitrarily assigned to the section on oxidase-positive, MacConkey-positive, gram-negative

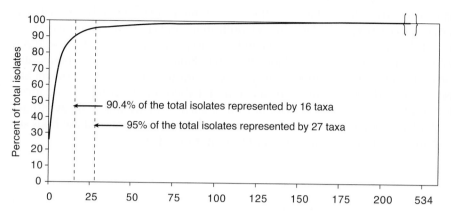

Figure 14-1   Data demonstrating that more than 500 different bacterial species or taxa are reported from clinical laboratories across the United States. However, 95% of the isolates reported are distributed among only 27 different taxa and more than 90% are represented by 16 different taxa. (Source: TSN Database — USA).

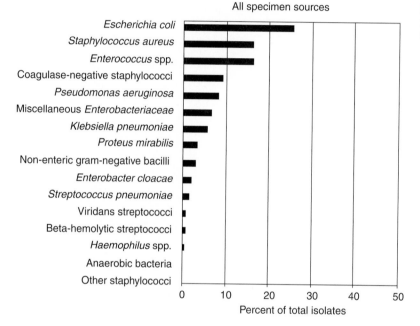

Figure 14-2   Relative frequency with which the most common bacterial species and taxa are reported by clinical laboratories.

bacilli and coccobacilli in this text, it is also included in the discussion of oxidase-positive, MacConkey-negative, gram-negative bacilli and coccobacilli. This example clearly demonstrates the limitations of solely depending on identification flow charts for the identification process.

The identification process often can be arduous and resource-draining so that laboratorians must make every effort to identify only those organisms that are most likely to be involved in the infection process. To that end, the chapters in Part III have also been designed to provide guidance for determining whether a clinical isolate is relevant and requires full identification. Furthermore, the clinical diagnosis and the source of the specimen also can aid in determining which group of organisms to consider. For example, if a patient has endocarditis or the specimen source is blood, and a small, gram-negative rod is observed on Gram stain, the microbiologist should consider a group of gram-negative

bacilli known as the *HACEK* (*Haemophilus aphrophilus, Actinobacillus actinomycetemcomitans, Cardiobacterium hominis, Eikenella corrodens,* and *Kingella* spp.), which are not commonly encountered in most clinical specimens. Similarly, if a patient has suffered an animal bite, the microbiologist should think of *Pasteurella multocida,* EF-4a, and EF-4b if the isolate is gram-negative and *Staphylococcus hyicus* and *S. intermedius* if the organism is gram-positive. Finally, in consideration of an isolate's clinical relevance, each chapter also provides information on whether antimicrobial susceptibility testing is indicated and, if needed, how it should be performed.

FUTURE TRENDS OF ORGANISM IDENTIFICATION

There are several dynamics that occur in clinical microbiology and infectious diseases that always have and will continue to impact and challenge bacterial identification practices. For instance, new species associated to some degree with human infections will continue to be discovered, and well-known species may change their characteristics in ways that impact the identification criteria that have been used historically for their identification. For these reasons identification schemes and strategies will always need to be updated for both conventional methods and commercial systems. Also, while the vast majority of identification schemes are based on the phenotypic characteristics of bacteria the use of molecular methods for bacterial detection, identification, and characterization will continue to expand and play a greater role in the diagnostic microbiology arena. Finally, as both phenotypic and molecular methods continue to be used, there will also be a greater trend for these methods to become incorporated into automated systems that will eventually replace most if not all of the conventional methods still used in most laboratories today.

Bacterial Identification Flow Charts and Schemes: A Guide to Part III

As discussed in Chapter 14, many factors and criteria must be taken into consideration during the process of bacterial identification. Identification flow charts can be useful tools to support the identification process. To that end, key flow charts outlined in this chapter provide a "map" to help guide students and readers to the appropriate chapters in Part III. However, as pointed out in Chapter 14, the amount of identification information that can be reasonably encompassed in the flow chart format is relatively limited. The in-depth details and information for each organism group is only found within the individual chapters.

For guidance to the appropriate chapters, the following four flow charts are provided for reference:

Figure 15-1. Flow chart guide for anaerobic bacteria.

Figure 15-2. Flow chart guide for aerobic and facultatively anaerobic gram-positive bacteria.

Figure 15-3. Flow chart guide for aerobic and facultatively anaerobic gram-negative bacteria.

Figure 15–4. Flow chart guide for mycobacteria and other bacteria not characterized by the Gram reaction.

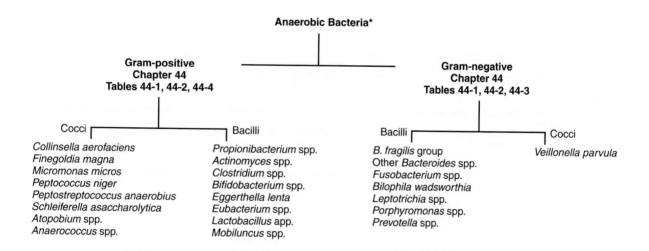

*Unable to grow in the presence of of oxygen. NOTE: any motile or curved organisms suspected of being anaerobes should be evaluated for aerotolerance and microaerophilic growth (these may be potential *Campylobacter* or *Suterella* spp.)

Figure 15-1   Flow chart guide for anaerobic bacteria.

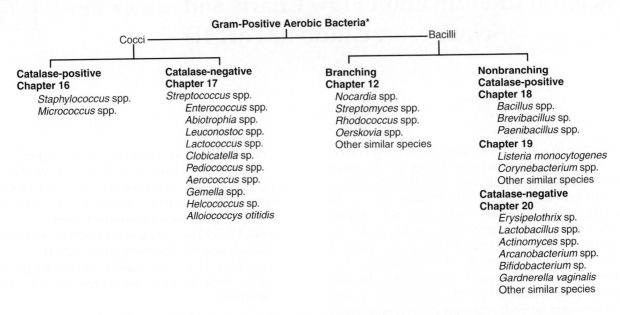

*Aerobic bacteria require oxygen to grow; facultative organisms are able to grow in the presence or absence of oxygen.

Figure 15-2   Flow chart guide for aerobic and facultatively anaerobic gram-positive bacteria.

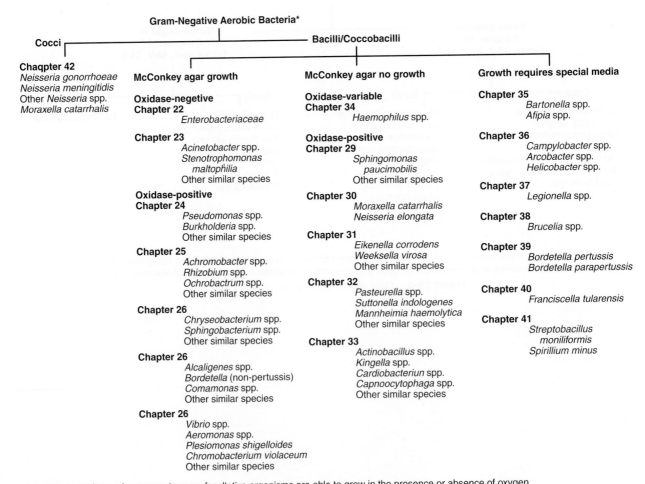

*Aerobic bacteria require oxygen to grow, facultative organisms are able to grow in the presence or absence of oxygen

Figure 15-3   Flow chart guide for aerobic and facultatively anaerobic gram-negative bacteria.

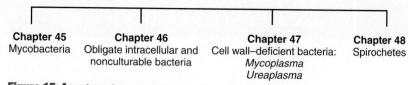

Chapter 45
Mycobacteria

Chapter 46
Obligate intracellular and
nonculturable bacteria

Chapter 47
Cell wall–deficient bacteria:
Mycoplasma
Ureaplasma

Chapter 48
Spirochetes

Figure 15-4   Flow chart guide for mycobacteria and other bacteria not
characterized by the Gram reaction.

Genera and Species to Be Considered

Staphylococcus aureus

Coagulase-negative staphylococci (most commonly encountered)
- *Staphylococcus epidermidis*
- *Staphylococcus haemolyticus*
- *Staphylococcus saprophyticus*
- *Staphylococcus lugdunensis*
- *Staphylococcus schleiferi*

Coagulase-negative staphylococci (less commonly encountered)
- *Staphylococcus capitis*
- *Staphylococcus caprae*
- *Staphylococcus warneri*
- *Staphylococcus hominis*
- *Staphylococcus auricularis*
- *Staphylococcus cohnii*
- *Staphylococcus xylosus*
- *Staphylococcus simulans*

Micrococcus spp. and related genera

GENERAL CHARACTERISTICS

The organisms described in this chapter are all catalase-positive, gram-positive cocci. However, only those belonging to the genus *Staphylococcus* are of primary clinical significance. Several of the coagulase-negative staphylococci (i.e., non–*Staphylococcus aureus*) species listed may be encountered in clinical specimens, but only those most prominently associated with infections in humans (i.e., *S. epidermidis, S. haemolyticus, S. saprophyticus, S. schleiferi,* and *S. lugdunensis*) are discussed in detail.

EPIDEMIOLOGY

As outlined in Table 16-1, the staphylococci that are associated with infections in humans are colonizers of various skin and mucosal surfaces. Because the carrier state is common among the human population, infections are frequently acquired when the colonizing strain gains entrance to a normally sterile site as a result of trauma or abrasion to the skin or mucosal surface. However, the traumatic event that allows entry of the organism often may be so minor that it goes unnoticed.

Staphylococci are also transmitted from person to person. Upon transmission, the organisms may become established as part of the recipient's normal flora and later be introduced to sterile sites by trauma or invasive procedures. Alternatively, the organism may be directly introduced into normally sterile sites, such as by a surgeon or nurse during surgery. Person-to-person spread of staphylococci, particularly those that have acquired antimicrobial resistance, most notably occurs in hospitals and presents substantial infection control problems. However, more recently serious *S. aureus* infections have been encountered in the community setting as well.

PATHOGENESIS AND SPECTRUM OF DISEASE

Without question, *S. aureus* is the most virulent species of staphylococci encountered. A wide spectrum of factors, not all of which are completely understood, contribute to this organism's ability to cause infections and disease. Several toxins and enzymes mediate tissue invasion and survival at the infection site (Table 16-2). Elaboration of these factors is chiefly responsible for the various skin, wound, and deep tissue infections commonly caused by *S. aureus*. Many of these infections can rapidly become life-threatening if not treated and managed appropriately.

Localized skin infections may involve hair follicles (i.e., folliculitis) and spread deeper to cause boils (i.e., furuncles). More serious, deeper infections result when the furuncles coalesce to form carbuncles. Impetigo, the *S. aureus* skin infection that involves the epidermis, is typified by the production of vesicles that rupture and crust over. Regardless of the initial site of infection, the invasive nature of this organism always presents a threat for deeper tissue invasion, bacteremia, and

Table 16-1   Epidemiology

Organism	Habitat (reservoir)	Mode of Transmission
Staphylococcus aureus	Normal flora of human anterior nares, nasopharynx, perineal area, and skin; can colonize various epithelial or mucosal surfaces	Spread of patient's endogenous strain to normally sterile site by traumatic introduction (e.g., surgical wound or microabrasions). Also may be transmitted person to person by fomites, air, or unwashed hands of health care workers, especially in the nosocomial setting. May be transmitted from infected skin lesion of health care worker to patient
Staphylococcus epidermidis	Normal flora of human skin and mucous membranes; distributed widely, often in large numbers, over body surface	Spread of patient's endogenous strain to normally sterile site, usually as a result of implantation of medical devices (e.g., shunts, prosthetic devices) during hospitalization. Person-to-person spread in hospitals can lead to patients becoming colonized and potentially infected with antibiotic-resistant strains
Staphylococcus haemolyticus and *Staphylococcus lugdunensis*	Normal human flora similar to *S. epidermidis* but present in fewer numbers	Probably the same as for *S. epidermidis*
Staphylococcus saprophyticus	Normal flora of human skin and mucosa of genitourinary tract	Introduction of endogenous flora into sterile urinary tract, notably in young, sexually active females. A community-acquired infection, not considered an agent of nosocomial infections
Micrococcus spp.	Normal flora of human skin, mucosa, and oropharynx	Uncertain; rarely implicated in infections. When infections occur, they likely involve endogenous strains

Table 16-2   Pathogenesis and Spectrum of Diseases

Organism	Virulence Factors	Spectrum of Diseases and Infections
Staphylococcus aureus	Produces and secretes toxins and enzymes that have a role in virulence; alpha, beta, gamma, and delta toxins act on host cell membranes and mediate cell destruction. Leucocidin (PVL) mediates destruction of phagocytes. Clumping factor, coagulase, and hyaluronidase enhance invasion and survival in tissues. Potent exotoxins include exfoliatins, toxic shock syndrome toxin (TSST-1), and enterotoxins. Biofilm production.	Infections generally involve intense suppuration and destruction (necrosis) of tissue. Infections can be generally grouped as localized skin infections such as folliculitis, furuncles (boils), carbuncles, and impetigo; various wound infections; deep infections that spread from skin to cause bacteremia (with and without endocarditis) and to involve bones, joints, deep organs, lungs and respiratory system, and tissues; scalded skin syndrome in neonates; toxic shock syndrome; and food poisoning.
Staphylococcus epidermidis	Certain factors facilitate initial attachment to implanted medical devices. Production of exopolysaccharide "slime" or biofilm enhances organism adhesion and provides conditions refractory to antibiotic action and host defense mechanisms. Propensity to acquire and disseminate antimicrobial resistance allows for survival in hospital setting.	Ubiquitous member of normal flora makes this species the most commonly encountered in clinical specimens, usually as a contaminant. Can be difficult to establish clinical significance. Most common infections include nosocomial bacteremia associated with indwelling vascular catheters; endocarditis involving prosthetic cardiac valves (rarely involves native valves); infection at intravascular catheter sites, frequently leading to bacteremia; and other infections associated with CSF shunts, prosthetic joints, vascular grafts, postsurgical ocular infections, and bacteremia in neonates under intensive care
S. haemolyticus and *S. lugdunensis*	Uncertain; probably similar to those described for *S. epidermidis*	Similar to those described for *S. epidermidis*
S. saprophyticus	Uncertain	Urinary tract infections in sexually active, young females; infections in sites outside urinary tract are not common
Micrococcus spp.	Unknown; probably of extremely low virulence	Usually considered contaminants of clinical specimens; rarely implicated as cause of infections in humans

spread to one or more internal organs including the respiratory tract. Furthermore, these serious infections have emerged more frequently among nonhospitalized patients and are associated with strains that produce the Panton-Valentine leukocidin (PVL) toxin. Also worrisome is that these serious "community-associated" infections are frequently mediated by methicillin-resistant *S. aureus* (community-acquired MRSA or CA-MRSA).

S. aureus can also produce toxin-mediated diseases, such as scalded skin syndrome and toxic shock syndrome. In these cases, the organisms may remain relatively localized, but production of potent toxins causes systemic or widespread effects. With scalded skin syndrome, which usually afflicts neonates, the exfoliatin toxins cause extensive sloughing of epidermis to produce a burnlike effect on the patient. The toxic shock syndrome toxin (TSST-1) has several systemic effects, including fever, desquamation, and hypotension potentially leading to shock and death.

The coagulase-negative staphylococci, among which *S. epidermidis* is the most commonly encountered, are substantially less virulent than *S. aureus* and are opportunistic pathogens. Their prevalence as nosocomial pathogens is as much, if not more, related to medical procedures and practices than to the organism's capacity to establish an infection. Infections with *S. epidermidis* and, less commonly, *S. haemolyticus* and *S. lugdunensis* usually involve implantation of medical devices (see Table 16-2).[5,20] This kind of medical intervention allows invasion by these normally noninvasive organisms. Two organism characteristics that do enhance the likelihood of infection include production of a slime layer or biofilm that facilitates attachment to implanted medical devices and the ability to acquire resistance to most of the antimicrobial agents used in hospital environments.[3,4,7]

Although most coagulase-negative staphylococci are primarily associated with nosocomial infections, urinary tract infections caused by *S. saprophyticus* are clear exceptions. This organism is most frequently associated with community-acquired urinary tract infections in young, sexually active females but is not commonly associated with hospital-acquired infections or any infections at non–urinary tract sites.

Because coagulase-negative staphylococci are ubiquitous colonizers, they are frequently found as contaminants in clinical specimens. This fact, coupled with the emergence of these organisms as nosocomial pathogens, complicates laboratory interpretation of their clinical significance. When these organisms are isolated from clinical specimens, every effort should be made to substantiate their clinical relevance in a particular patient so that unnecessary work and production of misleading information can be avoided.

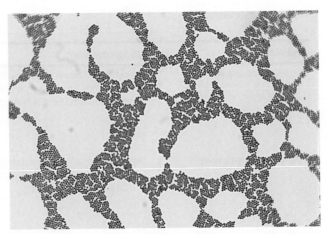

Figure 16-1 Gram stain of *Staphylococcus aureus* from blood agar.

What, if any, virulence factors are produced by *Micrococcus* is not known. Because these organisms are rarely associated with infection, they are probably of low virulence.

LABORATORY DIAGNOSIS

SPECIMEN COLLECTION AND TRANSPORT

No special considerations are required for specimen collection and transport of the organisms discussed in this chapter. Refer to Table 5-1 for general information on specimen collection and transport.

SPECIMEN PROCESSING

No special considerations are required for processing of the organisms discussed in this chapter. Refer to Table 5-1 for general information on specimen processing.

DIRECT DETECTION METHODS

Other than Gram stain of patient specimens, there are no specific procedures for the direct detection of these organisms in clinical material. All the *Micrococcaceae* (the family to which *Staphylococcus* and *Micrococcus* belong) produce spherical, gram-positive cells. During cell division, *Micrococcaceae* divide along both longitudinal and horizontal planes, forming pairs, tetrads, and, ultimately, irregular clusters (Figure 16-1). Gram stains should be performed on young cultures, because very old cells may lose their ability to retain crystal violet and may appear gram-variable or gram-negative. *Staphylococcus* appear as gram-positive cocci, usually in clusters. *Micrococcus* and related genera (i.e., *Kytococcus, Nesterenkonia, Dermacoccus, Arthrobacter,* and *Kocuria*[11,18,19]) resemble the staphylococci microscopically.

Table 16-3   Colonial Appearance and Characteristics on 5% Sheep Blood Agar

Organism	Appearance
Micrococcus spp. and related organisms*	Small to medium; opaque, convex; nonhemolytic; wide variety of pigments (white, tan, yellow, orange, pink)
Staphylococcus aureus	Medium to large; smooth, entire, slightly raised, translucent; most colonies pigmented creamy yellow; most colonies beta-hemolytic
S. epidermidis	Small to medium; translucent, gray-white colonies; most colonies nonhemolytic; slime-producing strains are extremely sticky and adhere to the agar surface
S. haemolyticus	Medium; smooth, butyrous and opaque; beta-hemolytic
S. hominis	Medium to large; smooth, butyrous and opaque; may be unpigmented or cream-yellow-orange
S. lugdunensis	Medium to large; smooth, glossy, entire edge with slightly domed center; unpigmented or cream to yellow-orange
S. warneri	Resembles *S. lugdunensis*
S. saprophyticus	Large; entire, very glossy, smooth, opaque, butyrous, convex; usually white but colonies can be yellow or orange
S. schleiferi	Medium to large; smooth, glossy, slightly convex with entire edges; unpigmented
S. intermedius	Large; slightly convex, entire, smooth, glossy, translucent; usually nonpigmented
S. hyicus	Large; slightly convex, entire, smooth, glossy, opaque; usually nonpigmented
S. capitis	Small to medium; smooth, slightly convex, glistening, entire, opaque; *S. capitis* subsp. *urealyticus* usually pigmented (yellow or yellow-orange); *S. capitis* subsp. *capitis* is nonpigmented
S. cohnii	Medium to large; convex, entire, circular, smooth, glistening, opaque. *S. cohnii* subsp. *urealyticum* usually pigmented (yellow or yellow-orange); *S. cohnii* subsp. *cohnii* is nonpigmented
S. simulans	Large; raised, circular, nonpigmented, entire, smooth, slightly glistening
S. auricularis	Small to medium; smooth, butyrous, convex, opaque, entire, slightly glistening; nonpigmented
S. xylosus	Large; raised to slightly convex, circular, smooth to rough, opaque, dull to glistening; some colonies pigmented yellow or yellow-orange
S. sciuri	Medium to large; raised, smooth, glistening, circular, opaque; most strains pigmented yellow in center of colonies
S. caprae	Small to medium; circular, entire, convex, opaque, glistening; nonpigmented

*Includes *Kytococcus*, *Nesterenkonia*, *Dermacoccus*, *Kocuria*, and *Arthrobacter*.[19]

CULTIVATION

Media of Choice

Micrococcaceae will grow on 5% sheep blood and chocolate agars but not MacConkey agar. They will also grow well in broth-blood culture systems and common nutrient broths, such as thioglycollate and brain-heart infusion.

Selective media can also be used to isolate staphylococci from clinical material. Mannitol salt agar is commonly used for this purpose. This agar contains a high concentration of salt (10%), the sugar mannitol, and phenol red as the pH indicator. On this medium, organisms such as *S. aureus* that can grow in the presence of salt and ferment mannitol produce colonies surrounded by a yellow halo.

Incubation Conditions and Duration

Visible growth on 5% sheep blood and chocolate agars incubated at 35° C in carbon dioxide (CO_2) or ambient air usually occurs within 24 hours of inoculation. Mannitol salt agar and other selective media may require incubation for at least 48 to 72 hours before growth is detected.

Colonial Appearance

Table 16-3 describes the colonial appearance and other distinguishing characteristics (e.g., hemolysis) of each genus and various staphylococcal species on 5% sheep blood agar. Growth on chocolate agar is similar. *S. aureus* yields colonies surrounded by a yellow halo on mannitol salt agar. However, other staphylococci (particularly *S. saprophyticus*) may also ferment mannitol and thus resemble *S. aureus* on this medium.

APPROACH TO IDENTIFICATION

The commercial systems for identification of *Staphylococcus* and *Micrococcus* are discussed in Chapter 13. Their accuracy in the identification of various staphylococcal species is highly variable and uncertain. Therefore, the ability of any commercial method to identify the

Table 16-4   Differentiation Among Gram-Positive, Catalase-Positive Cocci

Organism	Catalase	Microdase (modified oxidase)	Aerotolerance	Resistance to:		
				Bacitracin (0.04 U)[a]	Furazolidone (100 μg)[a]	Lysostaphin (200 μg/mL)
Staphylococcus	+[b]	−[c]	FA	R	S	S
Micrococcus (and related organisms)	+	+	A[d]	S	R	R[e]
Rothia	±	−	FA	R or S	R or S	R
Aerococcus	−[f]	−	FA[g]	S	S	R
Enterococcus	−[f]	−	FA	R	S	R

Data compiled from references 1, 18, 19.

+, ≥90% of species or strains positive; ±, ≥90% of species or strains weakly positive; −, ≥90% of species or strains negative; A, strict aerobe; FA, facultative anaerobe or microaerophile; R, resistant; S, sensitive.

[a]For bacitracin, susceptible ≥10 mm; for furazolidone, susceptible ≥15 mm.

[b]S. aureus subsp. anaerobius and S. saccharolyticus are catalase-negative and only grow anaerobically.

[c]S. sciuri, Macrococcus caseolyticus, S. lentus, and S. vitulus are microdase-positive.

[d]Kocuria (Micrococcus) kristinae is facultatively anaerobic.

[e]Some strains of Micrococcus, Arthrobacter (Micrococcus) agilis, and Kocuria are susceptible to lysostaphin.

[f]Some strains may show a pseudocatalase reaction.

[g]Grows best at reduced oxygen tension and may not grow anaerobically.

different species of coagulase-negative staphylococci should be validated before use. However, in general, commercially available methods do perform well in the identification of *S. epidermidis*, the most commonly encountered coagulase-negative staphylococcal species.

Table 16-4 shows how the catalase-positive, gram-positive cocci can be differentiated. Because they may show a pseudocatalase reaction, that is, they may appear to be catalase-positive, *Aerococcus* and *Enterococcus* are included in Table 16-4; *Rothia* (formerly *Stomatococcus*) is included for the same reason. Once an organism has been characterized as a gram-positive, catalase-positive, coccoid bacterium, complete identification may involve a series of tests, including (1) atmospheric requirements, (2) resistance to 0.04 U of bacitracin (Taxo A disk) and furazolidone, and (3) possession of cytochrome C as determined by the microdase (modified oxidase) test. However, in the busy setting of many clinical laboratories, microbiologists proceed immediately to a coagulase test based on recognition of a staphylococcal-like colony and a positive catalase test.

Microdase disks are available commercially (Remel, Inc., Lenexa, Kan). A visible amount of growth from an 18- to 24-hour-old culture is smeared on the disk; *Micrococcus* spp. turn blue within 2 minutes (see Figure 13-25).

Both for bacitracin and for furazolidone resistance, disk tests are used (Figure 16-2). A 0.04-U bacitracin-impregnated disk and a 100-μg furazolidone-impregnated disk, both available from Becton Dickinson and Company, are placed on the surface of a 5% sheep blood agar plate that has been previously streaked in

Figure 16-2   *Staphylococcus epidermis* screening plate showing resistance to bacitracin (taxo A disk) and susceptible to furazolidone (FX disk).

three directions with a cotton-tipped swab that has been dipped in a bacterial suspension prepared to match the turbidity of the 0.5 McFarland standard (i.e., the same as is used in preparing inocula for disk diffusion susceptibility tests as described in Chapter 12).

Comments Regarding Specific Organisms

Micrococcus spp. and related genera are (1) not lysed with lysostaphin, (2) resistant to the antibiotic furazolidone, (3) susceptible to 0.04 U of bacitracin, and (4) microdase-positive; they usually will only grow aerobically. In contrast, staphylococci are (1) lysed with lysostaphin, (2)

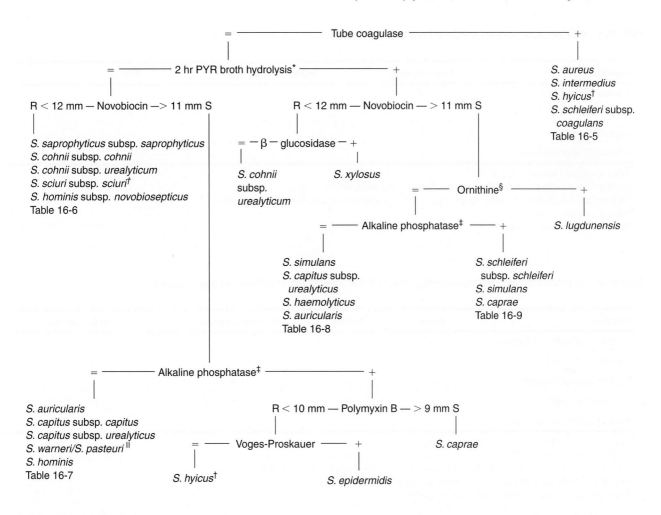

=, Signifies a negative result.
*Available commercially from Remel, Inc., Lenexa, Kan.
†Rarely involved in infections in humans.
‡Alkaline phosphatase available as a disk (Becton Dickinson and Company, Sparks, Md.) or a tablet (KEY Scientific Products, Round Rock, Texas).
§Moeller's decarboxylase medium.
‖rRNA gene restriction site polymorphism with pBA₂ as a probe may be required to separate these species.

Figure 16-3   Staphylococcal identification to species. (Based on the methods in Reference 4.)

resistant to 0.04 U of bacitracin, (3) susceptible to fura-zolidone, (4) microdase-negative, and (5) facultatively anaerobic.

Once an isolate is identified as, or strongly suspected to be, a species of staphylococci, a test for coagulase production is performed to separate *S. aureus* from the other species that are collectively referred to as *coagulase-negative staphylococci* (Figure 16-3).

The enzyme coagulase produced by *S. aureus* binds plasma fibrinogen and activates a cascade of reactions that cause plasma to clot. An organism can produce two types of coagulase, referred to as *bound* and *free* (see Procedure 13-11 for further information on coagulase tests). Detection of bound coagulase, or clumping factor, is accomplished using a rapid slide test (i.e., the slide coagulase test), in which a positive test is indicated when the organisms agglutinate on a glass slide when mixed with plasma (see Figure 13-11, *A*). Most, but not all, strains of *S. aureus* produce clumping factor and thus are readily detected by this slide test. Approximately 10% to 15% of strains may give a negative slide coagulase test.

Isolates suspected of being *S. aureus,* but failing to produce bound coagulase, must be tested for production of extracellular (i.e., free) coagulase because *S. lugdu-nensis* and *S. schleiferi* may give a positive slide coagulase test. This test, referred to as the *tube coagulase test,* is performed by inoculating a tube containing plasma and incubating at 35° C. Production of the enzyme results in a clot formation within 1 to 4 hours of inoculation (Figure 13-11, *B*). Some strains produce fibrolysin,

Table 16-5   Differentiation among Coagulase-Positive Staphylococci

Organism	Tube Coagulase	Voges-Proskauer Test	PYR*
S. aureus subsp. *aureus*	+	+	−
S. intermedius[†]	v	−	+
S. hyicus[†‡]	v	−	−
S. schleiferi subsp. *coagulans*	+	+	+

Data compiled from references 1, 2, 3, 10, 16.

+, >90% of strains positive; −, >90% of strains negative; *PYR,* pyrrolidonyl aminopeptidase; *v,* variable.

*Performed from disk (Becton Dickinson and Company, Sparks, Md.) or tablet (KEY Scientific Products, Round Rock, Texas).

[†]Primarily isolated from animals.

[‡]Rarely a cause of infections in humans.

Table 16-6   Differentiation among Coagulase-Negative, PYR-Negative, Novobiocin-Resistant Staphylococci

Organism	Urease	Oxidase	Alkaline Phosphatase*	Acid from D-Trehalose[†]
S. saprophyticus subsp. *saprophyticus*	+	−	−	+
S. cohnii subsp. *cohnii*	−	−	−	+
S. cohnii subsp. *urealyticum*	+	−	v	+
S. sciuri subsp. *sciuri*[‡]	−	+	+	(+)
S. hominis subsp. *novobiosepticus*	+	−	−	−

Data compiled from references 1, 6, 8, and 17.

+, >90% of strains positive; −, >90% of strains negative; (+), delayed positive; v, variable

*Performed from disk (Becton Dickinson and Company, Sparks, Md.) or tablet (KEY Scientific Products, Round Rock, Texas).

[†]Performed by the method of Kloos and Schleifer.[9] Results obtained by other methods may vary.

[‡]Primarily isolated from animals; rarely a cause of infections in humans.

dissolve the clot after 4 hours of incubation at 35° C, and may appear to be negative if not read at 4 hours. Because citrate-utilizing organisms may yield false-positive results, plasma containing EDTA (ethylenediaminetetraacetic) rather than citrate should be used.

Various commercial systems are available that substitute for the conventional coagulase tests just described. Latex agglutination procedures that detect clumping factor and protein A and passive hemagglutination tests that detect clumping factor are no longer in vogue because they often fail to detect methicillin-resistant *S. aureus* strains, which are being isolated from an increasing number of community-acquired infections. There is also a commercially available DNA probe for confirmation of *S. aureus* (Gen-Probe, Inc., San Diego, Calif).

Table 16-5 shows tests for differentiating among the coagulase-positive staphylococci; *S. intermedius* is an important agent of dogbite wound infections and may be misidentified as *S. aureus* if only coagulase testing is performed. Microbiologists may want to consider performing the additional tests in cases in which coagulase-positive staphylococci are isolated from dogbite infections. Otherwise, catalase-positive, gram-positive cocci in clusters from a white to yellow, creamy, opaque colony on blood agar that is slide coagulase-positive and tube coagulase-positive in 4 hours may be identified as *S. aureus.*[15]

Most laboratories do not identify the coagulase-negative staphylococci to species.[12] However, exceptions may include isolates from normally sterile sites (blood, joint fluid, or cerebrospinal fluid [CSF]), isolates from prosthetic devices, catheters, and shunts, and isolates from urinary tract infections that may be *S. saprophyticus.*

The coagulase-negative staphylococci may be identified based on the criteria shown in Figure 16-3 and Tables 16-5 through 16-9. Isolates not identified to species usually are simply reported as "coagulase-negative staphylococci."

It is particularly important to differentiate *S. lugdunensis* from other coagulase-negative staphylococci from sterile sites because there are different interpretive criteria for oxacillin for this organism. *S. lugdenensis* is

Table 16-7   Differentiation among Coagulase-Negative, PYR-Negative, Novobiocin-Susceptible, Alkaline Phosphatase-Negative Staphylococci

Organism	Urease	β-Glucosidase*	Anaerobic Growth	Acid from D-Trehalose‡
S. auricularis	–	–	(±)	(+)
S. capitis subsp. *capitis*	–	–	(+)	–
S. capitis subsp. *urealyticus*	+	–	(+)	–
S. warneri/S. pasteuri†	+	+	+	+
S. hominis	+	–	–	v

Data compiled from reference 1.
+, >90% of strains positive; (+), ?90% of strains delayed positive; –, >90% of strains negative; ±, 90% or more strains are weakly positive; ( ), reaction may be delayed; v, variable results.
*Performed from disk (Becton Dickinson and Company, Sparks, Md.) or tablet (KEY Scientific Products, Round Rock, Texas).
†rRNA gene restriction polymorphism with pBA₂ as a probe may be necessary to separate *S. warneri* from *S. pasteuri.*
‡Performed by the method of Kloos and Schleifer.[9] Results obtained by other methods may vary.

Table 16-8   Differentiation of Coagulase-Negative, PYR-Positive, Novobiocin-Susceptible, Alkaline Phosphatase-Negative Staphylococci

Organism	Urease	β-Glucuronidase*	β-Galactosidase*	Acid from Mannitol†
S. simulans	+	v	+	+
S. capitis subsp. *urealyticus*	+	–	–	+
S. haemolyticus‡	–	v	–	v
S. auricularis‡	–	–	(v)	–

Data compiled from references 1, 10, and 16.
+, >90% of strains positive; –, >90% of strains negative; (v), variable, positive reactions may be delayed.
*Performed from disk (Becton Dickinson Microbiology Systems, Sparks, Md) or tablet (KEY Scientific Products, Round Rock, Texas).
†Performed by the method of Kloos and Schleifer.[9] Results obtained by other methods may vary.
‡*S. haemolyticus* and *S. auricularis* are very difficult to separate; even fatty acid analysis does not work well.

positive for both the 2-hour PYR and ornithine decarboxylase tests.

SERODIAGNOSIS

Antibodies to teichoic acid, a major cell wall component of gram-positive bacteria, are usually produced in long-standing or deep-seated staphylococcal infections, such as osteomyelitis. This procedure, if required, is usually performed in reference laboratories. However, the clinical utility of performing this assay is, at best, uncertain.

ANTIMICROBIAL SUSCEPTIBILITY TESTING AND THERAPY

Antimicrobial therapy is vital to the management of patients suffering from staphylococcal infections (Table 16-10). Although a broad spectrum of agents may be used for therapy (see Table 12-6 for a detailed listing),

Table 16-9   Differentiation of Coagulase-Negative, PYR-Positive, Novobiocin-Susceptible, Alkaline Phosphatase–Positive Staphylococci

Organism	β-Galactosidase*	Urease
S. schleiferi subsp. *schleiferi*	(+)	–
S. simulans	+	+
S. caprae	–	+

Data compiled from references 1, 10, and 16.
+, >90% of strains positive; (+), >90% of strains delayed positive; –, >90% of strains negative.
*Performed from disk (Becton Dickinson Microbiology Systems, Sparks, Md) or tablet (KEY Scientific Products, Round Rock, Texas).

most staphylococci are capable of acquiring and using one or more of the resistance mechanisms presented in Chapter 11. The unpredictable nature of any clinical isolate's antimicrobial susceptibility requires that testing be done as a guide to therapy. As discussed in Chapter

Table 16-10   Antimicrobial Therapy and Susceptibility Testing

Organism	Therapeutic Options	Potential Resistance to Therapeutic Options	Validated Testing Methods*	Comments
Staphylococci	Several agents from each major class of antimicrobials, including aminoglycosides, beta-lactams, quinolones, and vancomycin. New agent available for use against MRSA includes linezolid, tigecycline daptomycin. See Table 12-6 for listing of specific agents that could be selected for testing and use. For many isolates a penicillinase-resistant penicillin (e.g., nafcillin, oxacillin, methicillin) is used; vancomycin is used when isolates resistant to these penicillin derivatives are encountered	Yes; resistance to every therapeutically useful antimicrobial has been described	As documented in Chapter 12: disk diffusion, broth dilution, agar dilution, and commercial systems	In vitro susceptibility testing results are important for guiding therapy. For species other than S. aureus clinical significance should be established before testing is done
Micrococcus spp.	No specific guidelines, because these species are rarely implicated in infections	Unknown	Not available	

*Validated testing methods include those standard methods recommended by the Clinical and Laboratory Standards Institute (CLSI) and those commercial methods approved by the Food and Drug Administration (FDA).

12, several standard methods and commercial systems have been developed for testing staphylococci.

Although a penicillinase-resistant penicillin, such as methicillin, nafcillin, or oxacillin, is the mainstay of antistaphylococcal therapy, resistance is common.[14] The primary mechanism for this resistance is production of an altered penicillin-binding protein (i.e., PBP 2a), which renders all currently available beta-lactams essentially ineffective. Vancomycin is the most commonly used cell wall–active agent that retains activity and is an alternative drug for resistant strains. High-level resistance to vancomycin (minimal inhibitory concentrations [MIC] > 8 µg/mL) has been described in six clinical *S. aureus* isolates and strains with MIC in the intermediate range (i.e., 8 to 16 µg/mL) have been encountered.[13,22] Two relatively newer agents available for use against such resistant strains include linezolid and daptomycin. Because of the substantial clinical and public health impact of vancomycin resistance emerging among staphylococci, laboratories should have a heightened awareness of this resistance pattern.

Because *Micrococcus* spp. are rarely encountered in infections in humans, therapeutic guidelines and standardized testing methods do not exist (see Table 16-10). However, in vitro results indicate that these organisms generally appear to be susceptible to most beta-lactam antimicrobials.[21]

PREVENTION

There are no approved antistaphylococcal vaccines. Health care workers identified as intranasal carriers of an epidemic strain of *S. aureus* are treated with topical mupirocin and, in some cases, with rifampin. Some physicians advocate the use of antibacterial substances such as gentian violet, acriflavine, chlorhexidine, or bacitracin to the umbilical cord stump to prevent staphylococcal disease in hospital nurseries. During epidemics, it is recommended that all full-term infants be bathed with 3% hexachlorophene as soon after birth as possible and daily thereafter until discharge.

Case Study 16-1

A teenage male had a history of colitis, most likely Crohn's disease. He had difficulty controlling his disease with medical management and had been treated with parenteral nutrition and pain medication. He attends high school and is socially adjusted, even though his illness has caused him to be small in stature. He lives with his mother, who works for a veterinarian. The reason for this admission was abdominal discomfort and erythema at the exit site and along the tunnel of his central line. Blood cultures were collected, and both the blood cultures and his catheter tip cultures grew a catalase-positive, gram-positive cocci. The coagulase tube test was positive, but the slide and latex test for coagulase were negative (See Procedure 13-11, Coagulase Test).

QUESTIONS

1. What further biochemical testing would you perform?

2. What additional test should always be performed from staphylococci that are PYR-positive from blood cultures?

3. Susceptibility testing for the penicillinase-resistant penicillins is problematic for coagulase-negative staphylococci, because they can be heteroresistant and express resistance poorly in vitro. This characteristic makes testing in the laboratory difficult, leading to reports of false susceptibility. What is the only reported mechanism of resistance to these agents?

4. Because of the difficulties in expression of the *mec*A gene product in staphylococci, studies have been done to determine which antimicrobial agent best induces the microorganism to produce PBP2a. After extensive studies with many challenge strains, which antimicrobial agent was found to best predict the susceptibility or resistance to the penicillinase-resistant penicillins?

REFERENCES

1. Bannerman TL: *Staphylococcus, Micrococcus,* and other catalase-positive cocci that grow aerobically. In Murray PR, Baron EJ, Pfaller MA, et al, editors: *Manual of clinical microbiology,* ed 8, Washington, DC, 2003, ASM Press.
2. Behme RJ, Shuttleworth R, McNabb A, et al: Identification of staphylococci with a self-educating system using fatty acid analysis and biochemical tests (published erratum appears in *J Clin Microbiol* 35:1043, 1997), *J Clin Microbiol* 34:2267, 1996.
3. Hébert GA: Hemolysin and other characteristics that help differentiate and biotype *Staphylococcus lugdunensis* and *Staphylococcus schleiferi, J Clin Microbiol* 28:2425, 1990.
4. Hébert GA, Crowder CG, Hancock GA, et al: Characteristics of coagulase-negative staphylococci that help differentiate these species and other members of the family Micrococcaceae, *J Clin Microbiol* 26:1939, 1988.
5. Isaac DW, Pearson TA, Hurwitz CA, et al: Clinical and microbiologic aspects of *Staphylococcus haemolyticus* infections, *Pediatr Infect J* 12:1018, 1993.
6. Kloos WE, Ballard DN, Webster JA, et al: Ribotype delineation and description of *Staphylococcus sciuri* subspecies and their potential as reservoirs of methicillin resistance and staphylolytic enzyme genes, *Int J Syst Bacteriol* 47:313, 1997.
7. Kloos WE, Bannerman TL: Update on clinical significance of coagulase-negative staphylococci, *Clin Microbiol Rev* 7:117, 1994.
8. Kloos WE, George CG, Olgiate JS, et al: *Staphylococcus hominis* subsp *novobiosepticus* subsp nov, a novel trehalose- and N-acetyl-D-glucosamine-negative, novobiocin- and multiple-antibiotic-resistant subspecies isolated from human blood cultures, *Int J Syst Bacteriol* 48:799, 1998

9. Kloos WE, Schleifer KH: Simplified scheme for routine identification of human *Staphylococcus* species, *J Clin Microbiol* 1:82, 1975.
10. Kloos WE, Wolfshohl JF: Identification of *Staphylococcus* species with API STAPH-IDENT System, *J Clin Microbiol* 16:509, 1982.
11. Koch C, Schumann P, Stackebrandt E: Reclassification of *Micrococcus agilis* (Ali-Cohen, 1889) to the genus *Arthrobacter* as *Arthrobacter agilis* comb nov and emendation of the genus *Arthrobacter, Int J Syst Bacteriol* 45:837, 1995.
12. Koontz F: Is there a clinical necessity to identify coagulase-negative staphylococci to the species level? "Okay, so I was wrong!" *Clin Microbiol Newsletter* 20:78, 1998.
13. Lyytikainen O, Vaara M, Jaarviluoma E, et al: Increased resistance among *Staphylococcus epidermidis* isolates in a large teaching hospital over a 12-year period, *Eur J Clin Microbiol Infect Dis* 15:133, 1996.
14. Mulligan ME, Murray-Leisure KA, Ribner BS, et al: Methicillin-resistant *Staphylococcus aureus:* a consensus review of the microbiology, pathogenesis, and epidemiology with implications for prevention and management, *Am J Med* 94:313, 1993.
15. National Committee for Clinical Laboratory Standards: *Abbreviated identification of bacteria and yeast;* approved guideline M35-A, Wayne, Pa, 2002, NCCLS.
16. Roberson JR, Fox LK, Hancock DD, et al: Evaluation of methods for differentiation of coagulase-positive staphylococci, *J Clin Microbiol* 30:3217, 1992.

17. Schleifer KH: Gram-positive cocci. In Sneath PH, Mair NS, Sharpe ME, et al, editors: *Bergey's manual of systematic bacteriology,* vol 2, Baltimore, 1986, Williams and Wilkins.

18. Schumann P, Spröer C, Burghardt J, et al: Reclassification of the species *Kocuria erythromyxa* (Brooks and Murray, 1981) as *Kocuria rosea* (Flügge, 1886), *Int J Syst Bacteriol* 49:393, 1999.

19. Stackerbrandt E, Koch C, Gvozdiak O, et al: Taxonomic dissection of the genus *Micrococcus: Kocuria* gen nov *Nesterenkonia* gen nov, *Kytococcus* gen nov, *Dermacoccus* gen nov, and *Micrococcus* (Cohn, 1872) gen emend, *Int J Syst Bacteriol* 45:682, 1995.

20. Vandenesch F, Etienne J, Reverdy ME, et al: Endocarditis due to *Staphylococcus lugdunensis:* report of 11 cases and review, *Clin Infect Dis* 17:871, 1993.

21. von Eiff C, Herrmann M, Peters G: Antimicrobial susceptibilities of *Stomatococcus mucilaginosus* and of *Micrococcus* spp, *Antimicrob Agents Chemother* 39:268, 1995.

22. Woolford N, Johnson AP, Morrison D, et al: Current perspectives on glycopeptide resistance, *Clin Microbiol Rev* 8:585, 1995.

ADDITIONAL READING

Archer GL, Climo MW: *Staphylococcus epidermidis* and other coagulase-negative staphylococci. In Mandell GL, Bennett JE, Dolin R, editors: *Principles and practice of infectious diseases,* Philadelphia, 2005, Elsevier Churchill Livingstone.

Moreillan P, Que YA, Glauser MP: *Staphylococcus aureus* (including Staphylococcal Toxic Shock). In Mandell GL, Bennett JE, Dolin R, editors: *Principles and practice of infectious diseases,* Philadelphia, 2005, Elsevier Churchill Livingstone.

Waldvogel FA: *Staphylococcus aureus* (including toxic shock syndrome). In Mandell GL, Bennett JE, Dolin R, editors: *Principles and practice of infectious diseases,* New York, 2000, Churchill Livingstone.

Streptococcus, Enterococcus, and Similar Organisms

CHAPTER **17**

Genera and Species to Be Considered

Beta-hemolytic streptococci
- *Streptococcus pyogenes* (group A beta-hemolytic streptococci)
- *Streptococcus agalactiae* (group B beta-hemolytic streptococci)

Groups C, F, and G beta-hemolytic streptococci

Streptococcus pneumoniae

Viridans (alpha-hemolytic) streptococci
- *Streptococcus mutans* group
- *Streptococcus salivarius* group
- *Streptococcus mitis* group
- *Streptococcus bovis* group
- *Streptococcus urinalis*
- *Streptococcus anginosus* group (also called *Streptococcus milleri* group)

Nutritionally variant streptococci
- *Abiotrophia defectiva*
- *Granulicatella adiacens* (formerly *Abiotrophia adiacens*)
- *Granulicatella balaenopterae*
- *Granulicatella elegans*

Enterococci (most commonly isolated)
- *Enterococcus faecalis*
- *Enterococcus faecium*
- Other *Enterococcus* spp. isolated from humans
- *E. durans*
- *E. mundtii*
- *E. dispar*
- *E. gallinarum*
- *E. avium*
- *E. hirae*
- *E. raffinosus*
- *E. casseliflavus*

Leuconostoc spp.

Lactococcus spp.

Globicatella sp.

Pediococcus spp.

Aerococcus spp.

Gemella spp.

Helcococcus sp.

Alloiococcus otitidis

Dolosicoccus paucivorans

Facklamia

Dolosigranulum pigrum

Ignavigranum ruoffiae

Tetragenococcus

GENERAL CHARACTERISTICS

The organisms discussed in this chapter are all catalase-negative, gram-positive cocci. *Alloiococcus,* which is catalase-negative only when tested on media devoid of whole blood (e.g., chocolate agar), is included here because it morphologically resembles the viridans streptococci. Some strains of *Enterococcus faecalis* produce a pseudocatalase when grown on blood-containing media and may appear weakly catalase-positive. Of the organisms considered in this chapter, those that are most commonly encountered in infections in humans include *Streptococcus pyogenes, S. agalactiae, S. pneumoniae,* viridans streptococci, and enterococci, usually *E. faecalis* or *E. faecium.* The other species listed in the tables are either rarely found in clinically relevant settings or are usually considered contaminants that can be mistaken for viridans streptococci or enterococci.

EPIDEMIOLOGY

Many of these organisms are commonly found as part of normal human flora and are encountered in clinical specimens as contaminants or as components of mixed cultures with minimal or unknown clinical significance (Table 17-1). However, when these organisms gain access to normally sterile sites, they can cause life-threatening infections. Other organisms, most notably *Streptococcus pneumoniae* and *Streptococcus pyogenes,* are notorious pathogens. Although *S. pneumoniae* can be found as part of the normal upper respiratory flora, this organism is also the leading cause of bacterial pneumonia and meningitis. Similarly, although *S. pyogenes* may be carried in the upper respiratory tract of humans, it is rarely considered as normal flora and should be deemed clinically important whenever it is encountered. At the other extreme, organisms such as *Leuconostoc* spp. and *Pediococcus* spp. usually are only capable of causing infections in severely compromised patients.

Many of the organisms listed in Table 17-1 are spread person to person by various means and subse-

Table 17-1 Epidemiology

Organism	Habitat (reservoir)	Mode of Transmission
Streptococcus pyogenes	Inhabits skin and upper respiratory tract of humans. Not considered part of normal flora but may be carried on nasal, pharyngeal, and, sometimes, anal mucosa. Presence in specimens is almost always considered clinically significant	Person to person by direct contact with mucosa or secretions, or by contaminated droplets produced by coughs or sneezes. Once exposed, recipient may become colonized, with subsequent development of infection
Streptococcus agalactiae	Normal flora of female genital tract and lower gastrointestinal tract; may occasionally colonize upper respiratory tract	Infections in fetuses and infants are acquired by person-to-person transmission from mother in utero or during delivery; may also be nosocomially transmitted by unwashed hands of mother or health care personnel. Mode of transmission for infections in adults is uncertain, but probably involves endogenous isolates gaining access to sterile site(s)
Groups C, F, and G beta-hemolytic streptococci	Normal flora of human skin, nasopharynx, gastrointestinal tract, and genital tract	Endogenous isolates gain access to sterile site, or by person-to-person transmission
Streptococcus pneumoniae	Colonizes nasopharynx of humans	Person-to-person spread by contact with contaminated respiratory secretions. Once exposed, recipient may become colonized, with potential for subsequent development of infection. For pneumonia, this occurs by aspiration of the organism into the lungs
Viridans streptococci	Normal flora of human oral cavity, gastrointestinal tract, and female genital tract	Usually by endogenous strains gaining access to normally sterile site; most notably results from dental manipulations with subsequent transient bacteremia
Enterococcus spp.	Found in soil, food, water, and as normal flora of animals, birds, and humans. Species most commonly associated with infections (i.e., *E. faecalis* and *E. faecium*) are normal flora of the human gastrointestinal tract and female genitourinary tract	Frequently by endogenous strains gaining access to sterile sites. Person-to-person transmission, directly or by contaminated medical equipment, allows nosocomial spread and colonization with multidrug-resistant strains. Once colonized, compromised patients are at risk of developing infections with resistant strains
Abiotrophia spp. (Nutritionally variant streptococci)	Normal flora of human oral cavity	Uncertain; probably by endogenous strains gaining access to normally sterile sites
Leuconostoc spp. *Lactococcus* spp. *Globicatella* sp. *Pediococcus* spp *Aerococcus* spp. *Gemella* spp. and *Helcococcus* sp.	Plants, vegetables, dairy products Foods and vegetation Uncertain Foods and vegetation Environmental; occasionally found on skin Normal flora of human oral cavity and upper respiratory tract Uncertain	Mode of transmission for the miscellaneous gram-positive cocci listed is unknown. Most are likely to transiently colonize the gastrointestinal tract after ingestion. From that site they gain access to sterile sites, usually in compromised patients. All are rarely associated with human infections
Alloiococcus otitidis	Occasionally isolated from human sources, but natural habitat is unknown	Uncertain; rarely implicated in infections

quently establish a state of colonization or carriage; infections may then develop when colonizing strains gain entrance to normally sterile sites. In some instances this may involve trauma (medically or nonmedically induced) to skin or mucosal surfaces or, as in the case of *S. pneumoniae* pneumonia, may result from aspiration into the lungs of organisms colonizing the upper respiratory tract.

PATHOGENESIS AND SPECTRUM OF DISEASE

The capacity of the organisms listed in Table 17-2 to produce disease and the spectrum of infections they cause vary widely with the different genera and species.

BETA-HEMOLYTIC STREPTOCOCCI

S. pyogenes produces several factors that contribute to its virulence; it is one of the most aggressive pathogens encountered in clinical microbiology laboratories. Among these factors are streptolysin O and S, which not only contribute to virulence but are also responsible for the beta-hemolytic pattern on blood agar plates that is used as a guide to identification of this species. The infections caused by *S. pyogenes* may be localized or systemic; other problems may arise as a result of the host's antibody

Table 17-2   Pathogenesis and Spectrum of Disease

Organism	Virulence Factors	Spectrum of Diseases and Infections
Streptococcus pyogenes	Protein F mediates epithelial cell attachment, and M protein is antiphagocytic; produces several enzymes and hemolysins that contribute to tissue invasion and destruction, including streptolysin O, streptolysin S, streptokinase, DNase, and hyaluronidase. Streptococcal pyrogenic exotoxins mediate production of rash (i.e., scarlet fever) or multisystem effects that may result in death	Acute pharyngitis, impetigo, erysipelas, necrotizing fasciitis and myositis, bacteremia with potential for infection in any of several organs, pneumonia, scarlet fever, streptococcal toxic shock syndrome
	Cross-reactions of antibodies produced against streptococcal antigens and human heart tissue	Rheumatic fever
	Deposition of antibody-streptococcal antigen complexes in kidney results in damage to glomeruli	Acute, poststreptococcal glomerulonephritis
Streptococcus agalactiae	Uncertain. Capsular material interferes with phagocytic activity and complement cascade activation	Infections most commonly involve neonates and infants, often preceded by premature rupture of mother's membranes. Infections often present as multisystem problems, including sepsis, fever, meningitis, respiratory distress, lethargy, and hypotension. Infections may be classified as early onset (occur within first 5 days of life) or late onset (occur 7 days to 3 months after birth). Infections in adults usually involve postpartum infections, such as endometritis, which can lead to pelvic abscesses and septic shock. Infections in other adults usually reflect compromised state of the patient and include bacteremia, pneumonia, endocarditis, arthritis, osteomyelitis, and skin and soft tissue infections
Groups C, F, and G beta-hemolytic streptococci	None have been definitively identified, but likely include factors similar to those produced by *S. pyogenes* and *S. agalactiae*	Cause similar types of acute infections in adults as described for *S. pyogenes* and *S. agalactiae,* but usually involve compromised patients. A notable proportion of infections caused by group G streptococci occur in patients with underlying malignancies. Group C organisms occasionally have been associated with acute pharyngitis
Streptococcus pneumoniae	Polysaccharide capsule that inhibits phagocytosis is primary virulence factor. Pneumolysin has various effects on host cells, and several other factors likely are involved in eliciting a strong cellular response by the host	A leading cause of meningitis and pneumonia with or without bacteremia; also causes sinusitis, and otitis media
Viridans streptococci	Generally considered to be of low virulence. Production of extracellular complex polysaccharides (e.g., glucans and dextrans) enhance attachment to host cell surfaces, such as cardiac endothelial cells or tooth surfaces in the case of dental caries	Slowly evolving (subacute) endocarditis, particularly in patients with previously damaged heart valves. Bacteremia and infections of other sterile sites do occur in immunocompromised patients. Meningitis can develop in patients suffering trauma or defects that allow upper respiratory flora to gain access to the central nervous system. *S. mutans* plays a key role in the development of dental caries
Enterococcus spp.	Little is known about virulence. Adhesions, cytolysins, and other metabolic capabilities may allow these organisms to proliferate as nosocomial pathogens. Multidrug resistance also contributes to proliferation	Most infections are nosocomial and include urinary tract infections, bacteremia, endocarditis, mixed infections of abdomen and pelvis, wounds, and occasionally, ocular infections. CNS and respiratory infections are rare
Abiotrophia spp. (nutritionally variant streptococci)	Unknown	Endocarditis; rarely encountered in infections of other sterile sites

Table 17-2   Pathogenesis and Spectrum of Disease—cont'd

Organism	Virulence Factors	Spectrum of Diseases and Infections
Leuconostoc spp., *Lactococcus* spp., *Globicatella* sp., *Pediococcus* spp., *Aerococcus* spp., *Gemella* spp., and *Helcococcus* sp. *Facklamia* spp. *Ignavigranum ruoffiae* *Dolosigranulum pigrum* *Dolosicoccus paucivorans*	Unknown; probably of low virulence. Opportunistic organisms that require impaired host defenses to establish infection. Intrinsic resistance to certain antimicrobial agents (e.g., *Leuconostoc* spp. and *Pediococcus* spp. resistant to vancomycin) may enhance survival of some species in the hospital setting	Whenever encountered in clinical specimens these organisms should first be considered as probable contaminants. *Aerococcus urinae* is notably associated with urinary tract infections
Alloiococcus sp.	Unknown	Chronic otitis media in children

response to the infections caused by these organisms. Localized infections include acute pharyngitis, for which *S. pyogenes* is the most common bacterial etiology, and skin infections, such as impetigo and erysipelas (see Chapter 60 for more information on skin and soft tissue infections).

S. pyogenes infections are prone to progression with involvement of deeper tissues and organs, a characteristic that has earned them the designation in lay publications of the "flesh-eating bacteria." Such systemic infections are life-threatening.[37] Additionally, even when infections remain localized, streptococcal pyrogenic exotoxins (SPE) may be released and produce scarlet fever, which occurs in association with streptococcal pharyngitis and is manifested by a rash of the face and upper trunk. Streptococcal toxic shock syndrome, typified by multisystem involvement that includes renal and respiratory failure, rash, and diarrhea, is another disease mediated by production of potent SPE.

Other complications that result from *S. pyogenes* infections are the poststreptococcal diseases rheumatic fever and acute glomerulonephritis. **Rheumatic fever,** which is manifested by fever, carditis (inflammation of heart muscle), subcutaneous nodules, and polyarthritis, usually follows respiratory tract infections and is thought to be mediated by antibodies produced against *S. pyogenes* that cross-react with human heart tissue. **Acute glomerulonephritis,** characterized by edema, hypertension, hematuria, and proteinuria, can follow respiratory or cutaneous infections and is mediated by antigen-antibody complexes that deposit in glomeruli, where they initiate damage.

S. agalactiae infections usually are associated with neonates and are acquired before or during the birthing process (see Table 17-2). Although the virulence factors associated with the other beta-hemolytic streptococci have not been definitively identified, groups C, G, and F streptococci cause infections similar to those associated with *S. pyogenes* (i.e., skin and soft tissue infections and bacteremia) but are less commonly encountered, often involve compromised patients, and do not produce postinfection sequelae.[39]

STREPTOCOCCUS PNEUMONIAE AND VIRIDANS STREPTOCOCCI

S. pneumoniae is a primary cause of bacterial pneumonia, meningitis, and otitis media, and the antiphagocytic properties of the polysaccharide capsule is the key to the organism's virulence. The organism may harmlessly inhabit the upper respiratory tract but may also gain access to the lungs by aspiration, where it may establish an acute suppurative pneumonia. In addition, this organism also accesses the bloodstream and the meninges to cause acute, purulent, and often life-threatening infections.

The viridans (greening) streptococci and *Abiotrophia* spp. (formally known as *nutritionally variant streptococci*) are generally considered to be opportunistic pathogens of low virulence. These organisms are not known to produce any factors that facilitate invasion of the host. However, when access is gained, a transient bacteremia occurs and endocarditis and infections at other sites in compromised patients may result.[3,5,6]

ENTEROCOCCI

Although virulence factors associated with enterococci are a topic of increasing research interest, very little is known about the characteristics that have allowed these organisms to become a prominent cause of nosocomial infections.[22,23,28] Compared with other clinically important gram-positive cocci, this genus is intrinsically more resistant to the antimicrobial agents commonly used in hospitals and is especially resistant to all currently available cephalosporins and aminoglycosides.[26] In addition, these organisms are capable of acquiring and exchanging genes that encode resistance to antimicrobial agents. This genus is the first clinically relevant group of gram-positive cocci to acquire and disseminate resistance to vancomycin, the single cell–wall active agent available for use against gram-positive organisms resistant to beta-lactams (e.g., methicillin-resistant staphylococci). Spread of this troublesome resistance marker from enterococci to other clinically relevant organisms is a serious public health concern and appears to have occurred with the emergence of vancomycin-resistant *S. aureus*.

A wide variety of enterococcal species have been isolated from human infections, but *E. faecalis* and *E. faecium* still clearly predominate as the species that are most commonly encountered. Between these two species, *E. faecalis* is the most commonly encountered, but the incidence of *E. faecium* infections is on the rise in many hospitals, which is probably related in some way to the acquisition of resistance to vancomycin and other antimicrobial agents.

MISCELLANEOUS OTHER GRAM-POSITIVE COCCI

The other genera listed in Table 17-2 are of low virulence and are almost exclusively associated with infections involving compromised hosts.[19] A possible exception is the association of *Alloiococcus otitidis* with chronic otitis media in children.[4] Certain intrinsic features, such as resistance to vancomycin among *Leuconostoc* spp. and pediococci, may contribute to the ability of these organisms to survive in the hospital environment. However, whenever they are encountered, strong consideration must be given to their clinical relevance and potential as contaminants. These organisms can also challenge many identification schemes used for gram-positive cocci, and they may be readily misidentified as viridans streptococci.

LABORATORY DIAGNOSIS

SPECIMEN COLLECTION AND TRANSPORT

No special considerations are required for specimen collection and transport of the organisms discussed in this chapter. Refer to Table 5-1 for general information on specimen collection and transport.

SPECIMEN PROCESSING

No special considerations are required for processing of the organisms discussed in this chapter. Refer to Table 5-1 for general information on specimen processing.

DIRECT DETECTION METHODS

Antigen Detection

Antigen detection screening methods are available for several streptococcal species. Detection of *S. pyogenes* antigen in throat specimens is possible using latex agglu-

tination, coagglutination, or enzyme-linked immuno-sorbent assay (ELISA) technologies. These commercial kits have been reported to be very specific, but false-negative results may occur if specimens contain low numbers of *S. pyogenes*. Sensitivity has ranged from approximately 60% to greater than 95% depending on the methodology and several other variables. Therefore, many microbiologists recommend collecting two throat swabs from each patient. If the first swab yields a positive result by a direct antigen method, the second swab can be discarded. However, for those specimens in which the rapid antigen test yielded a negative result, a blood agar plate or selective streptococcal blood agar plate should be inoculated with the second swab.

Several commercial antigen detection kits are available for diagnosis of neonatal sepsis and meningitis caused by group B streptococci. Developed for use with serum, urine, or cerebrospinal fluid (CSF), the best results have been achieved with CSF; false-positive results have been a problem using urine. Latex agglutination procedures appear to be the most sensitive and specific. Because neonates acquire *S. agalactiae* infection during passage through the colonized birth canal, direct detection of group B streptococcal antigen from vaginal swabs has also been attempted. However, direct extraction and latex particle agglutination have not been sensitive enough for use alone as a screening test.

Latex agglutination kits to detect the capsular polysaccharide antigen of the pneumococcus have also been developed for use with urine, serum, and CSF, although they are no longer commonly used in clinical microbiology laboratories.

Gram Stain

All the genera described in this chapter are gram-positive cocci. Microscopically, streptococci are round or oval-shaped, occasionally forming elongated cells that resemble pleomorphic corynebacteria or lactobacilli. They may appear gram-negative if cultures are old or if the patient has been treated with antibiotics. *Gemella haemolysans* is easily decolorized. *S. pneumoniae* is typically lancet-shaped and occurs singly, in pairs, or in short chains (see Figure 9-6, *B*).

Growth in broth should be used for determination of cellular morphology if there is a question regarding staining characteristics from solid media. In fact, the genera described in this chapter are subdivided on the basis of Gram stain based on whether they have a "strep"-like Gram stain or a "staph"-like Gram stain. For example, *Streptococcus* and *Abiotrophia* growing in broth form long chains of cocci (Figure 17-1), whereas *Aerococcus, Gemella,* and *Pediococcus* grow as large, spherical cocci arranged in tetrads or pairs or as individual cells. *Leuconostoc* may elongate to form coccobacilli, although cocci are the primary morphology. The cellular arrange-

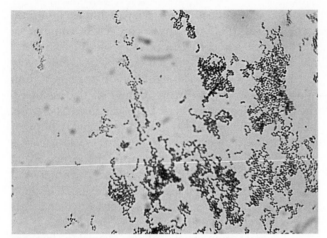

Figure 17-1   Chains of streptococci seen in Gram stain prepared from broth culture.

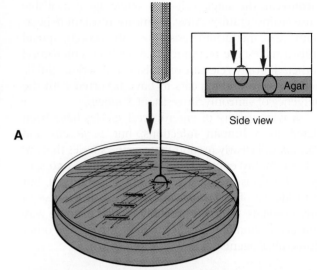

ments of the genera in this chapter are noted in Tables 17-3 and 17-4.

CULTIVATION

Media of Choice

Except for *Abiotrophia* and *Granulicatella,* the organisms discussed in this chapter will grow on standard laboratory media such as 5% sheep blood and chocolate agars. They will not grow on MacConkey agar but will grow on gram-positive selective media such as CNA (Columbia agar with colistin and nalidixic acid) and PEA (phenylethyl alcohol agar).

Abiotrophia and *Granulicatella* will not grow on blood or chocolate agars unless pyridoxal (vitamin B_6) is supplied either by placement of a pyridoxal disk, by cross-streaking with *Staphylococcus,* or by inoculation of vitamin B_6–supplemented culture media.

Blood culture media support the growth of all of these organisms, as do common nutrient broths, such as thioglycollate or brain-heart infusion. Blood cultures that appear positive and show chaining gram-positive cocci on Gram stain but do not grow on subculture should be resubcultured with a pyridoxal disk to cover for the possibility of *Abiotrophia* or *Granulicatella* bacteremia.

Other selective media are available for isolating certain species from clinical specimens. For isolating group A streptococci from throat swabs, the most common medium is 5% sheep blood agar supplemented with trimethoprim-sulfamethoxazole to suppress the growth of normal flora. However, this medium also inhibits growth of groups C, F, and G beta-hemolytic streptococci.

To detect genital carriage of group B streptococci during pregnancy, Todd-Hewitt broth with antimicrobials (gentamicin, nalidixic acid, or colistin and nalidixic

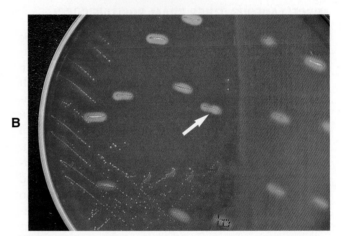

Figure 17-2   Stabbing the inoculating loop vertically into the agar after streaking the blood agar plate **(A)** allows subsurface colonies to display hemolysis caused by streptolysin O **(B)**.

acid) is used to suppress the growth of vaginal flora and allow growth of *S. agalactiae* following subculture to blood agar. LIM broth is one medium formulation used for this purpose (see Chapter 7).

Incubation Conditions and Duration

Most laboratories incubate blood or chocolate agar plates in 5% to 10% carbon dioxide. This is the preferred atmosphere for *S. pneumoniae* and is acceptable for all other genera discussed in this chapter. However, visualization of beta-hemolysis is enhanced by anaerobic conditions. Therefore, the blood agar plates should be inoculated by stabbing the inoculating loop into the agar several times (Figure 17-2, *A*). Colonies can then grow throughout the depth of the agar, producing subsurface oxygen-sensitive hemolysins (i.e., streptolysin O) (Figure 17-2, *B*). Most organisms will grow on agar media within 48 hours of inoculation.

Table 17-3 Differentiation of Catalase-Negative, Gram-Positive Coccoid Organisms Primarily in Chains

Organisms	Gram Stain from THIO Broth	Hemolysis α, β, or non[a]	Cytochrome[b]/ Catalase	Van	LAP	PYR	Gas in MRS Broth	Motility	on BE	in 6.5% NaCl Broth	Growth At 10°C	Growth At 45°C	Comments
Leuconostoc	cb, pr, ch	α, non	-/-	R	V	-	+	-	V	V	V	V	
Enterococcus Vancomycin R	c, ch	α, β, or non	-/-[c]	R	+	+	-	-	+	+	+	+	
Enterococcus Vancomycin S	c, ch	α, β, or non	-/-[c]	S	+	+	-	V[d]	+	+	+	+	
Streptococcus (all)	c, ch	α, β, or non	-/-	S	+[e]	V[f]	-	-	V[g]	V[h]	-	V	
S. agalactiae	c, ch	β, non	-/-	S	+	-	NT	-	-	V	NT	NT	
S. bovis	c, ch	α, non	-/-	S	+	-	-	-	+	-	-	+	
Viridans streptococci	c, ch	α, non	-/-	S	+	-	-	-	-[j]	-	-	V	
S. urinalis	c, pr, ch	non	-/-	S	+	+	-	-	+	+	-	+	
Abiotrophia	c, ch	α, non	-/-	S	V	V	-	-	-	-	-	V	Satellitism around *S. aureus*
Granulicatella	c, pr, ch	α	-/-	S	+	+	-	-	NT	-	-	V	Satellitism around *S. aureus*
Lactococcus	cb, ch	α, non	-/-	S	+	V	-	-	+	V	+	V[i]	
Dolosicoccus paucivorans	c, pr, ch	α	-/-	S	-	+	-	-	-	-	-	-	
Globicatella sanguinis	c, ch, pr	α, non	-/-	S	-	V	-	-	+	+	+	V	
Vagococcus	c, ch	α, non	-/-	S	+	+	-	+	+	V	+	V	
Lactobacillus	cb, ch	α, non	-/-	V	V	-	V	-	V	V	+	V	
Weissella confusa	Elongated bacilli[k]	α	-/-	R	-	NT	+	V	+	V	NT	+	Arginine positive

Data compiled from references 13, 16, 18-20, 32, 33, 36, 38, and 42.

α, alpha-hemolytic; β, beta-hemolytic; *c*, cocci; *cb*, coccobacilli; *ch*, chaining; *pr*, pairs; *LAP* leucine aminopeptidase; *MRS*, gas from glucose in Mann, Rogosa, Sharp *Lactobacillus* broth; *NT*, not tested; *PYR*, pyrrolidonyl arylamidase; *THIO*, thioglycollate broth; *Van*, vancomycin (30 μg) susceptible (*S*) or resistant (*R*); +, 90% or more of species or strains are positive; -, 90% or more of species or strains negative; *V*, variable reactions.

[a]Hemolysis tested on TSA with 5% sheep blood.
[b]Cytochrome enzymes as detected by the porphyrin broth test.
[c]Enterococci may produce a positive "pseudocatalase" effervescence. This occurs when *E. faecalis* strains grown on a blood-containing medium are tested for catalase production.
[d]*Vagococcus fluvialis* is negative for L-arabinose and raffinose, but the motile *Enterococcus gallinarum* is positive for both.
[e]The most common isolates are positive.
[f]*S. pyogenes, S. pneumoniae,* and *S. urinalis* are PYR positive.
[g]95% to 10% of viridans streptococci and *S. bovis* are bile esculin positive.
[h]Some beta streptococci grow in 6.5% salt broth.
[i]Occasional isolates are positive or give weakly positive reactions that are difficult to interpret.
[j]Majority of strains will not grow at 45°C in less than or equal to 48 hours.
[k]From blood agar, the organism resembles a gram-positive coccobacillus.

Table 17-4 Differentiation of Catalase-Negative, Gram-Positive, Coccoid Organisms Primarily in Clusters and/or Tetrads

Organisms	Gram Stain from THIO Broth[a]	Hemolysis α, β, or non[b]	Cytochrome[c]/Catalase	Van	LAP	PYR	Gas in MRS Broth	Motility	on BE	in 6.5% NaCl Broth	Growth At 10°C	Growth At 45°C	Comments
Alloiococcus[d]	c, pr, tet	non	-/+[k]	S	+	+	-	-	-	+[e]	-	-	Chronic otitis, no growth anaerobically at 72 hrs
Facklamia	c, pr, ch, cl	α, non	-/-	S	+	+	-	-	NT	+[f]	-	-	
Dolosigranulum pigrum	c, cl	non	-	S	+	+[wk]	-	-	-	+	-	-[g]	Enhanced growth around S. aureus; sauerkraut odor on SBA
Ignavigranum ruoffiae	c, pr, cl	α	-/-	S	+	+	-	-	-	+	-		
Rothia (formerly Stomatococcus mucilaginosa)	c, pr, cl	non	+/- or +[wk]	S	+	+[h]	-	-	NT	-	-	-	Strong adherence to agar surface
Gemella	c, pr, ch, cl, tet[i]	α, non	-/-	S[j]	V[k]	V[l]	-	-	-	V	-	V	
Pediococcus[m]	c, pr, tet, cl	α, non	-/-	R	+	-	-	-	+	V	-	V	
Tetragenococcus[n]	c, tet, cl	α	-/-	S	+	-	-	-	+	+	-	+	Rarely found in humans
Aerococcus													
Aerococcus urinae	c, pr, tet, cl	α	-/-	S	+	-	-	-	-	+	-	V[o]	
A. viridans	c, pr, tet, cl	α	-/+[wk]	S	-	+	-	-	V	+	V	V	
Helcococcus kunzii[p]	c, pr, ch, cl	non	-/-	S	-	+	-	-	-	+[q]	-	-	Lipophilic

Data compiled from references 2, 7-11, 14, 15, 17, 18, 19, 24, 25, 33, and 36.

α, alpha-hemolytic; β, beta-hemolytic; BE, bile esculin hydrolysis; c, cocci; cb, coccobacilli; ch, chaining; cl, clusters; pr, pairs; tet, tetrads; LAP, leucine aminopeptidase; MRS, gas from glucose in Mann, Rogosa, Sharp Lactobacillus broth; NT, not tested; PYR, pyrrolidonyl arylamidase; SBA, 5% sheep blood agar; THIO, thioglycollate broth; Van, vancomycin (30 μg) susceptible reactions. (S) or resistant (R); +, 90% or more of species or strains are positive; +[wk], strains or species may be weakly positive; −, 90% or more of species or strains negative; V, variable

[a]Alloiococcus will not grow in thioglycollate broth; Gram stain must be done from a solid medium. D. pigrum grows poorly in thioglycollate broth.

[b]Hemolysis tested on TSA with 5% sheep blood.

[c]Cytochrome enzymes as detected by the porphyrin broth test.

[d]No growth anaerobically. May be catalase negative when grown on non–blood-containing media.

[e]May take 2 to 7 days.

[f]Facklamia hominis, F. ignava, and F. languida are positive and F. sourekii is negative.

[g]Positive after 7 days.

[h]Most are positive.

[i]G. haemolysans easily decolorizes when Gram stained. They resemble Neisseria with adjacent flattened sides of pairs of cells.

[j]There is one literature report of a vancomycin-resistant Gemella haemolysans.

[k]G. haemolysans and G. sanguinis are LAP negative, and G. morbillorum and G. bergeri are positive.

[l]Weakly positive. Use a large inoculum.

[m]The most commonly isolated pediococci are arginine deaminase positive.

[n]Reactions are based on one isolate only.

[o]If inoculated too heavily, the organism will grow at 45°C.

[p]Lipophilic-growth stimulated on HIA (heart infusion agar) with 1% horse serum or 0.1% Tween.

[q]Because Helcococcus is lipophilic, the salt broth may appear to be negative unless supplemented with 1% horse serum or 0.1% Tween 80.

Table 17-5   Colonial Appearance and Characteristics on 5% Sheep Blood Agar

Organism	Appearance
Group A beta-hemolytic streptococci[a]	Grayish white, transparent to translucent, matte or glossy; large zone of beta hemolysis
Group B beta-hemolytic streptococci[b]	Larger than group A streptococci; translucent to opaque; flat, glossy; narrow zone of beta hemolysis; some strains nonhemolytic
Group C beta-hemolytic streptococci[c]	Grayish white, glistening; wide zone of beta hemolysis
Group F beta-hemolytic streptococci[d]	Grayish white, small, matte; narrow to wide zone of beta hemolysis
Group G beta-hemolytic streptococci[e]	Grayish white, matte; wide zone of beta hemolysis
S. pneumoniae	Small, gray, glistening; colonies tend to dip down in the center and resemble a doughnut (umbilicated) as they age; if organism has a polysaccharide capsule, colony may be mucoid; alpha-hemolytic
Viridans streptococci[f]	Minute to small, gray, domed, smooth or matte; alpha-hemolytic or nonhemolytic
Abiotrophia spp. and *Granulicatella* spp.[g]	Resemble viridans streptococci
Enterococcus spp.	Small, cream or white, smooth, entire; alpha-, beta-, or non-hemolytic
Leuconostoc, Aerococcus, Pediococcus, Gemella, Lactococcus, Globicatella, Helcococcus, Alloiococcus, Tetragenococcus, Dolosigranulum, Facklamia, Ignavigranum, Dolosicoccus, Vagococcus, and Weissella	Resemble viridans streptococci; see Tables 17-3 and 17-4 for hemolytic reactions

[a]Two colony sizes, that is, small (called large-colony and named *S. pyogenes*) and minute (called small-colony and named *S. anginosus* group).
[b]*S. agalactiae.*
[c]Two colony sizes, that is, small (called large-colony and named *S. dysgalactiae* subsp. *equisimilis*) and minute (called small-colony and named *S. anginosus* group).
[d]*S. anginosus* group.
[e]Two colony sizes, that is, small (called large-colony and named *S. dysgalactiae* subsp. *equisimilis*) and minute (called small-colony and named *S. anginosus* group).
[f]Includes *S. mutans* group, *S. salivarius* group, *S. anginosus* group, *S. bovis* and variants, and *S. urinalis* and *S. mitis* group.
[g]May satellite around staphylococcal colonies on 5% sheep blood or chocolate agars.

Colonial Appearance

Table 17-5 describes the colonial appearance and other distinguishing characteristics (e.g., hemolysis) of each genus on 5% sheep blood agar. The beta-hemolytic streptococci may have a distinctive buttery odor.

APPROACH TO IDENTIFICATION

The commercially available biochemical test systems for streptococci and enterococci are discussed in Chapter 13. None of them has been found to accurately identify all species of viridans streptococci or enterococci.

Comments Regarding Specific Organisms

Useful characteristics for differentiation among catalase-negative, gram-positive cocci are shown in Tables 17-3 and 17-4. Organisms that may be weakly catalase-positive, such as *Rothia mucilaginosa* (formerly *Stomatococcus mucilaginosus*), or coccobacillary, such as *Lactobacillus*, are included in these tables.

The cellular arrangement on Gram stain and the type of hemolysis are important considerations. If the presence of hemolysis is uncertain, the colony should be moved aside with a loop and the medium directly beneath the original colony should be examined by holding the plate in front of a light source.

A screening test for vancomycin susceptibility is often useful for differentiating among many alpha-hemolytic cocci. All streptococci, aerococci, gemellas, lactococci, and most enterococci are susceptible to vancomycin (any zone of inhibition), whereas pediococci, leuconostocs, and many lactobacilli are typically resistant (growth up to the disk). Other useful tests listed in Tables 17-3 and 17-4 include LAP (leucine aminopeptidase) and PYR (pyrrolidonyl arylamidase), which are commercially available as disks (see Chapter 13).

Leuconostoc produces gas from glucose in MRS broth; this distinguishes it from all other genera, except the lactobacilli. However, unlike *Leuconostoc* spp., lactobacilli appear as elongated bacilli when Gram stained from thioglycollate broth. Several organisms (e.g., *Leuconostoc, Pediococcus, Lactococcus, Helcococcus, Globicatella, Tetragenococcus, Streptococcus urinalis,* and *Aerococcus viridans*) will show growth on bile-esculin agar and in 6.5% salt broth; this is the reason these two tests no longer solely can be used to identify enterococci.

Table 17-6 Differentiation Among Clinically Relevant Beta-Hemolytic Streptococci

Species	Colony Size	Lancefield Group	Pyr	Vp	Hipp	Camp Test
S. pyogenes	Large	A	+	–	–	–
S. anginosus group[a]	Small	A	–	+	–	–
S. agalactiae		B	–	–[b]	+	+
S. dysgalactiae subsp. *equisimilis*	Large	C and G	–	–	–	–
S. anginosus group[a]	Small	C and G	–	+	–	–
S. anginosus group[a]	Small	F	–	+	–	–
S. anginosus group[a]	Small	Non-groupable	–	+	–	–

Data compiled from references 32, 40 and 41.
Hipp, Hydrolysis of hippurate; *PYR*, pyrrolidonyl arylamidase; *VP*, Voges-Proskauer test; +, >90% of strains positive; –, >90% of strains negative.
[a]Also called *S. milleri* group.
[b]Mixed reports of this result in the literature.

Serologic grouping of cell wall carbohydrates has classically been used to identify species of beta-hemolytic streptococci. The original Lancefield precipitin test is now rarely performed in clinical laboratories. It has been replaced by either latex agglutination or coagglutination procedures that are available as commercial kits. Serologic tests have the advantage of being rapid, confirmatory, and easily performed on one or two colonies. However, they are more expensive than biochemical screening tests.

The PYR and hippurate or CAMP tests can be used to identify group A and B streptococci, respectively. However, use of the 0.04-U bacitracin disk is no longer recommended for *S. pyogenes*, because groups C and G streptococci are also susceptible to this agent. *S. pyogenes* is the only species of beta-hemolytic streptococci that will give a positive PYR reaction.

S. agalactiae is able to hydrolyze hippurate and is positive in the CAMP test. The CAMP test detects production of a diffusible, extracellular protein that enhances the hemolysis of sheep erythrocytes by *Staphylococcus aureus*. A positive test is recognized by the appearance of an arrowhead shape at the juncture of the *S. agalactiae* and *S. aureus* streaks. Occasionally, non–beta-hemolytic strains of *Streptococcus agalactiae* may be encountered, but identification of such isolates can be accomplished using the serologic agglutination approach.

Table 17-6 shows the differentiation of the clinically relevant beta-hemolytic streptococci. Minute beta-hemolytic streptococci are all likely to be *S. anginosus* group; a positive Voges-Proskauer test and negative PYR test identify a beta-hemolytic streptococcal isolate as such.

Suspicious colonies (Figure 17-3) thought to be *S. pneumoniae* must be tested for either bile solubility or susceptibility to optochin (ethylhydrocupreine hydro-chloride). The bile solubility test is confirmatory and is based on the ability of bile salts to lyse *S. pneumoniae*. In the optochin test, which is presumptive, a filter paper disk ("P" disk) impregnated with optochin is placed on a blood agar plate previously streaked with a suspect organism. The plate is incubated at 35° C for 18 to 24 hours and read for inhibition. *S. pneumoniae* produce a zone of inhibition, whereas viridans streptococci grow up to the disk. A newly discovered organism, *Streptococcus pseudopneumoniae*, may interfere with appropriate interpretation of the optochin disk test.[2] *S. pseudopneumoniae* are resistant to optochin (zone ≤14 mm) when they are incubated under increased CO_2, but are susceptible to optochin (zone >14 mm) when they are incubated in ambient atmosphere. Therefore, optochin disk tests should be incubated under 5% CO_2 and all tests should be confirmed by a bile solubility test. Unfortunately, a commercial molecular probe for *S. pneumoniae*, AccuProbe Pneumococcus (Gen-Probe, San Diego, Calif), does not discriminate between *S. pneumoniae* and *S. pseudopneumoniae*. Because the pathogenic potential of *S. pseudopneumoniae* is currently unknown, it is important to differentiate it from *S. pneumoniae*, a known pathogen. Serologic identification of *S. pneumoniae* is also possible using coagglutination or latex agglutination test kits.

Once *S. pneumoniae* has been ruled out as a possibility for an alpha-hemolytic isolate, viridans streptococci and enterococci must be considered. Figure 17-3 outlines the key tests for differentiating among the viridans streptococci. Carbohydrate fermentation tests are performed in heart infusion broth with bromcresol purple indicator. Although alpha-hemolytic streptococci are not often identified to species, there are cases (i.e., endocarditis, isolation from multiple blood cultures) in which full identification is indicated. This is particularly

S
Optochin‡
R
S. pneumoniae
+
VP (Acetoin)
=
+
Arginine
=
+
Urease
=
+
PYR
=
S. vestibularis
S. mitis group
S. urinalis*
S. anginosus group
+
Urease
=
S. salivarius group
+
Mannitol
=
+
β-Gal
=
S. bovis variant†
S. salivarius†
S. mutans group
S. bovis

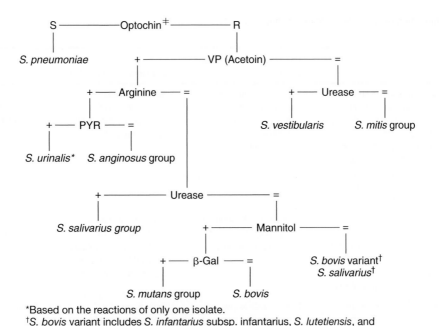

*Based on the reactions of only one isolate.
†*S. bovis* variant includes *S. infantarius* subsp. infantarius, *S. lutetiensis*, and *S. gallolyticus* subsp. *pasteurianus*. Most *S. bovis* variant strains will be positive for α-galactosidase and *S. salivarius* will be negative.
‡Optochin test must be performed in CO_2 to avoid misidentification with *S. pseudopneumoniae*.

Figure 17-3 Differentiation of clinically relevant viridans streptococcal groups. *S. mitis* group includes *S. mitis, S. sanguinis, S. parasanguinis, S. gordonii, S. oralis,* and *S. cristatus. S. mutans* group includes *S. mutans* and *S. sobrinus. S. anginosus* group includes *S. anginosus, S. constellatus* subsp. *constellatus,* and *S. intermedius. S. salivarius* group includes *S. salivarius* and *S. vestibularis. β-gal,* Beta-galactosidase; *PYR,* pyrrolidonyl arylamidase; *R,* resistant; *S,* sensitive; +, positive; =, negative. (Compiled from references 9, 30-32, and 34.)

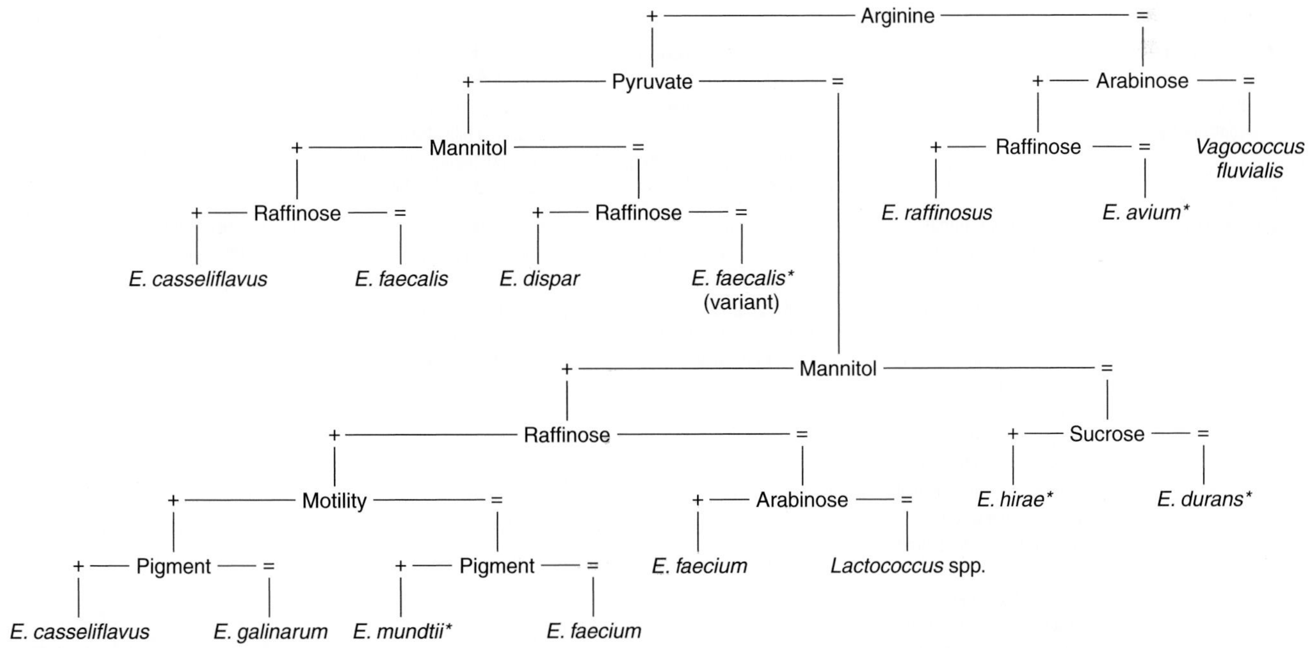

+
Arginine
=
+
Pyruvate
=
+
Arabinose
=
+
Mannitol
=
+
Raffinose
=
Vagococcus fluvialis
+
Raffinose
=
+
Raffinose
=
E. raffinosus
E. avium*
E. casseliflavus
E. faecalis
E. dispar
E. faecalis* (variant)
+
Mannitol
=
+
Raffinose
=
+
Sucrose
=
+
Motility
=
+
Arabinose
=
E. hirae*
E. durans*
+
Pigment
=
+
Pigment
=
E. faecium
Lactococcus spp.
E. casseliflavus
E. galinarum
E. mundtii*
E. faecium

*Rarely isolated from human sources.

Figure 17-4 Species identification of clinically relevant enterococcal and enterococcal-like isolates. =, Signifies a negative result. (Compiled from reference 38.)

Table 17-7   Antimicrobial Therapy and Susceptibility Testing

Organism	Therapeutic Options	Resistance to Therapeutic Options	Validated Testing Methods*	Comments
Streptococcus pyogenes	Penicillin is drug of choice; alternatives may include macrolides (e.g., azithromycin, clarithromycin, or erythromycin), telithromycin, and certain cephalosporins; vancomycin for penicillin-allergic patients with serious infections	No resistance to penicillin, cephalosporins, vancomycin known; resistance to macrolides does occur	As documented in Chapter 12: disk diffusion, broth dilution, and agar dilution	Testing to guide therapy is not routinely needed, unless a macrolide is being considered
Streptococcus agalactiae	Penicillin, with or without an aminoglycoside; ceftriaxone or cefotaxime may be used instead of penicillin; vancomycin is used for penicillin-allergic patients	No resistance to penicillins, cephalosporins, or vancomycin known	As documented in Chapter 12: disk diffusion, broth dilution, agar dilution, and some commercial methods	Testing to guide therapy is not routinely needed
Group C, F, and G beta-hemolytic streptococci	Penicillin; vancomycin for penicillin-allergic patients	No resistance known to penicillin or vancomycin	Same as used for *S. pyogenes* and *S. agalactiae*	Testing to guide therapy is not routinely needed
Streptococcus pneumoniae	Penicillin, ceftriaxone, or cefotaxime; telithromycin; macrolides, trimethoprim-sulfamethoxazole, and certain quinolones (levofloxacin, moxifloxacin, gemifloxacin)	Yes. Resistance to penicillin, cephalosporins, and macrolides is frequently encountered; vancomycin resistance has not been encountered. Fluoroquinolone resistance is rare	As documented in Chapter 12; disk diffusion, broth dilution, and certain commercial methods	In vitro susceptibility testing results are important for guiding therapy
Viridans streptococci	Penicillin or ceftriaxone, with or without an aminoglycoside; vancomycin is used in cases of penicillin allergies and beta-lactam resistance	Resistance to penicillin and cephalosporins is frequently encountered; vancomycin resistance has not been encountered	As documented in Chapter 12: disk diffusion, broth dilution, agar dilution, and some commercial methods	In vitro susceptibility testing results are important for guiding therapy
Abiotrophia spp. (nutritionally variant streptococci)	Penicillin, or vancomycin, plus an aminoglycoside	Resistance to penicillin in known, but impact on efficacy of combined penicillin and aminoglycoside therapy is not known	See CLSI document M45: methods for antimicrobial dilution and disk susceptibility testing of infrequently isolated or fastidious bacteria	Testing to guide therapy is not necessary
Enterococcus spp.	For systemic, life-threatening infections a cell wall–active agent (i.e. penicillin, ampicillin, or vancomycin) plus an aminoglycoside (gentamicin or streptomycin); newer agents such as linezdid and daptomycin may also be effective; occasionally, other agents such as chloramphenicol may be used when multidrug-resistant strains are encountered. For urinary tract isolates, ampicillin, nitrofurantoin, tetracycline, or quinolones may be effective	Resistance to every therapeutically useful antimicrobial agent, including vancomycin, linezolid, and daptomycin has been described	As documented in Chapter 12: disk diffusion, broth dilution, various screens, agar dilution, and commercial systems	In vitro susceptibility testing results are important for guiding therapy

Table 17-7   Antimicrobial Therapy and Susceptibility Testing (*cont'd*)

Organism	Therapeutic Options	Resistance to Therapeutic Options	Validated Testing Methods*	Comments
Leuconostoc spp., *Lactococcus* spp., *Globicatella* sp., *Pediococcus* spp., *Aerococcus* spp., *Gemella* spp., *Helcococcus* sp., and *Alloiococcus otitidis* and other miscellaneous opportunistic cocci	No definitive guidelines. Frequently susceptible to penicillins; effectiveness of cephalosporins is uncertain	Unknown. *Leuconostoc* and pediococci are intrinsically resistant to vancomycin	See CLSI document M45: Methods for Antimicrobial Dilution and Disk Susceptibility Testing of Infrequently Isolated or Fastidious Bacteria	Whenever isolated from clinical specimens, the potential of the isolate being a contaminant should be strongly considered

*Validated testing methods include those standard methods recommended by the Clinical and Laboratory Standards Institute (CLSI) and those commercial methods approved by the Food and Drug Administration (FDA).

true for blood culture isolates of *S. bovis* that have been associated with gastrointestinal malignancy and may be an early indicator of gastrointestinal cancer. *S. bovis* possesses group D antigen that may be detected using commercially available typing sera. However, this is not a definitive test, because other organisms (e.g., *Leuconostoc*) may also produce a positive result.

Except for species not usually isolated from humans (*E. saccharolyticus, E. cecorum, E. columbae,* and *E. pallens*), all enterococci hydrolyze PYR and possess group D antigen. A flow chart that may be used to identify enterococcal species is shown in Figure 17-4. Identifying the species of enterococcal isolates is important for understanding the epidemiology of antimicrobial resistance among isolates of this genus and for management of patients with enterococcal infections. Most clinical laboratories identify *Enterococcus* spp. presumptively by demonstrating that the isolate is PYR and LAP positive and that it grows at 45° C and in 6.5% NaCl. However, the recent discovery of *Streptococcus urinalis* presents a problem in this regard. *S. urinalis* and the commonly isolated *Enterococcus* spp. exhibit identical reactions in the four tests listed above and only differ in the ability to grow at 10° C (*S. urinalis* cannot).

SERODIAGNOSIS

Individuals with disease caused by *S. pyogenes* produce antibodies against various antigens. The most common are antistreptolysin O (ASO), anti-DNase B, antistreptokinase, and antihyaluronidase. Pharyngitis seems to be followed by rises in antibody titers against all antigens, whereas patients with pyoderma, an infection of the skin, only show a significant response to anti-DNase B. Use of serodiagnostic tests is most useful to demonstrate

prior streptococcal infection in patients from whom group A *Streptococcus* has not been cultured but who present with sequelae suggestive of rheumatic fever or acute glomerulonephritis. Serum obtained as long as 2 months after infection usually demonstrates increased antibodies. As with other serologic tests, an increasing titer over time is most useful for diagnosing previous streptococcal infection.

Commercial products are available for detection of antistreptococcal antibodies. Streptozyme (Wampole Laboratories, Princeton, NJ), which detects a mixture of antibodies, is a commonly used test. Unfortunately, no commercial system has been shown to accurately detect all streptococcal antibodies.

ANTIMICROBIAL SUSCEPTIBILITY TESTING AND THERAPY

For *S. pyogenes* and the other beta-hemolytic streptococci, penicillin is the drug of choice (Table 17-7). Because penicillin resistance has not been encountered among these organisms, susceptibility testing of clinical isolates for reasons other than resistance surveillance is not necessary.[21] However, if a macrolide such as erythromycin, is being considered for use, as is the case with patients who are allergic to penicillin, testing is needed to detect resistance that has emerged among these organisms. For serious infections caused by *S. agalactiae,* an aminoglycoside may be added to supplement beta-lactam therapy and enhance bacterial killing.

In contrast to beta-hemolytic streptococci, the emergence of resistance to a variety of different antimicrobial classes in *S. pneumoniae* and viridans streptococci dictates that clinically relevant isolates be subjected to in vitro susceptibility testing.[27] When testing is performed, methods that produce minimal inhibitory concentration (MIC) data for beta-lactams are preferred. The level of resistance (i.e., MIC in µg/mL) can provide important information regarding therapeutic management of the patient, particularly in cases of pneumococcal meningitis in which relatively slight increases in MIC can have substantial impact on clinical efficacy of penicillins and cephalosporins. Vancomycin resistance has not been described in either *S. pneumoniae* or in viridans streptococci.

Enterococci are intrinsically resistant to a wide array of antimicrobial agents, and they generally are resistant to killing by any of the single agents (e.g., ampicillin or vancomycin) that are bactericidal for most other gram-positive cocci. Therefore, effective bactericidal activity can only be achieved with the combination of a cell wall–active agent, such as ampicillin or vancomycin, and an aminoglycoside, such as gentamicin or streptomycin.

Unfortunately, many *E. faecalis* and *E. faecium* isolates have acquired resistance to one or more of these components of combination therapy. This resistance generally eliminates any contribution that the target antimicrobial agent could make to the synergistic killing of the organism. Therefore, performance of in vitro susceptibility testing with clinical isolates from systemic infections is critical for determining which combination of agents may still be effective therapeutic choices.

For uncomplicated urinary tract infections, bactericidal activity is usually not required for clinical efficacy, so that single agents such as ampicillin, nitrofurantoin, or a quinolone are often sufficient.

PREVENTION

A single-dose, 23-valent vaccine (Pneumovax, Merck & Co. Inc., West Point, Pa) to prevent infection by the most common serotypes of *S. pneumoniae* is available in the United States. Vaccination is recommended for individuals older than 65 years of age and for patients with (1) chronic pulmonary, cardiac, liver, or renal disease; (2) no spleen (asplenic); (3) sickle cell disease; (4) diabetes; (5) HIV infection; or (6) any other diseases that compromise the immune system. The vaccine is not effective in children younger than 2 years of age. A heptavalent (seven serotypes) vaccine (Prevnar, distributed by Wyeth-Ayerst Pharmaceuticals, Philadelphia, Pa) is available for children younger than 2 years of age. The seven serotypes in this vaccine account for the majority of cases of bacteremia, meningitis, and otitis media in children younger than 6 years of age. Moreover, 80% of penicillin-resistant strains are one of these seven serotypes.

Lifetime chemoprophylaxis with penicillin, given either monthly (intramuscular administration) or daily (oral administration), is recommended for patients with rheumatic heart disease to prevent development of bacterial endocarditis on a damaged heart valve. Likewise, penicillin may be indicated to control outbreaks of *S. pyogenes* in individuals in close physical contact, such as in households, military populations, or newborn nurseries.

Case Study

A 76-year-old man with atherosclerosis had been previously admitted for abdominal aneurysm and resection of the perirenal aorta. He had several follow-up admissions over the next year for postoperative wound infections, with accompanying bacteremia, alternating between *Pseudomonas aeruginosa*, vancomycin-resistant *Enterococcus faecium*, and *Candida glabrata*. On his final admission, blood cultures were positive, with numerous gram-positive cocci in pairs and chains in the smear, but subculture of the bottle showed no growth aerobically with increased CO_2 on blood agar or chocolate agar or anaerobically on *Brucella* agar.

QUESTIONS

1. What is this organism and what would you do to grow it?
2. To control infection, screening for vancomycin resistance in enterococci on selected hospitalized patients is important. What is a cost-effective screening method?
3. Many genera of gram-positive cocci are catalase-negative, but only a few are vancomycin-resistant. Name these genera, and indicate how they can be differentiated from vancomycin-resistant enterococci.

REFERENCES

1. Aguirre M, Morrison D, Cookson D, et al: Phenotypic and phylogenetic characterization of some *Gemella*-like organisms from human infections: description of *Dolosigranulum pigrum* gen nov, sp nov, *J Appl Bacteriol* 75:608, 1993.
2. Arbique JC, Poyart C, Trieu-Cuot P, et al: Accuracy of phenotypic and genotypic testing for identification of *Streptococcus pneumoniae* and description of *Streptococcus pseudopneumoniae* sp. nov, *J Clin Microbiol* 42:4686, 2004.
3. Bochud PY, Calandra T, Francioli P, et al: Bacteremia due to viridans streptococci in neutropenic patients: a review, *Am J Med* 97:256, 1994.

4. Bosley GS, Whitney AM, Prucker JM., et al: Characterization of ear fluid isolates of *Alloiococcus otitidis* from patients with recurrent otitis media, *J Clin Microbiol* 33:2876, 1995.

5. Bouvet A: Human endocarditis due to nutritionally variant streptococci: *Streptococcus adjacens* and *Streptococcus defectivus, Eur Heart J* 16(suppl B):24, 1995.

6. Carratala J, Alcaide F, Fernandez-Sevilla A, et al: Bacteremia due to viridans streptococci that are highly resistant to penicillin: increase among neutropenic patients with cancer, *Clin Infect Dis* 20:1169, 1995.

7. Christensen JJ, Vibits H, Ursing J, et al: *Aerococcus*-like organism: a newly recognized potential urinary tract pathogen, *J Clin Microbiol* 29:1049, 1991.

8. Collins MD, Falsen E, Lemozy J, et al: Phenotypic and phylogenetic characterization of some *Globicatella*-like organisms from human sources: description of *Facklamia hominis* gen nov, sp nov, *Int J Syst Bacteriol* 47:880, 1997.

9. Collins MD, Hutson RA, Falsen E, et al: An unusual *Streptococcus* from human urine, *Streptococcus urinalis* sp nov, *Int J Syst Evol Microbiol* 50:1173, 2000.

10. Collins MD, Hutson RA, Falsen E, et al: Description of *Gemella sanguinis* sp nov, isolated from human clinical specimen, *J Clin Microbiol* 36:3090, 1998.

11. Collins MD, Hutson RA, Falsen E, et al: *Facklamia sourekii* sp nov, isolated from human sources, *Int J Syst Bacteriol* 49:635, 1999.

12. Collins MD, Hutson RA, Falsen E, et al: *Gemella bergeriae* sp nov, isolated from human clinical specimens, *J Clin Microbiol* 36:1290, 1998.

13. Collins MD, Lawson PA: The genus *Abiotrophia* (Kawamura et al) is not monophyletic: proposal of *Granulicatella* gen nov, *Granulicatella adiacens* comb nov, *Granulicatella elegans* comb nov and *Granulicatella balaenopterae* comb nov, *Int J Syst Evol Microbiol* 50:365, 2000.

14. Collins MD, Lawson PA, Monasterio R, et al: *Facklamia ignava* sp nov, isolated from human clinical specimens, *J Clin Microbiol* 36:2146, 1998.

15. Collins MD, Lawson PA, Monasterio R, et al: *Ignavigranum ruoffiae* sp nov, isolated from human clinical specimens, *Int J Syst Bacteriol* 49:97, 1999.

16. Collins MD, Rodriguez Jovita M, Hutson RA, et al: *Dolosicoccus paucivorans* gen nov, sp nov, isolated from human blood, *Int J Syst Bacteriol* 49:1439, 1999.

17. Collins MD, Williams AM, Wallbanks S: The phylogeny of *Aerococcus* and *Pediococcus* as determined by 16S rRNA sequence analysis: description of *Tetragenococcus* gen nov, *FEMS Microbiol Lett* 70:255, 1990.

18. Facklam RR: Newly described, difficult-to-identify, catalase-negative, gram-positive cocci, *Clin Microbiol Newsl* 23:1, 2001.

19. Facklam R, Elliott JA: Identification, classification, and clinical relevance of catalase-negative, gram-positive cocci, excluding the streptococci and enterococci, *Clin Microbiol Rev* 8:479, 1995.

20. Facklam R, Hollis D, Collins MD: Identification of gram-positive coccal and coccobacillary vancomycin-resistant bacteria, *J Clin Microbiol* 27:724, 1989.

21. Gerber MA: Antibiotic resistance in group A streptococci, *Pediatr Clin North Am* 42:539, 1995.

22. Jett BD, Huycke MM, Gilmore MS: Virulence of enterococci, *Clin Microbiol Rev* 7:462, 1994.

23. Johnson AP: The pathogenicity of enterococci, *J Antimicrob Chemother* 33:1083, 1994.

24. LaClaire L, Facklam R: Antimicrobial susceptibility and clinical sources of *Dolosigranulum pigrum* cultures, *Antimicrob Agents Chemother* 44:2001, 2000.

25. Lawson PA, Collins MD, Falsen E, et al: *Facklamia languida* sp nov, isolated from human clinical specimen, *J Clin Microbiol* 37:1161, 1999.

26. Leclercq R: Epidemiology and control of multiresistant enterococci, *Drugs* 2:47, 1996.

27. McCracken GH: Emergence of resistant *Streptococcus pneumoniae*: a problem in pediatrics, *Pediatr Infect Dis J* 14:424, 1995.

28. Murray BE: The life and times of the enterococcus, *Clin Microbiol Rev* 3:46, 1990.

29. National Committee for Clinical Laboratory Standards: *Abbreviated identification of bacteria and yeast;* approved guideline M35-A, Wayne, Pa, 2002, NCCLS.

30. Poyart C, Quesne G, and Trieu-Cuot P: Taxonomic dissection of the *Streptococcus bovis* group by analysis of manganese-dependent superoxide dismutase gene (sodA) sequences: reclassification of "*Streptococcus infantarius* subsp. *coli*" as *Streptococcus lutetiensis* sp. nov. and of *Streptococcus bovis* biotype II.2 as *Streptococcus pasteurianus* sp. nov, *Int J Syst Evol Microbiol* 52:1247, 2002.

31. Ruoff KL, Miller SI, Garner CV, et al: Bacteremia with *Streptococcus bovis* and *Streptococcus salivarius*: clinical correlates of more accurate identification of isolates, *J Clin Microbiol* 27:305, 1989.

32. Ruoff KL, Whiley RA, Beighton B: *Streptococcus.* In Murray PR, Baron EJ, Jorgensen JH, et al, editors: *Manual of clinical microbiology*, ed 8, Washington, DC, 2003, ASM Press.

33. Ruoff KL: *Aerococcus, Abiotrophia,* and other infrequently isolated aerobic catalase-negative, gram-positive cocci. In Murray PR, Baron EJ, Jorgensen JH, et al, editors: *Manual of clinical microbiology*, ed 8, Washington, DC, 2003, ASM Press.

34. Schlegel L, Grimont F, Collins MD, et al: *Streptococcus infantarius* sp nov, *Streptococcus infantarius* subsp *infantarius* subsp nov, and *Streptococcus infantarius* subsp *coli* subsp nov, isolated from humans and food, *Int J Syst Evol Microbiol* 50:1425, 2000.

35. Schlegel L, Grimont F, Ageron E, et al: Reappraisal of the taxonomy of the *Streptococcus bovis/Streptococcus equinus* complex and related species: description of *Streptococcus gallolyticus* subsp. *gallolyticus* subsp. nov, *S. gallolyticus* subsp. *macedonicus* subsp. nov and *S. gallolyticus* subsp. *pasteurians* subsp. nov, *Int J Syst Evol Microbiol* 53:631, 2003.

36. Schlerfer KH: Gram-positive cocci. In Sneath PH, Mair NS, Sharpe ME, et al, editors: *Bergey's manual of systematic bacteriology,* vol 2, Baltimore, 1986, Williams and Wilkins.

37. Stevens DL: Invasive group A streptococcal disease, *Infect Agents Dis* 5:157, 1996.

38. Teixeira LM, Facklam RR: *Enterococcus.* In Murray PR, Baron EJ, Jorgensen JH, et al, editors: *Manual of clinical microbiology*, ed 8, Washington, DC, 2003, ASM Press.

39. Turner JC, Fox A, Fox K, et al: Role of group C beta-hemolytic streptococci in pharyngitis: epidemiologic study of clinical features associated with isolation of group C streptococci, *J Clin Microbiol* 31:808, 1993.

40. Vandamme P, Pot B, Falsen E, et al: Taxonomic study of Lancefield streptococcal groups C, G, and L (*Streptococcus dysgalactiae*) and proposal of *S. dysgalactiae* subsp *equisimilis* subsp nov, *Int J Syst Bacteriol* 46:774, 1996.

41. Whiley RA, Hall LM, Hardie JM, et al: A study of small-colony, β-hemolytic, Lancefield group C streptococci within the *anginosus* group: description of *Streptococcus constellatus* subsp *pharyngis* subsp nov, associated with the human throat and pharyngitis, *Int J Syst Bacteriol* 49:1443, 1999.

42. York MK, Ruoff KL, Clarridge J, et al: Identification of gram-positive bacteria. In Isenberg HD, editor: *Clinical microbiology procedures handbook,* ed 2, Washington, DC, 2004, ASM Press.

ADDITIONAL READING

Bisno AL, Ruoff KL: Classification of streptococci. In Mandell GL, Bennett JE, Dolin R, editors: *Principles and practice of infectious diseases,* ed 6, Philadelphia, 2005, Elsevier Churchill Livingstone.

Bisno AL, Stevens DL: *Streptococcus pyogenes.* In Mandell GL, Bennett JE, Dolin R, editors: *Principles and practice of infectious diseases,* ed 6, Philadelphia, 2005, Elsevier Churchill Livingstone.

Bisno AL: Non-suppurative poststreptococcal sequelae: rheumatic fever and glomerulonephritis. In Mandell GL, Bennett JE, Dolin R, editors: *Principles and practice of infectious diseases,* ed 6, Philadelphia, 2005, Elsevier Churchill Livingstone.

Edwards MS, Baker CJ: *Streptococcus agalactiae* (group B streptococcus). In Mandell GL, Bennett JE, Dolin R, editors: *Principles and practice of infectious diseases,* ed 6, Philadelphia, 2005, Elsevier Churchill Livingstone.

Johnson CC, Tunkel AR: *Viridans streptococci* and groups C and G streptococci and *Gemella morbillgrum.* In Mandell GL, Bennett JE, Dolin R, editors: *Principles and practice of infectious diseases,* ed 6, Philadelphia, 2005, Elsevier Churchill Livingstone.

Moellering RC Jr: *Enterococcus* species, *Streptococcus bovis,* and *Leuconostoc* species. In Mandell GL, Bennett JE, Dolin R, editors: *Principles and practice of infectious diseases,* ed 6, Philadelphia, 2005, Elsevier Churchill Livingstone.

Musher DM: *Streptococcus pneumoniae.* In Mandell GL, Bennett JE, Dolin R, editors: *Principles and practice of infectious diseases,* ed 6, Philadelphia, 2005, Elsevier Churchill Livingstone.

Mirzanejad Y, Stratton CW: In Mandell GL, Bennett JE, Dolin R, editors. *Principles and practice of infectious diseases,* ed 6, Philadelphia, 2005, Elsevier Churchill Livingstone.

Bacillus and Similar Organisms

CHAPTER 18

Genera and Species to Be Considered

- *Bacillus anthracis*
- *Bacillus cereus*
- *Bacillus mycoides*
- *Bacillus circulans*
- *Bacillus licheniformis*
- *Bacillus subtilis*
- *Bacillus megaterium*
- Other *Bacillus* spp.
- *Brevibacillus brevis*
- *Paenibacillus* spp.

GENERAL CHARACTERISTICS

Bacillus spp. and related genera *Brevibacillus* and *Paenibacillus* all are aerobic, gram-positive, spore-forming rods.[1,6] Although these genera contain many species, only those that are most commonly associated with human infections are considered here.

EPIDEMIOLOGY

Bacillus anthracis, the most notorious pathogen of this genus, inhabits the soil. Humans acquire infections when they are inoculated with the spores, either by inhalation during exposure to contaminated animal products, such as hides, or by traumatic introduction (Table 18-1). In addition, as of 2001, intentional use of *B. anthracis* as an agent of bioterrorism became a reality. All other *Bacillus* spp. are generally considered to be opportunistic pathogens of low virulence and are usually only associated with compromised patients exposed to contaminated materials.[3]

PATHOGENESIS AND SPECTRUM OF DISEASE

B. anthracis is the most highly virulent species for humans and is the causative agent of anthrax, of which

there are three forms (Table 18-2). Anthrax is a devastating disease that is rarely encountered in developed countries. *Bacillus anthracis* has become the centerpiece for counterterrorism planning efforts in that this organism has already been successfully used as an agent of bioterrorism.

Although other species, such as *B. cereus*, can cause serious infections, the relative virulence of this and other *Bacillus* spp. is trivial compared with that of *B. anthracis*. *B. cereus* can cause food poisoning and serious eye infections may result from trauma, use of contaminated needles during intravenous drug abuse, or inoculation of the eye with contaminated dust or dirt particles. Infections of other body sites also are known to occur in compromised hosts. However, because the spores of *Bacillus* spp. are ubiquitous in nature, contamination of various clinical specimens is common. Therefore, whenever *Bacillus* spp. are encountered, the clinical significance of the isolate should be carefully established before performing extensive identification procedures.

LABORATORY DIAGNOSIS

SPECIMEN COLLECTION AND TRANSPORT

No special considerations are required for specimen collection and transport of the organisms discussed in this chapter. Refer to Table 5-1 for general information on specimen collection and transport.

SPECIMEN PROCESSING

With few exceptions, special processing considerations are not required for these organisms. Refer to Table 5-1 for general information on specimen processing. The exceptions are processing procedures for foods implicated in *B. cereus* food poisoning outbreaks and animal hides or products, and environmental samples, for the isolation of *B. anthracis*. The organisms will be present as spores in these specimens, so initial processing will involve either heat or alcohol shock before plating on

Table 18-1 Epidemiology

Species	Habitat (reservoir)	Mode of Transmission
Bacillus anthracis	Soil, from where it is contracted by various herbivores. Causes disease in these animals. Not part of normal human flora.	Contact with infected animals or animal products. Infections occur by inhalation of spores or by their introduction through breaks in the skin or mucous membranes. Occasionally acquired by ingestion. Person-to-person transmission has not been documented. Also identified as agent for bioterrorism—intentional spread to humans.
Bacillus cereus, Bacillus circulans, Bacillus licheniformis, Bacillus subtilis, other *Bacillus* spp., *Brevibacillus* sp., and *Paenibacillus* spp.	Vegetative cells and spores of many species are widely distributed in nature. Not commonly considered part of normal flora but may transiently colonize skin or the gastrointestinal or respiratory tracts.	Traumatic introduction into normally sterile sites or by exposure to contaminated medical equipment or supplies. Infections often involve immunosuppressed patients. In cases of food poisoning, ingestion of food contaminated with *B. cereus* or toxins formed by this organism.

Table 18-2 Pathogenesis and Spectrum of Disease

Species	Virulence Factors	Spectrum of Diseases and Infections
Bacillus anthracis	Production of antiphagocytic capsule and potent exotoxins (i.e., edema toxin and lethal toxin) that mediate cell and tissue destruction	Causative agent of anthrax, of which there are three forms: • Cutaneous anthrax occurs at site of spore penetration 2 to 5 days after exposure and is manifested by progressive stages from an erythematous papule to ulceration and finally to formation of a black scar (i.e., eschar). May progress to toxemia and death. • Pulmonary anthrax, also known as "wool sorter's" disease, follows inhalation of spores and progresses from malaise with mild fever and nonproductive cough to respiratory distress, massive chest edema, cyanosis, and death. • Gastrointestinal anthrax may follow ingestion of spores and affects either the oropharyngeal or the abdominal area. Most patients die from toxemia and overwhelming sepsis.
Bacillus cereus	Produces enterotoxins and pyogenic toxin	Food poisoning of two types: diarrheal type, which is characterized by abdominal pain and watery diarrhea; and emetic type, which is manifested by profuse vomiting. *B. cereus* is the most commonly encountered species of *Bacillus* in opportunistic infections that include posttraumatic eye infections, endocarditis, and bacteremia. Infections of other sites are rare and usually involve intravenous drug abusers or immunocompromised patients.
Bacillus circulans, Bacillus licheniformis, Bacillus subtilis, other *Bacillus* spp., *Brevibacillus* sp., and *Paenibacillus* spp.	Virulence factors not known	Food poisoning has been associated with some species but is not common. These organisms may also be involved in opportunistic infections similar to those described for *B. cereus*.

solid media. The shock will allow only the spore-forming bacilli to survive; thus this is an enrichment and selection technique designed to increase the chance for laboratory isolation of these species. Despite its publicity as a potential agent of biologic warfare, *B. anthracis* is not highly contagious, so standard safety precautions are appropriate.[7]

DIRECT DETECTION METHODS

Other than Gram stain of patient specimens, there are no specific procedures for the direct detection of *Bacillus* spp. in clinical material. The organisms are gram-positive when stained from young cultures but become gram-variable or gram-negative with age (Figure 18-1).

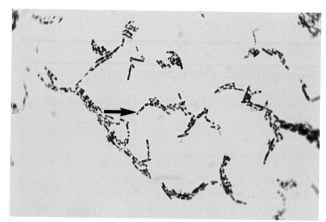

Figure 18-1   Gram stain of *Bacillus cereus*. The arrow is pointed at a spore, which is clear inside the gram-positive vegetative cell.

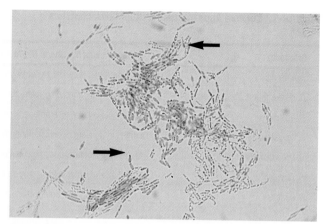

Figure 18-2   Spore stain of *Bacillus cereus*. The arrows are pointed at green spores in a pink vegetative cell.

A feature of *Bacillus* spp. that is unique from all other clinically relevant aerobic organisms is the ability to produce spores in the presence of oxygen. Although spores are not readily evident on all smears containing *Bacillus* spp., their presence confirms the genus identification.

To induce sporulation, organisms can be grown on triple sugar iron agar (TSI), urea agar, or nutrient agar containing 5 mg manganese sulfate per liter. On Gram-stained smears, spores appear clear because they do not retain the crystal violet or safranin. However, spores will stain with specific dyes such as malachite green, which is forced into the spore using heat; the vegetative cell is then counterstained with safranin (Figure 18-2).

CULTIVATION

Media of Choice
All *Bacillus* and related genera grow well on 5% sheep blood agar, chocolate agar, routine blood culture media, and commonly used nutrient broths. They will not grow on MacConkey agar, and those that are susceptible to nalidixic acid will not grow on CNA (Columbia agar with nalidixic acid and colistin). PEA (phenylethyl alcohol agar) is useful for the isolation of *Bacillus* spp. from contaminated specimens.

Two special media are used for isolation and identification of *B. anthracis*. PLET (polymyxin-lysozyme-EDTA-thallous acetate) can be used for selection and isolation of this species from contaminated specimens. As a means of identification, bicarbonate agar is used to induce *B. anthracis* capsule formation.

Incubation Conditions and Duration
Most species will produce detectable growth within 24 hours following incubation on media incubated at

35° C, in ambient air, or in 5% carbon dioxide (CO_2). Bicarbonate agar must be incubated in CO_2.

Colonial Appearance
Table 18-3 describes the colonial appearance and other distinguishing characteristics (e.g., hemolysis) of each species of *Bacillus* or related genera on blood agar. Colonies of *B. anthracis* growing on bicarbonate agar are large and mucoid.

APPROACH TO IDENTIFICATION

There are no commercial biochemical identification systems in routine use in clinical laboratories for *Bacillus* spp. Species differentiation within the genera *Bacillus*, *Brevibacillus*, and *Paenibacillus* is based on the size of the vegetative cell, whether the spore swells the vegetative cell, and biochemical tests (Table 18-4), including elaboration of the enzyme lecithinase (Figure 18-3).

Comments Regarding Specific Organisms
The vegetative cell width of *B. anthracis*, *B. cereus*, *B. mycoides*, *B. thuringiensis*, and *B. megaterium* is usually greater than 1 μm and the spores do not swell the cell. The vegetative cell width of *B. subtilis*, *B. pumilus*, and *B. licheniformis* is less than 1 μm, and the spores do not swell the cell. The cell width of *B. circulans*, *B. coagulans*, *B. sphaericus*, *B. brevis*, *P. macerans*, *P. alvei*, and *P. polymyxa* is less than 1 μm and the spores swell the cell. When determining cell width, only cells staining gram-positive should be measured. Organisms that do not retain the crystal violet appear narrower because the cell wall is not apparent.

It is important to be able to identify *B. anthracis*, if this pathogen is ever encountered. Clinical microbiologists are sentinels for recognition of a bioterrorist event, especially one involving *B. anthracis*. Even though

Table 18-3 Colonial Appearance and Other Characteristics

Organism	Appearance on 5% Sheep Blood Agar
B. anthracis	Medium-large, gray, flat, irregular with swirling projections ("Medusa head"); nonhemolytic
B. cereus and *B. thuringiensis*	Large, feathery, spreading; beta-hemolytic
B. mycoides	Rhizoid colony that resembles a fungus; weakly beta-hemolytic
B. megaterium	Large, convex, entire, moist; nonhemolytic
B. licheniformis	Large blister colony; becomes opaque with dull to rough surface with age; beta-hemolytic
B. pumilus	Large, moist, blister colony; may be beta-hemolytic
B. subtilis	Large, flat, dull, with ground-glass appearance; may be pigmented (pink, yellow, orange, or brown); may be beta-hemolytic
B. circulans	Large, entire, convex, butyrous; smooth, translucent surface; may be beta-hemolytic
B. coagulans	Medium-large, entire, raised, butyrous, creamy-buff; may be beta-hemolytic
B. sphaericus	Large, convex, smooth, opaque, butyrous; nonhemolytic
Brevibacillus brevis	Medium-large, convex, circular, granular; may be beta-hemolytic
Paenibacillus macerans	Large, convex, fine granular surface; nonhemolytic
P. alvei	Swarms over agar surface; discrete colonies are large, circular, convex, smooth, glistening, translucent or opaque; may be beta-hemolytic
P. polymyxa	Large, moist blister colony with "ameboid spreading" in young cultures; older colonies wrinkled; non- hemolytic

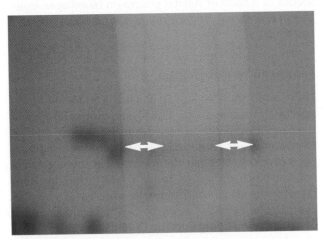

Figure 18-3 Lecithinase production *by Bacillus cereus* on egg yolk agar. The organism has been streaked down the center of the plate. The positive test for lecithinase is indicated by the opaque zone of precipitation around the bacterial growth *(arrows).*

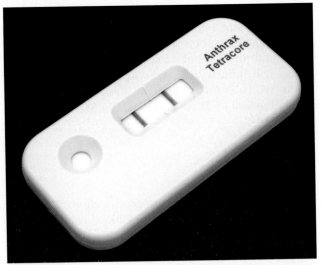

Figure 18-4 Red Line Alert Test. A red line appears on the cassette if the culture isolate is presumptive *Bacillus anthracis.* (Courtesy Tetracore, Inc., Gaithersburg, Md.)

this organism is rarely found, sentinel laboratory protocols require that each institution rule out the possibility of anthrax before reporting out any blood cultures in which a large gram-positive aerobic rod is isolated. The index case of anthrax was discovered in 2001 by an astute clinical microbiologist who saw large gram-positive rods in the patient's cerebrospinal fluid. *B. anthracis* should be suspected if typical nonhemolytic "Medusa head" colonies are observed on 5% sheep blood agar, and the organism is subsequently found to be nonmotile. The Red Line Alert Test (Tetracore, Inc., Gaithersburg, Md) is a Food and Drug Administration (FDA)-cleared immunochromatographic test that presumptively identifies *B. anthracis* from blood agar (Figure 18-4). The sentinel laboratory anthrax protocol was revised in 2005 to permit clinical microbiology laboratories who buy this or any other FDA-cleared test to perform this testing to rule out nonhemolytic,

Table 18-4 Differentiation of Clinically Relevant *Bacillus* spp., *Brevibacillus*, and *Paenibacillus*

Organism	Bacillary Body Width >1.0 μm	Wide Zone Lecithinase	Spores Swell Sporangium	Voges-Proskauer	Glucose with Gas	Fermentation of: Mannitol	Xylose	Anaerobic Growth	Citrate	Indole	Motility	Penicillin Sensitive	Parasporal Crystals
Bacillus anthracis	+	+	-	+	-	-	-	+	v	-	-	+	-
B. cereus	+	+	-	+	-	-	-	+	+	-	+	-	-
B. thuringiensis	+	+	-	+	-	-	-	+	+	-	+	-	+
B. mycoides	+	+	-	+	-	-	-	+	+	-	-	-	-
B. megaterium	+	-	-	-	-	+ or (+)	v	-	+	-			
B. licheniformis	-	-	-	+	-	+	+	+	+	-			
B. pumilus	-	-	-	+	-	+	+	-	+	-			
B. subtilis	-	-	-	+	-	+	+	-	+	-			
B. circulans	-	-	+	-	-	+	+	v	-	-			
B. coagulans	-	-	v	v	-	-	v	+	v	-			
B. sphaericus	-	-	+	-	-	-	-	-	v	-			
Brevibacillus brevis	-	-	+	-	-	+	-	-	v	-			
Paenibacillus macerans	-	-*	+	-	+	+	+	+	-	-			
P. alvei	-	-	+	+	-	-	-	+	-	+			
P. polymyxa	-	-*	+	+	+	v	+	+	-	-			

Compiled from references 2, 4, and 5.

+, 90% or more of species or strains are positive; –, 90% or more of species or strains are negative, v, variable reactions; ( ), reactions may be delayed.

*Weak lecithinase production only seen under the colonies.

Table 18-5 Antimicrobial Therapy and Susceptibility Testing

Species	Therapeutic Options	Resistance to Therapeutic Options	Validated Testing Methods*	Comments
Bacillus anthracis	Penicillin is drug of choice. Alternatives include ciprofloxacin, erythromycin, tetracycline, and chloramphenicol	None known	See CLSI document M100-S15. *Performed in approved reference laboratories only.*	
Other *Bacillus* spp., *Brevibacillus* sp., *Paenibacillus* spp.	No definitive guidelines. Vancomycin, ciprofloxacin, imipenem, and aminoglycosides may be effective	*B. cereus* frequently produces beta-lactamase	See CLSI document M45; Methods for antimicrobial dilution and disk susceptibility testing of infrequently isolated or fastidious bacteria.	Whenever isolated from clinical specimens, the potential for the isolate to be a contaminant must be strongly considered

*Validated testing methods include those standard methods recommended by the National Committee for Clinical Laboratory Standards (NCCLS) and those commercial methods approved by the Food and Drug Administration (FDA).

nonmotile *Bacillus* spp. that are not *B. anthracis*. Organisms that test positive, however, should be immediately forwarded to a reference laboratory for gamma bacteriophage testing or animal pathogenicity testing.

B. cereus is another important species to identify. It is penicillin-resistant, beta-hemolytic, and motile, and it produces a wide zone of lecithinase on egg yolk agar. *B. cereus* endophthalmitis is an emergency and should be reported to the physician immediately.

The other organisms discussed in this chapter are common environmental contaminants, so most laboratories only identify isolates from sterile sites (e.g., blood) or those found in large numbers in pure culture.

SERODIAGNOSIS

Indirect hemagglutination and enzyme-linked immunosorbent assays are available to detect antibodies to *B. anthracis*, but serodiagnostic methods are not used to diagnose infections caused by other *Bacillus* spp.

ANTIMICROBIAL SUSCEPTIBILITY TESTING AND THERAPY

Although penicillin has been established as the preferred therapy for anthrax, the infrequent nature with which other species are encountered limits recommendations concerning therapy (Table 18-5). Nonetheless, the threat of bioterrorism has spawned interest in the development of in vitro testing of antimicrobial agents against *B. anthracis*. To that end, the Clinical and Laboratory Standards Institute (CLSI) document M100 addresses the technical issues required for testing this organism. Most other *Bacillus* spp. will grow on the media under the conditions recommended for testing the more commonly encountered bacteria (see Chapter 12 for more information regarding validated testing methods), and technical information regarding the testing of these other species is provided in CLSI document M45 "Methods for Antimicrobial Dilution and Disk Susceptibility Testing of Infrequently Isolated or Fastidious Bacteria." Still, given that these other species are more frequent contaminants than etiologic agents, careful evaluation of their clinical significance must be established before extensive antimicrobial susceptibility testing efforts are undertaken.

PREVENTION

A cell-free inactivated vaccine, (BioThrax, Bioport Corp., Lansing, Mich) given in six doses (0, 2 weeks, 4 weeks, 6 months, 12 months, and 18 months) with annual boosters thereafter is available for immunizing high-risk adults (i.e., public health laboratory workers, workers handling potentially contaminated industrial raw materials, and military personnel) against anthrax. Chemoprophylaxis with ciprofloxacin (or doxycycline) for a minimum of 4 weeks is recommended following aerosol exposure to *B. anthracis* such as may follow a bioterrorist event.

Case Study

A 46-year-old male welder from central Louisiana was healthy until 5 days before admission, when he experienced cough, congestion, chills, and fever. The symptoms had resolved but he experienced a bout of hemoptysis and was referred to the emergency department by his local physician. His temperature was normal, pulse was 128 beats/min, respiratory rate was 26 breaths/min, and blood pressure was 170/98. Chest radiograph was markedly abnormal with a confluent alveolar infiltrate in the right lung, with only a small amount of aeration in the apex. The left lung had a confluent density in the midline suggestive of a mass. Oxygen therapy was started and the patient was placed on ciprofloxacin and cefotaxime. Shortly thereafter, the patient vomited coffee-colored emesis and had a cardiorespiratory arrest. Aerobic blood cultures were positive the next day for a large gram-positive rod with spores.

QUESTIONS

1. The spore-forming bacteria grew well aerobically. What characteristics of the colony would be useful in identifying this bacterium to the species level?
2. What tests can you do to confirm the identification?
3. If you were to isolate a nonhemolytic, nonmotile, aerobic spore-forming bacteria from a clinical specimen, what should you do as quickly as possible?

REFERENCES

1. Ash CF, Priest G, Collins MD: Molecular identification of rRNA group 3 bacilli (Ash, Farrow, Wallbanks, and Collins) using a PCR probe test, *Antonie van Leeuwenhoek* 64:253, 1993.
2. Claus D, Berkeley RC: Genus *Bacillus*. In Sneath PH, Mair NS, Sharpe ME, et al, editors: *Bergey's manual of systematic bacteriology*, vol 2, Baltimore, 1986, Williams and Wilkins.
3. Drobniewski FA: *Bacillus cereus* and related species, *Clin Microbiol Rev* 6:324, 1993.
4. Hollis DG, Weaver RE: *Gram-positive organisms: a guide to identification*, Atlanta, 1981, Centers for Disease Control.
5. Logan NA, Turnbull PC: *Bacillus* and other aerobic endospore-forming bacteria. In Murray PR, Baron EJ, Jorgensen JH, et al, editors: *Manual of clinical microbiology*, ed 8, Washington, DC, 2003, ASM Press.

6. Parry JM, Turnbull PC, Gibson JR: *A colour atlas of* Bacillus *species,* London, 1983, Wolf Medical Publications.

7. Shida O, Takagi H, Kadowaki K, et al: Proposal for two new genera, *Brevibacillus* gen nov and *Aneurinibacillus* gen nov, *Int J Syst Bacteriol* 46:939, 1996.

ADDITIONAL READING

Clinical and Laboratory Standards Institute: Methods for antimicrobial dilution and disk susceptibility testing of infrequently isolated or fastidious bacteria; M45, Wayne, Pa, CLSI.

Lucey D: *Bacillus anthracis* (anthrax). In Mandell GL, Bennett JE, Dolin R, editors: *Principles and practice of infectious diseases,* ed 6, Philadelphia, 2005, Elsevier Churchill Livingstone.

Pekete T: *Bacillus* species and related genera other than *Bacillus anthracis*. In Mandell GL, Bennett JE, Dolin R, editors: *Principles and practice of infectious diseases,* ed 6, Philadelphia, 2005, Elsevier Churchill Livingstone.

CHAPTER 19

Listeria, Corynebacterium, and Similar Organisms

Genera and Species to Be Considered

- *Arthrobacter* spp.
- *Brevibacterium* spp.
- *Cellulomonas* spp.
- *Cellulosimicrobium cellulans*
- *Corynebacterium amycolatum*
- *Corynebacterium auris*
- *Corynebacterium diphtheriae*
- *Corynebacterium jeikeium*
- *Corynebacterium minutissimum*
- *Corynebacterium pseudodiphtheriticum*
- *Corynebacterium pseudotuberculosis*
- *Corynebacterium striatum*
- *Corynebacterium ulcerans*
- *Corynebacterium urealyticum*
- *Corynebacterium xerosis*
- *Dermabacter hominis*
- *Exiguobacterium acetylicum*
- *Kurthia* spp.
- *Leifsonia aquatica* (formerly *C. aquaticum*)
- *Listeria monocytogenes*
- *Microbacterium* spp. (includes former genus *Aureobacterium*)
- *Oerskovia* spp.
- Other *Corynebacterium* spp. and CDC Coryneform groups
- *Rothia* spp.
- *Turicella otitidis*

GENERAL CHARACTERISTICS

The genera described in this chapter are all catalase-positive, gram-positive rods. They are not acid-fast, most do not branch, and they do not form spores. In previous editions of the text, *Rothia* and *Oerskovia* were included in the chapters on gram-positive cocci and *Norcardia*, respectively. They are included with the gram-positive rods in this edition because some species are rod-like. Furthermore, although *Oerskovia* exhibit extensive branching and vegetative hyphae, and they penetrate into the agar surface, they do not have aerial hyphae as *Nocardia* do.

EPIDEMIOLOGY

Most organisms listed in Table 19-1 are part of the normal human flora, are found in the environment, or are associated with various animals. The two most potent pathogens are *Listeria monocytogenes* and *Corynebacterium diphtheriae*.[30,32] However, these two species differ markedly in epidemiology. *L. monocytogenes* is widely distributed in nature and occasionally colonizes the human gastrointestinal tract. *C. diphtheriae* is only carried by humans but, rarely, is isolated from healthy individuals.

In contrast to these two organisms, *C. jeikeium* is commonly encountered in clinical specimens, mostly because it tends to proliferate as skin flora of hospitalized individuals. However, *C. jeikeium* is not considered to be highly virulent. The penetration of the patient's skin by intravascular devices is usually required for this organism to cause infection.

PATHOGENESIS AND SPECTRUM OF DISEASE

L. monocytogenes, by virtue of its ability to survive within phagocytes, and *C. diphtheriae*, by production of an extremely potent cytotoxic exotoxin, are the most virulent species listed in Table 19-2. Most of the other organisms are opportunistic and infection with them requires that patients be compromised.[22] For this reason, whenever *Corynebacterium* spp. or the other genera of gram-positive rods are encountered, careful consideration must be given to their role as infecting or contaminating agents.

LABORATORY DIAGNOSIS

SPECIMEN COLLECTION AND TRANSPORT

No special considerations are required for specimen collection and transport of the organisms discussed in this chapter. Refer to Table 5-1 for general information on specimen collection and transport.

SPECIMEN PROCESSING

No special considerations are required for processing of most of the organisms discussed in this chapter. Refer to Table 5-1 for general information on specimen processing. One exception is the isolation of *Listeria monocytogenes* from placental and other tissue. Because it may be difficult to isolate *Listeria* from these sources, the specimen may be placed in a nutrient broth at 4° C for several weeks to months. The broth is subcultured at

Table 19-1   Epidemiology

Organism	Habitat (reservoir)	Mode of Transmission
Listeria monocytogenes	Colonizes a wide variety of animals, soil, and vegetable matter; widespread in these environments; may colonize human gastrointestinal tract	Usually by ingestion of contaminated food, such as meat and dairy products. Colonized mothers may pass organism onto fetus. Portal of entry is probably from gastrointestinal tract to blood, and, in some instances, from blood to meninges
Corynebacterium diphtheriae	Inhabits human nasopharynx but only in carrier state; not considered part of normal flora. Isolation from healthy humans is not common	Person to person by exposure to contaminated respiratory droplets or direct contact with infected cutaneous lesions; may also be transferred by exposure to contaminated objects
Corynebacterium jeikeium	Skin flora of hospitalized patients, most commonly in the inguinal, axillary, and rectal sites	Uncertain; may be person to person or selection of endogenous resistant strains during antimicrobial therapy; introduced during placement or improper care of intravenous catheters
Corynebacterium ulcerans	Normal flora of humans and cattle	Uncertain; associated with close animal contact, especially during summer
Corynebacterium pseudotuberculosis	Associated with infections in animals such as sheep, goats, and horses	Uncertain; associated with close animal contact, but infections in humans are rare
Corynebacterium pseudodiphtheriticum	Normal human pharyngeal and, occasionally, skin flora	Uncertain; probably by access of patient's endogenous strain to normally sterile site
Corynebacterium minutissimum	Normal human skin flora	Uncertain; probably by access of patient's endogenous strain to normally sterile site
Corynebacterium urealyticum	Normal human skin flora	Uncertain; probably by access of patient's endogenous strain to normally sterile site
Leifsonia aquatica (formerly *Corynebacterium aquaticum*)	Fresh water	Uncertain
Corynebacterium xerosis	Normal flora of human conjunctiva, skin, and nasopharynx	Uncertain; probably by access of patient's endogenous strain to normally sterile site
Corynebacterium striatum	Normal human skin flora	Uncertain; probably by access of patient's endogenous strain to normally sterile site
Corynebacterium amycolatum	Normal flora of human conjunctiva, skin, and nasopharynx	Uncertain; probably by access of patient's endogenous strain to normally sterile site
Corynebacterium auris	Uncertain; probably part of normal human flora	Uncertain; rarely implicated in human infections
Kurthia spp.	Environmental; not part of normal human flora	Uncertain; rarely implicated in human infections
Brevibacterium spp.	Normal human flora and various foods	Uncertain; rarely implicated in human infections
Dermabacter hominis	Normal flora of human skin	Uncertain; rarely implicated in human infections
Turicella otitidis	Uncertain; probably part of normal human flora	Uncertain; rarely implicated in human infections
Arthrobacter spp., *Microbacterium* spp., *Cellulomonas* spp., and *Exiguobacterium* sp.	Uncertain; probably environmental; not part of normal human flora	Uncertain; rarely implicated in human infections

frequent intervals to enhance recovery; this procedure is called **cold enrichment.**

DIRECT DETECTION METHODS

Other than Gram stain of patient specimens, there are no specific procedures for the direct detection of these organisms in clinical material. Most of the genera in this chapter (except *Listeria, Rothia,* and *Oerskovia*) are classified as coryneform bacteria, that is, they are gram-positive, short or slightly curved rods with rounded ends; some have rudimentary branching. Cells are arranged singly, in "palisades" of parallel cells, or in pairs that remain connected after cell division to form V or L shapes. Groups of these morphologies seen together resemble and are often referred to as *Chinese letters* (Figure 19-1). The Gram stain morphologies of clinically important species are described in Table 19-3. *L. monocytogenes* is a short gram-positive rod that may occur singly or in short chains resembling streptococci.

Table 19-2   Pathogenesis and Spectrum of Diseases

Organism	Virulence Factors	Spectrum of Diseases and Infections
Listeria monocytogenes	Listeriolysin, a hemolytic and cytotoxic toxin that may allow for survival within phagocytes	Bacteremia, without any other known site of infection, and meningitis. Infections usually occur in neonates or immunosuppressed patients. Granulomatosis infantiseptica is an in utero infection that is disseminated systemically and causes stillbirth
Corynebacterium diphtheriae	Diphtheria toxin, a potent exotoxin that destroys host cells	Diphtheria, of which there are two forms. Respiratory diphtheria is a pharyngitis characterized by the development of an exudative membrane that covers the tonsils, uvula, palate, and pharyngeal wall; life-threatening if untreated, because respiratory obstruction develops and release of toxin in blood can damage various organs, including the heart. Cutaneous diphtheria is characterized by nonhealing ulcers and membrane formation
Corynebacterium jeikeium	Unknown; probably of low virulence. Multiple antibiotic resistance allows survival in hospital setting	Septicemia, wound infections and, rarely, endocarditis in compromised patients, especially those with intravenous catheters
Corynebacterium ulcerans	Unknown; probably of low virulence	Has been associated with diphtheria-like sore throat
Corynebacterium pseudotuberculosis	Unknown; probably of low virulence	Suppurative granulomatous lymphadenitis
Corynebacterium pseudodiphtheriticum	Unknown; probably of low virulence	Endocarditis, pneumonia, and lung abscesses
Corynebacterium minutissimum	Unknown; probably of low virulence	Superficial, pruritic skin infections known as *erythrasma;* rarely, causes septicemia and abscess formation
Corynebacterium urealyticum	Unknown; probably of low virulence. Multiple antibiotic resistance allows survival in hospital setting	Urinary tract infections and wound infections of compromised patients
Leifsonia aquatica (formerly *Corynebacterium aquaticum*)	Unknown; probably of low virulence	Bacteremia
Corynebacterium xerosis	Unknown; probably of low virulence	Endocarditis or septicemia in immunocompromised patients
Corynebacterium striatum	Unknown; probably of low virulence	Bacteremia, pneumonia, and lung abscesses in compromised patients
Corynebacterium amycolatum	Unknown; probably of low virulence	Endocarditis, septicemia, and pneumonia in immunocompromised patients
Corynebacterium auris	Unknown; probably of low virulence	Uncertain disease association but has been linked to otitis media
Kurthia spp., *Brevibacterium* spp., and *Dermabacter* sp.	Unknown; probably of low virulence	Rarely cause infections in humans; when infections do occur they usually involve bacteremia in compromised patients with indwelling catheters or penetrating injuries
Turicella sp.	Unknown; probably of low virulence	Uncertain disease association but has been linked to otitis media
Arthrobacter spp., *Microbacterium* spp., *Aureobacterium* spp., *Cellulomonas* spp., and *Exiguobacterium* sp.	Uncertain; probably of low virulence	Uncertain disease association

CULTIVATION

Media of Choice

All the genera described in this chapter usually will grow on 5% sheep blood and chocolate agars. Some coryneform bacteria will not grow on chocolate agar, and the **lipophilic** (lipid loving) species (e.g., *C.*

jeikeium, C. urealyticum, C. afermentans subsp. *lipophilum, C. accolens,* and *C. macginleyi*) produce much larger colonies if grown on 5% sheep blood agar supplemented with 1% Tween 80 (Figure 19-2).

Selective and differential media for *C. diphtheriae* should be used if diphtheria is suspected. The two media

Table 19-3   Gram Stain Morphology, Colonial Appearance, and Other Distinguishing Characteristics

Organism	Gram Stain	Appearance on 5% Sheep Blood Agar
Arthrobacter spp.	Typical coryneform gram-positive rods after 24 hrs, with "jointed ends" giving L and V forms, and coccoid cells after 72 hrs (i.e., rod-coccus cycle*)	Large colony; resembles *Brevibacterium* spp.
Brevibacterium spp.	Gram-positive rods; produce typical coryneform arrangements in young cultures (<24 hrs) and coccoid-to-coccobacillary forms that decolorize easily in older cultures (i.e., rod-coccus cycle*)	Medium to large; gray to white, convex, opaque, smooth, shiny; nonhemolytic; cheeselike odor
Cellulomonas spp.	Irregular, short, thin, branching gram-positive rods	Small to medium; two colony types, one starts out white and turns yellow within 3 days and the other starts out yellow
CDC coryneform group F-1	Typical coryneform gram-positive rods	Small, gray to white
CDC coryneform group G†	Typical coryneform gram-positive rods	Small, gray to white; nonhemolytic
Corynebacterium accolens	Resembles *C. jeikeium*	Resembles *C. jeikeium*
C. afermentans subsp. *afermentans*	Typical coryneform gram-positive rods	Medium; white; nonhemolytic; nonadherent
C. afermentans subsp. *lipophilum*	Typical coryneform gram-positive rods	Small; gray, glassy
C. amycolatum	Pleomorphic gram-positive rods with single cells, V forms, or Chinese letters	Small; white to gray, dry
C. argentoratense	Typical coryneform gram-positive rods	Medium; cream-colored; nonhemolytic
C. aurimucosum	Typical coryneform gram-positive rods	Slightly yellowish sticky colonies; some strains black-pigmented
C. auris	Typical coryneform gram-positive rods	Small to medium; dry, slightly adherent, become yellowish with time; nonhemolytic
C. coyleae	Typical coryneform gram-positive rods	Small, whitish and slightly glistening with entire edges; either creamy or sticky
C. diphtheriae group‡	Irregularly staining, pleomorphic gram-positive rods	Various biotypes of *C. diphtheriae* produce colonies ranging from small, gray, and translucent (biotype *intermedius*) to medium, white, and opaque (biotypes *mitis* and *gravis*); *C. diphtheriae* biotype *mitis* may be beta-hemolytic; *C. ulcerans* and *C. pseudotuberculosis* resemble *C. diphtheriae*
C. falsenii	Typical coryneform gram-positive rods	Small; whitish, circular with entire edges, convex, glistening, creamy; yellow pigment after 72 hrs
C. freneyi	Typical coryneform gram-positive rods	Whitish; dry; rough
C. glucuronolyticum	Typical coryneform gram-positive rods	Small; white to yellow, convex; nonhemolytic
C. jeikeium	Pleomorphic; occasionally, club-shaped gram-positive rods arranged in V forms or palisades	Small; gray to white, entire, convex; nonhemolytic
C. imitans	Typical coryneform gram-positive rods	Small, white to gray, glistening, circular, convex; creamy; entire edges
C. macginleyi	Typical coryneform gram-positive rods	Tiny colonies after 48 hrs; nonhemolytic
C. matruchotii	Gram-positive rods with whip-handle shape and branching filaments	Small; opaque, adherent
C. minutissimum	Typical coryneform gram-positive rods with single cells, V forms, palisading and Chinese letters	Small; convex, circular, shiny, and moist
C. mucifaciens	Typical coryneform gram-positive rods	Small, slightly yellow and mucoid; circular, convex, glistening
C. propinquum	Typical coryneform gram-positive rods	Small to medium with matted surface; nonhemolytic
C. pseudodiphtheriticum	Typical coryneform gram-positive rods	Small to medium; slightly dry
C. pseudotuberculosis	Typical coryneform gram-positive rods	Small, yellowish white, opaque, convex; matted surface

*Rod-coccus cycle means rods are apparent in young cultures; cocci are apparent in cultures >3 days old.
†Includes strains G-1 and G-2.
‡Includes *C. diphtheriae, C. ulcerans,* and *C. pseudotuberculosis.*

Continued

Table 19-3   Gram Stain Morphology, Colonial Appearance, and Other Distinguishing Characteristics—cont'd

Organism	Gram Stain	Appearance on 5% Sheep Blood Agar
C. riegelii	Typical coryneform gram-positive rods	Small, whitish, glistening, convex with entire edges; either creamy or sticky
C. simulans	Typical coryneform gram-positive rods	Grayish white; glistening; creamy
C. singulare	Typical coryneform gram-positive rods	Circular; slightly convex with entire margins; creamy
C. striatum	Regular medium to large gram-positive rods; can show banding	Small to medium; white, moist and smooth (resembles colonies of coagulase-negative staphylococci)
C. sundsvallense	Gram-positive rods, some with terminal bulges or knobs; some branching	Buff to slight yellow, sticky, adherent to agar
C. thomssenii	Typical coryneform gram-positive rods	Tiny after 24 hrs; whitish, circular, mucoid and sticky
C. ulcerans	Typical coryneform gram-positive rods	Small, dry, waxy, gray to white
C. urealyticum	Gram-positive coccobacilli arranged in V forms and palisades	Pinpoint (after 48 hrs); white, smooth, convex; nonhemolytic
C. xerosis	Regular medium to large gram-positive rods; can show banding	Small to medium; dry, yellowish, granular
Dermabacter hominis	Coccoid to short gram-positive rods	Small; gray to white, convex; distinctive pungent odor
Exiguobacterium acetylicum	Irregular, short, gram-positive rods arranged singly, in pairs, or short chains; (i.e., rod-coccus cycle*)	Golden yellow
Kurthia spp.	Regular gram-positive rods with parallel sides; coccoid cells in cultures >3 days old	Large, creamy or tan-yellow; nonhemolytic
Leifsonia aquatica	Irregular, slender, short gram-positive rods	Yellow
Listeria monocytogenes	Regular, short, gram-positive rods or coccobacilli occurring in pairs (resembles streptococci)	Small; white, smooth, translucent, moist; beta-hemolytic
Microbacterium spp.	Irregular, short, thin, gram-positive rods	Small to medium; yellow
Oerskovia spp.	Extensive branching; hyphae break up into coccoid to rod-shaped elements	Yellow-pigmented; convex; creamy colony grows into the agar; dense centers
Rothia spp.	Extremely pleomorphic; predominately coccoid and bacillary (in broth, Figure 19-3, A) to branched filaments (solid media, Figure 19-3, B)	Small, smooth to rough colonies; dry; whitish; raised
Turicella otitidis	Irregular, long, gram-positive rods	Small to medium; white to cream, circular, convex

Data compiled from references 2-7, 10-23, 33-36, and 42-44.

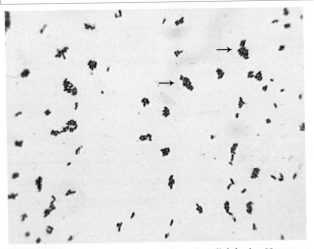

Figure 19-1   Gram stain of *Corynebacterium diphtheriae*. Note palisading and arrangements of cells in formations that resemble Chinese letters *(arrows)*.

commonly used for this purpose are cystine-tellurite blood agar and modified Tinsdale's agar. In addition, Loeffler medium, containing serum and egg, stimulates the growth of *C. diphtheriae* and the production of metachromatic granules within the cells. Primary inoculation of throat swabs to Loeffler serum slants is no longer recommended because of the inevitable over growth of other bacteria.

None of these organisms grows on MacConkey agar. They all should grow in routine blood culture broth and nutrient broths, such as thioglycollate or brain-heart infusion. The lipophilic coryneform bacteria will show better growth if broths are supplemented with rabbit serum.

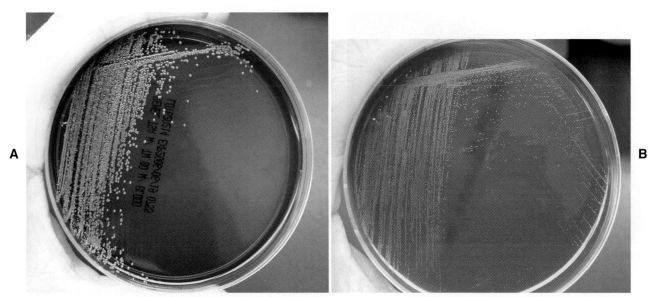

Figure 19-2 *Corynebacterium urealyticum* on blood agar with Tween 80 (**A**) and blood agar (**B**) at 48 hours. This organism is lipophilic and grows much better on the lipid-containing medium.

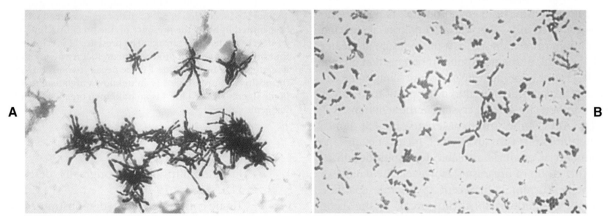

Figure 19-3 **A,** *Rothia dentocariosa* from broth. **B,** *R. dentocariosa* from solid media. (Courtesy Deanna Kiska, SUNY Upstate Medical University, Syracuse, NY.)

Incubation Conditions and Duration

Detectable growth on 5% sheep blood and chocolate agars incubated at 35° C in either ambient air or in 5% to 10% carbon dioxide should occur within 48 hours of inoculation. The lipophilic organisms grow more slowly, requiring 3 or more days to become visible on routine media. For growth of *C. diphtheriae,* cystine-tellurite blood agar and modified Tinsdale's agar should be incubated for at least 48 hours in ambient air.

Colonial Appearance

Table 19-3 describes the colonial appearance and other distinguishing characteristics (e.g., hemolysis and odor) of each clinically important genus or species on blood agar. Colonies of *C. diphtheriae* on cystine-tellurite blood agar appear black or gray, whereas those on

modified Tinsdale's agar are black with dark brown halos (Figure 19-4).

APPROACH TO IDENTIFICATION

Except for *Listeria monocytogenes* and a few *Corynebacterium* spp., the identification of the organisms in this chapter generally is complex and problematic. A multiphasic approach is required for definitive identification. This often requires biochemical testing, whole-cell fatty acid analysis, cell wall diamino acid analysis, or 16S rRNA gene sequencing. The last three methods are usually not available in routine clinical laboratories, so identification of isolates requires expertise available in reference laboratories. Further complicating the situation is the fact that coryneforms are present as normal flora throughout

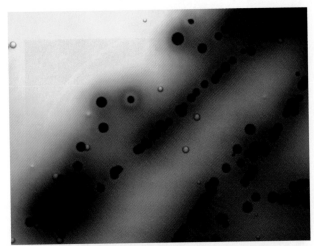

Figure 19-4 Colony of *Corynebacterium diphtheriae* on Tinsdale's agar. Note black colonies with brown halo.

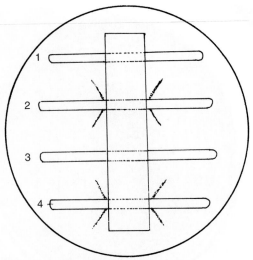

Figure 19-5 Diagram of an Elek plate for demonstration of toxin production by *Corynebacterium diphtheriae*. A filter paper strip impregnated with diphtheria antitoxin is buried just beneath the surface of a special agar plate before the agar hardens. Strains to be tested and known positive and negative toxigenic strains are streaked on the agar's surface in a line across the plate and at a right angle to the antitoxin paper strip. After 24 hours of incubation at 37° C, plates are examined with transmitted light for the presence of fine precipitin lines at a 45-degree angle to the streaks. The presence of precipitin lines indicates that the strain produced toxin that reacted with the homologous antitoxin. Line 1 is the negative control; Line 2 is the positive control. Line 3 is an unknown organism that is a nontoxigenic strain; Line 4 is an unknown organism that is a toxigenic strain.

the body. Thus, only clinically relevant isolates should be identified fully. Indicators of clinical relevance include isolation from normally sterile sites or multiple blood culture bottles, isolation in pure culture or as the predominant organism from symptomatic patients who have not yielded any other known etiologic agent, and isolation from urine if present as a pure culture at greater than 10^4 cfu/mL or the predominant organism at greater than 10^5 cfu/mL. Coryneforms are more likely to be the cause of a urinary tract infection if the pH of the urine is alkaline or there are struvite crystals in the sediment.

The API Coryne strip (bioMérieux, Inc., St. Louis, Mo) and the RapID CB Plus (Remel, Inc., Lenexa, Kan) are commercial products available for rapid identification of this group of organisms; however, the databases may not be current with recent taxonomic changes. Therefore, misidentifications can occur if the code generated using these kits is the exclusive criteria used for identification.

Table 19-4 shows the key tests needed to separate the genera discussed in this chapter. In addition to the features shown, the Gram stain and colonial morphology should be carefully noted.

Comments Regarding Specific Organisms

Two tests (halo on Tinsdale's agar and urea hydrolysis) can be used to separate *C. diphtheriae* from other corynebacteria. Definitive identification of a *C. diphtheriae* isolate as a true pathogen requires demonstration of toxin production by the isolate in question. A patient may be infected with several strains at once, so testing is performed using a pooled inoculum of at least 10 colonies. There are several methods by which toxin testing can be performed:

- Guinea pig lethality test to ascertain whether diphtheria antitoxin neutralizes the lethal effect of a cell-free suspension of the suspect organism

- Immunodiffusion test originally described by Elek (Figure 19-5)
- Tissue culture cell test to demonstrate toxicity of a cell-free suspension of the suspect organism in tissue culture cells and the neutralization of the cytopathic effect by diphtheria antitoxin
- Polymerase chain reaction (PCR) test to detect the toxin gene

Because the incidence of diphtheria in the United States is so low (fewer than 5 cases/year), it is not practical to perform these tests in routine clinical laboratories. Toxin testing is usually performed in reference laboratories.

Identification criteria for *Corynebacterium* spp. (including *C. diphtheriae*) are shown in Tables 19-5 through 19-9. Most clinically relevant strains are catalase-positive, nonmotile, nonpigmented, and esculin- and gelatin-negative. Therefore, isolation of an organism failing to demonstrate any of these characteristics provides a significant clue that another genus shown in Table 19-4 should be considered. In addition, an isolate that is an irregular gram-positive rod that is strictly aerobic and nonlipophilic and oxidizes or does not

Table 19-4 Catalase-Positive, Non–Acid-Fast, Gram-Positive Rods[a]

Organism	Metabolism[b]	Motility	Pigment[c]	Nitrate Reduction	Esculin	Glucose Fermentation	CAMP[d]	Mycolic Acids[e]	Cell Wall Diamino Acids[f]	Other Comments
Corynebacterium	F/O	–	n, w, y, bl	v	–[g]	v	v	+[h]	*meso*-DAP	Gelatin-positive
Arthrobacter	O	v[i]	w, g	v	v	⌐	–	–	L-lys	
Brevibacterium	O	–	w, g, sl y, t	v	–	⌐	–	–	*meso*-DAP	Gelatin- and casein-positive; cheese odor
Microbacterium[k]	F/O[m]	v[n]	y, o, y-o	v	v[o]	v	–[p]	–	L-lys, D-orn	Gelatin and casein variable
Turicella otitidis	O	–	w	–	–	–	+	–	*meso*-DAP	Isolated from ears
Dermabacter hominis	F	–	n, w	–	+	+		–	*meso*-DAP	Pungent odor; decarboxylates lysine and ornithine; gelatin positive
Cellulomonas	F[l]	v	sl y, y	+	+	+	–	–	L-orn	Gelatin positive; casein-negative
Leifsonia aquatica	O	+	y	v	v	–[q]	–	–	DAB	Gelatin- and casein-negative
Rhodococcus equi	O	–	p	v	–	–	+	+	*meso*-DAP	Usually mucoid; can be acid-fast; urease-positive
Cellulosimicrobium cellulans (formerly Oerskovia xanthineolytica)	F	–	y	+	+	+	NT	–	L-lys	Hydrolyzes xanthine; colonies pit agar
Oerskovia turbata	F	v	y	+	+	+	NT	–	L-lys	Does not hydrolyze xanthine
Listeria monocytogenes	F	+[r]	w	–	+	+	+	–	*meso*-DAP	Narrow zone of beta hemolysis on sheep blood agar; hippurate-positive
Kurthia	O	+[r]	n, c	–	–	–	NT	–	L-lys	Large, "Medusa-head" colony with rhizoid growth on yeast nutrient agar; may be H_2S-positive in TSI butt; gelatin-negative
Exiguobacterium acetylicum	F	+	Golden	v	+	+	NT	–	L-lys	Most are oxidase positive; casein- and gelatin-positive
Rothia dentocariosa	F	–	w	+	+	+	–	–	L-lys	If sticky, probable *R. mucilaginosa;* some strains are black-pigmented
Actinomyces neuii	F	–	n	v	–	+	+	–	NT	Nonhemolytic

Table 19-4 Catalase-Positive, Non–Acid-Fast, Gram-Positive Rods[a]—cont'd

Organism	Metabolism[b]	Motility	Pigment[c]	Nitrate Reduction	Esculin	Glucose Fermentation	CAMP[d]	Mycolic Acids[e]	Cell Wall Diamino Acids[f]	Other Comments
Actinomyces viscosus	F	–	n	+	–	+	–	–	NT	
Propionibacterium avidum/granulosum	F	–	w, n	–	v	v	+	–	NT	Beta-hemolytic, branching

Data compiled from references 1-3, 5-14, 20-25, 27, 29, 37, 40, 42, and 43.

NT, Not tested; *TSI*, triple sugar iron agar; *v*, variable reactions; +, ≥90% of species or strains positive; –, ≤90% of species or strains negative.

[a]The aerotolerant catalase-positive *Propionibacterium* spp. and *Actinomyces* spp. are also included in Table 23-4.

[b]F, fermentative; O, oxidative.

[c]c, cream; g, gray; n, nonpigmented; o, orange; sl, slightly; t, tan; w, white; y, yellow; y-o, yellowish-orange; p, pink; bl, black.

[d]CAMP test using a beta-lysin–producing strain of *Staphylococcus aureus*.

[e]Mycolic acids of various lengths are also present in the partially acid-fast *Nocardia*, *Gordona*, *Rhodococcus*, and *Tsukamurella* and the completely acid-fast *Mycobacterium* genera.

[f]DAB, diaminobutyric acid; D-orn, d-ornithine; L-lys, L-lysine; L-orn, L-ornithine; *meso*-DAP, *meso*-diaminopimelic acid.

[g]Of the significant clinical *Corynebacterium* isolates, only *C. matruchotii* and *C. glucuronolyticum* are esculin-positive.

[h]Of the significant *Corynebacterium* isolates, *Corynebacterium amycolatum* does not have mycolic acid as a lipid in the cell wall, as determined by high-performance liquid chromatography (HPLC) profiling methods.

[i]Rod forms of some species are motile.

[j]Glucose may be variably oxidized, but it is not fermented.

[k]*Microbacterium* spp. now includes the former *Aureobacterium* spp.

[l]Some grow poorly anaerobically.

[m]Slow and weak oxidative production of acid from some carbohydrates.

[n]Only the orange-pigmented species *M. imperiale* and *M. arborescens* are motile at 28° C.

[o]Positive reaction may be delayed.

[p]Some strains of *M. arborescens* are CAMP-positive.

[q]Glucose is usually oxidized, but it is not fermented.

[r]Motile at 20° to 25° C.

Table 19-5 Fermentative, Nonlipophilic, Tinsdale-Positive *Corynebacterium* spp.*

Organism	Urease[†]	Nitrate Reduction[†]	Esculin Hydrolysis[†]	Fermentation of Glycogen[†]	Lipophilic
C. diphtheria subsp. *gravis*	–	+	–	+	–
C. diphtheria subsp. *mitis*	–	+	–	–	–
C. diphtheria subsp. *belfanti*	–	–	–	–	–
C. diphtheria subsp. *intermedius*	–	+	–	–	+
C. ulcerans[‡,§]	+	–	–	+	–
C. pseudotuberculosis[‡,§]	+	v	–	–	–

Data compiled from references 5, 10, 24, 26.

+, ≥90% of species or strains positive; –, ≥90% of species or strains are negative; v, variable reactions.

*Separation of lipophilic and nonlipophilic species can be determined by comparing growth on sheep blood agar and sheep blood agar with 1% Tween 80 or growth in brain-heart infusion broth with and without 1 drop of Tween 80 or rabbit serum.

[†]Reactions from API Coryne.

[‡]Propionic acid produced as a product of glucose metabolism.

[§]Reverse CAMP positive.

utilize glucose, will likely be *Leifsonia aquatica, Arthrobacter, Brevibacterium,* or *Microbacterium.*

The enhancement of growth by lipids (e.g., Tween 80 or serum) by certain coryneform bacteria, such as *C. jeikeium* and *C. urealyticum,* is useful for preliminary identification. These two species are also resistant to several antibiotics commonly tested against grampositive bacteria, with vancomycin frequently being the only drug demonstrating inhibition of growth.

Listeria monocytogenes can be presumptively identified by observation of motility by direct wet mount. The organism exhibits characteristic end-over-end tumbling motility when incubated in nutrient broth at room temperature for 1 to 2 hours. Alternatively, characteristic motility can be seen by an umbrella-shaped pattern (Figure 19-6) that develops after overnight incubation at room temperature of a culture stabbed into a tube of semisolid agar. *L. monocytogenes* ferments glucose and is Voges-Proskauer–positive and esculin-positive. The isolation of a small gram-positive, catalase-positive rod with a narrow zone of beta-hemolysis isolated from blood or cerebrospinal fluid (CSF) should be used as strong presumptive evidence for listeriosis.

Listeria monocytogenes DNA in CSF and tissue (fresh or paraffin blocks) can be detected by molecular assays, although these are not available in most clinical laboratories.

SERODIAGNOSIS

Serodiagnostic techniques are not generally used for the laboratory diagnosis of infections caused by the organisms discussed in this chapter. Antilisteriolysin O IgG antibodies can be detected in cases of listeriosis, although IgM antibodies cannot. However, these tests are not commonly used for the clinical diagnosis.

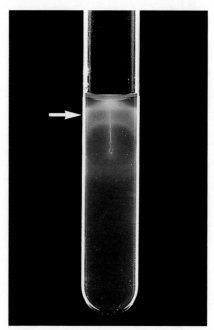

Figure 19-6 Umbrella motility of *Listeria monocytogenes* grown at room temperature.

ANTIMICROBIAL SUSCEPTIBILITY TESTING AND THERAPY

Definitive guidelines have been established regarding antimicrobial therapy for *L. monocytogenes* against certain antimicrobial agents, but because there is no resistance to the therapeutic agents of choice, antimicrobial susceptibility testing is not routinely necessary (Table 19-10).[28]

Although many of the other organisms listed in Table 19-10 will grow on the media and under the conditions recommended for testing the more commonly encountered bacteria (see Chapter 12 for more

Table 19-6   Fermentative, Nonlipophilic, Tinsdale-Negative Clinically Relevant *Corynebacterium* spp.*,†

Organism	Urea‡	Nitrate Reduction‡	Propionic Acid§	Motility	Esculin Hydrolysis‡	Fermentation of: Glucose‡	Maltose‡	Sucrose‡	Xylose‡	CAMP‖
C. amycolatum [a]	v	v	+	−	−	+	v	v	−	−
C. argentoratense	−	−	+	−	−	+	−	−	−	−
C. aurimucosum [b,c]	−	−	−	−	−	+	+	+	−	+
C. coyleae	−	−	ND	−	−	(+)	−	−	−	
C. falsenii [d]	(+)	v	ND	−	v	(+)	v	+	−	ND
C. freneyi [e]	−	v	−	−	−	+	+	+	v	+
C. glucuronolyticum	v	v	+	−	v	+	v	+	−	+
C. imitans	−	v	ND	−	−	+	+	(+)	−	−
C. matruchotii	−	+	+	−	v +	+	+	+	−	−
C. minutissimum [f]	+	−	−	−	−	+	(+)	v	−	−
C. riegelii	−	+	ND	−	−	+	−	+	−	−
C. simulans [g]	+	−	−	−	−	+	+	+	−	−
C. singulare	+	+	−	−	−	+	−	v	−	v
C. striatum	−	+	−	−	−	+	+	v	−	−
C. sundsvallense [b]	+	−	ND	−	−	(+)	+	+	+	−
C. thomssenii [b]	+	−	ND	−	−	(+)	+	+	−	−
C. xerosis	−	v		−	−	+	+	+	−	−

Data compiled from references 4, 5, 6, 10, 12, 16, 18, 19, 24, 34, 38, 41, and 44.

*Consider also *Dermabacter, Cellulomonas, Exiguobacterium,* and *Microbacterium* from Table 19-4 if the isolate is pigmented, motile, or esculin- or gelatin-positive. The aerotolerant catalase-positive *Propionibacterium* spp. and *Actinomyces* spp. included in Table 20-4 should also be considered in the differential with the organisms in this table.

†Separation of lipophilic and nonlipophilic species can be determined by comparing growth on sheep blood agar and sheep blood agar with 1% Tween 80 or growth in brain-heart infusion broth with and without 1 drop of Tween 80 or rabbit serum.

‡Reactions from API Coryne.

§Propionic acid as an end-product of glucose metabolism.

‖CAMP data using a beta-lysin-producing strain of *Staphylococcus aureus.*

ND, No data; *v,* variable reactions; *+,* ≥90% of species or strains are positive, *+*[w], weakly positive; *(+),* delayed positive reaction; *−,* ≥90% of species or strains are negative.

[a] Most frequently encountered species in human clinical material; frequently misidentified as *C. xerosis.*

[b] Sticky colonies.

[c] Yellow or black-pigmented; black-pigmented strains have been previously listed as *C. nigricans;* may be pathogenic from female genital tract.

[d] Yellow after 72 hours.

[e] Grows at 42° C; frequently misidentified as *C. xerosis.*

[f] DNase positive.

[g] Nitrite reduced.

Table 19-7 Strictly Aerobic, Nonlipophilic, Nonfermentative, Clinically Relevant *Corynebacterium* spp.[a,b]

Organism	Oxidation of Glucose	Nitrate Reduction[c]	Urease[c]	Esculin Hydrolysis[c,d]	Gelatin[c,d]	Camp[e]	Other Comments
C. afermentans subsp. *afermentens*	–	–	–	–	–	v	Isolated from blood; nonadherent colony
C. auris[f]	–	–	–	–	–	+	Isolated from ears; dry, usually adherent colony
C. mucifaciens	+	–	–	–	NT	–	Slightly yellow, mucoid colonies
C. pseudodiphtheriticum	–	+	+	–	–	–	
C. propinquum	–	+	–	–	–	–	

Data compiled from references 5, 10, and 24.

NT, Not tested; *v,* variable reactions; +, ≥90% of species or strains are positive; –, ≥90% of species or strains are negative.

[a]*Kurthia* is also a strictly aerobic, nonlipophilic, nonfermentative organism. However, as described in Table 19-3, the colonial and cellular morphology of *Kurthia* should easily distinguish it from the organisms in this table.

[b]Separation of lipophilic and nonlipophilic species can be determined by comparing growth on sheep blood agar and sheep blood agar with 1% Tween 80 or growth in brain-heart infusion broth with and without 1 drop of Tween 80 or rabbit serum.

[c]Reactions from API Coryne.

[d]Consider also *Brevibacterium, Microbacterium, Leifsonia aquatica,* and *Arthrobacter* from Table 19-4 in the differential if the isolate is gelatin- or esculin-positive.

[e]CAMP test using a beta-lysin–producing strain of *Staphylococcus aureus.*

[f]For isolates from the ear, consider also *Turicella otitidis,* which is nitrate- and urease-positive, from Table 19-4 in the differential.

Table 19-8 Strictly Aerobic, Lipophilic, Nonfermentative, Clinically Relevant *Corynebacterium* spp.*

Organism	Nitrate Reduction[†]	Urease[†]	Esculin Hydrolysis[†]	Oxidation of:	
				Glucose	Maltose
C. lipophiloflavum[‡]	–	–	–	–	–
C. jeikeium[§]	–	–	–	+	v
C. afermentens subsp. *lipophilum*	–	–	–	–	–
C. urealyticum[§]	–	+	–	–	–

Data compiled from references 5, 10, 24, and 33.

+, ≥90% of species or strains positive; –, ≥90% of species or strains negative; v, variable reactions.

*Separation of lipophilic and nonlipophilic species can be determined by comparing growth on sheep blood agar and sheep blood agar with 1% Tween 80 or growth in brain-heart infusion broth with and without one drop of Tween 80 or rabbit serum.

[†]Reactions from API Coryne.

[‡]Yellow.

[§]Isolates are usually multiply antimicrobial resistant.

Table 19-9 Lipophilic, Fermentative, Clinically Relevant *Corynebacterium* spp.*

Organism	Urease[†]	Esculin Hydrolysis[†]	Alkaline Phosphatase[†]	Pyrazinamidase[†]
C. kroppenstedtii[‡]	–	+	–	+
C. bovis	–	–	+	–
C. accolens[§]	–	–	–	v
C. macginleyi[§]	–	–	+	–
CDC coryneform group F-1	+	–	–	+
CDC coryneform group G	–	–	+	+

Data compiled from references 5, 10, 24 and 35.

+, ≥90% of species or strains positive; –, ≥90% of species or strains negative; v, variable reactions.

*Separation of lipophilic and nonlipophilic species can be determined by comparing growth on sheep blood agar and sheep blood agar with 1% Tween 80 or growth in brain-heart infusion broth with and without one drop of Tween 80 or rabbit serum.

[†]Reactions from API Coryne.

[‡]Propionic acid produced as a product of glucose metabolism.

[§]Nitrate reduced.

information regarding validated testing methods), the ability to grow and the ability to detect important antimicrobial resistances are not the same thing.[39] As shown in Table 19-10, CLSI document M45 does provide some guidelines for testing *Corynebacterium* spp. Chapter 12 should be reviewed for strategies that can be used to provide susceptibility information and data when warranted.

PREVENTION

The only effective control of diphtheria is through immunization with a multidose diphtheria toxoid prepared by inactivation of the toxin with formaldehyde. Immunization is usually initiated in infancy as part of a triple antigen vaccine (DPT) containing diphtheria toxoid, per-

tussis, and tetanus toxoid. Boosters are recommended every 10 years to maintain active protection and are given as part of a double-antigen vaccine with tetanus toxoid.

A single dose of intramuscular penicillin or a 7- to 10-day course of oral erythromycin is recommended for all individuals exposed to diphtheria, regardless of their immunization status. Follow-up throat cultures from individuals taking prophylaxis should be obtained at least 2 weeks after therapy; if the patient still harbors *C. diphtheriae*, an additional 10-day course of oral erythromycin should be given. Previously immunized contacts should receive a booster dose of diphtheria toxoid; nonimmunized contacts should begin the primary series of immunizations.

Patients who are immunocompromised and pregnant women should avoid eating soft cheeses such as

Table 19-10   Antimicrobial Therapy and Susceptibility Testing

Organism	Therapeutic Options	Resistance to Therapeutic Options	Validated Testing Methods*
Listeria monocytogenes	Ampicillin, or penicillin, with or without an aminoglycoside	Not to commonly used agents	Yes, but testing is rarely, if ever, needed to guide therapy
Corynebacterium diphtheriae	Antitoxin to neutralize diphtheria toxin plus penicillin or erythromycin to eradicate organism	Not to recommended agents	See CLSI document M45-A Methods for antimicrobial dilution and disk susceptibility testing of infrequently isolated or fastidious bacteria
Other *Corynebacterium* spp.	No definitive guidelines. All are susceptible to vancomycin	Resistance to penicillins and cephalosporins does occur	See CLSI document M45-A Methods for antimicrobial dilution and disk susceptibility testing of infrequently isolated or fastidious bacteria
Kurthia spp., *Brevibacterium* spp., *Dermabacter* sp., *Arthrobacter* spp., *Microbacterium* spp., *Cellulomonas* spp., and *Exiguobacterium* sp.	No definitive guidelines	Unknown	Not available

*Validated testing methods include those standard methods recommended by the Clinical and Laboratory Standards Institute (CLSI) and those commercial methods approved by the Food and Drug Administration (FDA).

Mexican-style cheese, feta, brie, Camembert, and blue-veined cheese to prevent foodborne listeriosis. Additionally, leftover or ready-to-eat foods such as hot dogs or cold cuts should be thoroughly heated before consumption.

Case Study

A 27-year-old man received a pancreas and kidney transplant and was admitted 3 months later for possible rejection of the organs. Five days earlier he developed fever, nausea, and dizziness. His creatinine was elevated, and he had white blood cells in his urine, but all other findings were normal. A biopsy did not demonstrate rejection. At this point, the laboratory reported greater than 100,000 gram-positive rods in the urine. The colonies were catalase-positive and beta-hemolytic. The following day, blood cultures were positive with the same organism.

QUESTIONS

1. A simple laboratory test indicated that the isolate was not a *Corynebacterium*. What was that test?
2. Which method is used for routine susceptibility testing for *Listeria*?
3. List the *Corynebacterium* spp. that are considered urinary tract pathogens. What test is helpful to screen for these pathogens?

REFERENCES

1. Barreau C, Bimet F, Kiredjian M, et al: Comparative chemo-taxonomic studies of mycolic acid-free coryneform bacteria of human origin, *J Clin Microbiol* 31:2085, 1993.
2. Bernard K, Bellefeuille M, Hollis DG, et al: Cellular fatty acid composition and phenotypic and cultural characterization of CDC fermentative coryneform groups 3 and 5, *J Clin Microbiol* 32:1217, 1994.
3. Billie J, Rocourt J, Swaminathan B: *Listeria* and *Erysipelothrix*. In Murray PR, Baron EJ, Jorgensen JH, et al, editors: *Manual of clinical microbiology*, ed 8, Washington, DC, 2003, ASM Press.
4. Collins MD, Bernard KA, Hutson RA, et al: *Corynebacterium sundsvallense* sp nov, from human clinical specimens, *Int J Syst Bacteriol* 49:361, 1999.
5. Coyle MB, Lipsky BA: Coryneform bacteria in infectious diseases: clinical and laboratory aspects, *Clin Microbiol Rev* 3:227, 1990.
6. Daneshvar MI, Hollis DG, Weyant RS, et al: Identification of some charcoal-black-pigmented CDC fermentative coryneform group 4 isolates as *Rothia dentocariosa* and some as *Corynebacterium aurimucosum*: proposal of *Rothia dentocariosa* emend. Georg and Brown 1967, *Corynebacterium aurimucosum* emend. Yassin et al. 2002, and *Corynebacterium nigricans* Shukla et al. 2003 pro synon. *Corynebacterium aurimucosum*, *J Clin Microbiol* 42:4189, 2004.
7. de Briel D, Coudere F, Riegel P, et al: High-performance liquid chromatography of corynomycolic acids as a tool in identification of *Corynebacterium* species and related organisms, *J Clin Microbiol* 30:1407, 1992.
8. Evtushenko LI, Dorofeeva LV, Subbotin SA, et al: *Leifsonia poae* gen nov, sp nov, isolated from nematode galls on *Poa annua*, and reclassification of "*Corynebacterium aquaticum*" (Leifson, 1962) as *Leifsonia aquatica* (ex Leifson, 1962) gen nov, nom rev, comb nov and *Clavibacter xyli* (Davis et al, 1984) with two subspecies as *Leifsonia xyli* (Davis et al, 1984) gen nov, comb nov, *Int J Syst Evol Microbiol* 50:371, 2000.
9. Farrow JA, Wallbanks S, Collins MD: Phylogenetic interrelationships of round-spore-forming bacilli containing cell walls based on lysine and the non-spore-forming genera *Caryophanon*, *Exiguobac-*

terium, Kurthia, and *Planococcus, Int J Syst Bacteriol* 44:74 (Erratum, 44:377), 1994.

10. Funke G, Bernard KA: Coryneform gram-positive rods. In Murray PR, Baron EJ, Jorgensen JH, et al, editors: *Manual of clinical microbiology,* ed 8, Washington, DC, 2003, ASM Press.

11. Funke G, Carlotti A: Differentiation of *Brevibacterium* spp encountered in clinical specimens, *J Clin Microbiol* 32:1729, 1994.

12. Funke G, Efstratiou A, Kuklinska D, et al: *Corynebacterium imitans* sp nov isolated from patients with suspected diphtheria, *J Clin Microbiol* 35:1978, 1997.

13. Funke G, Falsen E, Barreau C: Primary identification of *Microbacterium* spp encountered in clinical specimens as CDC coryneform group A-4 and A-5 bacteria, *J Clin Microbiol* 33:188, 1995.

14. Funke G, Hutson RA, Bernard KA, et al: Isolation of *Arthrobacter* spp. from clinical specimens and description of *Arthrobacter cumminsii* sp nov and *Arthrobacter woluwensis* sp nov, *J Clin Microbiol* 34:2356, 1996.

15. Funke G, Lawson PA, Collins MD: *Corynebacterium mucifaciens* sp nov, and unusual species from human clinical material, *Int J Syst Bacteriol* 47:952, 1997.

16. Funke G, Lawson PA, Collins MD: *Corynebacterium riegelii* sp nov: an unusual species isolated from female patients with urinary tract infections, *J Clin Microbiol* 36:624, 1998.

17. Funke G, Lawson PA, Collins MD: Heterogeneity within Centers for Disease Control and Prevention coryneform group ANF-1?like bacteria and description of *Corynebacterium auris* sp nov, *Int J Syst Bacteriol* 45:735, 1995.

18. Funke G, Lawson PA, Bernard KA, et al: Most *Corynebacterium xerosis* strains identified in the routine clinical laboratory correspond to *Corynebacterium amycolatum, J Clin Microbiol* 34:1124, 1996.

19. Funke G, Pascual Ramos C, Collins MD: *Corynebacterium coyleae* sp nov, isolated from human clinical specimens, *Int J Syst Bacteriol* 47:92, 1997.

20. Funke G, Pascual Ramos C, Collins MD: Identification of some clinical strains of CDC coryneform group A-3 and group A-4 bacteria as *Cellulomonas* species and proposal of *Cellulomonas hominis* sp nov for some group A-3 strains, *J Clin Microbiol* 33: 2091, 1995.

21. Funke G, Stubbs S, Altwegg M, et al: *Turicella otitidis* gen nov, sp nov: a coryneform bacterium isolated from patients with otitis media, *Int J Syst Bacteriol* 44:270, 1994.

22. Funke G, Stubbs S, Pfyffer GE, et al: Characteristics of CDC group 3 and group 5 coryneform bacteria isolated from clinical specimens and assignment to the genus *Dermabacter, J Clin Microbiol* 32:1223, 1994.

23. Funke G, von Graevenitz A, Weiss N: Primary identification of *Aureobacterium* spp isolated from clinical specimens as "*Corynebacterium aquaticum,*" *J Clin Microbiol* 32:2686, 1994.

24. Funke G, von Graevenitz A, Clarridge JE, et al: Clinical microbiology of coryneform bacteria, *Clin Microbiol Rev* 10:125, 1997.

25. Gruner E, Steigerwalt AG, Hollis DG, et al: Human infections caused by *Brevibacterium casei,* formerly CDC groups B-1 and B-3, *J Clin Microbiol* 32:1511, 1994.

26. Hollis DG, Weaver RE: *Gram-positive organisms: a guide to identification,* Special Bacteriology Section, Atlanta, 1981, Centers for Disease Control.

27. Jones D, Collins MD: Irregular, nonsporing gram-positive rods. In Sneath PH, Mair NS, Sharpe ME, editors: *Bergey's manual of systematic bacteriology,* vol 2, Baltimore, 1986, Williams and Wilkins.

28. Jones EM, MacGowan AP: Antimicrobial chemotherapy of human infection due to *Listeria monocytogenes, Eur J Clin Microbiol Infect Dis* 14:165, 1995.

29. Kandler D, Weiss N: Regular, nonsporing, gram-positive rods. In Sneath PH, Mair NS, Sharpe ME, editors: *Bergey's manual of systematic bacteriology,* vol 2, Baltimore, 1986, Williams and Wilkins.

30. Lorber B: *Listeriosis, Clin Infect Dis* 24:1, 1997.

31. McNeil MM, Brown JM: The medically important aerobic actinomycetes: epidemiology and microbiology, *Clin Microbiol Rev* 7:357, 1994.

32. Popovic T, Wharton M, Wenger JD, et al: Are we ready for diphtheria? A report from the diphtheria diagnostic workshop, Atlanta, July 11-12, 1994, *J Infect Dis* 171:765, 1995.

33. Riegel P, de Briel D, Prévost G, et al: Genomic diversity among *Corynebacterium jeikeium* strains and comparison with biochemical characteristics, *J Clin Microbiol* 32:1860, 1994.

34. Riegel P, Ruimy R, de Briel D, et al: *Corynebacterium argentoratense* sp nov, from the human throat, *Int J Syst Bacteriol* 45:533, 1995.

35. Riegel P, Ruimy R, de Briel D, et al: Genomic diversity and phylogenetic relationships among lipid-requiring diphtheroids from humans and characterization of *Corynebacterium macginleyi* sp nov, *Int J Syst Bacteriol* 45:128, 1995.

36. Shukla SK, Bernard KA, Harney M, et al: *Corynebacterium nigricans* sp. nov.: Proposed name for a black-pigmented *Corynebacterium* species recovered from the human female urogenital tract, *J Clin Microbiol* 41:4353, 2003.

37. Simonet M, deBriel D, Boucot I, et al: Coryneform bacteria isolated from middle ear fluid, *J Clin Microbiol* 31:1667, 1993.

38. Sjödén B, Funke A, Izquierdo E, et al: Description of some coryneform bacteria isolated from human clinical specimens as *Corynebacterium falsenii* sp nov, *Int J Syst Bacteriol* 48:69, 1998.

39. Soriano F, Zabardiel J, Nieto E: Antimicrobial susceptibilities of *Corynebacterium* species and other nonspore-forming gram-positive bacilli to 18 antimicrobial agents, *Antimicrob Agents Chemother* 39:208, 1995.

40. Takeuchi M, Hatano K: Union of the genera *Microbacterium* Orla-Jensen and *Aureobacterium* (Collins et al) in a redefined genus *Microbacterium, Int J Syst Bacteriol* 48:739, 1998.

41. Wauters G, Driessen A, Ageron E, et al: Propionic acid-producing strains previously designated as *Corynebacterium xerosis, C. minutissimum, C. striatum,* and CDC group I2 and group F2 coryneforms belonging to the species *Corynebacterium amycolatum, Int J Syst Bacteriol* 46:653, 1996.

42. Yokota A et al: Proposal of six new species in the genus *Aureobacterium* and transfer of *Flavobacterium esteraromaticum* Omelianski to the genus *Aureobacterium* as *Aureobacterium esteraromaticum* comb nov, *Int J Syst Bacteriol* 43:555, 1993.

43. Yokota A, Takeuchi M, Weiss N: Proposal of two new species in the genus *Microbacterium: Microbacterium dextranolyticum* sp nov and *Microbacterium aurum* sp nov, *Int J Syst Bacteriol* 43:549, 1993.

44. Zimmermann O, Spröer C, Kroppenstedt RM, et al: *Corynebacterium thomssenii* sp nov: a *Corynebacterium* with N-acetyl-β-glucosaminidase activity from human clinical specimens, *Int J Syst Bacteriol* 48:489, 1998.

ADDITIONAL READING

Lorber B: *Listeria monocytogenes.* In Mandell GL, Bennett JE, Dolin R, editors: *Principles and practice of infectious diseases,* ed 6, Philadelphia, 2005, Elsevier Churchill Livingstone.

MacGregor RR: *Corynebacterium diphtheriae.* In Mandell GL, Bennett JE, Dolin R, editors: *Principles and practice of infectious diseases,* ed 6, Philadelphia, 2005, Elsevier Churchill Livingstone.

Myer DK, Reboli AC: Coryneform bacteria and *Rhodococcus.* In Mandell GL, Bennett JE, Dolin R, editors: *Principles and practice of infectious diseases,* ed 6, Philadelphia, 2005, Elsevier Churchill Livingstone.

Erysipelothrix, Lactobacillus, and Similar Organisms

CHAPTER 20

GENERAL CHARACTERISTICS

The genera described in this chapter are all catalase-negative, non–spore-forming, gram-positive rods; some may exhibit rudimentary branching.

EPIDEMIOLOGY, PATHOGENESIS, AND SPECTRUM OF DISEASE

The organisms listed in Table 20-1 include those that are closely associated with animals and are contracted by humans through animal exposure (e.g., *Erysipelothrix rhusiopathiae* and *Arcanobacterium pyogenes*) and those that are part of the normal human flora (e.g., *Lactobacillus* spp. and *Gardnerella vaginalis*).

These species are not commonly encountered in infections in humans, except possibly for *G. vaginalis,* and when they are encountered the infections are rarely life-threatening (Table 20-2). Often the primary challenge is to determine the clinical relevance of these organisms when they are found in specimens from normally sterile sites.[4,15,16,19–21]

LABORATORY DIAGNOSIS

SPECIMEN COLLECTION AND TRANSPORT

Generally, no special considerations are required for specimen collection and transport of the organisms discussed in this chapter. Of note, skin lesions for *Erysipelothrix* should be collected by biopsy of the full thickness of skin at the leading edge of the discolored area. Refer to Table 5-1 for other general information on specimen collection and transport.

SPECIMEN PROCESSING

No special considerations are required for processing of the organisms discussed in this chapter. Refer to Table 5-1 for general information on specimen processing.

DIRECT DETECTION METHODS

Other than Gram stain of patient specimens, there are no specific procedures for the direct detection of these organisms in clinical material. Gram stain of *Arcanobacterium* spp. shows delicate, curved, gram-positive rods with pointed ends and occasional rudimentary branching. This branching is more pronounced after these organisms have been cultured anaerobically. *Arcanobacterium* spp. stain unevenly after 48 hours of growth on solid media and also exhibit coccal forms.

Lactobacillus is highly pleomorphic, occurring in long chaining rods, as well as in coccobacilli and spiral forms (Figure 20-1).

E. rhusiopathiae stains as both short rods and long filaments. These morphologies correspond to two colonial types, that is, (1) rough colonies that contain slender, filamentous gram-positive rods with a tendency to overdecolorize and become gram-negative and (2) smooth colonies that contain small, slender rods. This variability in staining and colonial morphology may be mistaken for a polymicrobial infection both on direct examination and culture.

Gardnerella are small, pleomorphic gram-variable or gram-negative coccobacilli and short rods. The wet mount and Gram stain of vaginal secretions are key tests for diagnosing bacterial vaginosis caused by *G. vaginalis.* A wet mount prepared in saline reveals the characteristic "clue cells," which are large, squamous epithelial cells with numerous attached small rods. A Gram-stained smear of the discharge shows the attached organisms to be gram-variable coccobacilli. Clue cells typically are present, and large numbers of gram-

Table 20-1 Epidemiology

Species	Habitat (reservoir)	Mode of Transmission
Erysipelothrix rhusiopathiae	Carried by and causes disease in animals. Not part of normal human flora	Abrasion or puncture wound of skin with animal exposure. Most often associated with persons who handle animals or animal products
Arcanobacterium haemolyticum	Normal flora of human skin and pharynx	Uncertain; infections probably caused by person's endogenous strains
Arcanobacterium pyogenes	Carried by and causes disease in animals. Not part of normal human flora	Uncertain; probably by abrasion or undetected wound during exposure to animals
Gardnerella vaginalis	Normal vaginal flora of humans; may also colonize distal urethra of males	Infections probably caused by person's endogenous strain
Lactobacillus spp.	Widely distributed in foods and nature. Normal flora of human mouth, gastrointestinal tract, and female genital tract	Infections are rare; when they do occur they probably are caused by person's endogenous strain

Table 20-2 Pathogenesis and Spectrum of Disease

Organisms	Virulence Factors	Spectrum of Diseases and Infections
Erysipelothrix rhusiopathiae	Uncertain; certain enzymes may promote virulence	Erysipeloid; a localized skin infection that is painful and may spread slowly; may cause diffuse skin infection with systemic symptoms; bacteremia may occur, but endocarditis is rare
Arcanobacterium haemolyticum	Unknown; probably of low virulence	Pharyngitis, cellulitis, and other skin infections
Arcanobacterium pyogenes	Unknown; probably of low virulence to humans	Rarely associated with human infection; when infections do occur, they generally are cutaneous and may be complicated by bacteremia
Gardnerella vaginalis	Uncertain; produces cell adherence factors and cytotoxin	Bacterial vaginosis; less commonly associated with urinary tract infections; bacteremia is extremely rare
Lactobacillus spp.	Uncertain; probably of low virulence	Most frequently encountered as contaminant. May cause bacteremia in immunocompromised patients

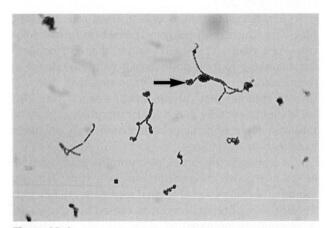

Figure 20-1 Gram stain of *Lactobacillus* spp. Note spiral forms *(arrow)*.

positive rods (i.e., lactobacilli) representing normal vaginal flora are absent or few in number.

Cultivation

Media of Choice. All the genera described in this chapter grow on 5% sheep blood and chocolate agars. They will not grow on MacConkey agar but will grow on Columbia colistin-nalidixic acid (CNA) agar. All genera, except *Gardnerella,* will grow in commercially available blood culture broths. *Gardnerella* is inhibited by sodium polyanetholsulfonate (SPS), which currently is used as an anticoagulant in most commercial blood culture media. An SPS-free medium or a medium with SPS that is supplemented with gelatin should be used when *G. vaginalis* sepsis is suspected.

Isolation of *G. vaginalis* from female genital tract specimens is best accomplished using the selective medium, human blood bilayer Tween agar (HBT).

Table 20-3   Colonial Appearance on 5% Sheep Blood Agar and Other Characteristics

Organism	Appearance
Arcanobacterium spp.	Small colonies with various appearances, including smooth, mucoid, and white and dry, friable, and gray; may be surrounded by narrow zone of beta-hemolysis
Erysipelothrix rhusiopathiae	Two colony types, that is, large and rough or small, smooth, and translucent; shows alpha-hemolysis after prolonged incubation
Gardnerella vaginalis	Pinpoint; nonhemolytic
Lactobacillus spp.	Multiple colonial morphologies, ranging from pinpoint, alpha-hemolytic colonies resembling streptococci to rough, gray colonies
Weissella confusa	Pinpoint; alpha-hemolytic

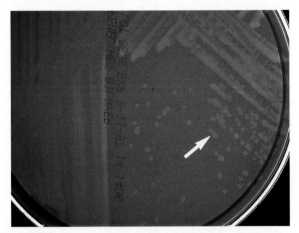

Figure 20-2   *Gardnerella vaginalis* on human blood bilayer Tween (HBT) agar. Note small colonies with diffuse zone of beta-hemolysis *(arrow)*.

Incubation Conditions and Duration

Detectable growth of these organisms should occur on 5% sheep blood and chocolate agars, CNA, and HBT incubated at 35° C in 5% to 10% carbon dioxide (CO_2) within 48 hours of inoculation.

Colonial Appearance

Table 20-3 describes the colonial appearance and other distinguishing characteristics (e.g., hemolysis) of each genus on sheep blood agar. *G. vaginalis* produces small, gray, opaque colonies surrounded by a diffuse zone of beta-hemolysis on HBT agar (Figure 20-2).

APPROACH TO IDENTIFICATION

The identification of the four genera described in this chapter must be considered along with that of *Actinomyces, Bifidobacterium,* and *Propionibacterium,* which are discussed in Chapter 44. Although the latter genera are usually considered with the anaerobic bacteria, they

will grow on routine laboratory media in 5% to 10% CO_2 and some are also catalase negative. Therefore, as shown in Table 20-4, these organisms must be considered together when a laboratory encounters catalase-negative, gram-positive, non–spore-forming rods.

Several commercial systems for fastidious gram-negative bacterial identifications will adequately identify *Gardnerella.* The HNID panel (*Haemophilus-Neisseria* identification panel, Dade MicroScan, Inc., West Sacramento, Calif) works particularly well. However, rapid identification panels usually only are used for isolates from extragenital sources (e.g., blood).

Comments Regarding Specific Organisms

A presumptive identification of *G. vaginalis* is sufficient for genital isolates, based on typical appearance on Gram stain, beta-hemolysis on HBT agar, and negative tests for oxidase and catalase. *Corynebacterium lipophiloflavum,* a bacteria isolated from a woman with bacterial vaginosis, is catalase-positive.[5]

The beta-hemolytic *Arcanobacterium* resemble the beta-hemolytic streptococci but can be differentiated from them by Gram stain morphology. *A. haemolyticum* and *A. pyogenes* can be differentiated based on liquefaction of gelatin; *A. pyogenes* is positive and *A. haemolyticum* is negative. *A. bernardiae* is nonhemolytic.

Erysipelothrix is the only catalase-negative, gram-positive non–spore-forming rod that produces hydrogen sulfide (H_2S) when inoculated into triple sugar iron (TSI) agar (Figure 20-3). Some *Bacillus* spp. also blacken the butt of TSI, but they are catalase-positive and produce spores.

Lactobacillus spp. are usually identified based on colony and Gram stain morphologies and catalase reaction (negative). Differentiation from viridans streptococci may be difficult, but the formation of chains of rods rather than cocci in thioglycollate broth is helpful. Alternatively, a Gram stain of growth just outside the zone of inhibition surrounding a 10-U penicillin disk placed on a blood agar plate inoculated with a lawn of

Table 20-4 Biochemical and Physiologic Characteristics of Catalase-Negative, Gram-Positive, Aerotolerant, Non–Spore-Forming Rods

	Urease	Nitrate Reduction	β-Hemolysis[b]	Glucose	Maltose	Mannitol	Sucrose	Xylose	CAMP[c]	GLC[d]	Other Comments:
				Fermentation[a] of:							
Actinomyces israelii	−	v	−	+	+	v	+	+	−	A, L, S	
A. odontolyticus	−	+	−[e]	+	v	−	+	v	−	A, S	Red pigment produced after 1 week on SBA
A. naeslundii	+	+	−	+	+	v	+	v	−	A, L, S	
A. radingae	−	−	−[w]	+	+	−	+	+	−	S?	Pyrazinamidase, beta-galactosidase–positive and esculin-positive
A. turicensis	−	−	−[w]	+	v	−	+	+	−	NT	Pyrazinamidase, beta-galactosidase–negative and esculin-negative
A. graevenitzii	−	−	−	+	+	−	+	−	ND	L > S	Beta-galactosidase–negative
Actinobaculum schaalii	−	−	−	+	+	−	v	+	+[w]	A, s	
Arcanobacterium haemolyticum	−	−	+	+	+	−	v	−	Reverse +[f]	A, L, S	Gelatin-negative at 48 hr; beta-hemolysis is stronger on agar containing human or rabbit blood
A. pyogenes	−	−	+[g]	+	v	v	v	+	−	A, L, S	Gelatin-positive at 48 hr; casein-positive
A. bernardiae	−	−	−	+	+	−	−	−	−	A, L, S	
Bifidobacterium adolescentis	−	−	−	+	+	−	+	+	ND	A > L (s)	
Erysipelothrix sp.	−	−	−	+[h]	−	−	−	−		A, L, S	H_2S-positive in TSI butt; vancomycin-resistant; alpha-hemolytic
Lactobacillus spp.	−	−	−	+	+	v	+	ND	ND	L (a s)	Some strains vancomycin-resistant; alpha-hemolytic
Propionibacterium acnes	−	+	−	+	−	−	−	−	+	A, P (iv L s)	Indole-positive; may show beta-hemolysis on rabbit blood agar
P. propionicum[i]	−	+	−	+	+	+	+	−	ND	A, P, S, (L)	Colony may show red fluorescence under long-wavelength UV light
Gardnerella vaginalis	−	−	−	+	+	−	v	−[k]	ND	A (l s)	Beta-hemolysis on HBT; usually hydrolyses hippurate

Table 20-4　Biochemical and Physiologic Characteristics of Catalase-Negative, Gram-Positive, Aerotolerant, Non–Spore-Forming Rods—cont'd

	Urease	Nitrate Reduction	β-Hemolysis[b]	Glucose	Maltose	Mannitol	Sucrose	Xylose	CAMP[c]	GLC[d]	Other Comments:
							Fermentation[a] of:				
Weissella spp.	NT	–	–	+	+	–	+	v	NT	L (as)	Vancomycin-resistant small, short rods; produces gas from MRS broth; alpha-hemolytic; esculin–positive; arginine–positive

Data compiled from references 3, 5, 8-14, 17, 18, 21-23.

HBT, Human blood bilayer Tween agar; *iv,* isovaleric acid; *ND,* not done, *NT,* not tested; *SBA,* 5% sheep blood agar; *TSI,* triple sugar iron agar; *v,* variable; *w,* weak; +, ≥90% of strains positive; –, ≥90% of strains negative; >, greater than.

[a]Fermentation is detected in peptone base with Andrade's indicator.

[b]On sheep blood agar.

[c]CAMP test using a beta-lysin–producing strain of *Staphylococcus aureus.*

[d]End products of glucose metabolism: *A,* Acetic acid; *L,* lactic acid; *P,* propionic acid; *S,* succinic acid; *(),* may or may not produce acid end product.

[e]May show beta-hemolysis on brain-heart infusion agar with sheep or human blood.

[f]Reverse CAMP test; *S. aureus* beta-lysins are inhibited by a diffusible substance produced by *A. haemolyticum* (Figure 20-4).

[g]May also show beta-hemolysis on brain-heart infusion agar with human blood.

[h]Reaction may be weak or delayed.

[i]Some strains are catalase-negative.

[j]Formerly *Arachnia propionica.*

[k]*Gardnerella vaginalis*–like organisms ferment xylose.

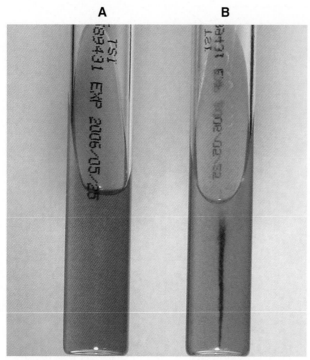

A B

Figure 20-3 H₂S production by *Erysipelothrix rhusiopathiae* in TSI **(B)**. A negative TSI **(A)** is included for comparison.

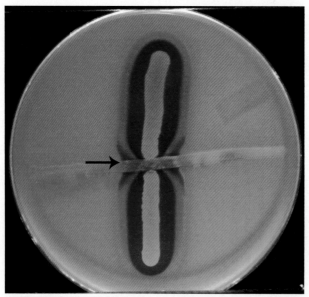

Figure 20-4 Reverse CAMP test. *Arcancobacterium haemolyticum* is struck on a blood agar plate. *Staphylococcus aureus* is then struck perpendicular to the *Arcanobacterium*. A positive reverse CAMP test is indicated by an arrow pointed to the *Staphylococcus.*

the organism should show long bacilli rather than coccoid forms if the organism is *Lactobacillus* spp.

SERODIAGNOSIS

Serodiagnostic techniques are not generally used for the laboratory diagnosis of infections caused by the organisms discussed in this chapter.

ANTIMICROBIAL SUSCEPTIBILITY TESTING AND THERAPY

The rarity with which most of these organisms are encountered as the cause of infection has made the development of validated in vitro susceptibility testing methods difficult (Table 20-5). However, most of the organisms are susceptible to the agents used to eradicate them so that in vitro testing is not usually necessary for guiding therapy.[2,13] *Lactobacillus* spp. can be resistant to various antimicrobial agents. Fortunately, these organisms are rarely implicated in infections. When they are encountered in specimens from normally sterile sites, careful evaluation of their clinical significance is warranted before any attempt is made at performing a nonstandardized susceptibility test.

Although some of these organisms may grow on the media and under the conditions recommended for testing other bacteria (see Chapter 12 for more information regarding validated testing methods), this does not necessarily mean that interpretable and reliable results will be produced. Chapter 12 should be reviewed for preferable strategies that can be used to provide susceptibility information when validated testing methods do not exist for a clinically important bacterial isolate.

PREVENTION

Because these organisms are ubiquitous in nature and many are part of normal human flora that are commonly encountered without deleterious effects on healthy human hosts, there are no recommended vaccination or prophylaxis protocols.

Table 20-5 Antimicrobial Therapy and Susceptibility Testing

Organism	Therapeutic Options	Resistance to Therapeutic Options	Validated Testing Methods*	Comments
Erysipelothrix rhusiopathiae	Susceptible to penicillins, cephalosporins, erythromycin, clindamycin, tetracycline, and ciprofloxacin	Not common	See CLSI document M45: methods for antimicrobial dilution and disk susceptibility testing of infrequently isolated or fastidious bacteria	Susceptibility testing not needed to guide therapy
Arcanobacterium haemolyticum	No definitive guidelines. Usually susceptible to penicillin, erythromycin, and clindamycin	Not known	Not available	Susceptibility testing not needed to guide therapy
Arcanobacterium pyogenes	No definitive guidelines. Usually susceptible to cephalosporins, penicillins, ciprofloxacin, and chloramphenicol	Not known	Not available	Susceptibility testing not needed to guide therapy
Gardnerella vaginalis	Metronidazole is the drug of choice; also susceptible to ampicillin	Not known	Not available	Susceptibility testing not needed to guide therapy
Lactobacillus spp.	No definitive guidelines. Systemic infections may require use of a penicillin with an aminoglycoside	Frequently resistant to cephalosporins; not killed by penicillin alone; frequently highly resistant to vancomycin	See CLSI document M45: methods for antimicrobial dilution and disk susceptibility testing of infrequently isolated or fastidious bacteria	Confirm that isolate is clinically relevant and not a contaminant

*Validated testing methods include those standard methods recommended by the Clinical and Laboratory Standards Institute (CLSI) and those commercial methods approved by the Food and Drug Administration (FDA).

Case Study

Police found an elderly alcoholic man unconscious near his fishing gear. He was taken to the hospital and blood cultures were collected; however, despite the efforts of the emergency department, they were not able to revive him. At autopsy, several vegetations were seen on both his aortic and mitral valves, and both valves were perforated. Blood cultures were reported positive with a gram-positive rod.

QUESTIONS

1. What test should the laboratory perform to confirm the genus of this gram-positive, catalase-negative rod, growing both aerobically and anaerobically?
2. Although susceptibility testing for this organism is not generally performed, what important information regarding the susceptibility of this organism is important to communicate to those caring for the patient?
3. What is the likely source of this patient's *Erysipelothrix*?

REFERENCES

1. Billie J, Rocourt J, Swaminathan B: *Listeria* and *Erysipelothrix*. In Murray PR, Baron EJ, Jorgensen JH, et al, editors: *Manual of clinical microbiology*, ed 8, Washington, DC, 2003, ASM Press.
2. Carlson P, Kontiainen S, Renkonen O: Antimicrobial susceptibility of *Arcanobacterium haemolyticum, Antimicrob Agents Chemother* 38: 142, 1994.
3. Coyle MB, Lipsky BA: Coryneform bacteria in infectious disease: clinical and laboratory aspects, *Clin Microbiol Rev* 3:227, 1990.
4. Drancourt M, Oules O, Bouche V, et al: Two cases of *Actinomyces pyogenes* infections in humans, *Eur J Clin Microbiol Infect Dis* 12:55, 1993.
5. Funke G, Bernard KA: Coryneform gram-positive rods. In Murray PR, Baron EJ, Jorgensen JH, et al, editors: *Manual of clinical microbiology*, ed 8, Washington, DC, 2003, ASM Press.
6. Funke G, Hutson RA, Hilleringmann M, et al: *Corynebacterium lipophiloflavum* sp nov isolated from a patient with bacterial vaginosis, *FEMS Microbiol Lett* 150:219, 1997.
7. Funke G, Martinett Lucchini G, Pfyffer GE, et al: Characteristics of CDC group 1 and group 1-like coryneform bacteria isolated from clinical specimens, *J Clin Microbiol* 31:2907, 1993.
8. Funke G, Pascual Ramos C, Fernandez-Garayzabal J, et al: Description of human-derived Centers for Disease Control coryneform group 2 bacteria as *Actinomyces bernardiae* sp nov, *Int J Syst Bacteriol* 45:57, 1995.
9. Funke G, von Graevenitz A, Clarridge JE, et al: Clinical microbiology of coryneform bacteria, *Clin Microbiol Rev* 10:125, 1997.
10. Hollis DG, Weaver RE: *Gram-positive organisms: a guide to identification*, Special Bacteriology Section, Atlanta, 1981, Centers for Disease Control.
11. Jones D, Collins MD: Irregular, non-spore-forming, gram-positive rods. In Sheath PH, Mair NS, Sharpe E, editors: *Bergey's manual of systematic bacteriology*, vol 2, Baltimore, 1986, Williams & Wilkins.
12. Kandler O, Weiss N: Regular, nonsporing gram-positive rods. In Sneath PH, Mair NS, Sharpe E, editors: *Bergey's manual of systematic bacteriology*, vol 2, Baltimore, 1986, Williams & Wilkins.
13. Kharsany AB, Hoosen AA, Van den Ende J: Antimicrobial susceptibilities of *Gardnerella vaginalis, Antimicrob Agents Chemother* 37: 2733, 1993.
14. Lawson A, Falsen E, Åkervall, et al: Characterization of some *Actinomyces*-like isolates from human clinical specimens: reclassification of *Actinomyces suis* (Soltys and Spratling) as *Actinobaculum suis* comb nov and description of *Actinobaculum schaalii* sp nov, *Int J Syst Bacteriol* 47:899, 1997.
15. Lidbeck A, Nord CE: Lactobacilli and the normal human anaerobic microflora, *Clin Infect Dis* 16(suppl 4):S181, 1993.
16. Mackenzie A, Fuite LA, Chan TH, et al: Incidence and pathogenicity of *Arcanobacterium haemolyticum* during a 2-year study in Ottawa, *Clin Infect Dis* 21:177, 1995.
17. Pascual Ramos C, Falsen E, Alvarez N, et al: *Actinomyces graevenitzii* sp nov, isolated from human clinical specimens, *Int J Syst Bacteriol* 47:885, 1997.
18. Pascual Ramos C, Foster G, Collins MD: Phylogenetic analysis of the genus *Actinomyces* based on 16S rRNA gene sequences: description of *Arcanobacterium phocae* sp nov, *Arcanobacterium bernardiae* comb nov, and *Arcanobacterium pyogenes* comb nov, *Int J Syst Bacteriol* 47:46, 1997.
19. Patel R, Cockerill FR, Porayko MK, et al: Lactobacillemia in liver transplant patients, *Clin Infect Dis* 18:207, 1994.
20. Schuster MG, Brennan PJ, Edelstein P: Persistent bacteremia with *Erysipelothrix rhusiopathiae* in a hospitalized patient, *Clin Infect Dis* 17:783, 1993.
21. Spiegel CA: Bacterial vaginosis, *Clin Microbiol Rev* 4:485, 1991.
22. Vandamme P, Falsen E, Vancanneyt M, et al: Characterization of *Actinomyces turicensis* and *Actinomyces radingae* strains from human clinical samples, *Int J Syst Bacteriol* 48:503, 1998.
23. van Esbroeck M, Vandamme P, Falsen E, et al: Polyphasic approach to the classification and identification of *Gardnerella vaginalis* and unidentified *Gardnerella vaginalis*–like coryneforms present in bacterial vaginosis, *Int J Syst Bacteriol* 46:675, 1996.

ADDITIONAL READING

Reboli AC, Farrar WE: *Erysipelothrix rhusiopathiae*. In Mandell GL, Bennett JE, Dolin R, editors: *Principles and practice of infectious diseases*, ed 6, Philadelphia, 2005, Elsevier Churchill Livingstone.

Steinberg JP, Del Rio C: Other gram-negative and gram-variable bacilli. In Mandell GL, Bennett JE, Dolin R, editors: *Principles and practice of infectious diseases*, ed 6, Philadelphia, 2005, Elsevier Churchill Livingstone.

Nocardia, Streptomyces, Rhodococcus, and Similar Organisms

CHAPTER 21

Genera and Species to Be Considered

- *Nocardia asteroides*
- *Nocardia nova*
- *Nocardia farcinica*
- *Nocardia brasiliensis*
- *Nocardia otitidiscaviarum*
- *Nocardia pseudobraziliensis*
- *Rhodococcus* spp.
- *Gordonia* spp.
- *Tsukamurella* spp.
- *Streptomyces somaliensis*
- *Streptomyces paraguayensis*
- *Streptomyces anulatus*
- *Actinomadura madurae*
- *Actinomadura pelletieri*
- *Dermatophilus congolensis*
- *Nocardiopsis dassonvillei*
- Thermophilic actinomycetes

The actinomycetes are a large and diverse group of gram-positive bacilli. For the most part, cells of all actinomycetes elongate to form branching, filamentous forms. The rate and extent of filament (hyphae) elongation with lateral branching is dependent on the strain of actinomycetes, the growth medium, and the temperature of incubation.[3] Some organisms form filaments, or hyphae, on the agar surface or into the agar, whereas others extend into the air.

These organisms are aerobic, facultatively anaerobic, or obligately anaerobic; only the aerobic actinomycetes are discussed in this chapter. In the previous edition of this text, *Oerskovia* and *Rothia* were also discussed; however, because of additional phylogenetic information, these genera are now reviewed in Chapter 19. Aerobic actinomycetes belong to the order Actinomycetales. There are more than 40 genera but only the clinically relevant aerobic actinomycetes genera are considered here (Table 21-1). In this chapter, only those aerobic actinomycetes that can exhibit branching and/or

partial acid-fastness are addressed. Although both belong to the order Actinomycetales, *Corynebacterium* spp. do not usually exhibit branching filaments or partial acid-fastness and *Mycobacterium* spp. do not exhibit branching and are strongly (acid-alcohol) acid-fast; for these reasons, the *Corynebacteriaceae* and *Mycobacteriaceae* are addressed in Chapters 19 and 45, respectively. Finally, another clinically significant aerobic actinomycete is *Tropheryma whipplei*; because this organism has not been cultured on artificial media, it is reviewed in Chapter 46. For purposes of discussion, the remaining genera of aerobic actinomycetes are divided into the following two large groups: those whose cell walls contain mycolic acid and are therefore partially acid-fast and those whose cell envelopes do not contain mycolic acid and therefore are non–acid-fast.

In general, the aerobic actinomycetes are not frequently isolated in the clinical laboratory; nevertheless, these organisms are causes of serious human disease. Not only are infections caused by these organisms difficult to recognize on clinical grounds, but they are also difficult to isolate. Further complicating matters from the laboratory's perspective is the difficulty in classifying, identifying, and performing antibiotic susceptibilities on aerobic actinomycetes isolated from clinical specimens. At the time of this writing, the taxonomy of the aerobic actinomycetes is complex and continues to evolve. New and reliable methods that can identify cell wall amino acids and sugars and characterize mycolic acid, menaquinones, and phospholipids in conjunction with nucleic acid phylogenetic studies are proving extremely useful for resolving the taxonomy of the actinomycetes.

GENERAL CHARACTERISTICS

The genera *Nocardia, Rhodococcus, Gordonia,* and *Tsukamurella* are partially acid-fast aerobic actinomycetes. *Nocardia* and *Rhodococcus* belong to the family *Nocardiacea* while *Gordonia* and *Tsukamurella* are in the *Gordoniaceae* and *Tsukamurellaceae* families, respectively. However, the

Table 21-1   Clinically Relevant Aerobic Actinomycetes*

Cell Wall Containing Mycolic Acid	Genus
Present	Nocardia Rhodococcus Gordonia Tsukamurella Corynebacterium
Absent	Streptomyces Actinomadura Dermatophilus Nocardiopsis

*The genera *Williamsia*, *Skermania*, and *Dietzia* are also aerobic actinomycetes but to date are not clinically relevant.

BOX 21-1   *Nocardia* spp. That Are Considered Human Pathogens or Have Been Implicated in Human Disease

- *Nocardia asteroides sensu stricto* type VI
- *N. nova*
- *N. farcinica*
- *N. brasiliensis*
- *N. otitidiscaviarum*
- *N. pseudobrasiliensis*

LESS COMMON OR PREVALENCE NOT ESTABLISHED

- *N. transvalensis*
- *N. brevicatena*
- *N. carnea*
- *N. abscessus*
- *N. africana*
- *N. paucivorans*
- *N. veterana*

---

presence of this property is variable, being dependent on the particular strain and culture conditions.

PARTIALLY ACID-FAST AEROBIC ACTINOMYCETES

Nocardia

Organisms belonging to the genus *Nocardia* are gram-positive (often with a beaded appearance), variably acid-fast, catalase-positive, and strictly aerobic. As they grow, *Nocardia* spp. form branched filaments that extend along the agar surface, that is, **substrate hyphae,** and into the air, that is, **aerial hyphae.** As the organisms age, nocardiae fragment into pleomorphic rods or coccoid elements.[15] Nocardiae also are characterized by the presence of mesodiaminopimelic acid (DAP) and the sugars arabinose and galactose in their cell wall peptidoglycan.

Currently, the taxonomy within the genus *Nocardia* is changing rapidly and the recognition and description of new species continues and remains controversial as to the number of validly described species; recent publications cite 22 to 30 valid species.[3, 5, 18] Of significance, Cloud and colleagues[6] reported that the most commonly identified species was *Nocardia cyriacigeorgica* and none were *N. asteroides* using partial 16S rRNA DNA sequencing, followed by *N. farcinica, N. nova, N. africana,* and *N. veterana.* The species that are considered human pathogens or have been implicated as human pathogens are listed in Box 21-1. *N. asteroides, N. nova, N. farcinica, N. brasiliensis, N. otitidiscaviarum* (formerly *N. caviae*), *N. pseudobrasiliensis,* and *N. transvalensis* account for most of the diseases in humans that are caused by *Nocardia* spp.

Rhodococcus, Gordonia, Tsukamurella

Organisms belonging to *Rhodococcus, Gordonia,* and *Tsukamurella* are similar to *Nocardia* spp. in that they are gram-positive, aerobic, catalase-positive, partially acid-

Table 21-2   Species Included in the Genera *Rhodococcus*, *Gordonia*, and *Tsukamurella*

Genus	Species
Rhodococcus	equi, erythropolis, rhodnii, rhodochrous, (other species of unknown significance include chlorophenicolus, coprophilus, fascians, globerulus, marinonascens, and ruber)
Gordonia	aichiensis, bronchialis, polyisoprenivorans, rubripertincta, sputi, terrae (remaining species isolated from environmental sources)
Tsukamurella	paurometabola, pulmonis, ichonensis, tyrosinosolvens, strandjordae (T. wratislaviensis isolated from nature)

Data compiled from Brown JM, et al: In Murray PR, Baron EJ, Pfaller MA, et al, editors: *Manual of clinical microbiology*, ed 8, Washington, DC, 2003, American Society for Microbiology; Goodfellow M, Chun J, Stubbs S, et al: *Lett Appl Microbiol* 19:401, 1994; Klatte S, Rainey FA, and Kroppenstedt RM: *Int J Syst Bacteriol* 44:769, 1994; Lasker BA, Brown JM, and McNeil MM: *Clin Infect Dis* 15:233, 1992; Maertens J, et al: *Clin Microbiol Infect* 4:51, 1998; Riegel P, et al: *J Clin Microbiol* 34:2045, 1996; Yassin AF, Rainey FA, Burrghardt J, et al: *Int J Syst Bacteriol* 47:607, 1997; Arenskötter M, et al: *Appl Environ Microbiol* 70:3195, 2004

fast, branching, filamentous bacteria that can fragment into rods and cocci. The extent of acid-fastness is dependent on the amount and complexity of mycolic acids in the organism's cell envelope and on culture conditions. The differentiation of these three genera, as well as species identification, is difficult.[3,9,12] In particular, the genus *Rhodococcus* consists of a very diverse group of organisms in terms of morphology, biochemical characteristics, and ability to cause disease.[13] As previously mentioned, the taxonomy of these organisms continues to evolve; species included in these three genera, as of this writing, are summarized in Table 21-2.

Table 21-3   Non–Acid-Fast Aerobic Actinomycetes Associated with Human Disease

Genus	Number of Species	Species Associated with Human Disease
Streptomyces	>3000	S. somaliensis S. paraguayensis S. anulatus
Actinomadura	27	A. madurae A. pelletieri A. latina
Dermatophilus	2	D. congolensis
Nocardiopsis	8	N. dassonvillei N. synnemataformans

NON–ACID-FAST AEROBIC ACTINOMYCETES: *STREPTOMYCES, ACTINOMADURA, DERMATOPHILUS, NOCARDIOPSIS,* AND THE THERMOPHILIC ACTINOMYCETES

These organisms are gram-positive, branching filaments that do not have mycolic acids present in their cell envelopes and are therefore non–acid-fast. This group of actinomycetes is heterogeneous and these organisms are encountered infrequently in the clinical laboratory. Only those non–acid-fast actinomycetes associated with human disease are addressed (Table 21-3). Another group of non–acid-fast actinomycetes, the thermophilic actinomycetes, are associated with infections in humans and include the genera *Thermoactinomyces, Saccharomonospora,* and *Saccharopolyspora,* which are medically important.

EPIDEMIOLOGY AND PATHOGENESIS

PARTIALLY ACID-FAST AEROBIC ACTINOMYCETES

Nocardia

Nocardia are normal inhabitants of soil and water, and are primarily responsible for the decomposition of plant material. Infections caused by *Nocardia* spp. are found worldwide. Because they are ubiquitous, the isolation of these organisms from clinical specimens does not always indicate infection. Rather, isolation possibly may indicate colonization of the skin and upper respiratory tract or laboratory contamination; laboratory contamination is rarely seen.[18] *Nocardia* infections can be acquired by either traumatic inoculation or inhalation. *N. asteroides sensu stricto* type VI is evenly distributed throughout the United States as is *N. farcinica.*[18] The prevalence of other species varies regionally, while *N.*

brasiliensis is associated with tropical climates and has a higher prevalence in the southwestern and southeastern United States.

Nocardia, particularly *N. asteroides,* are facultative intracellular pathogens that can grow in various human cells. The mechanisms of pathogenesis are complex and not completely understood. However, the virulence of *N. asteroides* appears to be associated with several factors, such as its stage of growth, resistance to intracellular killing, tropism for neuronal tissue, and ability to inhibit phagosome-lysosome fusion; other characteristics, such as production of large amounts of catalase and hemolysins, may also be associated with virulence.[1]

Rhodococcus, Gordonia, Tsukamurella

These organisms can be isolated from several environmental sources, especially soil and farm animals, as well as fresh and salt waters. It is believed that organisms are acquired primarily by inhalation.[3] For the most part, these aerobic actinomycetes are infrequently isolated from clinical specimens.

To date, *Rhodococcus equi* has been the organism most commonly associated with human disease, particularly in immunocompromised patients, such as those infected with human immunodeficiency virus (HIV). *R. equi* is a facultative intracellular organism that can persist and replicate within macrophages.[21] Determinants of virulence of *R. equi* are under investigation and may involve cell wall mycolic acids that play a role in intracellular survival, production of interleukin-4, and granuloma formation.[21] Although *Gordonia* spp. and *Tsukamurella* are able to cause opportunistic infections in humans, little is known regarding their pathogenic mechanisms.

NON–ACID-FAST AEROBIC ACTINOMYCETES: *STREPTOMYCES, ACTINOMADURA, DERMATOPHILUS, NOCARDIOPSIS,* AND THE THERMOPHILIC ACTINOMYCETES

Aspects regarding the epidemiology of the non–acid-fast aerobic actinomycetes are summarized in Table 21-4. Little is known about how these agents cause infection.

SPECTRUM OF DISEASE

PARTIALLY ACID-FAST AEROBIC ACTINOMYCETES

The partially acid-fast actinomycetes cause various infections in humans.

Nocardia

Infections caused by *Nocardia* spp. can occur in immunocompetent and immunocompromised individuals. *N.*

Table 21-4   Epidemiology of the Non–Acid-Fast Aerobic Actinomycetes

Organism	Habitat (reservoir)	Distribution	Routes of Primary Transmission
Streptomyces somaliensis	Sandy soil	Africa, Arabia, Mexico, South America	Penetrating wound/abrasions in the skin
S. anulatus	Soil	Most common isolate in United States	Penetrating wound/abrasions in the skin
Actinomadura madurae	Soil	Tropical and subtropical countries	Penetrating wound/abrasions in the skin
A. pelletieri, A. latina	Unknown, possibly soil	Tropical and subtropical countries	Penetrating wound/abrasions in the skin
Dermatophilus congolensis	Unknown; skin commensal or saprophyte in soil (?)[22]	Worldwide, but more prevalent in humid, tropical, and subtropical regions	Trauma to the epidermis caused by insect bites and thorns; contact with tissues of infected animals through abrasions in the skin
*Nocardiopsis dassonvillei**	Unknown	Unknown	Unknown
Thermophilic actinomycetes	Ubiquitous; water, air, soil, compost piles, dust, hay	Worldwide	Inhalation

*Only a few cases of infection in the world literature.

asteroides, N. brasiliensis, and *N. otitidiscaviarum* are the major causes of these infections, with *N. asteroides* causing greater than 80% of infections.

Nocardia spp. cause three types of skin infections in immunocompetent individuals:[14,15]

- Mycetoma, a chronic, localized, painless, subcutaneous infection
- Lymphocutaneous infections
- Skin abscesses or cellulites

Of note, *N. brasiliensis* is the predominant cause of these skin infections.

In individuals who are immunocompromised, *Nocardia* spp. can cause invasive pulmonary infections and disseminated infections. Patients receiving systemic immunosuppression, such as transplant recipients, individuals with impaired pulmonary immune defenses, and intravenous drug abusers, are examples of immunosuppressed patients at risk for these infections. Patients with pulmonary infections caused by *Nocardia* can exhibit a wide range of symptoms, from an acute to a more chronic presentation. Unfortunately, there are no specific signs to indicate pulmonary nocardiosis. Patients usually appear systemically ill, with fever, night sweats, weight loss, and a productive cough that may be bloody. Pulmonary infection can lead to complications such as pleural effusions, empyema, mediastinitis, and soft tissue infection. An acute inflammatory response follows infection, resulting in necrosis and abscess formation; granulomas are not usually formed.[3]

Nocardia can often spread hematogenously throughout the body from a primary pulmonary infection. Disseminated infection can result in lesions in the brain and skin; hematogenous dissemination involving the

Table 21-5   Infections Caused by *Rhodococcus, Gordonia*, and *Tsukamurella*

Organism	Clinical Manifestations
Rhodococcus	Pulmonary infections (pneumonia, lung abscess, pulmonary nodules) Bacteremia Skin, urinary tract, and wound infections Endophthalmitis Peritonitis Catheter-associates sepsis Abscesses: prostatic/splenic, thyroid, renal, brain, subcutaneous Osteomyelitis
Gordonia	Skin infections Chronic pulmonary disease Catheter-associated sepsis Wound infection—sternal Bacteremia
Tsukamurella	Peritonitis Catheter-associated sepsis Skin infection

central nervous system is particularly common, occurring in about 30% of patients. Disseminated nocardiosis has a very poor prognosis.

Rhodococcus, Gordonia, Tsukamurella

The types of infections caused by *Rhodococcus, Gordonia,* and *Tsukamurella* are listed in Table 21-5. For the most part, these organisms are considered opportunistic pathogens because the majority of infections are in immunocompromised individuals.

Table 21-6 Clinical Manifestations of Infections Caused by Non–Acid-Fast Aerobic Actinomycetes

Organism	Clinical Manifestations
Streptomyces spp. (*S. somaliensis* and other species such as *S. anulatus* and *S. albus*)	Actinomycetoma Other (rare): pericarditis, bacteremia, and brain abscess
Actinomadura spp. (*A. madurae, A. pelletieri,* and *A. latina*)	Actinomycetoma Other (rare): peritonitis, wound infection, pneumonia, and bacteremia
Dermatophilus congolensis	Exudative dermatitis with scab formation (dermatophilosis)
Nocardiopsis dassonvillei	Actinomycetoma and other skin infections

NON–ACID-FAST AEROBIC ACTINOMYCETES: *STREPTOMYCES, ACTINOMADURA, DERMATOPHILUS, NOCARDIOPSIS,* AND THE THERMOPHILIC ACTINOMYCETES

Infection caused by most of these non–acid-fast aerobic actinomycetes is usually chronic, granulomatous lesions of the skin referred to as *mycetoma*. Mycetoma is an infection of subcutaneous tissues resulting in tissue swelling and draining sinus tracts. These infections are acquired by traumatic inoculation of organisms (usually in the lower limbs) and are usually caused by fungi. If a mycetoma is caused by an actinomycete, the infection is then called *actinomycetoma*.

Except for the thermophilic actinomycetes, only rarely have most of these agents been associated with other types of infections (Table 21-6). These nonmycetomic infections have occurred in immunosuppressed patients, such as those infected with HIV.[16]

The thermophilic actinomycetes are responsible for hypersensitivity pneumonitis, an allergic reaction to these agents. This occupational disease occurs in farmers, factory workers, and others who are repeatedly exposed to these agents. There are acute and chronic forms of this disease. Patients with acute hypersensitivity pneumonitis experience malaise, sweats, chills, loss of appetite, chest tightness, cough, and fever within 4 to 6 hours after exposure; typically symptoms resolve within a day.[15] Under some circumstances involving continued exposure to the organisms, patients suffer from a chronic form of disease in which symptoms progressively worsen with subsequent development of irreversible lung fibrosis.

LABORATORY DIAGNOSIS

SPECIMEN COLLECTION, TRANSPORT, AND PROCESSING

Appropriate specimens should be collected aseptically from affected areas. For the most part, there are no special requirements for specimen collection, transport, or processing of the organisms discussed in this chapter; refer to Table 5-1 for general information. When nocardiosis is clinically suspected, multiple specimens should be submitted for culture, because smears and cultures are simultaneously positive in only a third of the cases. Also, the random isolation of *Nocardia* spp. from the respiratory tract is of questionable significance because these organisms are so widely distributed in nature.[14] Some of the actinomycetes tend to grow as a microcolony in tissues, leading to the formation of granules. Most commonly, these granules are formed in actinomycetomas, such as those caused by *Nocardia, Streptomyces, Nocardiopsis,* and *Actinomadura* spp. Therefore, material from draining sinus tracts is an excellent specimen for direct examination and culture.

DIRECT DETECTION METHODS

Direct microscopic examination of Gram-stained preparations of clinical specimens is of utmost importance in the diagnosis of infections caused by the aerobic actinomycetes. Often, the demonstration of gram-positive, branching or partially branching beaded filaments provides the first clue to the presence of an aerobic actinomycete (Figure 21-1). Unfortunately, the actinomycetes do not always exhibit such characteristic morphology; many times these organisms are not seen at all or appear as gram-positive cocci, rods, or short filaments. Nevertheless, if gram-positive, branching or partially branching organisms are observed, a modified acid-fast stain should be performed (i.e., 1% sulfuric acid rather than 3% hydrochloric acid as the decolorizing agent). Of note, the modified acid-fast stain is positive in only about half of these smears showing gram-positive beaded, branching filaments subsequently confirmed as *Nocardia*.[18] Histopathologic examination of tissue specimens using various histologic stains, such as Gomori's methenamine-silver (GMS) stain, can also detect the presence of actinomycetes.

Of importance, any biopsy or drainage material from actinomycetomas should be examined for the presence of granules. If observed, the granules are washed in saline, emulsified in 10% potassium hydroxide or crushed between two slides, Gram-stained, and examined microscopically for the presence of filaments.[3]

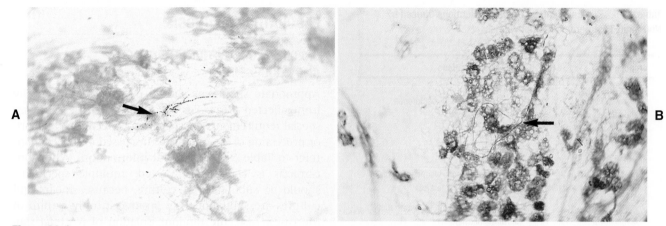

Figure 21-1 **A,** Gram stain of sputum obtained from patient with pulmonary nocardiosis caused by *Nocardia asteroides*. **B,** The same sputum stained using a modified acid-fast stain. The arrows indicate the organism.

CULTIVATION

The majority of aerobic actinomycetes do not have complex growth requirements and are able to grow on routine laboratory media, such as sheep blood, chocolate, Sabouraud dextrose, and brain-heart infusion agar. However, because many of the aerobic actinomycetes grow slowly, they may be overgrown by other normal flora present in contaminated specimens. This is particularly true for the nocardiae that require a minimum of 48 to 72 hours of incubation before colonies become visible. Because of their slow growth and the possibility of being overgrown because of contaminating flora, various selective media have been proposed to recover nocardiae. A solid medium that uses paraffin as the sole source of carbon has been effective for isolating *Nocardia* spp. and rapidly growing mycobacteria from contaminated clinical specimens.[19] Selective media formulated for the isolation of *Legionella* spp. from contaminated specimens, such as buffered charcoal-yeast extract medium with polymyxin, anisomycin, and vancomycin, have also allowed recovery of nocardiae from contaminated specimens; Martin Lewis and colistin nalidixic acid media have also been employed.[8,20] *Nocardia* spp. also grow well on Sabouraud dextrose agar as well as on fungal media containing cycloheximide such as Mycosel.[18] Occasional isolates are first seen on mycobacterial culture media, because *Nocardia* can survive the usual decontamination procedures.

If other aerobic actinomycetes are considered, a selective medium, such as brain-heart infusion agar with chloramphenicol and cycloheximide, is recommended in addition to routine agar to enhance their isolation from contaminated specimens.[3] Although most aerobic actinomycetes grow at 35° C, more isolates can be recovered at 30° C. Therefore, selective and nonselective agars should be incubated at 35° C and 30° C. Plates should be incubated for 2 to 3 weeks. The typical Gram-stain morphology and colonial appearance of the aerobic actinomycetes are summarized in Table 21-7. Examples of Gram stains and cultures of different aerobic actinomycetes are shown in Figures 21-2 and 21-3.

Clinical laboratories are rarely asked to diagnose hypersensitivity pneumonitis that is caused by the thermophilic actinomycetes. These organisms grow rapidly on trypticase soy agar with 1% yeast extract. The ability to grow at temperatures of 50° C or greater is a characteristic of all thermophilic actinomycetes. Differentiation of the various agents is based on microscopic and macroscopic morphologies; Hollick[10] and McNeil and Brown[15] provide practical identification schemes for this group of actinomycetes.

APPROACH TO IDENTIFICATION

If Gram-stain morphology or colonial morphology is suggestive of a possible actinomycetes (see Table 21-7), a Ziehl-Neelsen acid-fast stain should first be performed to rule out rapidly growing mycobacteria (see Chapter 45), followed by a modified acid-fast stain (Procedure 21-1). If the modified acid-fast stain results are positive, the isolate is a probable partially acid-fast aerobic actinomycete, that is, *Nocardia, Rhodococcus, Tsukamurella,* or *Gordonia.* (If acid-fast stain-negative, these organisms are still not completely ruled out because of the variability of acid-fastness among isolates belonging to this group.) Aerobic actinomycetes can be initially placed into major groupings by taking into consideration the following:

- Gram-stain morphology (see Figures 21-1 and Figure 21-2)
- Modified acid-fast stain results
- Presence or absence of aerial hyphae when grown on tap water agar

Table 21-7   Typical Gram-Stain Morphology and Colonial Appearance

Organism	Gram Stain*	Colonial Appearance on Routine Agar
Nocardia spp.	Branching, fine, delicate filaments with fragmentation	Extremely variable; adherent; some isolates are beta-hemolytic on sheep blood agar; wrinkled; often dry, chalky-white appearance to orange-tan pigment; crumbly
Rhodococcus	Diphtheroid-like with minimal branching or coccobacillary; colonial growth appears as coccobacilli in zigzag configuration	Nonhemolytic; round; often mucoid with orange to red, salmon-pink pigment developing within 4 to 7 days (pigment may vary widely)
Gordonia	Non-motile, short rods	Somewhat pigmented; *G. sputi* smooth, mucoid and adherent to media; *G. bronchialis* dry and raised
Tsukamurella	Mostly long rods that fragment, no spores or aerial hyphae	May have rhizoid edges, dry, white to creamy to orange
Streptomyces spp.	Extensive branching with chains and spores; does not fragment easily	Glabrous or waxy heaped colonies; variable morphology
Actinomadura spp.	Moderate, fine, intertwining branching with short chains of spores, fragmentation	White-to-pink pigment, mucoid, molar tooth appearance after 2 weeks' incubation
Dermatophilus sp.	Branched filaments divided in transverse and longitudinal planes; fine, tapered filaments	Round, adherent, gray-white colonies that later develop orange pigments; often beta-hemolytic
Nocardiopsis sp.	Branching with internal spores	Coarsely wrinkled and folded with well-developed aerial mycelium

Data compiled from Brown JH, Mcneil MM: In Murray PR, Baron EJ, Pfaller MA et al, editors: *Manual of clinical microbiology,* ed 8, Washington, DC, American Society for Microbiology, 2003; McNeil MM, Brown JM: *Clin Microbiol Rev* 7:357, 1994.
*Aerobic actinomycetes are gram-positive organisms that are often beaded in appearance.

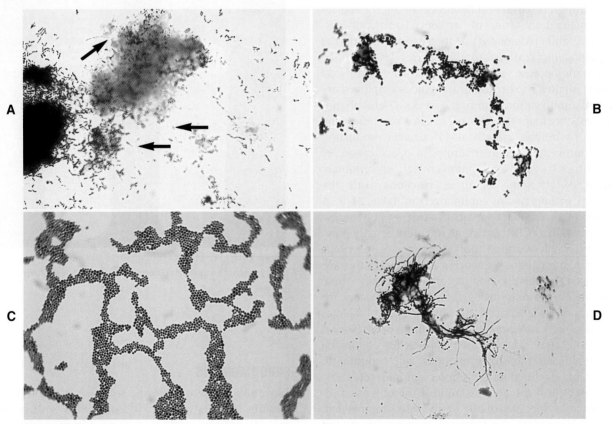

Figure 21-2   Gram stains of different aerobic actinomycetes. **A,** *Nocardia asteroides* grown on Löwenstein-Jensen media. The arrows indicate branching rods. **B,** *Rhodococcus equi* from broth. **C,** *R. equi* grown on chocolate agar. **D,** *Streptomyces* spp. on Sabouraud dextrose agar.

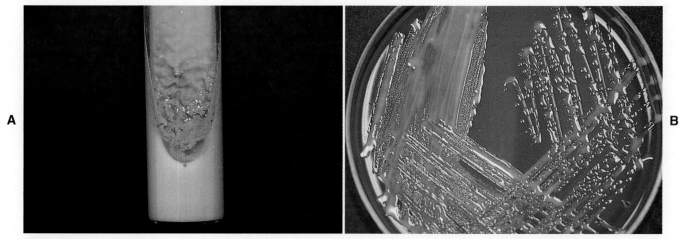

Figure 21-3   Aerobic actinomycetes grown on solid media. **A**, *Nocardia asteroides* grown on Löwenstein-Jensen media. **B**, *Rhodococcus equi* grown on chocolate agar.

● Growth or no growth in nutrient broth containing lysozyme (250 µg/mL; Figure 21-4, Procedure 21-2)

● Other tests: urea hydrolysis, nitrate reduction, and ability to grow anaerobically

Table 21-8 summarizes these key characteristics for aerobic actinomycetes.

Accurate identification of *Nocardia* to the species level is important, in that differences among the species have emerged in terms of virulence, antibiotic susceptibility, and epidemiology. However, identification of the pathogenic *Nocardia* to the species level can be problematic because no single method can identify all *Nocardia* isolates and because the methods employed are time consuming, often requiring 2 weeks. Useful phenotypic tests include the use of casein, xanthine, and tyrosine hydrolysis; growth at 45° C; acid production from rhamnose; gelatin hydrolysis; opacification of Middlebrook agar; and antimicrobial susceptibility patterns.[2,4,7,11,18] Some of these reactions with the nocardial pathogens are summarized in Table 21-9. A number of molecular approaches such as polymerase chain reaction (PCR)-restriction fragment length polymorphism analysis and PCR of a sequence of 16S rRNA gene with subsequent sequencing (see Chapter 8) have been used to rapidly and accurately identify these organisms.[8,17] Of note, using the MicroSeq System for identification (see Chapter 8), almost 15% of isolates were identified as *Nocardia* but no definitive species was given.[8] Therefore, identification approaches employing phenotypic and molecular methods will continue to evolve and be refined for the aerobic actinomycetes.

To confirm the identification of the other actinomycetes and speciate them, many tests are needed; these are beyond the capabilities of most routine clinical microbiology laboratories and should therefore be referred to a reference laboratory.

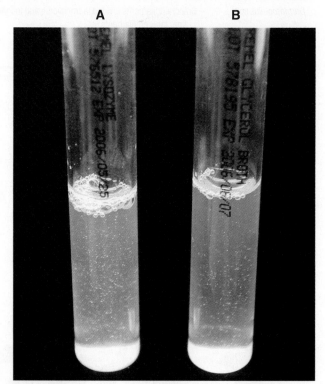

Figure 21-4   Lysozyme (A) and glycerol broths (B). The lysozyme broth is growing better, which is typical of *Nocardia asteroides*.

SERODIAGNOSIS

Currently, no serodiagnostic tests are available to help identify patients with active nocardiosis with certainty and can only be used to augment culture results. Infections caused by other aerobic actinomycetes cannot presently be diagnosed on a serologic basis.

Table 21-8   Preliminary Grouping of the Clinically Relevant Aerobic Actinomycetes

Characteristics	Nocardia	Rhodococcus	Gordonia	Tsukamurella	Streptomyces	Actinomadura	Dermatophilus	Nocardiopsis
Partially acid-fast	+	±	±	±	–	–	–	–
Appearance on tap water agar*: branching/aerial hyphae	Extensive/+	Minimal/–	Minimal/–	Minimal/–	Extensive/+	Variable/sparse	Branching	Extensive/+
Lysozyme resistance	+	±	–	+	–	–	–	–
Urea hydrolysis	+	±	+	+	±	–	+	+
Nitrate reduction	±	±	+	–	±	+	–	+
Growth anaerobically	–	–	–	–	–	–	–	–

Modified from Brown JM, McNeil MM: In Murray PR, Baron EJ, Pfaller MA et al, editors: *Manual of clinical microbiology,* ed 8, Washington, DC, 2003, American Society for Microbiology.
+, predominantly positive; –, predominantly negative; ±, mostly positive with some negative isolates.
*Tap water agar: Bacto agar (Difco Laboratories, Detroit, Mich.) is added to 100 mL of tap water, sterilized, and then poured into plates.[2] Plates are lightly inoculated using a single streak and incubated at 30° C for up to 7 days and examined daily.

TABLE 21-9   Key Tests for Differentiation of the Pathogenic Nocardia

Test	N. asteroides sensu stricto	N. farcinica*	N. nova	N. travalensis (N. asteroides type IV)	N. transvalensis sensu stricto	N. brasiliensis	N. otitidiscaviarum	N. pseudobrasiliensis
Hydrolysis of:								
Casein	–	–	–	–	–/+	+	–	+
Xanthine	–	–	–	–	–/+	–	–	–
Tyrosine	–	–	–	–	–/+	+	+	+
Growth at 42° C after 3 days	±	+	–	–	–	–	±	–
14-day arylsulfatase	–	–	+	NT	–	–	–	–
Acid from rhamnose	±	±	–	–	–	–	–	–
Gelatin hydrolysis	–	–	–	–	–	+	–	–
Opacification of Middlebrook agar[4]	–	+	–	–	–	–	–/+	–
Api 20C Assimilation:								
Galactose	–	–	–	+	+	+	–	+
Glycerol	+	+	–	+	+	+	+	+
Trehalose	–	–	–	+	+	+	+	+
Adonitol	–	–	–	–	+	–	–	–
Sensitivity by Kirby Bauer disk diffusion:[†]								
Gentamicin	S	R	S/R	R/S	R/S	S	S	S
Tobramycin	S/R	R	S/R	R	R	S	S/R	S
Amikacin	S	S	S	S/R	R	S	S	S
Erythromycin	R	S/R	S	R	R	R	R	R

NT, Not tested; +, predominantly positive; –, predominantly negative; ±, mostly positive with some negative isolates; –/+, mostly negative with some positive isolates.
*Cefotaxime resistant.
[†]Sensitive (S) or resistant (R) as determined by Kirby-Bauer disk diffusion.

Table 21-10 Primary Drugs of Choice for Infections Caused by Aerobic Actinomycetes

Organisms	Primary Drugs of Choice
Nocardia spp.	Sulfonamides Trimethoprim-sulfamethoxazole Other primary agents: amikacin, ceftriaxone, cefotaxime, linezolid, or imipenem Minocycline Combination of sulfa-containing agent and one of the primary agents is recommended for serious systemic disease
Rhodococcus, Gordonia, Tsukamurella	Erythromycin and rifampin Gentamicin, tobramycin, or ciprofloxacin Vancomycin and imipenem
Streptomyces spp.	Streptomycin and trimethoprim-sulfamethoxazole or dapsone
Actinomadura spp.	Streptomycin and trimethoprim-sulfamethoxazole or dapsone Amikacin and imipenem
Nocardiopsis dassonvillei	Trimethoprim-sulfamethoxazole
Dermatophilus congolensis	Highly variable susceptibilities; no specific drugs of choice

Procedure 21-1

PARTIALLY ACID-FAST STAIN FOR IDENTIFICATION OF *NOCARDIA*

PRINCIPLE

The nocardiae, because of unusual long-chain fatty acids in their cell walls, can retain carbolfuchsin dye during mild acid decolorization, whereas other aerobic branching bacilli cannot.

METHOD

1. Emulsify a very small amount of the organisms to be stained in a drop of distilled water on the slide. A known positive control and a negative control should be stained along with the unknown strain.
2. Allow to air dry and heat fix.
3. Flood the smear with Kinyoun carbolfuchsin and allow the stain to remain on the slide for 3 minutes.
4. Rinse with tap water, shake off excess water, and decolorize briefly with 1% sulfuric acid alcohol until no more red color rinses off the slides.
5. Counterstain with Kinyoun methylene blue for 30 seconds.
6. Rinse again with tap water. Allow the slide to air dry, and examine the unknown strain compared with the controls. Partially acid-fast organisms show reddish to purple filaments, whereas non–acid-fast organism are blue only.

ANTIMICROBIAL SUSCEPTIBILITY TESTING AND THERAPY

A standard for susceptibility testing by broth microdilution and with cation-supplemented Mueller-Hinton broth has been approved by the Clinical and Laboratory Standards Institute (formerly National Committee for Clinical and Laboratory Standards) along with interpretive guidelines.[22] Other methods, including modified disk diffusion, agar dilution, broth microdilution, E-test, and radiometric growth index have been used for antimicrobial susceptibility testing of *Nocardia* spp. However, although these methods demonstrate good interlaboratory and intralaboratory agreement and reproducibility, correlation of in vitro susceptibility testing results with clinical outcome has not been systematically performed at the time of this writing. Nevertheless, antimicrobial susceptibility testing should be performed on clinically significant isolates of *Nocardia*.

However, if required, the isolate should be sent to a reference laboratory. For all other actinomycetes, there are no standardized methods currently available. In some instances, susceptibility studies of *Rhodococcus* and *Gordonia* can be used as a guide for directing therapy. The primary drugs of choice against the aerobic actinomycetes are shown in Table 21-10; there is no effective antimicrobial therapy for hypersensitivity pneumonitis caused by the thermophilic actinomycetes.

PREVENTION

There are no vaccines available that are directed against any aerobic actinomycete, although some have been developed but with little success. With respect to hypersensitivity pneumonitis caused by the thermophilic actinomycetes, patients must prevent disease by avoiding exposure to these sensitizing microorganisms.

Procedure 21-2

LYSOZYME RESISTANCE FOR DIFFERENTIATING *NOCARDIA* FROM *STREPTOMYCES*

PRINCIPLE
The enzyme lysozyme, present in human tears and other secretions, can break down cell walls of certain microorganisms. Susceptibility to the action of lysozyme can differentiate certain morphologically similar genera and species.

METHOD
1. Prepare basal broth as follows:
   - Peptone (Difco Laboratories), 5 g
   - Beef extract (Difco), 3g
   - Glycerol (Difco), 70 mL
   - Distilled water, 1000 mL

Dispense 500 mL of this solution into 16- × 125-mm screw-cap glass test tubes, 5 mL per tube. Autoclave the test tubes and the remaining solution for 15 minutes at 120° C. Tighten caps and store the tubes in the refrigerator for a maximum of 2 months.

2. Prepare lysozyme solution as follows:
   - Lysozyme (Sigma Chemical Co.), 100 mg
   - HCl (0.01 N), 100 mL
   - Sterilize through a 0.45-mm membrane filter.

3. Add 5 mL lysozyme solution to 95 mL basal broth; mix gently, avoiding bubbles, and aseptically dispense in 5-mL amounts to sterile, screw-cap tubes as in step 1. Store refrigerated for a maximum of 2 weeks.

4. Place several bits of the colony to be tested into a tube of the basal glycerol broth without lysozyme (control) and into a tube of broth containing lysozyme.

5. Incubate at room temperature for up to 7 days. Observe for growth in the control tube. An organism that grows well in the control tube but not in the lysozyme tube is considered to be susceptible to lysozyme.

Case Study

A 60-year-old woman with a history of steroid-treated rheumatoid arthritis presented to the emergency department with increasing confusion, lethargy, and fever, which began approximately 2 weeks before admission. She had an elevated white blood cell (WBC) count. A spinal tap was performed, which showed elevated protein, low glucose, and a WBC count of 200 mm³. The cerebrospinal fluid was cultured and two colonies of nonhemolytic, catalase-positive, gram-positive rods grew on the second quadrant of the blood agar plate after 48 hours. The chocolate plate had no growth.

QUESTIONS

1. Because the colonies were nonhemolytic, *Listeria* was not in the differential. Before dismissing these colonies as plate contaminants, what testing should be performed?

2. Identification and susceptibility testing for such isolates is important but is usually left to reference laboratories. However, a simple disk test using an inoculum equivalent to a 0.5 McFarland showed that the isolate was resistant to gentamicin, erythromycin, and cefotaxime. Which species is characteristically resistant to these antimicrobial agents?

3. If the isolate was mucoid and salmon pink in color, what testing would be helpful to the identification?

REFERENCES

1. Beaman BL, Beaman L: *Nocardia* species: host parasite relationships, *Clin Microbiol Rev* 7:213, 1994.
2. Berd D: Laboratory identification of clinically important aerobic actinomycetes, *Appl Microbiol* 25:665, 1973.
3. Brown JM, McNeil MM: *Nocardia, Rhodococcus, Gordonia, Actinomadura, Streptomyces,* and other aerobic actinomycetes. In Murray PR, Baron EJ, Pfaller MA, et al, editors: *Manual of clinical microbiology,* ed 8, Washington, DC, 2003, American Society for Microbiology.
4. Carson M, Hellyar A: Opacification of Middlebrook agar as an aid in distinguishing *Nocardia farcinica* within the *Nocardia asteroides* complex, *J Clin Microbiol* 32:2270, 1994.
5. Chun J, Goodfellow M: A phylogenetic analysis of the genus *Nocardia* with 16S rRNA gene sequences, *Int J Syst Bacteriol* 45:240, 1995.
6. Cloud JL, Conville PS, Croft A, et al: Evaluation of partial 16S ribosomal DNA sequencing for identification of *Nocardia* species by using the MicroSeq 500 system with an expanded database, *J Clin Microbiol* 42:578, 2004.
7. Flores M, Desmond E: Opacification of Middlebrook agar as an aid in identification of *Nocardia farcinica, J Clin Microbiol* 31:3040, 1993.
8. Garrett MA, Holmes HT, Nolte FS: Selective buffered charcoal-yeast extract medium for isolation of nocardiae from mixed cultures, *J Clin Microbiol* 30:1891, 1992.
9. Goodfellow M, Chun J, Stubbs S, et al: Transfer of *Nocardia amarae* (Lechevalier and Lechevalier, 1974) to the genus *Gor-*

donia as *Gordonia amarae* comb nov, *Letters Appl Microbiol* 19:401, 1994.

10. Hollick G: Isolation and identification of aerobic actinomycetes, *Clin Microbiol Newsl* 17:25, 1995.

11. Kiska DL, Hicks K, Pettit DJ: Identification of medically relevant *Nocardia* species with an abbreviated battery of tests, *J Clin Microbiol* 40:1346, 2002.

12. Klatte S, Rainey FA, Kroppenstedt RM: Transfer of *Rhodococcus aichensis* (Tsukamura, 1982) and *Nocardia amarae* (Lechevalier and Lechevalier, 1974) to the genus *Gordonia* as *Gordonia aichiensis* comb nov and *Gordonia amarae* comb nov, *Int J Syst Bacteriol* 44: 769, 1994.

13. Lasker BA, Brown JM, McNeil MM: Identification and epidemiological typing of clinical and environmental isolates of the genus *Rhodococcus* with use of a digoxigenin-labeled rDNA gene probe, *Clin Infect Dis* 15:233, 1992.

14. Lerner PI: *Nocardiosis, Clin Infect Dis* 22:891, 1996.

15. McNeil MM, Brown JM: The medically important aerobic actinomycetes: epidemiology and microbiology, *Clin Microbiol Rev* 7:357, 1994.

16. McNeil MM, Brown JM, Scalise G, et al: Nonmycetomic *Actinomadura madulae* infection in a patient with AIDS, *J Clin Microbiol* 30:1008, 1992.

17. Roth A, Andrees S, Kroppenstedt RM, et al: Phylogeny of the genus *Nocardia* based on reassessed 16S rRNA gene sequences reveals underspeciation and division of strains classified as *Nocadia asteroides* into three established species and two unnamed taxons, *J Clin Microbiol* 41:851, 2003.

18. Saubolle MA, Sussland D: Nocardiosis: review of clinical laboratory experience, *J Clin Microbiol* 41:4497, 2003.

19. Shawar RM, Moore DG, LaRocco MT: Cultivation of *Nocardia* spp. on chemically defined media for selective recovery of isolates from clinical specimens, *J Clin Microbiol* 28:508, 1990.

20. Vickers RM, Rihs JD, Yu VL: Clinical demonstration of isolation of *Nocardia asteroides* on buffered charcoal-yeast extract media, *J Clin Microbiol* 30:227, 1992.

21. Weinstock DM, Brown AE: *Rhodococcus equi*: an emerging pathogen, *Clin Infect Dis* 34:1379, 2002.

22. Woods GL, Brown-Elliott BA, Desmond EP, et al: Susceptibility testing of Mycobacteria, *Nocardia*, and other actinomycetes; Approved Standard M24-A, vol 23, no. 18 NCCLS, Wayne, Pa, 2003.

Enterobacteriaceae CHAPTER 22

GENERA AND SPECIES TO BE CONSIDERED

Because of the large number and diversity of genera included in the *Enterobacteriaceae*, it is helpful to consider the bacteria of this family as belonging to one of two major groups. The first group comprises those species that either commonly colonize the human gastrointestinal tract or are most notably associated with human infections (Box 22-1). Although many *Enterobacteriaceae* that cause human infections are part of our normal gastrointestinal flora, there are exceptions, such as the plague bacillus, *Yersinia pestis*. The second group consists of genera that may colonize humans but are rarely associated with human infections or are most commonly recognized as environmental inhabitants or colonizers of other animals. For this reason, the discovery of these species in clinical specimens should alert laboratorians to possible identification errors; careful confirmation of both the laboratory results and the clinical significance of such isolates is warranted.

Although the organisms that compose this second group may have substantial significance outside diagnostic microbiology and infectious diseases, this chapter focuses on organisms of the first group, whose clinical significance is well recognized.

GENERAL CHARACTERISTICS

Enterobacteriaceae ferment glucose, are oxidase-negative, and, with rare exception, reduce nitrates to nitrites. Furthermore, except for *Shigella dysenteriae* type 1, all commonly isolated *Enterobacteriaceae* are catalase-positive.

EPIDEMIOLOGY

Enterobacteriaceae inhabit a wide variety of niches that include the human gastrointestinal tract, the gastrointestinal tract of other animals, and various environmental sites. Some are agents of zoonoses, causing infections in animal populations (Table 22-1). Just as the reservoirs for these organisms vary, so do their modes of transmission to humans.

For those species that normally colonize humans, infections may result when a patient's own bacterial strains (i.e., endogenous strains) establish infections in a normally sterile body site. These organisms can also be passed from one patient to another. Such infections often depend on the debilitated state of a hospitalized patient and are nosocomially acquired. However, this is not always the case. For example, although *Escherichia coli* is the most common cause of nosocomial infections, it is also the leading cause of urinary tract infections in nonhospitalized patients.

Other species, such as *Salmonella* spp., *Shigella* spp., and *Yersinia enterocolitica,* only inhabit the bowel at the time they are causing infection and are acquired by ingestion of contaminated foods or water. This is also the mode of transmission for the various types of *E. coli* that are known to cause gastrointestinal infections. In contrast, *Yersinia pestis* is unique among the Enterobacteriaceae that infect humans. This is the only species that is transmitted from animals by the bite of an insect (i.e., flea) vector.

PATHOGENESIS AND SPECTRUM OF DISEASES

The clinically relevant members of the *Enterobacteriaceae* can be considered as two groups: the opportunistic pathogens and the overt pathogens. *Salmonella typhi, Shigella* spp., and *Y. pestis* are among the latter and are the causative agents of typhoid fever, dysentery, and the "black" plague, respectively. Therefore, their discovery in clinical material should always be considered significant. These organisms, as well as other *Salmonella* spp., produce various potent virulence factors and are capable of producing life-threatening infections (Table 22-2).

The opportunistic pathogens most commonly include *Citrobacter* spp., *Enterobacter* spp., *Klebsiella* spp., *Proteus* spp., and *Serratia* spp. Although considered opportunistic pathogens, these organisms produce signifi-

BOX 22-1	Genera and Species of the Family *Enterobacteriaceae* That Commonly Colonize Humans or Are Associated with Human Infections

Citrobacter *freundii*
Citrobacter *(diversus) koseri*
Citrobacter *amalonaticus*
Edwardsiella *tarda*
Enterobacter *aerogenes*
Enterobacter *cloacae*
Enterobacter *agglomerans* group (*Pantoea agglomerans*)
Enterobacter *gergoviae*
Enterobacter *sakazakii*
Enterobacter *amnigenus*
Enterobacter *(cancerogenous) taylorae*
Escherichia *coli*
Hafnia *alvei*
Klebsiella *pneumoniae*
Klebsiella *oxytoca*
Klebsiella *ozaenae*
Morganella *morganii* subsp. *morganii*
Plesiomonas *shigelloides*
Proteus *mirabilis*
Proteus *vulgaris*
Proteus *penneri*
Providencia *rettgeri*
Providencia *stuartii*
Salmonella, all serotypes
Serratia *marcescens*
Serratia *liquefaciens* group
Shigella *dysenteriae* (group A)
Shigella *flexneri* (group B)
Shigella *boydii* (group C)
Shigella *sonnei* (group D)
Yersinia *pestis*
Yersinia *enterocolitica* subsp. *enterocolitica*
Yersinia *frederiksenii*
Yersinia *intermedia*
Yersinia *pseudotuberculosis*

cant virulence factors, such as endotoxins, that can mediate fatal infections. However, because they generally do not initiate disease in healthy, uncompromised human hosts, they are considered opportunistic.

Although *E. coli* is a normal bowel inhabitant, its pathogenic classification is somewhere between that of the overt pathogens and opportunistic organisms. Strains of this species, such as enterotoxigenic *E. coli* (ETEC), enteroinvasive *E. coli* (EIEC), and enteroaggregative *E. coli* (EAEC), express potent toxins and cause serious gastrointestinal infections. Additionally, in the case of enterohemorrhagic *E. coli* (EHEC), life-threatening systemic disease can result from infection. Furthermore, as the leading cause of nosocomial infections among *Enterobacteriaceae*, *E. coli* is likely to have greater virulence capabilities than the other species categorized as "opportunistic" *Enterobacteriaceae*.

LABORATORY DIAGNOSIS

SPECIMEN COLLECTION AND TRANSPORT

No special considerations are required for specimen collection and transport of the organisms discussed in this chapter. Refer to Table 5-1 for general information on specimen collection and transport.

SPECIMEN PROCESSING

No special considerations are required for processing of the organisms discussed in this chapter. Refer to Table 5-1 for general information on specimen processing.

DIRECT DETECTION METHODS

Other than Gram stain of patient specimens, there are no specific procedures for the direct detection of most *Enterobacteriaceae*. Microscopically the cells of these organisms generally appear as coccobacilli, or straight rods with rounded ends. *Yersinia pestis* resembles a closed safety pin when it is stained with methylene blue or Wayson stain[2]; this is a key characteristic for rapid diagnosis of plague.

Klebsiella granulomatis (formerly *Calymmatobacterium granulomatis*) can be visualized in scrapings of lesions stained with Wright's or Giemsa stain. Cultivation in vitro is very difficult, so the direct examination is important diagnostically. Groups of organisms are seen within mononuclear endothelial cells; this pathognomonic entity is known as a *Donovan body,* named after the physician who first visualized the organism in such a lesion. The organism stains as a blue rod with prominent polar granules, giving rise to a "safety pin" appearance, surrounded by a large, pink capsule. Subsurface infected cells must be present; surface epithelium is not an adequate specimen.

CULTIVATION

Media of Choice

Most *Enterobacteriaceae* grow well on routine laboratory media, such as 5% sheep blood, chocolate, and MacConkey agars. In addition to these media, selective agars, such as Hektoen enteric (HE) agar, xylose-lysine-deoxycholate (XLD) agar, and *Salmonella-Shigella* (SS) agar are commonly used to cultivate enteric pathogens from gastrointestinal specimens (see Chapter 7 and Table 7-1 for more information on the characteristics and appearance of these media; also see Chapter 59 for more information about laboratory procedures for the diagnosis of bacterial gastrointestinal infections). The broths used in blood culture systems, as well as thioglycollate and brain-heart infusion broths, all support the growth of *Enterobacteriaceae*.

Table 22-1 Epidemiology of Clinically Relevant *Enterobacteriaceae*

Organism	Habitat (reservoir)	Mode of Transmission
Escherichia coli	Normal bowel flora of humans and other animals; may also inhabit female genital tract	Varies with the type of infection. For nongastrointestinal infections, organisms may be endogenous or spread person to person, especially in the hospital setting; for gastrointestinal infections, transmission mode varies with the type of *E. coli* (see Table 22-2), and may involve fecal-oral spread between humans via contaminated food or water or consumption of undercooked beef or milk from colonized cattle
Shigella spp.	Only found in humans at times of infection; not part of normal bowel flora	Person-to-person spread by fecal-oral route, especially in overcrowded areas and areas with poor sanitary conditions
Salmonella typhi and *paratyphi*	Only found in humans but not part of normal bowel flora	Person-to-person spread by fecal-oral route by ingestion of food or water contaminated with human excreta
Other *Salmonella* spp.	Widely disseminated in nature and associated with various animals	Ingestion of contaminated food products processed from animals, frequently of poultry or dairy origin. Direct person-to-person transmission by fecal-oral route can occur in health care settings when handwashing guidelines are not followed
Edwardsiella tarda	Gastrointestinal tract of cold-blooded animals, such as reptiles	Uncertain; probably by ingestion of contaminated water or close contact with carrier animal
Yersinia pestis	Carried by urban and domestic rats and wild rodents, such as the ground squirrel, rock squirrel, and prairie dog	From rodents to humans by the bite of flea vectors, or by ingestion of contaminated animal tissues; during human epidemics of pneumonic (i.e., respiratory) disease the organism can be spread directly from human to human by inhalation of contaminated airborne droplets; rarely transmitted by handling or inhalation of infected animal tissues or fluids
Yersinia enterocolitica	Dogs, cats, rodents, rabbits, pigs, sheep, and cattle. Not part of normal human flora	Consumption of incompletely cooked food products (especially pork), dairy products such as milk, and, less commonly, by ingestion of contaminated water or by contact with infected animals
Yersinia pseudotuberculosis	Rodents, rabbits, deer, and birds. Not part of normal human flora	Ingestion of organism during contact with infected animal or by contaminated food or water
Citrobacter spp., *Enterobacter* spp., *Klebsiella* spp., *Morganella* spp., *Proteus* spp., *Providencia* spp., and *Serratia* spp.	Normal human gastrointestinal flora	Endogenous, or person-to-person spread, especially in hospitalized patients

Cefsulodin-irgasan-novobiocin (CIN) agar is a selective medium specifically used for the isolation of *Y. enterocolitica* from gastrointestinal specimens. Similarly, MacConkey-sorbitol agar (described in Table 7-1), is used to differentiate sorbitol-negative *E. coli* O157:H7 from other types of *E. coli* that are capable of fermenting this sugar alcohol.

Klebsiella granulomatis will not grow on routine agar media. Recently, the organism was cultured in human monocytes from biopsy specimens of genital ulcers of patients with donovanosis. Historically, the organism has also been cultivated on a special medium described by Dienst that contains growth factors found in egg yolk. In clinical practice, however, the diagnosis of granuloma inguinale is made solely on the basis of direct examination.

Incubation Conditions and Duration

Under normal circumstances, most *Enterobacteriaceae* produce detectable growth on commonly used broth and agar media within 24 hours of inoculation. For isolation, 5% sheep blood and chocolate agars may be incubated at 35° C in carbon dioxide or ambient air. However, MacConkey agar and other selective agars (e.g., SS, HE, XLD) should be incubated only in ambient air. Unlike most other *Enterobacteriaceae*, *Y. pestis* grows best at 25° to 30° C. Colonies of *Y. pestis* are pinpoint at 24 hours but resemble those of other *Enterobacteriaceae*

Table 22-2   Pathogenesis and Spectrum of Disease for Clinically Relevant *Enterobacteriaceae*

Organism	Virulence Factors	Spectrum of Disease and Infections
Escherichia coli (as a cause of extraintestinal infections)	Several, including endotoxin, capsule production, and pili that mediate attachment to host cells	Urinary tract infections, bacteremia, nosocomial infections of various body sites. Most common cause of gram-negative nosocomial infections
Enterotoxigenic *E. coli* (ETEC)	Pili that permit gastrointestinal colonization. Heat-labile (LT) and heat-stable (ST) enterotoxins that mediate secretion of water and electrolytes into the bowel lumen	Traveler's and childhood diarrhea characterized by profuse, watery stools. Transmitted by contaminated food and water
Enteroinvasive *E. coli* (EIEC)	Virulence factors uncertain, but organism invades enterocytes lining the large intestine in a manner nearly identical to *Shigella* spp.	Dysentery (i.e., necrosis, ulceration, and inflammation of large bowel); usually in young children living in areas of poor sanitation
Enteropathogenic *E. coli* (EPEC)	Bundle-forming pilus, intimin, and other factors that mediate organism attachment to mucosal cells of the small bowel, resulting in changes in cell surface (i.e., loss of microvilli)	Diarrhea in infants in developing, low-income nations; can cause a chronic diarrhea
Enterohemorrhagic *E. coli* (EHEC)	Toxin similar to Shiga toxin produced by *Shigella dysenteriae.* Most frequently associated with certain serotypes, such as *E. coli* O157:H7	Inflammation and bleeding of the mucosa of the large intestine (i.e., hemorrhagic colitis); can also lead to hemolytic-uremic syndrome resulting from toxin-mediated damage to kidneys. Transmitted by ingestion of undercooked ground beef or raw milk
Enteroaggregative *E. coli* (EAEC)	Probably involves binding by pili, ST-like, and hemolysin-like toxins; actual pathogenic mechanism not known	Watery diarrhea that, in some cases, can be prolonged. Mode of transmission is not well understood
Shigella spp.	Several factors involved to mediate adherence and invasion of mucosal cells, escape from phagocytic vesicles, intercellular spread, and inflammation. Shiga toxin role in disease is uncertain, but it does have various effects on host cells	Dysentery defined as acute inflammatory colitis and bloody diarrhea characterized by cramps, tenesmus, and bloody, mucoid stools. Infections with *S. sonnei* may produce only watery diarrhea
Salmonella spp.	Several factors serve to protect organisms from stomach acids, promote attachment and phagocytosis by intestinal mucosal cells, allow survival in and destruction of phagocytes, and facilitate dissemination to other tissues	Three general categories of infection: (1) Gastroenteritis and diarrhea caused by a wide variety of serotypes that produce infections limited to the mucosa and submucosa of the gastrointestinal tract. *S. typhimurium* and *S. enteritidis* are the serotypes most commonly associated with *Salmonella* gastroenteritis in the United States (2) Bacteremia and extraintestinal infections occur by spread from the gastrointestinal tract. These infections usually involve *S. choleraesius* or *S. dublin,* although any serotype may cause bacteremia (3) Enteric fever (typhoid fever, or typhoid) is characterized by prolonged fever and multisystem involvement, including lymph nodes, liver, and spleen. This life-threatening infection is most frequently caused by *S. typhi* or *S. paratyphi* strains
Yersinia pestis	Multiple factors play a role in the pathogenesis of this highly virulent organism. These include ability to adapt for intracellular survival and production of antiphagocytic capsule, exotoxins, endotoxins, coagulase, and fibrinolysin	Two major forms of infection are bubonic plague and pneumonic plague. Bubonic plague is characterized by high fever and painful inflammatory swelling of axilla and groin lymph nodes (i.e., the characteristic buboes); infection rapidly progresses to fulminant bacteremia that is frequently fatal if untreated. Pneumonic plague involves the lungs and is characterized by malaise and pulmonary signs; the respiratory infection can occur as a consequence of bacteremic spread associated with bubonic plague or can be acquired by the airborne route during close contact with other pneumonic plague victims; this form of plague is also rapidly fatal
Yersinia enterocolitica subsp. *enterocolitica*	Various factors allow the organism to attach to and invade the intestinal mucosa and spread to lymphatic tissue	Enterocolitis characterized by fever, diarrhea, and abdominal pain; also can cause acute mesenteric lymphadenitis, which may present clinically as appendicitis (i.e., pseudoappendicular syndrome). Bacteremia can occur with this organism but is uncommon

Table 22-2   Pathogenesis and Spectrum of Diseases for Clinically Relevant *Enterobacteriaceae*—cont'd

Organism	Virulence Factors	Spectrum of Disease and Infections
Y. pseudotuberculosis	Similar to those of *Y. enterocolitica*	Causes similar infections as described for *Y. enterocolitica* but much less common
Citrobacter spp., *Enterobacter* spp., *Klebsiella* spp., *Morganella* spp., *Proteus* spp., *Providencia* spp., and *Serratia* spp.	Several factors, including endotoxins, capsules, adhesion proteins, and resistance to multiple antimicrobial agents	Wide variety of nosocomial infections of respiratory tract, urinary tract, blood, and several other normally sterile sites; most frequently infect hospitalized and seriously debilitated patients

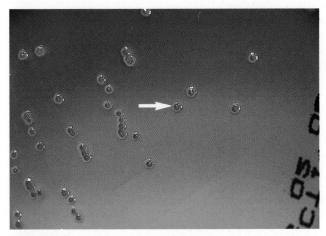

Figure 22-1 "Bulls-eye" *(arrow)* colony of *Yersinia enterocolitica* on cefsoludin-irgasan-novobiocin (CIN) agar.

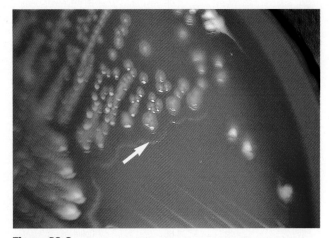

Figure 22-2 *Proteus mirabilis* swarming on blood agar *(arrow* at swarming edge).

after 48 hours. CIN agar, used for the isolation of *Y. enterocolitica*, should be incubated 48 hours to allow for the development of typical "bulls-eye" colonies (Figure 22-1).

Colonial Appearance

Table 22-3 describes the colonial appearance and other distinguishing characteristics (pigment and odor) of the most commonly isolated *Enterobacteriaceae* on MacConkey, HE, and XLD agars (see Figures 7-4, 7-6, and 7-9 for examples). All *Enterobacteriaceae* produce similar growth on blood and chocolate agars; colonies are large, gray, and smooth. Colonies of *Klebsiella* or *Enterobacter* may be mucoid because of their polysaccharide capsule. *Escherichia coli* is often beta-hemolytic on blood agar, but most other genera are nonhemolytic. As a result of motility, *Proteus mirabilis, P. penneri,* and *P. vulgaris* "swarm" on blood and chocolate agars. Swarming results in the production of a thin film of growth on the agar surface (Figure 22-2) as the motile organisms spread from the original site of inoculation.

Colonies of *Y. pestis* on 5% sheep blood agar are pinpoint at 24 hours but exhibit a rough, cauliflower appearance at 48 hours. Broth cultures of *Y. pestis* exhibit a characteristic "stalactite pattern" in which clumps of cells adhere to one side of the tube.

Y. enterocolitica produces bull's-eye colonies (dark red or burgundy centers surrounded by a translucent border; see Figure 22-1) on CIN at 48 hours. However, because most *Aeromonas* spp. produce similar colonies on CIN, it is important to perform an oxidase test to verify that the organisms are *Yersinia* spp. (oxidase-negative).

APPROACH TO IDENTIFICATION

In the early decades of the twentieth century, *Enterobacteriaceae* were identified using more than 50 biochemical tests in tubes; this method is still used today in reference and public health laboratories. Certain key tests such as indole, methyl red, Voges-Proskauer, and citrate, known by the acronym, IMViC, were routinely performed to group the most commonly isolated

Table 22-3 Colonial Appearance and Characteristics*

Organism	Medium	Appearance
Citrobacter spp.	Mac	Late lactose fermenter; therefore, NLF after 24 hours; LF after 48 hours; colonies are light pink after 48 hours
	HE	Colorless
	XLD	Red, yellow, or colorless colonies, with or without black centers (H_2S)
Edwardsiella spp.	Mac	NLF
	HE	Colorless
	XLD	Red, yellow, or colorless colonies, with or without black centers (H_2S)
Enterobacter spp.	Mac	LF; may be mucoid
	HE	Yellow
	XLD	Yellow
Escherichia coli	Mac	LF; flat, dry, pink colonies with a surrounding darker pink area of precipitated bile salts[†]
	HE	Yellow
	XLD	Yellow
Hafnia alvei	Mac	NLF
	HE	Colorless
	XLD	Red or yellow
Klebsiella spp.	Mac	LF; mucoid
	HE	Yellow
	XLD	Yellow
Morganella spp.	Mac	NLF
	HE	Colorless
	XLD	Red or colorless
Proteus spp.	Mac	NLF; may swarm depending on the amount of agar in the medium; characteristic foul smell
	HE	Colorless
	XLD	Yellow or colorless, with or without black centers
Providencia spp.	Mac	NLF
	HE	Colorless
	XLD	Yellow or colorless
Salmonella spp.	Mac	NLF
	HE	Green
	XLD	Red with black center
Serratia spp.	Mac	Late LF; S. marcescens may be red pigmented, especially if plate is left at 25° C (Figure 22-3)
	HE	Colorless
	XLD	Yellow or colorless
Shigella spp.	Mac	NLF; S. sonnei produces flat colonies with jagged edges
	HE	Green
	XLD	Colorless
Yersinia spp.	Mac	NLF; may be colorless to peach
	HE	Salmon
	XLD	Yellow or colorless

HE, Hektoen enteric agar; *LF*, lactose fermenter, pink colony; *Mac*, MacConkey agar; *NLF*, non–lactose-fermenter, colorless colony; *XLD*, xylose-lysine-deoxycholate agar.

*Most *Enterobacteriaceae* are indistinguishable on blood agar; see text for colonial description.

[†]Pink colonies on MacConkey agar with sorbitol are sorbitol fermenters; colorless colonies are nonsorbitol fermenters.

pathogens. Today, this type of conventional biochemical identification of enterics has become a historical footnote in most clinical and hospital laboratories.

In the latter part of the twentieth century, manufacturers began to produce panels of miniaturized tests for identification, first of enteric gram-negative rods and later of other groups of bacteria and yeast. Original panels were inoculated manually; these were followed by semiautomated and automated systems, the most sophisticated of which inoculate, incubate, read, and discard the panels. Practically any commercial identification system can be used to reliably identify the commonly isolated *Enterobacteriaceae*. Depending on the system, results are available within 4 hours or after

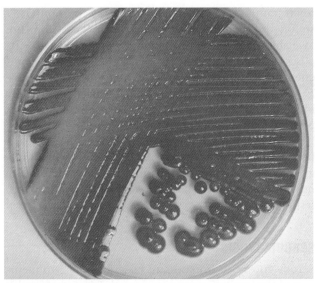

Figure 22-3 Red-pigmented *Serratia marcescens* on MacConkey agar.

overnight incubation. The extensive computer databases employed by these systems also include information on unusual biotypes. The number of organisms used to define individual databases is important and rarely isolated organisms or new microorganisms may be misidentified or not identified at all.

The definitive identification of enterics is now based on molecular methods, especially 16S ribosomal RNA (rRNA) sequencing and DNA-DNA hybridization. Using molecular methods, the genus *Plesiomonas*, composed of one species of oxidase-positive gram-negative rods, has now been included in the family *Enterobacteriaceae*.[2] *Plesiomonas* clusters with the genus *Proteus* within the family *Enterobacteriaceae* by 16S rRNA sequencing. *Proteus*, however, like all other *Enterobacteriaceae*, is oxidase-negative. The finding that an oxidase-positive genus and an oxidase-negative genus cluster together is a revolutionary concept in microbial taxonomy. Although *Plesiomonas* is included with the *Enterobacteriaceae* in the eighth edition of the *Manual of Clinical Microbiology*, this classification is still tentative and *Plesiomonas* is still included with the genus *Aeromonas* in this text (see Chapter 28).

In the interests of cost containment, many clinical laboratories use an abbreviated identification scheme for identification of commonly isolated enterics.[3] *E. coli*, the most commonly isolated enteric organism, can be identified using a positive spot indole test (see Procedure 13-39). To presumptively identify an organism as *E. coli*, the characteristic colonial appearance on MacConkey agar, as described in Table 22-3, is documented as is the positive spot indole test. A spot indole test can also be used to quickly separate swarming *Proteae*, such as *P.*

mirabilis and *P. penneri*, which are negative, from the indole-positive *P. vulgaris*.

Specific Considerations for Identifying Enteric Pathogens

Table 22-4 illustrates the use of biochemical profiles obtained with triple sugar iron (TSI) agar and lysine iron agar (LIA) to presumptively identify enteric pathogens (see Chapter 13 for information on the principles, performance, and interpretation of these tests). Organisms that exhibit the profiles shown in Table 22-4 require further biochemical profiling and, in the case of *Salmonella* spp. and *Shigella* spp., serotyping to establish a definitive identification. Bacterial species not considered capable of causing gastrointestinal infections give profiles other than those shown.

In most clinical laboratories, serotyping of *Enterobacteriaceae* is limited to the preliminary grouping of *Salmonella* spp., *Shigella* spp., and *E. coli* O157:H7. Typing should be performed from a non–sugar-containing medium, such as 5% sheep blood agar or LIA. Use of sugar-containing media, such as MacConkey or TSI agars, can cause the organisms to autoagglutinate.

Commercially available polyvalent antisera designated A, B, C_1, C_2, D, E, and Vi are commonly used to preliminarily group *Salmonella* spp. because 95% of isolates belong to groups A through E. The antisera A through E contain antibodies against somatic ("O") antigens, and the Vi antiserum is prepared against the capsular ("K") antigen of *S. typhi*. Typing is performed using a slide agglutination test. If an isolate agglutinates with the Vi antiserum and does not react with any of the "O" groups, then a saline suspension of the organism should be prepared and heated to 100° C for 10 minutes to inactivate the Vi antigen. The organism should then be retested. *S. typhi* is positive with Vi and group D. Complete typing of *Salmonella* spp., including the use of antisera against the flagellar ("H") antigens, is performed at reference laboratories.

Preliminary serologic grouping of *Shigella* spp. is also performed using commercially available polyvalent somatic ("O") antisera designated A, B, C, and D. As with *Salmonella* spp., *Shigella* spp. may produce a capsule and therefore heating may be required before typing is successful. Subtyping of *Shigella* spp. beyond the groups A, B, and C (*Shigella* group D only has one serotype) usually is performed by reference laboratories.

Plesiomonas shigelloides, a new member of the *Enterobacteriaceae* that can cause gastrointestinal infections (see Chapter 28), might cross-react with *Shigella* grouping antisera, particularly group D, and lead to misidentification. This mistake can be avoided by performing an oxidase test.

Sorbitol-negative *E. coli* can be serotyped using commercially available antisera to determine whether

Table 22-4 TSI and LIA Reactions Used to Screen for Enteropathogenic *Enterobacteriaceae**†

TSI Reactions‡	LIA Reactions‡	Possible Identification
K/Ⓐ or K/A H₂S +	K/K or K/NC H₂S +	*Salmonella* spp. *Edwardsiella* spp.
K/A H₂S+	K/K or K/NC H₂S +	*Salmonella* spp. (rare)
K/Ⓐ	K/K or K/NC	*Salmonella* spp. (rare)
K/A	K/K or K/NC H₂S +	*Salmonella typhi* (rare)
K/Ⓐ	K/K or K/NC	*Salmonella* spp. (rare)
K/Ⓐ	K/A H₂S +	*Salmonella paratyphi* A (usually H₂S−)
K/Ⓐ	K/A or A/A	*Escherichia coli* *Salmonella paratyphi* A *Shigella flexneri* 6 (uncommon) *Aeromonas* spp. (oxidase-positive)
K/A	K/K or K/NC	*Plesiomonas* sp. (oxidase-positive) *Salmonella typhi* (rare) *Vibrio* spp. (oxidase-positive)
K/A	K/A or A/A	*Escherichia coli* *Shigella* groups A-D *Yersinia* spp.
A/Ⓐ H₂S +	K/K or K/NC H₂S +	*Salmonella* spp. (rare)
A/Ⓐ	K/A or A/A	*Escherichia coli* (rare)
A/A	K/A or A/A	*Escherichia coli* *Yersinia* spp. *Aeromonas* spp. (oxidase-positive) *Vibrio cholerae* (rare, oxidase-positive)
A/A	K/K or K/NC	*Vibrio* spp. (oxidase-positive)

A, Acid; Ⓐ, acid and gas production; *H₂S*, hydrogen sulfide; *K*, alkaline; *LIA*, lysine iron agar; *NC*, no change; *TSI*, triple sugar iron agar.
*Vibrio spp. and *Aeromonas* spp. are included in this table because they grow on the same media as the *Enterobacteriaceae* and may be enteric pathogens; identification of these organisms is discussed in Chapter 28.
†TSI and LIA reactions described in this table are only screening tests. The identity of possible enteric pathogens must be confirmed by specific biochemical and serologic testing.
‡Details regarding the TSI and LIA procedures can be found in Chapter 13.

the somatic "O" antigen 157 and the flagellar "H" antigen 7 are present. Latex reagents and antisera are now also available for detecting some non-0157, sorbitol-fermenting Shiga toxin–producing strains of *E. coli* (Meridian Diagnostics Inc., Cincinnati, Ohio; Oxoid, Inc., Ogdensburg, NY). Some national reference laboratories are, therefore, simply performing tests for Shiga-

toxin rather than searching for O157 or non-O157 strains by culture. Unfortunately, isolates are not available then for strain typing for epidemiologic purposes. Laboratory tests to identify enteropathogenic, enterotoxigenic, enteroinvasive, and enteroaggregative *E. coli* that cause gastrointestinal infections usually involve animal, tissue culture, or molecular studies performed in reference laboratories.

Most commercial systems will identify *Y. pestis* if a heavy inoculum is used. *All* isolates biochemically grouped as a *Yersinia* spp. should be forwarded to the nearest public health laboratory for confirmatory testing.

SERODIAGNOSIS

Serodiagnostic techniques are used for only two members of the family *Enterobacteriaceae*, that is, *S. typhi* and *Y. pestis*. Agglutinating antibodies can be measured in the diagnosis of typhoid fever; a serologic test for *S. typhi* is part of the "febrile agglutinins" panel and is individually known as the Widal test. Because results obtained by using the Widal test are somewhat unreliable, this method is no longer widely used.

Serologic diagnosis of plague is possible using either a passive hemagglutination test or enzyme-linked immunosorbent assay; these tests are usually performed in reference laboratories.

ANTIMICROBIAL SUSCEPTIBILITY TESTING AND THERAPY

For many of the gastrointestinal infections caused by Enterobacteriaceae, the inclusion of antimicrobial agents as part of the therapeutic strategy is controversial or at least uncertain (Table 22-5).

For extraintestinal infections, antimicrobial therapy is a vital component of patient management (Table 22-6). Although a broad spectrum of agents may be used for therapy against Enterobacteriaceae (see Table 12-6 for a detailed listing), every clinically relevant species is capable of acquiring and using one or more of the resistance mechanisms discussed in Chapter 11. The unpredictable nature of any clinical isolate's antimicrobial susceptibility requires that testing be done as a guide to therapy. As discussed in Chapter 12, several standard methods and commercial systems have been developed for this purpose.

PREVENTION

Vaccines are available for typhoid fever and bubonic plague; however, neither is routinely recommended in

Table 22-5 Therapy for Gastrointestinal Infections Caused by *Enterobacteriaceae*

Organisms	Therapeutic Strategies
Enterotoxigenic *E. coli* (ETEC) Enteroinvasive *E. coli* (EIEC) Enteropathogenic *E. coli* (EPEC) Enterohemorrhagic *E. coli* (EHEC) Enteroaggregative *E. coli* (EAEC)	Supportive therapy, such as oral rehydration, is indicated in cases of severe diarrhea; for life-threatening infections, such as hemolytic-uremic syndrome associated with EHEC, transfusion and hemodialysis may be necessary. Antimicrobial therapy may shorten duration of gastrointestinal illness, but many of these infections will resolve without such therapy. Because these organisms may develop resistance (see Table 22-6), antimicrobial drug therapy for non–life-threatening infections may be contraindicated
Shigella spp.	Oral rehydration; antimicrobial drug therapy may be used to shorten the period of fecal excretion and perhaps limit the clinical course of the infection. However, because of the risk of resistance, using antimicrobial drug therapy for less serious infections may be questioned
Salmonella spp.	For enteric fevers (e.g., typhoid fever) and extraintestinal infections (e.g., bacteremia, etc.) antimicrobial agents play an important role in therapy. Potentially effective agents for typhoid include quinolones, chloramphenicol, trimethoprim/sulfamethoxazole, and advance-generation cephalosporins, such as ceftriaxone; however, first- and second-generation cephalosporins and aminoglycosides are not effective. For nontyphoidal *Salmonella* bacteremia, a third-generation cephalosporin (e.g., ceftriaxone) is frequently used. For gastroenteritis, replacement of fluids is most important. Antimicrobial therapy generally is not recommended for either treatment of the clinical infection or decreasing the time that a patient excretes the organism
Yersinia enterocolitica and *Yersinia pseudotuberculosis*	The need for antimicrobial therapy for enterocolitis and mesenteric lymphadenitis is not clear. In cases of bacteremia, piperacillin, third-generation cephalosporins, aminoglycosides, and trimethoprim/sulfamethoxazole are potentially effective agents. *Y. enterocolitica* is frequently resistant to ampicillin and first-generation cephalosporins, whereas *Y. pseudotuberculosis* isolates are generally susceptible

Table 22-6 Antimicrobial Therapy and Susceptibility Testing of Clinically Relevant *Enterobacteriaceae*

Organism	Therapeutic Options	Potential Resistance to Therapeutic Options	Testing Methods*	Comments
E. coli, Citrobacter spp., *Enterobacter* spp., *Morganella* spp., *Proteus* spp., *Providencia* spp., and *Serratia* spp.	Several agents from each major class of antimicrobials, including aminoglycosides, beta-lactams, and quinolones have activity. See Table 12-6 for listing of specific agents that should be selected for in vitro testing. For urinary tract infections, single agents may be used; for systemic infections, potent beta-lactams are used, frequently in combination with an aminoglycoside	Yes; every species is capable of expressing resistance to one or more antimicrobials belonging to each drug class	As documented in Chapter 12; disk diffusion, broth dilution, agar dilution, and commercial systems	In vitro susceptibility testing results are important for guiding therapy
Yersinia pestis	Streptomycin is the therapy of choice; tetracycline or chloramphenicol are effective alternatives	Yes, but rare	See CLSI document M100-515, testing must only be performed in a licensed reference laboratory	Manipulation of cultures for susceptibility testing is dangerous for laboratory personnel and is not necessary

*Validated testing methods include those standard methods recommended by the Clinical and Laboratory Standards Institute (CLSI) and those commercial methods approved by the Food and Drug Administration (FDA).

the United States. An oral, multiple-dose vaccine prepared against *S. typhi* strain Ty2la or a parenteral single-dose vaccine containing Vi antigen is available for people traveling to an endemic area or for household contacts of a documented *S. typhi* carrier.[1]

An inactivated multiple-dose, whole-cell bacterial vaccine is available for bubonic plague for people traveling to an endemic area. However, this vaccine will not provide protection against pneumonic plague.[1] Individuals exposed to pneumonic plague should be given chemoprophylaxis with doxycycline (adults) or trimethoprim/sulfamethoxazole (children younger than 8 years of age).[1]

Case Study

A 47-year-old woman who had undergone kidney transplantation 2 years earlier presented to the hospital with fever and confusion. Blood cultures obtained on admission were positive with a gram-negative rod. A direct identification strip was inoculated from the blood culture that keyed out as *Shigella* spp., with very few positive reactions. It did not type with *Shigella* antisera. The test was repeated from a colony the next day, with the same low number of positive reactions. However, the technologist noticed that the original strip had been incubating on the counter and was now positive for urease and a number of sugar fermentation reactions. A new code was determined adding the additional reactions, and the organism keyed out as *Yersinia pseudotuberculosis*. When the patient was questioned, she admitted that she had been eating unpasteurized imported goat cheese.

QUESTIONS

1. What tests would you do to confirm the identification of *Y. pseudotuberculosis*?
2. Why did the reactions change in the second incubation period?
3. Had this organism been *Yersinia pestis*, the agent of plague and an agent on the list of potential agents for biologic warfare, what reactions would have been different?
4. If the isolate were urease-negative and nonmotile, what should you do next?

REFERENCES

1. Committee on Infectious Diseases: *2006 Red book: report of the Committee on Infectious Diseases*, ed 27, Elk Grove, Ill, 2006, American Academy of Pediatrics.

2. Farmer JJ III: *Enterobacteriaceae*: introduction and identification. In Murray PR, Baron EJ, Jorgensen JH, et al, editors: *Manual of clinical microbiology*, ed 8, Washington, DC, 2003, ASM Press.

3. National Committee for Clinical Laboratory Standards: *Abbreviated identification of bacteria and yeast*; approved guideline M35-A, Wayne, Pa, 2002, NCCLS.

ADDITIONAL READING

Abbott SL: *Klebsiella, Enterobacter, Citrobacter, Serratia, Plesiomonas*, and other *Enterobacteriaceae*. In Murray PR, Baron EJ, Jorgensen JH, et al, editors: *Manual of clinical microbiology*, ed 8, Washington, DC, 2003, ASM Press.

Balows A, Truper HG, Dworkin M, et al, editors: *The prokaryotes. A handbook on the biology of bacteria: ecophysiology, isolation, identification, applications*, ed 2, New York, 1992, Springer-Verlag.

Bockemühl J, Wang JD: *Yersinia*. In Murray PR, Baron EJ, Jorgensen JH, et al, editors: *Manual of clinical microbiology*, ed 8, Washington, DC, 2003, ASM Press.

Boemare NE, Akhurst RJ, Mourant RG: DNA relatedness between *Xenorhabdus* spp (*Enterobacteriaceae*), symbiotic bacteria of entomopathogenic nematodes, and a proposal to transfer *Xenorhabdus luminescens* to a new genus, *Photorhabdus* gen nov, *Int J Syst Bacteriol* 43:249, 1993.

Bopp C, Brenner FW, Fields PI, et al: *Escherichia, Shigella*, and *Salmonella*. In Murray PR, Baron EJ, Jorgensen JH, et al, editors: *Manual of clinical microbiology*, ed 8, Washington, DC, 2003, ASM Press.

Brenner DJ, O'Hara CM, Grimont PA: Biochemical identification of *Citrobacter* species defined by DNA hybridization and description of *Citrobacter gillenii* sp nov (formerly *Citrobacter* genomospecies 10) and *Citrobacter murliniae* sp nov (formerly *Citrobacter* genomospecies 11), *J Clin Microbiol* 37:2619, 1999.

Butler T, Dennis DT: *Yersinia* species (including plague). In Mandell GL, Bennett JE, Dolin R, editors: *Principles and practice of infectious diseases*, ed 6, Philadelphia, 2005, Elsevier Churchill Livingstone.

Carter JS, Bowden FJ, Bastian I, et al: Phylogenetic evidence for reclassification of *Calymmatobacterium granulomatis* as *Klebsiella granulomatis* comb nov, *Int J Syst Bacteriol* 49:1695, 1999.

Donnenberg MS: *Enterobacteriaceae*. In Mandell GL, Bennett JE, Dolin R, editors: *Principles and practice of infectious diseases*, ed 6, Philadelphia, 2005, Elsevier Churchill Livingstone.

DuPont HL: *Shigella* species (bacillary dysentery). In Mandell GL, Bennett JE, Dolin R, editors: *Principles and practice of infectious diseases*, ed 6, Philadelphia, 2005, Elsevier Churchill Livingstone.

Kharsany AB, Hoosen AA, Kiepiela P, et al: Culture of *Calymmatobacterium granulomatis*, *Clin Infect Dis* 22:391, 1996.

Kosako Y, Tamura K, Sakazaki R, et al: *Enterobacter kobei* sp nov: a new species of the family *Enterobacteriaceae* resembling *Enterobacter cloacae*, *Curr Microbiol* 33:261, 1996.

Müller HE, Brenner DJ, Fanning GR: Emended description of *Buttiauxella agrestis* with recognition of six new species of *Buttiauxella* and two new species of *Kluyvera: Buttiauxella ferragutiae* sp nov, *Buttiauxella gaviniae* sp nov, *Buttiauxella brennerae* sp nov, *Buttiauxella izardii* sp nov, *Buttiauxella noackiae* sp nov, *Buttiauxella warmboldiae* sp nov, *Kluyvera cochleae* sp nov, and *Kluyvera georgiana* sp nov, *Int J Syst Bacteriol* 46:50, 1996.

Neubauer H, Aleksic S, Hensel A, et al: *Yersinia enterocolitica* 16S rRNA gene types belong to the same genospecies but form three homology groups, *Int J Med Microbiol* 290:61, 2000.

O'Hara CM, Brenner FW, Steigerwalt AG, et al: Classification of *Proteus vulgaris* biotype 3 with recognition of *Proteus hauseri* sp nov, nom rev and unnamed *Proteus* genomospecies 4, 5 and 6, *Int J Syst Evol Microbiol* 50:1869, 2000.

Pegues DA, Ohl ME, Miller SI: *Salmonella* species, including *Salmonella typhi.* In Mandell GL, Bennett JE, Dolin R, editors: *Principles and practice of infectious diseases,* ed 6, Philadelphia, 2005, Elsevier Churchill Livingstone.

Quan TJ: Plague. In Wentworth BB, editor: *Diagnostic procedures for bacterial infections,* ed 7, Washington, DC, 1987, American Public Health Association.

Ryan KJ, editor: *Sherris medical microbiology: an introduction to infectious diseases,* Norwalk, Conn, 1994, Appleton and Lange.

Salyers AA, Whitt DD: *Bacterial pathogenesis: a molecular approach,* Washington, DC, 1994, ASM Press.

Schaechter M, Medoff G, Eisenstein BI, editors: *Mechanisms of microbial disease,* ed 2, Baltimore, 1993, Williams & Wilkins.

Schonheyder HC, Jensen KT, Frederiksen W: Taxonomic notes: synonymy of *Enterobacter cancerogenus* (Urošević, 1966) Dickey and Zumoff, 1988 and *Enterobacter taylorae* Farmer et al, 1985 and resolution of an ambiguity in the biochemical profile, *Int J Syst Bacteriol* 44:586, 1994.

Validation List No 75, *Int J Syst Evol Microbiol* 50:1415, 2000.

Validation of the publication of new names and new combinations previously effectively published outside the IJSEM: List No 73, *Int J Syst Evol Microbiol* 50:423, 2000.

Genera and Species to Be Considered

Current Name	Previous Name
Acinetobacter spp.; saccharolytic, nonhemolytic	Acinetobacter baumanii, Acinetobacter calcoaceticus, Acinetobacter anitratus Acinetobacter calcoaceticus subsp. anitratus
Acinetobacter spp.; saccharolytic, hemolytic	Acinetobacter alcaligenes, Acinetobacter anitratus, Acinetobacter haemolyticus
Acinetobacter spp.; asaccharolytic, nonhemolytic	Acinetobacter calcoaceticus subsp. Iwoffi, Acinetobacter johnsonii, A. junii, A. Iwoffi
Acinetobacter spp.; asaccharolytic, hemolytic	
Bordetella holmesii	CDC group NO-2
Bordetella parapertussis	
Bordetella trematum	
Burkholderia gladioli	Pseudomonas gladioli, Pseudomonas marginata
CDC group NO-1	
Pseudomonas luteola	Chryseomonas luteola, CDC Group Ve-1
Pseudomonas oryzihabitans	Flavimonas oryzihabitans, CDC Group Ve-2
Stenotrophomonas maltophilia	Xanthomonas maltophilia, Pseudomonas maltophilia

GENERAL CHARACTERISTICS

The organisms discussed in this chapter are considered together because, except for CDC group NO-1, they are all oxidase-negative and grow on MacConkey agar, as do the *Enterobacteriaceae.* However, unlike the *Enterobacteriaceae,* which ferment glucose, these organisms either oxidize or do not utilize glucose. Although NO-1 is oxidase-negative and does not usually grow on MacConkey agar, it is included here because it must be distinguished from the asaccharolytic *Acinetobacter* spp. The specific morphologic and physiologic features of these organisms are considered later in this chapter in the discussion of laboratory diagnosis. Of note, only *Acinetobacter* and *Stenotrophomonas* are routinely found in clinical specimens. *Bordetella parapertussis* is included in Table 23-4 in this chapter but is discussed in Chapter 39.

EPIDEMIOLOGY

These organisms inhabit environmental niches, with *Acinetobacter* spp. and *S. maltophilia* being widely distributed in nature and hospital environments (Table 23-1). Although none of these organisms are commonly part of the human flora, the relatively high prevalence of *Acinetobacter* spp. and *S. maltophilia* in hospitals frequently results in colonization of the skin and respiratory tract of patients.[1] The prevalence of these organisms is evidenced by the fact that, excluding the *Enterobacteriaceae, Acinetobacter* spp. and *S. maltophilia* are the second and third most common gram-negative bacilli, respectively, encountered in clinical specimens. In contrast, *Pseudomonas luteola, Pseudomonas oryzihabitans,* and CDC group NO-1 are not commonly encountered in the clinical setting.[4]

PATHOGENESIS AND SPECTRUM OF DISEASE

All of the organisms listed in Table 23-2 are opportunistic pathogens for which no definitive virulence factors are known. Because *Acinetobacter* spp. and *S. maltophilia* are relatively common colonizers of hospitalized patients, their clinical significance when found in patient specimens can be difficult to establish. In fact, these organisms are more frequently isolated as colonizers than as infecting agents. When infection does occur, it usually involves debilitated patients, such as those in burn or intensive care units, who have undergone medical instrumentation and/or have received multiple antimicrobial agents.[1,15] Infections caused by *Acinetobacter* spp. and *S. maltophilia* usually involve the respiratory or genitourinary tract, bacteremia, and, occasionally, wound infections, but infections involving several other body sites have been described. Community-acquired infections with these organisms can occur, but the vast majority of infections are nosocomial.[1,2,11-15]

LABORATORY DIAGNOSIS

SPECIMEN COLLECTION AND TRANSPORT

No special considerations are required for specimen collection and transport of the organisms discussed in this chapter. Refer to Table 5-1 for general information on specimen collection and transport.

Table 23-1   Epidemiology

Species	Habitat (Reservoir)	Mode of Transmission
Acinetobacter spp.	Widely distributed in nature, including the hospital environment. May become established as part of skin and respiratory flora of patients hospitalized for prolonged periods	Colonization of hospitalized patients from environmental factors; medical instrumentation (e.g., intravenous or urinary catheters) introduces organism to normally sterile sites
Stenotrophomonas maltophilia	Widely distributed in nature, including moist hospital environments. May become established as part of respiratory flora of patients hospitalized for prolonged periods	Colonization of hospitalized patients from environmental factors; medical instrumentation introduces organism to normally sterile sites (similar to transmission of *Acinetobacter* spp.)
CDC group NO-1	Oropharynx of animals. Not part of human flora	Animal bite or scratch
B. gladioli	Environmental pathogen of plants; occasionally found in respiratory tract of patients with cystic fibrosis but not part of normal flora	Transmission to humans uncommon, mode of transmission not known
Pseudomonas luteola *Pseudamonas oryzihabitans*	Environmental including moist hospital environments (e.g. respiratory therapy equipment). Not part of normal human flora	Uncertain, probably involves exposure of debilitated hospital patients to contaminated fluids and medical equipment
Bordetella holmseii, B. trematum	Unknown or part of normal human flora	Unknown, rarely found in humans

Table 23-2   Pathogenesis and Spectrum of Diseases

Species	Virulence Factors	Spectrum of Disease and Infections
Acinetobacter spp.	Unknown	Clinical isolates are often colonizers. True infections are usually nosocomial, occur during warm seasons, and most commonly involve the genitourinary tract, respiratory tract, wounds, soft tissues, and bacteremia
Bordetella holmesii, Bordetella trematum	Unknown	Bacteremia is only type of infection described
Burkholderia gladioli	Unknown	Role in human disease is uncertain, occasionally found in sputa of patients with cystic fibrosis, but clinical significance in this setting is uncertain
Pseudomonas luteola, P. oryzihabitans	Unknown	Catheter-related infections, septicemia, and peritonitis, usually associated with continuous ambulatory peritoneal dialysis, and miscellaneous mixed infections of other body sites
Stenotrophomonas maltophilia	Unknown. Intrinsic resistance to almost every commonly used antibacterial agent supports the survival of this organism in hospital environment	Most infections are nosocomial and include catheter-related infections, bacteremia, wound infections, pneumonia, urinary tract infections, and miscellaneous infections of other body sites
CDC group NO-1	Unknown	Animal bite wound infections

SPECIMEN PROCESSING

No special considerations are required for processing of the organisms discussed in this chapter. Refer to Table 5-1 for general information on specimen processing.

DIRECT DETECTION METHODS

Other than Gram stain of patient specimens, there are no specific procedures for the direct detection of these organisms in clinical material. *Acinetobacter* spp. are plump coccobacilli that tend to resist alcohol decolorization; they may be mistaken for *Neisseria* spp. The *Bordetella* spp. are coccobacilli or short rods. *S. maltophilia, P. oryzihabitans,* and *P. luteola* are short to medium-size straight rods. CDC group NO-1 are coccoid to medium-size bacilli.

Table 23-3 Colonial Appearance and Characteristics

Organism	Medium	Appearance
Stenotrophomonas maltophilia	BA	Large, smooth, glistening colonies with uneven edges and lavender-green to light purple pigment; greenish discoloration underneath growth; ammonia smell
	Mac	NLF
Acinetobacter spp.	BA	Smooth, opaque, raised, creamy, and smaller than *Enterobacteriaceae;* some genospecies are beta-hemolytic
	Mac	NLF, but colonies exhibit a purplish hue that may cause the organism to be mistaken for a LF (Figure 23-1)
Burkholderia gladioli	BA	Yellow
	Mac	NLF
Bordetella parapertussis	BA	Smooth, opaque, beta-hemolytic
	Mac	NLF, delayed growth
Bordetella holmesii	BA	Punctate, semiopaque, convex, round with greening of blood usually accompanied by lysis
	Mac	NLF, delayed growth
Bordetella trematum	BA	Convex, circular, grayish cream to white
	Mac	NLF
Pseudomonas oryzihabitans	BA	Wrinkled and smooth, transparent, yellow
	Mac	NLF
Pseudomonas luteola	BA	Smooth, opaque, yellow
	Mac	NLF
CDC group NO-1	BA	Small colonies that can be transferred intact by an inoculating needle
	Mac	NLF, but only 20% of strains grow

BA, 5% sheep blood agar; *LF,* lactose fermenter; *Mac,* MacConkey agar; *NLF,* non–lactose-fermenter.

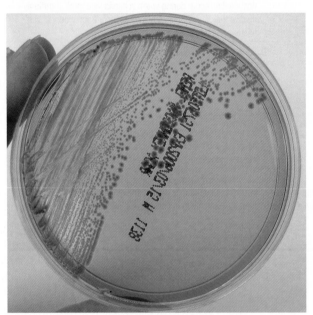

Figure 23-1 Colony of *Acinetobacter* spp. on MacConkey agar. Note purple color.

CULTIVATION

Media of Choice

In addition to their ability to grow on MacConkey agar, all of the genera described in this chapter grow well on 5% sheep blood and chocolate agars. These organisms also grow well in the broth of blood culture systems and in common nutrient broths such as thioglycollate and brain-heart infusion.

Incubation Conditions and Duration

These organisms generally produce detectable growth on 5% sheep blood and chocolate agars when incubated at 35° C in carbon dioxide or ambient air for a minimum of 24 hours. MacConkey agar should be incubated only in ambient air.

Colonial Appearance

Table 23-3 describes the colonial appearance and other distinguishing characteristics (e.g., hemolysis and odor) of each genus when grown on 5% sheep blood and MacConkey agars.

APPROACH TO IDENTIFICATION

Acinetobacter spp. and *S. maltophilia* are reliably identified by the API 20E system (bioMérieux Inc., St. Louis, Mo), although other commercial systems may not perform as well. The other organisms are most reliably identified using conventional biochemical and physiologic characteristics, such as those outlined in Table 23-4.

Table 23-4 Key Biochemical and Physiologic Characteristics

Organism	Growth on Mac Conkey	Motile	Oxidizes Glucose	Oxidizes Maltose	Esculin Hydrolysis	Lysine Decarboxylase	Nitrate Reduction	Urea, Christensen's
Stenotrophomonas maltophilia	+	+	+	+	V	+	V	–
Saccharolytic *Acinetobacter*	+	–	+	–	–	–	–	V
Asaccharolytic *Acinetobacter*	+	–	–	V	–	–	–	V
Burkholderia gladioli	+	+	+	–	–	–	V	V
Bordetella parapertussis	+	–	–	–	–	ND	–	+
Bordetella holmesii†	+ or (+)	–	–	–	–	–	–	–
Bordetella trematum	+	+	–	–	–	–	V	–
Pseudomonas oryzihabitans	+	+ p, 1-2	+	+	–	–	–	V
Pseudomonas luteola	+	+ p, >2	+	+	+	–	V	V
CDC group NO–1	V	–	–	–	–	–	+	–

Compiled from references 7, 14, 16, and 17.

V, Variable; +, >90% of strains are positive; –, >90% of strains are negative; ( ), delayed; *ND,* no data; *p,* polar flagella.

B. gladioli is included with the oxidase-negative organisms because oxidase reactions are frequently weak and may only be positive with Kovacs' method.

†Brown soluble pigment

Comments Regarding Specific Organisms

There are 25 genospecies or genomospecies in the genus *Acinetobacter.* Each genospecies comprises a distinct DNA hybridization group and is given a numeric designation, which has replaced previous species names. The genus is also divided into two groups; one contains the saccharolytic (glucose-oxidizing) species and the other contains the asaccharolytic (non–glucose-utilizing) species.

Most glucose-oxidizing, nonhemolytic strains were previously identified as *A. baumanii,* and most non–glucose-utilizing, nonhemolytic strains were designated as *A. lwoffi.* The majority of beta-hemolytic organisms were previously called *A. haemolyticus.* Nitrate-reducing strains of asaccharolytic *Acinetobacter* spp. are difficult to differentiate from CDC group NO-1. The *Acinetobacter* transformation test provides the most dependable criterion for this purpose, but this test is not commonly performed in clinical microbiology laboratories.

S. maltophilia can produce biochemical profiles similar to those of *Burkholderia cepacia,* but a negative oxidase test most often rules out the latter. *S. maltophilia* also oxidizes maltose faster than glucose (hence the species name, *maltophilia,* "maltose loving") and produces a brown pigment on heart infusion agar that contains tyrosine.

SERODIAGNOSIS

Serodiagnostic techniques are not generally used for the laboratory diagnosis of infections caused by the organisms discussed in this chapter.

ANTIMICROBIAL SUSCEPTIBILITY TESTING AND THERAPY

Acinetobacter spp. and *S. maltophilia* can exhibit resistance to a wide array of antimicrobial agents, making the selection of agents for optimal therapy difficult (Table 23-5). This underscores the importance of establishing clinical significance of individual isolates before antimicrobial testing is performed and results are reported (see Chapter 12 for a discussion of criteria used to establish significance). Not doing so could lead to the inappropriate treatment of patients with expensive and potentially toxic agents.

For urinary tract infections caused by *Acinetobacter* spp., single-drug therapy is usually sufficient. In contrast, more serious infections, such as pneumonia or bacteremia, may require the use of a beta-lactam agent in combination with an aminoglycoside. A new agent, tigecycline, also has potent activity against these organisms. Because this genus is able to acquire and express resistance to most antimicrobial agents, including imipenem, in vitro testing is recommended for clinically relevant isolates.[1,5] Methods outlined by the Clinical and Laboratory Standards Institute (CLSI) appear to be suitable for testing *Acinetobacter* spp., *S. maltophilia,* and other organisms listed in Table 23-5.[7-9]

S. maltophilia is notoriously resistant to most currently available antimicrobial agents, leaving trimethoprim-sulfamethoxazole as the primary drug of choice for infections caused by this species.[3,5,10] Although a few other agents, such as minocycline,

Table 23-5   Antimicrobial Therapy and Susceptibility Testing

Species	Therapeutic Options	Potential Resistance to Therapeutic Options	Validated Testing Methods*	Comments
Acinetobacter spp.	No definitive guidelines. Potentially active agents include beta-lactam/ beta-lactamase inhibitor combinations, ceftazidime, imipenem, ciprofloxacin, tigecycline, and aminoglycosides	Yes; resistance to beta-lactams, aminoglycosides, and quinolones	Disk diffusion, broth dilution, and agar dilution	In vitro susceptibility testing results important for guiding therapy
Bordetella holmesii	No definitive guidelines. Potentially active agents include penicillins, cephalosporins, and quinolones	Unknown	Not available	
Burkholderia. gladioli	No definitive guidelines. Potentially active agents include imipenem, piperacillin, and ciprofloxacin	Yes	See CLSI document M100, Performance Standards for Antimicrobial Susceptibility Testing	Rarely involved in human infections. Reliable therapeutic data are limited
Pseudomonas luteola, P. oryzihabitans	No definitive guidelines. Potentially active agents include cefotaxime, ceftriaxone, ceftazidime, imipenem, quinolones, and aminoglycosides	Yes	See CLSI document M100, Performance Standards for Antimicrobial Susceptibility Testing	Rarely involved in human infections
Stenotrophomonas maltophilia	Multiple resistance leaves few therapeutic choices. Therapy of choice is trimethoprim-sulfamethoxazole. Potential alternatives include minocycline, ticarcillin/clavulanic acid, and chloramphenicol	Yes, intrinsically resistant to most beta-lactams and aminoglycosides; frequently resistant to quinolones	See CLSI document M100, Performance Standards for Antimicrobial Susceptibility Testing	May be tested by various methods, but profiles obtained with beta-lactams can be seriously misleading
CDC group NO-1	No definitive guidelines. Appear susceptible to beta-lactam antibiotics	Unknown	Not available	

*Validated testing methods include those standard methods recommended by the Clinical and Laboratory Standards Institute (CLSI) and those commercial methods approved by the Food and Drug Administration (FDA).

ticarcillin/clavulanic acid, and chloramphenicol, often exhibit in vitro activity, clinical experience with these agents is not extensive. Therefore, trimethoprim-sulfamethoxazole remains the drug of choice.

The other agents should only be considered when trimethoprim-sulfamethoxazole–resistant strains are encountered. Even then, the potential efficacy of these other agents is suspect because of the ability of *S. maltophilia* to rapidly develop resistance.[3,6] As indicated in Table 23-5, there are CLSI guidelines available for testing several of the organisms listed in this chapter.[7,9]

PREVENTION

Because these organisms are ubiquitous in nature and are not generally a threat to human health, there are no recommended vaccination or prophylaxis protocols.

Hospital-acquired infections are best controlled by following appropriate sterile techniques and infection control guidelines and by implementing effective protocols for the sterilization and decontamination of medical supplies.

Case Study

A 3-month-old boy hospitalized in the intensive care nursery since birth was recovering from corrective congenital heart surgery. The child developed signs of sepsis. It was noted that a central line had been in place for some time, through which heparin had been given to reduce clot formation. A blood culture drawn through the line was positive. Subsequently, a second culture was collected from a peripheral stick. Both cultures grew a gram-negative rod. The child's diagnosis was catheter-related bacteremia; antibiotics and removal of the catheter were successful in clearing the infection.

QUESTIONS

1. A commercial system identified the gram-negative bacilli as *Acinetobacter,* but indicated that all the biochemical tests in the system, including utilization of glucose, were negative and the identification should be confirmed. What rapid biochemical tests would you do to confirm the identification?
2. The patient's isolate did not grow on MacConkey agar. Which test is needed to separate this genus from NO-1?
3. How did the patient acquire this microorganism?
4. What is the meaning of a glucose-oxidizing gram-negative rod?
5. What is the best method to distinguish an asaccharolytic microorganism from a fastidious gram-negative rod that is negative in OF medium because it is unable to grow?

REFERENCES

1. Bergogne-Berezin E, Towner KJ: *Acinetobacter* spp. as nosocomial pathogens: microbiological, clinical, and epidemiological features, *Clin Microbiol Rev* 9:148, 1996.
2. Clinical and Laboratory Standards Institute: *Methods for dilution antimicrobial tests for bacteria that grow aerobically,* M7-A6, Villanova, Pa, CLSI, 2005.
3. Clinical and Laboratory Standards Institute: *Performance standards for antimicrobial disk susceptibility tests,* M2-A8, Villanova, Pa, CLSI, 2005.
4. Clinical and Laboratory Standards Institute: *Performance standards for antimicrobial susceptibility testing,* M100-S15, Villanova, Pa, CLSI, 2005.
5. Esteban J, Valero-Moratalla ML, Alcazar R, et al: Infections due to *Flavimonas oryzihabitans:* case report and literature review, *Eur J Clin Microbiol Infect Dis* 12:797, 1993.
6. Garrison MW, Anderson DE, Campbell DM, et al: *Stenotrophomonas maltophilia:* emergence of multidrug-resistant strains during therapy and in an in vitro pharmacodynamic chamber model, *Antimicrob Agents Chemother* 40:2859, 1996.
7. Gilligan PH, Lum G, Vandamme PA, et al: *Burkholderia, Stenotrophomonas, Ralstonia, Brevundimonas, Comamonas, Delftia, Pandoraea,* and *Acidovorax.* In Murray PR, Baron EJ, Jorgensen JH, et al, editors: *Manual of clinical microbiology,* ed 8, Washington, DC, 2003, ASM Press.
8. Hollis DG, Moss CW, Daneshvar MI, et al: Characterization of Centers for Disease Control group NO-1, a fastidious, nonoxidative, gram-negative organism associated with dog and cat bites, *J Clin Microbiol* 31:746, 1993.
9. Livermore DM: β-Lactamases in laboratory and clinical resistance, *Clin Microbiol Rev* 8:557, 1995.
10. Metchock B, Thornsberry C: Susceptibility of *Xanthomonas (Pseudomonas) maltophilia* to antimicrobial agents, *Antimicrob Newsletter* 6:35, 1989.
11. Pankuch GA, Jacobs MR, Rittenhouse SF, et al: Susceptibilities of 123 strains of *Xanthomonas maltophilia* to eight β-lactams (including β-lactam–β-lactamase inhibitor combinations) and ciprofloxacin tested by five methods, *Antimicrob Agents Chemother* 38:2317, 1994.
12. Rahav G, Simhon A, Mattan Y, et al: Infections with *Chryseomonas luteola* (CDC group Ve-1) and *Flavimonas oryzihabitans* (CDC group Ve-2), *Medicine* 74:83, 1995.
13. Reed RP: *Flavimonas oryzihabitans* sepsis in children, *Clin Infect Dis* 22:733, 1996.
14. Schreckenberger PC, Daneshvar MI, Weyant RS, et al: *Acinetobacter, Achromobacter, Chryseobacterium, Moraxella,* and other nonfermentative gram-negative rods. In Murray PR, Baron EJ, Jorgensen JH, editors: *Manual of clinical microbiology,* ed 8, Washington, DC, 2003, ASM Press.
15. Seifert H, Strate A, Pulverer G: Nosocomial bacteremia due to *Acinetobacter baumanii:* clinical features, epidemiology, and predictors of mortality, *Medicine* 74:340, 1995.
16. Weyant RS, Hollis DG, Weaver RE, et al: *Bordetella holmesii* sp nov: a new gram-negative species associated with septicemia, *J Clin Microbiol* 33:1, 1995.
17. Weyant RS, Moss CW, Weaver RE, et al, editors: *Identification of unusual pathogenic gram-negative aerobic and facultatively anaerobic bacteria,* ed 2, Baltimore, 1996, Williams & Wilkins.

ADDITIONAL READING

Allen DM, Hartman BJ: *Acinetobacter* species. In Mandell GL, Bennett JE, Dolin R, editors: *Principles and practice of infectious diseases,* ed 6, Philadelphia, 2005, Elsevier Churchill Livingstone.
Hewlett EL: *Bordetella* species. In Mandell GL, Bennett JE, Dolin R, editors: *Principles and practice of infectious diseases,* ed 6, Philadelphia, 2005, Elsevier Churchill Livingstone.
Maschmeyer G, Göbel UB: *Stenotrophomonas maltophilia* and *Burkholderia cepacia.* In Mandell GL, Bennett JE, Dolin R, editors: *Principles and practice of infectious diseases,* ed 6, Philadelphia, 2005, Elsevier Churchill Livingstone.
Steinberg JP, Del Rio C: Other gram-negative and gram-variable bacilli. In Mandell GL, Bennett JE, Dolin R, editors: *Principles and practice of infectious diseases,* ed 6, Philadelphia, 2005, Elsevier Churchill Livingstone.

CHAPTER 24 *Pseudomonas, Burkholderia, and Similar Organisms*

Genera and Species to Be Considered

Current Name	Previous Name
Acidovorax delafieldii	*Pseudomonas delafieldii*
Acidovorax temperans	
Brevundimonas diminuta	*Pseudomonas diminuta*
Brevundimonas vesicularis	*Pseudomonas vesicularis*
Burkholderia cepacia complex	*Pseudomonas cepacia*
Burkholderia pseudomallei	*Pseudomonas pseudomallei*
Burkholderia mallei	*Pseudomonas mallei*
Pandoraea spp.	CDC group WO-2
Pseudomonas aeruginosa	
Pseudomonas fluorescens	
Pseudomonas mendocina	
Pseudomonas monteilii	
Pseudomonas putida	
Pseudomonas stutzeri (includes CDC group Vb-3)	CDC group IVd
Pseudomonas veronii	
Pseudomonas-like group 2	
CDC group Ic	
Ralstonia mannitolilytica	"*Pseudomonas thomasii*," *Ralstonia pickettii* biovar 3
Ralstonia pickettii	*Pseudomonas pickettii, Burkholderia pickettii*, Va-1, Va-2

GENERAL CHARACTERISTICS

At one time, most of the species belonging to the genera *Brevundimonas, Burkholderia, Ralstonia,* and *Acidovorax* were all members of the genus *Pseudomonas.* Organisms in these genera share many similar characteristics. They are aerobic, straight, and slender gram-negative bacilli whose cells range from 1 to 5 μm in length and 0.5 to 1 μm in width. All species except *B. mallei* are motile. Members of these genera use a variety of carbohydrate, alcohol, and amino acid substrates as carbon and energy sources. Although they are able to survive and possibly grow at relatively low temperatures (i.e., as low as 4° C),

the optimum temperature range for growth of most species is between 30° and 37° C; that is, they are mesophilic. *Burkholderia gladioli, Pseudomonas luteola,* and *Pseudomonas oryzihabitans* are oxidase-negative and are discussed in Chapter 23. *Pseudomonas alcaligenes, Pseudomonas pseudoalcaligenes, Ralstonia paucula, Ralstonia gilardii, Comamonas* spp. (including the former *Pseudomonas testosteroni*), and *Delftia acidovorans* (formerly *Pseudomonas acidovorans*) are not able to utilize glucose oxidatively or fermentively and are discussed in Chapter 27. *Acidovorax facilis* is MacConkey-negative. The organisms in this chapter are all oxidase-positive, grow on MacConkey agar, and utilize glucose oxidatively.

EPIDEMIOLOGY

BURKHOLDERIA SPP. AND RALSTONIA PICKETTII

Burkholderia spp. and *Ralstonia pickettii* are inhabitants of the environment and are not considered part of normal human flora (Table 24-1). As such, their transmission usually involves human contact with heavily contaminated medical devices or substances that are encountered in the hospital setting.

Among *Burkholderia* spp. found in the United States, *B. cepacia* is the one most commonly found in clinical specimens. Plants, soil, and water serve as reservoirs. This organism is also able to survive on or in medical devices and disinfectants. Intrinsic resistance to multiple antimicrobial agents also contributes to the organism's survival in hospitals. Human acquisition of *B. cepacia* resulting in colonization or infection usually involves direct contact with contaminated foods; medical solutions, including disinfectants; and devices such as respiratory equipment. Alternatively, person-to-person transmission also has been documented.[6]

B. pseudomallei also is an environmental inhabitant of niches similar to those described for *B. cepacia*, but it is geographically restricted to tropical and subtropical

Table 24-1   Epidemiology

Species	Habitat (reservoir)	Mode of Transmission
Burkholderia cepacia	Environmental (soil, water, plants); survives well in hospital environment; not part of normal human flora; may colonize respiratory tract of patients with cystic fibrosis	Exposure of medical devices and solutions contaminated from the environment; person-to-person transmission also documented
B. pseudomallei	Environmental (soil, streams, surface water such as rice paddies); limited to tropical and subtropical areas, notably Southeast Asia; not part of human flora	Inhalation or direct inoculation from environment through disrupted epithelial or mucosal surfaces
B. mallei	Causative agent of glanders in horses; mules, and donkeys; not part of human flora	Transmission to humans is extremely rare; associated with close animal contact and introduced through mucous membranes or broken skin.
Ralstonia pickettii	Environmental (multiple sources); found in variety of clinical specimens; not part of human flora	Mode of transmission is not known; likely involves exposure to contaminated medical devices and solutions
Pseudomonas aeruginosa	Environmental (soil, water, plants); survives well in domestic environments (e.g., hot tubs, whirlpools, contact lens solutions) and hospital environments (e.g., sinks, showers, respiratory equipment); rarely part of normal flora of healthy humans	Ingestion of contaminated food or water; exposure to contaminated medical devices and solutions; introduction by penetrating wounds; person-to-person transmission is assumed to occur
P. alcaligenes, P. pseudoalcaligenes, Pseudomonas sp. CDC group 1, "*P. denitrificans*," *Pseudomonas*-like group 2, and CDC group Ic	Environmental; not part of normal human flora	Uncertain. Rarely encountered in clinical specimens
P. fluorescens, P. putida, P. stutzeri, (including Vb-3), and *P. mendocina*	Environmental (soil and water); not part of normal human flora	Exposure to contaminated medical devices and solutions
Brevundimonas vesicularis and *B. diminuta*	Environmental; not part of normal human flora	Uncertain. Rarely encountered in clinical specimens
Acidovorax spp.	Environmental, soil; not part of human flora	Unknown. Rarely found in humans

areas of Australia and Southeast Asia. The organism is widely disseminated in soil, streams, ponds, and rice paddies. Human acquisition occurs through inhalation of contaminated debris or by direct inoculation through damaged skin or mucous membranes.[4]

Although *B. mallei* causes severe infections in horses and related animals, infections of humans are exceedingly rare. When transmission has occurred, it has been associated with close animal contact. *B. gladioli* is a plant pathogen that is only rarely found in sputa of patients with cystic fibrosis; the mode of transmission to humans and its clinical significance are unknown.[23]

Ralstonia pickettii is another environmental organism that is occasionally found in a variety of clinical specimens, such as blood, sputa of patients with cystic fibrosis, and urine. The mode of transmission is uncertain but is likely to involve exposure to contaminated materials.

PSEUDOMONAS SPP. AND BREVUNDIMONAS SPP.

These genera comprise several environmental species that rarely inhabit human skin or mucosal surfaces. In the clinical setting, *Pseudomonas aeruginosa* is the most commonly encountered gram-negative species that is not a member of the family *Enterobacteriaceae*. This organism survives in various environments in nature and in homes and hospitals (see Table 24-1). Because of the ubiquitous nature of *P. aeruginosa*, transmission of this organism to humans can occur in a variety of ways.

P. fluorescens, P. putida, and *P. stutzeri* also are environmental inhabitants, but they are much less commonly found in clinical specimens than is *P. aeruginosa*. The other pseudomonads and *Brevundimonas* spp. listed in Table 24-1 also are environmental organisms. Because they are rarely encountered in patient specimens, the mode of transmission to humans remains uncertain.

Table 24-2   Pathogenesis and Spectrum of Disease

Species	Virulence Factors	Spectrum of Disease and Infections
Burkholderia cepacia	Unknown. Binding of mucin from patients with cystic fibrosis may be involved. Intrinsic resistance to multiple antibiotics complicates therapy and may promote organism survival in hospital	Nonpathogenic to healthy human hosts; able to colonize and cause life-threatening infections in patients with cystic fibrosis or chronic granulomatous disease; other patients may suffer nonfatal infections of the urinary tract, respiratory tract, and other sterile body sites
B. pseudomallei	Unknown. Bacilli can survive within phagocytes	Wide spectrum from asymptomatic infection to melioidosis, of which there are several forms, including infections of the skin and respiratory tract, multisystem abscess formation, and bacteremia with septic shock
B. mallei	Unknown for human infections	Human disease is extremely rare. Infections range from localized acute or chronic suppurative infections of skin at site of inoculation to acute pulmonary infections and septicemia
Ralstonia pickettii	Unknown	Rarely encountered as cause of disease; non-pathogenic to healthy human host, but may be isolated from a variety of clinical specimens, including blood, sputum, and urine; when encountered, environmental contamination should be suspected
Pseudomonas aeruginosa	Exotoxin A, endotoxins, proteolytic enzymes, alginate, and pili; intrinsic resistance to many antimicrobial agents	Opportunistic pathogen that can cause community- or hospital-acquired infections Community-acquired infections: skin (folliculitis); external ear canal (otitis externa); eye, following trauma; bone (osteomyelitis), following trauma; heart (endocarditis) in IV drug abusers; and respiratory tract (patients with cystic fibrosis) Hospital acquired infections: respiratory tract, urinary tract, wounds, bloodstream (bacteremia), and central nervous system Key pathogen that infects lungs of cystic fibrosis patients
P. fluorescens, P. putida, and P. stutzeri (includes Vb-3)	Unknown. Infection usually requires patient with underlying disease to be exposed to contaminated medical devices or solutions	Uncommon cause of infection; have been associated with bacteremia, urinary tract infections, wound infections, and respiratory tract infections; when found in clinical specimen, significance should always be questioned
P. mendocina, P. alcaligenes, P. pseudoalcaligenes, Pseudomonas sp. CDC group 1, "P. denitrificans," Pseudomonas-like group 2, and CDC group Ic	Unknown	Not known to cause human infections
Brevundimonas vesicularis and B. diminuta	Unknown	Rarely associated with human infections. B. vesicularis is rare cause of bacteremia in patients suffering underlying disease
Acidovorax spp.	Unknown	Rarely isolated from clinical specimens. Not implicated in human infections.

PATHOGENESIS AND SPECTRUM OF DISEASE

BURKHOLDERIA SPP. AND RALSTONIA PICKETTII

Because these organisms are uncommon causes of infection in humans, very little is known about what, if any, virulence factors they exhibit.[14] Except for *B. pseudomallei*, the species listed in Table 24-2 generally are nonpathogenic for healthy human hosts.

The capacity of *B. cepacia* to survive in the hospital environment, which may be linked to the organism's intrinsic resistance to many antibiotics, provides the opportunity for this species to occasionally colonize and infect hospitalized patients.[8,9] In patients with cystic fibrosis or chronic granulomatous disease, the organism can cause fulminant lung infections and bacteremia, resulting in death.[17-20,22] In other types of patients, infections of the blood, urinary tract, and respiratory tract usually result from exposure to contaminated medical solutions or devices but are rarely fatal.[20]

Infections caused by *B. pseudomallei*, which can survive within human macrophages,[21] can range from being asymptomatic to severe. The disease is referred to as *melioidosis*; it has several forms, including formation of skin abscesses, sepsis and septic shock, abscess formation in several internal organs, and acute pulmonary disease.[4]

The remaining species listed in Table 24-2 are rarely encountered in human disease, and their clinical significance should be questioned when they are found in clinical specimens.[2]

PSEUDOMONAS SPP. AND *BREVUNDIMONAS* SPP.

Of the species in these two genera, *Pseudomonas aeruginosa* is the most thoroughly studied with regard to infections in humans. Although this organism is an environmental inhabitant, it is also a very successful opportunistic pathogen. Factors that contribute to pathogenicity include production of exotoxin A, which kills host cells by inhibiting protein synthesis, and production of several proteolytic enzymes and hemolysins that destroy cells and tissue. On the bacterial cell surface, pili may mediate attachment to host cells. Some strains produce alginate, a polysaccharide polymer that inhibits phagocytosis and contributes to infection potential in patients with cystic fibrosis. Additionally, *P. aeruginosa* can survive harsh environmental conditions and displays intrinsic resistance to a wide variety of antimicrobial agents that facilitate the organism's ability to survive in the hospital setting (see Table 24-2).

Even with the variety of potential virulence factors discussed, *P. aeruginosa* remains an opportunistic pathogen that requires compromised host defenses to establish infection. In normal, healthy hosts, infection is usually associated with events that disrupt or bypass protection provided by the epidermis (e.g., burns, puncture wounds, use of contaminated needles by intravenous drug abusers, eye trauma with contaminated contact lenses). The result is infections of the skin, bone, heart, or eye (see Table 24-2).

In patients with cystic fibrosis, *P. aeruginosa* has a predilection for infecting the respiratory tract. Although organisms rarely invade through respiratory tissue and into the bloodstream of these patients, the consequences of respiratory involvement alone are serious and life-threatening. In other patients, *P. aeruginosa* is a notable cause of nosocomial infections of the respiratory and urinary tracts, wounds, bloodstream, and even the central nervous system. For immunocompromised patients, such infections are often severe and frequently life-threatening. In some cases of bacteremia, the organism may invade and destroy walls of subcutaneous blood vessels, resulting in formation of cutaneous papules that become black and necrotic. This condition is known as *ecthyma gangrenosum.* Similarly, patients with diabetes may suffer a severe infection of the external ear canal (malignant otitis externa), which can progress to involve the underlying nerves and bones of the skull.

No known virulence factors have been associated with *P. fluorescens, P. putida,* or *P. stutzeri.* When infections do occur they usually involve a compromised patient exposed to contaminated medical materials.[15,16] Such exposure has been known to result in infections of the respiratory and urinary tracts, wounds, and bacteremia (see Table 24-2). However, because of their low virulence, whenever these species are encountered in clinical specimens, their significance should be highly suspect. Similar caution should be applied whenever the other *Pseudomonas* spp. or *Brevundimonas* spp. listed in Table 24-2 are encountered.[16]

LABORATORY DIAGNOSIS

SPECIMEN COLLECTION AND TRANSPORT

No special considerations are required for specimen collection and transport of organisms discussed in this chapter. Refer to Table 5-1 for general information on specimen collection and transport.

SPECIMEN PROCESSING

No special considerations are required for processing of the organisms discussed in this chapter. Refer to Table 5-1 for general information on specimen processing.

DIRECT DETECTION METHODS

Other than Gram stain of patient specimens, there are no specific procedures for the direct detection in clinical material of the organisms discussed in this chapter. These organisms usually appear as medium-size straight rods by Gram stain. Exceptions are *Brevundimonas diminuta,* which is a long straight rod; *Burkholderia mallei,* which is a coccobacillus; *Pseudomonas pseudomallei,* which is a small gram-negative rod with bipolar staining, making cells resemble "safety pins"; and CDC group Ic, which is a thin pleomorphic rod.

CULTIVATION

Media of Choice

Pseudomonas spp., *Brevundimonas* spp., *Burkholderia* spp., *Ralstonia pickettii,* and CDC group Ic grow well on routine laboratory media, such as 5% sheep blood agar and chocolate agar (Figure 24-1). Except for *Brevundimonas vesicularis,* all usually grow on MacConkey agar. All four genera also grow well in broth-blood culture systems and common nutrient broths, such as thioglycollate and brain-heart infusion. Specific selective media, such as *Pseudomonas cepacia* (PC) agar or oxidative-fermentative base–polymyxin B–bacitracin-lactose (OFPBL) agar may be used to isolate *Burkholderia cepacia* from respiratory

Table 24–3 Colonial Appearance and Other Characteristics of *Pseudomonas, Brevundimonas, Burkholderia, Ralstonia,* and Other Organisms

Organism	Medium	Appearance
Acidovorax delafieldii	BA Mac	No distinctive appearance NLF
A. temperans	BA Mac	No distinctive appearance NLF
Brevundimonas diminuta	BA Mac	Chalk white NLF
B. vesicularis	BA Mac	Orange pigment NLF, but only 66% grow
Burkholderia cepacia complex	BA Mac PC or OFPBL	Smooth and slightly raised; dirtlike odor NLF; colonies become dark pink to red due to oxidation of lactose after 4-7 days Smooth
B. pseudomallei	BA Mac Ashdown	Smooth and mucoid to dry and wrinkled (may resemble *P. stutzeri*) NLF Dry, wrinkled, violet–purple
B. mallei	BA Mac	No distinctive appearance NLF
Pandoraea spp.	BA Mac	No distinctive appearance NLF
Pseudomonas aeruginosa	BA Mac	Spreading and flat, serrated edges; confluent growth; often shows metallic sheen; bluish–green, red, or brown pigmentation; colonies often beta–hemolytic; grapelike or corn taco–like odor; mucoid colonies commonly seen in patients with cystic fibrosis NFL
P. fluorescens	BA Mac	No distinctive appearance NLF
P. mendocina	BA Mac	Smooth, nonwrinkled, flat, brownish–yellow pigment NLF
P. monteilii	BA Mac	No distinctive appearance NLF
P. putida	BA Mac	No distinctive appearance NLF
P. stutzeri and CDC group Vb–3	BA Mac	Dry, wrinkled, adherent, buff to brown NLF
P. veronii	BA Mac	No distinctive appearance NLF
Pseudomonas–like group 2	BA Mac	No distinctive appearance but colonies tend to stick to agar NLF
CDC group Ic	BA Mac	No distinctive appearance NLF
Ralstonia mannitolilytica	BA Mac	No distinctive appearance NLF
R. pickettii	BA Mac	No distinctive appearance but may take 72 hr to produce visible colonies NLF

BA, 5% sheep blood agar; *Mac,* MacConkey agar; *NLF,* non–lactose-fermenter; *OFPBL,* oxidative–fermentative base–polymyxin B–bacitracin–lactose; *PC, Pseudomonas cepacia* agar.

secretions of patients with cystic fibrosis (see Table 24-3). Ashdown medium is used to isolate *Burkholderia pseudomallei* when infections with this species are suspected.

Incubation Conditions and Duration

Detectable growth on 5% sheep blood and chocolate agars, incubated at 35° C in carbon dioxide or ambient air, generally occurs within 24 to 48 hours after inoculation. Growth on MacConkey agar incubated in ambient air at 35° C also is detectable within this same time. Selective media (PC or OFPBL) used for patients with cystic fibrosis may require incubation at 35° C in ambient air for up to 72 hours before growth is detected.

Colonial Appearance

Table 24-3 describes the colonial appearance and other distinguishing characteristics (e.g., hemolysis and odor) of each genus on common laboratory media.

APPROACH TO IDENTIFICATION

Most of the commercial systems available for identification of these organisms reliably identify *Pseudomonas aeruginosa* and *Burkholderia cepacia* complex, but their reliability for identification of other species is less certain.

Table 24-4 provides key phenotypic characteristics for identifying the species discussed in this chapter. These tests provide useful information for presumptive organism identification, but definitive identification often requires the use of a more extensive battery of tests usually performed by reference laboratories.

Comments Regarding Specific Organisms

A convenient and reliable identification scheme for *P. aeruginosa* involves the following conventional tests and characteristics:

- Oxidase-positive
- Triple sugar iron slant with an alkaline/no change (K/NC) reaction
- Good growth at 42° C
- Production of bright bluish-green, red, or brown diffusible pigment on Mueller-Hinton agar or trypticase soy agar (Figures 24-2 and 24-3)

P. aeruginosa, P. fluorescens, P. putida, P. veronii, and *P. monteilii* comprise the group known as the *fluorescent pseudomonads. P. aeruginosa* can be distinguished from the others in this group by its ability to grow at 42° C. Mucoid strains of *P. aeruginosa* from patients with cystic fibrosis may not exhibit the characteristic pigment and may react slower in biochemical tests than nonmucoid strains. Therefore, standard biochemicals should be held for the complete 7 days before being recorded as negative. This slow biochemical activity is often what prevents the identification of mucoid *P. aeruginosa* by commercial systems. *P. monteilii* can be distinguished from *P. putida* by its inability to oxidize xylose. Both can be distinguished from *P. fluorescens* by their inability to liquefy gelatin.

Burkholderia cepacia should be suspected whenever a nonfermentative organism that decarboxylates lysine

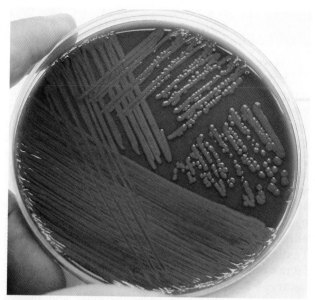

Figure 24-1 *Burkholderia cepacia* on chocolate agar. Note green pigment.

is encountered. Lysine decarboxylation is positive in 80% of strains. Correct identification of the occasional strains that are lysine-negative (20%), or oxidase-negative (14%), requires full biochemical profiling. *Pandoraea* spp. may be differentiated from *B. cepacia* by their failure to decarboxylate lysine and their inability to liquefy gelatin. They do not hydrolyze Tween 80, unlike *Ralstonia paucula,* which does.

The presumptive identification of other species in this chapter is fairly straightforward using the key characteristics given in Table 24-4. However, there are a few notable exceptions. First, when *B. cepacia* complex is identified by a commercial system in a cystic fibrosis patient, it should be reidentified by a combination of phenotypic and genotypic methods. This is also true if a rapid system identifies an organism as *B. gladioli* or *R. pickettii.* There are nine genomovars in the *B. cepacia* complex and appropriate speciation is very important.

SERODIAGNOSIS

Serodiagnostic techniques are not generally used for the laboratory diagnosis of infections caused by the organisms discussed in this chapter. An indirect hemagglutination assay is available in endemic areas in the Far East to diagnose infections caused by *Pseudomonas pseudomallei;* acute and convalescent sera are required. Cross-reactions with other organisms (i.e., *B. cepacia* complex) occurs and interpretation of any serology must include compatible clinical symptoms.

Case Study

A 31-year-old man was seen by his physician for low-grade fever and chronic cough with purulent sputum production. A radiograph showed diffuse shadowing of the upper lungs. These chronic respiratory symptoms had been present since youth, when the patient had been diagnosed with cystic fibrosis (CF). A sputum was sent for culture for CF pathogens, and the patient was admitted for antimicrobial therapy and supportive care. A smear of the sputum was not performed. However, several mucoid and nonmucoid morphologies of oxidase-positive, gram-negative, non-glucose-fermenting rods were isolated. The mucoid organism (Figure 24-4) grew at 42° C and had a grapelike odor but did not produce blue-green or fluorescent pigment (see Figure 24-2). Using the disk method, the isolates were found to be resistant to aminoglycosides and fluoroquinolone antibiotics. Growth was seen around the colistin disk on the plate, only from the nonmucoid strain.

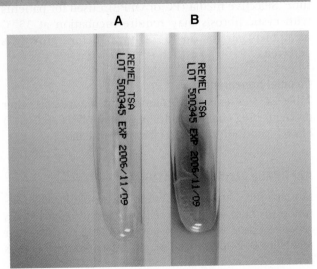

Figure 24-2 *Pseudomonas aeruginosa* on tryptic soy agar **(B)**. Note bluish-green color. Uninoculated tube **(A)** is shown for comparison.

QUESTIONS

1. What are the likely gram-negative agents found in cultures from CF patients?
2. What is the likely identification of the mucoid gram-negative rod? Why did the organism produce atypical reactions?
3. The Cystic Fibrosis Foundation recommends against using rapid methods to perform susceptibility testing on isolates from CF patients. What is the reason for this recommendation?
4. Give the reasons that the disk method is useful in testing for the pathogens of CF patients.
5. What is the likely identification of the colistin-resistant gram-negative rod? This organism might be confused with what other nonfermenting gram-negative rods?
6. Why do you think a smear was not useful for evaluation of the patient's infection?

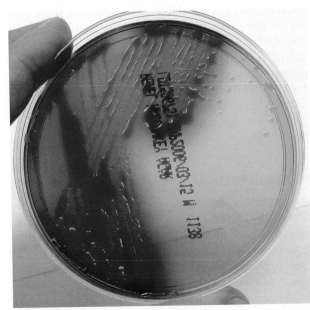

Figure 24-3 *Pseudomonas aeruginosa* on MacConkey agar. This stain produces a red pigment.

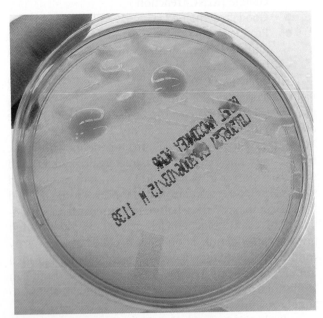

Figure 24-4 Mucoid stain of *Pseudomonas aeruginosa* on MacConkey agar.

Table 24-4   Biochemical and Physiologic Characteristics

Organisms	Growth at 42°C	Nitrate Reduction	Gas from Nitrate	Gelatin Liquefied	Arginine Dihydrolase	Lysine Decarboxylase	Urea Hydrolysis	Oxidizes Glucose	Oxidizes Lactose	Oxidizes Mannitol	Oxidizes Xylose
Acidovorax delafieldii	v	+	-	-	+	-	+	+	-	v	v
Acidovorax temperans	+	+	-	-	-	-	v	+	-	v	-
Brevundimonas diminuta	v	-	-	v	-	-	-	v	-	-	-
B. vesicularis	v	-	-	v	-	-	-	v	-	-	v
Burkholderia cepacia complex	v	v	-	v	+	v	v	+	v	+	v
B. pseudomallei	+	+	+	v	+	-	v	+	+	+	+
B. mallei	-	+	-	-	+	-	v	+	v	-	v
Pandoraea spp.	v	v	-	-	-	-	v	+w	-	-	-
Pseudomonas aeruginosa	+	+	+	v	+	-	v	+	-	v	+
P. fluorescens	-	-	-	+	+	-	v	+	v	v	+
P. mendocina	+	+	+	-	+	-	v	+	-	-	+
P. monteilii	-	-	-	-	+	-	v	+	-	-	-
P. putida	-	-	-	-	+	-	v	+	v	v	+
P. stutzeri	v	+	+	-	-	-	v	+	-	+	+
Pseudomonas veronii	-	+	-	v	+	ND	v	+	ND	+	+
Pseudomonas–like group 2	v	v	-	-	v	-	+	+	+	+	+
CDC group Ic	+	+	-	-	+	-	v	+	-	-	-
Ralstonia mannitolilytica	+	-	-	v	-	-	+	+	+	+	+
R. pickettii	v	+	v	v	-	-	+	+	v	-	+

Data compiled from references 1, 3, 5, 6, 7, 24, 29, 30, 31, 32.
ND, No data; *v*, variable; +, >90% of strains are positive; −, >90% of strains are negative; *w*, weak.
*Arginine-positive strains of *P. stutzeri* formerly classified as CDC group Vb-3.

Table 24-5   Antimicrobial Therapy and Susceptibility Testing

Species	Therapeutic Options	Potential Resistance to Therapeutic Options	Validated Testing Methods*	Comments
Burkholderia cepacia	No definitive guidelines. Potentially active agents include piperacillin, ceftazidime, imipenem, ciprofloxacin, chloramphenicol, and trimethoprim/sulfamethoxazole	Yes	Not available	Antimicrobial therapy rarely eradicates organism; will grow on susceptibility testing media, but standards for interpretation of results do not exist
B. pseudomallei	No definitive guidelines. Potentially active agents include ceftazidime, piperacillin/tazobactam, ticarcillin/clavulanate, amoxicillin/clavulanate, imipenem, trimethoprim/sulfamethoxazole, and chloramphenicol	Yes	Not available	Optimal therapy still controversial
B. mallei	No definitive guidelines. Potentially active agents may include those listed for *B. pseudomallei*	Yes	Not available	Rarely involved in human infections, so reliable therapeutic data are limited
Ralstonia pickettii	No definitive guidelines. Potentially active agents include those listed for *B. cepacia*	Yes	Not available	Rarely involved in human infections, so reliable therapeutic data are limited
Pseudomonas aeruginosa	An antipseudomonal beta-lactam (listed below) with or without an aminoglycoside; certain quinolones may also be used. Specific agents include piperacillin/tazobactam, ceftazidime, cefepime, aztreonam, imipenem, meropenem, gentamicin, tobramycin, amikacin, netilmicin, ciprofloxacin, and levofloxacin	Yes	Disk diffusion, broth dilution, agar dilution, and commercial systems	In vitro susceptibility testing results important for guiding therapy
P. fluorescens, P. putida, P. stutzeri (includes Vb-3), *P. mendocina, P. alcaligenes, P. pseudoalcaligenes, Pseudomonas* sp. CDC group 1, "*P. denitrificans,*" *Pseudomonas*-like group 2, and CDC group Ic	Because rarely implicated in human infections, there are no definitive guidelines; agents used for *P. aeruginosa* may be effective for these species	Yes	Not available	Most will grow on susceptibility testing media, but standards for interpretation of results do not exist
Pseudomonas luteola Pseudomonas oryzihabitans	No definitive guidelines. Potentially active agents include cefotaxime, ceftriaxone, ceftazidime, imipenem, quinolones, and aminoglycosides	Yes, activity of penicillins is variable; commonly resistant to first- and second-generation cephalosporins	Not available	
Brevundimonas vescularis and *B. diminuta*	Because rarely implicated in human infections, there are no definitive guidelines	Unknown	Not available	Rarely involved in human infection
Acidovorax spp.	No definitive guidelines	Unknown	Not available	No clinical experience

*Validated testing methods include those standard methods recommended by the Clinical and Laboratory Standards Institute (CLSI) and those commercial methods approved by the Food and Drug Administration (FDA).

ANTIMICROBIAL SUSCEPTIBILITY TESTING AND THERAPY

Except for *Pseudomonas aeruginosa,* validated susceptibility testing methods do not exist for the organisms discussed in this chapter. Therefore, when these organisms are isolated, the laboratory has a conflict between the urge to contribute in some way to patient management by providing data and the lack of confidence in producing interpretable and accurate information.

Although many of these organisms will grow on the media and under the conditions recommended for testing the more commonly encountered bacteria (see Chapter 12 for more information regarding validated testing methods), the ability to grow under test conditions does not guarantee reliable detection of important antimicrobial resistance. Therefore, even though testing can provide an answer, substantial potential to obtain the wrong answer exists. Chapter 12 should be reviewed for strategies that can be used to provide susceptibility information and data when validated testing methods do not exist for a clinically important bacterial isolate.

The infrequency with which *Burkholderia* spp. and *Ralstonia pickettii* are encountered in infections in humans and the lack of validated in vitro susceptibility testing methods do not allow definitive treatment and testing guidelines to be given (Table 24-5). Potential therapies for *B. cepacia* and *B. pseudomallei* are provided, but antimicrobial therapy rarely eradicates *B. cepacia,* especially from the respiratory tract of patients with cystic fibrosis, and the optimum therapy for melioidosis remains controversial.[8,24-28] *Burkholderia* spp. are capable of expressing resistance to various antibiotics, so devising effective treatment options can be problematic.[7,10,23] Additionally troublesome is that valid testing methods for these organisms do not exist. Because of these complex issues, the importance of establishing the clinical significance of these species when they are isolated in clinical specimens is emphasized.

Among *Pseudomonas* spp. and *Brevundimonas* spp., *P. aeruginosa* is the only species for which valid in vitro susceptibility testing methods exist[11-13] and for which there is extensive therapeutic experience (see Table 24-5; also see Chapter 12 for a discussion of available testing methods). Therapy usually involves the use of a beta-lactam developed for antipseudomonal activity and an aminoglycoside. The particular therapy used depends on several clinical factors and on the antimicrobial resistance profile that the laboratory reports for a particular *P. aeruginosa* isolate.

P. aeruginosa is intrinsically resistant to various antimicrobial agents; only those with potential activity are given in Table 24-5. However, *P. aeruginosa* also readily acquires resistance to the potentially active agents listed, necessitating that agents selected from the list be tested against each clinically relevant isolate.

Although antimicrobial resistance is also characteristic of the other *Pseudomonas* spp. and *Brevundimonas* spp.,[7,10,15,16,19] the fact that these organisms are rarely clinically significant and the lack of validated testing methods prohibit the provision of specific guidelines (see Table 24-5). Antimicrobial agents used for *P. aeruginosa* infections are often considered for use against the other species; however, before proceeding with the development of treatment strategies, the first critical step should be to establish the clinical significance of these organisms.

PREVENTION

Because these organisms are ubiquitous in nature and many are commonly encountered without deleterious effects on healthy human hosts, there are no recommended vaccination or prophylaxis protocols. Hospital-acquired infections are best controlled by following appropriate infection control guidelines and implementing effective protocols for the sterilization and decontamination of medical supplies.

REFERENCES

1. Anzai Y, Kudo Y, Oyaizu H: The phylogeny of the genera *Chryseomonas, Flavimonas* and *Pseudomonas* supports synonymy of these three genera, *Int J Syst Bacteriol* 47:249, 1997.
2. Christenson JC, Welch DF, Mukwaya G, et al: Recovery of *Pseudomonas gladioli* from respiratory tract specimens of patients with cystic fibrosis, *J Clin Microbiol* 27:270, 1989.
3. Coenye T, Falsen E, Hoste B, et al: Description of *Pandoraea* gen nov with *Pandoraea apista* sp nov, *Pandoraea pulmonicola* sp nov, *Pandoraea pnomenusa* sp nov, *Pandoraea sputorum* sp nov and *Pandoraea norimbergensis* comb nov, *Int J Syst Evol Microbiol* 50:887, 2000.
4. Dance DA: Melioidosis: the tip of the iceberg? *Clin Microbiol Rev* 4:52, 1991.
5. Daneshvar MI, Hollis DG, Steigerwalt AG, et al: Assignment of CDC Weak Oxidizer Group 2 (WO-2) to the genus *Pandoraea* and characterization of three new *Pandoraea* genomospecies, *J Clin Microbiol* 39:1819, 2001.
6. De Baere T, Steyaert S, Wauters G, Des Vos P, et al: Classification of *Ralstonia pickettii* biovar 3/'thomasii' strains (Pickett 1994) and of new isolates related to nosocomial recurrent meningitis as *Ralstonia mannitolytica* sp. nov, *J Syst Evol Microbiol* 51(pt 2):547, 2001.
7. Elomari M, Caroler L, Verhille S: *Pseudomonas monteilii* sp nov, isolated from clinical specimens, *Int J Syst Bacteriol* 47:846, 1997.
8. Gilligan PH: Microbiology of airway disease in patients with cystic fibrosis, *Clin Microbiol Rev* 4:35, 1991.
9. Godfrey AJ, Wong S, Dance DA, et al: *Pseudomonas pseudomallei* resistance to β-lactam antibiotics due to alterations in the chromosomally encoded β-lactamase, *Antimicrob Agent Chemother* 35:1635, 1991.
10. Gold R, Jin E, Levison H, et al: Ceftazidime alone and in combination in patients with cystic fibrosis: lack of efficacy in

treatment of severe respiratory infections caused by *Pseudomonas cepacia, J Antimicrob Chemother* 12(suppl A):331, 1983.

11. Lewin C, Doherty C, Govan J: In vitro activities of meropenem, PD12731, PD 131628, ceftazidime, chloramphenicol, co-trimoxazole, and ciprofloxacin against *Pseudomonas cepacia, Antimicrob Agents Chemother* 37:123, 1993.

12. Livermore D: β-lactamases in laboratory and clinical resistance, *Clin Microbiol Rev* 8:557, 1995.

13. National Committee for Clinical Laboratory Standards: *Methods for dilution antimicrobial susceptibility tests for bacteria that grow aerobically;* M7-A4, ed 4, Villanova, Pa, 1997, NCCLS.

14. National Committee for Clinical Laboratory Standards: *Performance standards for antimicrobial disk susceptibility tests;* M2-A6, ed 6, Villanova, Pa, 1997, NCCLS.

15. National Committee for Clinical Laboratory Standards: *Performance standards for antimicrobial susceptibility testing;* M100-S7, Villanova, Pa, 1997, NCCLS.

16. Nelson JW, et al: Virulence factors of *Burkholderia cepacia, FEMS Immunol Med Microbiol* 8:89, 1994.

17. Noble RC, Overman SB: *Pseudomonas stutzeri* infection: a review of hospital isolates and a review of the literature, *Diagn Microbiol Infect Dis* 19:51, 1994.

18. Oberhelman RA, Humbert JR, Santorelli FW: *Pseudomonas vesicularis* causing bacteremia in a child with sickle cell anemia, *South Med J* 87:821, 1994.

19. O'Neil KM, Herman JH, Modlin JF, et al: *Pseudomonas cepacia: an emerging pathogen in chronic granulomatous disease, J Pediatr* 108:940, 1986.

20. Pallent LJ, Hugo WB, Grant DJ, et al: *Pseudomonas cepacia* as a contaminant and infective agent, *J Hosp Infect* 4:9, 1983.

21. Papapetropoulou M, Iliopoulou J, Rodopoulou G, et al: Occurrence and antibiotic-resistance of *Pseudomonas* species isolated from drinking water in Southern Greece, *J Chemother* 6:111, 1994.

22. Pegues DA, Carson LA, Anderson RL, et al: Outbreak of *Pseudomonas cepacia* bacteremia in oncology patients, *Clin Infect Dis* 16: 407, 1993.

23. Pruksachartvuthi S, Aswapokee N, and Thankerngpol K: Survival of *Pseudomonas pseudomallei* in human phagocytes, *J Med Microbiol* 31:109, 1990.

24. Segers P, Vancanneyt M, Pot B, et al: Classification of *Pseudomonas diminuta* (Leifson and Hugh, 1954) and *Pseudomonas vesicularis* (Basing, Dîll, and Freytag, 1953) in *Brevundimonas* gen nov as *Brevundimonas diminuta* comb nov and *Brevundimonas vesicularis* comb nov, respectively, *Int J Syst Bacteriol* 44:499, 1994.

25. Simpson IN, Finlay J, Winstanleyet DJ, et al: Multi-resistance isolates possessing characteristics of both *Burkholderia (Pseudomonas) cepacia* and *Burkholderia gladioli* from patients with cystic fibrosis, *J Antimicrob Chemother* 34:353, 1994.

26. Smith MD, Wuthiekanum V, Walsh AL, et al: Susceptibility of *Pseudomonas pseudomallei* to some newer β-lactam antibiotics and antibiotic combinations using time-kill studies, *J Antimicrob Chemother* 33:145, 1994.

27. Sookpranee M, Boonma P, Susaengrat M, et al: Multicenter prospective randomized trial comparing ceftazidime plus co-trimoxazole with chloramphenicol plus doxycycline and co-trimoxazole for treatment of severe melioidosis, *Antimicrob Agents Chemother* 36:158, 1992.

28. Sokpranee T, Sookpranee M, Mellencamp MA, et al: *Pseudomonas pseudomallei:* a common pathogen in Thailand that is resistant to the bactericidal effects of many antibiotics, *Antimicrob Agents Chemother* 35:484, 1991.

29. Stryjewski ME, LiPuma JJ, Messier RH Jr, et al: Sepsis: multiple organ failure, and death due to *Pandoraea pnomenusa* infection after lung transplantation, *J Clin Microbiol* 41:2255, 2003.

30. Vandamme P, Goris J, Coenye T: Assessment of Centers for Disease Control group IVc-2 to the genus *Ralstonia* as *Ralstonia paucula* sp nov, *Int J Syst Bacteriol* 49:663, 1999.

31. Weyant RS, Moss CW, Weaver RE, et al, editors: *Identification of unusual pathogenic gram-negative aerobic and facultatively anaerobic bacteria,* ed 2, Baltimore, 1996, Williams & Wilkins.

32. Yabuuchi E, Kosako Y, Yano H, et al: Transfer of two *Burkholderia* and an *Alcaligenes* species to *Ralstonia* gen nov: proposal of *Ralstonia pickettii* (Ralston, Palleroni, and Doudoroff, 1973) comb nov, *Ralstonia solanacearum* (Smith, 1896) comb nov and *Ralstonia eutropha* (Davis, 1969) comb nov, *Microbiol Immunol* 39:897, 1995.

ADDITIONAL READING

Balows A, Truper HG, Dworkin M, et al, editors: *The prokaryotes. A handbook on the biology of bacteria: ecophysiology, isolation, identification, applications,* ed 2, New York, 1981, Springer-Verlag.

Fick RB Jr: Pseudomonas aeruginosa: *the opportunist,* Boca Raton, Fla, 1993, CRC Press.

Gilligan PH, Lum G, Vandamme PA, et al: *Burkholderia, Stenotrophomonas, Ralstonia, Brevundimonas, Comamonas, Delftia, Pandoraea,* and *Acidovorax.* In Murray PR, Baron EJ, Jorgensen JH, et al, editors: *Manual of clinical microbiology,* ed 8, Washington, DC, 2003, ASM Press.

Kiska DL, Gilligan PH: *Pseudomonas.* In Murray PR, Baron EJ, Jorgensen JH, et al, editors: *Manual of clinical microbiology,* ed 8, Washington, DC, 2003, ASM Press.

Pollack M: *Pseudomonas aeruginosa.* In Mandell GL, Bennett JE, Dolin R, editors: *Principles and practice of infectious diseases,* New York, 2000, Churchill Livingstone.

Achromobacter, Rhizobium, Ochrobactrum, and Similar Organisms

CHAPTER 25

Genera and Species to Be Considered

Current Name	Previous Name
"Achromobacter"* group	
Alcaligenes xylosoxidans	Achromobacter xylosoxidans, Alcalignes xylosoxidans ssp. xylosoxidans
CDC group EF-4b	CDC group EF-4
CDC group Ic	
CDC group O-3	
CDC group OFBA-1	
Ochrobactrum anthropi	CDC group Vd
Paracoccus yeei	CDC group EO-2
Psychrobacter immobilis (saccharolytic strains)	Part of CDC group EO-2
Rhizobium radiobacter	Agrobacterium radiobacter, CDC group Vd-3
Shewanella putrefaciens	Alteromonas putrefaciens, Achromobacter putrefaciens, CDC group Ib

*Quotation mark indicates a proposed organism name.

GENERAL CHARACTERISTICS

Most of the organisms discussed in this chapter exist in the environment, although CDC group EF-4b inhabits the upper respiratory tract of certain animals and *Ochrobactrum anthropi* may occasionally inhabit the human gastrointestinal tract. All are nonpigmented, oxidase-positive, and oxidize glucose; most grow on MacConkey agar. However, their specific morphologic and physiologic features are somewhat diverse and are considered later in this chapter in the discussion of laboratory diagnosis.

EPIDEMIOLOGY

As environmental organisms, these bacteria are rarely encountered in human specimens or infections. When they are encountered, they are found on contaminated medical devices or are isolated from immunocompromised or otherwise debilitated patients. Of the organisms listed in Table 25-1, *Rhizobium radiobacter* and

Ochrobactrum anthropi are the species most commonly encountered in the clinical setting. The other bacteria have rarely been discovered in clinical material, and several have never been established as the cause of human infection.[7-9,11,12]

R. radiobacter inhabits the soil, and human infections occur by exposure to contaminated medical devices usually involving already ill patients.[5]

The specific environmental niche of *O. anthropi* is unknown, but this organism is capable of survival in water, including moist areas within the hospital environment. The organism may also be a transient colonizer of the human gastrointestinal tract. Similar to *R. radiobacter*, human infections caused by *O. anthropi* are associated with implantation of intravenous catheters or other foreign bodies in patients with a debilitating illness. Acquisition by contaminated pharmaceuticals and by puncture wound has also been documented.[1-4,10]

The epidemiology of CDC group EF-4b is unlike that of the other bacteria discussed in this chapter. Animals, rather than the environment, are the reservoir, and transmission to humans occurs by dog or cat bites and scratches.

PATHOGENESIS AND SPECTRUM OF DISEASE

Because these organisms rarely cause human infections, little is known about what, if any, virulence factors they may produce to facilitate infectivity (Table 25-2). The fact that *R. radiobacter* and *O. anthropi* infections frequently involve contaminated medical materials and immunocompromised patients, and rarely if ever occur in healthy hosts, suggests that these bacteria have relatively low virulence. One report suggests that *R. radiobacter* is capable of capsule production,[5] and the ability of *O. anthropi* to adhere to the silicone material of catheters may contribute to this organism's propensity to cause catheter-related infections.[1] No known virulence factors have been described for CDC group EF-4b. Infection appears to require traumatic introduction by a puncture wound (i.e., bite) or scratch, which indicates that the organism itself does not express any invasive properties.

For both *R. radiobacter* and *O. anthropi*, bacteremia is the most common type of infection (see Table 25-2);

Table 25-1   Epidemiology

Species	Habitat (reservoir)	Mode of Transmission
"*Achromobacter*" group	Uncertain, probably environmental; not part of human flora	Unknown. Rarely found in humans
Rhizobium radiobacter	Environmental, soil and plants; not part of human flora	Contaminated medical devices such as intravenous and peritoneal catheters
CDC group EF-4b	Animal oral and respiratory flora; not part of human flora	Animal contact, particularly bites or scratches from dogs and cats
Paracoccus yeei	Environmental. Not part of human flora	Infections in humans not yet described
Psychrobacter immobilis	Environmental, particularly cold climates such as the Antarctic; not part of human flora	Unknown. Rarely found in humans. Has been found in fish, poultry, and meat products
CDC group OFBA-1	Uncertain, probably environmental; not part of human flora	Unknown. Rarely found in humans
Ochrobactrum anthropi	Uncertain, probably environmental; found in water and hospital environments; may also be part of human flora	Uncertain. Most likely involves contaminated medical devices, such as catheters or other foreign bodies, or contaminated pharmaceuticals. Also can be acquired in community by puncture wounds
Alcaligens xylosoxidans	Environment, not part of normal flora	Unknown, rarely found in humans
Shewanella putrefaciens	Environmental and foodstuffs, not part of human flora	Unknown, rarely found in humans

Table 25-2   Pathogenesis and Spectrum of Disease

Species	Virulence Factors	Spectrum of Disease and Infections
"*Achromobacter*" group	Unknown	Rarely isolated from humans. Patients with septicemia have been reported
Rhizobium radiobacter	Unknown. One blood isolate described as mucoid, suggestive of exopolysaccharide capsule production	Exposure of immunocompromised or debilitated patient to contaminated medical devices resulting in bacteremia, and less commonly, peritonitis, endocarditis, or urinary tract infection
CDC group EF-4b	Unknown	Infected bite wounds of fingers, hands, or arm leading to cellulitis or abscess formation. Systemic infections are rare
Paracoccus yeei	Unknown	No infections described in humans. Rarely encountered in clinical specimens
Psychrobacter immobilis	Unknown	Rare cause of infection in humans. Has been described in wound and catheter site infections, meningitis, and eye infections
CDC group OFBA-1	Unknown	Rarely isolated from clinical specimens; found in blood, respiratory, wound, and catheter specimens
Ochrobactrum anthropi	Unknown. Exhibits ability to adhere to silicone catheter material in a manner similar to staphylococci	Catheter and foreign body-associated bacteremia. May also cause pyogenic infections, community-acquired wound infections, and meningitis in tissue graft recipients. Patients are usually immunocompromised or otherwise debilitated
Alcaligens xylosoxidans	Unknown	Rare cause of human infection
Shewanella putrefaciens	Unknown	Clinical significance uncertain; often found in mixed cultures. Has been implicated in cellulites, otitis media, and septicemia; also may be found in respiratory tract, urine, feces and pleural fluid

peritonitis, endocarditis, meningitis, urinary tract, and pyogenic infections are much less commonly encountered.[2,5,6] Cellulitis and abscess formation typify the infections resulting from the traumatic introduction of CDC group EF-4b into the skin and subcutaneous tissue.

Although other species listed in Table 25-2 may be encountered in clinical specimens, their association with human infection is rare and their clinical significance in such encounters should be carefully analyzed.

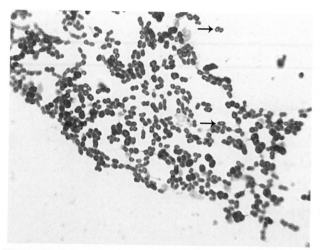

Figure 25-1   *Paracoccus yeei;* note doughnut-shaped organism on Gram stain *(arrows).*

LABORATORY DIAGNOSIS

SPECIMEN COLLECTION AND TRANSPORT

No special considerations are required for specimen collection and transport of the organisms discussed in this chapter. Refer to Table 5-1 for general information on specimen collection and transport.

SPECIMEN PROCESSING

No special considerations are required for processing the organisms discussed in this chapter. Refer to Table 5-1 for general information on specimen processing.

DIRECT DETECTION METHODS

Other than Gram stain of patient specimens, there are no specific procedures for the direct detection of these organisms in clinical material. *Achromobacter, Alcaligenes xylosoxidans, Ochrobactrum,* CDC group OFBA-1, and CDC group Ic are slender, short to long rods, and CDC group O-3 are thin, medium to slightly long curved rods with tapered ends resembling a sickle. *R. radiobacter* is a short, pleomorphic rod. *Psychrobacter immobilis,* CDC group EF-4b, and *Paracoccus yeei* are coccobacilli. *P. yeei* has a characteristic O appearance on Gram stain (Figure 25-1). *Shewanella putrefaciens* are long, short, or filamentous rods.

CULTIVATION

Media of Choice

Achromobacter, Alcaligenes xylosoxidans, Rhizobium, Paracoccus yeei, CDC group Ic, CDC group O-3, *S. putrefaciens,* CDC group EF-4b, *Ochrobactrum,* CDC group OFBA-1, and *Psychrobacter* spp. grow well on routine laboratory

media such as 5% sheep blood, chocolate, and MacConkey agars. These organisms also grow well in the broth of blood culture systems and in common nutrient broths such as thioglycollate and brain-heart infusion.

Incubation Conditions and Duration

These organisms will produce detectable growth on 5% sheep blood and chocolate agars in 5% CO_2 and MacConkey in ambient air, when incubated at 35° C for a minimum of 24 hours. *Psychrobacter* spp. is an exception in that it usually grows poorly at 35° C and grows best between 20° and 25° C.

Colonial Appearance

Table 25-3 describes the colonial appearance and other distinguishing characteristics (e.g., hemolysis and odor) of each genus when grown on 5% sheep blood or MacConkey agars.

APPROACH TO IDENTIFICATION

The ability of most commercial identification systems to accurately identify the organisms discussed in this chapter is limited or uncertain. Identification often requires the use of conventional biochemical profiles.

The key biochemical reactions that can be used to presumptively differentiate among the genera discussed in this chapter are provided in Table 25-4. However, definitive identification of these organisms often requires performing an extensive battery of biochemical tests not commonly available in many clinical microbiology laboratories. Therefore, full identification of clinically relevant isolates may require that they be sent to a reference laboratory.

Comments Regarding Specific Organisms

Although the EF portion of the CDC group EF-4b designation stands for *eugonic* (an organism that grows well on common laboratory media) *fermenter,* most CDC group EF-4b strains oxidize glucose, so the designation as a eugonic fermenter is a misnomer. *P. yeei,* formerly CDC group EO-2 (a *eugonic oxidizer*) has a biochemical profile very similar to that of the saccharolytic, nonhemolytic *Acinetobacter* spp. (see Chapter 23 for more information regarding this genus), except the latter is oxidase negative.

The notable characteristic of CDC group OFBA-1 is that it produces an acidlike reaction in the OF medium control tube, even though no carbohydrates are present. In contrast, *Rhizobium radiobacter* produces acid from various carbohydrates, but it does not acidify the OF control tube.[14]

O. anthropi and the *Achromobacter* group are phenotypically very similar so that cellular fatty acid

Table 25-3   Colonial Appearance and Characteristics

Organism	Medium	Appearance
"*Achromobacter*" group	BA	Smooth, glistening, entire
	Mac	NLF; biovar F does not grow
Alcaligenes xylosoxidans	BA	Colonies resemble *Acinetobacter*, that is, smooth, opaque, glistening, and entire
	Mac	NLF
CDC group EF-4b	BA	No distinctive appearance, but cultures smell like popcorn
	Mac	NLF
CDC group Ic	BA	No distinctive appearance
	Mac	NLF
CDC group O-3	BA	Circular, entire, translucent, very punctate
	Mac	NLF, may grow poorly or not at all
CDC group OFBA-1	BA	Beta-hemolytic
	Mac	NLF
Ochrobactrum anthropi	BA	Resemble colonies of *Enterobacteriaceae,* only smaller
	Mac	NLF
Paracoccus yeei	BA	Growth frequently mucoid
	Mac	NLF
Psychrobacter immobilis (saccharolytic strains)	BA	No distinctive appearance but usually does not grow well at 35° C; grows best at 20° C; cultures smell like roses
	Mac	NLF
Rhizobium radiobacter	BA	No distinctive appearance
	Mac	NLF
Shewanella putrefaciens	BA	Convex, circular, smooth; occasionally mucoid; lavender greening of blood; soluble brown to tan pigment
	Mac	NLF

BA, 5% sheep blood agar; *Mac,* MacConkey agar; *NLF,* non–lactose-fermenter.

Table 25-4   Key Biochemical and Physiologic Characteristics

Organism	Oxidizes Glucose	Oxidizes Xylose	Oxidizes Mannitol	Nitrate Reduction	Gas From Nitrate	Arginine Dihydrolase	Esculin Hydrolyzed	Growth on Cetrimide
"*Achromobacter*" group[a,b]	+	+	v	+	+	+	+	v
Alcaligenes xylosoxidans[b]	v	+	–	+	v	v	–	+
CDC group EF–4b	+	–	–	+	–	–	–	ND
CDC group Ic	+	–	–	+	–	+	–	+
CDC group O–3	+	+	–	–	–	–	+	ND
CDC group OFBA–1[c]	+	+	(+)	+	+	+	–	+
Ochrobactrum anthropi[b]	+	+	v	v	v	v	v	–
Paracoccus yeei	+	+	–	+	v	v	–	–
Psychrobacter immobilis[d]	(+)	(+)	–	v	–	v	–	–
Rhizobium radiobacter	+	+	+	v	–	–	+	–
Shewanella putrefaciens[e]	v	–	–	+	–	–	–	–

Compiled from data in references 15, 17, and 19.

ND, No data available; *v,* variable; +, >90% of strains are positive; –, >90% of strains are negative; (+), delayed.

[a]Includes biovars B, E, and F; F does not grow on MacConkey agar.

[b]Usually motile by peritrichous flagella.

[c]Oxidizes base.

[d]Saccharolytic variety; prefers growth at 25° C.

[e]H$_2$S in butt of triple sugar iron (TSI) agar.

Table 25-5   Antimicrobial Therapy and Susceptibility Testing

Species	Therapeutic Options	Potential Resistance to Therapeutic Options	Validated Testing Methods*	Comments
"*Achromobacter*" group	No definitive guidelines. Human infections rare	Unknown	Not available	No clinical experience
Rhizobium radiobacter	Optimal therapy uncertain. Treatment involves removal of foreign body. Potentially active agents include ceftriaxone, cefotaxime, imipenem, gentamicin, and ciprofloxacin	Yes	Not available	Will grow on susceptibility testing media, but standards for interpretation of results do not exist
CDC group EF-4b	No definitive guidelines. Potentially active agents include penicillin, ampicillin, ciprofloxacin, and ofloxacin	Unknown; some cephalosporins may be less active than the penicillins	Not available	Limited clinical experience
Paracoccus yeei	No definitive guidelines	Unknown	Not available	No clinical experience
Psychrobacter immobilis	No definitive guidelines	Unknown	Not available	Limited clinical experience
CDC group OFBA-1	No definitive guidelines	Unknown	Not available	No clinical experience
Ochrobactrum anthropi	Optimal therapy uncertain. Treatment involves removal of foreign body. Potentially active agents include trimethoprim/ sulfamethoxazole, ciprofloxacin, and imipenem; aminoglycoside activity variable	Commonly resistant to all penicillins and cephalosporins	Not available	Will grow on susceptibility testing media, but standards for interpretation of results do not exist
Alcaligens xylosoxidans	No definitive guidelines	Unknown	Not available	
Shewanella putrefaciens	No definitive guidelines. Generally susceptible to various antimicrobial agents	Often resistant to ampicillin and cephalothin	Not available	

*Validated testing methods include those standard methods recommended by the Clinical and Laboratory Standards Institute (CLSI) and those commercial methods approved by the Food and Drug Administration (FDA).

analysis is often required for their differentiation. *Psychrobacter* spp. can be either saccharolytic or asaccharolytic, although all members of this genus have in common an optimal growth temperature of less than 35° C.

Because *Alcaligenes xylosoxidans* may oxidize glucose and consistently oxidizes xylose, this organism could be considered along with the other organisms of this chapter. Because other *Alcaligenes* spp. usually do not utilize glucose, all *Alcaligenes* spp. are considered together in Chapter 27.

CDC group O-3 is often misidentified as *Campylobacter* spp. because of its curved shape on Gram stain.

Shewanella putrefaciens is notable for the production of H_2S in the butt of triple sugar iron (TSI) agar; this characteristic is rare among the nonfermentative gram-negative rods.

SERODIAGNOSIS

Serodiagnostic techniques are not generally used in the laboratory diagnosis of infections caused by the organisms discussed in this chapter.

ANTIMICROBIAL SUSCEPTIBILITY TESTING AND THERAPY

Validated susceptibility testing methods do not exist for the organisms discussed in this chapter. Although many of these organisms will grow on the media and under the conditions recommended for testing the more commonly encountered bacteria (see Chapter 12 for more information regarding validated testing methods), the ability to grow and the ability to detect important antimicrobial resistances are not the same thing. Therefore the lack of validated in vitro susceptibility testing methods does not allow definitive treatment and testing guidelines to be given for any of the organisms listed in Table 25-5. Although susceptibility data for some of these bacteria can be found in the literature, the lack of understanding of potential underlying resistance mechanisms prohibits the validation of such data. Review Chapter 12 for preferable strategies that can be used to provide susceptibility information and data when validated testing methods do not exist for a clinically important bacterial isolate.

Because *R. radiobacter* and *O. anthropi* infections are frequently associated with implanted medical devices, therapeutic management of the patient often involves removal of the contaminated material. Although definitive antimicrobial therapies for these infections have not been established, in vitro data suggest that certain agents could be more effective than others (see Table 25-5). For *R. radiobacter*, certain cephalosporins and aminoglycosides show in vitro activity. However, this organism is known to produce inactivating enzymes of both beta-lactams and aminoglycosides, so that resistance to potentially effective agents is a possibility.[13]

O. anthropi is commonly resistant to all currently available penicillins and cephalosporins but usually is susceptible to imipenem. This resistance profile is sufficiently consistent with the species that it may be useful in confirming the identification. The organism may also appear susceptible to trimethoprim-sulfamethoxazole and ciprofloxacin, but antimicrobial therapy without removal of the contaminated medical device may not successfully eradicate the organism.

PREVENTION

Because these organisms are ubiquitous in nature and are not generally a threat to human health, there are no recommended vaccination or prophylaxis protocols. Hospital-acquired infections are best controlled by following appropriate sterile techniques and infection control guidelines, and implementing effective protocols for the sterilization and decontamination of medical supplies.

Case Study

A 31-year-old female bartender, who is right-hand dominant, was bitten by a dog on her right second finger. She was treated at the emergency department later that day for pain and swelling of the finger, and flexor tenosynovitis and tendon laceration were diagnosed. She went to surgery the following day, where cultures were obtained. The patient did well following surgery and the physicians wanted to send her home on ampicillin/sulbactam. However, the laboratory had reported an unidentified gram-negative coccobacilli. The organism did not grow on MacConkey, but was an oxidase- and catalase-positive glucose oxidizer without any pigment production. The organism emitted the distinct odor of popcorn.

QUESTIONS

1. What is the most likely identification of this bacterium and how would you definitively identify it?
2. List the other common microorganisms that would be expected to be isolated from dog and cat bites.
3. This patient isolate had an unusual reaction in nitrate medium, which is characteristic of EF-4b and a very few other microorganisms. No gas was present in the Durham tube, no red color was detected with the addition of the reagents, and no pink color was present after addition of zinc. See Procedure 13-29, Nitrate Reduction. When no pink color is present after the addition of zinc, is the microorganism positive for nitrate reduction?

REFERENCES

1. Alnor D, Frimodt-Moller N, Espersen F, et al: Infections with the unusual human pathogens *Agrobacterium* species and *Ochrobactrum anthropi*, *Clin Infect Dis* 18:914, 1994.
2. Chang HJ, Christenson JC, Pavia AT, et al: *Ochrobactrum anthropi* meningitis in pediatric pericardial allograft transplant recipients, *J Infect Dis* 173:656, 1996.
3. Cieslak TJ, Drabick CJ, Robb ML: Pyogenic infections due to *Ochrobactrum anthropi*, *Clin Infect Dis* 22:845, 1996.
4. Cieslak TJ, Robb ML, Drabick CJ, et al: Catheter-associated sepsis caused by *Ochrobactrum anthropi*: report of a case and review of related non-fermentative bacteria, *Clin Infect Dis* 14:902, 1992.
5. Dunne WM, Tillman J, Murray JC: Recovery of a strain of *Agrobacterium radiobacter* with a mucoid phenotype from an immunocompromised child with bacteremia, *J Clin Microbiol* 31:2541, 1993.
6. Edmond MB, Riddler SA, Baxter CM, et al: *Agrobacterium radiobacter*: a recently recognized opportunistic pathogen, *Clin Infect Dis* 16:388, 1993.
7. Gini GA: Ocular infection caused by *Psychrobacter immobilis* acquired in the hospital, *J Clin Microbiol* 28:400, 1990.
8. Holmes B, Lewis R, Trevett A: Septicemia due to *Achromobacter* group B: a report of two cases, *Med Microbiol Lett* 1:177, 1992.
9. Hulse M, Johnson S, Ferrieri P: *Agrobacterium* infections in humans: experience at one hospital and review, *Clin Infect Dis* 16:112, 1993.

10. Kern WV, Oethinger M, Kaufhold A, et al: *Ochrobactrum anthropi* bacteremia: report of four cases and short review, *Infection* 21:306, 1993.

11. Lloyd-Puryear M, Wallace D, Baldwin T, et al: Meningitis caused by *Psychrobacter immobilis* in an infant, *J Clin Microbiol* 29:2041, 1991.

12. Lozano F, Florez C, Recio FJ, et al: Fatal *Psychrobacter immobilis* infection in a patient with AIDS, *AIDS* 8:1189, 1994.

13. Martinez JL, Martinez-Suarez J, Culebras E, et al: Antibiotic inactivating enzymes from a clinical isolate of *Agrobacterium radiobacter, J Antimicrob Chemother* 23:283, 1989.

14. Nozue H, Hayashi T, Hashimoto Y: Isolation and characterization of *Shewanella alga* from human clinical specimens and emendation of the description of *S. alga* Simidu et al, 1990, 335, *Int J Syst Bacteriol* 42:628, 1992.

15. Sawada H, Hiroyuki I, Oyaizu H, et al: Proposal for rejection of *Agrobacterium tumefaciens* and revised descriptions for the genus *Agrobacterium* and for *Agrobacterium radiobacter* and *Agrobacterium rhizogenes, Int J Syst Bacteriol* 43:694, 1993.

16. Shideh K, Janda JM: Biochemical and pathogenic properties of *Shewanella alga* and *Shewanella putrefaciens, J Clin Microbiol* 36:783, 1998.

17. Weyant RS, Moss CW, Weaver RE et al, editors: *Identification of unusual pathogenic gram-negative aerobic and facultatively anaerobic bacteria,* ed 2, Baltimore, 1996, Williams & Wilkins.

18. Validation of publication of new names and new combinations previously effectively published outside the IJSEM, *Int J Syst Evol Microbiol,* 2003, 53:935.

19. Young JM, Kuykendall LD, Martínez-Romero E, et al: A revision of *Rhizobium* Frank, 1889, with an emended description of the genus, and the inclusion of all species of *Agrobacterium* Conn, 1942 and *Allorhizobium undicola* de Lajudie et al, 1998 as new combinations: *Rhizobium radiobacter, R. rhizogenes, R. rubi, R. undicola* and *R. vitis, Int J Syst Evol Microbiol* 51:89, 2001.

ADDITIONAL READING

Schreckenberger PC , Daneshvar MI, Weyant RS, et al: *Acinetobacter, Achromobacter, Chryseobacterium, Moraxella,* and other nonfermentative gram-negative bacteria. In Murray PR, Baron EJ, Jorgensen JH, et al, editors: *Manual of clinical microbiology,* ed 8, Washington, DC, 2003, ASM Press.

Steinberg JP, Del Rio C: Other gram-negative and gram-variable bacilli. In Mandell GL, Bennett JE, Dolin R, editors: *Principles and practice of infectious diseases,* ed 6, Philadelphia, 2005, Elsevier Churchill Livingstone.

Genera and Species to Be Considered

Current Name	Previous Name
Agrobacterium yellow	
CDC group IIb*	*Flavobacterium* spp.(IIb)
CDC group EO-3	
CDC group EO-4	
CDC group O-1	
Chryseobacterium spp.†	
Elizabethkingia	*Chryseobacterium meningosepticum,*
meningoseptica	*Flavobacterium meningosepticum,*
	and CDC group IIa
Empedobacter brevis	*Flavobacterium breve*
Sphingobacterium multivorum	*Flavobacterium multivorum* and
	CDC group IIK-2
Sphingobacterium spiritivorum	*Flavobacterium spiritivorum* and
	CDC group IIK-3
Sphingobacterium thalpophilum	

---

*Includes clinical strains of *C. gleum* and *C. indologenes* other than the type strains.

†Includes type strain of *C. gleum* and *C. indologenes* (formerly *Flavobacterium gleum* and *F. indologenes).*

GENERAL CHARACTERISTICS

The organisms that constitute the genera discussed in this chapter are environmental inhabitants that are occasionally encountered in human specimens. They are considered together because they share similar physiologic and morphologic characteristics, that is, most are yellow-pigmented, oxidase-positive, glucose oxidizers that grow on MacConkey agar. *Sphingobacterium mizutaii* does not grow on MacConkey agar and is discussed in Chapter 29. At one time, many of the species presented were members of the *Flavobacterium* genus.

EPIDEMIOLOGY

As environmental inhabitants, these organisms may be found in various niches (Table 26-1). Most notable in terms of clinical relevance is their ability to survive in hospital environments, especially in moist areas. Although they are not considered part of normal human flora, these species can colonize a patient's respiratory tract during hospitalization, probably as a result of exposure to a contaminated water source or medical devices. Alternatively, transmission may occur directly by contaminated pharmaceutical solutions and, in the case of *Chryseobacterium meningosepticum,* from person to person.

Because of their ability to survive well in hospital environments, these organisms also have the potential to contaminate laboratory culture media and blood culture systems. Whenever these species are encountered, their clinical significance and the potential for contamination should be seriously considered.

PATHOGENESIS AND SPECTRUM OF DISEASE

As environmental organisms, no specific virulence factors have been identified for these species. However, the ability to survive in chlorinated tap water may give these organisms an edge in their ability to thrive in hospital water systems.

The development of infection basically requires exposure of debilitated patients to a contaminated source that results in respiratory colonization (Table 26-2). Depending on the health of the patient, subsequent infections such as bacteremia and pneumonia may develop. These infections are most frequently caused by *C. meningosepticum* or *Myroides odoratum.* In addition, infections of several other body sites, which may or may not be preceded by respiratory colonization, have been associated with the other species.[1-5,7-13]

Meningitis caused by *C. meningosepticum* is the most notable infection associated with the organisms listed in Table 26-2. This life-threatening infection, which may be accompanied by bacteremia, originally gained attention because it occurred in neonates. However, *C. meningosepticum* meningitis can also occur in compromised adults, and the organism has been implicated in hospital-based outbreaks of both meningitis and pneumonia.[14]

LABORATORY DIAGNOSIS

SPECIMEN COLLECTION AND TRANSPORT

No special considerations are required for specimen collection and transport of the organisms discussed in this chapter. Refer to Table 5-1 for general information on specimen collection and transport.

Table 26-1   Epidemiology

Species	Habitat (reservoir)	Mode of Transmission
Chryseobacterium meningosepticum, Chryseobacterium spp., *Empedobacter brevis, Sphingobacterium* spp.	Soil; plants; water; foodstuffs; hospital water sources, including incubators, sinks, faucets, tap water, hemodialysis systems, saline solutions, and other pharmaceuticals. Not part of human flora	Exposure of patients to contaminated medical devices or solutions, but source is not always known. May colonize upper respiratory tract. *C. meningosepticum* occasionally may be transmitted from birth canal to neonate

Table 26-2   Pathogenesis and Spectrum of Diseases

Species	Virulence Factors	Spectrum of Disease and Infections
Chryseobacterium meningosepticum, Chryseobacterium spp., *Empedobacter brevis, Sphingobacterium* spp.	Specific virulence factors are unknown. Able to survive chlorinated tap water. *C. meningosepticum,* the species most often associated with human infections, can be encapsulated and produces proteases and gelatinases that may be destructive for host cells and tissues	Bacteremia (often associated with implanted devices such as catheters or contaminated medical solutions). *C. meningosepticum,* particularly associated with meningitis in neonates and less commonly in adults. Other organisms associated with pneumonia, mixed infections of wounds, ocular and urinary tract infections, and occasionally other infections, including sinusitis, endocarditis, peritonitis, and fasciitis

SPECIMEN PROCESSING

No special considerations are required for processing the organisms discussed in this chapter. Refer to Table 5-1 for general specimen processing information.

DIRECT DETECTION METHODS

Other than Gram stain of patient specimens, there are no specific procedures for the direct detection of these organisms in clinical material. The *Chryseobacterium* spp., *Elizabethkingia meningoseptica,* and CDC group IIb are medium to long straight rods and often appear as "II-forms" (i.e., cells that appear thin in the center and thicker at the ends). *Empedobacter brevis* varies in being short to long rods. *Sphingobacterium* spp. are short straight rods, *S. thalpophilum* may exhibit II-forms. *Agrobacterium* yellow group are slender, medium to long gram-negative rods. CDC groups EO-3 and EO-4 are coccobacilli and CDC group O-1 are short gram-negative rods.

CULTIVATION

Media of Choice

All genera and CDC groups in this chapter grow well on routine laboratory media such as 5% sheep blood and chocolate agars. They also grow well in the broth of blood culture systems and in common nutrient broths such as thioglycollate and brain-heart infusion.

Incubation Conditions and Duration

These organisms will produce detectable growth on blood and chocolate agars when incubated at 35° C in either carbon dioxide or ambient air for a minimum of 24 hours. Growth on MacConkey agar is usually detectable within 24 hours of inoculation.

Colonial Appearance

Table 26-3 describes the colonial appearance and other distinguishing characteristics of each genus on 5% sheep blood and MacConkey agars.

APPROACH TO IDENTIFICATION

The ability of most commercial identification systems to accurately identify the organisms discussed in this chapter is limited or uncertain. The key biochemical reactions that can be used to presumptively differentiate among the genera discussed in this chapter are provided in Table 26-4. However, definitive identification of these organisms often requires performing a battery of biochemical tests not commonly available in many clinical microbiology laboratories. Therefore, full identification of clinically relevant isolates may require that they be sent to a reference laboratory.

Comments Regarding Specific Organisms

The growth of *S. spiritivorum* and *Chryseobacterium* spp., other than *E. meningoseptica,* is variable on MacConkey

Table 26-3 Colonial Appearance and Characteristics

Organism	Medium	Appearance
Agrobacterium yellow group	BA Mac	Yellow NLF
CDC group IIb	BA Mac	Yellow to orange pigment NLF, growth variable
CDC group EO-3	BA Mac	Yellow NLF
CDC group EO-4	BA Mac	Most strains yellow NLF
CDC group O-1	BA Mac	Yellow NLF, growth variable
Chryseobacterium spp.	BA	Circular, smooth, shiny with entire edge; light yellow to orange
Chryseobacterium meningosepticum	Mac	NLF
Elizabethkingia meningoseptica	BA Mac	Usually nonpigmented although may exhibit a slight yellow pigment; smooth, circular, large, shiny with entire edge NLF
Empedobacter brevis	BA Mac	Circular, smooth, shiny with entire edge; light yellow NLF, if growth
Sphingobacterium multivorum	BA Mac	Small, circular, convex, smooth, opaque with light yellow pigment after overnight incubation at room temperature NLF
Sphingobacterium spiritivorum	BA Mac	Small, circular, convex, smooth with pale yellow pigment NLF, if growth
Sphingobacterium thalpophilum	BA Mac	Pale yellow NLF

BA, 5% sheep blood agar; *Mac,* MacConkey agar; *NLF,* non–lactose-fermenter.

agar. Therefore, these organisms often need to be differentiated from yellow-pigmented, MacConkey-negative, oxidase-positive genera considered in Chapters 29 and 33.

Indole and urea hydrolysis are key biochemical tests for the separation of *Empedobacter brevis, Elizabethkingia meningoseptica,* and *Chryseobacterium* spp. from *Sphingobacterium* spp.

SERODIAGNOSIS

Serodiagnostic techniques are not generally used for the laboratory diagnosis of infections caused by the organisms discussed in this chapter.

ANTIMICROBIAL SUSCEPTIBILITY TESTING AND THERAPY

Validated susceptibility testing methods do not exist for these organisms. Although they will grow on the media and under the conditions recommended for testing the more commonly encountered bacteria (see Chapter 12 for more information regarding validated testing methods), the ability to grow and the ability to detect important antimicrobial resistances are not the same thing. Therefore, the lack of validated in vitro susceptibility testing methods does not allow definitive treatment and testing guidelines to be given for any of the organisms listed in Table 26-5.

Although susceptibility data for some of these bacteria can be found in the literature, the lack of understanding of potential underlying resistance mechanisms prohibits the validation of such data. Review Chapter 12 for preferable strategies that can be used to provide susceptibility information and data when validated testing methods do not exist for a clinically important bacterial isolate.

In general, the species considered in this chapter are frequently resistant to the penicillins (including carbapenems such as imipenem), cephalosporins, and aminoglycosides commonly used to treat infections caused by other gram-negative bacilli. However, the susceptibility data can vary substantially with the type of testing method used.[1,2,14] An unusual feature of many of these species is that they often appear susceptible to, and may be treated with, antimicrobial agents that are usually considered only effective against gram-positive bacteria; clindamycin, rifampin, and vancomycin are notable examples.[14]

PREVENTION

Because these organisms are ubiquitous in nature and are not generally a threat to human health, there are no recommended vaccination or prophylaxis protocols. Hospital-acquired infections are best controlled by following appropriate sterile techniques and infection control guidelines, and implementing effective protocols for the sterilization and decontamination of medical supplies.

Table 26-4   Key Biochemical and Physiologic Characteristics

Organism	Oxidizes Mannitol	Indole	Gelatin	Urea	Nitrate Reduction	Esculin Hydrolysis	Motility
Agrobacterium yellow group[a]	–	–	–	+	–	(+)	p,1–2
CDC group EO–3	(+)	–	–	(+)	–	–	nm
CDC group EO–4	–	–	–	+	–	–	nm
CDC group O–1	–	–	v	–	–	+	p, 1–2
Chryseobacterium spp.[b,d]	–	+	v	v	v	v	nm
Elizabethkingia meningoseptica [b,c]	+	+	+	–	–	+	nm
Empedobacter brevis [b,c]	–	+	+	–	–	–	nm
Sphingobacterium multivorum	–	–	–	+	–	+	nm
Sphingobacterium spiritivorum	+	–	v	+ or (+)	–	+	nm
Sphingobacterium thalpophilum	–	–	v	+	+	+	nm

Data compiled from references 6, 8, 12, 15, and 16.

v, Variable; +, >90% of strains are positive; –, >90% of strains are negative; *nm*, nonmotile; *p*, polar flagella, (+), reaction may be delayed.
[a]Only a positive 3-ketolactonate test differentiates this group from *Sphingomonas paucimobilis*
[b]Colonial pigmentation is critical to separate *Chryseobacterium* spp. and *Empedobacter brevis*.
[c]DNase positive.
[d]Includes *Chryseobacterium gleum*, *C. indologenes*, and CDC group IIb.

Table 26-5   Antimicrobial Therapy and Susceptibility Testing

Species	Therapeutic Options	Potential Resistance to Therapeutic Options	Validated Testing Methods*	Comments
Chryseobacterium indologenes, Chryseobacterium meningosepticum, Empedobacter brevis, Sphingobacterium spp.	No definitive guidelines. Potentially active agents include ciprofloxacin, rifampin, clindamycin, trimethoprim/ sulfamethoxazole, and vancomycin	Yes, most produce beta-lactamases and are also frequently resistant to aminoglycosides	Not available	In vitro susceptibility results with disk diffusion may be seriously misleading

*Validated testing methods include those standard methods recommended by the Clinical and Laboratory Standards Institute (CLSI) and those commercial methods approved by the Food and Drug Administration (FDA).

Case Study

A 48-year-old male with underlying acute myelogenous leukemia developed fever during a hospitalization for induction chemotherapy. At the time of infection, the patient was severely neutropenic. Blood drawn through an indwelling venous catheter was positive with an indole-positive and oxidase-positive yellow bacteria. Subsequent blood cultures were negative on therapy. Fourteen days into treatment, the patient developed progressive dyspnea (difficulty breathing), fever, and a pulmonary infiltrate, and the sputum grew the same yellow bacteria, now resistant to the therapy. The patient was treated with minocycline with resolution of the pneumonia.

QUESTIONS

1. List the bacteria that would be in the differential for indole-positive, glucose-nonfermenting gram-negative rods?

2. *C. meningosepticum* is the most significant pathogen in this group. How can this species be separated from the other indole-positive bacteria?

3. Although *C. meningosepticum* is a ubiquitous inhabitant of the aqueous environment, disease is rare. However, it can be present in the hospital environment most commonly, resulting in serious disease in the neonate. Ninety percent of meningitis caused by *C. meningosepticum* occurs in neonates, predominantly in the premature infant. Mortality is high with more than half of the infants succumbing to the disease. Because of its importance and rarity, which method should be used to perform susceptibility testing for this kind of bacteria?

REFERENCES

1. Blahovea J, Hupkova M, Krcmery V, et al: Resistance to and hydrolysis of imipenem in nosocomial strains of *Flavobacterium meningosepticum, Eur J Clin Microbiol Infect Dis* 13:833, 1994.
2. Fass RJ, Barnishan J: In vitro susceptibilities of nonfermentative gram-negative bacilli other than *Pseudomonas aeruginosa* to 32 antimicrobial agents, *Rev Infect Dis* 2:841, 1980.
3. Ferrer C, Jakob E, Pastorino G, et al: Right-sided bacterial endocarditis due to *Flavobacterium odoratum* in a patient on chronic hemodialysis, *Am J Nephrol* 15:82, 1995.
4. Hsueh P, Wu J, Hsiue T, et al: Bacteremic necrotizing fasciitis due to *Flavobacterium odoratum, Clin Infect Dis* 21:1337, 1995.
5. Jorgensen JH, Maher LA, Howell AW: Activity of meropenem against antibiotic-resistant or infrequently encountered gram-negative bacilli, *Antimicrob Agents Chemother* 35:2410, 1991.
6. Kim KK, Kim MK, Lim JH, et al: Transfer of *Chryseobacterium meningosepticum* and *Chryseobacterium miricola* to *Elizabethkingia meningoseptica* comb. nov. and *Elizabethkingia miricola* comb. nov., *Int J Syst Evol Microbiol* 55:1287, 2005.
7. Marnejon T, Watanakunakorn C: *Flavobacterium meningosepticum* septicemia and peritonitis complicating CAPD, *Clin Nephrol* 38:176, 1992.
8. Pickett MJ, Hollis DG, Bottone EJ: Miscellaneous gram- negative bacteria. In Balows A, Hausler WJ Jr, Herrmann KL, et al, editors: *Manual of clinical microbiology,* ed 5, Washington, DC, 1991, American Society for Microbiology.
9. Pokrywka M, Viazanko K, Medvick J, et al: A *Flavobacterium meningosepticum* outbreak among intensive care patients, *Am J Infect Control* 21:139, 1993.
10. Reina J, Borrell N, Figuerola J: *Sphingobacterium multivorum* isolated from a patient with cystic fibrosis, *Eur J Clin Microbiol Infect Dis* 11:81, 1992.
11. Sader HS, Jones RN, Pfaller MA: Relapse of catheter-related *Flavobacterium meningosepticum* bacteremia demonstrated by DNA macrorestriction analysis, *Clin Infect Dis* 21:997, 1995.
12. Schreckenberger PC, Daneshvar MI, Weyant RS, et al: *Acinetobacter, Achromobacter, Chryseobacterium, Moraxella,* and other nonfermentative gram-negative bacteria. In Murray PR, Baron EJ, Jorgensen JH, et al, editors: *Manual of clinical microbiology,* ed 8, Washington, DC, 2003, ASM Press.
13. Skapek SX, Jones WS, Hoffman KM, et al: Sinusitis and bacteremia caused by *Flavobacterium meningosepticum* in a sixteen-year-old with Shwachman Diamond syndrome, *Pediatr Infect Dis J* 11:411, 1992.
14. Tizer KB, Cervia JS, Dunn A, et al: Successful combination of vancomycin and rifampin therapy in a newborn with community acquired *Flavobacterium meningosepticum* neonatal meningitis, *Pediatr Infect Dis J* 14:916, 1995.
15. Vandamme P, et al: New perspectives in the classification of the flavobacteria: description of *Chryseobacterium* gen nov, *Bergeyella* gen nov, and *Empedobacter* nov rev, *Int J Syst Bacteriol* 44:827, 1994.
16. Weyant RS, Moss CW, Weaver RE et al, editors: *Identification of unusual pathogenic gram-negative aerobic and facultatively anaerobic bacteria,* ed 2, Baltimore, 1996, Williams & Wilkins.

ADDITIONAL READING

Balows A, Truper HG, Dworkin M et al, editors: *The prokaryotes. A handbook on the biology of bacteria: ecophysiology, isolation, identification, applications,* ed 2, New York, 1981, Springer-Verlag.
Steinberg JP, Del Rio C: Other gram-negative and gram-variable bacilli. In Mandell GL, Bennett JE, Dolin R, editors: *Principles and practice of infectious diseases,* ed 6, Philadelphia, 2005, Elsevier Churchill Livingstone.

Alcaligenes, Bordetella (Non-pertussis), Comamonas, and Similar Organisms

CHAPTER 27

Genera and Species to Be Considered

Current Name	Previous Name
Achromobacter denitrificans	Alcaligenes denitrificans, Achromobacter xylosoxidans subsp. denitrificans
Alcaligenes faecalis	Pseudomonas or Alcaligenes odorans
Alcaligenes piechaudii	Alcaligenes faecalis type I
Bordetella bronchiseptica	CDC group IVa
CDC Alcaligenes-like group 1	
CDC group IIg	
Comamonas spp.	
Cupriavidus pauculus	CDC group IVc-2, Wautersia paucula, Ralstonia paucula
Delftia acidovorans	Comamonas acidovorans, Pseudomonas acidovorans
Gilardi rod group 1	
Myroides spp.	Flavobacterium odoratum
Oligella ureolytica	CDC group IVe
Oligella urethralis	Moraxella urethralis, CDC group M-4
Pseudomonas alcaligenes	
Pseudomonas pseudoalcaligenes	
Psychrobacter spp. (asaccharolytic strains)	Moraxella phenylpyruvia
Psychrobacter phenylpyruvicus	Moraxella phenylpyruvica
Roseomonas spp.	

GENERAL CHARACTERISTICS

The genera discussed in this chapter are considered together because most of them are usually MacConkey-positive, oxidase-positive, nonglucose utilizers. They are a diverse group of organisms whose specific morphologic and physiologic features are presented later in this chapter in the discussion of laboratory diagnosis.

EPIDEMIOLOGY

The habitats of the species listed in Table 27-1 vary from the soil and water environment to the upper respiratory tract of various mammals. Certain species have been exclusively found in humans, whereas the natural habitat for other organisms remains unknown.

The diversity of the organisms' habitats is reflected in the various ways they are transmitted. For example, transmission of environmental isolates such as *Achromobacter denitrificans* frequently involves exposure of debilitated patients to contaminated fluids or medical solutions.[11] In contrast, *Bordetella bronchiseptica* transmission primarily occurs by close contact with animals, whereas *B. holmesii* has been detected only in human blood, and no niche or mode of transmission is known.[27]

PATHOGENESIS AND SPECTRUM OF DISEASE

Identifiable virulence factors are not known for most of the organisms listed in Table 27-2. However, because infections usually involve exposure of compromised patients to contaminated materials, most of these species are probably of low virulence.[14,15] Among the environmental organisms listed, *Achromobacter* is the organism most frequently associated with various infections. This organism also has been implicated in outbreaks of nosocomial infections.[4,6,9,11] Other organisms, such as *Oligella urethralis, O. ureolytica,* and *Roseomonas* spp., whose reservoirs are unknown, are rarely implicated in human infections and are likely to be of low virulence.[2,3,7,13,17-19]

LABORATORY DIAGNOSIS

SPECIMEN COLLECTION AND TRANSPORT

No special considerations are required for collection and transport of the organisms discussed in this chapter. Refer to Table 5-1 for general information on specimen collection and transport.

SPECIMEN PROCESSING

No special considerations are required for processing of the organisms discussed in this chapter. Refer to Table 5-1 for general information on specimen processing.

DIRECT DETECTION METHODS

Other than Gram stain of patient specimens, there are no specific procedures for the direct detection of

Table 27-1 Epidemiology

Species	Habitat (reservoir)	Mode of Transmission
Achromobacter denitrificans	Environment, including moist areas of hospital. Transient colonizer of human gastrointestinal or respiratory tract of patients with cystic fibrosis	Not often known. Usually involves exposure to contaminated fluids (e.g., intravenous fluids, hemodialysis fluids, irrigation fluids), soaps, and disinfectants
Alcaligenes faecalis	Environment; soil and water, including moist hospital environments. May transiently colonize the skin	Exposure to contaminated medical devices and solutions
Alcaligenes piechaudii	Environment	Unknown. Rarely found in humans
Bordetella bronchiseptica	Normal respiratory flora of several mammals, including dogs, cats, and rabbits. Not part of human flora	Probably by exposure to contaminated respiratory droplets during close contact with animals
CDC group IVc-2	Uncertain. Probably water sources, including those in the hospital setting. Not part of human flora	Usually involves contaminated dialysis systems or exposure of wounds to contaminated water
Delftia acidovorans *Comamonas testosteroni* *Comamonas* spp.	Environmental, soil and water; can be found in hospital environment. Not part of human flora	Uncertain. Rarely found in humans. Probably involves exposure to contaminated solutions or devices
Gilardi rod group 1	Unknown. Probably environmental. Not part of human flora	Unknown. Rarely found in humans
Oligella urethralis *Oligella ureolytica*	Unknown. May colonize distal urethra	Manipulation (e.g., catheterization) of urinary tract
Pyschrobacter spp.	Unknown	Unknown
Roseomonas spp.	Unknown	Unknown. Rarely found in humans

Table 27-2 Pathogenesis and Spectrum of Disease

Species	Virulence Factors	Spectrum of Disease and Infections
Achromobacter dentrificans	Unknown. Survival in hospital the result of inherent resistance to disinfectants and antimicrobial agents	Infections usually involve compromised patients and include bacteremia, urinary tract infections, meningitis, wound infections, pneumonia, and peritonitis, and occur in various body sites; can be involved in nosocomial outbreaks
Alcaligenes faecalis	Unknown	Infections usually involve compromised patients. Often a contaminant; clinical significance of isolates should be interpreted with caution. Has been isolated from blood, respiratory specimens, and urine.
Alcaligenes piechaudii	Unknown	Rare cause of human infection
Bordetella bronciseptica	Unknown for humans. Has several factors similar to *B. parapertussis*	Opportunistic infection in compromised patients with history of close animal contact. Infections are uncommon and include pneumonia, bacteremia, urinary tract infections, meningitis, and endocarditis
CDC group IVc-2	Unknown	Rare cause of human infection. Infections in compromised patients include bacteremia and peritonitis
Delftia acidovorans *Comamonas testosteroni* *Comamonas* spp.	Unknown	Isolated from respiratory tract, eye, and blood but rarely implicated as being clinically significant
Gilardi rod group 1	Unknown	Clinical significance uncertain, has been isolated from wounds, urine, and blood
Oligella urethralis	Unknown	Urinary tract infections, particularly in females
Oligella urealytica	Unknown	Also isolated from kidney, joint, and peritoneal fluid
Pyschrobacter spp.	Unknown	Rare cause of human infection
Roseomonas spp.	Unknown	Clinical significance uncertain. Most isolated from blood, wounds, or genitourinary tract of immunocompromised or debilitated patients

these organisms in clinical material. *B. bronchiseptica* is a medium-sized straight rod, whereas *Oligella urethralis, Psychrobacter* spp., *Roseomonas* spp., and *Moraxella* spp. are all coccobacilli, although *P. phenylpyruvicus* may appear as a broad rod, and some *Roseomonas* spp. may appear as short, straight rods. *Oligella ureolytica* is a short, straight rod, and the cells of Gilardi rod group 1 tend to be short and broad; *Myroides* spp. are pleomorphic rods and are either short or long and straight to slightly curved.

Alcaligenes and *Achromobacter* spp. are medium to long straight rods, as are CDC *Alcaligenes*-like group 1, *Cupriavidus pauculus, Delftia acidovorans, Pseudomonas alcaligenes,* and *Pseudomonas pseudoalcaligenes*. The *Comamonas* spp. are pleomorphic, and may appear as long, paired, curved rods or filaments. The cells of CDC group IIg appear as small coccoid-to-rod forms or occasionally as rods with long filaments.

CULTIVATION

Media of Choice

B. bronchiseptica, will grow on 5% sheep blood, chocolate, and MacConkey agars, usually within 1 to 2 days after inoculation. It should also grow in thioglycollate broth. *Psychrobacter* spp., Gilardi rod group 1, *Myroides* spp., *Oligella* spp., *Achromobacter* spp., *Delftia acidovorans, Alcaligenes* spp., CDC *Alcaligenes*-like group 1, *Comamonas* spp., *Roseomonas* spp., *Pseudomonas alcaligenes* and *P. pseudoalcaligenes, Cupriavidus pauculus,* and CDC group IIg all grow well on 5% sheep blood, chocolate, and MacConkey agars. Most of these genera should also grow well in the broth of blood culture systems, as well as in common nutrient broths such as thioglycollate and brain-heart infusion.

Incubation Conditions and Duration

Most of the organisms just mentioned will produce detectable growth on media incubated at 35° C in ambient air or 5% CO_2. *Psychrobacter* spp. usually grow better at 25° C than at 35° C.

Colonial Appearance

Table 27-3 describes the colonial appearance and other distinguishing characteristics (e.g., pigment and odor) of each genus on 5% sheep blood and MacConkey agars.

APPROACH TO IDENTIFICATION

The ability of most commercial identification systems to accurately identify the organisms discussed in this chapter is limited or uncertain. Strategies for identification of these genera therefore are based on the use of conventional biochemical tests and special staining for flagella. Although most clinical microbiology laboratories do not routinely perform flagella stains, motility

and flagella placement is the easiest way to separate these organisms.

Many microbiologists groan at the mere mention of having to perform a flagella stain, but the method described in Procedure 13-16 is a wet mount that is easy to perform. At the very least, a simple wet mount to observe cells for motility will help separate the motile and nonmotile genera. The pseudomonads, *Brevundimonas, Burkholderia,* and *Ralstonia* species described in Chapter 24 are motile by means of single or multiple polar flagella, and the motile organisms described in this chapter have peritrichous flagella (e.g., *B. bronchiseptica, Alcaligenes* spp., and *Achromobacter* spp.), or polar flagella (e.g., *Delftia, Comamonas*).

Organisms are first divided on the basis of Gram stain morphology, that is, coccoid (Table 27-4) or rod-shaped (Tables 27-5, 27-6, and 27-7). They are then further divided based on whether the organisms are nonmotile (see Table 27-5), peritrichously flagellated (see Table 27-6), or flagellated by polar tufts (see Table 27-7).

Comments Regarding Specific Organisms

B. bronchiseptica is oxidase-positive, motile, and rapidly urease-positive, sometimes in as little as 4 hours. This organism must be differentiated from *Cupriavidus pauculus* and *Oligella ureolytica.*

Urea hydrolysis is a key test for *Myroides* spp., which is also distinguished by production of a characteristic fruity odor. CDC group IIg is the only indole-positive, nonmotile species included in this chapter.

The genus *Oligella* includes one nonmotile species *(O. urethralis)* and one motile species *(O. ureolytica).* Urease hydrolysis is a key test for differentiating between these species; *O. ureolytica* often turns positive within minutes. *Psychrobacter phenylpyruvicus* is nonmotile and both urea-positive and phenylalanine deaminase–positive.

Achromobacter xylosoxidans subsp. *denitrificans* and *Alcaligenes piechaudii* reduce nitrate to nitrite, but only *A. xylosoxidans* subsp. *denitrificans* reduces nitrite to gas. *Alcaligenes faecalis* has a fruity odor and also reduces nitrite to gas. CDC *Alcaligenes*-like group 1 is similar to *Achromobacter xylosoxidans* subsp. *denitrificans* but is usually urea-positive.

Delftia acidovorans is unique in producing an orange color when Kovac's reagent is added to tryptone broth (indole test).

Roseomonas must be separated from other pink-pigmented gram-negative (e.g., *Methylobacterium*) and gram-positive (e.g., certain *Rhodococcus* spp. or *Bacillus* spp.) organisms. *Roseomonas* differs from *Rhodococcus* and *Bacillus* spp. by being resistant to vancomycin, as determined by using a 30-µg vancomycin disk on an inoculated 5% blood agar plate. Unlike *Methylobacterium, Roseomonas* will grow on MacConkey agar and at 42° C.

Table 27-3   Colonial Appearance and Characteristics

Organism	Medium	Appearance
Achromobacter denitrificans	BA	Small, convex, and glistening
	Mac	NLF
Alcaligenes faecalis	BA	Feather-edged colonies usually surrounded by zone of green discoloration; produces a highly characteristic, fruity odor resembling apples or strawberries
	Mac	NLF
Alcaligenes piechaudii	BA	Non pigmented, glistening, convex colonies surrounded by zone of greenish-brown discoloration
	Mac	NLF
Bordetella bronchiseptica	BA	Small, convex, round
	Mac	NLF
CDC *Alcaligenes*-like group 1	BA	Resembles *A. denitrificans*
	Mac	NLF
CDC group IIg	BA	No distinctive appearance
	Mac	NLF
Comamonas spp.	BA	No distinctive appearance
	Mac	NLF
Cupriavidus pauculus	BA	Small, yellow
	Mac	NLF
Delftia acidovorans	BA	No distinctive appearance
	Mac	NLF
Gilardi rod group 1	BA	No distinctive appearance
	Mac	NLF
Myroides spp.	BA	Most colonies are yellow, have a characteristic fruity odor, and tend to spread
	Mac	NLF
Oligella spp.	BA	Small, opaque, whitish
	Mac	NLF
Pseudomonas alcaligens	BA	No distinctive appearance
	Mac	NLF
Pseudomonas pseudoalcaligenes	BA	No distinctive appearance
	Mac	NLF
Psychrobacter spp. (asaccharolytic strains)	BA	Smooth, small, translucent to semiopaque
	Mac	NLF
Roseomonas spp.	BA	Pink-pigmented; some colonies may be mucoid
	Mac	NLF

BA, 5% sheep blood agar; *Mac,* MacConkey agar; *NLF,* non–lactose fermenter.

Table 27-4   Key Biochemical and Physiologic Characteristics for Coccoid Species

Organisms	Motility	Urea Hydrolysis	Nitrate Reduction	Nitrite Reduction
Oligella ureolytica	+ or (+)*	+	+	+
Oligella urethralis	nm	–	–	+
Psychrobacter phenylpyruvicus[†]	nm	+	v	–
Psychrobacter immobilis[‡] (asaccharolytic strains)	nm	v	v	ND

Data compiled from references 8, 16, 20, and 26.

ND, No data available; *v,* variable; +, >90% of strains are positive; –, >90% of strains are negative; *nm,* nonmotile; (+), positive delayed

*Petrichous flagella but motility may be delayed or difficult to demonstrate

[†]Deaminates phenylalanine

[‡]Best growth at 25° C

Table 27-5   Key Biochemical and Physiologic Characteristics for Rod-Shaped Nonmotile Species

Organisms	Insoluble Pigment	Indole	Urea Hydrolysis	Nitrite Reduction
CDC Group IIg	v, tan or salmon	+	−	+
Myroides spp.	v, yellow	−	+	v
Gilardi rod group 1	v, amber	−	−	−

Data compiled from references 8, 16, 20, 22, and 26.
v, Variable; +, >90% of strains are positive; −, >90% of strains are negative

Table 27-6   Key Biochemical and Physiologic Characteristics for Rod-Shaped Motile Species with Polar Flagella

Organism	Number of Flagella	Oxidizes Mannitol	Insoluble Pigment	Growth at 42° C	Nitrate Reduction
Delftia acidovorans	>2	+	−	v	+
Comamonas spp.	>2	−	−	v	+
Pseudomonas alcaligenes	1-2	−	v[c]	v	v
Pseudomonas pseudoalcaligenes[a]	1-2	−	−	+	+
Roseomonas spp.[b]	1-2[d]	v	pink	v	v

Data compiled from references 8, 16, 20, 21, 25, and 26.
v, Variable; +, >90% of strains are positive; −, >90% of strains are negative.
[a]Oxidizes fructose.
[b]Represents composite of several species and genomospecies.
[c]Some strains have a yellow-orange insoluble pigment.
[d]Genomospecies 5 is nonmotile.

Table 27-7   Key Biochemical and Physiologic Characteristics for Rod-Shaped Motile Species with Peritrichous Flagella

Organism	Urea Hydrolysis	Nitrate Reduction	Gas from Nitrate	Growth on Cetrimide	Jordan's Tartrate
Achromobacter denitrificans	−	+	+	v	+
Alcaligenes faecalis*	−	−	−	v	−
Alcaligenes piechaudii	−	+	−	+	+
CDC *Alcaligenes*-like group 1	v	+	+	−	−
Bordetella bronchiseptica	+	+	−	−	−
Cupriavidus pauculus	+	v	−	−	+

Data compiled from references 8, 16, 20, 23, 24, 26, and 28.
v, Variable; +, >90% of strains are positive; −, >90% of strains are negative.
*Reduces nitrite.

Table 27-8 Antimicrobial Therapy and Susceptibility Testing

Species	Therapeutic Options	Potential Resistance to Therapeutic Options	Validated Testing Methods*	Comments
Achromobacter denitrificans	No definitive guidelines. Potentially active agents include mezlocillin, piperacillin, ticarcillin/clavulanic acid, ceftazidime, imipenem, trimethoprim/sulfamethoxazole, and quinolones	Capable of beta-lactamase production	Not available	
Alcaligenes faecalis	No definitive guidelines	Capable of beta-lactamase production	Not available	
Alcaligenes piechaudii	No definitive guidelines	Unknown	Not available	
Bordetella bronchiseptica	No definitive guidelines. Potentially active agents include aminoglycosides, mezlocillin, piperacillin, ceftazidime, imipenem, and quinolones	Commonly resistant to ampicillin and several cephalosporins	Not available	
CDC group IVc-2	No definitive guidelines. Potentially active agents include cefotaxime, ceftazidime, ceftriaxone, and imipenem	Often resistant to penicillins, even with beta-lactamase inhibitor, and aminoglycosides	Not available	
Comamonas acidovorans, Comamonas testosteroni, Comamonas spp.	No definitive guidelines. Potentially active agents include ceftazidime, piperacillin, imipenem, and ciprofloxacin	Unknown	Not available	C. acidovorans tends to be more resistant than the other two species, especially to aminoglycosides
Gilardi rod group 1	No definitive guidelines	Unknown	Not available	Generally susceptible to various antimicrobial agents
Oligella urethralis, Oligella urealytica	No definitive guidelines. Potentially active agents include several penicillins, cephalosporins, and quinolones	Produces beta-lactamases; may develop resistance to quinolones	Not available	
Roseomonas spp.	No definitive guidelines. Potentially active agents include aminoglycosides, imipenem, and quinolones	Generally resistant to cephalosporins and penicillins	Not available	

*Validated testing methods include those standard methods recommended by the Clinical and Laboratory Standards Institute (CLSI) and those commercial methods approved by the Food and Drug Administration (FDA).

SERODIAGNOSIS

Serodiagnostic techniques are not generally used for the laboratory diagnosis of infections caused by the organisms discussed in this chapter.

ANTIMICROBIAL SUSCEPTIBILITY TESTING AND THERAPY

Validated susceptibility testing methods do not exist for these organisms. Although they will grow on the media and under the conditions recommended for testing the more commonly encountered bacteria (see Chapter 12 for more information regarding validated testing methods), this does not necessarily mean that interpretable and reliable results will be produced. Chapter 12 should be reviewed for preferable strategies that can be used to provide susceptibility information when validated testing methods do not exist for a clinically important bacterial isolate.

The lack of validated in vitro susceptibility testing methods does not allow definitive treatment and testing guidelines to be given for most organisms listed in Table 27-8. *B. parapertussis* is an exception; significant clinical experience indicates that erythromycin is the antimicrobial agent of choice for whooping cough caused by this organism (see Chapter 39 for more information regarding therapy for *B. pertussis* and *B. parapertussis* infections). Standardized testing methods do not exist for this species,[9] but the recent recognition of erythromycin resistance in *B. pertussis* indicates that development of such testing may be warranted for the causative agents of whooping cough.[12]

Even though standardized methods have not been established for the other species discussed in this chapter, in vitro susceptibility studies have been published and antimicrobial agents that have potential activity are noted, where appropriate, in Table 27-8.*

PREVENTION

Because the organisms may be encountered throughout nature and do not generally pose a threat to human health, there are no recommended vaccination or prophylaxis protocols. For those organisms occasionally associated with nosocomial infections, prevention of infection is best accomplished by following appropriate sterile techniques and infection control guidelines.

Case Study

An HIV-positive patient developed pneumonia while traveling in Europe. His cultures were reported as negative for pathogens and he did not respond to the usual treatment with cephalosporin therapy. The patient returned home and submitted a sputum culture to the local laboratory. The smear showed white blood cells but low numbers of bacteria. In 24 hours no pathogens were detected, but after 48 hours heavy growth of a gram-negative coccobacilli was observed. After identification of the cause of his pneumonia, the patient was successfully treated with ciprofloxacin. It was later found that he had a pet dog.

QUESTIONS

1. The organism was able to grow on MacConkey agar but the colonies were small and colorless. Oxidase and catalase were positive. Colonies on blood agar were without pigment. What rapid tests can be done to identify this bacterium?
2. In this case the indole and PDA were negative but the motility and urease were positive. Which genera are in the differential for the pathogen?
3. Had the organism been nonmotile, which serious pathogen should be considered?
4. Our patient was not diagnosed at another hospital. What do you suspect as the reason for the inability to detect the organism?

REFERENCES

1. Bizet C, Tekaia F, Philippon A: In vitro susceptibility of *Alcaligenes faecalis* compared with those of other *Alcaligenes* spp. to antimicrobial agents including seven β-lactams, *J Antimicrob Chemother* 32:907, 1993.
2. Bowman JP, Cavanagh J, Austin JJ, et al: Novel *Psychrobacter* species from Antarctic ornithogenic soils, *Int J Syst Bacteriol* 46:841, 1996.
3. Castagnola E, Tasso L, Conte M, et al: Central venous catheter-related infection due to *Comamonas acidovorans* in a child with non-Hodgkin's lymphoma, *Clin Infect Dis* 19:559, 1994.
4. Cheron M, Abachin E, Guerot E, et al: Investigation of hospital-acquired infections due to *Alcaligenes denitrificans* subsp. *xylosoxidans* by DNA restriction fragment length polymorphism, *J Clin Microbiol* 32:1023, 1994.
5. Decre D, Arlet G, Bergogne-Berezin E, et al: Identification of a carbenicillin-hydrolyzing β-lactamase in *Alcaligenes denitrificans* subsp. *xylosoxidans*, *Antimicrob Agents Chemother* 39:771, 1995.
6. Dunne WM, Maisch S: Epidemiological investigation of infections due to *Alcaligenes* species in children and patients with cystic fibrosis: use of repetitive-element–sequence polymerase chain reaction, *Clin Infect Dis* 20:836, 1995.
7. Hollis DG, Weaver RE, Moss CW, et al: Chemical and cultural characterization of CDC group WO-1, a weakly oxidative gram-negative group of organisms isolated from clinical sources, *J Clin Microbiol* 30:291, 1992.
8. Holt JG, Krieg NR, Sneath PH, et al, editors: *Bergey's manual of determinatative bacteriology*, ed 9, Baltimore, 1994, Williams & Wilkins.
9. Hoppe JE, Tschirner T: Comparison of media for agar dilution susceptibility testing of *Bordetella pertussis* and *Bordetella parapertussis*, *Eur J Clin Microbiol Infect Dis* 14:775, 1995.
10. Knippschild M, Schmid EN, Uppenkamp M, et al: Infection by *Alcaligenes xylosoxidans* subsp. *xylosoxidans* in neutropenic patients, *Oncology* 53:258, 1996.
11. Legrand C, Anaissie E: Bacteremia due to *Achromobacter xylosoxidans* in patients with cancer, *Clin Infect Dis* 14:479, 1992.
12. Lewis K, Saubolle MA, Tenover FC, et al: Pertussis caused by an erythromycin-resistant strain of *Bordetella pertussis*, *Pediatr Infect Dis J* 14:388, 1995.
13. Lindquist SW, Weber DJ, Mangum ME, et al: *Bordetella holmesii* sepsis in an asplenic adolescent, *Pediatr Infect Dis J* 14:813, 1995.
14. Moss CW, Daneshvar MI, Hollis DG: Biochemical characteristics and fatty acid composition of Gilardi rod group 1 bacteria, *J Clin Microbiol* 31:689, 1993.
15. Musso D, Drancourt M, Bardot J, et al: Human infection due to the CDC group IVc-2 bacterium: case report and review, *Clin Infect Dis* 18:482, 1994.
16. Pickett MJ, Hollis DG, Bottone EJ: Miscellaneous gram-negative bacteria. In Balows A, Hausler WJ Jr, Herrmann KL, et al, editors: *Manual of clinical microbiology*, ed 5, Washington, DC, 1991, American Society for Microbiology.
17. Pugliese A, Pacris B, Schoch PE, et al: *Oligella urethralis* urosepsis, *Clin Infect Dis* 17:1069, 1993.

18. Rihs JD, Brenner DJ, Weaver RE, et al: *Roseomonas*: a new genus associated with bacteremia and other human infections, *J Clin Microbiol* 31:3275, 1993.

19. Riley UBG, Bignardi G, Goldberg L, et al: Quinolone resistance in *Oligella urethralis*-associated chronic ambulatory peritoneal dialysis peritonitis, *J Infect* 32:155, 1996.

20. Schreckenberger PC, Daneshvar MI, Weyant RS, et al: *Acinetobacter, Achromobacter, Chryseobacterium, Moraxella,* and other nonfermentative gram-negative rods. In Murray PR, Baron EJ, Jorgensen JH, editors: *Manual of clinical microbiology,* ed 8, Washington, DC, 2003, ASM Press.

21. Validation of the publication of new names and new combinations previously effectively published outside the IJSB. Validation List No 65, *Int J Syst Bacteriol* 48:627, 1998.

22. Vancanneyt M, Segers P, Torck U, et al: Reclassification of *Flavobacterium odoratum* (Stutzer 1929) strains to a new genus, *Myroides,* as *Myroides odoratus* comb nov and *Myroides odoratimimus* sp nov, *Int J Syst Bacteriol* 46:926, 1996.

23. Vandamme P, Heyndrickx M, Vancanneyt M, et al: *Bordetella trematum* sp nov, isolated from wounds and ear infections in humans, and reassessment of *Alcaligenes denitrificans* (Rüger and Tan, 1983), *Int J Syst Bacteriol* 46:849, 1996.

24. Vandamme P, Goris J, Coenye T, et al: Assignment of Centers for Disease Control group IVc-2 to the genus *Ralstonia* as *Ralstonia paucula* sp. nov, *Int J Syst Bacteriol,* 49:663, 1999.

25. Wen A, Fegan M, Hayward C, et al: Phylogenetic relationships among members of the *Comamonadaceae,* and description of *Delftia acidovorans* (den Dooren de Jong, 1926 and Tamaoka et al, 1987) gen nov, comb nov, *Int J Syst Bacteriol* 49:567, 1999.

26. Weyant RS, Moss CW, Weaver RE, editors: *Identification of unusual pathogenic gram-negative aerobic and facultatively anaerobic bacteria,* ed 2, Baltimore, 1996, Williams & Wilkins.

27. Woolfrey BF, Moody JA: Human infections associated with *Bordetella bronchiseptica, Clin Microbiol Rev* 4:243, 1991.

28. Yabuuchi E, Kawamura Y, Kosako Y, et al: Emendation of the genus *Achromobacter* and *Achromobacter xylosoxidans* (Yabuuchi and Yano) and proposal of *Achromobacter ruhlandii* (Packer and Vishniac) comb nov, *Achromobacter piechaudii* (Kiredjian et al) comb nov, and *Achromobacter xylosoxidans* subsp *denitrificans* (Rüger and Tan) comb nov, *Microbiol Immunol* 42:429, 1998.

ADDITIONAL READING

Balows A, Truper HG, Dworkin M, et al, editors: *The prokaryotes. A handbook on the biology of bacteria: ecophysiology, isolation, identification, applications,* ed 2, New York, 1981, Springer-Verlag.

Loeffelholz MS: *Bordetella.* In Murray PR, Baron EJ, Jorgensen JH, et al, editors: *Manual of clinical microbiology,* ed 8, Washington, DC, 2003, ASM Press.

Steinberg JP, Del Rio C: Other gram-negative and gram-variable bacilli. In Mandell GL, Bennett JE, Dolin R, editors: *Principles and practice of infectious diseases,* ed 6, Philadelphia, 2005, Elsevier Churchill Livingstone.

von Graevenitz A: Ecology, clinical significance, and antimicrobial susceptibility of infrequently encountered glucose-nonfermenting gram-negative rods. In Gilardi GL, editor: *Nonfermentative gram-negative rods: laboratory identification and clinical aspects,* New York, 1985, Marcel Dekker.

Vibrio, Aeromonas, Plesiomonas, and *Chromobacterium*

Genera and Species to Be Considered

Current Name	Previous Name
Aeromonas caviae	
Aeromonas hydrophila	
Aeromonas jandaei	
Aeromonas schubertii	
Aeromonas veronii	
A. veronii biovar *sobria*	
A. veronii biovar *veronii*	
Chromobacterium	
violaceum	
Grimontia hollisae[18]	CDC group EF-13; *Vibrio hollisae*
Plesiomonas shigelloides	*Aeromonas* or *Pseudomonas shigelloides*
Vibrio alginolyticus	*Vibrio parahaemolyticus* biotype 2
Vibrio cholerae	
Vibrio cincinnatiensis	
Vibrio damsela	*Photobacterium damselae* subsp. *damselae*
Vibrio fluvialis	CDC group EF-6
Vibrio furnissii	
Vibrio harveyi	*Vibrio carchariae*
Vibrio metschnikovii	CDC enteric group 16
Vibrio mimicus	*Vibrio cholerae* (sucrose-negative)
Vibrio parahaemolyticus	*Pasteurella parahaemolyticus*
Vibrio vulnificus	CDC group EF-3

GENERAL CHARACTERISTICS

The organisms discussed in this chapter are considered together because they are all oxidase-positive, glucose-fermenting, gram-negative bacilli that grow on MacConkey agar. Their individual morphologic and physiologic features are presented later in this chapter in the discussion of laboratory diagnosis. There are other halophilic organisms, for example, *Halomonas venusta, Shewanella algae,* and CDC halophilic nonfermenter group 1, which require salt but do not ferment glucose as the halophilic vibrios do.

EPIDEMIOLOGY

The epidemiology of *Vibrio* spp., *Aeromonas* spp., *Plesiomonas shigelloides,* and *Chromobacterium violaceum* are similar in many aspects (Table 28-1). The primary habitat for most of these organisms is water, generally brackish or marine water for *Vibrio* spp. and fresh water for *Aeromonas* spp., *P. shigelloides,* and *C. violaceum.* None of these organisms are considered part of the normal human flora. Transmission to humans is by ingestion of contaminated water or seafood or by exposure of disrupted skin and mucosal surfaces to contaminated water.

The epidemiology of the most notable human pathogen in this chapter, *Vibrio cholerae,* is far from being fully understood.[5] This organism causes epidemics and pandemics (i.e., epidemics that span worldwide) of the diarrheal disease cholera. Since 1817 the world has witnessed seven cholera pandemics. During these outbreaks the organism is spread among people by the fecal-oral route, usually in environments with poor sanitation.

Of interest, the niche that *V. cholerae* inhabits between epidemics is uncertain. The form of the organism shed from infected humans is somewhat fragile and cannot survive long in the environment. However, evidence does suggest that survival, or dormant, stages of the bacillus exist to allow long-term survival in brackish water or saltwater environments during inter-epidemic periods.[4,10] Asymptomatic carriers of *V. cholerae* have been documented, but they are not thought to be a significant reservoir for maintaining the organism between outbreaks.

PATHOGENESIS AND SPECTRUM OF DISEASE

As a notorious pathogen, *V. cholerae* elaborates several toxins and factors that play important roles in the organism's virulence.[10] Cholera toxin (CT) is primarily responsible for the key features of cholera (Table 28-2). Release of this toxin causes mucosal cells to hypersecrete water and electrolytes into the lumen of the gastrointestinal tract. The result is profuse watery diarrhea, leading to dramatic fluid loss. The fluid loss results in severe dehydration and hypotension that, without medical intervention, frequently leads to death. This toxin-mediated disease does not require the organism to penetrate the mucosal barrier. Therefore, blood and the inflammatory cells typical of dysenteric stools are notably absent in cholera. Instead, "rice water stools," composed of fluids and mucous flecks, are the hallmark of cholera toxin activity.

Table 28-1 Epidemiology

Species	Habitat (reservoir)	Mode of Transmission
Vibrio cholerae	Niche outside of human gastrointestinal tract between occurrence of epidemics and pandemics is uncertain; may survive in a dormant state in brackish or salt water; human carriers also are known but are uncommon	Fecal-oral route, by ingestion of contaminated washing, swimming, cooking, or drinking water; also by ingestion of contaminated shellfish or other seafood
V. alginolyticus	Brackish or salt water	Uncertain; exposure to contaminated water
V. cincinnatiensis	Unknown	Unknown
Vibrio damsela	Brackish or salt water	Exposure of wound to contaminated water
V. fluvialis	Brackish or salt water	Ingestion of contaminated water or seafood
V. furnissii	Brackish or salt water	Ingestion of contaminated water or seafood
Grimontia hollisae	Brackish or salt water	Ingestion of contaminated water or seafood
V. metschnikovii	Unknown	Unknown
V. mimicus	Brackish or salt water	Ingestion of contaminated water or seafood
V. parahaemolyticus	Brackish or salt water	Ingestion of contaminated water or seafood
V. vulnificus	Brackish or salt water	Ingestion of contaminated water or seafood
Aeromonas caviae A. hydrophila, A. sobia, A. jandaei, A. schubertii, and A. veronii	Aquatic environments around the world, including fresh water, polluted or chlorinated water, brackish water, and, occasionally, marine water; may transiently colonize gastrointestinal tract; often infect various warm- and cold-blooded animal species	Ingestion of contaminated food (e.g., dairy, meat, produce) or water; exposure of disrupted skin or mucosal surfaces to contaminated water or soil; traumatic inoculation of fish fins or fishing hooks
Plesiomonas shigelloides	Fresh water, especially in warmer climates	Ingestion of contaminated water or seafood; exposure to cold-blooded animals, such as amphibia and reptiles
Chromobacterium violaceum	Environmental, soil and water of tropical and subtropical regions. Not part of human flora	Exposure of disrupted skin to contaminated soil or water

The somatic antigens O1 and O139 associated with the *V. cholerae* cell envelope are positive markers for strains capable of epidemic and pandemic spread of the disease. Strains carrying these markers almost always produce cholera toxin, whereas non-O1/non-O139 strains do not produce the toxin and hence do not produce cholera. Therefore, although these somatic antigens are not virulence factors per se, they are important virulence and epidemiologic markers that provide important information regarding *V. cholerae* isolates. The non-O1/non-O139 strains are associated with nonepidemic diarrhea and extraintestinal infections.

V. cholerae produces several other toxins and factors whose exact role in disease is still uncertain (see Table 28-2). For the organism to effectively release toxin, it first must infiltrate and distribute itself along the cells lining the mucosal surface of the gastrointestinal tract. Motility and chemotaxis mediate the distribution of organisms, and mucinase production allows for penetration of the mucous layer. TCP pili provide the means by which bacilli attach to mucosal cells for release of cholera toxin.

Depending on the species, other vibrios are variably involved in three types of infection: gastroenteritis, wound infections, and bacteremia. Although some of these organisms have not been definitively associated with human infections, others, such as *V. vulnificus,* are known to cause fatal septicemia, especially in patients suffering from an underlying liver disease.

Aeromonas spp. and *P. shigelloides* are similar to *Vibrio* spp. in terms of the types of infections they cause.[7-9,11,12,16] Although these organisms can cause gastroenteritis, most frequently in children, their role in intestinal infections is not always clear. Therefore, the significance of their isolation in stool specimens should be interpreted with caution. Although noncholera vibrios, *Aeromonas* spp., and *P. shigelloides* produce factors that may contribute to virulence, little is known about any specific virulence factors.

C. violaceum is not associated with gastrointestinal infections, but acquisition of this organism by contamination of wounds can lead to fulminant, life-threatening systemic infections.[19]

Table 28-2 Pathogenesis and Spectrum of Diseases

Species	Virulence Factors	Spectrum of Disease and Infections
Vibrio cholerae	Cholera toxin, Zot toxin, Ace toxin, O1 and O139 somatic antigens, hemolysin/cytoxins, motility, chemotaxis, mucinase, and TCP pili	Cholera: profuse, watery diarrhea leading to dehydration, hypotension, and, often, death; occurs in epidemics and pandemics that span the globe. May also cause nonepidemic diarrhea and, occasionally, extraintestinal infections of wounds, respiratory tract, urinary tract, and central nervous system
V. alginolyticus	Specific virulence factors for the non-*V. cholerae* species are uncertain	Ear infections, wound infections; rare cause of septicemia; involvement in gastroenteritis is uncertain
V. cincinnatiensis		Rare cause of septicemia
V. damsela		Wound infections and rare cause of septicemia
V. fluvialis		Gastroenteritis
V. furnissii		Rarely associated with human infections
Grimontia hollisae		Gastroenteritis; rare cause of septicemia
V. metschnikovii		Rare cause of septicemia; involvement in gastroenteritis is uncertain
V. mimicus		Gastroenteritis; rare cause of ear infection
V. vulnificus		Wound infections and septicemia; involvement in gastroenteritis is uncertain
Aeromonas caviae, A. hydrophila, A. sobia, A. jandaei, A. schubertii, and *A. veronii*	*Aeromonas* spp. produce various toxins and factors, but their specific role in virulence is uncertain	Gastroenteritis, wound infections, bacteremia, and miscellaneous other infections, including endocarditis, meningitis, pneumonia, conjunctivitis, and osteomyelitis
Plesiomonas shigelloides	Unknown	Gastroenteritis; septicemia in compromised adults and infants experiencing complicated delivery
Chromobacterium violaceum	Unknown	Rare but dangerous infection. Begins with cellulitis or lymphadenitis and can rapidly progress to systemic infections with abscess formation in various organs and septic shock

LABORATORY DIAGNOSIS

SPECIMEN COLLECTION AND TRANSPORT

Because there are no special considerations for isolation of these genera from extraintestinal sources, general specimen collection and transport information provided in Table 5-1 is applicable. However, stool specimens suspected of containing *Vibrio* spp. should be collected and transported only in Cary-Blair medium. Buffered glycerol saline is not acceptable, because glycerol is toxic for vibrios. Feces is preferable, but rectal swabs are acceptable during the acute phase of diarrheal illness.

SPECIMEN PROCESSING

No special considerations are required for processing the organisms discussed in this chapter. Refer to Table 5-1 for general information on specimen processing.

DIRECT DETECTION METHODS

V. cholerae toxin can be detected in stool using an enzyme-linked immunosorbent assay (ELISA) or a com-

mercially available latex agglutination test (Oxoid Inc., Odgensburg, NY), but these tests are not widely used in the United States.

Microscopically, vibrios are gram-negative, straight or slightly curved rods (Figure 28-1). When stool specimens from cholera patients are examined using dark-fifield microscopy, the bacilli exhibit characteristic rapid darting or shooting-star motility. However, direct microscopic examination of stools by any method is not commonly used for the laboratory diagnosis of enteric bacterial infections.

Aeromonas spp. are gram-negative, straight rods, whereas *P. shigelloides* cells tend to be pleomorphic gram-negative rods that occur singly, in pairs, in short chains, or even as long, filamentous forms. Cells of *C. violaceum* are slightly curved, medium to long, gram-negative rods with rounded ends.

CULTIVATION

Media of Choice

Stool cultures for *Vibrio* spp. are plated on the selective medium thiosulfate citrate bile salts sucrose (TCBS) agar.

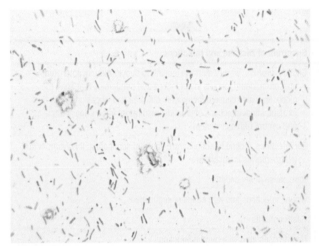

Figure 28-1   Gram stain of *Vibrio parahaemolyticus.*

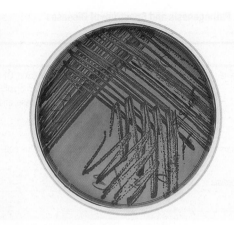

Figure 28-2   Colonies of *Chromobacterium violaceum* on DNase agar. Note violet pigment.

Although some *Vibrio* spp. grow very poorly on this medium, those that grow well produce either yellow or green colonies, depending on whether they are able to ferment sucrose (and produce yellow colonies). Alkaline peptone water (pH 8.4) may be used as an enrichment broth for obtaining growth of vibrios from stool. After inoculation, the broth is incubated for 5 to 8 hours at 35° C and then subcultured to TCBS.

Aeromonas spp. are indistinguishable from *Yersinia enterocolitica* on cefsulodin-irgasan-novobiocin (CIN) agar so it is important to perform an oxidase test to differentiate these two genera.

All of the genera considered in this chapter grow well on 5% sheep blood, chocolate, and MacConkey agars. They also grow well in the broth of blood culture systems and in thioglycollate or brain-heart infusion broths.

Incubation Conditions and Duration

These organisms produce detectable growth on 5% sheep blood and chocolate agars when incubated at 35° C in carbon dioxide or ambient air for a minimum of 24 hours. MacConkey and TCBS agars only should be incubated at 35° C in ambient air. The typical violet pigment of *C. violaceum* colonies (Figure 28-2) is optimally produced when cultures are incubated at room temperature (22° C).

Colonial Appearance

Table 28-3 describes the colonial appearance and other distinguishing characteristics (e.g., hemolysis and odor) of each genus on 5% sheep blood and MacConkey agars. The appearance of *Vibrio* spp. on TCBS is described in Table 28-4 and shown in Figure 28-3.

Table 28-3   Colonial Appearance and Characteristics

Organism	Medium	Appearance
Aeromonas spp.	BA	Large, round, raised, opaque; most pathogenic strains are beta-hemolytic, except *A. caviae,* which is usually nonhemolytic
	Mac	Both NLF and LF
Chromobacterium violaceum	BA	Round, smooth, convex, some strains beta-hemolytic; most colonies appear black or very dark purple; cultures smell of ammonium cyanide
	Mac	NLF
Plesiomonas shigelloides	BA	Shiny, opaque, smooth, nonhemolytic
	Mac	Both NLF and LF
Vibrio spp. and *Grimontia hollisae*	BA	Medium to large, smooth, opaque, iridescent with a greenish hue; *V. cholerae, V. fluvialis, V. mimicus,* and *V. damsela* can be beta-hemolytic
	Mac	NLF except *V. vulnificus,* which may be LF

BA, 5% sheep blood agar; *Mac,* MacConkey agar; *LF,* lactose fermenter, *NLF,* non–lactose-fermenter.

Table 28-4   Key Biochemical and Physiologic Characteristics of *Vibrio* spp. and *Grimontia hollisae*

Species	Oxidase	Gas from Glucose	Lactose	Sucrose	Lysine Decarboxylase[a]	Arginine Dihydrolase[a]	Ornithine Decarboxylase[a]	Growth in 0% NaCl[b]	Growth In 6% NaCl[b]	TCBS[c] Growth	Colony on TCBS[c]
Grimonti hollisae	+	–	–	–	–	–	–	–	+	Very poor	Green
Vibrio alginolyticus	+	–	–	+	+	–	v	–	+	Good	Yellow
Vibrio cholerae	+	–	v	+	+	–	+	+	v	Good	Yellow
Vibrio cincinnatiensis[g]	+	–	–	+	v	–	–	–	+	Very poor	Yellow
Vibrio damsela	+	–	–	–	v	+	–	–	+	Reduced at 36° C	Green[d]
Vibrio fluvialis	+	–	–	+	–	+	–	–	+	Good	Yellow
Vibrio furnissi	+	+	–	+	–	+	–	–	+	Good	Yellow
Vibrio harveyi	+	–	–	v	+	–	–	–	+	Good	Yellow
Vibrio metschnikovii	–	–	v	+	v	v	–	–	v	May be reduced	Yellow
Vibrio mimicus	+	–	v	–	+	–	+	+	v	Good	Green
Vibrio parahaemolyticus	+	–	–	–	+	–	+	–	+	Good	Green[e]
Vibrio vulnificus	+	–	(+)	–	+	–	+	–	+	Good	Green[f]

Data compiled from references 7 and 13.

v, Variable; +, >90% of strains are positive; –, >90% of strains are negative; (+), delayed.

[a]1% NaCl added to enhance growth.

[b]Nutrient broth with 0% or 6% NaCl added.

[c]Thiosulfate citrate bile salts-sucrose agar.

[d]5% yellow.

[e]1% yellow.

[f] 0% yellow.

[g]Ferments myo-inositol.

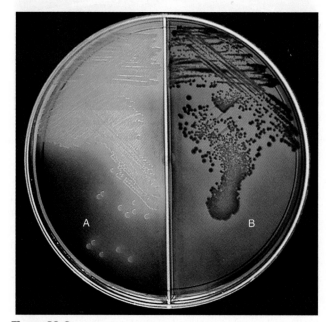

Figure 28-3   Colonies of *Vibrio cholerae* (**A**) and *V. parahaemolyticus* (**B**) on TCBS agar.

APPROACH TO IDENTIFICATION

The colonies of these genera resemble those of the family *Enterobacteriaceae* but notably differ by their positive oxidase test (except *V. metschnikovii,* which is oxidase-negative). The oxidase test must be performed from 5% sheep blood or another medium without a fermentable sugar (e.g., lactose in MacConkey agar or sucrose in TCBS). The reason for this is that fermentation of a carbohydrate results in acidification of the medium, and a false-negative oxidase test may result if the surrounding pH is below 5.1. Likewise, if the violet pigment of a suspected *C. violaceum* isolate interferes with performance of the oxidase test, the organism should be grown under anaerobic conditions (where it cannot produce pigment) and retested.

The reliability of commercial identification systems has not been widely validated for identification of these organisms, although most are listed in the databases of several systems. The API 20E system (bioMérieux, Inc., St. Louis, Mo) is one of the best for vibrios. Because the

Table 28-5 Key Biochemical and Physiologic Characteristics of *Aeromonas* spp., *P. shigelloides*, and *C. violaceum*

Species	Oxidase	Gas from Glucose	Esculin Hydrolysis	Fermentation of Sucrose	Lysine Decarboxylase	Arginine Dihydrolase	Ornithine Decarboxylase	Growth in 0% NaCl[a]	Growth in 6% NaCl[a]	TCBS[b] Growth
Aeromonas caviae	+	−	+	+	−	+	−	+	−	−
Aeromonas hydrophila	+	v	v	v	v	+	−	+	v	−
Aeromonas jandaei	+	+	−	−	+	+	−	+	−	−
Aeromonas schubertii	+	−	−	−	+	+	−	+	−	−
Aeromonas veronii biovar *sobria*	+	+	−	+	+	+	−	+	−	−
Aeromonas veronii biovar *veronii*	+	+	+	+	+	−	+	+	−	−
Chromobacterium violaceum[c]	v	−[d]	−	v	−	+	−	+	−	ND
Plesiomonas shigelloides	+	−	−	−	+	+	+	+	−	−

Data compiled from references 1, 20, and 22.
ND, No data; *v*, variable; +, >90% of strains are positive; −, >90% of strains are negative.
[a]Nutrient agar with 0% or 6% NaCl added.
[b]Thiosulfate citrate bile salts sucrose agar.
[c]91% produce an insoluble violet pigment; often, nonpigmented strains are indole-positive.
[d]Gas-producing strains have been described.

inoculum is prepared in 0.85% saline, the amount of salt often is enough to allow growth of the halophilic (salt-loving) organism.

The ability of most commercial identification systems to accurately identify *Aeromonas* spp. to the species level is limited and uncertain, and some kits have trouble separating *Aeromonas* spp. from *Vibrio* spp. Therefore, identification of potential pathogens should be confirmed using conventional biochemicals or serotyping. Tables 28-4 and 28-5 show several characteristics that can be used to presumptively group *Vibrio* spp., *Aeromonas* spp., *P. shigelloides*, and *C. violaceum*.

Comments Regarding Specific Organisms

V. cholerae and *V. mimicus* are the only *Vibrio* spp. that do not require salt for growth. Therefore, a key test in separating the halophilic species from *V. cholerae*, *V. mimicus*, *Aeromonas* spp., and *P. shigelloides* is growth in nutrient broth with 6% salt. Furthermore, the addition of 1% NaCl to conventional biochemicals is recommended to allow growth of halophilic species.

The string test can be used to separate *Vibrio* spp. from *Aeromonas* spp. and *P. shigelloides*. In this test, organisms are emulsified in 0.5% sodium deoxycholate, which lyses the vibrio cells, but not those of *Aeromonas* spp. and *P. shigelloides*. With cell lysis there is release of DNA, which can then be pulled up into a string using an inoculating loop (Figure 28-4).

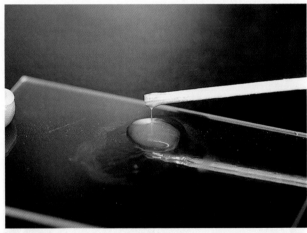

Figure 28-4 String test used to separate *Vibrio* spp. (positive) from *Aeromonas* spp. and *P. shigelloides* (negative).

A vibriostatic test using 0/129 (2,4-diamino-6, 7-diisopropylpteridine)–impregnated disks also has been used to separate vibrios (susceptible) from other oxidase-positive, glucose-fermenters (resistant) and to differentiate *V. cholerae* 01 and non-01 (susceptible) from other *Vibrio* spp. (resistant). However, recent strains of *V. cholerae* 0139 have been resistant to 0/129, so the dependability of this test may be waning.

Serotyping should be performed immediately to further characterize *V. cholerae* isolates. Toxigenic strains of serogroups 01 and 0139 can be involved in cholera

Table 28-6   Antimicrobial Therapy and Susceptibility Testing

Species	Therapeutic Options	Potential Resistance to Therapeutic Options	Validated Testing Methods*	Comments
Vibrio cholerae	Adequate rehydration plus antibiotics. Recommended agents include tetracycline or doxycycline; alternatives include trimethoprim/sulfamethoxazole, erythromycin, chloramphenicol, and quinolones	Yes; resistance to tetracycline, chloramphenicol, and trimethoprim/sulfamethoxazole is known	See CLSI document M100[†]	
Other *Vibrio* spp.	No definitive guidelines. For gastroenteritis therapy may not be needed; for wound infections and septicemia potentially active agents include tetracycline, chloramphenicol, nalidixic acid, most cephalosporins, and quinolones	Yes; similar to resistance reported for *V. cholerae*	See CLSI document M45[†]	
Aeromonas spp.	No definitive guidelines. For gastroenteritis therapy may not be needed; for soft tissue infections and septicemia potentially active agents include ceftriaxone, cefotaxime, ceftazidime, imipenem, aztreonam, amoxicillin-clavulanate, quinolones, and trimethoprim/sulfamethoxazole	Yes; capable of producing various beta-lactamases that mediate resistance to penicillins and certain cephalosporins	See CLSI document M45[†]	
Plesiomonas shigelloides	No definitive guidelines. For gastroenteritis therapy may not be needed; for septicemia potentially active agents include most cephalosporins, imipenem, aztreonam, beta-lactam/beta-lactamase-inhibitor combinations, and quinolones	Yes; capable of beta-lactamase production and resistance to ampicillin, mezlocillin, and piperacillin	See CLSI document M45[†]	
Chromobacterium violaceum	No definitive guidelines. Potentially active agents include cefotaxime, ceftazidime, imipenem, and aminoglycosides	Yes; activity of penicillins is variable; activity of first- and second-generation cephalosporins is poor	Not available	Will grow on Mueller-Hinton agar, but interpretive standards do not exist

*Validated testing methods include those standard methods recommended by the Clinical and Laboratory Standards Institute (CLSI) and those commercial methods approved by the Food and Drug Administration (FDA).
[†]Cited under Additional Reading.

epidemics. Strains that do not type in either antiserum are identified as non-01. Although typing sera are commercially available, isolates of *V. cholerae* are usually sent to a reference laboratory for serotyping.

Identification of *V. cholerae* or *V. vulnificus* should be reported immediately because of the life-threatening nature of these organisms.

Identification of *Aeromonas* spp., *P. shigelloides,* and *C. violaceum* can be accomplished using the characteristics shown in Table 28-5. *P. shigelloides* is unusual in being among the few species of clinically relevant bacteria that decarboxylate lysine, ornithine, and arginine.

Pigmented strains of *C. violaceum* are so distinctive that a presumptive identification can be made based on colonial appearance, oxidase, and Gram stain. Nonpigmented strains (approximately 9% of isolates) may be differentiated from *Pseudomonas, Burkholderia, Brevundimonas,* and *Ralstonia* based on glucose fermentation and a positive test for indole. Negative lysine and ornithine reactions are useful criteria for separating *C. violaceum* from *P. shigelloides*. In addition to the characteristics listed in Table 28-5, failure to ferment either maltose or mannitol also differentiates *C. violaceum* from *Aeromonas* spp.

SERODIAGNOSIS

Agglutination, vibriocidal, or antitoxin tests are available for diagnosing cholera using acute and convalescent sera. However, these methods are most commonly used for epidemiologic purposes. Serodiagnostic techniques are not generally used for the laboratory diagnosis of infections caused by the other organisms discussed in this chapter.

ANTIMICROBIAL SUSCEPTIBILITY TESTING AND THERAPY

Two components to the management of patients with cholera are rehydration of the patient and antimicrobial therapy (Table 28-6). Antimicrobials decrease the severity of illness and shorten the duration of organism shedding. The drugs of choice for cholera are tetracycline or doxycycline; however, resistance to these agents is known, and the use of other agents, such as chloramphenicol, ampicillin, or trimethoprim-sulfamethoxazole, may be necessary.[10] Clinical and Laboratory Standards Institute (CLSI) methods for testing *V. cholerae* do exist, and the CLSI document M100 should be consulted for this purpose.

The need for antimicrobial intervention for gastrointestinal infections caused by other *Vibrio* spp., *Aeromonas* spp., and *P. shigelloides* is less clear. In contrast, extraintestinal infections with these organisms and with *C. violaceum* can be life-threatening, and directed therapy is required.[7,8,12,14,15] Many of these organisms will grow on the media and under the conditions recommended for testing the more commonly encountered bacteria; CLSI document M45 addresses methods that may be applied to testing these organisms.

Antimicrobial agents that have potential activity are listed, where appropriate, in Table 28-6. Of importance is the capacity of these species to exhibit resistance to therapeutic agents; especially noteworthy is the capacity of *Aeromonas* spp. and *P. shigelloides* to produce various beta-lactamases.[2,6,17,21]

PREVENTION

No cholera vaccine is available in the United States. Two oral vaccines are available outside the United States, although the World Health Organization no longer recommends immunization for travel to or from cholera-infected areas. Individuals who have recently shared food and drink with a patient with cholera (e.g., household contacts) should be given chemoprophylaxis with tetracycline, doxycycline, or trimethoprim-sulfamethoxazole. However, mass chemoprophylaxis during epidemics is not indicated.[3] There are no approved vaccines or chemoprophylaxis for the other organisms discussed in this chapter.

Case Study

After vacationing in San Diego, a 21-year-old male surfer complained of severe left ear pain. He was afebrile but the auditory canal and tympanic membrane were erythematous. Amoxicillin was prescribed for presumed otitis media. Over the next 4 days the symptoms persisted and a bloody discharge developed. The patient returned to his physician, who cultured the drainage and prescribed gentamicin eardrops. His symptoms improved over the next 7 days. On culture, a non–lactose fermenter was isolated from MacConkey agar.

QUESTIONS

1. The isolate was indole- and oxidase-positive. A biochemical identification system had positive reactions for lysine and ornithine, but not arginine. What genus and species of bacteria are in the differential and how would you identify this microorganism?
2. How do you think the patient acquired this infection?
3. Commercial systems are known to misidentify the *Vibrio* as *Aeromonas* and vice versa. What is the reason for such a critical error?
4. Susceptibility testing using the disk method is not problematic for the *Vibrio* as long as which extra step is taken when doing testing?

REFERENCES

1. Abbott S: *Aeromonas*. In Murray PR, Baron EJ, Jorgensen JH, et al, editors: *Manual of clinical microbiology,* ed 8, Washington, DC, 2003, ASM Press.

2. Clark RB, Lister PD, Arneson-Rotert L, et al: In vitro susceptibilities of *Plesiomonas shigelloides* to 24 antibiotics and antibiotic-β-lactamase-inhibitor combinations, *Antimicrob Agents Chemother* 34:159, 1990.

3. Clinical and Laboratory Standards Institute: Methods for antimicrobial dilution and disk susceptibility testing of infrequently isolated or fastidious bacteria; M45, Villanova, Pa, 2006, CLSI.

4. Clinical and Laboratory Standards Institute: Performance standards for antimicrobial susceptibility testing; M-100 S16, Villanova, Pa, 2006, CLSI.

5. Colwell RR: Global climate and infectious disease: the cholera paradigm, *Science* 274:2025, 1996.

6. Committee on Infectious Diseases: *2006 Red book: report of the Committee on Infectious Diseases,* ed 27, Elk Grove, Ill, 2006, American Academy of Pediatrics.

7. Farmer JJ, Janda JM, Birkhead K: *Vibrio.* In Murray PR, Baron EJ, Jorgensen JH, et al, editors: *Manual of clinical microbiology,* ed 8, Washington, DC, 2003, ASM Press.

8. Hayes MV, Thomson CJ, Amyes SG: The "hidden" carbapenemase of *Aeromonas hydrophila, J Antimicrob Chemother* 37:33, 1996.

9. Janda JM: Recent advances in the study of the taxonomy, pathogenicity, and infectious syndromes associated with the genus *Aeromonas, Clin Microbiol Rev* 4:397, 1991.

10. Janda JM, Guthertz LS, Kokka RP, et al: *Aeromonas* species in septicemia: laboratory characteristics and clinical observations, *Clin Infect Dis* 19:77, 1994.

11. Jones BL, Wilcox MH: *Aeromonas* infections and their treatment, *J Antimicrob Chemother* 35:453, 1995.

12. Kaper JB, Morris JG, Levine MM: Cholera, *Clin Microbiol Rev* 8:48, 1995.

13. Kimura B, Hokimoto S, Takahasi H: *Photobacterium histaminum* Okuzumi et al, 1994 is a later subjective synonym *Photobacterium damselae* subsp *damselae* (Love et al, 1981) Smith et al, 1991, *Int J Syst Evol Microbiol* 50:1339, 2000.

14. Ko WC, Chuang YC: *Aeromonas bacteremia:* review of 59 episodes, *Clin Infect Dis* 20:1298, 1994.

15. Ko WC, Yu KW, Liu CY, et al: Increasing antibiotic resistance in clinical isolates of *Aeromonas* strains in Taiwan, *Antimicrob Agents Chemother* 40:1260, 1996.

16. Lee AC, Yuen KY, Ha SY: *Plesiomonas shigelloides*: case report and literature review, *Pediatr Hematol Oncol* 13:265, 1995.

17. Morita K, Watanabe N, Kurata S, et al: β-Lactam resistance of motile *Aeromonas* isolates from clinical and environmental sources, *Antimicrob Agents Chemother* 38:353, 1994.

18. Thompson FL, Hoste B, Vandemeulebroecke K, et al.: Reclassification of *Vibrio hollisae* as *Grimontia hollisae* gen nov, comb nov, *Int J Syst Evol Microbiol,* 53:1615, 2003.

19. Ti TY, Tan CW, Chong AP, et al: Nonfatal and fatal infections caused by *Chromobacterium violaceum, Clin Infect Dis* 17:505, 1993.

20. von Graevenitz A, Zbinden R, Mutters R: *Actinobacillus, Capnocytophaga, Eikenella, Kingella, Pasteurella,* and other fastidious or rarely encountered gram-negative rods. In Murray PR, Baron EJ, Jorgensen JH, et al, editors: *Manual of clinical microbiology,* ed 8, Washington, DC, 2003, ASM Press.

21. Walsh TR, Payne DJ, MacGowan AP, et al: A clinical isolate of *Aeromonas sobria* with three chromosomally mediated inducible β-lactamases: a cephalosporinase, a penicillinase, and a third enzyme, displaying carbapenemase activity, *J Antimicrob Chemother* 35:271, 1995.

22. Weyant RS, Moss CW, Weaver RE, et al, editors: *Identification of unusual pathogenic gram-negative aerobic and facultatively anaerobic bacteria,* ed 2, Baltimore, 1996, Williams & Wilkins.

ADDITIONAL READING

Farmer JJ III, Hickman-Brenner FW: The genera *Vibrio* and *Photobacterium.* In Balows A, Truper HG, Dworkin M, et al, editors: *The prokaryotes. A handbook on the biology of bacteria: ecophysiology, isolation, identification, applications,* ed 2, New York, 1992, Springer-Verlag.

Farmer JJ III, Arduino MJ, Hickman-Brenner FW: The genera *Aeromonas* and *Plesiomonas.* In Balows A, Truper HG, Dworkin M, et al, editors: *The prokaryotes. A handbook on the biology of bacteria: ecophysiology, isolation, identification, applications,* ed 2, New York, 1992, Springer-Verlag.

Neill MA, Carpenter CCJ: Other pathogenic vibrios. In Mandell GL, Bennett JE, Dolin R, editors: *Principles and practice of infectious diseases,* ed 6, Philadelphia, 2005, Elsevier Churchill Livingstone.

Seas C, Gatuzzo E: *Vibrio cholerae.* In Mandell GL, Bennett JE, Dolin R, editors: *Principles and practice of infectious diseases,* ed 6, Philadelphia, 2005, Elsevier Churchill Livingstone.

Steinberg JP, Del Rio C: Other gram-negative and gram-variable bacilli. In Mandell GL, Bennett JE, Dolin R, editors: *Principles and practice of infectious diseases,* ed 6, Philadelphia, 2005, Elsevier Churchill Livingstone.

CHAPTER **29** *Sphingomonas paucimobilis* and Similar Organisms

Organisms to Be Considered

Current Name	Previous Name
Acidovorax facilis	*Pseudomonas facilis*
CDC group IIc	
CDC group IIe	
CDC group IIh	
CDC group IIi	
CDC group O-2	
Sphingobacterium mizutaii	
Sphingomonas parapaucimobilis	
Sphingomonas paucimobilis	*Pseudomonas paucimobilis*, CDC group IIk-1

GENERAL CONSIDERATIONS

The organisms discussed in this chapter are considered together because they usually fail to grow on Mac-Conkey agar, they are oxidase-positive, and they oxidatively utilize glucose.

EPIDEMIOLOGY, SPECTRUM OF DISEASE, AND ANTIMICROBIAL THERAPY

As demonstrated in Table 29-1, these organisms are rarely isolated from human materials and they have an extremely limited role as agents of infection. Because they are rarely encountered in the clinical setting, little information exists regarding their epidemiology, capacity to cause human infections, and potential for antimicrobial resistance.[3-5] For example, even though the O-1 and O-2 organisms have been submitted to CDC after being isolated from clinical materials such as blood, cerebrospinal fluid, wounds, and pleural fluid, their natural habitat is unknown. When these organisms are encountered in clinical specimens, their clinical signi-

ficance and potential as contaminants should be seriously considered.

LABORATORY DIAGNOSIS

SPECIMEN COLLECTION AND TRANSPORT

No special considerations are required for specimen collection and transport of the organisms discussed in this chapter. Refer to Table 5-1 for general information on specimen collection and transport.

SPECIMEN PROCESSING

No special considerations are required for processing of the organisms discussed in this chapter. Refer to Table 5-1 for general information on specimen processing.

DIRECT DETECTION METHODS

There are no specific procedures for the direct detection of these organisms in clinical material. Microscopically, CDC groups IIc, IIe, IIh, IIi, and O-2 are all short, straight rods that may appear as "II forms," which are bacteria with thickened ends and thin centers. *Sphingobacterium mizutaii* exhibits II forms as well. *Sphingomonas paucimobilis* and *parapaucimobilis* are medium-sized straight rods. *Acidovorax facilis* are straight to slightly curved gram-negative rods that occur singly or in short chains.

CULTIVATION

Media of Choice

Sphingomonas spp., *Sphingobacterium mitzutaii*, *Acidovorax facilis*, and all CDC groups considered in this chapter grow well on routine laboratory media, such as 5% sheep blood and chocolate agars, but most fail to grow on MacConkey agar. They grow well in thioglycollate

Table 29-1 Epidemiology, Spectrum of Disease, and Antimicrobial Therapy

Organism	Epidemiology	Disease Spectrum	Antimicrobial Therapy
Sphingomonas paucimobilis	Exists in environmental niches, such as water, including hospital water systems. Not part of human flora. Mode of transmission is uncertain, probably involves patient exposure to contaminated medical devices or solutions	Virulence factors are unknown. Infections include catheter-related bacteremia, wound and urinary tract infections	No definitive guidelines; potentially active agents include trimethoprim/sulfamethoxazole, chloramphenicol, ciprofloxacin, and aminoglycosides; resistance to beta-lactams is known, but validated susceptibility testing methods do not exist
CDC group IIc, CDC group IIe, CDC group IIh, and CDC group IIi	Soil, plants, foodstuffs, and water, including moist areas in hospitals. Not part of human flora	Rarely found in clinical material and not substantiated as cause of human infections	No guidelines; little is known about antimicrobial resistance potential
CDC group O-1, O-2, and O-3	Unknown	Rarely found in clinical material and not implicated as cause of human infections	No guidelines; nothing is known about antimicrobial resistance potential

and brain-heart infusion broths and in broths used in blood culture systems.

Incubation Conditions and Duration
Within 24 hours of inoculation these organisms produce detectable growth on media incubated at 35°C in carbon dioxide or in ambient air.

Colonial Appearance
Table 29-2 describes the colonial appearance and other distinguishing characteristics (e.g., pigment) of each organism on 5% sheep blood agar. When these organisms (i.e., *Sphingomonas paucimobilis* and CDC groups O-2 and IIe) do grow on MacConkey agar, they appear as nonlactose fermenters.

APPROACH TO IDENTIFICATION
The ability of most commercial identification systems to accurately identify the organisms discussed in this chapter is limited or uncertain. Table 29-3 shows some conventional biochemical tests that can be used to presumptively differentiate between the various organisms.

Comments Regarding Specific Organisms
Strains of CDC groups IIe and IIh are similar to *Empedobacter brevis* (see Chapter 26) in that they oxidize glucose and maltose and produce indole. CDC group IIi resembles *Sphingobacterium multivorum,* but IIi produces indole. Similarly, *Sphingomonas paucimobilis* resembles

Table 29-2 Colonial Appearance and Characteristics

Organism	Medium	Appearance
Acidovorax facilis	BA	No distinctive appearance
CDC group IIc	BA	No distinctive appearance but colonies sticky
CDC group IIe	BA	No distinctive appearance
CDC group IIh	BA	No distinctive appearance
CDC group IIi	BA	No distinctive appearance
CDC group O-2	BA	Yellow pigment
Sphingobacterium mizutaii	BA	Yellow pigment
Sphingomonas spp.	BA	Small, circular, smooth, convex, bright yellow growth pigment

BA, 5% sheep blood agar.

CDC group O-1 but oxidizes more carbohydrates. CDC group O-2 will not oxidize xylose, mannitol, or lactose; this can help separate it from the other bright yellow–pigmented organisms discussed in this chapter.

SERODIAGNOSIS
Serodiagnostic techniques are not generally used for the laboratory diagnosis of infections caused by the organisms discussed in this chapter.

Table 29-3   Key Biochemical and Physiologic Characteristics

Organism	Insoluble Pigment	Glucose Oxidized	Xylose Oxidized	Sucrose Oxidized	Esculin Hydrolysis	Motility
Acidovorax facilis	–	+	(+)	–	–	+
CDC group IIc	Tan or buff	+	–	+	+	nm
CDC group IIe	–	+	–	–	–	nm
CDC group IIh	–	+	–	–	+	nm
CDC group IIi	Yellow	+	+	+	+	nm
CDC group O-2	Yellow to orange	v	–	+	v	v‡
Sphingobacterium mizutaii	v†	+	(+)	+	+	nm
Sphingomonas spp*	Yellow	+	+	+	+	+§

Data compiled from references 1, 2, and 6.
nm, Nonmotile; *v,* variable; +, >90% of strains are positive; –, >90% of strains are negative; (+), delayed.
*Includes *Sphingomonas paucimobilis* and *S. parapaucimobilis*.
†Yellow pigment production may be enhanced by incubation at room temperature.
‡Only 20% are motile; motility is only apparent upon wet mount or flagellar staining.
§Usually nonmotile in motility medium, but motility is present on wet mount.

PREVENTION

Because these organisms are rarely implicated in human infections, vaccines or prophylactic measures are not indicated.

Case Study

A 16-year-old acute lymphoblastic leukemia patient presented to his oncologist with pain and swelling of the left knee. He had recently received a course of chemotherapy and radiotherapy and was taking oral steroids. Straw-colored fluid with 400 white blood cells per microliter was aspirated from his knee. No microorganisms were seen on smear and none grew in culture. Unfortunately, only a few drops of the fluid were cultured on plate media. Over the next 6 months the patient was in and out of the hospital, receiving antibiotics and having more cultures obtained with no positive findings to explain his pain and swelling. He was admitted to the hospital where an arthroscopic procedure was performed to evaluate the problem. Widespread synovitis was seen. Cultures obtained from the surgery grew a yellow-pigmented gram-negative rod on blood agar but not MacConkey. Indole and urease testing were negative but the oxidase test and wet mount motility were positive.

QUESTIONS

1. Which microorganisms are in the differential for the correct identification of motile, yellow, gram-negative rods that do not grow on MacConkey? What tests can be done to distinguish among them?
2. The isolate was identified as *Sphingomonas paucimobilis*. What method should be used to test for susceptibility of this unusual pathogen?
3. Why do you think that many of the initial cultures from this patient failed to detect the pathogen?

Reference for case: Charity R, Foukas A: Osteomyelitis and secondary septic arthritis caused by *Sphingomonas paucimobilis, Infection* 33:93-95, 2005.

REFERENCES

1. Daneshvar MI, Hill B, Hollis DG, et al: CDC group O-3: Phenotypic characteristics, fatty acid composition, isoprenoid quinone content, and in vitro antimicrobic susceptibilities of an unusual gram-negative bacterium isolated from clinical specimens, *J Clin Microbiol* 36:1674, 1998.
2. Hollis DG, Moss CW, Daneshvar MI, et al: CDC group IIc phenotypic characteristics, fatty acid composition, and isoprenoid quinone content, *J Clin Microbiol* 34:2322, 1996.
3. Lemaitre D, Elaichouni A, Hundhausen M, et al: Tracheal colonization with *Sphingomonas paucimobilis* in mechanically ventilated neonates due to contaminated ventilator temperature probes, *J Hosp Infect* 32:199, 1996.
4. Reina J, Bassa A, Llompart I, et al: Infections with *Pseudomonas paucimobilis:* report of four cases and review, *Rev Infect Dis* 13:1072, 1990.
5. Salazar R, Martino R, Suredo A, et al: Catheter-related bacteremia due to *Pseudomonas paucimobilis* in neutropenic cancer patients: report of two cases, *Clin Infect Dis* 20:1573, 1995.
6. Weyant RS, Moss CW, Weaver RE, et al, editors: *Identification of unusual pathogenic gram-negative aerobic and facultatively anaerobic bacteria,* ed 2, Baltimore, 1996, Williams & Wilkins.

ADDITIONAL READING

Schreckenberger PC, Daneshvar MI, Weyant RS, et al: *Acineto-bacter, Achromobacter, Chryseobacterium, Moraxella,* and other nonfermentative gram-negative rods. In Murray PR, Baron EJ, Jorgensen JH, et al, editors: *Manual of clinical microbiology*, ed 8, Washington, DC, 2003, ASM Press.

Steinberg JP, Del Rio C: Other gram-negative and gram-variable bacilli. In Mandell GL, Bennett JE, Dolin R, editors: *Principles and practice of infectious diseases*, ed 6, Philadelphia, 2005, Elsevier Churchill Livingstone.

Genera and Species to Be Considered

Current Name	Previous Name
Moraxella atlantae	
Moraxella canis	
Moraxella lacunata	
Moraxella lincolnii	
Moraxella nonliquefaciens	
Moraxella osloensis	
Neisseria elongata subspecies *elongata*	CDC group M6
Neisseria elongata subspecies *glycolytica*	
Neisseria elongata subspecies *nitroreducens*	
Neisseria weaverii	CDC group M5

GENERAL CHARACTERISTICS

The organisms discussed in this chapter are either coccobacilli or short to medium-sized, gram-negative rods. Subinhibitory concentrations of penicillin, such as occurs in the presence of a 10-unit penicillin disk, cause the coccoid forms of these bacteria to elongate to a bacillus morphology. In contrast, true cocci, such as most *Neisseria* spp. and *Moraxella* (*Branhamella*) *catarrhalis,* with which these organisms may be confused, maintain their original coccus shape in the presence of penicillin. In addition, the organisms discussed in this chapter do not use glucose and most do not grow on MacConkey agar. Specific morphologic and physiologic features are presented later in this chapter in the discussion of laboratory diagnosis.

EPIDEMIOLOGY, SPECTRUM OF DISEASE, AND ANTIMICROBIAL THERAPY

Infections caused by *Moraxella* spp. and *Neisseria elongata* most likely result when a breakdown of the patient's mucosal or epidermal defensive barriers allows subsequent invasion of sterile sites by an organism that is part of the patient's normal flora (i.e., an endogenous strain; Table 30-1). The fact that these organisms rarely cause infection indicates that they have low virulence.[4-7] Whenever they are encountered in clinical specimens, the possibility that they are contaminants should be seriously considered (Table 30-2). This is especially the case when the specimen source may have come in contact with a mucosal surface.

Moraxella catarrhalis is the species most commonly associated with human infections, primarily of the respiratory tract. However, because the cellular morphology of this species is more similar to that of *Neisseria* spp. than that of the other *Moraxella* spp., details of this organism's characteristics are discussed in Chapter 42.

The rarity with which these organisms are encountered as the cause of infection, and the lack of validated in vitro susceptibility testing methods, does not allow definitive treatment guidelines to be given (Table 30-3). Although many of these organisms may grow on the media and under the conditions recommended for testing other bacteria (see Chapter 12 for more information regarding validated testing methods), this does not necessarily mean that interpretable and reliable results will be produced. Chapter 12 should be reviewed for preferable strategies that can be used to provide susceptibility information when validated testing methods do not exist for a clinically important bacterial isolate.

In general, beta-lactam antibiotics are thought to be effective against these species. However, some evidence suggests that beta-lactamase–mediated resistance may be capable of spreading among *Moraxella* spp.[9]

LABORATORY DIAGNOSIS

SPECIMEN COLLECTION AND TRANSPORT

No special considerations are required for specimen collection and transport of the organisms discussed in this chapter. Refer to Table 5-1 for general information on specimen collection and transport.

SPECIMEN PROCESSING

No special considerations are required for processing of the organisms discussed in this chapter. Refer to Table 5-1 for general information on specimen processing.

Table 30-1 Epidemiology

Organism	Habitat (reservoir)	Mode of Transmission
Moraxella nonliquefaciens, Moraxella lacunata, Moraxella osloensis, Moraxella lincolnii, Moraxella canis, and *Moraxella atlantae*	Normal human flora that inhabit mucous membranes covering the nose, throat, other parts of the upper respiratory tract, conjunctiva, and, for some species (i.e., *M. osloensis*), the urogenital tract. May also colonize the skin	Infections are rare. When they occur, they are probably caused by the patient's endogenous strains. Person-to-person transmission may be possible, but this has not been documented
Neisseria elongate	Normal flora of upper respiratory tract	When infections occur, they are probably caused by the patient's endogenous strains
Neisseria weaverii	Oral flora of dogs	Dog bite

Table 30-2 Pathogenesis and Spectrum of Disease

Organism	Virulence Factors	Spectrum of Disease and Infections
Moraxella nonliquefaciens, Moraxella lacunata, Moraxella osloensis, Moraxella lincolnii, Moraxella canis, and *Moraxella atlantae*	Unknown. Because they are rarely associated with infections, they are considered opportunistic organisms of low virulence	*M. lacunata* has historically been associated with eye infections, but these infections also may be caused by other *Moraxella* spp. Other infections include bacteremia, endocarditis, septic arthritis, and, possibly, respiratory infections
Neisseria elongata	Unknown. An opportunistic organism of low virulence	Rarely implicated in infections. Has been documented as a cause of bacteremia, endocarditis, and osteomyelitis
Neisseria weaverii	Unknown	Infections of dog bite wounds

Table 30-3 Antimicrobial Therapy and Susceptibility Testing

Organism	Therapeutic Options	Potential Resistance to Therapeutic Options	Validated Testing Methods*
Moraxella spp.	No definitive guidelines. Generally susceptible to penicillins and cephalosporins	β-lactamase–mediated resistance to penicillins common	Not available. Exception: See CLSI document M45 for testing guidelines for *M. catarrhalis* (see Chapter 42)
Neisseria elongata and *Neisseria weaverii*	No definitive guidelines. Generally susceptible to penicillins and cephalosporins	None known	Not available

*Validated testing methods include those standard methods recommended by the Clinical and Laboratory Standards Institute (CLSI) and those commercial methods approved by the Food and Drug Administration (FDA).

DIRECT DETECTION METHODS

Other than Gram stain of patient specimens, there are no specific procedures for the direct detection of these organisms in clinical material. *M. atlantae, M. nonliquefaciens,* and *M. osloensis* may appear as either coccobacilli or as short, broad rods that tend to resist decolorization and may appear gram-variable. This is also true for *M. canis,* which appears as cocci in pairs or short chains. *M. lacunata* is a coccobacillus or medium-sized rod, and *M. lincolnii* is a coccobacillus that may appear in chains. All subspecies of *Neisseria elongata* are either coccobacilli

or short, straight rods and *N. weaverii* is a medium-length, straight bacillus.

CULTIVATION

Media of Choice

Moraxella spp. and the elongated *Neisseria* spp. grow well on 5% sheep blood and chocolate agars. Most strains grow slowly on MacConkey agar and resemble the non-lactose–fermenting *Enterobacteriaceae*. Both genera also grow well in the broth of commercial blood culture sys-

Table 30-4   Colonial Appearance and Characteristics

Organism	Medium	Appearance
Moraxella atlantae	BA Mac	Small, pitting and spreading NLF
M. lacunata	BA Mac	Small colonies that pit the agar No growth
M. lincolnii	BA Mac	Smooth, translucent to semiopaque No growth
M. nonliquefaciens	BA Mac	Smooth, translucent to semiopaque; occasionally, colonies spread and pit agar NLF, if growth
M. osloensis	BA Mac	Smooth, translucent to semiopaque NLF, if growth
M. canis	BA Mac	Resemble colonies of *Enterobacteriaceae* NLF
Neisseria elongata (all subspecies)	BA Mac	Gray, translucent, smooth, glistening; may have dry, claylike consistency NLF, if growth
N. weaverii	BA Mac	Small, smooth, semiopaque NLF, if growth

BA, 5% sheep blood agar; *Mac*, MacConkey agar; *NLF*, non–lactose-fermenter.

tems and in common nutrient broths, such as thioglycollate and brain-heart infusion.

Incubation Conditions and Duration

Five percent sheep blood and chocolate agars should be incubated at 35° C in carbon dioxide or ambient air for a minimum of 48 hours. For those species that may grow on MacConkey agar, the medium should be incubated at 35° C in ambient air.

Colonial Appearance

Table 30-4 describes the colonial appearance and other distinguishing characteristics (e.g., pitting) of each species on 5% sheep blood and MacConkey agars.

APPROACH TO IDENTIFICATION

The ability of most commercial identification systems to accurately identify the organisms discussed in this chapter is limited or uncertain. Table 30-5 lists some conventional biochemical tests that can be used to presumptively differentiate the species in this chapter. This is a simplified scheme; clinically important isolates should be sent to a reference laboratory for definitive identification.

As just mentioned, these organisms can be difficult to differentiate from gram-negative diplococci (see Chapter 42 for more information about gram-negative diplococci). In addition, these organisms are relatively biochemically inert. Elongation in the presence of penicillin is a useful criterion for differentiating them from true cocci. The effect of penicillin is determined by streaking a blood agar plate and adding a 10-unit penicillin disk to the first quadrant before overnight incubation at 35° C. A Gram stain of the growth taken from around the edge of the zone of inhibition readily demonstrates whether the isolate in question is a true coccus or has elongated.

Comments Regarding Specific Organisms

M. nonliquefaciens and *M. osloensis,* the two most frequently isolated species, can be differentiated by the ability of *M. osloensis* to utilize acetate. *M. lacunata* is able to liquefy serum, so depressions are formed on the surface of Loeffler's serum agar slants. Most of the species considered in this chapter do not utilize glucose; *Neisseria elongata* subsp. *glycolytica,* which produces acid from glucose in the rapid sugar test used for *Neisseria* spp., is the only exception. Unlike *Oligella* spp. (see Chapter 27 for more information regarding this genus) none of the organisms considered here is motile.

SERODIAGNOSIS

Serodiagnostic techniques are not generally used for the laboratory diagnosis of infections caused by the organisms discussed in this chapter.

Table 30-5 Key Biochemical and Physiologic Characteristics

Organism	Growth on MacConkey	Catalase	Nitrate Reduction	Nitrite Reduction	DNase	Digests Loeffler's Slant	Sodium Acetate Utilization	Growth in Nutrient Broth
Moraxella atlantae	+	+	–	–	–	–	ND	–
M. lacunata	–	+	+	–	–	+	–	–
M. lincolnii	–	+	–	–*	–	–	–	–
M. nonliquefaciens	–	+	+	–	–	–	–	v
M. osloensis	v	+	v	–	–	–	+	+
M. canis	+	+	+	v	+	–	+	+
Neisseria elongata subsp. *elongata*	v	–	–	+	ND	ND	v	+
Neisseria elongata subsp. *glycolytica*	+	+	–	v	ND	ND	+	+
Neisseria elongata subsp. *nitroreducens*	v	–	+	+	ND	ND	v	v
N. weaverii	v	+	–	+	ND	ND	–	v

Data compiled from references 1-3, 6, 8, and 10.

ND, No data; *v*, variable; +, >90% of strains positive; –, >90% of strains negative.

*Nitrite-positive strains have been reported.

PREVENTION

Because these organisms do not generally pose a threat to human health, there are no recommended vaccination or prophylaxis protocols.

Case Study

A 44-year-old woman was rehospitalized following a gastrojejunostomy. She had increased white blood cells (21,000/μL) and was thought to have a post-surgical infection. X-rays showed a leakage from the gastrojejunostomy site into the left upper abdomen and communication with the large cavity. Aspiration of the fluid found by radiologic examination contained small gram-variable bacilli. Tiny, yellowish nonhemolytic colonies grew on blood agar that slightly pitted the agar. They were catalase-negative and oxidase-positive, but failed to grow on MacConkey.

QUESTIONS

1. Both rapid indole and hanging drop motility tests were negative for this bacillus. What microorganisms are in the differential and how would you approach this identification?

2. The nitrate test was negative and the organism did not ferment glucose. The OF glucose test was also negative, meaning that the organism was a glucose nonoxidizer. What is the most likely identification and how would you confirm it?

3. How is the nitrite reduction test performed when the nitrate test is negative?

REFERENCES

1. Grant PE, Brenner DJ, Steigerwalt AG, et al: *Neisseria elongata* subsp *nitroreducens* subsp nov, formerly CDC group M-6, a gram-negative bacterium associated with endocarditis, *J Clin Microbiol* 28:2591, 1990.

2. Jannes G, Vaneechoutte M, Lannoo M, et al: Polyphasic taxonomy leading to the proposal of *Moraxella canis* sp nov for *Moraxella catarrhalis*–like strains, *Int J Syst Bacteriol* 43:438, 1993.

3. Kodjo A, Richard Y, Tønjum T: *Moraxella boevrei* sp nov, a new *Moraxella* species found in goats, *Int J Syst Bacteriol* 47:115, 1997.

4. Montejo M, Ruiz-Irastorza G, Aguirrebengoa K, et al: Endocarditis due to *Neisseria elongata* subspecies *nitroreducens*, *Clin Infect Dis* 20:1431, 1995.

5. Mueleman P, Erard K, Herregods MC, et al: Bioprosthetic valve endocarditis caused by *Neisseria elongata* subspecies *nitroreducens*, *Infection* 24:258, 1996.

6. Schreckenberger PC, Daneshvar MI, Weyant RS, et al: *Acineto-bacter, Achromobacter, Chryseobacterium, Moraxella,* and other

nonfermentative gram-negative rods. In Murray PR, Baron EJ, Jorgensen JH, et al, editors: *Manual of clinical microbiology,* ed 8, Washington, DC, 2003, ASM Press.

7. Struuillou L, Raffi F, Barrier JH: Endocarditis caused by *Neisseria elongata* subspecies *nitroreducens:* case report and literature review, *Eur J Clin Microbiol Infect Dis* 12:625, 1993.

8. Vandamme P, Gillis M, Vancanneyt M, et al: *Moraxella lincolnii* sp nov, isolated from the human respiratory tract, and reevaluation of the taxonomic position of *Moraxella osloensis, Int J Syst Bacteriol* 43:474, 1993.

9. Wallace RJ, Steingrube DR, Nash DR, et al: BRO β-lactamases of *Branhamella catarrhalis* and *Moraxella* subgenus *Moraxella,* including evidence for chromosomal β-lactamase transfer by conjugation in *B. catarrhalis, M. nonliquefaciens,* and *M. lacunata, Antimicrob Agent Chemother* 30:1845, 1989.

10. Weyant RS, Moss CW, Weaver RE, et al, editors: *Identification of unusual pathogenic gram-negative aerobic and facultatively anaerobic bacteria,* ed 2, Baltimore, 1996, Williams & Wilkins.

ADDITIONAL READING

Doern GV: The *Moraxella* and *Branhamella* subgenera of the genus *Moraxella.* In Balows A, Truper HG, Dworkin M, et al, editors: *The prokaryotes. A handbook on the biology of bacteria: ecophysiology, isolation, identification, applications,* ed 2, New York, 1992, Springer-Verlag.

Janda JM, Knapp JS: *Neisseria* and *Moraxella catarrhalis.* In Murray PR, Baron EJ, Jorgensen JH, et al, editors: *Manual of clinical microbiology,* ed 8, Washington, DC, 2003, ASM Press.

Lyons RW: Ecology, clinical significance, and antimicrobial susceptibility of *Acinetobacter* and *Moraxella.* In Gilardi GL, editor: *Nonfermentative gram-negative rods: laboratory identification and clinical aspects,* New York, 1985, Marcel Dekker.

Murphy TF: *Moraxella (Branhamella) catarrhalis* and other gram-negative cocci. In Mandell GL, Bennett JE, Dolin R, editors: *Principles and practice of infectious diseases,* ed 6, Philadelphia, 2005, Elsevier Churchill Livingstone.

GENERAL CHARACTERISTICS

The organisms discussed in this chapter are considered together because they are all asaccharolytic, oxidase-positive bacilli that do not grow on MacConkey agar. Their individual morphologic and physiologic features are presented later in this chapter in the discussion of laboratory diagnosis.

EPIDEMIOLOGY, SPECTRUM OF DISEASE, AND ANTIMICROBIAL THERAPY

The organisms listed in Table 31-1 are not commonly associated with human infections, but they are occasionally encountered in clinical specimens.[1,3,5,6] Of those listed, *Eikenella corrodens* is the one most commonly encountered and is usually found in mixed infections resulting from bites or clenched-fist wounds (Table 31-2). This organism also is the "E," for *Eikenella,* among the HACEK group of bacteria known to cause subacute bacterial endocarditis (see Chapter 52 for more information regarding endocarditis and bloodstream infections).

The rarity with which these organisms are encountered as the cause of infection and the lack of validated in vitro susceptibility testing methods do not allow definitive treatment guidelines to be given for most of them (Table 31-3). Although beta-lactamase production has been described in *E. corrodens,* this species is usually susceptible to penicillins and many other beta-lactam antimicrobials so that therapy with these agents is frequently used.[2,4]

LABORATORY DIAGNOSIS

SPECIMEN COLLECTION AND TRANSPORT

No special considerations are required for specimen collection and transport for the organisms discussed in this chapter. Refer to Table 5-1 for general information on specimen collection and transport.

SPECIMEN PROCESSING

No special considerations are required for processing of the organisms discussed in this chapter. Refer to Table 5-1 for general information on specimen processing.

DIRECT DETECTION METHODS

Other than Gram stain and microscopic examination, there are no specific procedures for the direct detection of these organisms in clinical material. *E. corrodens* is a slender, medium-length straight rod, and *Methylobacterium* is a vacuolated, pale-staining, short to medium-length bacillus that may resist decolorization. *Weeksella virosa* and *Bergeyella zoohelcum* are medium to long rods with parallel sides and rounded ends that may form "II-forms" similar to the *Sphingobacterium* (see Chapter 26 for more information regarding this genus).

CULTIVATION

Media of Choice

E. corrodens and *Methylobacterium* do not grow well on either 5% sheep blood or chocolate agars, and *Weeksella* and *Bergeyella* only grow slowly on these media. None of these genera grow to a detectable extent on MacConkey agar, and all grow slowly in the broth media used in blood culture systems, thioglycollate broth, and in brain-heart infusion broth.

Incubation Conditions and Duration

To detect growth on 5% sheep blood and chocolate agars, incubation at 35° to 37° C in carbon dioxide for a minimum of 48 hours is required. In contrast to the other genera, *Methylobacterium* grows best at 25° C.

Table 31-1   Epidemiology

Organism	Habitat (reservoir)	Mode of Transmission
Eikenella corrodens	Normal human flora of mouth and gastrointestinal tract	Person to person involving trauma by human teeth incurred by bites or by clenched-fist wounds incurred by facial punches. Also, certain infections may be caused by patient's endogenous strains (e.g., endocarditis)
Methylobacterium spp.	Found on vegetation and occasionally in hospital environment. Not part of normal human flora	Uncertain; probably involves contaminated medical devices such as catheters
Weeksella virosa	Uncertain; probably environmental. Not part of normal human flora	Uncertain, rarely found in clinical material
Bergeyella zoohelcum	Oral flora of dogs and other animals. Not part of normal human flora	Bite or scratch of dog or cat

Table 31-2   Pathogenesis and Spectrum of Disease

Organism	Virulence Factors	Spectrum of Disease and Infections
Eikenella corrodens	Unknown; opportunistic organism usually requires trauma for introduction into normally sterile sites. Also may enter bloodstream to cause transient bacteremia or be introduced by intravenous drug abuse	Human bite wound infections, head and neck infections, and aspiration pneumonias as part of mixed infection. Can also cause endocarditis that is slow to develop and indolent (i.e., subacute). Less commonly associated with brain and intraabdominal abscesses
Methylobacterium spp.	Unknown; an opportunistic organism probably of low virulence that is an uncommon cause of infection	Bacteremia and peritonitis in patients undergoing chronic ambulatory peritoneal dialysis (CAPD)
Weeksella virosa	Unknown; role in human disease is uncertain	Asymptomatic bacteruria; also isolated from female genital tract
Bergeyella zoohelcum	Unknown; an opportunistic organism that requires traumatic introduction to normally sterile site	Dog and cat bite wound infections

Table 31-3   Antimicrobial Therapy and Susceptibility Testing

Organism	Therapeutic Options	Potential Resistance to Therapeutic Options	Validated Testing Methods*
Eikenella corrodens	Often susceptible to penicillins, quinolones, cephalosporins, and trimethoprim-sulfamethoxazole	May produce beta-lactamases. Usually resistant to clindamycin, metronidazole, and aminoglycosides	See CLSI document M45 section on "HACEK" organisms
Methylobacterium spp.	No guidelines	Unknown	Not available
Weeksella virosa and *Bergeyella zoohelcum*	No guidelines; potentially active agents include beta-lactams and quinolones	Susceptibility to tetracycline, aminoglycosides, and trimethoprim-sulfamethoxazole	Not available

*Validated testing methods include those standard methods recommended by the Clinical and Laboratory Standards Institute (CLSI) and those commercial methods approved by the Food and Drug Administration (FDA).

Table 31-4   Colonial Appearance and Characteristics

Organism	Medium*	Appearance
Bergeyella zoohelcum	BA	Colonies may be sticky; tan to yellow in color
Eikenella corrodens	BA	Colonies are tiny at 24 hours; mature colonies have moist, clear centers surrounded by flat, spreading growth; colonies may pit or corrode the agar surface; slight yellow pigmentation in older cultures; sharp odor of bleach
Methylobacterium spp.	BA	Pink to coral pigment; does not grow well on blood agar
Weeksella virosa	BA	Small colonies at 24 hours; mature colonies mucoid and adherent with a tan to brown pigment

BA, 5% sheep blood agar.
*These organisms usually do not grow on MacConkey agar; if breakthrough growth occurs, the organisms appear as non–lactose-fermenters.

Table 31-5   Key Biochemical and Physiologic Characteristics

Organism	Catalase	Oxidizes Xylose	Indole	Arginine Dihydrolase
Eikenella corrodens	–	–	–	–
Methylobacterium spp.*	+	+	–	ND
Weeksella virosa	+	–	+	–
Bergeyella zoohelcum	+	–	+	+

Data compiled from reference 7.
ND, No data; +, >90% of strains positive; –, >90% of strains negative.
*Colonies are pigmented pink and must be differentiated from *Roseomonas* spp. (see Chapter 27); *Roseomonas* spp. usually grow on MacConkey agar and will grow at 42° C.

Colonial Appearance

Table 31-4 describes the colonial appearance and other distinguishing characteristics (e.g., odor and pigment) of each genus on 5% sheep blood agar.

APPROACH TO IDENTIFICATION

The ability of most commercial identification systems to accurately identify the organisms discussed in this chapter is limited or, at best, uncertain. Therefore, strategies for identification of these genera are based on the use of conventional biochemical tests. Table 31-5 outlines basic criteria that are useful for differentiating among the genera in this chapter.

Comments Regarding Specific Organisms

Methylobacterium can be differentiated from other pink-pigmented, gram-negative rods (e.g., *Roseomonas* spp., discussed in Chapter 27) based on the criterion that *Methylobacterium* will utilize acetate whereas *Roseomonas* will not. Additionally, *Roseomonas* will grow at 42° C and *Methylobacterium* will not. Some strains of *Methylobacterium* weakly oxidize glucose, and many oxidize xylose.

E. corrodens can be recognized in culture by its bleachlike odor. The organism does not utilize glucose or other carbohydrates and is catalase-negative. However, it reduces nitrate to nitrite and hydrolyzes both ornithine and lysine.

Weeksella and *Bergeyella* are biochemically similar to each other, although *W. virosa* is urease-negative and *B. zoohelcum* is urease-positive. Both organisms are indole-positive, which is an unusual characteristic for most nonfermentative bacteria. *W. virosa* will grow on selective media for *Neisseria gonorrhoeae* but can be separated from the gonococci by its positive test for indole and its Gram stain morphology.

SERODIAGNOSIS

Serodiagnostic techniques are not generally used for the laboratory diagnosis of infections caused by the organisms discussed in this chapter.

PREVENTION

Because these organisms do not generally pose a threat to human health, there are no recommended vaccination or prophylaxis protocols.

Case Study

A 64-year-old Indonesian man was in good health until 3 months ago when he awoke with back pain localized to the upper thoracic area. His symptoms were not improved with physical therapy or acupuncture. A bone scan was positive for inflammation at C5 and C6 and he was treated with antibiotics for 2 weeks. His symptoms returned when antibiotics were discontinued. A vertebrectomy of C5 and C6 was performed and bone tissue was sent for culture. A nonhemolytic gram-negative rod was isolated that was catalase-negative and oxidase-positive, but did not grow on MacConkey. Colonies showed pits in the agar and exuded an odor of bleach (Figure 31-1). When questioned, the patient admitted to no dental procedures before the illness but did admit that he had previously sought medical attention for a swallowed fishbone, which was not successfully removed. Review of the radiology films did show a dense area that could have been the foreign body.

QUESTIONS

1. Because the bacterium is catalase-negative, it needs to be separated from the genera described in Chapters 30, 32, and 33. However, one unique positive biochemical reaction will definitely identify it from all others. What is that test?
2. Why do you think the physicians were so interested in whether dental work had been performed?
3. Why is this bacterium a member of the HACEK group?

REFERENCES

1. Chen CK, Wilson ME: *Eikenella corrodens* in human oral and non-oral infections: a review, *J Periodontol* 63:941, 1992.
2. Fass RJ, Barnishan J, Solomon MC, et al: In vitro activities of quinolones, β-lactams, tobramycin, and trimethoprim-sulfamethoxazole, against nonfermentative gram-negative bacilli, *Antimicrob Agents Chemother* 40:1412, 1996.
3. Kay KM, Macone S, Kazanjian PH: Catheter infections caused by *Methylobacterium* in immunocompromised hosts: report of three cases and review of the literature, *Clin Infect Dis* 14:1010, 1992.
4. Lacroix JM, Walker CB: Identification of a streptomycin resistance gene and a partial Tn3 transposon coding for a β-lactamase in a periodontal strain of *Eikenella corrodens, Antimicrob Agents Chemother* 36:740, 1992.
5. Reina J, Borell N: Leg abscess caused by *Weeksella zoohelcum* following a dog bite, *Clin Infect Dis* 14:1162, 1992.
6. Reina J, Gil J, Alomar P: Isolation of *Weeksella virosa* (formally CDC group IIf) from a vaginal sample, *Eur J Clin Microbiol Infect Dis* 8:569, 1989.
7. Weyant RS, Moss CW, Weaver RE, et al, editors: *Identification of unusual pathogenic gram-negative aerobic and facultatively anaerobic bacteria,* ed 2, Baltimore, 1996, Williams & Wilkins.

ADDITIONAL READING

Schreckenberger PC, Daneshvar MI, Weyant RS, et al: *Acinetobacter, Achromobacter, Chryseobacterium, Moraxella,* and other non-fermentive gram-negative rods. In Murray PR, Baron EJ, Jorgensen JH, et al, editors: *Manual of clinical microbiology,* ed 8, Washington, DC, 2003, ASM Press.

Steinberg JP, Del Rio C: Other gram-negative and gram-variable bacilli. In Mandell GL, Bennett JE, Dolin R., editors: *Principles and practice of infectious diseases,* ed 6, Philadelphia, 2005, Elsevier Churchill Livingstone.

von Graevenitz A, Zbinden R, Mutters R: *Actinobacillus, Capnocytophaga, Eikenella, Kingella, Pasteurella,* and other fastidious or rarely encountered gram-negative rods. In Murray PR, Baron, EJ, Jorgensen JH, et al, editors: *Manual of clinical microbiology,* ed 8, Washington, DC, 2003, ASM Press.

Genera and Species to Be Considered

Current Name	Previous Name
Bisgaard's taxon 16	
CDC group EF-4a	CDC group EF-4
Mannheimia haemolytica	*Pasteurella haemolytica*
Pasteurella aerogenes	
Pasteurella bettyae	CDC group HB-5
Pasteurella canis	
Pasteurella dagmatis	
Pasteurella multocida	
Pasteurella pneumotropica	
Pasteurella stomatis	
Suttonella indologenes	*Kingella indologenes*

GENERAL CHARACTERISTICS

The organisms discussed in this chapter are small, gram-negative, nonmotile, oxidase-positive bacilli that ferment glucose; most will not grow on MacConkey agar. Their individual morphologic and physiologic features are presented later in this chapter in the discussion of laboratory diagnosis.

EPIDEMIOLOGY, SPECTRUM OF DISEASE, AND ANTIMICROBIAL THERAPY

The majority of the organisms presented in this chapter are part of animal flora and are transmitted to humans during close animal contact, including bites. For most, virulence factors are not recognized and the organisms may be considered opportunistic pathogens that require mechanical disruption of the host's anatomic barriers, such as occurs with bite-induced wounds (Table 32-1). Of the organisms listed in Table 32-2, *Pasteurella multocida* is the species most commonly encountered in clinical specimens.

Because the mode of transmission often involves traumatic inoculation of the organism through human skin, it is expected that most infections caused by these bacteria involve wounds and soft tissues (see Table 32-2). However, in some settings involving compromised patients, infections of the respiratory tract and other body sites can occur.[2-4]

An unusual feature of the organisms considered in this chapter is that most are susceptible to penicillin. Although most other clinically relevant gram-negative bacilli are intrinsically resistant to penicillin, it is the drug of choice for infections involving *P. multocida* and several other species listed in Table 32-3.

The general therapeutic effectiveness of penicillin and the lack of resistance to this agent among *Pasteurella* spp. suggest that in vitro susceptibility testing is not usually necessary. However, if needed the Clinical and Laboratory Standards Institute (CLSI) document M45 does provide suggested methods and drugs for testing.

LABORATORY DIAGNOSIS

SPECIMEN COLLECTION AND TRANSPORT

No special considerations are required for specimen collection and transport of the organisms discussed in this chapter. Refer to Table 5-1 for general information on specimen collection and transport.

SPECIMEN PROCESSING

No special considerations are required for processing of the organisms discussed in this chapter. Refer to Table 5-1 for general information on specimen processing.

DIRECT DETECTION METHODS

Other than Gram stain of patient specimens, there are no specific procedures for the direct detection of these organisms in clinical material. *Pasteurella* spp. are typically short, straight bacilli, although *P. aerogenes* may also appear as coccobacilli. The cells of *P. bettyae* are usually thinner rods than those of the other species. *Mannheimia haemolytica* is a small rod or coccobacillus. *Suttonella indologenes* is a variable length, broad rod, whereas Bisgaard's taxon 16 is a short, straight bacillus. EF-4a appears as a coccobacillus or a short, straight rod.

CULTIVATION

Media of Choice

The bacteria described in this chapter grow well on routine laboratory media such as 5% sheep blood and chocolate agars; most strains do not grow on MacConkey

Table 32-1   Epidemiology

Organism	Habitat (reservoir)	Mode of Transmission
Pasteurella multocida, other *Pasteurella* spp., and *Mannheimia haemolytica*	Nasopharynx and gastrointestinal tract of wild and domestic animals. Part of an animal's normal flora; also associated with infections in various animals. May be found as part of upper respiratory flora of humans who have extensive exposure to animals (e.g., animal handlers)	Animal (usually cat or dog) bite or scratch. Infections may also be associated with non-bite exposure to animals. Less commonly, infections may occur without a history of animal exposure
Suttonella indologenes	Uncertain; rarely encountered in clinical specimens but may be part of human flora	Uncertain
Bisgaard's taxon 16	Uncertain; probably similar to that of *Pasteurella* spp.	Dog bite or close exposure to other animals
CDC group EF-4a	Animal oral and respiratory flora. Not part of human flora	Animal contact, particularly bites or scratches from dogs and cats

Table 32-2   Pathogenesis and Spectrum of Disease

Organism	Virulence Factors	Spectrum of Disease and Infections
Pasteurella multocida, other *Pasteurella* spp., and *Mannheimia haemolytica*	Endotoxin and antiphagocytic capsule associated with *P. multocida*	Focal soft tissue infections following bite or scratch. Chronic respiratory infection, usually in patients with preexisting chronic lung disease and heavy exposure to animals. Bacteremia with metastatic abscess formation may also occur, occasionally in patients with no history of animal exposure
Suttonella indologenes	Unknown	Associated with eye infections but rare
Bisgaard's taxon 16	Unknown	Associated with animal bite wounds
CDC group EF-4a	Unknown	Infected bite wounds of fingers, hands, or arm leading to cellulitis or abscess formation. Systemic infections are rare

agar. *Pasteurella, Suttonella, Mannheimia,* Bisgaard's taxon 16, and CDC group EF-4 also grow well in broth blood culture systems and common nutrient broths such as thioglycollate and brain-heart infusion.

Incubation Conditions and Duration
Five percent sheep blood and chocolate agars should be incubated at 35° C in carbon dioxide or ambient air for a minimum of 24 hours.

Colonial Appearance
Table 32-4 describes the colonial appearance and other distinguishing characteristics (e.g., hemolysis and odor) of each genus on blood agar.

APPROACH TO IDENTIFICATION
Some commercial biochemical identification systems are able to identify strains of *P. multocida*. However, the ability of most commercial identification systems to accurately identify the other species is limited or uncertain. Table 32-5 shows some conventional biochemical tests that can be used to presumptively differentiate among the organisms discussed in this chapter. These organisms closely resemble those described in Chapter 33. Therefore, organisms discussed in Chapters 32 and 33 all should be considered when evaluating an isolate in the clinical laboratory. Definitive identification of these fermentative organisms usually requires that the isolate be sent to a reference laboratory for a full battery of biochemical tests.

Comments Regarding Specific Organisms
The pasteurellae should all be oxidase-positive, based on the use of the tetramethyl-p-phenylenediamine dihydrochloride reagent; however, several subcultures may be necessary to obtain a positive reaction. Except for *P. bettyae*, these organisms are all catalase-positive,

Table 32-3   Antimicrobial Therapy and Susceptibility Testing

Organism	Therapeutic Options	Potential Resistance to Therapeutic Options	Validated Testing Methods*	Comments
Pasteurella multocida, other *Pasteurella* spp., and *Mannheimia haemolytica*	Cleaning, irrigation, and debridement of bite wound. Antimicrobial of choice is penicillin. Other active agents include mezlocillin, piperacillin-tazobactam, cefuroxime, cefotaxime, fluoroquinolones, tetracycline, and trimethoprim/ sulfamethoxazole	None known to penicillin	See CLSI document M45	Unusual for gram-negative bacilli to be susceptible to penicillin
Suttonella indologenes	No guidelines. Susceptible to penicillins, similar to *Pasteurella* spp.	Unknown	Not available	
Bisgaard's taxon 16	No guidelines	Unknown	Not available	
CDC group EF-4a	No definitive guidelines. Potentially active agents include penicillin, ampicillin, ciprofloxacin, and ofloxacin	Unknown; some cephalosporins may be less active than the penicillins	Not available	Limited clinical experience

*Validated testing methods include those standard methods recommended by the Clinical and Laboratory Standards Institute (CLSI) and those commercial methods approved by the Food and Drug Administration (FDA).

Table 32-4   Colonial Appearance and Characteristics on 5% Sheep Blood Agar

Organism	Appearance
Bisgaard's taxon 16*	No distinctive appearance
CDC group EF-4a	Some strains have a yellow to tan pigment; smells like popcorn
*Pasteurella aerogenes**	Convex, smooth, translucent, nonhemolytic†
P. bettyae‡	Convex, smooth, nonhemolytic
P. canis	Convex, smooth, nonhemolytic
P. dagmatis	Convex, smooth, nonhemolytic
P. multocida	Convex, smooth, gray, nonhemolytic; rough and mucoid variants can occur; may have a musty or mushroom smell
*P. pneumotropica**	Smooth, convex, nonhemolytic
P. stomatis	Smooth, convex, nonhemolytic
Suttonella indologenes	Resembles *Kingella* spp. (see Chapter 33)
*Mannheimia haemolytica**	Convex, smooth, grayish, beta-hemolytic (this feature may be lost on subculture)

*Breakthrough growth may occur on MacConkey agar; will appear as lactose fermenter.
†After 48 hours, colonies may be surrounded by a narrow green to brown halo.
‡Breakthrough growth may occur on MacConkey agar; will appear as non–lactose-fermenter.

and all species reduce nitrates to nitrites. *P. aerogenes* and some strains of *P. dagmatis* ferment glucose with the production of gas. *P. multocida,* the most frequently isolated species, can be separated from the other species based on positive reactions for ornithine decarboxylase and indole, and a negative reaction for urease.

Mannheimia haemolytica may be differentiated from the pasteurellae by its inability to produce indole or ferment mannose. *S. indologenes* can be separated from the pasteurellae by its negative nitrate test and differs from *Kingella* spp. (discussed in Chapter 33) in being both indole- and sucrose-positive. Bisgaard's taxon 16 can be separated from the pasteurellae by its fermentation of sucrose and maltose.

CDC group EF-4a is an eugonic-fermenter (EF), meaning that it grows well on routine laboratory media and ferments glucose. This distinguishes it from the dysgonic-fermenters (DF), which grow poorly on blood and chocolate agars (see Chapter 33). EF-4a cannot ferment any sugar except glucose, and is indole-negative and arginine dihydrolase-positive.

SERODIAGNOSIS

Serodiagnostic techniques are not generally used for the laboratory diagnosis of infections caused by the organisms discussed in this chapter.

Table 32-5 Key Biochemical and Physiologic Characteristics

Organism	Indole	Urea	Nitrate Reduction	Catalase	Ornithine Decarboxylase	Fermentation of: Mannitol	Sucrose	Maltose
Pasteurella aerogenes	−	(+)	(+)	+	v	−	+	+
P. bettyae	(+)	−	(+)	−	−	−	−	−
P. canis	+	−	+	+	(+)	−	(+)	−
P. dagmatis	(+)	(+)	(+)	+	−	−	+	(+)
P. multocida	(+)	−	(+)	+	(+)	+	+	−
P. pneumotropica	(+)	(+)*	(+)	+	(+)	−	+	+
P. stomatis	(+)	−	+	+	−	−	(+)	−
Suttonella indologenes	(+)	−	−	v	−	−	(+)	+or (+)
Mannheimia haemolytica	−	−	+	+	−	(+)	+	+
CDC group EF-4a	−	−	(+)†	+	−	−	−	−
Bisgaard's taxon 16	(+)	−	(+)	+	−	−	+	(+)

Data compiled from references 1 and 5.

+, >90% of strains positive; (+), >90% of strains positive but reaction may be delayed (i.e., 2-7 days); −, >90% of strains negative; v, variable.

*May require a drop of rabbit serum on the slant or a heavy inoculum.

†Gas may be produced during reduction of nitrate.

PREVENTION

Because these organisms do not generally pose a threat to human health, there are no recommended vaccination or prophylaxis protocols.

Case Study

A cat lover of many years was bitten on her right thumb. Over the next few days the thumb became swollen and red and did not improve. She went to the emergency department where the wound was debrided and drained. She was started on ampicillin therapy. The culture grew *Staphylococcus aureus* at 24 hours but a small gram-negative rod was seen on the blood agar plate at 48 hours. The rod was unable to grow on MacConkey agar but was oxidase- and indole-positive and fermented a variety of sugars. It was identified by the laboratory's gram-negative identification system.

QUESTIONS

1. What is the likely identification of this organism?
2. Had this bite been of human origin, what would be the most likely organism to cause an infection in the victim?
3. What is the most likely antimicrobial agent used to treat *Pasteurella multocida*?
4. The report of *S. aureus* isolated from this wound may be erroneous. Can you explain why?

REFERENCES

1. Angen O, Mutters R, Caugant DA, et al: Taxonomic relationships of the [*Pasteurella*] *haemolytica* complex as evaluated by DNA-DNA hybridization and 16S rRNA sequencing with proposal of *Mannheimia haemolytica* gen nov, comb nov, *Mannheimia granulomatis* comb nov, *Mannheimia glucosida* sp nov, *Mannheimia ruminalis* sp nov and *Mannheimia varigena* sp nov, *Int J Syst Bacteriol* 49:67, 1999.
2. Cuadrado-Gomez LM, Arranz-Caso JA, Cuadros-Gonzalez J, et al: *Pasteurella pneumotropica* pneumonia in a patient with AIDS, *Clin Infect Dis* 21:445, 1995.
3. Holst E, Roloff J, Larsson L, et al: Characterization and distribution of *Pasteurella* species recovered from infected humans, *J Clin Microbiol* 30:2984, 1992.
4. Shapiro DS, Brooks PE, Coffey DM, et al: Peripartum bacteremia with CDC group HB-5 (*Pasteurella bettyae*), *Clin Infect Dis* 22:1125, 1996.
5. Weyant RS, Moss CW, Weaver RE, et al, editors: *Identification of unusual pathogenic gram-negative aerobic and facultatively anaerobic bacteria*, ed 2, Baltimore, 1996, Williams & Wilkins.

ADDITIONAL READING

Clinical and Laboratory Standards Institute: Methods for antimicrobial dilution and disk susceptibility testing of infrequently isolated or fastidious bacteria; M45, Villanova, Pa, CLSI.

von Graevenitz A, Zbinden R, Mutters R: *Actinobacillus, Capnocytophaga, Eikenella, Kingella, Pasteurella*, and other fastidious or rarely encountered gram-negative rods. In Murray PR, Baron EJ, Jorgensen JH, et al, editors: *Manual of clinical microbiology*, ed 8, Washington, DC, 2003, ASM Press.

Zolvo JJ: *Pasteurella* species. In Mandell GL, Bennett JE, Dolin R, editors: *Principles and practice of infectious diseases*, ed 6, Philadelphia, 2005, Elsevier Churchill Livingstone.

Actinobacillus, Kingella, Cardiobacterium, *Capnocytophaga,* and Similar Organisms

CHAPTER **33**

Genera and Species to Be Considered

Current Name	Previous Name
Actinobacillus *actinomycetemcomitans*	
Other *Actinobacillus* spp., including	
A. ureae	
A. suis	
A. lignieresii	
A. hominis	
A. equuli	
Capnocytophaga gingivalis	CDC group DF-1
Capnocytophaga ochracea	CDC group DF-1
Capnocytophaga sputigena	CDC group DF-1
Capnocytophaga canimorsus	CDC group DF-2
Capnocytophaga cynodegmi	CDC group DF-2
Cardiobacterium hominis	
Dysgonomonas *capnocytophagoides*	CDC group DF-3
Kingella denitrificans	
Kingella kingae	

GENERAL CHARACTERISTICS

The organisms discussed in this chapter are dysgonic, that is, they grow slowly or poorly. Although they all ferment glucose, their fastidious nature requires that serum be added to the basal fermentation medium to enhance growth and detect fermentation byproducts. These bacteria are capnophiles, that is, they require additional carbon dioxide (CO_2) for growth, and most species will not grow on MacConkey agar. Their individual morphologic and physiologic features are presented later in this chapter in the discussion of laboratory diagnosis.

EPIDEMIOLOGY, SPECTRUM OF DISEASE, AND ANTIMICROBIAL THERAPY

The organisms listed in Table 33-1 are part of the normal flora of humans and other animals. As such, they generally are of low virulence and, except for those species associated with periodontal infections,

usually only cause infections in humans after introduction into sterile sites following trauma such as bites or manipulations in the oral cavity.[1,6-9]

The types of infections caused by these bacteria vary from periodontitis to endocarditis (Table 33-2). Three of these organisms, *Actinobacillus actinomycetemcomitans, Cardiobacterium hominis,* and *Kingella* spp., are the A, C, and K, respectively, of the HACEK group of organisms that cause slowly progressive (i.e., subacute) bacterial endocarditis. *Kingella* spp. can also be involved in other serious infections involving children.[3,11]

Among the species acquired from animals, *Capnocytophaga canimorsus* can cause fulminant, life-threatening infections following dog bites.[9]

Infections are frequently treated using beta-lactam antibiotics, occasionally in combination with an aminoglycoside (Table 33-3). Beta-lactamase production has been described in *Kingella* spp., but the impact of this resistance mechanism on clinical efficacy of beta-lactams is uncertain.[2,5] When in vitro susceptibility testing is required, Clinical and Laboratory Standards Institute (CLSI) document M45 does provide guidelines for testing *A. actinomycetemcomitans, Cardiobacterium* spp., and *Kingella* spp.

LABORATORY DIAGNOSIS

SPECIMEN COLLECTION AND TRANSPORT

No special considerations are required for specimen collection and transport of the organisms discussed in this chapter. Refer to Table 5-1 for general information on specimen collection and transport.

SPECIMEN PROCESSING

No special considerations are required for processing of the organisms discussed in this chapter. Refer to Table 5-1 for general information on specimen processing.

DIRECT DETECTION METHODS

Other than Gram stain of patient specimens, there are no specific procedures for the direct detection of these organisms in clinical material. *Actinobacillus* spp. are short to very short bacilli. They occur singly, in pairs,

397

Table 33-1 Epidemiology

Organism	Habitat (reservoir)	Mode of Transmission
Actinobacillus actinomycetemcomitans	Normal flora of human oral cavity	Endogenous; enters deeper tissues by minor trauma to mouth, such as during dental procedures
Other *Actinobacillus* spp.	Normal oral flora of animals such as cows, sheep, and pigs; not part of human flora	Rarely associated with human infection. Transmitted by bite wounds or contaminated of preexisting wounds during exposure to animals
Kingella denitrificans and *Kingella kingae*	Normal flora of human upper respiratory and genitourinary tracts	Infections probably caused by patient's endogenous strains
Cardiobacterium hominis	Normal flora of human upper respiratory tract	Infections probably caused by patient's endogenous strains
Capnocytophaga gingivalis, Capnocytophaga ochracea, and *Capnocytophaga sputigena*	Subgingival surfaces and other areas of human oral cavity	Infections probably caused by patient's endogenous strains
Capnocytophaga canimorsus and *Capnocytophaga cynodegmi*	Oral flora of dogs	Dog bite or nonbite, long exposure to dogs
Dysgonomonas capnocytophagoides	Uncertain; possibly part of human gastrointestinal flora	Uncertain; possibly endogenous

Table 33-2 Pathogenesis and Spectrum of Diseases

Organism	Virulence Factors	Spectrum of Diseases and Infections
Actinobacillus actinomycetemcomitans	Unknown; probably of low virulence; an opportunistic pathogen	Destructive periodontitis, including bone loss; endocarditis, often following dental manipulations; soft tissue and human bite infections, often mixed with anaerobic bacteria and *Actinomyces* spp.
Other *Actinobacillus* spp.	Unknown for human disease; probably of low virulence	Rarely cause infection in humans but may be found in animal bite wounds; association with other infections, such as meningitis or bacteremia, is extremely rare and involves compromised patients
Kingella denitrificans and *Kingella kingae*	Unknown; probably of low virulence; opportunistic pathogens	Endocarditis and infections in various other sites, especially in immunocompromised patients; *K. kingae* associated with blood, bone, and joint infections of young children
Cardiobacterium hominis	Unknown; probably of low virulence	Infections in humans are rare; most commonly associated with endocarditis, especially in persons with anatomic heart defects
Capnocytophaga gingivalis, Capnocytophaga ochracea, and *Capnocytophaga sputigena*	Unknown. Produce wide variety of enzymes that may mediate tissue destruction	Most commonly associated with periodontitis and other types of periodontal disease; less commonly associated with bacteremia in immunocompromised patients
Capnocytophaga canimorsus and *Capnocytophaga cynodegmi*	Unknown	Range from mild, local infection at bite site to bacteremia culminating in shock and disseminated intravascular coagulation. Most severe in splenectomized or otherwise debilitated (e.g., alcoholism) patients but can occur in healthy people. Miscellaneous other infections such as pneumonia, endocarditis, and meningitis may also occur
Dysgonomonas capnocytophagoides	Unknown; probably of low virulence	Role in disease is uncertain. May be associated with diarrheal disease in immunocompromised patients. Rarely isolated from other clinical specimens, such as urine, blood, and wounds

Table 33-3 Antimicrobial Therapy and Susceptibility Testing

Organism	Therapeutic Options	Potential Resistance to Therapeutic Options	Validated Testing Methods*
Actinobacillus actinomycetemcomitans	No definitive guidelines. For periodontitis, debridement of affected area; potential agents include ceftriaxone, ampicillin, amoxicillin-clavulanic acid, fluoroquinolone or trimethoprim-sulfamethoxazole. For endocarditis penicillin, ampicillin, or a cephalosporin (perhaps with an aminoglycoside) may be used	Some strains appear resistant to penicillin and ampicillin, but clinical relevance of resistance is unclear	See CLSI document M45
Other *Actinobacillus* spp.	No guidelines	Unknown	Not available
Kingella denitrificans, Kingella kingae	A beta-lactam with or without an aminoglycoside; other active agents include erythromycin, trimethoprim/sulfamethoxazole, and ciprofloxacin	Some strains produce beta-lactamase that mediates resistance to penicillin, ampicillin, ticarcillin, and cefazolin	See CLSI document M45
Cardiobacterium hominis	For endocarditis, penicillin with or without an aminoglycoside; usually susceptible to other beta-lactams, chloramphenicol, and tetracycline	Unknown	See CLSI document M45
Capnocytophaga gingivalis, Capnocytophaga ochracea, Capnocytophaga sputigena	No definitive guidelines. Generally susceptible to clindamycin, erythromycin, tetracyclines, chloramphenicol, imipenem, and other beta-lactams	Beta-lactamase–mediated resistance to penicillin	Not available
Capnocytophaga canimorsus, Capnocytophaga cynodegmi	Penicillin is drug of choice; also susceptible to penicillin derivatives, imipenem, and third-generation cephalosporins	Unknown	Not available
Dysgonomonas capnocytophagoides	No guidelines. Potentially effective agents include chloramphenicol, trimethoprim/sulfamethoxazole, tetracycline, and clindamycin	Often resistance to beta-lactams and ciprofloxacin	Not available

*Validated testing methods include those standard methods recommended by the Clinical and Laboratory Standards Institute (CLSI) and those commercial methods approved by the Food and Drug Administration (FDA).

and in chains, and they tend to exhibit bipolar staining. This staining morphology gives the overall appearance of the dots and dashes of Morse code.

Kingella spp. stain as short, plump coccobacilli with squared-off ends that may form chains. *Cardiobacterium hominis* is a pleomorphic gram-negative rod with one rounded end and one tapered end, giving the cells a teardrop appearance. *C. hominis* tends to form clusters, or rosettes, when Gram stains are prepared from 5% sheep blood agar.

Capnocytophaga spp. are gram-negative, fusiform-shaped bacilli with one rounded end and one tapered end and occasional filamentous forms; *C. cynodegmi* and *C. canimorsus* may be curved. *Dysgonomonas capnocytophagoides* stains as short gram-negative rods or coccobacilli.

CULTIVATION

Media of Choice

All genera described in this chapter grow on 5% sheep blood and chocolate agars. Most actinobacilli (except *A. actinomycetemcomitans* and *A. ureae*) show light growth on MacConkey agar, but the other genera will not grow on this medium. *Dysgonomonas capnocytophagoides* can be recovered from stool on CVA (cefoperazone-vancomycin-amphotericin B) agar. For recovery of *D. capnocytophagoides*, this medium, a *Campylobacter* selective agar, is incubated at 35° C instead of 42° C (preferred for *Campylobacter*).

These genera grow in the broths of commercial blood culture systems and in common nutrient broths such as thioglycollate and brain-heart infusion. Growth

Table 33-4   Colonial Appearance and Characteristics on 5% Sheep Blood Agar

Organism	Appearance
Actinobacillus actinomycetemcomitans	Pinpoint colonies after 24 hrs; rough, sticky, adherent colonies surrounded by a slight greenish tinge after 48 hrs; characteristic finding is presence of a four- to six- pointed starlike configuration in the center of a mature colony growing on a clear medium (e.g., brain-heart infusion agar), which can be visualized by examining the colony under low power (100×) of a standard light microscope
*A. equuli**	Small colonies at 24 hrs that are sticky, adherent, smooth or rough, and nonhemolytic
*A. lignieresii**	Resembles *A. equuli*
*A. suis**	Beta-hemolytic but otherwise resembles *A. equuli* and *A. lignieresii*
A. ureae	Resembles the pasteurellae (see Chapter 32)
Cardiobacterium hominis	After 48 hrs, colonies are small, slightly alpha-hemolytic, smooth, round, glistening and opaque; pitting may be produced
Capnocytophaga spp.	After 48-74 hrs, colonies are small- to medium-size, opaque, shiny; nonhemolytic; pale beige or yellowish color may not be apparent unless growth is scraped from the surface with a cotton swab; gliding motility may be observed as outgrowths from the colonies or as a haze on the surface of the agar, similar to swarming of *Proteus*
Dysgonomonas capnocytophagoides	Pinpoint colonies after 24 hrs; small, wet, gray-white colonies at 48-72 hrs; usually nonhemolytic, although some strains may produce a small zone of beta hemolysis; characteristic odor alternately described as fruity or bitter
Kingella denitrificans	Small, nonhemolytic; frequently pits agar; can grow on *Neisseria gonorrhoeae* selective agar (e.g., Thayer-Martin agar)
K. kingae	Small, with a small zone of beta hemolysis; may pit agar

*May grow on MacConkey agar as tiny lactose fermenters.

of *Actinobacillus* in broth media is often barely visible, with no turbidity produced. Microcolonies may be seen as tiny puffballs growing on the blood cell layer in blood culture bottles or as a film or tiny granules on the sides of a tube.

Incubation Conditions and Duration
The growth of all genera discussed in this chapter occurs best at 35° C and in the presence of increased CO_2. Therefore 5% sheep blood and chocolate agars should be incubated in a CO_2 incubator or candle jar. In addition, *Actinobacillus* and *Cardiobacterium* grow best in conditions of elevated moisture; a candle jar with a sterile gauze pad moistened with sterile water is ideal for this purpose. *Capnocytophaga* will grow anaerobically but will not grow in ambient air.

Even when optimum growth conditions are met, the organisms discussed here are all slow-growing; therefore inoculated plates should be held 2 to 7 days for colonies to achieve maximal growth.

Colonial Appearance
Table 33-4 describes the colonial appearance and other distinguishing characteristics (e.g., hemolysis and pigment) of each genus on 5% sheep blood agar. Most species will not grow on MacConkey agar; exceptions are noted in Table 33-4.

APPROACH TO IDENTIFICATION
None of the commercial identification systems will reliably identify any of the genera in this chapter. Table 33-5 outlines some conventional biochemical tests that are useful for differentiating among *Actinobacillus, Cardiobacterium, Kingella,* and *Haemophilus aphrophilus;* these are four of the five HACEK bacteria that cause subacute bacterial endocarditis. Table 33-6 shows key conventional biochemicals that can be used to differentiate *Capnocytophaga* spp., *Dysgonomonas capnocytophagoides,* and aerotolerant *Leptotrichia buccalis.*

Comments Regarding Specific Organisms
The genus *Actinobacillus* is very similar to *Pasteurella* (see Chapter 32), which must also be considered when a fastidious gram-negative rod requiring rabbit serum is isolated. *A. actinomycetemcomitans,* the most frequently isolated of the actinobacilli, can be distinguished from *H. aphrophilus* (see Chapter 34) by its positive test for catalase and negative test for lactose fermentation.

A. actinomycetemcomitans differs from *C. hominis* in being indole-negative and catalase-positive; catalase is also an important test for differentiating *Kingella* spp., which are catalase-negative, from *A. actinomycetemcomitans. C. hominis* is indole-positive following extraction with xylene and addition of Ehrlich's reagent; this is a key feature in differentiating it from *H. aphro-*

Table 33-5   Biochemical and Physiologic Characteristics of *Actinobacillus* spp. and Related Organisms

Organism	Catalase	Nitrate Reduction	Indole	Urea	Esculin Hydrolysis	Fermentation of:[†] Xylose	Lactose	Trehalose
Actinobacillus actinomycetemcomitans	+	+	−	−	−	v	−	−
A. equuli	v	+	−	(+)*	−	+	+	(+)
A. lignieresii	v	+	−	(+)*	−	+ or (+)	v	−
A. suis	v	+	−	(+)*	+	+	+ or (+)	+
A. ureae	v	+	−	(+)*	−	−	−	−
Cardiobacterium hominis	−	−	+	−	−	−	−	ND
Haemophilus aphrophilus	−	+	−	−	−	−	(+)	(+)
Kingella denitrificans	−	(+)[‡]	−	−	−	−	−	ND
K. kingae	−	−	−	−	−	−	−	ND

Data compiled from reference 10.
ND, No data; *v*, variable; +, >90% of strains positive; (+), >90% of strains positive but reaction may be delayed (i.e., 2 to 7 days); −, >90% of strains negative.
*May require a drop of rabbit serum on the slant or a heavy inoculum.
[†]May require the addition of 1 to 2 drops rabbit serum per 3 mL of fermentation broth to stimulate growth.
[‡]Nitrate is usually reduced to gas.

Table 33-6   Biochemical and Physiologic Characteristics of *Capnocytophaga* spp., *Dysgonomonas*, and Similar Organisms

Organism	Oxidase	Catalase	Esculin Hydrolysis	Indole	Nitrate Reduction	Xylose Fermentation
Capnocytophaga spp. (CDC group DF-1)*	−	−	(v)	−	v	−
C. canimorsus (CDC group DF-2)[†]	(+)	(+)	v	−	−	−[‡]
C. cynodegmi (CDC group DF-2-like)[†]	(+)	(+)	+ or (+)	−	−	−
*Leptotrichia buccalis**	−	−	v	−	−	−[‡]
*Dysgonomonas capnocytophagoides**	−	−	(+)	(v)	−	+or (+)[‡]
CDC group DF-3-like	−	v	v	(+)	−	−[‡]

Data compiled from references 4 and 10.
+, >90% of strains positive; (+), >90% of strains positive, but reaction may be delayed (i.e., 2 to 7 days); −, >90% of strains negative; *v*, variable; *(v)*, positive reactions may be delayed.
*Lactic acid is the major fermentation end product of glucose fermentation for *Leptotrichia buccalis*, and succinic acid and propionic is the major fermentation end product of glucose fermentation for *Capnocytophaga* spp. (CDC group DF-1) and *Dysgonomonas capnocytophagoides*.
[†]*C. canimorsus* does not ferment the sugars inulin, sucrose, or raffinose; *C. cynodegmi* will usually ferment one or all of these sugars.
[‡]May require the addition of 1 to 2 drops of rabbit serum per 3 mL of fermentation broth to stimulate growth.

philus, A. actinomycetemcomitans, and CDC group EF-4a. *C. hominis* is similar to *Suttonella indologenes* but can be distinguished by its ability to ferment mannitol and sorbitol.

Kingella spp. are catalase-negative, which helps to separate them from *Neisseria* spp. (see Chapter 42), with which they are sometimes confused. *K. denitrificans* may be mistaken for *Neisseria gonorrhoeae* when isolated from modified Thayer-Martin agar. Nitrate reduction is a key test in differentiating *K. denitrificans* from *N. gonorrhoeae,* which is nitrate-negative.

The species in the former CDC group DF-1, that is, *C. ochracea, C. sputigena,* and *C. gingivalis,* are catalase- and oxidase-negative; however, members of CDC group DF-1 cannot be separated by conventional biochemical tests. *C. canimorsus* and *C. cynodegmi,* formerly part of CDC group DF-2, are catalase- and oxidase-positive; these species are also difficult to differentiate from each other. However, for most clinical purposes, a presumptive identification to genus, that is, *Capnocytophaga,* is sufficiently informative and precludes the need to identify an isolate to the species level. Presumptive identification of an organism as *Capnocytophaga* spp. can be made when a yellow-pigmented, thin, gram-negative rod with tapered ends that exhibits gliding motility (see Table 33-4) and does not grow in ambient air is isolated.

Dysgonomonas capnocytophagoides, formerly CDC group DF-3, although similar to the other organisms in this chapter, are oxidase-negative. They are nonmotile, unlike the *Capnocytophaga,* which exhibit gliding motility. Gas-liquid chromatography is useful in separating *D. capnocytophagoides* and *Capnocytophaga* spp., but this technology is not commonly available in most clinical laboratories. *D. capnocytophagoides* produces succinic and propionic acid, whereas *Capnocytophaga* produces only succinic acid. Cellular fatty acid analysis can provide information necessary to distinguish *Capnocytophaga, D. capnocytophagoides,* and the aerotolerant strains of *Leptotrichia buccalis.*

SERODIAGNOSIS

Serodiagnostic techniques are not generally used for the laboratory diagnosis of infections caused by the organisms discussed in this chapter.

PREVENTION

Because the organisms discussed in this chapter do not generally pose a threat to human health, there are no recommended vaccination or prophylaxis protocols.

 Case Study

A 71-year-old woman with acute myeloid leukemia was being treated with immunosuppressive therapy and was neutropenic with 100 white blood cells per microliter. She had a low-grade fever and was not responding to treatment with a third-generation cephalosporin. Severe periodontal disease was noted. Her blood cultures became positive after 48 hours with a gram-negative fusiform rod that did not grow on MacConkey and was neither oxidase- nor catalase-positive. On blood agar, the organism was nonhemolytic but spread out from the initial colony, producing a haze on the agar. The laboratory reported that the organism was resistant to beta-lactam drugs. The patient's therapy was changed to ciprofloxacin, and she became afebrile within 24 hours.

QUESTIONS

1. What is the likely organism isolated from the blood culture, and what is the likely source of the organism?

2. Which tests will confirm the identification?
3. The patient was not responding to cephalosporin antimicrobial agents. What rapid testing can the laboratory perform to aid in the appropriate treatment of this organism?

REFERENCES

1. Gill VJ, Travis LB, Williams DY: Clinical and microbiological observations on CDC group DF-3, a gram-negative coccobacillus, *J Clin Microbiol* 29:1589, 1991.
2. Gordillo EM, Rendel M, Sood R, et al: Septicemia due to β-lactamase–*Kingella kingae, Clin Infect Dis* 17:818, 1993.
3. Hassan IJ, Hayek L: Endocarditis caused by *Kingella denitrificans, J Infect* 27:291, 1993.
4. Hofstad T, Olsen I, Eribe ER: *Dysgonomonas* gen nov to accommodate *Dysgonomonas gadei* sp nov, an organism isolated from human gall bladder, and *Dysgonomonas capnocytophagoides* (formerly CDC group DF-3), *Int J Syst Evol Microbiol* 50:2189, 2000.
5. Jensen KT, Schonheyder H, Thomsen VF: In-vitro activity of β-lactam and other antimicrobial agents against *Kingella kingae, J Antimicrob Chemother* 33:635, 1994.
6. Kaplan AH, Weber DJ, Oddone EZ, et al: Infection due to *Actinobacillus actinomycetemcomitans:* fifteen cases and review, *Rev Infect Dis* 11:46, 1989.
7. Minamoto GY, Sordillo EM: *Kingella denitrificans* as a cause of granulomatous disease in a patient with AIDS, *Clin Infect Dis* 15:1052, 1992.
8. Peel MM, Hornridge KA, Luppino M, et al: *Actinobacillus* spp. and related bacteria in infected wounds of humans bitten by horses and sheep, *J Clin Microbiol* 29:2535, 1991.
9. Pers C, Gahrn-Hansen B, Frederiksen W: *Capnocytophaga canimorsus* septicemia in Denmark, 1982-1995: review of 39 cases, *Clin Infect Dis* 23:71, 1996.
10. Weyant RS, Moss CW, Weaver RE, et al, editors: *Identification of unusual pathogenic gram-negative aerobic and facultatively anaerobic bacteria,* ed 2, Baltimore, 1996, Williams & Wilkins.
11. Yagupsky P, Dagan R: *Kingella kingae* bacteremia in children, *Pediatr Infect Dis J* 13:1148, 1994.

ADDITIONAL READING

Clinical and Laboratory Standards Institute: Methods for antimicrobial dilution and disk susceptibility testing of infrequently isolated or fastidious bacteria; M45, Villanova, Pa, CLSI.

Gill VJ: *Capnocytophaga.* In Mandell GL, Bennett JE, Dolin R, editors: *Principles and practice of infectious diseases,* ed 6, Philadelphia, 2005, Elsevier Churchill Livingstone.

Steinberg JP, Del Rio C: Other gram-negative and gram-variable bacilli. In Mandell GL, Bennett JE, Dolin R, editors: *Principles and practice of infectious diseases,* ed 6, Philadelphia, 2005, Elsevier Churchill Livingstone.

von Graevenitz A, Zbinden R, Mutters R: *Actinobacillus, Capnocytophaga, Eikenella, Kingella, Pasteurella,* and other fastidious or rarely encountered gram-negative rods. In Murray PR, Baron EJ, Jorgensen JH, et al, editors: *Manual of clinical microbiology,* ed 8, Washington, DC, 2003, ASM Press.

Haemophilus CHAPTER 34

Organisms to Be Considered

Current Name	Previous Name
Haemophilus influenzae	
Haemophilus influenzae biogroup *aegyptius*	*Haemophilus aegyptius*
Haemophilus ducreyi	
Other *Haemophilus* spp.	
H. parainfluenzae	
H. parahaemolyticus	
H. aphrophilus	
H. haemolyticus	
H. segnis	
H. paraphrophilus	

GENERAL CHARACTERISTICS

Species of the genus *Haemophilus*, except for *H. aphrophilus*, require hemin and nicotine adenine dinucleotide (NAD) for in vitro growth. The morphologic and physiologic features of individual species are presented later in this chapter in the discussion of laboratory diagnosis.

EPIDEMIOLOGY

As presented in Table 34-1, except for *Haemophilus ducreyi*, *Haemophilus* spp. normally inhabit the upper respiratory tract of humans.[5] Although *H. ducreyi* is only found in humans, the organism is not part of our normal flora and its presence in clinical specimens is indicative of infection.

Among *H. influenzae* strains, there are two broad categories: typeable and nontypeable. Strains are typed based on capsular characteristics. The capsule is composed of a sugar-alcohol phosphate (i.e., polyribitol phosphate) complex. Differences in this complex are the basis for separating encapsulated strains into one of six groups: type a, b, c, d, e, or f. Type b *H. influenzae* is most commonly encountered in serious infections in humans. Nontypeable strains do not produce a capsule[12] and are most commonly encountered as normal inhabitants of the upper respiratory tract.

Although person-to-person transmission plays a key role in infections caused by type b *H. influenzae* and *H. ducreyi*, infections caused by other *Haemophilus* strains and species likely arise endogenously as a person's own flora gains access to a normally sterile site. *H. ducreyi* is unique among the species considered here in that it is the agent of a sexually transmitted disease.

PATHOGENESIS AND SPECTRUM OF DISEASE

Production of a capsule and factors that mediate bacterial attachment to human epithelial cells are the primary virulence factors associated with *Haemophilus* spp.[8,12] In general, infections caused by type b *H. influenzae* are often systemic and life-threatening, whereas infections caused by nontypeable strains are usually localized (Table 34-2). Infections caused by other *Haemophilus* spp. occur much less frequently.[1,2,9] Also, the development and use of the anti–type b *H. influenzae* vaccine in children has greatly reduced the infection caused by this organism.

Chancroid is the sexually transmitted disease caused by *H. ducreyi* (see Table 34-2). Although small outbreaks of this disease have occurred in the United States, this disease is more commonly seen among socioeconomically disadvantaged populations that inhabit tropical environments.

LABORATORY DIAGNOSIS

SPECIMEN COLLECTION AND TRANSPORT

Haemophilus spp. can be isolated from most clinical specimens. The collection and transport of these specimens is outlined in Table 5-1, with emphasis on the following points. First, *Haemophilus* spp. are very susceptible to drying and temperature extremes.

Table 34-1 Epidemiology

Organism	Habitat (reservoir)	Mode of Transmission
Haemophilus influenzae (including biogroup *aegyptius*)	Normal flora of human upper respiratory tract	Person-to-person spread by contaminated respiratory droplets. Certain infections may be caused by person's endogenous strains
Haemophilus ducreyi	Not part of normal human flora; only found in humans during infection	Person-to-person spread by sexual contact (i.e., causative agent of a sexually transmitted disease known as chancroid)
Other *Haemophilus* spp.; species most commonly associated with infection include *H. parainfluenzae*, *H. parahaemolyticus*, and *H. paraphrophilus*	Normal flora of human upper respiratory tract	Spread of a patient's endogenous strain to sites outside the respiratory tract. Much less commonly involved than *H. influenzae* in infections in humans

Table 34-2 Pathogenesis and Spectrum of Diseases

Organism	Virulence Factors	Spectrum of Disease and Infections
Haemophilus influenzae	For encapsulated strains (of which type b is most common), the capsule is antiphagocytic and highly associated with virulence; other cell envelope factors also facilitate attachment to host cells For nonencapsulated strains, pili and other cell surface factors, not fully understood, play a role in attachment to host cells	Infections most commonly associated with encapsulated strains can be life-threatening and include meningitis, epiglottitis, cellulitis with bacteremia, septic arthritis, and pneumonia Nonencapsulated strains usually cause localized infections such as otitis media, sinusitis, conjunctivitis, and exacerbations of chronic bronchitis; pneumonia and bacteremia in adults with underlying medical conditions, such as malignancy or alcoholism, also can occur
Haemophilus influenzae biogroup *aegyptius*	Uncertain; probably similar to those of other *H. influenzae*	Purulent conjunctivitis; also one strain is the cause of Brazilian purpuric fever, which is a severe infection of high mortality in children between ages 1 and 4; infection includes purulent meningitis, bacteremia, high fever, vomiting, purpura (i.e., rash), and vascular collapse
Haemophilus ducreyi	Uncertain, but capsular factors, pili, and certain toxins are probably involved in attachment and penetration of host epithelial cells	Chancroid; genital lesions that progress from tender papules (i.e., small bumps) to painful ulcers with several satellite lesions. Regional lymphadenitis is common
Other *Haemophilus* spp.	Uncertain; probably of low virulence. Opportunistic pathogens	Associated with wide variety of infections similar to those caused by *H. influenzae* but much less commonly encountered in a clinically significant setting. *H. aphrophilus* is an uncommon cause of endocarditis and is the H member of the HACEK group of bacteria associated with slowly progressive (subacute) bacterial endocarditis

Therefore, specimens suspected of containing these organisms must be plated immediately, especially if they have not been submitted in a suitable transport medium.

Second, for recovery of *H. ducreyi* from genital ulcers, special measures are necessary because of the bacterium's fastidious nature. The ulcer should be cleaned with sterile gauze premoistened with sterile saline. A cotton swab premoistened with phosphate-buffered saline is then used to collect material from the base of the ulcer. To maximize the chance for recovering the organism, the swab must be plated to special selective media within 10 minutes of collection.

SPECIMEN PROCESSING

Other than the precautions just discussed for *H. ducreyi*, no special considerations are required for specimen processing of *Haemophilus* spp. Refer to Table 5-1 for general information on specimen processing.

DIRECT DETECTION METHODS

Direct Observation

Gram stain is generally used for the direct detection of *Haemophilus* in clinical material. However, in some instances the acridine orange stain (AO; see Chapter 6 for more information on this technique) is used to detect smaller numbers of organisms than may be detected by Gram stain.

To increase the sensitivity of direct Gram stain examination of body fluid specimens, especially cerebrospinal fluid (CSF), the specimens usually are centrifuged (2000 rpm for 10 minutes) and the smear is prepared from the resulting pellet. This concentration step can increase the sensitivity of direct microscopic examination by fivefold to tenfold. Moreover, cytocentrifugation of the specimen, in which clinical material is concentrated by centrifugation directly onto microscope slides, reportedly increases sensitivity of the Gram stain by as much as 100-fold (see Chapter 55 for information on infections of the central nervous system).[11]

Gram stains of the smears made with patient specimens must be examined carefully. *Haemophilus* spp. stain a pale pink and may be difficult to detect in the pink background of proteinaceous material that is often found in patient specimens.

H. influenzae appear as coccobacilli or small rods, whereas the cells of *H. influenzae* biotype *aegypticus* usually appear as long, slender rods. *H. haemolyticus* are small coccobacilli or short rods with occasional cells appearing as tangled filaments.

H. parainfluenzae produce either small pleomorphic rods or long filamentous forms, whereas *H. parahaemolyticus* usually are short to medium-length bacilli. *H. aphrophilus* and *H. paraphrophilus* are very short bacilli but occasionally are seen as filamentous forms. *H. segnis* are pleomorphic rods, and the cells of *H. ducreyi* may be either slender or coccobacillary. Traditionally, *H. ducreyi* cells are described as appearing as "schools of fish." However, this morphology is rarely seen in actual laboratory practice.

Antigen Detection

Detection of *H. influenzae* type b capsular polysaccharide directly in clinical specimens, such as CSF and urine, can be performed using commercially available particle agglutination assays (see Chapter 9). Because organisms in clinical infections are usually present at a sufficiently high concentration to be visualized by Gram stain, most clinical laboratories no longer perform the latex test. Historically, the latex tests have been very sensitive and specific for detection of *H. influenzae* type b, especially in

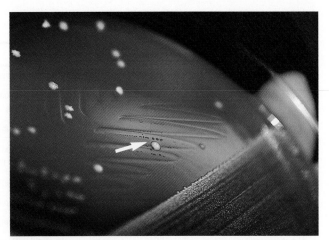

Figure 34-1 *Haemophilus influenzae* satelliting *(arrow)* around colonies of *Staphylococcus aureus*.

patients who have been on antimicrobial therapy before collection of the specimen. Unfortunately, these tests have been reported as positive in CSF and urine of patients who have recently received an *H. influenzae* type b vaccination,[5,6] so positive results of agglutination assays must be interpreted with caution.

CULTIVATION

Media of Choice

Chocolate agar provides the factors, that is, hemin (X factor) and NAD (V factor), necessary for the growth of *Haemophilus* spp. Most strains will not grow on 5% sheep blood agar, which contains hemin but not NAD. Several bacterial species, including *Staphylococcus aureus,* produce NAD as a metabolic byproduct. Therefore, tiny colonies of *Haemophilus* spp. may be seen growing on sheep blood agar very close to colonies of bacteria that can produce V factor; this phenomenon is known as *satelliting* (Figure 34-1). The phenomenon of satelliting has become important in this era of needing to rapidly identify potential agents of a bioterrorist attack. The protocol for ruling out *Francisella* or *Brucella* (see Chapter 65) is to first rule out *Haemophilus* by the lack of evidence of satelliting around *S. aureus*.

A selective medium, such as horse blood–bacitracin agar, may be occasionally used for isolation of *H. influenzae* from respiratory secretions of patients with cystic fibrosis. This medium is designed to prevent overgrowth of *H. influenzae* by mucoid *Pseudomonas aeruginosa. Haemophilus* spp. will not grow on MacConkey agar.

H. ducreyi requires special media to grow. Two such media are (1) Mueller-Hinton–based chocolate agar supplemented with 1% IsoVitaleX and 3 µg/mL vanco-

Table 34-3   Colonial Appearance and Characteristics

Organism	Medium	Appearance
Haemophilus aphrophilus	CA	Round; convex with opaque zone near center
H. ducreyi	Selective medium	Small, flat, smooth, and translucent to opaque at 48-72 hrs; colonies can be pushed intact across agar surface
H. haemolyticus	CA	Resembles *H. influenzae* except beta-hemolytic on rabbit or horse blood agar
H. influenzae	CA	Unencapsulated strains are small, smooth, and translucent at 24 hrs; encapsulated strains form larger, more mucoid colonies; mouse nest odor; nonhemolytic on rabbit or horse blood agar
H. influenzae biotype *aegyptius*	CA	Resembles *H. influenzae* except colonies are smaller at 48 hrs
H. parahaemolyticus	CA	Resembles *H. parainfluenzae* except beta-hemolytic on rabbit or horse blood agar
H. parainfluenzae	CA	Medium to large, smooth, and translucent; nonhemolytic on rabbit or horse blood agar
H. paraphrophilus	CA	Resembles *H. aphrophilus*
H. segnis	CA	Convex, grayish white, smooth or granular at 48 hrs

CA, Chocolate agar.

mycin and (2) heart infusion–based agar supplemented with 10% fetal bovine serum and 3 µg/mL vancomycin. The vancomycin inhibits gram-positive organisms normally colonizing the genital tract.

Haemophilus spp. will grow in the broths of commercial blood culture systems and in common nutrient broths such as thioglycollate and brain-heart infusion. However, they often produce only weakly turbid suspensions and may not be readily visible in broth cultures. For this reason, blind subcultures to chocolate agar or examination of smears by AO or Gram stain have been used to enhance detection. However, the productivity of this extra effort for detecting *Haemophilus* spp. in patient cultures has never been established, especially with the newer blood culture detection systems.

Rabbit or horse blood agars are commonly used for detecting hemolysis by hemolysin-producing strains of *Haemophilus* strains that will not grow on 5% sheep blood.

Incubation Conditions and Duration
Most strains of *Haemophilus* spp. are able to grow aerobically and anaerobically. Growth is stimulated by 5% to 10% carbon dioxide (CO_2), so that incubation in a candle extinction jar, CO_2 pouch, or CO_2 incubator is recommended. These organisms usually grow within 24 hours, but cultures are routinely held 72 hours before being discarded as negative. An exception is *H. ducreyi,* which may require as long as 7 days to grow.

Optimal growth of all *Haemophilus* spp., except *H. ducreyi,* occurs at 35° to 37° C. Cultures for *H. ducreyi* should be incubated at 33° C. In addition,

H. ducreyi requires high humidity, which may be established by placing a sterile gauze pad moistened with sterile water inside the candle jar or CO_2 pouch.

Colonial Appearance
Table 34-3 describes the colonial appearance and other distinguishing characteristics (e.g., odor and hemolysis) of each species.

APPROACH TO IDENTIFICATION
Commercial identification systems for *Haemophilus* spp. are summarized in Chapter 13. All of the systems incorporate several rapid enzymatic tests and generally work well for identifying these organisms.

Traditional identification criteria include hemolysis on horse or rabbit blood and the requirement for X and/or V factors for growth. To establish X and V factor requirements, disks impregnated with each factor are placed on unsupplemented media, usually Mueller-Hinton agar or trypticase soy agar, that has been inoculated with a light suspension of the organism (see Figure 13-42). After overnight incubation at 35° C in ambient air, the plate is examined for growth around each disk. Many X factor–requiring organisms are able to carry over enough factor from the primary medium to give false-negative results (i.e., growth occurs at such a distance from the X disk as to falsely indicate that the organism does not require the X factor).

The porphyrin test is another means for establishing an organism's X-factor requirements and eliminates the potential problem of carryover. This test detects the presence of enzymes that convert δ-aminolevulinic acid (ALA) into porphyrins or protoporphyrins.

Table 34-4 **Key Biochemical and Physiologic Characteristics of *Haemophilus* spp.**

Organism	X Factor	V Factor	Beta-Hemolytic on Rabbit Blood Agar	Lactose Fermentation[3]	Mannose Fermentation[3]
Haemophilus influenzae[4]	+	+	−	−	−
H. influenzae biotype *aegyptius*[4]	+	+	−	−	−
H. haemolyticus	+	+	+	−	−
H. parahaemolyticus	−	+	+	−	−
H. parainfluenzae	−	+	v[5]	−	+
H. paraphrophilus[1,2]	−	+	−	+	+
H. segnis	−	+	−	−	−
H. ducreyi[1]	+	−	−	−	−
H. aphrophilus[1]	−	−	−	+	+

Data compiled from references 7 and 14.
+, >90% of strains positive; −, >90% of strains negative.
[1]Catalase-negative.
[2]*H. aphrophilus* reactions are also included in Chapter 33 (Table 33-5).
[3]Test performed in rapid sugar fermentation medium.
[4]Cannot be differentiated biochemically.
[5]Biotypes 2 and 3 are β-hemolytic on rabbit blood agar.

The porphyrin test may be performed in broth, in agar, or on a disk.

Isolates from CSF or respiratory tract specimens that (1) are gram-negative rods or gram-negative coccobacilli, (2) grow on chocolate agar in CO_2 but not blood agar or satellite around other colonies on blood agar, and (3) are porphyrin-negative and nonhemolytic on rabbit or horse blood may be identified as *H. influenzae*.[10] *Haemophilus* isolates may also be identified to species using rapid sugar fermentation tests; an abbreviated identification scheme for the X- and/or V-requiring organisms is shown in Table 34-4.

Differentiation of *H. aphrophilus* from similar organisms (e.g., *A. actinomycetemcomitans,* Chapter 33) is shown in Table 33-5. *H. aphrophilus* does not require either X or V factors for growth. However, it is catalase-negative and ferments lactose or sucrose. *A. actinomycetemcomitans* yields the opposite reactions in these tests.

SEROTYPING

Although serologic typing of *H. influenzae* may be used to establish an isolate as being any one of the six serotypes (i.e., a, b, c, d, e, and f), it is usually only used to identify type b strains. All *H. influenzae* from cases of invasive infections should be serotyped to rule in/out type b infections. Testing can be performed using a slide agglutination test (see Chapter 9); a saline control should always be run to detect autoagglutination (i.e.,

the nonspecific agglutination of the test organism without homologous antiserum).

SERODIAGNOSIS

An enzyme-linked immunosorbent assay (ELISA) has been developed to detect antibodies to *H. ducreyi.* ELISA and radioimmunoassay (RIA) have been used to show seroconversion following *H. influenzae* type b vaccination. None of these assays are used commonly for diagnostic purposes.

ANTIMICROBIAL SUSCEPTIBILITY TESTING AND THERAPY

Standard methods have been established for performing in vitro susceptibility testing with clinically relevant isolates of *Haemophilus* spp. (see Chapter 12 for details on these methods). In addition, various agents may be considered for testing and therapeutic use (Table 34-5). Although resistance to ampicillin by production of beta-lactamase has become widespread among *H. influenzae,* resistance to cephalosporins that are not notably affected by the enzyme (i.e., ceftriaxone and cefotaxime) is rare. Therefore, routine susceptibility testing of clinical isolates as a guide to therapy may not be routinely necessary.

In addition to beta-lactamase production, beta-lactam resistance mediated by altered penicillin-binding proteins in *H. influenzae* has been well described and can

Table 34-5   Antimicrobial Therapy and Susceptibility Testing

Organism	Therapeutic Options	Potential Resistance to Therapeutic Options	Validated Testing Methods*	Comments
Haemophilus influenzae	Usually ceftriaxone or cefotaxime for life-threatening infections; for localized infections several cephalosporins, beta-lactam/beta-lactamase inhibitor combinations, macrolides, trimethoprim/sulfamethoxazole, and certain fluoroquinolones are effective	Beta-lactamase–mediated resistance to ampicillin is common; beta-lactam resistance by altered PBP target is rare (≤1% of strains)	As documented in Chapter 12: disk diffusion, broth dilution, and certain commercial systems	Resistance to third-generation cephalosporins has not been documented. Testing to guide therapy is not routinely needed
Haemophilus ducreyi	Erythromycin is the drug of choice; other potentially active agents include ceftriaxone and ciprofloxacin	Resistance to trimethoprim/sulfamethoxazole and tetracycline has emerged; beta-lactamase–mediated resistance to ampicillin and amoxicillin is also known	Not available	
Other *Haemophilus* spp.	Guidelines the same as for *H. influenzae*	Beta-lactamase–mediated resistance to ampicillin is known	As documented in Chapter 12: disk diffusion, broth dilution, and certain commercial systems. Also see CLSI document M45	Resistance to third-generation cephalosporins has not been documented. Testing to guide therapy is not routinely needed

*Validated testing methods include those standard methods recommended by the Clinical and Laboratory Standards Institute (CLSI) and those commercial methods approved by the Food and Drug Administration (FDA).

lead to decreased activity of both beta-lactam/beta-lactamase inhibitor combinations and of first- and second-generation cephalosporins. However, this mechanism of resistance is not common and its impact on the clinical efficacy of the antimicrobial agents affected is not fully known.

Although standardized testing guidelines for *H. ducreyi* have not been established by the Clinical and Laboratory Standards Institute (CLSI), erythromycin is widely accepted as an effective drug for treating chancroid. However, this organism can exhibit resistance to other therapeutic agents, so the potential exists for the emergence of significant antimicrobial resistance.[13]

PREVENTION

Several multiple-dose protein-polysaccharide conjugate vaccines are licensed in the United States for *H. influenzae* type b. These vaccines have substantially reduced the incidence of severe invasive infections caused by type b organisms, and vaccination of children starting at 2 months of age is strongly recommended.

Rifampin chemoprophylaxis is recommended for all household contacts of index cases of *H. influenzae* type b meningitis in which there is at least one unvaccinated contact younger than 4 years of age. Children and staff of daycare centers should also receive rifampin prophylaxis if at least two cases have occurred among the children.[3]

Case Study

A 20-year-old man presented to the emergency department with a history of temperature up to 103° F and mild respiratory distress. He reported that he had the worst sore throat of his life and was having difficulty swallowing. On physical examination, the patient was found to have a "cherry-red" epiglottis. Blood and throat cultures were obtained, and the patient was begun on treatment with cefotaxime. An endotracheal tube was placed for 48 hours until the inflammation of the epiglottis subsided. The throat culture grew normal respiratory microbiota, but a gram-negative rod was isolated from the blood culture in 24 hours only on chocolate agar.

QUESTIONS

1. What is the genus of the organism that was isolated from this patient's blood?
2. The organism grew on blood agar only around a colony of *Staphylococcus* (see Figure 34-1) but produced porphyrins from delta-aminolevulinic acid and fermented lactose. What is the species of this organism?
3. What is the importance of identification of *Haemophilus* to the species level from specimens from normally sterile sites?

REFERENCES

1. Chadwick PR, Malnick H, Ebizie AO: *Haemophilus paraphrophilus* infection: a pitfall in laboratory diagnosis, *J Infect* 30:67, 1995.
2. Coll-Vinent B, Suris X, Lopez-Soto A, et al: *Haemophilus paraphrophilus* endocarditis: case report and review, *Clin Infect Dis* 20:1381, 1995.
3. Committee on Infectious Diseases: 2006 *Red book: report of the Committee on Infectious Diseases*, ed 27, Elk Grove, Ill, 2006, American Academy of Pediatrics.
4. Darville T, Jacobs RF, Lucas RA, et al: Detection of *Haemophilus influenzae* type b antigen in cerebrospinal fluid after immunization, *Pediatr Infect Dis J* 11:243, 1992.
5. Foweraker JE, Cooke NJ, Hawkey PM: Ecology of *Haemophilus influenzae* and *Haemophilus parainfluenzae* in sputum and saliva and effects of antibiotics on their distribution in patients with lower respiratory tract infections, *Antimicrob Agents Chemother* 37:804, 1993.
6. Jones RG, Bass JW, Weisse ME, et al: Antigenuria after immunization with *Haemophilus influenzae* oligosaccharide CRM197 conjugate (H6OC) vaccine, *Pediatr Infect Dis J* 10:557, 1991.
7. Kilian M: *Haemophilus*. In Murray PR, Baron EJ, Jorgensen JH, et al, editors: *Manual of clinical microbiology*, ed 8, Washington, DC, 2003, ASM Press.
8. Lageragard T: *Haemophilus ducreyi*: pathogenesis and protective immunity, *Trends Microbiol* 3:87, 1995.
9. Merino D, Saavedra J, Pujol E, et al: *Haemophilus aphrophilus* as a rare cause of arthritis, *Clin Infect Dis* 19:320, 1994.
10. National Committee for Clinical Laboratory Standards: Abbreviated identification of bacteria and yeast; M35-A, Wayne, Pa, 2002, NCCLS.
11. Shanholtzer CJ, Schaper PJ, Peterson LR: Concentrated Gram-stained smears prepared with a cytospin centrifuge, *J Clin Microbiol* 16:1052, 1982.
12. St Geme JW III: Nontypeable *Haemophilus influenzae* disease: epidemiology, pathogenesis, and prospects for prevention, *Infect Agents Dis* 2:1, 1993.
13. Van Dyck E, Bogaerts J, Smet H, et al: Emergence of *Haemophilus ducreyi* resistance to trimethoprim-sulfamethoxazole in Rwanda, *Antimicrob Agents Chemother* 38:1647, 1994.
14. Weyant RS, Moss CW, Weaver RE, et al, editors: *Identification of unusual pathogenic gram-negative aerobic and facultatively anaerobic bacteria*, ed 2, Baltimore, 1996, Williams & Wilkins.

ADDITIONAL READING

Clinical and Laboratory Standards Institute: Methods for antimicrobial dilution and disk susceptibility testing of infrequently isolated or fastidious bacteria; M45, Villanova, Pa, CLSI.

Hand WL: *Haemophilus* species (including chancroid). In Mandell GL, Bennett JE, Dolin R, editors: *Principles and practice of infectious diseases*, New York, 2000, Churchill Livingstone.

CHAPTER 35 *Bartonella* and *Afipia*

Genera and Species to Be Considered

Bartonella bacilliformis
Other *Bartonella* spp., including
  B. quintana
  B. henselae
  B. elizabethae
  B. clarridgeiae
Afipia felis

The two genera, *Bartonella* and *Afipia,* are able to grow on chocolate agar and, albeit very slowly, on routine blood (trypticase soy agar with 5% sheep blood agar), typically appearing after 12 to 14 days and sometimes requiring as long as 45 days; neither organism grows on MacConkey agar. Presently, there is no optimal procedure for the isolation of these organisms from clinical specimens.[19] Because of these similarities and because two organisms, *Bartonella henselae* and *Afipia felis,* cause cat-scratch disease (CSD), these genera are addressed together in this chapter.

BARTONELLA

GENERAL CHARACTERISTICS

Bartonella spp. originally were grouped with members of the family *Rickettsiales.* However, because of extensive differences, the family *Bartonellaceae* was removed from this order.[2] As a result of phylogenetic studies using molecular biologic techniques, the genus *Bartonella* currently includes 16 species and three subspecies, most of which were reclassified from the genus *Rochalimeae* and from the genus *Grahamella.*[4,16] Only five species are currently recognized as major causes of disease in humans (Table 35-1), but other members of the genus have been found in animal reservoirs such as rodents, ruminants, and moles.[19] *Bartonella* spp. are most closely related to *Brucella abortus* and *Agrobacterium tumefaciens,*[20] and are short, gram-negative, rod-shaped, fastidious organisms that are oxidase-negative and grow best on blood-enriched media or cell co-culture systems.

EPIDEMIOLOGY AND PATHOGENESIS

Organisms belonging to the genus *Bartonella* cause numerous infections in humans; most of these infections are thought to be zoonoses.[7] Interest in these organisms has increased because of their recognition as causes of an expanding array of clinical syndromes in immunocompromised and immunocompetent patients. For example, *Bartonella* species have been recognized with increasing frequency over the past decade as a cause of culture-negative endocarditis. Humans acquire infection either naturally (infections caused by *Bartonella quintana* or *Bartonella bacilliformis*) or incidentally (other *Bartonella* species) via arthropod-borne transmission. Nevertheless, questions remain regarding the epidemiology of these infections; some epidemiologic information is summarized in Table 35-1.

Bartonella is a facultative intracellular bacterium that closely interacts with its host cells and has unique abilities to cause either acute or chronic infection as well as the proliferation of microvascular endothelial cells and angiogenesis (forming new capillaries from pre-existing ones) or suppurative manifestations. Three *Bartonella* species (*B. quintana, B. bacilliformis,* and *B. henselae*) are unique in their abilities to cause angiogenic lesions. Research has revealed that organisms belonging to this genus interact with host red blood cells, endothelial cells, and possibly bone marrow progenitor cells.[9] Colonization of vascular endothelium is considered a crucial step in the establishment and maintenance of *Bartonella*-triggered angioproliferative lesions. Within several hours following infection of cultured human umbilical vein endothelial cells, *Bartonella* species adhere to and enter these cells by an actin-dependent process that resembles other bacterial-directed phagocytosis or uptake into host cells.[4] Recent studies have also shown that *B. henselae* possess nine outer membrane proteins (OMP), one of which is able to bind to endothelial cells.

Table 35-1   Organisms Belonging to the Genus *Bartonella* and Recognized to Cause Disease in Humans*

Organism	Habitat (reservoir)	Mode of Transmission	Clinical Manifestation(s)
Bartonella bacilliformis	Uncertain; humans	Sand flies*	Carrión's disease[†]
B. quintana	Uncertain; possibly small rodents; humans	Human body louse*	Trench fever; chronic bacteremia; endocarditis; bacillary angiomatosis; chronic lymphadenopathy; pericarditis
B. henselae	Domestic cats	Domestic cat by bites and/or scratches; cat fleas*	Bacteremia; endocarditis; cat-scratch disease; bacillary angiomatosis; peliosis hepatitis; neuroretinitis
B. clarridgeiae	Domestic cats	Domestic cat by bites and/or scratches	Bacteremia; cat-scratch disease
B. elizabethae	Rats	Fleas	Endocarditis

*Other *Bartonella* species have caused incidental infections in humans, but only one or a few documented cases.
[†]Disease confined to a small endemic area in South America; characterized by a septicemic phase with anemia, malaise, fever, and enlarged lymph nodes, liver and spleen followed by a cutaneous phase with bright red cutaneous nodules that are usually self-limited.[3]

Typically, *Bartonella* species multiply and persist in the red blood cells in their reservoir host and share common persistence and dissemination strategies. In addition to angioproliferation, recent data indicate that bartonellae can inhibit endothelial cell apoptosis (programmed cell death); these organisms also activate monocyte and macrophage cells that are able to produce potent angiogenic factors.[4] Although much remains to be learned regarding the pathogenesis of infections caused by *Bartonella*, it is clear that these organisms possess unique pathogenic strategies to expand their bacterial habitat to sustain their survival within the human host. It is also evident that the pathologic response to these infections varies substantially with the status of the host immune system. For example, infection with the same *Bartonella* species, such as *B. henselae*, can cause a focal suppurative reaction (i.e., CSD), in immunocompetent patients, or a multifocal angioproliferative response (i.e., bacillary angiomatosis) in immunocompromised patients.[19]

SPECTRUM OF DISEASE

The diseases caused by *Bartonella* species are listed in Table 35-1. Because *B. quintana* and *B. henselae* are more common causes of infections in humans, these agents are addressed in greater depth.

Trench fever, caused by *B. quintana*, was largely considered a disease of the past that afflicted about 1 million soldiers. Clinical manifestations of trench fever range from a mild influenza-like headache, bone pain, splenomegaly (enlarged spleen), and a short-lived maculopapular rash. During the febrile stages of trench fever, infection may persist long after the disappearance

of all clinical signs; some patients may have six or more recurrences.[11] *B. quintana* has reemerged and has been reported to cause bacteremia, endocarditis, chronic lymphadenopathy, and bacillary angiomatosis primarily in homeless people in Europe and the United States, as well as in patients infected with the human immunodeficiency virus (HIV). **Bacillary angiomatosis** is a vascular proliferative disease most often involving the skin (other organs such as the liver, spleen, and lymph nodes may also be involved) and occurs in immunocompromised individuals such as organ transplant recipients and people who are HIV-positive. Prolonged bacteremias that occur in patients with *B. quintana* infections may be associated with the development of endocarditis and bacillary angiomatosis.

B. henselae can also cause bacteremia, endocarditis, and bacillary angiomatosis. Of note, recent observations indicate that *B. henselae* infections may be commonly subclinical and/or markedly underreported, as problems with current diagnostic approaches are recognized (see Laboratory Diagnosis). In addition, *B. henselae* causes CSD and peliosis hepatitis. About 24,000 cases of CSD occur annually in the United States; about 80% of these occur in children.[18] Usually this infection begins as a papule or pustule at the primary inoculation site; regional tender lymphadenopathy develops in 1 to 7 weeks. The spectrum of disease ranges from chronic, self-limited adenopathy to a severe systemic illness affecting multiple body organs. Although complications such as a suppurative (draining) lymph node or encephalitis are reported, fatalities are rare. Diagnosis of CSD traditionally required that a patient fulfill three of the four following criteria:

- History of animal contact plus site of primary inoculation (e.g., a scratch)
- Negative laboratory studies for other causes of lymphadenopathy
- Characteristic histopathology of the lesion
- A positive skin test using antigen prepared from heat-treated pus taken from another patient's lesion

Bartonella clarridgeiae is a recently described species that also causes CSD and bacteremia.[12,14]

Peliosis hepatitis caused by *B. henselae* can occur alone or develop with cutaneous bacillary angiomatosis or bacteremia. Patients with peliosis hepatitis present with gastrointestinal symptoms, fever, chills, and an enlarged liver and spleen containing blood-filled cavities. This systemic disease is seen in patients infected with HIV and other immunocompromised individuals.

LABORATORY DIAGNOSIS

Specimen Collection, Transport, and Processing

Clinical specimens frequently submitted to the laboratory for direct examination and culture include blood, which has been collected in a lysis-centrifugation blood culture tube (Isolator; Wampole Laboratories, Cranbury, NJ), as well as aspirates and/or tissue specimens (e.g., lymph node, spleen, or cutaneous biopsies). There are no special requirements for specimen collection, transport, or processing of the organisms discussed in this chapter. Refer to Table 5-1 for general information on specimen collection, transport, and processing.

Direct Detection Methods

The presence of *Bartonella* spp. can be detected on histopathologic examination of tissue biopsies stained with the Warthin-Starry silver stain; immunofluorescence and immunohistochemical techniques have also been used. Because of these organisms' fastidious nature and slow growth, the use of molecular methods to identify *Bartonella* species directly in clinical specimens is particularly attractive and numerous methods have been used. However, a word of caution regarding molecular detection of *Bartonella* species in patient samples is in order. Although polymerase chain reaction (PCR) targeting the 16S-23S rRNA gene intergenic transcribed spacer region has been proposed as a reliable method for the detection of *Bartonella* species DNA in clinical samples, a recent study revealed some potential limitations regarding the specificity of primers.[15] In addition, because of the expanding number of species belonging to this genus, the use of PCR with restriction fragment length polymorphism (see Chapter 8) as well as sequencing may require the targeting of not one but several genes and subsequent sequencing for accurate species identification.[1]

Cultivation

Although still being debated is the optimum conditions required for recovery of bartonellae from clinical specimens, two widely used methods are direct plating onto fresh chocolate agar plates (less than 2 weeks old) and co-cultivation in cell culture.[13,17] Fresh agar helps supply moisture necessary for growth. Lysed, centrifuged sediment of blood collected in an Isolator tube or minced tissue is directly inoculated to fresh chocolate agar plates and incubated at 35° C in a very humid atmosphere with 5% to 10% carbon dioxide (CO_2), examined daily for 3 days, and examined again after 2 weeks of incubation. One study indicated that collection of blood in EDTA and subsequent freezing may improve the sensitivity of recovering *B. henselae* from blood.[3] Biopsy material is co-cultivated with an endothelial cell culture system; co-cultures are incubated at 35° C in 5% to 10% CO_2 for 15 to 20 days. Blood-enriched agar, such as Columbia or heart infusion agar base with 5% sheep blood, can also be employed, but horse or rabbit blood is reported to be a more effective supplement than sheep blood.[14] Lymph node tissue, aspirates, or swabs can be inoculated onto laked horse blood agar slopes supplemented with hemin; plates are then sealed and incubated in 5% CO_2 atmosphere up to 6 weeks at 37° C with 85% humidity.[6]

Approach to Identification

Bartonella spp. should be suspected when colonies of small, gram-negative bacilli are recovered after prolonged incubation (Figure 35-1). Organisms are all oxidase-, urease-, nitrate reductase–, and catalase-negative. Various means can be used to confirm the identification of *Bartonella*. Species can be delineated using biochemical profiles when 100 mg/mL of hemin is added to the test medium, the MicroScan rapid or Rapid ANAII system (Innovative Diagnostic Systems, Norcross, Ga) anaerobe panels, polyvalent antisera, or a number of molecular biological methods.[5,6,17]

Serodiagnosis

Several serologic methods for detecting antibodies to *Bartonella* spp. have been developed. An indirect fluorescent antibody using antigen prepared from *Bartonella* spp. co-cultivated with Vero cells[18] and enzyme-linked immunoassays has been developed and evaluated. However, the sensitivity and specificity of these assays have been questioned. Cross-reactivity between *Bartonella* and *Chlamydia* spp. and *Coxiella burnettii* have

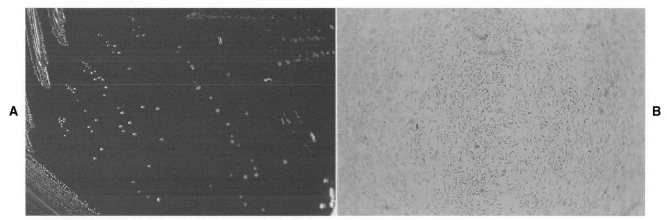

Figure 35-1 **A,** Colonies of *Bartonella henselae* on blood agar. **B,** Gram stain of a colony of *B. henselae* from blood agar.

been reported.[17] Of note, a significant antibody response to infection is not mounted in most HIV-infected patients, who as a group are susceptible to *Bartonella* infections.[10]

In conclusion, LaScola and colleagues in a 5-year study of various samples obtained for culture of *Bartonella* species demonstrated that the success of recovery or detection of *B. henselae* or *B. quintana* was dependent on several factors. These factors include the clinical form of the disease (i.e., endocarditis, bacteremia, bacillary angiomatosis, or CSD), previous antibiotic therapy, the type of clinical specimen (e.g., blood, heart valve, skin, or lymph node), and the type of laboratory diagnostic method employed (serology, PCR, shell vial cultures with human endothelial cell monolayers, direct plating of blood onto agar or broth blood cultures).[13]

ANTIMICROBIAL SUSCEPTIBILITY TESTING AND THERAPY

Treatment recommendations for *Bartonella* diseases, including CSD, depend on the specific disease presentation. The efficacy of various antibiotics for CSD is difficult to assess because symptoms are generally self-limiting over time, even without specific therapy. In addition to the clinical situation, the approach to treatment must be adapted to each species.[19] Although susceptibilities have been determined in the presence of eucaryotic cells or without cells (i.e., axenic media), these conditions are not standardized nor do they have interpretive criteria according to Clinical and Laboratory Standards Institute (CLSI). Moreover, results of in vitro testing do not always correlate with clinical efficacy; for example, penicillin is not effective therapy despite exquisite susceptibility in vitro.[19]

Recent experience with azithromycin indicates this antibiotic might shorten the time to resolution of adenopathy of CSD; however, it is presently unclear that antibiotic therapy is useful in immunocompetent patients. For patients with more severe CSD (about 5% to 14% of cases), other successful antibiotic regimens have included azithromycin or doxycycline in combination with rifampin or rifampin alone, doxycycline, or erythromycin. For bacillary angiomatosis and peliosis, doxycycline and erythromycin are considered the drugs of choice, whereas proposed therapy for endocarditis, suspected or documented, is gentamicin with or without doxycycline and gentamicin with doxycycline, respectively.[19]

PREVENTION

There are no vaccines available to prevent infections caused by *Bartonella* spp. Because of the link between cats and/or cat fleas and transmission of *B. henselae* to humans, it has been recommended that immunocompromised individuals avoid contact with cats, especially kittens, and control flea infestation.

AFIPIA FELIS

CSD was first reported in 1931; however, the causative agent was unknown for several decades. Finally a bacterial agent was isolated and characterized and given the name *Afipia felis*. However, the role of *A. felis* in the etiology of CSD was subsequently questioned because patients with CSD failed to mount an immune response to *A. felis* antigen and the organism was not detected from patients with CSD by either culture or PCR.

Subsequently, data were provided that patients with CSD mounted not only an immune response to *B. henselae* but *B. henselae* was also detected by culture, PCR, and immunocytochemistry; *B. henselae* was also detected in CSD skin test antigens, cats, and cat fleas. In light of all the data, *B. henselae* is now recognized as the primary causative agent of CSD, whereas *A. felis* plays a rare role in this disease.[8,17] Despite its rare isolation, indirect evidence suggests that *A. fipia* may be more commonly linked to CSD than is currently appreciated due to the lack of appropriate laboratory methods for detection.

Case Study

A 52-year-old male with a 25 pack-year smoking history had been living on the street for an unknown period of time. He sought medical attention because of overall poor health and was found to be anemic with weight loss. A spiculated mass was seen in his left middle lung lobe on chest film and a lobectomy was performed with the possible diagnosis of carcinoma. Numerous necrotizing granulomas and chronic inflammation were reported by the pathology department but no carcinoma was observed in the lung tissue. Gram and Warthin Starry stains were observed for "dark-staining gram-variable debris" but no definitive organisms. The patient had an uneventful recovery without antiinfective therapy. Routine bacterial and fungal cultures of the lung tissue were negative but the broth mycobacterial culture grew a gram-negative rod. The rod only grew on charcoal yeast extract agar (CYE), but not on blood or chocolate agars. It was oxidase- and urease-positive, motile, and beta-lactamase–positive. The catalase reaction was weak; nitrate was negative. It did not react with *Legionella* genus antiserum.

QUESTIONS

1. The significant characteristics of this bacterium are that it grows only in broth and on CYE plates. Most laboratories do not stock CYE. What should laboratory personnel do if they isolate a gram-negative rod in a culture from a normally sterile site that was sent to detect *Mycobacterium*?
2. Our isolate was identified as *Afipia broomeae* using DNA homology testing. According to Weyant and colleagues,[21] this bacterium is characterized for its growth on CYE and in broth, but not on other laboratory media. The species identification is based on a positive oxidase, catalase, urease, and xylose and a negative nitrate reduction. *A. felis* is identical except it is nitrate-positive. Although the CDC collection of

A. felis is mostly from lymph nodes, most of the *A. broomeae* were from respiratory specimens. How do you think our patient acquired the infection?
3. Because both *Afipia* and *Bartonella* are difficult to grow, should a laboratory attempt to provide culture services?

REFERENCES

1. Avidor B, Graidy M, Efrat G, et al: *Bartonella koehlerae*, a new cat-associated agent of culture-negative human endocarditis, *J Clin Microbiol* 42:3462, 2004.
2. Breitschwerdt EB, Kordick DL: *Bartonella* infections in animals: carriership, reservoir potential, and zoonotic potential for human infection, *Clin Microbiol Rev* 13:428, 2000.
3. Brenner SA, Rooney JA, Manzewitsch P, et al: Isolation of *Bartonella (Rochalimaea) henselae:* effects of methods of blood collection and handling, *J Clin Microbiol* 35:544, 1997.
4. Dehio C: Recent progress in understanding *Bartonella*-induced vascular proliferation, *Curr Opin Microbiol* 6:61, 2003.
5. Drancourt M, Raoult D: Proposed tests for the routine identification of *Rochalimaea* species, *Eur J Clin Microbiol Infect Dis* 12:710, 1993.
6. Fournier PE, Robson J, Zeaiter Z, et al: Improved culture from lymph nodes of patients with cat scratch disease and genotypic characterization of *Bartonella henselae* isolates in Australia, *J Clin Microbiol* 40: 3620, 2002.
7. Garcia-Caceres U, Garcia FU: Bartonellosis: an immunosuppressive disease and the life of Daniel Alcides Carrión, *Am J Clin Pathol* 95(suppl 1):S58, 1991.
8. Giladi M, Avidor B, Kletter Y, et al: Cat scratch disease: the rare role of *Afipia felis*, *J Clin Microbiol* 36:2499, 1998.
9. Greub G, Raoult D: *Bartonella*: new explanations for old diseases, *J Med Microbiol* 51:915, 2002.
10. Jacomo V, Raoult D: Human infections caused by *Bartonella* spp. Parts 1 and 2, *Clin Microbiol Newsletter* 22:1-5 and 9-13, 2000.
11. Jacomo V, Raoult D: Natural history of *Bartonella* infections (an exception to Koch's postulates), *Clin Diagn Lab Immunol* 9:8, 2002.
12. Kordick DL, Hilyard EJ, Hadfield TL, et al: *Bartonella clarridgeiae*: a newly recognized zoonotic pathogen causing inoculation papules, fever, and lymphadenopathy (cat-scratch disease), *J Clin Microbiol* 35:1813, 1997.
13. LaScola B, Raoult D: Culture of *Bartonella quintana* and *Bartonella henselae* from human samples: a 5-year experience (1993-1998), *J Clin Microbiol* 37:1899, 1999.
14. Lawson PA, Collins MD: Description of *Bartonella clarridgeiae* sp nov isolated from the cat of a patient with *Bartonella henselae* septicemia, *Med Microbiol Lett* 5:640, 1996.
15. Maggie RG, Breitschwerdt EB: Potential limitations of the 16S-23S rRNA intergenic region for molecular detection of *Bartonella* species, *J Clin Microbiol* 43:1171, 2005.
16. Manfredi R, Sabbatini S, Chiodo F: Bartonellosis: light and shadows in diagnostic and therapeutic issues, *Clin Microbiol Infect* 11:167, 2004.
17. Maurin M, Raoult D: *Bartonella (Rochalimaea) quintana* infections, *Clin Microbiol Rev* 9:273, 1996.
18. Midani S, Ayoub EM, Anderson B: Cat scratch disease, *Adv Pediatr* 43:397, 1996.

19. Rolain JM, Brouqui P, Koehler JE, et al: Recommendations for treatment of human infections caused by *Bartonella* species, *J Clin Microbiol* 48:1921, 2004.

20. Weisburg WG, Woese CR, Dobson ME, et al: A common origin of rickettsiae and certain plant pathogens, *Science* 230:556, 1985.

21. Weyant, RS, Moss CW, Weaver RE, et al, editors: *Identification of unusual pathogenic gram-negative aerobic and facultatively anaerobic bacteria*, ed 2, Baltimore, 1996, Williams & Wilkins.

Campylobacter, Arcobacter, and Helicobacter

CHAPTER 36

Genera and Species to Be Considered

Campylobacter coli
Campylobacter concisus
Campylobacter curvus
Campylobacter fetus subsp. *fetus*
Campylobacter fetus subsp. *venerealis*
Campylobacter gracilis
Campylobacter hyointestinalis subsp. *hyointestinalis*
Campylobacter jejuni subsp. *doylei*
Campylobacter jejuni subsp. *jejuni*
Campylobacter lari
Campylobacter rectus
Campylobacter showae
Campylobacter sputorum biovar. *sputorum*
Campylobacter upsaliensis
Arcobacter cryaerophilus
Arcobacter butzleri
Helicobacter pylori
Helicobacter cinaedi
Helicobacter fennelliae

Because of morphologic similarities and an inability to recover these organisms using routine laboratory media for primary isolation, the genera *Campylobacter, Arcobacter,* and *Helicobacter* are considered in this chapter. All organisms belonging to these genera are small, curved, motile, gram-negative bacilli. With few exceptions, most of these bacteria also have a requirement for a microaerophilic atmosphere. A recently described organism that morphologically resembles campylobacteria and has similar culture requirements is *Sutterella wadsworthiensis*. This organism has been isolated from human stool specimens as well as from patients with appendicitis, peritonitis, and rectal and perirectal abscesses. Because the pathogenic potential of *S. wadsworthiensis* requires further clarification, this organism is not discussed in this chapter.

CAMPYLOBACTER AND ARCOBACTER

GENERAL CHARACTERISTICS

The taxonomy of the genus *Campylobacter* has been extensively revised, particularly over the past 10 years. Being somewhat closely related to *Campylobacter* spp., *Arcobacter cryaerophilus* and *Arcobacter butzleri* belonged to this genus until 1991. Of significance, the number of campylobacters implicated in human disease has increased as methods of detection have improved. Currently, 17 species and various subspecies are recognized in the genus *Campylobacter. Campylobacter* and *Arcobacter* spp. are relatively slow-growing, fastidious, and, in general, asaccharolytic; those organisms known to cause disease in humans are listed in Table 36-1.

EPIDEMIOLOGY AND PATHOGENESIS

The vast majority of campylobacters appear to be pathogenic and are associated with a wide variety of diseases in humans and other animals. These organisms also demonstrate considerable ecologic diversity. *Campylobacter* spp. are microaerobic inhabitants of the gastrointestinal tracts of various animals, including poultry, dogs, cats, sheep, and cattle, as well as the reproductive organs of several animal species. When fecal samples from chicken carcasses chosen at random from butcher shops in the New York City area were tested for *Campylobacter,* 83% of the samples yielded more than 10^6 colony-forming units per gram of feces. In general, *Campylobacter* spp. produce three syndromes in humans: febrile systemic disease, periodontal disease, and, most commonly, gastroenteritis. *Arcobacter* species appear to be associated with gastroenteritis as well; in one recent study, *A. butzleri* was the fourth most common *Campylobacter*-like organism isolated from stool and was associated with a persistent, watery diarrhea.[23]

Within the genus *Campylobacter, C. jejuni* and *C. coli* are most often associated with infections in humans and are usually transmitted via contaminated food, milk, or water. Outbreaks have been associated with contaminated drinking water and improperly pasteurized milk, among other sources. In contrast to other agents of foodborne gastroenteritis, including *Salmonella* and staphylococci, *Campylobacter* does not multiply in food. Other campylobacters have been isolated from patients who drank untreated water, were compromised in some way, or were returning from international travel. *C. jejuni* subsp. *doylei* has been isolated from children with diarrhea and from gastric biopsies from adults. In developed countries, the majority of *C. jejuni* infections in humans is acquired during the preparation and eating of chicken.[1] Of note, person-to-person transmission of

Table 36-1   ***Campylobacter* and *Arcobacter* spp., Their Source, and Spectrum of Disease in Humans**

Organism	Source	Spectrum of Disease in Humans
C. concisus, C. curvus, C. rectus, C. showae	Humans	Periodontal disease; gastroenteritis (?)
C. gracilis	Humans	Deep-tissue infections of head, neck, and viscera; gingival crevices
C. coli	Pigs, poultry, sheep, bulls, birds	Gastroenteritis*; septicemia
C. jejuni subsp. jejuni	Poultry, pigs, bulls, dogs, cats, birds, and other animals	Gastroenteritis*; septicemia, meningitis; proctitis
C. jejuni subsp. doylei	Humans	Gastroenteritis*; gastritis, septicemia
C. lari	Birds, poultry, other animals; river and seawater	Gastroenteritis*; septicemia; prosthetic joint infection
C. hyointestinalis subsp. hyointestinalis	Pigs, cattle, hamsters, deer	Gastroenteritis
C. upsaliensis	Dogs, cats	Gastroenteritis; septicemia, abscesses
C. fetus subsp. fetus	Cattle, sheep	Septicemia; gastroenteritis; abortion; meningitis
C. fetus subsp. venerealis	Cattle	Septicemia
C. sputorum biovar sputorum	Humans, cattle, pigs	Abscesses, gastroenteritis
Arcobacter cryaerophilus	Pigs, bulls, and other animals	Gastroenteritis*; septicemia
A. butzleri	Pigs, bulls, humans, other animals; water	Gastroenteritis*; septicemia

*Most common clinical presentation.

Campylobacter infections plays only a minor role in the transmission of disease. There is a marked seasonality with the rates of *C. jejuni* infection in the United States; the highest rates of infection occur in late summer and early fall. *Campylobacter* spp., usually *C. jejuni,* have been recognized as the most common etiologic agent of gastroenteritis in the United States. However, this predominance may be a reflection of the laboratory methods used to detect campylobacters.

Motility contributes to the ability of campylobacters to colonize and infect intestinal mucosa. Although infection with *C. jejuni* results in an acute inflammatory enteritis that affects the small intestine and colon, the pathogenesis of infection remains unclear. However, multiplication of organisms in the intestine leads to cell damage and an inflammatory response. Blood and polymorphonuclear neutrophils are often observed in stool specimens from infected patients. Most strains of *C. jejuni* are susceptible to the nonspecific bactericidal activity of normal human serum; this susceptibility probably explains why *C. jejuni* bacteremia is uncommon.[1] Humoral immune responses are important in controlling *C. jejuni* infections and cell-mediated immunity probably plays a role as well.

SPECTRUM OF DISEASE

As previously mentioned, campylobacters can cause either gastrointestinal or extraintestinal infections. Extraintestinal disease, including meningitis, endocarditis, and septic arthritis, is being recognized increasingly, particularly in patients with acquired immunodeficiency syndrome (AIDS) and other immunocompromised individuals. The different campylobacters and the types of infections they cause are summarized in Table 36-1. Gastroenteritis caused by *Campylobacter* spp. is usually a self-limiting illness and does not require antibiotic therapy. Most recently, postinfectious complications following infection with *C. jejuni* have been recognized and include reactive arthritis and most notably, Guillain-Barré syndrome, an acute demyelination (removal of the myelin sheath from a nerve) of the peripheral nerves. Studies indicate that 20% to 40% of patients with this syndrome are infected with *C. jejuni* in the 1 to 3 weeks before the onset of neurologic symptoms.[3]

LABORATORY DIAGNOSIS

Specimen Collection, Transport, and Processing

There are no special requirements for the collection, transport, and processing of clinical specimens for the detection of campylobacters; the two most common specimens submitted to the laboratory are feces (rectal swabs are also acceptable for culture) and blood. If a delay of more than 2 hours is anticipated, stool should be placed either in Cary-Blair transport medium or in campy thio, a thioglycollate broth base with 0.16% agar and vancomycin (10 mg/L), trimethoprim (5 mg/L), cephalothin (15 mg/L), polymyxin B (2500 U/L), and amphotericin B (2 mg/L). Cary-Blair transport medium

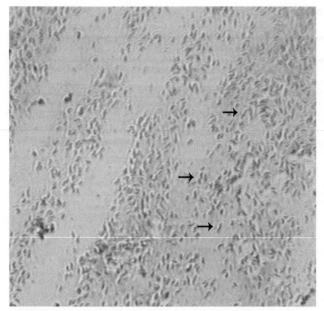

Figure 36-1   Gram stain appearance of *Campylobacter jejuni* subsp. *jejuni* from a colony on a primary isolation plate. Note seagull and curved forms *(arrows)*.

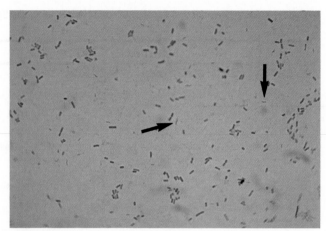

Figure 36-2   Appearance of *Campylobacter jejuni* subsp. *jejuni* in a direct Gram stain of stool obtained from a patient with campylobacteriosis. Arrows point to the "seagull" form.

is also suitable for other enteric pathogens; specimens received in this transport medium should be processed immediately or stored at 4° C until processed.

Direct Detection

Because of their characteristic microscopic morphology, that is, small, curved or seagull-winged, faintly staining, gram-negative rods (Figure 36-1), *Campylobacter* spp. can sometimes be detected by direct Gram stain examination of stool (Figure 36-2). Various molecular assays based on polymerase chain reaction (PCR) amplification may provide an alternative to culture methods for the detection of *Campylobacter* spp. directly in clinical specimens. The finding of *Campylobacter* DNA in stools from a large number of patients with diarrhea suggests that *Campylobacter* spp. other than *C. jejuni* and *C. coli* may account for a proportion of cases of acute gastroenteritis in which no etiologic agent is currently identified.[15]

Cultivation

Stool. To successfully isolate *Campylobacter* spp. from stool, selective media and optimum incubation conditions are critical; culture methods are biased toward the detection of *C. jejuni* and *C. coli* in feces. For optimum recovery, the inoculation of two selective agars is recommended.[7] Because *Campylobacter* and *Arcobacter* spp. have different optimum temperatures, two sets of selective plates should be incubated, one at 42° C and one at 37° C. Table 36-2 describes the selec-

tive plating media and incubation conditions required to recover *Campylobacter* spp. from stool.

A filtration method can also be used in conjunction with a nonselective medium to recover *Campylobacter* and *Arcobacter* spp. A filter (0.65-mm pore-size cellulose acetate) is placed on the agar surface, and a drop of stool is placed on the filter. The plate is incubated upright. After 60 minutes at 37° C, the filter is removed and the plates are reincubated in a microaerobic atmosphere. Because organisms are able to move through the filter, the organism is effectively removed from contaminating stool flora and colonies are produced on the agar surface. *C. concisus, A. butzleri, A. cryaerophilus, H. cinaedi* were isolated after 5 to 6 days of incubation by use of the filter technique.[8] Finally, an enrichment broth for the recovery of *Arcobacter* or *Campylobacter* species from stool has also been employed.[15,23]

Blood. *Campylobacter* spp. that cause septicemia grow in most blood culture media, although they may require as long as 2 weeks for growth to be detected. Subcultures from broths must be incubated in a microaerobic atmosphere or the organisms will not multiply. Turbidity is often not visible in blood culture media; therefore, blind subcultures or microscopic examination using acridine orange stain may be necessary. The presence of *Campylobacter* spp. in blood cultures is detected effectively by carbon dioxide (CO_2) monitoring. Isolation from sources other than blood or feces is extremely rare but is ideally accomplished by inoculating the material (minced tissue, wound exudate) to a nonselective blood or chocolate agar plate and incubating the plate at 37° C in a CO_2-enriched, microaerobic atmosphere. (Selective agars containing a cephalosporin, rifampin, and polymyxin B may inhibit growth of some strains and should not be used for isolation from normally sterile sites.)

Table 36-2   Selective Media and Incubation Conditions to Recover *Campylobacter* and *Arcobacter* spp. from Stool Specimens

Organism	Primary Plating Media	Incubation Conditions
C. jejuni *C. coli*	Modified Skirrow's media: Columbia blood agar base, 7% horse-lysed blood, and antibiotics (vancomycin, trimethoprim, and polymyxin B) Campy-BAP: *Brucella* agar base with antibiotics (trimethoprim, polymyxin B, cephalothin, vancomycin, and amphotericin B) and 10% sheep blood Blood-free, charcoal-based selective medium: Columbia base with charcoal, hemin, sodium pyruvate, and antibiotics (vancomycin, cefoperazone, and cyclohexamide) Modified charcoal cefoperazone deoxycholate agar (CCDA) Semisolid motility agar: Mueller-Hinton broth II, agar, cefoperazone, and trimethoprim lactate Campy-CVA: *Brucella* agar base with antibiotics (cefoperazone, vancomycin, and amphotericin B) and 5% sheep blood	42° C under microaerophilic conditions* for 72 hr
C. fetus subsp. *fetus*† *C. jejuni* subsp. *doylei* *C. upsaliensis* *C. lari* *C. hyointestinalis*	Modified Skirrow's media Blood-free charcoal-based selective media Campy-CVA CCDA Semisolid motility agar	37° C under microaerophilic conditions for at least 72 hr up to 7 days‡
A. cryaerophilus, A. butzleri	Campy-CVA	37° C under microaerophilic conditions§ for 72 hr

*Atmosphere can be generated in several ways, including commercially produced, gas-generating envelopes to be used with plastic bags or jars. Evacuation and replacement in plastic bags or anaerobic jars with an atmosphere of 10% CO_2, 5% O_2, and the balance of nitrogen (N_2) is the most cost-effective method, although it is labor intensive.
†All these organisms are susceptible to cephalothin.
‡*C. upsaliensis* will grow at 42° C but not on cephalothin-containing selective agar.
§*A. cryaerophilus* does not require microaerophilic conditions.

Approach to Identification

Plates should be examined for characteristic colonies, which are gray to pinkish or yellowish gray and slightly mucoid-looking; some colonies may exhibit a tailing effect along the streak line (Figure 36-3). However, other colony morphologies are also frequently seen, depending on the media used. Suspicious-looking colonies seen on selective media incubated at 42° C may be presumptively identified as *Campylobacter* spp., usually *C. jejuni* or *C. coli,* with a few simple tests. A wet preparation of the organism in broth may be examined for characteristic darting motility and curved forms on Gram stain. Almost all the pathogenic *Campylobacter* spp. are oxidase- and catalase-positive. For most laboratories, reporting of such isolates from feces as *"Campylobacter* spp." should suffice.

Most *Campylobacter* spp. are asaccharolytic, are unable to grow in 3.5% NaCl, although strains of *Arcobacter* appear more resistant to salt, and except for *Arcobacter cryaerophilus,* are unable to grow in air. Growth in 1% glycine is variable. Susceptibility to nalidixic acid and cephalothin, an important differential characteristic among species (Table 36-3), is determined by inoculating a 5% sheep blood or Mueller-Hinton agar plate with a McFarland 0.5 turbidity suspension of the organism as for agar disk diffusion susceptibility testing, placing 30-mg disks on the agar surface and incubating

Figure 36-3   Colonies of *Campylobacter jejuni* after 48 hours of incubation on selective medium in a microaerophilic atmosphere.

microaerobically at 37° C. Other tests useful for identifying these species are the rapid hippurate hydrolysis test, production of hydrogen sulfide (H_2S) in triple sugar iron agar butts, nitrate reduction, and hydrolysis of indoxyl acetate.[17] Indoxyl acetate disks are available commercially. Cellular fatty acid analysis can also help differentiate among species. This method is not available to most clinical microbiology laboratories. Several commercial products are available for species identification,

Table 36-3 Differential Characteristics of Clinically Relevant *Campylobacter*, *Arcobacter*, and *Helicobacter* spp.

Genus and Species	Growth at 25° C	Growth at 42° C	Hipprate Hydrolysis	Catalase	H₂S in Triple Sugar Iron Agar	Indoxyl Acetate Hydrolysis	Nitrate to Nitrite	Susceptible to 30-µg Disk Cephalothin	Nalidixic Acid
C. coli	−	+	−	+	−	+	+	−	+
C. concisus	−	+	−	−	+	−	+	−	−
*C. curvus**	−	+	−	−	+	+	+	ND	+
C. fetus subsp. *fetus*	+	−/+	−	+	−	+	+	+	−
C. hyointestinalis	+/−	−	−	+	+	−	+	+	+
C. jejuni subsp. *jejuni*	−	+	+	+	−	+	+	−	+
C. jejuni subsp. *doylei*	−	+/−	+	+/− or weak +	−	+	−	+	+
C. lari	−	+	−	+	−	−	+	−	−
*C. rectus**	−	Slight +	−	−	+	+	+	ND	+
C. sputorum	−	+	−	−/+	+	−	+	+	−/+
C. upsaliensis	−	+	−	−/weak +	−	+	+	+	+
A. butzleri†	+	−	−	−/weak +	−	+	+	−/+	+/−
A. cryaerophilus‡	+	−	−	+/−	−	+	+	−/+	+/−
H. cinaedi	−	−/+	−	+	−	−/+	+	+/−	+
H. fennelliae	−	−	−	+	−	+	−	+	+
H. pylori§	−	+	−	+	−	−	+/−	+	−

ND, test not done; +, most strains positive; −, most strains negative; +/−, variable (more often positive); −/+, variable (more often negative).

*Anaerobic, not microaerobic.

†Grows at 40° C.

‡Aerotolerant, not microaerobic; except for a few strains, *A. cryaerophilus* cannot grow on MacConkey agar, whereas *A. butzleri* grows on MacConkey agar.

§Strong and rapid positive urease.

including particle agglutination methods and nucleic acid probes.

Molecular assays using PCR-based amplification of the 16S rRNA gene and direct sequencing of the PCR product have also successfully identified the majority of *Campylobacter* species.[11] These assays discriminate closely related taxa such as *Arcobacter* or *Helicobacter* from *Campylobacter* species. Finally, another approach using 16S-23S PCR-based amplification with a DNA probe colorimetric membrane assay proved to rapidly detect and identify *Campylobacter* in stool specimens.[15]

Serodiagnosis

Serodiagnosis is not widely applicable for the diagnosis of infections caused by these organisms.

ANTIMICROBIAL SUSCEPTIBILITY TESTING AND THERAPY

Because susceptibility tests for *Campylobacter* spp. are not standardized, susceptibility testing of isolates is not routinely performed. *C. jejuni* and *C. coli* are susceptible to many antimicrobial agents, including macrolides, tetracyclines, aminoglycosides, and quinolones.[22] Erythromycin is the drug of choice for patients with more severe gastroenteritis (severe dehydration, bacteremia), with ciprofloxacin as an alternative drug.[1] Until a few years ago, if antibiotic therapy was indicated for *Campylobacter* infection, fluoroquinolones were considered the drugs of choice; however, a rapidly increasing proportion of *Campylobacter* strains worldwide have been found to be fluoroquinolone-resistant.[1,3] **Parenteral therapy** (not taken through the alimentary canal but by some other route such as intravenous) is used to treat systemic infections.

PREVENTION

No vaccines are available for *Campylobacter* spp. Because many infections caused by *Campylobacter* spp. are acquired by ingesting contaminated foodstuffs or water, all foods derived from animal sources, particularly poultry, should be thoroughly cooked. All milk should be pasteurized and drinking water chlorinated.

Table 36-4   Genes and Their Possible Role in Enhancing Virulence of *H. pylori*

Gene	Possible Role
vacA	Secreted product (VacA) that causes a variety of changes in epithelial cells such as vacuole formation, decreased apoptosis, and loosening of cell junctions.
cagA	Marker for pathogenicity island that encodes a type IV secretion system that transfers CagA proteins into host cells.
babA	Encodes outer membrane protein that mediates adherence to a specific blood group antigen on the surface of gastric epithelial cells.
iceA	Presence appears to be associated with peptic ulcer disease in some populations.

Care must be taken during food preparation to prevent cross-contamination from raw poultry to other food items.

HELICOBACTER

GENERAL CHARACTERISTICS

In 1983, spiral-shaped organisms resembling campylobacteria were isolated from the human stomach; these organisms were named *Campylobacter pylori*. Based on many studies, the genus *Helicobacter* was established in 1989 and *C. pylori* was renamed *Helicobacter pylori*. At least 22 species are now included in this genus, the majority of which colonize mammalian stomachs or intestines. The genus *Helicobacter* consists mostly of curved, microaerophilic, gram-negative rods, with most species having strong urease activity. *Helicobacter* spp. isolated from humans include *H. pylori, H. cinaedi, H. fennelliae, H. heilmannii* (formerly known as *Gastrospirillum hominis*), *H. westmeadii, H. canis, H. canadensis* sp. nov., *H. pullorum,* and "*H. rappini*" (formerly known as "*Flexispira rappini*"). Because *H. pylori, H. cinaedi,* and *H. fennelliae* are significant human pathogens, only these species are addressed.

EPIDEMIOLOGY AND PATHOGENESIS

Helicobacter pylori's primary habitat is the human gastric mucosa. The organism is found worldwide. Although acquired early in life in underdeveloped countries, its exact mode of transmission is unknown. An oral-oral, fecal-oral, or a common environmental source have been proposed as possible routes of transmission, with transmission of *H. pylori* primarily occurring in families.[12] Recent findings suggest that mother-to-child transmission is the most probable cause of intrafamilial spread.[14] In industrialized nations, antibody surveys show that approximately 50% of adults older than age 60 are infected. Gastritis incidence increases with age. *H. pylori* has occasionally been cultured from feces and dental plaque, thereby suggesting a fecal-oral or oral-oral transmission.[6]

The habitat for *H. cinaedi* and *H. fennelliae* appears to be the gastrointestinal tract of humans, and these organisms may be part of the resident flora; hamsters have also been proposed as a reservoir for *H. cinaedi*. Although the epidemiology of these organisms is not clearly delineated, these two bacterial agents are sexually transmitted among homosexual men.

H. pylori colonizes the mucous layer of the antrum and fundus of the stomach but does not invade the epithelium. By virtue of this ability to colonize the gastric mucosa, persist despite the host immune defense, and cause host tissue damage, *H. pylori* is an effective and significant bacterial pathogen. Motility allows *H. pylori* to escape the acidity of the stomach and burrow through and colonize the gastric mucosa in close association with the epithelium. The study of specific virulence determinants of *H. pylori* is ongoing. For example, a protein called *CagA* is injected by *H. pylori* into human gastric epithelial cells. The protein then subsequently affects host cell gene expression such that cytokine release, cell structure, and interactions with neighboring cells are altered thereby enabling *H. pylori* to more successfully colonize the gastric epithelium. It is clear that persons with serologic evidence of carrying *cag*-positive strains are at increased risk of developing both peptic ulcer disease and gastric carcinoma.[2,20] Urease enzyme likely plays a significant role in the survival and growth of *H. pylori* in the stomach by creating an alkaline microenvironment. Other possible virulence determinants include adhesins for colonization of mucosal surfaces, mediators of inflammation, and a vacuolating cytotoxin that causes damage to host cells (Table 36-4).[4,5,17,20] Although *H. pylori* is noninvasive, untreated infection lasts for the life of the host and persists despite a significant host immune response.

SPECTRUM OF DISEASE

H. cinaedi and *H. fennelliae* cause proctitis, enteritis, and sepsis in homosexual men. Septic shock caused by *H. fennelliae* was reported in a non–HIV-infected heterosexual immunocompromised patient. *H. cinaedi* has also been reported to cause septicemia, cellulitis, and meningitis in immunocompromised patients. *H. pylori* causes gastritis, peptic ulcer disease, and gastric cancer.[4,6] However, most individuals tolerate the presence of *H. pylori* for decades with few, if any, symptoms.

LABORATORY DIAGNOSIS

Specimen Collection, Transport, and Processing

There are no special requirements for the collection, transport, or processing of stool or blood specimens for *H. cinaedi* and *H. fennelliae*. Tissue biopsy material of the stomach for detection of *H. pylori* should be placed directly into transport media such as Stuart's transport medium to prevent drying. Specimens for biopsy may be refrigerated up to 24 hours before processing; tissues should be minced and gently homogenized.

Direct Detection

The Warthin-Starry or other silver stain and Giemsa stains are used by pathologists for examination of biopsy specimens. One potential problem is that of sampling error. Squash preparations of biopsy material can be Gram-stained with good results; the 0.1% basic fuchsin counterstain enhances recognition of the bacteria's typical morphology.

Presumptive evidence of the presence of *H. pylori* in biopsy material may be obtained by placing a portion of crushed tissue biopsy material directly into urease broth or onto commercially available urease agar kits. A positive test is considered indicative of the organism's presence. Another noninvasive indirect test to detect *H. pylori* is the urea breath test. This test relies on the presence of *H. pylori* urease. The patient ingests radioactively labeled (^{13}C) urea, and if infection is present, the urease produced by *H. pylori* hydrolyzes the urea to form ammonia and labeled bicarbonate that is exhaled as CO_2; the labeled CO_2 is detected by either a scintillation counter or a special spectrometer. This test has excellent sensitivity and specificity. Two enzyme immunoassay *H. pylori* stool antigen tests (Premier Platinum HpSA, Meridian Diagnostics, Inc., Cincinnati, Ohio; FemtoLab *H. pylori,* Connex, Martinsried, Germany) and a one-step immunochromatographic assay using monoclonal antibodies (Immunocard STAT! HpSA, Meridian Bioscience Europe) have been introduced to directly detect *H. pylori*. Finally, a variety of molecular methods have been developed to not only directly detect *H. pylori* in clinical specimens but also determine strains and host genotype characteristics, bacterial density in the stomach, and antimicrobial resistance.[21]

Cultivation

Stool specimens submitted for culture of *H. cinaedi* and *H. fennelliae* are plated to selective media used for campylobacter isolation but without cephalothin such as Campy-CVA. For the recovery of *H. pylori* from tissue biopsy specimens including gastric antral biopsies, nonselective agar media, including chocolate agar and *Brucella* agar with 5% sheep blood, have been useful. Selective agar, such as Skirrow's and modified Thayer-Martin agars, also support growth. Recently, the combination of a selective agar (Columbia agar with an egg yolk emulsion, supplements, and antibiotics) and a nonselective agar (modified chocolate agar with Columbia agar, 1% Vitox, and 5% sheep blood) was reported as the optimal combination for recovering *H. pylori* from antral biopsies.[19] Incubation up to 1 week in a humidified, microaerobic atmosphere at 35° to 37° C may be required before growth of this human pathogen is visible.

Approach to Identification

Colonies of *Helicobacter* spp. may require 4 to 7 days of incubation before small, translucent, circular colonies are observed. Organisms are identified presumptively as *Helicobacter pylori* by the typical cellular morphology and positive results for oxidase, catalase, and rapid urease tests. Definitive identification of *H. pylori, H. cinaedi,* and *H. fennelliae* is accomplished using a similar approach to that for *Campylobacter* spp. (see Table 36-3).

Serodiagnosis

Another approach to *H. pylori* diagnosis is serologic testing. Numerous serologic enzyme-linked immunoassays (EIAs) designed to detect immunoglobulin G (IgG) and immunoglobulin A (IgA) antibodies to *H. pylori* are commercially available. Reported performance of these assays varies as a result of the reference method used to confirm *H. pylori* infection, antigen source for the assay, and the population studied.[16] In addition to variability in assay performance, the clinical utility of these assays has not been defined and may not differentiate between active vs. past *H. pylori* infections in all cases. However, one study did confirm the role of seroconversion in determining a cure of *H. pylori* infection.[9]

ANTIMICROBIAL SUSCEPTIBILITY TESTING AND THERAPY

Except for metronidazole and clarithromycin, most laboratory susceptibility assays do not always predict clinical outcome. Routine testing of *H. pylori* isolates' susceptibility to metronidazole is recommended using agar or broth dilution methods or possibly the E test.[13,18]

Therapy for *H. pylori* infection is problematic. *H. pylori* readily becomes resistant if metronidazole, clarithromycin, azithromycin, rifampin, or ciprofloxacin is given as a single agent.[10] Current regimens usually use triple-drug therapy that usually includes metronidazole, a bismuth salt, and either amoxicillin or tetracycline. An alternative and more simple regimen for patients with metronidazole-resistant strains is omeprazole or lansoprazole (proton pump inhibitors that cause rapid symptom relief while working synergistically with the antibiotics) and amoxicillin or clarithromycin. Relapses occur often. *Helicobacter* spp. associated with enteritis and proctitis may respond to quinolones, but ideal therapy has not been established.

PREVENTION

No vaccines are currently available that are directed against *H. pylori*. However, several vaccines are being developed that use numerous strategies.

Case Study

A 10-year-old boy became ill a few days after a fourth of July picnic where fried chicken was served. He complained of diarrhea, abdominal pain, and fever. Symptoms continued over the next week and he was seen at the local clinic. Blood was found in his stool and cultures were ordered. He was treated with ampicillin but switched to azithromycin (a macrolide similar to erythromycin) for 5 days when the cultures results were reported.

QUESTIONS

1. At 42° C in a microaerophilic environment, water droplet–type colonies were seen that were oxidase-positive and strongly catalase-positive. A Gram stain showed gram-negative rods with "seagull" shapes. What rapid test can be done to confirm the identity of this bacterium?
2. What testing is needed if the hippurate is negative?
3. How do you think our patient became infected?
4. Why can't the nalidixic acid disk be used in the identification of *C. jejuni/coli*?

REFERENCES

1. Allos BM: *Campylobacter jejuni* infections: update on emerging issues and trends, *Clin Infect Dis* 32:1201, 2001.
2. Blaser MJ: The biology of *cag* in the *Helicobacter pylori*-human interaction, *Gastroenterology* 128:1512, 2005.
3. Butzler JP: *Campylobacter,* from obscurity to celebrity, *J Clin Microbiol Infect* 10:868, 2004.
4. Crowe SE: *Helicobacter* infection, chronic inflammation, and the development of malignancy, *Curr Opin Gastroenterol* 21:32, 2005.
5. Day AS, Jones NL, Lynetl JT, et al: *cagE* is a virulence factor associated with *Helicobacter pylori*-induced duodenal ulceration in children, *J Infect Dis* 181:1370, 2000.
6. Dunn BE, Cohen H, Blaser MJ: *Helicobacter pylori, Clin Microbiol Rev* 10:720, 1997.
7. Endtz HP, Ruijs GJ, van Klingeren B, et al: Comparison of six media, including a semisolid agar, for the isolation of various *Campylobacter* species from stool specimens, *J Clin Microbiol* 29:1007, 1991.
8. Engberg J, On SL, Harrington CS, et al: Prevalence of *Campylobacter, Arcobacter, Helicobacter* and *Sutterella* spp. in human fecal samples as estimated by a reevaluation of isolation methods for campylobacters, *J Clin Microbiol* 38:286, 2000.
9. Feldman M, Cryer B, Lee E, et al: Role of seroconversion in confirming cure of *Helicobacter pylori* infection, *JAMA* 280:363, 1998.
10. Goodwin CS: Antimicrobial treatment of *Helicobacter pylori* infection, *Clin Infect Dis* 25:1023, 1997.
11. Gorkiewicz G, Feierl G, Schober C, et al: Species-specific identification of campylobacters by partial 16S rRNA gene sequencing, *J Clin Microbiol* 41:2537, 2003.
12. Han S, Zschausch H, Meyer HW, et al: *Helicobacter pylori:* clonal population structure and restricted transmission within families revealed by molecular typing, *J Clin Microbiol* 38:3646, 2000.
13. Henriksen TH, Brorson Ö, Schöyen R, et al: A simple method for determining metronidazole resistance of *Helicobacter pylori, J Clin Microbiol* 35:1424, 1997.
14. Konno M, Fujii N, Yakota S, et al: Five year follow-up study of mother-to-child transmission of *Helicobacter pylori* infection detected by a random amplified polymorphic DNA fingerprinting method, *J Clin Microbiol* 43:2246, 2005.
15. Maher M, Finnegan C, Collins E, et al: Evaluation of culture methods and a DNA probe-based PCR assay for detection of *Campylobacter* species in clinical specimens of feces, *J Clin Microbiol* 41:2980, 2003.
16. Marchildon PA, Ciota LM, Zamaniyan FZ, et al: Evaluation of three commercial enzyme immunoassays compared with the ^{13}C urea breath test for detection of *Helicobacter pylori* infection, *J Clin Microbiol* 34:1147, 1996.
17. On SL: Identification methods for campylobacters, helicobacters, and related organisms, *Clin Microbiol* 9:405, 1996.
18. Pavicic MJ, Namavar F, Verboom T, et al: In vitro susceptibility of *Helicobacter pylori* to several antimicrobial combinations, *Antimicrob Agents Chemother* 37:1184, 1993.
19. Piccolomini R, Di Bonaventura G, Catamo G, et al: Optimal combination of media for primary isolation of *Helicobacter pylori* from gastric biopsy specimens, *J Clin Microbiol* 35:1541, 1997.
20. Rieder G, Fischer W, Haas R: Interaction of *Helicobacter pylori* with host cells: function of secreted and translocated molecules, *Curr Opin Microbiol* 8:67, 2005.
21. Simala-Grant J, Taylor DE: Molecular biology methods for the characterization of *Helicobacter pylori* infections and their diagnosis, *APMIS* 112:886, 2004.
22. Tajada P, Gomez-Graces JL, Alos JI, et al: Antimicrobial susceptibilities of *Campylobacter jejuni* and *Campylobacter coli* to 12 β-lactam agents and combinations with β-lactamase inhibitors, *Antimicrob Agents Chemother* 40:1924, 1996.
23. Vandenberg O, Dediste A, Houf K, et al: *Arcobacter* species in humans, *Emerg Infect Dis* 10:1863, 2004.

CHAPTER 37 *Legionella*

This chapter addresses organisms that will not grow on routine primary plating media and belong to the genus *Legionella*. *Legionella* is the causative agent of **legionnaires' disease,** a febrile and pneumonic illness with numerous clinical presentations. *Legionella* was discovered in 1976 by scientists at the Centers for Disease Control and Prevention (CDC) who were investigating an epidemic of pneumonia among Pennsylvania State American Legion members attending a convention in Philadelphia. There is retrospective serologic evidence of *Legionella* infection as far back as 1947.

GENERAL CHARACTERISTICS

There is only one genus, *Legionella*, within the family *Legionellaceae*. All members of this genus are faintly staining, thin, gram-negative bacilli that require a medium supplemented with iron and L-cysteine, and buffered to pH 6.9 for optimum growth. The overwhelming majority of *Legionella* spp. are motile. As of this writing, 45 species belong to this genus. Nevertheless, the organism *Legionella pneumophila* predominates as a human pathogen within the genus and consists of 15 serotypes. In approximately decreasing order of clinical importance are *L. pneumophila* serotype 1 (about 50% of the cases of legionnaires' disease), *L. pneumophila* serotype 6, *L. micdadei*, *L. dumoffii*, *L. anisa*, and *L. feeleii*.[14] Of note, many species of *Legionella* have only been isolated from the environment or recorded as individual cases. To date, 19 species of *Legionella* are documented as human pathogens in addition to *L. pneumophila*.[8] Box 37-1 is an abbreviated list of some of the species of *Legionella*.

EPIDEMIOLOGY AND PATHOGENESIS

EPIDEMIOLOGY

Legionellae are ubiquitous and widely distributed in the environment. As a result, most individuals are exposed to *Legionella* spp.; however, few develop symptoms. In nature, legionellae are found primarily in aquatic habitats and thrive at warmer temperatures; these bacteria are capable of surviving extreme ranges of environmental conditions for long periods; studies have shown that *L. pneumophila* can survive for up to 14 months in water with only a slight loss in viability. *Legionella* spp. have been isolated from the majority of natural water sources investigated, including lakes, rivers, and marine waters, as well as moist soil.[8] Organisms are also widely distributed in man-made facilities, including air-conditioning ducts and cooling towers; potable water; large, warm-water plumbing systems; humidifiers; whirlpools; and technical-medical equipment in hospitals.

Legionella infections are acquired exclusively from environmental sources; no person-to-person spread has been documented.[2] Inhalation of infectious aerosols (1 to 5 mm diameter) is considered the primary means of transmission. Exposure to these aerosols can occur in the workplace or in industrial or nosocomial settings; for example, nebulizers filled with tap water and showers have been implicated. Legionnaires' disease occurs in sporadic, endemic, and epidemic forms. The incidence of disease varies greatly and appears to depend on the geographic area, but it is estimated that *Legionella* spp. cause less than 1% to 5% of cases of pneumonia.

PATHOGENESIS

Legionella spp. can infect and multiply within some species of free-living amoebae (*Hartmannella*, *Acanthamoeba*, and *Naegleria* spp.), as well as within *Tetrahymena* spp., a ciliated protozoa, or within **biofilms** (well-organized microcolonies of bacteria usually enclosed in polymer matrices that are separated by water channels that remove wastes and deliver nutrients).[6] This ability of legionellae contributes to its survival in the environment. In addition, *L. pneumophila* exists in two well-defined, morphologically distinct forms in Hela cells: (1) a highly differentiated, cystlike form that is highly infectious, metabolically dormant, and resistant to antibiotics and detergent-mediated lysis and (2) a replicative intracellular form that is ultrastructurally similar to agar-grown bacteria.[4] The existence of this cystlike form may account for the ability of *L. pneumophila* to survive for long periods between hosts (amoebae or humans).

BOX 37-1	Some *Legionella* spp. Isolated from Humans and Environmental Sources

SPECIES ISOLATED FROM HUMANS

L. pneumophila, serotypes 1-15
L. micdadei
L. bozemanii
L. dumoffii
L. feelei
L. gormanii
L. hackeliae
L. longbeachae
L. oakridgensis
L. wadsworthii

SPECIES ISOLATED FROM ENVIRONMENT ONLY

L. cherrii
L. erythra
L. gratiana
L. jamestowniensis
L. brunensis
L. fairfieldensis
L. santicrucis

BOX 37-2	Examples of *L. pneumophila* Factors Crucial for Intracellular Infection

- Heat shock protein 60
- Outer membrane protein
- Macrophage infectivity potentiator
- Genes encoding for the type II secretion systems required for intracellular growth
- Type IV pili
- Flagella
- *Dot/icm* type IV secretion system

Although the exact mechanisms by which *L. pneumophila* causes disease are not totally delineated, its ability to avoid destruction by the host's phagocytic cells plays a significant role in the disease process. *L. pneumophila* is considered a facultative intracellular pathogen. Following infection, organisms are taken up by phagocytosis into alveolar macrophages, where they survive and replicate within a specialized, membrane-bound vacuole by resisting acidification and evading fusion with lysosomes; it is still unknown how *Legionella* prevent vacuole acidification.[11] This sequestering of legionellae also makes it difficult to deliver and accumulate antimicrobials within macrophages. Of significance, studies have shown that although certain antimicrobials can penetrate the macrophage and inhibit bacterial multiplication, *L. pneumophila* is not killed and, when drugs are removed, the organism resumes replicating.[1] Therefore a competent cell-mediated immune response is also important in recovery from legionella infections. Humoral immunity appears to play an insignificant role in the defense against this organism.

In eukaryotic cells, most proteins that are secreted or transported inside vesicles to other cellular compartments are synthesized at the endoplasmic reticulum (EPR). Many bacterial pathogens use secretion systems as a part of how they cause disease. *L. pneumophila* possess genes that are able to "trick" eukaryotic cells into transporting them to the EPR; these virulence genes are called *dot* (defective organelle trafficking) or *icm* (intracellular multiplication). This dot/icm secretion system in *L. pneumophila* consists of 23 genes and is a type IV secretion system.[5,11,12] Bacterial type IV secretion systems are bacterial devices that deliver macromolecular

molecules such as proteins across and into cells. After entry but before bacterial replication, *L. pneumophila*, residing in a membrane-bound vacuole, is surrounded by a ribosome-studded membrane derived from the host cell's EPR and mitochondria. Thus, by exploiting host cell functions, *L. pneumophila* is able to gain access to the lumen of the EPR, which supports its survival and replication because the organelle is rich in peptides. A number of additional bacterial factors have also been identified as crucial for intracellular infection; some of these are listed in Box 37-2.

Finally, several cellular components and extracellular products of *L. pneumophila,* such as an extracellular cytotoxin that impairs the ability of phagocytic cells to use oxygen and various enzymes (e.g., phospholipase C), have been purified and proposed as virulence factors. However, their exact role in the pathogenesis of legionella infections is not completely clear.

SPECTRUM OF DISEASE

Legionella spp. are associated with a spectrum of clinical presentations, ranging from asymptomatic infection to severe, life-threatening diseases. Serologic evidence exists for the presence of asymptomatic disease, because many healthy people surveyed possess antibodies to *Legionella* spp. The following are the three primary clinical manifestations[8]:

- Pneumonia with a case fatality rate of 10% to 20% (referred to as legionnaires' disease)
- Pontiac fever, which is a self-limited, nonfatal respiratory infection
- Other infection sites, such as wound abscesses, encephalitis, or endocarditis

Individuals at risk for pneumonia are those who are immunocompromised, older than age 60, or heavy smokers. The clinical manifestations following infection with a particular species are primarily caused by differences in the host's immune response and perhaps by inoculum size; the same *Legionella* spp. gives rise to different expressions of disease in different individuals.

There are a number of bacteria that grow only within amoebae and are closely related phylogenetically based on 16S rRNA gene sequencing to *Legionella* species; these organisms are referred to as "*Legionella*-like amoeba pathogens" (LLAPs). Several LLAPs have been assigned to the *Legionella* genus. One LLAP has been isolated from the sputum of a patient with pneumonia after the specimen was incubated with the amoeba *Acanthamoeba polyphaga*. Serologic surveys of patients with community-acquired pneumonia suggest LLAP may be occasional human pathogens.[7]

LABORATORY DIAGNOSIS

SPECIMEN COLLECTION AND TRANSPORT

Specimens from which *Legionella* can be isolated include respiratory tract secretions of all types, including sputum and pleural fluid; other sterile body fluids, such as blood; and lung, transbronchial, or other biopsy material. Because sputum from patients with Legionnaires' disease is usually nonpurulent and may appear bloody or watery, the grading system used for screening sputum for routine cultures is not applicable.[9] Patients with legionnaires' disease usually have detectable numbers of organisms in their respiratory secretions, even for some time after antibiotic therapy has been initiated. If the disease is present, the initial specimen is often likely to be positive. However, additional specimens should be processed if the first specimen is negative and suspicion of the disease persists. Pleural fluid has not yielded many positive cultures in studies performed in several laboratories, but it may contain organisms. Specimens should be transported without holding media, buffers, or saline, which may inhibit the growth of *Legionella*. The organisms are actually very hardy and are best preserved by maintaining specimens in a small, tightly closed container to prevent desiccation and transporting them to the laboratory within 30 minutes of collection. If a longer delay is anticipated, specimens should be refrigerated. If one cannot ensure that specimens will remain moist, 1 mL of sterile broth may be added.

SPECIMEN PROCESSING

All specimens for *Legionella* culture should be handled and processed in a class II biological safety cabinet (BSC). When specimens from nonsterile body sites are submitted for culture, selective media and/or treatment of the specimen to reduce the numbers of contaminating organisms is proposed. Brief treatment of sputum specimens with hydrochloric acid before culture has been shown to enhance the recovery of legionellae.[3] However, this technique is time-consuming and is now only recommended on specimens from patients with cystic fibrosis.[10] Respiratory secretions may be held for up to 48 hours at 5° C before culture; if culturing is delayed longer, then the specimen may be frozen.[10]

Tissues are homogenized before smears and cultures are performed, and clear, sterile body fluids are centrifuged for 30 minutes at 4000× g. The sediment is then vortexed and used for culture and smear preparation. Blood for culture of *Legionella* probably should be processed with the lysis-centrifugation tube system (Isolator; Wampole Laboratories, Cranbury, NJ) and plated directly to buffered charcoal-yeast extract (BCYE) agar. Specimens collected by bronchoalveolar lavage are quite dilute and therefore should be concentrated at least tenfold by centrifugation before culturing.

DIRECT DETECTION METHODS

Several laboratory methods are used to detect *Legionella* spp. directly in clinical specimens.

Stains

Because of their faint staining, *Legionella* spp. are not usually detectable directly in clinical material by Gram stain. Organisms can be observed on histologic examination of tissue sections using silver or Giemsa stains.

Antigens

One approach to direct detection of legionellae in clinical specimens is the direct immunofluorescent antibody (DFA) test of respiratory secretions. Polyclonal and monoclonal antisera conjugated with fluorescein are available from several commercial suppliers. Specimens are first tested with pools of antisera containing antibodies to several serotypes of *L. pneumophila* or several *Legionella* spp. Those that exhibit positive results are then reexamined with specific conjugated antisera. One reagent made by Genetic Systems (Seattle, Wash) is a monoclonal antibody directed against a cell wall protein common to *L. pneumophila*. The manufacturer's directions should be followed explicitly, and material from commercial systems should never be divided and used separately. Laboratories should decide which serotypes to test for routinely based on the prevalence of isolates in their geographic area. The sensitivity of the DFA test ranges from 25% to 75%, and its specificity is greater than 95%.[13] If positive, organisms appear as brightly fluorescent rods (Figure 37-1). Of importance, cultures always must be performed, because *Legionella* spp. or serotypes not included in the antisera pool can be recovered.

Rapid detection of *Legionella* antigen in urine and other body fluids has been accomplished by radio-immunoassay (RIA), enzyme immunoassay (EIA), and immunochromatography; the RIA is no longer manufactured. Antigen may be present in the prodromal

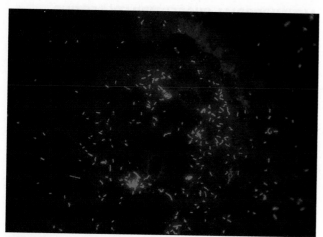

Figure 37-1 Fluorescent antibody-stained *Legionella pneumophila*.

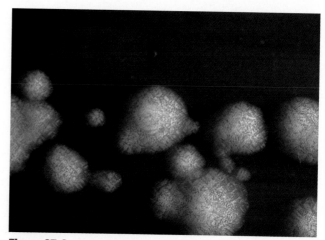

Figure 37-2 Colonies of *Legionella pneumophila* on buffered charcoal-yeast extract agar.

period and by 3 days after the onset of symptoms. A major drawback with the immunochromatographic urine antigen assay is that it only detects the presence of antigen of *L. pneumophila* serogroup 1. One *Legionella* EIA (Biotest, Dreieich, Germany) is marketed by the manufacturer as for the detection of all serotypes of *L. pneumophila* and some other species in that a broadly cross-reactive antibody is used. At this writing, the relative sensitivity of this test in detecting infections caused by organisms other than *L. pneumophila* serogroup 1 is unknown. Of note, a recent study demonstrated that the clinical utility for the diagnosis of legionnaires' disease of two EIAs differed depending on the category of infection being investigated. Sensitivity (about 45%) for both EIAs were significantly lower for nosocomial cases than for either community-acquired or travel-associated ones.[6] These assays have a sensitivity of 80% in their ability to detect infection caused by *L. pneumophila* serogroup 1 and are highly specific, although nonspecific false-positive results do occur. Boiling urine for 5 minutes and concentrating urine by centrifugation help increase assay specificity and sensitivity, respectively. Of importance, because bacterial antigen may persist in urine for days to weeks after initiation of antibiotic therapy, these assays may be positive when other diagnostic tests are negative.

Nucleic Acid

Although not yet commercially available, direct detection of *Legionella* nucleic acid by conventional and real-time polymerase chain reaction (PCR) has the potential to offer rapid results and increased sensitivity on respiratory samples over current methods[13]; of significance, PCR assays can detect all *Legionella* spp., not just *L. pneumophila*.

CULTIVATION

Specimens for culture should be inoculated to two agar plates for recovery of *Legionella*, at least one of which is BCYE without inhibitory agents. This medium contains charcoal to detoxify the medium, remove carbon dioxide (CO_2), and modify the surface tension to allow the organisms to proliferate more easily. BCYE is also prepared with ACES buffer and the growth supplements cysteine (required by *Legionella*), yeast extract, α-ketoglutarate, and iron. A second medium, BCYE base with polymyxin B, anisomycin (to inhibit fungi), and cefamandole, is recommended for specimens, such as sputum, that are likely to be contaminated with other flora. These media are commercially available. Several other media, including a selective agar containing vancomycin and a differential agar containing bromthymol blue and bromcresol purple, are also available from Remel (Lenexa, Kan) and others. Specimens obtained from sterile body sites may be plated to two media without selective agents and perhaps also inoculated into the special blood culture broth without SPS. (Specimens should always be plated to standard media for recovery of pathogens other than *Legionella* that may be responsible for the disease.)

Plates are incubated in a candle jar at 35° to 37° C in a humid atmosphere. Only growth of *L. gormanii* is stimulated by increased CO_2, so incubation in air is preferable to 5% to 10% CO_2, which may inhibit some legionellae. Within 3 to 4 days, colonies should be visible. Plates are held for a maximum of 2 weeks before discarding. Blood cultures in biphasic media should be held for 1 month. At 5 days, colonies are 3 to 4 mm in diameter, gray-white to blue-green, glistening, convex, and circular and may exhibit a cut-glass type of internal granular speckling (Figure 37-2). A Gram stain yields thin, gram-negative bacilli (Figure 37-3).

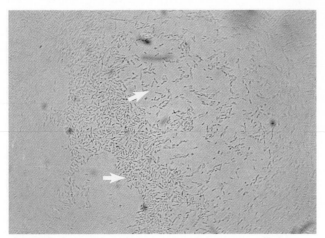

Figure 37-3   Gram stain of a colony of *Legionella pneumophila* showing thin, gram-negative bacilli *(arrows)*.

APPROACH TO IDENTIFICATION

Because *Legionella* spp. are biochemically inert and many tests produce equivocal results, extensive biochemical testing is of little use.[10,14] Definitive identification requires the facilities of a specialized reference laboratory. Identification of *L. pneumophila* spp. can be achieved, however, by a monoclonal immunofluorescent stain (Genetic Systems). Emulsions of organisms from isolated colonies are made in 10% neutral formalin, diluted 1:100 (to produce a very thin suspension), and placed on slides for fluorescent antibody staining. Clinical laboratories probably perform sufficient service to clinicians by indicating the presence of *Legionella* spp. in a specimen. If further identification is necessary, the isolate should be forwarded to an appropriate reference laboratory.

SERODIAGNOSIS

Most patients with legionellosis have been diagnosed retrospectively by detection of a fourfold rise in anti-*Legionella* antibody with an indirect fluorescent antibody (IFA) test. Serum specimens should be tested no closer than 2 weeks apart. Confirmation of disease is accomplished by a fourfold rise in titer to more than 128. A single serum with a titer of more than 256 and a characteristic clinical picture may be presumptive for legionellosis; however, because as many as 12% of healthy persons yield titers as high as 1:256, this practice is strongly discouraged.[9] Unfortunately, individuals with legionnaires' disease may not exhibit an increase in serologic titers until as long as 10 weeks after the primary illness or they may never display significant antibody titer increases. Commercially prepared antigen-impregnated slides for IFA testing are available from numerous suppliers.

ANTIMICROBIAL SUSCEPTIBILITY TESTING AND THERAPY

In vitro susceptibility studies are not predictive of clinical response and should not be performed for individual isolates of legionellae. Because newer agents such as fluoroquinolones and the newer macrolides (e.g., clarithromycin and azithromycin) are more active against *L. pneumophila*, erythromycin has been replaced. Alternative regimens include doxycycline and the combination of erythromycin and rifampin. Clinical response usually follows within 48 hours after the introduction of effective therapy. Penicillins, cephalosporins of all generations, and aminoglycosides are not effective and should not be used.

PREVENTION

Although under development, a vaccine against *Legionella* infections is not currently available. The effectiveness of other approaches to the prevention of legionella infections, such as the elimination of its presence from cooling towers and potable water, is uncertain.[8,11]

Case Study

A 6-month-old infant was diagnosed clinically with pneumonia. She was treated with intramuscular ceftriaxone followed by an oral cephalosporin for 3 days. The next day she was found to be unresponsive and rushed to the hospital. She was afebrile but tachypneic (increased breathing) and tachycardic (increased heart rate); she had an increased white blood cell (WBC) count predominated with lymphocytes. A bronchoalveolar lavage was collected and was positive for *Legionella* by direct fluorescent antibody. A culture grew *Legionella pneumophila* serogroup 6 after 8 days. Despite appropriate therapy with erythromycin and rifampin, the infant's pulmonary disease was fatal. No underlying disease was found in the baby.

QUESTIONS

1. The *Legionella* urine antigen test was negative in this baby. How do you explain this finding?
2. The baby appeared to be a normal infant. List as many risk factors as possible for acquiring *Legionella* pneumonia.
3. List the factors that hamper the laboratory diagnosis of *Legionella*.
4. Generally sputum sent to the laboratory to diagnose pneumonia will be purulent with mucus and increased polymorphonuclear WBCs. Describe the type of sputum seen with *Legionella* infection.

REFERENCES

1. Barker J, Brown MRW: Speculations on the influence of infecting phenotype on virulence and antibiotic susceptibility of *Legionella pneumophila, J Antimicrob Chemother* 36:7, 1995.

2. Breiman RF: Modes of transmission in epidemic and non-epidemic *Legionella* infections: directions of further study. In Barbaree JM, Breiman RF, and Dufour AP, editors: Legionella: *current status and emerging perspectives,* 1993, Washington, DC, American Society for Microbiology.

3. Buesching WJ, Brust RA, Ayers LW: Enhanced primary isolation of *Legionella pneumophila* from clinical specimens by low pH treatment, *J Clin Microbiol* 17:1153, 1983.

4. Garduño RA, Garduño E, Hiltz M et al: Intracellular growth of *Legionella pneumophila* gives rise to a differentiated form dissimilar to stationary-phase forms, *Infect Immun* 70:6273, 2002.

5. Harb OS, Kwaik YA: Interaction of *Legionella pneumophila* with protozoa provides lessons, *ASM News* 66:609, 2000.

6. Helbig JH, Uldum SA, Bernander S, et al: Clinical utility of urinary antigen detection for diagnosis of community-acquired, travel-associated, and nosocomial Legionnaires' disease, *J Clin Microbiol* 41:838, 2003.

7. Marrie TJ, Raoult D, LaScola B, et al: *Legionella*-like and other amoebal pathogens as agents of community-acquired pneumonia, *Emerg Infect Dis* 7:1026, 2001.

8. Muder RR, VL Yu: Infection due to *Legionella* species other than *Legionella pneumophila, Clin Infect Dis* 35:990, 2002.

9. Murdock DR: Diagnosis of *Legionella* infection, *Clin Infect Dis* 36:64, 2003.

10. Pasculle W: Update on *Legionella, Clin Microbiol Newsletter* 22:97, 2000.

11. Roy CR, Tilney LG: The road less traveled: transport of *Legionella* to the endoplasmic reticulum, *J Cell Biol* 158:415, 2002.

12. Salcedo SP, Holden DW: Bacterial interactions with the eukaryotic secretory pathway, *Curr Opin Microbiol* 8:92, 2005.

13. Shelhamer JH, Gill VJ, Quinn TC, et al: The laboratory evaluation of opportunistic pulmonary infections, *Ann Intern Med* 124:585, 1996.

14. Winn WC: *Legionella* and the clinical microbiologist, *Infect Dis Clin North Am* 7:377, 1993.

Genera and Species to Be Considered

Brucella abortus

Brucella melitensis

Brucella suis

Brucella canis

Although most isolates of *Brucella* grow on blood and chocolate agars (some isolates are also able to grow on MacConkey agar), more enriched agars and special incubation conditions are generally needed to achieve optimal recovery of these very fastidious organisms from clinical specimens.

GENERAL CHARACTERISTICS

Organisms belonging to the genus *Brucella* are small, nonmotile, aerobic, gram-negative coccobacilli or short rods that stain poorly by conventional Gram stain. Many isolates require supplementary carbon dioxide (CO_2) for growth, especially on primary isolation. *Brucella* spp. are closely related to *Bartonella, Rhizobium,* and *Agrobacterium.*[5] There are eight known species including two new species isolated from marine hosts, but only *Brucella abortus, B. melitensis, B. suis,* and *B. canis* are pathogenic for humans. Of note, *Brucella* spp. are considered as potential bioterrorism agents (see Chapter 65).

EPIDEMIOLOGY AND PATHOGENESIS

The disease **brucellosis** occurs worldwide, especially in Mediterranean and Persian Gulf countries, India, and parts of Mexico and Central and South America.[10] Brucellosis is a zoonosis and is recognized as a cause of devastating economic losses among domestic livestock. In the United States, 79 to 136 cases were reported to the Centers for Disease Control and Prevention (CDC) from 1996 to 2003, but it is estimated that these represent only 4% to 10% of all cases.[5] Each of the four *Brucella* spp. that are pathogenic for humans has a limited number of preferred animal hosts (Table 38-1). In their hosts, brucella organisms tend to localize in

tissues rich in erythritol (e.g., placental tissue), a four-carbon alcohol that enhances their growth.[6] Humans become infected by three primary routes[5]:

- Ingestion of infected unpasteurized animal milk products (most common means of transmission)
- Inhalation of infected aerosolized particles
- Direct contact with infected animal parts through ruptures of skin and mucous membranes

Persons considered at greatest risk for contracting brucellosis are dairy farmers, livestock handlers, slaughterhouse employees, veterinarians, and laboratory personnel.

Brucella spp. are facultative, intracellular parasites (able to exist in both intracellular and extracellular environments); only the following bacteria are classified as facultatively intracellular: *Salmonella, Shigella, Brucella, Yersinia, Listeria,* and *Francisella.*[2] Following infection, brucellae are ingested by neutrophils within which they replicate, resulting in cell lysis. Neutrophils containing viable organisms circulate in the bloodstream and are subsequently phagocytized by reticuloendothelial cells in the spleen, liver, and bone marrow. If left untreated, granulomas subsequently develop in these organs, with organisms surviving in monocytes and macrophages. Brucellae exhibit a tendency to invade and persist in the human host through inhibiting apoptosis (programmed cell death). Resolution of infection depends on the host's nutritional and immune status, the size of inoculum and route of infection, and the species of *Brucella* causing the infection; in general, *B. melitensis* and *B. abortus* are more virulent for humans.[5]

Survival and multiplication of *Brucella* in phagocytic cells are features essential to the establishment, development, and chronicity of the disease. The mechanisms by which brucellae avoid intracellular killing are not completely understood. *Brucella* spp. can change from smooth to rough colonial morphology based on the composition of their cell wall lipopolysaccharide (LPS); those with smooth LPS are more resistant to intracellular killing by neutrophils than those with rough LPS.[4,8] *Brucella* ensure their intracellular survival by avoiding fusion of their membrane-bound compartment with lysosomes in both macrophages and epithelial cells. In addition, *Brucella,* like *Legionella* (see

Table 38-1 *Brucella* spp. Pathogenic for Humans and Their Respective Natural Animal Host

Organism	Preferred Animal Host
B. abortus	Cattle
B. melitensis	Sheep or goats
B. suis	Swine
B. canis	Dogs

Chapter 37), use a type IV secretion system, VirB for intracellular survival and replication. Unlike *Legionella*, however, *Brucella* modulates phagosome transport to avoid being delivered to lysosomes.[7] Essentially, VirB is involved in controlling the maturation of the *Brucella* vacuole into an organelle that allows replication; if mutations occur in this gene region, *B. abortus* are unable to establish chronic infections in the mouse model.[1] Nevertheless, many questions remain unanswered regarding the pathogenesis of disease caused by *Brucella*.

SPECTRUM OF DISEASE

The clinical manifestations of brucellosis vary greatly, ranging from asymptomatic infection to serious, debilitating disease. For the most part, brucellosis is a systemic infection that can involve any organ of the body. Symptoms are nonspecific and include fever, chills, weight loss, sweats, headache, muscle aches, fatigue, and depression. Lymphadenopathy and splenomegaly are common physical findings. After an incubation period of about 2 to 3 weeks, the onset of disease is commonly insidious. Complications can occur, such as arthritis, spondylitis (inflammation of the spinal cord), and endocarditis.

LABORATORY DIAGNOSIS

SPECIMEN COLLECTION, TRANSPORT, AND PROCESSING

Definitive diagnosis of brucellosis requires isolation of the organisms in cultures of blood, bone marrow, or other tissues. For the following reasons it is essential that the clinical microbiology laboratory be notified whenever brucellosis is suspected:

- To ensure that specimens are cultivated in an appropriate manner for optimum recovery from clinical specimens

- To avoid accidental exposure of laboratory personnel handling the specimens, because *Brucella* spp. are considered class III pathogens

Blood for culture can be routinely collected (see Chapter 52) into most commercially available blood culture bottles and the lysis-centrifugation system (Isolator; Wampole Laboratories, Cranbury, NJ). There are no special requirements for specimen collection, transport, or processing for other clinical specimens.

DIRECT DETECTION METHODS

Direct stains of clinical specimens are not particularly useful for the diagnosis of brucellosis. Preliminary studies using conventional and real-time polymerase chain reaction assays indicate that these assays may prove to be a reliable, sensitive, and specific means to directly detect *Brucella* spp. in clinical specimens.

CULTIVATION

Commercial blood culture systems, such as the BacT/Alert, BACTEC, and lysis-centrifugation systems, have all successfully detected *Brucella* in blood.[9] A recent comparison of the BACTEC 9240 with the Isolator blood culture systems found the BACTEC system more sensitive and faster in its detection of *B. melitensis*.[11] Other blood culture bottles, such as those with brain-heart infusion and trypticase soy broth, also support the growth of brucellae if bottles are continuously vented and placed in a CO_2 incubator. Although the majority of isolates can be detected within 7 days using commercial systems, prolonged incubation of 30 days and periodic blind subcultures to blood and chocolate agar plates at 7, 14, and 30 days is recommended to maximize recovery of brucellae.[10] Culture bottles may not become turbid. All subculture plates should be held for a minimum of 7 days.

Although *Brucella* spp. grow on blood and chocolate agars, supplemented media such as *Brucella* agar or some type of infusion base are recommended for specimen types other than blood. The addition of 5% heated horse or rabbit serum enhances growth on all media. Cultures should be incubated in 5% to 10% CO_2 in a humidified atmosphere; inoculated plates are incubated for 3 weeks before being discarded as negative.

On culture, colonies appear small, convex, smooth, translucent, nonhemolytic, and slightly yellow and opalescent after at least 48 hours of incubation (Figure 38-1). The colonies may become brownish with age.

APPROACH TO IDENTIFICATION

Presumptive identification of *Brucella* can be made, but appropriate biohazard facilities must be used. On Gram

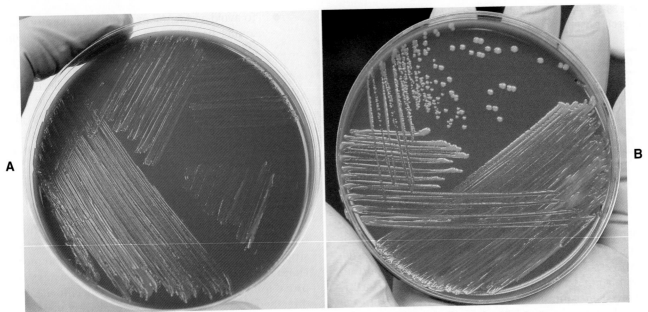

Figure 38-1   Growth of *Brucella* spp. on chocolate agar after 2 days (**A**) and 4 days (**B**) of incubation.

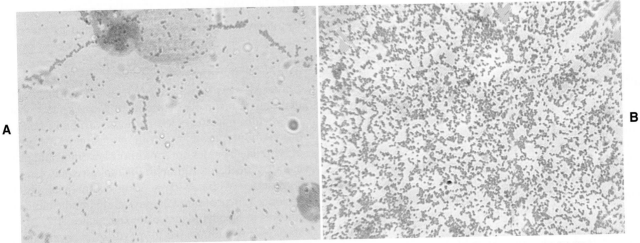

Figure 38-2   *Brucella melitensis* with traditional Gram stain (**A**) and Gram stain with 2-minute safranin counterstain (**B**) to allow for easier visualization of the organism.

stain, organisms are small coccobacilli that resemble fine grains of sand (Figure 38-2). Although organisms are catalase-positive, this test should not be done because organisms could become nebulized. Most strains are oxidase-positive. Other nonfermentative gram-negative coccobacilli that may be confused with *Brucella* are *Bordetella*, *Moraxella*, *Kingella*, and *Acinetobacter* spp. *Brucella* spp., however, are nonmotile, urease- and nitrate-positive, and strictly aerobic. The most rapid test for presumptive identification of *Brucella* is the particle agglutination test with antismooth *Brucella* serum (Difco Laboratories, Detroit, Mich). *Brucella* spp. are differentiated by the rapidity with which they hydrolyze urea, relative ability to produce H_2S, require-

ments for CO_2, and susceptibility to the aniline dyes thionine and basic fuchsin. Differential characteristics are listed in Table 38-2. For determination of CO_2 requirement, identical plates of *Brucella* agar or brain-heart infusion agar should be given equal inocula (e.g., with a calibrated loop) of a broth suspension of the organism to be tested. One plate should be incubated in a candle jar and the other plate in air within the same incubator. Most strains of *B. abortus* do not grow in air but show growth in the candle jar. Isolates of *Brucella* should be sent to state or other reference laboratories for confirmation or definitive identification, because most clinical laboratories lack the necessary media and containment facilities.

Table 38-2   Characteristics of *Brucella* spp. Pathogenic for Humans

Species	CO₂ Required for Growth	Time to Positive Urease	H₂S Produced	Inhibition by Dye Thionine*	Fuschin*
B. abortus	±	2 hr (rare 24 hr)	+ (most strains)	+	−
B. melitensis	−	2 hr (rare 24 hr)	+	−	−
B. suis	−	15 min	±	−	+ (most)
B. canis	−	15 min	−	−	+

+, >90% of strains positive; −, >90% of strains negative; ±, variable results.
*Dye tablets (Key Scientific Products, Inc., Round Rock, Texas).

SERODIAGNOSIS

Because of the difficulty of isolating the organism, a serologic test, the serum agglutination test (SAT), is widely used and detects antibodies to *Brucella abortus, B. melitensis,* and *B. suis;* SAT will not detect *B. canis* antibodies. A titer of 1:160 or greater in the SAT is considered diagnostic if this result fits the clinical and epidemiologic findings. The SAT can cross-react with class M immunoglobulins with a variety of bacteria such as *Francisella tularensis* and *Vibrio cholera.* Enzyme-linked immunosorbent assays (ELISAs) also have been developed, but further evaluation must be done before these assays replace the SAT. In patients with neurobrucellosis, ELISA offers significant diagnostic advantages over conventional agglutination methods.[5]

ANTIMICROBIAL SUSCEPTIBILITY TESTING AND THERAPY

Because of the fastidious nature of these organisms coupled with their intracellular localization, in vitro susceptibility testing is not reliable.[3] Prolonged treatment with antibiotics that can penetrate macrophages and act in the acidic intracellular environment (6 weeks) is given to patients with brucellosis to prevent relapse of infection. For initial therapy, doxycycline in combination with streptomycin or rifampin is recommended. Sometimes, surgical drainage is also required to treat localized foci of infection.

PREVENTION

Prevention of brucellosis in humans is dependent on the elimination of disease in domestic livestock. Vaccines directed against *B. abortus* or *B. melitensis* have been used in the appropriate animal host (i.e., cattle and goats/sheep, respectively) and are very successful in eradicating the disease in animals. An effective vaccine against *B. suis* and various human vaccines using killed brucellae fractions are under development.

 Case Study

A 67-year-old woman from the Middle East had total knee arthroplasty of the right knee in 1991 and left knee in 1994. She was seeking medical attention because of pain in her left knee. Her knee was aspirated and 3600 white blood cells per microliter were reported, but no organisms were seen on Gram stain. The culture grew coagulase-negative *Staphylococcus* from joint fluid cultured in blood culture bottles after 3 days of incubation. A few tiny, poorly staining gram-negative rods were present on the direct blood and chocolate plates after 5 days of incubation, but not in the blood culture. The rods were oxidase- and catalase-positive. Repeat culture 2 weeks later grew only the gram-negative rods.

Surgical debridement with appropriate antimicrobial therapy resulted in control of the infection.

QUESTIONS

1. When a fastidious gram-negative coccobacilli is isolated from a normally sterile site, what is the first step that should be performed in the laboratory?
2. What rapid test will expedite the identification of this fastidious coccobacillus?
3. How did this woman acquire the infection with this organism?
4. How is the diagnosis confirmed?

REFERENCES

1. Boschiroli ML, Ouahrani-Betlache S, Foulongne V, et al: Type IV secretion and *Brucella* virulence, *Vet Microbiol* 90:341, 2002.
2. Fortier AH, Green SJ, Polsinelli T, et al: Life and death of an intracellular pathogen: *Francisella tularensis* and the macrophage, *Immunol Series* 60:349, 1994.
3. Hall WH: Modern chemotherapy for brucellosis in humans, *Rev Infect Dis* 12:1060, 1990.
4. Maria-Pilar J, Dudal S, Jacques D, et al: Cellular bioterrorism: how *Brucella* corrupts macrophage physiology to promote invasion and proliferation, *Clin Immunol* 114:227, 2004.
5. Pappas G, Akritidis N, Bosilkovski M, et al: Brucellosis, *N Engl J Med* 352:2325, 2005.
6. Radolf JD: Brucellosis: don't let it get your goat! *Am J Med Sci* 307:64, 1994.
7. Roy CR: Exploitation of the endoplasmic reticulum by bacterial pathogens, *Trends Microbiol* 10:418, 2002.
8. Smith LD, Ficht TA: Pathogenesis of *Brucella*, *Crit Rev Microbiol* 17:209, 1990.
9. Yagupsky P: Detection of *Brucella* in blood cultures, *J Clin Microbiol* 37:3437, 1999.
10. Yagupsky P: Detection of *Brucella melitensis* by BACTEC NR660 blood culture system, *J Clin Microbiol* 32:1899, 1994.
11. Yagupsky P, Peled N, Press J, et al: Comparison of BACTEC 9240 Peds Plus medium and Isolator 1.5 microbial tube for detection of *Brucella melitensis* from blood cultures, *J Clin Microbiol* 35:1382, 1997.

The genus *Bordetella* contains three human pathogens: *Bordetella bronchiseptica*, *B. pertussis*, and *B. parapertussis*. *B. bronchiseptica* is reviewed in Chapter 27 because it grows on MacConkey agar. Although *B. parapertussis* can also grow on MacConkey agar, it is addressed with *B. pertussis* in this chapter for two reasons. First, *B. pertussis* and *B. parapertussis* both cause human upper respiratory tract infections with almost identical symptoms, epidemiology, and therapeutic management. Second, optimal recovery of both organisms from respiratory specimens requires the addition of blood and/or other suitable factors to culture media.

GENERAL CHARACTERISTICS

General features of *Bordetella* spp. other than *B. pertussis* and *B. parapertussis* are summarized in Chapter 27. In contrast to *B. bronchiseptica*, *B. pertussis* and *B. parapertussis* are nonmotile and only infect humans. In the evolutionary process these exclusive human pathogens have a close genetic relationship.[10]

EPIDEMIOLOGY AND PATHOGENESIS

EPIDEMIOLOGY

Before the introduction of vaccines and in populations in which immunization is not performed, **pertussis** (whooping cough) is an epidemic disease with cycles every 2 to 5 years. Pertussis is a highly contagious, acute infection of the upper respiratory tract caused primarily by *B. pertussis* and less commonly by *B. parapertussis*.[10] The latter agent generally has a less severe clinical presentation both in duration of symptoms and in the percentage of patients affected.[9] Recently, *Bordetella holmesii* was reported to cause a pertussis-like illness but as of this writing, little is known about the biology, virulence mechanisms, and pathogenic significance.[10,14] Pertussis was first described in the sixteenth century and occurs worldwide, totalling about 48.5 million cases

annually. Although the incidence has decreased significantly since the widespread use of vaccination, outbreaks of pertussis occur periodically. It appears that *B. pertussis* infections are endemic in adults and adolescents, most likely due to waning vaccine-induced immunity; therefore these infections might serve as the source of the epidemic cycles involving unvaccinated or partially immunized infants and children.[1,11] Infection is transmitted from person-to-person, presumably by airborne transmission from the cough of an infected person; humans are the only reservoir.

PATHOGENESIS

The mechanisms by which *B. pertussis*, the primary pathogen of whooping cough, overcome the immune defenses of healthy individuals are complex, involving the interplay of several virulence factors (Table 39-1). Some factors help establish infection, others are toxigenic to the host, and still others override specific components of the host's mucosal defense system. For example, once reaching the host's respiratory tract, *B. pertussis* attaches to respiratory ciliated epithelial cells by means of adhesins and paralyzes the beating cilia by elaborating a tracheal cytotoxin.[2,10,11] A major virulence factor, pertussis toxin (PT) is also produced by the attached organism. PT enters the bloodstream, subsequently binding to specific receptors on host cells. After binding, PT disrupts several host cell functions, such as initiation of host cell translation; the inability of host cells to receive signals from the environment causes a generalized toxicity. The center membrane of *B. pertussis* also blocks the access of the host's lysozyme to the bacterial cell wall via its outer membrane.[13] Of note, *B. pertussis* and *B. parapertussis* share a nearly identical virulence control system encoded by the *bvgAS* locus that is responsive to varied environmental conditions.[10] By virtue of this very complex system, it appears that *Bordetella* are able to change into different phenotypes that allow for their transmission, colonization, and survival.

SPECTRUM OF DISEASE

Several factors influence the clinical manifestations of *B. pertussis* and are listed in Box 39-1. Classic pertussis is usually a disease of children and can be divided into

Table 39-1   Major Virulence Determinants of *Bordetella pertussis*

Function	Factor/Structure
Adhesion	Fimbriae Filamentous hemagglutinin (FHA) Pertactin Tracheal colonization factor Brk A*
Toxicity	Pertussis toxin (an A/B toxin related to cholera toxin) Adenylate cyclase toxin (hemolyzes red cells and activates cyclic adenosine monophosphate, thereby inactivating several types of host immune cells) Dermonecrotic toxin (exact role unknown) Tracheal cytotoxin (ciliary dysfunction and damage) Endotoxin Type III secretion†
Overcome host defenses	Outer membrane (host lysozyme is inhibited) Siderophore production (host lactoferrin and transferrin are unable to limit iron)

*Plays a role in pathogenesis by conferring serum resistance.[3,10]

†Allows *Bordetella* to transport proteins directly into host cells and is required for persistent tracheal colonization.

<table>
<tr><td>BOX 39-1</td><td>Factors Known to Affect the Clinical Manifestation of B. pertussis Infection</td></tr>
</table>

- Patient age
- Previous immunization or infection
- Presence of passively acquired antibody
- Antibiotic treatment

---

three symptomatic stages: catarrhal, paroxysmal, and convalescent. During the **catarrhal stage,** symptoms are the same as for a mild cold with a runny nose and mild cough; this stage may last for several weeks.[10] Episodes of severe and violent coughing increase in number, marking the beginning of the **paroxysmal stage.** As many as 15 to 25 paroxysmal coughing episodes can occur in 24 hours and are associated with vomiting and "whooping" as air is rapidly inspired into the lungs past the swollen glottis. Lymphocytosis occurs, although typically there is no fever and there are no signs and symptoms of systemic illness. This stage may last from 1 to 4 weeks.

In addition to classic pertussis, *B. pertussis* can also cause mild illness and asymptomatic infection primarily in household contacts as well as in a number of unvaccinated and previously vaccinated children. Since the 1990s, a shift in the age distribution of pertussis cases to adolescence and adults has been observed in highly vaccinated populations.[5] Thus adults and adolescents are now recognized as a reservoir for transmitting infection to vulnerable infants. Among these immunized individuals, a prolonged cough may be the only manifestation of pertussis; scratchy throat, other pharyngeal symptoms, and episodes of sweating commonly occur in adults with pertussis.[6] A number of studies have documented that between 13% and 32% of adolescents and adults with an illness involving a cough of 6 days' duration or longer have serologic and/or culture evidence of infection with *B. pertussis.*

LABORATORY DIAGNOSIS

SPECIMEN COLLECTION, TRANSPORT, AND PROCESSING

Confirming the diagnosis of pertussis is challenging. Culture, which is most sensitive early in the illness has been the traditional diagnostic standard for pertussis with nearly 100% specificity but varied sensitivity. Organisms may become undetectable by culture 2 weeks after the start of paroxysms. Nasopharyngeal aspirates or a nasopharyngeal swab (calcium alginate or Dacron on a wire handle) are acceptable specimens because *B. pertussis* colonizes the ciliated epithelial cells of upper respiratory tract. Specimens obtained from throat, sputum, or anterior nose are unacceptable because these sites are not lined with ciliated epithelium. For collection, the swab is bent to conform to the nasal passage and held against the posterior aspect of the nasopharynx. If coughing does not occur, another swab is inserted into the other nostril to initiate the cough. The swab is left in place during the entire cough, removed, and immediately inoculated onto a selective medium at the bedside (Table 39-2).

A fluid transport medium may be used for swabs but must be held for less than 2 hours. Half-strength

Table 39-2   Examples of Selective Media for Primary Isolation of *B. pertussis* and *B. parapertussis*

Agar Media	Description
Bordet-Gengou	Potato infusion agar with glycerol and sheep blood with methicillin or cephalexin* (short shelf-life)
Modified Jones-Kendrick charcoal[12]	Charcoal agar with yeast extract, starch, and 40 μg cephalexin (2- to 3-month shelf-life but inferior to Regan-Lowe)
Regan-Lowe[†]	Charcoal agar with 10% horse blood and cephalexin (4- to 8-week shelf-life)

*Cephalexin is superior to methicillin or penicillin for inhibiting normal respiratory flora.[12]
[†]Regan-Lowe medium was found best for recovery of *B. pertussis* from nasopharyngeal swabs.[7]

Regan-Lowe agar enhances recovery when used as a transport and enrichment medium. Cold casein hydrolysate medium and casamino acid broth (available commercially) have been found to be effective transport media, particularly for preparation of slides for direct fluorescent antibody staining.

DIRECT DETECTION METHODS

A direct fluorescent antibody (DFA) stain using polyclonal antibodies against *B. pertussis* and *B. parapertussis* is commercially available for detection of *B. pertussis* in smears made from nasopharyngeal (NP) material (Becton Dickinson, Sparks, Md); an NP specimen DFA-positive for *B. pertussis* is shown in Figure 6-15, *B*. Although rapid, this DFA stain has limited sensitivity and variable specificity; therefore the DFA should always be used in conjunction with culture.[6] DFA monoclonal reagent is also commercially available with two antisera with different fluors to detect *B. pertussis* and *B. parapertussis* (Accu-Mab, Altachem Pharma, Edmonton, Canada). Because of the limitations with currently available diagnostic methods, significant effort has been put into developing nucleic acid amplification methods. To date, most studies employing direct detection of *B. pertussis* and *B. parapertussis* by various polymerase chain reaction (PCR) procedures, including real-time PCR, indicate that these assays have a diagnostic sensitivity at least comparable to and in most cases superior to that of culture. A word of caution: Positive results have been obtained with samples containing *Bordetella holmesii* and *Bordetella bronchiseptica* (see Chapter 27) depending on the sequence targeted in conventional and real time PCR assays. Nasopharyngeal swabs (rayon or Dacron swabs on plastic shafts) and aspirates are the two types of samples primarily used for pertussis PCR; calcium-alginate swabs are unacceptable because these inhibit PCR-based detection.

CULTIVATION

Plates are incubated at 35° C in a humidified atmosphere without elevated carbon dioxide for up to 12 days[8]; most

Figure 39-1   Growth of *Bordetella pertussis* on Regan-Lowe media.

isolates are detected in 3 to 5 days. Young colonies of *B. pertussis* and *B. parapertussis* are small and shiny, resembling mercury drops; colonies become whitish-gray with age (Figure 39-1). Sensitivity of culture approaches 60% in the best of hands and depends on the stage of illness at the time of specimen collection, the technique used for specimen collection, specimen adequacy and transport, and culture conditions.

APPROACH TO IDENTIFICATION

A Gram stain of the organism reveals minute, faintly staining coccobacilli singly or in pairs (Figure 39-2). The use of a 2-minute safranin "O" counterstain or a 0.2% aqueous basic fuchsin counterstain enhances their visibility. The DFA reagent is used to presumptively identify organisms. Whole-cell agglutination reactions in specific antiserum can be used for species identification.

SERODIAGNOSIS

Although there are several serologic tests available for the diagnosis of pertussis, including agglutination, com-

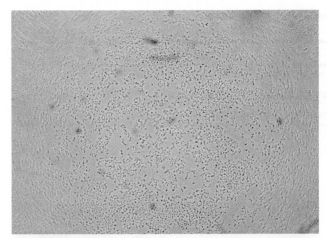

Figure 39-2   Typical Gram stain appearance of *Bordetella pertussis*.

plement fixation, and enzyme immunoassay, no single method can be recommended for serologic diagnosis at this time.

ANTIMICROBIAL SUSCEPTIBILITY TESTING AND THERAPY

Laboratories currently do not perform routine susceptibility testing of *B. pertussis* and *B. parapertussis* because erythromycin or the newer macrolides (clarithromycin, azithromycin) remain active and are the antibiotic of choice. However, because of three erythromycin-resistant isolates of *B. pertussis*, continued surveillance of *B. pertussis* is advised.[4,10]

PREVENTION

Whole-cell vaccines prepared from various *B. pertussis* preparations to prevent pertussis are manufactured in many countries and are efficacious in controlling epidemic pertussis. However, because of reactions to these vaccines and an apparent lack of long-term immunity, new acellular vaccines have replaced whole-cell vaccines in the United States and elsewhere. Also, some countries such as Germany, France, and Canada now recommend routine vaccination of adolescents.[6] Prompt recognition of clinical cases and treatment of contacts and cases are also very important in preventing the transmission of *B. pertussis* and *B. parapertussis;* viable organisms can be recovered from untreated patients for 3 weeks after the onset of cough. To prevent nosocomial outbreaks, patients with suspected or confirmed pertussis should be placed on droplet precautions.

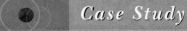

Case Study

A 36-year-old female surgeon was discharged from the hospital after an uncomplicated delivery of a healthy second child. Three days after arriving home she awoke with a fever, malaise, and nonproductive cough. An induced sputum grew *Pseudomonas aeruginosa*. She was admitted to the hospital and treated with ceftazidime and tobramycin. Her white blood cell count on admission was 13,500/mm³ and scattered coarse rhonchi were heard on deep inspiration. The rest of the family was in good health. The Infectious Disease Service (IDS) did not believe this woman had pseudomonas pneumonia and suggested a nasopharyngeal aspirate be collected for *Bordetella pertussis* PCR, even though the patient was feeling better. The laboratory performed the test with a positive result. Subsequent culture on Regan-Lowe medium was positive for the organism (see Figure 39-1).

QUESTIONS

1. Why was the IDS so interested in having the diagnosis correct in a patient whose disease was improving?
2. Pertussis in young and older adults is underdiagnosed. What are some of the reasons?
3. Why is the pertussis PCR assay so sensitive?
4. What biochemical tests uniquely identify *B. pertussis*?

REFERENCES

1. Cattaneo LA, Reed GW, Haase DH, et al: The seroepidemiology of *Bordetella pertussis* infections: a study of persons ages 1-65 years, *J Infect Dis* 173:1257, 1996.
2. Cherry JD: Historical review of pertussis and the classical vaccine, *J Infect Dis* 174(suppl 3): S259, 1996.
3. Fernandez RC, Weiss AA: Cloning and sequencing of a *Bordetella pertussis* serum resistance locus, *Infect Immunol* 62:4727, 1994.
4. Gordon KA, Fusco J, Biedenback DJ, et al: Antimicrobial susceptibility testing of clinical isolates of *Bordetella pertussis* from Northern California: report from the SENTRY antimicrobial surveillance program, *Antimicrob Agent Chemother* 45:3599, 2001.
5. Gurls D, Strebel PM, Bardenheier B, et al: Changing epidemiology of pertussis in the United States: increasing reported incidence among adolescents and adults, 1990-1996, *Clin Infect Dis* 28:1230, 1999.
6. Hewlett EL, Edwards KM: Pertussis—not just for kids, *N Engl J Med* 352:212, 2005.
7. Hoppe JE, Vogl R: Comparison of three media for cultures of *Bordetella pertussis*, *Eur J Clin Microbiol* 5:361, 1986.
8. Katzko C, Hofmeister M, Church D: Extended incubation of culture plates improves recovery of *Bordetella* spp., *J Clin Microbiol* 34:1563, 1996.
9. Mastrantonio P, Stefanelli P, Giuliano M, et al: *Bordetella parapertussis* infection in children: epidemiology, clinical symptoms, and molecular characteristics of isolates, *J Clin Microbiol* 36:999, 1998.

10. Mattoo S, Cherry JD: Molecular pathogenesis, epidemiology, and clinical manifestations of respiratory infections due to *Bordetella pertussis* and other *Bordetella* species, *Clin Microbiol Rev* 18: 326, 2005.

11. Rappuoli R: Pathogenicity mechanisms of *Bordetella, Curr Top Microbiol Immunol* 192:319, 1994.

12. Stauffer LR, Brown DR, Sandstrom RE: Cephalexin-supplemented Jones-Kendrick charcoal agar for selective isolation of *Bordetella pertussis:* comparison with previously described media, *J Clin Microbiol* 17:60, 1983.

13. Weiss A: Mucosal immune defenses and the response of *Bordetella pertussis, ASM News* 63:22, 1997.

14. Yih WK, Silva EA, Ida JH, et al: *Bordetella holmesii*-like organisms isolated from Massachusetts patients with pertussis-like symptoms, *Emerg Infect Dis* 5:441, 1999.

Genera and Species to Be Considered

Current Name

Franciscella tularensis
   subsp. *tularensis*
   subsp. *holartica*
   subsp. *mediasiatica*
   subsp. *novicida*
Franciscella philomiragia

Blood, chocolate, and MacConkey agars cannot be used for the primary isolation of organisms belonging to the genus *Francisella*. *Francisella* is a facultative, intracellular pathogen that requires cysteine and a source of iron for growth.[5] Because of this requirement for a complex medium for isolation and growth, these organisms are discussed in this chapter.

GENERAL CHARACTERISTICS

Organisms belonging to the genus *Francisella* are faintly staining, gram-negative coccobacilli that are nonmotile and obligately aerobic. The taxonomy of this genus continues to be in flux. The most current proposed taxonomy is summarized in Table 40-1. For the most part, different subspecies are associated with different geographic regions. In addition, based on the analysis of 14 human isolates initially classified as *Yersinia philomiragia*, these organisms were transferred to the genus *Francisella* as *F. philomiragia*.[7]

EPIDEMIOLOGY AND PATHOGENESIS

Francisella tularensis is the agent of human and animal tularemia. Worldwide in distribution, *F. tularensis* is carried by many species of wild rodents, rabbits, beavers, and muskrats in North America. Humans become infected by handling the carcasses or skin of infected animals, through insect vectors (primarily deerflies and ticks in the United States), by being bitten by carnivores that have themselves eaten infected animals, or by inhalation.[2] Evidence exists that *Francisella* can persist in waterways, possibly in association with amebae.[3]

The capsule of *F. tularensis* appears to be a necessary component for expression of full virulence, allowing the organism to avoid immediate destruction by polymorphonuclear neutrophils.[10] In addition to being extremely invasive, *F. tularensis* is an intracellular parasite that can survive in the cells of the reticuloendothelial system, where it resides after a bacteremic phase. Granulomatous lesions may develop in various organs. Humans are infected by less than 50 organisms by either aerosol or cutaneous routes; *F. tularensis* subsp. *tularensis* are the most virulent for humans with an infectious dose of less than 10 colony forming units.[3] *F. philomiragia* has been isolated from several patients, many of whom were immunocompromised or were victims of near-drowning incidents. The organism is present in animals and ground water.

SPECTRUM OF DISEASE

Following inoculation of *F. tularensis* through abrasions in the skin or arthropod bites, a lesion appears at the site and progresses to an ulcer; lymph nodes adjacent to the site of inoculation become enlarged and often necrotic. Once the organism enters the bloodstream, patients become systemically ill with high temperature, chills, headache, and generalized aching. Clinical manifestations of infection with *F. tularensis* can be glandular, ulceroglandular, oculoglandular, oropharyngeal, systemic, and pneumonic. These clinical presentations are briefly summarized in Table 40-2.

LABORATORY DIAGNOSIS

F. tularensis is a Biosafety Level 2 pathogen, a designation that requires technologists to wear gloves and to work within a biological safety cabinet (BSC) when handling clinical material that potentially harbors this agent. For cultures, the organism is designated Biosafety Level 3; a mask, recommended for handling all clinical specimens, is very important for preventing aerosol acquisition with *F. tularensis*. Because tularemia is one of the most common laboratory-acquired infections, most microbiologists do not attempt to work with infectious material from suspected patients. It is recommended that specimens be sent to reference laboratories or state or other

Table 40-1   Most Recent Taxonomy of the Genus *Francisella* and Key Characteristics

Organism	Primary Region	Disease in Humans
F. tularensis subsp. *tularensis*	North America	Most severe
F. tularensis subsp. *holartica*	Europe, former Soviet Union, Japan, North America	Least severe
F. tularensis subsp. *mediasiatica*	Kazakhstan, Uzbekistan	Severe
F. tularensis subsp. *novicida*	North America	Severe
F. philomiragia (formerly *Yersinia philomiragia*)	North America	Least severe: virulent only in immunocompromised individuals and near-drowning victims

Table 40-2   Clinical Manifestations of *Francisella tularensis* Infection

Types of Infection	Clinical Manifestations and Description
Ulceroglandular	Common; ulcer and lymphadenopathy; rarely fatal
Glandular	Common; lymphadenopathy; rarely fatal
Oculoglandular	Conjunctivitis, lymphadenopathy
Oropharyngeal	Ulceration in the oropharynx
Systemic tularemia	Acute illness with septicemia; 30% to 60% mortality rate; no ulcer or lymphadenopathy
Pneumonic tularemia	Acquired by inhalation of infectious aerosols or by dissemination from the bloodstream; pneumonia; most serious form of tularemia

public health laboratories that are equipped to handle *Francisella*.

SPECIMEN COLLECTION, TRANSPORT, AND PROCESSING

The most common specimens submitted to the laboratory are scrapings from infected ulcers, lymph node biopsies, and sputum. Because there are no special requirements for specimen collection, transport, or processing, except for specimen handling as just described, refer to Table 5-1 for general information. In light of recent events and concerns about bioterrorism, laboratories must keep in mind that isolation of *F. tularensis* from blood cultures might be considered as a potential bioterrorist attack and are considered Select Biological Agents of Human Disease (see Chapter 65).

DIRECT DETECTION METHODS

Gram stain of clinical material is of little use. Fluorescent antibody stains are commercially available for direct detection of the organism in lesion smears, but such procedures are best performed by reference laboratories. Conventional and real-time polymerase chain reaction (PCR) assays have been developed to detect *F. tularensis* directly in clinical specimens. Of significance, several

patients with clinically suspected tularemia with negative serology and culture had detectable DNA by PCR.

CULTIVATION

Isolation of *F. tularensis* is difficult. *F. tularensis* is strictly aerobic and requires enriched media (containing cysteine and cystine) for primary isolation. Commercial media for cultivation of the organism are available (glucose cystine agar, BBL Microbiology Systems, Sparks, Md; cystine-heart agar, Difco Laboratories, Detroit, Mich); both require the addition of 5% sheep or rabbit blood. *F. tularensis* also may grow on chocolate agar supplemented with IsoVitaleX, the nonselective buffered charcoal-yeast extract agar used for isolation of legionellae, or modified Mueller-Hinton broth.[1] Growth is not enhanced by carbon dioxide. These slow-growing organisms require 2 to 4 days for maximal colony formation and are weakly catalase-positive and oxidase-negative. Some strains may require up to 2 weeks to develop visible colonies. *F. philomiragia* is less fastidious than *F. tularensis*. Although *F. philomiragia* does not require cysteine or cystine for isolation, it resembles *F. tularensis* in that it is a small, coccobacillary rod that grows poorly or not at all on MacConkey agar. This organism grows well on heart infusion agar with 5% rabbit blood or buffered charcoal-yeast extract agar with or without cysteine.[7] *F. tularensis* can be detected in commercial blood culture systems within 2 to 5 days; because these organisms Gram stain poorly, an acridine orange stain may be required to visualize the organisms in a positive blood culture bottle.

APPROACH TO IDENTIFICATION

Colonies are transparent, mucoid, and easily emulsified. Although carbohydrates are fermented, isolates should be identified serologically (by agglutination) or by a fluorescent antibody stain. Ideally, isolates should be sent to a reference laboratory for characterization.

F. *philomiragia* differs from *F. tularensis* biochemically; *F. philomiragia* is oxidase-positive by Kovac's modification, and most strains produce hydrogen sulfide in triple sugar iron agar medium, hydrolyze gelatin, and grow in 6% sodium chloride (no strains of *F. tularensis* share these characteristics).

Of concern, two recent reports describe problems in identifying *Francisella* species isolated from clinical specimens.[6,9] In one study, 12 microbiology employees were exposed to *F. tularensis* despite bioterrorism procedures being in place; the organism had been isolated from blood, respiratory, and autopsy specimens and grew on chocolate agar.[9] In this situation, multiple cultures were worked up on open benches without any additional personal protective equipment for what had been thought to be most consistent with a *Haemophilus* species. As a result of these reports, microbiologists must be aware of not only the key characteristics of this group of organisms (Box 40-1) but also the possible pitfalls in their identification (e.g., some strains grow well on sheep blood agar; identification kits may incorrectly suggest an identification of *Actinobacillus actinomycetemcomitans*).

SERODIAGNOSIS

Because of the risk of infection to laboratory personnel and other inherent difficulties with culture, diagnosis of

---

BOX 40-1	**Indications of a Possible *Francisella* Species**

- Unusual Gram stain: small, poorly staining gram-negative rods seen mostly as single cells *or* amorphous gram-negative mass without distinct cell forms (*F. philomiragia*)
- Subcultures yield primarily pinpoint colonies on chocolate agar
- Oxidase-negative; weak or negative catalase test
- Negative satellite or X and V tests
- Small, gram-negative coccobacillus observed in a Gram-stained smear of a positive blood culture in which time to detection is longer than 24 hours
- Organism requires prolonged incubation on chocolate agar

---

tularemia is usually accomplished serologically by whole-cell agglutination (febrile agglutinins or newer enzyme-linked immunosorbent assay techniques).

ANTIMICROBIAL SUSCEPTIBILITY TESTING AND THERAPY

There is no standardized antimicrobial susceptibility test for *Francisella* spp. The organism is susceptible to aminoglycosides, and streptomycin is the drug of choice. Gentamicin is a possible alternative[4]; doxycycline and chloramphenicol also have been used, although these two agents have been associated with a higher rate of relapse after treatment. Fluoroquinolones appear promising for treatment of even severe tularemia.[3,8]

PREVENTION

The primary way to prevent tularemia is by reducing the possibility for exposure to the etiologic agent in nature, such as wearing protective clothing to prevent insect bites and not handling dead animals. An investigative live-attenuated vaccine is available.[3]

Case Study

A 36-year-old man with human immunodeficiency virus had been doing well on a prophylactic regimen. After camping in Yosemite, he presented to his physician for a nonhealing, erythematous 3-mm "cyst" on his neck. He had been treated with ampicillin-sulbactam without resolution, so a biopsy was performed. No organisms were seen on Gram stain, but the culture grew a tiny gram-negative rod only on chocolate agar after 3 days of incubation (Figure 40-1). The microbiologist found the organism to be oxidase- and urease-negative but weakly catalase-positive. It did not satellite around a staphylococcal dot on blood agar. A beta-lactamase test was positive. The patient was treated with a 4-week course of ciprofloxacin with resolution of his lesion.

QUESTIONS

1. The local health department identified the isolate as *Francisella tularensis* by PCR and by fluorescent stain. According to the state laboratory, what do you think is the most commonly misidentified genus and species submitted to them as *F. tularensis*? What test will separate these two organisms?
2. The isolate was oxidase- and urease-negative but weakly catalase-positive. After it was determined to be negative for satelliting around a staphylococcal dot on blood agar, what precautions should be taken when working with the culture?
3. Why was a beta-lactamase test performed?

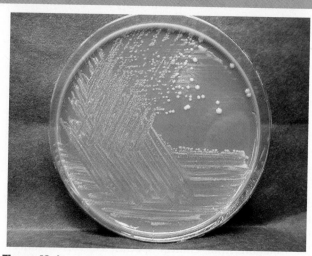

Figure 40-1   *F. tularensis* growing on chocolate agar after 72 hours of incubation. (Courtesy Mary K. York.)

REFERENCES

1. Chu MC, Weyant: *Francisella* and *Brucella*. In Murray PR, Baron EJ, Pfaller MA, et al, editors: *Manual of clinical microbiology,* ed 8, Washington, DC, 2003, American Society for Microbiology.
2. Craven R, Barnes AM: Plague and tularemia, *Infect Dis Clin North Am* 5:165, 1991.
3. Ellis J, Oyston PC, Green M, et al: Tularemia, *Clin Microbiol Rev* 15: 631, 2002.
4. Enderlin G, Morales L, Jacobs RF, et al: Streptomycin and alternative agents for the treatment of tularemia: review of the literature, *Clin Infect Dis* 19:42, 1994.
5. Fortier AH, Green SJ, Polsinelli T, et al: Life and death of an intracellular pathogen: *Francisella tularensis* and the macrophage, *Immunol Series* 60:349, 1994.
6. Friis-Møller, Lemming LE, Valerius NH, et al: Problems in identification of *Francisella philomiragia* associated with fatal bacteremia in a patient with chronic granulomatous disease, *J Clin Microbiol* 40:1840, 2004.
7. Hollis DG, Weaver RE, Steigerwalt AG, et al: *Francisella philomiragia* comb nov (formerly *Yersinia philomiragia*) and *Francisella tularensis* biogroup novicida (formerly *Francisella novicida*) associated with human disease, *J Clin Microbiol* 27:1601, 1989.
8. Limaye AP, Hooper CJ: Treatment of tularemia with fluoroquines: two cases and review, *Clin Infect Dis* 29:922, 1999.
9. Shapiro DS, Schwartz DR: Exposure of laboratory workers to *Francisella tularensis* despite a bioterrorism procedure, *J Clin Microbiol* 40:2278, 2002.
10. Sjöstedt A: Virulence determinants and protective antigens of *Francisella tularensis, Curr Opin Microbiol* 6:66, 2003.

Genera and Species to Be Considered

Streptobacillus moniliformis
Spirillum minus

Streptobacillus moniliformis is a gram-negative bacillus that requires media containing blood, serum, or ascites fluid as well as incubation under carbon dioxide (CO_2) for isolation from clinical specimens. This organism causes rat-bite fever and Haverhill fever in humans. *Spirillum minus* has never been grown in culture but, because both are causative agents of rat-bite fever, these organisms are considered in this chapter.

STREPTOBACILLUS MONILIFORMIS

GENERAL CHARACTERISTICS

There is only one species in the genus *Streptobacillus*: *S. moniliformis*. This facultative anaerobe is nonmotile and tends to be highly pleomorphic. With respect to taxonomy, it is somewhat related to members of the order *Mycoplasmatales* (see Chapter 47); however, the exact phylogenetic origin of *S. moniliformis* remains to be delineated.[8]

EPIDEMIOLOGY AND PATHOGENESIS

The natural habitat of *S. moniliformis* is the upper respiratory tract (nasopharynx, larynx, upper trachea, and middle ear) of wild and laboratory rats; only rarely has this organism been isolated from other animals, such as mice, guinea pigs, gerbils, and turkeys.[8] *S. moniliformis* is pathogenic for humans and is transmitted by two routes:

- By rat bite or possibly by direct contact with rats[8]
- By ingestion of contaminated food such as unpasteurized milk or milk products and, less frequently, water[4]

The incidence of *S. moniliformis* infections is unknown, but human infections appear to occur worldwide.

The pathogenic mechanisms of *S. moniliformis* are unknown. The organism is known to spontaneously develop L forms (bacteria without cell walls) that may allow its persistence in some sites.[2]

SPECTRUM OF DISEASE

Despite the different modes of transmission, the clinical manifestations of *S. moniliformis* infection are similar. When *S. moniliformis* is acquired by ingestion, the disease is called **Haverhill fever.**

Patients with rat-bite, or Haverhill, fever develop acute onset of chills, fever, headache, vomiting, and often, severe joint pains. Febrile episodes may persist for weeks or months.[5] Within the first few days of illness, patients develop a rash on the palms, soles of the feet, and other extremities. Complications can occur and include endocarditis, septic arthritis, pneumonia, pericarditis, brain abscess, amnionitis, prostatitis, and pancreatitis.[6,8]

LABORATORY DIAGNOSIS

Specimen Collection, Transport, and Processing

Unfortunately, the diagnosis of rat-bite fever caused by *S. moniliformis* is often delayed owing to lack of exposure history, atypical clinical presentation, and the unusual microbiologic characteristics of the organism. Organisms may be cultured from blood or aspirates from infected joints, lymph nodes, or lesions. There are no special requirements for the collection, transport, and processing of these specimens except for blood. Because recovering *S. moniliformis* from blood cultures is impeded by concentrations of sodium polyanethol sulfonate (SPS) used in blood culture bottles, an alternative to using most commercially available bottles must be employed.[3,7] After collection using routine procedures (described in Chapter 52), blood and joint fluids are mixed with equal volumes of 2.5% citrate to prevent clotting and are then inoculated to brain-heart infusion cysteine broth supplemented with heated horse serum and yeast extract, commercially available fastidious anaerobe broth without SPS, or thiol broth.[3,5,7]

Direct Detection Methods

Pus or exudates should be smeared, stained with Gram or Giemsa stain, and examined microscopically (Figure 41-1). Direct detection of *S. moniliformis* using polymerase chain reaction is under development.[8]

Cultivation

As previously mentioned, *S. moniliformis* requires the presence of blood, ascitic fluid, or serum for growth.

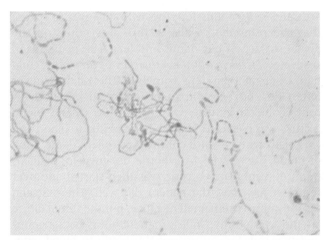

Figure 41-1 Gram stain of *Streptobacillus moniliformis* from growth in thioglycollate broth with 20% serum. (Courtesy Robert E. Weaver, Centers for Disease Control and Prevention, Atlanta, Ga.)

Growth occurs on blood agar, incubated in a very moist environment with 5% to 10% CO_2, usually after 48 hours of incubation at 37° C. Colonies are nonhemolytic. Addition of 10% to 30% ascitic fluid (available commercially from some media suppliers) or 20% horse serum should facilitate recovery of the organism. In broth cultures, the organism grows as "fluff balls" or "bread crumbs" near the bottom of the tube of broth or on the surface of the sedimented red blood cell layer in blood culture media. Colonies grown on brain-heart infusion agar supplemented with 20% horse serum are small, smooth, glistening, colorless or grayish, with irregular edges.

Colonies are embedded in the agar and may also exhibit a fried-egg appearance, with a dark center and a flattened, lacy edge. These colonies have undergone the spontaneous transformation to the L form. Stains of L-form colonies yield coccobacillary or bipolar staining coccoid forms; usually a special stain, such as the Dienes stain (performed by pathologists), is required. Acridine orange stain also reveals the bacteria when Gram stain fails because of lack of cell wall constituents.

Gram-stained organisms from standard colonies show extreme pleomorphism, with long, looped, filamentous forms, chains, and swollen cells. The club-shaped cells can be 2 to 5 times the diameter of the filament. The carbolfuchsin counterstain or the Giemsa stain may be necessary for visualization (see Figure 41-1).

Approach to Identification

S. moniliformis does not produce indole and is catalase-, oxidase-, and nitrate-negative, in contrast to organisms with which *Streptobacillus* may be confused, including *Actinobacillus, Haemophilus aphrophilus,* and *Cardiobacterium.* In addition, *S. moniliformis* is nonmotile and urea-and lysine decarboxylase–negative; H_2S is not produced in triple sugar iron agar but can be detected using lead acetate paper.[3]

Serodiagnosis

Serologic diagnosis of rat-bite fever is also useful; most patients develop agglutinating titers to the causative organism. The specialized serologic tests are performed only at national reference laboratories, because the disease is extremely rare in the United States. A titer of 1:80 is considered diagnostic, unless a fourfold rise in titer is demonstrated.[5]

ANTIMICROBIAL SUSCEPTIBILITY TESTING AND THERAPY

There are no standardized methods to determine *S. moniliformis* susceptibility to various antimicrobials. Different in vitro techniques, such as agar dilution and disk diffusion, have had similar results.[8] Although *S. moniliformis* is susceptible to a broad spectrum of antibiotics, penicillin is regarded as the drug of choice for human rat-bite fever. An aminoglycoside or tetracycline can be used to eliminate L forms or for patients allergic to penicillin.

PREVENTION

There are no vaccines available to prevent rat-bite fever. Disease is best prevented by avoiding contact with animals known to harbor the organism. Individuals with frequent animal contact should wear gloves, practice regular handwashing and avoid hand-to-mouth contact when handling rats or cleaning rat cages.

SPIRILLUM MINUS

GENERAL CHARACTERISTICS

Spirillum minus is a gram-negative, helical, strictly aerobic organism.

EPIDEMIOLOGY AND PATHOGENESIS

Little information is available regarding the epidemiology or pathogenesis of *S. minus,* but it is supposed to be similar in some regards to that of *S. moniliformis.* The mode of transmission of infection is by a rat bite.

SPECTRUM OF DISEASE

S. minus also causes rat-bite fever in humans and is referred to as **Sodoku.** The clinical signs and symptoms are similar to those caused by *S. moniliformis* except that arthritis is rarely seen in patients with Sodoku and swollen lymph nodes are prominent[1]; febrile episodes are also more predictable in Sodoku. Following the bite, the wound heals spontaneously but 1 to 4 weeks later reulcerates to form a granulomatous lesion at the same time that the patient develops constitutional symptoms of fever, headache, and a generalized, blotchy, purplish, maculopapular rash. Differentiation between rat-bite fever caused by *S. minus* and that caused by *S. moniliformis* is usually accomplished based on clinical presentations of the two infections and the isolation of the latter organism in culture. The incubation period for *S. minus* is much longer than that for streptobacillary rat-bite fever, which has occurred within 12 hours of the initial bite.

LABORATORY DIAGNOSIS

Specimen Collection, Transport, and Processing

Specimens commonly submitted for diagnosis of Sodoku include blood, exudate, or lymph node tissues. There are no requirements for specimen collection, transport, or processing of the organisms discussed in this chapter. Refer to Table 5-1 for general information on this subject.

Direction Detection Methods

Because *S. minus* cannot be grown on synthetic media, diagnosis relies on direct visualization of characteristic spirochetes in clinical specimens using Giemsa or Wright stains, or dark-field microscopy. *S. minus* appears as a thick, spiral, gram-negative organism with two or three coils and polytrichous polar flagella. Diagnosis is definitively made by injection of lesion material or blood into experimental white mice or guinea pigs and subsequent recovery 1 to 3 weeks after inoculation.

Serodiagnosis

There is no specific serologic test available for *S. minus* infection.

ANTIMICROBIAL SUSCEPTIBILITY TESTING AND THERAPY

Because this spirochete is nonculturable, routine antimicrobial susceptibility testing is not performed.

PREVENTION

No vaccines are available to prevent rat-bite fever. Disease is best prevented by avoiding contact with animals known to harbor the organism.

Case Study

An 8-year-old girl presented with a 7-day history of worsening flulike illness with fever, cough, and

arthralgias. By admission, the arthralgia was so severe that she refused to walk. A rash was noted on the dorsal surface of her hands and feet. The pediatrician suspected *Streptobacillus moniliformis* because the child had a pet rat that slept with her. The rat had never bitten her, but she did carry it around her neck. A routine blood culture was drawn. A second blood culture was collected in a tube and plated to blood and chocolate agars. A third culture was collected and 5 mL of blood was placed in a bottle designed for a maximum of 3 mL. Doxycycline was then started and over the next few days the patient did well. The first and second cultures remained negative, but the third culture attempt was positive in less than 24 hours with a gram-negative rod. Subcultures of the bottle to blood, chocolate, *Brucella* (aerobic and anaerobic), and *Legionella* selective agars incubated in CO_2 were negative at 48 hours.

QUESTIONS

1. Why did the laboratory set up a culture with more blood than recommended by the manufacturer?
2. Because *S. moniliformis* was suspected and there was no growth on any of the subcultures at 48 hours, what can the laboratory do to grow the organism on plated medium to identify it?
3. How is the organism definitively identified?

REFERENCES

1. Buranakitjaroen P, Nilganuwong S, Gherunpong V: Rat bite fever caused by *Streptobacillus moniliformis*, *Southeast Asian J Trop Med Public Health* 25:778, 1994.
2. Freundt EA: Experimental investigations into the pathogenicity of the L-phase variant of *Streptobacillus moniliformis*, *Acta Pathol Microbiol Scand* 38:246, 1956.
3. Lambe DW Jr, McPhedran AM, Mertz JA, et al: *Streptobacillus moniliformis* isolated from a case of Haverhill fever: biochemical characterization and inhibitory effect of sodium polyanethol sulfonate, *Am J Clin Pathol* 60:854, 1973.
4. McEvoy MB, Noah ND, Pilsworth R: Outbreak of fever caused by *Streptobacillus moniliformis*, *Lancet* ii:1361, 1987.
5. Rogosa M: *Streptobacillus moniliformis* and *Spirillum minus*. In Lennette EH, Balows A, Hausler WJ Jr, et al, editors: *Manual of clinical microbiology*, ed 4, Washington, DC, 1985, American Society for Microbiology.
6. Rupp ME: *Streptobacillus moniliformis* endocarditis: case report and review, *Clin Infect Dis* 14:769, 1992.
7. Shanson D, Pratt J, Green P: Comparison of media with and without "Panmede" for the isolation of *Streptobacillus moniliformis* from blood cultures and observations on the inhibitory effect of sodium polyanethol sulfonate, *J Med Microbiol* 19:181, 1985.
8. Wullenweber M: *Streptobacillus moniliformis*—a zoonotic pathogen: taxonomic considerations, host species, diagnosis, therapy, geographical distribution, *Lab Anim* 29:1, 1985.

CHAPTER 42 *Neisseria* and *Moraxella catarrhalis*

Genera and Species to Be Considered

Current Name	Previous Name
Moraxella catarrhalis	*Branhamella catarrhalis, Neisseria catarrhalis*
Neisseria gonorrhoeae	
Neisseria meningitidis	
Other *Neisseria* spp.	
N. cinerea	
N. lactamica	
N. polysaccharea	
N. subflava	*N. subflava, N. flava,* and *N. perflava*
N. sicca	
N. mucosa	
N. flavescens	

GENERAL CHARACTERISTICS

Species of the genus *Neisseria* discussed in this chapter and *Moraxella catarrhalis* are all oxidase-positive, gram-negative diplococci that do not elongate when exposed to subinhibitory concentrations of penicillin. The rodlike *Neisseria* spp. are described in Chapter 30.

EPIDEMIOLOGY

Except for *Neisseria gonorrhoeae*, the organisms considered in Table 42-1 are normal inhabitants of the upper respiratory tract of humans. *N. gonorrhoeae*, primarily a pathogen found in the urogenital tract, is never considered normal flora and is always considered clinically significant.

Transmission of the two pathogenic species of *Neisseria*, *N. gonorrhoeae* and *N. meningitidis*, is person to person. *N. gonorrhoeae* is sexually transmitted and *N. meningitidis* is spread by contaminated respiratory droplets. Infections caused by *M. catarrhalis* and the other *Neisseria* spp. usually involve a patient's endogenous strain.

PATHOGENESIS AND SPECTRUM OF DISEASE

As presented in Table 42-2, infections caused by *M. catarrhalis* are usually localized to the respiratory tract and rarely disseminate.[7,12]

N. gonorrhoeae is a leading cause of sexually transmitted disease, and infections caused by this organism usually are localized to the mucosal surfaces in the area of initial exposure to the organism (e.g., cervix, conjunctiva, pharyngeal surface, anorectal area, or urethra of males). Localized infections can be acute with a pronounced purulent response, or they may be asymptomatic. Not all infections remain localized, and dissemination from the initial infection site can lead to severe disseminated disease (see Table 42-2).

N. meningitidis is a leading cause of fatal bacterial meningitis.[9] However, the virulence factors responsible for the spread of this organism from a patient's upper respiratory tract to the bloodstream and meninges to cause life-threatening infections are not fully understood (see Table 42-2).

The other *Neisseria* spp. are not considered pathogens and are often referred to as the **saprophytic** *Neisseria*. Although they are most commonly encountered as contaminants in clinical specimens, they can occasionally be involved in bacteremia and endocarditis.[4]

LABORATORY DIAGNOSIS

SPECIMEN COLLECTION AND TRANSPORT

The pathogenic *Neisseria* spp. described in this chapter are very sensitive to drying and temperature extremes. In addition to general information on specimen collection and transport provided in Table 5-1, there are some special requirements for isolation of *N. gonorrhoeae* and *N. meningitidis*.

Swabs are acceptable for *N. gonorrhoeae* if the specimen will be plated within 6 hours. If cotton swabs are used, the transport medium should contain charcoal to inhibit toxic fatty acids present in the cotton fibers. Calcium alginate or rayon fibers are preferred, however.

Table 42-1   Epidemiology

Organism	Habitat (reservoir)	Mode of Transmission
Moraxella catarrhalis	Normal human flora of upper respiratory tract; occasionally colonizes female genital tract	Spread of patient's endogenous strain to normally sterile sites. Person-to-person nosocomial spread by contaminated respiratory droplets also can occur
Neisseria gonorrhoeae	Not part of normal human flora. Only found on mucous membranes of genitalia, anorectal area, oropharynx, or conjunctiva at time of infection	Person-to-person spread by sexual contact, including rectal intercourse and orogenital sex. May also be spread from infected mother to newborn during birth
Neisseria meningitidis	Colonizes oro- and nasopharyngeal mucous membranes of humans. Human carriage of the organism without symptoms is common	Person-to-person spread by contaminated respiratory droplets, usually in settings of close contact
Other Neisseria spp.	Normal human flora of the upper respiratory tract	Spread of patient's endogenous strain to normally sterile sites. Person-to-person spread may also be possible, but these species are not common causes of human infections

Table 42-2   Pathogenesis and Spectrum of Disease

Organism	Virulence Factors	Spectrum of Disease and Infections
Moraxella catarrhalis	Uncertain; factors associated with cell envelope probably facilitate attachment to respiratory epithelial cells	Most infections are localized to sites associated with the respiratory tract and include otitis media, sinusitis, and pneumonia. Lower respiratory tract infections often target elderly patients and those with chronic obstructive pulmonary disease. Rarely causes disseminated infections such as bacteremia or meningitis
Neisseria gonorrhoeae	Several surface factors, such as pili, mediate attachment to human mucosal cell surface, invasion of host cells, and survival in the presence of inflammatory cells	A leading cause of sexually transmitted diseases. Genital infections include purulent urethritis in males and cervicitis in females. These infections also may be asymptomatic. Other localized infections include pharyngitis, anorectal infections, and conjunctivitis (e.g., ophthalmia neonatorum of newborns acquired during birth from an infected mother). Disseminated infections result when the organism spreads from a local infection to cause pelvic inflammatory disease or disseminated gonococcal infection that includes bacteremia, arthritis, and metastatic infection at other body sites
Neisseria meningitidis	Surface structures, perhaps pili, facilitate attachment to mucosal epithelial cells and invasion to the submucosa. Once in the blood, survival is mediated by production of a polysaccharide capsule. Endotoxin release mediates many of the systemic manifestations of infection such as shock	Life-threatening, acute, purulent, meningitis. Meningitis may be accompanied by appearance of petechiae (i.e., rash) that is associated with meningococcal bacteremia (i.e., meningococcemia). Bacteremia leads to thrombocytopenia, disseminated intravascular coagulation, and shock. Disseminated disease is often fatal. Less common infections include conjunctivitis, pneumonia, and sinusitis
Other Neisseria spp.	Unknown; probably of low virulence	Rarely involved in human infections. When infections occur they can include bacteremia, endocarditis, and meningitis

The best method for culture and transport of *N. gonorrhoeae* is to inoculate the agar immediately after specimen collection and to place the medium in an atmosphere of increased carbon dioxide (CO_2) for transport. Specially packaged media consisting of selective agar in plastic trays that contain a CO_2-generating system are commercially available (JEMBEC plates) and widely used (Figure 42-1). The JEMBEC system is transported to the laboratory at room temperature. Upon receipt in the laboratory, the agar surface is cross-streaked to obtain isolated colonies, and the plate is incubated at 35° C in 3% to 5% CO_2.

The recovery of *N. gonorrhoeae* or *N. meningitidis* from normally sterile body fluids requires no special methods, blood cultures being a notable exception. Both organisms are sensitive to sodium polyanethol sulfonate (SPS), so the content of SPS in blood culture broths should not exceed 0.025%. In addition, if blood is first collected in Vacutainer tubes (Becton Dickinson and Company, Sparks, Md), which may

Figure 42-1 JEMBEC system. Plate contains modified Thayer-Martin medium. The CO_2-generating tablet is composed of sodium bicarbonate and citric acid. After inoculation, the tablet is placed in the well and the plate is closed and placed in the zip-lok plastic pouch. The moisture in the agar activates the tablet, generating a CO_2 atmosphere in the pouch.

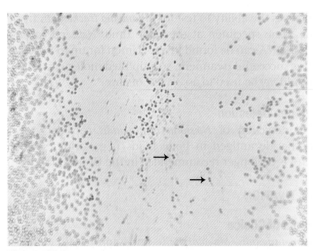

Figure 42-2 Gram stain of *Neisseria gonorrhoeae* showing gram-negative diplococci *(arrows)*.

contain concentrations of SPS toxic to gonococci and meningococci, the blood specimen must be transferred to the broth culture system within 1 hour of collection.

Nasopharyngeal swabs collected to detect *N. meningitidis* carriers should be plated immediately to the JEMBEC system, or they should be submitted on swabs placed in charcoal transport media.

SPECIMEN PROCESSING

The JEMBEC system should be incubated at 35° to 37° C as soon as the plate is received in the laboratory. Body fluids (e.g., joint or cerebrospinal fluid [CSF]) should be kept at room temperature or placed at 37° C before culturing because both the gonococci and meningococci are sensitive to cold.

Any volume of clear fluid greater than 1 mL suspected of containing either of these pathogens should be centrifuged at room temperature at 1500× *g* for 15 minutes. The supernatant fluid should then be removed and the sediment should be vortexed and inoculated onto the appropriate media (described later).

Any specimens or cultures in which *N. meningitidis* is a consideration should be handled in a biological safety cabinet to avoid laboratory-acquired infections.

DIRECT DETECTION METHODS

Gram Stain

Members of the genus *Neisseria* discussed in this chapter and *Moraxella catarrhalis* appear as gram-negative diplococci (Figure 42-2) with adjacent sides flattened. The direct Gram stain of urethral discharge from

symptomatic males with urethritis is an important test for gonococcal disease. The appearance of gram-negative diplococci inside polymorphonuclear leukocytes is diagnostic in this situation. However, because the normal vaginal and rectal flora are composed of gram-negative coccobacilli, which can resemble *Neisseria* spp., direct examination of endocervical secretions in symptomatic women is still only presumptive evidence of gonorrhea and the diagnosis must be confirmed by culture.

The direct Gram stain of body fluids for either *N. gonorrhoeae* or *N. meningitidis* is best accomplished using a cytocentrifuge, which can concentrate small numbers of organisms 100-fold.

Commercial Molecular Assays

Molecular assays have replaced old enzyme-linked immunosorbent assay systems for rapid diagnosis of *Neisseria gonorrhoeae*. The Food and Drug Administration has cleared a number of amplified and nonamplified tests. (For a discussion of molecular technology, see Chapter 8.) The nonamplified DNA probe assay (Gen-Probe, San Diego, Calif) has a chemiluminescent detection system for direct detection of gonococcal ribosomal RNA in genital and conjunctival specimens. This test performs well in high-risk patients, is rapid (results are available in 2 hours), and is suitable for screening many patients simultaneously.

Amplified assays are more sensitive than the nonamplified assay and are commercially available from Roche Diagnostic Systems (Branchburg, NJ), Gen-Probe (San Diego, Calif), and Becton Dickinson and Company (Sparks, Md). These tests are suitable for large-scale screening programs, but none are admissible as evidence in medicolegal cases. An advantage of all molecular assays is the ability to test for *Chlamydia trachomatis* from the same specimen at the same time. *Neisseria gonorrhoeae* DNA can be found in a specimen for up to 3 weeks after successful treatment, so amplified molecular assays should *not* be used to assess test of cure.

Antigen Detection

The detection of *Neisseria meningitidis* capsular polysaccharide antigen in body fluids (e.g., urine, serum, CSF) is no longer recommended in the United States.

CULTIVATION

Media of Choice

N. meningitidis, M. catarrhalis, and saprophytic *Neisseria* spp. grow well on 5% sheep blood and chocolate agars; *N. gonorrhoeae* is more fastidious and requires an enriched chocolate agar for growth on primary culture. Because gonococci, and sometimes meningococci, must be isolated from sites that contain large numbers of normal flora (e.g., genital or upper respiratory tracts), selective media have been developed to facilitate their recovery. The first of these was Thayer-Martin medium, a chocolate agar with an enrichment supplement (IsoVitaleX) and the antimicrobials colistin (to inhibit gramnegative bacilli), nystatin (to inhibit yeast), and vancomycin (to inhibit gram-positive bacteria). This original medium was subsequently modified to include trimethoprim (to inhibit swarming *Proteus*), and its name was changed to MTM (modified Thayer-Martin) medium. Martin Lewis (ML) medium is similar to MTM except that anisomycin, an antifungal agent, is substituted for nystatin and the concentration of vancomycin is increased.

A transparent medium containing lysed horse blood, horse plasma, yeast dialysate, and the same antibiotics as MTM, called **New York City (NYC) medium,** also has been used. The advantage of NYC medium is that genital mycoplasma (*Mycoplasma hominis* and *Ureaplasma urealyticum;* see Chapter 47 for more information regarding these organisms) will also grow on this agar. Some strains of *N. gonorrhoeae* are inhibited by the concentration of vancomycin in the selective media, so the addition of nonselective chocolate agar is recommended, especially in suspect cases that are culture-negative or for sterile specimens (e.g., joint fluid).

Unlike the pathogenic species, some of the saprophytic *Neisseria* spp. (*N. flavescens, N. mucosa, N. sicca,* and *N. subflava*) may grow on MacConkey agar, although poorly. *N. gonorrhoeae* and *N. meningitidis* will grow in most broth blood culture media but grow poorly in common nutrient broths such as thioglycollate and brain-heart infusion. *M. catarrhalis* and the other *Neisseria* spp. grow well in almost any broth medium.

Figure 42-3 Candle jar.

Incubation Conditions and Duration

Agar plates should be incubated at 35° to 37° C for 72 hours in a CO_2-enriched, humid atmosphere. The pathogenic neisseriae and *M. catarrhalis* grow best under conditions of increased CO_2 (3% to 7%). This atmosphere can be achieved using a candle jar, CO_2-generating pouch, or CO_2 incubator. Only white, unscented candles should be used in candle jars because other types may be toxic to *N. gonorrhoeae* and *N. meningitidis.*

Humidity can be provided by placing a pan with water in the bottom of a CO_2 incubator or by placing a sterile gauze pad soaked with sterile water in the bottom of a candle jar (Figure 42-3).

Colonial Appearance

Table 42-3 describes the colonial appearance and other distinguishing characteristics (e.g., pigment) on chocolate agar.

APPROACH TO IDENTIFICATION

Various commercial systems are available for the rapid identification of the coccoid *Neisseria* spp. and *M. catarrhalis.* Some of these systems are described briefly in Table 13-1. These systems employ biochemical or enzymatic substrates and work very well for the pathogenic species (*N. gonorrhoeae, N. meningitidis,* and *M. catarrhalis*). A heavy inoculum of the organism is required, but because these systems detect the activity of preformed enzymes, viability of the organisms in the inoculum is not essential. Manufacturers' instructions should be followed exactly; several systems have been developed only for strains isolated on selective media and should not be used to test other gram-negative diplococci.

Table 42-3   Colonial Appearance and Other Characteristics on Chocolate Agar*

Organism	Appearance
Moraxella catarrhalis	Large, nonpigmented or gray, opaque, smooth; friable "hockey puck" consistency; colony may be moved intact over surface of agar
Neisseria gonorrhoeae	Small, grayish white, convex, translucent, shiny colonies with either smooth or irregular margins; may be up to five different colony types on primary plates
N. meningitidis	Medium, smooth, round, moist, gray to white; encapsulated strains are mucoid; may be greenish cast in agar underneath colonies
N. cinerea	Small, grayish white; translucent; slightly granular
N. flavescens	Medium, yellow, opaque, smooth
N. lactamica	Small, nonpigmented or yellowish, smooth, transparent
N. mucosa	Large, grayish white to light yellow, translucent; mucoid because of capsule
N. polysaccharea	Small, grayish white to light yellow, translucent, raised
N. sicca	Large, nonpigmented, wrinkled, coarse and dry, adherent
N. subflava	Medium, greenish yellow to yellow, smooth, entire edge

*Appearance on blood agar is the same as on chocolate agar except for pigmentation; colonies are less opaque on blood agar.

Table 42-4   Biochemical and Physiologic Characteristics of *Moraxella catarrhalis* and Coccoid *Neisseria* spp.

Organism	Growth on: Modified Thayer-Martin*	Nutrient Agar at 35° C	Blood or Chocolate Agar at 25° C	Rapid Fermentation Sugars Glucose	Maltose	Lactose	Nitrate Reduction	Gas from Nitrate Reduction	0.1% Nitrite Reduction
*Moraxella catarrhalis***	v	+	+	−	−	−	+	−	v
Neisseria cinerea[†]	v	+	−	−‡	−	−	−	−	+
N. flavescens	−	+	+	−	−	−	−	−	+[§]
N. gonorrhoeae[ǁ]	+	−	−	+	−	−	−	−	−
N. lactamica	+	v	v	+	+	+	−	−	+
N. meningitidis	+	−	−	+	+	−	−	−	v
N. mucosa	−	+	+	+ or (+)	+	−	+	+	+
N. sicca[¶]	−	+	+	+ or (+)	+	−	−	−	+
N. subflava[¶]	−	+	+	v	+	−	−	−	+

Data compiled from references 5 and 13.

+, >90% of strains positive; (+), >90% of strains positive but reaction may be delayed (i.e., 2 to 7 days); −, >90% of strains negative; v, variable.

*Growth defined as >10 colonies.

** Butyrate and DNase positive.

[†]*Neisseria cinerea* may be differentiated from *N. flavescens* by a positive reaction with the amylosucrase test.

‡Some strains of *N. cinerea* may appear glucose-positive in some rapid systems and be mistaken for *N. gonorrhoeae*. However, *N. cinerea* grows on nutrient agar at 35° C and reduces nitrite, unlike the gonococcus.

§Only 2 of 10 strains were tested.

ǁ*Kingella denitrificans* may grow on modified Thayer-Martin agar and be mistaken for *N. gonorrhoeae* on microscopic examination. However, *K. denitrificans* can reduce nitrate and is catalase-negative, unlike the gonococcus.

¶*Neisseria subflava* produces a yellow pigment on Loeffler's agar; *N. sicca* does not.

Biochemical Identification

Table 42-4 shows some conventional biochemical tests that traditionally have been used to definitively identify these organisms. The extent to which identification of isolates is carried out depends on the source of the specimen and the suspected species of the organism involved.

An isolate from a child or a case of sexual abuse must be identified unequivocally, because of the medicolegal ramifications of these results. It is recommended that these organisms be identified using at least two different types of tests, that is, biochemical, immunologic, enzymatic, or the nonamplified DNA probe previously discussed. Isolates from normally sterile body fluids should also be completely identified. However, isolates from genital sites of adults at risk of sexually transmitted disease can be identified presumptively, that is, oxidase-positive, gram-negative diplococci growing on gonococcal selective agar. Likewise, an oxidase-positive, gram-negative diplococcus that hydrolyzes tributyrin from an eye or ear culture can be identified as *M. catarrhalis* (see Figure 13-6).[8]

Comments Regarding Specific Organisms

Determination of carbohydrate utilization patterns has historically been performed in cystine trypticase soy agar (CTA) with 1% dextrose, maltose, lactose, and sucrose. This medium is no longer widely used because it does not work well for oxidative *Neisseria* spp., specifically *N. gonorrhoeae* and *N. meningitidis*. Therefore, carbohydrate utilization patterns are currently determined by inoculating an extremely heavy suspension of the organism to be tested in a small volume of buffered, low-peptone substrate with the appropriate carbohydrate (see Chapter 7 for a detailed description of this test).

The saprophytic *Neisseria* spp. are not routinely identified in the clinical laboratory. *N. cinerea* may be misidentified as *N. gonorrhoeae* if the isolate produces a weak positive glucose reaction. However, it will grow on nutrient agar at 35° C, whereas the gonococcus will not. Moreover, *N. cinerea* is inhibited by colistin, whereas *N. gonorrhoeae* is not.

M. catarrhalis may be differentiated from the gonococci and meningococci based on its growth on blood agar at 22° C and on nutrient agar at 35° C, the reduction of nitrate to nitrite, its inability to utilize carbohydrates, and its production of DNase. *M. catarrhalis* is the only member of this group of organisms that hydrolyzes DNA.

Immunoserologic Identification

Particle agglutination methods are available for immunoserologic identification of *N. gonorrhoeae*. They include the Phadebact GC OMNI test (Karo Bio Diagnostics AB, Huddinge, Sweden) and the GonoGen II test (Becton Dickinson and Company, Sparks, Md). These tests can be performed from colonies growing on primary plates; isolates are typed with specific monoclonal antibodies.

Serotyping

Antisera are commercially available for identifying *N. meningitidis* serogroups A, B, C, Y, and W135, the types that most frequently cause systemic disease in the United States.

SERODIAGNOSIS

Serodiagnostic techniques are not generally used for the laboratory diagnosis of infections caused by the organisms discussed in this chapter.

ANTIMICROBIAL SUSCEPTIBILITY TESTING AND THERAPY

Although beta-lactamase production is common among *M. catarrhalis* isolates, many beta-lactam antibiotics maintain activity. Several other agents are also effective so that susceptibility testing to guide therapy is not routinely required (Table 42-5).

Standard methods have been established for performing in vitro susceptibility testing with *N. gonorrhoeae* and *N. meningitidis* (see Chapter 12 for details regarding these methods). In addition, various agents are available that may be considered for testing and therapeutic use. Although resistance to penicillin by production of beta-lactamase has become widespread among *N. gonorrhoeae*, resistance to ceftriaxone, which is not notably affected by the enzyme, has not been described.[10] Therefore, routine testing of isolates as a guide for using this therapeutic agent does not appear to be necessary. On the other hand, quinolones also are widely used to treat gonorrhea, but resistance to these agents is emerging and testing methods to detect and track this resistance are needed.[5,6,11]

Beta-lactamase production in *N. meningitidis* is extremely rare, but decreased susceptibility to penicillin mediated by altered penicillin-binding proteins is emerging.[1,2,14] However, optimum laboratory methods for detecting this relatively low level of resistance have not been established, and the impact of this resistance on the clinical efficacy of penicillin is not known.

PREVENTION

A single-dose vaccine to the polysaccharide capsular antigens of *N. meningitidis* groups A, C, Y, and W135 is available in the United States.[3] It is given to military recruits, patients without spleens who are older than age 2, and laboratory workers. Chemoprophylaxis with rifampin or ciprofloxacin (PO) or ceftriaxone (IM) is indicated for close contacts of patients with meningococcal meningitis. Household contacts, day care contacts, and health care workers who have given mouth-to-mouth resuscitation are at risk and should be treated within 24 hours. No chemoprophylaxis is necessary for asymptomatic carriers.

A single application of either a 2.5% solution of povidone-iodine, 1% tetracycline eye ointment, 0.5% erythromycin eye ointment, or 1% silver nitrate eye

Table 42-5 Antimicrobial Therapy and Susceptibility Testing

Species	Therapeutic Options	Potential Resistance to Therapeutic Options	Validated Testing Methods*	Comments
Moraxella catarrhalis	Several beta-lactams are effective, including: beta-lactam/beta-lactamase inhibitor combinations, cephalosporins, macrolides, quinolones, and trimethoprim-sulfamethoxazole	Commonly produce beta-lactamases that mediate resistance to ampicillin. Although not common, resistance to erythromycin, and trimethoprim-sulfamethoxazole may occur	See CLSI document M45	Testing to guide therapy is not routinely needed
Neisseria gonorrhoeae	Recommended therapy includes ceftriaxone or quinolones (e.g., ciprofloxacin, ofloxacin, levofloxacin). Macrolides also may be used	Penicillin resistance by beta-lactamase production is common. Resistance to quinolones is emerging	As documented in Chapter 12: disk diffusion, agar dilution, limited commercial methods	Testing by disk diffusion may not detect decrease in quinolone activity. No ceftriaxone resistance has been documented
Neisseria meningitidis	Supportive therapy for shock and antimicrobial therapy using penicillin, ceftriaxone, cefotaxime, or chloramphenicol	Subtle increases in beta-lactam resistance have been described, but clinical relevance is uncertain. Beta-lactamase production is extremely rare	As documented in Chapter 12: broth dilution, agar dilution	Testing to guide therapy is not routinely needed
Other *Neisseria* spp.	Usually susceptible to penicillin and other beta-lactams	Uncertain; potential for beta-lactamase production	Not available	

*Validated testing methods include those standard methods recommended by the Clinical and Laboratory Standards Institute (CLSI) and those commercial methods approved by the Food and Drug Administration (FDA).

drops is instilled in newborns within 1 hour of delivery to prevent gonococcal ophthalmia neonatorum.

Case Study

An elderly man had a history of chronic obstructive pulmonary disease (COPD) following 20 years of heavy smoking. He presented to the emergency department with shortness of breath, severe cough, and profuse, yellow sputum production. Crackles and wheezing could be heard on chest examination. A sputum culture was positive for many intracellular gram-positive, lancet-shaped diplococci. Gram-negative diplococci were also observed in the smear. The patient was placed on amoxicillin-clavulanic acid along with other supportive measures.

QUESTIONS

1. List the agents most often found to be involved in acute infections of COPD patients.
2. List the required tests to definitively and rapidly identify *Moraxella catarrhalis*.

3. The culture from this patient grew *Streptococcus pneumoniae* and *Moraxella catarrhalis*. The *S. pneumoniae* was susceptible to penicillin but the organism was still present in the sputum, even though the patient was being treated with amoxicillin before the acute episode that brought him to the emergency department. Can you explain this observation?

REFERENCES

1. Abadi FJ, Yakubu DE, Pennington TH: Antimicrobial susceptibility of penicillin-sensitive and penicillin-resistant meningococci, *J Antimicrob Chemother* 35:687, 1995.
2. Blondeau JM, Ashton FE, Isaacson M, et al: *Neisseria meningitidis* with decreased susceptibility to penicillin in Saskatchewan, Canada, *J Clin Microbiol* 33:1784, 1995.
3. Committee on Infectious Diseases: *2006 Red book: report of the Committee on Infectious Diseases,* ed 27, Elk Grove, Ill, 2006, American Academy of Pediatrics.
4. Heiddal S, Sverrisson JT, Yngvason FE, et al: Native valve endocarditis due to *Neisseria sicca*: case report and review, *Clin Infect Dis* 16:667, 1993.
5. Janda WM, Knapp JS: *Neisseria* and *Moraxella catarrhalis.* In Murray PR, Baron EJ, Jorgensen JH, et al, editors: *Manual of clinical microbiology,* ed 8, Washington, DC, 2003, ASM Press.

6. Kam KM, Wong PW, Cheung MM, et al: Detection of quinolone-resistant *Neisseria gonorrhoeae, J Clin Microbiol* 34:1462, 1996.

7. Myer GA, Shope TR, Waeker NJ, et al: *Moraxella (Branhamella) catarrhalis* bacteremia in children, *Clin Pediatr* 34:146, 1995.

8. National Committee for Clinical Laboratory Standards: Abbreviated identification of bacteria and yeast; M35-A, Wayne, Pa, 2002, NCCLS.

9. Riedo FX, Plikaytis BD, Broome CV: Epidemiology and prevention of meningococcal disease, *Pediatr Infect Dis J* 14:643, 1995.

10. Schwebke JR, Whittington W, Rice RJ, et al: Trends in susceptibility of *Neisseria gonorrhoeae* to ceftriaxone from 1985 through 1991, *Antimicrob Agents Chemother* 39:917, 1995.

11. Tanaka M, Matsumoto T, Kobayashi I, et al: Emergence of in vitro resistance to fluoroquinolones in *Neisseria gonorrhoeae* isolated in Japan, *Antimicrob Agents Chemother* 39:2367, 1995.

12. Verghese A, Berk SL: *Moraxella (Branhamella) catarrhalis, Infect Dis Clin North Am* 5:523, 1991.

13. Weyant RS, Moss CW, Weaver RE, et al, editors: *Identification of unusual pathogenic gram-negative aerobic and facultatively anaerobic bacteria,* ed 2, Baltimore, 1996, Williams & Wilkins.

14. Woods CR, Smith AL, Wasilauskas BL, et al: Invasive disease caused by *Neisseria meningitidis* relatively resistant to penicillin in North Carolina, *J Infect Dis* 170:453, 1994.

ADDITIONAL READING

Apicella MA: *Neisseria meningitidis.* In Mandell GL, Bennett JE, Dolin R, editors: *Principles and practice of infectious diseases,* ed 6, Philadelphia, 2005, Elsevier Churchill Livingstone.

Handsfield HH, Sparling PF: *Neisseria gonorrhoeae.* In Mandell GL, Bennett JE, Dolin R, editors: *Principles and practice of infectious diseases,* ed 6, Philadelphia, 2005, Elsevier Churchill Livingstone.

Murphy TF: *Moraxella (Branhamella) catarrhalis* and other gram-negative cocci. In Mandell GL, Bennett JE, Dolin R, editors: *Principles and practice of infectious diseases,* ed 6, Philadelphia, 2005, Elsevier Churchill Livingstone.

CHAPTER **43** **Overview and General Considerations**

GENERAL CHARACTERISTICS

The organisms described in this chapter and in Chapter 44 usually will not grow in the presence of oxygen. These chapters also include some microaerophilic organisms, such as *Actinomyces* spp., *Bifidobacterium* spp., and *Clostridium* spp., that can grow in the presence of either reduced or atmospheric oxygen but grow best under anaerobic conditions.

EPIDEMIOLOGY

Most of the anaerobic bacteria that cause infections in humans are also part of our normal flora. The ecology of these organisms is such that various species and genera exhibit preferences for the body sites that they inhabit (Table 43-1). Other pathogenic anaerobes (e.g., *Clostridium botulinum* and *Clostridium tetani*) are soil and environmental inhabitants and are not considered part of human flora.

The ways in which anaerobic infections are acquired are summarized in Table 43-2. Although person-to-person nosocomial spread of *Clostridium difficile* among hospitalized patients presents an enormous clinical and infection control dilemma, the majority of anaerobic infections occurs when a patient's normal flora gains access to a sterile site as a result of disruption of some anatomic barrier.[2]

Table 43-1 Incidence of Anaerobes as Normal Flora of Humans

Genus	Skin	Upper Respiratory Tract*	Intestine	External Genitalia	Urethra	Vagina
Gram-Negative Bacteria						
Bacteroides	0	±	2	±	±	±
Prevotella	0	2	2	1	±	1
Porphyromonas	0	1	1	U	U	±
Fusobacterium	0	2	1	U	U	±
Veillonella	0	2	1	0	U	1
Gram-Positive Bacteria						
Finegoldia magna	1	2	2	1	±	1
Micromonas micros	1	2	2	1	±	1
Clostridium	±	±	2	±	±	±
Actinomyces	0	1	1	0	0	±
Bifidobacterium	0	1	2	0	0	±
Eubacterium	0	1	2	U	U	1
Lactobacillus	0	1	1	0	±	2
Propionibacterium	2	1	±	±	±	1

Modified from Summanen PE, Baron EJ, Citron DM et al: *Wadsworth anaerobic bacteriology manual,* ed 5, Belmont, Calif, 1993, Star.
U, Unknown; *0,* not found or rare; *±,* irregular; *1,* usually present; *2,* present in large numbers.
*Includes nasal passages, nasopharynx, oropharynx, and tonsils.

Table 43-2 Acquisition of Anaerobic Infections and Diseases

Mode of Acquisition	Examples
Endogenous strains of normal flora gain access to normally sterile sites, usually as result of one or more predisposing factors that compromise normal anatomic barriers (e.g., surgery or accidental trauma) or alter other host defense mechanisms (e.g., malignancy, diabetes, burns, immunosuppressive therapy, aspiration, etc.)	Wide variety of infections involving several anatomic locations, including bacteremia, head and neck infections, dental and orofacial infections, pneumonia and other infections of the thoracic cavity, intraabdominal and obstetric and gynecologic infections, bite wound and other soft tissue infections, and gangrene (i.e., clostridial myonecrosis). Organisms most commonly encountered in these infections include *Bacteroides fragilis* group, *Prevotella* spp., *Porphyromonas* spp., *Fusobacterium nucleatum,* *Peptostreptococcus,* and *Clostridium perfringens*
Contamination of existing wound or puncture by objects contaminated with toxigenic *Clostridium* spp.	Tetanus *(Clostridium tetani),* gas gangrene *(Clostridium perfringens,* and, less commonly, *C. septicum, C. novyi,* and others)
Ingestion of preformed toxins in vegetable or meat-based foods	Botulism *(Clostridium botulinum),* and other clostridial food poisonings *(C. perfringens)*
Colonization of gastrointestinal tract with potent toxin-producing organism	Infant botulism *(C. botulinum)*
Person-to-person spread	Nosocomial spread of *Clostridium difficile*–induced diarrhea and pseudomembranous colitis; bite-wound infections caused by variety of anaerobic species

PATHOGENESIS AND SPECTRUM OF DISEASE

The types of infections and diseases in humans that are caused by anaerobic bacteria span a wide spectrum. Certain species, such as *C. botulinum* and *C. tetani,* produce some of the most potent toxins known. In contrast, specific virulence factors for the organisms most commonly encountered in infections (e.g., *B. fragilis* group, *C. difficile*) are not well understood (Table 43-3).

Bile
Biopsy of endometrial tissue obtained with an endometrial suction curette (Pipelle, Unimar, Inc., Wilton, Conn.)
Blood
Bone marrow
Bronchial washings obtained with double-lumen plugged catheter
Cerebrospinal fluid
Culdocentesis aspirate
Decubitus ulcer, if obtained from base of lesion after thorough debridement of surface debris
Fluid from normally sterile site (e.g., joint)
Material aspirated from abscesses (the best specimens are from loculated or walled-off lesions)
Percutaneous (direct) lung aspirate or biopsy
Peritoneal (ascitic) fluid
Sulfur granules from draining fistula
Suprapubic bladder aspirate
Thoracentesis (pleural) fluid
Tissue obtained at biopsy or autopsy
Transtracheal aspirate
Uterine contents, if collected using a protected swab

Bronchial washing or brush (except if collected with double-lumen plugged catheter)
Coughed (expectorated) sputum
Feces (except for *Clostridium difficile*)
Gastric or small-bowel contents (except in blind loop syndrome)
Ileostomy or colostomy drainage
Nasopharyngeal swab
Rectal swab
Secretions obtained by nasotracheal or orotracheal suction
Swab of superficial (open) skin lesion
Throat swab
Urethral swab
Vaginal or cervical swab
Voided or catheterized urine

Most anaerobic infections involve a mixture of anaerobic and facultatively anaerobic organisms (e.g., *Enterobacteriaceae*), so that it is problematic to establish the extent to which a particular anaerobic species is contributing to infection. In addition, as ubiquitous members of our normal flora, anaerobic organisms frequently contaminate clinical materials. For these reasons, assigning clinical significance to anaerobic bacteria isolated in the laboratory is important, although often difficult.

SPECIMEN COLLECTION AND TRANSPORT

The proper collection and transport of specimens for anaerobic culture cannot be overemphasized. Because indigenous anaerobes are often present in large numbers as normal flora on mucosal surfaces, even minimal contamination of a specimen can give misleading results. Box 43-1 shows the specimens acceptable for anaerobic culture; Box 43-2 shows the specimens that are likely to be contaminated and are unacceptable for anaerobic culture. In general, material for anaerobic culture is best obtained by tissue biopsy or by aspiration using a needle and syringe. Use of swabs is a poor alternative because of excessive exposure of the specimen to the deleterious effects of drying, the possibility of contamination during collection, and the easy retention of microorganisms within the fibers of the swab. If a swab must be used, it should be from an oxygen-free transport system.

There are special collection instructions for some clostridial illnesses, that is, foodborne *C. perfringens* and *C. botulinum*, *C. difficile* pseudomembranous enterocolitis, and *C. septicum* neutropenic enterocolitis. Food and fecal specimens must be sent to a public health laboratory for confirmation of *C. perfringens* food poisoning; these should be transported at 4° C. The clinical diagnosis of botulism is confirmed by demonstration of botulinum toxin in serum, feces, vomitus, or gastric contents, as well as by the recovery of the organism from the stool of patients. The Centers for Disease Control and Prevention maintains a 24-hour/day, 365-day/year hotline to provide emergency assistance in cases of botulism. Botulinum toxin is a potential bioweapon, so clinical microbiologists should be alert to numbers of potential cases of botulism. Specimens for infant botulism should include serum and stool; those for wound botulism should include serum, stool, and tissue biopsy. Feces for *C. difficile* culture and toxin assay should be liquid or unformed; solid, formed stool or rectal swabs are adequate to detect carriers but not to detect active cases of enterocolitis. Stools should be placed in an anaerobic transport container if culture is to be performed. Specimens for toxin assay may be collected in leak-proof containers and may be stored for up to 3 days at 4° C or frozen at –70° C. The specimens of choice for neutropenic enterocolitis involving *C. septicum* are three different blood cultures, stool, and lumen contents or tissue from the involved ileocecal area; a muscle biopsy should also be collected if myonecrosis (death of muscle tissue) is suspected.

A crucial factor in the final success of anaerobic cultures is the transport of the specimen; the lethal effect of atmospheric oxygen must be nullified until the specimen can be processed in the laboratory. Recapping a syringe and transporting the needle and syringe to the laboratory is no longer acceptable because of safety concerns involving needlestick injuries. Therefore, even aspirates must be injected into some type of oxygen-free

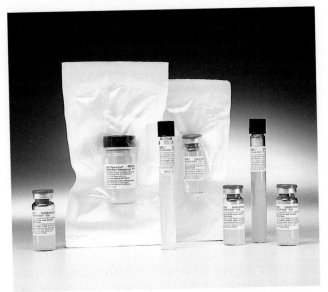

Figure 43-1 Anaerobic transport system for liquid specimens. Specimen is injected into tube through the rubber septum. Agar at the bottom contains oxygen tension indicator. (Courtesy BD Diagnostic Systems, Sparks, Md. Port-A-Cul is a trademark of Becton Dickinson and Company.)

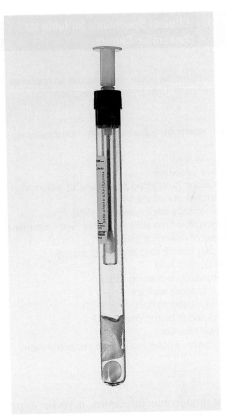

Figure 43-2 Anaerobic transport system for swab specimens. Vacutainer Anaerobic Specimen Collector, BD Diagnostic Systems (Sparks, Md). Sterile pack contains a sterile swab and an oxygen-free inner tube. When the specimen is collected, the swab is inserted back into the inner tube. Agar on the bottom of the outer tube contains an oxygen tension indicator.

transport tube or vial. Three different kinds of anaerobic transport systems are shown in Figures 43-1 to 43-3. Figure 43-1 is a rubber-stoppered collection vial containing an agar indicator system. The vial is gassed out with oxygen-free carbon dioxide (CO_2) or nitrogen. The specimen (pus, body fluid, or other liquid material) is injected through the rubber stopper after all air is expelled from the syringe and needle. If only a swab specimen can be obtained, a special collection device with an oxygen-free atmosphere is required (see Figure 43-2). When reinserting the swab, care must be taken not to tip the container, which would cause the oxygen-free CO_2 or nitrogen to spill out and be displaced by ambient air. Tissue can be placed in a small amount of liquid to keep it from drying out and then placed in an anaerobic pouch (see Figure 43-3). All specimens should be held at room temperature pending processing in the laboratory, because refrigeration can oxygenate the specimen.

SPECIMEN PROCESSING

Specimens for anaerobic culture may be processed on the open bench-top with incubation in anaerobic jars or pouches or in an anaerobic chamber. The roll tube method developed at the Virginia Polytechnic Institute is no longer widely used and is not discussed here.

Anaerobe Jars or Pouches

The most frequently used system for creating an anaerobic atmosphere is the anaerobe jar. Anaerobe jars

are available commercially from several companies, including Becton Dickinson and Company (Sparks, Md) (GasPak, Figure 43-4), EM Diagnostic Systems (Gibbstown, NJ), and Oxoid U.S.A. (Columbus, Md). These systems all use a clear, heavy plastic jar with a lid that is clamped down to make it airtight. Anaerobic conditions can be set up by two different methods. The easiest method uses a commercially available hydrogen and CO_2 generator envelope that is activated by either adding water (GasPak; Becton Dickinson and Company) or by the moisture from the agar plates (EM Diagnostic Systems; Oxoid). Production of heat within a few minutes (detected by touching the top of the jar) and subsequent development of moisture on the walls of the jar are indications that the catalyst and generator envelope are functioning properly. Reduced conditions are achieved in 1 to 2 hours, although the methylene blue or resazurin indicators take longer to decolorize. Alternatively, the "evacuation-replacement" method can be used. Air is removed from the sealed jar by drawing a

Figure 43-3 Anaerobic transport system for tissue specimens. Tissue is placed in a small amount of saline to keep it moist. It is inserted into a self-contained atmosphere-generating anaerobic bag for transportation. This system is called the GasPak Pouch. (Courtesy BD Diagnostic Systems, Sparks, Md.)

Figure 43-4 GasPak anaerobe jar (BD Diagnostic Systems, Sparks, Md). Inside the jar are inoculated plates, activated gas-generating envelope, and indicator strip. Palladium-coated alumina pellets that catalyze the reaction to remove oxygen are in a wire-mesh basket attached to the lid of the jar.

vacuum of 25 inches (62.5 cm) of mercury. This process is repeated two times, filling the jar with an oxygen-free gas, such as nitrogen, between evacuations. The final fill of the jar is made with a gas mixture containing 80% to 90% nitrogen, 5% to 10% hydrogen, and 5% to 10% CO_2. Many anaerobes require CO_2 for maximal growth. The atmosphere in the jars is monitored by including an indicator to check anaerobiosis. Anaerobe bags or pouches are useful for laboratories processing small numbers of anaerobic specimens. A widely used anaerobic pouch is shown in Figure 43-3. Besides specimen transport, the pouch also can be used to incubate one or two agar plates.

Holding Jars

If anaerobic jars or pouches are used for incubation, the use of holding jars is recommended during specimen processing and examination of cultures. Holding jars are anaerobic jars with loosely fitted lids that are attached by rubber tubing to nitrogen gas. Uninoculated plates are kept in holding jars pending use for culture setup, and inoculated plates are kept in holding jars pending incubation or examination; this minimizes exposure to oxygen.

Anaerobe Chamber

Anaerobic chambers, or glove boxes, are made of molded or flexible clear plastic. Specimens and other materials are placed in the chamber through an air lock. The technologist uses gloves (Forma Scientific, Marietta, Ohio) or sleeves (Sheldon Manufacturing, Inc., Cornelius, Ore) that form airtight seals around the arms to handle items inside the chamber (Figure 43-5). Media stored in the chamber are kept oxygen-free, and all work on a specimen from inoculation through workup is performed under anaerobic conditions. A gas mixture of 5% CO_2, 10% hydrogen, and 85% nitrogen and a palladium catalyst maintain the anaerobic environment inside the chamber.

ANAEROBIC MEDIA

Initial processing of anaerobic specimens involves inoculation of appropriate media. Table 43-4 lists commonly used anaerobic media. Primary plates should be freshly prepared or used within 2 weeks of preparation. Plates stored for longer periods accumulate peroxides and become dehydrated; this results in growth

Table 43-3   Pathogenesis and Spectrum of Disease for Anaerobic Bacteria

Organism	Virulence Factors	Spectrum of Disease and Infections
Clostridium perfringens	Produces several exotoxins; α-toxin is most important and mediates destruction of host cell membranes; enterotoxin inserts and disrupts membranes of mucosal cells	Gas gangrene, a life-threatening, toxin-mediated destruction of muscle and other tissues following traumatic introduction of the organism. Food poisoning caused by release of the toxin after ingestion of large quantities of organism. Disease is usually self-limiting and benign and is manifested by abdominal cramps, diarrhea, and vomiting
Clostridium tetani	Produces tetanospasmin, a neurotoxic exotoxin that disrupts nerve impulses to muscles	Tetanus (also commonly known as *lockjaw*). Organism establishes a wound infection and elaborates the potent toxin that mediates generalized muscle spasms. If untreated, spasms continue to be triggered by even minor stimuli, leading to exhaustion and, eventually, respiratory failure
Clostridium botulinum	Produces extremely potent neurotoxins	The disease botulism results from ingestion of preformed toxin in nonacidic vegetable or mushroom foodstuffs. Absorption of the toxin leads to nearly complete paralysis of respiratory and other essential muscle groups. Other forms of botulism can occur when the organism elaborates the toxin after it has colonized the gastrointestinal tract of infants (i.e., infant botulism). Wound botulism is more rare than the other forms, and occurs when *C. botulinum* produces the toxin from an infected wound site
Clostridium difficile	Produces toxin A, which is an enterotoxin that is thought to be primarily responsible for the gastrointestinal disease caused by this organism. Toxin B, a cytotoxin, has a less clear role in *C. difficile* infections	Organism requires diminution of normal gut flora by the activity of various antimicrobial agents to become established in the gut of hospitalized patients. Once established, elaboration of toxin(s) results in diarrhea (i.e., antibiotic-associated diarrhea) or potentially life-threatening inflammation of the colon. When the surface of the inflamed bowel is overlaid with a "pseudomembrane" composed of necrotic debris, white blood cells, and fibrin, the disease is referred to as *pseudo-membranous colitis*
Actinomyces spp., including *A. israelii A. meyeri* *A. naeslundii* *A. odontolyticus*	No well-characterized virulence factors. Infections usually require disruption of protective mucosal surface of the oral cavity, respiratory tract, gastrointestinal tract, and/or female genitourinary tract	Usually involved in mixed oral or cervicofacial, thoracic, pelvic, and abdominal infections caused by patient's endogenous strains; certain species (*A. viscosus* and *A. naeslundii*) also involved in periodontal disease and dental caries
Propionibacterium spp.	No definitive virulence factors known	Associated with inflammatory process in acne but only rarely implicated in infections of other body sites. As part of normal skin flora they are the most common anaerobic contaminants of blood cultures
Bifidobacterium spp.	No definitive virulence factors known	Not commonly found in clinical specimens. Usually encountered in mixed infections of pelvis or abdomen
Eubacterium spp.	No definitive virulence factors known	Usually associated with mixed infections of abdomen, pelvis, or genitourinary tract
Mobiluncus spp.	No definitive virulence factors known	Organisms are found in the vagina and have been associated with bacterial vaginosis, but their precise role in gynecologic infections is unclear. Rarely encountered in infections outside the female genital tract
Bacteroides fragilis group, other *Bacteroides* spp., *Bacteroides gracilis*, *Bacteroides ureolyticus*, *Prevotella* spp., *Porphyromonas* spp., *Fusobacterium nucleatum*, and other *Fusobacterium* spp.	These anaerobic gram-negative bacilli produce capsules, endotoxin, and succinic acid, which inhibit phagocytosis, and various enzymes that mediate tissue damage. Most infections still require some breach of mucosal integrity that allows the organisms to gain access to deeper tissues	Organisms most commonly encountered in anaerobic infections. Infections are often mixed with other anaerobic and facultatively anaerobic organisms. Infections occur throughout the body, usually as localized or enclosed abscesses, and may involve the cranium, periodontium, thorax, peritoneum, liver, and female genital tract. May also cause bacteremia, aspiration pneumonia, septic arthritis, chronic sinusitis, decubitus ulcers, and other soft tissue infections. The hallmark of most, but not all, infections is the production of a foul odor. In general, infections caused by *B. fragilis* group occur below the diaphragm; pigmented *Prevotella* spp., *Porphyromonas* spp., and *F. nucleatum* generally are involved in head and neck and pleuropulmonary infections
Finegoldia magna, *Micromonas micros*	No definitive virulence factors known	Most often found mixed with other anaerobic and facultatively anaerobic bacteria in cutaneous, respiratory, oral, or female pelvic infections
Veillonella spp.	No definitive virulence factors known	May be involved in mixed infections but rarely play a significant role

Table 43-4 Common Anaerobic Media

Medium	Components/Comments	Primary Purpose
Anaerobic blood agar	May be prepared with Columbia, Schaedler, CDC, *Brucella*, or brain-heart infusion base supplemented with 5% sheep blood, 0.5% yeast extract, hemin, L-cystine, and vitamin K_1	Nonselective medium for isolation of anaerobes and facultative anaerobes
Bacteroides bile esculin agar (BBE)	Trypticase soy agar base with ferric ammonium citrate and hemin; bile salts and gentamicin act as inhibitors	Selective and differential for *Bacteroides fragilis* group; good for presumptive identification
Laked kanamycin-vancomycin blood agar (LKV)	*Brucella* agar base with kanamycin (75 μg/ mL), vancomycin (7.5 μg/mL), vitamin K_1, (10 μg/mL), and 5% laked blood	Selective for isolation of *Prevotella* and *Bacteroides* spp.
Anaerobic phenylethyl alcohol agar (PEA)	Nutrient agar base, 5% blood, phenylethyl alcohol	Selective for inhibition of enteric gram-negative rods and swarming by some clostridia
Egg-yolk agar (EYA)	Egg yolk base	Nonselective for determination of lecithinase and lipase production by clostridia and fusobacteria
Cycloserine cefoxitin fructose agar (CCFA)	Egg yolk base with fructose, cycloserine (500 mg/L), and cefoxitin (16 mg/L); neutral red indicator	Selective for *Clostridium difficile*
Cooked meat (also called *chopped meat*) broth	Solid meat particles initiate growth of bacteria; reducing substances lower oxidation-reduction potential (Eh)	Nonselective for cultivation of anaerobic organisms; with addition of glucose, can be used for gas-liquid chromatography
Peptone-yeast extract glucose broth (PYG)	Peptone base, yeast extract, glucose, cysteine (reducing agent), resazurin (oxygen tension indicator), salts	Nonselective for cultivation of anaerobic bacteria for gas-liquid chromatography
Thioglycollate broth	Pancreatic digest of casein, soy broth, and glucose enrich growth of most bacteria. Thioglycollate and agar reduce Eh. May be supplemented with hemin and vitamin K_1	Nonselective for cultivation of anaerobes, as well as facultative anaerobes and aerobes

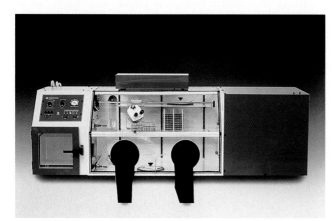

Figure 43-5 Gloveless anaerobe chamber. (Courtesy Anaerobe Systems, Morgan Hill, Calif.)

Figure 43-6 Prereduced, anaerobically sterilized (PRAS) plated media. (Courtesy Anaerobe Systems, Morgan Hill, Calif.)

inhibition. Reduction of media in an anaerobic environment eliminates dissolved oxygen but has no effect on the peroxides. Prereduced, anaerobically sterilized (PRAS) media are produced, packaged, shipped, and stored under anaerobic conditions. They are commercially available from Anaerobe Systems (Morgan Hill, Calif) (Figure 43-6) and have an extended shelf life of up to 6 months.

PREVENTION

A multiple-dose vaccine is available for prevention of tetanus.[1] The immunogen is adsorbed tetanus toxoid and is generally administered with diphtheria toxoid and pertussis vaccine as a triple antigen called DTP. Single boosters of diphtheria and tetanus (DT) or tetanus alone are recommended every 10 years. Immunoprophylaxis in wound management is based on the type of wound. Completely immunized individuals with minor and/or uncontaminated wounds do not need any specific treatment. However, completely immunized individuals with major and/or contaminated wounds should get a booster of tetanus toxoid if they have not had one in the previous 5 years. Finally, a partially immunized individual or one who has never been immunized should receive a dose of tetanus toxoid immediately. In addition, passive immunization with human tetanus immune globulin (TIG) should be given if the individual has a major wound or a wound contaminated with soil that contains animal feces.

Individuals who have eaten food suspected of containing botulinum toxin should be purged with cathartics (laxatives), have their stomach pumped, and be given high enemas.

REFERENCES

1. Committee on Infectious Diseases: *2006 Red book: report of the Committee on Infectious Diseases,* ed 27, Elk Grove, Ill, 2006, American Academy of Pediatrics.
2. Knoop FC, Owens M, and Crocker IC: *Clostridium difficile:* clinical disease and diagnosis, *Clin Microbiol Rev* 6:251, 1993.

ADDITIONAL READING

Allen SD, Emery CL, Lyerly DM: *Clostridium.* In Murray PR, Baron EJ, Jorgensen JH, et al, editors: *Manual of clinical microbiology,* ed 8, Washington, DC, 2003, ASM Press.

Bleck TP: *Clostridium botulinum.* In Mandell GL, Bennett JE, Dolin R, editors: *Principles and practice of infectious diseases,* ed 6, Philadelphia, 2005, Elsevier Churchill Livingstone.

Bleck TP: *Clostridium tetani,* In Mandell GL, Bennett JE, Dolin R, editors: *Principles and practice of infectious diseases,* ed 6, Philadelphia, 2005, Elsevier Churchill Livingstone.

Jousimies-Somer HR, Summanen P, Citron DM, et al: *Wadsworth-KTL anaerobic bacteriology manual,* ed 6, Belmont, Calif, 2002, Star.

Jousimies-Somer HR, Summanen PH, Wexler H: *Bacteroides, Porphyromonas, Prevotella, Fusobacterium,* and other anaerobic gram-negative bacteria. In Murray PR, Baron EJ, Pfaller Jorgensen JH, et al, editors: *Manual of clinical microbiology,* ed 8, Washington, DC, 2003, ASM Press.

Lorber B: *Bacteroides, Prevotella, Porphyromonas,* and *Fusobacterium* species (and other medically important anaerobic gram-negative bacilli). In Mandell GL, Bennett JE, Dolin R, editors: *Principles and practice of infectious diseases,* ed 6, Philadelphia, 2005, Elsevier Churchill Livingstone.

Lorber B: Gas gangrene and other *Clostridium*-associated diseases. In Mandell GL, Bennett JE, Dolin R, editors: *Principles and practice of infectious* diseases, ed 6, Philadelphia, 2005, Elsevier Churchill Livingston.

Mascini EM, Verdoef J: Anaerobic cocci. In Mandell GL, Bennett JE, Dolin R, editors: *Principles and practice of infectious diseases,* ed 6, Philadelphia, 2005, Elsevier Churchill Livingstone.

Moncla BJ, Hillier SL: *Peptostreptococcus, Propionibacterium, Lactobacillus, Actinomyces* and other non-sporeforming anaerobic gram-positive bacteria. In Murray PR, Baron EJ, Jorgensen JH et al, editors: *Manual of clinical microbiology,* ed 8, Washington DC, 2003, ASM Press.

Tzianabos AO, Kasper DL: Anaerobic infections: general concepts. In Mandell GL, Bennett JE, Dolin R, editors: *Principles and practice of infectious diseases,* ed 6, Philadelphia, 2005, Elsevier Churchill Livingstone.

MACROSCOPIC EXAMINATION OF SPECIMENS

Upon receipt in the laboratory, specimens should be inspected for characteristics that strongly indicate the presence of anaerobes, such as (1) foul odor; (2) sulfur granules associated with *Actinomyces* spp., *Propionibacterium* spp., or *Eubacterium nodatum;* or (3) brick-red fluorescence under long-wavelength ultraviolet (UV) light associated with pigmented *Prevotella* or *Porphyromonas.*

DIRECT DETECTION METHODS

ANTIGEN DETECTION

The cytotoxin (toxin B) of *Clostridium difficile* can be detected using a tissue culture assay. This assay, performed in various cell lines, is based on the neutralization of cytopathic effect if a cell-free fecal extract is adsorbed by either *Clostridium sordellii* or *C. difficile* antitoxins. Latex particle agglutination tests or enzyme-linked immunosorbent assays (ELISA) to detect toxins A and/or B are also available.

GRAM STAIN

The Gram stain is an important rapid tool for anaerobic bacteriology. Not only does it reveal the types and relative numbers of microorganisms and host cells present but also it serves as a quality control measure for the adequacy of anaerobic techniques. The absence of leukocytes does not rule out the presence of a serious anaerobic infection, however, because certain organisms, such as clostridia, produce necrotizing toxins that destroy white blood cells. A positive Gram stain with a negative culture may indicate (1) poor transport methods, (2) excessive exposure to air during specimen processing, (3) failure of the system (jar, pouch, or chamber) to achieve an anaerobic atmosphere, (4) inadequate types of media or old media, or (5) that microorganisms have been killed by antimicrobial therapy.

Standard Gram stain procedures and reagents are used, except that the safranin counterstain is left on for 3 to 5 minutes. Alternatively, 0.5% aqueous basic fuchsin can be used as the counterstain.

Table 44-1 indicates the cellular morphology on Gram stain of common anaerobes.

CULTIVATION.

MEDIA OF CHOICE

The primary plating media for inoculating anaerobic specimens can be picked from the media listed in Table 43-4. In general, they include a nonselective anaerobic blood agar and one or all of the following selective media: *Bacteroides* bile esculin agar (BBE), laked kanamycin-vancomycin blood agar (LKV), or anaerobic phenylethyl alcohol agar (PEA). In addition, aerobic 5% sheep blood agar, chocolate agar, and MacConkey agar are set up because most anaerobic infections are polymicrobic and may include aerobic or facultative anaerobic bacteria. A backup broth, usually thioglycollate, to enrich for small numbers of anaerobes may also be included on tissues and other sterile specimens. Most anaerobes grow well on any of the foregoing media.

Cultures for *C. difficile* are plated on a special selective medium, cycloserine cefoxitin fructose agar (CCFA). There are also selective media for certain groups of anaerobes, such as *Actinomyces,* although these are rarely used in the clinical laboratory.

Special anaerobic blood bottles containing various media, including thioglycollate broth, thiol broth, and Schaedler's broth, are commercially available. Although many anaerobes will grow in the aerobic blood culture bottle, it is better to use an unvented anaerobic broth when attempting to isolate these organisms from blood or bone marrow.

INCUBATION CONDITIONS AND DURATION

Inoculated plates should be immediately incubated under anaerobic conditions at 35° to 37° C for 48 hours. In general, cultures should not be exposed to oxygen until after 48 hours' incubation, because anaerobes are most sensitive to oxygen during their log phase of growth. Furthermore, colony morphology often changes dramatically between 24 and 48 hours. Plates incubated in an anaerobe chamber or bag can be examined at 24 hours without oxygen exposure for typical colonies of *B. fragilis* group or *C. perfringens.* Plates that show no growth at 48 hours should be incubated for at least 5 days before discarding.

The thioglycollate broth can be incubated anaerobically with the cap loose or aerobically with the cap tight. Broths should be inspected daily for 7 days.

Table 44-1   Gram-Stain Morphology, Colonial Appearance, and Other Distinguishing Features of Common Anaerobic Bacteria

Organism	Gram Stain*	Medium	Appearance
Actinomyces spp.	Gram-positive, branching, beaded or banded, thin, filamentous rods	Ana BA	Colonies of most species are small, smooth, flat, convex, gray-white, translucent, with entire margins; colonies of *A. israelii* and *A. gerencseriae* are white, opaque, and may resemble a "molar tooth"; *A. odontolyticus* turns red after several days in ambient air and may be beta-hemolytic
Anaerococcus spp.	Gram-positive cocci arranged in short chains or tetrads	Ana BA	Small, white, translucent, smooth
Atopobium spp.	Elongated gram-positive cocci occurring singly, in pairs, or in short chains	Ana BA	Resemble lactobacilli
Bacteroides distasonis	Gram-negative, straight rods with rounded ends; occur singly or in pairs	Ana BA BBE	Gray-white, circular, entire, convex, smooth, translucent to opaque; nonhemolytic At 48 hrs, colonies are >1 mm, circular, entire, raised, and either (1) low convex, dark gray, friable, and surrounded by a dark gray zone (esculin hydrolysis) and sometimes a precipitate (bile) or (2) glistening, convex, light to dark gray, and surrounded by a gray zone
B. fragilis	Gram-negative, pale-staining, pleomorphic rods with rounded ends; occur singly or in pairs; cells often described as resembling a safety pin (Figure 44-1)	Ana BA BBE	White to gray, circular, entire, convex, translucent to semiopaque; nonhemolytic (Figure 44-2) At 48 hrs, colonies are >1 mm, circular, entire, raised, and either (1) low convex, dark gray, friable, and surrounded by a dark gray zone (esculin hydrolysis) and sometimes a precipitate (bile) or (2) glistening, convex, light to dark gray, and surrounded by a gray zone (Figure 44-3)
B. ovatus	Gram-negative, ovoid rods with rounded ends; occur singly or in pairs	Ana BA BBE	Pale buff, circular, entire, convex, semiopaque; often mucoid; nonhemolytic At 48 hrs, colonies are >1 mm, circular, entire, raised, and either (1) low convex, dark gray, friable, and surrounded by a dark gray zone (esculin hydrolysis) and sometimes a precipitate (bile) or (2) glistening, convex, light to dark gray, and surrounded by a gray zone
B. thetaiotaomicron	Gram-negative, irregularly staining, pleomorphic rods with rounded ends; occur singly or in pairs	Ana BA BBE	White, circular, entire, convex, semiopaque, shiny, punctiform; nonhemolytic At 48 hrs, colonies are >1 mm, circular, entire, raised, and either (1) low convex, dark gray, friable, and surrounded by a dark gray zone (esculin hydrolysis) and sometimes a precipitate (bile) or (2) glistening, convex, light to dark gray, and surrounded by a gray zone
B. ureolyticus	Gram-negative, pale-staining, thin delicate rods with rounded ends; some curved	Ana BA	Small, translucent or transparent; may produce greening of agar on exposure to air; colonies corrode (pit) the agar (Figure 44-4); or may be smooth and convex or spreading
B. vulgatus	Gram-negative, pleomorphic rods with rounded ends; occur singly, in pairs, or in short chains; swellings or vacuoles may be seen	Ana BA BBE	Gray, circular, entire, convex, semiopaque; nonhemolytic At 48 hrs, colonies are >1 mm, circular, entire, raised, glistening, convex, light to dark gray but with no gray zone (esculin not hydrolyzed)
Bifidobacterium spp.	Gram-positive diphtheroid; coccoid or thin, pointed shape; or larger, highly irregular, curved rods with branching; rods terminate in clubs or thick, bifurcated (forked) ends ("dog bones")	Ana BA	Small, white, convex, shiny, with irregular edge
Bilophila wadsworthia	Gram-negative, pale-staining, delicate rods	Ana BA BBE	Small, translucent Grows at 3 to 5 days; colonies are usually gray with a black center because of production of hydrogen sulfide (H_2S); black center may disappear after exposure to air
Clostridium botulinum	Gram-positive, straight rods occurring singly or in pairs; spores usually subterminal and resemble a tennis racket	Ana BA	Gray-white; circular to irregular; usually beta-hemolytic

Table 44-1   Gram-Stain Morphology, Colonial Appearance, and Other Distinguishing Features of Common Anaerobic Bacteria—cont'd

Organism	Gram Stain*	Medium	Appearance
C. clostridioforme	Gram-positive rod that stains gram-negative; long, thin rods; spores usually not seen; elongated football shape with cells often in pairs	Ana BA	Small, convex, entire edge; nonhemolytic
C. difficile	Gram-positive straight rods; may produce chains of up to six cells aligned end to end; spores oval and subterminal	Ana BA CCFA	Large, white, circular, matte to glossy, convex, opaque; nonhemolytic; horse stable odor; fluoresces yellow-green Yellow, ground-glass colony (Figure 44-5)
C. perfringens	Gram-variable straight rods with blunt ends occurring singly or in pairs; spores seldom seen but if present are large and central to subterminal, oval, and swell cell; large boxcar shapes (Figure 44-6)	Ana BA	Gray to grayish yellow; circular, glossy, dome-shaped, entire, translucent; double zone of beta hemolysis (Figure 44-7)
C. ramosum	Gram-variable straight or curved rods; spores rarely seen but are round and terminal; more slender and longer than *C. perfringens*	Ana BA	Small, gray-white to colorless; circular to slightly irregular, smooth, translucent or semiopaque; nonhemolytic
C. septicum	Gram-positive in young cultures but becomes gram-negative with age; stains unevenly; straight or curved rods occurring singly or in pairs; spores subterminal, and oval, and swell cells	Ana BA	Gray; circular, glossy, translucent; markedly irregular to rhizoid margins resembling a "Medusa head"; beta-hemolytic; swarms over entire agar surface in less than 24 hrs
C. sordellii	Gram-positive rods; subterminal spores	Ana BA	Large colony with irregular edge
C. sporogenes	Gram-positive rods; subterminal spores	Ana BA	Colonies firmly adhere to agar; may swarm over agar surface
C. tertium	Gram-variable rods; terminal spores	Ana BA	Resembles *Lactobacillus* spp.
C. tetani	Gram-positive becoming gram-negative after 24 hrs' incubation; occur singly or in pairs; spores oval and terminal or subterminal with drumstick or tennis racket appearance	Ana BA	Gray; matte surface, irregular to rhizoid margin, translucent, flat; narrow zone of beta hemolysis; may swarm over agar surface
Collinsella aerofaciens[8]	Gram-positive chains of coccoid cells	Ana BA	Circular, entire, white center with translucent edge
Eggerthella lenta[9]	Gram-positive small, straight rod with rounded ends	Ana BA	Small, gray, translucent, circular, entire, convex
Eubacterium spp.	Gram-positive pleomorphic rods or coccobacilli occurring in pairs or short chains; *E. alactolyticum* has a seagull-wing shape similar to *Campylobacter*; *E. nodatum* is similar to *Actinomyces*, with beading, filaments, and branching	Ana BA	Small, gray, transparent to translucent, raised to convex; colonies of *E. nodatum* may resemble *A. israelii*
Finegoldia magna[11]	Gram-positive cocci with cells >0.6 μm in diameter; in pairs and clusters; resemble staphylococci	Ana BA	Tiny, gray, translucent; nonhemolytic
Fusobacterium mortiferum	Gram-negative, pale-staining, irregularly stained, highly pleomorphic rods with swollen areas, filaments, and large, bizarre, round bodies	Ana BA BBE	Circular; entire or irregular edge, convex or slightly umbonate, smooth, translucent; nonhemolytic > 1 mm in diameter, flat and irregular
F. necrophorum subsp. *necrophorum*[13]	Gram-negative, pleomorphic rods with round to tapered ends; may be filamentous or contain round bodies; becomes more pleomorphic with age	Ana BA	Circular, umbonate, ridged surface, translucent to opaque; fluoresces chartreuse; greening of agar on exposure to air; some strains beta-hemolytic
F. nucleatum subsp. *nucleatum*[2]	Gram-negative, pale-staining, long, slender, spindle-shaped with sharply pointed or tapered ends; occasionally the cells occur in pairs end to end; resembles *Capnocytophaga* (Figure 44-8)	Ana BA	Three colony types: bread crumblike (white; Figure 44-9), speckled, or smooth (gray to gray-white); greening of the agar on exposure to air; fluoresces chartreuse; usually nonhemolytic
F. varium	Gram-negative, unevenly staining, pleomorphic, coccoid, and rod shapes occurring singly or in pairs	Ana BA BBE	Gray-white center with colorless edge resembling a fried egg; circular, entire, convex, translucent; nonhemolytic > 1 mm in diameter, flat and irregular

Table 44-1 Gram-Stain Morphology, Colonial Appearance, and Other Distinguishing Features of Common Anaerobic Bacteria—cont'd

Organism	Gram Stain*	Medium	Appearance
Lactobacillus spp.	Gram-variable pleomorphic rods or coccobacilli; straight, uniform rods have rounded ends; short coccobacilli resemble streptococci	Ana BA	Resemble *Lactobacillus* spp. colonies on aerobic blood or chocolate agar, except colonies are usually larger when incubated anaerobically
Leptotrichia spp.	Gram-negative, large fusiform rods with one pointed end and one blunt end	Ana BA	Large, raspberry-like colonies
Micromonas micros[11]	Gram-positive cocci with cells <0.7 μm in diameter; occur in packets and short chains	Ana BA	Tiny, white, opaque; nonhemolytic
Mobiluncus spp.	Gram-variable, small, thin, curved rods; the two species can be divided based on length of the cell	Ana BA	Tiny colonies after 48 hrs' incubation; after 3 to 5 days colonies are small, low convex, and translucent
Peptococcus niger	Gram-positive, spherical cells occurring singly or in pairs, tetrads, and irregular masses	Ana BA	Tiny, black, convex, shiny, smooth, circular, entire edge; becomes light gray when exposed to air
Peptostreptococcus anaerobius	Gram-positive large coccobacillus; often in chains	Ana BA	Medium, gray-white, opaque; sweet, fetid odor; colonies usually larger than most anaerobic cocci (Figure 44-10)
Porphyromonas spp.	Gram-negative coccobacilli	Ana BA	Dark brown to black; more mucoid than *Prevotella;* except for *P. gingivalis,* fluoresces brick red (Figure 44-11)
Prevotella disiens	Gram-negative rods; occur in pairs or short chains	Ana BA LKV	White, circular, entire, convex, translucent to opaque, smooth, shiny; nonhemolytic; fluoresces brick red Black pigment (Figure 44-12)
P. melaninogenica	Gram-negative coccobacilli	Ana BA LKV	Dark center with gray to light brown edges; circular, entire, convex, smooth, shiny; nonhemolytic; fluoresces brick red Black pigment
Propionibacterium spp.	Gram-positive pleomorphic diphtheroid-like rod; club-shaped to palisade arrangements; called *anaerobic diphtheroids*	Ana BA	Young colonies are small and white to gray-white and become larger and more yellowish tan with age; *P. avidum* is beta-hemolytic
Schleiferella asaccharolytica	Gram-positive cocci occurring in pairs, tetrads, or irregular clusters	Ana BA	Small, gray, translucent
Veillonella parvula	Gram-negative, tiny diplococci in clusters, pairs, and short chains; unusually large cocci, especially in clusters, suggests *Megasphaera* or *Acidaminococcus*	Ana BA	Small, almost transparent; grayish white; smooth, entire, opaque, butyrous; may show red fluorescence under UV light (360 nm)

Ana BA, Anaerobic blood agar; *BBE, Bacteroides* bile esculin agar; *CCFA,* cycloserine cefoxitin fructose agar; *LKV,* laked kanamycin-vancomycin blood agar.
*Typical Gram stain appearance will be seen from broth (thioglycollate or peptone-yeast-glucose).

COLONIAL APPEARANCE

The colonial appearance and other characteristics (e.g., hemolysis and odor) of common anaerobes are described in Table 44-1.

APPROACH TO IDENTIFICATION

Complete identification of anaerobes can be costly, often requiring various biochemical tests, gas-liquid chromatography to analyze the metabolic end products of glucose fermentation, and/or gas chromatography for whole-cell long chain fatty acid methyl ester (FAME) analysis. Most clinical laboratories no longer perform complete identification of anaerobes, because presump-tive identification is just as useful in assisting the physician in determining appropriate therapy. There-fore, the approach to identification taken in this chapter emphasizes simple, rapid methods to identify commonly isolated anaerobic bacteria. Identification should pro-ceed in a stepwise fashion, beginning with examination of the primary plates.

EXAMINATION OF PRIMARY PLATES

Anaerobes are usually present in mixed culture with other anaerobes and facultative bacteria. The combi-nation of selective and differential agar plates yields information that suggests the presence and perhaps the types of anaerobe(s). Primary anaerobic plates should be

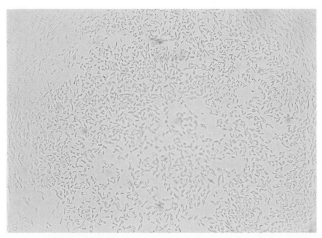

Figure 44-1 Gram stain of *Bacteroides fragilis*.

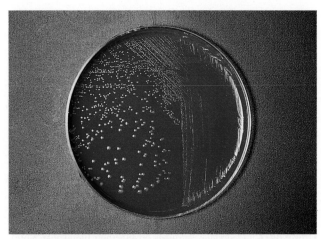

Figure 44-2 *Bacteroides fragilis* on anaerobic blood agar.

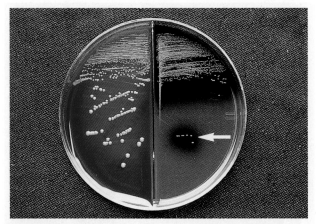

Figure 44-3 *Bacteroides fragilis* on *Bacteroides* bile esculin agar (BBE) *(arrow)*. (Courtesy Anaerobe Systems, Morgan Hill, Calif.)

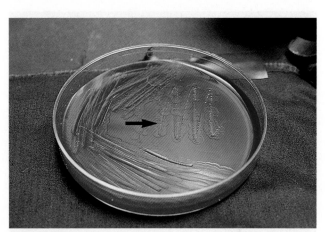

Figure 44-4 *Bacteroides ureolyticus* on anaerobic blood agar. Note pitting of agar *(arrow)*. (Courtesy Anaerobe Systems, Morgan Hill, Calif.)

Figure 44-5 *Clostridium difficile* on cycloserine cefoxitin fructose agar (CCFA). (Courtesy Anaerobe Systems, Morgan Hill, Calif.)

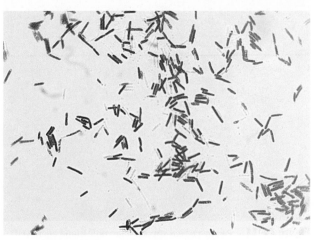

Figure 44-6 Gram stain of *Clostridium perfringens*.

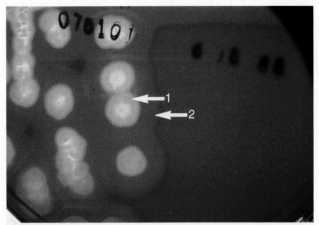

Figure 44-7   *Clostridium perfringens* on anaerobic blood agar. Note double zone of beta hemolysis. *1,* First zone; *2,* second zone. (Courtesy Anaerobe Systems, Morgan Hill, Calif.)

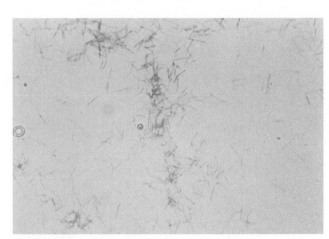

Figure 44-8   Gram stain of *Fusobacterium nucleatum* subsp. *nucleatum.* Note pointed ends.

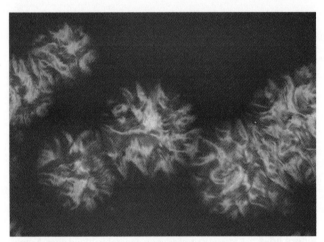

Figure 44-9   *Fusobacterium nucleatum* subsp. *nucleatum* on anaerobic blood agar. Note breadcrumb-like colonies and greening of agar.

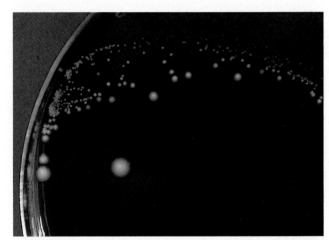

Figure 44-10   *Peptostreptococcus anaerobius* on anaerobic blood agar.

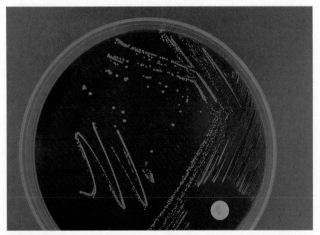

Figure 44-11   *Porphyromonas* spp. on anaerobic blood agar. Red fluorescence under ultraviolet light (365 nm).   (Courtesy Anaerobe Systems, Morgan Hill, Calif.)

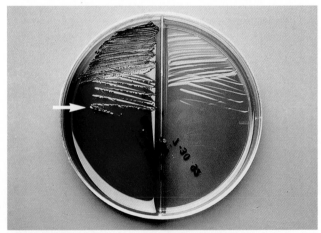

Figure 44-12   *Prevotella disiens* on laked kanamycin-vancomycin blood agar. Note black pigment *(arrow).*

examined with a hand lens (×8) or, preferably, a stereo-scopic microscope. Colonies should be described from the various media and semiquantitated.

All colony morphotypes from the nonselective anaerobic blood agar should be characterized and sub-cultured to purity plates, because facultative and obligately anaerobic bacteria frequently have similar colonial appearances. Colonies on the PEA (phenylethyl alcohol agar) are processed further only if they are different from colonies growing on the anaerobic blood agar or if colonies on the anaerobic blood agar are im-possible to subculture because of overgrowth by swarm-ing clostridia, *Proteus,* or other organisms.

The backup broth (e.g., thioglycollate) should be Gram-stained; if cellular types are seen that were not present on the primary plates, the broth should be subcultured.

SUBCULTURE OF ISOLATES

A single colony of each distinct morphotype is exam-ined microscopically using a Gram stain and is sub-cultured for **aerotolerance testing.** A sterile wooden stick or platinum loop should be used to subculture colonies to:

- A chocolate agar plate to be incubated in carbon dioxide (CO_2)
- An anaerobic blood agar plate to be incubated anae-robically (purity plate)

The chocolate agar plate should be inoculated first, so that if only the anaerobic blood agar plate grows, there is no question of not having enough organisms to initiate growth. The following antibiotic identification disks are placed on the first quadrant of the purity plate (Procedure 44-1):

Kanamycin, 1 mg

Colistin, 10 µg

Vancomycin, 5 µg

These disks aid in preliminary grouping of anae-robes and serve to verify the Gram stain, but they do not imply susceptibility of an organism for antibiotic therapy.

Three other disks may be added to the anaerobic blood agar plate at this time. A nitrate disk may be placed on the second quadrant for subsequent determi-nation of nitrate reduction. A sodium polyanethol-sulfonate (SPS) disk can be placed near the colistin disk for rapid presumptive identification of *Peptostreptococcus anaerobius* if gram-positive cocci are seen on Gram stain. A bile disk may be added to the second quadrant to detect bile inhibition if gram-negative rods are seen on Gram stain.

If processing is performed on the open bench, all plates should promptly be incubated anaerobically,

because some clinical isolates (e.g., *Fusobacterium necro-phorum* subsp. *necrophorum* and some *Prevotella* spp.) may die after relatively short exposure to oxygen. The primary plates are reincubated along with the purity plates for an additional 48 to 72 hours and are again inspected for slowly growing or pigmenting strains.

PRESUMPTIVE IDENTIFICATION OF ISOLATES

Information from the primary plates in conjunction with the atmospheric requirements, Gram stain, and colony morphology of a pure isolate provides prelimi-nary separation of many anaerobic organisms. Table 44-2 summarizes the extent to which isolates can be identified using this information. It is also useful to consider the specimen source and expected organisms from that site to aid in this process.

Presumptive identification of many clinically rele-vant anaerobic bacteria can be accomplished using a few simple tests. These tests are shown in Tables 44-3 and 44-4 and are discussed in detail in another text.[13]

The gram-negative organisms are divided into the following major categories, based on microscopic mor-phology, antibiotic identification disks pattern, or a few other simple tests.

Bacteroides fragilis Group

The gram-negative *Bacteroides fragilis* group grows in 20% bile and organisms are almost always resistant to all three special-potency antibiotic disks. Rare strains of *B. fragilis* are susceptible to colistin.

Nonpigmented *Prevotella* spp.

Most bile-sensitive, kanamycin-resistant, gram-negative rods belong in the genus *Prevotella*. Colistin susceptibility is variable, and almost all strains are catalase- and indole-negative.

Pigmented *Prevotella* and *Porphyromonas* spp.

Colonies that fluoresce brick red or produce brown to black pigment are placed among the pigmented *Pre-votella* and *Porphyromonas* spp. Some species are coccobacillary.

Bacteroides ureolyticus

Bacteroides ureolyticus reduces nitrate and requires for-mate and fumarate for growth in broth culture. Its disk pattern is the same as for the fusobacteria; however, colony morphology is different. *B. ureolyticus* forms small, translucent to transparent colonies that may corrode the agar, whereas the *Fusobacterium* colony is generally larger and more opaque. *B. ureolyticus* formerly was grouped with organisms that have been transferred to the genus *Campylobacter* (*C. gracilis, C. concisus, C. recta,* and *C. curva*).

Table 44-2   Preliminary Grouping of Anaerobic Bacteria Based on Minimal Criteria

Organism	Gram Reaction	Cell Shape	Gram-Stain Morphology	Aerotolerance	Distinguishing Characteristics
Bacteroides fragilis group	−	B	Can be pleomorphic with safety pin appearance	−	Grows on BBE; >1 mm in diameter; some strains hydrolyze esculin
Pigmented gram-negative bacilli	−	B, CB	Can be very coccoid or *Haemophilus*-like	−	Foul odor; black or brown pigment; some fluoresce brick red
Bacteroides ureolyticus	−	B	Thin; some curved	−	May pit agar or spread; transparent colony
Fusobacterium nucleatum	−	B	Slender cells with pointed ends	−	Foul odor; three colony types; bread crumblike, speckled, or smooth
Gram-negative bacillus	−	B		−	
Gram-negative coccus	−	C	*Veillonella* cells are tiny	−	
Gram-positive coccus	+	C, CB	Variable size	−	
Clostridium perfringens (presumptive)	+	B	Large; boxcar shape; no spores observed; may appear gram-negative	−	Double-zone beta hemolysis
Clostridium spp.	+	B	Spores usually observed; may appear gram-negative	−*	
Gram-positive bacillus	+	B, CB	No spores observed, no boxcar-shaped cells	−*	
Actinomyces-like	+	B	Branching cells	−*	Sulfur granules on direct examination; "molar tooth" colony

B, Bacillus; *BBE*, *Bacteroides* bile esculin agar; *C*, coccus; *CB*, coccobacillus; −, negative; + positive.
*Some strains are aerotolerant; these include *Clostridium tertium*, *C. histolyticum*, some bifidobacteria, some propionibacteria, and most *Actinomyces* spp.

The campylobacters, as well as a new genus, *Sutterella*, are all microaerophiles, not anaerobes, and are discussed in Chapter 36. Curved or motile organisms that grow anaerobically but not in 5% CO_2 should be retested in a microaerophilic atmosphere with approximately 6% oxygen.

Bilophila phenotypically resembles *B. ureolyticus* but is resistant to bile and is strongly catalase-positive.

Other *Bacteroides* or *Prevotella* spp.

The term **other *Bacteroides* or *Prevotella* spp.** is applied to gram-negative bacilli that do not fit the preceding categories or the *Fusobacterium* spp.

Fusobacterium spp.

The gram-negative *Fusobacterium* spp. are sensitive to kanamycin, and most strains fluoresce a chartreuse color. Different species have characteristic cell and colony morphology.

Leptotrichia

Leptotrichia are very large fusiform rods with one pointed end and one blunt end. Colonies are large, gray, and convoluted. They are most often isolated from the oral cavity or urogenital tract.

Anaerobic Gram-Negative Cocci

The category of anaerobic gram-negative cocci is based on Gram stain morphology and includes *Veillonella*, *Megasphaera*, and *Acidaminococcus*.

Anaerobic Gram-Positive Cocci

Among anaerobic gram-positive cocci, the genera of clinical importance are *Peptostreptococcus*, *Finegoldia*, *Micromonas*, *Peptococcus*, *Anaerococcus*, and *Schleiferella*. If a coccus is resistant to metronidazole, it probably is an anaerobic *Streptococcus* or *Staphylococcus*.

The gram-positive rods are divided into two major categories based on microscopic morphology and the presence or absence of spores.

Anaerobic, Gram-Positive, Spore-Forming Bacilli

The clostridia are the endospore-forming, anaerobic, gram-positive bacilli. If spores are not present on Gram stain, the ethanol shock spore or heat shock spore test will separate this group from the non–spore-forming anaerobic bacilli. Some strains of *C. perfringens*, *C. ramosum*, and *C. clostridioforme* may not produce spores or survive a spore test, so it is important to recognize these organisms using other characteristics. Some clostridia typically stain gram-negative, although they are sus-

Table 44-3 Abbreviated Identification of Gram-Negative Anaerobes

	Cell Shape	Slender Cells with Pointed Ends	Kanamycin (1 mg)	Vancomycin (5 µg)	Colistin (10 µg)	Growth in Bile	Spot Indole	Catalase	Pigmented Colony	Brick Red Fluorescence	Lipase	Pits the Agar	Requires Formate-Fumarate	Nitrate Reduction	Urease	Motile
Gram-Negative Rods																
Bacteroides fragilis group	B	−	R	R	R	+	V	V	−	−	−	−	−	−	−	−
Bacteroides ureolyticus	B	−	S	R	S	−	−	−	−	−	−	+⁻	+	+	+	−
Pigmented spp.																
Prevotella intermedia	B, CB	−	R	R	V	−	V	−	+*	V	V	−	−	−	−	−
Prevotella loescheii	B, CB	−	R	R	S	−	+	−	+	+	+	−	−	−	−	−
Other *Prevotella* spp.	B, CB	−	R	R	V	−	−	−	−⁺	+	−⁺	−	−	−	−	−
Prophyromonas spp.	B, CB	−	R	S‡	R	−	+	−	+	+§	−	−	−	−	−	−
Bilophila sp.	B	−	S	R	S	+	−	+	−	−	−	−	−	+	+⁻	−
Fusobacterium spp.	B	V	**S**	**R**	**S**	V	V	−	−	−	V	−	−	−	−	−
F. nucleatum subsp. *nucleatum*	B¶	+	S	R	S	−	+	−	−	−	−	−	−	−	−	−
F. necrophorum subsp. *necrophorum*	B	−	S	R	S	−⁺	+	−	−	−	+⁻	−	−	−	−	−
F. mortiferum-varium	B	−	S	R	S	+	V	−	−	−	−	−	−	−	−	−
Leptotrichia spp.	B	+**	S	R	S	V	−	−	−	−	−	−	−	−	−	−
Gram-Negative Cocci																
Veillonella	C	−	S	R	S	−	−	V	−	−⁺	−	−	−	+	−	−

Reactions in **bold type** are key tests; *B*, Bacillus; *C*, coccus; *CB*, coccobacillus; *R*, resistant; *S*, sensitive; *V*, variable; +, positive; −, negative; superscripts indicate reactions of occasional strains.
P. melaninogenica group often requires prolonged incubation before pigment is observed.
†*P. bivia* produces pigment on prolonged incubation.
‡Will not grow on LKV because of susceptibility to vancomycin.
§*P. gingivalis* does not fluoresce.
¶Thin, pointed fusiform cells.
**One pointed end, one blunt end.

Table 44-4   Abbreviated Identification of Gram-Positive Anaerobes

	Cell Shape	Spores Observed	Boxcar-Shaped Cells	Double-Zone Beta Hemolysis	Kanamycin (1 mg)	Vancomycin (5 µg)	Colistin (10 µg)	Spot Indole	Sodium Polyanethol Sulfonate	Catalase	Survives Ethnol Spore Test	Lecithinase	Nagler Test	Strong Reverse-Camp Test	Arginine Stimulation	Urease	Nitrate Reduction	Groun-Glass, Yellow Colonies on CCFA* Medium	Comment
Gram-Positive Cocci	C, CB	–	–	–	V	S	R	V	V	V	–	–	–	–	–		–/+	–	
Peptostreptococcus anaerobius	C, CB	–	–	–	R^s	S	R	–	S	–/+	–	–	–	–	–		–	–	Sweet, putrid odor; may chain
Schleiferella asaccharolytica	C	–	–	–	S	S	R	+	R	V	–	–	–	–	–	–	–	–	Indole odor
Finegoldia magna	C†	+	–	–	S	S	R	–	R	V	–	–	–	+	–	–	–	–	
Micromonas micros	C‡	–	–	–	S	S	R	–	V‖	–	–	–	–	–	–	–	–	–	
Peptococcus niger	C	+	–	–	S	S	R	–	R	–	–	–	–	–		–	–	–	Black to olive green colonies
Gram-Positive Rods Spore-Forming																			
Clostridium spp.	B	+/–	–/+	–	V	S	R	V	R	–/+	+/–	V	–		–	V	–/+	V	
Nagler-positive																			
C. perfringens	B	–	+	+/–	S	S	R	–	R	–/+	+/–	+	+	+	–	–	+/–	–	
C. baratii	B	+	–	–	S	S	R	–	R	–	+	+	+^w	–	–	–	V	–	
C. sordellii	B	+	–	–	S	S	R	+	R	–/+	+	+	+^w	–	–	–/+	–		Swarming with serpentine-edged colonies
C. bifermentans	B	+	–	–	S	S	R	+	R	–/+	+	+	+^w	–	–	–	–	–	
Nagler-negative																			
C. difficile	B	+/–	–	–	S	S	R	–	R	–/+	+	–	–	–	–	–	–	+	Horse stable odor; fluoresces chartreuse

Table 44-4 Abbreviated Identification of Gram-Positive Anaerobes—cont'd

	Cell Shape	Spores Observed	Boxcar-Shaped Cells	Double-Zone Beta Hemolysis	Kanamycin (1 mg)	Vancomycin (5 µg)	Colistin (10 µg)	Spot Indole	Sodium Polyanethol Sulfonate	Catalase	Survives Ethnol Spore Test	Nagler Test	Lecithinase	Strong Reverse-Camp Test	Arginine Stimulation	Urease	Nitrate Reduction	Groun-Glass, Yellow Colonies on CCFA* Medium	Comment
C. septicum	B	+	–	–	S	S	R	–		–⁺	⁺⁻	–	–	–	–	–	V	–	Smoothly swarming over agar surface
Non–Spore-Forming																			
Propionibacterium acnes	B, CB	–	–	–	V	S	R	**+**⁻		**+**⁺	–	–	V	–	V	V	**+**	–	May branch; diphtheroid
Eggerthella lenta	B	–	–	–	S	S	R	–		–⁺	–	–	–	–	**+**	–	**+**	–	Small rod
Bifidobacterium spp.	B§	–	–	–	S	S	R	V		–	–	–	–	–	–	–	–	–	Some strains are aerotolerant, e.g., *B. adolescentis*
Eubacterium spp.	B	–	–	–	S	S	R	V		–		–		–			–		

Data compiled from references 5-7 and 11.

Reactions in **bold type** are key tests; B, Bacillus; C, coccus; CB, coccobacillus; R, resistant; S, sensitive; V, variable; w, weak; +, positive; –, negative; superscripts indicate reactions of occasional strains.

*Cycloserine cefoxitin fructose agar.

†Cell size > 0.6 µm.

‡Cell size < 0.6 µm.

§Rods with or without one bifurcated end.

‖Some strains inhibited but zone usually < 12 mm.

Table 44-5   Antimicrobial Therapy and Susceptibility Testing of Anaerobic Bacteria

Organism Group	Therapeutic Options	Potential Resistance to Therapeutic Options	Validated Testing Methods*
Bacteroides fragilis group, other *Bacteroides* spp., *Porphyromonas* spp., *Prevotella* spp., and *Fusobacterium* spp.	Highly effective agents include most beta-lactam/beta-lactamase–inhibitor combinations, imipenem, metronidazole, and chloramphenicol	Beta-lactamase production does occur, but generally does not significantly affect imipenem or beta-lactamase–inhibitor combinations. However, isolates of *B. fragilis* are known to produce beta-lactamases capable of hydrolyzing imipenem. Metronidazole resistance is extremely rare; resistance to various cephalosporins or clindamycin does occur, and susceptibility to these agents cannot be assumed	Yes; see Table 44-6
Clostridium spp.	Penicillins, with or without beta-lactamase–inhibitor combinations and imipenem; metronidazole or vancomycin for *C. difficile*-induced gastrointestinal disease; for botulism and *C. perfringens* food poisoning antimicrobial therapy is not indicated	Resistance to therapeutic options is not common, but cephalosporins and clindamycin exhibit uncertain clinical efficacy	Yes; see Table 44-6
Actinomyces spp., *Propionibacterium* spp., *Bifidobacterium* spp., *Eubacterium* spp.	Penicillins, with or without beta-lactamase–inhibitor combinations, imipenem, cefotaxime, and ceftizoxime	Resistance to therapeutic options not common; generally resistant to many cephalosporins and metronidazole	Yes; see Table 44-6
Peptostreptococcus spp., and *Peptococcus niger*	Penicillins, most cephalosporins, imipenem, vancomycin, clindamycin, and chloramphenicol	Resistance to therapeutic options is not common	Yes; see Table 44-6

*Validated testing methods include those standard methods recommended by the Clinical and Laboratory Standards Institute (CLSI) and those commercial methods approved by the Food and Drug Administration (FDA).

ceptible to vancomycin on the disk test. Several species of clostridia grow aerobically (*C. tertium, C. carnis, C. histolyticum,* and occasional strains of *C. perfringens*), but they produce spores only under anaerobic conditions.

Clostridium botulinum is listed by the Centers for Disease Control and Prevention (CDC) as a potential agent of bioterrorism (see Chapter 65). Diagnosis of botulism is made by the demonstration of botulinum neurotoxin in serum, feces, gastric contents, vomitus, or suspect food (food poisoning) or environmental specimen (potential bioterrorism incident). For most hospital laboratories, that means knowing how to package and ship the specimen to their State Health Department or CDC. Isolation of *C. botulinum* is rarely seen in the clinical microbiology laboratory.

Anaerobic, Gram-Positive, Non–Spore-Forming Bacilli

The genera *Actinomyces, Bifidobacterium, Eubacterium, Eggerthella, Collinsella,* anaerobic *Lactobacillus, Atopobium,* and *Propionibacterium* are among the anaerobic, gram-positive, non–spore-forming bacilli. It is difficult to differentiate this group accurately to the genus level without end-product analysis, except for *P. acnes* and

Eggerthella lenta. Mobiluncus, a cause of bacterial vaginosis, is usually diagnosed on Gram stain of vaginal secretions by observation of gram-variable, curved rods with tapered ends. It is rarely isolated in the clinical laboratory as vaginal secretions are not acceptable specimens for anaerobic culture.

DEFINITIVE IDENTIFICATION

Various techniques are available for definitive identification of anaerobic bacteria. These methods may include the following:

- Pre-reduced anaerobically sterilized (PRAS) biochemicals
- Miniaturized biochemical systems (e.g., API 20A, bioMérieux, Inc., St. Louis, Mo)
- Rapid, preformed enzyme detection panels (e.g., AnIdent, bioMérieux, Inc.; RapID-ANA II, Remel, Inc., Lenexa, Kan; BBL Brand Crystal Anaerobe ID, Becton Dickinson and Company; Rapid Anaerobe Identification Panel, Dade MicroScan, West Sacramento, Calif; Vitek ANI card, bioMérieux, Inc.)

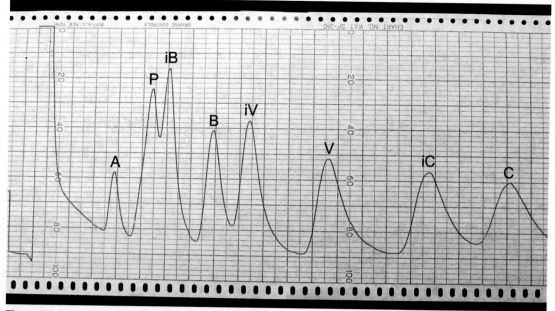

Figure 44-14   Actual chromatogram of volatile acid standard. *A,* Acetic acid; *P,* propionic acid; *iB,* isobutyric acid; *B,* butyric acid; *iV,* isovaleric acid; *V,* valeric acid; *iC,* isocaproic acid; *C,* caproic acid.

- Gas-liquid chromatography (GLC) for end products of glucose fermentation (Supelco, Bellefonte, Pa) GLC is used to separate and identify anaerobic metabolic end products (i.e., volatile fatty acids [Figure 44-14] and nonvolatile organic acids) of carbohydrate fermentation and amino acid degradation. Chromatograms produced with anaerobic bacteria can greatly facilitate identification of certain genera and species that are not readily identified based on other phenotypic characteristics.

- High-resolution gas-liquid chromatography (GLC) for cellular fatty acid analysis. GLC application has also been expanded for the analysis of longer-chain fatty acids (i.e., 9 to 20 carbons in length) to produce chromatograms for identifying organisms often without the need for other phenotypic information (e.g., Gram stain morphology, oxidase profile). One such commercial system (MIDI Microbial Identification System; Newark, Del.) has more than 600 bacteria in the chromatographic database. Although this approach may not be practical for identification of many commonly encountered bacterial species, it has great potential for use as a reference method for organisms difficult to identify by conventional methods.

None of the commercial identification systems reliably identify all anaerobic bacteria. Therefore, their high cost alone probably does not justify their use in most clinical laboratories. To ensure accurate identifi-

---

BOX 44-1   Indications for Performing Antimicrobial Susceptibility Testing with Anaerobic Bacteria

To establish patterns of susceptibility of anaerobes to new antimicrobial agents

To periodically monitor susceptibility patterns of anaerobic bacteria collected within and among specific geographic areas or particular health care institutions

To assist in the therapeutic management of patients, but only when such information may be critical because of the following:
- Known resistance of a particular species to most commonly used agents
- Therapeutic failures and/or persistence of organism at site of infection
- Lack of a precedence for therapeutic management of a particular infection
- Severity of the infection (e.g., brain abscess, osteomyelitis, infections of prosthetic devices, and refractory or recurrent bacteremia)

Modified from Clinical and Laboratory Standards Institute (CLSI) Document M11-A6.

---

cation, reference laboratories use a combination of PRAS biochemicals and GLC or high-resolution GLC. Several approaches to complete anaerobic identification are discussed in other texts.[1,3,4,13] Identification criteria for the aerotolerant *Actinomyces* spp. are presented in Table 20-4 and in Tables 44-2 and 44-4.

Table 44-6   Summary of Antimicrobial Susceptibility Testing Methods for Anaerobic Bacteria

	Test Methods	
Test Conditions	**Agar Dilution**	**Broth Microdilution and Macrodilution**
Medium	Brucella agar supplemented with hemin (5 μg/mL), vitamin K (1 μg/mL) and 5% (V/V) laked sheep blood.	Brucella broth supplemented with hemin (5 μg/mL), vitamin K (1 μg/mL) and lysed horse blood (5%)
Inoculum size	1×10^5 CFU/spot	1×10^6 CFU/mL
Incubation conditions	Anaerobic, 35°-37° C	Anaerobic, 35°-37° C
Incubation duration	48 hrs	48 hrs

From the Clinical and Laboratory Standards Institute (CLSI): Methods for antimicrobial susceptibility testing of anaerobic bacteria, M11-A6, Wayne, Pa, CLSI.

Procedure 44-1

ANTIBIOTIC IDENTIFICATION DISKS

PRINCIPLE
Most anaerobes have a characteristic susceptibility pattern to colistin (10 μg), vancomycin (5 μg), and kanamycin (1 mg) disks. The pattern generated will usually confirm a dubious Gram stain reaction (with few exceptions, most gram-positive anaerobes are susceptible to vancomycin) and aid in subdividing the anaerobic gram-negative bacilli into groups.

METHOD
1. Allow the three cartridges of disks to equilibrate to room temperature.
2. Transfer a portion of one colony to an anaerobic blood agar plate. Streak the first quadrant several times to produce a heavy lawn of growth, and then streak the other quadrants for isolation.
3. Place the colistin, kanamycin, and vancomycin disks in the first quadrant, well separated from each other (Figure 44-13).
4. Incubate the plates anaerobically for 48 hours at 35° C.

Figure 44-13
Special potency antibiotic and other disks. *Actinomyces odontolyticus.* Note red pigment. (Courtesy Anaerobe Systems, Morgan Hill, Calif.)

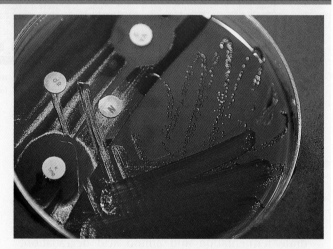

EXPECTED RESULTS
Observe for a zone of inhibition of growth. A zone of 10 mm or less indicates resistance, and a zone greater than 10 mm indicates susceptibility.

QUALITY CONTROL
1. Colistin
   Positive: *Fusobacterium necrophorum* subsp. *necrophorum*

   Negative: *Bacteroides fragilis*
2. Kanamycin
   Positive: *Clostridium perfringens*
   Negative: *Bacteroides fragilis*
3. Vancomycin
   Positive: *Clostridium perfringens*
   Negative: *Bacteroides fragilis*

ANTIMICROBIAL SUSCEPTIBILITY TESTING AND THERAPY

When mixed infections are encountered, definitive information regarding the identification of each species present usually will not affect therapeutic management. Because most clinically relevant anaerobes are susceptible to first-line antimicrobials (Table 44-5), knowledge of their presence and Gram-stain morphologies in mixed cultures is usually sufficient for guiding therapy. Therefore, definitive identification methods that follow

the schemes just outlined should be judiciously applied to clinical situations in which an anaerobic organism is isolated in pure culture from a normally sterile site (e.g., clostridial myonecrosis), when there is solid evidence for clinical relevance, or when antimicrobial susceptibility testing is being considered for the reasons listed in Box 44-1.

The therapeutic options listed for each of the major groups of anaerobic bacteria in Table 44-5 are the antimicrobial agents known to be effective against 95% or more of the organisms that constitute a particular organism group.[8] Therefore, therapeutic use of the antimicrobial agents listed generally precludes the need to routinely perform antimicrobial susceptibility testing with anaerobic isolates.

Although standard susceptibility testing methods have been established for testing anaerobic bacteria against various antimicrobial agents (Table 44-6), the fastidious nature of many species and the labor intensity of using these methods indicate that testing should only be done under special circumstances (see Box 44-1).

Although certain commercial methods (e.g., E-test and Spiral Biotech, Inc., described in Chapter 12) may facilitate anaerobic susceptibility testing in some way, the difficulty in assigning clinical significance to many anaerobic isolates and the availability of several highly effective empiric therapeutic choices significantly challenges a laboratory policy of routinely performing susceptibility testing with these organisms.

REFERENCES

1. Allen SD, Emery CL, Lyerly DM: *Clostridium.* In Murray PR, Baron EJ, Jorgensen JH, et al, editors: *Manual of clinical microbiology,* ed 8, Washington, DC, 2003, ASM Press.
2. Dowell VR, Hawkins TM: Laboratory methods in anaerobic bacteriology: CDC laboratory manual. Centers for Disease Control, DHHS Pub No (CDC) 81-8272, Atlanta, 1981, US Department of Health and Human Services.
3. Dzink JL, Sheenan MT, Socransky SS: Proposal of three subspecies of *Fusobacterium nucleatum* Knorr, 1922: *Fusobacterium nucleatum* subsp nov, comb nov; *Fusobacterium nucleatum* subsp *polymorphum* subsp nov, nom rev, comb nov; and *Fusobacterium nucleatum* subsp *vincentii* subsp nov, nom rev, comb nov, *Int J Syst Bacteriol* 40:74, 1990.
4. Holdeman LV, Cato EP, Moore WEC, editors: *Anaerobic laboratory manual,* ed 4, Blacksburg, Va, 1977, Virginia Polytechnic Institute and State University.
5. Holdeman LV, Cato EP, Moore WEC, editors: *Anaerobic laboratory manual: update,* ed 4, Blacksburg, Va, 1987, Virginia Polytechnic Institute and State University.
6. Jousimies-Somer HR, Summanen P, Citron DM, et al: *Wadsworth anaerobic bacteriology manual,* ed 6, Belmont, Calif, 2002, Star.
7. Jousimies-Somer HR, Summanen PH, Wexler H: *Bacteroides, Porphyromonas, Prevotella, Fusobacterium,* and other anaerobic gram-negative bacteria. In Murray PR, Baron EJ, Jorgensen JH, et al, editors: *Manual of clinical microbiology,* ed 8, Washington, DC, 2003, ASM Press.
8. Johnson CC: Susceptibility of anaerobic bacteria to β-lactam antibiotics in the United States, *Clin Infect Dis* 16(suppl 4):S371, 1993.
9. Kageyama A, Benno Y, Nakase T: Phylogenetic and phenotypic evidence for the transfer of *Eubacterium aerofaciens* to the genus *Collinsella* as *Collinsella aerofaciens* gen nov, comb nov, *Int J Syst Bacteriol* 49:557, 1999.
10. Kageyama A, Benno Y, Nakase T: Phylogenetic evidence for the transfer of *Eubacterium lentum* to the genus *Eggerthella* as *Eggerthella lenta* gen nov, comb nov, *Int J Syst Bacteriol* 49:1725, 1999.
11. Moncla BJ, Hillier SL: *Peptostreptococcus, Propionibacterium, Lactobacillus, Actinomyces* and other non–sporeforming anaerobic gram-positive bacteria. In Murray PR, Baron EJ, Jorgensen JH, et al, editors: *Manual of clinical microbiology,* ed 8, Washington DC, 2003, ASM Press.
12. Murdoch DA, Shah HN: Reclassification of *Peptostreptococcus magnus* (Prevot, 1933) Holdeman and Moore, 1972 as *Finegoldia magna* comb nov and *Peptostreptococcus micros* (Prevot, 1933) Smith, 1957 as *Micromonas micros* comb nov, *Anaerobe* 5:553, 1999.
13. National Committee for Clinical Laboratory Standards: Abbreviated identification of bacteria and yeast; Approved guideline, M35-A, Wayne, Pa, 2002, NCCLS.
14. Shinjo T, Fujisawa T, Mitsuoka T: Proposal of two subspecies of *Fusobacterium necrophorum* (Flügge) Moore and Holdeman: *Fusobacterium necrophorum* subsp *necrophorum* subsp nov, nom rev (ex Flügge, 1886), and *Fusobacterium necrophorum* subsp *funduliforme* subsp nov, nom rev (ex Halle, 1898), *Int J Syst Bacteriol* 41:395, 1991.

ADDITIONAL READING

Finegold SM: Anaerobic bacteria: general concepts. In Mandell GL, Bennett JE, Dolin R, editors: *Principles and practice of infectious diseases,* ed 5, New York, 2000, Churchill Livingstone.

CHAPTER 45 Mycobacteria

Genera and Species to Be Considered

Mycobacterium tuberculosis complex:
  Mycobacterium tuberculosis
  Mycobacterium bovis
  Mycobacterium bovis BCG
  Mycobacterium africanum
Nontuberculosis mycobacteria

Organisms belonging to the genus *Mycobacterium* are very thin, rod-shaped (0.2 to 0.4 × 2 to 10 μm), and nonmotile. This genus belongs in the *Mycobacteriaceae* family (the only genus in this family), Actinomycetales order, and *Actinomycetes* class. Genera that are closely related to members of the genus *Mycobacterium* include *Nocardia, Rhodococcus,* and *Corynebacterium. Mycobacterium* spp. have an unusual cell wall structure that contains *N*-glycolylmuramic acid in lieu of *N*-acetylmuramic acid and has a very high lipid content. Because of this cell wall structure, mycobacteria are difficult to stain with commonly used basic aniline dyes, such as those used in the Gram stain. However, these organisms resist decolorization by acidified alcohol (3% hydrochloric acid) after prolonged application of a basic fuchsin dye or with heating of this dye following its application. This important property of mycobacteria that is dependent on its cell wall is referred to as **acid-fastness.** By virtue of this acid-fast characteristic, mycobacteria can be distinguished from other genera. Another important feature of these organisms is that they grow more slowly than most other human pathogenic bacteria because of their hydrophobic cell surface. Because of this hydrophobicity, organisms tend to clump, so that nutrients are not easily allowed into the cell. Growth is slow or very slow, with colonies becoming visible in 2 to 60 days at optimum temperature.

Currently, there are more than 100 recognized or proposed species in the genus *Mycobacterium.*[23] These species produce a spectrum of infections in humans and animals ranging from localized lesions to disseminated disease. Although some species cause only human infections, others have been isolated from a wide variety of animals. Many species are also found in water and soil.

For the most part, mycobacteria can be divided into two major groups based on fundamental differences in epidemiology and association with disease: those belonging to the *Mycobacterium tuberculosis* complex *(M. tuberculosis, M. bovis, M. africanum)* and those referred to as **nontuberculous mycobacteria (NTMs).** For purposes of discussion, the mycobacteria are addressed using these two major groups (Box 45-1).

MYCOBACTERIUM TUBERCULOSIS COMPLEX

Tuberculosis was endemic in animals in the Paleolithic period, long before it ever affected humans. This disease, also known as *consumption,* has been known in all ages and climates. For example, tuberculosis was the subject of a hymn in a sacred text of India dating from 2500 BC, while DNA unique to *Mycobacterium tuberculosis* was identified in lesions from the lung of 1000-year-old human remains found in Peru.

GENERAL CHARACTERISTICS

The term **complex** is frequently used in the clinical microbiology laboratory setting to describe two or more species whose distinction is complicated and of little or no medical importance.[25] The mycobacterial species that occur in humans and belong to the **M. tuberculosis complex** include *M. tuberculosis, M. bovis, M. bovis* BCG, and *M. africanum;* all species are capable of causing tuberculosis. It should be noted that species identification may be required for epidemiologic and public health reasons. Organisms belonging to the *M. tuberculosis* complex are considered slow-growers, and colonies are nonpigmented.

BOX 45-1 Major Groupings of Organisms Belonging to the Genus *Mycobacterium**

MYCOBACTERIUM TUBERCULOSIS COMPLEX
M. tuberculosis
M. bovis
M. bovis BCG
M. africanum

NONTUBERCULOUS MYCOBACTERIA

Slow-Growing	**Rapid-Growing**
Nonphotochromogens:	**Potentially Pathogenic:**
M. avium complex	M. fortuitum
M. celatum	M. chelonae
M. ulcerans	M. abscessus
M. gastri	M. smegmatis
M. genavense	M. peregrinum
M. haemophilum	M. immunogenum
M. malmoense	M. mucogenicum
M. shimoidei	M. fortuitum third biovariant
M. simiae	complex, sorbitol-positive†
M. xenopi	M. fortuitum third biovariant
M. terrae complex	complex, sorbitol-negative†
M. heidelbergense	M. wolinskyi
M. branderi	M. goodiii
M. triplex	M. septicum
M. conspicuum	M. mageritense
Photochromogens:	M. canariasense
M. kansasii	M. alvei
M. asiaticum	M. novocastrense
M. marinum	**Rarely Pathogenic or Not Yet**
M. intermedium	**Associated with Infection:**
Scotochromogens:	M. agri, M. aichiense, M.
M. szulgai	austroafricanum, M. aurum,
M. scrofulaceum	M. brumae, M. chitae, M.
M. interjectum	chubuense, M. diernhoferi,
M. gordonae	M. duvalii, M. fallax, M.
M. cookii	flavescens, M. gadium,
M. hiberniae	M. gilvum, M. hassiacum,
M. lentiflavum	M. komossense, M.
M. conspicuum	moriokaense,
M. heckeshornense	M. murale, M. neoaurum,
M. tusciae	M. obuense,
M. kubicae	M. parafortuitum,
M. ulcerans	M. phlei, M. pulveris,
M. bohemicum	M. rhodesiae,
Noncultivatable:	M. senegalense,
M. leprae	M. sphagni,
	M. thermoresistible,
	M. tokaiense, M. vaccae

*This box is not inclusive, but rather lists only the prominent mycobacteria isolated from humans.
†Currently consists of genomospecies with new species names proposed of *M. bonikei, M. houstonense, M. neworleansense,* and *M. brisbanense.*

EPIDEMIOLOGY AND PATHOGENESIS

Epidemiology

All members belonging to *M. tuberculosis* complex cause tuberculous infections. *M. tuberculosis* is the cause of most cases of human tuberculosis, particularly in developed countries. An estimated 1.7 billion persons, one third of the world's population, are infected with *M. tuberculosis.* This reservoir of infected individuals results in 8 million new cases of tuberculosis and 2.9 million deaths annually. Tuberculosis continues to be a public health problem in the United States, with more than 20,000 cases reported annually; an estimated 10 million persons in this country are already infected.[4] Following the emergence of epidemic multidrug-resistant strains of *M. tuberculosis* in conjunction with the ending of the downward trend of reported cases of tuberculosis in the United States, the incidence of tuberculosis in the United States decreased 44% from 1993 to 2003 and is currently at a historic low level.[28] Although the organisms belonging to *M. tuberculosis* complex have numerous characteristics in common with one another, they do differ in certain aspects regarding their epidemiology (Table 45-1).

Pathogenesis

The pathogenesis of tuberculosis caused by organisms belonging to *M. tuberculosis* complex is discussed in Chapter 53. Inhalation of a single viable organism has been shown to lead to infection, although close contact is usually necessary for acquisition of infection. With regard to *M. tuberculosis,* 15% to 20% of persons who become infected develop disease. Disease usually occurs some years after the initial infection, when the patient's immune system breaks down for some reason other than the presence of tuberculosis bacilli within the lung. In a small percentage of infected hosts, the disease becomes systemic, affecting a variety of organs.

Following ingestion of milk from infected cows, *M. bovis* may penetrate the gastrointestinal mucosa or invade the lymphatic tissue of the oropharynx.[27] An attenuated strain of *M. bovis,* **bacille Calmette-Guérin (BCG),** has been used extensively in many parts of the world to immunize susceptible individuals against tuberculosis. BCG has recently been used as part of a controversial protocol to boost the nonspecific cellular immune response of certain immunologically deficient patients, particularly those with malignancies. Because mycobacteria are the classic examples of intracellular pathogens and the body's response to BCG hinges on cell-mediated immunoreactivity, immunized individuals are expected to react more aggressively against all antigens that elicit cell-mediated immunity. Rarely, the unfortunate individual's immune system will be so compromised that it cannot handle the BCG, and systemic BCG infection may develop.

SPECTRUM OF DISEASE

Tuberculosis may mimic other diseases such as pneumonia, neoplasm, or fungal infections. In addition, clinical manifestations of patients infected with *M. tuberculosis* complex may range from asymptomatic to

Table 45-1   Epidemiology of Organisms Belonging to the *M. tuberculosis* Complex That Cause Human Infections

Organism	Habitat	Primary Route of Transmission	Distribution
M. tuberculosis	Patients with cavitary disease are primary reservoir	Person to person by inhalation of droplet nuclei: droplet nuclei containing the organism (infectious aerosols, 1 to 5 μm) are produced when people with pulmonary tuberculosis cough, sneeze, speak, or sing; infectious aerosols may also be produced by manipulations of lesions or processing clinical specimens in the laboratory. Droplets are so small that air currents keep them airborne for long periods; once inhaled they are small enough to reach the lung's alveoli*	Worldwide
M. bovis	Humans and wide host range of animals, cattle, nonhuman primates, goats, cats, buffalo, badgers, possums, dogs, pigs, deer, etc.	Ingestion of contaminated milk from infected cows[†]; airborne transmission[‡]	Worldwide
M. africanum	Humans[§]	Inhalation of droplet nuclei	East and West Africa

*Infection can occasionally occur through the gastrointestinal tract or skin.
[†]Incidence is significantly decreased in developed countries since introducing universal pasteurization of milk and milk products, as well as effective control programs for cattle.
[‡]Can be transmitted human to human, animal to human, and human to animal.
[§]Infections in animals have not yet been totally excluded.

acutely symptomatic. Patients who are symptomatic can have systemic symptoms, pulmonary signs and symptoms, signs and symptoms related to other organ involvement (e.g., kidney), or a combination of these features. Of note, cases of pulmonary disease caused by *M. tuberculosis* complex organisms are clinically, radiologically, and pathologically indistinguishable.[21] Common presenting symptoms include low-grade fevers, night sweats, fatigue, anorexia (loss of appetite), and weight loss. If a patient presents with pulmonary tuberculosis, a productive cough is usually present, along with fevers, chills, myalgias (aches), and sweating, which are signs and symptoms similar for not only influenza but also acute bronchitis or pneumonia. As previously mentioned, other organs besides the lung can be involved following infection with *M. tuberculosis* complex organisms in a small percentage of patients and include the following:

- Genitourinary tract
- Lymph nodes
- Central nervous system (meningitis)
- Bone and joint (arthritis and osteomyelitis)
- Peritoneum
- Pericardium
- Larynx

Individuals infected with the human immunodeficiency virus (HIV) are particularly susceptible to development of active tuberculosis. These patients are likely to have rapidly progressive primary disease instead of a subclinical infection. Of further concern, the diagnosis of tuberculosis is more difficult in persons

infected with HIV, because chest radiographs of the pulmonary disease often lack specificity and frequently patients are anergic to **tuberculin skin testing,** a primary means to identify individuals infected with *M. tuberculosis*. The tuberculin skin test, or **PPD (purified protein derivative)** test, is based on the premise that following infection with *M. tuberculosis*, a patient will develop a delayed hypersensitivity cell-mediated immunity to certain antigenic components of the organism. To determine whether a person has been infected with *M. tuberculosis*, a culture extract of *M. tuberculosis* (i.e., PPD of tuberculin) is intracutaneously injected. After 48 to 72 hours a person who has been infected will exhibit a delayed hypersensitivity reaction to the PPD; this reaction is characterized by erythema (redness) and, most important, induration (firmness as a result of influx of immune cells). The diameter of induration is measured and then interpreted as to whether the patient has been infected with *M. tuberculosis*; different interpretative criteria exist for different patient populations (e.g., immunosuppressed persons, such as those infected with HIV). This test is not 100% sensitive or specific, and a positive reaction to the skin test does not necessarily signify the presence of disease. Because of these issues, a new Food and Drug Administration (FDA)-approved enzyme-linked immunosorbent assay called QuantiFERON-TB Gold (Cellestis Limited, Carnegie, Victoria, Australia) has become available. This test measures a component of the cell-mediated immune response to *M. tuberculosis* as a means to diagnose latent tuberculosis infection and tuberculosis disease. Based on the quantification of interferon-gamma released from sensitized lymphocytes in heparinized whole blood incubated overnight with a mixtures of synthetic peptides simulating two proteins in *M. tuberculosis*, this test assesses responses to multiple antigens, does not require more than one patient visit, and is less subject to reader bias and error. Importantly, results of the assay are unaffected by prior BCG vaccination. The Centers for Disease Control and Prevention (CDC) has published guidelines recommending that this assay may be used in all circumstances in which the tuberculin skin test is currently used (e.g., contact investigations, evaluation of recent immigrants) and has provided specific cautions for interpreting negative results in persons from selected populations.[14]

NONTUBERCULOUS MYCOBACTERIA

The NTMs include all other mycobacterial species that do not belong to *M. tuberculosis* complex. This large group of mycobacteria have been known by several names (Box 45-2). Significant geographic variability exists both in the prevalence and species responsible for NTM disease.[9,23] As previously mentioned, NTM are present everywhere in the environment and sometimes colonize healthy individuals in the skin and respiratory and gastrointestinal tracts. Little is known about how

BOX 45-2	Other Names That Have Been Used to Designate the Nontuberculous Mycobacteria

Anonymous
Atypical
Unclassified
Unknown
Tuberculoid
Environmental
Opportunistic
MOTT (mycobacteria other than tubercle bacilli)

From Debrunner M, et al: *Clin Infect Dis* 15:330, 1992.

infection is acquired, but some NTM diseases appear to be acquired by trauma, inhalation of infectious aerosols, or ingestion; a few diseases are nosocomially or iatrogenically acquired. In contrast to *M. tuberculosis* complex organisms, NTM are not usually transmitted from person to person nor does their isolation necessarily mean that they are associated with a disease process. Therefore, interpretation of a positive NTM culture is complicated because these organisms are widely distributed in nature, their pathogenic potential varies greatly from one another, and humans can be colonized by these mycobacteria without necessarily developing infection or disease. With few exceptions, little is known about the pathogenesis of infections caused by these bacterial agents.

In 1959, Runyon[24] classified NTM into four groups (Runyon groups I through IV) based on phenotypic characteristics of the various species, most notably growth rate and colonial pigmentation (Table 45-2). This large group of organisms is addressed by first discussing the slow-growing NTM (Runyon groups I to III) and then the rapid-growers (Runyon group IV). One other NTM, *M. leprae*, which is noncultivable, is also reviewed. (As with many classification schemes, the Runyon classification does not always hold true. For example, some NTM can be either a photochromogen or a nonphotochromogen.)

Because of the difficulty in determining the clinical significance of isolating an NTM from a clinical sample, several clinical classification schemes for the NTM have also been proposed. For example, one such scheme classifies NTM recovered from humans into four major groupings (pulmonary, lymphadenitis, cutaneous, disseminated) based on the clinical disease that they cause.[33] Other NTM classifications have been made based on the organism's pathogenic potential.

SLOW-GROWING NTM

The slow-growing NTM can be further divided into three groups based on phenotypic characteristics of the

Table 45-2   Runyon Classification of NTM

Runyon Group Number	Group Name	Description
I	Photochromogens	Colonies of NTM that develop pigment following exposure to light after being grown in the dark and take more than 7 days to appear on solid media
II	Scotochromogens	Colonies of NTM that develop pigment in the dark or light and take more than 7 days to appear on solid media
III	Nonphotochromogens	Colonies of NTM that are nonpigmented regardless of whether they are grown in the dark or light and take more than 7 days to appear on solid media
IV	Rapid-growers	Colonies of NTM that appear on solid media in less than 7 days

Table 45-3 Characteristics of the NTM Classified as Photochromogens

Organism	Epidemiology	Pathogenicity	Type of Infection
M. kansasii	Infection more common in white males. Natural reservoir is tap water with aerosols involved in transmission.	Potentially pathogenic	Chronic pulmonary disease; extrapulmonary diseases, such as cervical lymphadenitis and cutaneous disease
M. asiaticum	Not commonly encountered (primarily Australia)	Potentially pathogenic	Pulmonary disease
M. marinum	Natural reservoir is fresh water and saltwater as a result of contamination from infected fish and other marine life Transmission by contact with contaminated water and organism entry by trauma or small breaks in the skin; associated with aquatic activity usually involving fish	Potentially pathogenic	Cutaneous disease; bacteremia
M. intermedium	Unknown	Potentially pathogenic	Pulmonary disease
M. novocastrense	Unknown	Potentially pathogenic	Cutaneous disease

Table 45-4 Characteristics of the NTM Classified as Scotochromogens

Organism	Epidemiology/Habitat	Pathogenicity	Type of Infection
M. szulgai	Water and soil	Potentially pathogenic	Pulmonary disease, predominantly in middle-aged men; cervical adenitis; bursitis
M. scrofulaceum	Raw milk, soil, water, dairy products	Potentially pathogenic	Cervical adenitis in children; bacteremia; pulmonary disease; skin infections
M. interjectum	Unknown	Potentially pathogenic	Chronic lymphadenitis; pulmonary disease
M. heckeshornense	Unknown	Potentially pathogenic	Pulmonary disease (two cases)
M. tusciae	Unknown—isolated from tap water	Potentially pathogenic	Cervical lymphadenitis (one case)
M. kubicae	Unknown	Potentially pathogenic	Pulmonary disease
M. gordonae	Tap water, water, soil	Nonpathogenic*	NA
M. cookie	Sphagnum, surface waters in New Zealand	Nonpathogenic*	NA
M. hiberniae	Sphagnum, soil in Ireland	Nonpathogenic*	NA

NA, Not applicable.
*Rarely, if ever, causes disease.

various species. Species of mycobacteria synthesize **carotenoids** (a group of pigments that are yellow to red) in varying amounts and are categorized into three groups based on the production of these pigments. Some of these NTMs are considered potentially pathogenic for humans, whereas others are rarely associated with disease.

Photochromogens

The photochromogens are slow-growing NTM whose colonies become pigmented when exposed to light. Salient features of the photochromogens are summarized in Table 45-3.

Scotochromogens

The scotochromogens are slow-growing NTM whose colonies are pigmented when grown in the dark or the light. Salient features of the scotochromogens are summarized in Table 45-4. Of note, the epidemiology of the potentially pathogenic scotochromogens has not been definitively described. In contrast to potentially pathogenic nonphotochromogens (see later), these agents are rarely recovered in clinical laboratories.

Nonphotochromogens

The nonphotochromogens are slow-growing NTM whose colonies produce no pigment whether they are

Table 45-5   Characteristics of the NTM Classified as Nonphotochromogens and Considered as Potential Pathogens

Organism	Epidemiology	Type of Infection
M. avium complex	Environmental sources including natural waters; soil	Patients without AIDS: pulmonary infections in patients with preexisting pulmonary disease; cervical lymphadenitis; disseminated disease* in immunocompromised, HIV-negative patients Patients with AIDS: disseminated disease
M. xenopi[†]	Water, especially hot water taps in hospitals; believed to be transmitted in aerosols	Primarily pulmonary infections in adults. Less common: extrapulmonary infections (bone, lymph node, sinus tract) and disseminated disease
M. ulcerans	Stagnant tropical waters; also harbored in an aquatic insect's salivary glands; infections occur in tropical or temperate climates	Indolent cutaneous and subcutaneous infections (African Buruli ulcer or Australian Bairnsdale ulcer)
M. malmoense	Majority of cases from England, Wales, Sweden. Rarely isolated from patients infected with HIV. Little known about epidemiology; to date, isolated from only humans and captured armadillos	Chronic pulmonary infections primarily in patients with preexisting disease; cervical lymphadenitis in children; less common, infections of the skin or bursa
M. genavense	Isolated from pet birds and dogs. Mode of acquisition unknown	Disseminated disease in patients with AIDS (wasting disease characterized by fever, weight loss, hepatosplenomegaly, anemia)
M. haemophilum	Unknown	Disseminated disease; cutaneous infections in immunosuppressed adults. Mild and limited skin infections in preadolescence or early adolescence; cervical lymphadenitis in children
M. heidelbergense	Unknown	Lymphadenitis in children; also isolated from sputum, urine, and gastric aspirate
M. shimoidei	To date, has not been isolated from environmental sources Few case reports, but widespread geographically	Tuberculosis-like pulmonary infection, disseminated disease
M. simiae	Tap water and hospital water tanks; rarely isolated	Tuberculosis-like pulmonary infections

AIDS, Acquired immunodeficiency syndrome; *HIV*, human immunodeficiency virus.
*Disseminated disease: can involve multiple sites, such as bone marrow, lung, liver, lymph nodes.
[†]Can be either nonphotochromogenic or scotochromogenic.

grown in the dark or the light. Of the organisms classified in this group, those belonging to *M. terrae* complex (*M. terrae*, *M. triviale*, and *M. nonchromogenicum*) and *M. gastri* are considered nonpathogenic for humans. The other nonphotochromogens are considered potentially pathogenic (Table 45-5), and many are frequently encountered in the clinical laboratory. Because organisms belonging to *Mycobacterium avium* complex are frequently isolated in the clinical laboratory and are able to cause infection in the human host, these nonphotochromogens are discussed in greater detail.

Mycobacterium avium *Complex.* In large part as a result of increasing populations of immunosuppressed patients, such as individuals infected with HIV, the incidence of infections caused by organisms belonging to *M. avium* complex and their corresponding clinical significance has changed significantly since they were first recognized as human pathogens in the 1950s. With the introduction of highly active antiretroviral therapy, in-

fections caused by these organisms in patients with acquired immunodeficiency syndrome (AIDS) have been dramatically reduced.

General Characteristics. Taxonomically, the *M. avium* complex comprises *M. avium*, *M. intracellulare*, *M. paratuberculosis*, *M. lepraemurium*, and the "wood pigeon" bacillus. It has been proposed that *M. avium*, *M. paratuberculosis*, and the wood pigeon bacillus be placed in one species with three subspecies (*M. avium* subsp. *avium*, *M. avium* subsp. *paratuberculosis*, and *M. avium* subsp. *silvaticum* [wood pigeon bacillus])[30]; it has been suggested that *M. lepraemurium* be reduced to a subspecies of *M. avium* as well. Recently, it has been proposed that bird isolates of *M. avium* be designated as *M. avium* subsp. *avium* while pig and human isolates be designated as *M. avium* subsp. *homicissuis*.[15] Only the *M. avium* complex organisms that cause human infections are discussed and include *M. avium*, *M. intracellulare*, and *M. paratuberculosis*. Unfortunately, the nomenclature is somewhat confusing. Although *M. avium* and *M. intracellulare* are clearly different organisms, they so closely resemble each other that the distinction cannot be made by routine laboratory determinations or on clinical grounds. As a result, sometimes these organisms are referred to as *M. avium-intracellulare*. Furthermore, because the isolation of *M. paratuberculosis* in a routine laboratory setting is exceedingly rare, the term *M. avium* complex is most commonly used to report the isolation of *M. avium-intracellulare*. For purposes of this discussion, *M. avium* complex (MAC) is used and includes only *M. avium* and *M. intracellulare*.

Epidemiology and Pathogenesis. MAC is an important pathogen in immunocompromised and immunocompetent populations. This agent is among the most commonly isolated NTM species in the United States, particularly in its potentially pathogenic role in pulmonary infections in patients with AIDS and non–HIV-infected patients. The organisms are ubiquitous in the environment and have been isolated from natural water, soil, dairy products, pigs, chickens, cats, and dogs; through extensive studies it is generally accepted that natural waters serve as the major reservoir for most human infections. Infections caused by MAC are acquired by inhalation or ingestion. The pathogenesis of MAC infections is not clearly understood. However, these organisms as well as other environmental NTM have extraordinary starvation survival, persisting well over a year in tap water; MAC tolerate temperature extremes. In addition, similar to legionellae, *M. avium* can infect and replicate in protozoa with amoebae-grown *M. avium* being more invasive toward human epithelial and macrophage cells.[23] Cultures of MAC can exhibit opaque and translucent or transparent colony morphology. Studies suggest that transparent colonies are more virulent by virtue of being more drug resistant, are isolated more frequently from blood of patients with AIDS, and appear more virulent in macrophage and animal models.[25]

M. paratuberculosis (*M. avium* subsp. *paratuberculosis*) is known to cause an inflammatory bowel disease referred to as **Johne's disease** in cattle, sheep, and goats. This organism was also isolated from the bowel mucosa of patients with Crohn's disease, a chronic inflammatory bowel disease of humans. This organism is extremely fastidious, seems to require a growth factor (mycobactin, produced by other species of mycobacteria, such as *M. phlei*, a saprophytic strain) and may take as long as 6 to 18 months for primary isolation. Whether these and other mycobacteria actually contribute to development of Crohn's disease or are simply colonizing an environmental niche in the bowel of these patients remains to be elucidated.

Clinical Spectrum of Disease. The clinical manifestations of *M. avium* complex infections are summarized in Table 45-5.

Other Nonphotochromogens. There are several other mycobacterial species that are considered nonphotochromogens that are potentially pathogenic in humans. The epidemiology and spectrum of disease for these organisms are summarized in Table 45-5. In addition to the species in this table, other newer species of mycobacteria that are nonphotochromogens have been described, such as *M. celatum*[25] and *M. conspicuum*.[26] These newer agents appear to be potentially pathogenic in humans.

RAPIDLY GROWING NTM

Mycobacteria whose colonies appear on solid media in 7 days or less constitute the second major group of NTM.

General Characteristics

Although a large group of organisms, only eight taxonomic groups of potentially pathogenic, rapidly growing mycobacteria are currently recognized.[32] In contrast to the majority of other mycobacteria, most rapid-growers can grow on routine bacteriologic media and on media specific for cultivation of mycobacteria. On Gram stain, organisms appear as weakly gram-positive rods resembling diphtheroids. Only the potentially pathogenic, rapid-growing mycobacteria are considered.

Epidemiology and Pathogenesis

The rapidly growing mycobacteria considered as potentially pathogenic can cause disease in either healthy or immunocompromised patients. Like many other NTM, these organisms are ubiquitous in the environment and are present worldwide. They have been found in soil, marshes, rivers, municipal water supplies, and in marine and terrestrial life forms.[36] Of importance, infections caused by rapidly growing mycobacteria may be acquired in the community from environmental sources or nosocomially as a result of medical intervention. They may be commensals on the skin. Organisms gain entry into the host by inoculation into the skin and subcutaneous tissues during trauma, injections, or surgery or through animal contact; organisms can also cause disseminated cutaneous infections. The description of chronic pulmonary infections caused by rapidly growing mycobacteria suggests a possible respiratory route for acquisition of organisms present in the environment.[7] Of the potentially pathogenic, rapidly growing NTM, *M. fortuitum*, *M. chelonae*, and *M. abscessus* are commonly encountered; these three species account for approximately 90% of clinical disease.[32] To date, little is known about the pathogenesis of these organisms.

Table 45-6   Common Types of Infections Caused by Rapidly Growing Mycobacteria

Organism	Common Types of Infection
M. abscessus	Disseminated disease primarily in immunocompromised individuals, skin and soft tissue infections, pulmonary infections, postoperative infections
M. fortuitum	Postoperative infections in breast augmentation and median sternotomy; skin and soft tissue infections; pulmonary infections
M. chelonae	Skin and soft tissue infections, postoperative wound infections, keratitis
M. fortuitum third biovariant complex, sorbitol-positive or sorbitol-negative	Skin and soft tissue infections
M. peregrinum	Skin and soft tissue infections; bacteremia
M. mucogenicum	Posttraumatic wound infections, catheter-related sepsis
M. smegmatis	Skin or soft tissue infections

Spectrum of Disease

The spectrum of disease caused by the most commonly encountered rapid-growers is summarized in Table 45-6.

NONCULTIVATABLE NTM—*MYCOBACTERIUM LEPRAE*

Mycobacterium leprae is an NTM that is a close relative of *M. tuberculosis*. This organism causes **leprosy** (also called **Hansen's disease**). Leprosy is a chronic disease of the skin, mucous membranes, and nerve tissue.

General Characteristics

M. leprae has not yet been cultivated in vitro, although it can be cultivated in the armadillo and the footpads of mice. Largely through the application of molecular biologic techniques, information has been gained regarding this organism's genomic structure and its various genes and their products. Although polymerase chain reaction (PCR) has been used to detect and identify *M. leprae* in infected tissues, it has not proved so far to be as effective diagnostically as anticipated in indeterminate or paucibacillary (few organisms present) disease.[17] Thus, diagnosis of leprosy is accomplished based on distinct clinical manifestations such as hypopigmented skin lesion and peripheral nerve involvement in conjunction with a skin-smear positive for acid-fast bacilli.

Epidemiology and Pathogenesis

Understanding of the epidemiology and pathogenesis of the disease is hampered by our inability to grow the organism in culture. In tropical countries, where the disease is most prevalent, it may be acquired from infected humans; however, infectivity is very low. Prolonged close contact and host immunologic status play a role in infectivity.

Epidemiology. Although leprosy is rare in the United States and most Western countries, there are about 1 million cases worldwide, and the number of new cases has significantly increased since 1994.[17] The primary reservoir is infected humans. Transmission of leprosy occurs person to person through inhalation or contact with infected skin. It appears that inhalation of *M. leprae* discharged in nasal secretions of an infected individual is the more important mode of transmission.[20]

Pathogenesis. Although the host's immune response to *M. leprae* plays a key role in control of infection, the immune response is also responsible for the damage to skin and nerves; in other words, leprosy is both a bacterial and immunologic disease.[6] After acquisition of *M. leprae*, infection passes through many stages in the host that are characterized by various clinical and histopathologic features. Although there are many intermediate stages, the primary stages include a silent phase, during which leprosy bacilli multiply in the skin within macrophages, and an intermediate phase, in which the bacilli multiply in peripheral nerves and begin to cause sensory impairment. More severe disease states may follow. A patient may recover spontaneously at any stage.

Spectrum of Disease

Based on the host's response, the spectrum of disease caused by *M. leprae* ranges from subclinical infection to intermediate stages of disease to full-blown and serious clinical manifestations involving the skin, upper respiratory system, testes, and peripheral nerves. The two major forms of the disease are a localized form called *tuberculoid leprosy* and a more disseminated form called *lepromatous leprosy*. Patients with lepromatous leprosy are anergic to *M. leprae* because of a defect in their cell-mediated immunity. Because of unimpeded growth, individuals display extensive skin lesions containing numerous acid-fast bacilli; organisms can spill over into blood and disseminate. In contrast, individuals with tuberculoid leprosy do not have an immune defect, so the disease is localized to the skin and nerves; few organisms are observed in skin lesions. Most of the serious sequelae associated with leprosy are a result of this organism's tropism for peripheral nerves.

LABORATORY DIAGNOSIS OF MYCOBACTERIAL INFECTIONS

Specimens received by the laboratory for mycobacterial smear and culture must be handled in a safe manner. Tuberculosis ranks high among laboratory-acquired infections. Therefore, laboratory and hospital administrators must provide laboratory personnel with facilities, equipment, and supplies that will reduce this risk to a minimum. All tuberculin-negative personnel should be skin-tested at least every year. Biosafety Level 2 practices, containment equipment, and facilities for preparing acid-fast smears and culture are strongly recommended. Of great significance, all aerosol-generating procedures must be performed in a class II A or B or III biological safety cabinet (BSC). If *M. tuberculosis* is grown and then propagated and manipulated, Biosafety Level 3 practices are recommended.

SPECIMEN COLLECTION AND TRANSPORT

Acid-fast bacilli may infect almost any tissue or organ of the body. The successful isolation of the organism depends on the quality of the specimen obtained and the appropriate processing and culture techniques used by the mycobacteriology laboratory. In suspected mycobacterial disease, as in all other infectious diseases, the diagnostic procedure begins at the patient's bedside. Collection of proper clinical specimens requires careful attention to detail by health care professionals. Specimens should be collected in sterile, leak-proof, disposable, and appropriately labeled containers and placed into bags to contain leakage.

Pulmonary Specimens

Pulmonary secretions may be obtained by any of the following methods: spontaneously produced or induced sputum, gastric lavage, transtracheal aspiration, bronchoscopy, and laryngeal swabbing. Sputum, aerosol-induced sputum, bronchoscopic aspirations, and gastric lavage constitute the majority of specimens submitted for examination. Spontaneously produced sputum is the specimen of choice. To raise sputum, patients must be instructed to take a deep breath, hold it momentarily,

and then cough deeply and vigorously. Patients must also be instructed to cover their mouths carefully while coughing and to discard tissues in an appropriate receptacle. Saliva and nasal secretions are not to be collected nor is the patient to use oral antiseptics during the period of collection. Sputum specimens must be free of food particles, residues, and other extraneous matter.

The aerosol (saline) induction procedure can best be done on ambulatory patients who are able to follow instructions. Aerosol-induced sputum specimens have been collected from children as young as 5 years of age. This procedure should be performed only in an enclosed area with appropriate airflow and by operators wearing particulate respirators and taking all appropriate safety measures to avoid exposure. The patient is told that the procedure is being performed to induce coughing to raise sputum that the patient cannot raise spontaneously and that the salt solution is irritating. The patient is instructed to inhale slowly and deeply through the mouth and to cough at will, vigorously and deeply, coughing and expectorating into a collection tube. The procedure is discontinued if the patient fails to raise sputum after 10 minutes or feels any discomfort. Ten milliliters of sputum should be collected; if the patient continues to raise sputum, a second specimen should be collected and submitted. Specimens should be delivered promptly to the laboratory and refrigerated if processing is delayed.

Gastric Lavage Specimens

Gastric lavage is used to collect sputum from patients who may have swallowed sputum during the night. The procedure is limited to senile, nonambulatory patients, children younger than 3 years of age, and patients who fail to produce sputum by aerosol induction. The most desirable gastric lavage is collected at the patient's bedside before the patient arises and before exertion empties the stomach. Gastric lavage cannot be performed as an office or clinic procedure. A series of three specimens are collected within 3 days.

The collector should wear a cap, gown, and particulate respirator mask and stand beside (not in front of) the patient, who should sit up on the edge of the bed or in a chair, if possible. The Levine collection tube is inserted through a nostril, and the patient is instructed to swallow the tube. When the tube is fully inserted, a syringe is attached to the end of the tube and filtered distilled water is inserted through the tube. The syringe is then used to withdraw 20 to 25 mL of gastric secretions that are expelled slowly down the sides of the 50-mL conical collecting tube. The top of the collection tube is screwed on tightly, and the tube is held upright during prompt delivery to the laboratory.

Bronchial lavages, washings, and brushings are collected and submitted by medical personnel. These are the specimens of choice for detecting nontuberculous mycobacteria and other opportunistic pathogens in patients with immune dysfunctions.

Urine Specimens

The incidence of urogenital infections shows little evidence of decreasing. Whereas 2% to 3% of patients with pulmonary tuberculosis exhibit urinary tract involvement, 30% to 40% of patients with genitourinary disease have tuberculosis at some other site. The clinical manifestations of urinary tuberculosis are variable, including frequency of urination (most common), dysuria, hematuria, and flank pain. Definitive diagnosis requires recovery of acid-fast bacilli from the urine.

Early morning voided urine specimens in sterile containers should be submitted daily for at least 3 days. The procedure for collection is that used to collect a clean-catch midstream urine specimen, described in Chapter 57. Twenty-four-hour urine specimens are undesirable because of excessive dilution, higher contamination, and difficulty in concentrating.

Fecal Specimens

Acid-fast stain and/or culture of stool from patients with AIDS has been used to identify patients who may be at risk for developing disseminated *M. avium* complex disease. The clinical utility of this practice remains controversial[8,16]; however, if screening stains and/or cultures are positive, dissemination often follows.[10] Feces should be submitted in a clean, dry, wax-free container without preservative or diluent. Contamination with urine should be avoided.

Tissue and Body Fluid Specimens

Tuberculous meningitis is uncommon but still occurs in both immunocompetent and immunosuppressed patients. Sufficient quantity of specimen is most critical for isolation of acid-fast bacilli from cerebrospinal fluid (CSF). There may be very few organisms in the spinal fluid, which makes their detection difficult. At least 10 mL of CSF is recommended for recovery of mycobacteria. Similarly, as much as possible (10 to 15 mL minimum) of other body fluids, such as pleural, peritoneal, and pericardial fluids, should be collected in a sterile container or syringe with a Luer tip cap.

Blood Specimens

Immunocompromised patients, particularly those who are infected with HIV, can have disseminated mycobacterial infection; the majority of these infections are caused by *M. avium* complex. A blood culture positive for *M. avium* complex is always associated with clinical evidence of disease. Best recovery of mycobacteria is achieved by collecting the blood in either a broth such as the radiometric BACTEC 13A vial or the Isolator lysis-

centrifugation system (see Chapter 52). Some studies have indicated that the lysis-centrifugation system is advantageous, because quantitative data can be obtained with each blood culture; in patients with AIDS, quantitation of such organisms can be used to monitor therapy and determine prognosis. However, the necessity for doing quantitative blood cultures remains unclear. Blood for culture of mycobacteria should be collected in a manner as for routine blood cultures.

Wounds, Skin Lesions, and Aspirates

If attempting to culture a skin lesion or wound, an aspirate is the best type of specimen to collect. The skin should be cleansed with alcohol before aspiration of the material into a syringe. If the volume is insufficient for aspiration, pus and exudates may be obtained on a swab and then placed in transport medium, such as Amie's or Stuart's (dry swabs are unacceptable). However, a negative culture of a specimen obtained on a swab is not considered reliable, and this should be noted in the culture report.

SPECIMEN PROCESSING

Specimen processing for the recovery of acid-fast bacilli from clinical specimens involves several complex steps, each of which must be carried out with precision. Specimens from sterile sites can be inoculated directly to media (small volume) or concentrated to reduce volume. Other specimens require decontamination and concentration. A scheme for processing is depicted in Figure 45-1. The procedures are explored in detail in the following discussions.

Contaminated Specimens

The majority of specimens submitted for mycobacterial culture consist of organic debris, such as mucin, tissue, serum, and other proteinaceous material that is contaminated with organisms. A typical example of such a specimen is sputum. Laboratories must process these specimens so that contaminating bacteria that can rapidly outgrow mycobacteria are either killed or reduced in numbers, and mycobacteria are released from mucin and/or cells. After decontamination, mycobacteria are concentrated, usually by centrifugation, to enhance their detection by acid-fast stain and culture. Unfortunately, there is not one ideal method for decontaminating and digesting clinical specimens. Although continuously faced with the inherent limitations of various methods, laboratories must strive to maximize the survival and detection of mycobacteria on the one hand, while maximizing the elimination of contaminating organisms on the other. Of note, rapidly growing mycobacteria are especially susceptible to high or prolonged exposure to greater than or equal to 2% NaOH.

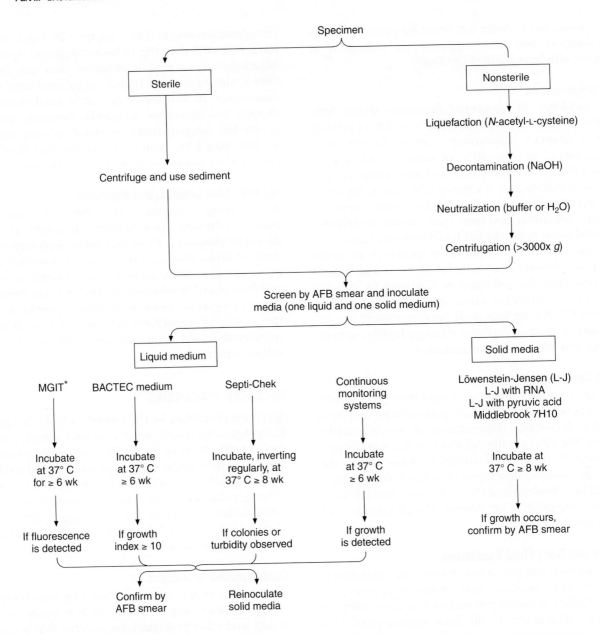

*Mycobacterium Growth Indicator Tube.

Figure 45-1 A flowchart for specimen processing for isolation of mycobacteria.

Digestion-decontamination procedures should be as gentle as possible, with no more than an overall contamination rate of 5%.[5]

Overview. Commonly used digestion-decontamination methods are the sodium hydroxide (NaOH) method, the Zephiran-trisodium phosphate method, and the N-acetyl-L-cysteine (NALC)-NaOH method. The NALC-NaOH method is detailed in Procedure 45-1. Of note, another decontaminating procedure using oxalic acid is very useful for treating specimens known to harbor gram-negative rods, particularly *Pseudomonas* and *Proteus*, which are extremely troublesome conta-

minants. NaOH is a commonly used decontaminant that is also mucolytic. Several agents can be used to liquefy a clinical specimen, including NALC, dithiothreitol (sputolysin), and enzymes. None of these agents are inhibitory to bacterial cells. In most procedures, liquefaction (release of the organisms from mucin or cells) is enhanced by vigorous mixing with a vortex type of mixer in a closed container. Following mixing, the container should be allowed to stand for 15 minutes before opening, to prevent the dispersion of fine aerosols generated during mixing. Of utmost importance during processing is strict adherence to processing and laboratory safety protocols. All of these procedures should be carried

Procedure 45-1

N-ACETYL-L-CYSTEINE-SODIUM HYDROXIDE METHOD FOR LIQUEFACTION AND DECONTAMINATION OF SPECIMENS

PRINCIPLE

Sodium hydroxide (NaOH), a decontaminating agent, also acts as an emulsifier. Because of its potential toxicity, NaOH should be used at the lowest concentration that effectively digests and decontaminates the specimen. The addition of a mucolytic agent, *N*-acetyl-L-cysteine (NALC), reduces the concentration of NaOH required and also shortens the time required for decontamination, thus aiding the optimal recovery of acid-fast bacilli.

METHOD

1. Reagent preparation:

   A. NALC-NaOH preparation:

   For each day's cultures, add up the total volume of specimens to be treated and prepare an equal volume of the digestant-decontamination mixture, as follows:

   - 1 N (4%) NaOH (50 mL)
   - 0.1 M (2.94%) trisodium citrate
   - 3H$_2$O (50 mL)
   - NALC powder (0.5 g)

   Use sterile distilled water for preparation of solutions to minimize chances of inadvertently adding acid-fast tap water contaminants to the specimens. Mix, sterilize, and store the NaOH and the citrate in sterile, screw-capped flasks for later use. This solution should be used within 24 hours after the NALC is added.

   B. 0.67 M phosphate buffer, pH 6.8 preparation:

   Make up the following solutions:

   - Solution A (0.067 M disodium phosphate):
     Sodium monohydrogen phosphate (anhydrous) 9.47 g
   Distilled water 1000 mL
   - Solution B (0.067 M monopotassium phosphate):
     Potassium dihydrophosphate 9.07 g
     Distilled water 1000 mL

   Add 50 mL of solution B to 50 mL of solution A and adjust pH to 6.8.

2. Work within a biological safety cabinet and wear protective clothing, gloves, and mask. Transfer a maximum of 10 mL of sputum, urine, or other fluid to be processed to a sterile, disposable, plastic 50-mL conical centrifuge tube with a leak-proof and aerosol-free plastic screw-cap. Tubes with easily visible volume indicator marks are best.

3. Add an equal volume of freshly prepared digestant to the tube, being very careful when pouring digestant not to touch the lip of the specimen container, which might inadvertently transfer positive material to a negative specimen. Tighten the cap completely.

4. Vortex the specimen for approximately 15 seconds or for a maximum of 30 seconds, being certain to create a vortex in the liquid and not to merely agitate the material. Check for homogeneity by inverting the tube. If clumps remain, vortex the specimen intermittently while the rest of the specimens are being digested. An extra pinch of NALC crystals may be necessary to liquefy mucoid sputa.

5. Start a 15-minute timer when the first specimen is finished being vortexed. Continue digesting the other specimens, noting the amount of time that the entire run takes. The

digestant should remain on the specimens for a maximum exposure of 20 minutes.

6. After 15 minutes of digestion, add enough phosphate buffer to reach within 1 cm of the top, screw the cap tightly closed, and invert the tube to mix the solutions and stop the digestion process. Addition of this solution also reduces the specific gravity of the specimen, aiding sedimentation of the bacilli during centrifugation.

7. Centrifuge all tubes at 3600× *g* for 15 minutes, using aerosol-free sealed centrifuge cups.

8. Carefully pour off the supernatant into a splash-proof container. To ensure that the specimen does not run down the outside of the tube after pouring, the lip of the tube may be wiped with an amphyl- or phenol-soaked gauze to absorb drips. Be careful not to touch the lip of any tube to another container. It is helpful to watch the sediment carefully as the supernatant is being decanted, because a very mucoid sediment may be loose and may pour out with the supernatant. If the sediment begins to slip, stop decanting and use a sterile capillary pipette to remove the supernatant without losing the sediment.

9. Resuspend the sediment in 1 to 2 mL phosphate buffer, pH 6.8 buffer (with bovine serum albumin [BSA]).

10. Inoculate the sediment to culture media and prepare slides.

out in a BSC. Following digestion and decontamination, specimens are concentrated by centrifugation at ≥3000× *g*.

Special Considerations. Many specimen types besides respiratory samples contain normal flora and require decontamination and concentration; handling procedures for such specimens are described briefly.

Aerosol-induced sputum should be treated as sputum. Gastric lavages should be processed within 4 hours of collection or neutralized with 10% sodium carbonate (check with pH paper to determine that the specimen is at neutral pH) and refrigerated until processed as for sputum. If more than 10 mL of watery-appearing aspirate was obtained, the specimen can be centrifuged at 3600× *g* for 30 minutes, the supernatant decanted, and the sediment processed as for sputum.

Urine specimens should be divided into a maximum of four 50-mL centrifuge tubes and centrifuged at 3600× g for 30 minutes. The supernatant should be decanted, leaving approximately 2 mL of sediment in each tube. The tubes are vortexed to suspend the sediments, and sediments are combined. If necessary, distilled water can be added to a total volume of 10 mL. This urine concentrate is then treated as for sputum or with the sputolysin-oxalic acid method.

For fecal specimens, approximately 0.2 g of stool (a portion about the size of a pea) is emulsified in 11 mL of sterile, filtered, distilled water.[38] The suspension is vortexed thoroughly, and particulate matter is allowed to settle for 15 minutes. Ten milliliters of the supernatant is then transferred to a 50-mL conical centrifuge tube and decontaminated using the oxalic acid or NALC-NaOH method.

Swabs and wound aspirates should be transferred to a sterile 50-mL conical centrifuge tube containing a liquid medium (Middlebrook 7H9, Dubos Tween albumin broth) at a ratio of 1 part specimen to 5 to 10 parts liquid medium. The specimen is vortexed vigorously and allowed to stand for 20 minutes. The swab is removed, and the resulting suspension is processed as for sputum.

Large pieces of tissue thought to be contaminated should be finely minced using a sterile scalpel and scissors. This material is then homogenized in a sterile tissue grinder with a small amount of sterile saline or sterile 0.2% bovine albumin; this suspension is then processed as for sputum. If not known to be sterile, tissue is homogenized and half is directly inoculated to solid and liquid media. The other half is processed as for sputum. If the tissue is collected aseptically and not thought to be contaminated, it may be processed without treatment with NALC-NaOH.

Specimens Not Requiring Decontamination

Tissues or body fluids collected aseptically usually do not require the digestion and decontamination methods used with contaminated specimens. Processing clinical specimens that do not routinely require decontamination for acid-fast culture is described here. If such a specimen appears contaminated because of color, cloudiness, or foul odor, perform a Gram stain to detect bacteria other than acid-fast bacilli. Specimens found to be contaminated should be processed as described in the previous section.

CSF should be handled aseptically and centrifuged for 30 minutes at 3600× g to concentrate the bacteria. The supernatant is decanted, and the sediment is vortexed thoroughly before preparing the smear and inoculating media. If insufficient quantity of spinal fluid is received, the specimen should be used directly for smear and culture. Because recovery of acid-fast bacilli from CSF is difficult, additional solid and liquid media should be inoculated if material is available.

Pleural fluid should be collected in sterile anticoagulant (1 mg/mL ethylenediaminetetraacetic acid [EDTA] or 0.1 mg/mL heparin). If the fluid becomes clotted, it should be liquefied with an equal volume of sputolysin and vigorously mixed. To lower the specific gravity and density of pleural fluid, transfer 20 mL to a sterile 50-mL centrifuge tube and dilute the specimen by filling the tube with distilled water. Invert several times to mix the suspension and centrifuge at 3600× g for 30 minutes. The supernatant should be removed, and the sediment should be resuspended for smear and culture.

Joint fluid and other sterile exudates can be handled aseptically and inoculated directly to media. Bone marrow aspirates may be injected into Wampole Pediatric Isolator tubes, which help to prevent clotting; the specimen can be removed with a needle and syringe for preparation of smears and cultures. As an alternative, these specimens are either inoculated directly to media or, if clotted, treated with sputolysin or glass beads and distilled water before concentration.

DIRECT DETECTION METHODS

Acid-Fast Stains

Mycobacteria possess cell walls that contain mycolic acids, which are long-chain, multiply cross-linked fatty acids. These long-chain mycolic acids probably serve to complex basic dyes, contributing to the characteristic of acid-fastness that distinguishes mycobacteria from other bacteria. Mycobacteria are not the only group with this unique feature. Species of *Nocardia* and *Rhodococcus* are also partially acid-fast; *Legionella micdadei*, a causative agent of pneumonia, is partially acid-fast in tissue. Cysts of the genera *Cryptosporidium* and *Isospora* are distinctly acid-fast. The mycolic acids and lipids in the mycobacterial cell wall probably account for the unusual resistance of these organisms to the effects of drying and harsh decontaminating agents, as well as for acid-fastness.

When Gram-stained, mycobacteria usually appear as slender, poorly stained, beaded gram-positive bacilli; sometimes they appear as "gram-neutral" or "gram-ghosts" by failing to take up either crystal violet or safranin. Acid-fastness is affected by age of colonies, medium on which growth occurs, and ultraviolet light. Rapidly growing species appear to be acid-fast–variable. Three types of staining procedures are used in the laboratory for rapid detection and confirmation of acid-fast bacilli: fluorochrome, Ziehl-Neelsen, and Kinyoun. Smears for all methods are prepared in the same way (Procedure 45-2).

Procedure 45-2

PREPARATION OF SMEARS FOR ACID-FAST STAIN FROM DIRECT OR CONCENTRATED SPECIMENS

METHOD

1. Vortex concentrated sediment, unconcentrated sputum, other purulent material, or stool. Aspirate 0.1 to 0.2 mL into a Pasteur pipette and place 2 to 3 drops on the slide. Place the end of the pipette or a sterile applicator stick parallel to the slide and slowly spread the liquid uniformly to make a thin smear.

2. For cerebrospinal fluid (CSF) sediment, vortex thoroughly and apply to the slide in heaped drops. A heaped drop is allowed to air dry, and a second application of sediment is placed on the same spot and allowed to dry. A minimum of three layers, applied to the same 1-cm diameter circle, should facilitate detection of small numbers of bacilli. (**NOTE:** some laboratories have stopped performing acid-fast stains on CSF because positive stains are extremely rare.)

3. Fix the smear at 80° C for 15 minutes or for 2 hours at 65° to 70° C on an electric hot plate.

NOTE: Survival of mycobacteria at this temperature has been reported; handle all specimens with proper precautions.

4. Stain slides by Ziehl-Neelsen or fluorochrome stain.

Procedure 45-3

AURAMINE-RHODAMINE FLUOROCHROME STAIN

PRINCIPLE

The fluorochrome dyes used in this stain complex to the mycolic acids in acid-fast cell walls. Detection of fluorescing cells is enhanced by the brightness against a dark background.

METHOD

1. Heat-fix slides at 80° C for at least 15 minutes or for 2 hours at 65° to 70° C.

2. Flood slides with auramine-rhodamine reagent and allow to stain for 15 to 20 minutes at room temperature.

3. Rinse with deionized water and tilt slide to drain.

4. Decolorize with 0.5% acid-alcohol (70% ethanol and 0.5% hydrochloric acid) for 2 to 3 minutes.

5. Rinse with deionized water and tilt slide to drain.

6. Flood slides with 0.5% potassium permanganate for 2 to 4 minutes.

7. Rinse with deionized water and air dry.

8. Examine under low power (250×) for fluorescence.

EXPECTED RESULTS

Mycobacterium spp. will fluoresce yellow to orange depending on the filter system used.

The visualization of acid-fast bacilli in sputum or other clinical material should be considered only presumptive evidence of tuberculosis, because stain does not specifically identify *M. tuberculosis*. The report form should indicate this. For example, *M. gordonae*, a nonpathogenic scotochromogen commonly found in tap water, has been a problem when tap water or deionized water has been used in the preparation of smears or even when patients have rinsed their mouths with tap water before the use of aerosolized saline solution for inducing sputum. However, the incidence of false-positive smears is very low when good quality control is maintained. Conversely, acid-fast stained smears of clinical specimens require at least 10^4 acid-fast bacilli per milliliter for detection from concentrated specimens.[14]

Methods

Fluorochrome Stain. Fluorochrome stain is the screening procedure recommended for those laboratories that possess a fluorescent (ultraviolet) microscope (Procedure 45-3). This stain is more sensitive than the conventional carbolfuchsin stains, because the fluorescent bacilli stand out brightly against the background (Figure 45-2); the smear can be initially examined at lower magnifications (250× to 400×), and therefore more fields can be visualized in a short period. In addition, a positive fluorescent smear may be restained by the conventional Ziehl-Neelsen or Kinyoun procedure, thereby saving the time needed to make a fresh smear. Screening of specimens with rhodamine or rhodamine-auramine will result in a higher yield of positive smears and will substantially reduce the amount of time needed for examining smears. One drawback associated with the fluorochrome stains is that many rapid-growers may not appear fluorescent with these reagents. It is recommended that all positive fluorescent smears be confirmed with a Ziehl-Neelsen stain or examination by another technologist. It is important to wipe the immersion oil from the objective lens after examining a positive smear, because stained bacilli can float off the slide into the oil and may possibly contribute to a false-positive reading for the next smear examined.

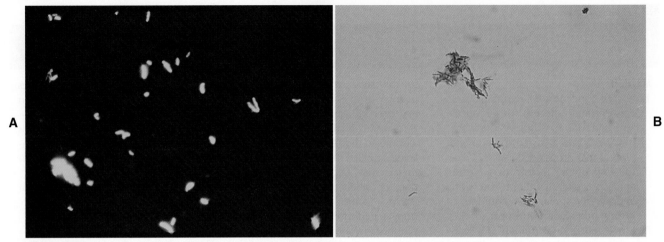

Figure 45-2 *M. tuberculosis* stained with **(A)** fluorochrome stain (400× magnification) and **(B)** Kinyoun acid-fast stain (1000× magnification).

Procedure 45-4

ZIEHL-NEELSEN ACID-FAST STAIN

PRINCIPLE

Heating the slide allows greater penetration of carbolfuchsin into the cell wall. Mycolic acids and waxes complex the basic dye, which then fails to wash out with mild acid decolorization.

METHOD

1. Heat-fix slides as previously described.
2. Flood smear with carbolfuchsin stain reagent and steam the slides gently for 1 minute by flaming from below the rack with a gas burner or by staining the slides directly on a special hot plate. Do not permit the slides to boil or dry out.
3. Allow the stain to remain on the slides for an additional 4 to 5 minutes without heat.
4. Rinse with deionized water and tilt slides to drain.
5. Decolorize with 3.0% acid-alcohol (95% ethanol and 3.0% hydrochloric acid) for 2 minutes. Rinse slides with deionized water and tilt to drain.
6. Flood slides with methylene blue reagent for 1 minute.
7. Rinse with deionized water and allow to air dry.
8. Examine under oil immersion (1000×) for presence of acid-fast bacilli.

EXPECTED RESULTS

Mycobacterium spp. will appear red or have a red-blue, beaded appearance, whereas nonmycobacteria will appear blue.

Fuchsin Acid-Fast Stains. The classic carbolfuchsin (Ziehl-Neelsen) stain requires heating the slide for better penetration of stain into the mycobacterial cell wall; hence it is also known as the *hot stain* procedure (Procedure 45-4). Procedure 45-5 describes the Kinyoun acid-fast stain. The method is similar to the Ziehl-Neelsen stain but without heat (see Figure 45-2); hence the term *cold stain*. If present, typical acid-fast bacilli appear as purple to red, slightly curved, short or long rods (2 to 8 μm); they may also appear beaded or banded (*M. kansasii*). For some nontuberculous species, such as *M. avium* complex, they appear pleomorphic, usually coccoid.

Examination, Interpretation, and Reporting of Smears.

Smears should be examined carefully by scanning at least 300 oil immersion fields (equivalent to three full horizontal sweeps of a smear that is 2 cm long and 1 cm wide) before reporting a smear as negative. Because the fluorescent stain can be examined using a lower magnification than for the fuchsin-stained smear, the equivalent number of fields can be examined in less time, thus making the fluorochrome stain the preferred method.

When acid-fast organisms are observed on a smear, results must be quantified to be meaningful. Because this quantitation estimates the number of bacilli being excreted, the extent of a patient's infectiousness can be assessed for clinical and epidemiologic purposes. The positive smear is reported by the laboratory, as well as the staining method and the quantity of organisms. The recommended interpretations and ways to report smear results are given in Table 45-7.

The overall sensitivity of an acid-fast smear ranges from 20% to 80%.[5,13] Factors such as specimen type, staining method, and culture method can influence the

Procedure 45-5

KINYOUN STAIN

PRINCIPLE
By increasing the concentration of basic fuchsin and phenol, the need for heating the slide is avoided.

METHOD
1. Heat-fix slides as previously described.
2. Flood slides with Kinyoun carbol-fuchsin reagent and allow to stain for 5 minutes at room temperature.
3. Rinse with deionized water and tilt slide to drain.
4. Decolorize with 3.0% acid-alcohol (95% ethanol and 0.5% hydrochloric acid) for 2 minutes.
5. Rinse with deionized water and drain standing water from slide surface by tipping slide.
6. Flood slide with methylene blue counterstain and allow to stain for 1 to 3 minutes.
7. Rinse with deionized water and allow to air dry.
8. Examine under oil immersion (1000×).

EXPECTED RESULTS
Mycobacterium spp. will appear red or have a red-blue beaded appearance, whereas nonmycobacteria will appear blue.

Table 45-7 Acid-Fast Smear Reporting

Number of AFB* Seen Fuchsin Stain (1000× Magnification)	Number of AFB Seen Fluorochrome Stain (450× Magnification)	Report
0	0	No AFB seen
1-2/300 fields	1-2/70 fields	Doubtful; request another specimen
1-9/100 fields	2-18/50 fields	1+
1-9/10 fields	4-36/10 fields	2+
1-9/field	4-36/field	3+
>9/field	>36/field	4+

Modified from Kent PT, Kubica GP: *Public health mycobacteriology: a guide for the level III laboratory,* US Department of Health and Human Service, Public Health Service, Centers for Disease Control, Atlanta, 1985.
*AFB, Acid-fast bacilli.

acid-fast smear sensitivity. In general, specificity of acid-fast smear examination is very high. However, cross-contamination of slides during the staining process and use of water contaminated with saprophytic mycobacteria can lead to false-positive results. Staining receptacles should not be used; acid-fast bacilli can also be transferred from one slide to another via immersion oil. For these reasons, it is best to confirm a positive result. Although not without some limitations, because of its simplicity and speed, the stained smear is an important and useful test, particularly for the detection of smear-positive patients ("infectious reservoirs"), who pose the greatest risk to others in their environment.

Antigen-Protein Detection

The detection of microbial products or components has been used in recent years to diagnose infections caused by *M. tuberculosis*. For example, tuberculostearic acid is a fatty acid that can be extracted from the cell wall of mycobacteria and then detected by gas chromatography/mass spectrometry in clinical samples containing few mycobacteria.[22] Because of the limited number of species that can cause meningitis and the fact that *M. tuberculosis* appears to be the only one of these species that releases tuberculostearic acid into the surrounding environment, the presence of this substance in CSF is thought to be diagnostic of tuberculous meningitis. Performance of this assay is limited to only a few laboratories. In addition, various immunoassays for antigen detection directly in clinical specimens, including sputum and CSF, have been evaluated and show some promise.

Adenosine deaminase is a host enzyme whose production is increased in certain infections caused by *M. tuberculosis*. For example, elevated levels of this enzyme were found in the majority of patients with tuberculous pleural effusions (98% sensitive); the test for the enzyme was determined to be highly specific as well (96% specificity).[3]

Nucleic Acid Amplification

Molecular techniques, such as conventional and real-time PCR, have been used to detect *M. tuberculosis* directly in clinical specimens; two kits are available commercially that are approved by the FDA. The

Amplicor *Mycobacterium tuberculosis* test (Roche Diagnostic Systems, Branchburg, NJ) uses PCR to detect *M. tuberculosis* directly in respiratory specimens; the Amplified *Mycobacterium tuberculosis* Direct Test (Gen-Probe, San Diego, Calif) is based on ribosomal RNA amplification. The Roche assay currently is approved for use on acid-fast smear-positive specimens only because numerous studies have demonstrated less than optimum sensitivity on smear-negative specimens. Because of subsequent kit modifications that improved sensitivity, Gen-Probe's assay is approved on both smear-positive and smear-negative specimens. Clinical laboratories also have developed their own PCR assays to detect *M. tuberculosis* directly in clinical specimens. In addition to in-house developed assays, there are three other non-FDA approved commercial amplification tests that are widely used outside the United States: AMPLICOR *Mycobacterium* system (Roche Diagnostic Systems), an automated format of the AMPLICOR *Mycobacterium tuberculosis* test; the ProbeTec Direct TB energy transfer system (Becton Dickinson, Sparks, Md); and the INNO-LiPA Rif TB assay (Innogenetics NV, Ghent, Belgium). These systems have comparable sensitivities and specificities to the FDA-approved amplification assays. The ultimate usefulness of these assays in the clinical laboratory is still being sorted out because of issues associated with amplification assays, such as expense and clinical utility.

CULTIVATION

A combination of different culture media is required to optimize recovery of mycobacteria from culture; at least one solid medium in addition to a liquid medium should be used.[18] The ideal media combination should support the most rapid and abundant growth of mycobacteria, allow for the study of colony morphology and pigment production, inhibit the growth of contaminants, and be economical.

Solid Media

Solid media, such as those listed in Box 45-3, are recommended because of the development of characteristic, reproducible colonial morphology, good growth from small inocula, and a low rate of contamination. Optimally, at least two solid media (a serum [albumin] agar base medium, e.g., Middlebrook 7H10, and an egg-potato base medium, e.g., L-J) should be used for each specimen (these media are available from commercial sources). All specimens must be processed appropriately before inoculation. It is imperative to inoculate test organisms to commercially available products for quality control (Procedure 45-6).

Cultures are incubated at 35° C in the dark in an atmosphere of 5% to 10% carbon dioxide (CO_2) and

BOX 45-3	**Suggested Media for Cultivation of Mycobacteria from Clinical Specimens***

SOLID
Agar-Based
Middlebrook 7H10 and Middlebrook 7H10 selective
Middlebrook 7H11 and Middlebrook 7H11 selective
Middlebrook biplate (7H10/7H11S agar)
Egg-Based
Löwenstein-Jensen (L-J)
L-J Gruft
L-J with pyruvic acid
L-J with iron
LIQUID
BACTEC 12B medium
Middlebrook 7H9 broth
Septi-Chek AFB
Mycobacteria Growth Indicator Tube
Commercially supplied broths for continuously monitoring systems for mycobacteria

*For optimal recovery of mycobacteria, a minimum combination of liquid medium and solid media is recommended.

high humidity. Tubed media is incubated in a slanted position with screw-caps loose for at least 1 week to allow for evaporation of excess fluid and the entry of CO_2; plated media are either placed in a CO_2-permeable plastic bag or wrapped with CO_2-permeable tape. If specimens are obtained from skin or superficial lesions suspected to contain *M. marinum* or *M. ulcerans*, an additional set of solid media should be inoculated and incubated at 25° to 30° C. In addition, a chocolate agar plate (or the placement of an X-factor [hemin] disk on conventional media) and incubation at 25° to 33° C is needed for recovery of *M. haemophilum* from these specimens.

Cultures are examined weekly for growth. Contaminated cultures are discarded and reported as "contaminated, unable to detect presence of mycobacteria"; additional specimens are also requested. If available, sediment may be recultured after enhanced decontamination or by inoculating the sediment to a more selective medium. Most isolates appear between 3 and 6 weeks; a few isolates appear after 7 or 8 weeks of incubation. When growth appears, the rate of growth, pigmentation, and colonial morphology are recorded. Typical colonial appearance of *M. tuberculosis* and other mycobacteria are shown in Figure 45-3. After 8 weeks of incubation, negative cultures (those showing no growth) are reported, and the cultures are discarded.

Because of the resurgence of tuberculosis in the United States in the late 1980s and early 1990s, significant effort has been put into developing methods to provide more rapid diagnosis of tuberculosis. Welch et al[35] refined a method that decreased the time to

Procedure 45-6

QUALITY CONTROL FOR MYCOBACTERIOLOGY

REAGENTS

1. Media: routine media used for cultivation of mycobacteria
2. Quality control organism: a recent isolate of *M. tuberculosis* or *M. tuberculosis* strain H37Rv
3. Other materials:
   Autoclaved sputum (AFB [acid-fast bacilli]-negative)
   7H9 liquid medium containing 15% glycerol
   Sterile buffer, pH 7.0
   50-mL plastic, conical centrifuge tubes

METHOD

1. Suspend several colonies of H37Rv in a tube containing 3 mL of 7H9 liquid medium and several plastic or glass beads. Mix vigorously on a test tube mixer; then allow large particles to settle for 15 minutes.
2. Prepare a dilution of approximately 10^6 organisms per mL by adding the above cell suspension drop by drop to 1 mL of buffer until a barely turbid suspension occurs. Transfer 0.5 mL of the 10^6 cells per mL dilution to 4.5 mL of glycerol broth to give a suspension of 10^5 cells per mL. Repeat the procedure to make a 10^4 per mL and a 10^3 per mL suspension.
3. Label fifteen 3-dram vials for each suspension (10^5, 10^4, and 10^3). Transfer 0.3 mL of the appropriate suspension to each vial. Store the vials at $-70°$ C to use for future quality control testing.
4. Thaw one vial of each of the three dilutions each time the quality control procedure is performed.
5. Add 2.7 mL of autoclaved sputum to each cell suspension to effect a tenfold dilution, and inoculate three sets of the media used for primary isolation with each of the three dilutions of sputum. Inoculate 0.1 mL of sputum per bottle.
6. Decontaminate and concentrate the remainder as with sputum specimens. Reconstitute the sediments with sterile buffer to 2.6 mL, resuspend vigorously, and inoculate a second set of media with 0.1 mL of each of the concentrated and resuspended samples.
7. Incubate at $35°$ C in 5% to 10% CO_2 for 21 days.

INTERPRETING AND RECORDING RESULTS

Egg media should have been inoculated with approximately 10^4, 10^3, and 10^2 organisms, respectively. The first dilution should produce semiconfluent growth, and the second and third dilutions should produce countable colonies in each bottle. Because of the retrospective nature of these determinations, close comparisons must be made between current and previous results to note trends or developing deficiencies. Failures may be the result of faulty media, lethal effects of decontamination and concentration procedures, improperly prepared reagents, or overexposure of specimens to these reagents. Should deficiencies become evident, techniques should be reviewed and attempts made to determine the source of the problem. New batches of media must be substituted for deficient media, and the latter rechecked to verify deficiencies. Personnel should be included in all discussions of problems and corrective measures. All deficiencies and corrective actions should be recorded in the appropriate section of the quality control records.

Example of Interpreting Quality Control Test Results of Decontamination and Concentration Procedure

	Sputum						
	Unprocessed			**Processed**			
Sputum Sample	**10^4**	**10^3**	**10^2**	**10^4**	**10^3**	**10^2**	**Interpretation**
1	3+	2+	50-100 colonies	2+	1+ 2+	Approximately 10 colonies	Media and decontamination procedures are acceptable
2	3+	2+	50-100 colonies	1+	0	0	Media acceptable; procedures too toxic
3	2+ or 1+	2+ or 1+	0	1+ or 0	1+ or 0	0	One or more of the media is not supporting growth of AFB adequately

detection of mycobacterial growth by half or more compared with conventional culture methods by using a thinly poured Middlebrook 7H11 plate. These plates are inoculated in a routine manner, sealed, incubated, and examined microscopically (40× magnification) at regular intervals for the appearance of microcolonies. Of note, presumptive identification of *M. tuberculosis* or *M. avium* complex could be made for about 83% of the isolates within 10 and 11 days after inoculation, respectively.

Liquid Media

In general, the use of a liquid media system reduces the turn-around time for isolation of acid-fast bacilli to approximately 10 days, compared with 17 days or longer for conventional solid media.[2,9,19] There are several different systems to culture and detect the growth of mycobacteria in liquid media. The most commonly used systems are summarized in Table 45-8. Growth of mycobacteria in liquid media, regardless of the type, requires 5% to 10% CO_2; CO_2 is either already

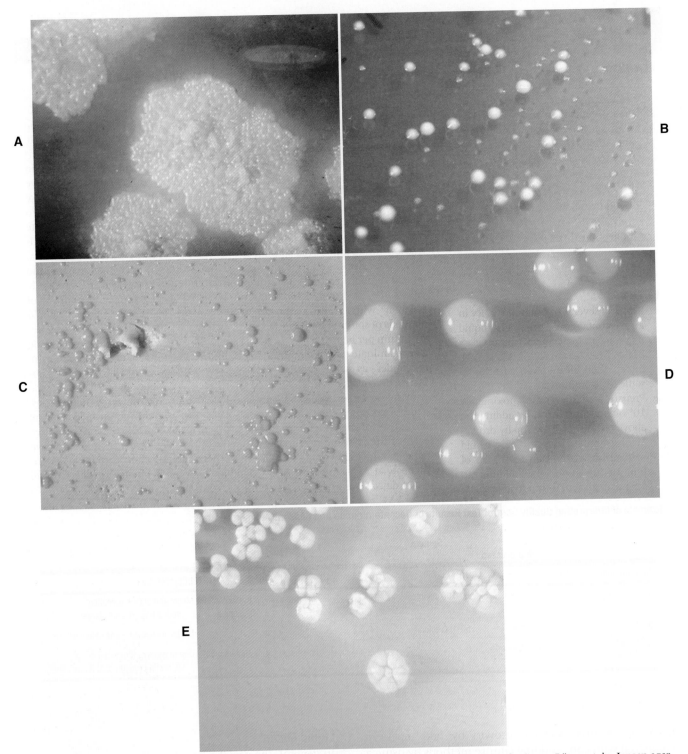

Figure 45-3 Typical appearance of some mycobacteria on solid agar medium. **A,** *M. tuberculosis* colonies on Löwenstein-Jensen agar after 8 weeks of incubation. **B,** Different colonial morphology seen on culture of one strain of *M. avium* complex. **C,** *M. kansasii* colonies exposed to light. **D,** Scotochromogen *M. gordonae* with yellow colonies. **E,** Smooth, multilobate colonies of *M. fortuitum* on Löwenstein-Jensen medium.

Table 45-8   Commonly Used Liquid Media Systems to Culture and Detect the Growth of Mycobacteria

System	Basic Principles of Detection
BACTEC 460 TB (Becton Dickinson Diagnostic Systems, Cockeysville, Md)	Culture media contains ^{14}C-labeled palmitic acid. If present in the broth, mycobacteria metabolize the ^{14}C-labeled substrates and release radioactively labeled $^{14}CO_2$ in the atmosphere, which collects above the broth in the bottle. The instrument withdraws this CO_2-containing atmosphere and measures the amount of radioactivity present. Bottles that yield a radioactive index, called a *growth index,* greater than or equal to 10 are considered positive.
Septi-Chek AFB System (Becton Dickinson)	Biphasic culture system made up of a modified Middlebrook 7H9 broth with a three-sided paddle containing chocolate, egg-based, and modified 7H11 solid agars. The bottle is inverted regularly to inoculate the solid media. Growth is detected by observing the three-sided paddle.
Mycobacteria Growth Indicator Tube (MGIT) (Becton Dickinson)	Culture tube contains Middlebrook 7H9 broth and a fluorescent compound embedded in a silicone sensor. Growth is detected visually using an ultraviolet light. Oxygen (O_2) diminishes the fluorescent output of the sensor; therefore, O_2 consumption by organisms present in the medium are detected as an increase in fluorescence.
Continuous Growth Monitoring Systems	
ESP Culture System II (TREK Diagnostic Systems, Cleveland, Ohio)	Organisms are cultured in a modified Middlebrook 7H9 broth with enrichment and a cellulose sponge to increase the culture's surface area. The instrument detects growth by monitoring pressure changes that occur as a result of O_2 consumption or gas production by the organisms as they grow.
BacT/Alert System (bioMérieux, Inc., Durham, NC)	Organisms are cultured in modified Middlebrook 7H9 broth. The instrument detects growth by monitoring CO_2 production by means of a colorimetric CO_2 sensor in each bottle.
BACTEC 9000 MB (Becton Dickinson)	Organisms are cultured in a modified Middlebrook 7H9 broth. The instrument detects growth by monitoring O_2 consumption by means of a fluorescent sensor.
BACTEC MGIT 960 (Becton Dickinson)	See above for basic principle (MGIT). The instrument detects growth by monitoring O_2 consumption by means of a fluorescent sensor.

provided in the culture vials or is added according to manufacturer's instructions. Once growth is detected in liquid medium, an acid-fast stain of a culture aliquot is performed to confirm the presence of acid-fast bacilli and subcultured to solid agar. A Gram stain can also be performed if contamination is suspected.

Interpretation

Although isolation of *M. tuberculosis* complex organisms represents infection, the clinician must determine the clinical significance of isolating an NTM in most cases; in other words, does the organism represent mere colonization or significant infection? Because these organisms vary greatly in their pathogenic potential, can colonize an individual without causing infection, and are ubiquitous in the environment, interpretation of a positive NTM culture is complicated. Therefore, the American Thoracic Society recommends diagnostic criteria for NTM disease to help physicians interpret culture results.[33]

APPROACH TO IDENTIFICATION

Regardless of the identification methods used, the first test that is always performed on organisms growing on solid or liquid mycobacterial media is an acid-fast stain to confirm that the organisms are indeed mycobacteria. Identification of other than *M. tuberculosis* complex and the more frequently isolated NTM (MAC, *M. avium*, *M. intracellulare*, *M. gordonae*, and *M. kansasii*) has become challenging for routine clinical microbiology laboratories, particularly in light of the ever-increasing number of new mycobacterial species; currently, there are 109 valid mycobacterial species. Traditional methods (i.e., phenotypic methods) for identifying mycobacteria and in particular the NTM, are based on growth parameters, biochemical characteristics, and the analysis of cell wall lipids, all of which are slow, cumbersome, and often inconclusive. Over the last decade, the rate of non-AIDS associated infections is also increasing and many of the newly identified NTM species have been associated with various diseases. As a result, it is vital to selecting effective antimicrobial therapy and deciding whether to perform susceptibility testing that NTM are accurately speciated. Of note, most of the newer species have been identified using nucleic acid sequencing with often only limited phenotypic characteristics published. Because of these issues and limitations with conventional phenotypic methods for identification, molecular

Table 45-9   Controls and Media Used for the Biochemical Identification of Mycobacteria

Biochemical Test	Control Organisms		Result		Medium Used and Amount	Duration	Incubation Conditions
	Positive	Negative	Positive	Negative			
Niacin	M. tuberculosis	M. intracellulare	Yellow	No color change	0.5 mL DH$_2$O	15-30 min	Room temperature
Nitrate	M. tuberculosis	M. intracellulare	Pink or red	No color change	0.3 mL DH$_2$O	2 hours	37° C bath
Urease	M. fortuitum	M. avium	Pink or red	No color change	Urea broth for AFB	1, 3, and 5 days	37° C incubator without CO$_2$
68° C Catalase	M. fortuitum or M. gordonae	M. tuberculosis	Bubbles	No bubbles	0.5 mL phosphate buffer (pH 7)	20 min	68° C bath
SQ Catalase	M. kansasii or M. gordonae	M. avium	>45 mm	<45 mm	Commercial medium	14 days	37° C incubator (with CO$_2$)
Tween 80	M. kansasii	M. intracellulare	Pink or red	No color change	1 mL DH$_2$O	5 or 10 days	37° C incubator (in the dark, without CO$_2$)
Tellurite	M. avium	M. tuberculosis	Smooth, fine, black precipitate (smokelike action)	Gray clumps (no smokelike action)	Middlebrook, 7H9 broth	7, then 3 additional days	37° C incubator (with CO$_2$)
Arylsulfatase	M. fortuitum	M. intracellulare	Pink or red	No color change	Wayne's arylsulfatase medium	3 days	37° C incubator (without CO$_2$)
5% NaCl	M. fortuitum	M. gordonae	Substantial growth	Little or no growth	Commercial slant with and without 5% NaCl	28 days	37° C incubator (with CO$_2$)
TCH	M. bovis	M. tuberculosis	No growth (i.e., susceptible)	Growth (i.e., resistant or ≥1% of colonies are resistant)	TCH slant	3 weeks	37° C incubator (with CO$_2$)

AFB, Acid-fast bacilli; *DH$_2$O*, distilled water; *SQ*, semiquantitative; *TCH*, thiophene-2-carboxylic acid hydrazide.

and genetic investigations are becoming indispensable to accurately identify the NTM. Thus, if timely and accurate identification of mycobacteria is to be accomplished, molecular approaches in conjunction with some phenotypic characteristics should be employed.

Regardless of whether molecular or phenotypic methods are employed, when growth is detected, broth subcultures of colonies growing on solid media (several colonies inoculated to Middlebrook 7H9 broth [5 mL] and incubated at 35°C for 5 to 7 days with daily agitation to enhance growth) or liquid media is then used for pigmentation and growth rate determinations as well as to inoculate all test media for biochemical tests, if used. In some instances, additional cultures may be inoculated and then incubated at different temperatures when more definitive identification is needed.

Conventional Phenotypic Tests

Growth Characteristics. The preliminary identification of mycobacterial isolates depends on their rate of growth, colonial morphology (see Figure 45-3), colonial texture, pigmentation, and, in some instances, the permissive incubation temperatures of mycobacteria. Despite the limitations of phenotypic tests, the growth characteristics of mycobacteria is helpful to determine a preliminary identification, for example, that an isolate appears as a rapidly growing mycobacteria. To perform identification procedures, quality control organisms should be tested along with unknowns, as listed in Table 45-9. The commonly used quality control organisms can be maintained in broth at room temperature and transferred monthly. In this way they will always be available for inoculation to test media

Procedure 45-7

DETERMINATION OF PIGMENT PRODUCTION AND GROWTH RATE

PRINCIPLE

Certain mycobacteria produce carotenoids, either dependently or independently of exposure to light. This characteristic, in addition to their doubling time under standard conditions, is useful for initial identification.

METHOD

1. After the broth culture has incubated for 5 to 7 days, adjust the turbidity to that of a McFarland 0.5 standard.
2. Dilute the broth (McFarland 0.5 turbidity) 10^4.
3. Inoculate 0.1 mL of the diluted broth to each of three tubes of Löwenstein-Jensen agar. Completely wrap two of

the tubes in aluminum foil to block all light. If the isolate was obtained from a skin lesion or the initial colony was yellow-pigmented (possible *M. szulgai*), six tubes should be inoculated. The second set of tubes, two of them also wrapped with aluminum foil, is incubated at 30° C, or at room temperature if a 30° C incubator is not available.

4. Examine the cultures after 5 and 7 days for the appearance of grossly visible colonies. Examine again at intervals of 3 days. Interpretation: rapid-growers produce visible colonies in less than 7 days; slow-growers require more than 7 days.

5. When colonies are mature, expose the growth from a foil-wrapped tube to a bright light, such as a desk lamp, for 2 hours. The cap must be loose during exposure, because pigment production is an oxygen-dependent reaction. The tube is rewrapped and returned to the incubator, and the cap is left loose.
6. The three tubes are examined 24 and 48 hours after light exposure. For tubes incubating at 30° C, pigment may require 72 hours for development.

EXPECTED RESULTS

Interpret as shown in Figure 45-4.

---

along with suspensions of the unknown mycobacteria being tested.

Growth Rate. The rate of growth is an important criterion for determining the initial category of an isolate. Rapid-growers usually produce colonies within 3 to 4 days after subculture. Even a rapid-grower, however, may take longer than 7 days to initially produce colonies because of inhibition by a harsh decontaminating procedure. Therefore, the growth rate (and pigment production) must be determined by subculture; the method is described in Procedure 45-7. The dilution of the organism used to assess growth rate is critical. Even slow-growing mycobacteria appear to produce colonies in less than 7 days if the inoculum is too heavy. One organism particularly likely to exhibit false-positive rapid growth is *M. flavescens*. This species therefore serves as an excellent quality control organism for this procedure.

Pigment Production. As previously discussed, mycobacteria may be categorized into three groups based on pigment production. Procedure 45-7 describes how to determine pigment production. To achieve optimum photochromogenicity, colonies should be young, actively metabolizing, isolated, and well-aerated.[34] Although some species, such as *M. kansasii*, turn yellow after a few hours of light exposure, others, such as *M. simiae*, may take prolonged exposure to light. Scotochromogens produce pigmented colonies even in the absence of light, and colonies often become darker with prolonged exposure to light (Figure 45-4). One member of this group, *M. szulgai*, is peculiar in that it is a scotochromogen at 35° C and nonpigmented when grown at

25° to 30° C. For this reason, all pigmented colonies should be subcultured to test for photoactivated pigment at both 35° C and 25° to 30° C. Nonchromogens are not affected by light.

Biochemical Testing. Once placed into a preliminary subgroup based on its growth characteristics, an organism must be definitively identified to species or complex level. Although conventional biochemical tests can be used for this purpose, new methods (discussed later in this section) have replaced biochemical tests for the identification of mycobacterial species because of the previously discussed limitations of phenotypic testing. Although key biochemical tests are still discussed in this edition, the reader must be aware that this approach to identification will be ultimately replaced with molecular methods. Table 45-10 summarizes distinctive properties of the more commonly cultivable mycobacteria isolated from clinical specimens; key biochemical tests for each of the major mycobacterial groupings, including *M. tuberculosis* complex, are listed in Table 45-11. Key biochemical tests are discussed in the following text; detailed procedures are given in other texts.[1,5]

Niacin. Niacin (nicotinic acid) plays an important role in the oxidation-reduction reactions that occur during mycobacterial metabolism. Although all species produce nicotinic acid, *M. tuberculosis* accumulates the largest amount. (*M. simiae* and some strains of *M. chelonae* also produce niacin.) Niacin therefore accumulates in the medium in which these organisms are growing. A positive niacin test is preliminary evidence that an

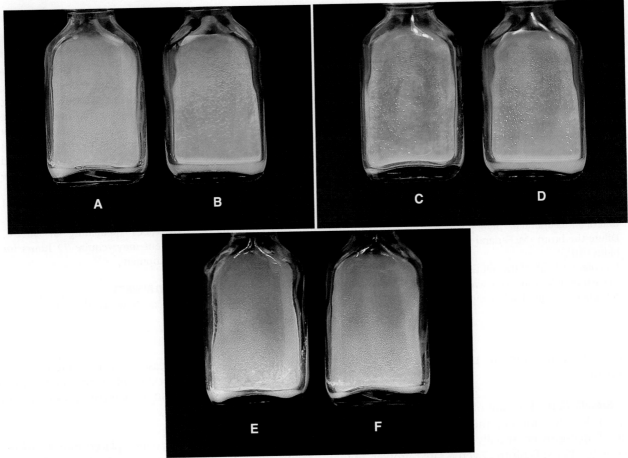

Figure 45-4 Initial grouping of mycobacteria based on pigment production before and after exposure to light. In one test system, subcultures of each isolate are grown on two agar slants. One tube is wrapped in aluminum foil to prevent exposure of the organism to light, and the other tube is allowed light exposure. After sufficient growth is present, the wrapped tube is unwrapped, and the tubes are examined together. Photochromogens are unpigmented when grown in the dark (tube A) and develop pigment after light exposure (tube B). Scotochromogens are pigmented in the dark (tube C); the color does not intensify after exposure to light (tube D). Nonphotochromogens are nonpigmented when grown in the dark (tube E) and remain so even after light exposure (tube F).

organism that exhibits a buff-colored, slow-growing, rough colony may be *M. tuberculosis* (Figure 45-5). The method is delineated in Procedure 45-8. This test is not sufficient, however, for confirmation of the identification. If sufficient growth is present on an initial L-J slant (the egg-base medium enhances accumulation of free niacin), a niacin test can be performed immediately. If growth on the initial culture is scanty, the subculture used for growth rate determination can be used. If this culture yields only rare colonies, the colonies should be spread around with a sterile cotton swab (after the growth rate has been determined) to distribute the inoculum over the entire slant. The slant is then reincubated until light growth over the surface of the medium is visible. For reliable results, the niacin test should be performed only from cultures on L-J that are at least 3 weeks old and show at least 50 colonies; other-

wise, enough niacin might not have been produced to be detected.

Nitrate Reduction. This test is valuable for the identification of *M. tuberculosis*, *M. kansasii*, *M. szulgai*, and *M. fortuitum*. The ability of acid-fast bacilli to reduce nitrate is influenced by age of the colonies, temperature, pH, and enzyme inhibitors. Although rapid-growers can be tested within 2 weeks, slow-growers should be tested after 3 to 4 weeks of luxuriant growth. Commercially available nitrate strips yield acceptable results only with strongly nitrate-positive organisms, such as *M. tuberculosis*. This test may be tried first because of its ease of performance. The *M. tuberculosis*–positive control must be strongly positive in the strip test or the test results will be unreliable. If the paper strip test is negative or if the control test result is not strongly positive, the chemical

Table 45-10   Distinctive Properties of Commonly Cultivable Mycobacteria Encountered in Clinical Specimens

Group/Complex	Species	Optimal Temp (°C)	Usual Colonial Morphology[a]	Niacin	Growth On TCH (10 μg/mL)[b]	Nitrate Reduction	Semiquantitative Catalase (>45 mm)	68°C Catalase	Tween Hydrolysis, 5 Days	Tellurite Reduction	Tolerance to 5% NaCl	Iron Uptake	Arylsulfatase, 3 Days	Growth on MacConkey Agar	Urease	Pyrazinamidase, 4 Days
TB	M. tuberculosis	37	R	+	+	+	-	-	-[c]	±	-	-	-	-	±	+
	M. bovis	37	Rt	-	-	-	-	-	-	±	-	-	-	-	±	-
	M. africanum	37	R	V	V	V	-	-	-	-	-	-	-	-	±	-
Photochromogens	M. marinum	30	S/SR	∓	+	-	+	-	+	±	-	-	∓[d]	-	-/+	+
	M. kansasii	35	SR/S	-	+	+	+	+	+	±	-	-	-	-	±	-
	M. simiae	37	S	∓	+	-	+	+	-	±	-	-	-	-	±	∓
	M. asiaticum	37	S	∓	+	-	+	+	+	-	-	-	-	-	-	-
Scotochromogens	M. scrofulaceum	37	S	-	+	-	+	+	∓[c]	±	-	-	>	-	>	∓
	M. szulgai	37	S or R	-	+	-	+	+	∓	±	-	-	>	-	+	+
	M. gordonae	37	S	-	+	-	+	+	+	+	-	-	>	-	>	∓
Nonphotochromogens	M. avium complex	35-37	St/R	-	+	-	±	±	-	+	-	-	-	-	-	+
	M. genavense[e]	37	St	-	+	-		+	+	±	-	-	-	∓	+	+
	M. gastri	35	S/SR/R	-	+	-	+	+	+	+	-	-	-	-	±	-
	M. malmoense	30	S	-	+	-	+	-	+	-	-	-	-	-	-	+
	M. haemophilum[f]	30	R	-	+	-	±	±	-	+	-	-	-	-	+	+
	M. shimoidei	37	R	-	+	-	+	+	+	-	-	-	-	-	-	+
	M. ulcerans	30	R	-	+	-	-	-	-	-	-	-	-	-	-	+
	M. flavescens[g]	37	S	-	+	+	+	+	+	±	+	-	-	-	±	-
	M. xenopi[h]	42	Sf	-	+	-	+	+	+	-	-	-	±	>	-	+
	M. terrae complex (M. terrae, M. triviale,[i] M. nonchromogenicum)	35	SR	-	-	+	+	+	+	-	-	-	-	-	-	>
Rapidly growing	M. fortuitum group	28-30	Sf/Rf	-/+	+	+	+	+	V	+	+	+	+	+	+	>
	M. chelonae	28-30	S/R	-/+	+	-	+	V	V	+	+	-	+	+	+	>
	M. abscessus	28-30	S/R	-	-	-	+	V	V	+	-	-	+	+	+	+
	M. smegmatis	28-30	R/S	-	+	+	+	+	V	+	+	+	-	±	+	+

Plus and minus signs indicate the presence or absence, respectively, of the feature; blank spaces indicate either that the information is not currently available or that the property is unimportant. V, Variable; ±, usually present; ∓, usually absent.

See Clinical microbiology procedures handbook, vol 2, The American Society for Microbiology, Washington, DC, 2004, and Vincent V, Brown-Elliott BA, Jost KC, Wallace RJ: In Murray PR, Baron EJ, Pfaller MA, et al: Manual of clinical microbiology, ed 8, The American Society for Microbiology, Washington, DC, 2003, for other mycobacterial species' biochemical reactions and additional biochemical reactions on the mycobacteria included in this table.

[a]R, Rough; S, smooth; SR, intermediate in roughness; t, thin or transparent; f, filamentous extensions.

[b]TCH, Thiophene-2-carboxylic acid hydrazide.

[c]Tween hydrolysis may be positive at 10 days.

[d]Arylsulfatase, 14 days, is positive.

[e]Requires mycobactin for growth on solid media.

[f]Requires hemin as a growth factor.

[g]Young cultures may be nonchromogenic or possess only pale pigment that may intensify with age.

[h]Strains of M. xenopi can be nonphotochromogenic or scotochromogenic.

[i]M. triviale is tolerant to 5% NaCl, and a rare isolate may grow on MacConkey agar.

Table 45-11   Key Biochemical Reactions to Help Distinguish Mycobacteria Belonging to the Same Mycobacterial Group

Mycobacterial Group	Key Biochemical Tests
M. tuberculosis complex	Niacin, nitrate reduction, susceptibility to TCH if *M. bovis* is suspected
Photochromogens	Tween 80 hydrolysis, nitrate reduction, pyrazinamidase, 14-day arylsulfatase, urease, niacin
Scotochromogens	Permissive growth temperature, Tween 80 hydrolysis, nitrate reduction, semiquantitative catalase, urease, 14-day arylsulfatase
Nonphotochromogens	Heat-resistant and semiquantitative catalase activity, nitrate reduction, Tween 80 hydrolysis, urease, 14-day arylsulfatase, tellurite reduction, acid phosphatase activity
Rapidly growing	Growth on MacConkey agar, nitrate reduction, Tween 80 hydrolysis, 3-day arylsulfatase, iron uptake

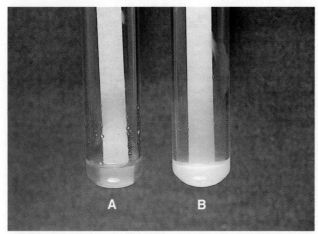

Figure 45-5   Niacin test performed with filter paper strips. The positive test (**A**) displays a yellow color. The negative result (**B**) remains milky white or clear.

procedure (Procedure 45-9) must be carried out using strong and weakly positive controls.

Catalase. Most species of mycobacteria, except for certain strains of *M. tuberculosis* complex (some isoniazid-resistant strains) and *M. gastri*, produce the intracellular enzyme catalase, which splits hydrogen peroxide into water and oxygen. Catalase is assessed in the following two ways:

1. By the relative activity of the enzyme, as determined by the height of a column of bubbles of oxygen (Figure 45-6) formed by the action of untreated enzyme produced by the organism (semiquantitative catalase test). Based on the semiquantitative catalase test, mycobacteria are divided into two groups: those producing less than 45 mm of bubbles and those producing more than 45 mm of bubbles.

 Procedure 45-8

NIACIN TEST WITH COMMERCIALLY AVAILABLE PAPER STRIPS*

PRINCIPLE
The accumulation of niacin in the medium caused by lack of an enzyme that converts niacin to another metabolite in the coenzyme pathway is characteristic for *M. tuberculosis* and a few other species. Niacin is measured by a colored end product.

METHOD
1. Add 1 mL of sterile distilled water to the surface of the egg-based medium on which the colonies to be tested are growing.
2. Lay the tube horizontally, so that the fluid is in contact with the entire

surface. Using a pipette, scratch or lightly poke through the surface of the agar; this allows niacin in the medium to dissolve in the water.
3. Allow the tube to sit for up to 30 minutes at room temperature. It can incubate longer to achieve a stronger reaction.
4. Remove 0.6 mL of the distilled water (which appears cloudy at this point) to a clean, 12- × 75-mm screw-cap or snap-top test tube. Insert a niacin test strip with the arrow down, following manufacturer's instructions.
5. Cap the tube tightly and incubate at room temperature, occasionally

shaking the tube to mix the fluid with the reagent on the bottom of the strip.
6. After 20 minutes, observe the color of the liquid against a white background (see Figure 45-5).

EXPECTED RESULTS
Yellow liquid indicates a positive test. The color of the strip should not be considered when evaluating results. If the liquid is clear, the test is negative. Discard the strip into alkaline disinfectant (10% NaOH) to neutralize the cyanogen bromide.

*Remel, Inc., Lenexa, Kan.

Procedure 45-9

NITRATE REDUCTION TEST USING CHEMICAL REAGENTS

PRINCIPLE

As in the conventional nitrate test, the presence of nitrite (product of the nitro-reductase enzyme) is detected by production of a red-colored product on the addition of several reagents. If the enzyme has reduced nitrate past nitrite to gas, then addition of zinc dust (which converts nitrate to nitrite) will detect the lack of nitrate in the reaction medium.

METHOD

1. Prepare the dry crystalline reagent as follows:

Sulfanilic acid (Sigma Chemical Co., St Louis, Mo)	1 part
N-(1-Naphthyl) ethylene-diamine dihydrochloride (Eastman Chemical Co., Rochester, NY)	1 part
l-Tartaric acid (Sigma Chemical Co.)	10 parts

These crystals can be measured with any small scoop or tiny spoon, because the proportions are by volume, not weight. The mixture should be ground in a mortar and pestle to ensure adequate mixing, because the crystals are of different textures. The reagent can be stored in a dark glass bottle at room temperature for at least 6 months.

2. Add 0.2 mL sterile distilled water to a 16- × 125-mm screw-cap tube. Emulsify two very large clumps of growth from a 4-week culture on Löwenstein-Jensen agar in the water. The suspension should be milky.

3. Add 2 mL nitrate substrate broth (Difco Laboratories or Remel, Lenexa, Kan) to the suspension and cap tightly. Shake gently and incubate upright for 2 hours in a 35° C water bath.

4. Remove from water bath and add a small amount of the crystalline reagent. A wooden stick or a small spatula can be used to add crystals; the amount is not critical. Examine immediately.

EXPECTED RESULTS

Development of a pink to red color indicates the presence of nitrite, demonstrating the ability of the organism to reduce nitrate to nitrite. If no color results, the organisms may have reduced nitrate beyond nitrite (as in the conventional nitrate test). Add a small amount of powdered zinc to the negative tube. The development of a red color indicates that unreduced nitrate was present in the tube and the organism was nitroreductase-negative.

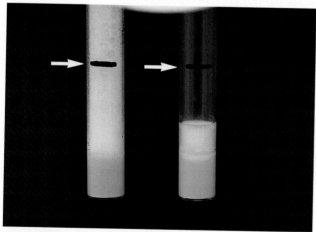

Figure 45-6 Semiquantitative catalase test. The tube on the left contains a column of bubbles that has risen past the line *(arrow)*, indicating 45-mm height (a positive test). The tube on the right is the negative control.

2. By the ability of the catalase enzyme to remain active after heating, a measure of the heat stability of the enzyme (heat-stable catalase test). When heated to 68° C for 20 minutes, the catalase of *M. tuberculosis*, *M. bovis*, *M. gastri*, and *M. haemophilum* becomes inactivated.

Tween 80 Hydrolysis. The commonly nonpathogenic, slow-growing scotochromogens and nonphotochromogens produce a lipase that is able to hydrolyze Tween 80 (the detergent polyoxyethylene sorbitan monooleate) into oleic acid and polyoxyethylated sorbitol, whereas pathogenic species do not. Tween 80 hydrolysis is useful for separating species of photochromogens, nonchromogens, and scotochromogens. Because laboratory-prepared media have a very short shelf life, the CDC recommends the use of a commercial Tween 80 hydrolysis substrate (Difco Laboratories or Remel Laboratories) that is stable for up to 1 year.

Tellurite Reduction. Some species of mycobacteria reduce potassium tellurite at variable rates. The ability to reduce tellurite in 3 to 4 days distinguishes members of *M. avium* complex from most other nonchromogenic species. All rapid-growers reduce tellurite in 3 days.

Arylsulfatase. The enzyme arylsulfatase is present in most mycobacteria. Test conditions can be varied to differentiate different forms of the enzyme. The rate by which this enzyme breaks down phenolphthalein disulfate into phenolphthalein (which forms a red color in the presence of sodium bicarbonate) and other salts helps to differentiate certain strains of mycobacteria. The

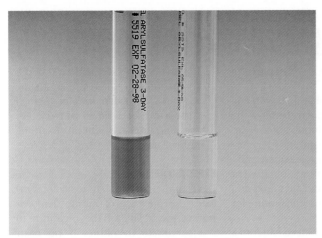

Figure 45-7   A positive arylsulfatase test is shown on the left; the tube containing the negative control is on the right.

3-day test is particularly useful for identifying the potentially pathogenic rapid-growers, *M. fortuitum* and *M. chelonae.* Slow-growing *M. marinum* and *M. szulgai* are positive in the 14-day test (Figure 45-7).

Growth Inhibition by Thiophene-2-Carboxylic Acid Hydrazide (TCH). This test is used to distinguish *M. bovis* from *M. tuberculosis,* because only *M. bovis* is unable to grow in the presence of 10 mg per mL TCH.

Other Tests. Other tests (see Table 45-11) are often performed to make more subtle distinctions between species. It is not cost effective for routine clinical microbiology laboratories to be able to perform all the procedures necessary for definitive identification of mycobacteria. Thus, those that require further testing can be forwarded to regional laboratories.

Molecular Identification

The genus *Mycobacterium* consists of a diverse group of acid-fast bacilli. As previously mentioned, conventional methods for identification of these organisms are well established and inexpensive, but these methods are time-consuming and labor-intensive, and they provide inconclusive results, often leading to species identifications based on a "best-fit" approach. The introduction of molecular methods has revolutionized the identification of mycobacteria by reducing time to identification as well as providing more accurate speciation.

DNA Hybridization. DNA hybridization is used to identify some of the more common mycobacterial species isolated on solid culture media or from broth culture. Of importance, these tests can be performed with sufficient growth from primary cultures. Non-isotopically labeled (i.e., acridinium ester–labeled) DNA probes specific for mycobacterial ribosomal RNA (rRNA)

sequences are commercially available (GenProbe, San Diego, Calif) and are listed in Box 45-4. rRNA is released from the mycobacteria by means of a lysing agent, sonication, and heat. The specific DNA probe is allowed to react with the extracted rRNA to form a stable DNA-RNA hybrid. Any unhybridized DNA-acridinium ester probes are chemically degraded. When an alkaline hydrogen peroxide solution is added to elicit chemiluminescence, only the hybrid-bound acridinium ester is available to emit light; the amount of light emitted is directly related to the amount of hybridized probe. The light produced is measured on a chemiluminometer. Numerous laboratories have incorporated these tests into their routine procedures.

Amplification and Amplification with Reverse Hybridization. Subsequent to the introduction of commercially available hybridization assays, commercially available and in-house developed nucleic acid amplification tests were successfully used for the early identification of *M. tuberculosis* complex grown in liquid cultures such as the BACTEC MGIT cultures. In addition, a commercially available system in which the 16S-23S rRNA spacer region of mycobacterial species (INNO-LiPA *Mycobacteria*; Innogenetics, Ghent, Belgium) has been successfully used to directly detect and identify eight of the most clinically relevant mycobacterial species in aliquots of positive liquid cultures (Box 45-5). A second commercial system, GenoType *Mycobacterium* (Hain Lifescience GmbH, Nehrin, Germany), using a similar format, has additional probes for *M. celatum, M. malmoense, M. peregrinum, M. phlei,* and two subgroups of

M. fortuitum as well as a supplemental kit that allows for 16 additional mycobacterial species.

Other Molecular Methods of Identification.

Amplification and Restriction Enzyme Analysis or DNA Sequencing. As previously mentioned, it has become increasingly important that identification of mycobacteria be accurate because organisms that are responsible for disease have important consequences for the proper selection of therapy. Thus, molecular methods have become increasingly important due to their rapidity and, in most cases, production of unequivocal results. Most of these methods are mainly based on 16S rRNA, 16S rDNA, the internal transcribed spacer (ITS) between 16S-23S rDNA and 16S rDNA, and the heat shock protein hsp65 gene of mycobacteria. PCR-based sequencing for mycobacterial identification consists of PCR amplification of mycobacterial DNA with genus-specific primers and sequencing of the amplicons. The organism is identified by comparison of the nucleotide sequence with reference sequences. The target most commonly used for this approach is the gene coding for the 16S ribosomal RNA that is present in all bacterial species. This region contains both conserved and variable regions, thus making it an ideal target for identification purposes. Despite its accuracy, problems remain in that the sequences in some databases are not accurate, there is no present consensus as to the quantitative definition of a genus or species based on 16S rRNA gene sequence data, and procedures are not standardized.

Another common noncommercial approach for identification is to amplify by PCR a highly conserved gene, such as the hsp65 gene, and perform restriction enzyme analysis on the PCR product. Following restriction enzyme digestion, usually using two different enzymes, digested products are separated by gel electrophoresis with subsequent sizing of resulting fragments The restriction fragment patterns are species-specific and are able to differentiate many species of NTM.

DNA Microarrays. DNA microarrays are also attractive for the rapid examination of large numbers of DNA sequences by a single hybridization step. This approach has been used to simultaneously identify mycobacterial species and detect mutations that confer rifampin resistance in mycobacteria. Fluorescent-labeled PCR amplicons generated from bacterial colonies are hybridized to a DNA array containing nucleotide probes. The bound amplicons emit a fluorescent signal that is detected with a scanner. Using this approach, 82 unique 16S ribosomal RNA sequences allow for the discrimination of 54 mycobacterial species and 51 sequences that contain unique *rpo*B gene mutations (mutations that confer resistance to rifampin).[31]

Chromatographic Analysis

The analysis of mycobacterial lipids by chromatographic methods, including thin-layer chromatography, gas-liquid chromatography (GLC), capillary gas chromographic methods, and reverse-phase high-performance liquid chromatography (HPLC), has been used to identify mycobacteria. In HPLC a liquid mobile phase is combined with various technical advances to separate large cellular metabolites and components. HPLC of extracted mycobacteria is a specific and rapid method for identification of species.[29] Many state health departments and the CDC now use this method routinely. The long-chain mycolic acids are separated better by HPLC than by GLC, because they do not withstand the high temperatures needed for GLC. The patterns produced by different species are very easily reproducible, and a typical identification requires only a few hours.

SERODIAGNOSIS

Immunodiagnostic methods based on the detection of antibodies to various mycobacterial antigens have not come into widespread clinical use compared with other infectious diseases.

ANTIMICROBIAL SUSCEPTIBILITY TESTING AND THERAPY

M. TUBERCULOSIS COMPLEX

In vitro drug susceptibility testing should be performed on the first isolate of *M. tuberculosis* from all patients.[18, 28] Susceptibility testing of *M. tuberculosis* requires meticulous care in the preparation of the medium, selection of adequate samples of colonies, standardization of the inoculum, use of appropriate controls, and interpretation of results. Laboratories that see very few positive cultures should consider sending isolates to a reference laboratory for testing. Isolates must be saved in sterile 10% skim milk in distilled water at −70° C for possible future additional studies (e.g., susceptibilities if the patient does not respond well to treatment).

Direct vs. Indirect Susceptibility Testing

Susceptibilities may be performed by either the direct or the indirect method. The direct method uses as the inoculum a smear-positive concentrate containing more than 50 acid-fast bacilli per 100 oil immersion fields; the indirect method uses a culture as the inoculum source. Although direct testing provides more rapid results, this method is less standardized and contamination may occur.

Table 45-12 Overview of Conventional Methods to Determine Susceptibility of *M. tuberculosis* Isolates to Antimycobacterial Agents

Method	Principle
Absolute concentration	For each drug tested, a standardized inoculum is inoculated to control (drug free) media and media containing several appropriate graded drug concentrations. Resistance is expressed as the lowest concentration of drug that inhibits all or almost all of the growth, that is, the minimum inhibitor concentration (MIC).
Resistance ratio	The resistance of the test organism is compared with that of a standard laboratory strain. Both strains are tested in parallel by inoculating a standard inoculum to media containing twofold serial dilutions of the drug. Resistance is expressed as the ratio of the MIC of the test strain divided by the MIC for the standard strain for each drug.
Proportion	For each drug tested, several dilutions of standardized inoculum are inoculated onto control and drug-containing agar medium. The extent of growth in the absence or presence of drug is compared and expressed as a percentage. If growth at the critical concentration of a drug is >1%, the isolate is considered clinically resistant.
Commercial systems cleared for use by the FDA: BACTEC 460TB (Becton Dickinson, Sparks, Md); BACTEC MGIT 960 (Becton (Dickinson); ESPII Culture System (Trek Diagnostic Systems, Cleveland, Ohio); MB/BacT Alert 3D (bioMérieux, Durham, NC)	Employing the principles of the agar proportion method, these methods use liquid medium. Growth is indicated by the amount of ^{14}C-labeled-carbon dioxide (CO_2) released, as measured by the BACTEC 460 instrument, or the amount of fluorescence or gas measured by the BACTEC MGIT 960, MB/BacT ALERT or the ESPII, respectively. For each drug tested, a standardized inoculum is inoculated into a drug-free and a drug-containing vial. The rate and amount of CO_2 produced in the absence or presence of drug is then compared.

Conventional Methods

Development of primary drug resistance in tuberculosis represents an increase in the proportion of resistant organisms. This increase in resistant organisms results from a spontaneous mutation and subsequent selection to predominance of these drug-resistant mutants by the action of a single or ineffective drug therapy. A poor clinical outcome is predicted with an agent when more than 1% of bacilli in the test population are resistant.[12] If an isolate is reported as resistant to a drug, then treatment failure will most likely occur if this drug is used for therapy. Drug resistance is defined for *M. tuberculosis* complex in terms of the critical concentration of the drug. The critical concentration of a drug is the amount of drug required to prevent growth above the 1% threshold of the test population of tubercle bacilli.

Four general methods are used throughout the world for determining susceptibility of isolates of *M. tuberculosis* to various antituberculous agents; these methods are summarized in Table 45-12. The proportion and BACTEC radiometric methods are most commonly used in the United States. Usually the initial isolate of *M. tuberculosis* is tested against five antimicrobials; these antimicrobials are referred to as **primary drugs** and are listed in Box 45-6. If any drug resistance is detected to any of the primary drugs, a second battery of agents is tested; these agents are also listed in Box 45-6.

New Approaches

Several new technologies recently have been introduced that are promising in terms of being faster, more reliable,

---

BOX 45-6 Antituberculous Agents Commonly Tested against *M. tuberculosis*

PRIMARY DRUGS
Streptomycin
Isoniazid
Rifampin
Ethambutol
Pyrazinamide

SECONDARY DRUGS
Ethionamide
Capreomycin
Ciprofloxacin
Ofloxacin
Kanamycin
Cycloserine
Rifabutin

---

and/or easier to perform than most conventional methods for susceptibility testing. For example, mutations leading to rifampin resistance have been detected using molecular methods. One molecular method, the line probe assay (INNO-LiPA Rif TB, Innogenetics, Ghent, Belgium), is a commercially available reverse hybridization-based probe assay for rapid detection of rifampin mutations leading to rifampin resistance in *M. tuberculosis*. Many different genotypic assays are currently available for drug susceptibility testing. Most are based on PCR amplification of a specific region of an *M. tuberculosis* gene followed by analysis of the amplicon for specific mutations associated with resistance to a particular drug. The presence or absence of mutations

Table 45-13   CLSI Recommendations for Susceptibility Testing of NTM

Organism	Isolates to Be Tested	Recommended Method	Drugs to Be Tested
MAC	• Clinically significant isolates from patients on prior macrolide therapy • Isolates from patients who become bacteremic while on macrolide preventive therapy • Isolates from patients who relapse while on macrolide therapy • Initial isolates from blood/tissue of patients with disseminated disease or respiratory samples from patients with pulmonary disease • Repeat testing after 3 months for patients with disseminated disease and 6 months for patients with chronic pulmonary disease	Broth-based method	Clarithromycin or azithromycin
M. kansasii	• All initial isolates • Repeat testing if cultures remain positive after 3 months of appropriate therapy.	Agar proportion Broth-based	Rifampin If rifampin-resistant, test rifabutin, ethambutol, streptomycin, clarithromycin, amikacin, ciprofloxacin, trimethoprim-sulfamethoxazole, gatifloxacin
M. marinum	Only if patient fails to responds clinically after several months of therapy and remains culture-positive	Agar proportion Agar disk elution Broth microdilution	Rifampin, ethambutol, clarithromycin, amikacin, sulfamethoxazole or trimethoprim-sulfamethoxazole
Rapidly growing mycobacteria	• Clinically significant isolates • Isolates: M. fortuitum group, M. chelonae, M. abscessus • Repeat testing if cultures remain positive after 6 months of appropriate therapy	Broth microdilution	Amikacin, cefoxitin, ciprofloxacin, clarithromycin, doxycycline, imipenem, linezolid, sulfamethoxazole, tobramycin

can then be detected by several methods such as automated sequencing. As previously mentioned, high-density DNA probe assays (see Chapter 8) have also been used to detect rifampin resistance as well as for mycobacterial species identification.[31] In addition, Jacobs et al[11] used an innovative approach to perform susceptibility testing by using a luciferase-reporter mycobacteriophage (bacterial viruses). The basis for this assay is simple—only viable mycobacteria can become infected with and replicate the mycobacteriophage; dead tubercle bacilli cannot. The mycobacteriophage was constructed to have the firefly luciferase gene next to a mycobacterial promoter; therefore, the presence and growth of the mycobacteriophage is detected by chemiluminescence. In brief, the isolate of *M. tuberculosis* to be tested is grown in the presence and absence of drug and the specially constructed mycobacteriophage is added. Following infection, luciferin, a substrate of luciferase, is added. If organisms are viable (i.e., thereby allowing infection of the bacteriophage and subsequent transcription and translation of the luciferase gene), the luciferin is broken down and light is emitted that can be measured; the amount of light emitted is directly propor-

tional to the number of viable *M. tuberculosis*. Therefore, if an organism is resistant to the drug, light will be emitted, whereas organisms susceptible to the drug will not emit light. Another commercially available assay employing mycobacteriophages is the FAST Plaque TB–RIF test. (Bio Tec Laboratories, Ipswich, UK). Of note, testing should be repeated if the patient remains culture-positive after 3 months of appropriate therapy or fails to respond clinically to therapy.

Therapy

Therapy directed against *M. tuberculosis* is dependent on the susceptibility of the isolate to various antimicrobial agents. To prevent the selection of resistant mutants, treatment of tuberculosis requires two or three drugs. The most common two-drug regimen is INH and rifampin administered for 9 months in cases of uncomplicated tuberculosis; if pyrazinamide is added to this regimen during the first 2 months, the total length of therapy can be shortened to 6 months. Finally, INH prophylaxis is recommended for those individuals with a recent skin-test conversion who are disease free.

NONTUBERCULOUS MYCOBACTERIA (NTM)

In general, treatment of patients with diseases caused by NTM require more individualization of therapy than patients with tuberculosis. This individualization is based on the species of mycobacteria recovered, site and severity of infection, antimicrobial drug susceptibility results, concurrent diseases, and the patient's general condition.[9] Currently, sufficient data exist to base general recommendations for susceptibility testing of MAC, *M. kansasii*, and *M. marinum*.[18,37] In addition, susceptibility testing should be performed on clinically significant, rapidly growing mycobacteria. Clinical Laboratory Standards Institute (CSLI; formerly National Committee for Clinical Laboratory Standards [NCCLS]) recommendations for susceptibility testing of NTM are summarized in Table 45-13.

PREVENTION

As previously mentioned, prophylactic chemotherapy with INH is used when known or suspected primary tuberculous infection causes a risk for clinical disease. At present, the BCG vaccine (named after Calmette and Guérin) is the only available vaccine against tuberculosis. The effectiveness of this live vaccine is controversial, because studies have demonstrated ineffectiveness to 80% protection. The greatest potential value for this vaccine is in developing countries that have high prevalence rates for tuberculosis. Nevertheless, at least four types of antituberculosis vaccines are currently being evaluated in experimental studies using animals for possible subsequent use in humans.

Case Study

A 40-year-old HIV-positive man on HAART (highly active antiretroviral therapy) presented with progressive encephalomyeloradicuopathy. He had no fever, cough or weakness, but had severe headaches. CSF was collected and contained 25 white blood cells (WBCs) per cubic millimeter (/mm³), low glucose, elevated protein, and no organisms on Gram stain or acid-fast stain. His studies were negative for cryptococcal antigen, toxoplasma by serology, and herpes simplex virus (HSV) by PCR. Routine bacterial culture was negative. Despite therapy for HSV and routine aerobic bacterial causes of meningitis, over the next 4 days he spiked fevers and a second CSF specimen showed 415 WBCs/mm³, with no diagnosis. A battery of viral encephalitis serology tests was done and all were negative. An in-house PCR of a third CSF specimen was positive for *Mycobacterium tuberculosis*, which grew in culture after 4 weeks.

QUESTIONS

1. Why were the acid-fast smears negative from all three of the specimens, but the PCR was positive?
2. How can *M. tuberculosis* be identified to the species level?
3. List the organisms present in the *Mycobacterium tuberculosis* complex.
4. Sometimes in processing for mycobacterial culture, an aerosol is created and one specimen splashes into another tube and contaminates it. If a physician questions that his patient does not appear to have tuberculosis, how can the laboratory confirm that the positive culture does not represent contamination?

REFERENCES

1. American Society for Microbiology: *Clinical microbiology procedures handbook*, vol 2, Washington, DC, 2004, American Society for Microbiology.
2. Badak FZ, Kiska DL, Setterquist S, et al: Comparison of mycobacteria growth indicator tube with BACTEC 460 for detection and recovery of mycobacteria from clinical specimens, *J Clin Microbiol* 34:2236, 1996.
3. Banales JL, Pineda PR, Fitzgerald JM, et al: Adenosine deaminase in the diagnosis of tuberculous pleural effusions: a report of 218 patients and review of the literature, *Chest* 99:355, 1991.
4. Centers for Disease Control and Prevention: *Core curriculum on tuberculosis*, ed 2, Atlanta, 1991, US Department of Health and Human Services, Public Health.
5. Centers for Disease Control and Prevention: In Kent PT, Kubica GP, editors: *Public health mycobacteriology: a guide for the level III laboratory*, Atlanta, 1995, US Department of Health and Human Services, Public Health Service.
6. Colston MJ: The microbiology of *Mycobacterium leprae*; progress in the last 30 years, *Trans R Soc Trop Med Hyg* 87:508, 1993.
7. Griffith DE, Girard WM, Wallace RJ: Clinical features of pulmonary disease caused by rapidly growing mycobacteria, *Am Rev Respir Dis* 147:1271, 1993.
8. Havlik JA Jr, Metchock B, Thompson SE III, et al: A prospective evaluation of *Mycobacterium avium* complex colonization of the respiratory and gastrointestinal tracts of persons with human immunodeficiency virus infection, *J Infect Dis* 168:1045, 1993.
9. Heifets L: Mycobacterial infections caused by nontuberculous mycobacteria, *Semin Respir Crit Care Med* 25: 283, 2004.
10. Horsburgh C Jr, Metchock BG, McGowan JE Jr, et al: Clinical implications of recovery of *Mycobacterium avium* complex from the stool or respiratory tract of HIV-infected individuals, *AIDS* 6:512, 1992.
11. Jacobs WR Jr, Barletta RG, Udani R, et al: Rapid assessment of drug susceptibilities of *Mycobacterium tuberculosis* by means of luciferase reporter phages, *Science* 260:819, 1993.
12. Kent PT, Kubica GP: *Public health mycobacteriology: a guide for the level III laboratory*, US Department of Health and Human Services, Public Health Service, Atlanta, 1985, Centers for Disease Control and Prevention.
13. Lipsky BJ, Gates J, Tenover FC, et al: Factors affecting the clinical value of microscopy for acid-fast bacilli, *Rev Infect Dis* 6:214, 1984.
14. Mazurek GH, Jereb J, LoBue P, et al: Guidelines for using QuantiFERON-TB Gold Test for detecting *Mycobacterium tuberculosis* infection, *MMWR* 54:15, 2005.

15. Mijs W, de Haas P, Rossau R, et al: Molecular evidence to support a proposal to reserve the designation *Mycobacterium avium* subsp. *avium* for bird-type isolates and *M. avium* subsp. *hominissuis* for the human/porcine type of *M. avium*, *Int J Syst Bacteriol* 52:1505, 2002.

16. Morris A, Reller LB, Salfinger M, et al: Mycobacteria in stool specimens: the nonvalue of smears for predicting culture results, *J Clin Microbiol* 31:1385, 1993.

17. Moschella SL: An update on the diagnosis and treatment of leprosy, *J Am Acad Dermatol* 51:417, 2004.

18. National Committee for Laboratory Standards: Susceptibility testing of mycobacteria, and other aerobic actinomycetes; Approved Standard, volume 23 (no. 18), Wayne, Pa, 2003, NCCLS.

19. Nolte FS, Metchock B: *Mycobacterium*. In Murray PR, Baron EJ, Pfaller MA, et al, editors: *Manual of clinical microbiology,* Washington, DC, 1995, American Society for Microbiology.

20. Noordeen SK: Epidemiology and control of leprosy: a review of progress over the last 30 years, *Trans R Soc Trop Med Hyg* 87:515, 1993.

21. O'Reilly LM, Daborn CJ: The epidemiology of *Mycobacterium bovis* infections in animals and man: a review, *Tubercle Lung Dis* 76(suppl 1):1, 1995.

22. Pfaller MF: Application of new technology to the detection, identification, and antimicrobial susceptibility testing of mycobacteria, *Am J Clin Pathol* 101:329, 1994.

23. Primm TP, Lucero CA, Falkinham JO: Health impacts of environmental mycobacteria, *Clin Microbiol Rev* 17: 98, 2004.

24. Runyon EH: Anonymous bacteria in pulmonary disease, *Med Clin North Am* 43:273, 1959.

25. Shinnick TM, Good RC: Mycobacterial taxonomy, *Eur J Clin Microbiol Infect Dis* 13:884, 1994.

26. Springer B, Tortoli E, Richter I, et al: *Mycobacterium conspicuum* sp nov, a new species isolated from patients with disseminated infections, *J Clin Microbiol* 33:2805, 1995.

27. Steele JH, Ranney AF: Animal tuberculosis, *Am Rev Tuberculosis* 77:908, 1958.

28. Taylor Z, Nolan CM, Blumberg HM: Controlling tuberculosis in the United States: recommendations from the American Thoracic Society, CDC, and the Infectious Diseases Society of America, *MMWR* 54:1-81, 2005.

29. Thibert L, Lapierre S: Routine application of high-performance liquid chromatography for identification of mycobacteria, *J Clin Microbiol* 31:1759, 1993.

30. Thorel MF, Krichevsky M, Levy-Frebault VV: Numerical taxonomy of mycobactin-dependent mycobacteria: emended description of *Mycobacterium avium*, and description of *Mycobacterium avium* subsp *avium* subsp nov, *Mycobacterium avium* subsp *paratuberculosis* subsp nov, and *Mycobacterium avium* subsp *silvaticum* subsp nov, *Int J Syst Bacteriol* 40:254, 1990.

31. Vernet G, Jay C, Rodrigue M, et al: Species differentiation and antibiotic susceptibility testing with DNA microarrays, *J Appl Microbiol* 96:59, 2004.

32. Wallace RJ: Recent changes in taxonomy and disease manifestations of the rapidly growing mycobacteria, *Eur J Clin Microbiol Infect Dis* 13:953, 1994.

33. Wallace RJ, et al: Diagnosis and treatment of disease caused by nontuberculous mycobacteria, *Am Rev Respir Dis* 142:940, 1990.

34. Wayne LG: The role of air in the photochromogenic behavior of *Mycobacterium kansasii*, *Am J Clin Pathol* 42:431, 1964.

35. Welch DF, Guruswamy AP, Sides SJ, et al: Timely culture for mycobacteria which utilizes a microcolony method, *J Clin Microbiol* 31:2178, 1993.

36. Wolinsky E: Mycobacterial diseases other than tuberculosis, *Clin Infect Dis* 15:1, 1992.

37. Woods GL: Mycobacterial susceptibility testing and reporting: when, how, and what to test, *Clin Microbiol Newsl* 27: 67, 2005.

38. Yajko DM, Nassos PS, Sanders CA, et al: Comparison of four decontamination methods for recovery of *Mycobacterium avium* complex from stools, *J Clin Microbiol* 31:302, 1993.

ADDITIONAL READING

Brown-Elliott BA, Griffith DE, Wallace RJ: Newly described or emerging human species of nontuberculous mycobacteria, *Infect Dis Clin North Am* 16:187, 2002.

Brown-Elliott BA, Griffith DE, Wallace RJ: Diagnosis of nontuberculous mycobacterial in fections, *Clin Lab Med* 22:911, 2002.

Goutzamanis JJ, Gilbert GL: *Mycobacterium ulcerans* infection in Australian children: report of eight cases and review, *Clin Infect Dis* 21:1186, 1995.

Hoffner SE: Pulmonary infections caused by less frequently encountered slowly growing mycobacteria, *Eur J Clin Microbiol Infect Dis* 13:925, 1994.

Inderlied CB: Antimycobacterial agents: in vitro susceptibility testing, spectra of activity, mechanisms of action and resistance, and assays for activity in biologic fluids. In Lorian V, editor: *Antibiotics in laboratory medicine,* Baltimore, 1996, Williams & Wilkins.

Tortoli E: Impact of gentotypic studies on mycobacterial taxonomy: the new mycobacteria of the 1990s, *Clin Microbiol Rev* 16:319, 2003.

CHAPTER 46

Obligate Intracellular and Nonculturable Bacterial Agents

Genera and Species to Be Considered

Current Name	Previous Name
Chlamydia trachomatis	
Chlamydophila psittaci	
Chlamydophila pneumoniae	
Rickettsia akari	
Rickettsia conorii	
Rickettsia rickettsii	
Rickettsia prowazekii	
Rickettsia typhi	
Orientia tsutsugamushi	
Ehrlichia chaffeensis	
Ehrlichia ewingii	
Anaplasma phagocytophilum	*Ehrlichia phagocytophila, Ehrlichia equi,* and human granulocytic ehrlichiosis agent
Neorickettsia sennetsu	*Ehrlichia sennetse*
Coxiella burnetii	
Tropheryma whipplei	*T. whippelii*
Calymmatobacterium granulomatis	

The organisms addressed in this chapter are obligate intracellular bacteria or are considered either extremely difficult to culture or unable to be cultured. Organisms of the genera *Chlamydia, Chlamydophila, Rickettsia, Orientia, Anaplasma,* and *Ehrlichia* are prokaryotes that differ from most other bacteria with respect to their very small size and obligate intracellular parasitism. Three other organisms, *Coxiella, Calymmatobacterium granulomatis,* and *Tropheryma whipplei,* are also reviewed here because they are difficult to cultivate or are noncultivable.

CHLAMYDIA

The members of the order Chlamydiales have been regrouped into two genera and nine species based on differences in phenotype, 16S rRNA, and 23S rRNA.[10] The genus *Chlamydia* comprises three species: *C. trachomatis, C. suis,* and *C. muridarum.* The other genus, *Chlamydophila,* consists of six species: *C. pneumoniae, C. psittaci, C. pecorum, C. felis, C. caviae,* and *C. abortus.*[10,11] Chlamydiae

possess a heat-stable, family-specific antigen that is an essential component of the cell membrane lipopolysaccharide[10]; species and type-specific protein antigens also exist. Members of the order Chlamydiales are obligate intracellular bacteria that were once regarded as viruses. These organisms require the biochemical resources of the eukaryotic host cell environment to fuel their metabolism for growth and replication because they are unable to produce high-energy compounds such as adenosine triphosphate. Chlamydiae have a unique developmental life cycle, with an intracellular growth, or replicative form, the reticulate body (RB), and an extracellular, metabolically inert, infective form, the elementary body (EB). Structurally, the chlamydial EB closely resembles a gram-negative bacillus; however, its cell wall lacks a peptidoglycan layer. In addition to this replicative cycle associated with acute chlamydial infections, multiple lines of evidence indicate that *Chlamydia* can also persist.[2,19] Following an unresolved infection, it is believed that *Chlamydia* persist in a viable but noncultivable growth stage referred to as a persistent body (PB), which results in a long-term, stable relationship with the infected host cell, that is, a chronic infection. Through an unknown stimulus, PBs can then presumably resume differentiation into EB and lead to active infection. The entire life cycle is illustrated in Figure 46-1. *C. trachomatis, C. pneumoniae,* and *C. psittaci* are important causes of human infection; *C. psittaci* and *C. pecorum* are common pathogens among animals. These three organisms that infect humans differ with respect to their antigens, host cell preference, antibiotic susceptibility, EB morphology, and inclusion morphology (Table 46-1).

CHLAMYDIA TRACHOMATIS

Over the past few decades, the importance of infections caused by *Chlamydia trachomatis* has been recognized. Of significance, the association of infertility and ectopic pregnancy with *C. trachomatis* infections and the realization that the majority of *C. trachomatis* infections are asymptomatic are well recognized.

General Characteristics

C. trachomatis infects humans almost exclusively and is responsible for various clinical syndromes. Based on major outer membrane protein (MOMP) antigenic dif-

Table 46-1   Differential Characteristics Among Chlamydiae That Cause Human Disease

Property	C. trachomatis	C. psittaci	C. pneumoniae
Host range	Humans (except one biovar that causes mouse pneumonitis)	Birds, lower mammals, humans (rare)	Humans
Elementary body morphology	Round	Round	Pear-shaped
Inclusion morphology	Round, vacuolar	Variable, dense	Round, dense
Glycogen-containing inclusions	Yes	No	No
Plasmid DNA	Yes	Yes	No
Susceptibility to sulfonamides	Yes	No	No

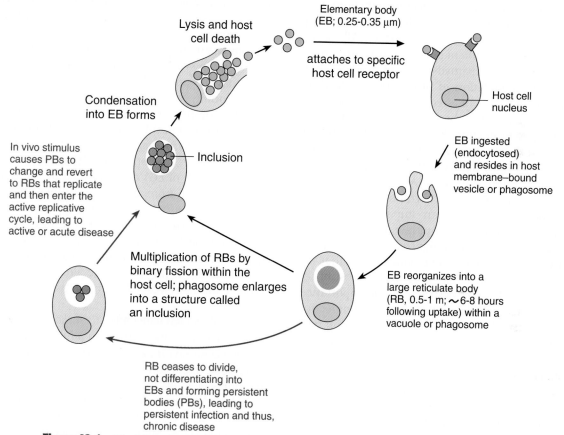

Figure 46-1   The life cycle of chlamydiae. The entire cycle takes approximately 48 to 72 hours.

ferences, *C. trachomatis* is divided into 17 different serovars that are associated with different primary clinical syndromes.[26]

Epidemiology and Pathogenesis

C. trachomatis causes significant infection and disease worldwide. In the United States, *C. trachomatis* is the most common sexually transmitted bacterial pathogen and a major cause of **pelvic inflammatory disease (PID)** (see Chapter 58 for more information on PID); an estimated 3 million *C. trachomatis* infections occur annually in the United States. Another disease caused by this organism, **ocular trachoma,** affects 500 million people, with 7 to 9 million of those infected becoming blind.

C. trachomatis infections are primarily spread from human to human by sexual transmission. Some infections, such as neonatal pneumonia or inclusion conjunctivitis, are transmitted from mother to infant during birth. The various routes of transmission for *C. trachomatis* infection are summarized in Table 46-2.

The natural habitat of *C. trachomatis* is humans. The mechanisms by which *C. trachomatis* cause inflammation and tissue destruction are not completely understood.

Table 46-2 Primary Syndromes Caused by *C. trachomatis*

Subtypes	Clinical Syndrome	Route(s) of Transmission
A, B, Ba, C	Endemic trachoma (multiple or persistent infections that ultimately lead to blindness)	Hand to eye from fomites, flies
L1, L2, L3	Lymphogranuloma venereum	Sexual
D-K	Urethritis, cervicitis, pelvic inflammatory disease, epididymitis, infant pneumonia and conjunctivitis (does not lead to blindness)	Sexual, hand to eye by autoinoculation of genital secretions; eye to eye by infected secretions; neonatal

Chlamydiae grow in cytoplasmic inclusions in susceptible host cells, which include mucosal epithelium, vascular endothelium, smooth muscle cells, circulating monocytes and tissue-specific macrophages. All species share a unique developmental cycle (see Figure 46-1). One mechanism that chlamydiae use to promote their survival is the modulation of programmed cell death pathways (apoptosis) in infected host cells.[2] These intracellular pathogens can either induce host cell death, thereby possibly facilitating their spread to neighboring host cells and down-regulating inflammation, or they can inhibit apoptosis, allowing for sustained survival within the infected host cell. As previously mentioned, it appears that chronic asymptomatic or persistent infections might play an important role. For example, the importance of multiple, recurrent infections with *C. trachomatis* is recognized for the development of ocular trachoma; some data suggest that the host's immune response may account for most of the inflammation and tissue destruction. Of importance, immunity provides little protection from reinfection and appears to be short-lived, following infection with *C. trachomatis*.

Spectrum of Disease

As previously mentioned, infection with *C. trachomatis* can lead to several different clinical syndromes. These infections are summarized in Table 46-2.

Trachoma. Trachoma is manifested by a chronic inflammation of the conjunctiva and remains a major cause of preventable blindness worldwide. As the infection progresses, the conjunctiva becomes scarred, which causes distortion of the eyelids such that the eyelashes become misdirected and turn in. The eyelashes then mechanically damage the cornea, resulting in ulceration, scarring, and visual loss.

Lymphogranuloma Venereum. Lymphogranuloma venereum (LGV) is a sexually transmitted disease that is unusual in Europe and North America but relatively frequent in Africa, Asia, and South America. The disease is characterized by a primary genital lesion at the initial infection site, which lasts a short time. This lesion is often small and may be unrecognized, especially by female patients. The second stage, acute lymphadenitis, often involves the inguinal lymph nodes, causing them to enlarge and become matted together, forming a large area of groin swelling, or bubo. During this stage, infection may become systemic and cause fever or may spread locally, causing granulomatous proctitis. In a few patients (more women than men), the disease progresses to a chronic third stage, causing the development of genital hyperplasia, rectal fistulas, rectal stricture, draining sinuses, and other manifestations.

Diagnosis of LGV is established by the isolation of an LGV strain from a bubo or other infected site. However, organism recovery rates of only 24% to 30% are reported. An intradermal skin test of LGV antigen, Frei's test, lacks sensitivity in early LGV and lacks specificity later, because Frei's antigen is only a genus-specific antigen. Moreover, the Frei's test can remain positive for many years, limiting its usefulness.

Oculogenital Infections. *C. trachomatis* can cause an acute inclusion conjunctivitis in adults. These infections are associated with a purulent discharge. In contrast to trachoma, inclusion conjunctivitis does not lead to blindness in adults (or newborns).

C. trachomatis infections have surpassed gonococcal infections as a cause of sexually transmitted disease in the United States. Similar to gonococci, *C. trachomatis* is a cause of urethritis, cervicitis, bartholinitis, proctitis, salpingitis, epididymitis, and acute urethral syndrome in women. In the United States, 60% of cases of nongonococcal urethritis are caused by chlamydiae. Both chlamydiae and gonococci are major causes of PID, contributing significantly to the rising rate of infertility and ectopic pregnancies in young women. After only one episode of PID, as many as 10% of women may become infertile because of tubal occlusion. The risk increases dramatically with each additional episode.

Many genital chlamydial infections in both sexes are asymptomatic or not easily recognized by clinical criteria; asymptomatic carriage in both men and women may persist, often for months. As many as 25% of men and 70% to 80% of women identified as having chlamydial genital tract infections have no symptoms. Of significance, these asymptomatic infected individuals serve as a large reservoir to sustain transmission within a community.

Perinatal Infections.
Approximately one fourth to half of infants born to women infected with *C. trachomatis* develop inclusion conjunctivitis. Usually, the incubation period is 5 to 12 days from birth, but may be as late as 6 weeks. Although most develop inclusion conjunctivitis, about 10% to 20% of infants develop pneumonia. Of significance, perinatally acquired *C. trachomatis* infection may persist in the nasopharynx, urogenital tract, or rectum for more than 2 years.

Laboratory Diagnosis
Diagnosis of *C. trachomatis* can be achieved by cytology, culture, direct detection of antigen or nucleic acid, and serologic testing.

Specimen Collection and Transport.
The organism can be recovered from or detected in infected cells of the urethra, cervix, conjunctiva, nasopharynx, and rectum and from material aspirated from the fallopian tubes and epididymis. For collecting specimens from the endocervix (preferred anatomic site to collect screening specimens from women), the specimen for *C. trachomatis* culture should be obtained after all other specimens (e.g., those for Gram-stained smear, *Neisseria gonorrhoeae* culture, or Papanicolaou [Pap] smear). A large swab should first be used to remove all secretions from the cervix. The appropriate swab (for nonculture tests, use the swab supplied or specified by the manufacturer) or endocervical brush is inserted 1 to 2 cm into the endocervical canal, rotated against the wall for 10 to 30 seconds, withdrawn without touching any vaginal surfaces, and then placed in the appropriate transport medium or swabbed onto a slide prepared for direct fluorescent antibody (DFA) testing.[3]

Urethral specimens should not be collected until 2 hours after the patient has voided. A urogenital swab (or one provided or specified by the manufacturer) is gently inserted into the urethra (females, 1 to 2 cm; males, 2 to 4 cm), rotated at least once for 5 seconds; and then withdrawn. Again, swabs should be placed into the appropriate transport medium or onto a slide prepared for DFA testing. Because chlamydiae are relatively labile, viability can be maintained by keeping specimens cold and minimizing transport time to the laboratory. For successful culture, specimens should be submitted in a chlamydial transport medium such as 2SP (0.2 M sucrose-phosphate transport medium with antibiotics); a number of commercial transport media are commercially available. Specimens should be refri-

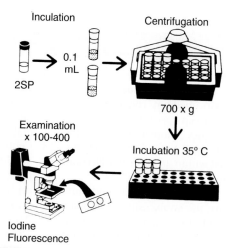

Figure 46-2   Processing of specimens for the cultivation of *C. trachomatis*. (From Smith TF: Role of the diagnostic virology laboratory in clinical microbiology: tests for *Chlamydia trachomatis* and enteric toxins in cell culture. In de la Maza LM, Peterson EM, editors: *Medical virology,* vol 1, New York, 1982, Elsevier Biomedical.)

gerated upon receipt, and if they cannot be processed for culture within 24 hours, they should be frozen at –70° C.

Cultivation.
Cultivation of *C. trachomatis* is discussed before methods for direct detection and serodiagnosis because all nonculture methods for the diagnosis of *C. trachomatis* are compared with culture.

Several different cell lines have been used to isolate *C. trachomatis* in cell culture, including McCoy, Hela, and monkey kidney cells; cycloheximide-treated McCoy cells are commonly used. After shaking the clinical specimens with 5-mm glass beads, centrifugation of the specimen onto the cell monolayer (usually growing on a coverslip in the bottom of a vial, the "shell vial") presumably facilitates adherence of elementary bodies. After 48 to 72 hours of incubation, monolayers are stained with an immunofluorescent stain and examined microscopically for inclusions; less specific inclusion-detection methods such as using iodine staining are not recommended.[3] Figure 46-2 is an overview of the basic steps in processing of specimens for chlamydial culture. Procedure 46-1 describes a method for isolation of chlamydiae. Although its specificity approaches 100%, the sensitivity of culture has been estimated at between 70% and 90% in experienced laboratories. Limitations of *Chlamydia* culture that contribute to this lack of sensitivity include prerequisites to maintain viability of patient specimens by either rapid or frozen transport and to ensure the quality of the specimen submitted for testing (i.e., endocervical specimens devoid of mucus and containing endocervical epithelial or metaplastic cells and/or urethral epithelial cells).[22] In addition to these issues is the requirement for a sensitive

Procedure 46-1

CELL CULTURE METHOD FOR ISOLATION OF CHLAMYDIAE

METHOD

1. Collect a swab or tissue in sucrose transport medium containing antibiotics. Keep specimens refrigerated at all times before transport to the laboratory. If the culture is not inoculated within 24 hours of collection, freeze the specimen at –70° C.

2. Vortex the specimen vigorously and remove the swab. (Mince the tissue and grind to make a cell suspension at a 1:10 dilution using transport medium.)

3. Use aseptic technique and aspirate the cell culture medium above the McCoy cell monolayer in the shell vial (commercially available), and add 0.2 mL of the patient specimen to the vial.

4. Centrifuge the vial for at least 1 hour at 2500 to 3000 × g at ambient temperature in a temperature-controlled centrifuge. Do not allow the temperature to increase past 40° C.

5. Aseptically remove the inoculum and add 1 mL fresh maintenance medium containing 1 µg/mL cycloheximide.

6. Recap the vials tightly and incubate at 35° C for 48 to 72 hours.

7. Remove medium from the vials, fix (according to manufacturer's instructions), and stain with iodine or fluorescein-conjugated chlamydial antibody.

8. Examine slides at 200 to 400× magnification for apple-green (fluorescent stain) intracytoplasmic inclusions (see Figure 46-3).

Table 46-3 Use of Different Laboratory Tests to Diagnose *C. trachomatis* Infections

Patient Population	Specimen Type	Acceptable Diagnostic Test
Prepubertal girls	Vaginal	Culture (if culture is unavailable, certain specialists accept AMPA)
Neonates and infants	Nasopharyngeal	Culture, DFA
	Rectal	Culture
	Conjunctiva	Culture, DFA, EIA, NAH
Women	Cervical	AMPA, culture, DFA, EIA, NAH
	Urethral	AMPA, culture, DFA, EIA, NAH
	Urine	AMPA
Children, women and men	Rectal	Culture, DFA
Men	Urethral	AMPA (DFA, EIA, NAH recommended when AMPA is unavailable)
	Urine†	AMPA

Modified from Centers for Disease Control and Prevention: Recommendations for the prevention and management of *Chlamydia trachomatis* infections, *MMWR* 42(RR-12):1-39, 1993; Centers for Disease Control and Prevention: Screening tests to detect *Chlamydia trachomatis* and *Neisseria gonorrhoeae* infections, *MMWR* 51(RR-15):1-27, 2002.

DFA, Direct fluorescent antibody staining; *EIA*, enzyme-linked immunosorbent assay; *NAH*, nucleic acid hybridization; *AMPA*, nucleic acid amplification assay.

*Must be confirmed in a population with a low prevalence (<5%) of *C. trachomatis* infection.

†EIA can be used on urine from symptomatic men but not on urine from older men. Also, a positive result must be confirmed in a population with a low prevalence of *C. trachomatis* infection.

cell culture system and the minimum delay of at least 2 days between specimen receipt and the availability of results. Despite these limitations, culture is still recommended as the test of choice for some situations (Table 46-3). As of this writing, only chlamydia cultures should be used in situations with legal implications (e.g., sexual abuse) when the possibility of a false-positive test is unacceptable.[3]

Direct Detection Methods

Cytologic Examination. Cytologic examination of cell scrapings from the conjunctiva of newborns or persons with ocular trachoma can be used to detect *C. trachomatis* inclusions, usually after Giemsa staining. Cytology has also been used to evaluate endocervical and urethral scrapings, including those obtained for Pap smears. However, this method is insensitive compared with culture or other methods discussed below.

Antigen Detection and Nucleic Acid Hybridization. To circumvent the shortcomings of cell culture, antigen detection methods are commercially available.

Direct fluorescent antibody (DFA) staining methods use fluorescein-isothiocyanate conjugated monoclonal antibodies to either outer membrane proteins or lipopolysaccharides of *C. trachomatis* to detect elementary bodies in smears of clinical material (Figure 46-3). The MicroTrak DFA (Syva Co., San Jose, Calif.) is one such commercial system.

Chlamydial antigen can also be detected by enzyme-linked immunosorbent assays (ELISA). Numerous U.S. Food and Drug Administration (FDA)-approved

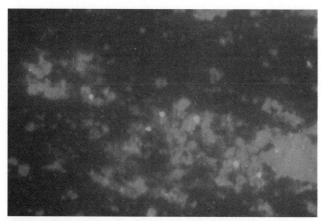

Figure 46-3 Appearance of fluorescein-conjugated, monoclonal antibody–stained elementary bodies in direct smear of urethral cell scraping from a patient with chlamydial urethritis. (Courtesy Syva Co, San Jose, Calif.)

kits are commercially available. These assays use polyclonal or monoclonal antibodies that detect chlamydial LPS. These tests are not species-specific for *C. trachomatis* and may cross-react with LPS of other bacterial species present in the vagina or urinary tract and thereby produce a false-positive result. Also available are nucleic acid hybridization tests that use a chemiluminescent type of DNA probe (PACE, Gen-Probe, San Diego, Calif.) that is complementary to a sequence of ribosomal RNA (rRNA) in the chlamydial genome. Once formed, the DNA-rRNA complex is absorbed onto a magnetic bead and detected by a luminometer. This assay is species-specific for *C. trachomatis.*

Based on numerous studies, these nonculture tests are more reliable in patients who are symptomatic and shedding large numbers of organisms than in those who are asymptomatic and most likely shedding fewer organisms. For the most part, these assays have sensitivities of greater than 70% and specificities of 97% to 99% in populations with prevalences of *C. trachomatis* infection of 5% or more.[3] In a low-prevalence population, that is, less than 5%, a significant proportion of positive tests will be falsely positive. Therefore, a positive result in a low-prevalence population should be handled with care, and a positive result should be verified.[3,4] Positive results can be validated by the following methods:

- Culture
- Performing a second nonculture test that identifies a *C. trachomatis* antigen or nucleic acid sequence that is different from that used in the screening test
- Using a blocking antibody or competitive probe that verifies a positive test result by preventing attachment of a labeled antibody or probe used in the standard assay

Amplification Assays. At the time of this writing, five commercial assays using nucleic acid amplification are FDA-approved for the laboratory diagnosis of *C. trachomatis* infection. These assays use different formats: polymerase chain reaction (PCR), ligase chain reaction, standard displacement amplification, and transcription-mediated amplification. The first three assay formats amplify or target DNA sequences present in the cryptic plasmid that is present in 7 to 10 copies in the chlamydial EB, whereas the last format amplifies ribosomal RNA sequences. Studies clearly indicate that these tests are more sensitive than culture and other nonnucleic acid amplification assays.[5,15,21] Because of the increased sensitivity of detection, first-voided urine specimens from symptomatic and asymptomatic men and women are acceptable specimens to detect *C. trachomatis,* thereby affording a noninvasive means of chlamydia testing.

Table 46-3 summarizes the various uses of the different *C. trachomatis* laboratory diagnostic tests. Unfortunately, there is currently no straightforward approach to the laboratory diagnosis of chlamydial infections. Each laboratory test format has its own advantages and limitations with regard not only to sensitivity, specificity, and cost, but also to the type of patient population tested (i.e., low- or high-prevalence populations and symptomatic vs. asymptomatic individuals).[21] Moreover, wide variations in sensitivities with these nonculture tests are reported in the literature. The Centers for Disease Control and Prevention developed new guidelines for *C. trachomatis* laboratory tests, recommending additional testing after a positive *C. trachomatis* screening test by nonnucleic acid amplification tests.[4] At the time of this writing, however, confirming a positive *C. trachomatis* result obtained by nucleic acid amplification has only received limited evaluation.

Serodiagnosis. Serologic testing has limited value for diagnosis of urogenital infections in adults. Most adults with chlamydial infection have had a previous exposure to *C. trachomatis* and are therefore seropositive. Serology can be used to diagnose LGV. Antibodies to a genus-specific antigen can be detected by complement fixation, and a single-point titer greater than 1:64 is indicative of LGV. This test is not useful in diagnosing trachoma, inclusion conjunctivitis, or neonatal infections. The microimmunofluorescence assay (micro-IF), a tedious and difficult test, is used for type-specific antibodies of *C. trachomatis* and can also be used to diagnose LGV. A high titer of IgM (1:32) suggests a recent infection; however, not all patients produce IgM. In contrast to CF, micro-IF may be used to diagnose trachoma and inclusion conjunctivitis using acute and convalescent phase sera. Detection of *C. trachomatis*–specific IgM is useful in the diagnosis of neonatal infections. Negative serology can reliably exclude chlamydial infection.

Antibiotic Susceptibility Testing and Therapy

Because *C. trachomatis* is an obligate intracellular bacterium, susceptibility testing is not practical in the routine clinical microbiology laboratory setting and is performed in only a few laboratories. In addition, a standardized in vitro assay as well as an understanding of the relationship between in vitro test results and clinical outcome after treatment are lacking. Antibiotics that have activity against *C. trachomatis* include erythromycin and other macrolide antibiotics, tetracyclines, and fluoroquinolones.

Prevention

Because no effective vaccines are available, strategies to prevent chlamydial urogenital infections focus on trying to manifest behavioral changes. By identifying and treating persons with genital chlamydia before infection is transmitted to sexual partners or in the case of pregnant women to babies, the risk of acquiring or transmitting infection may be significantly decreased.

CHLAMYDOPHILA PSITTACI

Although members of this chlamydial species are common in birds and domestic animals, infections in humans are relatively uncommon.

General Characteristics

C. psittaci differs from *C. trachomatis* in that it is not sulfonamide-sensitive and in the morphology of its EB and inclusion bodies (see Table 46-1).

Epidemiology and Pathogenesis

C. psittaci is an endemic pathogen of all bird species. Psittacine birds (e.g., parrots, parakeets) are a major reservoir for human disease, but outbreaks have occurred among turkey-processing workers and pigeon aficionados. The birds may show diarrheal illness or may be asymptomatic. Humans acquire the disease by inhalation of aerosols. The organisms are deposited in the alveoli; some are ingested by alveolar macrophages and then carried to regional lymph nodes. From there they are disseminated systemically, growing within cells of the reticuloendothelial system. Human-to-human transmission is rare, thus obviating the need for isolating patients if admitted to the hospital.

Spectrum of Disease

Disease usually begins after an incubation period of 5 to 15 days. Onset may be insidious or abrupt. Clinical findings associated with this infection are diverse and include pneumonia, severe headache, mental status changes, and hepatosplenomegaly. The severity of infection ranges from unapparent or mild disease to a life-threatening systemic illness with significant respiratory problems.

Laboratory Diagnosis

Diagnosis of psittacosis is almost always by serologic means. Because of hazards associated with working with the agent, only laboratories with Biosafety Level 3 biohazard containment facilities can culture *C. psittaci* safely. State health departments take an active role in consulting with clinicians about possible cases. Complement fixation has been the most frequently used serologic test to detect psittacosis infection. A more recent test, indirect microimmunofluorescence, is more sensitive but difficult to perform. For this new test, different strains of *C. psittaci* are grown in hens' egg yolk sac cultures. The cultures, rich with elementary bodies, are diluted and suspended in buffer. Next, a dot of each antigen suspension is placed in a geometric array on the surface of a glass slide. Several identical arrays, each containing one dot of every antigen, are prepared on a single slide. These slides are fixed and frozen until use. At the time of the test, serial dilutions of a patient's serum are placed over the antigen arrays. After incubation and washing steps, fluorescein-conjugated anti-human immunoglobulin (either IgG or IgM) is overlaid on the slide. The slide is finally read for fluorescence of particular antigen dots using a microscope. Either a fourfold rise in titer between acute and convalescent serum samples or a single IgM titer of 1:32 or greater in a patient with an appropriate illness is considered diagnostic of an infection.

Finally, amplification of rDNA sequences using a PCR assay followed by restriction fragment length polymorphism (RFLP) analysis was able to identify and distinguish all nine chlamydial species, including *C. psittaci*.[10]

Antibiotic Susceptibility Testing and Therapy

Because *C. psittaci* is an obligate intracellular pathogen and its incidence of infection is rare, susceptibility testing is not practical in the routine clinical microbiology laboratory. Tetracycline is the drug of choice for psittacosis. If left untreated, the fatality rate is about 20%.

Prevention

Prevention of disease is accomplished by treatment of infected birds and/or quarantining of imported birds for a month.

CHLAMYDOPHILA PNEUMONIAE (CHLAMYDIA PNEUMONIAE [TWAR])

The TWAR strain was first isolated from the conjunctiva of a child in Taiwan. It was initially considered to be a psittacosis strain, because the inclusions produced in cell culture resembled those of *C. psittaci*. The Taiwan isolate (TW-183) was shown to be serologically related to a pharyngeal isolate (AR-39) isolated from a college student in the United States, and thus the new strain

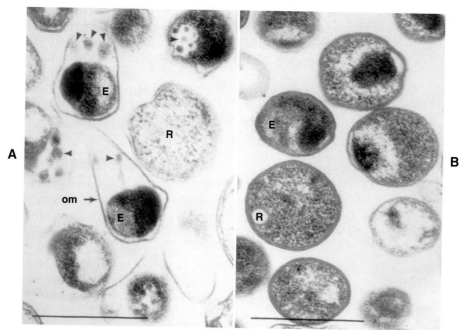

Figure 46-4   Electron micrograph of *C. pneumoniae* (**A**) and *C. trachomatis* (**B**) (bar = 50.5 μm). *E,* Elementary body; *om,* outer membrane; *R,* reticulate body; *arrowhead,* small electron-dense bodies of undetermined function.   (From Grayson JT, et al: *Int J Syst Bacteriol* 39:88, 1989.)

was called "TWAR," an acronym for TW and AR (acute respiratory). To date, only this one serotype of the new species, *C. pneumoniae,* has been identified.

General Characteristics

C. pneumoniae is considered more homogeneous than either *C. trachomatis* or *C. psittaci,* because all isolates tested are immunologically similar. One significant difference between *C. pneumoniae* and the other chlamydiae is the pear-shaped appearance of its EB (Figure 46-4).

Epidemiology and Pathogenesis

C. pneumoniae appears to be a human pathogen; no bird or animal reservoirs have been identified. The mode of transmission from person to person is by aerosolized droplets via the respiratory route.[12] The spread of infection is low. Antibody prevalence to *C. pneumoniae* starts to rise in school-age children and reaches 30% to 45% in adolescents; more than half of adults in the United States and in other countries have *C. pneumoniae* antibody. Of interest, *C. pneumoniae* infections are both endemic and epidemic. Unfortunately, little is known about the pathogenesis of *C. pneumoniae* infections.

Spectrum of Disease

C. pneumoniae has been associated with pneumonia, bronchitis, pharyngitis, sinusitis, and a flulike illness. Infection in young adults is usually mild to moderate; the microbiologic differential diagnosis primarily in-

cludes *Mycoplasma pneumoniae.* Severe pneumonia may occur in older or respiratory-compromised patients. Of note, asymptomatic infection or unrecognized, mildly symptomatic illnesses caused by *C. pneumoniae* are common. In addition, an association exists between *C. pneumoniae* infection and the development of asthmatic symptoms. Finally, an association between coronary artery disease and other atherosclerotic syndromes and *C. pneumoniae* infection has been suggested by sero-epidemiologic studies and the demonstration of the organism in atheromatous plaques (yellow deposits within arteries containing cholesterol and other lipid material). Such an etiologic role by this organism is currently under intense scrutiny.

Laboratory Diagnosis

The laboratory diagnosis of *C. pneumoniae* infections is accomplished by cell culture, serology, or nucleic acid.[14]

Direct Detection Methods.   To date, assays to directly detect *C. pneumoniae* antigens have poor sensitivity. A variety of nucleic acid amplification assays including conventional and real-time PCR assays have been developed to detect *C. pneumoniae* nucleic acid sequences in clinical specimens. Several of these amplification assays are commercially available. Using these methods, the organism has been detected in throat swabs and other specimens, such as nasopharyngeal, broncho-alveolar lavage fluids, and sputum.

Cultivation. Specimens for isolation are usually swabs of the oropharynx; techniques for isolation from sputum are unsatisfactory. Swabs should be placed into chlamydial transport media, transported on ice, and stored at 4° C; organisms are rapidly inactivated at room temperature or by rapid freezing or thawing. A cell culture procedure similar to that used for *C. trachomatis* but using the more sensitive HL or Hep-2 cell lines must be substituted for McCoy cells. Multiple blind passages might be necessary to improve recovery rates. *C. pneumoniae* species-specific monoclonal antibodies can detect the organism in cell culture.

Serodiagnosis. Diagnosis of *C. pneumoniae* infection can also be achieved by serology. However, serologic testing has had variable success and questionable validity. Complement fixation using a genus-specific antigen has been used, but it is not specific for *C. pneumoniae*. A microimmunofluorescence test using *C. pneumoniae* elementary bodies as antigen is more reliable but available only in specialized laboratories. A fourfold rise in either IgG or IgM is diagnostic, and a single IgM titer of 16 or greater or an IgG titer of 512 or greater is suggestive of recent infection. Commercial kits for diagnosing *C. pneumoniae* are not available yet because of the difficulty in developing species-specific detection methods.

Antibiotic Susceptibility Testing and Therapy

Methods for susceptibility testing of *C. pneumoniae* have been largely adapted from those used for *C. trachomatis*.[18] Similar to *C. trachomatis*, susceptibility testing is not practical for the clinical microbiology laboratory and, because methods are not yet standardized, the results can be influenced by several variables. Treatment with tetracycline and erythromycin has been successful.

Prevention

Little is known regarding effective ways to prevent *C. pneumoniae* infections.

RICKETTSIA, ORIENTIA, ANAPLASMA, AND EHRLICHIA

Agents that are responsible for human disease within the family *Rickettsiaceae* belong to five genera, *Rickettsia, Orientia, Ehrlichia, Anaplasma,* and *Neorickettsia. Orientia tsutsugamushi* (formerly called *Rickettsia tsutsugamushi*) was placed into its own genus primarily based on the likely absence of lipopolysaccharide (LPS) and peptidoglycan and the presence of a 54-58 kDa major surface protein. Until recently, two other genera causing human disease, *Coxiella* and *Bartonella,* were included in this family. However, based on phylogenetic differences,

these two genera were recently removed from the *Rickettsiaceae* family. *Bartonella* spp. can be cultured on standard bacteriologic media; therefore, this group of organisms is addressed in Chapter 35. Because *Coxiella burnetii* can survive extracellularly, unlike the rickettsiae, yet requires cultivation in cell culture similar to the rickettsiae, this organism is discussed separately in this chapter.

GENERAL CHARACTERISTICS

Rickettsiae are fastidious bacteria that are obligate, intracellular parasites. These bacterial agents survive only briefly outside of a host (reservoir or vector) and multiply only intracellularly. Organisms are small (0.3 μm × 1 to 2 μm), pleomorphic, gram-negative bacilli that multiply by binary fission in the cytoplasm of host cells; the release of mature rickettsiae results in the lysis of the host cell.

EPIDEMIOLOGY AND PATHOGENESIS

This group of organisms infects wild animals, with humans acting as accidental hosts in most cases. Most of these organisms are passed between animals by an insect vector. Similarly, humans become infected following the bite of an infected arthropod vector or by inhalation of infectious aerosols.[35] Characteristics, including the respective arthropod vector of the prominent species of *Rickettsia, Orientia, Anaplasma,* and *Ehrlichia,* are summarized in Table 46-4.

Organisms belonging to the genus *Rickettsia* do not undergo any type of intracellular developmental cycle.[17] Different species of *Rickettsia* not only share some antigenic properties and are genetically similar but also share a similar mechanism of pathogenesis. After being deposited directly into the bloodstream through the bite of an arthropod vector, these organisms induce the endothelial cells of the host's blood vessels to engulf them and are carried into the cell's cytoplasm within a vacuole. Following infection, organisms escape the vacuole, becoming free in the cytoplasm. *Rickettsia* spp. then multiply, causing cell injury and death. Subsequent vascular lesions caused by *Rickettsia*-induced damage to endothelial cells account for the changes that occur throughout the body, particularly in the skin, heart, brain, lung, and muscle. Rickettsiae also have numerous ways to evade human host defenses such as cell-to-cell spread, escaping from the phagosome and entering into a latent state (primarily *R. prowazekii*).[35]

In contrast to *Rickettsia* and *Orientia* spp., organisms belonging to the genus *Ehrlichia* undergo an intracellular developmental cycle following infection of circulating leukocytes. Similar to chlamydiae, *A. phagocytophilum, Ehrlichia* spp., and *N. sennetsu* cannot survive

Table 46-4   Characteristics of Prominent *Rickettsia* Orientia, Anaplasma*, and *Ehrlichia* spp.

Agent	Disease	Vector	Distribution	Diagnostic Tests
Spotted Fever Group				
Rickettsia akari	Rickettsialpox	Mites	Worldwide	Serology, immunohistology, PCR with sequencing
R. conorii	Mediterranean and Israeli spotted fevers; Indian tick typhus; Kenya tick typhus	Ticks	Southern Europe, Mideast, Africa	Serology, immunohistology, PCR with sequencing
R. rickettsii	Rocky Mountain spotted fever	Ticks	North and South America; particularly in southeastern states and Oklahoma in the United States	Serology, immunohistology, PCR with sequencing
Typhus Group				
R. prowazekii	Epidemic typhus Brill-Zinsser disease	Lice None; recrudescent disease	Worldwide Worldwide	Serology, PCR with sequencing Serology, PCR with sequencing
R. typhi	Murine typhis	Fleas	Worldwide	Serology, PCR with sequencing
Scrub Typhus Group				
O. tsutsugamushi	Scrub typhus	Chiggers	South and Southeast Asia, South Pacific	Serology, PCR with sequencing
Ehrlichia/ Anaplasma/ Neorickettsia				
Ehrlichia chaffeenis	Human monocytic ehrlichiosis	Ticks	Southeast, South Central, and mid-Atlantic United States	Serology, PCR, immunohistology, immunocytology
E. ewingii	Granulocytic ehrlichiosis	Ticks	United States—to date, most reported infections in immunocompromised patients	PCR with species-specific primers or DNA sequencing of amplicons
Anaplasma phagocytophilum	Human granulocytic anaplasmosis	Ticks	United States, Europe	Serology, PCR, immunohistology, peripheral blood smear, immunocytology
Neorickettsia sennetsu (formerly *E. sennetsu*)	Sennetsu fever	Ticks	Southeast Asia (primarily Japan)	Serology

*Other *Rickettsia* species recognized as emerging human pathogens include *R. africae, R. sibirica, R. japonica, R. honei, R. australis, R. slovaca, R. aeschilimannie, R. helvetica, R. heilongjianggensis,* and *R. parkeri*[29,30]; all belong to the spotted fever group.

outside host cells and once released, must rapidly induce signals for their own uptake into another host cell that is unique to each genus. How these organisms accomplish this entry, replicate in the host milieu, and then exit is largely unknown. *E. chaffeensis* primarily infects monocytes and causes human monocytic ehrlichiosis (HME), whereas *A. phagocytophilum* infects bone marrow–derived cells, primarily infecting neutrophils, causing human granulocytic ehrlichiosis (HGE).

SPECTRUM OF DISEASE

The genus *Rickettsia* encompasses the following three groups of bacteria: the spotted fever group, the typhus group, and the scrub typhus group *(O. tsutsugamushi),* based on the arthropod mode of transmission, clinical manifestations, rate of intracellular growth, rate of intracellular burden, and extent of intracellular growth (see Table 46-4). The triad of fever, headache, and rash is the primary clinical manifestation in patients with an exposure to insect vectors. Infections caused by these organisms may be severe and are sometimes fatal.

Although HME and HGE cause distinct infections, their clinical findings are similar. In general, patients with ehrlichial infections present with nonspecific symptoms such as fever, headache, and myalgias; rashes occur only rarely. The severity of illness can range from asymptomatic to mild to severe.

LABORATORY DIAGNOSIS

Because rickettsial and ehrlichial infections can be severe or even fatal, a timely diagnosis is essential.

Direct Detection Methods

Immunohistology and conventional and real-time PCR have been used to diagnose rickettsial[13,20] and ehrlichial infections[29,30,34], but these methods are available in only a limited number of laboratories. Biopsy of skin tissue from the rash caused by the spotted fever group rickettsiae is the preferred specimen to demonstrate their presence. Organisms are identified using polyclonal antibodies; detection is achieved by fluorescein-labeled antibodies or enzyme-labeled indirect procedures. The sensitivity of these techniques is about 70% and is dependent on correct tissue sampling, examination of multiple tissue levels, and biopsying before or during the first 24 hours of therapy.[9] In addition, PCR has been used to detect *Rickettsia* spp. in clinical specimens (see Table 46-4).[24,29,30]

Direct detection of *Ehrlichia* and *Anaplasma* from peripheral blood or cerebrospinal fluid (CSF) includes PCR amplification, direct microscopic examination of Giemsa-stained or Wright's stained specimens, or immunocytologic or immunohistologic stains with *E. chaffeensis* or *Anaplasma* species antibodies. Direct microscopic examination of Giemsa-stained or Diff-Quik–stained peripheral blood buffy-coat smears can detect **morulae** (cytoplasmic vacuoles containing enriched organisms) during the febrile stage of infection in ehrlichiosis; morulae-like structures also can be observed in CSF cells and tissues.[8,29] Finally, both *Ehrlichia* and *Anaplasma* have been detected directly in clinical specimens by conventional and real-time PCR.[29,34] Recent reports have described the development of rapid, species-specific real-time PCR assays to detect single or co-infections with *Anaplasma* species and/or *Ehrlichia* species in peripheral blood specimens.[7,34]

Cultivation

Although the rickettsiae can be cultured in embryonated eggs and in tissue culture, the risk of laboratory-acquired infection is extremely high, limiting availability of culture to a few specialized laboratories. Blood should be collected as early as possible in the course of disease

Table 46-5 Reaction of *Proteus* Strains in Weil-Felix Test

Disease	OX-19	OX-2	OX-K
Brill-Zinsser	V	V	−
Epidemic typhus	+	V	−
Murine typhus	+	V	−
Rickettsialpox	−	−	−
Rocky Mountain spotted fever	+	+	−
Scrub typhus	−	−	V
Q fever	−	−	−
Ehrlichiosis	−	−	−

+, >90% positive agglutination; −, >90% negative agglutination; *V,* variable results.

in a sterile, heparin-containing vial. Similarly, punch biopsies of skin or eschars are also acceptable, but must be collected early in the course of disease. These same specimens are also acceptable for PCR.

To date, culture of *Ehrlichia* and *Anaplasma* is limited and culture conditions are still being optimized. Currently, the preferred specimen for culture is peripheral blood obtained in a sterile, EDTA- or acid-citrate-dextrose-anticoagulated blood tube; if specimens must be transported, specimens should be transported overnight at approximately 4° C.

Serodiagnosis

Although it is not fast, the diagnosis of rickettsial disease and ehrlichiosis is primarily accomplished serologically. Serologic assays for the diagnosis of rickettsial infections include the indirect immunofluorescence assay (IFA), enzyme immunoassay (EIA), *Proteus vulgaris* OX-19 and OX-2 and *Proteus mirabilis* OX-K strain agglutination, line blot, and Western immunoblotting. The least specific but widely used test is the Weil-Felix reaction, the fortuitous agglutination of certain strains of *P. vulgaris* by serum from patients with rickettsial disease. Because false-positive and false-negative tests are a continuing problem, these tests should be replaced by more accurate serologic methods such as IFA. Table 46-5 outlines the positive reactions associated with each disease. This test is only presumptive, however, and a more specific serologic test such as IFA must be used for confirmation of disease. These tests are performed primarily by reference laboratories.

Except for latex agglutination, IFA, and DFAfl testing for diagnosing Rocky Mountain spotted fever, none of the serologic tests is useful for diagnosing disease in time to influence therapy. This lack of utility for serology is because antibodies to rickettsiae other than *R. rickettsii* cannot be reliably detected until at least 2 weeks after the patient has become ill. With newer immunologic recombinant reagents under development, the potential exists for new tests for all the rickettsial diseases.

To date, the sensitivity and specificity of serologic assays for ehrlichiosis is unknown but is presumed to be relatively high; indirect immunofluorescent antibody testing is available for *E. chaffeensis* or *A. phagophilum*. A fourfold or greater rise in antibody titer during the course of disease is considered significant.

ANTIBIOTIC SUSCEPTIBILITY TESTING AND THERAPY

Tetracyclines are the primary drugs of choice for treatment of most infections caused by *Rickettsia*, *Ehrlichia* or *Anaplasma* species. Depending on the specific species of *Rickettsia*, some fluoroquinolones may be used, as may chloramphenicol.

PREVENTION

The best means of prevention for rickettsial and ehrlichial infection is to avoid contact with the respective vectors.

COXIELLA

Coxiella burnetii is the causative agent of **Q fever,** an acute systemic infection that primarily affects the lung.

GENERAL CHARACTERISTICS

C. burnetii is smaller than *Rickettsia* spp. and is more resistant to various chemical and physical agents. Recent phylogenetic studies of this gram-negative coccobacillus have demonstrated that it is far removed from the rickettsiae. In contrast to the rickettsiae, *C. burnetii* can survive extracellularly; however, it can be grown only in lung cells. The organism has a sporelike life cycle and can exist in two antigenic states.[31] When isolated from animals, *C. burnetii* is in a phase I form and is highly infectious. In its phase II form, *C. burnetii* has been grown in cultured cell lines and is not infectious.

EPIDEMIOLOGY AND PATHOGENESIS

The most common animal reservoirs for the zoonotic disease caused by *C. burnetii* are cattle, sheep, and goats. In infected animals, organisms are shed in urine, feces, milk, and birth products. Usually, the infected animals are asymptomatic. Humans are infected by the inhalation of contaminated aerosols. Of significance, because

BOX 46-1	Clinical Manifestations of *C. burnetii* Infection[30]
Febrile, self-limited illness
Atypical pneumonia
Granulomatous hepatitis
Endocarditis
Neurologic manifestations (e.g., encephalitis, meningoencephalitis)
Osteomyelitis

of its resistance to desiccation and sunlight by virtue of forming spores, *C. burnetii* is able to withstand harsh environmental conditions. Q fever is endemic worldwide except in Scandinavia.

Following infection, *C. burnetii* is passively phagocytized by host cells and multiplies within vacuoles. The incubation period is about 2 weeks to 1 month. After infection and proliferation in the lungs, organisms are picked up by macrophages and carried to the lymph nodes, from which they then reach the bloodstream.

SPECTRUM OF DISEASE

After the incubation period, initial clinical manifestations of *C. burnetii* infections are systemic and non-specific: headache, fever, chills, myalgias. In contrast to rickettsial infections, a rash does not develop. Both acute and chronic forms of the disease are recognized. Possible clinical manifestations are listed in Box 46-1.

LABORATORY DIAGNOSIS

Because laboratory-acquired infections caused by *C. burnetii* have occurred, cultivation of the organism must be done in a biosafety level 3 containment facility. However, the use of a shell vial assay with human lung fibroblasts to isolate the organism from buffy coat and biopsy specimens has not resulted in any laboratory-acquired infections.[31] Once inoculated, cultures are incubated for 6 to 14 days at 37° C in carbon dioxide. The organism is detected using a direct immunofluorescent assay.[28]

Although organisms can be detected by nucleic acid amplification assays, serology is the most convenient and commonly used diagnostic tool. Three serologic techniques are available: indirect immunofluorescent antibody (IFA), complement fixation, and EIA. IFA is considered the reference method for both acute and chronic Q fever that is both highly specific and sensitive and is recommended for its reliability, cost effectiveness, and ease of performance.[31] Many reference and state health laboratories perform phase I and phase II IgG and IgM serologic assays.

ANTIBIOTIC SUSCEPTIBILITY TESTING AND THERAPY

Because *C. burnetii* does not multiply in bacteriologic culture media, susceptibility testing has been performed in only a very limited number of laboratories. Tetracyclines are recommended for the treatment of acute and chronic Q fever.

PREVENTION

The best way to prevent infection with *C. burnetii* is to avoid contact with infected animals. A vaccine is commercially available in Australia and Eastern European countries; a vaccine is being developed in the United States.[36]

TROPHERYMA WHIPPLEI

Although observed in diseased tissue, some organisms are nonculturable yet associated with specific disease processes, making the development of "traditional" diagnostic assays difficult (e.g., serology or antigen detection). However, with the ability to detect and classify bacteria using molecular techniques without culture, some of these agents have been detected and characterized. This strategy uses PCR to amplify ribosomal DNA sequences by PCR of the unknown agent followed by sequencing and phylogenetic analysis.[32] Using this approach, *Tropheryma whipplei* has been identified as the causative agent of Whipple's disease.[33]

GENERAL CHARACTERISTICS

Phylogenetic analysis shows that this organism is a gram-positive actinomycete that is not closely related to any other genus known to cause infection.

EPIDEMIOLOGY, PATHOGENESIS, AND SPECTRUM OF DISEASE

Whipple's disease, found primarily in middle-age men, is characterized by the presence of PAS-staining macrophages (indicating mucopolysaccharide or glycoprotein) in almost every organ system. The bacillus is seen in macrophages and affected tissues but it has never been cultured. Patients develop diarrhea, weight loss, arthralgia, lymphadenopathy, hyperpigmentation, often a long history of joint pain, and a distended and tender abdomen. Neurologic and sensory changes often occur. Although less common than intestinal or articular involvement, cardiac manifestations can also occur, including endocarditis.[25] It has been suggested that a cellular immune defect is involved in pathogenesis of this disease.

LABORATORY DIAGNOSIS

Detection of *T. whipplei* is performed by only a few laboratories using conventional and real-time PCR.

ANTIBIOTIC SUSCEPTIBILITY TESTING AND THERAPY

No susceptibility testing is performed, because this organism is nonculturable. Patients usually respond well to long-term therapy with antibacterial agents, including trimethoprim/sulfamethoxazole, tetracycline, and penicillin; tetracycline has been associated with serious relapses, however. Colchicine therapy appears to control symptoms.[27] Without treatment the disease is uniformly fatal.

PREVENTION

Little is known about the prevention of this disease.

CALYMMATOBACTERIUM GRANULOMATIS

Calymmatobacterium granulomatis is the etiologic agent of **granuloma inguinale,** or **donovanosis,** a sexually transmitted disease.

GENERAL CHARACTERISTICS

C. granulomatis is an encapsulated, pleomorphic, gram-negative bacillus that is usually observed in vacuoles in the cells of large mononuclear cells.

EPIDEMIOLOGY AND PATHOGENESIS

Granuloma inguinale is uncommon in the United States but is recognized as a major cause of genital ulcers in India, Papua New Guinea, the Caribbean, Australia, and parts of South America. The causative agent is sexually transmitted, although there is a possibility that it may be nonsexually transmitted as well. Infectivity of this bacillus must be low, because sexual partners of infected patients often do not become infected.

SPECTRUM OF DISEASE

Granuloma inguinale is characterized by subcutaneous nodules that enlarge and evolve to form beefy, erythematous, granulomatous, painless lesions that bleed easily. The lesions, which usually occur on the genitalia, have been mistaken for neoplasms. Patients often have inguinal lymphadenopathy.

LABORATORY DIAGNOSIS

The organism can be visualized in scrapings of lesions stained with Wright's or Giemsa stain. Subsurface infected cells must be present; surface epithelium is not an adequate specimen. Groups of organisms are seen within mononuclear endothelial cells; this pathognomonic entity is known as a *Donovan body,* named after the physician who first visualized the organism in such a lesion. The organism stains as a blue rod with prominent polar granules, giving rise to a "safety pin" appearance, surrounded by a large, pink capsule. The capsular polysaccharide shares several cross-reactive antigens with *Klebsiella* species, and there also appears to be a high degree of molecular homology, fueling speculation that *Calymmatobacterium* is closely related to *Klebsiella.*[1] Based on the DNA-DNA hybridization, it has been recently proposed that the name of this organism be changed to *Klebsiella granulomatis.*

Cultivation in vitro is difficult, but it can be done using media containing some of the growth factors found in egg yolk. A medium described by Dienst[6] has been used to culture *Calymmatobacterium* from aspirated bubo material.[16] More recently, this agent was cultured in human monocytes from biopsies of genital ulcers of patients with donovanosis.[23]

ANTIBIOTIC SUSCEPTIBILITY TESTING AND THERAPY

No antibiotic susceptibility testing is performed. Although gentamicin and chloramphenicol are the most effective drugs for the therapy of granuloma inguinale, tetracycline or ampicillin is the drug of first choice. Trimethoprim/sulfamethoxazole or erythromycin (in pregnancy) also provides effective treatment for granuloma inguinale.

Case Study

A 13-year-old girl was well until 9 days before admission, when she developed headache, fever, and myalgias. Amoxicillin therapy was started 4 days before admission. Over the next few days, the symptoms continued and she developed abdominal pain and disorientation. Laboratory findings revealed neutropenia and elevated liver enzymes. On admission she presented with photophobia, nuchal rigidity (stiff neck) and disseminated intravascular coagulation. She had recently traveled to Arkansas where she stayed on a farm, rode horses, and removed multiple ticks from her legs. A bone marrow aspiration was performed which yielded the diagnosis from the Wright's stain of the monocytes. After treatment with doxycycline, she had a complete recovery.

QUESTIONS

1. What is the etiologic agent of this disease and how was it detected.
2. Why is this disease called a zoonosis?
3. Diagnosis of this disease is difficult. What laboratory methods are available to the physician to diagnose a patient with this disease?

REFERENCES

1. Bastian I, Bowden FJ: Amplification of *Klebsiella*-like sequences from biopsy samples from patients with donovanosis, *Clin Infect Dis* 23:1328, 1996.
2. Byrne GI, Ojcius DM: Chlamydia and apoptosis: life and death decisions of an intracellular pathogen, *Nat Rev* 2: 802, 2004.
3. Centers for Disease Control and Prevention: Recommendations for the prevention and management of *Chlamydia trachomatis* infections, Atlanta, 1993, Centers for Disease Control.
4. Centers for Disease Control and Prevention: Sexually transmitted diseases treatment guidelines, *MMWR* 51:1, 2002.
5. Cook RL, Hutchison SL, Østergaard L, et al: Systematic review: noninvasive testing for *Chlamydia trachomatis* and *Neisseria gonorrhoeae,* *Ann Intern Med* 142:914, 2005.
6. Dienst RB, Brownell GH: Genus *Calymmatobacterium Aragao* and *Vianna* 1913. In Krieg NR, Holt JG, editors: *Bergey's manual of systematic bacteriology,* vol 1, Baltimore, 1984, Williams & Wilkins.
7. Doyle CK, Labruna MB, Breitschwerdt EB, et al: Detection of medically important *Ehrlichia* by quantitative multicolor TaqMan real-time polymerase chain reaction of the *dsb* gene, *J Molec Diagn* 7:504, 2005.
8. Dumler JS, Bakken JS: Ehrlichial diseases of human: emerging tick-borne infections, *Clin Infect Dis* 20:1102, 1994.
9. Dumler JS, Walker DH: Diagnostic tests for Rocky Mountain spotted fever and other rickettsial diseases, *Dermatol Clin* 12:25, 1994.
10. Everett KD, Andersen AA: Identification of nine species of the Chlamydiaceae using PCR-RFLP, *Int J Syst Bacteriol* 49:803, 1999.
11. Everett KD, Bush RM, Andersen AA: Emended description of the order Chlamydiales, proposal of Parachlamydiaceae fam nov and Simkaniaceae fam nov, each containing one monotypic genus, revised taxonomy of the family Chlamydiaceae, including a new genus and five new species, and standards for the identification of organisms, *Int J Syst Bacteriol* 49:415, 1999.
12. Falsey AR, Walsh EE: Transmission of *Chlamydia pneumoniae,* *J Infect Dis* 168:493, 1993.
13. Fournier PE, Raoult D: Suicide PCR on skin biopsy specimens for diagnosis of rickettsioses, *J Clin Microbiol* 42:3428, 2004.
14. Gaydos CA, Roblin PM, Hammerschlag MR, et al: Diagnostic utility of PCR-enzyme immunoassay, culture, and serology for detection of *Chlamydia pneumoniae* in symptomatic and asymptomatic patients, *J Clin Microbiol* 32:903, 1994.
15. Gaydos CA, Quinn TC: Urine nucleic acid amplification tests for the diagnosis of sexually transmitted infections in clinical practice, *Curr Opin Infect Dis* 18:55, 2005.
16. Goldberg J: Studies on granuloma inguinale. IV. Growth requirements of *Donovania granulomatis* and its relationship to the natural habitat of the organism, *Br J Ven Dis* 35:266, 1959.
17. Hackstadt T: The biology of rickettsiae, *Infect Agents Dis* 5:127, 1996.
18. Hammerschlag MR: Antimicrobial susceptibility and therapy of infections caused by *Chlamydia pneumoniae,* *Antimicrob Agents Chemother* 38:1873, 1994.

19. Hogan, RJ, Mathews SA, Mukhopadhyay S, et al: Chlamydial persistence: beyond the biphasic paradigm, *Infect Immun* 72: 1843, 2004.

20. Jensenius M, Fournier PE, Raoult D: Rickettsioses and the international traveler, *Clin Infect Dis* 39:1493, 2004.

21. Johnson RE, Green TA, Schachter J, et al: Evaluation of nucleic acid amplification tests as reference tests for *Chlamydia trachomatis* infections in asymptomatic men, *J Clin Microbiol* 38:4382, 2000.

22. Kellogg J: Impact of variation in endocervical specimen collection and testing techniques on frequency of false-positive and false-negative chlamydia detection results, *Am J Clin Pathol* 104:554, 1995.

23. Kharsany AB, Hoosen AA, Kiepiela P, et al: Culture of *Calymmatobacterium granulomatis, Clin Infect Dis* 22:391, 1996.

24. La Scola B, Raoult D: Laboratory diagnosis of rickettsioses: current approaches to diagnosis of old and new rickettsial diseases, *J Clin Microbiol* 35:2715, 1997.

25. Lepidi H, Fenollar F, Dumler JS, et al: Cardiac valves in patients with Whipple endocarditis: microbiological, molecular, quantitative histologic and immunohistochemical studies of 5 patients, *J Infect Dis* 190:935, 2004.

26. Martin DH: Chlamydial infections, *Med Clin North Am* 74:1367, 1990.

27. McMenemy A: Whipple's disease, a familial Mediterranean fever, adult-onset Still's disease, and enteropathic arthritis, *Curr Opin Rheumatol* 4:479, 1992.

28. Musso D, Raoult D: *Coxiella burnetii* blood cultures from acute and chronic Q-fever patients, *J Clin Microbiol* 33:3129, 1995.

29. Paddock CD, Childs JE: *Ehrlichia chaffeensis*: a prototypical emerging pathogen, *Clin Microbiol Rev* 16: 37, 2003.

30. Pariola P, Paddock C, Raoult D: Tick-borne rickettsioses around the world: emerging diseases challenging old concepts, *Clin Microbiol Rev* 18: 719, 2005.

31. Raoult D, Marrie T: Q fever, *Clin Infect Dis* 20:489, 1995.

32. Relman DA: The identification of uncultured microbial pathogens, *J Infect Dis* 168:1, 1983.

33. Relman DA, Schmidt TM, MacDermott RP, et al: Identification of the uncultured bacillus of Whipple's disease, *N Engl J Med* 327: 293, 1992.

34. Sirigireddy KR, Ganta RR: Multiplex detection of *Ehrlichia* and *Anaplasma* species pathogens in peripheral blood by real-time reverse transcriptase-polymerase chain reaction, *J Molec Diagn* 7:308, 2005.

35. Walker DH, Valbuena GA, Olano JP: Pathogenic mechanisms of diseases caused by *Rickettsia, Ann N Y Acad* Sci 990: 1, 2003.

36. Williams JC, Waag D: Antigens, virulence factors, and biological response modifiers of *Coxiella burnetii*: strategies for vaccine development. In Williams JC and Thompson HA, editors: *Q fever: the biology of* Coxiella burnetii, Boca Raton, Fla, 1991, CRC Press.

ADDITIONAL READING

Black CM: Current methods of laboratory diagnosis of *Chlamydia trachomatis* infections, *Clin Microbiol Rev* 10:160, 1997.

Centers for Disease Control and Prevention: Sexually transmitted diseases treatment guidelines, MMWR 51:1, 2002.

Dumler JS: Laboratory diagnosis of human rickettsial and ehrlichial infections, *Clin Microbiol Newsl* 18:57, 1996.

Evermann JF: *Chlamydia psittaci:* zoonotic potential worthy of concern, *Clin Microbiol Newsl* 9:1, 1987.

Kuo CC, Jackson LA, Campbell LA, et al: *Chlamydia pneumoniae* (TWAR), *Clin Microbiol Rev* 8:451, 1995.

Leber AL, Hall GS, LeBar WD: Nucleic acid amplification tests for detection of *Chlamydia trachomatis* and *Neisseria gonorrhoeae*. In Sharp SE, coordinating editor: *Cumitech 44*, Washington, DC, 2006, American Society for Microbiology.

Paddock CD, Childs JE: *Ehrlichia chaffeensis*: a prototypical emerging pathogen, *Clin Microbiol Rev* 16: 37, 2003.

Raoult D, Roux V: Rickettsioses as paradigms of new or emerging infectious diseases, *Clin Microbiol Rev* 10:694, 1997.

Cell Wall–Deficient Bacteria: *Mycoplasma* and *Ureaplasma*

CHAPTER 47

Genera and Species to Be Considered

Mycoplasma pneumoniae
Mycoplasma hominis
Mycoplasma genitalium
Ureaplasma urealyticum

This chapter addresses a group of bacteria, the mycoplasmas, which are the smallest known free-living forms; unlike all other bacteria, these prokaryotes do not have a cell wall. Although mycoplasmas are ubiquitous in the plant and animal kingdoms (more than 150 different species exist within this class), this chapter addresses only the most prominent varieties, that is, *Mycoplasma* spp. and *Ureaplasma urealyticum*, that colonize or infect humans and are not of animal origin.

GENERAL CHARACTERISTICS

Mycoplasmas belong to the class *Mollicutes* (Latin, meaning *soft skin*). This class comprises four orders, which, in turn, contain five families (Figure 47-1). The mycoplasmas that colonize and/or infect humans belong to the family *Mycoplasmataceae*; this family comprises two genera, *Mycoplasma* and *Ureaplasma*. Besides lacking a cell wall, these agents with their small cell size (0.3×0.8 μm) and small genome size, require sterols for membrane function and growth. Mollicutes appear most closely related to the gram-positive bacterial subgroup that includes bacilli, streptococci, and lactobacteria that diverged from the *Streptococcus* branch of gram-positive bacteria. With few exceptions, most mycoplasmas are aerobic and have fastidious growth requirements.

EPIDEMIOLOGY AND PATHOGENESIS

Mycoplasmas are part of the microbial flora of humans and are found mainly in the oropharynx, upper respiratory tract, and genitourinary tract. Besides those that are considered primarily as commensals, considerable evidence indicates the pathogenicity of some mycoplasmas; for others, a role in a particular disease is less clearly delineated.

EPIDEMIOLOGY

The mycoplasmas usually considered as commensals are listed along with their respective sites of colonization in Table 47-1. One species of *Acholeplasma* (these organisms are widely disseminated in animals), *Acholeplasma laidlawii*, has been isolated from the oral cavity of humans a limited number of times; however, the significance of these mycoplasmas and their colonization of humans remains uncertain.

Of the other mycoplasmas that have been isolated from humans, the possible roles that *M. pirum*, *M. penetrans*, and *M. fermentans* might play in human disease is uncertain at this time. *M. fermentans*, *M. pirum*, and *M. penetrans* have been isolated from patients infected with the human immunodeficiency virus (HIV).[1,2] It now appears that *M. genitalium* may account for as much as 15% to 20% of nongonococcal urethritis.[13,19] Of note, *M. genitalium* is not associated with the presence of other mycoplasmas and ureaplasmas. In women, this organism may also cause cervicitis and endometritis. The distribution, habitat, and transmission of these mycoplasmas are unknown.

Finally, the remaining three species of mycoplasmas that have been isolated from humans, *M. pneumoniae*, *U. urealyticum*, and *M. hominis*, have well-established roles in human infections. Both *U. urealyticum* and *M. hominis* have been isolated from the genitourinary tract of humans, and *M. pneumoniae* has been isolated from the respiratory tract.

Infants are commonly colonized with *U. urealyticum* and *M. hominis*. In general, this colonization does not persist beyond age 2.[5] Once an individual reaches puberty, colonization with these mycoplasmas can occur primarily as a result of sexual contact.[10,11] In situations in which these agents cause disease in neonates, organisms are transmitted from a colonized mother to her newborn infant by an ascending route from colonization of the mother's urogenital tract, by crossing the placenta from the mother's blood, by delivery through a colonized birth canal, or postnatally from mother to infant.[6,16]

M. pneumoniae is a cause of community-acquired pneumonia (see Chapter 53); infections caused by this agent are distributed worldwide, with an estimated 2 million cases per year in the United States. Of significance, the incidence of other respiratory infections caused by *M. pneumoniae* may be 10 to 20 times higher.

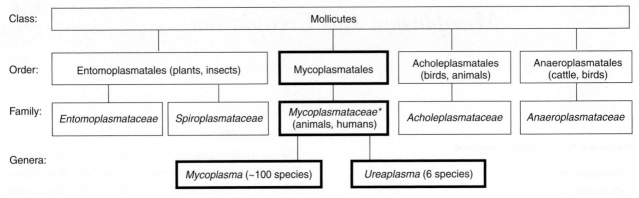

*Mollicutes that have been detected in humans.

Figure 47-1   Taxonomy of the class Mollicutes.

Table 47-1   Mycoplasmas That Are Considered Normal Flora of the Oropharynx or Genital Tract

Organism	Site of Colonization
M. orale	Oropharynx
M. salivarium	Oropharynx
M. buccale	Oropharynx
M. faucium	Oropharynx
M. lipophilum	Oropharynx
M. primatum	Genital tract
M. spermatophilum	Genital tract

Infections can occur singly or as outbreaks in closed populations such as families and military recruit camps. Besides respiratory infection, *M. pneumoniae* can cause extrapulmonary manifestations. This organism infects children and adults and is spread by respiratory droplets produced by coughing.

PATHOGENESIS

In general, mycoplasmas colonize mucosal surfaces of the respiratory and urogenital tracts. Except for those mycoplasmas noted, most rarely produce invasive disease except in immunocompromised hosts. Of the mycoplasmas that are established as causes of human infections, these agents reside extracellularly, attaching with great affinity to ciliated and nonciliated epithelial cells. *M. pneumoniae* has a complex and specialized attachment organelle to accomplish this process that includes a P1 adhesin protein that primarily interacts with host cells.[22] With respect to the mycoplasmas that are clearly able to cause disease, many of the disease processes are thought to be immunologically mediated.[15] In addition to adherence properties and possibly

immune-mediated injury, the ability to cause localized cell injury appears to contribute to their pathogenicity. Finally, *M. pneumoniae* may be able to establish a latent or chronic state to avoid the immune response.

Of interest, the mycoplasmas associated with patients with HIV (*M. fermentans, M. penetrans,* and *M. pirum*) are all capable of invading humans cells and modulate the immune system.[14] Based on these findings, some investigators have proposed that these mycoplasmas might play a role in certain disease processes in these patients.[1,2,9,14]

SPECTRUM OF DISEASE

The clinical manifestations of infections caused by *M. pneumoniae* and the pathogenic genital mycoplasmas, *U. urealyticum, M. hominis,* and *M. genitalium* are summarized in Table 47-2.

LABORATORY DIAGNOSIS

The laboratory diagnosis of mycoplasma infections is extremely challenging because of complex and time-consuming culture requirements and the lack of reliable, widely available rapid diagnostic tests. Accurate, rapid diagnosis for *M. pneumoniae* is highly desired, because penicillin and other beta-lactam agents are ineffective treatment. The laboratory diagnosis of the mycoplasmas well recognized as able to cause human disease (i.e., *M. pneumoniae, U. urealyticum, M. hominis,* and *M. genitalium*) is addressed.

SPECIMEN COLLECTION, TRANSPORT, AND PROCESSING

Various specimens are appropriate for the diagnosis of mycoplasma infections by culture or other means of

Table 47-2   **Clinical Manifestations of Mycoplasma Infections Caused by *Mycoplasma pneumoniae*, *Ureaplasma urealyticum*, *M. hominis*, and *M. genitalium***

Organism	Clinical Manifestations
Mycoplasma pneumoniae	Asymptomatic infection Upper respiratory tract infection in young children: Mild, nonspecific symptoms including runny nose, coryza, and cough; most without fever Lower respiratory tract infection in adults: Typically mild illness with nonproductive cough, fever, malaise, pharyngitis, myalgias; 3% to 13% of patients develop pneumonia; complications include rash, arthritis, encephalitis, myocarditis, pericarditis, and hemolytic anemia
Genital mycoplasmas: *U. urealyticum* and *M. hominis*	Systemic infections in neonates: Meningitis, abscess, and pneumonia. *U. urealyticum* is also associated with the development of chronic lung disease Invasive disease in immunosuppressed patients: Bacteremia, arthritis (particularly in patients with agammaglobulinemia), abscesses and other wound infections, pneumonia, peritonitis Urogenital tract infections: Prostatitis, pelvic inflammatory disease, bacterial vaginosis, amnionitis, nongonococcal urethritis
M. genitalium	Nongonococcal urethritis in men. Possible cause of cervicitis and endometritis in females.

detection. Acceptable specimens include body fluids (e.g., blood, joint fluid, amniotic fluid, urine, prostatic secretions, semen, pleural secretions, bronchoalveolar lavage specimens), tissues, and swabs of the throat, nasopharynx, urethra, cervix, or vagina. Blood for culture of genital mycoplasmas should be collected without anticoagulants and immediately inoculated into an appropriate broth culture medium (see later text). Swab specimens should be obtained without the application of any disinfectants, analgesics, or lubricant[18]; Dacron, polyester, or calcium alginate swabs on aluminum or plastic shafts should be used.

Because mycoplasmas have no cell wall, they are highly susceptible to drying and therefore transport media are necessary, particularly when specimens are collected on swabs. Transport and storage conditions of various types of specimens are summarized in Table 47-3.

DIRECT DETECTION METHODS

At present, no direct methods for identifying *M. pneumoniae, U. ureaplasma, or M. hominis* in clinical samples are recommended, although some methods have been described, such as polymerase chain reaction (PCR), immunoblotting, and indirect immunofluorescence.[8,22] Although PCR appears to hold the greatest promise for being a highly sensitive and specific means to detect *M. pneumoniae* directly in clinical specimens, the possibility for a long-term carrier state of *M. pneumoniae* in the respiratory tract has somewhat hampered its direct detection by PCR. Because it is unknown whether there is a specific threshold quantity of *M. pneumoniae* in respiratory tract specimens that can differentiate colonization from infection, a positive PCR result may overestimate its clinical importance. Thus, although PCR is an attractive means to detect *M. pneumoniae*, partic-

ularly in light of difficulties with culture (see later), PCR assays must be standardized and evaluated as to their clinical utility; quantitative assays may be required.[7,22] Multiplexed real-time PCR assays that detect *M. pneumoniae* as well as other atypical respiratory tract pathogens such as *Chlamydophila pneumonia* and *Legionella pneumoniae* have been developed and sold commercially.

Because there is no reliable medium for its isolation, *M. genitalium* has been directly detected by PCR targeting its attachment protein in urine and urethral swabs in men.[19] In women, vaginal or cervical swabs are used.

CULTIVATION

In general, the media for mycoplasma isolation contains a beef or soybean protein with serum, fresh yeast extract, and other factors. Of importance, because these organisms grow more slowly than most other bacteria, the media must be selective to prevent overgrowth of faster-growing organisms that may be present in a clinical sample. Culture media and incubation conditions for these organisms are summarized in Table 47-4. Culture methods for *M. pneumoniae, U. urealyticum,* and *M. hominis* are provided at the end of this chapter in procedures 47-1, 47-2, and 47-3, respectively. The quality control of the growth media with a fastidious isolate is of great importance.

For the most part, the different metabolic activity of the mycoplasmas for different substrates is used to detect their growth. Glucose (dextrose) is incorporated into media selective for *M. pneumoniae*, because this mycoplasma ferments glucose to lactic acid; the resulting pH change is then detected by a color change in a dye indicator. Similarly, urea

Table 47-3 Transport and Storage Conditions for *Mycoplasma pneumoniae, Ureaplasma urealyticum,* and *M. hominis*

Specimen Type	Transport Conditions	Transport Media (examples)	Storage	Processing
Body fluid or liquid specimens*	Within 1 hr of collection on ice or at 4° C	Not required	4° C up to 24 hr[†]	Concentrate by high-speed centrifugation and dilute (1:10 to 1:1000) in broth culture media to remove inhibitory substances and contaminating bacteria. Urine should be filtered through a 0.45-μm pore size filter
Swabs	Place immediately into transport media	0.5% albumin in trypticase soy broth Modified Stuart's 2SP (sugar-phosphate medium with 10% heat-inactivated fetal calf serum) Shepard's 10B broth for ureaplasmas[17] SP-4 broth[20] for other mycoplasmas and *M. pneumoniae*[‡] Mycoplasma transport medium (trypticase phosphate broth, 10% bovine serum albumin, 100,000 U of penicillin/milliliter)[3]	4°C up to 24 hr[†]	None
Tissue	Within 1 hr of collection on ice or at 4° C	Not required as long as prevented from drying out	4° C up to 24 hr[†]	Mince (not ground) and dilute (1:10 and 1:100) in transport media

*Except blood (see text).
[†]Can be stored indefinitely at -70° C if diluted in transport media following centrifugation.
[‡]SP-4 broth: sucrose phosphate buffer, 20% horse serum, *Mycoplasma* base, and neutral red.

Table 47-4 Cultivation of *Mycoplasma pneumoniae, Ureaplasma urealyticum,* and *M. hominis*

Organism	Media (examples)	Incubation Conditions
M. pneumoniae	Biphasic SP-4 (pH 7.4) Triphasic system (Mycotrim RS, Irvine Scientific, Irvine, Calif.) PPLO broth or agar with yeast extract and horse serum Modified New York City medium	Agar and broths: 37° C, ambient air for up to 4 wk
U. urealyticum*/*M. hominis*[†]	A7 or A8 agar medium[‡] New York City medium Modified New York City medium SP-4 glucose broth with arginine[§] SP-4 glucose broth with urea[‖] Triphasic system (Mycotrim GU, Irvine Scientific) Shepard's 10B broth (or Ureaplasma 10C broth)[‖]	Broths: 37° C, ambient air for up to 7 days Agars: 37° C in 5% CO_2 or anaerobically for 2 to 5 days

*Utilizes urea and requires acidic medium.
[†]Converts arginine to ornithine and grows over a broad pH range.
[‡]Commercially available.
[§]For *M. hominis* isolation.
[‖]For *U. urealyticum* isolation.

and/or arginine can be incorporated into media to detect *U. urealyticum* and *M. hominis,* respectively. If a color change, that is, a pH change, is detected, a 0.1- to 0.2-mL aliquot is immediately subcultured to fresh broth and agar media.

In some clinical situations, it may be necessary to provide quantitative information regarding the numbers of genital mycoplasmas in a clinical specimen. For example, quantitation of specimens taken at different stages during urination or after prostatic massage can

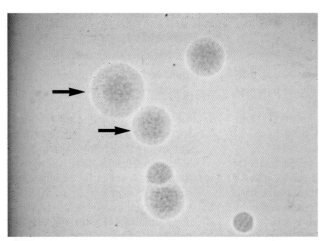

Figure 47-2   Colonies of *Mycoplasma pneumoniae* visualized under 100× magnification. Note the variation in the size of the colonies *(arrows)*. (Courtesy Clinical Microbiology Laboratory, SUNY Upstate Medical University, Syracuse, NY.)

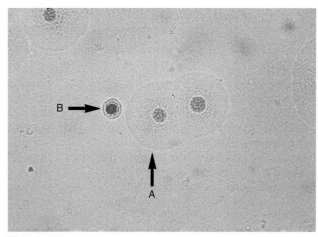

Figure 47-3   Isolation of *Mycoplasma hominis* and *Ureaplasma urealyticum* (100× magnification). Note the "fried egg" appearance of the large *M. hominis* colony *(arrow A)* and the relatively small size of the *U. urealyticum* colony *(arrow B)*. (Courtesy Clinical Microbiology Laboratory, SUNY Upstate Medical University, Syracuse, NY.)

help determine the location of mycoplasmal infection in the genitourinary tract.[12]

APPROACH TO IDENTIFICATION

On agar, *M. pneumoniae* will appear as spherical, grainy, yellowish forms that are embedded in the agar, with a thin outer layer similar to those shown in Figure 47-2. The agar surface is examined under 40× magnification every 24 to 72 hours. Because only *M. pneumoniae* and one serovar of *U. urealyticum* hemadsorb, definitive identification of *M. pneumoniae* is accomplished by overlaying suspicious colonies with 0.5% guinea pig erythrocytes in phosphate-buffered saline instead of water. After 20 to 30 minutes at room temperature, colonies are observed for adherence of red blood cells.

Cultures for the genital mycoplasmas are handled in a similar fashion, including culture examination and the requirement for subculturing. Colonies may be definitely identified on A8 agar as *U. urealyticum* by urease production in the presence of a calcium chloride indicator.[4,14,17] *U. urealyticum* colonies (15 to 60 μm in diameter) will appear as dark brownish clumps. Colonies that are typical in appearance for *U. urealyticum* are shown in Figure 47-3. *M. hominis* are large (about 20 to 300 μm in diameter) and are urease-negative (see Figure 47-3), with a characteristic "fried egg" appearance. On conventional blood agar, strains of *M. hominis,* but not of *U. urealyticum,* produce nonhemolytic, pinpoint colonies that do not stain with Gram stain. These colonies can be stained with the Dienes or acridine orange stains. Numerous transport and growth media systems for the detection, quantitation, identification,

and antimicrobial susceptibility testing of the genital mycoplasma are commercially available in the United States and Europe.

SERODIAGNOSIS

Laboratory diagnosis of *M. pneumoniae* is usually made serologically. Nonspecific production of cold agglutinins occurs in approximately half of patients with atypical pneumonia caused by this organism. The most widely used serologic tests today are enzyme-linked immunosorbent assay (ELISA), although newly developed indirect fluorescent antibody tests are being used with some success. IgM-specific tests such as the Immuno Card (Meridian Diagnostics, Cincinnati, Ohio) are commercially available, and a single positive result in children, adolescents, and young adults may be considered diagnostic in some cases.[21,22] In addition, there is a commercially available, membrane-based assay that simultaneously detects IgM and IgG against *M. pneumoniae* (Remel EIA, Lenexa, Kansas) with good sensitivity and specificity compared to other tests.

Although serologic tests such as indirect hemagglutination and metabolism inhibition for genital mycoplasmas are available, they are rarely used. Because of the antigenic complexity of the mycoplasmas, the development of a specific and useful serologic assay is a challenge.

SUSCEPTIBILITY TESTING AND THERAPY

Although agar and broth dilution methods may be used to determine antibiotic susceptibilities, the complex

Procedure 47-1

ISOLATION OF *MYCOPLASMA PNEUMONIAE*

PRINCIPLE

Mycoplasma pneumoniae is primarily a respiratory tract pathogen that is extremely fastidious. Because this bacterial agent lacks a cell wall and has strict nutritional requirements, the recovery of this organism from clinical specimens requires special conditions both in terms of culture media and incubation conditions.

METHOD

1. Prepare biphasic SP-4 culture media as follows:

   A. Combine the ingredients for the media base.

   Mycoplasma broth base (BD), 3.5 g
   Tryptone (Difco), 10 g
   Bacto-Peptone (Difco), 5 g
   (50% aqueous solution, filter sterilized), 10 mL
   Distilled water, 615 mL

   B. Stir the solids into boiling water to dissolve them and adjust the pH to 7.4 to 7.6. Autoclave according to the manufacturer's instructions. Cool to 56° C before adding the following supplements for each 625 mL of base to make a final volume of 1 L:

   CMRL 1066 tissue culture medium with glutamine, 10× (GIBCO), 50 mL
   Aqueous yeast extract (prepared as described in New York City Medium), 35 mL
   Yeastolate (Difco; 2% solution), 100 mL

Fetal bovine serum (heat inactivated at 56° C for 30 minutes), 170 mL
Penicillin G sodium, 1000 1U/mL
Amphotericin B, 0.5 g
Polymyxin B, 500,000 U

   C. To prepare the agar necessary for making biphasic media, add 8.5 g Noble agar (Difco) to the basal medium ingredients before adding the supplements.

   D. Dispense 1 mL of SP-4 agar aseptically into the bottom of sterile 4-mL screw-capped vials. Allow the agar to set and dispense 2 mL of SP-4 broth above the agar layer in each vial. Seal caps tightly and store at −20° C.

2. Place 0.1 to 0.2 mL of liquid specimen or dip and twirl a specimen received on a swab in a vial of biphasic SP-4 culture medium. After expressing as much fluid as possible from the swab, remove the swab to prevent contamination.

3. Seal the vial tightly and incubate in air at 35° C for up to 3 weeks.

4. Inspect the vial daily. During the first 5 days, a change in pH, indicated by a color shift from orange to yellow or violet, or increased turbidity is a sign that the culture is contaminated and should be discarded.

5. If either a slight acid pH shift (yellow color) with no increase in turbidity or no change occurs after 7 days' incubation, subculture several

drops of the broth culture to agar. Continue to incubate the original broth.

6. If broth that exhibited no changes at 7 days shows a slight acid pH shift at any time, subculture to agar as above. Broths that show no change at 3 weeks are blindly subcultured to agar.

7. Incubate the agar plates in a very moist atmosphere with 5% CO_2 at 35° C for 7 days.

8. Observe the agar surface under 40× magnification after 5 days for colonies, which appear as spherical, grainy, yellowish forms, embedded in the agar, with a thin outer layer (see Figure 47-2).

9. Definitive identification of *M. pneumoniae* is accomplished by overlaying agar plates showing suspicious colonies with 5% sheep or guinea pig erythrocytes in 1% agar prepared in physiologic saline (0.85% NaCl) instead of water. The 1% agar is melted and cooled to 50° C, the blood cells are added, and a thin layer is poured over the original agar surface.

10. Reincubate the plate for 24 hours, and observe for beta hemolysis around colonies of *M. pneumoniae* caused by production of hydrogen peroxide. Additional incubation at room temperature overnight enhances the hemolysis. No other species of *Mycoplasma* produces this reaction.

Modified from Clyde WA Jr, Kenny GE, Schachter J: *Cumitech 19*: *Laboratory diagnosis of chlamydial and mycoplasmal infections*, Drew ML, coordinating editor, Washington, DC, 1984, American Society for Microbiology; Waites KB, Bébéar CM, Robertson JA, et al: *Laboratory diagnosis of mycoplasmal infections*, Nolte FS, coordinating editor, Washington, DC, 2001, American Society for Microbiology.

growth requirements of mycoplasmas have restricted their performance to only a few laboratories. Further complicating matters, the lack of standardized methods and interpretive criteria have contributed to problems with susceptibility tests for the mycoplasmas. To date, the Mycoplasma Chemotherapy Working Team of the International Research Program on Comparative Mycoplasmology has formulated agar and broth dilution methods while the Clinical Laboratory Standards Institute develops standardized methods.[23] Most mycoplasmal infections are treated empirically.

Most *M. pneumoniae* infections are self-limited and usually do not require treatment. However, treatment can markedly shorten the illness, although complete eradication of the organism takes a long time, even after therapy. Because of the lack of a cell wall, *M. pneumoniae* as well as the other mollicutes are innately resistant to all beta-lactams. *M. pneumoniae* is usually susceptible to the macrolides, tetracycline, ketolides and fluoroquinolones antibiotics. Of the quinolones, the newer quinolones such as sparfloxacin appear to have activity against *M. pneumoniae*.[22]

Porcedure 47-2

ISOLATION OF *UREAPLASMA UREALYTICUM*

PRINCIPLE

Similar to other mycoplasmas, *Ureaplasma urealyticum* requires special cultural conditions to allow recovery of this bacterial agent from clinical specimens.

METHOD

1. Inoculate one *Ureaplasma* agar plate and one *Ureaplasma* broth each with 0.1 mL specimen from transport medium.

2. Incubate broth in tightly sealed test tubes for 5 days. Observe twice daily for a color change in the broth to red, with no increase in turbidity. If color change occurs, immediately transfer one loopful to a *Ureaplasma* agar plate and streak for isolated colonies.

3. Agar plates are incubated in a candle jar or, optimally, in an anaerobic jar at 35° C. Colonies appear on agar within 48 hours. Plates are inspected in the same way as described for *M. pneumoniae* (see Procedure 47-1). *Ureaplasma* colonies appear as small, granular, yellowish spheres.

4. To identify definitively colonies on *Ureaplasma* agar after 48 hours' incubation, pour a solution of 1% urea and 0.8% $MnCl_2$ in distilled water over the agar surface. *U. urealyticum* stains dark brown because of production of urease (see Figure 47-3).

Porcedure 47-3

ISOLATION OF *MYCOPLASMA HOMINIS*

PRINCIPLE

Similar to the other mycoplasmas, the successful recovery of *M. hominis* by culture from clinical specimens requires special conditions.

METHOD A

1. Inoculate one *M. hominis* agar plate and two *M. hominis* broth tubes, one broth containing phenol red indicator and one without the possibly inhibitory phenol red, each with 0.1-mL specimen from transport media.

2. Incubate broths in tightly sealed test tubes for 5 days. If the phenol red–containing broth changes color to red or violet, both broths are subcultured to *M. hominis* agar. After 48 hours of incubation, transfer 0.1 mL or a loopful of broth from tubes that exhibited no change or only a slight increase in turbidity to *M. hominis* agar and streak for isolated colonies.

3. *M. hominis* agar plates are incubated in the same manner as *Ureaplasma* cultures. Plates should be observed daily for up to 5 days for colonies.

METHOD B (ALTERNATIVE METHOD)

1. Inoculate specimen onto a prereduced colistin-nalidixic acid (CNA) sheep blood agar or anaerobic blood plate and incubate anaerobically for 48 to 72 hours.

2. Examine for pinpoint colonies that show no bacteria on Gram stain.

3. Streak suspicious colonies to *M. hominis* agar and incubate as in step 3 above.

Case Study

A 29-year-old previously healthy female presented with a productive cough, fever to 102° F, and severe headache. She had cervical adenopathy (swollen glands), although she had a nonerythematous throat with no exudate. Chest examination showed crackles bilaterally at the lung base with decreased breath sounds diffusely. This finding was confirmed by chest film that showed bilateral multifocal areas of patchy consolidation. Her neck was not stiff, but because of the severity of the headache, she was admitted to the neurologic service. A spinal fluid was obtained and was negative for bacteria, cryptococcus, and acid-fast smear. Blood cultures and sputum cultures did not produce a pathogen. The patient did not improve on ceftriaxone. On day 3 she was started on erythromycin. On day 4, cold agglutinins were done and were positive. The patient gradually improved although the headache, photophobia, and cough continued for some time.

QUESTIONS

1. What is the agent of this disease? Explain how the diagnosis was quickly made.

2. Can you explain why the bacterial cultures were negative?

3. Why is erythromycin an effective therapy for *M. pneumoniae,* but ceftriaxone is not?

Unfortunately, the susceptibility of *M. hominis* and *U. urealyticum* to various agents is not as predictable. For the most part, the tetracyclines are the drugs of choice for these agents, although resistance has been reported.

PREVENTION

As of this writing, no vaccines have been developed for the mycoplasmas.

REFERENCES

1. Ainsworth JG, Katseni V, Hourshid S, et al: *Mycoplasma fermentans* and HIV-associated nephropathy, *J Infect* 29:323, 1994.
2. Bauer FA, Wear DJ, Angritt P, et al: *Mycoplasma fermentans* (incognitus strain) infection in the kidneys of patients with acquired immunodeficiency syndrome and associated nephropathy: a light microscopic, immunohistochemical and ultrastructural study, *Human Pathol* 22:63, 1991.
3. Clegg A, Passey M, Yoannes M, et al: High rates of genital mycoplasma infection in highland Papua New Guinea determined both by culture and by a commercial detection kit, *J Clin Microbiol* 35:197, 1997.
4. Clyde WA, Kenny GE, Schachter J: *Cumitech 19: laboratory diagnosis of chlamydial and mycoplasmal infections*, Drew ML, coordinating editor, Washington, DC, 1984, American Society for Microbiology.
5. Foy HM, Kenny GE, Leninsohn EM, et al: Acquisition of mycoplasmata and T-strains during infancy, *J Infect Dis* 121:579, 1970.
6. Goldenberg RL, Thompson BS: The infectious origins of stillbirth, *Am J Obstet Gynecol* 189:861-873, 2003.
7. Loens K, Ursi D, Goosens H, et al: Molecular diagnosis of *Mycoplasma pneumoniae* respiratory tract infections, *J Clin Microbiol* 4:4915, 2003.
8. Lo L-C: New understandings of mycoplasmal infections and disease, *Clin Microbiol Newsletter* 17:169, 1995.
9. Lo S-C, et al: Identification of *Mycoplasma incognitus* infection in patients with AIDS: an immunohistochemical, in situ hybridization and ultrastructural study, *Am J Trop Med Hyg* 41:601, 1989.
10. McCormack WM, Almedia PC, Bailey PE, et al: Sexual activity and vaginal colonization with genital mycoplasmas, *JAMA* 221:1375, 1972.
11. McCormack WM, Lee YH, Zinner SH: Sexual experience and urethral colonization with genital mycoplasma: a study in normal men, *Ann Intern Med* 78:696, 1973.
12. Meares EM, Stamey TA: Bacteriologic localization patterns in bacterial prostatitis and urethritis, *Invest Urol* 5:492, 1968.
13. Mena L, et al: *Mycoplasma genitalium* infections in asymptomatic men and men with urethritis attending a sexually transmitted diseases clinic in New Orleans, *J Clin Infect Dis* 35:1167, 2001.
14. Montagnier L, Blanchard A: *Mycoplasmas* as cofactors in infection due to the human immunodeficiency virus, *Clin Infect Dis* 17 (suppl 1):S309, 1993.
15. Razin S, Yogev D, Naot V: Molecular biology and pathogenicity of mycoplasmas, *Microbiol Mol Biol Rev* 62:1094, 1998.
16. Sanchez P: Perinatal transmission of *Ureaplasma urealyticum:* current concepts based on review of the literature, *Clin Infect Dis* 17(suppl):S107, 1993.
17. Shepard MC, Lunceford CD: Serological typing of *Ureaplasma urealyticum* isolates from urethritis patients by an agar growth inhibition method, *J Clin Microbiol* 8:566, 1978.
18. Taylor-Robinson D, Furr PM: Recovery and identification of human genital tract mycoplasmas, *Isr J Med Sci* 17:648, 1981.
19. Totten PA, et al: Association of *Mycoplasma genitalium* with nongonococcal urethritis in heterosexual men, *J Infect Dis* 183:269, 2001.
20. Tully JG, Whitcomb RF, Clark HF, et al: Pathogenic mycoplasmas: cultivation and vertebrate pathogenicity of a new spiroplasma, *Science* 195:892, 1977.
21. Waites KB, Rikihisa Y, Taylor-Robinson D: *Mycoplasma* and *Ureaplasma*. In Murray PR, et al, editors: *Manual of clinical microbiology*, ed 8, Washington, DC, 2003, ASM Press.
22. Waites KB, Talkington DF: *Mycoplasma pneumoniae* and its role as a human pathogen, *Clin Microbiol Rev* 17:697, 2004.
23. Waites KB, Bebear CM, Robertson JA, et al: Cumitech 34, Laboratory diagnosis of mycoplasmal infections. *Am Society Microbiology*, Washington DC, 2001.

ADDITIONAL READING

Lo S-C: New understandings of mycoplasmal infections and disease, *Clin Microbiol Newsl* 17:169, 1995.

Taylor-Robinson D: Infections due to species of *Mycoplasma* and *Ureaplasma:* an update, *Clin Infect Dis* 23:671, 1996.

Tully JG: Current status of the mollicute flora of humans, *Clin Infect Dis* 17(suppl 1):S2, 1993.

Waites KB, Bébéar CM, Robertson JA et al: Cumitech 34, Laboratory diagnosis of mycoplasmal infections. *Amer Society Microbiology*, Washington DC, 2001.

Waites KB, Talkington DF: *Mycoplasma pneumoniae* and its role as a human pathogen, *Clin Microbiol Rev* 17:697, 2004.

Genera and Species to Be Considered

Treponema pallidum subsp. *pallidum*

Treponema pallidum subsp. *pertenue*

Treponema pallidum subsp. *endemicum*

Treponema carateum

Borrelia recurrentis

Borrelia burgdorferi sensu stricto

Borrelia garinii

Borrelia afzelii

Borrelia valaisiana

Leptospira interrogans

This chapter addresses the bacteria that belong in the order Spirochaetales. Although there are at least eight genera in this family, only the genera *Treponema, Borrelia,* and *Leptospira,* which contain organisms pathogenic for humans, are addressed in this chapter. Although there are two other human intestinal spirochetes identified to date, *Brachyspira (Serpulina) pilosicoli* and *Brachyspira aalborgi,* that have been isolated from biopsy material from patients with intestinal disease, a clear association between their presence and intestinal disease has not been established. A few cases of intestinal spirochetosis with septicemia and dissemination have been reported thereby providing some support as to a pathogenic role for these organisms.[8]

The spirochetes are all long, slender, helically curved, gram-negative bacilli, with the unusual morphologic features of axial fibrils and an outer sheath. These fibrils, or axial filaments, are flagella-like organelles that wrap around the bacteria's cell wall, are enclosed within the outer sheath, and facilitate motility of the organisms. The fibrils are attached within the cell wall by platelike structures, called *insertion disks,* located near the ends of the cells. The protoplasmic cylinder gyrates around the fibrils, causing bacterial movement to appear as a corkscrew-like winding. Differentiation of genera within the family *Spirochaetaceae* is based on the number of axial fibrils, the number of insertion disks present (Table 48-1), and biochemical and metabolic features. The spirochetes also fall into genera based loosely on their morphology (Figure 48-1): *Treponema* are slender with tight coils; *Borrelia* are somewhat thicker with fewer and looser coils; and *Leptospira* resemble *Borrelia* except for their hooked ends.

TREPONEMA

GENERAL CHARACTERISTICS

The major pathogens in the genus *Treponema*—*T. pallidum* subsp. *pallidum, T. pallidum* subsp. *pertenue, T. pallidum* subsp. *endemicum* and *T. carateum*—infect only humans and have not been cultivated for more than one passage in vitro. Most species stain poorly with Gram's or Giemsa's methods and are best observed with the use of dark-field or phase-contrast microscopy. These organisms are considered to be microaerophilic.

Other treponemes such as *T. vincentii, T. denticola, T. refringens, T. socranskii,* and *T. pectinovorum* are normal inhabitants of the oral cavity or the human genital tract. These organisms are cultivable anaerobically on artificial media. Acute necrotizing ulcerative gingivitis, also known as **Vincent's disease,** is a destructive lesion of the gums. Methylene blue–stained material from the lesions of patients with Vincent's disease show certain morphologic types of bacteria. Observed morphologies include spirochetes and fusiforms; oral spirochetes, particularly an unusually large one, may be important in this disease, along with other anaerobes.

EPIDEMIOLOGY AND PATHOGENESIS

Key features of the epidemiology of diseases caused by the pathogenic treponemes are summarized in Table 48-2. In general, these organisms enter the host by either penetrating intact mucous membranes (as is the case for *T. pallidum* subsp. *pallidum*—hereafter referred to as *T. pallidum*) or entering through breaks in the skin. After penetration, *T. pallidum* subsequently invades the bloodstream and spreads to other body sites. Although the mechanisms by which damage is done to the host are unclear, *T. pallidum* has a remarkable tropism (attraction) to arterioles; infection ultimately leads to endarteritis (inflammation of the lining of arteries) and subsequent progressive tissue destruction.

Table 48-1 Spirochetes Pathogenic for Humans

Genus	Axial Filaments	Insertion Disks
Treponema	6 to 10	1
Borrelia	30 to 40	2
Leptospira	2	3 to 5

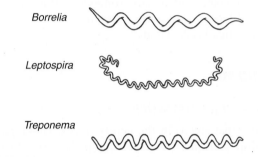

Figure 48-1 Species designation of spirochetes based on morphology.

SPECTRUM OF DISEASE

Treponema pallidum causes venereal (transmitted through sexual contact) syphilis. The clinical presentation of venereal syphilis is varied and complex, often mimicking many other diseases. This disease is divided into stages: incubating, primary, secondary, early latent, latent, and tertiary.[16] **Primary syphilis** is characterized by the appearance of a **chancre** (a painless ulcer) usually at the site of inoculation, most commonly the genitalia. Within 3 to 6 weeks, the chancre heals. Dissemination of the organism occurs during this primary stage; once the organism has reached a sufficient number (usually 2 to 24 weeks are required), clinical manifestations of **secondary syphilis** become apparent. During this phase the patient is ill and seeks medical attention. Systemic symptoms such as fever, weight loss, malaise, and loss of appetite are present in about half of the patients. The skin is the organ most commonly affected in secondary syphilis, with patients having a widespread rash. After the secondary phase, the disease becomes subclinical but not necessarily dormant (inactive); during this latent period diagnosis can be made only by serologic tests. **Late,** or **tertiary syphilis,** is the tissue-destructive phase that appears 10 to 25 years after the initial infection in up to 35% of untreated patients.[16] Complications of syphilis at this stage include central nervous disease, cardiovascular abnormalities, eye disease, and granuloma-like lesions, called *gummas,* in any organ.

The other pathogenic treponemes are major health concerns in developing countries. Although morphologically and antigenically similar, these agents differ epidemiologically and with respect to their clinical presentation from *T. pallidum.* The diseases caused by these treponemes are summarized in Table 48-2.

LABORATORY DIAGNOSIS

Direct Detection

Treponemes can be detected in material taken from skin lesions by dark-field examination or fluorescent antibody staining and microscopic examination. Material for microscopic examination is collected from suspicious lesions by first cleansing the area around the lesion with

Table 48-2 Epidemiology and Spectrum of Disease of the Treponemes Pathogenic for Humans

Agent	Transmission	Geographic Location	Disease	Clinical Manifestations*	Age Group
T. pallidum subsp. pallidum	Sexual contact or congenital (mother to fetus)	Worldwide	Venereal syphilis†	Refer to text above, this page	All ages
T. pallidum subsp. pertenue	Traumatized skin comes in contact with an infected lesion	Humid, warm climates: Africa, South and Central America, Pacific Islands	Yaws	Skin—papules,† nodules, ulcers	Children
T. pallidum subsp. endemicum	Mouth to mouth by utensils	Arid, warm climates: North Africa, Southeast Asia, Middle East	Endemic nonvenereal syphilis	Skin/mucous patches, papules, macules, ulcers, scars†	Children
T. carateum	Traumatized skin comes in contact with an infected lesion	Semiarid, warm climates: Central and South America, Mexico	Pinta	Skin papules, macules	All ages but primarily children and adolescents

*All diseases have a relapsing clinical course and prominent cutaneous manifestations.[12]
†If untreated, organisms can disseminate to other parts of the body such as bone.

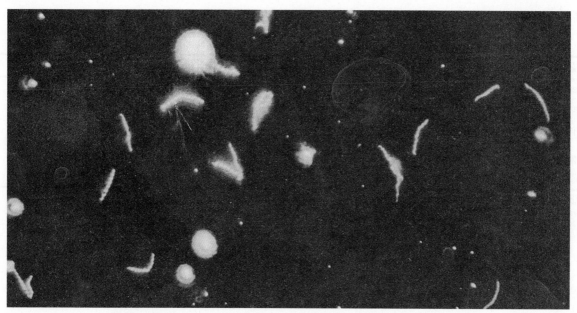

Figure 48-2   Appearance of *Treponema pallidum* in dark-field preparation.

a sterile gauze pad moistened in saline. The surface of the ulcer is then abraded until some blood is expressed. After blotting the lesion until there is no further bleeding, the area is squeezed until serous fluid is expressed. The surface of a clean glass slide is touched to the exudate, allowed to air-dry, and transported in a dust-free container for fluorescent antibody staining. A *T. pallidum* fluorescein-labeled antibody is commercially available for staining (Viro Stat, Portland, Maine). For dark-field examination, the expressed fluid is aspirated using a sterile pipette, dropped onto a clean glass slide, and coverslipped. The slide containing material for dark-field examination must be transported to the laboratory immediately. Because positive lesions may be teeming with viable spirochetes that are highly infectious, all supplies and patient specimens must be handled with extreme caution and carefully discarded as required for contaminated materials. Gloves should always be worn.

Material for dark-field examination is examined immediately under 400× high-dry magnification for the presence of motile spirochetes. Treponemes are long (8 to 10 μm, slightly larger than a red blood cell) and consist of 8 to 14 tightly coiled, even spirals (Figure 48-2). Once seen, characteristic forms should be verified by examination under oil immersion magnification (1000×). Although the dark-field examination depends greatly on technical expertise and the numbers of organisms in the lesion, it can be highly specific when performed on genital lesions.

Direct detection of *T. pallidum* in clinical material, such as serous exudate, has been accomplished using polymerase chain reaction (PCR) technology but is not presently widely available. Optimal clinical specimens for

PCR analysis have not been clearly delineated. However, PCR appears useful in the diagnosis of congenital syphilis.

Serodiagnosis

Class serologic tests for treponematosis measure the presence of two types of antibodies: treponemal and nontreponemal. **Treponemal antibodies** are produced against antigens of the organisms themselves, whereas **nontreponemal antibodies,** often called **reaginic antibodies,** are produced by infected patients against components of mammalian cells. Reaginic antibodies, although almost always produced by patients with syphilis, are also produced by patients with other infectious diseases such as leprosy, tuberculosis, chancroid, leptospirosis, malaria, rickettsial disease, trypanosomiasis, lymphogranuloma venereum (LGV), measles, chickenpox, hepatitis, and infectious mononucleosis; noninfectious conditions such as drug addiction; autoimmune disorders, including rheumatoid disease; and factors such as old age, pregnancy, and recent immunization.

The two most widely used nontreponemal serologic tests are the **VDRL** (Venereal Disease Research Laboratory) tests and **RPR** (rapid plasma reagin) test. Each of these tests is a flocculation (or agglutination) test, in which soluble antigen particles are coalesced to form larger particles that are visible as clumps when they are aggregated by antibody.

Specific treponemal serologic tests include the **FTA-ABS** (fluorescent treponemal antibody absorption) test and the **TP-PA** (*T. pallidum* particle agglutination) test. Once positive, their usefulness is limited because these tests tend to yield positive results through-

Table 48-3 Sensitivity of Commonly Used Serologic Tests for Syphilis

Method	Stage*		
	Primary	Secondary	Late
Nontreponemal (reaginic tests)—Screening			
Venereal Disease Research Laboratory (reaginic) test (VDRL)	70%	99%	0% to 1%†
Rapid plasma-reagin (RPR) card test and automated reagin test (ART)	80%	99%	0% to 1%†
Specific Treponemal Tests—Confirmatory			
Fluorescent treponemal antibody absorption test (FTA-ABS)	85%	100%	98%†

Modified from Tramont EC: *Clin Infect Dis* 21:1361, 1995.
*Percentage of patients with positive serologic tests in treated or untreated primary or secondary syphilis.
†Treated late syphilis.

out the patient's life. The FTA-ABS test is performed by overlaying whole treponemes that are fixed to a slide with serum from patients suspected of having syphilis because of a previous positive VDRL or RPR test. The patient's serum is first absorbed with non-*T. pallidum* treponemal antigens (sorbent) to reduce nonspecific cross-reactivity. Fluorescein-conjugated anti-human antibody reagent is then applied as a marker for specific antitreponemal antibodies in the patient's serum. This test should not be used as a primary screening procedure. TP-PA (Fujirebio America, Fairfield, NJ) tests utilize gelatin particles sensitized with *T. pallidum* subsp. *pallidum* antigens. Serum samples are diluted in a microtiter plate and sensitized gelatin particles are added. The presence of specific antibody causes the gelatin particles to agglutinate and form a flat mat across the bottom of the microdilution well in which the test is performed. The nontreponemal serologic tests for syphilis can be used to determine quantitative titers of antibody, which are useful for following response to therapy. The relative sensitivity of each test is shown in Table 48-3 to confirm that a positive nontreponemal test result is due to syphilis rather than to one of the other infections or biologic false-positive conditions previously mentioned.

ANTIMICROBIAL SUSCEPTIBILITY TESTING AND THERAPY

Because the treponemes are noncultivable, susceptibility testing is not performed. For all treponemal infections, penicillin G is the drug of choice. Treatment varies depending on the stage of disease and the host (e.g., children or adults, HIV-infected or congenital syphilis).[6]

PREVENTION

No vaccines are available for the treponematoses. Prevention is best accomplished by early and appropriate treatment, thereby preventing person-to-person spread.

BORRELIA

GENERAL CHARACTERISTICS

Organisms belonging to the genus *Borrelia* are composed of 3 to 10 loose coils (see Figure 48-1) and are actively motile. In contrast to the treponemes, *Borrelia* spp. stain well with Giemsa's stain. Species that have been grown in vitro are microaerophilic.

EPIDEMIOLOGY AND PATHOGENESIS

Although pathogens for mammals and birds, borreliae are the causative agents of tickborne and louseborne relapsing fever and tickborne Lyme disease in humans.

Relapsing Fever

Human relapsing fever is caused by more than 15 species of *Borrelia* and is transmitted to humans by the bite of a louse or tick. *B. recurrentis* is responsible for louseborne or epidemic relapsing fever. This spirochete is transmitted by the louse *Pediculus humanus* subsp. *humanus* and disease is found worldwide; humans are the only reservoir for *B. recurrentis*. All other borreliae that cause disease in the United States are transmitted via tick bites and are named after the species of tick, usually of the genus *Ornithodoros* (soft tick), from which they are recovered. Common species in the United States include *B. hermsii*, *B. turicatae*, *B. parkeri*, and *B. mazzottii*. Depending on the organisms and the disease, their reservoir is either humans or rodents in most cases. Although their pathogenic mechanisms are unclear, these spirochetes exhibit antigenic variability that may account for the cyclic fever patterns associated with this disease.

Lyme Disease

Although there are currently at least 10 different *Borrelia* species within the *B. burgdorferi* sensu lato complex, only *Borrelia burgdorferi* sensu stricto (strict sense of *B. burgdorferi*) as well as *B. garinii*, *B. afzelii*, and *B. valaisiana* are agents of **Lyme disease** and are transmitted by the bite of Ixodes ticks.[17] Lyme disease is the most common vector-borne disease in North America and Europe and is an emerging problem in northern Asia.[10] Hard ticks, belonging primarily to the genus *Ixodes*, act as vectors in the United States, including *Ixodes pacificus* in California and *I. scapularis* in other areas. The ticks' natural hosts are deer and rodents, although they will attach to pets as well as to humans; all stages of ticks—larva, nymph, and adult—can harbor the spirochete and transmit disease. The nymphal form of the tick is most likely to transmit disease because it is active in the spring and summer when people are dressed lightly and in the woods. Because at this stage the tick is the size of a pinhead, the initial tick bite may be overlooked. Ticks require a period of attachment of at least 24 hours before they transmit disease. Endemic areas of disease have been identified in many states, including Massachusetts, Connecticut, Maryland, Minnesota, Oregon, and California, as well as in Europe, Russia, Japan, and Australia. Direct invasion of tissues by the organism is responsible for the clinical manifestations. However, IgM antibodies are produced continually months to years after initial infection as the spirochete changes its antigens. *B. burgdorferi*'s potential ability to induce an autoimmune process in the host because of cross-reactive antigens may contribute to the pathology associated with Lyme disease.[3] Moreover, by virtue of its ability to vary its surface antigens (e.g., outer surface protein [Osp] A to G) as well as avoid complement attack, *B. burgdorferi* is able to avoid the human host response. The pathologic findings associated with Lyme disease are also believed to be due to the release of host cytokines initiated by the presence of the organism.

SPECTRUM OF DISEASE

Relapsing Fever

Two to 15 days after infection, patients have an abrupt onset of fever, headache, and myalgia that lasts for 4 to 10 days. As the host produces a specific antibody in response to the agent, organisms disappear from the bloodstream, becoming sequestered (hidden) in different body organs during the afebrile period. Subsequently, organisms reemerge with newly modified antigens and multiply, resulting in another febrile period. Subsequent relapses are usually milder and of shorter duration. Generally, there are more relapses associated with cases of untreated tickborne relapsing fever, but louse-borne relapsing fevers tend to be more severe.

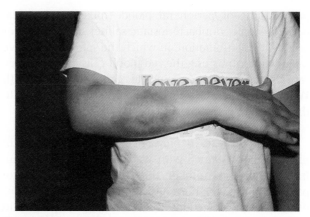

Figure 48-3   Appearance of the classic erythema migrans lesion of acute Lyme disease.

Lyme Disease

Lyme disease is characterized by three stages, not all of which occur in any given patient. The first stage, erythema migrans (EM), is the characteristic red, ring-shaped skin lesion with a central clearing that first appears at the site of the tick bite but may develop at distant sites as well (Figure 48-3). Patients may experience headache, fever, muscle and joint pain, and malaise during this stage. The second stage, beginning weeks to months after infection, may include arthritis, but the most important features are neurologic disorders (i.e., meningitis, neurologic deficits) and carditis. The third stage is usually characterized by chronic arthritis and may continue for years. Of note, there is an association between *Borrelia* species and distinct clinical manifestations.[17,18] For example, *B. garinii* has been associated with up to 72% of European cases of neuroborreliosis.[17,18]

LABORATORY DIAGNOSIS

Specimen Collection, Transport, and Processing

Peripheral blood is the specimen of choice for direct detection of borreliae that cause relapsing fever. *Borrelia burgdorferi* can be visualized and cultured, although serology has, so far, been the best means to diagnose Lyme disease. Specimens submitted for stain and/or culture include blood, biopsy specimens, and body fluids including joint and cerebrospinal fluids; there are no special requirements for collection, transport, or processing of these specimens.

Direct Detection Methods

Relapsing Fever. Clinical laboratories rely on direct observation of the organism in peripheral blood from patients for diagnosis. Organisms can be found in 70% of cases when blood specimens from febrile patients are examined. The organisms can be seen directly in wet preparations of peripheral blood (mixed with equal parts of sterile, nonbacteriostatic saline) under dark- or bright-field illumination, in which the spirochetes move rapidly, often pushing the red blood cells around. They can also be seen by staining thick and thin films with Wright's or Giemsa stains using procedures similar to those used to detect malaria.

Lyme Disease. *B. burgdorferi* may be visualized in tissue sections stained with Warthin-Starry silver stain. In general, the number of spirochetes in blood of patients with Lyme borreliosis is below the lower limits of microscopic detection. PCR has become important in diagnosing Lyme disease. PCR has detected *B. burgdorferi* DNA in clinical specimens from patients with early and late clinical manifestations; optimal specimens include urine, synovial tissue, synovial fluid, and skin biopsies from patients with EM.[1,15,18] A variety of molecular methods have been used by laboratories to increase sensitivity and specificity and decrease turnaround time for diagnosing Lyme borreliosis. PCR has confirmed EM with an overall sensitivity and specificity of 68% and 100%, respectively. The ability to detect spirochetes in blood or plasma by PCR is dependent on the stage of illness (from 40% of patients with secondary EM to only 9.5% of patients with primary EM); PCR also does relatively well in detecting *B. burgdorferi* sensu lato in synovial fluids.[7,14] In contrast, variable results using PCR have been achieved in cerebrospinal fluid (CSF) specimens obtained from patients with peripheral or central nervous system involvement with Lyme borreliosis; overall sensitivity is only in the range of 20%.

Cultivation

Although the organisms that cause relapsing fever can be cultured in nutritionally rich media under microaerobic conditions, the procedures are cumbersome and unreliable and are used primarily as research tools. Similarly, the culture of *B. burgdorferi* may be attempted, although the yield is low. The periphery of the ring lesion of EM or synovial tissue provide the best specimens for culture in untreated patients. CSFfl and blood or plasma (greater than 9 mL) in general are of low diagnostic yield by culture or PCR.[14,18] The resuspended plasma from blood, spinal fluid sediment, or macerated tissue biopsy is inoculated into a tube of modified Kelly's medium (BSK II, BSK-H, or Preac-Mursic), and incubated at 30° to 34° C for up to 12 weeks under microaerophilic conditions.[1] Blind subcultures (0.1 mL) are performed weekly from the lower portion of the broth to fresh media, and the cultures are examined by dark-field microscopy or by fluorescence microscopy after staining with acridine orange for the presence of spirochetes.

Serodiagnosis

Relapsing Fever. Serologic tests for relapsing fever have not proved reliable for diagnosis because of the many antigenic shifts *Borrelia* organisms undergo during the course of disease. Patients may exhibit increased titers to *Proteus* OX K antigens (up to 1:80), but other cross-reacting antibodies are rare. Certain reference laboratories, such as those at the Centers for Disease Control and Prevention (CDC), may perform special serologic procedures on sera from selected patients.

Lyme Disease. Despite its inadequacies, serology continues to be the standard for the diagnosis of Lyme disease. *B. burgdorferi* has numerous immunogenic lipids, proteins, lipoproteins, and carbohydrate antigens on its surface and outer membrane. Numerous serologic tests are commercially available; however, standardization of these tests is not yet accomplished and their performance characteristics vary greatly. The most common of these tests are the indirect immunofluorescence assay (IFA), the enzyme-linked immunosorbent assay (ELISA), and Western blot. Measuring antibody by ELISA is the primary screening method because it is quick, reproducible, and relatively inexpensive. However, false-positive rates are high, mainly as a result of cross-reactivity. Patients with syphilis, HIV infection, leptospirosis, mononucleosis, parvovirus infection, rheumatoid arthritis, and other autoimmune diseases commonly show positive results.[1,3,13,14,18] For the United States, the CDC recommends a two-step approach to the serologic diagnosis of Lyme disease.[5] The first step is to use a sensitive screening test such as an ELISA or IFA; if this test is positive or equivocal, the result must be confirmed by immunoblotting. In certain clinical situations, results of serologic tests must be interpreted with caution. For example, patients with Lyme arthritis frequently remain antibody-positive despite treatment but do not necessarily have persistent infection. Conversely, patients with only a localized EM may be seronegative. Because of these limitations and others, the Food and Drug Administration (FDA) and the American College of Physicians[2] have published guidelines regarding the use of laboratory tests for Lyme disease diagnosis. Of paramount importance is the clinician's determination before ordering serologic tests of the pretest probability of Lyme disease based on clinical symptoms and the incidence of Lyme disease in the population represented by the patient.

ANTIBIOTIC SUSCEPTIBILITY TESTING AND THERAPY

Because there are no standardized methods and borreliae are difficult to culture, antimicrobial susceptibility testing is not routinely performed.

Several antibiotics, including tetracycline, are effective in treating relapsing fever. Doxycycline, amoxicillin, or cefuroxime and parenteral cephalosporins are drugs of choice during the first stage of Lyme disease. Broad-spectrum cephalosporins, particularly ceftriaxone or cefotaxime, have been used successfully with patients who either fail initial treatment or present in later stages of the disease. Symptomatic treatment failures, particularly in patients with chronic Lyme disease, have been reported.

PREVENTION

A recombinant outer surface protein A vaccine has been licensed for use in humans against Lyme disease caused by infection with organisms belonging to the *B. burgdorferi* complex. Although many issues surrounding vaccine use remain unsettled, the vaccine is expected to be in widespread use in endemic areas. Infection is also prevented by avoiding tick-infested areas; wearing protective clothing; checking your clothing, body, and pets for ticks; and removing them promptly. There are no vaccines against infections caused by other *Borrelia* spp.

LEPTOSPIRA

GENERAL CHARACTERISTICS

The leptospires include both free-living and parasitic forms. Pathogenic *Leptospira* species are currently classified in seven species based on DNA relatedness, with *Leptospira interrogans* sensu stricto being the main species associated with human leptospirosis; in France, this organism is responsible for about 60% of human cases. The pathogens include about 80 serologically defined types that were formerly designated as species and are now known as serovars, or serotypes, of *L. interrogans* sensu stricto. Each serovar is usually associated with a particular animal host and therefore serovar identification is important for epidemiology studies and prevention strategies.

EPIDEMIOLOGY AND PATHOGENESIS

Leptospirosis, a zoonosis, has a worldwide distribution but is most common in developing countries and warm climates where contact with infected animals or water contaminated with their urine is likely to occur. *L. interrogans* can infect most mammals throughout the world, as well as reptiles, amphibians, fish, birds, and invertebrates.[9] The organism is maintained in nature by virtue of persistent colonization of renal tubules of carrier animals.[4] Humans become infected through direct or indirect contact with the urine or blood of infected animals. Leptospires enter the human host

BOX 48-1	Potential Virulence Factors of *Leptospira*

Hemolipins
Sphingomyelinases C and H
Fibronectin-binding protein for adhesion and invasion
Lipopolysaccharide and outer membrane proteins

through breaks in the skin, mucous membranes, or conjunctivae. Infection can be acquired in home and recreational settings (e.g., swimming, hunting, canoeing) or in certain occupational settings (e.g., farmers, ranchers, abattoir workers, trappers, veterinarians).[4,9]

Pathogenic leptospires rapidly invade the bloodstream after entry and spread throughout all sites in the body such as the central nervous system and kidneys. Virulent strains show chemotaxis toward hemoglobin as well as the ability to migrate through host tissues. A number of potential virulence factors that might facilitate this process are shown in Box 48-1. Precisely how *L. interrogans* causes disease is not completely understood, but it appears that the presence of endotoxin and other toxins may play a role in which hemostasis pathways are activated as is an autoimmune response in the human host.

SPECTRUM OF DISEASE

Symptoms begin abruptly 2 to 20 days after infection and include fever, headache, and myalgia. The most common clinical syndrome is **anicteric leptospirosis,** which is a self-limiting illness consisting of a septicemic stage, with high fever and severe headache that lasts 3 to 7 days, followed by the immune stage. Symptoms associated with the immune stage (onset coincides with the appearance of IgM) are varied, but in general are milder than the septicemic stage. The hallmark of the immune stage is aseptic meningitis. Weil's disease, or **icteric leptospirosis,** is generally the most severe illness, with symptoms caused by liver, kidney, and/or vascular dysfunction with lethal pulmonary hemorrhage; death can occur in up to 10% of cases. Unfortunately, the clinical presentations of leptospirosis mimic those of many other diseases.

LABORATORY DIAGNOSIS

Specimen Collection, Transport, and Processing

During the first 10 days of illness, leptospires are present in the blood and CSFfl. Urine specimens can be obtained beginning in the second week of illness and up to 30 days after the onset of symptoms. Other than the time of collection of an appropriate specimen and avoiding collection in citrate anticoagulants, there are no other special requirements for specimen collection, transport, or processing. Specimens should be transported at room temperature and inoculated for culture within 24 hours.[4]

Direct Detection

Blood, CSF, and urine may be examined directly by dark-field microscopy examination. Detection of motile leptospires in these specimens is optimized by centrifuging at 1500× *g* for 30 minutes; sodium oxalate or heparin-treated blood is initially spun at 500× *g* for 15 minutes to remove blood cells. Other techniques, such as fluorescent antibody staining and hybridization techniques using leptospira-specific DNA probes have also detected leptospires in clinical specimens, but only limited studies have been performed. Conventional and real-time PCR assays have been used to detect leptospires in clinical and environmental samples.

Cultivation

Currently, albeit insensitive, the definitive method for laboratory diagnosis of leptospirosis is to culture the organisms from blood, CSF, or urine. A few drops of heparinized or sodium oxalate–anticoagulated blood are inoculated into tubes of semisolid media enriched with rabbit serum (Fletcher's or Stuart's) or bovine serum albumin. Urine should be inoculated soon after collection, because acidity (diluted out in the broth medium) may harm the spirochetes. One or 2 drops of undiluted urine and a 1:10 dilution of urine are added to 5 mL of medium. The addition of 200 μg/mL of 5-fluorouracil (an anticancer drug) may prevent contamination by other bacteria without harming the leptospires. Tissue specimens, especially from the liver and kidney, may be aseptically macerated and inoculated in dilutions of 1:1, 1:10, and 1:100 as for urine cultures.

All cultures are incubated at room temperature or 30° C in the dark for up to 6 to 8 weeks. Because organisms grow below the surface, material collected from a few centimeters below the surface of broth cultures should be examined weekly for the presence of growth using a direct wet preparation under dark-field illumination. Leptospires exhibit corkscrew-like motility.

Approach to Identification

Based on the number of coils and hooked ends, leptospires can be distinguished from other spirochetes. Physiologically, the saprophytes can be differentiated from pathogens by their ability to grow to 10° C and lower, or at least 5° C lower than the growth temperature of pathogenic leptospires.[11]

Serodiagnosis

Serodiagnosis of leptospirosis requires a fourfold or greater rise in titer of agglutinating antibodies. The microscopic agglutination (MA) test using live cells is the standard serologic procedure used. Serologic diagnosis of leptospirosis is best performed using pools of bacterial antigens containing many serotypes in each pool. Positive results of the MA test are visualized by examining for the presence of agglutination under dark-field examination. However, a macroscopic agglutination procedure is more readily accessible to routine clinical laboratories. Reagents are available commercially. Indirect hemagglutination and an ELISA test for IgM antibody are also available; IgM-detection assays are primarily used because these antibodies become detectable during the first week of illness.

ANTIBIOTIC SUSCEPTIBILITY AND THERAPY

Standard procedures have not been developed for testing leptospires against various drugs. Treatment of leptospirosis is supportive management and use of appropriate antibiotics. Ceftriaxone, penicillin, amoxicillin, doxycycline, and tetracycline are recommended for treatment of leptospirosis.

PREVENTION

General preventive measures include the vaccination of domestic livestock and pet dogs. In addition, protective clothing, rodent control measures, and preventing recreational exposures, such as avoiding freshwater ponds, are indicated in preventing leptospirosis.

Case Study

A 43-year-old woman was referred to the Infectious Disease Clinic for recurrent symptoms beginning about 4 months after noticing a painful, swollen spot on her leg after removal of a tick. She had been treated with doxycycline for 14 days for presumed Lyme disease, documented by a positive IFA titer and presentation with migratory arthralgias, which were worse in the small joints of the hands. She also complained of fatigue, poor mentation, and occasional headaches. Because her symptoms recurred 2 months after treatment, she was unable to continue employment. Six months after a course of amoxicillin and later two courses of ceftriaxone, she again became symptomatic. A Western blot and PCR of her serum were ordered. The Western blot result was equivocal, but the PCR was positive for the agent of Lyme disease.

QUESTIONS

1. How did this patient acquire Lyme disease?
2. Why did the physician order further testing to diagnose Lyme disease in this patient?

3. Our patient did not respond to therapy for *B. burgdorferi*. Can you explain why this can happen?

REFERENCES

1. Aguero-Rosenfeld ME, Wang G, Schwartz I, et al: Diagnosis of Lyme borreliosis, *Clin Microbiol Rev* 18:484, 2005.
2. American College of Physicians: Guidelines for laboratory evaluation in the diagnosis of Lyme disease, *Ann Intern Med* 127:1106, 1997.
3. Athreya BH and Rose CD: Lyme disease, *Curr Probl Pediatr* 26:189, 1996.
4. Bharti AR, Nally JE, Ricaldi JN, et al: Leptospirosis: a zoonotic disease of global importance, *Lancet Infect Dis* 3:757, 2003.
5. Centers for Disease Control and Prevention: Recommendations for test performance and interpretation from the second national conference on serologic diagnosis of Lyme disease, *Morb Mortal Wkly Rep* 44:590, 1995.
6. Centers for Disease Control and Prevention: Sexually transmitted diseases treatment guidelines, *Morb Mortal Wkly Rep* 51:18, 2002.
7. Dummler JS: Molecular diagnosis of Lyme disease, *Mol Diagn* 6:1, 2001.
8. Erlandson KM and Klingler ET: Intestinal spirochitosis: epidemiology, microbiology and clinical significance, *Clin Microbiol Newsl* 27:91, 2005.
9. Farr RW: Leptospirosis, *Clin Infect Dis* 21:1, 1995.
10. Gomes-Solecki M, Dunn JJ, Luft BJ, et al: Recombinant chimeric *Borrelia* proteins for diagnosis of Lyme disease, *J Clin Microbiol* 38:2530, 2000.
11. Johnson RC and Harris VG: Differentiation of pathogenic and saprophytic leptospires. I. Growth at low temperature, *J Bacteriol* 94:27, 1967.
12. Koff AB and Rosen T: Nonvenereal treponematoses: yaws, endemic syphilis, and pinta, *J Am Acad Dermatol* 29:519, 1993.
13. Raoult D, Hechemy KE, and Baranton G: Cross-reaction with *Borrelia burgdorferi* antigen of sera from patients with human immunodeficiency virus infection, syphilis, and leptospirosis, *J Clin Microbiol* 27:2152, 1989.
14. Reed KD: Laboratory testing for Lyme disease: possibilities and practicalities, *J Clin Microbiol* 40:319, 2002.
15. Schmidt B, Muellegger RR, Stockenhuber C, et al: Detection of *Borrelia burgdorferi*-specific DNA in urine specimens from patients with erythema migrans before and after antibiotic therapy, *J Clin Microbiol* 34:1359, 1996.
16. Tramont EC: Syphilis in adults: from Christopher Columbus to Sir Alexander Fleming to AIDS, *Clin Infect Dis* 21:1361, 1995.
17. Wang G, van Dam A, Schwartz I, et al: Molecular typing of *Borrelia burgdorferi* sensu lato: taxonomic, epidemiological, and clinical manifestations, *Clin Microbiol Rev* 12:633, 1999.
18. Wilske B: Diagnosis of Lyme borreliosis in Europe, *Vector-Borne Zoonotic Dis* 3:215, 2003.

IV PART

Parasitology

Parasites to Be Considered

Protozoa

Amebae (Intestinal)
- *Entamoeba histolytica*
- *Entamoeba dispar**
- *Entamoeba hartmanni*
- *Entamoeba coli*
- *Entamoeba polecki*
- *Endolimax nana*
- *Iodamoeba bütschlii*
- *Blastocystis hominis*

Flagellates (Intestinal)
- *Giardia lamblia*[†]
- *Chilomastix mesnili*
- *Dientamoeba fragilis*
- *Pentatrichomonas hominis*
- *Enteromonas hominis*
- *Retortamonas intestinalis*

Ciliates (Intestinal)
- *Balantidium coli*

Coccidia, Microsporidia (Intestinal)
- *Cryptosporidium* spp.
- *Cyclospora cayetanensis*
- *Isospora belli*
- *Sarcocystis hominis*
- *Sarcocystis suihominis*
- *Sarcocystis* "lindemanni"

Microsporidia
- *Enterocytozoon bieneusi*
- *Encephalitozoon* spp.

Amebae, Flagellates (Other Body Sites)

Amebae
- *Naegleria fowleri*
- *Acanthamoeba* spp.
- *Hartmannella* spp.
- *Balamuthia mandrillaris (Leptomyxid ameba)*

Protozoa (cont'd)

- *Entamoeba gingivalis*

Flagellates
- *Trichomonas vaginalis*
- *Trichomonas tenax*

Coccidia and Microsporidia (Other Body Sites)

Coccidia
- *Toxoplasma gondii*

Microsporidia
- *Nosema*
- *Brachiola*
- *Vittaforma*
- *Pleistophora*
- *Trachipleistophora*
- *Encephalitozoon*
- *Enterocytozoon*
- "Microsporidium"*

Helminths

Nematodes (Roundworms)

Intestinal
- *Ascaris lumbricoides*
- *Enterobius vermicularis*
- *Ancylostoma duodenale*
- *Necator americanus*
- *Strongyloides stercoralis*
- *Trichostrongylus* spp.
- *Trichuris trichiura*
- *Capillaria philippinensis*

Tissue
- *Trichinella spiralis*
- Visceral larva migrans (*Toxocara canis* or *Toxocara cati*)
- Ocular larva migrans (*Toxocara canis* or *Toxocara cati*)
- Cutaneous larva migrans (*Ancylostoma braziliense* or *Ancylostoma caninum*)
- *Dracunculus medinensis*
- *Angiostrongylus cantonensis*
- *Angiostrongylus costaricensis*
- *Gnathostoma spinigerum*
- *Anisakis* spp. (larvae from saltwater fish)
- *Phocanema* spp. (larvae from saltwater fish)
- *Contracaecum* spp. (larvae from saltwater fish)

---

*Entamoeba histolytica is being used to designate true pathogens, whereas *E. dispar* is now being used to designate nonpathogens. Unless trophozoites containing ingested red blood cells (*E. histolytica*) are seen, the two organisms cannot be differentiated based on morphology seen in permanent stained smears of fecal specimens. Reagents are currently available for identifying the *E. histolytica/E. dispar* group and for differentiating *E. histolytica* from *E. dispar.*

†Although some individuals have changed the species designation for the genus *Giardia* to *G. intestinalis* or *G. duodenalis,* there is no general agreement. Therefore, for this listing, we retain the name *Giardia lamblia.*[29]

---

*This designation is not a true genus, but a "catch-all" for those organisms that have not been (or may never be) identified to the genus and/or species level.

    Capillaria hepatica

    Thelazia spp.

Cestodes (Tapeworms)

Intestinal

    Diphyllobothrium latum

    Dipylidium caninum

    Hymenolepis nana

    Hymenolepis diminuta

    Taenia solium

    Taenia saginata

Tissue (Larval Forms)

    Taenia solium

    Echinococcus granulosus

    Echinococcus multilocularis

    Multiceps multiceps

    Spirometra mansonoides

    Diphyllobothrium spp.

Trematodes (Flukes)

Intestinal

    Fasciolopsis buski

    Echinostoma ilocanum

    Heterophyes heterophyes

    Metagonimus yokogawai

Liver/Lung

    Clonorchis (Opisthorchis) sinensis

    Opisthorchis viverrini

    Fasciola hepatica

    Paragonimus westermani

    Paragonimus mexicanus

Protozoa

Protozoa

  Sporozoa, Flagellates (Blood, Tissue)

  Sporozoa (Malaria and Babesiosis)

    Plasmodium vivax

    Plasmodium ovale

    Plasmodium malariae

    Plasmodium falciparum

    Babesia spp.

  Flagellates (Leishmaniae, Trypanosomes)

    Leishmania tropica complex

    Leishmania mexicana complex

    Leishmania braziliensis complex

    Leishmania donovani complex

    Leishmania peruviana

    Trypanosoma brucei gambiense

    Trypanosoma brucei rhodesiense

    Trypanosoma cruzi

    Trypanosoma rangeli

Nematodes

Blood and Tissues (Filarial Worms)

    Wuchereria bancrofti

    Brugia malayi

    Brugia timori

    Loa loa

    Onchocerca volvulus

    Mansonella ozzardi

    Mansonella streptocerca

    Mansonella perstans

    Dirofilaria immitis (usually lung lesion; in dogs, heartworm)

    Dirofilaria spp. (may be found in subcutaneous nodules)

Trematodes

Blood

    Schistosoma mansoni

    Schistosoma haematobium

    Schistosoma japonicum

    Schistosoma intercalatum

    Schistosoma mekongi

GENERAL CHARACTERISTICS

The field of parasitology is often associated with tropical areas; however, many parasitic organisms that infect humans are worldwide in distribution and occur with some frequency in the temperate zones. In addition to the normal population, an increase in the number of compromised patients, particularly those who are immunodeficient or immunosuppressed, has led to increased interest in the field of parasitology. These persons are greatly at risk for certain parasitic infections. Parasites of humans are classified in six major divisions. These include the Protozoa (amebae, flagellates, ciliates, sporozoans, coccidia, microsporidia); the Nematoda, or roundworms; the Platyhelminthes, or flatworms (cestodes, trematodes); the Pentastomids, or tongue worms; the Acanthocephala, or thorny-headed worms; and the Arthropoda (e.g., insects, spiders, mites, ticks). The identification of parasitic organisms depends on morphologic criteria; these criteria, in turn, depend on correct specimen collection and adequate fixation. Improperly submitted specimens may result in failure to find the organisms or in their misidentification. Tables 49-1 through 49-3 contain information on the various groups of parasites, the possible parasites recovered from different body sites, the most frequently used specimen collection approaches, and appropriate processing methods.

EPIDEMIOLOGY

Parasites are usually restricted to specialized environments inside and outside their hosts. A **zoonosis** is a disease of wild or domestic animals that can occur in humans as a result of parasitic infection. Animals that are potential sources of infection for humans are called

Table 49-1 Description of the More Common Groups of Parasites That Infect Humans

Parasite Group	Description
Protozoa, Intestinal	
Amebae	These single-celled organisms are characterized by having pseudopods (motility) and trophozoite and cyst stages in the life cycle. However, there are some exceptions in which a cyst form has not been identified. Amebae are usually acquired by humans via fecal-oral transmission of the infective cyst. *Entamoeba histolytica* causes amebiasis and is the most significant organism within this group
Flagellates	These protozoa move by means of flagella and are acquired through fecal-oral transmission. Except for *Dientamoeba fragilis* (internal flagella) and those in the genus *Pentatrichomonas,* they have both the trophozoite and cyst stages in the life cycle. Reproduction is by longitudinal binary fission; *Giardia lamblia* and *D. fragilis* are pathogens within this group
Ciliates	These single-celled protozoa move by means of cilia and are acquired through fecal-oral transmission, *Balantidium coli* being the only human pathogen in the group. The cilia beat in a coordinated rhythmic pattern and the trophozoite moves in a spiral path. They have both the trophozoite and cyst stages in the life cycle, and both contain a large macronucleus and a smaller micronucleus. These organisms are considerably larger than most intestinal protozoa and can be mistaken for debris or junk when seen in a permanent stained smear
Coccidia	These protozoa are acquired by ingestion of various meats or through fecal-oral transmission via contaminated food and/or water. In some cases, these organisms may disseminate to other body sites, particularly in the severely compromised patient. These protozoa have both asexual and sexual cycles, the most common infective stage being the oocyst, containing sporocysts and/or sporozoites, all of which can be acquired through fecal-oral transmission. Representatives within this group include *Cryptosporidium* spp., *Cyclospora cayetanensis, Isospora belli,* and *Sarcocystis* spp.
Microsporidia	The most difficult intestinal protozoa to diagnose are the microsporidia (size range 1 to 2.5 μm); these organisms disseminate to other body sites, including the kidneys and lungs. Routine parasitology stains are not useful; modified trichrome stains have been developed specifically for these organisms. The infective form is called the *spore;* each spore contains a polar tubule that is used to penetrate new host cells, thus initiating or continuing the life cycle. Infections are acquired through ingestion, inhalation, or direct inoculation of spores from the environment. Currently there are nine genera that cause disease in humans; the two most important are *Encephalitozoon* and *Enterocytozoon*
Protozoa, Other Sites	
Amebae	Except for *Entamoeba gingivalis* (found in the mouth), these amebae are pathogenic, free-living organisms that may be associated with warm, freshwater environments. They have been found in the central nervous system (CNS), the eye, and other body sites. Those amebae that invade the CNS *(Naegleria fowleri)* can cause severe, life-threatening infection that often ends in death within a few days. Other amebae in this group can cause more chronic CNS disease (*Acanthamoeba* spp., *Balamuthia mandrillaris,* particularly in the immunocompromised patient); they can also cause keratitis
Flagellates	*Trichomonas vaginalis* is found in the genitourinary system and is usually acquired by sexual transmission. *Trichomonas tenax* can be found in the mouth and is considered to be nonpathogenic
Coccidia	Coccidian parasites are particularly important in the compromised patient and can cause life-threatening disease. These organisms can disseminate from the intestinal tract to other body sites, including the lung. In the compromised patient, sequelae may be serious and life-threatening (*Cryptosporidium* spp.)
Microsporidia	These organisms are the most difficult protozoa to diagnose (size range 1 to 2.5 μm). Dissemination from the intestine to other body sites has been well documented. Modified trichrome stains have been developed specifically for these organisms, since routine parasitology stains for fecal specimens are not that effective for the microsporidial spores. Optical brightening agents are also recommended; although they are very sensitive, they are nonspecific. Specific genera include *Encephalitozoon, Pleistophora, Trachipleistophora,* and *Brachiola*
Protozoa, Blood, Tissue	
Malaria, Babesiosis	All of these organisms are arthropod-borne. The genus *Plasmodium* includes parasites that undergo exoerythrocytic and pigment-producing erythrocytic schizogony in vertebrates and a sexual stage followed by sporogony in mosquitoes. *Babesia* spp. is tickborne and can cause severe disease in patients who have been splenectomized or are otherwise immunologically compromised. Diagnosis may be somewhat more difficult than that of the intestinal protozoa, particularly if automated blood differential systems are used; the microscopic examination of both thick and thin blood films is recommended
Flagellates (Leishmaniae)	The leishmaniae have undergone extensive classification revisions. However, from a clinical perspective, recovery and identification of the organisms are still related to body site. Recovery of leishmanial amastigotes is limited to the site of the lesion in infections other than those caused by the *Leishmania donovani* complex (visceral leishmaniasis). These protozoa have both amastigote (mammalian host) and promastigote (sand fly) stages in the life cycle. Reproduction in both forms occurs by binary longitudinal division

Table 49-1   Description of the More Common Groups of Parasites That Infect Humans—cont'd

Parasite Group	Description
Flagellates (Trypanosomes)	The trypanosomes are normally identified to the species level based on geographic exposure history and clinical symptoms. These protozoa have at some time in the life cycle the trypomastigote form with the typical undulating membrane and free flagellum at the anterior end. The longer the duration of the infection, the more difficult it may be to confirm the diagnosis. Those organisms that cause African sleeping sickness *(T. b. gambiense, T.b. rhodesiense)* generally cause different disease entities, one of which tends to be chronic and more typically the patient appears to have sleeping sickness *(T.b. gambiense)*, and the other causing a more fulminant disease often leading to death before typical sleeping sickness symptoms *(T.b. rhodesiense)*. The etiologic agent of American trypanosomiasis is *T. cruzi*, which contains amastigote and trypomastigote stages (mammalian host) and the epimastigote form in the arthropod host
Nematodes, Intestinal	The largest number of helminthic parasites of humans belongs to the roundworm group. Nematodes are elongate-cylindric and bilaterally symmetric, with a tri-radiate symmetry at the anterior end. Nematodes have an outer cuticle layer, no circular muscles, and a pseudocele containing all systems (digestive, excretory, nervous, reproductive). These organisms are normally acquired by egg ingestion or skin penetration of larval forms from the soil. Representative roundworms include *Ascaris, Enterobius, Trichuris, Strongyloides,* and hookworm
Nematodes, Tissue	Many of these organisms are less commonly seen within the United States; however, some are more important and are found worldwide. Diagnosis may be difficult if the only specimens are obtained through biopsy or autopsy, and interpretation must be based on examination of histologic preparations. Examples include *Trichinella,* visceral larva migrans, ocular larva migrans, cutaneous larva migrans, all of which are associated with eosinophilia
Nematodes, Filarial	These nematodes are arthropod-borne. The adult worms tend to live in the tissues or lymphatics of the vertebrate host. Diagnosis is based on the recovery and identification of the larval worms (microfilariae) in the blood, other body fluids, or skin. While circulating in peripheral blood or cutaneous tissues, the microfilariae can be ingested by blood-sucking insects. After the larvae mature, they can escape into the vertebrate host's skin when the arthropod takes its next blood meal. Severity of disease with these nematodes varies; however, elephantiasis may be associated with some of the filarial worms. Examples include *Wuchereria, Brugia, Loa,* and *Onchocerca*
Cestodes, Intestinal	The adult form of the tapeworm is acquired through ingestion of the larval forms contained in poorly cooked or raw meats or freshwater fish. In the case of *Dipylidium caninum,* infection is acquired by the accidental ingestion of dog fleas. Both *Hymenolepis nana* and *H. diminuta* are transmitted via ingestion of certain arthropods (fleas, beetles). Also, *H. nana* can be transmitted through egg ingestion (life cycle can bypass the intermediate beetle host). The adult tapeworm consists of a chain of egg-producing units called *proglottids,* which develop from the neck region of the attachment organ, the scolex. Food is absorbed through the worm's integument. The intermediate host contains the larval forms that are acquired through ingestion of the adult tapeworm eggs. Humans can serve as both the intermediate and definitive hosts in *H. nana* and *Taenia solium* infections
Cestodes, Tissue	The ingestion of certain tapeworm eggs or accidental contact with certain larval forms can lead to tissue infection with *Taenia solium, Echinococcus granulosus,* and several others. The human serves as the accidental intermediate host
Trematodes, Intestinal	Trematodes are flatworms and are exclusively parasitic. Except for the schistosomes (blood flukes), flukes are hermaphroditic. They may be flattened; most have oral and ventral suckers. All of the intestinal trematodes require a freshwater snail to serve as an intermediate host; these infections are foodborne (freshwater fish, mollusks, or plants) and are emerging as a major public health problem (>40 million people infected with intestinal and liver/lung trematodes). An example is *Fasciolopsis buski,* the giant intestinal fluke
Trematodes, Liver, Lung	These trematodes also require a freshwater snail to serve as an intermediate host; these infections are foodborne (freshwater fish, crayfish or crabs, or plants). Public health concerns include cholangiocarcinoma associated with *Clonorchis* and *Opisthorchis* infections, severe liver disease associated with *Fasciola* infections, and the misdiagnosis of tuberculosis in those infected with *Paragonimus*
Trematodes, Blood	The sexes of these trematodes (schistosomes) are separate, and infection is acquired by skin penetration by the cercarial forms that are released from freshwater snails. The males are characterized by having an infolded body that forms the gynecophoral canal in which the female worm is held during copulation and oviposition. The adult worms reside in the blood vessels over the small intestine, large intestine, or bladder. Although these parasites are not endemic within the United States, patients are seen who may have acquired schistosomiasis elsewhere. Examples are *Schistosoma mansoni, S. haematobium,* and *S. japonicum*

reservoir hosts. The host specificity of any particular parasite influences factors associated with both transmission and control. Where humans are the only host for a parasite or any stage of its development, control options are relatively easy to define. However, if an infection is a zoonosis, then control measures can become complex because of the existence of one or more reservoir animals. Some organisms are free-living during certain portions of their life cycles and do not depend on the human host for their survival. In some cases, the human

Table 49-2   Body Sites and Possible Parasites Recovered (Trophozoites, Cysts, Oocysts, Spores, Adults, Larvae, Eggs, Amastigotes, Trypomastigotes)*

Site	Parasites	Site	Parasites
Blood		Intestinal (cont'd)	*Strongyloides stercoralis*
Red cells	*Plasmodium* spp.		*Trichuris trichiura*
	Babesia spp.		*Hymenolepis nana*
White cells	*Leishmania donovani*		*Hymenolepis diminuta*
	Toxoplasma gondii		*Taenia saginata*
Whole blood/plasma	*Trypanosoma* spp.		*Taenia solium*
	Microfilariae		*Diphyllobothrium latum*
Bone marrow	*Leishmania donovani*		*Clonorchis sinensis*
			(*Opisthorchis*)
Central Nervous System			*Paragonimus* spp.
Cutaneous Ulcers	*Leishmania* spp.		*Schistosoma* spp.
	Acanthamoeba spp.		*Fasciolopsis buski*
	Taenia solium (cysticerci)		*Fasciola hepatica*
	Echinococcus spp.		*Metagonimus yokogawai*
	Naegleria fowleri		*Heterophyes heterophyes*
	Acanthamoeba spp.	Liver, Spleen	*Echinococcus* spp.
	Balamuthia mandrillaris		*Entamoeba histolytica*
	Toxoplasma gondii		*Leishmania donovani*
	Microsporidia		Microsporidia
	Trypanosoma spp.	Lung	*Cryptosporidium* spp.[†]
Intestinal Tract	*Entamoeba histolytica*		*Echinococcus* spp.
	Entamoeba dispar		*Paragonimus* spp.
	Entamoeba coli		Microsporidia
	Entamoeba hartmanni	Muscle	*Taenia solium* (cysticerci)
	Endolimax nana		*Trichinella spiralis*
	Iodamoeba bütschlii		*Onchocerca volvulus* (nodules)
	Blastocystis hominis		*Trypanosoma cruzi*
	Giardia lamblia		Microsporidia
	Chilomastix mesnili	Skin	*Leishmania* spp.
	Dientamoeba fragilis		*Onchocerca volvulus*
	Pentatrichomonas hominis		Microfilariae
	Balantidium coli	Urogenital system	*Trichomonas vaginalis*
	Cryptosporidium spp.		*Schistosoma* spp.
	Isospora belli		Microsporidia
	Microsporidia		Microfilariae
	Ascaris lumbricoides	Eye	*Acanthamoeba* spp.
	Enterobius vermicularis		*Toxoplasma gondii*
	Hookworm		*Loa loa*
			Microsporidia

*This table does not include every possible parasite that could be found in a particular body site. However, the most likely organisms have been listed.
[†]Disseminated in severely immunosuppressed individuals.

becomes an accidental host or even an accidental intermediate host. Parasites are transmitted from host to host through sexual means (venereal transmission) *(Trichomonas vaginalis),* from ingestion of infective forms in food or water *(Giardia lamblia, Cryptosporidium* spp., *Ascaris lumbricoides),* through skin penetration of infective larvae *(Strongyloides stercoralis,* hookworm), or through the bites of various arthropods *(Plasmodium, Trypanosoma, Leishmania).* Specific examples are given in Table 49-4.

PATHOGENESIS AND SPECTRUM OF DISEASE

Although a number of parasites can cause serious and life-threatening disease, particularly in the compromised patient, many organisms reach a certain "status quo" with the host and produce no detectable pathogenic effect. Disease may not be the ultimate outcome of infection. Depending on the parasite, one, some, or multiple body sites may be infected with no or few symptoms or, at the other extreme, death. Also, some

Table 49-3   Specimens and/or Body Site, Specimen Options, Collection and Transport Methods and Specimen Processing

Specimens and/or Body Site	Specimen Options	Collection and Transport Methods	Specimen Processing	Comments
Stool for O&P examination	Fresh stool Preserved stool*	½-pint waxed container; 30 min if liquid, 60 min if semi-formed, 24 hrs if formed delivery to laboratory 5% or 10% formalin, MIF, SAF, Schaudinn's, PVA, modified PVA, single-vial systems	Direct wet smear (not on formed specimen), concentration, permanent stained smear Concentration, permanent stained smear Depending on specimen (fresh or preserved) and patient's clinical history, immunoassays may also be performed.	Stool specimens containing barium are unacceptable; intestinal protozoa may be undetectable for 5 to 10 days after barium is given to the patient. Certain substances and medications will also interfere with the detection of intestinal protozoa: mineral oil, bismuth, antibiotics, antimalarial agents, and nonabsorbable antidiarrheal preparations. After administration of any of these compounds, parasitic organisms may not be recovered for a week to several weeks. Specimen collection should be delayed after barium or antibiotics are administered for 5 to 10 days or at least 2 weeks, respectively.
Stool for culture of nematodes	Fresh stool, entire stool specimen	½-pint waxed container; immediate delivery to laboratory	Filter paper strip, Petri dish, agar plate, charcoal cultures all available	Fresh stool (do not refrigerate) is required for these procedures.
Stool for recovery of tapeworm scolex	Preserved stool, entire stool specimen Adult worms, worm segments	5% or 10% formalin (10% recommended) Saline, 70% alcohol	Using a series of mesh screens, the stool is filtered, looking for the very small tapeworm scolex (proof of therapy efficacy) and/or proglottids (uncommon procedure, but an option).	After treatment for tapeworm removal, the patient should be instructed to take a saline cathartic and to collect all stool material passed for the next 24 hrs. The stool should be immediately placed in 10% formalin and thoroughly broken up and mixed with the preservative (1-gallon [3.8-liter] plastic jars are recommended, half full of 10% formalin).
Cellophane tape preparation for pinworm	Surface sample from perianal skin; anal impression smear	Cellophane tape preparation or commercial sampling paddle or swab	Tape is lifted from slide, a drop of xylene-substitute is added, the tape is replaced, and the specimen is ready for examination under the microscope.	Specimens should be collected late at night after the person has been asleep for several hours or first thing in the morning before going to the bathroom or taking a shower. A minimum of 4 to 6 consecutive negative tapes is required to rule out the infection.
Sigmoid colon	Sigmoidoscopy material, prepared as smears	Fresh, PVA, or Schaudinn's smears; specimen taken with spatula rather than cotton-tipped swabs; transported as smears in preservative	Direct wet smear, permanent stained smears.	Material from the mucosal surface should be aspirated or scraped and should not be obtained with cotton-tipped swabs. At least six representative areas of the mucosa should be sampled and examined (six samples, six slides). It is recommended that a parasitology specimen tray (containing Schaudinn's fixative, PVA, and 5% or 10% formalin) be provided or a trained technologist be available at the time of sigmoidoscopy to prepare the

Table 49-3   Specimens and/or Body Site, Specimen Options, Collection and Transport Methods and Specimen Processing—cont'd

Specimens and/or Body Site	Specimen Options	Collection and Transport Methods	Specimen Processing	Comments
Sigmoid colon (cont'd)				slides. The examination of sigmoidoscopy specimens does not take the place of routine O&P examinations. If the amount of material is limited, the use of PVA fixative is highly recommended.
Duodenum	Duodenal contents	Entero-Test or aspirates; string in Petri dish or tube; immediate transport to laboratory	The specimen may be centrifuged (10 min at 500 × g) and should be examined immediately as a wet mount for motile organisms. Iodine can be used as well. Direct wet smear of mucus; permanent stained smears can also be prepared.	A fresh specimen is required; the amount may vary from <0.5 mL to several milliliters of fluid. If the specimen cannot be completely examined within 2 hrs after it is taken, any remaining material should be preserved in 5% to 10% formalin.
Entero-Test capsule (string)	Duodenal contents	Entero-Test (string test) in Petri dish (fresh) or preserved in PVA	Bilestained mucus clinging to the yarn should be scraped off (mucus can also be removed by pulling the yarn between thumb and finger) and collected in a small Petri dish; disposable gloves are recommended. Usually 4 or 5 drops of material are obtained. The specimen should be examined immediately as a wet mount for motile organisms (iodine may be added later to facilitate identification of any organisms present). Organism motility will be like that described above for duodenal drainage. The pH of the terminal end of the yarn should be checked to ensure adequate passage into the duodenum (a very low pH means that it never left the stomach). The terminal end of the yarn should be yellowgreen, indicating it was in the duodenum (the bile duct drains into the intestine at this point). Permanent stained smears can also be prepared.	If the specimen cannot be completely examined within an hour after the yarn has been removed, the material should be preserved in 5% to 10% formalin or PVA mucus smears should be prepared.
Urogenital tract	Vaginal discharge Urethral discharge Prostatic secretions Urine	Saline swab, transport swab (no charcoal), culture medium, plastic envelope culture Air-dried smear for FA Single unpreserved specimen, 24-hr unpreserved specimen, early morning	Direct wet smear; fluorescence; urine must be centrifuged before examination. Examination of urinary sediment may be indicated in certain filarial infections. Administration of the drug diethylcarbamazine (Hetrazan) has been reported to enhance the recovery of microfilariae from the urine. The triple-concentration technique is recommended for the recovery of	Fresh specimens are required; air-dried smear may be an option for fluorescence. Do not refrigerate swabs and/or culture containers at any time; motility and/or ability to grow will probably be lost.

Table 49-3   Specimens and/or Body Site, Specimen Options, Collection and Transport Methods and Specimen Processing—cont'd

Specimens and/or Body Site	Specimen Options	Collection and Transport Methods	Specimen Processing	Comments
Urogenital tract (cont'd)			microfilariae. The membrane filtration technique can also be used with urine for the recovery of microfilariae. A membrane filter technique for the recovery of *Schistosoma haematobium* eggs has also been useful.	
Sputum	Sputum Induced sputum Bronchoalveolar lavage (BAL)	True sputum (not saliva) No preservative (10% formalin, if time delay) Sterile, immediate delivery to laboratory	Direct wet smear; permanent stained smears; fluorescence also available. Sputum is usually examined as a wet mount (saline or iodine), using low and high dry power (×100 and ×400). The specimen is not concentrated before preparation of the wet mount. If the sputum is thick, an equal amount of 3% sodium hydroxide (NaOH) (or undiluted chlorine bleach) can be added; the specimen is thoroughly mixed and then centrifuged. NaOH should not be used if one is looking for *Entamoeba* spp. or *T. tenax*. After centrifugation, the supernatant fluid is discarded, and the sediment can be examined as a wet mount with saline or iodine. If examination has to be delayed for any reason, the sputum should be fixed in 5% or 10% formalin to preserve helminth eggs or larvae or in PVA fixative to be stained later for protozoa.	True sputum is required; all specimens, especially induced specimens and BALs, should be delivered immediately to the laboratory (do not refrigerate).

Aspirates	Bone marrow
	
	Cutaneous ulcers
	Liver, spleen
	
	Lung
	  Transbronchial aspirate
	  Tracheobronchial aspirate
		Sterile, immediate delivery to laboratory
		Sterile plus air-dried smears
		Sterile, collected in four separate aliquots (liver)
		
		Air-dried smears
		
		Air-dried smears
			Permanent stained smears; cultures can also be set (specifically designed for the recovery of blood parasites)	All aspirates for culture must be collected using sterile conditions and containers; this is mandatory for the culture isolation of leishmaniae and trypanosomes.

Central nervous system	CSF	Sterile	Direct wet smear, permanent stained smears; culture for free-living amebae *(Naegleria, Acanthamoeba)*	All specimens must be transported immediately to the laboratory (stat procedure).
Biopsy	Intestinal tract Cutaneous ulcers Eye Scrapings Liver, spleen Lung	Routine histology Sterile, nonsterile to histopathology (formalin acceptable) Sterile (in saline), nonsterile to histopathology Sterile (in saline) Sterile, nonsterile to histopathology	Direct wet smears, permanent stained smears; specimens to histology for routine processing.	The more material that is collected and tested, the more likely the organism will be isolated and subsequently identified. Sterile collection required for all specimens that will be cultured; bacterial and/or fungal contamination will prevent isolation of parasites in culture.

Table 49-3   Specimens and/or Body Site, Specimen Options, Collection and Transport Methods and Specimen Processing—cont'd

Specimens and/or Body Site	Specimen Options	Collection and Transport Methods	Specimen Processing	Comments
Biopsy (cont'd)	Brush biopsy Open lung biopsy Muscle Skin biopsy Scrapings Skin snip	Air-dried smears Air-dried smears Fresh, squash preparation, nonsterile to histopathology Nonsterile to histopathology (formalin acceptable) Sterile (in saline), nonsterile to histopathology Aseptic, smear or vial No preservative		
Blood	Smears of whole blood Anticoagulated blood	Thick and thin films, immediate delivery to laboratory Fresh (first choice) Anticoagulant (second choice) EDTA (first choice) Heparin (second choice)	Thick and thin films, specialized concentrations and/or screening methods Thick and thin films, specialized concentrations and/or rapid methods QBC Microhematocrit Centrifugation Method (Becton Dickinson, Tropical Disease Diagnostics, Sparks, Md.)	Examination of blood films (particularly for malaria) is considered a stat procedure; immediate delivery to the laboratory is mandatory. Delivery to the laboratory within 30 min or less; if these time frames are not met, typical parasite morphology may not be seen in blood collected using anticoagulants. Knott Concentration: The Knott concentration procedure is used primarily to detect the presence of microfilariae in the blood, especially when a light infection is suspected. The disadvantage of the procedure is that the microfilariae are killed by the formalin and are therefore not seen as motile organisms. Membrane Filtration Technique: The membrane filtration technique using Nuclepore filters has proved highly efficient in demonstrating filarial infections when microfilaremias are of low density. It has also been successfully used in field surveys.

Modified from Garcia LS: *Diagnostic medical parasitology*, ed 5, Washington, DC, 2007, ASM Press.

CSF, Cerebrospinal fluid; *EDTA*, ethylenediaminetetraacetic acid; *FA*, fluorescent antibody; *MIF*, merthiolate-iodine-formalin; *PVA*, polyvinyl alcohol; *SAF*, sodium acetate-acetic acid-formalin.

*A number of new stool fixatives are now available; some use a zinc sulfate base rather than mercuric chloride. Some collection vials can be used as a single-vial system; both the concentration and permanent stained smear can be performed from the preserved stool. However, not all single-vial systems (proprietary formulas) provide material that can be used for fecal immunoassay procedures.

parasites will multiply within the human body, whereas others will mature, but not increase in numbers. These life cycle differences play important roles in pathogenicity and disease outcome. It is also important to remember that any patient who is debilitated or in any way immunocompromised (including the very young and the very old), may react very differently to the presence of a parasitic infection than an immunocompetent person. Certainly, it is not to the parasite's advantage to damage the host to the extent that severe illness or even death occurs; this makes survival of the parasite difficult, at best, and transmission from host to host becomes a critical issue. It is important to understand the life cycle, both in terms of potential control and prevention, and regarding infectivity and disease outcomes in both the normal and immunosuppressed host (Table 49-5). Table 49-6 lists mechanisms of pathogenesis and the spectrum of parasitic diseases.

Table 49-4 Epidemiology of the More Common Groups of Parasites That Infect Humans

Parasite Group	Habitat (reservoir)	Mode of Transmission	Prevention
Protozoa, Intestinal			
Amebae	These single-celled organisms are generally found in humans. Although certain animals harbor some of these organisms, they are not considered important reservoir hosts	Humans acquire these infections by the ingestion of food and water contaminated with fecal material containing the resistant, infective cyst stage of the protozoa. Various sexual practices have also been documented in transmission	Preventive measures involve increased attention to personal hygiene and sanitation measures; elimination of sexual activities where the potential for fecal-oral contact is possible
Flagellates	The flagellates are generally found in humans. Although certain animals harbor some of these organisms, they are not considered important reservoir hosts. One exception may be animals such as the beaver that harbor *Giardia lamblia* and contaminate water supplies	Humans acquire these infections by the ingestion of food and water contaminated with fecal material containing the resistant, infective cyst stage of the protozoa; in some cases (*Dientamoeba fragilis*), no cyst stage has been identified; the trophozoite forms may be transmitted from person to person within certain helminth eggs	Preventive measures involve increased attention to personal hygiene and sanitation measures; elimination of sexual activities where the potential for fecal-oral contact is possible. Adequate water treatment (including filtration) is required; awareness of environmental sources of infection
Ciliates	*Balantidium coli* is generally found in humans but is also found in pigs. In some areas of the world, pigs are considered important reservoir hosts	Humans acquire these infections by the ingestion of food and water contaminated with fecal material containing the resistant, infective cyst stage of the protozoa	Preventive measures involve increased attention to personal hygiene and sanitation measures; elimination of sexual activities where the potential for fecal-oral contact is possible
Coccidia	The coccidia are found in humans; in some cases such as cryptosporidiosis, animal reservoirs (cattle) can serve as important hosts as well. The muscle of various animals may contain sarcocysts that are infective for humans on consumption of raw or poorly cooked meats. Numerous waterborne outbreaks with *Cryptosporidium* have been reported throughout the world. Coccidian oocysts are extremely resistant to environmental conditions, particularly if they are kept moist	These protozoa are acquired by ingestion of various meats or through fecal-oral transmission via contaminated food and/or water. The infective forms are called *oocysts* (*Cryptosporidium* spp., *Cyclospora cayetanensis*), or sarcocysts (*Sarcocystis* spp.) contained in infected meats. *Cryptosporidium* has also been implicated in nosocomial infections	Preventive measures involve increased attention to personal hygiene and sanitation measures; elimination of sexual activities where the potential for fecal-oral contact is possible. Adequate water treatment (including filtration) is mandatory; awareness of environmental sources of infection
Microsporidia	Microsporidia have the potential to infect every living animal, some of which probably serve as reservoir hosts for human infection. However, host specificity issues have not been well defined to date. The spores are environmentally resistant and can survive years if kept moist	Infection with microsporidial spores usually occurs through ingestion; however, inhalation of spores, as well as direct inoculation from the environment, almost certainly occurs as well	Preventive measures involve increased attention to personal hygiene and sanitation measures; increased awareness of environmental exposure possibilities and adequate water treatment
Protozoa, Other Sites			
Amebae	Free-living amebae are associated with warm, freshwater environments; they are also found in soil. Although humans can harbor these organisms, person-to-person transfer is thought to be rare. Environmental sources are the primary link to human infections. Contaminated eye care solutions have been linked to those organisms that cause keratitis	Infection occurs through contact with contaminated water; organisms enter through the nasal mucosa and may travel via the olfactory nerve to the brain. Disease can be severe and life-threatening; keratitis is also caused by these organisms and infection can be linked to blindness or severe corneal damage. Eye infections can be linked to contaminated lens solutions or direct, accidental inoculation of the eye from environmental water and/or soil sources	Avoidance of contaminated environmental water and soil sources; adequate care of contact lens systems

Table 49-4   Epidemiology of the More Common Groups of Parasites that Infect Humans—cont'd

Parasite Group	Habitat (reservoir)	Mode of Transmission	Prevention
Flagellates	*Trichomas vaginalis* infection is found in a large percentage of humans, many of whom will be asymptomatic. Human-to-human transfer is common; reinfection is also common, particularly if sexual partners are not treated	*Trichomonas vaginalis* is found in the genitourinary system and is usually acquired by sexual transmission	Awareness of sexual transmission; treatment of all partners when infection is diagnosed in an individual patient
Protozoa, Blood, Tissue			
Malaria, babesiosis	Humans harbor the four most common species of malaria: *Plasmodium vivax, P. ovale, P. malariae,* and *P. falciparum.* Other animals can carry *Babesia* spp., and animal reservoir hosts play a large role in human transmission	These organisms are arthropod-borne, malaria by the female anopheline mosquito and *Babesia* by one or more genera of ticks. These infections can also be transmitted transplacentally, via shared needles, through blood transfusions, and from organ transplants	Vector control, awareness of transmission through blood transfusions, shared drug needles, congenital infections, organ transplants. Careful monitoring of blood supply. Malaria prophylaxis if traveling to endemic areas
Flagellates (Leishmaniae)	Some strains of leishmaniae have reservoir hosts, including dogs for the Mediterranean strain of *Leishmania donovani* and wild rodents for the African strains of *L. donovani. L. tropica* has also been linked to the same two types of animal reservoirs	Transmission is via the bite of infected sandflies. Infection can also occur from person to person (cutaneous lesions), from blood transfusion, shared needles, and organ transplants	Vector control, avoiding environmental sources (e.g., dogs, wild rodents); careful handling of all clinical specimens from infected patients
Flagellates (Trypanosomes)	Humans are the only known hosts for *Trypanosoma brucei gambiense* (West African trypanosomiasis), whereas *Trypanosoma brucei rhodesiense* (East African trypanosomiasis) infections are found in a number of antelope and other ungulates serving as reservoir hosts. Chickens are reservoir hosts for *Trypanosoma cruzi,* as are certain rodents	Transmission is through the bite of the infected tsetse fly, through blood transfusion, shared needles, and organ transplants. Transmission of *T. cruzi* is through the infected feces of the triatomid bug; the bug takes a blood meal, then defecates, and the human host scratches the infected feces into the bite site; bug saliva contains irritant	Vector control, awareness of potential exposure/infection from blood sources (transfusions, shared needles, organ transplants). Laboratory accidents have been reported when handling infected blood
Nematodes, Intestinal	These roundworms generally do not have animal reservoirs that are relevant to human infection. One minor exception is the pig ascarid; a few human infections have been reported. These worms are found worldwide and *Ascaris lumbricoides* is probably the most common parasite of humans, although some would argue that *Enterobius vermicularis* is number one. *Strongyloides stercoralis* is particularly important as causing severe disease in the compromised host	*Ascaris lumbricoides* and *Trichuris trichiura* eggs must undergo development in the soil before they are infective; thus children who play in the dirt are a particularly high-risk group. Ingestion of food and water contaminated with infective eggs is the primary route of infection. Hookworm and *Strongyloides stercoralis* infections are initiated by larval penetration of the skin from contaminated soil. Pinworm infection *(Enterobius vermicularis)* is acquired through the ingestion of infective eggs from the environment (hands to mouth situation)	Avoid ingestion of contaminated soil and/or frequenting soil contaminated with hookworm eggs (pets, soil, water, warmth—warm weather); treatment for pinworm is recommended; however, reinfection is common

Table 49-4 Epidemiology of the More Common Groups of Parasites that Infect Humans—cont'd

Parasite Group	Habitat (reservoir)	Mode of Transmission	Prevention
Nematodes, Tissue	*Trichinella* spp. has a number of animal reservoir hosts, including bear, walrus, pigs, rodents, and other animals. Dog and cat hookworms cause cutaneous larva migrans (CLM), and the dog and cat ascarid, *Toxocara* spp., causes visceral and ocular larva migrans (VLM, OLM). These infections can be serious and cause severe disease if not treated	*Trichinella* is acquired through the ingestion of raw or poorly cooked infected meats. CLM is caused by skin penetration of infective larvae from the soil; a combination to avoid is children, sandboxes, and dogs and cats that defecate in the sand. Larval migration is limited to the skin. VLM and OLM are caused by the accidental ingestion of *Toxocara* spp. eggs from contaminated soil; larval migration is throughout the body, including the eyes	Adequate cooking of infected meats; awareness of possibility of contaminated soils for dog and cat hookworms and/or ascarids; covering of all sand boxes where pets have access to defecation and children play
Nematodes, Filarial	*Wuchereria bancrofti, Loa loa,* and *Onchocerca volvulus* have no animal reservoirs and are found only in humans, whereas *Brugia* can also be found in cats and monkeys. *Dracunculus medinensis* can infect dogs, cats, and monkeys, as well as humans	Filarial nematodes are transmitted through the bite of a blood-sucking arthropod (midges, mosquitoes, flies). *Dracunculus* infections are acquired through the ingestion of water contaminated with small crustaceans, *Cyclops,* which contain infective larvae. This infection has been targeted for world eradication	Vector control; protection of well water sources
Cestodes, Intestinal	The human serves as the definitive host for the beef (*Taenia saginata*) and pork (*Taenia solium*) tapeworms, whereas cows/camels and pigs serve as intermediate hosts, respectively. The human also serves as the intermediate host for *T. solium* (cysticercosis). *Diphyllobothrium latum* adult tapeworms can also be found in a number of wild animals, the most important being dogs, bears, seals, and walrus, that serve as reservoir hosts; the human serves as the definitive host. *Hymenolepis nana* (dwarf tapeworm) can also occur in rodents; the human can serve as both the intermediate and definitive host, with development from the egg to adult worm occurring in the human intestine	Human infection with the adult worm occurs through the ingestion of raw or poorly cooked meat (beef, camel, pork) containing the intermediate forms, the cysticerci. The human becomes the accidental intermediate host when eggs from an adult *T. solium* tapeworm are ingested. The cysticerci will then develop in the muscle and tissues of the human, rather than the pig. Infection with adult *D. latum* tapeworm occurs via ingestion of poorly cooked freshwater fish containing the sparganum or plerocercoid larval form. Infection with *Hymenolepis nana* is primarily acquired from accidental ingestion of eggs from an adult tapeworm	Adequate cooking of infected meats; treatment of patients harboring adult tapeworms (accidental ingestion of eggs can lead to infection)
Cestodes, Tissue	The adult worms are found in a variety of animals; the human becomes the accidental intermediate host after ingestion of eggs from the adult worms. Reservoir hosts include dogs, cats, and rodents	The ingestion of certain tapeworm eggs or accidental contact with certain larval forms can lead to tissue infection with *Taenia solium, Echinococcus* spp., and several others	Preventive measures involve increased attention to personal hygiene and sanitation measures
Trematodes, Intestinal	Fish-eating wild and domestic animals serve as reservoir hosts. The definitive host of *Fasciolopsis buski* is the pig	Ingestion of water chestnut and caltrop (raw, peeled with the teeth) is the source of infection; metacercariae are encysted on the plant material. Pig feces are used to fertilize the water plant crops	Refrain from eating raw water plants that may contain encysted larval forms of the flukes; adequate waste disposal of farm animal feces (pigs)
Trematodes, Liver, Lung	Cats, dogs, and wild fish-eating mammals can serve as reservoir hosts for *Opisthorchis* spp., *Clonorchis sinensis,* and *Paragonimus* spp. *Fasciola hepatica* is normally a parasite of sheep and *F. gigantica,* is a parasite of cattle, whereas humans are accidental hosts	Infection occurs through the ingestion of raw or poorly cooked fish, crabs, crayfish, and certain plants in or on which metacercariae are encysted. Infection with *Fasciola* spp. is not that easily acquired (parasite not that well adapted to the human host)	Thorough cooking of potentially infected fish, crabs, crayfish; refrain from eating raw water plants that may contain encysted metacercariae

Table 49-4   Epidemiology of the More Common Groups of Parasites that Infect Humans—cont'd

Parasite Group	Habitat (reservoir)	Mode of Transmission	Prevention
Trematodes, Blood	*Schistosoma mansoni* and *S. haematobium* appear to be restricted to the human host, whereas *S. japonicum* can be found in cattle, deer, dogs, and rodents. The worms mature in the blood vessels, and eggs make their way outside of the body in stool and/or urine. A freshwater snail is a mandatory part of the life cycle (contains developmental forms of schistosome)	Infection occurs through skin penetration by infected cercariae released from a freshwater snail containing the intermediate stages within the schistosome life cycle. Cercariae can be released from the snail intermediate host singly or in groups	Protection from potentially contaminated water sources; awareness of mode of transmission; proper handling of human waste containing eggs (continued infection of snail intermediate hosts)

Table 49-5   Parasitic Infections: Clinical Findings in Normal and Compromised Hosts

Organism	Normal Host	Compromised Host
Entamoeba histolytica	Asymptomatic to chronic-acute colitis; extraintestinal disease may also occur (primary site: right upper lobe of liver)	Diminished immune capacity may lead to extraintestinal disease
Free-living amebae	Patients tend to have eye infections with *Acanthamoeba* spp.; linked to poor lens care	Primary amebic meningoencephalitis (PAM); granulomatous amebic encephalitis (GAE)
Giardia lamblia	Asymptomatic to malabsorption syndrome	Certain immunodeficiencies tend to predispose an individual to infection
Toxoplasma gondii	Approximately 50% of individuals have antibody and organisms in tissue but are asymptomatic	Disease in compromised host tends to involve the central nervous system with various neurologic symptoms
Cryptosporidium spp.	Self-limiting infection with diarrhea and abdominal pain	Because of autoinfective nature of life cycle, will not be self-limiting; may produce fluid loss of >10 L/day and there may be multisystem involvement. No known totally effective therapy
Cyclospora cayetanensis	Self-limiting infection with diarrhea (3-4 days); relapses common	Diarrhea may persist for 12 weeks or more; biliary disease has also been reported in this group, particularly those with acquired immunodeficiency syndrome (AIDS)
Isospora belli	Self-limiting infection with mild diarrhea or no symptoms	May lead to severe diarrhea, abdominal pain, and possible death (rare case reports); diagnosis may occasionally be missed because of nonrecognition of the oocyst stage; will not be seen when concentrated from polyvinyl alcohol fixative
Sarcocystis spp.	Self-limiting infection with diarrhea, or mild symptoms	Symptoms may be more severe and last for a longer time
Microsporidia *Nosema* *Brachiola* *Vittaforma* *Encephalitozoon* *Enterocytozoon* *Pleistophora* *Trachipleistophora* "*Microsporidium*"	Little is known about these infections in the normal host. Most infections have been identified as eye infections (*Nosema, Vittaforma, Encephalitozoon*)	Can infect various parts of the body; diagnosis often depends on histologic examination of tissues; routine examination of clinical specimens (stool, urine, etc., becoming more common); can probably cause death; dissemination to multiple body sites occurs
Leishmania spp.	Asymptomatic to mild disease	More serious manifestations of visceral leishmaniasis; some cutaneous species manifest visceral disease; difficult to treat and manage; definite coinfection with AIDS
Strongyloides stercoralis[13]	Asymptomatic to mild abdominal complaints; can remain latent for many years because of low-level infection maintained by internal autoinfective life cycle	Can result in disseminated disease (hyperinfection syndrome caused by infective nature of life cycle); abdominal pain, pneumonitis, sepsis-meningitis with gram-negative bacilli, eosinophilia; distinct link to certain leukemias/lymphomas; can be fatal
Crusted (Norwegian) scabies	Infections can range from asymptomatic to moderate itching	Severe infection with reduced itching response; hundreds of thousands of mites on body; infection easily transferred to others; secondary infection common

Table 49-6 Pathogenesis and Spectrum of Parasitic Diseases

Parasite Group	Pathogenesis	Spectrum of Disease
Protozoa, Intestinal		
Amebae	Pathogens can cause severe disease; however, exposure does not always lead to disease; infection may be self-limiting; disease more likely in the compromised host	Nonpathogens cause no disease, patients are asymptomatic; *Entamoeba histolytica* causes intestinal symptoms (bloody diarrhea), as well as the potential for amebic liver abscess; other tissues may be involved, especially in the immunocompromised patient
Flagellates	Not all patients become infected upon exposure; disease spectrum varies; some patients who harbor *Giardia* may remain asymptomatic; if nonexposed patients become infected, symptoms are much more likely to occur	Nonpathogens cause no disease, patients are asymptomatic; *Giardia lamblia* (malabsorption syndrome) and *Dientamoeba fragilis* cause intestinal symptoms ranging from "indigestion" to nonbloody diarrhea, cramping, gas, etc.
Ciliates	*Balantidium coli* is a rare infection within the United States; people who have common contact with pigs are much more likely to get infections; wide range of symptoms	*Balantidium coli* causes intestinal symptoms, including severe watery diarrhea, similar to that seen with coccidial and microsporidial infections
Coccidia	All of the coccidia infective to humans can cause severe disease, particularly in the compromised patient; infections in the immunocompetent patient tend to be self-limiting; *Cryptosporidium* spp. can maintain the infective cycle within the patient because of the autoinfective portion of the life cycle (compromised patient cannot produce antibody to limit this autoinfective cycle); huge waterborne outbreaks documented for *Cryptosporidium* spp.; infecting dose is low for *Cryptosporidium* spp.	*Cryptosporidium* spp., *Cyclospora cayetanensis*, and *Isospora belli* can all cause intestinal symptoms, including severe watery diarrhea; infections are more severe in immunocompromised patients. Life-threatening infections can be seen with *Cryptosporidium* spp.; the organisms can also disseminate to other body tissues, primarily the lung. *Sarcocystis* can also cause intestinal symptoms and/or muscle pain, etc., depending on the mode of infection (ingestion of oocysts or infected meats)
Microsporidia	Multiple genera are pathogenic for humans, as well as for animals; wide range of body sites; disease varies, depending on immune status of patient; disease outcome complicated by lack of treatment for some genera; albendazole effective for *Encephalitozoon* spp.	Every body tissue can be infected; *Enterocytozoon bieneusi* and *Encephalitozoon (Septata) intestinalis* are the most common and are found in the intestinal tract, the latter can also disseminate to other tissues, including the kidneys. Eye infections have been seen in both normal and compromised patients—severe corneal infections seen
Protozoa, Other Sites		
Amebae	Pathogenic for humans; disease ranging from acute meningoencephalitis to chronic encephalitis to cutaneous infections to keratitis, and the potential for other body sites; disease spectrum depends on patient's immune capacity and the organism involved; disease can be mild (*Acanthamoeba* spp.) to fatal (*Naegleria fowleri*)	Infection occurs through contact with contaminated water; organisms enter through the nasal mucosa and may travel via the olfactory nerve to the brain. Disease caused by *Naegleria fowleri* can be severe and life-threatening (primary amebic meningoencephalitis—PAM); chronic granulomatous amebic encephalitis (GAE) can be caused by *Acanthamoeba* spp. and *Balamuthia mandrillaris*; keratitis is also caused by these organisms, and infection can be linked to blindness or severe corneal damage. Eye infections can be linked to contaminated lens solutions or direct, accidental inoculation of the eye from environmental water and/or soil sources
Flagellates	*Trichomonas vaginalis* causes genitourinary disease, depending on vaginal pH, presence or absence of other organisms, sexual practices, etc. The disease can vary from mild to severe	*Trichomonas vaginalis* is found in the genitourinary system and is usually acquired by sexual transmission. Disease can be asymptomatic in the male, but can cause pain, itching, and discharge in females; some strains of drug-resistant *Trichomonas* have been documented

Table 49-6   Pathogenesis and Spectrum of Parasitic Diseases—cont'd

Parasite Group	Pathogenesis	Spectrum of Disease
Protozoa, Blood, Tissue		
Malaria, Babesiosis	*Plasmodium vivax, P. falciparum, P. ovale,* and *P. malariae* are definitely pathogenic for humans, malaria with *P. falciparum* being one of the leading causes of death in endemic areas; although the host can develop antibody, protection is strain specific and short-lived	Malaria can cause a range of symptoms, with life-threatening illness caused by *P. falciparum;* symptoms include fever, chills, nausea, central nervous system (CNS) symptoms, etc.; *Babesia* infections mimic those seen with malaria, however, there is no fever periodicity
Flagellates (Leishmaniae)	*Leishmania donovani* that invades the spleen, liver, and bone marrow can cause serious disease, particularly in the compromised host; more severe disease seen in all leishmanial species in immunosuppressed patients	Leishmaniasis can infect the skin and mucous membranes, and the organs of the reticuloendothelial system; symptoms can be mild to life-threatening, particularly in the immunocompromised patient
Flagellates (Trypanosomes)	Humans are the only known hosts for *Trypanosoma brucei gambiense* (West African trypanosomiasis), whereas *Trypanosoma brucei rhodesiense* (East African trypanosomiasis) infections are found in a number of antelope and other hoofed mammals serving as reservoir hosts. *Trypanosoma cruzi* (American trypanosomiasis) can be found in rodents as well as in chickens	*T.b. gambiense* and *T.b. rhodesiense* cause African sleeping sickness, with eventual invasion of the CNS, leading to coma and death; Chagas' disease *(T. cruzi)* causes acute to chronic problems, primarily linked to cardiac disease and diminished cardiac capacity; the muscles of the gastrointestinal (GI) tract are also infected, leading to loss of function in terms of movement of food through the GI tract
Nematodes, Intestinal	These worms can cause mild to severe disease, depending on the worm burden (original number of eggs ingested or infective larvae penetrating the skin); young children and debilitated patients more likely to be symptomatic; severe infections seen in hyperinfections caused by *Strongyloides stercoralis* (autoinfective life cycle and immunocompromised patients); outcome varies tremendously from patient to patient and depends on the original infective dose	*Ascaris lumbricoides, Trichuris trichiura,* and hookworm symptoms range from none to diarrhea, pain, etc., depending on the worm burden; anemia may be seen with severe hookworm infection; *Strongyloides stercoralis* infections can involve many body tissues (disseminated disease) in the compromised patient and can cause death; pinworm infection *(Enterobius vermicularis)* symptoms range from none to anal itching, irritability, loss of sleep, etc.
Nematodes, Tissue	Again, depending on the original infective dose, *Trichinella* spp. can cause mild to severe disease; both cutaneous larva migrans (CLM) (dog/cat hookworm larvae), and toxocariasis (VLM, OLM) (ingestion of dog/cat ascarid eggs) cause serious disease if not treated; often CLM, VLM, and OLM are seen in children more often than adults	*Trichinella* can cause eosinophilia, muscle aches and pains and death, depending on the worm burden; CLM can cause severe itching and eosinophilia due to larval migration in the skin; VLM and OLM are caused by larval migration throughout the body, including the eyes (will mimic retinoblastoma)
Nematodes, Filarial	*Wuchereria bancrofti, Loa loa,* and *Onchocerca volvulus* cause human disease; however, some filarial infections are not that well adapted to humans and require many years of exposure before disease is manifest; some infections not evident, some cause multiple disease manifestations	Filarial nematode symptoms range from none to elephantiasis, blindness, skin changes, lymphadenitis, lymphangitis; in some cases, Loeffler's syndrome may be seen as well
Cestodes, Intestinal	The beef *(Taenia saginata),* pork *(Taenia solium),* and freshwater fish *(Diphyllobothrium latum)* tapeworms infect humans and are generally found in the intestine as a single worm. In the case of cysticercosis, the ingestion of *T. solium* eggs can cause mild to severe disease, depending on the infecting dose and body site (muscle, CNS). *Hymenolepis nana* (dwarf tapeworm) infection can lead to many worms in the intestinal tract (autoinfective cycle—also can go from egg to larval form to adult within the human host)	Human infection with the adult tapeworms can cause no symptoms or mild intestinal symptoms may occur. When the human becomes the accidental intermediate host for *T. solium,* CNS symptoms may occur, including epileptic seizures. Infection with adult *D. latum* tapeworm can also cause intestinal symptoms such as pain, diarrhea, etc., but the patient may also be asymptomatic; a vitamin B_{12} deficiency may be seen. Infection with *Hymenolepis nana* is primarily acquired from accidental ingestion of eggs from an adult tapeworm; symptoms may be absent or diarrhea may be present

Table 49-6   Pathogenesis and Spectrum of Parasitic Diseases—cont'd

Parasite Group	Pathogenesis	Spectrum of Disease
Cestodes, Tissue	*Echinococcus* spp. can cause severe disease, depending on the original infecting dose of tapeworm eggs; multiple organs can be involved, including brain, liver, lung, bone; some cysts grow like a metastatic tumor; surgical removal can be difficult, if not impossible	Depending on the body site, hydatid cysts can cause pain, anaphylactic shock (fluid leakage), or CNS symptoms. The patient may be unaware of infection until cyst begins to press on other body organs or a large fluid leak occurs
Trematodes, Intestinal	Many genera and species pathogenic for humans; disease severity depends on original infective dose of metacercariae; some patients may be unaware of infection	Intestinal trematodes can cause pain, diarrhea; intestinal toxicity can sometimes be seen in heavy infections with *F. buski*
Trematodes, Liver, Lung	Many genera and species pathogenic for humans; disease severity depends on original infective dose of metacercariae; some patients may be unaware of infection	*Paragonimus* spp. infection in the lungs can be severe, with coughing, shortness of breath, etc; liver fluke infection can also involve the bile ducts and gall bladder; symptoms depend on worm burden
Trematodes, Blood	Schistosomes are definitely pathogenic for humans; however, loading dose of cercariae from infected water sources determines outcome of disease; light infections may not produce symptoms; heavy infections can lead to death	Symptoms may range from none in light infections to severe organ failure caused by deposition of eggs with subsequent formation of granuloma within the tissues; "pipe-stem" fibrosis seen in blood vessels; collateral circulation may develop; severe disease can cause death

LABORATORY DIAGNOSIS

The ability to detect and identify human parasites is directly linked to the quality of the clinical specimen, submission of the appropriate specimen(s), relevant diagnostic test orders, and the experience and training of those who examine the specimens and interpret the results. A summary of human parasites and applicable clinical specimens, diagnostic tests, what constitutes a positive finding, and comments can be found in Table 49-7.

SPECIMEN COLLECTION AND TRANSPORT

Depending on its stage of development (adult, larvae, eggs, trophozoites, cysts, oocysts, spores) within the clinical specimen, a particular parasite may or may not be able to survive for any length of time outside of the host. For this reason, clinical specimens should be transported immediately to the laboratory to increase the likelihood of finding intact organisms. Because there is often a lag time between when the specimen is collected and when the laboratory receives it, many facilities routinely use preservatives for collection and transport. This approach ensures that any parasites present will maintain their morphology and can be identified after appropriate processing and examination. Of course, this assumes the use of appropriate fixatives, immediate fixation after obtaining the specimen, and adequate mixing between the fixative and specimen (Table 49-8). It is mandatory that clearly defined specimen collection guidelines are available for health care personnel and that all clients recognize the importance of following such guidelines. Specimen rejection criteria must be included as a part of these guidelines; they must be followed and enforced to reduce wasted personnel time and to limit the possibility of providing misleading or incorrect results. Detailed specimens and/or body sites and collection and transport information can be found in Table 49-3.

SPECIMEN PROCESSING

Diagnostic parasitology includes laboratory procedures that are designed to detect organisms within clinical specimens using morphologic criteria and visual identification, rather than culture, biochemical tests, and/or physical growth characteristics. Many clinical specimens, such as those from the intestinal tract, contain numerous artifacts that complicate the differentiation of parasites from surrounding debris. Specimen preparation often requires concentration of some type, all of which are designed to increase the chances of finding the organism(s). Microscopic examination requires review of the prepared clinical specimen using multiple magnifications and different time frames; organism identification also depends on the skill of the microbiologist. Final identification is usually based on microscopic examination of stained preparations, often using high magnification such as oil immersion (×1000). See Table 49-3 for specific details on specimen processing.

Table 49-7 Common Human Parasites, Diagnostic Specimens, Tests, and Positive Findings

Organism	Acquire Infection	Location in Host	Diagnostic Specimen	Diagnostic Test*	Positive Specimen	Comments
Intestinal Amebae *Entamoeba histolytica* *Entamoeba dispar* *Entamoeba hartmanni* *Entamoeba coli* *Endolimax nana* *Iodamoeba bütschlii* *Blastocystis hominis*	Ingestion of food or water contaminated with infective cysts; fecal-oral transmission	Intestinal tract; *E. histolytica* infection may disseminate to the liver (extraintestinal amebiasis)	Stool; sigmoidoscopy specimens	O&P examination; stained sigmoidoscopy slides; stool immunoassays	Trophozoites and/or cysts	Many of the protozoa can look very much alike; see diagnostic tables for details; immunoassays for *E. histolytica/E. dispar* group and *E. histolytica* (fresh, frozen stool required)
Free-Living Amebae *Naegleria fowleri* *Acanthamoeba* spp. *Balamuthia mandrillaris*	Contaminated water or soil; dust, contaminated eye solutions; organisms may enter through nasal mucosa; travel to brain via olfactory nerve	CNS (PAM, GAE); eye	CSF, corneal scrapings, biopsy, eye care solutions; CSF examination Stat request	Stains, culture, FA, biopsy; *Balamuthia* cannot be grown on agar culture, whereas *Naegleria* and *Acanthamoeba* can	Trophozoites or cysts	CNS disease life-threatening with *Naegleria*; other CNS infections more chronic; keratitis can lead to blindness
Intestinal Flagellates *Giardia lamblia* *Dientamoeba fragilis* *Chilomastix mesnili* *Pentatrichomonas hominis*	Ingestion of food or water contaminated with infective cysts or trophozoites (*D. fragilis, B. hominis*); fecal-oral transmission	Intestinal tract	Stool; duodenal specimens or Entero-Test capsule (string test) for *Giardia*	O&P examination; wet preparations or stains of duodenal material; stool immunoassays (can use fresh or formalin-fixed specimens—no PVA)	Trophozoites or cysts	*G. lamblia* is difficult to recover; stool immunoassays more sensitive than routine O&P examinations; *D. fragilis* requires permanent stain for identification
Urogenital Flagellates *Trichomonas vaginalis*	Sexually transmitted; wet towels less likely but possible	Urinary tract; genital system; males may be asymptomatic	Vaginal secretions, prostatic fluid, often recovered in urine sediment	Wet preparations, culture, immunoassays	Trophozoites	Often diagnosed by motility in urine sediment or wet preparations
Intestinal Ciliate *Balantidium coli*	Ingestion of food or water contaminated with infective cysts; fecal-oral transmission	Intestinal tract	Stool	O&P examination; wet preparations better than permanent stained smear	Trophozoites and/or cysts	Not common in the United States; associated with pigs; seen in proficiency testing specimens

Table 49-7 Common Human Parasites, Diagnostic Specimens, Tests, and Positive Findings—cont'd

Organism	Acquire Infection	Location in Host	Diagnostic Specimen	Diagnostic Test*	Positive Specimen	Comments
Intestinal Coccidia *Cryptosporidium* spp. *Cyclospora cayetanensis* *Isospora belli*	Ingestion of food or water contaminated with infective oocysts; fecal-oral transmission	Intestinal tract; *Cryptosporidium* can disseminate to other tissues in compromised host (lung, gall bladder)	Stool; biopsy; duodenal specimen; sputum	Modified acid-fast stains; stool immunoassays; concentration wet prep for *I. belli*	Oocysts in stool or scrapings; other developmental stages in tissues	*Cryptosporidium* cause of severe diarrhea in compromised patient; nosocomial transmission
Intestinal Microsporidia *Enterocytozoon bieneusi* *Encephalitozoon (Septata) intestinalis*	Ingestion of food or water contaminated with infective spores; fecal-oral transmission	Intestinal tract; *E. intestinalis*, *E. bieneusi* can disseminate to other body sites (kidney)	Stool; biopsy	Modified trichrome stains; optical brightening agents; experimental immunoassays; biopsy and histology (tissue Gram stains)	Spores in stool; other developmental stages in tissues	Can cause serious diarrhea in compromised host; less known about infections in normal host
Microsporidia, Other Body Sites *Encephalitozoon* spp. *Brachiola vesicularum* *Microsporidium* *Nosema* *Pleistophora* *Trachipleistophora* *Vittaforma*	Ingestion of food or water contaminated with infective spores; fecal-oral transmission; inhalation; direct environmental contact to eyes; probably hands to eyes	All tissues	All body fluids and/or tissues relevant, depending on the body site	Modified trichrome stains; optical brightening agents; experimental immunoassays; biopsy and histology (tissue Gram stains)	Spores in stool, urine, other body fluids; developmental stages in tissues	Can cause serious diarrhea in compromised host; less known about infections in normal host; a number of eye infections documented in immunocompetent patients
Tissue Protozoa *Toxoplasma gondii*	*Toxoplasma:* ingestion of raw meats, oocysts from cat feces	Eye, CNS in compromised patient (*T. gondii*)	Biopsies (any tissue), CSF	Serology, tissue culture; recovery from CSF	Positive serology; recovery of trophozoites in CSF	Many people have positive serologies for *T. gondii*; infection serious in compromised patients
Intestinal Nematodes *Enterobius vermicularis* *Trichuris trichiura* *Ascaris lumbricoides* Hookworm *Strongyloides stercoralis*	Ingestion of food or water contaminated with infective eggs; penetration of skin by infective larvae in soil	Intestine; *Strongyloides* may disseminate (hyperinfection), primarily in compromised patients	Stool; duodenal contents (*Strongyloides*); cellophane tape preps or paddles for *Enterobius*	O&P examination; special concentrates and cultures; examination of tapes for *Enterobius*	Adult worms, eggs and/or larvae, depending on the roundworm involved	Review direct and indirect life cycles (migration through heart, lung, trachea to intestine); 4-6 consecutive tapes required to "rule out" *Enterobius* infection

Table 49-7 Common Human Parasites, Diagnostic Specimens, Tests, and Positive Findings—cont'd

Organism	Acquire Infection	Location in Host	Diagnostic Specimen	Diagnostic Test*	Positive Specimen	Comments
Tissue Nematodes						
VLM, OLM (Toxocara)	Ingestion of infective eggs	Migration through tissues	Serum	Serology	Positive serology	Human is accidental host for VLM, OLM, and CLM
CLM (dog/cat hookworm)	Skin penetration of larvae	Skin tracks, migration	Visual inspection	Presence of tracks/skin	Eosinophilia, visual tracks	Outbreaks still occur
Trichinella spp.	Ingestion of raw pork, bear	Muscle	Serum, muscle biopsy	Serology, squash prep	Positive serology, larvae	Sometimes ID only after surgical removal
Anisakis, etc.	Ingestion of raw marine fish	Intestine	Submission of larvae	ID of larvae	Positive larval ID	
Intestinal Cestodes						
Taenia saginata (beef)	Ingestion of: Raw beef	Intestine	Stool and/or proglottids	O&P, India ink proglottids	Eggs, proglottid branches	Both *Taenia* eggs look alike; need gravid proglottid or scolex to make the ID
Taenia solium (pork)	Raw pork		Stool and/or proglottids	O&P, India ink proglottids	Eggs, proglottid branches	
Diphyllobothrium latum	Raw freshwater fish		Stool and/or proglottids	O&P	Eggs, proglottid shape	
Hymenolepis nana	Tapeworm eggs		Stool	O&P	Eggs	
Hymenolepis diminuta	Grain beetles		Stool	O&P	Eggs	
Dipylidium caninum	Fleas from dogs/cats		Stool and/or proglottids	O&P	Eggs, proglottid shape	
Tissue Cestodes						
Echinococcus granulosus	Ingestion of: Eggs from dog tapeworm	Liver, lung, etc.	Serum, hydatid cyst aspirate; biopsy	Serology, centrifugation of fluid; histology	Positive serology; hydatid sand, tapeworm tissue	*E. granulosus* (enclosed cyst)
E. multilocularis	Eggs from fox tapeworm	Liver, lung, etc.				*E. multilocularis* (cyst develops throughout tissue)
Taenia solium (pork)	Eggs from human tapeworm	CNS, subcutaneous tissues	Serum, scans, biopsy	Serology, films, histology	Positive serology, positive scans, tapeworm tissue	Small, enclosed cysticerci (cysticercosis)
Intestinal Trematodes						
Fasciolopsis buski	Ingestion of metacercariae: On water chestnuts	Intestine	Stool	O&P examination; *M. yokogawai*, *H. heterophyes* eggs tiny; use high dry power	Eggs in stool	Eggs of *F. buski* look identical to that of the liver fluke, *Fasciola hepatica*
Metagonimus yokogawai	In raw fish					
Heterophyes heterophyes						
Liver and Lung Trematodes						
Fasciola hepatica	Ingestion of metacercariae: On watercress	Liver	Stool	O&P	Eggs in stool	Eggs of *Fasciola* look almost identical to those of *Fasciolopsis*; lung fluke eggs in sputum look like "brown metal filings"
Clonorchis sinensis	In raw fish	Liver, bile ducts	Stool, duodenal drainage	O&P	Eggs in stool or duodenal drainage	
Paragonimus spp.	In raw crabs	Lung	Stool, sputum	O&P	Eggs in stool and/or sputum	
Blood Trematodes						
Schistosoma mansoni	Skin penetration of cercariae released from the freshwater snail intermediate host	Veins over the: Large intestine	Because the adult worms may become located in the "incorrect" veins, both urine and stool (unpreserved) should be examined	O&P; hatching test for egg viability (all specimens collected with no preservatives); concentrates performed with saline, not water	Eggs in stool and/or urine	When schistosomiasis suspected, stool, spot-urine, and 24-hr urine (collected with no preservatives)
S. haematobium		Bladder				
S. japonicum		Small intestine				

Table 49-7 Common Human Parasites, Diagnostic Specimens, Tests, and Positive Findings—cont'd

Organism	Acquire Infection	Location in Host	Diagnostic Specimen	Diagnostic Test*	Positive Specimen	Comments
Malaria *Plasmodium vivax* *P. ovale* *P. malariae* *P. falciparum*	Infection through mosquito bite, blood transfusion, shared drug needles, transplacental	Preerythrocytic: Liver Blood Blood Blood Blood plus capillaries of deep tissues (spleen, liver, bone marrow)	Drawn immediately: **Stat request** Blood; draw every 6 hours until confirmed as positive or negative (another option: draw daily for 3 days)	Thick, thin blood films; rapid immunoassay methods (not yet FDA-approved in the United States); concentration methods	Parasites present	*P. falciparum* infections are *medical emergencies;* complete patient history mandatory (travel, prophylaxis, prior history); Giemsa stain recommended
Babesiosis *Babesia* spp.	Tickborne; transfusion; organ transplants	Blood	Blood	Thick, thin blood films	Parasites present	Can mimic ring forms of *P. falciparum;* patient may have no travel history outside of the United States
Trypanosomes *Trypanosoma brucei gambiense* *T. b. rhodesiense* *Trypanosoma cruzi*	Bite of: Tsetse fly Tsetse fly Feces of triatomid bug (kissing bug) (bug feces scratched into bite site)	Blood, lymph nodes, CNS Blood, lymph nodes, CNS Blood, striated muscle (heart, GI tract, etc.)	Blood, node aspirate, CSF Blood, cardiac changes, muscle biopsy	Thick, thin films Thick, thin films Thick, thin films, histology; culture	Trypomastigotes Trypomastigotes Trypomastigotes in blood, amastigotes in tissue	African sleeping sickness more common with *T.b. gambiense* Chagas' disease (American trypanosomiasis) (xenodiagnosis an option)
Leishmaniae *Leishmania tropica* or *L. mexicana* complex (cutaneous) *L. braziliensis* complex (mucocutaneous) *L. donovani* complex (visceral)	Bite of the sand fly	Macrophages of: Skin Skin, mucous membranes Spleen, liver, bone marrow (RE system)	Skin biopsy Skin, membrane biopsy Blood, bone marrow, liver/spleen biopsy	Stained smears, cultures Stained smears, cultures Thick, thin blood films; stained smears, cultures	Amastigotes within the clinical specimens indicated	Animal inoculation is rarely used; some research labs now using PCR
Filarial Nematodes *Wuchereria bancrofti* (S) *Onchocerca volvulus* (NS)	Bite of: Mosquito Black fly	Lymphatics (adults), blood (microfilariae) Nodules (adults), skin, eye (microfilariae)	Blood Skin snips, blood; biopsy nodule	Thick, thin films; various concentrations Biopsy, tease skin apart in water; thick, thin films	Microfilariae Microfilariae; adult in tissue nodules	Elephantiasis possible; periodicity a factor in finding microfilariae; some microfilariae sheathed (S), some not (NS)

Table 49-7   Common Human Parasites, Diagnostic Specimens, Tests, and Positive Findings—cont'd

Organism	Acquire Infection	Location in Host	Diagnostic Specimen	Diagnostic Test*	Positive Specimen	Comments
Less Common						
Loa loa (S)	Black gnat	Eye (adults)	Blood	Thick, thin films; various concentrations	Microfilariae; adult worm	"African eye worm"
Brugia malayi (S)	Mosquito	Lymphatics (adults)			Microfilariae	
Mansonella spp. (NS)	Mosquito	blood (microfilariae) for all three				

BAL, Bronchoalveolar lavage; *CLM,* cutaneous larva migrans; *CNS,* central nervous system; *CSF,* cerebrospinal fluid; *FA,* fluorescent antibody; *FDA,* Food and Drug Administration; *GAE,* granulomatous amebic encephalitis; *GI,* gastrointestinal tract; *ID,* identification; *NS,* not sheathed; *O&P,* ova and parasite; *OLM,* ocular larva migrans; *PAM,* primary amebic meningoencephalitis; *PCR,* polymerase chain reaction; *PVA,* polyvinyl alcohol; *RE,* reticuloendothelium; *S,* sheathed; *TTA,* transtracheal aspiration; *VLM,* visceral larva migrans.

*Although serologic tests are not always mentioned, they are available for a number of parasitic infections. Unfortunately, most are not routinely available. Contact your state public health laboratory or the Centers for Disease Control and Prevention in Atlanta, Ga.

Table 49-8   Preservatives Used in Diagnostic Parasitology (Intestinal Tract Specimens)

Preservative	Concentration	Permanent Stained Smear	Immunoassays (*Giardia lamblia, Cryptosporidium* spp.)	Comments
5% or 10% formalin	Yes	No	Yes	EIA, FA, cartridge
5% or 10% buffered formalin	Yes	No	Yes	EIA, FA, cartridge
MIF	Yes	Polychrome IV stain	ND	No published data
SAF	Yes	Iron hematoxylin	Yes	EIA, FA, cartridge
PVA*	Yes	Trichrome or iron hematoxylin	No	PVA may interfere with immunoassays
PVA-Modified†	Yes	Trichrome or iron hematoxylin	No	PVA may interfere with immunoassays
PVA-Modified‡	Yes	Trichrome or iron hematoxylin	Some, but not all	PVA may interfere with immunoassays
Single Vial Systems§	Yes	Trichrome or iron hematoxylin	Some, but not all	Check with the manufacturer
Schaudinn's (without PVA)*	No	Trichrome or iron hematoxylin	No	Mercury interferes with immunoassays

EIA, Enzyme immunoassay; *FA*, fluorescent antibody; *MIF*, merthiolate-iodine-formalin; *ND*, no data; *PVA*, polyvinyl alcohol; *SAF*, sodium acetate-acetic acid-formalin.

*These two fixatives use the mercuric chloride base in the Schaudinn's fluid; this formulation is still considered to be the "gold standard" against which all other fixatives are evaluated (organism morphology after permanent staining). Additional fixatives prepared with nonmercuric chloride-based compounds continue to be developed and tested.

†This modification uses a copper sulfate base rather than mercuric chloride.

‡This modification uses a zinc base rather than mercuric chloride and apparently works well with both trichrome and iron-hematoxylin stains.

§These modifications use a combination of ingredients (including zinc) but are prepared from proprietary formulas. The aim is to provide a fixative that can be used for the fecal concentration, permanent stained smear, and available immunoassays for *Giardia lamblia, Cryptosporidium* spp., and *Entamoeba histolytica* (or the *Entamoeba histolytica/E. dispar* group). Currently, most immunoassays for *Entamoeba histolytica* or the *E. histolytica/E. dispar* group require fresh or frozen stool specimens; specimens in preservative are not acceptable.

APPROACH TO IDENTIFICATION

Protozoa are small and tend to range from 1.5 μm (microsporidia) to approximately 80 μm (*Balantidium coli*, a ciliate). Some are intracellular and require multiple isolation and staining methods for identification. Helminth infections are usually diagnosed by finding eggs, larvae, and/or adult worms in various clinical specimens, primarily those from the intestinal tract. Identification to the species level may require microscopic examination of the specimen. The recovery and identification of blood parasites can require concentration, culture, and microscopy. Confirmation of suspected parasitic infections depends on the proper collection, processing, and examination of clinical specimens; often, multiple specimens must be submitted and examined to find and confirm the suspected organism(s) (see Table 49-7).

MICROSCOPIC EXAMINATION

Good, clean microscopes and light sources are mandatory when examining specimens for parasites. Organism identification depends on morphologic differences, most of which must be seen using stereoscopic (magnification ≤50×) or regular microscopes at low (100×), high dry (400×), and oil immersion (1000×) magnifications. The use of a 50× or 60× oil immersion objective for scanning can be very helpful, particularly when the 50× oil and 100× oil immersion objectives are placed side by side. This arrangement on the microscope can help avoid accidentally getting oil on the 40× high dry objective.

A stereoscopic microscope is recommended for larger specimens (e.g., arthropods, tapeworm proglottids, various artifacts). The total magnification usually varies from approximately 10× to 45×, either with a zoom capacity or with fixed objectives (0.66×, 1.3×, 3×) that can be used with 5× or 10× oculars. Depending on the density of the specimen or object being examined, the light source must be directed either from under the stage or onto the top of the stage.

Intestinal Tract

Stool is the most common specimen submitted to the diagnostic laboratory; the most commonly performed procedure in parasitology is the ova and parasite (O&P) examination. Several other diagnostic techniques are available for the recovery and identification of parasitic organisms from the intestinal tract. Most laboratories do not routinely offer all of these techniques, but many are relatively simple and inexpensive to perform. The clinician should be aware of the possibilities and the clinical relevance of information obtained from using such techniques. Rarely, it is necessary to examine stool specimens for the presence of scolices and proglottids of cestodes and adult nematodes and trematodes to

BOX 49-1	Direct Smear—Review

PRINCIPLE:
To assess worm burden of patient, to provide quick diagnosis of heavily infected specimen, to check organism motility (primarily protozoan trophozoites), and to diagnose organisms that might not be seen from the permanent stain methods; seeing organism motility is the primary objective

SPECIMEN:
Any fresh liquid or very soft stool specimen that has not been refrigerated or frozen

REAGENTS:
0.85% NaCl; Lugol's or D'Antoni's iodine

EXAMINATION:
Low-power examination (×100) of entire 22- by 22-mm coverslip preparation (both saline and iodine); high dry power examination (×400) of at least one third of the coverslip area (both saline and iodine)

RESULTS AND LABORATORY REPORTS:
Results from the direct smear examination should often be considered presumptive; however, some organisms could be definitively identified (*Giardia lamblia* cysts and *Entamoeba coli* cysts, helminth eggs, and larvae, *Isospora belli* oocysts). These reports should be considered "preliminary," while the final report would be available after the results of the concentration and permanent stained smear were available

PROCEDURE NOTES AND LIMITATIONS:
Once iodine is added to the preparation, the organisms will be killed and motility will be lost. Specimens submitted in stool preservatives or fresh, formed specimens should not be examined using this procedure; the concentration and permanent stained smear should be performed instead. Oil immersion examination (×1000) is not recommended (organism morphology not that clear)

BOX 49-2	Concentration—Review

PRINCIPLE:
To concentrate the parasites present, either through sedimentation or by flotation. The concentration is specifically designed to allow recovery of protozoan cysts, coccidian oocysts, microsporidian spores, and helminth eggs and larvae

SPECIMEN:
Any stool specimen that is fresh or preserved in formalin (most common), PVA (mercury- or non-mercury–based), SAF, MIF, or the newer single-vial-system fixatives

REAGENTS:
5% or 10% formalin, ethyl acetate, zinc sulfate (specific gravity, 1.18 for fresh stool and 1.20 for preserved stool); 0.85% NaCl; Lugol's or D'Antoni's iodine.

EXAMINATION:
Low-power examination (×100) of entire 22- by 22-mm coverslip preparation (iodine recommended, but optional); high dry power examination (×400) of at least one third of the coverslip area (both saline and iodine)

RESULTS AND LABORATORY REPORTS:
Often, results from the concentration examination should be considered presumptive; however, some organisms could be definitively identified (*Giardia lamblia* cysts and *Entamoeba coli* cysts, helminth eggs, and larvae, *Isospora belli* oocysts). These reports should be considered "preliminary," while the final report would be available after the results of the permanent stained smear were available

PROCEDURE NOTES AND LIMITATIONS:
Formalin-ethyl acetate sedimentation concentration is the most commonly used. Zinc sulfate flotation will not detect operculated or heavy eggs (*Clonorchis* eggs, unfertilized *Ascaris* eggs); both the surface film and sediment will have to be examined before a negative result is reported. Smears prepared from concentrated stool are normally examined at low (×100) and high dry (×400) power; oil immersion examination (×1000) is not recommended (organism morphology not that clear). The addition of too much iodine may obscure helminth eggs (will mimic debris)

confirm the diagnosis and/or for identification to the species level.

Other specimens from the intestinal tract, such as duodenal aspirates or drainage, mucus from the Entero-Test Capsule technique, and sigmoidoscopy material, can also be examined as wet preparations and as permanent stained smears after processing with either trichrome or iron-hematoxylin staining. Although not all laboratories perform these procedures, they are included to give some idea of the possibilities for diagnostic testing.

O&P EXAMINATION. The O&P examination comprises three separate protocols: the direct wet mount, the concentration, and the permanent stained smear.*

The direct wet mount requires fresh stool, is designed to allow detection of motile protozoan trophozoites, and is examined microscopically at low and high dry magnifications (×100, entire 22- × 22-mm coverslip; ×400, ¹⁄₃ to¹⁄₂ 22- × 22-mm coverslip) (Box 49-1). However, because of potential problems resulting from the lag time between specimen passage and receipt in the laboratory, the direct wet examination has been

eliminated from the routine O&P in favor of receipt of specimens collected in stool preservatives. The direct wet preparation is not performed for specimens received in the laboratory in stool collection preservatives. Specific fixatives can be found in Table 49-8.

The second part of the O&P is the concentration, which is designed to facilitate recovery of protozoan cysts, coccidian oocysts, microsporidial spores, and helminth eggs and larvae (Box 49-2). Both flotation and sedimentation methods are available; the most common procedure is the formalin-ethyl acetate sedimentation method (formerly called the formalin-ether method). The concentrated specimen is examined as a wet preparation, with or without iodine, using low and high dry magnifications (×100, ×400) as indicated for the direct wet smear examination.

The third part of the O&P examination is the permanent stained smear, which is designed to facilitate identification of intestinal protozoa (Box 49-3). Several staining methods are available; the two most common methods are the Wheatley modification of the Gomori

---

*References 6, 7, 11, 12, 17, 19, 20, 26, 28.

BOX 49-3 Permanent Stained Smear—Review

PRINCIPLE:

To provide contrasting colors for the background debris and parasites present; designed to allow examination and recognition of detailed organism morphology under oil immersion examination (100× objective for a total magnification of ×1000). Designed primarily to allow recovery and identification of the intestinal protozoa

SPECIMEN:

Any stool specimen that is fresh or preserved in PVA (mercury- or non-mercury–based), SAF, MIF, or the newer single-vial-system fixatives

REAGENTS:

Trichrome, iron-hematoxylin, modified iron-hematoxylin, polychrome IV, chlorazol black E stains, and their associated solutions; dehydrating solutions (alcohols and xylenes or xylene substitutes); mounting fluid optional. Absolute alcohol (100% ethanol) is preferred, rather than using the 95%/5% alcohol option for absolute alcohol

EXAMINATION:

Oil immersion examination (×1000) of at least 300 fields; additional fields may be required if suspect organisms have been seen in the wet preparations from the concentrated specimen

RESULTS AND LABORATORY REPORTS:

The majority of the suspect protozoa and/or human cells could be confirmed by the permanent stained smear. These reports should be categorized as "final" and would be reported as such (the direct wet smear and the concentration examination would provide "preliminary" results)

PROCEDURE NOTES AND LIMITATIONS:

The most commonly used stains include trichrome and iron-hematoxylin. Unfortunately, helminth eggs and larvae will generally take up too much stain and may not be identified from the permanent stained smear. Also, coccidian oocysts and microsporidian spores require other staining methods for identification. Permanent stained smears are normally examined under oil immersion examination (×1000), and low or high dry power is not recommended. The slide may be screened using the newer 50× or 60× oil immersion objectives but should not be reported until examination has been completed using the 100× oil immersion lens. Confirmation of the intestinal protozoa (both trophozoites and cysts) is the primary purpose of this technique

IMPORTANT REMINDER:

When using non-mercury fixatives or one of the single-vial options (usually zinc-based, proprietary formula), the iodine-alcohol step can be eliminated. After drying, the slides can be placed directly into the stain (trichrome or hematoxylin). However, when material arrives for proficiency testing (e.g., College of American Pathologists, American Association of Bioanalysts), these fecal specimens may have been preserved using mercury-based fixatives. Therefore, you must include the iodine-alcohol step and subsequent rinse steps for mercury removal in the routine staining protocol. Some laboratories have decided to leave the staining protocol as is; the use of the iodine-alcohol step will not harm smears preserved using non-mercury–based fixatives

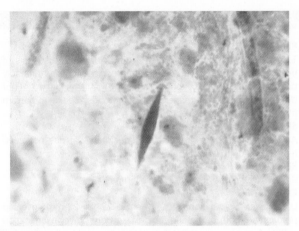

Figure 49-1 Charcot-Leyden crystals; stool material stained with Wheatley's trichrome stain.

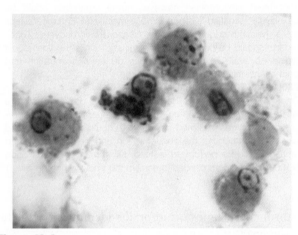

Figure 49-2 Polymorphonuclear leukocytes; stool material stained with Wheatley's trichrome stain.

important procedure performed for the identification of intestinal protozoan infections; the permanent stained smears are examined using oil immersion objectives (×600 for screening, ×1000 for final review of 300 or more oil immersion fields).

Modified acid-fast stains are recommended for the intestinal coccidia (Box 49-4), and modified trichrome stains are recommended for the intestinal microsporidia (Box 49-5). These stains are specifically designed to allow identification of the coccidian oocysts and microsporidian spores, respectively.[5,6]

Recovery of Tapeworm Scolex. Because the medication used for treatment of tapeworms is usually very effective, the procedure for recovery of the tapeworm scolex is rarely requested and is no longer clinically relevant. However, stool specimens may have to be examined for the presence of scolices and gravid proglottids of cestodes for proper species identification. This

tissue trichrome and the iron-hematoxylin stains (Figures 49-1 to 49-3). This part of the O&P examination is critical for the confirmation of suspicious objects seen in the wet examination and identification of protozoa that might not have been seen in the wet preparation. The permanent stained smear is the most

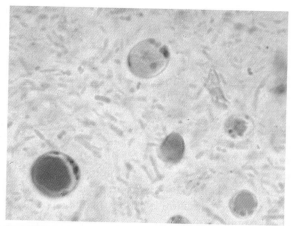

Figure 49-3 *Blastocystis hominis* central body forms (larger objects) and yeast cells (smaller, more homogeneous objects); stool material stained with Wheatley's trichrome stain.

procedure requires mixing a small amount of feces with water and straining the mixture through a series of wire screens (graduated from coarse to fine mesh) to look for scolices and proglottids. Appearance of scolices after therapy is an indication of successful treatment. If the scolex has not been passed, it may still be attached to the mucosa; the parasite is capable of producing more segments from the neck region of the scolex, and the infection continues. If this occurs, then the patient can be retreated when proglottids begin to reappear in the stool.[6]

Examination for Pinworm.
A roundworm parasite that has worldwide distribution and is commonly found in children is *Enterobius vermicularis,* known as pinworm or seatworm. The adult female worm migrates out of the anus, usually at night, and deposits her eggs on the perianal area. The adult female (8 to 13 mm long) may occasionally be found on the surface of a stool specimen or on the perianal skin. Because the eggs are usually deposited around the anus, they are not commonly found in feces and must be detected by other diagnostic techniques. Diagnosis of pinworm infection is usually based on the recovery of typical eggs, which are described as thick-shelled, football-shaped eggs with one slightly flattened side. Each egg often contains a fully developed embryo and will be infective within a few hours after being deposited (Figure 49-4).[6,7,17]

Sigmoidoscopy Material.
Material obtained from sigmoidoscopy can be helpful in the diagnosis of amebiasis that has not been detected by routine fecal examinations; however, a series of at least three routine stool examinations for parasites should be performed for each patient before sigmoidoscopy examination is done. The specimen should be processed immediately. Three

---

BOX 49-4   Modified Acid-Fast Permanent Stained Smear—Review

PRINCIPLE:

To provide contrasting colors for the background debris and parasites present; designed to allow examination and recognition of acid-fast characteristic of the organisms under high dry examination (40× objective for a total magnification of ×400). Designed primarily to allow recovery and identification of intestinal coccidian oocysts. Internal morphology (sporozoites) will be seen in some *Cryptosporidium* oocysts under oil immersion (×1000 magnification), while *Cyclospora* oocysts contain no specific internal morphology

SPECIMEN:

Any stool specimen that is fresh or preserved in formalin, SAF, or the newer single-vial-system fixatives

REAGENTS:

Kinyoun's acid-fast stain, modified Ziehl-Neelsen stain and their associated solutions; dehydrating solutions (alcohols and xylenes or xylene substitutes); mounting fluid optional; the decolorizing agents are less intense than the routine acid-alcohol used in routine acid-fast staining (this fact is what makes these procedures "modified" acid-fast procedures). The recommended decolorizer is a 1% to 3% sulfuric acid, no stronger; many laboratories are using 1% so the *Cyclospora* oocysts retain more of the color

EXAMINATION:

High dry examination (×400) of at least 300 fields; additional fields may be required if suspect organisms have been seen but are not clearly acid-fast

RESULTS AND LABORATORY REPORTS:

The identification of *Cryptosporidium* and *Isospora* oocysts should be possible; *Cyclospora* oocysts, which are twice the size of *Cryptosporidium* oocysts, should be visible but tend to be more acid-fast variable. Although microsporidia are acid-fast, their small size will make their recognition difficult. Final laboratory results would depend heavily on the appearance of the quality control (QC) slides and comparison with positive patient specimens

PROCEDURE NOTES AND LIMITATIONS:

Both the cold and hot modified acid-fast methods are excellent for the staining of coccidian oocysts. Some believe that the hot method may result in better stain penetration, but the differences are probably minimal. Procedure limitations are related to specimen handling (proper centrifugation speeds and time, use of no more than two layers of wet gauze for filtration, and complete understanding of the difficulties in recognizing microsporidial spores). There is also some controversy concerning whether the organisms lose the ability to take up acid-fast stains after long-term storage in 10% formalin. The organisms will be more difficult to find in specimens from patients who do not have the typical, watery diarrhea (more formed stool contains more artifact material)

---

methods of examination can be performed. All three are recommended; however, depending on the availability of trained personnel, proper fixatives, or the amount of specimen obtained, one or two procedures may be used. Many physicians who perform sigmoidoscopy procedures do not realize the importance of selecting the proper fixative for material to be examined for parasites. Even the most thorough examination will be meaningless if the specimen has been improperly prepared.[6]

PRINCIPLE:
To provide contrasting colors for the background debris and parasites present; designed to allow examination and recognition of organism morphology under oil immersion examination (100× objective for a total magnification of ×1000). Designed primarily to allow recovery and identification of intestinal microsporidial spores. Internal morphology (horizontal or diagonal "stripes" may be seen in some spores under oil immersion)

SPECIMEN:
Any stool specimen that is fresh or preserved in formalin, SAF, or the newer single-vial-system fixatives

REAGENTS:
Modified trichrome stain (using high-dye-content chromotrope 2R) and associated solutions; dehydrating solutions (alcohols, xylenes, or xylene substitutes); mounting fluid optional

EXAMINATION:
Oil immersion examination (×1000) of at least 300 fields; additional fields may be required if suspect organisms have been seen but are not clearly identified

RESULTS AND LABORATORY REPORTS:
The identification of microsporidial spores may be possible; however, their small size will make recognition difficult. Final laboratory results would depend heavily on the appearance of the QC slides and comparison with positive patient specimens

PROCEDURE NOTES AND LIMITATIONS:
Because of the difficulty in getting dye to penetrate the spore wall, this staining approach can be helpful. Procedure limitations are related to specimen handling (proper centrifugation speeds and time, use of no more than two layers of wet gauze for filtration, and complete understanding of the difficulties in recognizing microsporidial spores because of their small size [1 to 3.0 μm])

IMPORTANT QUESTIONS FOR COMMERCIAL SUPPLIERS:
Make sure to ask about specific fixatives and whether the fecal material can be stained with the modified trichrome stains and modified acid-fast stains. Also, ask whether the fixatives prevent the use of any of the newer fecal immunoassay methods now available for several of the intestinal amebae, flagellates, coccidia, and microsporidia

Duodenal Drainage. In infections with *G. lamblia* or *Strongyloides stercoralis,* routine stool examinations may not reveal the organisms. Duodenal drainage material can be submitted for examination, a technique that may reveal the parasites. The "falling leaf" motility often described for *Giardia* trophozoites is rarely seen in these preparations. The organisms may be caught in mucus strands, and the movement of the flagella on the *Giardia* trophozoites may be the only subtle motility seen for these flagellates. *Strongyloides* larvae will usually be very motile. Remember to keep the light intensity low.

The duodenal fluid may contain mucus, and the organisms tend to be found within the mucus. Therefore, centrifugation of the specimen is important, and the sedimented mucus should be examined. Fluorescent antibody or various immunoassay detection kits (*Cryptosporidium* or *Giardia*) can also be used with fresh or formalinized material.

If a presumptive diagnosis of giardiasis is obtained based on the wet preparation examination, the coverslip can be removed and the specimen can be fixed with either Schaudinn's fluid or polyvinyl alcohol (PVA) for subsequent staining with either trichrome or iron hematoxylin. If the amount of duodenal material submitted is very small, then permanent stains can be prepared rather than using any of the specimen for a wet smear examination. Some believe that this approach provides a more permanent record, and the potential problems with unstained organisms, very minimal motility, and a lower-power examination can be avoided by using oil immersion examination of the stained specimen at ×1000 magnification.[6]

Duodenal Capsule Technique (Entero-Test). The duodenal capsule technique is a simple and convenient method of sampling duodenal contents that eliminates the need for intestinal intubation. The technique involves using a length of nylon cord coiled inside a gelatin capsule. The cord protrudes through one end of the capsule and is taped to the side of the patient's face. The capsule is then swallowed. The gelatin dissolves in the stomach, and the weighted cord is carried by peristalsis into the duodenum. The cord is attached to the weight by a slipping mechanism; the weight is released and passes out in the stool when the cord is retrieved after 4 hours. The mucus collected on the cord is then examined for the presence of parasites, including *Strongyloides stercoralis, Giardia lamblia, Cryptosporidium* spp., microsporidia, and the eggs of *Clonorchis sinensis.*[6]

Urogenital Tract Specimens

The identification of *Trichomonas vaginalis* is usually based on the examination of wet preparations of vaginal and urethral discharges and prostatic secretions or urine sediment. Multiple specimens may need to be examined to detect the organisms. These specimens are diluted with a drop of saline and examined under low power (×100) and reduced illumination for the presence of actively motile organisms; as the jerky motility begins to diminish, it may be possible to observe the undulating membrane, particularly under high dry power (×400) (Figure 49-5). Stained smears are usually not necessary for the identification of this organism. Many times the number of false-positive and false-negative results reported based on stained smears strongly suggests the value of confirmation by observation of motile organisms from the direct mount, from appropriate culture media or more sensitive direct antigen detection methods such as the OSOM Trichomonas Rapid Test (Figure 49-6).[6,25]

Sputum

Although not one of the more common specimens, expectorated sputum may be submitted for examination

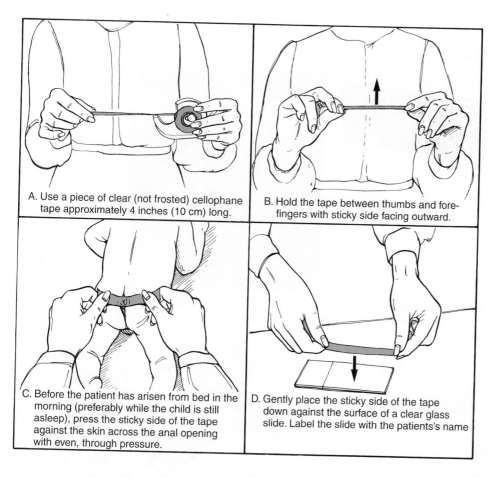

A. Use a piece of clear (not frosted) cellophane tape approximately 4 inches (10 cm) long.

B. Hold the tape between thumbs and forefingers with sticky side facing outward.

C. Before the patient has arisen from bed in the morning (preferably while the child is still asleep), press the sticky side of the tape against the skin across the anal opening with even, through pressure.

D. Gently place the sticky side of the tape down against the surface of a clear glass slide. Label the slide with the patients's name

Figure 49-4   Method for collection of a cellophane (Scotch) tape preparation for pinworm diagnosis. This method dispenses with the tongue depressor, requiring only tape and a glass microscope slide. The tape must be pressed deep into the anal crack.

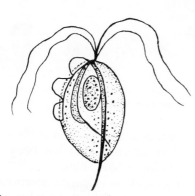

Figure 49-5   *Trichomonas vaginalis* trophozoite. (Illustration by Nobuko Kitamura.)

for parasites. Organisms in sputum that may be detected and may cause pneumonia, pneumonitis, or Loeffler's syndrome include the migrating larval stages of *Ascaris lumbricoides, Strongyloides stercoralis,* and hookworm; the eggs of *Paragonimus westermani; Echinococcus granulosus* hooklets; and the protozoa *Entamoeba histolytica, Entamoeba gingivalis, Trichomonas tenax, Cryptosporidium* spp., and possibly the microsporidia. Some of the smaller organisms must be differentiated from fungi such as *Candida* spp. and *Histoplasma capsulatum.* In a *Parago-*

nimus infection, the sputum may be viscous and tinged with brownish flecks, which are clusters of eggs ("iron filings"), and may be streaked with blood.

Induced sputa are collected after patients have used appropriate cleansing procedures to reduce oral contamination. The induction protocol is critical for the success of the procedure, and well-trained individuals are necessary to recover the organisms.

Aspirates

The examination of aspirated material for the diagnosis of parasitic infections may be extremely valuable, particularly when routine testing methods have failed to demonstrate the organisms.[6] These specimens should be transported to the laboratory immediately after collection. Aspirates include liquid specimens collected from a variety of sites where organisms might be found. Aspirates most commonly processed in the parasitology laboratory include fine-needle aspirates and duodenal aspirates. Fluid specimens collected by bronchoscopy include bronchoalveolar lavages and bronchial washings.

Fine-needle aspirates may be submitted for slide preparation and/or culture. Aspirates of cysts and abscesses for amebae may require concentration by centrifugation, digestion, microscopic examination for motile

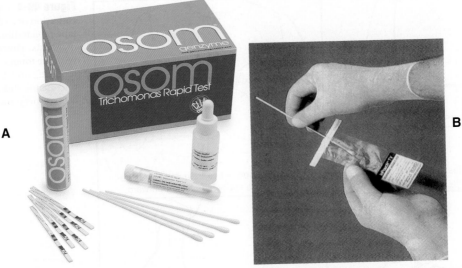

A

B

Figure 49-6 **A,** Rapid identification kit for *Trichomonas vaginalis*. **B,** Culture system for *T. vaginalis*. (**A** Courtesy Genzyme Diagnostics, Cambridge, Mass.)

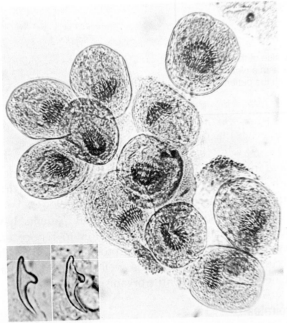

Figure 49-7 *Echinococcus granulosus,* hydatid sand (300×). *Inset,* Two individual hooklets (1000×).

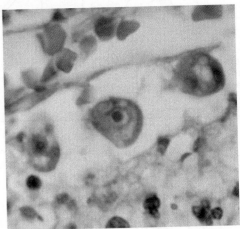

Figure 49-8 *Naegleria fowleri* in brain tissue. Hematoxylin and eosin stain. Note the large karyosome.

organisms in direct preparations, and cultures and microscopic evaluation of stained preparations. Aspiration of cyst material, usually liver or lung, for the diagnosis of hydatid disease is usually performed when open surgical techniques are used for cyst removal. The aspirated fluid is submitted to the laboratory and examined for the presence of hydatid sand (scolices) or hooklets; the absence of this material does not rule out the possibility of hydatid disease, in that some cysts are sterile (Figure 49-7).

Bone marrow aspirates for *Leishmania* amastigotes, *Trypanosoma cruzi* amastigotes, or *Plasmodium* spp. re-

quire Giemsa staining. Examination of these specimens may confirm an infection that has been missed by examination of routine blood films.

Cases of primary amebic meningoencephalitis are rare, but the examination of spinal fluid may reveal the causative agent, *Naegleria fowleri,* one of the free-living amebae (Figure 49-8).

Biopsy Specimens

Biopsy specimens are recommended for the diagnosis of tissue parasites.[6,17] The following procedures may be used for this purpose in addition to standard histologic preparations: impression smears and teased and "squash" preparations of biopsy tissue from skin, muscle, cornea, intestine, liver, lung, and brain. Tissue to be examined by permanent sections or electron microscopy should be fixed as specified by the laboratories that will process the tissue. In certain cases, a biopsy may be the only means of confirming a suspected parasitic problem. Specimens that are going to be examined as fresh material rather than as tissue sections should be kept moist in saline and submitted to the laboratory immediately.

Detection of parasites in tissue depends in part on specimen collection and having sufficient material to perform the recommended diagnostic procedures. Biopsy specimens are usually small and may not be representative of the diseased tissue. Multiple tissue samples often improve diagnostic results. To optimize the yield from any tissue specimen, examine all areas and use as many procedures as possible. Tissues are obtained from invasive procedures, many of which are very expensive and lengthy; consequently, these specimens deserve the most comprehensive procedures possible. A muscle biopsy can be obtained for the diagnosis of infection with *Trichinella* spp.; the specimen can be processed as a routine histology slide or can be examined as a "squash" preparation (Figure 49-9); the life cycle of this human tissue nematode can be seen in Figure 49-10.

Tissue submitted in a sterile container on a sterile sponge dampened with saline may be used for cultures of protozoa after mounts for direct examination or impression smears for staining have been prepared. If cultures for parasites will be made, use sterile slides for smear and mount preparation. Examination of tissue impression smears is detailed in Table 49-9.

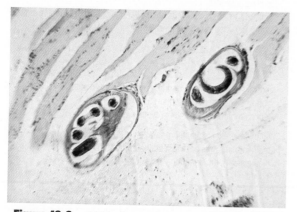

Figure 49-9   *Trichinella* spp. larvae encysted in muscle.

Blood

Depending on the life cycle, a number of parasites may be recovered in a blood specimen, either whole blood, buffy coat preparations, or various types of concentrations. Although some organisms may be motile in fresh, whole blood, normally species identification is accomplished from the examination of permanent stained blood films, both thick and thin films.* Blood films can be prepared from fresh, whole blood collected with no anticoagulants, anticoagulated blood, or sediment from the various concentration procedures. The recommended stain of choice is Giemsa stain; however, the parasites can also be seen on blood films stained with Wright's stain. Delafield's hematoxylin stain is often used to stain the microfilarial sheath; in some cases, Giemsa stain does not provide sufficient stain quality to allow differentiation of the microfilariae.

Thin Blood Films. In any examination of thin blood films for parasitic organisms, the initial screen should be carried out with the low-power objective (10×) of a microscope. Microfilariae may be missed if the entire thin film is not examined. Microfilariae are rarely present in large numbers, and frequently only a few organisms occur in each thin film preparation. Microfilariae are commonly found at the edges of the thin film or at the feathered end of the film because they are carried to these sites when the blood is spread. The feathered end of the film where the red blood cells (RBCs) are drawn out into one single, distinctive layer of cells should be examined for the presence of malaria parasites and trypanosomes. In these areas, the morphology and size of the infected RBCs are most clearly seen (Box 49-6).

Depending on the training and experience of the microscopist, examination of the thin film usually takes 15 to 20 minutes (≥300 oil immersion fields) for the thin film at a magnification of ×1000. Although some people use a 50× or 60× oil immersion objective to screen stained blood films, there is some concern that small parasites such as plasmodia, *Babesia* spp., or *L. donovani*

*References 6, 11, 12, 17, 19, 20.

Humans
Ingestion of undercooked pork
containing encysted larvae

Larvae carried via
bloodstream to muscles
(penetrate and encyst)

Larvae liberated
when cyst digested

Female worms
penetrate mucosa
and liberate larvae

Parasites mature
and mate in
upper intestine

Figure 49-10   Life cycle of *Trichinella spiralis.*

Table 49-9 Examination of Impression Smears

Tissue	Possible Parasite	Stain
Lung	Microsporidia	Modified trichrome, acid-fast stain, Giemsa, optical brightening agent (calcofluor), methenamine silver, EM*
	Toxoplasma gondii	Giemsa, immune-specific reagent
	Cryptosporidium spp.	Modified acid-fast stain, immune-specific reagent
	Entamoeba histolytica	Giemsa, trichrome
Liver	*Toxoplasma gondii*	Giemsa
	Leishmania donovani	Giemsa
	Cryptosporidium spp.	Modified acid-fast stain, immune-specific reagent
	Entamoeba histolytica	Giemsa, trichrome
Brain	*Naegleria fowleri*	Giemsa, trichrome
	Acanthamoeba spp.	Giemsa, trichrome
	Balamuthia mandrillaris (Leptomyxid ameba)	Giemsa, trichrome
	Entamoeba histolytica	Giemsa, trichrome
	Toxoplasma gondii	Giemsa, immune-specific reagent
	Microsporidia	Modified trichrome, acid-fast stain, Giemsa, optical brightening agent (calcofluor), methenamine silver, EM
	Encephalitozoon spp.	
Skin	*Leishmania* spp.	Giemsa
	Onchocerca volvulus	Giemsa
	Mansonella streptocerca	Giemsa
	Acanthamoeba spp.	Giemsa, trichrome
Nasopharynx, sinus cavities	Microsporidia	Modified trichrome, acid-fast stain, Giemsa, optical brightening agent (calcofluor), methenamine silver, EM
	Acanthamoeba spp.	Giemsa, trichrome
	Naegleria fowleri	Giemsa, trichrome
Intestine		
Small intestine	*Cryptosporidium* spp (both small and large intestine)	Modified acid-fast, immune-specific reagent
	Cyclospora cayetanensis	Modified acid-fast
Jejunum	Microsporidia	Modified trichrome, acid-fast stain, Giemsa, optical brightening agent (calcofluor), methenamine silver, EM
	Enterocytozoon bieneusi	
	Encephalitozoon (Septata) intestinalis	
Duodenum	*Giardia lamblia*	Giemsa, trichrome
Colon	*Entamoeba histolytica*	Giemsa, trichrome
Cornea, conjunctiva	Various genera of microsporidia	Acid-fast stain, Giemsa, modified trichrome, methenamine silver, optical brightening agent (calcofluor), EM
	Acanthamoeba spp.	Giemsa, trichrome, calcofluor (cysts)
Muscle	*Trichinella spiralis*	Wet examination, squash preparation
	Microsporidia	Modified trichrome, acid-fast stain, Giemsa, optical brightening agent (calcofluor), methenamine silver, EM
	Pleistophora sp., *Brachiola* sp., *Trachipleistophora* sp.	

*EM, Electron microscopy

may be missed at this smaller total magnification (×500 or ×600) compared with the ×1000 total magnification using the more traditional 100× oil immersion objective. Because people tend to scan blood films at different rates, it is important to examine a minimum number of fields. If something suspicious has been seen in the thick film, often the number of fields examined on the thin film may be considerably more than 300. The request for blood film examination should always be considered a STAT procedure, with all results (negative as well as positive) reported by telephone to the physician as soon as possible. If positive, notification of appropriate governmental agencies (local, state, and federal) should be done within a reasonable time frame in accordance with guidelines and laws.

Both malaria and *Babesia* infections have been missed with automated differential instruments, and therapy was delayed. Although these instruments are not designed to detect intracellular blood parasites, the inability of the automated systems to discriminate between uninfected RBCs and those infected with parasites may pose serious diagnostic problems.

BOX 49-6   Thin Blood Films—Review

PRINCIPLE:

To provide contrasting colors for the background debris and parasites present, either outside of the red blood cells (RBCs) or within the RBCs; designed to allow examination and recognition of detailed organism morphology under oil immersion examination (100× objective for a total magnification of ×1000). Designed primarily to allow recovery and identification of *Plasmodium* spp., *Babesia* spp., *Trypanosoma* spp., *Leishmania donovani,* and filarial blood parasites. The thin blood film is routinely used for specific parasite identification, although the number of organisms per field is significantly reduced compared with the thick blood film. The primary purpose is to allow malarial parasites to be seen within the RBCs and to assess the size of the infected RBCs compared with the uninfected RBCs. RBC morphology is preserved using this method

SPECIMEN:

Finger stick blood, whole blood, or anticoagulated blood (EDTA recommended)

REAGENTS:

Giemsa (films will have to be pre-fixed with absolute methanol before staining), Wright (fixative contained in the stain itself), or Wright/Giemsa stains and their associated solutions; mounting fluid optional

EXAMINATION:

Oil immersion examination (×1000) of at least 300 fields; additional fields may be required if suspect organisms have been seen in the thick blood film. The slide may be screened using the newer 50× or 60× oil immersion objectives but should not be reported until examination has been completed using the 100× oil immersion lens. A blood film must be examined totally at a lower power to rule out the presence of microfilariae; this is particularly important for proficiency testing specimens

RESULTS AND LABORATORY REPORTS:

The thin blood film is routinely used for parasite identification to the species level (*Plasmodium* spp.). Both the thick and thin films should be examined before reporting the final result

PROCEDURE NOTES AND LIMITATIONS:

The thin blood film is prepared exactly as one used for a differential count. A well-prepared film is thick at one end and thin at the other. It is mandatory to use clean, grease-free slides; long streamers of blood indicate the slide used as a spreader was dirty or chipped. Streaks in the film are usually caused by dirt; holes in the film indicate grease on the slide. Although Giemsa stain is the stain of choice, blood parasites can be seen using other stains; however, the parasite morphology/color may not be consistent with that described for Giemsa-stained organisms. Giemsa stain will not stain the sheath of *Wuchereria bancrofti;* hematoxylin-based stains are recommended (Delafield's hematoxylin)

BOX 49-7   Thick Blood Films—Review

PRINCIPLE:

To provide contrasting colors for the background debris and parasites present; designed to allow examination and recognition of detailed organism morphology under oil immersion examination (100× objective for a total magnification of ×1000). Designed primarily to allow recovery and identification of *Plasmodium* spp., *Babesia* spp., *Trypanosoma* spp., *Leishmania donovani,* and filarial blood parasites. The thick blood film is routinely used for the detection of parasites, since the number of organisms per field is much greater compared with the thin blood film. The primary purpose is to allow the examination of a larger volume of blood than that seen with the thin blood film. RBC morphology is not preserved using this method. The blood films must be laked before or during staining (rupture of all RBCs); the only structures that are left on the blood film are white blood cells, platelets, and parasites

SPECIMEN:

Finger stick blood, whole blood, or anticoagulated blood (EDTA recommended)

REAGENTS:

Giemsa (water-based stain, laking of the RBCs occurs during staining process), Wright (thick blood films must be laked before staining), or Wright/Giemsa stains and their associated solutions; mounting fluid optional

EXAMINATION:

Oil immersion examination (×1000) of at least 300 fields; additional fields may be required if suspect organisms have been seen in the thin blood film. The slide may be screened using the newer 50× or 60× oil immersion objectives but should not be reported until examination has been completed using the 100× oil immersion lens. A blood film must be examined totally at a lower power to rule out the presence of microfilariae. This is particularly true for proficiency testing specimens

RESULTS AND LABORATORY REPORTS:

The thick blood film is routinely used to detect the presence of parasites; final identification may require examination of the thin blood film. Both should be examined before reporting the final result

PROCEDURE NOTES AND LIMITATIONS:

The thick blood film is prepared by spreading a few drops of blood (using a circular motion) over an area approximately 2 cm in diameter. If using whole blood, continue stirring about 30 seconds to prevent the formation of fibrin strands. It is mandatory to use clean, grease-free slides. Allow the film to air dry at room temperature (never apply heat to these films). After the thick films are thoroughly dry, they can be laked to remove the hemoglobin. Although Giemsa stain is the stain of choice, blood parasites can be seen using other stains; however, the parasite morphology/color may not be consistent with that described for Giemsa-stained organisms. Giemsa stain will not stain the sheath of *Wuchereria bancrofti;* hematoxylin-based stains are recommended (Delafield's hematoxylin)

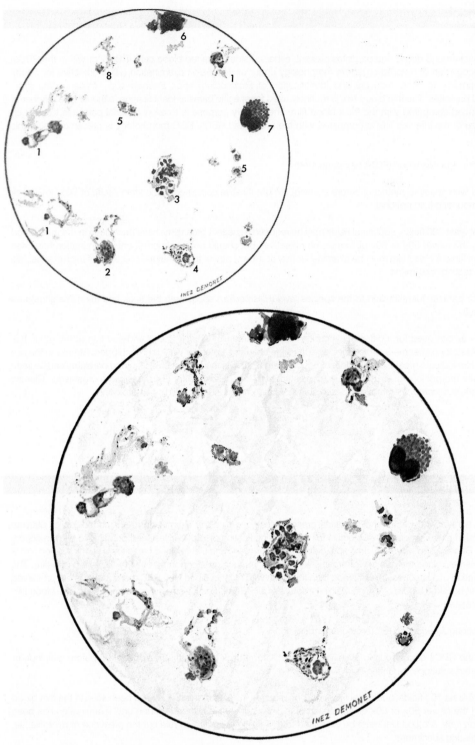

Figure 49-11 *Plasmodium vivax* in thick smear. **1,** Ameboid trophozoites. **2,** Schizont, two divisions of chromatin. **3,** Mature schizont. **4,** Microgametocyte. **5,** Blood platelets. **6,** Nucleus of neutrophil. **7,** Eosinophil. **8,** Blood platelet associated with cellular remains of young erythrocytes. (From Wilcox A: *Manual for the microscopical diagnosis of malaria in man,* Washington, DC, 1960, U.S. Public Health Service.)

Thick Blood Films. In the preparation of a thick blood film, the greatest concentration of blood cells is in the center of the film. The examination should be performed at low magnification to detect microfilariae more readily. Examination of a thick film usually requires 5 to 10 minutes (approximately 100 oil immersion fields). Search for malarial organisms and try-panosomes is best done under oil immersion (total magnification of ×1000). Intact RBCs are frequently seen at the very periphery of the thick film; such cells, if infected, may prove useful in malaria diagnosis because they may demonstrate the characteristic morphology necessary to identify the organisms to the species level (Box 49-7 and Figure 49-11).

Blood Film Stains. For accurate identification of blood parasites, a laboratory should develop proficiency in the use of at least one good staining method. It is better to select one method that will provide reproducible results than to do several on a hit-or-miss basis. Blood films should be stained as soon as possible, because prolonged storage may result in stain retention. Failure to stain positive malarial smears within a month may result in failure to demonstrate typical staining characteristics for individual species.

The most common stains are of two types. Wright's stain has the fixative in combination with the staining solution, so that both fixation and staining occur at the same time; therefore, the thick film must be laked before staining. In Giemsa stain, the fixative and stain are separate; thus, the thin film must be fixed with absolute methanol before staining.

Buffy Coat Films. *L. donovani,* trypanosomes, and *Histoplasma capsulatum* (a fungus with intracellular elements resembling those of *L. donovani*) may occasionally be detected in the peripheral blood. The parasite or fungus will be found in the large mononuclear cells that are found in the buffy coat (a layer of white blood cells resulting from centrifugation of whole citrated blood). The nuclear material will stain dark red-purple, and the cytoplasm will be light blue *(L. donovani).* *H. capsulatum* will appear as a large dot of nuclear material (dark red-purple) surrounded by a clear halo area. Trypanosomes in the peripheral blood will also concentrate with the buffy coat cells.

DIRECT DETECTION METHODS

Progress has been made in the development and application of molecular methods for diagnosis, including the use of purified or recombinant antigens and nucleic acid probes.[6,8,9] The detection of parasite-specific antigen is more indicative of current disease. Many of the assays were originally developed with polyclonal antibodies that were targeted to unpurified antigens that markedly decreased the sensitivity and specificity of the tests. Nucleic acid–based diagnostic tests for parasitology are primarily available only in specialized research or reference centers. Polymerase chain reaction (PCR) and other nucleic acid probe tests have been reported for almost all species of parasites. The only nucleic acid–based probe test commercially available is for the detection of *T. vaginalis.* As the costs of these tests decrease and the various steps necessary to perform the tests become automated, demand will increase for commercially available reagents.

Intestinal Parasites

Immunoassays are generally simple to perform and allow for many tests to be performed simultaneously, thereby reducing overall costs. A major disadvantage of antigen detection in stool specimens is that the method can detect only one or two pathogens at one time. A routine O&P examination must be performed to detect other parasitic pathogens. The current commercially available antigen tests (DFA, EIA, IFA, or the cartridge formats) have excellent sensitivity and specificity compared with routine microscopy.[6,8,9] Available antigen detection tests are listed in Table 49-10. Currently, the most common immunoassays are designed to confirm infection with *Entamoeba histolytica,* the *Entamoeba histolytica/Entamoeba dispar* group, *Giardia lamblia,* and *Cryptosporidium* spp. Test formats include fluorescence, enzyme immunoassays, and immunochromatographic test strips, often used in test cartridges.[6,8,9]

Blood Parasites

Several new blood parasite antigen detection systems are available and have been found to be very effective in field trials. Most of these procedures are based on an antigen capture approach and have been incorporated in a dipstick format. These kits are listed in Table 49-11.[6,21]

CULTIVATION

Nematode infections giving rise to larval stages that hatch in soil or in tissues may be diagnosed by using certain fecal culture methods to concentrate the larvae. *Strongyloides stercoralis* larvae are generally the most common larvae found in stool specimens. Depending on the fecal transit time through the intestine and the patient's condition, rhabditiform and rarely filariform larvae may be present. Also, if stool examination is delayed, then embryonated ova as well as larvae of hookworm may be present. Culture of feces for larvae is useful to (1) reveal their presence when they are too scanty to be detected by concentration methods, (2) distinguish whether the infection is due to *S. stercoralis* or hookworm based on rhabditiform larval morphology by allowing hookworm egg hatching to occur, releasing first-stage larvae, and (3) allow development of larvae into the filariform stage for further differentiation.[6,11,20]

Very few clinical laboratories offer specific culture techniques for protozoan parasites. The methods for in vitro culture are often complex, whereas quality control is difficult and not really feasible for the routine diagnostic laboratory. In certain institutions, some techniques may be available, particularly where consultative services are provided and for research.[6]

Cultures of parasites grown in association with an unknown flora are referred to as *xenic cultures.* A good example of this type of culture is stool specimens cultured for *Entamoeba histolytica.* If the parasites are grown with a single known bacterium, the culture is referred to as *monoxenic.* An example of this type of

Table 49-10 Commercially Available Kits for Immunodetection of Parasitic Organisms or Antigens in Stool Samples

Organism and Kit Name*	Manufacturer and/or Distributor	Type of Test	Comments[†]
Cryptosporidium spp.			The tests detect *C. hominis*, different *C. parvum* genotypes, and other species depending on intensity of the infection
ProSpecT *Cryptosporidium* Microplate Assay	Remel	EIA	Can be used with fresh, frozen, or formalin-preserved stool www.remel.com
XPECT *Cryptosporidium* Kit	Remel	Cartridge	
		EIA	Contact manufacturer; www.med-chem.com
Cryptosporidium Antigen ELISA	Medical Chemical	EIA	Can be used with fresh, frozen, or formalin-preserved stool
Crypto-CELISA	Cellabs		www.cellabs.com
Crypto Cel		DFA	
Cryptosporidium Test	TechLab		
Cryptosporidium Test	Wampole	EIA	www.techlabinc.com
		EIA	www.wampolelabs.com
Entamoeba histolytica			These tests will differentiate between *E. histolytica* and *E. dispar*
ProSpecT *Entamoeba histolytica* Microplate Assay	Remel	EIA	
Entamoeba histolytica II Test	TechLab	EIA	Requires fresh or frozen stool
Entamoeba CELISA Path	Cellabs	EIA	Contact manufacturer
Entamoeba histolytica Test	Wampole	EIA	Contact manufacturer
Giardia lamblia			
ProSpecT *Giardia* Microplate Assay	Remel	EIA	Different EIA formats—contact manufacturer
GiardEIA	Antibodies, Inc.	EIA	www.antibodiesinc.com
Giardia Antigen ELISA	Medical Chemical	EIA	Contact manufacturer
Giardia-CELISA	Cellabs	DFA	Contact manufacturer
Giardia-Cel	Cellabs	EIA	
Giardia	Wampole	EIA	See TechLab
SIMPLE-READ *Giardia*	Medical Chemical	Cartridge device, lateral flow	Can be used with fresh, frozen, or formalin-preserved stool
XPECT *Giardia*	Remel	Cartridge device, lateral flow	Can be used with fresh, frozen, or formalin-preserved stool
Giardia II	TechLab	EIA	
Giardia Test	Wampole	EIA	
Combination tests: *Cryptosporidium* and *Giardia*			
ProSpecT *Giardia/Cryptosporidium* Microplate Assay	Remel	EIA	Can be used with fresh, frozen, or formalin-preserved stool
MERIFLUOR *Cryptosporidium/Giardia*	Meridian Bioscience	DFA	www.meridianbioscience.com
Cryptosporidium/Giardia DFA	Medical Chemical	DFA	Contact manufacturer
Crypto/Giardia-Cel	Cellabs	DFA	Contact manufacturer
ColorPAC *Giardia/Cryptosporidium* Rapid Assay	Becton-Dickinson	Cartridge device, lateral flow	Can be used with fresh, frozen, or formalin-preserved stool www.bd.com
ImmunoCard STAT! *Cryptosporidium/Giardia*	Meridian Bioscience	Cartridge device, lateral flow	Can be used with fresh, frozen, or formalin-preserved stool
Xpect *Giardia/Cryptosporidium*	Remel	Cartridge	Can be used with fresh, frozen, or formalin-preserved stool;
Combination tests: *Cryptosporidium*, *Giardia*, and *Entamoeba*			
Triage Parasite Panel	Biosite Diagnostics, Inc.	Cartridge device, lateral flow	Requires fresh or frozen stool; combination test with *Giardia* and *E. histolytica/E. dispar* group; will not differentiate between *E. histolytica/E. dispar* www.biosite.com

*A number of the kits are manufactured by a single manufacturer but are labeled under different company names; consequently, some of the data for sensitivity and specificity may be identical to kits with another name/company.

[†]Website addresses are given only the first time the company appears in the table.

Table 49-11 Commercially Available Test Kits for Immunodetection or Molecular Detection of Parasitic Organisms or Antigens in Serum, Plasma, Blood, or Vaginal Discharge*

Manufacturer	Test	Company Websites	Usable Specimen
Malaria			
ACON Laboratories	Malaria P.f. Rapid Test	www.aconlabs.com	Whole blood
AmeriTek, Inc.	One Step Malaria Test	www.ameritek.org	Whole blood
Binax	NOW Malaria P.f./P.v.	www.binax.com	Whole blood
Bio-Quant, Inc	One Step Malaria Test	www.bio-quant.com	Whole blood
Core Diagnostics Ltd.	CORE Malaria (Pf, Pf/Pv, Pan Pf)	www.corediag.com	Whole blood
Cortez Diagnostics, Inc.	OneStep RapiCard InstaTest	www.rapidtest.com	Whole blood
DiaMed SA	OptiMAL Rapid Malaria Test, Optimal –IT Rapid Malaria Test	www.diamed.com	Whole blood
Genix Technology	Malaria (P. fal.) Ag	www.genixtech.com	Whole blood
MP Biomedicals Asia Pacific Pte. Ltd.	ASSURE Malaria P.f. Rapid Test	www.genelabs.com.sg	Whole blood
Global eMed, LLC	Smart Strip Malaria Test	www.globalemed.com	Whole blood
International Immuno-Diagnostics	One-Step Malaria (PF/PV)	www.intlimmunodiagnostics.com	Whole blood
Kat-Medical	KAT-Quick Malaria test	www.katmedical.com	Whole blood
Mega Diagnostics, Inc.	MegaKwik Malaria (Pf) Card Test	www.mega-dx.com	Whole blood
Orchid Biomedical Systems	Paracheck Pf	www.tulipgroup.com/Orchid/	Plasma, whole blood
Premier Medical Corporation	First Response Malaria P.f / P.v Antigen Strips	www.premiermedcorp.com	Whole blood
Princeton BioMeditech Corporation	BioSign Malaria	www.pbmc.com	Whole blood
SPAN Diagnostics	ParaHIT f	www.span.co.in	Whole blood
Standard Diagnostics Inc.	Malaria P.f/P.v Antigen Test	www.standardia.com	Whole blood, serum
Trinity Biotech	Rapid Uni-Gold Malaria (Pf.)	www.trinitybiotech.com	Whole blood
Vencor International Inc.	DxStrip Malaria Combo	www.vencorinternational.com	Whole blood
Filariasis			
Cellabs Pty Ltd	CELISA	www.cellabs.com.au	Whole blood, serum
Binax Inc.	NOW Filariasis Test	www.binax.com	Whole blood, plasma, serum
Trichomonas vaginalis			
Genzyme Diagnostics	OSOM Trichomonas Rapid Test	www.genzymediagnostics.com	Genital swab
Xenotope Diagnostics Inc.	XenoStrip-Tv rapid test	www.xenotope.com	Genital swab
Thermo Electron corp.	Acceava Trichomonas	www.thermo.com	Genital swab
Kalon Biological limited	Kalon Tvaginalis latex kit	www.kalonbio.co.uk	Genital swab
Chemicon LTD	Light Diagnostics *T. vaginalis* DFA Kit	www.chemicon.com	Genital swab
Becton-Dickinson DNA Probe	BD AffirmVP III	www.bd.com	Genital swab

*For detailed information, see www.rapid-diagnostics.org. It is very important to verify FDA approval within the United States for any product being considered for purchase; not all products listed are FDA approved.

culture is the clinical specimen (corneal biopsy) cultured with *Escherichia coli* as a means of recovering species of *Acanthamoeba* and *Naegleria*. If parasites are grown as pure culture without any bacterial associate, the culture is referred to as *axenic*. An example of this type of culture is the use of media for the isolation of *Leishmania* spp. or *Trypanosoma cruzi*.

Larval Stage Nematodes

The use of certain fecal culture methods (sometimes referred to as *coproculture*) is especially helpful to detect light infections of hookworm, *S. stercoralis,* and *Trichostrongylus* spp. and for specific identification of parasites. The rearing of infective-stage nematode larvae also helps in the specific diagnosis of hookworm and trichostrongyle infections because the eggs of many of these species are identical and specific identifications are based

on larval morphology. Additionally, such techniques are useful for obtaining numerous infective-stage larvae for research purposes. Diagnostic methods that are available include the Harada-Mori filter paper strip culture, the Petri dish filter paper culture, the agar plate method, the charcoal culture, and the Baermann concentration.[6,11,20]

Protozoa

Few parasites can be routinely cultured, and the only procedures that are in general use are for *Entamoeba histolytica, Naegleria fowleri, Acanthamoeba* spp., *Trichomonas vaginalis, Toxoplasma gondii, Trypanosoma cruzi,* and the leishmanias.* These procedures are usually available only after consultation with the laboratory and on special request. Any laboratory providing these types of

*References 3, 6, 14, 16, 24, 25.

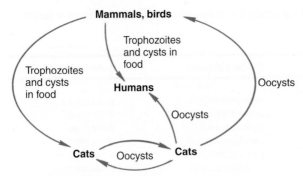

Figure 49-12 Life cycle of *Toxoplasma gondii.*

cultures must also be willing to maintain stock quality control (QC) cultures of specific organisms, often obtained from the American Type Culture Collection (ATCC).[6] The relevant QC organism would be cultured at the same time as the patient specimen, thus providing some assurance that the culture system was performing properly.

SERODIAGNOSIS

Although parasites and their by-products are immunogenic for the host, the host immune response is usually not protective. Any immunity that does develop is usually species specific and even strain or stage specific. Human parasites are generally divided into two groups: (1) those that multiply within the host (e.g., protozoa) and (2) those that mature within the host but never multiply (e.g., schistosomes, *Ascaris*). In infections caused by protozoa that multiply within the host, there is continuous antigenic stimulation of the host's immune system as the infection progresses. In these instances, a positive correlation usually exists between clinical symptoms and serologic test results. *Toxoplasma gondii,* one of the tissue protozoa, is acquired by humans via ingestion of stages contained in different types of meat, although the life cycle does include stages within animal hosts. The diagnosis of toxoplasmosis is primarily through serologic means, rather than from the identification of organisms in human clinical specimens (Figure 49-12).

In contrast to the protozoa, helminths often migrate through the body and pass through a number of developmental stages before becoming mature adults. These infections are often difficult to confirm serologically, probably because of a limited antigenic response by the host or failure to use the appropriate antigen in the test system. Most parasitic antigens used in serologic procedures are heterogeneous mixtures that are not well defined. Test results using such antigens may represent cross-reactions or poor sensitivity.

Serologic procedures have been available for many years; however, they are not routinely offered by most clinical laboratories because of high cost, lack of trained personnel, low number of test orders, and problems with sensitivity, specificity, and interpretation. Standard techniques that have been used include complement fixation (CF), indirect hemagglutination (IHA), indirect fluorescent antibody (IFA), soluble antigen fluorescent antibody, bentonite flocculation, latex agglutination (LA), double diffusion, counterelectrophoresis, immunoelectrophoresis, radioimmunoassay, and intradermal tests.[30]

The Centers for Disease Control and Prevention (CDC) offers a number of serologic procedures for diagnostic purposes, some of which are not available elsewhere. Because regulations regarding submission of specimens may vary from state to state, each laboratory should check with its own county or state department of public health for appropriate instructions. Additional information on procedures, availability of skin test antigens, and interpretation of test results may be obtained directly from the CDC by writing or calling:

Serology Unit Parasitology Diseases Branch
Building 4 Room 1009
Mail Stop F13
Centers for Disease Control and Prevention
4770 Buford Highway
Atlanta, GA 30034
Serology (770) 488-7760
Chagas' Disease and Leishmaniasis (770) 488-4474
Malaria (770) 488-7765

ORGANISM IDENTIFICATION

INTESTINAL, OTHER BODY SITE PROTOZOA

The protozoa are unicellular organisms, most of which are microscopic. They possess a number of specialized organelles, which are responsible for life functions and which allow further division of the group into classes.

The important characteristics of the intestinal protozoa are found in Tables 49-12 to 49-19. The clinically important intestinal protozoa are generally considered to be *Entamoeba histolytica, Dientamoeba fragilis, Giardia lamblia, Balantidium coli, Isospora belli, Cryptosporidium* spp., *Cyclospora cayetanensis,* and the microsporidia.

Amebae

The class Sarcodina, or amebae, contains the organisms that move by means of cytoplasmic protrusions called *pseudopodia.* Included in this group are free-living organisms, as well as nonpathogenic and pathogenic organisms found in the intestinal tract and other areas of the body (see Tables 49-12 and 49-13).

Table 49-12 Intestinal Protozoa: Trophozoites of Common Amebae

Characteristic	Entamoeba histolytica	Entamoeba dispar	Entamoeba hartmanni	Entamoeba coli	Endolimax nana	Iodamoeba bütschlii
Size* (diameter or length)	12-60 µm; usual range: 15-20 µm; invasive forms may be larger than 20 µm	Same size range as Entamoeba histolytica	5-12 µm; usual range: 8-10 µm	15-50 µm; usual range: 20-25 µm	6-12 µm; usual range: 8-10 µm	8-20 µm; usual range: 12-15 µm
Motility	Progressive, with hyaline, fingerlike pseudopodia; motility may be rapid	Same motility as E. histolytica	Usually nonprogressive	Sluggish nondirectional, with blunt, granular pseudopodia	Sluggish, usually nonprogressive	Sluggish, usually nonprogressive
Nucleus: number and visibility	Difficult to see in unstained preparations; 1 nucleus	Difficult to see in unstained preparations; 1 nucleus	Usually not seen in unstained preparations; 1 nucleus	Often visible in unstained preparation; 1 nucleus	Occasionally visible in unstained preparations; 1 nucleus	Usually not visible in unstained preparations; 1 nucleus
Peripheral chromatin (stained)	Fine granules, uniform in size and usually evenly distributed; may have beaded appearance	Fine granules, uniform in size and usually evenly distributed; may have beaded appearance	Nucleus may stain more darkly than E. histolytica although morphology is similar; chromatin may appear as solid ring rather than beaded (trichrome)	May be clumped and unevenly arranged on the membrane; may also appear as solid, dark ring with no beads or clumps	Usually no peripheral chromatin; nuclear chromatin may be variable	Usually no peripheral chromatin
Karyosome (stained)	Small, usually compact; centrally located but may also be eccentric	Small, usually compact; centrally located but may also be eccentric	Usually small and compact; may be centrally located or eccentric	Large, not compact; may or may not be eccentric; may be diffuse and darkly stained	Large, irregularly shaped; may appear "blotlike"; many nuclear variations are common; may mimic E. hartmanni or Dientamoeba fragilis	Large; may be surrounded by refractile granules that are difficult to see ("basket nucleus")
Cytoplasm appearance (stained)	Finely granular, "ground glass" appearance; clear differentiation of ectoplasm and endoplasm; if present, vacuoles are usually small	Finely granular, "ground glass" appearance; clear differentiation of ectoplasm and endoplasm; if present, vacuoles are usually small	Finely granular	Granular, with little differentiation into ectoplasm and endoplasm; usually vacuolated	Granular, vacuolated	Granular, may be heavily vacuolated
Inclusions (stained)	Noninvasive organism may contain bacteria; presence of red blood cells (RBCs) diagnostic; the presence of RBCs is the only morphologic characteristic that allows differentiation between pathogenic E. histolytica and nonpathogenic E. dispar	Organisms usually contain bacteria; RBCs not present in cytoplasm	May contain bacteria; no RBCs	Bacteria, yeast, other debris	Bacteria	Bacteria

*These sizes refer to wet preparation measurements. Organisms on a permanent stained smear may be 1 to 1.5 µm smaller as a result of artificial shrinkage.

Table 49-13 Intestinal Protozoa: Cysts of Common Amebae

Characteristic	Entamoeba histolytica/ Entamoeba dispar	Entamoeba hartmanni	Entamoeba coli	Endolimax nana	Iodamoeba bütschlii
Size* (diameter or length)	10-20 μm; usual range: 12-15 μm	5-10 μm; usual range: 6-8 μm	10-35 μm; usual range: 15-25 μm	5-10 μm; usual range: 6-8 μm	5-20 μm; usual range: 10-12 μm
Shape	Usually spherical	Usually spherical	Usually spherical; may be oval, triangular, or other shapes; may be distorted on permanent stained slide because of inadequate fixative penetration	Usually oval, may be round	May vary from oval to round; cyst may collapse as a result of large glycogen vacuole space
Nucleus: number and visibility	Mature cyst: 4; immature: 1 or 2 nuclei; nuclear characteristics difficult to see on wet preparation	Mature cyst: 4; immature: 1 or 2 nuclei; two nucleated cysts very common	Mature cyst: 8; occasionally, 16 or more nuclei may be seen; immature cysts with 2 or more nuclei are occasionally seen	Mature cyst: 4; immature cysts: 2; very rarely seen and may resemble cysts of Enteromonas hominis	Mature cyst: 1
Peripheral chromatin (stained)	Peripheral chromatin present; fine, uniform granules, evenly distributed; nuclear characteristics may not be as clearly visible as in trophozoite	Fine granules evenly distributed on the membrane; nuclear characteristics may be difficult to see	Coarsely granular; may be clumped and unevenly arranged on membrane; nuclear characteristics not as clearly defined as in trophozoite; may resemble E. histolytica	No peripheral chromatin	No peripheral chromatin
Karyosome (stained)	Small, compact, usually centrally located but occasionally may be eccentric	Small, compact, usually centrally located	Large, may or may not be compact and/or eccentric; occasionally may be centrally located	Smaller than karyosome seen in trophozoites, but generally larger than those of the genus Entamoeba	Larger, usually eccentric refractile granules may be on one side of karyosome ("basket nucleus")
Cytoplasm, chromatoidal bodies (stained)	May be present; bodies usually elongate, with blunt, rounded, smooth edges; may be round or oval	Usually present; bodies usually elongate with blunt, rounded, smooth edges; may be round or oval	May be present (less frequently than E. histolytica); splinter shaped with rough, pointed ends	Rare chromatoidal bodies present; occasionally, small granule or inclusions seen; fine linear chromatoidals may be faintly visible on well-stained smears	No chromatoidal bodies present; occasionally, small granules may be present
Glycogen (stained with iodine)	May be diffuse or absent in mature cyst; clumped chromatin mass may be present in early cysts (stains reddish brown with iodine)	May or may not be present as in E. histolytica/E. dispar	May be diffuse or absent in mature cyst; clumped mass occasionally seen in mature cysts (stains reddish brown with iodine)	Usually diffuse if present (will stain reddish brown with iodine)	Large, compact, well-defined mass (will stain reddish brown with iodine)

*Wet preparation measurements (permanent stains: organisms usually measure 1-2 μm less).

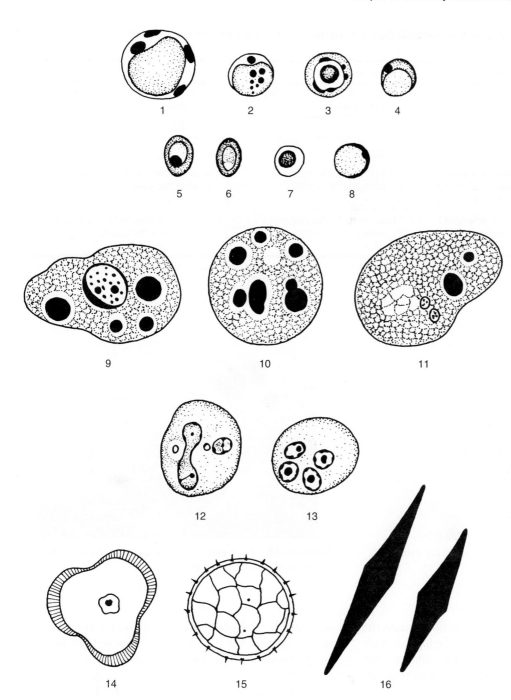

Figure 49-13   Various structures that may be seen in stool preparations. **1, 2,** and **4,** *Blastocystis hominis.* **3** and **5** to **8,** Various yeast cells. **9,** Macrophage with nucleus. **10** and **11,** Deteriorated macrophage without nucleus. **12** and **13,** Polymorphonuclear leukocytes. **14** and **15,** Pollen grains. **16,** Charcot-Leyden crystals. (Modified from Markell EK, Voge M: *Medical parasitology,* ed 5, Philadelphia, 1981, WB Saunders. Illustration by Nobuko Kitamura.)

Occasionally, when fresh stool material is examined as a direct wet mount, motile trophozoites may be seen, as well as other nonparasitic structures (Figure 49-13). The true pathogen, *E. histolytica,* has directional and progressive motility, whereas the other amebae tend to move more slowly and at random. Also, the cytoplasm is generally more finely granular and the presence of RBCs in the cytoplasm is considered to be diagnostic for *E. histolytica* (Figure 49-14). However, these characteristics are rarely seen in wet smear preparations;

morphology is best seen on the permanent stained smear. *E. histolytica* is one of the most important pathogens and may invade other tissues of the body, resulting in severe symptoms and possible death.[4,6,28]

Recent evidence from molecular studies confirms the differentiation of pathogenic *E. histolytica* from nonpathogenic *E. dispar* (Figure 49-15) as two distinct species.[4,6] *E. histolytica* is considered to be the etiologic agent of amebic colitis and extraintestinal abscesses (amebic liver abscess), whereas nonpathogenic *E. dispar*

Table 49-14   Comparison of Free-Living Amebae: *Naegleria fowleri, Acanthamoeba* spp., and *Balamuthia mandrillaris*

Characteristic	*Naegleria fowleri*	*Acanthamoeba* spp.	*Balamuthia mandrillaris*
Trophozoite	Biphasic (amebic) and flagellate forms; 8-15 μm; lobate pseudopodia (amebic form)	Large (15-25 μm): no flagella; filiform pseudopodia	Large (15-60 μm); no flagella; branched pseudopods
Cysts	Not present in tissue; small, smooth, rounded	Present in tissue; large with wrinkled double wall	Present in tissue; large (15-30 μm); outer wall thick and irregular
Growth on media	Require living cells (bacteria or cell culture); do not grow with >0.4% NaCl	May grow without bacteria; not affected by 0.85% NaCl	Does not grow well on bacteria-seeded nonnutrient agar plates
Appearance in tissue*	Smaller than *Acanthamoeba;* dense endoplasm; less distinct nuclear staining	Large; rounded; less endoplasm; nucleus more distinct	Very similar to *Acanthamoeba;* may occasionally see two nuclei in the trophozoites

Entamoeba histolytica has a delicate nuclear membrane and a small, pale-staining nucleolus. Freshwater amebae have a distinct nuclear membrane and a large, deep-staining nucleolus.

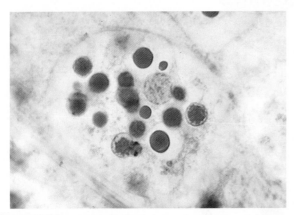

Figure 49-14   *Entamoeba histolytica* trophozoite containing ingested red blood cells.

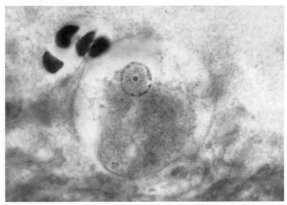

Figure 49-15   *Entamoeba histolytica/E. dispar* trophozoite; no ingested red blood cells are present.

produces no intestinal symptoms and is not invasive in humans. Nuclear and cytoplasmic morphology of both the trophozoite and cyst stages of the amebae are important criteria for organism identification (Figures 49-14 to 49-28).

Another organism, *Blastocystis hominis,* (see Figures 49-3 and 49-13), is considered to be potentially pathogenic for humans. Although usually listed with the amebae, its classification is still under review and there are continuing reports of symptomatic infection in humans.[6,22,27] However, others believe this organism may be an incidental finding and symptoms are really caused by other proven pathogens found in very low numbers. Although for a number of years the true role of this organism in terms of colonization or disease was still somewhat controversial, it is now generally considered a causative agent of intestinal disease. It has also been suggested that ribodemes I, III, and VI may be responsible for gastrointestinal symptoms,[22] which would account for why some patients tend to become symptomatic with this infection, while others remain asymptomatic.

The current recommendation is to report the presence of *B. hominis* and quantitate it from the permanent stained smear (i.e., rare, few, moderate, many, packed); this information may be valuable in helping assess the pathogenicity of the organism.

Members of the free-living amebae, *Acanthamoeba* spp. and *Naegleria fowleri,* may be identified from stained smears of culture material (see Table 49-14). Key characteristics include the pseudopods, *Naegleria* having lobed pseudopods and *Acanthamoeba* having spiky pseudopodia. Also, when organisms from culture plates are placed in distilled water, *Naegleria fowleri* undergoes transformation within a few hours to a pear-shaped flagellate, usually with two flagella. The flagellate stage

Table 49-15   Intestinal Protozoa: Trophozoites of Flagellates

Characteristic	Shape and Size	Motility	Number of Nuclei and Visibility	Number of Flagella (usually difficult to see)	Other Features
Dientamoeba fragilis	Shaped like amebae. 5-15 μm; usual range: 9-12 μm	Usually nonprogressive; pseudopodia are angular, serrated, or broad lobed and almost transparent	Percentage may vary, but 40% of organisms have 1 nucleus and 60% have 2 nuclei; not visible in unstained preparations; no peripheral chromatin; karyosome is composed of a cluster of 4-8 granules	No visible flagella	Cytoplasm finely granular and may be vacuolated with ingested bacteria, yeasts, and other debris; may be great variation in size and shape on a single smear
Giardia lamblia	Pear-shaped, 10-20 μm; width, 5-15 μm	"Falling leaf" motility may be difficult to see if organism in mucus	2; not visible in unstained mounts	4 lateral, 2 ventral, 2 caudal	Sucking disc occupying ½-¾ of ventral surface; pear-shaped front view; spoon-shaped side view
Chilomastix mesnili	Pear-shaped, 6-24 μm; usual range: 10-15 μm; width, 4-8 μm	Stiff, rotary	1; not visible in unstained mounts	3 anterior, 1 in cytostome	Prominent cytostome extending ⅓-½ length of body; spiral groove across ventral surface
Pentatrichomonas hominis	Pear-shaped, 5-15 μm; usual range: 7-9 μm; width 7-10 μm	Jerky, rapid	1; not visible in unstained mounts	3-5 anterior, 1 posterior	Undulating membrane extends length of the body; posterior flagellum extends free beyond end of body
Trichomonas tenax	Pear shaped, 5-12 μm; average of 6.5-7.5 μm; width, 7-9 μm	Jerky, rapid	1; not visible in unstained mounts	4 anterior, 1 posterior	Seen only in preparations from mouth; axostyle (slender rod) protrudes beyond the posterior end and may be visible; posterior flagellum extends only halfway down body and there is no free end
Enteromonas hominis	Oval, 4-10 μm; usual range 8-9 μm; width, 5-6 μm	Jerky	1; not visible in unstained mounts	3 anterior, 1 posterior	One side of body flattened; posterior flagellum extends free posteriorly or laterally
Retortamonas intestinalis	Pear-shaped or oval, 4-9 μm; usual range: 6-7 μm; width, 3-4 μm	Jerky	1; not visible in unstained mounts	1 anterior, 1 posterior	Prominent cytostome extending approximately ½ length of body

Table 49-16   Intestinal Protozoa: Cysts of Flagellates

Species	Size	Shape	Number of Nuclei	Other Features
Dientamoeba fragilis, Pentatrichomonas hominis, Trichomonas tenax	No cyst stage			
Giardia lamblia	8-19 μm; usual range: 11-14 μm; width, 7-10 μm	Oval, ellipsoidal, or may appear round	4; not distinct in unstained preparations; usually located at one end	Longitudinal fibers in cysts may be visible in unstained preparations; deep staining median bodies usually lie across the longitudinal fibers; there is often shrinkage and the cytoplasm pulls away from the cyst wall; there may also be a "halo" effect around the outside of the cyst wall as a result of shrinkage caused by dehydrating reagents
Chilomastix mesnili	6-10 μm; usual range: 7-9 μm; width, 4-6 μm	Lemon shaped with anterior hyaline knob	1; not distinct in unstained preparations	Cytostome with supporting fibrils, usually visible in stained preparation; curved fibril along side of cytostome usually referred to as "shepherd's crook"
Enteromonas hominis	4-10 μm; usual range: 6-8 μm; width, 4-6 μm	Elongate or oval	1-4; usually 2 lying at opposite ends of cyst; not visible in unstained mounts	Resembles *E. nana* cyst; fibrils or flagella usually not seen
Retortamonas intestinalis	4-9 μm; usual range: 4-7 μm; width, 5 μm	Pear shaped or slightly lemon shaped	1; not visible in unstained mounts	Resembles *Chilomastix* cyst; shadow outline of cytostome with supporting fibrils extends above nucleus; bird beak fibril arrangement

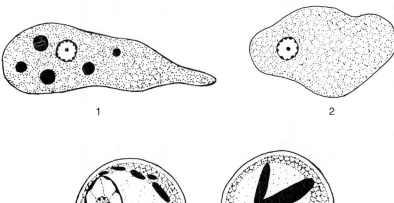
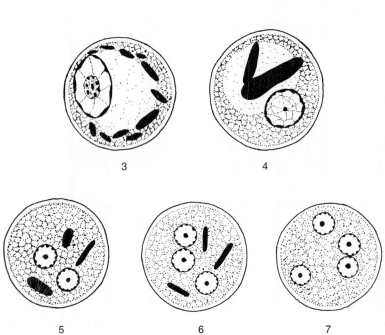

Figure 49-16   **1,** Trophozoite of *Entamoeba histolytica* (note ingested red blood cells). **2,** Trophozoite of *Entamoeba histolytica/Entamoeba dispar* (morphology does not allow differentiation between the two species). **3** and **4,** Early cysts of *E. histolytica/E. dispar.* **5** to **7,** Cysts of *E. histolytica/E. dispar.*

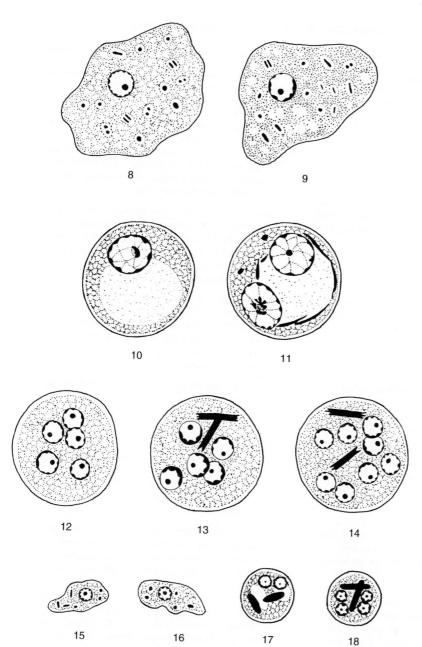

Figure 49-16   (*cont'd*) **8** and **9**, Trophozoites of *Entamoeba coli*. **10** and **11**, Early cysts of *E. coli*. **12** to **14**, Cysts of *E. coli*. **15** and **16**, Trophozoites of *E. hartmanni*. **17** and **18**, Cysts of *E. hartmanni* (From Garcia LS: *Diagnostic medical parasitology*, ed 5, Washington, DC, 2007, ASM Press. Illustrations 4 and 11 by Nobuko Kitamura.)

Table 49-17   Intestinal Protozoa: Ciliates

Species	Shape and Size	Motility	Number of Nuclei	Other Features
Balantidium coli Trophozoite	Ovoid with tapering anterior end; 50-100 μm in length; 40-70 μm in width; usual range; 40-50 μm	Ciliates: rotary, boring; may be rapid	1 large, kidney-shaped macronucleus; 1 small, round micronucleus, which is difficult to see even in the stained smear; macronucleus may be visible in unstained preparation	Body covered with cilia, which tend to be longer near cytostome; cytoplasm may be vacuolated
Cyst	Spherical or oval, 50-70 μm; usual range: 50-55 μm		1 large macronucleus visible in unstained preparation; micronucleus difficult to see	Macronucleus and contractile vacuole are visible in young cysts; in older cysts, internal structure appears granular; cilia difficult to see within the cyst wall

Table 49-18   Morphologic Criteria Used to Identify Intestinal Protozoa (Coccidia, Microsporidia, *Blastocystis hominis*)

Species	Shape and Size	Other Features
Cryptosporidium spp.	Oocyst generally round, 4-6 μm, each mature oocyst containing 4 sporozoites	Oocyst, diagnostic stage in stool, sporozoites occasionally visible within oocyst wall; acid-fast positive using modified acid-fast stains; various other stages in life cycle can be seen in biopsy specimens taken from gastrointestinal tract (brush border of epithelial cells) and other tissues; disseminated infection well documented in compromised host; oocysts immediately infective (in both formed and/or watery specimens); nosocomial infections documented; use enteric precautions for inpatients
Cyclospora cayetanensis	Oocyst generally round, 8-10 μm, oocysts are not mature, no visible internal structure; oocysts may appear wrinkled	Oocyst, diagnostic stage in stool; acid-fast variable using modified acid-fast stains; color range from clear to deep purple (tremendous variation); best results obtained with decolorizer of 1% acid, 3% maximum; oocysts may appear wrinkled (like crumpled cellophane); mimic *Cryptosporidium* oocysts but are twice as large
Isospora belli	Ellipsoidal oocyst; range 20-30 μm in length, 10-19 μm in width; sporocysts rarely seen broken out of oocysts, but measure 9-11 μm	Mature oocyst contains 2 sporocysts with 4 sporozoites each; usual diagnostic stage in feces is immature oocyst containing spherical mass of protoplasm (intestinal tract); oocysts are modified acid-fast positive; whole oocyst may stain pink if immature, but just the internal sporocysts stain if the oocyst is mature
Sarcocystis hominis *S. suihominis* *S. bovihominis*	Oocyst thin-walled and contains 2 mature sporocysts, each containing 4 sporozoites; frequently, thin oocyst wall ruptures; ovoidal sporocysts each measure 10-16 μm in length and 7.5-12 μm in width	Thin-walled oocyst or ovoid sporocysts occur in stool (intestinal tract)
S. "lindemanni"	Shapes and sizes of skeletal and cardiac muscle sarcocysts vary considerably	Sarcocysts contain from several hundred to several thousand trophozoites, each of which measures from 4-9 μm in width and 12-16 μm in length. The sarcocysts may also be divided into compartments by septa, not seen in *Toxoplasma* cysts (tissue/muscle)
Microsporidia *Encephalitozoon* spp. *Enterocytozoon bieneusi* *Brachiola* spp. *Pleistophora* spp. *Trachipleistophora hominis* *T. anthropophthera* *Nosema connori* *Vittaforma corneae* *Microsporidium* spp.	Small, oval spores (1-4 μm, most are 1-2.5 μm) can be found in routine histologic sections; electron microscopy is very specific but less sensitive; spores in fecal specimens can appear round; diagonal or horizontal stripe in modified trichrome stains often visible in some spores and represents polar tubule structure within the spore; some will also contain a terminal vacuole; the spore wall will fluoresce with optical brightening agents	Spores shed from enterocytes can now be identified in stool specimens. In addition to the various modified trichrome procedures, preliminary results indicate monoclonal antibodies may provide a more sensitive detection method; however, these are not yet available commercially; optical brightening agents (calcofluor, FungiFluor, Uvitex 2B) are nonspecific staining agents for microsporidial spores; specimens from other body sites are also acceptable and have been found to be positive, depending on the genus involved
Blastocystis hominis	Organisms are generally round, measure approximately 6-40 μm, and are usually characterized by a large, central body (looks like a large vacuole); this stage has been called the *central body form*	The more amebic form can be seen in diarrheal fluid but will be difficult to identify; the central body forms vary tremendously in size, even on a single fecal smear; this is the most common form seen; routine fecal examinations may indicate a positive rate much higher than other protozoa; some laboratories report figures of 20% and higher

Table 49-19 Microsporidia That Cause Human Infection

Species	Immunocompromised	Immunocompetent	Comments[a]
Common			
Enterocytozoon bieneusi	Chronic diarrhea; wasting syndrome, cholangitis, acalculous cholecystitis, chronic sinusitis, chronic cough, pneumonitis; cause of diarrhea in organ transplant recipients	Self-limiting diarrhea in adults and children; traveler's diarrhea; asymptomatic carriers	Short-term culture only; three strains identified, but not named; AIDS patients with chronic diarrhea (present in 5% to 30% of patients when CD4 lymphocyte counts very low); pigs, nonhuman primates
Encephalitozoon hellem	Disseminated infection; keratoconjunctivitis; sinusitis, bronchitis, pneumonia, nephritis, ureteritis, cystitis, prostatitis, urethritis	Possibly diarrhea	Cultured in vitro; detected in persons with traveler's diarrhea and co-infection with Enterocytozoon bieneusi; pathogenicity unclear; spores not reported yet from stool; Psittacine birds
Encephalitozoon intestinalis	Chronic diarrhea, cholangiopathy; sinusitis, bronchitis, pneumonitis; nephritis, bone infection, nodular cutaneous lesions	Self-limiting diarrhea; asymptomatic carriers	Cultured in vitro; formerly Septata intestinalis; AIDS patients with chronic diarrhea; dogs, donkeys, pigs, cows, goats
Encephalitozoon cuniculi	Disseminated infection; keratoconjunctivitis, sinusitis, bronchitis, pneumonia; nephritis; hepatitis, peritonitis, symptomatic and asymptomatic intestinal infection; encephalitis	Not described. Two HIV-seronegative children with seizure disorder (suspect E. cuniculi infection) presumably were immunocompromised	Cultured in vitro; wide mammalian host range
Uncommon			
Pleistophora sp.	Myositis (skeletal muscle)	Not described	Tend to infect fish
Pleistophora ronneafiei	Myositis	Not described	
Trachipleistophora hominis	Myositis; keratoconjunctivitis; sinusitis	Keratitis	Cultured in vitro; AIDS patients
Trachipleistophora anthropophthera	Disseminated infection	Not described	AIDS patients
Brachiola connori	Disseminated infection	Not described	Formerly Nosema connori; often infect insects; disseminated in infant with SCID
Brachiola vesicularum	Myositis	Not described	
Brachiola algerae	Myositis; nodular cutaneous lesions	Keratitis	Cultured in vitro; formerly Nosema algerae; skin nodules in boy with acute lymphocytic leukemia; found in arthropods
Nosema ocularum	Not described	Keratitis	HIV-seronegative individual
Vittaforma corneae	Disseminated infection	Keratitis	Cultured in vitro; formerly Nosema corneum; non-HIV patient
Microsporidium ceylonensis*	Not described	Corneal ulcer, keratitis	HIV seronegative individual, autopsy
Microsporidium africanum*	Not described	Corneal ulcer, keratitis	HIV seronegative individual, autopsy
Microsporidia (not classified)		Keratoconjunctivitis in a contact lens wearer	

AIDS, Acquired immunodeficiency syndrome; HIV, human immunodeficiency virus; SCID, severe combined immunodeficiency.

*Microsporidium is a collective generic name for microsporidia that cannot be classified because available information is insufficient.

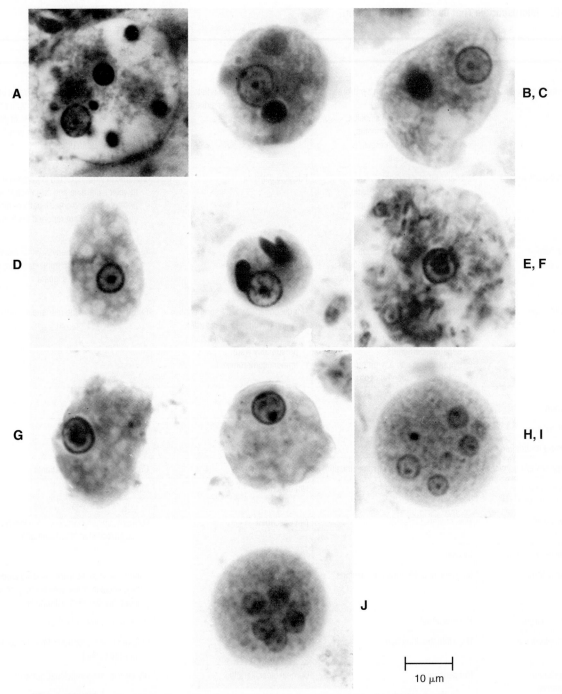

Figure 49-17   **A** to **C,** Trophozoites of *Entamoeba histolytica* (note ingested red blood cells).
D, Trophozoite of *E. histolytica/E. dispar.* **E,** Early cyst of *E. histolytica/E. dispar.* **F** to **H,** Trophozoites of
Entamoeba coli. **I** and **J,** Cysts of *E. coli.*

is a temporary, nonfeeding stage and usually reverts
back to the trophozoite stage. Both genera of free-living
amebae are also characterized by having the typical
hexagonal, double-walled cyst that can be seen using
calcofluor white stain. *Balamuthia mandrillaris,* like
Acanthamoeba spp., the cause of granulomatous amebic
encephalitis (GAE), looks much like *Acanthamoeba,* but

does not grow well on the *Escherichia coli*–seeded
nonnutrient agar plates. The trophozoites are characterized by extensive branching, and the cysts have the
typical double wall. Unlike the acute infection primary
amebic meningoencephalitis (PAM) caused by *Naegleria
fowleri,* GAE is a more chronic infection like that seen
with the *Acanthamoeba* spp.

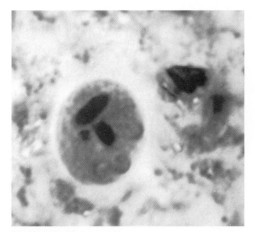

Figure 49-18 *Entamoeba histolytica/E. dispar* cyst.

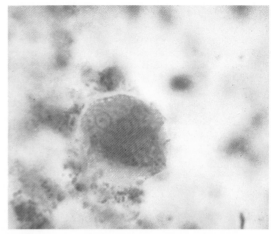

Figure 49-21 *Entamoeba coli* cyst, trichrome stain (poor preservation—typical appearance of some *E. coli* cysts).

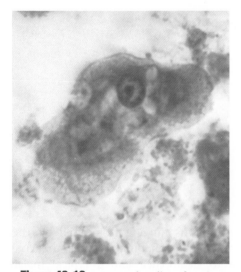

Figure 49-19 *Entamoeba coli* trophozoite.

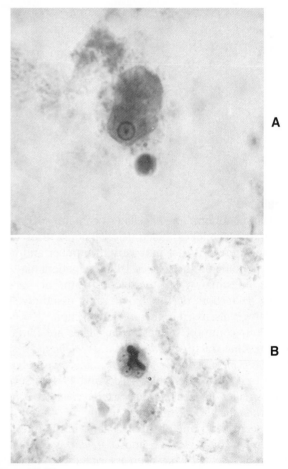

Figure 49-22 **A,** *Entamoeba hartmanni trophozoite.* **B,** *E. hartmanni* cyst.

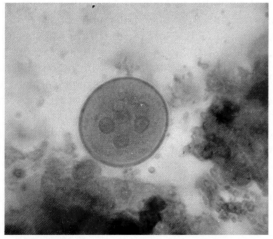

Figure 49-20 *Entamoeba coli* cyst, iodine stain.

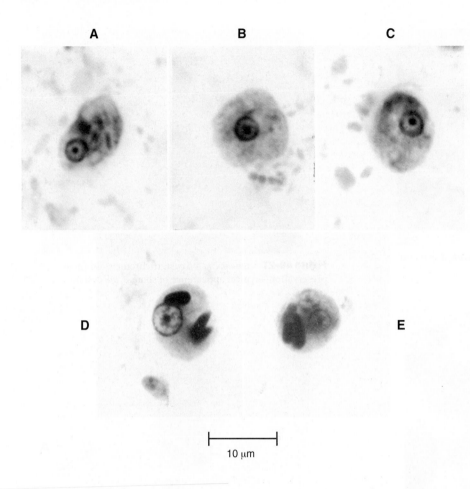

Figure 49-23   **A** to **C**, Trophozoites of *Entamoeba hartmanni*. **D** and **E**, Cysts of *E. hartmanni*.

10 µm

Flagellates

The Mastigophora, or flagellates, contain specialized locomotor organelles called *flagella;* long, thin cytoplasmic extensions that may vary in number and position depending on the species. Different genera may live in the intestinal tract, the bloodstream, or various tissues. Detection of the blood- and tissue-dwelling flagellates is discussed in the previous section.

Four common species of flagellates are found in the intestinal tract: *Giardia lamblia, Dientamoeba fragilis, Chilomastix mesnili,* and *Pentatrichomonas hominis* (Figures 49-29 to 49-36) (see Tables 49-15 and 49-16). Several other smaller flagellates, such as *Enteromonas hominis* and *Retortamonas intestinalis* (see Figure 49-29), are rarely seen, and none of the flagellates in the intestinal tract, except for *G. lamblia* and *D. fragilis,* is considered pathogenic. *D. fragilis* has been associated with diarrhea, nausea, vomiting, and other nonspecific intestinal complaints. *Trichomonas vaginalis* is pathogenic but occurs in the urogenital tract. *Trichomonas tenax* is occasionally found in the mouth and may be associated with poor oral hygiene.

G. lamblia is probably the most common protozoan organism found in persons within the United States and causes symptoms ranging from mild diarrhea, flatulence, and vague abdominal pains to acute, severe diarrhea to steatorrhea and a typical malabsorption syndrome. Various documented waterborne and foodborne outbreaks have occurred during the past several years, and various animals may serve as reservoir animal hosts for *G. lamblia.* A number of immunoassays are now commercially available for the detection of antigen or organisms in clinical specimens (see Table 49-10).

Most flagellate trophozoites are pear shaped and have different numbers and arrangements of flagella. The sucking disk and axonemes of *Giardia,* the cytostome and spiral groove of *Chilomastix,* and the undulating membrane of *Trichomonas* are all distinctive criteria for identification (see Figures 49-29 to 49-33).

Dientamoeba has no known cyst stage and the trophozoite has one or two nuclei, which have no peripheral chromatin and which have four to eight chromatin granules in a central mass. This organism varies tremendously in size and shape and may contain ingested bacteria and other debris; it will be overlooked unless the permanent stained smear is performed as a routine part of the O&P examination (see Figures 49-34

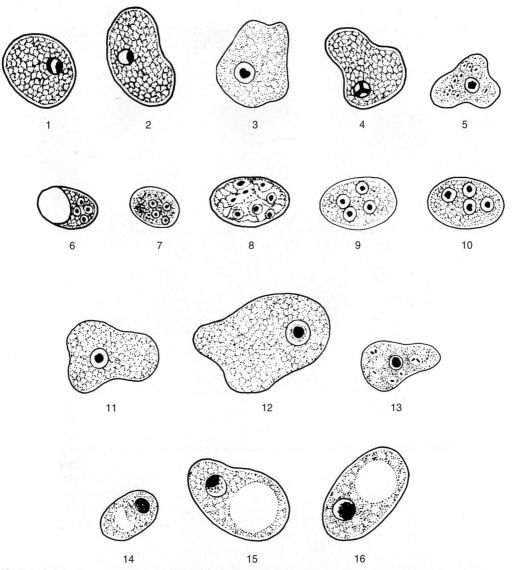

Figure 49-24   1 to 5, Trophozoites of *Endolimax nana.* 6 to 10, Cysts of *E. nana.* 11 to 13, Trophozoites of *Iodamoeba bütschlii.* 14 to 16, Cysts of *I. bütschlii.* (From Garcia LS: *Diagnostic medical parasitology,* ed 5, Washington, DC, 2007, ASM Press.)

to 49-36). Symptoms include intermittent diarrhea, abdominal pain, nausea, anorexia, malaise, fatigue, poor weight gain, and unexplained eosinophilia.

Ciliates

The class Ciliata, or ciliates, contains species that move by means of cilia, short extensions of cytoplasm that cover the surface of the organism. This group contains only one organism that infects humans: *Balantidium coli* infects the intestinal tract and may produce severe symptoms. *Balantidium coli* is the largest protozoan and the only ciliate that infects humans (Figures 49-37 and 49-38) (see Table 49-17). The trophozoites have a

rapid, rotatory, boring motion and the surface is covered with short cilia. The cytoplasm contains both a large, bean-shaped macronucleus and a smaller, round micronucleus that is often difficult to see. Although this infection is uncommon within the United States, the organisms can be seen in proficiency testing specimens.

Coccidia

Members of the class Sporozoa are found in the blood and other tissues and have a complex life cycle that involves both sexual and asexual generations. The four species of *Plasmodium,* the cause of malaria, are found in

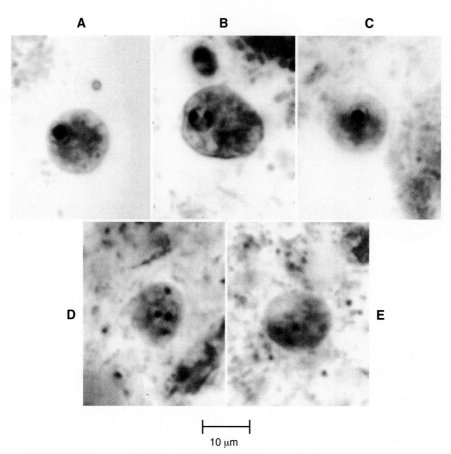

10 μm

Figure 49-25 **A** to **C**, Trophozoites of *Endolimax nana.* **D** and **E**, Cysts of *E. nana.*

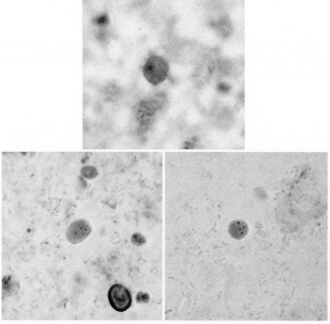

Figure 49-26 *Top, Endolimax nana* trophozoite. *Bottom left, E. nana* cyst. *Bottom right, E. nana* cyst.

A   B   C

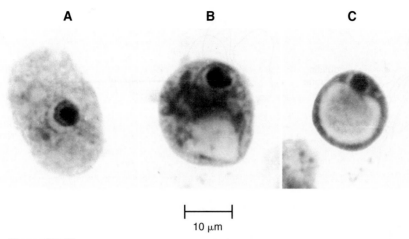

10 μm

Figure 49-27   **A,** Trophozoites of *Iodamoeba bütschlii.* **B** and **C,** Cysts of *I. bütschlii.*

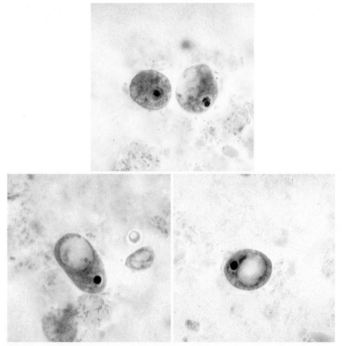

Figure 49-28   *Top, Iodamoeba bütschlii* trophozoites. *Bottom left, Iodamoeba bütschlii* cyst. *Bottom right, I. bütschlii* cyst.

this group; their diagnosis is discussed in the following section. Members of the coccidian genera—*Isospora, Cryptosporidium,* and *Cyclospora*—can be found in the intestinal mucosa and other tissues. These organisms have been seen with increasing frequency in specimens from immunosuppressed patients, particularly those with AIDS (see Table 49-18).

Isospora, Cryptosporidium, and *Cyclospora* parasites are passed in the stool as oocysts; the other members of the protozoa exist in the intestinal tract in the trophozoite or cyst stages. Humans are generally infected with the coccidian parasite, *Toxoplasma gondii,* from the ingestion of infected meats, although its life cycle includes stages in animal hosts within the cat family and oocyst ingestion is also a potential route of infection (see Figure 49-12). The organisms can be seen in routine histologic tissue preparations, and the disease itself is normally diagnosed using serologic tests for antibody.

Isospora belli infects humans and diagnosis is based on finding the oocysts in the stool, primarily in the concentration sediment. These oocysts are immature when passed and will stain using modified acid-fast

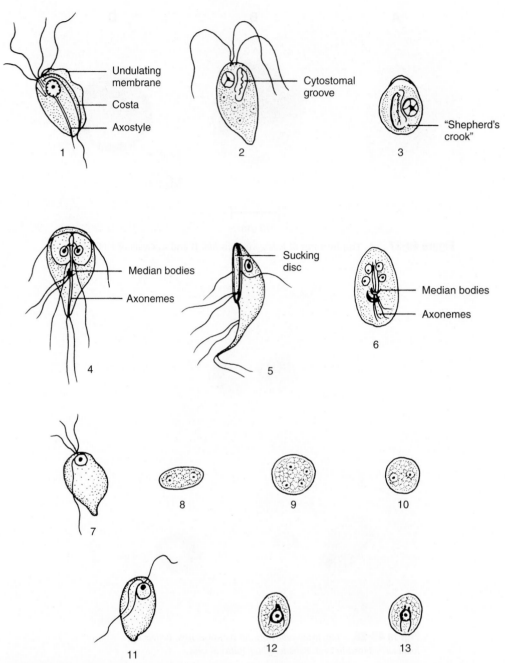

Figure 49-29   1, Trophozoite of *Trichomonas hominis*. 2, Trophozoite of *Chilomastix mesnili*. 3, Cyst of *C. mesnili*. 4, Trophozoite of *Giardia lamblia* (front view). 5, Trophozoite of *G. lamblia* (side view). 6, Cyst of *G. lamblia*. 7, Trophozoite of *Enteromonas hominis*. 8 to 10, Cysts of *E. hominis*. 11, Trophozoite of *Retortamonas intestinalis*. 12 to 13, Cysts of *R. intestinalis*.   (Adapted from Garcia LS, Bruckner DA: *Diagnostic medical parasitology,* Washington, DC, 1993, ASM Press. Illustration 5 by Nobuko Kitamura. Illustrations 7 to 13 modified from Markell EK, Voge M: *Medical parasitology,* ed 5, Philadelphia, 1981, WB Saunders.)

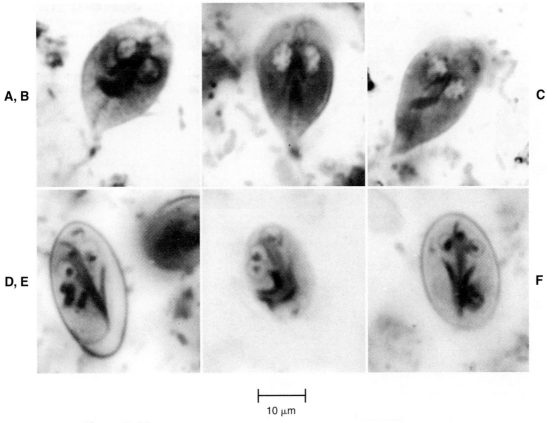

Figure 49-30   **A** to **C,** Trophozoites of *Giardia lamblia.* **D** to **F,** Cysts of *G. lamblia.*

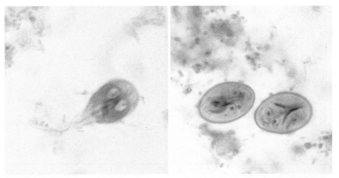

Figure 49-31   *Left, Giardia lamblia* trophozoite. *Right, G. lamblia* cysts.

stains (Figure 49-39). These infections are seen more frequently in compromised patients, such as those with AIDS.

Of the coccidia, *C. parvum* is the most important; this organism has been reported in numerous waterborne outbreaks, some of which have involved thousands of people.[2,3,31] Outbreaks have been reported worldwide; these infections tend to be self-limiting in the immunocompetent host but can be life-threatening in the immunocompromised host. Of 15 named species of *Cryptosporidium* infectious for nonhuman vertebrate hosts *C. baileyi, C. canis, C. felis, C. hominis, C. meleagridis, C. muris,* and *C. parvum* have been reported to also infect humans. Humans are the primary hosts for *C. hominis,* and except for *C. parvum,* which is widespread among nonhuman hosts and is the most frequently reported zoonotic species, the remaining species have been reported primarily in immunocompromised humans. The

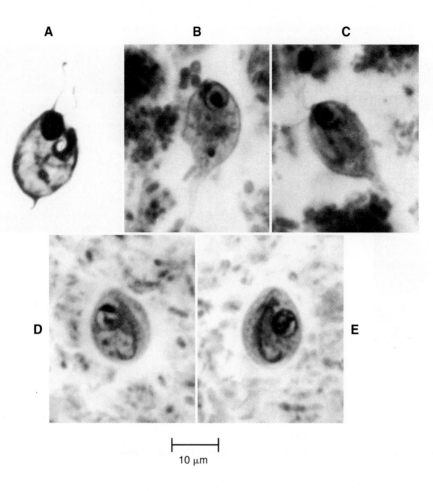

A **B** **C**

D **E**

10 µm

Figure 49-32 **A** to **C,** Trophozoites of *Chilomastix mesnili* (**A,** silver stain). **D** and **E,** Cysts of *C. mesnili.*

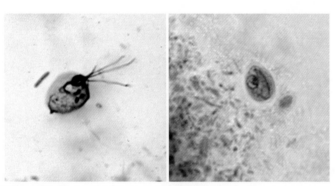

Figure 49-33 *Left, Chilomastix mesnili* trophozoite, silver stain. *Right, C. mesnili* cyst.

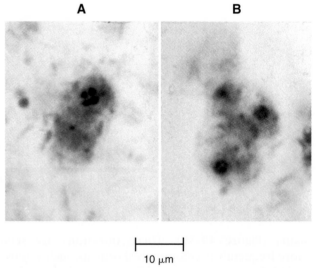

A **B**

10 µm

Figure 49-35 **A** and **B,** Trophozoites of *Dientamoeba fragilis.*

Figure 49-34 Trophozoites of *Dientamoeba fragilis.*

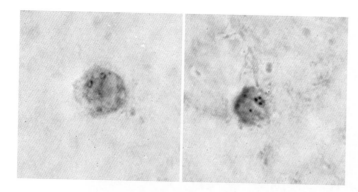

Figure 49-36 *Left, Dientamoeba fragilis,* two nuclei. *Right, D. fragilis,* one nucleus.

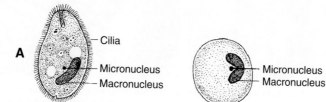

Figure 49-37 **A,** Trophozoite of *Balantidium coli.* **B,** Cyst of *B. coli.* (From Garcia LS: *Diagnostic medical parasitology,* ed 5, Washington, DC, 2007, ASM Press.)

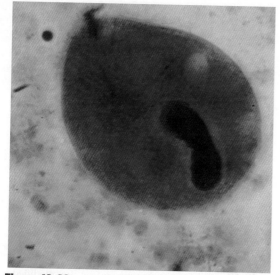

Figure 49-38 *Balantidium coli* trophozoite, iodine stain.

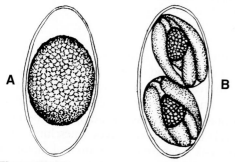

Figure 49-39 **A,** Immature concept of *Isospora belli.* **B,** Mature oocyst of *I. Belli.* (Illustration by Nobuko Kitamura.)

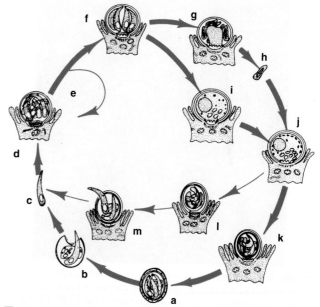

Figure 49-40 Life cycle of *Cryptosporidium. (a)* Sporulated oocyst in feces. *(b)* Excystation in intestine. *(c)* Free sporozoite in intestine. *(d)* Type I meront (six or eight merozoites). *(e)* Recycling of type I merozoite. *(f)* Type II meront (four merozoites). *(g)* Microgametocyte with approximately 16 microgametes. *(h)* Microgamete fertilizes macrogamete *(i)* to form zygote *(j).* Approximately 80% of the zygotes form thick-walled oocysts *(k),* which sporulate within the host cell. About 20% of the zygotes do not form an oocyst wall; their sporozoites are surrounded only by a unit membrane *(l).* Sporozoites within "autoinfective," thin-walled oocysts *(l)* are released into the intestinal lumen *(m)* and reinitiate the endogenous cycle (at *c).* (Life cycle from William L. Current, Lilly Research Laboratories, Greenfield, Ind.)

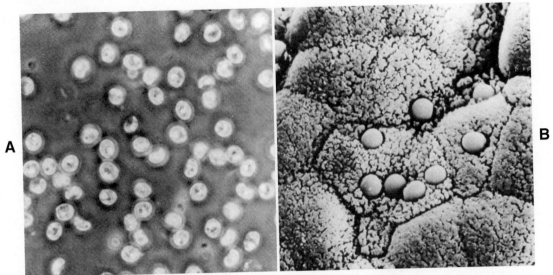

Figure 49-41 *Cryptosporidium.* **A,** Oocysts recovered from a Sheather's sugar flotation; organisms measure 4 to 6 μm. **B,** Scanning electron microscopy view of organisms at brush border of epithelial cells. (From Garcia LS: *Diagnostic medical parasitology,* ed 5, Washington, DC, 2007, ASM Press.)

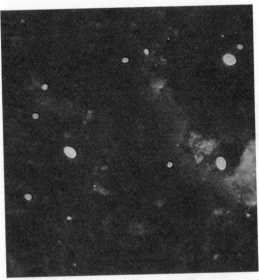

Figure 49-42 *Cryptosporidium* oocysts and *Giardia* cysts stained with monoclonal antibody-conjugated fluorescent reagent. (Courtesy Merifluor, Meridian Diagnostics, Cincinnati, Ohio.)

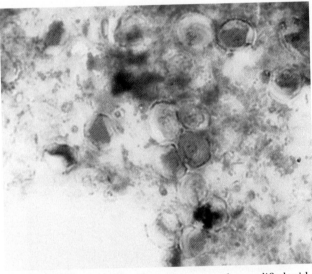

Figure 49-43 *Cyclospora cayetanensis* oocysts after modified acid-fast staining. Note the variability in the intensity of stain. These oocysts measure 8 to 10 μm—twice the size of *Cryptosporidium* spp. Photographed using oil immersion (×1000). (Courtesy Charles R. Sterling, University of Arizona.)

presence of autoinfective oocysts may explain why a small inoculum can lead to an overwhelming infection in compromised patients and why they may have persistent, life-threatening infections in the absence of documentation of repeated exposure to oocysts (Figure 49-40). In the immunocompromised individual, the infection is not always limited to the gastrointestinal tract but can disseminate to other tissues, including the respiratory tract. Oocysts range from 4 to 6 μm and can

be identified using modified acid-fast stains and immunoassays (Figures 49-41 and 49-42).

Several outbreaks have been reported with *Cyclospora cayetanensis* and have been linked to the ingestion of imported contaminated foods, including strawberries, raspberries, basil, baby lettuce leaves (mesclun), and peas.[6,26] The oocysts measure 8 to 10 μm and modified acid-fast methods stain the oocysts from light pink to deep purple (Figures 49-43 and 49-44). However, the

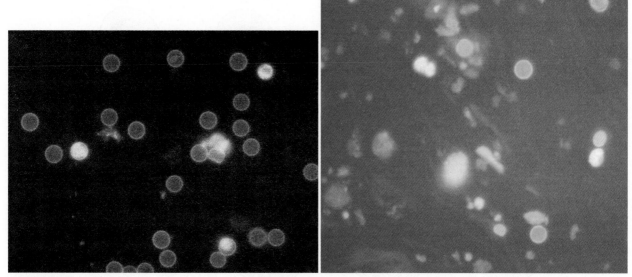

Figure 49-44   *Cyclospora cayetanensis* oocysts exhibiting autofluorescence. Photographed using high dry (×400). (*Left photograph* courtesy Charles R. Sterling, University of Arizona; *right photograph* courtesy E. Long, Centers for Disease Control and Prevention, Atlanta, Ga.)

oocysts tend to be very acid-fast variable and some of the oocysts may resemble clear, wrinkled cellophane.

Sarcocystis spp. appear in Table 49-18 but are not discussed in detail. According to the literature, extra-intestinal human sarcocystosis is rare, with a much lower incidence than that seen with the intestinal infection.

Microsporidia

The microsporidia are obligate intracellular parasites that can infect both animals and humans, through ingestion, inhalation, or direct inoculation of the infective spores, which are environmentally resistant (Figures 49-45 and 49-46) (see Table 49-19). To date, eight genera have been recognized in humans: *Brachiola, Encephalitozoon, Enterocytozoon, Microsporidium, Nosema, Pleistophora, Trachipleistophora,* and *Vittaforma.* Although these infections are found primarily in the compromised host, there is documented evidence that immuno-competent hosts can also become infected. Classification criteria include spore size, configuration of the nuclei within the spores and developing forms, the number of polar tubule coils within the spore, and the relationship between the organism and host cell. The life cycle includes merogony (repeated divisions), schizogony (multiple fission), and sporogony (spore production).

With *Enterocytozoon* and *Encephalitozoon,* two of the genera of microsporidia that can infect humans, spore stages in the life cycle are passed in the stool or urine in a disseminated infection. Within the past few years, diagnostic methods have become available for the identification of microsporidial spores in stool specimens.

However, considering the spore size (approximately 1 to 4 μm), the clinical laboratory may have to wait until immunoassays for antigen or organism detection are available commercially for spore detection. These two genera cause chronic intractable diarrhea accompanied by fever, malaise, and weight loss; dissemination to other body sites such as the kidneys can also occur and is primarily seen in infections with *Encephalitozoon* spp.

Modified trichrome stains are recommended and should be used with concentrated stool sediment, urine, or other body site specimens (Figure 49-47). Routine histologic stains can also be used on tissues and include Giemsa, trichrome, tissue Gram stains, silver stains, and PAS (Figures 49-48 and 49-49).[6]

BLOOD PROTOZOA

Malaria

Malaria is caused by four species of the protozoan genus *Plasmodium: P. vivax, P. falciparum, P. ovale,* and *P. malariae* (Table 49-20; see also Tables 49-4 to 49-7) (Figures 49-50 to 49-52). The life cycle can be seen in Figure 49-51. Humans become infected when the sporozoites are introduced into the blood from the salivary secretion of the infected mosquito when the mosquito takes a blood meal. These sporozoites then enter the parenchymal cells of the liver, where they undergo asexual multi-plication, called the **preerythrocytic cycle.** Timing of this cycle depends on the species; however, the schi-zonts eventually rupture, releasing thousands of mero-zoites into the bloodstream, where they invade the erythrocytes (red blood cells [RBCs]), called the **eryth-**

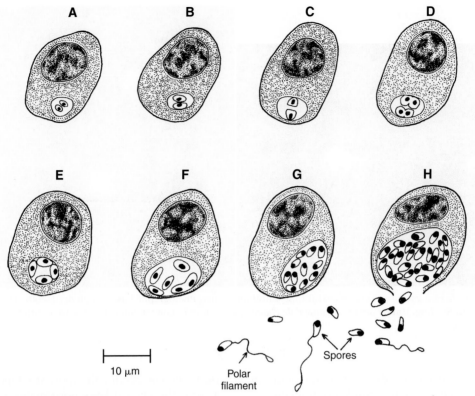

Figure 49-45   Life cycle diagram of the microsporidia. **A** to **G**, Asexual development of sporoblasts. **H**, Release of spores. (Modified from Gardiner CH, Fayer R, Dubey JP: *An atlas of protozoan parasites in animal tissues*, US Department of Agriculture, Agriculture Handbook No 651, 1988. Illustration by Sharon Belkin.)

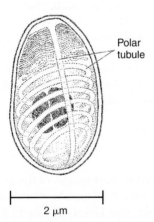

Figure 49-46   Diagram illustrating the polar tubule within a microsporidian spore. (Modified from Bryan RT, Cali A, Owen RL, Spencer HC. In Sun T, editor: *Progress in clinical parasitology*, vol II, Field and Wood Medical Publishers, distributed by WW Norton, New York, 1991. Illustration by Sharon Belkin.)

rocytic cycle. The early forms in the RBCs are called **ring forms,** or young trophozoites (see Figure 49-52). As the parasites continue to grow and feed, they become actively ameboid within the RBC. They feed on hemoglobin, and the residue that is left is called *malarial pigment.*

During the next phase of the cycle, the nuclear chromatin and cytoplasm begin to divide **(schizogony),** leading to the development of individual merozoites. This infected RBC then ruptures, releasing merozoites that infect other RBCs and metabolic products into the bloodstream. If many of the RBCs rupture simultaneously, a malarial paroxysm may result from the toxic materials released into the bloodstream. In the early stages of a malarial infection, the RBC rupture is not synchronized and patients may exhibit low-grade symptoms; after several days, a 48- or 72-hour periodicity is usually established.

After several erythrocytic cycles, **gametocytes** are formed and are infective for the mosquito vector. After ingestion of gametocytes, the cycle continues in the mosquito, with eventual production of **sporozoites,** which are infective for humans when the mosquito takes the next blood meal.

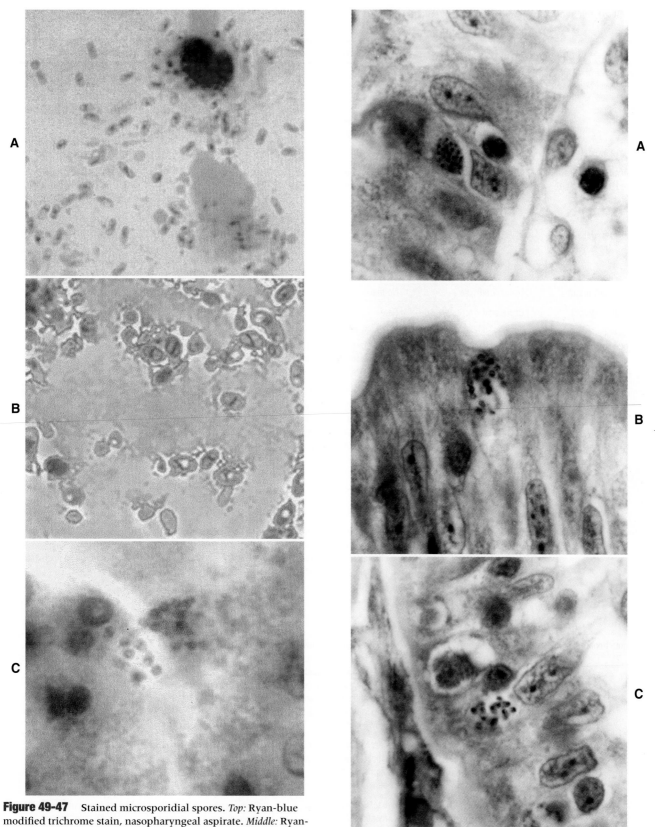

Figure 49-47   Stained microsporidial spores. *Top:* Ryan-blue modified trichrome stain, nasopharyngeal aspirate. *Middle:* Ryan-blue modified trichrome stain, stool (enlarged image). *Bottom:* Ryan-blue modified trichrome stain, urine.

Figure 49-48   Routine histology micrograph of microsporidian spores in enterocytes, stained with Giemsa stain. *Top:* Note the small size. *Middle:* The spores are more easily seen; note the position between the nucleus and the brush border of the cell. *Bottom:* In these spores, the granule is easily seen.

Table 49-20 Plasmodia in Giemsa-Stained Thin Blood Films

	Plasmodium vivax	*Plasmodium malariae*	*Plasmodium falciparum*	*Plasmodium ovale*
Persistence of exoerythrocytic cycle	Yes	No	No	Yes
Relapses	Yes	No, but long-term recrudescences are recognized	No long-term relapses	Possible, but usually spontaneous recovery
Time of cycle	44-48 hr	72 hr	36-48 hr	48 hr
Appearance of parasitized red blood cells; size and shape	1½ to 2 times larger than normal; oval to normal; may be normal size until ring fills ½ of cell	Normal shape; size may be normal or slightly smaller	Both normal	60% of cells larger than normal and oval; 20% have irregular, frayed edges
Schüffner's dots (eosinophilic stippling)	Usually present in all cells except early ring forms	None	None; occasionally, comma-like red dots are present (Maurer's dots)	Present in all stages, including early ring forms; dots may be larger and darker than in *P. vivax*
Color of cytoplasm	Decolorized, pale	Normal	Normal, bluish tinge at times	Decolorized, pale
Multiple rings/cell	Occasional	Rare	Common	Occasional
All developmental stages present in peripheral blood	All stages present	Ring forms few, as ring stage brief; mostly growing and mature trophozoites and schizonts	Young ring forms and no older stages; few gametocytes; rare mature schizonts	All stages present
Appearance of parasite; young trophozoite (early ring form)	Ring is ⅓ diameter of cell, cytoplasmic circle around vacuole; heavy chromatin dot	Ring often smaller than in *P. vivax*, occupying ⅛ of cell; heavy chromatin dot; vacuole at times "filled in"; pigment forms early	Delicate, small ring with small chromatin dot (frequently 2)*; scanty cytoplasm around small vacuoles; sometimes at edge of red blood cell (appliqué form) or filamentous slender form; may have multiple rings per cell	Ring is larger and more ameboid than in *P. vivax*, otherwise similar to *P. vivax*
Growing trophozoite	Multishaped, irregular ameboid parasite; streamers of cytoplasm close to large chromatin dot; vacuole retained until close to maturity; increasing amounts of brown pigment	Non-ameboid rounded or band-shaped solid forms; chromatin may be hidden by coarse, dark brown pigment	Heavy ring forms; fine pigment grains	Ring shape maintained until late in development; non-ameboid compared to *P. vivax*.
Mature trophozoite	Irregular ameboid mass; 1 or more small vacuoles retained until schizont stage; fills almost entire cell; fine, brown pigment	Vacuoles disappear early; cytoplasm compact, oval, band shaped, or nearly round almost filling cell; chromatin may be hidden by peripheral coarse, dark brown pigment	Not seen in peripheral blood (except in severe infections); development of all phases following ring form occurs in capillaries of viscera	Compact; vacuoles disappear; pigment dark brown, less than in *P. malariae*
Schizont (presegmenter)	Progressive chromatin division; cytoplastic bands containing clumps of brown pigment	Similar to *P. vivax* except smaller; darker, larger pigment granules peripheral or central	Not seen in peripheral blood (see above)	Smaller and more compact than *P. vivax*
Mature schizont	Merozoites, 16 (12 to 24) each with chromatin and cytoplasm, filling entire red blood cell, which can hardly be seen	8 (6 to 12) merozoites in rosettes or irregular clusters filling normal-sized cells, which can hardly be seen; central arrangement of brown-green pigment	Not seen in peripheral blood; in rare cases may be seen	¾ of cells occupied by 8 (8 to 12) merozoites in rosettes or irregular clusters

*If there is a lag time between blood collection (EDTA anticoagulant) and blood film preparation, *P. falciparum* rings may appear larger than normal.

Table 49-20 Plasmodia in Giemsa-Stained Thin Blood Films—cont'd

	Plasmodium vivax	*Plasmodium malariae*	*Plasmodium falciparum*	*Plasmodium ovale*
Macrogametocyte	Rounded or oval homogeneous cytoplasm; diffuse, delicate, light brown pigment throughout parasite; eccentric compact chromatin	Similar to *P. vivax*, but fewer in number, pigment darker and more coarse	Sex differentiation difficult; "crescent" or "sausage" shapes characteristic; may appear in "showers," black pigment near chromatin dot, which is often central	Smaller than *P. vivax*
Microgametocyte	Large pink to purple chromatin mass surrounded by pale or colorless halo; evenly distributed pigment	Similar to *P. vivax*, but fewer in number, pigment darker and more coarse	Same as macrogametocyte (described above)	Smaller than *P. vivax*
Main criteria	Large, pale red blood cell; trophozoite irregular; pigment usually present; Schüffner's dots not always present; several phases of growth seen in one smear; gametocytes appear as early as third day	Red blood cell normal in size and color; trophozoites compact, stain usually intense, band forms not always seen; coarse pigment; no stippling of red blood cells; gametocytes appear after a few weeks	Development following ring stage takes place in blood vessels of internal organs; delicate ring forms and crescent-shaped gametocytes are only forms normally seen in peripheral blood; gametocytes appear after 7-10 days	Red blood cell enlarged, oval, with fimbriated edges; Schüffner's dots seen in all stages; gametocytes appear after 4 days or as late as 18 days

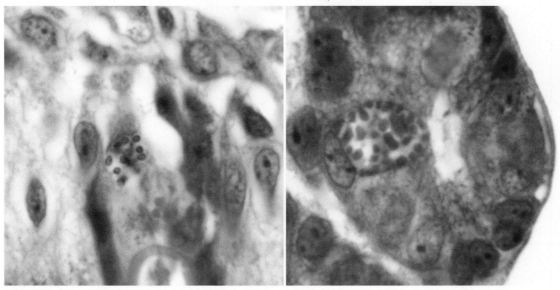

Figure 49-49   Routine histology micrograph of microsporidian spores in enterocytes, stained with Giemsa stain. *Left:* Note the fully formed spores. *Right:* These spores are not fully mature.

The asexual and sexual forms circulate in the human bloodstream in three species. However in *P. falciparum* infections, as the parasite continues to grow, the RBC membrane becomes sticky, and the cells tend to adhere to the endothelial lining of the capillaries of internal organs. Interference with normal blood flow in these vessels gives rise to additional problems, which are responsible for the different clinical manifestations of this type of malaria.

A number of genetic factors provide some resistance to the various *Plasmodium* spp. Hemoglobin S, thalassemia, and glucose-6-phosphate dehydrogenase deficiency are associated with increased resistance to *P. falciparum*. Apparently, the Duffy-negative RBCs convey

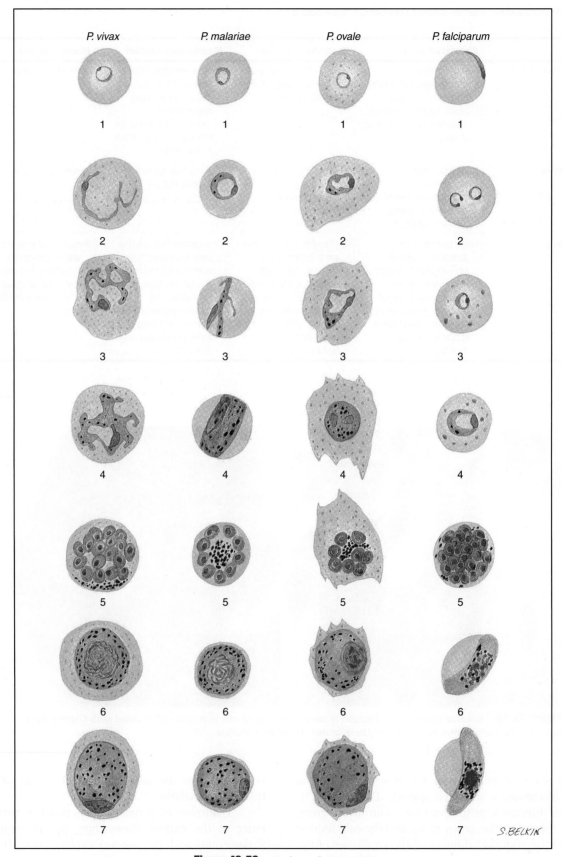

Figure 49-50   For legend see p. 605

Figure 49-50   The morphology of malaria parasites. *Plasmodium vivax 1*, Early trophozoite (ring form). *2*, Late trophozoite with Schüffner's dots (note enlarged red blood cell). *3*, Late trophozoite with ameboid cytoplasm (very typical of *P. vivax*). *4*, Late trophozoite with ameboid cytoplasm. *5*, Mature schizont with merozoites (18) and clumped pigment. *6*, Microgametocyte with dispersed chromatin. *7*, Macrogametocyte with compact chromatin. *Plasmodium malariae 1*, Early trophozoite (ring form). *2*, Early trophozoite with thick cytoplasm. *3*, Early trophozoite (band form). *4*, Late trophozoite (band form) with heavy pigment. *5*, Mature schizont with merozoites (9) arranged in rosette. *6*, Microgametocyte with dispersed chromatin. *7*, Macrogametocyte with compact chromatin. *Plasmodium ovale 1*, Early trophozoite (ring form) with Schüffner's dots. *2*, Early trophozoite (note enlarged red blood cell). *3*, Late trophozoite in red blood cell with fimbriated edges. *4*, Developing schizont with irregularly shaped red blood cell. *5*, Mature schizont with merozoites (8) arranged irregularly. *6*, Microgametocyte with dispersed chromatin. *7*, Macrogametocyte with compact chromatin. *Plasmodium falciparum 1*, Early trophozoite (accolé or appliqué form). *2*, Early trophozoite (one ring is in headphone configuration/double chromatin dots). *3*, Early trophozoite with Maurer's dots. *4*, Late trophozoite with larger ring and Maurer's dots. *5*, Mature schizont with merozoites (24). *6*, Microgametocyte with dispersed chromatin. *7*, Macrogametocyte with compact chromatin. NOTE: Without the appliqué form, Schüffner's dots, multiple rings/cell, and other developing stages, differentiation among the species can be difficult. It is obvious that the early rings of all four species can mimic one another very easily. *Remember that one set of negative blood films cannot rule out a malarial infection.* (Garcia LS: *Diagnostic medical parasitology,* ed 5, Washington, DC, 2007, Copyright by American Society for Microbiology.)

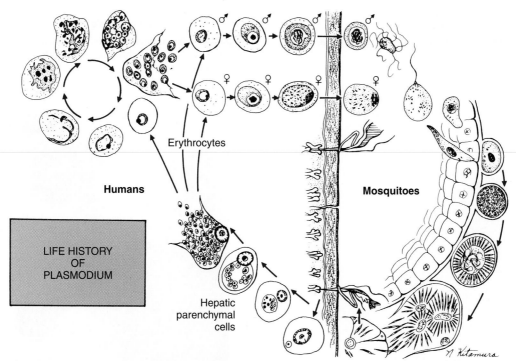

Figure 49-51   Life cycle of *Plasmodium*. (Modified from Wilcox A: *Manual for the microscopical diagnosis of malaria in man,* U.S. Public Health Service, Washington, DC, 1960. Illustration by Nobuko Kitamura.)

increased resistance to infection with *P. vivax*. *Plasmodium* spp. can be transmitted from shared needles, blood transfusions, and congenital infections. These former infections, as well as mosquito vectors, are still seen within the United States.[1,10,18,23] *The request for examination of blood films for parasites is always considered a STAT request.*[6,21]

Babesiosis

Babesia are tickborne (they can also be transmitted via a blood transfusion or organ transplantation) sporozoan parasites that can cause disease in humans, especially in those who have been splenectomized or tend to be compromised for other reasons. However, immunocompetent persons can also become infected; several different *Babesia* species can be found within the United States.[6,15] *Babesia* organisms infect the RBCs and appear as pleomorphic, ringlike structures when stained with any of the recommended stains used for blood films (Figure 49-53) (see also Tables 49-4 to 49-7). They may be confused with the ring forms in *Plasmodium* infections; however, in a *Babesia* infection, four or five rings

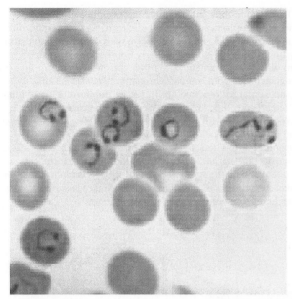

Figure 49-52 *Plasmodium falciparum,* early ring forms.

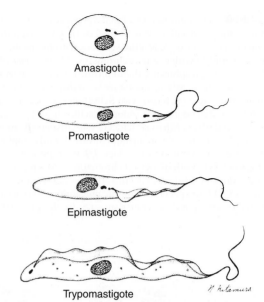

Figure 49-54 Characteristic stages of species of *Leishmania* and *Trypanosoma* in human and insect hosts. (Illustration by Nobuko Kitamura.)

Amastigote

Promastigote

Epimastigote

Trypomastigote

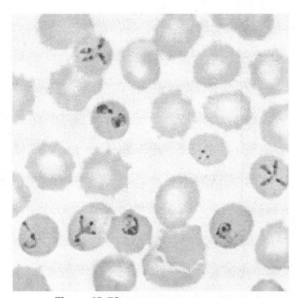

Figure 49-53 *Babesia* in red blood cells

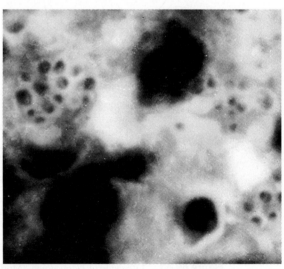

Figure 49-55 *Leishmania donovani* parasites in Küpffer cells of liver (2000×).

may be seen per RBC, and the individual rings are small and pleomorphic compared with those found in malarial infections.

Hemoflagellates

Hemoflagellates are blood and tissue flagellates, two genera of which are medically important for humans: *Leishmania* and *Trypanosoma* (Figures 49-54 to 49-59) (see also Tables 49-4 to 49-7). Some species may circulate in the bloodstream or at times may be present in lymph nodes or muscle. Other species tend to parasitize the reticuloendothelial cells of the hematopoietic

organs. The hemoflagellates of humans have four morphologic types (see Figure 49-54): amastigote (leishmanial form, or Leishman-Donovan [L-D] body), promastigote (leptomonal form), epimastigote (crithidial form), and trypomastigote (trypanosomal) form.

The amastigote form is an intracellular parasite in the cells of the reticuloendothelial system and is oval, measuring 1.5 to 5 µm, and contains a nucleus and kinetoplast. *Leishmania* spp. exist as the amastigote form in humans and as the promastigote form in the insect host. The life cycle is essentially the same for all species; however, clinical symptoms vary. As the vector takes a

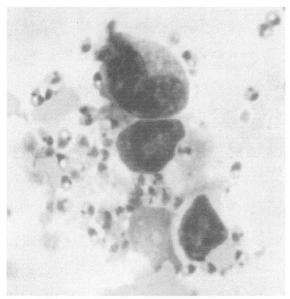

Figure 49-56 *Leishmania donovani* amastigotes.

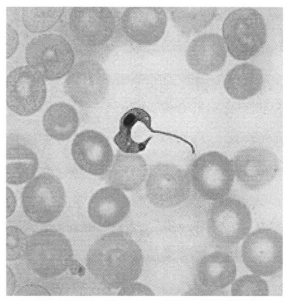

Figure 49-58 *Trypanosoma cruzi* trypomastigote.

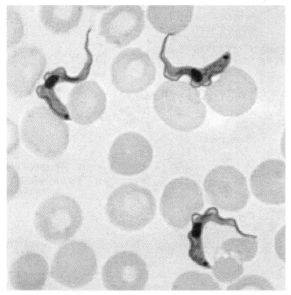

Figure 49-57 *Trypanosoma gambiense* in blood film.

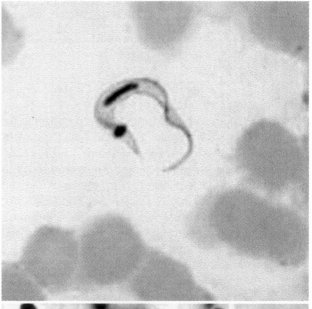

A

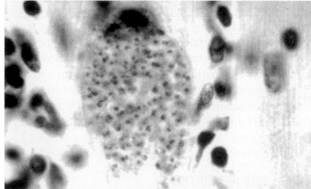

B

blood meal, promastigotes are introduced into the human host, thus initiating the infection. Depending on the species, the parasites then move from the bite site to the organs within the reticuloendothelial system (bone marrow, spleen, liver) or to the macrophages of the skin or mucous membranes (see Figure 49-55).

Many strains make up the three main groups. *L. tropica* causes oriental sore or cutaneous leishmaniasis of the Old World, *L. braziliensis* causes mucocutaneous leishmaniasis of the New World, and *L. donovani* causes visceral leishmaniasis (kala-azar) (see Figure 49-56).[6,17] Diagnosis of leishmanial organisms is based on the

Figure 49-59 **A,** *Trypanosoma cruzi* in blood film (1600×). **B,** *Trypanosoma cruzi* parasites in cardiac muscle (2500×).

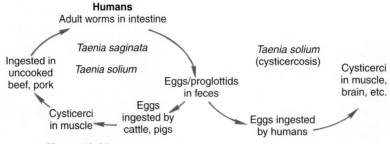

Figure 49-60   Life cycle of *Taenia saginata* and *Taenia solium*.

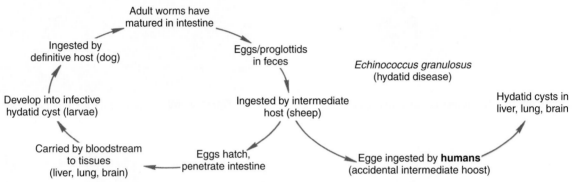

Figure 49-61   Life cycle of *Echinococcus granulosus* (hydatid disease).

demonstration of the L-D bodies or the recovery of promastigotes in culture.

Three species of trypanosomes are pathogenic for humans: *Trypanosoma brucei gambiense* (see Figure 49-57) causes West African sleeping sickness, *T. b. rhodesiense* causes East African sleeping sickness (can result in death within 1 year), and *T. cruzi* causes American trypanosomiasis, or Chagas' disease (see Figure 49-58). The two species that cause African sleeping sickness are morphologically similar and cause illness characterized by acute (organisms found in blood) and chronic stages (organisms found in central nervous system [comatose stage, or "sleeping sickness"]).

In the early stages of infection with *T. cruzi*, the trypomastigote forms appear in the blood but do not multiply (see Figure 49-59). They then invade the tissue cells, most often striated muscle (cardiac), and begin to divide, producing many L-D bodies. When these forms are liberated into the blood, they transform into the trypomastigote forms, which are carried to other sites, where tissue invasion again occurs. Diagnosis is confirmed by finding the parasites in blood films and/or tissue biopsies.

INTESTINAL HELMINTHS

The intestinal helminths that infect humans belong to two phyla: the Nematoda, or roundworms, and the

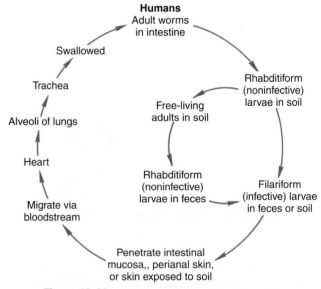

Figure 49-62   Life cycle of *Strongyloides stercoralis*.

platyhelminths, or flatworms (Figures 49-60 to 49-80) (see also Tables 49-4 to 49-7). The platyhelminths, most of which are hermaphroditic, have a flat, bilaterally symmetric body. The two classes, Trematoda (flukes) and Cestoda (tapeworms), contain organisms that are parasitic for humans.

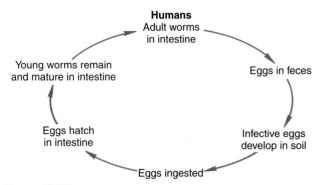

Figure 49-63   Life cycle of *Enterobius vermicularis* and *Trichuris trichiura* (direct type of cycle).

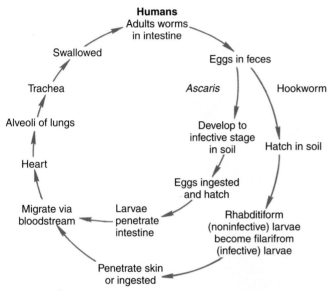

Figure 49-64   Life cycle of *Ascaris lumbricoides* and hookworms (indirect type of cycle).

Nematodes

Nematodes are elongate, cylindric worms with a well-developed digestive tract, and have separate sexes, with the male being smaller than the female. Most nematodes are diagnosed by finding the characteristic eggs in the stool (see Figures 49-62 to 49-69). The eggs of *Ancylostoma duodenale* and *Necator americanus* are essentially identical, so an infection with either species is reported as "hookworm eggs present."

Strongyloides stercoralis is passed in the feces as the noninfective rhabditiform larva (see Figure 49-67). Although hookworm eggs are normally passed in the stool, these eggs will continue to develop and hatch if the stool is left at room temperature for several days. These larvae may be mistaken for those of *Strongyloides*. Figure 49-68 illustrates the morphologic differences between the rhabditiform larvae of hookworm and *Strongyloides*. Recovery of *Strongyloides* larvae in duodenal contents is discussed earlier in this chapter.

The technique for recovery of *Enterobius vermicularis* (pinworm) eggs can be seen in Figures 49-4 and 49-69. Eggs of other nematodes are relatively easy to find and differentiate one from another.

Cestodes

The tapeworms have a long, segmented, ribbonlike body that has a special attachment portion, or scolex, at the anterior end (see Figures 49-70 to 49-74). Adult worms inhabit the small intestine; however, humans may be host to either the adult or the larval forms, depending on the species. Cestodes generally require one or more intermediate hosts for the completion of their life cycle (see Figures 49-60 and 49-61). *Taenia* spp. tapeworm proglottids tend to be longer than wide and contain uterine branches; more than 12 branches per side is *T. saginata*; less than 12 is *T. solium* (see Figure 49-71).

Except for *Diphyllobothrium latum*, tapeworm eggs are embryonated and contain a six-hooked oncosphere (Table 49-21; see also Figure 49-70). *Taenia saginata* and *T. solium* cannot be identified to species based on egg morphology; gravid proglottids (see Figure 49-71) or the scolices must be examined.

Hymenolepis nana has an unusual life cycle; ingestion of the egg can lead to the adult worm in humans, thus bypassing the need for an intermediate host (see Figure 49-72). The eggs of *H. nana* and *H. diminuta* are very similar; however, *H. nana* eggs are smaller and have polar filaments, which are present in the space between the oncosphere and the eggshell (see Figure 49-70).

The freshwater fish tapeworm, *Diphyllobothrium latum*, does not have embryonated eggs. The eggs are operculated, similar to trematode eggs (see Figure 49-70). The proglottids from the adult worm are much wider than long (opposite that seen with the *Taenia* spp. tapeworms). While the adult worms of *H. nana* and *H. diminuta* are relatively small, *Taenia* spp. range from about 12 to 15 feet, and *D. latum* can reach 30 feet in length.

Trematodes

Humans acquire most fluke infections by ingesting the encysted metacercariae (see Figure 49-75). Most trematodes have operculated eggs, which are best recovered by the sedimentation concentration technique rather than the flotation method. Often, careful measurements may be required for egg identification (see Figure 49-76). Some of the smaller eggs may be missed unless the high dry objective (40×) is used for microscopic examination.

Paragonimus spp. eggs are found not only in sputum but also in stool (sputum swallowed). These eggs are similar in size and shape to those of *D. latum*.

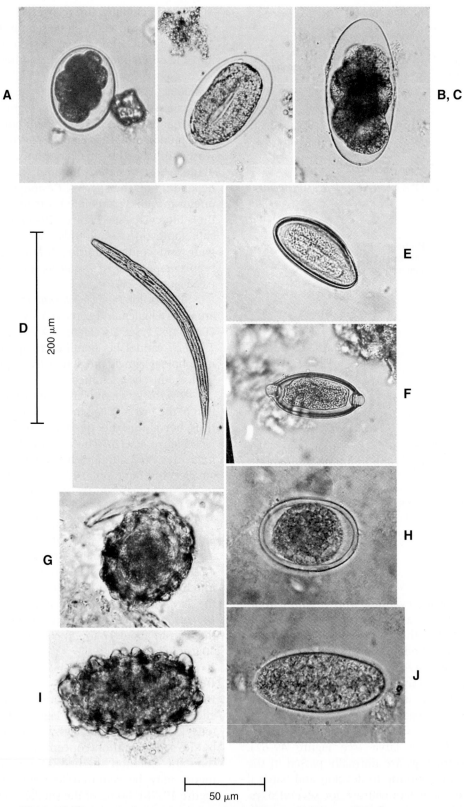

Figure 49-65 **A,** Immature hookworm egg. **B,** Embryonated hookworm egg.
C, *Trichostrongylus orientalis,* immature egg. **D,** *Strongyloides stercoralis,* rhabditiform
larva (200 µm). **E,** *Enterobius vermicularis* egg. **F,** *Trichuris trichiura* egg. **G,** *Ascaris
lumbricoides,* fertilized egg. **H,** *A. lumbricoides,* fertilized egg, decorticate. **I,** *A.
lumbricoides,* unfertilized egg. **J,** *A. lumbricoides,* unfertilized egg, decorticate.

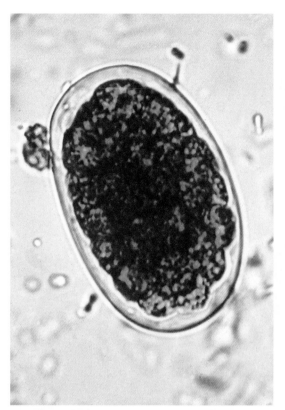

Figure 49-66 Hookworm egg, iodine stain.

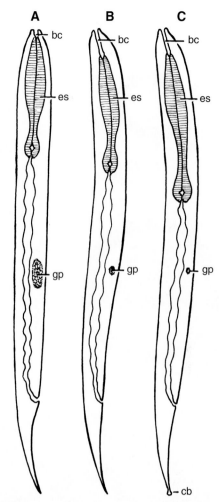

Figure 49-68 Rhabditiform larvae. **A,** *Strongyloides.* **B,** Hookworm. **C,** *Trichostrongylus. bc,* Buccal cavity; *cb,* beadlike swelling of caudal tip; *es,* esophagus; *gp,* genital primordia. (Illustration by Nobuko Kitamura.)

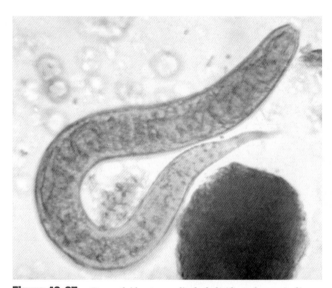

Figure 49-67 *Strongyloides stercoralis* rhabditiform larva, iodine stain.

Schistosome eggs are relatively easy to identify and have terminal *(Schistosoma haematobium),* lateral *(S. mansoni),* or small lateral *(S. japonicum)* spines (see Figure 49-76). These eggs are nonoperculated; specific procedures for egg recovery and identification are found in the earlier section on the egg hatching procedure.

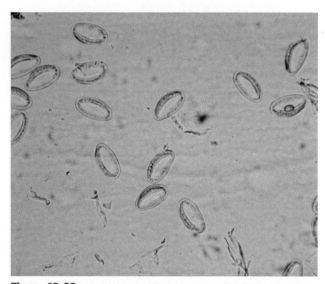

Figure 49-69 *Enterobius vermicularis* eggs (cellophane [Scotch] tape preparation).

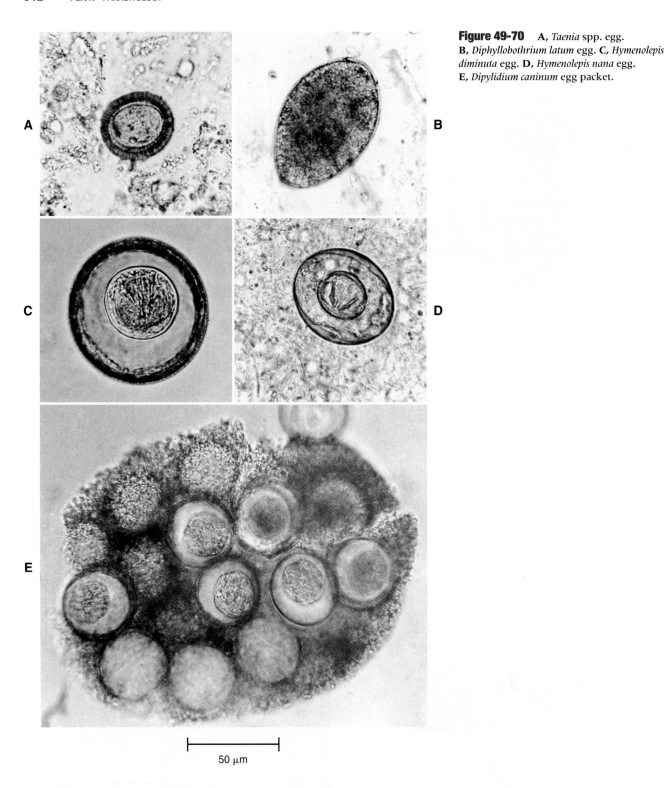

Figure 49-70   **A,** *Taenia* spp. egg.
B, *Diphyllobothrium latum* egg. **C,** *Hymenolepis diminuta* egg. **D,** *Hymenolepis nana* egg.
E, *Dipylidium caninum* egg packet.

50 μm

Within the life cycle (see Figure 49-77), free-swimming cercariae attach to and penetrate unbroken skin of the human host. In temperate zones, such as the Great Lakes area, cercariae of other bird or animal trematodes can penetrate human skin and cause a strong local reaction (swimmer's itch), although they do not complete the life cycle.

BLOOD HELMINTHS

Filarial Nematodes
The filarial worms are long, thin nematodes that inhabit the lymphatic system and the subcutaneous and deep connective tissues (see Figure 49-78). Most species produce larval worms, the microfilariae, which can be

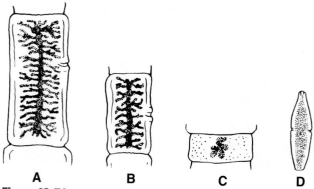

Figure 49-71   Gravid proglottids. **A,** *Taenia saginata.* **B,** *Taenia solium.* **C,** *Diphyllobothrium latum.* **D,** *Dipylidium caninum.* (From Garcia LS: *Diagnostic medical parasitology,* ed 5, Washington, DC, 2007, ASM Press.)

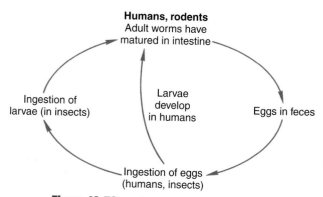

Figure 49-72   Life cycle of *Hymenolepis nana.*

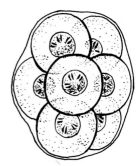

Figure 49-73   *Dipylidium caninum* egg packet.   (Illustration by Nobuko Kitamura.)

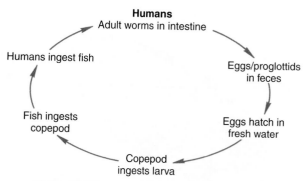

Figure 49-74   Life cycle of *Diphyllobothrium latum.*

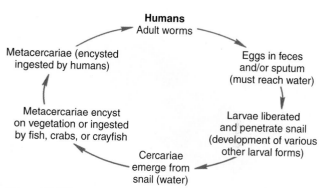

Figure 49-75   Life cycle of trematodes acquired by humans through ingestion of raw fish, crabs, or crayfish and vegetation.

found in the peripheral blood; two species, *Onchocerca volvulus* and *Dipetalonema streptocerca,* produce microfilariae found in the subcutaneous tissues and dermis (see Figure 49-79).

Diagnosis is usually based on clinical findings, but demonstration of the parasite is the only accurate way of confirming the diagnosis. Fresh, thin, and thick blood films can be prepared, as well as the Knott and mem-brane filtration concentration techniques. Because some of the infections are periodic and microfilariae are found in the peripheral blood at certain times only, the time the blood is drawn may be critical in demonstrating the parasites. Differentiation of the species depends on (1) the presence or absence of the sheath and (2) the distribution of nuclei in the tail region of the microfilaria (see Figure 49-80).

ANTIMICROBIAL SUSCEPTIBILITY TESTING AND THERAPY

Antimicrobial susceptibility testing methods are available for limited parasites but are performed in the research laboratory setting only. Routine testing may be available in the future, but probably not until automated or semiautomated systems are available. Certainly this type of testing has greater relevance in other areas of the world in which certain parasites are known to be resistant to specific therapeutic agents.

Chemotherapy plays an important role not only in reducing patient morbidity and mortality but also in reducing the transmission of the parasitic infection. Many of the drugs used to treat parasitic infections have

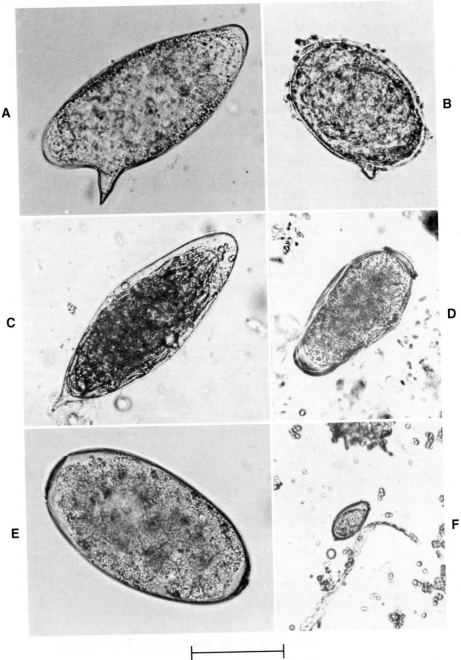

50 μm

Figure 49-76 **A,** *Schistosoma mansoni* egg. **B,** *Schistosoma japonicum* egg. **C,** *Schistosoma haematobium* egg. **D,** *Paragonimus westermani* egg. **E,** *Fasciola hepatica* egg. **F,** *Clonorchis (Opisthorchis) sinensis* egg.

serious side effects; therefore, before initiation of therapy, the following factors must be considered: health of the patient, parasite drug resistance, accuracy of the original dose, potential drug toxicity, and need for follow-up examinations to monitor therapy[6,17] (Table 49-22). The mechanisms of action for most antiparasitic drugs are not well known, including potential drug toxicity to the patient. Because of the limited resources of developing nations where a majority of parasitic infections occur, commercial incentive is low for developing effective therapeutics or vaccines. For some drugs, such as triclabendazole and tryparsamide used extensively in veterinary practices, information is scant regarding adverse side effects or toxicity to humans. Specific information can be obtained from the Centers for Disease Control and Prevention, CDC Drug Service, in Atlanta, Georgia (day [404] 639-3670 or evenings, weekends, and holidays [404] 639-2888). Drugs that

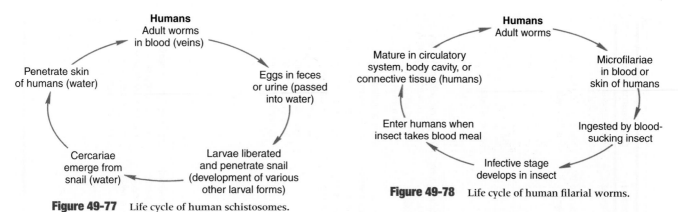

Humans
Adult worms
in blood (veins)

Penetrate skin
of humans (water)

Eggs in feces
or urine (passed
into water)

Cercariae
emerge from
snail (water)

Larvae liberated
and penetrate snail
(development of various
other larval forms)

Figure 49-77   Life cycle of human schistosomes.

Humans
Adult worms

Mature in circulatory
system, body cavity, or
connective tissue (humans)

Microfilariae
in blood or
skin of humans

Enter humans when
insect takes blood meal

Ingested by blood-
sucking insect

Infective stage
develops in insect

Figure 49-78   Life cycle of human filarial worms.

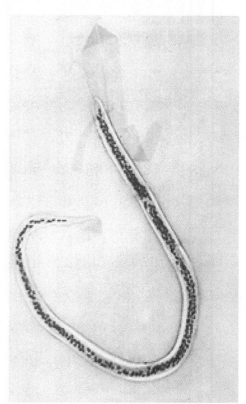

Figure 49-80   Microfilaria of *Wuchereria bancrofti* in thick blood film.

A   B   C   D   E   F   G

N. Kitamura

Figure 49-79   Anterior and posterior ends of microfilariae found in humans. **A,** *Wuchereria bancrofti.* **B,** *Brugia malayi.* **C,** *Loa loa.* **D,** *Onchocerca volvulus.* **E,** *Mansonella perstans.* **F,** *Mansonella streptocerca.* **G,** *Mansonella ozzardi.*

are not commercially available in the United States can often be obtained from the CDC.

PREVENTION

Prevention of human parasitic infections is directly linked to the various organism life cycles and modes of infection (see Table 49-4). Preventive measures involve increased attention to personal hygiene and sanitation measures and elimination of sexual activities in which the potential for fecal-oral contact is possible. Adequate

water treatment (including filtration) may be required, as well as an overall awareness of environmental sources of infection. In some cases, avoidance of contaminated environmental water and soil sources may be important; this is mandatory when dealing with lens care systems and potential infections with free-living amebae.

Chemoprophylactic agents for the prevention of clinical symptoms are given to patients who plan to travel to areas in which malaria is endemic. These drugs for prophylaxis are effective against the erythrocytic forms only and do not prevent the person from getting

Table 49-21   Cestode Parasites of Humans (Intestinal)

	Diphyllobothrium latum	Taenia saginata	Taenia solium	Hymenolepis nana	Hymenolepis diminuta	Dipylidium caninum
Intermediate hosts (common)	Two: copepods and fish	One: cattle	One: pig	One: various arthropods (beetles, fleas); or none	One: various arthropods (beetles, fleas)	One: various arthropods (fleas, dog lice)
Mode of infection	Ingestion of plerocercoid (sparganum) in flesh of infected fish	Ingestion of cysticercus in infected beef	Ingestion of cysticercus in infected pork	Ingestion of cysticercoid in infected arthropod or by direct ingestion of egg; autoinfection may also occur	Ingestion of cysticercoid in infected arthropod	Ingestion of cysticercoid in fleas, lice
Prepatent period	3-5 weeks	10-12 weeks	5-12 weeks	2-3 weeks	~3 weeks	3-4 weeks
Normal life span	Up to 25 years	Up to 25 years	Up to 25 years	Perhaps many years as a result of autoinfection	Usually <1 year	Usually <1 year
Length	4-10 m	4-12 m	1.5-8 m	2.5-4.0 cm	20-60 cm	10-70 cm
Scolex	Spatulate, 3 × 1 mm; no rostellum or hooklets; has 2 shallow grooves (bothria)	Quadrate, 1- to 2-mm diameter; no rostellum or hooklets; 4 suckers	Quadrate, 1-mm diameter, has rostellum and hooklets, 4 suckers	Knoblike but not usually seen; has rostellum and hooklets; 4 suckers	Knoblike but not usually seen; has rostellum but no hooklets; 4 suckers	0.2-0.5 mm in diameter; has conical/retractile rostellum armed with 4-7 rows of small hooklets; 4 suckers
Usual means of diagnosis	Ovoid, operculate yellow-brown eggs (58-75 μm by 40-50 μm) in feces; egg usually has small knob at abopercular end; proglottids may be passed, usually in chain of segments (few cm to ½ m long); proglottids wider than long (3 × 11 mm) and have rosette-shaped central uterus	Gravid proglottids in feces; they are longer than wide (19 × 17 mm) and have 15-20 lateral branches on each side of central uterine stem; they usually appear singly; spheroidal yellow-brown, thick-shelled eggs (31-43 μm in diameter) containing an oncosphere may be found in feces	Gravid proglottids in feces; they are longer than wide (11 × 5 mm) and have 7-13 lateral branches on each side of central uterine stem; usually appear in chain of 5-6 segments, spheroidal, yellow-brown, thick-shelled eggs (31-43 μm) containing an oncosphere may be found in feces	Nearly spheroidal, pale, thin-shelled eggs (30-47 μm in diameter) in feces; oncosphere surrounded by rigid membrane, which has two polar thickenings from which 4-8 filaments extend into the space between the oncosphere and thin, outer shell	Large, ovoid, yellowish, moderately thick-shelled eggs (70-85 μm by 60-80 μm) in feces; egg contains oncosphere	Gravid proglottids (8-23 μm long) containing compartmented cluster of eggs in feces; proglottids have genital pores at both lateral margins; occasionally may see individual oncospheres (20-33 mm in diameter) in feces
Diagnostic problems or notes	Eggs are sometimes confused with eggs of Paragonimus; eggs are unembryonated when passed in feces	Eggs are identical to those of Taenia solium; ordinarily can distinguish between species only by examination of gravid proglottids; eggs can be confused with pollen grains (handle all proglottids with extreme care)	Eggs are identical to those of T. saginata; one is less likely to find eggs in feces than with T. saginata (handle all proglottids with extreme care since T. solium eggs are infective to humans)	Sometimes confused with eggs of Hymenolepis diminuta; rodents serve as reservoir hosts	Should not be confused with H. nana since eggs lack polar filaments; rodents serve as reservoir hosts	Gravid proglottids resemble rice grains (dry) or cucumber seeds (moist); dogs and cats serve as reservoir hosts

Table 49-22   Therapy for Parasitic Infections[15]

Parasite Group	Therapy*†	Comments
Protozoa, Intestinal		
Amebae		
Entamoeba histolytica	Iodoquinol	For asymptomatic cyst passers
	Metronidazole	Mild to moderate intestinal disease
	Metronidazole	Invasive amebiasis
Blastocystis hominis	Metronidazole	Clinical significance is controversial; symptomatic patients may be treated
Flagellates		
Dientamoeba fragilis	Iodoquinol	Fecal examination 2-4 weeks after therapy must include permanent stained smear
Giardia lamblia	Metronidazole	Immunoassays can be used for "test of cure" (as well as O&P examinations); some patients who harbor *Giardia* may remain asymptomatic
Ciliates		
Balantidium coli	Tetracycline	Infection relatively rare in the United States; at least 3 negative fecal examinations 1 month after therapy
Coccidia		
Cryptosporidium spp.	Paromomycin	Therapy marginal; may be combined with azithromycin
Cyclospora cayetanensis	TMP/SMX	Monitor results using modified acid-fast stains
Isospora belli	TMP/SMX	Fecal examinations 1-2 weeks after therapy
Microsporidia	Albendazole	Multiple genera are pathogenic for humans, as well as for animals; wide range of body sites; disease varies, depending of immune status of patient; disease outcome complicated by lack of treatment for some genera; albendazole effective for *Encephalitozoon* spp.
Protozoa, Other Sites		
Amebae		
Free-living amebae		Disease ranging from acute meningoencephalitis to chronic encephalitis to cutaneous infections to keratitis, and the potential for other body sites; disease spectrum depends on patient's immune capacity and the organism involved; disease can be mild (*Acanthamoeba* spp.) to fatal *(Naegleria fowleri)*
Acanthamoeba spp.	Amphotericin B, sulfadiazine (meningoencephalitis)	
	Polyhexamethylene biguanide, propamidine isethionate (keratitis)	
Balamuthia mandrillaris	Pentamidine isethionate	
Naegleria fowleri	Amphotericin B	
Flagellates		
Trichomonas vaginalis	Metronidazole	The disease can vary from mild to severe. Drug resistance does occur; however, continued positive clinical specimens can also be due to reinfection unless all sexual partners are also treated
Coccidia		
Toxoplasma gondii	Pyrimethamine plus sulfadiazine	Those with clinically active disease, diagnosed congenital disease, and disease in symptomatic compromised patients should be treated. Prophylactic treatment is often recommended for the newborn until it can be demonstrated that IgM antibody is not present
Protozoa, Blood, Tissue		
Babesia spp.	Clindamycin plus quinine	If symptoms persist, repeat blood film examination
Plasmodium spp.	Chloroquine phosphate	Chemoprophylaxis (chloroquine-sensitive endemic areas)
	Mefloquine	Chemoprophylaxis (chloroquine-resistant endemic areas)
	Chloroquine phosphate	All species except chloroquine-resistant *P. falciparum* and *P. vivax*
		Treatment regimens vary tremendously, depending on endemic area, potential species of *Plasmodium* spp., and severity of disease[6]
Flagellates (Leishmaniae)	Stibogluconate sodium (meglumine antimonate, amphotericin B)	*Leishmania donovani* that invades the spleen, liver, and bone marrow can cause serious disease, particularly in the compromised host; more severe disease seen in all leishmanial species in immunosuppressed patients

Table 49-22   Therapy for Parasitic Infections[15]—cont'd

Parasite Group	Therapy*†	Comments

Flagellates
  Trypanosoma brucei gambiense
 	
  Trypanosoma brucei rhodesiense
 	
  Trypanosoma cruzi
	Pentamidine isethionate (early stages)
	Melarsoprol (late stages)
	Suramin (early stages)
	Melarsoprol (late stages)
	Benznidazole
		T.b. gambiense and *T.b. rhodesiense* cause African sleeping sickness, with eventual invasion of the CNS, leading to coma and death. Drugs are relatively toxic and side effects common
		
		Chagas' disease *(T. cruzi)* causes acute to chronic problems that are primarily linked to cardiac disease and diminished cardiac capacity. Monitor serology and ECG; alcohol should be avoided during treatment; other contraindications present

Nematodes, Intestinal		
Ascaris lumbricoides *Trichuris trichiura* Hookworm *Strongyloides stercoralis* *Enterobius vermicularis*	Albendazole or mebendazole Ivermectin Pyrantel pamoate	Therapy for these infections is usually effective
Nematodes, Tissue		
Trichinella spp. *Ancylostoma* spp. *Toxocara* spp.	Mebendazole Albendazole Albendazole	*Trichinella* can cause eosinophilia, muscle aches and pains, and death, depending on the worm burden; corticosteroids can be used with severe symptoms
Nematodes, Filarial		
Brugia malayi *Loa loa* *Wuchereria bancrofti* *Mansonella* spp. *Onchocerca volvulus*	Diethylcarbamazine Species-dependent[15] Ivermectin	Some filarial infections are not that well adapted to humans and require many years of exposure before disease is manifest; some infections not evident, some cause multiple disease manifestations
Cestodes, Intestinal		
Taenia spp. *Hymenolepis* spp. *Diphyllobothrium latum* *Dipylidium caninum*	Praziquantel (niclosamide)	Human infection with the adult tapeworms can cause no symptoms or mild intestinal symptoms may occur. When the human becomes the accidental intermediate host for *T. solium,* CNS symptoms may occur, including epileptic seizures. Infection with adult *D. latum* tapeworm can also cause intestinal symptoms such as pain, diarrhea, etc., but the patient may also be asymptomatic; a vitamin B_{12} deficiency may be seen. Infection with *Hymenolepis nana* is primarily acquired from accidental ingestion of eggs from an adult tapeworm; symptoms may be absent or diarrhea may be present
Cestodes, Tissue		
Echinococcus spp. *Taenia solium* (cysticercosis)	Albendazole or surgery Albendazole	Monitor with CT, ultrasonography, and radionucleotide scans. Depending on the body site, hydatid cysts can cause pain, anaphylactic shock (fluid leakage), or CNS symptoms. The patient may be unaware of infection until cyst begins to press on other body organs or a large fluid leak occurs
Trematodes, Intestinal		
Fasciolopsis buski *Heterophyes heterophyes* *Metagonimus yokogawai*	Praziquantel	Intestinal trematodes can cause pain, diarrhea; intestinal toxicity can sometimes be seen in heavy infections with *F. buski*

Table 49-22   Therapy for Parasitic Infections[15]—cont'd

Parasite Group	Therapy*†	Comments
Trematodes, Liver, Lung		
Clonorchis sinensis Fasciola hepatica Metorchis conjunctus Opisthorchis viverrini Paragonimus spp.	Praziquantel Bithionol Praziquantel	Many genera and species pathogenic for humans; symptoms depend on worm burden. Metorchis sp. is the North American liver fluke
Trematodes, Blood		
Schistosoma spp.	Praziquantel	Symptoms may range from none in light infections to severe organ failure as a result of deposition of eggs with subsequent formation of granuloma within the tissues; severe disease can cause death

CNS, Central nervous system; CT, computed tomography; ECG, electrocardiogram; O&P, ova and parasite; TMP/SMX, trimethoprim/sulfamethoxazole.

*There are a number of alternative therapies for many parasitic infections; one should refer to more extensive references regarding treatment options. Not all drugs are available within the United States. Additional information can be obtained from the Centers for Disease Control and Prevention, CDC Drug Service, Atlanta, Ga. (day, [404] 639-3670; evenings, weekends, and holidays, [404] 639-2888).

†The mechanisms of action for most antiparasitic drugs are not well known, including those involving potential drug toxicity. Many of the drugs used to treat parasitic infections have serious side effects; therefore, before initiation of therapy, it is important to consider the following: health of the patient, parasite drug resistance, accuracy of the original dose, potential drug toxicity, and the need for follow-up examinations to monitor therapy.

Procedure 49-1

FORMALIN-ETHER (FORMALIN-ETHYL ACETATE) SEDIMENTATION TECHNIQUES

PRINCIPLE
Formalin fixes the eggs, larvae, oocysts, and spores, so that they are no longer infectious, as well as preserves their morphology. Fecal debris is extracted into the ethyl acetate phase of the solution, freeing the sedimented parasitic elements from at least some of the artifact material in stool. Numerous ether substitutes are available; the term *ethyl acetate* used throughout this chapter is used in the general sense (ether substitute)

METHOD
1. Transfer $1/4$ to $1/2$ teaspoon of fresh stool into 10 mL of 5% or 10% formalin in a 15-mL shell vial, unwaxed paper cup, or 16- × 125-mm tube (container may vary depending on individual preferences) and comminute thoroughly. Let stand 30 minutes for adequate fixation.
2. Filter this material (funnel or pointed paper cup with end cut off) through two layers of gauze into a 15-mL centrifuge tube.
3. Add physiologic saline or 5% or 10% formalin to within $1/2$ inch (1.5 cm) of the top and centrifuge for 10 minutes at 500× g.

4. Decant, resuspend the sediment (should have 0.5 to 1 mL sediment) in saline to within $1/2$ inch (1.5 cm) of the top, and centrifuge again for 10 minutes at 500× g. This second wash may be eliminated if the supernatant fluid after the first wash is light tan or clear.
5. Decant and resuspend the sediment in 5% or 10% formalin (fill the tube only half full). If the amount of sediment left in the bottom of the tube is very small, do not add ethyl acetate in step 6; merely add the formalin, then spin, decant, and examine the remaining sediment.
6. Add approximately 3 mL of ethyl acetate, stopper, and shake vigorously for 30 seconds. Hold the tube so that the stopper is directed away from your face; remove stopper carefully to prevent spraying of material caused by pressure within the tube.
7. Centrifuge for 10 minutes at 500× g. Four layers should result: a small amount of sediment in the bottom of the tube, containing the parasites; a layer of formalin; a plug of fecal debris on top of the formalin layer; and a layer of ether substitute at the top.
8. Free the plug of debris by ringing

with an applicator stick, and decant all the fluid. After proper decanting, a drop or two of fluid remaining on the side of the tube will drain down to the sediment. Mix the fluid with the sediment and prepare a wet mount for examination.

The formalin-ethyl acetate sedimentation procedure may be used on PVA-preserved material. Steps 1 and 2 differ as follows:
1. Fixation time with PVA should be at least 30 minutes. Mix contents of PVA bottle (stool-PVA mixture: 1 part stool to 2 or 3 parts PVA) with applicator sticks. Immediately after mixing, pour approximately 2 to 5 mL (amount will vary depending on the viscosity and density of the mixture) of the stool-PVA mixture into a 15-mL shell vial, 16- × 125-mm tube, or such, and add approximately 10 mL physiologic saline or 5% or 10% formalin.
2. Filter this material (funnel or paper cup with pointed end cut off) through two layers of gauze into a 15-mL centrifuge tube.

Steps 3 through 8 are the same for both fresh and PVA-preserved material.

FORMALIN-ETHER (FORMALIN-ETHYL ACETATE) SEDIMENTATION TECHNIQUES—CONT'D

NOTE: Tap water may be substituted for physiologic saline throughout this procedure; however, saline is recommended. Some workers prefer to use 5% or 10% formalin for all the rinses (steps 3 and 4).

Note: The introduction of ethyl acetate (or Hemo-De, or other substitute) as a substitute for diethyl ether (ether) in the formalin-ether sedimentation concentration procedure provides a much safer chemical for the clinical laboratory. Tests comparing the use of these two compounds on formalin-preserved and PVA-preserved material indicate that the differences in organism recovery and identification are minimal and pro-

bably do not reflect clinically relevant differences.

When examining the sediment in the bottom of the tube:

1. Prepare a saline mount (1 drop of sediment and 1 drop of saline solution mixed together), and scan the whole 22- × 22-mm coverslip under low power for helminth eggs or larvae.
2. Iodine may then be added to aid in the detection of protozoan cysts and should be examined under high dry power. If iodine is added before low-power scanning, be certain that the iodine is not too strong; otherwise, some of the helminth eggs will stain

so darkly that they will be mistaken for debris.

3. Occasionally a precipitate is formed when iodine is added to the sediment obtained from a concentration procedure with PVA-preserved material. The precipitate is formed from the reaction between the iodine and excess mercuric chloride that has not been thoroughly rinsed from the PVA-preserved material. The sediment can be rinsed again to remove any remaining mercuric chloride, or the sediment can be examined as a saline mount without the addition of iodine.

TRICHROME STAIN

PRINCIPLE

The internal elements that distinguish among cysts and trophozoites can best be visualized with a stain that enhances the morphologic features. In addition, such a stained smear provides a permanent record of the results.

REAGENTS

A. FORMULA

Chromotrope 2R	0.6 g
Light green SF	0.3 g
Phosphotungstic acid	0.7 g
Acetic acid (glacial)	1 mL
Distilled water	100 mL

B. STAIN PREPARATION
1. Add 1 mL of glacial acetic acid to the dry components.
2. Allow the mixture to stand for 15 to 30 minutes to "ripen"; then add 100 mL of distilled water. This preparation gives a highly uniform and reproducible stain; the stain should be purple. Store in Coplin jars.

METHOD
1. Prepare fresh fecal smears or PVA smears as described.
2. Place in 70% ethanol for 5 minutes. (This step may be eliminated for PVA smears.)

3. Place in 70% ethanol plus D'Antoni's iodine (dark reddish brown) for 2 to 5 minutes.
4. Place in two changes of 70% ethanol—one for 5 minutes* and one for 2 to 5 minutes.
5. Place in trichrome stain solution for 10 minutes.
6. Place in 90% ethanol, acidified (1% acetic acid) for up to 3 seconds (*do not leave the slides in this solution any longer*).
7. Dip once in 100% ethanol.
8. Place in two changes of 100% ethanol for 2 to 5 minutes each.*
9. Place in two changes of xylene or toluene for 2 to 5 minutes each.*
10. Mount in Permount or some other mounting medium; use a No. 1 thickness coverglass.

EXPECTED RESULTS

Background debris will be green and protozoa will show blue-green to purple cytoplasm. The nuclei and inclusions will be red or purple-red and sharply delineated from background.

NOTE

If you are currently using one of the stool fixatives that contains a mercuric chloride substitute (e.g., zinc sulfate), remember that the proficiency testing specimens you receive for permanent staining may have been preserved in PVA using the mercuric chloride fixative

base. If you use the trichrome staining method for your *mercuric chloride substitute fixatives,* you may have eliminated the 70% alcohol/iodine step and the following 70% alcohol rinse steps from your method. However, when you stain the proficiency testing fecal smears, you will need to incorporate the iodine step plus the next 70% alcohol rinse steps back into your staining protocol before placing your slides into the trichrome stain. These two steps are designed to remove the mercury from the smear and then to remove the iodine; when your slide is placed into the trichrome stain, both the mercury and iodine are no longer present in the fecal smear. If you fail to incorporate these two steps into your staining protocol, the quality of your proficiency testing stained smears will be poor.

Although you may be using mercuric chloride substitute fixatives, both the iodine/70% alcohol and subsequent 70% alcohol rinse steps before the trichrome stain can be used with no damage to the slides. However, until notified differently, some proficiency testing specimens have been fixed using mercuric chloride and will require the iodine and subsequent alcohol rinses before the trichrome staining step. All steps after the trichrome stain would remain the same for either type of fixative.

---

*At this stage, slides can be held several hours or overnight.

PRINCIPLE

The internal elements that distinguish among cysts and trophozoites can best be visualized with a stain that enhances the morphologic features. Also, the stained smear provides a permanent record of the results.

REAGENTS

A. MAYER'S ALBUMIN

Add an equal quantity of glycerine to a fresh egg white. Mix gently and thoroughly. Store at 4° C and indicate an expiration date of 3 months. Mayer's albumin from commercial suppliers can normally be stored at 25° C for 1 year (e.g., Produce #756, E.M. Diagnostic Systems, Inc., 480 Democrat Road, Gibbstown, NJ 08027; [800] 443-3637).

B. STOCK SOLUTION OF HEMATOXYLIN STAIN

Hematoxylin powder	10 g
Ethanol (95% or 100%)	1000 mL

1. Mix well until dissolved.
2. Store in a clear glass bottle, in a light area. Allow to ripen for 14 days before use.
3. Store at room temp with an expiration date of 1 year.

C. MORDANT

Ferrous ammonium sulfate [Fe(NH$_4$)$_2$(SO$_4$)$_2$ • 6H$_2$O]	10 g
Ferric ammonium sulfate [FeNH$_4$(SO$_4$)$_2$ • 12 H$_2$O]	10 g
Hydrochloric acid (HCl) (concentrated)	10 mL
Add distilled water to 1000 mL	

D. WORKING SOLUTION OF HEMATOXYLIN STAIN

1. Mix equal quantities of stock solution of stain and mordant.
2. Allow mixture to cool thoroughly before use (prepare at least 2 hours before use). The working solution should be made fresh every week.

E. PICRIC ACID

Mix equal quantities of distilled water and an aqueous saturated solution of picric acid to make a 50% saturated solution.

F. ACID-ALCOHOL DECOLORIZER

Hydrochloric acid (HCl) (concentrated)	30 mL
Alcohol to 1000 mL	

G. 70% ALCOHOL AND AMMONIA

70% alcohol	50 mL
Ammonia	0.5-1 mL
Add enough ammonia to bring the pH to approximately 8.0.	

H. CARBOLFUCHSIN

Basic fuchsin (solution A): Dissolve 0.3 g basic fuchsin in 10 mL of 95% ethanol.

Phenol (solution B): Dissolve 5 g of phenol crystals in 100 mL distilled water. (Gentle heat may be needed.)

1. Mix solution A with solution B.
2. Store at room temperature. Stable for 1 year.

METHOD

1. Slide preparation
   a. Place 1 drop of Mayer's albumin on a labeled slide.
   b. Mix the sediment from the SAF concentration well with an applicator stick.
   c. Add approximately 1 drop of the fecal concentrate to the albumin and spread the mixture over the slide.
2. Allow slide to air dry at room temperature (smear will appear opaque when dry).
3. Place slide in 70% alcohol for 5 minutes.
4. Wash in container (not running water) of tap water for 2 minutes.
5. Place slide in Kinyoun stain for 5 minutes.
6. Wash slide in running tap water (constant stream of water into container) for 1 minute.
7. Place slide in acid/alcohol decolorizer for 4 minutes.*
8. Wash slide in running tap water (constant stream of water into container) for 1 minute.
9. Place slide in iron hematoxylin working solution for 8 minutes.
10. Wash slide in distilled water (in container) for 1 minute.
11. Place slide in picric acid solution for 3 to 5 minutes.
12. Wash slide in running tap water (constant stream of water into container) for 10 minutes.
13. Place slide in 70% alcohol plus ammonia for 3 minutes.
14. Place slide in 95% alcohol for 5 minutes.
15. Place slide in 100% alcohol for 5 minutes.
16. Place slide in two changes of xylene for 5 minutes.

PROCEDURE NOTES

1. The first 70% alcohol step acts with the Mayer's albumin to "glue" the specimen to the glass slide. The specimen may wash off if insufficient albumin is used or if the slides are not completely dry before staining.
2. The working hematoxylin stain should be checked each day of use by adding a drop of stain to alkaline tap water. If a blue color does not develop, prepare fresh working stain solution.
3. The picric acid differentiates the hematoxylin stain by removing more stain from fecal debris than from the protozoa and removing more stain from the organism cytoplasm than the nucleus. When properly stained, the background should be various shades of gray-blue and protozoa should be easily seen with medium-blue cytoplasm and dark blue-black nuclei.

---

*This step can also be performed as follows:
a. Place slide in acid/alcohol decolorizer for 2 minutes.
b. Wash slide in running tap water (constant stream of water into container) for 1 minute.
c. Place slide in acid/alcohol decolorizer for 2 minutes.
d. Wash slide in running tap water (constant stream of water into container) for 1 minute.
e. Continue staining sequence with step 9 above (iron hematoxylin working solution).

Procedure 49-4

MODIFIED ACID-FAST STAIN FOR COCCIDIA

REAGENTS

A. CARBOLFUCHSIN

Basic fuchsin	4 g
Phenol	8 mL
Alcohol (95%)	20 mL
Distilled water	100 mL

Dissolve the basic fuchsin in the alcohol, and add the water slowly while shaking. Melt the phenol in a 56° C water bath, and add 8 mL to the stain, using a pipette with a rubber bulb.

B. DECOLORIZER

Ethanol (95%)	97 mL
Concentrated HCl	3 mL

Add the hydrochloric acid to the alcohol slowly, working under a chemical fume hood.

C. COUNTERSTAIN

Methylene blue	0.3 g
Distilled water	100 mL

"COLD" MODIFIED ACID-FAST STAIN METHOD (KINYOUN)

1. Spin an aliquot of 10% formalinized stool for 10 minutes at 500× *g*.
2. Remove upper layer of sediment with pipette and place a thin layer onto a microscope slide.
   NOTE: If the stool specimen contains a lot of mucus, 10 drops of 10% KOH can be added to the sediment (step 2), vortexed, rinsed with 10% formalin, and respun before smear preparation. Some laboratories use this approach routinely before smear preparation.
3. Heat fix the smear at 70° C for 10 minutes.
4. Stain the fixed smear for 3 to 5 minutes (no heat necessary).
5. Wash in distilled, filtered water and shake off excess water.
6. Flood with decolorizer for approximately 1 minute. Check to see that no more red color runs when the slide is tipped. Add a bit more decolorizer for very thick slides or those that continue to bleed red dye.
7. Wash thoroughly with filtered water as above and shake off excess.
8. Flood with counterstain for approximately 1 minute.
9. Wash with distilled water and drain by standing slides upright. Do not blot dry.
   By the addition of a detergent or wetting agent the staining of acid-fast organisms may be accelerated. Tergitol No. 7 (Sigma Chemical Co.) may be used. Add 1 drop of Tergitol No. 7 to every 30 to 40 mL of the Kinyoun carbolfuchsin stain.
   Acid-fast bacteria stain red with carbolfuchsin stains. The background color is dependent on the counterstain; methylene blue imparts a blue color to non–acid-fast material, whereas brilliant green results in green background, and picric acid results in yellow.

"HOT" MODIFIED ACID-FAST STAIN METHOD

1. Spin an aliquot of 10% formalinized stool for 10 minutes at 500× *g*.
2. Remove upper layer of sediment with pipette and place a thin layer onto a microscope slide.
   NOTE: If the stool specimen contains a lot of mucus, 10 drops of 10% KOH can be added to the sediment (step 2), vortexed, rinsed with 10% formalin, and respun before smear preparation. Some laboratories use this approach routinely before smear preparation.
3. Heat fix the smear at 70° C for 10 minutes.
4. Place slide on staining rack and flood with carbolfuchsin.
5. Heat to steaming and allow to stain for 5 minutes. If the slide begins to dry, more stain is added without additional heating.
6. Rinse the smear with tap or distilled water.
7. Decolorize with 5% aqueous sulfuric acid for 30 seconds (thicker smears may require a longer time).
8. Rinse smear with tap or distilled water, drain, and flood smear with methylene blue counterstain for 1 minute.
9. Rinse with tap or distilled water, drain, and air dry.

Kinyoun: *Am J Pub Health* 5:867, 1915.

Procedure 49-5

MODIFIED TRICHROME STAIN FOR THE MICROSPORIDIA (WEBER-GREEN)

PRINCIPLE

The oval shape, spore wall, and diagonal or horizontal "stripe" that distinguish microsporidia spores can best be visualized with a stain that enhances the morphologic features. In addition, such a stained smear provides a permanent record of the results.

REAGENTS

A. MODIFIED TRICHROME STAIN

Chromotrope 2R	6 g*
Fast green	0.15 g
Phosphotungstic acid	0.7 g
Acetic acid (glacial)	3 mL
Distilled water	100 mL

1. Prepare the stain by adding 3 mL of acetic acid to the dry ingredients. Allow the mixture to stand (ripen) for 30 minutes at room temperature.
2. Add 100 mL of distilled water. Properly prepared stain will be dark purple.

---

*10 times the normal trichrome stain formula.

3. Store in a glass or plastic bottle at room temperature. The shelf life is at least 24 months.

B. ACID-ALCOHOL

90% ethyl alcohol	995.5 mL
Acetic acid (glacial)	4.5 mL

Prepare by combining the two solutions.

METHOD

1. Using a 10-μL aliquot of unconcentrated, preserved liquid stool (5% or 10% formalin or SAF), prepare the smear by spreading the material over an area of 45 × 25 mm.
2. Allow the smear to air dry.
3. Place the smear in absolute methanol for 5 minutes.
4. Allow the smear to air dry.
5. Place in trichrome stain for 90 minutes.
6. Rinse in acid-alcohol for no more than 10 seconds.
7. Dip slides several times in 95% alcohol. Use this step as a rinse.
8. Place in 95% alcohol for 5 minutes.

9. Place in 100% alcohol for 10 minutes.
10. Place in xylene substitute for 10 minutes.
11. Mount with coverslip (No. 1 thickness), using mounting medium. Check the specimen for adherence to the slide.
12. Examine smears under oil immersion (1000×) and read at least 100 fields; the examination time will probably be at least 10 minutes per slide.

EXPECTED RESULTS

Known parasites should be detected readily. When the smear is thoroughly fixed and the stain is performed correctly, the spores will be ovoid and refractile, with the spore wall being bright pinkish red. Occasionally the polar tube can be seen either as a stripe or as a diagonal line across the spore. The majority of the bacteria and other debris will tend to stain green. However, there will still be some bacteria and debris that will stain red.

Procedure 49-6

MODIFIED TRICHROME STAIN FOR THE MICROSPORIDIA (RYAN-BLUE)

PRINCIPLE

The oval shape, spore wall, and diagonal or horizontal "stripe" that distinguish microsporidia spores can best be visualized with a stain that enhances the morphologic features. In addition, such a stained smear provides a permanent record of the results.

Numerous variations to the modified trichrome (Weber-Green) were tried in an attempt to improve the contrast between the color of the spores and the background staining. Optimal staining was achieved by modifying the composition of the trichrome solution. This stain is also available commercially from several suppliers. The specimen can be fresh stool or stool that has been preserved in 5% or 10% formalin, SAF, or some of the newer single-vial system fixatives. Actually, any specimen other than tissue thought to contain microsporidia can be stained by this method.

REAGENTS

A. TRICHROME STAIN (MODIFIED FOR MICROSPORIDIA) (RYAN-BLUE)

Chromotrope 2R	6 g*
Aniline blue	0.5 g
Phosphotungstic acid	0.25 g
Acetic acid (glacial)	3 mL
Distilled water	100 mL

1. Prepare the stain by adding 3 mL of acetic acid to the dry ingredients. Allow the mixture to stand (ripen) for 30 minutes at room temperature.
2. Add 100 mL of distilled water and adjust the pH to 2.5 with 1M HCl. Properly prepared stain will be dark purple. The staining solution should be protected from light.
3. Store in a glass or plastic bottle at room temperature. The shelf life is at least 24 months.

B. ACID-ALCOHOL (see procedure 49-5)

METHOD

1. Using a 10-μL aliquot of unconcentrated, preserved liquid stool (5% or 10% formalin SAF), prepare the smear by spreading the material over an area of 45 × 25 mm.
2. Allow the smear to air dry.
3. Place the smear in absolute methanol for 5 or 10 minutes.
4. Allow the smear to air dry.
5. Place in trichrome stain for 90 minutes.
6. Rinse in acid-alcohol for no more than 10 seconds
7. Dip slides several times in 95% alcohol. Use this step as a rinse (no more than 10 seconds).
8. Place in 95% alcohol for 5 minutes.
9. Place in 95% alcohol for 5 minutes.
10. Place in 100% alcohol for 10 minutes.
11. Place in xylene substitute for 10 minutes.
12. Mount with coverslip (No. 1 thickness), using mounting medium.
13. Examine smears under oil immersion (1000×) and read at least 100 fields; the examination time will probably be at least 10 minutes per slide.

EXPECTED RESULTS

Known parasites should be detected readily. When the smear is thoroughly fixed and the stain is performed correctly, the spores will be ovoid and refractile, with the spore wall being bright pinkish red. Occasionally the polar tube can be seen either as a stripe or as a diagonal line across the spore. The majority of the bacteria and other debris will tend to stain blue. However, there will still be some bacteria and debris that will stain red.

PROCEDURE NOTES FOR MODIFIED TRICHROME STAINING METHODS (WEBER OR RYAN)

1. It is mandatory that positive control smears be stained and examined each time patient specimens are stained and examined.
2. Because of the difficulty in getting stain penetration through the spore wall, prepare thin smears and do not reduce the staining time in trichrome. Also, make sure the slides are not left too long in the decolorizing agent (acid-alcohol). If the control organisms are too light, leave them in the trichrome longer and shorten the time to two dips in the acid-alcohol solution. Also, remember that the 95% alcohol rinse after the acid-alcohol should be performed quickly to prevent additional destaining from the acid-alcohol reagent.
3. When you purchase the chromotrope 2R, obtain the highest dye content available. Two sources are Harleco (Gibbstown, NJ) and Sigma Chemical Co. (St. Louis, Mo.) (dye content among the highest [85%]). Fast green and aniline blue can be obtained from Allied Chemical and Dye, New York.
4. In the final stages of dehydration, the 100% ethanol and the xylenes (or xylene substitutes) should be kept as free from water as possible. Coplin jars must have tight-fitting caps to prevent both evaporation of reagents and absorption of moisture. If the xylene becomes cloudy after addition of slides from 100% alcohol, return the slides to 100% alcohol and replace the xylene with fresh stock.

PROCEDURE LIMITATIONS FOR MODIFIED TRICHROME STAINING METHODS (WEBER OR RYAN)

1. Although this staining method will stain the microsporidia, the range of stain intensity and the small size of the spores will cause some difficulty in identifying these organisms. Because this procedure will result in many other organisms or objects staining in stool specimens, differentiation of the microsporidia from surrounding material will still be difficult. There also tends to be some slight size variation among the spores.
2. If the patient has severe watery diarrhea, there will be less artifact material in the stool to confuse with the microsporidial spores; however, if the stool is semiformed or formed, the amount of artifact material will be much greater; thus, the spores will be much harder to detect and

---

*10 times the normal trichrome stain formula.

Procedure 49-6

MODIFIED TRICHROME STAIN FOR THE MICROSPORIDIA (RYAN-BLUE)—CONT'D

identify. Also, remember that the number of spores will vary according to the stool consistency (the more diarrhetic, the more spores that will be present).

3. Those who developed some of these procedures believe that concentration procedures result in an actual loss of microsporidial spores; thus there is a strong recommendation to use unconcentrated, formalinized stool. However, there are no data indicating what centrifugation

speeds, etc., were used in the study.
4. In the UCLA Clinical Microbiology Laboratory, we have generated data (unpublished) to indicate that centrifugation at 500× g for 10 minutes increases dramatically the number of microsporidial spores available for staining (from the concentrate sediment). This is the same protocol we use for centrifugation of all stool specimens, regardless of the suspected organism.
5. Avoid the use of wet gauze filtration

(an old, standardized method of filtering stool before centrifugation) with too many layers of gauze that may trap organisms and prevent them from flowing into the fluid to be concentrated. It is recommended that no more than two layers of gauze be used. Another option is to use the commercially available concentration systems that use metal or plastic screens for filtration.

Procedure 49-7

MODIFIED TRICHROME STAIN FOR THE MICROSPORIDIA (KOKOSKIN—HOT METHOD)

PRINCIPLE

The oval shape, spore wall, and diagonal or horizontal "stripe" that distinguish microsporidia spores can best be visualized with a stain that enhances the morphologic features. Changes in temperature from room temperature to 50° C and the staining time from 90 to 10 minutes have been recommended as improvements for the modified trichrome staining methods. In addition, such a stained smear provides a permanent record of the results.

METHOD

1. Using a 10-μL aliquot of unconcentrated, preserved liquid stool

(5% or 10% formalin or SAF), prepare the smear by spreading the material over an area of 45 × 25 mm.
2. Allow the smear to air dry.
3. Place the smear in absolute methanol for 5 minutes.
4. Allow the smear to air dry.
5. Place in trichrome stain for 10 minutes at a temperature of 50° C.
6. Rinse in acid-alcohol for no more than 10 seconds.
7. Dip slides several times in 95%, alcohol. Use this step as a rinse (no more than 10 seconds).
8. Place in 95% alcohol for 5 minutes.
9. Place in 100% alcohol for 10 minutes.

10. Place in xylene substitute for 10 minutes.
11. Mount with coverslip (No. 1 thickness), using mounting medium.
12. Examine smears under oil immersion (1000×) and read at least 100 fields; the examination time will probably be at least 10 minutes per slide.

Procedure 49-8

STAINING THIN FILMS: GIEMSA STAIN

PRINCIPLE

By spreading the blood cells in a thin layer, the size of red cells, inclusions, and extracellular forms can be more easily visualized.

METHOD

1. Fix blood films in absolute methanol (acetone-free) for 30 seconds.
2. Allow slides to air dry.
3. Immerse slides in a solution of 1 part Giemsa stock (commercial liquid stain or stock prepared from powder) to 10 to 50 parts of Triton-buffered water (pH 7.0 to 7.2). Stain 10 to 60

minutes (see note below). Fresh working stain should be prepared from stock solution each day.
4. Dip slides briefly in Triton X-100 buffered water.
5. Drain thoroughly in vertical position and allow to air dry.

NOTE: A good general rule for stain dilution vs. staining time is that if dilution is 1:20, stain for 20 minutes; if 1:30, stain for 30 minutes; and so forth. However, a series of stain dilutions and staining times should be tried to determine the best dilution/time for each batch of stock stain.

EXPECTED RESULTS

Giemsa stain colors the components of blood as follows: erythrocytes, pale gray-blue; nuclei of white blood cells, purple and pale purple cytoplasm; eosinophilic granules, bright purple-red; neutrophilic granules, deep pink-purple. Parasitic forms are blue to purple, with reddish nuclei. Their characteristic morphologies are used for differentiation. Inexperienced workers may confuse platelets with parasites.

Procedure 49-9

STAINING THICK FILMS: GIEMSA STAIN

PRINCIPLE

A large amount of blood can be examined for parasitic forms by lysing the red blood cells and staining for parasites. The lack of methanol fixation allows lysis of red blood cells by the aqueous stain solution. Although parasites can be found in the larger volume of blood, definitive morphologic criteria necessary for specific organism identification may be more difficult to see.

METHOD

The procedure to be followed for thick films is the same as for thin films, except

that the first two steps are omitted. If the slide has a thick film at one end and a thin film at the other, fix only the thin portion and then stain both parts of the film simultaneously.

Note: Although Giemsa stain has been used for many years, a number of blood stains can be used for the blood parasites (Giemsa, Wright, Wright-Giemsa combination, rapid blood stains). The use of different stains will produce somewhat different colors for the cell nuclei and cytoplasm; however, color variability is also seen using Giemsa stain. Regardless of the stain used, the quality

control is built into the slide; the color seen in any parasites present will mimic those seen in the white blood cells (WBCs). Therefore the WBCs serve as the quality control organisms. It is not mandatory to use actual blood parasites as the quality control organisms.

---

malaria, that is, the drugs do not prevent sporozoites from entering the liver and beginning the preerythrocytic developmental cycle. In general, chloroquine is still the drug of choice, although different regimens may be used in areas with chloroquine-resistant malaria.

Vector control, awareness of transmission through blood transfusions, shared drug needles, congenital infections, and organ transplants all are important considerations for the prevention of human disease caused by parasites. Careful monitoring of the blood supply is required for parasites transmitted via this route. This is particularly important in areas of the world in which bloodborne parasites play a large role in human disease.

Adequate cooking of potentially infected meats is also important; cultural habits may influence the handling and/or eating of raw or poorly cooked foods. Prevention depends on a thorough understanding of the life cycle and epidemiology of all parasites involved in causing human disease. This information is critical for the prevention of human disease caused by parasites limited to the human host, as well as those that can cause disease in humans, as well as other animal hosts.

Case Study 49-1

A patient is diagnosed as having diarrhea and specimens are submitted to the laboratory. When examining the permanent stained smears, trophozoites (18 μm) with evenly arranged nuclear chromatin and compact central karyosomes are seen. However, these trophozoites contain no red blood cells (RBCs).

QUESTIONS

1. How should these findings be reported?
2. Why is the correct report important to the physician?
3. Is it mandatory that immunoassay testing be performed to differentiate nonpathogenic *Entamoeba dispar* from pathogenic *Entamoeba histolytica*?

Case Study 49-2

A patient has come to the emergency department at 11:00 pm with an unexplained fever; the person had just returned from a trip to Africa. Blood is sent to microbiology for malaria smears and to hematology for routine testing.

QUESTIONS

1. Should these requests be handled as stat requests or as routine orders?
2. How should the blood films be examined and why is it important to identify any organisms present to the species level?
3. What are the chances that a malarial infection could be a "mixed infection"?
4. Can malaria parasites be detected using automated hematology instruments?

REFERENCES

1. Chen LH, Keystone JS: New strategies for the prevention of malaria in travelers, *Infect Dis Clin North Am* 19:185, 2005.

2. Clark DP: New insights into human cryptosporidiosis, *Clin Microbiol Rev* 12:554, 1999.

3. Desjeux P: Leishmaniasis: current situation and new perspectives, *Comp Immunol Microbiol Infect Dis* 27:305, 2004.

4. Espinosa-Cantellano M, Martinez-Paloma A: Pathogenesis of intestinal amebiasis: from molecules to disease, *Clin Microbiol Rev* 13:318, 2000.

5. Fayer R: *Cryptosporidium*: a water-borne zoonotic parasite, *Vet Parasitol* 126:37, 2004.

6. Garcia LS: *Diagnostic medical parasitology,* ed 5, Washington, DC, 2007, ASM Press.

7. Garcia LS: *Practical guide to diagnostic parasitology,* Washington, DC, 1999, ASM Press.

8. Garcia LS, Shimizu RY: Detection of *Giardia lamblia* and *Cryptosporidium parvum* antigens in human fecal specimens using the ColorPAC combination rapid solid-phase qualitative immunochromatographic assay, *J Clin Microbiol* 38:1267, 2000.

9. Garcia LS, Shimizu RY, Bernard CN: Detection of *Giardia lamblia, Entamoeba histolytica/E. dispar,* and *Cryptosporidium parvum* antigens in human fecal specimens using the EIA Triage Parasite Panel, *J Clin Microbiol* 38:3337, 2000.

10. Greenwood BM, Bojang K, Whitty CJ, et al: Malaria, *Lancet* 365:1487, 2005.

11. Isenberg HD, editor: *Clinical microbiology procedures handbook,* vol 1, 2, and 3, ed 2, Washington, DC, 2005, ASM Press.

12. Isenberg HD, editor: *Essential procedures for clinical microbiology,* Washington, DC, 1995, ASM Press.

13. Keiser PB, Nutman TB: *Strongyloides stercoralis* in the immunocompromised population, *Clin Microbiol Rev* 17:208, 2004.

14. Kerr SF, McHugh CP, Merkelz R: Short report: a focus of *Leishmania mexicana* near Tucson, Arizona, *Am J Trop Med Hyg* 61:378, 1999.

15. Krause PJ: Babesiosis diagnosis and treatment, *Vector Borne Zoonotic Dis* 3:45, 2003.

16. Magill AJ: Cutaneous leishmaniasis in the returning traveler, *Infect Dis Clin North Am* 19:241, 2005.

17. Markell EK, Voge M, John DT: *Medical parasitology,* ed 7, Philadelphia, 1992, WB Saunders.

18. Martens P, Hall L: Malaria on the move: human population movement and malaria transmission, *Emerg Infect Dis* 6:103, 2000.

19. Melvin DM, Brooke MM: Laboratory procedures for the diagnosis of intestinal parasites, US Department of Health, Education, and Welfare Pub No (CDC) 85-8282, Washington, DC, 1985, U.S. Government Printing Office.

20. National Committee for Clinical Laboratory Standards/Clinical and Laboratory Standards Institute: Procedures for the recovery and identification of parasites from the intestinal tract; Approved guideline M28-A2, Villanova, Pa, 2005, Clinical and Laboratory Standards Institute.

21. National Committee for Clinical Laboratory Standards/Clinical and Laboratory Standards Institute: Laboratory Diagnosis of Blood-borne Parasitic Diseases; Approved guideline M15-A, Wayne, Pa, 2000, Clinical and Laboratory Standards Institute.

22. Noel C, Dufernez F, Gerbod D, et al: Molecular phylogenies of *Blastocystis* isolates from different hosts: implications for genetic diversity, identification of species, and zoonosis, *J Clin Microbiol* 43:348, 2005.

23. Pasvol G: Management of severe malaria: interventions and controversies, *Infect Dis Clin North Am* 19:211, 2005.

24. Paulinoa M, Iribarnea F, Dubinb M, et al: The chemotherapy of Chagas' disease: an overview, *Mini Rev Med Chem* 5:499, 2005.

25. Schwebke JR, Burgess D: Trichomoniasis, *Clin Microbiol Rev* 17:794, 2004.

26. Shields JM, Olson BH: *Cyclospora cayetanensis*: a review of an emerging parasitic coccidian, *Int J Parasitol* 33:371, 2003.

27. Tan KS: *Blastocystis* in humans and animals: new insights using modern methodologies, *Vet Parasitol* 126:121, 2004.

28. Tanyuksel M, Petri WA Jr: Laboratory diagnosis of amebiasis, *Clin Microbiol Rev* 16:713, 2003.

29. Thompson RC, Monis PT: Variation in *Giardia*: implications for taxonomy and epidemiology, *Adv Parasitol* 58:69, 2004.

30. Wilson M, Schantz PM: Parasitic immunodiagnosis. In Strickland GT, editor: *Hunter's tropical medicine and emerging infectious diseases,* ed 8, Philadelphia, 2000, WB Saunders.

31. Xiao L, Ryan UM: Cryptosporidiosis: an update in molecular epidemiology, *Curr Opin Infect Dis* 17:483, 2004.

V PART

Mycology

Historically, the fungi were regarded as relatively insignificant causes of infection. However, these microorganisms began to be recognized as important causes of disease in the early to mid-twentieth century, particularly because of changes in patient profiles, which continues today. In recognition of the endemic systemic mycoses, which may cause disease in healthy hosts, a number of fungal species that are normally found in the environment have been recognized as important causes of human disease, particularly in the immunocompromised host.[108,145] The modern clinical laboratory, therefore, needs to provide methods for isolating and identifying the common causes of mycologic disease. Susceptibility testing of these isolates is also often necessary. Some clinical microbiology laboratories have kept pace with changing times and have offered somewhat complete mycologic services. However, because of continued economic restraints in the health care environment, some laboratories are unable to offer these services. In such instances, diagnostic clinical mycology is performed by reference laboratories that have varying degrees of experience. The lack of experience in clinical mycology has been influenced by a shortage of trained individuals, lack of quality educational programs, and the inability of clinical laboratories to support the cost of sending personnel to training courses. Commonly, individuals with experience who retire or leave their position are replaced by someone with considerably less experience. Training and continuing education programs are needed to assist in the development of such individuals, if quality laboratory services are to be offered. A real concern is that the changing health care environment and implementation of cost containment measures without continuing education will prevent future generations from being well trained in diagnostic clinical mycology.

This chapter is designed to assist technologists or microbiologists with the basics of diagnostic clinical mycology with the hope that this information will allow some laboratories to offer clinical mycology services.

OVERVIEW OF CLINICAL MYCOLOGY

More than 50,000 valid species of fungi exist, but only 100 to 150 species are generally recognized as causes of human disease and approximately 25 species cause the majority of human disease. Most of these organisms normally live a **saprophytic** (living on dead or decayed organic matter) existence in nature, enriched by nitrogenous matter, but they are capable of maintaining a separate existence as an opportunistic pathogen in humans or animals. Fungal infections are generally not communicable in the usual sense from person-to-person transmission. Animal-to-person transfer is also rare, with most such transmissions involving certain **dermatophytes** (a particular group of fungi that infect the skin and grow primarily in the keratin layer). Humans become an accidental host for fungi by inhaling spores or by the introduction of fungal elements into tissue by trauma. Except for disease caused by the dimorphic fungi, humans are relatively resistant to infections caused by the fungi. The major factors responsible for the increase in the number of fungal infections have been alterations in the host, particularly in the immune system. Whether caused by immunosuppressive agents or serious underlying diseases, these alterations may lead to infection by organisms that are normally nonpathogenic or are part of the patient's normal microbiota (i.e., normal flora). These infections may occur in patients with debilitating diseases such as progressive infection with the human immunodeficiency virus (HIV) or diabetes mellitus, or in patients with impaired immunologic function resulting from corticosteroid or antimetabolite chemotherapy.[108] Other common predisposing factors also include long-term intravenous cannulation, complex surgical procedures, and antibacterial therapy. It is because of this ability of normally saprophytic fungi to cause disease in the immunocompromised patient that laboratories now must be able to identify and report a wide array of fungi. No fungus species can be considered completely innocuous. It is therefore difficult at times for the mycologist at the bench to determine the significance of an organism recovered from a clinical specimen. In some instances, clinical and histopathologic correlations are the only means by which usually saprophytic fungi can be proven to represent true pathogens.

GENERAL FEATURES OF THE FUNGI

Fungi seen in the clinical laboratory can generally be separated into two groups based on the appearance of the colonies formed. The **yeasts** produce moist, creamy, opaque or pasty colonies on media, whereas the **filamentous fungi** or **molds** produce fluffy, cottony, woolly, or powdery colonies. Several systemic fungal pathogens exhibit either a yeast (or yeastlike) phase and filamentous forms are referred to as being **dimorphic.** When dimorphism is temperature dependent, the fungi are designated as **thermally dimorphic.** In general, these fungi produce a mold form at 25° to 30° C and a yeast form at 35° to 37° C under certain circumstances. The medically important dimorphic fungi include *Histoplasma capsulatum, Blastomyces dermatitidis, Coccidioides immitis, Paracoccidioides brasiliensis, Sporothrix schenckii,* and *Penicillium marneffei. Coccidioides immitis* is not thermally dimorphic. Additionally, some of the medically important yeasts, particularly the *Candida* species, may produce yeasts forms, pseudohyphae, and/or true hyphae. The polymorphic features of this group of organisms are not temperature dependent.

YEASTS

Yeasts are unicellular organisms that are round to oval and range in size from 2 to 60 μm. The microscopic morphologic features have limited utility in helping differentiate or identify these organisms. The microscopic morphology on cornmeal agar is most useful when used in conjunction with the **biophysical profile** (i.e., a combination of the biochemical and physical characteristics used in the identification of a microorganism) obtained using a commercial system. The differentiation of yeasts in direct microscopic and histopathologic examination of clinical specimens is often impossible, but sometimes particular characteristics are seen that suggest the identification or are **pathognomonic** (i.e., unique) for a particular organism. Important morphologic characteristics that are useful in differentiating yeasts include the size of the yeasts, the presence or absence of a capsule, and broad-based versus narrow-necked budding. For example, variability in size with evidence of a capsule and narrow-necked budding are features that can be helpful for separating *Cryptococcus* species from *Candida* species. The medically important yeasts and yeastlike organisms belong to different taxonomic groups, including the Ascomycota, the Basidiomycota, or Deuteromycota.

In general, the yeasts reproduce asexually by **blastoconidia** formation **(budding)** (Figure 50-1) and sexually by the production of ascospores or basidiospores. The process of budding begins with a weakening and subsequent outpouching of the yeast cell wall. This process continues until the bud or daughter cell is completely formed. The cytoplasm of the bud is contiguous with the cytoplasm for the original cell. Finally, a cell wall septum is created between mother and daugh-

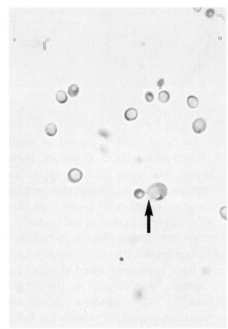

Figure 50-1 Blastoconidia (budding cells *[arrow]*) characteristic of the yeasts (430×).

ter yeast cells. The daughter cell often eventually detaches from the mother cell and a residual defect occurs at the site of the budding (i.e., a **bud scar**).

Two terms need explanation: germ tube and pseudohyphae. An outpouching of the cell wall that becomes tubular and does not have a constriction at its base is termed a **germ tube;** it represents the initial stage of true hyphae formation (Figure 50-2). Alternatively, if buds elongate, fail to dissociate, and form subsequent buds then **pseudohyphae** are formed; some believe these have a "links of sausage" appearance (Figure 50-3). Pseudohyphae have cell wall constrictions rather than true intracellular septations delineating the fungal cell borders. Far less common, some yeasts reproduce asexually through binary fission. These cells undergo cell wall expansion followed by mitosis. Following mitosis, a new cell wall septum is created between the two newly formed nuclei. The two cells may either separate after the septum is formed or remain attached through subsequent rounds of mitosis. Binary fission is a characteristic of a parasitic (infective) form of *Penicillium marneffei* (Figure 50-4).

MOLDS

The basic structural units of the molds are tubelike projections known as **hyphae.** As the hyphae grow, they become intertwined to form a loose network, the **mycelium,** which penetrates the substrate from which it obtains the necessary nutrients for growth. **Vegetative hyphae** comprise the body of the fungus, in

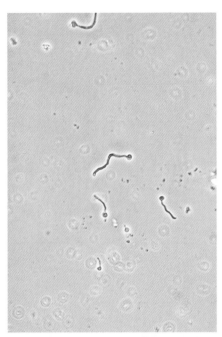

Figure 50-2   *C. albicans,* germ-tube test, showing yeast cells with germ tubes present (430×).

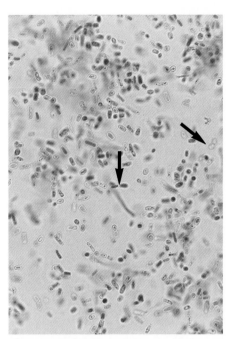

Figure 50-4   *Penicillium marneffei* and binary fission *(arrows)* (500×).

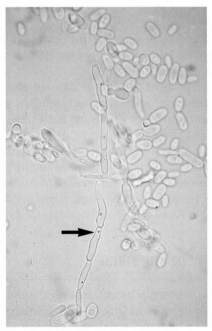

Figure 50-3   Pseudohyphae consisting of elongated cells *(arrow)* with constrictions where attached (430×).

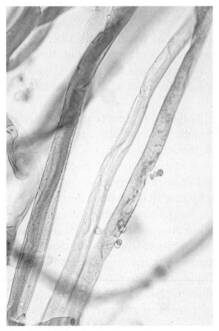

Figure 50-5   Hyaline hyphae that have rare or no discernible septations (aseptate or pauciseptate) (430×).

contrast to specialized **reproductive hyphae.** The nutrient-absorbing and water-exchanging portion of the fungus is called a **vegetative mycelium.** The portion extending above the substrate surface is known as **aerial mycelium;** aerial mycelia often give rise to fruiting bodies from which asexual spores are borne.

Three types of hyphae exist in the medically important fungi: the **coenocytic (sparsely septate) hyphae** of the Zygomycetes (Figure 50-5), the dark and pigmented septate hyphae of the dematiaceous fungi (Figure 50-6), and the septate, nonpigmented hyphae of the **hyaline** molds (Figure 50-7). The terms

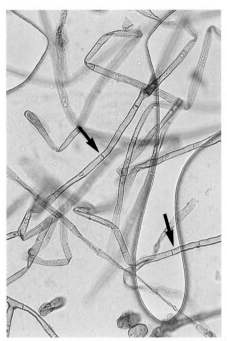

Figure 50-6   Dematiaceous hyphae showing pigmentation and septations *(arrows)* (430×).

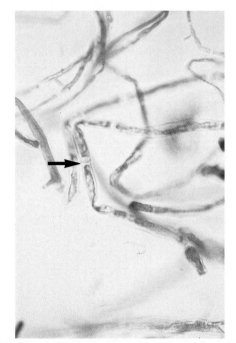

Figure 50-7   Hyaline hyphae showing septations *(arrow)* (430×).

dematiaceous and **hyaline** describe the presence or absence, respectively, of pigmentation within the hyphae of molds. Hyphal pigmentation is a useful feature to differentiate some fungi and is discussed later in this chapter.

The hyphae of Zygomycetes are wider in diameter compared with those of the fungi producing septate hyaline or dematiaceous hyphae. The branching of the Zygomycetes often occurs at angles greater than 45 degrees and up to 90 degrees, in contrast to the acutely branching hyphae of the septate dematiaceous and hyaline molds. A limited sampling of hyphae of the Zygomycetes often appear nonseptate or aseptate; however, a few septa are usually present on close inspection and vary depending on the organism and the age of the culture. These septa are often located near specialized reproductive hyphae. In these pauciseptate hyphae, nuclei and cytoplasm are free to flow throughout the length of the hyphal element. This has important implications when a commonly employed technique of tissue grinding is used to process all clinical specimens. When such processing is used, the hyphae of Zygomycetes may be destroyed, because the pauciseptate hyphae are not as compartmentalized by septa as they are with the hyaline and dematiaceous septate molds. Any disruption of any area of the hyphal strand in the Zygomycetes causes cytoplasmic leakage and eventually death of the organism. Conversely, the septate hyphae of dematiaceous and hyaline molds will survive such tissue processing because of the excessive compartmentalization of the hyphae by septa and the viability of individual vegetative fungal cells.

The vegetative hyphae of the dematiaceous and hyaline molds are morphologically similar, apart from the differentiating pigmentation. Both have true septate hyphae that branch at acute angles. The cell walls of the dematiaceous fungi contain melanin-related compounds, which make the hyphae darkly pigmented when observed by light microscopy. In histopathologic sections, these pigments may be accentuated by the use of the special stains (e.g., Fontana-Masson stain).

TAXONOMY OF THE FUNGI

Fungi are composed of a vast array of organisms that are unique compared with plants and animals. Included among these are the mushrooms, rusts and smuts, molds and mildews, and the yeasts. Despite the great variation in morphologic features of the fungi, most or all share the following characteristics:

- The presence of chitin in the cell wall
- The presence of ergosterol in the cell membrane
- Reproduction by means of spores, produced either asexually or sexually
- The lack of chlorophyll
- Lack of susceptibility to antibacterial antibiotics

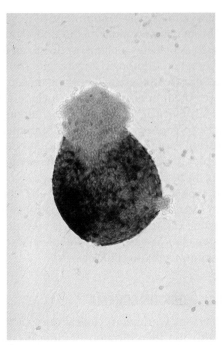

Figure 50-8   A cleistothecium of *Pseudallescheria boydii* that has opened and is releasing numerous ascospores (750×).

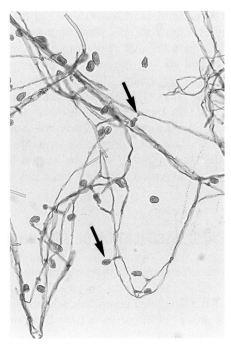

Figure 50-9   *Scedosporium apiospermum* showing asexually produced conidia borne singly on conidiophores (anellophores [*arrows*]) (430×).

● Their heterotopic (derive nutrition from organic materials) nature

Traditionally, the fungi have been categorized into four well-established phyla: the Zygomycota, Ascomycota, Basidiomycota, and Deuteromycota. The phylum Zygomycota includes those organisms that produce sparsely septate hyphae and exhibit asexual reproduction by sporangiospores and sexual reproduction by the production of zygospores. Some of the clinically important genera in this phylum include *Rhizopus, Mucor, Rhizomucor, Absidia, Cunninghamella,* and *Saksenaea*

The Ascomycota include many fungi that reproduce asexually by the formation of **conidia** (asexual spores) and sexually by the production of ascospores. The filamentous ascomycetes are ubiquitous in nature, and all produce true septate hyphae. All exhibit a sexual form **(teleomorph)** but also exist in an asexual form **(anamorph).** In general, the anamorphic form correlates well with the teleomorphic classification. However, different anamorphic forms may have the same teleomorphic form. For example, *Pseudallescheria boydii* (Figure 50-8), in addition to having the *Scedosporium apiospermum* anamorph (Figure 50-9) may exhibit a *Graphium* anamorph (Figure 50-10). This latter anamorph may be seen with several other fungi. Examples of other clinically important fungi that belong to the phylum Ascomycota include *Histoplasma capsulatum* and *Blastomyces dermatitidis,* which have a teleomorph designated as *Ajellomyces.* Some species of *Aspergillus* have a

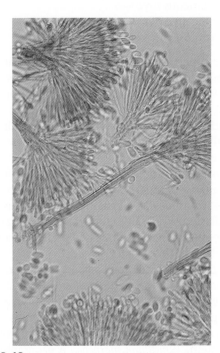

Figure 50-10   *Graphium* anamorph of *P. boydii* (500×).

teleomorph, *Eurotium.* Additionally, numerous yeast species belong to the Ascomycota and include *Saccharomyces* and some species of *Candida.*

The phylum Basidiomycota includes those fungi that reproduce sexually by the formation of basidiospores on a specialized structure called the *basidia*. The basidiomycetes are generally plant pathogens or environmental organisms that rarely cause disease in humans. Included are the smuts, rusts, mushrooms, and the *Cryptococcus neoformans*. The teleomorphic form of *Cryptococcus neoformans* is *Filobasidiella neoformans*.

The phylum Deuteromycota includes those fungi that lack a sexual reproductive cycle and are characterized by their asexual reproductive structures, primarily conidia. It is possible that the sexual forms of organisms in this group exist but have not yet been described.

PRACTICAL CLASSIFICATION OF THE FUNGI

The botanic taxonomic schema for grouping the fungi has little value in a clinical microbiology laboratory. Table 50-1 is a simplified taxonomic schema illustrating the major groups of fungi; these have been previously described within this chapter.

Clinicians find value in categorizing the fungi into four categories of mycoses:

- Superficial or cutaneous mycoses
- Subcutaneous mycoses
- Systemic mycoses
- Opportunistic mycoses

The superficial, or cutaneous, mycoses are fungal infections that involve the hair, skin, or nails without direct invasion of the deeper tissue. The fungi in this category include the dermatophytes (agents of ringworm, athlete's foot, and so on) and agents of infections such as tinea, tinea nigra, and piedra. All of these infect keratinized tissues.

Some fungi produce infections that are confined to the subcutaneous tissue without dissemination to distant sites. Examples of subcutaneous infections include chromoblastomycosis, mycetoma, and the phaeohyphomycotic cysts, which are discussed under subcutaneous mycoses later in this chapter.

As traditionally defined, agents of systemic fungal infections contain the genera *Blastomyces, Coccidioides, Histoplasma,* and *Paracoccidioides*. Infections caused by these organisms primarily involve the lungs but also may become widely disseminated and involve any organ system. *Penicillium marneffei,* a geographically limited cause of systemic mycosis in a select patient population may also be considered a part of this group.

Any of the fungi could be considered an opportunistic pathogen in the appropriate clinical setting. The list of uncommon fungi found to cause disease in humans expands every year. Fungi previously thought to be nonpathogenic may be the cause of infections. The infections these organisms cause occur primarily in

patients with some type of compromise to their immune system. This may be secondary to an underlying disease process, such as diabetes mellitus, or due to an immunosuppressive agent. Although any fungus could potentially cause disease in these patients, the most commonly encountered organisms in this group include *Aspergillus,* Zygomycetes, *Candida,* and *Cryptococcus,* among others. All of these organisms may cause disseminated (systemic) disease. Some of the dematiaceous fungi may cause deeply invasive phaeohyphomycoses in this patient population.

This type of classification allows the clinician to attempt to categorize organisms in a logical fashion into groups having clinical relevance. Table 50-2 presents an example of a clinical classification of infections and their etiologic agents useful to clinicians.

PRACTICAL WORKING SCHEMA

To assist persons working in clinical microbiology laboratories with the identification of clinically important fungi, Koneman and Roberts[70] have suggested the use of a practical working schema designed to:

- Assist in the recognition of fungi most commonly encountered in clinical specimens
- Assist with the recognition of fungi recovered on culture media that are strictly pathogenic fungi
- Provide a pathway that is easy to follow and that allows an identification to be made based on a few colonial and microscopic features

Table 50-3 presents these features. However, the table includes only those organisms commonly seen in the clinical laboratory. With practice, most laboratorians should be able to recognize these on a day-to-day basis. The identification of other, less commonly encountered fungi will require the use of various texts containing photomicrographs useful for their identification.

The use of this schema requires that one first examine the culture for the presence, absence, and number of septa. If the hyphae appear to be broad and predominantly nonseptate, Zygomycetes should be considered. If the hyphae are septate, they must be examined further for the presence or absence of pigmentation. If a dark pigment is present in the hyphae, the organism is considered to be dematiaceous and the conidia are then examined for their morphologic features and their arrangement on the hyphae. If the hyphae are nonpigmented, they are considered to be hyaline. They are then examined for the type and the arrangement of the conidia produced. The molds are identified by recognition of their characteristic microscopic features (see Table 50-3). Murray has developed an expanded morphologic classification of medically important fungi based on general microscopic

Table 50-1   Phylogenetic Position of Medically Significant Fungi

Class	Order	Genus/Species
Phylum Zygomycota Zygomycetes	Entomophthorales Mucorales	Basidiobolus‡ Conidiobolus Absidia‡ Cunninghamella‡ Mucor† Rhizopus† Syncephalastrum
Phylum Ascomycota Hemiascomycetes Loculoascomycetes Plectomycetes	Endomycetales Myriangiales Microascales Eurotiales Onygerales	Endomyces (Geotrichum sp.)* Kluyveromyces (Candida pseudotropicalis)* Piedraia hortae Pseudallescheria boydii† Scedosporium prolificans‡ Emericella (Aspergillus nidulans)* Ajellomyces (Histoplasma capsulatum,† Blastomyces dermatitidis†)* Arthroderma (Trichophyton sp.† and Microsporum sp.†)*
Phylum Basidiomycota Teliomycetes	Filobasidiales	Filobasidiella (Cryptococcus neoformans†)*
Phylum Deutromycota Blastomycetes Hyphomycetes	Moniliales	Candida† Acremonium‡ Aspergillus† Blastomyces† Chrysosporium Coccidioides† Epidermophyton† Geotrichum Gliocladium Histoplasma† Microsporum† Paecilomyces‡ Paracoccidioides† Penicillium‡ Sepedonium Scopulariopsis Sporothrix† Trichoderma Trichophyton† Alternaria Aureobasidium Bipolaris‡ Cladosporium Curvularia‡ Drechslera Exophiala‡ Exserohilum Fonsecaea‡ Helminthosporium Madurella‡ Nigrospora Phialophora‡ Rhinocladiella Stemphylium Ulocladium Cladophialophora‡ Epicoccum Fusarium† Phoma

Modified from Kwon-Chung KJ, Bennett JE: *Medical mycology*, Philadelphia, 1992, Lea & Febiger.
*When the sexual form is known.
†Most commonly encountered as causes of infection.
‡Infrequently cause of infection.

Table 50-2 General Clinical Classification of Pathogenic Fungi

Cutaneous	Subcutaneous	Opportunistic	Systemic

Superficial mycoses
    Tinea
    Piedra
    Candidosis
	Chromoblastomycosis
	Sporotrichosis
	Mycetoma (eumycotic)
	Phaeohyphomycosis
		Aspergillosis
		Candidosis
		Cryptococcosis
		Geotrichosis
			Aspergillosis
			Blastomycosis
			Candidosis
			Coccidioidomycosis

Dermatophytosis		Zygomycosis Fusariosis Trichosporonosis Others*	Histoplasmosis Cryptococcosis Geotrichosis Paracoccidioidomycosis Zygomycosis Fusariosis Trichosporonosis

*Virtually any fungus may cause disease in the profoundly immunocompromised host.

features and colonial morphology. The color pigmentation of colonies is presented as a useful diagnostic feature[91] (Box 50-1).

VIRULENCE FACTORS OF THE MEDICALLY IMPORTANT FUNGI

The increase in the number of opportunistic fungal infections in humans is due in large part to the immunocompromised nature of the host. However, there are certain factors, virulence factors, which make it easier for them to invade tissues and cause disease. Some virulence factors have been known for years and include:

● The size of an organism (Disease that begins by inhalation requires that the organism be small enough to reach the alveoli.)
● The ability of an organism to grow at 37° C at a neutral pH
● Conversion of the dimorphic fungi from the mycelial form to the corresponding yeast or spherule form within the host
● Possibly, toxin production

Most of the fungi have a place in the environment where they live out a saprophytic existence (Table 50-4). Perhaps the fungi that cause disease in humans have developed various mechanisms over time that have allowed them to establish disease in a somewhat foreign setting (the human host). A review by Hogan, Klein, and Levitz[57] describes the known or speculative virulence factors for certain of the fungi known to be pathogenic for humans. Table 50-5 presents a summary of those virulence factors. Particularly important virulence factors associated with these organisms are highlighted here.

ASPERGILLUS SPECIES

It is well documented that there is an increased risk of aspergillosis in individuals who have received high doses of corticosteroids and other immunosuppressive regimens. This is thought to be the result of the suppression of cell-mediated immunity and neutropenia. The association of neutropenia and neutrophil dysfunction with invasive aspergillosis was known before the advent of bone marrow transplantation, particularly in patients having chronic granulomatous disease of childhood. Because the ability of *Aspergillus* to establish disease in humans is directly related to the immunocompromised host, it is difficult to assess whether specific fungal virulence factors play a role in the development of the disease.

Aspergillus fumigatus is known to produce two elastases, including a serum protease and a metalloprotease,[69,83] which may act on elastin that compromises approximately 30% of the lung tissue. However, no definitive evidence has been shown that elastases are related to invasive pulmonary infection. *Aspergillus* species also are known to produce catalase that may be a virulence factor contributing to aspergillosis associated with chronic granulomatous disease of childhood.[22] The association of catalase production with any other clinical presentations of aspergillosis is uncertain. Most fungi are not thought to cause disease through the elaboration of toxins. *Aspergillus flavus,* however, is a notable exception. The aflatoxin produced by this fungus is a known carcinogenic hepatotoxin.

BLASTOMYCES DERMATITIDIS

Little is known about the pathogenesis or virulence mechanisms of *Blastomyces dermatitidis*. It is assumed that either conidia or hyphal fragments are the infectious

Table 50-3   Most Commonly Encountered Fungi of Clinical Laboratory Importance: A Practical Working Schema

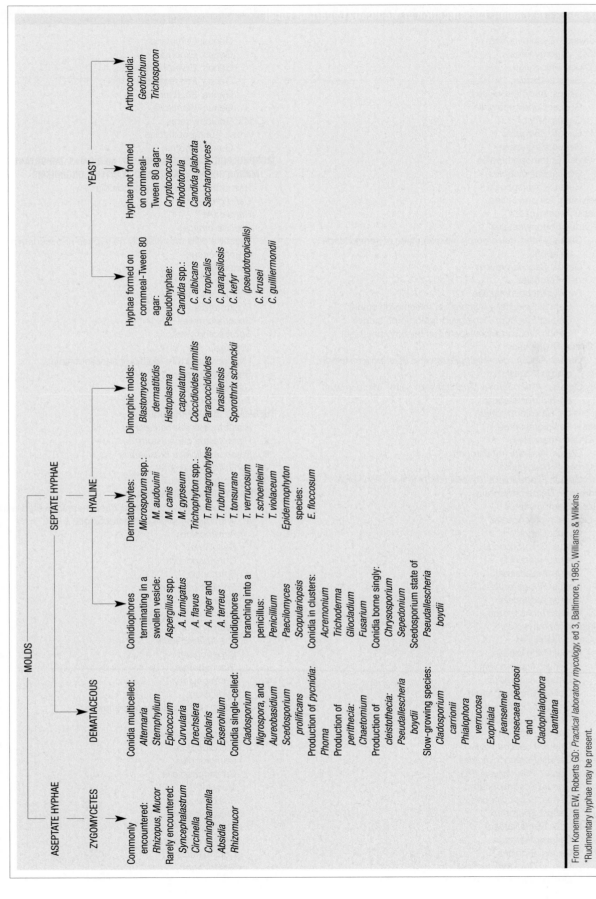

From Koneman EW, Roberts GD: *Practical laboratory mycology*, ed 3, Baltimore, 1985, Williams & Wilkins.
*Rudimentary hyphae may be present.

BOX 50-1   Taxonomic Classification of Medically Important Fungi

Subdivision: Zygomycotina
  Class: Zygomycetes
    Order: Mucorales
      Genus: *Absidia*
      Genus: *Apophysomyces*
      Genus: *Cunninghamella*
      Genus: *Mucor*
      Genus: *Rhizopus*
      Genus: *Saksenaea*
    Order: Entomophthorales
      Genus: *Basidiobolus*
      Genus: *Conidiobolus*
Subdivision: Ascomycotina
  Class: Ascomycetes
    Order: Endomycetales
      Genus: *Pichia* (teleomorph [sexual] stage of some *Candida* spp.)
      Genus: *Saccharomyces*
    Order: Eurotiales
      Family: *Trichocomaceae*
        Genus: *Emericella* (*Aspergillus* teleomorph stage)
        Genus: *Eurotium* (*Aspergillus* teleomorph stage)
        Genus: *Neosartorya* (*Aspergillus* teleomorph stage)
    Order: Onygenales
      Genus: *Ajellomyces* (*Histoplasma* and *Blastomyces* teleomorph stages)
      Genus: *Arthroderma* (*Trichophyton* and *Microsporum* teleomorph stages)
Subdivision: Basidiomycotina
  Class: Basidiomycetes
    Order: Agaricales
      Genus: *Amanita* (mushroom)
    Order: Filobasidiales
      Genus: *Filobasidiella* (*Cryptococcus* teleomorph stage)
Subdivision: Deuteromycotina
  Class: Blastomycetes
    Order: Cryptococcales
      Genus: *Candida*
      Genus: *Cryptococcus*
      Genus: *Hansenula*
      Genus: *Malassezia*
      Genus: *Rhodotorula*
      Genus: *Torulopsis*
      Genus: *Trichosporon*
  Class: Hyphomycetes
    Order: Moniliales
      Family: *Moniliaceae*
        Genus: *Acremonium*
        Genus: *Aspergillus*
        Genus: *Chrysosporium*
        Genus: *Coccidioides*
        Genus: *Epidermophyton*
        Genus: *Fusarium*
        Genus: *Paecilomyces*
        Genus: *Paracoccidioides*
        Genus: *Pseudallescheria*
        Genus: *Scedosporium*
        Genus: *Scopulariopsis*
        Genus: *Sporothrix*
      Family: *Dematiaceae*
        Genus: *Alternaria*
        Genus: *Bipolaris*
        Genus: *Cladophialophora*
        Genus: *Cladosporium*

        Genus: *Curvularia*
        Genus: *Exophiala*
        Genus: *Exserohilum*
        Genus: *Fonsecaea*
        Genus: *Phialophora*
        Genus: *Wangiella*
  Class: Coelomycetes
    Order: Sphaeropsidales
      Genus: *Phoma*

MORPHOLOGIC CLASSIFICATION OF MEDICALLY IMPORTANT FUNGI, MONOMORPHIC YEASTS, AND YEASTLIKE ORGANISMS

1. Pseudohyphae with blastoconidia
   Candida spp.
   Hansenula
   Saccharomyces
2. Yeastlike cells only (usually no hyphae or pseudohyphae)
   Cryptococcus
   Hansenula
   Malassezia
   Prototheca
   Rhodotorula
   Saccharomyces
   Sporobolomyces
   Ustilago
3. Hyphae and arthroconidia or annelloconidia
   Blastoschizomyces
   Geotrichum
   Trichosporon

THERMALLY DIMORPHIC FUNGI

1. *Blastomyces dermatitidis*
2. *Histoplasma capsulatum*
3. *Paracoccidioides brasiliensis*
4. *Penicillium marneffei*
5. *Sporothrix schenckii*

THERMALLY MONOMORPHIC MOLDS

1. White, cream, or light gray surface; nonpigmented reverse
   a. With microconidia or macroconidia
      Acremonium
      Beauveria
      Chrysosporium
      Emmonsia
      Epidermophyton
      Fusarium
      Graphium
      Microsporum
      Pseudallescheria
      Sepedonium
      Stachybotrys
      Trichophyton
      Verticillium
   b. Having sporangia or sporangiola
      Absidia
      Apophysomyces
      Basidiobolus
      Conidiobolus
      Cunninghamella
      Mucor
      Rhizomucor
      Rhizopus
      Saksenaea
   c. Having arthroconidia
      Coccidioides
      Geotrichum

d. Having only hyphae with chlamydoconidia
Epidermophyton
Microsporum
Trichophyton
2. White, cream, beige, or light gray surface; yellow, orange, or reddish reverse
Acremonium
Chaetomium
Microsporum
Trichophyton
3. White cream, beige, or light gray surface; red to purple reverse
Microsporum
Penicillium
Trichophyton
4. White, cream, beige, or light gray surface; brown reverse
Chaetomium
Chrysosporium
Cokeromyces
Emmonsia
Madurella
Microsporum
Scopulariopsis
Sporotrichum
Trichophyton
5. White, cream, beige, or light gray surface; black reverse
Chaetomium
Graphium
Nigrospora
Phoma
Pseudallescheria
Scedosporium
Trichophyton
6. Tan to brown surface
   a. Having small conidia
   Aspergillus
   Botrytis
   Chrysosporium
   Cladosporium
   Emmonsia
   Ochroconis
   Paecilomyces
   Phialophora
   Scedosporium
   Scopulariopsis
   Sporotrichum
   Trichophyton
   Verticillium
   b. Having large conidia or sporangia
   Alternaria
   Apophysomyces
   Basidiobolus
   Bipolaris
   Botrytis
   Cokeromyces
   Conidiobolus
   Curvularia
   Epicoccum
   Epidermophyton
   Fusarium
   Microsporum
   Rhizomucor
   Rhizopus

   Stemphylium
   Trichophyton
   Ulocladium
   c. Having miscellaneous microscopic morphology
   Chaetomium
   Coccidioides
   Madurella
   Phoma
   Ustilago
7. Yellow to orange surface
Aspergillus
Chrysosporium
Epicoccum
Epidermophyton
Microsporum
Monilia
Penicillium
Sepedonium
Sporotrichum
Trichophyton
Trichothecium
Verticillium
8. Pink to violet surface
Acremonium
Aspergillus
Beauveria
Chrysosporium
Fusarium
Gliocladium
Microsporum
Monilia
Paecilomyces
Sporotrichum
Trichophyton
Trichothecium
Verticillium
9. Green surface; light reverse
Aspergillus
Epidermophyton
Gliocladium
Penicillium
Trichoderma
Verticillium
10. Dark gray or black surface; light reverse
Aspergillus
Syncephalastrum
11. Green, dark gray, or black surface; dark reverse
   a. Having small conidia
   Aureobasidium
   Botrytis
   Cladosporium/Cladophialophora
   Exophiala
   Fonsecaea
   Hortaea
   Phialophora
   Pseudallescheria
   Scedosporium
   b. Having large conidia
   Alternaria
   Bipolaris
   Curvularia
   Dactylaria
   Epicoccum

Helminthosporium	c. Having only hyphae (with or without chlamydoconidia)
Nigrospora	*Hortaea*
Pithomyces	*Madurella*
Stachybotrys	d. Having large fruiting bodies
Stemphylium	*Chaetomium*
Ulocladium	*Phoma*

From Murray PR: *ASM pocket guide to clinical microbiology*, Washington, DC, 1996, ASM Press.

Table 50-4   Summary of Common Pathogens

Organism	Natural Habitat	Infectious Form	Mode of Transmission	Common Sites of Infection	Clinical Form
Aspergillus spp.	Ubiquitous, plants	Conidia	Inhalation	Lung, eye, skin, nails	Hyphae
Blastomyces dermatitidis	Unknown? soil/wood	Probably conidia	Usually inhalation	Lungs, skin, long bones	Yeast
Candida spp.	Human flora	Yeast, pseudo- and true hyphae	Direct invasion/ dissemination	GI and GU tract, nails, viscera, blood	Yeast, Pseudo- and true hyphae
Coccidioides immitis	Soil of many arid regions	Arthroconidia	Inhalation	Lungs, skin, meninges	Spherules, endospores
Cryptococcus neoformans	Bird feces, soil	Yeast*	Inhalation	Lungs, skin, meninges	Yeast
Histoplasma capsulatum	Bat and bird feces	Conidia	Inhalation	Lungs, bone marrow, blood	Yeast
Paracoccidioides brasiliensis	?Soil, plants	Conidia	Inhalation/trauma	Lungs, skin, mucous membranes	Yeast
Sporothrix schenckii	Soil, plants	Conidia/hyphae	Trauma/rarely inhalation	Skin and lymphatics, lungs, meninges	Yeast
Dermatophytes	Human disease, animals, soil	Conidia/hyphae	Contact	Skin, hair, or nails	Hyphae

*Possibly the conidia of the teleomorphic stage *(Filobasidiella neoformans)*.

particles responsible for establishing disease. Thermal dimorphism plays a role in the conversion from the mold to the yeast form that occurs in patients with active disease. It has also been shown that decreased virulence is associated with an increase in the amount of α-1, 3-glucan content present in *Blastomyces dermatitidis*.[24] It appears that the ability of *Blastomyces dermatitidis* to produce a granulomatous response is directly related to the amount of phospholipid component found in an alkali-soluble fraction of this organism.

CANDIDA SPECIES

The pathogenesis of *Candida* infections is extremely complex and probably varies with each species. Adhesion of *Candida* to the epithelium of the gastrointestinal or urinary tract is critically important. *Candida* spp.

commonly colonize the mucosal surfaces, and their ability to invade and cause infection is first dependent on binding.[60] Fibronectin, a component of the host extracellular matrix, may play a role in the initiation and dissemination of *Candida albicans* infections.[157] Three distinct aspartyl proteases have been described in *C. albicans*, and strains with high levels of proteases have been shown to have an increased ability to cause disease in experimental animal models.[94] Hydrophobic molecules present on the surface of *Candida* spp. also appear to be important in pathogenesis, and there is a strong correlation between adhesion and surface hydrophobicity.[37] Also, high levels of phospholipase, found in strains of *C. albicans*, has been correlated with a higher mortality rate for experimental animals, compared with experimental infections caused by strains that produce a lower level of phospholipase.[63] Furthermore, the pheno-

Table 50-5   Virulence Factors of Medically Important Fungi

Fungal Pathogen	Putative Virulence Factor

Aspergillus spp.	Elastase-serine protease
	    Proteases
	    Toxins
	    Other
	 Elastase-metalloprotease
	 Aspartic acid proteinase
	 Aflatoxin
	 Catalase
	 Lysine biosynthesis
	 p-aminobenzoic acid synthesis

Blastomyces dermatitidis	Cell wall α-1,3-glucan
Coccidioides immitis	Extracellular proteinases
Cryptococcus neoformans	Capsule Phenoloxidase melanin synthesis Varietal differences
Dematiaceous fungi	Phenoloxidase melanin synthesis
Histoplasma capsulatum	Cell wall α-1,3-glucan Intracellular growth Thermotolerance

Paracoccidioides brasiliensis	Estrogen-binding proteins
	 Cell wall components
	    β-glucan
	    α-1,3-glucan

Sporothrix schenckii	Thermotolerance Extracellular enzymes

Modified from Hogan LH, Klein BS, Levitz SM: Virulence factors of medically important fungi, *Clin Microbiol Rev* 9:469, 1996.

typic switching (i.e., ability to produce pseudohyphae and hyphae) seen in *Candida albicans* may also play a role in the pathogenesis.[143] Finally, the immunosuppressive activity of yeast mannans also may contribute to virulence of these yeasts.[26]

COCCIDIOIDES IMMITIS

The arthroconidia of *Coccidioides immitis* are 2 to 5 μm in size and are easily deposited within the alveoli of the lung. From the arthroconidia, spherules develop and ultimately segment into numerous endospores that are released from the spherule cell wall as it ruptures. This provides for local and hematologic dissemination of the organism. It has been noted that spherules are coated with an extracellular matrix that is inhibitory to polymorphonuclear leukocytes (PMN) and may be responsible for the resistance of spherules to killing by PMNs.[41]

The dissemination of *Coccidioides immitis* from the lungs and the marked tissue damage caused by the organism has led to speculation that an extracellular protease may be responsible. It has been shown that the protease, found in hyphae and spherules, promotes dissolution of the wall of the cell matrix as the endospores become mature. Concurrently, the action of the protease on immunoglobulins and against elastin in the lungs may be a virulence factor that promotes tissue damage when the ruptured spherule releases endospores and any residual protease.[157]

Men are several times more likely than women to develop disseminated infection. Pregnancy paradoxically reverses the resistance of women to dissemination and with each increasing term of pregnancy the risk of dissemination also increases. In fact, numerous hormones during pregnancy have been studied for their effects on *C. immitis*. Several were shown to stimulate growth in vitro by accelerating the rate of spherule development and endospore release. It is speculated that estrogen-binding proteins found in the cytosols of *C. immitis* may play a role in controlling its growth in vitro.[41]

CRYPTOCOCCUS NEOFORMANS

Cryptococcosis is most often seen in immunocompromised patients with impaired T cell–mediated immunity.[30] Important risk factors include acquired immunodeficiency syndrome (AIDS), corticosteroid therapy, lymphoma, and T-cell dysfunction. *Cryptococcus neoformans* is the only pathogenic yeast known to have a polysaccharide capsule. The capsule collapses and protects the yeast from desiccation under drying conditions.[21] The cell size reduction, resulting from capsular collapse, places the yeast in the ideal size range for alveolar deposition.[95] In addition, the polysaccharide capsule contains compounds that are not recognized by phagocytes.[72] The so-called acapsular strains of *C. neoformans*, which actually just have a very reduced capsule, are more easily phagocytosed. In some instances, *C. neoformans* elicits minimal tissue response in infected individuals, particularly severely immunocompromised patients.

Phenoloxidase, an enzyme found in *C. neoformans*, is responsible for melanin production. It has been speculated that melanin might act as a virulence factor by making the organism resistant to leukocyte attack.[148] Evidence has also been presented to show that increased melanin production can decrease lymphocyte proliferation and tumor necrosis factor production.[61] Whether phenoloxidase is truly a virulence factor has yet to be determined, but studies of dematiaceous fungi, such as *Exophiala* species support these findings. It is interesting to hypothesize, however, that interactions of substance in the brain, which are known to react with phenoloxidase, may play a role in the affinity displayed by *C. neoformans* and certain neurotropic dematiaceous fungi to invade the central nervous system.

HISTOPLASMA CAPSULATUM

Histoplasma capsulatum, one of the thermally dimorphic fungi, produces microconidia and/or small hyphal fragments are 2 to 5 µm in size and ideal for alveolar deposition. Conversion of the mold to the yeast form is a prerequisite for the organism to cause infection. *Histoplasma capsulatum* is an intracellular parasite and can survive within macrophages for an extended period.[156] Its ability to survive within the macrophages is a key virulence factor. It is speculated that the organism modulates the pH within the phagolysosomes.[59]

PARACOCCIDIOIDES BRASILIENSIS

The virulence of *Paracoccidioides brasiliensis,* another of the thermally dimorphic fungi, is also influenced by its ability to convert from the mold to the yeast form. Infection is initiated by inhalation of conidia, followed by rapid conversion to the yeast form within the lungs.[85] As with *Coccidioides immitis, Paracoccidioides brasiliensis* is 13 to 80 times more common in men than in women.[130] It is speculated that an estrogen-binding protein present in the fungus inhibits the conversion of inhaled conidia to the yeast form in women.[16] In addition, a beta-glucan has been implicated as an important immunomodulator in initiation of the host inflammatory response.[14]

SPOROTHRIX SCHENCKII

Conversion of this mold to the yeast form is also necessary before an infection with *Sporothrix schenckii* can be established, because the mold form exists in the environment and the yeast form is found in infected tissues. The organism most often causes infection after it is acquired by traumatic implantation into the skin and subcutaneous tissues. Conidia and/or small hyphal fragments are capable of deposition within the alveoli when inhaled, but this route of infection is rare. Kwon-Chung and Bennett reported that several clinical isolates of *Sporothrix schenckii* failed to grow at 37° C and that they caused infection only in the skin and testes of infected rats, suggesting that they required a lower temperature to produce infection.[74] The effect of thermotolerance on the virulence of this organism is not well documented. Extracellular enzymes such as acid phosphatases have also been speculated to play a role in the interaction of *Sporothrix schenckii* yeast forms within macrophages.

DERMATOPHYTES

The dermatophytes, which infect only the skin, hair, or nails, secrete extracellular enzymes that likely aid in the colonization of keratinous tissues.[15] These extracellular enzymes that may be associated with virulence include keratinase, elastase, and lipase.[78]

GENERAL CONSIDERATIONS FOR THE LABORATORY DIAGNOSIS OF FUNGAL INFECTIONS

COLLECTION, TRANSPORT, AND CULTURING OF CLINICAL SPECIMENS

The diagnosis of fungal infections is dependent entirely on the selection and collection of an appropriate clinical specimen for culture. Many fungal infections are similar clinically to mycobacterial infections, and often the same specimen is cultured for both fungi and mycobacteria. Many infections have a primary focus in the lungs; respiratory tract secretions are almost always included among the specimens selected for culture. It should be emphasized that dissemination to distant body sites may occur, and fungi may be recovered from nonrespiratory sites. The proper collection of specimens and their rapid transport to the clinical laboratory are of major importance for the recovery of fungi. In many instances, specimens not only contain the etiologic agent but also contain contaminating bacteria or fungi that will rapidly overgrow some of the slower-growing pathogenic fungi. It is common for many yeasts, such as *Candida* spp., to be recovered on routine bacteriology media and fungal culture media. A few specific comments concerning specimen collection and culturing are included in this chapter.

Respiratory Tract Secretions

Respiratory tract secretions (sputum, induced sputum, bronchial washings, bronchoalveolar lavage, and tracheal aspirations) are perhaps the most commonly submitted specimens for fungal culture. To ensure the optimal recovery of fungi and prevent overgrowth by contaminants, antibacterial antibiotics should be included in the battery of media to be used. Cyclohexi-mide, an antifungal agent that prevents overgrowth by rapidly growing molds, should be included in at least one of the culture media used. As much specimen as possible (0.5 mL) should be used to inoculate each medium.

Cerebrospinal Fluid

Cerebrospinal fluid (CSF) collected for culture should be filtered through a 0.45-µm membrane filter attached to a sterile syringe. After filtration, the filter is removed and is placed onto the surface of an appropriate culture medium, with the inoculum side down. Cultures should be examined daily and the filter moved to another location every other day. If less than 1.0 mL of specimen is submitted for culture, it should be centrifuged and 1-drop aliquots of the sediment should be placed onto several areas on the agar surface. Media used for the recovery of fungi from CSF should contain no antibacterial or antifungal agents. Once submitted to the laboratory, CSF specimens should be processed promptly. If prompt processing is not possible, samples should be kept at room temperature or placed in a 30° C incubator, because most organisms will continue to replicate in this environment.

Blood

Disseminated fungal infections are more prevalent than previously recognized, and blood cultures provide an accurate method for determining their etiology in many instances. Only a few manual fungal blood culture systems have been available over past years, and most are not used routinely by clinical microbiology laboratories. Currently, several automated blood culture systems, including the BACTEC (Becton Dickinson, Sparks, Md), BacT/ALERT (bioMérieux, Durham, NC), and ESP (Trek Diagnostics, Westlake, Ohio), are adequate systems for the recovery of yeasts. However, those laboratories that have a high incidence of dimorphic fungi recovered from blood are encouraged to use the lysis-centrifugation system, the Isolator (Wampole Laboratories, Princeton, NJ). The Isolator has been shown to be optimal for the recovery of *H. capsulatum* and other filamentous fungi.[12,13] Using this system, red blood cells and white blood cells, which may contain the microorganisms, are lysed, and centrifugation concentrates the organisms before culturing. The concentrate is inoculated onto the surface of appropriate culture media, and most fungi are detected within the first 4 days of incubation. However, occasional isolates of *H. capsulatum* may require approximately 10 to 14 days for recovery. The optimal temperature for fungal blood cultures is 30° C, and the suggested incubation time is 21 days.

Hair, Skin, and Nail Scrapings

Specimens of hair, skin scrapings or biopsies, and nail clippings are usually submitted for dermatophyte culture and are contaminated with bacteria and/or rapidly growing fungi. Samples collected from lesions may be obtained by scraping the skin or nails with a scalpel blade or microscope slide; infected hairs are removed by plucking them with forceps. These specimens should be placed in a sterile container; they should not be refrigerated. Mycosel agar, which contains chloramphenicol and cycloheximide, is satisfactory for the recovery of dermatophytes. Cultures should be incubated for a minimum of 21 days at 30° C before being reported as negative.

Urine

Urine samples collected for fungal culture should be processed as soon after collection as possible. Twenty-four-hour urine samples are unacceptable for culture. The usefulness of quantitative cultures is undetermined, and it should not be done. All urine samples should be centrifuged and the sediment cultured using a loop to provide adequate isolation of colonies. Because urine is often contaminated with gram-negative bacteria, it is necessary to use media containing antibacterial agents to ensure the recovery of fungi.

Tissue, Bone Marrow, and Sterile Body Fluids

All tissues should be processed before culturing by mincing or grinding or placement in a Stomacher (Tekmar, Cincinnati, Ohio). The Stomacher expresses the cytoplasmic contents of cells by pressure exerted from the action of rapidly moving metal paddles against the tissue in a broth suspension. Large portions of tissue should be cut into smaller pieces if the Stomacher is to be used for processing. After processing, at least 1.0 mL of specimen should be spread onto the surface of appropriate culture media, and incubation should be at 30° C for 21 days; cultures incubation may be extended if there is a high clinical suspicion of a mycotic disease. If tissue was submitted, it is important to be sure that portions of the tissue are inoculated onto the agar surface (i.e., not just the broth used to assist in the dissolution of the tissue in the Stomacher).

Bone marrow may be placed directly onto the surface of appropriate culture media and incubated in the manner previously mentioned. Sterile body fluids should be concentrated by centrifugation before culturing, and at least 1 mL of specimen should be placed onto the surface of appropriate culture media. An alternative is to place bone marrow and other body fluids in an Isolator tube and process it as a blood culture. All specimens should be cultured as soon as they are received by the laboratory to ensure the recovery of fungi from these important sources.

CULTURE MEDIA AND INCUBATION REQUIREMENTS

Any of a number of fungal culture media are satisfactory for use in the clinical microbiology laboratory. Most are adequate for the recovery of fungi, and selection is usually left up to each individual laboratory director. Table 50-6 lists various media and the indications for their use. For optimal recovery a battery of media should be used, and the following are recommended:

1. Media with and without cycloheximide
2. Media with and without an antibacterial agent. This is for specimens that are likely to contain contaminating bacteria. It is not necessary to use these on specimens from sterile sites.

Agar plates or screw-capped agar tubes are satisfactory for the recovery of fungi, but plates are pre-

Table 50-6   Fungal Culture Media: Indications for Use

Media	Indications for Use
Primary Recovery Media	
Brain-heart infusion agar	Primary recovery of saprobic and pathogenic fungi
Brain-heart infusion agar with antibiotics	Primary recovery of pathogenic fungi exclusive of dermatophytes
Brain-heart infusion biphasic blood culture bottles	Recovery of fungi from blood
Dermatophyte test medium	Primary recovery of dermatophytes, recommended as screening medium only
Inhibitory mold agar	Primary recovery of pathogenic fungi exclusive of dermatophytes
Potato flake agar	Primary recovery of saprobic and pathogenic fungi
Mycosel	Primary recovery of dermatophytes
SABHI agar	Primary recovery of saprobic and pathogenic fungi
Yeast-extract phosphate agar	Primary recovery of pathogenic fungi exclusive of dermatophytes
Differential Test Media	
Ascospore agar	Detection of ascospores in ascosporogenous yeasts such as *Saccharomyces* spp.
Cornmeal agar with Tween 80 and trypan blue	Identification of *C. albicans* by chlamydospore production; identification of *Candida* by microscopic morphology
Cottonseed conversion agar	Conversion of dimorphic fungus *B. dermatitidis* from mold to yeast form
Czapek's agar	Differential identification of *Aspergillus* spp.
Niger seed agar	Identification of *C. neoformans*
Nitrate reduction medium	Detection of nitrate reduction in confirmation of *Cryptococcus* spp.
Potato dextrose agar	Demonstration of pigment production by *T. rubrum;* preparation of microslide cultures and sporulation of dermatophytes
Rice medium	Identification of *M. audouinii*
Trichophyton agars 1-7	Identification of members of *Trichophyton* genus
Urea agar	Detection of *Cryptococcus* spp.; differentiate *T. mentagrophytes* from *T. rubrum;* detection of *Trichosporon* spp.
Yeast fermentation broth	Identification of yeasts by determining fermentation
Yeast nitrogen base agar	Identification of yeasts by determining carbohydrate assimilation

From Koneman EW, Roberts GD: *Practical laboratory mycology,* ed 3, Baltimore, 1985, Williams & Wilkins.

ferred, because they provide better aeration of cultures, a large surface area for better isolation of colonies, and greater ease of handling by technologists making microscopic preparations for examination. Agar tends to dehydrate during the extended incubation period required for fungal recovery; however, this problem can be minimized by using culture dishes containing at least 40 mL of agar and placing them in a humid field incubator. Dishes should be opened and examined only within a certified biological safety cabinet. Many laboratories discourage the use of culture dishes because of safety considerations; however, the advantages of using them outweigh the disadvantages.

Compared with agar plates, screw-capped culture tubes are more easily stored, require less space for incubation, and are more easily handled. In addition, they have a lower dehydration rate, and laboratory workers believe that cultures are less hazardous to handle when in tubes. However, disadvantages, such as relatively poor isolation of colonies, a reduced surface area for culturing, and a tendency to promote anaerobiosis, discourage their routine use in most clinical microbiology laboratories. If culture tubes are used, the tube should be as large as possible to provide an adequate surface area for isolation. After inoculation, tubes should be placed in a horizontal position for at least 1 to 2 hours to allow the specimen to absorb to the agar surface and avoid settling at the bottom of the tube. Cotton-plugged tubes are unsatisfactory for fungal cultures.

Cultures should be incubated at room temperature, or preferably at 30° C, for 21 to 30 days before reporting as negative. A relative humidity in the range of 40% to 50% can be achieved by placing an open pan of water in the incubator. Cultures should be examined at least three times weekly during incubation.

As previously mentioned, some clinical specimens are contaminated with bacteria and/or rapidly growing fungi. The need for antifungal and antibacterial agents is obvious. The addition of 0.5 mg/mL of cycloheximide and 16 μg/mL of chloramphenicol to media has been traditionally advocated to inhibit the growth of contaminating molds and bacteria, respectively. Better results have been achieved using a combination of 5 μg/mL of gentamicin and 16 μg/mL of chloramphenicol as antibacterial agents. Ciprofloxacin at a concentration of 5 μg/mL may be used.

Cycloheximide may be added to any of the media that contain or lack antibacterial antibiotics. However, if cycloheximide is included as a part of the battery of culture media used, a medium lacking this ingredient should also be included. Certain of the pathogenic fungi, such as *Cryptococcus neoformans, Candida krusei,* other *Candida* spp., *Trichosporon cutaneum, Pseudallescheria boydii,* and *Aspergillus* spp., are partially or completely inhibited by cycloheximide.

Although the use of antibiotics in fungal culture media is necessary for the optimal recovery of organisms, the use of decontamination and concentration methods advocated for the recovery of mycobacteria are not appropriate because many fungi are killed by sodium hydroxide treatment.[115]

DIRECT MICROSCOPIC EXAMINATION OF CLINICAL SPECIMENS

Direct microscopic examination of clinical specimens has been used for many years; however, its usefulness should be reemphasized.[88] Because the mission of a clinical microbiology laboratory is to provide a rapid and accurate diagnosis, the mycology laboratory can provide this service in many instances by direct examination (particularly the Gram stain) of the clinical specimen submitted for culture. Microbiologists are encouraged to become familiar with the diagnostic features of fungi commonly encountered in clinical specimens and to recognize them when stained by various dyes. This important procedure can often provide the first microbiologic proof of the etiology of disease in patients with fungal infection. This is the most rapid method currently available.

Tables 50-7 and 50-8 present the methods available for the direct microscopic detection of fungi in clinical specimens and a summary of the characteristic microscopic features of each. Figures 50-11 to 50-28 present photomicrographs of some of the fungi commonly seen in clinical specimens.

Traditionally, the potassium hydroxide preparation has been the recommended method for the direct microscopic examination of specimens.[114] However, now it is believed that the calcofluor white stain is superior[51] (Procedure 50-1). Slides prepared by this method may be observed using fluorescent or bright-field microscopy as used for the potassium hydroxide preparation; the former is optimal in that fungal cells will fluoresce.

Molecular detection methods are becoming popular in all areas of clinical microbiology; however, none has been accepted as a routine diagnostic tool in clinical mycology. Ideally, a panel of primers specific for the detection of fungi in clinical specimens would include the most common organisms known to cause disease in immunocompromised patients (including the dimorphic fungi and *P. jiroveci*). However, currently, no commercial methods are available to the clinical laboratory and reports in the literature deal predominantly with selected organisms. The large number of fungi may limit the development of a screening method that is cost effective. Studies by Makimura, Murayama, and Yamaguchi,[81] Hopfer,[58,59] Sandhu et al,[118] and Lu et al[79] present examples of what has been done with molecular methods in mycology for the detection of fungi in clinical specimens.

EXTENT OF IDENTIFICATION OF FUNGI RECOVERED FROM CLINICAL SPECIMENS

The question of when and how far to go with the identification of fungi recovered from clinical specimens presents an interesting challenge. The current emphasis on cost containment and the ever-increasing number of opportunistic fungi causing infection in compromised patients prompts one to consider whether all fungi recovered from clinical specimens should be thoroughly identified and reported. Murray, Van Scoy, and Roberts[92] were concerned with the time and expense associated with the identification of yeasts from respiratory tract specimens. Because these are the specimens most commonly submitted for fungal culture, they questioned whether it was important to identify every organism recovered. After evaluating the clinical usefulness of information provided by the identification of yeast recovered from respiratory tract specimens, they suggested the following:

- The routine identification of yeasts recovered in culture from respiratory secretions is not warranted, but all yeasts should be screened for the presence of *Cryptococcus neoformans.*
- All respiratory secretions submitted for fungal culture, regardless of the presence or absence of oro-

Table 50-7   Summary of Methods Available for Direct Microscopic Detection of Fungi in Clinical Specimens

Method	Use	Time Required	Advantages	Disadvantages
Acid-fast stain and partial acid fast stain	Detection of mycobacteria and *Nocardia*, respectively	12 min	Detects *Nocardia** and some isolates of *B. dermatitidis*	Tissue homogenates are difficult to observe because of background staining
Calcofluor white	Detection of fungi	1 min	Can be mixed with KOH: detects fungi rapidly because of bright fluorescence	Requires use of a fluorescence microscope; background fluorescence prominent, but fungi exhibit more intense fluorescence; vaginal secretions are difficult to interpret
Gram stain	Detection of bacteria	3 min	Is commonly performed on most clinical specimens submitted for bacteriology and will detect most fungi, if present	Some fungi stain well; however, others (e.g., *Cryptococcus* spp.) show only stippling and stain weakly in some instances; some isolates of *Nocardia* fail to stain or stain weakly
India ink	Detection of *C. neoformans* in CSF	1 min	When positive in CSF, is diagnostic of meningitis	Positive in less than 50% of cases of meningitis; not sensitive in non–HIV-infected patients
Potassium hydroxide (KOH)	Clearing of specimen to make fungi more readily visible	5 min; if clearing is not complete, an additional 5-10 min is necessary	Rapid detection of fungal elements	Experience required since background artifacts are often confusing; clearing of some specimens may require an extended time
Methenamine silver stain	Detection of fungi in histologic section	1 hr	Best stain to detect fungal elements	Requires a specialized staining method that is not usually readily available to microbiology laboratories
Papanicolaou stain	Examination of secretions for presence of malignant cells	30 min	Cytotechnologist can detect fungal elements	Fungal elements stain pink to blue
Periodic acid- Schiff (PAS) stain	Detection of fungi	20 min; 5 min additional if counterstain is used	Stains fungal elements well; hyphae of molds and yeasts can be readily distinguished	*Nocardia* spp. do not stain well
Wright stain	Examination of bone marrow or peripheral blood smears	7 min	Detects *H. capsulatum* and *C. neoformans*	Detection is most commonly of *H. capsulatum* and *C. neoformans* in disseminated disease.

From Merz WG, Roberts GD: Detection and recovery of fungi from clinical specimens. In Murray PR, Baron EJ, Pfaller MA, et al, editors: *Manual of clinical microbiology,* ed 6, Washington, DC, 1995, ASM Press.
CSF, Cerebrospinal fluid; *HIV,* human immunodeficiency virus.
*Partially acid-fast bacterium.

pharyngeal contamination, should be cultured because common pathogens such as *Histoplasma capsulatum, Blastomyces dermatitidis, Coccidioides immitis,* and *Sporothrix schenckii* may be recovered.
● The routine identification of yeasts in respiratory secretions is of little or no value to the clinician and probably represents "normal flora" except for *C. neoformans.*

The extent of identification of yeasts from other specimen sources is discussed under yeast identification later in this chapter. The utility of the identification and susceptibility testing of non-*Cryptococcus* yeast isolates has recently been studied by Barenfenger.[8] She found that compared with only superficial characterization (i.e., "Yeast present, not *C. neoformans*"), the rendering of identification and susceptibility testing of *Candida* isolates from respiratory secretions lead to unnecessary treatment and increased costs in the latter group. There was no statistical difference in the mortality of these groups.

When and how far to proceed with an identification of the molds is a much more difficult question to

Table 50-8   Summary of Characteristic Features of Fungi Seen in Direct Examination of Clinical Specimens

Morphologic Form Found in Specimens	Organism(s)	Size Range (diameter, μm)	Characteristic Features
Yeastlike	*Histoplasma capsulatum*	2-5	Small; oval to round budding cells; often found clustered within histiocytes; difficult to detect when present in small numbers
	Sporothrix schenckii	2-6	Small; oval to round to cigar-shaped; single or multiple buds present; uncommonly seen in clinical specimens
	Cryptococcus neoformans	2-15	Cells exhibit great variation in size; usually spherical but may be football-shaped; buds single or multiple and "pinched off"; capsule may or may not be evident; occasionally, pseudohyphal forms with or without a capsule may be seen in exudates of CSF
	Malassezia furfur (in fungemia)	1.5-4.5	Small; bottle-shaped cells, buds separated from parent cell by a septum; emerge from a small collar
	Blastomyces dermatitidis	8-15	Cells are usually large, double refractile when present; buds usually single; however, several may remain attached to parent cells; buds connected by a broad base
	Paracoccidioides brasiliensis	5-60	Cells are usually large and are surrounded by smaller buds around the periphery ("mariner's wheel appearance"); smaller cells may be present (2-5 μm) and resemble *H. capsulatum;* buds have "pinched-off" appearance
Spherules	*Coccidioides immitis*	10-200	Spherules vary in size; some may contain endospores, others may be empty; adjacent spherules may resemble *B. dermatitidis;* endospores may resemble *H. capsulatum* but show no evidence of budding; spherules may produce multiple germ tubes if a direct preparation is kept in a moist chamber for ≥24 hrs
	Rhinosporidium seeberi	6-300	Large, thick-walled sporangia containing sporangiospores are present; mature sporangia are larger than spherules of *C. immitis;* hyphae may be found in cavitary lesions
Yeast and pseudohyphae or hyphae	*Candida* spp. except *C. glabrata*	5-10 (pseudo hyphae)	Cells usually exhibit single budding; pseudohyphae, when present, are constricted at the ends and remain attached like links of sausage; hyphae, when present, are septate
	M. furfur (in tinea versicolor)	3-8 (yeast) 2.5-4 (hyphae)	Short, curved hyphal elements are usually present along with round yeast cells that retain their spherical shape in compacted clusters
Pauciseptate hyphae	Zygomycetes; *Mucor, Rhizopus,* and other genera	10-30	Hyphae are large, ribbonlike, often fractured or twisted; occasional septa may be present; smaller hyphae are confused with those of *Aspergillus* spp., particularly *A. flavus*
Hyaline septate hyphae	Dermatophytes Skin and nails	3-15	Hyaline, septate hyphae are commonly seen; chains of arthroconidia may be present
	Hair	3-15	Arthroconidia on periphery of hair shaft producing a sheath are indicative of ectothrix infection; arthroconidia formed by fragmentation of hyphae within the hair shaft are indicative of endothrix infection
		3-15	Long hyphal filaments or channels within the hair shaft are indicative of favus hair infection
	Aspergillus spp.	3-12	Hyphae are septate and exhibit dichotomous, 45-degree angle branching; larger hyphae, often disturbed, may resemble those of Zygomycetes
	Geotrichum spp.	4-12	Hyphae and rectangular arthroconidia are present and are sometimes rounded; irregular forms may be present
	Trichosporon spp.	2-4 by 8	Hyphae and rectangular arthroconidia are present and sometimes rounded; occasionally, blastoconidia may be present

Table 50-8   Summary of Characteristic Features of Fungi Seen in Direct Examination of Clinical Specimens—cont'd

Morphologic Form Found in Specimens	Organism(s)	Size Range (diameter, μm)	Characteristic Features
Dematiaceous septate hyphae	*Bipolaris* spp. *Cladosporium* spp. *Curvularia* spp. *Drechslera* spp. *Exophiala* spp. *Exserohilum* spp. *Hortaea werneckii* *Phialophora* spp.	2-6	Dematiaceous polymorphous hyphae are seen; budding cells with single septa and chains of swollen rounded cells are often present; occasionally, aggregates may be present in infection caused by *Phialophora* and *Exophiala* spp.
	Wangiella dermatitidis	1.5-5	Usually large numbers of frequently branched hyphae are present along with budding cells
Sclerotic bodies	*Cladosporium carrionii* *Fonsecaea compacta* *Fonsecaea pedrosoi* *Phialophora verrucosa* *Rhinocladiella aquaspersa*	5-20	Brown, round to pleomorphic, thick-walled cells with transverse septations; commonly, cells contain two fission planes that form a tetrad of cells (sclerotic bodies)

Granules	*Acremonium*
	  A. falciforme
	  A. kiliense
	  A. recifei
		200-300	White, soft granules without a cementlike matrix

	Aspergillus *A. nidulans*	500-1000	Black, hard grains with a cementlike matrix at periphery

	Curvularia
	  C. geniculata
	  C. lunata
		65-160	White, soft granule without a cementlike matrix

	Exophiala *E. jeanselmei*	200-300	Black, soft granules, vacuolated, without a cementlike matrix, made of dark hyphae and swollen cells

	Fusarium
	  F. moniliforme
	  F. solani
		200-500
		
		300-600
			White, soft granules without a cementlike matrix

	Leptosphaeria *L. senegalensis*	400-600	Black, hard granules with cementlike matrix present
	L. tompkinsii	500-1000	Periphery composed of polygonal swollen cells and center of a hyphal network
	Madurella *M. grisea*	350-500	Black, soft granules without a cementlike matrix, periphery composed of polygonal swollen cells and center of a hyphal network
	M. mycetomatis	200-900	Black to brown, hard granules of two types: (1) Rust brown, compact, and filled with cementlike matrix (2) Deep brown, filled with numerous vesicles, 6-14 μm in diameter, cementlike matrix in periphery, and central area of light-colored hyphae
	Neotestudina *N. rosatii*	300-600	White, soft granules with cementlike matrix present at periphery
	Pseudallescheria *P. boydii*	200-300	White, soft granules composed of hyphae and swollen cells at periphery in cementlike matrix
	Pyrenochaeta *P. romeroi*	300-600	Black, soft granules composed of polygonal swollen cells at periphery, center is network of hyphae, no cementlike matrix present

From Merz WG, Roberts GD: Detection and recovery of fungi in clinical specimens. In Murray PR, Baron EJ, Pfaller MA, et al, editors: *Manual of clinical microbiology,* ed 6, Washington, DC, 1995, ASM Press.
CSF, Cerebrospinal fluid.

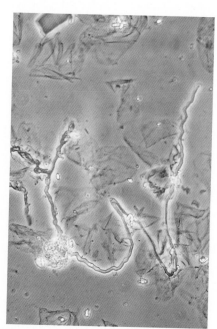

Figure 50-11   This potassium hydroxide preparation of a skin scraping from a patient with a dermatophyte infection shows septate hyphae intertwined among epithelial cells. Phase-contrast microscopy (500×).

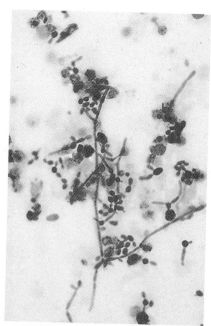

Figure 50-13   This periodic acid-Schiff stain of urine demonstrates the blastoconidia and pseudohyphae of *C. albicans*.

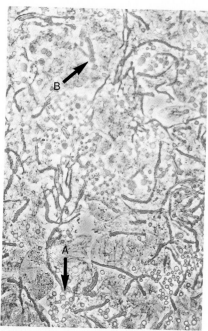

Figure 50-12   This potassium hydroxide preparation a skin scraping from a patient with tinea versicolor demonstrates spherical yeast cells (**A**) and short hyphal fragments (**B**) of *M. furfur*. Phase-contrast microscopy (500×).

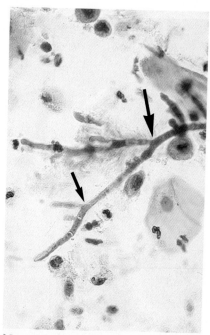

Figure 50-14   A Papanicolaou stain of sputum shows the dichotomously branching septate hyphae *(arrows)* of *A. fumigatus.*

answer. Except for obvious plate contaminants, all commonly encountered molds should be identified and reported if recovered from patients at risk for invasive fungal disease. Immunocompromised patients may have serious or even fatal disease by fungi that were once thought not to be of clinical significance.[108] Those organisms that fail to sporulate after a reasonable time should be reported as being present but

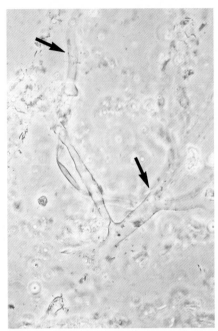

Figure 50-15 This potassium hydroxide preparation of sputum shows the fragmented portions *(arrows)* of broad, predominately nonseptate hyphae of *Rhizopus* spp. Phase-contrast microscopy.

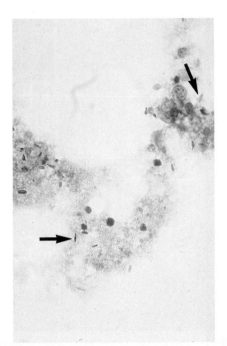

Figure 50-17 This periodic acid-Schiff stain of exudate shows the cigar-to-oval-shaped yeast cells *(arrows)* of *S. schenckii.*

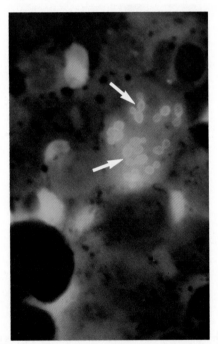

Figure 50-16 This calcofluor white stain of sputum shows the intracellular yeast cells *(arrows)* of *H. capsulatum,* which are 2 to 5 μm in diameter.

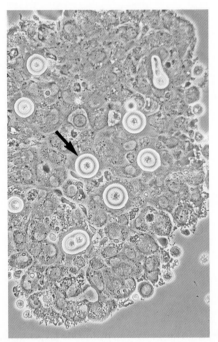

Figure 50-18 This potassium hydroxide preparation of pleural fluid shows the encapsulated, variably sized, spherical yeast cells *(arrow)* of *C. neoformans.* Phase-contrast microscopy.

the identification need not be attempted if the dimorphic fungi have been ruled out or if the clinician believes that the organism is not clinically significant. Ideally, all laboratories should identify all fungi recovered from clinical specimens; however, the limits of practicality and economic considerations play a definite role in making a decision as to how far to go with their identification. Each individual laboratory director, in consul-

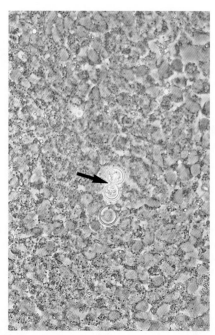

Figure 50-19 This potassium hydroxide preparation of exudate shows a large budding yeast cell with a distinct broad base *(arrow)* between the cells, which is characteristic of *B. dermatitidis*. Phase-contrast microscopy.

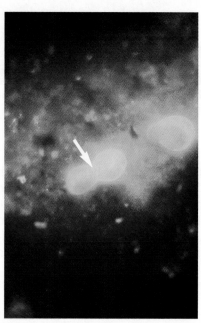

Figure 50-20 This auramine-rhodamine preparation of specimen material from a bone lesion demonstrates the characteristic broad-based budding yeast cell *(arrow)* of *Blastomyces dermatitidis*.

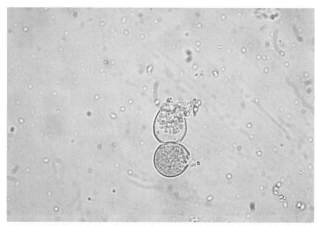

Figure 50-21 This potassium hydroxide preparation of sputum demonstrates two spherules of *C. immitis* filled with endospores. When these lie adjacent to each other they may be mistaken for *B. dermatitidis*. Bright field microscopy.

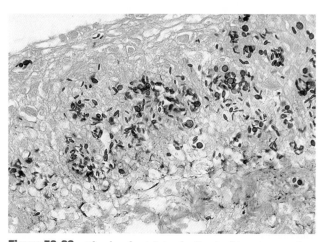

Figure 50-22 The deeply staining bodies in this mouse testis are the yeast forms of *Sporothrix schenckii*.

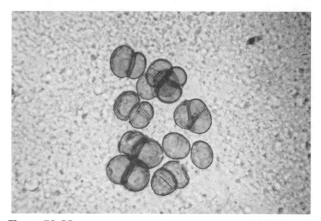

Figure 50-23 Sclerotic bodies from the tissue of a patient with chromoblastomycosis (400×). (From Velasquez LF, Restrepo A: *Sabouraudia* 13:1, 1975.)

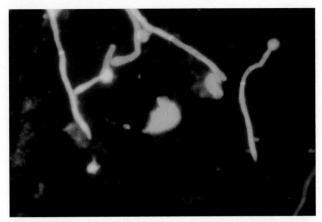

Figure 50-24 This calcofluor white stain of urine demonstrates *Candida albicans*.

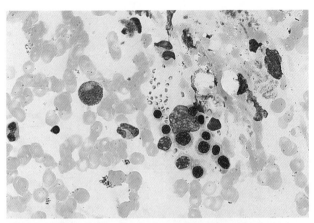

Figure 50-27 *Histoplasma capsulatum* is often detected by the hematopathologist in disseminated histoplasmosis. The small yeasts (just above center in the photomicrograph) of *H. capsulatum* are present in this bone marrow aspirate. Wright stain (1000×).

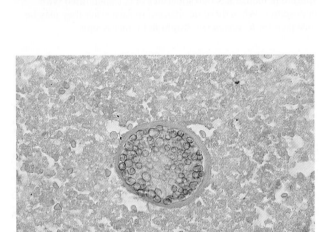

Figure 50-25 This histologic section demonstrates a well-developed spherule of *Coccidioides immitis* that is filled with endospores.

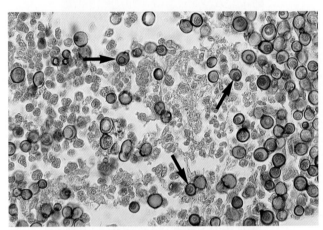

Figure 50-28 *Blastomyces dermatitidis (arrows)* in tissue. Methenamine silver stain (430×).

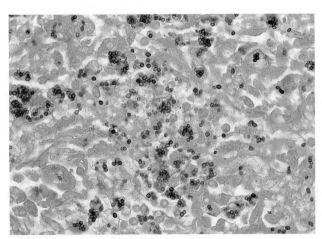

Figure 50-26 The deeply staining small uniform yeast cells in this histologic section of lung tissue are typical of *Histoplasma capsulatum*. Methenamine silver stain (430×).

tation with the clinicians being served, will have to make this decision after considering the patient population, laboratory practice, and economic impact.

As shown in Table 50-9, an increasing number of fungi not discussed thus far are seen in the clinical microbiology laboratory. They are considered to be environmental flora, but in reality they must be regarded as potential pathogens because infections with a number of these organisms have been reported. Less commonly encountered fungal pathogens that have been shown to cause human infections include but are not limited to *Pseudallescheria boydii*,[10] *Scedosporium prolificans*,[5,129] *Bipolaris*,[2,32,86] *Exserohilum*,[2,32,86] *Trichosporon*,[45,80,135] *Aureobasidium*,[18] and others.[93] It is necessary for the laboratory to identify and report all organisms recovered from clinical specimens so that their clinical significance can be determined. In many instances, the presence of environmental fungi is unimportant; how-

Table 50-9   Fungi Most Commonly Recovered from Clinical Specimens

Blood	Cerebrospinal Fluid	Genitourinary Tract	Respiratory Tract	Skin
Candida albicans	Cryptococcus neoformans	Candida albicans	Yeast, not Cryptococcus	Trichophyton rubrum
Candida tropicalis	Candida albicans	Candida glabrata	Penicillium spp.	Trichophyton mentagrophytes
Candida parapsilosis	Candida parapsilosis	Candida tropicalis	Aspergillus spp.	Alternaria spp.
Cryptococcus neoformans	Candida tropicalis	Candida parapsilosis	Aspergillus fumigatus	Candida albicans
Histoplasma capsulatum	Coccidioides immitis	Penicillium spp.	Cladosporium spp.	Penicillium spp.
Candida lusitaniae	Histoplasma capsulatum	Candida krusei	Alternaria spp.	Scopulariopsis spp.
Candida krusei		Cryptococcus neoformans	Aspergillus niger	Epidermophyton floccosum
Saccharomyces spp.		Saccharomyces spp.	Geotrichum candidum	Candida parapsilosis
Candida kefyr		Histoplasma capsulatum	Fusarium spp.	Aspergillus spp.
Candida zeylanoides		Cladosporium spp.	Aspergillus versicolor	Acremonium spp.
Trichosporon spp.		Aspergillus spp.	Aspergillus flavus	Aspergillus versicolor
Coccidioides immitis		Trichosporon spp.	Acremonium spp.	Cladosporium spp.
Candida guilliermondii		Alternaria spp.	Scopulariopsis spp.	Fusarium spp.
Malassezia furfur			Beauveria spp.	Trichosporon spp.
			Trichosporon spp.	Phialophora spp.

ever, that is not always the case. The following section presents the molds most commonly recovered from clinical specimens and a brief description of their colonial and morphologic features. In addition, Tables 50-10 and 50-11 present the molds and yeasts implicated in causing human infection, the time required for their identification, most likely site for their recovery, and the clinical implications of each.

GENERAL CONSIDERATIONS FOR THE IDENTIFICATION OF MOLDS

The identification of molds is made using a combination of the following:

● The growth rate
● Colonial morphologic features
● Microscopic morphologic features

In most instances the latter provides the most definitive means for identification. The determination of the growth rate can be the most helpful when examining a mold culture. However, this may be of limited value because the growth rate of certain fungi is variable, depending on the amount of inoculum present in a clinical specimen. In general, the growth rate for the dimorphic fungi, *B. dermatitidis, H. capsulatum,* and *P. brasiliensis,* is slow; 1 to 4 weeks are usually required before colonies become visible. The growth of *C. immitis* is often rapid and is hazardous to bacteriologists. In some instances, however, cultures of *B. dermatitidis* and *H. capsulatum* may be detected within 3 to 5 days. This is a somewhat uncommon circumstance and is encountered only when large numbers of the organism are present in the specimen. Colonies of the Zygomycetes may appear within 24 hours, whereas the other hyaline

and dematiaceous fungi often exhibit growth within 1 to 5 days. The growth rate of an organism, therefore, is important, but it must be used in combination with other features before a definitive identification can be made.

The colonial morphologic features may be of limited value in identifying the molds, because of natural variation among isolates and colonies grown on different culture media. Although it may be possible to recognize common organisms recovered repeatedly in the laboratory, colonial morphology is an unreliable criterion and should be used only to supplement the microscopic morphologic features of the organism. Incubation conditions and culture media must be considered. For example, *H. capsulatum,* which appears as a white-to-tan fluffy mold on brain-heart infusion agar, may appear yeastlike in appearance when grown on the same medium containing blood enrichment.

In general, the microscopic morphologic features of the molds are stable and exhibit minimal variation. The definitive identification is based on the characteristic shape, method of reproduction, and arrangement of spores; however, the size of the hyphae also provides helpful information. The large, ribbonlike, pauciseptate hyphae of the Zygomycetes are easily recognized, whereas small hyphae, approximately 2 μm in diameter, may suggest the presence of one of the dimorphic fungi or a dermatophyte.

The fungi may be prepared for microscopic observation using several techniques. The procedure traditionally used by most laboratories is the adhesive tape preparation (Procedure 50-2, Figure 50-29). It can be prepared easily and quickly and often is sufficient to make the identification for most of the fungi. However, the wet mount (Procedure 50-3, Figure 50-30) is preferred by some laboratories, and a microslide culture

Table 50-10   Common Filamentous Fungi Implicated in Human Mycotic Infections

Etiologic Agent	Time Required for Identification	Probable Recovery Sites	Clinical Implication
Acremonium spp.	2-6 days	Skin, nails, respiratory secretions, cornea, vagina, gastric washings, blood	Skin and nail infections, mycotic keratitis, mycetoma
Alternaria spp.	2-6 days	Skin, nails, conjunctiva, respiratory secretions, subcutaneous tissue	Skin and nail infections, sinusitis, conjunctivitis, hypersensitivity pneumonitis, skin abscess
Aspergillus flavus	1-4 days	Skin, respiratory secretions, gastric washings, nasal sinuses, lung	Skin infections, allergic bronchopulmonary infection, sinusitis, myocarditis, disseminated infection, renal infection, subcutaneous mycetoma
Aspergillus fumigatus	2-6 days	Respiratory secretions, skin, ear, cornea, gastric washings, nasal sinuses, lung	Allergic bronchopulmonary infection, fungus ball, invasive pulmonary infection, skin and nail infections, external otomycosis, mycotic keratitis, sinusitis, myocarditis, renal infection
Aspergillus niger	1-4 days	Respiratory secretions, gastric washings, ear, skin	Fungus ball, pulmonary infection, external otomycosis, mycotic keratitis
Aspergillus terreus	2-6 days	Respiratory secretions, skin, gastric washings, nails, lung	Pulmonary infection, disseminated infection, endocarditis, onychomycosis, allergic bronchopulmonary infection
Bipolaris spp.	2-6 days	Respiratory secretions, skin, nose, bone, sinuses	Sinusitis, brain abscess, peritonitis, subcutaneous abscess, pulmonary infection, osteomyelitis, encephalitis
Blastomyces dermatitidis	6-21 days (recovery time) (additional 1-2 days required for confirmatory identification)	Respiratory secretions, skin, oropharyngeal ulcer, bone, prostate, lung	Pulmonary infection, skin infection, oropharyngeal ulceration, osteomyelitis, prostatitis, arthritis, central nervous system (CNS) infection, disseminated infection
Cladosporium spp.	6-10 days	Respiratory secretions, skin, nails, nose, cornea	Skin and nail infections, mycotic keratitis, chromoblastomycosis caused by *Cladophialophora carrionii*
Coccidioides immitis	3-21 days	Respiratory secretions, skin, bone, cerebrospinal fluid, synovial fluid, urine, gastric washings, blood	Pulmonary infection, skin infection, osteomyelitis, meningitis, arthritis, disseminated infection
Curvularia spp.	2-6 days	Respiratory secretions, cornea, brain, skin, nasal sinuses	Pulmonary infection, disseminated infection, mycotic keratitis, brain abscess, mycetoma, endocarditis
Drechslera spp.	2-6 days	Respiratory secretions, skin, peritoneal fluid (following dialysis)	Pulmonary infection (rare)
Epidermophyton floccosum	7-10 days	Skin, nails	Tinea cruris, tinea pedis, tinea corporis, onychomycosis
Exserohilum spp.	2-6 days	Eye, skin, nose, bone	Keratitis, subcutaneous abscess, sinusitis, endocarditis, osteomyelitis
Fusarium spp.	2-6 days	Skin, respiratory secretions, cornea, nails, blood	Mycotic keratitis, skin infection (in burn patients), disseminated infection, endophthalmitis
Geotrichum spp.	2-6 days	Respiratory secretions, urine, skin, stool, vagina, conjunctiva, gastric washings, throat	Bronchitis, skin infection, colitis, conjunctivitis, thrush, wound infection
Histoplasma capsulatum	≤10-45 days (recovery time) (additional 1-2 days required for confirmatory identification)	Respiratory secretions, bone marrow, blood, urine, adrenals, skin, cerebrospinal fluid, eye, pleural fluid, liver, spleen, oropharyngeal lesions, vagina, gastric washings, larynx	Pulmonary infection, oropharyngeal lesions, CNS infection, skin infection (rare), uveitis, peritonitis, endocarditis, brain abscess, disseminated infection

Table 50-10   Common Filamentous Fungi Implicated in Human Mycotic Infections—cont'd

Etiologic Agent	Time Required for Identification	Probable Recovery Sites	Clinical Implication
Microsporum audouinii	10-14 days (recovery time) (additional 14-21 days required for confirmatory identification)	Hair (scalp)	Tinea capitis
Microsporum canis	5-7 days	Hair, skin	Tinea corporis, tinea capitis, tinea barbae, tinea manuum
Microsporum gypseum	3-6 days	Hair, skin	Tinea capitis, tinea corporis
Mucor spp.	1-5 days	Respiratory secretions, skin, nose, brain, stool, orbit, cornea, vitreous humor, gastric washings, wounds, ear, lung	Rhinocerebral infection, pulmonary infection, gastrointestinal infection, mycotic keratitis, intraocular infection, external otomycosis, orbital cellulitis, disseminated infection
Penicillium spp.	2-6 days	Respiratory secretions, gastric washings, skin, urine, ear, cornea	Allergic; human infections rare except with *P. marneffei.*
Phialophora spp.	6-21 days	Respiratory secretions, gastric washings, skin, cornea, conjunctiva	Some species produce chromoblastomycosis or mycetoma; mycotic keratitis, conjunctivitis, intraocular infection
Pseudallescheria boydii	2-6 days	Respiratory secretions, gastric washings, skin, cornea	Pulmonary fungus ball, mycetoma, mycotic keratitis, endocarditis, disseminated infection, brain abscess
Rhizopus spp.	1-5 days	Respiratory secretions, skin, nose, brain, stool, orbit, cornea, vitreous humor, gastric washings, wounds, ear, lung	Rhinocerebral infection, pulmonary infection, mycotic keratitis, intraocular infection, orbital cellulitis, external otomycosis, disseminated infection
Scedosporium prolificans	2-6 days	Respiratory secretions, skin, nasal sinuses, bone	Arthritis, osteomyelitis, sinusitis, endocarditis
Scopulariopsis spp.	2-6 days	Respiratory secretions, gastric washings, nails, skin, vitreous humor, ear	Pulmonary infection, nail infection, skin infection, intraocular infection, external otomycosis
Sporothrix schenckii	3-12 days (recovery time) (additional 2-10 days required for confirmatory identification)	Respiratory secretions, skin, subcutaneous tissue, maxillary sinuses, synovial fluid, bone marrow, bone, cerebrospinal fluid, ear, conjunctiva	Pulmonary infection, lymphocutaneous infection, sinusitis, arthritis, osteomyelitis, meningitis, external otomycosis, conjunctivitis, disseminated infection
Trichophyton mentagrophytes	7-10 days	Hair, skin, nails	Tinea barbae, tinea capitis, tinea corporis, tinea cruris, tinea pedis, onychomycosis
Trichophyton rubrum	10-14 days	Hair, skin, nails	Tinea pedis, onychomycosis, tinea corporis, tinea cruris
Trichophyton tonsurans	10-14 days	Hair, skin, nails	Tinea capitis, tinea corporis, onychomycosis, tinea pedis
Trichophyton verrucosum	10-18 days	Hair, skin, nails	Tinea capitis, tinea corporis, tinea barbae
Trichophyton violaceum	14-18 days	Hair, skin, nails	Tinea capitis, tinea corporis, onychomycosis
Exophiala dermatitidis	5-21 days	Respiratory secretions, skin, eye	Phaeohyphomycosis, endophthalmitis, pneumonia

From Koneman EW, Roberts GD: Mycotic disease. In Henry JB, editor: *Clinical diagnosis and management by laboratory methods,* ed 18, Philadelphia, 1991, Saunders.

Table 50-11   Common Yeastlike Organisms Implicated in Human Infection*

Etiologic Agent	Probable Recovery Sites	Clinical Implication
Candida albicans	Respiratory secretions, vagina, urine, skin, oropharynx, gastric washings, blood, stool, transtracheal aspiration, cornea, nails, cerebrospinal fluid, bone, peritoneal fluid	Pulmonary infection, vaginitis, urinary tract infection, dermatitis, fungemia, mycotic keratitis, onychomycosis, meningitis, osteomyelitis, peritonitis, myocarditis, endocarditis, endophthalmitis, disseminated infection, thrush, arthritis
Candida glabrata	Respiratory secretions, urine, vagina, gastric washings, blood, skin, oropharynx, transtracheal aspiration, stool, bone marrow, skin (rare)	Pulmonary infection, urinary tract infection, vaginitis, fungemia, disseminated infection, endocarditis
Candida tropicalis	Respiratory secretions, urine, gastric washings, vagina, blood, skin, oropharynx, transtracheal aspiration, stool, pleural fluid, peritoneal fluid, cornea	Pulmonary infection, vaginitis, thrush, endophthalmitis, endocarditis, arthritis, peritonitis, mycotic keratitis, fungemia
Candida parapsilosis	Respiratory secretions, urine, gastric washings, blood, vagina, oropharynx, skin, transtracheal aspiration, stool, pleural fluid, ear, nails	Endophthalmitis, endocarditis, vaginitis, mycotic keratitis, external otomycosis, paronychia, fungemia
Saccharomyces spp.	Respiratory secretions, urine, gastric washings, vagina, skin, oropharynx, transtracheal aspiration, stool, blood	Pulmonary infection (rare), endocarditis
Candida krusei	Respiratory secretions, urine, gastric washings, vagina, skin, oropharynx, blood, transtracheal aspiration, stool, cornea	Endocarditis, vaginitis, urinary tract infection, mycotic keratitis
Candida guilliermondii	Respiratory secretions, gastric washings, vagina, skin, nails, oropharynx, blood, cornea, bone, urine	Endocarditis, fungemia, dermatitis, onychomycosis, mycotic keratitis, osteomyelitis, urinary tract infection
Rhodotorula spp.	Respiratory secretions, urine, gastric washings, blood, vagina, skin, oropharynx, stool, cerebrospinal fluid, cornea	Fungemia, endocarditis, mycotic keratitis
Trichosporon spp.	Respiratory secretions, blood, skin, oropharynx, stool	Pulmonary infection, brain abscess, disseminated infection, piedra
Cryptococcus neoformans	Respiratory secretions, cerebrospinal fluid, bone, blood, bone marrow, urine, skin, pleural fluid, gastric washings, transtracheal aspiration, cornea, orbit, vitreous humor	Pulmonary infection, meningitis, osteomyelitis, fungemia, disseminated infection, endocarditis, skin infection, mycotic keratitis, orbital cellulitis, endophthalmic infection
Cryptococcus albidus subsp. *albidus*	Respiratory secretions, skin, gastric washings, urine, cornea	Meningitis, pulmonary infection
Candida kefyr (pseudotropicalis)	Respiratory secretions, vagina, urine, gastric washings, oropharynx	Vaginitis, urinary tract infection
Cryptococcus luteolus	Respiratory secretions, skin, nose	Not commonly implicated in human infection
Cryptococcus laurentii	Respiratory secretions, cerebrospinal fluid, skin, oropharynx, stool	Not commonly implicated in human infection
Cryptococcus albidus subsp. *diffluens*	Respiratory secretions, urine, cerebrospinal fluid, gastric washings, skin	Not commonly implicated in human infection
Cryptococcus terreus	Respiratory secretions, skin, nose	Not commonly implicated in human infection

From Koneman EW, Roberts GD: Mycotic disease. In Henry JB, editor: *Clinical diagnosis and management by laboratory methods*, ed 18, Philadelphia, 1991, Saunders.
*Arranged in order of occurrence in the clinical laboratory.

method (Procedure 50-4, Figure 50-31) may be used when greater detail of the morphologic features is required for identification.

LABORATORY SAFETY CONSIDERATIONS

Although there are risks associated with the handling of fungi recovered from clinical specimens, a common sense approach concerning their handling protects the laboratory from contamination and workers from becoming infected.

All mold cultures and clinical specimens must be handled in a class II biological safety cabinet (BSC), *with no exceptions.* Some laboratory directors believe that it is necessary to handle mold cultures within an enclosed BSC equipped with gloves, but this is not necessary if a

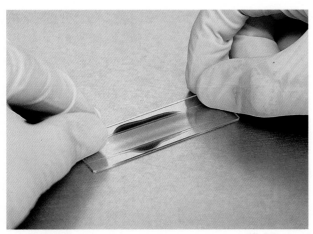

Figure 50-29   Cellophane tape preparation showing placement of tape onto slide containing lactophenol cotton or aniline blue.

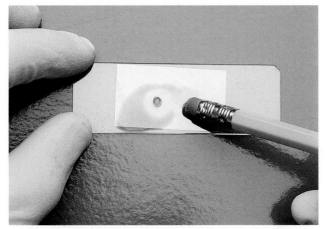

Figure 50-30   Performance of a wet mount showing agar positioned under coverslip before using pressure to disperse growth.

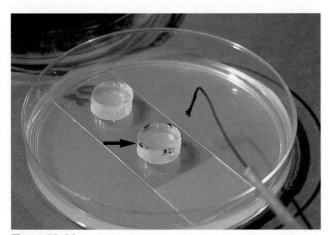

Figure 50-31   Microslide culture showing inoculation of agar plug *(arrow).*

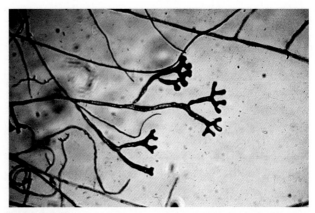

Figure 50-32   Antler hyphae showing swollen hyphal tips resembling antlers, with lateral and terminal branching (favic chandeliers) (500×).

laminar flow BSC is used. It is permissible, however, to handle yeast cultures on the bench-top; they must be treated as infectious agents. The use of an electric incinerator or a gas flame is suitable for the decontamination of a loop used for transfer of yeast cultures. Cultures of organisms suspected of being pathogens should be sealed with tape to prevent laboratory contamination and should be autoclaved as soon as the definitive identification is made. Few problems with laboratory contamination or acquired infection by laboratory personnel are seen if common safety precautions are used.

GENERAL MORPHOLOGIC FEATURES OF THE MOLDS

Specialized types of vegetative hyphae may be helpful in placing an organism into a certain group. For example, dermatophytes often produce several types of hyphae, including antler hyphae that are curved, freely branching, and "antlerlike" in appearance (Figure 50-32). Racquet hyphae are enlarged, club-shaped structures (Figure 50-33). In addition, certain of the dermatophytes produce spiral hyphae that are coiled or exhibit corkscrewlike turns seen within the hyphal strand (Figure 50-34). These structures are not characteristic for any certain group; however, they are found most commonly in dermatophytes.

Some species of fungi produce sexual spores in a large saclike structure called an **ascocarp** (Figure 50-35). This, in turn, contains smaller sacs called **asci,** each of which contains four to eight **ascospores.** This type of sexual reproduction is uncommonly seen in the fungi recovered in the clinical microbiology laboratory; most exhibit only asexual reproduction. It is possible that all fungi possess a sexual form, but this form may not have

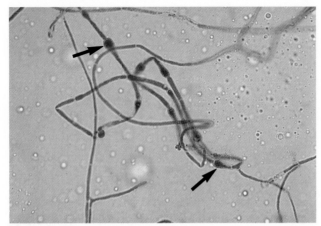

Figure 50-33 Racquet hyphae showing swollen areas *(arrows)* resembling a tennis racquet.

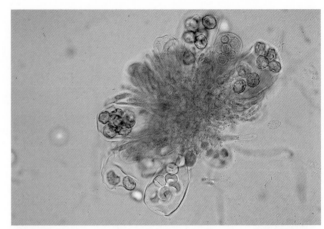

Figure 50-35 Ascocarp showing dark-appearing ascospores (430×).

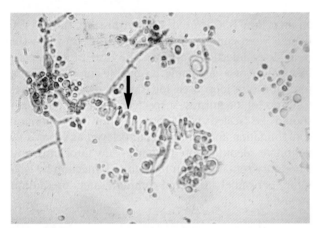

Figure 50-34 Spiral hyphae *(arrow)* exhibiting corkscrewlike turns (430×).

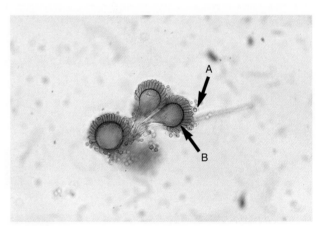

Figure 50-36 Conidia (asexual spores [**A**]) produced on specialized structures (conidiophores [**B**]) of *Aspergillus* (430×).

been observed yet on artificial culture media. Conidia are produced by most fungi and represent the asexual reproductive cycle (Figure 50-36). The type of conidia, their morphology, and arrangement are important criteria for establishing the definitive identification of an organism.

The simplest type of sporulation is the development of a spore directly from the vegetative hyphae. **Arthroconidia** are formed directly from the hyphae by fragmentation through the points of septation (Figure 50-37). When mature, they appear as square, rectangular, or barrel-shaped, thick-walled cells. These result from the simple fragmentation of the hyphae into spores, which are easily dislodged and disseminated into the environment. **Chlamydoconidia (chlamydospores)** are round, thick-walled spores formed directly from the differentiation of hyphae in which there is a concentration of protoplasm and nutrient material (Figure 50-38). These appear to be resistant resting spores produced by the rounding up and enlargement of

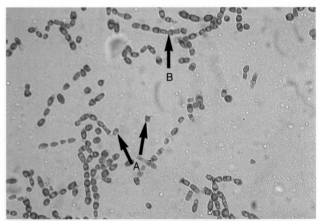

Figure 50-37 Arthroconidia formation (**A**) produced by the breaking down of a hyphal strand (**B**) into individual rectangular units (430×).

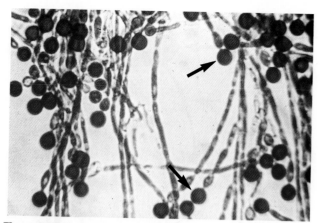

Figure 50-38 Chlamydoconidia composed of thick-walled spherical cells *(arrows)* (430×).

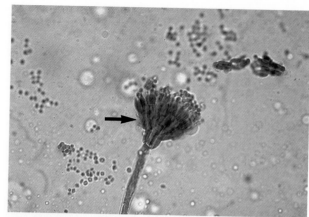

Figure 50-40 Complex method of sporulation in which conidia are borne on phialides produced on secondary branches (metulae *[arrow]*) characteristic of *Penicillium* (430×).

Figure 50-39 Simple tubular phialide with a cluster of conidia at its tip *(arrow)* characteristic of *Acremonium* (430×).

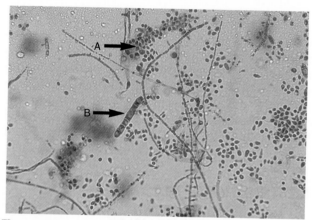

Figure 50-41 In this preparation of a *Trichophyton* species, the numerous small, spherical microconidia **(A)** are contrasted with a large, elongated macroconidium **(B)** (430×).

the cells of the hyphae. Chlamydoconidia may be **intercalary** (within the hyphae) or **terminal** (on the end of hyphae).

A variety of other types of spores occur with many species of fungi. Conidia are asexual spores produced singly or in groups by specialized hyphal strands, **conidiophores.** In some instances, the conidia are freed from their point of attachment by pinching off, or abstriction. Some conidiophores terminate in a swollen vesicle. From the surface of the vesicle are formed secondary small, flask-shaped phialides, which, in turn, give rise to long chains of conidia. This type of fruiting structure is characteristic of the aspergilli. A single, simple, slender, tubular conidiophore **(phialide)** that produces a cluster of conidia, held together as a gelatinous mass, is characteristic of certain fungi, including the genus *Acremonium* (Figure 50-39). In other instances, conidiophores form a branching structure

termed a **"penicillus,"** in which each branch terminates in secondary branches **(metulae),** and phialides, from which chains of conidia are borne (Figure 50-40). Species of *Penicillium* and *Paecilomyces* are representative of this type of sporulation. In other instances, fungi may produce conidia of two sizes: **microconidia** that are small, unicellular, round, elliptical, or pyriform in shape or **macroconidia** that are large, usually multiseptate, and club- or spindle-shaped (Figure 50-41). Microconidia may be borne directly on the side of a hyphal strand or at the end of a conidiophore. Macroconidia are usually borne on a short to long conidiophore and may be smooth or rough-walled. Microconidia and macroconidia are seen in some fungal species and are not specific, except as they are used to differentiate a limited number of genera.

The hyphae of the Zygomycetes are sparsely septate. Sporulation takes place by progressive cleavage

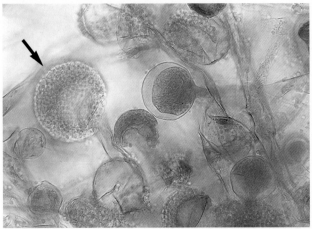

Figure 50-42   Large saclike sporangia that contain sporangiospores *(arrow)* characteristic of the zygomycetes (250×).

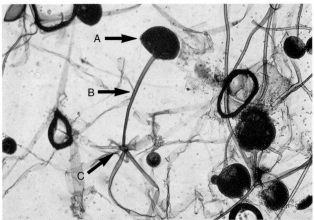

Figure 50-43   *Rhizopus* spp. showing sporangium **(A)** on long sporangiophore **(B)** arising from pauciseptate hyphae. Note presence of characteristic rhizoids **(C)** at the base of the sporangiophore (250×).

during maturation within the sporangium, a saclike structure produced at the tip of a long stalk **(sporangiophore).** Sporangiospores, spores produced within the sporangium, are produced and released by the rupture of the sporangial wall (Figure 50-42). Rarely, some isolates may produce **zygospores,** rough-walled spores produced by the union of two matching types of a zygomycete; this is an example of sexual reproduction.

IDENTIFICATION OF MOLDS

HYALINE, PAUCISEPTATE MOLDS: THE ZYGOMYCETES

Genera and Species to Be Considered—Zygomycetes (*Rhizopus, Mucor, Absidia*)

GENERAL CHARACTERISTICS

The Zygomycetes characteristically produce large, ribbonlike hyphae that are irregular in diameter and contain occasional septa. The septa may not be apparent in some preparations, which has resulted in the characterization of this group as aseptate. The specific identification of these organisms is confirmed by observing the characteristic saclike fruiting structures **(sporangia),** which produce internally spherical, yellow or brown spores **(sporangiospores)** (Figure 50-43). Each sporangium is formed at the tip of a supporting structure **(sporangiophore).** During maturation, the sporangium becomes fractured and sporangiospores are released into the environment. Sporangiophores are usually connected to one another by occasionally septate hyphae called **stolons,** which attach at contact points where

rootlike structures **(rhizoids)** may appear and anchor the organism to the agar surface. The identification of the three most common Zygomycetes—*Mucor, Rhizopus,* and *Absidia*—is, in part, based on the presence or absence of rhizoids and the position of the rhizoids in relation to the sporangiophores.

EPIDEMIOLOGY AND PATHOGENESIS

Although the Zygomycetes (*Rhizopus, Mucor, Absidia, Syncephalastrum, Cunninghamella,* and others) are a less common cause of infection compared with the aspergilli, they are an important cause of morbidity and mortality in immunocompromised patients, particularly patients with diabetes mellitus. The organisms involved have a worldwide distribution and are commonly found on decaying vegetable matter or old bread (they are a common bread mold), or in soil. The infection is generally acquired by inhalation of spores followed by subsequent development of infection. Once established, infection is rapidly progressive, particularly in patients with diabetes mellitus who have infections that involve the sinuses.

SPECTRUM OF DISEASE

Immunocompromised patients, particularly those who have uncontrolled diabetes mellitus and those who are receiving prolonged corticosteroid, antibiotic, or cytotoxic therapy, are at greatest risk.[71] The organisms involved in causing **zygomycosis** (infection caused by a zygomycete) have a marked propensity for vascular invasion and rapidly produce thrombosis and necrosis of tissue. One of the most common presentations is the rhinocerebral form wherein the nasal mucosa, palate, sinuses, orbit, face, and brain are involved; each shows massive necrosis with vascular invasion and infarction. Perineural invasion also occurs in zygomycoses and is a potential means of retroorbital spread (i.e., invasion into the brain). Other types of infection involve the lungs and gastrointestinal tract; some patients develop disseminated infection. The Zygomycetes have also caused skin infections of patients with severe burns and in infections of subcutaneous tissue of patients who have undergone surgery.

LABORATORY DIAGNOSIS

Specimen Collection and Transport
See General Considerations for the Laboratory Diagnosis of Fungal Infections.

Specimen Processing
See General Considerations for the Laboratory Diagnosis of Fungal Infections.

Direct Detection Methods

Stains.   The rapid diagnosis of zygomycosis may be made by examination of tissue specimens or exudate from infected lesions in a calcofluor white or potassium hydroxide preparation. Branching, broad-diameter, predominantly nonseptate hyphae are observed (see Figure 50-15). It is important that the laboratory notify the clinician of these findings because Zygomycetes grow rapidly and vascular invasion occurs at a rapid rate.

Antigen-Protein.   Antigen-protein–based assays are not used for the diagnosis of zygomycosis.

Nucleic Acid Amplification   Nucleic acid testing is not routinely used for the diagnosis of zygomycosis. These assays may be available in research settings.

Cultivation.   The colonial morphologic features of the Zygomycetes allow one to immediately suspect organisms belonging to this group. Colonies characteristically produce a fluffy, white to gray or brown hyphal growth that diffusely covers the surface of the agar within 24 to 96 hours (Figure 50-44). The hyphae appear to be coarse and fills the entire culture dish or tube rapidly with loose, grayish hyphae dotted with brown or black sporangia. It is impossible to distinguish between the different genera and species of Zygomycetes based on their colonial morphologic features, because most are identical in appearance.

Approach to Identification
Rhizopus has unbranched sporangiophores with rhizoids that appear at the point where the stolon arises, at the

Figure 50-44   *Rhizopus* colony.

Figure 50-45   *Mucor* spp., showing numerous sporangia without rhizoids (430×).

base of the sporangiophore (see Figure 50-43). In contrast, *Mucor* is characterized by sporangiophores that are singularly produced or branched and have at their tip a round sporangium filled with sporangiospores. It does not have rhizoids or stolons; this distinguishes it from the other genera of Zygomycetes (Figure 50-45). *Absidia,* an uncommon isolate in the clinical laboratory, is characterized by the presence of rhizoids that originate between sporangiophores (Figure 50-46). The sporangia of *Absidia* are pyriform and have a funnel-shaped area **(apophysis)** at the junction of the sporangium and the sporangiophore. Usually a septum is formed in the sporangiophore just below the sporangium. Other genera of Zygomycetes that are encountered much less frequently in the clinical laboratory include *Rhizomucor, Saksenaea, Cunninghamella, Apophysomyces,*[87] *Conidiobolus,* and *Basidiobolus.*[111]

Serodiagnosis
Serology is not useful for the diagnosis of zygomycosis.

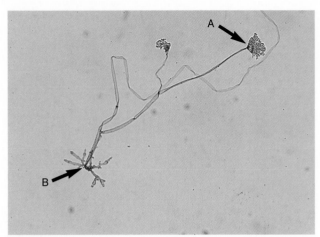

Figure 50-46   *Absidia* spp. **(A)** showing sporangia on long sporangiophores arising from pauciseptate hyphae **(B)**. Note that rhizoids are produced between sporangiophores and not at their bases (250×).

HYALINE, SEPTATE, MONOMORPHIC MOLDS: THE DERMATOPHYTES

Genera and Species to Be Considered—*Trichophyton, Microsporum,* and *Epidermophyton*

GENERAL CHARACTERISTICS

The dermatophytes produce infections that involve the superficial areas of the body, including the hair, skin, and nails (Dermatomycoses). The genera *Trichophyton, Microsporum,* and *Epidermophyton* are the principle etiologic agents of the dermatomycoses.

EPIDEMIOLOGY AND PATHOGENESIS

The dermatophytes break down and utilize keratin as a source of nitrogen but are usually incapable of penetrating the subcutaneous tissue, unless the host is immunocompromised; even in this situation, however, penetration into the subcutis is rare. The genus *Trichophyton* is capable of invading the hair, skin, and nails, whereas the genus *Microsporum* involves only the hair and skin; the genus *Epidermophyton* involves the skin and nails. Common species of dermatophytes recovered from clinical specimens, in order of frequency, include *Trichophyton rubrum, Trichophyton mentagrophytes, Epidermophyton floccosum, Trichophyton tonsurans, Microsporum canis,* and *Trichophyton verrucosum*.[151] The frequency of recovery of these species may differ by geographic local. Other geographically limited species are described elsewhere.

SPECTRUM OF DISEASE

Cutaneous mycoses are perhaps the most common fungal infections of humans and are usually referred to as **tinea** (Latin for "worm" or "ringworm"). The gross appearance of the lesion is that of an outer ring of the active, progressing infection, with central healing within the ring. These infections may be characterized by another Latin noun to designate the area of the body involved. Examples include *Tinea corporis,* which is ringworm of the body; *Tinea cruris,* which is ringworm of the groin (i.e., "jock itch"); *Tinea capitis,* which is ringworm of the scalp and hair; *Tinea barbae* which is ringworm of the beard; and *Tinea unguium,* which is ringworm of the nail.

Trichophyton

Members of the genus *Trichophyton* are widely distributed and are the most important and common causes of infections of the feet and nails; they may be responsible for *tinea corporis, tinea capitis, tinea unguium,* and *tinea barbae.* They are most commonly seen in adult infections, which vary in their clinical manifestations. Most cosmopolitan species are **anthropophilic,** or "human-loving"; few are **zoophilic,** primarily infecting animals.

Generally, hairs infected with members of the genus *Trichophyton* do not fluoresce under the ultraviolet (UV) light of a Wood's lamp; the demonstration of fungal elements inside, surrounding, and penetrating the hair shaft, or within a skin scraping, is necessary to diagnose a dermatophyte infection by direct examination. The recovery and identification of the causative organism are necessary for confirmation.

LABORATORY DIAGNOSIS

Specimen Collection and Transport
See General Considerations for the Laboratory Diagnosis of Fungal Infections.

Specimen Processing
See General Considerations for the Laboratory Diagnosis of Fungal Infections.

Direct Detection Methods

Stains. Calcofluor white or potassium hydroxide preparations will reveal the presence of hyaline septate hyphae and/or arthroconidia (see Figures 50-11 and 50-16). The direct microscopic examination of infected hairs may reveal the hair shaft to be filled with masses of large (4 to 7 μm) arthroconidia in chains, characteristic of an **endothrix** type of invasion. In other instances, the hair will show external masses of spores that ensheath the hair shaft; this is characteristic of the

ectothrix type of hair invasion. Hairs infected with *Trichophyton schoenleinii* reveal hyphae and airspaces within the shaft.

Antigen-Protein. Antigen-protein—based assays are not useful for the detection or identification of dermatophytes.

Nucleic Acid Amplification. Nucleic acid amplification assays for dermatophytes are not routine; these are available in research settings.

Cultivation. Because the dermatophytes generally present a similar microscopic appearance within infected hair, skin, or nails, the final identification is typically made by culture. A summary of the colonial and microscopic morphologic features of these fungi is presented in Table 50-12. Figure 50-47 presents an identification schema that will be of use to the clinical laboratory for the identification of commonly encountered dermatophytes. Figure 50-47 begins with the microscopic features of the dermatophytes as they might be observed in an initial examination of the culture. In many instances, the primary recovery medium fails to function as well as a sporulation medium. It is commonly necessary to subculture the initial growth onto cornmeal agar or potato dextrose agar so that sporulation will occur.

Approach to Identification

Trichophyton. Microscopically, the genus *Trichophyton* is characterized by smooth, club-shaped, thin-walled macroconidia with 3 to 8 septa ranging in size from 4 × 8 μm to 8 × 15 μm. The macroconidia are borne singly at the terminal ends of hyphae or on short conidiophores; the microconidia predominate and are usually spherical, pyriform (teardrop-shaped), or clavate (club-shaped), and 2 to 4 μm (Figure 50-48). Only the common species of *Trichophyton* are described here.

Trichophyton rubrum and *T. mentagrophytes* are the most common species recovered in the clinical laboratory. *Trichophyton rubrum* is a slow-growing organism that produces a flat or heaped-up colony that is generally white to reddish with a cottony or velvety surface. The characteristic cherry-red color is best observed on the reverse side of the colony; however, this is produced only after 3 to 4 weeks of incubation. Occasional strains may lack the deep red pigmentation on primary isolation. Colonies may be of two types: fluffy or granular. Microconidia are uncommon in most of the fluffy strains but are more common in the granular strains and occur as small, teardrop-shaped conidia often borne laterally along the sides of the hyphae (see Figure 50-48). Macroconidia are seen less commonly, although they are

sometimes found in the granular strains, where they appear as thin-walled, smooth-walled, multicelled, cigar-shaped conidia with three to eight septa. *Trichophyton rubrum* has no specific nutritional requirements. It does not perforate hair in vitro (Procedure 50-5) or produce urease.

Trichophyton mentagrophytes produces two distinct colonial forms: the downy variety recovered from patients with *tinea pedis* and the granular variety recovered from lesions acquired by contact with animals. The rapidly growing colonies may appear as white, cottony or downy colonies to cream-colored or yellow colonies that are coarsely granular to powdery. Granular colonies may show evidence of red pigmentation. The reverse side of the colony is usually rose-brown, occasionally orange to deep red, and may be confused with *T. rubrum*. The white, downy colonies produce only a few spherical microconidia; the granular colonies sporulate freely, with numerous small, spherical microconidia produced in grapelike clusters and thin-walled, smooth-walled, cigar-shaped macroconidia measuring 6 × 20 μm to 8 × 50 μm, with two to five septa (Figure 50-49). Macroconidia characteristically exhibit a definite narrow attachment to their base. Spiral hyphae may be found in one third of the isolates recovered.

Trichophyton mentagrophytes produces urease within 2 to 3 days after inoculation onto Christensen's urea agar. In contrast to *T. rubrum, T. mentagrophytes* perforates hair (Figure 50-50). This latter criterion may be used when distinguishing between these two species is difficult.

Trichophyton tonsurans is responsible for an epidemic form of *tinea capitis* that occurs most commonly in children but also occurs occasionally in adults. It has displaced *M. audouinii* as a primary cause of *tinea capitis* in most of the United States. The fungus causes a low-grade superficial lesion of varying severity and produces circular, scaly patches of alopecia (loss of hair). The stubs of hair remain in the epidermis of the scalp after the brittle hairs have broken off and may give the typical "black dot" ringworm appearance. Because the infected hairs do not fluoresce under a Wood's lamp, a careful search for the embedded stub should be carried out by the physician with the use of a bright light.

Cultures of *T. tonsurans* develop slowly and are typically buff to brown, wrinkled and suedelike in appearance. The colony surface shows radial folds and often develops a craterlike depression in the center, with deep fissures. The reverse side of the colony is yellowish to reddish brown. Microscopically, numerous microconidia with flat bases are produced on the sides of hyphae. With age, the microconidia tend to become pleomorphic, are swollen to elongated, and are referred to as balloon forms (Figure 50-51). Chlamydoconidia are abundant in old cultures; swollen and fragmented

Table 50-12 Characteristics of Dermatophytes Commonly Recovered in the Clinical Laboratory

Dermatophyte	Colonial Morphology	Growth Rate	Microscopic Identification
Microsporum audouinii*	Downy white to salmon-pink colony: reverse tan to salmon-pink	2 wks	Sterile hyphae: terminal chlamydoconidia, favic chandeliers, and pectinate bodies; macroconidia rarely seen—bizarre-shaped if seen; microconidia rare or absent
Microsporum canis	Colony usually membranous with feathery periphery; center of colony white to buff over orange-yellow; lemon-yellow or yellow-orange apron and reverse	1 wk	Thick-walled, spindle-shaped, multiseptate, rough-walled macroconidia, some with a curved tip; microconidia rarely seen
Microsporum gypseum	Cinnamon-colored, powdery colony; reverse light tan	1 wk	Thick-walled, rough, elliptical, multiseptate macroconidia; microconidia few or absent
Epidermophyton floccosum	Center of colony tends to be folded and is khaki green; periphery is yellow; reverse yellowish brown with observable folds	1 wk	Macroconidia large, smooth-walled, multiseptate, clavate, and borne singly or in clusters of two or three; microconidia not formed by this species
Trichophyton mentagrophytes	Different colonial types; white granular and fluffy varieties; occasional light yellow periphery in younger cultures; reverse buff to reddish brown	7-10 days	Many round to globose microconidia most commonly borne in grapelike clusters or laterally along the hyphae; spiral hyphae in 30% of isolates; macroconidia are thin-walled, smooth, club-shaped, and multiseptate; numerous or rare depending upon strain
Trichophyton rubrum	Colonial types vary from white downy to pink granular; rugal folds are common; reverse yellow when colony is young; however, wine red color commonly develops with age	2 wks	Microconidia usually teardrop, most commonly borne along sides of the hyphae; macroconidia usually absent, but when present are smooth, thin-walled, and pencil-shaped
Trichophyton tonsurans	White, tan to yellow or rust, suedelike to powdery; wrinkled with heaped or sunken center; reverse yellow to tan to rust red	7-14 days	Microconidia are teardrop or club-shaped with flat bottoms; vary in size but usually larger than other dermatophytes; macroconidia rare and balloon forms found when present
Trichophyton schoenleinii*	Irregularly heaped, smooth white to cream colony with radiating grooves; reverse white	2-3 wks	Hyphae usually sterile; many antler-type hyphae seen (favic chandeliers)
Trichophyton violaceum*	Port wine to deep violet colony, may be heaped or flat with waxy-glabrous surface; pigment may be lost on subculture	2-3 wks	Branched, tortuous hyphae that are sterile; chlamydoconidia commonly aligned in chains
Trichophyton verrucosum	Glabrous to velvety white colonies; rare strains produce yellow-brown color; rugal folds with tendency to skin into agar surface	2-3 wks	Microconidia rare; large and teardrop when seen; macroconidia extremely rare, but form characteristic "rat-tail" types when seen; many chlamydoconidia seen in chains, particularly when colony is incubated at 37° C

From Koneman EW, Roberts GD: *Practical laboratory mycology*, ed 3, Baltimore, 1985, Williams & Wilkins.
*These organisms are uncommonly seen in the United States.

hyphal cells resembling arthroconidia may be seen. *Trichophyton tonsurans* grows poorly on media lacking enrichments (casein agar); however, growth is greatly enhanced by the presence of thiamine or inositol in casein agar.

Trichophyton verrucosum causes a variety of lesions in cattle and in humans; it is most often seen in farmers, who acquire their infection from cattle. The lesions are found chiefly on the beard, neck, wrist, and back of the hands; they are deep, pustular, and inflammatory. With pressure, short stubs of hair may be recovered from the purulent lesion. Direct examination of the hair shaft reveals sheaths of isolated chains of large (5 to 10 μm) spores surrounding the hair shaft **(ectothrix),** and hyphae within the hair **(endothrix).** Masses of these conidia may also be seen in exudate from the lesions.

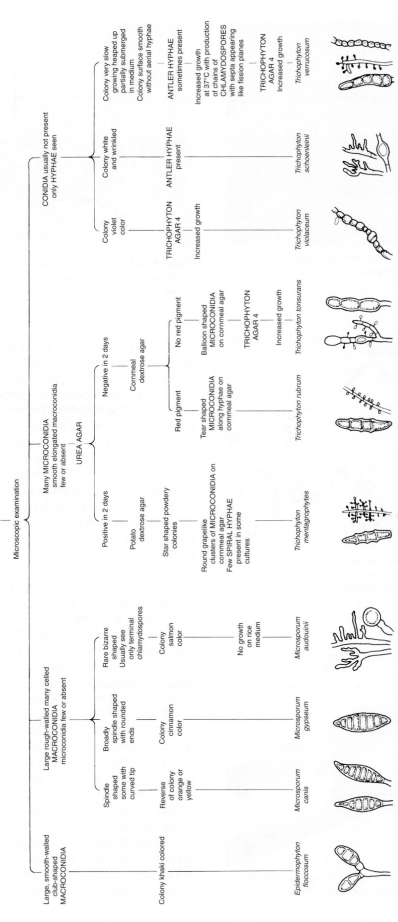

Figure 50–47   Dermatophyte identification schema. (From Koneman EW, Roberts GD: *Practical laboratory mycology*, ed 3, Baltimore, 1985, Williams & Wilkins.)

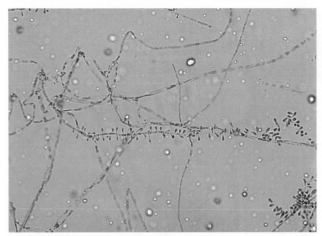

Figure 50-48 *Trichophyton rubrum* showing numerous pyriform microconidia borne singly on hyphae (750×).

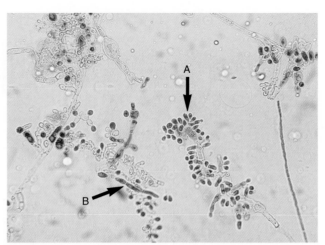

Figure 50-51 *Trichophyton tonsurans* showing numerous microconidia (**A**) that are borne singly or in clusters. A single macroconidium (**B**) (rare) is also present (600×).

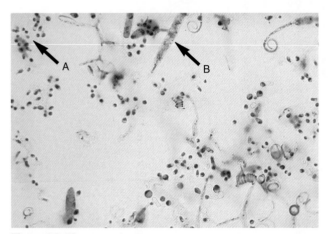

Figure 50-49 *Trichophyton mentagrophytes* showing numerous microconidia in grapelike clusters (**A**). There are also several thin-walled macroconidia present (**B**) 500×.

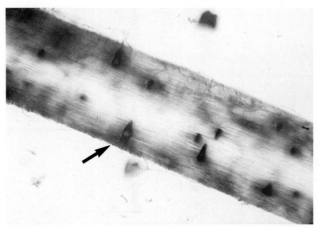

Figure 50-50 Hair perforation by *T. mentagrophytes.* Wedge-shaped areas *(arrow)* illustrate hair perforation (100×).

Trichophyton verrucosum grows slowly (14 to 30 days), and growth is enhanced at 35° to 37° C and also on media enriched with thiamine and inositol. *Trichophyton verrucosum* may be suspected when slowly growing colonies appear to embed themselves into the agar surface. Kane and Smitka[64] described a medium for the early detection and identification of *T. verrucosum.* The ingredients for this medium are 4% casein and 0.5% yeast extract. The organism is recognized by its early hydrolysis of casein and very slow growth rate. Chains of chlamydoconidia are formed regularly at 37° C. The early detection of hydrolysis, formation of characteristic chains of chlamydoconidia, and the restrictive slow growth rate of *T. verrucosum* differentiate it from *T. schoenleinii,* another slowly growing organism. Colonies are small, heaped, and folded, occasionally flat and disk-shaped. At first they are glabrous and waxy, with a short aerial mycelium. Colonies range from gray and waxlike to a bright ochre. The reverse of the colony is most often nonpigmented but may be yellow.

Microscopically, chlamydoconidia in chains and antler hyphae may be the only structures observed microscopically in cultures of *T. verrucosum* (see Figures 50-32 and 50-38). Chlamydoconidia may be abundant at 35° to 37° C. Microconidia may be produced by some cultures if the medium is enriched with yeast extract or a vitamin (Figure 50-52). Conidia, when present, are borne laterally from the hyphae and are large and clavate. Macroconidia are rarely formed, vary considerably in size and shape, and are referred to as "rat tail" in appearance.

Trichophyton schoenleinii causes a severe type of infection called *favus.* It is characterized by the formation of yellowish cup-shaped crusts or scutulae, considerable scarring of the scalp, and sometimes permanent alo-

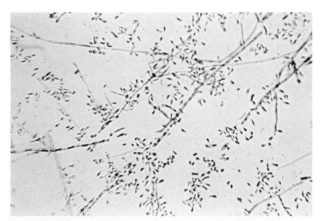

Figure 50-52   *Trichophyton verrucosum,* showing microconidia, which are rarely seen (500×).

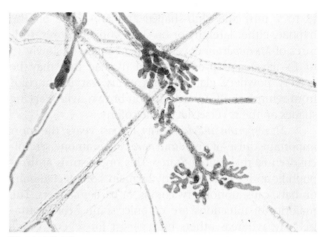

Figure 50-53   *Trichophyton schoenleinii,* showing swollen hyphal tips with lateral and terminal branching (favic chandeliers). Microconidia and macroconidia are absent (500×).

Figure 50-54   Large, rough-walled macroconidia of *M. canis* (430×).

pecia. Infections are common among members of the same family. A distinctive invasion of the infected hair, the favic type, is demonstrated by the presence of large, inverted cones of hyphae and arthroconidia at the base of the hair follicle and branching hyphae throughout the length of the hair shaft. Longitudinal tunnels or empty spaces appear in the hair shaft where the hyphae have disintegrated. In calcofluor white or potassium hydroxide preparations, these tunnels are readily filled with fluid; air bubbles may also be seen in these tunnels.

Trichophyton schoenleinii is a slowly growing organism (30 days or longer) and produces a white to light gray colony that has a waxy surface. Colonies have an irregular border that consists mostly of submerged hyphae and that tends to crack the agar. The surface of the colony is usually nonpigmented or tan, furrowed, and irregularly folded. The reverse side of the colony is usually tan or nonpigmented. Microscopically, conidia are not formed commonly. The hyphae tend to become knobby and club-shaped at the terminal ends, with the production of many short lateral and terminal branches (Figure 50-53). Chlamydoconidia are generally numerous. All strains of *T. schoenleinii* may be grown in a vitamin-free medium and grow equally well at room temperature or at 35° to 37° C.

Trichophyton violaceum produces an infection of the scalp and body and is seen primarily in persons living in the Mediterranean region, the Middle and Far East, and Africa. Hair invasion is of the endothrix type; the typical "black dot" type of *tinea capitis* is observed clinically. Direct microscopic examination of the calcofluor white or potassium hydroxide preparation of the nonfluorescing hairs shows dark, thick hairs filled with masses of arthroconidia arranged in chains, similar to those seen in *T. tonsurans* infections.

Colonies of *T. violaceum* are very slow growing, beginning as cone-shaped, cream-colored, glabrous colonies. Later, these become heaped up, verrucous

(warty), violet to purple, and waxy in consistency. Colonies may often be described as being "port wine" in color. The reverse side of the colony is purple or nonpigmented. Older cultures may develop a velvety area of mycelium and sometimes lose their pigmentation. Microscopically, microconidia and macroconidia are generally not present; only sterile, distorted hyphae and chlamydoconidia are found. In some instances, however, swollen hyphae that contain cytoplasmic granules may be seen. The growth of *T. violaceum* is enhanced on media containing thiamine.

Microsporum. The genus *Microsporum* is immediately recognized by the presence of large (8 to 15 × 35 to 150 µm), spindle-shaped, rough-walled macroconidia with thick (up to 4 µm) walls that contain 4 or more septa (Figure 50-54). The exception is *Microsporum nanum*, which characteristically produces macroconidia having two cells. Microconidia, when present, are small

(3 to 7 µm) and club-shaped and are borne on the hyphae, either laterally or on short conidiophores. Cultures of *Microsporum* develop either rapidly or slowly (5 to 14 days) and produce aerial hyphae that may be velvety, powdery, glabrous, or cottony, varying in color from whitish, buff, to a cinnamon brown, with varying shades on the reverse side of the colony.

Microsporum audouinii was, in past years, the most important cause of epidemic *tinea capitis* among schoolchildren in the United States. This organism is anthropophilic and is spread directly by means of infected hairs on hats, caps, upholstery, combs, or barber clippers. The majority of infections are chronic; some heal spontaneously, whereas others may persist for several years. Infected hair shafts fluoresce yellow-green using a Wood's lamp. Colonies of *M. audouinii* generally grow more slowly than other members of the genus *Microsporum* (10 to 21 days), and they produce a velvety aerial mycelium that is colorless to light gray to tan. The reverse side often appears salmon-pink to reddish brown. Colonies of *M. audouinii* do not usually sporulate in culture. The addition of yeast extract may stimulate growth and the production of macroconidia in some instances. Most commonly, atypical vegetative forms such as terminal chlamydoconidia and antler and racquet hyphae are the only clues to the identification of this organism. It is common to identify *M. audouinii* by exclusion of all the other dermatophytes as a cause of infection.

Microsporum canis is primarily a pathogen of animals (zoophilic); it is the most common cause of ringworm infection in dogs and cats in the United States. Children and adults acquire the disease through contact with infected animals, particularly puppies and kittens, although human-to-human transfer has been reported. Hairs infected with *M. canis* fluoresce a bright yellow-green using a Wood's lamp, which is a useful tool for screening pets as possible sources of human infection. Direct examination of a calcofluor white or potassium hydroxide preparation of infected hairs reveals small spores (2 to 3 µm) outside the hair. Culture must be performed to provide the specific identification.

Colonies of *M. canis* grow rapidly, are granular or fluffy with a feathery border, are white to buff, and characteristically have a lemon-yellow or yellow-orange fringe at the periphery. On aging, the colony becomes dense and cottony and a deeper brownish-yellow or orange and frequently shows an area of heavy growth in the center. The reverse side of the colony is bright yellow, becoming orange or reddish-brown with age. Rarely, strains are recovered that show no reverse side pigment. Microscopically, *M. canis* shows an abundance of large (15 to 20 × 60 to 125 µm), spindle-shaped, multisegmented (four to eight) macroconidia with curved ends (Figure 50-55). These are thick-walled

Figure 50-55 *Microsporum canis,* showing several spindle-shaped, thick-walled, multicelled macroconidia (500×).

with spiny **(echinulate)** projections on their surfaces. Microconidia are usually few in number; however, large numbers may occasionally be seen.

Microsporum gypseum, a free-living organism of the soil **(geophilic)** that only rarely causes human or animal infection, may be occasionally seen in the clinical laboratory. Infected hairs generally do not fluoresce using a Wood's lamp. However, microscopic examination of the infected hairs shows them to be irregularly covered with clusters of spores (5 to 8 µm), some in chains. These arthroconidia of the ectothrix type are considerably larger than those of other *Microsporum* species.

Microsporum gypseum grows rapidly as a flat, irregularly fringed colony with a coarse, powdery surface that appears to be buff or cinnamon color. The underside of the colony is conspicuously orange to brownish. Microscopically, macroconidia are seen in large numbers and are characteristically large, ellipsoidal, have rounded ends, and are multisegmented (three to nine) with echinulated surfaces (Figure 50-56). Although they are spindle-shaped, these macroconidia are not as pointed at the distal ends as those of *M. canis.* The appearance of the colonial and microscopic morphologic features is sufficient to make the distinction between *M. gypseum* and *M. canis.*

Epidermophyton. *Epidermophyton floccosum,* the only member of the genus *Epidermophyton,* is a common cause of *tinea cruris* and *tinea pedis.* This organism is susceptible to cold temperatures and for this reason it is recommended that specimens submitted for dermatophyte culture not be refrigerated before culture and that cultures not be stored at 4° C. In direct examination of skin scrapings using the calcofluor white or potassium hydroxide preparation, the fungus is seen as fine branching hyphae. *Epidermophyton floccosum* grows slow-

Figure 50-56   *Microsporum gypseum* showing ellipsoidal, multicelled macroconidia (750×).

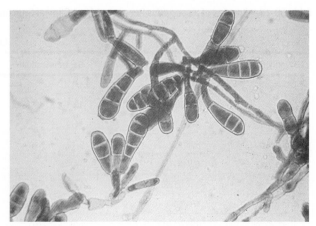

Figure 50-57   *Epidermophyton floccosum* showing numerous smooth, multiseptate, thin-walled macroconidia that appear club-shaped (1000×).

ly, and growth appears as an olive-green to khaki color, with the periphery surrounded by a dull orange-brown color. After several weeks, colonies develop a cottony white aerial mycelium that completely overgrows the colony and is sterile, and remains so, even after subculture. Microscopically, numerous smooth, thin-walled, club-shaped, multiseptate (2 to 4 μm) macroconidia are seen (Figure 50-57). They are rounded at the tip and are borne singly on a conidiophore or in groups of two or three. Microconidia are absent, spiral hyphae are rare, and chlamydoconidia are usually numerous. The absence of microconidia is useful in differentiating this organism from *Trichophyton* species, whereas the morphology of the macroconidia (smooth and thin-walled) is useful in differentiating it from *Microsporum* species.

Serodiagnosis
Serology is not useful for the diagnosis of disease caused by dermatophytes.

HYALINE, SEPTATE, MONOMORPHIC MOLDS: THE OPPORTUNISTIC MYCOSES

Genera and Species to Be Considered—*Aspergillus, Fusarium, Geotrichum, Acremonium, Penicillium, Paecilomyces,* and *Scopulariopsis*

GENERAL CHARACTERISTICS
The tissue invasive opportunistic mycoses are a group of fungal infections that occur almost exclusively in immunocompromised patients. The type of patient who acquires an opportunistic fungal infection is one who is compromised by some underlying disease process such as lymphoma, leukemia, diabetes mellitus, or another

defect of the immune system. Many patients, particularly those who undergo some type of transplantation, are often placed on treatment with corticosteroids, cytotoxic drugs, or other immunosuppressive agents to control rejection of the transplanted organ. Many fungi previously thought to be nonpathogenic are now recognized as etiologic agents of opportunistic fungal infections. Because most of the organisms known to cause infection in this group of patients are commonly encountered in the clinical laboratory as **saprobes** (saprophytic fungi), it may be impossible for the laboratorian to determine the clinical significance of these isolates recovered from clinical specimens. It is necessary, therefore, that the laboratory identify and report completely the presence of all fungi recovered, because each is a potential pathogen. Many of the organisms associated with opportunistic infections are acquired during construction, demolition, or remodeling of buildings, or are hospital-acquired. Other information regarding the specific clinical aspects of the opportunistic fungal infections is discussed with each individual organism.

EPIDEMIOLOGY AND PATHOGENESIS

Aspergillus
Several species of aspergilli are among the most frequently encountered fungi in the clinical laboratory; any are potentially pathogenic in the immunocompromised host, but some species are more frequently associated with disease than others. The aspergilli are widespread in the environment, where they colonize grain, leaves, soil, and living plants. Conidia of the aspergilli are easily dispersed into the environment, and humans become infected by inhaling them. It may be difficult to assess the significance of members of the genus *Aspergillus* in a clinical specimen. They are found frequently in cultures of respiratory secretions, skin

Table 50-13   Species of *Aspergillus* Recovered from Clinical Specimens During a 10-Year Period at Mayo Clinic

Organism	Clinical Specimen Source				
	Respiratory Secretions	Gastrointestinal	Genitourinary	Skin, Subcutaneous Tissue	Blood, Bone, CNS, etc.
A. clavatus	*97/93*	1/1	—	1/1	—
A. flavus	1298/740	10/10	11/11	177/131	2/2
A. fumigatus	3247/2656	11/9	14/14	175/137	8/8
A. glaucus	503/307	1/1	—	8/8	1/1
A. nidulans	52/48	—	—	5/3	—
A. niger	1484/1376	18/18	17/17	151/124	11/11
A. terreus	164/146	—	—	23/21	3/3
A. versicolor	1237/1202	6/6	24/22	226/224	16/16
Other species of *Aspergillus*	3463/3418	18/14	32/32	319/314	16/16

CNS, Central nervous system.
Numerator, Number of cultures; *denominator*, number of patients.

scrapings, and other specimens. Table 50-13 presents the Mayo Clinic experience with the recovery of species of *Aspergillus* from clinical specimens. It has been reported that *Aspergillus* is significant in only 10% of cases[132]; however, this depends on the hospital setting and type of patient population seen. Table 50-13 also illustrates the diversity of aspergilli seen in clinical specimens.

PATHOGENESIS AND SPECTRUM OF DISEASE

Aspergillus

Aspergillus species are capable not only of causing disseminated infection, as is seen in immunocompromised patients, but also of causing a wide variety of other types of infections, including a pulmonary or sinus fungus ball, allergic bronchopulmonary aspergillosis, external otomycosis (a fungus ball of the external auditory canal), mycotic keratitis, **onychomycosis** (infection of nail and surrounding tissue), sinusitis, endocarditis, and central nervous system infection. Most often, immunocompromised patients acquire a primary pulmonary infection that becomes rapidly progressive and may disseminate to virtually any organ.

Fusarium and Other Hyaline Septate Opportunistic Molds

Infection caused by species of *Fusarium* and other hyaline septate monomorphic molds is becoming more common, particularly in immunocompromised patients.[42,96,140,145] These organisms are common environmental flora and have long been known to cause mycotic keratitis after traumatic implantation into the cornea. Disseminated fusariosis is commonly accompanied by fungemia,

which are detected by routine blood culture systems. This is in distinct contrast to the aspergilli, which are rarely recovered from blood culture, even in instances of endovascular infections. Necrotic skin lesions are common with disseminated fusariosis. Other types of infection caused by *Fusarium* include sinusitis, wound (burn) infection, allergic fungal sinusitis, and endophthalmitis. *Fusarium* spp. are commonly recovered from respiratory tract secretions, skin, and other specimens of patients having no evidence of infection. Interpretation of culture results that contain these rests with the clinician, and is often assisted by correlating with the histopathology. *Geotrichum* is an uncommon cause of infection but has been shown to cause wound infections and oral thrush; it is an opportunistic pathogen in the immunocompromised host.[56] *Acremonium* species are also recognized as important pathogens in immunocompromised hosts; these have been associated with disseminated infection, fungemia, subcutaneous lesions, and esophagitis.[40,48,120] *Penicillium* is one of the most common organisms recovered by the clinical laboratory. In North America, these are rarely associated with invasive fungal disease. They may, however, be a cause of allergic bronchopulmonary penicilliosis or chronic allergic sinusitis. One species, *Penicillium marneffei*, however, is an important and emerging pathogen in Southeast Asia and is discussed further in the section considering dimorphic pathogens.[122,158] Of the *Paecilomyces* species, *P. lilacinus* appears to be the most pathogenic species and has been associated with endophthalmitis, cutaneous infections, and arthritis.[19,50] *Paecilomyces variotti* has also been shown to be an important pathogen, causing endocarditis, fungemia, and invasive disease.[147]

A variety of other saprobic fungi not discussed here may be encountered in the clinical laboratory but are seen less commonly. Other references are recommended for further information regarding the identification of these organisms.[27,67,121,126]

LABORATORY DIAGNOSIS

Specimen Collection and Transport

See General Considerations for the Laboratory Diagnosis of Fungal Infections.

Specimen Processing

See General Considerations for the Laboratory Diagnosis of Fungal Infections.

Direct Detection Methods

Stains. Specimens submitted for direct microscopic examination that contain organisms in this group demonstrate septate hyphae that usually shows evidence of **dichotomous branching**, often of 45-degrees (see Figure 50-14). In addition, some hyphae may show the presence of rounded, thick-walled cells. Although often considered to represent an *Aspergillus* species, these cannot be reliably distinguished from hyphae of *Fusarium* spp., *Pseudallescheria boydii,* or other hyaline molds.

Antigen-Protein. Antigen-protein–based assays are beginning to be used to monitor patients who are at high risk for developing invasive fungal infections. One of these assays, the galactomannan assay, specifically targets antigens of *Aspergillus* species, because this is the most common cause of invasive fungal infections caused by the hyaline septate molds (i.e., hyalohyphomycosis).[107,153] Conversely, the beta-glucan assay is designed to detect antigens common to all clinically important fungi.[97,99,153] How these tests will be used in the future and how they will compare with nucleic acid amplification tests remain to be determined.

Nucleic Acid Amplification. Nucleic acid amplification assays are not commonly performed for the detection or identification of these fungi. However, a variety of both broad-range (i.e., those that detect all fungi) and species-specific assays have been developed and in specialized centers may be used for patient care.

Cultivation. Because aspergilli are recovered frequently, it is imperative that the organism be demonstrated in the direct microscopic examination of fresh clinical specimens and/or that it be recovered repeatedly from patients having a compatible clinical picture to ensure that the organism is clinically significant. Correlation with biopsy results is the best way of establishing the significance of an isolate. Most species of *Aspergillus* are susceptible to cycloheximide. Therefore, specimens submitted for its recovery or subcultures should be inoculated onto media that lacks this ingredient.

Aspergillus fumigatus is the most commonly recovered species from immunocompromised patients; moreover, it is the species most often seen in the clinical laboratory. In addition, *Aspergillus flavus* is sometimes recovered from immunocompromised patients and represents a frequent isolate in the clinical microbiology laboratory. The recovery of *A. fumigatus* or *A. flavus* from surveillance (nasal) cultures has been correlated with subsequent invasive aspergillosis.[141] The absence of a positive nasal culture does not preclude infection, however. *Aspergillus niger* is seen commonly in the clinical laboratory, but its association with clinical disease is somewhat limited; this organism is a cause of fungus ball and otitis externa. *Aspergillus terreus* is a significant cause of infection in immunocompromised patients, but its frequency of recovery is much lower than the previously mentioned species. It is, however, important to correctly identify this species, because it is innately resistant to ampicillin B.

Approach to Identification

Aspergillus. *Aspergillus fumigatus* is a rapidly growing mold (2 to 6 days) that produces a fluffy to granular, white to blue-green colony. Mature sporulating colonies most often exhibit the blue-green powdery appearance. Microscopically, *A. fumigatus* is characterized by the presence of septate hyphae and short or long conidiophores having a characteristic "foot cell" at their base. The tip of the conidiophore expands into a large, dome-shaped vesicle that has bottle-shaped phialides covering the upper half or two thirds of its surface. Long chains of small (2 to 3 μm in diameter), spherical, rough-walled, green conidia form a columnar mass on the vesicle (Figure 50-58). Cultures of *A. fumigatus* are thermotolerant and are able to withstand temperatures up to 45° C.

Aspergillus flavus is a somewhat more rapidly growing species (1 to 5 days) and produces a yellow-green colony. Microscopically, vesicles are globose and phialides are produced directly from the vesicle surface **(uniserate)** or from a primary row of cells termed metulae **(biserate)**. The phialides give rise to short chains of yellow-orange elliptical or spherical conidia that become roughened on the surface with age (Figure 50-59). The conidiophore of *A. flavus* is also rough.

Aspergillus niger produces darkly pigmented roughened spores, but the hyaline septate hyphae like other aspergilli (i.e., it is not dematiaceous). *Aspergillus niger* produces mature colonies within 2 to 6 days. Growth

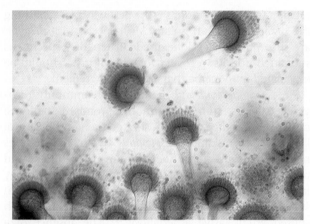

Figure 50-58 *Aspergillus fumigatus* conidiophore and conidia (400×).

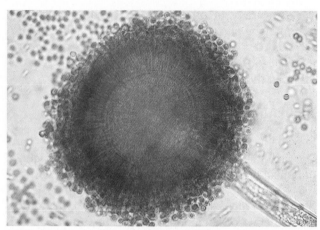

Figure 50-60 *Aspergillus niger* showing larger spherical vesicle that gives rise to metulae, phialides, and conidia (750×).

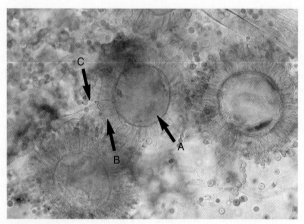

Figure 50-59 *Aspergillus flavus* showing spherical vesicles **(A)** that give rise to metulae **(B)** and phialides **(C)** that produce chains of conidia (750×).

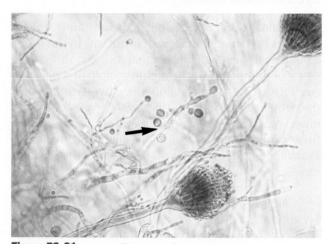

Figure 50-61 *Aspergillus terreus* showing typical head of aspergillus and aleurioconidia *(arrow)* found on submerged hyphae of this species (500×).

begins initially as a yellow colony that soon develops a black, dotted surface as conidia are produced. With age, the colony becomes jet black and powdery while the reverse remains buff or cream color; this occurs on any culture medium. Microscopically, *A. niger* exhibits septate hyphae, long conidiophores that support spherical vesicles that give rise to large metulae, and smaller phialides (biseriate), from which long chains of brown to black, rough-walled conidia are produced (Figure 50-60). The entire surface of the vesicle is involved in sporulation.

Aspergillus terreus is less commonly seen in the clinical laboratory; it produces colonies that are tan and resemble cinnamon. Vesicles are hemispherical, as seen microscopically, and phialides cover the entire surface and are produced from a primary row of metulae (biserate). Phialides produce globose to elliptical conidia arranged in chains. This species produces larger cells, **aleurioconidia,** which are found on submerged

hyphae (Figure 50-61). Refer to the work by Kennedy and Sigler[67] for further information regarding these and other species.

Fusarium. Colonies of *Fusarium* grow rapidly, within 2 to 5 days, and are fluffy to cottony and may be pink, purple, yellow, green, or other colors, depending on the species. Microscopically the hyphae are small and septate and give rise to phialides that produce either single-celled microconidia, usually borne in gelatinous heads similar to those seen in *Acremonium* (see Figure 50-39) or large multicelled macroconidia that are sickle- or boat-shaped and contain numerous septations (Figure 50-62). It is common to find numerous chlamydoconidia produced by some cultures of *Fusarium*. The most common medium used to induce sporulation is cornmeal agar. Keys for identification of species by *Fusarium* are based on growth on potato dextrose agar.

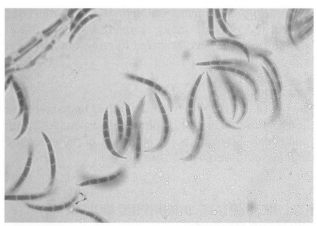

Figure 50-62   *Fusarium* spp. showing characteristic multicelled, sickle-shaped macroconidia (500×).

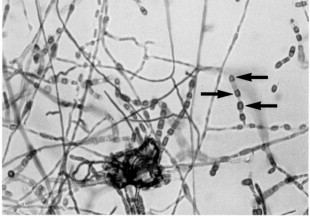

Figure 50-64   The mycelial form of *Coccidioides immitis* showing numerous thick-walled, rectangular or barrel-shaped *(arrows)* alternate arthroconidia (500×).

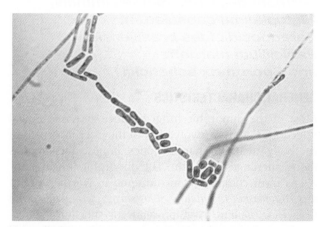

Figure 50-63   *Geotrichum candidum* showing numerous arthroconidia. Note that arthroconidia do not alternate with a clear (dysjunctor) cell as in the case of *C. immitis* (430×).

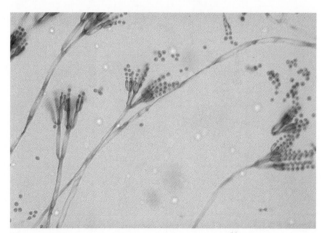

Figure 50-65   *Penicillium* spp. showing typical brushlike conidiophores (penicillus) (430×).

Geotrichum.   *Geotrichum* often initially appears as a white to cream yeastlike colony; some isolates may appear as white, powdery molds. Hyphae are septate and produce numerous rectangular to cylindrical to barrel-shaped arthroconidia (Figure 50-63). Arthrocondia do not alternate, but are contiguous as contrasted with *Coccidioides immitis* (Figure 50-64). Blastoconidia are not produced.

Acremonium.   Colonies of *Acremonium* are rapidly growing and also may appear yeastlike when initial growth is observed. Mature colonies become white to gray to rose or reddish orange in color. Microscopically, small septate hyphae that produce single, unbranched, tubelike phialides are observed. Phialides give rise to clusters of elliptical, single-celled conidia contained in a gelatinous cluster at the tip of the phialide (see Figure 50-39).

Penicillium.   Colonies of *Penicillium* are most commonly shades of green or blue-green, but pink, white, or other colors may be seen. The surface of the colonies may be velvety to powdery because of the presence of conidia. Microscopically, hyphae are hyaline and septate and produce brushlike conidiophores (i.e., a penicillus). Conidiophores produce metulae from which phialides producing chains of conidia arise (Figure 50-65). *Penicillium marneffei*, a particularly virulent species in their genus, is discussed in the section on hyaline, septate, dimorphic molds.

Paecilomyces.   Colonies of *Paecilomyces* spp. are often velvety, tan to olive brown, and somewhat powdery. Colonies of *P. lilacinus* exhibit shades of lavender to pink. Microscopically, *Paecilomyces* resemble a *Penicillium* in that a penicillus is formed. However, the phialides of *Paecilomyces* are long,

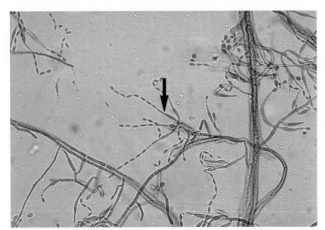

Figure 50-66   *Paecilomyces* spp. showing long, tapering, delicate phialides *(arrow).*

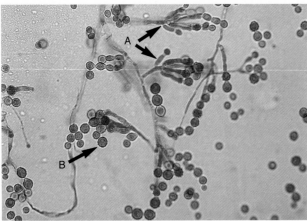

Figure 50-67   *Scopulariopsis* spp. showing a large penicilllus **(A)** with echinulate conidia **(B)** (430×).

delicate, and tapering (Figure 50-66) in contrast to the more blunted phialides of *Penicillium*. The penicillus produces numerous chains of small, oval conidia that are easily dislodged. Single phialides producing chains of conidia may also be present.

   Scopulariopsis. *Scopulariopsis* has been associated with onychomycosis, pulmonary infection, fungus ball, and more recently invasive fungal disease in the immunocompromised host. Colonies of *Scopulariopsis* initially appear white but later become light brown and powdery in appearance. Colonies often resemble those of *M. gypseum*. Microscopically, *Scopulariopsis* resembles a large *Penicillium* at first glance, in that a rudimentary penicillus is produced. Conidia are produced from annellophores and may also be produced singly directly from the hyphae. Conidia are large, have a flat base, and are rough-walled (Figure 50-67). The hyaline and septate species is *S.*

brevicaulis. *Scopulariopsis brumptii* is a dematiaceous species and is occasionally recovered in the clinical laboratory; it has been reported to cause brain abscess in a liver transplant patient.[104]

Serodiagnosis

The use of *Aspergillus* serology is limited to assistance in the diagnosis of bronchopulmonary aspergillosis and fungus ball. Serology is currently not of value for the diagnosis of disseminated aspergillosis.

HYALINE, SEPTATE, DIMORPHIC MOLDS: SYSTEMIC MYCOSES

Genera and Species to Be Considered: *Blastomyces dermatitidis, Coccidioides immitis, Histoplasma capsulatum, Paracoccidioides brasiliensis, Penicillium marneffei, and Sporothrix schenckii*

GENERAL CHARACTERISTICS

Most of the dimorphic fungi produce systemic fungal infections that may involve any of the internal organs of the body, including lymph nodes, bone, subcutaneous tissue, meninges, and skin. The dimorphic fungal pathogens most commonly encountered in North America are *Histoplasma capsulatum, Blastomyces dermatitidis,* and *Coccidioides immitis.* Asymptomatic or subclinical infection is common with *H. capsulatum* and *C. immits,* and may go unrecognized clinically. These patients may be detected only by serologic testing or following histopathologic review of tissues removed because of lesions found during a roentgenographic examination. Symptomatic infections may present signs of only a mild or more severe, but self-limited disease, with positive supportive evidence from cultural or immunologic findings. Patients with disseminated or progressive infection have severe symptoms, with spread of the initial disease, often from a pulmonary locus, to several, distant organs. However, some cases of disseminated infection may exhibit little in the way of signs or symptoms of disease for long periods, only to exacerbate later. Immunocompromised patients, particularly those infected with advanced HIV infection (i.e., AIDS), or those receiving long-term corticosteroid therapy, most often present with disseminated infection.[74]

   The classic terminology of systemic mycoses to refer to the dimorphic fungi is somewhat misleading, in that other fungi, including *C. neoformans* and species of *Candida* and *Aspergillus,* among others, may also cause disseminated systemic infections.

EPIDEMIOLOGY

Blastomyces dermatitidis

Blastomyces dermatitidis commonly produces a chronic infection that contains a mixture of suppurative and granulomatous inflammation. The disease **(blastomycosis)** is most commonly found in North America and extends southward from Canada to the Mississippi, Ohio, and Missouri River valleys, Mexico, and Central America. Some isolated cases have also been reported from Africa. The largest number of cases occur in the Mississippi, Ohio, and Missouri River valley regions. The exact ecologic niche for this organism in nature has not been determined; however, patients with blastomycosis often have a history of exposure to soil or wood. Several outbreaks have been reported and have been related to a common exposure. Blastomycosis is more common in men than women, and seems to be associated with outdoor occupations or activities. Disease also occurs in dogs.

Coccidioides immitis

Coccidioides immitis is found primarily in the desert portion of the southwestern United States as well as in the semiarid regions of Mexico and Central and South America. Although the geographic distribution of the organism is well defined, infection may be seen in any part of the world because of the ease of travel. The infection **(coccidioidomycosis)** is acquired by inhalation of the infective arthroconidia of *C. immitis*.

Histoplasma capsulatum

Outbreaks of **histoplasmosis** have been associated with activities that disperse aerosolized conidia or small hyphal fragments. Infection is acquired via inhalation of these infective structures from the environment. The severity of the disease is generally related directly to the inoculum size and the immunologic status of the host. Numerous cases of histoplasmosis have been reported in persons cleaning an old chicken house or barn that has been undisturbed for long periods or from working in or cleaning those areas that have served as roosting places for starlings and similar birds. Spelunkers (i.e., cave explorers) are commonly exposed to the organism when it is aerosolized from bat guano in caves. It is estimated that 500,000 persons are infected annually by *H. capsulatum*. Although it is perhaps one of the most common systemic fungal infections seen in the midwestern and southern parts of the United States, history of exposure is often impossible to document.

Paracoccidioides brasiliensis

Infection caused by this organism is most commonly found in South America, with the highest prevalence in Brazil, Venezuela, and Colombia. It also has been seen in many other areas, including Mexico, Central America, and Africa. Occasional imported cases are seen in the United States and Europe. The exact mechanism by which **paracoccidioidomycosis** is acquired is unclear; however, it is speculated that its origin is pulmonary and that it is acquired by inhalation of the organism from the environment. Because mucosal lesions are an integral part of the disease process, it is also speculated that the infection may be acquired through trauma to the oropharynx caused by vegetation that is commonly chewed by some residents of the endemic areas. The specific ecologic niche of the organism in nature is not known.

Penicillium marneffei

Penicillium marneffei is an emerging dimorphic pathogenic fungus endemic to Southeast Asia, particularly in the Guangxi Zhuang Autonomous Region of the People's Republic of China.[76] *Penicillium marneffei* has been associated with the bamboo rat *(Rhizomys pruinosus)*. In one study, the internal organs from 18 of 19 bamboo rats were culture positive for *P. marneffei*.[29] This fungus has also been associated with the Vietnamese bamboo rat *(Rhizomys sinensis)*.

Sporothrix schenckii

Sporothrix schenckii has a worldwide distribution, and its natural habitat is living or dead vegetation. Humans acquire the infection **(sporotrichosis)** through trauma (thorns, splinters), usually to the hand, arm, or leg. The infection is an occupational hazard for farmers, nursery workers, gardeners, florists, and miners. It is commonly known as "rose gardener's" disease. Rarely, pulmonary sporotrichosis will occur secondary to inhalation of spores.

PATHOGENESIS AND SPECTRUM OF DISEASE

Traditionally, the systemic mycoses have included only blastomycosis, coccidioidomycosis, histoplasmosis, and paracoccidioidomycosis. Of these, only *H. capsulatum* and *B. dermatitidis* are genetically related. Although these fungi are morphologically dissimilar, have one characteristic in common, dimorphism. Most of these organisms, except for *C. immitis*, are thermally dimorphic. The dimorphic fungi exist in nature as the mold form, which is distinct from the parasitic or invasive form, sometimes called the tissue form. Distinct morphologic differences may be observed with the dimorphic fungi both in vivo and in vitro as discussed later in this chapter.

Blastomyces dermatitidis

Blastomyces dermatitidis commonly produces an acute or chronic suppurative and granulomatous infection. Blastomycosis begins as a respiratory infection and is probably acquired by inhalation of the conidia or hyphal fragments of the organism. The infection may spread and involve the secondary sites of infection in the lungs, long bones, soft tissue, and skin.

Coccidioides immitis

Approximately 60% of infected patients with coccidioidomycosis are asymptomatic and have self-limited respiratory tract infections. The infection, however, may become disseminated, with extension to visceral organs, meninges, bone, skin, lymph nodes, and subcutaneous tissue. Fewer than 1% of people who acquire coccidioidomycosis ever become seriously ill; dissemination does, however, occur most frequently in persons of dark-skinned races. Pregnancy also appears to predispose women to disseminated infection. This infection has been known to occur in epidemic proportions. In 1992, an epidemic occurred in northern California, with more than 4000 cases seen in Kern County near Bakersfield. Persons who visit endemic areas and return to a distant location may present to their local physician; therefore the endemic mycoses should be considered in the differential diagnosis, if the patient has the appropriate travel history. All laboratories should be prepared to deal with the laboratory diagnosis of coccidioidomycosis. *Coccidioides immitis* is listed as a "select agent," which means proper documentation of the isolation and destruction of this organism is required by federal law, within 10 days of identification. Additionally, this documentation must be submitted to the Centers for Disease Control and Prevention. A special license is required to maintain stock cultures of *C. immitis*.

Histoplasma capsulatum

Histoplasma capsulatum most commonly produces a chronic, granulomatous infection (histoplasmosis) that is primary and begins in the lung and eventually invades the reticuloendothelial system. Approximately 95% of cases are asymptomatic and self-limited. Chronic pulmonary infections, however, do occur. Dissemination through the reticuloendothelial system occurs, with the primary sites of dissemination being the lymph nodes, liver, spleen, and bone marrow. Infections of the kidneys and meninges are also possible. Resolution of disseminated infection is the rule in immunocompetent hosts, but progressive disease is more common in immunocompromised patients (e.g., patients with AIDS). Ulcerative lesions of the upper respiratory tract may occur in both immunocompetent and immunocompromised hosts.

Paracoccidioides brasiliensis

Paracoccidioides brasiliensis produces a chronic granulomatous infection (paracoccidioidomycosis) that begins as a primary pulmonary infection. It is often asymptomatic and then disseminates to produce ulcerative lesions of the mucous membranes. Ulcerative lesions are commonly present in the nasal and oral mucosa, gingivae, and less commonly in the conjunctivae. Lesions occur most commonly on the face in association with oral mucous membrane infection. The lesions are characteristically ulcerative, with a **serpiginous** (snakelike) active border and a crusted surface. Lymph node involvement in the cervical area is common. Pulmonary infection is seen most often, and progressive chronic pulmonary infection is found in approximately 50% of cases. Dissemination to other anatomic sites, including the lymphatic system, spleen, intestines, liver, brain, meninges, and adrenal glands, occurs in some patients.

Penicillium marneffei

Penicillium marneffei is an emerging pathogen that most commonly infects immunosuppressed individuals. *Penicillium marneffei* causes either a focal cutaneous or mucocutaneous infection, or may produce a progressive disseminated and frequently fatal infection. Granulomatous, suppurative, and necrotizing inflammatory responses have been demonstrated.[28] The mode of transmission and the primary source in the environment is unknown, but the bamboo rat has been implicated.

Sporothrix schenckii

Sporothrix schenckii, also a dimorphic fungus, is most often associated with chronic subcutaneous infection. The primary lesion begins as a small, nonhealing ulcer, commonly of the index finger or the back of the hand. With time, the infection is characterized by the development of nodular lesions of the skin or subcutaneous tissues at the point of contact and later involves the lymphatic channels and lymph nodes draining the region. The subcutaneous nodules ulcerate to form an infection that becomes chronic. Only rarely is the disease disseminated. Pulmonary infection may be seen in patients that inhale the spores of *Sporothrix schenckii*.

LABORATORY DIAGNOSIS

Specimen Collection and Transport
See General Considerations for the Laboratory Diagnosis of Fungal Infections.

Specimen Processing
See General Considerations for the Laboratory Diagnosis of Fungal Infections.

Direct Detection Methods

Stains. The microscopic morphologic features of the tissue forms, or what has been termed the parasitic

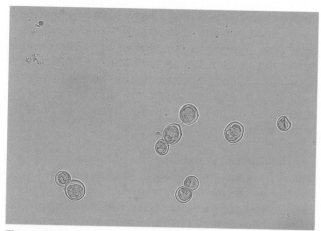

Figure 50-68 *Blastomyces dermatitidis,* yeast form showing thick-walled, oval to round, single-budding, yeastlike cells (500×).

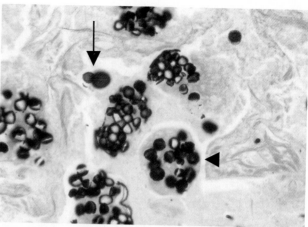

Figure 50-69 The tissue form of *Coccidioides immitis* (i.e., the spherule). The external wall of the spherule does not stain with the silver stain, whereas the internal endospores do stain *(arrowhead)*. Also note how the juxtaposed endospores, which have been released from a spherule that has burst, resemble a budding yeast *(arrow)*. GMS stain (400×).

forms, of the dimorphic fungi vary with the genus and are described for each below.

Blastomyces dermatitidis. The diagnosis of blastomycosis may easily be made when a clinical specimen is observed by direct microscopy. *Blastomyces dermatitidis* appears as large, spherical, thick-walled yeast cells 8 to 15 μm in diameter, usually with a single bud that is connected to the parent cell by a broad base (Figure 50-68; also see Figures 50-19 and 50-20). A smaller form (2-8 μm) may be rarely seen.

Coccidioides immitis. In direct microscopic examinations of sputum or other body fluids, *C. immitis* appears as a nonbudding, thick-walled spherule, 20 to 200 μm in diameter, containing either granular material or numerous small (2 to 5 μm in diameter), nonbudding endospores (Figure 50-69; also see Figures 50-21 and 50-25). The endospores are freed by rupture of the spherule wall; therefore, empty and collapsed "ghost" spherules may also be present. Small, immature spherules measuring 5 to 20 μm may be confused with *H. capsulatum* or *B. dermatitidis.* When two endospores or immature spherules are lying adjacent to one another, it may appear that budding yeast is present. When identification of *C. immitis* is questionable, a wet preparation of the clinical specimen may be made using sterile saline, and the edges of the coverglass may be sealed with petrolatum and incubated overnight. When spherules are present, multiple hyphal strands will be produced from the endospores.

Histoplasma capsulatum. The direct microscopic examination of respiratory tract specimens and other similar specimens often fails to reveal the presence of *H. capsulatum.* The organism, however, may be detected by an astute laboratorian when examining Wright- or

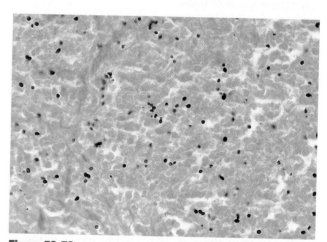

Figure 50-70 The small oval yeast cells, which are relatively uniform in size, are characteristic of *Histoplasma capsulatum* (2000×).

Giemsa-stained specimens of bone marrow and rarely in peripheral blood. *Histoplasma capsulatum* is found intracellularly within mononuclear cells as small, round to oval yeast cells 2 to 5 μm in diameter (Figure 50-70; also see Figure 50-16).

Paracoccidioides brasiliensis. Specimens submitted for direct microscopic examinations are important for the diagnosis of paracoccidioidomycosis. Large, round or oval, multiply budding yeast cells (8 to 40 μm in diameter) are usually recognized in sputum, mucosal biopsy, and other exudates. Characteristic multiply budding yeast forms resemble a "mariner's wheel" (Figure 50-71). The yeast cells surrounding the periphery of the parent cell range from 8 to 15 μm in

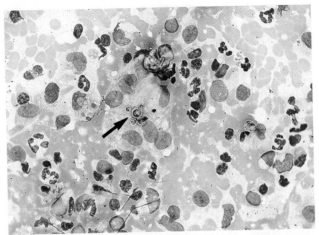

Figure 50-71 *Paracoccidioides brasiliensis* in a bone marrow aspirate shows a yeast cell with multiple buds *(arrow).*

diameter. Some cells may be as small as 2 to 5 μm but still exhibit multiple buds.

Penicillium marneffei. The direct examination of infected tissues and exudates that contain *Penicillium marneffei* produces small (2 to 6 μm), yeastlike cells that have internal crosswalls; no budding cells are produced (see Figure 50-4). Like *Histoplasma capsulatum*, *P. marneffei* may also be detected in peripheral blood smears in disseminated disease.

Sporothrix schenckii. Exudate aspirated from unopened subcutaneous nodules or from open draining lesions is often submitted for culture and direct microscopic examination. Direct examination of this material is usually of little diagnostic value because it is difficult to demonstrate the rare characteristic yeast forms. If identified, *S. schenckii* usually appears as small (2 to 5 μm), round to oval to cigar-shaped yeast cells (see Figure 50-22). If stained with the periodic acid-Schiff (PAS) method in histologic section, an amorphous pink material may be seen surrounding the yeast cells (see Figure 50-17).

Antigen-Protein. Immunodiffusion methods (the exoantigen test) may be used to identify isolates of these organisms based on precipitation bands of identity between specific antibodies and fungal antigen extracts. These assays, however, have been largely replaced by the more rapid nucleic acid hybridization reactions described below.

Nucleic Acid Amplification. Nucleic acid amplification assays are not routinely performed but are available in some reference laboratories and in research settings. Real-time or homogeneous, rapid cycle poly-

merase chain reaction (PCR) assays have been described for *H. capsulatum* and *C. immitis*.[11,84] These assays have been demonstrated to be suitable for isolate identification, but the potential to use them on DNA extracts from direct clinical specimens also exists. This latter application is exciting, given the slow growth of *H. capsulatum* and the severe disease that may be cause by both of these dimorphic pathogens.

Cultivation. Commonly, the dimorphic fungi are regarded as slow-growing organisms that require 7 to 21 days for visible growth to appear at 25° to 30° C. However, exceptions to this rule occur with some frequency. Occasionally cultures of *B. dermatitidis* and *H. capsulatum* are recovered in as short a time as 2 to 5 days when many organisms are present in the clinical specimen. In contrast, when small numbers of colonies of *B. dermatitidis* and *H. capsulatum* are present, sometimes 21 to 30 days of incubation are required before they are detected. *Coccidioides immitis* is consistently recovered within 3 to 5 days of incubation, but when many organisms are present, colonies may be detected within 48 hours. Cultures of *P. brasiliensis* are commonly recovered within 5 to 25 days, with a usual incubation period of 10 to 15 days. As one can see, the growth rate, if slow, might lead one to suspect the presence of a dimorphic fungus; however, considerable variation in the time for recovery exists. The exceptions to this slow growth are *C. immitis* and *P. marneffei*, which may be recovered within 3 to 5 days.

Textbooks present descriptions for the dimorphic fungi that the reader assumes are typical for each particular organism. As is true in other areas of microbiology, variation in the colonial morphologic features also occurs depending on the strain and the type of medium used. One must be aware of this variation and must not rely heavily on colonial morphologic features for the identification of members of this group of fungi.

The pigmentation of colonies is sometimes helpful but also varies widely; colonies of *B. dermatitidis* and *H. capsulatum* are described as being fluffy white, with a change in color to tan or buff with age. Some isolates initially appear darkly pigmented, with colors ranging from gray or dark brown to red.[62] On media containing blood enrichment, these organisms may appear heaped, wrinkled, glabrous, neutral in color, and yeastlike in appearance; often tufts of aerial hyphae project from the top of the colony. Some colonies may appear pink to red, possibly because of the adsorption of hemoglobin from the blood in the medium. *Coccidioides immitis* is described as being fluffy white, with scattered areas of hyphae that are adherent to the agar surface so as to give an overall "cobweb" appearance to the colony. However, numerous morphologic forms, including textures ranging from woolly to powdery and pigmentation ranging from pink-lavender or yellow to brown or buff, have been reported.

The definitive identification of a dimorphic fungus has traditionally been made by observing both the mold and tissue or parasitic forms of the organism. In general 25° to 30° C is the optimal temperature for the recovery and identification of the dimorphic fungi from clinical specimens. Temperature (35° to 37° C), certain nutritional factors, and stimulation of growth in tissue independent of temperature are among the factors necessary to initiate the transformation of the mold form to the tissue form.[47,136] Previously, the definitive identification of *B. dermatitidis* and *H. capsulatum* was made by the in vitro conversion of a mold form to the corresponding yeast form by in vitro conversion on a blood-enriched medium incubated at 35° to 37° C, and for *C. immitis* conversion to the spherule form by animal inoculation. The conversion of dimorphic molds to the yeast form (except for *C. immitis*) can be accomplished with some difficulty, as outlined in Procedure 50-6. Some laboratories use the exoantigen test (Procedure 50-6) to identify the dimorphic pathogens. However, this test requires an extended incubation before cultures may be identified.

Perhaps the most significant advance in clinical mycology in the last few decades was the development of specific nucleic acid probes for the identification of some of the dimorphic fungi (Procedure 50-8). DNA probes are commercially available from (Gen-Probe Inc., San Diego, Calif) that are complementary to species-specific ribosomal RNA. Fungal cells are heat-killed, disrupted by a lysing agent and sonication, and the nucleic acid is exposed to a species-specific DNA probe, which has been labeled with a chemiluminescent tag (acridinium ester). The labeled DNA probe combines with an ribosomal RNA of the organism to form a stable DNA:RNA hybrid. All unbound DNA probes are "quenched" and light generated from the DNA:RNA hybrids are measured in a luminometer. Total time for testing is less than 1 hour, and young colonies may be tested.[131]

Nucleic acid probe identification is sensitive, specific, and rapid. Colonies that are contaminated with bacteria or other fungi may be tested; however, results from colonies recovered on a blood-enriched media must be interpreted with caution because hemin may cause false-positive chemiluminescence. The major disadvantage of nucleic acid probe identification relates to the cost per test. It is recommended that nucleic acid probes be used whenever possible, however, to confirm the identification of an organism suspected of being *H. capsulatum, B. dermatitidis,* or *C. immitis.* A brief consideration of the colonial morphology of the dimorphic systemic pathogens follows.

Blastomyces dermatitidis. *Blastomyces dermatitidis* commonly requires 5 days to 4 weeks or longer for growth to be detected but may be detected in as short a time as 2 to 3 days. On enriched culture media, the mold form develops initially as a glabrous or waxy appearing colony is off-white to white. With age, the aerial hyphae often turn gray to brown. The more waxy, yeastlike appearance is typified on media enriched with blood. Tufts of hyphae often project upward from the colonies and have been referred to as the "prickly state" of the organism. Some isolates, however, appear fluffy on primary recovery and remain so throughout the incubation period.

Coccidioides immitis. Cultures of *C. immitis* represent a biohazard to laboratory workers, and strict safety precautions must be followed when examining cultures. Mature colonies may appear within 3 to 5 days of incubation and may be present on most media, including those used in bacteriology. Procop and colleagues reviewed 1270 fungal isolates recovered from 176,144 Isolator blood cultures and compared them with the recovery fungi on standard bacteriologic media; all the isolates of *C. immitis* present were recovered on standard bacteriologic media.[110] ***Laboratory workers are cautioned not to open cultures of fluffy white molds unless they are placed inside of a biological safety cabinet.*** Colonies of *C. immitis* often appear as delicate, cobweblike growth after 3 to 21 days of incubation. Some portions of the colony will exhibit aerial hyphae, whereas in others the hyphae will be adherent to the agar surface. Most isolates appear fluffy white; however, colonies of varying colors ranging from pink to yellow to purple and black have been reported.[62] On blood agar, some colonies exhibit a greenish discoloration, whereas others appear yeastlike, smooth, wrinkled, and tan.

Histoplasma capsulatum. *Histoplasma capsulatum* is easily cultured from clinical specimens; however, it may be overgrown by bacteria or rapidly growing molds. A procedure useful for the recovery of *H. capsulatum, B. dermatitidis,* and *C. immitis* from contaminated specimens (e.g., sputa) utilizes a yeast extract phosphate medium[14] and a drop of concentrated ammonium hydroxide (NH$_4$OH) placed on one side of the inoculated plate of medium.[123] In the past, it has been recommended that specimens not be kept at room temperature before culture, because *H. capsulatum* would not survive. The organism will survive transit in the mail for as long as 16 days.[53] It is, however, recommended that specimens be cultured as soon as possible to ensure the optimal recovery of *H. capsulatum* and other dimorphic fungi. The above-described method works well with specimens shipped via mail.

Histoplasma capsulatum is usually considered to be a slow-growing mold at 25° to 30° C and commonly requires 2 to 4 weeks or more for colonies to appear. The organism may, however, be recovered in 5 days or less if many yeast cells are present in the clinical specimen. Isolates of *H. capsulatum* have been reported to be recovered from blood cultures with the Isolator within a mean time of 8 days.[12] Textbooks describe the colonial morphology of *H. capsulatum* as being a white, fluffy mold that turns brown to buff with age. Some isolates ranging from gray to red have also been reported. The organism also may produces wrinkle, moist, heaped, yeastlike colonies that are soft and cream, tan, or pink. Tufts of hyphae often project upward from the colonies as described with *B. dermatitidis*. It is not possible to differentiate *H. capsulatum* and *B. dermatitidis* from each other using colonial morphologic features.

Paracoccidioides brasiliensis.

Colonies of *P. brasiliensis* grow very slowly (21 to 28 days) and are heaped, wrinkled, moist, and yeastlike. With age, colonies may become covered with a short aerial mycelium and turn tan to brown. The surface of colonies is often heaped with crater formations.

Penicillium marneffei.

At 25° C, *Penicillium marneffei* grows rapidly and produces blue-green to yellowish colonies on Sabouraud's agar. A soluble, red to maroon pigment that diffuses into the agar and is often best observed by viewing the reverse of the colony is suggestive of *P. marneffei*. Although the growth rate and colonial morphologic features may help one to recognize the possibility of the presence of a dimorphic fungus, they should be used in combination with the microscopic morphologic features used to make the identification. *Penicillium marneffei* cannot be definitively identified by morphologic features alone; thermal conversion studies or nucleic acid–based testing is needed to confirm the identification of this pathogen.

Sporothrix schenckii.

Colonies of *S. schenckii* grow rapidly (3 to 5 days) and are initially usually small, moist, and white to cream-colored. On further incubation, these become membranous, wrinkled, and coarsely matted, with the color becoming irregularly dark brown or black and the colony becoming leathery in consistency. It is not uncommon for the clinical microbiology laboratory to mistake a young culture of *S. schenckii* for that of a yeast until the microscopic features are observed.

Approach to Identification

Blastomyces dermatitidis.

Microscopically, hyphae of the mold form of *Blastomyces dermatitidis* are

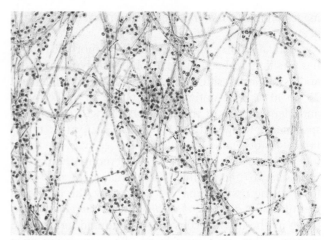

Figure 50-72   The mycelial form of *Blastomyces dermatitidis* shows oval conidia borne laterally on branching hyphae (1000×).

septate and delicate, and measure approximately 2 μm in diameter. Commonly, ropelike strands of hyphae are seen; however, these are found with most of the dimorphic fungi. The characteristic microscopic morphologic features are single, circular-to-pyriform conidia produced on short conidiophores that resemble lollipops (Figure 50-72); less commonly, the conidiophores may be elongated. The production of conidia in some isolates is minimal or absent, particularly on a medium containing blood enrichment.

When incubated at 37° C, colonies of the yeast form develop within 7 days and appear waxy and wrinkled, and cream to tan. Microscopically, large, thick-walled yeast cells (8 to 15 μm) with buds attached by a broad base are seen (see Figure 50-68). Some strains may produce yeast cells as small as 2 to 5 μm, which have been termed *microforms*. These small forms may resemble *C. neoformans* or *H. capsulatum*. Although, these microforms may be present, a thorough search should reveal more typical yeast forms. During conversion, swollen hyphal forms and immature cells with rudimentary buds will likely also be present. Because the conversion of *B. dermatitidis* is easily accomplished, this is feasible in the clinical laboratory; however, this is the most appropriate instance where mold-to-yeast conversion should be attempted. *Blastomyces dermatitidis* may also be identified by the presence of a specific (i.e., A) band in the exoantigen test[97] or by nucleic acid probe testing. In some instances, *H. capsulatum*, *P. boydii*, or *T. rubrum* might be confused microscopically with *B. dermatitidis*. The relatively slow growth rate of *B. dermatitidis* and careful examination of the microscopic morphologic features will usually differentiate it from these fungi.

Coccidioides immitis.

Microscopically, some cultures show small septate hyphae that often exhibit right-angle branches and racquet forms. With age, the hyphae form arthroconidia that are characteristically rectangular-to-barrel-shaped. The arthroconidia are larger than the hyphae from which they were produced and stain darkly with lactophenol cotton or aniline blue. The arthroconidia are separated from one another by clear or lighter staining nonviable cells (**disjunctor cells**). These types of conidia are referred to as **alternate arthroconidia** (see Figure 50-64). Arthroconidia have been reported to range from 1.5 to 7.5 μm in width and 1.5 to 30 μm in length, whereas most are 3 to 4.5 μm in width and 3 μm in length. Variation has been reported in the shape of arthroconidia and ranges from rounded to square or rectangular to curved; however, most are barrel-shaped. Even if alternate arthroconidia are observed microscopically, the definitive identification should be made using nucleic acid probe testing. If a culture is suspected of being **C. immitis,** *it should be sealed with tape to prevent chances of laboratory-acquired infection.* Because *C. immitis* is the most infectious of all the fungi, extreme caution should be used when handling cultures of this organism.

Safety precautions include the following:

1. If culture dishes are used, they should be handled only in a biological safety cabinet. Cultures should be sealed with tape if the specimen is suspected of containing *C. immitis.*
2. The use of cotton plug test tubes is discouraged, and screw-capped tubes should be used if culture tubes are preferred. All handling of cultures of *C. immitis* in screw-capped tubes should be performed inside a biological safety cabinet.
3. All microscopic preparations for examination should be prepared in a biological safety cabinet.
4. Cultures should be autoclaved as soon as the final identification of *C. immitis* is made.

Other, usually nonvirulent fungi that resemble *C. immitis* microscopically may be found in the environment. Some molds, such as *Malbranchea* sp., also produce alternate arthroconidia, but these tend to be more rectangular, and it is necessary to consider them when making the identification. *Geotrichum candidum* and *Trichosporon* spp. produce hyphae that disassociate into contiguous arthroconidia; these should not be confused with *C. immitis* (Figure 50-73 and see Figure 50-63). The colonial morphologic features of older cultures of these fungi may resemble *C. immitis*, but, as noted, the arthroconidia are not alternate. It is also important to remember that if confusion in identification does arise, or when occasional strains of *C. immitis* that fail to sporulate are encountered, identification by exoantigen or nucleic acid probe testing may be performed.

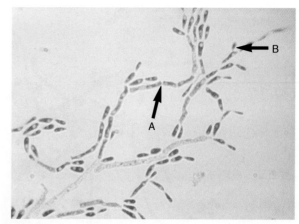

Figure 50-73   *Trichosporon* spp. produce arthroconidia (**A**) and an occasional blastoconidium (**B**).

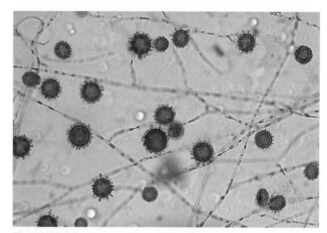

Figure 50-74   The mycelial form of *H. capsulatum* produces characteristic tuberculate macroconidia (1000×).

Histoplasma capsulatum. Microscopically, the hyphae of *H. capsulatum* are small (approximately 2 μm in diameter) and are often intertwined to form ropelike strands. Commonly, large (8 to 14 μm in diameter) spherical or pyriform, smooth-walled macroconidia are seen in young cultures. With age, the macroconidia become roughened or tuberculate and provide enough evidence to make a tentative identification (Figure 50-74). The macroconidia are produced either on short or long conidiospores. Some isolates produce round to pyriform, smooth microconidia (2 to 4 μm in diameter), in addition to the characteristic tuberculate macroconidia. Some isolates of *H. capsulatum* fail to sporulate despite numerous attempts to induce sporulation.

Conversion of the mold to the yeast form is usually difficult and is not recommended. Microscopically, a mixture of swollen hyphae and small budding yeast cells that are 2 to 5 μm should be observed. These are similar to the intracellular yeast cells seen in mononuclear cells

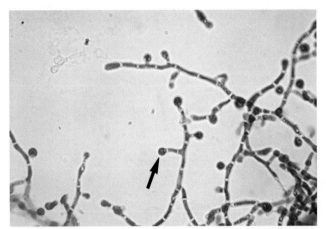

Figure 50-75 The mycelial form of *Paracoccidioides brasiliensis* is demonstrated here, which shows septate hyphae and pyriform conidia singly borne *(arrow)* (430×)

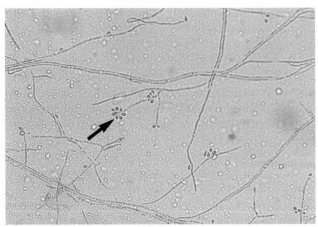

Figure 50-76 The mycelial form of *Sporothrix schenckii* shows pyriform-to-ovoid microconidia in a flowerette morphology at the tip of the conidiophore *(arrow)* (750×).

in infected tissue. The yeast form of *H. capsulatum* cannot be recognized unless the corresponding mold form is present on another culture or unless the yeast is converted directly to the mold form by incubation at 25° to 30° C after yeast cells have been observed. The exoantigen test can be used for identification, but nucleic acid probe testing is now recommended as the most user-friendly, definitive means of providing a rapid identification of this organism. *Sepedonium,* an environmental organism that grows on mushrooms, is always mentioned as being confused with *H. capsulatum,* because it produces similar tuberculate macroconidia. This organism, however, is almost never recovered from clinical specimens, does not have a yeast form, fails to produce characteristic bands in the exoantigen test with *H. capsulatum* antiserum, and does not react in nucleic acid probe tests.

Paracoccidioides brasiliensis. Microscopically, the mold form is similar to that seen with *B. dermatitidis.* Small hyphae (approximately 2 μm in diameter) are seen, along with numerous chlamydoconidia. Small (3 to 4 μm), delicate, globose or pyriform conidia may be seen arising from the sides of the hyphae or on very short conidiophores (Figure 50-75). Most often, cultures reveal only fine septate hyphae and numerous chlamydoconidia.

After temperature-based conversion on a blood-enriched medium, the colonial morphology of the yeast form is characterized by smooth, soft-wrinkled, yeastlike colonies that are cream to tan. Microscopically, the colonies are composed of yeast cells 10 to 40 μm in diameter that are surrounded by narrow-necked yeast cells around the periphery, as previously described (see Figure 50-71). If in vitro conversion to the yeast form is unsuccessful, the exoantigen test[127] (Procedure 50-6)

should be used to make the definitive identification of *P. brasiliensis.* Nucleic acid probe testing is not available for this organism. However, this organism is known to cross-react with the *Blastomyces dermatitidis.* This cross-reaction, in conjunction with microscopic and colonial morphology, epidemiologic data, and clinical features, may be used for the definitive identification of this fungus.

Penicillium marneffei. At 25° C, *Penicillium marneffei* grows rapidly and produces blue-green to yellowish colonies. A soluble red to maroon pigment, which diffuses into the agar, is highly suggestive of *P. marneffei.* At 37° C, the conversion of mycelium-to-the infective, yeastlike form occurs in approximately 2 weeks. Oval, yeastlike cells (2 to 6 μm) with septa are seen; abortive, extensively branched, and highly septate hyphae may also be present (see Figure 50-4).

Sporothrix schenckii. Microscopically, hyphae are delicate (approximately 2 μm in diameter), septate, and branching. Single-celled conidia that are 2 to 5 μm in diameter are borne in clusters from the tips of single conidiophores (flowerette arrangement). Each conidium is attached to the conidiophore by an individual, delicate, threadlike structure **(denticle)** that may require examination under oil immersion to be visible. As the culture ages, single-celled, thick-walled, black-pigmented conidia may also be produced along the sides of the hyphae, simulating the arrangement of microconidia produced by *T. rubrum* (sleeve arrangement) (Figure 50-76).

Because of similar morphologic features, saprophytic species of the genus *Sporotrichum* may be confused with *S. schenckii,* and it is necessary to distinguish between them. During incubation of a culture at 37° C,

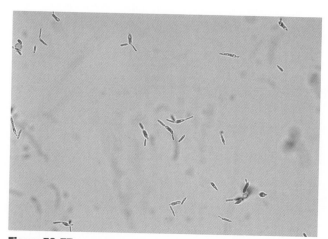

Figure 50-77   The yeast form of *S. schenckii* consists of cigar-shaped and oval budding cells (500×).

the colony of *S. schenckii* transforms to a soft, cream-colored to white, yeastlike colony. Microscopically, singly or multiply budding, spherical, oval, or elongate, cigar-shaped yeast cells are observed without difficulty (Figure 50-77). Conversion from the mold form to the yeast form is easily accomplished and usually occurs within 1 to 5 days after transfer of the culture to a medium containing blood enrichment; most isolates of *S. schenckii* are converted to the yeast form within 12 to 48 hours at 37° C. *Sporotrichum* does not produce a yeast form.

Table 50-14 presents a summary of the colonial and microscopic morphologic features of the dimorphic fungi in addition to other organisms previously discussed.

Serodiagnosis

Fungal serologies are rapid and useful tests that may aid in the diagnosis of systemic fungal infections caused by *B. dermatitidis, H. capsulatum,* and *C. immitis.* These tests have also been useful to study the epidemiology of these fungal infections, because even individuals with historically distant, asymptomatic, or subclinical infections will have often developed an antibody response to the infecting pathogen. Unfortunately, these tests require detailed preparation and technical expertise. False-negative reactions may occur if serologies are drawn in immunocompromised individuals who are unable to produce an antibody response. False-positive reactions may occur because of cross-reactivity with other fungi. For example, because the antigens of *H. capsulatum* are similar to those of *B. dermatitidis,* occasionally a patient with histoplasmosis will demonstrate positive reactions in serologic tests for *B. dermatitidis.*

Two assays, complement fixation and immunodiffusion, should be used together for the detection of antibodies directed toward *B. dermatitidis, H. capsulatum,* and *C. immitis.* In the complement fixation assay, titers of

1:8 to 1:16 are suggestive of active infection by *B. dermatitidis* and *H. capsulatum,* while titers of 1:32 or greater are indicative of active disease. Titers as low as 1:2 to 1:4 have been seen in patients with coccidioidomycosis. Titers greater than 1:16 are usually indicative of active disease. Bands of identity form in the immunodiffusion test between known antisera, known fungal antigen, and the antibodies present in the patient's serum. Specific bands of identity are used for the serologic detection of particular fungi, whereas nonspecific bands suggest the possibility of an infection by another fungal pathogen. One or two bands of identity, the H and M bands, may occur in patients with histoplasmosis. The presence of both bands is indicative of active infection. The presence of an M band may indicate early or chronic infection.

SEPTATE, DEMATIACEOUS MOLDS

Genera and Species to Be Considered—Agents of:

Superficial infections: Tinea nigra (*Hortaea werneckii*) and black piedra (*Piedraia hortae*)

Mycetoma: *Pseudallescheria boydii* and *Acremonium* (agents of white grain mycetoma); *Exophiala jeanselmei, Curvularia,* and *Madurella mycetomatis* (agents of black grain mycetoma).

Chromoblastomycosis: *Cladosporium (Cladophialophora), Phialophora,* and *Fonsecaea*

Phaeohyphomycosis: *Alternaria, Bipolaris* (and *Drechslera), Curvularia,* and *Exophiala jeanselmei, Exophiala dermatitidis,* and *Exserohilum*

GENERAL CHARACTERISTICS

The dematiaceous fungi are known agents of superficial and subcutaneous mycoses that involve the skin and subcutaneous tissues; less commonly, deeply invasive or disseminated disease may be caused by these fungi. These organisms are ubiquitous in nature and exist as saprophytes and plant pathogens. The etiologic agents are found in several unrelated fungal genera. Humans and animals serve as accidental hosts after traumatic inoculation of the organism into cutaneous and subcutaneous tissues. In the mycology laboratory, these are often initially separated by growth rate into the slow-growing dematiaceous molds, which may require 7 to 10 days to grow, and rapid-growing dematiaceous molds, which usually grow in less than 7 days. When nonsterile body sites are cultured, it is difficult or impossible for the laboratorian to determine the significance of these organisms. If colonies of common saprophytic molds occur near the edge of the plate and

Table 50-14 Summary of the Characteristic Features of Fungi Known to Be Common Causes of Selected Fungal Infection in Humans

Infection	Etiologic Agent	Cultural Characteristics at 30° C				Microscopic Morphologic Features			Confirmatory Tests for Identification
		Growth Rate (days)	Blood-Enriched Medium	Medium Lacking Blood Enrichment	Blood-Enriched Medium	Nonblood-Enriched Medium	Microscopic Recommended Screening Tests	Recommended Morphologic Features of Tissue Form	
Blastomycosis	*Blastomyces dermatitidis*	2-30	Colonies are cream to tan, soft, moist, wrinkled, waxy, flat to heaped, and yeastlike; "tufts" of hyphae often project upward from colonies	Colonies are white to cream to tan, some with drops of exudate present, fluffy to glabrous, and adherent to the agar surface	Hyphae 1-2 μm in diameter are present; some are aggregated in ropelike clusters; sporulation is rare	Hyphae 1-2 μm in diameter are present; single pyriform conidia are produced on short to long conidiophores; some cultures produce few conidia	Not available	8-15 μm, broad-based budding cells with double-contoured walls are seen; cytoplasmic granulation is often obvious	1. Specific nucleic acid probe 2. Broad-based budding cells may be seen after in vitro conversion on cottonseed agar 3. Exoantigen test
Histoplasmosis	*Histoplasma capsulatum*	3-45	Colonies are heaped, moist, wrinkled, yeastlike, soft, and cream, tan, or pink in color; "tufts" of hyphae often project upward from colonies	Colonies are white, cream, tan, or gray, fluffy to glabrous; some colonies appear yeastlike and adherent to the agar surface; many variations in colonial morphology occur	Hyphae 1-2 μm in diameter are present; some are aggregated in ropelike clusters; sporulation is rare	Young cultures usually have a predominance of smooth-walled macroconidia that become tuberculate with age; macroconidia may be pyriform or spherical; some isolates produce small pyriform microconidia in the presence or absence of macroconidia	Not available	2-5 μm, small, oval to spherical budding cells often seen inside of mononuclear cells	1. Specific nucleic acid probe 2. Exoantigen test

Table 50-14   Summary of the Characteristic Features of Fungi Known to Be Common Causes of Selected Fungal Infection in Humans—cont'd

Infection	Etiologic Agent	Growth Rate (days)	Cultural Characteristics at 30° C — Blood-Enriched Medium	Medium Lacking Blood Enrichment	Microscopic Morphologic Features — Blood-Enriched Medium	Nonblood-Enriched Medium	Microscopic Recommended Screening Tests	Recommended Morphologic Features of Tissue Form	Confirmatory Tests for Identification
Paracoccidioidomycosis	*Paracoccidioides brasiliensis*	21-28	Colonies are heaped, wrinkled, moist, and yeastlike; with age, colonies may become covered with short aerial mycelium and may turn brown		Hyphae 1-2 μm in diameter are present; some isolates produce conidia similar to those of *B. dermatitidis*; chlamydoconidia may be numerous, and multiple budding yeast cells 10-25 μm in diameter may be present		Not available	10-25 μm, multiple budding cells (buds 1-2 μm) resembling a mariner's wheel may be present; buds are attached to the parent cell by a narrow neck	Exoantigen test
Zygomycosis	*Rhizopus* spp., *Mucor* spp., and other Zygomycetes	1-3	Colonies are extremely fast growing, woolly, and gray to brown to gray-black		1. *Rhizopus* spp.—rhizoids are produced at the base of sporangiophore 2. *Mucor* spp.—no rhizoids are produced		Not available	Large, ribbonlike (10-30 μm), twisted, often distorted pieces of aseptate hyphae may be present; septa occasionally may be seen	Identification is based on characteristic morphologic features
Aspergillosis	*Aspergillus fumigatus* *Aspergillus flavus* *Aspergillus niger* *Aspergillus terreus* *Aspergillus* spp.	3-5	Colonies of *A. fumigatus* are usually blue-green to gray-green, whereas those of *A. flavus* and *A. niger* are yellow-green and black, respectively; colonies of *A. terreus* resemble powdered cinnamon; other species of *Aspergillus* exhibit a wide range of colors; blood-enriched media usually have little effect on the colonial morphologic features		1. *A. fumigatus*—uniserate heads with phialides covering the upper half to two thirds of the vesicle 2. *A. flavus*—uniserate or biserate or both with phialides covering the entire surface of a spherical vesicle 3. *A. niger*—biserate with phialides covering the entire surface of a spherical vesicle; conidia are black 4. *A. terreus*—biserate with phialides covering the entire surface of a hemispherical vesicle; aleurioconidia are formed on submerged hyphae		Not available	Septate hyphae 5-10 μm in diameter that exhibit dichotomous branching	Identification is based on microscopic morphologic features and colonial morphology; *A. fumigatus* can tolerate elevated temperatures ≥45° C

Table 50-14 Summary of the Characteristic Features of Fungi Known to Be Common Causes of Selected Fungal Infection in Humans—cont'd

			Cultural Characteristics at 30° C		Microscopic Morphologic Features			
Infection	Etiologic Agent	Growth Rate (days)	Blood-Enriched Medium	Medium Lacking Blood Enrichment	Blood-Enriched Medium	Nonblood-Enriched Medium	Microscopic Recommended Screening Tests	Recommended Morphologic Features of Tissue Form	Confirmatory Tests for Identification
Coccidioidomycosis	*Coccidioides immitis*	2-21	Colonies may be white and fluffy to greenish on blood-enriched media; some isolates are yeastlike, heaped, wrinkled, and membranous	Colonies usually are fluffy white but may be pigmented gray, orange, brown, or yellow; mycelium is adherent to the agar surface in some portions of the colony	Chains of alternate, barrel-shaped arthroconidia are characteristic; some arthroconidia may be elongated; hyphae are small and often arranged in ropelike strands, and racquet forms are seen in young cultures		Not available	Round spherules 30-60 μm in diameter containing 2-5 μm endospores are characteristic Empty spherules are commonly seen	1. Specific nucleic acid probe 2. Exoantigen test

are clearly away from the inoculum, they should be considered contaminants, unless there is additional evidence of infection.

EPIDEMIOLOGY AND PATHOGENESIS

Superficial Infections (Tinea Nigra and Black Piedra)

Tinea nigra is a superficial skin infection caused by *Hortaea werneckii*, manifested by blackish brown macular patches on the palm of the hand or sole of the foot. Lesions have been compared with silver nitrate staining of the skin. Black piedra is a fungal infection of the hair of the scalp and rarely of axillary and pubic hair, caused by the dematiaceous fungus *Piedraia hortae*. These diseases occur primarily in tropical areas of the world, with cases reported from Africa, Asia, and Latin America.

Mycetoma

A **mycetoma** is a chronic granulomatous infection that usually involves the lower extremities but may occur in any part of the body. The infection is characterized by swelling, purplish discoloration, tumorlike deformities of the subcutaneous tissue, and multiple sinus tracts that drain pus containing yellow, white, red, or black granules. The color of the granules is in part due to the type of infecting organism. The infection gradually progresses to involve the bone, muscle, or other contiguous tissue and ultimately requires amputation in most progressive cases. Rarely there is dissemination of the organism, but this is uncommon. Mycetoma is usually seen among persons who live in tropical and subtropical regions of the world, whose outdoor occupations and failure to wear protective clothing predispose them to trauma.

Two types of mycetomas are described. Actinomycotic (bacterial) mycetoma is caused by species of the aerobic actinomycetes, including *Nocardia, Actinomadura,* and *Streptomyces*. Eumycotic (fungal) mycetoma is caused by a heterogeneous group of fungi having septate hyphae.[100] Eumycotic mycetomas are, in turn, separated into white- or black-grain mycetomas, a distinction that is usually determined by the pigmentation of the hyphae of the infecting agent. Although some hyaline septate molds may cause mycetomas, this disease is covered in this section because many of the etiologic agents are dematiaceous fungi. Etiologic agents of eumycotic mycetoma to be discussed include *Pseudallescheria boydii* and *Acremonium* spp. as causative agents of white grain mycetomas, and *Exophiala jeanselmei, Curvularia,* and *Madurella mycetomatis* as causative agents of black grain mycetomas.

Most patients with mycetomas live in tropical locales, but infections in temperate zones do occur. The most common etiologic agent of white grain mycetoma in the United States is *P. boydii,* a member of the Ascomycota. The organism is commonly found in soil, standing water, and sewage; humans acquire the infection by traumatic implantation of the organism into the skin and subcutaneous tissues

Chromoblastomycosis

Chromoblastomycosis is a chronic fungal infection acquired via traumatic inoculation of an organism, primarily into the skin and subcutaneous tissue. The infection is characterized by the development of a papule at the site of the traumatic insult that slowly spreads to form warty or tumorlike lesions characterized as cauliflower-like. There may be secondary infection and ulceration. The lesions are usually confined to the feet and legs but may involve the head, face, neck, and other body surfaces. Histologic examination of lesion tissue reveals characteristic **sclerotic bodies**, which are copper-colored, septate cells that appear to be dividing by binary fission and thought by some to resemble copper pennies (see Figure 50-23). These infections cause hyperplasia of the epidermis of the skin, which may be mistaken for squamous cell carcinoma. Fungal brain abscess, known in the past as *cerebral chromoblastomycosis,* may be caused by the dematiaceous fungi, but is more appropriately considered a type of phaeohyphomycosis and is discussed with this disease below.[101] Chromoblastomycosis is widely distributed, but most cases occur in tropical and subtropical areas of the world. Occasional cases are reported from temperate zones, including the United States. The infection is seen most often in areas in which agricultural workers do not wear protective clothing and suffer thorn or splinter puncture wounds through which an organism enters from the soil.

Phaeohyphomycosis

Phaeohyphomycosis is a general term used to describe any infection caused by a dematiaceous organism, except those described above. These infections may be subcutaneous, localized, or systemic infections, and may be caused by any of a number of dematiaceous fungi. Infections include the phaeohyphomycotic cyst, progressive soft tissue infection, brain abscess, sinusitis, endocarditis, mycotic keratitis, pulmonary infection, and systemic infection.[2,32,86,152]

PATHOGENESIS AND SPECTRUM OF DISEASE

The spectrum of disease caused by the dematiaceous fungi ranges from superficial (e.g., skin and hair) infections to emergent, rapidly progressive, and often fatal disease (e.g., brain abscess). The list below is not

Table 50-15 Dematiaceous (Dm) Fungi

Organism	Disease	Site	Tissue Form
Slow-Growing Species			
Cladosporium spp.	Chromoblastomycosis Phaeohyphomycosis	Subcutaneous Subcutaneous, brain	Sclerotic bodies Septate, dematiaceous hyphae
Ochroconis galloparvum	Phaeohyphomycosis	Brain, subcutaneous, lungs	Septate, dematiaceous hyphae
Exophiala dermatitidis	Phaeohyphomycosis Pneumonial	Brain, eye, subcutaneous, and dissemination Lungs	Dematiaceous hyphal fragments and budding yeasts
Hortaea jeanselmei	Mycetoma Phaeomycotic cyst	Subcutaneous Subcutaneous	Dematiaceous hyphal fragments and budding yeasts
Hortaea werneckii	Tinea nigra	Skin	Dematiaceous hyphal fragments and budding yeast
Fonsecaea spp.	Chromoblastomycosis Phaeohyphomycosis Cavitary lung disease	Subcutaneous Brain Lungs	Sclerotic bodies Septate, dematiaceous hyphae Septate, dematiaceous hyphae
Phialophora spp.	Chromoblastomycosis Phaeohyphomycosis Septic arthritis	Subcutaneous Subcutaneous Joints	Sclerotic bodies Septate, dematiaceous hyphae Septate, dematiaceous hyphae
Piedraia hortae	Black piedra	Hair	Asci containing nodules cemented to hair shafts
Madurella mycetomatis	Mycetoma	Subcutaneous	Dematiaceous hyphal fragments
Rapid-Growing Species			
Alternaria spp.	Phaeohyphomycosis Sinusitis Nasal septal erosion Ulcers and onychomycosis	Subcutaneous Sinuses Nasal septum Skin, Nails	Septate, dematiaceous hyphae Septate, dematiaceous hyphae; possible fungus-ball Septate, dematiaceous hyphae Septate, dematiaceous hyphae
Bipolaris spp.	Phaeohyphomycosis Sinusitis, fungus-ball	Subcutaneous, brain, eye, bones Sinuses	Septate, dematiaceous hyphae Septate, dematiaceous hyphae; possible fungus-ball
Curvularia spp.	Sinusitis Phaeohyphomycosis	Sinuses Subcutaneous, heart valves, eye, and lungs	Septate, dematiaceous hyphae; possible fungus-ball Septate, dematiaceous hyphae
Drechslera spp.	Phaeohyphomycosis Sinusitis	Subcutaneous and brain Sinuses	Septate, dematiaceous hyphae Septate, dematiaceous hyphae
Exserohilum spp.	Phaeohyphomycosis	Subcutaneous	Septate, dematiaceous hyphae
Pseudallescheria boydii	Mycetoma Phaeohyphomycosis	Subcutaneous Subcutaneous, skin, joints, bones, brain, and lungs	Granules of hyaline hyphae Septate, hyaline hyphae
Cladophialophora bantiana	Phaeohyphomycosis	Brain	Septate, dematiaceous hyphae

comprehensive but provides the common etiologic agents of diseases that may be caused by dematiaceous fungi (Table 50-15).

- Common etiologic agents of subcutaneous infection include the following:
- Mycetoma common etiologic agents
  - Bacterial: *Nocardia, Actinomadura,* and *Streptomyces*
  - White grain mycetoma: *Pseudallescheria boydii, Acremonium,* and *Fusarium* spp.
  - Black grain mycetoma: *Madurella mycetomatis, Exophiala jeanselmei,* and *Curvularia*

- Chromoblastomycosis common etiologic agents
  - *Cladosporium/Cladophialophora, Phialophora,* and *Fonsecaea*
- Phaeohyphomycosis common etiologic agents
  - *Exophiala jeanselmei, E. dermatitidis, Curvularia, Bipolaris, Alternaria,* and *Exserohilum*
- Sinusitis common agents
- *Alternaria, Bipolaris, Exserohilum,* and *Curvularia*
- Brain abscess common etiologic agents
  - *Cladophialophora bantiana, Exophiala dermatitidis,* and *Bipolaris*
- Mycotic keratitis and endophthalmitis
- *Exophiala dermatitidis, Bipolaris,* and *Curvularia*

LABORATORY DIAGNOSIS

Specimen Collection and Transport

See General Considerations for the Laboratory Diagnosis of Fungal Infections.

Specimen Processing

See General Considerations for the Laboratory Diagnosis of Fungal Infections.

Direct Detection Method

Stains. In general, dematiaceous fungal hyphae are seen in clinical specimens by direct microscopic examination or by histopathologic examination of tissue obtained at surgery or autopsy. The dematiaceous character of the hyphae may not be appreciated if the examination is performed using calcofluor white/fluorescent microscopy alone, without observing the hyphae using traditional transmitted light microscopy.

Superficial Infections. Direct microscopic examination of a clinical specimen from a patient with tinea nigra may show dematiaceous hyphae and small budding yeast cells and/or hyphal fragments. Portions of hairs from a patient with black piedra are examined, in wet mounts using potassium hydroxide that is gently heated, for the presence of nodules composed of cemented mycelium. When mature nodules are crushed, oval asci, containing two to eight aseptate ascospores, 19 to 55 μm long by 4 to 8 μm in diameter, are seen. The asci are spindle-shaped and have a filament at each pole.

Chromoblastomycosis. The laboratory diagnosis of chromoblastomycosis is made easily. Scrapings from crusted lesions added to 10% potassium hydroxide show the presence of the sclerotic bodies, which appear rounded, brown, and 4 to 10 μm in diameter and resemble "copper pennies" having fission planes (see Figure 50-23).

Mycetoma and Phaeohyphomycosis. The direct examination of clinical specimens from patients with a eumycotic mycetoma or phaeohyphomycosis demonstrate yellowish brown, septate to moniliform hyphae, with or without budding yeast cells present. The presence of dematiaceous yeasts depends on the fungus. Dematiaceous yeast are commonly seen in the direct examination of clinical specimens from patients with infections caused by *Exophiala* spp. Macroscopic examination of granules from lesions of mycetoma caused by *P. boydii* reveal them to be white to yellow, and 0.2 to 2 mm in diameter. Microscopically, the granules of *P. boydii* consist of loosely arranged, intertwined septate hyaline hyphae that are cemented together.

The observation of pigmented hyphae in hematoxylin-eosin or unstained histopathologic sections will provide the diagnosis of dematiaceous fungal disease. The methenamine silver stain used for the detection of fungal elements in tissues stains fungi black, thereby making it impossible to determine whether they are hyaline septate or dematiaceous septate molds. The Fontana-Masson stain, which stains the melanin and melanin-like pigments in the cell walls of these organisms, may be used to confirm the presence of pigmented hyphae in histologic sections. Culture of the specific etiologic agent is necessary for final confirmation.

Antigen-Protein. Antigen-protein–based assays are not used for the diagnosis of disease caused by these organisms.

Nucleic Acid Amplification. Nucleic acid amplification assays are not routinely used for the detection or identification of these organisms, but may be available in research settings.

Cultivation. Although dematiaceous molds recovered in the clinical mycology laboratory may represent true pathogens, more often they represent transient flora, inhaled spores, or contaminants. Cultures from sterile body sites, if aseptically obtained, should not contain these molds. Cultures should be interpreted in conjunction with the results of the direct examination for fungal elements, corresponding histopathology, and discussion with the clinician to most effectively establish the diagnosis of mycotic infection caused by these organisms.

Superficial Infections. *Hortaea werneckii*, the causative agent of tinea nigra, may be recovered on common fungal media but grows very slowly. Initial colonies of *H. werneckii* may be olive to black, shiny, and yeastlike in appearance (Figure 50-78), and usually grow within 2 to 3 weeks. As the culture ages, colonies become filamentous, with velvety-gray aerial hyphae. *Piedraia hortae*, the causative agent of black piedra, is easily cultured on any fungal culture medium lacking cycloheximide. Colonies of this organism are also very slow growing, appear dark brown to black, and also produce aerial mycelium. A red to brown diffusible pigment may be produced by some isolates.

Mycetoma

WHITE GRAIN MYCETOMA. *Pseudallescheria boydii* is an organism that grows rapidly (5 to 10 days) on common laboratory media. Initial growth begins as a white, fluffy colony that changes in several weeks to brownish gray (the so-called mousy grey) colony; the reverse of

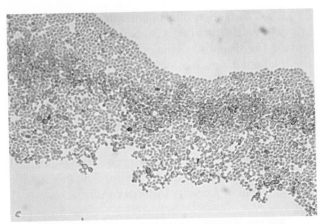

Figure 50-78   The yeast forms of *Hortaea werneckii.*

the colony progresses from tan to dark brown. *Acremonium* spp. that cause mycetomas, such as *A. falciforme,* grow slowly and produce gray to brown colonies.

BLACK GRAIN MYCETOMA. Colonies of *M. mycetomatis* and *E. jeanselmei* are slow growing, unlike colonies of *Curvularia.* Colonies of *M. mycetomatis* vary from white, during the early phases of growth, to olive-brown; a brown diffusible pigment is characteristic of this fungus. Colonies of *E. jeanselmei* appear yeastlike and darkly pigmented olive to black, but in time develop a more velvety appearance with the production of aerial hyphae. *Curvularia* produces a fluffy or downy, olive-gray to black colony and growth is rapid.

Chromoblastomycosis. The fungi known to cause chromoblastomycosis, *Cladosporium (Cladophialophora), Phialophora,* and *Fonsecaea,* are all dematiaceous. These fungi are slow growing and produce heaped-up and slightly folded, darkly pigmented colonies with a gray to olive to black and velvety or suedelike appearance. The reverse side of the colonies is jet black. Microscopic examination is necessary to definitively identify the pathogenic agent.

Phaeohyphomycosis. The colonies of many of the rapidly growing dematiaceous molds are similar; therefore, identification relies on microscopic examination. The colonies of *Alternaria* are rapidly growing and appear to be fluffy and gray to gray-brown or gray-green in color. *Curvularia* produces colonies that are rapidly growing and resemble those of *Alternaria.* *Bipolaris* produces colonies that are gray-green to dark brown and slightly powdery, as does *Drechslera* and *Exserohilum.* *Drechslera* is a dematiaceous organism that has been described erroneously as *Helminthosporium* or *Bipolaris* in many older textbooks.

The colonies of many of the slow-growing dematiaceous molds are also similar to one another and require identification based on microscopic morphology. *Exophiala jeanselmei* and *Exophiala dermatitidis* are dematiaceous organisms that grow slowly (7 to 21 days) and initially produce shiny black, yeastlike colonies. *Exophiala dermatitidis* is often more mucoid and may be brown, compared with *Exophiala jeanselmei,* but both are very similar in appearance. Colonies become filamentous and velvety with age as a result of the production of mycelium. The colonial morphology of other slowly growing dematiaceous fungi, such as *Fonsecaea,* are described above.

Approach to Identification

Superficial Infections. *Hortaea werneckii* is a dematiaceous fungus that produces yeast-like cells that may be one- or two-celled. Conidia produced by this organism are produced by annellophores, which bear successive rings **(annellides),** which are difficult to see microscopically. The biophysical profile is used to differentiate this fungus from other *Exophiala* species. In contrast, *Piedraia hortae* usually does not sporulate on routine mycologic media but demonstrates only highly septate dematiaceous hyphae and swollen intercalary cells.

Mycetoma. It is not possible to predict the specific etiologic agent of a eumycotic mycetoma without culture. Culture media containing antibiotics should not be used as the sole medium for culturing clinical specimens from a mycetoma, because species of the aerobic actinomycetes are susceptible to antibacterial antibiotics and may be inhibited by these agents. For further information on other less common fungi involved in mycetoma, the reader is referred to the following references.[1,3,100,103,117]

White Grain Mycetoma: Pseudallescheria boydii and Acremonium. As previously mentioned, these fungi are hyaline molds that produce septate hyphae and could have been in covered above with *Aspergillus* and *Fusarium* as causes of hyalohyphomycosis. The features described here are useful for identification regardless of the disease process (i.e., mycetoma or hyalohyphomycosis). *Pseudallescheria boydii* is also involved in causing a variety of infections elsewhere in the body. These include infections of the nasal sinuses and septum, meningitis, arthritis, endocarditis, mycotic keratitis, external otomycosis, brain abscess or disseminated invasive infection. Most of these more serious infections occur primarily in immunocompromised patients.

Pseudallescheria boydii is an example of an organism that exhibits both asexual and sexual reproduction. The telomorphic or sexual form of this fungus, which is evidenced by the production of cleistothecia, is *P. boydii*, whereas if asexual reproductive structures alone are observed, it may be termed *Scedosporium apiospermum*. The asexually produced conidia of *Pseudallescheria boydii/ Scedosporium apiospermum* are golden brown, elliptical-to-pyriform, single-celled conidia borne singly from the tips of long or short conidiophores (annellophores) (see Figure 50-9). This anamorph predominates in cultures from clinical specimens. Another anamorphic form, the *Graphium* stage of *P. boydii*, may be seen less commonly that consists of clusters of conidiophores with conidia produced at the ends; this has also been referred to as *coremia* (see Figure 50-10). The teleomorphic (sexual) form of the organism produces brown to black cleistothecia, which are pseudoparenychatous saclike structures that contain asci and ascospores. When the latter are fully developed, the large (50 to 200 μm), thick-walled cleistothecia rupture and liberate the asci and ascospores (see Figure 50-8). The ascospores are oval and delicately pointed at each end. Isolates of *P. boydii* may be induced to form cleistothecia by culturing on plain water agar; however, they are seldom found on primary recovery of a culture from a clinical specimen.

It is important to recognize *P. boydii*, because it is resistant to amphotericin B, a commonly used antifungal agent for systemic infections. Another species of *Scedosporium, S. prolificans*, has been associated with infections other than mycetoma, such as arthritis or invasive disease in immunocompromised patients.[124] It differs from *S. apiospermum* by producing inflated annellophores that appear flask-shaped. The obsolete or previous name for this organism was *S. inflatum*, which more accurately reflects the morphology of the conidiophore. This organism is important to recognize, because it is resistant to most, if not all, of the commonly used antifungal agents. *Acremonium* develops hyaline hyphae and produces simple, unbranched, erect conidiophores. Single-celled conidia are produced loosely or in gelatinous masses at the tip of the conidiophore (see Figure 50-39). Intercalary and terminal chlamydoconidia may also be produced.

Black Grain Mycetoma: Exophiala jeanselmei, Curvularia, and Madurella mycetomatis.

Sterile hyphae are produced when *Madurella mycetomatis* is grown on rich fungal media. Nutritionally poor media may be used to induce sporulation. Long, tapering phialides with collarettes and sclerotia may be seen. Temperature tolerance, and biochemical hydrolysis and assimilation studies may be used to differentiate *M. mycetomatis* from *M. grisea*. See Phaeohyphomycosis later for the description of *Exophiala jeanselmei* and *Curvularia*.

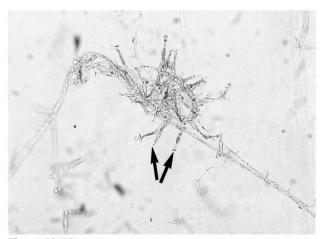

Figure 50-79   *Phialophora richardsiae* showing phialides having prominent saucerlike collarette *(arrows)* (500×).

Chromoblastomycosis: Cladosporium, Phialophora, and Fonsecaea.

The taxonomy of the organisms that cause chromoblastomycosis is complex.[121] Their identification is based on somewhat distinct microscopic morphologic features. These are polymorphic fungi that may produce more than one type of conidiation. The genus *Cladosporium* includes those species that produce long chains of budding, often fusiform, conidia (blastoconidia) that have a dark septal scar present. The reclassification of some of these organisms into the genus *Cladophialophora* is because some of these organisms also produce phialides[139]; for the most part, the genus *Cladosporium* is retained in this chapter. The genus *Phialophora* includes those species that produce short, flask-shaped to tubular phialides, each having a well-developed collarette. Clusters of conidia are produced by the phialides through an apical pore and often remain aggregated near the opening in a gelatinous mass. *Phialophora* species produce colonies that are woolly and olive-brown to brownish gray; some strains may appear to have concentric zones of color. Microscopically, hyphae are dematiaceous and sporulation is common. *Philaophora richardsiae* produces phialides with distinct flattened or saucerlike collarettes are readily observed (Figure 50-79). In contrast, *Phialophora verrucosa* produces deeper more cup or flask-shaped phialides. Pleomorphic phialides may also be seen with these species; however, all produce either or both hyaline elliptical conidia or brown elliptical conidia within the phialides. The genus *Fonsecaea* includes those organisms that exhibit a mixed type of sporulation. The genus produces a distinct *Fonsecaea*-type conidiophore, which somewhat resembles truncated *Cladosporium*-type sporulation. It may also produce a *Rhinocladiella*-type of sporulation, wherein single-celled conidia are produced on denticles that arise from all sides of conidiophores **(sympodially).** A mixture of the *Fonsecaea, Rhinocladiella* and

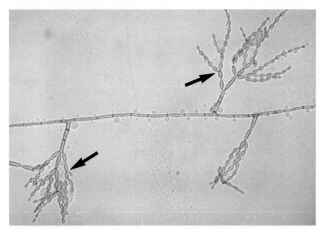

Figure 50-80   *Cladosporium* spp. showing *Cladosporium* type of sporulation *(arrows)* with chains of elliptical conidia (430×).

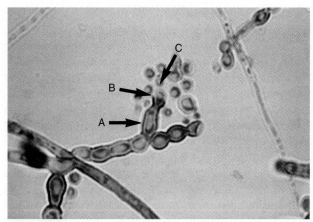

Figure 50-81   *Phialophora verrucosa* showing flask-shaped phialide (**A**) with distinct collarette (**B**) and conidia (**C**) near its tip (750×).

Cladosporium types may occur; moreover, phialides with collarettes or *Phialophora* type sporulation may also be present.

The diagnostic features of the three genera are summarized as follows:

1. *Cladosporium (Cladosporium carrionii): Cladosporium* type of sporulation with long chains of elliptical conidia (2 to 3 μm × 4 to 5 μm) borne from erect, tall, branching conidiophores (Figure 50-80)
2. *Phialophora* species: *Phialophora verrucosa* produces phialides, each with a distinct cup or flask-shaped collarette (Figure 50- 81), whereas *P. richardsiae* produces phialides with a flattened collarette (see Figure 50-79). Conidia are produced endogenously and occur in clusters at the tip of the phialide.
3. *Fonsecaea* species*:* Conidial heads with sympodial arrangement of conidia, with primary conidia giving rise to secondary conidia (Figure 50-82). *Cladosporium*-type, *Phialophora*-type, and/or *Rhinocladiella* types of sporulation may also occur.

Phaeohyphomycosis: *Alternaria, Bipolaris, Clados-porium, Curvularia, Drechslera, Exophiala, Exserohilum,* and *Phialophora*.   A useful approach to the identification of dematiaceous molds is first to determine whether single-celled or multicelled conidia are produced. If conidia are produced singly, determine whether they are produced individually or in chains such as *Cladosporium* species. In cellophane tape preparations, the chains of conidia produced by *Cladosporium* are easily disrupted. If multicellular conidia are produced by the dematiaceous mold, then it is useful to examine the septations within the conidium. Multicellular conidia with septations that occur only in the horizontal axis of the conidium (i.e., the axis perpendicular to the longitudinal axis of the conidium) are characteristic of certain

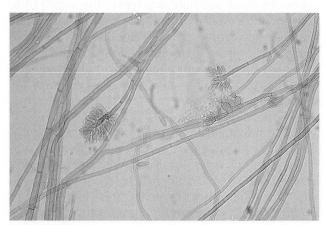

Figure 50-82   Both the *Rhinocladiella*-type and the *Phialophora*-type of sporulation may be produced by *Fonsecaea pedrosoi* and are demonstrated here (430×).

organisms, such as in *Bipolaris, Curvularia,* and *Drechslera,* whereas conidia with septations present in both the longitudinal and horizontal axis of the conidium are characteristic of other fungi, such as with *Alternaria.*

Alternaria: Microscopically, hyphae are septate and golden brown–pigmented; conidiophores are simple but sometimes branched. Conidiophores bear a chain of large brown conidia resembling a drumstick and contain both horizontal and longitudinal septa (Figure 50-83). It is sometimes difficult to observe chains of conidia, because they may be dislodged as the culture mount is prepared.

Bipolaris: Hyphae are dematiaceous and septate. Conidiophores, however, are characteristically bent (**geniculate**) at the locations where conidia are attached; conidia are arranged sympodially and are oblong to fusoid. The hilum protrudes only slightly (Figure 50-84). Germ tubes are formed at one or both ends, parallel to the long axis of the conidium, when the fungus is

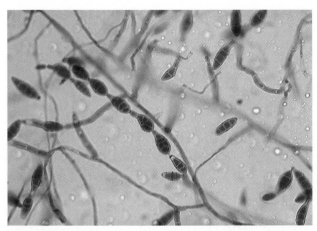

Figure 50-83 *Alternaria* spp. showing chaining muriform dematiaceous conidia with horizontal and longitudinal septa.

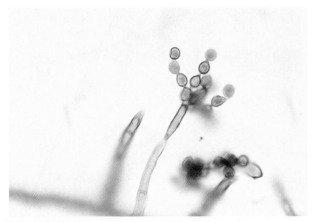

Figure 50-85 *Cladosporium* spp. showing branching chains of dematiaceous blastoconidia that are easily dislodged during the preparation of a microscopic mount (430×).

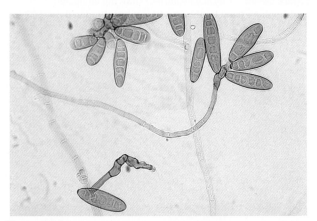

Figure 50-84 *Bipolaris* spp. showing dematiaceous multicelled conidia produced sympodially from geniculate conidiophores (430×).

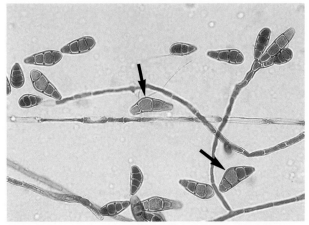

Figure 50-86 *Curvularia* spp. showing twisted conidiophore and curved conidia having a swollen central cell *(arrows)* (500×).

incubated in water at 25° C for up to 24 hours (i.e., from both poles, thus the name *Bipolaris*).

Cladosporium: Microscopically, hyphae are septate and brown in color. Conidiophores are long and branched, and give rise to branching chains of darkly pigmented budding conidia. Conidia are usually single-celled and exhibit prominent attachment scars (disjunctors). The cells that produce the branch points are often referred to as "shield" cells (Figure 50-85). This organism often fails to reveal chains of conidia on wet mounts because conidia are so easily dislodged.

Curvularia: Hyphae are dematiaceous and septate, as observed microscopically. Conidiophores are geniculate (i.e., bent where conidia are attached); conidia are arranged sympodially, are golden brown, multicelled, and curved, with a central swollen cell (Figure 50-86). The end cells are lighter in color compared with the swollen cell.

Drechslera: Microscopically, the hyphae are septate and darkly pigmented, and conidiophores are geniculate.

Conidia are produced sympodially (Figure 50-87). However, sporulation is generally sparse with this organism and is not commonly seen. The conidia of *Drechslera* are impossible to differentiate from those of *Bipolaris* based on morphologic criteria alone. The germ-tube test may be used to differentiate these organisms. Conidia are placed in a drop of water, coverslipped, and observed after an incubation period of at least 24 hours. Any of the cells of the conidia belonging to *Drechslera* may germinate, and these may grow perpendicular to the long axis of the conidium. In contrast, only the end cells of the conidia belonging to *Bipolaris* germinate, and these grow predominantly parallel to the longitudinal axis of the conidium.

Exophilia species: Only the *Exophiala* species *E. jeanselmei* and *E. dermatitidis* are considered here; although other species exist, they are recovered far less commonly in the clinical laboratory. *Exophiala jeanselmei:* The microscopic features of young colonies of these fungi exhibit dematiaceous yeastlike cells (Figure 50-88). Although

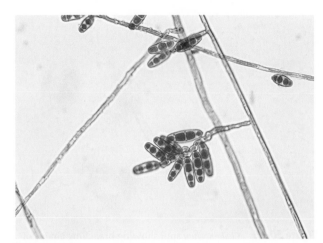

Figure 50-87 *Drechslera* spp. showing dematiaceous multicelled conidia. Most isolates produce only a few conidia.

Figure 50-89 *Exophiala jeanselmei* showing elongated conidiophore (annellophore) with a narrow, tapered tip (500×).

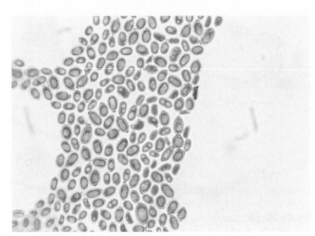

Figure 50-88 *Exophiala dermatitidis* showing dematiaceous yeastlike cells from a young culture. These forms asexually reproduce via annellides rather than true budding (blastoconidiation) (500×).

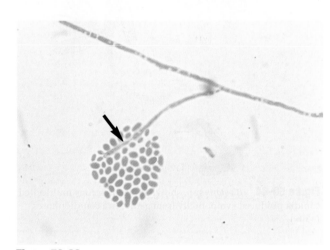

Figure 50-90 *Exophiala dermatitidis* showing elongated tubular annellophores *(arrow)*; morphologically very similar to *E. jeanselmei* (500×).

these may appear to be budding, a close inspection may disclose that the daughter cells are produced by annellides rather than true buds. Feltlike, filamentous colonies produce dematiaceous hyphae and conidiophores that are cylindrical and have a tapered tip. Annellations may be visible at the tip, and clusters of oval to round conidia are apparent (Figures 50-89 and 50-90). Potassium nitrate is utilized by *E. jeanselmei* but not *E. dermatitidis*. Temperature studies are also useful for differentiating the most common *Exophilia* species. Both *E. jeanselmei* and *E. dermatitidis* grow at 37° C, whereas only *E. dermatitidis* can grow at 40° to 42° C.

Exserohilum: Hyphae are septate and dematiaceous. Conidiophores are geniculate, and conidia are produced sympodially. Conidia are elongate, ellipsoid to fusoid, and exhibit a prominent hilum that is truncated and protruding (Figure 50-91). The conidia are multi-

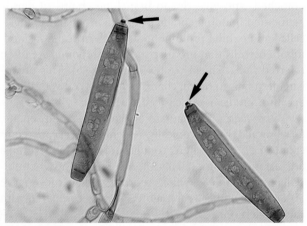

Figure 50-91 *Exserohilum* spp. showing elongated, multicelled conidia with prominent hila *(arrows)*.

cellular with only perpendicular septa and usually contain 5 to 9 septa.

Phialophora: Phialophora richardsiae is considered to be a cause of phaeohyphomycosis. See Chromoblastomycosis on p. 691 for a description of *P. richardsiae*.

Serodiagnosis

Some serologic and skin testing may be useful for the diagnosis of allergy to dematiaceous fungi. Serology is not useful for the diagnosis of invasive dematiaceous fungal disease.

PNEUMOCYSTIS JIROVECI (AN ATYPICAL FUNGUS)

Genera and Species to Be Considered: *Pneumocystis jiroveci (P. carinii)*

GENERAL CHARACTERISTICS

Pneumocystis jiroveci is an opportunistic atypical fungus that infects immunocompromised hosts.[146] Originally thought to be a trypanosome, the precise taxonomic categorization of *P. jirovecii* has remained challenging.[20] The clinical response to antiprotozoal drugs and the lack of response to antifungal drugs in patients with pneumocystosis further supported the notion that *P. carinii* was a protozoan parasite.[133] The inability to successfully maintain and propagate this organism in routine culture has further limited its characterization, but cultivation is possible under special conditions.

Although *Pneumocystis jiroveci* has been shown to be a fungus, it differs from other fungi in various aspects. The cell membrane of *P. jiroveci* contains cholesterol rather than ergosterol.[65] A flexible-walled trophozoite form exists that is susceptible to osmotic disturbances. Additionally, this organism contains only one to two copies of the small ribosomal subunit gene, whereas most other fungi contain numerous copies of this gene.[46]

DNA sequence analysis of the small ribosomal subunit gene has been performed on *P. jiroveci*. These comparative studies have disclosed a greater sequence homology with the fungi than with the protozoa.[34,35,134] Two independent analyses that compared DNA sequences of *P. jiroveci* with those of other fungi confirm the position of *P. jiroveci* within the fungal kingdom and place it somewhere between the ascomycetes and the basidiomycetes.[17,142]

EPIDEMIOLOGY AND PATHOGENESIS

Pneumocystis jiroveci has a worldwide distribution and most commonly presents as pneumonia in an immunocompromised host. Infection appears to be species specific, with *P. carinii* causing disease in rodents and *P. jiroveci* causing human disease.

SPECTRUM OF DISEASE

After inhalation, the organism is thought to adhere to type I pneumocytes.[146] The organisms exist and replicate extracellularly while bathed in alveolar lining fluid. Successful replication of the organism results in alveolar spaces being filled with foamy material. These changes result in impaired oxygen diffusing capacity and hypoxemia. A predominantly interstitial mononuclear inflammatory response is associated with this type of pneumonia.

LABORATORY DIAGNOSIS

Specimen Collection and Transport

Respiratory specimens that reflect the deep portions of the lung, such as bronchoalveolar lavage fluid, are best for the detection of this organism. If a sputum specimen is submitted for direct examination, it should be an induced sputum obtained by a trained respiratory therapist or the false-negative test result rate may be unacceptably high.

Specimen Processing

See Direct Detection Methods below.

Direct Detection Methods

Stains. The diagnosis of *P. jiroveci* pneumonia is currently based on the clinical presentation, radiographic studies, and direct and/or pathologic examination of bronchoalveolar lavage fluid or biopsy material. The flexible-walled trophozoite forms are the predominant form of the organism, but these are difficult to visualize.[133] These forms are somewhat discernible in Giemsa-stained material, but their pleomorphic appearance makes this form of the organism difficult to identify. A firm-walled cystic form also exists. The cysts are outnumbered by the trophozoites 1:10.[133] The cystic forms are more easily recognizable than the trophozoites and may be definitively identified using a variety of stains, such as calcofluor white, methenamine silver, or immunofluorescent staining (Figure 50-92). The cysts of these organisms are spherical to concave, uniform in size (4 to 7 μm in diameter), do not bud, and contain distinctive intracystic bodies. A comparison of these four most common staining method have demonstrated that the calcofluor white and methenamine silver methods likely represent the best balance between sensitivity and specificity, and have the best overall positive and negative predictive values.[109]

Antigen-Protein. Commercial kits, which employ monoclonal antibodies directed against *P. jiroveci*, stain both cysts and trophozoites. It is a highly sensitive microscopic method of detection, but is expensive and

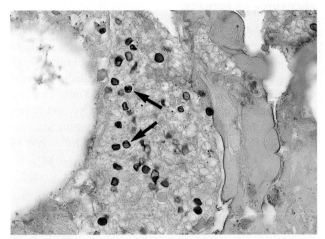

Figure 50-92 The cystic forms of *Pneumocystis jiroveci* (arrows) stain well with methenamine silver (500×).

nonspecific staining, which may be thought to represent the pleomorphic trophozoite forms, may limit the specificity of this assay.

Nucleic Acid Amplification. A variety of nucleic acid amplification assays for *P. jiroveci* have been developed, including most recently real-time PCR methods. These are currently migrating from research settings to the clinical molecular microbiology laboratory, but commercial kits are not yet available.

Cultivation. This organism does not grow in routine culture.

Approach to Identification. See Direct Detection Methods above.

Serodiagnosis

Serology is not useful for the diagnosis of pneumocystosis.

THE YEASTS

Genera and Species to Be Considered: *Candida*, *Cryptococcus*, *Trichosporon*, and *Malassezia*

GENERAL CHARACTERISTICS

There has been a significant increase in the number of fungal infections caused by yeasts and yeastlike fungi during recent years. These are most commonly due to infection with various *Candida* species, but other yeasts also cause significant disease, particularly in the immunocompromised host. In addition causing disease in immunocompromised patients, infections are also common in postsurgical patients, trauma patients, and patients with long-term indwelling venous catheters.

Some of these yeasts are resistant to commonly used antifungal agents, which denotes the need for prompt, appropriate identification and, in some instances, antifungal susceptibility testing.[31,49,77,138]

Just how far the laboratory should go with a complete identification of all yeast species is of question. It is recommended that:

- All yeasts recovered from sterile body fluids, including cerebrospinal fluid, blood, urine, paracentesis, and other fluids, should be identified.
- Yeasts from all seriously ill or immunocompromised patients or in whom a mycotic infection is suspected should be identified.
- Yeasts from respiratory secretions should not be identified on a routine basis; however, they should be screened for the presence of *Cryptococcus neoformans*.
- Yeasts recovered from several successive specimens, except respiratory secretions, should be identified. Each laboratory director must decide how much time, effort, and expense is to be spent on the identification of yeasts in the laboratory.

The development of commercially available yeast identification systems provides laboratories of all sizes with the capability of using accurate and standardized methods, but these methods should be used in conjunction with cornmeal agar morphology to avoid misidentifications. Some of these systems have extensive computer databases that include biochemical profiles of large numbers of yeasts. Variations in the reactions of carbohydrates and other substrates utilized are considered in the identification of yeasts provided by these systems. Commercially available systems are recommended for all laboratories. These may be used in conjunction with some less expensive and rapid screening tests that will provide the presumptive identification of *C. neoformans* and a definitive identification of *C. albicans*. More recent diagnostic tools that have been introduced for the more rapid characterization of yeasts include CHROMagar Candida (BD Diagnostics, Sparks, Md) and *C. albicans* PNA FISH (AdvanDX, Woburn, Mass).* Some laboratories might prefer the use of conventional systems; therefore, the information presented within this section discusses rapid screening methods for the presumptive identification of yeasts, commercially available systems, and a conventional schema that will provide for the identification of commonly encountered species of yeast.

EPIDEMIOLOGY

Candida

Candida spp. are responsible for the most frequently encountered opportunistic fungal infections. *Candida*

---

*References 4, 7, 89, 90, 98, 113, 137, 154.

infections are caused by a variety of species, but *Candida albicans* is the most frequent etiologic agent. The next three most commonly isolated *Candida* species are *Candida tropicalis, Candida parapsilosis,* and *Candida glabrata.* The frequency with which these are isolated varies by institution. A number of other species have been shown to cause human infections, but the rate of infection is not as high as these four species. *Candida albicans* and other *Candida* species are a part of the our microbiota (i.e., normal flora), but these have also become endemic in most hospitals. Infections may be caused by endogenous yeasts or may be nosocomial (hospital acquired).

Cryptococcus

Cryptococcus neoformans exists as a saprobe in nature. It is most often found associated with avian excreta, particularly that of pigeons. The hypothesis that pigeon habitats serve as reservoirs for human infection is substantiated by numerous reports; the pigeon manure apparently serves as an enrichment for *C. neoformans* because of its composition. It is believed that *C. neoformans* is widely distributed in nature, and aerosolization is a prerequisite to most infections. Alternatively, the conidia of the telemorphic form of *C. neoformans, Filobasidiella neoformans,* could be the aerosolized form that initiates the infection.

Trichosporon and *Malassezia*
See Pathogenesis and Spectrum of Disease below.

PATHOGENESIS AND SPECTRUM OF DISEASE

Candida

Candida may be recovered from the oropharynx, gastrointestinal tract, genitourinary tract, and skin. *Candida* spp. are responsible for several different types of infections in healthy and immunocompromised patients. Included are intertriginous candidiasis, in which skin folds are involved; paronychia; onychomycosis; perlèche; vulvovaginitis; thrush; pulmonary infection; eye infection; endocarditis; meningitis; fungemia; and disseminated infection. The latter two types of infection are most commonly seen in seriously ill and/or immunocompromised patients, whereas the others may occur in healthy hosts. Esophagitis produced by *C. albicans* is common in patients with AIDS.

The clinical significance of *Candida* recovered from respiratory tract secretions is difficult to determine, in that it is considered to be part of the normal oropharyngeal flora of humans. A study at Mayo Clinic evaluated the clinical significance of yeasts recovered from respiratory secretions, except for *Cryptococcus neoformans,* and concluded that they are part of the normal flora and that their routine identification is unneces-

sary.[92] Similarly, Barenfanger and colleagues have demonstrated that the routine identification of yeasts from respiratory specimens result in unnecessary antifungal therapy, an extended hospital stay, and increased healthcare costs without demonstrable benefit.[8] The simultaneous recovery of the same species of yeast from several body sites, including urine, is a good indicator of disseminated infection and fungemia.

Cryptococcus

Cryptococcosis, specifically caused by *C. neoformans,* is an acute, subacute or chronic fungal infection that has several manifestations. In the immunocompromised patient, it is common to see disseminated disease with meningitis. Patients that have a moderately compromised immune system or early in the disease process cryptococcal fungemia may be seen without concomitant meningitis. Disseminated cryptococcosis and cryptococcal meningitis became well recognized in patients with AIDS and they remain an important cause of morbidity and mortality in these patients in resource-poor countries who do not have access to highly active antiretroviral therapy.[25] It has become infrequent in countries wherein patients have ready access to highly active antiretroviral therapy. Patients with disseminated infection may exhibit painless papular skin lesions that may ulcerate. Other less common manifestations of cryptococcosis include endocarditis, hepatitis, renal infection, and pleural effusion. It is of interest to note that a review of patient records at the Mayo Clinic revealed more than 100 immunocompetent patients with colonization of the respiratory tract with *C. neoformans* who did not develop subsequent infection. Follow-up on these patients was as long as 6 years; none in this group were considered to be immunocompromised. This makes the clinical significance of *C. neoformans* somewhat difficult to assess; however, given the severity of disease its presence in clinical specimens should be considered significant. In many instances, the clinical symptoms are suppressed by corticosteroid therapy, which is a risk factor for disease, and culture or serologic evidence (detection of cryptococcal antigens) provides the earliest proof of infection. There is a strong association of cryptococcal infection with such debilitating diseases as leukemia, lymphoma, and the immunosuppressive therapy that may be required for these and other underlying diseases. The presence of *C. neoformans* in clinical specimens in some instances precedes the symptoms of an underlying disease.

Four serotypes of *C. neoformans* have been described (A, B, C, and D), with somewhat different geographic distribution.[74] The organism is one of the few basidiomycetes that infect humans; its teleomorph has three varieties, *Filobasidiella neoformans* var. *neoformans* (serotype D), *F. neoformans* var. *gattii* (serotypes B and C), and

Filobasidiella neoformans var. *grubii* (serotype A).[42] *Cryptococcus* will continue to be used as the genus name, because it is well-established by tradition, and because the yeast anamorphic form is the form recovered in the laboratory. Despite the discovery of three varieties of *Cryptococcus,* there is no difference in disease produced or in the response to chemotherapy among the three.

Trichosporon

Trichosporonosis is caused by a variety of *Trichosporon* species, which have undergone changes in nomenclature based on DNA sequence comparisons. The yeast-like fungus causes disease almost exclusively in immunocompromised patients, particularly those who have leukemia.[6] Disseminated trichosporonosis is the most common clinical manifestation. Skin lesions accompanied by fungemia are frequently seen. Endocarditis, endophthalmitis, and brain abscess have been reported. *Trichosporon* is occasionally recovered from respiratory tract secretions, skin, the oropharynx, and the stool of patients who have no evidence of infection and may represent transient fungal colonization of those individuals.

White piedra is an uncommon fungal infection of immunocompetent patients and is found in both tropical and temperate regions of the world. It is characterized by the development of soft, yellow or pale brown aggregations around hair shafts in the axillary, facial, genital, and scalp regions of the body. The *Trichosporon* species that cause this disease frequently invades the cortex of the hair and causes damage.

Malassezia

Malassezia furfur causes **tinea versicolor,** a skin infection that is characterized by superficial brownish scaly areas on light-skinned persons and lighter areas on dark-skinned persons. The lesions occur on the smooth surfaces of the body, namely, the trunk, arms, shoulders, and face. It has a worldwide distribution. *Malassezia furfur* is also a cause of disseminated infection in infants and young children and even adults given lipid replacement therapy.[82] *Malassezia pachydermatis,* another species, may be recovered from skin lesions of patients. Rarely, it may cause of fungemia in immunocompromised patients.

LABORATORY DIAGNOSIS

Specimen Collection and Transport

See General Considerations for the Laboratory Diagnosis of Fungal Infections.

Specimen Processing

See General Considerations for the Laboratory Diagnosis of Fungal Infections.

Direct Detection Methods

Stains

Candida. The direct microscopic examination of clinical specimens containing *Candida* reveals budding yeast cells **(blastoconidia)** 2 to 4 μm in diameter and/or pseudohyphae (see Figure 50-13) showing regular points of constriction, resembling links of sausage. True septate hyphae may also be produced. The blastoconidia, hyphae, and pseudohyphae are strongly gram-positive. It is advisable to report the approximate number of such forms, in that the presence of large numbers in a fresh clinical specimen may be of diagnostic significance.

Cryptococcus. Traditionally, the India ink preparation has been the most widely used method for the rapid detection of *C. neoformans* in clinical specimens. This method is still used as a rapid and inexpensive assessment tool in many institutions and is of considerable diagnostic value in resource poor settings. This method delineates the large capsule of *C. neoformans,* because the ink particles cannot penetrate the capsular polysaccharide material. Although this test is useful, many laboratories have replaced this test with the more sensitive cryptococcal latex agglutination test that detects cryptococcal antigen. The cryptococcal antigen detection (CAD) test is described in a subsequent section. The India ink preparation is commonly positive in specimens from patients who have AIDS. Laboratories examining many specimens from these patients may wish to retain the use of this procedure in combination with CAD and culture.

The microscopic examination of other clinical specimens, including respiratory secretions, can also be of value in making a diagnosis of cryptococcosis. *Cryptococcus neoformans* appears as a spherical, single or multiple budding, thick-walled yeast that is 2 to 15 μm in diameter. It is usually surrounded by a wide, refractile polysaccharide capsule (see Figure 50-18). Perhaps the most important characteristic of *C. neoformans* is the extreme variation in the size of the yeast cells; this is unrelated to the amount of polysaccharide capsule present. It is important to remember that not all isolates of *C. neoformans* exhibit a discernible capsule.

Trichosporon. The microscopic examination of clinical specimens reveals hyaline hyphae, numerous round to rectangular arthroconidia, and occasionally a few blastoconidia. Usually, hyphae and arthroconidia predominate.

In white piedra, white nodules are removed and observed using the potassium hydroxide preparation after applying light pressure to the coverslip so that crushing of the nodule occurs. Hyaline hyphae, 2 to 4 μm in width, and arthroconidia are found in the preparation of the cementlike material that bind the hyphae together. The organism may be identified in culture by the presence of true hyphae, blastoconidia, and arthroconidia, in conjunction with urease positivity (see Figure 50-73). Although *Trichosporon asahii* may be distinguished from other species in the genus by its biophysical profile (carbohydrates and substrate utilization), these are likely best distinguished at the species level with molecular tools, such as DNA sequencing.

Malassezia. Most often, the detection of *M. furfur* is made by the direct microscopic examination of skin scrapings. Here the organism will easily be recognized as oval- or bottle-shaped cells that exhibit monopolar budding in the presence of a cell wall with a septum at the site of the bud scar. In addition, small hyphal fragments are observed (see Figure 50-12). In cases of fungemia, the morphologic form seen in the direct examination of the blood cultures are small yeasts without the presence of pseudohyphae.

Antigen-Protein. The cryptococcal latex test for antigen of *Cryptococcus neoformans* may be performed on cerebrospinal fluid or serum. This assay has replaced the use of India ink to screen for *C. neoformans* in many laboratories. It should be noted that *Trichosporon* shares an antigen that is similar to that produced by *C. neoformans*. Sera from patients who have trichosporonosis often yield false-positive cryptococcal antigen tests when latex agglutination methods are used.

Nucleic Acid Amplification. Nucleic acid amplification assays have been developed for a variety of yeast species, but are usually performed in research settings.

Cultivation.

Candida. The colonial and microscopic morphologic features of *Candida* spp. are of little value in making a definitive identification. Most *Candida* species produce creamy white colonies, but some produce flatter, drier colonies.

Cryptococcus. *Cryptococcus neoformans* is easily cultured on routine fungal culture media without cycloheximide. The organism is inhibited by the presence of cycloheximide at 25° to 30° C. For the optimal recovery of *C. neoformans* from cerebrospinal fluid, it is recommended that a 0.45-μm membrane filter be used with a sterile syringe. The filter is placed on the surface of the culture medium and is removed at daily intervals so that growth under the filter can be visualized. An alternative

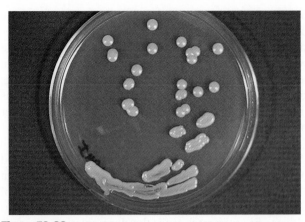

Figure 50-93   The colonies of *C. neoformans* appear shiny and mucoid because of the presence of a polysaccharide capsule.

to the membrane filter technique culture following centrifugation.

Colonies of *C. neoformans* usually appear on culture media within 1 to 5 days and begin as a smooth white to tan colony that may be mucoid-to-creamy (Figure 50-93). It is important to recognize the colonial morphology on different culture media since variation does occur; for example, on inhibitory mold agar, *C. neoformans* appears as a golden yellow, nonmucoid colony. Textbooks typically characterize the colonial morphology as being *Klebsiella*-like because of the large amount of polysaccharide capsule material present (see Figure 50-93). In reality, most isolates of *C. neoformans* do not have large capsules and may not have the typical mucoid appearance.

Trichosporon. Colonies of *Trichosporon* species vary in their morphology; however, most are cream-colored, heaped, dry to moist, and wrinkled. Some may appear white, dry, and powdery.

Malassezia. *Malassezia furfur* is infrequently cultured in the clinical laboratory. Recovery of the organism is not required to establish a diagnosis (in skin infections), and it is seldom attempted for this purpose. Cultivation, for example from a positive blood culture, requires an agar medium overlaid with a long-chain fatty acid (olive oil). The colonies are small when compared with the colonies of *C. albicans*, and are creamy and white-to-off white.

Approach to Identification

The general approach to yeast identification consists of evaluating the carbohydrate and substrate utilization profile with a commercial system, and observing the morphology in a cornmeal preparation. This latter aspect is particularly important in discovering errors in identification that are made by the commercial system prior to the release of an erroneous identification to the clinician. For example, if a commercial system designated an isolate as *Candida glabrata* but pseudohyphae were seen in the cornmeal preparation, then additional testing needs to be done to correctly identify the isolate, since *C. glabrata* does not make pseudohyphae. This traditional approach could be modified to use newer methods of confirmation, such a CHROMagar Candida.

***Candida* species** *Candida albicans* may be identified by the production of a germ tubes or chlamydoconidia (Figure 50-94; see also Figure 50-2). Other species of *Candida* are most commonly identified by the utilization of specific substrates and the fermentation or assimilation of particular carbohydrates. The morphologic features of yeast on cornmeal agar containing Tween 80 often allow for the tentative identification of selected

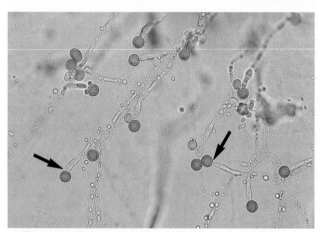

Figure 50-94   The chlamydoconidia of *C. albicans (arrows).*

species, and importantly may detect misidentifications by the commercial systems (Table 50-16).

Germ-Tube Test. The germ-tube test (Procedure 50-9) is the most generally accepted and economical method used in the clinical laboratory for the identification of yeasts. Approximately 75% of the yeasts recovered from clinical specimens are *C. albicans,* and the germ-tube test usually provides a sufficient identification of this organism within 3 hours.

Germ tubes appear as early hyphal-like extensions of yeast cells that are produced without a constriction at the point of origin from the yeast cell (see Figure 50-2). Another *Candida* species, *C. dubliniensis,* has been shown to also produce true germ tubes, but this species is infrequently encountered. *Candida tropicalis* produces what has been termed "pseudo-germ tubes," which are constricted at the base or point of germ tube origin from the yeast cell. Unless this is recognized and skills developed to differentiate between true and pseudo-germ tubes then *C. tropicalis* isolates will be misidentified as *C. albicans*

Another method of identification of *C. albicans* is based on the presence of chlamydoconidia (see Figure 50-94) on cornmeal agar containing 1% Tween 80 and trypan blue incubated at room temperature for 24 to 48 hours. Many of the finer points of yeast identification are discussed in *The Yeasts: A Taxonomic Study,* written by Kreger-Van Rij.[73]

The search for a more rapid, less subjective methods of identifying *C. albicans* and other *Candida* species continues. *Candida albicans* produces β-glactose aminidase and ʟ-proline aminopeptidase. Other species of *Candida* may produce one enzyme, but not both. Assays

Table 50-16   Characteristic Microscopic Features of Commonly Encountered Yeasts on Cornmeal Tween 80 Agar

Organism	Arthroconidia	Blastoconidia	Pseudohyphae or Hyphae
Candida albicans	—	Spherical clusters at regular intervals on pseudohyphae	Chlamydoconidia present on hyphae
C. glabrata	—	Small, spherical, and tightly compacted	None produced
C. krusei	—	Elongated; clusters occur at septa of pseudohyphae	Branched pseudohyphae
C. parapsilosis	—	Present but not characteristic	Sagebrush appearance; large (giant) hyphae present
C. kefyr (pseudotropicalis)	—	Elongated, lie parallel to pseudohyphae	Pseudohyphae present but not characteristic
C. tropicalis	—	Produced randomly along hyphae or pseudohyphae	Pseudohyphae present but not characteristic
Cryptococcus	—	Round to oval, vary in size, separated by a capsule	Rare; usually not seen
Saccharomyces	—	Large and spherical	Rudimentary hyphae sometimes present
Trichosporon	Numerous; resemble *Geotrichum*	May be present but difficult to find	Septate hyphae present

that include the BactiCard Candida (REMEL Laboratories, Lenexa, Kan), Murex *C. albicans*-50 (Murex Diagnostics, Norcross Ga), and Albicans-sure (Clinical Standards Laboratories, Rancho Dominguez, Calif) were designed to detect these enzymes.

Heelan, Siliezar, and Coon[55] compared the germ-tube test to BactiCard, Murex C. albicans-50, Albicans-sure, and the API 20C AUX yeast identification system. All rapid enzymatic screening methods were sensitive and specific for the rapid identification of *C. albicans*. Compared with the germ-tube test, all required less time (5 to 30 minutes), were more expensive, and required some additional equipment. Overall, all methods provided rapid and objective alternatives to the germ-tube test. CHROMagar Candida is also a product that utilizes enzymatic reactions to differentiate *C. albicans* and several other yeast species.[7,89,90,106,137] More recently *C. albicans* PNA FISH has been release for the detection and differentiation of *C. albicans* from non-albicans *Candida* species directly in positive blood cultures that contain yeast.[4,98,113,154] The use of any new or additional testing for the identification of yeasts should be submitted to a financial impact and outcomes analysis and be compared with the traditional methods of identification. Thereafter, the medical director, in conjunction with the medical staff may decide what is the optimal approach for the laboratory.

Cryptococcus neoformans. The microscopic examination of colonies of *C. neoformans* may be of help in providing a tentative identification of *C. neoformans,* because the cells are spherical and exhibit wide variation in size. A presumptive identification of *C. neoformans* may be based on rapid urease production and failure to utilize an inorganic nitrate substrate. The final identification of *C. neoformans* is usually based on typical substrate utilization patterns and, in some laboratories, pigment production on niger seed agar (Figure 50-95). A discussion of tests useful for the identification of *C. neoformans* and other species of cryptococci is presented in a subsequent section.

Rapid Uease Test. The rapid urease test (Procedure 50-10) is a most useful tool for screening for urease-producing yeasts recovered from respiratory secretions and other clinical specimens. Alternatives to this method include the heavy inoculum of the tip of a slant of Christensen's urea agar and subsequent incubation at 35° to 37° C. In many instances, a positive reaction will occur within several hours; however, 1 to 2 days of incubation may be required. It is of interest to note that strains of *Rhodotorula,* some *Candida* species, and *Trichosporon* species hydrolyze urea with time, so a distinction should be made between a traditional urease test, which takes hours, and the rapid urease test. In

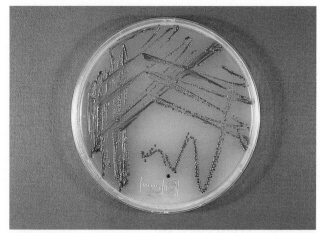

Figure 50-95   The colonies *C. neoformans* are brown when grown on Niger seed agar.

addition, the microscopic morphologic features of the yeast in question are helpful in interpreting the usefulness of the traditional urease test. An alternative to traditional urease testing is the rapid selective urease test (Procedure 50-11). This method appears to be useful in rapidly detecting *C. neoformans*. It is helpful to use these screening tests to make a presumptive identification of *C. neoformans*. In instances in which inoculum is limited, the laboratorian must use tests that can be performed and then prepare a subculture so that additional tests can be performed later. It is often just as fast to inoculate the organism onto the surface of a plate of niger seed agar; results may be obtained during the same day of incubation at 25° C. This is discussed further within this section.

All of the tests mentioned provide a tentative identification of *C. neoformans;* however, they must be supplemented with additional tests (usually the results of a commercial system in conjunction with cornmeal agar morphology) before a final identification can be reported. Additional tests useful in the identification of potential cryptococci are the nitrate reduction test and the detection of phenol oxidase production.

Trichosporon species. The presence of contiguous arthroconidia that are rectangular, often with rounded ends, predominate along with septate hyaline hyphae raise the possibility of *Trichosporon*. Blastoconidia are sometimes present but are not seen in all cultures. Urease production is helpful in differentiating *Trichosporon* species, which are positive, from *Blastoschizomyces* and *Geotrichum* species, which are negative. The final identification is based on the characteristic substrate utilization profile.

Malassezia. *Malassezia furfur* may be recovered from the blood of patients who have fungemia. In most instances, the residual lipid (from lipid replacement therapy) is adequate to support the primary growth of the organism in the blood culture. However, subculture onto additional media requires overlayment of the inoculum by olive oil or another source of long-chain fatty acids. These findings, in conjunction with the "bowling pin" or "pop bottle" morphology, are sufficient for identification. Other *Malassezia* species do not require long-chain fatty acids and are traditionally identified using substrate utilization analysis in conjunction with cornmeal agar morphology.

COMMERCIALLY AVAILABLE YEAST IDENTIFICATION SYSTEMS

Commercially available yeast identification systems have provided laboratories of all sizes with standardized identification methods. The methods, for the most part, are rapid and provide results within 24 to 72 hours. The major advantage is that the systems provide an identification based on a database of thousands of yeast biotypes that considers a number of variations and substrate utilization patterns. Another advantage is that manufacturers of these products provide computer consultation services to help the laboratorian with the identification of isolates that give an atypical result. Although these are powerful tools, it is not recommended that these be used as the sole method of identification; traditionally, these are used most effectively in conjunction with yeast morphology on cornmeal agar.

API-20C AUX YEAST SYSTEM

The API-20C AUX yeast identification system (bio-Mérieux, Durham, NC) has perhaps the most extensive computer-based data set of all commercial systems available. The system consists of a strip that contains 20 microcupules, 19 of which contain dehydrated substrates for determining utilization profiles of yeasts. Reactions are compared with growth in the first cupule, which lacks a carbohydrate substrate. Reactions are read and results are converted to a seven-digit biotype profile number. Most of the yeasts are identified within 48 hours; however, species of *Cryptococcus* and *Trichosporon* may require up to 72 hours. The API-20C AUX yeast identification system, as well as all the other commercially available products, requires that the microscopic morphologic features of yeast grown on cornmeal agar containing 1% Tween 80 and trypan blue be used in conjunction with the substrate utilization patterns. This is particularly helpful when more than one possibility for an identification is provided and the microscopic morphologic features can be used to distinguish bet-

ween the possibilities given by the profile register. Several evaluations of the API-20C AUX yeast identification system have been made, and results have all been favorable. This system is limited in that it cannot identify unusual species; however, most of those seen in the clinical laboratory are accurately identified to the species level.

UNI-YEAST TEK SYSTEM

The Uni-Yeast Tek yeast identification system (Remel Laboratories, Lexena, Kan) consists of a sealed, multi-compartment plate containing media used to indicate carbohydrate utilization, nitrate utilization, urease production, and cornmeal agar morphology. Past evaluations of this product have shown that it is generally satisfactory for the identification of commonly encountered yeasts[149]; however, it is not widely used currently.

MICROSCAN YEAST IDENTIFICATION PANEL

The MicroScan Yeast Identification Panel (Baxter-MicroScan, West Sacramento, Calif) is a 96-well microtiter plate containing 27 dehydrated substrates and was introduced as an alternative to the API-20C AUX yeast identification system. It utilizes chromogenic substrates to assess specific enzyme activity detected within 4 hours. Specific enzyme profiles have been generated for many of the yeasts commonly encountered in the clinical microbiology laboratory. The most recent evaluation of the method showed that it was moderately accurate within 4 hours using no supplementary tests. When supplementary tests were used, the sensitivity was excellent compared with the API-20C AUX yeast identification system. The accuracy for the identification of common yeasts was high, while uncommon yeasts were identified in most instances.[125] An evaluation of the MicroScan Yeast Identification Panel used with the Baxter MicroScan WalkAway was reported by Riddle and colleagues[112] and was found to be less accurate than when the panel was used in conjunction with microscopic morphology on cornmeal agar. It is recommended that Cornmeal Tween-80 agar be used in conjunction with the Yeast Identification Panel.

VITEK BIOCHEMICAL CARDS

The Yeast Biochemical Card (bioMérieux, Durham, NC) is a 30-well disposable plastic card that contains conventional biochemical tests and negative controls. The Yeast Biochemical Card is used with the automated Vitek Legacy System, which is used for bacterial identification and susceptibility testing in many laboratories. The most recent evaluation of this system showed an overall accuracy of identification near 100% compared with API-20C AUX. Less than one fourth of the yeasts required supplemental biochemical or morphologic features to confirm their identification. Of all correctly identified yeasts, more than half were reported after 24 hours of incubation.[36] The accuracy of identification of common and uncommon species was satisfactory. It is not necessary to identify germ tube–positive yeasts using this system. For laboratories already using this system, accurate and reliable identification of most commonly encountered yeasts can be accomplished. Another evaluation by Dooley, Beckius, and Jeffrey[31] showed that the Vitek YBC system correctly identified 93% of common yeasts; however, only 55% of uncommon yeasts were correctly identified. The latter included isolates such as *Candida guilliermondii, Candida krusei, Candida lambica, Candida lusitaniae, Candida rugosa,* species of *Cryptococcus* (non-*neoformans*), *Geotrichum candidum, Rhodotorula, Saccharomyces,* and *Trichosporon.* The YBC on the Vitek Legacy system is being replaced by the Yeast Identification Card on the Vitek II system.[7] The YBC on the Vitek II has an expanded database. A yeast susceptibility card is also in evaluations for this newer, more automated system.

Interest in commercially available yeast identification systems has taken precedence over the more cumbersome and labor intensive conventional yeast identification methods. Currently the rapid identification methods are financially feasible and provide the capability of identifying yeasts to laboratories of all sizes. Commercially available systems are recommended and provide accurate and rapid identification of yeasts and yeast-like organisms. These systems are recommended for all laboratories. In general, they are easy to use, easy to interpret, and relatively inexpensive compared with conventional methods. In most instances, they are faster than conventional systems, provide more standardized results, and require less technical skill to perform. As with any system, uncommon identifications should be scrutinized to avoid misidentifications.

CHROMAGAR

CHROMagar Candida (BD Diagnostics) is a differential medium useful for the recovery, isolation of colonies, and differentiation of species of *Candida* found in clinical specimens.[7,89,90,106,137] Different yeast species react with chromogenic substrates to yield a characteristic colony color. When used with colonial morphologic features, it can provide a presumptive identification. Sand-Millan, Ribacoda, and Ponton[119] reported an evaluation of 1537 isolates of yeast, which after 48 hours of incubation at 37° C showed that CHROMagar had a sensitivity and specificity near 100% for *C. albicans, C. tropicalis,* and

Candida krusei Another evaluation by Pfaller, Houston, and Coffmann[106] showed that more than 95% of stock and clinical isolates of *C. albicans, C. tropicalis,* and *C. krusei* were correctly identified. Furthermore, a similar sensitivity was observed for *Candida glabrata.* In addition, CHROMagar was evaluated as a recovery medium and was found to detect mixed cultures of *Candida* species. Considering that the previously mentioned species account for approximately 90% of the yeast recovered in the clinical laboratory, CHROMagar appears to be a suitable alternative to the other yeast identification systems.

RaPID YEAST PLUS SYSTEM

The RapID Yeast Plus System (Innovative Diagnostic Systems, Norcross, Ga) is a qualitative micromethod using conventional and chromogenic substrates to identify the medically important yeasts. Kitch and colleagues[68] showed that the system correctly identified more than 90% of 286 strains tested within 5 hours, without the need for additional tests. It appears that the RapID Yeast Plus System can be a useful alternative to other commercial systems.

CONVENTIONAL YEAST IDENTIFICATION METHODS[54,149]

A few laboratories still prefer to use conventional methods for the identification of yeasts. Regardless of the type of identification system used, the germ-tube test is often the first step in screening a large number of isolates, unless another screening test for *C. albicans* (e.g., PNA FISH) has been used. As previously mentioned, approximately 75% of yeasts recovered in the clinical laboratory can be identified using the germ-tube test.

CORNMEAL AGAR MORPHOLOGY

The second major step using this practical identification schema is to use cornmeal agar morphology as a means to determine if the yeast produces blastoconidia, arthroconidia, pseudohyphae, true hyphae, and/or chlamydoconidia (Procedure 50-12). Cornmeal agar morphology may be used successfully for the detection of characteristic chlamydoconidia produced by *C. albicans.* This method is currently satisfactory for the definitive identification of *C. albicans* when the germ-tube test is negative. In other instances, microscopic morphologic features on cornmeal agar help to differentiate the genera *Cryptococcus, Saccharomyces, Candida, Geotrichum,* and *Trichosporon.* Previously, it was believed that the morphologic features of the common species of *Candida* were distinct enough to provide a presumptive identification. This can be accomplished for *C. albicans, C. glabrata, C. krusei, C. parapsilosis, C. tropicalis,* and *C. kefyr* if one keeps in mind that there are numerous other species, uncommonly recovered in the clinical laboratory, that might resemble microscopically any of the previously mentioned species. In general, this method performs well since the previously mentioned genera and species are more commonly seen in clinical laboratories. For the uncommonly encountered isolates, cornmeal agar morphology has less value. It is recommended as an adjunct test for use with most commercially available yeast identification systems. It aids in the differentiation of yeasts hat yield similar biochemical profiles, and helps avoid misidentifications, particularly of less commonly encountered species that may not be well represented in the commercial database.

CARBOHYDRATE UTILIZATION

Carbohydrate utilization patterns are the most commonly used conventional methods for the definitive identification of yeast recovered in a clinical laboratory. Various methods have been advocated for use in determining carbohydrate utilization patterns by clinically important yeast, and all work equally well. Procedure 50-13 outlines the method previously found to be most useful by the Mayo Clinic Mycology Laboratory; however, this method is not commonly used in clinical laboratories. Most use commercially available methods.

Once the carbohydrate utilization profile is obtained, reactions may be compared with those listed in tables in most mycology laboratory manuals.[73] In most instances, carbohydrate utilization tests provide the definitive identification of an organism, and additional tests are unnecessary. Carbohydrate fermentation tests are preferred by some laboratories and are simply performed using purple broth containing different carbohydrate substrates. In general, carbohydrate fermentation tests are unnecessary and are not recommended for routine use.

PHENOLOXIDASE DETECTION USING NIGER SEED AGAR (PROCEDURE 50-14)

A simplified *Guizotia abyssinica* medium (niger seed medium) is a definitive method for detection of phenoloxidase production by yeasts.[66,102] Most isolates of *C. neoformans* readily produce phenoloxidase; however, some do not. In addition, in some instances, cultures of *C. neoformans* have been shown to contain both phenoloxidase-producing and phenoloxidase-deficient colonies within the same culture. If conventional methods are used, it is necessary to use all the criteria, including urease production, carbohydrate utilization, and the phenoloxidase test, before making a final identification of *C. neoformans*.

ANTIMICROBIAL SUSCEPTIBILITY TESTING AND THERAPY

ANTIFUNGAL AGENTS

Numerous antifungal agents have been developed, and newer agents are on the horizon. The increasing numbers of immunosuppressed patients and the expansion of drug resistance of microorganisms makes the development and appropriate use of antimicrobial agents one of the most important areas in microbiology and infectious diseases. This section is meant only to introduce the reader to the more commonly used antifungal agents and is by no means comprehensive. This section is not to be used as a guide for therapy. Therapeutic guidelines may be found in the referenced texts.

Polyene Macrolide Antifungals

Polyene macrolide antifungal agents consist of a group of complex organic molecules, most of which contain multiple, conjugated double bond and one to three ring structures. This group includes many of the most commonly used antifungal agents, such as amphotericin B, the colloidal and liposomal preparations of amphotericin B, nystatin, and griseofulvin.

Amphotericin B. Amphotericin B is produced by the actinomycete *Streptomyces nodosus*. It is commonly infused intravenously to treat deep-seated fungal infections (e.g., invasive aspergillosis), as well as those caused by *Candida* spp., *Cryptococcus,* and members of the Zygomycetes. Amphotericin B binds the ergosterol component of the fungal cell membrane and alters the selective permeability of this membrane. However, other sterols, including those present in mammalian cell membranes, are also bound. The most significant adverse reaction associated with amphotericin B therapy is renal insufficiency. The liposomal amphotericin B compounds reportedly diminish this adverse reaction. Although amphotericin B is active against a wide variety of fungi, there are resistant organisms that are important for the laboratory to be able to identify. Fungi resistant to amphotericin B include *Pseudallescheria boydii, Aspergillus terreus, Trichosporon* species, and in most instances *Fusarium* species.[75,105,144,155]

Nystatin. Nystatin, an antifungal antibiotic produced by *Streptomyces noursei,* is not absorbed by the gastrointestinal tract. It is principally used locally to treat

Procedure 50-1

CALCOFLUOR WHITE-POTASSIUM HYDROXIDE PREPARATION

METHOD

1. Place a drop of calcofluor white (CW) reagent and a drop of 10% potassium hydroxide (KOH) glycerin in the center of a microscope slide.
2. Add a portion of the clinical specimen to the CW-KOH solution and apply a coverslip.
3. If necessary, dissociate particles by applying gentle pressure to the coverslip with a pencil eraser. Allow to stand for 5 minutes. If particles do not dissociate, repeat this step.
4. Examination for fungal elements is performed using a fluorescent microscope with a 400- to 500-nm exciter filter and a 500- to 520-nm barrier filter. Slides are scanned at 10× magnification for fluorescent fungal elements. The presence and nature of fungal elements are discerned using 40× magnification.

Reagents

KOH reagent: (10 g KOH, 10 mL glycerin, and 80 mL distilled water)

CW reagent: (0.05 g CW, 0.02 g Evans blue, and 50 mL distilled water)

Calcofluor white is an industrial textile brightener, which nonspecifically binds to chitin and other elements in the fungal cell wall. Calcofluor white fluorescence occurs maximally at an excitation wavelength of 440 nm. Under these conditions, fungal elements will fluoresce blue-white. Because CW is a nonspecific stain, an appreciation for fungal element morphology on direct examination is crucial for adequate specimen interpretation. The presence of KOH in the solution dissolves human cellular elements and debris, which allows for easier visualization of fungal elements.

Procedure 50-2

ADHESIVE (SCOTCH) TAPE PREPARATION

METHOD

1. Touch the adhesive side of a small length of transparent tape to the surface of the colony.
2. Adhere the length of tape to the surface of a microscope slide to which a drop of lactophenol cotton or aniline blue has been added (see Figure 50-29).
3. Observe microscopically for the characteristic shape and arrangement of the spores.

The transparent adhesive tape preparation allows one to observe the organism microscopically approximately the way it sporulates in culture. The relationship of the spores, spore-producing structures (e.g., conidiophores), and the body of the fungus are usually intact, and the microscopic identification of an organism can be made easily. If the tape is not pressed firmly enough to the surface of the colony, the sample may consist only of conidia and not be adequate for an identification. When spores are not observed, a wet mount should be made. In some cases, the macroconidia of *H. capsulatum* were seen in wet mount preparations when the adhesive tape preparation revealed only hyphal fragments. On the contrary, in some instances, cultures have sporulated heavily and revealed only the presence of conidia when the adhesive tape preparation was observed. In this type of situation, a second adhesive tape preparation should be made from the periphery of the colony where sporulation is not as prominent.

Some laboratories prefer to use the microslide culture (Procedure 50-4) for making the microscopic identification of an organism. This method might appear to be the most suitable because it allows one to observe microscopically the fungus growing directly underneath the coverslip. Microscopic features should be easily discerned, structures should be intact, and many representative areas of growth are available for observation.

Procedure 50-3

WET MOUNT

METHOD

1. With a wire bent at a 90-degree angle, cut out a small portion of an isolated colony. The portion should be removed from a point intermediate between the center and the periphery. The portion removed should contain a small amount of the supporting agar.
2. Place the portion onto a slide to which a drop of lactophenol cotton or aniline blue has been added (see Figure 50-30).
3. Place a coverslip into position and apply gentle pressure with a pencil eraser or other suitable object to disperse the growth and the agar. Examine microscopically.

The major disadvantage of the wet mount is that the characteristic arrangement of spores is disrupted when pressure is applied to the coverslip. This method is suitable in many instances because characteristic spores are often seen but their arrangement cannot be determined. In some instances, it is not adequate to make a definitive identification.

Procedure 50-4

MICROSLIDE CULTURE

METHOD

1. Cut a small block of a suitable agar medium that has been previously poured into a culture dish to a depth of approximately 2 mm. The block may be cut using a sterile scalpel blade or with a sterile test tube that has no lip (which produces a round block).
2. Place a sterile microscope slide onto the surface of a culture dish containing sterile 2% agar. Alternatively, place a round piece of filter paper or paper towel into a sterile culture dish, add two applicator sticks, and position the microscope slide on top.
3. Add the agar block to the surface of the sterile microscope slide.
4. With a right-angle wire, inoculate the four quadrants of the agar plug with the organism (see Figure 50-31).
5. Apply a sterile coverslip onto the surface of the agar plug.
6. If the filter paper applicator stick method is used, add a small amount of sterile water to the bottom of the culture dish. Replace the lid of the culture and allow it to incubate at 30° C.
7. After a suitable incubation period, remove the coverslip (working inside of a biological safety cabinet) and place it on a microscope slide containing a drop of lactophenol cotton or aniline blue. Some suggest placing the coverslip near the opening of an incinerator-burner to allow rapid drying of the organism on the coverslip to occur before adding it to the stain.
8. Observe microscopically for the characteristic shape and arrangement of spores.
9. The remaining agar block may be used later (if the microslide culture is unsatisfactory for the microscopic identification) if it is allowed to incubate further. The agar plug is then removed and discarded, and a drop of lactophenol cotton or aniline blue is placed on the area of growth and a coverslip is positioned into place. Many laboratorians like to make two cultures on the same slide so that if characteristic microscopic features are not observed on examination of the first culture, the second will be available after an additional incubation period.

Although this method is ideal for making a definitive identification of an organism, it is the least practical of all the methods described. It should be reserved for those instances in which an identification cannot be made based on an adhesive tape preparation or wet mount.

CAUTION: Do not make slide cultures of slow-growing organisms suspected to be dimorphic pathogens such as *H. capsulatum, B. dermatitidis, C. immitis, P. brasiliensis,* or *S. schenckii.* Microslide cultures must be observed only after a coverslip has been removed from the agar plug and not while it is in position on top of the agar plug. The latter method of observation is very dangerous, because it could cause a laboratory-acquired infection.

Procedure 50-5

HAIR PERFORATION TEST

METHOD

1. Place a filter paper disk into the bottom of a sterile culture dish.
2. Cover the surface of the paper disk with sterile distilled water.
3. Add a small portion of sterilized hair into the water.
4. Inoculate a portion of the colony to be studied directly onto the hair.
5. Incubate at 25° C for 10 to 14 days.
6. Observe hairs regularly by placing them into a drop of water on a microscope slide. Position a coverslip and examine microscopically for the presence of conical perforations of the hair shaft (see Figure 50-50).

---

oral or vulvovaginal candidiasis. Toxicity of this drug is prohibitive to parenteral use.

Griseofulvin. Griseofulvin is an antifungal antibiotic produced by a species of *Penicillium.* Its mechanism of action consists of binding microtubular proteins, which are required for mitosis. Griseofulvin is an oral agent used to treat dermatophytoses, which are not responsive to azole antifungal therapy. Headache, gastrointestinal disturbances, and photosensitivity are a few of the adverse reactions, which limit the usefulness of this drug.

Antimetabolite

5-Fluorocytosine (Flucytosine). Flucytosine is a pyrimidine base, which is fluorinated in the fifth position. Flucytosine is metabolized to 5-fluorouracil, which is incorporated into fungal RNA. This subsequently inhibits protein synthesis. Flucytosine is also metabolized into fluorodeoxyuridine monophosphate, which is a potent inhibitor of DNA synthesis. Flucytosine and amphotericin B act synergistically and have been used in combination therapy for treating infections by *Candida* spp. and *C. neoformans.* Side effects and the

Procedure 50-6

IN VITRO CONVERSION OF DIMORPHIC MOLDS[43]

PRINCIPLE

Dimorphic molds exist in the yeast or spherule form in infected tissue. Proof that a mold is actually one of the systemic dimorphic fungi can be achieved by simulating the environment of the host and converting the mold to the yeast form. This method is recommended only for the identification of *B. dermatitis*, *P. marneffei*, or *S. schenckii*. Conversion of *C. immitis* to its spherule form requires special media or animal inoculation, and is not recommended for clinical laboratories.[148] *Histoplasma capsulatum* most often will not convert from the mold to the yeast form in vitro or will do so only after extended incubation at 35° to 37° C, so for these reasons it is not recommended.

METHOD

1. Transfer a large inoculum of the mold form of the culture onto the surface of a fresh, moist slant of brain-heart infusion agar containing 5% to 10% sheep blood. If *B. dermatitidis* is suspected, a tube of cottonseed conversion medium should be inoculated.
2. Add a few drops of sterile distilled water to provide moisture if the surface of the culture medium appears to be dry.

3. Leave the cap of the screw-capped tube slightly loose to allow the culture to have adequate oxygen exchange.
4. Incubate cultures at 35° to 37° C for several days; observe for the appearance of yeastlike portions of the colony. It may be necessary to make several subcultures of any growth that appears, because several transfers are often required to accomplish the conversion of many isolates. Cultures of *B. dermatitidis*, however, are usually easily converted and require 24 to 48 hours on cottonseed agar medium. *Coccidioides immitis* may be converted in vitro to the spherule form using a chemically defined medium; however, this method is of little use to the clinical laboratory and it should not be attempted. Genetic probe hybridization, DNA sequencing, species-specific PCR, or exoantigen detection are recommended methods of definitively identifying isolates suspected to be *Coccidioides*.

QUALITY CONTROL

Because of their hazardous nature, it is not recommended that stock cultures be tested routinely. An extract of control strains can be used as a positive control. An older test, exoantigen test (Proce-

dure 50-7), as well as more recently described methods can be used to identify the dimorphic fungi. Because the conversion of the dimorphic molds to the corresponding yeast or spherule forms is technically cumbersome to perform and long delays are often experienced, the effort to convert the dimorphic fungi is not recommended in the routine mycology laboratory. The exoantigen technique has been used in many laboratories to make a definitive identification of *B. dermatitidis*, *C. immitis*, *H. capsulatum*, and *P. brasiliensis*. The exoantigen test relies on the principle that soluble antigens are produced and can be extracted from fungi; they are concentrated and subsequently reacted with serum known to contain antibodies directed against the specific antigenic components of the organism being tested. Reagents and materials for the exoantigen test are available commercially.

---

emergence of the resistance when this drug is used alone has limited its utility.

Azole Antifungal Drugs

The azole group of antifungal agents consists of the imidazoles and the triazoles. These compounds contain six carbon ring structures with conjugated double bonds, chloride residues, and five carbon ring structures that contain at least two nitrogen molecules. Traditionally used agents included in this group include clotrimazole, miconazole, fluconazole, itraconazole, and ketoconazole. Newer triazoles consist of voriconazole, posaconazole, and most recently ravuconazole. Of the newer triazoles, only voriconazole is discussed further below, because, of these newer agents, it was the first released and has been the most thoroughly reviewed. These antifungal agents disrupt the integrity of the fungal cell membrane by interfering with the synthesis of ergosterol.[9]

Clotrimazole and Miconazole. The synthetic imidazoles clotrimazole and miconazole are covered together because of their many similarities. These agents are available for topical or intravaginal applications. They are useful in mild cases of dermatophytosis, including tinea versicolor. Adverse reactions are generally limited to burning, itching, and/or skin irritation.

Fluconazole. Fluconazole is a triazole, which is exceptionally soluble in water. The solubility of this compound allows either oral or intravenous administration. Fluconazole has excellent activity against most *Candida* spp. and *Cryptococcus neoformans*; therapeutic levels are easily reached in the central nervous system. Side effects of fluconazole therapy are usually minimal. The susceptibility of *C. glabrata* to fluconazole is not predictable. Isolates of *C. glabrata* may be susceptible, dose-dependant susceptible, or resistant to fluconazole. Other notable yeasts or yeast-like fungi

Procedure 50-7

EXOANTIGEN TEST[89,137]

PRINCIPLE

Specific antibodies developed against particular mycelial antigens will react in a gel immunodiffusion precipitin test. The mold forms of the dimorphic fungi can be identified definitively by an antigen-antibody reaction, negating the need for conversion to the yeast form.

METHOD

1. A mature fungus culture on a Sabouraud dextrose agar slant is covered with an aqueous solution of merthiolate (1:5000 final concentration), which is allowed to remain in contact with the culture for 24 hours at 25° C. It is necessary that the entire surface of the colony be covered so that effective killing of the organism is ensured and solubilization of the exoantigen is maximized.

2. Filter the aqueous solution, which overlays the culture, through a 0.45-μm size membrane filter. This should be performed inside a biological safety cabinet.

3. Five milliliters of this solution is concentrated using a Minicon Macrosolute B-15 Concentrator (Millipore, Billerica, Mass). The solution is concentrated 50× when testing with *H. capsulatum* and *B. dermatitidis* antiserum and 5× and 25× for reaction with *C. immitis* antiserum.

4. The concentrated supernatant is used in the microdiffusion test. The supernatant is placed into wells punched into a plate of buffered, phenolized agar adjacent to the control antigen well and is tested against positive control antiserum obtained from commercial sources (e.g., Immunomycologics, Washington, Okla; Meridian Diagnostics, Cincinnati, Ohio; and Gibson Laboratories, Lexington, Ky).

5. The immunodiffusion test is allowed to react for 24 hours at 25° C, and the plate is observed for the presence of precipitin bands of identity with the reference reagents. The sensitivity of the exoantigen test for the identification of *B. dermatitidis* may be increased by incubating the immunodiffusion plates at 37° C for 48 hours; however, bands appear sharper at 25° C after 24 hours. It is recommended that any culture suspected of being *B. dermatitidis* be incubated at both temperatures.

6. *Coccidioides immitis* may be identified for the presence of the CF, TP, or HL antigens, while *H. capsulatum* may be identified by the presence of H or M bands, or both, and *B. dermatitidis* may be identified by the A band. Detailed instructions for the performance and interpretation of the tests are included with the manufacturers' package inserts.

QUALITY CONTROL

Extracts from known fungi are tested each time the test is performed. Lines of identity with the unknown strain are necessary for identification.

The exoantigen test has been largely replaced by specific nucleic acid probes that provide for the rapid and specific identification of *H. capsulatum, B. dermatitidis,* and *C. immitis.*

---

that are resistant to fluconazole are *C. krusei* and *Rhodotorula* species.[105]

Ketoconazole. Ketoconazole is an imidazole, which is either taken orally or applied topically. It is useful in mild cases of paracoccidioidomycosis and is an alternative to amphotericin B for infections caused by *Blastomyces* or *Histoplasma*. Ketoconazole may be used if prolonged oral therapy for chronic mucocutaneous candidiasis is needed. One group has reported some success in the treatment of *P. boydii* infections with ketoconazole.[43] In vivo, ketoconazole is fungistatic, because fungicidal levels are not achievable with therapeutic concentrations. Adverse reactions include transient elevations in liver enzymes, nausea, and dose-related gynecomastia, decreased libido, and oligospermia in males.

Itraconazole. Itraconazole is a triazole with a spectrum of activity that encompasses that of ketoconazole. Additionally, itraconazole has been shown to be effective in cases of aspergillosis, sporotrichosis, cryptococcosis, and onchymycosis. Adverse reactions principally include gastrointestinal disturbances; however, vestibular disturbances, edema, and skin irritations have been reported.

Voriconazole. Voriconizole is one of the new triazoles. It has a expanded spectrum of activity compared with itraconazole. In addition to used described above for itraconazole. Voriconazole demonstrates useful activity against some *Fusarium* strains, and fluconazole resistant yeasts, such as *C. krusei* and *C. glabrata.* Significantly, the zygomycetes are resistant to voriconizole.[105] Transient visual disturbances, which may significantly alarm the patient if not forewarned, may occur, as may elevated liver enzymes.

Echinocandins

The echinocandins are glucan synthesis inhibitors. More specifically, they inhibit 1, 3 beta-glucan synthase, an enzyme important in fungal cell wall synthesis. The presence of these drugs leads to cellular osmotic instability. There are three echinocandins: caspofungin, micafungin, and anidulafungin. Caspofungin was the

Case Study 50-1

A 42-year-old airline pilot presented to the local hospital with increasing shortness of breath, fever, and chills. The patient was thought to have pneumonia and was treated with ceftriaxone, after blood and sputum cultures were collected. Because the patient had recently traveled to a malaria-endemic region, blood was sent for parasite examination. The thick and thin smears for *Plasmodium* species were negative. The patient's condition continued to decline, and he was referred to a tertiary care medical center. The blood cultures were negative. The sputum cultures revealed only normal respiratory flora. The patient underwent bronchoscopy with bronchoalveolar lavage (BAL). Direct examination of the BAL fluid using a calcofluor white/potassium hydroxide (KOH) preparation revealed moderate-sized yeast with broad-based buds. The patient was started immediately on amphotericin B.

QUESTIONS

1. What is the suspected organism?
2. How does the laboratory confirm the identity of this pathogen?
3. Where did the patient likely acquire his infection?

Case Study 50-2

A 36-year-old man with acute myelocytic leukemia has undergone bone marrow transplantation and was profoundly neutropenic. He developed a fever and punctate skin lesions during his hospitalization. Blood cultures and biopsy cultures of the skin lesions were taken. The biopsy of one of the skin lesions revealed hyaline septate hyphae with some acute angle branching and was diagnosed as "invasive aspergillosis" in pathology. The patient was started on amphotericin B but failed to improve. The Gram stain of a positive blood culture bottle revealed hyphae and curious sickle-shaped or canoe-shaped fungal structures. The culture from the biopsy grew a hyaline septate mold that produces similar canoe-shaped, multicellular macroconidia.

QUESTIONS

1. What is the etiologic agent of infection and how is the identification made?
2. Are patients such as the one described here at increased risk for fungal infections?
3. What was wrong with the histopathologic diagnosis?
4. Why did this patient not respond to amphotericin B?

REFERENCES

1. Abd El-Bagi ME, et al: Mycetoma of the hand, *Saudi Med J* 25: 352, 2004.
2. Adam RD, et al: Phaeohyphomycosis caused by the fungal genera *Bipolaris* and *Exserohilum*. A report of 9 cases and review of the literature, *Medicine (Baltimore)* 65:203, 1986.
3. Ahmed AO, et al: Mycetoma caused by *Madurella mycetomatis*: a neglected infectious burden, *Lancet Infect Dis.* 4:566, 2004.
4. Alexander BD, et al: Cost savings with implementation of PNA FISH testing for identification of *Candida albicans* in blood cultures, *Diagn Microbiol Infect Dis* 54:277, 2006.
5. Alvarez M, et al: Nosocomial outbreak caused by Scedosporium prolificans (inflatum): four fatal cases in leukemic patients, *J Clin Microbiol* 33:3290, 1995.
6. Anaissie E, et al: Azole therapy for trichosporonosis: clinical evaluation of eight patients, experimental therapy for murine infection, and review, *Clin Infect Dis* 15:781, 1992.
7. Aubertine CL, et al: Comparative study of the new colorimetric VITEK 2 yeast identification card versus the older fluorometric card and of CHROMagar *Candida* as a source medium with the new card, *J Clin Microbiol* 44:227, 2006.
8. Barenfanger J, et al: Improved outcomes associated with limiting identification of *Candida* spp. in respiratory secretions, *J Clin Microbiol* 41:5645, 2003.
9. Beggs W, Andrews F, Sarosi G: Actions of imidazole-containing antifungal drugs, *Life Sci* 28:111, 1981.
10. Bernstein EF, et al: Disseminated cutaneous *Pseudallescheria boydii*, *Br J Dermatol* 132:456, 1995.
11. Bialek R, et al: PCR assays for identification of *Coccidioides posadasii* based on the nucleotide sequence of the antigen 2/proline-rich antigen, *J Clin Microbiol* 42:778, 2004.
12. Bille J, et al: Evaluation of a lysis-centrifugation system for recovery of yeasts and filamentous fungi from blood, *J Clin Microbiol* 18:469, 1983.
13. Bille J, Edson RS, Roberts GD: Clinical evaluation of the lysis-centrifugation blood culture system for the detection of fungemia and comparison with a conventional biphasic broth blood culture system, *J Clin Microbiol* 19:126, 1984.
14. Bradsher R, McDonnell R: *Blastomyces dermatitidis* and *Paracoccidioides brasiliensis*. In Chmel H, Bendinelli M, and Friedman I, editors: *Pulmonary infections and immunity*, New York, 1990, Plenum Press.
15. Brasch J: [Pathogens and pathogenesis of dermatophytoses], *Hautarzt* 41:9, 1990.
16. Brummer E, Castaneda E, Restrepo A: Paracoccidioidomycosis: an update, *Clin Microbiol Rev* 6:89, 1993.
17. Bruns TD, et al: Evolutionary relationships within the fungi: analyses of nuclear small subunit rRNA sequences, *Mol Phylogenet Evol* 1:231, 1992.
18. Caporale NE, et al: Peritoneal catheter colonization and peritonitis with *Aureobasidium pullulans*, *Perit Dial Int* 16:97, 1996.
19. Carey J, et al: *Paecilomyces lilacinus* vaginitis in an immuno-competent patient, *Emerg Infect Dis* 9:1155, 2003.

20. Chagas C: Nova trypanosomiaze humana, *Mem Inst Oswaldo Cruz Rio J* 1:159, 1909.

21. Cherniak R, Sundstrom JB: Polysaccharide antigens of the capsule of *Cryptococcus neoformans, Infect Immun* 62:1507, 1994.

22. Cohen M, et al: Fungal infection in chronic granulomatous disease: the importance of the phagocyte in defense against fungi, *Am J Med* 71:59, 1981.

23. Converse J: Growth of spherules of *Coccidioides immitis* in a chemically defined medium, *Proc Soc Exp Biol Med* 90:709, 1956.

24. Cox R, Best G: Cell wall composition of two strains of *Blastomyces dermatitidis* exhibiting differences in virulence for mice, *Infect Immun* 5:449, 1972.

25. Currie B, Casadevall A: Estimation of the prevalence of cryptococcal infection among patients infected with the human immunodeficiency virus in New York City, *Clin Infect Dis* 19:1029, 1994.

26. Cutler J: Putative virulence factors of *Candida albicans, Annu Rev Microbiol* 45:187, 1991.

27. de Hoog G, Guarro J et al: *Atlas of clinical fungi*, t.N.a.R. Centraal bureau voor Schimell cultures/Universita Rovira i Virgili. Utrecht and Reus, Spain.

28. Deng Z, et al: Infections caused by *Penicillium marneffei* in China and Southeast Asia: review of eighteen published cases and report of four more Chinese cases, *Rev Infect Dis* 10:640, 1988.

29. Denz Z, Yun M, Ajello L: Human penicilliosis marneffei and its relation to the bamboo rat (*Rhizomys x pruinosus*), *J Med Vet Mycol* 24:383, 1986.

30. Diamond R: *Cryptococcus neoformans*. In Mandell G, Bennett J, Dolin R, editors: *Mandell, Douglas, and Bennett's principles and practice of infectious diseases*, New York, 1995, Churchill Livingstone.

31. Dooley D, Beckius M, Jeffrey B: Misidentification of clinical yeast isolates by using updated Vitek Yeast Biochemical Card, *J Clin Microbiol* 32:2889, 1994.

32. Douer D, et al: Human *Exserohilum* and *Bipolaris* infections: report of *Exserohilum* nasal infection in a neutropenic patient with acute leukemia and review of the literature, *J Med Vet Mycol* 25:235, 1987.

33. Drutz D, et al: Human sex hormones stimulate the growth and maturation of *Coccidioides immitis, Infect Immun* 32:897, 1981.

34. Edman J, et al: Ribosomal RNA sequence shows *Pneumocystis* to be a member of the fungi, *Nature* 334:519, 1988.

35. Edman J, et al: Ribosomal RNA genes of *Pneumocystis carinii, J Protozool* 36:18S, 1989.

36. El-Zaatari M, et al: Evaluation of the updated Vitek Yeast Identification database, *J Clin Microbiol* 28:1938, 1990.

37. Ener B, Douglas L: Correlation between cell-surface hydrophobicity of *Candida albicans* and adhesion to buccal epithelial cells, *FEMS Microbiol Lett* 78:37, 1992.

38. Espinel-Ingroff A: Antifungal susceptibility testing, *Clin Microbiol Newsl* 184:161, 1996.

39. Espinel-Ingroff A, Pfaller M: Antifungal agents and susceptibility testing. In Murray P, et al, editors: *Manual of clinical microbiology*, Washington, DC, 1995, ASM Press.

40. Fleming RV, Walsh TJ, Anaissie EJ: Emerging and less common fungal pathogens, *Infect Dis Clin North Am* 16:915, 2002.

41. Frey C, Drutz D: Influence of fungal surface components on the interaction of *Coccidioides immitis* with polymorphonuclear neutrophils, *J Infect Dis* 153:933, 1986.

42. Friedank H: Hyalohyphomycoses due to *Fusarium* spp: two case reports and review of the literature, *Mycoses* 38:69, 1995.

43. Galgiani J, et al: *Pseudallescheria boydii* infections treated with ketoconazole: clinical evaluations of seven patients and in vitro susceptibility results, *Chest* 86:219, 1984.

44. Galgiani J, et al: Reference method for broth dilution antifungal susceptibility testing of yeasts; Approved standard M27-A, Wayne, Pa, 1996, National Committee for Clinical Laboratory Standards.

45. Girmenia C, et al: Invasive infections caused by *Trichosporon* species and *Geotrichum capitatum* in patients with hematological malignancies: a retrospective multicenter study from Italy and review of the literature, *J Clin Microbiol* 43:1818, 2005.

46. Giuntuli D, Stringer S, Stringer J: Extraordinary low number of ribosomal RNA genes in *P. carinii, J Eukaryot Microbiol* 41:88S, 1994.

47. Gray L, Roberts G: Laboratory diagnosis of systemic fungal disease, *Infect Dis Clin North Am* 2:779, 1988.

48. Guarro J, et al: *Acremonium* species: new emerging fungal opportunists—in vitro antifungal susceptibilities and review, *Clin Infect Dis* 25:1222, 1997.

49. Guinet R, et al: Fatal septicemia due to amphotericin B-resistant *Candida lusitaniae, J Clin Microbiol* 18:443, 1983.

50. Gutierrez-Rodero F, et al: Cutaneous hyalohyphomycosis caused by *Paecilomyces lilacinus* in an immunocompetent host successfully treated with itraconazole: case report and review, *Eur J Clin Microbiol Infect Dis* 18:814, 1999.

51. Hageage G, Harrington B: Use of calcofluor white in clinical mycology, *Lab Med* 15:109, 1984.

52. Hall G, Pratt-Rippin K, Washington J: Evaluation of a chemiluminescent probe assay for identification of *Histoplasma capsulatum* isolates, *J Clin Microbiol* 30:3003, 1992.

53. Harari A, et al: Effects of time lapse between sputum collection and culturing on isolation of clinically significant fungi, *J Clin Microbiol* 15:425, 1982.

54. Hazen K: New and emerging yeast pathogens, *Clin Microbiol Rev* 8:462, 1995.

55. Heelan J, Siliezar D, Coon K: Comparison of rapid testing methods for enzyme production with the germ tube method for presumptive identification of *Candida albicans, J Clin Microbiol* 34:2847, 1996.

56. Heinic G, et al: Oral *Geotrichum candidum* infection associated with HIV infection: a case report, *Oral Surg Oral Med Oral Pathol* 73:726, 1992.

57. Hogan L, Klein B, Levitz S: Virulence factors of medically important fungi, *Clin Microbiol Rev* 9:469, 1996.

58. Hopfer R, Walden P, Setterquist S: Detection and differentiation of fungi in clinical specimens using polymerase chain reaction (PCR) amplification and restriction enzyme analysis, *J Med Vet Mycol* 31:65, 1993.

59. Hopfer R: Use of molecular biological techniques in the diagnostic laboratory for detecting and differentiating fungi, *Arch Med Res* 26:287, 1995.

60. Hostetter M: Adhesions and ligands involved in the interaction of *Candida* spp with epithelial and endothelial surfaces, *Clin Microbiol Rev* 7:29, 1994.

61. Huffnagle G, Chen G, Curtis J: Down-regulation of the afferent phase of T cell-mediated pulmonary inflammation and immunity by a high melanin-producing strain of *Cryptococcus neoformans, J Immunol* 155:3607, 1995.

62. Huppert M, Sun S, Bailey J: Natural variability in *Coccidioides immitis*. In Ajello L, editor: *Coccidioidomycosis*, Tucson, 1967, University of Arizona Press.

63. Ibrahim A, et al: Evidence implicating phospholipase as a virulence factor of *Candida albicans, Infect Immun* 63:1993, 1995.

64. Kane J, Smitka C: Early detection and identification of *Trichophyton verrucosum, J Clin Microbiol* 8:740, 1978.

65. Kaneshiro E, et al: Evidence for the presence of "metabolic sterols" in *Pneumocystis*: identification and initial characterization of *Pneumocystis carinii* sterols, *J Eukaryot Microbiol* 41:78, 1994.

66. Kaufman C, Merz W: Two rapid pigmentation tests for identification of *Cryptococcus neoformans, J Clin Microbiol* 15:339, 1982.

67. Kennedy M, Sigler L: *Aspergillus, Fusarium* and other moniliaceous fungi. In Murray P, Baron E, Pfaller M, editors: *Manual of clinical microbiology*, Washington, DC, 1995, American Society for Microbiology Press.

68. Kitch T, et al: Ability of Rapid Yeast Plus system to identify 304 clinically significant yeasts within 5 hours, *J Clin Microbiol* 34:1069, 1996.

69. Kolattukudy PE, et al: Evidence for possible involvement of an elastolytic serine protease in aspergillosis, *Infect Immun* 61:2357, 1993.

70. Koneman E, Roberts G: *Practical laboratory mycology*, ed 3, Baltimore, 1985, Williams & Wilkins.

71. Kontoyiannis DP, et al: Zygomycosis in the 1990s in a tertiary-care cancer center, *Clin Infect Dis* 30:851, 2000.

72. Kozel TR, et al: Role of the capsule in phagocytosis of *Cryptococcus neoformans*, *Rev Infect Dis* 10:S436, 1988.

73. Kreger-Van Rij N: *The yeasts: A taxonomic study*, New York, 1984, Elsevier Science Publishing.

74. Kwon-Chung K, Bennet J: *Medical mycology*, Philadelphia, 1992, Lea & Febiger.

75. Lass-Florl C, et al: Epidemiology and outcome of infections due to *Aspergillus terreus*: 10-year single centre experience, *Br J Haematol* 131:201, 2005.

76. Li JS, et al: Disseminated penicilliosis marneffei in China. Report of three cases, *Chin Med J (Engl)* 104:247, 1991.

77. Libertin CR, Wilson WR, Roberts GD: *Candida lusitaniae*—an opportunistic pathogen, *Diagn Microbiol Infect Dis* 3:69, 1985.

78. Lopez-Martinez R, et al: [Exoenzymes of dermatophytes isolated from acute and chronic tinea], *Rev Latinoam Microbiol* 36:17, 1994.

79. Lu JJ, et al: Comparison of six different PCR methods for detection of *Pneumocystis carinii*, *J Clin Microbiol* 33:2785, 1995.

80. Madariaga MG, Tenorio A, Proia L: *Trichosporon inkin* peritonitis treated with caspofungin, *J Clin Microbiol* 41:5827, 2003.

81. Makimura K, Murayama SY, Yamaguchi H: Detection of a wide range of medically important fungi by the polymerase chain reaction, *J Med Microbiol* 40:358, 1994.

82. Marcon MJ, Powell DA: Epidemiology, diagnosis, and management of *Malassezia furfur* systemic infection, *Diagn Microbiol Infect Dis* 7:161, 1987.

83. Markaryan A, et al: Purification and characterization of an elastinolytic metalloprotease from *Aspergillus fumigatus* and immunoelectron microscopic evidence of secretion of this enzyme by the fungus invading the murine lung, *Infect Immun* 62:2149, 1994.

84. Martagon-Villamil J, et al: Identification of *Histoplasma capsulatum* from culture extracts by real-time PCR, *J Clin Microbiol* 41:1295, 2003.

85. McEwen JG, et al: Experimental murine paracoccidioidomycosis induced by the inhalation of conidia, *J Med Vet Mycol* 25:165, 1987.

86. McGinnis MR, Rinaldi MG, Winn RE: Emerging agents of phaeohyphomycosis: pathogenic species of *Bipolaris* and *Exserohilum*, *J Clin Microbiol* 24:250, 1986.

87. Meis JF, et al: Severe osteomyelitis due to the zygomycete *Apophysomyces elegans*, *J Clin Microbiol* 32: 3078, 1994.

88. Merz W, Roberts G: Detection and recovery of fungi from clinical specimens. In Murray P, et al, editors: *Manual of clinical microbiology*, Washington, DC, 1995, ASM Press.

89. Murray CK, et al: Use of chromogenic medium in the isolation of yeasts from clinical specimens, *J Med Microbiol* 54:981, 2005.

90. Murray MP, Zinchuk R, Larone DH: CHROMagar Candida as the sole primary medium for isolation of yeasts and as a source medium for the rapid-assimilation-of-trehalose test, *J Clin Microbiol* 43:1210, 2005.

91. Murray P: *ASM pocket guide to clinical microbiology*, Washington, DC, 1996, ASM Press.

92. Murray P, Van Scoy R, Roberts GD: Should yeasts in respiratory secretions be identified? *Mayo Clin Proc* 52:42, 1977.

93. Musial CE, Cockerill FR 3rd, Roberts GD: Fungal infections of the immunocompromised host: clinical and laboratory aspects, *Clin Microbiol Rev* 1:349, 1988.

94. Neely AN, Orloff MM, Holder IA: *Candida albicans* growth studies: a hypothesis for the pathogenesis of *Candida* infections in burns, *J Burn Care Rehabil* 13:323, 1992.

95. Neilson JB, Fromtling RA, Bulmer GS: *Cryptococcus neoformans*: size range of infectious particles from aerosolized soil, *Infect Immun* 17:634, 1977.

96. Nucci M: Emerging moulds: *Fusarium, Scedosporium* and *Zygomycetes* in transplant recipients, *Curr Opin Infect Dis* 16:607, 2003.

97. Odabasi Z, et al: Beta-D-glucan as a diagnostic adjunct for invasive fungal infections: validation, cutoff development, and performance in patients with acute myelogenous leukemia and myelodysplastic syndrome, *Clin Infect Dis* 39:199, 2004.

98. Oliveira K, et al: Differentiation of *Candida albicans* and *Candida dubliniensis* by fluorescent in situ hybridization with peptide nucleic acid probes, *J Clin Microbiol* 39:4138, 2001.

99. Ostrosky-Zeichner L, et al: Multicenter clinical evaluation of the (1-3) beta-D-glucan assay as an aid to diagnosis of fungal infections in humans, *Clin Infect Dis* 41:654, 2005.

100. Padhye A: Fungi causing eumycotic mycetoma. In Murray P, et al, editors: *Manual of clinical microbiology*, Washington, DC, 1995, ASM Press.

101. Palaoglu S, et al: Cerebral phaeohyphomycosis, *Neurosurgery* 33:894, 1993.

102. Paliwal DK, Randhawa HS: Evaluation of a simplified Guizotia abyssinica seed medium for differentiation of *Cryptococcus neoformans*, *J Clin Microbiol* 7:346, 1978.

103. Pang KR, et al: Subcutaneous fungal infections, *Dermatol Ther* 17:523, 2004.

104. Patel R, et al: Phaeohyphomycosis due to *Scopulariopsis brumptii* in a liver transplant recipient, *Clin Infect Dis* 19:198, 1994.

105. Pfaller MA, Diekema DJ: Rare and emerging opportunistic fungal pathogens: concern for resistance beyond *Candida albicans* and *Aspergillus fumigatus*, *J Clin Microbiol* 42):4419, 2004.

106. Pfaller MA, Houston A, Coffmann S: Application of CHROMagar Candida for rapid screening of clinical specimens for *Candida albicans, Candida tropicalis, Candida krusei*, and *Candida (Torulopsis) glabrata*, *J Clin Microbiol* 34:58, 1996.

107. Pfeiffer CD, Fine JP, Safdar N: Diagnosis of invasive aspergillosis using a galactomannan assay: a meta-analysis, *Clin Infect Dis* 42:1417, 2006.

108. Procop GW, Roberts GD: Emerging fungal diseases: the importance of the host, *Clin Lab Med* 24:691, 2004.

109. Procop GW, et al: Detection of *Pneumocystis jiroveci* in respiratory specimens by four staining methods, *J Clin Microbiol* 42:3333, 2004.

110. Procop GW, et al: Performance of five agar media for recovery of fungi from isolator blood cultures, *J Clin Microbiol* 38:3827, 2000.

111. Richardson M, Shankland G: *Rhizopus, Rhizomucor, Absidia*, and other agents of subcutaneous zygomycetes. In Murray P, et al, editors: *Manual of clinical microbiology*, Washington, DC, 1995, ASM Press.

112. Riddle DL, et al: Clinical comparison of the Baxter MicroScan Yeast Identification Panel and the Vitek Yeast Biochemical Card, *Am J Clin Pathol* 101:438, 1994.

113. Rigby S, et al: Fluorescence in situ hybridization with peptide nucleic acid probes for rapid identification of *Candida albicans* directly from blood culture bottles, *J Clin Microbiol* 40:2182, 2002.

114. Roberts GD: Detection of fungi in clinical specimens by phase-contrast microscopy, *J Clin Microbiol* 2:261, 1975.

115. Roberts GD, Karlson AG, DeYoung DR: Recovery of pathogenic fungi from clinical specimens submitted for mycobacteriological culture, *J Clin Microbiol* 3:47, 1976.

116. Roberts GD, et al: Rapid urea broth test for yeasts, *J Clin Microbiol* 7:584, 1978.

117. Sakayama K, et al: Mycetoma of foot: a rare case report and review of the literature, *Foot Ankle Int* 25:763, 2004.

118. Sandhu GS, et al: Molecular probes for diagnosis of fungal infections, *J Clin Microbiol* 33:2913, 1995.

119. Sand-Millan R, et al: Evaluation of a commercial medium for identification of *Candida* sp., *Eur J Clin Microbiol Infect Dis* 15:153, 1996.

120. Schell WA, Perfect JR: Fatal, disseminated *Acremonium strictum* infection in a neutropenic host, *J Clin Microbiol* 34:1333, 1996.

121. Schell W, et al: *Bipolaris, Exophiala, Scedosporium, Sporothrix* and other dematiaceous fungi. In Murray P, et al, editors: *Manual of clinical microbiology*, Washington, DC, 1995, ASM Press.

122. Skoulidis F, Morgan MS, MacLeod KM: *Penicillium marneffei*: a pathogen on our doorstep? *J R Soc Med* 97:394, 2004.

123. Smith C, Goodman N: Improved culture method for the isolation of *Histoplasma capsulatum* and *Blastomyces dermatitidis* from contaminated specimens, *Am J Clin Pathol* 68:276, 1975.

124. Spielberger RT, et al: Fatal *Scedosporium prolificans (S. inflatum)* fungemia following allogeneic bone marrow transplantation: report of a case in the United States, *Clin Infect Dis* 21:1067, 1995.

125. St. Germain G, Beauchesne D: Evaluation of the MicroScan Rapid Yeast Identification panel, *J Clin Microbiol* 29:2296, 1991.

126. St. Germain G, Summerbell R: *Identifying filamentous fungi: a clinical handbook*, Belmont, Calif, 1996, Star Publishing.

127. Standard PG, Kaufman L: A rapid and specific method for the immunological identification of mycelial form cultures of *Paracoccidioides brasiliensis, Curr Microbiol* 4:297, 1980.

128. Standard P, Kaufman L: Safety considerations in handling exoantigen extracts from pathogenic fungi, *J Clin Microbiol* 15:663, 1982.

129. Steinbach WJ, et al: *Scedosporium prolificans* osteomyelitis in an immunocompetent child treated with voriconazole and caspofungin, as well as locally applied polyhexamethylene biguanide, *J Clin Microbiol* 41:3981, 2003.

130. Stevens DA: The interface of mycology and endocrinology, *J Med Vet Mycol* 27:133, 1989.

131. Stockman L, et al: Evaluation of commercially available acridinium ester-labeled chemiluminescent DNA probes for culture identification of *Blastomyces dermatitidis, Coccidioides immitis, Cryptococcus neoformans*, and *Histoplasma capsulatum, J Clin Microbiol* 31:845, 1993.

132. Strimlan CV, et al: Respiratory tract *Aspergillus*: clinical significance, *Minn Med* 63:25, 1980.

133. Stringer JR: *Pneumocystis carinii*: what is it, exactly? *Clin Microbiol Rev* 9:489, 1996.

134. Stringer SL, et al: *Pneumocystis carinii*: sequence from ribosomal RNA implies a close relationship with fungi, *Exp Parasitol* 68:450, 1989.

135. Sugita T, et al: Taxonomic position of deep-seated, mucosa-associated, and superficial isolates of *Trichosporon cutaneum* from trichosporonosis patients, *J Clin Microbiol* 33:1368, 1995.

136. Sun SH, Huppert M, Vukovich KR: Rapid in vitro conversion and identification of *Coccidioides immitis, J Clin Microbiol* 3:186, 1976.

137. Tan GL, Peterson EM: CHROMagar Candida medium for direct susceptibility testing of yeast from blood cultures, *J Clin Microbiol* 43:1727, 2005.

138. Terreni AA, Strohecker JS, Dowda H Jr: *Candida lusitaniae* septicemia in a patient on extended home intravenous hyperalimentation, *J Med Vet Mycol* 25:63, 1987.

139. Tintelnot K, et al: Systemic mycosis caused by a new *Cladophialophora* species, *J Med Vet Mycol* 33:349, 1995.

140. Torres HA, Raad II, Kontoyiannis DP: Infections caused by *Fusarium* species, *J Chemother* 15(suppl 2):28, 2003.

141. Treger TR, et al: Diagnosis of pulmonary infection caused by *Aspergillus*: usefulness of respiratory cultures, *J Infect Dis* 152:572, 1985.

142. Van der Peer Y, et al: Evolution of basidiomycetous yeasts as deduced from small ribosomal subunit RNA sequences, *Syst Appl Microbiol* 15:250, 1992.

143. Vartivarian SE: Virulence properties and nonimmune pathogenetic mechanisms of fungi, *Clin Infect Dis* 14:S30, 1992.

144. Walsh M, et al: Fungal *Pseudoallescheria boydii* lung infiltrates unresponsive to amphotericin B in leukaemic patients, *Aust N Z J Med* 22:265, 1992.

145. Walsh TJ, et al: Infections due to emerging and uncommon medically important fungal pathogens, *Clin Microbiol Infect* 10(Suppl 1):48, 2004.

146. Walzer P: *Pneumocystis carinii*. In Mandell G, Bennett J, and Dolin R, editors: *Principles and practice of infectious disease*, New York, 1995, Churchill Livingstone.

147. Wang SM, Shieh CC, Liu CC, Successful treatment of *Paecilomyces variotii* splenic abscesses: a rare complication in a previously unrecognized chronic granulomatous disease child. *Diagn Microbiol Infect Dis* 53:149, 2005.

148. Wang Y, Aisen P, Casadevall A: *Cryptococcus neoformans* melanin and virulence: mechanism of action, *Infect Immun* 63:3131, 1995.

149. Warren N, Hazen K: *Candida, Cryptococcus* and other yeasts of medical importance. In Murray P, et al, editors: *Manual of clinical microbiology*, Washington, DC, 1995, ASM Press.

150. Weeks R: A rapid simplified medium for converting the mycelial phase of *Blastomyces dermatitidis* to the yeast phase, *Mycopathologia* 22:153, 1964.

151. Weitzman I, Kane J, Summerbell R: *Trichophyton, Microsporon, Epidermophyton* and agents of superficial mycoses. In Murray P, et al, editors: *Manual of clinical microbiology*, Washington, DC, 1995, ASM Press.

152. Whittle DL., Kominos S: Use of itraconazole for treating subcutaneous phaeohyphomycosis caused by *Exophiala jeanselmei, Clin Infect Dis* 21:1068, 1995.

153. Willinger B: Laboratory diagnosis and therapy of invasive fungal infections, *Curr Drug Targets* 7:513, 2006.

154. Wilson DA, et al: Multicenter evaluation of a *Candida albicans* peptide nucleic acid fluorescent in situ hybridization probe for characterization of yeast isolates from blood cultures, *J Clin Microbiol* 43:2909, 2005.

155. Wolf DG, et al: Multidrug-resistant *Trichosporon asahii* infection of nongranulocytopenic patients in three intensive care units, *J Clin Microbiol* 39:4420, 2001.

156. Wu-Hsieh B, Howard D: Histoplasmosis. In Murphy J, Friedman H, Bendinelli M, editors: *Fungal infections and immune responses*, New York, 1993, Plenum Press.

157. Yuan L, Cole G, Sun S: Possible role of a proteinase endosporulation of *Coccidioides immitis, Infect Immun* 56:1551, 1988.

158. Zhiyong Z, Mei K, Yanbin L: Disseminated *Penicillium marneffei* infection with fungemia and endobronchial disease in an AIDS patient in China, *Med Princ Pract* 15:235, 2006.

159. Zimmer BL, Roberts GD: Rapid selective urease test for presumptive identification of *Cryptococcus neoformans, J Clin Microbiol* 10:380, 1979.

PART VI

Virology

Viruses infect humans, lower animals, insects, plants, bacteria, and fungi. Classification and nomenclature are standardized by the International Committee on Taxonomy of Viruses through reports published periodically. Viruses are divided into genera and species, as are bacteria; however, most are referred to by common names that have been used for decades. For example, the genera *Simplexvirus* and *Pneumovirus* contain herpes simplex and respiratory syncytial viruses, respectively. Viruses of medical importance to humans comprise seven families of DNA viruses and 14 families of RNA viruses (Table 51-1).

GENERAL CHARACTERISTICS

VIRAL STRUCTURE

Viruses are composed of a nucleic acid genome surrounded by a protein coat called a **capsid.** Together the genome and capsid are referred to as the **nucleocapsid** (Figure 51-1). Genomes are either RNA or DNA. Viral capsids are composed of many individual subunits called **capsomeres.** Capsomeres assemble into an icosahedral or irregular-shaped capsid. Irregular-shaped capsids usually assume a helical form. Icosahedral-shaped capsids are cubical with 20 flat sides, whereas helical capsids are spiral shaped. Some of the larger viruses have a lipid-containing envelope that surrounds the capsid. In addition, many viruses have glycoprotein spikes that extend from the surface of the virus, acting as attachment projections or as enzymes. The entire virus, including nucleic acid, capsid, envelope, and glycoprotein spikes, is called the **virion,** or viral particle.

Viruses that cause disease in humans range in size from approximately 20 to 300 nm. Even the largest viruses, such as the poxviruses, cannot be detected with a light microscope, since they are less than one fourth the size of a staphylococcal cell (Figure 51-2).

VIRAL REPLICATION

Viruses are strict intracellular parasites, reproducing or replicating only within a host cell. The steps in virus replication, called the *infectious cycle,* include attachment, penetration, uncoating, macromolecular synthesis, assembly, and release (Figure 51-3).

To initiate the infectious cycle, a virus must first recognize and bind to a suitable host cell, referred to as *attachment.* Typically, glycoprotein spikes bind to host cell carbohydrate receptors. Viruses recognize and attach to a limited number of host cell types, allowing infection of some tissues but not others. This is referred to as viral **tropism.**

The process by which viruses enter the host cell is called *penetration.* One mechanism of penetration involves fusion of the viral envelope with the host cell membrane. Not only does this method internalize the virus but also it can lead to fusion between this and other host cells nearby, forming multinucleated cells called **syncytia.** The detection of syncytia can be used to identify the presence of virus in cell cultures or stained smears of clinical specimens.

Uncoating occurs once the virus is internalized. Uncoating is necessary to release viral genome before the viral DNA or RNA is delivered to its intracellular site of replication in the nucleus or cytoplasm.

Macromolecular synthesis includes the production of nucleic acid and protein polymers. Viral transcription leads to the synthesis of messenger RNA (mRNA), which encodes early and late viral proteins. Early proteins are nonstructural elements, such as enzymes, and late proteins are structural components. Rapid identification of virus in cell culture can be accomplished by detecting early viral proteins in infected cells using immunofluorescent staining techniques. Replication of viral nucleic acid is necessary to provide genomes for progeny virions.

During viral *assembly*, structural proteins, genomes, and, in some cases, viral enzymes are assembled into virions. Envelopes are acquired during viral "budding" from a host cell membrane. Nuclear and cytoplasmic membranes are common areas for budding. Acquisition of an envelope is the final step in viral assembly.

Release of intact virions occurs following cell lysis or by budding from cytoplasmic membranes. Release by budding may not result in rapid host cell death as does release by lysis. Detection of virus in cell culture is facilitated by recognition of areas of cell lysis. Detection of virus released by budding is more difficult, because the cell monolayer remains intact. Influenza viruses, which are released by budding with minimal cell destruction, can be detected in cell culture by an alter-

Table 51-1   List of DNA and RNA Viruses of Human Importance

DNA Viruses	
Family	**Viral Members**
Adenoviridae	Human adenoviruses
Hepadnaviridae	Hepatitis B virus
Herpesviridae	Herpes simplex virus 1 and 2, varicella-zoster virus, cytomegalovirus, Epstein-Barr virus, human herpes-viruses 6, 7, and 8
Papillomaviridae	Human papilloma viruses
Parvoviridae	Parvovirus B-19
Polyomaviridae	BK and JC polyomaviruses
Poxviridae	Variola, vaccinia, orf, molluscum contagiosum, monkeypox
RNA Viruses	
Family	**Viral Members**
Arenaviridae	Lymphocytic choriomeningitis virus, Lassa fever virus
Astroviridae	Gastroenteritis-causing astroviruses
Bunyaviridae	Arboviruses including California encephalitis and Lacrosse viruses; nonarboviruses including sin nombre and related hantaviruses
Caliciviridae	Noroviruses and hepatitis E virus
Coronaviridae	Coronaviruses, including SARS coronavirus
Filoviridae	Ebola and Marburg hemorrhagic fever viruses
Flaviviridae	Arboviruses including yellow fever, dengue, West Nile, Japanese encephalitis, and St. Louis encephalitis viruses; nonarboviruses including hepatitis C virus
Orthomyxoviridae	Influenza A, B, and C viruses
Paramyxoviridae	Parainfluenza viruses, mumps virus, measles virus, respiratory syncytial virus, metapneumovirus, Nipah virus
Picornaviridae	Polio viruses, coxsackie A viruses, coxsackie B viruses, echoviruses, enteroviruses 68-71, enterovirus 72 (hepatitis A virus), rhinoviruses
Reoviridae	Rotavirus, Colorado tick fever virus
Retroviridae	Human immunodeficiency viruses (HIV-1 and HIV-2), human T-lymphotropic viruses (HTLV-1 and HTLV-2)
Rhabdoviridae	Rabies virus
Togaviridae	Eastern, Western and Venezuela equine encephalitis viruses, rubella virus

native technique called **hemadsorption.** Influenza virus–infected cells contain virally encoded glycoprotein hemagglutinins that have inserted in the host cell cytoplasmic membrane, preparing for inclusion in the viral envelope at the time of release by cytoplasmic budding. Red blood cells (RBCs) added to the culture medium will adsorb to the outer membranes of infected cells, but not to uninfected cells.

Each infected host cell results in as many as 100,000 virions; however, as few as 1% of these may be infectious or "viable" in the practical sense. Noninfectious virions result from errors or mutations occurring during the infectious cycle.

CLASSIFICATION OF VIRUSES

Viruses can be classified according to morphology, type of genome, or means of replication. Morphology in-

cludes type of capsid, such as icosahedral or irregular. Type of genome includes RNA or DNA, and whether it is single- or double-stranded. Means of replication refers to the strategy each virus uses to duplicate its genome. For example, enteroviruses have single-stranded RNA genomes that synthesize additional strands of RNA directly, whereas retroviruses make RNA in a two-step process by first synthesizing DNA, which subsequently makes RNA. Clinical virologists generally categorize viruses as containing DNA or RNA and further organize by family and common names (see Table 51-1).

VIRAL PATHOGENESIS

Viruses are transmitted from person to person by respiratory, fecal-oral, and sexual contact routes, by trauma or injection with contaminated objects or needles, by tissue transplants (including blood transfusions), by

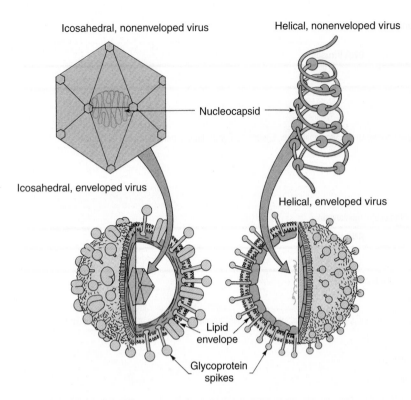

Figure 51-1 Illustration of viral particle. Enveloped and nonenveloped virions have icosahedral or irregular (usually helical) shape. (Modified from Murray PR, Drew WL, Kobayashi GS, et al, editors: *Medical microbiology*, St Louis, 1990, Mosby.)

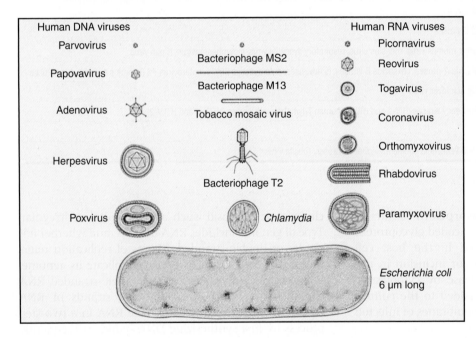

Figure 51-2 Relative sizes of representative viruses, bacteriophage (bacterial viruses), and bacteria, including chlamydia. (From Murray PR, Drew WL, Kobayashi GS, et al, editors: *Medical microbiology*, St Louis, 1990, Mosby.)

arthropod or animal bites, and during gestation (transplacental). Once introduced into a host, the virus infects susceptible cells, frequently in the upper respiratory tract. Local infection leads to a **viremia** (viruses in the blood), which inoculates secondary target tissue distant from the primary site and releases mediators of human immune cell functions. Symptomatic disease ensues. Disease resolves when specific antibody and cell-mediated immune mechanisms halt continued replication of the virus. Tissue is damaged by lysis of virus-infected cells or by immunopathologic mechanisms directed against the virus but which is also destructive to neighboring tissue. Most DNA-containing viruses, such as those in the herpes group, remain **latent** in host tissue with no observable clinical impact. Reactivation may occur accompanying immune suppression,

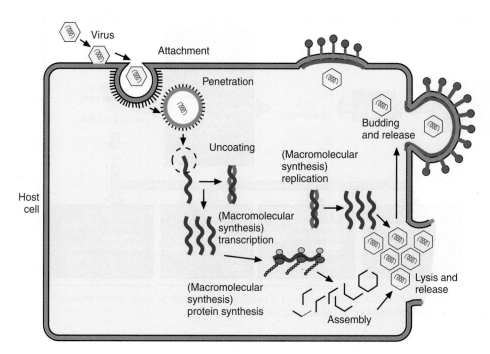

Figure 51-3   Illustration of viral infectious cycle. (Modified from Murray PR, Drew WL, Kobayashi GS, et al, editors: *Medical microbiology*, St Louis, 1990, Mosby.)

resulting in recurrence of clinically apparent disease. Occasionally, pathogenic viruses stimulate an immune reaction that cross reacts with related human tissue, resulting in damage to host function. This is termed *autoimmune pathogenesis* and, when present, occurs well after the acute viral infection has resolved. Rare viral infection promotes transformation or immortalization of host cells resulting in uncontrolled cell growth. Viruses with the ability to stimulate uncontrolled growth of host cells are referred to as **oncogenic** viruses. Some papilloma viruses (wart viruses) are oncogenic, giving rise to human cervical cancer.

Examples of the variety of pathogenic mechanisms shown during viral infection are illustrated by diseases caused by the measles virus. Following replication in the upper respiratory tract and subsequent viremia, the virus infects many susceptible cells throughout the body including endothelial cells in capillaries of the skin. This is accompanied by local inflammation and results in the characteristic rash of measles. Immunocompetent individuals eradicate the virus, resolving their infection, and have lifelong immunity. In some, antibody produced in response to the measles infection cross-reacts with tissue in the central nervous system (CNS), causing a post-infectious encephalitis. In others, slow but continuing replication of damaged virus in the brain gives rise to subacute sclerosing panencephalitis. In those who are severely immunocompromised, ongoing primary infection is not aborted by usual immune mechanisms, resulting in a fatal outcome (Figure 51-4). Measles virus is not an oncogenic virus, thus no cancers result from prolonged infection.

ANTIVIRAL AGENTS

Approximately 40 antiviral drugs are formally licensed for clinical use, with about one half used in the treatment of human immunodeficiency virus (HIV) infections. Antivirals are categorized based on their modes of action (Table 51-2). In addition to being used for HIV, antivirals are also used in the treatment of herpes viruses (herpes simplex virus [HSV], varicella zoster virus [VZV], and cytomegalovirus [CMV]), hepatitis B virus (HBV), hepatitis C virus (HCV), respiratory syncytial virus (RSV), and influenza viruses.

VIRUSES THAT CAUSE HUMAN DISEASES

Hundreds of viruses cause disease in humans. Complicating the understanding of viral disease in humans is the fact that individual viruses may cause multiple different diseases and, conversely, many viruses may cause the same disease. For example, viruses that can cause encephalitis includes HSV, many arboviruses, rabies virus, HIV, and measles virus. On the other hand, HSV can cause pharyngitis, genital infection, conjunctivitis, and encephalitis. Organ systems and the viruses that are important pathogens in each are summarized in Table 51-3.

Retroviruses and most DNA viruses establish a latent state following primary infection. During the latent state, viral genome is integrated into host cell chromosome and no viral replication occurs. **Latent viruses** can silently reactivate, resulting in viral replication and shedding but no clinical symptoms, or

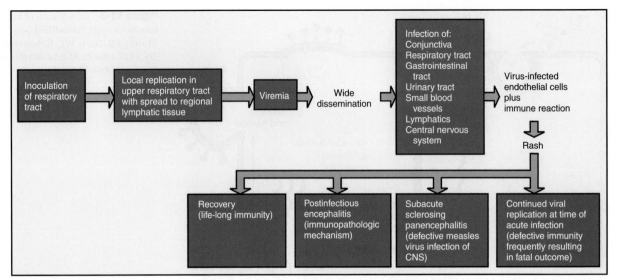

Figure 51-4   Viral pathogenesis illustrated by the mechanisms of the spread of measles virus within the body. (From Murray PR, Drew WL, Kobayashi GS, et al, editors: *Medical microbiology*, St Louis, 1990, Mosby.)

Table 51-2   Antiviral Agents

Virus	Mode of Action	Target	Common Drugs (examples)
CMV	Nucleoside analog	Viral DNA	Gancyclovir
HIV	Nucleoside analog	Viral DNA	Zidovudine, etc.
	Nucleotide analog	Viral DNA	Tenofovir disoproxil fumarate
	Nonnucleoside analog	Reverse transcriptase	Nevirapine, etc.
	Protease inhibitor	Viral protease	Saquinavir
	Fusion inhibitor	Virus, host cell membrane	Enfuvirtide
HSV/VZV	Nucleoside analog	Viral DNA	Acyclovir, etc.
	Pyrophosphate analog	DNA polymerase	Foscarnet
Hepatitis B	Nucleoside analog	Reverse transcriptase	Lamivudine
	Nucleotide analog	DNA polymerase	Adefovir dipivoxil
Influenza A	Inhibit penetration and uncoating of virus	Host cell membrane	Amantadine, imantadine
Influenza A and B	Prevent release of virus	Neuraminidase inhibitors	Zanamivir, seltamivir
RSV	Inhibit expression of viral mRNA and protein synthesis	Viral mRNA	Ribavirin
HCV	Inhibit expression of viral mRNA/increase resistance to virus	Viral mRNA/neighboring host cells	Ribavirin + interferon-alpha
Picornaviruses (enteroviruses and rhinoviruses)	Inhibit attachment and uncoating of virus	Binds to virus	Pleconaril

CMV, Cytomegalovirus; *HIV*, human immunodeficiency virus; *HSV*, herpes simplex virus; *VZV*, varicella zoster virus; *RSV*, respiratory syncytial virus; *HCV*, hepatitis C virus.

they can reactivate, causing symptomatic, even fatal, disease.

ADENOVIRUSES

Adenoviruses (Table 51-4) were first isolated from human adenoid tissues. At present, approximately 50 serotypes of human adenoviruses have been described; however, most disease is associated with only one third of these types. Adenoviruses cause less than 5% of all acute respiratory disease in the general population. In addition, adenovirus serotypes 40 and 41 cause gastroenteritis in infants and young children, and other serotypes cause conjunctivitis and keratitis. Compromised hosts may develop disseminated, multiorgan system disease. Laboratory diagnosis is accomplished most commonly by conventional or rapid cell culture using HEp-2 cells.

Table 51-3   List of Viral Syndromes and Common Viral Pathogens

Viral Syndrome	Viral Pathogens
Infants and Children	
Upper respiratory tract infection	Rhinovirus, coronavirus, parainfluenza, adenovirus, respiratory syncytial, influenza
Pharyngitis	Adenovirus, coxsackie A, herpes simplex, Epstein-Barr virus, rhinovirus, parainfluenza, influenza
Croup	Parainfluenza, respiratory syncytial, metapneumovirus
Bronchitis	Parainfluenza, respiratory syncytial, metapneumovirus
Bronchiolitis	Respiratory syncytial, parainfluenza, metapneumovirus
Pneumonia	Respiratory syncytial, adenovirus, influenza, parainfluenza
Gastroenteritis	Rotavirus, adenovirus 40-41, calicivirus, astrovirus
Congenital and neonatal disease	HSV-2, echovirus, and other enteroviruses, CMV, parvovirus B19, VZV, HIV, hepatitis viruses
Adults	
Upper respiratory tract infection	Rhinovirus, coronavirus, adenovirus, influenza, parainfluenza, Epstein-Barr
Pneumonia	Influenza, adenovirus, sin nombre virus (hantavirus), SARS coronavirus
Pleurodynia	Coxsackie B
Gastroenteritis	Noroviruses
All Patients	
Parotitis	Mumps, parainfluenza
Myocarditis/pericarditis	Coxsackie B and echoviruses
Keratitis/conjunctivitis	Herpes simplex, varicella-zoster, adenovirus, enterovirus 70
Pleurodynia	Coxsackie B
Herpangina	Coxsackie A
Febrile illness with rash	Echoviruses and coxsackie viruses
Infectious mononucleosis	Epstein-Barr virus and cytomegalovirus
Meningitis	Echoviruses and coxsackie viruses, mumps, lymphocytic choriomeningitis, HSV-2
Encephalitis	HSV-1, togaviruses, bunyaviruses, flaviviruses, rabies, enteroviruses, measles, HIV, JC virus
Hepatitis	Hepatitis A, B, C, D (delta agent), E, and non-A, B, C, D, E
Hemorrhagic cystitis	Adenovirus, BK virus
Cutaneous infection with or without rash	HSV 1 and 2, varicella-zoster, enteroviruses, measles, rubella, parvovirus B-19, human herpesvirus 6 and 7, HPV, poxviruses including smallpox, monkeypox, molluscum contagiosum, and orf
Hemorrhagic fever	Ebola, Marburg, Lassa, yellow fever, dengue, and other viruses
Generalized, no specific target organ	HIV-1, HIV-2, HTLV-1

Table 51-4   Adenoviruses

Family: *Adenoviridae*	Common Name: Adenovirus

Virus: Adenovirus
   Characteristics: Double-stranded DNA genome; icosahedral capsid, no envelope; approximately 50 human serotypes
   Transmission: Respiratory, fecal-oral, and direct contact (eye)
   Site of latency: Replication in oropharynx
   Disease: Pharyngitis, pharyngoconjunctival fever, keratoconjunctivitis, pneumonia, hemorrhagic cystitis, disseminated disease, and gastroenteritis in children
   Diagnosis: Cell culture (HEp-2 and other continuous human epithelial lines), EIA for gastroenteritis serotypes 40-41
   Treatment: Supportive
   Prevention: Vaccine (adenovirus serotypes 4 and 7) for military recruits

Table 51-5 Arenaviruses

Family: *Arenaviridae*   Common Name: Arenavirus
Virus: Lymphocytic choriomeningitis (LCM) and Lassa fever (Lassa, Nigeria) viruses Characteristics: Enveloped, irregular-shaped capsid containing a two-segmented (each segment is circular), single-stranded RNA genome Transmission: From rodent to human through contamination of human environment with rodent urine; virus enters through skin abrasions or inhalation Disease: LCM causes asymptomatic to influenza-like to aseptic meningitis–type disease; Lassa fever virus causes influenza-like disease to severe hemorrhagic fever Diagnosis: Serology, polymerase chain reaction Treatment: Supportive for LCM; ribavirin and immune plasma for Lassa fever Prevention: Avoid contact with virus, rodent control; isolation and barrier nursing prevent nosocomial spread

Table 51-6 Bunyaviruses

Family: *Bunyaviridae*   Common Name: Bunyavirus
Virus: Arboviruses* including the California encephalitis group containing LaCrosse virus, and nonarthropod-borne virus including hantaviruses (containing sin nombre virus) Characteristics: Segmented, single-stranded, RNA genome. Spherical or pleomorphic capsid with envelope Transmission: Mosquito, tick, and sandfly vectors, except for hantaviruses, which are zoonoses transmitted by contact with rodent host Disease: Encephalitis for arboviruses; pneumonia or hemorrhagic fever for hantaviruses Diagnosis: Serology and antibody detection in cerebrospinal fluid, RT-PCR for hantaviruses Treatment: Supportive Prevention: Avoid contact with arthropod vector. Vector control programs; hantaviruses, avoid rodent urine and feces
*Arthropod-borne viruses (arboviruses) are taxonomically heterogeneous but were once grouped together because of their common mode of transmission. Viruses adapted to arthropod vectors occur in several taxonomic families, including the *Togaviridae*, *Flaviviridae*, and *Bunyaviridae*. The virus group within *Togaviridae* that includes arboviruses is the alphavirus group. Common arboviruses are referred to as *bunyaviruses*, *flaviviruses*, and *alphaviruses*.

ARENAVIRUSES

Arenaviruses (Table 51-5) causing disease in humans include lymphocytic choriomeningitis (LCM) virus and Lassa fever virus (first detected in Lassa, Nigeria). Both are zoonoses transmitted from rodents to humans through aerosol or skin abrasion. LCM causes aseptic meningitis in Europe and the Americas. Lassa causes hemorrhagic fever, shock, and death in 5% to 15% of symptomatic patients (80% of cases are asymptomatic). Other, less commonly reported arenaviruses also cause hemorrhagic fever. All cause lifelong, persistent, inapparent infections in a natural rodent host. Diagnosis is accomplished by serology or reverse transcriptase polymerase chain reaction (RT-PCR) testing for viral nucleic acid. Viral isolation is not routinely recommended because of the risk of laboratory acquisition. LCM virus requires biosafety level 3 and Lassa fever virus requires biosaftey level 4 laboratory facilities.

BUNYAVIRUSES

Bunyaviruses (Table 51-6) comprise a large (approximately 300 total members with 12 human pathogens), diverse group of viruses, most of which are transmitted by mosquitoes. The family name is a reflection of where the viruses were first detected in Bunyamwera, Uganda. Clinical disease caused by arborviruses in the California serogroup of Bunyaviruses, including California encephalitis and La Crosse viruses, occur primarily in Minnesota, Wisconsin, Iowa, Illinois, Indiana, and Ohio. Disease is typically mild and self-limited; however, severe, even fatal, encephalitis ensues in approximately 2% of patients. Bunyaviruses belonging to the *Hantavirus* genus are not arboviruses, because they are rodent-borne and transmitted through aerosolized rodent excreta. Hantavirus pulmonary syndrome, originally reported in the four corners area of the southwestern United States, is caused by a hantavirus called *sin nombre* (no name) virus and is characterized by hemorrhagic fever, kidney disease, and acute respiratory failure. Laboratory diagnosis of arbovirus disease requires the detection of specific antibody in acute and convalescent serum specimens or in cerebrospinal fluid (CSF), or for hantavirus pulmonary syndrome the detection of viral RNA genome by RT-PCR.

CALICIVIRUSES

Caliciviruses (Table 51-7) have been known for years by veterinary virologists as major animal pathogens. Not

Table 51-7 Caliciviruses

Family: *Caliciviridae*	Common Name: Calicivirus

Virus: Noroviruses and hepatitis E* virus
  Characteristics: Nonenveloped, icosahedral capsid surrounding single-stranded RNA genome
  Transmission: Fecal-oral
  Disease: Nausea, vomiting, and diarrhea (Norovirus), hepatitis similar to that caused by hepatitis A virus except for extraordinarily high case fatality rate (10% to
    20%) among pregnant women
  Diagnosis: EM, RT-PCR, EIA for Noroviruses; serology for hepatitis E
  Treatment: Supportive
  Prevention: Avoid contact with virus

*Hepatitis E virus is being reclassified into its own family, *Hepeviridae*.

Table 51-8 Coronaviruses

Family: *Coronaviridae*	Common Name: Coronaviruses

Virus: Coronavirus
  Characteristics: Single-stranded, RNA genome. Helical capsid with envelope
  Transmission: Unknown, probably direct contact or aerosol
  Disease: Common cold; possibly gastroenteritis, especially in children; severe acute respiratory syndrome (SARS)
  Diagnosis: Electron microscopy and RT-PCR
  Treatment: Supportive
  Prevention: Avoid contact with virus

until the 1990s did the taxonomic status of noroviruses (formerly known as Norwalk-like viruses named after Norwalk, Ohio) and hepatitis E virus result in classification in the family *Caliciviridae*. Hepatitis E virus is being removed from the calicivirus family and included in a newly proposed family, the *Hepeviridae*. Noroviruses are representative of a number of small, noncultivatable viruses causing diarrhea in adults and children. The prototype norovirus is Norwalk virus. Following a 1- to 2-day incubation period, nausea, vomiting, and diarrhea last 1 to 3 days. In recent years noroviruses have been implicated in large outbreaks of disease on cruise ships and in nursing homes. Noroviruses are easily spread by water, direct person-to-person contact, or airborne droplets of vomitus. Virus persists in water despite efforts at disinfection. Diagnosis is by clinical criteria, electron microscopy, or more recently by RT-PCR if laboratory confirmation is needed. Infection does not confer long-lasting immunity. Hepatitis E virus was discovered in Asia by a Russian virologist who volunteered to drink stool filtrates from a patient with an unidentified form of hepatitis. Also uncultivable, this virus is well established in developing countries as a cause of hepatitis clinically similar to that caused by hepatitis A, except the case-fatality rate is exceptionally high (10% to 20%) in pregnant women. Diagnosis is made using serologic testing. There is no chronic disease or carrier state. Hepatitis E is not endemic in the United States and other developed areas of the world.

CORONAVIRUSES

Many coronaviruses of humans (Table 51-8) and other animals exist. The prefix *corona-* is used because of the crownlike surface projections that are seen when the virus is examined by electron microscopy. Human respiratory coronaviruses cause colds and occasionally pneumonia in adults. Coronaviruses are thought to cause diarrhea in infants based on the presence (using electron microscopy) of coronavirus-like particles in stool of symptomatic patients. No practical diagnostic methods, short of electron microscopy or RT-PCR, are available for laboratory diagnosis.

Severe acute respiratory syndrome (SARS) is a new infectious disease that first emerged in Guangdong province, China, in November 2002. SARS is caused by a novel coronavirus (SARS coronavirus) of animal origin. Within months, more than 8000 patients worldwide (with approximately 700 deaths) were affected. The virus has not been detected in humans in subsequent years, but the animal reservoir and live animal markets in China are still present, allowing animal-to-human transfer to once again happen. Low levels of

Table 51-9 Filoviruses

Family: *Filoviridae*	Common Name: Filovirus

Virus: Ebola (aka Ebola-Reston) and Marburg viruses
Characteristics: Enveloped, long, filamentous and irregular capsid forms with single-stranded RNA
Transmission: Transmissible to humans from monkeys and, presumably, other wild animals; human-to-human transmission via body fluids and respiratory droplets
Disease: Severe hemorrhages and liver necrosis; mortality as high as 90%
Diagnosis: Electron microscopy, cell culture in monkey kidney cells; Biosafety Level 4 required
Treatment: Supportive
Prevention: Avoid contact with virus; export prohibitions on wild-caught monkeys

Table 51-10 Flaviviruses

Family: *Flaviviridae*	Common Name: Flavivirus

Characteristics: Single-stranded RNA genome surrounded by spherical and icosahedral capsid with envelope
Virus: Arboviruses* including yellow fever, dengue, West Nile, Japanese encephalitis, and St. Louis encephalitis viruses
Transmission: Arthropod vector, usually mosquito
Disease: St. Louis and West Nile encephalitis, dengue and yellow fever
Diagnosis: Serology and antibody detection in cerebrospinal fluid; RT-PCR for dengue and yellow fever
Treatment: Supportive
Prevention: Avoid contact with vector; vector control programs
Virus: Hepatitis C virus
Transmission: Parenteral or sexual
Disease: Acute and chronic hepatitis; strong correlation between chronic HCV infection and hepatocellular carcinoma
Diagnosis: Serology, RT-PCR and viral genotyping
Treatment: Supportive, interferon
Prevention: Avoid contact with virus; blood supply screened for antibody to hepatitis C virus

*See footnote for Table 51-6.

virus in the respiratory tract during early disease provide a diagnostic challenge and make molecular testing by RT-PCR the best test for laboratory diagnosis.

FILOVIRUSES

Filoviruses (Table 51-9) are the most pathogenic of the hemorrhagic fever viruses. The name *filo* means thread-like, referring to their long filamentous morphology as seen with electron microscopy. Infections with Marburg or Ebola viruses, endemic in Africa, result in severe hemorrhages, vomiting, abdominal pain, myalgia, pharyngitis, conjunctivitis, and proteinuria. Case-fatality rates of 90% are expected with Ebola virus outbreaks, somewhat less with Marburg. Electron microscopy and RT-PCR are used for diagnosis.

FLAVIVIRUSES

The flaviviruses (Latin *flavus,* meaning yellow) include viruses that cause arbovirus diseases such as yellow fever, dengue and West Nile viruses, and Japanese and St. Louis encephalitis viruses. In addition, a non-arbovirus, HCV, is a flavivirus (Table 51-10).

Yellow fever is one of the great plagues to have occurred throughout history. As a result of thousands dying during the construction of the Panama Canal, an army physician, Dr. Walter Reed, uncovered the epidemiology of yellow fever in 1900. In the jungle habitat, monkeys serve as the reservoir and the mosquito as the vector. In urban settings, the human can serve as the reservoir during outbreaks as long as the mosquito vector is present. Yellow fever virus infects primarily the liver, resulting in fever, jaundice, and hemorrhage. A minority of cases progress to severe jaundice and massive hemorrhages, with mortality rates of up to 50% in this group. Yellow fever is effectively prevented with a vaccine.

Dengue virus, which is endemic in tropical and subtropical countries including Latin America and Mexico, has four serotypes that cause nonlethal fever, arthritis, and rash. Infection with one serotype confers immunity to this but not other serotypes. Subsequent infection with one of the three remaining serotypes

Table 51-11 Hepadnaviruses

Family: *Hepadnaviridae*	Common Name: Hepadnavirus

Virus: Hepatitis B virus (HBV)
  Characteristics: Partly double-stranded DNA genome; icosahedral capsid with envelope; virion also called *Dane particle;* surface antigen originally termed *Australia antigen*
  Transmission: Humans are reservoir and vector; spread by direct contact including exchange of body secretions, recipient of contaminated blood products, percutaneous injection of virus, and perinatal exposure
  Site of latency: Liver
  Disease: Acute infection with resolution (90%); fulminant hepatitis most co-infected with delta virus (1%); chronic hepatitis, persistence of HB_SAg (9%) followed by resolution (disappearance of HB_SAg), asymptomatic carrier state, chronic persistent (systemic disease without progressive liver disease), or chronic active disease (progressive liver damage)
  Diagnosis: Serology, viral antigen detection, and PCR
  Oncogenic: Liver cancer
  Treatment: Antivirals and liver transplant for fulminant disease
  Prevention: HBV vaccine; hepatitis B immune globulin

results in immune-enhanced disease in the form of severe hemorrhagic fever or dengue shock syndrome. Humans are the main reservoir for the dengue virus. Transmission occurs by a mosquito vector that is well adapted to urban settings. Dengue virus is the most common cause of arbovirus disease in the world, causing an estimated 100 million cases of dengue fever, 250,000 cases of dengue hemorrhagic fever, and 25,000 deaths per year. A vaccine is not available. Laboratory diagnosis is based on the presence of virus-specific IgM antibody, a fourfold rise in specific IgG antibody, or a positive RT-PCR test for dengue genomic sequences.

Scores of other arthropod-borne flaviviruses, most transmitted by mosquitoes or ticks, cause encephalitis, hemorrhagic fever, or milder disease characterized by fever, arthralgia, and rash. One such infection is caused by West Nile virus (first isolated in the West Nile district of Uganda), a flavivirus closely related to Japanese and St. Louis encephalitis viruses. West Nile virus, endemic for decades in Africa, Israel, and Europe, has been endemic since 1999 in the United States. West Nile virus is maintained in a bird-mosquito cycle. Amplification of virus during warm months results in the death of bird hosts, most commonly crows, ravens, and jays. Bridge mosquitoes, those that bite both humans and birds, are responsible for transmission to humans as viral populations in birds are amplified. Laboratory diagnosis is provided most efficiently by the detection of IgM antibody to West Nile virus in serum or cerebrospinal fluid.

HCV causes hepatitis with 170 million HCV carriers worldwide, about 4 million in the United States. Acute infection with HCV progresses to a chronic infection of 50% to 90% of infected individuals. Chronic infection with HCV is an important cause of liver disease and is associated with the development of end-stage liver disease and hepatocellular carcinoma. The virus is transmitted predominantly by means of exposure to infected blood, such as intravenous drug use and blood products. The screening of blood products for HCV beginning in 1990 has nearly eliminated transmission by this means. Less efficient modes of transmission included sexual contact with infected partners, acupuncture, tattooing, and sharing razors. HCV disease is detected using screening antibody tests, confirmatory antibody testing, and RT-PCR. In addition, RT-PCR is used to quantitate virus in blood to provide a measure of response to therapy. Finally, viral genotyping by molecular techniques is used to identify genotypes that do not respond to therapy. The full benefits of modern laboratory testing, including the application of molecular methods, has impacted significantly the recognition, monitoring, and treatment of HCV disease.

HEPADNAVIRUSES

HBV (Table 51-11) is the prototype virus found in the Hepadnaviridae family (*hepa* from hepatitis and *dna* from the genome type). Other mammalian and avian hepadnaviruses are known to exist. Disease in humans, transmitted via perinatal, sexual, and parenteral routes, is usually asymptomatic but may result in acute or chronic hepatitis with self-limited or fatal outcomes. Fatal disease is most likely to occur in people co-infected with hepatitis D virus (delta agent), a deficient RNA virus that replicates only in cells already infected with HBV. Chronic HBV infection is a significant worldwide cause of liver cirrhosis and hepatocellular carcinoma, despite the availability of an effective vaccine. Diagnosis is provided by the detection of a battery of antibodies and viral antigens. The best indication of active viral replication and a high state of infectivity is the presence of HBV DNA in the serum, which is detected by a number of molecular tests including PCR. Additionally,

Table 51-12 Herpesviruses

Family: *Herpesviridae*	Common name: Herpesvirus

Characteristics: Double-stranded DNA genome; icosahedral capsid with envelope; at least 8 human herpes viruses known: HSV-1, HSV-2, VZV, EBV, CMV, HH6, HH7, and HH8

Virus: Herpes simplex virus types 1 and 2 (HSV-1 and -2)
  Transmission: Direct contact with infected secretions
  Site of latency: Sensory nerve ganglia
  Disease: Predominant virus in parentheses. Gingivostomatitis (HSV-1), pharyngitis (HSV-1), herpes labialis (HSV-1), genital infection (HSV-2), conjunctivitis (HSV-1), keratitis (HSV-1), herpetic whitlow (HSV-1 and -2), encephalitis (HSV-1 in adults), disseminated disease (HSV-1 or HSV-2 in neonates)
  Detection: Cell culture (HDF, others), EIA, FA stain, PCR
  Treatment: Acyclovir
  Prevention: Avoid contact

Virus: Varicella-zoster virus (VZV)
  Transmission: Close personal contact, especially respiratory
  Site of latency: Dorsal root ganglia
  Disease: Chicken pox (varicella), shingles (zoster)
  Detection: FA stain, cell culture (HDF), shell vial culture, PCR
  Treatment: Acyclovir and famciclovir
  Prevention: Vaccine

Virus: Epstein-Barr virus (EBV)
  Transmission: Close contact with infected saliva
  Site of latency: B lymphocytes
  Disease: Infectious mononucleosis, progressive lymphoreticular disease, oral hairy leukoplakia in HIV- infected patients
  Detection: Serology, PCR
  Oncogenic: Burkitt's lymphoma, nasopharyngeal carcinoma
  Treatment: Supportive
  Prevention: Avoid contact

Virus: Cytomegalovirus (CMV)
  Transmission: Close contact with infected secretions, blood transfusions (WBCs), organ transplants, transplacental
  Site of latency: White blood cells, endothelial cells, cells in a variety of organs
  Disease: Asymptomatic infection, congenital disease of newborn, symptomatic disease of immunocompromised host, heterophile-negative infectious mononucleosis
  Diagnosis: Cell culture (HDF), shell vial culture, CMV antigenemia, FA stain, PCR
  Treatment: Supportive, decrease immune suppression, ganciclovir and foscarnet
  Prevention: Use CMV antibody-negative blood and tissue for transfusion and transplantation, respectively

Virus: Human herpesviruses 6 and 7 (HHV-6 and HHV-7)
  Transmission: Most likely close contact via respiratory route; almost all children infected by age 2 to 3 years
  Site of latency: T lymphocytes (CD4 cells)
  Disease: Roseola (exanthem subitum), fever, malaise, rash, leukopenia, and interstitial pneumonitis in organ transplant patients
  Detection: Detection of virus in peripheral blood specimens by PCR, cell culture using lymphocyte lines
  Treatment: Susceptible to ganciclovir and foscarnet
  Prevention: None practical

Virus: Human herpesvirus 8 (HHV-8)
  Transmission: Not known; much less widely disseminated than other herpesviruses
  Site of latency: Viral genome found in Kaposi's tumor cells, endothelial cells, and tumor-infiltrating leukocytes
  Disease: Kaposi's sarcoma
  Detection: PCR or in situ by hybridization
  Treatment: None known
  Prevention: Avoid contact with virus

detection of HBV DNA in serum is used to resolve questionable serologic results, and quantitation is helpful for predicting treatment response.

HERPES VIRUSES

There are eight known human herpes group viruses (Table 51-12). Herpes viruses also are widely disseminated among most other animal species and do not infect humans, except for herpes B virus from non-human primates (not counted among the eight human herpes viruses), which causes severe, usually fatal encephalitis in humans. Human herpes viruses include herpes simplex virus types 1 and 2 (HSV 1 and HSV 2), VZV, Epstein-Barr virus (EBV), and CMV. In addition, more recently detected herpes viruses include human herpes viruses 6 (HHV-6) and 7 (HHV-7), which are lymphotropic viruses acquired early in life, and HHV-8,

Table 51-13   Orthomyxoviruses

Family: *Orthomyxoviridae*   Common Name: Orthomyxovirus

Characteristics: Segmented (eight separate molecules), single-stranded, RNA genome; helical capsid with envelope; three major antigenic types, influenza A, B, and C. Types A and B cause nearly all human disease
Virus: Influenza A
  Transmission: Contact with respiratory secretions
  Disease: Influenza (fever, malaise, headache, myalgia, cough); primary influenza pneumonia; in children, bronchiolitis, croup, otitis media
  Detection: Cell culture (PMK), EIA, FA stain, RT-PCR
  Epidemiology: Viral subtypes based on hemagglutinin and neuraminidase glycoproteins abbreviated "H" and "N," respectively (e.g., H1N1 or H3N2); infects
      humans and other animals; antigenic drift, resulting in minor antigenic change, causes local outbreaks of influenza every 1-3 years; antigenic shift, resulting in
      major antigenic change, causes periodic worldwide outbreaks
  Treatment: Supportive; antivirals amantadine and rimantidine (influenza A only), and zanamivir and oseltamivir influenza A and B
  Prevention: Influenza vaccine or antiviral prophylaxis
Virus: Influenza B
  Transmission: Contact with respiratory secretions
  Disease: Similar to "mild" influenza
  Detection: Cell culture (PMK), EIA, FA stain, RT-PCR
  Epidemiology: Antigenic drift only, resulting in local outbreaks every 1-3 years
  Treatment: Supportive; antivirals zanamivir and oseltamivir
  Prevention: Influenza vaccine or antiviral prophylaxis

Kaposi's sarcoma–associated herpes virus (KSHV). HHV-6 and HHV-7 are associated with the childhood disease roseola (exanthem subitum), characterized by short-lasting fever and skin rash. HHV-8 is the primary and necessary factor for development of Kaposi's sarcoma. Herpes viruses are prototypical latent DNA viruses, with the lifelong integration of viral DNA into human host cells. Recurrence of viral replication at subsequent times results in disease, modified by host immune response. HSV-1 may reactivate, causing mucous membrane disease or life-threatening encephalitis. HSV-2 reactivates, causing mucous membrane vesicles or aseptic meningitis. VZV recurs with localized lesions as a disease called *shingles*. EBV reactivates, causing asymptomatic shedding of virus in the oropharynx or as disseminated disease in immunocompromised patients. CMV, like EBV, recurs symptomatically in compromised hosts as a pathogen in many tissues (e.g., heart, gastrointestinal tract, lung, brain). HHV-6 and -7 cause reactivation disease also in compromised hosts. Herpes viruses are detected by cell culture, direct antigen, or nucleic acid amplification systems (PCR), and by serologic testing of acute and convalescent serum specimens.

ORTHOMYXOVIRUSES

The *Orthomyxoviridae* family (Table 51-13) includes the influenza viruses. Influenza viruses are characterized by their hemagglutinin (H) and neuraminidase (N) antigens. Novel or altered H and N antigens represent viruses that will cause human influenza outbreaks. Influenza viruses are unique because of their ability to alter antigenic (H and N) composition, thus reinfecting

"nonimmune," susceptible hosts. Antigenic drift, caused by mutation during genome replication, results in minor antigenic change and relatively mild influenza outbreaks every 1 to 3 years. Antigenic shift, caused by reassortment or mixing of the segmented viral genome during co-infection in nonhuman animals, results in major antigenic change and periodic worldwide outbreaks (pandemics). Co-infection occurs when human virus infects an animal or human cell at the same time as an animal influenza virus. In 1917, 20 to 50 million people died worldwide as a result of an influenza pandemic. Fears of a similar pandemic stemming from the introduction of novel viruses, such as the avian strains circulating in domestic chickens and wild birds, relates to viruses recombining to produce novel strains with recognized pathogenic factors, such as those present in the 1917 pandemic strain. Influenza virus disease is detected in the laboratory by cell culture, fluorescent antibody (FA) staining or other antigen detection methods, RT-PCR, and serologic testing. RT-PCR is emerging as the test of choice because of its speed and sensitivity.

PAPILLOMAVIRUSES

The *Papillomaviridae* family (Table 51-14) includes the human papilloma viruses (HPVs). HPVs cause human warts. They have not been cultivated in cell culture, thus preventing the production of type-specific antigens and corresponding typing antisera. HPVs have been divided into more than 200 genotypes based on DNA sequences, approximately 80 of which have been well characterized. Genotypes have differing cellular tro-

Table 51-14 Papillomaviruses

Family: *Papovaviridae*	Common Name: Papillomavirus

Characteristics: Double-stranded DNA genome; icosahedral capsid, no envelope; includes papilloma viruses
Virus: Human papilloma virus (HPV)
  Characteristics: Contains more than 200 DNA types
  Transmission: Direct contact, sexual contact for genital warts
  Site of latency: Epithelial tissue
  Disease: Skin and genital warts, benign head and neck tumors, anogenital warts
  Diagnosis: Cytology, DNA probes
  Oncogenic: Cervical and penile cancer (especially HPV types 16 and 18)
  Treatment: Spontaneous disappearance the rule; surgical or chemical removal may be necessary
  Prevention: Avoid contact with infected tissue

pisms, resulting in associations of HPV types with specific clinical types of warts. For example, HPV-1 is associated with plantar warts, HPV-2 and HPV-4 are associated with common warts of the hands, and HPV-6, HPV-11, and others are associated with genital warts. Importantly, 15 to 20 types of HPV cause virtually all cases of cervical cancer, with types 16 and 18 causing more than 60% of cases. HPV is the most common sexually transmitted disease. Infection is detected by histopathologic or cytologic examination of cutaneous biopsy or cells, respectively, and by DNA probe assays used to identify specific genotypes within infected epithelial cells.

PARAMYXOVIRUSES

The *Paramyxoviridae* family (Table 51-15) includes many pathogenic viruses, especially for young children, including measles, mumps, parainfluenza viruses, and RSVs. Recently, human metapneumovirus and Nipah virus (Nipah is the area in Malaysia where the first virus was isolated) have been recognized as paramyxoviruses that cause human disease. Paramyxoviruses do not have a segmented genome like the orthomyxoviruses and therefore do not undergo antigenic shift. Disease caused by measles and mumps viruses can be eliminated by vaccination, which is required for all children in the United States. Human parainfluenza viruses are important pathogens in children, causing croup and other respiratory disease and resulting in more than 250,000 emergency department room visits per year in the United States. RSV causes bronchiolitis in young children. The virus contains a surface protein called *F* (fusion) protein. F protein mediates host cell fusion into syncytial cells, which are a hallmark of RSV infection. RSV immune serum prevents severe RSV bronchiolitis during susceptible early months of life in newborns with RSV disease and underlying medical conditions, especially premature children with developing lung function. Parainfluenza, measles, and mumps viruses hemadsorb guinea pig red blood cells. Laboratory detection is performed using cell culture with hemadsorption, FA staining, or enzyme immunoassay. PCR successfully detects virus in respiratory secretions, but offers little advantage over fluorescent or enzymatic antigen detection because young children have large quantities of virus in respiratory secretions.

Metapneumovirus, a newly discovered virus that is closely related to RSV, has caused disease, presumably, throughout human history but has avoided detection because it is difficult to grow in cell culture from clinical specimens. In children, infection appears to be less common than RSV but more common than parainfluenza virus, making it an important medical problem. It causes bronchiolitis and pneumonia in infants and, most likely, lower respiratory tract disease in older adults. Like RSV, metapneumovirus has winter epidemics with variation in severity from year to year.

Nipah virus is another recently discovered paramyxovirus that causes respiratory disease in pigs and acute, febrile encephalitis in humans. In 1999, the first human outbreak was described. It resulted from direct contact with diseased pigs and accounted for 265 human cases with 108 deaths. Multiple outbreaks have been described in subsequent years. The reservoir for Nipah virus is presumed to be fruit bats, with pigs and other animals being intermediate hosts.

PARVOVIRUSES

Parvoviruses (Latin *parvus* means small) (Table 51-16) have wide distribution among warm-blooded animals. Parvovirus B19 represents the one human pathogen in the family. Its replication in human cells is largely restricted to erythroid progenitor cells, making adult bone marrow and fetal liver, the site of erythropoiesis during fetal development, the major sites of viral replication. Important diseases caused by the B19 virus include fifth disease (the fifth of the childhood exanthems), aplastic crisis in patients with underlying hemoglobinopathies,

Table 51-15 Paramyxoviruses

Family: *Paramyxoviridae*	Common Name: Paramyxoviruses

Characteristics: Single-stranded, RNA genome; helical capsid with envelope; no segmented genome like orthomyxoviruses

Virus: Measles virus
  Transmission: Contact with respiratory secretions; extremely contagious
  Disease: Measles, atypical measles (occurs in those with waning "vaccine" immunity), and subacute sclerosing panencephalitis
  Detection: Cell culture (PMK) and serology
  Treatment: Supportive; immunocompromised patients can be treated with immune serum globulin
  Prevention: Measles vaccine

Virus: Mumps
  Transmission: Person-to-person contact, presumably respiratory droplets
  Disease: Mumps
  Detection: Cell culture (PMK) and serology
  Treatment: Supportive
  Prevention: Mumps vaccine

Virus: Parainfluenza virus
  Transmission: Contact with respiratory secretions
  Disease: Adults: upper respiratory, rarely pneumonia. Children: respiratory including croup, bronchiolitis, and pneumonia
  Detection: Cell culture (PMK), shell vial culture, and FA stain
  Epidemiology: Four serotypes, disease occurs year-round
  Treatment: Supportive
  Prevention: Avoid contact with virus

Virus: Respiratory syncytial virus (RSV)
  Transmission: Person-to-person by hand and respiratory contact
  Disease: Primarily in infants and children. Infants: bronchiolitis, pneumonia, and croup. Children: upper respiratory
  Detection: Cell culture (HEp-2), EIA, and FA stain
  Epidemiology: Disease occurs annually late fall through early spring; nosocomial transmission can occur readily
  Treatment: Supportive; treat severe disease in compromised infants with ribavirin
  Prevention: Avoid contact with virus. Immune globulin for infants with underlying lung disease; prevent nosocomial transmission with isolation and cohorting

Virus: Metapneumovirus
  Transmission: Person-to-person
  Disease: Primarily in infants and children; bronchiolitis and pneumonia
  Detection: RT-PCR
  Epidemiology: Winter epidemics, severity varies from year to year
  Treatment: Supportive
  Prevention: Avoid contact with virus

Table 51-16 Parvoviruses

Family: *Parvoviridae*	Common Name: Parvovirus

Virus: Parvovirus B-19
  Characteristics: Single-stranded DNA virus; icosahedral capsid, no envelope; parvovirus B-19 is the only known human parvovirus
  Transmission: Close contact, probably respiratory
  Disease: Erythema infectiosum (fifth disease), aplastic crises in patients with chronic hemolytic anemias, and fetal infection and stillbirth
  Detection: Serology, PCR, histology
  Treatment: Supportive
  Prevention: Avoid contact

and fetal infection (hydrops fetalis) resulting from transplacental inoculation. Parvovirus causes a biphasic illness in humans. The first phase comprises fever, malaise, myalgia, and chills, which corresponds to peak levels of virus and destruction of erythroblasts. This phase, when mild, may be overlooked or considered nonspecific viral disease. The second phase includes rash and arthralgia occurring after the virus has disappeared but at a time when parvovirus-specific antibody can be detected, consistent with the conclusion that the rash is caused by immune complexes in the capillaries of the skin. Laboratory diagnosis is accomplished using parvovirus-specific IgM or virus-specific IgG antibody testing with acute and convalescent sera, or by

Table 51-17   Picornaviruses

Family: *Picornaviridae*	Common Name: Picornaviruses

Characteristics: Single-stranded RNA genome; icosahedral capsid with no envelope
Virus: Enteroviruses
  Poliovirus (3 types)
  Coxsackievirus, group A (23 types)
  Coxsackievirus, group B (6 types)
  Echovirus (31 types)
  Enteroviruses (5 types)
  Transmission: Fecal-oral
  Disease: Predominant virus in parentheses; polio (poliovirus), herpangina (coxsackie A), pleurodynia (coxsackie B), aseptic meningitis (many enterovirus types), hand-foot-mouth disease (coxsackie A), pericarditis and myocarditis (coxsackie B), acute hemorrhagic conjunctivitis (enterovirus 70), and fever, myalgia, summer "flu" (many enterovirus types), neonatal disease (echo- and coxsackieviruses)
  Detection: Cell culture (PMK and HDF), PCR, and serology
  Treatment: Supportive, pleconaril in development
  Prevention: Avoid contact with virus Vaccination for polio
Virus: Hepatitis A virus (enterovirus type 72)
  Transmission: Fecal-oral
  Disease: Hepatitis with short incubation, abrupt onset, and low mortality; no carrier state
  Detection: Serology
  Treatment: Supportive
  Prevention: Vaccine; prevent clinical illness with serum immunoglobulin
Virus: Rhinovirus (common cold virus)
  Characteristics: Approximately 100 serotypes
  Transmission: Contact with respiratory secretions
  Disease: Common cold
  Detection: Cell culture (usually not clinically necessary), RT-PCR
  Treatment: Supportive
  Prevention: Avoid contact with virus

detection of viral DNA using PCR. Parvovirus cannot be cultivated in usual cells available in clinical virology laboratories.

PICORNAVIRUSES

Picornaviruses (Table 51-17) are small (from Italian *piccolo* meaning small) RNA viruses including the enteroviruses, rhinoviruses, and hepatitis A virus (HAV). The enteroviruses were originally divided into poliovirus, coxsackie virus, and echovirus groups based on similarity of characteristics in cell culture and disease in humans. Genetic diversity among these viruses, recognized by application of modern molecular techniques, dictates that newly characterized strains be given enterovirus-type designations rather than serotype status in one of the three original groups, because of the inexactness of the old disease- and phenotype-based classification system. The rhinoviruses grow best or exclusively at lower temperatures (30° C), making their detection in clinical virology laboratories, where incubation is at 35° to 37° C, unlikely. The application of PCR for the detection of rhinoviruses has expanded our understanding of the range of diseases caused by this group. Considerably more lower respiratory tract disease in adults is caused by rhinovirus than was previously

known. Most enteroviruses grow in cell culture; however, the most sensitive method for detection is by RT-PCR. Disease caused by HAV (enterovirus 72) is diagnosed using serology tests.

POLYOMAVIRUSES

The *Polyomaviridae* family (Table 51-18) includes JC and BK viruses, named with the initials of the persons from whom the viruses were first isolated. Both have latent states in the kidney and can result in symptomatic reactivation during periods of immune suppression. JC virus reactivates to cause disease in the CNS, whereas BK causes a hemorrhagic cystitis. JC virus is detected using PCR with CSF or electron microscopy of brain tissue, whereas BK virus is detected using PCR or cytologic examination of urine.

POXVIRUSES

The poxviruses (Table 51-19) are the largest and most complex of all viruses. Diseases caused by poxviruses include smallpox, a devastating and frequently fatal disease of historical importance (eliminated from the world in 1977), but feared as a possible biologic weapon if reintroduced; molluscum contagiosum, an exclusively

Table 51-18   Polyomaviruses

Family: *Polyomaviridae*   Common Name: Polyomavirus

Virus: Polyomavirus (BK and JC viruses infect humans)
  Characteristics: Double-stranded DNA genome; icosahedral capsid, no envelope; includes BK and JC polyomaviruses
  Transmission: Probably direct contact with infected respiratory secretions; both viruses are ubiquitous in humans
  Site of latency: Kidney
  Disease: Mild or asymptomatic primary infection, virus remains dormant in kidneys; reactivation in immunocompromised patients causes hemorrhagic cystitis (BKV) or progressive multifocal leukoencephalopathy (JCV)
  Detection: JC virus by PCR (cerebrospinal fluid) or EM (brain tissue); BK virus by PCR or cytology (urine)
  Treatment: Supportive. Decrease immune suppression
  Prevention: Avoid contact with virus; prevention of acquisition of virus unlikely

Table 51-19   Poxviruses

Family: *Poxviridae*   Common Name: Poxvirus

Virus: Smallpox, molluscum contagiosum, orf and monkeypox viruses
  Characteristics: Largest and most complex of all viruses; brick-shaped virion with nonconforming symmetry referred to as *complex;* double-stranded DNA genome
  Transmission: Respiratory droplets (smallpox); direct contact (molluscum, orf and monkeypox)
  Disease: All diseases of the skin; smallpox is a generalized infection with pustular rash (10%-25% fatal); molluscum manifests as benign nodules of skin; orf manifests as localized papules/vesicles of skin; monkeypox as generalized infection including skin
  Detection: Electron microscopy of material from skin lesion, PCR
  Epidemiology: Smallpox eradicated from world in 1977; smallpox and molluscum are limited to humans; orf and monkeypox are zoonoses
  Treatment: Supportive
  Prevention: Vaccine for smallpox; avoid contact for all viruses

Table 51-20   Reoviruses

Family: *Reoviridae*   Common Name: Reovirus

Virus: Rotavirus
  Characteristics: Segmented, double-stranded, RNA genome; icosahedral capsid with no envelope
  Transmission: Fecal-oral; survives well on inanimate objects
  Disease: Gastroenteritis in infants and children 6 months to 2 years
  Detection: EIA, latex agglutination
  Epidemiology: Winter-spring seasonality in temperate climates; nosocomial transmission can occur easily
  Treatment: Supportive, especially fluid replacement
  Prevention: Avoid contact with virus; vaccine in underdeveloped countries

human disease of the skin that is self-limited but may recur during periods of severe immunosuppression; orf, a localized infection of the skin caused by viruses responsible for dermatitis in sheep, goats, and related animals; and monkey pox, characterized by fever, headache, fatigue, and rash with adenopathy (adenopathy helps distinguish monkey pox from smallpox). Rarely, patients develop encephalitis. The source of monkeypox is wild rodents imported from Africa where the virus is endemic.

REOVIRUSES

The reoviruses (Table 51-20) were first isolated from respiratory and enteric specimens from which they gained their name, *respiratory-enteric-orphan* viruses (reoviruses). *Orphan* designated the absence of an associated disease when the viruses were first described. Although reoviruses infect most mammalian species and are readily detected in water contaminated with animal feces, they generally are not agents of serious disease. In humans, rotaviruses and the agent of Colorado tick fever are pathogens. Reoviruses are difficult to grow, necessitating enzyme immunoassays, electron microscopy, or RT-PCR for detection.

RETROVIRUSES

The retrovirus family (Table 51-21) constitutes a large group of viruses that primarily infect vertebrates.

Table 51-21   Retroviruses

Family: *Retroviridae*	Common Name: Retroviruses

Characteristics: Single-stranded, RNA genome; icosahedral capsid with envelope; reverse transcriptase converts genomic RNA into DNA
Virus: Human immunodeficiency virus types 1 and 2 (HIV-1 and HIV-2)
  Transmission: Sexual contact, blood and blood product exposure, and perinatal exposure
Site of latency: CD4 T lymphocytes
Disease: Most disease in humans caused by HIV-1; infected cells include CD4 (helper) T lymphocytes, monocytes, and some cells of the central nervous system; asymptomatic infection, acute flu-like disease, AIDS related complex, and AIDS, AIDS-associated infections and malignancies
Detection: Serology, antigen detection, RT-PCR
Epidemiology: Those at risk of infection are homosexual or bisexual males, intravenous drug abusers, sexual contacts of HIV-infected individuals, and infants of infected mothers
Treatment: Many including nucleoside reverse transcriptase inhibitors, nonnucleoside reverse transcriptase inhibitors, protease inhibitors, and inhibitors of viral entry into host cells. Treat infections resulting from immunosuppression
Prevention: Avoid contact with infected blood/blood products and secretions. Blood for transfusion is screened for antibody to HIV-1 and -2
Virus: Human T-lymphotropic viruses (HTLV-1 and HTLV-2)
Transmission: Known means of transmission are similar to HIV
Disease: T-cell leukemia and lymphoma, and tropical spastic paraparesis for HTLV-1; no known disease associations for HTLV-2
Detection: Serology
Epidemiology: HTLV-1 present in 0.025% of volunteer blood donors in United States. Blood is screened for antibody to HTLV-1 and -2; rates of HTLV-1 in areas of Japan and the Caribbean are considerably higher
Oncogenic: T-cell lymphoma (HTLV-1)
Treatment: Supportive
Prevention: Avoid contact with virus

Table 51-22   Rhabdoviruses

Family: *Rhabdoviridae*	Common Name: Rhabdovirus

Virus: Rabies virus
  Characteristics: Single-stranded, RNA genome; helical capsid with envelope, bullet-shaped
  Transmission: Bite of rabid animal most common; 20% of human rabies cases have no known exposure to rabid animal
  Disease: Rabies
  Detection: FA staining, PCR
  Treatment: Supportive
  Prevention: Avoid contact with rabid animals; vaccinate domestic animals; postexposure prophylaxis with hyperimmune antirabies globulin and immunization with rabies vaccine

Human retroviruses include the human immunodeficiency viruses (HIV-1 and HIV-2) and the human T-cell lymphoma viruses (HTLV-1 and HTLV-2). Retroviruses are unique because they possess the enzyme reverse transcriptase. Reverse transcriptase allows the viral RNA genome to be replicated into DNA and then RNA, rather than directly into RNA. Amino acid sequencing of the reverse transcriptase protein divides the retrovirus family into its many groups. The laboratory detection of HIV and monitoring of antiviral therapy is done using serologic testing, antigen detection, and qualitative and quantitative testing for RNA genome by the RT-PCR test. In addition, genome sequencing is used to establish susceptibility to some antiviral agents.

RHABDOVIRUSES

Rhabdoviruses (Table 51-22) infect plants and animals. Rabies virus is the one member that causes disease in

humans. The virus is present in saliva and is readily transmitted by animal bite. Local replication at the site of the bite wound is followed by invasion of the peripheral and central nervous systems. Once in the brain, rabies virus spreads to various tissues and organs, including salivary gland, kidney, heart, hair follicles, and cornea. The bullet-shaped virion is readily identifiable by electron microscopy; however, diagnosis is performed best by FA staining or RT-PCR of infected cells or tissue.

TOGAVIRUSES

The *Togaviridae* family (Table 51-23) includes rubella virus and the alpha viruses, a large group of mosquito-borne arboviruses. Rubella is a benign disease, causing rash and fever, unless it occurs in a pregnant patient, in whom it can infect the developing fetus and result in multiple congenital anomalies. Fetal infection can be

Table 51-23   Togaviruses

Family: *Togaviridae*	Common Name: Togaviruses

Characteristics: Single-stranded RNA genome and icosahedral capsid with envelope; family contains arboviruses and nonarthropod-borne rubella virus

Virus: Rubella virus
　Transmission: Respiratory, transplacental
　Disease: Rubella (mild exanthematous disease), congenital rubella
　Detection: Serology
　Treatment: Supportive
　Prevention: Rubella vaccine

Virus: Arboviruses referred to as alphaviruses*
　Transmission: Arthropod vector, usually mosquito
　Disease: Eastern, Western, and Venezuelan equine encephalitis
　Detection: Serology and antibody detection in CSF
　Treatment: Supportive
　Prevention: Avoid contact with vector; vector control programs

*See footnote for Table 51-6.

prevented by vaccinating all women before they become pregnant. In arbovirus infections, mosquitoes infect a vertebrate host (such as birds and rodents), the virus multiplies (amplifies) in this host, and is picked up and passed during subsequent mosquito bites. Humans are infected incidentally and are not amplifiers of the virus; rather, they are dead-end hosts. Human disease varies from asymptomatic infection to fatal encephalitis and includes Eastern, Western, and Venezuelan equine encephalitides. Togavirus disease is diagnosed by detecting specific serum IgG and IgM antibodies. Virus isolation is not practical in clinical laboratories.

MISCELLANEOUS VIRUSES

Additional viruses detected in humans include the astroviruses and potential agents of hepatitis, transfusion-transmitted virus (TTV), and hepatitis G virus (HGV). The astrovirus is a single-stranded RNA virus found in the gastrointestinal tract of many animals, including humans. Human astroviruses are ubiquitous in children, causing a minority of childhood diarrheas. Astroviruses are detected by electron microscopy.

TTV and HGV are DNA and RNA viruses, respectively, that are detected commonly in human blood specimens but have not yet been associated with any disease. HGV is a flavivirus, similar to HCV. TTV resembles a new group of animal viruses called circoviruses.

LABORATORY DIAGNOSIS OF VIRAL INFECTION

SETTING UP A CLINICAL VIROLOGY LABORATORY

The demand for clinical virology laboratory services has skyrocketed during the past two decades. This growth has resulted from the introduction of virus-specific antiviral drugs, the commercial availability of reagents, and development of rapid diagnostic techniques by conventional methods such as fluorescence microscopy and enzyme immunoassays, the ready availability of cell lines for cell culture procedures, and the introduction of real-time PCR for the detection of viral genomes. Additionally, improved medical care, in the forms of organ transplantation and immune suppression accompanying cancer therapy, have expanded greatly the number of patients acquiring viral disease. Combining these with the appearance of new viral diseases threatening local and world populations (e.g., SARS, avian influenza, monkeypox), makes the laboratory diagnosis of viral infection more important and achievable than in previous years. In deciding which virology tests to offer, those in each clinical laboratory should determine whether the test is required for the appropriate care of their patient population and whether techniques are available that provide an accurate and cost-effective test result. Viral diseases that require laboratory diagnosis include sexually transmitted diseases, diarrhea, respiratory disease in adults and children, aseptic meningitis, arbovirus encephalitides, congenital diseases, hepatitis, and infections in immunocompromised hosts. Viruses detected in a community clinical virology laboratory are listed in Table 51-24.

Those working in a clinical virology laboratory must be familiar with cell culture, enzyme immunoassay, immunofluorescence methods, and molecular methods such as PCR, in addition to other common laboratory techniques. Large equipment needed for a full-service virology laboratory include a laminar flow biological safety cabinet, fluorescence microscope, inverted bright-field microscope, refrigerated centrifuge, incubator, refrigerator and freezer, roller drum for holding cell culture tubes during incubation, and enzyme or molecular testing instrumentation (Figures 51-5 through 51-7).

Standard precautions and biosafety level 2 conditions are needed for community and most non-retroviral laboratories. Requirements include standard microbiologic practices, training in biosafety, protective clothing and gloves, limited access, decontamination of all infectious waste, and a class I or II biosafety cabinet. Recall that some viruses should not be propagated in biosafety level 2 laboratories. These include H5N1 influenza viruses, SARS coronavirus, hemorrhagic fever viruses, and smallpox if it were reintroduced. A virology laboratory floor plan that includes safety equipment and work areas is shown in Figure 51-8.

SPECIMEN SELECTION AND COLLECTION

General Principles

Specimen selection depends on the specific disease syndrome, viral etiologies suspected, and time of year. Selection of specimen based on disease is confusing, in that most viruses enter via the upper respiratory tract, yet they infect tissues and result in symptoms distant from the primary, respiratory tract site. For example, aseptic meningitis, caused by many enterovirus types, is identified by detecting virus in throat, rectal swab, or

Table 51-24 Viruses Detected by Culture, PCR, or Assay for Antigen in a Community Hospital Virology Laboratory

Virus	Number of Viruses From Adults and Chidlren
Adenovirus	25
Cytomegalovirus	30
Enterovirus	50
Herpes simplex virus	206
Influenza virus	426
Parainfluenza virus	41
Respiratory syncytial virus	151
Rotavirus	163
Varicella-zoster virus	38
TOTAL	1130

Data representing 1 year of testing from Evanston Northwestern Healthcare, Evanston, Ill.

Figure 51-5 Roller drums used to hold cell culture tubes during incubation. Slow rotation continually bathes cells in medium. (Courtesy Children's Hospital Medical Center of Akron, Akron, Ohio.)

Figure 51-6 Inverted microscope used to examine cell monolayers growing attached to the inside surface beneath the liquid medium. Note the objective is under the glass test tube facilitating observation of the cell monolayer.

Figure 51-7   Class II biological safety cabinets used in clinical virology laboratory. The cabinet on the left is used for "contaminated" work, including specimen inoculation and working with positive cell cultures. The cabinet on the right is used for "clean" work, such as preparation and maintenance of uninoculated cell cultures.

CSF specimen. Pharyngitis and gastrointestinal symptoms may not be included in the patient's complaints.

Specimen selection based on virus suspected is complicated by the fact that similar clinical syndromes can be caused by many different viruses. By collecting only specimens needed to detect a specific virus, other important etiologies may be missed. For example, testing smears of nasal secretions from an infant by fluorescence staining or enzyme immunoassay for the detection of RSV does not allow for diagnosis of similar disease caused by influenza, parainfluenza, and metapneumoviruses.

Appropriate specimen selection dictates that the specimen type and viruses suspected should be indicated on the requisition. *The laboratory should always be notified if rare agents representing a danger to laboratory workers, such as SARS coronavirus, H5N1 avian influenza virus, hemorrhagic fever viruses, and the like, are suspected.*

Serum for serologic testing may be necessary, and some viral disease need only to be considered during certain months because appearance is seasonal. Table 51-25 suggests specimens for the diagnosis of viral diseases, noting seasonality where important.

Specimens for the detection of virus should be collected as early as possible following the onset of symptomatic disease. Virus may no longer be present as early as 2 days after the appearance of symptoms. Recommendations for collection of common specimens are summarized below.

Throat, Nasopharyngeal Swab or Aspirate

Nasopharyngeal aspirates are superior, in general, to throat or nasopharyngeal swabs for recovering viruses, but swabs are considerably more convenient. Throat swabs are acceptable for recovering enteroviruses, adenoviruses, and HSV, whereas nasopharyngeal swab or aspirate specimens are preferred for the detection of respiratory syncytial, influenza, and parainfluenza viruses. Rhinovirus detection requires a nasal specimen. Throat specimens are collected by rubbing inflamed, vesiculated, or purulent areas of the posterior pharynx with a dry, sterile swab. Avoid touching the tongue, buccal mucosa, or teeth and gums. Nasopharyngeal secretion specimens are collected by inserting a swab with flexible shaft through the nostril to the nasopharynx or by using a bulb syringe with 3 to 7 mL of buffered saline. The saline is squirted into the nose by squeezing the bulb and aspirated back by releasing the bulb or using small tubing with suction inserted in the other nostril.

Bronchial and Bronchoalveolar Washes
Wash and lavage fluid collected during bronchoscopy are excellent specimens for the detection of viruses that infect the lower respiratory tract, especially influenza and adenoviruses.

Rectal Swabs and Stool Specimens
Stool and rectal swabs of fecal specimens are used to detect rotavirus, enteric adenoviruses (serotypes 40 and 41), and enteroviruses. Many agents of viral gastroenteritis do not grow in cell culture and require electron microscopy for detection (see section regarding detection of viruses in patient specimens). In general, stool specimens are preferable to rectal swabs and should be required for rotavirus and enteric adenovirus testing. Rectal swabs are acceptable for detecting enteroviruses in patients suspected of having enteroviral disease such as aseptic meningitis. Rectal swabs are collected by inserting a swab 3 to 5 cm into the rectum to obtain feces. Five to 10 mL of freshly passed diarrheal stool or stool collected in a diaper from young infants is sufficient for rotavirus and enteric adenovirus detection.

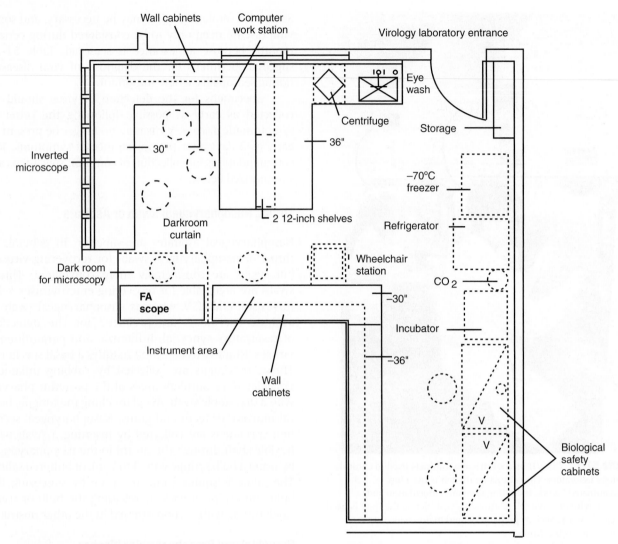

Figure 51-8 Floor plan of a clinical virology laboratory that includes biological safety cabinets for specimen processing and cell culture handling, darkroom area for fluorescence microscopy, stand-up and sit-down counter space, computer station, storage areas, incubator, refrigerator, and freezer. *V,* Vacuum.

Urine

CMV, mumps, rubella, measles, polyomaviruses, and adenoviruses can be detected in urine. Virus recovery may be increased by processing multiple (two to three) specimens because virus can be shed intermittently or in low numbers. The best specimen is at least 10 mL of a clean-voided first-morning urine.

Skin and Mucous Membrane Lesions

Enteroviruses, HSV, VZV, and, rarely, CMV or pox viruses can be detected in vesicular lesions of skin and mucous membranes. Once the vesicle has ulcerated or crusted, detection of virus is difficult.

Collection of specimen from cutaneous vesicles for detection of HSV or VZV may require a Tzanck smear if PCR testing is not available. Tzanck smears are prepared

by carefully unroofing the vesicle. If a tuberculin syringe is used, the small "drop" of vesicle fluid should be aspirated first to be used in case viral or bacterial culture is needed. Flush needle with viral transport medium and add rinse to viral transport tube. With the roof of the vesicle folded back, carefully remove excess fluid by dabbing with a sterile gauze. Press a clean glass microscope slide against the base of the ulcer. Lift, move slide slightly, and press again. Cells from the base of the ulcer will stick to the slide, making an "impression smear" of infected and noninfected cells. Additional smears can be made from other vesicles. Send slides to the laboratory for fixation and staining. Alternatively, vesicle fluid and cells scraped from the base of an unroofed vesicle can be added to 2 to 3 mL of viral transport medium. Smears can be prepared in the laboratory by cytocentrifugation

Table 51-25   Specimens for the Diagnosis of Viral Diseases*

Disease Categories and Probable Viral Agent	Season of Most Common Occurence	Throat/ Nasopharynx	Stool	CSF	Urine	Other
Respiratory						
Adenoviruses	Y	++++				
Influenza virus	W	++++				
Parainfluenza virus	Y	++++				
Respiratory syncytial virus (RSV)	W	++++				
Metapneumovirus	W	++++				
Rhinoviruses	Y					Nasal (+++)
SARS Coronavirus	W	++++				
Sin nombre virus	SP, S					Serum for antibody detection
Dermatologic and Mucous Membrane						
Vesicular						
Enterovirus	S, F	++	+++			Vesicle fluid or scraping
Herpes simplex[†]	Y					Vesicle fluid or scraping
Varicella-zoster[†]	Y	++				Vesicle fluid or scraping
Monkeypox	Y					Vesicle fluid or scraping
Exanthematous						
Enterovirus	S, F	+++	++			
Measles	Y	++			++	Serum for antibody detection
Rubella	Y				++	Serum for antibody detection
Parvovirus	Y					Serum for antibody detection, amniotic fluid (PCR)
Pustular/Nodular						
Molluscum contagiosum, orf	Y					Tissue
Warts	Y					Tissue/cells
Meningoencephalitis/Encephalitis						
Arboviruses	S, F					CSF/serum for antibody detection
Enteroviruses	S, F	+++		++	++++	
Herpes simplex	Y			++++		Brain biopsy (PCR)
Lymphocytic choriomeningitis	Y					Serum for antibody detection
Mumps virus	Y					Serum for antibody detection
HIV	Y					Brain biopsy (culture/PCR)
Polyomavirus (JC virus)	Y					Brain biopsy (EM/PCR)
Rabies virus	Y					Corneal cells, brain
Gastrointestinal Disease						
Adenoviruses (serotypes 40-41)	Y		++++			Stool (EIA or EM)
Noroviruses	S		++++			Stool (EM)
Rotavirus	W, SP		++++			Stool (EIA, Latex)

Continued

Table 51-25 Specimens for the Diagnosis of Viral Diseases* (cont'd)

Dermatologic and Mucous Membrane						
Congenital and Perinatal						
Cytomegalovirus	Y				+++	Serum for antibody (IgM) detection
Enteroviruses	S, F	+++		+++	+++	
Herpes simplex virus	Y					Vesicle fluid
Parvovirus	Y					Amniotic fluid, liver tissue
Rubella	Y				++	Serum for antibody (IgM) detection
Eye (Ocular Disease)						
Adenoviruses	Y	++				Conjunctival swab or scraping
Herpes simplex virus	Y					Conjunctival swab or scraping
Varicella-zoster virus	Y					Conjunctival swab or scraping
Posttransplantation Syndrome						
Cytomegalovirus	Y				++	Blood (++++) shell vial and/or antigenemia; tissue (++++)
Epstein-Barr virus	Y					Serology, tissue (PCR) (EBV)
Human herpesvirus-6 (HH6)	Y					Serology, blood (PCR)
Herpes simplex	Y					Tissue (+++) virus
BK virus	Y				++++	
Myocarditis, Pericarditis, and Pleurodynia						
Coxsackie B	S, F	+++		++		Pericardial fluid (++++)
Hemorrhagic Fevers						
Ebola/Marburg viruses	Y					Tissue, respiratory secretions, serum for antibody detection
Lassa fever virus		+++			+	Serum/throat washes for viral detection. Serum for antibody detection

Y, Year-round; *SP,* spring; *S,* summer; *F,* fall; *W,* winter; *CSF,* cerebrospinal fluid; *EIA,* enzyme immunoassay; *EM,* electron microscopy; *PCR,* polymerase chain reaction.

*Specimens indicated beside specific viruses should be obtained if that specific virus is suspected (++++, most appropriate; + least appropriate).

†Direct fluorescent antibody studies are available for herpes simplex virus and varicella-zoster virus.

of fluid medium or PCR can be performed using the specimen in viral transport medium.

Sterile Body Fluids Other Than Blood

Sterile body fluids, especially cerebrospinal, pericardial, and pleural fluids, may contain enteroviruses, HSV, VZV, influenza viruses, or CMV. These specimens are collected aseptically by the physician.

Blood

Viral culture of blood is used primarily to detect CMV; however, HSV, VZV, enteroviruses, and adenovirus may occasionally be encountered. Five to 10 mL of anti-coagulated blood collected in a Vacutainer tube is needed. Heparinized, citrated, or EDTA anticoagulated blood is acceptable for CMV detection. Citrated blood should be used in instances in which other viruses are being sought.

Bone Marrow

Bone marrow for viral detection should be added to a sterile tube with anticoagulant. Heparin, citrate, or EDTA anticoagulants are acceptable. Specimens are collected by aspiration. Most viruses, other than parvovirus

B19, are detected more readily from sites other than bone marrow.

Tissue

Tissue specimens are especially useful for detecting viruses that commonly infect lung (CMV, influenza virus, adenovirus, sin nombre virus), brain (HSV), and gastrointestinal tract (CMV). Specimens are collected during surgical procedures.

Serum for Antibody Testing

Acute and convalescent serum specimens may be needed to detect antibody to specific viruses. Acute specimens should be collected as soon as possible after the appearance of symptoms. The convalescent specimen is collected a minimum of 2 to 3 weeks after the acute specimen. In both cases, an appropriate specimen is 3 to 5 mL of serum collected by venipuncture.

SPECIMEN TRANSPORT AND STORAGE

Ideally, all specimens collected for detection of virus should be processed by the laboratory immediately. Although inoculation of specimens into cell culture at the bedside has been recommended in the past, potential biohazards, sophisticated processing steps, and necessary quality control make this practice impractical. Specimens for viral isolation should not be allowed to sit at room or higher temperature. Specimens should be placed in ice and transported to the laboratory at once. If a delay is unavoidable, the specimen should be refrigerated, *not frozen,* until processing occurs. Every attempt should be made to process the specimen within 12 to 24 hours of collection. Under unusual circumstances, specimens may need to be held for days before processing. For storage up to 5 days, hold specimen at 4° C. Storage for 6 or more days should be at –20° or preferably at –70° C. Specimens for freezing should first be diluted or emulsified in viral transport medium. Significant loss of viral infectivity may occur during prolonged storage, regardless of conditions, especially for the more labile enveloped viruses.

If a commercial kit is being used, specimens for molecular testing should be transported and stored according to the manufacturer's instructions. Specimens for processing using commercial reagents that are not approved by the Food and Drug Administration (FDA) (e.g., analyte specific reagents) or assays that have been created and validated in the user's laboratory are transported and stored at refrigeration temperatures. Storage for longer than 2 to 3 days should be at –70° C.

Many types of specimens for the detection of virus can be collected using a swab. Most types of synthetic swab material, such as rayon and Dacron, are acceptable. Once collected, it is recommended that specimens on swabs be emulsified in viral transport medium before transporting to the laboratory, especially if transport will occur at room temperature and require more than 1 hour. It has been shown, however, that transport of specimens containing HSV can be accomplished using a Culturette (Becton-Dickinson Microbiology Products, Cockeysville, Md.; modified Stuart's medium). Calcium alginate is not acceptable for the detection of HSV because it may inactivate HSV. If calcium alginate must be used, the specimen should be transferred into viral transport medium by swirling the swab vigorously and then the swab should be discarded immediately.

Viral transport media are used to transport small volumes of fluid specimens, small tissues and scrapings, and swab specimens, especially when contamination with microbial flora is expected. Transport media contain protein, such as serum, albumin, or gelatin, to stabilize virus and antimicrobials to prevent overgrowth of bacteria and fungi. Penicillin (500 units/mL) and streptomycin (500 to 1000 mcg/mL) have been used traditionally; however, a more potent mixture includes vancomycin (20 mcg/mL), gentamicin (50 mcg/mL), and amphotericin (10 mcg/mL). If serum is added as the protein source, fetal calf serum is recommended because it is less likely to contain inhibitors, such as antibody. Examples of successful transport media include Stuart's medium, Amie's medium, Leibovitz-Emory medium, Hanks balanced salt solution (HBSS), and Eagle' tissue culture medium. Respiratory and rectal/stool specimens can be maintained in modified Stuart, modified Hanks balanced salt solution, or Leibovitz-Emory medium containing antimicrobials.

Blood for viral culture, transported in a sterile tube containing anticoagulant, must be kept at refrigeration temperature (4° C) until processing.

Blood for viral serology testing should be transported to the laboratory in the sterile tube in which it was collected. Serum should be separated from the clot as soon as possible. Serum can be stored for hours or days at 4° C or for weeks or months at –20° C or below before testing. Testing for virus-specific IgM should be done before freeing whenever possible, because IgM may form insoluble aggregates upon thawing, giving a false-negative result.

SPECIMEN PROCESSING

General Principles

Specimens for viral culture should be processed immediately upon receipt in the laboratory. This may be accomplished best by combining bacteriology and virology processing responsibilities. Although the threat of cell culture contamination in the past dictated separation of virology procedures, the addition of broad-spectrum antimicrobials to cell cultures significantly

Table 51-26 Laboratory Processing of Viral Specimens

Source	Specimen	Processing*	Cells for detection of common viruses
Blood	Anticoagulated blood	Separate leukocytes (see Procedure 51-1)	PMK, HDF, HEp-2
Cerebrospinal fluid (CSF)	1 mL CSF	Inoculate directly	PMK, HDF, HEp-2
Stool or rectal swab	Pea-sized aliquot of feces	Place in 2 mL of viral transport medium vortex. Centrifuge at 1000× g for 15 min and use supernatant fluid for inoculum	PMK, HDF, HEp-2
Genital, skin	Vesicle fluid or scraping	Emulsify in viral transport medium	HDF
Miscellaneous	Swab, fluids	Emulsify in viral transport medium Fluid, inoculate directly	PMK, HDF, HEp-2
Respiratory tract	Nasopharyngeal secretions, throat swab, respiratory tract washings, sputum	Dilute with viral transport medium	PMK, HDF, HEp-2
Tissue	Tissue in sterile container	Mince with sterile scalpel and scissors and gently grind. Prepare 20% suspension in viral transport medium. Centrifuge at 1000× g for 15 min and use supernatant fluid for inoculum	PMK, HDF, HEp-2
Urine	Mid-stream urine	Inoculate directly if clear. Turbid specimens should be centrifuged at 1000× g for 15 min and use supernatant fluid for inoculum	HDF, HEp-2 (if adenovirus suspected)

HDF, Human diploid fibroblast; *HEp-2*, human epidermoid; *PMK*, primary monkey kidney.
*All inoculum into tissue culture tubes are 0.25 mL volumes.

decreased the possibility of cross contamination with bacteria and fungi. In most laboratories, processing with other microbiology specimens allows viral cultures to be processed 7 days per week. If delays must occur, specimens should be stored in viral transport medium at 4° C as described previously. Fluid specimens that have not been added to viral transport medium and that must be stored for days before processing should be diluted in transport medium (1:2 to 1:5) before storage.

Each specimen for virus isolation should be accompanied by a requisition that provides the following information, in addition to patient identification and demographics: source of specimen, clinical history or viruses suspected, and date and time of specimen collection. If this information is not available, a call for additional details should be made to the requesting physician or to the person caring for the patient.

Processing viral specimens should occur in a biological safety cabinet whenever possible (see Figure 51-7). This protects specimens from contamination by the processing technologist and protects those in the laboratory from infectious aerosols created when specimens are manipulated. If processing cannot be performed in a biological safety cabinet, it should occur behind a protective Plexiglas shield on the countertop. Latex gloves and a laboratory coat should be worn during manipulation of all patient specimens. Vortexing, pipetting, and centrifugation can all create dangerous aerosols. Vortexing should be done in a tightly capped tube behind a shield. After vortexing, the tube should be opened in a biological safety cabinet. Pipetting also should be performed behind a protective shield. Pipettes must be discarded into a disinfectant fluid so that the disinfectant reaches the inside of the pipette or into a leakproof biosafety bag for autoclaving or incineration. More than one patient specimen or series of cell culture tubes for that specimen should not be opened for inoculation or feeding (exchange of cell culture medium) at one time. Aerosols and "micro" splashes contribute to cross contamination of cultures, especially during viral respiratory season when a high percentage of specimens are positive for influenza virus, RSV, and other viruses. This pertains to specimens processed in a biological safety cabinet or behind a shield on the benchtop.

Processing virology specimens is not complicated (Table 51-26). In general, any specimen that might be contaminated with bacteria or fungi, or any swab specimen should be added to viral transport medium. Fluid specimens that are normally sterile can be inoculated directly to cell culture. Viral transport medium or fluid specimens not in transport medium should be vortexed just before inoculation to break up virus-

Table 51-27   Menu of Virus Detection or Quantitation Tests

Test	Frequency of Request (%)*
Culture/Antigen Detection	
Blood culture (CMV shell vial/antigenemia)	2
Bronchial secretions culture	5
Enterovirus culture	4
Herpes simplex virus culture	5
Influenza culture/antigen detection	12
Pediatric respiratory viruses culture	13
Rotavirus antigen detection	8
Respiratory syncytial virus detection	24
Urine culture	2
Viral culture (tissue and fluids)	12
Varicella zoster virus detection	13
Total tests (N = 4776)	100
Molecular Detection/Quantitation	
Cytomegalovirus detection	5
Enterovirus detection	< 1
Hepatitis C virus quantitation	8
Human immunodeficiency virus detection and quantitation	7
Human papilloma virus detection	54
Herpes simplex virus detection	4
Influenza virus detection	22
Parvovirus detection	< 1
Total tests (N = 4842)	100

Data representing 1 year of testing from Evanston Northwestern Healthcare, Evanston, Ill.

*Frequency of request expressed as percentage of all tests.

Table 51-28   Menu of Viral Serology Tests

Test	Frequency of Request (%)*
Cytomegalovirus IgM and/or IgG	3
Enterovirus antibody	< 1
Epstein-Barr virus panel or individual antibody	1
Hepatitis panel	8
Hepatitis A IgM and/or IgG	1
Hepatitis B panel or individual antigen/antibody test	28
Hepatitis C antibody	7
Human immunodeficiency virus antibody or antigen	34
HTLV-1 antibody	< 1
Influenza A or B antibody	< 1
Measles antibody	5
Mumps antibody	1
Parvovirus antibody	< 1
Rubella antibody	6
Varicella-zoster antibody	6
Total tests (N = 24,071)	100%

Data representing 1 year of testing from Evanston Northwestern Healthcare, Evanston, Ill.

*Frequency of request expressed as percentage of all tests.

containing cells and resuspend the inoculum. Sterile glass beads added to the transport medium help break up cell clumps and release virus from cell aggregates. Grossly contaminated or potentially toxic specimens, such as minced or ground tissue, can be centrifuged (1000× *g*, 15 minutes) and the virus-containing supernatant used as inoculum. Each viral cell culture tube is inoculated with 0.25 mL of specimen. If insufficient specimen is available, dilute with viral transport medium to increase volume. Excess specimen can be saved at 4° C in case the culture is contaminated. Contaminated specimens can be reprocessed with antibiotic-containing viral transport medium, if they were not originally handled in this manner, or they can be filtered using a disposable 0.22-µm filter and the filtrate recultured. In practice, most specimens requiring reprocessing because of contamination fail to have virus detected in culture. Blood for viral culture requires special processing to isolate leukocytes, which are then inoculated to cell culture tubes (Procedure 51-1). Rapid shell vial cell cultures are used to detect many viruses. Handling and examining cell cultures once they are inoculated with specimen are discussed later in this chapter.

Processing Based on Specimen Type

Virology laboratories should have a menu of individual virus detection and serology tests, and an algorithm for their use, rather than one comprehensive viral detection or serology test battery for use in all situations. Tables 51-27 and 51-28 represent virus detection and serology tests that are useful in a community clinical virology laboratory, based on physician orders. It is important to note the common and increasing use of molecular detection tests for viruses. In the author's setting, half of all virus detection tests performed were molecular based (e.g., PCR or DNA probe detection). An algorithm for the use of virus detection tests should be based on the type of specimen received or the specific virus suspected. Most laboratories receive specimens with little or no clinical data and no specific virus indicated. The algorithm in Figure 51-9 is designed for use with these specimens.

Lip and Genital. All lip and genital specimens should be tested only for HSV by culture or molecular methods. Disease by other viruses at lip and external genital sites is unusual and detection of these agents,

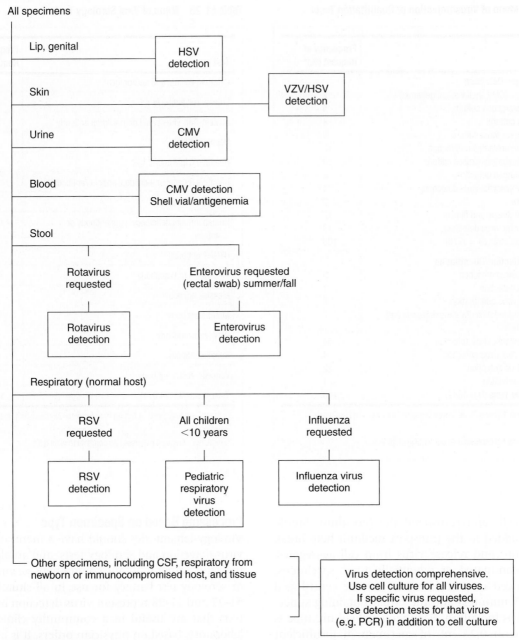

All specimens

Lip, genital — HSV detection

Skin — VZV/HSV detection

Urine — CMV detection

Blood — CMV detection Shell vial/antigenemia

Stool
- Rotavirus requested → Rotavirus detection
- Enterovirus requested (rectal swab) summer/fall → Enterovirus detection

Respiratory (normal host)
- RSV requested → RSV detection
- All children <10 years → Pediatric respiratory virus detection
- Influenza requested → Influenza virus detection

Other specimens, including CSF, respiratory from newborn or immunocompromised host, and tissue → Virus detection comprehensive. Use cell culture for all viruses. If specific virus requested, use detection tests for that virus (e.g. PCR) in addition to cell culture

Figure 51-9 Algorithm for processing viral specimens based on specimen type and virus suspected. Virus detection implies viral culture, antigen detection, or molecular testing (e.g., PCR).

such as VZV or enterovirus, should be prompted by a special request.

Urine. Urine specimens require a CMV detection test. This is done best using the shell vial cell culture or molecular assay.

Stool. Stool specimens from infants (5 years of age and younger) in North America should be tested for rotavirus during the fall, winter, and spring. Enteric adenoviruses (serotypes 40 and 41) cause diarrhea in young children and infants throughout the year. Adenovirus gastroenteritis appears to be more common in some geographic areas. Routine testing is only necessary in locations in which disease is known to occur. Stool or rectal swabs from adults, children, and infants should be tested for enterovirus in summer and fall as an aid in the diagnosis of aseptic meningitis. Stool for enterovirus should be tested in conjunction with throat and CSF specimens when possible.

Specimens from the Respiratory Tract. Respiratory specimens should be divided according to the patient's age and underlying medical condition. Immunocompromised patients require a comprehensive virus detection test consisting of cell culture or corresponding molecular tests. Immunocompetent adults should have an influenza virus test (culture or PCR) during influenza season, which is November to April in most areas. Children younger than 10 years of age are susceptible to serious infection caused by influenza virus, parainfluenza virus, RSV, and adenoviruses and need full respiratory virus detection. Infants younger than 2 years of age are especially vulnerable to RSV bronchiolitis, which may require hospitalization and comprehensive supportive care. A rapid, nonculture RSV detection test, such as FA staining or enzyme immunoassay, is appropriate in these situations.

Specimens from Neonatal Patients. Specimens from newborns with the possibility of congenital or perinatal viral disease should receive a comprehensive virus culture, rapid shell vial cell culture for CMV, and appropriate molecular tests (e.g., enteroviruses) when appropriate. Consultation with the pediatrician or neonatologist is necessary to match test orders to specific clinical conditions.

Cerebrospinal Fluid. CSF specimens can contain many different viral etiologies, although HSV, enteroviruses, HIV and arboviruses are the most likely to be detected. HSV, HIV and enteroviruses are detected best using molecular assays, while arboviruses require antibody testing. Testing for other, less frequently detected viruses, such as CMV, VZV, JC virus and many more, should be included following consultation with the patient's physician.

Blood. Blood for viral culture might contain CMV, VZV, adenovirus, or enteroviruses; however, CMV is by far the most common and the one agent whose detection and quantitation is clinically useful information. Therefore, blood for viral detection should be processed for CMV only, unless other agents are mentioned by the requesting physician. All specimens from immunocompromised hosts and tissues or fluids from presumably sterile sites should be processed for comprehensive virus detection. Processing of specimens (see Figure 51-9) should be modified to match the needs of local physicians and endemic viral diseases.

Processing Based on Requests for Specific Viruses

Arboviruses. Serologic tests for **arboviruses** are offered by state public health laboratories or by some commercial laboratories. Diagnosis of arbovirus ence-

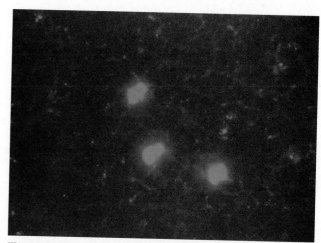

Figure 51-10   Typical fluorescing white blood cells containing CMV antigen as seen in the CMV antigenemia stain.

phalitis, such as Eastern, Western, Venezuelan, and St. Louis, and that caused by California encephalitis, LaCrosse, and West Nile viruses, requires detection of virus-specific IgM in serum or a rise in IgG antibody titer in acute and convalescent serum specimens. Detection of virus-specific IgM in CSF is available for most agents. Culture of arboviruses for diagnostic purposes is not practical. PCR for some agents is available through state public health laboratories but may be less sensitive than serodiagnosis because the virus and the viral nucleic acid detected by molecular methods are present for short periods only.

Cytomegalovirus. CMV can be detected in clinical specimens using conventional cell culture, shell vial assay, antigenemia immunoassay, or molecular methods. CMV produces cytopathic effects (CPE) in diploid fibroblast cells in 3 to 28 days, averaging 7 days. Shell vial for CMV has a sensitivity equivalent to conventional cell culture but takes only 16 hours to complete. The antigenemia immunoassay uses monoclonal antibody in an indirect immunoperoxidase or indirect immunofluorescent stain to detect CMV protein (pp65) in peripheral blood leukocytes. The antigenemia assay requires only 3 to 5 hours to complete and includes separation of leukocytes by sedimentation, counting of leukocytes, smear preparation with a standard density of cells, and staining and counting of infected (fluorescing) cells (Procedure 51-2). Results are reported as number of positive leukocytes per total number of leukocytes in the smear (Figure 51-10). Quantitative CMV PCR and a commercially available CMV hybrid capture assay (Digene Corp., Gaithersburg, Md) are also available for the detection and quantitation of CMV viremia.

Enteroviruses. Enteroviruses can be detected by conventional cell culture and PCR. Although most

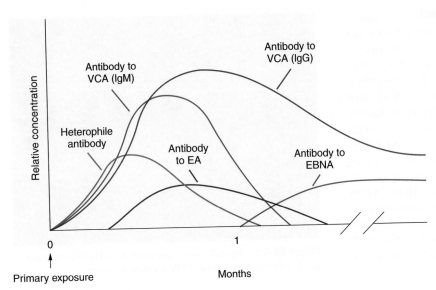

enteroviruses grow in primary monkey kidney cells, some strains grow faster in diploid fibroblast, buffalo green monkey kidney, or rhabdomyosarcoma cell lines. Presumptive diagnosis is based on CPE. Confirmation or definitive diagnosis is accomplished using commercially available FA stains. PCR is the test of choice for use with CSF to diagnose aseptic meningitis caused by enterovirus serotypes. Most enteroviruses are detected from June to December. Mean time to detection in cell culture is 4 days.

Epstein-Barr Virus. Serology tests are used to help diagnose EBV-associated diseases, including infectious mononucleosis. Isolation of EBV (in cultured B lymphocytes) is not routinely performed in clinical laboratories. Antibodies that are ordered appear at specific times following primary EBV infection (Figure 51-11).

Hepatitis Viruses. Disease or asymptomatic carriage caused by hepatitis A, B, C, D, and E viruses is detected using serology, antigen detection, or PCR tests (Table 51-29).

Herpes Simplex Virus. HSV grows rapidly in most cell lines. MRC-5 or mink lung fibroblast cell lines are recommended. Fifty percent of genital HSV isolates are detected within 1 day and 100% within 3 to 5 days. Cultures should be examined daily and finalized if negative after 5 days of incubation. Real-time PCR detects HSV within hours with a sensitivity equal to or greater than cell culture.

Human Immunodeficiency Virus and Other Retroviruses. HIV-1 is detected by antibody, antigen, and RT-PCR tests (Figure 51-12). HIV-1 ELISA antibody tests

Table 51-29 Serology Tests for Hepatitis Viruses

Disease	Virus	Diagnostic Tests
Hepatitis A	Enterovirus 72	Antibody to Hepatitis A virus (IgG and IgM)
Hepatitis B	Hepadnavirus	Hepatitis B surface-antigen (HBsAg) Hepatitis B early-antigen (HBeAg) Anti-HBsAg Anti-HBeAg Anti-HB core antigen
Hepatitis C	Flavivirus	Antibody to hepatitis C virus
Hepatitis D	Delta agent (hepatitis D virus)	Antibody to delta agent
Hepatitis E	Calicivirus-like (hepevirus)	Antibody to hepatitis E virus

detect antibody to both HIV-1 and HIV-2. Confirmation of the ELISA screening test is accomplished with an HIV-1–specific Western blot test or with an ELISA test for HIV-2 followed by an HIV-2–specific Western blot test. Recently infected patients who have not seroconverted or newborn babies with maternal antibody can be identified as HIV infected using sensitive RT-PCR assays. HIV infection is monitored in those receiving antiviral therapy using quantitative molecular testing with serum specimens. Successful antiviral therapy should reduce the HIV serum viral load to undetectable levels.

Blood for transfusion is screened for antibody indicative of HIV-1 and HIV-2, and HTLV-1 and HTLV-2 infection. HTLV-1 ELISA screening tests detect antibody to both HTLV-1 and HTLV-2. In addition, an HIV antigen (p24) test is performed to detect donors who may be recently infected and have not produced HIV antibody.

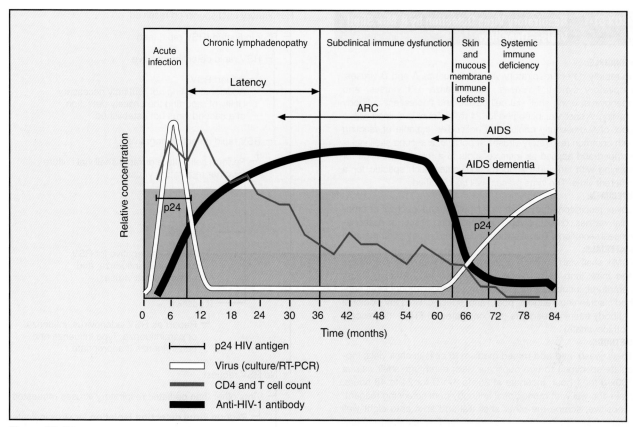

Figure 51-12 Illustration of the usual time-course of immune response, viremia, and disease resulting from untreated HIV-1 infection. (Redrawn from Murray PR, Kobayashi GS, Pfaller MA, et al, editors: *Medical microbiology*, ed 2, St Louis, 1994, Mosby.)

Units containing antigen or antibody are discarded because of the risk of transferring latent virus to the recipient.

Influenza A and B Viruses. Influenza A and B viruses are detected by using conventional cell culture, shell vial culture, membrane EIA (enzyme immunoassay), direct staining of respiratory tract secretions using FA methods, and PCR. PCR is superior to other methods. When cell culture is performed, primary monkey kidney cells are superior to other cell lines. Median time to detection, using hemadsorption at day 2 or 3, is 3 days. Fluorescent antibody staining is used for confirmation and typing of isolates as A or B. Nearly all influenza viruses are detected after 1 week of incubation.

Pediatric Respiratory Viruses. Influenza, parainfluenza, respiratory syncytial, and adenoviruses should be sought in specimens from hospitalized infants and children younger than 10 years of age with suspected viral lower respiratory tract disease. All viruses can be detected by fluorescent staining of respiratory secretions or rapid cell culture (shell vial). If direct fluorescent staining is used, cell culture confirmation of all negatives is needed for children suspected to have viruses other than RSV. Many laboratories use R-Mix Cells in a rapid shell vial format to detect respiratory viruses (Box 51-1). This approach mixes two cell lines (human lung carcinoma A549 and mink lung fibroblast Mv1Lu cells from Diagnostic Hybrids, Inc. Athens, Ohio) in one shell vial tube. Two R-Mix shell vial tubes are inoculated for each specimen. After an 18- to 24-hour incubation, the cell mixture from one tube is stained with a pooled antibody reagent designed to detect all common respiratory viruses. Positive (fluorescent) specimens have the second tube scraped, spotted onto eight-well slides, and stained with individual antibody reagents to identify the specific virus present. If conventional cell culture is used, influenza and parainfluenza viruses are detected in PMK cells by CPE or hemadsorption. Fluorescent staining is used for confirmation and typing. Adenovirus and RSV are detected in HEp-2 cell culture and confirmed, if necessary, by fluorescence staining. Specimens from infants or young children sent for RSV detection should be tested by a rapid, nonculture RSV test. Fluorescent antibody staining, conducted by experienced personnel, is equivalent to culture in sensitivity and is best used for single specimens or small batches. Conventional ELISA is also very accurate and is recommended for large

---

BOX 51-1 **Respiratory Virus Detection by R-Mix Shell Vials—Overview**

PRINCIPLE:

To rapidly detect respiratory viruses (influenza A and B viruses, respiratory syncytial viruses, parainfluenza 1-3 viruses, and adenovirus) using shell vial cell culture and fluorescent antibody staining. A shell vial incubated for 24 to 48 hours is stained with a pool of fluorescently conjugated antibodies capable of reacting with common respiratory viruses. If positive, a second shell vial is scraped and applied as multiple spots to a microscope slide for staining with individual antibody conjugates each specific for a different virus. The virus present will be identified

SPECIMEN:

Lower respiratory tract secretions (sputum, endotracheal or bronchial washes, bronchoalveolar lavages), lung tissue, or nasopharyngeal secretions. Throat swabs/specimens are not recommended

MATERIALS:

R-Mix shell vials containing a mixture of human lung carcinoma and mink lung cells (Diagnostic Hybrids, Inc., Athens, Ohio); centrifuge; incubator; fluorescent microscope; monoclonal antibody screening reagent pool, and specific virus monoclonal antibody staining reagents (all conjugated to FITC—fluorescein isothiocyanate)

METHODS:

Thaw, wash, and add refeed medium to cells in shell vials; inoculate specimen to two duplicate vials; centrifuge shell vials at 700× g for 1 hour; incubate at 35° to 37° C for 24 to 48 hours; stain one vial with monoclonal antibody pool screening reagent; if positive, scrape the other shell vial and spot onto eight-well slide; stain with specific monoclonal staining reagent to detect specific virus present

INTERPRETATION:

Report specific virus detected with specific monoclonal staining reagent. If no fluorescence is detected with the monoclonal pool screening reagent, report as "No Respiratory Viruses Detected"

PROCEDURE NOTES:

R-mix shell vials should be screened and stained at 24 hours for influenza A and B viruses, but respiratory syncytial viruses, parainfluenza viruses 1-3, and adenovirus require 48 hours of incubation for maximum sensitivity

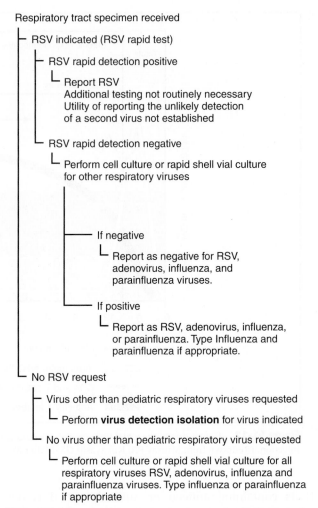

Figure 51-13 Flowchart for the detection and identification of pediatric respiratory viruses.

---

batches of specimens. Membrane ELISA and related testing methods are less sensitive than culture but rapid (less than 1 hour). They are convenient for stat testing. Figure 51-13 describes a comprehensive approach to the detection of pediatric respiratory viruses.

Gastroenteritis Viruses. Electron microscopy (EM) can be used to identify viral agents known to cause gastroenteritis (Table 51-30). EM, however, is labor intensive and not broadly available in clinical virology laboratories. Immunoassays for rotaviruses and enteric adenovirus types 40 and 41 are commercially available. Other viruses, such as noroviruses and astroviruses, do not cause life-threatening diarrheal disease and are not detected by clinical laboratories routinely at this time.

TORCH Viruses. TORCH is an acronym for *Toxo*plasma, *r*ubella, *c*ytomegalovirus, and *h*erpes simplex

virus. Testing for these agents and for other viral etiologies of infection in newborns is appropriate during pregnancy because transplacental infection followed by congenital defects can occur, and postnatally because some congenital disease of newborns can be treated to prevent serious, permanent damage to the infant. A blanket request for TORCH tests should be avoided when possible, especially in specimens from newborns. Clinical presentation in the newborn may be characteristic for one or two of the viral agents, and tests for these etiologies only should be pursued. Table 51-31 suggests laboratory tests for the diagnosis of the viral diseases in the newborn.

Varicella-Zoster Virus. VZV causes chicken pox (varicella) and shingles (zoster). Varicella (a vesicular eruption) represents the clinical consequences of a primary VZV infection. VZV, a DNA-containing virus, establishes latency in a dorsal nerve root ganglion. Months to years later, during periods of relative immune

Table 51-30   Human Gastroenteritis Viruses

Virus	Relative Medical Importance	Epidemiology	Diagnostic Tests
Rotavirus	++++	Major cause of diarrhea in infants	Enzyme immunoassay (EIA), latex agglutination
Enteric adenoviruses	++	Diarrhea in infants and young children	EIA, Electron microscopy (EM) (especially types 40-41)
Noroviruses (caliciviruses)	+++	Epidemics in children and adults	EM, RT-PCR
Astroviruses	+	Diarrhea in children	EM

Table 51-31   Laboratory Diagnosis of Viral Diseases in the Newborn

Virus	Specimen	Diagnostic Tests
Rubella virus	Serum	Serology
CMV	Urine Tissue	Cell culture (shell vial), PCR Cell culture (shell vial), histopathology
HSV	Cutaneous lesion CSF	Cell culture, PCR PCR, cell culture
HIV	Blood, tissue Serum	RT-PCR, cell culture Serology
HBV	Blood	PCR, serology
VZV	Cutaneous lesion Tissue, fluid, or secretions	FA (Tzanck preparation), PCR Cell culture (shell vial), PCR
Enteroviruses	Cutaneous lesion, tissue, CSF	Cell culture, RT-PCR

FA, Fluorescent antibody; *PCR,* polymerase chain reaction; *RT-PCR,* reverse transcriptase PCR.

suppression, VZV reactivates to cause zoster. Zoster is a modified or limited form of varicella, localized to a specific dermatome, the cutaneous area served by the infected nerve ganglion. Virus is present in the vesicular fluid and in the cells at the base of the vesicle. Material for virus detection should be collected from newly formed vesicles. Once the vesicle has opened and crusted over, detection is unlikely. Virus can be detected by staining cells from the base of the vesicles, by culturing cells and vesicular fluid, or by PCR testing of fluid and cells. The Tzank test, which is a smear of cells from the base of the vesicle stained by the Giemsa, Papanicolaou (Pap), or other suitable cytologic staining method, detects typical multinucleated giant cells and inclusions (Figure 51-14, *A*). The FA stain also can be used to detect VZV in Tzanck smears. Traditionally, diploid fibroblast cell culture (e.g., MRC-5) has been used to detect VZV, which requires up to 28 days before producing visible CPE. The shell vial assay reduces detection time to 48 hours and significantly increases sensitivity, identifying virus that fails to produce CPE in conventional cell culture. A comparison of FA staining of Tzanck smears, conventional cell culture, rapid shell vial culture, and PCR testing shows PCR to be the most sensitive detection method (Box 51-2). In laboratories

where PCR testing is not available, FA staining is the best and recommended method for diagnosis.

VIRUS DETECTION METHODS

Cytology and Histology

A readily available technique for the detection of virus is cytologic or histologic examination for the presence of characteristic viral inclusions. This involves the morphologic study of cells or tissue, respectively. Viral inclusions are intracellular structures formed by aggregates of virus or viral components within an infected cell or abnormal accumulations of cellular materials resulting from viral-induced metabolic disruption. Inclusions occur in single or syncytial cells. Syncytial cells are aggregates of cells fused to form one large cell with multiple nuclei. Pap- or Giemsa-stained cytologic smears are examined for inclusions or syncytia. Cytology is most frequently used to detect infections with VZV and HSV (see Figure 51-14, *A*). A stained smear of cells from the base of a skin vesicle used to detect VZV or HSV inclusions is called a Tzanck test. Inclusions resulting from infection with CMV, adenovirus, parvovirus, papilloma virus, and molluscum contagiosum virus are detected by histologic examination of tissue stained with hematoxylin and

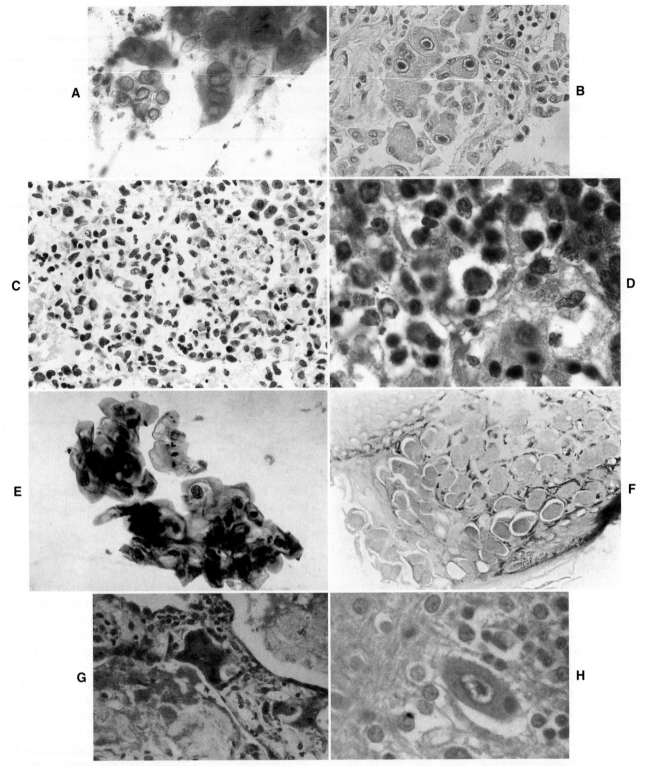

Figure 51-14 Viral inclusions. **A,** Pap-stained smear showing multinucleated giant cells typical of herpes simplex or varicella zoster viruses. **B,** Hematoxylin and eosin–stained (HE) lung tissue containing intranuclear inclusion within enlarged CMV infected cells. **C,** HE-stained lung tissue containing epithelial cells with intranuclear inclusions characteristic of adenovirus. **D,** HE-stained liver from stillborn fetus showing intranuclear inclusions in erythroblasts (extramedullary hematopoiesis) resulting from parvovirus infection. **E,** Pap stain of exfoliated cervicovaginal epithelial cells showing perinuclear vacuolization and nuclear enlargement characteristic of human papillomavirus infection. **F,** HE-stained epidermis filled with molluscum bodies, which are large eosinophilic cytoplasmic inclusions resulting from infection with molluscum contagiosum virus. **G,** HE-cells infected with measles virus. **H,** HE-stained brain tissue showing oval, eosinophilic rabies cytoplasmic inclusion (Negri body). (*E* and *F* from from Murray PR, Kobayashi GS, Pfaller MA, et al, editors: *Medical microbiology,* ed 2, St Louis, 1994, Mosby.)

BOX 51-2 Varicella Zoster Virus Detection by Polymerase Chain Reaction Assay

PRINCIPLE:
Detection of VZV in dermal lesions from patients with chicken-pox (varicella) or shingles (zoster). Dermal swab specimens can be used. Real-time PCR detects VZV DNA in a 2-hour assay with greater sensitivity than conventional cell culture, shell vial cell culture, VZV fluorescent antibody staining, or Papanicolaou smear.

SPECIMEN:
Dermal swab, vesicular scraping, or vesicular fluid

MATERIALS:
Nucleic acid extraction reagents or instrument (e.g., QIAamp DNA Mini Kit, Qiagen, Inc., Valencia, Calif); real-time PCR instrument (e.g., LightCycler System, Roche Diagnostics, Corp., Indianapolis, Ind); Molecular Laboratory for performance of PCR; PCR reaction mixture containing polymerase, primers, and deoxynucleoside triphosphates

METHODS:
Extract DNA; perform real-time PCR; computer analysis after each amplicon production cycle

INTERPRETATION:
Presence of VZV-specific amplicons detected by fluorescent signal using crossover plot (see Figure 51-20) to signify a positive test

PROCEDURE NOTES:
PCR provides 1.9-fold increase in positive findings compared to shell vial culture. Assay is uniformly negative when other viruses, not VZV, are present in the specimen. LightCycler is a closed system nearly eliminating carryover contamination. Confirmation of amplicon identify using melt curve analysis (see Figure 51-20)

---

eosin or Pap (see Figure 51-14, *B* through *F*). Less commonly, inclusions characteristic of measles and rabies viruses are detected by examining stained tissues (see Figure 51-14, *G* and *H*). Rabies virus inclusions in brain tissue are called *Negri bodies*. Cytology and histology are less sensitive than culture but are especially helpful for those viruses that are difficult or dangerous to isolate in the laboratory, such as parvovirus and rabies virus, respectively.

Electron Microscopy

Very few laboratories use EM to detect viruses because it is labor intensive and relatively insensitive. EM is most helpful for the detection of viruses that do not grow readily in cell culture and works best if the titer of virus is at least 10^6 to 10^7 particles per milliliter. Immune EM allows visualization of virus particles present in numbers too small for easy direct detection. The addition of specific antiserum to the test suspension causes the virus particles to form antibody-bound aggregates, which are more easily detected than are single virus particles. In the clinical virology laboratory, electron microscopy is most useful for the detection of gastroenteritis viruses that cannot be detected by other methods (noroviruses, coronaviruses, and astroviruses) and viruses causing encephalitis that are not detected with cell culture (HSV,

measles virus, and JC polyomavirus) (Figure 51-15). In addition, the etiology of newly recognized viral syndromes can be recognized rapidly by identifying characteristic viral morphology by EM in infected tissue. This was exemplified by the early recognition of Ebola virus as the cause of an outbreak of viral hemorrhagic fever in Africa in the 1970s and sin nombre virus as the cause of fatal pneumonia in the four corners area of the United States in the 1990s.

Immunodiagnosis (Antigen Detection)

High-quality, commercially available viral antibody reagents have led to the development of fluorescent antibody, enzyme immunoassay, radioimmunoassay, latex agglutination, and immunoperoxidase tests that detect viral antigen in patient specimens.

Direct and indirect immunofluorescent methods are used. Direct immunofluorescent testing involves use of a labeled antiviral antibody; the label is usually fluorescein isothiocyanate (FITC), which is layered over specimen suspected of containing homologous virus. The indirect immunofluorescent procedure is a two-step test in which unlabeled antiviral antibody is added to the slide first, followed by a labeled (FITC) antiglobulin that binds to the first-step antibody bound to virus in the specimen. Direct immunofluorescence is generally more rapid and specific than indirect immunofluorescence, but less sensitive. Increased sensitivity of the indirect test results from signal amplification that occurs with the addition of the second antibody. Signal amplification decreases specificity by increasing nonspecific background fluorescence. Direct immunofluorescence is best suited to situations in which large quantities of virus are suspected or when high-quality, concentrated monoclonal antibodies are used, such as for the detection of RSV in patient specimen or the identification of viruses growing in cell culture. Indirect immunofluorescence should be used when lower quantities of virus are suspected, such as detection of respiratory viruses in specimens from adult patients. High-quality monoclonal antibodies improve the sensitivity and specificity of immunofluorescence testing. Whenever possible the direct test should be used because it is faster without the second antibody step.

Strict criteria for the interpretation of fluorescent patterns must be used. This includes standard interpretation of fluorescent intensity (Table 51-32) and recognition of viral inclusion morphology. Nuclear and cytoplasmic staining patterns are typical for influenza virus, adenovirus, and the herpes viruses; cytoplasmic staining only is typical for respiratory syncytial, parainfluenza, and mumps viruses; and staining within multinucleated giant cells is typical of measles virus (Figure 51-16). False-positive staining can occur with specimens containing yeasts, certain bacteria, mucus, or leukocytes.

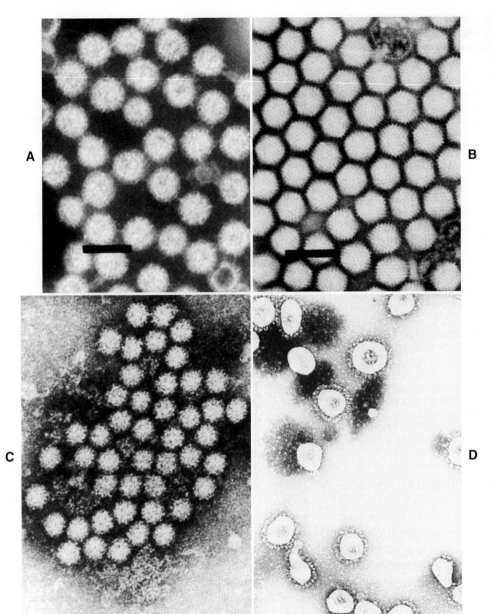

Figure 51-15 Electron micrographs of viruses. **A,** Rotavirus. **B,** Adenovirus. **C,** Norwalk agent virus. **D,** Coronavirus.

Table 51-32 Interpretation of Fluorescence Intensity

Intensity	Interpretation
Negative	No apple-green fluorescence
1+	Faint, yet unequivocal apple-green fluorescence
2+	Apple-green fluorescence
3+	Bright, apple-green fluorescence
4+	Brilliant, apple-green fluorescence

Leukocytes, which contain Fc receptors for antibody, can also cause nonspecific binding of antibody conjugates. To verify one's ability to interpret FA tests, every laboratory should perform viral culture or some alternative detection method in parallel with immunofluorescence until in-house performance has been established.

Immunofluorescent stains that are most useful in the clinical virology laboratory are those for RSV, influenza and parainfluenza viruses, adenovirus, HSV, VZV, and CMV. A pool of antibodies can be used to screen a specimen for multiple viruses. A positive screen is then tested with each individual reagent to identify the exact virus that is present. Screening pools have been used successfully for the detection of respiratory viruses

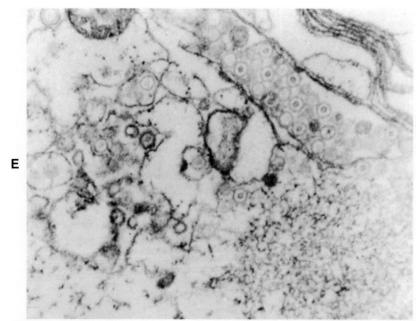

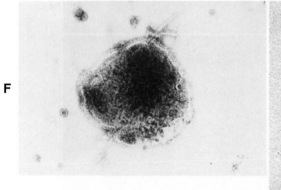

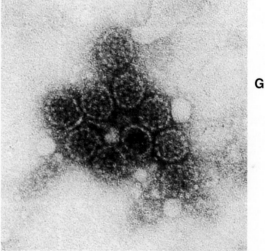

Figure 51-15, cont'd   E, Herpes simplex virus. F, Measles virus. G, Negatively stained preparation of JC virus in brain tissue. (*C* from Howard BJ, Klaas J, Rubin SJ, et al: *Clinical and pathogenic microbiology*, St Louis, 1987, Mosby; *D* and *F* from US Department of Health, Education, and Welfare, Public Health Service, Centers for Disease Control, Atlanta; *G* courtesy Dr. Gabriele M. ZuRhein.)

in specimens from children. Such pools are less sensitive when used with specimens from adults because of lower numbers of viral particles in the specimens.

Enzyme immunoassay methods used most in clinical virology are the solid-phase enzyme-linked immunosorbent assay (solid-phase ELISA) and the membrane bound enzyme-linked immunosorbent assay (membrane ELISA). Solid-phase ELISA is performed in a small test tube or microtiter tray. Breakaway strips of microtiter wells are available for low-volume test runs. The remaining, unused wells can be saved for future testing. A solid-phase ELISA used for the detection of rotavirus in stool specimens is illustrated in Figure 51-17. Membrane ELISA tests have been developed for low-volume testing and where rapid results are needed. They can be performed by those with minimum training and usually require less than 30 minutes to complete. The membrane method uses a handheld reaction chamber with a cellulose-like membrane. Specimen and reagents are applied to the membrane. Following a short incubation time, a chromogenic (color) reaction occurs on the surface of the membrane and is read visually. Built-in controls on the same membrane provide convenient monitoring of test procedures. Figure 51-18 illustrates a membrane ELISA used to detect rotavirus. The most used enzyme immunoassays for antigen detection are those for RSV (solid-phase and membrane), rotavirus (solid-phase and membrane), and influenza viruses (membrane).

Advantages of enzyme immunoassays are the use of nonradioactive, relatively stable reagents and results that can be interpreted qualitatively (positive or negative) or quantitatively (titer or degree of positive reaction). It is important to note that enzyme immunoassays

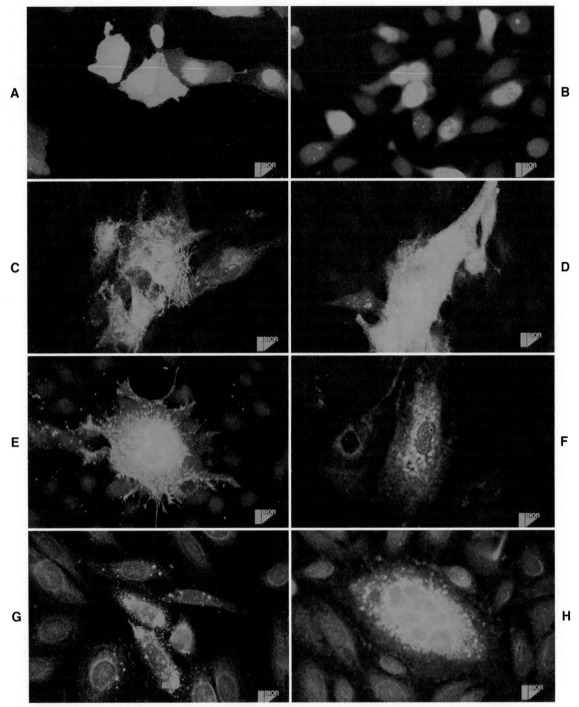

Figure 51-16 Fluorescent antibody staining of virus-infected cells. **A,** Influenza virus. **B,** Adenovirus.
C, Varicella zoster virus. **D,** Herpes simplex virus. **E,** Respiratory syncytial virus. **F,** Parainfluenza virus.
G, Mumps virus. **H,** Measles virus. (Courtesy Bion Enterprises Ltd., Park Ridge, Ill.)

frequently have an indeterminate or borderline interpretative category. This result implies that low levels of
viral antigen or background interference has prevented
a clear-cut positive or negative result. Such results
usually require testing of a second specimen to avoid
interference or to detect a rise in antigen level. ELISAs

are sensitive and simple to perform and can be easily
automated, but a drawback is that specimen quality
cannot be evaluated, that is, the number of cells cannot
be assessed, as can be done microscopically with fluorescent immunoassays. Acellular specimens are potentially
inferior, in that specimen collection could have been

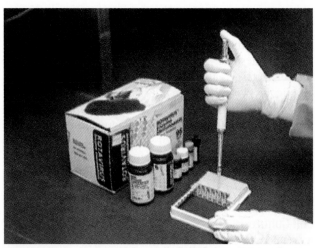

Figure 51-17   Solid-phase enzyme immunoassay for detection of rotavirus with breakaway strips of microtiter wells for small batch testing. (Courtesy Children's Hospital Medical Center of Akron, Akron, Ohio.)

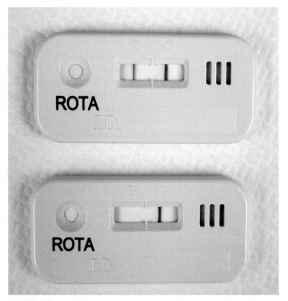

Figure 51-18   Positive (top) and negative membrane ELISA tests for the detection of rotavirus. The red line in the reaction area on the left represents a positive test. A red line in the reaction area on the right represents an internal test control ensuring that the test has been carried out correctly. If the test control line is not present, the test is invalid and must be repeated.

inadequate and cell-associated virus would not be present.

Radioimmunoassay (RIA), immunoperoxidase staining, and latex agglutination are additional techniques used to detect viral antigen. RIA has been largely replaced by ELISA because of the expensive equipment and disposal procedures needed for radioactive materials. Immunoperoxidase staining is commonly used to stain histologic sections for virus but is less popular than

immunofluorescence staining in clinical virology laboratories. Latex agglutination is an easy and inexpensive method but lacks sensitivity compared with ELISA and fluorescent immunoassays.

Enzyme-Linked Virus-Inducible System

The enzyme-linked virus-inducible system (ELVIS) uses a BHK (baby hamster kidney) cell culture system with a cloned (added) beta-galactosidase gene that is expressed only when cells are infected with a virus. In the ELVIS-HSV test system (Diagnostic Hybrids, Inc., Athens, Ohio), the genetically engineered BHK cells are sold in multiwelled microtiter plates. Following inoculation of specimens and overnight incubation, growth of HSV results in production of the beta-galactosidase enzyme by the BHK cells. Beta-galactosidase serves as the "reporter" molecule. When cells are fixed and stained for galactosidase activity, positive staining indicates the presence of HSV-1 or HSV-2. Wells not containing HSV show no staining.

Molecular Detection Using Nucleic Acid Probes and Polymerase Chain Reaction Assays

During the past decade, the introduction of nucleic acid detection techniques into the clinical virology laboratory has resulted in a major shift in testing strategy. With the use of both nucleic acid detection and amplification-based systems, in conjunction with automated nucleic acid isolation techniques for sample preparation, nearly all virology laboratories have access to commercial or in-house molecular assays. These technologic improvements make it possible to generate results within 2 to 3 hours. Nucleic acid detection can be accomplished using nucleic acid probes, which are short segments of DNA that hybridize with complementary viral DNA or RNA segments. The probe is labeled with a fluorescent, chromogenic, or radioactive tag that allows detection if hybridization occurs. The probe reaction can occur in situ, such as in a tissue thin section, in liquid, or on a reaction vessel surface or membrane. A DNA probe test used to detect papillomavirus DNA in a smear of cervical cells is illustrated in Figure 51-19. Nucleic acid probes are most useful when the amount of virus is relatively abundant, viral culture is slow or not possible, and immunoassays lack sensitivity or specificity. DNA target fragments that are too few in number in the original specimen to be detected by probes can be amplified using molecular techniques such as PCR, which is a method that duplicates short DNA targets thousands to a million-fold. This provides enough target, referred to as *amplicons*, to readily identify the presence of a specific virus. The PCR reaction with ensuing amplicon identification has been automated and made very rapid. Rapid testing, referred to as *real-time PCR*, is illustrated in Figure 51-20. The PCR test also can be used to amplify

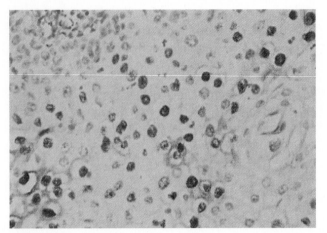

Figure 51-19 Smear of cervical cells stained with probe for papillomavirus DNA. Dark-staining cells contain virus DNA. (Courtesy Children's Hospital Medical Center of Akron, Akron, Ohio.)

and detect RNA viruses by using the enzyme reverse transcriptase (RT). The first step in RT-PCR includes making a complementary DNA strand of the RNA segment in question. The usual PCR steps used to multiply the DNA target are then performed, leading to DNA amplicons whose identification signifies the presence of the original RNA sequence. The rapid appearance and broad application of molecular diagnostics will require the introduction and use of standardized materials and external quality control programs. In addition, the use of universal internal controls throughout the procedure will further ensure accuracy.

Cell Culture

Conventional Cell Culture. Viruses are strict intracellular parasites, requiring a living cell for multiplication and spread. To detect virus using living cells, suitable host cells, cell culture media, and techniques in cell culture maintenance are necessary. Host cells, re-

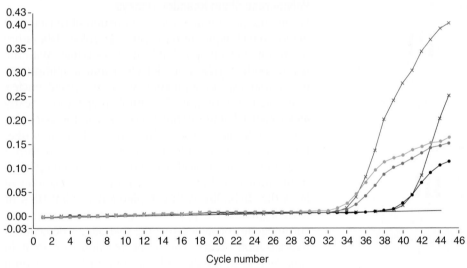

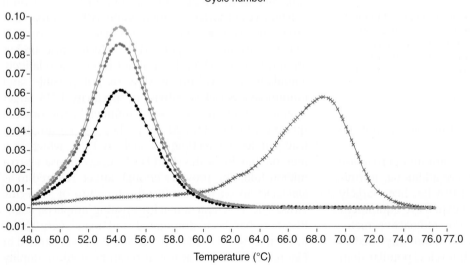

Figure 51-20 Real-time PCR detection of herpes simplex virus (HSV). Black, red, and light green lines represent three different HSV-1 viruses. Pink and dark green lines represent two different HSV-2 viruses. **A,** Cycle crossover detection of HSV-1 and HSV-2 amplicons, with all viruses detected between cycles 34 and 40. **B,** Melt curve confirmation of the presence of HSV-1 and HSV-2 viruses. HSV-1 amplicons melt at approximately 54° C (three HSV-1 viruses confirmed) and HSV-2 amplicons melt at approximately 68° C (one HSV-2 virus confirmed).

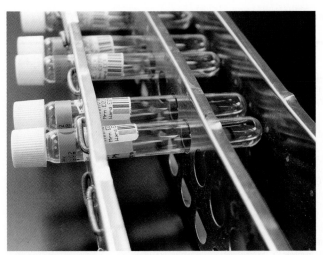

Figure 51-21   Cell culture tubes incubating on their sides in a stationary rack. Tubes are oriented with the same glass surface facing downward because an emblem printed on one side of the glass near the neck is used for correct positioning in the rack.

ferred to as *cell cultures* (referred to by some as *tissue cultures*), originate as a few cells and grow into a monolayer on the sides of glass or plastic test tubes. Cells are kept moist and supplied with nutrients by keeping them continuously immersed in cell culture medium (Figure 51-21). Cell cultures are routinely incubated in a roller drum that holds cell culture test tubes tilted 5 to 7 degrees while they slowly revolve ($\frac{1}{2}$ to 1 rpm) at 35° to 37° C (see Figure 51-5). Incubation of cell culture tubes in a stationary rack can be used in place of a roller drum. Rapidly growing viruses, such as HSV, appear to be detected equivalently by both methods. Comparative studies are not available for most viruses.

Metabolism of growing cells in a closed tube results in the production of carbon dioxide and acidification of the growth liquid. To counteract the pH decrease, a bicarbonate buffering system is used in the culture medium to keep the cells at physiologic pH (pH 7.2). Phenol red—a pH indicator that is red at physiologic, yellow at acidic, and purple at alkaline pH—is added to monitor adverse pH changes. Once inoculated with specimen, cell cultures are incubated for 1 to 4 weeks, depending on the viruses suspected. Periodically the cells are inspected microscopically for the presence of virus, indicated by areas of dead or dying cells called **cytopathic effect (CPE).**

Two kinds of media, growth medium and maintenance medium, are used for cell culture. Both are prepared with Eagle's minimum essential medium (EMEM) in Earle's balanced salt solution (EBSS) and include antimicrobials to prevent bacterial contamination. Usual antimicrobials added are vancomycin (10 µg/mL), gentamicin (20 µg/mL), and amphotericin (2.5 µg/mL). Growth medium is a serum-rich (10% fetal, newborn,

or agammaglobulinemic calf serum) nutrient medium designed to support rapid cell growth. This medium is used for initiating growth of cells in a tube when cell cultures are being prepared in-house or for feeding tubes of purchased cell cultures that have incomplete cell monolayers. *Feeding* refers to the removal of old medium followed by the addition of fresh culture medium. Maintenance medium is similar to growth medium but contains less serum (0% to 2%) and is used to keep cells in a steady state of metabolism. Fetal, newborn, or agammaglobulinemic calf serum is used to avoid inhibitors, such as specific antibody, and to be free of mycoplasmas present in serum from older animals.

Several kinds of cell cultures are routinely used for isolation of viruses. A cell culture becomes a **cell line** once it has been passed or subcultured in vitro. Cell lines are classified as primary, low passage, or continuous. **Primary cell lines** are those that have been passed only once or twice since harvesting, such as primary monkey kidney cells. Further passage of primary cells results in a decreased receptivity to viral infection. **Low passage cell lines** are those that remain virus-sensitive through 20 to 50 passages. Human diploid fibroblast cells, such as lung fibroblasts, are a commonly used low-passage cell line. **Continuous cell lines** such as HEp-2 cells (human epidermoid carcinoma cells), can be passed and remain sensitive to virus infections indefinitely. Unfortunately, most viruses do not grow well in continuous cell lines. The majority of clinically significant viruses can be recovered using one cell culture type from each group. A combination frequently used by clinical laboratories is Rhesus monkey kidney cells, MRC-5 lung fibroblast cells, and HEp-2 cells (Table 51-33).

Inoculated cell cultures should be incubated immediately at 35° C. After allowing virus to adsorb to the cell monolayer for 12 to 24 hours, it is common practice to remove the remaining inoculum and culture medium and replace it with fresh maintenance medium. This avoids most inoculum-induced cell culture toxicity and improves virus recovery. Incubation should be continued for 5 to 28 days depending on viruses suspected (see Table 51-33). Maintenance medium should be changed periodically (usually 1 to 2 times weekly) to provide fresh nutrients to the cells.

Blind passage refers to passing cells and fluid to a second cell culture tube. Blind passage is used to detect viruses that may not produce CPE in the initial culture tube but will when the "beefed-up" inoculum is passed to a second tube. Cell cultures that show nonspecific or ambiguous CPE are also passed to additional cell culture tubes. Toxicity, which causes ambiguous CPE, is diluted during passage and should not appear in the second cell culture tube. In both instances, passage is performed by scraping the monolayer off the sides of the tube with a pipette or disrupting the monolayer by vortexing with

Table 51-33 Isolation and Identification of Common, Clinically Encountered Viruses

Virus	PMK	HEp-2	HDF	CPE Description	Rate of Growth (days)	Identification and Comments
Adenovirus	*++*	+++	++	Rounding and aggregation of infected cells in grapelike clusters	2-10	Confirm by FA Serotype by cell culture neutralization
Cytomegalovirus (CMV)	—	—	++++	Discrete, small foci of rounded cells	5-28	Distinct CPE sufficient to identify Confirm by FA
Enterovirus	++++	+	++	Characteristic refractile angular or tear-shaped CPE; progresses to involve entire monolayer	2-8	Confirm by FA Stable at pH 3
Herpes simplex (HSV)	+	++++	++++	Rounded, swollen refractile cells. Occasional syncytia, especially with type 2. Rapidly involves entire monolayer	1-3 (may take up to 7)	Distinct CPE sufficient to identify. Confirm by FA
Influenza	++++	—	±	Destructive degeneration with swollen, vacuolated cells	2-10	Detect by hemadsorption or hemagglutination with guinea pig RBCs. Identify by FA
Mumps	+++	±	±	CPE, usually absent. Occasionally, syncytia are seen	5-10	Detect by hemadsorption with guinea pig RBCs. Confirm by FA
Parainfluenza	+++	—	—	CPE, usually minimal or absent	4-10	Detect by hemadsorption with guinea pig RBCs. Identify by FA
Respiratory syncytial virus (RSV)	+	+++	+	Syncytia in HEp-2	3-10	Distinct CPE in HEp-2 is sufficient for presumptive identification. Confirm by FA
Rhinovirus	++	—	+++	Characteristic refractile rounding of cells. In PMK, CPE is identical to that produced by enteroviruses	4-10	Labile at pH 3. Growth optimal at 32°-33° C
Varicella-zoster	—	—	++	Discrete foci of rounded, swollen, refractile cells. Slowly involves entire monolayer	5-28	Confirm by FA

CPE, Cytopathic effects; *FA*, fluorescent antibody; *PMK*, primary monkey kidney; *RBCs*, red blood cells.
*Relative sensitivity of cell cultures for recovering the virus: —, None recovered; ±, rare strains recovered; +, few strains recovered; ++++, ≥80% of strains recovered.

sterile glass beads added to the culture tube, followed by inoculating 0.25 mL of the resulting suspension into new cell cultures. Blind passage is less frequently used today because the added time and expense do not justify detection of a few additional isolates after extended incubation in two cell culture tubes.

Shell Vial Cell Culture. The shell vial cell culture is a rapid modification of conventional cell culture. Virus is detected more quickly using the shell vial technique because the infected cell monolayer is stained for viral antigens produced soon after infection, before the development of CPE. Viruses that normally take days to weeks to produce CPE can be detected within 1 to 2 days by detecting early produced viral antigens. A shell vial culture tube, a 15- × 45-mm 1-dram vial, is pre-

pared by adding a round coverslip to the bottom, covering this with growth medium, and adding appropriate cells (Figure 51-22). During incubation, a cell monolayer forms on top of the coverslip. Shell vials should be used 5 to 9 days after cells have been inoculated. Shell vials with the monolayer already formed can be purchased. Specimens are inoculated onto the shell vial cell monolayer by low-speed centrifugation. This enhances viral infectivity for reasons that are not well understood. Coverslips are stained using virus-specific immunofluorescent conjugates. Typical fluorescing inclusions confirm the presence of virus (Figure 51-23). The shell vial procedure for detecting CMV is presented in detail in Procedure 51-3.

The shell vial culture technique can be used to detect most viruses that grow in conventional cell cul-

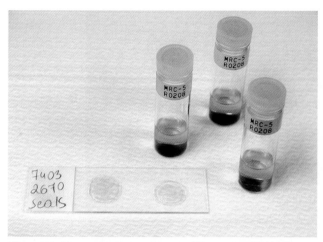

Figure 51-22 Shell vial cell culture tubes and stained coverslips. At the bottom of each shell vial tube under culture medium is a round coverslip with a cell monolayer on the top surface. After incubation, the coverslip is removed, stained, and placed on a microscope slide for fluorescence viewing. Note that two stained coverslips are on the glass slide.

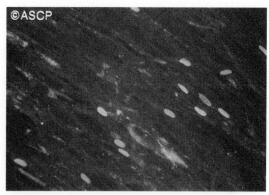

Figure 51-23 Typical fluorescing nuclei of human diploid fibroblast cells infected with cytomegalovirus as seen in the shell vial assay. (Courtesy Bostick CC: *Laboratory detection of CMV*, 1992, Microbiology Tech Sample No MB-3.)

ture. It is best used for viruses that require relatively long incubation before producing CPE, such as CMV and VZV. The advantage of shell vial is its speed; most viruses are detected within 24 hours. The disadvantage is that only one type of virus can be detected per shell vial. For example, a specimen that might contain influenza A or B, or adenovirus, would need to be inoculated to three separate shell vials so that each vial could be stained with a separate virus-specific conjugate. Other strategies pool antibody for detection of many viruses with a single vial. Additional vials from positive specimens are then stained with individual conjugates to identify the specific virus present. The shell vial procedure with mixed cell types used to detect seven different respiratory viruses is outlined in Box 51-1.

Identification of Viruses Detected in Cell Culture.

Viruses are most often detected in cell culture by the recognition of CPE. Virus-infected cells change their usual morphology and eventually lyse or detach from the glass surface while dying. Viruses have distinct CPEs, just as colonies of bacteria on agar plates have unique morphologies (Figure 51-24). CPE may be quantitated as indicated in Table 51-34. Preliminary identification of a virus can be made frequently based on the cell line that supports viral replication, how quickly the virus produced CPE, and a description of the CPE (see Table 51-33). Experienced virologists are able to presumptively identify most viruses isolated in clinical laboratories based on these criteria. When confirmation or definitive identification is required, additional testing can be performed. Fluorescent-labeled antisera, available for most viruses, are used for confirmation. In addition,

acid lability is used to differentiate enteroviruses from rhinoviruses, and neutralization is used to identify viruses with many serotypes for which fluorescent-labeled antisera are not available. Some viruses, such as influenza, parainfluenza, and mumps, which produce little or no CPE, can be detected by **hemadsorption**, because infected cells contain viral hemadsorbing glycoproteins in their outer membranes. The addition of guinea pig red blood cells (RBCs) to the cell culture tube, followed by a wash to remove nonadsorbed RBCs, results in a ring of RBCs around infected cells (see Figure 51-24, *G*). Cell cultures demonstrating hemadsorption can be stained with fluorescent-labeled antisera to identify the specific hemadsorbing virus present. Detailed procedures for culture confirmation by FA staining and hemadsorption for the detection of influenza and parainfluenza viruses are presented in Procedures 51-4 and 51-5.

VIRAL SEROLOGY

General Principles

Serology was the primary means for the laboratory diagnosis of viral infections until the mid-1970s. At that time, culture and detection of viral antigen became more widely available because of commercially available reagents, such as cell cultures, and a broad range of immunodiagnostic test kits for the detection of viral antigen directly in patient specimens. The use of viral serology tests has changed to complement this growing menu of diagnostic tests. Viral serology is now used primarily to detect immune status and to make the diagnosis of infections in situations in which the virus cannot be cultivated in cell culture, or detected readily by immunoassay or molecular assays.

In most viral infections, IgM is undetectable 1 to 4 months after acute infection resolves but detectable

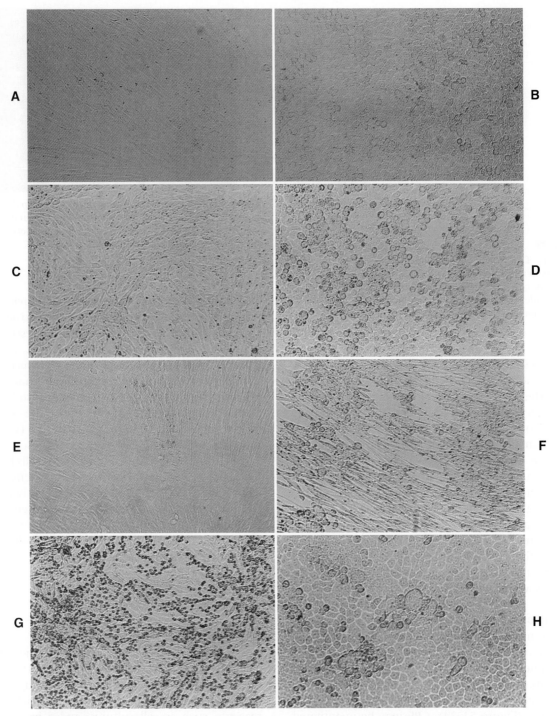

Figure 51-24 Cell culture morphology and viral CPE. **A,** Normal human diploid lung fibroblast cells (HDF). **B,** Normal HEp-2 cells. **C,** Normal primary monkey kidney cells (PMK). **D,** HEp-2 cells infected with adenovirus. **E,** HDF cells infected with cytomegalovirus. **F,** HDF cells infected with herpes simplex virus. **G,** PMK cells infected with hemadsorbing virus, such as influenza, parainfluenza, or mumps, plus guinea pig erythrocytes. **H,** HEp-2 cells infected with respiratory syncytial virus.

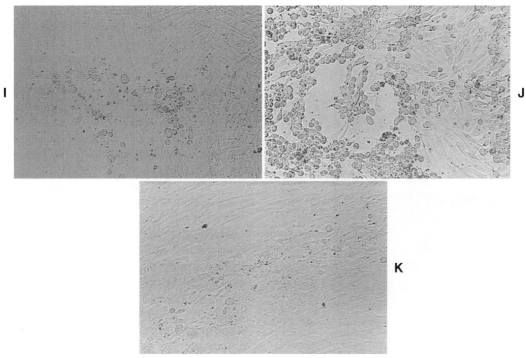

Figure 51-24, cont'd   **I,** HDF cells infected with rhinovirus. **J,** PMK cells infected with echovirus. **K,** HDF cells infected with varicella zoster virus. (From US Department of Health, Education, and Welfare, Public Health Service, Centers for Disease Control, Atlanta.)

Table 51-34   Quantitation of Cell Culture Cytopathic Effects (CPE)

Quantitation	Interpretation
Negative	Uninfected monolayer
Equivocal (±)	Atypical alteration of monolayer involving few cells
1+	1%-25% of monolayer exhibits CPE
2+	25%-50% of monolayer exhibits CPE
3+	50%-75% of monolayer exhibits CPE
4+	76%-100% of monolayer exhibits CPE

levels of IgG remain for the life of the patient. If a patient is infected by an antigenically similar virus or the original strain has remained latent and reactivates at a later time, these virus-specific IgG and IgM antibody levels may again rise. The secondary IgM response may be difficult to detect; however, a significant (fourfold) IgG titer rise is readily apparent in immunocompetent patients.

An immune status test measures whether a particular virus has previously infected a patient. A positive result with a sensitive, virus-specific IgG test indicates past infection. Some immune status tests include methods that detect both IgG and IgM, to detect recent or active infections. To help diagnose active disease, two

approaches are helpful. Detection of virus-specific IgM in an acute-phase specimen collected at least 7 to 14 days after the onset of infection indicates current, or very recent, disease. Detection of a fourfold (or equivalent increase if twofold dilutions are not tested) antibody titer rise between **acute** and **convalescent sera** also indicates current or recent disease. Acute-phase serum should be collected as soon as possible after onset of symptoms. The convalescent specimen should be collected 2 to 3 weeks after the acute. If a single post-acute serum, collected between acute and convalescent times, or convalescent specimen is all that is available for testing, an extremely high, virus-specific IgG titer may be suggestive of infection. The exact titer specific for active disease, if known at all, varies with each testing method and virus. In general, titers high enough to be diagnostic are unusual and testing of single specimens should not be performed. A reasonable policy would include using IgM tests, where available, and performing IgG testing only on paired, acute and convalescent specimens. No IgG testing need be performed on the first, acute specimen until receipt of the convalescent. This eliminates useless testing of single specimens in those instances in which a second sample is never submitted for analysis.

Many serologic methods are or have been routinely used to detect antiviral antibody. Prominent among these are **complement fixation (CF), enzyme-**

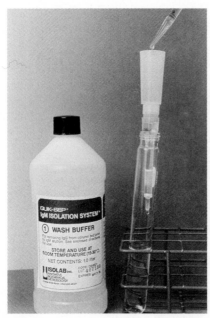

Figure 51-25   Separation of IgM from human serum by passing the serum through an ion-exchange column.

linked immunosorbent assay (ELISA), indirect immunofluorescence, anticomplement immuno-fluorescence (ACIF), and **Western immunoblotting.** Complement fixation is a labor-intensive, technically demanding method best fitted to batch testing. The major advantage of CF is the years of experience that have accumulated for interpreting results. As less demanding, easily automated techniques for batch testing are developed, such as ELISA, the need for CF testing is disappearing. Additional advantages of ELISA are that it can be used to detect IgM-specific antibodies free of common interfering factors, in particular through use of an antibody-capture technique. Indirect immuno-fluorescence is best used for individual specimens or small batch testing.

Immunofluorescence can also be used to detect virus-specific IgM but requires prior separation and elimination of the IgG fraction that, if present, can result in both false-positive and false-negative results. IgM and IgG can be separated by either ion exchange chroma-tography or immune precipitation (Figure 51-25).

Immunoglobulin G indirect FA testing is subject to false-positive results resulting from antibody-Fc receptors that occur in cells infected with virus. IFA testing is performed using virus-infected substrate cells fixed to a microscope slide. When the substrate cells are overlaid with patient serum, the Fc portion of the antibody molecule binds to these receptors. Fluorescent-labeled antiglobulin will attach to both homologous antibody (that bound to viral antigen) and to Fc-bound antibody. Subsequent fluorescence of Fc-bound antibody results in a false-positive or falsely elevated reading. To avoid this complication, the ACIF test can be used. Because fluorescent-labeled complement only binds to antigen-antibody complexes, the nonspecific antibody attached by Fc receptors, which is complement free, does not fluoresce. Western immunoblotting is also used for viral antibody detection. Because complex antigens are separated into individual components during the Western blot procedure, and positive or negative reactions are observed with each of these components, the Western blot provides a more specific result than other serologic tests such as EIA.

False-positive and false-negative results can occur when testing for virus-specific IgM antibodies. False-positive results occur when rheumatoid factor, an anti-IgG IgM-type globulin, combines with homologous or virus-specific IgG present in the patient specimen. Labeled anti-IgM combines with either bound virus-specific IgM or rheumatoid factor, causing falsely positive fluorescence. False-negative IgM tests occur when high levels of strongly binding homologous IgG antibodies prevent binding of IgM molecules, decreasing or eliminating IgM-specific fluorescence. Both problems can be eliminated by testing the IgG-free serum fraction.

Immune Status Testing

Immune status tests (Table 51-35) are used to detect patients who have been infected (or vaccinated) by a virus in the past, conferring lifelong immunity to re-infection. Rubella antibody immune status testing is used with women of childbearing age. A positive result (presence of antibody) indicates past infection or immunization and implies that congenital infection will not occur during subsequent pregnancies. Absence of antibody implies susceptibility to infection and should prompt rubella vaccination if the woman is not pregnant. Varicella and measles immune status assays are used most commonly to test health care workers. Those with no antibody must avoid diseased patients. CMV immune status is useful for organ transplant donors and recipients, and premature babies hospitalized in newborn intensive care nurseries who are likely to receive blood transfusions. Transplant recipients are susceptible to life-threatening CMV infection. Knowing the CMV status of the donor and recipient enables the physician to better diagnose and treat this disease. Newborns whose mothers were never infected with CMV are susceptible to serious, primary CMV infection that can be transmitted in white blood cells during blood transfusion. CMV-negative babies should receive only CMV-negative blood.

Serology Panels

Special situations exist in which testing for antibody to an individual virus is less helpful than using a battery of antigens to test for antibody to many viruses. The use of

Table 51-35   Serology Panels and Immune Status Tests for Common Viral Syndromes

Situation	Viruses under Consideration
AIDS	Human immunodeficiency virus (HIV)

Central nervous system (CNS) infections	Arboviruses especially; Eastern and Western equine encephalitis viruses
	    St. Louis encephalitis virus
	    La Crosse virus
	    West Nile virus
	Lymphocytic choriomeningitis virus
	Enteroviruses
	Measles
	Mumps
	Herpes simplex virus (HSV) and other herpesviruses
	Rabies

Exanthems	Measles Rubella Parvovirus
Vesicular	HSV Varicella-zoster virus
Hepatitis A	Hepatitis A virus
Hepatitis B	Hepatitis B virus
Heterophile-negative infectious mononucleosis syndromes	Cytomegalovirus Epstein-Barr virus
Myocarditis-pericarditis	Group B coxsackievirus, types 1-5 Influenza A and B Cytomegalovirus
Respiratory	Influenza A and B Respiratory syncytial virus Parainfluenza types 1-3 Adenovirus
Serology needed to determine immune status	Rubella Hepatitis B Varicella-zoster CMV Measles

a combination of serologic tests to diagnose a clinical syndrome (see Table 51-35) may be useful when the viruses under consideration cannot be cultured, specimens of infected tissue are not available (e.g., brain tissue), antiviral agents have been administered, or the patient is convalescing and isolation of virus is unlikely. In most cases, some but not all viruses in the battery require testing. Consultation with patient clinician is required to narrow the list of potential etiologies.

INTERPRETATION OF LABORATORY TEST RESULTS

Interpretation of laboratory test results must be based on knowledge of the normal viral flora in the site sampled, the clinical findings, and the epidemiology of viruses. Serologic testing, in addition to virus detection tests, may be needed to support or refute the association of a virus isolate with a disease state.

Viruses in Tissue and Body Fluids

In general, the detection of any virus in host tissues, CSF, blood, or vesicular fluid is significant. Recovery of adenovirus or mumps virus in urine is usually diagnostic of disease, as opposed to the detection of CMV, which may merely reflect asymptomatic reactivation. Occasionally, enteroviruses are detected in urine as a result of fecal contamination, or HSV from symptomatic or asymptomatic infection of the external urogenital tract is recovered in urine. Interpretation of these culture results requires correlation with clinical data. CMV viruria during the first 2 weeks of life establishes a diagnosis of congenital CMV infection, whereas detection at 4 weeks or after suggests intrapartum or postpartum acquisition.

Viruses in the Respiratory Tract

Detection of measles, mumps, influenza, parainfluenza, and respiratory syncytial viruses is significant because asymptomatic carriage and prolonged shedding is unusual. Conversely, HSV, CMV, and adenoviruses can be shed in the absence of symptoms for periods ranging from a few days to many months. Adenoviruses are detected commonly from asymptomatic infants and young children. Simultaneous detection of this virus from both throat and feces in febrile patients with respiratory syndromes increases the probability of association with illness. Isolation from throat but not feces has a lesser probability of association, and isolation from feces alone has the least diagnostic significance.

Viruses in the Eye

Detection of adenoviruses, HSV, VZV, and some enteroviruses from diseased cornea and conjunctiva usually establishes the etiology of the infection. Enterovirus type 70 is known to cause a particularly contagious form of viral conjunctivitis.

Detection of Epstein-Barr Virus

Disease caused by EBV is established by detecting antibody to multiple antigens. The time-course of appearance of antibody is depicted in Figure 51-11. Detection of antibody to viral capsid antigen, early antigen, and Epstein-Barr nuclear antigen is interpreted as indicated in Table 51-36.

Detection of Enteroviruses

Enteroviruses are most commonly found in asymptomatic infants and children, particularly during the late

Table 51-36 Interpretation of Epstein-Barr Virus Serology Test Results

Clinical Situation	Heterophile Antibody	IgG-VCA	IgM-VCA	EA	EBNA
No past infection	Usually −	−	−	−	−
Acute infection	Usually +	+	+	+	−
Convalescence phase	+/−	+	+or −	+ or −	+
Past infection	Usually −	+	−	− or W+	+
Chronic or reactivation	Not useful	+	−	+	+

EA, Early antigen; *EBNA*, Epstein-Barr nuclear antigen; *VCA*, viral capsid antigen; *W+*, weakly positive.

summer and early fall. Knowledge of the relative frequency of virus shedding is extremely helpful in assessing significance of results of throat or stool cultures. Prevalence of enteroviruses in the stools of infants and toddlers may approach 30% during peak periods. Shedding of enteroviruses in the throat is relatively transient, usually 1 to 2 weeks, whereas fecal shedding may last 4 to 16 weeks. Thus, isolation of an enterovirus from the throat supports etiology of a clinically compatible illness more than isolation from the feces alone. If live attenuated oral poliovirus vaccine is used, vaccine strains can be detected in stools of recently vaccinated children and their contacts, for example, siblings. Typing identifies the enteroviruses as a poliovirus serotype and, in absence of clinical findings that suggest polio along with a setting of recent vaccination, the isolate can be considered "normal."

Detection of Hepatitis Viruses

Disease caused by HAV is detected using serology tests specific for viral-induced IgM and IgG (Figure 51-26). In addition to clinical findings consistent with disease, the presence of hepatitis A–specific IgM is diagnostic of current, active disease. HBV requires detection of antigen and antibody to multiple antigens to classify disease type (Figure 51-27 and Table 51-37). Hepatitis D co-infection with HBV relies on the detection of anti-HDV antibodies. Diagnostic tests for HCV include RT-PCR and antibody detection. PCR is used to detect early acute hepatitis C disease, because antibody detection tests may be negative. Testing for antibody by ELISA is used to detect chronic hepatitis C disease (Figure 51-28). HCV RNA levels in serum, detected by a number of highly sensitive molecular biopsy methods, are used to differentiate patients who are likely to respond to therapy from those with lower response rates. All patients with sustained response to therapy became negative for HCV RNA within 6 months. HCV genotyping also is used to identify genotypes more or less likely to respond to therapy. For example, patients with HCV genotype 1 have significantly lower response rates to therapy.

Antibody tests are used to detect patients infected with hepatitis E; however, disease is rare in the United States.

Detection of Varicella Zoster Virus and Herpes Simplex Virus

The detection of VZV is always significant. Asymptomatic shedding does not appear to occur with this virus as it does with other herpes viruses. The detection of HSV from cutaneous or mucocutaneous vesicles is also significant, implying primary or reactivation disease. HSV may be detected in respiratory secretions during asymptomatic "stress" reactivation, unless typical vesicles or ulcers are present. HSV in stool usually represents either severe disseminated infection or infection of the anus or perianal areas. Isolation from any newborn infant specimen suggests potentially severe infection.

Detection of Cytomegalovirus

Interpretation of results of specimens containing CMV is most difficult. Primary CMV infection is usually asymptomatic and is commonly followed by silent reactivation of the latent virus throughout the patient's remaining life. CMV disease in immunocompromised patients can be life threatening, and antiviral therapy may be warranted. Detection of CMV in urine or respiratory secretions, however, is not diagnostic of significant disease. Detection of CMV in tissue, such as lung, by culture or histopathology, or in blood collected by venipuncture, suggests an active role in disease. The detection of CMV antigenemia or DNA in the blood, by a molecular method, is highly suggestive of active disease, and quantitative results can be used to evaluate therapeutic intervention and patient prognosis. Interpretation of the CMV antigenemia assay is dependent on the patient population and laboratory expertise. In general, CMV disease is accompanied by detectable virus in peripheral leukocytes. Disease severity is roughly proportional to quantity of virus, that is, number of fluorescing cells. As disease is treated and resolves, the number of positive cells decreases. Antigenemia levels should decrease to zero as the patient's immune func-

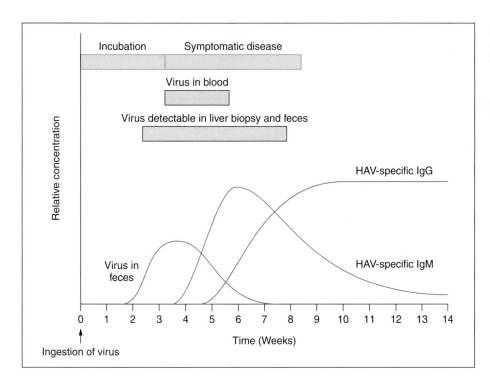

Figure 51-26   Illustration of time-course of disease and immune response to hepatitis A virus. (Modified from Murray PR, Kobayashi GS, Pfaller MA, et al, editors: *Medical microbiology*, ed 2, St Louis, 1994, Mosby.)

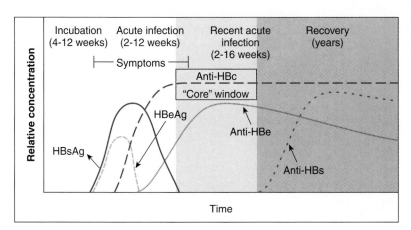

Figure 51-27   Illustration of time-course of antigenemia and immune response in a patient who recovers from acute hepatitis B infection.

Table 51-37   Serologic Profiles Following Typical Hepatitis B Virus (HBV) Infection

HBsAg	Anti-HBsAg	HBeAg	Anti-HBeAg	Anti-HBc	Most Likely Interpretation
–	–	–	–	–	No (or very early) exposure to HBV
+	–	+/–	–	–	Early acute hepatitis B (HB)
+	–	+	–	+	Acute or chronic HB
+	–	–	+	+	Chronic HBV carrier state
–	–	–	+	+	Early recovery phase from acute HB
–	+	–	+	+	Recovery from HB with immunity
–	+	–	–	–	Distant HBV infection or HB vaccine

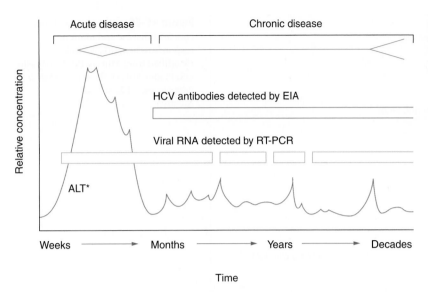

Figure 51-28   Illustration of time-course of immune response and disease caused by hepatitis C virus.

*Alanine aminotransferase marker for liver necrosis.

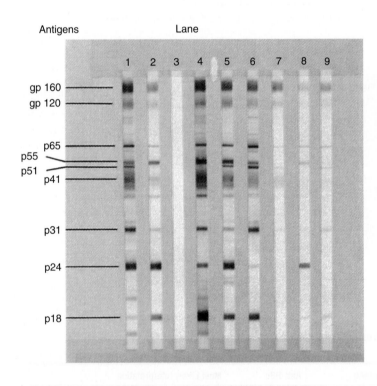

Figure 51-29   Western blot detecting specific HIV antibody. Lane 1 is the high-positive control; lane 2 is the low-positive control; lane 3 is the negative control; lanes 4 through 8 are positive sera; lane 9 is an indeterminate serum. Numbers at left refer to approximate molecular weights of HIV antigens. (Courtesy Hodinka RL: Children's Hospital of Philadelphia, Philadelphia, Pa.)

tion is restored and antiviral therapy is introduced. The presence of virus-specific IgM or a fourfold increase in IgG antibodies may indicate disease. However, positive serology results must be interpreted with caution. False-positive IgM results have been attributed to infections by other viruses, such as EBV, and rises of both IgM and IgG may result from transfusions or the use immune globulin therapy.

Detection of Human Immunodeficiency Virus

The HIV-1 Western blot test identifies antibody specific for several HIV antigens (Figure 51-29). Most often, antibody to HIV p24 and either gp41 or gp160 confirms HIV-1 infection. HIV-1 p24 antigen testing is used to detect acutely infected patients before the appearance of antibody. PCR testing for HIV is useful for newborns, who may have maternal HIV antibody confounding interpretation of serology tests, and for all patients who may not produce detectable antibody for months following primary infection. The quantitative plasma RNA test (viral load) is used to measure the amount of HIV in blood. As many as 10 billion new HIV virions are produced each day in untreated patients. Viral load testing has become an essential parameter in guiding decisions to begin or change antiviral therapy. Plasma HIV RNA can be quantified using various FDA-approved assays, including the Roche Monitor RT-PCR (Roche Molecular Diagnostics), Bayer Versant HIV-1 (bDNA) Assay (Bayer Diagnostics), and NucliSense EasyQ HIV-1 (NASBA) Assay (bioMérieux, Inc.). Viral load testing is performed at the time of diagnosis of HIV infection and periodically thereafter. Successful antiretroviral therapy should decrease plasma RNA to undetectable levels (less than 50 copies/mL).

PRESERVATION AND STORAGE OF VIRUSES

Clinical virology laboratories must have a method for storing and retrieving viruses. Isolates should be kept as control strains and, in rare instances, for epidemiologic investigations. Public health laboratories may use current enterovirus or influenza virus strains for typing. Viruses can be stored by freezing at –70° C or in liquid nitrogen. Freezing at –70° C is more practical for clinical laboratories. A method for the preservation and storage of viruses by freezing is described in Procedure 51-6.

ANTIVIRAL THERAPY AND IN VITRO ANTIVIRAL SUSCEPTIBILITY TESTING

The availability of specific antiviral chemotherapeutic agents has added importance and urgency to the need for the laboratory diagnosis of viral infections (see Table 51-2). Detection of virus to establish an exact etiologic diagnosis is usually needed to justify use of these expensive and sometimes toxic agents. Resistance is known to develop to all agents and may be detected in vitro using antiviral susceptibility testing. Phenotypic susceptibility assays measure viral replication in the presence of antiviral agent. These methods are not well standardized, require weeks to perform since virus must first be isolated in cell culture, and in vitro results do not always correlate with in vivo response (patients with susceptible strains in the laboratory have failed therapy). The Clinical Laboratory Standards Institute (Wayne, Pa) offers a consensus standard (document M33-A) for HSV antiviral susceptibility testing by a phenotypic assay (plaque reduction). Genotypic susceptibility assays use PCR to detect genes known to be responsible for resistance and sequencing to determine whether genome alterations associated with resistance have occurred. Genotypic assays are important for HIV and HCV. These

Table 51-38   Vaccination Against Viral Disease

Disease	Vaccine
Yellow fever	Attenuated—Live
Poliomyelitis	Attenuated—Live and inactivated
Measles	Attenuated—Live
Mumps	Attenuated—Live
Rubella	Attenuated—Live
Hepatitis B	Inactivated
Influenza	Inactivated
Smallpox	Attenuated—Live
Chickenpox	Attenuated—Live
Hepatitis A	Inactivated
Rabies	Inactivated
Rotavirus	Attenuated—Live

assays are rapid because isolation of virus in culture is not necessary for testing. Response to antiviral agent is also measured by quantitative monitoring of viral load in the patient's blood. Such testing is common in patients infected with HBV, HCV, and CMV. The viral load should significantly diminish following the addition of antiviral agent to which the virus is susceptible. Using molecular testing, such as quantitative PCR, to measure the amount of virus present in serum is a surrogate test for resistance to antiviral agents. The viral load rises quickly when resistance appears.

PREVENTION OF VIRAL INFECTION

VACCINATION

Control of many viral diseases has been accomplished by vaccination. Since Jenner developed the first vaccine against smallpox 200 years ago, attenuated-live or inactivated-dead viral vaccines have been used successively to prevent yellow fever, poliomyelitis, measles, mumps, rubella, hepatitis B, and influenza (Table 51-38). Smallpox was eliminated in 1977 by an effective vaccination program. Additional vaccines continue to appear. New smallpox vaccines with fewer side effects are being developed to prevent outbreaks in the event of bioterrorism. A live attenuated varicella (chickenpox) vaccine is now recommended for all children, and an inactivated hepatitis A vaccine is available for travelers or others entering areas of higher endemicity. Influenza vaccination is needed annually because of antigenic drift, or occasional major antigenic shift, of circulating virus. A new rotavirus vaccine is now FDA-approved and available.

IMMUNE PROPHYLAXIS AND THERAPY

Immune prophylaxis is used to prevent serious viral infection in immune or functionally compromised patients. In place of actively immunizing with an antiviral vaccine, it is possible to confer limited protection by the intramuscular inoculation of human immunoglobulin. Pooled human immunoglobulin contains antibody against all common viruses. Specific high-titered immunoglobulin can be collected from patients recovering from a specific infection to ensure maximal antibody levels. Immune prophylaxis should be considered an emergency procedure. Table 51-39 lists immune prophylaxis available for viral infections. Passive immunoprophylaxis of RSV disease in infants younger than 2 years, who have underlying lung disease resulting from premature birth or congenital heart disease, is particularly effective at preventing life-threatening bronchiolitis and pneumonia in this patient group.

Passive immunization is occasionally effective as therapy for viral infection (see Table 51-39). Therapy

Table 51-39   Immune Prophylaxis or Therapy for Viral Diseases

Disease	Situation
Prophylaxis	
Hepatitis A	Traveler to developing country
Hepatitis B	Newborns of infected mothers or unimmunized laboratory worker following needlestick
Rabies	Following bite by potentially rabid animal
Measles	Unimmunized close contact of patient
Varicella	Newborns of infected mothers at time of delivery
Respiratory syncytial virus	Infants younger than 2 yrs with underlying lung disease
Therapy	
Lassa fever	At time of disease to decrease severity

Procedure 51-1

PROCESSING BLOOD FOR VIRAL CULTURE: LEUKOCYTE SEPARATION BY POLYMORPHPREP

PRINCIPLE

Detection of most viruses in blood specimens is best accomplished by separating and culturing leukocytes. An exception is enterovirus, which, when infecting newborns, may be found free in serum. Leukocytes were originally added to cell culture by harvesting the buffy coat following concentration by centrifugation (100× *g*, 15 minutes). This method was found to be inferior to white blood cell concentration using density gradient centrifugation and sedimentation with Ficoll-Hypaque/Macrodex. Other density gradient methods have been employed, including those which use Plasmagel, LeucoPREP, Sepracell-MN, Mono-Poly Resolving Medium, and Polymorphprep. Leukocyte preparation using Polymorphprep, a mixture of sodium metrizoate and dextran, emerges as one of the best, because mononuclear and polymorphonuclear cells are separated from red blood cells in a one-step procedure.

SPECIMEN

Three to 5 mL of anticoagulated blood is optimal. EDTA-, citrate-, or heparin anticoagulant–containing tubes can be used. As little as 2 mL of specimen is acceptable for pediatric patients. Clotted

specimens are unacceptable. Specimens should be processed immediately. When processing delay is unavoidable, hold specimen at 2° to 8° C. Processing must occur within 12 to 24 hours. Loss of virus may occur during any processing delay.

MATERIALS

1. Polymorphprep. Axis-Shield PoC AS (U.S. supplier Fischer Scientific), Oslo, Norway. Store at or below 20° C
2. Phosphate buffered saline (PBS)—10×. Store at room temperature
3. Sterile phosphate buffered saline—0.5×. Dilute 10× PBS 1:20 with distilled water. Filter sterilize through 0.22-μm filter. Store at 2° to 8° C
4. Eagle's minimal essential medium (2% fetal bovine serum—FBS). Store at 2° to 8° C
5. Sterile sodium chloride (0.9%). Store at 2° to 8° C
6. Sterile conical centrifuge tubes
7. Sterile pipettes
8. Centrifuge

METHODS

1. Wear latex gloves and laboratory coat when handling patient speci-

mens. Use biological safety cabinet whenever tube containing specimen is opened. Use biocontainment safety covers during centrifugation procedures.
2. Allow blood specimen and all reagents to come to room temperature.
3. Mix blood specimen well by inverting five times. Using a 5- or 10-mL pipette, transfer specimen to a sterile conical centrifuge tube.
4. Centrifuge at 500× *g* for 10 minutes at room temperature (18° to 22° C).
5. Remove plasma with a pipette.
6. Measure approximate volume of remaining cells. Using a 5- or 10-mL pipette, dilute cells with the same approximate volume of 0.9% sodium chloride. Mix well.
7. Add 3.5 mL of Polymorphprep to a 15-mL conical centrifuge tube with a 5-mL pipette.
8. Carefully layer the entire volume of diluted blood cells over the 3.5 mL of Polymorphprep in the centrifuge tube. Do not mix the blood with the Polymorphprep.
9. Centrifuge the tube containing blood and Polymorphprep at 450 to 500× *g* for 30 minutes at room temperature (18° to 22° C).

Procedure 51-1

PROCESSING BLOOD FOR VIRAL CULTURE: LEUKOCYTE SEPARATION BY POLYMORPHPREP—CONT'D

10. Two leukocyte bands should be visible after centrifugation. The top band at the blood sample/Polymorphprep interface consists of mononuclear cells. The lower band contains polymorphonuclear cells. Erythrocytes are pelleted at the bottom of the tube.

11. The mononuclear and polymorphonuclear bands are collected and transferred to a 15-mL conical centrifuge tube with a Pasteur pipette.

12. Add 5 mL of 0.5× sterile PBS to the centrifuge tube containing the harvested cells. Mix well by aspirating and expelling repeatedly with a pipette.

13. Centrifuge cells at 400× g for 10 minutes at room temperature (18° to 22° C).

14. Aspirate supernatant PBS.

15. Add 5 mL of EMEM with 2% FBS to pelleted cells. Mix well by aspirating and expelling repeatedly with a pipette.

16. Centrifuge as in step 13.

17. Aspirate supernatant EMEM and resuspend in 2 mL of fresh EMEM with 2% FBS.

18. Specimen is ready for inoculation to cell culture.

TROUBLESHOOTING

1. If all cells are at the top of the Polymorphprep after centrifugation, remix and overlay on top of a new Polymorphprep layer. It may be necessary to increase the volume of Polymorphprep so proportion is still approximately 1:1 to 1:1.5. Centrifuge at either increased force of gravity for the same length of time or at the same force of gravity for a longer time.

2. If buffy coat is on top of the erythrocyte layer, take buffy coat and erythrocytes and relayer over a new gradient. Centrifuge for a shorter time at the same force of gravity or at a lesser force of gravity for the same time.

Procedure 51-2

CYTOMEGALOVIRUS ANTIGENEMIA STAIN

PRINCIPLE

The CMV antigenemia stain is used for the immunologic detection of cytomegalovirus lower matrix pp65 antigen in peripheral blood leukocytes. The assay is semiquantitative in that white blood cells are counted and a standard number are placed on the slide for staining. Experience with the CMV antigenemia stain suggests that positive results can serve as an early indicator of active CMV disease.

SPECIMEN

A total of 5 to 7 mL of anticoagulated blood is optimal. Heparin- or EDTA anticoagulant–containing tubes can be used. Specimens should be transported to the laboratory at room temperature immediately after collection. Clotted specimens are unacceptable. Specimens should be processed immediately. When processing delay is unavoidable, hold specimen at 2° to 8° C. Processing should occur within 8 hours.

MATERIALS

1. CMV Brite Kit (IQ Products, The Netherlands, distributed by Biotest Diagnostic Corporation, Denville, NJ), which includes dextran solution, RBC lysing solution, fixative solution, permeabilization solution, fetal calf serum, monoclonal antibodies, and control slides. Store at 2° to 8° C

2. Sterile pipettes and micropipettes

3. Refrigerated centrifuge and 15 mL conical bottom centrifuge tubes

4. Cytocentrifuge (Shandon, Inc., Pittsburgh, Pa.)

5. Phosphate buffered saline (PBS), calcium- and magnesium-free deionized water (dH$_2$O)

6. Coplin staining jars, moisture chamber, 37° C incubator, microscope slides for cytospin, coverslips and mounting medium

7. Coulter counter for WBC count (hematology laboratory)

8. Fluorescent microscope

METHODS

1. Wear latex gloves and laboratory coat when handling patient specimens. Use biological safety cabinet whenever tube containing specimen is open. Use biocontainment safety covers during centrifugation procedures.

2. Leukocyte isolation from peripheral blood
   a. Allow kit reagents to come to room temperature (20° to 25° C).
   b. Mix 5 to 7 mL of anticoagulated blood with 1.5 mL of dextran reagent (reagent A) in a 15-mL conical centrifuge tube. If a greater volume of blood is used, the dextran-to-blood ratio of 1:4 must be maintained.
   c. Incubate tube containing blood-dextran mixture for 20 minutes at 37° C at a 45-degree angle. Leave cap loose.
   d. Remove the top, leukocyte-rich layer to a 15-mL conical centrifuge tube using a Pasteur pipette. Centrifuge for 10 minutes at 300× g.

Continued

Procedure 51-2

CYTOMEGALOVIRUS ANTIGENEMIA STAIN—CONT'D

e. Dilute RBC lysing reagent (reagent B) 1:10 in dH$_2$O and allow to cool to 4° C.

f. Discard supernatant following 10-minute centrifugation. Resuspend cell pellet in 5 mL of diluted reagent B. Vortex and incubate for 5 minutes at 4° C.

g. Add 5 mL of PBS and centrifuge for 10 minutes at 300× *g*.

h. Wash cells in 5 mL PBS, centrifuge as above, and discard supernatant.

3. Count cells and prepare cell dilution.
   a. Resuspend cell pellet in 1.5 mL PBS.
   b. Aliquot 0.4 mL into a 1.5-mL microcentrifuge tube. This aliquot is used for a leukocyte count.
   c. Dilute cell suspension with PBS to make 1 mL of 2 × 10^6 cell/mL in a 1.5-mL microcentrifuge tube. Calculate the amount of cell suspension to use as follows:

$$\frac{\text{Desired number cells/mL}}{\text{Actual number cells/mL}} = \text{suspension required}$$

For example:

$$\frac{2 \times 10^6 \text{ cells/mL}}{5 \times 10^6 \text{ cells/mL}} = \frac{2}{5} = 0.4 \text{ mL cell suspension}$$

Add enough PBS to increase the volume to 1 mL.

For example: If the amount of cell suspension needed is 0.4 mL, add 0.6 mL of PBS to make 1 mL.

4. Cytospin preparation of slides
   a. Prepare two cytospin slides for testing.
   b. Resuspend 1 mL cell suspension by vortexing tube.
   c. Add 100 µL of the cell suspension in cytospin assembly.
   d. Centrifuge at 800 rpm for 4 minutes.
   e. Remove slides and air dry.

5. Fixation and permeabilization
   a. Dilute the fixation reagent (reagent C) 1:5 in dH$_2$O in a fume hood (5 mL reagent C, 20 mL dH$_2$O).

b. Flood slides with diluted reagent C and incubate at room temperature in a fume hood for 10 minutes.

c. Prepare 100 mL of 1% washing solution by diluting reagent E 1:100 in PBS (1 mL reagent E, 99 mL PBS). Fill 1 Coplin jar with 1% washing solution.

d. Dip slides three times in 1% washing solution and place slides in same Coplin jar for 5 minutes.

e. Dilute permeabilization solution (reagent D) 1:5 in dH$_2$O (5 mL reagent D, 20 mL dH$_2$O).

f. Immerse slides in dilute permeabilization solution for 5 minutes.

g. Fill Coplin jar with fresh 1% wash solution. Dip slides three times in 1% wash solution and place slides in same Coplin jar for 5 minutes.

h. Rinse slides in dH$_2$O for 15 seconds, then let slides dry. When dry, slides are ready for immediate staining. If not stained immediately, store at 2° to 8° C for 24 hours or up to 24 hours at –20° C. Wrap slides in foil and store in a desiccated environment for extended storage.

6. Immunofluorescent staining
   a. Allow slides, including control slide, to come to room temperature. Mark a small circle around the cell area (cytospin area) on the slide to contain the antibody solution during staining.
   b. Immerse slides in Coplin jar containing PBS for 5 minutes. Once slides have been immersed, do not allow specimen area to dry through remainder of staining procedure.
   c. Working with one slide at a time, remove slide from Coplin jar and carefully dry the area around the specimen with a sterile cotton swab. Place slide in moisture chamber and apply 35 µL of monoclonal antibody (reagent F).
   d. Incubate slides for 30 minutes at 37° C.
   e. Remove slides from staining chambers and rinse in a Coplin jar with fresh PBS for 3 minutes.

f. Again, working with one slide at a time, remove slide from Coplin jar and dry the area around the specimen with a sterile cotton swab. Place slide in moisture chamber and apply 35 µL of FITC-conjugated antibody to the slide.

g. Incubate slides for 30 minutes at 37° C.

h. Wash slides in PBS and rinse with dH$_2$O. Drain excess water and air dry. Do not blot.

i. Add mounting medium and coverslip.

7. Reading and reporting
   a. Examine using fluorescent microscope and 200× magnification (10× eyepiece, 20× objective). Read soon after staining (same day), because fluorescent intensity fades. Scan the whole cell spot on both slides.
   b. Positive cells show homogenous apple-green nuclear staining as demonstrated in the positive control. Negative cells show no apple-green staining. Nonspecific staining, which appears as cytoplasmic or dull yellowish staining, should be reported as negative. Atypical greenish staining of cells on the periphery of the cell inoculum is due to drying artifact and should be reported as negative.
   c. 200,000 cells should be present on each slide, 400,000 per two slides.
      (1) Report positive stains as "Positive: (number) CMV positive cells seen per 400,000 cells."
      (2) Report negative stains as "No CMV positive cells seen per 400,000 cells analyzed."

QUALITY CONTROL

1. The negative control should have no CMV-infected cells showing fluorescence.

2. The positive control should have CMV-infected cells showing homogenous apple-green nuclear staining.

3. If controls are not as above, results are unacceptable and test must be repeated.

Procedure 51-3

SHELL VIAL CULTURE FOR CYTOMEGALOVIRUS

PRINCIPLE

The shell vial culture for CMV is used to rapidly detect this virus in clinical specimens. Use of the shell vial culture with urine, blood, and bronchoalveolar lavage fluid specimens has demonstrated equivalent or superior sensitivity compared with conventional cell culture.

SPECIMEN

Acceptable specimens include urine, blood, respiratory secretions and washes, tissue, and cerebrospinal fluid. Specimens are collected, transported, and stored according to standard protocols.

MATERIALS

1. Shell vials, 15- × 45-mm, 1 dram vial with plastic closure
2. Circular coverslips, No. 1 12-mm diameter
3. MRC-5 cells growing in shell vial (5 to 9 days old)
4. In place of steps 1 to 3, use MRC-5 shell vials purchased from commercial source such as Biowhittaker Inc., Walkersville, Md., or Viromed. Laboratories, Inc., Minneapolis, Minn.
5. Centrifuge; sterile 15-mL conical centrifuge tubes
6. Sterile 1- and 5-mL pipettes and cotton-plugged sterile Pasteur pipettes
7. pH paper (3 to 9 pH)
8. Maintenance medium
9. CMV positive control (strain AD 169)
10. Negative control (use maintenance medium with no virus)
11. Anti-CMV early nuclear protein monoclonal antibody
12. FITC-labeled anti-mouse IgG conjugate
13. Phosphate buffered saline (PBS)
14. Methanol, forceps, buffered glycerol mounting medium, and glass microscope slides

METHODS

1. Wear latex gloves and laboratory coat when handling patient specimens. Use biological safety cabinet whenever tube containing specimen is open. Use biocontainment safety covers during centrifugation procedures.
2. Specimen preparation
   a. Urines
      (1) Using a 5-mL pipette, pipette 2 mL of urine into sterile centrifuge tube.
      (2) Using a 1-mL pipette, add 0.5 mL of viral transport medium containing antimicrobials.
      (3) Centrifuge at 1000× *g* for 10 minutes at 2° to 8° C.
      (4) Aspirate supernatant with cotton-plugged sterile Pasteur pipette and transfer to a sterile test tube.
      (5) Adjust pH to 7.0 with sterile cold 0.1N HCl or 0.1 N NaOH. Determine pH with pH paper.
      (6) Store at 2° to 8° C until inoculation into shell vial cultures.
   b. Blood: Refer to leukocyte separation by Polymorphprep (see Procedure 51-1). Store leukocytes at 2° to 8° C until inoculation into shell vial cultures.
   c. Respiratory secretions and washes
      (1) Process as for urines above. pH adjustment is not necessary.
      (2) Specimens received in viral transport medium do not require further preparation.
      (3) Store all specimens at 2° to 8° C until inoculation into shell vial culture.
   d. Tissue
      (1) Mince and grind 5 mm³ piece of tissue gently in approximately 5 mL of viral transport medium. This results in a 10% suspension.
      (2) Centrifuge at 1000× *g* for 10 minutes at 2° to 8° C.
      (3) Aspirate supernatant with cotton-plugged sterile Pasteur pipette and transfer to small sterile test tube.
      (4) Store at 2° to 8° C until inoculation into shell vial culture.
   e. Cerebrospinal fluid
      (1) CSF and other body fluids are inoculated into shell vial culture undiluted.
      (2) Specimens with less than sufficient volume for proper inoculation (≤0.5 mL) should be diluted with enough viral transport medium to inoculate shell vial cultures (0.5 to 0.6 mL).
      (3) Specimens received in viral transport medium are ready to inoculate.
      (4) Store at 2° to 8° C until inoculation into shell vial culture.
3. Inoculation and incubation of shell vial cultures
   a. Aspirate maintenance medium from sufficient shell vials for all specimens, including one positive and one negative control.
   b. Using a new sterile 1-mL pipette for each specimen, inoculate shell vials as follows:

Specimen	No. Vials	Inoculum/ Vial
Urine	1	0.25 mL
Respiratory	1	0.1 mL
Blood	3	all 0.25 mL
All other specimens	2	0.1 and 0.25 mL
Negative control	1	0.1 mL
Positive control	1	0.1 mL

Number of vials and volume of inoculum may vary from laboratory to laboratory depending on local experience. Inoculate all specimens before inoculating controls. A negative and positive control are used with each batch of shell vial cultures. Inoculate negative control before positive.

   c. Cap vials with sterile plastic lids and label.
   d. Centrifuge vials at 700× *g* for 40 minutes at 35° C.
   e. Using a 5-mL pipette, add 2 mL of maintenance medium to each shell vial.
   f. Recap vials and incubate at 36° C ± 1° C for 16 to 24 hours.
4. Staining shell vial coverslips
   a. Wear latex gloves and perform test in laminar flow hood.

Continued

Procedure 51-3

SHELL VIAL CULTURE FOR CYTOMEGALOVIRUS—CONT'D

b. Remove lids and aspirate media from shell vials.

c. Wash coverslips with 1 mL of PBS. Aspirate. Wash again with 1 mL PBS and let stand for 5 minutes. Aspirate. Additional PBS washes may be necessary for blood cultures, to remove blood cells from cell sheet. If necessary, check cell monolayer using the microscope.

d. Fix coverslips in 1 mL cold methanol for 10 minutes.

e. Aspirate methanol. Coverslips should be allowed to dry, but not overdry. A white clearing across coverslips is apparent at the right amount of dryness. Overdrying causes nonspecific fluorescence around the edge. Underdrying causes uneven staining.

f. Add 150 µL monoclonal antibody, at current working dilution, to each vial by dropping antibody into the middle of the coverslip. Replace shell vial lids and gently rotate to spread stain over entire coverslip. Incubate at

35° to 37° C for 30 minutes.

g. After staining, add 1 mL PBS to each vial (do not remove antibody first). Mix. Aspirate entire volume of fluid from vial.

h. Wash twice with 1 mL PBS. Let the second wash remain on coverslip for 5 minutes. Aspirate PBS; do not allow coverslips to dry.

i. Add 150 µL of antiimmunoglobulin conjugate (FITC), at current working dilution, to each vial by dropping antibody onto the middle of the coverslip. Spread as above in "f." Incubate at 35° to 37° C for 30 minutes.

j. Repeat wash steps "g" and "h." Wash coverslips an additional time with 1 mL distilled/deionized water for 2 minutes. A brief rinse may be sufficient for monoclonal antibody conjugates. The more thorough washing is needed for polyclonal reagents.

k. Using small forceps, remove coverslips from vials and place cell side down onto drop of mounting medium. Wipe tip of forceps

with alcohol-soaked gauze between each coverslip.

5. Reading and interpretation of shell vial coverslips: Read entire coverslip using the 20× fluorescence microscope objective. The nuclei of CMV-infected cells exhibit an apple-green fluorescence that may vary in intensity (see Figure 51-23). Infected nuclei may be distinguished from artifacts by their regular oval shape and evenly stained, slightly granular appearance. Occasionally the nuclei may appear rounded. The background should stain a dull red or a pale green (dull red if Evans blue counterstain is used with conjugate, pale green if no counterstain is used).

QUALITY CONTROL

1. The negative control coverslip should have no fluorescing nuclei typical of a CMV-infected cell.

2. The positive control coverslip should have multiple fluorescing nuclei typical of CMV-infected cells.

Procedure 51-4

CULTURE CONFIRMATION BY FLUORESCENT ANTIBODY STAINING

PRINCIPLE

Many viruses can be identified presumptively in cell culture by observing CPE. Definitive identification, or in some cases, typing, of a virus in cell culture can be accomplished by staining a smear of infected cells with a virus-specific antibody conjugate. Results are easier to interpret if there is at least 2+ CPE before collecting cells for staining.

MATERIALS

1. Virus-specific antibody conjugate
2. Phosphate buffered saline (PBS)
3. Glass slides with circled staining areas and coverslips
4. 25-µL pipettes and tips
5. Control slides with homologous virus for positive staining and nonhomologous virus for negative staining

6. Mounting fluid (25% buffered glycerol in PBS)
7. Sterile glass beads
8. Centrifuge
9. Pasteur pipettes

SPECIMEN

1. Cells can be collected from the surface of the glass test tube by scraping with a 1-mL pipette, adding three to four sterile glass beads and disrupting by vortexing, or detaching cells with trypsin. Scraping is best to use if there is less than 2+ CPE. Areas of CPE can be circled on the outside of the culture tube with a marker. If only these areas are scraped on the inside, the relative number of virus-infected cells will be greater and detection will be easier in the stained smear. Al-

though trypsin treatment is the traditional method for releasing cells, use of glass beads during vortexing is simple and reliable. The glass bead method is described.

a. Add three to four sterile glass beads to the cell culture tube. Vortex vigorously for 15 to 30 seconds. Examine microscopically to ensure that the cells are detached from the glass. Vortex a second time if cells remain on glass.

b. Pellet cells by centrifugation at 900× g for 10 minutes. Carefully remove supernatant.

c. Wash cells by resuspending the pellet in PBS and repeat centrifugation.

d. Remove supernatant and resuspend in 0.05 to 0.1 mL of PBS.

Procedure 51-4

e. Place 25-μL aliquot of the cell suspension onto two wells on clean, labeled slides. Prepare enough smears to test with antisera for all viruses under consideration.

f. Air dry at temperatures no higher than 30° C and fix in acetone for 10 minutes. Slides may be stored at −70° C.

2. Determine working dilution of fluorescent conjugate if not supplied by manufacturer at proper working dilution.

a. Prepare serial doubling dilutions of the stock conjugate in PBS.

b. Test each conjugate dilution for its ability to stain smears of homologous and nonhomologous virus-infected cells, as well as smears of normal uninfected cells.

c. The working dilution of the antiserum is the highest dilution producing 3 to 4+ fluorescence of homologous virus, ≤1+ fluorescence of nonhomologous virus, and no fluorescence of uninfected cells.

METHOD

1. Apply working dilution of conjugate to circled areas on slide. Cover entire circled area (approximately 25 μL of conjugate per circle).

2. Place slides in a moist chamber and cover the chamber to minimize evaporation. Incubate slides at 35° C for 20 minutes. Be careful not to tilt slides, because conjugates will run together. Do not allow conjugate to dry on the slide. This will cause nonspecific staining.

3. Tip slides to drain conjugate from smear. Remove excess reagent with running PBS, followed by a 5-minute PBS wash in a Coplin jar. Agitate slide occasionally during wash. A brief rinse with no agitation may be sufficient for monoclonal antibody conjugates. The more thorough washing is needed for polyclonal reagents.

4. Rinse slides in distilled water to remove PBS salt crystals.

5. Allow slides to air dry at room temperature. Protect slides from bright light.

6. Apply mounting fluid and coverslip.

7. Examine with a fluorescence microscope containing appropriate filters for FITC staining (490 nm UV needed). Use 20× and 40× objectives.

INTERPRETATION OF RESULTS

1. Positive staining cells are characterized by ≥1+ apple green fluorescent intranuclear and/or cytoplasmic granules (see Figure 51-16).

2. Negative staining cells are characterized by no or <1+ fluorescence. Uninfected cells may stain red if Evans blue counterstain is used.

QUALITY CONTROL

1. Positive and negative controls should be used each time the stain is performed.

2. The positive control (homologous virus) should exhibit ≥2+ staining.

3. The negative control (nonhomologous virus) should exhibit <1+ staining.

4. If control results are different than expected, stain should be repeated.

Procedure 51-5

PRINCIPLE

Hemadsorption of cell cultures is used to detect or confirm the presence of orthomyxoviruses (influenza A, B, and C) and some paramyxoviruses (parainfluenza 1 to 4 and mumps). Respiratory syncytial virus is not detected by this method. Cell cultures infected with these viruses may not show CPE; however, infected cells do acquire the ability to hemadsorb guinea pig red blood cells to their outer cell membranes. In addition, the presence of viral hemagglutinin in the culture medium, after release from the cell, can also cause agglutination of the RBCs in the fluid medium.

Hemadsorption is performed on various days during culture incubation depending on the time of year and the viruses suspected. Influenza virus detec-

tion is performed on specimens from adult and pediatric patients during the influenza season (November to March). Early hemadsorption (day 2 or 3) decreases time to detection of positive influenza cultures. Detection of parainfluenza viruses from pediatric patient specimens is performed year-round. Hemadsorption produced by parainfluenza viruses generally occurs later than that produced by influenza. Hemadsorption to detect mumps virus is performed, in most clinical settings, only if this virus is suspected by the clinician.

SPECIMEN

1. A culture exhibiting CPE should be hemadsorbed as soon as possible to expedite reporting of a positive result.

2. Perform hemadsorption on CPE-negative cell cultures of respiratory

tract specimens or of specimens for mumps virus detection according to the following schedule:

CULTURE TYPE	PERFORM HEMADSORPTION ON DAY:
Influenza virus suspected	2 and 7
Parainfluenza virus suspected	3 and 10
Mumps virus suspected	3 and 10

MATERIALS

1. Guinea pig red blood cells. Store in buffer in which cells were received at 2° to 8° C. Cells expire 10 days after date drawn from animal

2. Sterile, 10× phosphate buffered saline. Prepare working 1× phosphate

Continued

Procedure 51-5

HEMADSORPTION OF PRIMARY MONKEY KIDNEY MONOLAYERS TO DETECT INFLUENZA, PARAINFLUENZA, AND MUMPS VIRUSES—CONT'D

buffered saline (PBS) by diluting with distilled water. Filter (0.22 μm) sterilize. Store at 2° to 8° C.

3. Viral culture maintenance medium with antibiotics
4. Positive control virus (parainfluenza virus) in tube of current lot of PMK cells
5. Negative control is uninoculated tube of current lot of PMK cells.
6. Sterile, conical centrifuge tubes
7. Sterile pipettes
8. Repeater Eppendorf pipettes
9. Centrifuge

PROCEDURE

1. Preparation of guinea pig RBCs
   a. Transfer guinea pig cells to 15-mL conical centrifuge tube. Centrifuge for 5 minutes at 500× *g* (2° to 8° C). Aspirate supernatant and buffy coat.
   b. Resuspend RBCs in cold (2° to 8° C) PBS. Wash in cold PBS until supernate is clear (two to three times).
   c. Aspirate PBS after last wash. Measure remaining volume of cells using graduations on centrifuge tube. Add a sufficient amount of PBS to make a 10% cell solution.
   d. Hemadsorption test is performed using a 0.4% guinea pig cell solution prepared with maintenance medium. Depending on the number of tubes to be tested (0.2 mL/tube), make as follows:
   0.2 mL 10% RBCs + 4.8 mL medium (25 tubes)

or

0.4 mL 10% RBCs + 9.6 mL medium (50 tubes)

2. Hemadsorption
   a. Wear latex gloves and laboratory coat during procedure. Perform in biological safety cabinet.
   b. Aspirate maintenance medium from all patient and control tubes to be tested.
   c. Add 0.2 mL of the 0.4% RBC suspension to each tube. Be careful not to cross-contaminate tubes. Positive and negative controls should be tested last.
   d. Gently tilt stationary rack from side to side to ensure RBC contact with cell monolayer. Place stationary rack in refrigerator (2° to 8° C) for 30 minutes. Tubes should be placed horizontally so that the erythrocyte suspension covers the monolayer.
   e. Invert tubes quickly, following incubation, to dislodge RBCs lying on the cell sheet. Examine cell culture tubes microscopically with 4× objective for RBCs that adhere to the monolayer (see Figure 51-24, *G*). Culture fluids should be inspected for hemagglutination.

INTERPRETATION

1. Negative: No hemadsorption of guinea pig cells and no hemagglutination of cells in fluid medium. RBCs float freely in medium. Positive: Hemadsorption of guinea pig cells to infected PMK cells or hemagglutina-

tion of RBCs in fluid medium. Hemadsorption is graded on a scale of 1 to 4+ as follows:
   1+ = <25% of cell sheet showing hemadsorption
   2+ = 25% to 50% of cell sheet showing hemadsorption
   3+ = 51% to 75% of cell sheet showing hemadsorption
   4+ = 76% to 100% of cell sheet showing hemadsorption

2. Viruses that both hemadsorb and hemagglutinate guinea pig erythrocytes are usually influenza virus. Mumps and parainfluenza viruses generally produce only hemadsorption. Identification of virus causing hemadsorption should be confirmed using virus-specific antibody conjugates. Certain simian viruses that may contaminate the PMK cells hemadsorb at 4° C (see Quality Control). Examine all tubes as soon as possible after removal from the refrigerator because the neuraminidase of the myxoviruses is active at room temperature and will destroy hemagglutinins if tubes are left for extended periods at this temperature.

QUALITY CONTROL

1. Positive control: Parainfluenza virus should exhibit ≥2+ hemadsorption.
2. Negative control: Uninoculated PMK cells should exhibit no hemadsorption. If hemadsorption attributed to an endogenous simian virus occurs, the patient must exhibit more hemadsorption than the negative control to be considered positive.

---

with immune serum for some hemorrhagic fevers, such as Lassa fever, has reduced mortality.

ERADICATION

Global eradication of a viral disease has only occurred with smallpox. The eradication of any viral disease requires no animal reservoir, lack of recurrent infectivity, one or few stable serotypes, and an effective vaccine. Viral diseases currently considered candidates for eradication include measles and poliomyelitis.

Acknowledgment

The author acknowledges the extensive contributions made by those with whom he has worked at both Evanston Northwestern Healthcare in Evanston, Illinois, and Children's Hospital Medical Center of Akron in Akron, Ohio. In particular, the recent help and knowledge gained from Madeline Mainzer and Susan Hart are appreciated.

Procedure 51-6

PRESERVATION AND STORAGE OF VIRUSES BY FREEZING

PRINCIPLE

Viral isolates should be saved for use as quality control strains or for epidemiologic investigation. Maximal recovery of virus is accomplished by collecting cells and fluid from infected culture tubes.

MATERIALS

1. Sterile glass beads
2. Dimethyl sulfoxide (DMSO)
3. Dry ice and acetone
4. Freezer tubes

METHODS

1. Viruses other than CMV and VZV
   a. Select tubes showing ≥2+ CPE or hemadsorption. RBCs may be eluted following hemadsorption by placing tubes in a 37° C water bath for 15 minutes. Wash to remove RBCs. In all steps, open cell culture tubes in biological safety cabinet.
   b. Add three sterile glass beads to cell culture tube. Tighten cap. Vortex vigorously behind shield for 30 seconds. Examine microscopically to ensure that cells have been disrupted and removed from glass surface.
   c. Add cells and medium to freezer tube. Snap freeze by swirling the tube in a slurry of dry ice and acetone.
   d. Store at –70° C.

2. CMV and VZV: Use tubes showing ≥3+ CPE. Harvest as above except add 10% DMSO and glass beads. For example, add 0.2 mL of DMSO to 2 mL of medium in cell culture tube. Mix and proceed as described above.
3. Thaw in a 37° C water bath. Inoculate to appropriate cell culture.

Case Study

A previously healthy 19-year-old college student presented to the hospital emergency department with symptoms of respiratory tract infection. The patient experienced severe shortness of breath, was transferred to the intensive care unit, and was intubated and placed on a ventilator. A chest radiograph revealed left lung infiltrates.

QUESTIONS

1. What viruses cause serious respiratory tract disease in young adults?
2. What specimens are recommended to help diagnose viral respiratory tract infection?
3. What viral diagnostic tests would be appropriate for the detection of respiratory tract viruses?

ADDITIONAL READING

Balfour H: Drug therapy, *N Engl J Med* 340:1255, 1999.

Carr J, Gyorfi T: Human papillomavirus, *Clin Lab Med* 20:235, 2000.

Cockerill FR: Application of rapid-cycle real-time polymerase chain reaction for diagnostic testing in the clinical microbiology laboratory, *Arch Pathol Lab Med* 127:1112, 2003.

Constantine N, Zhao R: Molecular-based laboratory testing and monitoring for human immunodeficiency virus infections, *Clin Lab Sci* 18:263, 2005.

Debiasi RL, Tyler KL: Molecular methods for diagnosis of viral encephalitis, *Clin Microbiol Rev* 17:903, 2004.

De Clercq E: Antiviral drugs in current clinical use, *J Clin Virol* 30:115, 2004.

Espy MJ, Uhl JR, Sloan M, et al: Real-time PCR in clinical microbiology: applications for routine laboratory testing, *Clin Microbiol Rev* 19:165, 2006.

Fields BN, Howley PN, Griffin DN, et al, editors: *Virology,* ed 3, Philadelphia, 2001, Lippincott Williams and Wilkins.

Forbes BA: Introducing a molecular test into the clinical microbiology laboratory: development, evaluation, and validation, *Arch Pathol Lab Med* 127:1106, 2003.

Gavin PJ, Thomson RB: Review of rapid diagnostic tests fro influenza, *Clin Appl Immunol Rev* 4:151, 2003.

Harris KR, Dighe AS: Laboratory testing for viral hepatitis, *Am J Clin Pathol* 118(Suppl 1):S18, 2002.

Hodinka RL: What clinicians need to know about antiviral drugs and viral resistance, *Infect Dis Clin North Am* 11:945, 1997.

Isenberg HD, editor: *Clinical microbiology procedures handbook,* ed 2, Washington, DC, 2004, American Society for Microbiology.

Johnson FB: Transport of viral specimens, *Clin Microbiol Rev* 3:120, 1990.

Lauer GM, Walker BD: Hepatitis C virus infection, *N Engl J Med* 345:41, 2001.

Lee W: Hepatitis B virus infection, *N Engl J Med* 337:1733, 1997.

Lennette EH, Schmidt NJ, editors: *Diagnostic procedures for viral, rickettsial, and chlamydial infections,* ed 5, Washington, DC, 1979, American Public Health Association.

Lesprit P, Scieux C, Lemann M, et al: Use of the cytomegalovirus (CMV) antigenemia assay for the rapid diagnosis of primary CMV infection in hospitalized adults, *Clin Infect Dis* 26:646, 1998.

Levine AJ: *Viruses,* New York, 1991, Scientific American Library.

Liang TJ, Rehermann B, Seeff L, et al: Pathogenesis, natural history, treatment, and prevention of hepatitis C, *Ann Intern Med* 132:296, 2000.

McIntosh K, McAdam AJ: Human metapneumovirus—an important new respiratory virus, *N Engl J Med* 350:431, 2004.

Miller S: Diagnostic virology by electron microscopy, *ASM News* 54:475, 1988.

Minnich LL, Smith TF, Ray CG: *Cumitech* 24: *Rapid detection of viruses by immunofluorescence,* Washington, DC, 1988, American Society for Microbiology.

Murray PR, Baron EJ, Jorgensen JH, Pfaller MA, et al, editors: *Manual of clinical microbiology,* ed 8, Washington, DC, 2003, American Society for Microbiology.

Murray PR, Pfaller MA, Rosenthal KS: *Medical microbiology,* ed 5, St Louis, 2005, Mosby Elsevier.

Niesters HG: Molecular and diagnostic clinical virology in real-time, *Clin Microbiol Infect* 10:5, 2004.

Nolte F: Impact of viral load testing on patient care, *Arch Pathol Lab Med* 123:1011, 1999.

Petersen LR, Marfin AA: West Nile virus: a primer for the clinician, *Ann Intern Med* 137:173, 2002.

Pigott DC: Hemorrhagic fever viruses, *Crit Care Clin* 21:765, 2005.

Poon LL, Guan Y, Nicholls JM, et al: The aetiology, origins, and diagnosis of severe acute respiratory syndrome, *Lancet Infect Dis* 4:663, 2004.

Schiffman M, Castle PE: Human papillomavirus epidemiology and public health, *Arch Pathol Lab Med* 127:930, 2003.

Schmaljohn C and Hjelle B: Hantaviruses: a global disease problem, *Emerg Infect Dis* 3:95, 1997.

Sejvar JJ, Chowdary Y, Schomogyi M, et al: Human monkey-pox infections: a family cluster in the Midwestern United States, *J Infect Dis* 190:1833, 2004.

Sherlock CH, Brandt CJ, Middleton PJ, et al: *Cumitech* 26: *Laboratory diagnosis of viral infections producing enteritis,* Washington, DC, 1989, American Society for Microbiology.

Storch GA: Diagnostic virology, *Clin Infect Dis* 31:739, 2000.

Thomson RB, Bertram H: Laboratory diagnosis of central nervous system infections, *Infect Dis Clin North Am* 15:1047, 2001.

Van Helvoort T: When did virology start? *ASM News* 62:142, 1996.

Warrell MJ, Warrell DA: Rabies and other lyssavirus diseases, *Lancet* 363:959, 2004.

Wilder-Smith A, Schwartz E: Dengue in travelers, *N Engl J Med* 353:924, 2005.

Writing Committee of the World Health Organization Consultation on Human Influenza A/H5: Avian influenza A (H5N1) infection in humans, *N Engl J Med* 353:1374, 2005.

PART VII

Diagnosis by Organ System

Microorganisms present in the circulating blood, whether continuously, intermittently, or transiently, are a threat to every organ in the body. Microbial invasion of the bloodstream can have serious immediate consequences, including shock, multiple organ failure, disseminated intravascular coagulation (DIC), and death. Approximately 200,000 cases of bacteremia and fungemia occur annually, with mortality rates ranging from 20% to 50%. Thus, invasion of the bloodstream by microorganisms constitutes one of the most serious situations in infectious disease, and as a result, timely detection and identification of blood-borne pathogens is one of the most important functions of the microbiology laboratory. Pathogens of all four major groups of microbes—bacteria, fungi, viruses, and parasites—may be found circulating in blood during the course of many diseases. Positive blood cultures may help provide a clinical diagnosis, as well as a specific etiologic diagnosis.

GENERAL CONSIDERATIONS

The successful recovery of microorganisms from blood by the laboratory is dependent on many, often complex, factors: the possible types of bacteremia (presence of bacteria in blood), specimen collection methods, blood volumes, the number and timing of blood cultures, interpretation of results, and the type of patient population being served by the laboratory. If each of these factors or issues is not addressed by the laboratory in the development of their blood culture protocols, the detection and recovery of microorganisms may be severely compromised.

CAUSES

As previously mentioned, all major groups of microbes can be present in the bloodstream during the course of many diseases.

Bacteria

The organisms most commonly isolated from blood are gram-positive cocci, including coagulase-negative staphylococci, *Staphylococcus aureus,* and *Enterococcus* spp., and other organisms likely to be inhabitants of the hospital environment that colonize the skin, oropharyngeal area, and gastrointestinal tract of patients. Some of the most common, clinically significant bacteria isolated from blood cultures are listed in Box 52-1. In general, the number of fungi and coagulase-negative staphylococci has increased while the number of clinically significant anaerobic isolates has decreased over the past decade.[6]

Of importance, the laboratory isolation of certain bacterial species from blood can indicate the presence of an underlying, occult, or undiagnosed neoplasm.[4] Alterations in local conditions at the site of the neoplasm that allow bacteria to proliferate and seed the bloodstream have been suggested as a possible mechanism for this association between bacteremia and cancer. Reduced killing of bacterial cells by the host phagocytes is a second possible mechanism. Organisms associated with neoplastic disease include *Clostridium septicum* and other uncommonly isolated clostridial species, *Streptococcus bovis, Aeromonas hydrophila, Plesiomonas shigelloides,* and *Campylobacter* spp. Finally, if *Streptococcus anginosis* group bacteria are isolated from blood, the possibility of an abscess should be considered.

Fungi

Fungemia (presence of fungi in blood) is usually a serious condition, occurring primarily in immunosuppressed patients and in those with serious or terminal illness. *Candida albicans* is by far the most common species, but *Malassezia furfur* can often be isolated in patients, particularly neonates, receiving lipid-supplemented parenteral nutrition. *Candida* spp. account for approximately 8% to 10% of all nosocomial bloodstream infections.[15]

Except for *Histoplasma*, which multiply in leukocytes (white blood cells), fungi do not invade blood cells, but their presence in the blood usually indicates a focus of infection elsewhere in the body. Fungi in the bloodstream can **disseminate** (be carried) to all organs of the host, where they may grow, invade normal tissue, and produce toxic products. Fungi gain entrance to the circulatory system via loss of integrity of the gastrointestinal or other mucosa; through damaged skin; from primary sites of infection, such as the lung or other organs; or by means of intravascular catheters.

Systemic fungal infections that begin as pneumonia disseminate from the lungs, which serve as the portal of entry. Arthroconidia of *Coccidioides immitis* and microconidia of *Histoplasma capsulatum* and *Blastomyces*

BOX 52-1	**Organisms Commonly Isolated from Blood Cultures**

Staphylococcus aureus
Escherichia coli
Coagulase-negative staphylococci
Enterococcus spp.
Candida albicans
Pseudomonas aeruginosa
Klebsiella pneumoniae
Viridans streptococci
Streptococcus pneumoniae
Enterobacter cloacae
Proteus spp.
Beta-hemolytic streptococci
Anaerobic bacteria—*Bacteroides* and *Clostridium* spp.

---

dermatitidis are ingested by alveolar macrophages in the lung. These macrophages carry the fungi to nearby lymph nodes, usually the hilar nodes. The fungi multiply within the node tissue and ultimately are released into the circulating blood, from which they go on to seed other organs or are destroyed by the body's defenses. The large size and sterol-containing cell walls of molds make them particularly insensitive to the primary host defenses, that is, antibody and phagocytic cells.

Parasites

Eukaryotic parasites may be found transiently in the bloodstream as they migrate to other tissues or organs. Their presence, however, cannot be considered consistent with a state of good health. For example, tachyzoites of the parasite *Toxoplasma gondii* may be found in circulating blood. They invade cells within lymph nodes and other organs, including the lungs, liver, heart, brain, and eyes. The resulting cellular destruction accounts for the manifestations of toxoplasmosis. Also, microfilariae are seen in peripheral blood during infection with *Dipetalonema, Mansonella, Loa loa, Wuchereria,* or *Brugia.*

Malarial parasites invade host erythrocytes and hepatic parenchymal cells. The significant anemia and subsequent tissue hypoxia (reduction in oxygen levels) may result from destruction of red blood cells by the parasite. Vascular trapping of normal erythrocytes by the infected red cells, which are less flexible and tend to clog small capillaries, is a major cause of morbidity. The host's immunologic response is to remove the parasites and damaged red blood cells; the immune response may also have deleterious effects.

Parasites in the bloodstream are usually detected by direct visualization. Those parasites for which diagnosis is dependent on observation of the organism in peripheral blood smears include *Plasmodium, Trypanosoma,* and *Babesia.* Patients with malaria or filariasis may

display a periodicity in their episodes of fever that allows the physician to time the collection of blood for microscopic examination for optimal detection.

Viruses

Although many viruses do circulate in the peripheral blood at some stage of disease, the primary pathology relates to infection of the target organ or cells. Those viruses that preferentially infect blood cells are Epstein-Barr virus (invades lymphocytes), cytomegalovirus (invades monocytes, polymorphonuclear cells, and lymphocytes), and human immunodeficiency virus (HIV) (involves only certain T lymphocytes and perhaps macrophages) and other human retroviruses that attack lymphocytes. The pathogenesis of viral diseases of the blood is the same as that of viral diseases of any organ; by diverting the cellular machinery to create new viral components or by other means, the virus may prevent the host cell from performing its normal function. The cell may be destroyed or damaged by viral replication, and immunologic responses of the host may also contribute to the pathogenesis.

Although many viral diseases have a viremic stage, recovery of virus particles or detection of circulating viruses is used in the diagnosis of only a few diseases. Chapter 51 discusses recovery of viruses from blood in greater detail.

TYPES OF BACTEREMIA

Bacteremia may be **transient, continuous,** or **intermittent.** Transient, incidental bacteremia may occur spontaneously or with such minor events as brushing teeth or chewing food. Other conditions in which bacteria are only transiently present in the bloodstream include manipulation of infected tissues, instrumentation of contaminated mucosal surfaces, and surgery involving nonsterile sites. These circumstances may also lead to significant septicemia.

In septic shock, bacterial endocarditis, and other endovascular infections, organisms are released into the bloodstream at a fairly constant rate (continuous bacteremia). Also, during the early stages of specific infections, including typhoid fever, brucellosis, and leptospirosis, bacteria are continuously present in the bloodstream.

In most other infections, such as in patients with undrained abscesses, bacteria can be found intermittently in the bloodstream. Of note, the causative agents of meningitis, pneumonia, pyogenic arthritis, and osteomyelitis are often recovered from blood during the early course of these diseases. In the case of transient seeding of the blood from a sequestered focus of infection, such as an abscess, bacteria are released into the blood approximately 45 minutes before a febrile episode.

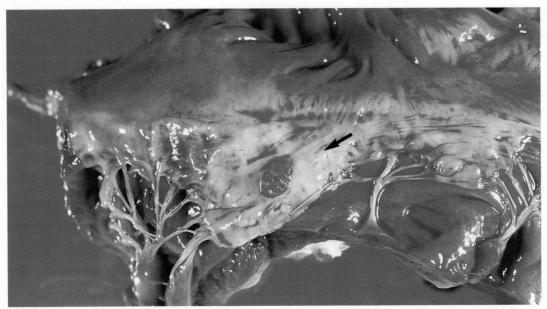

Figure 52-1   Vegetations of bacterial endocarditis. Arrow indicates the vegetations. (Courtesy Celeste N. Powers, MD, PhD, Virginia Commonwealth University Medical Center, Medical College of Virginia Campus, Richmond, Va.)

TYPES OF BLOODSTREAM INFECTIONS

The two major categories of bloodstream infections are **intravascular** (those that originate within the cardiovascular system) and **extravascular** (those that result from bacteria entering the blood circulation through the lymphatic system from another site of infection). Of note, other organisms, such as fungi, may also cause intravascular or extravascular infections. However, because bacteria account for the majority of significant vascular infections, these types of bloodstream infections are addressed. Factors that contribute to the initiation of bloodstream infections are immunosuppressive agents, widespread use of broad-spectrum antibiotics that suppress the normal flora and allow the emergence of resistant strains of bacteria, invasive procedures that allow bacteria access to the interior of the host, more extensive surgical procedures, and prolonged survival of debilitated and seriously ill patients.

Intravascular Infections

Intravascular infections include infective endocarditis, mycotic aneurysm, suppurative thrombophlebitis, and intravenous (IV), catheter-associated bacteremia. Because these infections are within the vascular system, organisms are present in the bloodstream at a fairly constant rate (i.e., a continuous bacteremia). These infections in the cardiovascular system are extremely serious and are considered life-threatening.

Infective Endocarditis. The development of **infective endocarditis** (infection of the endocardium most commonly caused by bacteria) is believed to involve several independent events. Cardiac abnormalities, such as congenital valvular diseases that lead to turbulence in blood flow or direct trauma from IV catheters, can damage cardiac endothelium. This damage to the endothelial surface results in the deposition of platelets and fibrin. If bacteria transiently gain access to the bloodstream (this can occur after an innocuous procedure such as brushing the teeth) after alteration of the capillary endothelial cells, the organisms may stick to and then colonize the damaged cardiac endothelial cell surface. After colonization, the surface will rapidly be covered with a protective layer of fibrin and platelets. This protective environment is favorable to further bacterial multiplication. This web of platelets, fibrin, inflammatory cells, and entrapped organisms is called a **vegetation** (Figure 52-1). The resulting vegetations ultimately seed bacteria into the blood at a slow but constant rate.

The primary causes of infective endocarditis are the viridans streptococci, comprising several species (Box 52-2). These organisms are normal inhabitants primarily of the oral cavity, often gaining entrance to the bloodstream as a result of gingivitis, periodontitis, or dental manipulation. Heart valves, especially those that have been previously damaged, present convenient surfaces for attachment of these bacteria. *Streptococcus sanguis* and *Streptococcus mutans* are most frequently isolated in streptococcal endocarditis.

With the ever-increasing use of IV catheters, intra-arterial lines, and vascular prostheses, organisms found

Viridans streptococci*
Nutritionally deficient streptococci (*Abiotrophia* spp. and *Granu-licatella* spp.)
Enterococci*
Streptococcus bovis
*Staphylococcus aureus**
Staphylococci (coagulase-negative)
Enterobacteriaceae
Pseudomonas spp. (usually in drug users)
Haemophilus spp., particularly *H. aphrophilus*
Unusual gram-negative bacilli (e.g., *Actinobacillus, Cardiobacterium, Eikenella, Coxiella burnetii*)
Yeast
Other (including polymicrobial infectious endocarditis)

*Most common organisms associated with native valve endocarditis in non–drug-using adults.

as normal or hospital-acquired inhabitants of the human skin are able to gain access to the bloodstream and find a surface on which to grow, including heart valves and vascular endothelium. In such a setting, *Staphylococcus epidermidis* and other coagulase-negative staphylococci have been increasingly implicated as causes of infection. *S. epidermidis* is the most common etiologic agent of prosthetic valve endocarditis, with *S. aureus* being the second most common. *S. aureus* is an important cause of septicemia without endocarditis and is found in association with other foci, such as abscesses, wound infections, and pneumonia, as well as sepsis related to indwelling intravascular catheters.

Mycotic Aneurysm and Suppurative Thrombophlebitis. Two other intravascular infections, mycotic aneurysms and suppurative thrombophlebitis, result from damage to the endothelial cells lining blood vessels. With respect to **mycotic aneurysm,** an infection causes inflammatory damage and weakening of an arterial wall; this weakening causes a bulging of the arterial wall (i.e., aneurysm) that can eventually rupture. The etiologic agents are similar to those that cause endocarditis.

Suppurative thrombophlebitis is an inflammation of a vein wall. The pathogenesis of this intravascular infection involves an alteration in a vein's endothelial lining that is followed by clot formation. This site is then seeded with organisms, thereby establishing a primary site of infection. Suppurative thrombophlebitis represents a frequent complication of hospitalized patients that is caused by the increasing use of IV catheters.

Intravenous Catheter–Associated Bacteremia. IV catheters are an integral part of the care of hospitalized patients; more than 3 million central venous catheters are used annually in the United States. For example,

central venous catheters are used to administer fluids, blood products, medications, antibiotics, and nutrition, and for hemodynamic monitoring. A short-term, triple-lumen (channel opening within a tube) central venous catheter is shown in Figure 52-2. Unfortunately, one major consequence of these medical devices is colonization of the catheter by either bacteria or fungi, which can lead to catheter infection and serious bloodstream infection. This consequence is a major nosocomial source of illness and even death.

IV catheter–associated bacteremia (or fungemia) is believed to occur primarily by two routes (Figure 52-3).[16] The first route involves the movement of organisms from the catheter skin entry site down the external surface of the catheter to the catheter tip within the bloodstream. After arriving at the tip, the organisms multiply and may cause a bacteremia. The second way that IV catheter–associated bacteremia may occur is by migration of organisms along the inside of the catheter (the lumen) to the catheter tip. The catheter's hub, where tubing connects into the IV catheter, is considered the site at which organisms can gain access to the patient's bloodstream through the catheter lumen. The most common etiologic agents for IV catheter–associated bloodstream infections, regardless of the route of infection, are organisms found on the skin (Box 52-3). Certain strains of *S. epidermidis* appear to be uniquely suited for causing catheter-related infections because of their ability to produce a biofilm or "slime" that consists of complex sugars (polysaccharides) that are believed to help the organism adhere to the catheter's surface. The initial attachment of *S. epidermidis* to the catheter's polystyrene surface is related to a cell surface protein. Once attached, the organism then proliferates, subsequently forming a biofilm. Uncommon routes of IV catheter–tip infection include contaminated fluids or blood-borne seeding from another infection site.

Extravascular Infections

Except for intravascular infections, bacteria usually enter the circulation through the lymphatic system. Most cases of clinically significant bacteremia are a result of extravascular infection. When organisms multiply at a local site of infection such as the lung, they can be drained by the lymphatics and reach the bloodstream. In most individuals, organisms in the bloodstream are effectively and rapidly removed by the reticuloendothelial system in the liver, spleen, and bone marrow and by circulating phagocytic cells. But, depending on the extent of local control of the infection, the organism may be circulated more widely, thereby causing a bacteremia or fungemia.

The most common **portals of entry** for bacteremia are the genitourinary tract (25%), respiratory tract (20%), abscesses (10%), surgical wound infections

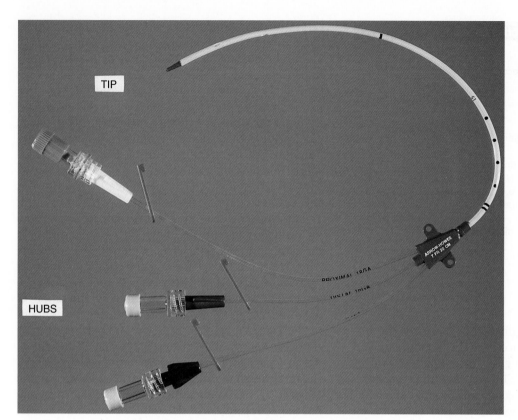

Figure 52-2   Short-term, triple-lumen central venous catheter. The ends from which the catheter is accessed are usually referred to as the hubs. After the catheter is inserted, the tip resides within the bloodstream.

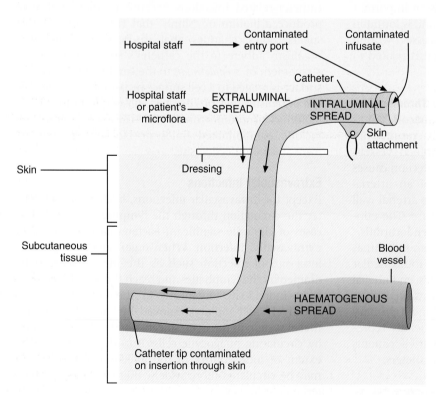

Figure 52-3   Possible routes by which microorganisms gain access to the bloodstream to cause intravenous catheter–associated bacteremias.   (Modified from Elliott TS: PHLS Communicable disease report: line-associated bacteraemias, *CDR Review* 3:R91, 1993.)

BOX 52-3	Common Agents of IV Catheter-Associated Bacteremia

Staphylococcus epidermidis
Other coagulase-negative staphylococci
Staphylococcus aureus
Enterobacteriaceae
Pseudomonas aeruginosa
Candida spp.
Corynebacterium spp.
Other gram-negative rods

Table 52-1   Organisms Commonly Associated with Bloodstream Invasion from Extravascular Sites of Infection

Organism	Extravascular Site of Infection
Haemophilus influenzae type b	Meninges, epiglottis, periorbital region
Streptococcus pneumoniae	Meninges, sometimes the lung
Neisseria meningitidis	Meninges
Brucella spp.	Reticuloendothelial system
Salmonella typhi	Small intestine, regional lymph nodes of the intestine, reticuloendothelial system
Listeria	Meninges

(5%), biliary tract (5%), miscellaneous sites (10%), and uncertain sites (25%). For the most part, the probability of bacteremia occurring from an extravascular site is dependent on the site of infection, its severity, and the organism. For example, any organism producing meningitis is likely to produce bacteremia at the same time. Of importance, certain organisms causing extravascular infections commonly invade the bloodstream; some of these organisms are listed in Table 52-1. In addition to these organisms, a large number of other bacteria and fungi that cause extravascular infections also invade the bloodstream but not as frequently. Whether these organisms invade the bloodstream depends on the host's ability to control the local infection and the organism's pathogenic potential. Some of the organisms that can frequently seed the bloodstream from a localized site of infection, if conditions allow, include members of the family *Enterobacteriaceae, Streptococcus pneumoniae, Staphylococcus aureus, Neisseria gonorrhoeae,* anaerobic cocci, *Bacteroides, Clostridium,* beta-hemolytic streptococci, and *Pseudomonas.* These are only some of the organisms frequently isolated from blood and should not be considered inclusive. Almost every known bacterial species and many fungal species have been implicated in extravascular bloodstream infections.

CLINICAL MANIFESTATIONS

As previously discussed, bacteremia may indicate the presence of a focus of disease, such as intravascular infection, pneumonia, or liver abscess, or it may merely represent transient release of bacteria into the bloodstream. **Septicemia** or sepsis indicates a situation in which bacteria or their products (toxins) are causing harm to the host. Unfortunately, clinicians often use the terms *bacteremia* and *septicemia* interchangeably. Signs and symptoms of septicemia may include fever or hypothermia (low body temperature), chills, hyperventilation (abnormally increased breathing that leads to excess loss of carbon dioxide from the body) and subsequent respiratory alkalosis (condition caused by the loss of acid leading to an increase in pH), skin lesions, change in mental status, and diarrhea. More serious manifestations include hypotension or shock, DIC, and major organ system failure. The syndrome, known as **septic shock,** characterized by fever, acute respiratory distress, shock, renal failure, intravascular coagulation, and tissue destruction, can be initiated by either exotoxins or endotoxins. Septic shock is mediated by activated mononuclear cells producing cytokines, such as tumor necrosis factor and interleukins.

Shock is the gravest complication of septicemia. In septic shock, the presence of bacterial products and the host's responding defensive components act to shut down major host physiologic systems. Manifestations include a drop in blood pressure, increase in heart rate, impairment of function of vital organs (brain, kidney, liver, and lungs), acid-base problems, and bleeding problems. Gram-negative bacteria contain a substance in their cell walls, called **endotoxin,** which has a strong effect on several physiologic functions. This substance, a lipopolysaccharide (LPS) comprising part of the cell wall structure (see Chapter 2), may be released during the normal growth cycles of bacteria or after the destruction of bacteria by host defenses. Endotoxin (or the core of the LPS, lipid A) has been shown to mediate numerous systemic reactions, including a febrile (producing fever) response, and the activation of complement and certain blood-clotting factors. Although most gram-positive bacteria do not contain endotoxin, many produce exotoxins, and the effects of their presence in the bloodstream may be equally devastating to the patient.

Disseminated intravascular coagulation (DIC) is a disastrous complication of sepsis. DIC is characterized by numerous small blood vessels becoming clogged with clotted blood and bleeding as a result of the depletion of coagulation factors. DIC can occur with septicemia involving any circulating pathogen, including parasites, viruses, and fungi, although it is most often a consequence of gram-negative bacterial sepsis.

IMMUNOCOMPROMISED PATIENTS

One of the greatest challenges facing microbiologists is the handling of blood cultures from immunocompromised patients. The number of immunocompromised patients has steadily increased in recent years in large part as the result of advances in medicine. People undergoing organ transplantation, elderly persons, individuals with malignant disease (e.g., malignancies and cancer), and those receiving therapy for the malignancy are examples of immunosuppressed patients. Acquired immunodeficiency syndrome (AIDS) has also contributed to the increase in the number of immunocompromised individuals. The marked immunosuppression brought about by infection with the human immunodeficiency virus (HIV) in patients with AIDS is a result of this virus' profound impairment of cellular immunity. Patients with AIDS have the greatest diversity of pathogens recovered from blood, including mycobacterial species, *Bartonella henselae*, *Corynebacterium jeikeium*, *Shigella flexneri*, unusual *Salmonella* species, *Histoplasma capsulatum*, *Cryptococcus neoformans*, and cytomegalovirus.

As is typically observed in other hospitalized patients, organisms such as gram-positive aerobic bacteria (e.g., *Staphylococcus aureus*, *Enterococcus*) and gram-negative aerobic bacteria (e.g., Enterobacteriaceae, *Pseudomonas aeruginosa*) are common causes of bloodstream infections in immunocompromised patients. In addition, bloodstream infections in immunocompromised patients are frequently caused by either unusual pathogens whose recovery from blood requires special techniques or by organisms that are normally considered contaminants when isolated from blood cultures. Therefore, microbiologists must be aware of the potential pathogenicity of some organisms in immunosuppressed patients that are generally regarded as probable blood culture contaminants. Without this knowledge, aerobic gram-positive rods isolated from blood cultures may be dismissed as contaminating diphtheroids, when, in fact, the organism is *C. jeikeium*, which is known to cause bacteremia in immunosuppressed patients. By the same token, microbiologists must be familiar with the unusual pathogens isolated from blood cultures obtained from immunocompromised patients and in particular, those organisms that require special techniques for isolation (some of the special considerations are covered later in this chapter).

DETECTION OF BACTEREMIA

Mortality rates associated with bloodstream infection range from 20% to 50%. Because bacteremia frequently portends life-threatening infection, the prompt detection and recovery of microorganisms from blood is of paramount importance.

To detect bloodstream infections, a patient's blood must be obtained by aseptic venipuncture and then incubated in culture media. Bacterial growth can be detected using techniques ranging from manual to totally automated methods. Once growth is detected, the organism is isolated, identified, and tested for its susceptibility to various antimicrobial agents when appropriate.

SPECIMEN COLLECTION

Preparation of the Site

Because blood culture media have been developed as enrichment broths to encourage the multiplication of even one bacterium, it follows that these media will enhance the growth of any stray contaminating bacterium, such as a normal inhabitant of human skin. Therefore, careful skin preparation before collecting the blood sample is of paramount importance to reduce the risk of introducing contaminants into blood culture media.

The vein from which the blood is to be drawn must be chosen before the skin is disinfected. If a patient has an existing IV line, the blood should be drawn below the existing line; blood drawn above the line will be diluted with fluid being infused. It is less desirable to draw blood through a vascular shunt or catheter, because these prosthetic devices are difficult to decontaminate completely.

Antisepsis. Once a vein is selected, the skin site is defatted with 70% isopropyl alcohol and an antiseptic is then applied to kill surface and subsurface bacteria. Regardless of the antiseptic used, it is critical that the manufacturer's recommendation be followed as to length of time the antiseptic is allowed to remain on the skin. Available data indicate that iodine tincture (iodine in alcohol) and chlorhexidine are equivalent for skin preparation before drawing blood cultures.[2,3,5] The steps necessary for drawing blood for culture are given in Procedure 52-1 at the end of this chapter. A standardized blood culture collection system is commercially available, BACTEC Blood Culture Procedural Trays (Becton Dickinson Biosciences, Sparks, Md), which contains all components for blood collection (Figure 52-4).

As part of ongoing quality assurance, laboratories should determine the rate of blood culture contamination by clinically evaluating patients' conditions in conjunction with the organism isolated from culture. Laboratories that recover contaminants at rates greater than 3% should suspect improper phlebotomy techniques and should institute measures to reeducate the phlebotomists in proper skin preparation methods.

Precautions. Standard precautions require that phlebotomists wear gloves for blood drawing. Because blood for culture must be obtained aseptically, it is important that any contaminated surfaces that might come in contact with the disinfected venipuncture site be disinfected. For example, if the site must be touched after preparation, the phlebotomist must disinfect the gloved fingers used for palpation in identical fashion. Also, if the rubber stopper or septum of the container into which blood is to be inoculated (e.g., test tubes or commercial culture bottles) is potentially contaminated, the phlebotomist must disinfect the septum.

Specimen Volume

Adults. For many years, it has been recognized that most bacteremias in adults have a low number of colony-forming units (CFU) per milliliter (mL) of blood. For example, in several studies, fewer than 30 CFU per

mL of blood were commonly found in patients with clinically significant bacteremias. Therefore, a sufficient sample volume is critical for the successful detection of bacteremia.

Because there is a direct relationship between the volume of blood and the yield, it follows that the more blood that is cultured, the greater the chance of isolating the organism. Therefore, collection of 10 to 20 mL of blood per culture is strongly recommended for adults. To illustrate, Cockerill and colleagues[6] reported that in patients without infective endocarditis, volumes of 20 mL increased the yield by 30% compared with 10-mL volumes. Unfortunately, a study confirmed that it is common practice to underinoculate blood culture bottles; findings from this study suggested that the yield increases by 3.2% for each milliliter of blood cultured.[12]

Children. It is not safe to take large samples of blood from children, particularly infants; also, the optimal volume of blood that should be obtained from infants and children has not been clearly delineated. Similar to adults, this patient population also has low level bacteremia. In light of low-level bacteremia in infants and children and based on the premise that it is safe to obtain as much as 4% to 4.5% of a patient's known total blood volume for culture and on the known relationship between blood volume and patient weight, recommendations for blood volumes for cultures from infants and children (Table 52-2) have been made by Baron and colleagues.[3] For infants and small children, only 1 to 5 mL of blood can usually be drawn for bacterial culture. Blood culture bottles are available that have been designed specifically for the pediatric patient. Because blood specimens from septic children may yield fewer than 5 CFU/mL of the organism, quantities less than 1 mL may not be adequate to detect pathogens. Nevertheless, smaller volumes should still be cultured because high levels of bacteremia (more than 1000 CFU/mL of blood) are detected in some infants.

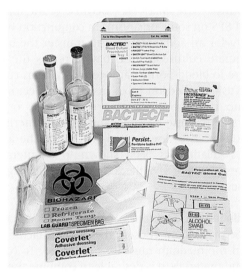

Figure 52-4 Standardized blood collection and blood culturing system (Becton Dickinson). Blood culture procedural tray consists of instructions, gauze pads, alcohol prep pad, latex-free tourniquet, Vacutainer brand standard needle holder and Safety-lok blood collection set, blood culture media, latex-free gloves, and Persist Skin Prep Swab.

Table 52-2   Blood Volumes Suggested for Cultures from Infants and Children

Weight of patient		Recommended volume of blood for culture (mL)				
kg	lb	Total blood volume (mL)	Culture No. 1	Culture No. 2	Total volume for culture (mL)	% of total blood volume
≤1	≤2.2	50-99	2		2	4
1.1-2	2.2-4.4	100-200	2	2	4	4
2.1-12.7	4.5-27	>200	4	2	6	3
12.8-36.3	28-80	>800	10	10	20	2.5
>36.3	>80	>2200	20-30	20-30	40-60	1.8-2.7

From Baron EJ, Weinstein MP, Dunne WM, et al: Blood cultures IV, Baron EJ, coordinating editor, *Cumitech* 1C, Washington, DC, 2005, American Society for Microbiology, reprinted with permission.

Number of Blood Cultures

Because periodicity of microorganisms in the bloodstream may be characteristic for some diseases, continuous for some and random in others, these patterns of bacteremia must be considered in establishing standards for the timing and number of blood cultures. If the volume of blood is adequate, usually two or three blood cultures are sufficient to achieve the optimum blood culture sensitivity. In patients with endocarditis who have not received antibiotics, a single blood culture is positive in 90% to 95% of the cases, whereas a second blood culture establishes the diagnosis in at least 98% of patients, depending on the study. For patients who have received prior antibiotic therapy, three separate blood collections of 16 to 20 mL each, and an additional blood culture or two taken on the second day, if necessary, detects most etiologic agents of endocarditis. This presumes use of a culture system that is adequate for growth of the organism involved, which often entails extending the incubation period. Similarly, for patients without infective endocarditis, 65.1% are detected in the first culture, 80% by the first two cultures, and 95.7% were detected with the first 3 blood cultures.[6]

Timing of Collection

The timing of cultures is not as important as other factors in patients with intravascular infections because organisms are released into the bloodstream at a fairly constant rate. Because the timing of intermittent bacteremia is unpredictable, it is generally accepted that two or three blood cultures be spaced an hour apart. However, a study found no significant difference in the yield between multiple blood cultures obtained simultaneously or those obtained at intervals.[11] The authors concluded that the overall volume of blood cultured was more critical to increasing organism yield than was timing.

When a patient's condition requires that therapy be initiated as rapidly as possible, little time is available to collect multiple blood culture samples over a timed interval. An acceptable compromise is to collect 40 mL of blood at one time, 20 mL from each of two separate venipuncture sites, using two separate needles and syringes before the patient is given antimicrobial therapy. Regardless, blood should be transported immediately to the laboratory and placed into the incubator or instrument as soon as possible. With blood culture instrumentation, a delay beyond 2 hours can delay the detection of positive cultures.

Miscellaneous Matters

Anticoagulation. Blood drawn for culture must not be allowed to clot. If bacteria become entrapped within a clot, their presence may go undetected. Thus, blood drawn for culture may be either inoculated directly into the blood culture broth media or into a sterile blood collection tube containing an anticoagulant for transport to the laboratory for subsequent inoculation. Heparin, ethylenediaminetetraacetic acid (EDTA), and citrate inhibit numerous organisms and are not recommended for use. Sodium polyanethol sulfonate (SPS, Liquoid) in concentrations of 0.025% to 0.03% is the best anticoagulant for blood. As a result, the most commonly used preparation in blood culture media today is 0.025% to 0.05% SPS. In addition to its anticoagulant properties, SPS is also anticomplementary and antiphagocytic, and interferes with the activity of some antimicrobial agents, notably aminoglycosides. SPS, however, may inhibit the growth of a few microorganisms, such as some strains of *Neisseria* spp., *Gardnerella vaginalis, Streptobacillus moniliformis,* and all strains of *Peptostreptococcus anaerobius.* Because of the inhibitory effect of SPS on some organisms in conjunction with the necessity for an additional step to transfer the blood to the ultimate culture bottles that increases the risk of exposure to blood-borne pathogens as well as contamination, using collection tubes instead of direct inoculation into culture bottles may compromise organism recovery. For these reasons, the use of intermediate collection tubes is discouraged.[3] Although the addition of 1.2% gelatin has been shown to counteract this inhibitory action of SPS, the recovery of other organisms decreases.

Dilution. In addition to volume of blood cultured and type of medium chosen, the dilution factor for the blood in the medium must be considered. To conserve space and materials, it is desirable to combine the largest feasible amount of blood from the patient (usually 10 mL) with the smallest amount of medium that will still encourage the growth of bacteria and dilute out or inactivate the antibacterial components of the blood. For this purpose, a 1:5 ratio of blood to unmodified medium has been found to be adequate in conventional blood cultures. All commercial blood culture systems (discussed later in this chapter) specify the appropriate dilution.

Blood Culture Media. The diversity of bacteria that is recovered from blood requires an equally diverse and large number of media to enhance the growth of these bacteria. Basic blood culture media contain a nutrient broth and an anticoagulant. Several different broth formulations are available, including those that can be prepared by the laboratory in-house and those that are commercially prepared. Most blood culture bottles available commercially contain trypticase soy broth, brain-heart infusion broth, supplemented peptone, or thioglycolate broth. More specialized broth bases include Columbia or *Brucella* broth.

Additives. Growth of cell wall–deficient bacteria (e.g., antibiotic-damaged organisms or *Mycoplasma pneumoniae*) may be enhanced by adding osmotic stabilizers, such as sucrose, mannitol, or sorbose, to create a hyperosmotic (hypertonic) medium. Media without osmotic additives are known as **isotonic.** The hypertonic bottles are difficult to inspect visually for evidence of bacterial growth, because red blood cells in the media become partially lysed and no longer settle as sediment to the bottom, giving a muddy appearance to the broth. Hypertonic media should be used only for specific problems, because numerous species of bacteria are inhibited in such media.

The addition of penicillinase to blood culture media for inactivation of penicillin has been largely superseded in recent years by the availability of a resin-containing medium that inactivates most antibiotics nonselectively by adsorbing them to the surface of the resin particles. Resin-containing media may enhance isolation of staphylococci, particularly when patients are receiving bacteriostatic drugs. The BACTEC system (Becton Dickinson Microbiology Systems, Sparks, Md) offers several resin-containing media. In addition to resin-containing media, BacT/ALERT has a blood culture bottle with supplemented BHI broth containing activated charcoal particles that also significantly increases the yield of microorganisms over standard blood culture media.

CULTURE TECHNIQUES

In general, each blood culture set includes a blood culture bottle designated for aerobic recovery and one for anaerobic recovery of bacteria. Because of the decline over the past 15 years in the proportion of positive blood cultures that yield anaerobic bacteria coupled with the increasing pressure for laboratories to be cost effective, some investigators have recommended that laboratories discard this routine practice of processing all blood samples aerobically and anaerobically.[12,13] Rather, it has been proposed that anaerobic cultures should be selectively performed and, in place of the anaerobic blood culture, a second aerobic bottle be included. Because this is a controversial proposal, laboratories must deal with conflicting recommendations as they attempt to provide clinically useful blood culture results. Also, depending on the patient population served by the laboratory, numbers of blood cultures submitted, and personnel and financial resources, the laboratory may have one or more methods available to ensure detection of the broadest range of organisms in the least possible time.

Conventional Blood Cultures

Incubation Conditions. The atmosphere in commercially prepared blood culture bottles is usually at a low oxidation-reduction potential, permitting the growth of most facultative and some anaerobic organisms. To encourage the growth of obligate (strict) aerobes, such as yeast and *Pseudomonas aeruginosa*, transient venting of the bottles with a sterile, cotton-plugged needle may be necessary. Constant agitation of the bottles during the first 24 hours of incubation also enhances the growth of most aerobic bacteria.

Detecting Growth. After 6 to 18 hours of incubation, most bacteria responsible for clinically significant infections are present in numbers large enough to recover by blind subculture or detect by acridine orange stain. In addition to daily visual examination, blind subcultures from conventional bottles after the first 6 to 12 hours of incubation are performed by aseptically removing a few drops of the well-mixed medium and spreading this inoculum onto a chocolate blood agar plate. The plate is incubated in 5% to 10% CO_2 at 35° C for 48 hours. Culture-negative bottles are then reincubated for 5 to 7 days unless the patient's condition requires special consideration, as is discussed later. Growth of anaerobic bacteria can be detected in stationary bottles by visual inspection with such success that blind anaerobic subcultures are not recommended. However, if growth is detected in the anaerobic bottles, subcultures are made and incubated both anaerobically and aerobically. After 48 hours of incubation, a second blind subculture or acridine orange stain may be performed.

Self-Contained Subculture System

A recent modification of the biphasic blood culture medium is the BD Septi-Chek system (Becton Dickinson Microbiology Systems, Sparks, Md) (Figure 52-5) consisting of a conventional blood culture broth bottle with an attached chamber containing a slide coated with agar or several types of agars. Special media for isolation of fungi and mycobacteria are also available. To subculture, the entire broth contents are allowed to contact the agar surface by inverting the bottle, a simple procedure that does not require opening the bottle or using needles. The large volume of broth that is subcultured and the ease of subculture allow faster detection time for many organisms than is possible with conventional systems. The Septi-Chek system appears to enhance the recovery of *Streptococcus pneumoniae*, but such biphasic systems do not efficiently recover anaerobic isolates.

Lysis Centrifugation

The lysis centrifugation system commercially available is the Isolator (Wampole Laboratories, Cranbury, NJ). The

Figure 52-5   Becton Dickinson Septi-Chek pediatric-size biphasic blood culture bottle. The medium-containing base bottle is inoculated with blood, and the top piece containing agar paddles is added in the laboratory. The agar is inoculated by tipping the bottle to allow the blood-containing medium to flow over the agar.

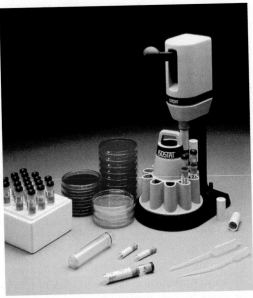

Figure 52-6   Lysis centrifugation blood culture (Isolator System, Wampole Laboratories) uses vacuum-draw collection tubes with a lysing agent and special apparatus (Isostat Press) to facilitate removal of the supernatant without use of needles. (Courtesy Wampole Laboratories, Cranbury, NJ.)

Isolator consists of a stoppered tube containing saponin to lyse blood cells, polypropylene glycol to decrease foaming, SPS as an anticoagulant, EDTA to chelate calcium ions and thus inhibit the complement cascade and coagulation, and a small amount of an inert fluorochemical (Fluorinert, 3M Co., St. Paul, Minn) to cushion and concentrate the microorganisms during 30-minute centrifugation at 3000× *g* (Figure 52-6). After centrifugation, the supernatant is discarded, the sediment containing the pathogen is vigorously vortexed, and the entire sediment is plated to solid agar. Benefits of this system include the more rapid and greater recovery of filamentous fungi, the presence of actual colonies for direct identification and susceptibility testing after initial incubation, the ability to quantify the colony-forming units present in the blood, rapid detection of polymicrobial bacteremia, dispensing with the need for a separate antibiotic-removal step, the ability to choose special media for initial culture setup based on clinical impression (e.g., direct plating onto media supportive of *Legionella* spp. or *Mycobacterium* spp.), and possible greater recovery of intracellular microorganisms caused by lysis of host cells. Possible limitations of the system seem to be a relatively high rate of plate contamination and a decreased ability to detect certain bacteria, such as *Streptococcus pneumoniae*, *Listeria monocytogenes*, *Haemophilus influenzae*, and anaerobic bacteria, compared with conventional systems.

Instrument-Based Systems

Conventional blood culture techniques are labor intensive and time consuming. During these times of cost constraints in health care and a corresponding requirement for clinically relevant care, the development of instrumentation for blood cultures has been accomplished. Instruments can rapidly and accurately detect organisms in blood specimens. By using such instrumentation, laboratories that process a lot of blood cultures can also provide results cost effectively. The decision to purchase a blood culture instrument is difficult and must take into account such matters as volume, patient population, and cost. More than half of all hospital microbiology laboratories use an automated blood culture system.

BACTEC Systems.   Many laboratories use the BACTEC system (Becton Dickinson Microbiology Systems, Sparks, Md), which measures the production of carbon dioxide (CO_2) by metabolizing organisms. Blood or sterile body fluid for routine culture is inoculated into bottles that contain the substrates.

The first BACTEC systems were semiautomated. Vials, containing ^{14}C-labeled substrates (glucose, amino acids, and alcohols) were incubated and often agitated on a rotary shaker. At predetermined time intervals thereafter, the bottles were placed into the monitoring module, where they were automatically moved to a detector. The detector inserted two needles through a rubber septum seal at the top of each bottle and withdrew the gas that had accumulated above the liquid medium and replaced it with fresh gas of the same mixture (aerobic or anaerobic). Any amount of radiolabeled CO_2, the final end product of metabolism of the ^{14}C-labeled substrates (above a preset baseline level), was considered to be suspicious for microbial growth. Microbiologists retrieved suspicious bottles and worked them up for possible microbial growth.

Subsequent modifications further automated the incubation and measuring device, and detection was accomplished by nonradioactive means. The BACTEC NR-860 is fully automated with the incubator, shaker, and detector all in one instrument and measures CO_2 produced by microbial metabolism by infrared spectrophotometry. The BACTEC 9240 and BACTEC 9120 were subsequently introduced. These fully automated blood culture systems use fluorescence to measure CO_2 that is released; a gas-permeable fluorescent sensor is on the bottom of each vial (Figure 52-7). As CO_2 diffuses into the sensor and dissolves in water present in the sensor matrix, hydrogen (H^+) ions are generated. These H^+ ions cause a decrease in pH, which, in turn, increases the fluorescent output of the sensor. These systems differ from all other BACTEC instruments in that there is continuous monitoring of each bottle and detection is external to the bottle. Of importance, the noninvasion of the blood culture bottle eliminates the potential for cross-contamination of cultures during repeated measurements, the need for a separate gas supply, and the use of ^{14}C-labeled substrates.

BacT/ALERT Microbial Detection System.
Other laboratories use the BacT/Alert System (bioMérieux, Durham, NC), which measures CO_2-derived pH changes by a colorimetric sensor in the bottom of each bottle (see Figure 52-7). The sensor is separated from the broth medium by a membrane that is only permeable to CO_2. As organisms grow, they release CO_2, which diffuses across the membrane and is dissolved in water present in the matrix of the sensor. As CO_2 is dissolved, free hydrogen ions are generated. These free hydrogen ions cause a color change in the sensor (blue to light green to yellow as the pH decreases); this color change is read by the instrument.

ESP System.
The ESP Culture System II (Trek Diagnostic Systems, Inc., Cleveland, Ohio) differs from the other previously discussed systems in that microbial growth is detected by the consumption and/or production of gases as organisms metabolize nutrients in the culture medium (see Figure 52-7). The consumption and/or production of gases is detected by monitoring changes in head space pressure by a sensitive detector that is attached to the blood culture bottles. Like the other systems, this is also a continuously monitoring instrument. Table 52-3 summarizes salient characteristics of the continuous monitoring blood culture instruments.

Techniques to Detect IV Catheter–Associated Infections

The insertion of an IV catheter during hospitalization is common practice. Infection, either locally at the catheter insertion site or bacteremia, is one of the most common complications of catheter placement. Because the skin of all patients is colonized with microorganisms that are also common pathogens in catheters, techniques used to diagnose catheter-related infections attempt to quantitate bacterial growth. Diagnosis of an IV catheter–related bacteremia (or fungemia) is difficult, because there are often no signs of infection at the catheter insertion site and the typical signs and symptoms of sepsis can overlap with other clinical manifestations; even the finding of a positive blood culture does not identify the catheter as the source. To date, various methods, such as semiquantitative cultures, Gram stains of the skin entry site, and culture of IV catheter tips following catheter removal, have been described to identify these infections. Many of these methods involve some type of quantitation in an attempt to differentiate colonization of the catheter from probable infection. Two major approaches to the diagnosis of catheter-related infection (CRI) in which the catheter remains in place are based on the premise that a greater number of organisms will be present in the intravascular catheter compared to the number found in blood specimens obtained from distant peripheral veins. The first approach, differential quantitative cultures, involves drawing two blood cultures—one from a peripheral site and the other from the suspected infected line. Quantitative cultures are processed for each specimen in which the same volume of blood is inoculated to standard microbiology media and colonies counted the following day. A colony count ratio greater than 4 to 10:1 between the central venous blood and a peripheral vein blood specimens indicates a probable CRI with a sensitivity of 78% to 94% and a specificity of 99% to 100%. The second approach involves the comparison of the differential time to positivity of blood specimens obtained from a peripheral and intravascular site; a differential time to positivity greater than 2 hours between bottles inoculated with blood from the catheter and those from a peripheral vein indicates a probable CRI.[3] Unfortunately, no single method has demonstrated a clear clinical benefit in diagnosing CRI and remains unsettled.

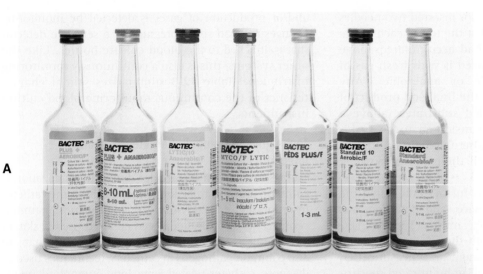

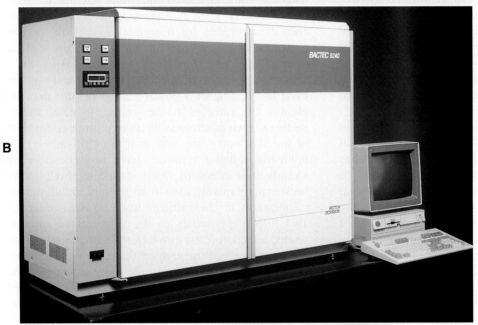

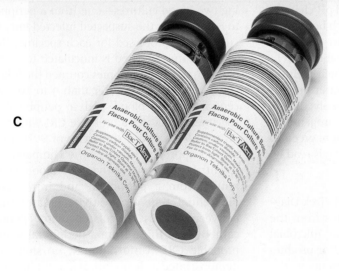

Figure 52-7 **A,** Blood culture bottles for the BACTEC 9240, 9120, and 9050 continuous monitoring instruments. **B,** The BACTEC 9240 continuous monitoring blood culture system. **C,** Blood culture bottles for the BacT/ALERT continuous monitoring blood culture instruments.

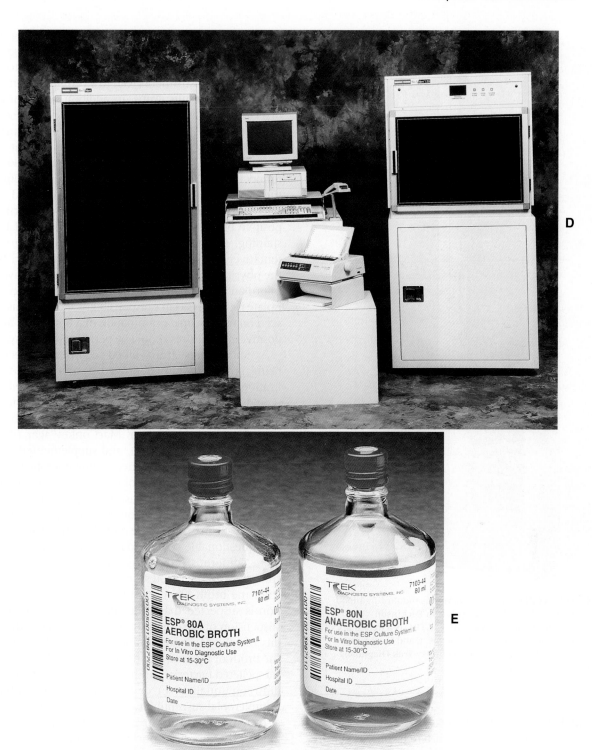

Figure 52-7—cont'd **D,** The BacT/ALERT continuously monitoring blood culture system. **E,** Blood culture bottles for Trek Diagnostic Systems, Inc., ESP Culture System II continuous monitoring instrument.

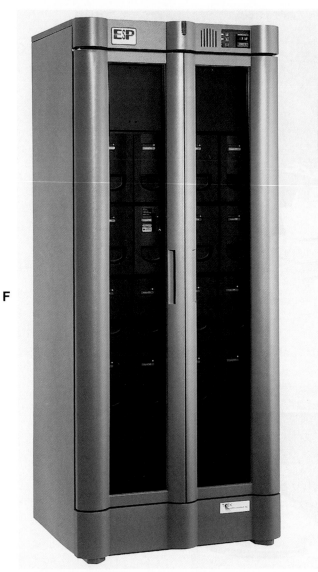

F

Figure 52-7—cont'd   F, ESP continuous monitoring blood culture system. (*A* and *B* courtesy Becton Dickinson Microbiology Systems, Sparks, Md; BACTEC is a trademark of Becton Dickinson Microbiology Systems. *C* and *D* courtesy bioMérieux, Durham, NC. *E* and *F* courtesy Trek Diagnostic Systems, Inc., Cleveland, Ohio.)

Handling Positive Blood Cultures

Most laboratories use a broth-based blood culture method. Bottles should be examined visually at least daily. Growth is usually indicated by hemolysis of the red blood cells, gas bubbles in the medium, turbidity, or the appearance of small aggregates of bacterial or fungal growth in the broth, on the surface of the sedimented red cell layer, or occasionally along the walls of the bottle. When macroscopic evidence of growth is apparent, a gram-stained smear of an air-dried drop of medium should be performed. Methanol fixation of the smear preserves bacterial and cellular morphology,

which may be especially valuable for detecting gram-negative bacteria among red cell debris. Designed to maximize sensitivity, detection algorithms of automated blood culture instruments lead to a certain percentage of false-positive results. Thus, in addition to performance of a Gram stain using methanol fixation, acridine orange (AO) staining is also useful for those blood culture bottles flagged by the instrument as positive but Gram stain-negative for organisms. Adler and colleagues[1] found that AO staining proved particularly helpful in the early detection of candidemia—one third of all microorganisms missed by Gram stain of instrument-positive bottles were yeasts that were detected by AO staining. As soon as a morphologic description can be tentatively assigned to an organism detected in blood, the physician should be contacted and given all available information. Determining the clinical significance of an isolate is the physician's responsibility. If no organisms are seen on microscopic examination of a bottle that appears positive, subcultures should be performed anyway.

Subcultures from blood cultures suspected of being positive, whether proved by microscopic visualization or not, should be made to various media that would support the growth of most bacteria, including anaerobes. Initial subculture may include chocolate agar, 5% sheep blood agar, MacConkey agar (if gram-negative bacteria are seen), and supplemented anaerobic blood agar. The incidence of polymicrobial bacteremia or fungemia ranges from 3% to 20% of all positive blood cultures. For this reason, samples must be resubcultured for isolated colonies. Subculturing positive bottles a second time at the end of the incubation period is of questionable value and is not recommended.

Numerous rapid tests for identification and presumptive antimicrobial susceptibilities can be performed from the broth blood culture if a monomicrobic infection is suspected (based on microscopic evaluation). A suspension of the organism that approximates the turbidity of a 0.5 McFarland standard, obtained directly from the broth or by centrifuging the broth and resuspending the pelleted bacteria, can be used to perform either disk diffusion (qualitative) or broth dilution (quantitative) antimicrobial susceptibility tests. These suspensions may also be used to perform preliminary tests such as coagulase, thermostable nuclease, esculin hydrolysis, bile solubility, antigen detection by fluorescent-antibody stain or agglutination procedures for gram-positive bacteria, oxidase, and commercially available rapid identification kits for gram-negative bacteria. Presumptive results must be verified with conventional procedures using pure cultures. In addition to these approaches, the introduction of a number of molecular methods, including conventional and peptide nucleic acid hybridization assays using specific probes, conventional and real-time polymerase chain reaction assays and micro-

Table 52-3 Summary Characteristics of the More Commonly Used Continuous-Monitoring Blood Culture Systems

System	Bottles Available*	Inoculum Volume (mL)	Blood: Broth	Bottle Capacity/ System Unit	Detection
BacT/ALERT	SA Aerobic	5-10	1:4	240 (max[†] = 2160 bottles)[‡]	Colorimetric detection of CO_2
	SN Anaerobic	5-10	1:4		
	FA, FN (aerobic and anaerobic bottles)	5-10	1:4		
	PF Pediatric	1-4	1:5		
	MB (mycobacteria whole blood)	3-5	~1:5		
	MP (mycobacteria processed specimen or body fluid other than blood)	0.50	—		
BACTEC	Standard aerobic/F	3-10	1:4	240 (max = 1200 bottles)[§]	Fluorescent detection of CO_2
	Standard anaerobic/F	5-7	1:4		
	Plus aerobic/F	3-10	1:2.5		
	Plus anaerobic/F	3-10	1:2.5		
	Peds Plus/F	1-5	1:8		
	Lytic/10 anaerobic/F	3-10	1:4		
	Myco/F Lytic medium (for fungi and mycobacteria)	1-5	1:8		
ESP	REDOX (aerobic)	10	1:9	384 (max = 1920 bottles)[¶]	Detection of O_2 consumption and/or CO_2, H_2, and/or N_2 production
	REDOX (anaerobic)	10	1:9		
	EZ Draw REDOX 1 aerobic	5	1:9		
	EZ Draw REDOX 2 anaerobic	5	1:9		

*No venting required on any bottles listed.

[†]*Max*, Maximum number of bottles that can be processed by assembling multiple system units.

[‡]Smaller instruments available: BacT/ALERT 120 or BACTEC 9120, with a total capacity of 120 bottles; BACTEC 9050, with a total capacity of 50 bottles.

[§]In addition to core computer system, there are the Vision and Epicenter computer systems, which allow for a maximum 4800 and 12,000 bottles, respectively.

[¶]Smaller instrument available: ESP128 with a total capacity of 128 bottles; ESP 256 with a total capacity of 256 bottles.

arrays have been used to directly identify microorganisms in blood culture bottles.

In the event of possible future studies (e.g., additional susceptibility testing), all isolates from blood cultures should be stored for a minimum of 6 months by freezing at −70° C in 10% skim milk. Storing an agar slant of the isolate under sterile mineral oil at room temperature is a good alternative to freezing. It is often necessary to compare separate isolates from the same patient or isolates of the same species from different patients, sometimes even months after the bacteria were isolated.

Interpretation of Blood Culture Results

Because of the increasing incidence of blood/vascular infection caused by bacteria that are normally considered nonvirulent, indigenous microflora of a healthy human host, interpretation of the significance of growth of such bacteria in blood cultures has become increasingly difficult. On one hand, contaminants may lead to unnecessary antibiotic therapy, additional testing and consultation, and increased length of hospital stay. Costs related to false-positive blood culture results (i.e., contaminants) are associated with 40% higher charges for IV antibiotics and microbiology testing. On the other hand, failure to recognize and appropriately treat indi-

genous microflora can have dire consequences. Guidelines that can assist in distinguishing probable pathogens from contaminants are as follows:

- Probable contaminant
  - Growth of *Bacillus* spp., *Corynebacterium* spp., *Propionibacterium acnes,* or coagulase-negative staphylococci in only one of several cultures

NOTE: *Bacillus anthracis* must be ruled out before dismissing *Bacillus* species as a probable contaminant.

  - Growth of multiple organisms from only one of several cultures (polymicrobial bacteremia is uncommon)
  - The clinical presentation and/or course is not consistent with sepsis (physician-based, not laboratory-based criteria)
  - The organism causing the infection at a primary site of infection is not the same as that isolated from the blood culture
- Probable pathogen
  - Growth of the same organism in repeated cultures obtained either at different times or from different anatomic sites
  - Growth of certain organisms in cultures obtained from patients suspected of endocarditis,

such as enterococci, or gram-negative rods in patients with clinical gram-negative sepsis
- Growth of certain organisms such as members of *Enterobacteriaceae*, *Streptococcus pneumoniae*, gram-negative anaerobes, and *Streptococcus pyogenes*
- Isolation of commensal microbial flora from blood cultures obtained from patients suspected to be bacteremic (e.g., immunosuppressed patients or those having prosthetic devices)

SPECIAL CONSIDERATIONS FOR OTHER RELEVANT ORGANISMS ISOLATED FROM BLOOD

The organisms discussed in this section require somewhat different conditions for their successful recovery from blood culture samples. Most of these organisms are infrequently isolated from blood. Therefore, it is important for the physician to notify the laboratory of remarkable patient history, such as travel abroad, that is leading him or her to suspect these agents. In light of recent events and concerns about bioterrorism, it is also important that the laboratory be aware of organisms isolated from blood cultures that are considered potential agents for bioterrorist attacks. These bacteria include *Bacillus anthracis*, *Francisella tularensis*, *Brucella* spp., and *Yersinia pestis*. Finally, in addition to the organisms discussed below with different conditions for their successful isolation from blood, there are a number of organisms that do not grow on artificial media and are best diagnosed by alternative methods such as serology or molecular amplification assays; these organisms are listed in Box 52-4.

HACEK BACTERIA

The term *HACEK* represents a group of fastidious, gram-negative bacilli that include *Haemophilus aphrophilus*, *Actinobacillus actinomycetemcomitans*, *Cardiobacterium hominis*, *Eikenella corrodens*, and *Kingella kingae*. Recovery of these organisms from blood cultures is usually associated with infective endocarditis. Other fastidious organisms, such as *Capnocytophaga* spp., *Rothia dentocariosa*,

Flavobacterium spp., and *Chromobacterium* spp., may also be isolated from blood cultures. In the past, if these less common organisms were suspected by the clinician, the laboratory held the blood cultures for an extended period beyond the first week and made blind subcultures to several enriched media, including more supportive media such as buffered charcoal-yeast extract. However, recent studies using continuous monitoring blood culture systems have showed that almost all bloodstream infections, including endocarditis as well as other bloodstream infections, were detected within 5 days of incubation.

CAMPYLOBACTER AND HELICOBACTER

Several species of *Campylobacter* and *Helicobacter* can be occasionally isolated from blood cultures, usually growing within the 5-day incubation protocol. However, these organisms are small, thin, curved gram-negative rods which may only be visualized using an AO stain following detection by continuous monitoring instruments. Because of the fastidious nature of these organisms, appropriate media and atmospheric conditions for subculture from blood culture bottles must be employed (see Chapter 36).

FUNGI

Even though all disseminated fungal disease is preceded by fungemia, recovery of the microorganisms from blood cultures has not been accomplished until recently. One reason for this may be that cultures were not often collected during the fungemia stage, because clinical symptoms had not yet developed. However, introduction of better methods for isolating fungi from blood, including the lysis centrifugation system, has resulted in greater recovery of fungi from peripheral blood and greater physician awareness to order fungal blood cultures.

Many fungi, particularly yeast, can be recovered in standard blood culture media, if the bottle is incubated at the appropriate temperature and has been vented and agitated to allow sufficient oxygen in the atmosphere for fungal growth. However, some fungi may grow slowly and poorly in these media, which best support bacterial growth. Optimal isolation of fungi in blood cultures is achieved with either agitated incubation of a recently introduced commercial biphasic system, such as the Septi-Chek, or by using the lysis centrifugation system. Manufacturers of media for automated blood culture systems have developed specific media for fungal isolation. These new formulas have dramatically increased the numbers of fungi isolated from patients with fungemia and have shortened the incubation time required for detection of the fungi. Blood specimens for detecting

Table 52-4   Newer Methods Used to Detect Mycobacteria in Blood

System	Type of System	Manufacturer
BACTEC 460/BACTEC 13A vial*	Semiautomated, radiometric	Becton Dickinson
Isolator system†	Manual	Wampole Laboratories
MB/Bact system (BacT/Alert)‡	Continuous monitor, fully automated	bioMérieux
ESP‡	Continuous monitor, fully automated	TREK Diagnostic Systems
Mycobacterial Growth Indicator Tube (MGIT; and manual)	Manual and fully automated system	Becton Dickinson
BACTEC 9000 Myco/F Lytic Medium	Continuous monitor, fully automated	Becton Dickinson

*Media specifically for the culture of mycobacteria from blood specimens. Bottles can be directly inoculated with blood.
†Blood can be held in Isolator tubes for prolonged periods without compromising the recovery of *M. avium* complex.
‡Media specifically for the culture of mycobacteria from blood specimens. Bottles can be directly inoculated with blood.

fungemia are collected in the same manner as for bacterial culture.

MYCOBACTERIA

Patients with HIV infection can have disseminated infection with species of nontuberculous mycobacteria, predominantly *Mycobacterium avium* complex. Isolation of *M. tuberculosis* from the blood of these patients also occurs; as many as 42% of HIV-positive patients with tuberculosis have positive blood cultures.

In the past, the use of special media was recommended, such as Middlebrook 7H9 broth with 0.05% SPS or brain-heart infusion broth with 0.5% polysorbate 80, with or without a Middlebrook 7H11 agar slant. However, newer methods of detection (Table 52-4) have increased sensitivity in their ability to detect mycobacteria present in blood specimens and have significantly shortened the time required for mycobacterial blood cultures to become positive.

BRUCELLA

Brucellosis is a common disease in many developing countries but is uncommon in developed countries. Because brucellosis may be in the differential diagnosis of many infections, microbiologists should be prepared to process blood cultures suspected of having *Brucella;* blood cultures are positive in 70% to 90% of patients with brucellosis. Septicemia occurs primarily during the first 3 weeks of illness. Special handling may be required for recovering *Brucella* spp. from blood because these organisms are fastidious, often slow-growing, intracellular parasites. Best recovery is obtained with *Brucella* or trypticase soy broth. The use of biphasic media may enhance growth, or the Isolator system may allow release of intracellular bacteria. The bottles should be continuously vented and incubated in 10% CO_2 at 37° C

for at least 4 weeks. Blind subcultures should be performed at 4 days and weekly thereafter onto *Brucella* blood agar plates incubated as described. *Brucella* spp. may grow slowly, so cultures must be incubated for a minimum of 3 weeks. Cultures should be handled in a biological safety cabinet.

The use of continuous monitoring systems enhances recovery of *Brucella* spp. and reduces the need for prolonged incubation—for example, the use of the BACTEC 9000 instruments makes possible the diagnosis of more than 95% of positive cultures within a 7-day period without routine subcultures of negative vials.[17]

SPIROCHETES

Borrelia

Visualization in direct preparations is diagnostic for 70% of cases of relapsing fever, a febrile disease caused by *Borrelia recurrentis.* Organisms may be seen in direct wet preparations of a drop of anticoagulated blood diluted in saline as long, thin, unevenly coiled spirochetes that seem to push the red blood cells around as they move. Thick and thin smears of blood, prepared as for malaria testing and stained with Wright's or Giemsa stain, are also sensitive for the detection of *Borrelia.*

Leptospira

Leptospirosis can be diagnosed by isolating the causative spirochete from blood during the first 4 to 7 days of illness. Media with up to 14% (vol/vol) rabbit serum, such as Fletcher's or polysorbate 80 (Tween 80)-albumin are recommended. After adding 1 to 3 drops of fresh or SPS-anticoagulated blood to each of several tubes with 10 mL of culture medium, the cultures are incubated for up to 13 weeks at 28° to 30° C in air in the dark; cultures are examined weekly by dark-field microscopy. Leptospires will grow 1 to 3 cm below the surface, usually within 2 weeks. The organisms remain viable in blood with SPS for 11 days, allowing for transport of specimens from distant locations. Direct dark-field examination of peripheral blood is not recommended because many artifacts are present that may resemble spirochetes. If blood must be shipped to a reference laboratory for culture, blood may be collected into heparin, oxalate, or citrate tubes and maintained at ambient temperature.[3] (Further information about *Borrelia* and *Leptospira* is provided in Chapter 48.)

VITAMIN B₆-DEPENDENT STREPTOCOCCI

Granulicatella spp. and *Abiotrophia* spp. are unable to multiply without the addition of 0.001% pyridoxal hydrochloride (also called thiol or vitamin B_6). These streptococci are known as "nutritionally variant" or "satelliting" streptococci and have been associated with bacteremia and endocarditis. Although human blood introduced into the blood culture medium provides enough of the pyridoxal to allow the organisms to multiply in the bottle, standard sheep blood agar plates may not support their growth. Subculturing the broth to a 5% sheep blood agar plate and either overlaying a streak of *Staphylococcus aureus* or dropping a pyridoxal disk to produce the supplement generally demonstrates colonies of the streptococci growing as tiny satellites next to the streak. Some commercial media may be supplemented with enough pyridoxal (0.001%) to support growth of nutritionally variant streptococci.

MYCOPLASMA HOMINIS

Mycoplasma hominis can be recovered during postabortal or postpartum fever, following gynecologic or urologic procedures, or in patients who were immunocompromised.[7,10,13] Isolates can be recovered from both non-automated and automated blood culture systems. However, because so few clinical isolates have been recovered to date, it has not been determined which blood culture system is optimal for recovering *M. hominis*. Although some studies report that *M. hominis* can produce sufficient CO_2 to be detected by instrumentation, the majority of isolates have been recovered only by subculture; in some cases, 7 days of incubation were required before growth was detected. It should be noted that *M. hominis* should be suspected if there are colonies on subculture yet no organisms seen on Gram stain. Thus, if *M. hominis* bacteremia is suspected, routine blind and terminal subcultures to special media to support the growth of *M. hominis* (e.g., arginine broth) and at least 7 days of incubation are recommended.

BARTONELLA

Based on phenotypic and genotypic characteristics, bacteria of the genus *Rochalimaea* were reclassified into the genus *Bartonella*. *Bartonella* previously contained only a single species, *B. bacilliformis*, the agent of verruga peruana and a septicemic, hemolytic disease known as Oroya fever (see Chapter 35). New species such as *Bartonella henselae* and *B. elizabethae*, as well as *Bartonella quintana*, have been reported to cause bacteremia and endocarditis in both immunocompetent and immunocompromised patients. *B. henselae* has also been linked to cat-scratch disease, a common infectious disease in the United States. Cat-scratch disease is characterized by a persistent necrotizing inflammation of the lymph nodes. For the most part, the most reliable method for diagnosis of *Bartonella* bacteremia is serology.

Because experience in successful primary isolation of *Bartonella* from blood using either broth-based or biphasic blood culture systems is limited to date, use of

---

Procedure 52-1

DRAWING BLOOD FOR CULTURE

PRINCIPLE

Organisms found in circulating blood can be enriched in culture for isolation and further studies. Blood for culture must be obtained aseptically. Once removed from the circulation, unclotted blood must be diluted in growth media.

METHOD

NOTE: Universal precautions require that phlebotomists wear gloves for this procedure.

1. Choose the vein to be drawn by touching the skin before it has been disinfected.

2. Using 70% alcohol, cleanse the skin over the venipuncture site in a circle approximately 5 cm in diameter, rubbing vigorously. Allow to air dry.

3. Starting in the center of the circle, apply 2% tincture of iodine (or povidone-iodine) in ever-widening circles until the entire circle has been saturated with iodine. Allow the iodine to dry on the skin for at least 1 minute. The timing is critical; a watch or timer should be used.

4. If the site must be touched by the phlebotomist after preparation, the phlebotomist must disinfect the gloved fingers used for palpation in identical fashion.

5. Insert the needle into the vein and withdraw blood. Do not change needles before injecting the blood into the culture bottle.[8,9]

6. After the needle has been removed, the site should be cleansed with 70% alcohol again, because many patients are sensitive to iodine.

the Isolator system is recommended. Of importance, use of the Isolator overrides the inhibition of *B. henselae* growth by SPS concentrations present in broth-based systems. Once processed, blood is plated onto enriched (chocolate or blood-containing) media, incubated at 35° to 37° C under elevated CO_2 and humidity. For optimal growth, media should be freshly prepared; some manufacturers will provide media fulfilling these criteria. Plates can be sealed with either Parafilm or Shrink seals after the first 24 hours of incubation and incubated up to 30 days.

Case Study

A college student was admitted to the hospital with fever and chills. He also appeared disoriented and complained of chest pain. Blood cultures were collected on admission, and he was started on cefazolin therapy. His condition worsened and within 12 hours the blood cultures were reported positive for a gram-positive cocci, suggestive of *Staphylococcus* spp. Two hours later the laboratory confirmed the identification of *Staphylococcus aureus*. An echocardiogram was performed that showed multiple vegetations on his tricuspid and aortic valves (see Figure 52-1). Surgery was planned to repair the valves, but the patient died within 48 hours of admission. It was discovered that the student had been injecting cocaine with his friends.

QUESTIONS

1. How did the laboratory identify the organism so quickly?
2. How many blood cultures should be submitted to diagnose bacteremia?
3. If the isolate had been coagulase-negative, what would indicate that this was a true infection with the organism?

REFERENCES

1. Adler H, Baumlin N, Frie R: Evaluation of acridine orange staining as a replacement of subcultures for BacT/ALERT-positive, Gram stain-negative blood cultures, *J Clin Microbiol* 41:5238, 2003.
2. Barenfanger J, Drake C, Lawhorn J, et al: Comparison of chlorhexidine and tincture of iodine for skin antisepsis in preparation for blood sample collection, *J Clin Microbiol* 42:2216, 2004.
3. Baron EJ, Weinstein MP, Dunne WM, et al: Blood cultures IV. Baron EJ, coordinating editor: *Cumitech* 1C, Washington, DC, 2005, American Society for Microbiology.
4. Beebe JL, Koneman EL: Recovery of uncommon bacteria from blood: association with neoplastic disease, *Clin Microbiol Rev* 8:336, 1995.
5. Calfee DP, Farr BM: Comparison of four antiseptic preparations for skin in the prevention of contamination of percutaneously drawn blood cultures: a randomized trial, *J Clin Microbiol* 40:1660, 2002.
6. Cockerill FR, Wilson JW, Vetter EA, et al: Optimal testing parameters for blood cultures, *Clin Infect Dis* 38:1724, 2004.
7. Fernandez-Guerrero ML, Ramos J, and Soriano F: *Mycoplasma hominis* bacteraemia not associated with genital infections, *J Infect* 39:91, 1999.
8. Isaacman DJ, Karasic RB: Lack of effect of changing needles on contamination of blood cultures, *Pediatr Infect Dis J* 9:274, 1990.
9. Krumholz HM, Cummings S, York M: Blood culture phlebotomy: switching needles does not prevent contamination, *Ann Intern Med* 113:290, 1990.
10. Lamey JR, Eschenbach DA, Mitchell SH, et al: Isolation of mycoplasmas and bacteria from the blood of postpartum women, *Am J Obstet Gynecol* 143:104, 1982.
11. Li J, Plorde JL, Carlson LG: Effects of volume and periodicity on blood cultures, *J Clin Microbiol* 32:2829, 1994.
12. Mermel LA, Maki DG: Detection of bacteremia in adults: consequences of culturing an inadequate volume of blood, *Ann Intern Med* 119:270, 1993.
13. Meyer RD, Clough W: Extragenital *Mycoplasma hominis* infections in adults: emphasis on immunosuppression, *Clin Infect Dis* 17(Suppl 1):243, 1993.
14. Morris AJ, Wilson ML, Mirrett S, et al: Rationale for selective use of anaerobic blood cultures, *J Clin Microbiol* 31:2110, 1993.
15. Pfaller MA, Diekema DJ: Twelve years of fluconazole in clinical practice: global trends in species distribution and fluconazole susceptibility of bloodstream isolates of *Candida*, *Eur Soc Clin Microbiol Infect Dis* 10(Suppl 1):11, 2004.
16. Salzman MB, Rubin LG: Intravenous catheter-related infections, *Adv Pediatr Infect Dis* 10:337, 1995.
17. Yagupsky P: Detection of *Brucellae* in blood cultures, *J Clin Microbiol* 37:3437, 1999.

GENERAL CONSIDERATIONS

ANATOMY

The respiratory system can be divided into upper and lower tracts (Figure 53-1). For purposes of study and discussion, the lower respiratory tract comprises structures including the trachea, bronchi, and bronchioles. The respiratory and gastrointestinal tracts are the two major connections between the interior of the body and the outside environment. The respiratory tract is the pathway through which the body acquires fresh oxygen and removes unneeded carbon dioxide. It begins with the nasal and oral passages, which serve to humidify inspired air, and extends past the nasopharynx and oropharynx to the trachea and then into the lungs. The trachea divides into bronchi, which subdivide into bronchioles, the smallest branches that terminate in the alveoli. Some 300 million alveoli are estimated to be present in the lungs; these are the primary, microscopic gas exchange structures of the respiratory tract.

One must be familiar with the anatomic structure of the thoracic cavity, so that specimens collected from various sites in the lower respiratory tract are appropriately processed by the laboratory. The thoracic cavity, which contains the heart and lungs, has three partitions that are separated from one another by pleura (see Figure 53-1). The lungs occupy the right and left pleural cavities while the mediastinum (space between the lungs) is occupied mainly by the esophagus, trachea, large blood vessels, and heart.

PATHOGENESIS OF THE RESPIRATORY TRACT: BASIC CONCEPTS

Microorganisms primarily cause disease by a limited number of pathogenic mechanisms (see Chapter 3). Because these mechanisms relate to respiratory tract infections, they are discussed briefly. Encounters between the human body and microorganisms occur many times each day. However, establishment of infection after such contact tends to be the exception rather than the rule. Whether an organism is successful in establishing an infection is dependent not only on the organism's ability to cause disease (pathogenicity) but also on the human host's ability to prevent the infection.

Host Factors

The human host has several mechanisms that nonspecifically protect the respiratory tract from infection: the nasal hairs, convoluted passages, and the mucous lining of the nasal turbinates; secretory IgA and nonspecific antibacterial substances (lysozyme) in respiratory secretions; the cilia and mucous lining of the trachea; and reflexes such as coughing, sneezing, and swallowing. These mechanisms prevent foreign objects or organisms from entering the bronchi and gaining access to the lungs, which remain sterile in the healthy host. Of note, aspiration of minor amounts of oropharyngeal material, as occurs often during sleep, plays an important role in the pathogenesis of many types of pneumonia. Once particles that have escaped the mucociliary sweeping activity enter the alveoli, alveolar macrophages ingest them and carry them to the lymphatics.

In addition to these nonspecific host defenses, normal flora of the nasopharynx and oropharynx help prevent colonization of the upper respiratory tract. Some of the bacteria that can be isolated as part of the indigenous flora of healthy hosts, as well as many species that may cause disease under certain circumstances but that are often isolated from the respiratory tracts of healthy persons, are listed in Box 53-1. Under certain circumstances and for unknown reasons, these colonizing organisms can cause disease—perhaps because of previous damage by a viral infection, loss of some host immunity, or physical damage to the respiratory epithelium (e.g., from smoking). (Organisms isolated from normally sterile sites in the respiratory tract by methods that avoid contamination with normal flora should be definitively identified and reported to the clinician.)

Microorganism Factors

Organisms possess either certain traits and/or produce certain products that promote colonization and subsequent infection in the host.

Adherence. For any organism to cause disease, it must first gain a foothold within the respiratory tract to grow to sufficient numbers to produce symptoms. Therefore, most etiologic agents of respiratory tract disease must first adhere to the mucosa of the respiratory tract. The presence of normal flora and the overall state of the

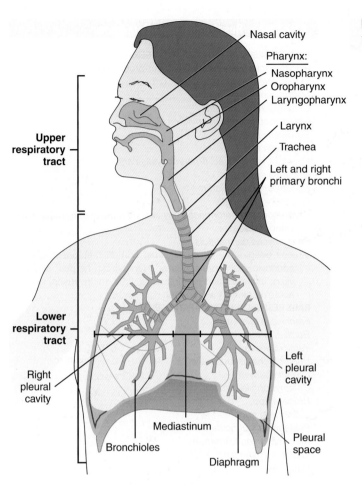

Nasal cavity

Pharynx:
- Nasopharynx
- Oropharynx
- Laryngopharynx

Larynx

Trachea

Left and right primary bronchi

Upper respiratory tract

Lower respiratory tract

Right pleural cavity

Left pleural cavity

Bronchioles

Mediastinum

Diaphragm

Pleural space

Figure 53-1   Anatomy of the respiratory tract, including upper and lower respiratory tract regions.

host affect the ability of microorganisms to adhere. Surviving or growing on host tissue without causing overt harmful effects is called **colonization.** Except for those microorganisms that are breathed directly into the lungs, all etiologic agents of disease must first colonize the respiratory tract to some degree before they can cause harm.

Streptococcus pyogenes possess specific adherence factors and its gram-positive cell wall contains lipoteichoic acids and certain proteins (M protein and others) visible as a thin layer of fuzz surrounding the bacteria. Other bacteria that possess lipoteichoic acid adherence complexes are *Staphylococcus aureus* and certain viridans streptococci. Many gram-negative bacteria (which do not have lipoteichoic acids), including *Enterobacteriaceae, Legionella* spp., *Pseudomonas* spp., *Bordetella pertussis,* and *Haemophilus* spp., adhere by means of proteinaceous fingerlike surface structures called **fimbriae.**

Fimbriae are also called **pili,** although this is technically the term for a similar structure used in sexual interaction rather than just adherence (see Chapter 2). Viruses possess either a hemagglutinin (influenza and parainfluenza viruses) or other proteins that mediate their epithelial attachment.

Toxins.   Certain microorganisms are almost always considered to be etiologic agents of disease if they are present in any numbers in the respiratory tract because they possess virulence factors that are expressed in every host. These organisms are listed in Box 53-2. The production of extracellular toxin was one of the first pathogenic mechanisms discovered among bacteria. *Corynebacterium diphtheriae* is a classic example of a bacterium that produces disease through the action of an extracellular toxin. Once the organism colonizes the upper respiratory epithelium, it produces a toxin that is disseminated systemically, adhering preferentially to central nervous system cells and muscle cells of the heart. Systemic disease is characterized by myocarditis, peripheral neuritis, and local disease that can lead to respiratory distress. Growth of *C. diphtheriae* causes necrosis and sloughing of the epithelial mucosa, producing a "diphtheritic (pseudo) membrane," which may extend from the anterior nasal mucosa to the bronchi, or may be limited to any area between—most often the

---

BOX 53-1 Organisms Present in the Nasopharynx and Oropharynx of Healthy Humans

POSSIBLE PATHOGENS
Acinetobacter spp.
Viridans streptococci, including *Streptococcus anginosus* group
Beta-hemolytic streptococci
Streptococcus pneumoniae
Staphylococcus aureus
Neisseria meningitidis
Mycoplasma spp.
Haemophilus influenzae
Haemophilus parainfluenzae
Moraxella catarrhalis
Candida albicans
Herpes simplex virus
Enterobacteriaceae
Mycobacterium spp.
Pseudomonas spp.
Burkholderia cepacia
Filamentous fungi
Klebsiella ozaenae
Eikenella corrodens
Bacteroides spp.
Peptostreptococcus spp.
Actinomyces spp.
Capnocytophaga spp.
Actinobacillus spp., *A. actinomycetemcomitans*
Haemophilus aphrophilus
Entamoeba gingivalis
Trichomonas tenax
RARELY PATHOGENS
Nonhemolytic streptococci
Staphylococci
Micrococci
Corynebacterium spp.
Coagulase-negative staphylococci
Neisseria spp., other than *N. gonorrhoeae* and *N. meningitidis*
Lactobacillus spp.
Veillonella spp.
Spirochetes
Rothia dentocariosa
Leptotrichia buccalis
Selenomonas
Wolinella
Stomatococcus mucilaginosus
Campylobacter spp.

BOX 53-2 Respiratory Tract Pathogens

DEFINITE RESPIRATORY TRACT PATHOGENS
Corynebacterium diphtheriae (toxin-producing)
Mycobacterium tuberculosis
Mycoplasma pneumoniae
Chlamydia trachomatis
Chlamydophila pneumoniae
Bordetella pertussis
Legionella spp.
Pneumocystis jiroveci (Pneumocystis carinii)
Nocardia spp.
Histoplasma capsulatum
Coccidioides immitis
Cryptococcus neoformans (may also be recovered from patients without disease)
Blastomyces dermatitidis
Viruses (respiratory syncytial virus, human metapneumovirus, adenoviruses, enteroviruses, Hantavirus, herpes simplex virus, influenza and parainfluenza virus, rhinoviruses, severe acute respiratory syndrome)
RARE RESPIRATORY TRACT PATHOGENS
Francisella tularensis
Bacillus anthracis
Yersinia pestis
Burkholderia pseudomallei
Coxiella burnetti
Chlamydophila psittaci
Brucella spp.
Salmonella spp.
Pasteurella multocida
Klebsiella rhinoscleromatis
Varicella-zoster virus (VZV)
Parasites

---

tonsillar and peritonsillar areas. The membrane may cause sore throat and interfere with respiration and swallowing. Although nontoxic strains of *C. diphtheriae* can cause local disease, it is much milder than the version mediated by toxin.

Some strains of *Pseudomonas aeruginosa* produce a toxin similar to diphtheria toxin. Whether this toxin actually contributes to the pathogenesis of respiratory tract infection with *P. aeruginosa* has not been established. *Bordetella pertussis,* the agent of whooping cough, also produces toxins. The role of these toxins in production of disease is not clear. They may act to inhibit the activity of phagocytic cells or to damage cells of the

respiratory tract. *Staphylococcus aureus* and beta-hemolytic streptococci produce extracellular enzymes that act to damage host cells or tissues. Extracellular products of staphylococci aid in production of tissue necrosis and destruction of phagocytic cells, contributing to the commonly seen phenomenon of abscess formation associated with infection caused by this organism. Although *S. aureus* can be recovered from throat specimens, it has not been proved to cause pharyngitis. Enzymes of streptococci, including hyaluronidase, allow rapid dissemination of the bacteria. Many other etiologic agents of respiratory tract infection also produce extracellular enzymes and toxins.

Microorganism Growth. In addition to adherence and toxin production, pathogens cause disease by merely growing in host tissue, interfering with normal tissue function, and attracting host immune effectors, such as neutrophils and macrophages. Once these cells begin to attack the invading pathogens and repair the damaged host tissue, an expanding reaction ensues with more nonspecific and immunologic factors being attracted to the area, increasing the amount of host tissue damage. Respiratory viral infections usually progress in this manner, as do many types of pneumonias, such as those caused by *Streptococcus pneumoniae, S. pyogenes, Staphylococcus aureus, Haemophilus influenzae, Neisseria meningitidis, Moraxella catarrhalis, Mycoplasma pneumoniae, Mycobacterium tuberculosis,* and most gram-negative bacilli.

Avoiding the Host Response. Another virulence mechanism that certain respiratory tract pathogens possess is the ability to evade host defense mechanisms. *S. pneumoniae, N. meningitidis, H. influenzae, Klebsiella pneumoniae,* mucoid *P. aeruginosa, Cryptococcus neoformans,* and others possess polysaccharide capsules that serve both to prevent engulfment by phagocytic host cells and to protect somatic antigens from being exposed to host immunoglobulins. The capsular material is produced in such abundance by certain bacteria, such as pneumococci, that soluble polysaccharide antigen particles can bind host antibodies, blocking them from serving as opsonins. Proof that the capsular polysaccharide is a major virulence mechanism of *H. influenzae, S. pneumoniae,* and *N. meningitidis* was established when vaccines consisting of capsular antigens alone were shown to protect individuals from disease.

Some respiratory pathogens evade the host immune system by multiplying within host cells. *Chlamydia trachomatis, Chlamydophila psittaci (Chlamydia psittaci),* and all viruses replicate within host cells. They have evolved methods for being taken in by the "nonprofessional" phagocytic cells of the host to achieve their required environment. Once within these cells, the organism is protected from host humoral immune factors and other phagocytic cells. This protection lasts until the host cell becomes sufficiently damaged that the organism is then recognized as foreign by the host and is attacked. A second group of organisms that cause respiratory tract disease comprises those that are able to be taken up by phagocytic host cells (usually macrophages). Once within the phagocytic cell, these respiratory tract pathogens are able to multiply. *Legionella, Pneumocystis jiroveci (Pneumocystis carinii),* and *Histoplasma capsulatum* are some of these more common intracellular pathogens.

Mycobacterium tuberculosis is the classic representative of an intracellular pathogen. In primary tuberculosis the organism is carried to an alveolus in a droplet nucleus, a tiny aerosol particle containing a few tubercle bacilli (the minimum infective dose is small). Once phagocytized by alveolar macrophages, organisms are carried to the nearest lymph node, usually in the hilar or other mediastinal chains. In the lymph node, the organisms slowly multiply within macrophages. Ultimately, *M. tuberculosis* destroys the macrophage and is subsequently taken up by other phagocytic cells. Tubercle bacilli multiply to a critical mass within the protected environment of the macrophages, which are prevented from accomplishing lysosomal fusion by the bacteria. Having reached a critical mass, the organisms spill out of the destroyed macrophages, through the lymphatics, and into the bloodstream, producing mycobacteremia and carrying tubercle bacilli to many parts of the body. In most cases, the host immune system reacts sufficiently at this point to kill the bacilli; however, a small reservoir of live bacteria may be left in areas of normally high oxygen concentration, such as the apical (top) portion of the lung. These bacilli are walled off, and years later, an insult to the host, either immunologic or physical, may cause breakdown of the focus of latent tubercle bacilli, allowing active multiplication and disease (secondary tuberculosis). In certain patients with primary immune defects, the initial bacteremia seeds bacteria throughout a host that is unable to control them, leading to disseminated or miliary tuberculosis. Growth of the bacteria within host macrophages and histiocytes in the lung causes an influx of more effector cells, including lymphocytes, neutrophils, and histiocytes, eventually resulting in granuloma formation, then tissue destruction and cavity formation. The lesion is characteristically a semisolid, amorphous tissue mass resembling semisoft cheese, from which it received the name **caseating necrosis** (death of cells or tissues). The infection can extend into bronchioles and bronchi from which bacteria are disseminated via respiratory secretions by coughing. Aerosol droplets produced by coughing and containing organisms are then inhaled by the next victim. Other portions of the patient's own lungs may become infected as well through **aspiration** (inhalation of a fluid or solid).

DISEASES OF THE LOWER RESPIRATORY TRACT

BRONCHITIS

Acute

Acute bronchitis is characterized by acute inflammation of the tracheobronchial tree. This condition may be part of, or preceded by, an upper respiratory tract infection such as influenza (the "flu") or the common cold. Most infections occur during the winter when acute respiratory tract infections are common.

The pathogenesis of acute bronchitis has not been studied for all of the causative agents but appears to be a mixture of viral cytopathic events and the host-related inflammatory response. Regardless of the cause, the protective functions of the bronchial epithelium are disturbed and excessive fluid accumulates in the bronchi. Depending on the etiology, destruction of the bronchial epithelium is either extensive (e.g., influenza virus) or minimal (e.g., rhinovirus colds).

Clinically, bronchitis is characterized by cough, variable fever, and sputum production. Sputum (matter

Table 53-1 Major Causes of Acute Bronchitis

Bacteria	Viruses
Bordetella pertussis, B. parapertussis, Mycoplasma pneumoniae, Chlamydophila pneumoniae	Influenza virus, adenovirus, rhinovirus, coronavirus (other less common viruses: respiratory syncytial virus, human metapneumovirus, coxsackie A21 virus)

BOX 53-3   Viral Agents That Cause Bronchiolitis

Respiratory syncytial virus
Parainfluenza viruses, types 1-3
Rhinoviruses
Adenoviruses
Influenza viruses
Enteroviruses
Human metapneumovirus

ejected from the trachea, bronchi, and lungs through the mouth) is often clear at the onset but may become purulent as the illness persists. Bronchitis may manifest as croup (a clinical condition marked by a barking cough and/or hoarseness).

The value of microbiologic studies to determine the cause of acute bronchitis in otherwise healthy individuals has not been established. Acute bronchitis is usually caused by viral agents, but a key bacteriologic consideration in infants and preschool children is *Bordetella pertussis* (Table 53-1). The best specimen for diagnosis of pertussis is a deep nasopharyngeal swab (see Chapter 39).

Chronic

Chronic bronchitis is a common condition affecting about 10% to 25% of adults. This disease is defined by clinical symptoms in which excessive mucus production leads to coughing up sputum on most days during at least 3 consecutive months for more than 2 successive years.[24] Cigarette smoking, infection, and inhalation of dust or fumes are important contributing factors.

Patients with chronic bronchitis can suffer from acute flare-ups of infection, but determination of the cause of the infection is difficult. Potentially pathogenic bacteria, such as nonencapsulated strains of *Haemophilus influenzae*, *Streptococcus pneumoniae*, and *Moraxella catarrhalis*, are frequently cultured from the bronchi of these patients. Because of chronic colonization, it is difficult to incriminate one of these organisms as the specific cause of an acute infection in patients with chronic bronchitis. Although the role of bacteria in acute infections in these patients is questionable, viruses are frequent causes.

BRONCHIOLITIS

Bronchiolitis, the inflammation of the smaller diameter bronchiolar epithelial surfaces, is an acute viral lower respiratory tract infection that primarily occurs during the first 2 years of life.[9] Characteristic clinical manifestations include an acute onset of wheezing and hyperinflation as well as cough, **rhinorrhea** (runny nose), **tachypnea** (rapid breathing), and respiratory distress. The disease is primarily caused by viruses including a

recently discovered virus, human metapneumovirus.[3] Respiratory syncytial virus (RSV) accounts for 40% to 80% of cases of bronchiolitis as well as demonstrating a marked seasonality; the etiologic agents of bronchiolitis are listed in Box 53-3. Like other viral infections, bronchiolitis shows a marked seasonality in temperate climates with a yearly increase in cases during winter to early spring.[9]

Initially, the virus replicates in the epithelium of the upper respiratory tract but in the young infant it rapidly spreads to the lower tract airways. Early inflammation of the bronchial epithelium progresses to necrosis. Symptoms such as wheezing are believed to be related to the type of inflammatory response to the virus as well as other host factors. For the most part, patients are managed based on clinical parameters, with the laboratory having a role in cases that require hospitalization; a specific viral etiology can be identified in a large number of infants by viral isolation from respiratory secretions, preferably from a nasal wash (see Chapter 51).

PNEUMONIA

Pneumonia (inflammation of the lower respiratory tract involving the lung's airways and supporting structures) is a major cause of illness and death. There are two major categories of pneumonias: those that are considered community-acquired (patients are believed to have acquired their infection outside the hospital setting) and those that include hospital- or ventilator-associated (patients are believed to have acquired their infection within the hospital setting, usually at least 2 days following admission) or health care–associated pneumonia (includes only patients hospitalized in an acute care hospital for 2 or more days within 90 days of infection from a long-term care facility, or received recent intravenous antibiotic therapy, chemotherapy, or wound care within the past 30 days of the current infection, or attended a hospital or hemolysis clinic).[1] Nevertheless, once a microorganism has successfully invaded the lung, disease can follow that includes the alveolar spaces and their supporting structure, the interstitium, and the terminal bronchioles.

Pathogenesis

Organisms can cause infection of the lung by four possible routes: by upper airway colonization or infection that subsequently extends into the lung; by aspiration of organisms (thereby avoiding the upper airway defenses); by inhalation of airborne droplets containing the organism; or by seeding of the lung via the blood from a distant site of infection. Viruses cause primary infections of the respiratory tract, as well as inhibit host defenses that, in turn, can lead to a secondary bacterial infection. For example, viruses may destroy respiratory epithelium and disrupt normal ciliary activity. Presumably, growth of viruses in host cells disrupts the function of the latter and encourages the influx of nonspecific immune effector cells that exacerbate the damage. Damage to host epithelial tissue by virus infection is known to predispose patients to secondary bacterial infection.

Aspiration of oropharyngeal contents, often not overt, is important in the pathogenesis of many types of pneumonia. Aspiration happens most often during a loss of consciousness such as might occur with anesthesia or a seizure, or after alcohol or drug abuse, but other individuals, particularly geriatric patients, may also develop aspiration pneumonia. Neurologic disease or esophageal pathology and periodontal disease or gingivitis are other important risk factors. Aided by gravity and often by loss of some host nonspecific protective mechanisms, organisms reach lung tissue, where they multiply and attract host inflammatory cells. Other mechanisms include inhalation of aerosolized material and hematogenous seeding. The buildup of cell debris and fluid contributes to the loss of lung function and thus to the pathology.

Furthermore, regarding the pathogenesis of hospital-associated, health care–associated, and ventilator-associated pneumonias, health care devices, the environment, and the transfer between the patient and staff or other patients can serve as sources of pathogens that cause pneumonia.[1] Also, the primary routes of bacterial entry into the lower respiratory tract in this category of pneumonia are by aspiration of oropharyngeal organisms or leakage of secretions containing bacteria around the endotracheal tube. For these reasons, intubation and mechanical ventilation significantly increase the risk of pneumonia (6- to 21-fold). Finally, it may be that the infected biofilm in the endotracheal tube with subsequent spread to distal airways may be important in the pathogenesis of ventilator-associated pneumonia.

Clinical Manifestations

The symptoms that suggest pneumonia are fever, chills, chest pain, and cough. In the past, pneumonias were classified into two major groups: (1) typical or acute pneumonias (e.g., *Streptococcus pneumoniae*) and (2) aty-pical pneumonias, based on whether the cough was productive or nonproductive of mucoid sputum. However, analysis of symptoms of pneumonia caused by the atypical pneumonia pathogens (*Mycoplasma pneumoniae, Legionella pneumophila,* and *Chlamydophila pneumoniae*) has revealed no significant differences from those symptoms of patients with typical bacterial pneumonias.[8,18] Because of this overlap in symptoms, many clinicians believe that this distinction is no longer clinically useful.

Some patients with pneumonia exhibit no signs or symptoms related to their respiratory tract (i.e., some only have fever). Therefore, physical examination of the patient, chest radiograph findings, patient history, and clinical laboratory findings are important. In addition to respiratory symptoms, 10% to 30% of patients with pneumonia complain of headache, nausea, vomiting, abdominal pain, diarrhea, and myalgias.

Epidemiology/Etiologic Agents

As previously mentioned, there are two major categories of pneumonias: those that are considered community-acquired and those considered hospital-, ventilator- or healthcare–associated. Because the epidemiology and etiologies can somewhat differ, these two major categories are discussed separately. Because of the many potential etiologic agents, pneumonia in the immuno-compromised patient is addressed separately in this chapter.

Community-Acquired Pneumonia. In the United States, pneumonia is the sixth leading cause of death and the number one cause of death from infectious diseases. It is estimated that as many as 2 to 3 million cases of community-acquired pneumonia occur annually and roughly one fifth of these require hospitalization; 45,000 deaths occur in the United States each year.[2] The etiology of acute pneumonias is strongly dependent on age. More than 80% of pneumonias in infants and children are caused by viruses, whereas less than 10% to 20% of pneumonias in adults are viral.

Children. Community-acquired pneumonia is a common and potentially serious infection that afflicts children. The annual incidence of pneumonia in children younger than 5 years of age is 34 to 40 cases per 1000 in Europe and North America.[18] Determining the cause of pneumonia is challenging because the lung is rarely sampled directly and sputum can rarely be obtained from children. Among previously healthy patients 2 months to 5 years old, RSV, human metapneumovirus, parainfluenza, influenza, and adenoviruses are the most common etiologic agents of lower respiratory tract disease. Children suffer less commonly from bacterial pneumonia, usually caused by *H. influen-*

zae, S. pneumoniae, or *S. aureus.* Neonates may acquire lower respiratory tract infections with *C. trachomatis* or *P. jiroveci* (which likely indicates an immature immune system or an underlying immune defect).

M. pneumoniae and *C. pneumoniae* are the most common causes of bacterial pneumonia in school-age children (5 to 14 years of age). Recently, a study confirmed the high prevalence of viral pneumonia in this age group[30]; the most common viruses, rhinovirus, adenovirus, parainfluenza viruses, influenza virus, and RSV were detected in 45%, 12%, 8%, 7%, and 3% of patients, respectively. Of note, mixed infections were documented in 35% of patients, with the majority of these (81%) being mixed viral-bacterial infections. However, it is not clear what these multiple microbial associations mean.

Young Adults. The most common etiologic agent of lower respiratory tract infection among adults younger than 30 years of age is *Mycoplasma pneumoniae,* which is transmitted via close contact. Contact with secretions seems to be more important than inhalation of aerosols for becoming infected. After contact with respiratory mucosa, *Mycoplasma* organisms are able to adhere to and colonize respiratory mucosal cells. Both a protein adherence factor and gliding motility may be virulence determinants. Once situated in their preferred site between the cilia of respiratory mucosal cells, *Mycoplasma* organisms multiply and somehow destroy ciliary function. Cytotoxins produced by *Mycoplasma* organisms may account for the cell damage they inflict. *Chlamydophila pneumoniae* (originally called *Chlamydia* TWAR or *Chlamydia pneumoniae*) is the third most common agent of lower respiratory tract infection in young adults, after mycoplasmas and influenza viruses; it also affects older individuals.[12,13] Chlamydiae are intracellular pathogens, thus their ability to disrupt cellular function and cause respiratory disease is similar to that of viruses.

Adults. In recent years, the epidemiology and treatment of community-acquired pneumonia has changed.[2,21] From 1979 through 1994, the overall rates of death due to pneumonia and influenza increased by 59%.[21,22] Pneumonia is increasing among older patients and those with underlying diseases such as chronic obstructive lung disease and diabetes mellitus. These patients may become infected with various organisms, including newly identified or previously unrecognized pathogens. Factors such as decreased mucociliary function, decreased cough reflex, decreased level of consciousness, periodontal disease, and decreased general mobility probably contribute to a greater incidence of pneumonia in older patients. Such patients have been found to be more frequently colonized with gram-negative bacilli than younger people, perhaps because of poor oral hygiene, decreased saliva, or decreased epithelial cell turnover.

Although mortality in the outpatient setting remains low (about 1%), mortality rates for community-acquired pneumonia approximate 25% if the patient requires hospitalization.[2,16,17] In addition, the epidemiology and distribution of etiologic agents in patients with severe community-acquired pneumonia who require hospitalization differs somewhat from patients with pneumonia who do not require hospitalization. According to the Infectious Diseases Society of America (IDSA), the decision to hospitalize a patient or to treat them as an outpatient is possibly the single most important clinical decision made by physicians during the course of illness.[2,15] This decision in turn impacts the subsequent site of treatment (home, hospital floor, or hospital intensive care unit), intensity of laboratory evaluation, antibiotic therapy, and cost. Thus, management guidelines for community-acquired pneumonia in adults were developed by the IDSA based on a three-step process: (1) assessment of preexisting conditions that might compromise safety of home care, (2) quantification of short-term mortality (referred to as the pneumonia port severity index [PSI] and based on a prediction rule derived from more than 14,000 patients) with subsequent assignment of patients to five risk classes (classes I through V), and (3) clinical judgment. In general, patients stratified to risk classes IV and V usually require hospitalization. Other guidelines have been set forth that differ from one another.[15] Future guidelines are being investigated in a joint effort by the IDSA and the American Thoracic Society.

Community-acquired pneumonia in adults is most commonly due to bacterial infections. Regardless of age or coexisting illness, *Streptococcus pneumoniae* is most prevalent, causing 15% to 80% of all community-acquired bacterial pneumonia. Additional common etiologies for those patients who are not hospitalized and those who are hospitalized are listed in Tables 53-2 and 53-3, respectively.

Pneumonia secondary to aspiration of gastric or oral secretions is common and occurs in the community setting. The most common agents are primarily the oral anaerobes such as black-pigmented *Prevotella* and *Porphyromonas* spp., *Prevotella oris, P. buccae, P. disiens, Bacteroides gracilis,* fusobacteria, and anaerobic and microaerophilic streptococci. The anaerobic agents possess many factors, such as extracellular enzymes and capsules that may enhance their ability to produce disease. It is their presence, however, in an abnormal site within the host producing lowered oxidation-reduction potential secondary to tissue damage that contributes most to their pathogenicity. *Staphylococcus aureus,* various Enterobacteriaceae, and *Pseudomonas* may also be acquired by aspiration; *Haemophilus influenzae, Legionella* spp.,

Table 53-2 Most Common Etiologies of Community-Acquired Pneumonia in Adults Who Are Not Hospitalized

Age	Coexisting Illness	Most Common Etiologies*
≤60 yrs	No	*M. pneumoniae*, respiratory viruses, *C. pneumoniae*, *H. influenzae*
>60 yrs	Yes	Respiratory viruses, *H. influenzae*, aerobic gram-negative rods, *Staphylococcus aureus*

*Streptococcus pneumoniae is the most common etiology for *all* categories of adult patients with pneumonia.

Table 53-3 Most Common Etiologies of Community-Acquired Pneumonia in Adults Who Are Hospitalized

Generally Requiring Intensive Care Unit	Most Common Etiologies*
No	*H. influenzae;* polymicrobial, aerobic gram-negative bacilli; *Legionella* spp.; *Staphylococcus aureus; C. pneumoniae;* respiratory viruses
Yes	*Legionella* spp., aerobic gram-negative bacilli, *M. pneumoniae*, respiratory viruses

*Streptococcus pneumoniae is the most common etiology for *all* categories of adult patients hospitalized with pneumonia.

Acinetobacter, Moraxella catarrhalis, Chlamydophila pneumoniae, meningococci, and other agents may also be implicated.

Adults also may suffer from viral pneumonia caused by influenza, adenovirus, cytomegalovirus, parainfluenza, varicella, rubeola, or RSV, particularly during epidemics. After viral pneumonia, especially influenza, secondary bacterial disease caused by beta-hemolytic streptococci, pneumococci, *Staphylococcus aureus, Moraxella catarrhalis, Haemophilus influenzae*, and *Chlamydophila pneumoniae* is more likely. Other agents that might be considered depending on geographic location and clinical presentation are viruses in the *Hantavirus* genus, the most common of which is sin nombre virus as well as severe acute respiratory syndrome (SARS) (see Chapter 51).

Unusual causes of acute lower respiratory tract infection in adults include *Actinomyces* and *Nocardia* spp. Other agents may rarely be recovered from sputum and include the agents of plague, tularemia, melioidosis (*Burkholderia pseudomallei*), *Brucella, Salmonella, Coxiella burnetti* (Q fever), *Bacillus anthracis, Pasteurella multocida*, and certain parasitic agents such as *Paragonimus westermani, Entamoeba histolytica, Ascaris lumbricoides*, and

Strongyloides spp. (the latter may cause fatal disease in immunosuppressed patients). A high index of suspicion by the clinician is usually a prerequisite to a diagnosis of parasitic pneumonia in the United States. Psittacosis should be ruled out as a cause of acute lower respiratory tract infection in patients who have had recent contact with birds. Among the fungal etiologies, *Histoplasma capsulatum, Blastomyces dermatitidis, Paracoccidioides brasiliensis, Coccidioides immitis, Cryptococcus neoformans*, and, occasionally, *Aspergillus fumigatus* may cause acute pneumonia. Therefore, occupational history and history of exposure to animals are important in suggesting specific potential infectious agents.

Hospital-, Ventilator-, and Healthcare–Associated Pneumonia. Pneumonia is the leading cause of death among patients with nosocomial infections (as high as 50% mortality among patients in intensive care units).[28] Some of these pneumonias are secondary to sepsis, and some are related to contaminated inhalation therapy equipment.

Nosocomial pneumonia is a risk for any hospitalized patient, particularly for intubated patients. Organisms associated with these infections can be hospital-specific, but overall the most common include *P. aeruginosa, Enterobacter* spp., *Klebsiella* spp., other *Enterobacteriaceae, S. aureus* (with methicillin-resistant strains rapidly emerging), *Acinetobacter* spp., *S. pneumoniae*, anaerobes, *Legionella*, and *H. influenzae*.[1,7] Other agents have been associated with nosocomial outbreaks, including influenza virus. Viruses, such as RSV, adenovirus, and influenza A, are often implicated as causes of nosocomial pneumonia among hospitalized children. The time of onset of hospital- or ventilator-associated pneumonia is an important epidemiologic variable and risk factor: early-onset pneumonia (defined as occurring within the first 4 days of hospitalization) usually carries a better prognosis, being more likely to be caused by antibiotic-sensitive bacteria, whereas late-onset pneumonia (5 days or more) is more likely to be caused by multidrug-resistant organisms and is associated with increased patient morbidity and mortality.[1]

Aspiration pneumonia with infection caused by gram-negative bacilli or staphylococci is probably the major type of hospital-acquired pneumonia, followed by pneumococcal disease. *Legionella* has been implicated in a number of hospital outbreaks, but the problem is typically specific to a given institution.

Chronic Lower Respiratory Tract Infections. *Mycobacterium tuberculosis* is the most likely etiologic agent of chronic lower respiratory tract infection, but fungal infection and anaerobic pleuropulmonary infection may also run a subacute or chronic course. Mycobacteria other than *M. tuberculosis* may also cause such disease, particularly *M. avium-intracellulare* and *M. kansasii.* Although possible causes of acute, community-acquired lower respiratory tract infections, fungi and parasites are more commonly isolated from patients with chronic disease. *Actinomyces* and *Nocardia* may also be associated with gradual onset of symptoms. *Actinomyces* is usually associated with an infection of the pleura or chest wall, and *Nocardia* may be isolated along with an infection caused by *M. tuberculosis.* The pathogenesis of many of the infections caused by agents of chronic lower respiratory tract disease is characterized by the requirement for some breakdown of cell-mediated immunity in the host or the ability of these agents to avoid being destroyed by host cell-mediated immune mechanisms. This may be caused by an effect on macrophages, the ability to mask foreign antigens, sheer size, or some other factor, allowing microbes to grow within host tissues without eliciting an overwhelming local immune reaction.

Cystic fibrosis (CF) is a genetic disorder that leads to persistent bacterial infection in the lung, causing airway wall damage and chronic obstructive lung disease. Eventually, a combination of airway secretions and damage leads to poor gas exchange in the lungs, cardiac malfunction, and subsequent death. Patients with CF may present as young adults with chronic respiratory tract disease, or the more common presentation is in children who may also present with gastrointestinal problems and stunted growth. A very mucoid *Pseudomonas*, characterized by production of copious amounts of extracellular capsular polysaccharide, can be isolated from the sputum of almost all patients with CF older than 18 years of age, becoming more prevalent with increasing age after 5 years. Even if CF has not been diagnosed, isolation of a mucoid *Pseudomonas aeruginosa* from sputum should alert the clinician to the possibility of such an underlying disease. Microbiologists should always report this unusual morphologic feature if it is encountered. In addition to mucoid *Pseudomonas*, patients with CF are likely to harbor *Staphylococcus aureus*, *Haemophilus influenzae*, and *Burkholderia cepacia* complex. RSV, influenza A, nontuberculous mycobacteria, and *Aspergillus* are also important pathogens in this population.

Lung abscess is usually a complication of acute or chronic pneumonia. In these circumstances, organisms infecting the lung cause localized destruction of the lung parenchyma (functional elements of the lung). Symptoms associated with lung abscess are similar to those of acute and chronic pneumonia except symptoms fail to resolve with treatment.

Immunocompromised Patients

Patients with Neoplasms. Patients with cancer are at high risk to become infected because of either gran-ulocytopenia and/or other defects in phagocytic defenses, cellular and/or humoral immune dysfunction, damage to mucosal surfaces and the skin, and various medical procedures such as blood product transfusion. In these patients, the nature of the malignancy often determines the etiology (Table 53-4) and pneumonia is a frequent clinical manifestation.

Transplant Recipients. For successful organ transplantation, the recipient's immune system must be suppressed in some fashion. For this reason, these patients are predisposed to infection. Regardless of the type of organ transplant (heart, renal, bone marrow, heart/lung, liver, pancreas), most infections occur within 4 months following transplantation. Major infections can occur within the first month but are usually associated with infections carried over from the pretransplant period. Pulmonary infections are of great importance in this patient population. Some of the most common causes of pneumonia include *Streptococcus pneumoniae*, *Haemophilus influenzae*, *Pneumocystis jiroveci*, and cytomegalovirus. Of importance, fungi, such as *Cryptococcus neoformans*, *Aspergillus* spp., *Candida* spp., and zygomycetes, can cause life-threatening pulmonary infection.

HIV-Infected Patients. Patients who are infected with human immunodeficiency virus (HIV) are at high risk for developing pneumonia. As discussed in the previous chapter, opportunistic infections as a result of severe immunodeficiency are a major cause of illness and death among these patients. In the United States, the most common opportunistic infection among patients with acquired immunodeficiency syndrome is *Pneumocystis jiroveci* pneumonia. Although *P. jiroveci* remains a major pulmonary pathogen, other organisms must be considered in this patient population, including *Mycobacterium tuberculosis* and *Mycobacterium-avium* complex, as well as common bacterial pathogens such as *Streptococcus pneumoniae* and *Haemophilus influenzae*. In addition to these common pathogens, many other organisms can cause lower respiratory tract infections, including *Nocardia* spp., *Rhodococcus equi* (a gram-positive, aerobic, pleomorphic organism), and *Legionella* spp.

PLEURAL INFECTIONS

As a result of an organism infecting the lung and subsequently gaining access to the pleural space via an abnormal passage (fistula), the patient may develop an empyema (pus in a body cavity such as the pleural cavity). Symptoms in these patients are insidious because early in the course of disease symptoms are related to the primary infection in the lung. Once enough purulent exudate is formed, typical physical and

Table 53-4   Infectious Agents Frequently Associated with Certain Malignancies

Malignancy (site and type of infections)	Pathogens
Acute nonlymphocytic leukemia (pneumonia, oral lesions, cutaneous lesions, urinary tract infections, hepatitis, most often sepsis without obvious focus)	*Enterobacteriaceae* *Pseudomonas* Staphylococci *Corynebacterium jeikeium* *Candida* *Aspergillus* *Mucor* Hepatitis C and other non-A, non-B
Acute lymphocytic leukemia (pneumonia, cutaneous lesions, pharyngitis, disseminated disease)	Streptococci (all types) *Pneumocystis jiroveci (P. carinii)* Herpes simplex virus Cytomegalovirus Varicella zoster virus
Lymphoma (disseminated disease, pneumonia, urinary tract, sepsis, cutaneous lesions)	*Brucella* *Candida* (mucocutaneous) *Cryptococcus neoformans* Herpes simplex virus (cutaneous) Varicella zoster virus Cytomegalovirus *Pneumocystis jiroveci (P. carinii)* *Toxoplasma gondii* *Listeria monocytogenes* Mycobacteria *Nocardia* *Salmonella* Staphylococci *Enterobacteriaceae* *Pseudomonas* *Strongyloides stercoralis*
Multiple myeloma (pneumonia, cutaneous lesions, sepsis)	*Haemophilus influenzae* *Streptococcus pneumoniae* *Neisseria meningitidis* *Enterobacteriaceae* *Pseudomonas* Varicella zoster virus *Candida* *Aspergillus*

radiographic findings indicative of an empyema are produced.

LABORATORY DIAGNOSIS OF LOWER RESPIRATORY TRACT INFECTIONS

SPECIMEN COLLECTION AND TRANSPORT

Although rapid determination of the etiologic agent is of paramount importance in managing pneumonia, the responsible pathogen is not determined in as many as 50% of patients, despite extensive diagnostic testing. Unfortunately, no single test is available that can identify all potential pathogens that cause lower respiratory tract infections. Refer to Table 5-1 for an overview of the collection, transport, and processing of specimens from the lower respiratory tract.

Sputum

Expectorated. The examination of expectorated sputum has been the primary means of determining the causes of bacterial pneumonia. However, lower respiratory tract secretions will be contaminated with upper respiratory tract secretions, especially saliva, unless they are collected using an invasive technique. For this reason, sputum is among the least clinically relevant specimens received for culture in microbiology laboratories, even though it is one of the most numerous and time-consuming specimens.

Good sputum samples depend on thorough health care worker education and patient understanding throughout all phases of the collection process. Food should not have been ingested for 1 to 2 hours before expectoration and the mouth should be rinsed with saline or water just before expectoration. Patients should be instructed to provide a deep-coughed specimen. The material should be expelled into a sterile container, with an attempt to minimize contamination by saliva. Specimens should be transported to the laboratory immediately, because even a moderate amount of time at room temperature can result in loss of some infectious agents.

Induced. Patients who are unable to produce sputum may be assisted by respiratory therapy technicians, who use postural drainage and thoracic percussion to stimulate production of acceptable sputum. As an alternative, an aerosol-induced specimen may be collected that is useful for isolating the agents of mycobacterial or fungal disease. Induced sputum is recognized for its high diagnostic yield in cases of *Pneumocystis jiroveci* pneumonia as well.[4] Aerosol-induced specimens are collected by allowing the patient to breathe aerosolized droplets of a solution containing 15% sodium chloride and 10% glycerin for approximately 10 minutes, or until a strong cough reflex is initiated. Lower respiratory secretions obtained in this way appear watery, resembling saliva, although they often contain material directly from alveolar spaces. These specimens are usually adequate for culture and should be accepted in the laboratory without prescreening. Obtaining such a specimen may obviate the need for a more invasive procedure, such as bronchoscopy or needle aspiration, in many cases.

The gastric aspirate is used exclusively for isolation of acid-fast bacilli and may be collected from patients who are unable to produce sputum, particularly young children. Before the patient wakes up in the morning, a nasogastric tube is inserted into the stomach and contents are withdrawn (on the assumption that acid-fast bacilli from the respiratory tract were swallowed during the night and will be present in the stomach). The relative resistance of mycobacteria to acidity allows them to remain viable for a short period. Gastric aspirate specimens must be delivered to the laboratory immediately so that the acidity can be neutralized. Specimens can be first neutralized and then transported if immediate delivery is not possible.

Endotracheal or Tracheostomy Suction Specimens

Patients with tracheostomies are unable to produce sputum in the normal fashion, but lower respiratory tract secretions can easily be collected in a Lukens trap (Figure 53-2). Tracheostomy aspirates or tracheostomy

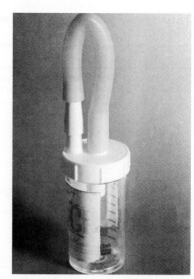

Figure 53-2   Tracheal secretions received in the laboratory in a Lukens trap.

suction specimens should be treated as sputum by the laboratory. Patients with tracheostomies rapidly become colonized with gram-negative bacilli and other nosocomial pathogens. Such colonization per se is not clinically relevant, but these organisms may be aspirated into the lungs and cause pneumonia. Thus, there can be a great deal of confusion for microbiologists and clinicians trying to ascertain the etiologic agent of pneumonia in these patients.

Bronchoscopy. The diagnosis of pneumonia, particularly in HIV-infected and other immunocompromised patients, often necessitates the use of more invasive procedures. Fiberoptic bronchoscopy has dramatically affected the evaluation and management of these infections. With this method, the bronchial mucosa can be directly visualized and collected for biopsy, and the lung tissue can be sent for transbronchial biopsy to evaluate lung cancer and other lung diseases. Although transbronchial biopsy is important, the procedure is often associated with significant complications such as bleeding.

During bronchoscopy, physicians can obtain bronchial washings or aspirates, bronchoalveolar lavage (BAL) samples, protected bronchial brush samples, or specimens for transbronchial biopsy. Bronchial washings or aspirates are obtained by instilling a small amount of sterile physiologic saline into the bronchial tree and withdrawing the fluid when purulent secretions are not visualized. Such specimens will still be contaminated with upper respiratory tract flora such as viridans streptococci and *Neisseria* spp. Recovery of potentially pathogenic organisms from bronchial washings should be attempted, because such specimens may be more diagnostically relevant than sputa.

A deeper sampling of desquamated host cells and secretions can also be obtained via bronchoscopy by BAL. Lavages are especially suitable for detecting *Pneumocystis* cysts and fungal elements.[5,23] During this procedure, a high volume of saline (100 to 300 mL) is infused into a lung segment through the bronchoscope to obtain cells and protein of the pulmonary interstitium and alveolar spaces. It is estimated that more than 1 million alveoli are sampled during this process. The value of this technique in conjunction with quantitative culture for the diagnosis of most major respiratory tract pathogens, including bacterial pneumonia, has been documented.[6,11] Scientists have found significant correlation between acute bacterial pneumonia and greater than 10^3 to 10^4 bacterial colonies per milliliter of BAL fluid. BAL has been shown to be a safe and practical method for diagnosing opportunistic pulmonary infections in immunosuppressed patients. Recently, a bedside, nonbronchoscopic "mini BAL" using a Metras catheter has been introduced; typically 20 mL or less of saline is instilled.[29]

Another type of respiratory specimen is obtained via a protected catheter bronchial brush as part of a bronchoscopy examination.[4] Specimens obtained by this moderately invasive collection procedure are best suited for microbiologic studies, particularly in aspiration pneumonia. Protected specimen brush bristles collect from 0.001 to 0.01 mL of material. An overview of the collection process is shown in Figure 53-3. Upon receipt, contents of the bronchial brush may be suspended in 1 mL of broth solution with vigorous vortexing and then inoculated onto culture media using a 0.01 mL calibrated inoculating loop. Some researchers have stated that specimens obtained via double-lumen–protected catheters are suitable for both anaerobic and aerobic cultures. Colony counts of greater than or equal to 1000 organisms per milliliter in the broth diluent (or 10^6/mL in the original specimen) have been considered to correlate with infection. All facets of the bronchoscopic procedure, such as order of sampling, use of anesthetic, and rapidity of plating, should be rigorously standardized.[4,25]

Transtracheal Aspirates. Percutaneous transtracheal aspirates (TTA) are obtained by inserting a small plastic catheter into the trachea via a needle previously inserted through the skin and cricothyroid membrane. This invasive procedure, although somewhat uncomfortable for the patient and not suitable for all patients (it cannot be used in uncooperative patients, in patients with bleeding tendency, or in patients with poor oxygenation), reduces the likelihood that a specimen will be contaminated by upper respiratory tract flora and diluted by added fluids, provided that care is taken to keep the catheter from being coughed back up into the pharynx. Although this technique is rarely used anymore, anaerobes, such as *Actinomyces* and those associated with aspiration pneumonia, can be isolated from TTA specimens.

Other Invasive Procedures. When pleural empyema is present, thoracentesis may be used to obtain infected fluid for direct examination and culture. This constitutes an excellent specimen that accurately reflects the bacteriology of an associated pneumonia. Laboratory examination of such material is discussed in Chapter 61. Blood cultures, of course, should always be obtained from patients with pneumonia, because they will be positive in about 20% of patients requiring hospitalization.

For patients with pneumonia, a thin needle aspiration of material from the involved area of the lung may be performed percutaneously. If no material is withdrawn into the syringe after the first try, approximately 3 mL of sterile saline can be injected and then withdrawn into the syringe. Patients with emphysema, uremia, thrombocytopenia, or pulmonary hypertension may be at increased risk of complication (primarily pneumothorax [air in the pleural space] or bleeding) from this procedure. The specimens obtained are very small in volume, and protection from aeration is usually impossible. This technique is more frequently used in children than in adults.

The most invasive procedure for obtaining respiratory tract specimens is the open lung biopsy. Performed by surgeons, this method is used to procure a wedge of lung tissue. Biopsy specimens are extremely helpful for diagnosing severe viral infections, such as herpes simplex pneumonia, for rapid diagnosis of *Pneumocystis* pneumonia, and for other hard-to-diagnose or life-threatening pneumonias. Ramifications of this and all other specimen collection techniques are discussed in *Cumitech 7B*, titled, "Laboratory Diagnosis of Lower Respiratory Tract Infections."[29]

SPECIMEN PROCESSING

Direct Visual Examination

Lower respiratory tract specimens can be examined by direct wet preparation for some parasites and by special procedures for *Pneumocystis*. Fungal elements can be visualized under phase microscopy with 10% potassium hydroxide, under ultraviolet light with calcofluor white, or using periodic acid-Schiff–stained smears.

For most other evaluations, the specimen must be fixed and stained. Bacteria and yeasts can be recognized on Gram stain. One of the most important uses of the Gram stain, however, is to evaluate the quality of ex-

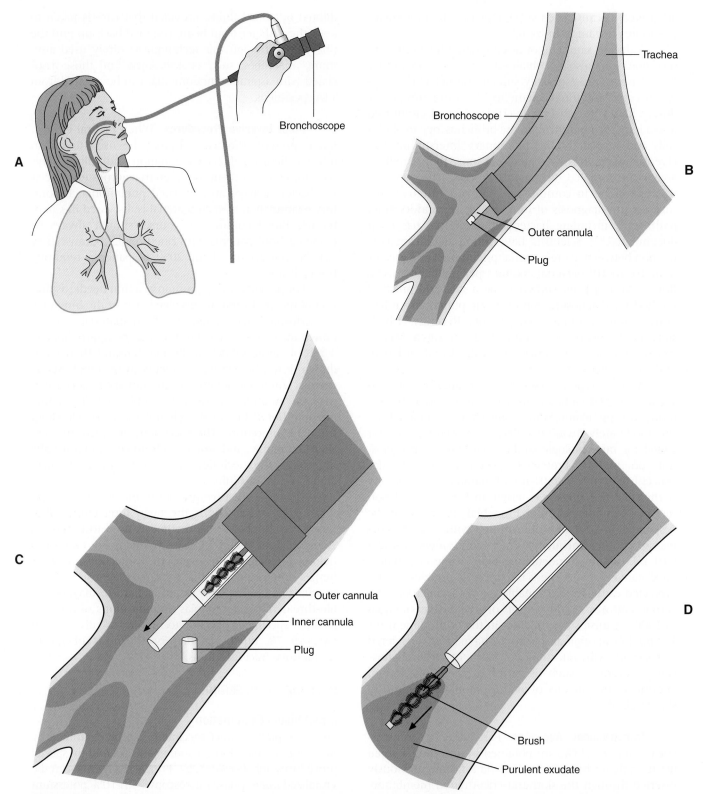

Figure 53-3 Overview of obtaining a protected catheter bronchial brush during a bronchoscopy examination. **A,** The bronchoscope is introduced into the nose and advanced through the nasopharyngeal passage into the trachea. The bronchoscope is then inserted into the lung area of interest. **B,** A small brush that holds 0.01 to 0.001 mL of secretions is placed within a double cannula. The end of the outermost tube or cannula is closed with a displaceable plug made of absorbable gel. The cannula is inserted to the proper area. **C,** Once in the correct area, the inner cannula is pushed out, dislodging the protective plug as it is extruded. **D,** The brush is then extended beyond the inner cannula, and the specimen is collected by "brushing" the involved area. The brush is withdrawn into the inner cannula, which is withdrawn into the outer cannula to prevent contamination by upper airway organisms as it is removed.

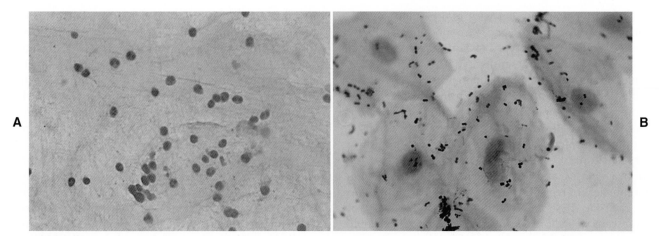

Figure 53-4   Gram stain of sputum specimens. **A,** This specimen contains numerous polymorphonuclear leukocytes and no visible squamous epithelial cells, indicating that the specimen is acceptable for routine bacteriologic culture. **B,** This specimen contains numerous squamous epithelial cells and rare polymorphonuclear leukocytes, indicating an inadequate specimen for routine sputum culture.

pectorated sputum received for routine bacteriologic culture.[13,14,29] A portion of the specimen consisting of purulent material is chosen for the stain. Of note, the smear can be evaluated adequately even before it is stained, thus negating the need for Gram stain of specimens later judged unacceptable. An acceptable specimen yields less than 10 squamous epithelial cells per low-power field (100×). The number of white blood cells may not be relevant, because many patients are severely neutropenic and specimens from these patients will not show white blood cells on Gram stain examination. On the other hand, the presence of 25 or more polymorphonuclear leukocytes per 100× field, together with few squamous epithelial cells, implies an excellent specimen (Figure 53-4). Until recently, only expectorated sputa were suitable for rejection based on microscopic screening. However, endotracheal aspirates (ETAs) from mechanically ventilated adult patients can now also be screened by Gram stain. Criteria used to reject ETAs from adult patients include greater than 10 squamous epithelial cells per low-power field or no organisms seen under oil immersion (1000×).[19] In *Legionella* pneumonia, sputum may be scant and watery, with few or no host cells. Such specimens may be positive by direct fluorescent antibody stain and culture, and should not be subjected to screening procedures. Conversely, sputum from patients with CF should be screened.[27] A throat swab is an acceptable specimen from patients with CF in selected clinical settings and should be processed in a similar manner as CF sputum.[27,29]

Respiratory secretions may need to be concentrated before staining. The cytocentrifuge instrument has been used successfully for this purpose, concentrating the cellular material in an easily examined monolayer on a glass slide. As an alternative, specimens are centrifuged, and the sediment is used for visual examinations and cultures. For screening purposes, the presence of ciliated columnar bronchial epithelial cells, goblet cells, or pulmonary macrophages in specimens obtained by bronchoscopy or BAL indicates a specimen from the lower respiratory tract.

In addition to the Gram stain, respiratory specimens may be stained for acid-fast bacilli with either the classic Ziehl-Neelsen or the Kinyoun carbolfuchsin stain. Auramine or auramine-rhodamine is also used to detect acid-fast organisms. Because they are fluorescent, these stains are more sensitive than the carbolfuchsin formulas and are preferable for rapid screening. Slides may be restained with the classic stains directly over the fluorochrome stains as long as all of the immersion oil has been removed carefully with xylene. All of the acid-fast stains will reveal *Cryptosporidium* spp. if they are present in the respiratory tract, as may occur in immunosuppressed patients. These patients are often at risk of infection with *P. jiroveci.* Although the modified Gomori methenamine silver stain has been used traditionally to recognize *Nocardia, Actinomyces,* fungi, and parasites, it takes approximately 1 hour of the technologist's time to perform, is technically demanding, and is not suitable as an emergency procedure. A fairly rapid stain, toluidine blue O, has been used in many laboratories with some success. Toluidine blue O stains not only *Pneumocystis* but also *Nocardia asteroides* and some fungi. A monoclonal antibody stain is the optimum stain for *Pneumocystis* (see Chapter 50) for less invasive specimens such as BAL and induced sputa.

Direct fluorescent antibody (DFA) staining has been used to detect *Legionella* spp. in lower respiratory tract specimens. Sputum, pleural fluid, aspirated material, and tissue are all suitable specimens. Because there are so many different serotypes of legionellae, poly-clonal antibody reagents and a monoclonal antibody directed against all serotypes of *Legionella pneumophila* are used. Because of low sensitivity (50% to 75%), DFA results should not be relied upon in lieu of culture. Rather, *Legionella* culture, DFA or urinary antigen, and serology should be performed for optimum sensitivity.

Commercially available DFA reagents are also used to detect antigens of numerous viruses, including herpes simplex, cytomegalovirus, adenovirus, influenza viruses, and RSV (see Chapter 51). Commercial suppliers of reagents provide procedure information for each of these tests. Monoclonal and polyclonal fluorescent stains for *Chlamydia trachomatis* are available and may be useful for staining respiratory secretions of infants with pneumonia. A number of molecular amplification techniques (see Chapter 8) for the direct detection of respiratory pathogens have been described; however, the sensitivity and specificity of these assays vary greatly from one study to another. To date, Federal Drug Administration (FDA)-approved amplification assays include those for the direct detection of *Mycobacterium tuberculosis* on smear-positive specimens (see Chapter 45).

Recently, a rapid urinary antigen test (Binax NOW *S. pneumoniae* urinary antigen test; Binax, Portland, Maine) for presumptive diagnosis of pneumococcal pneumonia has been approved by the FDA. This non-invasive procedure detects the C-polysaccharide antigen from the cell wall of *S. pneumoniae;* results can be made available within 15 minutes and are unaffected by the previous use of antibiotics. Studies indicate that the test's sensitivity in adult patients with high-risk pneumonia is significantly greater than for patients without high-risk pneumonia.[26] In addition, this assay appears to have limited use in pediatric patients for discriminating between children with and without pneumococcal pneumonia.[20]

Routine Culture

Most of the commonly sought etiologic agents of lower respiratory tract infection will be isolated on routinely used media: 5% sheep blood agar, MacConkey agar for the isolation and differentiation of gram-negative bacilli, and chocolate agar for *Haemophilus* and *Neisseria* spp. Because of contaminating oral flora, sputum specimens, specimens obtained by bronchial washing and lavage, tracheal aspirates, and tracheostomy or endotracheal tube aspirates are not inoculated to enrichment broth or incubated anaerobically. Only specimens obtained by percutaneous aspiration (including transtracheal aspiration) and by protected bronchial brush are suitable for anaerobic culture; the latter must be done quantitatively for proper interpretation (refer to prior discussion). Transtracheal and percutaneous lung aspiration material

may be inoculated to enriched thioglycollate as well as to solid media. For suspected cases of legionnaires' disease, buffered charcoal-yeast extract (BCYE) agar and selective BCYE are inoculated. Plates should be streaked in four quadrants to provide a basis for objective semiquantitation to define the amount of growth. After 24 to 48 hours of incubation, the numbers and types of colonies are recorded. For *Legionella* cultures, colonies form on the selective agar after 3 to 5 days at 35° C.

Sputum specimens from patients known to have CF should be inoculated to selective agar, such as mannitol salt, for recovery of *S. aureus* and selective horse blood–bacitracin, incubated anaerobically and aerobically, for recovery of *H. influenzae* that may be obscured by the mucoid *Pseudomonas* on routine media.[10] The use of a selective medium for *B. cepacia,* such as PC or OFPBL agars, is also necessary.[10]

For interpretation of culture results on those specimens contaminated by normal oropharyngeal flora (e.g., expectorated and induced sputum, bronchial washings), growth of the predominant aerobic and facultatively anaerobic bacteria is reported. To ensure optimum culture reporting, conditions must be well defined in terms of an objective grading system for streaked plates. Finally, the clinical significance of culture findings depends not only on standardized and appropriate laboratory methods being followed but also on how specimens are collected and transported, other laboratory data, and the patient's clinical presentation.

Numerous bacterial agents that cause lower respiratory tract infections are not detected by routine bacteriologic culture. Mycobacteria, *Chlamydia, Nocardia, Bordetella pertussis, Legionella,* and *Mycoplasma pneumoniae* require special procedures for detection; this also applies to viruses and fungi. Refer to the appropriate chapter section dealing with these organisms. Finally, one must keep in mind those potential agents for bioterrorist attack, such as *Bacillus anthracis, Francisella tularensis,* and *Yersinia pestis,* that might be recovered from respiratory specimens (see Chapter 65).

Case Study

A 16-month-old boy was admitted with fever, lethargy, and trouble breathing. A diagnosis of pneumonia was made by physical examination. The child had recently been to Panama and at that time was treated with ceftriaxone for cough and fever. His fever continued, despite treatment. On admission, he was given erythromycin therapy. Tracheal aspirate and blood cultures were obtained, but the respiratory specimen contained numerous epithelial cells and yielded normal respiratory flora on culture. A pleural aspirate and blood cultures were positive with *Streptococcus pneumoniae,* which was resistant to erythromycin and penicillin and intermediate in susceptibility to ceftriaxone. The patient was given high doses of ceftriaxone and vancomycin and responded to this therapy.

QUESTIONS

1. What criteria are used in laboratories to reject sputum and tracheal aspirates for culture?
2. If greater than 10 squamous epithelial cells per low-power field are seen in a Gram stain but the smear also has numerous white blood cells (greater than 25 per low-power field), should the specimen be rejected for culture?
3. In cases of pneumococcal pneumonia, what percentage of blood and sputum cultures is positive for *S. pneumoniae*?
4. The organism was reported as resistant to penicillin (MIC of 4 µg/mL) and intermediate in susceptibility to third-generation cephalosporins (minimum inhibitory concentration to ceftriaxone of 2 µg/mL). How does the laboratory perform testing for this organism?

REFERENCES

1. American Thoracic Society and the Infectious Diseases Society of America: Guidelines for the management of adults with hospital-acquired, ventilator-associated, and healthcare-associated pneumonia, *Am J Respir Crit Care Med* 171:388, 2005.
2. Bartlett JG, Dowell SF, Mandell LA, et al: Practice guidelines for the management of community-acquired pneumonia in adults, *Clin Infect Dis* 31:347, 2000.
3. Boivin G, Abed Y, Pelletier G, et al: Virological features and clinical manifestations associated with human metapneumovirus: a new paramyxovirus responsible for acute respiratory tract infections in all age groups, *J Infect Dis* 186:1330, 2002.
4. Broughton WA, Middleton RM III, Kirkpatrick MB, et al: Bronchoscopic protected specimen brush and bronchoalveolar lavage in the diagnosis of bacterial pneumonia, *Infect Dis Clin North Am* 5:437, 1991.
5. Caliendo AM: Enhanced diagnosis of *Pneumocystis carinii:* promises and problems, *Clin Microbiol Newsletter* 18:113, 1996.
6. Cantral DE, Tape TG, Reed EC, et al: Quantitative culture of bronchoalveolar lavage fluid for the diagnosis of bacterial pneumonia, *Am J Med* 95:601, 1993.
7. Carroll KA: Laboratory diagnosis of lower respiratory tract infections: controversy and conundrums, *J Clin Microbiol* 40:3115, 2002.
8. Current topics: atypical pneumonia agents are joining the mainstream, *ASM News* 61:621, 1995.
9. Denny F, Clyde WJ: Acute lower respiratory tract infections in non-hospitalized children, *J Pediatr* 108:635, 1989.
10. Gilligan P: Report on the consensus document for microbiology and infectious diseases in cystic fibrosis, *Clin Microbiol Newsletter* 18:11, 1996.
11. Kahn FW, Jones JM: Analysis of bronchoalveolar lavage specimens from immunocompromised patients with a protocol applicable in the microbiology laboratory, *J Clin Microbiol* 26:1150, 1988.
12. Kauppinen M, Saikku P: Pneumonia due to *Chlamydia pneumoniae:* prevalence, clinical features, diagnosis, and treatment, *Clin Infect Dis* 21:244, 1995.
13. Kuo CC, Jackson LA, Campbell LA, et al: *Chlamydia pneumoniae* (TWAR), *Clin Microbiol Rev* 8:451, 1995.
14. Lentino JR: The nonvalue of unscreened sputum specimens in the diagnosis of pneumonia, *Clin Microbiol Newsletter* 9:70, 1987.
15. Mandell LA, Bartlett JG, Dowell SF, et al: Update of practice guidelines for the management of community-acquired pneumonia in immunocompetent adults, *Clin Infect Dis* 37:1405, 2003.
16. Marrie TJ: Community-acquired pneumonia, *Clin Infect Dis* 18:501, 1994.
17. Marrie TJ, Durant H, Bates L: Community-acquired pneumonia requiring hospitalization: a 5-year prospective study, *Rev Infect Dis* 11:586, 1989.
18. McIntosh K: Community-acquired pneumonia in children, *N Engl J Med* 346: 429, 2002.
19. Morris AJ, Tanner DC, Reller RB: Rejection criteria for endotracheal aspirates from adults, *J Clin Microbiol* 31:1027, 1993.
20. Navarro D, Garcia-Maset L, Gimenao C, et al: Performance of the Binax NOW *Streptococcus pneumoniae* urinary antigen assay for diagnosis of pneumonia in children with underlying pulmonary diseases in the absence of acute pneumococcal infection, *J Clin Microbiol* 42:4853, 2004.
21. Niederman MS, Bass JB Jr, Campbell GD, et al: Guidelines for the initial management of adults with community-acquired pneumonia: diagnosis, assessment of severity, and initial antimicrobial therapy, *Am Rev Respir Dis* 148:1418, 1993.
22. Pinner RW, Teutsch SM, Simonsen I, et al: Trends in infectious diseases mortality in the United States, *JAMA* 275:189, 1996.
23. Pisani RJ, Wright AJ: Clinical utility of bronchoalveolar lavage in immunocompromised hosts, *Mayo Clin Proc* 67:221, 1992.
24. Poe R: Management of lower respiratory tract infections, *Guthrie J* 65:40, 1996.
25. Pollock HM, Hawkins EL, Bonner JR, et al: Diagnosis of bacterial pulmonary infections with quantitative protected catheter cultures obtained during bronchoscopy, *J Clin Microbiol* 17:255, 1983.
26. Roson B, Fernandez-Sabe N, Carratala J, et al: Contribution of a urinary antigen assay (Binax NOW) to the early diagnosis of pneumococcal pneumonia, *Clin Infect Dis* 38:222, 2004.
27. Sadeghi E, Matlow A, MacLusky I, et al: Utility of Gram stain in evaluation of sputa from patients with cystic fibrosis, *J Clin Microbiol* 32:54, 1994.
28. Salemi C, Morgan J, Padilla S, et al: Association between severity of illness and mortality from nosocomial infection, *Am J Infect Control* 23:188, 1995.
29. Sharp SE, Robinson A, Saubolle M, et al: Lower respiratory tract infections. In Sharp SE, coordinating editor: *Cumitech 7B,* Washington, DC, 2004, ASM Press.
30. Tsolia MN, Psarras S, Bossios A, et al: Etiology of community-acquired pneumonia in hospitalized school-age children: evidence for high prevalence of viral infections, *Clin Infect Dis* 39: 681, 2004.

CHAPTER 54 Upper Respiratory Tract Infections and Other Infections of the Oral Cavity and Neck

GENERAL CONSIDERATIONS

ANATOMY

The upper respiratory tract includes the epiglottis and surrounding tissues, larynx, nasal cavity, and the pharynx (throat). These anatomic structures are shown in Figure 54-1.

The pharynx is a tubelike structure that extends from the base of the skull to the esophagus (see Figure 54-1). Made of muscle, this structure is divided into three parts:

- Nasopharynx (portion of the pharynx above the soft palate)
- Oropharynx (portion of the pharynx between the soft palate and epiglottis)
- Laryngopharynx (portion of the pharynx below the epiglottis that opens into the larynx)

The oropharynx and nasopharynx are lined with stratified squamous epithelial cells that are teeming with microbial flora. The tonsils are contained within the oropharynx; the larynx is located between the root of the tongue and the upper end of the trachea.

PATHOGENESIS

An overview of the pathogenesis of respiratory tract infections is presented in Chapter 53. It is important to keep in mind that upper respiratory tract infections may spread and become more serious because the mucosa (mucous membrane) of the upper tract is continuous with the mucosal lining of the sinuses, eustachian tube, middle ear, and lower respiratory tract.

DISEASES OF THE UPPER RESPIRATORY TRACT, ORAL CAVITY, AND NECK

UPPER RESPIRATORY TRACT

Diseases of the upper respiratory tract are named according to the anatomic sites involved. Most of these infections are self-limiting and most are caused by viruses.

Laryngitis

Acute laryngitis is usually associated with the common cold or influenza syndromes. Characteristically, patients complain of hoarseness and lowering or deepening of the voice. Acute laryngitis is generally a benign illness.

Acute laryngitis is caused almost exclusively by viruses. Although numerous viruses can cause laryngitis, influenza and parainfluenza viruses, rhinoviruses, adenoviruses, coronavirus, and human metapneumovirus are the most common etiologic agents. If examination of the larynx reveals an exudate or membrane on the pharyngeal or laryngeal mucosa, streptococcal infection, mononucleosis, or diphtheria should be suspected (see discussion about miscellaneous infections caused by other agents later in this chapter).

Laryngotracheobronchitis

Another clinical syndrome closely related to laryngitis is acute laryngotracheobronchitis, or croup. Croup is a relatively common illness in young children, primarily those younger than 3 years of age. Of significance, croup can represent a potentially more serious disease if the infection extends downward from the larynx to involve the trachea or even the bronchi. Illness is characterized by variable fever, inspiratory stridor (difficulty in moving enough air through the larynx), hoarseness, and a harsh, barking, nonproductive cough. These symptoms last for 3 to 4 days, although the cough may persist for a longer period. In young infants, severe respiratory distress and fever are common symptoms.

Similar to the etiologic agents of laryngitis, viruses are a primary cause of croup; parainfluenza viruses are the major etiologic agents. In addition to parainfluenza viruses, influenza viruses, respiratory syncytial virus, and adenoviruses can also cause croup. *Mycoplasma pneumoniae*, rhinoviruses, and enteroviruses can cause a few cases as well.

Epiglottitis

Epiglottitis is an infection of the epiglottis and other soft tissues above the vocal cords. Infection of the epiglottis can lead to significant edema (swelling) and inflammation. Most commonly, children between the ages of 2 and 6 years of age are infected. These children typically present with fever, difficulty in swallowing because of pain, drooling, and respiratory obstruction with inspiratory stridor. Epiglottitis is a potentially life-threatening disease because the patient's airway can become completely obstructed (blocked) if not treated.

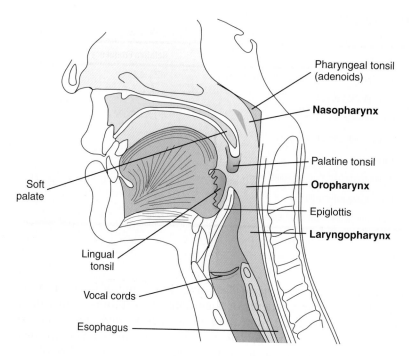

Figure 54-1   The pharynx, including its three divisions and nearby structures.

In contrast to laryngitis, epiglottitis is usually caused by bacteria. In the past, 2- to 4-year-old children were typically infected with *Haemophilus influenzae* type b as the primary cause of epiglottitis. However, as the use of *Haemophilus influenzae* type b conjugated vaccine increases, the typical patient is becoming an adult with sore throat. Other organisms occasionally implicated are streptococci and staphylococci. Diagnosis is established on clinical grounds, including the visualization of the epiglottis, which appears swollen and bright red in color. Bacteriologic culture of the epiglottis is contraindicated because swabbing of the epiglottis may lead to respiratory obstruction. Of importance, *H. influenzae* bacteremia usually occurs in children with epiglottitis caused by this organism.

Pharyngitis, Tonsillitis, and Peritonsillar Abscesses

Pharyngitis and Tonsillitis. Pharyngitis (sore throat) and tonsillitis are common upper respiratory tract infections affecting both children and adults.

Clinical Manifestations. Infection of the pharynx is associated with pharyngeal pain. Visualization of the pharynx reveals that affected tissues are erythematous (red) and swollen. Depending on the causative microorganism, either inflammatory exudates (fluid with protein, inflammatory cells, and cellular debris), vesicles (small blisterlike sacs containing liquid) and mucosal ulceration, or nasopharyngeal lymphoid hyperplasia (swollen lymph nodes) may be observed.

Pathogenesis. Pathogenic mechanisms differ and are dependent on the organism causing the pharyngitis. For example, some organisms directly invade the pharyngeal mucosa (e.g., *Arcanobacterium haemolyticum*), others elaborate toxins and other virulence factors at the site (e.g., *Corynebacterium diphtheriae*), and still others invade the pharyngeal mucosa and elaborate toxins and other virulence factors (e.g., group A streptococci [*Streptococcus pyogenes*]). Pathogenic mechanisms are reviewed in Part III according to various organism groups.

Epidemiology/Etiologic Agents. Most cases of pharyngitis occur during the colder months and often accompany other infections, primarily those caused by viruses. Patients with respiratory tract infections caused by influenza types A and B, parainfluenza, coxsackie A, rhinoviruses, or coronaviruses frequently complain of a sore throat. Pharyngitis, often with ulceration, is also commonly found in patients with infectious mononucleosis caused by either Epstein-Barr virus or cytomegalovirus. Although less common, pharyngitis caused by adenovirus or herpes simplex virus is clinically severe. Finally, acute retroviral syndrome caused by human immunodeficiency virus 1 (HIV-1) is associated with acute pharyngitis.

Although different bacteria can cause pharyngitis and/or tonsillitis, the primary cause of bacterial pharyngitis is *Streptococcus pyogenes* (or group A beta-hemolytic streptococci). Viral pharyngitis or other causes of pharyngitis/tonsillitis must be differentiated from that caused by *S. pyogenes* because pharyngitis resulting from

Table 54-1   Bacteria That Can Cause Acute Pharyngitis and/or Tonsillitis

Organism	Disease	Relative Frequency
Streptococcus pyogenes	Pharyngitis/tonsillitis/rheumatic fever/scarlet fever	15% to 35%
Group C and G beta-hemolytic streptococci	Pharyngitis/tonsillitis	< 3% to 11%
Arcanobacterium (Corynebacterium) haemolyticum[2,11]	Pharyngitis/tonsillitis/rash	< 1% to 10%
Neisseria gonorrhoeae	Pharyngitis/disseminated disease	Rare*
Corynebacterium ulcerans	Pharyngitis	Rare
Mycoplasma pneumoniae	Pneumonia/bronchitis/pharyngitis	Rare
Yersinia enterocolitica	Pharyngitis/enterocolitis	Rare
Human immunodeficiency virus-1	Pharyngitis/acute retroviral disease	Rare

*Less than 1%.

S. pyogenes is treatable with penicillin whereas viral infections are not. In addition, treatment is of particular importance because infection with *S. pyogenes* can lead to complications such as acute rheumatic fever and glomerulonephritis. These complications are referred to as *poststreptococcal sequelae* (diseases that follow a streptococcal infection) and are primarily immunologically mediated; these sequelae are discussed in greater detail in Chapter 17. *S. pyogenes* may also cause pyogenic infections (suppurations) of the tonsils, sinuses, and middle ear, or cellulitis as secondary pyogenic sequelae after an episode of pharyngitis. Accordingly, streptococcal pharyngitis is usually treated to prevent both the suppurative and nonsuppurative sequelae, as well as to decrease morbidity.

Cases of pharyngitis caused by groups C and G and by nonhemolytic members of these groups of streptococci (including group A) have been reported.[1,6,9,11,15-17] Although bacteria other than group A streptococci may cause pharyngitis, this occurs less often. Large colony isolates of group C (classified as *Streptococcus dysagalactiae* subsp. *equisimilus*) and G streptococci are pyogenic streptococci with similar virulence traits as *S. pyogenes*; symptoms of pharyngitis caused by these agents are also similar to *S. pyogenes*. In contrast to *S. pyogenes*, these agents have only been rarely associated with poststreptococcal sequelae, namely glomerulonephritis and possibly rheumatic fever.[12] Recent studies have demonstrated that these streptococci can exchange genetic information with *S. pyogenes* and thus potentially obtain virulence factors usually associated with *S. pyogenes* such as M proteins, streptolysin O, and superantigen genes.[14] *Arcanobacterium haemolyticum* is also a cause of pharyngitis among adolescents.[4] The agents that can cause pharyngitis or tonsillitis are listed in Table 54-1.

Although *H. influenzae, S. aureus,* and *S. pneumoniae* are frequently isolated from nasopharyngeal and throat cultures, they have not been shown to cause pharyngitis. Carriage of any of these organisms, as well as

Neisseria meningitidis, may have clinical importance for some patients or their contacts. Cultures of specimens obtained from the anterior nares often yield *S. aureus.* The carriage rate for this organism is especially high among health care workers, and even 20% of the general population can be colonized with this microbe.

Vincent's angina, also called acute necrotizing ulcerative gingivitis, or trench mouth, is a mixed bacterial-spirochetal infection of the gingival edge. The infection is relatively rare today, but it is considered a serious disease because it is often complicated by septic jugular thrombophlebitis, bacteremia, and widespread metastatic infection. Adults are more often affected than children; poor oral hygiene is a predisposing factor. Multiple anaerobes, especially *Fusobacterium necrophorum,* are implicated in this syndrome. Although Gram stain of a throat specimen is usually not predictive, in those patients with symptoms suggestive of Vincent's angina, Gram stain reveals numerous fusiform, gram-negative bacilli and spirochetes.

Peritonsillar Abscesses. Peritonsillar abscesses are generally thought of as a complication of tonsillitis. This infection is most common in children older than 5 years of age and in young adults. It is important to treat these infections because they can spread to adjacent tissues, as well as erode into the carotid artery to cause an acute hemorrhage. The predominant organisms in peritonsillar abscesses are non–spore-forming anaerobes, including *Fusobacterium* (especially *F. necrophorum*), *Bacteroides* (including the *B. fragilis* group), and anaerobic cocci. *Streptococcus pyogenes* and viridans streptococci may also be involved.

Rhinitis

Rhinitis (common cold) is an inflammation of the nasal mucous membrane or lining. Depending on the host response and the etiologic agent, rhinitis is characterized by variable fever, increased mucous secretions, inflammatory edema of the nasal mucosa, sneezing, and watery eyes. With rare exceptions, rhinitis is caused by viruses; some of these agents are listed in Box 54-1. Rhinitis is common because of the large number of different causative viruses, and reinfections may occur with same virus type.

Miscellaneous Infections Caused by Other Agents

Corynebacterium diphtheriae. Although much less common than streptococcal pharyngitis, *C. diphtheriae* can still be isolated from patients with sore throat, as well as more serious systemic disease. After an incubation period of 2 to 4 days, diphtheria usually presents as pharyngitis or tonsillitis. Patients are often febrile and complain of sore throat and malaise (body discomfort). The hallmark for diphtheria is the presence of an exudate or membrane that is usually on the tonsils or pharyngeal wall. The gray-white membrane is a result of the action of diphtheria toxin on the epithelium at the site of infection. Complications occur frequently with diphtheria and are usually seen during the last stage of the disease (paroxysmal stage). The most feared complications are those involving the central nervous system such as seizures, coma, or blindness. Information as to how this organism causes disease is discussed in Chapter 53. Additional specifics regarding this organism are also provided in Chapter 19.

Bordetella pertussis. Although mass immunization programs have greatly reduced the incidence of pertussis, enough cases (because of outbreaks and regional epidemics) still occur that laboratories should be able to either detect, isolate, and identify the organism or refer the specimen to a reference laboratory.

Characteristically, pertussis, or whooping cough, is a prolonged disease (lasting as long as 6 to 8 weeks) that is marked by paroxysmal (sudden and/or intense) coughing.

Following an incubation period of 7 to 13 days, the patient with symptomatic infection develops upper respiratory symptoms, including a dry cough, fever, runny nose, and sneezing. After about 2 weeks, this may progress to spells of paroxysmal coughing. As these

episodes worsen, the characteristic whoop, caused by attempted inspiration through an epiglottis undergoing spasm, begins. Vomiting may occur and usually a lymphocytosis is present. This phase of the illness may last as long as 6 weeks. Additional information about how *B. pertussis* causes this disease and how the organism is detected is provided in Chapter 39.

***Klebsiella* spp.** Rhinoscleroma is a rare form of chronic, granulomatous infection of the nasal passages, including the sinuses and occasionally the pharynx and larynx. Associated with *Klebsiella rhinoscleromatis,* the disease is characterized by nasal obstruction appearing over a long period, caused by tumorlike growth with local extension. Another species, *Klebsiella ozaenae,* can also be recovered from upper respiratory tract infections. This organism may contribute to another infrequent condition called **ozena,** characterized by a chronic, mucopurulent nasal discharge that is often foul-smelling. It is caused by secondary, low-grade anaerobic infection.

ORAL CAVITY

Stomatitis

Stomatitis is an inflammation of the mucous membranes of the oral cavity. Herpes simplex virus is the primary agent of this disease, in which multiple ulcerative lesions are seen on the oral mucosa. These lesions are painful and can be found not only in the mouth but also in the oropharynx. Herpetic infections of the oral cavity are prevalent among immunosuppressed patients.

Thrush

Candida spp. can also invade the oral mucosa. Immunosuppressed patients, including very young infants, may develop oral candidiasis, called *thrush.* Oral thrush can extend to produce pharyngitis and/or esophagitis, a common finding in patients with acquired immunodeficiency syndrome and in other immunosuppressed patients. Thrush is suspected if whitish patches of exudate on an area of inflammation are observed on the buccal (cheek) mucosa, tongue, or oropharynx. Oral mucositis or pharyngitis in the granulocytopenic patient may be caused by *Enterobacteriaceae, S. aureus,* or *Candida* spp. and is manifested by erythema, sore throat, and possibly exudate or ulceration.

Periodontal Infections

Types. The three dental problems for which help may be requested from clinical laboratories are (1) root canal infections, with or without periapical abscess; (2) orofacial odontogenic infections, with or without osteomyelitis (inflammation of a bone) of the jaw; and (3) perimandibular space infections. Oral bacteria are clearly important in other dental processes, such as caries (destruction of the mineralized tissues of the tooth; a cavity), periodontal (tissues in, around, and supporting the tooth) disease, and localized juvenile periodontitis, but clinical laboratories are not involved in culturing in such cases.

Etiologic Agents. The bacteriology is similar in all of these infections and involves primarily anaerobic bacteria and streptococci except for perimandibular space infections, which may also involve staphylococci and *Eikenella corrodens* in about 15% of patients. The streptococci are microaerobic or facultative and are usually alpha-hemolytic (particularly the *Streptococcus anginosus* group—see Chapter 17); they are usually found in 20% to 30% of dental infections of the preceding types.

Members of the *Bacteroides fragilis* group are found in root canal infections, orofacial odontogenic infections, and bacteremia secondary to dental extraction in 5% to 10% of patients. Anaerobic cocci (both *Peptostreptococcus* and *Veillonella*), pigmented *Prevotella* and *Porphyromonas*, the *Prevotella oralis* group, and *Fusobacterium* are found in about 20% to 50% of the three conditions mentioned, as well as in postextraction bacteremia. Infection with *Actinomyces israelii* may complicate oral surgery.

Salivary Gland Infections

Acute suppurative parotitis (inflammation of the salivary glands located under the cheek in front of and below the external ear) is seen in very ill patients, especially those who are dehydrated, malnourished, elderly, or recovering from surgery. It is associated with painful, tender swelling of the parotid gland; purulent drainage may be evident at the opening of the duct of the gland in the mouth. *Staphylococcus aureus* is the major pathogen but on occasion *Enterobacteriaceae*, other gram-negative bacilli, and oral anaerobes may play a role. A chronic bacterial parotitis has been described that usually involves *Staphylococcus aureus*. Less often, other salivary glands may be involved with a bacterial infection, usually because of ductal obstruction.

The mumps virus is traditionally the major viral agent involved in parotitis; however, since the advent of childhood vaccination, infection with mumps virus is rarely diagnosed. Influenza virus and enteroviruses may also cause this syndrome. Diagnosis of viral parotitis is usually done serologically. Infrequently, *Mycobacterium tuberculosis* may involve the parotid gland in conjunction with pulmonary tuberculosis.

NECK

Infections of the deep spaces of the neck are potentially serious because they may spread to critical structures such as major vessels of the neck or to the mediastinum, leading to mediastinitis, purulent pericarditis, and pleural empyema. The oral flora are responsible for these infections. Accordingly, the predominant organisms are anaerobes, primarily *Peptostreptococcus*, various *Bacteroides*, *Prevotella*, *Porphyromonas*, *Fusobacterium* spp., and *Actinomyces*. Streptococci, chiefly of the viridans variety, are also important. *Staphylococcus aureus* and various aerobic, gram-negative bacilli may be recovered, particularly from patients developing these problems in the hospital.

DIAGNOSIS OF UPPER RESPIRATORY TRACT INFECTIONS

COLLECTION AND TRANSPORT OF SPECIMENS

Cotton-, Dacron-, or calcium alginate–tipped swabs are suitable for collecting most upper respiratory tract microorganisms. If the swab remains moist, no further precautions need to be taken for specimens that are cultured within 4 hours of collection. After that period, some kind of transport medium to maintain viability and prevent overgrowth of contaminating organisms should be used. Swabs for detection of group A streptococci (*Streptococcus pyogenes*) are the only exception. This organism is highly resistant to desiccation and remains viable on a dry swab for as long as 48 to 72 hours. Throat swabs of this type can be placed in glassine paper envelopes for mailing or transport to a distant laboratory. Throat swabs are also adequate for recovery of adenoviruses and herpesviruses, *Corynebacterium diphtheriae*, *Mycoplasma*, *Chlamydia*, and *Candida* spp. Recovery of *C. diphtheriae* is enhanced by culturing both the throat and nasopharynx.

Nasopharyngeal swabs are better suited for recovery of respiratory syncytial virus, parainfluenza virus, *Bordetella pertussis*, *Neisseria* spp., and the other viruses causing rhinitis. Optimum conditions for the collection and transport of specimens for viral detection or culture are described in Chapter 51. Although swabs made of calcium alginate are commonly used to collect nasopharyngeal specimens (excluding those specimens for chlamydia or viral culture), nasopharyngeal secretions collected by either aspiration and/or washing are the best specimens for *Bordetella pertussis* because a larger amount of material is obtained.[8,13] The type of swab used for collection is very important. For example, cotton swabs should never be used for culture because fibers contain fatty acids on the surface that kill *Bordetella*. Calcium alginate or Dacron swabs are acceptable for obtaining nasopharyngeal swab specimens with calcium alginate being the best for culture. However, if polymerase chain reaction (PCR) is to be performed, Dacron or rayon swabs on plastic shafts are preferred.[7,13]

Specimens for *B. pertussis* ideally should be inoculated directly to fresh culture media at the patient's bedside. If this is not possible, transport for less than 2 hours in 1% casamino acid medium at room temperature is acceptable. If specimens are plated on the day of collection, Amies transport medium with charcoal is acceptable. If specimens are plated more than 24 hours after collection, Regan-Lowe or Jones-Kendrick transport medium is optimal; both contain charcoal, starch, and nutrients as well as cephalexin. If lengthy delays in transport are expected, transport of specimens in Regan-Lowe medium at 4° C is recommended.[5,13]

DIRECT VISUAL EXAMINATION OR DETECTION

A Gram stain of material obtained from upper respiratory secretions or lesions can do very little to help with diagnosis. Yeastlike cells can be identified, helpful in identifying thrush, and the characteristic pattern of fusiforms and spirochetes of Vincent's angina may be visualized. Plain Gram's crystal violet (allowed to remain on the slide for 1 minute before rinsing with tap water) and the Gram stain can be used to identify the agents of Vincent's angina. However, if only crystal violet is used, the smear should be very thin because everything will be intensely Gram positive, making a thick smear difficult to read.

For causes of pharyngitis, Gram stains are unreliable. Direct smears of exudate from membranelike lesions to differentiate diphtheria from other causes of such membranes are also not reliable or recommended.

Fungal elements, including yeast cells and pseudohyphae, may be visualized with either a 10% potassium hydroxide (KOH) preparation, calcofluor white fluorescent stain, or periodic acid-Schiff (PAS) stain. Direct examination of material obtained from the nasopharynx of suspected cases of whooping cough using a fluorescent antibody stain (see Chapter 39) has been shown to yield some early positive results for detection of *B. pertussis*. However, direct fluorescent antibody (DFA) staining of nasopharyngeal secretions suffers from both insensitivity and poor specificity depending on the antibody used. Numerous studies have demonstrated that PCR-based assays of *B. pertussis* in nasopharyngeal secretions are superior to both DFA and culture. Various methods, including fluorescent antibody stain reagents and enzyme immunoassays, are also commercially available to detect numerous viral agents (see Chapter 51).

One improvement in selected areas of clinical microbiology in recent years has been the development of rapid methods for detection of group A streptococcal antigen or nucleic acid that in most instances obviates the need for culture of pharyngeal specimens. At least 40 commercial products are available to identify group

A streptococcal antigens using membrane enzyme immunoassays or liposomal and optical immunoassay techniques. Although the specific procedures vary with the products, several generalizations can be made. Throat swabs are incubated in an acid reagent or enzyme to extract the group A–specific carbohydrate antigen. Dacron swabs seem to be most efficient at releasing antigen, although other types of swabs may yield acceptable results. In laboratory comparisons between a rapid antigen method and conventional culture methods for detecting the presence of group A streptococci in throat swabs, the commercial kits have shown relative acceptable (62% to more than 90%) sensitivity and specificity. Doubts have been expressed, however, about the ability of the commercial kits to detect all clinically important streptococcal infections (see Chapter 17). For this reason, specimens with a negative direct antigen test for group A streptococci should be cultured or confirmed using a nucleic acid method. Group A streptococci can also be directly detected from pharyngeal specimens by nucleic acid testing using two different molecular assay formats. The first is a commercially available assay (Gen-Probe Group A Strep Direct Test, GenProbe, Inc., San Diego, Calif) that employs a nonisotopic, chemiluminescent, single-stranded DNA probe that is complementary to the rRNA target of the group A *Streptococcus*; a 4-hour incubation step in selective media is required before testing. Gamma-irradiated Dacron swabs or rayon or Dacron swabs that are ethylene oxide–sterilized are acceptable for use with this assay; exposure of rayon swabs to gamma irradiation causes false-positive results.[2] Of note, use of a dry Dacron swab performed better than wet swabs in this assay.[3] Sensitivities of the Gen-Probe Group A Strep Direct Test range from 88% to almost 95% when compared with culture, and specificity is greater than 97%.[2] A rapid-cycle real-time PCR method, the LightCycler Strep-A (Roche Applied Science, Indianapolis, Ind) also detects *S. pyogenes* directly from throat swabs. Using this technology, 32 samples (including controls) can be tested per run in about 1.5 hours. Currently, assessment of the true performance characteristics of this assay is difficult; as previously discussed regarding other nucleic acid amplification methods (see Chapter 8), the sensitivity of the new test may exceed that of the culture "gold standard."

CULTURE

Streptococcus pyogenes (Beta-Hemolytic Group A Streptococci)

Because the primary cause of bacterial pharyngitis in North America is *Streptococcus pyogenes,* most laboratories routinely screen throat cultures for this organism. Group A streptococci are usually beta-hemolytic, with

Table 54-2 Medium and Atmosphere for Incubation of Cultures to Recover Group A Streptococci from Pharyngeal Specimens

Media	Atmosphere of Incubation
Sheep blood agar	Anaerobic
Sheep blood agar with coverslip over the primary area of inoculation	Aerobic
Sheep blood agar with trimethoprim-sulfamethoxazole	5%-10% CO_2 or anaerobic

less than 1% being nonhemolytic. Three variables must be taken into consideration regarding successful culture of group A streptococci from pharyngeal specimens: medium, atmosphere, and duration of incubation. Kellogg recommended four combinations of media and atmosphere of incubation for throat specimens; these are listed in Table 54-2.[10] Regardless of the medium and atmosphere of incubation employed, culture plates should be incubated for at least 48 hours before reporting as negative for group A streptococci. In addition, the incubation of sheep blood agar in 5% to 10% CO_2 was strongly discouraged.

Drawbacks to culture are that both 24 to 48 hours are required for colony formation and further manipulations of the beta-hemolytic organisms are necessary for definitive identification (see Chapter 17). If sufficient numbers of pure colonies are not available for identification, a subculture requiring additional incubation is necessary. By placing a 0.04-unit differential bacitracin filter paper disk (Taxo A, BBL, or Bacto Bacitracin, Difco Laboratories, Detroit, Mich) directly on the area of initial inoculation, presumptive identification of *S. pyogenes* can be made after overnight incubation (all of group A and a very small percentage of group B streptococci are susceptible). However, use of the bacitracin disk in the primary area of inoculation reduces the sensitivity and specificity of culture and identification of *S. pyogenes*. Sometimes growth of too few beta-hemolytic colonies or overgrowth of other organisms makes interpretation difficult. Therefore, using the bacitracin disk as the only method of identification of *S. pyogenes* is not recommended. New selective agars, such as streptococcal selective agar, have been developed that suppress the growth of almost all normal flora and beta-hemolytic streptococci except for groups A and B and *Arcanobacterium haemolyticum*. Direct antigen or nucleic detection tests or the PYR test (see Chapter 17) can also be carried out on the more isolated beta-hemolytic colonies that appear after overnight incubation on selective agar.

Corynebacterium diphtheriae

If diphtheria is suspected, the physician must communicate this information to the clinical laboratory. Because streptococcal pharyngitis is in the differential diagnosis of diphtheria and because dual infections do occur, cultures for *Corynebacterium diphtheriae* should be plated onto sheep blood agar or streptococcal selective agar, as well as onto special media for recovery of this agent. These special media include a Loeffler's agar slant and a cystine-tellurite agar plate. Chapter 19 discusses the identification of the organism. Recovery of this organism is enhanced by culturing specimens from both the throat and nasopharynx of potentially infected patients.

Bordetella pertussis

Freshly prepared Bordet-Gengou agar was the first medium developed for isolation of *Bordetella pertussis*. However, because it was inconvenient to use, other media were subsequently developed (see Chapter 39). Today, Regan-Lowe or charcoal horse blood agar is recommended for use in diagnostic laboratories. Because the organisms are extremely delicate, specimens should be plated directly onto media, if possible. The yield of positive isolations from clinical cases of pertussis seems to vary from 20% to 98% depending on the stage of disease, previous treatment of the patient, age of the patient, and laboratory techniques.

Neisseria

Specimens received in the laboratory for isolation of *Neisseria meningitidis* (for detection of carriers) or *N. gonorrhoeae* should be plated to a selective medium, either modified Thayer-Martin or Martin-Lewis agar. After 24 to 48 hours of incubation in 5% to 10% carbon dioxide, typical colonies of *Neisseria* spp. will be visible (see Chapter 42).

Epiglottitis

Clinical specimens from cases of epiglottitis (swabs obtained by a physician) should be plated to sheep blood agar, chocolate agar (for recovery of *Haemophilus* spp.), and a streptococcal selective medium if desired. Because *Staphylococcus aureus*, *Streptococcus pneumoniae*, and beta-hemolytic streptococci may cause this disease, albeit rarely, their presence should be sought. Refer to Table 5-1 for an overview of collection, transport, and processing of different specimens from the upper respiratory tract.

DIAGNOSIS OF INFECTIONS IN THE ORAL CAVITY AND NECK

COLLECTION AND TRANSPORT

A problem in collecting oral and dental infection material is to avoid or minimize contamination with oral flora. For collection of material from root canal infection, the tooth is isolated by means of a rubber dam. A sterile field is established, the tooth is swabbed with 70% alcohol, and after the root canal is exposed, a sterile paper point is inserted, removed, and placed into semisolid, nonnutritive, anaerobic transport medium. Alternatively, needle aspiration can be used if sufficient purulent material is present. Completely defining the flora of such infections is beyond the scope of routine clinical microbiology laboratories.

Specimens from neck space infections can usually be obtained with a syringe and needle or by biopsy during a procedure by the surgeon. Transport must be under anaerobic conditions.

DIRECT VISUAL EXAMINATION

All material submitted for culture should be smeared and examined by Gram stain and other appropriate techniques for fungi (i.e., calcofluor white, KOH, or PAS stains), if requested.

CULTURE

Infections such as peritonsillar abscesses, oral and dental infections, and neck space infections usually involve anaerobic bacteria. The anaerobes involved typically originate in the oral cavity and are often more delicate than anaerobes isolated from other clinical material. Very careful attention must be paid to providing optimal techniques for anaerobic cultivation, as well as collection and transport in order to recover and identify the etiologic agents.

Case Study

A 2-year-old girl presented to her physician with a sore throat and fever. On examination, her tonsils were enlarged and inflamed. A rapid test was performed for group A streptococci; the test result was negative. The physician decided to treat with amoxicillin regardless of the test results and asked that a culture be performed. The next day the laboratory reported that moderate growth of beta-hemolytic group A streptococcus was present.

QUESTIONS

1. List the tests that rapidly identify group A streptococcus (*Streptococcus pyogenes*).
2. Not all group A streptococci are *S. pyogenes*. How can the nonpathogenic group A streptococci be differentiated from the pathogenic strains?
3. Not all *S. pyogenes* are beta-hemolytic. What is the reason for this phenomenon, and how can the laboratory ensure that it detects the nonhemolytic strains?
4. What is the sensitivity of rapid diagnostic tests to detect group A streptococcal antigen?

REFERENCES

1. Benjamin JT, Perriello VA: Pharyngitis due to group C hemolytic streptococci in children, *J Pediatr* 89:254, 1976.
2. Bourbeau PP: Role of the microbiology laboratory in diagnosis and management of pharyngitis, *J Clin Microbiol* 41:3467, 2003.
3. Bourbeau PP, Heiter BJ: Use of swabs without transport media for the Gen-Probe Group A Strep Direct Test, *J Clin Microbiol* 42:3207, 2004.
4. Cambier M, Janssens M, Wauters G: Isolation of *Arcanobacterium haemolyticum* from patients with pharyngitis in Belgium, *Acta Clin Belg* 47:303, 1992.
5. Cassiday PK, Sanden GN, Kane CT, et al: Viability in *Bordetella pertussis* in four suspending solutions at three temperatures, *J Clin Microbiol* 32:1550, 1994.
6. Chretien JH, McGinniss CG, Thompson J, et al: Group B beta-hemolytic streptococci causing pharyngitis, *J Clin Microbiol* 10:63, 1979.
7. Cloud JL, Hymas W, Carroll KC: Impact of nasopharyngeal swab types on detection of *Bordetella pertussis* by PCR and culture, *J Clin Microbiol* 40:3838, 2002.
8. Hallander HO, Reizenstein E, Renemar B, et al: Comparison of nasopharyngeal aspirates with swabs for culture of *Bordetella pertussis*, *J Clin Microbiol* 31:50, 1993.
9. Hill HR, Caldwell GG, Wilson E, et al: Epidemic of pharyngitis due to streptococci of Lancefield group G, *Lancet* 2:371, 1969.
10. Kellogg JA: Suitability of throat culture procedures for detection of group A streptococci and as reference standards for evaluation of streptococcal antigen kits, *J Clin Microb* 28:165, 1990.
11. McCue JD: Group G streptococcal pharyngitis: analysis of an outbreak at a college, *JAMA* 248:1333, 1982.
12. Mc Donald M, Currie BJ, Carapetis JR: Acute rheumatic fever: a chink in the chain that links the heart to the throat? *Lancet* 4:240, 2004.
13. McGowan KL: Diagnostic tests for pertussis: culture vs. DFA vs. PCR, *Clin Microbiol Newsl* 24:143, 2002.
14. Sachse S, Seidel P, Gerlach D, et al: Superantigen-like gene(s) in human pathogenic *Streptococcus dysgalactiae*, subsp. *equisimilis*: genomic localization of the gene encoding streptococcal pyrogenic exotoxin G (spe G^{dys}), *FEMS Immunol Med Microbiol* 34:159, 2002.
15. Sellen TJ, Long DA: *Arcanobacterium haemolyticum* pharyngitis and tonsillitis, *Clin Microbiol Newsl* 18:30, 1996.
16. Turner JC, Fox A, Addy C, et al: Role of group C beta-hemolytic streptococci in pharyngitis: epidemiologic study of clinical features associated with isolation of group C streptococci, *J Clin Microbiol* 31:808, 1993.
17. Turner JC, Hayden GF, Kiselica D, et al: Association of group C beta-hemolytic streptococci with endemic pharyngitis among college students, *JAMA* 264:2644, 1990.

ADDITIONAL READING

Waites KB, Saubolle MA, Talkington DF, et al: Laboratory diagnosis of upper respiratory tract infections. In Sharp SE, coordinating editor: *Cumitech 10A*, Washington, DC, 2006, ASM Press.

CHAPTER 55

Meningitis and Other Infections of the Central Nervous System

GENERAL CONSIDERATIONS

ANATOMY

Diagnosis of an infection involving the central nervous system (CNS) is of critical importance. Most clinicians consider infection in the CNS to be one of the medical emergencies relating to infectious diseases. Therefore, an understanding of the basic anatomy and physiology of the CNS is helpful for the microbiologist to ensure appropriate specimen processing and interpretation of laboratory results.

Coverings and Spaces of the CNS

Because of the vital and essential role of the CNS in the body's regulatory processes, the brain and spinal cord have two protective coverings: an outer covering consisting of bone and an inner covering of membranes called the **meninges.** The outer bone covering encases the brain (i.e., cranial bones or skull) and spinal cord (i.e., the vertebrae). The meninges is a collective term for three distinct layers surrounding the brain and spinal column:

- Dura mater (outermost membrane layer)
- Arachnoid
- Pia mater (innermost membrane layer)

The pia mater and the arachnoid membrane are collectively called the **leptomeninges.** The portion of the arachnoid that covers the top of the brain contains arachnoid villi, which are special structures that absorb the spinal fluid and allow it to pass into the blood.

Between and around the meninges are spaces that include the epidural, subdural, and subarachnoid spaces. The relative location of the meninges and spaces to one another in the brain is shown in Figure 55-1. The location and nature of the meninges and spaces are summarized in Table 55-1.

Cerebrospinal Fluid

Cerebrospinal fluid (CSF) envelops the brain and spinal cord and has several functions. By cushioning and providing buoyancy for the bulk of the brain, the effective weight of the brain is reduced by a factor of 30. Of importance, CSF carries essential metabolites into the neural tissue and cleanses the tissues of wastes as it circulates around the entire brain, ventricles, and spinal cord. Every 3 to 4 hours, the entire volume of CSF is exchanged. In addition to these functions, CSF provides a means by which the brain monitors changes in the internal environment.

CSF is found in the subarachnoid space (see Table 55-1) and within cavities and canals of the brain and spinal cord. There are four large, fluid-filled spaces within the brain referred to as **ventricles.** Specialized secretory cells called the **choroid plexus** produce CSF. The choroid plexus is located centrally within the brain in the third and fourth ventricles; about 23 mL of CSF are contained within the ventricles. The fluid travels around the outside areas of the brain within the subarachnoid space, driven primarily by the pressure produced initially at the choroid plexus (Figure 55-2). By virtue of its circulation, chemical and cellular changes in the CSF may provide valuable information about infections within the subarachnoid space.

ROUTES OF INFECTION

Organisms may gain access to the CNS by several primary routes:

- Hematogenous spread: followed by entry into the subarachnoid space through the choroid plexus or through other blood vessels of the brain. This is the most common way that the CNS becomes infected
- Direct spread from an infected site: the extension of an infection close to or contiguous with the CNS can occasionally occur; examples of such infections include otitis media (infection of middle ear), sinusitis, and mastoiditis
- Anatomic defects in CNS structures: anatomic defects as a result of surgery, trauma, or congenital abnormalities can allow microorganisms easy and ready access to the CNS
- Travel along nerves leading to the brain (direct intraneural): the least common route of CNS infection caused by organisms such as rabies virus, which travels along peripheral sensory nerves, and herpes simplex virus

DISEASES OF THE CENTRAL NERVOUS SYSTEM

Meningitis

Infection within the subarachnoid space or throughout the leptomeninges is called *meningitis.* Based on the host's response to the invading microorganism, meningitis is

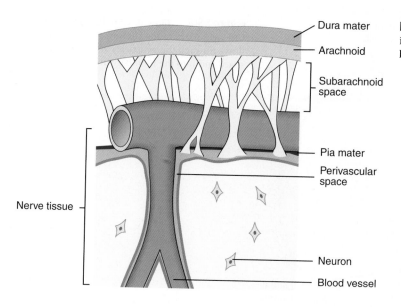

Figure 55-1   Cross section of the brain shows the important membrane coverings and spacings and other key structures.

Table 55-1   Inner Coverings (Meninges) of the Brain, Spinal Cord, and Surrounding Spaces

Anatomic Structure	Relative Location	Key Features
Epidural space	Outside the dura mater yet inside the skull	Cushion of fat and connective tissues
Dura mater	Outermost membrane	Membrane that adheres to the skull; white fibrous tissue
Subdural space	Between the dura mater and the arachnoid membrane	Cushion of lubricating serous fluid
Arachnoid membrane	Between the dura mater and pia mater	Delicate, cobweblike membrane covering the brain and spinal cord
Subarachnoid space	Beneath the arachnoid membrane	Contains a significant amount of CSF in an adult (~125-150 mL)
Pia mater	Beneath the subarachnoid space	Adheres to the outer surface of the brain and spinal cord; contains blood vessels

divided into two major categories: purulent and aseptic meningitis.

Purulent Meningitis. A patient with purulent meningitis typically has a marked, acute inflammatory exudate with large numbers of polymorphonuclear cells (PMNs). Frequently, the underlying CNS tissue, in particular the ventricles, may be involved. If the ventricles become involved, this process is referred to as **ventriculitis.** Bacteria usually cause these infections.

Pathogenesis. As discussed with other organ systems, the outcome of a host-microbe interaction depends on the characteristics of both the host and microorganism. An important host defense mechanism of the CNS is the blood-brain barrier; the choroid plexus, arachnoid membrane, and the cerebral microvascular endothelium are the key structures. Because of the unique structural properties of the vascular endothe-lium, such as continuous intercellular tight junctions, this barrier minimizes the passage of infectious agents into CSF in addition to regulating the transport of plasma proteins, glucose, and electrolytes.

Age of the host and other underlying host factors also contribute to whether an individual is predisposed to develop meningitis. Neonates have the highest prevalence of meningitis. This higher prevalence is probably due to the immature immune system of the newborn, the organisms present in the colonized female vaginal tract, and the increased permeability of the blood-brain barrier of newborns. Lack of demonstrable humoral antibody against *Haemophilus influenzae* type b in children has been associated with an increased incidence of meningitis. Before widespread vaccination, most children developed measurable antibody by age 5. The importance of antibody is also a factor in adults, because military recruits without antibody to *Neisseria meningitidis* are more likely to develop disease. *N. meningitidis*

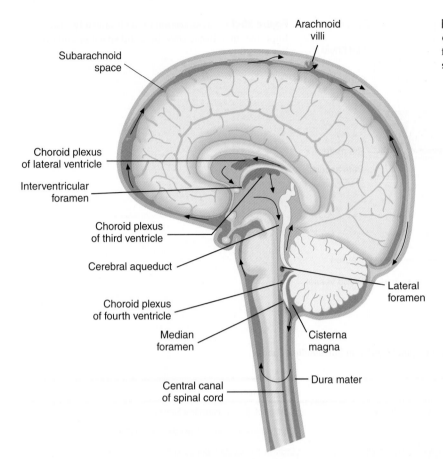

Figure 55-2   Flow of CSF through the brain. CSF originates in the choroid plexus, then flows through the ventricles and subarachnoid space and into the bloodstream.

Arachnoid villi

Subarachnoid space

Choroid plexus of lateral ventricle

Interventricular foramen

Choroid plexus of third ventricle

Cerebral aqueduct

Choroid plexus of fourth ventricle

Median foramen

Central canal of spinal cord

Lateral foramen

Cisterna magna

Dura mater

has been associated with epidemic meningitis among young adults in crowded conditions (e.g., military recruits and college dormitory mates).

Because the respiratory tract is the primary portal of entry for many etiologic agents of meningitis, factors that predispose adults to meningitis are often the same factors that increase the likelihood that the adult will develop pneumonia or other respiratory tract colonization or infection. Alcoholism, splenectomy, diabetes mellitus, prosthetic devices, and immunosuppression contribute to increased risk. Finally, patients with prosthetic devices, particularly CNS shunts, are also at increased risk of developing meningitis.

For organisms to reach the CNS (primarily by the blood-borne route), host defense mechanisms must be overcome. Most cases of meningitis caused by bacteria share a similar pathogenesis. A successful meningeal pathogen must first sequentially colonize and cross host mucosal epithelium and then enter and survive in the bloodstream. The most common causes of meningitis possess the ability to evade host defenses at each of these levels. For example, clinical isolates of *Streptococcus pneumoniae* and *N. meningitidis* secrete IgA proteases that destroy the action of the host's secretory IgA, thereby facilitating bacterial attachment to the epithelium.[8] In

addition, all of the most common etiologic agents of bacterial meningitis possess an antiphagocytic capsule that helps the organisms evade destruction by the host immune system.

Organisms appear to enter the CNS by interacting and subsequently breaking down the blood-brain barrier at the level of microvascular endothelium. To date, one of the least understood processes in the pathogenesis of meningitis is how organisms cross this barrier into the subarachnoid space. Nevertheless, there appear to be specific bacterial surface components, such as pili, polysaccharide capsules, and lipoteichoic acids, that facilitate adhesion of the organisms to the microvascular endothelial cells and subsequent penetration into the CSF.[7,14,16] Organisms can enter through (1) loss of capillary integrity by disrupting tight junctions of the blood-brain barrier, (2) transport within circulating phagocytic cells, or (3) by crossing the endothelial cell lining within endothelial cell vacuoles. After gaining access, the organism multiplies within the CSF, a site initially free of antimicrobial antibodies or phagocytic cells.

Clinical Manifestations. Meningitis can be either acute or chronic in onset and progression of disease.

Table 55-2 Guidelines for Interpretation of Results Following Hematologic and Chemical Analysis of Cerebrospinal Fluid (CSF) from Children and Adults (Excluding Neonates)

Clinical Setting	Leukocytes/mm³	Predominant Cell Type	Protein	Glucose*
Normal	0-5	None	15-50 mg/dL	45-100 mg/dL
Viral infection	2-2000 (mean of 80)	Mononuclear†	Slightly elevated (50-100 mg/dL) or normal	Normal
Purulent infection	5-20,000 (mean of 800)	PMN	Elevated (>100 mg/dL)	Low (<45 mg/dL), but may be normal early in the course of disease
Tuberculosis and fungi	5-2000 (mean of 100)	Mononuclear	Elevated (>50 mg/dL)	Normal or often low (>45 mg/dL)

*Must consider CSF glucose level in relation to blood glucose level. Normally, CSF glucose serum ratio is 0.6, or 50% to 70% of the blood glucose normal value.
†About 20% to 75% of cases may have PMN leukocytosis early in the course of infection.

ACUTE. Cases of acute meningitis are characterized by fever, stiff neck, headache, nausea and vomiting, neurologic abnormalities, and change in mental status.

With acute bacterial meningitis, CSF usually contains large numbers of inflammatory cells (>1000/mm³), primarily polymorphonuclear neutrophils. The CSF shows a decreased glucose level relative to the serum glucose level (the normal ratio of CSF to serum glucose is approximately 0.6), and shows an increased protein concentration (normal protein is 15 to 50 mg/dL in adults and as high as 170 mg/dL, with an average of 90 mg/dL, in newborns).

The sequelae of acute bacterial meningitis in children are frequent and serious. Seizures occur in 20% to 30% of patients seen at large urban hospitals, and other neurologic changes are common. Acute sequelae include cerebral edema, hydrocephalus, cerebral herniation, and focal neurologic changes. Permanent deafness can occur in 10% of children who recover from bacterial meningitis. More subtle physiologic and psychologic sequelae may follow an episode of acute bacterial meningitis.

CHRONIC. Chronic meningitis often occurs in patients who are immunocompromised, although this is not always the case. Patients experience an insidious onset of disease, with some or all of the following: fever, headache, stiff neck, nausea and vomiting, lethargy, confusion, and mental deterioration. Symptoms may persist for a month or longer before treatment is sought. The CSF usually manifests an abnormal number of cells (usually lymphocytic), elevated protein, and some decrease in glucose content (Table 55-2). The pathogenesis of chronic meningitis is similar to that of acute disease.[16]

Epidemiology/Etiologic Agents. The etiology of acute meningitis is dependent on the age of the patient.

Most cases in the United States occur in children younger than 5 years of age. Before 1985, *H. influenzae* type b (Hib) was the most common agent in children in the United States between 1 month and 6 years of age. Ninety-five percent of all cases were due to Hib, *Neisseria meningitidis,* and *Streptococcus pneumoniae.* In 1985, the first Hib vaccine, a polysaccharide vaccine, was licensed for use in children 18 months of age or older but was not efficacious in children younger than 18 months. However, the widespread use of conjugate vaccine, Hib polysaccharide-protein conjugate, in children as young as 2 months of age has significantly affected the incidence of invasive *H. influenzae* type b disease; the total number of annual cases of Hib disease in the United States have been reduced by 55% and the number of cases of *H. influenzae* meningitis by 94%.[10] To date the risks of meningococcal and pneumococcal diseases have remained level. After 6 years of age, patients are less likely to develop meningitis.

As previously mentioned, neonates have the highest prevalence of meningitis, with a concomitant increased mortality rate (as high as 20%). Organisms causing disease in the newborn are different from those that affect other age groups; many of them are acquired by the newborn during passage through the mother's vaginal vault. Neonates are likely to be infected with, in order of incidence, group B streptococci, *Escherichia coli,* other gram-negative bacilli, and *Listeria monocytogenes;* occasionally other organisms may be involved. For example, *Chryseobacterium meningosepticum (Flavobacterium meningosepticum)* has been associated with nursery outbreaks of meningitis. This organism is a normal inhabitant of water in the environment and is presumably acquired nosocomially.

Important causes of meningitis in the adult, in addition to the meningococcus in young adults, include pneumococci, *Listeria monocytogenes,* and, less commonly, *Staphylococcus aureus* and various gram-negative bacilli.

BOX 55-1 **Etiologic Agents of Chronic Meningitis**

Mycobacterium tuberculosis
Cryptococcus neoformans
Coccidioides immitis
Histoplasma capsulatum
Blastomyces dermatitidis
Candida spp.
Miscellaneous other fungi
Nocardia
Actinomyces
Treponema pallidum
Brucella
Borrelia burgdorferi
Sporothrix schenckii
Rare parasites—*Toxoplasma gondii*, cysticercus, *Pargonimus westermani*, *Trichinella spiralis*, *Schistosoma* spp., *Acanthamoeba*

Meningitis caused by the latter organisms results from hematogenous seeding from various sources, including urinary tract infections. Of note, the percentage of adults with nosocomial bacterial meningitis at large urban hospitals has increased.[13] The various etiologic agents of chronic meningitis are listed in Box 55-1.

Aseptic Meningitis. Aseptic meningitis is characterized by an increase of lymphocytes and other mononuclear cells (pleocytosis) in the CSF (in contrast to purulence, the PMN response characteristic of bacterial meningitis) and negative bacterial and fungal cultures. Patients may have fever, headache, a stiff neck, and nausea and vomiting.

Aseptic meningitis is commonly associated with viral infections and is usually a self-limiting infection. Aseptic meningitis is also a component of syphilis and some other spirochetal diseases (e.g., leptospirosis and Lyme borreliosis). Stiff neck and CSF pleocytosis may also be produced by various processes, such as malignancy, that are not CNS infections.

ENCEPHALITIS/MENINGOENCEPHALITIS

Encephalitis is an inflammation of the brain parenchyma and is usually a result of viral infection. Concomitant meningitis that occurs with encephalitis is known as **meningoencephalitis,** and the cellular infiltrate is more likely to be lymphocytic in this situation.

The host response to these CNS infections can differ somewhat from those associated with purulent or aseptic meningitis. Early in the course of viral encephalitis, or when considerable tissue damage occurs as a part of encephalitis, the nature of the inflammatory cells found in the CSF may be no different from that associated with bacterial meningitis; cell counts, however, are typically much lower.

Viral

Viral encephalitis, which cannot always be distinguished clinically from meningitis, is common in the warmer months. The primary agents are enteroviruses (coxsackieviruses A and B, echoviruses), mumps virus, herpes simplex virus, and arboviruses (West Nile virus, togavirus, bunyavirus, equine encephalitis, St. Louis encephalitis, and other encephalitis viruses). Other viruses, such as measles, cytomegalovirus, lymphocytic choriomeningitis, Epstein-Barr virus, hepatitis, varicella-zoster virus, rabies virus, myxoviruses, and paramyxoviruses, are less commonly encountered. Any preceding viral illness and exposure history are important considerations in establishing a cause by clinical means.

Involvement of the nervous system in patients who are infected with the human immunodeficiency virus (HIV) is common. HIV is a neurotropic (attracted to nerve cells) virus that enters the CNS by macrophage transport and causes various neurologic syndromes. As HIV-infected individuals become progressively more immunosuppressed, the CNS becomes a target for opportunistic pathogens, such as cytomegalovirus, BK virus, and JC virus, which can produce meningitis and/or encephalitis.[3]

Parasitic

Parasites can cause meningoencephalitis, brain abscess (see the following discussion), or other CNS infection via two routes. The free-living amebae, *Naegleria fowleri* and *Acanthamoeba* spp., invade the brain via direct extension from the nasal mucosa. These organisms are acquired by swimming or diving in natural, stagnating freshwater ponds and lakes.

Other parasites reach the brain via hematogenous spread. Toxoplasmosis, caused by a parasite that grows intracellularly and destroys brain parenchyma, is a common CNS affliction in HIV-infected patients with acquired immunodeficiency syndrome (AIDS). *Entamoeba histolytica* and *Strongyloides stercoralis* have been visualized in brain tissue, and the larval form of *Taenia solium* (the pork tapeworm), called a cysticercus, can travel to the brain via the bloodstream and encyst in that site. Amebic brain infection and cysticercosis cause changes in the CSF that mimic meningitis.[17]

BRAIN ABSCESS

Brain abscesses (localized collections of pus in a cavity formed by the breakdown of tissue) may occasionally cause changes in the CSF and clinical symptoms that mimic meningitis. Brain abscesses may also rupture into the subarachnoid space, producing a severe meningitis with high mortality. If anaerobic organisms or viridans streptococci are recovered from CSF cultures, the diagnosis of brain abscess must be entertained; however,

CSF culture is typically negative in brain abscess. Patients who are immunosuppressed or who have diabetes with ketoacidosis may show a rapidly progressive fungal infection (phycomycosis) of the nasal sinuses or palatal region that travels directly to the brain.

LABORATORY DIAGNOSIS OF CENTRAL NERVOUS SYSTEM INFECTIONS

MENINGITIS

Except in unusual circumstances, a lumbar puncture (spinal tap) is one of the first steps in the workup of a patient with suspected CNS infection, in particular, meningitis. Refer to Table 5-1 to review collection, transport, and processing of specimens obtained from the central nervous system.

Specimen Collection and Transport

CSF is collected by aseptically inserting a needle into the subarachnoid space, usually at the level of the lumbar spine. Three or four tubes of CSF should be collected and immediately labeled with the patient's name. Tube 3 or 4 is used for cell count and differential. If a small capillary blood vessel is inadvertently broken during the spinal tap, blood cells picked up from this source will usually be absent from the last tube collected; comparison of counts between tubes 1 and 4 is occasionally needed if such bleeding is suspected. The other tubes can be used for both microbiologic and chemical studies, including protein and glucose levels. In this way, a larger proportion of the total fluid can be concentrated, facilitating detection of infectious agents present in low numbers, and the supernatant can still be used for other required studies. The volume of CSF is critical for detection of certain microorganisms, such as mycobacteria and fungi. A minimum of 5 to 10 mL is recommended for detection of these agents by centrifugation and subsequent culture. When an inadequate volume of CSF is received, the physician should be consulted regarding the order of priority of laboratory studies. Processing too little specimen lowers the sensitivity of the testing, which leads to false-negative results. This is potentially more harmful to patient care than performing an additional lumbar puncture to obtain the necessary or required amount of sample.

CSF should be hand-delivered immediately to the laboratory. Specimens should never be refrigerated. Certain agents, such as *Streptococcus pneumoniae*, may not be detectable after an hour or longer unless antigen detection methods are used. If not rapidly processed, CSF should be incubated (35° C) or left at room temperature. One exception to this rule involves CSF for viral studies. These specimens may be refrigerated for as long as 23 hours after collection or frozen at −70° C if a longer delay is anticipated until they are inoculated. CSF for viral studies should never be frozen at temperatures above −70° C.

CSF is one of the few specimens handled by the laboratory for which information promptly relayed to the clinician can directly affect therapeutic outcome. Such specimens should be processed immediately upon receipt in the laboratory and all results reported to the physician.

Initial Processing

Initial processing of CSF for bacterial, fungal, or parasitic studies includes centrifugation of all specimens greater than 1 mL in volume for at least 15 minutes at $1500 \times g$. Specimens in which cryptococci or mycobacteria are suspected must be handled differently. (Discussions of techniques for culturing CSF for mycobacteria and fungi are found in Chapters 45 and 50, respectively.) The supernatant is removed to a sterile tube, leaving approximately 0.5 mL of fluid in which to suspend the sediment before visual examination or culture. Mixing of the sediment after the supernatant has been removed is critical. Forcefully aspirating the sediment up and down into a sterile pipette several times will adequately disperse the organisms that remained adherent to the bottom of the tube after centrifugation. Laboratories that use a sterile pipette to remove portions of the sediment from underneath the supernatant will miss a significant number of positive specimens. The supernatant can be used to test for the presence of antigens or for chemistry evaluations (e.g., protein, glucose, lactate, C-reactive protein). As a safeguard, keep the supernatant even if it has no immediate use.

CSF Findings

As previously mentioned, CSF is also removed for analysis of cells, protein, and glucose. Ideally, the glucose content of the peripheral blood is determined simultaneously for comparison to that in CSF. General guidelines for interpretation of results are shown in Table 55-2.

Because the results of hematologic and chemical tests directly relate to the probability of infection, communication between the physician and the microbiology laboratory is essential. Among 555 cerebrospinal fluid samples from patients older than 4 months of age tested at the University of California–Los Angeles, only 2 showed normal cell count and protein in the presence of bacterial meningitis.[4] Thus, the diagnosis of acute bacterial meningitis can be excluded in patients with normal fluid parameters in almost all cases, precluding further expensive and labor-intensive microbiologic processing beyond a standard smear and culture (which must be included in all cases). Similar criteria have been used to exclude performance of smear and culture for tuberculosis, as well as syphilis serology, on CSF specimens.[1,2]

Visual Detection of Etiologic Agents

Following centrifugation, the resulting CSF sediment may be visually examined for the presence of cells and organisms.

Stained Smear of Sediment. Gram stain must be performed on all CSF sediments. False-positive smears have resulted from inadvertent use of contaminated slides. Therefore, use of alcohol-dipped and flamed, or autoclaved slides is recommended. After thoroughly mixing the sediment, a heaped drop is placed on the surface of a sterile or alcohol-cleaned slide. The sediment should never be spread out on the slide surface, because this increases the difficulty of finding small numbers of microorganisms. The drop of sediment is allowed to air dry, is heat- or methanol-fixed, and is stained by either Gram (Figure 55-3) or acridine orange stains. The acridine orange fluorochrome stain may

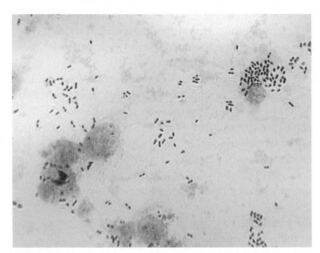

Figure 55-3   Gram stain of cerebrospinal fluid showing white blood cells and many gram-positive diplococci. This specimen subsequently grew *Streptococcus pneumoniae*.

allow faster examination of the slide under high-power magnification (400×) and thus a more thorough examination. The brightly fluorescing bacteria will be easily visible. All suspicious smears can be restained by the Gram stain (directly over the acridine orange stain) to confirm the presence and morphology of any organisms seen.

Using a cytospin centrifuge to prepare slides for staining has also been found to be an excellent alternative procedure. Use of this method for preparing smears for staining concentrates cellular material and bacterial cells up to a 1000-fold. By centrifugation, a small amount of CSF (or other body fluid) is concentrated onto a circular area of a microscopic slide (Figure 55-4), fixed, stained, and then examined.

The presence or absence of bacteria, inflammatory cells, and erythrocytes should be reported following examination. Based on demographic and clinical patient data and Gram stain morphology, the etiology of the majority of cases of bacterial meningitis can be presumptively determined within the first 30 minutes after receiving the specimen.

Wet Preparation. Amoebas are best observed by examining thoroughly mixed sediment as a wet preparation under phase-contrast microscopy. If a phase-contrast microscope is not available, observing under light microscopy with the condenser closed slightly is an alternative technique. Amoebas are recognized by typical slow, methodical movement in one direction by advancing pseudopodia. (The organisms may require a little time under the warm light of the microscope before they begin to move.) Organisms must be distinguished from motile macrophages, which occasionally occur in CSF. Following a suspicious wet preparation, a trichrome stain can help differentiate amoebas from somatic cells. The pathogenic amoebas can be cultured on a lawn of *Klebsiella pneumoniae* or *Escherichia coli* (see Chapter 49).

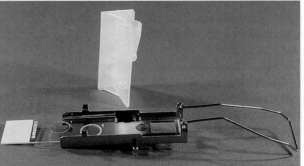

Figure 55-4   **A,** Cytocentrifuge. **B,** Device used to prepare the concentrated smears of material from body fluid specimens such as CSF by cytocentrifugation. (**A** Courtesy Cytospin 2, Shandon, Inc., Pittsburgh, Pa.)

India Ink Stain. The large polysaccharide capsule of *Cryptococcus neoformans* allows these organisms to be visualized by the India ink stain. However, latex agglutination testing for capsular antigen is more sensitive and extremely specific so that antigen testing is recommended for use in place of an India ink stain. Furthermore, strains of *C. neoformans* that infect patients with AIDS may not possess detectable capsules, so culture is also essential. To perform the India ink preparation, a drop of CSF sediment is mixed with one-third volume of India ink (Pelikan Drawing Ink, Block, Gunther, and Wagner; available at art supply stores). The India ink can be protected against contamination by adding 0.05 mL thimerosal (Merthiolate, Sigma Chemical Co., St. Louis, Mo) to the bottle when first opened. After mixing the CSF and ink to make a smooth suspension, a coverslip is applied to the drop and the preparation is examined under high-power magnification (400×) for characteristic encapsulated yeast cells, which can be confirmed by examination under oil immersion. The inexperienced microscopist must be careful not to confuse white blood cells with yeast. The presence of encapsulated buds, smaller than the mother cell, is diagnostic.

Direct Detection of Etiologic Agents

Antigen. Reagents and complete systems for the rapid detection of antigen in CSF are available.

Bacteria. Rapid antigen detection from CSF has been largely accomplished by the techniques of latex agglutination and coagglutination (see Chapter 9). All commercial agglutination systems use the principle of an antibody-coated particle that will bind to specific antigen, resulting in macroscopically visible agglutination. The soluble capsular polysaccharide produced by the most common etiologic agents of meningitis, including the group B streptococcal polysaccharide, are well suited to serve as bridging antigens. The systems differ in that certain antibodies are polyclonal and others are monoclonal, and not all systems detect all antigens. These reagents should only be used as an adjunct to standard procedures.

In general, the commercial systems have been developed for use with CSF, urine, or serum, although results with serum have not been as useful diagnostically as those with CSF. Soluble antigens from *Streptococcus agalactiae* and *Haemophilus influenzae* may concentrate in the urine. Urine, however, seems to produce a higher incidence of nonspecific reactions than either serum or CSF. The manufacturers' directions must be followed for performance of antigen detection test systems for different specimen types. Although some of the systems require pretreatment of samples (usually heating for 5 minutes), not all manufacturers recommend such a

step. The reagents, however, may yield false-positive or cross-reactions unless specimen pretreatment is performed. Interference by rheumatoid factor and other substances, more often present in body fluids other than CSF, has also been reported. The method of Smith and colleagues[11] has been shown to effectively reduce a substantial portion of nonspecific and false-positive reactions, at least for tests performed with latex particle reagents. This pretreatment, called **rapid extraction of antigen procedure** (**REAP**; Procedure 55-1), is recommended for laboratories that use commercial body fluid antigen detection kits. Certain commercial systems have such an extraction procedure included in their protocols.

Based on the findings of several studies, only a limited number of situations exist in which bacterial antigen testing (BAT) may be clinically useful and is warranted. Some of these situations include CSF specimens from previously treated patients and Gram stain–negative CSF specimens with abnormal parameters (elevated protein, decreased glucose, and/or an abnormal white blood cell count). Of significance, these systems are not substitutes for properly performed smears and cultures because they provide less than 100% sensitivity and specificity. In light of these limitations, practice guidelines for the diagnosis and management of bacterial meningitis do not recommend routine use of BAT.[12]

Cryptococcus neoformans. Reagents for the detection of the polysaccharide capsular antigen of *Cryptococcus neoformans* are available commercially. CSF specimens that yield positive results for cryptococcal antigen should be tested with a second latex agglutination test for rheumatoid factor. The commercial test systems incorporate rheumatoid factor testing in their protocol. A positive rheumatoid factor test renders the cryptococcal latex test uninterpretable, and the results should be reported as such, unless the rheumatoid factor antibodies have been inactivated. Undiluted specimens that contain large amounts of capsular antigen may yield a false-negative reaction caused by a prozone phenomenon. Patients with AIDS may have an antigen titer in excess of 100,000, requiring many dilutions to reach an end point. Such dilutions are usually done because the test is used for following a patient's response to treatment, as well as for initial diagnosis.

Molecular Methods. With the introduction of amplification technologies, such as polymerase chain reaction, many reports in the literature detail the application of these technologies to diagnose CNS infections caused by various microorganisms. Published data indicate that many of these assays are more sensitive and specific compared with presently available techniques, particularly of CNS infections caused by herpes simplex virus and enteroviruses.[5,6] As of this writing, reagents for some of these amplification assays are commercially available as analyte specific reagents (ASRs) for both conventional and real-time PCR assays.[9]

Miscellaneous Tests

Other tests, such as the limulus lysate test, CSF lactate determinations, C-reactive protein, mass spectrometry, and gas-liquid chromatography, have been evaluated for use in the diagnosis of CNS infections. However, the utility and value of these tests are either controversial, remain to be defined, or are impractical for routine use in the clinical laboratory.

Culture

Because the majority of cases of bacterial meningitis are usually caused by only one organism, a minimal number of media types are required.

Bacteria and Fungi. Routine bacteriologic media should include a chocolate agar plate, 5% sheep blood agar plate, and an enrichment broth, usually thioglycolate without indicator. The chocolate agar plate is needed to recover fastidious organisms, most notably *H. influenzae* and isolates of *N. meningitidis* that do not grow on blood agar plates; the use of the blood agar plate also aids in the recognition of *S. pneumoniae*. After vortexing the sediment and preparing smears, several drops of the sediment should be inoculated to each medium. Plates should be incubated at 37° C in 5% to 10% carbon dioxide (CO_2) for at least 72 hours. If a CO_2 incubator is not available, a candle jar can be used. The broth should be incubated in air at 37° C for at least 5 days. The broth cap must be loose to allow free exchange of air. If organisms morphologically resembling anaerobic bacteria are seen on the Gram stain or if a brain abscess is suspected,

an anaerobic blood agar plate may also be inoculated. These media will support the growth of almost all bacterial pathogens and several fungi.

The symptoms of chronic meningitis that prompt a physician to request fungal cultures are the same as those for tuberculous meningitis, which should always be sought by the laboratory if chronic or fungal meningitis is suspected. Cultures for mycobacteria are addressed in Chapter 45. For CSF fungal cultures, two drops of the well-mixed sediment should be inoculated onto Sabouraud dextrose agar or other non–blood-containing medium and brain-heart infusion with 5% sheep blood. Fungal media should be incubated in air at 30° C for 4 weeks. If possible, two sets of media should be inoculated, with one set incubated at 30° C and the other at 35° C.

Parasites and Viruses. Conditions for the culture of free-living amoebae and viral agents are discussed in Chapters 49 and 51, respectively. The physician must notify the laboratory to culture these agents.

Brain Abscess/Biopsies
Specimen Collection, Transport and Processing. Whenever possible, biopsy specimens or aspirates from brain abscesses should be submitted to the laboratory under anaerobic conditions. Several devices are commercially available to transport biopsy specimens under anaerobic conditions. Swabs are not considered an optimum specimen but if used to collect abscess material they should be sent in a transport device that maintains an anaerobic environment.

Biopsy specimens should be homogenized in sterile saline before plating and smear preparation. This processing should be kept to a minimum to reduce oxygenation.

◼ *Procedure 55-1*

RAPID EXTRACTION OF ANTIGEN PROCEDURE (REAP)

PRINCIPLE
Removal of nonspecific cross-reactive material can improve the specificity of direct antigen detection particle agglutination tests. Ethylene-diaminetetraacetic acid (EDTA) forms complexes with cross-reactive materials, and they are removed from the reaction mixture by centrifugation.

METHOD
1. Pipette 0.05 mL fluid to be tested (CSF, serum, or urine) into a 1.5-mL plastic, conical microcentrifuge tube.

2. Add 0.15 mL of 0.1 M EDTA (Sigma Chemical Co.) to the microcentrifuge tube, close the cap tightly, and vortex the tube.

3. Heat in a dry bath (available from instrument supply companies) for 3 minutes at 100° C.

4. Centrifuge the tubes for 5 minutes at 13,000× *g* in a tabletop microcentrifuge. Be certain that the instrument achieves the required centrifugal force.

5. Remove the supernatant with a capillary pipette and use 1 drop of this

solution as the test sample in the antigen detection test, following the manufacturer's instructions for performance of the test.

EXPECTED RESULTS
Nonspecific agglutination should not occur.

Culture. Abscess and biopsy specimens submitted for culture should be inoculated onto 5% sheep blood and chocolate agar plates. Plates should be incubated in 5% to 10% CO_2 for 72 hours at 35° C. In addition, an anaerobic agar plate and broth with an anaerobic indicator, vitamin K, and hemin should be inoculated and incubated in an anaerobic environment at 35° C. Anaerobic culture plates are incubated for a minimum of 72 hours but are examined after 48 hours of incubation. Anaerobic broths should be incubated for a minimum of 5 days. If a fungal etiology is suspected, fungal media, such as brain-heart infusion with blood and antibiotics or inhibitory mold agar, should be inoculated as well.

Case Study

A 2-year-old girl presented at midnight to the hospital emergency department with a temperature of 104° F. She was diagnosed with bilateral otitis. Her parents were given a prescription for amoxicillin/clavulanic acid, which they were unable to have filled that night. In the morning they found the child lethargic and having seizures. They returned to the hospital where the child presented with fever of 105° C, purpura, and nuchal rigidity. CSF cultures were collected. The child was started on ceftriaxone. The next day the laboratory reported growth of a gram-negative diplococcus.

QUESTIONS

1. What is the suspected organism of this infection and how can the laboratory rapidly identify it?
2. Is it recommended that laboratories do susceptibility testing for *N. meningitidis*?
3. How can the laboratory improve the speed with which it detects this organism in CSF?
4. What measures are taken to prevent the spread of infection among health care workers who are exposed to patients with *N. meningitidis*?

REFERENCES

1. Albright RE, Christenson RH, Emlet JL, et al: Issues in cerebrospinal fluid management: acid-fast bacillus smear and culture, *Am J Clin Pathol* 95:48, 1991.
2. Albright RE, Graham CB, Christenson RH, et al: Issues in cerebrospinal fluid management: CSF venereal disease research laboratory testing, *Am J Clin Pathol* 95:387, 1991.
3. Gradon JD, Timpone JG, Schnittman SM: Emergence of unusual opportunistic pathogens in AIDS: a review, *Clin Infect Dis* 15:134, 1992.
4. Hayward RA, Shapiro MF, Oye RK: Laboratory testing on cerebrospinal fluid: a reappraisal, *Lancet* 1:1, 1987.
5. Huang C, Morse D, Slater B, et al: Multiple-year experience in the diagnosis of viral central nervous system infections with a panel of polymerase chain reaction assays for detection of 11 viruses, *Clin Infect Dis* 39:630, 2004.
6. Korimbocus J, Scaramossino N, Lacroix B, et al: DNA probe array for the simultaneous identification of herpesviruses, enteroviruses, and flaviviruses, *J Clin Micrbiol* 43: 3779, 2005.
7. Parkkinen J, Korhonen TK, Pere A, et al: Binding sites in the rat brain for *Escherichia coli* S fimbriae associated with neonatal meningitis, *J Clin Invest* 81:860, 1988.
8. Plaut AG: The IgA1 proteases of pathogenic bacteria, *Ann Rev Microbiol* 37:603, 1983.
9. Poppert S, Essig A, Stoehr B, et al: Rapid diagnosis of bacterial meningitis by real-time PCR and fluorescence in situ hybridization, *J Clin Microbiol* 43:3390, 2005.
10. Schwartz MN: Bacterial meningitis—a view of the past 90 years, *N Engl J Med* 351:1826, 2004.
11. Smith LP, Hunter KW Jr, Hemming VG, et al: Improved detection of bacterial antigens by latex agglutination after rapid extraction from body fluids, *J Clin Microbiol* 20:981, 1984.
12. Tunkel AR, Hartman BJ, Kaplan SL, et al: Practice guidelines for the management of bacterial meningitis, *Clin Infect Dis* 39:1267, 2004.
13. van de Beek D, de Gan J, Spanjaard L, et al: Clinical features and prognostic factors in adults with bacterial meningitis, *N Engl J Med* 351:1849, 2004.
14. Virji M, Alexandrescu C, Ferguson DJ, et al: Variations in the expression of pili: the effect on adherence of *Neisseria meningitidis* to human epithelial and endothelial cells, *Mol Microbiol* 6:1271, 1992.
15. Virji M, Kayhty H, Ferguson DJ, et al: The role of pili in the interactions of pathogenic *Neisseria* with cultured human endothelial cells, *Mol Microbiol* 5:1831, 1991.
16. Wilhelm C and Ellner JJ: Chronic meningitis, *Neurol Clin* 4:115, 1986.

EYES

ANATOMY

Only a small portion of the eye is exposed to the environment; about five sixths of the eyeball is enclosed within bony orbits that are shaped like four-sided pyramids. In general, ocular (eye) infections can be divided into those that involve the exposed, or external, structures and those that involve internal sites.

The external structures of the eye—eyelids, conjunctiva, sclera, and cornea—are shown in Figure 56-1. The eyeball comprises three layers. From the outside in, these tissues are the sclera, choroid, and retina. The sclera is a tough, white, fibrous tissue (i.e., "white" of the eye). The anterior (toward the front) portion of the sclera is the cornea, which is transparent and has no blood vessels. A mucous membrane, called the **conjunctiva,** lines each eyelid and extends onto the surface of the eye itself.

The large interior space of the eyeball is divided into two sections: the anterior and posterior cavities (see Figure 56-1). The anterior cavity is filled with a clear and watery substance called **aqueous humor;** the posterior cavity is filled with a soft, gelatin-like substance called **vitreous humor.**

Infections can also occur in the eyes' lacrimal (pertaining to tears) system. The major components of the lacrimal apparatus include the lacrimal gland, lacrimal canaliculi (short channel), and lacrimal sac.

RESIDENT MICROBIAL FLORA

Rather sparse indigenous flora exist in the conjunctival sac. *Staphylococcus epidermidis* and *Lactobacillus* spp. are the most frequently encountered organisms; *Propionibacterium acnes* may also be present. *Staphylococcus aureus* is found in less than 30% of people, and *Haemophilus influenzae* colonizes 0.4% to 25%. *Moraxella catarrhalis,* various *Enterobacteriaceae,* and various streptococci (*Streptococcus pyogenes, Streptococcus pneumoniae,* other alpha-hemolytic and gamma-hemolytic forms) are found in a very small percentage of people.

DISEASES

The eye and its associated structures are uniquely predisposed to infection by various microorganisms. The major infections of the eye are listed in Table 56-1 along with a brief description of the disease.

PATHOGENESIS

The eye has a number of defense mechanisms. The eyelashes prevent entry of foreign material into the eye. The lids blink 15 to 20 times per minute, during which time secretions of the lacrimal glands and goblet cells wash away bacteria and foreign matter. Lysozyme and immunoglobulin A (IgA) are secreted locally and serve as part of the eye's natural defense mechanisms. Also, the eyes themselves are enclosed within the bony orbits. The delicate intraocular structures are enveloped in a tough collagenous coat (sclera and cornea). If these barriers are broken by a penetrating injury or ulceration, infection may occur. Infection can also reach the eye via the bloodstream from another site of infection. Finally, because three of the four walls of the orbit are contiguous with the paranasal (facial) sinuses, sinus infections may extend directly to the periocular orbital structures.

EPIDEMIOLOGY AND ETIOLOGY OF DISEASE

Blepharitis

Bacteria, viruses, and occasionally, lice can cause blepharitis. Although occasionally isolated from surfaces surrounding the healthy eye, *Staphylococcus aureus* and *S. epidermidis* are the most common infectious cause of blepharitis in developed countries. Symptoms include burning, itching, sensation of a foreign body, and crusting of the eyelids.

Viruses can also cause a vesicular (blisterlike) eruption of the eyelids. Herpes simplex virus (HSV) produces vesicles on the eyelids that typically crust and heal with scarring over 2 weeks. Unfortunately, once this vesicular stage has resolved, the lesions can be confused with bacterial blepharitis.

Finally, the pubic louse *Phthirus pubis* has a predilection for eyelash hair. Presence of this organism produces irritation, itch, and swelling of the lid margins (edges).

Conjunctivitis

Bacterial conjunctivitis is the most common type of ocular infection. The principal causes of acute conjunctivitis

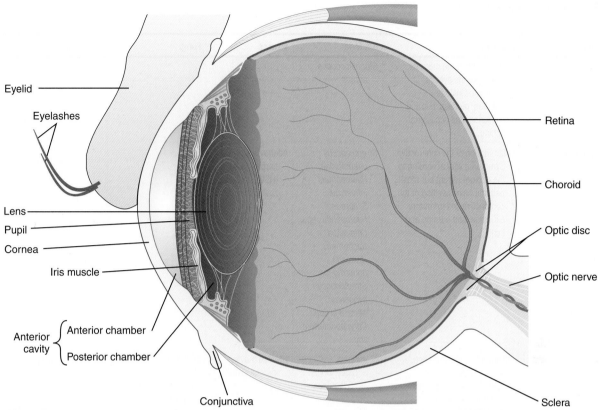

Figure 56-1   Key anatomic structures of the eye. (Modified from Thibodeau GA, Patton KT: *Anatomy and physiology,* ed 2, St Louis, 1993, Mosby.)

in the normal host are listed in Table 56-1. Age-related factors are key determinants of the etiologic agent. In neonates, neisserial and chlamydial infections are frequent and are acquired during passage through an infected vaginal canal. With the common practice of instilling antibiotic drops into the eyes of newborns in the United States, the incidence of gonococcal and chlamydial conjunctivitis has dropped dramatically. However, *Chlamydia trachomatis* is responsible for one of the most important types of conjunctivitis, trachoma, one of the leading causes of blindness in the world, primarily in underdeveloped countries.[7]

In children the most common causes of bacterial conjunctivitis are *Haemophilus influenzae, S. pneumoniae,* and perhaps *S. aureus. S. pneumoniae* and *H. influenzae* (especially subsp. *aegyptius*) have been responsible for epidemics of conjunctivitis.

Numerous other bacteria may also cause conjunctivitis. For example, diphtheritic conjunctivitis may occur in conjunction with diphtheria elsewhere in the body. *Moraxella lacunata* produces a localized conjunctivitis with little discharge from the eye. Distinctive clinical pictures may also occur with conjunctivitis caused by *Mycobacterium tuberculosis, Francisella tularensis, Treponema pallidum,* and *Yersinia enterocolitica.*

Fungi may be responsible for this type of infection as well, often in association with a foreign body in the eye or an underlying immunologic problem. However, these infections are infrequently encountered.

Viruses are an important cause of conjunctivitis; 20% of such infections in children resulted from adenoviruses in one large U.S. study and 14% of infections in adult patients in another study.[4,8] Adenoviruses types 4, 3, and 7A are common. Most viral conjunctivitis is self-limited but is highly contagious, with potential to cause major outbreaks. Worldwide, enterovirus 70 and coxsackievirus A24 are responsible for outbreaks and epidemics of acute hemorrhagic conjunctivitis.

Keratitis

Keratitis (corneal infection) may be caused by various infectious agents, usually only after some type of trauma produces a defect in the ocular surface. Keratitis should be regarded as an emergency, because corneal perforation and loss of the eye can occur within 24 hours when organisms such as *Pseudomonas aeruginosa, Staphylococcus aureus,* or HSV are involved. Bacteria account for 65% to 90% of corneal infections.

In the United States, *S. aureus, S. pneumoniae,* and *P. aeruginosa* account for more than 80% of all bacterial

Table 56-1   Major Infections of the Eye

Infection	Description	Bacteria	Viruses	Fungi	Parasites
Blepharitis	Inflammation of the margins (edges) of the eyelids. Symptoms include irritation, redness, burning sensation, and occasional itching.	*Staphylococcus aureus*			
Conjunctivitis	Inflammation of the conjunctiva. Symptoms vary according to the etiologic agent but most patients have swelling of the conjunctiva, inflammatory exudates, and burning and itching.	*Streptococcus pneumoniae; Haemophilus influenzae; S. aureus; Haemophilus* spp., *Chlamydia trachomatis; Neisseria gonorrhoeae; Streptococcus pyogenes; Moraxella* spp.; *Corynebacterium* spp.	Adenoviruses; herpes simplex (HSV); varicella zoster		
Keratitis	Inflammation of the cornea. Although there are no specific clinical signs to confirm infection, most patients complain of pain and usually some decrease in vision, with or without discharge from the eye.	*S. aureus; S. pneumoniae; Pseudomonas aeruginosa; Moraxella lacunata; Bacillus* spp.	HSV; adenoviruses; varicella zoster	*Fusarium solani; Aspergillus* spp. *Candida* spp.; *Acremonium; Curvularia*	*Acanthamoeba* spp.
Keratoconjunctivitis	Infection involving both the conjunctiva and cornea. *Ophthalmia neonatorum* is an acute conjunctivitis or keratoconjunctivitis of the newborn caused by either *N. gonorrhoeae* or *C. trachomatis.*	Refer to agents for keratitis/ conjunctivitis	Refer to agents for keratitis/ conjunctivitis	Refer to agents for keratitis	*Toxoplasma gondii; Toxocara*
Chorioretinitis and uveitis	Inflammation of the retina and underlying choroid or the uvea. Infection can result in loss of vision.	*Mycobacterium tuberculosis; Treponema pallidum; Borrelia burgdorferi*	Cytomegalovirus; HSV	*Candida* spp.	*Toxoplasma gondii; Toxocara*
Endophthalmitis	Infection of the aqueous or vitreous humor. This infection is usually caused by bacteria or fungi, is rare, develops suddenly and progresses rapidly, often leading to blindness. Pain, especially while moving the eye, and decreased vision, are prominent features.	*S. aureus; S. epidermidis; S. pneumoniae;* other streptococcal spp.; *P. aeruginosa;* other gram-negative organisms	HSV	*Candida* spp.; *Aspergillus* spp.; *Volutella* spp.; *Acremonium* spp.	*Toxocara; Onchocerca volvulus*
Lacrimal infections; canaliculitis	A rare, chronic inflammation of the lacrimal canals in which the eyelid swells and there is a thick, mucopurulent discharge.	*Actinomyces; Propionibacterium propionicum*			

Table 56-1   Major Infections of the Eye—cont'd

Infection	Description	Bacteria	Viruses	Fungi	Parasites
Dacryocystis	Inflammation of the lacrimal sac that is accompanied by pain, swelling, and tenderness of the soft tissue in the medial canthal region.	S. pneumoniae; S. aureus; S. pyogenes; Haemophilus influenzae		C. albicans; Aspergillus spp.	
Dacryoadenitis	Acute infection of the lacrimal gland. These infections are rare and can be accompanied by pain, redness, and swelling of the upper eyelid, and conjunctival discharge.	S. pneumoniae; S. aureus; S. pyogenes			

corneal ulcers. Many culture-positive cases are now being recognized as polymicrobial.[15] A toxic factor known as exopeptidase has been implicated in the pathogenesis of corneal ulcer produced by *S. pneumoniae*. With *P. aeruginosa*, proteolytic enzymes are responsible for the corneal destruction. The gonococcus may cause keratitis in the course of inadequately treated conjunctivitis. *Acinetobacter*, which may look identical microscopically to the gonococcus and is resistant to penicillin and many other antimicrobial agents, can cause corneal perforation. Many other bacteria, several viruses other than HSV, and many fungi, may cause keratitis. Fungal keratitis is usually a complication of trauma.

Although still unusual, a previously rare etiologic agent of corneal infections has become more common in users of soft and extended-wear contact lenses. *Acanthamoeba* spp., free living amebae, can survive in improperly sterilized cleaning fluids and be introduced into the eye with the contact lens. Other bacterial and fungal causes of infections in such patients have also been traced to inadequate cleaning of lenses.

Endophthalmitis

Surgical trauma, nonsurgical trauma (infrequently), and hematogenous spread from distant sites of infection are the background factors in endophthalmitis. The infection may be limited to specific tissues within the eye or may involve all of the intraocular contents. Bacteria are the most common infectious agents responsible for endophthalmitis.

After surgery or trauma, evidence of the disease is usually found within 24 to 48 hours but can be delayed for several days. Postoperative infection involves primarily bacteria from the ocular surface microflora. Although *Staphylococcus epidermidis* and *S. aureus* are responsible for the majority of cases of endophthalmitis after cataract removal, any bacterium, including those considered to be primarily saprophytic, may cause endophthalmitis. In hematogenous endophthalmitis, a septic

focus elsewhere is usually evident before onset of the intraocular infection. *Bacillus cereus* has caused endophthalmitis in people addicted to narcotics and after transfusion with contaminated blood. Endophthalmitis associated with meningitis may involve various organisms, including *Haemophilus influenzae*, streptococci, and *Neisseria meningitidis*. *Nocardia* endophthalmitis may follow pulmonary infection with this organism.

Mycotic infection of the eye has increased significantly over the past three decades because of increased use of antibiotics, corticosteroids, antineoplastic chemotherapy, addictive drugs, and hyperalimentation. Fungi generally considered to be saprophytic are important causes of postoperative endophthalmitis (see Table 56-1). Endogenous mycotic endophthalmitis is most often caused by *Candida albicans*. Patients with diabetes and underlying disease are most at risk. Other causes of hematogenous ocular infection include *Aspergillus, Cryptococcus, Coccidioides, Sporothrix*, and *Blastomyces*.

Viral causes of endophthalmitis include HSV, varicella (herpes) zoster virus (VZV), cytomegalovirus, and measles viruses. The most common parasitic cause is *Toxocara. Toxoplasma gondii* is a well-known cause of chorioretinitis. Thirteen percent of patients with cysticercosis have ocular involvement. *Onchocerca* usually produces keratitis, but intraocular infection also occurs.

Periocular

Canaliculitis, one of three infections of the lacrimal apparatus (see Table 56-1), is an inflammation of the lacrimal canal and is usually caused by *Actinomyces* or *Propionibacterium propionicum* (formerly *Arachnia*). Infection of the lacrimal sac (dacryocystitis) may involve numerous bacterial and fungal agents; the major causes are listed in Table 56-1. Dacryoadenitis is an uncommon infection of the lacrimal gland characterized by pain of the upper eyelid with erythema and often involves pyogenic bacteria such as *S. aureus* and streptococci. Chronic infections of the lacrimal gland occur in tuberculosis, syphilis, leprosy, and schistosomiasis. Acute inflammation of the gland may occur in the course of mumps and infectious mononucleosis.

Orbital cellulitis is an acute infection of the orbital contents and is most often caused by bacteria. This is a potentially serious infection because it may spread posteriorly to produce central nervous system complications.[2] Most cases involve spread from contiguous sources such as the paranasal sinuses. In children, blood-borne bacteria, notably *Haemophilus influenzae*, may lead to orbital cellulitis. *S. aureus* is the most common etiologic agent; *Streptococcus pyogenes* and *S. pneumoniae* are also common. Anaerobes may cause a cellulitis secondary to chronic sinusitis, primarily in adults. Mucormycosis of the orbit is a serious, invasive fungal infection seen particularly in patients with diabetes who have poor control of their disease, patients with acidosis from other causes, and patients with malignant disease receiving cytotoxic and immunosuppressive therapy. *Aspergillus* may produce a similar infection in the same settings but also can cause mild, chronic infections of the orbit.

Newer surgical techniques involving the ocular implantation of prosthetic or donor lenses have resulted in increasing numbers of iatrogenic (resulting from the activities of a physician) infections. Isolation of *Propionibacterium acnes* may have clinical significance in such situations, in contrast to many other sites in which it is usually considered to be a contaminant.

Other Infections

Opportunistic infections in human immunodeficiency virus (HIV)-infected individuals can involve the eye. Ocular manifestations were previously reported in up to 70% of HIV-infected patients.[3] Systemic infections that involve the eye included cytomegalovirus, *Pneumocystis jiroveci*, *Cryptococcus neoformans*, *Mycobacterium avium* complex, and *Candida* spp. Most often the retina, choroid, and optic nerve are involved with these agents, resulting in significant visual morbidity (unhealthy condition) if left untreated. However, because widespread use of highly active antiretroviral therapy that restores the immune system and lowers the viral load in patients with HIV infection, the incidence of acquired immunodeficiency syndrome (AIDS) and related ophthalmic infections have declined sharply.[11]

LABORATORY DIAGNOSIS

Specimen Collection and Transport

Purulent material from the surface of the lower conjunctival sac and inner canthus (angle) of the eye is collected on a sterile swab for cultures of conjunctivitis. Both eyes should be cultured separately. Chlamydial cultures are taken with a dry calcium alginate swab and placed in 2-SP transport medium. An additional swab may be rolled across the surface of a slide, fixed with methanol, and sent if direct fluorescent antibody (DFA) chlamydia stains are used for detection.

In the patient with keratitis, an ophthalmologist should obtain scrapings of the cornea with a heat-sterilized platinum spatula. Multiple inoculations with the spatula are made to blood agar, chocolate agar, an agar for fungi, thioglycollate broth, and an anaerobic blood agar plate. Other special media may be used if indicated. For culture of HSV and adenovirus, corneal material is transferred to viral transport media. Recently, the collection of only two corneal scrapes (one used for Gram stain and the other transported in brain heart infusion medium and used for culture) was determined to provide a simple method for diagnosis of bacterial keratitis.[6]

Cultures of endophthalmitis specimens are inoculated with material obtained by the ophthalmologist from the anterior and posterior chambers of the eye, wound abscesses, and wound dehiscences (splitting open). Lid infection material is collected on a swab in a conventional manner. For microbiologic studies of canaliculitis, material from the lacrimal canal should be transported under anaerobic conditions. Aspiration of fluid from the orbit is contraindicated in patients with orbital cellulitis. Because sinusitis is the most common background factor, an otolaryngologist's assistance in obtaining material from the maxillary sinus by antral puncture is helpful. Blood cultures should also be obtained. Tissue biopsy is essential for microbiologic diagnosis of mucormycosis. Because cultures are usually negative, the diagnosis is made by histologic examination.

Direct Visual Examination

All material submitted for culture should always be smeared and examined directly by Gram stain or other appropriate techniques. In bacterial conjunctivitis, polymorphonuclear leukocytes predominate; in viral infection, the host cells are primarily lymphocytes and monocytes. Specimens in which chlamydia is suspected can be stained immediately with monoclonal antibody conjugated to fluorescein for detection of elementary bodies or inclusions. Using histologic stains, basophilic intracytoplasmic inclusion bodies are seen in epithelial cells. Cytologists and anatomic pathologists usually perform these tests. Direct examination of conjunctivitis specimens using histologic methods (Tzanck smear) may reveal multinucleated epithelial cells typical of herpes group viral infections. However, DFA stains available for both HSV and VZV are most reliable for rapid diagnosis of these viral infections. In patients with keratitis, scrapings are examined by Gram, Giemsa, periodic acid-Schiff (PAS), and methenamine silver stains. If *Acanthamoeba* or other amebae are suspected, a direct wet preparation should be examined for motile trophozoites, and a trichrome stain should be added to the regimen. For this diagnosis, however, culture is by far the most sensitive detection method. In patients with endophthalmitis, material is also examined by Gram, Giemsa, PAS, and methenamine silver stains. When submitted in large volumes of fluid, ophthalmic specimens must be concentrated by centrifugation before additional studies are performed.

Culture

Because of the constant washing action of the tears, the number of organisms recovered from cultures of certain eye infections may be relatively low. Unless the clinical specimen is obviously purulent, using a relatively large inoculum and various media are recommended to ensure recovery of an etiologic agent. Conjunctival scrapings placed directly onto media yield the best results. At a minimum, one should use blood and chocolate agar plates incubated under increased carbon dioxide tension (5% to 10% CO_2). Because potential pathogens may be present in an eye without causing infection, it may be very helpful to the clinician, when only one eye is infected, to culture both eyes. If a potential pathogen grows in cultures of the infected and the uninfected eye, the organism may not be causing the infection; however, if the organism only grows in culture from the infected eye, it is most likely the causative agent. When *Moraxella lacunata* is suspected, Loeffler's medium may prove useful; the growth of the organism often leads to proteolysis and pitting of the medium, although nonproteolytic strains may be found. If diphtheritic conjunctivitis is suspected, Loeffler's or cystine-tellurite medium should be used. For more serious eye infections, such as keratitis, endophthalmitis, and orbital cellulitis, one should always include, in addition to the media just noted, a reduced anaerobic blood agar plate, a medium for fungi, and a liquid medium such as thioglycolate broth. Blood cultures are also important in serious eye infections.

Cultures of material for chlamydiae and viruses should be inoculated to appropriate media from transport broth. Cycloheximide-treated McCoy cells for chlamydiae isolation and human embryonic kidney, primary monkey kidney, and Hep-2 cell lines for virus isolation should be inoculated.

Nonculture Methods

Although acute and convalescent serologic tests for viral agents might be used in the event of epidemic conjunctivitis, they typically are not performed because the infections are self-limited. Enzyme-linked immunosorbent assay (ELISA) tests and DFA staining are now available for detection of *Chlamydia trachomatis*. It is anticipated that the direct antigen tests should perform well, particularly because so many eyes have been partially treated before culture. An ELISA test of aqueous humor is available for diagnosis of *Toxocara* infection. Finally, single and multiplex polymerase chain reaction (PCR) assays including both conventional and real-time formats have been used to diagnose viral and chlamydial keratoconjunctivitis and other ophthalmic infections including uveitis.

EARS

ANATOMY

The ear is divided into three anatomic parts: the external, middle, and inner ear. Important anatomic structures are shown in Figure 56-2.

The middle ear is part of a continuous system that includes the nares, nasopharynx, auditory tube, and the mastoid air spaces. These structures are lined with respiratory epithelium (e.g., ciliated cells, mucus-secreting goblet cells).

RESIDENT MICROBIAL FLORA

The normal flora of the external ear canal are rather sparse, similar to flora of the conjunctival sac qualitatively except that pneumococci, *Propionibacterium acnes*, *Staphylococcus aureus*, and *Enterobacteriaceae* are encountered somewhat more often. *Pseudomonas aeruginosa* is found on occasion. *Candida* spp. (non-*C. albicans*) are also common.

DISEASES, EPIDEMIOLOGY, AND ETIOLOGY OF DISEASE

Otitis Externa (External Ear Infections)

Otitis externa is similar to skin and soft tissue infections elsewhere. Two major types of external otitis exist: acute or chronic. Acute external otitis may be localized or diffuse. Acute localized disease occurs in the form of a pustule or furuncle and typically results from *Staphylococcus aureus*. Erysipelas caused by group A streptococci may involve the external ear canal and the soft tissue of the ear itself. Acute diffuse otitis externa (swimmer's ear) is related to maceration (softening of tissue) of the ear from swimming and/or hot, humid weather. Gram-negative bacilli, particularly *Pseudomonas aeruginosa*, play an important role. A severe, hemorrhagic external otitis caused by *P. aeruginosa* is difficult to treat and has occasionally been related to hot tub use.

Chronic otitis externa results from the irritation of drainage from the middle ear in patients with chronic, suppurative otitis media and a perforated eardrum. Ma-

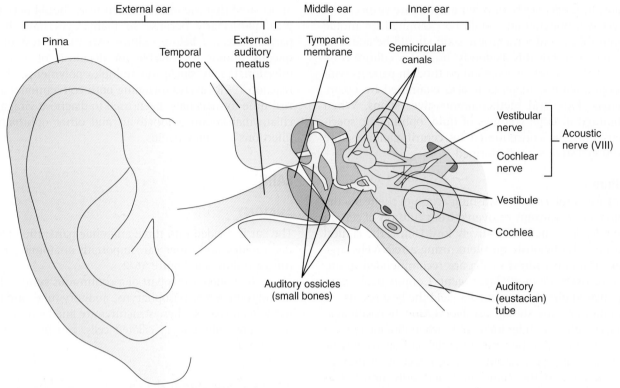

Figure 56-2 The ear. (Modified from Thibodeau GA, Patton KT: *Anatomy and physiology*, ed 2, St Louis, 1993, Mosby.)

lignant otitis externa is a necrotizing infection that spreads to adjacent areas of soft tissue, cartilage, and bone. If allowed to progress and spread into the central nervous system or vascular channel, a life-threatening situation may develop. *P. aeruginosa*, in particular, and anaerobes are frequently associated with this process. Malignant otitis media is seen in patients with diabetes who have blood vessel disease of the tissues overlying the temporal bone in which the poor local perfusion of tissues results in a milieu for invasion by bacteria. On occasion, external otitis can extend into the cartilage of the ear, usually requiring surgical intervention. Certain viruses may infect the external auditory canal, the soft tissue of the ear, or the tympanic membrane; influenza A virus is a suspected, but not an established, cause. VZV may cause painful vesicles within the soft tissue of the ear and the ear canal. *Mycoplasma pneumoniae* is a cause of bullous myringitis (a painful infection of the eardrum with hemorrhagic bullae); the ear canal itself may be involved as well.

Otitis Media (Middle Ear Infections)

In children (in whom otitis media is most common), pneumococci (33% of cases) and *Haemophilus influenzae* (20%) are the usual etiologic agents in acute disease. Group A streptococci *(Streptococcus pyogenes)* are the third most frequently encountered agents, found in 8% of cases. Other organisms, encountered in 1% to 6% of

cases, include *Moraxella catarrhalis, Staphylococcus aureus,* gram-negative enteric bacilli, and anaerobes; in one recent study, *M. catarrhalis, S. pneumoniae,* and *H. influenzae* were the most common bacterial pathogens.[9] Viruses, chiefly respiratory syncytial virus (RSV) and influenza virus, have been recovered from the middle ear fluid of 4% of children with acute or chronic otitis media. *Chlamydia trachomatis* and *Mycoplasma pneumoniae* have occasionally been isolated from middle ear aspirates. Otitis media with effusion (fluid) is considered a chronic sequela of acute otitis media. A slowly growing organism, *Alloiococcus otitidis* is a potential pathogen that is found solely in patients with otitis media with effusion.[1,5]

Chronic otitis media yields a predominantly anaerobic flora, with *Peptostreptococcus* spp., *Bacteroides fragilis* group, *Prevotella melaninogenica* (pigmented, anaerobic, gram-negative rods), and *Porphyromonas,* other *Prevotella* spp., and *Fusobacterium nucleatum* as the principal pathogens; less frequently present are *S. aureus, Pseudomonas aeruginosa, Proteus* spp., and other gram-negative facultative bacilli. Table 56-2 summarizes the major causes of ear infections.

The mastoid is a portion of the temporal bone (lower sides of the skull) that contains the mastoid sinuses (cavities). Mastoiditis is a complication of chronic otitis media in which organisms find their way into the mastoid sinuses.

Table 56-2 **Major Infectious Causes of Ear Disease**

Disease	Common Causes
Otitis externa	Acute: *Staphylococcus aureus, Streptococcus pyogenes, Pseudomonas aeruginosa;* other gram-negative bacilli Chronic: *P. aeruginosa;* anaerobes
Otitis media	Acute *Streptococcus pneumoniae; Haemophilus influenzae; Moraxella catarrhalis; S. pyogenes;* respiratory syncytial virus; influenza virus Chronic: Anaerobes

PATHOGENESIS

Local trauma, the presence of foreign bodies, or excessive moisture can lead to otitis externa (external ear infections). Infrequently, an infection from the middle ear can extend by purulent drainage to the external ear.

Anatomic or physiologic abnormalities of the auditory tube can predispose individuals to develop otitis media. The auditory tube is responsible for protecting the middle ear from nasopharyngeal secretions, draining secretions produced in the middle ear into the nasopharynx, and ventilating the middle ear so that air pressure is equilibrated with that in the external ear canal. If any of these functions becomes compromised and fluid develops in the middle ear, infection may occur. To illustrate, if a person has a viral upper respiratory infection, the auditory tube becomes inflamed and swollen. This inflammation and swelling may, in turn, compromise the auditory tube's ventilating function, thereby resulting in a negative, rather than a positive, pressure in the middle ear. This change in pressure can then allow for potentially pathogenic bacteria present in the nasopharynx to enter the middle ear.

LABORATORY DIAGNOSIS

Specimen Collection and Transport

For the laboratory diagnosis of external otitis, the external ear should be cleansed with a mild germicide such as 1:1000 aqueous solution of benzalkonium chloride to reduce the contaminating skin flora before obtaining the culture. Material from the ear, especially that obtained after spontaneous perforation of the eardrum or by needle aspiration of middle ear fluid (tympanocentesis), should be collected by an otolaryngologist, using sterile equipment. Cultures from the mastoid are generally taken on swabs during surgery, although actual bone is preferred. Specimens should be transported anaerobically.

Direct Visual Examination

Material aspirated from the middle ear or mastoid is also examined directly for bacteria and fungi. The calcofluor

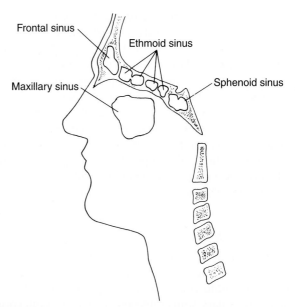

Figure 56-3 Location of the paranasal sinuses. (From Milliken ME, Campbell G: *Essential competencies for patient care,* St Louis, 1985, Mosby.)

white or PAS stains can reveal fungal elements. Methenamine silver stains have the added efficiency of staining most bacterial, fungal, and several parasitic species.

Culture and Nonculture Methods

Ear specimens submitted for culture should be inoculated to blood, MacConkey, and chocolate agars. Anaerobic cultures should also be set up on those specimens obtained by tympanocentesis or those obtained from patients with chronic otitis media or mastoiditis. Because cultures of middle ear effusions are culture-positive for only 20% to 30% of patients, conventional and real-time PCR assays have been used to detect the common middle ear pathogens.

SINUSES

ANATOMY

The sinuses, like the mastoids, are unique, air-filled cavities within the head (Figure 56-3). The sinuses are normally sterile. These structures, as well as the eustachian tube, the middle ear, and the respiratory portion of the pharynx, are lined by respiratory epithelium. The clearance of secretions and contaminants depends on normal ciliary activity and mucous flow.

DISEASES

Acute sinusitis usually develops during the course of a cold or influenzal illness and tends to be self-limited, lasting 1 to 3 weeks. Acute sinusitis is often difficult to distinguish from the primary illness. Symptoms include purulent nasal and postnasal discharge, a feeling of pressure over the sinus areas of the face, cough, and a nasal quality to the voice. Fever is sometimes present.

Occasionally, acute sinusitis persists and reaches a chronic state in which bacterial colonization occurs and the condition no longer responds to antibiotic treatment. Ordinarily, surgery or drainage is required for successful management. Patients with chronic sinusitis may have acute exacerbations (flare-ups). Other complications include local extension into the orbit, skull, meninges, or brain, and development of chronic sinusitis.

PATHOGENESIS

Most cases of acute sinusitis are believed to be bacterial complications of viral colds. The exact mechanisms involved are unknown. Of significance, about 5% to 10% of acute maxillary sinus infections result from infection originating from a dental source. The maxillary sinuses are close to the roots of the upper teeth so that dental infections can extend into these sinuses. The primary problems believed to be associated with chronic sinusitis are inadequate drainage, impaired mucociliary clearance, and mucosal damage.

EPIDEMIOLOGY AND ETIOLOGY OF DISEASE

Although difficult to access, the actual incidence of acute sinusitis parallels that of acute upper respiratory tract infections (i.e., being most prevalent in the fall through spring).

Most studies of the microbiology of acute sinusitis have dealt with maxillary sinusitis because it is the most common type and the only one really accessible for puncture and aspiration. Acute viral sinusitis is one of the most common causes of respiratory tract infection and in most cases resolves without treatment. However, published estimates indicate that 0.5% to 2% of cases of acute viral sinusitis in adults are complicated by bacterial sinusitis. This scenario is even more common in children.[12] Bacterial cultures are positive in about three fourths of patients. Studies over the past 2 decades have indicated that *Streptococcus pneumoniae* and *Haemophilus influenzae* are the major bacterial pathogens in adults with acute sinusitis; other species such as beta-hemolytic and alpha-hemolytic streptococci, *Staphylococcus aureus*, and anaerobes have also been cultured but less frequently.[10]

Among children, *S. pneumoniae*, *H. influenzae*, and *M. catarrhalis* are most common.[14] Rhinovirus is found in 15% of patients, influenza virus in 5%, parainfluenza virus in 3%, and adenovirus in less than 1%. The major causes of acute sinusitis are summarized in Table 56-3. A wider variety of bacteria, particularly anaerobes, are

Table 56-3   Major Infectious Causes of Acute Sinusitis

Age Group	Common Causes
Young adults	*Haemophilus influenzae, Streptococcus pneumoniae, Streptococcus pyogenes, Moraxella catarrhalis*
Children	*S. pneumoniae, H. influenzae, M. catarrhalis,* rhinovirus

more frequently involved in chronic sinusitis in adults. One report has shown *M. catarrhalis* to be an important agent in chronic sinusitis in children.[13]

LABORATORY DIAGNOSIS

In most cases, a diagnosis can be made on the basis of physical findings, history, radiograph studies, and other imaging techniques such as magnetic resonance imaging. However, if a laboratory diagnosis is needed, an otolaryngologist obtains the material from the maxillary sinus by puncture and aspiration or during surgery. Sinus drainage is unacceptable for smear or culture because this material will be contaminated with aerobic and anaerobic normal respiratory flora. Once received by the laboratory, Gram-stained smears and aerobic and anaerobic cultures should be performed. Aerobic culture media should include blood, chocolate, and MacConkey agars.

Case Study

A 12-year-old boy complained of severe ear pain of 4 days' duration. He was afebrile, but his tympanic membrane was erythematous and bleeding. The boy had been swimming a few days earlier in a local lake. His physician took cultures, which grew a gram-negative rod with a pleasant odor. The patient was given antibiotic eardrops and did well, with resolution of his symptoms.

QUESTIONS

1. What organism caused this infection?
2. How can this isolate be identified rapidly?
3. If the characteristic odor is lacking, what characteristics of the organism make it easy to identify?
4. Name the other fluorescent *Pseudomonas* spp.

REFERENCES

1. Bosley GS, Whitney AM, Pruckler JM, et al: Characterization of ear fluid isolates of *Alloiococcus otitidis* from patients with recurrent otitis media, *J Clin Microbiol* 33:2876, 1995.

2. Chow AW: Life-threatening infections of the head and neck, *Clin Infect Dis* 14:991, 1992.

3. Dugel PV, Rao NA: Ocular infections in the acquired immunodeficiency syndrome, *Int Ophthalmol Clin* 33:103, 1993.

4. Gigliotti F, Williams WT, Hayden FG, et al: Etiology of acute conjunctivitis in children, *J Pediatr* 98:531, 1981.

5. Hendolin PH, Paulin L, Ylikoski J: Clinically applicable multiplex PCR for four middle ear pathogens, *J Clin Microbiol* 38:125, 2000.

6. Kaye SB, Rao PG, Smith G, et al: Simplifying collection of corneal specimens in cases of suspected bacterial keratitis, *J Clin Microbiol* 41:3192, 2003.

7. Leibowitz HM, Pratt MV, Flagstad IJ, et al: Human conjunctivitis: a diagnostic evaluation, *Arch Ophthalmol* 94:1747, 1976.

8. Palmu AA, Herva E, Savolainen H, et al: Association of clinical signs and symptoms with bacterial findings in acute otitis media, *Clin Infect Dis* 38:234, 2004.

9. Piccirillo JF: Acute bacterial sinusitis, *N Engl J Med* 351:902, 2004.

10. Roels P: Ocular infections of AIDS: new considerations for patients using highly active anti-retroviral therapy (HAART), *Optometry* 75:624, 2004.

11. Sande M, Gwaltney JM: Acute community-acquired bacterial sinusitis: continuing challenges and current management, *Clin Infect Dis* 39:S151, 2004.

12. Solomon AW, Peeling RW, Foster A, et al: Diagnosis and assessment of trachoma, *Clin Microbiol Rev* 17:982, 2004.

13. Tinkelman DG, Silk HJ: Clinical and bacteriologic features of chronic sinusitis in children, *Am J Dis Child* 143:938, 1989.

14. Wald ER: Microbiology of acute and chronic sinusitis in children and adults, *Am J Med Sci* 316:13, 1998.

15. Wilhelmus KR, Liesegang TJ, Osato MS, et al: Laboratory diagnosis of ocular infections. In Spector SC, editor: *Cumitech* 13*A*, Washington, DC, 1994, American Society for Microbiology.

GENERAL CONSIDERATIONS

ANATOMY

The urinary tract consists of the kidneys, ureters, bladder, and urethra (Figure 57-1). Often, urinary tract infections (UTIs) are characterized as being either upper or lower based primarily on the anatomic location of the infection: the **lower urinary tract** encompasses the bladder and urethra, and the **upper urinary tract** encompasses the ureters and kidneys.

The anatomy of the female urethra is of particular importance to the pathogenesis of UTIs. The female urethra is relatively short compared with the male urethra and also lies in close proximity to the warm, moist, perirectal region, which is teeming with microorganisms. Because of the shorter urethra, bacteria can reach the bladder more easily in the female host.

RESIDENT MICROORGANISMS OF THE URINARY TRACT

The urethra has resident microflora that colonize its epithelium in the distal portion. Some of these organisms are listed in Box 57-1. Potential pathogens, including gram-negative aerobic bacilli (primarily *Enterobacteriaceae*) and occasional yeasts, are also present as transient colonizers. All areas of the urinary tract above the urethra in a healthy human are sterile. Urine is typically sterile, but noninvasive methods for collecting urine must rely on a specimen that has passed through a contaminated milieu. Therefore, quantitative cultures for diagnosis of UTIs have been used to discriminate between contamination, colonization, and infection.

INFECTIONS OF THE URINARY TRACT

EPIDEMIOLOGY

UTIs are among the most common bacterial infections that lead patients to seek medical care. It has been estimated that more than 7 million outpatient visits, 1 million visits to the emergency department, and 100,000 hospital stays every year in the United States are due to UTIs.[14,19,23] Approximately 10% of humans will have a UTI at some time during their lives. Of note, UTIs are also the most common hospital-acquired infection, accounting for as many as 35% of nosocomial infections.[21]

The exact prevalence of UTIs is age and sex dependent. During the first year of life, UTIs are less than 2% in males and females. The incidence of UTIs among males remains relatively low after 1 year of age and until approximately 60 years of age when enlargement of the prostate interferes with emptying of the bladder. Therefore, UTI is predominantly a disease of females. Extensive studies have shown that the incidence of bacteriuria (presence of bacteria in urine) among girls 5 through 17 years of age is 1% to 3%. The prevalence of bacteriuria in females increases gradually with time to as high as 10% to 20% in older women. In women between 20 and 40 years of age who have had UTIs, as many as 50% may become reinfected within 1 year. The association of UTIs with sexual intercourse may also contribute to this increased incidence because sexual activity increases the chances of bacterial contamination of the female urethra. Finally, as a result of anatomic and hormonal changes that favor development of UTIs, the incidence of bacteriuria increases during pregnancy. These infections can lead to serious infections in both mother and fetus.

UTIs are important complications of diabetes, renal disease, renal transplantation, and structural and neurologic abnormalities that interfere with urine flow. In 40% to 60% of renal transplant recipients, the urinary tract is the source of bacteremia and in these patients, the recurrence rate is about 40%.[4] In addition, UTIs are a leading cause of gram-negative sepsis in hospitalized patients and are the origin for about half of all nosocomial infections caused by urinary catheters.

ETIOLOGIC AGENTS

Community-Acquired

Escherichia coli is by far the most frequent cause of uncomplicated community-acquired UTIs. At the molecular level, the *E. coli* that causes UTIs is sufficiently different from other types of *E. coli* so as to be designated uropathogenic *E. coli* (UPEC). Other bacteria frequently isolated from patients with UTIs are *Klebsiella* spp., other *Enterobacteriaceae*, *Staphylococcus saprophyticus*, and enterococci. In more complicated UTIs, particularly in recurrent infections, the relative frequency of infection caused by *Proteus*, *Pseudomonas*, *Klebsiella*, and *Enterobacter* spp. increases.

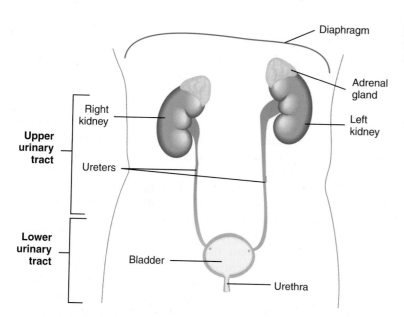

Figure 57-1   Overview of the anatomy of the urinary tract. (From Potter PH, Perry AG: *Fundamentals of nursing,* St Louis, 1985, Mosby.)

BOX 57-1   Resident Microflora of the Urethra

Coagulase-negative staphylococci (excluding *S. saprophyticus*)
Viridans and non-hemolytic streptococci
Lactobacilli
Diphtheroids (*Corynebacterium* spp.)
Nonpathogenic (saprobic) *Neisseria* spp.
Anaerobic cocci
Propionibacterium spp.
Anaerobic gram-negative bacilli
Commensal *Mycobacterium* spp.
Commensal *Mycoplasma* spp.

Hospital-Acquired

The hospital environment plays an important role in determining the organisms involved in UTIs. Hospitalized patients are most likely to be infected by *E. coli, Klebsiella* spp., *Proteus* spp., staphylococci, other *Enterobacteriaceae, Pseudomonas aeruginosa,* enterococci, and *Candida* spp. The introduction of a foreign body into the urinary tract, especially one that remains in place for a time (e.g., Foley catheter), carries a substantial risk of infection, particularly if obstruction is present. As many as 20% of all hospitalized patients who receive short-term catheterization develop a UTI. Consequently, UTI is the most common nosocomial infection in the United States, and the infected urinary tract is the most frequent source of bacteremia.

Miscellaneous

Other less frequently isolated agents are other gram-negative bacilli, such as *Acinetobacter* and *Alcaligenes* spp., other *Pseudomonas* spp., *Citrobacter* spp., *Gardnerella vaginalis, Aerococcus urinae,* and beta-hemolytic streptococci. Bacteria such as mycobacteria, *Chlamydia trachomatis,*

Ureaplasma urealyticum, Campylobacter spp., *Haemophilus influenzae, Leptospira,* and certain *Corynebacterium* spp. (e.g., *C. renale*) are rarely recovered from urine. Because renal transplant recipients are immunosuppressed, these patients not only suffer from common uropathogens but are also susceptible to opportunistic infections with unusual pathogens.[4] Of note, a recent study involving renal transplant recipients showed that for culture-negative urine, amplification of regions in bacterial 16S rRNA and subsequent analysis by a type of high-performance liquid chromatography detected the presence of a number of known uropathogens as well as unusual agents. *Salmonella* spp. may be recovered during the early stages of typhoid fever; their presence should be immediately reported to the physician. If anaerobes are suspected, the physician should perform a percutaneous bladder tap unless urine can be obtained from the upper urinary tract by another means (e.g., from a nephrostomy tube). Communication by the clinician to the laboratory that such an agent is suspected is most important for detecting such agents. However, the laboratory can exert some initiative as well. In patients with "sterile pyuria," Gram stain may reveal unusual organisms with distinctive morphology (e.g., *H. influenzae,* anaerobes). The presence of any organisms on smear that do not grow in culture is an important clue to the cause of the infection. The laboratory can then take the action necessary to optimize chances for recovery.

In general, viruses and parasites are not usually considered urinary tract pathogens. *Trichomonas vaginalis* may occasionally be observed in urinary sediment, and *Schistosoma haematobium* can lodge in the urinary tract and release eggs into the urine. Adenoviruses types 11 and 21 have been implicated as causative agents in hemorrhagic cystitis in children.

PATHOGENESIS

Routes of Infection

Bacteria can invade and cause a UTI via two major routes: ascending and hematogenous pathways.[9,19] Although the **ascending route** is the most common route of infection in females, ascent in association with instrumentation (e.g., urinary catheterization, cystoscopy) is the most common cause of hospital-acquired UTIs in both sexes. For UTIs to occur by the ascending pathway, enteric gram-negative bacteria and other microorganisms that originate in the gastrointestinal tract must be able to colonize the vaginal cavity and/or the periurethral area.[6] Once these organisms gain access to the bladder, they may multiply and then pass up the ureters to the kidneys. UTIs occur more often in women than men, at least partially because of the short female urethra and its proximity to the anus. As previously mentioned, sexual activity can increase chances of bacterial contamination of the female urethra.

In most hospitalized patients, UTI is preceded by urinary catheterization or other manipulation of the urinary tract.[5] The pathogenesis of catheter-associated UTI is not fully understood. It is certain that soon after hospitalization, patients become colonized with bacteria endemic to the institution, often gram-negative aerobic and facultative bacilli carrying resistance markers. These bacteria colonize the patient's skin, gastrointestinal tract, and mucous membranes, including the anterior urethra. With insertion of a catheter, the bacteria may be pushed along the urethra into the bladder or, with an indwelling catheter, may migrate along the track between the catheter and the urethral mucosa, gaining access to the bladder. It is estimated that approximately 10% to 30% of catheterized patients will develop **bacteriuria** (presence of bacteria in urine).

UTIs may also occur by the **hematogenous, or blood-borne, route.**[9] Hematogenous spread usually occurs as a result of bacteremia. Any systemic infection can lead to seeding of the kidney, but certain organisms, such as *Staphylococcus aureus* or *Salmonella* spp., are particularly invasive. Although most infections involving the kidneys are acquired by the ascending route, yeast (usually *Candida albicans*), *Mycobacterium tuberculosis*, *Salmonella* spp., *Leptospira* spp., or *Staphylococcus aureus* in the urine often indicates pyelonephritis acquired via hematogenous spread, or the **descending route.** Hematogenous spread accounts for less than 5% of UTIs.

The Host-Parasite Relationship

Many individuals, women in particular, are colonized in the vaginal and/or periurethral area with organisms originating from the gastrointestinal tract, yet they do not develop urinary infections. Whether an organism is able to colonize and then cause a UTI is determined in large part by a complex interplay of host and microbial factors.

In most cases, the host defense mechanisms are able to eliminate the organisms. Urine itself is inhibitory to some of the urethral flora such as anaerobes. In addition, if urine has a low pH, high or low osmolality, high urea concentration, or high organic acid content, even organisms that can grow in urine may be inhibited. Of importance, if bacteria do gain access to the bladder, the constant flushing of contaminated urine from the body either eliminates bacteria or maintains their numbers at low levels. Clearly, any interference with the act of normal voiding, such as mechanical obstruction resulting from kidney stones or strictures, will promote the development of UTI. Also, the bladder mucosal surface has antibacterial properties. If the infection is not eradicated, the site of infection remains in the superficial mucosa; deep layers of the bladder are rarely involved.

In addition to the previously described host defenses, a valvelike mechanism at the junction of the ureter and bladder prevents the **reflux** (backward flow) of urine from the bladder to the upper urinary tract. Therefore, if the function of these valves is inhibited or compromised in any way, such as by obstruction or congenital abnormalities, urine reflux provides a direct route for organisms to reach the kidney. Hormonal changes associated with pregnancy and their effects on the urinary tract increase the chance for urine reflux to the upper urinary tract.

Activation of the host immune response by uropathogens also plays a key role in fending off infection. For example, bacterial contact with urothelial cells initiates an immune response via a variety of signaling pathways.[8] Bacterial lipopolysaccharide (LPS; see Chapter 2) activates host cells to ultimately release cytokines such as tumor necrosis factor and interferon-gamma. In addition, bacteria can activate the complement cascade, leading to the production of biologically active components such as opsonins as well as augment the host's adaptive immune response. Recently, important advances in defining additional host factors that lead to host susceptibility or resistance to uropathogens have been made. For example, a glycoprotein synthesized exclusively by epithelial cells in a specific anatomic location in the kidney, referred to as *Tamm-Horsfall protein* or *uromucoid*, serves as an antiadherence factor by binding to *E. coli*–expressing type 1 fimbriae (see below). **Defensins,** a group of small antimicrobial peptides, are produced by a variety of host cells such as macrophages, neutrophils, and cells in the urinary tract and attach to the bacterial cell, eventually causing its death.

Although many microorganisms can cause UTIs, most cases are caused by only a few organisms. To illustrate, only a limited number of serogroups of *E. coli* cause a significant proportion of UTIs.[19] Numerous in-

---

BOX 57-2	Examples of Probable Virulence Factors of Uropathogenic *E. coli*

Type 1 fimbriae that bind to uroepithelial cells
Type P fimbriae that recognize kidney glycosphingolipids
Siderophores that help gather iron from the host
Alpha- and beta hemolysins that lyse host erythrocytes
Capsules
Sat protein that acts as a proteolytic toxin

BOX 57-3	Risk Factors Associated with Complicated Urinary Tract Infections

Underlying diseases that predispose the kidney to infection (e.g., diabetes, sickle cell anemia)
Kidney stones
Structural or functional abnormalities of the urinary tract (e.g., a tipped bladder)
Indwelling urinary catheters

---

vestigations indicate that UPEC possess certain virulence factors that enhance their ability to colonize and invade the urinary tract. Some of these virulence factors include increased adherence to vaginal and uroepithelial cells by bacterial surface structures (adhesins, in particular, pili), alpha-hemolysin production, and resistance to serum-killing activity (Box 57-2). Also, genome sequences of some UPEC strains have been determined that show that several potential virulence factor genes associated with the acquisition and development of UTIs are encoded on **pathogenicity islands** (e.g., hemolysins and P fimbriae). By definition, pathogenicity islands (see Chapter 3) contain genes that are associated with virulence and are absent from avirulent or less virulent strains of the same species.[15]

The importance of adherence in the pathogenesis of UTIs has also been demonstrated with other species of bacteria. Once introduced into the urinary tract, *Proteus* strains appear to be uniquely suited to cause significant disease in the urinary tract. Data indicate that these strains are able to facilitate their adherence to the mucosa of kidneys. Also, *Proteus* is able to hydrolyze urea via urease production. Hydrolysis of urea results in an increase in urine pH that is directly toxic to kidney cells and also stimulates the formation of kidney stones. Similar findings have been made with *Klebsiella* spp. *Staphylococcus saprophyticus* also adheres better to uroepithelial cells than does *S. aureus* or *S. epidermidis*.

Other bacterial characteristics may be important in the pathogenesis of UTIs. Motility may be important for organisms to ascend to the upper urinary tract against the flow of urine and cause pyelonephritis. Some strains demonstrate greater production of K antigen; this antigen protects bacteria from being phagocytosed.

Finally, despite numerous host defenses and even antibiotic treatments that can effectively sterilize the urine, a significant proportion of patients have recurrent UTIs. Current knowledge cannot account for this recurrence, despite repeated and long-term antibiotic treatment. However, recent studies show that uropathogens can invade superficial epithelial cells in the bladder and replicate, forming large foci of intracellular *E. coli*.[1,8] This invasion of bladder epithelial cells triggers the host immune response, which in turn causes the superficial cells to exfoliate within hours following infection. Although this exfoliation is considered a host defense mechanism by eliminating infected cells into the urine, intracellular organisms are able to reemerge from the bladder epithelial cells and invade the underlying, new superficial layer of epithelial cells, consequently persisting within the urinary tract. Recently, Anderson and colleagues[1] reported that these intracellular bacteria mature into numerous, large protrusions on the bladder surface that they called "pods." This bacterial organization in which the intracellular bacteria are embedded in a fibrous, polysaccharide-rich matrix resembling that of a biofilm, may help further explain the persistence of bladder infections despite strong host defenses.

TYPES OF INFECTION AND THEIR CLINICAL MANIFESTATIONS

UTI encompasses a broad range of clinical entities that differ in terms of clinical presentation, degree of tissue invasion, epidemiologic setting, and requirements for antibiotic therapy. There are five major types of UTIs: urethritis, asymptomatic bacteriuria, cystitis, the urethral syndrome, and pyelonephritis. Sometimes UTIs are classified as uncomplicated or complicated. Uncomplicated infections occur primarily in otherwise healthy females and occasionally in male infants and adolescent and adult males. Most uncomplicated infections respond readily to antibiotic agents to which the etiologic agent is susceptible. Complicated infections occur in both sexes. In general, individuals who develop complicated infections often have certain risk factors. Some of these risk factors are listed in Box 57-3. In general, complicated infections are more difficult to treat and have greater morbidity (e.g., kidney damage, bacteremia) and mortality compared with uncomplicated infections.

The clinical presentation of UTIs may vary, ranging from asymptomatic infection to full-blown pyelonephritis (infection of the kidney and its pelvis). Some UTI symptoms may be nonspecific and frequently symptoms overlap considerably in patients with lower UTIs and in those with upper UTIs.

Urethritis

Symptoms associated with **urethritis** (infection of the urethra), **dysuria** (painful or difficult urination) and frequency, are similar to those associated with lower UTIs. Urethritis is a common infection. Because *Chlamydia trachomatis*, *Neisseria gonorrhoeae*, and *Trichomonas vaginalis* are common causes of urethritis and are considered to be sexually transmitted, urethritis is discussed as a sexually transmitted disease in Chapter 58.

Asymptomatic Bacteriuria

Asymptomatic bacteriuria or asymptomatic UTI is the isolation of a specified quantitative count of bacteria in an appropriately collected urine specimen obtained from a person without symptoms or signs of urinary infection. Asymptomatic bacteriuria is common but its prevalence varies widely with age, gender, and the presence of genitourinary abnormalities or underlying diseases. For example, the prevalence of bacteriuria increases with age in healthy women from as low as about 1% among school girls to greater than or equal to 20% among women 80 years of age or older living in the community while bacteriuria is rare in healthy young men.[13] Because its clinical significance was controversial (asymptomatic bacteriuria precedes UTI but does not always lead to asymptomatic infection), guidelines were recently published for the diagnosis and treatment of asymptomatic bacteriuria in adults older than 18 years of age.[13] The foundation of these guidelines rests on the premise that screening of asymptomatic subjects for bacteriuria is appropriate if bacteriuria has adverse outcomes that can be prevented by antimicrobial therapy. Thus, screening and treatment for asymptomatic bacteriuria was recommended for pregnant women (because the risk of progression to severe symptomatic UTI and possible harm to the fetus), males undergoing transurethral resection of the prostate, and individuals undergoing urologic procedures for which mucosal bleeding is anticipated. In contrast, screening for or treatment of asymptomatic bacteriuria was not recommended for premenopausal, nonpregnant women, diabetic women, older persons living in the community, older institutionalized subjects, persons with spinal cord injury, or catheterized patients while the catheter is in place.

Cystitis

Typically, patients with **cystitis** (infection of the bladder) complain of dysuria, frequency, and urgency (compelling need to urinate). These symptoms are due not only to inflammation of the bladder but also to multiplication of bacteria in the urine and urethra. Often, there is tenderness and pain over the area of the bladder. In some individuals, the urine is grossly bloody. The patient may note urine cloudiness and a bad odor. Because cystitis is a localized infection, fever and other signs of a systemic (affecting the body as a whole) illness are usually not present.

Acute Urethral Syndrome

Another UTI is **acute urethral syndrome.** Patients with this syndrome are primarily young, sexually active women, who experience dysuria, frequency, and urgency but yield fewer organisms than 10^5 colony-forming units of bacteria per milliliter (CFU/mL) urine on culture.[7,10,17,18] (The classic criterion of greater than 10^5 CFU/mL of urine is highly indicative of infection in most patients with UTIs.) Almost 50% of all women who seek medical attention for complaints of symptoms of acute cystitis fall into this group. Although *Chlamydia trachomatis* and *N. gonorrhoeae* urethritis, anaerobic infection, genital herpes, and vaginitis account for some cases of acute urethral syndrome, most of these women are infected with organisms identical to those that cause cystitis but in numbers less than 10^5 CFU/mL urine. One must use a cutoff of 10^2 CFU/mL, rather than 10^5 CFU/mL, for this group of patients but must insist on concomitant pyuria (presence of 8 or more leukocytes per cubic millimeter on microscopic examination of uncentrifuged urine). Approximately 90% of these women have pyuria, an important discriminatory feature of infection.

Pyelonephritis

Pyelonephritis refers to inflammation of the kidney parenchyma, calices (cup-shaped division of the renal pelvis), and pelvis (upper end of the ureter that is located inside the kidney), and is usually caused by bacterial infection. The typical clinical presentation of an upper urinary tract infection includes fever and flank (lower back) pain and, frequently, lower tract symptoms (frequency, urgency, and dysuria). Patients can also exhibit systemic signs of infection such as vomiting, diarrhea, chills, increased heart rate, and lower abdominal pain. Of significance, 40% of patients with acute pyelonephritis are bacteremic.

LABORATORY DIAGNOSIS OF URINARY TRACT INFECTIONS

As previously mentioned, because noninvasive methods for collecting urine must rely on a specimen that has passed through a contaminated milieu, quantitative cultures for the diagnosis of UTI are used to discriminate between contamination, colonization, and infection. Refer to Table 5-1 for a quick reference for collecting, transporting, and processing urinary tract specimens.

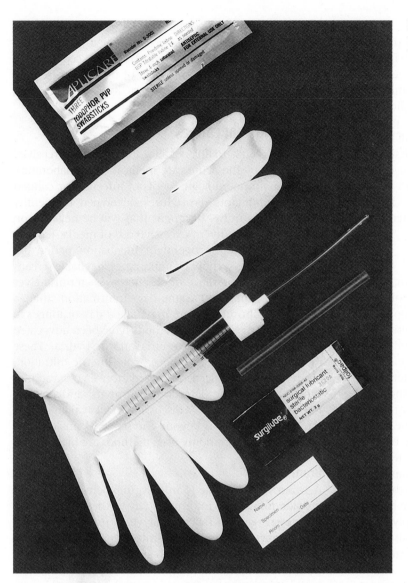

Figure 57-2   Collection device to obtain a urine by "in and out" or "straight," catheterization. (Courtesy Tristate Hospital Supply Corp., Howell, Mich.)

SPECIMEN COLLECTION

Prevention of contamination by normal vaginal, perineal, and anterior urethral flora is the most important consideration for collection of a clinically relevant urine specimen.

Clean-Catch Midstream Urine

The least invasive procedure, the clean-catch midstream urine specimen collection must be performed carefully for optimal results, especially in females. Good patient education is essential. Guidelines for proper specimen collection should be prepared on a printed card (bilingual, if necessary), with the procedure clearly described and preferably illustrated to help ensure patient compliance. The patient should be instructed to clean the periurethral area well with a mild detergent to avoid contamination. Of importance, the patient should also

be instructed to rinse well because the detergent may be bacteriostatic. Once cleansing is completed, the patient should retract the labial folds or glans penis, begin to void, and then collect a midstream urine sample. Studies showed that uncleansed, first-void specimens from males were as sensitive as (but less specific than) midstream urine specimens.

Straight Catheterized Urine

Although slightly more invasive, urinary catheterization may allow collection of bladder urine with less urethral contamination. Either a physician or other trained health professional performs this procedure. Risk exists, however, that urethral organisms will be introduced into the bladder with the catheter. An example of a collection device to obtain a "straight," or "in and out," catheterized urine is shown in Figure 57-2.

Suprapubic Bladder Aspiration

With suprapubic bladder aspiration, urine is withdrawn directly into a syringe through a percutaneously inserted needle, thereby ensuring a contamination-free specimen. The bladder must be full before performing the procedure. This collection technique may be indicated in certain clinical situations, such as pediatric practice, when urine is difficult to obtain. In brief, the full bladder is punctured using a needle and syringe and sampled following proper skin preparation (antisepsis). If good aseptic techniques are used, this procedure can be performed with little risk in premature infants, infants, small children, and pregnant women and other adults with full bladders.

Indwelling Catheter

The number of patients in hospitals and nursing homes with long-term, indwelling urinary catheters continues to increase. These patients ultimately develop bacteriuria, which predisposes them to more severe infections.[5] Specimen collection from patients with indwelling catheters requires scrupulous aseptic technique. Health care workers who manipulate a urinary catheter in any way should wear gloves. The catheter tubing should be clamped off above the port to allow collection of freshly voided urine. The catheter port or wall of the tubing should then be cleaned vigorously with 70% ethanol, and urine aspirated via a needle and syringe; the integrity of the closed drainage system must be maintained to prevent the introduction of organisms into the bladder. Specimens obtained from the collection bag are inappropriate, because organisms can multiply there, obscuring the true relative numbers. Cultures should be obtained when patients are ill; routine monitoring does not yield clinically relevant data.

SPECIMEN TRANSPORT

Because it is an excellent supportive medium for growth of most bacteria, urine must be immediately refrigerated or preserved. Bacterial counts in refrigerated (4° C) urine remain constant for as long as 24 hours. Urine transport tubes (B-D Urine Culture Kit [Becton Dickinson Vacutainer Kits, Rutherford, NJ]) containing boric acid, glycerol, and sodium formate have been shown to preserve bacteria without refrigeration for as long as 24 hours when greater than 10^5 CFU/mL (100,000 organisms per milliliter) were present in the initial urine specimen. The system may inhibit the growth of certain organisms, and it must be used with a minimum of 3 mL of urine. Another preservative system (Starplex Scientific, Inc., Etobicoke, Ontario, Calif) is also available. Both boric acid products preserve bacterial viability in urine for 24 hours in the absence of antibiotics. For patients from whom colony counts of organisms of less than 100,000/mL might be clinically significant, plating within 2 hours of collection is recommended. None of the kits has any advantage over refrigeration, except perhaps for convenience or for transport of urine from remote areas where refrigeration is not practical.

SCREENING PROCEDURES

As many as 60% to 80% of all urine specimens received for culture by the acute care medical center laboratory may contain no etiologic agents of infection or contain only contaminants. Procedures developed to identify quickly those urine specimens that will be negative on culture and circumvent excessive use of media, technologist time, and the overnight incubation period are discussed in this section. Of importance, a reliable screening test for the presence or absence of bacteriuria gives physicians important same-day information that a conventional urine culture may take a day or longer to provide. Many screening methods have been advocated for use in detecting bacteriuria and/or pyuria. These include microscopic methods, colorimetric filtration, bioluminescence, electrical impedance, enzymatic methods, photometric detection of growth, and enzyme immunoassay. Because a discussion of all available urine screening methods is beyond the scope of this chapter, only the more commonly used methods are highlighted.

Gram Stain

A Gram stain of urine is an easy, inexpensive means to provide immediate information as to the nature of the infecting organism (bacteria or yeast) to guide empiric therapy. After a drop of well-mixed urine is allowed to air-dry, the smear is fixed, stained, and examined under oil immersion (1000×) for the presence of ≥ 1 or 5 bacteria per oil immersion field (OIF). The performance characteristics of the urine Gram stain are not well defined in that different criteria have been used to define a positive result (≥1 or 5 bacteria per OIF).[22] Using either ≥1 or 5 bacteria/OIF has a sensitivity of 96% and 95%, respectively, and a specificity of 91% when correlated with significant bacteriuria ($>10^5$ CFU/mL). The Gram stain should not be relied on for detecting polymorphonuclear leukocytes in urine because leukocytes deteriorate quickly in urine that is not fresh or not adequately preserved. Many microbiologists have not adopted Gram stain examination of urine specimens because of its unreliability in detecting lower yet clinically significant numbers of organisms and because of its labor intensity. If employed, urine Gram stain should be limited to patients with acute pyelonephritis, patients with invasive UTIs, or other patients for whom immediate information is necessary for appropriate clinical management.[22]

Pyuria

Pyuria is the hallmark of inflammation, and the presence of polymorphonuclear neutrophils (PMNs) can be detected and enumerated in uncentrifuged specimens. This method of screening urine correlates fairly well with the number of PMNs excreted per hour, the best indicator of the host's state. Patients with more than 400,000 PMNs excreted into the urine per hour are likely to be infected, and the presence of more than 8 PMNs/mm^3 correlates well with this excretion rate and with infection.[22] This test can be performed using a hemocytometer, but it is not easily incorporated into the work flow of most microbiology laboratories. The standard urinalysis (usually done in hematology or chemistry sections) includes an examination of the centrifuged sediment of urine for enumeration of PMNs, results of which do not correlate well with either the PMN excretion rate or the presence of infection. Pyuria also can be associated with other clinical diseases, such as vaginitis, and therefore is not specific for UTIs.

Indirect Indices

Frequently, screening tests detect bacteriuria or pyuria by examining for the presence of bacterial enzymes and/or PMN enzymes rather than the organisms or PMNs themselves.

Nitrate Reductase (Greiss) Test.   This screening procedure looks for the presence of urinary nitrite, an indicator of UTI. Nitrate-reducing enzymes that are produced by the most common urinary tract pathogens reduce nitrate to nitrite. This test has been incorporated onto a paper strip that also tests for leukocyte esterase, an enzyme produced by PMNs (see below).

Leukocyte Esterase Test.   As previously mentioned, evidence of a host response to infection is the presence of PMNs in the urine. Because inflammatory cells produce leukocyte esterase, a simple, inexpensive, and rapid method that measures this enzyme has been developed. Studies have shown that leukocyte esterase activity correlates with hemocytometer chamber counts. The nitrate reductase and leukocyte esterase tests have been incorporated into a paper strip. Numerous manufacturers sell these strips commercially, and the strips are one of the most widely used enzymatic tests. Although the sensitivity of the combination strip is higher than either test alone, the sensitivity of this combination screening is not great enough to recommend its use as a stand-alone test in most circumstances. Of note, the leukocyte esterase test is not sensitive enough for determining pyuria in patients with acute urethral syndrome.

Catalase.   The Uriscreen (Diatech Diagnostics, Inc., Rehovet, Israel) is another rapid urine screening system based on the detection of catalase present in somatic (pertaining to the body) cells and in most bacterial species commonly causing UTIs except for streptococci and enterococci. Approximately 1.5 to 2 mL of urine are added to a tube containing dehydrated substrate. Hydrogen peroxide is added to the urine, and the solution is mixed gently. The formation of bubbles above the liquid surface is interpreted as a positive test. Some studies have reported that this system did not offer significant advantages over the leukocyte esterase-nitrite strip

Automated and Semiautomated Systems

Automated screening systems offer the promise of a large throughput with minimal labor and a rapid turn-around time compared with conventional cultures. However, these advantages may be offset by a substantial cost for the instrumentation. Often these costs can be justified only in laboratories that receive many specimens.

Several automated or semiautomated urine screening systems that are either bacterial growth independent or dependent are commercially available. By examining images of uncentrifuged urine samples using a video camera, the IRIS 939 UDx system (International Remote Imaging Systems, Inc., Chatsworth, Calif) and the Sysmex UF-100 (TOA Medical Electronics; Kobe, Japan) are able to recognize many cellular structures, including leukocytes and bacteria. A walk-away robotic instrument, Cellenium-160US (Combact Diagnostic Systems Ltd., Hertzliya, Israel) has been introduced for urine screening using fluorescent probes to stain a monolayer of bacteria from urine on a membrane. Following staining, the membrane is then examined at high magnification using computerized fluorescent microscopy imaging technology. Although the need to culture negative urine specimens is eliminated, only limited clinical performance data have been published. Finally, another semiautomated instrument, the Coral UTI Screen system (Coral Biotechnology, San Diego, Calif) uses a somatic-cell release agent to first release and destroy the adenosine triphosphate (ATP) in somatic cells while bacterial ATP remains protected within the bacterial cell. Bacterial ATP is then liberated and detected by the instrument. Studies to date have demonstrated this instrument to have a sensitivity and specificity of 86% and 75.5%, respectively.[3] Although a number of other automated urine screens are in various stages of development, to date these screens have not been widely accepted by clinical microbiology laboratories.[16,20,21]

General Comments Regarding Screening Procedures

In general, screening methods are insensitive at levels below 10^5 CFU/mL. Therefore, they are not acceptable for urine specimens collected by suprapubic aspiration, catheterization, or cystoscopy. Screening methods may also fail to detect a significant number of infections in symptomatic patients with low colony counts (10^2 to 10^3 CFU/mL) such as young, sexually active females with acute urethral syndrome. Further complicating the laboratory's decision whether to adopt a screening method is whether screening results will be used to rule out infection in asymptomatic patients. Under these circumstances, testing for pyuria is essential.

Therefore, given the importance of the 10^2 CFU/mL count and the PMN count, no screening test should be used indiscriminately. Selecting a screening method largely depends on the laboratory and the patient population being served by the laboratory. For example, there will be a cost advantage in screening urine in laboratories that receive many culture-negative specimens. On the other hand, urine from patients with symptoms of UTI plus a selected group expected to have asymptomatic bacteriuria should be cultured. For example, patients in their first trimester of pregnancy should be cultured because these women might appear asymptomatic but have a covert infection and become symptomatic later; UTIs in pregnant women may lead to pyelonephritis and the likelihood of a premature birth. Other situations in which patients with no symptoms of UTI might be cultured include the following:

- Bacteremia of unknown source
- Urinary tract obstruction
- Follow-up after removal of an indwelling catheter
- Follow-up of previous therapy

Other factors that must be considered when selecting a rapid urine screen include accuracy, ease of test performance, reproducibility, turn-around time, and whether bacteriuria and/or pyuria are detected.

URINE CULTURE

Inoculation and Incubation of Urine Cultures

Once it has been determined that a urine specimen should be cultured for isolation of the common agents of UTI, a measured amount of urine is inoculated to each of the appropriate media. The urine should be mixed thoroughly before plating. The plates can be inoculated using disposable sterile plastic tips with a displacement pipetting device calibrated to deliver a constant amount, but this method is somewhat cumbersome. Most often, microbiologists use a calibrated loop designed to deliver a known volume, either 0.01 or 0.001 mL of urine. These loops, made of platinum, plastic, or other material, can be obtained from laboratory supply companies.

The calibrated loop that delivers the larger volume of urine (0.01 mL) is recommended to detect lower numbers of organisms in certain specimens. For example, urine collected from catheterization, nephrostomies,

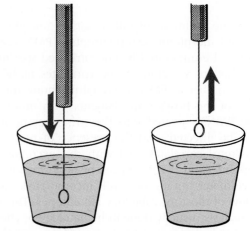

Figure 57-3 Method for inserting a calibrated loop into urine to ensure that the proper amount of specimen adheres to the loop.

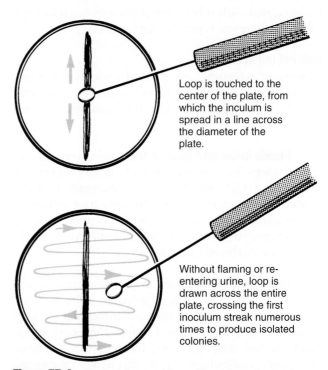

Loop is touched to the center of the plate, from which the inoculum is spread in a line across the diameter of the plate.

Without flaming or re-entering urine, loop is drawn across the entire plate, crossing the first inoculum streak numerous times to produce isolated colonies.

Figure 57-4 Method for streaking with calibrated urine loop to produce isolated colonies and countable colony-forming units.

ileal conduits, and suprapubic aspirates should be plated with the larger calibrated loop. The communication of pertinent clinical history to the laboratory is essential so that appropriate processing can be performed.

The choice of which media to inoculate depends on the patient population served and the microbiologist's preference. The use of a 5% sheep blood agar plate and a MacConkey agar plate allows detection of most gram-negative bacilli, staphylococci, streptococci, and enterococci. To save cost and somewhat streamline

Table 57-1 Overview of Bacteriologic Culture Systems for Urine

Culture System	Principle	Comment(s)
Bactercult (Wampole Laboratories, Princeton, NJ.)	Sterile, disposable tube coated with culture medium; contains phenol red indicator for presumptive ID of common uropathogens	Colonies counted
Bullseye Urine Plate (HealthLink Diagnostics, Southeast VetLab Supply, Miami, Fla.)	Five-chambered plate containing media for isolation and ID of common uropathogens and their AST patterns	Colonies counted; direct AST
Diaslide (Diatech Diagnostics, Inc., Boston, Mass.)	Transparent-hinged plastic casing contains CLED and MacConkey agars for detection and presumptive ID of common UTI pathogens	Growth is compared with reference photographs for quantitation, morphology, and color
DIP N COUNT (Starplex Scientific, Etobiocoke, Ontario, Canada)	Paddle contains CLED and MacConkey or EMB agar	Colonies are compared with a color density chart; color chart is used for ID
onSite Urine Culture Device (TREK Diagnostic Systems, Westlake, Ohio)	Transparent hinged casing with 5% blood, Columbia CNA and CLED agars. Kit contains a plastic sampler for inoculation.	Colonies counted.
Rainbow (Biolog, Inc., Hayward, Calif.)	Rainbow agar CP-8 used with eight confirmation spot tests for ID of eight microorganisms causing UTI; medium contains chromogenic substrates that color microorganisms	Interpretation may occur within 3 to 6 hr for rapidly growing organisms
URI-CHECK (Troy Biologicals, Inc., Troy, Mich.)	Dipslide culture for the enumeration and ID of uropathogens; dipslide contains CLED and MacConkey or EMB agar; similar to DIP N COUNT	Growth density is compared with chart
Uri-Kit/Uri-Three (Culture Kits, Inc., Norwich, NY.)	Agar plate systems used to detect common uropathogens; Uri-Kit is plastic-hinged case containing CLED medium; Uri-Three triplate contains 5% blood and MacConkey and CLED agars	Growth density is compared with a colony density chart
Uricult Trio (Orion Diagnostica, Espoo, Finland)	Three-medium dipslide containing CLED, MacConkey, and a β-glucuronidase substrate for enumeration of microorganisms and presumptive identification of *Escherichia coli*	*E. coli* appears as brown colonies

From Clarridge JE, Johnson JR, Pezzlo MT: Laboratory diagnosis of urinary tract infections. In Weissfeld AS, coordinating editor: *Cumitech 2B*, Washington, DC, 1998, American Society for Microbiology.

AST, Antimicrobial susceptibility testing; *ID*, identification; *UTI*, urinary tract infection.

culture processing, many laboratories use an agar plate split in half (biplate); one side contains 5% sheep blood agar and the other half contains MacConkey agar.

In some circumstances, enterococci and other streptococci may be obscured by heavy growth of *Enterobacteriaceae*. Because of this possibility, some laboratories add a selective plate for gram-positive organisms, such as Columbia colistin-nalidixic acid agar (CNA) or phenylethyl alcohol agar. Although some discriminatory capability may be added, cost is also added to the procedure. In addition to increased cost, inclusion of plated media selective for gram-positive organisms generally provides no or limited additional information. Many European laboratories use cystine-lactose electrolyte-deficient (CLED) agar. In recent years, chromogenic media has been introduced and become commercially available from a number of manufacturers that allows for more specific direct detection and differentiation of urinary tract pathogens on primary plates. This media uses enzymatic reactions to identify *E. coli* and *Enterococcus* without additional confirmatory testing from urine specimens as well as providing presumptive identification of *S. saprophyticus*, *Streptococcus agalactiae*, *Klebsiella-Enterobacter-Serratia* and the *Proteus-Morganella-Providentia* groups.

Before inoculation, urine is mixed thoroughly and the top of the container is then removed. The calibrated loop is inserted vertically into the urine in a cup. Otherwise, more than the desired volume of urine will be taken up, potentially affecting the quantitative culture result (Figure 57-3). A widely used method is described in Procedure 57-1. If the urine is in a small-diameter tube, the surface tension will alter the amount of specimen picked up by the loop. A quantitative

Table 57-2 Criteria for Classification of Urinary Tract Infections by Clinical Syndrome

Category	Criteria Clinical	Criteria Laboratory
Acute, uncomplicated UTI in women	Dysuria, urgency, frequency, suprapubic pain No urinary symptoms in last 4 wk before current episode No fever or flank pain	$\geq$10 WBC/mm^3 $\geq$10^3 CFU/mL uropathogens* in CCMS urine
Acute, uncomplicated pyelonephritis	Fever, chills Flank pain on examination Other diagnoses excluded No history or clinical evidence of urologic abnormalities	$\geq$10 WBC/mm^3 $\geq$10^4 CFU/mL uropathogens in CCMS urine
Complicated UTI and UTI in men	Any combination of symptoms listed above One or more factors associated with complicated UTI†	$\geq$10 WBC/mm^3 $\geq$10^5 CFU/mL uropathogens in CCMS urine
Asymptomatic bacteriuria	No urinary symptoms	$\pm$>10 WBC/mm^3 $\geq$10^5 CFU/mL in two CCMS cultures >24 hrs apart

From Stamm WE: *Infection* 20(suppl 3):S151, 1992.
*Uropathogens: Organisms that commonly cause UTIs.
†Factors associated with complicated UTI include any UTI in a male, indwelling or intermittent urinary catheter, >100 mL of postvoid residual urine, obstructive uropathy, urologic abnormalities, azotemia (excess urea in the blood, even without structural abnormalities), and renal transplantation.
UTI, Urinary tract infection; *WBC*, white blood cells; *CFU*, colony-forming unit; *CCMS*, clean-catch midstream urine.

pipette should be considered if the urine cannot be transferred to a larger container. Once inoculated, the plates are streaked to obtain isolated colonies (Figure 57-4).

Once plated, urine cultures are incubated overnight at 35° C. For the most part, incubation for a minimum of 24 hours is necessary to detect uropathogens.[12] Thus, some specimens inoculated late in the day cannot be read accurately the next morning. These cultures should either be reincubated until the next day or possibly interpreted later in the day when a full 24-hour incubation has been completed.

During the past several years, a number of self-contained bacteriologic culture systems for urine have been introduced. For larger diagnostic laboratories, these systems increase efficiency while decreasing costs and turn-around time. Although the review of each of these systems is beyond the scope of this chapter, Table 57-1 provides an overview of the more commonly used systems; the reader is referred to the list of additional reading at the end of this chapter for more information.

Interpretation of Urine Cultures

As previously mentioned, UTIs may be completely asymptomatic, produce mild symptoms, or cause life-threatening infections. Of importance, the criteria most useful for microbiologic assessment of urine specimens is dependent not only on the type of urine submitted (e.g., voided, straight catheterization) but the clinical history of the patient (e.g., age, sex, symptoms, antibiotic therapy).

One major problem in interpreting urine cultures arises because urine cultures collected by the voided technique may be contaminated with normal flora, in-

cluding *Enterobacteriaceae*. Determining what colony count represents true infection from contamination is of utmost importance and is related to the patient's clinical presentation. A number of studies have proposed the use of different cutoffs in colony counts based on clinical presentation; an example of one such set of guidelines is given in Table 57-2.

Ideally, the clinician caring for the patient should provide the laboratory with enough clinical information to allow specimens from different patient populations to be identified.[2,11] These specimens could then be selectively processed using the guidelines in Table 57-2. However, because microbiology laboratories frequently receive little or no clinical information about patients, questions have been raised as to whether these cutoffs are practical and realistic for routine laboratory use. Further complicating urine culture interpretation is the increasing difficulty in distinguishing between infection and contamination as the criterion for a positive culture is lowered from 10^5 CFU/mL to 10^2 CFU/mL. Because of these issues, many laboratories establish their own interpretative criteria for urine cultures based on the type of urine submitted (e.g., clean-catch midstream, catheterized, and surgically obtained specimens such as suprapubic aspirates). Variations in interpretative guidelines occur from one laboratory to another but some generalities can be made; these are listed in Table 57-3. Some examples of urine culture results are shown in Figure 57-5 to illustrate some of these interpretations. (For delineation of complete urine protocols, refer to sources listed in the Additional Reading at the end of this chapter.)

Table 57-3 General Interpretative Guidelines for Urine Cultures

Result	Specific Specimen Type/Associated Clinical Condition, if Known	Workup
$\geq10^4$ CFU/mL of a single potential pathogen or for each of two potential pathogens	CCMS urine/pyelonephritis, acute cystitis, asymptomatic bacteriuria, **or** catheterized urines	Complete*
$\geq10^3$ CFU/mL of a single potential pathogen	CCMS urine/symptomatic males **or** catheterized urines **or** acute urethral syndrome	Complete
$\geq$Three organism types with no predominating organism	CCMS urine **or** catheterized urines	None. Because of possible contamination, ask for another specimen
Either two or three organism types with predominant growth of one organism type and <10^4 CFU/mL of the other organism type(s)	CCMS urine	Complete workup for the predominating† organism(s); description of the other organism(s)
$\geq10^2$ CFU/mL of any number of organism types (set up with a 0.001- and 0.01-mL calibrated loop)	Suprapubic aspirates, any other surgically obtained urines (including ileal conduits, cystoscopy specimens)	Complete

*A complete workup includes identification of the organism and appropriate susceptibility testing.
†Predominant growth =10^4 to $\geq10^5$ CFU/mL.
CFU, Colony-forming unit; *CCMS,* clean-catch midstream urine.

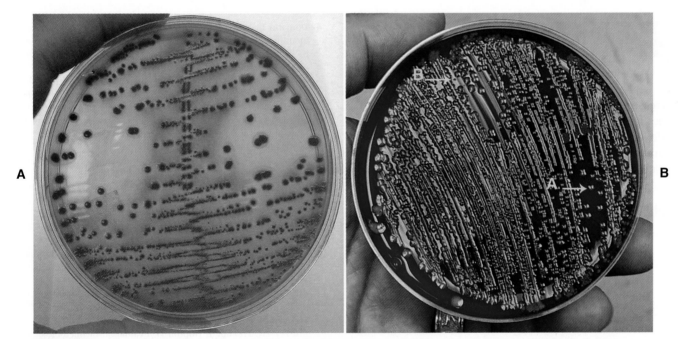

Figure 57-5 Culture results illustrating some of the various interpretative guidelines. **A,** Growth of $\geq10^5$ CFU/mL of a lactose-fermenting gram-negative rod in a clean-catch midstream (CCMS) urine from a patient with pyelonephritis; complete workup would be done. **B,** Growth of $\geq10^5$ CFU/mL of a lactose-fermenting gram-negative rod *(arrow A)* and <10^4 CFU/mL of another organism type *(arrow B)* from a CCMS urine; only the organism with a colony count of >10^4 to 10^5 CFU/mL would be worked up completely.

In addition to the previously described guidelines, a pure culture of *S. aureus* is considered to be significant regardless of the number of CFUs, and antimicrobial susceptibility tests are performed. The presence of yeast in any number is reported to physicians, and pure cultures of a yeast may be identified to the species level. In all urine, regardless of the extent of final workup, all isolates should be enumerated (e.g., three different organisms present at 10^3 CFU/mL), and those present in numbers greater than 10^4 CFU/mL should be described morphologically (e.g., non–lactose-fermenting gram-negative rods).

Procedure 57-1

INOCULATING URINE WITH A CALIBRATED LOOP

PRINCIPLE

The number of microorganisms per milliliter recovered on urine culture can aid in the differential diagnosis of UTI. Plastic or wire loops, available commercially, have been calibrated to deliver a known volume of liquid when handled correctly, thus enabling the microbiologist to estimate numbers of organisms in the original specimen based on CFU of growth on cultures.

METHOD

1. Flame a calibrated wire inoculating loop and allow it to cool without touching any surface. Alternatively, aseptically remove a plastic calibrated loop from its package.
2. Mix the urine thoroughly and remove the top of the container. If the urine is in a small-diameter tube, the surface tension will alter the amount of specimen picked up by the loop. A quantitative pipette should be considered if the urine cannot be transferred to a larger container.
3. Insert the loop vertically into the urine (see Figure 57-3) to allow urine to adhere to the loop.
4. Spread the loopful of urine over the surface of the agar plate, as shown in Figure 57-4. A standard quadrant streaking technique is also acceptable.
5. Without reflaming, insert the loop vertically into the urine again for transfer of a loopful to a second plate. Repeat for each plate.
6. Incubate plates for at least 24 hours at 35° to 37° C in air. Colonies are counted on each plate. The number of CFUs is multiplied by 1000 (if a 0.001-mL loop was used) or by 100 (if a 0.01-mL loop was used) to determine the number of microorganisms per milliliter in the original specimen.
7. Because antimicrobial treatment or other factors may inhibit initial growth, reincubate plates with no growth or tiny colonies for an additional 24 hours before discarding plates.
8. To store the inoculating loop, place (handle down) in a test tube taped to the wall, rather than flat on the bench, to prevent bending, which would destroy the calibration.

Case Study

A 25-year-old woman was seen urgently by her physician for severe suprapubic pain, which increased on urination. Her physician diagnosed this as her fourth episode of bladder infection in the past year. He decided to order a urinalysis and a culture. The urinalysis was positive for leucocyte esterase, and the patient was treated with nitrofurantoin. The culture subsequently grew 10^3 staphylococci/mL.

QUESTIONS

1. Is this organism significant?
2. What would you do to identify it to species?
3. Is susceptibility testing indicated for this organism?
4. Would a rapid urine screen be helpful in diagnosing the infection in this young woman?

REFERENCES

1. Anderson AC, Martin SM, Hultgren SJ: Host subversion by formation of intracellular bacterial communities in the urinary tract, *Microbes Infect* 6:1094, 2004.
2. Carroll KC, Hale DC, Von Boerum DH, et al: Laboratory evaluation of urinary tract infections in an ambulatory clinic, *Am J Clin Pathol* 10:100, 1994.
3. Churchill D, Gregson D: Screening urine samples for significant bacteriuria in the clinical microbiology laboratory, *Clin Microbiol Newsl* 26:179, 2004.
4. Domann E, Hong G, Imirzalioglu C, et al: Culture-independent identification of pathogenic bacteria and polymicrobial infections in the genitourinary tract of renal transplant recipients, *J Clin Microbiol* 41:5500, 2003.
5. Falkiner FR: The insertion and management of indwelling urethral catheters: minimizing the risk of infection, *J Hosp Infect* 25:79, 1993.
6. Foxman B, Brown P: Epidemiology of urinary tract infections—transmission and risk factors, incidence, and costs, *Infect Dis Clin North Am* 17:227, 2003.
7. Hamilton-Miller JM: The urethral syndrome and its management, *J Antimicrob Chemother* 33(suppl A):63, 1994.
8. Kucheria R, Dagsupta P, Sacks SH, et al: Urinary tract infections: new insights into a common problem, *Postgrad Med J* 81:83, 2004.
9. Kunin CM: Urinary tract infections in females, *Clin Infect Dis* 18:1, 1994.
10. Kunin CM, White LV, Hua TH: A reassessment of the importance of "low-count" bacteriuria in young women with acute urinary symptoms, *Ann Intern Med* 119:454, 1993.
11. Morgan MG, McHenzie H: Controversies in the laboratory diagnosis of community-acquired urinary tract infection, *Eur J Clin Microbiol Infect Dis* 12:491, 1993.
12. Murray PR, Traynor P, Hopson D: Evaluation of microbiological processing of urine specimens: comparison of overnight versus two-day incubation, *J Clin Microbiol* 30:1600, 1992.
13. Nicolle LE, Bradley S, Colgan R, et al: Infectious Diseases Society of America guidelines for the diagnosis and treatment of asymptomatic bacteriuria in adults, *Clin Infect Dis* 40:643, 2005.
14. Palac DM: Urinary tract infections in women: a physician's perspective, *Lab Med* 17:25, 1986.
15. Parham NJ, Pollard SJ, Chaudhuri RR, et al: Prevalence of pathogenicity isolate II_{CFT073} genes among extraintestinal clinical isolates of *Escherichia coli*, *J Clin Microbiol* 43:2425, 2005.
16. Smith TK, Hudson AJ, Spencer RC: Evaluation of six screening methods for detecting significant bacteriuria, *J Clin Pathol* 41:904, 1988.

17. Stamm WE: Criteria for the diagnosis of urinary tract infection and for the assessment of therapeutic effectiveness, *Infection* 20(suppl 3):S151, 1992.

18. Stamm WE: Protocol for the diagnosis of urinary tract infection: reconsidering the criterion for significant bacteriuria, *Urology* 32(suppl):6, 1988.

19. Stamm WE, Hooton TM, Johnson JR, et al: Urinary tract infections: from pathogenesis to treatment, *J Infect Dis* 159:400, 1989.

20. Stevens M: Evaluation of Questor urine screening for bacteriuria and pyuria, *J Clin Pathol* 46:817, 1993.

21. Stevens M: Screening urines for bacteriuria, *Med Lab Sci* 46:194, 1989.

22. Wilson ML, Gaido L: Laboratory diagnosis of urinary tract infections in adult patients, *Clin Infect Dis* 38:1150, 2004.

23. Wong ES: Guideline to prevention of catheter-associated urinary tract infections, *Am J Infect Control* 11:28, 1983.

ADDITIONAL READING

Clarridge JE, Johnson JR, Pezzlo MT: Laboratory diagnosis of urinary tract infections. In Weissfeld AS, coordinating editor: *Cumitech 2B*, Washington, DC, 1998, American Society for Microbiology.

Pezzlo M: Detection of urinary tract infections by rapid methods, *Clin Microbiol Rev* 1:268, 1988.

Pezzlo M, York MK: Urine cultures. Section 3.12. In Isenberg HD, editor: *Clinical microbiology procedures handbook*, vol 1, Washington, DC, 2004, American Society for Microbiology.

Sobel JD: Pathogenesis of urinary tract infections: host defenses, *Infect Dis Clin North Am* 1:751, 1987.

CHAPTER 58

Genital Tract Infections

GENERAL CONSIDERATIONS

ANATOMY

Familiarity with the anatomic structures is important for appropriate processing of specimens from genital tract sites and interpretation of microbiologic laboratory results. The key anatomic structures for the female and male genital tract in relation to other important structures are shown in Figure 58-1.

RESIDENT MICROBIAL FLORA

The lining of the normal human genital tract is a mucosal layer made up of transitional, columnar, and squamous epithelial cells. Various species of commensal bacteria colonize these surfaces, causing no harm to the host except under abnormal circumstances and helping to prevent the adherence of pathogenic organisms. Normal urethral flora include coagulase-negative staphylococci and corynebacteria, as well as various anaerobes. The vulva and penis, especially the area underneath the prepuce (foreskin) of the uncircumcised male, may harbor *Mycobacterium smegmatis* along with other gram-positive bacteria.

The flora of the female genital tract varies with the pH and estrogen concentration of the mucosa, which depends on the host's age. Prepubescent and postmenopausal women harbor primarily staphylococci and corynebacteria (the same flora present on surface epithelium), whereas women of reproductive age may harbor large numbers of facultative bacteria such as *Enterobacteriaceae*, streptococci, and staphylococci, as well as anaerobes such as lactobacilli, anaerobic non–spore-forming bacilli and cocci, and clostridia. Lactobacilli are the predominant organisms in secretions from normal, healthy vaginas. Recent studies have shown that hydrogen peroxide–producing lactobacilli are most associated with a healthy state.[3,4] The numbers of anaerobic organisms remain constant throughout the monthly cycle. Many women carry group B beta-hemolytic streptococci (*Streptococcus agalactiae*), which may be transmitted to the neonate. Although yeasts (acquired from the gastrointestinal tract) may be transiently recovered from the female vaginal tract, they are not considered normal flora.

SEXUALLY TRANSMITTED DISEASES AND OTHER GENITAL TRACT INFECTIONS

Genital tract infections may be classified as endogenous or exogenous. **Exogenous infections** may be acquired as people engage in sexual activity, and these infections are referred to as **sexually transmitted diseases (STDs).** In contrast, **endogenous infections** result from organisms that are members of the patient's normal genital flora.

In the female, genital tract infections can be divided between lower tract (vulva, vagina, and cervix) and upper tract (uterus, fallopian tubes, ovaries, and abdominal cavity) infections.[7] **Lower tract infections** are commonly acquired by sexual or direct contact. Although the organisms that cause lower tract infections are not usually part of the normal genital tract flora, some organisms that are normally present in very low numbers can increase sufficiently to cause disease. **Upper tract infections** are frequently an extension of a lower tract infection in which organisms from the vagina or cervix are believed to travel into the uterine cavity and on through the endometrium to the fallopian tubes and ovaries. Similarly, an organism can spread along contiguous mucosal surfaces in the male from a lower genital tract site of infection (i.e., urethra) and cause infection in a reproductive organ such as the epididymis.

GENITAL TRACT INFECTIONS

SEXUALLY TRANSMITTED DISEASES AND OTHER LOWER GENITAL TRACT INFECTIONS

Lower genital tract infections may be acquired either through sexual contact with an infected partner or through nonsexual means. These infections are some of the most common infectious diseases.

Epidemiology/Etiologic Agents

STDs are major public health problems in all populations and socioeconomic groups worldwide. An estimated 4 million genital infections caused by *Chlamydia trachomatis* occur annually in the United States alone. The incidence and spread of STDs are greatly influenced by numerous factors such as the availability of multiple

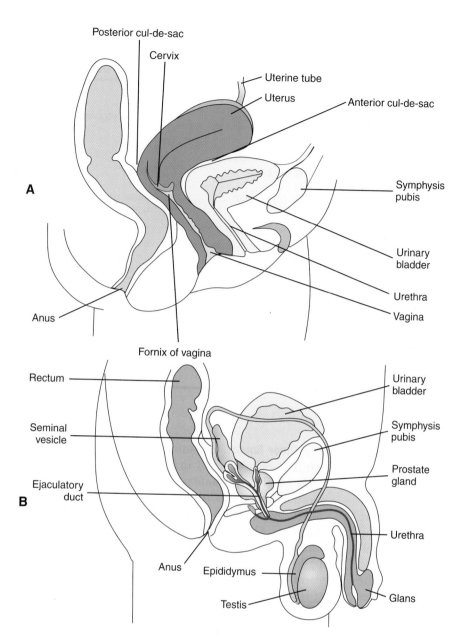

Figure 58-1   Location of key anatomic structures of the female (**A**) and male (**B**) genital tracts in relation to other major anatomic structures.

sexual partners, the presence of asymptomatic infection, the frequent movement of people within populations, and increasing affluence.[9]

The number of microorganisms that can cause genital tract infections is large. Of importance, these organisms are diverse, representing all four major groups of microorganisms (bacteria, viruses, fungi, and parasites). The major causes of genital tract infections are listed in Table 58-1.

Routes of Transmission
Although genital tract infections can be caused by members of the patient's genital flora (endogenous infections), the overwhelming majority of lower genital tract infections is sexually transmitted.

Sexually Transmitted.   *Chlamydia trachomatis, Neisseria gonorrhoeae, Trichomonas vaginalis,* human immunodeficiency virus (HIV), *Treponema pallidum, Ureaplasma urealyticum, Mycoplasma hominis,* other mycoplasmas, herpes simplex virus (HSV), and others may be acquired as people engage in sexual activity. In addition, other agents that cause genital tract disease and may be sexually transmitted include adenovirus, coxsackievirus, molluscum contagiosum virus (a member of the poxvirus group), the human papillomaviruses (HPVs) of genital warts (condylomata acuminata; types 6, 11, and others) and those associated with cervical carcinoma (predominantly types 16 and 18, but numerous others are also implicated), *Calymmatobacterium granulomatis,* and ectoparasites such as scabies and lice. Some of these

Table 58-1 Major Causes of Genital Tract Infections and Sexually Transmitted Diseases

Frequency	Disease	Agent	Organism Group
More common	Genital and anal warts (condyloma); cervical dysplasia; cancer	Human papillomavirus	Viruses
	Vaginitis	Gardnerella/Mobiluncus, Trichomonas vaginalis, Candida albicans	Bacteria, parasites, fungi
	Urethritis/cervicitis (also acute salpingitis, acute perihepatitis, urethritis, pharyngitis)	Neisseria gonorrhoeae, Chlamydia trachomatis, Ureaplasma urealyticum	Bacteria
	Herpes genitalis (genital/skin ulcers)	Herpes simples virus type 2 (less commonly type 1)	Viruses
	AIDS	Human immunodeficiency virus (HIV)	Viruses
	Hepatitis (acute and chronic infection)	Hepatitis B virus	Viruses
Less common	Lymphogranuloma venereum	C. trachomatis (L-1, L-2, L-3 serovars)	Bacteria
	Granuloma inguinale	Calymmatobacterium granulomatis (Donovania)	Bacteria
	Syphilis	Treponema pallidum	Bacteria
	Chancroid	Haemophilus ducreyi	Bacteria
	Scabies, mites	Sarcoptes scabei	Ectoparasites
	Pediculosis pubis, "crabs" infestation	Phthirius pubis	Ectoparasites
	Enteritis (homosexuals/proctitis)	Giardia lamblia, Entamoeba histolytica, Shigella spp., Salmonella spp. Enterobius vermicularis, Campylobacter spp., Helicobacter spp.	Bacteria, parasites
	Molluscum contagiosum	Poxlike virus	Viruses
	Heterophile-negative mononucleosis, congenital infections	Cytomegalovirus	Viruses

agents are not routinely isolated from clinical specimens. Infections with more than one agent may occur and therefore dual or concurrent infections should always be considered.

An individual's sexual habits and practices dictate potential sites of infection. Homosexual practices and increasingly common heterosexual practices of anal-genital or oral-genital intercourse allow for transmission of a genital tract infection to other body sites such as the pharynx or anorectal region. In addition, these practices have required that other gastrointestinal and systemic pathogens also be considered etiologic agents of STDs. The intestinal protozoa *Giardia lamblia, Entamoeba histolytica,* and *Cryptosporidium* spp. are significant causes of STDs, especially among homosexual populations. In the same group of patients, fecal pathogens, such as *Salmonella, Shigella, Campylobacter,* and *Microsporidium,* are often transmitted sexually. Oral-genital practices probably allow *N. meningitidis* to colonize and infect the genital tract. Viruses shed in secretions or present in blood (cytomegalovirus [CMV]; hepatitis B, possibly C and E; other non-A, non-B hepatitis viruses; human T-cell lymphotropic virus type I [HTLV-I]; and HIV) are spread by sexual practices.

Certain infections that are sexually transmitted occur on the surface epithelium of or near the lower genital tract. The major pathogens of these types of infections include HSV, *Haemophilus ducreyi,* and *T. pallidum.*

Other Routes. Organisms may also be introduced into the genital tract by instrumentation, presence of a foreign body, or irritation and can subsequently cause infection. Infections transmitted in this way are often caused by the same organisms that cause skin/wound infections. Of great significance, infection can also be transmitted from mother to infant either in vivo (within the living body) or during delivery. For example, transplacental infection may occur with syphilis, HIV, CMV, or HSV. Infection in the newborn can also be acquired during delivery by direct contact with an infectious lesion or discharge in the mother and a susceptible area in the infant (e.g., the eye). STDs, such as HSV, *C. trachomatis,* and *N. gonorrhoeae,* may be transmitted from mother to newborn in this manner. Other organisms, such as group B streptococci, *Escherichia coli,* and *Listeria monocytogenes,* that originate from the mother may also be transmitted to the infant before, during, or after birth. (Infections in the fetus and newborn are discussed later in this chapter.)

Clinical Manifestations

Clinical manifestations of lower genital tract infections are as varied and diverse as the etiologies.

Asymptomatic. Although symptoms of genital tract infections generally cause the patient to seek medical attention, a patient with an STD, especially a female, may be free of symptoms (i.e., asymptomatic). For

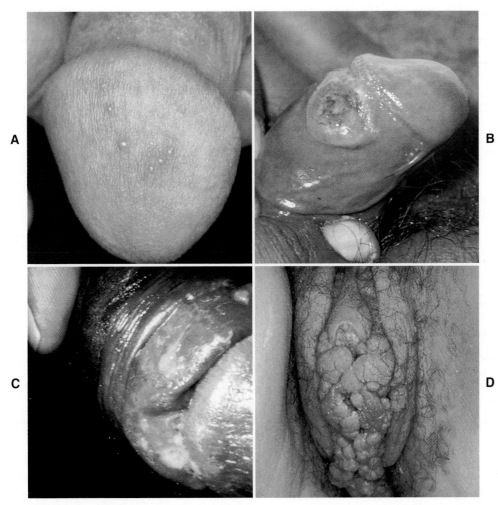

Figure 58-2 Genital lesions of the skin and mucous membranes that are sexually transmitted. **A,** Genital herpes showing vesicular lesions. **B,** Typical chancre of primary syphilis. **C,** Early chancroid lesion of the penis. **D,** *Condyloma acuminatum.* (All photographs from Farrar WE, Wood MJ, Innes JA, Hubbs H: *Infectious diseases text and color atlas,* ed 2, London, 1992, Gower Medical Publishing.)

example, gonorrhea or chlamydia infection in the male is usually obvious because of a urethral discharge, yet females with either or both of these infections may have either minimal symptoms or no symptoms at all. Also, the primary lesion of syphilis (chancre) can be unremarkable and go unnoticed by the patient. Therefore, the lack of symptoms does not guarantee the absence of disease. Unfortunately, these asymptomatic individuals can then serve as reservoirs for infection and unknowingly spread the pathogen to other individuals. Also, as in the case for asymptomatic infections in the female caused by *N. gonorrhoeae* or *C. trachomatis,* untreated infections can lead to serious sequelae such as pelvic inflammatory disease or infertility.

Dysuria. Although a frequent presenting symptom associated with urinary tract infection, dysuria can commonly result from an STD caused by organisms such as *N. gonorrhoeae, C. trachomatis,* and HSV.

Urethral Discharge. The presence of an inflammatory exudate at the tip of the urethral meatus is gene-

rally only observed in males; the symptoms of urethral infection in females are infrequently localized. Most males complain of discomfort at the penile tip as well as dysuria. Urethritis may be gonococcal, caused by *N. gonorrhoeae,* or nongonococcal. Nongonococcal urethritis can be caused by *C. trachomatis, Trichomonas vaginalis* (less frequently), and genital mycoplasmas such as *Mycoplasma hominis, M. genitalium,* and *Ureaplasma urealyticum.*

Lesions of the Skin and Mucous Membranes. Numerous organisms can cause genital lesions that are diverse in both their appearance and their associated symptoms (Figure 58-2). The agents and their features of infection are summarized in Table 58-2. Some of these infections, such as genital herpes (caused by HSV) or genital warts (caused by HPVs and discussed in Chapter 51), are common, whereas others, such as lymphogranuloma venereum and granuloma inguinale, are uncommon in the United States. Of note, specific HPVs, referred to as *genotypes,* infect mucosal cells in the cervix and can cause a progressive spectrum of abnormalities classified as low-grade and high-grade squamous intra-

Table 58-2 Summary of Common Causes of Genital Lesions of the Skin and Mucous Membranes

Agent	Disease	Lesion	Major Associated Symptoms
Herpes simplex virus	Genital herpes	Papules, vesicles (blisters), pustules, and/or ulcers	Multiple lesions that are usually painful and tender, can recur (see Figure 58-2, *A*)
Treponema pallidum	Primary syphilis	Genital ulcer (chancre)	Usually a single lesion, painless; lesion has even edges, represents the first of three stages of syphilis (see Figure 58-2, *B*)
Haemophilus ducreyi	Chancroid	Papule that becomes pustular and ulcerates (chancroid); multiple ulcers may develop	Ulcer is deeply invasive, tender, painful, and purulent in appearance; edges of lesion are ragged (see Figure 58-2, *C*)
Chlamydia trachomatis serotype L1, L2, and L3	Lymphogranuloma venereum	Small ulcer or vesicle that heals spontaneously without leaving a scar	After lesion heals, painful, swollen lymph nodes (lymphadenopathy) develop 2-6 weeks later; fever and chills; severe lymphatic obstruction and lymphedema can develop
Calymmatobacterium granulomatis	Granuloma inguinale	Single or multiple subcutaneous nodules	Indolent and chronic course; nodules enlarge and erode through the skin, producing a deep red, sharply defined ulcer that is painless
Human papillomavirus	Condylomata acuminate (primary genotypes 6 and 11) Condylomata planum (primary genotypes 16, 18, 31, 33)	Genital warts Flat, genital warts	Warts have a cauliflower-like appearance; usually multiple lesions that can be flat or elevated; usually asymptomatic apart from physical presence (see Figure 58-2, *D*) Cervical warts that must be visualized by using a magnifying lens after the application of acetic acid (called colposcopy). Infections can cause neoplasias that in some cases can progress to cervical cancer.

epithelial **neoplasia** (process of rapid cell growth that is faster than normal and continues to grow, i.e., a tumor), and in some cases, progress to invasive cervical cancer.

Vaginitis. Inflammation of the vaginal mucosa, called **vaginitis,** is a common clinical syndrome accounting for approximately 10 million office visits each year. Women who present with vaginal symptoms often complain of an abnormal discharge and possibly other symptoms such as an offensive odor or itching. The three most common causes of vaginitis in premenopausal women are vaginal candidiasis, bacterial vaginosis, and trichomoniasis.

Candida albicans causes about 80% to 90% of cases of vaginal candidiasis; other species of *Candida* account for the remaining cases. Yeast can be carried vaginally in small numbers and produce no symptoms. However, if conditions in the vagina change so as to give the yeast an advantage over other normal vaginal flora, candidiasis can result. Most patients complain of perivaginal itching, often with little or no discharge. Irritative symptoms such as erythema are also associated with candidiasis. Frequently, candidal discharge is classically thick and "cheesy" in appearance.

Vaginal infection with *T. vaginalis,* a protozoan parasite, produces a profuse, slightly offensive, yellow-green discharge; patients frequently complain of itching. About 25% of women carrying trichomonads are asymptomatic. The World Health Organization has ranked trichomoniasis as the most prevalent, nonviral, sexually transmitted disease in the world with an estimated 172 million new cases a year.

In addition to vaginitis caused by these two organisms, there is a third type of vaginitis referred to as **bacterial vaginosis (BV).** Initially, BV was believed to be caused by *Gardnerella vaginalis,* but *G. vaginalis* was isolated from 40% of women without vaginitis. Current understanding is that BV is polymicrobial in etiology, involving *G. vaginalis* and other facultative and anaerobic organisms. A recent study using three different molecular methods, including broad-range polymerase chain reaction (PCR) amplification of the 16S rDNA gene, confirms the bacterial diversity of organisms involved in this infection; 35 unique bacterial species were detected including many newly recognized species in women with BV.[3] This study also confirmed the loss of vaginal lactobacilli and concomitant overgrowth of anaerobic and facultative bacteria. The exact mechanism for the onset of BV is unknown, although it appears to be associated with a reduction in lactobacilli and hydrogen peroxide production, a rise in the vaginal pH, and the overgrowth of BV-associated organisms.[17] Syner-

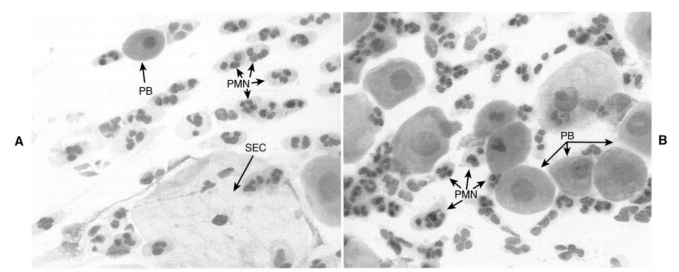

Figure 58-3   Gram stain of vaginal secretions from a patient with desquamative inflammatory vaginitis. **A,** Numerous polymorphonuclear cells *(PMNs)*, a squamous epithelial cell *(SEC)*, a parabasal cell *(PB)*, and the absence of lactobacilli are observed. **B,** Numerous PMNs, several PBs, and the absence of lactobacilli are observed.

gistic activity of various anaerobic organisms, including *Prevotella* spp., *Porphyromonas* spp., peptostreptococci, *Mobiluncus* spp. (curved, motile rods), and mycoplasmas, as well as *G. vaginalis,* seems to contribute to the pathology of BV. BV is characterized by perivaginal irritation that is considerably milder than trichomoniasis or candidiasis and is usually associated with a foul-smelling discharge that is often described as having a "fishy" odor. This odor is a result of products of bacterial metabolism (polyamines) being volatilized by vaginal fluids. Some patients also complain of abdominal discomfort. Of significance, it appears that BV and trichomoniasis frequently coexist.[12] Because BV can recur in the absence of sexual reexposure and other settings (e.g., nonsexually active women, virgins), BV is not exclusively sexually transmitted. BV also increases a woman's risk of acquiring HIV, is associated with increased complications in pregnancy, and may be involved in the pathogenesis of pelvic inflammatory disease. Moreover, although treatment trials report cure rates of 80% to 90% at 1 week, recurrence rates of BV were 15% to 30% within 3 months.[17]

Although uncommon, there are other infectious causes of vaginitis. Three are briefly mentioned here because Gram stain of vaginal secretions may be helpful. First, Sobel[12] described a number of perimenopausal patients with a diffuse, exudative vaginitis with massive vaginal cell exfoliation, purulent vaginal discharge, and an occasional vaginal and cervical spotted rash. Laboratory findings included elevated pH of vaginal secretions. Also, numerous polymorphonuclear cells, an increased number of parabasal cells, the absence of gram-positive bacilli, and their replacement by occasional gram-positive cocci are observed on direct Gram stain (Figure 58-3). Because of the extensive exfoliation of epithelial cells, basal cells then appear. This clinical syndrome is referred to as **desquamative inflammatory vaginitis.** Symptoms associated with another disorder, lactobacillosis, resemble those of candidiasis and often follow antifungal therapy. Gram stain or wet mount reveal a large number of very long lactobacilli. These predominately anaerobic lactobacilli are 40 to 75 μ in length and thus are quite longer than the average lactobacillus (5 to 15 microns) that are considered as normal flora in the vagina.[5] Finally, preexisting lesions due to other diseases may become secondarily infected with a mixed anaerobic flora of fusobacteria and spirochetes. This is referred to as fusiform-spirochetal disease; this infection can progress rapidly. Gram stain examination reveals inflammatory cells in conjunction with gram-negative, fusiform bacterial morphotypes and spirochetes.

Cervicitis.   Polymorphonuclear neutrophils (PMNs) are normally present in the endocervix; however, an abnormally increased number of PMNs may be associated with cervicitis (inflammation of the cervix). Therefore, a purulent discharge from the endocervix can be observed in some cases of cervicitis. The endocervix is the site from which *N. gonorrhoeae* is most frequently isolated in women with gonococcal infections. In patients presenting with cervicitis, *C. trachomatis* can also be isolated; chlamydia have not been associated with vaginitis. Frequently, patients are infected with both pathogens. Because most women with cervicitis caused by gonococci or chlamydia are asymptomatic and cervical abnormalities are either subtle or absent in these women, an appropriate laboratory diagnosis to detect these organisms must be performed.

HSV and human papillomavirus (HPV) can also infect the cervix. In women with herpes cervicitis, the cervix is friable (bleeds easily) and may have ulcers. Affected patients may also have lower abdominal pain.

Anorectal Lesions. As previously mentioned, because of the homosexual practice and increasingly common heterosexual practice of anal-genital intercourse, sites of infections in addition to those in the genital tract must be considered. The anorectum and pharynx are commonly infected with the classic STDs, including anal warts caused by HPV, as well as other viruses and parasites. Patients with symptoms of proctitis (inflammation of the rectum) caused by *N. gonorrhoeae* or *C. trachomatis* complain of itching, mucopurulent anal discharge, anal pain, bleeding, and tenesmus (painful straining during a bowel movement). Anorectal infections caused by HSV are associated with severe anal pain, rectal discharge, tenesmus, and systemic signs and symptoms such as fever, chills, and headaches.

In HIV-infected individuals and other immunocompromised patients, these infections tend to last longer, be more severe, and are more difficult to treat compared with infection in immunocompetent individuals. Anorectal lesions are common in HIV-infected patients and include anal condylomata, anal abscesses, and ulcers.[16] In this patient population, anal abscesses and ulcers can be due to various organisms, including CMV, *Mycobacterium avium* complex, HSV, *Campylobacter* spp., and *Shigella,* as well as traditional etiologic agents of STDs.

Bartholinitis. In adult women, the Bartholin's gland is a 1-cm mucus-producing gland on each side of the vaginal orifice. Each gland has a 2-cm duct that opens on the inner surface of the labia minora. If infected, this duct can become blocked and result in a Bartholin's gland abscess. Although *N. gonorrhoeae* and *C. trachomatis* can cause infection, anaerobic and polymicrobic infections originating from normal genital flora are more common.

Infections of the Reproductive Organs and Other Upper Tract Infections

Besides the lower genital tract, infections can occur in the reproductive organs of both males and females.

Females. Infection of the female reproductive organs (i.e., uterus, fallopian tubes, ovaries, and even the abdominal cavity) can occur. The organisms are believed to be frequently acquired as they ascend from lower-tract sites of infection. Organisms may also be introduced to the reproductive organs by surgery, instrumentation, or during childbirth.

Pelvic Inflammatory Disease. Pelvic inflammatory disease (PID) is an infection that results when cervical microorganisms travel upward to the endometrium, fallopian tubes, and other pelvic structures. This infection can produce one or more of the following inflammatory conditions: endometritis, salpingitis (inflammation of the salpinges), localized or generalized peritonitis, or abscesses involving the fallopian tubes or ovaries. Patients with PID often have intermittent abdominal pain and tenderness, vaginal discharge, dysuria, and possibly systemic symptoms such as fever, weight loss, and headache. Serious complications, such as permanent scarring of the fallopian tubes and infertility, can arise if PID is untreated.

Infection with *N. gonorrhoeae* and/or *C. trachomatis* in the lower genital tract can lead to PID if a woman is not adequately treated. Other organisms, such as anaerobes, gram-negative rods, streptococci, and mycoplasmas, may ascend through the cervix, particularly after parturition (childbirth), dilation of the cervix, or abortion. The presence of an intrauterine device (IUD) is associated with a slightly higher rate of PID. Such infections caused by *Actinomyces* have been associated with the use of IUDs.

Infections after Gynecologic Surgery. Following gynecologic surgery, such as a vaginal hysterectomy, women frequently develop postoperative infections that include pelvic cellulitis or abscesses. For the most part, these infections arise from the patient's own vaginal flora. Therefore, the major pathogens mirror the normal flora organisms: aerobic gram-positive cocci, gram-negative bacilli, anaerobes such as *Peptostreptococcus* spp., and genital mycoplasmas.

Infections Associated with Pregnancy. Infections can also occur during pregnancy (prenatal) or after birth (postpartum) in the mother. Of further significance, these infections may, in turn, be transmitted to the infant. Thus, these infections not only can compromise the mother's health but also the health of the developing fetus or neonate.

While developing within the uterus, the fetus is protected from most environmental factors, including infectious agents. The human immune system does not become fully competent until several months after birth. Immunoglobulins that cross the placental barrier, primarily immunoglobulin G (IgG), serve to protect the newborn from many infections until the infant begins to produce immunoglobulins of his or her own in response to antigenic stimuli. This unique environmental niche, however, does expose the vulnerable fetus to pathogens present in the mother.

Prenatal infections (those that occur anytime before birth) may be acquired by the blood-borne or as-

Table 58-3   Common Etiologic Agents of Prenatal and Neonatal Infections

Time of Infection*	Route of Infection	Common Agents
Prenatal	Transplacental	**Bacteria:** *Listeria monocytogenes, Treponema pallidum, Borrelia burgdorferi* **Viruses:** Cytomegalovirus (CMV), rubella, HIV, parvovirus B19, enteroviruses **Parasites:** *Toxoplasma gondii, Plasmodium* spp.
	Ascending	**Bacteria:** Group B streptococci, *Escherichia coli, L. monocytogenes, Chlamydia trachomatis,* genital mycoplasmas **Viruses:** CMV, herpes simplex virus (HSV)
Natal	Passing through the birth canal	**Bacteria:** Group B streptococci, *E. coli, L. monocytogenes, N. gonorrhoeae, C. trachomatis* **Viruses:** CMV, HSV, enteroviruses, hepatitis B virus, HIV
Postnatal	All of the above routes, from the nursery environment, or from maternal contact (e.g., breastfeeding)	All agents listed above and in addition, various organisms from the nursery environment, including gram-negative bacteria and viruses such as respiratory syncytial virus

*Some newborns develop infections during the first 4 weeks of postnatal life. Infections may be delayed manifestations of earlier prenatal (before birth), natal, or postnatal (after birth) acquisition of pathogens.

cending routes from mother to infant. If the mother has a bloodstream infection, organisms can reach and cross the placenta, with possible spread of infection to the developing fetus. Organisms that can cross the placenta are listed in Table 58-3. Alternatively, organisms can also infect the fetus by the ascending route from the vagina through torn or ruptured fetal membranes. **Chorioamnionitis** is an infection of the uterus and its contents during pregnancy. This infection is commonly acquired by organisms ascending from the vagina or cervix after premature or prolonged rupture of the membranes or labor. Organisms that are commonly isolated from amniotic fluid are listed in Box 58-1. Other maternal infections that are associated with adverse pregnancy outcomes that are not generally sexually transmitted include parvovirus B19, rubella, and *Listeria monocytogenes*.

Males. Infections in male reproductive organs can also occur and include epididymitis, prostatitis, and orchitis. **Epididymitis,** an inflammation of the epididymis, is commonly seen in sexually active men. Patients complain of fever and pain and swelling of the testicle. *N. gonorrhoeae* and *C. trachomatis* are common causes of epididymitis. However, enterics and coagulase-negative staphylococci can also cause infection in men older than 35 years of age and in homosexual men; these infections are often associated with obstruction by the prostate gland.

Prostatitis is a term to clinically describe adult male patients who have perineal, lower back, or lower abdominal pain, urinary discomfort, or ejaculatory complaints. Prostatitis is caused by both infectious and noninfectious means. Bacteria can cause an acute or chronic prostatitis. Patients with acute bacterial pro-

BOX 58-1	Organisms Frequently Isolated in Chorioamnionitis

Anaerobic bacteria
Genital mycoplasmas
Group B streptococci
Escherichia coli

statitis have dysuria and urinary frequency, symptoms that are associated with lower urinary tract infection. Frequently, these patients have systemic signs of illness such as fever. Chronic bacterial prostatitis is an important cause of persistent bacteriuria in the male that leads to recurrent bacterial urinary tract infections. The common causes of these infections are similar to the bacterial causes of lower urinary tract infections such as *Escherichia coli* and other enterics.

Finally, inflammation of the testicles, **orchitis,** is uncommon and generally acquired by blood-borne dissemination of viruses. Mumps is associated with most cases. Patients exhibit testicular pain and swelling following infection. Infections range from mild to severe.

LABORATORY DIAGNOSIS OF GENITAL TRACT INFECTIONS

LOWER GENITAL TRACT INFECTIONS

Urethritis, Cervicitis, and Vaginitis

Specimen Collection. This discussion focuses only on those specimens submitted for culture and/or direct examination. Procedures for the collection and transport of specimens for detection of agents by other noncultural methods (e.g., detection of *Chlamydia tracho-*

matis by amplification) should be followed according to the respective manufacturer's instructions. Refer to Table 5-1 for a review of collection, transport, and processing of genital tract specimens.

Urethral. Urethral discharge may occur in both males and females infected with pathogens such as *Neisseria gonorrhoeae* and *Trichomonas vaginalis*. The presence of infection is more likely to be asymptomatic in females because the discharge is usually less profuse and may be masked by normal vaginal secretions. *Ureaplasma urealyticum* can also be isolated from male urethral discharge.

A urogenital swab designed expressly for collection of such specimens should be used. These swabs are made of cotton or rayon that has been treated with charcoal to adsorb material toxic to gonococci and wrapped tightly over one end of a thin wire shaft. Cotton- or rayon-tipped swabs on a thin wire may also be used to collect specimens for isolation of mycoplasmas and chlamydiae. Calcium alginate swabs are generally more toxic for HSV, gonococci, chlamydiae, and mycoplasmas than are treated cotton swabs. Because Dacron swabs are least toxic, they are recommended for viral specimens. Dacron-tipped swabs on plastic shafts are also acceptable for chlamydiae and genital mycoplasmas.

To obtain a urethral specimen, a swab is inserted approximately 2 cm into the urethra and rotated gently before withdrawing. Because chlamydiae are intracellular pathogens, it is important to remove epithelial cells (with the swab) from the urethral mucosa. Separate swabs for cultivation of gonococci, chlamydiae, and ureaplasma are required. When profuse urethral discharge is present, particularly in males, the discharge may be collected externally without inserting a sampling device into the urethra. However, a urethral swab for chlamydia must be collected on males. A few drops of first-voided urine has also been used successfully to detect gonococci in males.

Because *T. vaginalis* may be present in urethral discharge, material for culture should be collected by swab as just described and another specimen collected on a swab and placed into a tube containing 0.5 mL of sterile physiologic saline. This specimen should be hand delivered to the laboratory immediately. Direct wet mounts and cultures for *T. vaginalis* can be performed from this second specimen. Commercial media for culture of *Trichomonas* are available. The first few drops of voided urine may also be a suitable specimen for recovery of *Trichomonas* from infected males, if it is inoculated into culture media immediately. Alternatively, material may be smeared onto a slide for later performance of a fluorescent antibody stain. Plastic envelopes for direct examination and subsequent culture are also available

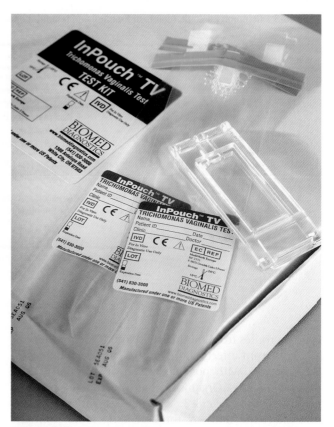

Figure 58-4 InPouch TV diagnostic system for wet mount examination and culture of *Trichomonas vaginalis*. The swab collected from the patient is inserted into liquid medium in the upper chamber of the plastic pouch and swirled. The top of the pouch is folded over and then sealed with the tabs. Once received in the laboratory, the upper chamber is examined for motile organisms; if no motile organisms are seen by microscopy, the upper chamber material is inoculated into the lower chamber and then incubated for up to 5 days. The lower chamber is similarly examined daily by microscopy for the presence of motile *T. vaginalis*. (Courtesy BioMed Diagnostics, White City, Ore.)

(InPouch TV, BIOMED, White City, Ore); sensitivity of this system is superior to other available methods and organism viability is maintained up to 48 hours (Figure 58-4). In addition, several other techniques are available including enzyme immunoassay, latex agglutination tests, and the Affirm VPIII probe (Becton Dickinson, Cockeysville, Md); polymerase chain reaction (PCR) has also been used to detect *T. vaginalis* directly in clinical specimens.

Cervical/Vaginal. Organisms that cause purulent vaginal discharge (vaginitis) include *T. vaginalis*, gonococci, and, rarely, beta-hemolytic streptococci. The same organisms that cause purulent infections in the urethra may also infect the epithelial cells in the cervical opening (os), as can HSV. Mucus is removed by gently rubbing the area with a cotton ball. The urethral swab just described is

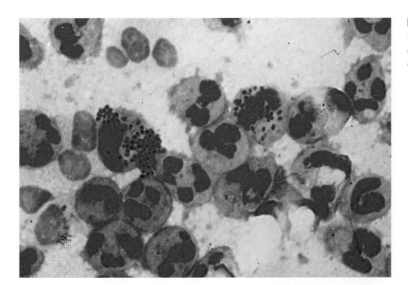

Figure 58-5   Gram-negative intracellular diplococci, which are diagnostic for gonorrhea in urethral discharge and presumptive for gonorrhea in vaginal discharge.

inserted into the cervical canal and rotated and moved from side to side for 30 seconds before removal.

Swabs are handled as described for urethral swabs for isolation of *Trichomonas* and gonococci. Chlamydiae cause a mucopurulent cervicitis with discharge. Endocervical specimens are obtained after the cervix has been exposed with a speculum, which allows visualization of vaginal and cervical architecture, and after ectocervical mucus has been adequately removed. The speculum is moistened with warm water, because many lubricants contain antibacterial agents. Because normal vaginal secretions contain great quantities of bacteria, care must be taken to avoid or minimize contaminating swabs for culture by contact with these secretions. A small, nylon-bristled cytology brush, or cytobrush, may be used to ensure that cellular material is collected, but its use is associated with discomfort and bleeding. Some controversy exists over whether the cytobrush results in better specimens, at least for detection of *Chlamydia trachomatis.*[7]

In addition to cervical specimens, which are particularly useful for isolating herpes, gonococci, mycoplasmas, and chlamydiae, vaginal discharge specimens may be collected. Organisms likely to cause vaginal discharge include *Trichomonas*, yeast, and the agents of BV. Swabs for diagnosis of BV are dipped into the fluid that collects in the posterior fornix of the vagina.

Genital tract infections caused by sexually transmitted agents in children (preadolescents) are most often the result of sexual abuse. Because of medico-legal implications, the laboratory should treat specimens from such patients with extreme care, carefully identifying and documenting all isolates. At the time of this writing, cultures should be obtained, especially for *Chlamydia trachomatis*, because amplification detection methods are still not admissible evidence in most states.

Because it is impossible to exclude contamination

with vaginal flora, obtaining swabs of Bartholin gland exudate are not recommended. Infected Bartholin glands should be aspirated with needle and syringe after careful skin preparation, and cultures should be evaluated for anaerobes and aerobes.

Transport. Swabs collected for isolation of gonococci may be transported to the laboratory in modified Stuart's or Amie's charcoal transport media and held at room temperature until inoculated to culture media. Good recovery of gonococci is possible if swabs are cultured within 12 hours of collection. Material that must be held longer than 12 hours should be inoculated directly to one of the commercial systems designed for recovery of gonococci, described later in this chapter.

Swabs for isolation of chlamydiae and mycoplasmas are best transported in specific transport media that have antibiotics and other essential components. Specimens for chlamydiae culture should be transported on ice. (Specimens transported at room temperature should be inoculated within 15 minutes of collection.) Specimens can be stored at 4° C for up to 24 hours. If culture inoculation will be delayed more than 24 hours, specimens should be quick-frozen in a dry ice and 95% ethanol bath and stored at −70° C until cultured. If collected and transported in specific transport media, specimens for genital mycoplasma culture may be transported on ice or at room temperature. If not in genital mycoplasma transport media, specimens should be transported on ice to suppress the growth of contaminating flora.

Direct Microscopic Examination. In addition to culture, urethral discharge may be examined by Gram stain for the presence of gram-negative intracellular diplococci (Figure 58-5), usually indicative of gonor-

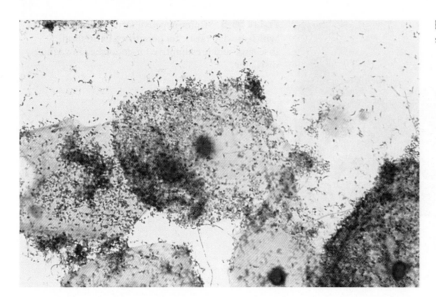

Figure 58-6   Clue cells in vaginal discharge suggestive of bacterial vaginosis.

rhea in males. After inoculation to culture media, the swab is rolled over the surface of a glass slide, covering an area of at least 1 cm². If the Gram stain is characteristic, cultures of urethral discharge need not be performed. Urethral smears from females may also be examined. However, presumptive diagnosis of gonorrhea from these smears is reliable only if the microscopist is experienced, because normal vaginal flora, such as *Veillonella* or occasional gram-negative coccobacilli, may resemble gonococci. If extracellular organisms resembling *N. gonorrhoeae* are seen, the microscopist should continue to examine the smear for intracellular diplococci. Presumptive diagnosis can be useful when decisions are to be made regarding immediate therapy, but confirmatory cultures or an alternative nonculture method should always be performed on specimens from females. Some strains of *N. gonorrhoeae* are sensitive to the amount of vancomycin present in selective media. If suspicious organisms seen on smear fail to grow in culture, reculture on chocolate agar without antibiotics may be warranted.

Fluorescein-conjugated monoclonal antibody reagents are sensitive and specific for visualization of the inclusions of *Chlamydia trachomatis* in cell cultures or elementary bodies in urethral and cervical specimens containing cells. Reagents for direct staining of specimens are available commercially in complete collection and test systems, but the relatively greater technologist time required for this method limits its usefulness for laboratories that receive many specimens, except as a confirmatory test for other antigen detection systems with borderline results. In some studies, the sensitivity of visual detection of chlamydiae with these newer reagents has been similar to that of culture, although such comparative results are obtained only by techno-

logists experienced in fluorescent techniques and when the slides are examined thoroughly. False-positive results should not occur if at least 10 morphologically compatible fluorescing elementary bodies are seen on the entire smear. No direct visual methods exist for detection of mycoplasmas at this time, but molecular assays have been evaluated.

Direct microscopic examination of a wet preparation of vaginal discharge provides the simplest rapid diagnostic test for *Trichomonas vaginalis* when such a specimen is available and can be examined immediately. The plastic envelope method combines direct visualization with culture. Motile trophozoites of *Trichomonas* can be visualized in a routine wet preparation performed by a proficient technologist in two thirds of cases or a direct fluorescent antibody (DFA) stain (Meridian Diagnostics, Cincinnati, Ohio) may be used.

Budding cells and pseudohyphae of yeast can also be easily identified in wet preparations by adding 10% potassium hydroxide (KOH) to a separate preparation, thereby dissolving host cell protein and enhancing the visibility of fungal elements.

BV, characterized by a foul-smelling discharge, can be diagnosed microscopically or clinically. The discharge is primarily sloughed epithelial cells, many of which are completely covered by tiny, gram-variable rods and coccobacilli. These cells are called **clue cells** (Figure 58-6). The absence of inflammatory cells in the vaginal discharge is another sign of BV. Although *Gardnerella vaginalis* has been historically associated with the syndrome and can be cultured on a human blood bilayer plate, culture is not recommended for diagnosis of BV.[13] A clinical diagnosis of BV is best made using three or more of the following criteria: homogeneous, gray discharge; clue cells seen on wet mount or Gram stain; a pH greater

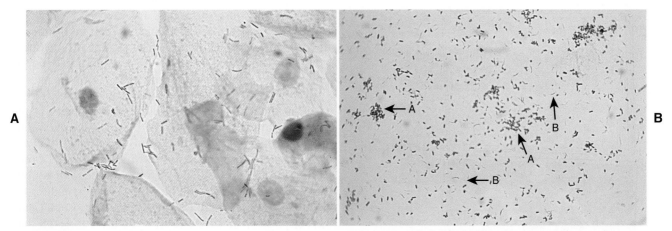

Figure 58-7   **A**, Predominance of lactobacilli in Gram stain from healthy vagina. **B**, Absence of lactobacilli and presence of *Gardnerella vaginalis (A arrows)* and *Mobiluncus* spp. *(B arrows)* morphologies.

than 4.5; and an amine or fishy odor elicited by the addition of a drop of 10% KOH to the discharge on a slide or on the speculum.[12]

Probably the best way to differentiate BV from other vaginal infections is by Gram stain (Figure 58-7). Nugent and colleagues has developed a grading system for Gram stains of vaginal discharge (Procedure 58-1).[8] This system is based on the presence or absence of certain bacterial morphologies. Typically, in patients with BV, lactobacilli are either absent or few in number, whereas curved, gram-variable rods (*Mobiluncus* spp.) and/or *G. vaginalis* and *Bacteroides* morphotypes predominate. The Gram stain is more sensitive and specific than either the wet mount for detection of clue cells or culture for *G. vaginalis,* and the smear can be saved and reexamined later.

Culture.   Samples for isolation of gonococci may be inoculated directly to culture media, obviating the need for transport medium. Commercially produced systems have been developed for this purpose, and many clinicians inoculate standard plates directly if convenient access to an incubator is available. Modified Thayer-Martin medium is most often used, although New York City (NYC) medium has the added advantage of supporting the growth of mycoplasmas and gonococci. Excellent recovery of gonococci is the rule when specimens are inoculated directly to any of these media in self-contained incubation systems such as JEMBEC plates (Figure 58-8). The specimen swab containing material is rolled across the agar with constant turning to expose all surfaces to the medium. The JEMBEC plate, which generates its own increased carbon dioxide atmosphere by means of a sodium bicarbonate tablet, is inoculated in a W pattern. The plate may be cross-streaked with a sterile loop in the laboratory (Figure 58-9).

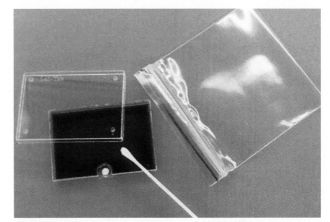

Figure 58-8   JEMBEC plate containing modified Thayer-Martin medium in a plastic, snap-top box with a self-contained CO_2-generating tablet, all sealed inside a Zip-lock plastic envelope after inoculation.

Specimens must be inoculated to additional media for isolation of yeast, streptococci, and mycoplasmas. Yeast grows well on Columbia agar base with 5% sheep blood and colistin and nalidixic acid (CNA), although more selective media are available. Most yeast and streptococci also grow on standard blood agar; thus, adding special fungal media such as Sabouraud brain-heart infusion agar (SABHI) is unwarranted.

A specimen from the lower vagina followed by the rectum using the same swab at 35 to 37 weeks' gestation reliably predicts the presence of group B streptococci at delivery.[11] The swab should be transported to the laboratory in a nonnutritive transport medium such as Amies or Stuart's without charcoal and then inoculated into a recommended selective broth medium such as Todd-Hewitt broth supplemented with either gentamicin and nalidixic acid or with colistin and nalidixic acid.

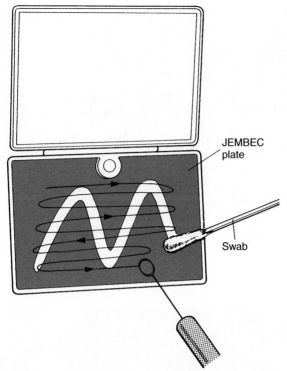

JEMBEC plate

Swab

Figure 58-9   Method of cross-streaking JEMBEC plate after original specimen has been inoculated by rolling the swab over the surface of the agar in a W pattern.

Selective enrichment broths are subcultured to agar the next day to isolate and identify group B streptococci. In addition, the presence of group B streptococci in urine in any concentration from a pregnant woman is a marker for heavy genital tract colonization.[11] Thus, any quantity of group B streptococci in urine from pregnant women should be worked up in the laboratory (see Chapter 57).

T. vaginalis may be cultured in Diamond's medium (available commercially) or plastic envelopes inoculated with discharge material. Culture techniques are most sensitive. A commercially available biphasic genital mycoplasma culture system (Mycotrim-GU, Irvine Scientific, Santa Ana, Calif) can be used to culture *Mycoplasma hominis* and *Ureaplasma urealyticum,* although commercially prepared media are not as sensitive as fresh media.[18] *Mycoplasma genitalium* may not grow on commercial media because of the presence of thallium acetate.[14]

Nonculture Methods.   Various nonculture methods may be used to diagnose genital tract diseases, including serology, latex agglutination, nucleic acid hybridization and amplification assays, and enzyme immunoassays. Most assays detect a single or possibly two genital tract pathogens, and most are commercially available.[6,15] These methods are described in more detail in chapters relating to individual pathogens.

As previously discussed, BV involves several organisms. Besides the Gram stain, BV can be diagnosed by using the Amsel criteria, which include pH measurement, performance of an amine test, and wet mount microscopy of vaginal secretions.[12] However, this approach has been considered unreliable because of the lack of microscopy-related skills and availability of pH paper in most doctors' offices.[1,12] Although the Gram stain offers high sensitivity and specificity, it is not immediately available. Currently, commercial laboratory tests are available to aid in the diagnosis of BV but are not all available in the United States; a test for sialidase (OSOM BVBLUE, Genzyme Diagnostics, Cambridge, Mass) in conjunction with measuring pH has been reported to be a rapid, highly sensitive and specific means to diagnose BV.[12] (Sialidases are secreted from anaerobic gram-negative rods such as *Bacteroides* and *Prevotella* as well as *Gardnerella* and play a role in bacterial nutrition, cellular interactions, and immune response evasion, which in turn improves the ability of bacteria to adhere, invade, and destroy mucosal tissue.) Of note, a hybridization assay (Affirm VP III Microbial Identification Test; Becton Dickinson Microbiology Systems, Sparks, Md) is commercially available to diagnose BV, as well as genital tract infections caused by *Candida* spp. and *Trichomonas vaginalis.* Once the appropriate reagents and specimen are added to special trays, the entire hybridization assays are then performed by instrumentation (Figure 58-10). Evaluations indicate this system is sensitive and specific.

Genital Skin and Mucous Membrane Lesions

External genital lesions are usually either vesicular or ulcerative. Causes of lesions can be determined by physical examination, histologic/cytologic examination, or microscopic examination and/or culture of exudate. Because any genital lesion may be highly contagious, clinicians should wear gloves when carrying out all manipulations of lesion material.

Vesicles in the genital area are almost always attributable to viruses, and herpes simplex is the most common cause. Epithelial cells from the base of a vesicle may be spread onto the surface of a slide and examined for the typical multinucleated giant cells of HSV or stained by immunofluorescent antibody stains for viral antigens. Additionally or alternatively, the material may be transported for culture of the virus, as outlined in Procedure 58-2.

Several commercial fluorescein-conjugated monoclonal and polyclonal antibodies directed against herpetic antigens of either type 1 or 2 are available. When fluorescent-antibody–stained lesion material containing enough cells is viewed under ultraviolet light, the diagnosis can be made in 70% to 90% of patients. Laboratories that routinely process genital material for herpes

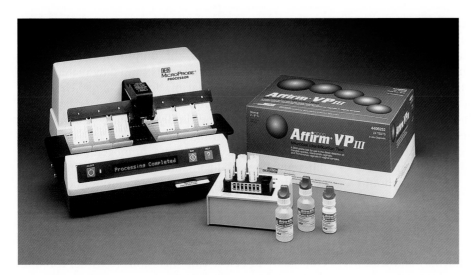

Figure 58-10 Affirm VP III Microbial Identification Test used to differentiate the three major causes of vaginitis/bacterial vaginosis from a single sample within 1 hour. (Courtesy Becton Dickinson Microbiology Systems. Affirm is a trademark of Becton Dickinson and Company.)

should be using immunofluorescent staining reagents when a rapid answer is desired; otherwise, culture, which is generally positive in 2 days, is the method of choice. Nonfluorescent markers, such as biotin-avidin-horseradish peroxidase or alkaline phosphatase, have also been conjugated to these specific antibodies, often allowing earlier detection of herpes-infected cells in tissue culture monolayers. Such reagents have been developed for use directly on clinical material, although their sensitivity is not great enough to forego culture if a definitive diagnosis is necessary.

Material from lesions suggestive of syphilis should be examined by dark-field or fluorescent microscopy. These procedures are described in Chapter 48.

All lesions suspected of infectious etiology may be Gram stained in addition to the procedures described. The smear of lesion material from a patient with chancroid may show many small, pleomorphic, gram-negative rods and coccobacilli arranged in chains and groups, characteristic of *H. ducreyi*. However, culture has been shown to be more sensitive for diagnosis of this agent. Material collected on cotton or Dacron swabs may be transported in modified Stuart's medium. Specimens should be inoculated to culture media within 1 hour of collection. A special agar, consisting of chocolate agar enriched with 1% IsoVitaleX (BBL Microbiology Systems) and vancomycin (3 mg/mL) has yielded good isolation if cultures are incubated in 5% to 7% carbon dioxide in a moist atmosphere, such as a candle jar. *H. ducreyi* grows best at 33° C.[2,10]

Granuloma inguinale is diagnosed by staining a crushed preparation of a small piece of biopsy tissue obtained from the edge of the base of the ulcer with Wright's or Giemsa stain and finding characteristic Donovan bodies (bipolar staining rods intracellularly within macrophages). Cytologists or pathologists usually examine such specimens rather than microbiologists. No acceptable media for isolation of *C. granulomatis* are available.

Bubo

Buboes, swollen lymph glands that occur in the inguinal (pelvic) region, are often evidence of a genital tract infection. Buboes are common in patients with primary syphilis, genital herpes, lymphogranuloma venereum, and chancroid. Patients with AIDS may show generalized lymphadenopathy. Other diseases that are not sexually transmitted, such as plague, tularemia, and lymphoma, can also produce buboes. Material from buboes may be aspirated for microscopic examination and culture.

INFECTIONS OF THE REPRODUCTIVE ORGANS

Pelvic Inflammatory Disease

Pelvic inflammatory disease is often caused by the same organisms that cause cervicitis or by organisms that make up the normal flora of the vaginal mucosa. Because of the profuse normal flora of the vaginal tract, specimens must be collected in such a way as to prevent vaginal flora contamination. Aspirated material collected by needle and syringe represents the best specimen. If this cannot be obtained at the time of surgery or laparoscopy, collection of intrauterine contents using a protected suction curetting device or double-lumen sampling device inserted through the cervix is also acceptable. Culdocentesis (aspiration of fluid in the cul-de-sac), after decontamination of the vagina by povidone-iodine, is satisfactory but rarely practiced today.

Aspirated material should be placed into an anaerobic transport container. The presence of either mixed anaerobic flora, gonococci, or both can be rapidly detected from a Gram stain. Direct examination with fluorescent monoclonal antibody stain may also detect chlamydiae. All such specimens should be inoculated to media that allow the recovery of anaerobic, facultative, and aerobic bacteria, gonococci, fungi, mycoplasmas, and chlamydiae. All material collected from normally sterile body sites in the genital tract should be inoculated to chocolate agar and placed into a suitable broth, such as chopped meat medium or thioglycollate, in addition to the other types of media noted. If only specimens obtained on routine swabs inserted through the cervix are available, cultures should be performed only for detection of gonococci and chlamydiae.

Miscellaneous Infections

Infections of the male prostate, epididymis, and testes are usually bacterial. In younger men, chlamydiae predominate as the cause of epididymitis and possibly of prostatitis. Urine or discharge collected via the urethra is the specimen of choice unless an abscess is drained surgically or by needle and syringe. The first few milliliters of voided urine may be collected before and after prostatic massage to try to pinpoint the anatomic site of the infection. Cultures are inoculated to support the growth of anaerobic, facultative, and aerobic bacteria, as well as gonococci.

Infections of Neonates and Human Products of Conception

Suspected infections acquired by the fetus as a result of a maternal infection that crosses the placenta (congenital infection) can be diagnosed culturally or serologically in the newborn. Because maternal IgG crosses the placenta, serologic tests are often difficult to interpret (see Chapter 10). For culturable agents, the most definitive diagnoses involve recovery of the pathogen in culture. HSV, varicella-zoster virus (VZV), enteroviruses, and cytomegalovirus (CMV) can be cultured easily, as can most bacterial agents. Rubella and parvovirus B19 are more difficult to culture. Nasal and urine specimens offer the greatest yield for viral isolation, whereas blood, cerebrospinal fluid, and material from a lesion can also be productive. Systemic neonatal herpes without lesions may be difficult to diagnose unless tissue biopsy material is examined, because the viruses may not be present in cerebrospinal fluid or blood. Bacteria and fungi can be isolated from lesions, blood, and other normally sterile sites.

Determining the presence of fetal immunoglobulin M (IgM) directed against the agent in question estab-

lishes the serologic diagnosis of congenital infection. Until recently, ultracentrifugation was required for separation of IgM from IgG, the only definitive means of preventing false-positive results caused by maternal IgG or fetal rheumatoid factor. Ion-exchange chromatography columns, antihuman IgG, and bacterial proteins that bind to IgG specifically are now commercially available for removing cross-reactive IgG and rheumatoid factor to obtain more homogeneous IgM for differentiation of fetal antibody. Indirect fluorescent antibody and enzyme-linked immunosorbent assay (ELISA) test systems are commercially available to detect IgM against *T. gondii,* rubella, CMV, HSV, and VZV. Interference by rheumatoid factor is still a consideration in most commercial IgM test systems (see Chapter 9). Our ability to detect viral inclusions in tissue, conjunctival scrapings, and vesicular lesions, traditionally performed with Giemsa stain, has been improved as a result of monoclonal and polyclonal fluorescent antiody reagents, which are described in the chapters that discuss individual agents.

Infections that infants can acquire as they pass through an infected birth canal or are related to difficult labor, premature birth, premature rupture of the membranes, or other events include the following:

● HSV and CMV infections
● Gonorrhea
● Group B streptococcal sepsis
● Chlamydial conjunctivitis and pneumonia
● *Escherichia coli* or other neonatal meningitis

In the laboratory, these infections are diagnosed by direct detection or culturing for the agents when possible, or by performing serologic tests. The appropriate specimens (e.g., cerebrospinal fluid, serum, pus, tracheal aspirate) should be examined and inoculated immediately. Routine body surface cultures of infants in intensive care have not been shown to be helpful for predicting subsequent disease.

Finally, certain infectious agents are known to cause fetal infection and even abortion. For example, *Listeria monocytogenes,* although usually causing only mild flulike symptoms in the mother, can cause extensive disease and abortion of the fetus if infection occurs late in the pregnancy. Therefore, isolation of the organism from the placenta and from tissues of the fetus is important.

Procedure 58-1

PREPARING AND SCORING VAGINAL GRAM STAINS FOR BACTERIAL VAGINOSIS

Organism Morphotype	Number/Oil Immersion Field	Score
Lactobacillus-like (parallel-sided, gram-positive rods)	>30	0
	5-30	1
	1-4	2
	<1	3
	0	4
Mobiluncus-like (curved, gram-negative rods)	>5	2
	<1-4	1
	0	0
Gardnerella/Bacteroides-like (tiny, gram-variable coccobacilli and rounded, pleomorphic, gram-negative rods with vacuoles)	>30	4
	5-30	3
	1-4	2
	<1	1
	0	0

Score	Interpretation
0-3	Normal
4-6	Intermediate, repeat test later
7-10	Bacterial vaginosis

Smear Preparation:
1. Roll the swab of vaginal discharge over the surface of a slide.
2. Allow the smear to air dry, fix with methanol, and Gram stain.
3. Assign scores; refer to box above.
4. Add up total score and interpret as above:

From Nugent RP, Krohn MA, Hillier SL: *J Clin Microbiol* 29:297, 1991.

Procedure 58-2

COLLECTION OF MATERIAL FROM SUSPECTED HERPETIC LESIONS

PRINCIPLE

Herpesvirus is best recovered from the base of active lesions in the vesicular stage. The older the lesion, the less likely it will yield viable virus.

METHOD

1. Open the vesicles with a small-gauge needle or Dacron-tipped swab.
2. Rub the base of the lesion vigorously with a small cotton-tipped or Dacron-tipped swab to recover infected cells.
3. Place the swab into viral transport medium (see Chapter 51) and refrigerate until inoculated to culture media. Specimens in media may be stored at −70° C for extended periods without loss of viral yield.
4. If large vesicles are present, material for culture may be aspirated directly by needle and syringe.
5. Material from another lesion can be applied directly to a glass slide for a Tzanck preparation (cytology) with Wright-Giemsa stain for detection of multinucleated giant cells, or for fluorescent antibody stain for detection of viral antigens.

Case Study

A 19-year-old woman presented to the clinic with low-grade fever and abdominal pain of 2 days' duration. She had a vaginal discharge for the previous 4 days and was sexually active with multiple partners in the past 2 months. She used no contraceptives. Urine was collected for an amplified probe assay for *Chlamydia trachomatis* and for *Neisseria gonorrhoeae*. Routine genital culture of the cervix was also collected. She was given an injection of ceftriaxone and placed on doxycycline for 2 weeks.

QUESTIONS

1. DNA probe assays are routinely performed from urine specimens from teenagers, because they are easily obtained. Why did the physician also do a culture in this woman?

2. This woman likely has pelvic inflammatory disease (PID). If left untreated, many cases result in infertility. What are the causes of this infection?

3. In this case, the culture was positive for an oxidase-positive, gram-negative diplococci. What is the likely agent?

REFERENCES

1. Anderson MR, Klink K, Cohrssen A: Evaluation of vaginal complaints, *JAMA* 291: 1368, 2004.

2. Clarridge JE, Shawar R, Simon B: *Haemophilus ducreyi* and chancroid: practical aspects for the clinical microbiology laboratory, *Clin Microbiol Newsl* 12:137, 1990.

3. Fredricks DN, Fiedler TL, Marrazzo JM: Molecular identification of bacteria associated with bacterial vaginosis, *N Engl J Med* 353: 1899, 2005.

4. Hillier SL, Krohn MA, Rabe LK, et al: Normal vaginal flora, H_2O_2-producing lactobacilli and bacterial vaginosis in pregnant women, *Clin Infect Dis* 16(suppl 4):S273, 1993.

5. Horowitz BJ, Mårdh P-A, Nagy E, et al: Vaginal lactobacillosis, *Am J Obstet Gynecol* 170:857, 1994.

6. Johnson RE, Newhall WJ, Rapp JR, et al: Screening tests to detect *Chlamydia trachomatis* and *Neisseria gonorrhoeae* infections-2002, *MMWR* 51(RR-15):1, 2002.

7. Kellogg JA, Seiple JW, Klinedinst JL, et al: Comparison of cytobrushes with swabs for recovery of endocervical cells and for Chlamydiazyme detection of *Chlamydia trachomatis*, *J Clin Microbiol* 30:2988, 1992.

8. Nugent RP, Krohn MA, Hillier SL: Reliability of diagnosing bacterial vaginosis is improved by a standardized method of Gram stain interpretation, *J Clin Microbiol* 29:297, 1991.

9. Romanowski B, Harris JR: Sexually transmitted diseases, *Clin Symp* 36:1, 1984.

10. Schmid GP, Faur YC, Valu JA, et al: Enhanced recovery of *Haemophilus ducreyi* from clinical specimens by incubation at 33° versus 35° C, *J Clin Microbiol* 33:3257, 1995.

11. Schrag S, Gorwitz R, Fultz-Butts K, et al: Prevention of perinatal group B streptococcal disease, revised guidelines from CDC, *MMWR* 51(RR-11):1, 2002.

12. Sobel JD: What's new in bacterial vaginosis and trichomoniasis? *Infect Dis Clin North Am* 19:387, 2005.

13. Spiegel CA: Bacterial vaginosis: changes in laboratory practice, *Clin Microbiol Newsl* 21:33, 1999.

14. Taylor-Robinson D, Bebear C: Antibiotic susceptibilities of mycoplasmas and treatment of mycoplasmal infections, *J Antimicrob Chemo* 40:622, 1997.

15. Warford A, Chernesky M, Peterson EM: Laboratory diagnosis of *Chlamydia trachomatis* infections. In Gleaves CA, coordinating editor, *Cumitech 19A*, Washington, DC, 1999, American Society for Microbiology.

16. Weiss EG, Wexner SD: Surgery for anal lesions in HIV-infected patients, *Ann Med* 27:467, 1995.

17. Wilson J: Managing recurrent bacterial vaginosis, *Sex Transm Infect* 80:8, 2004.

18. Wood JC, Lu RM, Peterson EM, et al: Evaluation of Mycotrim-GU for isolation of *Mycoplasma* species and *Ureaplasma urealyticum*, *J Clin Microbiol* 22:789, 1985.

GENERAL CONSIDERATIONS

ANATOMY

We are all connected to the external environment through our gastrointestinal (GI) tract (Figure 59-1). What we swallow enters the GI tract and passes through the esophagus into the stomach, through the small and large intestines, and finally to the anus. During passage, fluids and other components are added to this material as secretory products of individual cells and as enzymatic secretions of glands and organs, and are removed from this material by absorption through the gut epithelium.

The major components of the tract are listed in Box 59-1. The nature of the epithelial cells lining the GI tract varies with each portion. The lining of the GI tract is called the **mucosa.** Because of the differing nature of the mucosal surfaces of various segments of the bowel, specific infectious disease processes tend to occur in each segment.

The wall of the small intestine has folds that have millions of tiny, hairlike projections called **villi.** Each villus contains an arteriole, venule, and lymph vessel (Figure 59-2). The function of villi is to absorb fluids and nutrients from the intestinal contents. Epithelial cells lining the surface of villi have a surface that resembles a fine brush and is referred to as a **brush border.** The brush border is formed by nearly 2000 microvilli per epithelial cell. Intestinal digestive enzymes are produced in brush border cells toward the top of the villi. Villi and microvilli help make the small intestine the primary site of digestion and absorption by significantly increasing the surface area; more than 90% of physiologic net fluid absorption occurs here. Mucus-secreting goblet cells are found in large numbers of villi and intestinal crypts.

Similar to the small intestine, the large intestine is composed of several segments (see Box 59-1). The wall of the large intestine consists of columnar epithelial cells, many of which are mucus-producing goblet cells. In contrast to the small intestine, there are no villous projections into the lumen. The remaining excess fluid within the GI tract is resorbed by the cells lining the large intestine before waste is finally discharged through the rectum.

In addition to the previously discussed components of the GI tract, numerous other organs and structures are either located in the main digestive organs or open into them. These accessory organs and structures include the salivary glands, tongue, teeth, liver, gallbladder, and pancreas. Except for the teeth and salivary glands, these organs are illustrated in Figure 59-1.

RESIDENT MICROBIAL FLORA

The GI tract contains vast, diverse normal flora. Although the acidity of the stomach prevents any significant colonization in a normal host under most circumstances, many species can survive passage through the stomach to become resident within the lower intestinal tract. Normally, the upper small intestine contains only sparse flora (bacteria, primarily streptococci; lactobacilli; and yeasts; 10^1 to 10^3/mL), but in the distal ileum, counts are about 10^6 to 10^7/mL, with *Enterobacteriaceae* and *Bacteroides* spp. predominately present.

Infants usually are colonized by normal human epithelial flora, such as staphylococci, *Corynebacterium* spp., and other gram-positive organisms (bifidobacteria, clostridia, lactobacilli, streptococci), within a few hours of birth. Over time, the content of the intestinal flora changes. The normal flora of the adult large bowel (colon) is established relatively early in life and consists predominantly of anaerobic species, including *Bacteroides, Clostridium, Peptostreptococcus, Bifidobacterium,* and *Eubacterium.*

Aerobes, including *Escherichia coli,* other *Enterobacteriaceae,* enterococci, and streptococci are outnumbered by anaerobes 1000:1. The number of bacteria per gram of stool within the bowel lumen increases steadily as material approaches the sigmoid colon (the last segment). Eighty percent of the dry weight of feces from a healthy human consists of bacteria, which can be present in numbers as high as 10^{11} to 10^{12} colony-forming units (CFU)/g of stool.

GASTROENTERITIS

SCOPE OF THE PROBLEM

Worldwide, diarrheal diseases are the second leading cause of death; about 25 million enteric infections occur each year. These infections cause significant morbidity

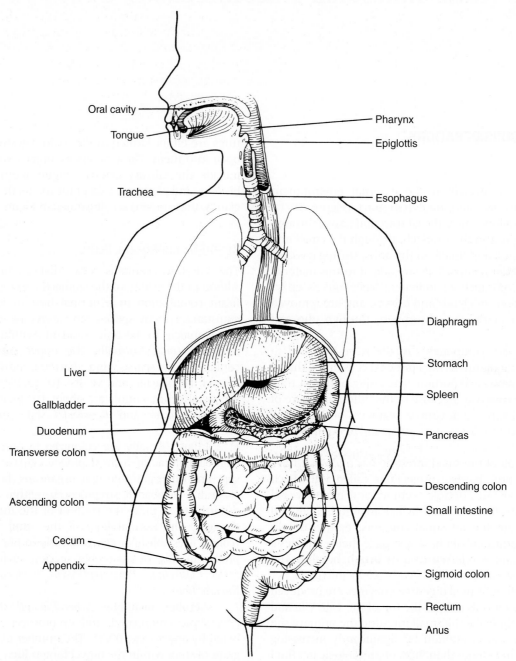

Figure 59-1   General anatomy of the gastrointestinal tract. (From Broadwell DC, Jackson BS: *Principles of ostomy care,* 1982, St Louis, Mosby.)

and death, particularly in elderly people and children younger than 5 years of age. It has been estimated that 4 to 6 million children die each year of diarrheal diseases, particularly in developing countries in Asia and Africa. Even in developed countries, significant morbidity occurs as a result of diarrheal illness. Although acute diarrheal syndromes are usually self-limited, some persons with infectious diarrhea require diagnostic studies and treatment.

PATHOGENESIS

Similar to the pathogenesis of urinary tract infections, the host and the invading microorganism possess key features that determine whether an enteric pathogen is able to cause microbial diarrhea.

Host Factors

The human host has numerous defenses that normally prevent or control disease produced by enteric

Mouth
Oropharynx
Esophagus
Stomach
- Fundus: enlarged portion of the stomach to the left and above the opening of the esophagus into the stomach
- Body: central part of the stomach
- Pylorus: lower portion of the stomach

Small intestine
- Duodenum: uppermost division; attached to pyloric end of the stomach
- Jejunum: midsection of the small intestine
- Ileum: lower portion of the small intestine

Large intestine
- Cecum
- Colon

  Ascending colon: lies on the right side of the abdomen and extends up to the lower portion of the liver; the ileum joins the large intestine at the junction of the cecum and the ascending colon

  Transverse colon: passes horizontally across the abdomen

  Descending colon: lies on the left side of the abdomen in a vertical position

  Sigmoid colon: extends downward, subsequently joining the rectum
- Rectum
- Anal canal

pathogens. For example, the acidity of the stomach effectively restricts the number and types of organisms that enter the lower GI tract. Normal peristalsis helps move organisms toward the rectum, interfering with their ability to adhere to the mucosa. The mucous layer coating the epithelium entraps microorganisms and helps propel them through the gut. The normal flora prevents colonization by potential pathogens.

Mucous membranes line the GI tract, as well as the respiratory and urogenital tracts. Although technically inside the body, some of these membranes can be considered as outside in the sense that they are exposed to the external environment in the form of food, water, and air. These membranes contain multiple cell types; some are secreting or absorbing cells that perform physiologic functions of the membrane while others serve a protective function. For example, sets of specialized cells called **follicles** are part of the mucous membrane lining the GI tract and serve a protective function. Collections of follicles are called **Peyer's patches.** Follicles contain M cells, macrophages, and B and T cells. As a result of the collective action of the follicle components following uptake and processing of the bacteria or antigens, **secretory immunoglobulin A (sIgA)** is released. Phagocytic cells and sIgA within the gut help destroy etiologic agents of disease, as do eosinophils, which are particularly active against parasites. Follicles and Peyer's patches are found in the small and large intestines.

Other factors that come into play include the host's personal hygiene and age. An initial step in the pathogenesis of enteric infections is ingestion of the pathogen. The majority of enteric pathogens, including bacteria, viruses, and parasites, is acquired by the fecal-oral route. Enteric infections can be spread by contamination of food products or drinking water and then subsequent ingestion. The age of the host also plays a role in whether disease is established. For example, diarrheal infections caused by rotavirus or enteropathogenic *Escherichia coli* tend to affect young children and not adults.

Finally, the normal intestinal flora is an important factor in the host response to the introduction of a potentially harmful microorganism. Whenever a reduction in normal flora occurs because of antibiotic treatment or some host factor, resistance to GI infection is significantly reduced. The most common example of the protective effect of normal flora is the development of the syndrome pseudomembranous colitis (PMC). This inflammatory disease of the large bowel is caused by the toxins of the anaerobic organism *Clostridium difficile* and occasionally other clostridia and perhaps even *Staphylococcus aureus,* and seldom occurs except after antimicrobial or antimetabolite treatment has altered the normal flora. Almost every antimicrobial agent and several cancer agents have been associated with the development of PMC. *C. difficile,* usually acquired from the hospital environment, is suppressed by normal flora. When normal flora are reduced, *C. difficile* is able to multiply and produce its toxins. This syndrome is also known as *antibiotic-associated colitis.* Other microorganisms that may gain a foothold when released from selective pressure of normal flora include *Candida* spp., staphylococci, *Pseudomonas* spp., and various *Enterobacteriaceae.*

Microbial Factors

The ability of an organism to cause GI infection depends not only on the susceptibility of the human host to the invading organism but also on the organism's virulence traits. To cause GI infection, a microorganism must possess one or more factors that allow it to overcome host defenses or it must enter the host at a time when one or more of the innate defense systems is inactive. For example, certain stool pathogens are able to survive gastric acidity only if the acidity has been reduced by bicarbonate, other buffers, or by medications for ulcers (e.g., cimetidine, ranitidine, H_2 blockers). Pathogens ingested with milk have a better chance of survival, because milk neutralizes stomach acidity. Organisms such as *Mycobacterium tuberculosis, Shigella, E. coli* O157:H7, and *C. difficile* (a spore-forming *Clostridium* spp.) are able to withstand exposure to gastric acids and thus require much smaller infectious inocula than do acid-sensitive organisms such as *Salmonella.*

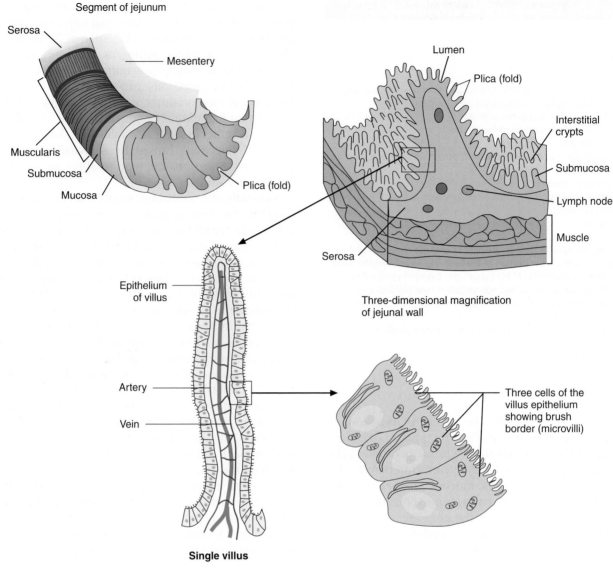

Figure 59-2   Wall of the small intestine. Villi cover the folds of the mucosal layer; in turn, each villus is covered with epithelial cells.

Primary Pathogenic Mechanisms. Because the normal adult GI tract receives up to 8 L of ingested fluid daily, plus the secretions of the various glands that contribute to digestion (salivary glands, pancreas, gallbladder, stomach), of which all but a small amount must be resorbed, any disruption of the normal flow or resorption of fluid will profoundly affect the host. Depending on how they interact with the human host, enteric pathogens may cause disease in one or more of the following three ways:

- By changing the delicate balance of water and electrolytes in the small bowel, resulting in massive fluid secretion. In many cases, this process is mediated by enterotoxin production. This is a non-inflammatory process

- By causing cell destruction and/or a marked inflammatory response following invasion of host cells and possible cytotoxin production, usually in the colon
- By penetrating the intestinal mucosa with subsequent spread to and multiplication in lymphatic or reticuloendothelial cells outside of the bowel; these infections are considered systemic infections

Examples of microorganisms for each of these pathogenic mechanisms are listed in Table 59-1.

Toxins

ENTEROTOXINS. Enterotoxins alter the metabolic activity of intestinal epithelial cells, resulting in an outpouring of electrolytes and fluid into the lumen. They act primarily in the jejunum and upper ileum, where

Table 59-1   Examples of Microorganisms That Cause GI Infection for Each Primary Pathogenic Mechanism

Mechanism	Examples of Microorganisms
Toxin Production	
Enterotoxin	*Vibrio cholerae* Noncholera vibrios *Shigella dysenteriae* type 1 Enterotoxigenic *Escherichia coli* *Salmonella* spp. *Clostridium difficile* (toxin A) *Aeromonas* *Campylobacter jejuni*
Cytotoxin	*Shigella* spp. *Clostridium difficile* (toxin B) Enterohemorrhagic *Escherichia coli*
Neurotoxin	*Clostridium botulinum* *Staphylococcus aureus* *Bacillus cereus*
Attachment Within or Close to Mucosal Cells/Adherence	Enteropathogenic *Escherichia coli* Enterohemorrhagic *Escherichia coli* *Cryptosporidium parvum* *Isospora belli* Rotavirus Hepatitis A, B, C Norwalk virus
Invasion	*Shigella* spp. Enteroinvasive *Escherichia coli* *Entamoeba histolytica* *Balantidium coli* *Campylobacter jejuni* *Plesiomonas shigelloides* *Yersinia enterocolitica* *Edwardsiella tarda*

most fluid transport takes place. The stool of patients with enterotoxic diarrheal disease involving the small bowel is profuse and watery, and blood or polymorphonuclear neutrophils are not prominent features.

The classic example of an enterotoxin is that of *Vibrio cholerae* (Figure 59-3). This toxin consists of two subunits, A and B.[21] The A subunit is composed of one molecule of A_1, the toxic moiety, and one molecule of A_2, which binds an A_1 subunit to five B subunits. The B subunits bind the toxin to a receptor (a ganglioside, an acidic glycolipid) on the intestinal cell membrane. Once bound, the toxin acts on adenylate cyclase enzyme, which catalyzes the transformation of adenosine triphosphate (ATP) to cyclic adenosine monophosphate (cAMP). Increased levels of cAMP stimulate the cell to actively secrete ions into the intestinal lumen. To maintain osmotic stabilization, the cells then secrete fluid into the lumen. The fluid is drawn from the intravascular fluid store of the body. Patients therefore can become dehydrated and hypotensive rapidly. *V. cholerae*

inhabits sea and stagnant water and is spread in contaminated water. The organsims have been isolated from coastal waters of several states, and sporadic cases of cholera occur in the United States. Additional information about *V. cholerae* is provided in Chapter 28.

Other organisms also produce a cholera-like enterotoxin. A group of vibrios similar to *V. cholerae* but serologically different, known as the **noncholera vibrios,** produce disease clinically identical to cholera, effected by a very similar toxin. The heat-labile toxin (LT) elaborated by certain strains of *E. coli,* called **enterotoxigenic *E. coli* (ETEC),** is similar to cholera toxin, sharing cross-reactive antigenic determinants. The enterotoxins of some *Salmonella* spp. (including *S. arizonae*), *Vibrio parahaemolyticus,* the *Campylobacter jejuni* group, *Clostridium perfringens, Clostridium difficile, Bacillus cereus, Aeromonas, Shigella dysenteriae,* and many other *Enterobacteriaceae* also cause positive reactions in at least one of the tests for enterotoxin (discussed below). The exact contribution of these enterotoxins to the pathogenicity of most stool pathogens remains to be elucidated.

Certain strains of *E. coli,* in addition to producing a heat-labile toxin similar to cholera toxin (LT), also produce a heat-stable toxin (ST) with other properties. Although ST also promotes fluid secretion into the intestinal lumen, its effect is mediated by activation of guanylate cyclase, resulting in increased levels of cyclic guanylate monophosphate (GMP), which yields the same net effect as increased cAMP. Tests for ST include enzyme-linked immunosorbent assay (ELISA), immunodiffusion, cell culture, and the classic suckling mouse assay, in which culture filtrate is placed into the stomach of a suckling mouse, with the intestinal contents later measured for fluid volume increase. Molecular techniques, including the use of DNA probes as well as several amplification assays, have been used to identify ETEC directly in clinical samples or isolated bacterial colonies.

Although several tests are available for the detection of enterotoxin, they are not performed routinely in diagnostic microbiology laboratories. These tests include the ligated rabbit ileal loop test, the Chinese hamster ovary cell assay, and the Y-1 adrenal cell assay. Because many enterotoxins are antigenic, homologous antibodies can be used to identify them specifically. Immunodiffusion, ELISA, and latex agglutination tests are all available to identify specific toxins. Molecular probes and amplification assays for toxin detection are also available primarily for research use.

CYTOTOXINS. The second category of toxins, cytotoxins, acts to disrupt the structure of individual intestinal epithelial cells. When destroyed, these cells slough from the surface of the mucosa, leaving it raw and unprotected. The secretory or absorptive functions

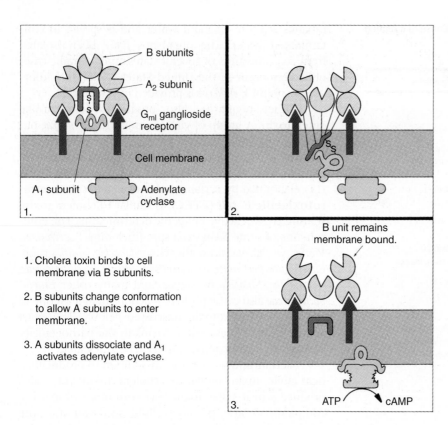

Figure 59-3   Diagrammatic representation of the structure and action of cholera toxin.

1. Cholera toxin binds to cell membrane via B subunits.

2. B subunits change conformation to allow A subunits to enter membrane.

3. A subunits dissociate and A_1 activates adenylate cyclase.

of the cells are no longer performed. The damaged tissue evokes a strong inflammatory response from the host, further inflicting tissue damage. Numerous polymorphonuclear neutrophils and blood are often seen in the stool, and pain, cramps, and tenesmus (painful straining during a bowel movement) are common symptoms. The term **dysentery** refers to this destructive disease of the mucosa, almost exclusively occurring in the colon. Cytotoxin has not yet been shown to be the sole virulence factor for any etiologic agent of GI disease, because most agents produce a cytotoxin in conjunction with another factor.

E. coli strains seem to possess virulence mechanisms of many types.[15,20] Some strains produce a cytotoxin that destroys epithelial cells and blood cells. Certain strains produce a cytotoxin that affects Vero cells (African green monkey kidney cells) and resemble the cytotoxin produced by *Shigella dysenteriae* (Shiga toxin); such strains of *E. coli* are associated with hemorrhagic colitis and the sequelae following infection of hemolytic-uremic syndrome (HUS) and thrombotic thrombocytopenia purpura (TTP).[8,9,15] These strains of *E. coli* are referred to as **enterohemorrhagic *E. coli* (EHEC).** Table 59-2 summarizes the key pathogenic features of the primary groups of diarrheagenic *E. coli*.

C. difficile produces a cytotoxin, the presence of which is a most useful marker for diagnosis of PMC. *S. dysenteriae, Staphylococcus aureus, C. perfringens,* and

V. parahaemolyticus produce cytotoxins that probably contribute to the pathogenesis of diarrhea, although they may not be essential for initiation of disease. Other vibrios, *Aeromonas hydrophila* (a relatively newly described agent of GI disease), and *Campylobacter jejuni,* the most common cause of GI disease in many areas of the United States, have been shown to produce cytotoxins. The role that these toxins play in the pathogenesis of the disease syndromes is not yet completely delineated.

NEUROTOXINS.   **Food poisoning,** or intoxication, may occur as a result of ingesting toxins produced by microorganisms. The microorganisms usually produce their toxins in foodstuffs before they are ingested; thus the patient ingests preformed toxin. Strictly speaking, these syndromes are not GI infections but rather intoxications; because they are acquired by ingestion of microorganisms or their products, they are considered in this chapter. Particularly in staphylococcal food poisoning and botulism, the causative organisms may not be present in the patient's bowel at all.

Bacterial agents of food poisoning that produce neurotoxins include *Staphylococcus aureus* and *Bacillus cereus.* Toxins produced by these organisms cause vomiting, independent of other actions on the gut mucosa. Staphylococcal food poisoning is one of the most frequently reported categories of food-borne disease. The organisms grow in warm food, primarily meat or dairy

Table 59-2   Overview of the Primary Groups of *E. coli* That Cause Diarrhea in Humans

Type	Primary Mode of Pathogenesis	Other Comments
Enterotoxigenic (ETEC)	Produces heat-labile (LT) and/or heat stable (ST) enterotoxins; genes of both toxins reside on a plasmid. LTs are closely related in structure and function to cholera toxin. STs result in net intestinal fluid secretion by stimulating guanylate cyclase.	Common cause of traveler's diarrhea; infects all ages
Enteroaggregative (EAEC)	Binds to small intestine cells via fimbriae encoded by a large molecular weight plasmid, forming small clumps of bacteria on the cell surface. Other plasmid-borne virulence factors include structured pilin, a heat-stable enterotoxin, novel anti-aggregative protein, and a heat-labile enterotoxin, all believed to be the cause of the associated diarrhea.	Infects primarily young children
Enteroinvasive (EIEC)	Pathogenesis has yet to be totally elucidated. Studies suggest that mechanisms by which diarrhea results are virtually identical to those of *Shigella* spp.	Very difficult to distinguish from *Shigella* spp. and other *E. coli* strains
Enteropathogenic (EPEC)	Initially attaches in the colon and small intestine and then becomes intimately adhered to intestinal epithelial cells, subsequently causing the loss of enterocyte microvilli (effacement). Genes for attachment/effacement reside in a cluster on the bacterial chromosome (i.e. pathogenicity island).	Diarrhea in infants, particularly in large urban hospitals
Enterohemorrhagic (EHEC)	Attaches to and effaces gut epithelial cells in a similar manner as EPEC. In addition, EHEC elaborates Shiga toxins.	Although many outbreaks are caused by *E. coli* O157:H7, other serotypes have been implicated in outbreaks and sporadic cases

products, and produce the toxin. Onset of disease is usually within 2 to 6 hours of ingestion. *B. cereus* produces two toxins, one of which is preformed, called the *emetic toxin,* because it produces vomiting. The second type, probably involving several enterotoxins, causes diarrhea. Often acquired from eating rice, *B. cereus* has also been associated with cooked meat, poultry, vegetables, and desserts.

Perhaps the most common cause of food poisoning is from type A *Clostridium perfringens,* which produces toxin in the host after ingestion. As a result, a relatively mild, self-limited (usually 24-hour) gastroenteritis occurs often in outbreaks in hospitals. Meats and gravies are typical offending foods.

One of the most potent neurotoxins known is produced by the anaerobic organism *Clostridium botulinum.* This toxin prevents the release of the neurotransmitter acetylcholine at the cholinergic nerve junctions, causing flaccid paralysis. The toxin acts primarily on the peripheral nerves but also on the autonomic nervous system. Patients exhibit descending symmetric paralysis and ultimately die of respiratory paralysis unless they are mechanically ventilated. In most cases, adult patients who develop botulism have ingested the preformed toxin in food (home-canned tomato products and canned, cream-based foods are often implicated), and the disease is considered to be an intoxication, although *C. botulinum* has been recovered from the stools of many adult patients. A relatively recently recognized syndrome, *infant botulism,* is a true GI infection. In adults, the normal flora probably prevents colonization by *C. botulinum,* whereas the organism is able to multiply and produce toxin in the infant bowel. Infant botulism is not an infrequent condition; babies acquire the organism by ingestion, although the source of the bacterium is not always clear. Because an association has been found with honey and corn syrup, infants younger than 9 months of age should not be fed honey. The effect of the toxin is the same, whether ingested in food or produced by growing organisms within the bowel.

Attachment. An organism's ability to cause disease can also depend on its ability to colonize and adhere to a relevant region of the bowel. To illustrate, ETEC must be able to adhere to and colonize the small intestine, as well as produce an enterotoxin. These organisms produce an adherence antigen, called *colonization factor antigen (CFA),* that gives the organism this adherence capacity. Certain strains of *E. coli* referred to as the **enteropathogenic *E. coli* (EPEC),** attach and then adhere to the intestinal brush border. This localized adherence is mediated by the production of pili. Subsequent to attaching, EPEC disrupts normal cell function by effacing the brush epithelium, thereby causing diarrheal disease. This complete process is referred to as **attachment and effacement.** Genes responsible for the initial adherence of ETEC, EHEC, and EPEC to intestinal epithelial cells reside on a transmissible plasmid.[15,20] Of note, EHEC

Origins of EHEC

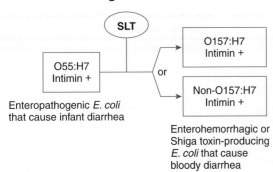

Figure 59-4 It appears that the presence of EHEC strain O157:H7 has actually increased in recent years and was not simply overlooked before 1982. *E. coli* O157:H7 strains are closely related to a Shiga toxin–negative EPEC strain O55:H7. It is proposed that this EPEC strain O55:H7 became infected by a bacteriophage that encoded Shiga toxin (SLT); it is now recognized that more than 100 different *E. coli* serotypes can express Shiga toxin.[9,20]

have the same ability to attach to intestinal epithelial cells and cause effacement. In addition, EHEC produces a Shiga toxin that spreads to the bloodstream, causing systemic damage to vascular endothelial cells of various organs, including kidney, colon, small intestine, and lung. EHEC is believed to have arisen as a result of an EPEC strain having become infected with a bacteriophage that carried the Shiga toxin gene (Figure 59-4).

Giardia lamblia, a parasite, has increasingly become more common as an etiologic agent of GI disease in the United States. Excreted into fresh water by natural animal hosts such as the beaver, the organism can be acquired by drinking stream water or even city water in some localities, particularly in the Rocky Mountain states, as well as throughout the world. The organism, a flagellated protozoan, adheres to the intestinal mucosa of the small bowel, possibly by means of a ventral sucker, destroying the mucosal cells' ability to participate in normal secretion and absorption. No evidence indicates invasion or toxin production.

Cryptosporidium and *Isospora* spp., parasitic etiologic agents of diarrhea in animals and poultry and more recently recognized as causing human disease, probably also act by adhering to intestinal mucosa and disrupting function. Cryptosporidia are often seen in the diarrhea of patients with acquired immunodeficiency syndrome (AIDS), as well as in travelers' diarrhea, day care epidemics, and diarrhea in people with animal exposure; cryptosporidia and *Isospora* spp. may cause severe, protracted diarrhea in AIDS patients. Other coccidian parasites, such as microsporidia, produce diarrhea by destroying intestinal cell function.

Invasion. Following initial and essential adherence to GI mucosal cells, some enteric pathogens are

able to gain access to the intracellular environment. Invasion allows the organism to reach deeper tissues, access nutrients for growth, and possibly avoid the host immune system.

In the case of diarrhea caused by *Shigella,* the primary mechanism of disease production consists of (1) the triggering and directing by *Shigella* of its own entry into colonic epithelial cells by genes located on a plasmid, and once internalized, (2) the rapid multiplication of *Shigella* in the submucosa and lamina propria and its intracellular and extracellular spread to other adjacent colonic epithelial cells.[19] Once in the host cell cytoplasm, *Shigella* spp. cause apoptosis and release of the cytokines interleukin (IL)-1 and IL-8.[20] The inflammatory response to these cytokines damages the colonic mucosa and exacerbates (aggravates) the infection. Of note, the genes for invasiveness are located on a large invasion plasmid. These activities lead to extensive superficial tissue destruction. If these two steps do not occur, one does not get the clinical presentation of classic dysentery (Table 59-3). The entry process is illustrated in Figure 59-5.

Salmonellae interact with the apical (top) microvilli of colonic epithelial cells, disrupting the brush border. Similar to *Shigella, Salmonella* spp. also stimulate the host cell to internalize them through rearrangements of host actin filaments and other cytoskeleton proteins.[3,4,6] Once the whole bacteria are internalized within endocytic vesicles of the host epithelial cell, organisms begin to multiply within the vacuoles. In contrast to *Shigella* spp. that use the colonic mucosal epithelium as a site of multiplication, certain serotypes of *Salmonella,* such as *S. typhi* and *S. choleraesuis,* use the colonic epithelium as a route to gain access to the submucosal layers, mesenteric lymph nodes, and subsequently the bloodstream. The entry of *Salmonella* is a complex process that involves several essential genes, as well as particular environmental conditions of the host cell; this process is still being delineated. Many virulence factors for invasion of salmonellae into nonphagocytic cells as well as their ability to cause systemic infections by surviving in phagocytic cells and replicating within the *Salmonella*-containing vesicle in a variety of eukaryotic cells are determined by chromosomal genes, many of which are located within pathogenicity islands. Invasiveness is also thought to contribute to the pathogenesis of disease associated with species of vibrios, campylobacters, *Yersinia enterocolitica, Plesiomonas shigelloides,* and *Edwardsiella tarda.*

Certain parasites, particularly *Entamoeba histolytica* and *Balantidium coli,* invade the intestinal epithelium of the colon as a primary site of infection. The ensuing amebic dysentery is characterized by blood and numerous white blood cells, and the patient experiences cramping and tenesmus. Other parasites that are

Table 59-3   Types of Enteric Infections

Pathogenic Mechanism	Major Symptoms	Examples of Etiologic Agents
Upsetting of fluid and electrolyte balance/noninflammatory	Watery diarrhea No fecal leukocytes No fever	*Vibrio cholerae* Rotavirus Norwalk virus Enterotoxigenic *Escherichia coli* *Giardia lamblia* *Bacillus cereus*
Invasion and possible cytotoxin production/ inflammatory (dysentery)	Dysenteric-like diarrhea (mucus, blood, white cells) Fever Fecal leukocytes	*Shigella* spp. Enteroinvasive *E. coli* *Salmonella enteritidis* *Entamoeba histolytica*
Penetration with subsequent access to the bloodstream (enteric fever)	Signs of systemic infection (headache, malaise, sore throat) Fever	*Salmonella typhi* *Yersinia enterocolitica*

acquired by ingestion, such as *Trichinella,* may cause transient bloody diarrhea and pain during migration through the intestinal mucosa to their preferred sites within the host.

Other organisms selectively destroy absorptive cells (e.g., villus tip cells) in the mucosa, disrupting their normal cell function and thereby causing diarrhea. Rotaviruses and Norwalk-like viruses are both visualized by electron microscopy within the absorptive cells at the ends of the intestinal villi, where they multiply and destroy cellular function. As a result, the villi become shortened, and inflammatory cells infiltrate the mucosa, further contributing to the pathologic condition. In addition to these viral agents, hepatitis A, B, and C and occasionally enteric adenoviruses have been associated with diarrheal symptoms in infected patients.

Miscellaneous Virulence Factors. Other virulence traits appear to be involved in the development of GI infections and include characteristics such as motility, chemotaxis, and mucinase production. Also, the possession of certain antigens, such as the Vi antigen of *Salmonella typhi* and certain cell wall components, are also associated with virulence.

CLINICAL MANIFESTATIONS

The clinical symptoms experienced by a patient are largely dependent on how the enteric pathogen causes disease. To illustrate, patients infected with an enteric pathogen that upsets fluid and electrolyte balance have no fecal leukocytes present in the stool and complain of watery diarrhea; fever is usually absent or mild. Although nausea, vomiting, and abdominal pain may also be present, the dominant feature is intestinal fluid loss. In contrast, patients infected with an enteric pathogen that causes significant cell destruction and inflam-

mation have fecal leukocytes present in the stool (Figure 59-6). Their diarrhea is often characterized by the presence of mucus and possibly blood; in many of these patients, fever is a prominent component of their disease, as well as abdominal pain, cramps, and tenesmus. Finally, patients who become infected with a pathogen that is able to penetrate the intestinal mucosa of the small intestine without producing enterocolitis and then subsequently spread and multiply at other sites will present with signs and symptoms of a systemic illness such as headache, sore throat, malaise, and fever; diarrhea in these patients is not a prominent feature and is absent or mild in many cases. Features of these three types of enteric infections are summarized in Table 59-3.

EPIDEMIOLOGY

Gastrointestinal infections occur in numerous epidemiologic settings. Awareness of these different settings is important because knowledge of a particular epidemiologic setting can help provide a basis for the diagnosis and clues to possible etiologies. When this knowledge is combined with clinical findings, the etiology of the infection can often be narrowed to three or four organisms.

Institutional Settings

Diarrheal illness can be a major problem in institutional settings such as day care centers, hospitals, and nursing homes. Because individual hygiene is often difficult to maintain in these settings, coupled with several organisms with relatively low infecting doses such as *Shigella* and *Giardia lamblia*, numerous outbreaks of diarrheal illness caused by various organisms have been reported. Organisms such as *Shigella, Campylobacter jejuni, Giardia lamblia, Cryptosporidium,* and rotaviruses have been reported to cause outbreaks in day care centers. Of significance, these infections can be spread to family

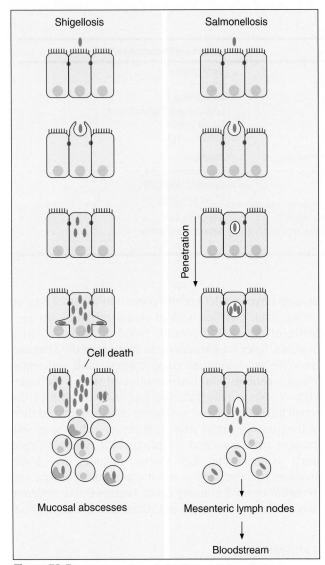

Figure 59-5 The invasion of *Shigella* and *Salmonella* into intestinal epithelial cells. (Modified from Sansonetti PJ: Genetic and molecular basis of epithelial invasion by *Shigella* species, *Rev Infect Dis* 13[suppl 4]:S282, 1991, University of Chicago Press.)

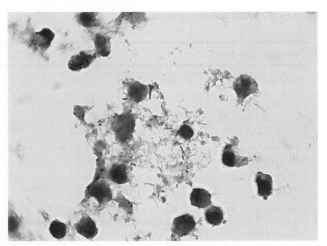

Figure 59-6 Wright's stain of stool from a patient with shigellosis showing moderate numbers of polymorphonuclear cells.

members. Similarly, outbreaks caused by these organisms, as well as hemorrhagic *E. coli* O157:H7, have been reported in nursing homes and other extended care facilities.

Nosocomial diarrheal illness is also a problem for hospital patients and personnel. Of importance, rotaviruses, adenoviruses, and coxsackie viruses are also nosocomially transmitted. In addition to these organisms, *Clostridium difficile* is a major nosocomial enteric pathogen in hospitals and other settings, including nursing homes and extended-care facilities. This organism is a hardy pathogen that readily survives on fomites (inanimate objects) such as floors, bed rails, call buttons, and doorknobs, and on the hands of hospital personnel caring for the patient. Of great concern has been the recent emergence of a new strain of *C. difficile* with increased virulence and fluoroquinolone resistance.[13] By virtue of partial deletions in a toxin regulatory gene, *tcdC*, these isolates are able to produce 16- to 23-fold more toxin A and B. In addition, a separate binary toxin has been described that is encoded by *cdtA* and *cdtB* genes; *cdtB* mediates cell surface binding and cellular translocation, whereas *cdtA* disrupts the assembly of the actin filament causing cell death.[11] These strains have emerged as a cause of geographically dispersed outbreaks of *C. difficile*–associated disease. Of significance, many of the reported cases caused by these strains were in otherwise healthy patients with minimal or no exposure to a health care setting.

Traveler's Diarrhea

Individuals who travel into developing geographic areas that have poor sanitation are at particularly high risk for developing diarrhea if they do not pay attention to their eating and drinking habits. In areas with poor sanitation, enteric pathogens heavily contaminate the water and food. Although many types of enteric pathogens can cause diarrhea in travelers, enterotoxigenic *E. coli* is a leading cause in Asia, Africa, and Latin America, accounting for about 50% of cases. Salmonellae, shigellae, *Campylobacter* spp., vibrios, rotavirus, and Norwalk virus can also cause diarrhea in travelers, depending on the area or country they visit.

Food- and Water-Borne Outbreaks

The Centers for Disease Control and Prevention report that more than 12,000 cases of food-borne illness occur in the United States each year. Because most of these illnesses are not reported, some estimate that millions of cases occur annually. Eating raw or undercooked fish, shellfish, or meats, and drinking unpasteurized milk increases the risks of certain bacterial, parasitic, and viral infections. Many food-borne outbreaks can be traced to poor hygienic practices of food handlers such as not washing hands after using the toilet; hepatitis A, Norwalk virus, and *Salmonella* are a few examples of organisms that have contaminated food during preparation by a food handler and causing diarrheal disease. Since 1968, the number of cases of salmonellosis has gradually increased, with many of these infections associated with eating raw or undercooked eggs. Also, the potential for widespread dissemination of food-borne pathogens has increased because of factors such as the tendency to eat outside the home, the export and import of food sources worldwide, and travel.

In addition to food-borne outbreaks of GI tract infections, water-borne outbreaks of diarrheal disease caused by *Giardia lamblia* and *Cryptosporidium* have been traced to inadequately filtered surface water. Recreational waters, including swimming pools, can also become contaminated with enteric pathogens such as *Shigella* and *G. lamblia* because of poor toilet facilities or practices.

Immunocompromised Hosts

GI tract infections in individuals infected with human immunodeficiency virus (HIV) and other patients who are immunosuppressed, such as organ transplant recipients or individuals receiving chemotherapy, are a diagnostic challenge for the clinician and microbiologist. For example, cytotoxic chemotherapy and/or antibiotic therapy may predispose patients to develop *C. difficile* colitis.

Diarrhea is a common clinical manifestation of infection with HIV, developing in about 30% to 80% of cases. Numerous pathogens and opportunistic pathogens have been identified and are believed to cause recurrent or chronic diarrhea. Commonly reported etiologic agents are:

- Species of *Salmonella, Shigella,* and *Campylobacter*
- Cytomegalovirus
- *Cryptosporidia, Isospora belli*
- *Microsporidia*
- *Entamoeba histolytica*
- *Mycobacterial* spp.
- *Giardia lamblia*

ETIOLOGIC AGENTS

Many microorganisms are able to cause enteric infections. A discussion of each organism is beyond the scope of this chapter. Rather, these organisms are addressed in Parts III through VI of the textbook. Table 59-4 summarizes the general characteristics of the more common agents of enteric infections.

OTHER INFECTIONS OF THE GASTROINTESTINAL TRACT

Besides causing disease in the small and large intestine, microorganisms can also infect other sites of the GI tract, as well as the GI tract's accessory organs.

ESOPHAGITIS

Infections of the mucosa of the esophagus **(esophagitis)** can cause painful or difficult swallowing, and/or the sensation that something is lodged in the throat while swallowing. Individuals who have esophagitis usually have local or systemic underlying illnesses such as hematologic malignancies or HIV infection, or are receiving immunosuppressive therapy. The most common etiologic agents are *Candida* spp. (primarily *C. albicans*), herpes simplex virus, and cytomegalovirus.

GASTRITIS

Gastritis refers to inflammation of the gastric mucosa. This illness is associated with nausea and upper abdominal pain; vomiting, burping, and fever may also be present. A curved organism called *Helicobacter pylori* is seen on the surface of gastric epithelial cells of patients with gastritis. The organism is recovered from gastric biopsy material obtained endoscopically but not from stool. Following acute infection, *H. pylori* can persist for years in most individuals, with most remaining asymptomatic. *H. pylori* is also the causative agent of peptic ulcer disease and a significant risk factor for stomach cancer.

PROCTITIS

Proctitis is the inflammation of the rectum (distal portion of the large intestine). Common symptoms associated with proctitis are itching and a mucous discharge from the rectum; if the infection progresses, ulcers and abscesses may form in the rectum. The majority of infections are sexually transmitted through anal intercourse. *Chlamydia trachomatis,* herpes simplex, syphilis, and gonorrhea are the most common etiologic agents.

MISCELLANEOUS

Unusual agents and those that have not been cultured, such as the mycobacteria that may be associated with Crohn's disease and the bacterium associated with Whipple's disease, identified by molecular methods as a new agent, *Trophyrema whipplei,* are also candidates as etiologic agents of GI disease. Occasionally, stool cultures from patients with diarrheal disease yield heavy growth of organisms such as enterococci, *Pseudomonas* spp., or

Table 59-4   General Characteristics of the Common Agents of Enteric Infections

Organism	Common Sources or Predisposing Condition	Distribution	Clinical Presentation	Predominant Pathogenic Mechanism	Fecal Leukocytes
Bacillus cereus	Meats, vegetables, rice	Worldwide	Intoxication: vomiting or watery diarrhea	Ingestion of preformed toxin (food poisoning)	–
Clostridium botulinum	Improperly preserved vegetables, meat, fish	Worldwide	Neuromuscular paralysis	Ingestion of preformed toxin (food poisoning)	–
Staphylococcus aureus	Meats, salads, dairy products	Worldwide	Intoxication: vomiting	Ingestion of preformed toxin (food poisoning)	–
Clostridium perfringens	Meats, poultry	Worldwide	Watery diarrhea	Ingestion of organism followed by toxin production	–
Aeromonas	Water	Worldwide	Watery diarrhea or dysentery	? Enterotoxin ? Cytotoxin	–
Campylobacter spp.	Water, poultry, milk	Worldwide	Dysentery	? Invasion ? Cytotoxins	+
Clostridium difficile	Antimicrobial therapy	Worldwide	Dysentery	Enterotoxin and cytotoxin	+/–
Diarrheagenic *Escherichia coli*:					
Enteropathogenic	?	Worldwide	Watery diarrhea	Adherence/? invasion without multiplication	–
Enterotoxigenic	Food, water	Worldwide—more prevalent in developing countries	Watery diarrhea	Enterotoxin	–
Enteroinvasive	Food	Worldwide	Dysentery	Invasion, enterotoxin	+
Enterohemorrhagic	Meats	Worldwide	Watery, often bloody diarrhea	Cytotoxin	–/+
Plesiomonas shigelloides	Fresh water, shellfish	Worldwide	? Dysentery	Unknown ? Enterotoxin	+/–
Salmonella spp. (nontyphoidal)	Food, water	Worldwide	Dysentery	Invasion	+
Salmonella typhi	Food, water	Tropical, developing countries	Enteric fever	Penetration	+ (monocytes, *not* PMNs)
Shigella spp.	Food, water	Worldwide	Dysentery	Invasion	+
Shigella dysenteriae	Water	Tropical, developing countries	Dysentery	Invasion, cytotoxin	+
Vibrio cholerae	Water, shellfish	Asia, Africa, Middle East, South and North American (along coastal areas)	Watery diarrhea	? Enterotoxin Cytotoxin	–/+
Yersinia enterocolitica	Milk, pork, water	Worldwide	Watery diarrhea and/or enteric fever	? Invasion ? Penetration	–
Giardia lamblia	Food, water	Worldwide	Watery diarrhea	Unknown-impaired absorption	–
Cryptospordium parvum	Animals, water	Worldwide	Watery diarrhea	? Adherence	–
Entamoeba histolytica	Food, water	Worldwide (more common in developing countries)	Dysentery	Invasion, cytotoxin	–/+ (amoeba destroy the white cells)

Table 59-4   General Characteristics of the Common Agents of Enteric Infections (cont'd)

Organism	Common Sources or Predisposing Condition	Distribution	Clinical Presentation	Predominant Pathogenic Mechanism	Fecal Leukocytes
Rotavirus	?	Worldwide	Watery diarrhea	Mucosal damage leading to impaired absorption in small intestine	–
Norwalk viruses	Shellfish, salads	Worldwide	Watery diarrhea	Mucosal damage leading to impaired absorption in small intestine	–

?, Questionable or uncertain; +, positive; –, negative; +/–, more frequently positive; +/–, more frequently negative.

Klebsiella pneumoniae, not usually found in such numbers as normal flora. Only anecdotal evidence suggests that these organisms actually contribute to the pathogenesis of the diarrhea. Agents of sexually transmitted disease may cause GI symptoms when they are introduced into the colon via sexual intercourse. *Mycobacterium avium-intracellulare* complex may be transmitted in this way, going on to cause systemic disease in patients with AIDS. The pathogenesis of infections resulting from *Blastocystis hominis* (a possible coccidian etiologic agent of human diarrheal disease) is not well documented, although these organisms are associated with GI symptoms.

LABORATORY DIAGNOSIS OF GASTROINTESTINAL TRACT INFECTIONS

SPECIMEN COLLECTION AND TRANSPORT

If enteric pathogens are to be detected by the laboratory, adherence to appropriate guidelines for specimen collection and transport is imperative (see Table 5-1 for a quick guide to specimen collection, transport, and processing). If an etiologic agent is not isolated with the first culture or visual examination, two additional specimens should be submitted to the laboratory over the next few days. Because organisms may be shed intermittently, collection of specimens at different times over several days enhances recovery. Certain infectious agents, such as *Giardia,* may be difficult to detect, requiring the processing of multiple specimens over weeks, duodenal aspirates (in the case of *Giardia*), or additional alternative methods.

General Comments

Specimens that can be delivered to the laboratory within 30 minutes may be collected in a clean, waxed cardboard or plastic container. Stool for direct wet-mount examination, *Clostridium difficile* toxin assay, immunoelectron microscopy for detection of viruses, and ELISA or latex agglutination test for rotavirus must be sent to the laboratory without any added preservatives or liquids. Volume of a liquid stool at least equal to 1 teaspoon (5 mL) or a pea-sized piece of formed stool is necessary for most procedures.

Stool Specimens for Bacterial Culture

If a delay longer than 2 hours is anticipated for stools for bacterial culture, the specimen should be placed in transport medium. Cary-Blair transport medium best preserves the viability of intestinal bacterial pathogens, including *Campylobacter* and *Vibrio* spp. However, the media produced by different manufacturers can vary.[14] Most workers recommend reducing the agar content of Cary-Blair medium from 0.5% to 0.16% (modified) for maintenance of *Campylobacter* spp. Buffered glycerol transport medium does not maintain these bacteria. Several manufacturers produce a small vial of Cary-Blair with a self-contained plastic scoop suitable for collecting samples.

Because *Shigella* spp. are delicate, a transport medium of equal parts of glycerol and 0.033 M phosphate buffer (pH 7.0) supports viability of *Shigella* better than does Cary-Blair. For this purpose, maintaining the glycerol transport medium at refrigerator or freezer temperatures yields better results.

If stool is unavailable, a rectal swab may be substituted as a specimen for bacterial or viral culture, but it is not as good, particularly for diagnosis in adults. For suspected intestinal infection with *Campylobacter,* the swab must be placed in Cary-Blair transport medium immediately to avoid drying. Swabs are not acceptable for the detection of parasites, toxins, or viral antigens.

Stool Specimens for Ova and Parasites

For detection of ova and parasites, specimen preservation with a fixative is recommended for visual examination (see Chapter 49).

Stool Specimens for Viruses

Stools for virus culture must be refrigerated if they are not inoculated onto media into cell cultures within 2 hours. A rectal swab, transported in modified Stuart's transport medium or another viral transport medium, is adequate for recovery of most viruses from feces. See Chapter 51 for more information regarding collection and transport of specimens for viral culture.

Miscellaneous Specimen Types

Other specimens that may be obtained for diagnosis of GI tract infection include duodenal aspirates (usually for detection of *Giardia* or *Strongyloides*), which should be examined immediately by direct microscopy for the presence of motile protozoan trophozoites, cultured for bacteria, and placed into polyvinyl alcohol (PVA) fixative for subsequent parasitic examination. The laboratory should be informed in advance that such a specimen is going to be collected so that the specimen can be processed and examined efficiently.

The string test has proved useful for diagnosing duodenal parasites, such as *Giardia,* and for isolating *Salmonella typhi* from carriers and patients with acute typhoid fever. The patient swallows a weighted gelatin capsule containing a tightly wound length of string, which is left protruding from the mouth and taped to the cheek. After a predetermined period, during which the capsule reaches the duodenum and dissolves, the string, now covered with duodenal contents, is retracted and delivered immediately to the laboratory. There the technologist, using sterile-gloved fingers, strips the mucus and secretions attached to the string and deposits some material on slides for direct examination and some material into fixative for preparation of permanent stained mounts. The technologist also inoculates some material to appropriate media for isolation of bacteria.

DIRECT DETECTION OF AGENTS OF GASTROENTERITIS IN FECES

Wet Mounts

A direct wet mount of fecal material, particularly with liquid or unformed stool, is the fastest method for detection of motile trophozoites of *Dientamoeba fragilis, Entamoeba, Giardia,* and other intestinal parasites that may not contribute to disease but may alert the microbiologist to the possibility of finding other parasites, such as *Entamoeba coli, Endolimax nana, Chilomastix mesnili,* and *Trichomonas hominis.* Occasionally the larvae or adult worms of other parasites may be visualized. Experienced observers can also see the refractile forms of cryptosporidia and many types of cysts on the direct wet mount, including *Cyclospora cayetanensis,* a parasite that is associated with the consumption of contaminated food such as raspberries.[7] If present in sufficient numbers, the ova of intestinal parasites can be seen.

Examination of a direct wet mount of fecal material taken from an area with blood or mucus, with the addition of an equal portion of Loeffler's methylene blue, is helpful for detection of leukocytes, which occasionally aids in differentiating among the various types of diarrheal syndromes. Under phase-contrast and dark-field microscopy, the darting motility and curved forms of *Campylobacter* may be observed in a warm sample. Water, which will immobilize *Campylobacter,* should not be used. However, for practical reasons most laboratories do not use a wet mount. Trained observers working in endemic areas can recognize the characteristic appearance and motility of *Vibrio cholerae.*

Stains

Feces may be Gram stained for detection of certain etiologic agents. For example, many thin, comma-shaped, gram-negative bacilli may indicate *Campylobacter* infection (if vibrios have been ruled out). In addition, polymorphonuclear cells may also be detected. Of importance, an acid-fast stain can be used to detect *Cryptosporidium* spp., mycobacteria, and *Isospora* spp. Examination of fixed fecal material for parasites by trichrome or other stains is covered in Chapter 49. A permanent stained preparation should be made from all stool specimens received for detection of parasites.

Antigen Detection

An accurate, sensitive, indirect fluorescent antibody stain for giardiasis and cryptosporidiosis is commercially available. These organisms can be visualized easily and unequivocally with a monoclonal antibody fluorescent stain (Meridian Diagnostics, Cincinnati, Ohio). Park and colleagues described a simple and rapid screening procedure using a direct fluorescent antibody stain for *E. coli* O157:H7.[18]

Enzyme immunoassays (EIAs) or latex agglutination can detect numerous microorganisms that cause GI tract infections. For example, EIAs are commercially available to detect *E. coli* O157:H7, the presence of the Shiga toxins produced by EHEC, or the presence of *C. difficile* toxins A or A and B.[5,10,12,15] In addition, rotavirus is detected using a solid-phase EIA procedure or a latex agglutination test. EIA methods are also available for detection of antigens of *Cryptosporidium* and *Giardia lamblia* as well as *E. histolytica.* EIA methods have also been evaluated for detection of certain bacterial pathogens.[17]

Molecular Biological Techniques

The development of amplification techniques has led to numerous publications for the direct detection of many enteric pathogens, including all major organism groups—bacteria, viruses, and parasites. Probes for *Salmonella*, *Shigella*, EHEC, and *Yersinia* are being evaluated. A disadvantage with probe technology is that the organism itself is not available for susceptibility testing, which is important for certain bacterial pathogens (e.g., *Shigella*) for which susceptibility patterns vary.

CULTURE OF FECAL MATERIAL FOR ISOLATION OF ETIOLOGIC AGENTS

Bacteria

Fecal specimens for culture should be inoculated to several media for maximal yield, including solid agar and broth. The choice of media is arbitrary and based on the particular requirements of the clinician and the laboratory. Recommendations for selection of media are given in this section.

Organisms for Routine Culture. Stools received for routine culture in most clinical laboratories in the United States should be examined for the presence of *Campylobacter*, *Salmonella*, and *Shigella* spp. under all circumstances. Detection of *Aeromonas* and *Plesiomonas* spp. should be incorporated into routine stool culture procedures. The cost of doing a stool examination on every patient for all potential enteric pathogens is prohibitive. The decision as to what other bacteria are routinely cultured should take into account the incidence of GI tract infections caused by particular etiologic agents in the area served by the laboratory. For example, if the incidence of *Yersinia enterocolitica* gastroenteritis is high enough in the area served by the laboratory, then this agent should also be sought routinely. Similarly, because of the increasing prevalence of disease caused by *Vibrio* spp. in individuals living in high-risk areas of the United States (seacoast), laboratories in these localities may routinely look for these organisms. Conversely, unless a patient has a significant travel history, a laboratory located in the Midwestern part of the United States should not routinely look for these organisms except by special request. Protocols for culture of enterohemorrhagic *E. coli* (e.g., *E. coli* O157:H7) vary greatly[1,15,16]; based on incidence of disease, some laboratories routinely culture for this organism while others carry out culture on request only. Other laboratories set up cultures routinely on bloody stool specimens from children only. Selective or screening media to detect *E. coli* O157:H7 also vary greatly such as using a 1% sorbitol-containing medium (most O157:H7 *E. coli* are sorbitol-negative), a specific trypticase blood agar (Unipath

GmbH, Wesel, Germany) or Rainbow Agar O157 (Biolog, Inc., Hayward, Calif.).[10,15,16]

Routine Culture Methods. An in-depth discussion regarding culture of all enteric pathogens is beyond the scope of this chapter. Because U.S. laboratories should routinely examine stools for the presence of *Salmonella*, *Shigella*, and *Campylobacter* spp., culture of these organisms is addressed. Culture conditions for all other pathogens, including viruses, are covered in Parts III, IV, and VI. Specimens received for detection of the most frequently isolated *Enterobacteriaceae* and *Salmonella* and *Shigella* spp. should be plated to a supportive medium, a slightly selective and differential medium, and a moderately selective medium. A highly selective medium does not seem to be cost effective for most microbiology laboratories.

Blood agar (tryptic soy agar with 5% sheep blood) is an excellent general supportive medium. Blood agar medium allows growth of yeast species, staphylococci, and enterococci, in addition to gram-negative bacilli. Of importance, the absence of normal gram-negative fecal flora and/or the presence of significant quantities of organisms such as *Staphylococcus aureus*, yeasts, and *Pseudomonas aeruginosa* can be evaluated. Another benefit of blood agar is that it allows oxidase testing of colonies. Several colonies that do not resemble *Pseudomonas* from the third or fourth quadrant should be routinely screened for production of cytochrome oxidase. If many are present, *Aeromonas*, *Vibrio*, or *Plesiomonas* spp. should be suspected.

The moderately selective agar should support growth of most *Enterobacteriaceae*, vibrios, and other possible pathogens; MacConkey agar works well. Some laboratories use eosin-methylene blue (EMB), which is slightly more inhibitory. All lactose-negative colonies should be tested further, ensuring adequate detection of most vibrios and most pathogenic *Enterobacteriaceae*. Lactose-positive vibrios (*V. vulnificus*), pathogenic *E. coli*, some *Aeromonas* spp., and *Plesiomonas* spp. may not be distinctive on MacConkey agar.

Salmonella/Shigella. The specimen should also be inoculated to a moderately selective agar such as Hektoen enteric (HE) or xylose-lysine desoxycholate (XLD) media. These media inhibit growth of most *Enterobacteriaceae*, allowing *Salmonella* and *Shigella* spp. to be detected. Colony morphologies of lactose-negative, lactose-positive, and H_2S-producing organisms are illustrated in Figure 59-7. Other highly selective enteric media, such as salmonella-shigella, bismuth sulfite, deoxycholate, or brilliant green, may inhibit some strains of *Salmonella* or *Shigella*. All these media are incubated at 35° to 37° C in air and examined at 24 and 48 hours for suspicious colonies.

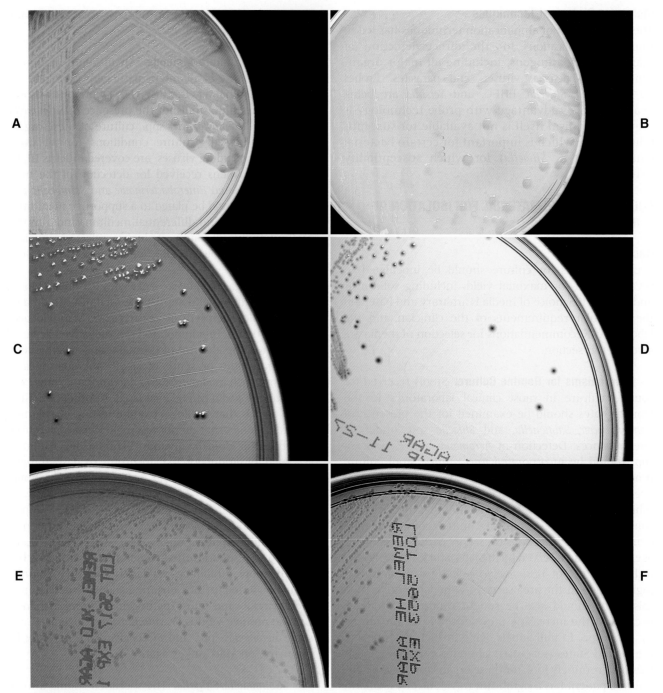

Figure 59-7   Colonies of a lactose-positive organism growing on xylose-lysine deoxycholate (XLD) agar **(A)** and Hektoen enteric (HE) agar **(B)**. Colonies of *Salmonella enteritidis* (lactose-negative) growing on XLD **(C)** and HE agar **(D)**. (Note how both agars detect H_2S production.) Colonies of *Shigella* (lactose-negative) growing on XLD **(E)** and HE agar **(F)**.

Campylobacter. Cultures for isolation of *Campylobacter jejuni* and *Campylobacter coli* should be inoculated to a selective agar containing antimicrobial agents that suppress the growth of normal flora but not of *Campylobacter* spp. The introduction of a blood-free, charcoal-containing medium that has more selective antibiotic components has resulted in better recovery of most enteropathogenic *Campylobacter* spp. compared with earlier media.[2] Brucella broth base has yielded less satisfactory recovery of *Campylobacter* spp. Commercially produced agar plates for isolation of campylobacters are available from several manufacturers. These plates are incubated in a microaerophilic atmosphere at 42° C and examined at 24 and 48 hours for suspicious colonies. Culture methods for other campylobacters that are associated with GI disease, such as *C. hyointestinalis* and *C. fetus* subsp. *fetus,* are provided in Chapter 36.

Enrichment Broths. Enrichment broths are sometimes used for enhanced recovery of *Salmonella, Shigella, Campy*lobacter, and *Y. enterocolitica* although *Shigella* usually does not survive enrichment. Gram-negative broth (Hajna GN) or selenite F broth yields good recovery. Enrichment broths for *Enterobacteriaceae* should be incubated in air at 35° C for 6 to 8 hours and then several drops should be subcultured to at least two selective medium plates. A commercial system that allows such broth to be tested for antigen of *Salmonella* or *Shigella* directly has been described; however, the reported sensitivity is lower than desired. Stool would be inoculated to broth only initially; those broths that tested negative could be discarded without subculturing. Campy-thioglycollate enrichment broth increases the yields of positive cultures for *Campylobacter* spp., although it is not necessary for routine use. Enrichment broth for *Campylobacter* is refrigerated overnight or for a minimum of 8 hours before a few drops are plated to *Campylobacter* agar and incubated at 42° C in a microaerophilic atmosphere.

LABORATORY DIAGNOSIS OF *CLOSTRIDIUM DIFFICILE*–ASSOCIATED DIARRHEA

The definitive diagnosis of *C. difficile*–associated diarrhea is based on clinical criteria combined with laboratory testing. Visualization of a characteristic pseudomembrane or plaque on endoscopy is diagnostic for pseudomembranous colitis and, with the appropriate history of prior antibiotic use, meets the criteria for diagnosis of antibiotic-associated pseudomembranous colitis. No single laboratory test will establish the diagnosis unequivocally. Three tests are currently available for routine use: culture, detection of cytotoxin by tissue culture, and antigen detection assays (e.g., enzyme immunoassay, latex agglutination) for *C. difficile* toxin. The laboratory diagnosis of *C. difficile*–associated diarrhea is reviewed in Chapter 44.

Case Study

A 30-year-old man developed diarrhea with severe abdominal cramping 3 days after eating in a local restaurant. He became febrile and weak and went to his physician. A stool specimen was collected and immediately hand-carried to the laboratory. Numerous WBCs and bacteria were seen in a wet mount,

but the bacteria were noteworthy in that they were nonmotile. The patient was treated with ciprofloxacin. A culture was performed that grew a non–lactose-fermenting gram-negative rod that was identified as *Shigella sonnei.*

QUESTIONS

1. What agent of diarrhea becomes nonviable in a stool that has not been cultured within 30 minutes of collection?
2. If the culture is going to be delayed in transit, what preservative and storage temperature is best for preservation of the specimen?
3. What method is used to identify *Shigella* to the species level, and what is the importance of such identifications?

REFERENCES

1. Abbott SL: Laboratory aspects of non-O157 toxigenic *E. coli, Clin Microbiol Newsl* 19:105, 1997.
2. Endtz HP, Ruijs GJ, Zwinderman AH, et al: Comparison of six media, including a semisolid agar, for the isolation of various *Campylobacter* species from stool specimens, *J Clin Microbiol* 29:1007, 1991.
3. Finlay BB, Falkow S: *Salmonella* interactions with polarized human intestinal Caco-2 epithelial cells, *J Infect Dis* 162:1096, 1990.
4. Galán JE, Ginocchio C, Costeas P: Molecular and functional characterization of the *Salmonella typhimurium* invasion gene invA: homology of invA to members of a new protein family, *J Bacteriol* 17:4338, 1992.
5. Gavin PJ, Thomson RB: Diagnosis of enterohemorrhagic *Escherichia coli* infection by detection of Shiga toxins, *Clin Microbiol Newsl* 26:49, 2004.
6. Ginocchio C, Pace J, Galan JE: Identification and molecular characterization of a *Salmonella typhimurium* gene involved in triggering the internalization of *Salmonellae* into cultured epithelial cells, *Proc Natl Acad Sci U S A* 89:5976, 1992.
7. Herwaldt BL, Beach MJ: The return of *Cyclospora* in 1997: another outbreak of cyclosporiasis in North America associated with imported raspberries, *Ann Intern Med* 130:210, 1999.
8. Kaplan BS: Commentary on the relationships between HUS and TTP. In Kaplan BS, et al, editors: *Hemolytic-uremic syndrome and thrombotic thrombocytopenic purpura,* New York, 1992, Marcel Dekker.
9. Kaye SA, Obrig TG: Pathogenesis of *E. coli* hemolytic-uremic syndrome, *Clin Microbiol Newsl* 18:49, 1996.
10. Kehl SC: Role of the laboratory in the diagnosis of enterohemorrhagic *Escherichia coli* infections, *J Clin Microbiol* 40:2711, 2002.
11. Loo VG, Poirier L, Miller MA, et al: A predominantly clonal multi-institutional outbreak of *Clostridium difficile*-associated diarrhea with high morbidity and mortality, *N Engl J Med* 353:2442, 2005.
12. MacKenzie AM, Orrbine E, Hyde L, et al: Performance of the ImmunoCard STAT! *E. coli* O157:H7 test for detection of *Escherichia coli* O157:H7 in stools, *J Clin Microbiol* 38:1866, 2000.
13. McDonald LC, Killgore GE, Thompson A, et al: An epidemic, toxin gene-variant of *Clostridium difficile, N Engl J Med* 353: 2433, 2005.
14. Mundy LS, Shanholtzer CJ, Willard KE, et al: An evaluation of three commercial fecal transport systems for the recovery of enteric pathogens, *Am J Clin Pathol* 96:364, 1991.

15. Nataro JP, Kasper JB: Diarrheagenic *Escherichia coli, Clin Microbiol Rev* 11:142, 1998.

16. Novicki TJ, Daly JA, Mottice SL, et al: Comparison of sorbitol MacConkey agar and a two-step method which utilizes enzyme-linked immunosorbent assay toxin testing and a chromogenic agar to detect and isolate enterohemorrhagic *Escherichia coli, J Clin Microbiol* 38:547, 2000.

17. Pal T, Pacsa AS, Emody L, et al: Modified enzyme-linked immunosorbent assay for detecting enteroinvasive *Escherichia coli* and virulent *Shigella* strains, *J Clin Microbiol* 26:948, 1985.

18. Park CH, Hixon DL, Morrison WL, Cook CB: Rapid diagnosis of enterohemorrhagic *Escherichia coli* O157:H7 directly from fecal specimens using immunofluorescence stain, *Am J Clin Pathol* 101:91, 1994.

19. Sansonetti PJ: Genetic and molecular basis of epithelial cell invasion by *Shigella* species, *Rev Infect Dis* 13(suppl 4):S282, 1991.

20. Schmidt H, Hensel M: Pathogenicity islands in bacterial pathogenesis, *Clin Microbiol Rev* 17:14, 2004.

21. Sears CL, Kaper JB: Enteric bacterial toxins: mechanisms of action and linkage to intestinal secretion, *Microbiol Rev* 60:167, 1996.

ADDITIONAL READING

Gilligan PH, Janda JM, Miller JM, et al: Laboratory diagnosis of bacterial diarrhea, *Cumitech 12A:1,* Washington, DC, 1992, American Society for Microbiology.

Gradus MS: Public health criteria for the diagnosis of foodborne illness, *Clin Microbiol Newsl* 8:85, 1986.

Guerrant RL: Bacterial and protozoal gastroenteritis, *N Engl J Med* 325:327, 1991.

Guerrant RL: Principles and syndromes of enteric infection. In Mandell GL, Bennett JE, and Dolin R, editors: *Principles and practice of infectious diseases,* ed 6, Philadelphia, 2005, Churchill Livingstone.

Guerrant RL and Hughes JM: Nausea, vomiting, and noninflammatory diarrhea. In Mandell GL, Bennett JE, and Dolin R, editors: *Principles and practice of infectious diseases,* ed 6, Philadelphia, 2005, Churchill Livingstone.

Sears CL and Kaper JB: Enteric bacterial toxins: mechanisms of action and linkage to intestinal secretion, *Microbiol Rev* 60:167, 1996.

Voth DE and Ballard JD: *Clostridium difficile* toxins: mechanisms of action and role in disease, *Clin Microbiol Rev* 18: 247, 2005.

Wilkins TD and Bartlett JG: *Clostridium difficile* testing: after 20 years, still challenging, *J Clin Microbiol* 41:531, 2003.

York M: Fecal culture for aerobic pathogens, In Isenberg H, editor in chief: *Clinical procedures handbook,* vol 1, ed 2, Washington, DC, 2004, ASM Press.

GENERAL CONSIDERATIONS

The skin serves as a barrier between the internal organs and the external environment. Skin is not only subjected to frequent trauma and thereby is at frequent risk of infection, but also it can reflect internal disease.

ANATOMY OF THE SKIN

From inside out the skin is divided into three distinct layers: the subcutaneous tissue, the dermis, and the epidermis (Figure 60-1). The subcutaneous tissue lies beneath the dermis and is rich in fat. Deeper hair follicles and sweat glands originate in this layer. Below the subcutaneous layer are thin fascial membranes (sheets or bands of fibrous tissue) that cover muscles, ligaments, and other connective tissues. Of importance, the fascia serves as a barrier to infection for the deeper tissues and organs of the body.

Above the subcutaneous tissue and fascial membranes lies the dermis, which comprises dense connective tissue that is rich in blood and nerve supply. Shorter hair follicles and sebaceous (oil-producing) glands originate in the dermis. Finally, the epidermis, which is the outermost layer of skin, is made of layered squamous epithelium. Hair follicles, sebaceous glands, and sweat glands open to the skin surface, through the epidermis.

FUNCTION OF THE SKIN

The skin is the body's largest and thinnest organ. It forms a self-repairing and protective boundary between the body's internal environment and an often hostile external environment. Skin plays a crucial role in the control of body temperature, excretion of water and salts, synthesis of important chemicals and hormones, and as a sensory receptor. Of significance, the skin has an important protective function by virtue of the outermost epithelial layer of the epidermis, which comprises cells containing keratin, a water-repellent protein. The skin's normal microbial flora, pH, and chemical defenses also help prevent colonization by many pathogens. The resident microbial flora is listed in Box 60-1.

INCIDENCE, ETIOLOGIC AGENTS, AND PATHOGENESIS

Approximately 15% of all patients who seek medical attention have either some skin disease or skin lesion, many of which are infectious. Many different bacteria, fungi, and viruses may be involved. Also, these infections can result from one or several causative agents. Because of this great diversity of etiologic agents and the complexity of these infections, only the most common infections involving the skin and subcutaneous tissues are addressed.

Skin infections can arise from the invasion of certain organisms from the external environment through breaks in the skin or from organisms that reach the skin through the blood as part of a systemic disease. In some infections, such as staphylococcal scalded-skin syndrome, toxins produced by the bacteria cause skin lesions. In others, lesions can also result from the host's immune response to microbial antigens.

Because of the diversity of etiologic agents, clinicians will often describe the appearance of skin lesions to microbiologists for possible input as to appropriate culture techniques. The physical characteristics of the lesions can indicate the need for smear, culture, biopsy, or surgery, for instance. Some of the terms most frequently used to describe manifestations of skin infections are provided in Table 60-1. Figure 60-2 shows examples of these skin lesions.

SKIN AND SOFT TISSUE INFECTIONS

SKIN

Numerous infections of the skin may occur. Several of the most common are discussed here.

Infections in or around Hair Follicles

Folliculitis, furunculosis, and carbuncles are localized abscesses either in or around hair follicles. These infections are distinguishable from one another based on size and the extent of involvement in subcutaneous tissues. Table 60-2 summarizes each infection's respective clinical features. For the most part, these infections are precipitated by blockage of the hair follicle with skin oils (sebum), or minor trauma resulting from friction

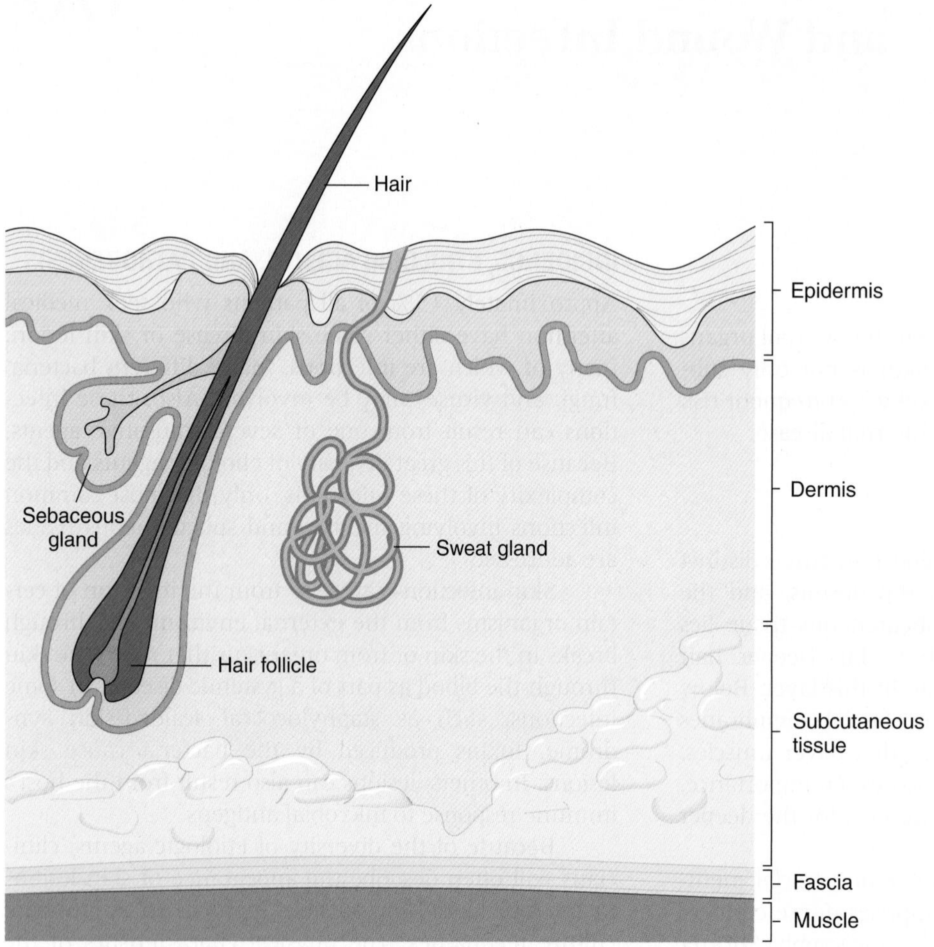

Figure 60-1   Diagram of skin.

BOX 60-1 | Resident Microbial Flora of the Skin

Diphtheroids
Staphylococcus epidermidis
Other coagulase-negative staphylococci
Propionibacterium acnes

such as that caused by clothes rubbing the skin. *Staphylococcus aureus* is the most common etiologic agent for all three infections. Members of the family *Enterobacteriaceae* can also cause folliculitis, albeit much less commonly. Also, outbreaks of folliculitis caused by *Pseudomonas aeruginosa* and associated with the use of whirlpools, swimming pools, and hot tubs have been reported.[9]

Infections in the Keratinized Layer of the Epidermis
Because of their ability to utilize keratin in the cells of the epidermis, the dermatophyte fungi are significant and well-suited pathogens for this site. Unlike the previously discussed infections, dermatophytes cannot invade the deeper layers of skin. Because keratin is also

present in hair and nails, these fungi may also cause superficial infections at these sites (see Chapter 50 for more information).

Infections in the Deeper Layers of the Epidermis and the Dermis
Most infections in the deeper layers of the epidermis and dermis result from the inoculation of microorganisms by traumatic breaks in the skin. These superficial skin infections usually do not require surgical intervention. Table 60-3 summarizes these infections. In most instances, these infections resolve with local care such as heat application. Antibiotics are only occasionally required.

Cutaneous ulcers usually involve a loss of epidermal and part of the dermal tissues. In contrast, nodules are inflammatory foci in which the epidermal and dermal layers remain largely intact. Various bacteria and fungi can cause ulcerative and/or nodular skin lesions following direct inoculation. Examples of these etiologies include *Corynebacterium diphtheriae*, *Bacillus anthracis*, *Nocardia* spp., *Mycobacterium marinum*, and *Sporothrix schenckii*.

Table 60-1   Manifestations of Skin Infections

Term	Description	Possible Etiologic Agents
Macule	A circumscribed (limited), flat discoloration of the skin	Dermatophytes *Treponema pallidum* (infection—secondary syphilis) Viruses such as enteroviruses (viral exanthems [rashes])
Papule	An elevated, solid lesion up to 0.5 cm in diameter. Multiple papules may become confluent and form plaques	Pox virus (infection—Molluscum contagiosum) *Sarcoptes scabei* (infection—scabies) Human papillomavirus types 3 and 10 (infection—flat warts) *S. aureus, P. aeruginosa*, etc. (infection—folliculitis)
Nodule	A circumscribed, raised, solid lesion greater than 0.5 cm in diameter	*Sporothrix schenckii* *S. aureus* (infection—furuncle) Miscellaneous fungi (infection—subcutaneous mycoses) *Mycobacterium marinum* *Nocardia* spp. *Corynebacterium diphtheriae*
Pustule	A circumscribed collection of leukocytes and fluid that varies in size	*Candida* spp. Dermatophytes *S. aureus* (infection—folliculitis) Herpes simplex virus (HSV) Varicella zoster virus (VZV) *Neisseria gonorrhoeae* (infection—gonococcemia) *S. aureus* or group A streptococci (*Streptococcus pyogenes* [infection—impetigo])
Vesicle	A circumscribed collection of fluid up to 0.5 cm in diameter (blisterlike)	HSV VZV Herpes zoster
Scales	Excess dead epidermal cells	Dermatophytes (infection—tinea) Group A streptococci (*Streptococcus pyogenes* [infection—scarlet fever])
Ulcer	A loss of epidermis and dermis	*Haemophilus ducreyi* (infection—chancroid) Bowel flora (infection—decubiti) *T. pallidum* (infection—chancre of primary syphilis) *Bacillus anthracis* (infection—anthrax)
Bulla	A circumscribed lesion of fluid more than 0.5 cm in diameter	Clostridial species (infection—necrotizing gas gangrene) *Staphylococcus aureus* (infection—staphylococcal scalded skin syndrome; bullous impetigo) HSV *Vibrio vulnificus* and other vibrios Group A streptococci (*Streptococcus pyogenes*) Other gram-negative rods

From reference 10.

INFECTIONS OF THE SUBCUTANEOUS TISSUES

Infections of the subcutaneous tissues may manifest as abscesses, ulcers, and boils. *Staphylococcus aureus* is the most common etiologic agent of subcutaneous abscesses in healthy individuals. Many subcutaneous abscesses contain mixed bacteria. To a large degree, the organisms isolated from subcutaneous abscesses depend on the site of infection. For example, anaerobes are commonly isolated from abscesses of the perineal, inguinal, and buttock area, whereas nonperineal infections are caused by mixed facultative aerobic organisms.

Chronic undermining ulcer, or **Meleney's ulcer,** is a slowly progressive infection of the subcutaneous tissue with associated ulceration of portions of the overlying skin. The causative organism is classically a microaerobic streptococcus, but anaerobic streptococci and, occasionally, other organisms may be involved.

In many instances, infections of the epidermis and dermis extend and can become subcutaneous infections and even reach the fascia and/or muscle. For example, erysipelas (Figure 60-3) can become a subcutaneous cellulitis and thereafter a streptococcal necrotizing fasciitis. Similarly, folliculitis can readily become a subcutaneous abscess or a carbuncle that can extend to the fascia. Cellulitis also can frequently extend to the subcutaneous tissues (Figure 60-4). Anaerobic cellulitis is

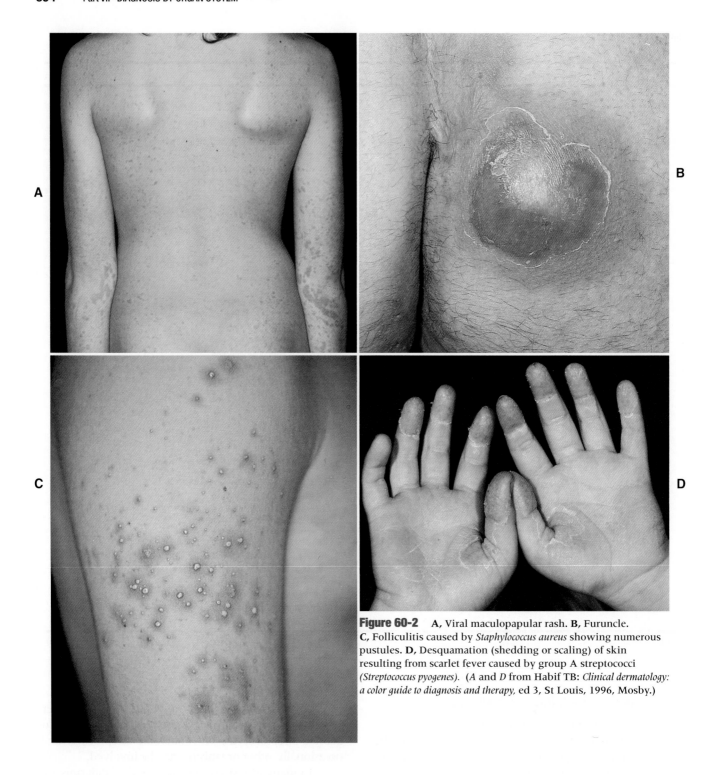

Figure 60-2 **A,** Viral maculopapular rash. **B,** Furuncle. **C,** Folliculitis caused by *Staphylococcus aureus* showing numerous pustules. **D,** Desquamation (shedding or scaling) of skin resulting from scarlet fever caused by group A streptococci *(Streptococcus pyogenes).* *(A and D from Habif TB: Clinical dermatology: a color guide to diagnosis and therapy, ed 3, St Louis, 1996, Mosby.)*

associated with considerable amounts of gas from organisms that are usually present in the subcutaneous tissue. This infection is most often found in the extremities and is particularly common among patients with diabetes and may involve the neck, abdominal wall, perineum, or connective tissue in other areas. Anaerobic cellulitis also may occur as a postoperative problem. Although the onset and spread of this lesion are not usually rapid and patients do not show impressive systemic effects at first, it is not an illness to be taken

Table 60-2 Infections Involving Hair Follicles

Infection	Skin Manifestations
Folliculitis—minor infection of hair follicles	Papules or pustules that are pierced by a hair and surrounded with redness
Furuncle (boil)	Abscess that begins as a red nodule in a hair follicle that ultimately becomes painful and full of pus
Carbuncle	Furuncles that spread more deeply to the dermis and subcutaneous tissues; these infections may be associated with fever and malaise

Table 60-3 Infections of the Epidermal and Dermal Layers of the Skin

Infection	Key Features of Infection	Etiologies	Other Comments
Erysipelas	Primarily involves the dermis and most superficial parts of the subcutaneous tissue. Lesions are painful, red, swollen, and indurated. Patients are febrile, and regional lymphadenopathy (swollen glands) is often present. Lesion has a marked, well-demarcated, raised border (see Figure 60-3)	Group A streptococci (*Streptococcus pyogenes* [sometimes groups B, C, or G streptococci])	Infants, children, and elderly individuals are most affected Primarily a clinical diagnosis
Erythrasma	Chronic infection of the keratinized layer of the epidermis. Lesions are dry, scaly, itchy and reddish brown.	*Corynebacterium minutissimum*—possible cause	Common in diabetics Resembles dermatophyte infection
Erysipeloid	Purplish-red, nonvesiculated skin lesion with an irregular, raised border. The lesions itch and burn. Fever and other systemic symptoms are uncommon	*Erysipelothrix rhusiopathiae*	Uncommon. Considered an occupational disease
Impetigo	Erythematous (red) lesions that may be bullous (less common) or nonbullous	Nonbullous—group A streptococci (*Streptococcus pyogenes*) Bullous—*Staphylococcus aureus*	
Cellulitis	Diffuse, spreading infection involving the deeper layers of the dermis. Lesions are ill-defined, flat, painful, red, and swollen. Patients have fever, chills, and regional lymphadenopathy (see Figure 60-4)	Group A streptococci *Staphylococcus aureus* Less common: *Aeromonas, Vibrio* spp., and *Hemophilus influenzae*—typically affects young children	Primarily a clinical diagnosis
Dermatophytoses	Superficial infections of the skin and its appendages (i.e., ringworm, athlete's foot, jock itch, as well as infections of nails and hair)	*Microsporum, Trichophyton,* and *Epidermophyton* spp.	
Hidradenitis	Chronic infection of obstructed apocrine (sweat) glands in the axillas, genital, or perianal areas with intermittent discharge of often foul-smelling pus	*S. aureus, Streptococcus anginosus* group, anaerobic streptococci, and *Bacteroides* spp.	
Infected pilonidal tuft cyst or hairs	Pain and swelling, redness	Anaerobes, including *Bacteroides fragilis* group, *Prevotella, Fusobacterium,* anaerobic gram-positive cocci, and *Clostridium* spp.	

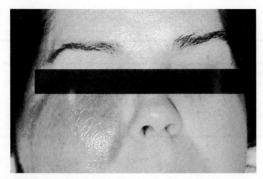

Figure 60-3 Erysipelas caused by group A streptococci *(Streptococcus pyogenes).*

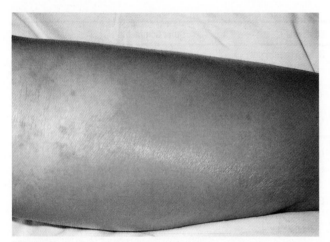

Figure 60-4 Cellulitis. (From Farrar WE et al: *Infectious diseases: text and color atlas,* ed 2, London, 1992, Mosby-Wolfe.)

lightly. The organisms are almost always a mixture of aerobes and anaerobes. The aerobes include *Escherichia coli,* alpha-hemolytic or nonhemolytic streptococci, and *S. aureus* predominantly, but group A streptococci *(Streptococcus pyogenes)* and other *Enterobacteriaceae* are encountered as well. The anaerobes are typically found in greater numbers and in more variety than the aerobes; *Peptostreptococcus* spp., *B. fragilis* group strains, *Prevotella, Porphyromonas,* other anaerobic gram-negative bacilli, and clostridia are seen. Bacteremia is not usually present.

INFECTIONS OF THE MUSCLE FASCIA AND MUSCLES

There are several rare, yet serious or potentially serious, forms of deep and sometimes extensive soft tissue and skin infections.

Necrotizing Fasciitis

Necrotizing fasciitis is a serious infection that occurs relatively infrequently. The basic pathology is infection of the fascia overlying muscle groups, often with involvement of the overlying soft tissue. At the fascial level, no barrier exists to spread of infection, so fas-

ciitis may extend widely and rapidly to involve huge areas of the body in short periods. This process, once known as *hospital gangrene,* typically involves group A streptococci or *S. aureus.* Necrotizing fasciitis also frequently involves anaerobic bacteria, especially *Bacteroides* and *Clostridium* spp.

Progressive Bacterial Synergistic Gangrene

Progressive bacterial synergistic gangrene is usually a chronic gangrenous condition of the skin most often encountered as a postoperative complication, particularly after abdominal or thoracic surgery. The lesions may be extensive and, with involvement of the abdominal wall, may lead to evisceration (extrusion of the internal organs). As the name suggests, this is typically a mixed infection with microaerobic streptococci and *S. aureus.* At times other organisms may be present, including anaerobic streptococci, *Proteus,* and other facultative and anaerobic bacteria. This infection occurs infrequently. Cultures should be taken from the advancing outer edge of the lesion (not the central portion of the wound) or the microaerobic streptococcus will be missed.

Myositis

Myositis (involvement of muscle) is caused by various infectious agents. The nature of the pathologic process is variable, sometimes involving extensive necrosis of muscle, as in gas gangrene or clostridial myonecrosis, necrotizing cutaneous myositis or synergistic nonclostridial anaerobic myonecrosis, anaerobic streptococcal myonecrosis, myonecrosis caused by *Bacillus* spp., or myonecrosis caused by *Aeromonas.* Focal collections of suppuration in muscle (staphylococcal or other pyomyositis [purulent myositis]) are sometimes seen.[6,8] Abscess in the psoas muscle (muscle arising from the lumbar vertebrae and extending to the lesser trochanter of the leg's femur) may involve *Mycobacterium tuberculosis, S. aureus,* or various facultative or anaerobic gram-negative bacilli. Serious vascular problems resulting from loss of blood supply may lead to death of muscle; such muscle may become secondarily infected (vascular gangrene). Organisms that produce myositis or other muscle pathology are listed in Box 60-2.

WOUND INFECTIONS

Besides skin and soft tissue infections that occur primarily as a result of a break in the skin surface, wound infections can occur as complications of surgery, trauma, and bites or diseases that interrupt a mucosal or skin surface.

Postoperative Infections

Sources of surgical wound infections can include the patient's own normal flora or organisms present in the

BOX 60-2	Organisms Producing Myositis or Other Muscle Pathology

Clostridium perfringens
C. novyi
C. septicum
C. bifermentans
C. histolyticum
C. sordellii
C. sporogenes
Bacillus spp.
Aeromonas spp.
Peptostreptococcus spp.
Microaerobic streptococci
Bacteroides spp.
Enterobacteriaceae
Staphylococcus aureus
Group A streptococci (*Streptococcus pyogenes*)
Pseudomonas mallei
P. pseudomallei
Vibrio vulnificus
Mycobacterium tuberculosis
Salmonella typhi
Legionella spp.
Rickettsia spp.
Viruses
Trichinella
Taenia solium
Toxoplasma

BOX 60-3	Organisms Encountered in Postoperative Wound Infections

Staphylococcus aureus
Coagulase-negative staphylococci
Streptococcus pyogenes
Streptococcus anginosus group streptococci (*S. anginosus, S. constellatus, S. intermedius*)
Microaerobic streptococci
Enterococci
Proteus, Morganella, Providencia
Other Enterobacteriaceae
Escherichia coli
Pseudomonas spp.
Candida spp.
Bacteroides spp.
Prevotella and *Porphyromonas* spp.
Fusobacterium spp.
Clostridium spp.
Peptostreptococcus spp.
Non–spore-forming, anaerobic, gram-positive rods

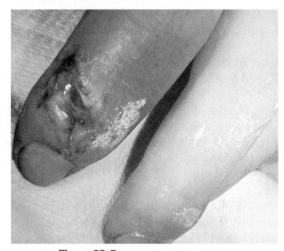

Figure 60-5   Human bite infection.

hospital environment that are introduced to the patient by medical procedures and/or a specific underlying disease or trauma (e.g., burns) that may interrupt a mucosal or skin surface. The nature of the infecting flora depends on the underlying problem and the location of the process. In the case of wound infections following appendectomy or other lower bowel surgery, indigenous flora of the lower gastrointestinal tract are involved, primarily *E. coli*, streptococci, *Bacteroides fragilis* group, and other anaerobic gram-negative rods, including *Bilophila* spp., *Peptostreptococcus* spp., anaerobic non–spore-forming rods, and *Clostridium* spp.[3,7] Principal pathogens are listed in Box 60-3. *Mycobacterium chelonae* and *Mycobacterium fortuitum* may also cause infections following cardiac surgery, mammoplasty, and other clean surgeries.[7]

Bites

Human bites (Figure 60-5) and clenched-fist injuries yield, in order of frequency, *Streptococcus anginosis*, alpha-hemolytic streptococci, *S. aureus*, *Eikenella corrodens*, and group A streptococci (*S. pyogenes*). Anaerobes that may be involved include, in order of frequency, *Prevotella* spp., *Fusobacterium* spp., *Veillonella* spp., and *Peptostreptococcus* spp. In a recent study, 54% of patients presenting with infected human bites were hospitalized; the median number of isolates per wound culture was 4

with both aerobic and anaerobic organisms isolated from 54% of cultures.[13]

Dog and cat bites introduce organisms commonly found in their oral and nasal fluids, including *Pasteurella* spp., Centers for Disease Control (CDC) group EF-4a, *Weeksella* spp., and *Staphylococcus intermedius,* with much smaller numbers of *S. aureus. Simonsiella steedae* is found in the oral cavity of most dogs and is also found in the oral cavity of humans and cats and other animals. Similar to human bites, approximately half of animal bite wound cultures yielded aerobes and anaerobes. The oral flora of snakes includes various gram-negative bacilli, including *Pseudomonas, Klebsiella, Proteus,* and *E. coli. Clostridia* may also be recovered from snakebite wounds. In infected animal bite wounds (Figure 60-6), the most frequently encountered aerobic and facultative bacteria are alpha-hemolytic streptococci, *S. aureus,*

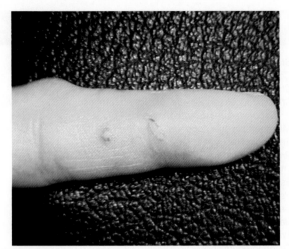

Figure 60-6 Animal bite infection caused by *Pasteurella* spp.

Pasteurella spp., and *Enterobacter cloacae*. Predominant anaerobes in animal bites are anaerobic gram-positive cocci, *Fusobacterium* spp., and anaerobic gram-negative rods such as *Bacteroides, Porphyromonas,* and *Prevotella* spp.[1,8,14]

Capnocytophaga canimorsus (former CDC group DF-2) and *Dysgonomonas capnocytophagoides* (former CDC group DF-3) have been responsible for several types of serious infections, including bacteremia, endocarditis, and meningitis. Most patients had a history of dog bite, and most had underlying diseases that impair host defense mechanisms.

Burns

Infection of burn wounds may be associated with bacteremia, carry a significant mortality, and interfere with the acceptance of skin grafts. Many organisms are capable of infecting the eschar (scab) of a burn. Those most often encountered are various streptococci, *Staphylococcus aureus, S. epidermidis, Enterobacteriaceae, Pseudomonas* spp., other gram-negative bacilli, *Candida,* and *Aspergillus.* Anaerobes, including clostridia and *Bacteroides,* have been recovered occasionally but are probably more frequently involved than has been appreciated to date.

SPECIAL CIRCUMSTANCES REGARDING SKIN AND SOFT TISSUE INFECTIONS

In addition to the infections previously discussed, other circumstances can cause the skin and underlying soft tissue to become infected. Some of these infections are associated with the host with compromised defenses; others are manifestations of systemic infection.

Infections Related to Vascular and Neurologic Problems

Classically, a patient with one of these common infections has diabetes mellitus, poor arterial circulation

(often both large-vessel and small-vessel disease), and peripheral neuropathy (neurologic problems, such as numbness). Because of a loss of sensation resulting from the neuropathy, these individuals traumatize their feet readily (often just by virtue of wearing a new pair of shoes) without being aware of it. The traumatized area develops an ulcer that does not heal readily because of the poor vascular supply and that often becomes infected.[2,12] The infections tend to be chronic and difficult to heal, particularly because these patients may also have poor vision and therefore may not recognize the problem and may not seek medical attention until the process has gone on for some time.

Foot infections in diabetic patients can accelerate dramatically with devastating consequences if appropriate treatment is not given promptly. Therefore, appropriate techniques used to obtain a microbiologic sample are critical. Culture of aspirated fluid or pus, not surface swabbing, is more likely to yield a causative agent, particularly if taken from a deep pocket within the wound. In addition, culture of debrided infected tissue is a good method for diagnosing foot ulcers[15]; it is essential, however, that superficial debris be removed before sampling. With respect to etiologic agents, *S. aureus* is a key pathogen in diabetic foot infections, even when it is not the only isolate.[12] Serious infections are often caused by three to five bacterial species including both aerobes and anaerobes, whereas gram-negative organisms are found in many patients with chronic or previously treated infections. These infections primarily have purulent discharge and necrotic tissue at the base of the ulcer, often with a foul odor. Extension to the underlying bone, producing a difficult-to-manage osteomyelitis (infection of the bone), occurs often. Definitive diagnosis requires a specimen of bone that is obtained by open or percutaneous biopsy; to avoid contamination, specimens need to be obtained without crossing an open wound.[12] Periodically, an acute cellulitis and lymphangitis may be associated with chronic, low-grade infection, thereby making control of the patient's diabetes difficult. Peripheral vascular disease unrelated to diabetes mellitus may also predispose a patient to skin or soft tissue infections, but usually these infections are less difficult to manage because no associated neuropathy is present.

Venous insufficiency also predisposes individuals to infection, again primarily of the lower extremities (in this case, often in the area of the calf or lower leg rather than the foot). Infections related to poor blood supply often involve *S. aureus* and group A streptococci (*Streptococcus pyogenes*). Those with open ulcers often become colonized with *Enterobacteriaceae* and *P. aeruginosa*, which may or may not play a role in the infection. Although less well appreciated, anaerobes are frequently involved in infection, particularly in patients with diabetes or

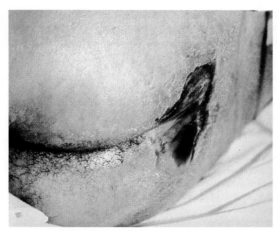

Figure 60-7   Sacral decubitus ulcer.

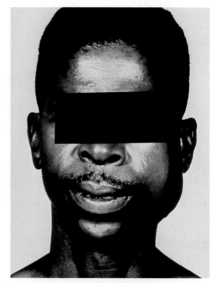

Figure 60-8   Actinomycosis. Note "lumpy jaw."

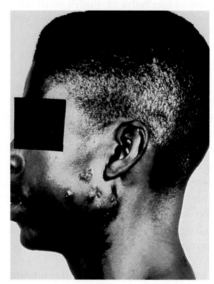

Figure 60-9   Actinomycosis, side view. Note sinuses in skin of face and neck.

peripheral vascular disease. The poor blood supply contributes to anaerobic conditions. Various anaerobes may be recovered, including the *Bacteroides fragilis* group, *Prevotella* and *Porphyromonas*, *Peptostreptococcus* spp., and less frequently *Clostridium* spp.

Another common type of infection in this general category, especially in the older or very ill, bedridden patient, is infected decubitus ulcer (pressure sore [Figure 60-7]). Anaerobic conditions are present in such lesions because of tissue necrosis. Because most of these lesions are located near the anus or on the lower extremities and because so many of these patients are relatively helpless, the ulcers become contaminated with bowel flora, which leads to chronic infection. This contributes to further death of tissue and extension of the decubitus ulcer. Bacteremia is a possible complication, with *B. fragilis* group often being involved, along with clostridia and enteric bacteria. The ulcers yield various anaerobes and aerobes characteristic of the colonic flora; nosocomial pathogens such as *S. aureus* and *P. aeruginosa* may also be recovered.

Sinus Tract and Fistulas

Sometimes, a deep-seated infection beneath the skin and subcutaneous soft tissue spontaneously drains itself externally by way of a sinus (channel or cavity) to the skin's surface. Draining sinus tracts are most often associated with a chronic osteomyelitis. Unfortunately, this type of drainage does not usually cure the underlying process, and such sinuses themselves tend to be chronic. The organisms most often involved in sinuses with an underlying osteomyelitis are *S. aureus*, various *Enterobacteriaceae*, *P. aeruginosa*, anaerobic gram-negative bacilli, anaerobic gram-positive cocci, and occasionally other anaerobes. In the case of actinomycosis (with or without bone involvement [Figures 60-8 and 60-9]), one would expect to recover *Actinomyces* spp., *Propionibacterium propionicum*, *Prevotella* or *Porphyromonas* and

other non–spore-forming anaerobes, and *Actinobacillus actinomycetemcomitans*. With other types of draining sinuses, the organisms involved depend on the nature of the underlying process.

Chronic draining sinuses are also found in patients with tuberculosis and atypical mycobacterial infection, *Nocardia* infection, and certain infections associated with implanted foreign bodies. Curettings or biopsy from the debrided, cleansed sinus is the best specimen.

Abnormal passages or communications between two organs or leading from an internal organ to the body surface, known as **fistulas,** are difficult management problems. They also often pose insurmountable problems in terms of collection of meaningful specimens,

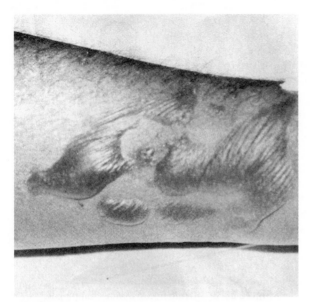

Figure 60-10 Bullae on the arm of a patient with *Vibrio vulnificus* sepsis. (From Pollak SJ, Parrish EF III, Barrett TJ, et al: *Arch Intern Med* 143:837, 1983. Copyright 1983, American Medical Association.)

BOX 60-4	Organisms Involved in Systemic Infection with Cutaneous Lesions

Viridans streptococci
Staphylococcus aureus
Enterococci
Group A and other beta-hemolytic streptococci
Neisseria gonorrhoeae
N. meningitidis
Haemophilus influenzae
Pseudomonas aeruginosa
P. mallei
P. pseudomallei
Listeria monocytogenes
Vibrio vulnificus
Salmonella typhi
Mycobacterium tuberculosis
M. leprae
Treponema pallidum
Leptospira
Streptobacillus moniliformis
Bartonella bacilliformis
Bartonella (Rochalimaea) henselae
Rickettsia
Candida spp.
Cryptococcus neoformans
Blastomyces dermatitidis
Coccidioides immitis
Histoplasma capsulatum

because the organ that has the abnormal communication to the skin surface often has its own profuse indigenous flora. Examples are perirectal fistulas from the small bowel to the skin in association with Crohn's disease or chronic intraabdominal infection. When the bowel is involved, only cultures for specific key organisms, such as mycobacteria or *Actinomyces,* are meaningful. Always attempt to rule out specific underlying causes such as tuberculosis, actinomycosis, and malignancy. Biopsy should be performed in such situations.

Systemic Infections with Skin Manifestations

Cutaneous manifestations of systemic infection, such as bacteremia or endocarditis, may be important clues for the clinician, and they present an opportunity for direct or cultural demonstration of the presence of a particular organism. For example, one may be able to scrape petechiae (a tiny red spot caused by the escape of a small amount of blood) from patients with meningococcemia and demonstrate gram-negative diplococci. In other patients, the skin lesion represents a more impressive type of metastatic infection. In *Vibrio vulnificus* sepsis, dramatic-appearing cutaneous ulcers with necrotizing vasculitis or bullae may be found (Figure 60-10). In some patients, skin lesions may actually represent a noninfectious complication of a local or systemic infection such as scarlet fever or toxic shock syndrome. Various organisms that may be involved in systemic infection with cutaneous lesions are listed in Box 60-4.

LABORATORY DIAGNOSTIC PROCEDURES

INFECTIONS OF THE EPIDERMIS AND DERMIS

For many of the infections of the epidermis and dermis, such as impetigo, folliculitis, cellulitis, and erysipelas, diagnosis is generally made on a clinical basis. Table 60-3 provides the key features and etiologic agents of these infections. Of importance, awareness of the common bacterial pathogens can help guide empiric therapy.

Erysipeloid

For the most part, Gram stain or culture of superficial wound drainage is usually negative. However, culture of a full-thickness skin biopsy taken at the margin of the lesion can confirm the clinical diagnosis.

Superficial Mycoses and Erythrasma

If a dermatophyte infection is suspected, the lesion is cleaned and scrapings are obtained from the active border of the lesion. These scrapings should be suspended in 10% potassium hydroxide, examined for the presence of hyphae, and may also be cultured (see Chapter 50).

A Wood's lamp examination of skin lesions may reveal golden-yellow fluorescent lesions of tinea versicolor. Wood's light examination may also reveal a coral red fluorescence that is characteristic for erythrasma. *Corynebacterium minutissimum* produces porphyrin that accounts for the red fluorescence and may be a possible cause. Skin scrapings may be cultured in media containing serum, but imprint smears of the lesion should reveal gram-positive pleomorphic rods, precluding the need for culture.

Erysipelas and Cellulitis

As previously mentioned, diagnosis of erysipelas and cellulitis can generally be made on a clinical basis. The value of needle aspiration for the bacteriologic diagnosis of these infections has not been clearly demonstrated, particularly in adults.[11] Of note, a higher percentage of positive cultures of soft tissue aspirates along the advancing margin of erythema using a 22- or 23-gauge needle attached to a 3- or 5-mL syringe may be achieved in children with cellulitis.[5] If received by the laboratory, aspirates should be inoculated onto blood and chocolate agars, as well as a broth such as trypticase soy broth.

Vesicles and Bullae

These fluid-filled lesions characteristically involve certain organisms (see Table 60-1), so if the laboratory is aware of the nature of the lesion, the flora and use of appropriate techniques to ensure recovery of the agent may be anticipated. Material in the blisterlike lesion varies from serous (resembling serum) fluid to serosanguineous (composed of serum and blood) or hemorrhagic (bloody) fluid. Large bullae permit withdrawal of 0.5 to 1 mL of fluid by needle and syringe aspiration. Some vesicles are tiny, so a swab must be used for specimen collection. The clinician can usually anticipate whether the lesion is viral or bacterial in nature and may even be able to suspect a particular organism. Recovery of the agent is facilitated if the microbiologist is provided with this information. Bullous (blisterlike) lesions are caused by bacteria and are often associated with sepsis, so blood cultures are mandatory. One may see bronzed skin with bullous lesions in gas gangrene, an entity with a distinctive clinical picture; *Clostridium perfringens* and other clostridia are the key pathogens. Gram stain of the fluid from such lesions typically reveals the etiologic agent to be a gram-positive bacillus and provides the clinician with additional valuable information on which to base initial therapy.

Depending on clinical presentation, specimens would be submitted for either viral or bacterial culture. Bacteriologic diagnosis can generally be made by culturing the fluid from the lesions aerobically or in carbon dioxide on blood and MacConkey agar at 35° C.

INFECTIONS OF THE SUBCUTANEOUS TISSUES

Proper collection and transport of specimens are important factors in the laboratory diagnosis of infections of the subcutaneous tissues. Specimen collection is particularly difficult because many of these lesions are open and therefore readily colonized by nosocomial pathogens that may not be involved in the infection. The most reliable specimens for determining the etiology of ulcers and nodules are those obtained from the base of the ulcer or nodule following removal of overlying debris or by surgical biopsy of deep tissues without contact with the superficial layers of the lesion. A Gram stain of the specimen should be performed and material aerobically cultured on blood agar and MacConkey agar. If fungi, *Nocardia* spp., or mycobacterial infection is considered likely, appropriate fungal media (e.g., Sabouraud dextrose medium with and without chloramphenicol and cycloheximide) and mycobacterial media (e.g., Middlebrook 7H9 or 7H13, Lowenstein-Jensen) should be used. These culture methods are addressed in greater detail in Chapters 50 and 45, respectively.

Similar problems are faced when trying to collect material for culture of sinus tracts. Material for culture should be obtained from the deepest portion of the sinus tract. If systemic symptoms such as fever are present, blood cultures should also be obtained. Again, a Gram stain should be routinely performed. Cultures should be inoculated to recover both facultative and anaerobic bacteria in the same manner as described for surgical wounds later in this chapter.

INFECTIONS OF THE MUSCLE FASCIA AND MUSCLES

Blood cultures should always be drawn from patients with significant myonecrosis. Transport of material (tissue is always better than pus, which is, in turn, better than a swab) should be under anaerobic conditions. Gram stains should be routinely performed. Cultures should be inoculated to recover both facultative and anaerobic bacteria as described for surgical wounds in the following section.

WOUND INFECTIONS

Postoperative

Because anaerobic bacteria are involved in many infections, specimen collection so as to avoid indigenous flora and specimen transport under good anaerobic conditions (which will not interfere with recovery of even obligate aerobes) are particularly important. Unusual organisms associated with postsurgical wound infections, such as *Mycoplasma hominis*, *Mycobacterium chelonae*, *Mycobacterium fortuitum*, fungi, and even *Legionella* spp., should not be overlooked. A Gram-stained smear of material submitted for culture should be examined. Exudates from superficial wounds should be inoculated to blood agar, MacConkey agar, a colistin-nalidixic acid agar plate, and a broth. Material from deep wounds should be set up for anaerobic and aerobic cultures.

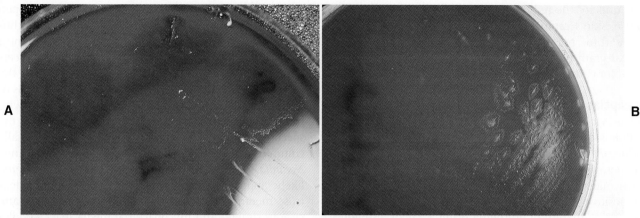

Figure 60-11   Demonstration of the swarming film of growth of *Clostridium septicum* at 24 hours (**A**) and *Clostridium sporogenes* (**B**) for comparison.

 Procedure 60-1

SEMIQUANTITATIVE BACTERIOLOGIC CULTURE OF TISSUE

PRINCIPLE
The degree or extent of bacterial wound contamination is directly related to the risk of wound sepsis. Because of this relationship, physicians use the results of a quantitative culture (the number of colony-forming units [CFUs] per gram of the eschar biopsy) in their management of severely burned patients.

METHOD
1. Cut a piece of tissue, measuring several cubic millimeters, aseptically onto a small, preweighed, sterile urine cup.
2. Determine the weight of the tissue by subtracting the weight of the aluminum foil from the total weight.
3. Place the specimen and 2 mL of sterile nutrient broth in a sterile tissue grinder; macerate the specimen.

4. Inoculate 0.1 mL of sample to a blood agar plate, in duplicate, and an anaerobic blood agar plate (if indicated), in duplicate. In addition, inoculate 0.01 mL of sample using a calibrated loop to a blood agar plate, in duplicate. Spread the inoculum on the plates with a sterile glass spreading rod or a loop.
5. Incubate plates in 5% to 10% carbon dioxide overnight, and count the colonies of bacteria on the plate that contain 30 to 300 CFUs. If more than 300 colonies are obtained on both plated dilutions, the factor 300 is used as *N* for calculations and the result is considered greater than the value.
6. Calculate the number of CFUs per gram of tissue with the following formula:

Number of CFUs counted × Reciprocal of volume of homogenate inoculated (10^{-1} or 10^{-2}) × 2 (volume of diluent used for tissue homogenization) ÷ weight of tissue

For example, for a tissue that weighed 0.002 g, 68 CFUs were observed on the plate that received the 10^{-2} dilution of suspension:

$$\frac{68 \times 10^2 \times 2}{0.002} = \frac{136 \times 10^2}{2 \times 10^{-3}} = 6.8 \times 10^6 \text{ CFU/g}$$

Modified from a method published by Buchanan and colleagues.[4]

More detailed information regarding the processing of specimens for anaerobic cultures is presented in Chapters 43 and 44.

Bites
Bite wound infections usually involve relatively small lesions and minimal exudate, so a swab technique with anaerobic transport usually is needed. Surrounding skin should be thoroughly disinfected before the specimen is obtained. The best material for culture is pus that is aspirated from the depth of the wound or samples obtained during surgery for incision and drainage or debridement, which is the removal of all dead and necrotic tissue. Again, Gram-stained smears should be prepared and examined. For aerobic cultures blood agar, MacConkey agar, and chocolate agar should be inoculated.

Burns
For burn patients, it is important to ascertain the number of organisms present per gram of tissue. Greater than 10^5 colony-forming units (CFUs) per gram of tissue is considered by some clinicians to be indicative of infection, whereas less than that number may indicate only colonization. A laboratory may be asked to perform quantitative cultures of tissue, albeit rarely. Because

conventional quantitative cultures are labor-intensive and expensive, Buchanan and colleagues[4] cultured biopsies by a semiquantitative technique that involved the inoculation of 0.1- and 0.01-mL samples to blood agar, in duplicate (Procedure 60-1). This modified procedure, which requires fewer manipulations in specimen processing, is comparable in terms of sensitivity and specificity to more involved conventional quantitative cultures.

Case Study

A 50-year-old man presented with extreme pain in his left thigh. He had an elevated measurement of creatine phosphokinase (CPK), an enzyme found predominantly in the heart, brain, and skeletal muscle. When the CPK is elevated, it usually indicates injury or stress to one or more of these areas. His heart muscle CPK fraction was normal. A biopsy specimen was collected by fine-needle aspiration. Gram stain showed gram-positive rods with subterminal spores. The aerobic culture was sterile, but the anaerobic culture grew a pure culture of bacteria with the same Gram stain morphology as seen in the direct smear. The colony had irregular edges like a medusa-head; a film of growth swarmed over the entire plate by 24 hours (Figure 60-11).

QUESTIONS

1. The isolate was indole-negative. Given this information, what is the genus and species of the bacteria?
2. What does the positive test for CPK indicate in this patient?
3. Infections with *C. septicum* are an indication of what underlying diseases?

REFERENCES

1. Abrahamian FM: Dog bites: bacteriology, management, and prevention, *Curr Infect Dis Rep* 2:446, 2000.
2. Amin N: Infected diabetic foot ulcers, *Am Fam Physician* 37:283, 1988.
3. Baron EJ, Bennion R, Thompson J, et al: A microbiological comparison between acute and complicated appendicitis, *Clin Infect Dis* 14:227, 1992.
4. Buchanan K, Heimbach DM, Minshew BH, et al: Comparison of quantitative and semiquantitative culture techniques for burn biopsy, *J Clin Microbiol* 23:258, 1986.
5. Fleisher G, Ludwig S, Campos J: Cellulitis: bacterial etiology, clinical features and laboratory findings, *J Pediatr* 97:591, 1980.
6. George WL: Other infections of skin, soft tissue, and muscle. In Finegold SM, George WL, editors: *Anaerobic infections in humans*, San Diego, 1989, Academic Press.
7. Giacometti A, Circone O, Schimizzi AM, et al: Epidemiology and microbiology of surgical wound infections, *J Clin Microbiol* 38:918, 2000.
8. Goldstein EJC: Bite wounds and infection, *Clin Infect Dis* 14:633, 1992.
9. Gustafson TL, Band JD, Hutcheson RH Jr, et al: *Pseudomonas* folliculitis: an outbreak and review, *Rev Infect Dis* 5:1, 1983.
10. Habif TP: Principles of diagnosis and anatomy. In *Clinical dermatology: a color guide to diagnosis and therapy*, ed 4, St Louis, 2004, Mosby.
11. Hook EW III, Hooton TM, Horton CA, et al: Microbiologic evaluation of cutaneous cellulitis in adults, *Arch Intern Med* 146:295, 1986.
12. Lipsky BA: Medical treatment of diabetic foot infections. *Clin Infect Dis* 39: S104, 2004.
13. Talan DA, Abrahamian FM, Moran GJ, et al: Clinical presentation and bacteriologic analysis of infected human bites in patients presenting to emergency departments, *Clin Infect Dis* 37:1481, 2003.
14. Talan DA, Citron DM, Abrahamian FM, et al: Bacteriologic analysis of infected dog and cat bites, *N Engl J Med* 340:85, 1999.
15. Williams DT, Hilton JR, Harding KG: Diagnosing foot infection in diabetes, *Clin Infect Dis* 39:583, 2004.

Any body tissue or sterile body fluid site can be invaded and infected with etiologic agents of disease from all four categories of microbes: bacteria, fungi, viruses, and parasites. Although from different areas of the body, all specimens discussed in this chapter are considered normally sterile. Therefore, even one colony of a potentially pathogenic microorganism may be significant. (Refer to Table 5-1 for a quick guide regarding collection, transport, and processing of specimens from sterile body sites.)

SPECIMENS FROM STERILE BODY SITES

FLUIDS

In response to infection, fluid may accumulate in any body cavity. Infected solid tissue often presents as cellulitis or with abscess formation. Areas of the body from which fluids are typically sent for microbiologic studies (in addition to blood and cerebrospinal fluid [see Chapters 52 and 55]) include those in Table 61-1.

Pleural Fluid

The parietal pleura, a serous membrane of the thoracic cavity (see Chapter 53), lines the entire thoracic cavity. The outer surface of each lung is also covered by the visceral pleura (Figure 61-1). Pleural fluid is a collection of fluid in the pleural space, normally found between the lung and the chest wall (see Figure 61-1). The fluid usually contains few or no cells and has a consistency similar to that of serum but with a lower protein content. When excess amounts of this fluid are present, it is called an **effusion,** or transudate, and is often the result of cardiac, hepatic, or renal disease. Pleural fluid that contains numerous white blood cells and other evidence of an inflammatory response (an exudate) is usually caused by infection, but malignancy, pulmonary infarction, or autoimmune diseases in which an antigen-antibody reaction initiates an inflammatory response may also be responsible. The material collected from the patient by needle aspiration (thoracentesis) is submitted to the laboratory as pleural fluid, thoracentesis fluid, or empyema fluid. Exudative pleural effusions that contain numerous polymorphonuclear neutrophils, particularly those that are grossly purulent, are called **empyema** fluids. Empyema usually occurs secondary to pneumonia, but other infections near the lung (e.g., subdia-phragmatic infection) may seed microorganisms into the pleural cavity.

Peritoneal Fluid

The **peritoneum** is a large, moist, continuous sheet of serous membrane that lines the walls of the abdominal-pelvic cavity and the outer coat of the organs contained within the cavity (Figure 61-2). In the abdomen, these two membrane linings are separated by a space called the **peritoneal cavity,** which contains or abuts the liver, pancreas, spleen, stomach and intestinal tract, bladder, and fallopian tubes and ovaries. The kidneys occupy a retroperitoneal (behind the peritoneum) position. Within the healthy human peritoneal cavity is a small amount of fluid that maintains moistness of the surface of the peritoneum. Normal peritoneal fluid may contain as many as 300 white blood cells per milliliter, but the protein content and specific gravity of the fluid are low. During an infectious or inflammatory process, increased amounts of fluid accumulate in the peritoneal cavity, a condition called **ascites.** The fluid, often called **ascitic fluid,** contains an increased number of inflammatory cells and an elevated protein level.

Agents of infection gain access to the peritoneum through a perforation of the bowel, through infection within abdominal viscera, by way of the bloodstream, or by external inoculation (as in surgery or trauma). On occasion, as in pelvic inflammatory disease (PID), organisms travel through the natural channels of the fallopian tubes into the peritoneal cavity.

Primary Peritonitis. There are two major types of infections in the peritoneal cavity: primary and secondary peritonitis. In **primary peritonitis,** no apparent focus of infection is evident. The organisms likely to be recovered from specimens from patients with primary peritonitis vary with the patient's age. The most common etiologic agents in children are *Streptococcus pneumoniae* and group A streptococci, *Enterobacteriaceae,* other gram-negative bacilli, and staphylococci. In adults, *Escherichia coli* is the most common bacterium, followed by *S. pneumoniae* and group A streptococci. Polymicrobial peritonitis is unusual in the absence of bowel perforation or rupture. Among sexually active young women, *Neisseria gonorrhoeae* and *Chlamydia trachomatis* are common etiologic agents of peritoneal infection, often in the form of a perihepatitis (inflammation of the surface of the liver, called *Fitz-Hugh-Curtis syndrome*). Tuberculous peritonitis occurs infrequently in the United States and is more likely to be found among persons recently arrived from South America, Southeast Asia, or Africa. Fungal causes of peritonitis are not common, but *Candida* spp. may be recovered from immunosuppressed patients and patients receiving prolonged antibacterial therapy.

Secondary Peritonitis.   **Secondary peritonitis** is a sequel to a perforated viscus (organ), surgery, traumatic injury, loss of bowel wall integrity because of destructive disease (e.g., ulcerative colitis, ruptured appendix, carcinoma), obstruction, or a preceding infection (liver abscess, salpingitis, septicemia). The nature, location, and etiology of the underlying process govern the agents

to be recovered from peritoneal fluid. With PID as the background, gonococci, anaerobes, or chlamydia are isolated. With peritonitis or intraabdominal abscess, anaerobes generally are found in peritoneal fluid, usually together with Enterobacteriaceae and enterococci or other streptococci. In patients whose bowel flora have been altered by antimicrobial agents, more resistant gram-negative bacilli and *Staphylococcus aureus* may be encountered. Because anaerobes outnumber aerobes in the bowel by 1000-fold, it is not surprising that anaerobic organisms play a prominent role in intraabdominal infection, perhaps acting synergistically with facultative bacteria. The organisms likely to be recovered include *E. coli*, the *Bacteroides fragilis* group, enterococci and other streptococci, *Bilophila* spp., other anaerobic gram-negative bacilli, anaerobic gram-positive cocci, and clostridia.

Table 61-1   Areas of the Body from Which Fluids Are Submitted to the Microbiology Laboratory

Body Area	Fluid Name(s)
Thorax	Thoracentesis or pleural or empyema fluid
Abdominal cavity	Paracentesis or ascitic or peritoneal fluid
Joint	Synovial fluid
Pericardium	Pericardial fluid

Peritoneal Dialysis Fluid

More than 13,000 patients with end-stage renal disease are maintained on continuous ambulatory peritoneal dialysis (CAPD). In this treatment, fluid is injected into the peritoneal cavity and subsequently removed, which allows exchange of salts and water and removal of various wastes in the absence of kidney function. Because the dialysate fluid is injected into the peritoneal cavity via a catheter, this break in the skin barrier places

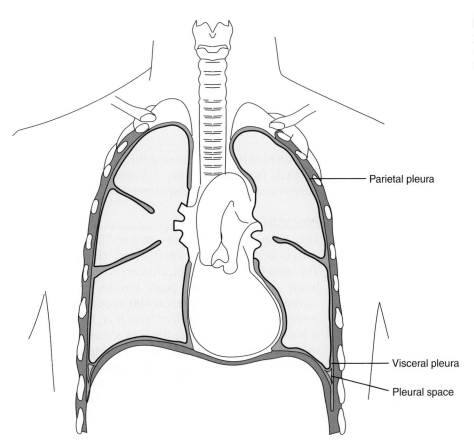

Figure 61-1   The location of the pleural space in relation to the parietal and visceral pleura and the rest of the respiratory tract.

Parietal pleura

Visceral pleura

Pleural space

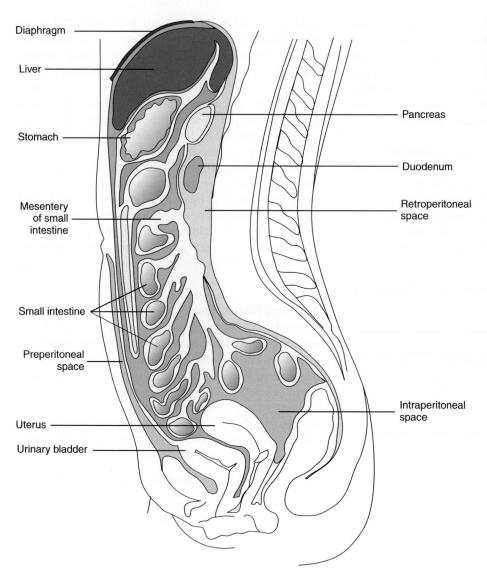

Diaphragm

Liver

Stomach

Mesentery
of small
intestine

Small intestine

Preperitoneal
space

Uterus

Urinary bladder

Pancreas

Duodenum

Retroperitoneal
space

Intraperitoneal
space

Figure 61-2   The abdominal
cavity. The retroperitoneal and
preperitoneal spaces are considered
as extraperitoneal (outside) spaces.
(Modified from Thibodeau GA:
Anatomy and physiology, St Louis,
1993, Mosby.)

the dialysis patient at significant risk for infection. The
average incidence of peritonitis in these patients is up to
two episodes per year per patient. Peritonitis is best
diagnosed clinically by the presence of two of the following: cloudy dialysate, abdominal pain, or a positive
culture from dialysate.[13] Although white blood cells
are usually plentiful (a value of leukocytes > 100/mL
is usually indicative of infection), the number of
organisms is usually too low for detection on Gram
stain of the peritoneal fluid sediment unless a concentrating technique is used; fungi are more readily
detected.

Most infections originate from the patient's own
skin flora; *Staphylococcus epidermidis* and *S. aureus* are the
most common etiologic agents, followed by streptococci,
aerobic or facultative gram-negative bacilli, *Candida* spp.,
Corynebacterium spp., and others. The oxygen content of
peritoneal dialysate is usually too high for development

of anaerobic infection. Among the gram-negative bacilli
isolated, *Pseudomonas* spp., *Acinetobacter* spp., and the
Enterobacteriaceae are frequently seen.

Pericardial Fluid

The heart and contiguous major blood vessels are
surrounded by a protective tissue, the pericardium.
The area between the epicardium, which is the membrane surrounding the heart muscle, and the pericardium is called the **pericardial space** and normally
contains 15 to 20 mL of clear fluid. If an infectious agent
is present within the fluid, the pericardium may become
distended and tight, and eventually **tamponade** (interference with cardiac function and circulation) can ensue.

Agents of **pericarditis** (inflammation of the pericardium) are usually viruses. Parasites, bacteria, certain
fungi, and noninfectious causes are also associated with
this disease.

BOX 61-1 **Common Etiologic Agents of Pericarditis and Myocarditis**

VIRUSES
  Enteroviruses (primary coxsackie A and B and, less frequently, echoviruses)
  Adenoviruses
  Influenza viruses
BACTERIA (RELATIVELY UNCOMMON)
  Mycoplasma pneumoniae
  Chlamydia trachomatis
  Mycobacterium tuberculosis
  Staphylococcus aureus
  Streptococcus pneumoniae
  Enterobacteriaceae and other gram-negative bacilli
FUNGI (RELATIVELY UNCOMMON)
  Coccidioides immitis
  Aspergillus spp.
  Candida spp.
  Cryptococcus neoformans
  Histoplasma capsulatum
PARASITES (RELATIVELY UNCOMMON)
  Entamoeba histolytica
  Toxoplasma gondii

Myocarditis (inflammation of the heart muscle itself) may accompany pericarditis. The pathogenesis of disease involves the host inflammatory response contributing to fluid buildup and cell and tissue damage. The most common etiologic agents of pericarditis and myocarditis are listed in Box 61-1. Other bacteria, fungi, and parasitic agents have been recovered from pericardial effusions, and therefore all agents should be sought.

Patients who develop pericarditis resulting from agents other than viruses are often compromised in some way. An example is infective endocarditis, in which a myocardial abscess develops and then ruptures into the pericardial space.

Joint Fluid

Infectious arthritis may involve any joint in the body. Infection of the joint usually occurs secondary to hematogenous spread of bacteria or, less often, fungi, or as a direct extension of infection of the bone. It may also occur after injection of material, especially corticosteroids, into joints or after insertion of prosthetic material (e.g., total hip replacement). Although infectious arthritis usually occurs at only one site (monoarticular), a preexisting bacteremia or fungemia may seed more than one joint to establish polyarticular infection, particularly when multiple joints are diseased, such as in rheumatoid arthritis. Knees and hips are the most frequently affected joints.

In addition to active infections associated with viable microorganisms within the joint, sterile, self-limited arthritis caused by antigen-antibody interactions may follow an episode of infection, such as meningococcal meningitis. When an etiologic agent cannot be isolated from an inflamed joint fluid specimen, either the absence of viable agents or inadequate transport or culturing procedures can be blamed. For example, even under the best circumstances, *Borrelia burgdorferi* is isolated from the joints of fewer than 20% of patients with Lyme disease. Nonspecific test results, such as increased white blood cell count, decreased glucose, or elevated protein, may seem to implicate an infectious agent but are not conclusive. A role has been postulated for the persistence of bacterial L-forms (cell-wall–deficient forms) in joint fluid after systemic infection, but such theories have not been proved.

Overall, *Staphylococcus aureus* is the most common etiologic agent of septic arthritis, accounting for approximately 70% of all such infections. In adults younger than 30 years of age, however, *Neisseria gonorrhoeae* is isolated most frequently. *Haemophilus influenzae* has been the most common agent of bacteremia in children younger than 2 years of age, and consequently it has been the most frequent cause of infectious arthritis in these patients, followed by *S. aureus*. The widespread use of *H. influenzae* type B vaccine should contribute to a change in this pattern. Streptococci, including groups A (*Streptococcus pyogenes*) and B (*Streptococcus agalactiae*), pneumococci, and viridans streptococci, are prominent among bacterial agents associated with infectious arthritis in patients of all ages. Among anaerobic bacteria, *Bacteroides*, including *B. fragilis*, may be recovered, as may *Fusobacterium necrophorum*, which usually involves more than one joint in the course of sepsis. Among people living in certain endemic areas of the United States and Europe, infectious arthritis is a prominent feature of Lyme disease. Some of the more frequently encountered etiologic agents of infectious arthritis are listed in Box 61-2.

These agents act to stimulate a host inflammatory response, which is initially responsible for the pathology of the infection. Arthritis is also a symptom associated with infectious diseases caused by certain agents, such as *Neisseria meningitidis*, group A streptococci (rheumatic fever), and *Streptobacillus moniliformis*, in which the agent cannot be recovered from joint fluid. Presumably, antigen-antibody complexes formed during active infection accumulate in a joint, initiating an inflammatory response that is responsible for the ensuing damage.

Infections in prosthetic joints are usually associated with somewhat different etiologic agents than those in natural joints. After insertion of the prosthesis, organisms that gained access during the surgical procedure slowly multiply until they reach a critical mass and produce a host response. This may occur long after the initial surgery; approximately half of all prosthetic joint infections occur more than 1 year after surgery.

BOX 61-2 | **Most Frequently Encountered Etiologic Agents of Infectious Arthritis**

BACTERIAL
Staphylococcus aureus
Beta-hemolytic streptococci
Streptococci (other)
Haemophilus influenzae
Haemophilus spp. (other)
Bacteroides spp.
Fusobacterium spp.
Neisseria gonorrhoeae
Pseudomonas spp.
Salmonella spp.
Pasteurella multocida
Moraxella osloensis
Kingella kingae
Moraxella catarrhalis
Capnocytophaga spp.
Corynebacterium spp.
Clostridium spp.
Peptostreptococcus spp.
Eikenella corrodens
Actinomyces spp.
Mycobacterium spp.
Mycoplasma spp.
Ureaplasma urealyticum
Borrelia burgdorferi

FUNGAL
Candida spp.
Cryptococcus neoformans
Coccidioides immitis
Sporothrix schenckii

VIRAL
Hepatitis B
Mumps
Rubella
Other viruses (rarely)

Skin flora are the most common etiologic agents, with *Staphylococcus epidermidis,* other coagulase-negative staphylococci, *Corynebacterium* spp., and *Propionibacterium* spp. predominating. However, *Staphylococcus aureus* is also a major pathogen in this infectious disease. Alternatively, organisms may reach joints during hematogenous spread from distant, infected sites.[9]

BONE

Bone Marrow Aspiration or Biopsy

Diagnosis of certain diseases, including brucellosis, histoplasmosis, blastomycosis, tuberculosis, and leishmaniasis, can sometimes be made only by detection of the organisms in the bone marrow. *Brucella* spp. can be isolated on culture, as can fungi, but parasitic agents must be visualized in smears or sections made from bone marrow material. Of importance, many of the etiologic agents associated with disseminated infections in patients infected with the human immunodeficiency virus (HIV) may be visualized or isolated from the bone marrow. Some of these organisms include cytomegalovirus, *Cryptococcus neoformans,* and *Mycobacterium avium* complex.

Bone Biopsy

A small piece of infected bone is occasionally sent to the microbiology laboratory for determination of the etiologic agent of **osteomyelitis** (infection of bone). Patients develop osteomyelitis from hematogenous spread of an infectious agent, invasion of bone tissue from an adjacent site of infection (e.g., joint infection, dental infection), breakdown of tissue caused by trauma or surgery, or lack of adequate circulation followed by colonization of a skin ulceration with microorganisms. Once established, infections in bone may tend to progress toward chronicity, particularly if effective blood supply to the affected area is lacking.

Staphylococcus aureus, seeded during bacteremia, is the most common etiologic agent of osteomyelitis among people of all age groups. The toxins and enzymes produced by this bacterium, as well as its ability to adhere to smooth surfaces and produce a protective glycocalyx coating, seem to contribute to its pathogenicity. Among young persons, osteomyelitis is usually associated with a single agent. Such infections are usually of hematogenous origin. Other organisms that have been recovered from hematogenously acquired osteomyelitis include *Salmonella* spp., *Haemophilus* spp., Enterobacteriaceae, *Pseudomonas* spp., *Fusobacterium necrophorum,* and yeasts. *S. aureus* or *P. aeruginosa* is often recovered from drug addicts. Parasites or viruses are rarely, if ever, etiologic agents of osteomyelitis.

Bone biopsies from infections that have spread to a bone from a contiguous source or that are associated with poor circulation, especially in patients with diabetes, are likely to yield multiple isolates. Gram-negative bacilli are increasingly common among hospitalized patients; a break in the skin (surgery or intravenous line) may precede establishment of gram-negative osteomyelitis. Breaks in skin from other causes, such as a bite wound or trauma, also may be the initial event that leads to underlying bone infection. For example, a human bite may lead to infection with *Eikenella corrodens,* whereas an animal bite may lead to *Pasteurella multocida* osteomyelitis. Poor oral hygiene may lead to osteomyelitis of the jaw with *Actinomyces* spp., *Capnocytophaga* spp., and other oral flora, particularly anaerobes. Pigmented *Prevotella* (formerly *Bacteroides melaninogenicus* group) and *Porphyromonas, Fusobacterium,* and *Peptostreptococcus* spp. are often involved. Pelvic infection in the female may lead to mixed aerobic and anaerobic osteomyelitis of the pubic bone.

Patients with neuropathy (pathologic changes in the peripheral nervous system) in the extremities,

BOX 61-3	Infectious Agents in Tissue Requiring Special Media

Actinomyces spp.
Brucella spp.
Legionella spp.
Bartonella (Rochalimaea) henselae (cat-scratch disease bacilli)
Systemic fungi
Mycoplasma
Mycobacteria
Viruses

notably patients with diabetes, who may have poor circulation as well, are subjected to trauma that they cannot feel simply by walking. They develop ulcers on the feet that do not heal, become infected, and may eventually progress to involve underlying bone. These infections are usually polymicrobial, involving anaerobic and aerobic bacteria.[3] *Prevotella* or *Porphyromonas*, other gram-negative anaerobes, including the *Bacteroides fragilis* group, *Peptostreptococcus* spp., *Staphylococcus aureus*, and group A and other streptococci are frequently encountered.

SOLID TISSUES

Pieces of tissue are removed from patients during surgical or needle biopsy procedures or may be collected at autopsy. Any agent of infection may cause disease in tissue, and laboratory practices should be adequate to recover bacteria, fungi, and viruses and to detect the presence of parasites. Fastidious organisms (e.g., *Brucella* spp.) and agents of chronic disease (e.g., systemic fungi and mycobacteria) may require special media and long incubation periods for isolation. Some agents that require special supportive or selective media are listed in Box 61-3.

LABORATORY DIAGNOSTIC PROCEDURES

SPECIMEN COLLECTION AND TRANSPORT

Because of the numerous specimen types from normally sterile body sites that can be submitted to the laboratory, requirements for collection and transport vary.

Fluids and Aspirates

Most specimens (pleural, peritoneal, pericardial, and synovial fluids) are collected by aspiration with a needle and syringe.

Collecting pericardial fluid is obviously hazardous because the sample is immediately adjacent to the beating heart. Collection is performed by needle aspiration with electrocardiographic monitoring or as a surgical procedure. Laboratory personnel should be alerted in advance so that the appropriate media, tissue culture media, and stain procedures are available immediately.

Body fluids from sterile sites should be transported to the laboratory in a sterile tube or vial that excludes oxygen. From 1 to 5 mL of specimen is adequate for isolation of most bacteria, but the larger the specimen, the better, particularly for isolation of *M. tuberculosis* and fungi; at least 5 mL should be submitted for each of these latter two cultures. Ten milliliters of fluid is recommended for the diagnosis of peritonitis. Anaerobic transport vials are available from several sources. These vials are prepared in an oxygen-free atmosphere and are sealed with a rubber septum or short stopper through which the fluid is injected. Fluid should never be transported in a syringe capped with a sterile rubber stopper because this method is unsafe. The use of syringes should be curtailed whenever possible because recapping and removal of needles is not permitted under standard precautions requirements for hospital safety. Most clinically significant anaerobic bacteria survive adequately in nonanaerobic transport containers (e.g., sterile, screw-capped tubes) for short periods if the specimen is frankly purulent and of adequate volume. Specimens received in anaerobic transport vials should be inoculated to routine aerobic (an enriched broth, blood, chocolate, and sometimes MacConkey agar plates) and anaerobic media as quickly as possible. Specimens for recovery of only fungi or mycobacteria may be transported in sterile, screw-capped tubes. At least 5 to 10 mL of fluid are required for adequate recovery of small numbers of organisms. If gonococci or chlamydiae are suspected, additional aliquots should be sent to the laboratory for smears and appropriate cultures.

Percutaneous catheters are placed during many surgical procedures to prevent the accumulation of exudate and blood at the operative site. Often, the laboratory receives drainage fluids from these catheters for culture when signs and symptoms suggest infection. However, culture of such fluid is potentially misleading when the fluid becomes contaminated within the catheter or collection device, or when the fluid does not originate from a site of the infection. Everts and colleagues[5] confirmed that direct aspiration of potentially infected fluid collections rather than catheter drainage fluid should be submitted for culture for the assessment of deep tissue infections in patients.

With respect to pericardial, pleural, synovial, and peritoneal fluids, the inoculation of blood culture broth bottles at the bedside or in the laboratory may be beneficial.[2,10,11,14] One must always send some of the specimen to the laboratory in a container other than a blood culture bottle, because putting the sample into a blood culture bottle dilutes it, making the preparation of a smear for Gram stain useless. The specimen in the blood culture bottle is processed as a blood culture, facilitating the recovery of small numbers of organisms and diluting out the effects of antibiotics. Citrate or sodium polyanetholesulfonate (SPS) may be used as an anticoagulant. Of importance, specimens collected by percutaneous needle aspiration (paracentesis) or at the time of surgery should be inoculated into aerobic and anaerobic blood culture broth bottles immediately at the bedside. Any delay results in decreased detection of true positive cultures.[11] In general, if sufficient specimen is available, cultures should be inoculated with the same volumes specified by the manufacturer for blood specimens. If the volume is insufficient to follow the manufacturer's instructions, as little as 0.1 mL can be inoculated.

Fluid from CAPD patients can be submitted to the laboratory in a sterile tube, urine cup, or in the original bag. The bag is entered only once with a sterile needle and syringe to withdraw fluid for culture. Fluid should be directly inoculated into blood culture bottles (at least 20 mL [10 mL in each of two culture bottles] and cultured). Numerous other studies indicate that in addition to blood culture bottles, an adult Isolator tube is a sensitive and specific method for culture.

Bone

Bone marrow is typically aspirated from the interstitium of the iliac crest. Usually, this material is not processed for routine bacteria, because blood cultures are equally useful for these microbes, and false-positive cultures for skin bacteria *(Staphylococcus epidermidis)* are frequent. Some laboratories report good recovery from bone marrow material that has been injected into a pediatric Isolator tube as a collection and transport device. The lytic agents within the Isolator tube are thought to lyse cellular components, presumably freeing intracellular bacteria for enhanced recovery. Bone removed at surgery or by percutaneous biopsy is sent to the laboratory in a sterile container.

Tissue

Tissue specimens are obtained after careful preparation of the skin site. It is critical that biopsy specimens be collected aseptically and submitted to the microbiology laboratory in a sterile container. A wide-mouthed, screw-capped bottle or plastic container is recommended. Anaerobic organisms survive within infected tissue long enough to be recovered from culture. A small amount of sterile, nonbacteriostatic saline may be added to keep the specimen moist. Because homogenizing with a tissue grinder can destroy some organisms by the shearing forces generated during grinding, it is often best to use a sterile scissors and forceps to mince larger tissue specimens into small pieces suitable for culturing (Figure 61-3). Note: *Legionella* spp. may be inhibited by saline; a

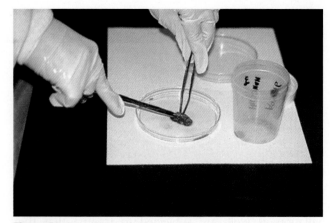

Figure 61-3 Mincing a piece of tissue for culture using a sterile forceps and scissors. Note: Perform this procedure in a biosafety cabinet.

section of lung should be submitted without saline for *Legionella* isolation.

If anaerobic organisms are of concern, a small amount of tissue can be placed into a loosely capped, wide-mouthed plastic tube and sealed into an anaerobic pouch system, which also seals in moisture enough for survival of organisms in tissue until the specimen is plated. The surgeon should take responsibility for seeing that a second specimen is submitted to anatomic pathology for histologic studies. Formaldehyde-fixed tissue is not useful for recovery of viable microorganisms, although some organisms can be recovered after very short periods. Therefore, an attempt may rarely be made to subculture from tissue in formalin if that is the only specimen available. Material from draining sinus tracts should include a portion of the tract's wall obtained by deep curettage. Tissue from infective endocarditis should contain a portion of the valve and vegetation if the patient is undergoing valve replacement.

In some instances, contaminated material may be submitted for microbiologic examination. Specimens, such as tonsils or autopsy tissue, may be surface cauterized with a heated spatula or blanched by immersing in boiling water for 5 to 10 seconds to reduce surface contamination. The specimen may then be dissected with sterile instruments to permit culturing of the specimen's center, which will not be affected by the heating. Alternatively, larger tissues may be cut in half with a sterile scissors or blade and the interior portion cultured for microbes.

Because surgical specimens are obtained at great risk and expense to the patient, and because supplementary specimens cannot be obtained easily, it is important that the laboratory save a portion of the original tissue (if enough material is available) in a small amount of sterile broth in the refrigerator and at −70° C (or, if necessary, at −20° C) for at least 4 weeks in case additional studies are indicated. If the entire tissue must be ground up for culture, a small amount of the suspension should be placed into a sterile tube and refrigerated.

SPECIMEN PROCESSING, DIRECT EXAMINATION, AND CULTURE

Fluids and Aspirates

Techniques for laboratory processing of all sterile body fluids are similar except for those previously discussed body fluids that are directly inoculated into blood culture bottles. Clear fluids may be concentrated by centrifugation or filtration, whereas purulent material can be inoculated directly to media. Any body fluid received in the laboratory that is already clotted must be homogenized to release trapped bacteria and minced or cut to release fungal cells. Either processing such specimens in a motorized tissue homogenizer or grinding them manually in a mortar and pestle or glass tissue grinder allows better recovery of bacteria. Hand grinding is often preferred, because motorized grinding can generate considerable heat and thereby kill microorganisms in the specimen. Grinding may lyse fungal elements; therefore, it is not recommended with specimens processed for fungi. Small amounts of whole material from a clot should be aseptically cut with a scalpel and placed directly onto media for isolation of fungi.

All fluids should be processed for direct microscopic examination. In general, if one organism is seen per oil immersion field, at least 10^5 organisms per milliliter of specimen are present. In such cases, often only a few organisms are present in normally sterile body fluids. Therefore, organisms must be concentrated in body fluids. For microscopic examination, cytocentrifugation (see Figure 55-4) should be used to prepare Gram-stained smears because organisms can be further concentrated up to a 1000-fold.[4] Body fluids for culture should be concentrated by either filtration or high-speed centrifugation. Once the sample is concentrated, the supernatant is aseptically decanted or aspirated with a sterile pipette, leaving approximately 1 mL liquid in which to mix the sediment thoroughly. Vigorous vortexing or drawing the sediment up and down into a pipette several times adequately resuspends the sediment. This procedure should be done in a biological safety cabinet. The suspension is used to inoculate media. Direct potassium hydroxide (KOH) or calcofluor white preparations for fungi and acid-fast stain for mycobacteria can also be performed.

Specimens for fungi should be examined by direct wet preparation or by preparing a separate smear for periodic acid-Schiff (PAS) staining in addition to Gram stain. Either 10% KOH or calcofluor white is recommended for visualization of fungal elements from a wet preparation. In addition to hyphal forms, material from the thoracic cavity may contain spherules of *Coccidioides* or budding yeast cells.

Lysis of leukocytes before concentration of CAPD effluents can significantly enhance recovery of organisms.[8] Filtration of CAPD fluid through a 0.45-mm poresize membrane filter allows a greater volume of fluid to be processed and usually yields better results. Because the numbers of infecting organisms may be low (fewer than 1 organism per 10 mL of fluid), a large quantity of fluid must be processed. Sediment obtained from at least 50 mL of fluid has been recommended.[2] If the specimen is filtered, the filter should be cut aseptically into three pieces, one of which is placed on chocolate agar for incubation in 5% carbon dioxide, one on MacConkey agar, and the other on a blood agar plate for anaerobic incubation.

If fluids have been concentrated by centrifugation, the resulting sediment should be inoculated to an enrichment broth and blood and chocolate agars. Because these specimens are from normally sterile sites, selective media are inadvisable because they may inhibit the growth of some of the organisms being sought. Appropriate procedures for the isolation of anaerobes, mycobacteria, fungi, *Chlamydia* spp., and viruses should be used when such cultures are clinically indicated.

Bone

Clotted bone marrow aspirates or biopsies must be homogenized or ground to release trapped microorganisms. Specimens are inoculated to the same media as for other sterile body fluids. A special medium for enhancement of growth of *Brucella* spp. and incubation under 10% carbon dioxide may be needed. A portion of the specimens may be inoculated directly to fungal media. Sections are also made from biopsy material (bone) for fixation, staining, and examination (usually by anatomic pathologists) for the presence of mycobacterial, fungal, or parasitic agents. With respect to obtaining specimens from patients suspected of having osteomyelitis, cultures taken from open wound sites above infected bone or material taken from a draining sinus leading to an area of osteomyelitis may not reflect the actual etiologic agent of the underlying osteomyelitis. Cultures of samples of bone obtained during wound debridement surgery appears to be more useful for directing antibiotic therapy for better clinical outcome.[7]

Diagnosis of prosthetic (artificial) joint infections is often difficult. Unfortunately, there is no universally accepted definition for the diagnosis of infection in the absence of microbiologic proof because clinical symptoms such as pain do not differentiate infection from mechanical joint failure. And, as of this writing, there is no standardized approach to the laboratory diagnosis of these infections and published data are conflicting. Further complicating the diagnosis is that the most common bacteria causing prosthesis infections are common skin contaminants such as coagulase-negative staphylococci. Some studies have reported that culture is relatively insensitive possibly due to organisms residing in biofilms, whereas polymerase chain reaction (PCR) assays were able to detect a majority of prosthetic joint infections. Atkins and colleagues[1] recommended that five or six operative bone specimens be submitted for culture and that the cutoff for a definite diagnosis of infection be three or more of these specimens yielding the same organism. However, a recent study using PCR and culture using multiple media types and prolonged incubation found that appropriate culture was adequate to exclude bacterial infection in hip prostheses and PCR did not enhance diagnostic sensitivity for infection.[6]

Normal bone is difficult to break up; however, most infected bone is soft and necrotic. Therefore, grinding the specimen in a mortar and pestle may break off some pieces. Small shavings from the most necrotic-looking areas of the bone specimen may sometimes be scraped off aseptically and inoculated to media. Pieces should be placed directly into media for recovery of fungi. Small bits of bone can be ground with sterile broth to form a suspension for bacteriologic and mycobacterial cultures. If anaerobes are to be recovered, all manipulations are best performed in an anaerobic chamber. If such an environment is unavailable, microbiologists should work quickly within a biosafety cabinet to inoculate prereduced anaerobic plates and broth with material from the bone.

Solid Tissue

Tissue should be manipulated within a laminar flow biological safety cabinet by an operator wearing gloves. Processing tissue within an anaerobic chamber is even better. The microbiologist should cut through the infected area (which is often discolored) with a sterile scalpel blade. Half of the specimen can then be used for fungal cultures and the other half for bacterial cultures. Both types of microbial agents should be sought in all tissue specimens. Some sample should also be sent to surgical pathology for histologic examination. Specimens should be cultured for viruses or acid-fast bacilli when such tests are requested. Material that is to be cultured for parasites should be finely minced or teased before inoculation into broth. Direct examination of stained tissue for parasites is performed by anatomic pathologists. Imprint cultures of tissues may yield bacteriologic results identical to homogenates and may help differentiate microbial infection within the tissue's center from surface colonization (growth only at the edge) when specimens are cut in half before processing. Additional media can be inoculated for incubation at lower temperatures, which may facilitate recovery of certain systemic fungi and mycobacteria.

Tissue may also be inoculated to virus tissue culture cells for isolation of viruses. Brain, lung, spinal fluid, and blood are most useful. Tissue may be examined by immunofluorescence for the presence of herpes simplex virus, varicella-zoster virus, cytomegalovirus, or rabies viral particles. Lung tissue should be examined by direct fluorescent antibody test for *Legionella* spp.

The tissues of all fetuses, premature infants, and young babies who have died of an infectious process should be cultured for *Listeria*. Specimens of the brain, spinal fluid, blood, liver, and spleen are most likely to contain the organism. The isolation procedure is given in detail by Seeliger and Cherry.[12]

Case Study

A 79-year-old woman had left knee arthroplasty to replace an arthritic joint with a prosthetic joint. The surgery was complicated and lasted more than 1 hour. She received antibiotics at the time of the surgery but not postoperatively, in accordance with usual protocols for this type of surgery. She did well postoperatively and went home in 7 days. After several weeks, she complained of low-grade fevers and pain in the joint. Her physician aspirated 50 mL of fluid from the knee and inoculated 10 mL into each of both aerobic and anaerobic blood culture bottles. Some fluid was also sent to the laboratory, where numerous white blood cells were found but no organisms were seen on Gram stain. The anaerobic blood culture bottle turned positive at 48 hours with a gram-positive cocci. The aerobic bottle remained negative.

QUESTIONS

1. What is the likely genus of the organism in the blood culture bottle?
2. What is the likely source of the infection in this patient?
3. The physician wanted the laboratory to be sure that this organism was not isolated because of poor technique in collection of the joint fluid specimen. Because the diagnosis of a septic joint means that the patient must have more surgery and long-term therapy, the physician wanted to be certain of the diagnosis. How can the laboratory ascertain that the organism caused the infection?

REFERENCES

1. Atkins BL, Athanasou N, Deeks JL, et al: Prospective evaluation of criteria for microbiological diagnosis of prosthetic-joint infection at revision arthroplasty, *J Clin Microbiol* 36:2932, 1998.

2. Bobadilla M, Sifuentes J, Garcia-Tsao G: Improved method for the bacteriological diagnosis of spontaneous bacterial peritonitis, *J Clin Microbiol* 27:2145, 1989.

3. Centers for Disease Control Department of Health and Human Services: *The prevention and treatment of complication of diabetes: A guide for primary care practitioners,* Atlanta, 1996, Public Health Service.

4. Chapin-Robertson K, Dahlberg SE, Edberg SC: Clinical and laboratory analyses of Cytospin-prepared Gram stains for recovery and diagnosis of bacteria from sterile body fluids, *J Clin Microbiol* 30:377, 1992.

5. Everts RJ, Heneghan JP, Adholla PO, et al: Validity of cultures of fluid collected through drainage catheters versus those obtained by direct aspiration, *J Clin Microbiol* 39:66, 2001.

6. Ince A, Rupp J, Frommelt L, et al: Is "aseptic" loosening of the prosthetic cup after total hip replacement due to nonculturable bacterial pathogens in patients with low-grade infection? *Clin Infect Dis* 39:1599, 2004.

7. Khatri G, Wagner DK, Sohnle PG: Effect of bone biopsy in guiding antimicrobial therapy for osteomyelitis complicating open wounds, *Am J Med Sci* 321:367, 2001.

8. Ludlam HA, Price TN, Berry AJ, et al: Laboratory diagnosis of peritonitis in patients on continuous ambulatory peritoneal dialysis, *J Clin Microbiol* 26:1757, 1988.

9. Maderazo EG, Judson S, Pasternak H: Late infections of total joint prostheses, *Clin Orthop* 229:131, 1988.

10. Reinhold CE, Nickolai DJ, Piccinini TE, et al: Evaluation of broth media for routine culture of cerebrospinal fluid and joint fluid specimens, *Am J Clin Pathol* 89:671, 1988.

11. Runyon B, Antillon MR, Akriviadis EA, et al: Bedside inoculation of blood culture bottles with ascitic fluid is superior to delayed inoculation in the detection of spontaneous bacterial peritonitis, *J Clin Microbiol* 28:2811, 1990.

12. Seeliger HPR, Cherry WB: *Human listeriosis: its nature and diagnosis,* Washington, DC, 1957, US Government Printing Office.

13. Teitelbaum I, Burkart J: Peritoneal dialysis, *Am J Kid Dis* 42:1082, 2003.

14. Von Essen R, Holtta A: Improved method of isolating bacteria from joint fluids by the use of blood culture bottle, *Ann Rheum Dis* 45:454, 1986.

PART VIII

Clinical Laboratory Management

Laboratory Physical Design, Management, and Organization

Microbiology laboratories operate in various venues. During the 1990s the trend changed from centralized facilities that operate from large hospital, public health, or independent laboratories to decentralized facilities that operate outside of a main laboratory but may still be within the physical hospital facility. Decentralized testing is also known as **alternate-site testing**. Alternate-site testing is performed in physician office laboratories (POLs), in ancillary outpatient hospital clinics, or at the hospital patient's bedside. Bedside testing is also known as **point of care testing** (POCT). POCT is the most radical departure from previous tenets involved in laboratory testing. The most common reason for POCT is enhanced turnaround time, and most POCT involves nonmicrobiologic testing (e.g., glucose, fecal occult blood, urinalysis, hemoglobin, hematocrit, arterial blood gases, and sodium or potassium levels). POCT is most often performed by nurses, perfusionists (who operate heart-lung machines during open-heart surgery), respiratory therapists, and physicians themselves. The most common microbiology bedside test is the rapid group A streptococcal antigen test for sore throats; this test is also popular in POLs or outpatient pediatric clinics where support personnel can perform the test and report the results to the physician before the patient goes home. Microbiologists manage testing in each of these venues. This chapter introduces management concepts that are applicable in a wide range of circumstances. Whole books have been written on the subject of laboratory management.[1] It is beyond the scope of this chapter to be more than an introduction to the concepts involved in the business of providing laboratory services.

SPACE REQUIREMENTS AND ORGANIZATION OF WORK FLOW

No matter where microbiology testing is performed a physical space will be set aside for this purpose. Depending on the complexity of testing and the type of pathogens sought, the laboratory space is organized based on the following considerations:

- Organization and staffing
- Functions (tests) that must be performed
- Equipment that will be used
- Special electrical, plumbing, or air handling requirements
- Safety equipment (e.g., shower/eye wash) and emergency systems (e.g., fire alarm or sprinkler system) requirements
- Waste-treatment requirements
- Containment requirements

All of these features are used to determine the space requirements or square footage. Although architects, engineers, and contractors are hired to design and construct the space, the microbiologist should take an active role in ensuring that the design and layout optimize workflow and efficiency. In the typical hospital or independent laboratory, space is provided for each of the following:

- Specimen receiving
- Specimen accessioning and processing
- Staining and light microscopy
- Dark-field and fluorescent microscopy, if applicable
- Waste disposal
- Media preparation and glassware washing
- An isolation room for processing acid-fast bacilli (AFB) and systemic fungi, if applicable
- Separate rooms for specialized molecular-based tests
- Open benches for routine specimen workup

There may also be separate areas for offices, record storage, restrooms, lunchrooms, and library or conference rooms. Laboratory design consultants at the Centers for Disease Control and Prevention (CDC) recommend 200 square feet to accommodate two to three technologists; others have recommended 8 to 10 linear feet per technologist. Closely related functions (waste treatment and glassware washing) should be adjacent to each other or in close proximity. Bubble diagrams (floor plans) (Figure 62-1) allow visualization of spatial relationships and traffic flow patterns. Several rules of thumb should be considered when designing spaces. Module design is recommended for maximum flexibility. The administrative (office) area should be located near the front entrance to the laboratory away from potentially contaminated and odor-producing activities. Likewise, the areas where work with the most potentially infectious organisms is performed should be located in the most secluded space. Corridors should be at least 5 feet wide to allow people to pass each other,

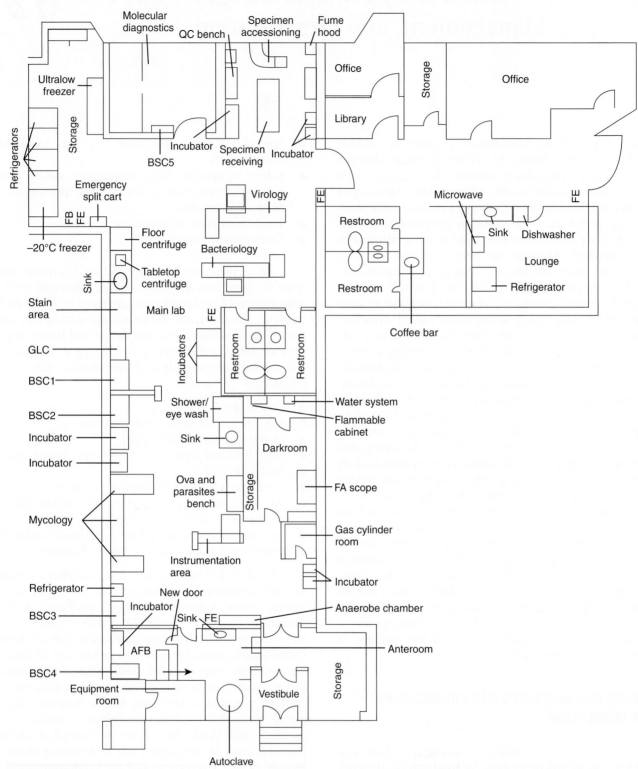

Figure 62-1   Bubble diagram for space layout. *AFB*, Acid-fast bacilli; *BSC*, biological safety cabinet; *FB*, fire blanket; *FE*, fire extinguisher; *GLC*, gas liquid chromatograph; *QC*, quality control.

and total floor space should include reasonable projections for expansion.

LABORATORY DESIGN WITH RESPECT TO SAFETY

Laboratory safety is discussed in depth in Chapter 4. However, a few points are worth repeating here since they pertain to laboratory design. The emergency shower and eye wash stations should be centrally located. The standard for the location of these units is the same, that is, within 10 seconds and 100 feet of each work area. Cold water is used for emergency flushing of eyes and skin because it will do the following:

- Slow the reaction rate of the splashed chemical
- Constrict blood vessels and minimize circulation of an absorbed chemical
- Slow cellular metabolism and enzyme reaction rates
- Help reduce the pain of chemical contact

The National Fire Protection Association (NFPA) can be consulted regarding the proper storage of chemicals to prevent unintentional mixing of chemicals from leaking or broken containers. Sulfuric acid and sodium hydroxide, which are both corrosives, for example, should not be stored in the same area. Flammables should be stored in a special cabinet and alkalines, oxidizers, and carcinogens should be stored separately.

Fire extinguishers and blankets should be readily available throughout the work area. A spill cart should be centrally located and should contain first aid supplies and kits to clean up radioactive, acid, alkali, and corrosive spills. Installation of an automatic fire detection and alarm system, as well as sprinklers (to extinguish fire), is part of the building code in many locations. Moreover, two means of exit should be planned in case of fire. All electrical outlets should be grounded.

The windows in the laboratory should be closed and sealed, and laboratory doors should be self-closing. There may be special requirements for storage and disposal of radionuclides, although ^{14}C (carbon-14) and ^{3}H (tritium), which are commonly used in the microbiology laboratory, are usually under a general license; state radiation safety offices should be consulted, if necessary. As stated in Chapter 4, the CDC and NIH (National Institutes of Health) address biosafety design criteria. Except for those facilities in which culturing of *Mycobacterium tuberculosis* (Biosafety Level 3) is conducted, clinical microbiology laboratories usually operate at Biosafety Level 2. Special or maximum containment facilities require a separate building, sealed openings into the laboratory, airlocks or liquid disinfectant barriers, dressing rooms and shower contiguous to the laboratory, double-door entry into the laboratory, biowaste treatment system, separate ventilation system, and a treatment system to decontaminate exhaust air.

DESIGN OF AIR HANDLING SYSTEM

HVAC (heating, ventilation, and air conditioning) systems must maintain a constant year-round temperature of 68° to 72° F and a relative humidity of 40% to 60%. The large amount of heat produced by instruments must be taken into account when planning an HVAC system. The American Society of Heating, Refrigeration and Air-Conditioning Engineers (ASHRAE) publishes standards for design of HVAC systems that take these environmental conditions into account. The supply air system provides fresh air to the various locations through adjustable vane diffusers mounted in the ceiling; it may condition the air (filter out mold spores and pollens, for instance) before its delivery (e.g., in laminar flow rooms). The supply air ductwork is generally constructed of galvanized sheet metal with external insulation to facilitate cleaning of the interior of the ducts in the case of aerosolization of pathogens following an accident. The exhaust air system is also externally insulated sheet metal; activated charcoal absorption filters and HEPA filters may be used to remove contaminants before exhaust air is discharged to the atmosphere. However, except in cases of Biosafety 3 or 4 facilities, the exhaust air can be discharged to the outside without being treated (HEPA-filtered).

The laboratory areas should maintain negative pressure with respect to the administrative areas to prevent toxic or pathogenic materials used in laboratory work areas from escaping and injuring humans or contaminating the environment. To set up negative pressure in the laboratory (or positive pressure in the offices), laboratory HVAC systems are designed so that a unidirectional airflow draws supply air into the laboratory (from the administrative areas) and discharges exhaust air directly to the outside away from occupied areas and air intakes. The amount of supply air being provided to a laboratory should equal 85% of the air being exhausted from the space. The other 15% of the makeup, or supply air, is provided from adjacent corridors through doorways to maintain the negative pressure. The number of air exchanges per hour is usually 4 to 6 for office spaces and 10 to 15 for biohazard rooms.

DESIGN OF MECHANICAL SYSTEMS

Many municipal or state government agencies have established utility specifications. These include electrical, plumbing, communications, and lighting. Electrical wall outlets should be located every 4 to 5 feet along walls and peninsular benches and should be positioned high enough to clear the top of a stand-up bench. The sum of the electrical requirements for all equipment, lighting, and air conditioning is used in calculating the total number of amps required. Dedicated circuits (with only one outlet) should be planned for computers or other instruments sensitive to power surges. Care should be taken to provide 110- and 220-volt circuitry for instruments, as needed. Telephones should be located for ease of use as should computer terminals, facsimile machines, and teleprinters. Commonly used gases such as carbon dioxide (CO_2) for incubators are often piped into the laboratory work areas to prevent having to transport compressed gas tanks throughout the laboratory. Sinks should be located at the end of benches or peninsulas. Typical plumbing services include hot and cold water and deionized or distilled water. Local building codes often call for plumbing effluent (discharge) to be diluted in a neutralization tank before it joins the building's sanitary sewer system. Light fixtures should be surface-mounted on or recessed into the ceiling and centered over the front edge of laboratory benches. The number of foot-candles required for illumination of work surfaces is usually between 160 and 200, depending on local ordinances. Because they cause the least disturbance to the ceiling, surface-mounted fixtures are preferred in microbiology laboratories. Recessed lighting should be tightly sealed to make the ceiling airtight.

WALLS, FLOORS, CEILINGS, AND FURNITURE

Sheetrock walls should be painted with epoxy so that they may be easily cleaned in the event of a spill. Square-foot, vinyl composition tiles are the most economical and effective floor covering for a laboratory. Seamless flooring is difficult to repair following spills, because the spilled material may dissolve the vinyl. Acoustic tile ceilings are permissible for Biosafety Level 1 or 2 facilities. However, sealed sheetrock or plaster ceilings with three coats of epoxy paint are recommended for Biosafety Level 3 areas; this type of ceiling will prevent airborne contamination from penetrating the space above the ceiling and can be washed with disinfectant as necessary. Flexible furniture systems that can be disassembled and moved as technologic changes occur or new equipment is purchased are preferred to permanently installed built-ins. They allow entire sections of the laboratory to be modified with minimal cost and inconvenience. Bench-tops should be impervious to water and resistant to acids, alkalis, organic solvents, and moderate heat. Tops are made of cast resin, quarried stone, particleboard or wood cores with acid-resistant plastic laminate surfaces, and stainless steel. Plastic laminates are usually acceptable for most uses and are the least expensive and easiest to replace. Standard benches have a depth of 30 inches.

INSTRUMENTATION

Instruments commonly found in the microbiology laboratory include:

- Bacterial detection devices (e.g., blood cultures)
- Bacterial identification and susceptibility testing devices
- Gas liquid chromatographs (GLC)
- Microscopes (light, fluorescent, phase, dark-field)
- Centrifuges
- Autoclave
- Dry-heat oven
- Incubators (radiant heat, CO_2)
- Water baths
- Heat baths
- Vortex mixers
- Anaerobe chamber
- Refrigerators/freezers
- Ultra-low freezer
- Biological safety cabinet (BSC)
- Fume hood
- Thermal cyclers

The life expectancy of major laboratory equipment is 5 years. Some equipment, (e.g., biological safety cabinets or fume hoods) must be carefully placed in the space when designing the floor plan. This equipment in particular should be located away from the main traffic areas or doorways to minimize air turbulence and the potential for air spillage. Autoclaves should be in a separate, back area, if possible, to reduce the noise, odors, and heat generated by this equipment.

The decision to buy any instrument should be analyzed based on a number of factors outlined later in this chapter, including an analysis of how the instrument will assist the microbiologist in turning out quality results. For example, will it decrease turn-around time (e.g., same-day bacterial identification and susceptibility instruments), will it allow a single person to handle more work by automating various parts of a process, or will it help save money by allowing testing of large volumes of specimens. Manufacturers' representatives are good resources for justifying instrument purchases because they know, for example, the minimum number of tests that a laboratory should perform to make the instrument cost effective and they usually have compared their instrument to others or to conventional methodology. Microbiology still remains mainly a manual discipline, although many semiautomated instruments are currently marketed.

REGULATION OF THE MICROBIOLOGY LABORATORY

The regulatory environment in which microbiologists currently work has evolved in the past 30 years and began after the television news magazine show *60 Minutes* aired a program about sink-testing. Hidden cameras recorded a laboratorian pouring a patient's urine down the sink and then fabricating the results of the urine culture and susceptibility. The public outcry that followed this broadcast led Congress to enact the Clinical Laboratory Improvement Act of 1967, now known as **CLIA '67.** The **Centers for Medicare and Medicaid Services (CMS),** formerly known as The Health Care Financing Administration (HCFA), was created to oversee the enforcement of the guidelines published in CLIA '67; CMS is part of the United States Department of Health and Human Services (DHHS). CMS is also responsible for overseeing the Medicare and Medicaid programs, which were instituted to provide health care for elderly and poor persons under the Social Security Act. CLIA '67 mandated the first quality control, personnel, and proficiency testing standards for clinical laboratories but applied only to laboratories engaged in interstate commerce; this represented less than 10% of all clinical laboratories. Therefore, in 1988, following media publicity regarding deaths from uterine and ovarian cancer in women whose Pap smears had been misread, the authority of CMS under CLIA '67 was expanded under the Clinical Laboratory Improvement Amendments of 1988, known as **CLIA '88.**

CLIA '67 and CLIA '88 are federal regulations; in addition, many state health departments (e.g., New York, Florida, Maryland, and California) have established their own, sometimes more stringent, regulations. A microbiology manager who lives in a state with specific laboratory guidelines must ensure that the laboratory meets all federal and state requirements and that the technical personnel are licensed by the state. The laboratory branch of the state Public Health Department is a good source for regulatory information.

A number of other federal agencies more indirectly regulate the practice of clinical microbiology. **The Centers for Disease Control and Prevention (CDC)** write regulations for the enforcement of CLIA. A Clinical Laboratory Improvement Advisory Committee (CLIAC) was established to advise the DHHS and the CDC as laws are turned into enforceable regulations. The CLIAC is composed of representatives of the laboratory industry, the manufacturing sector, health care administrators, a consumer advocate, and representatives of the CDC, FDA, and CMS. **The Food and Drug Administration (FDA)** clears the products bought by the laboratory for use in diagnostic testing. Labels of products that the FDA has approved state, "For Diagnostic Testing." Those that have not been cleared must be labeled "For Investigational Use." Laboratories may use devices or instruments that are not cleared by the FDA, but they must be subjected to rigorous validation and quality control testing. FDA clearance implies that the manufacturer has already verified the accuracy, precision, and reproducibility of the product, device, or

instrument, and the clinical laboratory may do less rigorous in-house testing. The FDA also designates each cleared test as waived, moderate, or high complexity (see Test Complexity Model on p. 920). The **Environmental Protection Agency (EPA)** regulates the disposal of toxic chemicals and biohazardous waste (see Chapter 4). The **Occupational Safety and Health Administration (OSHA)** regulates employee safety in the workplace (see Chapter 4). The **Office of the Inspector General (OIG)** of the **DHHS** monitors laboratories for fraud and abuse in billing Medicare and Medicaid. The OIG published a compliance document for clinical laboratories in 1997; this document explains the steps necessary to ensure that each laboratory conducts testing and bills under the highest ethical standards.

The Health Information Portability and Accountability Act (HIPAA) of 1996 legislates the need to maintain security and privacy of information found in patient records. This means having secure phone and fax lines for transmission of laboratory data and limiting access to patient data to only those individuals needing to know the results. Biosecurity concerns addressed in the Select Agent Rule are addressed in Chapter 65.

ACCREDITING AGENCIES

All laboratories must be accredited (licensed) under CLIA '88. This means that every laboratory must register with CMS for every analyte (test) it performs and must meet certain minimum quality standards. Except for laboratories performing only waived testing, abidance with CLIA is monitored by routine on-site inspections; even waived laboratories are subject to random inspections at the discretion of CMS. The inspections include a review of a laboratory's procedures, quality control program, documentation, and patient management. Several private agencies also inspect and accredit clinical laboratories for a fee. Laboratories inspected by private agencies with deemed status do not have to be reinspected by CMS because CMS has determined that their accreditation program is at least as stringent as that specified under CLIA '88. A list of private accrediting agencies is shown in Table 62-1. Most of the private agencies have their own rules and checklists and do not accept inspections by others. Therefore, microbiologists must be aware of their requirements, as well as CLIA and state requirements, if applicable. CMS reinspects approximately 10% of privately accredited laboratories each year to verify that agencies with deemed status are upholding the CLIA guidelines.

CLIA '88

The CLIA '88 regulations combined and replaced former laboratory standards with a single set of requirements

Table 62-1   Private Accrediting Agencies

Agency	Accredits	Deemed Status	Comments
College of American Pathologists (CAP)	Laboratories only	Yes	Oldest private accrediting agency
Joint Commission on the Accreditation of Health Care Organizations (JCAHO)	Entire hospitals or clinics, including laboratory	Yes	
Commission of Laboratory Accreditation (COLA)	Licenses laboratories in physician offices	Yes	
American Association of Bioanalysts (AAB)	Laboratories in free-standing facilities	No	

that apply to clinical laboratory testing wherever it is performed, that is, freestanding facility, POL (physician office laboratory), or hospital; this concept has come to be known as *site neutrality.* Uniform, minimum standards have been published in the federal register regarding laboratory personnel, quality control, quality assurance, and proficiency testing (PT) issues. The regulations are based on a test complexity model that, in turn, is based on the potential for harm to a patient. CLIA '88 also sets up enforcement procedures and defines sanctions to be applied if laboratories fail to meet minimum standards.

Test Complexity Model

CLIA '88 defined four categories of testing as follows:

- Waived tests
- Practitioner-performed microscopy
- Moderate complexity
- High complexity

Waived tests are those that are so simple to perform that there is virtually no risk to the patient even if the test is not performed correctly. Inspections and PT are waived for laboratories performing only these tests. The majority of microbiology waived tests include a variety of group A *Streptococcus* and influenza virus direct antigen tests. There is also an HIV antibody test performed from human saliva that was approved as a waived test with certain quality control and personnel restrictions. **Practitioner-performed microscopy** (PPM) is composed of microscopy performed at the point-of-care by a physician, physician's assistant, nurse, midwife, or nurse practitioner. The PPM category used to be called physician-performed microscopy but was subsequently expanded to include mid-level specialists. Laboratories holding a PPM certificate are not subject to on-site inspections but must still meet quality assurance (QA) and quality control (QC) standards. Microbiology tests in this category include wet mounts of vaginal and cervical secretions and skin, potassium hydroxide (KOH) preparations, and pinworm examinations. The third

category of testing is **moderate complexity.** Microbiology procedures in this category include urethral and cervical Gram stains and presumptive reporting of group A *Streptococcus (Streptococcus pyogenes)* using hemolysis and inhibition by bacitracin. The fourth category of testing is **high complexity.** Most microbiology cultures that require isolation, identification, and susceptibility testing of organisms are considered high complexity. Moderate and high complexity testing laboratories are both subject to on-site inspections and must comply with the same requirements for QC, QA, and PT. However, personnel requirements are different for each level.

PERSONNEL STANDARDS

Appropriately trained and experienced personnel are essential for the performance of quality laboratory testing. Therefore, a major part of CLIA '88 details the requirements for laboratory directors, supervisors, consultants, and testing personnel. Although the federal government emphasizes that the CLIA personnel standards represent minimum standards, education and experience requirements were lowered from those required in CLIA '67. For example, personnel in moderate testing laboratories may now be high school graduates if they have documented on-the-job training. Previously, individuals who performed microbiology procedures were required to have at least an associate's degree in medical laboratory technology or laboratory science and most held baccalaureate degrees in a biologic, chemical, or physical science or medical technology. Similarly, supervisory personnel who used to have baccalaureate degrees and 6 years of experience are now merely required to have an associate's degree in medical technology or laboratory science plus 2 years of clinical laboratory training and 2 years of supervisory experience. Laboratory directors who are doctoral scientists are now required to be board-certified from organizations such as the ABMM (American Board of Medical Microbiology) or ABMLI (American Board of Medical Laboratory Immunology).

PROFICIENCY TESTING STANDARDS

Proficiency testing (PT) was originally mandated in CLIA '67, but its scope was expanded in CLIA '88 to include sanctions (penalties) for poor performance. PT is a method whereby laboratories are sent unknown samples to test as they would test patient specimens. It is discussed in more detail in Chapter 63. CLIA '88 mandates that at least three times a year each laboratory participate in a program comparable in scope to the level of testing it performs. Public and private accrediting agencies send laboratories samples to test and return critiques with the answers. Critiques are summaries of how the laboratory performed in relation to its peers, and they serve as a method by which a laboratory can compare its test systems with that of others. Laboratories that fail a particular analyte two times consecutively must submit a plan of corrective action. If the laboratory continues to fail that analyte, CLIA '88 specifies sanctions, including the withdrawal of a laboratory's permit for that particular analyte. Both moderate and high complexity laboratories must participate in PT.

QUALITY ASSURANCE STANDARDS

Quality assurance (QA) and quality control (QC) standards are discussed in depth in Chapter 63. Both represent the procedures that laboratories establish to ensure that the patient results reported are accurate and timely, and assist the clinician in treating the patient. CLIA '88 mandates auditing (studying) the preanalytic, analytic, and postanalytic phases of a laboratory test to ensure that each step led to a favorable outcome for the patient. Laboratories must also keep problem logs showing complaints from clinicians or patients and detailing the investigation and any corrective action indicated.

SELECTION OF DIAGNOSTIC TESTS

One of the jobs of the clinical microbiologist is to decide what testing should be offered on-site and to select the best method of doing any test. The decision may be as basic as deciding whether to offer a particular test or as complex as deciding between different test methodologies. Although there are several aspects to consider when deciding to add a particular test, the most important is the clinical need for the test. The test should improve the quality of care delivered to patients by offering improved turn-around time or accessibility. The more serious the disease, the more a diagnostic test is indicated, especially if treatment is available but expensive. A good example of this is the decision by many facilities to offer on-site **STAT** (immediate) testing for

respiratory syncytial virus (RSV). Laboratories began offering antigen tests for RSV when treatment (ribavirin) became available. Ribavirin therapy is expensive, and infants with RSV can be very sick, so it makes sense to offer this test in order to make a rapid, specific diagnosis. Moreover, prompt therapy of infected infants may prevent spread of RSV in hospital nurseries.

Several other factors must be considered when deciding whether to add a diagnostic test:

- Projected test volume: Will it be worth the money required to verify this new procedure and perform ongoing quality control?
- Projected test cost: Will the laboratory be able to make money or save money relative to the current test?
- Medical staff needs: Is a screening test enough or should the test be confirmatory (diagnostic)?
- Should a culture be performed or is a noncultural test adequate?
- Disease prevalence: Analysis of population demographics (composition) will help determine whether your facility will see patients with a particular illness. In the example of the RSV antigen test, you might examine the incidence of emergency department admissions of children younger than 3 years of age who are in respiratory distress, and who subsequently receive treatment. However, if treatment will not be administered, then it would not be practical to offer a stat test.
- Availability of technically skilled personnel to perform the test, especially if it will be offered on all shifts.
- Availability of reagents cleared by the FDA.
- Availability of instrumentation and separate physical space, if needed. This would be an important consideration, for example, if the laboratory were considering a molecular technique, such as polymerase chain reaction (PCR), in which a thermal cycler must be used and separate work areas are needed.
- Extent of quality control required.
- Availability of proficiency test materials.
- Specimen requirements and storage conditions.
- Availability of a reference laboratory offering the test. Will it offer the service at a lower cost?

Once the decision has been made to offer the test, the different test methods should be studied. For example, if a laboratory is considering offering a test for chlamydia, the microbiologist would have to decide whether to do (1) a culture, (2) a direct fluorescent antibody test, (3) an enzyme-linked immunosorbent assay, (4) a nonamplified DNA probe, or (5) an amplified DNA probe. Although not all testing decisions are as complex as this, often a decision must be made among several methods.

ANALYSIS OF TESTS

Quantitative test methods must be validated based on accuracy, precision, sensitivity, specificity, predictive value, and efficiency before a decision is made. To understand how to quantitatively analyze the performance of various tests, some basic definitions are required.

Definitions

Accuracy (efficiency) is the ability of the test under study to match the results of a standard test commonly known as the "gold standard." For example, a non-cultural test may be compared with culture that is the "gold standard."

Precision is the reproducibility of a test when it is run (repeated) several times. Precision can be assessed both intralaboratory (within the same laboratory) and interlaboratory (between different laboratories). Precision does not imply accuracy.

Sensitivity is the percentage of individuals with the particular disease for which the test is used in whom positive test results are found. These are also called **true-positives.** Tests with very high sensitivity (≥99%) can, if negative, be used to exclude the presence of disease. For example, if an RSV direct antigen test is negative and the test sensitivity is 99.9%, then the infant is unlikely to have RSV. Most tests will show some **false-positives;** these represent cases in which individuals without the particular disease test positive by a certain test.

Specificity is the percentage of individuals who do not have the particular disease being tested and in whom negative test results are found. These are also called **true-negatives.** Tests with very high specificity (≥99%) should be negative both in healthy individuals and in individuals with symptoms similar to that of the disease being tested but not caused by the same organism. For example, in the infant with symptoms suggestive of RSV with a negative RSV direct antigen test, the physician would consider *Bordetella pertussis* or parainfluenza virus as the possible etiologic agent. Most tests will show some **false-negatives;** these represent cases in which individuals with the particular disease test negative by a certain test.

Prevalence is the frequency of disease in the population at a given time. It is based on the incidence of new cases of disease per year, usually calculated per 100,000 population. Prevalence is important in comparing the predictive value of a positive or negative test. The prevalence of chlamydiae in women attending a clinic for a sexually transmitted disease (STD), for example, is higher than in nuns.

Predictive value positive (PVP) is the percentage of true-positive test results based on the prevalence of the disease in the population studied. PVP measures the probability that a positive result indicates the presence of disease.

Predictive value negative (PVN) is the percentage of true-negative test results based on the prevalence of the disease in the population studied. PVN measures the probability that a negative result indicates the absence of disease.

Efficiency (accuracy) is the percentage of test results that are correctly identified by the test, that is, true positives and true negatives.

Mathematic Formulas

To calculate sensitivity, specificity, prevalence, PVP, PVN, and efficiency, the following formulas can be used. These formulas can be easily understood by remembering a simple table that contains all the critical information:

		Gold Standard	
		POSITIVES	NEGATIVES
	Positives	a	b
New		(true positives)	(false-positives)
test			
	Negatives	c	d
		(false-negatives)	(true negatives)

$$\text{Sensitivity} = \frac{a}{a+c} \times 100$$

$$\text{Specificity} = \frac{d}{b+d} \times 100$$

$$\text{PVP} = \frac{a}{a+b} \times 100$$

$$\text{PVN} = \frac{d}{c+d} \times 100$$

$$\text{Efficiency} = \frac{a+d}{a+b+c+d} \times 100$$

$$\text{Prevalence} = \frac{a+c}{a+b+c+d} \times 100$$

ASSESSING THE SENSITIVITY, SPECIFICITY, PVP, PVN, AND EFFICIENCY OF A TEST

All FDA-cleared products for diagnostic use must undergo clinical trials. During the trials, laboratories from different geographic regions with high and low prevalence of the disease compare the new test to a "gold standard." In most cases, culture (if available) is considered the "gold standard," although this is not always true. For example, because RSV, a very fastidious virus, does not easily grow in cell culture, a new direct antigen assay might be compared with an approved direct antigen test already on the market. The package insert

---

BOX 62-1   Use of Test Verification Data to Analyze Two Microbiology Tests

BACKGROUND INFORMATION:
A laboratory that serves a pediatric outpatient clinic wants to determine if there is any difference in performance between summer and winter months for two rotavirus tests, that is, a latex test and an enzyme-linked immunosorbent assay (ELISA). The prevalence of rotavirus in winter in infants with diarrhea is high; in summer the prevalence is low.

PURPOSE:
The laboratory has been using an ELISA test that is considered the gold standard. However, because the latex test is less expensive, easier to perform, and has a faster turnaround time, the laboratory wants to switch to this test.

STUDY DESIGN
One hundred diarrheal stool specimens were tested at random during January and July.

RESULTS:

JANUARY

	ELISA POSITIVES	ELISA NEGATIVES
Latex Positives	45	1
Latex Negatives	45	9

Prevalence = 90%

$Sensitivity = \frac{45}{90} \times 100 = 50\%$

$Specificity = \frac{9}{10} \times 100 = 90\%$

$PVP = \frac{45}{46} \times 100 = 98\%$

$PVN = \frac{9}{54} \times 100 = 17\%$

$Efficiency = \frac{54}{100} \times 100 = 54\%$

JULY

	ELISA POSITIVES	ELISA NEGATIVES
Latex Positives	1	10
Latex Negatives	1	88

Prevalence = 2%

$Sensitivity = \frac{1}{2} \times 100 = 50\%$

$Specificity = \frac{88}{98} \times 100 = 90\%$

$PVP = \frac{1}{11} \times 100 = 9\%$

$PVN = \frac{88}{89} \times 100 = 99\%$

$Efficiency = \frac{89}{100} \times 100 = 89\%$

CONCLUSIONS:
The latex test would perform well in the summer to screen out negative patients because the prevalence of disease is low and the predictive value of a negative test is 99%. However, when the disease prevalence is high, in winter, the laboratory would want to keep the ELISA test because the sensitivity of the latex test is low (50%), the predictive value of a negative latex test is only 17%, and the accuracy (efficiency) of the latex test is only 54%.

(instructions) for each FDA-cleared kit, reagent, or medium contains data from the clinical trials. This information is an indicator of the characteristics of the new test and can be used for comparison when in-house testing is performed.

Box 62-1 is an example of how to perform the calculations and make subsequent decisions regarding test verification.

TEST VERIFICATION AND VALIDATION

CLIA '88 mandates that all tests used by a laboratory after September 1, 1992, be **verified** before being used. Verification is a one-time process. Thereafter, the test must be **validated** as part of an ongoing process of quality assurance. Validation involves documentation of quality control, proficiency testing, and continued satisfactory performance of the test; it is discussed in more detail in Chapter 63. The comparability of different tests may be verified by one of several methods:

- Test patient samples in parallel with an established test using positive and negative specimens.
- Test specimens with known potency, for example, low-positive and high-positive sera.
- Determine the precision of the test by establishing the 95% confidence interval.

The manner of testing is left to the laboratory director's discretion as long as the full range of testing is evaluated. Currently, many laboratories perform concurrent testing of 20 to 30 samples using the new test and the standard test; others test 40 to 70 samples. At least 25% to 30% of the specimens tested should yield a positive result, however.

The confidence interval (CI) at the 95% level can be calculated as a way of determining the precision (or imprecision) of a test when a small sample size (e.g., 20 to 30 samples) is used. Wide confidence intervals point out the unreliability of a test in studies where a small number of samples are used. The

---

BOX 62-2 **Determination of Test Reliability at the 95% Confidence Level**

DATA:

Test	Number of Tests	Sensitivity
Current	30	90
New	30	95

STANDARD ERROR:

Current test: $SE = \sqrt{0.90 \times \dfrac{(1-0.90)}{30}} = \sqrt{0.003} = 0.05$

New test: $SE = \sqrt{0.95 \times \dfrac{(1-0.95)}{30}} = \sqrt{0.0016} = 0.04$

95% CONFIDENCE INTERVAL:

Current Test	New Test
0.90 − (1.96 × 0.05) to 0.90 + (1.96 × 0.05)	0.95 − (1.96 × 0.04) to 0.95 + (1.96 × 0.04)
0.90 − 0.10 to 0.90 + 0.10	0.95 − 0.08 to 0.95 + 0.08
0.80 to 1.00	0.87 to 1.03

CONCLUSION:
The new test would be acceptable (validated) at the 95% CI.

---

standard error (SE) and sensitivity of the tests must be known (Box 62-2).

The standard error can be calculated using the following formula:

$$SE = \sqrt{p \times \frac{(1-p)}{n}}$$

where   p   = sensitivity of the test
where   n   = sample size
where $\sqrt{}$ = square root

The example in Box 62-2 shows data collected by a laboratory that is validating a new test in parallel with the test being used; the 95% CI has been calculated for each test using the following formula:

$$p = (1.96 \times SE) \text{ to } p + (1.96 \times SE)$$

where   p   = sensitivity of the test
           1.96   = normal distribution value for 95% CI and is found in statistical tables; this number will change if the CI changes (e.g., 99% rather than 95%).

COST ACCOUNTING[2]

The preceding section on selection of diagnostic tests explains how a manager selects a test based on scientific accuracy. Once the test is selected, the cost must be determined. Cost accounting is the process by which managers determine what it costs to do a test. Cost is the amount of money spent for supplies (consumables), labor, and overhead. The amount of money related directly to test performance, supplies, and labor is also called **direct costs. Overhead,** also called **indirect costs,** applies to items necessary to run the laboratory or hospital facility, which are not test specific. Because overhead represents actual dollars spent to run the laboratory, however, a portion of it must be included in the price the laboratory charges for each test. Examples of indirect costs or overhead include the following:

- Labor to supervise performance of a test
- Quality control necessary to ensure test accuracy
- Maintenance and repairs to equipment or the physical facility
- Service contracts on equipment
- Equipment lease or rental costs
- Continuing education programs
- Travel to professional meetings
- Utilities (electricity, water, phone, vacuum, gases)
- Building security (alarm system, on-site guards)
- PT program
- Insurance, including fire, general, and professional liability; workers' compensation and medical or disability
- Property taxes
- Fees for licenses or certification
- Subscriptions to professional journals or books.

Depreciation, another indirect cost, is the amount of money the Internal Revenue Service (IRS) allows the institution to deduct over the useful life (usually 5 to 7 years) of the **capital equipment** (instruments) it buys. Depreciation takes into account that over time the equipment will need to be replaced because of wear, obsolescence, or because of procedural changes. Finally,

profit is the amount of money made per test that exceeds the total direct and indirect costs.

In calculating how much to charge for a test, the microbiology laboratory manager must consider the cost of supplies and labor. Supplies are usually placed on **bid** to several vendors so that the best price is obtained; tax and freight costs must be included, if applicable. Labor costs are most accurately determined by performing studies, that is, timing a technologist using a stopwatch as he or she performs all parts of the testing. All the steps performed *by the microbiology laboratory* must be considered in pricing the labor component. This includes the three segments of each clinical test, that is, preanalytic, analytic, and postanalytic.

The **preanalytic** stage involves specimen collection, accessioning, and preparation of a workcard. The **analytic** stage involves all steps in performing the test. The **postanalytic** stage involves result reporting, telephoning positive results, supervisory review of the written report, and delivery of the written report. The contribution of **fringe benefits** (paid time off such as holiday, vacation, and sick leave; medical insurance; employer contribution to a pension plan) must also be considered. Typically, benefits add an additional 25% to 30% to the employee's base salary. Realistically, in today's fast-paced laboratory, it is almost impossible for the manager to tally up all labor costs. Therefore, because labor is usually shown to represent 40% to 60% of the total cost of a test, many individuals estimate the cost of labor by doubling the cost of supplies. Box 62-3 illustrates how to determine the cost of a test, in this case a herpes culture. In the example, the cost of a positive and negative test are the same, although in most instances positive tests cost more than negative tests because of the additional costs associated with biochemical or immunologic confirmation.

The cost of the test with a 5% profit should be $33.46. If the laboratory runs a courier service to pick up specimens, this cost should also be added into the price of each test. Courier costs per test can be estimated by averaging the courier's salary over a 6- or 12-month period and then dividing by the total number of tests during that period.

REIMBURSEMENT AND CODING

The OIG compliance program requires the microbiologist to be a **gatekeeper**, that is, to assess the need for complex testing and to direct the clinician to the least expensive method to diagnose a patient's illness. It is also necessary for microbiologists to participate fully with their business office in charging the testing to third party payers, including Medicare and Medicaid. While inpatient reimbursement is covered under a lump-sum payment by **diagnosis-related groupings** based on

BOX 62-3	Pricing a Rapid Herpes Simplex Virus Culture

1. Cost of Supplies

Transport and Collection

Collection kit plastic bag	$ 0.17
Transport medium	1.28
Dual pack Dacron swabs (male and female)	0.34
Requisition	0.19
Collection instructions	0.03
	$2.01

Processing

Workcard	$ 0.03
Sheet of specimen labels	0.13
Cryovial	0.33
2 MRC5 shell vials	2.51
1 mL viral refeed medium	0.17
2 mL acetone	0.08
2 mL PBS (wash)	0.26
75-µL anti-HSV1 FITC monoclonal antibody	2.51
75-µL anti-HSV2 FITC monoclonal antibody	2.51
6 Pasteur pipettes	0.36
1 glass slide	0.09
	$ 8.98
TOTAL COST OF SUPPLIES	**$ 10.99**

Note: Cost of running QC is not included here because it is included as part of overhead.

2. Cost of Labor

Labor cost, estimated to be equal to the total cost of supplies	$ 10.99

3. Total Direct Cost of Test

Supplies + labor	$ 21.98

4. Overhead (see text)

Overhead typically runs between 40% and 50% of tests costs.
In this example, we shall estimate overhead to be 45% of total test cost.

Total direct cost × overhead rate ($21.98 × 45%)	$ 9.89
TOTAL TEST COST WITHOUT PROFIT	**$ 31.87**

5. Profit

Profit is determined by the institution and usually is between 3% and 10%.
In this example, we shall calculate a 5% profit.

Total text cost × profit rate ($31.87 × 5%)	$ 1.59
TOTAL TEST COST WITH PROFIT	**$ 33.46**

the reason for a specific hospitalization, outpatient reimbursement is based on a prospective payment system in which the patient's testing is coded using the CPT-4 system and the medical necessity of the test is justified using ICD9-CM codes. Table 62-2 indicates how to code a positive herpes culture. The ICD9-CM code, the reason for performing the test, indicates a genital lesion in this case. The clinician determines the ICD9-CM code. Medicare requires that its charge be the same as the lowest amount charged to the laboratory's most favored customer so as a general rule of thumb, charges should never be lower than what Medicare will reimburse in the schedules published annually. Compliance with all

Table 62-2 **Charging a Herpes Virus Culture**

Procedure	CPT Code	ICD9-CM	National Limitation Amount
Herpes virus culture (HSV1 and HSV2)	87254*	054.10 (herpes genitalis)	27.32 × 2 (HSV1 and HSV2) = 54.64

*Virus isolation, centrifuged enhanced (shell vial) technique, includes identification with immunofluorescence stain, each virus

billing rules is critical to assure ethical charging. Microbiologists responsible for this aspect of laboratory testing should take advantage of symposia and audio conferences on reimbursement and coding sponsored by professional organizations including the American Society for Microbiology.

BUDGETING

Budgeting for a laboratory is just like budgeting personal expenses. The manager looks at each line item, such as labor, supplies, travel, and subscriptions, in the previous year's ledger and estimates expenses for the next year. The laboratory is both a cost and revenue center. A **cost center** spends money to perform patient testing. A **revenue center** brings in money as it bills for these patient tests. The fiscally sound laboratory generates more dollars in revenue than it spends.

The institution's business office usually posts (writes) expenses to the **general ledger** once a month. The general ledger is the accounting record of the laboratory's expenditures and is taken from actual payroll figures and expenses. This ledger also includes revenues posted from tests billed during the month. Many budget reports have a column that shows the amount the microbiology laboratory manager budgeted for a particular line item in addition to the actual amount spent. The budget report may also show a column indicating the variance, or difference, between the budgeted and actual costs.

Figure 62-2 is a typical monthly budget for a small microbiology laboratory in a rural hospital. The laboratory anticipates revenues of $135,000 and expenses of $129,650 for a **gross profit** (before taxes) of $5350. Figure 62-3 shows the income statement for April 2005. The laboratory shows a gross profit of only $935 even though it expected to make $5350. The budget variance analysis (Figure 62-4) shows where the differences are between the estimated budget figures and the actual income and expenses. Numbers in parentheses indicate more money was spent than was budgeted. In this example, the laboratory made more money than expected, that is, $752. It also spent more money on salaries, laboratory supplies, and printing; however, this was offset by the fact that less money was spent on outside laboratory services, repairs and maintenance, and

Revenue	
Test income (inpatients)	$118,000
Test income (outpatients)	17,000
Total Revenue	**$135,000**

Expenses	
Salaries	$64,000
Payroll taxes	5600
Health insurance	4000
Workers' compensation insurance	1300
Employee medical testing	200
Laboratory supplies	34,000
Office supplies	1000
Printing	1000
Shipping	800
Outside laboratory services	1500
Licenses and permits	100
Proficiency tests	200
Rent	3100
Equipment lease/rental	250
Repairs and maintenance	2500
Telephone	1500
Utilities	2500
Travel	2000
Professional and general liability insurance	2000
Depreciation	2100
Total Expenses	**$129,650**

Figure 62-2 Monthly budget for a small hospital.

travel. The overall effect on the "bottom line" was the small gross profit. At this point, the prudent manager would check out the reasons for increased salaries and laboratory supplies; for example, increased overtime may be due to the fact that someone was on maternity leave or purchase of extra supplies may have been necessary to initiate a special project.

INVENTORY CONTROL

One of the best ways to control costs is to keep inventory at a level at which supplies neither run out nor stay on the shelf long enough to reach their expiration date. Most laboratories have lists of all reagents, media, and disposable supplies (pipettes, paper towels, slides) that they use. Past usage helps managers to establish minimum and maximum amounts to keep on hand. An

Revenue	
Test income (inpatients)	$118,562
Test income (outpatients)	17,190
Total Revenue	**$135,752**

Expenses	
Salaries	$67,945
Payroll taxes	5959
Health insurance	4231
Workers' compensation insurance	1322
Employee medical testing	56
Laboratory supplies	37,343
Office supplies	600
Printing	2509
Shipping	967
Outside laboratory services	308
Licenses and permits	125
Proficiency tests	0
Rent	3100
Equipment lease/rental	250
Repairs and maintenance	1716
Telephone	1709
Utilities	1954
Travel	680
Professional and general liability insurance	1935
Depreciation	2108
Total Expenses	**$134,817**

Figure 62-3 Monthly income statement for a small hospital.

inventory card system is commonly used to keep track of the status of all supplies. This card includes the following: (1) name of the item, (2) name of the vendor (supplier), (3) catalog number of the item, (4) packaging quantity (standard unit) of the item (package, gross), (5) cost per unit, (6) minimum and maximum volumes to keep on hand, (7) actual amount of the product on hand and the date the inventory was counted, (8) date and amount ordered and the **purchase order (PO)** number used (except in large facilities, where purchasing may be handled by a separate materials management department), (9) date and amount received, (10) amount **backordered,** and (11) date the backorder is filled. A backordered product is one that is not currently in stock. If this product will not be released soon, the manager may choose to make arrangements to purchase a substitute product. Some microbiology computer software packages also contain an inventory component.

A physical inventory is performed (i.e., stock is counted) weekly or monthly. Expired media or reagents are usually discarded at this time, and stock is rotated so that the product with the earliest expiration date is in front and used first. Sometimes, products are put on **standing order;** standing orders are products that arrive at a predetermined interval. For example, in a virology laboratory, tissue culture cells used for growing viruses arrive once a week as prearranged with the vendor. Standing orders are usually set up for 6- or 12-month intervals before reordering is necessary.

INTERVIEWING AND HIRING EMPLOYEES

Hiring good employees is one of the hardest jobs a manager faces. Interviewing prospective candidates is an art; managers must quickly determine not only whether the individual has a working knowledge of microbiology but also whether he or she has a good work ethic and will get along with other employees. The most important points to remember during an interview are to put the applicant at ease and not to ask personal questions. For example, the manager should not ask individuals about plans to marry or have children. Neither should the manager ask whether someone is divorced, separated, or cohabiting; whether a spouse is likely to be transferred; the applicant's religion or age; whether the applicant has been dishonorably discharged from the military; the applicant's sexual orientation; or whether the applicant has ever been arrested.

The interviewer should not ask about health problems unless a specific illness would preclude a candidate from safely performing his or her job. For example, a person who has received an organ transplant and is receiving immunosuppressive therapy would not be a good candidate to work with infectious materials in a microbiology laboratory. The manager must not discriminate against people with disabilities if reasonable accommodations can mean that they would be qualified for the job. An example of a reasonable accommodation would be to install a special telephone for a hearing-impaired microbiologist who would then be able to communicate with clinicians. Persons with mental and physical disabilities are protected under the Americans with Disabilities Act.

Before starting an interview, the manager should obtain the job description of the position for which the candidate is applying. The applicant should be asked if he or she can perform all functions outlined. It is permissible to ask why the candidate is leaving (or has left) his or her previous position, and about his or her career goals. Questions should be directed toward helping to determine whether the applicant has the experience, training, and interests necessary to perform the job. The interviewer should fully describe the job's functions and how it relates to the overall functioning of the laboratory; he or she should also describe any special requirements such as rotation to the second or third shifts, phlebotomy duties, or weekend and holiday rotations. The applicant should be allowed to ask questions, and the interview should only be closed when the manager is satisfied that enough information was gathered to evaluate how the applicant would perform.

	Budget	Actual	Variance*
Revenue			
Test income (inpatients)	$118,000	$118,562	$562
Test income (outpatients)	17,000	17,190	$190
Total Revenue	$135,000	$135,752	$752
Expenses	$64,000	$67,945	($3945)
Salaries	5600	5959	($359)
Payroll taxes	4000	4231	($231)
Workers' compensation insurance	1300	1322	($22)
Employee medical testing	200	56	$144
Laboratory supplies	34,000	37,343	($3343)
Office supplies	1000	600	$400
Printing	1000	2509	($1509)
Shipping	800	967	($167)
Outside laboratory services	1500	308	$1192
Licenses and permits	100	125	($25)
Proficiency tests	200	0	$200
Rent	3100	3100	$0
Equipment lease/rental	250	250	$0
Repairs and maintenance	2500	1716	$784
Telephone	1500	1709	($209)
Utilities	2500	1954	$546
Travel	2000	680	$1320
Professional and general liability insurance	2000	1935	$65
Depreciation	2100	2108	($8)
Total Expenses	$129,650	$134,817	($5167)

*Numbers in parentheses indicate more money was spent than budgeted.

Figure 62-4 Monthly budget variance analysis for a small hospital based on data presented in Figures 62-2 and 62-3.

EMPLOYEE JOB PERFORMANCE STANDARDS AND APPRAISALS

All employees should know what they are supposed to do, and every laboratory should have written job descriptions to provide to them. Standards of performance should be developed so that the employee knows how he or she will be judged when rated on job performance **(job appraisal).** Annual merit raises are often linked to job performance, and employees should get feedback throughout the year on how they are doing so that they have an opportunity to correct poor performance.

PERSONNEL RECORDS

Personnel records should be well organized and contain all the information required by regulatory bodies. This includes the following:

- Name of employee
- Application for employment, including:
  - Colleges or universities attended
  - Specialized laboratory training
  - Laboratory certification, if applicable
  - Previous experience

- Job description, signed by employee
- Documentation of credential review by the laboratory director and level under CLIA '88 personnel standards
- Record of job training at present job
- Record of continuing education
- Competency review (which shows that the employee meets minimum standards for performing tasks for which he or she is qualified)
- Performance appraisals (evaluation)
- Record of safety training
- Records of medical testing, for example, annual PPDs should be kept in a separate locked file to ensure their confidentiality.

ORGANIZATION OF THE MICROBIOLOGY LABORATORY

The organization of the microbiology laboratory is dependent on the site (public health hospital, POL, or independent laboratory) and the complexity of testing. However, some general guidelines may be applied to any situation, and they are discussed in the following section.

DIVISION OF WORK

Clinical microbiology comprises essentially eight sub-specialty areas: aerobic and anaerobic bacteriology, mycology, mycobacteriology (AFB), parasitology, virology, serology, and molecular diagnostics (PCR and other DNA probe technology). Cross-training (educating microbiologists in more than one of these areas) is now common practice in many laboratories. Because many of the disciplines require long, specialized training to demonstrate proficiency, rotations in mycology, AFB, parasitology, virology, serology, or molecular diagnostics may span as much as 6 months to 1 year. The same may also be the case in the anaerobe laboratory if the volume is high and special identification methods are used. On the other hand, aerobic bacteriology is usually broken down in one of two ways. Many laboratories separate specimens by source (e.g., blood, respiratory, stool, urine, and exudate [or miscellaneous] cultures.) The advantage of this system is that microbiologists working up the cultures only have to remember rules for that particular type of culture and work usually proceeds faster and more efficiently. Other laboratories split aerobic clinical specimens using the first letter of the patient's last name. In this system, one technologist works up all aerobic cultures on a single patient. The advantage of this system is that a single person has an overview of the results of all specimens on one patient. For example, if Mrs. Smith has a urine, wound, and stool culture, the same microbiologist would work up all three. In large laboratories, specimen setup and accessioning is usually handled by someone other than the one who will work up the culture. In smaller laboratories, the same individual may handle specimen setup and workup, although many times one technologist reads the 24-hour cultures and another reads the biochemicals and susceptibility tests. Specimens for AFB, virology, parasitology, serology, and molecular diagnostics are usually batched and set up once a day. The reason for this is that the processing protocol can be lengthy and can involve setting up positive and negative controls with each run, so it is not cost effective to set up specimens one at a time. Clinical bacteriology and mycology specimens are usually set up as they arrive unless specimens have been submitted in transport media that will allow them to be batched (see Chapter 5).

DESIGN OF LABORATORY HANDBOOK FOR CLINICAL STAFF

It is the microbiologist's responsibility to provide a laboratory handbook for the clinical practitioners or to make any applicable information available electronically on a computer screen. This handbook should be present on all hospital nursing units or given out to clients (by an independent laboratory). The handbook should list all the tests included in the test menu, with an individual description of how to collect and transport all specimen types. Handbooks are usually written in tabular form and include the following information:

- Test names
- CPT-4 test code number (which corresponds to numbers assigned by insurance carriers, including Medicare and Medicaid) based on standardized codes developed by the American Medical Association and updated annually
- Internal (billing) test code number if different from the CPT code
- The appropriate specimen to submit (e.g., thick and thin blood smears for malaria, nasopharyngeal swab for pertussis)
- The minimum specimen requirements (e.g., 1 mL serum, 0.5 mL cerebrospinal fluid)
- The methodology used (e.g., enzyme-linked immunosorbent assay [ELISA], complement fixation [CF], direct fluorescent antibody [DFA], culture, DNA probe)
- Appropriate container or transport medium (e.g., formalin and PVA [polyvinyl alcohol] for ova and parasite [O&P] examination, modified Cary-Blair medium for stool culture)
- Special collection instructions, including patient preparation if applicable (e.g., no O&P examination for 10 days after a barium enema, collect 3 to 5 early morning sputum specimens for AFB)
- Reference ranges, if applicable (e.g., titer greater than 1:8 for cryptococcal antigen test is a positive test)
- A comment section indicating turn-around times or other pertinent information such as whether testing is batched or sent to a reference laboratory

Tests are usually listed in alphabetic order for easy reference and cross-referenced, if possible (e.g., *Brucella* culture may be listed under *Brucella* culture, blood culture, and bone marrow culture). The microbiologist should use the handbook as a method of communicating laboratory information.

DESIGN OF LABORATORY REQUISITION FORM

The laboratory requisition form should be designed for ease of use and should be OIG compliant. It commonly includes several parts. The laboratory's name, address, and telephone and fax numbers, as well as its CLIA number and other applicable accreditation or licensure numbers, should be displayed. For reference laboratories, there should be a place for the client's name, address, and telephone number. Patient demographic information (full name, birth date or age, sex) and ordering physician are included as a minimum. Patient information is often stamped onto the requisition using a plastic card similar to a credit card; in the hospital, unique patient identification numbers are also stamped on the cards. Information about the source or type of specimen and date and time of collection is also required. Individuals who set up specimens in the laboratory should always check this information carefully to verify that the specimen has been received in a timely manner and is in an appropriate transport medium. Furthermore, because some specimens are unacceptable for particular types of cultures (e.g., vaginal specimens for anaerobic culture), checking the source of specimen against the test information is important. Certain medical information (current antimicrobial therapy, immunization history, and clinical syndrome or suspect agent) can be critical to guiding specimen setup or workup, and a space for it should be included on the requisition. The requisition often includes a list of commonly ordered tests so that clinical personnel can simply check off the tests they want to order. Finally, billing information should include the patient's address, telephone number, social security number, insurance information, the ordering physician's UPIN (unique person identification number), and the appropriate ICD-9-CM number (diagnosis code).

With the addition of computers, many laboratories have instituted electronic ordering. In many cases, a series of video screens are used to prompt the clinicians to enter the same information that can also be collected on sheets of paper. Whatever the format, however, the basic information requested is the same.

DESIGN OF LABORATORY WORKCARD

A workcard is the legal document that can be used to reconstruct the testing process. The record of the work performed on a specimen may be handwritten on the back of the laboratory requisition, or on a workcard designed especially for this purpose. Alternatively, the technologist can record the details of the specimen workup electronically, that is, directly into the computer; this is called a "paperless workcard."

The workcard should include the following information:

- Patient name
- Specimen source
- Laboratory number
- Date and time inoculated
- Initials of microbiologist setting up or working up the culture
- All notes made by the microbiologist during specimen workup, including test results

- Record of any telephone calls or faxes to clinicians
- Instrument printouts, if applicable; printouts on thermal paper should be copied onto xerographic paper, since thermal paper disintegrates over time

DESIGN OF LABORATORY REPORT FORM

The laboratory report form is the written means of communicating patient information to the clinician. It is a confidential record about the patient. The report should contain the following information:

- Name, address, and telephone number of laboratory performing test
- Laboratory accreditation numbers, as applicable (including CLIA, CAP [College of American Pathologists], Medicare and Medicaid, and state licensure)
- Patient name and other demographic information
- Specimen source
- Laboratory number
- Ordering physician
- Test ordered (e.g., aerobic culture, herpes culture)
- Date ordered
- Date and time inoculated
- Initials of microbiologist who set up specimen
- Type of report (direct examination, preliminary report, final report)
- Test results
- Initials of microbiologist performing the test and date of report
- Initials of supervisor reviewing report and date

The report should be legible if handwritten. Computer-generated reports should be laid out in an easy-to-read format so that the clinician can readily retrieve information.

WRITING A PROCEDURE MANUAL

The laboratory's procedure manual (also known as the SOPM [Standard Operating Procedure Manual]) is a compilation of all the tests performed by the microbiologist. It should be written in a step-by-step format. The validation of a well-written SOPM is that a microbiologist from another facility is able to read it and perform any procedure done by your facility. SOPMs are standardized in the sense that they are all written in a format outlined by the Clinical and Laboratory Standards Institute (CLSI) (formerly the National Committee for Clinical Laboratory Standards [NCCLS]).

Individual procedures should contain the following sections:

General Hospital

ANTIBIOTIC SUSCEPTIBILITY TABLES
JANUARY – DECEMBER 2005
Department of Pathology - Microbiology/Immunology

Table 1. Activity of selected antibiotics against gram-positive cocci

Organism (number tested)		Ampicillin	Ceftriaxone	Erythromycin[b]	Gentamicin	Levofloxacin	Oxacillin[c]	Penicillin	Tetracycline	TMP/SMX	Vancomycin[e]
MIC (μg/ml)		(MIC ≤ 8)	(MIC ≤ 1)		(MIC ≤ 4)	(MIC ≤ 2)	(MIC ≤ 0.06)	(MIC ≤ 2)	(MIC ≤ 2/38)		
Staphylococcus aureus	(1703)			42	98		54		95	97	100
Coag-negative staphylococci	(691)			26	75		28		60		100
Enterococcus species	(822)	76									76
Streptococcus pneumoniae	(144)	84[a]		72		99		60[d]	83		100

a 79 total isolates tested; 0% resistant; 16% intermediately resistant (MIC=2 μg/ml). **Note:** CLSI non-meningitis interpretive guidelines.

b Gram-positive bacteria resistant to erythromycin are also resistant to azithromycin and clarithromycin. For Staphylococci, susceptible if MIC ≤ 0.5 μg/ml; for *Streptococcus pneumoniae*, susceptible if MIC ≤ 0.25 μg/ml.

c Staphylococci resistant to oxacillin (methicillin) are also resistant to nafcillin, penicillin, ampicillin, cefazolin, cefoxitin, ceftriaxone, imipenem and all other beta-lactam antibiotics.

d 17% resistant (MIC ≥ 2 μg/ml), 23% intermediately resistant (MIC 0.12-1 μg/ml).

e For staphylococci and enterococci, susceptible if MIC ≤ 4 μg/ml; for *Streptococcus pneumoniae*, susceptible if MIC ≤ 1 μg/ml.

Table 2. Activity of selected antibiotics against gram-negative bacilli

Organism (number tested)		Ampicillin	Cefazolin	Cefepime		Ceftriaxone	Ciprofloxacin	Gentamicin	Imipenem	Levofloxacin	Piperacillin/Tazobactem	TMP/SMX
MIC (μg/ml)		(MIC ≤ 8)	(MIC ≤ 8)	(MIC ≤ 8)	(n)	(MIC ≤ 8)	(MIC ≤ 1)	(MIC ≤ 4)	(MIC ≤ 4)	(MIC ≤ 2)	(MIC ≤ 16/4)	(MIC ≤ 2/38)
Acinetobacter species	(115)			(100)	64		52	68	86	52	68	72
Citrobacter koseri (diversus)	(65)	0	97	(53)	98	98	100	100	100	100	100	100
Citrobacter freundii complex	(52)	4	2	(43)	98	83	77	88	100	77	88	85
Enterobacter aerogenes	(100)	0	1	(87)	100	92*	99	100	100	99	88	99
Enterobacter cloacae	(200)	0	0	(170)	99	85*	90	94	100	90	84	88
Escherichia coli	(2148)	63	93	(1750)	100	99	87	94	100	87	98	81
Klebsiella oxytoca	(66)	0	61	(55)	96	97	97	97	100	97	97	94
Klebsiella pneumoniae	(509)	0	97	(413)	99	99	97	98	100	97	95	95
Morganella morganii	(43)	2	0	(37)	100	86	60	86	98	60	95	65
Proteus mirabilis[a]	(411)	88	88	(349)	99	100	72	98	100	72	97	88
Pseudomonas aeruginosa	(374)			(291)	84			77	83	94	78[b]	94[c]
Serratia marcescens	(97)	0	0	(78)	100	99	99	99	100	99	97	99

* Use of cephalosporins not recommended for *Enterobacter* species because resistance develops rapidly. Imipenem, a quinolone, or TMP/SMX recommended.

a Other *Proteus* species are more resistant (similar to *Morganella*).

b For *Pseudomonas aeruginosa*, increase dosage to 750 mg daily.

c For *Pseudomonas aeruginosa*, susceptible if MIC ≤ 64/4 μg/ml; higher dose is necessary, 4.5 gm q6 for normal renal function; dose adjustment may be necessary for renal failure.

Data collected by the Clinical Microbiology Laboratory, Department of Pathology.

Figure 62-5   Annual cumulative antibiogram.

- Title (name of procedure)
- Principle (reason for performing the test)
- Preferred specimen patient preparation (if required)
- Transport container (need for anticoagulant, preservative, or holding medium)
- Transportation conditions (wet ice, room temperature)
- Specimen storage in laboratory (room temperature, 4° C, –20° C, –70° C)
- Criteria for unacceptable specimen (delay in transport, leaking container, presence of barium)
- Special safety precautions (tape plates for AFB or brucellae)
- Reagents or media required and incubation conditions
- Examination of cultures
- Guidelines for identification and susceptibility testing by culture type (respiratory, urine, blood, stool)
- Required quality control
- Methods for reporting positive, negative, and unsatisfactory results
- Technical notes, including possible sources of error and helpful hints
- References, including manufacturer's package inserts, textbooks, CLSI procedures, and research papers

The laboratory director is responsible for ensuring that a current SOPM is present, and he or she must review it at least annually. Each microbiologist should be required to review the procedures applicable to the work he or she is performing; many laboratories have their technologists and technicians initial the procedure when they review it. When procedures are updated, old procedures should be maintained for at least 2 years. Manufacturer's instrument manuals that conform to specifications of CLSI publication GP2-A4 are acceptable to both CLIA and CAP accreditation programs. However, CAP does not accept product package inserts or manufacturer's technical manuals in place of an individually written laboratory procedure.

PRODUCTION OF STATISTICAL REPORTS

One of the microbiologist's most important jobs is to publish a cumulative susceptibility report that tracks susceptibility or resistance of commonly isolated organisms to commonly administered antimicrobials in a particular hospital or geographic area. This report is called an **antibiogram.** Figure 62-5 illustrates an example of such a report. Cumulative antibiograms should be distributed to the medical staff so that physicians can plan empiric therapy. Empiric therapy (see Chapter 5) is basically therapy begun before actual culture and susceptibility data are available; it is based on the clinician's best guess as to the most probable organism causing the infection and the drug most likely to cure it. Many microbiology laboratories include the cost of drugs, provided by the pharmacy, along with the antibiogram, so clinicians will pick the most cost-effective antimicrobial.

From time to time, microbiologists are also asked to provide statistical information on the percentage of positive cultures (also known as positivity rate) for comparison with other laboratories or validation of testing methodologies. For example, no more than 3% to 5% of AFB cultures should be contaminated with normal upper respiratory flora to validate the decontamination procedure used in any facility. Quality assurance testing based on analysis of positive and unsatisfactory cultures is discussed in more detail in Chapter 63.

REFERENCES

1. American Institute of Architects Academy of Architecture for Health and the Facilities Guidelines Institute: *Guidelines for design and construction of hospital and healthcare facilities,* Washington, DC, 2001, American Institute of Architects.
2. Garcia LS, editor: *Clinical laboratory management,* Washington, DC, 2004, ASM Press.

ADDITIONAL READING

Barker JH: *Laboratory facilities planning and design,* Atlanta, 1985, Centers for Disease Control.

Barker JH, Blank CH, Steere NV: *Designing a laboratory,* Washington, DC, 1989, American Public Health Association.

Bickford G: Decentralized testing in the '90s, *Clin Lab Management Rev* 8:327, 1993.

College of American Pathologists: *Manual for laboratory planning and design,* Skokie, Ill, 1985, College of American Pathologists.

Elder BL, Hansen SA, Kellogg JA, et al: Verification and validation of procedures in the clinical microbiology laboratory. In McCurdy BW, coordinating editor: *Cumitech 31,* Washington, DC, 1997, American Society for Microbiology.

Elin RJ, Robertson EA, Sever G: Workload, space, and personnel of microbiology laboratories in teaching hospitals, *Am J Clin Pathol* 82:78, 1984.

Federal Register: Clinical Laboratories Improvement Act of 1967, *Federal Register* 33:15297, 1968.

Health Care Financing Administration: Clinical laboratory improvement amendments of 1988; final rule, *Federal Register* 57:7137, 1992.

Health Care Financing Administration: Medicare, Medicaid, and CLIA programs; revision of the clinical laboratory regulations for the Medicare, Medicaid, and Clinical Laboratories Improvement Act of 1967 programs, *Federal Register* 53:29590, 1988.

Ilstrup DM: Statistical method in microbiology, *Clin Microbiol Rev* 3:219, 1990.

McPherson BS, Needham CA: Method evaluation and test selection. In Wentworth BB, editor: *Diagnostic procedures for bacterial infections,* ed 7, Washington, DC, 1987, American Public Health Association.

National Committee for Clinical Laboratory Standards: Inventory control systems for laboratory supplies; Approved guideline GP6-A, Wayne, Pa, 1994, National Committee for Clinical Laboratory Standards.

National Committee for Clinical Laboratory Standards: Training verification for laboratory personnel, Approved guideline GP21-A, Wayne, Pa, 1995, National Committee for Clinical Laboratory Standards.

National Committee for Clinical Laboratory Standards: Cost accounting in the clinical laboratory; Approved guideline GP11-A, Wayne, Pa, 1998, National Committee for Clinical Laboratory Standards.

National Committee for Clinical Laboratory Standards: Laboratory design; Approved guideline GP18-A, Wayne, Pa, 1998, National Committee for Clinical Laboratory Standards.

National Committee for Clinical Laboratory Standards: Clinical laboratory technical procedure manuals, ed 4; Approved guideline GP2-A4, 2002, Wayne, Pa, National Committee for Clinical Laboratory Standards.

National Committee for Clinical Laboratory Standards: Analysis and presentation of cumulative antimicrobial susceptibility test data; Approved guideline M39-A, Wayne, Pa, 2002, National Committee for Clinical Laboratory Standards.

Stuart J and Hicks JM: Good laboratory management: an Anglo-American perspective, *J Clin Pathol* 44:793, 1991.

US Department of Health and Human Services: Model compliance plan for clinical laboratories, January 24, Office of the Inspector General, Washington, DC, 1997, US Department of Health and Human Services.

Washington JA: Confidence intervals: an important component of data presentations, *Clin Microbiol Newsl* 12:109, 1990.

Weissfeld AS, Baselski VS: Procedure coding, reimbursement, and billing compliance. In Isenberg HD, editor: *Clinical microbiology procedures handbook*, ed 2, Washington, DC, 2004, ASM Press.

CHAPTER 63

Quality in the Clinical Microbiology Laboratory

The issue of laboratory quality has evolved over more than three decades since the first recommendations for quality control (QC) were published in 1965. Just as microbial taxonomy has changed over the years, the approach to quality has evolved as well. QC is now seen as only one part of a total laboratory quality program. Quality now also includes total quality management (TQM), continuous quality improvement (CQI) or performance improvement (PI), and quality assurance (QA). TQM, CQI, and PI are umbrella terms, encompassing the whole institution's quality program. TQM evolved as an activity to improve patient care by having the laboratory monitor its work to detect deficiencies and subsequently correct them. CQI and PI went a step further by seeking to improve patient care by placing the emphasis on not making mistakes in the first place; CQI and PI advocate continuous training to guard against having to correct deficiencies. QC is now associated with the internal activities that ensure diagnostic test accuracy. QA is associated with those external activities that ensure positive patient outcomes. Positive patient outcomes in the microbiology laboratory are:

- Reduced length of stay
- Reduced cost of stay
- Reduced turn-around time for diagnosis of infection
- Change to appropriate antimicrobial therapy
- Customer (physician or patient) satisfaction

CQI or PI through well thought-out programs of QC and QA are part of the requirements for laboratory accreditation under CLIA '88 (see Chapter 62).

QC PROGRAM

The laboratory director is primarily responsible for the QC and QA programs. However, all laboratory personnel must actively participate in both programs. Federal guidelines (CLIA '88) are considered minimum standards and are superseded by higher standards imposed by states or private certifying agencies such as the College of American Pathologists (CAP), the Joint Commission on the Accreditation of Health Care Organizations (JCAHO), or the Commission of Laboratory Accreditation (COLA). The basic elements of a QC program are described in the following sections.

SPECIMEN COLLECTION AND TRANSPORT

The laboratory is responsible for providing instructions for the proper collection and transport of specimens. These instructions should be available to the clinical staff for use when specimens are collected. The written collection instructions should include:

- Test purpose and limitations
- Patient selection criteria
- Timing of specimen collection (e.g., before antimicrobials are administered)
- Optimal specimen collection sites
- Approved specimen collection methods
- Specimen transport medium
- Specimen transport time and temperature
- Specimen holding instructions if it cannot be transported immediately (e.g., hold at 4° C for 24 hours)
- Availability of test (on-site or sent to reference laboratory)
- Hours test performed (daily or batched)
- Turn-around time
- Result reporting procedures

The collection instructions should include information on how the requisition should be filled out, and the laboratory must include a statement indicating that the requisition must be filled out entirely. In addition to standard information, such as patient name, hospital or laboratory number, and ordering physician, other critical information includes (1) whether the patient is receiving antimicrobial therapy, (2) suspect agent or syndrome, and (3) immunization history (if applicable). The laboratory should also establish criteria for unacceptable specimens. Examples of unacceptable specimens include the following:

- Unlabeled or mislabeled specimens—these specimens should be identified by the collector or be recollected
- Use of improper transport medium such as stool for ova and parasites not submitted in preservative(s)
- Use of improper swab such as use of wooden shaft or calcium alginate tip for viruses
- Excessive transport time
- Improper temperature during transport or storage
- Improper collection site for test requested such as stool for respiratory syncytial virus

- Specimen leakage out of transport container
- Sera that are excessively hemolyzed, lipemic, or contaminated with bacteria

Sometimes, even though the specimen is not acceptable, the physician asks that it be processed anyway. If this happens, a disclaimer should be put on the final report, indicating that the specimen was not collected properly and that the results should be interpreted with caution.

STANDARD OPERATING PROCEDURE MANUAL (SOPM)

The requirement for a Standard Operating Procedure Manual (SOPM) is considered part of the QC program. The SOPM, discussed in detail in Chapter 62, should define test performance, tolerance limits, reagent preparation, required quality control, result reporting, and references. The SOPM should be written in Clinical and Laboratory Standards Institute (CLSI), formerly the National Committee for Clinical Laboratory Standards (NCCLS) format and must be reviewed and signed annually by the microbiology laboratory supervisor or director; in addition, all changes must be approved and dated by the laboratory director. The SOPM should be available in the work areas. It is the definitive laboratory reference and is used often for questions relating to individual tests. Any obsolete procedure should be dated when removed from the SOPM and retained for at least 2 years.

PERSONNEL

It is the laboratory director's responsibility to employ sufficient qualified personnel for the volume and complexity of the work performed. For example, published studies regarding staffing of virology laboratories suggest one technologist per 500 to 1000 specimens per year. Technical on-the-job training must be documented, and the employee's competency must be assessed twice each year. Figure 63-1 is an example of a personnel competency assessment. Continuing education programs should be provided, and documentation of attendance should be maintained in the employee's personnel file.

REFERENCE LABORATORIES

A laboratory should use only accredited or licensed reference laboratories. The referral laboratory's name, address, and licensure numbers should be included on the patient's final report.

PATIENT REPORTS

The laboratory should establish a system for supervisory review of all laboratory reports. This review should involve checking the specimen workup to verify that the correct conclusions were drawn and no clerical errors were made in reporting results. Reports should be given only to individuals authorized by law to receive them (physicians and various mid-level practitioners). Clinicians should be notified about "panic values" immediately. Panic values are potential life-threatening results, for example, positive Gram stain for cerebrospinal fluid (CSF) or a positive blood culture. Normal ranges are included on the report where appropriate, for example, serology tests. All patient records should be maintained for at least 2 years. In reality, records should be maintained for at least 10 years because they may be needed to support medical necessity in the event of a post-payment billing audit by the Centers for Medicare and Medicaid Services.

PROFICIENCY TESTING (PT)

Laboratories are required to participate in an external proficiency testing (PT) program for each analyte (test) for which a program is available; the laboratory must maintain an average score of 80% to maintain licensure in any subspecialty area. The federal government no longer maintains a PT program, but some states, such as New York, as well as several private accrediting agencies, such as CAP and the American Association of Bioanalysts (AAB) send out "blind unknowns." These unknowns are to be treated exactly as patient specimens, from accessioning into the laboratory computer or manual logbook through workup and reporting of results. The testing personnel and laboratory director are required to sign a statement when the PT is completed attesting to the fact that the specimen was handled exactly like a patient specimen. In this way, PT specimens serve to establish the accuracy and reproducibility of a laboratory's day-to-day performance. The laboratory's procedures, reagents, equipment, and personnel are all checked in the process. Furthermore, errors on PT help point out deficiencies, and the subsequent education of the staff can lead to overall improvements in laboratory quality. When grades (evaluations) come back, critiques (summaries) accompanying them should be discussed with the entire technical staff. Evidence of corrective action in the event of problems should be documented, including changes in procedures, retraining of personnel, or the purchase of alternative media and reagents.

Some laboratories have a system of internal PT in addition to those received from external agencies. When

Employee's Name: _____   Bench: _____

Evaluator's Name: _____   Date of Evaluation: _____

Numerical Parameters

5 = No errors for each task

4 = 1 error for each task

3 = 2 errors for each task

2 = 3 errors for each task

1 = 4 errors for each task

N/A = not applicable to individual's job description

Appearance

	Neatness					Legibility				
Workcards	1	2	3	4	5	1	2	3	4	5
Logbook	1	2	3	4	5	1	2	3	4	5
Reports	1	2	3	4	5	1	2	3	4	5

Specimen Setup and Transport

						N/A
Were specimens transported at correct temperature?	1	2	3	4	5	N/A
Were specimens preserved correctly for transport?	1	2	3	4	5	N/A
Was client inventory maintained current and at appropriate levels?	1	2	3	4	5	N/A
Were specimens received in timely manner?	1	2	3	4	5	N/A
Were specimen requests date stamped?	1	2	3	4	5	N/A
Did specimen name labels match request form?	1	2	3	4	5	N/A
Were specimens logged correctly?	1	2	3	4	5	N/A
Were shared, stat or special instructions noted on specimens label?	1	2	3	4	5	N/A
Were safety precautions for transportation and check-in of specimens followed?	1	2	3	4	5	N/A
Were all tests set up as ordered?	1	2	3	4	5	N/A
Were specimens inoculated onto appropriate media?	1	2	3	4	5	N/A
Were inoculated specimens incubated appropriately?	1	2	3	4	5	N/A
Were left over specimens saved appropriately?	1	2	3	4	5	N/A

Specimen Workup

						N/A
Were workcards easy to follow?	1	2	3	4	5	N/A
Were the correct biochemicals set up?	1	2	3	4	5	N/A
Did decision-making process yield proper results and follow accepted protocol?	1	2	3	4	5	N/A
Was each day's work dated?	1	2	3	4	5	N/A
Were all biochemical results recorded?	1	2	3	4	5	N/A
Were biochemicals inoculated properly?	1	2	3	4	5	N/A
Were isolates frozen when appropriate?	1	2	3	4	5	N/A
Was all work completed in a timely manner?	1	2	3	4	5	N/A
Were specimen stains read correctly?	1	2	3	4	5	N/A
Were specimens overworked?	1	2	3	4	5	N/A
Were safe work practices observed?	1	2	3	4	5	N/A
Were susceptibilities performed in a timely manner?	1	2	3	4	5	N/A
Were stat requests done in a timely manner?	1	2	3	4	5	N/A

Figure 63-1 Employee competency evaluation.

Quality Control						N/A
Were check plates done when appropriate?	1	2	3	4	5	N/A
Were check plates mixed?	1	2	3	4	5	N/A
Were necessary controls performed?	1	2	3	4	5	N/A
Were periodic QC controls performed?	1	2	3	4	5	N/A

Reporting						N/A
Were verbal reports given and documented?	1	2	3	4	5	N/A
Were final reports correct?	1	2	3	4	5	N/A
Were reports given in a timely manner?	1	2	3	4	5	N/A
Were correct susceptibilities reported?	1	2	3	4	5	N/A
Were all reports logged out?	1	2	3	4	5	N/A
Was supervisory review appropriate?	1	2	3	4	5	N/A

Conclusion						
Overall performance of employee at this bench (average of all tasks evaluated):	1	2	3	4	5	N/A
Is retraining necessary?		Yes			No	

NOTE: Overall satisfactory performance does not mean retraining is not necessary. Each task is evaluated individually for retraining purposes. Retraining is required on each task with a score less than 3.

Additional Comments:

Acknowledgment

Employee's signature: _____   Date: _____

Evaluator's signature: _____   Date: _____

Follow-up

Retraining completed.

Employee's signature: _____   Date: _____

Evaluator's signature: _____   Date: _____

Figure 63-1—cont'd

external audit is not available for a particular test method, laboratories are required to set up an internal program to revalidate the test at least semiannually. Internal PT samples can be set up by (1) seeding a simulated specimen and labeling it as an autopsy specimen so that no one panics if a pathogen is recovered, (2) splitting a routine specimen for workup by two different technologists, or (3) sending part of a specimen to a reference laboratory to confirm the laboratory's result.

PERFORMANCE CHECKS

INSTRUMENTS

Equipment logs should contain the following information:

- Instrument name, serial number, and date put in use
- Procedure and periodicity (daily, weekly, monthly, quarterly) for routine function checks
- Acceptable performance ranges
- Instrument function failures, including specific details of steps taken to correct the problems (**corrective action**)
- Date and time of service requests and response
- Date of routine preventive maintenance (PM), which should follow manufacturer's recommendations

Maintenance records should be retained in the laboratory for the life of the instrument. Specific guidelines regarding periodicity of testing for autoclaves, biological safety cabinets, centrifuges, incubators, microscopes, refrigerators, freezers, water baths, heat blocks, and other microbiology laboratory equipment can be found in a number of the references listed at the end of this chapter.

COMMERCIALLY PREPARED MEDIA EXEMPT FROM QC

The CSLI Subcommittee on Media Quality Control collected data over several years regarding the incidence of QC failures of commonly used microbiology media. Based on its findings, the subcommittee published a list of media that did not require retesting in the user's laboratory if purchased from a manufacturer who follows CSLI guidelines. The list of media is published in CSLI Standard M22-A3. The laboratory must inspect each shipment for cracked media or Petri dishes, hemolysis, freezing, unequal filling, excessive bubbles, clarity, and visible contamination. The manufacturer must supply written assurance that CSLI standards were followed; this verification must be maintained along with the laboratory's QC protocol.

USER-PREPARED AND NONEXEMPT, COMMERCIALLY PREPARED MEDIA

QC forms for user-prepared media should contain the amount prepared, the source of each ingredient, the lot number, the sterilization method, the preparation date, the expiration date (usually 1 month for agar plates and 6 months for tubed media), and the name of the preparer. Both user-prepared and nonexempt, commercially prepared media should be checked for proper color, consistency, depth, smoothness, hemolysis, excessive bubbles, and contamination. A representative sample of the lot should be tested for sterility; 5% of any lot is tested when a batch of 100 or fewer units is received, and a maximum of 10 units are tested in larger batches. A batch is any one shipment of a product with the same lot number; if a separate shipment of the same lot number of product is received, then it is considered a different batch and needs to be tested separately.

Sterility is routinely checked by incubating the medium for 48 hours at the temperature at which it will be used. Both user-prepared and nonexempt, commercially prepared media should also be tested with QC organisms of known physiologic and biochemical properties. Tables listing specific organisms to test for various media can be found in a number of the references listed at the end of this chapter.

ANTIMICROBIAL SUSCEPTIBILITY TESTS

The goal of quality control testing of antimicrobial susceptibility tests (ASTs) is to ensure the precision and accuracy of the supplies and microbiologists performing the test. Criteria regarding frequency of testing are the same regardless of the methodology, such as minimum inhibitory concentration (MIC) broth dilution or Kirby-Bauer (see Chapter 12). Each new shipment of microdilution trays or Mueller-Hinton plates should be tested with CSLI-approved American Type Culture Collection (ATCC [Rockville, Md]) reference strains.

Reference strains for MIC testing are selected for genetic stability and give MICs that are in the midrange of each antimicrobial agent tested. Reference strains for Kirby-Bauer testing have clearly defined mean diameters of their respective zone of inhibition for each antimicrobial tested. ATCC numbers of reference strains are different for various AST methods. Moreover, quality control MICs and zone diameters are updated annually by the CSLI Subcommittee on Antimicrobial Susceptibility Testing, so new tables should be obtained from the CSLI regularly.

Each susceptibility test system must also be tested with use (usually daily) for 20 consecutive days. If three or fewer MICs or zone of inhibition diameters per drug-reference strain combination are outside the reference range during the 20-day testing period, laboratories may switch to weekly QC testing. Thereafter, aberrant results obtained during the weekly testing must be vigorously investigated. If a source of error, such as contamination, incorrect reference strain used, or incorrect atmosphere of incubation, is found, quality control testing may simply be repeated. However, if no source of error is uncovered, 5 consecutive days of retesting must be performed. If accuracy and precision are again acceptable, weekly QC testing may resume; if the problem drug/organism combination(s) are still outside the reference ranges, 20 days of consecutive testing must be reinitiated before weekly testing can be reinstated. Under no circumstances should any drug/bug combination be reported for a patient isolate if QC testing has failed.

STAINS AND REAGENTS

Containers of stains and reagents should be labeled as to contents, concentration, storage requirements, date prepared (or received), date placed in service (commonly called the date opened), expiration date, source (commercial manufacturer or user prepared), and lot number. All stains and reagents should be stored according to manufacturer's recommendations and tested with positive and negative controls before use. Tables listing specific organisms to test for various stains or reagents can be found in a number of the references at the end of this chapter. Outdated materials or reagents that fail QC even after retesting with fresh organisms should be discarded immediately. Patient specimens should not be tested using the lot number in question until the problem is solved; in the case of a repeat failure, an alternative method should be used or the patient's specimen should be sent to a reference laboratory.

ANTISERA

The lot number, date received, condition received, and expiration date must be recorded for all shipments of antisera. In addition, the antisera should be dated when opened. New lots must be tested concurrently with previous lots, and testing must include positive and negative controls. Periodicity of testing thereafter should follow the requirements of agencies that inspect an individual laboratory (including the Centers for Medicare and Medicaid Services) and may include, with use, monthly or semiannual checks.

KITS

Kits that have been approved by the U.S. Food and Drug Administration need to be tested as specified in the manufacturer's package insert. Each shipment of kits must be tested even if it is the same lot number as a previously tested lot, because temperature changes during shipment may affect the performance.

MAINTENANCE OF QC RECORDS

All QC results should be recorded on an appropriate QC form. Corrective action should be noted on this form. If temperature is adjusted or a biochemical test is repeated, the new readings within the tolerance limits should be listed. In many laboratories, the supervisor reviews and initials all forms weekly and the director then reviews each one monthly. QC records should be maintained for at least 2 years except those on equipment, which must be saved for the life of each instrument.

MAINTENANCE OF REFERENCE QC STOCKS

Stock organisms may be obtained from the ATCC, commercial vendors, or PT programs; well-defined clinical isolates may also be used. The laboratory should have enough organisms on hand to cover the full range of testing of all necessary materials such as media, kits, and reagents.

BACTERIOLOGY

Nonfastidious (rapidly growing), aerobic bacterial organisms can be saved up to 1 year on trypticase soy agar (TSA) slants. Long-term storage (less than 1 year) of aerobes or anaerobes can be accomplished either by lyophilization (freeze drying) or freezing at −70° C. Frozen, nonfastidious organisms should be thawed, reisolated, and refrozen every 5 years; fastidious organisms should be thawed, reisolated, and refrozen every 3 years. Stock isolates may be maintained by freezing them in 10% skim milk, trypticase soy broth (TSB) with 15% glycerol, or 10% horse blood in sterile, screw-cap vials.

MYCOLOGY

Yeasts may be treated as nonfastidious bacterial organisms for maintaining stock cultures. Molds can be stored on potato dextrose agar (PDA) slants at 4° C for 6 months to 1 year. For longer-term storage, PDA slants may be overlaid with sterile mineral oil and stored at room temperature. Alternatively, sterile water can be added to an actively sporulating culture on PDA, the conidia (spores) can be teased apart to dislodge them from the agar surface and the water can then be dispensed to sterile, screw-top vials. These vials should be capped tightly and stored at room temperature.

MYCOBACTERIOLOGY

Acid-fast bacilli (AFB) may be kept on Lowenstein-Jenson (LJ) agar slants at 4° C for up to 1 year. They may also be frozen at –70° C in 7H9 broth with glycerol.

VIROLOGY

Viruses may be stored indefinitely at –70° C in a solution containing a cryoprotectant, such as 10% dimethyl sulfoxide (DMSO) or fetal bovine serum.

PARASITOLOGY

Slides and photographs must be available for QC purposes. Trichrome and other permanent slides may be purchased from commercial vendors. Clinical slides may be preserved indefinitely by adding a drop of Permount and a coverslip.

QA PROGRAM

Because QA is the method by which the overall process of infectious disease diagnosis is reviewed, any of the steps involved in the diagnosis of an infectious disease may be studied. These steps include:

- Preanalytic
  - Ordering of test by the clinician
  - Processing of test request by the clerical staff
  - Collection of specimen by nursing personnel or patient
  - Transport of specimen to the laboratory
  - Initial processing of specimen in laboratory, including specimen accessioning
- Analytic
  - Examination and workup of culture by the microbiologist
  - Interpretation of specimen results by the microbiologist
- Postanalytic
  - Formulation of a written report by the microbiologist
  - Communication of the microbiologist's conclusions to the clinician in written format
  - Interpretation of report by the clinician
  - Institution of appropriate therapy by the clinician

Analytic testing (the work actually done in the microbiology laboratory) is now seen as only one part of a continuing spectrum of steps that begins when the physician orders the test and ends when he or she receives the results and treats the patient.

QA audits (studies) are planned and conducted by examining the three stages of testing. The goal is to look at the proficiency with which the patient is served by the whole facility, including the laboratory. The outcome is to look at the consequences to the patient of the work that has been performed. QA audits involve the analysis of how the system works and how it can be improved.

Q-PROBES

A process is selected for audit in a number of ways. One is to subscribe to the Q-Probes program, which is a national interlaboratory QA program developed and administered by CAP. CAP selects topics to be audited and provides instructions and worksheets for collection of data as well as data entry forms. Data are collected for a specified period and then returned to CAP for analysis. CAP returns a summary of the institution's performance as well as a comparison with other facilities of similar size and scope of service. That way, an individual facility can compare its results with those of its peers, a process called **benchmarking.** Q-Probes are designed for all areas of laboratory medicine so that every one will not be directed toward microbiology. Since inception of the program, microbiology Q-Probe audits have included areas such as (1) blood culture utilization, (2) nosocomial (hospital-acquired) infections, (3) cumulative susceptibility results, (4) antibiotic usage, (5) turn-around time of Gram stains of CSF, (6) viral hepatitis test utilization, (7) laboratory diagnosis of tuberculosis, (8) blood culture contamination rates, (9) appropriateness of the ordering of stools for microbiology testing, and (10) sputum quality. Several generic (laboratory-wide) audits are also applicable to the microbiology laboratory, including error reporting, quality of reference laboratories, and effects of laboratory computer down time.

IN-HOUSE QA AUDITS

A facility that does not subscribe to the Q-Probe program may select topics for audits through suggestions from the medical, nursing, or pharmacy staffs; complaints from the medical or nursing staff; or deficiency or observation noted in the laboratory.

Physicians may suggest an audit to measure the transcription accuracy of their orders by nursing unit clerical personnel. Nursing administrators may suggest an audit of contaminated urine cultures to access the compliance of the nursing staff in instructing patients about proper urine culture collection techniques. Pharmacists may notice improper antibiotic utilization by the clinical staff, for example, a patient was not placed on the appropriate therapy after the pathogen was reported or the patient remains on antibiotic therapy to which his or her organism is resistant after the susceptibility report

Complaint Identification

Date:	Facility:
Time	Contact:
Recipient:	Phone:

Notes

Corrective Action

Effect on Patient Care

Follow-up

Reviewed by:_____   Date: _____

Figure 63-2   Client complaint form.

Problem

Steps Taken to Solve the Problem

Corrective Action Taken to Avoid This Problem in Future

Effect on Patient Care

Follow-up

Reviewed by:_____   Date: _____

Figure 63-3   Laboratory problem report.

BACKGROUND:
Following a complaint regarding turn-around time for STAT RSV direct antigen tests one winter, the microbiology laboratory at General Hospital has decided to audit its turn-around time. The medical staff indicated that it would like to turn the test around in 2.5 hours (150 minutes) from the time of collection to the time the physician is notified; the medical staff feels that this will insure the maximum patient benefits.

STUDY DESIGN:
All RSV requests for direct antigen testing were evaluated for a 3-month period to determine if this turn-around time was being met by laboratory personnel.

RESULTS:

	Reports Given in <150 Min		Report Time Exceeding 150 Min		Combined Averages	
Month	# Specimens	Average time	# Specimens	Average time	# Specimens	Average time
December	57	114 min	15	195 min	72	130 min
January	114	108 min	14	179 min	128	116 min
February	70	114 min	3	165 min	73	116 min

ANALYSIS:
Two hundred and seventy-three reports were reviewed. The average reporting time for the 3-month period was under the acceptable 150 minutes. In December and January 15 and 14 specimens, respectively, had turn-around times that exceeded 150 minutes, with an average of 195 minutes in December and 179 minutes in January; February had 3 reports exceeding 150 minutes.

CONCLUSIONS:
The overall (combined) average reporting time, while remaining within 150 minutes, could be improved. There was a dramatic drop in February after the medical staff complaint. This was undoubtedly a result of in-services given to courier and clerical staff regarding the need to transport and accession the stat specimens quickly.

RECOMMENDATIONS:
Because hospital-wide systems have been improved, appropriate follow-up would be to audit STAT turn-around time for another test, for example, Gram stain of CSF, in 3 to 6 months to verify that they also meet the 150-minute turn-around time requirement.

has been charted. Complaints from the medical or nursing staff can involve failure of the laboratory to conduct all the tests requested on the requisition, performance of the wrong test, or prolonged delay in turn-around time of test results. All complaints to the laboratory must be documented; Figure 63-2 shows a form designed for this purpose. Corrective action and follow-up with the laboratory, medical, and nursing staffs must also be documented.

Deficiencies or problems in the laboratory should also be documented (Figure 63-3). If, for example, the laboratory notices a dramatic rise in the number of positive respiratory syncytial virus (RSV) direct antigen tests in the summer (not RSV season) and the problem is traced back to a quality control problem that was not recognized by a new employee, a QA audit might be indicated to study the outcome of the patients, including inappropriate treatment for RSV and failure to institute treatment for the true causative agent. Alternatively, microbiologists may notice they are receiving many ova and parasite (O&P) examinations and stool cultures on patients hospitalized for more than 3 days. Because current cost containment guidelines suggest that this is inappropriate, the microbiology laboratory personnel could undertake a study to determine the percentage of positive results and the number of patients who tested positive for *Clostridium difficile* cytotoxin, which is the more likely cause of diarrhea in patients hospitalized for more than 3 days. If the audit showed that none of the stool cultures or O&P examinations were positive and

no stools were analyzed for *C. difficile* cytotoxin, then these findings would be presented to the medical staff. Some months following the medical staff in-service, the number of stool culture and O&P requests on patients hospitalized longer than 3 days would be reevaluated. It is hoped this would result in a dramatic decrease in numbers of inappropriate specimens.

CONDUCTING A QA AUDIT

Box 63-1 is an example of how an in-house QA audit may be conducted.

CONTINUOUS DAILY MONITORING

Daily activities of microbiologists and supervisory personnel ensure that patients get the best quality care. These activities include (1) comparing results of morphotypes seen on direct examinations with what grows on the culture to ensure that all organisms have been recovered, (2) checking antimicrobial susceptibility reports to verify that profiles match those expected from a particular species, and (3) studying culture and susceptibility reports for clusters of patients with unusual infections or multiple-drug–resistant organisms. These and many other processes result in continual improvement to all test systems that will ultimately benefit the patient.

ADDITIONAL READING

Anderson NL, Noble MA, Weissfeld AS, et al: Quality systems in the clinical microbiology laboratory. In Sewell DL, coordinating editor: *Cumitech 3B,* Washington, DC, 2005, ASM Press.

Bartlett RC, Mazens-Sullivan M, Tetreault JZ, et al: Evolving approaches to management of quality in clinical microbiology, *Clin Microbiol Rev* 7:55, 1994.

Clinical and Laboratory Standards Institute: Methods for dilution antimicrobial susceptibility tests for bacteria that grow aerobically; Approved standard M7-A7, Wayne, Pa, 2006, Clinical and Laboratory Standards Institute.

Clinical and Laboratory Standards Institute: Performance standards for antimicrobial disk susceptibility tests; Approved standard M2-A9, Wayne, Pa, 2006, Committee for Clinical and Laboratory Standards Institute.

International Organization for Standardization: Medical laboratories—particular requirements for quality and competence; ISO 15189:2003(E), Geneva, Switzerland, 2003, International Organization for Standardization.

Jenkins SG: Quality assurance, quality control, laboratory records, and water quality. In Isenberg HD, editor: *Clinical microbiology procedure handbook,* ed 2, Washington, DC, 2004, ASM Press.

LaRocco ML: Quality and productivity in the microbiology laboratory: continuous quality improvement, *Clin Microbiol Newsletter* 17:129, 1995.

National Committee for Clinical Laboratory Standards: Application of a quality management system model for laboratory services; Approved guideline GP26-A3, Wayne, Pa, 2004, National Committee for Clinical Laboratory Standards.

National Committee for Clinical Laboratory Standards: Assessment of laboratory tests when proficiency testing is not available; Approved guideline GP29-A, Wayne, Pa, 2002, National Committee for Clinical Laboratory Standards.

National Committee for Clinical Laboratory Standards: Clinical laboratory technical procedure manuals; Approved guideline GP2-A3, Wayne, Pa, 1996, National Committee for Clinical Laboratory Standards.

National Committee for Clinical Laboratory Standards: Continuous quality improvement: integrating five key quality system components; Approved guideline, GP22-A2, Wayne Pa, 2004, National Committee for Clinical Laboratory Standards.

National Committee for Clinical Laboratory Standards: Development of in vitro susceptibility testing criteria and quality control parameters; Approved guideline M23-A2, Wayne, Pa, 2001, National Committee for Clinical Laboratory Standards.

National Committee for Clinical Laboratory Standards: Methods for antimicrobial susceptibility testing of anaerobic bacteria; Approved standard M11-A6, Wayne, Pa, 2004, National Committee for Clinical Laboratory Standards.

National Committee for Clinical Laboratory Standards: Quality assurance for commercially prepared microbiological culture media; Approved standard M22-A3, Wayne, Pa, 2004, National Committee for Clinical Laboratory Standards.

National Committee for Clinical Laboratory Standards: Quality control of microbiological transport systems; Approved standard M40-A, Wayne, Pa, 2003, National Committee for Clinical Laboratory Standards.

National Committee for Clinical Laboratory Standards: Selecting and evaluating a referral laboratory; Approved guideline GP9-A, Wayne, Pa, 1998, National Committee for Clinical Laboratory Standards.

National Committee for Clinical Laboratory Standards: Training and competence assessment; Approved guideline GP21-A2, Wayne, Pa, 2004, National Committee for Clinical Laboratory Standards.

National Committee for Clinical Laboratory Standards: Using proficiency testing (PT) to improve the clinical laboratory; Approved guideline GP27-A, Wayne, Pa, 1999, National Committee for Clinical Laboratory Standards.

Warford A: Quality assurance in clinical virology. In Specter S, Hodinka RL, Young SA, editors: *Clinical virology manual,* ed 3, Washington, DC, 2000, ASM Press.

Every year, between 1.75 and 3 million (5% to 10%) of the 35 million patients admitted to acute-care hospitals in the United States acquire an infection that was neither present nor in the prodromal (incubation) stage when they entered the hospital. These infections are called **nosocomial,** or hospital-acquired, infections. Treatment of nosocomial infections is estimated to add between $4.5 and $15 billion annually to the cost of health care and represents an enormous economic problem in today's environment of cost containment. In addition, many of these infections lead to the death of hospitalized patients (patient mortality) or, at minimum, additional complications (patient morbidity) and antimicrobial chemotherapy.

Some of the earliest efforts to control infection followed the recognition in the nineteenth century that women were dying in childbirth from bloodstream infections caused by group A *Streptococcus (Streptococcus pyogenes)* because physicians were spreading the organism by not washing their hands between examinations of different patients. Hand washing is still the cornerstone of modern infection control programs. Moreover, the first recommendations for isolation precautions in U.S. hospitals were published in the late 1800s when guidelines appeared advocating placement of patients with infectious diseases in separate hospital facilities. By the late 1950s, the advent of nosocomial infections caused by *Staphylococcus aureus* finally ushered in the modern age of infection control. In the past four decades, we have learned that, in addition to hospitalized patients acquiring infections, health care workers are also at risk of acquiring infections from patients. Thus, present-day infection control programs have evolved to prevent the acquisition of infection by patients and caregivers.

INCIDENCE OF NOSOCOMIAL INFECTIONS

The Centers for Disease Control and Prevention (CDC) has established the **National Nosocomial Infections Surveillance** (NNIS) program to monitor the incidence of nosocomial infections in the United States. Regardless of a hospital's size or medical school affiliation, the rates of infections at each body site are consistent across institutions. The most common nosocomial infections

are urinary tract infections (33%), followed by pneumonia (15%), surgical site infections (15%), and bloodstream infections (13%). A companion CDC program, **Study of the Efficacy of Nosocomial Infection Control** (SENIC), keeps statistics on morbidity and mortality of hospital-acquired infections. Each nosocomial infection adds 5 to 10 days to the affected patient's hospital stay. Of individuals with hospital-acquired bloodstream or lung infections, 40% to 60% die each year. Likewise, patients with indwelling (Foley) catheters have a threefold increased chance of dying from urosepsis, a bloodstream infection that is a complication of a urinary tract infection, than those who do not have one.

Attack rates vary according to the type of hospital. Large, tertiary-care hospitals that treat the most seriously ill patients often have higher rates of nosocomial infection than do small, acute-care community hospitals; large medical school–affiliated (teaching) hospitals have higher infection rates than do small teaching hospitals. This difference in the risk of infection is probably related to several factors, including the severity of illness, the frequency of invasive diagnostic and therapeutic procedures, and variation in the effectiveness of infection control programs. Within hospitals, the surgical and medical services have the highest rates of infection; the pediatric and nursery services have the lowest. Moreover, within services, the predominant type of infections varies, that is, surgical site infections are the most common on the surgical service whereas urinary tract or bloodstream infections are the most common on medical services or in the nursery.

TYPES OF NOSOCOMIAL INFECTIONS

The majority of nosocomial infections are endogenous in origin, that is, they involve the patient's own microbial flora. Three principal factors determine the likelihood that a given patient will acquire a nosocomial infection:

- Susceptibility of the patient to the infection
- The virulence of the infecting organism
- The nature of the patient's exposure to the infecting organism

In general, hospitalized individuals have increased susceptibility to infection. Corticosteroids, cancer chemotherapeutic agents, and antimicrobial agents all contribute to the likelihood of nosocomial infection by suppressing the immune system or altering the host's normal flora to that of resistant (hospital) microbes. Likewise, foreign objects, such as urinary or intravenous catheters, break the body's natural barriers to infection. Nonetheless, these medications or devices are necessary to cure the patient's primary medical condition. Finally, exerting influence over the virulence of the pathogens is not possible because it is not possible to immunize patients against nosocomial infections. Patients with serious community-acquired infections are frequently admitted to the hospital, and the disease may spread nosocomially by either direct contact; contact with contaminated food, water, medications, or medical devices (fomites); or by airborne transmission. Thus, nosocomial infections may never be completely eliminated, only controlled.

URINARY TRACT INFECTIONS

Gram-negative rods cause the majority of hospital-acquired urinary tract infections, and *Escherichia coli* is the number one organism involved. Gram-positive organisms, *Candida* spp., and other fungi cause the remainder of the infections. The risk factors that predispose patients to acquire a nosocomial urinary tract infection include advanced age, female gender, severe underlying disease, and the placement of indwelling urinary catheters.

LUNG INFECTIONS

The most common nosocomial pathogens causing pneumonia are gram-negative rods, *S. aureus,* and *Moraxella catarrhalis. Streptococcus pneumoniae* and *Haemophilus influenzae,* which cause the majority of community-acquired pneumonias, are not important etiologic agents in hospital-acquired infections except very early during the hospital course (first 2 to 5 days); these infections probably represent infections that were already incubating at the time of the hospital admission. The risk factors that predispose patients to acquire a nosocomial lung infection include (1) advanced age, (2) chronic lung disease, (3) large-volume aspiration (the microorganisms in the upper respiratory tract are coughed up and lodge in the lungs instead of being spit out or swallowed), (4) chest surgery, (5) monitoring intracranial pressure (in which a catheter inserted through the skull measures the amount of fluid on the brain), (6) hospitalization in intensive care units, and (7) intubation (placement of a breathing tube down a patient's throat) or attachment to a mechanical ventilator (which controls breathing).

SURGICAL SITE INFECTIONS

Approximately 4% of surgical patients develop surgical site infections; 50% of these infections develop after the patient has left the hospital so this number may be an underestimate. Gram-positive organisms (*S. aureus,* coagulase-negative staphylococci and enterococci) cause the majority of these infections, followed by gram-negative rods and *Candida* spp. The risk factors that predispose patients to acquire a nosocomial wound infection include (1) advanced age, (2) obesity, (3) infection at a remote site (that spreads through the bloodstream), (4) malnutrition, (5) diabetes, (6) extended preoperative hospital stay, (7) greater than 12 hours between preoperative shaving of site and surgery, (8) extended time of surgery, and (9) inappropriate timing of prophylactic antibiotics (given to prevent common infections before they seed the surgical site). Surgical wounds are classified as either clean, clean-contaminated, contaminated, or dirty depending on the number of contaminating organisms at the site. Bowel surgery is considered dirty, for example, whereas surgery for a total hip replacement is considered clean.

BLOODSTREAM INFECTIONS

The overall rate of nosocomial bloodstream infections increased in all NNIS hospitals between 1980 and 1989 when the incidence of infections with coagulase-negative staphylococci, enterococci, *S. aureus,* and *Candida* spp. increased. The risk factors that predispose patients to acquire a nosocomial bloodstream infection include (1) age 1 year of age or younger or 60 years of age and older, (2) malnutrition, (3) immunosuppressive chemotherapy, (4) loss of skin integrity (e.g., burn or decubiti [bedsore]), (5) severe underlying illness, (6) indwelling device (e.g., catheter), (7) intensive care unit stay, and (8) prolonged hospital stay.

EMERGENCE OF ANTIBIOTIC-RESISTANT MICROORGANISMS

The organisms that cause nosocomial infections have changed over the years because of selective pressures from the use (and overuse) of antibiotics (see Chapter 11). Risk factors for the acquisition of highly resistant organisms include prolonged hospitalization and prior treatment with antibiotics. In the preantibiotic era, most hospital-acquired infections were caused by *S. pneumoniae* and group A *Streptococcus (Streptococcus pyogenes).* In the 1940s and 1950s, with the advent of treatment of patients with penicillin and sulfonamides, resistant strains of *S. aureus* appeared. Then, in the 1970s, treatment of patients with narrow-spectrum cephalosporins and aminoglycosides led to the emergence of resistant aerobic gram-negative rods, such as *Klebsiella, Enterobacter, Serratia,* and *Pseudomonas.* During the late 1970s and early 1980s, the use of more potent cephalosporins played a role in the emergence of antibiotic-resistant, coagulase-negative staphylococci, enterococci, methicillin-resistant *S. aureus* (MRSA), and *Candida* spp. The 1990s witnessed the emergence of beta-lactamase–producing, high-level gentamicin-resistant, and vancomycin-resistant enterococci (VRE). The twenty-first century has seen the emergence of vancomycin-resistant *Staphylococcus aureus*; preliminary work at the CDC suggests that, at least in one case, the plasmid coding for resistance to vancomycin was transferred from vancomycin-resistant *Enterococcus.*[1]

Patients' normal flora changes quickly after hospitalization from viridans streptococci, saprophytic *Neisseria* spp., and diphtheroids to potentially resistant microorganisms found in the hospital environment. Then, their colonized nares, skin, gastrointestinal tract, or genitourinary tract can serve as reservoirs for endogenously acquired infections. Moreover, if patients colonized with resistant microorganisms return to nursing homes in the community harboring these organisms, they can also transfer them to other patients. This further increases the pool of patients who harbor multidrug-resistant organisms when they, in turn, are hospitalized. These new patients recontaminate the hospital environment and serve as potential reservoirs for spread to additional patients.

HOSPITAL INFECTION CONTROL PROGRAMS

Hospital infection control programs are designed to detect and monitor hospital-acquired infections and to prevent or control their spread. The Infection Control Committee is multidisciplinary and should include the microbiologist, the infection control practitioner (often a nurse with special training), the hospital epidemiologist (usually an infectious disease physician), and a pharmacist. The infection control practitioner collects and analyzes surveillance data, monitors patient care practices, and participates in epidemiologic investigations. Daily review of charts of patients with fever or positive microbiology cultures allows the infection control practitioner to recognize problems with hospital-acquired infections and to detect outbreaks as early as possible. The infection control practitioner is also responsible for education of health care providers in techniques, such as hand washing and isolation precautions, that minimize the acquisition of infections.

It is the infection control practitioner's job to identify all cases of an outbreak. The investigation of the cluster of cases during a particular outbreak involves its characterization in terms of commonalities, such as loca-

tion in the hospital (nursery, intensive care unit), same caregiver, or prior respiratory or physical therapy. Risk factors, including underlying diseases, current or prior antimicrobial therapy, and placement of a Foley catheter, are also assessed. This information helps the infection control committee determine (1) the reservoir of the organism in the hospital, that is, the place where it exists, and (2) the means of transmission of the organism from its reservoir to the patient.

Microorganisms are spread in hospitals through several modes:

- **Direct contact,** for example, in contaminated food or intravenous solutions
- **Indirect contact,** for example, from patient to patient on the hands of health care workers (MRSA, rotavirus)
- **Droplet contact,** for example, inhalation of droplets (>5 μm in diameter) that cannot travel more than 3 feet (pertussis)
- **Airborne contact,** for example, inhalation of droplets (≤5 μm) that can travel large distances on air currents (tuberculosis)
- **Vector-borne contact,** for example, disease spread by vectors, such as mosquitoes (malaria) or rats (rat-bite fever); this mode of transmission is rare in hospitals in developed countries
- Once the reservoir is known, the infection control practitioner can implement control measures, such as reeducation regarding hand washing (in the case of spread by health care workers) or hyperchlorination of cooling towers in the case of nosocomial legionellosis.

ROLE OF THE MICROBIOLOGY LABORATORY

The microbiology laboratory supplies the data on organism identification and antimicrobial susceptibility profile that the infection control practitioner reviews daily for evidence of nosocomial infection. Thus, the laboratory must be able to detect potential microbial pathogens and then accurately identify them to species level and perform susceptibility testing. The microbiology laboratory should also monitor multidrug-resistant organisms by tabulating data on antimicrobial susceptibilities of common isolates and studying trends indicating emerging resistance. Significant findings should be immediately reported to the infection control practitioner. If an outbreak is identified, the laboratory works in tandem with the infection control committee by (1) saving all isolates, (2) culturing possible reservoirs (patients, personnel, or the environment), and (3) performing typing of strains to establish relatedness between isolates of the same species. Microbiology laboratories are also obligated by law to report certain isolates or syndromes to public health authorities. For example, Table 64-1 lists organisms to be reported to state health authorities in Texas. Other states have similar criteria.

CHARACTERIZING STRAINS INVOLVED IN AN OUTBREAK

The ideal system for typing microbial strains involved in outbreaks should be standardized, reproducible, sensitive, stable, readily available, inexpensive, applicable to a wide range of microorganisms, and field tested in other epidemiologic investigations. Although no such perfect system is currently available, a number of methods are used to aid in typing epidemic strains. There are two major ways to type strains using either phenotypic traits or molecular typing methods.

Classic phenotypic techniques include **biotyping** (analyzing unique biologic or biochemical characteristics), **antibiograms** (analyzing antimicrobial susceptibility patterns), and **serotyping** (serologic typing of bacterial or viral antigens, such as bacterial cell wall [O] antigens). **Bacteriocin typing,** which examines an organism's susceptibility to bacterial peptides (proteins), and **bacteriophage typing,** which examines the ability of bacteriophages (viruses capable of infecting and lysing bacterial cells) to attack certain strains, have been useful for typing *Pseudomonas aeruginosa* and *S. aureus,* respectively; these techniques, however, are not widely available.

Genotypic, or molecular, methods have largely replaced phenotypic methods as a means of confirming the relatedness of strains involved in an outbreak. Plasmid analysis and restriction endonuclease analysis of chromosomal DNA are widely used. Plasmids are extrachromosomal pieces of genetic material (nucleic acids) that self-replicate (reproduce). Plasmids may be transferred from one bacterial cell to another by conjugation or transduction (see Chapter 2). Plasmid analysis has often been used to explain the occurrence of unusual or multiple-antibiotic resistance patterns. It has been shown that plasmids or R factors (resistance genes carried on plasmids) can cause outbreaks when a specific plasmid is transmitted from one genus of bacteria to another. Plasmid profiles, patterns created when plasmids are separated based on molecular weight by agarose gel electrophoresis, can also be used to characterize the similarity of bacterial strains. Relatedness of strains is based on the number and size of plasmids, with strains from identical sources showing identical plasmid profiles. Plasmids themselves or chromosomal DNA may also be typed by means of restriction endonuclease digestion patterns. Restriction enzymes recognize specific nucleotide sequences in DNA and produce double-stranded cleavages that break the DNA into smaller fragments. The fragments of various sizes are separated using gel electrophoresis based on molecular weight. The specific recognition sequence and cleavage site have been defined for a great many of these enzymes.

Modifications of the basic restriction endonuclease technique have been developed to reduce the number of bands generated to less than 20 in an attempt to make the gels easier to interpret. These include pulsed-field gel electrophoresis (PFGE) and hybridization of ribosomal RNA with short fragments of DNA. Plasmid restriction digests have been used to type *S. aureus* and coagulase-negative staphylococci, and PFGE is the preferred method for typing enterococci, enteric gram-negative rods, and other gram-negative rods.

Other molecular methods, such as PCR (polymerase chain reaction), are used in conjunction with these methods for strain typing. All molecular methods are discussed in more detail in Chapter 8.

PREVENTING NOSOCOMIAL INFECTIONS

The CDC published guidelines in the 1970s specifying isolation precautions in hospitals. Techniques for isolation precautions included (1) health care workers washing their hands between caring for different patients; (2) segregation of infected patients in private rooms or cohorting of patients (placing patients with the same clinical syndrome in semiprivate rooms) if private rooms are not available; (3) wearing of masks, gowns, and gloves when caring for infected patients; (4) bagging of contaminated articles, such as bed linens, when removed from the room; (5) cleaning of all isolation rooms after the patient is discharged; and (6) placement of cards on the patient's door specifying the type of isolation and instructions for visitors and nursing staff. Categories of isolation were also established and included (1) strict isolation for highly contagious diseases such as chicken pox, pneumonic plague, and Lassa fever; (2) respiratory isolation for diseases such as measles or *Haemophilus influenzae* or *Neisseria meningitidis;* (3) enteric precautions for diseases such as amebic dysentery, *Salmonella,* and *Shigella;* (4) contact isolation for patients infected with multidrug-resistant bacteria; (5) AFB (tuberculosis) isolation for persons with *M. tuberculosis;* (6) drainage and secretion precautions for persons with conjunctivitis and burns; and (7) blood and body fluid precautions for individuals with acquired immunodeficiency syndrome (AIDS). Over time, a system of disease-specific precautions was added to the category-specific ones, and hospitals were given the option of using one of the two systems. Disease-specific precautions were more cost effective, in that only those

Table 64-1   Notifiable Infectious Conditions in Texas*

Diseases to Be Reported Immediately by Telephone/Fax[†]	Diseases to Be Reported within 1 Working Day	Diseases to Be Reported within 1 Week	Diseases to Be Reported Quarterly
Anthrax	Brucellosis	Acquired immunodeficiency syndrome (AIDS)	Vancomycin-resistant *Enterococcus* (VRE)
Botulism, Food-borne	Hepatitis A (acute)	Amebiasis	Penicillin-resistant *Streptococcus pneumoniae*
Diphtheria	Q fever	Botulism, infant	
H. influenzae, type b invasive infections	Rubella (including congenital)	Campylobacteriosis	
Measles (rubeola)	Tuberculosis	Chancroid	
Meningococcal infections, invasive	Tularemia	*Chlamydia trachomatis* infections	
Pertussis	*Vibrio* infection, including cholera	Creutzfeldt-Jacob disease	
Plague		*Cryptosporidium* infections	
Poliomyelitis, acute paralytic		Cyclospora	
Rabies in humans		Dengue	
Severe acute respiratory syndrome (SARS)		Encephalitis (specify etiology)	
Smallpox		Ehrlichiosis	
Viral hemorrhagic fevers		*Escherichia coli* O157:H7	
Yellow fever		Gonorrhea	
Vancomycin-resistant *Staphylococcus aureus* (VRSA)		Hansen's disease (leprosy)	
Vancomycin-resistant coagulase-negative *Staphylococcus* spp.		Hantavirus infection Hemolytic-uremic syndrome (HUS) Hepatitis B, D, E, and unspecified (acute) Hepatitis B (chronic) identified prenatally or at delivery Hepatitis C (newly diagnosed infection) Human immunodeficiency virus (HIV) infection Legionellosis Listeriosis Lyme disease Malaria Meningitis (specify type) Mumps Relapsing fever Salmonellosis, including typhoid fever Shigellosis Spotted fever group rickettsioses Streptococcal disease, invasive (group A or B or *S. pneumoniae)* Syphilis Tetanus Trichinosis Typhus Varicella (chickenpox) Yersiniosis	

*In addition to individual case reports, any outbreak, exotic disease, or unusual group expression of disease that may be of public health concern should be reported by the most expeditious means.

[†]Report even if only suspected; waiting for confirmation may hamper public health intervention activities.

BOX 64-1	Infection Control Measures for Standard Precautions

- Health care workers (HCWs) should wash hands frequently using a plain soap except in special circumstances, for example, preoperatively or after handling dressings from patients on contact isolation.
- HCWs should wear gloves when touching blood, body fluids, secretions, excretions, and contaminated items.
- HCWs should wear a mask, gown, eye protection, or face shield as appropriate.
- Each hospital should ensure that it has adequate procedures for routine care and cleaning and disinfection of environmental surfaces, beds, bed rails, and bedside equipment.
- Hospitals should handle, transport, and launder used linen soiled with blood, body fluids, secretions, and excretions in a manner that prevents skin and mucous membrane exposure and contamination of clothing, and that avoids transfer of microorganisms to other patients or the environment.
- HCWs should take care to prevent injuries when using needles, scalpels, and other sharp instruments or devices.
- HCWs should use equipment, such as mouthpieces and resuscitation bags, instead of mouth-to-mouth resuscitation.
- HCWs should refrain from handling patient care equipment if they have exudative lesions or weeping dermatitis.
- Hospitals should place incontinent or nonhygienic patients in a private room.
- Hospitals should ensure that reusable equipment is properly sterilized.
- Hospitals should ensure that single-use items are discarded properly.

Modified from Hospital Infection Control Practices Advisory Committee, 1996.

precautions specifically necessary were used to interrupt transmission of that one disease.

In 1996 the CDC developed a new system of **Standard Precautions** synthesizing the features of universal precautions (described in Chapter 4) and body substance isolation. Standard precautions are used in the care of all patients and apply to blood; all body fluids, secretions, and excretions *except sweat,* regardless of whether they contain visible blood; nonintact skin; and mucous membranes.

In addition, **transmission-based precautions** are used for patients known (or suspected) to be infected with pathogens spread by airborne or droplet transmission or by contact with dry skin or fomites. Box 64-1 lists infection control measures for standard precautions. Table 64-2 lists the infectious agents or syndromes along with the respective infection control measures for each transmission-based precaution. Many infection control practitioners find these guidelines a lot less cumbersome to implement than the old category- and disease-specific measures. Hospitals, however, may modify these guidelines to fit their individual situations as long as their number of nosocomial infections remains low.

Some of the potential agents of bioterrorism can be transmitted person-to-person (smallpox, pneumonic plague, and viral hemorrhagic fevers) and some cannot (anthrax). The ones that can be easily transmitted have specific transmission-based precautions, that is, airborne precautions for smallpox, droplet precautions for patients with pneumonic plague, and contact precautions for individuals with one of the viral hemorrhagic fevers (Ebola, Marburg).

SURVEILLANCE CULTURES

Most routine environmental cultures in the hospital are now considered to be of little use and should not be performed unless there are specific epidemiologic reasons. The decision to perform these cultures should be determined by the microbiologist, infection control practitioner, and hospital epidemiologist. However, certain surveillance cultures are still performed as a method of limiting outbreaks. These include culturing cooling towers or hot water sources for *Legionella* spp., culture of water and dialysis fluids for hemodialysis as well as endotoxin testing, culture of blood bank products, especially platelets, and surveillance cultures for vancomycin-resistant enterococci, methicillin (or oxacillin)-resistant *S. aureus* and vancomycin-resistant *S. aureus* using rectal and oropharyngeal swabs. Physical rehabilitation centers often culture hydrotherapy equipment (whirlpools) quarterly to verify that cleaning methods are adequate; some centers culture more frequently.

Routine surveillance of air handlers, food utensils, food equipment surfaces, and respiratory therapy equipment is no longer recommended; neither is monitoring infant formulas prepared in-house nor items purchased as sterile. A better approach is for the infection control team to monitor patients for the development of nosocomial infections that might be related to the use of contaminated commercial products. In the event of an outbreak or an incident related to suspected contamination, a microbiologic study would be indicated. However, most often, such infections are actually caused by in-use contamination, rather than contamination during the manufacturing process. Suspect lots of fluid and catheter trays should be saved, and the U.S. Food and Drug Administration should be notified if contamination of an unopened product is suspected.

Although some institutions still require preemployment stool cultures and ova and parasite examinations on food handlers, most now recognize that this is of limited value. It is much more important for food handlers to submit specimens for these tests if they develop diarrhea. Similarly, most hospitals no longer screen personnel routinely for nasal carriage of *S. aureus.* Although a significant percentage of the general

Table 64-2   Transmission-Based Precautions

Type of Precaution	Specific Etiologic Agents or Syndromes	Infection Control Measure to Be Undertaken by Hospital
Airborne	Measles Varicella Tuberculosis Smallpox	1. Place patient in private room that has monitored negative air pressure, 6-12 air changes per hour, and appropriate discharge of air outdoors *or* monitored HEPA filtration of room air before air is circulated to other areas of the hospital *or* cohorting of patients, that is, placing patients with the same infection in the same room, if private rooms are not available 2. Health care workers (HCWs) to wear respiratory protection when entering room of patient with known or suspected tuberculosis and, if not immune, for patients with measles or varicella as well 3. Transport patients out of their room only after placement of a surgical mask
Droplet	Invasive *Haemophilus influenzae* type b infection, including meningitis, pneumonia, epiglottitis, and sepsis Invasive *Neisseria meningitidis* infection, including meningitis, pneumonia, and sepsis Diphtheria (pharyngeal) *Mycoplasma pneumoniae* Pertussis Pneumonic plague Streptococcal pharyngitis, pneumonia, or scarlet fever in infants and young children Adenovirus, influenza virus Mumps Parvovirus B19 Rubella	1. Place patient in private room without special air handling or ventilation *or* cohort patients 2. HCWs should wear mask when working within 3 feet of patient 3. Transfer patients out of their room only after placement of a surgical mask
Contact	Gastrointestinal, respiratory, skin, or wound infections, or colonization with multidrug-resistant bacteria *Clostridium difficile* For diapered or incontinent patients: *Escherichia coli* 0157:H7, *Shigella,* hepatitis A virus, or rotavirus Respiratory syncytial virus, parainfluenza virus, and enterovirus infections in infants and young children Skin infections such as diphtheria (cutaneous), herpes simplex virus (neonatal or mucocutaneous), impetigo, major abscesses, cellulitis, or decubiti, pediculosis (lice infestation), scabies (mite infestation), staphylococci furunculosis (boils) in infants and young children, zoster (disseminated or in the immunocompromised host) Viral hemorrhagic infections (Ebola, Lassa, or Marburg)	1. Place patient in private room without special air handling or ventilation *or* cohort patients 2. HCWs should wear gloves when entering patient's room 3. HCWs should wash hands with a special antimicrobial agent or a waterless antiseptic agent 4. HCWs should wear a mask and eye protection during activities that are likely to generate splashes of blood, body fluids, secretions, and excretions 5. HCWs should wear a gown during procedures likely to generate splashes 6. HCWs should ensure reusable equipment is properly sterilized 7. HCWs should ensure that single-use items are properly discarded

Modified from Hospital Infection Control Practices Advisory Committee, 1996.

population, including hospital personnel, are known to carry this organism, most individuals rarely shed enough organism to pose a hazard and there is no simple way to predict which nasal carriers will disseminate staphylococci.

All steam and dry-heat sterilizers and ethylene oxide gas sterilizers should be checked at least once each week with a liquid spore suspension.

Hospitals that perform bone marrow transplantation or treat hematologic malignancies may also conduct surveillance cultures of severely immunocom-

promised patients who occupy laminar flow rooms. In these instances, isolation of specific organisms may have predictive value for subsequent systemic infection. Air sampling for fungi during construction is also indicated, especially if patients are immunocompromised and are being treated near the construction site.

Recently, the US Pharmacopeia published requirements for monitoring of sterile compounding in hospital pharmacies. The laminar flow hoods, biological safety cabinets, clean rooms, and donning areas must be monitored weekly or monthly so that intravenous or intrathecal products and drugs used in the operating room are made (compounded) under sterile conditions.

REFERENCES

1. Centers for Disease Control and Prevention. Public Health Dispatch; Vancomycin-resistant *Staphylococcus aureus*. Pennsylvania, 2002, *MMWR* 51:902, 2002.

ADDITIONAL READING

Banerjee SN, Emori TG, Culver DH, et al, and the National Nosocomial Infection Surveillance System: Secular trends in nosocomial primary bloodstream infections in the United States, 1980-1989, *Am J Med* 91(suppl 3B):86S, 1991.

Centers for Disease Control and Prevention: Public health focus: surveillance, prevention and control of nosocomial infections, *MMWR* 41:783, 1992.

Craven DE, Steger KA, Barber TW: Preventing nosocomial pneumonia: state of the art and perspectives for the 1990s, *Am J Med* 91(suppl 3B):44S, 1991.

Emori TG, Gaynes RP: An overview of nosocomial infections, including the role of the microbiology laboratory, *Clin Microbiol Rev* 6:428, 1993.

Garibaldi RA, Cushing D, Lerer T: Risk factors for postoperative infection, *Am J Med* 91(suppl 3B):158S, 1991.

Garner JS, Favero MS: Guideline for hand washing and hospital environmental control, 1985, PB85-923404, Atlanta, 1985, Centers for Disease Control.

Garner JS, Simmons BP: CDC guideline for isolation precautions in hospitals, PB85-923401, Atlanta, 1983, Centers for Disease Control.

Hospital Infection Control Practices Advisory Committee: Guideline for infection control in health care personnel, *Am J Infect Control* 26:289, 1998.

Hospital Infection Control Practices Advisory Committee: Guideline for isolation precaution in hospitals, *Infect Control Hosp Epidemiol* 17:53, 1996.

Hospital Infection Control Practices Advisory Committee: Guideline for prevention of intravascular device-related infections, *Am J Infect Control* 24:262, 1996.

Hospital Infection Control Practices Advisory Committee: Guideline for prevention of nosocomial pneumonia, PB95-176970, Atlanta, 1994, Centers for Disease Control and Prevention.

Hospital Infection Control Practices Advisory Committee: Guideline for prevention of surgical site infection, *Infect Control Hosp Epidemiol* 20:247, 1999.

Hospital Infection Control Practices Advisory Committee: Recommendations for preventing the spread of vancomycin resistance, *Infect Control Hosp Epidemiol* 16:105, 1995.

Javis WR: Infection control and changing health-care deliver systems, *Emerg Infect Dis* 7:170, 2001.

Jewett JK, et al: Childbed fever: a continuing entity, *JAMA* 206:344, 1968.

McGowan JE Jr, Weinstein RA: The role of the laboratory in control of nosocomial infection. In Bennett JV, Brachman PS, editors: *Hospital infections,* ed 3, Boston, 1992, Little, Brown, & Co.

Miller JM, Bell M, editors: Epidemiologic and infection control microbiology. In Isenberg HD, editor: *Clinical microbiology procedures handbook,* ed 2, Washington, DC, 2004, ASM Press.

Nichols RL: Surgical wound infection, *Am J Med* 91(suppl 3B): 54S, 1991.

Stamm WE: Catheter-associated urinary tract infections: epidemiology, pathogenesis, and prevention, *Am J Med* 91 (suppl 3B):65S, 1991.

US Pharmacopeial Convention, Inc: Pharmaceutical compounding—sterile preparations. In *United States pharmacopeia* 27, pp 2350, Rockville, Md, 2004, US Pharmacopeial Convention, Inc.

US Pharmacopeial Convention, Inc: Pharmaceutical compounding—sterile preparations. In *United States pharmacopeia* 27, Supplement 1, pp 3121, Rockville, Md, 2004, US Pharmacopeial Convention, Inc.

Wenzel RP, editor: *Prevention and control of nosocomial infections,* ed 3, Baltimore, 1997, Williams & Wilkins.

Wong ES, Hooton TM: Guideline for prevention of catheter-associated urinary tract infections, PB84-923402, Atlanta, 1982, Centers for Disease Control.

Sentinel Laboratory Response to Bioterrorism

CHAPTER 65

GENERAL CONSIDERATIONS

The practice of clinical microbiology changed completely in October 2001. Prior to the index case of inhalation anthrax in a Florida journalist, there were only two times that microorganisms had been used to intentionally harm the civilian population in the United States.

The first incident was a large community outbreak of salmonellosis caused by intentional contamination of restaurant salad bars in 1984 in The Dalles, Oregon.[5] In this incident, a cult leader, Baghwan Sri Rajneesh, set out to influence the outcome of a local election by incapacitating voters on election day. Cultures of *Salmonella typhimurium* were grown at a laboratory at the cult's compound. Ultimately, 750 individuals fell ill; luckily there were no deaths.

The second incident was an outbreak of *Shigella dysenteriae* type 2 among laboratory workers caused by intentional contamination of muffins and donuts by a microbiology technologist in Dallas, Texas, in 1996.[3] Forty-five of the laboratory workers developed gastrenteritis; four individuals were sick enough to be hospitalized.

The third incident in October 2001[2] stunned the country. Although there had previously been sporadic instances of letters filled with a white powder delivered to an abortion clinic with a note indicating that the person opening the envelope had been exposed to anthrax spores, all previous letters had been hoaxes. The October 2001 outbreak resulted from intentional delivery of anthrax spores in mailed letters or packages; ultimately there were 11 cases of inhalational anthrax and 11 of cutaneous disease. The unknown attacker terrorized the East Coast states of Florida, New Jersey, New York, and Connecticut as well as Washington, DC. Five individuals died. The attacks were a wake-up call for institutions everywhere to develop bioterrorism readiness plans.[1] The United States government also stepped in to regulate biosecurity in laboratories and create new regulations governing security to try to prevent another incident.

BIOCRIME

A bioterrorism event (biocrime) is either overt (announced) or covert (unannounced). The letters sent to Senators Daschle and Leahy in 2001 are examples of an overt event; a note inside each envelope announced that the individual opening it had been exposed to *Bacillus anthracis* spores. The package sent to the journalist at American Media Inc. is an example of a covert event; an environmental investigation of his office uncovered the anthrax spores following his death and the illness of a coworker.

GOVERNMENT LAWS AND REGULATIONS

The bombings at the federal building in Oklahoma City and the World Trade Center led Congress to pass the Antiterrorism and Effective Death Penalty Act of 1996. Section 511 (d) restricts the possession and use of materials capable of producing catastrophic damage in the hands of terrorists by requiring their registration. A companion law, the Uniting and Strengthening America by Providing Appropriate Tools Required to Intercept and Obstruct Terrorism (USA PATRIOT) Act of 2001 prohibits any person to knowingly possess any biologic agent, toxin or delivery system of a type or in a quantity that, under the circumstances, is not reasonably justified by prophylactic, protective, bona fide research, or other peaceful purpose. Later, the Public Health Security and Bioterrorism Preparedness and Response Act of 2002 required institutions to notify the Department of Health and Human Services (DHHS) or the United States Department of Agriculture (USDA) of the possession of specific pathogens or toxins called *select agents*. Laboratories possessing any select agents must register with the Centers for Disease Control and Prevention (CDC). Violation of any of these statutes carries criminal penalties. The animal and plant pathogens or toxins classified as select agents are listed in Box 65-1.

BIOSECURITY

Appendix F of the fourth edition of the CDC and the National Institutes of Health (NIH) manual *Biosafety in Microbiological and Biomedical Laboratories* (BMBL) was the first to address laboratory security concerns.[6] Laboratory managers and microbiology laboratory directors have now added biosecurity to the list of issues (including biosafety) that they need to know about. Laboratories must conduct a risk assessment and threat analysis

BOX 65-1 List of Select Agents*

VIRUSES

Crimean-Congo hemorrhagic fever virus
Eastern equine encephalitis virus
Ebola viruses
Equine Morbillivirus
Lassa fever virus
Marburg virus
Rift Valley fever virus
South American haemorrhagic fever viruses (Junin, Machupo, Sabia, Flexal, Guanarito)
Tick-borne encephalitis complex viruses
Variola major virus (smallpox virus)
Venezuelan equine encephalitis virus
Viruses causing hantavirus pulmonary syndrome
Yellow fever virus
Exemptions: Vaccine strains

BACTERIA

Bacillus anthracis
Brucella abortus, B. melitensis, B. suis
Burkholderia (Pseudomonas) mallei
Burkholderia (Pseudomonas) pseudomallei
Clostridium botulinum
Francisella tularensis
Yersinia pestis
Exemptions: vaccine strains

RICKETTSIAE

Coxiella burnetii
Rickettsia prowazekii
Rickettsia rickettsii

FUNGI

Coccidioides immitis

TOXINS

Abrin
Aflatoxins
Botulinum toxins
Clostridium perfringens epsilon toxin
Conotoxins
Diacetoxyscirpenol
Ricin
Saxitoxin
Shigatoxin
Staphylococcal enterotoxins
Tetrodotoxin
T-2 toxin
Exemptions: Toxins for medical use, inactivated for use as vaccines, or toxin preparations for biomedical research; national standard toxins required for biologic potency testing

RECOMBINANT ORGANISMS/MOLECULES

Genetically modified microorganisms or genetic elements from organisms in this box, shown to produce or encode for a factor associated with a disease

Genetically modified microorganisms or genetic elements that contain nucleic acid sequences coding for any of the toxins listed in this box, or their toxic subunits

OTHER RESTRICTIONS

The deliberate transfer of a drug resistance trait to microorganisms listed in this box that are not known to acquire the trait naturally is prohibited by NIH *Guidelines for Research Involving Recombinant DNA Molecules*, if such acquisition could compromise the use of the drug to control these disease agents in humans or veterinary medicine.

*Appendix A, Part 72 of Title 42 CFR (Code of Federal Regulations)

in order to write their security plan. This plan not only includes physical security (e.g., electronic card key access and locked freezers and refrigerators where select agents are stored), but also data system (laboratory information system) security and security policies for personnel. Individuals with access to select agents must be cleared by the U.S. Department of Justice (DOJ); this involves fingerprinting and background checks by the Federal Bureau of Investigation (FBI).

Most hospital clinical laboratories have made a decision not to store any select agents. Commercial laboratories, on the other hand, who must store *Coccidiodes immitis* in order to use it as a positive control in molecular assays for confirmation of coccidioidomycosis, have to register to possess and store this organism. This involves writing standard operating procedures for (1) accessing select agents, (2) specimen accountability, (3) receipt of select agents into the laboratory, (4) transfer or shipping of select agents from the laboratory, (5) reporting of incidents, injuries, and breeches of security, and (6) an emergency response plan if security is breached or the isolate is unintentionally released during an accident.

Each clinical laboratory should have a bioterrorism response plan.[4] All potential bioterrorism agents are classified as select agents and are, therefore, regulated by the federal government. For most hospital clinical laboratories, that means that none of the select agents may be kept by the laboratory unless the laboratory has previously registered to keep this agent. Therefore, CDC must be immediately informed whenever a select agent is recovered from a clinical specimen (telephone: (404) 498-2255, facsimile: (404) 498-2265). The select agent must either be sent to a public health laboratory or destroyed within 7 days of identification. If the agent is autoclaved, its destruction must be documented using APHIS/CDC Form 4, which can be downloaded at www.selectagents.gov/formsOverview.htm for this purpose.

LABORATORY RESPONSE NETWORK

Because communication between clinical and public health laboratories is so critical in the post-2001 era, CDC created an organizational structure whereby isolates and lines of communication flow freely among members.[4] The Laboratory Response Network (LRN) was originally designed as a four-tier system. Clinical (Level A) laboratories were at the bottom of a triangle, sending specimens to a Level B (local public health laboratory) or Level C (state public health laboratory) for confirmation. Isolates were ultimately sent to CDC or USAMRIID (United States Army Medical Research Institute for Infectious Diseases) (Level D) for archiving

Table 65-1 Algorithm for Sentinel Laboratories for Critical Agents*

Agent	Sentinel Lab Workup	Comments
Bacillus anthracis	Colony: large, **nonhemolytic,** stands up like beaten egg (Figure 65-1) Gram stain: large, gram-positive rods (Figure 65-2) Catalase: positive Motility: nonmotile Optional: use FDA-cleared Red Line Alert Test (Tetracore, Inc.) to rule out *B. anthracis* (see Chapter 18 for a fuller discussion of this test)	May be mistaken for *Bacillus megaterium*
Brucella spp.	Colony: small, nonhemolytic Gram stain: lightly staining tiny gram-negative coccobacilli Oxidase: positive Urease: positive	May be mistaken for *Haemophilus* or *Francisella*
Francisella tularensis	Colony: pinpoint growth after 48 hours Gram stain: pleomorphic, minute, faintly staining gram-negative coccobacilli Oxidase: negative Urease: negative β-lactamase: positive	May be mistaken for *Haemophilus* and *Actinobacillus*
Yersinia pestis	Colony: pinpoint growth on blood agar after 24 hours Gram stain: gram-negative rods exhibiting bipolar staining Oxidase: negative Urease: negative Indole: negative	Rapid systems may misidentify as *Shigella* spp., H$_2$S-negative *Salmonella* spp., *Acinetobacter* spp., and *Yersinia pseudotuberculosis*
Clostridium botulinum	None	Send all specimens to reference laboratory; patient must get antitoxin immediately
Smallpox and hemorrhagic fever viruses	None	Smallpox can be mistaken for herpes virus if inoculated into routine tissue culture cells

*See individual chapters for a more detailed discussion of each organism.

and sophisticated molecular testing. This original system has now been changed to a three-tier system in which former Level A laboratories are now called *sentinel laboratories*; Level B and C laboratories are now called *reference laboratories*; and Level D laboratories are now called *national laboratories*. Reference and national laboratories have access to special assays and reagents developed by the CDC that are used to confirm the identification of critical agents.

ROLE OF THE SENTINEL LABORATORY

The main role of sentinel microbiology laboratories is to raise suspicion when a targeted agent is suspected in a human specimen. Detection and recognition of a possible bioterrorism event will depend on:

- A laboratory having an active microbial surveillance and monitoring program
- Vigilant technologists looking for a disease that (1) does not occur naturally in a particular geographic region (e.g., plague in New York City); (2) is transmitted by an aerosol route of infection; and (3) is a single case of disease caused by an unusual agent

(e.g., *Burkholderia mallei* usually only seen in the Far East).
- Good communication with infection control practitioners, infectious disease physicians, and local or regional public health laboratories.

Sentinel laboratories must have a Class II biological safety cabinet, copies of Level A protocols containing the algorithms for ruling out suspicious microorganisms (Table 65-1), and participate in an applicable proficiency testing program such as the College of American Pathologist's Laboratory Preparedness Survey. Because sentinel laboratories rule out and refer microorganisms, proper knowledge of appropriate packaging and shipping is critical (see Chapter 4); all specimens must be classified as infectious. Sentinel laboratories should *never* accept nonhuman specimens such as those from animals or the environment. This will prevent the shutdown of the entire facility if a nonhuman sample is explosive or contains a volatile toxic chemical or radioactive substance.

Rapid communication between LRN sentinel members and their reference public health laboratories is essential. Sentinel laboratories, however, do *not* make

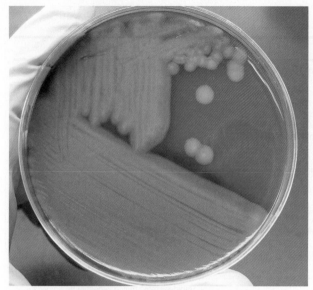

Figure 65-1 Colony of *Bacillus anthracis*.

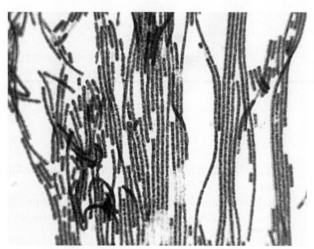

Figure 65-2 Gram stain of *Bacillus anthracis*.

BOX 65-2   Targeted Critical Agents Likely to Be Used in a Bioterrorist Event*

BACTERIA
- *Bacillus anthracis*
- *Yersinia pestis*
- *Francisella tularensis*
- *Coxiella burnetii*
- *Brucella* spp.
- *Burkholderia mallei*
- *Burkholderia pseudomallei*

VIRUSES
- Variola major virus (smallpox)
- Ebola virus
- Marburg virus
- Lassa virus
- Junin virus
- Venezuelan equine encephalitis virus (VEE)
- Western equine encephalitis virus (WEE)
- Eastern equine encephalitis virus (EEE)

TOXINS
- *Clostridium botulinum*
- *Staphylococcus aureus* enterotoxin B
- *Ricinus communis* (castor bean)

*Adapted from www.bt.cdc.gov/Agent/agentlist.asp

BOX 65-3   Sentinel (Level A) Laboratory Guidelines*

Introduction to Level A Laboratory Protocols
Clinical Laboratory Bioterrorism Readiness Plan
Anthrax
Botulinum toxin
Brucella
Plague
Tularemia
Coxiella burnetii
Burkholderia mallei and *Burkholderia pseudomallei*
Staphylococcal enterotoxin B
Unknown viruses
Packing and Shipping

*www.asm.org/Policy/index.asp?bid=6342

the determination that a bioterrorist event has occurred and do *not* notify law enforcement. Each sentinel laboratory *must* know how to contact public health officials 24 hours/day. The FBI has primary responsibility when a bioterrorism event occurs as outlined in Presidential Decision Directive 39. A bioterrorist event is first and foremost a criminal investigation. FEMA (the Federal Emergency Management Agency) has the lead role in consequence management. FEMA is, in turn, assisted by DOD (Department of Defense), DOE (Department of Energy), USDA, DOT (Department of Transportation), DHHS, and EPA (Environmental Protection Agency). FEMA, for example, calls for the deployment of the National Pharmaceutical Stockpile by CDC so victims may be appropriately treated. Early recognition is the

key to saving lives and sentinel laboratorians are on the front lines in the fight against bioterrorism.

Because sentinel laboratories are charged with ruling out possible bioterrorism agents and referring suspicious isolates to reference laboratories for confirmatory testing, each sentinel laboratory's bioterrorism response plan must include a telephone and pager number for their reference laboratory.

Therefore, a sentinel laboratory's key responsibility is to be familiar with likely agents involved in a biocrime; it must have standard operating procedures (SOPs) to accomplish this task. In order to standardize the process nationwide, the American Society for Microbiology (ASM) has put together a series of algorithms designed for the detection of targeted critical agents.

These agents are listed in Box 65-2. Agents are categorized based on their ability to:

- Be easily disseminated usually by aerosolization
- Be transmitted from person-to-person
- Cause significant morbidity and mortality. Algorithms can be accessed on the ASM website at www.asm.org/Policy/index.asp?bid=6342 and are given in Box 65-3.

REFERENCES

1. English JF: Overview of bioterrorism readiness plan: a template for healthcare facilities, *Am J Infect Cont* 27:468, 1999.
2. Jernigan JA, Stephens DS, Ashford DA, et al: Bioterrorism-related inhalational anthrax: the first 10 cases reported in the United States, *Emerg Infect Dis* 7:933, 2001.
3. Kolavic SA, Kimura A, Simons SL, et al: An outbreak of *Shigella dysenteriae* type 2 among laboratory workers due to intentional food contamination, *JAMA* 278:396, 1997.
4. Snyder JW: Role of the hospital-based microbiology laboratory in preparation for and response to a bioterrorism event, *J Clin Microbiol* 41:1, 2003.
5. Torok TJ, Tauxe RV, Wise RP, et al: A large community outbreak of salmonellosis caused by intentional contamination of restaurant salad bars, *JAMA* 278:389, 1997.
6. US Department of Health and Human Services/CDC and National Institutes of Health, Richmond JY and McKinney RW, editors: Biosafety in microbiological and biomedical laboratories (BMBL), ed 4, Washington, DC, 1999, US Department of Health and Human Services.

ADDITIONAL READING

Centers for Disease Control and Prevention: Biological and chemical terrorism: strategic plan for preparedness and response, *MMWR* 49(RR-4):1, 2000.

Centers for Disease Control and Prevention: Laboratory security and emergency response guidance for laboratories working with select agents, *MMWR* 51(RR19):1, 2002.

Christopher GW, Cieslak TJ, Pavlin JA, et al: Biological warfare: a historical perspective, *JAMA* 278:412, 1997.

Franz DR, Jahrling PB, Friedlancder AM, et al: Clinical recognition and management of patients exposed to biological warfare agents, *JAMA* 278:399, 1997.

Gilchrist MJR, McKinney WP, Miller JM, et al: Laboratory safety, management and diagnosis of biological agents associated with bioterrorism. In Snyder JW, coordinating editor: *Cumitech 33*, Washington DC, 2000, ASM Press.

Hawley RJ, Eitzen EM Jr: Biological weapons—a primer for microbiologists, *Ann Rev Microbiol* 55:235, 2001.

Klietmann WF, Ruoff KL: Bioterrorism: implications for the clinical microbiologist, *Clin Microbiol Rev* 14:364, 2001.

Morse SA: Bioterrorism: laboratory security, *Lab Med* 32:303, 2001.

Sewell DL: Laboratory safety practices associated with potential agents of biocrime or bioterrorism, *J Clin Microbiol* 41:2801, 2003.

Snyder JW, Weissfeld AS: Laboratory detection of potential agents of bioterrorism. In Murray PR, Baron EJ, Jorgensen JH, et al, editors: *Manual of clinical microbiology*, ed 8, Washington, DC, 2003, ASM Press.

CHAPTER 16

STAPHYLOCOCCUS, MICROCOCCUS, AND SIMILAR ORGANISMS

1. PYR, β-galactosidase, and VP will differentiate among the tube coagulase-positive staphylococci (see Table 16-5). *S. intermedius* is VP-negative and PYR-positive and is a pathogen in dogs. This isolate was *S. intermedius*, which was probably acquired by the mother from her occupation.

2. An ornithine decarboxylase test should be performed on PYR-positive staphylococci from significant normally sterile sites, to identify *Staphylococcus lugdunensis*, which is among the few staphylococci that have a positive reaction for this test. *S. lugdunensis* can be problematic to identify, because it can be slide coagulase-positive, but is always tube coagulase-negative. Fortunately it does not have the typical beta-hemolysis of *S. aureus*. The identification of *S. lugdunensis* is important, because this organism is highly associated with infectious endocarditis, a complication of bacteremia. Treatment must be long and aggressive. Because more than half of *S. lugdunensis* are β-lactamase–negative, the laboratory should include this result in its report, so that penicillin can be considered for therapy.

3. The only known mechanism of resistance is mediated by the *mec*A gene in staphylococci, which produces penicillin binding protein 2a (PBP2a). This protein does not bind well to either penicillin or to the penicillinase-resistant penicillins. Thus, it is available to the cell to complete the formation of the cell wall. Direct detection of the gene or its protein product is an excellent method to detect resistance in coagulase-negative staphylococci.

4. Surprisingly, the agent is cefoxitin, a drug that would not generally be considered to treat staphylococcal infections. The disk test is recommended by the CLSI as the best predictor of both susceptibility and resistance, having a high sensitivity and specificity.[1]

---

[1]Clinical Laboratory Standards Institute: Performance standards for antimicrobial susceptibility testing; M100-S16, Wayne, Pa, 2006, CLSI.

CHAPTER 17

STREPTOCOCCUS, ENTEROCOCCUS, AND SIMILAR ORGANISMS

1. The organism is an enterococci that has become vancomycin-dependent. To isolate it, place a vancomycin disk on the blood plate in the area of the inoculum from the blood culture bottle (Figure A-1).

2. A brain-heart infusion agar plate with 6 μg/mL can be used. Inoculate 1 to 10 mL of a 0.5 suspension of the isolate equal to a 0.5 McFarland tubidity standard and incubate for 24 hours. Greater than one colony of growth is considered presumptive for resistance.

3. Other than enterococci, two genera are recognized to be intrinsically resistant to vancomycin. *Leuconostoc* are gram-positive cocci in chains; they are PYR-variable and are identified by their characteristic production of gas in MRS broth (Procedure 13-27). *Pediococcus* are gram-positive cocci in clusters and tetrads; they are PYR-negative and do not produce gas in MRS broth. Although *Weissella confusa* are also vancomycin-resistant and produce gas in MRS broth, they are generally rod-shaped. The best way to demonstrate their rod shape is to prepare the Gram stain from growth in broth. They are also arginine decarboxylase–positive, unlike *Leuconostoc* or *Pediococcus*.

CHAPTER 18

BACILLUS AND SIMILAR ORGANISMS

1. The beta-hemolytic spore-formers are easily separated from all of the other species of *Bacillus* and related genera by their colony morphology. *Bacillus mycoides* displays a rhizoid colony that looks like a fungus. *Bacillus thurengiensis* is not a human pathogen, but is pathogenic to insects and would be unlikely to cause the demise of this unfortunate man. That leaves only *Bacillus cereus* in the differential, which is known to have virulence factors.

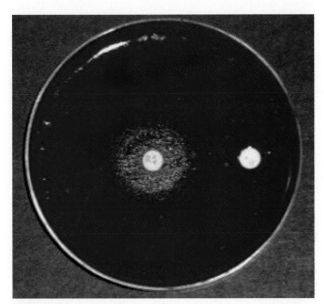

Figure A-1 Growth only around the vancomycin disk of an enterococci that has become vancomycin-dependent. *(Courtesy Mary K. York.)*

2. *Bacillus cereus* has a β-lactamase, which is more reliably detected by the observation of growth up to the edge of a 10-unit penicillin disk. β-lactamase testing is less reliable. The organism is also motile, which separates it from *Bacillus anthracis*, the other cause of serious human infections with *Bacillus* spp. However, *B. anthracis* is not hemolytic. A positive lecithinase test (see Figure 18-3) unequivocally confirms the identification, although this is rarely needed if the colony is typical.

3. There is the potential that this organism could be *B. anthracis*. Such an isolate could provide proof that a bioterrorist event is occurring. All plates and tubes should immediately be placed in a biological safety cabinet. Then the laboratory supervisor should be notified to contact the physician, the local epidemiologist, and the local public health department immediately. They will instruct you what action to take and how to handle the cultures. They will want you to carefully examine the Gram stain. The bacillary vegetative cells should be greater than 1 μm wide and the spores should not be wider than the vegetative cells. *B. anthracis* is nonmotile, whereas most other bacilli are motile. Old cultures of saprophytic *Bacillus* spp. tend to be nonmotile. Preparation of a fresh culture plate that is observed microscopically in a wet mount after a few hours at 35° C may be all that is needed to rule out *B. anthracis*. There is no danger in doing this test, because spores will not form in such a short time.

CHAPTER 19

LISTERIA, CORYNEBACTERIUM, AND SIMILAR ORGANISMS

1. The technologist performed a wet mount motility test, which was positive for the characteristic "tumbling motility" of *Listeria*. In the confirmatory tube motility test at 20° to 25° C, an umbrella-shaped growth of the organism was formed at the top of the tube (See Figure 19-2). A bile-esculin test was also performed, which was positive. Both of these tests indicated that the isolate was *Listeria monocytogenes*. Laboratory identification protocols should require checking for hemolysis on all gram-positive rods and include both motility and bile-esculin tests for isolates that are hemolytic.

2. Susceptibility testing is not routinely recommended because these organisms are predictably susceptible to ampicillin. Treatment is generally with ampicillin and an aminoglycoside.

3. *Corynebacterium* that are pathogens in the urinary tract are urease-positive. They include *C. urealyticum, C. glucuronolyticum,* and *C. riegelii*. Only *C. glucuronolyticum* is CAMP test–positive. *C. urealyticum* is lypophilic. All of these species grow slowly and might not be detected if the urine cultures are incubated for less than 24 hours.

CHAPTER 20

ERYSIPELOTHRIX, LACTOBACILLUS, AND SIMILAR ORGANISMS

1. *Erysipelothrix rhusiopathiae* is the only catalase-negative rod to produce hydrogen sulfide, visible on either triple sugar iron agar (TSI), Kligler's agar, or sulfide indole motility agar (SIM). This test should be part of the laboratory protocol for identification of catalase-negative, gram-positive rods from wounds and blood cultures. Because the organism grows slowly, the test may take up to 3 days for a positive result. *E. rhusiopathiae* is a known, but rare, cause of endocarditis, with a 38% fatality rate. The disease is indolent, and the patient probably had been infected for a long time.

2. *E. rhusiopathiae* is intrinsically resistant to vancomycin, which might be considered for therapy, because this drug is often used as empiric therapy for infections with gram-positive bacteria. The organism is susceptible to penicillin. Lactobacillus is the only other catalase-negative gram-positive rod that is vancomycin-resistant. These two genera are often confused.

3. *Erysipelothrix* is an animal pathogen. It is likely that the patient became infected from exposure to the

marine environment, probably through an open wound originating from a fishhook.

CHAPTER 21

NOCARDIA, STREPTOMYCES, RHODOCOCCUS, AND SIMILAR ORGANISMS

1. A penicillin disk will separate most of the contaminating coryneforms, which will be penicillin susceptible, from the more pathogenic actinomycetes. If the isolate is penicillin resistant, a positive partial acid-fast stain will separate *Nocardia* spp. from saprophytic actinomycetes and coryneforms. This test is difficult to do and should always have controls on the same slide as the unknown culture. Young colonies may give a false-negative result. Definitive identification takes much longer for the demonstration of microscopic aerial hyphae on tap water agar and a positive lysozyme test. It is important that the microbiologist not wait for these latter tests to turn positive to alert the physician of the possibility of the presence of this serious infection.

2. *Nocardia farcinica* is resistant to all three classes of agents, making it very difficult to treat. Treatment with antimicrobial agents is usually for 6 to 12 months. Generally sulfamethoxazole-trimethoprim is one of the drugs used. In our case, this woman probably had multiple brain abscesses that spilled into the CSF. The reason the chocolate agar was negative probably had to due with the low numbers of cells in the specimen and the fact that enriched media can be inhibitory to *Nocardia*.

3. When mucoid, pink colonies of gram-positive rods are isolated, *Rhodococcus equi* is suspected. The characteristic colony plus the ability to be CAMP test–positive and urease-positive identifies this species. *R. equi* is a serious pathogen in immunocompromised patients.

CHAPTER 22

ENTEROBACTERIACEAE

1. *Yersinia* organisms are characterized by having negative reactions at 35° C that are positive at 25° C. Both motility and the VP reaction are positive at 25° C for this organism but are negative at 35° C. However, many strains are not positive for either reaction and require serologic tests to identify.

2. *Yersinia* species are very slow-growing organisms, and the additional incubation allowed expression of enzymes that changed the reactions. In addition, *Yersinia* organisms prefer to grow at low temperatures and produce more enzymes at 25° C.

3. The plague bacillus is not motile and is urease-negative. *Y. pseudotuberculosis* is motile at 25° C and is urease-positive; however, not all strains produce positive reactions.

4. It is imperative to notify the physician, the local epidemiologist, and the local health department immediately. Even if the gram-negative rod is not ultimately diagnosed as the plague bacillus, the life-threatening nature of this infection prompts the institution of appropriate antimicrobial therapy and infection control action to prevent human-to-human transmission.

CHAPTER 23

ACINETOBACTER, STENOTROPHOMONAS, AND SIMILAR ORGANISMS

1. A positive catalase test and negative tests for oxidase and hanging drop motility (Procedure 13-26) will separate this species from a number of other genera of asaccharolytic gram-negative rods (discussed in later chapters). If the isolate had been able to oxidize glucose, these tests would be useful to separate the organism from *Pseudomonas luteola* and *P. oryzihabitans,* which, unlike *Acinetobacter,* have a yellow pigment.

2. The nitrate test will separate NO-1, which is positive, from asaccharolytic *Acinetobacter,* which do not convert nitrate to nitrite. However, in this case, the test is not necessary, because NO-1 is found in dogs and cats and our patient has not left the hospital. Some species of *Bordetella* (see Chapter 27) are also asaccharolytica and oxidase- and nitrate-negative, but they are either motile or urease-positive. The commercial system would have detected the positive urease. *Bordetella holmesii,* previously called NO-2, should also be separated from asaccharolytic *Acinetobacter.* It grows poorly on MacConkey, is often catalase-negative or weakly positive, and has a brown soluble pigment on Mueller Hinton agar.

3. *Acinetobacter* organisms are endogenous microbiota of the skin. Our patient probably acquired the organism from either his own skin or that of his caregivers. *Acinetobacter baumanii,* a glucose-oxidizing member of this genus, is the third most common cause of gram-negative infections in infants and has been responsible for serious nosocomial spread of resistant strains in neonatal intensive care units. Its association with catheter infections is rare.

4. An organism that can produce acid from glucose only in the presence of oxygen is a glucose-oxidizing gram-negative rod. If it can produce acid from glucose without the presence of oxygen, it is a

glucose-fermenting gram-negative rod. This test is performed by inoculating two tubes of oxidation-fermentation medium, and overlaying one with oil to prevent the absorption of oxygen in the medium as demonstrated in Figure 13-34.

5. Serious infections are caused by the HACEK bacteria (see Chapter 33 for list). These bacteria are glucose-fermenters, do not grow in OF medium, and are frequently misidentified by commercial identification systems. They often need rabbit serum or other enrichment to grow in tube medium, giving rise to falsely negative nitrate and glucose fermentation results. One way to detect glucose fermentation in these bacteria is to add a drop of rabbit serum to Andrade's glucose (Procedure 13-15). Another is to use a rapid commercial system designed to detect preformed enzymes of these fastidious microorganisms (see Table 13-1).

CHAPTER 24

PSEUDOMONAS, BURKHOLDERIA, AND SIMILAR ORGANISMS

1. *Pseudomonas aeruginosa*—mucoid strains, *Alcaligenes xylosoxidans, Burkholderia cepacia,* and *Stenotrophomonas maltophilia* are frequently found in respiratory specimens from CF patients.

2. *Pseudomonas aeruginosa* with mucoid morphology is characteristic of patients with CF. It is not clear what lung factors in the patient promote the expression of the mucoid variant, but it is a hallmark of the disease. Often these strains are unable to produce the characteristic pigment of *P. aeruginosa.*

3. The agents of infection are very slow growing. In a rapid test, their growth could be inhibited, masking their true resistance.

4. The disk test can be incubated for 24 hours without loss of sensitivity. It allows testing of drugs that may not routinely be tested in the laboratory but are needed to treat these patients who have acquired very drug-resistant strains. The disk test can aid in detection of mixed cultures, which are often seen in this patient population.

5. *Burkholderia cepacia* are colistin-resistant; *S. maltophilia* can also be resistant. They both are lysine-positive and ferment maltose, but only *B. cepacia* ferments mannitol. *S. maltophilia* hydrolyses DNA (see Figure 13-13) and is usually susceptible to sulfamethoxazole-trimethoprim but is intrinsically resistant to imipenem. *Ralstonia pickettii* and other *Burkholderia* organisms can be confused with *B. cepacia.* Because isolation of *B. cepacia* labels a CF patient as contagious to other CF patients, it is important that the identification be confirmed.

6. Because CF patients are uniformly infected with *Pseudomonas,* the smear will show gram-negative rods. The important information for the physician caring for the patient is the susceptibility of the gram-negative isolates, not the rapid detection of their presence in a smear.

CHAPTER 25

ACHROMOBACTER, RHIZOBIUM, OCHROBACTRUM, AND SIMILAR ORGANISMS

1. The organism is EF-4b, which has a characteristic odor but no yellow pigment. A negative indole reaction is also useful to differentiate it from other organisms found in cats and dogs.

2. *Pasteurella multocida* (indole-positive; see Chapter 32) is the most common bite organism. EF-4a and Capnocytophaga (DF-2) are glucose-fermenting rods. Others include the asaccharolytic *Neisseria weaverii* and NO-1. EF-4b is the only glucose-oxidizer on the list.

3. This reaction is a positive reaction. The organism reduced nitrate without gas production. The negative result after addition of the reagents merely indicted that there was no nitrite in the tube. The reagents produce a red color only when nitrite is present. Even after the addition of zinc (which changes nitrate to nitrite), no color was present, because the organism had converted nitrate to nitrite and nitrite to amines and other nongaseous end-products.

CHAPTER 26

CHRYSEOBACTERIUM, SPHINGOBACTERIUM, AND SIMILAR ORGANISMS

1. In the differential are *Weeksiella virosa* (a normal inhabitant of the genitourinary tract, primarily of females), *Bergeyella zoohelcum,* a number of CDC group II species, *Chryseobacterium* spp., and *Empedobacterium brevis.* All of these genera are oxidase- and catalase-positive. Some have yellow insoluble pigments.

2. Oxidation of mannitol in OF medium will definitively separate *C. meningosepticum* from the other *Chryseobacterium* spp, and the related species in the differential. Colonies frequently lack the distinct yellow pigment, or, if present, it can be pale yellow. *W. virosa* and *B. zoohelcum* are asaccharolytic and thus unable to produce acid even from glucose. Our patient's isolate was identified as *C. meningosepticum.*

3. It is thought that the failure of antimicrobial agents to eradicate the bacteria is the bacteria's resistance

to usual agents. Conflicting results are found in the literature, especially when disk methods are used. Testing should only be performed by qualified persons in a reference laboratory that uses a CLSI MIC method. Drugs tested should include trimethoprim-sulfamethoxazole, rifampin, quinolones, vancomycin, erythromycin, and minocycline. Many of these agents are not appropriate for neonates, making treatment difficult.

CHAPTER 27

ALCALIGENES, BORDETELLA (NON-PERTUSSIS), COMAMONAS, AND SIMILAR ORGANISMS

1. Spot indole, hanging drop motility, and 2-hour urease-phenylalanine deaminase (PDA) tests are helpful. The urease-PDA test is purchased as a disk. Heavy growth of the organism is emulsified in 1 mL of saline and the disk is added. After 2 hours of incubation, the tube is observed for the pink color of urea hydrolysis. If it is positive, a drop of 10% ferric chloride is added to detect the dark green color of a positive PDA test.

2. *Myroides* and *Psychrobacter immobilis* are nonmotile. *Oligella ureolytica* and *Psychrobacter phenylpyruvicus* are PDA-positive. *Bordetella bronchiseptica* and *Cupriavidus pauculus* (formerly CDC IVc-2; see Chapter 24) are motile and urease-positive but PDA-negative. *C. pauculus* is PYR-positive. This isolate was negative for PYR and was identified as *B. bronchiseptica*, a pathogen that causes "kennel cough" in dogs. As further confirmation of the rapid identification, tube biochemical tests were inoculated. *B. bronchiseptica* is asaccharolytic, does not produce H$_2$S, and is nitrate-positive with no gas.

3. *Brucella* are oxidase-, catalase-, and urease-positive. The genus does not grow well on Mac-Conkey. The importance of performing rapid tests to screen for this agent cannot be overemphasized. Whenever all three of these tests are positive and the isolate is not growing on Mac-Conkey, the microbiologist should begin to perform all testing in a biological safety cabinet. Laboratory-acquired infections with this pathogen have been documented both from the respiratory route and from direct skin contact. This pathogen is also on the list of potential bioterrorist agents.

4. Because the isolate took 48 hours to form colonies, the laboratory could have missed the identification, either because they did not incubate the plates for 48 hours or because they did not recognize this tiny colony as different from the normal oral microbiota that was on the plate.

CHAPTER 28

VIBRIO, AEROMONAS, PLESIOMONAS AND CHROMOBACTERIUM

1. This isolate could be either *Aeromonas* or *Vibrio*. The string test (see Figure 28-3), growth on TCBS, O129 disk susceptibility, or salt tolerance can be used to separate these genera. The organism produced yellow colonies on TCBS because it was positive for sucrose. It was accurately identified by a commercial system as *Vibrio alginolyticus*.

2. The most common *Vibrio* found in the ocean is *V. alginolyticus*. *Pseudomonas aeruginosa* is most often responsible for external ear infections from fresh water (called "swimmer's ear"), and *V. alginolyticus* is the common pathogen responsible for ear infections from swimming in salt water. See Chapter 56 Case Study.

3. Both genera share a number of biochemical reactions. *V. fluvialis* is a pathogen with serious consequences but shares a biochemical profile similar to *Aeromonas caviae*, a common species found in stool cultures. Without performing the tests that separate these genera, which are not found in commercial systems, misidentifications can easily be made.

4. To successfully perform the disk test, the initial inoculum equivalent to 0.5 McFarland turbidity standard should be made in saline, not water. This step will assure that the organism will grow well and demonstrate its resistance. *V. alginolyticus* is resistant to polymyxin B, which is commonly found in eardrops used to treat "swimmer's ear." Identification of the pathogen with susceptibility testing is needed for proper treatment.

CHAPTER 29

SPHINGOMONAS PAUCIMOBILIS AND SIMILAR ORGANISMS

1. *Sphingomonas* spp. and CDC groups O-1 and O-2 are motile, yellow-pigmented rods. Of the three, only *Sphingomonas* oxidizes xylose. Some *Pseudomonas* species, such as *P. stutzeri* and *P. mendocina,* are also yellow and oxidase-positive, but they grow on Mac-Conkey and have distinct colony morphologies.

2. Only an overnight MIC method is reliable to test these slow-growing pathogens. Disk testing and rapid commercial systems are not standardized for such a rare species.

3. The joint fluid from this patient probably had less than 10 organisms per milliliter of fluid. The infection was low-grade and not detected by putting a few drops onto a blood agar plate. Joint fluid is very viscous; consequently centrifugation is of little

value in concentrating the bacteria. A number of researchers have shown that the sensitivity of culture can be increased by culturing large amounts of fluid, much like the method used for blood cultures. Up to 10 mL can be placed in each blood culture bottle, increasing the chance of detecting the agent of the disease.

CHAPTER 30

MORAXELLA

1. It is possible from the description that this could be one the microorganisms discussed in Chapter 33, known as the HACEK group. However, most members of the HACEK group are glucose-fermenters. *Eikenella corrodens* is the exception. We have already discussed the use of Andrade's glucose medium supplemented with rabbit serum or plasma to grow these bacteria and demonstrate their fermentation properties (see Chapter 23 Case Study). When this test is set up, a nitrate reduction test is also important to separate *Eikenella corrodens* from *Neisseria elongata*.

2. The most likely identification is that of *Neisseria elongata* subsp. *elongata*. Because the catalase test was negative, the isolate is not *Neisseria weaverii*. Because the nitrate was negative, it is not *Eikenella corrodens* (see Chapter 31). However, this identification is based on completely negative results, except for the positive oxidase reaction, colony morphology, and Gram stain. Confirmation with at least one positive test should be done. The test used to confirm the identification is a positive nitrite reduction test.

3. To do the test, nitrite must be the substrate. When the nitrate A and B reagents are added to the uninoculated nitrite tube, a red color is produced from the reaction of the reagents with nitrite. If the organism is inoculated to a tube of nitrite and nitrite is reduced, no red color will be seen when the reagents are added. Incubation may have to be extended, because the red color will continue to appear until the organism reduces all the nitrite in the tube. Most laboratories do not keep this medium on hand, in which case, confirmation should be done at a reference laboratory.

CHAPTER 31

EIKENELLA AND SIMILAR ORGANISMS

1. The organism in the culture was ornithine-positive. That combined with its unique odor identifies it as *Eikenella corrodens*. For completeness, as with all isolates of gram-negative rods, the spot indole was performed to be sure of a negative reaction.

2. The physicians were interested in finding a source of the infection. *E. corrodens*, like other members of the HACEK group (see Chapter 33), are found as normal microbiota of the mouth and generally cause infections only when there is a contact of the mouth with the vascular system. Most infections occur because of bleeding during dental procedures.

3. This bacterium can cause infective endocarditis, as can the other members of the HACEK group. These bacteria travel from the mouth through the bloodstream and can lodge in the heart. Blood cultures are positive in most cases. Without treatment, usually with a penicillin, the prognosis is poor. Because of the importance of detection, the laboratory plays a critical role in the diagnosis of these infections.

CHAPTER 32

PASTEURELLA AND SIMILAR ORGANISMS

1. This organism likely is *Pasteurella multocida*, a common part of the normal oral flora of dogs and cats. Even though it does not grow on MacConkey, most laboratory identification systems will correctly identify it.

2. Human bite infections are often due to *S. aureus* but also include viridans streptococci, *Eikenella corrodens*, and a variety of anaerobic microbiota. *Pasteurella* is not associated with these infections.

3. These organisms are susceptible to many antibiotics, including ampicillin. As yet, beta-lactamases have not been found in *Pasteurella*. Because these organisms can grow slowly, overnight methods work best for susceptibility testing. Generally a beta-lactamase test is sufficient to guide therapy.

4. Both *S. aureus* and *S. intermedius* are coagulase-positive staphylococci. The latter is predominantly found in cats and dogs. Occasionally, it is difficult to demonstrate the coagulase for *S. intermedius*. These species are separated by other biochemical tests. See the case study in Chapter 16. If only the coagulase test is performed, the isolate can be erroneously called *S. aureus*.

CHAPTER 33

ACTINOBACILLUS, KINGELLA, CARDIOBACTERIUM, CAPNOCYTOPHAGA, AND SIMILAR ORGANISMS

1. The organism is a *Capnocytophaga* spp. (DF-1 and related species) that is part of the normal microbiota of the mouth of humans. The patient is at risk

for bacteremia because of her low white blood cell count and her periodontal disease.

2. Few human pathogenic gram-negative rods are both catalase- and oxidase-negative and grow in 5% CO_2 under aerobic conditions. Among these, only *Capnocytophaga* is a fusiform rod. Confirmation of the identification includes its characteristics of being indole-negative, nonhemolytic, and often esculin-positive. Identification to the species level is difficult and is not clinically important.

3. *Capnocytophaga* has historically been treated with beta-lactam antibiotics, but now more than 75% of the organisms have been found to produce a beta-lactamase. Testing for the enzyme with the chromogenic cephalosporin test is sufficient to inform the physician that treatment with a beta-lactam drug will not be effective.

CHAPTER 34

HAEMOPHILUS

1. The organism is in the genus *Haemophilus*, which is the most common cause of epiglottitis. In 5% of the cases, endocarditis develops as a complication.

2. The organism is *H. paraphrophilus* and is differentiated from *H. parainfluenzae* by its ability to ferment lactose. For specimens from blood cultures, fermentation testing is important to accurately identify *Haemophilus* to the species level.

3. Because the organism was not *H. influenzae*, another cause of epiglottitis, family contacts did not have to receive prophylaxis with rifampin or other agents. Isolation of *H. paraphrophilus*, as well as *H. influenzae*, in blood cultures is generally considered an indication of epiglottitis and warrants aggressive treatment. Isolation of *H. parainfluenzae* and *H. parahaemolyticus* are less often associated with a serious disease.

CHAPTER 35

BARTONELLA AND AFIPIA

1. Laboratories should have a policy that if a bacterium is isolated from a normally sterile site, even if the test ordered is for a specific agent, the bacterium should be identified. If the laboratory is unable to grow the organism, the broth bottle should be submitted to a reference laboratory.

2. The patient was homeless and slept on the ground where stray cats and dogs are prevalent. Possibly he breathed in the bacteria and, with his poor health, was not able to clear it from his lungs.

3. Culture is rarely necessary because both diseases are self-limiting in healthy persons. For immuno-

compromised, especially HIV-positive, patients it is probably advisable to offer either culture or PCR or arrange to have specimens sent to a reference laboratory for testing.

CHAPTER 36

CAMPYLOBACTER, ARCOBACTER, AND HELICOBACTER

1. Only *Campylobacter jejuni* is positive for hippurate among the strongly catalase-positive campylobacters.

2. Both disk testing and the indoxyl acetate test are needed to identify species that are hippurate-negative. *C. coli* is positive and *C. lari* is negative for indoxyl acetate. However, many other species are positive in this test. Only *C. coli* and *C. jejuni* are positive and cephalothin resistant. Only *C. lari* is indoxyl acetate negative and cephalothin resistant. Thus the combination of both tests will identify these three species, the most common stool pathogens.

3. It is likely that the chicken was the source of the *Campylobacter*. However, cooking does kill the bacterium. If the cooked meat had been placed on the same area that was used to clean the uncooked meat, it could have become contaminated after cooking. Chickens harbor large numbers of campylobacter.

4. From 20% to 50% of *C. jejuni* and *C. coli* isolates in the United States are resistant to ciprofloxacin. None were resistant when the bacterium was discovered. At that time, part of the identification included susceptibility to nalidixic acid. Because nalidixic acid is in the same family of drugs as ciprofloxacin, all ciprofloxacin-resistant strains are resistant to nalidixic acid, making the test unusable in the identification. A disk test for ciprofloxacin susceptibility has not been standardized yet, but the lack of a zone around either erythromycin or ciprofloxacin disks is an indication that the isolate is resistant to the antimicrobial agent. Such a result should be communicated to the physician, because both agents are drugs of choice to treat serious infection.

CHAPTER 37

LEGIONELLA

1. The urinary antigen test effectively detects *Legionella pneumophila* type 1, but is less often positive in infections with the other *L. pneumophila* types.

2. Legionella is an opportunistic disease, meaning that there is generally an underlying factor in the host to account for the ability of the bacteria to cause infection. Risk factors include lowered immune system either from corticosteroids, malignancy, or age. Infants and the elderly naturally have a

lowered immune system. Surprisingly patients with AIDS do not get *Legionella* commonly, despite their lowered immune system. Other risk factors would include a heavy exposure to water containing the *Legionella* bacteria.

3. The diagnosis is hampered by lack of sputum production by the patient to provide a good specimen. Cultures are limited because of the long incubation time for growth, the need for special media with a rather short outdate, the need to keep the cultures moist to grow the bacteria, the need to inhibit overgrowth of other bacteria by using antibiotics in the media or by acid treatment of the specimen, the inhibitory effect of yeast and of CO_2 on the organism, and the lack of biochemical tests to identify the organism once cultured. These factors, plus the fact that a good PCR test is not available, probably results in underdiagnosis and treatment of this infection.

4. The body's immune response to *Legionella* does not include the usual febrile response with polymorphonuclear cells. Sputum, if produced at all, will be watery and lack an increase in WBCs. Sometimes a more invasive specimen is needed to make the diagnosis.

CHAPTER 38

BRUCELLA

1. Because airborne transmission of *Brucella* and *Francisella* has led to laboratory-acquired infections, the cultures should be handled in a biological safety cabinet whenever these genera are in the differential.

2. A positive rapid urease test will presumptively identify *Brucella* spp., one of the few coccobacilli that are both oxidase- and catalase-positive and found in normally sterile site specimens. Although several other coccobacilli are urease-positive, including *Bordetella* spp. and *Oligella urealytica*, they are either motile, grow on MacConkey, or would be unlikely in a joint specimen.

3. Pasteurization is not always used for cheese and milk products in Middle Eastern countries. *Brucella* infection is chronic in nature and had probably been present in this woman for some time. In 40% of untreated cases, the infection is present in the joints. The foreign body in her knee makes her more susceptible to infections in that location.

4. The diagnosis can be confirmed by serologic testing for *Brucella* agglutinins. A titer of 1:160 is considered positive, but a negative titer does not rule out the disease. The organism identification can also be confirmed with serologic reagents. Without appro-

priate diagnosis and treatment, the organism will not be eradicated. Laboratory susceptibility testing is unreliable, leading to erroneous results. If the identification is not correct, inappropriate treatment can result. Usually the identification is confirmed by the local health department. In fact, *Brucella* disease is reportable and is on the list of potential bioterrorist organisms.

CHAPTER 39

BORDETELLA PERTUSSIS AND BORDETELLA PARAPERTUSSIS

1. This woman had a child at home that had no antibodies to *B. pertussis*. Pertussis in the newborn is often fatal. The ID service wanted to be certain that the mother was not infectious to the baby. Treatment of the mother for 5 days with erythromycin or a similar macrolide should protect the baby. Because the number of pertussis cases in the United States is increasing, a vaccine was licensed in 2005 for use in teens and adults. Hopefully use of the vaccine will decrease the number of days lost in school and the workplace and decrease the number of infants exposed to *B. pertussis*.

2. Some reasons include the following: Specimen collection is usually not adequate, the organism is quite labile in transport, the culture takes 5 days for colonies, and special media must be kept on hand. A PCR test for this bacterium has increased the ability to detect cases.

3. It is based on a gene that is present in multiple copies in the bacteria. While this gene is present in other species of *Bordetella*, the combination of a positive result with symptoms of cough greater than 2 weeks makes the definitive diagnosis.

4. The organism is a tiny coccobacilli that is catalase- and oxidase-positive, nonmotile, and urease-negative. It is unable to grow on MacConkey or blood agar, but grows on Regan-Lowe medium.

CHAPTER 40

FRANCISELLA

1. The most commonly misidentified microorganism submitted as *F. tularensis* is *Haemophilus influenzae*. Microbiologists do not expect *F. tularensis* to grow on chocolate agar, but today the commercial vendors make a very high quality medium that allows this fastidious organism to grow. Lack of growth on sheep blood agar even around the staphylococcus dot, which supplies NAD to the organism, rules out *Haemophilus*.

2. The culture should only be handled with biosafety level 3 precautions, the same precautions used to handle *Mycobacterium tuberculosis*. However at this point, it should be sent to the state health department and the physician notified of the possible identification of *F. tularensis*. Also, the local epidemiologist should be contacted immediately if you cannot rule out a bioterrorist event associated with isolation of the organism.

3. This test is part of the recommended protocol for the identification of *F. tularensis* for sentinel laboratories. *F. tularensis* has a beta-lactamase that is active against all penicillins and cephalosporins, including imipenem. A positive test alerts the microbiologist that this in not a fastidious HACEK bacteria (see Chapter 33) that does not grow on blood agar. HACEK bacteria are usually beta-lactamase–negative. Because it is difficult to do susceptibility testing on *F. tularensis*, the positive beta-lactamase test can be a guide to therapy.

CHAPTER 41

STREPTOBACILLUS MONILIFORMIS AND *SPIRILLUM MINUS*

1. *S. moniliformis* is susceptible to SPS, the anticoagulant used in blood culture bottles. By adding more blood, all of the SPS would be bound up, and the inhibitory effect of the anticoagulant will be minimized. This was successful and the organisms grew. The effort to grow it from the tube failed because there was too much SPS in the tube used to draw the blood.

2. *S. moniliformis* will not grow on laboratory media unless there is sufficient blood, ascitic fluid, or horse serum in the medium. Some strains require 8% to 10% CO_2. In the case of this patient, tiny colonies finally appeared on the blood agar plate after 72 hours in 8% CO_2.

3. Once colonies are apparent, the identification is based on negative results for oxidase, catalase, nitrate, and indole. One must be certain that there is growth in the nitrate tube; it may be necessary to add serum to the broth. In addition, the Gram stain is critical to the identification. L-forms or long rods with thick knobs are characteristic (see Figure 41-1), whereas coccoid forms or rods with tapered ends are more likely *Capnocytophaga*. Another genus, *Leptotrichia*, is similar, but does not usually grow on laboratory media. In addition to biochemical testing, it is important to communicate with the physician, to be sure that the identification fits with the clinical picture.

CHAPTER 44

LABORATORY CONSIDERATIONS

1. *Clostridium botulinum* produces three toxins, all of which can be fatal even in small doses. In addition, this strict anaerobe is difficult to isolate. The diagnosis is made by testing tissue or patient's serum in a mouse bioassay. Mice are inoculated with the specimen in combination with various antitoxins. Persons who work with the organism are given botulism toxoid vaccination so that they will be immune to the effects of the toxin. Our patient was given antitoxin against all three botulism toxins: A, B, and C. Routine clinical laboratories should not attempt to grow the organism, because the staff has not been vaccinated.

2. Because the organism is a strict anaerobe, it is important to keep oxygen from getting into the specimen. One good way to do this is to collect the specimen with a needle and syringe and to inject the specimen into a Vacutainer tube without any anticoagulant. Other choices include special tubes with agar as an oxygen acceptor. The specimen should then be transported at room temperature. Extremes of temperature (either cold or hot) should be avoided during transportation.

3. The patient was an intravenous drug user and probably acquired spores of *C. botulinum* from the impure drugs he injected subcutaneously into his arms and leg. There were 17 cases of this outbreak reported to the state health department. An investigation was performed, and heroin use was found to be common to all cases. While the drug is briefly boiled with a solvent before injection, this boiling is not sufficient to kill the spores. Spores that are injected subcutaneously can germinate in vivo in the anaerobic atmosphere. Here they produce the toxin resulting in the manifestations of the disease.

4. All *Clostridium* species are catalase-negative rods that do not branch. This is helpful but does not rule out all other anaerobic gram-positive rods. Many species of *Clostridium* produce spores, but not all of them will do so. Some will even stain gram-negative, making their identification even more difficult. A few are indole-positive, which is helpful. In summary, identifications may be difficult and require special RNA and DNA analysis.

CHAPTER 45

MYCOBACTERIA

1. In meningitis caused by *M. tuberculosis*, there are very few organisms. Even with concentration of the CSF, smears can only detect 10^4 bacteria/mL.

Smears are difficult also because the mycobacteria are lipophilic and do not easily stick to the slides. Cultures have a better chance, especially if 10 mL is inoculated into a broth culture. PCR has the advantage that it can detect lower numbers of microorganisms and even nonviable microorganisms.

2. *M. tuberculosis* is identified as the only species with a positive niacin test, provided that the colony is buff colored, slow growing, and rough. In addition, there are commercial FDA-cleared probe assays that are specific for the *M. tuberculosis* complex. These tests are more rapid than the niacin test and can be done before colonies are present on solid media.

3. This complex includes *M. tuberculosis, M. africanum,* and *M. bovis,* including *M. bovis* BCG. There are other members of the complex, but these are rarely seen in humans.

4. The laboratory can determine whether any other positive specimens were processed on the same day as the patient in question. If so, all of the positive isolates from that day should be sent for a test called DNA fingerprinting. It is unlikely that two strains have the same fingerprint. If they do, it is possible that one specimen contaminated another.

CHAPTER 46

OBLIGATE INTRACELLULAR AND NONCULTURABLE BACTERIAL AGENTS

1. The patient has *Ehrlichia chaffeensis,* which was detected in the bone marrow smear by the presence of morulae (variable-sized basophilic inclusions).

2. A zoonosis is a disease that humans get from association with animals or their excretions. Humans are usually the accidental host in such diseases. In this case, the vector of the disease is the tick and the usual host is the horse.

3. If the morula is not visualized in the peripheral smear, PCR and serology are the currently available methods to make the diagnosis. In our patient, serology was performed and a single titer of 1:512 was obtained. Culture is difficult in that these organisms have an intracellular life cycle that is limited to cell lines that are difficult to grow in vitro.

CHAPTER 47

CELL WALL–DEFICIENT BACTERIA: *MYCOPLASMA* AND *UREAPLASMA*

1. The patient had an infection with *Mycoplasma pneumoniae,* which was diagnosed by a positive test for cold agglutinins. For this test, blood is collected on the patient and incubated at 4° C. The red cells clump but the clumps disappear at 37° C. While this is not specific for *Mycoplasma pneumoniae,* it is positive in about 70% of such infections and is the first test that is positive in the disease. These are oligoclonal IgM autoantibodies directed against an altered I antigen on the surface of the RBCs from infected patients. A titer of 1:128 against type O Rh-negative erythrocytes is diagnostic. Culture takes several days and serum for detection of IgM antibodies is not positive for about 10 days.

2. Generally bacterial cultures do not include the media needed to detect *M. pneumoniae.* Colonies, even on appropriate media, are only visible microscopically. PCR holds promise but is not offered routinely by most laboratories.

3. *M. pneumoniae* does not have a cell wall, hence it lacks the typical peptidoglycan in the wall of bacteria. All penicillins and cephalosporins act on the cell wall of the bacteria to inhibit cross-linking of the peptidoglycan. Erythromycin inhibits protein synthesis inside the bacterial cell. Although *M. pneumoniae* lacks the cell wall of bacteria, it does use the same method of protein synthesis shared by other bacteria. Thus erythromycin and other macrolide antibiotics are effective treatment for infections with this bacterium.

CHAPTER 48

THE SPIROCHETES

1. She acquired the disease most likely from the bite of the nymphal form of the Ixodes tick infected with *Borrelia burgdorferi.* Even though symptoms were not manifest for a few weeks, her description of the initial lesion is consistent with erythema migrans (see Figure 48-3).

2. IFA tests are very sensitive to detect antibodies to *B. burgdorferi,* but they are known to have a high false-positive rate. Initially the patient's presentation combined with the positive IFA seemed to be enough to treat the patient effectively. However, when she did not respond, the physician wanted to be certain that the diagnosis was correct. Western blot tests look for a number of antibodies to the spirochete and are considered positive only if multiple bands are present. Our patient may not have a good serologic response, but the positive PCR was evidence that the organism was still present.

3. There are many reasons why patients do not respond to therapy. First, the patient could have another disease that is similar in symptoms but does not respond to the therapy for *Borrelia.* Second, the

patient may have a strain that has become resistant to the antibiotics used treat the infection. Another reason could be that the dose of antibiotic was not sufficient to kill the organism. She may also not have an immune system that is effective in mounting a response to the spirochete. In fact, all of these reasons could be true in this patient, who may have a poor prognosis.

CHAPTER 49

LABORATORY METHODS FOR DIAGNOSIS OF PARASITIC INFECTIONS

Case Study 49-1

1. If no RBCs are seen in the cytoplasm and the trophozoites measure ≥ 12 μm, *Entamoeba histolytica* (true pathogen) cannot be differentiated from nonpathogenic *Entamoeba dispar*. Currently, there are immunoassays available that will confirm organisms in the *Entamoeba histolytica/E. dispar* group in the specimen, as well as one immunoassay that will confirm the presence of the pathogen *E. histolytica*.

2. It is important for the physician to recognize that a report of *E. histolytica/E. dispar* does not rule out the possibility of infection with a true pathogen *(E. histolytica);* however, if ingested RBCs are not present within the trophozoites, the organisms are more likely to be nonpathogenic *E. dispar*.

3. If immunoassays are not available or if the physician does not feel this procedure is required, the patient will be treated symptomatically. Therapy may or may not be given, depending on the presence or absence of symptoms.

Case Study 49-2

1. Patients with malaria may appear for diagnostic blood work when least expected. Laboratory personnel should be aware of the "STAT" nature of such requests and the importance of obtaining some specific patient history information. The typical textbook description of malarial parasites in an established infection may not match the organism morphology seen by the technologist who is examining a smear during an early infection. A request for malarial smears *should always be treated as a STAT request.*

2. Blood smears should be prepared upon admission of the patient. A fever pattern may not be apparent early in the course of the infection. Both thick and thin blood smears should be prepared. At least 200 to 300 oil immersion fields should be examined before the smears are considered negative. Wright's, Wright-Giemsa, or Giemsa stain can be used, although Giemsa is the stain of choice for blood parasites. The number of oil immersion fields examined may have to be increased if the patient has had any prophylactic medication during the past 48 hours (the number of infected cells may be decreased). One negative set of blood smears does not rule out malaria. The most important thing to remember is that even though a low parasitemia may be present on the blood smears, the patient may still be faced with a serious, life-threatening disease.

3. Species of malaria may overlap in terms of geographic areas, thus it is always important to try to "rule out" an infection with *Plasmodium falciparum*, the only species that often results in death.

4. Malarial parasites may be missed with use of automated differential instruments. Even with technologist review of the smears, a light parasitemia is very likely to be missed.

CHAPTER 50

LABORATORY METHODS IN BASIC MYCOLOGY

1. The organism is *Blastomyces dermatitidis*. This organism may be rapidly and accurately identified in the specimen upon direct examination, by the experienced microscopist. The characteristic features are the moderate size and the pathognomonic "broad-base bud." The calcofluor white/KOH is a rapid and useful method for detecting and, in some instances, such as this one, identifying fungi in clinical specimens.

2. This organism is one of the systemic dimorphic molds. These organisms have a yeast or a yeastlike "parasitic" form that occurs in tissue at 37° C but grows as a mold in the mycology laboratory on agar media at 25° C. The mold forms of these organisms have characteristic conidiation. Unfortunately, some saprophytic molds have similar conidiation. Therefore, confirmatory testing is necessary. In North America, the other systemic dimorphic molds most frequently encountered are *Coccidioides immitis* and *Histoplasma capsulatum*. *Blastomyces dermatitidis* and *Histoplasma capsulatum* are temperature dimorphic molds, and their identity may be confirmed by effecting a mycelial-to-yeast conversion, but this is time consuming. The identity of fungi suspected to be one of these pathogens is most effectively made using commercially available nucleic acid probes.

3. The extensive travel history of the patient suggested the possibility of an "exotic" infection. He had, however, recently been camping with his family in Wisconsin, where this fungus is endemic. He likely contracted the disease in the woods associated with decaying vegetable matter or possibly near a beaver's dam.

CHAPTER 51

LABORATORY METHODS IN BASIC VIROLOGY

1. Common viruses that cause disease this serious in young adults include influenza A and adenovirus. Other viruses that should be considered if implicated by epidemiologic investigation are the hantaviruses (e.g., sin nombre virus).

2. Sputum and upper respiratory tract secretions are excellent specimens for the detection of influenza and adenoviruses; however, adenoviruses may be present in the absence of overt disease. Specimens collected by invasive techniques, such as bronchoalveolar lavage and lung biopsy, contain virus present in the lung and lower airways and are more likely to preclude upper respiratory tract viruses of no clinical significance.

3. Rapid detection of influenza and adenovirus can be accomplished using fluorescent antibody staining of respiratory secretions. Negative stains can be confirmed with cell culture.

Resolution of case: Specimen collection occurred during bronchoscopy that was performed at the time of intubation. Fluorescent staining for adenovirus antigen was positive. The subsequent viral culture confirmed the presence of adenovirus. The patient was treated with ribavirin and eventually returned to normal health after a 1-month hospitalization.

CHAPTER 52

BLOODSTREAM INFECTIONS

1. The technologist read the Gram stain and saw gram-positive cocci in clusters. Because of this finding, 2 drops of the blood culture were inoculated into 0.5 mL of coagulase plasma. After 2 hours of incubation at 35° C, a clot formed, presumptively identifying the organism as *S. aureus.*

2. Generally, two blood cultures collected consecutively are sufficient, but in cases of endocarditis three cultures may be indicated. In adults a minimum of 20 mL should be collected for each culture.

3. Coagulase-negative staphylococci in blood cultures are most likely skin contaminants due to improper collection of the blood for culture. If a coagulase-negative staphylococcus with the same antibiogram is isolated from more than one blood culture collection, there is a higher likelihood that the isolate is causing an infection. To be certain, laboratories may perform a pulse-field gel electrophoresis on the DNA from the isolates. Only such testing confirms that two isolates are the truly the same. Laboratories should monitor their rate of contamination and attempt to keep it below 3%.

CHAPTER 53

INFECTIONS OF THE LOWER RESPIRATORY TRACT

1. Greater than 10 epithelial cells per low-power field is considered a reasonable rejection criterion; however, in children tracheal aspirates should not be rejected for culture.

2. Such a specimen is contaminated with mouth flora but probably was collected from the lower respiratory tract. Although not ideal, such a specimen could yield the causative agent of the pneumonia and should be cultured. The Gram stain should be reported from the areas with WBCs.

3. The sensitivity of Gram stain for the diagnosis of pneumococcal pneumonia is 62%, but that of culture is reported to be about 50%. In fact, only 60% of patients with pneumonia even produce sputum. Blood cultures are positive in less than 16% of cases of community-acquired pneumonia overall but are diagnostic indicators of the agent of pneumonia if they are positive.

4. As documented in Chapter 12, susceptibility testing for the pneumococcus is difficult. In this case, an MIC method must be used to accurately detect resistance. Because the pneumococcus requires blood to grow, the medium must contain some form of blood. Lysed horse blood is used in MIC trays; however, gradient strips can be used for testing on Mueller-Hinton agar supplemented with sheep blood (Figure A-2).

CHAPTER 54

UPPER RESPIRATORY TRACT INFECTIONS AND OTHER INFECTIONS OF THE ORAL CAVITY AND NECK

1. This organism is a gram-positive cocci in chains and is catalase-negative. The identification can be confirmed with a positive PYR test from a beta-hemolytic colony. Alternatively, agglutination with specific streptococcal grouping reagents can be used to definitively identify the organism. Bacitracin susceptibility has proved to be an unreliable indicator for identification.[*]

2. The pathogenic strains are PYR-positive, but the less virulent strains that are part of normal respiratory microbiota are PYR-negative.

3. *S. pyogenes* has two types of hemolysins-streptolysin O and streptolysin S. Streptolysin O requires oxygen to be active. If only streptolysin O is present, the colonies will not be hemolytic unless the agar surface is stabbed to allow oxygen to penetrate the

---

[*]York MK, Gibbs L, Brooks GF: Characterization of antimicrobial resistance in *Streptococcus pyogenes* from San Francisco Bay Area, *J Clin Microbiol* 37:1727, 1999.

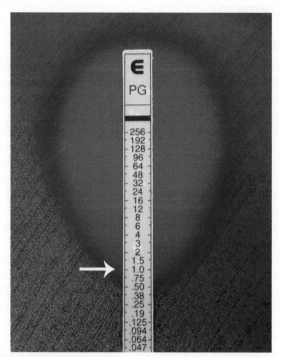

Figure A-2   A penicilin E-test strip showing an intermediately resistant *S. pneumoniae.*

subsurface colonies or the plate is incubated anaerobically (see Figure 17-2).

4. Most of these assays have a sensitivity of about 80% compared with culture on selective medium. It has not been shown that those that are missed by rapid testing are less significant. In fact, the same number of immune responses to infection are reported from both the patients who have positive antigen tests and those who have negative antigen test results. Consequently, it is the standard of care to follow up with culture on negative rapid antigen test results in children.

CHAPTER 55

MENINGITIS AND OTHER INFECTIONS OF THE CENTRAL NERVOUS SYSTEM

1. This organism is most likely *Neisseria meningitidis,* which is oxidase-positive. To confirm the identification, numerous commercial rapid tests are available. They take advantage of the fact that this species ferments maltose and glucose but not lactose. Other members of the genus have different patterns of fermentation. Some systems rely on specific enzyme production by the organism to identify *Neisseria* to the species level (see Table 13-1). Because of the importance of correctly diagnosing this infection, laboratories should confirm the results with more than one identification method.

2. Because resistance to penicillin is rare in the United States, testing is not recommended. Resistance has been reported occasionally in other countries, but the mechanism is not that of a beta-lactamase. Thus, beta-lactamase testing is not helpful for detection of resistance. If testing is performed, it should be done by an MIC method.

3. Using a cytocentrifuge, CSF can be concentrated 10 to 100-fold onto a glass slide (see Figure 55-3). Detection of 10^4 organisms/mL of fluid by Gram stain can be achieved with this preparation.

4. Rifampin or ciprofloxacin is usually given as prophylaxis for persons who have been exposed to this organism. Recently, several cases of laboratory-acquired infection with *N. meningitidis* have been reported. Laboratory workers should avoid creating aerosols when working with the agent and should use precautions to prevent inhalation of the organism. A vaccine is available that includes four or five of the capsular types; laboratory workers are encouraged to get vaccinated.

CHAPTER 56

INFECTIONS OF THE EYES, EARS, AND SINUSES

1. This patient has "swimmer's ear," which is caused by *Pseudomonas aeruginosa.* Pseudomonads are commonly found in fresh water, which is the likely source of the infecting organism.

2. When a beta-hemolytic dark colony with rough edges produces the characteristic grapelike odor, it is presumptively identified as *Pseudomonas aeruginosa.* If the isolate is oxidase-positive and indole-negative, the identification is definitive.

3. *Pseudomonas aeruginosa* has a characteristic blue-green pigment and a fluorescent green pigment, which can be enhanced by inoculation to special media (see Figure 24-1). Some other *Pseudomonas* spp. produce the fluorescent pigment, but these species do not grow at 42° C. By incubating the enhancing medium at 42° C, the organism can easily be separated from the other less common *Pseudomonas* spp.

4. *P. putida,* which has a foul odor; *P. monteilii,* and *P. fluorescens* are the other fluorescent *Pseudomonas* spp. *P. monteilii* is a newly identified member of the fluorescent group that, like *P. putida,* is unable to liquefy gelatin.

CHAPTER 57

INFECTIONS OF THE URINARY TRACT

1. Although the count is low, the presence of a uropathogen in pure culture in a young female can be significant.

2. The isolate was a catalase-positive, gram-positive cocci in clusters, which identified it as a *Staphylococcus* spp. It is important to identify *S. aureus* and *S. saprophyticus* in urinary specimens. *S. aureus* is coagulase-positive, and *S. saprophyticus* is coagulase-negative. Care must be used if one performs a Latex test for coagulase, rather than a test with rabbit plasma, because *S. saprophyticus* can give a false-positive result. *S. saprophyticus* is resistant to novobiocin. Testing for resistance is done using a disk impregnated with 5 μg/mL of novobiocin. Any zone less than 11 mm is considered resistant. In this case, the isolate was resistant. Even though there are other coagulase-negative staphylococci that are novobiocin-resistant (see Table 16-5), the isolate is presumed to be *S. saprophyticus*, because the other species are not known to cause urinary infections.

3. Susceptibility testing is not indicated, since these organisms are generally always susceptible to the usual drugs used to treat urinary tract infections, such as sulfamethoxazole-trimethoprim and nitrofurantoin.

4. Rapid tests tend to be falsely negative for sexually active females, because they do not detect low counts of bacteria, which are often found in infections in this age group. Also, they tend to be falsely negative when the infecting organism is gram-positive.

CHAPTER 58

GENITAL TRACT INFECTIONS

1. The physician wanted to be sure he had tested for other causes of infection. He also increased the sensitivity of detection of *Neisseria gonorrhoeae* by using two methods. The sensitivity of detection of *N. gonorrhoeae* in urine in females is not as high as that of appropriately collected culture of the cervix.

2. Usually, *N. gonorrhoeae* and *Chlamydia trachomatis* cause most of these infections. However, both enteric flora and anaerobic vaginal flora are also implicated.

3. This likely is *N. gonorrhoeae*. The organism identification can be confirmed with rapid sugar or enzyme assays specific for the organism.

CHAPTER 59

GASTROINTESTINAL TRACT INFECTIONS

1. *Shigella* is susceptible to exposure to air without a pH-buffering preservative. In addition, the WBCs also deteriorate in the stool in a short time.

2. A Cary-Blair type of preservative with a buffering capacity or buffered glycerol saline keeps the stool at neutral pH, which is essential to preserve the agents of diarrhea. Buffered glycerol saline is preferred for *Shigella* but is inhibitory to *Campylobacter*. Specimens in preservative should be stored at 4° C before inoculation to inhibit growth of normal fecal flora. Tests for WBCs cannot be done from stools in preservative.

3. *Shigella* is usually identified by agglutination of the bacteria with antisera to one of the groups A, B, C, and D. *Shigella sonnei* (group D) has a unique biochemical profile. It is ornithine decarboxylase–positive and ONPG-positive. Identification to species can be important in the determination of the spread of disease in the community. Many states require the laboratory to report the isolation of agents of gastroenteritis to their local health department.

CHAPTER 60

SKIN, SOFT TISSUE, AND WOUND INFECTIONS

1. Because the isolate has spores and only grows anaerobically, it is a *Clostridium*. This genus identification can be confirmed with a catalase test; *Clostridium* organisms are catalase-negative and *Bacillus*, the only other genus to have spores, is catalase-positive. There are two species of *Clostridium* common to humans that swarm and that have subterminal spores and are indole-negative. One is *C. sporogenes* and the other is *C. septicum*. Only *C. septicum* swarms over the entire plate; *C. sporogenes* is also adherent to the agar surface. This isolate is *Clostridium septicum*.

2. This patient had damage to the muscle probably caused by the *Clostridium*. This is known as myositis. Because the infection is anaerobic, cure will likely depend on debridement of the infected tissue with exposure to air as well as treatment with antibiotics.

3. Because this organism is present in the colon of normal individuals, its presence extraintestinally usually indicates a break in the integrity of the colon. This could be from colitis, from a colonic malignancy, and occasionally from infections with other bacteria, such as *E. coli* O157. The physician should do an investigation for these other diseases.

CHAPTER 61

NORMALLY STERILE BODY FLUIDS, BONE AND BONE MARROW, AND SOLID TISSUES

1. This organism likely is *Peptostreptococcus* spp., because it did not grow aerobically. This genus has been subdivided into a number of new genera. However, most laboratories use the general genus name

Peptostreptococcus spp., because it is not clinically important to separate the anaerobic gram-positive cocci further.

2. The patient is likely infected with her own skin flora, which was introduced at the time of the surgery. Because of the placement of a foreign object,

her body was unable to mount a sufficient response to clear the organism.

3. Because collection of joint fluid specimens is easily performed, ask the physician to aspirate fluid a second time. If the same organism is isolated, it is likely that the organism caused the infection.

Abscess Localized collection of pus.

Accessioning Receipt and recording of specimens delivered to the laboratory.

Accuracy Ability of a test under evaluation to match the results of an accepted standard test (i.e., "gold standard").

Acid-fast Characteristic of certain bacteria, such as mycobacteria, that involves resistance to decolorization by acids when stained by an aniline dye, such as carbolfuchsin.

Acquired immunity The specific response of the host to the infecting organism.

Acquired immunodeficiency syndrome (AIDS) Severe immune deficiency disease caused by human immunodeficiency virus (HIV-1) infection of T cells, characterized by opportunistic infections and other complications.

Acute serum Serum collected for antibody determination early in the course of an illness when little or no antibody would have been produced.

Acute urethral syndrome Lower urinary tract infection that may be difficult to differentiate from cystitis; seen most commonly in younger, sexually active females and caused by *Escherichia coli* (counts as low as 100 per milliliter may be significant in this situation), *Chlamydia*, and other organisms.

Aerobe, obligate Microorganism that lives and grows freely in air and cannot grow anaerobically.

Aerosol Atomized particles suspended in air; in context of this textbook, microorganisms suspended in air.

Aerotolerant Ability of an anaerobic microorganism to grow in air, usually poorly, especially after initial anaerobic isolation.

Agarose gel electrophoresis Separation of proteins based on molecular weight by electrical current–stimulated movement through a semisolid gel matrix.

Agglutination Aggregation or clumping of particles, such as bacteria when exposed to specific antibody.

AIDS Acquired immunodeficiency syndrome.

Aminoglycosides Group of related antibiotics including streptomycin, kanamycin, neomycin, tobramycin, gentamicin, and amikacin.

Amniotic Pertaining to the innermost fetal membrane forming a fluid-filled sac.

Amplicon Amplified nucleic acid product.

Anaerobe, obligate Microorganism that grows only in complete or nearly complete absence of air or molecular oxygen.

Anamnestic response More rapid production of antibodies in response to exposure to an antigen previously encountered.

Anamorph Asexual fungal form.

Anergy Absence of reaction to antigens or allergens.

Angiogenesis Forming new capillaries from preexisting ones.

Antibiogram A cumulative susceptibility report that tracks resistance or susceptibility of commonly isolated organisms to commonly administered antimicrobials.

Antibiotic Substance produced by a microorganism that inhibits or kills other microorganisms; a broad-spectrum antibiotic is therapeutically effective against a wide range of bacteria.

Antibody Substance (immunoglobulin) formed in blood or tissues that interacts only with antigen that induced its synthesis (e.g., agglutinin).

Anticoagulant Substance used to prevent clotting of specimens such as blood, bone marrow, and synovial fluid so that organisms do not become bound up in the clot.

Antigen Molecular structure that is capable of stimulating production of antibody.

Antimicrobial Chemical substance produced either by a microorganism or by synthetic means that is capable of killing or suppressing growth of microorganisms.

Antiseptic Compound that stops or inhibits growth of bacteria without necessarily killing them.

Apical Top.

Apoptosis Programmed cell death.

Arboviruses Arthropod-borne viruses.

Arthritis, septic Infection of synovial tissue and joint fluid of one or more joints; characterized by joint pain, stiffness, swelling, and fever.

Arthroconidium Spore formed by septation of a hypha and subsequent separation of septa.

Ascites Condition in which amounts of fluids increase and accumulate in the peritoneal cavity during infection or an inflammatory process.

Ascitic fluid Serous fluid in peritoneal cavity.

Ascocarp A large, saclike structure that contains sexual spores of fungi.

Aseptic meningitis Meningitidis characterized by an increase in lymphocytes and other mononuclear cells in the cerebrospinal fluid and negative bacterial and fungal cultures.

Aspiration Inhalation of a fluid or solid.

ASR Analyte specific reagent.

Assimilation Utilization of nutrients. Assimilation tests are used to determine whether yeasts are able to grow with only a single carbohydrate or nitrate; these tests are useful for classification of yeasts.

ATCC American Type Culture Collection.

Autotroph Organism that can utilize inorganic carbon sources (CO_2).

Auxotroph Differing from the wild strain (prototroph) by an additional nutritional requirement.

Avid The property of binding strongly, such as an antibody that strongly binds to an antigen.

Avidity Firmness of union of two substances; used commonly to describe union of antibody to antigen.

B cells Lymphocytes involved in antibody production.

Bacteremia Presence of viable organisms in blood.

Bacterial vaginosis Noninflammatory condition in vagina characterized by foul-smelling vaginal discharge and presence of mixed bacteria.

Bactericidal Term used to describe a drug that kills microorganisms.

Bacteriophage Virus that infects a bacterial cell, sometimes bringing about its lysis.

Bacteriostatic Term used to describe a drug that inhibits growth of an organism without killing it.

Bacteriuria Presence of bacteria in urine.

BCG Bacille Calmette-Guérin, an attenuated strain of *Mycobacterium tuberculosis* used for immunization.

Beta-lactamases Enzymes that destroy penicillins and/or cephalosporins and are produced by a variety of bacteria.

Bifurcated Divided into two branches.

Biofilm Well-organized microcolonies of bacteria usually enclosed in polymer matrices that are separated by water channels that remove wastes and deliver nutrients.

Biological safety cabinet Enclosure in which one can work with relatively dangerous organisms without risk of acquiring or spreading infection caused by them. These cabinets, also called biosafety hoods, vary in design according to the nature of the agents to be worked with. The simpler ones maintain a negative pressure within the work area and a laminar air curtain, both of which operate to prevent escape of organisms from the interior of hood. Air that is exhausted may be passed through a high-efficiency bacterial filter that traps all microorganisms that are anticipated or may be passed through a furnace that incinerates any organisms.

Bioluminescence Light generation by living organisms.

Biopsy Removal of tissue from a living body for diagnostic purposes (e.g., lymph node biopsy).

Biotin Small vitamin with two binding sites, one of which can bind covalently with nucleic acid, leaving the other free to form a strong bond with the protein avidin, which, in turn, can be bound to enzymes. The system is used as a label for nucleic acid probe detection.

Biotype   Biologic or biochemical type of an organism. Organisms of the same biotype display identical biologic or biochemical characteristics. Certain key markers are used to define and recognize biotypes in tracing the spread of organisms in the environment and in epidemics or outbreaks.

Blastoconidium   A spore formed by budding, as in yeasts.

Blepharitis   Inflammation of eyelids.

B lymphocytes (or B cells)   Bursa-derived lymphocytes important in humoral immunity.

Breakpoint   Level of an antibacterial drug achievable in serum; organisms inhibited by this level of drug are considered susceptible. In certain situations, clinicians strive to achieve serum or body fluid levels several times that of the breakpoint.

Bright-field microscopy   Conventional microscopy in which the object to be viewed is illuminated from below.

Bronchial lavage   Similar to bronchial washings but this term implies instillation of a larger volume of fluid before aspiration. Alveolar organisms may be present in the lavage.

Bronchial washings   Fluid that may be aspirated from bronchial tree during bronchoscopy.

Bronchitis   Inflammation of mucous membranes of bronchi; often caused by infectious agents, viruses in particular.

Bronchoscopy   Examination of bronchi through a bronchoscope, a tubular, illuminated instrument introduced through the trachea (windpipe).

BSC   Biological safety cabinet.

Bubo   Inflammatory enlargement of lymph node, usually in the groin or axilla.

Buffy coat   Layer of white blood cells and platelets above red blood cell mass when blood is sedimented.

Bullae   Large blebs or blisters, filled with fluid, in or just beneath the epidermal layer of skin.

Bursitis   Inflammation of a bursa, which is a small sac lined with synovial membrane and filled with fluid interposed between parts that move on each other.

Butt   Lower portion of medium in a tube in which the medium is dispensed such that the lower portion fills the tube entirely (i.e., the butt) while the upper portion is distributed in the form of a slanted surface, leaving an air space between the slant and the opposite wall of the tube.

Butyrous   Butterlike consistency.

Calibrated loop   Bacteriologic loop that is carefully calibrated to deliver a specified volume of fluid, as long as directions are followed carefully and the loop has not been damaged; used as a simple means of quantitating the number of organisms present, especially for urine culture.

CAMP   A diffusible extracellular protein named after Christie, Atkins, and Munch-Peterson that is produced by certain organisms (e.g., group B streptococci) and acts synergistically with the beta lysin of *S. aureus* to cause enhanced lysis of red blood cells.

Candle jar   A jar with a lid providing a gas-tight seal in which a small white candle is placed and lit after the culture plates have been placed inside. Candle will burn only until the oxygen concentration has been lowered to the point at which it will no longer support the flame. Atmosphere of such a jar has a lower oxygen content than room air and a carbon dioxide content of about 3%.

Cannula   An artificial tube for insertion into a tube or cavity of the body.

CAPD   Chronic ambulatory peritoneal dialysis.

Capnophilic   Term used to describe microorganisms that prefer an incubation atmosphere with increased carbon dioxide concentration.

Capsid   Protein layer or coat surrounding viral nucleic acid core.

Capsomere   Protein subunits that serve as components of the viral capsid.

Capsule   Gelatinous material surrounding bacterial cell wall, usually of polysaccharide nature.

Carrier   One who harbors a pathogenic organism but is not affected by it.

Catalase   Bacterial enzyme that breaks down peroxides with liberation of free oxygen.

Catheter   Flexible tubular (rubber or plastic) instrument used for withdrawing fluids from (or introducing fluids into) a body cavity or vessel (e.g., urinary bladder catheter).

Cation   A positive ion.

Cell line   A cell culture that has been passed (subcultured) in vitro.

Cell line, continuous   Line of tissue cells that is maintained by serial culture of an established cell line.

Cell line, primary   Line of tissue cells established by cutting up fresh tissue, often kidney, into tiny pieces, trypsinizing, and putting in a flask with appropriate medium.

Cell-mediated immunity   Human specific immune response carried out by special lymphocytes of the T (thymus-derived) class.

Cellulitis   Inflammation of subcutaneous tissue.

Cerebriform   With brainlike folds.

Cervical   Pertaining either to the neck or to the cervix of the uterus.

CF   Complement fixation.

CFU   Colony-forming unit (i.e., colony count).

Charcot-Leyden crystals   Slender crystals shaped like a double pyramid with pointed ends, formed from the breakdown products of eosinophils and found in feces, sputum, and tissues; indicative of an immune response that may have parasitic or nonparasitic causes.

Chemotherapeutic   Chemical agent used to treat infections (e.g., sulfonamides).

Chlamydospore   Thick-walled spore formed from a vegetative cell.

Chorioamnionitis   Infection of the uterus and its contents during pregnancy.

Chromatography   Method of chemical analysis by which a mixture of substances is separated by fractional extraction or adsorption or ion exchange on a porous solid.

Chromogen   Bacterial species whose colonial growth is pigmented (e.g., *Flavobacterium* spp., yellow).

Chromogenic   Giving rise to color, as chromogenic substrates for colored products of biochemical reactions or chromogenic bacteria that produce pigmented colonies.

Clavate   Club-shaped.

Clone   Group of microorganisms of identical genetic makeup derived from a single common ancestor.

CMI   Cell-mediated immunity.

CNS   Central nervous system.

Coagglutination   Agglutination of protein A–containing cells of *Staphylococcus aureus* coated with antibody molecules when exposed to corresponding antigen.

Coenocytic hyphae   Sparsely septated.

Colitis   Inflammation of mucosa of colon.

Colony   Macroscopically visible growth of a microorganism on a solid culture medium.

Commensal   Microorganism living on or in a host but causing the host no harm.

Community-acquired   Pertaining to outside the hospital, such as a community-acquired infection.

Complement fixation test   Antigen-antibody test based on fixation of complement in the presence of both elements and use of an indicator system to determine whether complement has been fixed.

Congenital   Existing before or at birth.

Conidia   Asexual spores.

Conjugation   Passing genetic information between bacteria by transferring chromosomal material, often via pili.

Conjunctivitis   Inflammation of the conjunctivae or membranes of the eye and eyelid.

Convalescent serum   Serum collected later in the course of an illness than the acute serum, usually at least 2 weeks after initial collection.

CPE   Cytopathogenic (cytopathic) effect; visual effect of virus infection on cell culture.

Creutzfeldt-Jakob disease   Debilitating prion-caused disease characterized by dementia, ataxia, delirium, stupor, coma, and death; has been transmitted by organ transplant.

Crossing point   See "threshold cycle."

Croup   Inflammation of upper airways (larynx, trachea) with respiratory obstruction, often caused by virus infections in children.

CSF   Cerebrospinal fluid.

Culdocentesis   Aspiration of fluid from the cul-de-sac by puncture of the vaginal vault.

Cystitis   Inflammation of urinary bladder, most often caused by bacterial infection

Cytokine   Group of biochemicals that is a key component of inflammation.

Cytopathic effect (CPE)   Alteration in cell morphology resulting from viral infection of a cell culture monolayer.

Cytotoxin   Toxin that produces cytopathic effects in vivo or in a tissue culture system.

Dark-field microscopy   Technique used to visualize very small microorganisms or their characteristics by a system that permits light to be reflected or refracted from the surface of objects being viewed.

Debridement   Surgical or other removal of nonviable tissue.

Decontamination Process of rendering an object or area safe for unprotected people by removing or making harmless biologic or chemical agents.

Decubitus ulcer A craterlike defect in skin and subcutaneous tissue caused by prolonged pressure on the area. This occurs primarily over bony prominences of the lower back and hips in individuals who are unable to care for themselves well and unable to roll or move periodically; also known as a *pressure sore* or *bedsore*.

Definitive host Host in which the sexual reproduction of a parasite occurs.

Dematiaceous Presence of pigmentation in fungal hyphae or spores.

Denaturation Process in which double-stranded DNA becomes single-stranded by heating or chemical means; also referred to as *melting*.

Dermatophyte A fungus parasitic on skin, hair, or nails.

Dermis Layer of skin beneath the epidermis that is composed of dense connective tissue rich in blood and nerve supply.

Desquamation Shedding or scaling of skin or mucous membrane.

DFA Direct fluorescent antibody test.

DIC Disseminated intravascular coagulation.

Dichotomous Branching in two directions.

Diluent Fluid used to dilute a substance.

Dimorphic fungi Fungi with both a mold phase and a yeast phase.

Direct wet mount A preparation from clinical material suspended in sterile saline or other liquid medium on a glass slide and covered with a coverslip; used for microscopic examination to detect microorganisms in clinical material and, in particular, to detect motility directly.

Disinfectant Agent that destroys or inhibits microorganisms that cause disease.

Disseminated intravascular coagulation (DIC) Disastrous complication of sepsis.

DNA Deoxyribonucleic acid, the lipoprotein molecule that contains the genetic code for most living things.

DNA minor groove A location on double-stranded DNA in which the strand backbones are closer together on one side the helix than on the other.

Droplet nucleus A tiny aerosolized particle that, because of its lack of mass, may stay suspended in air for extended periods.

Duplex Two nucleic acid strands that have complementary base sequences that have specifically bonded with each other and formed a double-stranded molecule.

Dx Diagnosis.

Dysentery Inflammation of the intestinal tract, particularly the colon, with frequent bloody stools (e.g., bacillary dysentery).

Dysgonic Growing poorly (bacterial cultures).

Dysuria Painful or difficult urination.

Ectoparasite Organism that lives on or within skin.

Ectothrix Outside of hair shafts.

Edema Excessive accumulation of fluid in tissue spaces.

Effusion Fluid escaping into a body space or tissue (e.g., pleural effusion).

Eh Oxidation-reduction potential.

Elementary body The infectious stage of *Chlamydia* or a cellular inclusion body of a viral disease.

ELISA Enzyme-linked immunosorbent assay.

Elution Process of extraction by means of a solvent.

EMB Eosin-methylene blue (agar plate).

Empyema Accumulation of pus in a body cavity, particularly empyema of the thorax or chest.

Encephalitis Inflammation of the brain.

Endocarditis A serious infection of the endothelium of the heart, usually involving leaflets of the heart valves where destruction of valves or distortion of them by formation of vegetations may lead to serious physiologic disturbances and death; also, an inflammation of the endocardial surface (much less common).

Endocervix Mucous membrane of the cervical canal.

Endogenous Developing from within the body.

Endoparasite Parasite that lives within the body.

Endophthalmitis Inflammation of internal tissues of eye; may rapidly destroy the eye.

Endothelium Squamous epithelium lining blood vessels.

Endothrix Within the hair shaft.

Endotoxin Substance containing lipopolysaccharide complexes found in the cell wall of bacteria, principally gram-negative bacteria; believed to play an important role in many of the complications of sepsis such as shock, DIC, and thrombocytopenia.

Enteric fever Typhoid fever; paratyphoid fever.

Enteroinvasive Capable of invading the mucosal surface and sometimes the deeper tissues of the bowel.

Enterotoxin Toxin affecting the cells of the intestinal mucosa.

Enzyme-linked immunosorbent assay (ELISA) An immunologic assay that uses an enzyme conjugated to antibodies to produce a visible endpoint.

Epidemiology The study of the occurrence and distribution of disease and factors that control presence or absence of disease.

Epidermis Outermost layer of skin made of layered squamous epithelial cells.

Epididymitis Inflammation of the epididymis characterized by fever and pain on one side of the scrotum; seen as a complication of prostatitis and cystitis.

Epiglottitis Inflammation of the epiglottis, a structure that prevents aspirating swallowed food and fluids into the tracheobronchial tree; a serious infection because the swollen epiglottis may block the airway.

Epithelium Tissue composed of contiguous cells that forms the epidermis and lines hollow organs and all passages of the respiratory, digestive, and genitourinary systems.

Erysipelas An acute cellulitis caused by group A streptococci.

Erythema Redness of the skin from various causes.

Erythrasma A minor, superficial skin infection caused by *Corynebacterium minutissimum*.

Eschar A dry scar, particularly one related to a burn.

Etiology Cause or causative agent.

Eugonic Growing luxuriantly (bacterial cultures).

Eukaryotic Organisms with a true nucleus, in contrast to bacteria and viruses.

Exanthem Skin eruption as a symptom of an acute disease, usually viral.

Exoantigen test In vitro immunodiffusion test method for identifying fungal hyphae as *Histoplasma*, *Blastomyces*, or *Coccidioides*.

Exoerythrocytic cycle Portion of the malarial life cycle occurring in the vertebrate host in which sporozoites, introduced by infected mosquitoes, penetrate the parenchymal liver cells and undergo schizogony, producing merozoites, which then initiate the erythrocytic cycle.

Exogenous From outside the body.

Exotoxin A toxin produced by a microorganism that is released into the surrounding environment.

Exudate Fluid that has passed out of blood vessels into adjacent tissues or spaces; high protein content.

Facultative anaerobe Microorganism that grows under either anaerobic or aerobic conditions.

Fascia Membranous covering of muscle.

Fermentation Anaerobic decomposition of carbohydrate.

Filamentous Threadlike.

Fimbriae Proteinaceous fingerlike surface structures of bacteria that provide for adherence to host surfaces.

Fistula Abnormal communication between two surfaces or between a viscus or other hollow structure and the exterior.

Flagella Complex structures mostly composed of the protein flagellin that are responsible for bacterial motility.

Floccose Cottony, in tufts.

Flocculation test Antigen-antibody test in which a precipitin end product forms macroscopically or microscopically visible clumps.

Fluorescent Emission of light by a substance (or a microscopic preparation) while acted on by radiant energy, such as ultraviolet rays, as in the immunofluorescent procedure.

Fluorochrome A dye that becomes fluorescent or self-luminous after exposure to ultraviolet light.

Fluorophore A fluorescent molecule that can absorb light energy and then is elevated to an excited state that is released as fluorescence in the absence of a quencher.

Fomite Any inanimate object that may be contaminated with disease-causing microorganisms and thus serves to transmit disease.

FTA Fluorescent treponemal antibody.

FTA-ABS Fluorescent treponemal antigen-antibody absorption test; indirect fluorescent antibody stain used to detect antibodies directed against whole-cell antigens of *Treponema pallidum* (syphilis bacillus).

Fungemia Presence of viable fungi in blood.

FUO Fever of unknown origin.

Fusiform Spindle-shaped, as in the anaerobe *Fusobacterium nucleatum*.

Gamma hemolysis No hemolysis of red blood cells.

Gangrene Death of a part of tissue resulting from disease, injury, or failure of blood supply.

Gas-liquid chromatography (GLC)   A method for separating substances by allowing their volatile phase to flow through a heated column with a carrier gas and measuring the time required to detect their presence at the distal end of the column.

Gastric aspirate   Fluid that may be aspirated from the stomach via a tube placed in the stomach by way of the nose or mouth.

Gastroenteritis   Inflammation of the mucosa of the stomach and intestines.

GC   *Gonococcus*.

Genotype   Related to characteristics of an organism's genetic makeup, that is, genus and constituent nucleic acids.

Germicide   An agent that destroys germs; disinfectant.

Germ tube   Tubelike process, produced by a germinating spore, that develops into mycelium.

Glabrous   Smooth.

GLC   Gas-liquid chromatography.

Granulocytopenia   Reduced number of granulocytic white blood cells in the blood.

Granuloma   Aggregation and proliferation of macrophages to form small (usually microscopic) nodules.

HAI   Hemagglutination inhibition.

Halophilic   Preferring high halide (salt) content.

Hansen's disease   Leprosy, a disease cause by *Mycobacterium leprae*.

Hemadsorption   Ability of certain virally infected cells to bind erythrocytes; mediated by glycopeptide adherence molecules (induced by viral activities within the cell) on the cell's surface.

Hemagglutination   Agglutination of red blood cells caused by certain antibodies, virus particles, or high molecular weight polysaccharides.

Hematogenous   Disseminated by the bloodstream.

Hemolysis, alpha   Partial destruction of, or enzymatic damage to, red blood cells in a blood agar plate, leading to greenish discoloration about the colony of the organism producing the alpha hemolysin.

Hemolysis, beta   Total lysis of red blood cells about a colony on a blood agar plate, leading to a completely clear zone surrounding the colony.

Hemolysis, gamma   No hemolysis is seen with organisms classed as gamma-hemolytic; nonhemolytic would be a better designation.

HEPA   High-efficiency particulate air filter; used in biological safety cabinets to trap pathogenic microorganisms.

Herpes   Inflammation of the skin characterized by clusters of small vesicles (e.g., caused by herpes simplex); disease caused by herpes simplex virus.

Heterotroph   Organism that requires an organic carbon source.

High-pressure liquid chromatography (HPLC)   Similar to GLC but capable of higher resolution because of increased pressure of liquid carrier that runs through the column.

HPLC   High-pressure (or performance) liquid chromatography.

Humoral immunity   Immunity affected by antibody.

Hyaline   Colorless, transparent.

Hybridoma   The product of fusion of an antibody-producing cell and an immortal malignant antibody-producing cell.

Hydrolysis   Breakdown of a substrate by an enzyme that adds the components of water to key bonds within the substrate molecule.

Hyperalimentation   Process by which nutrition (literally "extra nutrition") is provided; typically administered intravenously in subjects who are not able to absorb foods well from the gut because of disease of the bowel, in subjects in whom it is desirable to put the bowel "at rest" to promote healing, and in malnourished individuals to improve their nutritional status (e.g., before surgery); usually done over an extended period and requires the use of a special-access intravenous catheter such as a Hickman catheter.

Hyperemia   Increased blood in a part, resulting in distention of blood vessels.

Hypertonic   Hyperosmotic.

Hypertrophy   Increased size of an organ resulting from enlargement of individual cells.

Hypha   Tubular cell making up the vegetative portion of mycelium of fungi.

Hypoxia   Decreased oxygen content of tissues.

IFA   Indirect fluorescent antibody; test that detects antibody by allowing an antibody to react with its substrate and adding a second fluorescein dye–labeled antibody that will bind to the first.

Ig, IgG, IgM, etc.   Immunoglobulin, immunoglobulin G, immunoglobulin M, etc.

Immunodiffusion   Detection of antigen or antibody by observing the precipitin line formed in a semisolid gel matrix when homologous antigens and antibodies are allowed to diffuse toward each other and react.

Immunofluorescence   Microscopic method of determining the presence or location of an antigen (or antibody) by demonstrating fluorescence when the preparation is exposed to a fluorescein-tagged antibody (or antigen) using ultraviolet radiation.

Immunoglobulin   Synonymous with antibody; five distinct classes have been isolated: IgG, IgM, IgA, IgE, and IgD.

Immunoperoxidase stain   Combination of an enzyme that catalyzes production of a colored product with an antibody to facilitate detection of certain antigens, particularly viral antigens.

Immunosuppression   Depression of the immune response caused by disease, irradiation, or administration of antimetabolites, antilymphocyte serum, or corticosteroids.

Impetigo   Acute inflammatory skin disease, caused by streptococci or staphylococci, characterized by vesicles and bullae that rupture and form yellow crusts.

Inclusion bodies   Microscopic bodies, usually within body cells; thought to be virus particles in morphogenesis.

Indigenous flora   Normal or resident flora.

Induration   Abnormal hardness of a tissue or part resulting from hyperemia or inflammation, as in a reactive tuberculin skin test.

Infection   Invasion by and multiplication of microorganisms in body tissue resulting in disease.

Inhibitory quotient   Ratio of the average peak achievable level of antibiotic in a body fluid from which the organism was isolated to the MIC of that organism.

Insertion sequence   Transposable element containing genes that encode the information required to move among plasmids and chromosomes.

In situ hybridization   Detection of nucleic acid of a pathogenic organism in tissue sections by separating the DNA into single-stranded molecules and allowing a labeled strand of homologous DNA to bind to the target. The target is visualized by developing the label (i.e., enzymatic precipitate, fluorescence, or radiolabel).

Inspissation   Process of making a liquid or semisolid medium thick by evaporation or absorption of fluid.

Interfacing   Communicating.

Intermediate host   Required host in the life cycle in which essential larval development must occur before a parasite is infective to its definitive host or to additional intermediate hosts.

Intramuscular (intraperitoneal, intravenous)   Within the muscle (peritoneum, vein), as in intramuscular injection.

In vitro   Literally, within glass (i.e., in a test tube, culture plate, or other nonliving material).

In vivo   Within the living body.

Involution forms   Abnormally shaped bacterial cells occurring in an aging culture population.

Iodine tincture   Iodine in alcohol.

Ion-exchange chromatography   Separation of components of a solution by chromatography based on the reversible exchange of ions in the solution with ions present in or on an external matrix.

Isotonic   Of the same osmolality of body tissues, red blood cells, bacteria, etc.

Keratitis   Inflammation of the cornea.

KIA   Kligler's iron agar (tube).

KOH   Potassium hydroxide.

Lag phase   Period of slow microbial growth that occurs following inoculation of the culture medium.

Laked blood   Hemolyzed blood; hemolysis may be affected in various ways, but alternate freezing and thawing is a simple method.

Laminar flow   Nonturbulent flow of air in layers (flowing in a vertical direction in the case of a biosafety hood).

Latent   Not manifest; potential.

Latex agglutination   Agglutination of latex particles coated with antibody molecules when exposed to the corresponding antigen.

LCR   Ligase chain reaction.

Lectin   Naturally produced proteins or glycoproteins that can bind with carbohydrates or sugars to form stable complexes.

Legionnaires' disease   Febrile and pneumonic illness caused by *Legionella* species.

Leishman-Donovan (L-D) body   Small, round intracellular form (called amastigote or leishmanial stage) of *Leishmania* spp. and *Trypanosoma cruzi*.

Leukocytosis   Elevated white blood cell count.

Leukopenia   Low white blood cell count.

LGV   Lymphogranuloma venereum; the name for certain strains of *Chlamydia trachomatis* that cause a systemically expressed sexually transmitted disease.

Lipopolysaccharide   Carbohydrate-lipid complex; integral substance in gram-negative cell walls. Also known as endotoxin.

Liposome Small, closed vesicle consisting of a single lipid bilayer.

Logarithmic phase Period of maximal growth rate of a microorganism in a culture medium.

LPS Lipopolysaccharide; see endotoxin.

Lysis Disintegration or dissolution of bacteria or cells.

Lysogeny Process by which a viral genome is integrated into that of its host bacterium.

MAC *Mycobacterium avium* complex.

Macroconidia Large, usually multiseptate, club- or spindle-shaped fungal spores.

Mass spectrometry Method for determining composition of a substance by observing its volatile products during disintegration and comparing them with known standards.

MBC Minimum bactericidal concentration.

Media, differential Media that permit ready recognition of a particular organism or group of organisms by virtue of facilitating recognition of a natural product of the organism being sought or by incorporating an appropriate substrate and indicator system so that organisms possessing certain enzymes are readily recognized.

Media, enrichment Media, usually liquid, that favor the growth of one or more organisms while suppressing most of the competing flora in a specimen with a mixture of organisms.

Media, selective Culture media that contain inhibitory substances or unique growth factors such that one particular organism or group of organisms is conferred a real advantage over other organisms that may be found in a mixture. Efficient selective media selects out only the organism or organisms being sought, with little or no growth of other types of organisms.

Mediastinum Space in the middle of the chest between the medial surfaces of the two pleurae.

Melioidosis Disease caused by *Burkholderia pseudomallei.*

Melting temperature (Tm) The temperature at which 50% of double-stranded DNA becomes single-stranded.

Meningitis Inflammation of the meninges, the membranes that cover the brain and spinal cord (e.g., bacterial meningitis).

Meningoencephalitis Concomitant meningitis that occurs with encephalitis (inflammation of the brain parenchyma).

Merozoite Product of schizogonic cycle in malaria that invades red blood cells.

Mesenteric adenitis Inflammation of mesenteric lymph nodes.

Mesentery A fold of the peritoneum that connects the intestine with the posterior abdominal wall.

Metastatic Spread of an infectious (or other) process from a primary focus to a distant one via the bloodstream or lymphatic system.

MHA-TP Microhemagglutination test for antibody to *Treponema pallidum.*

MIC Minimum inhibitory concentration.

Microaerobic Requiring a partial pressure of oxygen less than that of atmospheric oxygen for growth. New term for "microaerophilic."

Microaerophile, obligate Microorganism that grows only under reduced oxygen tension and cannot grow aerobically or anaerobically.

Microaerophilic See "microaerobic."

Microconidia Small, single-celled fungal spores.

Microfilaria Embryos produced by filarial worms and found in the blood or tissues of individuals with filariasis.

Miliary Of the size of a millet seed (0.5 to 1.0 mm); characterized by the formation of numerous lesions of the above size distributed rather uniformly throughout one or more organs.

Minimum bactericidal concentration (MBC) The minimum concentration of antimicrobial agent needed to yield a 99.9% reduction in viable colony-forming units of a bacterial or fungal suspension.

Minimum inhibitory concentration (MIC) The minimum concentration of antimicrobial agent needed to prevent visually discernible growth of a bacterial or fungal suspension.

Mixed culture (pure culture) More than one organism growing in or on the same culture medium, as opposed to a single organism in pure culture.

Monoclonal antibody Antibody that is derived from a single cell producing one antibody molecule type that reacts with a single epitope.

Monolayer A confluent layer of tissue culture cells one cell thick.

MOTT Mycobacteria other than *Mycobacterium tuberculosis.*

Mucopurulent Term used to describe material containing both mucus and pus (e.g., mucopurulent sputum).

Mucosa A mucous membrane.

Multiple myeloma Malignancy involving antibody-producing plasma cells.

Multiplex PCR A PCR reaction with more than one primer pair in the reaction mixture.

Mutation Change in the original nucleotide sequence of a gene or genes.

Mycelium Mass of hyphae making up a colony of a fungus.

Mycetoma Chronic infection, usually of feet, caused by various fungi or by *Nocardia* or *Streptomyces,* resulting in swelling and sinus tracts; pulmonary mycetoma is a mass of fungal hyphae ("fungus ball") growing in a cavity formed during previous tuberculosis infection or other pathologic condition.

Mycoses Diseases caused by fungi (e.g., dermatomycosis, fungal infection of the superficial skin).

Mycotic aneurysm Bacterial infection causing inflammatory damage and weakening of an arterial wall.

Myocarditis Inflammation of the heart muscle.

Myositis Inflammation of a muscle, sometimes caused by infection as in pyomyositis; an infection caused by *Staphylococcus aureus* that leads to small abscesses within the muscle substance.

Nares External openings of nose (i.e., nostrils).

Nasopharyngeal Pertaining to the part of the pharynx above the level of the soft palate.

Necrosis Pathologic death of a cell or group of cells.

Necrotizing fasciitis A very serious, painful infection involving the fascia (membranous covering) of one or more muscles; may spread widely in short periods since there is no anatomic barrier to spread in this type of infection.

Neonatal First 4 weeks after birth.

Nested PCR A PCR assay that involves the sequential use of two primer sets.

Neurotrophic Having a selective affinity for nerve tissue. Rabies is caused by a neurotrophic virus.

NGU Nongonococcal urethritis.

Nick translation Use of enzymes to break DNA and repolymerize small sections of the molecule, usually to label the DNA with a radioactive nucleotide.

Nomenclature Naming of microorganisms according to established rules and guidelines.

Nonphotochromogens Slow-growing, nonpigmented mycobacteria.

Nontuberculous mycobacteria (NTM) All species of mycobacteria that do not belong to *M. tuberculosis* complex.

Nosocomial Pertaining to or originating in a hospital, for example, nosocomial infection.

Nucleic acid hybridization Process by which the single-stranded probe unites with complementary DNA.

Nucleic acid probe Piece of labeled single-stranded DNA used to detect complementary DNA in clinical material or a culture that specifically identifies the presence in these materials of an organism identical to that used to make the probe.

Nucleocapsid Name of viral particle that includes virus nucleic acid core enclosed in the protein capsid coat.

O&P Ova and parasites.

O-F Oxidation-fermentation medium.

Octal numbers Numbers used in computer databases to identify biochemical profiles of organisms and thus their identification.

Oil immersion microscopy Use of immersion oil to fill the space between the slide being studied and the special objective of the microscope; this keeps the light rays from dispersing and provides good resolution at high magnification (total magnification of 1000×).

Oncogenic Possessing the potential to cause normal cells to become malignant; causing cancer.

ONPG o-nitrophenol-β-galactopyranoside (β-galactosidase test).

Operculated ova Ova possessing a cap or lid.

Opsonize To facilitate destruction of pathogens by phagocytic ingestion or lysis by complement through the action of adherent antibodies.

Optical density A measurement of turbidity.

Osteomyelitis Inflammation of the bone and the marrow.

Otitis Inflammation of the ear from a variety of causes, including bacterial infection; otitis media is inflammation of the middle ear.

Oxidation A metabolic pathway of the microorganism that involves use of oxygen as a terminal electron acceptor. This type of reaction occurs in air.

Oxidation-reduction potential Electromotive force exerted by a nonreacting electrode in a solution containing the oxidized and reduced forms of a chemical, relative to a standard hydrogen electrode; the more negative the value, the more anaerobic conditions are.

Pandemic   Epidemic over a wide geographic area, or even worldwide.

Paracentesis   Surgical transcutaneous puncture of the abdominal cavity to aspirate peritoneal fluid.

Parasite   Organism that lives on or within and at the expense of another organism.

Parenteral   Route of administration of a drug other than by mouth; includes intramuscular and intravenous administration.

Parotitis   Inflammation of the parotid gland, the largest of the salivary glands; mumps is the most common cause of this.

Paroxysm   Rapid onset (or return) of symptoms; term usually applies to cyclic recurrence of malaria symptoms, which are chills, fever, and sweating.

Pathogen   Microorganism that causes infection and/or disease.

Pathogenic   Producing disease.

Pathogenicity islands   Stretches of DNA that contain genes that are associated with bacterial virulence and are absent in from avirulent or less virulent strains of the same species.

Pathologic   Caused by or involving a morbid condition, as a pathologic state.

PCR   Polymerase chain reaction.

Penicillin-binding protein   Enzymes essential for bacterial cell wall production.

Penicillinase (β-lactamase I)   Enzyme produced by some bacterial species that inactivates the antimicrobial activity of certain penicillins (e.g., penicillin G).

Peptidoglycan   Bacterial cell wall or murein layer that gives the bacterial cell shape and strength to withstand changes in environmental osmotic pressures.

Percutaneous   Performed through the skin (e.g., percutaneous bladder aspiration).

Pericarditis   Inflammation of the covering of the heart (pericardium).

Perineum   The portion of the body bound by the pubic bone anteriorly, the coccyx posteriorly, and the bony prominences (tuberosities) of the ileum on both sides.

Peritoneal cavity   Space between the visceral and parietal layers of the peritoneum.

Peritoneum   Large, moist, continuous sheet of serous membrane lining the abdominal pelvic cavity and the outer coat of the organs contained within the cavity.

Peritonitis   Inflammation of the peritoneal cavity, most often caused by bacterial infection.

Pertussis   Upper respiratory infection caused by *Bordetella pertussis*.

Petechiae   Tiny hemorrhagic spots in the skin or mucous membranes.

PFGE   Pulsed-field gel electrophoresis.

Phaeohyphomycosis   Term used to describe any infection caused by a dematiaceous organism.

Phase-contrast microscopy   Technique for direct observation of unstained material in which light beams pass through the object to be visualized and are partially deflected by the different densities of the object. These light beams are deflected again when they impinge on a special objective lens, increasing in brightness when aligned in phase.

Phenotype   Related to characteristics of an organism beyond the genetic level and include readily observable features.

Photochromogens   Mycobacteria that produce pigment after exposure to light but whose colonies remain buff-colored in the dark.

Phycomycosis   Serious infection involving fungi of the zygomycete group, often beginning with necrotic lesions in the nasal mucus or palate but rapidly spreading to involve other tissues. Seen in immunocompromised patients.

PID   Pelvic inflammatory disease.

Pili   Structures in bacteria similar to fimbriae that participate in bacterial conjugation and transfer of genetic material.

Plasma   Fluid portion of blood; obtained by centrifuging anticoagulated blood.

Plasmids   Extrachromosomal DNA elements of bacteria carrying a variety of determinants that may permit survival in an adverse environment or successful competition with other microorganisms of the same or different species.

Pleomorphic   Having more than one form, usually widely different forms, as in pleomorphic bacteria.

Pleura   The serous membrane enveloping the lung and lining the internal surface of the thoracic cavity.

PMN   Polymorphonuclear leukocyte or neutrophil.

Pneumonia   Inflammation of the lungs, primarily caused by infectious agents.

Pneumothorax   Introduction of air (usually inadvertently) into the pleural space, leading to collapse of the lung on that side.

Polymerase chain reaction (PCR)   A method for expanding small discrete sections of DNA by binding DNA primers to sections at the ends of the DNA to be expanded and using cycles of heat (to create single-stranded DNA) and cooler temperatures (to allow a DNA polymerase enzyme to create new sections of DNA between the primer ends).

PPD   Purified protein derivative (skin test antigen for tuberculosis).

Precipitin test   Detection of antigen by allowing specific antibody to diffuse through liquid or gel until an antigen-antibody complex forms; this complex is visualized as a line of precipitated material.

Precision   Reproducibility of a test when run several times.

Prevalence   Frequency of disease in a population at a given time.

Prion   Proteinaceous infectious agent associated with Creutzfeldt-Jakob disease and perhaps other chronic, debilitating central nervous system diseases.

Proctitis   Inflammation of the rectum.

Prodromal   Early manifestations of a disease before specific symptoms become evident.

Proglottid   Segments of the tapeworm containing male and female reproductive systems; may be immature, mature, or gravid.

Prognosis   Forecast as to the possible outcome of a disease.

Prokaryotic   Organisms without a true nucleus.

Prophylaxis   Preventive treatment (e.g., the use of drugs to prevent infection).

Prostatitis   Inflammation of the prostate gland, usually caused by infection, characterized by fever, low back or perineal pain, and at times urinary frequency and urgency; a common background factor for recurrent cystitis in males.

Prosthesis   An artificial part such as a hip joint or eye.

Protein A   A protein on the cell wall of strains of *Staphylococcus aureus* (Cowan strain) that binds the Fc portion of antibodies.

Prototroph   Naturally occurring or wild strain.

Pseudomembrane   Necrosis of mucosal surface simulating a membrane.

Pseudomembranous colitis (PMC)   Syndrome in the large bowel characterized by a layer of necrotic tissue and dead inflammatory cells often caused by the toxin of *Clostridium difficile*.

Psychrophilic   Cold-loving (e.g., microorganisms that grow best at low [4° C] temperatures).

Purulent   Consisting of pus.

Pus   Product of inflammation, consisting of fluid and many white blood cells; often bacteria and cellular debris are also present.

Pyelonephritis   Infection of the kidney and renal pelvis and the late effects of such infection.

Pyocin   Pigment produced by a bacterium that has antibacterial properties against other strains or species of bacteria.

Pyogenic   Pus-producing.

PYR test   The enzyme, L-pyroglutamyl-amino peptidase, hydrolyzes L-pyrolidonyl-β-naphthylamide (PYR) to produce β-naphthylamine. When the β-naphthylamine combines with cinnamaldehyde reagent, a red color is produced.

Pyuria   Presence of eight or more leukocytes per cubic millimeter on microscopic examination of uncentrifuged urine.

QC   Quality control.

QNS   Quantity not sufficient.

Quencher   Molecule that can accept energy from a fluorophore and then dissipate the energy so that no fluorescence results.

Radioisotope   Unstable molecule that emits detectable radiation (e.g., gamma rays, X-rays) for a known period (half-life). Can be incorporated into other compounds as a label for later detection by radiographic film exposure or by measurement in a scintillation counting instrument.

Reagin   An antibody that reacts in various serologic tests for syphilis.

Reservoir   Source from which an infectious agent may be disseminated; for example, humans are the only reservoir for *Mycobacterium tuberculosis*.

Resin   Plant product composed largely of esters and ethers of organic acids and acid anhydrides.

Restriction endonuclease   Enzyme that breaks nucleic acid (usually DNA) at only one specific sequence of nucleotides.

Reticulate body   The metabolically more active form of elementary bodies of *Chlamydia* spp.

Reticuloendothelial system   Macrophage system, which includes all the phagocytic cells of the body except for the granulocytic leukocytes.

Reverse transcription   Synthesis of DNA from RNA by using the enzyme reverse transcriptase.

RFLP   Restriction fragment length polymorphism.

Rheumatoid factor IgM antibodies produced by some patients against their own IgG.

Rhinorrhea Runny nose.

RNA Ribonucleic acid.

RPR Rapid plasma reagin, nontreponemal test for antibodies developed in response to syphilis infection.

Saccharolytic Capable of breaking down sugars.

Saprophytic Nonpathogenic.

Schizogony Stage in the asexual cycle of the malaria parasite that takes place in the red blood cells of humans.

Schlichter test Synonym for the serum bactericidal level test.

Sclerotic Hard, indurated.

Scolex (pl., scolices) Head portion of a tapeworm; may attach to the intestinal wall by suckers or hooklets.

Scotochromogens Mycobacteria that are pigmented even in the absence of exposure to light.

Sensitivity Ability of a test to detect all true cases of the condition being tested for; absence of false-negative results. (Also see "specificity.")

Septate Having cross walls.

Septic shock Acute circulatory failure caused by toxins of microorganisms; often leads to multiple organ failure and is associated with a relatively high mortality.

Septicemia (sepsis) Systemic disease associated with presence of pathogenic microorganisms or their toxins in the blood.

Sereny test Test for bacterial invasiveness; involves applying a suspension of the organism to the conjunctiva of a small mammal and observing for development of conjunctivitis.

Serosanguineous Like serous but with some blood present grossly.

Serous Like serum.

Serum Cell- and fibrinogen-free fluid remaining after whole blood clots.

Serum bactericidal level Lowest dilution of a patient's serum that kills a standard inoculum of an organism isolated from that patient; it is related to antibiotic level achieved in the patient's serum and the bactericidal activity of the drug being used.

Sinus Suppurating tract; paranasal sinus, hollows, or cavities near the nose (e.g., frontal and maxillary sinuses).

Slant See definition of "butt." The slant is the upper surface of the medium in the tube described. It is exposed to air in the tube.

Solid-phase immunosorbent assay (SPIA) ELISA test in which the capture antigen or antibody is attached to the inside of a plastic tube, microwell, or to the outside of a plastic bead, in a filter matrix, or some other solid support. Allows faster interaction between reactants and more concentrated visual end products than ELISA tests performed in liquid.

Somatic Pertaining to the body (of a cell) (e.g., the somatic antigens of *Salmonella* spp.).

Southern blot Identification of specific genetic sequences by separating DNA fragments by gel electrophoresis and transferring them to membrane filters in situ. Labeled complementary DNA applied to the filter binds to homologous fragments, which can then be identified by detecting the presence of the labeled DNA in association with bands of certain molecular size. Named after its discoverer, E.M. Southern.

Specificity Ability of a test to correctly yield a negative result when the condition being detected is absent; absence of false-positive results. (Also see "sensitivity.")

Spore Reproductive cell of bacteria, fungi, or protozoa; in bacteria, may be inactive, resistant forms within the cell.

Sporozoite Slender, spindle-shaped organism that is the infective stage of the malarial parasite; it is inoculated into humans by an infected mosquito and is the result of the sexual cycle of the malarial parasite in the mosquito.

Sputum Material discharged from the surface of the lower respiratory tract air passages and expectorated (or swallowed).

Standard precautions Infection control guidelines used in the care of all patients; they apply to blood, body fluids, and secretions and excretions except sweat.

Stat Statim (Latin); immediately.

Stationary phase Stage in the growth cycle of a bacterial culture in which the vegetative cell population equals the dying population.

STD Sexually transmitted disease.

Sterile (sterility) Free of living microorganisms (the state of being sterile).

Substrate A substance on which an enzyme acts.

Sulfur granule Small colony of organisms with surrounding clublike material; yellow-brown; resembles grain of sulfur.

Superantigen Molecules produced by microbes (viruses, bacteria, and perhaps parasites) that act independently to stimulate T-cell activities, including cytokine release. Among the most potent T-cell mitogens, superantigen stimulation can result in anergy, or alternatively, systemic immune system activation.

Superinfection Strictly speaking, superinfection refers to a new infection superimposed on another being treated with an antimicrobial agent. The new infecting agent is resistant to the therapy initially used and thus survives and causes persistence of the infection (now resistant to the treatment) or a new infection at a different site. The term is also used to indicate persistence or colonization with a new organism without any evidence of resulting infection.

Suppuration Formation of pus.

Suppurative thrombophlebitis Inflammation of a vein wall.

Syncytia Structure resulting from fusion of cell membranes of several cells to form a multinucleated cellular structure; usually the result of viral infection of the cells.

Syndrome Set of symptoms occurring together (e.g., nephrotic syndrome).

Synergism Combined effect of two or more agents that is greater than the sum of their individual effects.

Synovial fluid Viscid fluid secreted by the synovial membrane; formed in joint cavities, bursae, and so forth.

Tachypnea Rapid breathing

T cells Lymphocytes involved in cellular immunity.

Teichoic acids Glycerol or ribitol phosphate polymers combined with various sugars, amino acids, and amino sugars that are in the cell wall of gram-positive bacteria.

Teleomorph Sexual fungal form.

TEM Transmission electron micrograph.

Tenesmus Painful, unsuccessful straining in an attempt to empty the bowels.

Therapy, antimicrobial Treatment of a patient to combat an infectious disease.

Thermolabile Adversely affected by heat (as opposed to thermostable, not affected by heat).

Thoracentesis Drainage of fluid from the pleural space.

Thoracic Pertaining to the chest cavity.

Threshold cycle (C_T) The amplification cycle number in which the fluorescent signal rises above background; also referred to as the crossing point.

Thrush A form of *Candida* infection that typically produces white plaquelike lesions in the oral cavity.

Tinea Dermatophyte infection (tinea capitis, tinea of scalp; tinea corporis, tinea of the smooth skin of the body; tinea cruris, tinea of the groin; tinea pedis, tinea of the foot).

Titer Level of a substance such as antibody or toxin present in material such as serum; reciprocal of the highest dilution at which the substance can still be detected.

T lymphocytes (or T cells) Thymus-derived lymphocytes important in cell-mediated immunity.

Tm See "melting temperature."

Tolerance A form of resistance to antimicrobial drugs; of uncertain clinical importance. See "tolerant."

Tolerant Characteristic of an organism that requires a great deal more antimicrobial agent to kill it than to inhibit its growth.

TPI *Treponema pallidum* immobilization test, a test for antibodies against the agent of syphilis that uses live treponemes.

Trachoma Serious eye infection caused by *Chlamydia trachomatis*; often leads to blindness.

Transduction Moving genetic material from one prokaryote to another via a bacteriophage or viral vector.

Transformation Process in which an organism takes up free DNA that is released into the environment when another organism dies and then lyses.

Transient bacteremia Incidental and brief presence of bacteria in the bloodstream.

Transmission-based precautions Infection control guidelines used for patients known or suspected to be infected with pathogens spread by airborne or droplet transmission or by contact with dry skin or fomites.

Transposon Genetic material that can move from one genetic element to another (i.e., between plasmids or from a plasmid to a chromosome); so-called jumping genes.

Transtracheal aspiration Passage of needle and plastic catheter into the trachea for obtaining lower respiratory tract secretions free of oral contamination.

Transudate Similar to exudate but with low protein content.

Trophozoite Feeding, motile stage of protozoa.

Tropism Preferred environment or destination. In viral infection, preference for a particular tissue site (rabies viruses have a tropism for neural tissue).

Glossary

TSI   Triple sugar iron (agar tube).

TTA   Transtracheal aspiration.

Type III secretion system   Found in many gram-negative pathogens and is responsible for secretion and injection of virulence-associated factors into the cytoplasm of host cells.

Type IV secretion systems (bacterial)   Bacterial devices that deliver macromolecular molecules such as proteins across and into cells.

Typing   Methods of grouping organisms, primarily for epidemiologic purposes (e.g., biotyping, serotyping, bacteriophage typing, and the antibiogram).

Tzanck test   Stained smear of cells from the base of a vesicle examined for inclusions produced by herpes simplex virus or varicella-zoster virus.

Urethritis   Inflammation of urethra, the canal through which urine is discharged (e.g., gonococcal urethritis).

UTI   Urinary tract infection.

VD   Venereal disease.

VDRL   Veneral Disease Research Laboratory; classic nontreponemal serologic test for syphilis antibodies. Uses cardiolipin, lecithin, and cholesterol as cross-reactive antigen that flocculates in the presence of "reaginic" antibodies produced by patients with syphilis. Best test for cerebrospinal fluid in cases of neurosyphilis.

Vector   An arthropod or other agent that carries microorganisms from one infected individual to another.

Vegetation   In endocarditis, the aggregates of fibrin and microorganisms on the heart valves or other endocardium.

Vesicle   A small bulla or blister containing clear fluid.

Villi   Minute, elongated projections from the surface of intestinal mucosa that are important in absorption.

Vincent's angina   An old term, seldom used currently, referring to anaerobic tonsillitis.

Viremia   Presence of viruses in the bloodstream.

Virion   The whole viral particle, including nucleocapsid, outer membrane or envelope, and all adherence structures.

Virulence   Degree of pathogenicity or disease-producing ability of a microorganism.

Viscus (pl., viscera)   Any of the organs within one of the four great body cavities (cranium, thorax, abdomen, and pelvis).

V-P   Voges-Proskauer.

Western blot   Similar to Southern blot, except that antigenic proteins of an organism are separated by gel electrophoresis and transferred to membrane filters. Antiserum is allowed to react with the filters, and specific antibody bound to its homologous antigen is detected using labeled anti-antibody detectors.

Zoonosis   A disease of lower animals transmissible to humans (e.g., tularemia).

Zygomycetes   Group of fungi with nonseptate hyphae and spores produced within a sporangium.